ENCYCLOPEDIA OF TELEVISION

SERIES, PILOTS AND SPECIALS

THE INDEX: WHO'S WHO IN TELEVISION 1937–1984

VINCENT TERRACE

A **BASE**LINE Production

New York Zoetrope

Encyclopedia of Television
Series, Pilots and Specials 1937-1984
Volume III

Library of Congress Cataloging in Publication Number: 86-43211

ISBN 0-918432-71-5

New York Zoetrope
838 Broadway
New York NY 10003

The Encyclopedia of Television was set in H-P Times Roman. The book was printed and bound by the Maple-Vail Book Manufacturing Group at their York PA plant.

The Encyclopedia of Television is a product of BASELINE, the comprehensive data source for the film and television industry. More information about BASELINE—and about becoming a Baseline subscriber—can be obtained from BASELINE, 838 Broadway, New York NY 10003. (212) 254-8235.

Printed in the United States of America
First Printing December, 1986
5 4 3 2 1

This is the third volume of the ENCYCLOPEDIA OF TELEVISION: SERIES, PILOTS AND SPECIALS. This volume serves as both an index to Volumes I and II, and as a 'who's who' in television from 1937 to 1984. The index contains the names of over 18,000 performers, 5000 producers, 5000 writers, and 3500 directors. Alphabetically listed, each entry is supplemented with a list of that person's lifetime credits. Each program is followed by the volume number where that title can be found, and the corresponding entry number in that volume. In volumes I and II, each program is detailed with cast, credits, plot synopsis, running time, air date, and network. Together, the three volumes of this massive undertaking constitute the largest reference work on television.

A NOTE ON THE FORMAT:

Some performers have used more than one name in their credits. For example, a 'Herb' may also have called himself 'Herbert' at one time. The index lists names as they appeared on the actual credits, and at times, the same actor will be listed under both names. In addition, a last name is sometimes comprised of two words, such as 'De Luca'. In these cases, a 'De' with a space following it will be listed before a 'De' with a letter following. Hence, the name 'De Luca' will precede the name 'Dean'.

Vincent Terrace
New York City
November 1986

Contents

Performers

AAKER, LEE
THE ADVENTURES OF RIN TIN TIN (I,73); YOUR JEWELER'S SHOWCASE (I,4953)

AALDA, MARIANN
THE EDGE OF NIGHT (II,760)

AAMES, ANGELA
AUTOMAN (II,120); B.J. AND THE BEAR (II,125); CHEERS (II,494); THE CRACKER BROTHERS (II,599); LOVE AT FIRST SIGHT (II,1530)

AAMES, WILLIE
BATTLE OF THE NETWORK STARS (II,169); BATTLE OF THE NETWORK STARS (II,171); BOB HOPE SPECIAL: BOB HOPE'S ALL-STAR LOOK AT TV'S PRIME TIME WARS (II,303); CHARLES IN CHARGE (II,482); DOCTOR DAN (II,684); DUNGEONS AND DRAGONS (II,744); EIGHT IS ENOUGH (II,762); FAMILY (II,813); FRANKENSTEIN (I,1671); THE ODD COUPLE (II,1875); RICH MAN, POOR MAN—BOOK I (II,2161); SWISS FAMILY ROBINSON (II,2517); THE TOM SWIFT AND LINDA CRAIG MYSTERY HOUR (II,2633); WE'LL GET BY (II,2753); WE'LL GET BY (II,2754); WE'RE MOVIN' (II,2756)

ABBATE, NANCY
THE MICKEY MOUSE CLUB (I,3025)

ABBOTT, BRUCE
THE BLUE AND THE GRAY (II,273)

ABBOTT, BUD
THE ABBOTT AND COSTELLO CARTOON SHOW (I,1); THE ABBOTT AND COSTELLO SHOW (I,2); THE COLGATE COMEDY HOUR (I,997)

ABBOTT, DAVID
FOR RICHER, FOR POORER (II,906); LOVERS AND FRIENDS (II,1549)

ABBOTT, DOROTHY
DRAGNET (I,1369); LEAVE IT TO BEAVER (I,2648)

ABBOTT, GEORGE
ROYAL SHOWCASE (I,3864); SKIN OF OUR TEETH (I,4073); U.S. ROYAL SHOWCASE (I,4688)

ABBOTT, JOHN
GREAT BIBLE ADVENTURES (I,1872); SHANGRI-LA (I,3996)

ABBOTT, PHILIP
CAMPO 44 (I,813); THE COPS AND ROBIN (II,573); THE FBI (I,1551); THE HOUSE ON HIGH STREET (I,2141); KILROY (I,2541); RICH MAN, POOR MAN—BOOK II (II,2162); SEARCH FOR TOMORROW (II,2284)

ABBOTT, STEVE
CAMP WILDERNESS (II,419)

ABE, TORU
SHOGUN (II,2339)

ABEL, WALTER
FAITH BALDWIN'S THEATER OF ROMANCE (I,1515); ON TRIAL (I,3363); PHILCO TELEVISION PLAYHOUSE (I,3583); PRUDENTIAL FAMILY PLAYHOUSE (I,3683); SUSPICION (I,4309); TALES OF TOMORROW (I,4352)

ABELES, SARA
TALES OF THE APPLE DUMPLING GANG (II,2536)

ABELEW, ALAN
LUCAS TANNER (II,1555); LUCAS TANNER (II,1556)

ABELLIRA, REMI
BIG HAWAII (II,238)

ABELS, GREG
WHERE THE HEART IS (I,4831)

ABELSON, ARTHUR
OPERATION GREASEPAINT (I,3402)

ABERCROMBIE, IAN
FANTASY ISLAND (II,830)

ABERCROMBIE, JUDY
ALL-STAR ANYTHING GOES (II,50)

ABERG, SIV
THE GONG SHOW (II,1026); THE GONG SHOW (II,1027); THE NEW TREASURE HUNT (I,3275); THE NEW TREASURE HUNT (II,1831); ONCE UPON A MATTRESS (I,3371)

ABERLIN, BETTY
MR. ROGERS' NEIGHBORHOOD (I,3150); THE SMOTHERS BROTHERS SHOW (II,2384)

ABINERI, JOHN
MIRROR OF DECEPTION (I,3060)

ABINERI, SEBASTIAN
FLAMBARDS (II,869)

ABLE, GEORGE
IT'S A HIT (I,2257)

ABLE, WILL
HANSEL AND GRETEL (I,1934); SUPER CIRCUS (I,4287)

ABRAHAM, DAWN
PRISONERS OF THE LOST UNIVERSE (II,2086)

ABRAHAM, F MURRAY
A.E.S. HUDSON STREET (II,17); HOW TO SURVIVE A MARRIAGE (II,1198); MARCO POLO (II,1626); NIGHTSIDE (I,3296)

ABRAHMS, LYDIA SUE
THE UNCLE FLOYD SHOW (II,2703)

ABRELL, SARAH
O'MALLEY (II,1872); W*A*L*T*E*R (II,2731)

ABSALOM, CHRISTINE
GREAT EXPECTATIONS (II,1055)

ACAVONE, JAY
PAROLE (II,1960)

ACCENTS, THE
THE MICKIE FINN SPECIAL (I,3028)

ACE TRUCKING COMPANY, THE
THE TOM JONES SPECIAL (I,4545)

ACE, GOODMAN
EASY ACES (I,1398)

ACE, JANE
EASY ACES (I,1398)

ACE, ROSEMARY
THE MISS AND THE MISSILES (I,3062)

ACKER, CINDY
ALEX AND THE DOBERMAN GANG (II,32)

ACKER, SANDY
ALEX AND THE DOBERMAN GANG (II,32)

ACKER, SHARON
BATTLES: THE MURDER THAT WOULDN'T DIE (II,180); THE CENTURY TURNS (I,890); EXECUTIVE SUITE (II,796); GALACTICA 1980 (II,953); HEC RAMSEY (II,1121); THE NEW ADVENTURES OF PERRY MASON (I,3252); OUR MAN FLINT:)DEAD ON TARGET (II,1932); THE SENATOR (I,3976); THE STRANGER (I,4251); TEXAS (II,2566)

ACKERMAN, BETTYE
BEN CASEY (I,394); BRACKEN'S WORLD (I,703); DOCTORS' PRIVATE LIVES (II,693); DYNASTY (II,746); THE FEATHER AND FATHER GANG: NEVER CON A KILLER (II,846); HEAT OF ANGER (I,1992); RETURN TO PEYTON PLACE (I,3779)

ACKERMAN, LESLIE
THE CITY (II,541); THE LAST HURRAH (I,2626); MISSING PIECES (II,1706); ROXY PAGE (II,2222); SKAG (II,2374); STUDS LONIGAN (II,2484)

ACKERMAN, PETER
GIDGET'S SUMMER REUNION (I,1799)

ACKROYD, DAVID
THE ACADEMY (II,3); THE ACADEMY II (II,4); AFTERMASH (II,23); ANOTHER WORLD (II,97); EXO-MAN (II,797); LITTLE WOMEN (II,1498); THE SECRET STORM (I,3969); TOM AND JOANN (II,2630); WOMEN IN WHITE (II,2823); THE WORD (II,2833); THE YEAGERS (II,2843)

ACOSTA, RODOLFO
THE HIGH CHAPARRAL (I,2051)

ACROMANIACS, THE
THE ALL-STAR SUMMER REVUE (I,139)

ACTION FACTION DANCERS, THE
SOMETHING ELSE (I,4117)

ACTMAN, JANE
BLACK BEAUTY (II,261); THE CABOT CONNECTION (II,408); THE CHADWICK FAMILY (II,478); THE PAUL LYNDE SHOW (I,3489); ROOM 222 (I,3843); YOUR PLACE OR MINE? (II,2874)

ACUFF, ROY
THE CONCRETE COWBOYS (II,564); COUNTRY COMES HOME (II,585); COUNTRY COMES HOME (II,586); GEORGE BURNS IN NASHVILLE?? (II,969); JOHNNY CASH: THE FIRST 25 YEARS (II,1336); ROY ACUFF—50 YEARS THE KING OF COUNTRY MUSIC (II,2223)

ADAIR, DEBORAH
DYNASTY (II,746); FINDER OF LOST LOVES (II,857); THE YOUNG AND THE RESTLESS (II,2862)

ADAIR, EARL
THE BENNY HILL SHOW (II,207)

ADAIR, PATTY
COUNTRY STYLE (I,1072) COUNTRY STYLE (I,1073); SAY IT WITH ACTING (I,3928)

ADAM, RONALD
THE ADVENTURES OF A JUNGLE BOY (I,49); CASANOVA (II,451)

ADAMLE, MIKE
GAMES PEOPLE PLAY (II,956)

ADAMS, ALICE
SANDY DREAMS (I,3909)

ADAMS, ARTHUR
BRADDOCK (I,704); LANDON, LANDON & LANDON (II,1426)

ADAMS, ASHBY
THE DOCTORS (II,694)

ADAMS, BETTY
YOUR SHOW TIME (I,4958)

ADAMS, BEVERLY
CAMP RUNAMUCK (I,811); GIDGET (I,1795); THE GOOD OLD DAYS (I,1851)

ADAMS, BILL
AUNT JENNY'S REAL LIFE STORIES (I,314)

ADAMS, BROOKE
BLACK BART (II,260); THE DAUGHTERS OF JOSHUA CABE RETURN (I,1170); FLATBUSH/AVENUE J (II,878); LACE (II,1418); NERO WOLFE (II,1805); O.K. CRACKERBY (I,3348); THE TONY RANDALL SHOW (II,2640)

ADAMS, CAITLIN
HOW TO SURVIVE THE 70S AND MAYBE EVEN BUMP INTO HAPPINESS (II,1199); SQUARE PEGS (II,2431)

ADAMS, CASEY
G.E. SUMMMER ORIGINALS (I,1747)

ADAMS, CEDRIC
PRIZE PERFORMANCE (I,3673)

ADAMS, DALLAS
NANCY ASTOR (II,1788); RUNNING BLIND (II,2232); TERROR FROM WITHIN (I,4404)

ADAMS, DAVID
S.W.A.T. (II,2236)

ADAMS, DESMOND
DAY OF THE TRIFFIDS (II,627)

ADAMS, DON
THE BILL DANA SHOW (I,446); THE BOB HOPE SHOW (I,619); CELEBRATION: THE AMERICAN SPIRIT (II,461); CHANGING SCENE (I,899); CLASS OF '67 (I,970); A COUPLE OF DONS (I,1077); THE DON ADAMS SCREEN TEST (II,705); THE DON ADAMS SPECIAL: HOORAY FOR HOLLYWOOD (I,1329); DON RICKLES—ALIVE AND KICKING (I,1340); GET SMART (II,983); INSPECTOR GADGET (II,1236); A LAST LAUGH AT THE 60'S (I,2627); THE LOVE BOAT I (II,1532); THE MANY SIDES OF DON RICKLES (I,2898); MILTON BERLE'S MAD MAD MAD WORLD OF COMEDY (II,1701); MURDER AT NBC (A BOB HOPE SPECIAL) (I,3159); THE PARTNERS (I,3462); THE PERRY COMO CHRISTMAS SHOW (I,3529); THE PERRY COMO SHOW (I,3532); RICKLES (II,2170); SAGA OF SONORA (I,3884); SHOOT-IN AT NBC (A BOB HOPE SPECIAL) (I,4021); SOLDIERS IN GREASEPAINT (I,4110); TENNESSEE TUXEDO AND HIS TALES (I,4403); THREE TIMES DALEY (II,2610); WHERE THE GIRLS ARE (I,4830); THE WONDERFUL WORLD OF BURLESQUE II (I,4896)

ADAMS, EDIE
AMERICAN COWBOY (I,169); APPOINTMENT WITH ADVENTURE (I,246); THE BING CROSBY SPRINGTIME SPECIAL (I,473); THE BOB HOPE SHOW (I,594); THE BOB HOPE SHOW (I,643); THE BOB HOPE SHOW (I,649); BOSOM BUDDIES (II,355); THE CHEVY SUMMER SHOW (I,929); THE COLGATE COMEDY HOUR (I,998); THE DON ADAMS SPECIAL: HOORAY FOR HOLLYWOOD (I,1329); ERNIE IN KOVACSLAND (I,1454); THE ERNIE KOVACS SHOW (I,1455); EVIL ROY SLADE (I,1485); HERE'S EDIE (I,2030); IT'S TIME FOR ERNIE (I,2277); THE JACK PAAR SHOW (I,2316); KOVACS ON THE CORNER (I,2582); KOVACS UNLIMITED (I,2583); PRIVATE EYE, PRIVATE EYE (I,3671); THE ROCK N' ROLL SHOW (I,3820); THE SEEKERS (II,2298); SHOOTING STARS (II,2341); THE SPIRAL STAIRCASE (I,4160); TAKE A GOOD LOOK (II,2529); VERSATILE VARIETIES (I,4713)

ADAMS, HANK
MR. MOON'S MAGIC CIRCUS (II,1753)

ADAMS, HARRY
THE PHIL SILVERS SHOW (I,3580)

ADAMS, HARVEY
THE ADVENTURES OF LONG JOHN SILVER (I,67)

ADAMS, INGE
LIGHT'S OUT (I,2699); THREE STEPS TO HEAVEN (I,4485)

ADAMS, JANE
THE AMOS AND ANDY SHOW (I,179); SOLDIERS IN GREASEPAINT (I,4110)

ADAMS, JANET
BUFFALO BILL (II,387)

ADAMS, JEB
THE BLACK SHEEP SQUADRON (II,262)

ADAMS, JILL
THE FLYING DOCTOR (I,1610)

ADAMS, JOEY
BACK THE FACT (I,327); FUN FOR 51 (I,1707)

ADAMS, JONATHAN
CORONATION STREET (I,1054)

ADAMS, JULIE
330 INDEPENDENCE S.W. (I,4486); ALCOA/GOODYEAR THEATER (I,107); ALFRED HITCHCOCK PRESENTS (I,115); THE ANDY GRIFFITH SHOW (I,192); CAPITOL (II,426); CODE RED (II,552); CODE RED (II,553); THE DICK POWELL SHOW (I,1269); DOC HOLLIDAY (I,1307); GENERAL HOSPITAL (II,964); THE JIMMY STEWART SHOW (I,2391); KRAFT SUSPENSE THEATER (I,2591); THE LUX VIDEO THEATER (I,2795); PLAYHOUSE 90 (I,3623)

ADAMS, LIONEL
THE VALIANT (I,4697)

ADAMS, LYNN
THE GUIDING LIGHT (II,1064); THE SECRET STORM (I,3969)

ADAMS, MARLA

CAPITOL (II,426); CONCRETE BEAT (II,562); HAPPY DAYS (II,1084); THE SECRET STORM (I,3969); TRAUMA CENTER (II,2655); THE YOUNG AND THE RESTLESS (II,2862)

ADAMS, MARY
DOCTOR MIKE (I,1316); IT'S A MAN'S WORLD (I,2258); MICHAEL SHAYNE, DETECTIVE (I,3022); ONE MAN'S FAMILY (I,3383)

ADAMS, MASON
FLAMINGO ROAD (II,871); THE KID WITH THE BROKEN HALO (II,1387); LOU GRANT (II,1526); MURDER CAN HURT YOU! (II,1768); PEKING ENCOUNTER (II,1991); WHERE THE HEART IS (I,4831)

ADAMS, MAUD
BATTLE OF THE NETWORK STARS (II,173); BIG BOB JOHNSON AND HIS FANTASTIC SPEED CIRCUS (II,231); CHICAGO STORY (II,506); EMERALD POINT, N.A.S. (II,773); WOMEN WHO RATE A "10" (II,2825)

ADAMS, NEILE
THE BOB HOPE SHOW (I,579)

ADAMS, NICK
AMOS BURKE: WHO KILLED JULIE GREER? (I,181); THE FIRST HUNDRED YEARS (I,1582); THE GENERAL ELECTRIC THEATER (I,1753); THE OUTER LIMITS (I,3426); THE REBEL (I,3746); SAINTS AND SINNERS (I,3887); SAVAGE SUNDAY (I,3926)

ADAMS, NOELLE
THE JONATHAN WINTERS SPECIAL (I,2445); THE KEEFE BRASSELLE SHOW (I,2514)

ADAMS, PETER
BARRIER REEF (I,362)

ADAMS, POLLY
THE EDGE OF NIGHT (II,760)

ADAMS, RICHARD C
LUCAN (II,1553)

ADAMS, SONNY
GOLDEN WINDOWS (I,1841)

ADAMS, STANLEY
THE APARTMENT HOUSE (I,241); THE BOSTON TERRIER (I,697); DEATH OF A SALESMAN (I,1208); THE NIGHT STALKER (I,3292); THE NORLISS TAPES (I,3305); THE PIED PIPER OF HAMELIN (I,3595); PISTOLS 'N' PETTICOATS (I,3608); THE RED SKELTON SHOW (I,3755)

ADAMS, SUSAN
CRISIS IN SUN VALLEY (II,604); THE DOCTORS (II,694); FATHER KNOWS BEST: HOME FOR CHRISTMAS (II,839); FATHER KNOWS BEST: THE FATHER KNOWS BEST REUNION (II,840)

ADAMS, TREVOR
THE FALL AND RISE OF REGINALD PERRIN (II,810)

ADAMSON, PETER
CORONATION STREET (I,1054)

ADDAMS, DAWN
THE ALAN YOUNG SHOW (I,104); FATHER DEAR FATHER (II,837); STAR MAIDENS (II,2435)

ADDID WILLIAMS PUPPETS, THE
PIXANNE (II,2048)

ADDISON, JOHN
NO—HONESTLY (II,1857)

ADDISON, NANCY
THE DAIN CURSE (II,613); THE GUIDING LIGHT (II,1064); RYAN'S HOPE (II,2234)

ADDOTTA, KIP
THE LOU RAWLS SPECIAL (II,1527)

ADDY, WESLEY
THE DOCTOR (I,1320); LOVING (II,1552); RAGE OF ANGELS (II,2107); SHORT, SHORT DRAMA (I,4023)

ADELE SCOTT TRIO, THE
THE SKIP FARRELL SHOW (I,4074)

ADELL, ILLUNGA
THAT'S MY MAMA (II,2580)

ADELMAN, STEVE
THE WORLD OF PEOPLE (II,2839)

ADERMAN, JOHN
NOWHERE TO HIDE (II,1868)

ADIARTE, PATRICK
M*A*S*H (II,1569)

ADLER, BILL
CRAZY TIMES (II,602); KID TALK (I,2533)

ADLER, CLYDE
THE NEW SOUPY SALES SHOW (II,1829); THE SOUPY SALES SHOW (I,4137)

ADLER, CYNTHIA
THE CONEHEADS (II,567)

ADLER, JEFF
POLICE WOMAN: THE GAMBLE (II,2064)

ADLER, LUTHER
D.H.O. (I,1249); MEETING AT APALACHIN (I,2994); THE PSYCHIATRIST (I,3686); THE PSYCHIATRIST: GOD BLESS THE CHILDREN (I,3685)

ADLER, WILLIAM
NOT FOR PUBLICATION (I,3309)

ADORF, MARIO
MARCO POLO (II,1626); SMILEY'S PEOPLE (II,2383)

ADRIAN, IRIS
THE JACK BENNY PROGRAM (I,2294); MURDER CAN HURT YOU! (II,1768); THE TED KNIGHT SHOW (II,2550)

ADRIAN, JANE
THE MARSHAL OF GUNSIGHT PASS (I,2925)

ADRIAN, MAX
TWELFTH NIGHT (I,4636)

AFF, JAMES
RIVKIN: BOUNTY HUNTER (II,2183)

AGUILAR, TOMMY
BATTLES: THE MURDER THAT WOULDN'T DIE (II,180); FAME (II,812)

AGUTTER, JENNY
BEULAH LAND (II,226); THE SAVAGE CURSE (I,3925); THE SNOW GOOSE (I,4103)

AHEARNE, TOM
OUR FIVE DAUGHTERS (I,3413)

AHERN, GLADYS
JOHNNY CARSON PRESENTS THE SUN CITY SCANDALS (I,2420)

AHERN, WILL
JOHNNY CARSON PRESENTS THE SUN CITY SCANDALS (I,2420)

AHERNE, BRIAN
PULITZER PRIZE PLAYHOUSE (I,3692); REUNION IN VIENNA (I,3780)

AHN, PHILIP
ALCOA/GOODYEAR THEATER (I,107); FOUR STAR PLAYHOUSE (I,1652); THE KILLER WHO WOULDN'T DIE (II,1393); KUNG FU (II,1416); MR. GARLUND (I,3138); NAVY LOG (I,3233); TV READERS DIGEST (I,4630)

AHRENS, JIMMY
A CHARLIE BROWN THANKSGIVING (I,909); IT'S A MYSTERY, CHARLIE BROWN (I,2259); IT'S THE EASTER BEAGLE, CHARLIE BROWN (I,2275); THERE'S NO TIME FOR LOVE, CHARLIE BROWN (I,4438); YOU'RE A GOOD SPORT, CHARLIE BROWN (I,4963)

AIDMAN, BETTY
COPS (I,1047)

AIDMAN, CHARLES
THE BARBARY COAST (II,147); ONE STEP BEYOND (I,3388); THE PICTURE OF DORIAN GRAY (I,3592); THE SOUND OF ANGER (I,4133); STAR TONIGHT (I,4199); THRILLER (I,4492); THE WILD WILD WEST (I,4863)

AIELLO, DANNY
CAR WASH (II,437)

AIKEN, CHARLES
FRIENDS (II,927); THE WILDS OF TEN THOUSAND ISLANDS (II,2803)

AIKEN, DAVID
AMAHL AND THE NIGHT VISITORS (I,148)

AINLEY, ANTHONY
DOCTOR WHO (II,689); DOCTOR WHO—THE FIVE DOCTORS (II,690)

AINSLEY, JEAN
MASTER OF THE GAME (II,1647)

AINSLEY, PAUL
THREE'S COMPANY (II,2614)

AIO, GABRIEL
THE RETURN OF LUTHER GILLIS (II,2137)

AIR SUPPLY
STATE FAIR USA (II,2449)

AIRD, HOLLY
THE FLAME TREES OF THIKA (II,870); THE TALE OF BEATRIX POTTER (II,2533)

AJAYE, FRANKLYN
THE CHEAP DETECTIVE (II,490); COTTON CLUB '75 (II,580); KEEP ON TRUCKIN' (II,1372); NATIONAL LAMPOON'S HOT FLASHES (II,1795)

AKERLING, MYA
JENNIFER SLEPT HERE (II,1278); LITTLE SHOTS (II,1495)

AKERS, ANDRA
FLAMINGO ROAD (II,872); HARDCASE (II,1088); JOE FORRESTER (II,1303); MARY HARTMAN, MARY HARTMAN (II,1638); PEN 'N' INC. (II,1992)

AKINS, CLAUDE
THE ALL-STAR SALUTE TO MOTHER'S DAY (II,51); B.J. AND THE BEAR (II,125); B.J. AND THE BEAR (II,127); BELLE STARR (I,391); BOB HOPE SPECIAL: BOB HOPE'S ALL-STAR LOOK AT TV'S PRIME TIME WARS (II,303); BUS STOP (II,399); CELEBRITY (II,463); THE CONCRETE COWBOYS (II,564); THE COUNTRY MUSIC MURDERS (II,590); EBONY, IVORY AND JADE (II,755); ERIC (I,1451); FIRESIDE THEATER (I,1580); THE HAT OF SERGEANT MARTIN (I,1963); IN TANDEM (II,1228); JANE WYMAN PRESENTS THE FIRESIDE THEATER (I,2345); KISS ME, KILL ME (II,1400); LAREDO (I,2615); LEGMEN (II,1458); LOBO (II,1504); LOCK, STOCK, AND BARREL (I,2736); THE MISADVENTURES OF SHERIFF LOBO (II,1704); MOVIN' ON (II,1743); MURDER, SHE WROTE (II,1770); NASHVILLE 99 (II,1791); THE NIGHT STALKER (I,3292); THE NORLISS TAPES (I,3305); OUTPOST (I,3428); THE RHINEMANN EXCHANGE (II,2150); SAM HILL (I,3895); THE WAY THEY WERE (II,2748); YOU ARE THERE (I,4929); ZANE GREY THEATER (I,4979)

AKINS, TOM
SHERLOCK HOLMES (II,2327)

AKIRA
MODESTY BLAISE (II,1719)

AKUNE, SHUKO
E/R (II,748)

ALABAMA
CHRISTMAS LEGEND OF NASHVILLE (II,522); COUNTRY COMES HOME (II,586)

ALADDIN
THE LAWRENCE WELK SHOW (I,2643); MEMORIES WITH LAWRENCE WELK (II,1681)

ALAIMO, MARC
ARCHER—FUGITIVE FROM THE EMPIRE (II,103); THE BLUE KNIGHT (II,277); MR. & MS. AND THE BANDSTAND MURDERS (II,1746); NO MAN'S LAND (II,1855); SOMERSET (I,4115)

ALAIMO, MICHAEL
BILLY (II,247) BLUE JEANS (II,275)

ALAIMO, STEVE
WHERE THE ACTION IS (I,4828)

ALAIO, ROSE
THE GUIDING LIGHT (II,1064); RYAN'S HOPE (II,2234)

ALAMO, TONY
SO YOU WANT TO LEAD A BAND (I,4106)

ALAN COPELAND SINGERS, THE
THE RED SKELTON SHOW (I,3755); TOP OF THE MONTH (I,4578)

ALAN JOHNSON DANCERS, THE
JACK LEMMON IN 'S WONDERFUL, 'S MARVELOUS, 'S GERSHWIN (I,2313); JACK LEMMON—GET HAPPY (I,2312)

ALAN, BUDDY
BUCK OWENS TV RANCH (I,750)

ALAN, DON
THE MAGIC RANCH (I,2819)

ALAN, RICO
ELFEGO BACA (I,1427)

ALANN, LLOYD
JOE AND VALERIE (II,1299)

ALANSU, JOHN
ALL THE RIVERS RUN (II,43)

ALBEE, DENNY
THE EDGE OF NIGHT (II,760); ONE LIFE TO LIVE (II,1907)

ALBEE, JOSHUA
CODE RED (II,553); LASSIE (I,2624); SEALAB 2020 (I,3949)

ALBERG, SOMER
THE DEVIL'S DISCIPLE (I,1247)

ALBERGHETTI, ANNA MARIA
THE BOB HOPE SHOW (I,512); THE BOB HOPE SHOW (I,563); THE DESILU PLAYHOUSE (I,1237); DUPONT SHOW OF THE MONTH (I,1387); FORD TELEVISION THEATER (I,1634); THE JAZZ SINGER (I,2351); KISMET (I,2565); ROBERTA (I,3815); SCHLITZ PLAYHOUSE OF STARS (I,3936); TWELVE STAR SALUTE (I,4638)

ALBERONI, SHERRY
BRIGHT PROMISE (I,727); DOBIE GILLIS (I,1302); THE ED WYNN SHOW (I,1404); FAMILY AFFAIR (I,1519); INTERTECT (I,2228); JOSIE AND THE PUSSYCATS (I,2453); JOSIE AND THE PUSSYCATS IN OUTER SPACE (I,2454); LOST IN SPACE (I,2759); MCNAB'S LAB (I,2972); THE MOUSEKETEERS REUNION (II,1742); PARTRIDGE FAMILY: 2200 A.D. (II,1963); SUPER FRIENDS (II,2497); THE TOM EWELL SHOW (I,4538)

ALBERS, DICK
THE BOB HOPE SHOW (I,629)

ALBERT, DARLENE
THE GEORGE BURNS AND GRACIE ALLEN SHOW (I,1763)

ALBERT, EDDIE
ALCOA PREMIERE (I,109); THE BALLAD OF LOUIE THE LOUSE (I,337); BEULAH LAND (II,226); BEYOND WITCH MOUNTAIN (II,228); THE BORROWERS (I,693); CAROL (I,846); THE CHOCOLATE SOLDIER (I,944); CHRYSLER MEDALLION THEATER (I,951); A CONNECTICUT YANKEE (I,1038); DADDY'S GIRL (I,1124); THE DAVID NIVEN THEATER (I,1178); THE EDDIE ALBERT SHOW (I,1406); FORD THEATER HOUR (I,1635); FRONT ROW CENTER (I,1694); GOLIATH AWAITS (II,1025); GREEN ACRES (I,1884); HIPPODROME (I,2060); HOLLYWOOD SINGS (I,2091); HOWDY (I,2150); JOHNNY BELINDA (I,2415); KRAFT SUSPENSE THEATER (I,2591); LEAVE IT TO LARRY (I,2649); LI'L ABNER (I,2702); LIGHT'S OUT (I,2699); LIVING IN PARADISE (II,1503); THE LORETTA YOUNG THEATER (I,2756); THE MOTOROLA TELEVISION HOUR (I,3112); THE NIGHT OF CHRISTMAS (I,3290); NOTHING BUT THE BEST (I,3312); ON YOUR ACCOUNT (I,3364); THE OUTER LIMITS (I,3426); PARADE OF STARS (II,1954); REVLON MIRROR THEATER (I,3781); ROOSTER (II,2210); THE SATURDAY NIGHT REVUE (I,3921); SIEGFRIED AND ROY (II,2350); THE SPIRAL STAIRCASE (I,4160); SWITCH (II,2519); SWITCH (II,2520); TELLER OF TALES (I,4388); TROUBLE IN HIGH TIMBER COUNTRY (II,2661); THE WORD (II,2833); ZANE GREY THEATER (I,4979)

ALBERT, EDWARD
BATTLE OF THE NETWORK STARS (II,177); BLACK BEAUTY (II,261); THE LAST CONVERTIBLE (II,1435); THE MILLIONAIRE (II,1700); THE YELLOW ROSE (II,2847)

ALBERT, TRIGGER
THE MATT DENNIS SHOW (I,2948)

ALBERT, WIL
THE DUMPLINGS (II,743); LANDON, LANDON & LANDON (II,1426)

ALBERTS, BILLY
ACROBAT RANCH (I,20)

ALBERTS, VALERIE
ACROBAT RANCH (I,20); JUNIOR RODEO (I,2494)

ALBERTSON, FRANK
THE HOUSE NEXT DOOR (I,2139); PETER HUNTER, PRIVATE EYE (I,3563); SHORT, SHORT DRAMA (I,4023)

ALBERTSON, GRACE
KAREN (I,2505); OCTAVIUS AND ME (I,3327); THE TYCOON (I,4660)

ALBERTSON, JACK
ALCOA/GOODYEAR THEATER (I,107); BOB HOPE SPECIAL: BOB HOPE'S ALL-STAR COMEDY TRIBUTE TO VAUDEVILLE (II,302); CHICO AND THE MAN (II,508); DOBIE GILLIS (I,1302); DOCTOR SIMON LOCKE (I,1317); ELVIS REMEMBERED: NASHVILLE TO HOLLYWOOD (II,772); ENSIGN O'TOOLE (I,1448); THE FIRST 50 YEARS (II,862); GRANDPA GOES TO WASHINGTON (II,1050); HEY, LANDLORD (I,2039); THE INNER SANCTUM (I,2216); JACK CARTER AND COMPANY (I,2305); LOCK, STOCK, AND BARREL (I,2736); MARRIAGE IS ALIVE AND WELL (II,1633); MITZI AND A HUNDRED GUYS (II,1710); THE MONK (I,3087); MR. ED (I,3137); THE OATH: THE SAD AND LONELY SUNDAYS (II,1874); ONCE UPON A DEAD MAN (I,3369); THE PAUL LYNDE COMEDY HOUR (II,1977); ROOM FOR ONE MORE (I,3842); THE SLOWEST GUN IN THE WEST (I,4088); THE STAR MAKER (I,4193); TERROR AT ALCATRAZ (II,2563); THE THIN MAN (I,4446); UPTOWN (II,2710)

ALBERTSON, MABEL
ACCIDENTAL FAMILY (I,17); THE ANDY GRIFFITH SHOW (I,192); BEWITCHED (I,418); PETE 'N' TILLIE (II,2027); THAT GIRL (II,2570); THAT'S MY BOY (I,4427); THOSE WHITING GIRLS (I,4471); THE TOM EWELL SHOW (I,4538)

ALBIN, ABBY
THE BOB NEWHART SHOW (I,666)

ALBIN, DOLORES
THICKER THAN WATER (I,4445)

ALBRIGHT, LOLA
ABC'S MATINEE TODAY (I,7); ALFRED HITCHCOCK PRESENTS (I,115); THE ARMSTRONG CIRCLE THEATER (I,260); ARROYO (I,266); BRANDED (I,707); DELTA COUNTY, U.S.A. (II,657); THE GENERAL ELECTRIC THEATER (I,1753); KRAFT MYSTERY THEATER (I,2589); MIGHTY O (I,3036); NO WARNING (I,3301); PEPSI-COLA PLAYHOUSE (I,3523); PETER GUNN (I,3562); PEYTON PLACE (I,3574); READY AND WILLING (II,2115); SCREEN DIRECTOR'S PLAYHOUSE (I,3946); TALES OF TOMORROW (I,4352)

ALBRITTON, LOUISE
CONCERNING MISS MARLOWE (I,1032); ROBERT MONTGOMERY PRESENTS YOUR LUCKY STRIKE THEATER (I,3809); SILVER THEATER (I,4051); STAGE DOOR (I,4179)

ALCALDE, MARIO
MCCLOUD: WHO KILLED MISS U.S.A.? (I,2965)

ALCROFT, JAMIE
ALL NIGHT RADIO (II,40); ROCK-N-AMERICA (II,2195)

ALDA, ALAN
6 RMS RIV VU (II,2371); ANNIE AND THE HOODS (II,91); CBS: ON THE AIR (II,460); THE FOUR SEASONS (II,915); HIGHER AND HIGHER, ATTORNEYS AT LAW (I,2057); HOTEL 90 (I,2130); LILY (I,2705); M*A*S*H (II,1569); MARLO THOMAS AND FRIENDS IN FREE TO BE. . . YOU AND ME (II,1632); WHERE'S EVERETT? (I,4835)

ALDA, ANTONY
BUNGLE ABBEY (II,396); HOME ROOM (II,1172); THREE COINS IN THE FOUNTAIN (I,4473)

ALDA, BEATRICE
THE FOUR SEASONS (II,915)

ALDA, ELIZABETH
THE FOUR SEASONS (II,915)

ALDA, GENE
GULF PLAYHOUSE (I,1904)

ALDA, ROBERT
THE BOB HOPE SHOW (I,520); BY POPULAR DEMAND (I,784); CAN DO (I,815); CODE RED (II,553); THE FACTS OF LIFE (II,805); FAITH BALDWIN'S THEATER OF ROMANCE (I,1515); FAME (I,1517); LOVE OF LIFE (I,2774); LUCY MOVES TO NBC (II,1562); M*A*S*H (II,1569); PERSONALITY PUZZLE (I,3554); RHODA (II,2151); THE ROCK RAINBOW (II,2194); SAY IT WITH ACTING (I,3928); SECRET FILE, U.S.A. (I,3963); THE SECRET STORM (I,3969); SUPERTRAIN (II,2504); SUPERTRAIN (II,2505); WHAT'S YOUR BID? (I,4825); A YEAR AT THE TOP (II,2845)

ALDEN, HORTENSE
THE CAT AND THE CANARY (I,868)

ALDEN, JANE
COUNTERATTACK: CRIME IN AMERICA (II,581); HILL STREET BLUES (II,1154)

ALDEN, LYNN
YOU'RE ONLY YOUNG ONCE (I,4969)

ALDEN, NORMAN
330 INDEPENDENCE S.W. (I,4486); DEVLIN (II,663); ELECTRA WOMAN AND DYNA GIRL (II,764); FAY (II,844); THE FESS PARKER SHOW (II,850); FLAMINGO ROAD (II,871); THE GREATEST AMERICAN HERO (II,1060); JERRY (II,1283); MARY HARTMAN, MARY HARTMAN (II,1638); MURDOCK'S GANG (I,3164); MY THREE SONS (I,3205); THE NEW SUPER FRIENDS HOUR (II,1830); PANIC! (I,3447); THE PIGEON (I,3596); THE PLANT FAMILY (II,2050); THE PSYCHIATRIST: GOD BLESS THE CHILDREN (I,3685); RANGO (I,3721); SAMURAI (II,2249); SUPER FRIENDS (II,2497)

ALDEN, ROBERT
NANCY ASTOR (II,1788)

ALDERMAN, JANE
THE AWAKENING LAND (II,122)

ALDERSON, BROOKE
CONDO (II,565); HARRY'S BATTLES (II,1100)

ALDERSON, JOHN
LAST STAGECOACH WEST (I,2629); SECRET AGENT (I,3961)

ALDON, MARI
TONIGHT IN HAVANA (I,4557)

ALDRED, JOEL
MAN BEHIND THE BADGE (I,2857)

ALDREDGE, TOM
HENRY WINKLER MEETS WILLIAM SHAKESPEARE (II,1129)

ALDRICH, RHONDA
THE BOYS IN BLUE (II,359)

ALDRIDGE, MICHAEL
TINKER, TAILOR, SOLDIER, SPY (II,2621)

ALEC HOUSTON'S WILDCATS
THE JIMMY DEAN SHOW (I,2384)

ALEJANDRO, MIGUEL
YUMA (I,4976)

ALEKSANDER, GRANT
THE GUIDING LIGHT (II,1064)

ALENDE, TOMMY
CAP'N AHAB (I,824)

ALEONG, AKI
V: THE SERIES (II,2715)

ALETTER, FRANK
THE BANANA SPLITS ADVENTURE HOUR (I,340); THE BIG BRAIN (I,424); BRINGING UP BUDDY (I,731); THE CARA WILLIAMS SHOW (I,843); HUNTER (II,1204); IT'S ABOUT TIME (I,2263); NANCY (I,3221); RICH MAN, POOR MAN—BOOK I (II,2161)

ALETTER, KYLE
CIRCUS OF THE STARS (II,538)

ALETTER, LESLIE
CIRCUS OF THE STARS (II,533); CIRCUS OF THE STARS (II,536)

ALEXANDER, BEN
ABOUT FACES (I,12); DRAGNET (I,1369); THE FELONY SQUAD (I,1563); PARTY TIME AT CLUB ROMA (I,3465); PEOPLE (I,3517)

ALEXANDER, CHERISH
DOUG HENNING'S WORLD OF MAGIC V (II,729)

ALEXANDER, CRIS
WONDERFUL TOWN (I,4893)

ALEXANDER, DAVID
THE BEST CHRISTMAS PAGEANT EVER (II,212)

ALEXANDER, DENISE
DAYS OF OUR LIVES (II,629); GENERAL HOSPITAL (II,964); LIGHT'S OUT (I,2699); TACK REYNOLDS (I,4329)

ALEXANDER, HARRY
BEHIND THE SCREEN (II,200)

ALEXANDER, JANE
GEORGE WASHINGTON (II,978); MIRACLE ON 34TH STREET (I,3058)

ALEXANDER, JASON
E/R (II,748)

ALEXANDER, JEAN
CORONATION STREET (I,1054)

ALEXANDER, JOAN
THE DOCTORS (II,694); THE NEW ADVENTURES OF SUPERMAN (I,3255)

ALEXANDER, JOHN
ARSENIC AND OLD LACE (I,268); THE RIGHT MAN (I,3790)

ALEXANDER, KATHERINE
THE NASH AIRFLYTE THEATER (I,3227)

ALEXANDER, KEITH
U.F.O. (I,4662)

ALEXANDER, KIRK
ALL THE RIVERS RUN (II,43)

ALEXANDER, LAMAR
JOHNNY CASH: THE FIRST 25 YEARS (II,1336)

ALEXANDER, MILLETTE
THE EDGE OF NIGHT (II,760); FROM THESE ROOTS (I,1688); THE GUIDING LIGHT (II,1064); THE MILLION DOLLAR INCIDENT (I,3045)

ALEXANDER, NICK
THE GREEN FELT JUNGLE (I,1885)

ALEXANDER, NORMAN
THE JERRY REED WHEN YOU'RE HOT, YOU'RE HOT HOUR (I,2372)

ALEXANDER, ROD
BABES IN TOYLAND (I,318); THE BIG TIME (I,434); THE CHOCOLATE SOLDIER (I,944); GOOD TIMES (I,1853); HOLIDAY (I,2074); KALEIDOSCOPE (I,2504); LADY IN THE DARK (I,2602); MAX LIEBMAN PRESENTS (I,2958); THE MERRY WIDOW (I,3012); NAUGHTY MARIETTA (I,3232); PANORAMA (I,3448); STEP ON THE GAS (I,4215); TEXACO COMMAND PERFORMANCE (I,4407); VICTOR BORGE'S

COMEDY IN MUSIC III (I,4725); YOUR SHOW OF SHOWS (I,4957); YOUR SHOW OF SHOWS (II,2875)

ALEXANDER, SANDRA
DOMESTIC LIFE (II,703)

ALEXANDER, SHARON
BARBARA MANDRELL AND THE MANDRELL SISTERS (II,143)

ALEXANDER, TERENCE
ANOTHER WORLD (II,97); BERGERAC (II,209); THE FORSYTE SAGA (I,1645); IKE (II,1223); OUR FAMILY BUSINESS (II,1931); THE PALLISERS (II,1944); THE SEVEN DIALS MYSTERY (II,2308)

ALEXANDRA, TIANA
FLY AWAY HOME (II,893); PEARL (II,1986)

ALFASA, JOE
MORE WILD WILD WEST (II,1733)

ALFORD, KRISTI
GEORGE WASHINGTON (II,978)

ALFORD, PHIL
THE TYREES OF CAPITOL HILL (I,4661)

ALFORD, ROBERT
AS THE WORLD TURNS (II,110)

ALI, MUHAMMAD
I AM THE GREATEST: THE ADVENTURES OF MUHAMMAD ALI (II,1211); LOLA (II,1508); THE MUHAMMAD ALI VARIETY SPECIAL (II,1762)

ALICE, MARY
JOSHUA'S WORLD (II,1344); SANFORD AND SON (II,2256)

ALICIA, ANA
BATTLE OF THE NETWORK STARS (II,177); CONDOMINIUM (II,566); FALCON CREST (II,808); ROUGHNECKS (II,2219)

ALINDER, DALLAS
QUICK AND QUIET (II,2099); THE RAINBOW GIRL (II,2108)

ALISON, DOROTHY
A TOWN LIKE ALICE (II,2652)

ALLAN DAVIES DANCERS, THE
THE PEARL BAILEY SHOW (I,3501)

ALLAN, ANDREA
U.F.O. (I,4662)

ALLAN, JED
BRENDA STARR (II,372); CELEBRITY BOWLING (I,883); CONSPIRACY OF TERROR (II,569); DAYS OF OUR LIVES (II,629); GOING PLACES (I,1834); LASSIE (I,2622); RANSOM FOR A DEAD MAN (I,3722)

ALLAN, SIAN BARBARA
ERIC (I,1451)

ALLDREDGE, MICHAEL
FAMILY TIES (II,819); FLY AWAY HOME (II,893); HARD KNOCKS (II,1086); LOVE IN A COLD CLIMATE (II,1539)

ALLEGRETTI, COSMO
MR. MAYOR (I,3143)

ALLEGRETTI, COSMO F
CAPTAIN KANGAROO (II,434); GOOD EVENING, CAPTAIN (II,1031); WAKE UP (II,2736)

ALLEGRO, QUINTETTO
THE BOB HOPE SHOW (I,538)

ALLEN III, VERRETT
THE TRAVELS OF JAIMIE MCPHEETERS (I,4596)

ALLEN SISTERS, THE
THE TOMMY HUNTER SHOW (I,4552)

ALLEN, ADRIENNE
TELLER OF TALES (I,4388); YOU ASKED FOR IT (II,2855)

ALLEN, ARTHUR
BORDER COLLIE (I,688)

ALLEN, BILLIE
THE EDGE OF NIGHT (II,760)

ALLEN, BOB
THE GREATEST GIFT (I,1880)

ALLEN, BYRON
BATTLE OF THE NETWORK STARS (II,171); HOLLYWOOD STARS' SCREEN TESTS (II,1165); REAL PEOPLE (II,2118); SALUTE TO LADY LIBERTY (II,2243)

ALLEN, CASEY
KITTY FOYLE (I,2571)

ALLEN, CHARLOTTE
MISTRAL'S DAUGHTER (II,1708)

ALLEN, CHET
AMAHL AND THE NIGHT VISITORS (I,148); BONINO (I,685); THE TROUBLESHOOTERS (I,4614)

ALLEN, CHRIS
ADVENTURES OF HOPPITY HOPPER FROM FOGGY BOGG (I,62); THE BUFFALO BILLY SHOW (I,756); JACK AND THE BEANSTALK (I,2288); SPACE KIDDETTES (I,4143)

ALLEN, COREY
MEN INTO SPACE (I,3007)

ALLEN, DAVE
DAVE ALLEN AT LARGE (II,621)

ALLEN, DAVID
THE CHISHOLMS (II,512)

ALLEN, DAYTON
THE ADVENTURES OF OKY DOKY (I,70); THE ASTRONAUT SHOW (I,307); CBS CARTOON THEATER (I,878); DEPUTY DAWG (I,1233); HOWDY DOODY (I,2151); THE MILTON THE MONSTER CARTOON SHOW (I,3053); THE NEW HOWDY DOODY SHOW (II,1818); THE STEVE ALLEN SHOW (I,4219); WINKY DINK AND YOU (I,4879); WINKY DINK AND YOU (I,4880)

ALLEN, DEBBIE
BATTLE OF THE NETWORK STARS (II,174); BATTLE OF THE NETWORK STARS (II,175); BEN VEREEN—HIS ROOTS (II,202); CELEBRITY (II,463); EBONY, IVORY AND JADE (II,755); FAME (II,812); THE JIM STAFFORD SHOW (II,1291); JOHN SCHNEIDER'S CHRISTMAS HOLIDAY (II,1320); THE KIDS FROM FAME (II,1392); LORETTA LYNN IN THE BIG APPLE (II,1520); PARADE OF STARS (II,1954); ROOTS: THE NEXT GENERATIONS (II,2213); TEXACO STAR THEATER: OPENING NIGHT (II,2565); THREE GIRLS THREE (II,2608)

ALLEN, DENNIS
ROWAN AND MARTIN'S LAUGH-IN (I,3856); WE INTERRUPT THIS SEASON (I,4780); WHAT'S IT ALL ABOUT WORLD? (I,4816)

ALLEN, ELIZABETH
BATTLE OF THE NETWORK STARS (II,164); BRACKEN'S WORLD (I,703); C.P.O. SHARKEY (II,407); JACKIE GLEASON AND HIS AMERICAN SCENE MAGAZINE (I,2321); THE PAUL LYNDE SHOW (I,3489); TEXAS (II,2566); THRILLER (I,4492)

ALLEN, FRED
DATELINE (I,1165); ENTERTAINMENT —1955 (I,1450); JUDGE FOR YOURSELF (I,2466); SOUND OFF TIME (I,4135); TWO FOR THE MONEY (I,4654)

ALLEN, GARY
HARPER VALLEY PTA (II,1095); THE NEW MAVERICK (II,1822); THE NIGHT RIDER (II,1848); NOW WE'RE COOKIN' (II,1866); WHEN THE WHISTLE BLOWS (II,2781)

ALLEN, GEORGE
BIZARRE (II,258); FOREST RANGERS (I,1642)

ALLEN, GRACIE
THE GEORGE BURNS AND GRACIE ALLEN SHOW (I,1763)

ALLEN, HENRY 'RED'
TIMEX ALL-STAR JAZZ SHOW II (I,4516)

ALLEN, HERB
HAIL THE CHAMP (I,1918)

ALLEN, JEFF
THE DOCTORS (II,694)

ALLEN, JIMMY
KOBB'S CORNER (I,2577)

ALLEN, JOHN
A TOWN LIKE ALICE (II,2652)

ALLEN, JONELLE
BATTLE OF THE NETWORK STARS (II,170); CAGNEY AND LACEY (II,409); COTTON CLUB '75 (II,580); OPRYLAND USA—1975 (II,1920); PALMERSTOWN U.S.A. (II,1945); SPARROW (II,2411)

ALLEN, JOSEPH
MR. AND MRS. NORTH (I,3124)

ALLEN, KAREN
EAST OF EDEN (II,752); KNOTS LANDING (II,1404)

ALLEN, KEITH
FAT ALBERT AND THE COSBY KIDS (II,835)

ALLEN, LEE
THE BELLE OF 14TH STREET (I,390); THE CLIFF DWELLERS (I,974)

ALLEN, LINDA LOU
THE MARTIAN CHRONICLES (II,1635)

ALLEN, LIZ
ED MCMAHON AND HIS FRIENDS. . . DISCOVER WET AT CYPRESS GARDENS (I,1400)

ALLEN, LYNN
THINGS WE DID LAST SUMMER (II,2589)

ALLEN, MARILYN
DYNASTY (II,746)

ALLEN, MARK
THE TRAVELS OF JAIMIE MCPHEETERS (I,4596)

ALLEN, MARTY
BENNY AND BARNEY: LAS VEGAS UNDERCOVER (II,206); THE BOB HOPE SHOW (I,633); CIRCUS OF THE STARS (II,530); CIRCUS OF THE STARS (II,531); CIRCUS OF THE STARS (II,532); CIRCUS OF THE STARS (II,533); CIRCUS OF THE STARS (II,534); CIRCUS OF THE STARS (II,535); CIRCUS OF THE STARS (II,536); CIRCUS OF THE STARS (II,537); HELLO DERE (I,2010); MISTER JERICO (I,3070); MITZI AND A HUNDRED GUYS (II,1710); MURDER CAN HURT YOU! (II,1768); STEVE MARTIN: COMEDY IS NOT PRETTY (II,2462)

ALLEN, MCKENZIE
RYAN'S HOPE (II,2234)

ALLEN, MEL
LET'S CELEBRATE (I,2662)

ALLEN, MICHAEL
THE GUIDING LIGHT (II,1064)

ALLEN, PATRICK
GLENCANNON (I,1822); SIGN IT DEATH (I,4046); U.F.O. (I,4662)

ALLEN, PATRIE
WKRP IN CINCINNATI (II,2814)

ALLEN, PETER
LORETTA LYNN IN THE BIG APPLE (II,1520)

ALLEN, PHILIP R
THE BAD NEWS BEARS (II,134); SHEILA (II,2325); THE SIX OF US (II,2369); SNAFU (II,2386); WASHINGTON: BEHIND CLOSED DOORS (II,2744)

ALLEN, RAE
ACE (II,5); MARLO THOMAS IN ACTS OF LOVE—AND OTHER COMEDIES (I,2919); PHYL AND MIKHY (II,2037); THE RAINBOW GIRL (II,2108)

ALLEN, RAYMOND
SANFORD AND SON (II,2256); SANFORD ARMS (II,2257)

ALLEN, REX
FIVE STAR JUBILEE (I,1589); FRONTIER DOCTOR (I,1698); WHEN THE WEST WAS FUN: A WESTERN REUNION (II,2780)

ALLEN JR, REX
THE CBS NEWCOMERS (I,879); MUSIC CITY NEWS TOP COUNTRY HITS OF THE YEAR (II,1772); OPERATION GREASEPAINT (I,3402)

ALLEN, RICKY
PETE KELLY'S BLUES (I,3559); TIMMY AND LASSIE (I,4520); WHERE THERE'S SMOKEY (I,4832)

ALLEN, RONALD
ROMEO AND JULIET (I,3837)

ALLEN, SCOTT ARTHUR
THE WILD WOMEN OF CHASTITY GULCH (II,2801)

ALLEN, SETH
THE BLUE KNIGHT (II,277); MASQUERADE (I,2940); NIGHTSIDE (I,3296)

ALLEN, SHEILA
THE ASSASSINATION RUN (II,112); THE WALTONS (II,2740)

ALLEN, SHERRY
THE MICKEY MOUSE CLUB (I,3025)

ALLEN, STEVE
THE APARTMENT HOUSE (I,241); THE ARTHUR GODFREY SPECIAL (II,109); ARTHUR GODFREY'S TALENT SCOUTS (I,291); THE BIG SHOW (II,243); THE BOB HOPE SHOW (I,549); THE BOB HOPE SHOW (I,551); THE BOB HOPE SHOW (I,588); THE BOB HOPE SHOW (I,633); THE COMEDY ZONE (II,559); DANGER (I,1134); EVENING AT THE IMPROV (II,786); FANFARE (I,1526); THE GOLDDIGGERS (I,1838); GOOD TIMES (I,1853); HAVE I GOT A CHRISTMAS FOR YOU (II,1105); I'VE GOT A SECRET (I,2283); I'VE GOT A SECRET (I,2284); I'VE HAD IT UP TO HERE (II,1221); THE JONATHAN WINTERS SHOW (I,2442); THE JUNE ALLYSON SHOW (I,2488); LIFE'S MOST EMBARRASSING MOMENTS II (II,1471); LIFE'S MOST EMBARRASSING MOMENTS III (II,1472); LIFE'S MOST EMBARRASSING MOMENTS IV (II,1473); THE LUCILLE BALL SPECIAL (II,1559); THE MAD MAD MAD MAD WORLD OF THE SUPER BOWL (II,1586); MITZI AND A HUNDRED GUYS (II,1710); THE NEW STEVE ALLEN SHOW (I,3271); ONE MAN SHOW (I,3381); PLIMPTON! DID YOU HEAR THE ONE ABOUT. . .? (I,3629); RICH MAN, POOR MAN—BOOK I (II,2161); ROMP (I,3838); SHOOT-IN AT NBC (A BOB HOPE SPECIAL) (I,4021); SOLID GOLD (II,2395); SOLID GOLD '79 (II,2396); SONGS FOR SALE (I,4124); THE STEVE ALLEN COMEDY HOUR (I,4218); THE STEVE ALLEN COMEDY HOUR (II,2454); THE STEVE ALLEN SHOW (I,4219); THE STEVE ALLEN SHOW (I,4220); THE STEVE ALLEN SHOW (I,4221); THE STEVE ALLEN SHOW (I,4222); STEVE ALLEN'S LAUGH-BACK (II,2455); STEVE MARTIN: COMEDY IS NOT PRETTY (II,2462); STONE (II,2470); SUNDAY IN TOWN (I,4285); TALENT PATROL (I,4340); TEXACO COMMAND PERFORMANCE (I,4407); TEXACO STAR THEATER: OPENING NIGHT (II,2565); TIMEX ALL-STAR JAZZ SHOW I (I,4515); TONIGHT PREVIEW (I,4560); THE TONIGHT SHOW (I,4561); THE TONIGHT SHOW (I,4562); VAUDEVILLE (II,2722); WARNING SHOT (I,4770)

ALLEN, VALERIE
ELFEGO BACA (I,1427)

ALLEN, VERA
ANOTHER WORLD (II,97); FROM THESE ROOTS (I,1688); THE O'NEILLS (I,3392); SEARCH FOR TOMORROW (II,2284)

ALLEN, WOODY
THE BEST ON RECORD (I,409); HIPPODROME (I,2060); HOT DOG (I,2122); PLIMPTON! DID YOU HEAR THE ONE ABOUT. . .? (I,3629); THE SENSATIONAL, SHOCKING, WONDERFUL, WACKY 70S (II,2302); THE WOODY ALLEN SPECIAL (I,4904)

ALLENBY, PEGGY
ABC-POWERS CHARM SCHOOL (I,5); COSMOPOLITAN THEATER (I,1057); FIRST LOVE (I,1583); HERB SHRINER TIME (I,2020); SHORT, SHORT DRAMA (I,4023)

ALLENDE, FERNANDO
FLAMINGO ROAD (II,872); MASTER OF THE GAME (II,1647); THE PHOENIX (II,2035)

ALLEY, KIRSTIE
HIGHWAY HONEYS (II,1151); MASQUERADE (II,1644)

ALLIN, HENRY
CALL TO DANGER (I,803)

ALLINSON, MICHAEL
AFTER HOURS: GETTING TO KNOW US (II,21); ELIZABETH THE QUEEN (I,1431); GEORGE WASHINGTON (II,978); LOVE OF LIFE (I,2774)

ALLISON, BETSI
THREE ABOUT TOWN (I,4472)

ALLISON, FRAN
DAMN YANKEES (I,1128); DON MCNEILL'S TV CLUB (I,1337); KUKLA, FRAN, AND OLLIE (I,2594); THE KUKLAPOLITAN EASTER SHOW (I,2595); LET'S DANCE (I,2664); PINOCCHIO (I,3602)

ALLISON, JONE
THE GUIDING LIGHT (II,1064)

ALLISON, KEITH
HAPPENING '68 (I,1936)

ALLISON, LISA
WONDER GIRL (I,4891)

ALLISON, PATRICIA
OUR FIVE DAUGHTERS (I,3413)

ALLMAN, ELVIA
THE ABBOTT AND COSTELLO SHOW (I,2); THE ADDAMS FAMILY (II,13); BLONDIE (I,486); THE CLAUDETTE COLBERT SHOW (I,971); THE DICK VAN DYKE SHOW (I,1275); THE FIRST HUNDRED YEARS (I,1582); THE FREEWHEELERS (I,1683); THE GEORGE BURNS AND GRACIE ALLEN SHOW (I,1763); GIMME A BREAK (II,995); I MARRIED JOAN (I,2174); THE ODD COUPLE (II,1875); THE PEOPLE'S CHOICE (I,3521); PETTICOAT JUNCTION (I,3571); PINE LAKE LODGE (I,3597); STARSTRUCK (II,2446); TOPPER (I,4582)

ALLMAN, FRIEDA
SEARCH FOR TOMORROW (II,2284)

ALLMAN, SHELDON
ADAMS OF EAGLE LAKE (II,10); THE C.B. BEARS (II,406); HARRIS AGAINST THE WORLD (I,1956)

ALLPORT, CHRISTOPHER
ANOTHER WORLD (II,97); THE CHISHOLMS (II,512); A GIRL'S LIFE (II,1000); HECK'S ANGELS (II,1122); LOVE, NATALIE (II,1545)

ALLRUD, ROMOLA ROBB
LOVE OF LIFE (I,2774)

ALLSTAR, LEONARD
THE NEW ERNIE KOVACS SHOW (I,3264)

ALLWINE, WAYNE
THE NEW MICKEY MOUSE CLUB (II,1823)

ALLYSON, JUNE
20TH CENTURY FOLLIES (I,4641); THE DICK POWELL SHOW (I,1269); THE GENERAL MOTORS 50TH ANNIVERSARY SHOW (I,1758); THE GREAT MYSTERIES OF HOLLYWOOD (II,1058); THE JUNE ALLYSON SHOW (I,2488); THE KID WITH THE BROKEN HALO (II,1387); LETTERS FROM THREE LOVERS (I,2671); THE PERRY COMO SPECIAL (I,3534); VEGAS (II,2723); ZANE GREY THEATER (I,4979)

ALMAGOR, GILLA
A WOMAN CALLED GOLDA (II,2818)

ALMANZAR, JAMES
DOCTORS HOSPITAL (II,691); THE HIGH CHAPARRAL (I,2051)

ALMEDIA, LAURINDO
THE SHIRLEY BASSEY SHOW (I,4015)

ALOMAR, CARLOS
DAVID BOWIE—SERIOUS MOONLIGHT (II,622)

ALPERN, SUSAN
THE BELLE OF 14TH STREET (I,390); GRANDPA MAX (II,1051)

ALPERT, ARTHUR
WE INTERRUPT THIS SEASON (I,4780)

ALPERT, DAVID
THE ALAN YOUNG SHOW (I,104)

ALPERT, HERB
BEAT OF THE BRASS (I,376); THE BRASS ARE COMING (I,708); HERB ALBERT AND THE TIJUANA BRASS (I,2019); HERB ALPERT AND THE TIJUANA BRASS (II,1130)

ALPERT, WILL
STEVE MARTIN: COMEDY IS NOT PRETTY (II,2462)

ALRICH, RANDY
EMERGENCY PLUS FOUR (II,774)

ALSTON, BARBARA
ROSETTI AND RYAN: MEN WHO LOVE WOMEN (II,2217)

ALT, JOAN
THESE ARE MY CHILDREN (I,4440)

ALTAY, DERIN
THE AWAKENING LAND (II,122); THE BAXTERS (II,183)

ALTERMAN, STEVE
KIDD VIDEO (II,1388)

ALTHOFF, CHARLIE
VILLAGE BARN (I,4733)

ALTIERI, ANN
CHARLIE BROWN'S ALL STARS (I,910); YOU'RE IN LOVE, CHARLIE BROWN (I,4964)

ALTMAN, FRIEDA
THE TRAP (I,4593)

ALTMAN, JEFF
ALL'S FAIR (II,46); ANSON AND LORRIE (II,98); BULBA (II,392); COS (II,576); THE DUKES OF HAZZARD (II,742); GOSSIP (II,1043); GOSSIP (II,1044); PINK LADY (II,2041); SCARED SILLY (II,2269); THE STARLAND VOCAL BAND (II,2441)

ALTO, BOBBY
GOOD PENNY (II,1036)

ALTON, JOHN
GEORGE WASHINGTON (II,978)

ALTON, KENNETH
MACKENZIE'S RAIDERS (I,2803)

ALTON, ZOHRA
THE CRADLE SONG (I,1088); A TIME TO LIVE (I,4510)

ALVARADO, GINA
CUTTER TO HOUSTON (II,612); MISSING PIECES (II,1706)

ALVAREZ, ABRAHAM
ARCHIE BUNKER'S PLACE (II,105)

ALVAREZ, ANITA
THROUGH THE CRYSTAL BALL (I,4494)

ALVILA, CHRISTINA
THE LAZARUS SYNDROME (II,1451)

ALZADO, LYLE
SHE'S WITH ME (II,2321)

ALZAMORA, ARMAND
THE LEATHERNECKS (I,2647); THE TEAHOUSE OF THE AUGUST MOON (I,4368)

AMATEAU, ROD
HIGHWAY HONEYS (II,1151)

AMATO, JULIE
20TH CENTURY FOLLIES (I,4641); HALF THE GEORGE KIRBY COMEDY HOUR (I,1920)

AMATO, TONY
THE BOBBY DARIN AMUSEMENT COMPANY (I,672)

AMBER, LEE
BRACKEN'S WORLD (I,703)

AMECHE, DON
BOSTON AND KILBRIDE (II,356); COKE TIME WITH EDDIE FISHER (I,996); THE DON AMECHE THEATER (I,1331); DON'S MUSICAL PLAYHOUSE (I,1348); THE FRANCES LANGFORD SHOW (I,1657); THE FRANCES LANGFORD-DON AMECHE SHOW (I,1655); GIDGET GETS MARRIED (I,1796); HIGH BUTTON SHOES (I,2050); HOLIDAY HOTEL (I,2075); INTERNATIONAL SHOWTIME (I,2225); THE JACK CARSON SHOW (I,2304); JUNIOR MISS (I,2493); NOT IN FRONT OF THE KIDS (II,1861); SHEPHERD'S FLOCK (I,4007); STARTIME (I,4212); TAKE A CHANCE (I,4331); TOO YOUNG TO GO STEADY (I,4574)

AMECHE, JIM
FESTIVAL OF STARS (I,1568)

AMENDOLIA, DON
MAMA MALONE (II,1609)

AMERICAN FOLLIES BERGERE GIRLS, THE
GENE KELLY'S WONDERFUL WORLD OF GIRLS (I,1752)

AMERICAN STRING QUARTET, THE
STEVE MARTIN'S BEST SHOW EVER (II,2460)

AMES BROTHERS, THE
THE BIG PARTY FOR REVLON (I,427); THE RED SKELTON REVUE (I,3754)

AMES, AMANDA
THE RIFLEMAN (I,3789)

AMES, BARBARA
FORD THEATER HOUR (I,1635); MAGNAVOX THEATER (I,2827)

AMES, BETSY
KENNEDY (II,1377)

AMES, ED
THE AMES BROTHERS SHOW (I,178); ANDROCLES AND THE LION (I,189); DANIEL BOONE (I,1142)

AMES, FLORENZ
THE ADVENTURES OF ELLERY QUEEN (I,59); BLONDIE (I,486); THE HARDY BOYS AND THE MYSTERY OF THE APPLEGATE TREASURE (I,1951); THE LIFE OF VERNON HATHAWAY (I,2687)

AMES, GENE
THE AMES BROTHERS SHOW (I,178)

AMES, JOE
THE AMES BROTHERS SHOW (I,178)

AMES, JOYCE
DEAN MARTIN PRESENTS THE GOLDDIGGERS (I,1194); THE FUNNY SIDE (I,1714)

AMES, JUDITH
IF YOU KNEW TOMORROW (I,2187)

AMES, LEON
THE BARBARA STANWYCK THEATER (I,350); BEWITCHED (I,418); FATHER OF THE BRIDE (I,1543); FRONTIER JUDGE (I,1699); THE JEFFERSONS (II,1276); LIFE WITH FATHER (I,2690); MAGGIE (I,2810); MR. ED (I,3137); NO WARNING (I,3301); SCREEN DIRECTOR'S PLAYHOUSE (I,3946); STARS OVER HOLLYWOOD (I,4211); THE WIDE OPEN DOOR (I,4859)

AMES, NANCY
THE BOB HOPE SHOW (I,628); MUSIC BY COLE PORTER (I,3166); THE PERRY COMO WINTER SHOW (I,3542); THAT WAS THE WEEK THAT WAS (I,4422); THAT WAS THE WEEK THAT WAS (I,4423); TIN PAN ALLEY TODAY (I,4521)

AMES, RACHEL
GENERAL ELECTRIC TRUE (I,1754); GENERAL HOSPITAL (II,964); THE LINE-UP (I,2707); MIKE AND THE MERMAID (I,3039)

AMES, SHELLY
FATHER OF THE BRIDE (I,1543)

AMES, TEAL
THE EDGE OF NIGHT (II,760)

AMES, TRUDI
KAREN (I,2505)

AMES, VIC
THE AMES BROTHERS SHOW (I,178)

AMEY, MARLENA
BOSTON AND KILBRIDE (II,356)

AMOR, CHRISTINE
PRISONER: CELL BLOCK H (II,2085)

AMOROSA, JOHNNY
SAMMY KAYE'S MUSIC FROM MANHATTAN (I,3902); THE VINCENT LOPEZ SHOW (I,4735)

AMOS, JOHN
THE COPS AND ROBIN (II,573); THE FUNNY SIDE (I,1714); THE FUNNY SIDE (I,1715); FUTURE COP (II,945); FUTURE COP (II,946); GOOD TIMES (II,1038); HUNTER (II,1206); KEEPING UP WITH THE JONESES (II,1374); THE MARY TYLER MOORE SHOW (II,1640); MAUDE (II,1655); ROOTS (II,2211); TWO'S COMPANY (I,4659)

AMOUR, JOHN
DAYS OF OUR LIVES (II,629)

AMSBERRY, BOB
THE HARDY BOYS AND THE MYSTERY OF GHOST FARM (I,1950); THE MICKEY MOUSE CLUB (I,3025)

AMSTERDAM, MOREY
BATTLE OF THE AGES (I,367); BROADWAY OPEN HOUSE (I,736); CAN YOU TOP THIS? (I,817); THE DICK VAN DYKE SHOW (I,1275); HERE IT IS, BURLESQUE! (II,1133); HOLLYWOOD STAR REVUE (I,2093); HONEYMOON SUITE (I,2108); MIXED NUTS (II,1714); THE MOREY AMSTERDAM SHOW (I,3100); THE MOREY AMSTERDAM SHOW (I,3101); ONE MAN SHOW (I,3381); RUDOLPH'S SHINY NEW YEAR (II,2228)

AMTEER, BARBARA
CAROL FOR ANOTHER CHRISTMAS (I,852)

AMYES, ISABELLE
LOVE IN A COLD CLIMATE (II,1539)

ANDELIN, JIM
THE BENNETS (I,397); HAIL THE CHAMP (I,1918); A TIME TO LIVE (I,4510)

ANDERMAN, MAUREEN
COCAINE AND BLUE EYES (II,548); EVERY STRAY DOG AND KID (II,792)

ANDERS, BETTY
MOTHERS DAY (I,3109)

ANDERS, IAIN
PENMARRIC (II,1993)

ANDERS, JOE
THE BIG TOP (I,435)

ANDERS, KAREN
THE BOB HOPE SHOW (I,661)

ANDERS, LAUREN
THE GARY COLEMAN SHOW (II,958)

ANDERS, LAURIE
THE KEN MURRAY SHOW (I,2525)

ANDERS, LOTT
THE BIG TOP (I,435)

ANDERS, LUANA
EVIL ROY SLADE (I,1485); MY LUCKY PENNY (I,3196); ONE STEP BEYOND (I,3388)

ANDERS, MARGOT
YOUNG DR. MALONE (I,4938)

ANDERS, MERRY
FBI CODE 98 (I,1550); HOW TO MARRY A MILLIONAIRE (I,2147); INNOCENT JONES (I,2217); IT'S ALWAYS JAN (I,2264); THE LORETTA YOUNG THEATER (I,2756); MICHAEL SHAYNE, DETECTIVE (I,3022); NEVER TOO YOUNG (I,3248); TROUBLE WITH FATHER (I,4610); TV READERS DIGEST (I,4630)

ANDERS, RUDOLPH
SPACE PATROL (I,4144)

ANDERSEN, MOANA
HAWAIIAN HEAT (II,1111)

ANDERSON, ALBERT
KOSTA AND HIS FAMILY (I,2581)

ANDERSON, BARBARA
DOCTORS' PRIVATE LIVES (II,692); DOCTORS' PRIVATE LIVES (II,693); IRONSIDE (I,2240); IRONSIDE (II,1246); MISSION: IMPOSSIBLE (I,3067); NIGHT GALLERY (I,3287)

ANDERSON, BILL
THE BETTER SEX (II,223); THE BILL ANDERSON SHOW (I,440); COUNTRY COMES HOME (II,585); COUNTRY NIGHT OF STARS (II,592); JOHNNY CASH: THE FIRST 25 YEARS (II,1336); ROY ACUFF—50 YEARS THE KING OF COUNTRY MUSIC (II,2223)

ANDERSON, BOB
WICHITA TOWN (I,4857)

ANDERSON, BRAD
THE FANTASTIC FUNNIES (II,825)

ANDERSON, BRIDGETTE
GUN SHY (II,1068)

ANDERSON, BROOKE
THE CHICAGO STORY (II,507)

ANDERSON, CARL
THAT'S TV (II,2581)

ANDERSON, CAROL
CAMP RUNAMUCK (I,811)

ANDERSON, CASEY
FOLK SOUND U.S.A. (I,1614)

ANDERSON, CAT
TIMEX ALL-STAR JAZZ SHOW IV (I,4518)

ANDERSON, CHERYL
MCCLAIN'S LAW (II,1659)

ANDERSON, DAME JUDITH
THE ELGIN HOUR (I,1428); HENRY FONDA PRESENTS THE STAR AND THE STORY (I,2016); MACBETH (I,2800); RHEINGOLD THEATER (I,3783); SANTA BARBARA (II,2258); TELEPHONE TIME (I,4379)

ANDERSON, DANIEL
IT'S YOUR FIRST KISS, CHARLIE BROWN (I,2280); YOU'RE THE GREATEST, CHARLIE BROWN (I,4973)

ANDERSON, DARYL
LOU GRANT (II,1526); THE PHOENIX (II,2035)

ANDERSON, DAVID
GROWING PAYNES (I,1891)

ANDERSON, DONNA
THE TRAVELS OF JAIMIE MCPHEETERS (I,4596)

ANDERSON, DOUGLAS
MAGIC MIDWAY (I,2818)

ANDERSON, EDDIE 'ROCHESTER'
GO! (I,1830); THE GREEN PASTURES (I,1887); THE HARLEM GLOBETROTTERS (I,1953); THE JACK BENNY PROGRAM (I,2294); JACK BENNY'S 20TH ANNIVERSARY TV SPECIAL (I,2303); JACK BENNY'S NEW LOOK (I,2301); LAST OF THE PRIVATE EYES (I,2628)

ANDERSON, ERIC
THE CLAUDETTE COLBERT SHOW (I,971)

ANDERSON, GEORGINE
THE PRIME OF MISS JEAN BRODIE (II,2082); THE WOMAN IN WHITE (II,2819)

ANDERSON, HARRY
CHEERS (II,494); DOCTOR STRANGE (II,688); MAGIC WITH THE STARS (II,1599); NIGHT COURT (II,1844); TWILIGHT THEATER (II,2685)

ANDERSON, HERBERT
DENNIS THE MENACE (I,1231); DOBIE GILLIS (I,1302); RYAN'S HOPE (II,2234); STRATEGIC AIR COMMAND (I,4252)

ANDERSON, INGRID
CONCRETE BEAT (II,562); COVER UP (II,597); MICKEY SPILLANE'S MIKE HAMMER: MORE THAN MURDER (II,1693); RIPTIDE (II,2178)

ANDERSON, JAMES
SWAMP FOX (I,4311)

ANDERSON, JANE
THE BILLY CRYSTAL COMEDY HOUR (II,248); P.O.P. (II,1939)

ANDERSON, JEAN
LITTLE WOMEN (I,2723)

ANDERSON, JOAN
OUR FIVE DAUGHTERS (I,3413)

ANDERSON, JOHN
BACKSTAIRS AT THE WHITE HOUSE (II,133); BEN CASEY (I,394); BOOTS AND SADDLES: THE STORY OF THE FIFTH CAVALRY (I,686); BRIDGER (II,376); BROCK'S LAST CASE (I,742); CALL TO DANGER (I,804); CHECKING IN (II,492); DEAD MAN ON THE RUN (II,631); EGAN (I,1419); HITCHED (I,2065); JOHNNY CASH: COWBOY HEROES (II,1334); THE LAST HURRAH (I,2626); THE LIFE AND LEGEND OF WYATT EARP (I,2681); THE LOG OF THE BLACK PEARL (II,1506); MRS R.—DEATH AMONG FRIENDS (II,1759); ONCE AN EAGLE (II,1897); THE OUTER LIMITS (I,3426); RICH MAN, POOR MAN—BOOK II (II,2162); THE RIFLEMAN (I,3789); SCALPLOCK (I,3930); SECOND CHANCE (I,3957); SMILE JENNY, YOU'RE DEAD (II,2382); WON'T IT EVER BE MORNING? (I,4902)

ANDERSON, JUDITH
THE BORROWERS (I,693); CAESAR AND CLEOPATRA (I,788); THE CRADLE SONG (I,1088); ELIZABETH THE QUEEN (I,1431); THE FILE ON DEVLIN (I,1575); LIGHT'S DIAMOND JUBILEE (I,2698); MACBETH (I,2801); THE MOON AND SIXPENCE

(I,3098); A SALUTE TO TELEVISION'S 25TH ANNIVERSARY (I,3893)

ANDERSON, LARRY
BROTHERS AND SISTERS (II,382); DRIBBLE (II,736)

ANDERSON, LEW
HOWDY DOODY (I,2151); HOWDY DOODY AND FRIENDS (I,2152); THE NEW HOWDY DOODY SHOW (II,1818)

ANDERSON, LONI
BOB HOPE SPECIAL: BOB HOPE'S ALL-STAR COMEDY BIRTHDAY PARTY (II,297); BOB HOPE SPECIAL: BOB HOPE'S ALL-STAR COMEDY LOOK AT THE FALL SEASON: IT'S STILL FREE AND WORTH IT! (II,299); BOB HOPE SPECIAL: BOB HOPE'S ALL-STAR LOOK AT TV'S PRIME TIME WARS (II,303); BOB HOPE SPECIAL: BOB HOPE'S SPRING FLING OF COMEDY AND GLAMOUR (II,315); BOB HOPE SPECIAL: BOB HOPE'S WICKI-WACKY SPECIAL FROM WAIKIKI (II,321); BOB HOPE SPECIAL: THE BOB HOPE CHRISTMAS SPECIAL (II,328); BOB HOPE SPECIAL: THE BOB HOPE CHRISTMAS SPECIAL (II,329); BOB HOPE SPECIAL: THE BOB HOPE CHRISTMAS SPECIAL (II,331); THE CANDID CAMERA SPECIAL (II,422); CIRCUS OF THE STARS (II,533); THE FANTASTIC FUNNIES (II,825); HARRY O (II,1099); MAGIC WITH THE STARS (II,1599); MERRY CHRISTMAS FROM THE GRAND OLE OPRY (II,1686); PARTNERS IN CRIME (II,1961); SIEGFRIED AND ROY (II,2350); SUNDAY FUNNIES (II,2493); TOM SNYDER'S CELEBRITY SPOTLIGHT (II,2632); WINNER TAKE ALL (II,2808); WKRP IN CINCINNATI (II,2814)

ANDERSON, LYNN
BERT CONVY SPECIAL—THERE'S A MEETING HERE TONIGHT (II,210); BOB HOPE SPECIAL: HAPPY BIRTHDAY, BOB! (II,326); A COUNTRY CHRISTMAS (II,584); COUNTRY GALAXY OF STARS (II,587); COUNTRY MUSIC HIT PARADE (I,1069); COUNTRY STARS OF THE 70S (II,594); DEAN MARTIN PRESENTS MUSIC COUNTRY, U.S.A. (I,1192); DEAN MARTIN'S CHRISTMAS AT SEA WORLD (II,644); THE LAWRENCE WELK SHOW (I,2643); LUCY COMES TO NASHVILLE (II,1561); MEMORIES WITH LAWRENCE WELK (II,1681); SKINFLINT (II,2377); TENNESSEE ERNIE FORD'S WHITE CHRISTMAS (I,4402)

ANDERSON, MARIAN
AMERICA PAUSES FOR THE MERRY MONTH OF MAY (I,167); CHRISTMAS STARTIME WITH LEONARD BERNSTEIN (I,949); DATELINE (I,1165); FESTIVAL OF MUSIC (I,1567); THE FORD 50TH ANNIVERSARY SHOW (I,1630)

ANDERSON, MARY
STAGE DOOR (I,4177)

ANDERSON, MELISSA SUE
ADVICE TO THE LOVELORN (II,16); THE ALL-STAR SALUTE TO MOTHER'S DAY (II,51); BATTLE OF THE NETWORK STARS (II,163); CIRCUS LIONS, TIGERS AND MELISSAS TOO (II,528); JAMES AT 15 (II,1271); LITTLE HOUSE ON THE PRAIRIE (II,1487); THE LOVE BOAT (II,1535); PRINCESS (II,2084); THAT'S TV (II,2581)

ANDERSON, MELODY
HIGH SCHOOL, U.S.A. (II,1149); MANIMAL (II,1622); PLEASURE COVE (II,2058)

ANDERSON JR, MICHAEL
THE DAUGHTERS OF JOSHUA CABE (I,1169); ESPIONAGE (I,1464); EVEL KNIEVEL (II,785); HERE COME THE DOUBLE DECKERS (I,2025); KISS ME, KILL ME (II,1400); THE MAGNIFICENT SIX AND A HALF (I,2828); THE MARTIAN CHRONICLES (II,1635); THE MONROES (I,3089)

ANDERSON, MICKEY
THE INA RAY HUTTON SHOW (I,2207)

ANDERSON, MILDRED
FOLK SOUND U.S.A. (I,1614)

ANDERSON, PAT
THE DOCTORS (I,1323)

ANDERSON, RANDY
GEORGE WASHINGTON (II,978)

ANDERSON, RENEE
GENERAL HOSPITAL (II,964)

ANDERSON, RICHARD
THE BIONIC WOMAN (II,255); BUS STOP (I,778); CONDOMINIUM (II,566); COVER UP (II,597); DAN AUGUST (I,1130); THE FRENCH ATLANTIC AFFAIR (II,924); THE FUGITIVE (I,1701); GHOSTBREAKER (I,1789); THE HARDY BOYS (I,1948); JARRETT (I,2349); KRAFT SUSPENSE THEATER (I,2591); THE LONELY WIZARD (I,2743); THE NIGHT STRANGLER (I,3293); PEARL (II,1986); PERRY MASON (II,2025); THE RIFLEMAN (I,3789); THE SIX MILLION DOLLAR MAN (I,4067); THE SIX-MILLION-DOLLAR MAN (II,2372); ZORRO (I,4982)

ANDERSON, RICHARD DEAN
BATTLE OF THE NETWORK STARS (II,178); GENERAL HOSPITAL (II,964); THE PARKERS (II,1959); SEVEN BRIDES FOR SEVEN BROTHERS (II,2307)

ANDERSON, RICK
GENERAL HOSPITAL (II,964)

ANDERSON, SAM
JOE DANCER: MURDER ONE, DANCER 0 (II,1302); MAMA MALONE (II,1609); WKRP IN CINCINNATI (II,2814)

ANDERSON, SARA
LEAVE IT TO BEAVER (I,2648); SURE AS FATE (I,4297)

ANDERSON, SHARON
THE WORLD OF PEOPLE (II,2839)

ANDERSON, SHEILA
THE NEW ODD COUPLE (II,1825)

ANDERSON, STEVE
AFTER GEORGE (II,19); THE NEW LORENZO MUSIC SHOW (II,1821); ONE DAY AT A TIME (II,1900)

ANDERSON, SYLVIA
CAPTAIN SCARLET AND THE MYSTERONS (I,837); FIREBALL XL-5 (I,1578); SUPERCAR (I,4294); THUNDERBIRDS (I,4497)

ANDERSON, VASS
SMILEY'S PEOPLE (II,2383)

ANDERSON, WARNER
THE DOCTOR (I,1320); GIDGET GROWS UP (I,1797); THE LINE-UP (I,2707); PEYTON PLACE (I,3574)

ANDES, KEITH
BIRDMAN (II,256); BIRDMAN AND THE GALAXY TRIO (I,478); BLOOMER GIRL (I,488); DOCTOR MIKE (I,1316); THE DOCTOR WAS A LADY (I,1318); GENERAL ELECTRIC TRUE (I,1754); GLYNIS (I,1829); THE GREAT WALTZ (I,1878); HIDE AND SEEK (I,2047); HOLIDAY (I,2074); JANE WYMAN PRESENTS THE FIRESIDE THEATER (I,2345); THE LORETTA YOUNG THEATER (I,2756); PARADISE BAY (I,3454); THIS MAN DAWSON (I,4463); THE ULTIMATE IMPOSTER (II,2700)

ANDES, OLIVER
BONINO (I,685)

ANDON, KURT
FOG (II,897)

ANDOR, PAUL
THE DIARY OF ANNE FRANK (I,1262)

ANDRE TAYER DANCERS, THE
THE JONATHAN WINTERS SHOW (I,2443)

ANDRE THE CIRCUS GIANT
THE SIX-MILLION-DOLLAR MAN (II,2372)

ANDRE, ANNETTE
MY PARTNER THE GHOST (I,3199)

ANDRE, E J
LOGAN'S RUN (II,1507)

ANDREAS, LUKE
POTTSVILLE (II,2072)

ANDREI, FREDERIC
FACTS OF LIFE: THE FACTS OF LIFE GOES TO PARIS (II,806)

ANDRESS, URSULA
THE BOB HOPE SHOW (I,638); THE BOB HOPE SHOW (I,646); THRILLER (I,4492)

ANDREW, MICHAEL
THE 25TH MAN (MS) (II,2678)

ANDREWS TWINS, THE
THE BENNY RUBIN SHOW (I,398)

ANDREWS, ANDY
RED GOOSE KID'S SPECTACULAR (I,3751)

ANDREWS, ANTHONY
BRIDESHEAD REVISITED (II,375)

ANDREWS, BEN
RETURN TO PEYTON PLACE (I,3779)

ANDREWS, CARMEN
STORY FOR AMERICANS (I,4238)

ANDREWS, CHRIS
OH, BOY! (I,3341)

ANDREWS, DANA
THE BOB HOPE SHOW (I,582); BRIGHT PROMISE (I,727); THE HORROR OF IT ALL (II,1181); IKE (II,1223); THE LAST HURRAH (I,2626); A SHADOW IN THE STREETS (II,2312)

ANDREWS, EAMONN
TOP OF THE WORLD (II,2646)

ANDREWS, EDWARD
ACCIDENTAL FAMILY (I,17); ACRES AND PAINS (I,19); BELL, BOOK AND CANDLE (II,201); BROADSIDE (I,732); CALL TO DANGER (I,802); DEATH OF A SALESMAN (I,1208); THE DON RICKLES SHOW (I,1342); THE DORIS DAY SHOW (I,1355); EDDIE (I,1405); HANDS OF MURDER (I,1929); THE HARDY BOYS (I,1948); INSIDE O.U.T. (II,1235); KEEP U.S. BEAUTIFUL (I,2519); THE KOWBOYS (I,2584); LACY AND THE MISSISSIPPI QUEEN (II,1419); MAKE MORE ROOM FOR DADDY (I,2842); THE MAN FROM GALVESTON (I,2865); THE MAN WHO CAME TO DINNER (I,2881); NOTORIOUS (I,3314); O'CONNOR'S OCEAN (I,3326); POOR MR. CAMPBELL (I,3647); THE STREETS OF SAN FRANCISCO (I,4260); STUDIO ONE (I,4268); SUPERTRAIN (II,2504); SUPERTRAIN (II,2505); TRAVIS LOGAN, D.A. (I,4598); WHERE'S THE FIRE? (II,2785)

ANDREWS, HARRY
THE SEVEN DIALS MYSTERY (II,2308); VALLEY FORGE (I,4698)

ANDREWS, JOHNNY
HI MOM (I,2043); SONGS AT TWILIGHT (I,4123)

ANDREWS, JULIE
BOB HOPE SPECIAL: BOB HOPE'S PINK PANTHER THANKSGIVING GALA (II,312); DISNEY WORLD —A GALA OPENING—DISNEYLAND EAST (I,1292); AN EVENING WITH JULIE ANDREWS AND HARRY BELAFONTE (I,1476); THE FABULOUS 50S (I,1501); HIGH TOR (I,2056); THE JACK BENNY HOUR (I,2290); JULIE AND CAROL AT CARNEGIE HALL (I,2478); JULIE AND CAROL AT LINCOLN CENTER (I,2479); JULIE AND DICK IN COVENT GARDEN (II,1352); THE JULIE ANDREWS HOUR (I,2480); THE JULIE ANDREWS SHOW (I,2481); THE JULIE ANDREWS SPECIAL (I,2482); THE JULIE ANDREWS SPECIAL (I,2483); JULIE ANDREWS: ONE STEP INTO SPRING (II,1353); JULIE ON SESAME STREET (I,2485); JULIE! (I,2477); JULIE—MY FAVORITE THINGS (II,1355); MERRY CHRISTMAS. . . WITH LOVE, JULIE (II,1687); A WORLD OF LOVE (I,4914)

ANDREWS, MARK
THE EDGE OF NIGHT (II,760); SKIP TAYLOR (I,4075)

ANDREWS, MICHAEL A
MICKEY SPILLANE'S MIKE HAMMER: MURDER ME, MURDER YOU (II,1694)

ANDREWS, NANCY
HAPPY ENDINGS (II,1085); KANGAROOS IN THE KITCHEN (II,1362); KIBBE HATES FINCH (I,2530)

ANDREWS, PATTI
ONE MORE TIME (I,3386)

ANDREWS, STANLEY
ANNIE OAKLEY (I,225); DEATH VALLEY DAYS (I,1211); STARS OVER HOLLYWOOD (I,4211)

ANDREWS, TIGE
THE DETECTIVES (I,1245); JET FIGHTER (I,2374); MITZI AND A HUNDRED GUYS (II,1710); THE MOD SQUAD (I,3078); THE PHIL SILVERS SHOW (I,3580); THE RETURN OF THE MOD SQUAD (I,3776); SHEEHY AND THE SUPREME MACHINE (II,2322)

ANDREWS, TINA
THE CONTENDER (II,571); DAYS OF OUR LIVES (II,629); SANFORD ARMS (II,2257); TOP OF THE MONTH (I,4578)

ANDREWS, TOD
BRIGHT PROMISE (I,727); CAMEO THEATER (I,808); COUNTERTHRUST (I,1061); THE DOCTOR (I,1320); FIRST LOVE (I,1583); THE GRAY GHOST (I,1867)

ANDROSKY, CAROL
DIANA (I,1259); THE NOONDAY SHOW (II,1859); SQUARE PEGS (II,2431)

ANDRUSCO, GENE
THE AMAZING CHAN AND THE CHAN CLAN (I,157); THE BARKLEYS (I,356)

ANGAROLA, RICHARD
HOW THE WEST WAS WON (II,1196); IN TANDEM (II,1228); MR. AND MRS. COP (II,1748); TRAVIS LOGAN, D.A. (I,4598)

ANGEL, HEATHER
FAMILY AFFAIR (I,1519); MR. NOVAK (I,3145); PEYTON PLACE (I,3574)

ANGEL, JACK
THE SMURFS (II,2385); THE WORLD'S GREATEST SUPER HEROES (II,2842)

ANGELA, JUNE
MR. T. AND TINA (II,1758)

ANGELIQUE
GET SMART (II,983)

ANGELIS, PAUL
IF IT'S A MAN, HANG UP (I,2186)

ANGELO, RONNIE
CAPITOL (II,426)

ANGELOU, MAYA
THE RICHARD PRYOR SPECIAL (II,2164); THE RICHARD PRYOR SPECIAL? (II,2165); ROOTS (II,2211)

ANGLIM, PHILIP
THE ADAMS CHRONICLES (II,8); THE THORN BIRDS (II,2600)

ANGOLO, EDITH
SVENGALI AND THE BLONDE (I,4310)

ANGUS, TERRY
FRAGGLE ROCK (II,916)

ANGUSTAIN, IRA
CELEBRITY CHALLENGE OF THE SEXES 5 (II,469)

ANHALT, EDWARD
NOWHERE TO HIDE (II,1868)

ANHOLT, TONY
APPOINTMENT WITH A KILLER (I,245); THE LAST DAYS OF POMPEII (II,1436); THE PROTECTORS (I,3681); SPACE: 1999 (II,2409); THE STRAUSS FAMILY (I,4253)

ANISTON, JOHN
SEARCH FOR TOMORROW (II,2284)

ANITA KERR SINGERS, THE
THE SMOTHERS BROTHERS COMEDY HOUR (I,4095)

ANITA MANN DANCERS, THE
THE KEANE BROTHERS SHOW (II,1371); TOP OF THE MONTH (I,4578)

ANKA, PAUL
BING!. . . A 50TH ANNIVERSARY GALA (II,254); HAPPY BIRTHDAY, AMERICA (II,1083); LAS VEGAS PALACE OF STARS (II,1431); LILY—SOLD OUT (II,1480); LINDSAY WAGNER—ANOTHER SIDE OF ME (II,1483); PAUL ANKA IN MONTE CARLO (II,1971); THE PAUL ANKA SHOW (II,1972); PAUL ANKA. . . MUSIC MY WAY (II,1973); THE ROOTS OF ROCK 'N' ROLL (II,2212); SINATRA—THE FIRST 40 YEARS (II,2359); THE TOM JONES SPECIAL (I,4540)

ANKERS, EVELYN
STORY THEATER (I,4240); YOUR SHOW TIME (I,4958)

ANKRUM, DAVID
THE EYES OF TEXAS (II,799); TABITHA (II,2527); TABITHA (II,2528)

ANKRUM, MORRIS
FOR THE DEFENSE (I,1624); MACKENZIE'S RAIDERS (I,2803); MAN OF THE COMSTOCK (I,2873); PEPSI-COLA PLAYHOUSE (I,3523)

ANN-MARGRET
THE ANDY WILLIAMS SPECIAL (I,211); ANN-MARGRET OLSSON (II,84); THE ANN-MARGRET SHOW (I,218); ANN-MARGRET SMITH (II,85); ANN-MARGRET'S HOLLYWOOD MOVIE GIRLS (II,86); ANN-MARGRET. . . RHINESTONE COWGIRL (II,87); ANN-MARGRET: FROM HOLLYWOOD WITH LOVE (I,217); ANN-MARGRET—WHEN YOU'RE SMILING (I,219); THE BOB HOPE SHOW (I,629); THE BOB HOPE SHOW (I,642); THE BOB HOPE SHOW (I,662); BOB HOPE SPECIAL: BOB HOPE'S 30TH ANNIVERSARY TV SPECIAL (II,293); BOB HOPE SPECIAL: BOB HOPE'S ALL-STAR COMEDY SPECTACULAR FROM LAKE TAHOE (II,301); BOB HOPE SPECIAL: HAPPY BIRTHDAY, BOB! (II,326); FROM HOLLYWOOD WITH LOVE: THE ANN-MARGRET SPECIAL (I,1687); GEORGE BURNS CELEBRATES 80 YEARS IN SHOW BUSINESS (II,967); GEORGE BURNS EARLY, EARLY, EARLY, CHRISTMAS SHOW (II,968); HOLLYWOOD'S PRIVATE HOME MOVIES (II,1167); JACK BENNY'S BIRTHDAY SPECIAL (I,2299); LAS VEGAS PALACE OF STARS (II,1431); PERRY COMO IN LAS VEGAS (II,2003); PERRY COMO'S CHRISTMAS IN ENGLAND (II,2008); SWING OUT, SWEET

LAND (II,2515); THE WAY THEY WERE (II,2748)

ANNA-LISA
BLACK SADDLE (I,480); THE LLOYD BRIDGES SHOW (I,2733)

ANNABELL
DOCTOR DOLITTLE (I,1309)

ANNENBERG, WALTER
BOB HOPE SPECIAL: BOB HOPE PRESENTS A CELEBRATION WITH STARS OF COMEDY AND MUSIC (II,340)

ANNIS, FRANCESCA
SIGN IT DEATH (I,4046); WHY DIDN'T THEY ASK EVANS? (II,2795)

ANONYMOUS
THE INVISIBLE MAN (I,2232)

ANSARA, EDWARD
EVEL KNIEVEL (II,785)

ANSARA, MICHAEL
THE ADAM MACKENZIE STORY (I,30); THE BARBARA STANWYCK THEATER (I,350); THE BARBARY COAST (II,147); BROKEN ARROW (I,744); BUCK ROGERS IN THE 25TH CENTURY (II,383); CALL TO DANGER (I,804); CENTENNIAL (II,477); DOCTOR STRANGE (II,688); I DREAM OF JEANNIE (I,2167); THE INDIAN (I,2210); LAW OF THE PLAINSMAN (I,2639); THE OUTER LIMITS (I,3426); POWDERKEG (I,3657); WHEN THE WEST WAS FUN: A WESTERN REUNION (II,2780)

ANSBORO, GEORGE
DOCTOR I.Q. (I,1314)

ANSONIA, DINO
BIG CITY COMEDY (II,233)

ANSPACH, SUSAN
ROSETTI AND RYAN: MEN WHO LOVE WOMEN (II,2217); THE YELLOW ROSE (II,2847)

ANTEBI, CARY
THE MAGIC GARDEN (II,1591); A MAGIC GARDEN CHRISTMAS (II,1590)

ANTES, JERRY
THE ALAN YOUNG SHOW (I,104); THE DESILU REVUE (I,1238)

ANTHONY, BOB
FATHER MURPHY (II,841)

ANTHONY, GERALD
ONE LIFE TO LIVE (II,1907)

ANTHONY, JANE
TO SIR, WITH LOVE (II,2624)

ANTHONY, LEE
PALMS PRECINCT (II,1946)

ANTHONY, LUKE
PANDORA AND FRIEND (I,3445)

ANTHONY, MICHAEL
IN THE BEGINNING (II,1229)

ANTHONY, PHILIP
SPELL OF EVIL (I,4152)

ANTHONY, RAY
HOLLYWOOD'S PRIVATE HOME MOVIES II (II,1168); THE RAY ANTHONY SHOW (I,3728); THE RAY ANTHONY SHOW (I,3729); THE SATURDAY NIGHT DANCE PARTY (I,3919); THE SWINGING SCENE OF RAY ANTHONY (I,4323)

ANTHONY, ROGER
OPPENHEIMER (II,1919)

ANTIE, STEPHANIE
AND HERE'S THE SHOW (I,187); HERE'S THE SHOW (I,2033)

ANTON, MATTHEW
ALL MY CHILDREN (II,39); MO AND JO (II,1715)

ANTON, SUSAN
BATTLE OF THE BEAT (II,161); BOB HOPE SPECIAL: BOB HOPE'S STAND UP AND CHEER FOR THE NATIONAL FOOTBALL LEAGUE'S 60TH YEAR (II,316); THE MAGIC OF DAVID COPPERFIELD (II,1594); MEL AND SUSAN TOGETHER (II,1677); PRESENTING SUSAN ANTON (II,2077); THE SHAPE OF THINGS (II,2315); STOP SUSAN WILLIAMS (II,2473); THE SUZANNE SOMERS SPECIAL (II,2511); THAT'S TV (II,2581)

ANTONACCI, GREG
BUSTING LOOSE (II,402); MAKIN' IT (II,1603)

ANTONIO, JIM
THE BASTARD/KENT FAMILY CHRONICLES (II,159); DELTA COUNTY, U.S.A. (II,657); DOC ELLIOT (I,1305); LANIGAN'S RABBI (II,1427); LASSIE: THE NEW BEGINNING (II,1433); PLANET EARTH (I,3615); TERROR AT ALCATRAZ (II,2563)

ANTONIO, LOU
DOG AND CAT (II,695); DOG AND CAT (II,696); MAKIN' IT (II,1603); PARTNERS IN CRIME (I,3461); THE SNOOP SISTERS (II,2389)

APLON, BORIS
STAND BY FOR CRIME (I,4186)

APOSTL'E, STEVE
OUR FAMILY BUSINESS (II,1931)

APPLEBAUM III, LEROY
MICKEY SPILLANE'S MIKE HAMMER: MORE THAN MURDER (II,1693)

APPLEBY, JAMES
COLOR HIM DEAD (II,554)

APPLEGATE, CHARLIE
THE MILTON BERLE SHOW (I,3047)

APPLEGATE, EDDIE
THE LUCY SHOW (I,2791); NANCY (I,3221); THE PATTY DUKE SHOW (I,3483)

APPLEGATE, ROY
AT EASE (II,116); BRENDA STARR (II,372); HOME COOKIN' (II,1170); THE JIMMY STEWART SHOW (I,2391)

APPLEGATE, ROYCE D
THE BLUE AND THE GRAY (II,273); THERE'S ALWAYS ROOM (II,2583); THE TOM SWIFT AND LINDA CRAIG MYSTERY HOUR (II,2633)

APPLETON, CINDY
GILLIGAN'S ISLAND: THE HARLEM GLOBETROTTERS ON GILLIGAN'S ISLAND (II,993)

APPLEWHITE, CHARLIE
ALL IN FUN (I,131); BEST FOOT FORWARD (I,401); COUNTRY STYLE, U.S.A. (I,1074)

APREA, JOHN
ANOTHER WORLD (II,97); CRAZY TIMES (II,602); MATT HOUSTON (II,1654); THE MONTEFUSCOS (II,1727); THE ROCK RAINBOW (II,2194)

AQUA MAIDS, THE
AQUA VARIETIES (I,248)

AQUINO, BUTZ
SAVAGE: IN THE ORIENT (II,2264)

ARAGNO, ANNA
LITTLE WOMEN (I,2724)

ARAGON, MARLENE
THE WORLD'S GREATEST SUPER HEROES (II,2842)

ARBOGAST, BOB
HOT WHEELS (I,2126); SKYHAWKS (I,4079)

ARBUS, ALLAN
THE FOUR SEASONS (II,915); THE GANGSTER CHRONICLES (II,957); M*A*S*H (II,1569); WORKING STIFFS (II,2834)

ARCHARD, BERNARD
SEPARATE TABLES (II,2304)

ARCHDEATON, SALLY
THE THORNTON SHOW (I,4468)

ARCHER, ANNE
BOB & CAROL & TED & ALICE (I,495); THE CELEBRITY FOOTBALL CLASSIC (II,474); THE FAMILY TREE (II,820); THE LOG OF THE BLACK PEARL (II,1506); SEVENTH AVENUE (II,2309)

ARCHER, BEVERLY
THE ADVENTURES OF POLLYANNA (II,14); LITTLE LULU (II,1493); THE NANCY WALKER SHOW (II,1790); SPENCER (II,2420); WE'VE GOT EACH OTHER (II,2757)

ARCHER, GLENN
CAPITOL CAPERS (I,823)

ARCHER, HARRY
THE RAY KNIGHT REVUE (I,3733)

ARCHER, JOHN
LASSIE (I,2622); TV READERS DIGEST (I,4630); YOUR SHOW TIME (I,4958)

ARCHER, OSCEOLA
PANAMA HATTIE (I,3444); THE TEAHOUSE OF THE AUGUST MOON (I,4368)

ARCHERD, ARMY
CELEBRITY DAREDEVILS (II,473); GUEST SHOT (I,1899); JACQUELINE SUSANN'S VALLEY OF THE DOLLS, 1981 (II,1268); THE MOVIE GAME (I,3114)

ARCHERD, SELMA
MALIBU (II,1607)

ARCHIBALD, DEBBIE
THE SHAPE OF THINGS (II,2315)

ARCHIBALD, DONNIE
EVENING AT THE IMPROV (II,786)

ARCHIE TAYIR DANCERS, THE
THE ANDY WILLIAMS SHOW (I,210)

ARDAN, JAN

ROBERT Q'S MATINEE (I,3812)

ARDEN, EVE
CBS: ON THE AIR (II,460); THE EVE ARDEN SHOW (I,1470); THE EVE ARDEN SHOW (I,1471); THE EYES OF TEXAS II (II,800); HARRY AND MAGGIE (II,1097); MEET CYD CHARISSE (I,2982); THE MOTHERS-IN-LAW (I,3110); NUTS AND BOLTS (II,1871); OUR MISS BROOKS (I,3415); THE ROYAL FOLLIES OF 1933 (I,3862); A VERY MISSING PERSON (I,4715)

ARDEN, ROBERT
FOREIGN INTRIGUE: DATELINE EUROPE (I,1638); MASTER OF THE GAME (II,1647); SABER OF LONDON (I,3879)

ARDEN, SUZI
JUBILEE U.S.A. (I,2461); NANCY ASTOR (II,1788)

ARDEN, TONI
GUY LOMBARDO AND HIS ROYAL CANADIANS (I,1910); THE MUSIC OF GERSHWIN (I,3174); THIS IS SHOW BUSINESS (I,4459)

ARDISSON, GIORGIO
HERCULES (I,2022)

ARDITO, GINO
FROM HERE TO ETERNITY (II,933)

ARENA, FORTUNATO
S*H*E (II,2235)

ARENO, LOIS
QUICK AND QUIET (II,2099); THE ROPERS (II,2214); TOPPER (II,2649); WOMEN WHO RATE A "10" (II,2825)

ARESCO, JOEY
THE BLACK SHEEP SQUADRON (II,262); SNAFU (II,2386); SUPERTRAIN (II,2505)

ARGENZIANO, CARMEN
THE LAST NINJA (II,1438)

ARGO, ALLISON
CANNON: THE RETURN OF FRANK CANNON (II,425); LADIES' MAN (II,1423); WKRP IN CINCINNATI (II,2814)

ARGO, VICTOR
FORCE FIVE (II,907)

ARGOUD, KARIN
MAMA'S FAMILY (II,1610); THE RENEGADES (II,2132)

ARGUE, MICHELLE
PRISONER: CELL BLOCK H (II,2085)

ARGUETTE, CLIFF
THE JACK PAAR SPECIAL (I,2318)

ARI, BOB
ALICE (II,33); ALMOST AMERICAN (II,55)

ARIFF, ZAIN
A TOWN LIKE ALICE (II,2652)

ARIS, BEN
THE ASSASSINATION RUN (II,112)

ARKIN, ADAM
BETWEEN THE LINES (II,225); BUSTING LOOSE (II,402); MO AND JO (II,1715); PEARL (II,1986); TEACHERS ONLY (II,2547)

ARKIN, ALAN
ABC STAGE '67 (I,6); LOVE, LIFE, LIBERTY & LUNCH (II,1544); THE TROUBLE WITH PEOPLE (I,4611); TWO GUYS FROM MUCK (II,2690)

ARKIN, DAVID
MEN AT LAW (I,3004); THE STOREFRONT LAWYERS (I,4233)

ARKIN, LINDA
ALIVE AND WELL (II,36)

ARLAND, ROXANNE
ADVENTURES OF A MODEL (I,50)

ARLEN, DAVID
THE CBS NEWCOMERS (I,879)

ARLEN, RICHARD
ALARM (I,105); MATINEE THEATER (I,2947); THE NASH AIRFLYTE THEATER (I,3227)

ARLISS, DIMITRA
THE ART OF CRIME (II,108); RICH MAN, POOR MAN—BOOK II (II,2162)

ARLT, LEWIS
PIAF (II,2039); SEARCH FOR TOMORROW (II,2284)

ARMEN, KAY
JOHNNY OLSEN'S RUMPUS ROOM (I,2433); THE RAY BOLGER SHOW (I,3731); STOP THE MUSIC (I,4232); WASHINGTON SQUARE (I,4772)

ARMENDARIZ JR, PEDRO
ALL THAT GLITTERS (II,41); KILLER BY NIGHT (I,2537); THE LOG OF THE BLACK PEARL (II,1506)

ARMS, RUSSELL
54TH STREET REVUE (I,1573); AMERICA PAUSES FOR THE MERRY MONTH OF MAY (I,167); CHANCE OF A LIFETIME (I,898); HIDDEN TREASURE (I,2045); JOSEPHINE LITTLE: DRAGON BY THE TAIL (I,2451); A SALUTE TO TELEVISION'S 25TH ANNIVERSARY (I,3893); SVENGALI AND THE BLONDE (I,4310); THIS IS SHOW BUSINESS (I,4459)

ARMSTRONG, ALUN
THE EYES HAVE IT (I,1496); NICHOLAS NICKLEBY (II,1840)

ARMSTRONG, BENJAMIN LEIGHTON
CELEBRITY (II,463)

ARMSTRONG, BESS
BAREFOOT IN THE PARK (II,151); LACE (II,1418); ON OUR OWN (II,1892); THIS GIRL FOR HIRE (II,2596)

ARMSTRONG, BILL
THE LIAR'S CLUB (II,1466)

ARMSTRONG, DAVID
MARKHAM (I,2916)

ARMSTRONG, DEBBIE
WHY DIDN'T THEY ASK EVANS? (II,2795)

ARMSTRONG, HERB
YES, VIRGINIA, THERE IS A SANTA CLAUS (II,2849)

ARMSTRONG, HUGH
U.F.O. (I,4662)

ARMSTRONG, JOHN
HAPPILY EVER AFTER (I,1937)

ARMSTRONG, KERRY
PRISONER: CELL BLOCK H (II,2085)

ARMSTRONG, KEVIN
CROSSROADS (I,1105)

ARMSTRONG, LEE
FRAGGLE ROCK (II,916)

ARMSTRONG, LOUIS
THE BEST ON RECORD (I,409); THE BING CROSBY SPECIAL (I,472); CRESCENDO (I,1094); THE EDSEL SHOW (I,1417); AN HOUR WITH DANNY KAYE (I,2136); JOHNNY CARSON PRESENTS THE SUN CITY SCANDALS (I,2420); MUSIC '55 (I,3168); TIMEX ALL-STAR JAZZ SHOW I (I,4515); TIMEX ALL-STAR JAZZ SHOW II (I,4516); TIMEX ALL-STAR JAZZ SHOW III (I,4517); TIMEX ALL-STAR JAZZ SHOW IV (I,4518); YOU'RE THE TOP (I,4974)

ARMSTRONG, MARY
CELEBRITY (II,463); ME AND MRS. C. (II,1673)

ARMSTRONG, MICHAEL
THE ADVENTURES OF BLACK BEAUTY (I,51)

ARMSTRONG, R G
THE CENTURY TURNS (I,890); KINGSTON: THE POWER PLAY (II,1399); THE LEGEND OF THE GOLDEN GUN (II,1455); MANHUNTER (II,1620); THE RIFLEMAN (I,3789); THE SHARPSHOOTER (I,4001); SKAG (II,2374); THE STOCKERS (II,2469); T.H.E. CAT (I,4430); TEXAS JOHN SLAUGHTER (I,4414)

ARMSTRONG, ROBERT
THE FIRST HUNDRED YEARS (I,1581)

ARMSTRONG, VALERIE
THE SANDY DUNCAN SHOW (II,2253); SANTA BARBARA (II,2258); THE STOCKARD CHANNING SHOW (II,2468); SUPERTRAIN (II,2504)

ARMSTRONG, VALORIE
FUNNY FACE (I,1711)

ARMSTRONG, VAUGHN
ROGER AND HARRY (II,2200); WILDER AND WILDER (II,2802)

ARNAUTS, THE
CAVALCADE OF STARS (I,877)

ARNAZ JR, DESI
ADVICE TO THE LOVELORN (II,16); AUTOMAN (II,120); HAVING BABIES I (II,1107); HERE'S LUCY (II,1135); I LOVE LIBERTY (II,1215); PEOPLE DO THE CRAZIEST THINGS (II,1997); WHACKED OUT (II,2767)

ARNAZ, DESI
THE BOB HOPE SHOW (I,550); THE DESILU PLAYHOUSE (I,1237); THE DESILU REVUE (I,1238); I LOVE LUCY (I,2171); THE LUCILLE BALL-DESI ARNAZ SHOW (I,2784); THE MILTON BERLE SPECIAL (I,3052); THE MOTHERS-IN-LAW (I,3110); RIDDLE AT 24000 (II,2171); SHOW OF THE YEAR (I,4031); THE UNTOUCHABLES (I,4681)

ARNAZ, LUCIE
BOB HOPE SPECIAL: BOB HOPE'S ALL-STAR BIRTHDAY PARTY (II,295); CIRCUS OF THE STARS (II,531); CLOWNAROUND (I,980); HERE'S LUCY (II,1135); ONE MORE TRY (II,1908)

ARNE, PETER
THE FAR PAVILIONS (II,832); HORNBLOWER (I,2121); QUILLER: NIGHT OF THE FATHER (II,2100)

ARNER, GWEN
THE NEW LAND (II,1820); STICKIN' TOGETHER (II,2465)

ARNESS, JAMES
THE CHEVROLET GOLDEN ANNIVERSARY SHOW (I,924); FRONT ROW CENTER (I,1694); GUNSMOKE (II,1069); HOW THE WEST WAS WON (II,1196); THE MACAHANS (II,1583); MCCLAIN'S LAW (II,1659); THE RED SKELTON CHEVY SPECIAL (I,3753); A SALUTE TO TELEVISION'S 25TH ANNIVERSARY (I,3893)

ARNETTE, JEANNETTE
THE BOUNDER (II,357); BROTHERS (II,381); CAMP GRIZZLY (II,418); STEPHANIE (II,2453)

ARNGRIM, ALISON
LITTLE HOUSE ON THE PRAIRIE (II,1487)

ARNGRIM, STEFAN
LAND OF THE GIANTS (I,2611)

ARNO, SIG
MY FRIEND IRMA (I,3191); ROSALINDA (I,3846); TIME REMEMBERED (I,4509)

ARNOLD
GREEN ACRES (I,1884)

ARNOLD HOLOP ENSEMBLE, THE
THE ALAN DALE SHOW (I,96)

ARNOLD, BETTY
THOSE ENDEARING YOUNG CHARMS (I,4469)

ARNOLD, BEVERLY
THE MILTON THE MONSTER CARTOON SHOW (I,3053)

ARNOLD, DANNY
TENNESSEE ERNIE FORD MEETS KING ARTHUR (I,4397)

ARNOLD, ED
NO SOAP, RADIO (II,1856)

ARNOLD, EDDY
THE ARTHUR GODFREY SPECIAL (II,109); THE BEST ON RECORD (I,409); CAROL CHANNING AND 101 MEN (I,849); COUNTRY MUSIC HIT PARADE (I,1069); COUNTRY NIGHT OF STARS II (II,593); CRESCENDO (I,1094); DANNY THOMAS GOES COUNTRY AND WESTERN (I,1146); THE EDDY ARNOLD SHOW (I,1411); EDDY ARNOLD TIME (I,1412); THE PERRY COMO WINTER SHOW (I,3542); THE R.C.A. THANKSGIVING SHOW (I,3736); ROY ACUFF—50 YEARS THE KING OF COUNTRY MUSIC (II,2223)

ARNOLD, EDWARD
THE EDWARD ARNOLD THEATER (I,1418); THE ETHEL BARRYMORE THEATER (I,1467); STRANGE STORIES (I,4249); YOUR STAR SHOWCASE (I,4959)

ARNOLD, JEANNE
THE CARA WILLIAMS SHOW (I,843); THE GUIDING LIGHT (II,1064); THE JONES BOYS (I,2446); MAKING IT (II,1605); MUNSTER, GO HOME! (I,3157)

ARNOLD, JOHN
PRISONER: CELL BLOCK H (II,2085)

ARNOLD, KAY
MS. FIXER UPPER (I,3156)

ARNOLD, MADISON
MICKEY SPILLANE'S MIKE HAMMER: MURDER ME, MURDER YOU (II,1694)

ARNOLD, MARK
THE EDGE OF NIGHT (II,760)

ARNOLD, MARY
ROLLIN' ON THE RIVER (I,3834)

ARNOLD, MONROE
OFF WE GO (I,3338)

ARNOLD, MURRY
THE HAZEL BISHOP SHOW (I,1983)

ARNOLD, OCTAVIA
GEORGE WASHINGTON (II,978)

ARNOLD, PAUL
33 1/3 REVOLUTIONS PER MONKEE (I,4451); AMERICAN SONG (I,172); MIDWESTERN HAYRIDE (I,3033); THE PAUL ARNOLD SHOW (I,3484)

ARNOLD, PHIL
ADVENTURES OF A MODEL (I,50)

ARNOLD, SEAN
BERGERAC (II,209)

ARNOLD, VICE
THE BEST OF EVERYTHING (I,406)

ARNOLD, VICTOR
THE EDGE OF NIGHT (II,760)

ARNOLDO DANCERS, THE
BROADWAY GOES LATIN (I,734)

ARNONE, LEE
THE RETURN OF MARCUS WELBY, M.D (II,2138)

ARNOTT, DAVID
SHEEHY AND THE SUPREME MACHINE (II,2322)

ARONE, JAMES
MICKEY SPILLANE'S MIKE HAMMER: MURDER ME, MURDER YOU (II,1694)

ARONSON, JUDIE
THINGS ARE LOOKING UP (II,2588)

ARQUETTE, CLIFF
THE ARTHUR MURRAY PARTY FOR BOB HOPE (I,292); DAVE AND CHARLEY (I,1171); DO IT YOURSELF (I,1299); F TROOP (I,1499); HOBBY LOBBY (I,2068); THE HOLLYWOOD SQUARES (II,1164); JACK PAAR PRESENTS (I,2315); THE JONATHAN WINTERS SHOW (I,2443); MCNAB'S LAB (I,2972); THE R.C.A. VICTOR SHOW (I,3737); THE ROY ROGERS AND DALE EVANS SHOW (I,3858); STARS OVER HOLLYWOOD (I,4211); THE TONIGHT SHOW (I,4564)

ARQUETTE, HARRY
STRANGER IN OUR HOUSE (II,2477)

ARQUETTE, KERRY
STRANGER IN OUR HOUSE (II,2477)

ARQUETTE, LEWIS
DRIBBLE (II,736); HARPER VALLEY PTA (II,1095); IT ONLY HURTS WHEN YOU LAUGH (II,1251); THE MARILYN MCCOO AND BILLY DAVIS JR. SHOW (II,1630); THE NEW LORENZO MUSIC SHOW (II,1821); PRIME TIMES (II,2083); THE WALTONS (II,2740)

ARQUETTE, ROSANNA
HAVING BABIES II (II,1108); JOHNNY BELINDA (I,2418); SHIRLEY (II,2330)

ARRANGA, IRENE
HOME ROOM (II,1172); SECRETS OF MIDLAND HEIGHTS (II,2296); WELCOME BACK, KOTTER (II,2761)

ARRANTS, ROD
FOR RICHER, FOR POORER (II,906); LOVERS AND FRIENDS (II,1549); SEARCH FOR TOMORROW (II,2284)

ARRENDONDO, RODRIGO
THE BANANA SPLITS ADVENTURE HOUR (I,340)

ART REYNOLDS SINGERS, THE
BILL COSBY DOES HIS OWN THING (I,441)

ART VAN DAMME DANCERS
CHICAGO JAZZ (I,933)

ART VAN DAMME QUINTET, THE
CLUB 60 (I,987); THE MUSIC OF GERSHWIN (I,3174)

ARTHUR MURRAY DANCERS, THE
ARTHUR MURRAY'S DANCE PARTY (I,293)

ARTHUR, BEATRICE
ALL IN THE FAMILY (II,38); AMANDA'S (II,62); THE BEATRICE ARTHUR SPECIAL (II,196); THE BEST OF ANYTHING (I,404); BOB HOPE SPECIAL: BOB HOPE—HOPE, WOMEN AND SONG (II,324); CBS: ON THE AIR (II,460); THE CONNIE FRANCIS SHOW (I,1039); THE DEAN MARTIN CELEBRITY ROAST (II,638); MAUDE (II,1655); P.O.P. (II,1939)

ARTHUR, CAROL
THE BRADY GIRLS GET MARRIED (II,364); DOM DELUISE AND FRIENDS, PART 2 (II,702); THE DOM DELUISE SHOW (I,1328); VENICE MEDICAL (II,2726)

ARTHUR, INDUS
GENERAL HOSPITAL (II,964); THE GREEN FELT JUNGLE (I,1885)

ARTHUR, JEAN
THE JEAN ARTHUR SHOW (I,2352)

ARTHUR, MAUREEN
THE BOB HOPE SHOW (I,632); CHARGE ACCOUNT (I,903); THE DUCK FACTORY (II,738); EMPIRE (II,777); THE HERO (I,2035); HOLIDAY LODGE (I,2077); JOE AND SONS (II,1297); LOOK OUT WORLD (II,1516); THE STOCKARD

CHANNING SHOW (II,2468); THE TONIGHT SHOW (I,4563); WHAT'S IT ALL ABOUT WORLD? (I,4816); YOU'RE JUST LIKE YOUR FATHER (II,2861)

ARTIE MALVIN CHORUS, THE
THE PAT BOONE SHOW (I,3470); PERRY COMO PRESENTS THE JULIUS LAROSA SHOW (I,3530)

ARTIE MALVIN SINGERS, THE
THE STEVE LAWRENCE AND EYDIE GORME SHOW (I,4226)

ARVAN, JAN
A COUPLE OF DONS (I,1077); THE DONALD O'CONNOR SHOW (I,1344); IT'S ABOUT TIME (I,2263); MAUDE (II,1655); THE RED SKELTON SHOW (I,3756); ZORRO (I,4982)

ARVIN, WILLIAM
MORNING STAR (I,3104)

ASH, GLEN
A COUNTRY HAPPENING (I,1065); HIGHWAY HONEYS (II,1151); THE NEW ANDY GRIFFITH SHOW (I,3256)

ASHBURN, GREG
GEORGE WASHINGTON (II,978)

ASHBY, CHRISPIN
A TOWN LIKE ALICE (II,2652)

ASHDOWN, ISA
ANNIE OAKLEY (I,225)

ASHE, EVE BRENT
BEST OF THE WEST (II,218)

ASHE, JENNIFER
LOVING (II,1552)

ASHE, MARTIN
GILLIGAN'S ISLAND: RESCUE FROM GILLIGAN'S ISLAND (II,991)

ASHER, PETER
THE ADVENTURES OF ROBIN HOOD (I,74); THINGS WE DID LAST SUMMER (II,2589)

ASHERSON, RENEE
THE LILLI PALMER THEATER (I,2704); A MAN CALLED INTREPID (II,1612)

ASHFORD, MATTHEW
ONE LIFE TO LIVE (II,1907)

ASHKENAZI, IRVIN
DAVY CROCKETT (I,1181)

ASHLAND, CAMILIA
GENERAL HOSPITAL (II,964); V (II,2713); V: THE FINAL BATTLE (II,2714)

ASHLEY, ED
GENESIS II (I,1760)

ASHLEY, ELIZABETH
BLONDES VS. BRUNETTES (II,271); DUPONT SHOW OF THE MONTH (I,1387); THE FILE ON DEVLIN (I,1575); GHOST STORY (I,1788); THE MAGICIAN (I,2824); SANDBURG'S LINCOLN (I,3907); SECOND CHANCE (I,3958); TOM AND JOANN (II,2630)

ASHLEY, JOHN
STRAIGHTAWAY (I,4243)

ASHMORE, FRANK
MITCHELL AND WOODS (II,1709); V (II,2713); V: THE FINAL BATTLE (II,2714); V: THE SERIES (II,2715)

ASHROW, DAVID
THE KID WITH THE BROKEN HALO (II,1387)

ASHTON, FREDERICK
CINDERELLA (I,955); SLEEPING BEAUTY (I,4083)

ASHTON, JOHN
THE TOM SWIFT AND LINDA CRAIG MYSTERY HOUR (II,2633); THE WILDS OF TEN THOUSAND ISLANDS (II,2803)

ASHTON, JUNE
LITTLE WOMEN (I,2721)

ASHTON, PAT
THE BENNY HILL SHOW (II,207)

ASHTON, SOPHIE
THE TALE OF BEATRIX POTTER (II,2533)

ASHTON, STEVE
PERRY COMO PRESENTS THE JULIUS LAROSA SHOW (I,3530)

ASIA BOYS, THE
THE JACK CARSON SHOW (I,2304)

ASING, NANI
CATALINA C-LAB (II,458)

ASKEW, LEW
MANHUNTER (II,1620)

ASKEW, LUKE
NIGHT GAMES (II,1845); THE QUEST (II,2096)

ASKIN, LEON
ALFRED OF THE AMAZON (I,116); THE CHARLIE FARRELL SHOW (I,913); GENESIS II (I,1760); HOGAN'S HEROES (I,2069); RUSSIAN ROULETTE (I,3875)

ASLAN, GREGOIRE
THE KILLER WHO WOULDN'T DIE (II,1393)

ASNER, EDWARD
BATTLE OF THE NETWORK STARS (II,169); CELEBRITY CHALLENGE OF THE SEXES 1 (II,465); CIRCUS OF THE STARS (II,530); DECOY (I,1217); DINAH IN SEARCH OF THE IDEAL MAN (I,1278); THE HOUSE ON GREENAPPLE ROAD (I,2140); THE IMPOSTER (II,1225); LOU GRANT (II,1526); THE MARY TYLER MOORE SHOW (II,1640); MONTY HALL'S VARIETY HOUR (II,1728); NAKED CITY (I,3216); OFF THE RACK (II,1878); PARADE OF STARS (II,1954); THE POLICE STORY (I,3638); PROFILES IN COURAGE (I,3678); RICH MAN, POOR MAN—BOOK I (II,2161); ROOTS (II,2211); THE ROWAN AND MARTIN SPECIAL (I,3855); THE SEEKERS (I,3974); SLATTERY'S PEOPLE (I,4082); TED KNIGHT MUSICAL COMEDY VARIETY SPECIAL SPECIAL (II,2549); THEY CALL IT MURDER (I,4441); TWIGS (II,2683)

ASPEL, MICHAEL
THE GOODIES (II,1040)

ASSA, RENE
THE MAN WITH THE POWER (II,1616)

ASSALONE, TOM
GEORGE WASHINGTON (II,978)

ASSANG, GEORGE
BARRIER REEF (I,362)

ASSANTE, ARMAND
THE DOCTORS (II,694); HOW TO SURVIVE A MARRIAGE (II,1198); HUMAN FEELINGS (II,1203); RAGE OF ANGELS (II,2107)

ASSOCIATION, THE
CAROL CHANNING AND 101 MEN (I,849); SCROOGE'S ROCK 'N' ROLL CHRISTMAS (II,2279); WHERE THE GIRLS ARE (I,4830)

AST, PAT
BEST OF THE WEST (II,218); THE DONNA SUMMER SPECIAL (II,711)

ASTAIRE, FRED
ALCOA PREMIERE (I,109); ANOTHER EVENING WITH FRED ASTAIRE (I,229); ASTAIRE TIME (I,305); THE EASTER BUNNY IS COMIN' TO TOWN (II,753); AN EVENING WITH FRED ASTAIRE (I,1474); THE FRED ASTAIRE SHOW (I,1677); THE FRED ASTAIRE SPECIAL (I,1678); IT TAKES A THIEF (I,2250); JACK LEMMON IN 'S WONDERFUL, 'S MARVELOUS, 'S GERSHWIN (I,2313); MAKE MINE RED, WHITE, AND BLUE (I,2841); MERRY CHRISTMAS FROM THE CROSBYS (II,1685); OF THIS TIME, OF THAT PLACE (I,3335); THE OVER-THE-HILL GANG RIDES AGAIN (I,3432); SANTA CLAUS IS COMIN' TO TOWN (II,2259); SEVEN AGAINST THE SEA (I,3983); THINK PRETTY (I,4448)

ASTAR, BEN
OPERATION GREASEPAINT (I,3402); THE WINDS OF WAR (II,2807)

ASTER, DAVID
YOU CAN'T DO THAT ON TELEVISION (I,4933)

ASTIN, JOHN
THE ADDAMS FAMILY (II,11); THE ADDAMS FAMILY (II,13); BATMAN (I,366); EVIL ROY SLADE (I,1485); I'M DICKENS...HE'S FENSTER (I,2193); THE MOUSE FACTORY (I,3113); OPERATION PETTICOAT (II,1916); OPERATION PETTICOAT (II,1917); PHILLIP AND BARBARA (II,2034); THE PRUITTS OF SOUTHAMPTON (I,3684); SHERIFF WHO? (I,4009); SIMON AND SIMON (II,2357)

ASTIN, PATTY DUKE
see **DUKE, PATTY**

ASTOR, DAVE
RUN JACK RUN (I,3872)

ASTOR, MARY
THE PHILADELPHIA STORY (I,3581); THE PHILADELPHIA STORY (I,3582); THE WOMEN (I,4890)

ASTOR, SUZANNE
GRANDPA MAX (II,1051)

ASTREDO, HUMBERT A
DARK SHADOWS (I,1157)

ATCHER, BOB
JUNIOR RODEO (I,2494)

ATES, ROSCOE
ANNIE OAKLEY (I,225); THE MARSHAL OF GUNSIGHT PASS (I,2925)

ATHERTON, WILLIAM
MALIBU (II,1607)

ATKIN, HARVEY
CAGNEY AND LACEY (II,409); CAGNEY AND LACEY (II,410)

ATKINE, FEODOR
LACE (II,1418)

ATKINS, CHET
ANN-MARGRET. . . RHINESTONE COWGIRL (II,87); THE BOOTS RANDOLPH SHOW (I,687); COUNTRY COMES HOME (II,585); COUNTRY COMES HOME (II,586); THE EDDY ARNOLD SHOW (I,1411); AN EVENING WITH THE STATLER BROTHERS (II,790); FIFTY YEARS OF COUNTRY MUSIC (II,853); THE GRAND OLE OPRY (I,1866); JOHNNY CASH: THE FIRST 25 YEARS (II,1336); MUSIC CITY NEWS TOP COUNTRY HITS OF THE YEAR (II,1772); ROY ACUFF—50 YEARS THE KING OF COUNTRY MUSIC (II,2223)

ATKINS, CHRISTOPHER
CELEBRITY DAREDEVILS (II,473); DALLAS (II,614); I LOVE LIBERTY (II,1215)

ATKINS, DAVE
MASTER OF THE GAME (II,1647)

ATKINS, EILEEN
SMILEY'S PEOPLE (II,2383)

ATKINS, ENNIS
CHIEFS (II,509)

ATKINS, JANE
NORMA RAE (II,1860)

ATKINS, JERRY
THE PERRY COMO WINTER SHOW (I,3542)

ATKINS, NORMAN
THE YEOMAN OF THE GUARD (I,4926)

ATKINS, TOM
MICKEY SPILLANE'S MIKE HAMMER: MURDER ME, MURDER YOU (II,1694); THE ROCKFORD FILES (II,2197); SERPICO (II,2306); SKEEZER (II,2376)

ATKINSON, BARBARA
CROWN MATRIMONIAL (I,1106)

ATKINSON, BEVERLY HOPE
DOWN HOME (II,731)

ATKINSON, BRUCE
MAGNUM, P.I. (II,1601)

ATKINSON, DAVID
THE CHOCOLATE SOLDIER (I,944); THE MARCH OF DIMES BENEFIT SHOW (I,2900)

ATKINSON, JOHN
THE SAVAGE CURSE (I,3925)

ATKINSON, ROWAN
NOT THE NINE O'CLOCK NEWS (II,1864)

ATMORE, ANN
POLICE STORY (I,3636)

ATTELL, ANTOINETTE
LAUGH-IN (II,1444)

ATTERBURY, MALCOLM
APPLE'S WAY (II,101); BELLE STARR (I,391); THE FABULOUS SYCAMORES (I,1505); THE HARDY BOYS (I,1948); HICKEY VS. ANYBODY (II,1142); SUDDEN SILENCE (I,4275); THICKER THAN WATER (I,4445)

ATTMORE, WILLIAM
KELLY'S KIDS (I,2523); THE NEW MICKEY MOUSE CLUB (II,1823)

ATWATER, BARRY
FRONTIER (I,1695); NIGHT GALLERY (I,3288); THE NIGHT STALKER (I,3292); ONE STEP BEYOND (I,3388); YOU ARE THERE (I,4929)

ATWATER, EDITH
FAMILY TIES (II,819); THE HARDY BOYS MYSTERIES (II,1090); KAZ (II,1370); LOVE ON A ROOFTOP (I,2775)

ATWELL, JUDITH
THE WINDS OF WAR (II,2807)

ATWILL, ROY
THE ADMIRAL BROADWAY REVUE (I,37)

AUBERJONOIS, RENE
BENSON (II,208); MORE WILD WILD WEST (II,1733); ONCE UPON A DEAD MAN (I,3369); THE RHINEMANN EXCHANGE (II,2150); SCALPELS (II,2266); THE SCOOBY-DOO AND SCRAPPY-DOO SHOW (II,2275); THE TV TV SHOW (II,2674); THE WILD WILD WEST REVISITED (II,2800)

AUBREY, GRACE
IN THE STEPS OF A DEAD MAN (I,2205)

AUBREY, SKYE
THE CITY (I,969); ELLERY QUEEN: DON'T LOOK BEHIND YOU (I,1434); A VERY MISSING PERSON (I,4715)

AUBRY, DANIELE
THE ELISABETH MCQUEENEY STORY (I,1429)

AUBUCHON, JACQUES
CAFE DE PARIS (I,791); CYRANO DE BERGERAC (I,1119); THE GOOD OLD DAYS (I,1851); JOHNNY BELINDA (I,2417); MCHALE'S NAVY (I,2969); PARIS 7000 (I,3459)

AUDLEY, ELEANOR
THE DOCTOR WAS A LADY (I,1318); THE FABULOUS SYCAMORES (I,1505); GREEN ACRES (I,1884); MCNAB'S LAB (I,2972); MY THREE SONS (I,3205); PISTOLS 'N' PETTICOATS (I,3608)

AUDLEY, MAXINE
CROWN MATRIMONIAL (I,1106)

AUDRAN, STEPHANIE
BRIDESHEAD REVISITED (II,375); MISTRAL'S DAUGHTER (II,1708)

AUER, CHRIS
ANOTHER LIFE (II,95)

AUER, MISCHA
THE HAPPY TIME (I,1943); NINOTCHKA (I,3299)

AUERBACH, ARTIE
THE JACK BENNY PROGRAM (I,2294)

AUGER, BRIAN
33 1/3 REVOLUTIONS PER MONKEE (I,4451)

AUGER, CLAUDINE
THE DANNY THOMAS SPECIAL (I,1150); DANNY THOMAS: THE ROAD TO LEBANON (I,1153)

AUGUSTAIN, IRA
THE WHITE SHADOW (II,2788)

AUGUSTINE, CRAIG
SEARCH FOR TOMORROW (II,2284)

AUL, JANE
MARINELAND CARNIVAL (I,2913)

AULDTTO, TEDDY
SAMMY KAYE'S MUSIC FROM MANHATTAN (I,3902)

AUMONT, JEAN-PIERRE
CIRCUS OF THE STARS (II,530); THE ERROL FLYNN THEATER (I,1458); THE FRENCH ATLANTIC AFFAIR (II,924); INTERMEZZO (I,2223); THE LORETTA YOUNG THEATER (I,2756); SO HELP ME, APHRODITE (I,4104); THE SPIRAL STAIRCASE (I,4160); STUDIO '57 (I,4267)

AUNT GRACE
HAPPY BIRTHDAY (I,1940)

AUSTEN, FEATHER
I'D RATHER BE CALM (II,1216)

AUSTEN, JERRY
WINNER TAKE ALL (I,4881)

AUSTIN, AL
THE INVESTIGATORS (I,2231)

AUSTIN, BOB
COUNTRY STYLE (I,1072)

AUSTIN, JAMES
THE WORLD OF DARKNESS (II,2836)

AUSTIN, KAREN
CELEBRITY (II,463); LONDON AND DAVIS IN NEW YORK (II,1512); NIGHT COURT (II,1844); THE QUEST (II,2098); ST. ELSEWHERE (II,2432)

AUSTIN, NANCY
THE JIMMIE RODGERS SHOW (I,2383)

AUSTIN, PAMELA
EVIL ROY SLADE (I,1485); ROWAN AND MARTIN'S LAUGH-IN (I,3857)

AUSTIN, RAY
SPACE: 1999 (II,2409)

AUSTIN, VIRGINIA
ON THE CORNER (I,3359)

AUTRE, EMIL
THE NEW TREASURE HUNT (II,1831)

AUTRY, ALAN
LONE STAR (II,1513)

AUTRY, GENE
THE GENE AUTRY SHOW (I,1748); GENE AUTRY'S MELODY RANCH (I,1749); ROY ACUFF—50 YEARS THE KING OF COUNTRY MUSIC (II,2223); THE SINGING COWBOYS RIDE AGAIN (II,2363)

AUTTERSON, GAY
FLINTSTONE FAMILY ADVENTURES (II,882); THE FLINTSTONE FUNNIES (II,883); THE FLINTSTONES (II,885); FRED AND BARNEY MEET THE THING (II,919)

AVALON, FRANKIE
THE BIG BEAT (I,423); DICK CLARK'S GOOD OL' DAYS: FROM BOBBY SOX TO BIKINIS (II,671); EASY DOES IT. . . STARRING FRANKIE AVALON (II,754); FRANKIE AND ANNETTE: THE SECOND TIME AROUND (II,917); FRANKIE AVALON'S EASTER SPECIAL (I,1673); THE ROOTS

OF ROCK 'N' ROLL (II,2212); SAGA OF SONORA (I,3884); WHO'S AFRAID OF MOTHER GOOSE? (I,4852)

AVALOS, LUIS
THE BOYS IN BLUE (II,359); CONDO (II,565); E/R (II,748); HIGHCLIFFE MANOR (II,1150); PALS (II,1947); SIDE BY SIDE (II,2347)

AVERBACK, HY
MEET CORLISS ARCHER (I,2980); NBC COMEDY HOUR (I,3237); OUR MISS BROOKS (I,3415); THE SATURDAY NIGHT REVUE (I,3921)

AVERDON, DOE
BIG TOWN (I,436)

AVERY, BRIAN
A BELL FOR ADANO (I,388)

AVERY, JAMES
GOING BANANAS (II,1015)

AVERY, MARGARET
A.E.S. HUDSON STREET (II,17)

AVERY, PHYLLIS
CLEAR HORIZON (I,973); THE DANNY KAYE SHOW (I,1144); THE DOCTOR (I,1320); THE GEORGE GOBEL SHOW (I,1768); THE HUMAN COMEDY (I,2158); I LOVE MY DOCTOR (I,2172); MEET MR. MCNUTLEY (I,2989); MR. NOVAK (I,3145); THE RAY MILLAND SHOW (I,3734)

AVERY, TOL
333 MONTGOMERY (I,4487); JUSTICE OF THE PEACE (I,2500); SECOND LOOK (I,3960); SLATTERY'S PEOPLE (I,4082); THE THIN MAN (I,4446)

AVERY, VAL
THE DANGEROUS DAYS OF KIOWA JONES (I,1140); FIREHOUSE (I,1579); I LOVE A MYSTERY (I,2170); STONESTREET: WHO KILLED THE CENTERFOLD MODEL? (II,2472); THE STREETS (II,2479); THERE SHALL BE NO NIGHT (I,4437)

AVILA, CHRISTINE
THE TARZAN/LONE RANGER/ZORRO ADVENTURE HOUR (II,2543); TRUE LIFE STORIES (II,2665)

AVILA, CYNTHIA
BORN TO THE WIND (II,354); NIGHT PARTNERS (II,1847); THE RETURN OF MARCUS WELBY, M.D (II,2138)

AVILES, RICK
THE DAY THE WOMEN GOT EVEN (II,628); MR. & MRS. DRACULA (II,1745)

AVIS-KRAUSS, CHRISTINA
JOHNNY GARAGE (II,1337)

AVON, ROGER
THE EVE ARDEN SHOW (I,1471)

AVONNE, MICHELLE
MICKEY SPILLANE'S MIKE HAMMER: MURDER ME, MURDER YOU (II,1694)

AVRAMO, PETER
THE GREATEST GIFT (I,1880)

AVRUCH, FRANK
BOZO THE CLOWN (I,702)

AWITI, MILDRED
BUSH DOCTOR (II,400)

AWTREY, DONNA MARIE
JAKE'S WAY (II,1269)

AXELROD, NINA
HARD KNOCKS (II,1086)

AXTON, HOYT
THE ALL-STAR SALUTE TO MOTHER'S DAY (II,51); DIFF'RENT STROKES (II,674); DOMESTIC LIFE (II,703); THE HOYT AXTON SHOW (II,1200); THE ROUSTERS (II,2220); SEVEN BRIDES FOR SEVEN BROTHERS (II,2307); SKINFLINT (II,2377)

AYER, HAROLD
CONDOMINIUM (II,566)

AYKROYD, DAN
THE BEACH BOYS SPECIAL (II,190); THE CONEHEADS (II,567); SATURDAY NIGHT LIVE (II,2261); STEVE MARTIN'S BEST SHOW EVER (II,2460); THINGS WE DID LAST SUMMER (II,2589)

AYLEN, RICHARD
POSSESSION (I,3655); THE TALE OF BEATRIX POTTER (II,2533)

AYLER, ETHEL
THE COSBY SHOW (II,578)

AYLMER, FELIX
CAPTAIN BRASSBOUND'S CONVERSION (I,828); THE CHAMPIONS (I,896); MACBETH (I,2801); VICTORIA REGINA (I,4728)

AYLWARD, DEREK
THE ADVENTURES OF SIR LANCELOT (I,76)

AYR, MICHAEL
PIAF (II,2039)

AYRES, CATHERINE
MEMBER OF THE WEDDING (I,3002)

AYRES, CURT
HOME ROOM (II,1172)

AYRES, JACK
WESLEY (I,4795)

AYRES, JERRY
GENERAL HOSPITAL (II,964)

AYRES, LEAH
9 TO 5 (II,1852); THE EDGE OF NIGHT (II,760); MOTHER AND ME, M.D. (II,1740); VELVET (II,2725)

AYRES, LEW
ALCOA/GOODYEAR THEATER (I,107); THE BARBARA STANWYCK THEATER (I,350); FRONTIER JUSTICE (I,1700); HAWAII FIVE-O (I,1972); JOHNNY RISK (I,2435); JOSEPHINE LITTLE: ADVENTURES IN HAPPINESS (I,2450); THE JUNE ALLYSON SHOW (I,2488); MARCUS WELBY, M.D. (I,2905); THE MARY TYLER MOORE SHOW (II,1640); SAVAGE: IN THE ORIENT (II,2264); THE STRANGER (I,4251); ZANE GREY THEATER (I,4979)

AYRES, ROBERT
THE CHEATERS (I,918); FOG (II,897)

AYRES, ROSALIND
PENMARRIC (II,1993)

AYRES-ALLEN, PHYLICIA
THE COSBY SHOW (II,578)

AYSHEA
U.F.O. (I,4662)

AZAROW, MARTIN
IN TROUBLE (II,1230); MISSING PIECES (II,1706); THE NEW OPERATION PETTICOAT (II,1826); THE OUTLAWS (II,1936)

AZNAVOUR, CHARLES
THE SAMMY DAVIS JR. SPECIAL (I,3900)

AZZARA, CANDICE
ANOTHER MAN'S SHOES (II,96); CALUCCI'S DEPARTMENT (I,805); EDDIE AND HERBERT (II,757); THE GRADY NUTT SHOW (II,1048); HOUSE CALLS (II,1194); HOW TO SURVIVE THE 70S AND MAYBE EVEN BUMP INTO HAPPINESS (II,1199); THE LOVE BOAT II (II,1533); MAMA MALONE (II,1609); THE RAINBOW GIRL (II,2108); RHODA (II,2151); SOAP (II,2392); TENSPEED AND BROWN SHOE (II,2562); THE TWO OF US (II,2693); WIVES (II,2812)

BABBITT, HARRY
GLAMOUR GIRL (I,1817)

BABCOCK, BARBARA
BENSON (II,208); THE BIG EASY (II,234); BLISS (II,269); DALLAS (II,614); DOBIE GILLIS (I,1302); THE FOUR SEASONS (II,915); HILL STREET BLUES (II,1154); OPERATING ROOM (II,1915)

BABILONIA, TAI
BOB HOPE SPECIAL: BOB HOPE'S ALL-STAR COMEDY BIRTHDAY PARTY (II,297)

BABSON, THOMAS
240-ROBERT (II,2688); SCARECROW AND MRS. KING (II,2268)

BACALL, LAUREN
APPLAUSE (I,244); BLITHE SPIRIT (I,484); CIRCUS OF THE STARS (II,532); THE DUPONT SHOW OF THE WEEK (I,1388); HAPPY ENDINGS (II,1085); THE LIGHT FANTASTIC, OR HOW TO TELL YOUR PAST, PRESENT AND MAYBE YOUR FUTURE THROUGH SOCIAL DANCING (I,2696); LIGHT'S DIAMOND JUBILEE (I,2698); PARADE OF STARS (II,1954); THE PETRIFIED FOREST (I,3570); THE WAYNE NEWTON SPECIAL (II,2750)

BACALLA, DONNA
GENERAL HOSPITAL (II,964); THE NEW PEOPLE (I,3268); THE SURVIVORS (I,4301)

BACCOLONI, SALVATORE
THE DESERT SONG (I,1236)

BACH, CATHERINE
BATTLE OF THE NETWORK STARS (II,168); BATTLE OF THE NETWORK STARS (II,170); BATTLE OF THE NETWORK STARS (II,174); BLONDES VS. BRUNETTES (II,271); BOB HOPE SPECIAL: BOB HOPE'S MERRY CHRISTMAS SHOW (II,310); CELEBRITY CHALLENGE OF THE SEXES 4 (II,468); CELEBRITY CHALLENGE OF THE SEXES 5 (II,469); CIRCUS OF THE STARS (II,535); THE DUKES (II,741); THE DUKES OF HAZZARD (II,742); GEORGE BURNS' HOW TO LIVE TO BE 100 (II,972); THE MAGIC OF DAVID COPPERFIELD (II,1594);

MATT HELM (II,1652); THE NASHVILLE PALACE (II,1793)

BACH, ROBIN
NOT IN FRONT OF THE KIDS (II,1861)

BACHARACH, BURT
ANDY WILLIAMS KALEIDOSCOPE COMPANY (I,201); BURT BACHARACH IN SHANGRI-LA (I,773); THE BURT BACHARACH SPECIAL (I,775); THE BURT BACHARACH SPECIAL (I,776); BURT BACHARACH! (I,771); BURT BACHARACH: CLOSE TO YOU (I,772); BURT BACHARACH—OPUS NO. 3 (I,774); THE DIONNE WARWICK SPECIAL (I,1289); FELICIANO—VERY SPECIAL (I,1560); THE MAGICAL MUSIC OF BURT BACHARACH (I,2822); MOVIN' (I,3117)

BACHER, ROBERT
OPPENHEIMER (II,1919)

BACKER, BRIAN
TOO GOOD TO BE TRUE (II,2643)

BACKES, ALICE
BACHELOR PARTY (I,324); THE DANNY KAYE SHOW (I,1144); THE MUNSTERS (I,3158); WHATEVER HAPPENED TO DOBIE GILLIS? (II,2774)

BACKLINE, SUSAN
CATALINA C-LAB (II,458)

BACKPORCH MAJORITY, THE
THE BUDDY GRECO SHOW (I,754); THE BURL IVES THANKSGIVING SPECIAL (I,763)

BACKUS, ALICE
BACHELOR FATHER (I,323); HAZEL (I,1982)

BACKUS, HENNY
ACCIDENTAL FAMILY (I,17); BLONDIE (I,487)

BACKUS, JIM
ACCIDENTAL FAMILY (I,17); ALCOA/GOODYEAR THEATER (I,107); BLONDIE (I,487); THE BRADY BUNCH (II,362); CONFLICT (I,1036); CONTINENTAL SHOWCASE (I,1041); DAMN YANKEES (I,1128); ESPIONAGE (I,1464); THE FAMOUS ADVENTURES OF MR. MAGOO (I,1524); THE FEATHER AND FATHER GANG: NEVER CON A KILLER (II,846); GILLIGAN'S ISLAND (II,990); GILLIGAN'S ISLAND: RESCUE FROM GILLIGAN'S ISLAND (II,991); GILLIGAN'S ISLAND: THE CASTAWAYS ON GILLIGAN'S ISLAND (II,992); GILLIGAN'S ISLAND: THE HARLEM GLOBETROTTERS ON GILLIGAN'S ISLAND (II,993); GILLIGAN'S PLANET (II,994); THE GOOD GUYS (I,1847); HAPPY BIRTHDAY, AMERICA (II,1083); HAVE I GOT A CHRISTMAS FOR YOU (II,1105); HOLLYWOOD TALENT SCOUTS (I,2094); I MARRIED JOAN (I,2174); THE JIM BACKUS SHOW—HOT OFF THE WIRE (I,2379); MATINEE THEATER (I,2947); MIRACLE ON 34TH STREET (I,3058); MR. MAGOO (I,3142); THE NEW ADVENTURES OF GILLIGAN (II,1813); OF THEE I SING (I,3334); THE PIED PIPER OF HAMELIN (I,3595); STAR STAGE (I,4197); WHAT'S NEW MR. MAGOO? (II,2770); YES, VIRGINIA, THERE IS A SANTA CLAUS (II,2849)

BACKUS, RICHARD
ANOTHER WORLD (II,97); CAMELOT (II,416); FOR RICHER, FOR POORER (II,906); LOVERS AND FRIENDS (II,1549); RYAN'S HOPE (II,2234)

BACON, CATHERINE
LOVE OF LIFE (I,2774)

BACON, KEVIN
THE GUIDING LIGHT (II,1064); MR. ROBERTS (II,1754)

BADDELEY, ANGELA
UPSTAIRS, DOWNSTAIRS (II,2709)

BADDELEY, HERMIONE
CALL HOLME (II,412); CAMP RUNAMUCK (I,811); THE CARA WILLIAMS SHOW (I,843); DEAN MARTIN'S RED HOT SCANDALS OF 1926 (II,648); DEAN MARTIN'S RED HOT SCANDALS PART 2 (II,649); GETTING THERE (II,985); THE GOOD LIFE (I,1848); THE LAST LEAF (II,1437); MAUDE (II,1655); THIS GIRL FOR HIRE (II,2596)

BADE, TOM
O'MALLEY (II,1872)

BADEL, ALAN
THE SECRET OF CHARLES DICKENS (II,2293); SHOGUN (II,2339); THE WOMAN IN WHITE (II,2819)

BADEL, SARAH
THE PALLISERS (II,1944)

BADLAND, ANNETTE
LACE (II,1418)

BADLER, JANE
THE DOCTORS (II,694); ONE LIFE TO LIVE (II,1907); V (II,2713); V: THE FINAL BATTLE (II,2714); V: THE SERIES (II,2715)

BADSEN, MICHAEL
DINER (II,679)

BAER, BUDDY
TALES OF THE VIKINGS (I,4351)

BAER, CYNTHIA
DOCTOR CHRISTIAN (I,1308)

BAER, JOHN
ARROYO (I,266); THE HARDY BOYS AND THE MYSTERY OF GHOST FARM (I,1950); TERRY AND THE PIRATES (I,4405)

BAER, MAX
THE ASPHALT COWBOY (II,111); THE BEVERLY HILLBILLIES (I,417); THE CIRCLE FAMILY (II,527); MARINELAND CARNIVAL (I,2912)

BAER, PARLEY
THE ADDAMS FAMILY (II,11); THE ADVENTURES OF OZZIE AND HARRIET (I,71); THE ANDY GRIFFITH SHOW (I,192); CAROLYN (I,854); DOC (I,1303); THE DON RICKLES SHOW (I,1342); THE DOUBLE LIFE OF HENRY PHYFE (I,1361); GALLAGHER (I,1727); HELLO, LARRY (II,1128); THE MARRIAGE BROKER (I,2920) THE OVER-THE-HILL GANG RIDES AGAIN (I,3432); PUNCH AND JODY (II,2091); THE SLOWEST GUN IN THE WEST (I,4088); SNEAK PREVIEW (I,4100); TRUE GRIT (II,2664)

BAEZ, JOAN
FOLK SOUND U.S.A. (I,1614); JOAN BAEZ (II,1294)

BAFALOUKOS, TED
DINER (II,679)

BAGDASARIAN, CAROL
ACES UP (II,7)

BAGDASARIAN, ROSS
ALVIN AND THE CHIPMUNKS (II,61); THE ALVIN SHOW (I,146); SECRET AGENT (I,3961)

BAGGETTA, VINCENT
CHICAGO STORY (II,506); THE CHICAGO STORY (II,507); THE EDDIE CAPRA MYSTERIES (II,758); GENERAL HOSPITAL (II,964); LOU GRANT (II,1526)

BAGLEY, FUDDLE
RETURN OF THE WORLD'S GREATEST DETECTIVE (II,2142)

BAGOT, MEGAN
SEARCH FOR TOMORROW (II,2284)

BAGUL, DOMINIC
FACTS OF LIFE: THE FACTS OF LIFE GOES TO PARIS (II,806)

BAHAM, TOM
NIGHT GALLERY (I,3288)

BAIL, CHUCK
BOBBY PARKER AND COMPANY (II,344)

BAILE, DEBBIE
A TOWN LIKE ALICE (II,2652)

BAILEY, BILL
CACTUS JIM (I,786); THE M AND M CANDY CARNIVAL (I,2797); MASTER OF THE GAME (II,1647)

BAILEY, BILL 'BOOM BOOM'
MAGIC MIDWAY (I,2818)

BAILEY, CHUCK
A CHRISTMAS CAROL (II,517)

BAILEY, DAVID
ANOTHER WORLD (II,97); MCNAB'S LAB (I,2972)

BAILEY, F LEE
GOOD COMPANY (I,1844); LIE DETECTOR (II,1467); WHODUNIT? (II,2794)

BAILEY, G W
THE EARTHLINGS (II,751); FOG (II,897); GOODNIGHT, BEANTOWN (II,1041); HARDCASE (II,1088); M*A*S*H (II,1569); THE RUNAWAY BARGE (II,2230); SECOND EDITION (II,2288); ST. ELSEWHERE (II,2432); THERE GOES THE NEIGHBORHOOD (II,2582)

BAILEY, GILLIAN
HERE COME THE DOUBLE DECKERS (I,2025)

BAILEY, HILLARY
AS THE WORLD TURNS (II,110); THE DOCTORS (II,694); FAMILY BUSINESS (II,815); NO SOAP, RADIO (II,1856)

BAILEY, JACK
QUEEN FOR A DAY (I,3705); TRUTH OR CONSEQUENCES (I,4619)

BAILEY, JEFF
THERE'S NO TIME FOR LOVE, CHARLIE BROWN (I,4438)

BAILEY, JOEL
FLAMINGO ROAD (II,872); JOE DANCER: MURDER ONE, DANCER 0 (II,1302); THIS IS KATE BENNETT (II,2597)

BAILEY, JOHN ANTHONY
CAR WASH (II,437); HAPPY DAYS (II,1084); THE PLASTICMAN COMEDY/ADVENTURE SHOW (II,2051); WONDERBUG (II,2830)

BAILEY, JONATHAN
ADVENTURES OF CLINT AND MAC (I,55)

BAILEY, PEARL
THE BIG PARTY FOR REVLON (I,427); BING CROSBY AND HIS FRIENDS (I,456); BING CROSBY AND HIS FRIENDS (II,251); BING!. . . A 50TH ANNIVERSARY GALA (II,254); THE BOB HOPE SHOW (I,547); THE BOB HOPE SHOW (I,548); THE BOB HOPE SHOW (I,621); BOB HOPE SPECIAL: BOB HOPE PRESENTS A CELEBRATION WITH STARS OF COMEDY AND MUSIC (II,340); BOB HOPE SPECIAL: HAPPY BIRTHDAY, BOB! (II,326); CAROL CHANNING AND PEARL BAILEY ON BROADWAY (I,850); THE LOVE BOAT (II,1535); THE MEMBER OF THE WEDDING (II,1680); MIKE AND PEARL (I,3038); ONE MORE TIME (I,3386); THE PEARL BAILEY SHOW (I,3501); SILVER SPOONS (II,2355); SOMETHING SPECIAL (I,4118); STAND UP AND CHEER (I,4188)

BAILEY, PHILIP
EARTH, WIND & FIRE IN CONCERT (II,750)

BAILEY, RAYMOND
THE BEVERLY HILLBILLIES (I,417); DEAR MOM, DEAR DAD (I,1202); DOBIE GILLIS (I,1302); THE LIFE OF VERNON HATHAWAY (I,2687); MY SISTER EILEEN (I,3200)

BAILEY, ROBIN
THE GATHERING STORM (I,1741)

BAIN, BARBARA
GILLIGAN'S ISLAND: THE HARLEM GLOBETROTTERS ON GILLIGAN'S ISLAND (II,993); MISSION: IMPOSSIBLE (I,3067); RICHARD DIAMOND, PRIVATE DETECTIVE (I,3785); SAVAGE (I,3924); SPACE: 1999 (II,2409); YOUNG IN HEART (I,4939)

BAIN, COLLINS
ACADEMY THEATER (I,14)

BAIN, CONRAD
ALMOST AMERICAN (II,55); DIFF'RENT STROKES (II,674); THE JIMMY MCNICHOL SPECIAL (II,1292); MAUDE (II,1655); SEARCH FOR TOMORROW (II,2284); TV FUNNIES (II,2672); TWIGS (II,2683)

BAIN, CYNDI
EIGHT IS ENOUGH (II,762)

BAIN, MARY
MEET CORLISS ARCHER (I,2980)

BAIN, SHERRY
WILD AND WOOLEY (II,2798)

BAINTER, FAY
THE DOCTOR (I,1320); EYE WITNESS (I,1494)

BAIO, JIMMY
THE ACADEMY (II,3); THE ACADEMY II (II,4); FAMILY BUSINESS (II,815); FREEMAN (II,923); JOE AND SONS (II,1296); THE LOVE BOAT I (II,1532); SHEEHY AND THE SUPREME MACHINE (II,2322); SOAP (II,2392)

BAIO, JOEY
THE HERO (I,2035)

BAIO, SCOTT
BATTLE OF THE NETWORK STARS (II,168); BATTLE OF THE NETWORK STARS (II,170); BATTLE OF THE NETWORK STARS (II,171); BATTLE OF THE NETWORK STARS (II,172); BATTLE OF THE NETWORK STARS (II,173); BATTLE OF THE NETWORK STARS (II,176); BATTLE OF THE NETWORK STARS (II,178); BLANSKY'S BEAUTIES (II,267); CELEBRITY CHALLENGE OF THE SEXES 3 (II,467); CELEBRITY CHALLENGE OF THE SEXES 5 (II,469); CHARLES IN CHARGE (II,482); CIRCUS OF THE STARS (II,534); CIRCUS OF THE STARS (II,536); DOM DELUISE AND FRIENDS, PART 2 (II,702); HAPPY DAYS (II,1084); JOANIE LOVES CHACHI (II,1295); LEGS (II,1459); LILY FOR PRESIDENT (II,1478); MAGIC WITH THE STARS (II,1599); WE'RE MOVIN' (II,2756); WHO'S WATCHING THE KIDS? (II,2793)

BAIO, STEVEN
HAPPY DAYS (II,1084)

BAIRD, BIL
ART CARNEY MEETS PETER AND THE WOLF (I,272); ART CARNEY MEETS THE SORCERER'S APPRENTICE (I,273); ARTHUR MURRAY'S DANCE PARTY (I,293); THE BIL BAIRD SHOW (I,439); LIFE WITH SNARKY PARKER (I,2693); THE WHISTLING WIZARD (I,4844)

BAIRD, CORA
ART CARNEY MEETS PETER AND THE WOLF (I,272); ART CARNEY MEETS THE SORCERER'S APPRENTICE (I,273); ARTHUR MURRAY'S DANCE PARTY (I,293); THE BIL BAIRD SHOW (I,439); LIFE WITH SNARKY PARKER (I,2693); THE WHISTLING WIZARD (I,4844)

BAIRD, EUGENIE
THE BABY SITTER (I,321)

BAIRD, HARRY
U.F.O. (I,4662)

BAIRD, JIMMY
FURY (I,1721); SUDDEN SILENCE (I,4275)

BAIRD, SHARON
ANNETTE (I,222); THE BAY CITY ROLLERS SHOW (II,187); THE BUGALOOS (I,757); DONNY AND MARIE (II,712); LAND OF THE LOST (II,1425); LIDSVILLE (I,2679); THE MICKEY MOUSE CLUB (I,3025); THE MOUSEKETEERS REUNION (II,1742); THE NEW ZOO REVUE (I,3277); SIGMUND AND THE SEA MONSTERS (II,2352)

BAKALYAN, DICK
THE BOBBY DARIN AMUSEMENT COMPANY (I,672); BORDER PALS (II,352); BUNCO (II,395); PINE CANYON IS BURNING (II,2040); SHOOTING STARS (II,2341)

BAKEN, BONNIE
MIXED DOUBLES (I,3075)

BAKER, ANN
MEET CORLISS ARCHER (I,2980)

BAKER, ART
END OF THE RAINBOW (I,1443); YOU ASKED FOR IT (I,4931)

BAKER, BENNY
F TROOP (I,1499); THE JERK, TOO (II,1282); MCKEEVER AND THE COLONEL (I,2970)

BAKER, CARROLL
ANNE MURRAY'S LADIES' NIGHT (II,89); THE BOB HOPE SHOW (I,625); DANGER (I,1134); THE NEXT VICTIM (I,3281)

BAKER, CATHY
HEE HAW (II,1123)

BAKER, COLIN
THE CITADEL (II,539); DOCTOR WHO (II,689); WAR AND PEACE (I,4765)

BAKER, DAVID LAIN
CALL TO GLORY (II,413)

BAKER, DIANE
THE BLUE AND THE GRAY (II,273); THE D.A.: MURDER ONE (I,1122); THE DANGEROUS DAYS OF KIOWA JONES (I,1140); HERE WE GO AGAIN (I,2028); INHERIT THE WIND (I,2213); KILLER BY NIGHT (I,2537); KOJAK (II,1406); THE LLOYD BRIDGES SHOW (I,2733); LOVE STORY (I,2778); THE POLICE STORY (I,3638); SARGE: THE BADGE OR THE CROSS (I,3915); A WOMAN OF SUBSTANCE (II,2820)

BAKER, DICK
THE WHITE SHADOW (II,2788)

BAKER, DON
PRISONER: CELL BLOCK H (II,2085)

BAKER, FAY
FIRESIDE THEATER (I,1580); MAGGIE (I,2810); PANDORA AND FRIEND (I,3445); STORY THEATER (I,4240)

BAKER, GEORGE
A WOMAN OF SUBSTANCE (II,2820)

BAKER, JERRY
DINAH'S PLACE (II,678)

BAKER, JIM B
FLO (II,891)

BAKER, JOBY
THE CITY (II,541); GOOD MORNING WORLD (I,1850); THE SIX O'CLOCK FOLLIES (II,2368); STONE (II,2470); STONE (II,2471)

BAKER, JOE
THE BIG SHOW (II,243); THE CHEAP SHOW (II,491); THE DES O'CONNOR SHOW (I,1235); FRED AND BARNEY MEET THE THING (II,919); THE KOPYCATS (I,2580); NO SOAP, RADIO (II,1856); THE PLASTICMAN COMEDY/ADVENTURE SHOW (II,2051); THE RICH

LITTLE SHOW (II,2153); THE SKATEBIRDS (II,2375); THE STEVE ALLEN COMEDY HOUR (II,2454); THE THREE ROBONIC STOOGES (II,2609)

BAKER, JOE DON
DOC ELLIOT (I,1305); EISCHIED (II,763)

BAKER, LENNY
THE RUBBER GUN SQUAD (II,2225); SPARROW (II,2410)

BAKER, LESLIE
PRISONER: CELL BLOCK H (II,2085)

BAKER, MELANIE
VALLEY OF THE DINOSAURS (II,2717)

BAKER, PHIL
WHO'S WHOSE? (I,4855)

BAKER, RACHEL
THUNDARR THE BARBARIAN (II,2616)

BAKER, RYAN
COMPUTERCIDE (II,561)

BAKER, STANLEY
CODE NAME: HERACLITUS (I,990); WHO HAS SEEN THE WIND? (I,4847)

BAKER, STUART
HARDCASE (II,1088)

BAKER, SUSAN
DAVE ALLEN AT LARGE (II,621)

BAKER, SYBIL
AH!, WILDERNESS (I,91); THE WOMEN (I,4890)

BAKER, TOM
DOCTOR WHO (II,689); DOCTOR WHO—THE FIVE DOCTORS (II,690)

BAKEWELL, WILLIAM
DAVY CROCKETT (I,1181); DO NOT DISTURB (I,1300); THE PINKY LEE SHOW (I,3600)

BAKEY, ED
THE GEORGIA PEACHES (II,979)

BAKLHAM, ADAM
THE FAR PAVILIONS (II,832)

BAL, JEANNE
BACHELOR FATHER (I,323); MR. NOVAK (I,3145); NBC PLAYHOUSE (I,3243); SID CAESAR INVITES YOU (I,4040)

BALASKI, BELINDA
FORCE FIVE (II,907)

BALBRINA
THE ADVENTURES OF THE SCARLET PIMPERNEL (I,80)

BALCON, JILL
THE STRAUSS FAMILY (I,4253)

BALDARO, BARRY
OUT OF OUR MINDS (II,1933)

BALDAVIN, BARBARA
ADAM-12 (I,31); MEDICAL CENTER (II,1675); THE NEW HEALERS (I,3266); STAR TREK (II,2440); TERROR AT ALCATRAZ (II,2563)

BALDERSON, ROBERT
LITTLE HOUSE: LOOK BACK TO YESTERDAY (II,1490); NIGHT PARTNERS (II,1847); THE OUTLAWS (II,1936)

BALDING, REBECCA
THE DEADLY GAME (II,633); FAMILY TIES (II,819); THE FRENCH ATLANTIC AFFAIR (II,924); LOU GRANT (II,1526); MAKIN' IT (II,1603); MR. & MRS. & MR. (II,1744); SOAP (II,2392)

BALDUZZI, DICK
COCAINE AND BLUE EYES (II,548); HEAVEN ON EARTH (II,1119); THE MARY TYLER MOORE SHOW (II,1640); PETE 'N' TILLIE (II,2027); THE SAN PEDRO BUMS (II,2251); A SHADOW IN THE STREETS (II,2312); SIDE BY SIDE (II,2346); SPIDER-MAN (II,2424); SUSAN AND SAM (II,2506)

BALDWIN, ALEC
CUTTER TO HOUSTON (II,612); THE DOCTORS (II,694); THE SHERIFF AND THE ASTRONAUT (II,2326)

BALDWIN, BARBARA
THE DUDLEY DO-RIGHT SHOW (I,1378); GEORGE OF THE JUNGLE (I,1773)

BALDWIN, BILL
THE ANDY GRIFFITH SHOW (I,193); LEAVE IT TO BEAVER (I,2648); MAYOR OF HOLLYWOOD (I,2962); SAWYER AND FINN (II,2265)

BALDWIN, FAITH
FAITH BALDWIN'S THEATER OF ROMANCE (I,1515)

BALDWIN, JANIT
DOG AND CAT (II,695)

BALDWIN, JUDITH
THE DELPHI BUREAU (I,1224); GILLIGAN'S ISLAND: RESCUE FROM GILLIGAN'S ISLAND (II,991); GILLIGAN'S ISLAND: THE CASTAWAYS ON GILLIGAN'S ISLAND (II,992); TALES OF THE APPLE DUMPLING GANG (II,2536)

BALDWIN, ROY
ALL THE RIVERS RUN (II,43)

BALDWIN, STACEY
GENERAL HOSPITAL (II,964)

BALDWIN, VICTOR
GREEN ACRES (I,1884)

BALENDA, CARLA
THE ADVENTURES OF DR. FU MANCHU (I,58); HEY MULLIGAN (I,2040)

BALENDER, TODD
SUNDAY IN TOWN (I,4285)

BALES, BARBARA
IT'S A GREAT LIFE (I,2256)

BALFOUR, MICHAEL
SABER OF LONDON (I,3879)

BALICANA IVANKO TROUPE, THE
TEXACO STAR THEATER (I,4411)

BALIK, SHELBY
ADAMS HOUSE (II,9); FULL HOUSE (II,937)

BALIN, INA
CALL TO DANGER (I,804); DANGER IN PARADISE (II,617); THE DUPONT SHOW OF THE WEEK (I,1388); THE POLICE STORY (I,3638)

BALIN, RICHARD
CONDOMINIUM (II,566)

BALK, FAIRUZA
THE BEST CHRISTMAS PAGEANT EVER (II,212)

BALKIN, BARRY
MOBY DICK AND THE MIGHTY MIGHTOR (I,3077)

BALL, DEE DEE
THE INA RAY HUTTON SHOW (I,2207)

BALL, KENNY
THE MORECAMBE AND WISE SHOW (II,1734)

BALL, LUCILLE
ANN-MARGRET: FROM HOLLYWOOD WITH LOVE (I,217); THE BOB HOPE SHOW (I,506); THE BOB HOPE SHOW (I,550); THE BOB HOPE SHOW (I,582); THE BOB HOPE SHOW (I,591); THE BOB HOPE SHOW (I,595); THE BOB HOPE SHOW (I,611); THE BOB HOPE SHOW (I,644); THE BOB HOPE SHOW (I,664); BOB HOPE SPECIAL: BOB HOPE IN WHO MAKES THE WORLD LAUGH, PART 2 (II,286); BOB HOPE SPECIAL: BOB HOPE'S 30TH ANNIVERSARY TV SPECIAL (II,293); BOB HOPE SPECIAL: BOB HOPE'S ALL-STAR COMEDY TRIBUTE TO VAUDEVILLE (II,302); BOB HOPE SPECIAL: BOB HOPE'S ROAD TO HOLLYWOOD (II,313); BOB HOPE SPECIAL: BOB HOPE'S WORLD OF COMEDY (II,323); BOB HOPE SPECIAL: HAPPY BIRTHDAY, BOB! (II,325); BOB HOPE SPECIAL: HAPPY BIRTHDAY, BOB! (II,326); BOB HOPE SPECIAL: BOB HOPE'S HILARIOUS UNREHEARSED ANTICS OF THE STARS (II,341); CAROL + 2 (I,853); CBS SALUTES LUCY—THE FIRST 25 YEARS (II,459); CBS: ON THE AIR (II,460); CHER AND OTHER FANTASIES (II,497); CIRCUS OF THE STARS (II,531); THE DESILU PLAYHOUSE (I,1237); THE DESILU REVUE (I,1238); THE DINAH SHORE SPECIAL—LIKE HEP (I,1282); EVERYTHING YOU ALWAYS WANTED TO KNOW ABOUT JACK BENNY AND WERE AFRAID TO ASK (I,1482); FROM HOLLYWOOD WITH LOVE: THE ANN-MARGRET SPECIAL (I,1687); GENE KELLY. . . AN AMERICAN IN PASADENA (II,962); GENERAL ELECTRIC'S ALL-STAR ANNIVERSARY (II,963); HAPPY ANNIVERSARY AND GOODBYE (II,1082); HAVE GIRLS—WILL TRAVEL (I,1967); HEDDA HOPPER'S HOLLYWOOD (I,2002); HERE'S LUCY (II,1135); I LOVE LUCY (I,2171); THE JACK BENNY SPECIAL (I,2296); JACK BENNY'S 20TH ANNIVERSARY TV SPECIAL (I,2303); JACK BENNY'S BIRTHDAY SPECIAL (I,2299); THE LUCILLE BALL COMEDY HOUR (I,2783); A LUCILLE BALL SPECIAL STARRING LUCILLE BALL AND DEAN MARTIN (II,1557); A LUCILLE BALL SPECIAL STARRING LUCILLE BALL AND JACKIE GLEASON (II,1558); A LUCILLE BALL SPECIAL: WHAT NOW CATHERINE CURTIS? (II,1560); THE LUCILLE BALL SPECIAL (II,1559); THE LUCILLE BALL-DESI ARNAZ SHOW (I,2784); LUCY COMES TO NASHVILLE (II,1561); LUCY IN LONDON (I,2789); LUCY MOVES TO NBC (II,1562); THE LUCY SHOW (I,2791); THE MILTON BERLE SPECIAL (I,3052); THE MUSIC MART (II,1774); A SALUTE TO STAN LAUREL (I,3892); A SALUTE TO TELEVISION'S 25TH ANNIVERSARY (I,3893); SHOW BUSINESS SALUTE TO MILTON BERLE (I,4029); SHOW OF THE YEAR (I,4031);

SINATRA—THE FIRST 40 YEARS (II,2359); STEVE AND EYDIE. . . ON STAGE (I,4223); SUPER COMEDY BOWL 1 (I,4288); SWING OUT, SWEET LAND (II,2515); A TRIBUTE TO "MR. TELEVISION" MILTON BERLE (II,2658); TWELVE STAR SALUTE (I,4638); THE WONDERFUL WORLD OF BURLESQUE I (I,4895)

BALL, ROBERT E
DUFFY (II,739); THE HARVEY KORMAN SHOW (II,1103)

BALLANTINE, CARL
THE ALL-STAR SUMMER REVUE (I,139); THE ANDY WILLIAMS SHOW (I,206); CAMP GRIZZLY (II,418); CAR 54, WHERE ARE YOU? (I,842); MCHALE'S NAVY (I,2969); ONE IN A MILLION (II,1906); OUT OF THE BLUE (I,3422); THE QUEEN AND I (I,3704); SAGA OF SONORA (I,3884); WHERE'S THE FIRE? (II,2785)

BALLANTINE, SARA
THE JOHN FORSYTHE SHOW (I,2410)

BALLANTYNE, ELSPETH
PRISONER: CELL BLOCK H (II,2085)

BALLANTYNE, PAUL
FAMILY HONOR (I,1522) THE TEMPEST (I,4391)

BALLARD, DAVE
CAPTAIN VIDEO AND HIS VIDEO RANGERS (I,838); COLONEL STOOPNAGLE'S STOOP (I,1006)

BALLARD, KAYE
THE BOB HOPE SHOW (I,643); THE COLGATE COMEDY HOUR (I,998); THE DORIS DAY SHOW (I,1355); THE ENGELBERT HUMPERDINCK SHOW (I,1445); HENRY MORGAN'S GREAT TALENT HUNT (I,2017); IRENE (II,1245); THE MEL TORME SHOW (I,2995); THE MOTHERS-IN-LAW (I,3110); THE PERRY COMO SHOW (I,3532); THE STEVE ALLEN COMEDY HOUR (II,2454)

BALLARD, MICHAEL
ALICE (II,33)

BALLEN, TONY
GOLIATH AWAITS (II,1025); NICHOLS AND DYMES (II,1841)

BALLERINAS DIANA, THE
THE MUSIC OF GERSHWIN (I,3174)

BALLESTEROS, CARLOS
TEN WHO DARED (II,2559)

BALLINGER, ART
ADAM-12 (I,31); DRAGNET (I,1371); EMERGENCY! (II,775)

BALSAM, MARTIN
ALCOA/GOODYEAR THEATER (I,107); ALFRED HITCHCOCK PRESENTS (I,115); ARCHIE BUNKER'S PLACE (II,105); THE DEFENDER (I,1219); DUPONT SHOW OF THE MONTH (I,1387); THE GREATEST GIFT (I,1880); THE MILLIONAIRE (II,1700); MRS R.—DEATH AMONG FRIENDS (II,1759); THE TIME ELEMENT (I,4504); THE TWILIGHT ZONE (I,4651); WINTERSET (I,4882)

BALSAM, TALIA
CRAZY TIMES (II,602); FIT FOR A KING (II,866); PUNKY BREWSTER (II,2092); STICKIN' TOGETHER (II,2465); TAXI (II,2546); WHEN THE WHISTLE BLOWS (II,2781)

BALSON, ALLISON
GOLDIE AND KIDS: LISTEN TO ME (II,1021); THE LIFE AND TIMES OF EDDIE ROBERTS (II,1468); LITTLE HOUSE ON THE PRAIRIE (II,1487); LITTLE HOUSE:(A NEW BEGINNING (II,1488); LITTLE HOUSE: BLESS ALL THE DEAR CHILDREN (II,1489); LITTLE HOUSE: LOOK BACK TO YESTERDAY (II,1490)

BALSON, PAUL JOHN
FANTASY ISLAND (II,829)

BALTZELL, DEBORAH
I'M A BIG GIRL NOW (II,1218); LOVE AT FIRST SIGHT (II,1530); MARRIAGE IS ALIVE AND WELL (II,1633)

BAMBER, JUDY
ANYBODY CAN PLAY (I,234)

BAMFORD, FREDA
THE SNOW GOOSE (I,4103)

BAMFORD, GEORGE
ANOTHER WORLD (II,97); SEARCH FOR TOMORROW (II,2284)

BANACUDA, JOHNNY
ALL ABOUT MUSIC (I,127)

BANAS, CARL
THE RELUCTANT DRAGON AND MR. TOAD (I,3762)

BANCROFT, ANNE
ABC STAGE '67 (I,6); THE ALCOA HOUR (I,108); ANNIE AND THE HOODS (II,91); ANNIE, THE WOMAN IN THE LIFE OF A MAN (I,226); THE BOB HOPE SHOW (I,599); THE BOB HOPE SHOW (I,622); CLIMAX! (I,976); THE LUX VIDEO THEATER (I,2795); MARCO POLO (II,1626); THE PERRY COMO SPECIAL (I,3535); PLAYHOUSE 90 (I,3623)

BANFIELD, BEVER-LEIGH
THE CURSE OF DRACULA (II,611); EMERALD POINT, N.A.S. (II,773); OPEN ALL NIGHT (II,1914); ROOTS: THE NEXT GENERATIONS (II,2213); THIS IS KATE BENNETT (II,2597)

BANG, JOY
THE KOWBOYS (I,2584); THE PSYCHIATRIST: GOD BLESS THE CHILDREN (I,3685)

BANGERT, JOHNNY
MARGIE (I,2909)

BANGHAM, ROSEMARY
RENDEZVOUS WITH MUSIC (I,3768)

BANGHART, KENNETH
15 WITH FAYE (I,1570)

BANHAM, RUSS
JOE'S WORLD (II,1305)

BANIEL, NOA
A WOMAN CALLED GOLDA (II,2818)

BANK, FRANK
LEAVE IT TO BEAVER (I,2648); STILL THE BEAVER (II,2466)

BANKE, RICHARD
DANIEL BOONE (I,1141)

BANKHEAD, TALLULAH
THE ALL-STAR REVUE (I,138); BATMAN (I,366); THE BIG PARTY FOR REVLON (I,427); THE UNITED STATES STEEL HOUR (I,4677)

BANKS JR, MONTY
CONCERNING MISS MARLOWE (I,1032); SAY IT WITH ACTING (I,3928)

BANKS, C TILLY
HARRIS AND COMPANY (II,1096)

BANKS, DAVID
DOCTOR WHO—THE FIVE DOCTORS (II,690)

BANKS, EMILY
ALL THAT GLITTERS (II,41); BRILLIANT BENJAMIN BOGGS (I,730); THE TIM CONWAY SHOW (I,4502); YOUNG LIVES (II,2866)

BANKS, JOAN
DATE WITH THE ANGELS (I,1164); DOBIE GILLIS (I,1302); PRIVATE SECRETARY (I,3672)

BANKS, JONATHAN
THE BOYS IN BLUE (II,359); THE FIGHTING NIGHTINGALES (II,855); G.I.'S (II,949); THE GANGSTER CHRONICLES (II,957); THE GIRL IN THE EMPTY GRAVE (II,997); MICKEY SPILLANE'S MIKE HAMMER: MURDER ME, MURDER YOU (II,1694); REPORT TO MURPHY (II,2134)

BANKS, TOMMY
CELEBRITY REVUE (II,475)

BANKS, TYLER
DALLAS (II,614)

BANNEN, IAN
DEATH IN DEEP WATER (I,1206); THE GATHERING STORM (I,1741); JANE EYRE (I,2339); JOHNNY BELINDA (I,2417); MACBETH (I,2801); TERROR FROM WITHIN (I,4404)

BANNER, JILL
AMANDA FALLON (I,151)

BANNER, JOHN
THE CHICAGO TEDDY BEARS (I,934); HOGAN'S HEROES (I,2069)

BANNISTER, HARRY
FIRESIDE THEATER (I,1580); THE GIRLS (I,1811)

BANNISTER, JEFF
240-ROBERT (II,2689)

BANNISTER, PAUL
THE BRIGHTER DAY (I,728)

BANNON, JACK
LOU GRANT (II,1526); MAUREEN (II,1656); SUSAN AND SAM (II,2506); TRAUMA CENTER (II,2655)

BANNON, JIM
THE ADVENTURES OF CHAMPION (I,54); HAWKINS FALLS, POPULATION 6200 (I,1977)

BAR-YOTAM, REUBEN
CASABLANCA (II,450); MASADA (II,1642)

BARA, FAUTSO
THE RENEGADES (II,2133)

BARA, NINA
SPACE PATROL (I,4144)

BARAB, NIRA
THE FUZZ BROTHERS (I,1722)

BARAGREY, JOHN
LITTLE WOMEN (I,2720); TONIGHT AT 8:30 (I,4556)

BARAL, EILEEN
NANNY AND THE PROFESSOR (I,3225); OCCASIONAL WIFE (I,3325)

BARALEY, JOHN
JENNIE: LADY RANDOLPH CHURCHILL (II,1277)

BARASH, OLIVIA
DAUGHTERS (II,620); IN THE BEGINNING (II,1229); OUT OF THE BLUE (II,1934); THROUGH THE MAGIC PYRAMID (II,2615)

BARAT, MAXINE
AND EVERYTHING NICE (I,186)

BARBAR, BOBBY
THE ABBOTT AND COSTELLO SHOW (I,2)

BARBEAU, ADRIENNE
BATTLE OF THE NETWORK STARS (II,163); BATTLE OF THE NETWORK STARS (II,165); THE CELEBRITY FOOTBALL CLASSIC (II,474); THE FIGHTING NIGHTINGALES (II,855); HAVE I GOT A CHRISTMAS FOR YOU (II,1105); HAVING BABIES I (II,1107); MAUDE (II,1655); RETURN TO FANTASY ISLAND (II,2144); TOP OF THE HILL (II,2645); TOURIST (II,2651); VALENTINE MAGIC ON LOVE ISLAND (II,2716)

BARBEE, TODD
A CHARLIE BROWN THANKSGIVING (I,909); IT'S A MYSTERY, CHARLIE BROWN (I,2259); IT'S THE EASTER BEAGLE, CHARLIE BROWN (I,2275); THERE'S NO TIME FOR LOVE, CHARLIE BROWN (I,4438); YOU'RE NOT ELECTED, CHARLIE BROWN (I,4967)

BARBER, AVA
THE LAWRENCE WELK SHOW (I,2644); MEMORIES WITH LAWRENCE WELK (II,1681)

BARBER, ELLEN
AS THE WORLD TURNS (II,110); RYAN'S HOPE (II,2234); THE SECRET STORM (I,3969)

BARBER, FRAN
THE KATE SMITH HOUR (I,2508)

BARBER, GILLIAN
ALICE IN WONDERLAND (I,120)

BARBER, GLYNIS
BLAKE'S SEVEN (II,264)

BARBER, OLLIE
THE ETHEL WATERS SHOW (I,1468)

BARBERA, MICHAEL
I DREAM OF JEANNIE (I,2167)

BARBI, VINCENT
SIMON LASH (I,4052)

BARBOUR, JOHN
THE GONG SHOW (II,1026); ON STAGE AMERICA (II,1893); REAL PEOPLE (II,2118)

BARBOUR, JOYCE
THE LILLI PALMER THEATER (I,2704)

BARBOUR, KEITH
FITZ AND BONES (II,867); TERROR AT ALCATRAZ (II,2563)

BARBUTTI, PETE
THE GARRY MOORE SHOW (I,1740); THE LAS VEGAS SHOW (I,2619); PEGGY FLEMING AT SUN VALLEY (I,3507)

BARCLAY, DON
THE WALT DISNEY CHRISTMAS SHOW (I,4755)

BARCLAY, JERRY
THE BARRETTS OF WIMPOLE STREET (I,361); HIS MODEL WIFE (I,2064)

BARCLAY, MARY
ADVENTURES OF CLINT AND MAC (I,55)

BARCLAY, SIMON
CASANOVA (II,451)

BARCROFT, JUDITH
ALL MY CHILDREN (II,39); ANOTHER WORLD (II,97); AS THE WORLD TURNS (II,110); RYAN'S HOPE (II,2234); A WORLD APART (I,4910)

BARCROFT, ROY
DOCTOR MIKE (I,1316); THE FURTHER ADVENTURES OF SPIN AND MARTY (I,1720); LARAMIE (I,2614); LAST STAGECOACH WEST (I,2629); THE NEW ADVENTURES OF SPIN AND MARTY (I,3254); SPIN AND MARTY (I,4158)

BARD, KATHERINE
THE DOCTOR (I,1320); FRONT ROW CENTER (I,1694); JOHNNY BELINDA (I,2415); LEAVE IT TO LARRY (I,2649); REVLON MIRROR THEATER (I,3781)

BARD, RACHEL
IF I LOVE YOU, AM I TRAPPED FOREVER? (II,1222)

BARDETTE, TREVOR
THE LIFE AND LEGEND OF WYATT EARP (I,2681); SUDDEN SILENCE (I,4275)

BARDFORD, ANDREW
SMILEY'S PEOPLE (II,2383)

BARDOT, BRIGITTE
BRIGITTE BARDOT (I,729)

BARE, BOBBY
CHRISTMAS LEGEND OF NASHVILLE (II,522); COUNTRY COMES HOME (II,585); JOHNNY CASH: THE FIRST 25 YEARS (II,1336)

BARGY, JEAN
BLUES BY BARGY (I,494)

BARI, LENNY
FISH (II,864); FISHERMAN'S WHARF (II,865)

BARI, LYNN
ARROYO (I,266); BOSS LADY (I,694); THE DETECTIVE'S WIFE (I,1244); THE LUX VIDEO THEATER (I,2795); PEPSI-COLA PLAYHOUSE (I,3523); SCHAEFER CENTURY THEATER (I,3935); SCREEN DIRECTOR'S PLAYHOUSE (I,3946)

BARID, JIMMY
MR. NOVAK (I,3145)

BARING, VICTOR
CASANOVA (II,451)

BARKER, BOB
CIRCUS OF THE STARS (II,533); END OF THE RAINBOW (I,1443); THE FAMILY GAME (I,1520); THE NEW PRICE IS RIGHT (I,3270); THE PRICE IS RIGHT (I,3665); THE PRICE IS RIGHT (II,2079); THAT'S MY LINE (II,2579); TRUTH OR CONSEQUENCES (I,4619); TRUTH OR CONSEQUENCES (I,4620); TV'S FUNNIEST GAME SHOW MOMENTS (II,2676)

BARKER, DON
HOMICIDE (I,2101)

BARKER, EDDIE
RETURN OF THE MAN FROM U.N.C.L.E.: THE 15 YEARS LATER AFFAIR (II,2140)

BARKER, RONNIE
OPEN ALL HOURS (II,1913); THE TWO RONNIES (II,2694)

BARKER, WILLIAM
MR. ROGERS' NEIGHBORHOOD (I,3150)

BARKIN, ELLEN
MURDER INK (II,1769); PAROLE (II,1960)

BARKIN, MARCIE
PLEASE STAND BY (II,2057); THE STOCKARD CHANNING SHOW (II,2468); TOP TEN (II,2648)

BARKIN, SANDRA
THE GAY COED (I,1742)

BARKLEY, LUCILLE
FOUR STAR PLAYHOUSE (I,1652)

BARKLEY, ROGER
BEDTIME STORIES (II,199); THE NAMEDROPPERS (I,3219)

BARKO, STEPHEN
COCAINE AND BLUE EYES (II,548)

BARKWORTH, PETER
CROWN MATRIMONIAL (I,1106)

BARLOW, THELMA
CORONATION STREET (I,1054)

BARNARD, HENRY
THE GREATEST GIFT (I,1880)

BARNE, PATRICIA
HAWAII FIVE-O (II,1110)

BARNES, ANN
BLONDIE (I,486)

BARNES, BILLY
PURE GOLDIE (I,3696)

BARNES, CHRISTOPHER
BIG CITY BOYS (II,232); RISE AND SHINE (II,2179); THROUGH THE MAGIC PYRAMID (II,2615); YOU'RE GONNA LOVE IT HERE (II,2860)

BARNES, CHUCK
THE MAGIC LAND OF ALLAKAZAM (I,2817)

BARNES, DERYCK
THE HANDS OF CORMAC JOYCE (I,1927)

BARNES, EMILY
COUNTRY STYLE (I,1072); STAGE 13 (I,4183)

BARNES, GEORGENE
THE JONATHAN WINTERS SHOW (I,2443)

BARNES, JOANNA
21 BEACON STREET (I,4646); 333 MONTGOMERY (I,4487); ANYTHING FOR MONEY (I,236); THE BETTY WHITE SHOW (II,224); DATELINE: HOLLYWOOD (I,1167); EDDIE (I,1405); EXECUTIVE SUITE (II,796); FORD TELEVISION THEATER (I,1634); PATRICK STONE (I,3476); TARZAN THE APE MAN (I,4363); THE TRIALS OF O'BRIEN (I,4605)

BARNES, MAE
KITTY FOYLE (I,2571)

BARNES, MARJORIE
FEELING GOOD (I,1559)

BARNES, PAUL
HOLLYWOOD JUNIOR CIRCUS (I,2084)

BARNES, PRISCILLA
THE AMERICAN GIRLS (II,72); THE FUNNIEST JOKE I EVER HEARD (II,940); SCRUPLES (II,2281); THREE'S COMPANY (II,2614); THE WILD WOMEN OF CHASTITY GULCH (II,2801)

BARNES, RAYFORD
THE WILD WOMEN OF CHASTITY GULCH (II,2801)

BARNES, RAYMOND
DYNASTY (II,747)

BARNES, STEVE
THE BOOK OF LISTS (II,350)

BARNES, SUSAN
LAVERNE AND SHIRLEY (II,1446); SHE'S IN THE ARMY NOW (II,2320)

BARNES, WALTER
MOVIN' ON (I,3118); TALES OF THE VIKINGS (I,4351); WALKING TALL (II,2738)

BARNET, CHARLIE
ONE MORE TIME (I,3385)

BARNETT, EILEEN
DAYS OF OUR LIVES (II,629)

BARNETT, MARK
OPRYLAND: NIGHT OF STARS AND FUTURE STARS (II,1921)

BARNETT, MARY LOU
DOCTOR DAN (II,684)

BARNETT, ROBERT
SUNDAY IN TOWN (I,4285)

BARNETT, VINCE
GIRL ON THE RUN (I,1809)

BARNEY, JAY
FIRST LOVE (I,1583)

BARNSTABLE, CYB
OPERATING ROOM (II,1915); QUARK (II,2095)

BARNSTABLE, TRISH
OPERATING ROOM (II,1915); QUARK (II,2095)

BARON, DAVID
THE LONG DAYS OF SUMMER (II,1514)

BARON, JOAN-CARROLL
RAGE OF ANGELS (II,2107)

BARON, JOANNE
LOVERS AND OTHER STRANGERS (II,1550)

BARON, SANDY
DELLA (I,1222); HEY, LANDLORD (I,2039); THE POLICE STORY (I,3638)

BARONSKI, CHRISTINE
MURDER INK (II,1769)

BARR, DOUGLAS
BATTLE OF THE NETWORK STARS (II,173); BATTLE OF THE NETWORK STARS (II,174); CIRCUS OF THE STARS (II,537); THE FALL GUY (II,811); HOW DO I KILL A THIEF—LET ME COUNT THE WAYS (II,1195); SEMI-TOUGH (II,2299); WHEN THE WHISTLE BLOWS (II,2781)

BARR, JULIA
THE ADAMS CHRONICLES (II,8); ALL MY CHILDREN (II,39)

BARR, LEONARD
MARY (II,1637); SZYSZNYK (II,2523)

BARR, SHARON
ARCHER—FUGITIVE FROM THE EMPIRE (II,103)

BARRA, ELENA
IT HAPPENS IN SPAIN (I,2247)

BARRA, VANDA
A LUCILLE BALL SPECIAL STARRING LUCILLE BALL AND DEAN MARTIN (II,1557)

BARRETT, BARBARA
TOP 40 VIDEOS (II,2644)

BARRETT, CLAUDIA
STUDIO '57 (I,4267)

BARRETT, DOROTHY
THE THORNTON SHOW (I,4468)

BARRETT, HELEN
THOSE ENDEARING YOUNG CHARMS (I,4469)

BARRETT, JOE
HECK'S ANGELS (II,1122)

BARRETT, JUNE
DOCTOR STRANGE (II,688)

BARRETT, LAURINDA
FOR RICHER, FOR POORER (II,906); THE PATRIOTS (I,3477); THE TURN OF THE SCREW (I,4624)

BARRETT, LYNNE
MELODY, HARMONY, RHYTHM (I,2998)

BARRETT, MAJEL
THE DESILU REVUE (I,1238); GENESIS II (I,1760); LEAVE IT TO BEAVER (I,2648); PLANET EARTH (I,3615); STAR TREK (II,2439); STAR TREK (II,2440)

BARRETT, MAXINE
WINDOW SHADE REVUE (I,4874)

BARRETT, NANCY
DARK SHADOWS (I,1157); THE DOCTORS (II,694); ONE LIFE TO LIVE (II,1907); RYAN'S HOPE (II,2234)

BARRETT, RAY
STINGRAY (I,4228); THUNDERBIRDS (I,4497)

BARRETT, RONA
ALAN KING LOOKS BACK IN ANGER—A REVIEW OF 1972 (I,99); DATELINE: HOLLYWOOD (I,1167); THE MORNING SHOW (I,3103) THE OLIVIA NEWTON-JOHN SHOW (II,1885); PEOPLE TO PEOPLE (II,2000); RONA BARRETT'S HOLLYWOOD (I,3839); RONA LOOKS AT JAMES, MICHAEL, ELLIOTT, AND BURT (II,2205); RONA LOOKS AT KATE, PENNY, TONI, AND CINDY (II,2206); RONA LOOKS AT RAQUEL, LIZA, CHER AND ANN-MARGRET (II,2207); A SPECIAL OLIVIA NEWTON-JOHN (II,2418); TELEVISION INSIDE AND OUT (II,2552); THE TOMORROW SHOW (II,2635)

BARRETT, VICKIE
THE COLLEGE BOWL (I,1001)

BARRIE, BARBARA
79 PARK AVENUE (II,2310); ABC'S MATINEE TODAY (I,7); ALL TOGETHER NOW (II,45); BACKSTAIRS AT THE WHITE HOUSE (II,133); BAREFOOT IN THE PARK (II,151); BARNEY MILLER (II,154); BREAKING AWAY (II,371); DECOY (I,1217); DIANA (I,1259); DOUBLE TROUBLE (II,724); FULL MOON OVER BROOKLYN (I,1703); KOSTA AND HIS FAMILY (I,2581); THE PHIL SILVERS SHOW (I,3580); PRIVATE BENJAMIN (II,2087); REGGIE (II,2127); TUCKER'S WITCH (II,2667)

BARRIE, ELLEN
BABES IN TOYLAND (I,318)

BARRIE, WENDY
THE ADVENTURES OF OKY DOKY (I,70); STARLIGHT THEATER (I,4201); STARS IN KHAKI AND BLUE (I,4209); THROUGH WENDY'S WINDOW (I,4495)

BARRINGTON, JOSEPHINE
PAUL BERNARD—PSYCHIATRIST (I,3485)

BARRIS, CHUCK
THE CHUCK BARRIS RAH-RAH SHOW (II,525); THE GONG SHOW (II,1026); THE GONG SHOW (II,1027)

BARRIS, MARTY
COMEDY TONIGHT (I,1027); THE LATE FALL, EARLY SUMMER BERT CONVY SHOW (II,1441)

BARRIS, NORMAN
A CRY OF ANGELS (I,1111)

BARRON, ALLISON
GIDGET'S SUMMER REUNION (I,1799)

BARRON, BARBARA
NIGHT PARTNERS (II,1847)

BARRON, BOB
THE DESILU REVUE (I,1238)

BARRON, GERALDINE
TESTIMONY OF TWO MEN (II,2564)

BARRON, JEANNA
GIMME A BREAK (II,995)

BARRON, JEFFREY
ALAN KING'S THIRD ANNUAL FINAL WARNING!! (II,31); MARIE (II,1629); TV FUNNIES (II,2672)

BARRON, JOHN
THE FALL AND RISE OF REGINALD PERRIN (II,810); TYCOON: THE STORY OF A WOMAN (II,2697)

BARRON, KEITH
MIRROR OF DECEPTION (I,3060)

BARROW, BERNARD
THE GAY COED (I,1742); RYAN'S HOPE (II,2234); THE SECRET STORM (I,3969)

BARROWS, DAN
MIXED NUTS (II,1714)

BARRY, DON 'RED'

SAWYER AND FINN (II,2265)

BARRY, DONALD
MR. NOVAK (I,3145); PUNCH AND JODY (II,2091); SECOND CHANCE (I,3957); SURFSIDE 6 (I,4299)

BARRY, GENE
THE ADVENTURER (I,43); ALFRED HITCHCOCK PRESENTS (I,115); AMOS BURKE, SECRET AGENT (I,180); APPOINTMENT WITH ADVENTURE (I,246); BAT MASTERSON (I,363); BURKE'S LAW (I,762); THE CLOCK (I,978); FORD TELEVISION THEATER (I,1634); THE GIRL, THE GOLD WATCH AND DYNAMITE (II,1001); HONEY WEST: WHO KILLED THE JACKPOT? (I,2106); JANE WYMAN PRESENTS THE FIRESIDE THEATER (I,2345); THE LORETTA YOUNG THEATER (I,2756); THE NAME OF THE GAME (I,3217); OUR MISS BROOKS (I,3415); PRESCRIPTION: MURDER (I,3660); RANSOM FOR ALICE (II,2113); TIPTOE THROUGH TV (I,4524); TV READERS DIGEST (I,4630); VARIETY: THE WORLD OF SHOW BIZ (I,4704); WAR CORRESPONDENT (I,4767)

BARRY, GUERIN
FITZ AND BONES (II,867); TERROR AT ALCATRAZ (II,2563)

BARRY, HAROLD
VERSATILE VARIETIES (I,4713)

BARRY, IVOR
BRIDGET LOVES BERNIE (I,722); THE CABOT CONNECTION (II,408); CALL HOLME (II,412); DANIEL BOONE (I,1142); THE GOSSIP COLUMNIST (II,1045); MOMENT OF TRUTH (I,3084); MR. DEEDS GOES TO TOWN (I,3134); SCARLETT HILL (I,3933)

BARRY, J J
THE CORNER BAR (I,1051); THE CORNER BAR (I,1052); HAPPY DAYS (II,1084); MADHOUSE 90 (I,2807); MASQUERADE (I,2940); PEEPING TIMES (II,1990); THREE'S COMPANY (II,2614); TWO GUYS FROM MUCK (II,2690)

BARRY, JACK
THE $100,000 BIG SURPRISE (I,3378); THE BIG SURPRISE (I,433); BREAK THE BANK (II,369); CONCENTRATION (I,1031); THE GENERATION GAP (I,1759); HIGH LOW QUIZ (I,2053); THE JOE DIMAGGIO SHOW (I,2399); JOKER! JOKER!! JOKER!!! (II,1340); THE JOKER'S WILD (II,1341); JUVENILE JURY (I,2501); LIFE BEGINS AT EIGHTY (I,2682); OH, BABY! (I,3340); THE REEL GAME (I,3759); TIC TAC DOUGH (I,4498); TWENTY-ONE (I,4645); WINKY DINK AND YOU (I,4879)

BARRY, JUNE
THE FORSYTE SAGA (I,1645)

BARRY, MATTHEW
IVAN THE TERRIBLE (II,1260)

BARRY, PATRICIA
ALL MY CHILDREN (II,39); DAYS OF OUR LIVES (II,629); THE DUPONT SHOW OF THE WEEK (I,1388); EDDIE (I,1405); FIRST LOVE (I,1583); FOR RICHER, FOR POORER (II,906); THE FREEWHEELERS (I,1683); HARRIS AGAINST THE WORLD (I,1956); THE JERK, TOO (II,1282); STUDIO ONE (I,4268); THE TEXAS RANGERS (II,2567)

BARRY, ROBERT
THE ALDRICH FAMILY (I,111)

BARRYMORE, ETHEL
THE ETHEL BARRYMORE THEATER (I,1467); HOLLYWOOD OPENING NIGHT (I,2087); SVENGALI AND THE BLONDE (I,4310); TEXACO COMMAND PERFORMANCE (I,4408)

BART, TEDDY
MUSIC CITY U.S.A. (I,3167)

BARTELL, HARRY
MOBILE TWO (II,1718)

BARTELL, JACK
BELLE STARR (I,391)

BARTELL, RICHARD
SAM HILL (I,3895)

BARTER, SYLVIA
SEPARATE TABLES (II,2304)

BARTH, ED
ALAN KING'S FINAL WARNING (II,29); ETHEL IS AN ELEPHANT (II,783); HOW TO SURVIVE THE 70S AND MAYBE EVEN BUMP INTO HAPPINESS (II,1199); HUSBANDS AND WIVES (II,1208); HUSBANDS, WIVES AND LOVERS (II,1209); NUMBER 96 (II,1869); THE ORPHAN AND THE DUDE (II,1924); RICH MAN, POOR MAN—BOOK I (II,2161); SHAFT (I,3993); SIMON AND SIMON (II,2357); SUNDAY FUNNIES (II,2493)

BARTH, PETER
BOZO THE CLOWN (I,702)

BARTHOLOMEW, FREDDIE
CAMEO THEATER (I,808) FORD THEATER HOUR (I,1635)

BARTLE, BARRY
HOW DO I KILL A THIEF—LET ME COUNT THE WAYS (II,1195)

BARTLETT, BONNIE
CELEBRITY (II,463); IKE (II,1223); LITTLE HOUSE ON THE PRAIRIE (II,1487); LOVE OF LIFE (I,2774); ST. ELSEWHERE (II,2432); V (II,2713)

BARTLETT, DEBORAH
DREAM HOUSE (II,734); STRIPPERS (II,2481)

BARTLETT, MARTINE
BEN JERROD (I,396); SANDBURG'S LINCOLN (I,3907)

BARTLETT, ROBIN
RYAN'S HOPE (II,2234)

BARTLETT, TONY
WELCOME TRAVELERS (I,4791)

BARTO, BETTY
TELE-VARIETIES (I,4377)

BARTO, DOMINIC
THE GANGSTER CHRONICLES (II,957)

BARTOK, EVA
ON TRIAL (I,3363)

BARTOLD, NORMAN
ADAM'S RIB (I,32); LANDON, LANDON & LANDON (II,1426); LAVERNE AND SHIRLEY (II,1446); TEACHERS ONLY (II,2547)

BARTON, ANN
DOCTOR MIKE (I,1316); LEAVE IT TO BEAVER (I,2648)

BARTON, DAN
CAPTAIN AMERICA (II,427); DAN RAVEN (I,1131); IF YOU KNEW TOMORROW (I,2187)

BARTON, EARL
BROADWAY TO HOLLYWOOD (I,739)

BARTON, EILEEN
THE BILL GOODWIN SHOW (I,447); PANORAMA (I,3448); THE SWIFT SHOW (I,4316); VIDEO VILLAGE (I,4731)

BARTON, GARY
EVEL KNIEVEL (II,785)

BARTON, JOAN
A COUPLE OF JOES (I,1078)

BARTON, LAURA
THE PATTY DUKE SHOW (I,3483)

BARTON, MIKE
THE BRIGHTER DAY (I,728)

BARTON, NICKY
13 QUEENS BOULEVARD (II,2591)

BARTON, PETER
BATTLE OF THE NETWORK STARS (II,176); LEADFOOT (II,1452); THE POWERS OF MATTHEW STAR (II,2075); SHIRLEY (II,2330); THREE EYES (II,2605)

BARTY, BILLY
ACE CRAWFORD, PRIVATE EYE (II,6); THE BAY CITY ROLLERS SHOW (II,187); BOB HOPE SPECIAL: BOB HOPES'S STAR-STUDDED SPOOF OF THE NEW TV SEASON— G RATED—WITH GLAMOUR, GLITTER & GAGS (I,317); THE BUGALOOS (I,757); THE CAPTAIN AND TENNILLE (II,429); CHERYL LADD: SCENES FROM A SPECIAL (II,503); CIRCUS BOY (I,959); CIRCUS OF THE STARS (II,530); CLOWN ALLEY (I,979); CLUB OASIS (I,984); CLUB OASIS (I,985); DOCTOR SHRINKER (II,686); DON'T CALL US (II,709); FORD FESTIVAL (I,1629); GREAT DAY (II,1053); THE LIFE AND TIMES OF EDDIE ROBERTS (II,1468); PETER GUNN (I,3562); PUNCH AND JODY (II,2091); THE RED SKELTON SHOW (I,3755); SIGMUND AND THE SEA MONSTERS (II,2352); THE SPIKE JONES SHOW (I,4156); TWIN DETECTIVES (II,2687)

BARUCH, ANDRE
MASTERS OF MAGIC (I,2943)

BARYON, JAMES
JANE WYMAN PRESENTS THE FIRESIDE THEATER (I,2345)

BARYSHNIKOV, MIKHAIL
BARYSHNIKOV IN HOLLYWOOD (II,157); BARYSHNIKOV ON BROADWAY (II,158); BOB HOPE SPECIAL: BOB HOPE ON THE ROAD TO CHINA (II,291)

BARZMAN, ALAN
CARLTON YOUR DOORMAN (II,440)

BAS, CARLOS
THE JACKIE GLEASON SHOW (I,2323); THE JACKIE GLEASON SPECIAL (I,2324)

BASARABA, MARY LOU
SUPER PAY CARDS (II,2500)

BASCH, HARRY
FALCON CREST (II,808); MARY HARTMAN, MARY HARTMAN (II,1638)

BASCOMBE, JERRY
THE MIGHTY HERCULES (I,3034)

BASEHART, RICHARD
ASSIGNMENT: MUNICH (I,302); CITY BENEATH THE SEA (I,965); DUPONT SHOW OF THE MONTH (I,1387); HOW THE WEST WAS WON (II,1196); THE JUDGE (I,2468); KNIGHT RIDER (II,1402); THE PARADINE CASE (I,3453); THE REBELS (II,2121); SHANGRI-LA (I,3996); STONESTREET: WHO KILLED THE CENTERFOLD MODEL? (II,2472); VALLEY FORGE (I,4698); VOYAGE TO THE BOTTOM OF THE SEA (I,4743); W.E.B. (II,2732)

BASELEON, MICHAEL
COVER GIRLS (II,596); THE DOCTORS (I,1323); FLAMINGO ROAD (II,872); FOUR EYES (II,912); HARD KNOX (II,1087); LUCAS TANNER (II,1555); MAN ON A STRING (I,2877)

BASHAM, LARRY
SAWYER AND FINN (II,2265)

BASIE, COUNT
ALL-STAR SWING FESTIVAL (II,53); ASTAIRE TIME (I,305); THE BIG BAND AND ALL THAT JAZZ (I,422); DUKE ELLINGTON. . . WE LOVE YOU MADLY (I,1380); RODGERS AND HART TODAY (I,3829); THE SWINGIN' YEARS (I,4321)

BASILE, LOUIS
THE SUPER (I,4293)

BASINGER, KIM
CHARLIE'S ANGELS (II,486); DOG AND CAT (II,695); DOG AND CAT (II,696); FROM HERE TO ETERNITY (II,932); FROM HERE TO ETERNITY (II,933)

BASS, EMORY
ANGIE (II,80); MIXED NUTS (II,1714); RENDEZVOUS HOTEL (II,2131); SECOND EDITION (II,2288)

BASS, TODD
MY SISTER HANK (I,3201); ROWAN AND MARTIN'S LAUGH-IN (I,3856)

BASSETT, WILLIAM H
BEYOND WITCH MOUNTAIN (II,228); THE FEATHER AND FATHER GANG (II,845); NANCY (I,3221); THE TEMPEST (I,4391)

BASSEY, SHIRLEY
THE SHIRLEY BASSEY SHOW (I,4014); THE SHIRLEY BASSEY SHOW (I,4015); THE SHIRLEY BASSEY SPECIAL (II,2331)

BASTEDO, ALEXANDRA
THE CHAMPIONS (I,896)

BATANIDES, ARTHUR
JOHNNY MIDNIGHT (I,2431); MEETING AT APALACHIN (I,2994); SECOND LOOK (I,3960)

BATCHELOR, LUKE
THE SECRET OF CHARLES DICKENS (II,2293)

BATCHELOR, RUTH
PERSONAL AND CONFIDENTIAL (II,2026)

BATE, ANTHONY
FANNY BY GASLIGHT (II,823); SMILEY'S PEOPLE (II,2383); TINKER, TAILOR, SOLDIER, SPY (II,2621); A WOMAN CALLED GOLDA (II,2818)

BATEMAN, CHARLES
FOR RICHER, FOR POORER (II,906); HAZEL (I,1982); TWO FACES WEST (I,4653)

BATEMAN, JASON
IT'S YOUR MOVE (II,1259); LITTLE HOUSE ON THE PRAIRIE (II,1487); SILVER SPOONS (II,2355)

BATEMAN, JUSTINE
FAMILY TIES (II,819)

BATES, ALAN
SEPARATE TABLES (II,2304)

BATES, BRIDGET
CASANOVA (II,451)

BATES, JEANNE
BEN CASEY (I,394); DAYS OF OUR LIVES (II,629); HARD CASE (I,1947); PEPSI-COLA PLAYHOUSE (I,3523); THE SLOWEST GUN IN THE WEST (I,4088); TOPPER RETURNS (I,4583)

BATES, JENNIFER
THE GOOD LIFE (II,1033)

BATES, JIMMY
DEAR MOM, DEAR DAD (I,1202); FATHER KNOWS BEST (II,838)

BATES, JOHN TERRY
AMAHL AND THE NIGHT VISITORS (I,150)

BATES, JOLYON
CAESAR AND CLEOPATRA (I,789)

BATES, LULU
GAY NINETIES REVUE (I,1743)

BATES, RALPH
MURDER MOTEL (I,3161); PENMARRIC (II,1993)

BATES, RHONDA
BATTLE OF THE NETWORK STARS (II,166); BLANSKY'S BEAUTIES (II,267); KEEP ON TRUCKIN' (II,1372); PLEASURE COVE (II,2058); THE ROLLERGIRLS (II,2203); THE SHAPE OF THINGS (II,2315); SPEAK UP AMERICA (II,2412)

BATESON, TIMOTHY
THE TALE OF BEATRIX POTTER (II,2533); THERESE RAQUIN (II,2584)

BATHURST, PETER
DOCTOR IN THE HOUSE (I,1313) TEN LITTLE INDIANS (I,4394)

BATINKOFF, RANDALL
ONE MORE TRY (II,1908)

BATSON, SUSAN
GIDGET GROWS UP (I,1797) PALMERSTOWN U.S.A. (II,1945)

BATT, SHELLY
THE OUTLAWS (II,1936)

BATTAGLIA, ANTHONY
MR. & MRS. DRACULA (II,1745)

BATTEN, MARY
THE NEW TEMPERATURES RISING SHOW (I,3273); WILD WILD WORLD OF ANIMALS (I,4864)

BATTEN, THOMAS
HOW TO SUCCEED IN BUSINESS WITHOUT REALLY TRYING (II,1197)

BATTISTA, LLOYD
LOVE OF LIFE (I,2774)

BATTLES, MARJORIE
DOMINIC'S DREAM (II,704); MARY HARTMAN, MARY HARTMAN (II,1638)

BAUER, BELINDA
AIRWOLF (II,27); ARCHER—FUGITIVE FROM THE EMPIRE (II,103)

BAUER, BRUCE
TONI'S BOYS (II,2637)

BAUER, CHARITA
THE ALDRICH FAMILY (I,111); THE DRUNKARD (II,737); THE GUIDING LIGHT (II,1064)

BAUER, JAMIE LYN
BARE ESSENCE (II,150); BOSTON AND KILBRIDE (II,356); CIRCUS OF THE STARS (II,535); CIRCUS OF THE STARS (II,536); THE YOUNG AND THE RESTLESS (II,2862)

BAUER, ROCKY
NICHOLS AND DYMES (II,1841); SHE'S IN THE ARMY NOW (II,2320)

BAUERSMTIH, PAULA
THE WOMEN (I,4890)

BAUM, BRUCE
THE STOCKARD CHANNING SHOW (II,2468)

BAUM, CHARLES
A TIME TO LIVE (I,4510)

BAUM, JANINE
HARDCASE (II,1088)

BAUM, LAL
STRANDED (II,2475)

BAUMAN, JON
ALL NIGHT RADIO (II,40); BOWZER (II,358); THE MATCH GAME—HOLLYWOOD SQUARES HOUR (II,1650); THE POP 'N' ROCKER GAME (II,2066); WE DARE YOU! (II,2751)

BAUMANN, KATHERINE
BORDER PALS (II,352); HARRY O (II,1099); KEEP ON TRUCKIN' (II,1372); MRS R.—DEATH AMONG FRIENDS (II,1759); ROOSTER (II,2210); TERROR AT ALCATRAZ (II,2563)

BAUR, ELIZABETH
IRONSIDE (II,1246); LANCER (I,2610)

BAUTISTA, ROLAND
EARTH, WIND & FIRE IN CONCERT (II,750)

BAVAAR, RONY
CLUB 7 (I,986)

BAVIER, FRANCES

THE ANDY GRIFFITH SHOW (I,192); THE ANDY GRIFFITH SHOW (I,193); THE EVE ARDEN SHOW (I,1470); IF YOU KNEW TOMORROW (I,2187); IT'S A GREAT LIFE (I,2256); MAYBERRY R.F.D. (I,2961)

BAXLEY, BARBARA
ALL THAT GLITTERS (II,42); THE IMPOSTER (II,1225); THE INNER SANCTUM (I,2216); SEARCH FOR TOMORROW (II,2284); SENSE OF HUMOR (II,2303); WHERE THE HEART IS (I,4831)

BAXTER, ALAN
TELEPHONE TIME (I,4379)

BAXTER, ANNE
BATMAN (I,366); THE CATCHER (I,872); THE DEAN MARTIN CELEBRITY ROAST (II,638); EAST OF EDEN (II,752); HOLLYWOOD STARS' SCREEN TESTS (II,1165); HOTEL (II,1192); THE JUNE ALLYSON SHOW (I,2488); LISA, BRIGHT AND DARK (I,2711); LOVE STORY (I,2778); MARCUS WELBY, M.D. (I,2905); MARCUS WELBY, M.D. (II,1627); THE MONEYCHANGERS (II,1724); NERO WOLFE (II,1805); ZANE GREY THEATER (I,4979)

BAXTER, CAROL
THE A-TEAM (II,119); THE CURSE OF DRACULA (II,611); LITTLE WOMEN (II,1497)

BAXTER, CHARLES
ANOTHER WORLD (II,97); LOVE OF LIFE (I,2774); THE SECRET STORM (I,3969)

BAXTER, CHARLOTTE
PARADISE BAY (I,3454)

BAXTER, DR FRANK
FAIR WINDS TO ADVENTURE (I,1513); STAR FOR TODAY (I,4192); TELEPHONE TIME (I,4379)

BAXTER, GERRY
THE MISS AND THE MISSILES (I,3062)

BAXTER, KEITH
MELODY OF HATE (I,2999)

BAXTER, LULU
LITTLE LULU (II,1493)

BAXTER, LYNSEY
PETER PAN (II,2030); THE PRIME OF MISS JEAN BRODIE (II,2082)

BAXTER, MEREDITH *see* **BIRNEY, MEREDITH BAXTER**

BAXTER, STANLEY
BING CROSBY'S MERRIE OLDE CHRISTMAS (II,252)

BAXTER, TREVOR
THE KILLING GAME (I,2540)

BAY CITY ROLLERS
THE BAY CITY ROLLERS SHOW (II,187)

BAY, SUSAN
ALONE AT LAST (II,60); AMERICA, YOU'RE ON (II,69); LOOK OUT WORLD (II,1516)

BAYES, JOANNE
CRUNCH AND DES (I,1108)

BAYLOR, HAL
MY SISTER EILEEN (I,3200)

BAYLY, LORRAINE
1915 (II,1853); THE SULLIVANS (II,2490)

BAZARRE HARPERS
ROMP (I,3838); TIN PAN ALLEY TODAY (I,4521); THE WONDERFUL WORLD OF PIZZAZZ (I,4901)

BAZLEN, BRIGID
DAYS OF OUR LIVES (II,629); TOO YOUNG TO GO STEADY (I,4574)

BAZLEY, SALLY
FATHER DEAR FATHER (II,837)

BEACH BOYS, THE
THE BEACH BOYS SPECIAL (II,190); DICK CLARK'S GOOD OL' DAYS: FROM BOBBY SOX TO BIKINIS (II,671); GOOD VIBRATIONS FROM CENTRAL PARK (I,1854) THE JACK BENNY SPECIAL (I,2295)

BEACH, DYLAN
IT'S ARBOR DAY, CHARLIE BROWN (I,2267)

BEACH, JAMES
TENSPEED AND BROWN SHOE (II,2562)

BEACH, SARAH
IT'S ARBOR DAY, CHARLIE BROWN (I,2267)

BEACH, SCOTT
SHE'S A GOOD SKATE, CHARLIE BROWN (I,4012); YOU'RE THE GREATEST, CHARLIE BROWN (I,4973)

BEAIRD, BARBARA
FIBBER MCGEE AND MOLLY (I,1569); MAKE ROOM FOR DADDY (I,2843)

BEAIRD, BETTY
JULIA (I,2476); THE LILY TOMLIN SPECIAL (II,1479); THROUGH THE MAGIC PYRAMID (II,2615)

BEAIRD, PAMELA
LEAVE IT TO BEAVER (I,2648); MY FRIEND FLICKA (I,3190)

BEAL, JOHN
THE ADAMS CHRONICLES (II,8); ANOTHER WORLD (II,97); FREEDOM RINGS (I,1682); THE NURSES (I,3318); THE NURSES (I,3319); ROAD TO REALITY (I,3800)

BEAL, ROYAL
THE DEVIL AND DANIEL WEBSTER (I,1246); YOUNG DR. MALONE (I,4938)

BEALE, RICHARD
A HORSEMAN RIDING BY (II,1182)

BEALL, CHARLES
CELEBRITY (II,463)

BEALS, DICK
BIRDMAN (II,256); BIRDMAN AND THE GALAXY TRIO (I,478); FRANKENSTEIN JR. AND THE IMPOSSIBLES (I,1672); JACK AND THE BEANSTALK (I,2288); THE RICHIE RICH SHOW (II,2169)

BEALS, JACK
THE GUIDING LIGHT (II,1064)

BEAN, DAVID
PETER PAN (I,3566)

BEAN, ORSON
ARSENIC AND OLD LACE (I,268); THE ARTHUR GODFREY SHOW (I,287); THE BEAN SHOW (I,373); THE BLUE ANGEL (I,489); THE CHAMBER MUSIC SOCIETY OF LOWER BASIN STREET (I,894); THE DUNNINGER SHOW (I,1385); THE ELGIN HOUR (I,1428); FOREVER FERNWOOD (II,909); THE MAN IN THE DOG SUIT (I,2869); THE MELTING POT (II,1678); MIRACLE ON 34TH STREET (I,3057); MR. BEVIS (I,3128); STUDIO ONE (I,4268); TO TELL THE TRUTH (II,2625); THE TWILIGHT ZONE (I,4651)

BEANBLOSSOM, BILLIE JEAN
THE MICKEY MOUSE CLUB (I,3025); THE MOUSEKETEERS REUNION (II,1742)

BEANE, HILARY
CAR WASH (II,437); KLEIN TIME (II,1401) WINDOWS, DOORS AND KEYHOLES (II,2806)

BEANE, REGINALD
ONCE UPON A TUNE (I,3372)

BEARD, RENEE
CAPTAIN MIDNIGHT (I,833)

BEARD, STYMIE
BACKSTAIRS AT THE WHITE HOUSE (II,133); EAST OF EDEN (II,752); GOOD TIMES (II,1038)

BEASLEY, ALLYCE
CHEERS (II,494)

BEATRICE KROFT DANCERS, THE
STAR OF THE FAMILY (I,4194)

BEATTIE, LYNDA
STUNT SEVEN (II,2486); YOUNG LIVES (II,2866)

BEATTY, LEONARD
FOOLS, FEMALES AND FUN: WHAT ABOUT THAT ONE? (II,901)

BEATTY, NED
ALL THE WAY HOME (II,44); CELEBRITY (II,463); HUNTER (II,1204); THE LAST DAYS OF POMPEII (II,1436); LUCAN (II,1553); OUR TOWN (I,3421); SZYSZNYK (II,2523); A WOMAN CALLED GOLDA (II,2818)

BEATTY, ROBERT
DIAL 999 (I,1257); DOUGLAS FAIRBANKS JR. PRESENTS THE RHEINGOLD THEATER (I,1364); SLEEPWALKER (I,4084)

BEATTY, WARREN
DOBIE GILLIS (I,1302); LOVE OF LIFE (I,2774); SUSPICION (I,4309)

BEAUCHAMP, RICHARD
ANGIE (II,80); C.P.O. SHARKEY (II,407); THE CHEAP DETECTIVE (II,490); JACKIE AND DARLENE (II,1265); ZORRO AND SON (II,2878)

BEAUDIN, TANGIE
CASTLE ROCK (II,456)

BEAUDINE, DEKA
THE PAPER CHASE (II,1949)

BEAULIEU, IVAN
SEEING THINGS (II,2297)

BEAULIEU, PHILIPPINE LEROY
MISTRAL'S DAUGHTER (II,1708)

BEAUMONT, CHRIS
HERE WE GO AGAIN (I,2028)

BEAUMONT, HUGH
ALIAS MIKE HERCULES (I,117); FIRESIDE THEATER (I,1580); G.E. SUMMMER ORIGINALS (I,1747); LEAVE IT TO BEAVER (I,2648); THE LORETTA YOUNG THEATER (I,2756); MEDIC (I,2976); STUDIO '57 (I,4267)

BEAUMONT, KATHRYN
ONE HOUR IN WONDERLAND (I,3376)

BEAUMONT, RICHARD
ONLY A SCREAM AWAY (I,3393)

BEAUVY, NICK
ANDERSON AND COMPANY (I,188)

BEAVANS, PHILIPPA
THE KING AND MRS. CANDLE (I,2544)

BEAVER, TERRY
SIX PACK (II,2370)

BEAVERS, LOUISE
BEULAH (I,416); MAKE ROOM FOR DADDY (I,2843); SWAMP FOX (I,4311)

BECAUD, GILBERT
MONTE CARLO, C'EST LA ROSE (I,3095)

BECHER, JOHN C
ETHEL IS AN ELEPHANT (II,783)

BECHET, SIDNEY
FLOOR SHOW (I,1605)

BECK, BILLY
THE CHEAP SHOW (II,491); LOU GRANT (II,1526); ONCE UPON A MATTRESS (I,3371); STRANGER IN OUR HOUSE (II,2477); TWILIGHT THEATER II (II,2686)

BECK, JACKSON
JET FIGHTER (I,2374); KING LEONARDO AND HIS SHORT SUBJECTS (I,2557); POPEYE THE SAILOR (I,3648); POPEYE THE SAILOR (I,3649); TENNESSEE TUXEDO AND HIS TALES (I,4403)

BECK, JAMES
THE CLIFF DWELLERS (I,974)

BECK, JENNY
V: THE FINAL BATTLE (II,2714); V: THE SERIES (II,2715)

BECK, JIM
TEXAS JOHN SLAUGHTER (I,4414)

BECK, JOHN
BATTLE OF THE NETWORK STARS (II,171); BATTLE OF THE NETWORK STARS (II,176); THE BUFFALO SOLDIERS (II,388); DALLAS (II,614); FLAMINGO ROAD (II,871); FLAMINGO ROAD (II,872); LOCK, STOCK, AND BARREL (I,2736); NICHOLS (I,3283); SIDEKICKS (II,2348); WHEELS (II,2777)

BECK, KIMBERLY
EIGHT IS ENOUGH (II,762); FANTASY ISLAND (II,829); LUCAS TANNER (II,1556); PEYTON PLACE (I,3574); RICH MAN, POOR MAN—BOOK II (II,2162); SCALPELS (II,2266); STARTING FRESH (II,2447); THE WESTWIND (II,2766)

BECK, MICHAEL
CELEBRITY (II,463); FLY AWAY HOME (II,893); THE LAST NINJA (II,1438); THE STREETS (II,2479)

BECK, VINCENT
THE IMMORTAL (I,2196)

BECK-HILTON, KIMBERLY
CAPITOL (II,426)

BECKEL, GRAHAM
LOVE, SIDNEY (II,1547)

BECKER, ANTOINE
DYNASTY (II,746)

BECKER, BARBARA
CLUB 60 (I,987); ROAD OF LIFE (I,3798); THE WAYNE KING SHOW (I,4779)

BECKER, GEORGE
THE DINAH SHORE SHOW (I,1280)

BECKER, JOHN KLAUS
GEORGE WASHINGTON (II,978)

BECKER, SANDY
THE GO GO GOPHERS (I,1831); UNDERDOG (I,4673); WIN WITH A WINNER (I,4870)

BECKER, TERRY
VOYAGE TO THE BOTTOM OF THE SEA (I,4743)

BECKER, TONY
FOR LOVE AND HONOR (II,903); THE KILLER WHO WOULDN'T DIE (II,1393); KUDZU (II,1415); THE OREGON TRAIL (II,1922); THE OREGON TRAIL (II,1923); THE TEXAS WHEELERS (II,2568); THE WALTONS: A DAY FOR THANKS ON WALTON'S MOUNTAIN (EPISODE 3) (II,2741); THE WALTONS: MOTHER'S DAY ON WALTON'S MOUNTAIN (EPISODE 2) (II,2742); THE WALTONS: WEDDING ON WALTON'S MOUNTAIN (EPISODE 1) (II,2743)

BECKERMAN, TERI
THE INVISIBLE WOMAN (II,1244); THE OUTLAWS (II,1936)

BECKET, SIDNEY
THE EDDIE CONDON SHOW (I,1408)

BECKETT, ELAINE
ROWAN AND MARTIN'S LAUGH-IN (I,3856)

BECKETT, SCOTT
BACKBONE OF AMERICA (I,328); ROCKY JONES, SPACE RANGER (I,3823)

BECKLEY, BARBARA
SKEEZER (II,2376)

BECKLEY, WILLIAM
DYNASTY (II,746); HIGH RISK (II,1146)

BECKMAN, HENRY
BROCK'S LAST CASE (I,742); BRONK (II,379); THE FAMILY HOLVAK (II,817); FUNNY FACE (I,1711); HERE COME THE BRIDES (I,2024); I'M DICKENS...HE'S FENSTER (I,2193); MCHALE'S NAVY (I,2969); MY LIVING DOLL (I,3195); OWEN MARSHALL: COUNSELOR AT LAW (I,3435); PEYTON PLACE (I,3574)

BEDARD, ROLLAND
FOREST RANGERS (I,1642)

BEDDOE, DON
MIRACLE ON 34TH STREET (I,3056); MR. O'MALLEY (I,3146); THE SECOND HUNDRED YEARS (I,3959)

BEDELIA, BONNIE
HAWKINS ON MURDER (I,1978); LOVE OF LIFE (I,2774); THE NEW LAND (II,1820); THEN CAME BRONSON (I,4436); TOURIST (II,2651)

BEDFORD, BRIAN
ANDROCLES AND THE LION (I,189); THE HOLY TERROR (I,2098)

BEDI, KABIR
ARCHER—FUGITIVE FROM THE EMPIRE (II,103)

BEDLIN, JANET
SUSPENSE THEATER ON THE AIR (II,2507)

BEE, MOLLY
AMERICA PAUSES FOR THE MERRY MONTH OF MAY (I,167); THE BOB HOPE SHOW (I,566); THE JACK JONES SPECIAL (I,2310); THE JIMMY DEAN SHOW (I,2385); SWINGING COUNTRY (I,4322); THE TENNESSEE ERNIE FORD SHOW (I,4398)

BEECROFT, GREGORY
THE GUIDING LIGHT (II,1064)

BEEMS, FRED
PETE KELLY'S BLUES (I,3559)

BEENY, CHRISTOPHER
UPSTAIRS, DOWNSTAIRS (II,2709)

BEER, JACQUELINE
77 SUNSET STRIP (I,3988); THE FRENCH ATLANTIC AFFAIR (II,924)

BEERS, FRANCINE
6 RMS RIV VU (II,2371); ALONE AT LAST (II,59); THE EDGE OF NIGHT (II,760) ONE OF THE BOYS (II,1911)

BEERY JR, NOAH
THE ASPHALT COWBOY (II,111); THE BASTARD/KENT FAMILY CHRONICLES (II,159); BEYOND WITCH MOUNTAIN (II,228); CIRCUS BOY (I,959); DOC ELLIOT (I,1306); HONDO (I,2103); THE MURDOCKS AND THE MCCLAYS (I,3163); THE MYSTERIOUS TWO (II,1783); THE QUEST (II,2098); REVENGE OF THE GRAY GANG (II,2146); RIVERBOAT (I,3797); THE ROCKFORD FILES (II,2197); SIDEKICKS (II,2348); THE YELLOW ROSE (II,2847)

BEGG, JIM
THE CATTANOOGA CATS (I,874); INSIDE O.U.T. (II,1235); OZZIE'S GIRLS (I,3440); THE PIRATES OF FLOUNDER BAY (I,3607)

BEGGS, HAGAN
I LOVE A MYSTERY (I,2170); STAR TREK (II,2440)

BEGLEY JR, ED
BATTLESTAR GALACTICA (II,181); BOBBY JO AND THE BIG APPLE GOODTIME BAND (I,674); MAKING IT (II,1605); MARY HARTMAN, MARY HARTMAN (II,1638); MIXED NUTS (II,1714); ROLL OUT! (I,3833); ST. ELSEWHERE (II,2432); STILL THE BEAVER (II,2466); TALES OF THE APPLE DUMPLING GANG (II,2536); WONDER WOMAN (SERIES 2) (II,2829)

BEGLEY, ED

THE ARMSTRONG CIRCLE THEATER (I,260); CAMEO THEATER (I,808); THE DICK POWELL SHOW (I,1269); INHERIT THE WIND (I,2213); KRAFT TELEVISION THEATER (I,2592); LEAVE IT TO LARRY (I,2649); LIGHT'S OUT (I,2699); NEITHER ARE WE ENEMIES (I,3247); ROBERT MONTGOMERY PRESENTS YOUR LUCKY STRIKE THEATER (I,3809); WARNING SHOT (I,4770)

BEHAR, JOY
KIDS ARE PEOPLE TOO (II,1390)

BEHAR, PHYLLIS
ONE LIFE TO LIVE (II,1907)

BEHETS, BRIONY
PRISONER: CELL BLOCK H (II,2085)

BEHRENS, FRANK
CALL TO DANGER (I,802); MEETING AT APALACHIN (I,2994)

BEIGER, PETER
THE DIARY OF ANNE FRANK (I,1262)

BEIR, FRED
ANOTHER WORLD (II,97)

BEITZEL, ROBERT
GENERAL HOSPITAL (II,964)

BEKASSY, STEVE
SKYFIGHTERS (I,4078)

BEKINS, RICHARD
ANOTHER WORLD (II,97)

BEL GEDDES, BARBARA
ALFRED HITCHCOCK PRESENTS (I,115); THE BURNING COURT (I,766); DALLAS (II,614); JOURNEY TO THE UNKNOWN (I,2457); OUR TOWN (I,3421)

BELACK, DORIS
ANOTHER WORLD (II,97); BAKER'S DOZEN (II,136); THE EDGE OF NIGHT (II,760); ONE LIFE TO LIVE (II,1907)

BELAFONTE, HARRY
ABC STAGE '67 (I,6); BELAFONTE, NEW YORK (I,386); THE DIAHANN CARROLL SHOW (I,1252); THE DINAH SHORE SPECIAL (I,1284); AN EVENING WITH JULIE ANDREWS AND HARRY BELAFONTE (I,1476); HARRY AND LENA (I,1957); THE JULIE ANDREWS SPECIAL (I,2483); MARLO THOMAS AND FRIENDS IN FREE TO BE. . . YOU AND ME (II,1632); PETULA (I,3572); THE ROWAN AND MARTIN SPECIAL (I,3855); SUGAR HILL TIMES (I,4276); TONIGHT WITH BELAFONTE (I,4566); A WORLD OF LOVE (I,4914)

BELAFONTE-HARPER, SHARI
BATTLE OF THE NETWORK STARS (II,177); BATTLE OF THE NETWORK STARS (II,178); HOTEL (II,1192); VELVET (II,2725)

BELAH, RICHARD
DAVY AND GOLIATH (I,1180)

BELAND, HELLANE
DOUGLAS FAIRBANKS JR. PRESENTS THE RHEINGOLD THEATER (I,1364)

BELANEY, TOM
THE SEVEN DIALS MYSTERY (II,2308)

BELASCO, LEON
THE JERRY LESTER SHOW (I,2360); MY SISTER EILEEN (I,3200); NINOTCHKA (I,3299); RUSSIAN ROULETTE (I,3875)

BELCHER, DAVID
SPELL OF EVIL (I,4152)

BELFORD, CHRISTINE
100 CENTRE STREET (II,1901); BANACEK (II,138); BANACEK: DETOUR TO NOWHERE (I,339); COLORADO C.I. (II,555); DYNASTY (II,746); EMPIRE (II,777); IT'S NOT EASY (II,1256); KATE MCSHANE (II,1369); MARRIED: THE FIRST YEAR (II,1634); THE MASK OF MARCELLA (I,2937); PERSONAL AND CONFIDENTIAL (II,2026); SILVER SPOONS (II,2355); SUSAN AND SAM (II,2506)

BELL SISTERS, THE
THE BOB HOPE SHOW (I,513); OPERATION ENTERTAINMENT (I,3400)

BELL, E E
HERNDON AND ME (II,1139)

BELL, EDWARD
HUNTER (II,1204); THE SECRET WAR OF JACKIE'S GIRLS (II,2294)

BELL, IVAN
F TROOP (I,1499)

BELL, JAMES
CONFIDENTIALLY YOURS (I,1035); DENNIS THE MENACE (I,1231)

BELL, JOHN
MASADA (II,1642); TEN WHO DARED (II,2559)

BELL, LUCY
A TOWN LIKE ALICE (II,2652)

BELL, MICHAEL
THE 25TH MAN (MS) (II,2678); AMERICA, YOU'RE ON (II,69); THE BARKLEYS (I,356); BATMAN AND THE SUPER SEVEN (II,160); THE C.B. BEARS (II,406); CHARLIE'S ANGELS (II,486); DALLAS (II,614); DEVLIN (II,663); EGAN (I,1419); FLO'S PLACE (II,892); GO WEST YOUNG GIRL (II,1013); GODZILLA (II,1014); THE KWICKY KOALA SHOW (II,1417); LOST IN SPACE (I,2759); MADHOUSE 90 (I,2807); MORNING STAR (I,3104); THE NEW SUPER FRIENDS HOUR (II,1830); PETROCELLI (II,2031); THE PLASTICMAN COMEDY/ADVENTURE SHOW (II,2051); THE SMURFS (II,2385); SPACE STARS (II,2408); SPEED BUGGY (II,2419); SUPER FRIENDS (II,2497); THE SURVIVORS (I,4301); TROLLKINS (II,2660); THE WORLD'S GREATEST SUPER HEROES (II,2842)

BELL, NOVA
SUMMER (II,2491)

BELL, RALPH
THE JIMMY DURANTE SHOW (I,2389); THE PATTY DUKE SHOW (I,3483); YOUNG DR. MALONE (I,4938)

BELL, REX
COWBOYS AND INJUNS (I,1087)

BELL, RODNEY
THE BROTHERS (I,748)

BELL, STEVEN
DOCTOR KILDARE (I,1315)

BELL, TITA
HAPPY DAYS (II,1084)

BELLAMY, ANNE
THE NEW ADVENTURES OF HUCKLEBERRY FINN (I,3250)

BELLAMY, RALPH
AMOS BURKE: WHO KILLED JULIE GREER? (I,181); THE BARBARA STANWYCK THEATER (I,350); THE BILLION DOLLAR THREAT (II,246); BRIEF ENCOUNTER (I,723); CHARLIE COBB: NICE NIGHT FOR A HANGING (II,485); CONDOMINIUM (II,566); THE DEFENDER (I,1219); THE DEVIL'S DISCIPLE (I,1247); THE ELEVENTH HOUR (I,1425); FRONTIER JUSTICE (I,1700); HUNTER (II,1205); THE IMMORTAL (I,2196); THE IMMORTAL (I,2197); JOSEPHINE LITTLE: THE MIRACULOUS JOURNEY OFTADPOLE CHAN (I,2452); THE LOG OF THE BLACK PEARL (II,1506); THE MAN AGAINST CRIME (I,2853); THE MILLIONAIRE (II,1700); THE MONEYCHANGERS (II,1724); THE MOST DEADLY GAME (I,3106); PHILCO TELEVISION PLAYHOUSE (I,3583); SATURDAY'S CHILDREN (I,3923); THE SURVIVORS (I,4301); TESTIMONY OF TWO MEN (II,2564); TEXACO COMMAND PERFORMANCE (I,4407); THE UNITED STATES STEEL HOUR (I,4677); WHEELS (II,2777); THE WINDS OF WAR (II,2807)

BELLAN, JOE
HEAR NO EVIL (II,1116)

BELLAND AND SOMERVILLE
THE TIM CONWAY COMEDY HOUR (I,4501)

BELLAND, TRACY
YES, VIRGINIA, THERE IS A SANTA CLAUS (II,2849)

BELLARAN, RAY
SEARCH FOR TOMORROW (II,2284)

BELLAVER, HARRY
ABE LINCOLN IN ILLINOIS (I,10); ANNIE GET YOUR GUN (I,224); ANOTHER WORLD (II,97); THE BILLY BEAN SHOW (I,449); THE COUNTRY MUSIC MURDERS (II,590); HAPPY BIRTHDAY (I,1941); MR. BELVEDERE (I,3127); NAKED CITY (I,3216); RIVKIN: BOUNTY HUNTER (II,2183)

BELLER, KATHLEEN
AT EASE (II,116); THE BLUE AND THE GRAY (II,273); DYNASTY (II,746); HAVING BABIES III (II,1109); THE MANIONS OF AMERICA (II,1623); SEARCH FOR TOMORROW (II,2284)

BELLER, MARY LYNN
THE BRIGHTER DAY (I,728); A DATE WITH JUDY (I,1162)

BELLFLOWER, NELLIE
EAST OF EDEN (II,752); GOLDIE GOLD AND ACTION JACK (II,1023); THE KELLY MONTEITH SHOW (II,1376); NIGHT PARTNERS (II,1847); RENDEZVOUS HOTEL (II,2131); THUNDARR THE BARBARIAN (II,2616); YOU'RE JUST LIKE YOUR FATHER (II,2861)

BELLI, MELVIN

GUILTY OR INNOCENT (II,1065)

BELLING, STEVIE
CAMP WILDERNESS (II,419)

BELLINGTON, LYNDA
MACKENZIE (II,1584)

BELLINI, CAL
THE ASPHALT COWBOY (II,111); DIAGNOSIS: UNKNOWN (I,1251); IN THE DEAD OF NIGHT (I,2202); KATE MCSHANE (II,1369); MATT HOUSTON (II,1654); WAIKIKI (II,2734)

BELLSON, LOUIS
ONE MORE TIME (I,3385)

BELLWOOD, PAMELA
BATTLE OF THE NETWORK STARS (II,178); BLONDES VS. BRUNETTES (II,271); DYNASTY (II,746); EMILY, EMILY (I,1438); W.E.B. (II,2732); THE WILD WOMEN OF CHASTITY GULCH (II,2801)

BELMORE, DAISY
JANE EYRE (I,2337)

BELTRAN, ALMA
BIG JOHN (II,239); THE HARVEY KORMAN SHOW (II,1103); ST. ELSEWHERE (II,2432)

BELUSHI, JIM
THE BEST LEGS IN 8TH GRADE (II,214); SATURDAY NIGHT LIVE (II,2261); WHO'S WATCHING THE KIDS? (II,2793); WORKING STIFFS (II,2834)

BELUSHI, JOHN
THE BEACH BOYS SPECIAL (II,190); THE RICHARD PRYOR SPECIAL (II,2164); THE RICHARD PRYOR SPECIAL? (II,2165); SATURDAY NIGHT LIVE (II,2261); STEVE MARTIN'S BEST SHOW EVER (II,2460); THINGS WE DID LAST SUMMER (II,2589)

BELVADERES, THE
THE RAY ANTHONY SHOW (I,3728)

BELZER, RICHARD
THE FACTS (II,804); THICKE OF THE NIGHT (II,2587)

BENADERET, BEA
ALL ABOUT BARBARA (I,125); THE BEVERLY HILLBILLIES (I,417); THE FLINTSTONES (II,884); THE GEORGE BURNS AND GRACIE ALLEN SHOW (I,1763); THE GEORGE BURNS SHOW (I,1765); THE LIFE OF RILEY (I,2686); PETER LOVES MARY (I,3565); PETTICOAT JUNCTION (I,3571)

BENARD, FRANCOIS-MARIE
BARE ESSENCE (II,149); SCRUPLES (II,2280)

BENATAR, PAT
MUSIC CENTRAL (II,1771); PAT BENATAR IN CONCERT (II,1967)

BENBEN, BRIAN
THE GANGSTER CHRONICLES (II,957); LOVERS AND OTHER STRANGERS (II,1550)

BENCH, JOHNNY
THE BOB HOPE SHOW (I,638); THE BOB HOPE SHOW (I,646); GAMES PEOPLE PLAY (II,956)

BENCHLEY, NAT
KENNEDY (II,1377)

BENDA, KENNETH
NO—HONESTLY (II,1857)

BENDEL, ROBERT
ANOTHER LIFE (II,95)

BENDER, JERRY
BOLD VENTURE (I,682)

BENDER, JIM
FAMOUS JURY TRIALS (I,1525)

BENDER, RUSS
BROKEN ARROW (I,744)

BENDETT, KATHY
THE RITA MORENO SHOW (II,2181)

BENDIX, LARRAINE
THE LIFE OF RILEY (I,2686)

BENDIX, WILLIAM
330 INDEPENDENCE S.W. (I,4486); THE BOB HOPE SHOW (I,511); THE DICK POWELL SHOW (I,1269); FIRESIDE THEATER (I,1580); HOLLYWOOD OPENING NIGHT (I,2087); IVY LEAGUE (I,2286); LAST OF THE PRIVATE EYES (I,2628); THE LIFE OF RILEY (I,2686); LIGHT'S OUT (I,2699); OVERLAND TRAIL (I,3433); PINE LAKE LODGE (I,3597); PLAYHOUSE 90 (I,3623); SCHLITZ PLAYHOUSE OF STARS (I,3936); THE TIME ELEMENT (I,4504)

BENDIXSEN, MIA
MY WIVES JANE (I,3207)

BENDYKE, BETTY
THE GOLDBERGS (I,1836)

BENE, CARMEN
THE BENNY HILL SHOW (II,207)

BENEDICK, NICK
THE YOUNG AND THE RESTLESS (II,2862)

BENEDICT, BILLY
THE BLUE KNIGHT (II,276)

BENEDICT, DIRK
THE A-TEAM (II,119); BATTLESTAR GALACTICA (II,181); THE CABOT CONNECTION (II,408); CELEBRITY CHALLENGE OF THE SEXES 3 (II,467); CHOPPER ONE (II,515); FAMILY IN BLUE (II,818); THE GEORGIA PEACHES (II,979); SCRUPLES (II,2281)

BENEDICT, GREG
NO TIME FOR SERGEANTS (I,3300)

BENEDICT, LAWRENCE
GOLIATH AWAITS (II,1025)

BENEDICT, NICHOLAS
ALL MY CHILDREN (II,39)

BENEDICT, PAUL
THE BLUE AND THE GRAY (II,273); THE JEFFERSONS (II,1276); MAMA MALONE (II,1609)

BENEDICT, ROBERT
OPEN HOUSE (I,3394)

BENEDICT, WILLIAM
CALL HER MOM (I,796)

BENEKE, TEX
THE ANITA BRYANT SPECTACULAR (II,82); PETER MARSHALL SALUTES THE BIG BANDS (II,2028)

BENESCH, LYNN
ONE LIFE TO LIVE (II,1907)

BENET, BRENDA
DAYS OF OUR LIVES (II,629); TO ROME WITH LOVE (I,4526); WEDNESDAY NIGHT OUT (II,2760); THE YOUNG MARRIEDS (I,4942)

BENET, VICKI
THE MORT SAHL SPECIAL (I,3105)

BENGELL, NORMA
T.H.E. CAT (I,4430)

BENHAM, JOAN
DOCTOR IN THE HOUSE (I,1313)

BENIADES, TED
THE ANDROS TARGETS (II,76); SATURDAY'S CHILDREN (I,3923); WHERE THE HEART IS (I,4831)

BENISH, LYNN
ONE LIFE TO LIVE (II,1907)

BENJAMIN, CHRISTOPHER
BAFFLED (I,332); THE STRAUSS FAMILY (I,4253); THERESE RAQUIN (II,2584)

BENJAMIN, JULIA
HAZEL (I,1982)

BENJAMIN, PAUL
MR. INSIDE/MR. OUTSIDE (II,1750)

BENJAMIN, RICHARD
BATTLE OF THE NETWORK STARS (II,166); FAME (I,1517); HE AND SHE (I,1985); A LAST LAUGH AT THE 60'S (I,2627); MY LUCKY PENNY (I,3196); QUARK (II,2095); THE WAY THEY WERE (II,2748)

BENJAMIN, SUSAN
ACCIDENTAL FAMILY (I,17)

BENJI
BENJI, ZAX AND THE ALIEN PRINCE (II,205)

BENNET, JULIE
THE FUNKY PHANTOM (I,1709)

BENNETT, BRUCE
DAMON RUNYON THEATER (I,1129); MAN OF THE COMSTOCK (I,2873); NO WARNING (I,3301)

BENNETT, CONSTANCE
ALWAYS APRIL (I,147); THE BIG PAYOFF (I,428); CAMEO THEATER (I,808); FAITH BALDWIN'S THEATER OF ROMANCE (I,1515); ROBERT MONTGOMERY PRESENTS YOUR LUCKY STRIKE THEATER (I,3809); SUSPENSE (I,4305); TELLER OF TALES (I,4388)

BENNETT, EVELYN
THE GAY COED (I,1742)

BENNETT, FRANCES
COLOR HIM DEAD (I,1007); COLOR HIM DEAD (II,554)

BENNETT, FRANK
THE ASSASSINATION RUN (II,112)

BENNETT, GABRIELLE
KIDD VIDEO (II,1388)

BENNETT, HYWELL
TINKER, TAILOR, SOLDIER, SPY (II,2621)

BENNETT, JOAN
DANGER (I,1134); DARK SHADOWS (I,1157); THE EYES OF CHARLES SAND (I,1497); FORD TELEVISION THEATER (I,1634); GIDGET GETS

MARRIED (I,1796); JUNIOR MISS (I,2493); THE MAN WHO CAME TO DINNER (I,2880); THE NASH AIRFLYTE THEATER (I,3227); PLAYHOUSE 90 (I,3623); PURSUIT (I,3700); TELLER OF TALES (I,4388); TOO YOUNG TO GO STEADY (I,4574)

BENNETT, JOE
FAME (I,1517); THE PRIMARY ENGLISH CLASS (II,2081)

BENNETT, JOHN
THE FORSYTE SAGA (I,1645)

BENNETT, JULIE
THE ALL-NEW POPEYE HOUR (II,49); THE BANANA SPLITS ADVENTURE HOUR (I,340); THE CATTANOOGA CATS (I,874); THE CLOCK (I,978); THE FAMOUS ADVENTURES OF MR. MAGOO (I,1524); GOLIATH AWAITS (II,1025); THE NEW ADVENTURES OF HUCKLEBERRY FINN (I,3250); SCOOBY'S ALL-STAR LAFF-A-LYMPICS (II,2273)

BENNETT, KELLY
RITUALS (II,2182)

BENNETT, LEE
CARTUNE-O (I,859)

BENNETT, LINDA
THE BOB HOPE SHOW (I,629); JACK LEMMON IN 'S WONDERFUL, 'S MARVELOUS, 'S GERSHWIN (I,2313); STAR STAGE (I,4197)

BENNETT, MARJORIE
BIG EDDIE (II,235); DOBIE GILLIS (I,1302); THE ELISABETH MCQUEENEY STORY (I,1429); THE GOOD LIFE (II,1033); LOVE THAT BOB (I,2779); THE MAN FROM GALVESTON (I,2865); MY DARLING JUDGE (I,3186); PAPA SAID NO (I,3452); PINE LAKE LODGE (I,3597)

BENNETT, MARY
15 WITH FAYE (I,1570)

BENNETT, MATT
DOG AND CAT (II,695)

BENNETT, MEG
AFTER HOURS: GETTING TO KNOW US (II,21); SEARCH FOR TOMORROW (II,2284); THE YOUNG AND THE RESTLESS (II,2862)

BENNETT, MICHELLE
THE YELLOW ROSE (II,2847)

BENNETT, NORMAN
CELEBRITY (II,463)

BENNETT, PETER
THE ADVENTURES OF SIR LANCELOT (I,76)

BENNETT, RAMONA
THIS IS MUSIC (I,4458)

BENNETT, SUE
KAY KYSER'S KOLLEGE OF MUSICAL KNOWLEDGE (I,2512)

BENNETT, TONY
THE ARTHUR GODFREY SHOW (I,287); THE ARTHUR MURRAY PARTY FOR BOB HOPE (I,292); THE BEST ON RECORD (I,409); THE BOB HOPE SHOW (I,614); THE JACK JONES SPECIAL (I,2310); THE JERRY LEWIS SHOW (I,2368); MAKE ROOM FOR DADDY (I,2843); THE MUSIC OF GERSHWIN (I,3174); ON PARADE (I,3356); ON THE TOWN WITH TONY BENNETT (II,1895); PERRY PRESENTS (I,3549); SINATRA AND FRIENDS (II,2358); SINATRA—THE FIRST 40 YEARS (II,2359); STAR SEARCH (II,2437); TONY BENNETT IN LONDON (I,4567); TONY BENNETT IN WAIKIKI (I,4568); THE TONY BENNETT SHOW (I,4569); THE TONY BENNETT SHOW (I,4570); THE TONY BENNETT SPECIAL (I,4571); UPBEAT (I,4684); WITH LOVE, SOPHIA (I,4884)

BENNY GOODMAN ORCHESTRA
THE KATE SMITH SHOW (I,2509)

BENNY GOODMAN SEXTET, THE
STARTIME (I,4212)

BENNY, JACK
THE ANN-MARGRET SHOW (I,218); ANNIE AND THE HOODS (II,91); THE BING CROSBY SPECIAL (I,470); THE BOB HOPE SHOW (I,530); THE BOB HOPE SHOW (I,531); THE BOB HOPE SHOW (I,533); THE BOB HOPE SHOW (I,570); THE BOB HOPE SHOW (I,592); THE BOB HOPE SHOW (I,607); THE BOB HOPE SHOW (I,645); THE BOB HOPE SHOW (I,650); THE BOB HOPE SHOW (I,659); BURLESQUE (I,765); EVERYTHING YOU ALWAYS WANTED TO KNOW ABOUT JACK BENNY AND WERE AFRAID TO ASK (I,1482); FRANK SINATRA JR. WITH FAMILY AND FRIENDS (I,1663); THE GENERAL FOODS 25TH ANNIVERSARY SHOW (I,1756); GEORGE BURNS IN THE BIG TIME (I,1764); HAVE GIRLS—WILL TRAVEL (I,1967); HOW TO HANDLE A WOMAN (I,2146); THE JACK BENNY CHRISTMAS SPECIAL (I,2289); THE JACK BENNY HOUR (I,2290); THE JACK BENNY HOUR (I,2291); THE JACK BENNY HOUR (I,2292); THE JACK BENNY HOUR (I,2293); THE JACK BENNY PROGRAM (I,2294); THE JACK BENNY SPECIAL (I,2295); THE JACK BENNY SPECIAL (I,2296); JACK BENNY WITH GUEST STARS (I,2298); JACK BENNY'S 20TH ANNIVERSARY TV SPECIAL (I,2303); JACK BENNY'S BIRTHDAY SPECIAL (I,2299); JACK BENNY'S FIRST FAREWELL SHOW (I,2300); JACK BENNY'S NEW LOOK (I,2301); JACK BENNY'S SECOND FAREWELL SHOW (I,2302); THE MILTON BERLE SPECIAL (I,3050); REMEMBER HOW GREAT? (I,3764); SHOW BUSINESS SALUTE TO MILTON BERLE (I,4029); THE SLOWEST GUN IN THE WEST (I,4088); SWING OUT, SWEET LAND (II,2515); TIME OUT FOR GINGER (I,4508)

BENNY, JOAN
THE JACK BENNY PROGRAM (I,2294)

BENOIT, PATRICIA
AS THE WORLD TURNS (II,110); THE DOCTOR (I,1320); MR. PEEPERS (I,3147); TODAY IS OURS (I,4531)

BENSON, BARBARA
SO YOU WANT TO LEAD A BAND (I,4106)

BENSON, BOBBY
SEARCH FOR TOMORROW (II,2284)

BENSON, COURT
THE DOCTORS (II,694); FIRST LOVE (I,1583)

BENSON, DEBORAH
THE FRENCH ATLANTIC AFFAIR (II,924)

BENSON, GEORGE
CASANOVA (II,451); LYNDA CARTER: STREET LIGHTS (II,1567); MUSIC CENTRAL (II,1771)

BENSON, IRVING
THE MILTON BERLE SHOW (I,3049)

BENSON, LUCILLE
THE ADVENTURES OF POLLYANNA (II,14); BOSOM BUDDIES (II,355); BUNCO (II,395); THE CONCRETE COWBOYS (II,564); THE COUNTRY MUSIC MURDERS (II,590); EBONY, IVORY AND JADE (II,755); HARRY AND MAGGIE (II,1097); THE KILLER WHO WOULDN'T DIE (II,1393); LASSIE: THE NEW BEGINNING (II,1433); NASHVILLE 99 (II,1791); THE ROPERS (II,2214); THE RUNAWAY BARGE (II,2230); THE TOM SWIFT AND LINDA CRAIG MYSTERY HOUR (II,2633)

BENSON, MARTIN
THE HITCHHIKER'S GUIDE TO THE GALAXY (II,1158); THE SWORD OF FREEDOM (I,4326)

BENSON, PAT
ALICE (II,33); JOE AND VALERIE (II,1298)

BENSON, RED
NAME THAT TUNE (I,3218)

BENSON, REX
RODNEY DANGERFIELD SPECIAL: IT'S NOT EASY BEIN' ME (II,2199)

BENSON, ROBBY
OUR TOWN (I,3421)

BENTINE, MICHAEL
THE JONATHAN WINTERS SHOW (I,2441); THE JONATHAN WINTERS SHOW (I,2444)

BENTLEY, BEVERLY
BEAT THE CLOCK (I,377); THE PRICE IS RIGHT (I,3664)

BENTLEY, DAVID
MOLL FLANDERS (II,1720)

BENTLEY, JOHN
AFRICAN PATROL (I,88); THE PENDULUM (I,3511); THE TRAP (I,4593)

BENTLEY, JORDAN
WONDERFUL TOWN (I,4893)

BENTLEY, MARC
THE YOUNG AND THE RESTLESS (II,2862)

BENTON, BARBI
BARBI BENTON SPECIAL: A BARBI DOLL FOR CHRISTMAS (II,148); BOWZER (II,358); CELEBRITY CHALLENGE OF THE SEXES 4 (II,468); CELEBRITY DAREDEVILS (II,473); CIRCUS OF THE STARS (II,533); CIRCUS OF THE STARS (II,534); CIRCUS OF THE STARS (II,536); A COUNTRY CHRISTMAS (II,582); DEAN MARTIN AT THE WILD ANIMAL PARK (II,636); DOUG HENNING'S WORLD OF MAGIC III (II,727); FANTASY ISLAND (II,829); HEE HAW (II,1123); MAGIC WITH THE

STARS (II,1599); MEN WHO RATE A "10" (II,1684); PLAYBOY AFTER DARK (I,3620); SUGAR TIME (II,2489)

BENTON, EDDIE
DOCTOR STRANGE (II,688); DOCTORS' PRIVATE LIVES (II,692); RAFFERTY (II,2105)

BENTON, JAMES GRANT
HAWAIIAN HEAT (II,1111)

BENTON, JESSICA
THE ONEDIN LINE (II,1912)

BENTON, LEE
MICKEY SPILLANE'S MIKE HAMMER (II,1692); SCROOGE'S ROCK 'N' ROLL CHRISTMAS (II,2279)

BENTZEN, JAYNE
THE EDGE OF NIGHT (II,760)

BENZ, DONNA KEI
FORCE SEVEN (II,908) MURDOCK'S GANG (I,3164)

BENZELL, MIMI
TEXACO COMMAND PERFORMANCE (I,4407)

BERADINO, JOHN
GENERAL HOSPITAL (II,964); I LED THREE LIVES (I,2169); THE NEW BREED (I,3258); A TALE OF WELLS FARGO (I,4338)

BERAIL, ROXANNE
COOGAN'S REWARD (I,1044)

BERCOVICI, KAREN
SORORITY '62 (II,2405)

BERCOVICI, LUCA
THE RENEGADES (II,2132)

BERDIS, BERT
AMERICA ALIVE! (II,67); THE TIM CONWAY SHOW (II,2619)

BEREGI, OSCAR
LUXURY LINER (I,2796); THE RELUCTANT SPY (I,3763); SAFARI (I,3882)

BERENGER, TOM
ONE LIFE TO LIVE (II,1907)

BERENSON, BERRY
SCRUPLES (II,2280)

BERENSON, MARISA
TOURIST (II,2651)

BERG, GERTRUDE
THE ELGIN HOUR (I,1428); THE GOLDBERGS (I,1835); THE GOLDBERGS (I,1836); THE KATE SMITH SHOW (I,2509); MRS. G. GOES TO COLLEGE (I,3155); THE PERRY COMO THANKSGIVING SHOW (I,3539); THE UNITED STATES STEEL HOUR (I,4677)

BERG, GREG
THE BOOK OF LISTS (II,350); JIM HENSON'S MUPPET BABIES (II,1289)

BERG, NANCY
THE DOCTORS (II,694)

BERG, STORMY
JACKIE GLEASON AND HIS AMERICAN SCENE MAGAZINE (I,2321)

BERGAN, JUDITH-MARIE
DOMESTIC LIFE (II,703); MAGGIE (II,1589); SOAP (II,2392); THE SON-IN-LAW (II,2398); WHERE'S POPPA? (II,2784)

BERGANSKY, CHUCK
THE MARY TYLER MOORE SHOW (II,1640)

BERGEN, BILL
THE POLLY BERGEN SHOW (I,3643)

BERGEN, CANDICE
THE WAY THEY WERE (II,2748); THE WOODY ALLEN SPECIAL (I,4904)

BERGEN, EDGAR
20TH CENTURY FOLLIES (I,4641); AMOS BURKE: WHO KILLED JULIE GREER? (I,181); THE DICK POWELL SHOW (I,1269); FRANCES LANGFORD PRESENTS (I,1656); THE GENERAL FOODS 25TH ANNIVERSARY SHOW (I,1756); THE GREATEST SHOW ON EARTH (I,1883); THE JUNE ALLYSON SHOW (I,2488); THE KATE SMITH SHOW (I,2509); THE KRAFT 75TH ANNIVERSARY SPECIAL (II,1409); MY SISTER HANK (I,3201); ONE HOUR IN WONDERLAND (I,3376); THE STRAWBERRY BLONDE (I,4254); VAUDEVILLE (II,2722); THE WALTONS (II,2740); WHO DO YOU TRUST? (I,4846)

BERGEN, FRANCES
THE DOCTOR WAS A LADY (I,1318); FOUR STAR PLAYHOUSE (I,1652); JANE WYMAN PRESENTS THE FIRESIDE THEATER (I,2345); YANCY DERRINGER (I,4925)

BERGEN, JERRY
BUZZY WUZZY (I,782)

BERGEN, POLLY
79 PARK AVENUE (II,2310); THE ALAN YOUNG SHOW (I,104); APPOINTMENT WITH ADVENTURE (I,246); THE BEST IN MYSTERY (I,402); THE BLUE ANGEL (I,489); THE BOB HOPE SHOW (I,580); DANNY THOMAS: AMERICA I LOVE YOU (I,1145); THE DICK POWELL SHOW (I,1269); AN ECHO OF THERESA (I,1399); THE ELGIN HOUR (I,1428); THE GUY MITCHELL SHOW (I,1912); THE JACK BENNY HOUR (I,2292); JUST POLLY AND ME (I,2497); THE LUX VIDEO THEATER (I,2795); THE MAURICE CHEVALIER SHOW (I,2954); NOT FOR WOMEN ONLY (I,3310); PEPSI-COLA PLAYHOUSE (I,3523); THE POLLY BERGEN SHOW (I,3643); THE POLLY BERGEN SPECIAL (I,3644); THE ROSALIND RUSSELL SHOW (I,3845); A SPECIAL HOUR WITH DINAH SHORE (I,4149); SPRING HOLIDAY (I,4168); VELVET (II,2725); THE WINDS OF WAR (II,2807); YVES MONTAND ON BROADWAY (I,4978)

BERGER, ANNA
SEVENTH AVENUE (II,2309)

BERGER, ELIZABETH
THE NEW PEOPLE (I,3268)

BERGER, GREGG
G.I.'S (II,949); I'D RATHER BE CALM (II,1216); NO SOAP, RADIO (II,1856); THE SHERIFF AND THE ASTRONAUT (II,2326)

BERGER, HELMUT
DYNASTY (II,746)

BERGER, ION
THE MAGNIFICENT YANKEE (I,2829)

BERGER, IRA
A TIME FOR US (I,4507)

BERGER, JERRY
JACKIE GLEASON AND HIS AMERICAN SCENE MAGAZINE (I,2321)

BERGER, LAUREE
SILVER SPOONS (II,2355)

BERGER, MEL
THROUGH THE MAGIC PYRAMID (II,2615)

BERGER, ROD
IF I LOVE YOU, AM I TRAPPED FOREVER? (II,1222)

BERGER, SARAH
THE CITADEL (II,539); Q. E. D. (II,2094)

BERGER, SENTA
PERRY COMO'S CHRISTMAS IN AUSTRIA (II,2007); THE POPPY IS ALSO A FLOWER (I,3650)

BERGERAC, JACQUES
THE FREEWHEELERS (I,1683)

BERGERE, LEE
DYNASTY (II,746); HOT L BALTIMORE (II,1187); THE SIX MILLION DOLLAR MAN (I,4067); THE SLIGHTLY FALLEN ANGEL (I,4087); SUSAN AND SAM (II,2506)

BERGERSON, BEV
THE MAGIC LAND OF ALLAKAZAM (I,2817)

BERGHOF, HERBERT
REUNION IN VIENNA (I,3780)

BERGLAS, RON
A WOMAN CALLED GOLDA (II,2818)

BERGMAN, ALAN
OUR FIVE DAUGHTERS (I,3413)

BERGMAN, INGRID
THE BOB HOPE SHOW (I,654); BOGART (I,679); HEDDA GABLER (I,2001); HOLLYWOOD: THE SELZNICK YEARS (I,2095); THE HUMAN VOICE (I,2160); THE TURN OF THE SCREW (I,4624); TWENTY-FOUR HOURS IN A WOMAN'S LIFE (I,4644); A WOMAN CALLED GOLDA (II,2818)

BERGMAN, PETER
ALL MY CHILDREN (II,39); THE STARLAND VOCAL BAND (II,2441)

BERGMAN, RICHARD
DAYS OF OUR LIVES (II,629); FATHER MURPHY (II,841)

BERGMAN, TRACY
CIRCUS OF THE STARS (II,536); DAYS OF OUR LIVES (II,629) THE YOUNG AND THE RESTLESS (II,2862)

BERGSTROM, CATHERINE
THE NATIONAL SNOOP (II,1797)

BERITO, JENNIFER
THIS IS KATE BENNETT (II,2597)

BERJER, BARBARA
AS THE WORLD TURNS (II,110); FROM THESE ROOTS (I,1688); THE GUIDING LIGHT (II,1064)

BERK, RICHARD
RYAN'S HOPE (II,2234)

BERKELEY, BALLARD
FAWLTY TOWERS (II,843)

BERKELEY, GEORGE
CAPTAINS AND THE KINGS (II,435)

BERKSON, MURIEL
CONCERNING MISS MARLOWE (I,1032)

BERLAND, TERRI
HOUSE CALLS (II,1194)

BERLE GIRLS, THE
THE MILTON BERLE SHOW (I,3049)

BERLE, JACK
THE BOYS (II,360)

BERLE, MILTON
20TH CENTURY FOLLIES (I,4641); THE BARBARA STANWYCK THEATER (I,350); BATMAN (I,366); THE BIG TIME (I,434); THE BOB HOPE CHRYSLER THEATER (I,502); THE BOB HOPE SHOW (I,551); THE BOB HOPE SHOW (I,570); THE BOB HOPE SHOW (I,661); BOB HOPE SPECIAL: BOB HOPE'S 30TH ANNIVERSARY TV SPECIAL (II,293); BOB HOPE SPECIAL: BOB HOPE'S HILARIOUS UNREHEARSED ANTICS OF THE STARS (II,341); THE CHEVY CHASE NATIONAL HUMOR TEST (II,504); THE CRACKER BROTHERS (II,599); THE DEAN MARTIN CELEBRITY ROAST (II,638); DEAN MARTIN'S CELEBRITY ROAST (II,643); THE DICK POWELL SHOW (I,1269); EVIL ROY SLADE (I,1485); THE FIRST 50 YEARS (II,862); GEORGE BURNS CELEBRATES 80 YEARS IN SHOW BUSINESS (II,967); GEORGE BURNS' 100TH BIRTHDAY PARTY (II,971); HAVE I GOT A CHRISTMAS FOR YOU (II,1105); THE JACK JONES SPECIAL (I,2310); THE KRAFT 75TH ANNIVERSARY SPECIAL (II,1409); THE KRAFT MUSIC HALL (I,2587); KRAFT TELEVISION THEATER (I,2592); THE MANY FACES OF COMEDY (I,2894); THE MILTON BERLE SHOW (I,3047); THE MILTON BERLE SHOW (I,3048); THE MILTON BERLE SHOW (I,3049); THE MILTON BERLE SPECIAL (I,3050); THE MILTON BERLE SPECIAL (I,3051); THE MILTON BERLE SPECIAL (I,3052); MILTON BERLE'S MAD MAD MAD WORLD OF COMEDY (II,1701); MURDER AT NBC (A BOB HOPE SPECIAL) (I,3159); PARADE OF STARS (II,1954); PLIMPTON! DID YOU HEAR THE ONE ABOUT. . .? (I,3629); A SALUTE TO TELEVISION'S 25TH ANNIVERSARY (I,3893); SHEILA (II,2325); SHOW BUSINESS SALUTE TO MILTON BERLE (I,4029); SHOW OF THE YEAR (I,4031); SINATRA—THE FIRST 40 YEARS (II,2359); TEXACO STAR THEATER (I,4411); A TRIBUTE TO "MR. TELEVISION" MILTON BERLE (II,2658); VAUDEVILLE (II,2722)

BERLIN, THOMAS
THE ULTIMATE IMPOSTER (II,2700)

BERLINGER, WARREN
THE ALCOA HOUR (I,108); CRASH ISLAND (II,600); ELLERY QUEEN: TOO MANY SUSPECTS (II,767); THE FUNNY SIDE (I,1714); THE FUNNY SIDE (I,1715); THE HELLCATS (I,2009); THE JOEY BISHOP SHOW (I,2403); KEEPING UP WITH THE JONESES (II,1374); KILROY (I,2541); LILY FOR PRESIDENT (II,1478); MEET CORLISS ARCHER (I,2981); THE NEW OPERATION PETTICOAT (II,1826); QUICK AND QUIET (II,2099); THE SECRET STORM (I,3969); SHAUGHNESSEY (II,2318); SMALL AND FRYE (II,2380); A TOUCH OF GRACE (I,4585)

BERLOSOVA, SVETLANA
CINDERELLA (I,955)

BERMAN, DAVID
CONCRETE BEAT (II,562)

BERMAN, SHELLY
THE BOB HOPE SHOW (I,633); BRENDA STARR, REPORTER (II,373); THE FABULOUS 50S (I,1501); FOREVER FERNWOOD (II,909); JACK PAAR PRESENTS (I,2315); THE JACK PAAR SPECIAL (I,2318); THAT'S LIFE (I,4426)

BERNARD, ARTHUR
THE GIRL, THE GOLD WATCH AND EVERYTHING (II,1002)

BERNARD, CRYSTAL
HAPPY DAYS (II,1084); HIGH SCHOOL, U.S.A. (II,1148); HIGH SCHOOL, U.S.A. (II,1149)

BERNARD, DOROTHY
THE BARRETTS OF WIMPOLE STREET (I,361)

BERNARD, ED
COOL MILLION (I,1046); MURDOCK'S GANG (I,3164); POLICE WOMAN (II,2063); POLICE WOMAN: THE GAMBLE (II,2064); THAT'S MY MAMA (II,2580); THE WHITE SHADOW (II,2788)

BERNARD, GIL
CLUB OASIS (I,985)

BERNARD, HENRY
LITTLE WOMEN (I,2720)

BERNARD, IAN
TINKER, TAILOR, SOLDIER, SPY (II,2621)

BERNARD, JASON
FLATFOOTS (II,879); HIGH PERFORMANCE (II,1145); V (II,2713); V: THE FINAL BATTLE (II,2714); THE WHITE SHADOW (II,2788)

BERNARD, JOSEPH
ABC'S MATINEE TODAY (I,7); THE IMMORTAL (I,2196)

BERNARD, LILLIAN
THE BENNY RUBIN SHOW (I,398)

BERNARD, LOIS
THE BENNY RUBIN SHOW (I,398)

BERNARD, SUE
GENERAL HOSPITAL (II,964)

BERNARD, TOMMY
THE RUGGLES (I,3868); TOM AND JERRY (I,4533)

BERNARDI, HERSCHEL
ARNIE (I,261); A HATFUL OF RAIN (I,1964); HOORAY FOR HOLLYWOOD (I,2114); NEWMAN'S DRUGSTORE (II,1837); PETER GUNN (I,3562); PROFILES IN COURAGE (I,3678); SEVENTH AVENUE (II,2309) SQUADRON (I,4170)

BERNARDI, JACK
THINK PRETTY (I,4448); THE WINDS OF WAR (II,2807)

BERNARDO, AL
SEEING THINGS (II,2297); THE WORLD OF DARKNESS (II,2836)

BERNAU, CHRISTOPHER
DARK SHADOWS (I,1157); THE GUIDING LIGHT (II,1064)

BERNEAU, LAURA
BARE ESSENCE (II,150)

BERNER, SARA
THE HANK MCCUNE SHOW (I,1932)

BERNEY, BERYL
TOM CORBETT, SPACE CADET (I,4535)

BERNHARDT, MEREDITH
ROWAN AND MARTIN'S LAUGH-IN (I,3856)

BERNHEIM, SHIRL
THE CHEAP SHOW (II,491)

BERNIE, AL
54TH STREET REVUE (I,1573)

BERNIE, HELEN
HAWKINS FALLS, POPULATION 6200 (I,1977)

BERNIER, DAISEY
THE FRED WARING SHOW (I,1679)

BERNOUY, BENJAMIN
THE MEMBER OF THE WEDDING (II,1680); ONE MORE TRY (II,1908)

BERNS, BILL
TELEVISION SCREEN MAGAZINE (I,4383)

BERNSEN, CORBIN
ALLISON SIDNEY HARRISON (II,54)

BERNSTEIN, FRED
RYAN'S HOPE (II,2234)

BERNSTEIN, JAY
MICKEY SPILLANE'S MIKE HAMMER: MORE THAN MURDER (II,1693)

BERNSTEIN, LEONARD
CHRISTMAS STARTIME WITH LEONARD BERNSTEIN (I,949)

BERNSTEIN, SCOTT
HAPPY DAYS (II,1084)

BERRELL, LLOYD
THE ADVENTURES OF LONG JOHN SILVER (I,67)

BERRIDGE, ELIZABETH
ONE OF THE BOYS (II,1911)

BERRIDGE, ROBBIE
THE GUIDING LIGHT (II,1064)

BERRY, ALICE
ROGER DOESN'T LIVE HERE ANYMORE (II,2201)

BERRY, CHUCK
THE BIG BEAT (I,423); THE WOLFMAN JACK RADIO SHOW (II,2816)

BERRY, ERIC
BAREFOOT IN ATHENS (I,354)

BERRY, FRED
BATTLE OF THE NETWORK STARS (II,165); WHAT'S HAPPENING!! (II,2769)

BERRY, JUNE
W*A*L*T*E*R (II,2731)

BERRY, KEN
THE ANDY GRIFFITH SHOW (I,192); THE ANN SOTHERN SHOW (I,220); ARTHUR GODFREY'S PORTABLE

ELECTRIC MEDICINE SHOW (I,290); THE BOB NEWHART SHOW (I,666); CAROL AND COMPANY (I,847); DOCTOR KILDARE (I,1315); EUNICE (II,784); F TROOP (I,1499); THE FABULOUS FUNNIES (II,801); FANTASY ISLAND (II,829); FEATHERSTONE'S NEST (II,847); THE FIRST NINE MONTHS ARE THE HARDEST (I,1584); KELLY'S KIDS (I,2523); THE KEN BERRY WOW SHOW (I,2524); LETTERS FROM THREE LOVERS (I,2671); LI'L ABNER (I,2702); THE LOVE BOAT II (II,1533); MAMA'S FAMILY (II,1610); MAYBERRY R.F.D. (I,2961); MITZI AND A HUNDRED GUYS (II,1710); MITZI. . . ROARIN' IN THE 20S (II,1711); MITZI. . . THE FIRST TIME (I,3073); ONCE UPON A MATTRESS (I,3371); OVER AND OUT (II,1938); ROWAN AND MARTIN'S LAUGH-IN (I,3857); THE ROYAL FOLLIES OF 1933 (I,3862); TEXACO STAR THEATER: OPENING NIGHT (II,2565)

BERRYMAN, JOE
HIGHWAY HONEYS (II,1151)

BERSELL, MIKE
ALL MY CHILDREN (II,39); WHERE THE HEART IS (I,4831)

BERTHEAU, STACEY
VALLEY OF THE DINOSAURS (II,2717)

BERTHRONG, DEIRDRE
JAMES AT 15 (II,1270); JAMES AT 15 (II,1271)

BERTI, DEHL
BORN TO THE WIND (II,354); OPERATION: NEPTUNE (I,3404)

BERTI, MARTINA
MOSES THE LAWGIVER (II,1737); THE THREE MUSKETEERS (I,4480)

BERTINELLI, VALERIE
BATTLE OF THE NETWORK STARS (II,165); BATTLE OF THE NETWORK STARS (II,167); BATTLE OF THE NETWORK STARS (II,168); BATTLE OF THE NETWORK STARS (II,169); CELEBRITY CHALLENGE OF THE SEXES 3 (II,467); ONE DAY AT A TIME (II,1900); THE SECRET OF CHARLES DICKENS (II,2293)

BERTINO, JIM
THE RETURN OF LUTHER GILLIS (II,2137)

BERTISH, SUZANNE
NICHOLAS NICKLEBY (II,1840)

BERTRAND, COLETTE
CHARLIE'S ANGELS (II,487)

BERTRAND, JACQUELINE
EAGLE IN A CAGE (I,1392)

BERWICK, BRAD
WINDOW ON MAIN STREET (I,4873)

BERWICK, JAMES
APPLAUSE (I,244); THE KILLING GAME (I,2540); NO—HONESTLY (II,1857)

BERWICK, JOHN
THE KID SUPER POWER HOUR WITH SHAZAM (II,1386)

BERWICK, VIOLA
THE BENNETS (I,397); HAWKINS FALLS, POPULATION 6200 (I,1977); A TIME TO LIVE (I,4510)

BESCH, BIBI
BACKSTAIRS AT THE WHITE HOUSE (II,133); THE HAMPTONS (II,1076); SECRETS OF MIDLAND HEIGHTS (II,2296); SKYWARD CHRISTMAS (II,2378); SOMERSET (I,4115); STEELTOWN (II,2452); THREE TIMES DALEY (II,2610); TOM AND JOANN (II,2630)

BESS, ARDON
KING OF KENSINGTON (II,1395)

BESSELL, TED
BOBBY PARKER AND COMPANY (II,344); GOMER PYLE, U.S.M.C. (I,1843); GOOD TIME HARRY (II,1037); GREAT ADVENTURE (I,1868); THE GREATEST SHOW ON EARTH (I,1883); IT'S A MAN'S WORLD (I,2258); THE MARY TYLER MOORE SHOW (II,1640); ME AND THE CHIMP (I,2974); THE TED BESSELL SHOW (I,4369); THAT GIRL (II,2570)

BESSER, JOE
THE ABBOTT AND COSTELLO SHOW (I,2); BING CROSBY'S CHRISTMAS SHOW (I,474); THE GALAXY GOOFUPS (II,954); THE HOUNDCATS (I,2132); THE JACK BENNY PROGRAM (I,2294); JEANNIE (II,1275); THE JOEY BISHOP SHOW (I,2404); THE KEN MURRAY SHOW (I,2525); THE MONK (I,3087); SCOOBY'S ALL-STAR LAFF-A-LYMPICS (II,2273); YOGI'S SPACE RACE (II,2853)

BEST, EDNA
BERKELEY SQUARE (I,399); PULITZER PRIZE PLAYHOUSE (I,3692)

BEST, JAMES
THE ANDY GRIFFITH SHOW (I,192); CENTENNIAL (II,477); THE CODE OF JONATHAN WEST (I,992); THE DAVID NIVEN THEATER (I,1178); THE DUKES (II,741); THE DUKES OF HAZZARD (II,742); GENERAL ELECTRIC TRUE (I,1754); GENTRY'S PEOPLE (I,1762); MCLAREN'S RIDERS (II,1663); THE RUNAWAY BARGE (II,2230); SKIP TAYLOR (I,4075)

BEST, JANEEN
YOUNG HEARTS (II,2865)

BEST, LARRY
GRANDPA MAX (II,1051); THE MILTON THE MONSTER CARTOON SHOW (I,3053)

BEST, MARY
HEAVENS TO BETSY (I,1997)

BEST, WILLIE
MY LITTLE MARGIE (I,3194); TROUBLE WITH FATHER (I,4610); WATERFRONT (I,4774)

BESWICK, MARTINE
THE HORROR OF IT ALL (II,1181); LONGSTREET (I,2750); WINNER TAKE ALL (II,2808)

BETHENCOURT, FRANCIS
TONIGHT AT 8:30 (I,4556)

BETHUNE, IVY
COUNTERATTACK: CRIME IN AMERICA (II,581); DAYS OF OUR LIVES (II,629); FATHER MURPHY (II,841)

BETHUNE, JOHN
MOMENT OF TRUTH (I,3084)

BETHUNE, ZINA
THE GUIDING LIGHT (II,1064); LITTLE WOMEN (I,2722); LOVE OF LIFE (I,2774); THE NURSES (I,3318); YOUNG DR. MALONE (I,4938)

BETTEN, MARY
OUR FAMILY BUSINESS (II,1931)

BETTGER, LYLE
AUNT JENNY'S REAL LIFE STORIES (I,314); BROTHER RAT (I,746); THE COURT OF LAST RESORT (I,1082); THE GRAND JURY (I,1865); M STATION: HAWAII (II,1568); TV READERS DIGEST (I,4630)

BETTI, LAURA
THE WORD (II,2833)

BETTIS, VALERIE
THE WOMEN (I,4890)

BETTON, JACQUELINE
A WALK IN THE NIGHT (I,4750)

BETTS, JACK
CHECKMATE (I,919); ONE LIFE TO LIVE (II,1907)

BETZ, CARL
CRISIS (I,1102); THE DAUGHTERS OF JOSHUA CABE RETURN (I,1170); THE DONNA REED SHOW (I,1347); JUDD, FOR THE DEFENSE (I,2463); LOVE OF LIFE (I,2774); MITZI AND A HUNDRED GUYS (II,1710); THE MONK (I,3087)

BEUTEL, JACK
JUDGE ROY BEAN (I,2467)

BEVANS, CLEM
DAVY CROCKETT (I,1181)

BEVERLY
WHERE'S HUDDLES (I,4836)

BEXLEY, DON
BEANE'S OF BOSTON (II,194); SANFORD AND SON (II,2256); SANFORD ARMS (II,2257); UPTOWN SATURDAY NIGHT (II,2711)

BEY, RICHARD
SOAP WORLD (II,2394)

BEY, SALOME
ANNE MURRAY'S LADIES' NIGHT (II,89)

BEYERS, BILL
CAPITOL (II,426); JOE AND VALERIE (II,1298)

BEYMER, RICHARD
GUILTY OR NOT GUILTY (I,1903); PAPER DOLLS (II,1952)

BEZIC, VAL
DOROTHY HAMILL IN ROMEO & JULIET ON ICE (II,718)

BEZICOS, SANDRA
THE MAGIC PLANET (II,1598)

BEZICOS, VAL
THE MAGIC PLANET (II,1598)

BIBBY, CHARLES KING
CHIEFS (II,509)

BIBERMAN, ABNER
KODIAK (II,1405)

BICKARD, SANFORD

BUCK ROGERS IN THE 25TH CENTURY (I,751)

BICKETT, SARA
OPPENHEIMER (II,1919)

BICKFORD, CHARLES
ADAMSBURG, U.S.A. (I,33); THE BARBARA STANWYCK THEATER (I,350); THE FARMER'S DAUGHTER (I,1534); MAN BEHIND THE BADGE (I,2857); THE VIRGINIAN (I,4738); WINTERSET (I,4882)

BICKLE, TED
THE SONNY AND CHER COMEDY HOUR (II,2400)

BICKNELL, HEATHER
LOVE OF LIFE (I,2774)

BICTAK, WILLY
TO EUROPE WITH LOVE (I,4525)

BIDMEAD, STEPHANIE
LITTLE WOMEN (I,2723); THE ONEDIN LINE (II,1912)

BIEG, GLYNIS
THE LEPRECHAUN'S CHRISTMAS GOLD (II,1461)

BIEHN, MICHAEL
OPERATION: RUNAWAY (II,1918); THE RUNAWAYS (II,2231); STEELTOWN (II,2452)

BIENDIE, IRENE
ANOTHER WORLD (II,97)

BIENER, TOM
THE DONNA FARGO SHOW (II,710); THE JIM STAFFORD SHOW (II,1291); THE PAUL LYNDE COMEDY HOUR (II,1978)

BIERI, RAMON
BRET MAVERICK (II,374); CROSSFIRE (II,606); HOW THE WEST WAS WON (II,1196); JOE'S WORLD (II,1305); LUCAS TANNER (II,1555); OWEN MARSHALL: COUNSELOR AT LAW (I,3436); THE RHINEMANN EXCHANGE (II,2150); THE SAN PEDRO BUMS (II,2251); SARGE (I,3914); SARGE: THE BADGE OR THE CROSS (I,3915); SOUTHERN FRIED (I,4139); TRUE GRIT (II,2664)

BIG, GUY
THE HILARIOUS HOUSE OF FRIGHTENSTEIN (II,1153)

BIGELOW, SUSAN
LOVE AT FIRST SIGHT (II,1530); LOVE AT FIRST SIGHT (II,1531); THE ROCK RAINBOW (II,2194); WILD AND WOOLEY (II,2798)

BIGGINS, CHRISTOPHER
MASADA (II,1642)

BIGGS, CASEY
FLYING HIGH (II,895)

BIGGS, JERRY
HIGHWAY HONEYS (II,1151)

BIHELLER, ROBERT
HERE COME THE BRIDES (I,2024)

BIKEL, THEODORE
ALFRED HITCHCOCK PRESENTS (I,115); APPOINTMENT WITH ADVENTURE (I,246); THE DIARY OF ANNE FRANK (I,1262); THE ELGIN HOUR (I,1428); EXPOSE (I,1491); THE GENERAL ELECTRIC THEATER (I,1753); KILLER BY NIGHT (I,2537); THE KING AND MRS. CANDLE (I,2544); LOOSE CHANGE (II,1518); ST. JOAN (I,4175); TESTIMONY OF TWO MEN (II,2564); THERE SHALL BE NO NIGHT (I,4437); WHO HAS SEEN THE WIND? (I,4847)

BIL AND CORA BAIRD PUPPETS, THE
THE BIG TIME (I,434); MAX LIEBMAN PRESENTS (I,2958); PANORAMA (I,3448); THE TONIGHT SHOW (I,4564); VARIETY (I,4703); WASHINGTON SQUARE (I,4772)

BILL CALLAHAN DANCE TROUPE, THE
THE SATURDAY NIGHT REVUE (I,3921)

BILL FOSTER DANCERS, THE
THE KRAFT MUSIC HALL (I,2587); KRAFT MUSIC HALL PRESENTS THE DAVE KING SHOW (I,2586)

BILL THOMPSON SINGERS, THE
ANOTHER EVENING WITH FRED ASTAIRE (I,229); AN EVENING WITH FRED ASTAIRE (I,1474)

BILL WIMBERLY AND HIS COUNTRY RHYTHM BOYS
JUBILEE U.S.A. (I,2461)

BILL, FRANCESCA
BIG CITY BOYS (II,232)

BILL, TONY
HAVING BABIES II (II,1108); WASHINGTON: BEHIND CLOSED DOORS (II,2744); WHAT REALLY HAPPENED TO THE CLASS OF '65? (II,2768)

BILLINGS, EARL
WHAT'S HAPPENING!! (II,2769)

BILLINGSLEY, BARBARA
THE BROTHERS (I,748); CAVALCADE OF AMERICA (I,875); COUNTERPOINT (I,1059); FIRESIDE THEATER (I,1580); HIGH SCHOOL, U.S.A. (II,1149); IT'S A SMALL WORLD (I,2260); JIM HENSON'S MUPPET BABIES (II,1289); LEAVE IT TO BEAVER (I,2648); MATINEE THEATER (I,2947); PANIC! (I,3447); PROFESSIONAL FATHER (I,3677); SCHLITZ PLAYHOUSE OF STARS (I,3936); STILL THE BEAVER (II,2466)

BILLINGSLEY, DICK
DAYS OF OUR LIVES (II,629)

BILLINGSLEY, JENNIFER
GENERAL HOSPITAL (II,964); I AND CLAUDIE (I,2164); THE ROOKIES (I,3840)

BILLINGSLEY, NEIL
FATHER MURPHY (II,841); SEARCH FOR TOMORROW (II,2284)

BILLINGSLEY, PATRICK
THE LAST LEAF (II,1437)

BILLINGSLEY, PETER
MASSARATI AND THE BRAIN (II,1646); REAL KIDS (II,2116); REAL PEOPLE (II,2118)

BILLINGSLEY, SHERMAN
THE STORK CLUB (I,4236)

BILLINGTON, MICHAEL
THE QUEST (II,2098); U.F.O. (I,4662)

BILLY WILLIAMS QUARTET, THE
YOUR SHOW OF SHOWS (I,4957)

BILLY WILLIAMS QUINTET
THE KATE SMITH SHOW (I,2509)

BILTON, MICHAEL
BRIDESHEAD REVISITED (II,375)

BINCHY, KATE
LITTLE LORD FAUNTLEROY (II,1492)

BINDER, BRAMMER
BROTHER RAT (I,746)

BINDER, MICHAEL
DINER (II,679); G.I.'S (II,949); HOT (II,1184); THAT'S TV (II,2581)

BINDIGER, EMILY
THE GREAT SPACE COASTER (II,1059)

BINGHAM, BING
DEAR ALEX AND ANNIE (II,651)

BINGO THE CHIMP
THE ABBOTT AND COSTELLO SHOW (I,2)

BINKLEY, LANE
ALEX AND THE DOBERMAN GANG (II,32); DUFFY (II,739); INSTITUTE FOR REVENGE (II,1239); ROOTS (II,2211)

BINNEY, JEFFREY
BATTLESTAR GALACTICA (II,181)

BINNS, EDWARD
ADAMSBURG, U.S.A. (I,33); BATTLES: THE MURDER THAT WOULDN'T DIE (II,180); BRENNER (I,719); THE DOCTOR (I,1320); DOWN HOME (II,731); THE INNER SANCTUM (I,2216); IT TAKES A THIEF (I,2250); THE JUDGE (I,2468); THE NURSES (I,3318); PENTAGON U.S.A. (I,3514); THE POWER WITHIN (II,2073); STUBBY PRINGLE'S CHRISTMAS (II,2483)

BIPNEY, FRANK
THE JERK, TOO (II,1282)

BIRCH, JOHNNY
THE RED BUTTONS SHOW (I,3750)

BIRCH, PAUL
CANNONBALL (I,822); THE COURT OF LAST RESORT (I,1082); MEDIC (I,2976); YOU ARE THERE (I,4929)

BIRCH, PETER
PENMARRIC (II,1993)

BIRD, BILLIE
BENSON (II,208); IT TAKES TWO (II,1252); THREE'S A CROWD (II,2613)

BIRD, DAVID
MARY HARTMAN, MARY HARTMAN (II,1638)

BIRK, RAYE
THE EARTHLINGS (II,751); MR. ROBERTS (II,1754)

BIRKETT, BERNADETTE

THE BOOK OF LISTS (II,350)

BIRMAN, LEN
CAPTAIN AMERICA (II,427); CAPTAIN AMERICA (II,428); DOCTOR SIMON LOCKE (I,1317); ESCAPADE (II,781); POLICE SURGEON (I,3639); THE RHINEMANN EXCHANGE (II,2150)

BIRNBAUM, STAN
RYAN'S HOPE (II,2234)

BIRNEY, DAVID
THE ADAMS CHRONICLES (II,8); BATTLE OF THE NETWORK STARS (II,176); BRIDGET LOVES BERNIE (I,722); GLITTER (II,1009); JACQUELINE SUSANN'S VALLEY OF THE DOLLS, 1981 (II,1268); MASTER OF THE GAME (II,1647); SERPICO (II,2306); ST. ELSEWHERE (II,2432); ST. JOAN (I,4175); TESTIMONY OF TWO MEN (II,2564)

BIRNEY, MEREDITH BAXTER
BATTLE OF THE NETWORK STARS (II,176); BEULAH LAND (II,226); BRIDGET LOVES BERNIE (I,722); FAMILY (II,813); FAMILY TIES (II,819); THE IMPOSTER (II,1225); LITTLE WOMEN (II,1497); VANITIES (II,2720)

BIRO, BARNEY
NIGHT COURT (I,3285)

BISCAILUZ, EUGENE W
CODE 3 (I,993)

BISCARDI, JESSICA
COCAINE AND BLUE EYES (II,548)

BISHOP, ED
CAPTAIN SCARLET AND THE MYSTERONS (I,837); THE DEVIL'S WEB (I,1248); MASTER OF THE GAME (II,1647); QUILLER: PRICE OF VIOLENCE (II,2101); U.F.O. (I,4662)

BISHOP, JENNIFER
HEE HAW (II,1123)

BISHOP, JOEY
THE ANDY WILLIAMS SPECIAL (I,212); CELEBRITY SWEEPSTAKES (II,476); ESTHER WILLIAMS AT CYPRESS GARDENS (I,1465); EVERYTHING HAPPENS TO ME (I,1481); THE FRANK SINATRA TIMEX SHOW (I,1670); THE JOEY BISHOP SHOW (I,2403); THE JOEY BISHOP SHOW (I,2404); THE JOEY BISHOP SHOW (I,2405); THE LIAR'S CLUB (II,1466); ROMP (I,3838); SORORITY '62 (II,2405); A TRIBUTE TO "MR. TELEVISION" MILTON BERLE (II,2658)

BISHOP, JOHN
THE LIFE AND TIMES OF GRIZZLY ADAMS (II,1469)

BISHOP, JULIE
THE ETHEL BARRYMORE THEATER (I,1467); FIRESIDE THEATER (I,1580); MY HERO (I,3193); TV READERS DIGEST (I,4630)

BISHOP, KELLY
ADVICE TO THE LOVELORN (II,16); LOVE AND LEARN (II,1529); A YEAR AT THE TOP (II,2844); A YEAR AT THE TOP (II,2845)

BISHOP, LARRY
BEANE'S OF BOSTON (II,194); CONDOMINIUM (II,566)

BISHOP, LOANNE
GENERAL HOSPITAL (II,964)

BISHOP, MEL
THE RICH LITTLE SHOW (II,2153)

BISHOP, PAT
PRISONER: CELL BLOCK H (II,2085)

BISHOP, RONALD
CANDIDA (II,423)

BISHOP, STEPHEN
THE NATALIE COLE SPECIAL (II,1794)

BISHOP, SUE
THE BENNY HILL SHOW (II,207)

BISHOP, WILLIAM
ALL ABOUT BARBARA (I,125); FIRESIDE THEATER (I,1580); IT'S A GREAT LIFE (I,2256); THE MARRIAGE BROKER (I,2920)

BISNEY, STEVE
A TOWN LIKE ALICE (II,2652)

BISOGLIO, VAL
FLYING HIGH (II,895); INSIDE O.U.T. (II,1235); JOHNNY GARAGE (II,1337); MATT HELM (II,1652); POLICE WOMAN (II,2063); QUINCY, M. E. (II,2102); ROLL OUT! (I,3833); WORKING STIFFS (II,2834)

BISSELL, WHIT
ARCHIE (II,104); BACHELOR FATHER (I,323); BACHELOR PARTY (I,324); CITY BENEATH THE SEA (I,965); THE DOCTOR (I,1320); FIRESIDE THEATER (I,1580); THE GOLDEN AGE OF TELEVISION (II,1018); MIRACLE ON 34TH STREET (I,3056); MR. TUTT (I,3153); NICK AND NORA (II,1842); NO WARNING (I,3301); ONE STEP BEYOND (I,3388); SANDBURG'S LINCOLN (I,3907); THE TIME TUNNEL (I,4511); THE UNEXPLAINED (I,4675); THE WEB (I,4784); YOU ARE THERE (I,4929)

BISSET, JACQUELINE
PAVAROTTI AND FRIENDS (II,1985)

BISSET, SHARON
20 MINUTE WORKOUT (II,2680)

BITTER END SINGERS, THE
FAVORITE SONGS (I,1545); NBC FOLLIES OF 1965 (I,3240)

BITTIS, ELEANOR
JANE EYRE (I,2337)

BITTNER, JACK
ABE LINCOLN IN ILLINOIS (I,11)

BIXBY, BILL
THE BARBARY COAST (II,147); THE BOOK OF LISTS (II,350); THE COURTSHIP OF EDDIE'S FATHER (I,1083); ELVIS REMEMBERED: NASHVILLE TO HOLLYWOOD (II,772); FANTASY ISLAND (II,830); GOODNIGHT, BEANTOWN (II,1041); I'VE HAD IT UP TO HERE (II,1221); THE INCREDIBLE HULK (II,1232); THE MAGICIAN (I,2823); THE MAGICIAN (I,2824); MASQUERADE PARTY (II,1645); MITZI AND A HUNDRED GUYS (II,1710); MY FAVORITE MARTIAN (I,3189); THE NATURAL LOOK (II,1799); NIGHT GALLERY (I,3287); OF MEN OF WOMEN (I,3331); REX HARRISON PRESENTS SHORT STORIES OF LOVE (II,2148); RICH MAN, POOR MAN—BOOK I (II,2161); THE SENSATIONAL, SHOCKING, WONDERFUL, WACKY 70S (II,2302); SPENCER'S PILOTS (II,2421); THAT THING ON ABC (II,2574); WHATEVER BECAME OF. . .? (II,2773)

BJOERLING
FESTIVAL OF MUSIC (I,1567)

BJORN, ANNE
ALMOST AMERICAN (II,55)

BJURMAN, SUSAN
THE ADAMS CHRONICLES (II,8)

BLACK, KAREN
CIRCUS OF THE STARS (II,530); E/R (II,748); GHOST STORY (I,1788); THE SHAMEFUL SECRETS OF HASTINGS CORNERS (I,3994)

BLACK, MARIANNE
SUGAR TIME (II,2489); THREE'S COMPANY (II,2614)

BLACKBURN, CLARICE
DARK SHADOWS (I,1157); THE DOCTORS (II,694); STUDIO ONE (I,4268); WHERE THE HEART IS (I,4831)

BLACKBURN, DOROTHY
ANOTHER WORLD (II,97); THE DOCTORS (II,694); HARVEY (I,1962); INTERMEZZO (I,2223); MONTGOMERY'S SUMMER STOCK (I,3096)

BLACKBURN, GRETA
MASSARATI AND THE BRAIN (II,1646); V: THE FINAL BATTLE (II,2714)

BLACKBURN, HARRIET
THE INA RAY HUTTON SHOW (I,2207)

BLACKBURN, RICHARD
RETURN TO THE PLANET OF THE APES (II,2145)

BLACKBURN, ROBERT
KENNEDY (II,1377)

BLACKERBY, KATHLEEN
MATINEE AT THE BIJOU (II,1651)

BLACKETT, TONY
RETURN TO EDEN (II,2143)

BLACKMAN, HONOR
THE AVENGERS (II,121); KRAFT MYSTERY THEATER (I,2589); LACE (II,1418); ROBIN'S NEST (II,2187); THE WIDE OPEN DOOR (I,4859)

BLACKMAN, JOAN
PEYTON PLACE (I,3574)

BLACKMAN, JOHN
THE PAUL HOGAN SHOW (II,1975)

BLACKMER, SIDNEY
HOWDY (I,2150); THE LITTLE FOXES (I,2712); THE SHARPSHOOTER (I,4001)

BLACKMORE, STEPHANIE
BEYOND WITCH MOUNTAIN (II,228); MICKEY SPILLANE'S MIKE HAMMER: MORE THAN MURDER (II,1693); THIS IS

KATE BENNETT (II,2597)

BLACKSTONE JR, HARRY
MANDRAKE (II,1617)

BLACKSTONE SR, HARRY
BLACKSTONE MAGIC SPOTS (I,481)

BLACKTON, JENNY
THE INTERNS (I,2226)

BLACKWELL, CHARLES T
CHIEFS (II,509)

BLACKWELL, MR
THE MORNING SHOW (I,3103)

BLACKWELL, WILLIAM
DAVY CROCKETT (I,1181)

BLACQUE, TAUREAN
ALEX AND THE DOBERMAN GANG (II,32); FRANKIE AND ANNETTE: THE SECOND TIME AROUND (II,917); HILL STREET BLUES (II,1154)

BLAIN, JAMIE
RETURN TO EDEN (II,2143)

BLAINE, JIMMY
THE BILLY DANIELS SHOW (I,451); THE DIANE DOXEE-JIMMY BLAINE SHOW (I,1261); HOLD THAT CAMERA (I,2072); JERRY MAHONEY'S CLUB HOUSE (I,2371); MUSIC FROM THE MEADOWBROOK (I,3171); THE PAUL WINCHELL AND JERRY MAHONEY SHOW (I,3492); STOP THE MUSIC (I,4232)

BLAINE, MARTIN
GENERAL HOSPITAL (II,964); TODAY IS OURS (I,4531); YOUNG DR. MALONE (I,4938)

BLAINE, SALLY
ON TRIAL (I,3362)

BLAINE, VIVIAN
THE BOB HOPE SHOW (I,538); THE BOB HOPE SHOW (I,546); DREAM GIRL (I,1374); G.E. SUMMMER ORIGINALS (I,1747); HEREAFTER (II,1138); MARY HARTMAN, MARY HARTMAN (II,1638); THE RAY BOLGER SHOW (I,3731); THOSE TWO (I,4470); A YEAR AT THE TOP (II,2845)

BLAIR, BETSY
FORD THEATER HOUR (I,1635); MAGNAVOX THEATER (I,2827)

BLAIR, JANET
ALCOA/GOODYEAR THEATER (I,107); THE ARMSTRONG CIRCLE THEATER (I,260); THE BOB HOPE CHRYSLER THEATER (I,502); CAESAR'S HOUR (I,790); CHEVROLET ON BROADWAY (I,925); THE CHEVROLET TELE-THEATER (I,926); THE CHEVY SUMMER SHOW (I,929); THE CHEVY SUMMER SHOW (I,930); A CONNECTICUT YANKEE (I,1038); THE ELGIN HOUR (I,1428); FORD TELEVISION THEATER (I,1634); FORD THEATER HOUR (I,1635); HEAVEN WILL PROTECT THE WORKING GIRL (I,1996); ONE TOUCH OF VENUS (I,3389); SHIRLEY TEMPLE'S STORYBOOK (I,4017); THE SMITH FAMILY (I,4092); THE STRAWBERRY BLONDE (I,4254); TEXACO COMMAND PERFORMANCE (I,4407)

BLAIR, JOHN
GEORGE WASHINGTON (II,978)

BLAIR, JOYCE
THE LAST DAYS OF POMPEII (II,1436)

BLAIR, JUNE
THE ADVENTURES OF OZZIE AND HARRIET (I,71); TWO FACES WEST (I,4653)

BLAIR, KEVIN
RITUALS (II,2182)

BLAIR, LINDA
CELEBRITY CHALLENGE OF THE SEXES 2 (II,466); CELEBRITY DAREDEVILS (II,473); CIRCUS OF THE STARS (II,536); CIRCUS OF THE STARS (II,537); ROCK COMEDY (II,2191); STRANGER IN OUR HOUSE (II,2477)

BLAIR, PAMELA
LOVING (II,1552)

BLAIR, PATRICIA
DANIEL BOONE (I,1142); THE RIFLEMAN (I,3789); YOU'RE ONLY YOUNG ONCE (I,4969)

BLAIRE AND DEANE
FOR YOUR PLEASURE (I,1627)

BLAISDAL, BRAD
THREE'S COMPANY (II,2614)

BLAISDELL, NESBITT
KENNEDY (II,1377)

BLAKE, AMANDA
THE BEST LITTLE SPECIAL IN TEXAS (II,215); CLOWN ALLEY (I,979); FIRESIDE THEATER (I,1580); FOUR STAR PLAYHOUSE (I,1652); THE GENERAL ELECTRIC THEATER (I,1753); GUNSMOKE (II,1069); STUDIO ONE (I,4268)

BLAKE, CARTER
TOM CORBETT, SPACE CADET (I,4535)

BLAKE, ELLEN
THE DEADLY GAME (II,633); THE FRENCH ATLANTIC AFFAIR (II,924)

BLAKE, GEOFFREY
OFF THE RACK (II,1878)

BLAKE, JANET
PHILCO TELEVISION PLAYHOUSE (I,3583)

BLAKE, JEAN
BEWITCHED (I,418); THE CASE OF THE DANGEROUS ROBIN (I,862)

BLAKE, LARRY
CAP'N AHAB (I,824); THAT'S MY BOY (I,4427)

BLAKE, MADGE
BABY CRAZY (I,319); BATMAN (I,366); BEULAH (I,416); DEAR MOM, DEAR DAD (I,1202); EVERYTHING HAPPENS TO ME (I,1481); THE JOEY BISHOP SHOW (I,2403); LEAVE IT TO BEAVER (I,2648); MONA MCCLUSKEY (I,3085); THE REAL MCCOYS (I,3741); THE SHAMEFUL SECRETS OF HASTINGS CORNERS (I,3994); THE SOFT TOUCH (I,4107)

BLAKE, OLIVER
THE BROTHERS (I,748)

BLAKE, PETER
PENMARRIC (II,1993)

BLAKE, ROBERT
BARETTA (II,152); FIRESIDE THEATER (I,1580); JOE DANCER: THE BIG BLACK PILL (II,1300); JOE DANCER: THE MONKEY MISSION (II,1301); JOE DANCER: MURDER ONE, DANCER 0 (II,1302); ONE STEP BEYOND (I,3388); THE RICHARD BOONE SHOW (I,3784)

BLAKE, ROBIN
GENERAL HOSPITAL (II,964)

BLAKE, SONDRA
IN TANDEM (II,1228); JOE DANCER: THE BIG BLACK PILL (II,1300); JOE DANCER: THE MONKEY MISSION (II,1301); JOE DANCER: MURDER ONE, DANCER 0 (II,1302); NIGHTSIDE (II,1851); TURNOVER SMITH (II,2671)

BLAKE, STUART
DOCTOR WHO—THE FIVE DOCTORS (II,690)

BLAKE, TIMOTHY
MAUREEN (II,1656); PINOCCHIO'S CHRISTMAS (II,2045); THE ROPERS (II,2214); TABITHA (II,2527)

BLAKE, WHITNEY
HAZEL (I,1982); KRAFT MYSTERY THEATER (I,2589); MEDIC (I,2976)

BLAKELY, GENE
BEST FOOT FORWARD (I,401)

BLAKELY, GEORGE
ALL THE RIVERS RUN (II,43)

BLAKELY, SUSAN
THE HITCHHIKER (II,1157); RICH MAN, POOR MAN—BOOK I (II,2161)

BLAKEMOR, KORRY
CAMP WILDERNESS (II,419)

BLAKENEY, OLIVE
DOCTOR HUDSON'S SECRET JOURNAL (I,1312); MR. TUTT (I,3153)

BLAKISTON, CAROLINE
THE FORSYTE SAGA (I,1645)

BLAKSTAD, MATTHEW
PENMARRIC (II,1993)

BLANC, CHRISTOPHER
HOUSE CALLS (II,1194)

BLANC, DOMINIQUE
LACE (II,1418)

BLANC, MEL
THE ALVIN SHOW (I,146); ARTHUR GODFREY LOVES ANIMALS (I,282); THE ATOM ANT/SECRET SQUIRREL SHOW (I,311); BUCK ROGERS IN THE 25TH CENTURY (II,383); BUCK ROGERS IN THE 25TH CENTURY (II,384); THE BUGS BUNNY SHOW (II,390); THE BUGS BUNNY/ROAD RUNNER SHOW (II,391); CAPTAIN CAVEMAN AND THE TEEN ANGELS (II,433); CURIOSITY SHOP (I,1114); THE DICK TRACY SHOW (I,1272); THE FLINTSTONE COMEDY HOUR (II,881); FLINTSTONE FAMILY ADVENTURES (II,882); THE FLINTSTONE FUNNIES (II,883); THE FLINTSTONES (II,884); THE FLINTSTONES (II,885); FRED AND BARNEY MEET THE THING (II,919); THE GALAXY GOOFUPS (II,954); THE HEATHCLIFF AND MARMADUKE SHOW (II,1118); THE JACK BENNY

PROGRAM (I,2294); JACK BENNY'S 20TH ANNIVERSARY TV SPECIAL (I,2303); THE JETSONS (I,2376); LIPPY THE LION (I,2710); THE MAGILLA GORILLA SHOW (I,2825); MR. O'MALLEY (I,3146); MURDER CAN HURT YOU! (II,1768); THE NEW FRED AND BARNEY SHOW (II,1817); THE PETER POTAMUS SHOW (I,3568); PORKY PIG AND FRIENDS (I,3651); THE ROAD RUNNER SHOW (I,3799); SCOOBY'S ALL-STAR LAFF-A-LYMPICS (II,2273); SPEED BUGGY (II,2419); THE TOM AND JERRY SHOW (I,4534); THE WACKY RACES (I,4745); WHERE'S HUDDLES (I,4836); YOGI'S SPACE RACE (II,2853)

BLANC, SHELLEY
LOVE OF LIFE (I,2774)

BLANCH, JEWEL
BAFFLED (I,332)

BLANCHARD, JERRI
HOTEL BROADWAY (I,2127)

BLANCHARD, MARIE
KLONDIKE (I,2573); TERRY AND THE PIRATES (I,4405)

BLANCHARD, NINA
YOUR NEW DAY (II,2873)

BLANCHARD, SUSAN
ALL MY CHILDREN (II,39); BEACON HILL (II,193); HOW TO SUCCEED IN BUSINESS WITHOUT REALLY TRYING (II,1197); MAGNUM, P.I. (II,1601); MR. T. AND TINA (II,1758); THE NEW MAVERICK (II,1822); SAINT PETER (II,2238); SHE'S IN THE ARMY NOW (II,2320); YOUNG MAVERICK (II,2867)

BLANDE, CHRISTOPHER
MAKING THE GRADE (II,1606); THE TOM SWIFT AND LINDA CRAIG MYSTERY HOUR (II,2633)

BLANKFIELD, MARK
FRIDAYS (II,926); THE JERK, TOO (II,1282); TAXI (II,2546)

BLANKS, MARY LYN
AS THE WORLD TURNS (II,110)

BLANTON, ARELL
ALL THAT GLITTERS (II,41)

BLANSHARD, JOBY
A HORSEMAN RIDING BY (II,1182)

BLAUNER, STEVE
THE COMEDY OF ERNIE KOVACS (I,1022)

BLAVAT, JERRY
THE DISCOPHONIC SCENE (I,1291)

BLAYLOCK, TOM
DOC (II,682)

BLAZER, JUDITH
AS THE WORLD TURNS (II,110)

BLAZO, JOHN
RYAN'S HOPE (II,2234)

BLEDSOE, TEMPEST
THE COSBY SHOW (II,578)

BLEDSOE, WILL
HIGHWAY HONEYS (II,1151)

BLEE, DEBRA
T.J. HOOKER (II,2524)

BLEETH, YASMINE
RYAN'S HOPE (II,2234)

BLEIER, NANCY
BENDER (II,204)

BLEILER, WELDON BOYCE
TWIGS (II,2684)

BLEIWESS, NANCY
LAUGH-IN (II,1444)

BLENDERS, THE
THE LAWRENCE WELK SHOW (I,2643); MEMORIES WITH LAWRENCE WELK (II,1681)

BLESSED, BRIAN
APPOINTMENT WITH A KILLER (I,245); THE LAST DAYS OF POMPEII (II,1436)

BLESSING, JACK
SMALL AND FRYE (II,2380)

BLETCHER, BILLY
LI'L ABNER (I,2702)

BLIGH, BETTY
CLUB CELEBRITY (I,981)

BLINCOE, BRENDON
MR. MOM (II,1752)

BLISS, BRADLEY
ANOTHER WORLD (II,97)

BLISS, HELENA
THE MERRY WIDOW (I,3012)

BLISS, JUDITH
HAY FEVER (I,1979)

BLISS, LELA
BLONDIE (I,486); MY LITTLE MARGIE (I,3194)

BLISS, LUCILLE
CRUSADER RABBIT (I,1109); THE SMURFS (II,2385); SPACE KIDDETTES (I,4143)

BLOCH, ANDREW
ARCHER—FUGITIVE FROM THE EMPIRE (II,103); NUMBER 96 (II,1869); THE OUTLAWS (II,1936); SHIPSHAPE (II,2329)

BLOCH, ROBERT
THE HORROR OF IT ALL (II,1181)

BLOCK, BARBARA
ADVICE TO THE LOVELORN (II,16); SHE'S IN THE ARMY NOW (II,2320)

BLOCK, DAVID A
MASADA (II,1642)

BLOCK, HAL
TAG THE GAG (I,4330)

BLOCK, LARRY
THE SECRET STORM (I,3969); SESAME STREET (I,3982); SPACE FORCE (II,2407)

BLOCKER, DAN
BONANZA (II,347); CIMARRON CITY (I,953); THE DESILU PLAYHOUSE (I,1237); HENRY FONDA AND THE FAMILY (I,2015); JACK BENNY'S BIRTHDAY SPECIAL (I,2299); SAM HILL (I,3895); SWING OUT, SWEET LAND (II,2515)

BLOCKER, DIRK
THE BLACK SHEEP SQUADRON (II,262); BRIDGER (II,376); CIRCUS OF THE STARS (II,532); RYAN'S FOUR (II,2233)

BLODGETT, MICHAEL
NEVER TOO YOUNG (I,3248); ROMP (I,3838)

BLONDELL, GLORIA
THE FORTUNE HUNTER (I,1646); THE LIFE OF RILEY (I,2686)

BLONDELL, JOAN
BABY CRAZY (I,319); BANYON (I,344); THE BARBARA STANWYCK THEATER (I,350); BOBBY PARKER AND COMPANY (II,344); BURLESQUE (I,765); THE GREATEST SHOW ON EARTH (I,1883); HERE COME THE BRIDES (I,2024); HOORAY FOR HOLLYWOOD (I,2114); THE NASH AIRFLYTE THEATER (I,3227); PLAYHOUSE 90 (I,3623); THE REAL MCCOYS (I,3741); THE REBELS (II,2121); TALES OF TOMORROW (I,4352); THREE FOR THE GIRLS (I,4477); THE TWILIGHT ZONE (I,4651)

BLONIGAN, COLETTE
THE NIGHTENGALES (II,1850)

BLOOM, ANNE
EVERYDAY (II,793); INSPECTOR PEREZ (II,1237); NOT NECESSARILY THE NEWS (II,1862); NOT NECESSARILY THE NEWS (II,1863)

BLOOM, CHARLES
IN TROUBLE (II,1230); MORK AND MINDY (II,1735); NUMBER 96 (II,1869); THE WAVERLY WONDERS (II,2746)

BLOOM, CLAIRE
BACKSTAIRS AT THE WHITE HOUSE (II,133); BRIDESHEAD REVISITED (II,375); CAESAR AND CLEOPATRA (I,788); CYRANO DE BERGERAC (I,1119); ELLIS ISLAND (II,768); PLAYHOUSE 90 (I,3623); ROMEO AND JULIET (I,3837); SEPARATE TABLES (II,2304); SHIRLEY TEMPLE'S STORYBOOK (I,4017); SOLDIER IN LOVE (I,4108)

BLOOM, JOHN
THE BOYS IN BLUE (II,359)

BLOOM, LINDSAY
CIRCUS OF THE STARS (II,536); CIRCUS OF THE STARS (II,538); DALLAS (II,614); THE DUKES OF HAZZARD (II,742); GOOBER AND THE TRUCKERS' PARADISE (II,1029); MICKEY SPILLANE'S MIKE HAMMER (II,1692); MICKEY SPILLANE'S MIKE HAMMER: MORE THAN MURDER (II,1693)

BLOOM, VERNA
THE BLUE KNIGHT (II,277); DOC ELLIOT (I,1305); RIVKIN: BOUNTY HUNTER (II,2183)

BLOOMFIELD, AL
VAN DYKE AND COMPANY (II,2719)

BLOSSOM, JUNE
THE MILKY WAY (I,3044)

BLOSSOM, ROBERTS
ANOTHER WORLD (II,97); JOHNNY BELINDA (I,2418)

BLOSSOMS, THE
...AND BEAUTIFUL (I,184); ELVIS (I,1435); FELICIANO—VERY SPECIAL (I,1560)

BLOUNT, LISA
MICKEY SPILLANE'S MIKE HAMMER: MURDER ME, MURDER YOU (II,1694)

BLU, SUSAN
ARCHIE (II,104); THE ARCHIE SITUATION COMEDY MUSICAL VARIETY SHOW (II,106); FANG FACE (II,822);

THE PLASTICMAN COMEDY/ADVENTURE SHOW (II,2051); THE WILD WILD WEST REVISITED (II,2800)

BLUE FAMILY, THE
THE FRANK SINATRA SHOW (I,1664)

BLUE, BEN
ACCIDENTAL FAMILY (I,17); BEN BLUE'S BROTHERS (I,393); FIVE STAR COMEDY (I,1588); THE FRANK SINATRA SHOW (I,1664); FUN FOR 51 (I,1707); THE JACK BENNY SPECIAL (I,2296); THE SATURDAY NIGHT REVUE (I,3921); SHIRLEY TEMPLE'S STORYBOOK (I,4017); STARTIME (I,4212); WE'LL TAKE MANHATTAN (I,4792)

BLUESTEIN, STEVE
PLAYBOY'S PLAYMATE PARTY (II,2055)

BLUESTONE, ED
LAUGH-IN (II,1444)

BLUM, MARK
THINGS ARE LOOKING UP (II,2588)

BLUNT, ERIN
STRANDED (II,2475)

BLUTHAL, JOHN
FIREBALL XL-5 (I,1578); SUPERCAR (I,4294)

BLYDEN, LARRY
ABC STAGE '67 (I,6); THE AD-LIBBERS (I,29); AMERICA PAUSES FOR THE MERRY MONTH OF MAY (I,167); CALL TO DANGER (I,802); GHOSTBREAKER (I,1789); HARRY'S GIRLS (I,1959); HARVEY (I,1961); JOE AND MABEL (I,2398); THE LORETTA YOUNG THEATER (I,2756); THE MOVIE GAME (I,3114); PERSONALITY (I,3553); WHAT'S MY LINE? (I,4820); YOU'RE PUTTING ME ON (I,4972)

BLYE, MAGGIE
GOLDEN GATE (II,1020); HANK (I,1931); KODIAK (II,1405); MELVIN PURVIS: G-MAN (II,1679)

BLYTH, ANN
THE JUNE ALLYSON SHOW (I,2488); KRAFT SUSPENSE THEATER (I,2591); SAVAGE SUNDAY (I,3926); THE SECRET WORLD OF KIDS (I,3970); THE TWILIGHT ZONE (I,4651)

BLYTHE, ROBYN
FIRST TIME, SECOND TIME (II,863)

BLYTHWOOD, REGGIE
ANOTHER WORLD (II,97)

BORATTO, CATHERINA
THE FAR PAVILIONS (II,832)

BOA, BRUCE
LACE (II,1418); MIRROR OF DECEPTION (I,3060); MISTER JERICO (I,3070); A WOMAN CALLED GOLDA (II,2818)

BOARD, CAROL
DINAH'S PLACE (II,678)

BOARDMAN, ERIC
THE TIM CONWAY SHOW (II,2619); THE YESTERDAY SHOW (II,2850)

BOARDMAN, TRUE
THE BURNS AND SCHREIBER COMEDY HOUR (I,768)

BOATANEERS, THE
THE NAT KING COLE SHOW (I,3230)

BOAX, CHARLES
MACKENZIE'S RAIDERS (I,2803)

BOB BANAS DANCERS, THE
MALIBU U. (I,2849)

BOB BARKER MARIONETTES, THE
CURIOSITY SHOP (I,1114)

BOB CROSBY'S WILDCATS
TIMEX ALL-STAR JAZZ SHOW III (I,4517)

BOB HAMILTON DANCERS, THE
THE GARRY MOORE SHOW (I,1740)

BOB HAMILTON TRIO, THE
THE MUSIC OF GERSHWIN (I,3174); YOUR SHOW OF SHOWS (I,4957)

BOB SIDNEY DANCERS, THE
PERRY COMO'S WINTER SHOW (I,3546)

BOB TURK ICE DANCERS, THE
THE ICE PALACE (I,2182)

BOB WILLIAMS AND HIS DOG LOUIE
HELLZAPOPPIN (I,2011)

BOBATOON, STAR-SHEMAH
PALMERSTOWN U.S.A. (II,1945)

BOBBITT, BETTY
PRISONER: CELL BLOCK H (II,2085)

BOBBY LORD AND HIS TIMBER JACK TRIO
JUBILEE U.S.A. (I,2461)

BOBO
MR. SMITH (II,1755)

BOBO, WILLIE
COS (II,576)

BOBSIN, THOMAS W
CASINO (II,452)

BOCHEY, WILLIS
THE GREAT GILDERSLEEVE (I,1875)

BOCHNER, HART
CALLAHAN (II,414); EAST OF EDEN (II,752)

BOCHNER, LLOYD
ARENA (I,256); BRADDOCK (I,704); DYNASTY (II,746); THE EYES OF TEXAS II (II,800); GENERAL ELECTRIC TRUE (I,1754); HONG KONG (I,2112); REBECCA (I,3745); REX HARRISON PRESENTS SHORT STORIES OF LOVE (II,2148); THE RICHARD BOONE SHOW (I,3784); RICHIE BROCKELMAN: MISSING 24 HOURS (II,2168); SANTA BARBARA (II,2258); SCALPLOCK (I,3930); STAR TONIGHT (I,4199); THEY CALL IT MURDER (I,4441); TWELFTH NIGHT (I,4636)

BODS, LOU
DAMN YANKEES (I,1128)

BODWELL, BOYD
DOWN HOME (II,731)

BOELSEN, JIM
ARCHIE (II,104); THE ARCHIE SITUATION COMEDY MUSICAL VARIETY SHOW (II,106)

BOEN, EARL
ADAMS HOUSE (II,9); FIRST & TEN (II,861); FOR MEMBERS ONLY (II,905); LOVE AND LEARN (II,1529); MAKING A LIVING (II,1604); SHIPSHAPE (II,2329); SILVER SPOONS (II,2355); WHO'S THE BOSS? (II,2792)

BOGARDE, DIRK
BLITHE SPIRIT (I,485); LITTLE MOON OF ALBAN (I,2715)

BOGART, HUMPHREY
THE PETRIFIED FOREST (I,3570)

BOGART, LINDA SCRUGGS
THE DEADLY TRIANGLE (II,634)

BOGAZIANOS, VASILI
ALL MY CHILDREN (II,39); THE EDGE OF NIGHT (II,760)

BOGERT, WILLIAM
CENTENNIAL (II,477); THE FACTS OF LIFE (II,805); THE GREATEST AMERICAN HERO (II,1060); MISS WINSLOW AND SON (II,1705); THIS IS KATE BENNETT (II,2597)

BOGGS, BILL
ALL-STAR ANYTHING GOES (II,50); TELL ME ON A SUNDAY (II,2557)

BOGUE, MERWYN
KAY KYSER'S KOLLEGE OF MUSICAL KNOWLEDGE (I,2512)

BOHAN, DENNIS
THE FASHION STORY (I,1537)

BOHAY, HEIDI
CALIFORNIA FEVER (II,411); HOTEL (II,1192)

BOHRER, CORINNE
E/R (II,748)

BOLAND, BONNIE
ARNOLD'S CLOSET REVIEW (I,262); BARNEY AND ME (I,358); CHICO AND THE MAN (II,508); THE TIM CONWAY COMEDY HOUR (I,4501); TURN ON (I,4625)

BOLAND, JOE
ASK ME ANOTHER (I,296)

BOLAND, MARY
FORD THEATER HOUR (I,1635); THE GUARDSMAN (I,1893); THE WOMEN (I,4890)

BOLAND, NORMA
CELEBRITY (II,463)

BOLENDER, CHARLEY
JACKIE GLEASON AND HIS AMERICAN SCENE MAGAZINE (I,2321)

BOLES, BARBARA
PARADISE BAY (I,3454)

BOLES, JIM
CONCERNING MISS MARLOWE (I,1032); ONE MAN'S FAMILY (I,3383)

BOLGER, RAY
THE BIG TIME (I,434); THE BOB HOPE SHOW (I,640); CAPTAINS AND THE KINGS (II,435); GIVE MY REGARDS TO BROADWAY (I,1815); OPERATION ENTERTAINMENT (I,3400); THE PARTRIDGE FAMILY (II,1962); THE RAY BOLGER SHOW (I,3731); WASHINGTON

SQUARE (I,4772); WHERE'S RAYMOND? (I,4837)

BOLGER, ROBERT
ACADEMY THEATER (I,14); BRIEF PAUSE FOR MURDER (I,725); THE JUDGE (I,2468)

BOLIN, SHANNON
SOUNDS OF HOME (I,4136)

BOLKAN, FLORINDA
THE WORD (II,2833)

BOLKE, BRADLEY
OZMOE (I,3437); TENNESSEE TUXEDO AND HIS TALES (I,4403); THE YEAR WITHOUT A SANTA CLAUS (II,2846)

BOLL, RICHARD
ARCHER—FUGITIVE FROM THE EMPIRE (II,103)

BOLLER, NORMAN
THE PICTURE OF DORIAN GRAY (I,3591)

BOLLING, ANGIE
BENJI, ZAX AND THE ALIEN PRINCE (II,205)

BOLLING, TIFFANY
ELECTRA WOMAN AND DYNA GIRL (II,764); KEY WEST (I,2528); THE NEW PEOPLE (I,3268)

BOLOGNA, JOSEPH
BEDROOMS (II,198); THE JOE PISCOPO SPECIAL (II,1304); A LUCILLE BALL SPECIAL: WHAT NOW CATHERINE CURTIS? (II,1560); PARADISE (II,1955)

BOLSTER, STEPHEN
ANOTHER WORLD (II,97); ONE LIFE TO LIVE (II,1907); YOUNG DR. MALONE (I,4938)

BOLTON, BARBARA
THE HUNTER (I,2162); STAGE 13 (I,4183)

BOLTON, ELAINE
BLANSKY'S BEAUTIES (II,267); LEGS (II,1459); WHO'S WATCHING THE KIDS? (II,2793)

BOLTON, LOIS
CONCERNING MISS MARLOWE (I,1032)

BOMBECK, ERMA
BOB HOPE SPECIAL: BOB HOPE IN WHO MAKES THE WORLD LAUGH? (II,287); ERMA BOMBECK (I,1452); HERE'S RICHARD (II,1136); JOHN DENVER AND THE LADIES (II,1312); MAGGIE (II,1589)

BONACORSO, DANNY
WONDER GIRL (I,4891)

BONADUCE, DANNY
CALL HOLME (II,412); GOOBER AND THE GHOST CHASERS (II,1028); HONEST AL'S A-OK USED CAR AND TRAILER RENTAL TIGERS (II,1174); A KNIGHT IN SHINING ARMOUR (I,2575); THE PARTRIDGE FAMILY (II,1962); PARTRIDGE FAMILY: 2200 A.D. (II,1963); THANKSGIVING REUNION WITH THE PARTRIDGE FAMILY AND MY THREE SONS (II,2569)

BONANOVA, FORTUNIO
THE COUNT OF MONTE CRISTO (I,1058)

BONAR, IVAN
THE ADVENTURES OF OZZIE AND HARRIET (I,71); GENERAL HOSPITAL (II,964); THE GREEN FELT JUNGLE (I,1885); HAWKINS (I,1976); WALKING TALL (II,2738)

BONASS, CLARE
THE WOMAN IN WHITE (II,2819)

BOND III, JAMES
BIG BOB JOHNSON AND HIS FANTASTIC SPEED CIRCUS (II,231); THE RED HAND GANG (II,2122)

BOND, DAVID
DOBIE GILLIS (I,1302); SKINFLINT (II,2377)

BOND, DEREK
THE EVE ARDEN SHOW (I,1471); NOT GUILTY! (I,3311)

BOND, FORD
CITIES SERVICE BAND OF AMERICA (I,962)

BOND, J BLASINGAME
SAMMY KAYE'S MUSIC FROM MANHATTAN (I,3902)

BOND, JULIAN
SIX PACK (II,2370)

BOND, MISCHA
DOROTHY IN THE LAND OF OZ (II,722); YOUNG LIVES (II,2866)

BOND, PHILIP
THE ONEDIN LINE (II,1912)

BOND, RALEIGH
FAMILY TIES (II,819); THE GRADY NUTT SHOW (II,1048); LAVERNE AND SHIRLEY (II,1446); SALVAGE (II,2244)

BOND, RUDY
THE GOLDEN AGE OF TELEVISION (II,1019)

BOND, SHEILA
INSIDE U.S.A. WITH CHEVROLET (I,2220)

BOND, SHERI
FIRST & TEN (II,861)

BOND, STEVE
GENERAL HOSPITAL (II,964)

BOND, SUDIE
BENSON (II,208); FLO (II,891); THE FOUR OF US (II,914); THE GUIDING LIGHT (II,1064); LANDON, LANDON & LANDON (II,1426); MASQUERADE (I,2940); THE NEW TEMPERATURES RISING SHOW (I,3273); THE RAG BUSINESS (II,2106); SANCTUARY OF FEAR (II,2252); WE INTERRUPT THIS SEASON (I,4780); ZERO HOUR (I,4980)

BOND, SUE
THE BENNY HILL SHOW (II,207); CASANOVA (II,451)

BOND, WARD
CAVALCADE OF AMERICA (I,875); THE ELISABETH MCQUEENEY STORY (I,1429); FORD TELEVISION THEATER (I,1634); GULF PLAYHOUSE (I,1904); SILVER THEATER (I,4051); STAR STAGE (I,4197); SUSPENSE (I,4305); WAGON TRAIN (I,4747)

BOND, WILLIAM
THE BARRETTS OF WIMPOLE STREET (I,361)

BOND-OWEN, NICHOLAS
GEORGE AND MILDRED (II,965)

BONDI, BEULAH
CHRYSLER MEDALLION THEATER (I,951); THE DOCTOR (I,1320); THE GENERAL ELECTRIC THEATER (I,1753); ON BORROWED TIME (I,3354); SANDBURG'S LINCOLN (I,3907)

BONERZ, PETER
9 TO 5 (II,1852); THE BASTARD/KENT FAMILY CHRONICLES (II,159); THE BOB NEWHART SHOW (II,342); ELKE (I,1432); THE PIRATES OF FLOUNDER BAY (I,3607); A STORM IN SUMMER (I,4237); STORY THEATER (I,4241)

BONET, LISA
THE COSBY SHOW (II,578)

BONILLA-GIANNINI, ROXANNA
AT EASE (II,116)

BONINO, STEVE
THE KIDS FROM C.A.P.E.R. (II,1391)

BONKETTES CHORUS, THE
BONKERS! (II,348)

BONNAR, IVAN
THE COPS AND ROBIN (II,573)

BONNELL, VIVIAN
AMERICA, YOU'RE ON (II,69); THE CAROL BURNETT SHOW (II,443); MADHOUSE 90 (I,2807); NICK AND THE DOBERMANS (II,1843); ST. ELSEWHERE (II,2432)

BONNER, FRANK
FACTS OF LIFE: THE FACTS OF LIFE GOES TO PARIS (II,806); NO MAN'S LAND (II,1855); SUTTERS BAY (II,2508); WKRP IN CINCINNATI (II,2814)

BONNER, TONY
SKIPPY, THE BUSH KANGAROO (I,4076)

BONNIE BIRDS PLUS TWO, THE
THE DES O'CONNOR SHOW (I,1235)

BONO, CHASTITY
THE SONNY AND CHER COMEDY HOUR (II,2400)

BONO, SONNY
BATTLE OF THE NETWORK STARS (II,164); THE COUNTRY MUSIC MURDERS (II,590); THE FIRST NINE MONTHS ARE THE HARDEST (I,1584); HOW TO HANDLE A WOMAN (I,2146); THE SENSATIONAL, SHOCKING, WONDERFUL, WACKY 70S (II,2302); THE SONNY AND CHER COMEDY HOUR (II,2400); THE SONNY AND CHER SHOW (II,2401); THE SONNY COMEDY REVUE (II,2403); TOP OF THE HILL (II,2645)

BONO, SUSIE
ON STAGE AMERICA (II,1893)

BOOCOCK, MANDY
A TOWN LIKE ALICE (II,2652)

BOOKE, SORRELL
BRENDA STARR (II,372); THE DUKES (II,741); THE DUKES OF HAZZARD (II,742); GREAT BIBLE ADVENTURES (I,1872); JOE AND SONS (II,1296); OWEN MARSHALL: COUNSELOR AT LAW (I,3436); RICH MAN, POOR MAN—BOOK II (II,2162); SOAP (II,2392); VICTORIA REGINA

(I,4728); ZERO INTELLIGENCE (II,2876)

BOOKENDS, THE
THE RAY ANTHONY SHOW (I,3728); THE RAY ANTHONY SHOW (I,3729); THE SWINGING SCENE OF RAY ANTHONY (I,4323)

BOOKWALTER, DEVEREL
RYAN'S HOPE (II,2234)

BOOMER
HERE'S BOOMER (II,1134)

BOOMER, LINWOOD
LITTLE HOUSE ON THE PRAIRIE (II,1487)

BOON, ROBERT
ROLL OUT! (I,3833); SQUADRON (I,4170)

BOONE, BARBARA
AN APARTMENT IN ROME (I,242)

BOONE, BRENDON
GARRISON'S GORILLAS (I,1735)

BOONE, CHERRY
PAT BOONE AND FAMILY (II,1968)

BOONE, DEBBY
BATTLE OF THE NETWORK STARS (II,166); BATTLE OF THE NETWORK STARS (II,167); BATTLE OF THE NETWORK STARS (II,178); BOB HOPE SPECIAL: BOB HOPE—HOPE, WOMEN AND SONG (II,324); BOB HOPE SPECIAL: BOB HOPE PRESENTS A CELEBRATION WITH STARS OF COMEDY AND MUSIC (II,340); BOB HOPE SPECIAL: THE BOB HOPE SPECIAL (II,336); A COUNTRY CHRISTMAS (II,584); DEBBY BOONE. . . ONE STEP CLOSER (II,655); DEBBY BOONE. . . THE SAME OLD BRAND NEW ME (II,656); GEORGE BURNS' 100TH BIRTHDAY PARTY (II,971); OPRYLAND: NIGHT OF STARS AND FUTURE STARS (II,1921); PAT BOONE AND FAMILY (II,1968); PERRY COMO'S EASTER BY THE SEA (II,2014); PERRY COMO'S FRENCH-CANADIAN CHRISTMAS (II,2016)

BOONE, LAURY
PAT BOONE AND FAMILY (II,1968)

BOONE, LINDA
PAT BOONE AND FAMILY (II,1968)

BOONE, MARY JANE
RENDEZVOUS WITH MUSIC (I,3768)

BOONE, PAT
THE ANITA BRYANT SPECTACULAR (II,82); ARTHUR GODFREY AND FRIENDS (I,278); BOB HOPE SPECIAL: THE BOB HOPE SPECIAL (II,336); CHRISTMAS LEGEND OF NASHVILLE (II,522); THE GENERAL MOTORS 50TH ANNIVERSARY SHOW (I,1758); NIGHT GALLERY (I,3287); ONE MORE TIME (I,3386); PAT BOONE AND FAMILY (II,1968); THE PAT BOONE AND FAMILY CHRISTMAS SPECIAL (II,1969); THE PAT BOONE AND FAMILY EASTER SPECIAL (II,1970); PAT BOONE IN HOLLYWOOD (I,3469); THE PAT BOONE SHOW (I,3470); THE PAT BOONE SHOW (I,3471); PERRY COMO'S SPRINGTIME SPECIAL (II,2023); THE PIGEON (I,3596); ROCK 'N' ROLL: THE FIRST 25 YEARS (II,2189); STAR SEARCH (II,2437)

BOONE, RANDY
CIMARRON STRIP (I,954); IT'S A MAN'S WORLD (I,2258); THE VIRGINIAN (I,4738)

BOONE, RICHARD
ARENA (I,256); THE CENTURY TURNS (I,890); FRONTIER (I,1695); HAVE GUN—WILL TRAVEL (I,1968); HEC RAMSEY (II,1121); MATINEE THEATER (I,2947); MEDIC (I,2976); MR. I. MAGINATION (I,3140); PLAYHOUSE 90 (I,3623); THE RICHARD BOONE SHOW (I,3784); THE RIGHT MAN (I,3790); STUDIO ONE (I,4268)

BOONE, SHIRLEY
MY SISTER EILEEN (I,3200); PAT BOONE AND FAMILY (II,1968)

BOOSLER, ELAYNE
THE SHAPE OF THINGS (II,2315)

BOOTH, BILLY
DENNIS THE MENACE (I,1231)

BOOTH, CONNIE
FAWLTY TOWERS (II,843); WHY DIDN'T THEY ASK EVANS? (II,2795)

BOOTH, JAMES
WHEELS (II,2777)

BOOTH, KAREN
STAGE 7 (I,4181)

BOOTH, L WILLIAM
KENNEDY (II,1377)

BOOTH, NESDON
SAM HILL (I,3895)

BOOTH, SHIRLEY
THE GLASS MENAGERIE (I,1819); HAZEL (I,1982); THE MARCH OF DIMES BENEFIT SHOW (I,2900); PLAYHOUSE 90 (I,3623); A TOUCH OF GRACE (I,4585); THE YEAR WITHOUT A SANTA CLAUS (II,2846)

BOOTH, STEVE
HOBBIES IN ACTION (I,2067)

BOOTHBY, VICTORIA
ALL MY CHILDREN (II,39)

BOOTHE, POWERS
PHILIP MARLOWE, PRIVATE EYE (II,2033); SKAG (II,2374)

BORATO, MARINA
FELLINI: A DIRECTOR'S NOTEBOOK (I,1562)

BORDEN, ALICE
THE BOB NEWHART SHOW (II,342); THAT GIRL (II,2570)

BORDEN, GARY
BACKSTAIRS AT THE WHITE HOUSE (II,133)

BORDEN, LYNN
CENTENNIAL (II,477); HAZEL (I,1982)

BORDEN, MARSHALL
A CHRISTMAS CAROL (II,517); ONE LIFE TO LIVE (II,1907)

BORDERS, JAMES
HARDCASE (II,1088)

BORELLI, CARLA
BANYON (I,345); THE BETTY WHITE SHOW (II,224); DAYS OF OUR LIVES (II,629); TEXAS (II,2566)

BORG, RICHARD
THE EDGE OF NIGHT (II,760)

BORG, VEDA ANN
THE ABBOTT AND COSTELLO SHOW (I,2); DUFFY'S TAVERN (I,1379); THE LIFE OF RILEY (I,2686); NAVY LOG (I,3233); THE UNEXPECTED (I,4674)

BORGE, VICTOR
THE ALL-STAR REVUE (I,138); ARTHUR MURRAY'S DANCE PARTY (I,293); THE CARPENTERS (II,444); THE FUNNIEST JOKE I EVER HEARD (II,940); THE MICKEY ROONEY SHOW (I,3027); MR. BELVEDERE (I,3127); THE PERRY COMO SPECIAL (I,3535); THE VICTOR BORGE SHOW (I,4719); THE VICTOR BORGE SHOW (I,4720); THE VICTOR BORGE SHOW (I,4721); THE VICTOR BORGE SPECIAL (I,4722); VICTOR BORGE'S 20TH ANNIVERSARY SHOW (I,4727); VICTOR BORGE'S COMEDY IN MUSIC I (I,4723); VICTOR BORGE'S COMEDY IN MUSIC II (I,4724); VICTOR BORGE'S COMEDY IN MUSIC III (I,4725); VICTOR BORGE'S COMEDY IN MUSIC IV (I,4726)

BORGESON, LINDA
ANOTHER WORLD (II,97)

BORGH, TANYA
FOUR STAR PLAYHOUSE (I,1652)

BORGNINE, ERNEST
AIRWOLF (II,27); ALCOA PREMIERE (I,109); THE ANDY WILLIAMS SHOW (I,206); THE BOB HOPE CHRYSLER THEATER (I,502); THE BOB HOPE SHOW (I,619); CAPTAIN VIDEO AND HIS VIDEO RANGERS (I,838); THE COPS AND ROBIN (II,573); FIRESIDE THEATER (I,1580); THE FUNNIEST JOKE I EVER HEARD (II,940); FUTURE COP (II,945); FUTURE COP (II,946); THE GENERAL ELECTRIC THEATER (I,1753); THE GENERAL MOTORS 50TH ANNIVERSARY SHOW (I,1758); THE LAST DAYS OF POMPEII (II,1436); MCHALE'S NAVY (I,2969); NAVY LOG (I,3233); THE ROWAN AND MARTIN SPECIAL (I,3855); SAM HILL: WHO KILLED THE MYSTERIOUS MR. FOSTER? (I,3896); SANDY IN DISNEYLAND (II,2254); SEVEN AGAINST THE SEA (I,3983); TAKE ONE STARRING JONATHAN WINTERS (II,2532); WHAT'S UP, AMERICA? (I,4824); ZANE GREY THEATER (I,4979)

BORLEK, BOB
THE MAGIC SLATE (I,2820)

BORN, DAVID
HAPPY (I,1938)

BORN, ROSCOE
PAPER DOLLS (II,1952); RYAN'S HOPE (II,2234)

BORN, STEVEN
HAPPY (I,1938)

BORREGO, JESSE
FAME (II,812)

BORUFF, JOHN
FIRST LOVE (I,1583); ROD BROWN OF THE ROCKET RANGERS (I,3825)

BOSCHETTI, DESIREE
MY WIFE NEXT DOOR (II,1781)

BOSLEY, TOM
ALICE IN WONDERLAND (I,120); ARSENIC AND OLD LACE (I,269); THE BASTARD/KENT FAMILY CHRONICLES (II,159); BOBBY JO AND THE BIG APPLE GOODTIME BAND (I,674); THE DEAN MARTIN SHOW (I,1201); THE DEBBIE REYNOLDS SHOW (I,1214); THE DRUNKARD (II,737); EVENING AT THE IMPROV (II,786); GILLIGAN'S ISLAND: THE CASTAWAYS ON GILLIGAN'S ISLAND (II,992); HAPPY DAYS (II,1084); THE LOVE BOAT I (II,1532); MARCUS WELBY, M.D. (I,2905); MIRACLE ON 34TH STREET (I,3058); MITZI AND A HUNDRED GUYS (II,1710); MURDER, SHE WROTE (II,1770); NIGHT GALLERY (I,3288); THE OLIVIA NEWTON-JOHN SHOW (II,1885); PROFILES IN COURAGE (I,3678); THE REBELS (II,2121); THE RETURN OF THE MOD SQUAD (I,3776); RICH LITTLE'S WASHINGTON FOLLIES (II,2159); THE RIGHT MAN (I,3790); THE SANDY DUNCAN SHOW (I,3910); A SPECIAL OLIVIA NEWTON-JOHN (II,2418); THE STREETS OF SAN FRANCISCO (I,4260); TESTIMONY OF TWO MEN (II,2564); WAIT TIL YOUR FATHER GETS HOME (I,4748); WHAT'S UP, AMERICA? (I,4824); WHAT'S UP? (I,4823)

BOSSON, BARBARA
HILL STREET BLUES (II,1154); RICHIE BROCKELMAN, PRIVATE EYE (II,2167); RICHIE BROCKELMAN: MISSING 24 HOURS (II,2168); SUNSHINE (II,2495)

BOSTICK, CYNTHIA
FOR RICHER, FOR POORER (II,906)

BOSTOCK, BARBARA
THE FARMER'S DAUGHTER (I,1533); THE FIRST HUNDRED YEARS (I,1582); LOVE ON A ROOFTOP (I,2775)

BOSTON, MATT
CAMP WILDERNESS (II,419)

BOSTWICK, BARRY
THE CHADWICK FAMILY (II,478); FOUL PLAY (II,910); GEORGE WASHINGTON (II,978); SCRUPLES (II,2280); SLITHER (II,2379); A WOMAN OF SUBSTANCE (II,2820); YOU CAN'T TAKE IT WITH YOU (II,2857); YOUNG GUY CHRISTIAN (II,2864)

BOSTWICK, JACKSON
THE RED SKELTON SHOW (I,3756); SHAZAM! (II,2319)

BOSWALL, JOHN
LADY KILLER (I,2603)

BOSWELL, CONNEE
HOLLYWOOD STAR REVUE (I,2093); PETE KELLY'S BLUES (I,3559)

BOSWORTH, PATRICIA
CONCERNING MISS MARLOWE (I,1032); YOUNG DR. MALONE (I,4938)

BOTTOMS, JOSEPH
BATTLE OF THE NETWORK STARS (II,167); CELEBRITY (II,463); RETURN ENGAGEMENT (I,3775); WISHMAN (II,2811)

BOTTOMS, SAM
EAST OF EDEN (II,752)

BOTTOMS, TIMOTHY
EAST OF EDEN (II,752); THE MONEYCHANGERS (II,1724)

BOTWINICK, AMY
LOBO (II,1504); ROOSTER (II,2210)

BOUCHER, SHERRY
LASSIE (I,2624)

BOUCHET, BARBARA
THE MASK OF MARCELLA (I,2937)

BOUCHEY, WILLIS
DOBIE GILLIS (I,1302); FOREST RANGER (I,1641); LAST STAGECOACH WEST (I,2629)

BOUDROT, MEL
ALL MY CHILDREN (II,39)

BOULD, BRUCE
THE FALL AND RISE OF REGINALD PERRIN (II,810)

BOULER, RICHARD
DIAL "M" FOR MURDER (I,1256)

BOURNE, PETER
THE WINDS OF WAR (II,2807)

BOURNEUF, PHILIP
FRANKENSTEIN (I,1671); TAMING OF THE SHREW (I,4356); TONIGHT AT 8:30 (I,4556)

BOUTON, JIM
BALL FOUR (II,137)

BOUTROS, MAHER
TEXAS (II,2566)

BOUTSIKARIS, DENNIS
NURSE (II,1870)

BOUVIER, CHARLES
HEAR NO EVIL (II,1116)

BOUVIER, LEE
LAURA (I,2636)

BOVA, JOE
THE GOOD OLD DAYS (I,1851); OFF CAMPUS (II,1877)

BOVASSO, JULIE
DAUGHTERS (II,620); FROM THESE ROOTS (I,1688)

BOVE, LINDA
SESAME STREET (I,3982)

BOVEN, TANG
MARCO POLO (II,1626)

BOW, MICHAEL
MCCLOUD: WHO KILLED MISS U.S.A.? (I,2965)

BOW, SIMMY
SCRUPLES (II,2281)

BOWAN, PATRICIA
THE PATRICIA BOWAN SHOW (I,3475)

BOWAN, SYBIL
TIME REMEMBERED (I,4509)

BOWBESVLO, CAROLYN
CASANOVA (II,451)

BOWEN, DEBBIE
APPLAUSE (I,244)

BOWEN, DENNIS
ARCHIE (II,104); THE ARCHIE SITUATION COMEDY MUSICAL VARIETY SHOW (II,106); WELCOME BACK, KOTTER (II,2761)

BOWEN, JO
MELODY STREET (I,3000)

BOWEN, PHILIP
THERESE RAQUIN (II,2584)

BOWEN, ROGER
ARNIE (I,261); AT EASE (II,115); THE BASTARD/KENT FAMILY CHRONICLES (II,159); BATTLES: THE MURDER THAT WOULDN'T DIE (II,180); BULBA (II,392); DEADLOCK (I,1190); DUFFY (II,739); GOOD PENNY (II,1036); HOUSE CALLS (II,1194); THE LITTLE PEOPLE (I,2716); THE MONEYCHANGERS (II,1724); THE RANGERS (II,2112); STOP ME IF YOU HEARD THIS ONE (I,4231) SUZANNE PLESHETTE IS MAGGIE BRIGGS (II,2509); WHERE'S THE FIRE? (II,2785)

BOWER, ANTOINETTE
BORDER TOWN (I,689); HAWKINS ON MURDER (I,1978); KRAFT MYSTERY THEATER (I,2589); MISSION: IMPOSSIBLE (I,3067); RAPTURE AT TWO-FORTY (I,3724); THE THORN BIRDS (II,2600)

BOWER, TOM
THE DAIN CURSE (II,613); TALES OF THE APPLE DUMPLING GANG (II,2536); THE WALTONS (II,2740)

BOWERS, CHUCK
JUBILEE U.S.A. (I,2461)

BOWERS, JANE
THE CRACKER BROTHERS (II,599)

BOWERS, RAYMOND
THE CITADEL (II,539)

BOWIE, DAVID
BING CROSBY'S MERRIE OLDE CHRISTMAS (II,252); DAVID BOWIE—SERIOUS MOONLIGHT (II,622)

BOWIE, SANDRA
POWERHOUSE (II,2074)

BOWKER, JUDI
THE ADVENTURES OF BLACK BEAUTY (I,51); DR JEKYLL AND MR. HYDE (I,1368); ELLIS ISLAND (II,768)

BOWLES, JIMMY
BATTLES: THE MURDER THAT WOULDN'T DIE (II,180)

BOWLES, PETER
THE DOUBLE KILL (I,1360); TO THE MANOR BORN (II,2627)

BOWLINE, FRANK
TESTIMONY OF TWO MEN (II,2564)

BOWLING, ROGER
MUSIC CITY NEWS TOP COUNTRY HITS OF THE YEAR (II,1772)

BOWMAN, CHUCK
ADAM-12 (I,31)

BOWMAN, GIL
ONE NIGHT BAND (II,1909)

BOWMAN, GLEN
THE BOYS IN BLUE (II,359)

BOWMAN, JON
CELEBRITY CHARADES (II,470)

BOWMAN, LEE
THE ADVENTURES OF ELLERY QUEEN (I,59); EYE WITNESS (I,1494); FAME IS THE NAME OF THE GAME (I,1518); LOVE STORY (I,2776);

MIAMI UNDERCOVER (I,3020); STAGE 7 (I,4181); WHAT'S GOING ON? (I,4814)

BOWMAN, PAMELA
THE CRACKER BROTHERS (II,599); THE HITCHHIKER (II,1157)

BOWMAN, RUSSELL
THE WOMAN IN WHITE (II,2819)

BOWREN, DON
PAPER DOLLS (II,1952)

BOXLEITNER, BRUCE
BARE ESSENCE (II,149); BATTLE OF THE NETWORK STARS (II,175); BRING 'EM BACK ALIVE (II,377); THE CHADWICK FAMILY (II,478); EAST OF EDEN (II,752); FLY AWAY HOME (II,893); HOW THE WEST WAS WON (II,1196); KISS ME, KILL ME (II,1400); THE LAST CONVERTIBLE (II,1435); THE MACAHANS (II,1583); SCARECROW AND MRS. KING (II,2268); WILD TIMES (II,2799)

BOY GEORGE
CULTURE CLUB IN CONCERT (II,610)

BOYCE, TOMMY
THE SOUPY SALES SHOW (I,4138)

BOYD, GUY
LOOSE CHANGE (II,1518); REMINGTON STEELE (II,2130); RYAN'S HOPE (II,2234)

BOYD, JIMMY
BACHELOR FATHER (I,323); BROADSIDE (I,732); THE HOUSE NEXT DOOR (I,2139); SPACE FORCE (II,2407)

BOYD, JOHN
HARD KNOX (II,1087)

BOYD, STEPHEN
THE HANDS OF CORMAC JOYCE (I,1927); HEDDA HOPPER'S HOLLYWOOD (I,2002); KEY WEST (I,2528); THE LIVES OF JENNY DOLAN (II,1502); OF MEN OF WOMEN (I,3332); THE POPPY IS ALSO A FLOWER (I,3650)

BOYD, TANYA
BIZARRE (II,258); GOOD TIMES (II,1038); MCCLAIN'S LAW (II,1659); MURDER CAN HURT YOU! (II,1768); ROOTS (II,2211); THE TED KNIGHT SHOW (II,2550)

BOYD, WILLIAM
HOPALONG CASSIDY (I,2118)

BOYDEN, PETER
I LOVE HER ANYWAY! (II,1214); KENNEDY (II,1377)

BOYDEN, SALLY
LASSIE: THE NEW BEGINNING (II,1433)

BOYER, CHARLES
THE CHARLES BOYER THEATER (I,904); FOUR STAR PLAYHOUSE (I,1652); THE ROGUES (I,3832); THERE SHALL BE NO NIGHT (I,4437); TURN OF FATE (I,4623)

BOYETT, BILL
MOBILE TWO (II,1718)

BOYETT, WILLIAM
ADAM-12 (I,31); DRAGNET (I,1371); HIGHWAY PATROL (I,2058)

BOYINGTON, GREGORY 'PAPPY'
DANGER ZONE (I,1137)

BOYLAN, BARBARA
THE LAWRENCE WELK SHOW (I,2643); MEMORIES WITH LAWRENCE WELK (II,1681)

BOYLE, BOB
THE DOCTORS (II,694)

BOYLE, DAVID E
CONCRETE BEAT (II,562)

BOYLE, DAWN MARIE
ALL MY CHILDREN (II,39)

BOYLE, KATIE
THE KENNY EVERETT VIDEO SHOW (II,1379)

BOYLE, MICHAEL
WHIZ KIDS (II,2790)

BOYLE, PETER
COMEDY TONIGHT (I,1027); FROM HERE TO ETERNITY (II,932); KLEIN TIME (II,1401); KOVACS ON THE CORNER (I,2582)

BOYS, THE BEACH
GOOD VIBRATIONS FROM CENTRAL PARK (I,1854)

BRACKEN, EDDIE
ARCHY AND MEHITABEL (I,254); THE DAVID NIVEN THEATER (I,1178); THE GENERAL MOTORS 50TH ANNIVERSARY SHOW (I,1758); GULF PLAYHOUSE (I,1904); MASQUERADE PARTY (I,2941); STAR STAGE (I,4197); THE STRAWBERRY BLONDE (I,4254); WILLIE WONDERFUL (I,4867)

BRACKETT-ZIKA, CHRISTIAN
DOMESTIC LIFE (II,703)

BRADBURY, LANE
BOSTON AND KILBRIDE (II,356); DIAL HOT LINE (I,1254)

BRADDOCK, MICKEY
CIRCUS BOY (I,959)

BRADEN, JOHN
ANOTHER WORLD (II,97)

BRADEN, KIM
LAUGH-IN (II,1444); THE MAN WHO CAME TO DINNER (I,2881); PENMARRIC (II,1993)

BRADFORD, GREG
THE FACTS OF LIFE (II,805)

BRADFORD, HANK
IT ONLY HURTS WHEN YOU LAUGH (II,1251)

BRADFORD, JOHNNY
THE BABY SITTER (I,321); THE RANSOM SHERMAN SHOW (I,3723); TELE PUN (I,4376)

BRADFORD, RICHARD
BIG JOHN (II,239); MAN IN A SUITCASE (I,2868); MEDICAL CENTER (II,1675); ROARING CAMP (I,3804); U.M.C. (I,4666)

BRADLEY TRIO
JOHN CONTE'S LITTLE SHOW (I,2408)

BRADLEY, ANNE H
COLORADO C.I. (II,555)

BRADLEY, BART
I WAS A BLOODHOUND (I,2180)

BRADLEY, BRUCE
ALL ABOUT MUSIC (I,127)

BRADLEY, LESLEY
TARZAN AND THE TRAPPERS (I,4362)

BRADLEY, LESLIE
THE COUNT OF MONTE CRISTO (I,1058)

BRADLEY, TRUMAN
SCIENCE FICTION THEATER (I,3938)

BRADSHAW, BROOKER
GOLDIE GOLD AND ACTION JACK (II,1023)

BRADSHAW, TERRY
THE STOCKERS (II,2469)

BRADY, DANA
ROLL OUT! (I,3833)

BRADY, MONICA
THE OMEGA FACTOR (II,1888)

BRADY, PAT
THE ROY ROGERS AND DALE EVANS SHOW (I,3858); THE ROY ROGERS SHOW (I,3859)

BRADY, RONNIE
CARRY ON LAUGHING (II,448)

BRADY, RUTH
THE JACK BENNY PROGRAM (I,2294)

BRADY, SCOTT
CONFLICT (I,1036); THE D.A.: MURDER ONE (I,1122); DAMON RUNYON THEATER (I,1129); THE DAVID NIVEN THEATER (I,1178); FLOYD GIBBONS, REPORTER (I,1608); GOOD PENNY (II,1036); JOHNNY NIGHTHAWK (I,2432); LITTLE LEATHERNECK (I,2713); THE NIGHT STRANGLER (I,3293); PAPA SAID NO (I,3452); SHADOW OF SAM PENNY (II,2313); SHOTGUN SLADE (I,4026); THIS GIRL FOR HIRE (II,2596); WHEELS (II,2777); THE WINDS OF WAR (II,2807); WOMEN IN WHITE (II,2823)

BRAEDEN, ERIC
THE ALIENS ARE COMING (II,35); CODE NAME: DIAMOND HEAD (II,549); HOW THE WEST WAS WON (II,1196); INTERTECT (I,2228); THE JUDGE AND JAKE WYLER (I,2464); THE SIX MILLION DOLLAR MAN (I,4067); WONDER WOMAN (PILOT 2) (II,2827); THE YOUNG AND THE RESTLESS (II,2862)

BRAFF, RUBY
TONY BENNETT IN WAIKIKI (I,4568)

BRAHA, HERB
CALLAHAN (II,414)

BRAILSFORD, PAULINE
CHICAGO STORY (II,506); THE CHICAGO STORY (II,507)

BRAKE, PATRICIA
THE UGLIEST GIRL IN TOWN (I,4663)

BRAMLEY, RAYMOND
THE BACHELOR (I,325); DOORWAY TO DANGER (I,1353); HARVEY (I,1961); THE KING AND MRS. CANDLE (I,2544)

BRAMLEY, WILLIAM
READY FOR THE PEOPLE (I,3739); SEVEN AGAINST THE SEA (I,3983)

BRAMMER, SHELLY

BREAKING AWAY (II,371)

BRAND, CLIFF
SIX PACK (II,2370)

BRAND, JACK
ACTION AUTOGRAPHS (I,24)

BRAND, JOLENE
THE COMEDY OF ERNIE KOVACS (I,1022); THE ERNIE KOVACS SPECIAL (I,1456); GUESTWARD HO! (I,1900); THE HOOFER (I,2113); THE NEW ERNIE KOVACS SHOW (I,3264); ZORRO (I,4982)

BRAND, NEVILLE
THE ADVENTURES OF NICK CARTER (I,68); APPOINTMENT WITH ADVENTURE (I,246); ARROYO (I,266); THE BARBARY COAST (II,147); CAPTAINS AND THE KINGS (II,435); HITCHED (I,2065); JANE WYMAN PRESENTS THE FIRESIDE THEATER (I,2345); LAREDO (I,2615); LOCK, STOCK, AND BARREL (I,2736); PURSUIT (I,3700); THE QUEST (II,2096); SCREEN DIRECTOR'S PLAYHOUSE (I,3946); THE SEEKERS (II,2298); STAGE 7 (I,4181); TWO FOR THE MONEY (I,4655); THE UNTOUCHABLES (I,4681); THE UNTOUCHABLES (I,4682); WHEN THE WEST WAS FUN: A WESTERN REUNION (II,2780); WINCHESTER (I,4872)

BRAND, OSCAR
DRAW ME A LAUGH (I,1372)

BRANDE, TONY
SUPERCOPS (II,2502)

BRANDO, JOCELYN
ACTOR'S STUDIO (I,28); THE MISS AND THE MISSILES (I,3062) ONE STEP BEYOND (I,3388)

BRANDO, KEVIN
IS THIS GOODBYE, CHARLIE BROWN? (I,2242); THE LAST NINJA (II,1438); STARSTRUCK (II,2446)

BRANDO, MARLON
ROOTS: THE NEXT GENERATIONS (II,2213)

BRANDON, CLARK
THE FACTS OF LIFE (II,805); THE FITZPATRICKS (II,868); MR. MERLIN (II,1751); OUT OF THE BLUE (II,1934); WHEN, JENNY? WHEN (II,2783)

BRANDON, GAVIN
MARCUS WELBY, M.D. (II,1627)

BRANDON, HENRY
CAPTAIN BRASSBOUND'S CONVERSION (I,828); LITTLE HOUSE: LOOK BACK TO YESTERDAY (II,1490)

BRANDON, JANE ALICE
ADVICE TO THE LOVELORN (II,16); THE BEST OF EVERYTHING (I,406); THE DAUGHTERS OF JOSHUA CABE RETURN (I,1170); ONE LIFE TO LIVE (II,1907)

BRANDON, JOHN
GOLIATH AWAITS (II,1025)

BRANDON, MICHAEL
EMERALD POINT, N.A.S. (II,773); LINDSAY WAGNER—ANOTHER SIDE OF ME (II,1483); MAN IN THE MIDDLE (I,2870); VENICE MEDICAL (II,2726)

BRANDON, PETER
AS THE WORLD TURNS (II,110); A BELL FOR ADANO (I,388); COMPUTERCIDE (II,561); HOW TO SURVIVE A MARRIAGE (II,1198); THE TRAP (I,4593); YOUNG DR. MALONE (I,4938)

BRANDS, X
BRIDGER (II,376); WHEN THE WEST WAS FUN: A WESTERN REUNION (II,2780); YANCY DERRINGER (I,4925)

BRANDT, HANK
THE ALIENS ARE COMING (II,35); CANNON: THE RETURN OF FRANK CANNON (II,425); DYNASTY (II,746); JOE AND SONS (II,1297); JULIA (I,2476); MANDRAKE (II,1617); MOONLIGHT (II,1730); RANSOM FOR A DEAD MAN (I,3722)

BRANDT, JANE
DYNASTY (II,746)

BRANDT, JANET
THE SUPER (I,4293)

BRANDT, MEL
FARAWAY HILL (I,1532); MODERN ROMANCES (I,3079); STORIES IN ONE CAMERA (I,4234)

BRANDT, VICTOR
ALLAN (I,142); NATIONAL LAMPOON'S TWO REELERS (II,1796); NOBODY'S PERFECT (II,1858)

BRANIGAN, LAURA
STAR SEARCH (II,2437)

BRANNEN, RALPH
HOW THE WEST WAS WON (II,1196)

BRANNUM, HUGH
THE FRED WARING SHOW (I,1679)

BRANNUM, LUMPY
CAPTAIN KANGAROO (II,434); GOOD EVENING, CAPTAIN (II,1031); WAKE UP (II,2736)

BRASFIELD, ROD
COUNTRY MUSIC CARAVAN (I,1066)

BRASH, MARIAN
THE BOB AND RAY SHOW (I,496); SEARCH FOR TOMORROW (II,2284); TOM CORBETT, SPACE CADET (I,4535)

BRASSELLE, KEEFE
BE OUR GUEST (I,370); THE DAVID NIVEN THEATER (I,1178); GENTRY'S PEOPLE (I,1762); HENRY FONDA PRESENTS THE STAR AND THE STORY (I,2016); THE KEEFE BRASSELLE SHOW (I,2514); KEEFE BRASSELLE'S VARIETY GARDEN (I,2515); KEEP IT IN THE FAMILY (I,2516); THE LORETTA YOUNG THEATER (I,2756)

BRATCHER, JOE
BENDER (II,204)

BRAUM, BOB
THE DOTTY MACK SHOW (I,1358)

BRAUN, IRIS
THE SECRET STORM (I,3969)

BRAUN, JUDITH
LITTLE WOMEN (I,2721); ROAD TO REALITY (I,3800)

BRAUNER, ASHER
B.A.D. CATS (II,123); GENERAL HOSPITAL (II,964); MICKEY SPILLANE'S MARGIN FOR MURDER (II,1691)

BRAVERMAN, BART
FANG FACE (II,822); THE HARVEY KORMAN SHOW (II,1103); THE KROFFT KOMEDY HOUR (II,1411); LADIES IN BLUE (II,1422); LOOK OUT WORLD (II,1516); MAGIC MONGO (II,1592); THE NEW ODD COUPLE (II,1825); THE PLASTICMAN COMEDY/ADVENTURE SHOW (II,2051); VEGAS (II,2723); VEGAS (II,2724)

BRAVERMAN, MARVIN
THE FURTHER ADVENTURES OF WALLY BROWN (II,944); IT ONLY HURTS WHEN YOU LAUGH (II,1251); THE NEW ODD COUPLE (II,1825); THE ROBERT KLEIN SHOW (II,2185)

BRAVO, DANNY
THE ADAM MACKENZIE STORY (I,30); THE ADVENTURES OF JONNY QUEST (I,64)

BRAXTON, STEPHANIE
ALL MY CHILDREN (II,39); KING'S CROSSING (II,1397); MARLO THOMAS IN ACTS OF LOVE—AND OTHER COMEDIES (I,2919); THE SECRET STORM (I,3969)

BRAY, ROBERT
LASSIE (I,2622); STAGECOACH WEST (I,4185)

BRAY, THOM
BREAKING AWAY (II,371); CONCRETE BEAT (II,562); FOUR EYES (II,912); ONE DAY AT A TIME (II,1900); PRIME TIMES (II,2083); RIPTIDE (II,2178)

BRAZZI, ROSANNO
DOUGLAS FAIRBANKS JR. PRESENTS THE RHEINGOLD THEATER (I,1364) THE FAR PAVILIONS (II,832); THE SURVIVORS (I,4301)

BREAUX, MARC
THE MORT SAHL SPECIAL (I,3105)

BRECHT, SUSAN
DOROTHY (II,717)

BRECK, PETER
THE BIG VALLEY (I,437); BLACK BEAUTY (II,261); BLACK SADDLE (I,480); THE OUTER LIMITS (I,3426); THE SECRET EMPIRE (II,2291)

BRECKMAN, ANDY
HOT HERO SANDWICH (II,1186)

BREE, JAMES
THE TALE OF BEATRIX POTTER (II,2533)

BREEDING, LARRY
THE LAST RESORT (II,1440); SHEILA (II,2325); THIS IS KATE BENNETT (II,2597); WHO'S WATCHING THE KIDS? (II,2793)

BREEN, BRIDGET
THE DOCTORS (II,694)

BREEN, DANNY
NOT NECESSARILY THE NEWS (II,1863)

BREEN, KAREN
BUCKSHOT (II,385)

BREEN, PAULETTE

ALL MY CHILDREN (II,39)

BREESE, SIDNEY
CHICAGOLAND MYSTERY PLAYERS (I,936)

BREGMAN, BUDDY
THE MUSIC SHOP (I,3177)

BREMEN, LENNIE
IT'S A SMALL WORLD (I,2260); MODESTY BLAISE (II,1719); NEW GIRL IN HIS LIFE (I,3265)

BRENDEL, EL
DOWN HOME (I,1366); THE SOFT TOUCH (I,4107)

BRENLIN, GEORGE
GENERAL HOSPITAL (II,964); TOP SECRET (II,2647)

BRENNAN, CAROL
YOUR SHOW TIME (I,4958)

BRENNAN, CLAIRE
FORCE FIVE (II,907); KINGSTON: THE POWER PLAY (II,1399); PLANET EARTH (I,3615)

BRENNAN, EILEEN
13 QUEENS BOULEVARD (II,2591); BLACK BEAUTY (II,261); KRAFT SALUTES WALT DISNEY WORLD'S 10TH ANNIVERSARY (II,1410); LILY FOR PRESIDENT (II,1478); A NEW KIND OF FAMILY (II,1819); OFF THE RACK (II,1878); PRIVATE BENJAMIN (II,2087); ROWAN AND MARTIN'S LAUGH-IN (I,3856); THE SHAPE OF THINGS (II,2315); TAXI (II,2546)

BRENNAN, KEVIN
COME OUT, COME OUT, WHEREVER YOU ARE (I,1014)

BRENNAN, KIRK
FATHER MURPHY (II,841); KIDS 2 KIDS (II,1389)

BRENNAN, MISSY
SANTA BARBARA (II,2258)

BRENNAN, PATRICK
NIGHT PARTNERS (II,1847)

BRENNAN, TOM
KENNEDY (II,1377)

BRENNAN, WALTER
ALIAS SMITH AND JONES (I,118); CAVALCADE OF AMERICA (I,875); A DATE WITH DEBBIE (I,1161); THE ETHEL BARRYMORE THEATER (I,1467); FORD TELEVISION THEATER (I,1634); THE GUNS OF WILL SONNETT (I,1907); LIGHT'S DIAMOND JUBILEE (I,2698); MR. TUTT (I,3153); THE OVER-THE-HILL GANG RIDES AGAIN (I,3432); THE REAL MCCOYS (I,3741); SCHLITZ PLAYHOUSE OF STARS (I,3936); SCREEN DIRECTOR'S PLAYHOUSE (I,3946); TO ROME WITH LOVE (I,4526); TWO FOR THE MONEY (I,4655); THE TYCOON (I,4660); ZANE GREY THEATER (I,4979)

BRENNER, DAVID
EBONY, IVORY AND JADE (II,755); NEIL SEDAKA STEPPIN' OUT (II,1804); STEVE MARTIN'S THE WINDS OF WHOOPIE (II,2461)

BRENNER, DORI
BROTHERS (II,381); CASSIE AND COMPANY (II,455); FRIENDS (II,928); HILL STREET BLUES (II,1154); THE OATH: THE SAD AND LONELY SUNDAYS (II,1874); SEVENTH AVENUE (II,2309); SHEILA (II,2325); SPARROW (II,2410)

BRENNER, MAURICE
THE PHIL SILVERS SHOW (I,3580)

BRENT, BIX
THE BERT PARKS SHOW (I,400); THE JULIUS LAROSA SHOW (I,2486)

BRENT, EVE
STRATEGIC AIR COMMAND (I,4252); TARZAN AND THE TRAPPERS (I,4362)

BRENT, GEORGE
EL COYOTE (I,1423); WIRE SERVICE (I,4883)

BRENT, ROMMEY
EYE WITNESS (I,1494); TEN LITTLE INDIANS (I,4394)

BRESH, THOM
THE RICH LITTLE SPECIAL (II,2155)

BRESLIN, JIMMY
REAL PEOPLE (II,2118)

BRESLIN, PATRICIA
ALCOA/GOODYEAR THEATER (I,107); ALFRED HITCHCOCK PRESENTS (I,115); APPOINTMENT WITH ADVENTURE (I,246); THE ARMSTRONG CIRCLE THEATER (I,260); ARSENIC AND OLD LACE (I,268); CAMEO THEATER (I,808); THE CAMPBELL TELEVISION SOUNDSTAGE (I,812); GENERAL HOSPITAL (II,964); LOVE STORY (I,2776); MATINEE THEATER (I,2947); THE PEOPLE'S CHOICE (I,3521); PEYTON PLACE (I,3574); PHILCO TELEVISION PLAYHOUSE (I,3583); SCHLITZ PLAYHOUSE OF STARS (I,3936); TATE (I,4364); THRILLER (I,4492)

BRESMETH, LLOYD
THE DOCTORS (II,694)

BRESSLER, BRIAN
ROWAN AND MARTIN'S LAUGH-IN (I,3856)

BRESTOFF, RICHARD
THE BRADY GIRLS GET MARRIED (II,364); THE NEW OPERATION PETTICOAT (II,1826); OPERATION PETTICOAT (II,1916); OPERATION PETTICOAT (II,1917); POTTSVILLE (II,2072)

BRETT, JEREMY
MACBETH (I,2801); ONE DEADLY OWNER (I,3373); ROMEO AND JULIET (I,3837)

BRETT, SUSSANA
BARRIER REEF (I,362)

BREWER, BETTY
THE BILL CULLEN SHOW (I,445); CAPTAIN BILLY'S MISSISSIPPI MUSIC HALL (I,826); DON'S MUSICAL PLAYHOUSE (I,1348); HOLIDAY HOTEL (I,2075)

BREWER, CHARLES
DRAGNET (I,1371); THE LAWYERS (I,2646)

BREWER, MILLICENT
GOLDEN WINDOWS (I,1841)

BREWER, TERESA
ARTHUR MURRAY'S DANCE PARTY (I,293); THE GEORGE SEGAL SHOW (II,977); PERRY PRESENTS (I,3549); SUMMERTIME U.S.A. (I,4283); UPBEAT (I,4684)

BREWSTER, DIANE
THE FUGITIVE (I,1701); THE GENERAL ELECTRIC THEATER (I,1753); THE ISLANDERS (I,2243); IT'S A SMALL WORLD (I,2260); THE LONELY WIZARD (I,2743); MAVERICK (II,1657); MICHAEL SHAYNE, DETECTIVE (I,3022); STILL THE BEAVER (II,2466)

BRIAN, DAVID
MR. DISTRICT ATTORNEY (I,3136)

BRIAN, EDDIE
CONCERNING MISS MARLOWE (I,1032)

BRIANT, SHANE
THE PICTURE OF DORIAN GRAY (I,3592)

BRICE, MONTY
THE BOB HOPE SHOW (I,585)

BRICE, PAUL B
SESAME STREET (I,3982)

BRICE, PIERRE
STAR MAIDENS (II,2435)

BRICK, MICHAEL
BOSTON AND KILBRIDE (II,356); GALACTICA 1980 (II,953)

BRICKELL, BETH
BROCK'S LAST CASE (I,742); GENTLE BEN (I,1761); SAN FRANCISCO INTERNATIONAL AIRPORT (I,3906); STONE (II,2471)

BRIDGE, LOIE
OCTAVIUS AND ME (I,3327)

BRIDGES, BEAU
THE DOROTHY HAMILL WINTER CARNIVAL SPECIAL (II,720); ENSIGN O'TOOLE (I,1448); FRANK MERRIWELL (I,1659); THE LLOYD BRIDGES SHOW (I,2733); MR. NOVAK (I,3145); ROBERT YOUNG AND THE FAMILY (I,3813); SEA HUNT (I,3948); STUBBY PRINGLE'S CHRISTMAS (II,2483); UNITED STATES (II,2708)

BRIDGES, BETTY
BACKSTAIRS AT THE WHITE HOUSE (II,133); CONDOMINIUM (II,566); ROOSTER (II,2210)

BRIDGES, CINDY
THE LLOYD BRIDGES SHOW (I,2733)

BRIDGES, DON
ALL THE RIVERS RUN (II,43)

BRIDGES, JEFF
THE LLOYD BRIDGES SHOW (I,2733)

BRIDGES, JIM
MACKENZIE'S RAIDERS (I,2803)

BRIDGES, KEN
YOUR HIT PARADE (I,4952)

BRIDGES, LLOYD
ALCOA PREMIERE (I,109); THE BLUE AND THE GRAY (II,273); CELEBRITY CHALLENGE OF THE SEXES 1 (II,465); CIRCUS OF THE STARS (II,534); CLIMAX! (I,976); THE CRIME CLUB (I,1096); THE DESILU PLAYHOUSE (I,1237); THE DICK POWELL SHOW (I,1269); EAST OF EDEN (II,752); FRONT ROW CENTER (I,1694); GEORGE WASHINGTON (II,978); GREAT ADVENTURE (I,1868); HOW THE WEST WAS WON

(II,1196); JOE FORRESTER (II,1303); THE JUNE ALLYSON SHOW (I,2488); THE LLOYD BRIDGES SHOW (I,2733); LLOYD BRIDGES WATER WORLD (I,2734); THE LONER (I,2745); PAPER DOLLS (II,1952); THE PEOPLE NEXT DOOR (I,3519); ROBERT MONTGOMERY PRESENTS YOUR LUCKY STRIKE THEATER (I,3809); ROOTS (II,2211); SAN FRANCISCO INTERNATIONAL AIRPORT (I,3905); SEA HUNT (I,3948); THE TYREES OF CAPITOL HILL (I,4661)

BRIDGES, RAND
BLACK BART (II,260); THE YOUNG PIONEERS' CHRISTMAS (II,2870)

BRIDGES, TODD
ALMOST AMERICAN (II,55); BATTLE OF THE NETWORK STARS (II,168); CIRCUS OF THE STARS (II,535); CIRCUS OF THE STARS (II,536); DIFF'RENT STROKES (II,674); FISH (II,864); GOOD EVENING, CAPTAIN (II,1031); HIGH SCHOOL, U.S.A. (II,1148); KOMEDY TONITE (II,1407); THE ORPHAN AND THE DUDE (II,1924); THE RETURN OF THE MOD SQUAD (I,3776); ROOTS (II,2211)

BRIDGEWATER, DEE DEE
NIGHT PARTNERS (II,1847)

BRIERLEY, DAVID
DOCTOR WHO (II,689)

BRIERS, RICHARD
GOOD NEIGHBORS (II,1034)

BRIGGS, BRUCE
GILLIGAN'S ISLAND: THE HARLEM GLOBETROTTERS ON GILLIGAN'S ISLAND (II,993)

BRIGGS, BUNNY
UPTOWN (II,2710)

BRIGGS, CHARLES
ROLL OF THUNDER, HEAR MY CRY (II,2202); YOU'RE ONLY YOUNG ONCE (I,4969)

BRIGGS, DONALD
STAGE 13 (I,4183)

BRIGGS, JOHN
THE JOEY BISHOP SHOW (I,2403)

BRIGGS, KALEB
ROLL OF THUNDER, HEAR MY CRY (II,2202)

BRIGGS, RICHARD
RYAN'S HOPE (II,2234)

BRIGGS, RUTH
THE STEVE ALLEN COMEDY HOUR (I,4218)

BRIGGS, TRACI LEE
THE KID WITH THE BROKEN HALO (II,1387)

BRIGHT, JACK
TRY AND DO IT (I,4621)

BRIGHT, PATRICIA
'TWAS THE NIGHT BEFORE CHRISTMAS (II,2677); 54TH STREET REVUE (I,1573); CAR 54, WHERE ARE YOU? (I,842); IT'S ALWAYS JAN (I,2264); MOVIELAND QUIZ (I,3116); OPEN HOUSE (I,3394); THE PAUL WINCHELL AND JERRY MAHONEY SHOW (I,3492); PINOCCHIO'S CHRISTMAS (II,2045)

BRIGHT, RICHARD
THE COPS AND ROBIN (II,573); FROM HERE TO ETERNITY (II,932); SKAG (II,2374)

BRILES, CHARLES
THE BIG VALLEY (I,437)

BRILL, CHARLIE
RHYME AND REASON (II,2152); ROWAN AND MARTIN'S LAUGH-IN (I,3856); SUPERTRAIN (II,2504)

BRILL, FRAN
HOW TO SURVIVE A MARRIAGE (II,1198)

BRILL, KEN
BEULAH LAND (II,226)

BRILL, MARTIN
HOW TO HANDLE A WOMAN (I,2146); MR. BELVEDERE (I,3127)

BRILL, MARTY
ALL IN THE FAMILY (II,38); ARCHIE BUNKER'S PLACE (II,105); THE NEW DICK VAN DYKE SHOW (I,3262); THE NEW SOUPY SALES SHOW (II,1829); SERGEANT T.K. YU (II,2305)

BRIMBLE, NICK
PENMARRIC (II,1993)

BRIMBLE, VINCENT
PENMARRIC (II,1993)

BRIMLEY, WILFRED
THE FIRM (II,860); JOE DANCER: THE BIG BLACK PILL (II,1300); ROUGHNECKS (II,2219)

BRINCKERHOFF, BURT
ABE LINCOLN IN ILLINOIS (I,11); INHERIT THE WIND (I,2213); OF THIS TIME, OF THAT PLACE (I,3335)

BRINDLE, EUGENE
DAVY CROCKETT (I,1181)

BRINEGAR, PAUL
CRISIS IN SUN VALLEY (II,604); LANCER (I,2610); THE LIFE AND LEGEND OF WYATT EARP (I,2681); MATT HOUSTON (II,1654); THE TEXAS RANGERS (II,2567); THE WILD WOMEN OF CHASTITY GULCH (II,2801)

BRINKLEY, CHRISTIE
BOB HOPE SPECIAL: BOB HOPE'S ALL-STAR BIRTHDAY AT ANNAPOLIS (II,294); BOB HOPE SPECIAL: HAPPY BIRTHDAY, BOB! (II,325)

BRINKLEY, DAVID
THE FIRST 50 YEARS (II,862)

BRION, FRANCOISE
MISTRAL'S DAUGHTER (II,1708)

BRISCOE, DON
DARK SHADOWS (I,1157)

BRISCOE, LAURIE
20 MINUTE WORKOUT (II,2680)

BRISEBOIS, DANIELLE
ALL IN THE FAMILY (II,38); ARCHIE BUNKER'S PLACE (II,105); BATTLE OF THE NETWORK STARS (II,172); BATTLE OF THE NETWORK STARS (II,174); BATTLE OF THE NETWORK STARS (II,176); CIRCUS OF THE STARS (II,535); CIRCUS OF THE STARS (II,536); GLORIA COMES HOME (II,1011); KNOTS LANDING (II,1404)

BRITT, ELTON
THE GEORGE HAMILTON IV SHOW (I,1769); SATURDAY NIGHT JAMBOREE (I,3920)

BRITT, LEO
THE DEVIL'S DISCIPLE (I,1247)

BRITT, MARY
THE BOB HOPE SHOW (I,573)

BRITT, MELENDY
BATMAN AND THE SUPER SEVEN (II,160); FLASH GORDON (II,874); FLASH GORDON—THE GREATEST ADVENTURE OF ALL (II,875); THE NEW ADVENTURES OF BATMAN (II,1812); THE PLASTICMAN COMEDY/ADVENTURE SHOW (II,2051); THEY ONLY COME OUT AT NIGHT (II,2586); YOU CAN'T DO THAT ON TELEVISION (I,4933)

BRITT, RUTH
THE BOYS IN BLUE (II,359); THE SHERIFF AND THE ASTRONAUT (II,2326)

BRITTANY, MORGAN
DALLAS (II,614); DELTA COUNTY, U.S.A. (II,657); THE FUNNIEST JOKE I EVER HEARD (II,940); GLITTER (II,1009); HOLLYWOOD STARS' SCREEN TESTS (II,1165); THE RICHARD BOONE SHOW (I,3784); SAMURAI (II,2249); STUNT SEVEN (II,2486); THE WILD WOMEN OF CHASTITY GULCH (II,2801)

BRITTON, BARBARA
APPOINTMENT WITH ADVENTURE (I,246); THE BIG PARTY FOR REVLON (I,427); CLIMAX! (I,976); A DATE WITH LIFE (I,1163); THE FABULOUS SYCAMORES (I,1505); HEAD OF THE FAMILY (I,1988); I'M SOOO UGLY (II,1219); LIGHT'S OUT (I,2699); MR. AND MRS. NORTH (I,3125); THE REVLON REVUE (I,3782); ROBERT MONTGOMERY PRESENTS YOUR LUCKY STRIKE THEATER (I,3809)

BRITTON, PAMELA
BLONDIE (I,486); THE LIFE OF RILEY (I,2686); MY FAVORITE MARTIAN (I,3189); SILVER THEATER (I,4051)

BRITTON, TONY
FATHER DEAR FATHER (II,837); ROBIN'S NEST (II,2187); THE SCARECROW OF ROMNEY MARSH (I,3931)

BRITZ, WALLACE
ROUGHNECKS (II,2219)

BRIZARD, PHILIP
FACTS OF LIFE: THE FACTS OF LIFE GOES TO PARIS (II,806)

BROADHEAD, JAMES E
TONI'S BOYS (II,2637)

BROADIN, ANDRE
THE CLIFFWOOD AVENUE KIDS (II,543)

BROCCO, PETER
ALIAS SMITH AND JONES (I,119); COMMANDO CODY (I,1030); NIGHT PARTNERS (II,1847); THE RAINBOW GIRL (II,2108); THE SHAMEFUL SECRETS OF HASTINGS CORNERS (I,3994); THE WINDS OF WAR (II,2807)

BROCK, ALAN
OUR FIVE DAUGHTERS (I,3413)

BROCK, JIMMY
THE COLLEGE BOWL (I,1001)

BROCK, STANLEY
13 THIRTEENTH AVENUE (II,2592); THE BUREAU (II,397); EVERY STRAY DOG AND KID (II,792); NO MAN'S LAND (II,1855); PALMS PRECINCT (II,1946)

BROCKETT, DON
MR. ROGERS' NEIGHBORHOOD (I,3150)

BROCKMAN, JANE
WINDY CITY JAMBOREE (I,4877)

BROCKSMITH, ROY
STARSTRUCK (II,2446)

BROCKWELL, LEONARD
TO SIR, WITH LOVE (II,2624)

BRODERICK, JAMES
ABE LINCOLN IN ILLINOIS (I,11); BRENNER (I,719); FAMILY (II,813); ROOTS: THE NEXT GENERATIONS (II,2213)

BRODERICK, MALCOLM
FREEDOM RINGS (I,1682)

BRODERICK, PETER
SEARCH FOR TOMORROW (II,2284)

BRODERICK, SUSAN
JANE EYRE (II,1273)

BRODIE, KEVIN
MIKE AND THE MERMAID (I,3039)

BRODIE, STEVE
GRUEN GUILD PLAYHOUSE (I,1892); THE LIFE AND LEGEND OF WYATT EARP (I,2681); MAGGIE MALONE (I,2812); NAVY LOG (I,3233)

BRODRICK, MALCOLM
THE MARRIAGE (I,2922)

BRODSKY, BRIAN
STUNT SEVEN (II,2486)

BRODY, RONNIE
THE BENNY HILL SHOW (II,207); DAVE ALLEN AT LARGE (II,621)

BROEKMAN, JANE
JAMBOREE (I,2331)

BROGAN, JAMES
OUT OF THE BLUE (II,1934)

BROLIN, JAMES
HOTEL (II,1192); MARCUS WELBY, M.D. (I,2905); MARCUS WELBY, M.D. (II,1627); THE MONROES (I,3089)

BROMFIELD, JOHN
THE DESILU REVUE (I,1238); SHERIFF OF COCHISE (I,4008); U.S. MARSHAL (I,4686); WHEN THE WEST WAS FUN: A WESTERN REUNION (II,2780)

BROMFIELD, LOIS
THE FACTS (II,804)

BROMFIELD, VALRI
ANGIE (II,80); BEST OF THE WEST (II,218); BOBBIE GENTRY'S HAPPINESS HOUR (I,671); BOBBIE GENTRY'S HAPPINESS HOUR (II,343); THE DAVID LETTERMAN SHOW (II,624); LILY (II,1477); THE LILY TOMLIN SPECIAL (II,1479); THE NEW SHOW (II,1828)

BROMILOW, PETER
DANIEL BOONE (I,1142)

BROMLEY, SHEILA
BACHELOR FATHER (I,323); FIRESIDE THEATER (I,1580); HANK (I,1931); I MARRIED JOAN (I,2174); IVY LEAGUE (I,2286); THE LIFE OF RILEY (I,2686); MORNING STAR (I,3104); THREE ON AN ISLAND (I,4483); WENDY AND ME (I,4793)

BRONDER, WILLIAM
WHITE AND RENO (II,2787)

BRONGA, JOHNNY
THE RED HAND GANG (II,2122)

BRONSON, CHARLES
CAVALCADE OF AMERICA (I,875); THE DOCTOR (I,1320); EMPIRE (I,1439); THE GENERAL ELECTRIC THEATER (I,1753); LUKE AND THE TENDERFOOT (I,2792); MAN WITH A CAMERA (I,2883); MEDIC (I,2976); PEPSI-COLA PLAYHOUSE (I,3523); PLAYHOUSE 90 (I,3623); STAGE 7 (I,4181); THE TRAVELS OF JAIMIE MCPHEETERS (I,4596); THE TWILIGHT ZONE (I,4651)

BRONSON, LILLIAN
ADAM-12 (I,31); KING'S ROW (I,2563); THE OVER-THE-HILL GANG RIDES AGAIN (I,3432)

BRONSTEIN, MAX
HOLLYWOOD JUNIOR CIRCUS (I,2084)

BROOK, FAITH
CLAUDIA: THE STORY OF A MARRIAGE (I,972); IN THE STEPS OF A DEAD MAN (I,2205); WAR AND PEACE (I,4765)

BROOKE, HILLARY
THE ABBOTT AND COSTELLO SHOW (I,2); FORD TELEVISION THEATER (I,1634); FOUR STAR PLAYHOUSE (I,1652); MY LITTLE MARGIE (I,3194); PEPSI-COLA PLAYHOUSE (I,3523); SCREEN DIRECTOR'S PLAYHOUSE (I,3946)

BROOKE, MERIEL
AN ECHO OF THERESA (I,1399)

BROOKE, WALTER
ADVICE TO THE LOVELORN (II,16); THE BLUE AND THE GRAY (II,273); THE GREEN HORNET (I,1886); HOTEL COSMOPOLITAN (I,2128); THE LAWYERS (I,2646); ONE MAN'S FAMILY (I,3383); OWEN MARSHALL: COUNSELOR AT LAW (I,3436); PARADISE BAY (I,3454); SCRUPLES (II,2281); THE WALTONS (II,2740)

BROOKE-TAYLOR, TIM
CAMBRIDGE CIRCUS (I,807); THE GOODIES (II,1040)

BROOKES, JACQUELINE
ANOTHER WORLD (II,97); ONE OF OUR OWN (II,1910); RYAN'S HOPE (II,2234); THE SECRET STORM (I,3969); A TIME FOR US (I,4507)

BROOKS, ALBERT
DEAN MARTIN PRESENTS THE GOLDDIGGERS (I,1194); GENERAL ELECTRIC'S ALL-STAR ANNIVERSARY (II,963); HOT WHEELS (I,2126); MILTON BERLE'S MAD MAD MAD WORLD OF COMEDY (II,1701)

BROOKS, ANNROSE
ANOTHER WORLD (II,97)

BROOKS, ARTHUR
THE DOCTORS (II,694)

BROOKS, BARRY
WON'T IT EVER BE MORNING? (I,4902)

BROOKS, BOB
SPACE: 1999 (II,2409)

BROOKS, BONNIE
THE TROUBLE WITH TRACY (I,4613)

BROOKS, CLAUDE
MOMMA THE DETECTIVE (II,1721)

BROOKS, DAVID
THE EDGE OF NIGHT (II,760)

BROOKS, ELISABETH
DAYS OF OUR LIVES (II,629); DOCTORS HOSPITAL (II,691)

BROOKS, FOSTER
THE ALL-STAR SALUTE TO MOTHER'S DAY (II,51); BOB HOPE SPECIAL: BOB HOPE PRESENTS A CELEBRATION WITH STARS OF COMEDY AND MUSIC (II,340); CIRCUS OF THE STARS (II,532); CIRCUS OF THE STARS (II,537); COUNTRY GALAXY OF STARS (II,587); DEAN MARTIN'S CELEBRITY ROAST (II,643); DEAN'S PLACE (II,650); THE MAD MAD MAD MAD WORLD OF THE SUPER BOWL (II,1586); THE MICKIE FINNS FINALLY PRESENT HOW THE WEST WAS LOST (II,1695); MORK AND MINDY (II,1735); THE NEW BILL COSBY SHOW (I,3257); THE REAL TOM KENNEDY SHOW (I,3743)

BROOKS, GERALDINE
ABC'S MATINEE TODAY (I,7); BONANZA (II,347); COLOSSUS (I,1009); THE DUMPLINGS (II,743); EXECUTIVE SUITE (II,796); FARADAY AND COMPANY (II,833); FORD THEATER HOUR (I,1635); THE GENERAL ELECTRIC THEATER (I,1753); IRONSIDE (I,2240); THE KAISER ALUMINUM HOUR (I,2503); KRAFT MYSTERY THEATER (I,2589); KRAFT SUSPENSE THEATER (I,2591); LOVE OF LIFE (I,2774); MAGNAVOX THEATER (I,2827); THE MUSIC MAKER (I,3173); SILVER THEATER (I,4051); STARLIGHT THEATER (I,4201)

BROOKS, HELEN
THE STEVE ALLEN COMEDY HOUR (II,2454)

BROOKS, HERB
SPEAK UP AMERICA (II,2412)

BROOKS, HILDY
HART TO HART (II,1102); THE NIGHT RIDER (II,1848)

BROOKS, JAN
THE LINE-UP (I,2707)

BROOKS, JEFF
ONE MORE TRY (II,1908)

BROOKS, JOE
F TROOP (I,1499)

BROOKS, JOEL
AFTER GEORGE (II,19); THE FACTS OF LIFE (II,805); PRIVATE BENJAMIN (II,2087);

TEACHERS ONLY (II,2548)

BROOKS, MARTIN E
THE BIONIC WOMAN (II,255); THE CAMPBELL TELEVISION SOUNDSTAGE (I,812); MEDICAL CENTER (II,1675); SEARCH FOR TOMORROW (II,2284); THE SIX-MILLION-DOLLAR MAN (II,2372)

BROOKS, MEL
THE 2,000 YEAR OLD MAN (II,2696); THE ALL-STAR COMEDY SHOW (I,137); ANNIE AND THE HOODS (II,91); THE COLGATE COMEDY HOUR (I,998); DOM DELUISE AND FRIENDS (II,701); MARLO THOMAS AND FRIENDS IN FREE TO BE. . . YOU AND ME (II,1632)

BROOKS, PATRICIA
GOLDEN CHILD (I,1839)

BROOKS, PETER
SWINGIN' TOGETHER (I,4320)

BROOKS, RAND
THE ADVENTURES OF RIN TIN TIN (I,73); THE ROY ROGERS SHOW (I,3859)

BROOKS, RANDI
HERNDON AND ME (II,1139); MICKEY SPILLANE'S MIKE HAMMER: MURDER ME, MURDER YOU (II,1694); RETURN OF THE MAN FROM U.N.C.L.E.: THE 15 YEARS LATER AFFAIR (II,2140); WIZARDS AND WARRIORS (II,2813)

BROOKS, RANDY
BROTHERS AND SISTERS (II,382); FLY AWAY HOME (II,893); THE RENEGADES (II,2132); THE RENEGADES (II,2133)

BROOKS, ROXANNE
RICHARD DIAMOND, PRIVATE DETECTIVE (I,3785)

BROOKS, SHELTON
JOHNNY CARSON PRESENTS THE SUN CITY SCANDALS '72 (I,2421)

BROOKS, STEPHEN
THE FBI (I,1551); THE INTERNS (I,2226); LITTLE MOON OF ALBAN (I,2715); THE NURSES (I,3318); TWO FOR THE MONEY (I,4655)

BROOKS, WALTER
THREE STEPS TO HEAVEN (I,4485)

BROOKSHIRE, TOM
CELEBRITY CHALLENGE OF THE SEXES (II,464); CELEBRITY CHALLENGE OF THE SEXES 3 (II,467); CELEBRITY CHALLENGE OF THE SEXES 4 (II,468); CELEBRITY CHALLENGE OF THE SEXES 5 (II,469)

BROOKSMITH, JANE
THESE ARE MY CHILDREN (I,4440)

BROOME JR, HEYWIN
THE PHIL SILVERS SHOW (I,3580)

BROPHY, KEVIN
LUCAN (II,1553); LUCAN (II,1554); TROUBLE IN HIGH TIMBER COUNTRY (II,2661); THE YEAGERS (II,2843)

BROPHY, SALLIE
BUCKSKIN (I,753); EYE WITNESS (I,1494); FOLLOW YOUR HEART (I,1618); FRONTIER (I,1695); THE LUX VIDEO THEATER (I,2795); ON TRIAL (I,3362); ONE STEP BEYOND (I,3388)

BROSNAN, PIERCE
THE MANIONS OF AMERICA (II,1623); NANCY ASTOR (II,1788); REMINGTON STEELE (II,2130)

BROSTROM, GUNEL
NUMBER 13 DEMON STREET (I,3316)

BROTHERS, DR JOYCE
THE ALL-STAR COMEDY SHOW (I,137); THE ALL-STAR SALUTE TO MOTHER'S DAY (II,51); BLONDES VS. BRUNETTES (II,271); THE BOB HOPE SHOW (I,649); BOB HOPE SPECIAL: BOB HOPE'S FUNNY VALENTINE (II,309); CAPTAIN KANGAROO (II,434); THE CHEVY CHASE NATIONAL HUMOR TEST (II,504); THE CHEVY CHASE SHOW (II,505); DICK CLARK'S GOOD OL' DAYS: FROM BOBBY SOX TO BIKINIS (II,671); GEORGE BURNS' HOW TO LIVE TO BE 100 (II,972); LIVING EASY (I,2730); THE LOVE REPORT (II,1542); MORE WILD WILD WEST (II,1733); THE NATIONAL SNOOP (II,1797); ONE LIFE TO LIVE (II,1907); TELL ME, DR. BROTHERS (I,4387); THAT SECOND THING ON ABC (II,2572)

BROTHERS, JAMIE
OF THIS TIME, OF THAT PLACE (I,3335)

BROTMAN, STUART
HAPPY ANNIVERSARY, CHARLIE BROWN (I,1939); IT'S ARBOR DAY, CHARLIE BROWN (I,2267); YOU'RE A GOOD SPORT, CHARLIE BROWN (I,4963)

BROUGH, CANDI
B.J. AND THE BEAR (II,126); BRANAGAN AND MAPES (II,367); MORE WILD WILD WEST (II,1733); WOMEN WHO RATE A "10" (II,2825)

BROUGH, RANDI
B.J. AND THE BEAR (II,126); BRANAGAN AND MAPES (II,367); MORE WILD WILD WEST (II,1733); WOMEN WHO RATE A "10" (II,2825)

BROUN, HEYWOOD HALE
BLOOMER GIRL (I,488)

BROUWER, PETER
AS THE WORLD TURNS (II,110); LOVE OF LIFE (I,2774); ONE LIFE TO LIVE (II,1907)

BROWEN, PHILIP
THE ASSASSINATION RUN (II,112)

BROWN FERRY FOUR, THE
MIDWESTERN HAYRIDE (I,3033)

BROWN JR, JOE
THREE STEPS TO HEAVEN (I,4485)

BROWN JR, LES
THE BAILEYS OF BALBOA (I,334); POLICE STORY (I,3636); THE YOUNG MARRIEDS (I,4942)

BROWN JR, RAY
FIRST LOVE (I,1583)

BROWN, ANDREA
MASTER OF THE GAME (II,1647)

BROWN, ANGELA
THE GHOST SQUAD (I,1787)

BROWN, BARRY
TESTIMONY OF TWO MEN (II,2564)

BROWN, BERNARD
THE PALLISERS (II,1944)

BROWN, BILL
MR. CHIPS (I,3132)

BROWN, BLAIR
CAPTAINS AND THE KINGS (II,435); CHARLIE COBB: NICE NIGHT FOR A HANGING (II,485); KENNEDY (II,1377); THE OREGON TRAIL (II,1922); WHEELS (II,2777)

BROWN, BOB
THE BANANA COMPANY (II,139); NANCY ASTOR (II,1788)

BROWN, BRYAN
AGAINST THE WIND (II,24); THE THORN BIRDS (II,2600); A TOWN LIKE ALICE (II,2652)

BROWN, CANDIDA
ONLY A SCREAM AWAY (I,3393)

BROWN, CANDY ANN
THE ROLLERGIRLS (II,2203)

BROWN, CARLOS
SEMI-TOUGH (II,2300)

BROWN, CHARITY
ANNE MURRAY'S LADIES' NIGHT (II,89)

BROWN, CHARLES
THE ROSEY GRIER SHOW (I,3849); TODAY'S FBI (II,2628)

BROWN, CHELSEA
DIAL HOT LINE (I,1254); MATT LINCOLN (I,2949); ROWAN AND MARTIN'S LAUGH-IN (I,3856)

BROWN, CHERYL LYNN
THE GUIDING LIGHT (II,1064)

BROWN, CHRISTOPHER J
ANOTHER WORLD (II,97); HEAVEN ON EARTH (II,1119); OPERATION PETTICOAT (II,1916); OPERATION PETTICOAT (II,1917)

BROWN, D W
JO'S COUSINS (II,1293)

BROWN, DONNA
KAY KYSER'S KOLLEGE OF MUSICAL KNOWLEDGE (I,2512)

BROWN, DORIS
FOODINI THE GREAT (I,1619); LUCKY PUP (I,2788)

BROWN, DWIER
THE MEMBER OF THE WEDDING (II,1680); THE THORN BIRDS (II,2600)

BROWN, EARL
THE DINAH SHORE SHOW (I,1280); THAT GOOD OLD NASHVILLE MUSIC (II,2571)

BROWN, ERIC
MAMA'S FAMILY (II,1610)

BROWN, ERIN NICOLE
LITTLE SHOTS (II,1495); SCAMPS (II,2267)

BROWN, GAIL
ANOTHER WORLD (II,97)

BROWN, GARRETT
THE DOOLEY BROTHERS (II,715)

BROWN, GEORG STANFORD

THE KID WITH THE BROKEN HALO (II,1387); POLICE SQUAD! (II,2061); THE ROOKIES (I,3840); THE ROOKIES (II,2208); ROOTS (II,2211); ROOTS: THE NEXT GENERATIONS (II,2213); THE YOUNG LAWYERS (I,4940)

BROWN, GEORGIA
THE ROADS TO FREEDOM (I,3802); THE TRINI LOPEZ SHOW (I,4606)

BROWN, GILLIAN
CASANOVA (II,451)

BROWN, GREGORY
FIT FOR A KING (II,866)

BROWN, HELEN GURLEY
OUTRAGEOUS OPINIONS (I,3429)

BROWN, HENRY
STAT! (I,4213)

BROWN, JACK
SING ALONG WITH MITCH (I,4057)

BROWN, JAMES L
THE ADVENTURES OF RIN TIN TIN (I,73)

BROWN, JAMES MASON
TONIGHT ON BROADWAY (I,4559)

BROWN, JEANNINE
HELP! IT'S THE HAIR BEAR BUNCH (I,2012)

BROWN, JERRY
ALAN KING LOOKS BACK IN ANGER—A REVIEW OF 1972 (I,99)

BROWN, JIM ED
THE COUNTRY PLACE (I,1071)

BROWN, JIMMY
THE MISSUS GOES A SHOPPING (I,3069); THE PINKY LEE SHOW (I,3601)

BROWN, JO GIESE
WOMAN'S PAGE (II,2822)

BROWN, JOE E
ARTHUR GODFREY'S TALENT SCOUTS (I,291); BUICK CIRCUS HOUR (I,759); FIVE'S A FAMILY (I,1591); G.E. SUMMMER ORIGINALS (I,1747); THE GREATEST SHOW ON EARTH (I,1883); HANDS OF MURDER (I,1929); SCREEN DIRECTOR'S PLAYHOUSE (I,3946)

BROWN, JOHN
THE GEORGE BURNS AND GRACIE ALLEN SHOW (I,1763); THE LIFE OF RILEY (I,2685)

BROWN, JOHN MASON
AMERICANA (I,175)

BROWN, JOHNNY
ALVIN AND THE CHIPMUNKS (II,61); THE BILL COSBY SPECIAL, OR? (I,443); GOOD TIMES (II,1038); THE LESLIE UGGAMS SHOW (I,2660); THE MOUSE FACTORY (I,3113); THE PLASTICMAN COMEDY/ADVENTURE SHOW (II,2051); ROWAN AND MARTIN'S LAUGH-IN (I,3856); SAMMY AND COMPANY (II,2248); THE TV SHOW (II,2673); WHERE'S THE FIRE? (II,2785)

BROWN, JOSEPHINE
HANDS OF MURDER (I,1929)

BROWN, JUDITH
NICK AND NORA (II,1842); OLIVIA NEWTON-JOHN IN CONCERT (II,1884)

BROWN, JULIE
EVENING AT THE IMPROV (II,786); TV FUNNIES (II,2672)

BROWN, JUNE
LACE (II,1418)

BROWN, KATHIE
BONANZA (II,347); HONDO (I,2103); SLATTERY'S PEOPLE (I,4082)

BROWN, KELLY
MR. PORTER OF INDIANA (I,3148)

BROWN, LES
ONE MORE TIME (I,3385)

BROWN, LEW
INSPECTOR PEREZ (II,1237); TALES OF THE APPLE DUMPLING GANG (II,2536)

BROWN, LISA
THE GUIDING LIGHT (II,1064)

BROWN, LOU
THE BILL ANDERSON SHOW (I,440)

BROWN, LUTHER
KENNEDY (II,1377)

BROWN, LYNDA
OPEN ALL HOURS (II,1913)

BROWN, MAGGI
THE COMEDY OF ERNIE KOVACS (I,1022); THE ERNIE KOVACS SPECIAL (I,1456); THE NEW ERNIE KOVACS SHOW (I,3264)

BROWN, MARIE
THE LIFE OF RILEY (I,2686)

BROWN, MARK ROBERT
THE WONDERFUL WORLD OF PHILIP MALLEY (II,2831)

BROWN, MARSHA
ENIGMA (II,778)

BROWN, MITCH
THE COWBOYS (II,598); JOE AND SONS (II,1297)

BROWN, NICHOLAS
ALL THE RIVERS RUN (II,43)

BROWN, OLIVIA
MIAMI VICE (II,1689)

BROWN, PAMELA
THE ADMIRABLE CRICHTON (I,36); THE ETHEL BARRYMORE THEATER (I,1467); THE LILLI PALMER THEATER (I,2704); ONE STEP BEYOND (I,3388); VICTORIA REGINA (I,4728)

BROWN, PAUL
CONDOMINIUM (II,566); THE GOSSIP COLUMNIST (II,1045)

BROWN, PENDLETON
MAMA MALONE (II,1609)

BROWN, PEPE
FAT ALBERT AND THE COSBY KIDS (II,835); PAT PAULSEN'S HALF A COMEDY HOUR (I,3473)

BROWN, PETER
CAR CARE CENTRAL (II,436); DAYS OF OUR LIVES (II,629); LAREDO (I,2615); LAWMAN (I,2642); WHEN THE WEST WAS FUN: A WESTERN REUNION (II,2780)

BROWN, PHILIP
THE DORIS DAY SHOW (I,1355); A HORSEMAN RIDING BY (II,1182); MASTER OF THE GAME (II,1647); WHEN THE WHISTLE BLOWS (II,2781)

BROWN, PHYLLIS
MIDWESTERN HAYRIDE (I,3033)

BROWN, R G
THE GLEN CAMPBELL GOODTIME HOUR (I,1820); I'VE HAD IT UP TO HERE (II,1221); THE JOHN BYNER COMEDY HOUR (I,2407); KING OF THE ROAD (II,1396); THE PAUL LYNDE COMEDY HOUR (II,1978); THE RICH LITTLE SHOW (II,2153)

BROWN, REB
CAPTAIN AMERICA (II,427); CAPTAIN AMERICA (II,428); CENTENNIAL (II,477)

BROWN, RENEE
HARRIS AND COMPANY (II,1096)

BROWN, ROBERT
COLOSSUS (I,1009); HERE COME THE BRIDES (I,2024); IVANHOE (I,2282); THE LAST HURRAH (I,2626); PRIMUS (I,3668)

BROWN, ROGER AARON
DAYS OF OUR LIVES (II,629)

BROWN, RUSS
THE LAW AND MR. JONES (I,2637); THE LOSERS (I,2757); THE TWO OF US (I,4658)

BROWN, RUTH
CHECKING IN (II,492); HELLO, LARRY (II,1128)

BROWN, SHARON
SALT AND PEPE (II,2240)

BROWN, SUSAN
BRIGHT PROMISE (I,727); FROM THESE ROOTS (I,1688); GENERAL HOSPITAL (II,964); PUNCH AND JODY (II,2091); THE YOUNG MARRIEDS (I,4942)

BROWN, TALLY
THE ART OF CRIME (II,108)

BROWN, TED
ACROSS THE BOARD (I,21); THE GREATEST MAN ON EARTH (I,1882); HAPPY BIRTHDAY (I,1940); HOWDY DOODY (I,2151); THE JOE DIMAGGIO SHOW (I,2399); STOP ME IF YOU HEARD THIS ONE (I,4231)

BROWN, TIMOTHY
FAMILY IN BLUE (II,818); M*A*S*H (II,1569)

BROWN, TOD
GUNSMOKE (II,1069)

BROWN, TOM
GENERAL HOSPITAL (II,964); MR. LUCKY (I,3141)

BROWN, TOM HENRY
THE GEORGE BURNS AND GRACIE ALLEN SHOW (I,1763)

BROWN, TRACY
ANOTHER WORLD (II,97)

BROWN, VANESSA
ALL THAT GLITTERS (II,42); GENERAL HOSPITAL (II,964); MY FAVORITE HUSBAND (I,3188); ONE STEP BEYOND (I,3388); STAGE 7 (I,4181); WALTER FORTUNE (I,4758)

BROWN, VIVIAN
FACTS OF LIFE: THE FACTS OF LIFE GOES TO PARIS (II,806)

BROWN, WALLY

CIMARRON CITY (I,953); MY DARLING JUDGE (I,3186); THE ROARING TWENTIES (I,3805)

BROWN, WALTER
ADVENTURES OF THE SEASPRAY (I,82)

BROWN, WOODY
BATTLE OF THE NETWORK STARS (II,172); FLAMINGO ROAD (II,871); FLAMINGO ROAD (II,872); LOVE OF LIFE (I,2774); WELCOME TO PARADISE (II,2762)

BROWNE, CORAL
TIME EXPRESS (II,2620)

BROWNE, DIK
THE FANTASTIC FUNNIES (II,825)

BROWNE, JUNE
PENTHOUSE SONATA (I,3516); U.S. TREASURY SALUTES (I,4689)

BROWNE, KALE
MR. & MRS. & MR. (II,1744); SCRUPLES (II,2281)

BROWNE, KATHIE
THE SLOWEST GUN IN THE WEST (I,4088)

BROWNE, ROBERT ALAN
THE RAINBOW GIRL (II,2108); SANTA BARBARA (II,2258)

BROWNE, ROSCOE LEE
DOCTOR SCORPION (II,685); HIGH FIVE (II,1143); MCCOY (II,1661); MISS WINSLOW AND SON (II,1705); REX HARRISON PRESENTS SHORT STORIES OF LOVE (II,2148); SOAP (II,2392)

BROWNELL, BARBARA
ACE (II,5); M*A*S*H (II,1569); PRESENTING SUSAN ANTON (II,2077); WHIZ KIDS (II,2790)

BROWNING, ALAN
CORONATION STREET (I,1054); THE FEAR IS SPREADING (I,1553)

BROWNING, DOUG
Q.E.D. (I,3701)

BROWNING, MAXWELL
MISTRAL'S DAUGHTER (II,1708)

BROWNING, ROD
GILLIGAN'S ISLAND: THE CASTAWAYS ON GILLIGAN'S ISLAND (II,992); MR. & MS. AND THE MAGIC STUDIO MYSTERY (II,1747)

BROWNING, SUSAN
LET'S CELEBRATE (I,2663); MARY HARTMAN, MARY HARTMAN (II,1638)

BROWNSTEIN, JEFF
STORY THEATER (I,4241)

BROZAK, EDITH
SUNDAY IN TOWN (I,4285)

BRUBAKER, ROBERT
DAYS OF OUR LIVES (II,629); GUNSMOKE (II,1069); MICHAEL SHAYNE, DETECTIVE (I,3022); U.S. MARSHAL (I,4686)

BRUBECK, DAVE
SEE AMERICA WITH ED SULLIVAN (I,3972); TIMEX ALL-STAR JAZZ SHOW I (I,4515)

BRUCE, ALAN
THE PATTY DUKE SHOW (I,3483)

BRUCE, BETTY
THE BOB HOPE SHOW (I,508); THE SATURDAY NIGHT REVUE (I,3921); THE WONDERFUL WORLD OF JACK PAAR (I,4899)

BRUCE, BRENDA
ALL CREATURES GREAT AND SMALL (I,129)

BRUCE, CAROL
CURTAIN CALL THEATER (I,1115); HOTPOINT HOLIDAY (I,2131); SILVER THEATER (I,4051); SOUNDS OF HOME (I,4136); WKRP IN CINCINNATI (II,2814)

BRUCE, DAVID
BEULAH (I,416)

BRUCE, DERRICK
FAME (II,812)

BRUCE, DOROTHY
FIRESIDE THEATER (I,1580)

BRUCE, ED
BRET MAVERICK (II,374)

BRUCE, EDWIN
MAGGIE (I,2810)

BRUCE, EVE
FAR OUT SPACE NUTS (II,831)

BRUCE, JEAN SCOTT
MAGNUM, P.I. (II,1601)

BRUCE, LAURALEE
KANGAROOS IN THE KITCHEN (II,1362)

BRUCE, LYDIA
THE DOCTORS (II,694)

BRUCE, MARIAN
FRONT ROW CENTER (I,1693)

BRUCE, NIGEL
FOUR STAR PLAYHOUSE (I,1652)

BRUCK, BELLA
ALAN KING'S FINAL WARNING (II,29); I AND CLAUDIE (I,2164); THE SUNSHINE BOYS (II,2496); TAMMY (I,4357); THE UGILY FAMILY (II,2699)

BRUCK, KARL
THE YOUNG AND THE RESTLESS (II,2862)

BRUCKER, BO
RIVKIN: BOUNTY HUNTER (II,2183)

BRUDER, PATRICIA
AS THE WORLD TURNS (II,110)

BRULL, PAMELA
OFF THE RACK (II,1878); THE SECRET EMPIRE (II,2291); T.J. HOOKER (II,2524)

BRUNDIN, BO
CENTENNIAL (II,477); RICH MAN, POOR MAN—BOOK I (II,2161); VELVET (II,2725); THE WORD (II,2833)

BRUNER, NATALIE
WALLY'S WORKSHOP (I,4753)

BRUNER, WALLY
WALLY'S WORKSHOP (I,4753); WHAT'S MY LINE? (I,4820)

BRUNNER, BOB
LAWRENCE WELK'S TOP TUNES AND NEW TALENT (I,2645)

BRUNO, CATHERINE
THE EDGE OF NIGHT (II,760)

BRUNS, PHILIP
BUT MOTHER! (II,403); THE CHOPPED LIVER BROTHERS (II,514); FOREVER FERNWOOD (II,909); GREAT DAY (II,1054); JACKIE GLEASON AND HIS AMERICAN SCENE MAGAZINE (I,2321); MARY HARTMAN, MARY HARTMAN (II,1638); MASQUERADE (I,2940); MR. INSIDE/MR. OUTSIDE (II,1750); SHADOW OF SAM PENNY (II,2313)

BRUSSET, ANTOINE
MASTER OF THE GAME (II,1647)

BRUZAC, ANNE
THE BENNY HILL SHOW (II,207)

BRY, ELLEN
THE AMAZING SPIDER-MAN (II,63); BATTLE OF THE NETWORK STARS (II,178); ST. ELSEWHERE (II,2432)

BRYAN, ARTHUR Q
THE HALLS OF IVY (I,1923); THE HANK MCCUNE SHOW (I,1932); MEET THE GOVERNOR (I,2991); MOVIELAND QUIZ (I,3116); PROFESSIONAL FATHER (I,3677); TOM AND JERRY (I,4533)

BRYAN, WARREN
BRIEF PAUSE FOR MURDER (I,725)

BRYANT, ANITA
THE ANITA BRYANT SPECTACULAR (II,82); THE BOB HOPE SHOW (I,581); THE BOB HOPE SHOW (I,587); THE BOB HOPE SHOW (I,598); THE BOB HOPE SHOW (I,609); THE BOB HOPE SHOW (I,613); NBC SATURDAY PROM (I,3245); STARS AND STRIPES SHOW (II,2442); THE STARS AND STRIPES SHOW (I,4208); THE STARS AND STRIPES SHOW (II,2443)

BRYANT, BEN
SWINGIN' TOGETHER (I,4320)

BRYANT, JOHN
BELLE STARR (I,391); DOBIE GILLIS (I,1302); MEET THE GIRLS: THE SHAPE, THE FACE, AND THE BRAIN (I,2990); PINE LAKE LODGE (I,3597)

BRYANT, JOSHUA
BEHIND THE SCREEN (II,200); THE MAN FROM ATLANTIS (II,1614); MARRIED: THE FIRST YEAR (II,1634); SCENE OF THE CRIME (II,2271); WAIT UNTIL DARK (II,2735)

BRYANT, LEE
LASSIE: THE NEW BEGINNING (II,1433); THE MANY LOVES OF ARTHUR (II,1625); THE RITA MORENO SHOW (II,2181); T.J. HOOKER (II,2524)

BRYANT, MARDI
THE TED STEELE SHOW (I,4373)

BRYANT, MARGOT
CORONATION STREET (I,1054)

BRYANT, MEL
ARCHIE BUNKER'S PLACE (II,105)

BRYANT, MICHAEL
CAESAR AND CLEOPATRA (I,789); THE ROADS TO FREEDOM (I,3802)

BRYANT, NANA
THE FIRST HUNDRED YEARS (I,1581); MAKE ROOM FOR DADDY (I,2843); OUR MISS BROOKS (I,3415)

BRYANT, NEA
A GIRL'S LIFE (II,1000)

BRYANT, NICOLA
DOCTOR WHO (II,689)

BRYANT, SUZANNE
SCARLETT HILL (I,3933)

BRYANT, WILLIAM
THE BILLION DOLLAR THREAT (II,246); BRANDED (I,707); COMBAT (I,1011); GENERAL HOSPITAL (II,964); THE LEGEND OF THE GOLDEN GUN (II,1455); THE NEW HEALERS (I,3266); ROOSTER (II,2210); SWITCH (II,2519); WAR CORRESPONDENT (I,4767)

BRYANT, WILLIE
SHOWTIME AT THE APOLLO (I,4037); SUGAR HILL TIMES (I,4276)

BRYAR, CLAUDE
THE DAUGHTERS OF JOSHUA CABE (I,1169);

BRYAR, CLAUDIA
LEAVE IT TO BEAVER (I,2648); MANHUNTER (II,1620); MANHUNTER (II,1621)

BRYAR, PAUL
THE HUSTLER OF MUSCLE BEACH (II,1210); THE LONG HOT SUMMER (I,2747)

BRYCE, ED
ANOTHER WORLD (II,97); THE GUIDING LIGHT (II,1064); SOMERSET (I,4115)

BRYCE, EDWARD
TOM CORBETT, SPACE CADET (I,4535)

BRYCE, SCOTT
AS THE WORLD TURNS (II,110)

BRYDON, W B
THE ADAMS CHRONICLES (II,8); THE FABULOUS DR. FABLE (I,1500)

BRYGGMAN, LARRY
AS THE WORLD TURNS (II,110)

BRYNE, BARBARA
LOVE, SIDNEY (II,1547)

BRYNNER, YUL
ANNA AND THE KING (I,221); FIRESIDE THEATER (I,1580); THE GENERAL FOODS 25TH ANNIVERSARY SHOW (I,1756); THE POPPY IS ALSO A FLOWER (I,3650)

BUA, GENE
HOW TO SURVIVE A MARRIAGE (II,1198); LOVE OF LIFE (I,2774)

BUA, TONI BULL
LOVE OF LIFE (I,2774)

BUBBLES, JOHN
THE BELLE OF 14TH STREET (I,390); THE BOB HOPE SHOW (I,598); THE PETER LIND HAYES SHOW (I,3564)

BUBKIN, CELIA
THE O'NEILLS (I,3392)

BUCHANAN, EDDIE
THE BENNY HILL SHOW (II,207)

BUCHANAN, EDGAR
CADE'S COUNTY (I,787); CROSSROADS (I,1105); CROWN THEATER WITH GLORIA SWANSON (I,1107); GREEN ACRES (I,1884); HOPALONG CASSIDY (I,2118); JUDGE ROY BEAN (I,2467); LEAVE IT TO BEAVER (I,2648); THE LLOYD BRIDGES SHOW (I,2733); LUKE AND THE TENDERFOOT (I,2792); NATIONAL VELVET (I,3231); THE OVER-THE-HILL GANG RIDES AGAIN (I,3432); PETTICOAT JUNCTION (I,3571); THE RIFLEMAN (I,3789); SAM HILL (I,3895); SCREEN DIRECTOR'S PLAYHOUSE (I,3946); TELEPHONE TIME (I,4379); THE TYREES OF CAPITOL HILL (I,4661); YUMA (I,4976)

BUCHANAN, JACK
THE BOB HOPE SHOW (I,518); SPOTLIGHT (I,4162)

BUCHANAN, MORRIS
B.A.D. CATS (II,123)

BUCHANON, KIRBY
THE ROY ROGERS AND DALE EVANS SHOW (I,3858)

BUCHHOLZ, HORST
THE FRENCH ATLANTIC AFFAIR (II,924); RETURN TO FANTASY ISLAND (II,2144)

BUCHWALD, ART
THE ENTERTAINERS (I,1449); THE SENSATIONAL, SHOCKING, WONDERFUL, WACKY 70S (II,2302); STANDARD OIL ANNIVERSARY SHOW (I,4189) THAT WAS THE YEAR THAT WAS (I,4424); THE WORLD OF SOPHIA LOREN (I,4918)

BUCK
TOPPER (I,4582)

BUCK AND BUBBLES
NIGHT CLUB (I,3284)

BUCK, DAVID
HORNBLOWER (I,2121); THE SCARECROW OF ROMNEY MARSH (I,3931)

BUCK, GEORGIA
THE ETHEL WATERS SHOW (I,1468)

BUCK, LISA
SEARCH FOR TOMORROW (II,2284)

BUCK, RICHARD
THE EDGE OF NIGHT (II,760)

BUCKAROOS, THE
HEE HAW (II,1123)

BUCKERT, SANDY
TROPIC HOLIDAY (I,4607)

BUCKINGHAM, LINDSAY
FLEETWOOD MAC IN CONCERT (II,880)

BUCKINGHAM, STEVE L
THE OUTLAWS (II,1936)

BUCKLEY JR, WILLIAM F
BRIDESHEAD REVISITED (II,375); THAT WAS THE YEAR THAT WAS (I,4424)

BUCKLEY, BETTY
EIGHT IS ENOUGH (II,762); THE RUBBER GUN SQUAD (II,2225); SALUTE TO LADY LIBERTY (II,2243)

BUCKLEY, DENISE
IN THE STEPS OF A DEAD MAN (I,2205)

BUCKLEY, FLOYD
POPEYE THE SAILOR (I,3648)

BUCKLEY, HAL
MARCO POLO (II,1626); O.K. CRACKERBY (I,3348)

BUCKLEY, JEANNE
PEYTON PLACE (I,3574)

BUCKLEY, KEITH
SEARCH FOR THE NILE (I,3954)

BUCKMAN, TARA
LOBO (II,1504); THE MASTER (II,1648); STONE (II,2470)

BUCKNER, SUSAN
FRIENDS (II,928); THE NANCY DREW MYSTERIES (II,1789); WHEN THE WHISTLE BLOWS (II,2781)

BUDDY FOSTER DANCERS, THE
THE KEEFE BRASSELLE SHOW (I,2514)

BUDDY MILES EXPRESS, THE
33 1/3 REVOLUTIONS PER MONKEE (I,4451)

BUDDY RICH ORCHESTRA
AWAY WE GO (I,317)

BUDDY SCHWAB DANCERS, THE
THE HOLLYWOOD PALACE (I,2088)

BUENO, DOLORA
FLIGHT TO RHYTHM (I,1598)

BUERHLEN, JUSTIN
THE WINDS OF WAR (II,2807)

BUFANO, VINCENT
THE BEST LEGS IN 8TH GRADE (II,214); EISCHIED (II,763); FLATBUSH (II,877)

BUFFANO, JULES
THE JIMMY DURANTE SHOW (I,2388); TEXACO STAR THEATER (I,4412)

BUFFERT, KENNY
THE COLLEGE BOWL (I,1001)

BUFFINGTON, SAM
WHISPERING SMITH (I,4842)

BUFORD, IVAN
DAVE ALLEN AT LARGE (II,621)

BUGARD, MEDALINE
ONE MAN'S FAMILY (I,3383)

BUGGY, NIALL
THE CITADEL (II,539)

BUICK BELLS, THE
FIREBALL FUN FOR ALL (I,1577)

BUJOLD, GENEVIEVE
CAESAR AND CLEOPATRA (I,789); ST. JOAN (I,4175)

BUKTENICA, RAY
BUMPERS (II,394); COUSINS (II,595); HOUSE CALLS (II,1194); RHODA (II,2151); W*A*L*T*E*R (II,2731)

BULA, TAMMI
THE MARY TYLER MOORE SHOW (II,1640)

BULARD, NIGEL
THE WORLD OF PEOPLE (II,2839)

BULIFANT, JOYCE
ARNOLD'S CLOSET REVIEW (I,262); ARTHUR MURRAY'S DANCE PARTY (I,293); THE BAD NEWS BEARS (II,134);

BIG JOHN, LITTLE JOHN (II,240); THE BILL COSBY SHOW (I,442); CHARLEY'S AUNT (II,483); THE FIRST HUNDRED YEARS (I,1582); FLO (II,891); LITTLE WOMEN (II,1497); LOVE THY NEIGHBOR (I,2781); THE MARY TYLER MOORE SHOW (II,1640); THE MICHELE LEE SHOW (II,1690); SPORT BILLY (II,2429); TOM, DICK, AND MARY (I,4537)

BULL, PETER
TALES OF THE VIKINGS (I,4351)

BULL, RICHARD
LITTLE HOUSE ON THE PRAIRIE (II,1487); LITTLE HOUSE: A NEW BEGINNING (II,1488); LITTLE HOUSE: BLESS ALL THE DEAR CHILDREN (II,1489); LITTLE HOUSE: LOOK BACK TO YESTERDAY (II,1490); LITTLE HOUSE: THE LAST FAREWELL (II,1491); NO WARNING (I,3301); VOYAGE TO THE BOTTOM OF THE SEA (I,4743)

BULL, TONI
A HATFUL OF RAIN (I,1964)

BULLARD, FRANKLYN
KENNEDY (II,1377)

BULLOCH, JEREMY
HORNBLOWER (I,2121); ONLY A SCREAM AWAY (I,3393)

BULLOCK, EARL
BRANAGAN AND MAPES (II,367)

BULLOCK, JM J
FAMILY BUSINESS (II,815); TOO CLOSE FOR COMFORT (II,2642)

BULLOCK, OSMUND
PRIDE AND PREJUDICE (II,2080)

BULOFF, JOSEPH
TWO GIRLS NAMED SMITH (I,4656); WONDERFUL TOWN (I,4893)

BULTAND, NIGEL
COCAINE AND BLUE EYES (II,548)

BUMILLER, WILLIAM
FLATFOOTS (II,879)

BUMPASS, RODGER
NATIONAL LAMPOON'S HOT FLASHES (II,1795); NATIONAL LAMPOON'S TWO REELERS (II,1796)

BUMSTEAD, J P
BORDER PALS (II,352); CELEBRITY (II,463); GABE AND WALKER (II,950); MOONLIGHT (II,1730); THINGS ARE LOOKING UP (II,2588); THIS GIRL FOR HIRE (II,2596); USED CARS (II,2712)

BUNARO, DIEDRA
ONE LIFE TO LIVE (II,1907)

BUNCE, ALAN
ETHEL AND ALBERT (I,1466); THE KATE SMITH HOUR (I,2508); THE RIGHT MAN (I,3790)

BUNCH, MARIANNE
FLYING HIGH (II,895)

BUNDY, BROOKE
THE ADVENTURES OF NICK CARTER (I,68); THE BEST YEARS (I,410); DAYS OF OUR LIVES (II,629); FOOLS, FEMALES AND FUN: WHAT ABOUT THAT ONE? (II,901); GENERAL HOSPITAL (II,964); HOLLOWAY'S DAUGHTERS (I,2079); KELLY'S KIDS (I,2523); MCCLAIN'S LAW (II,1659); MR. NOVAK (I,3145); TRAVIS LOGAN, D.A. (I,4598)

BUNETA, BILL
CELEBRITY BOWLING (I,883)

BUNIN, HOPE
FEARLESS FOSDICK (I,1554); FOODINI THE GREAT (I,1619); LUCKY PUP (I,2788)

BUNIN, MORY
FEARLESS FOSDICK (I,1554); FOODINI THE GREAT (I,1619); LUCKY PUP (I,2788)

BUNNAGE, AVIS
LITTLE LORD FAUNTLEROY (II,1492)

BUNTROCK, BOBBY
HAZEL (I,1982)

BUONO, CARA
SESAME STREET (I,3982)

BUONO, VICTOR
BACKSTAIRS AT THE WHITE HOUSE (II,133); BATMAN (I,366); BRENDA STARR (II,372); THE CRIME CLUB (I,1096); GENERAL ELECTRIC TRUE (I,1754); HIGH RISK (II,1146); JUDGEMENT DAY (II,1348); THE MAN FROM ATLANTIS (II,1614); THE MAN FROM ATLANTIS (II,1615); MORE WILD WILD WEST (II,1733); MURDER CAN HURT YOU! (II,1768); NIGHT GALLERY (I,3287); THE RETURN OF THE MOD SQUAD (I,3776); THE RITA MORENO SHOW (II,2181); RUSSIAN ROULETTE (I,3875); TAXI (II,2546); TWO GUYS FROM MUCK (II,2690); VEGAS (II,2724); THE WILD WILD WEST (I,4863)

BURDETT, WINSTON
YOU ARE THERE (I,4929)

BURDICK, HAL
NIGHT EDITOR (I,3286)

BURGAN, JOANNE
H.M.S. PINAFORE (I,2066)

BURGE, JAMES
CASTLE ROCK (II,456); KENNEDY (II,1377)

BURGE, WENDY
THE MAGIC PLANET (II,1598)

BURGESS, BARBARA
ELVIS (I,1435)

BURGESS, BOBBY
THE LAWRENCE WELK SHOW (I,2643); THE LAWRENCE WELK SHOW (I,2644); MEMORIES WITH LAWRENCE WELK (II,1681); THE MICKEY MOUSE CLUB (I,3025); THE MOUSEKETEERS REUNION (II,1742)

BURGESS, DENNIS
FATHER BROWN (II,836); TEN WHO DARED (II,2559); TRUMAN AT POTSDAM (I,4618)

BURGESS, PATRICIA
THE GALLOPING GOURMET (I,1730)

BURGESS, SCOTT
1915 (II,1853)

BURGESS, VIVIENNE
THE ROCK FOLLIES (II,2192)

BURGETT, GINGER
ANOTHER LIFE (II,95)

BURGETTE, BERNADETTE
THE SHOW MUST GO ON (II,2344)

BURGHARDT, ARTHUR
DIFF'RENT STROKES (II,674); ONE LIFE TO LIVE (II,1907)

BURGHOFF, GARY
BATTLE OF THE NETWORK STARS (II,163); CASINO (II,452); THE DON KNOTTS SHOW (I,1334); M*A*S*H (II,1569); TWIGS (II,2683); W*A*L*T*E*R (II,2731)

BURGOYNE, MARCELLA
THE SULLIVANS (II,2490)

BURGOYNE, VICTORIA
NANCY ASTOR (II,1788)

BURGUNDY STREET SINGERS, THE
ED MCMAHON AND HIS FRIENDS. . . DISCOVER WET AT CYPRESS GARDENS (I,1400); FRANKIE AVALON'S EASTER SPECIAL (I,1673); THE JIMMIE RODGERS SHOW (I,2383); THE RED SKELTON SHOW (I,3756)

BURINER, JENNIFER
THE GOLDDIGGERS (I,1838)

BURKE, BILLIE
ARSENIC AND OLD LACE (I,268); DOC CORKLE (I,1304); THE EDDIE CANTOR COMEDY THEATER (I,1407); LIGHT'S OUT (I,2699); TEXACO COMMAND PERFORMANCE (I,4407)

BURKE, CECILY
THE AD-LIBBERS (I,29)

BURKE, CLEM
BLONDIE (II,272)

BURKE, CONSUELA
VANITY FAIR (I,4701)

BURKE, DAVID
EIGHT IS ENOUGH (II,762)

BURKE, DELTA
BATTLE OF THE NETWORK STARS (II,175); THE CHISHOLMS (II,513); FILTHY RICH (II,856); FIRST & TEN (II,861); THE HOME FRONT (II,1171); JOHNNY BLUE (II,1321); MICKEY SPILLANE'S MIKE HAMMER: MURDER ME, MURDER YOU (II,1694); ROOSTER (II,2210); THE SEEKERS (II,2298)

BURKE, GEORGIA
THE LITTLE FOXES (I,2712)

BURKE, JAMES
MARK SABER (I,2914); STAGECOACH WEST (I,4185)

BURKE, JERRY
THE LAWRENCE WELK SHOW (I,2643); MEMORIES WITH LAWRENCE WELK (II,1681)

BURKE, JOE
BENDER (II,204)

BURKE, JOSEPH
THE CONCRETE COWBOYS (II,564)

BURKE, MAURICE
THE FORTUNE HUNTER (I,1646)

BURKE, MICHELLE
WONDERFUL TOWN (I,4893)

BURKE, PATRICK
LITTLE VIC (II,1496)

BURKE, PAUL

ADVICE TO THE LOVELORN (II,16); BEACH PATROL (II,192); THE CRIME CLUB (I,1096); DYNASTY (II,746); AN ECHO OF THERESA (I,1399); FIVE FINGERS (I,1586); HARBOURMASTER (I,1946); NAKED CITY (I,3216); NOAH'S ARK (I,3302); PANIC! (I,3447); THE ROOKIES (I,3840); STAGE 7 (I,4181); TWELVE O'CLOCK HIGH (I,4637); WILD AND WOOLEY (II,2798)

BURKE, ROBERT
THE GANGSTER CHRONICLES (II,957)

BURKE, WALTER
THE HUSTLER OF MUSCLE BEACH (II,1210); MAGGIE BROWN (I,2811); MURDOCK'S GANG (I,3164); THE OVER-THE-HILL GANG RIDES AGAIN (I,3432)

BURKETT, BILL
CELEBRITY (II,463)

BURKHARDT, BOB
QUADRANGLE (I,3703)

BURKLEY, DENNIS
B.J. AND THE BEAR (II,125); THE BOOK OF LISTS (II,350); DELTA COUNTY, U.S.A. (II,657); FOREVER FERNWOOD (II,909); HANGING IN (II,1078); MARY HARTMAN, MARY HARTMAN (II,1638); MR. DUGAN (II,1749); SANFORD (II,2255); SUTTERS BAY (II,2508); THE TEXAS WHEELERS (II,2568)

BURKLEY, JACQUELINE
A TIME TO LIVE (I,4510)

BURL, ROBERT
ANASTASIA (I,182)

BURLAT, KEVIN
LITTLE SHOTS (II,1495)

BURLEIGH, STEPHEN
THE DOCTORS (II,694); SEARCH FOR TOMORROW (II,2284)

BURLESON, DON
JEREMIAH OF JACOB'S NECK (II,1281)

BURMESTER, LEO
CHIEFS (II,509); A FINE ROMANCE (II,858); FLO (II,891); GEORGE WASHINGTON (II,978); RAGE OF ANGELS (II,2107)

BURN, JONATHAN
THE FORSYTE SAGA (I,1645); SMILEY'S PEOPLE (II,2383)

BURNARD, SANDY
AFTERMASH (II,23)

BURNELL, PETER
THE DOCTORS (II,694)

BURNETT, CAROL
6 RMS RIV VU (II,2371); ALL MY CHILDREN (II,39); AMERICAN COWBOY (I,169); BERT CONVY SPECIAL—THERE'S A MEETING HERE TONIGHT (II,210); BING CROSBY AND CAROL BURNETT—TOGETHER AGAIN FOR THE FIRST TIME (I,454); BING CROSBY AND HIS FRIENDS (I,456); BURNETT "DISCOVERS" DOMINGO (II,398); BURT AND THE GIRLS (I,770); BURT BACHARACH: CLOSE TO YOU (I,772); CALAMITY JANE (I,793); CAROL + 2 (I,853); CAROL AND COMPANY (I,847); CAROL BURNETT AND COMPANY (II,441); THE CAROL BURNETT SHOW (II,443); CAROL CHANNING PROUDLY PRESENTS THE SEVEN DEADLY SINS (I,851); CBS: ON THE AIR (II,460); CHERYL LADD: SCENES FROM A SPECIAL (II,503); THE DUPONT SHOW OF THE WEEK (I,1388); THE ENTERTAINERS (I,1449); EUNICE (II,784); THE GARRY MOORE SHOW (I,1739); THE GENERAL MOTORS 50TH ANNIVERSARY SHOW (I,1758); GIRL FRIENDS AND NABORS (I,1807); JULIE AND CAROL AT CARNEGIE HALL (I,2478); JULIE AND CAROL AT LINCOLN CENTER (I,2479); KEEP U.S. BEAUTIFUL (I,2519); A LAST LAUGH AT THE 60'S (I,2627); LI'L ABNER (I,2702); MAMA'S FAMILY (II,1610); ONCE UPON A MATTRESS (I,3371); PANTOMIME QUIZ (I,3449); THE PERRY COMO CHRISTMAS SHOW (I,3529); ROWAN AND MARTIN BITE THE HAND THAT FEEDS THEM (I,3853); THE SENSATIONAL, SHOCKING, WONDERFUL, WACKY 70S (II,2302); SHIRLEY MACLAINE: IF THEY COULD SEE ME NOW (II,2335); SILLS AND BURNETT AT THE MET (II,2354); STANLEY (I,4190); STEVE AND EYDIE CELEBRATE IRVING BERLIN (II,2456); SUPER COMEDY BOWL 1 (I,4288); TEXACO STAR THEATER: OPENING NIGHT (II,2565); THE TIM CONWAY SPECIAL (I,4503); TWIGS (II,2683)

BURNETT, DON
NORTHWEST PASSAGE (I,3306)

BURNETT, LARISSA
1915 (II,1853)

BURNETT, SADI
MUSIC HALL AMERICA (II,1773)

BURNETT, SHEILA
CANDID CAMERA (II,420)

BURNETTE, SMILEY
JUBILEE U.S.A. (I,2461); PETTICOAT JUNCTION (I,3571); RANCH PARTY (I,3719)

BURNHAM, TERRY
I LOVE MY DOCTOR (I,2172)

BURNIER, JANINE
KEEP ON TRUCKIN' (II,1372); THE POWDER ROOM (I,3656)

BURNS, BART
THE BLUE KNIGHT (II,277); MIKE HAMMER, DETECTIVE (I,3042)

BURNS, BOB
THE GHOST BUSTERS (II,986)

BURNS, CAROL
PRISONER: CELL BLOCK H (II,2085)

BURNS, CATHERINE
THE CATCHER (I,872); A CHRISTMAS CAROL (II,517); SANDBURG'S LINCOLN (I,3907); TWO FOR THE MONEY (I,4655); THE WORD (II,2833)

BURNS, DANIELLE
ANOTHER WORLD (II,97)

BURNS, DAVID
THE IMOGENE COCA SHOW (I,2198); MIRROR, MIRROR, OFF THE WALL (I,3059); THE PRICE (I,3666); THE TRIALS OF O'BRIEN (I,4605)

BURNS, GEORGE
100 YEARS OF AMERICA'S POPULAR MUSIC (II,1904); ANN-MARGRET'S HOLLYWOOD MOVIE GIRLS (II,86); ANN-MARGRET—WHEN YOU'RE SMILING (I,219); THE BOB HOPE SHOW (I,630); THE BOB HOPE SHOW (I,644); BOB HOPE SPECIAL: BOB HOPE IN WHO MAKES THE WORLD LAUGH, PART 2 (II,286); BOB HOPE SPECIAL: BOB HOPE LAUGHS WITH THE MOVIE AWARDS (II,288); BOB HOPE SPECIAL: BOB HOPE'S 30TH ANNIVERSARY TV SPECIAL (II,293); BOB HOPE SPECIAL: BOB HOPE'S ALL-STAR TRIBUTE TO THE PALACE THEATER (II,305); BOB HOPE SPECIAL: BOB HOPE'S ROAD TO HOLLYWOOD (II,313); BOB HOPE SPECIAL: HAPPY BIRTHDAY, BOB! (II,325); THE BRASS ARE COMING (I,708); CBS: ON THE AIR (II,460); CIRCUS OF THE STARS (II,531); AN EVENING WITH CAROL CHANNING (I,1473); EVERYTHING YOU ALWAYS WANTED TO KNOW ABOUT JACK BENNY AND WERE AFRAID TO ASK (I,1482); THE FUNNIEST JOKE I EVER HEARD (II,939); THE GEORGE BURNS AND GRACIE ALLEN SHOW (I,1763); GEORGE BURNS AND OTHER SEX SYMBOLS (II,966); GEORGE BURNS CELEBRATES 80 YEARS IN SHOW BUSINESS (II,967); GEORGE BURNS EARLY, EARLY, EARLY, CHRISTMAS SHOW (II,968); GEORGE BURNS IN NASHVILLE?? (II,969); GEORGE BURNS IN THE BIG TIME (I,1764); THE GEORGE BURNS SHOW (I,1765); THE GEORGE BURNS SPECIAL (II,970); GEORGE BURNS' 100TH BIRTHDAY PARTY (II,971); GEORGE BURNS' HOW TO LIVE TO BE 100 (II,972); GEORGE BURNS: AN HOUR OF JOKES AND SONGS (II,973); THE GOLDIE HAWN SPECIAL (II,1024); JACK BENNY'S NEW LOOK (I,2301); JACK BENNY'S SECOND FAREWELL SHOW (I,2302); THE KIDS FROM FAME (II,1392); A LOVE LETTER TO JACK BENNY (II,1540); THE MANY FACES OF COMEDY (I,2894); ONE MORE TIME (I,3386); THE OSMOND BROTHERS SHOW (I,3409); THE OTHER BROADWAY (II,1930); PARADE OF STARS (II,1954); THE PERRY COMO SPECIAL (I,3534); THE RICH LITTLE SHOW (II,2154); TIN PAN ALLEY TODAY (I,4521); TWO OF A KIND: GEORGE BURNS AND JOHN DENVER (II,2692); WENDY AND ME (I,4793)

BURNS, JACK
THE ANDY GRIFFITH SHOW (I,192); BONKERS! (II,348); THE BURNS AND SCHREIBER COMEDY HOUR (I,768); THE BURNS AND SCHREIBER COMEDY HOUR (I,769); CLASS OF '67 (I,970); THE ENTERTAINERS (I,1449); THE FABULOUS FUNNIES (II,801); GETTING TOGETHER (I,1784); THE JUD STRUNK SHOW (I,2462); OPERATION GREASEPAINT (I,3402); OUR PLACE (I,3416); THE PERRY

COMO VALENTINE SPECIAL (I,3541); THE PERRY COMO WINTER SHOW (I,3544); PERRY COMO'S SUMMER SHOW (I,3545); PERRY COMO'S WINTER SHOW (I,3548); THE SUMMER SMOTHERS BROTHERS SHOW (I,4281); THAT WAS THE YEAR THAT WAS (I,4424); THIS WILL BE THE YEAR THAT WILL BE (I,4466); WAIT TIL YOUR FATHER GETS HOME (I,4748)

BURNS, MICHAEL
BROCK'S LAST CASE (I,742); GIDGET GETS MARRIED (I,1796); HARRY'S BUSINESS (I,1958); IT'S A MAN'S WORLD (I,2258); OFF WE GO (I,3338); WAGON TRAIN (I,4747)

BURNS, MONA
ANOTHER WORLD (II,97); THE BRIGHTER DAY (I,728); GREEN ACRES (I,1884); ONE MAN'S FAMILY (I,3383); THREE STEPS TO HEAVEN (I,4485)

BURNS, PAUL
JOHNNY GUITAR (I,2427)

BURNS, RONNIE
FRONT ROW CENTER (I,1694); THE GEORGE BURNS AND GRACIE ALLEN SHOW (I,1763); THE GEORGE BURNS SHOW (I,1765); HAPPY (I,1938); NO WARNING (I,3301); TIME OUT FOR GINGER (I,4508)

BURNS, SANDRA
THE GEORGE BURNS AND GRACIE ALLEN SHOW (I,1763)

BURNS, STEPHAN
240-ROBERT (II,2689)

BURNS, STEPHEN
THE THORN BIRDS (II,2600)

BURNS, STEVE
THE WESTWIND (II,2766)

BURNS, TIMOTHY
TALES OF THE APPLE DUMPLING GANG (II,2536)

BURR, ANN
THE DOCTOR (I,1320)

BURR, ANNE
AS THE WORLD TURNS (II,110); CITY HOSPITAL (I,967); THE GREATEST GIFT (I,1880)

BURR, DONALD
ANNIE GET YOUR GUN (I,223)

BURR, FRITZI
HOLMES AND YOYO (II,1169); NEWMAN'S DRUGSTORE (II,1837)

BURR, LONNY
THE MICKEY MOUSE CLUB (I,3025); THE MOUSEKETEERS REUNION (II,1742)

BURR, RAYMOND
79 PARK AVENUE (II,2310); CENTENNIAL (II,477); CLIMAX! (I,976); COMMAND PERFORMANCE (I,1029); COUNTERPOINT (I,1059); EISCHIED (II,763); FORD TELEVISION THEATER (I,1634); GRUEN GUILD PLAYHOUSE (I,1892); IRONSIDE (I,2240); IRONSIDE (II,1246); JACK BENNY WITH GUEST STARS (I,2298); THE JORDAN CHANCE (II,1343); KEEP U.S. BEAUTIFUL (I,2519); KINGSTON: CONFIDENTIAL (II,1398); KINGSTON: THE POWER PLAY (II,1399); THE LUX VIDEO THEATER (I,2795); MALLORY: CIRCUMSTANTIAL EVIDENCE (II,1608); PERRY MASON (II,2025); PLAYHOUSE 90 (I,3623); RIDDLE AT 24000 (II,2171); STARS OVER HOLLYWOOD (I,4211); TALES OF TOMORROW (I,4352); THE UNEXPECTED (I,4674)

BURR, ROBERT
LOVE OF LIFE (I,2774); SEARCH FOR TOMORROW (II,2284); SOMERSET (I,4115)

BURRELL, JIMMY
AMERICAN MINSTRELS (I,171)

BURRELL, MARYEDITH
FAMILY TIES (II,819); FRIDAYS (II,926); REMINGTON STEELE (II,2130)

BURRELL, RUSTY
THE PEOPLE'S COURT (II,2001)

BURRELL, SHEILA
THE SIX WIVES OF HENRY VIII (I,4068)

BURRIS, NEAL
PEE WEE KING SHOW (I,3504)

BURROUGHS, JACKIE
THE PSYCHIATRIST: GOD BLESS THE CHILDREN (I,3685)

BURROWS, ABE
ABE BURROWS ALMANAC (I,9); FUN FOR 51 (I,1707); GIANT IN A HURRY (I,1790)

BURRUD, BILL
ANIMALS ARE THE FUNNIEST PEOPLE (II,81); HIGH ROAD TO ADVENTURE (I,2055); ISLANDS IN THE SUN (I,2244); THRILL HUNTERS (I,4490); TREASURE (I,4599); TRUE ADVENTURE (I,4616); WANDERLUST (I,4762)

BURSKY, ALAN
HOW TO SUCCEED IN BUSINESS WITHOUT REALLY TRYING (II,1197); THE PARTRIDGE FAMILY (II,1962)

BURTIN, MICHELLE
THE GIRL, THE GOLD WATCH AND DYNAMITE (II,1001); THE GIRL, THE GOLD WATCH AND EVERYTHING (II,1002)

BURTIS, ERNIE
THE RAY KNIGHT REVUE (I,3733)

BURTIS, MARILYN
YOU BET YOUR LIFE (I,4932)

BURTON, CHRIS
BENJI, ZAX AND THE ALIEN PRINCE (II,205)

BURTON, DONALD
THE ROADS TO FREEDOM (I,3802); WAR AND PEACE (I,4765)

BURTON, ENID
CASANOVA (II,451)

BURTON, IRV
THAT SECOND THING ON ABC (II,2572)

BURTON, JEFF
ROLL OUT! (I,3833)

BURTON, KATE
ELLIS ISLAND (II,768)

BURTON, LEVAR
BATTLE OF THE NETWORK STARS (II,164); BATTLE OF THE NETWORK STARS (II,167); CELEBRITY CHALLENGE OF THE SEXES 2 (II,466); CELEBRITY CHALLENGE OF THE SEXES 4 (II,468); THE CELEBRITY FOOTBALL CLASSIC (II,474); I LOVE LIBERTY (II,1215); THE PAUL LYNDE COMEDY HOUR (II,1978); ROOTS (II,2211)

BURTON, NORMANN
FORCE FIVE (II,907); MORNING STAR (I,3104); SANDBURG'S LINCOLN (I,3907); THE TED KNIGHT SHOW (II,2550); THE ULTIMATE IMPOSTER (II,2700); WONDER WOMAN (SERIES 2) (II,2829)

BURTON, RICHARD
BOB HOPE SPECIAL: THE BOB HOPE SPECIAL (II,338); BRIEF ENCOUNTER (I,724); THE BROADWAY OF LERNER AND LOEWE (I,735); CELANESE THEATER (I,881); DUPONT SHOW OF THE MONTH (I,1387); ELLIS ISLAND (II,768); THE GATHERING STORM (I,1741); THE TEMPEST (I,4391); A WORLD OF LOVE (I,4914)

BURTON, ROBERT
AMOS BURKE: WHO KILLED JULIE GREER? (I,181); AS THE WORLD TURNS (II,110); THE DAUGHTERS OF JOSHUA CABE RETURN (I,1170); FOR RICHER, FOR POORER (II,906); KING'S ROW (I,2563); STONESTREET: WHO KILLED THE CENTERFOLD MODEL? (II,2472); TEXAS (II,2566); THE VIRGINIAN (I,4737)

BURTON, ROD
THE FACTS (II,804)

BURTON, SARAH
CONCERNING MISS MARLOWE (I,1032); THE EDGE OF NIGHT (II,760); FROM THESE ROOTS (I,1688)

BURTON, SHELLY
CAR 54, WHERE ARE YOU? (I,842)

BURTON, SKIP
LASSIE (I,2624)

BURTON, STEVE
HEAR NO EVIL (II,1116)

BURTON, TOD
THE LINE-UP (I,2707)

BURTON, TONY
MITCHELL AND WOODS (II,1709)

BURTON, WARREN
ALL MY CHILDREN (II,39); ANOTHER WORLD (II,97); LILY FOR PRESIDENT (II,1478); THREE GIRLS THREE (II,2608); WILDER AND WILDER (II,2802)

BURTON, WENDELL
EAST OF EDEN (II,752); YOU'RE A GOOD MAN, CHARLIE BROWN (I,4962)

BUSCH, ROBERT
THE STAR MAKER (I,4193)

BUSE, KATHLEEN
THE HAMPTONS (II,1076)

BUSEY, GARY
THE TEXAS WHEELERS (II,2568)

BUSFIELD, TIMOTHY
REGGIE (II,2127); TRAPPER JOHN, M.D. (II,2654)

BUSH, GEORGE
BOB HOPE SPECIAL: BOB HOPE PRESENTS A CELEBRATION WITH STARS

OF COMEDY AND MUSIC (II,340)

BUSH, KAY
THE KENNY EVERETT VIDEO SHOW (II,1379)

BUSH, NAT
TOP SECRET (II,2647)

BUSH, OWEN
HERNDON AND ME (II,1139); NATIONAL LAMPOON'S TWO REELERS (II,1796); SIROTA'S COURT (II,2365)

BUSH, PAU
SKYFIGHTERS (I,4078)

BUSHKIN, JOE
THE BING CROSBY SPECIAL (I,472); A COUPLE OF JOES (I,1078); THE EDDIE CONDON SHOW (I,1408); FLOOR SHOW (I,1605)

BUSHMAN, FRANCIS X
THE BOB HOPE SHOW (I,558); HEDDA HOPPER'S HOLLYWOOD (I,2002)

BUSIA, AKOSHU
YOUNG LIVES (II,2866)

BUSINI, CHRISTINE
STRIPPERS (II,2481)

BUSSERT, MEG
CAMELOT (II,416)

BUSSY, CLAUDE
THE GENE KELLY PONTIAC SPECIAL (I,1750)

BUSWELL, NEVILLE
CORONATION STREET (I,1054)

BUTCH, DANNY
HAPPY DAYS (II,1084)

BUTCHER, JOSEPH
BIG FOOT AND WILD BOY (II,237)

BUTENUTH, CLAUDIA
THE LEGEND OF SILENT NIGHT (I,2655)

BUTKO, NATALIE
SANDY (I,3908)

BUTKUS, DICK
BLUE THUNDER (II,278); CASS MALLOY (II,454); THE DEAN MARTIN CELEBRITY ROAST (II,637); RICH MAN, POOR MAN—BOOK I (II,2161); SUNDAY FUNNIES (II,2493)

BUTLER, ALFRED
THE ALFRED AND JOSEPHINE BUTLER SHOW (I,114)

BUTLER, DAVID
ANOTHER WORLD (II,97)

BUTLER, DAWS
THE ALL-NEW POPEYE HOUR (II,49); THE BANANA SPLITS ADVENTURE HOUR (I,340); THE BEANY AND CECIL SHOW (I,374); THE BULLWINKLE SHOW (I,761); THE C.B. BEARS (II,406); THE FUNKY PHANTOM (I,1709); THE GALAXY GOOFUPS (II,954); HELP! IT'S THE HAIR BEAR BUNCH (I,2012); THE HOUNDCATS (I,2132); THE HUCKLEBERRY HOUND SHOW (I,2155); THE JETSONS (I,2376); LIPPY THE LION (I,2710); MR. MAGOO (I,3142); THE PETER POTAMUS SHOW (I,3568); THE QUICK DRAW MCGRAW SHOW (I,3710); THE ROMAN HOLIDAYS (I,3835); THE RUFF AND REDDY SHOW (I,3866); SCOOBY'S ALL-STAR LAFF-A-LYMPICS (II,2273); SPACE KIDDETTES (I,4143); SUPER PRESIDENT (I,4291); THE SUPER SIX (I,4292); TIME FOR BEANY (I,4505); THE WACKY RACES (I,4745); WALLY GATOR (I,4751); THE WOODY WOODPECKER SHOW (I,4906); YOGI BEAR (I,4928); YOGI'S GANG (II,2852); YOGI'S SPACE RACE (II,2853)

BUTLER, DEAN
BATTLE OF THE NETWORK STARS (II,175); CIRCUS OF THE STARS (II,537); GIDGET'S SUMMER REUNION (I,1799); LITTLE HOUSE ON THE PRAIRIE (II,1487); LITTLE HOUSE: A NEW BEGINNING (II,1488); LITTLE HOUSE: BLESS ALL THE DEAR CHILDREN (II,1489); LITTLE HOUSE: LOOK BACK TO YESTERDAY (II,1490); LITTLE HOUSE: THE LAST FAREWELL (II,1491)

BUTLER, DUKE
BUCK ROGERS IN THE 25TH CENTURY (II,383)

BUTLER, ED
THE SQUARE WORLD OF ED BUTLER (I,4171)

BUTLER, EUGENE
THE NIGHTENGALES (II,1850)

BUTLER, FRANCIS
BONINO (I,685)

BUTLER, HOLLY
20 MINUTE WORKOUT (II,2680); VELVET (II,2725)

BUTLER, JERRY
...AND BEAUTIFUL (I,184)

BUTLER, JOSEPHINE
THE ALFRED AND JOSEPHINE BUTLER SHOW (I,114)

BUTLER, LOIS
THE R.C.A. VICTOR SHOW (I,3737)

BUTLER, LOU
THE R.C.A. VICTOR SHOW (I,3737)

BUTLER, RICHARD
MATT HELM (II,1652); MICKEY SPILLANE'S MIKE HAMMER: MORE THAN MURDER (II,1693); REVENGE OF THE GRAY GANG (II,2146)

BUTLIN, JAN
THE BENNY HILL SHOW (II,207)

BUTLIN, LOIS
MASTER OF THE GAME (II,1647)

BUTRICK, MERRITT
SQUARE PEGS (II,2431)

BUTTERFIELD, BILLY
THE EDDIE CONDON SHOW (I,1408); FLOOR SHOW (I,1605)

BUTTERFIELD, HERB
THE CLAUDETTE COLBERT SHOW (I,971); THE HALLS OF IVY (I,1923)

BUTTERFIELD, MAGGIE
OUT OF OUR MINDS (II,1933)

BUTTERWORTH, DONNA
LITTLE LEATHERNECK (I,2713)

BUTTERWORTH, PETER
CARRY ON LAUGHING (II,448)

BUTTERWORTH, SHANE
THE BAD NEWS BEARS (II,134)

BUTTERWORTH, TYLER
MARRIED ALIVE (I,2923)

BUTTON, DICK
HANS BRINKER OR THE SILVER SKATES (I,1933); RAINBOW OF STARS (I,3717)

BUTTON, DOLORES
AH!, WILDERNESS (I,91)

BUTTONS, RED
THE BOB HOPE SHOW (I,633); THE BOB HOPE SHOW (I,659); BOB HOPE SPECIAL: BOB HOPE'S SUPER BIRTHDAY SPECIAL (II,319); THE DEAN MARTIN CELEBRITY ROAST (II,638); THE DEAN MARTIN CELEBRITY ROAST (II,639); DEAN MARTIN'S CELEBRITY ROAST (II,643); THE DOUBLE LIFE OF HENRY PHYFE (I,1361); FLANNERY AND QUILT (II,873); GEORGE BURNS CELEBRATES 80 YEARS IN SHOW BUSINESS (II,967); GEORGE M! (I,1772); GEORGE M! (II,976); HANSEL AND GRETEL (I,1934); THE LOVE BOAT (II,1535); MURDER AT NBC (A BOB HOPE SPECIAL) (I,3159); THE RED BUTTONS SHOW (I,3750); THE SUNSHINE BOYS (II,2496); SUSPENSE (I,4305); VAUDEVILLE (II,2722); VEGAS (II,2723); WONDER WOMAN (PILOT 2) (II,2827)

BUTTRAM, PAT
ARTHUR GODFREY IN HOLLYWOOD (I,281); DANNY THOMAS GOES COUNTRY AND WESTERN (I,1146); DOWN HOME (I,1366); EVIL ROY SLADE (I,1485); THE GENE AUTRY SHOW (I,1748); GREEN ACRES (I,1884); HOWDY (I,2150); THE MOUSE FACTORY (I,3113); WHEN THE WEST WAS FUN: A WESTERN REUNION (II,2780)

BUTTS, DOROTHY
BUSTING LOOSE (II,402)

BUTTS, K C
OUR MAN HIGGINS (I,3414)

BUX, KUDA
KUDA BUX, HINDU MYSTIC (I,2593)

BUXTON, FRANK
BATFINK (I,365); GET THE MESSAGE (I,1782)

BUXTON, SARAH
TOO GOOD TO BE TRUE (II,2643)

BUZBY, ZANE
BUMPERS (II,394)

BUZZ MILLER DANCERS, THE
THE MORT SAHL SPECIAL (I,3105)

BUZZI, RUTH
ALICE (II,33); ALVIN AND THE CHIPMUNKS (II,61); BAGGY PANTS AND THE NITWITS (II,135); THE BOB HOPE SHOW (I,643); DEAN MARTIN'S CELEBRITY ROAST (II,643); DOM DELUISE AND FRIENDS, PART 2 (II,702); THE FLIP WILSON COMEDY SPECIAL (II,886); GENE KELLY'S WONDERFUL WORLD OF GIRLS (I,1752); KEEP U.S. BEAUTIFUL (I,2519); THE LOST SAUCER (II,1524); THE MAD MAD MAD MAD WORLD OF THE SUPER BOWL (II,1586); PARADISE

(II,1955); ROWAN AND MARTIN'S LAUGH-IN (I,3856); ROWAN AND MARTIN'S LAUGH-IN (I,3857); SANDY IN DISNEYLAND (II,2254); THE SHAPE OF THINGS (II,2315); SINGLES (I,4062); THAT GIRL (II,2570); THE WONDERFUL WORLD OF GIRLS (I,4898)

BYAN, PAUL
BROADSIDE (I,732)

BYINGTON, SPRING
DECEMBER BRIDE (I,1215); THE DESILU REVUE (I,1238); I DREAM OF JEANNIE (I,2167); LARAMIE (I,2614); PULITZER PRIZE PLAYHOUSE (I,3692)

BYNER, JOHN
THE BEST OF SULLIVAN (II,217); BING CROSBY'S SUN VALLEY CHRISTMAS SHOW (I,475); BIZARRE (II,259); THE CAPTAIN AND TENNILLE IN NEW ORLEANS (II,431); THE GARRY MOORE SHOW (I,1740); THE JOHN BYNER COMEDY HOUR (I,2407); A LAST LAUGH AT THE 60'S (I,2627); LIFE'S MOST EMBARRASSING MOMENTS IV (II,1473); MCNAMARA'S BAND (II,1668); MCNAMARA'S BAND (II,1669); MURDER CAN HURT YOU! (II,1768); THE NANCY DUSSAULT SHOW (I,3222); THE PAT BOONE AND FAMILY EASTER SPECIAL (II,1970); THE PRACTICE (II,2076); SHEEHY AND THE SUPREME MACHINE (II,2322); THE SINGERS (I,4060); SINGLES (I,4062); SOAP (II,2392); SOMETHING ELSE (I,4117)

BYRD, CARL
CATALINA C-LAB (II,458)

BYRD, CAROLYN
THE FIRM (II,860)

BYRD, DAVID
MISSING PIECES (II,1706)

BYRD, JULIAN
FEEL THE HEAT (II,848)

BYRD, RALPH
DICK TRACY (I,1271); FIRESIDE THEATER (I,1580)

BYRD, RICHARD
THREE GIRLS THREE (II,2608)

BYRD, THOMAS
BOONE (II,351)

BYRD, TOM
ALL TOGETHER NOW (II,45)

BYRD-NETHERY, MIRIAM
MR. T. AND TINA (II,1758); PALMERSTOWN U.S.A. (II,1945)

BYRDS, THE
WHERE THE GIRLS ARE (I,4830)

BYRNE, GEORGE
INSPECTOR PEREZ (II,1237)

BYRNE, JACKIE
RETURN TO EDEN (II,2143)

BYRNE, MARTHA
THE HAMPTONS (II,1076); I DO, I DON'T (II,1212)

BYRNE, MICHAEL
ELLIS ISLAND (II,768); IF IT'S A MAN, HANG UP (I,2186); SMILEY'S PEOPLE (II,2383)

BYRNES, BURKE
GIDGET GETS MARRIED (I,1796); ONCE UPON A SPY (II,1898)

BYRNES, EDD
77 SUNSET STRIP (I,3988); GIRL ON THE RUN (I,1809); THE KILLING GAME (I,2540); MOBILE TWO (II,1718); SWEEPSTAKES (II,2514); VEGAS (II,2723)

BYRON, CAROL
OH, THOSE BELLS! (I,3345); WINDOW ON MAIN STREET (I,4873)

BYRON, JAMES
WINDOW ON MAIN STREET (I,4873)

BYRON, JEAN
BATMAN (I,366); THE BRADY GIRLS GET MARRIED (II,364); DOBIE GILLIS (I,1302); FULL CIRCLE (I,1702); HENRY FONDA PRESENTS THE STAR AND THE STORY (I,2016); MAYOR OF THE TOWN (I,2963); PAT PAULSEN'S HALF A COMEDY HOUR (I,3473); THE PATTY DUKE SHOW (I,3483); PEPSI-COLA PLAYHOUSE (I,3523); RANSOM FOR A DEAD MAN (I,3722); STUDIO '57 (I,4267); TV READERS DIGEST (I,4630)

CAAN, JAMES
CELEBRATION: THE AMERICAN SPIRIT (II,461); PLAYBOY'S 25TH ANNIVERSARY CELEBRATION (II,2054); RICKLES (II,2170)

CABAL, ROBERT
RAWHIDE (I,3727)

CABOT, BRUCE
GRUEN GUILD PLAYHOUSE (I,1892); HAVE GIRLS—WILL TRAVEL (I,1967); THE SLOWEST GUN IN THE WEST (I,4088); STARS OVER HOLLYWOOD (I,4211); TALES OF TOMORROW (I,4352)

CABOT, CEIL
BARNEY AND ME (I,358); I'VE HAD IT UP TO HERE (II,1221); KATE LOVES A MYSTERY (II,1367); THE NUT HOUSE (I,3320)

CABOT, SEBASTIAN
THE BEACHCOMBER (I,371); BRAVO DUKE (I,711); CHECKMATE (I,919); FAMILY AFFAIR (I,1519); GHOST STORY (I,1788); JACK THE RIPPER (II,1264); JANE WYMAN PRESENTS THE FIRESIDE THEATER (I,2345); LAST OF THE PRIVATE EYES (I,2628); MIRACLE ON 34TH STREET (I,3058); STUMP THE STARS (I,4273); SUSPENSE (I,4306); THE THREE MUSKETEERS (I,4480)

CADELL, AVA
THE JERK, TOO (II,1282)

CADELL, JEAN
DOUGLAS FAIRBANKS JR. PRESENTS THE RHEINGOLD THEATER (I,1364)

CADILLAC, FRANK
WONDER GIRL (I,4891)

CADOGAN, ALICE
AFTERMASH (II,23)

CADORETTE, MARY
THREE'S A CROWD (II,2613)

CADY, FRANK
THE ADVENTURES OF OZZIE AND HARRIET (I,71); THE ANDY GRIFFITH SHOW (I,193); GREEN ACRES (I,1884); NEW GIRL IN HIS LIFE (I,3265); PETTICOAT JUNCTION (I,3571); SUTTERS BAY (II,2508); THESE ARE THE DAYS (II,2585)

CAESAR, HARRY
THE BLUE KNIGHT (II,277); ROLL OUT! (I,3833)

CAESAR, IRVING
JOHNNY CARSON PRESENTS THE SUN CITY SCANDALS '72 (I,2421)

CAESAR, JIMMY
THE KEANE BROTHERS SHOW (II,1371)

CAESAR, SID
THE ADMIRAL BROADWAY REVUE (I,37); AMERICA 2100 (II,66); ANN-MARGRET SMITH (II,85); AS CAESAR SEES IT (I,294); THE BOB HOPE SHOW (I,509); THE BOB HOPE SHOW (I,633); CAESAR'S HOUR (I,790); DATELINE (I,1165); DOROTHY IN THE LAND OF OZ (II,722); GABRIEL KAPLAN PRESENTS THE SMALL EVENT (II,952); IT ONLY HURTS WHEN YOU LAUGH (II,1251); MARRIAGE—HANDLE WITH CARE (I,2921); THE MUNSTERS' REVENGE (II,1765); PERRY COMO'S CHRISTMAS IN AUSTRIA (II,2007); PINK LADY (II,2041); A SALUTE TO TELEVISION'S 25TH ANNIVERSARY (I,3893); THE SATURDAY NIGHT REVUE (I,3921); SID CAESAR INVITES YOU (I,4040); THE SID CAESAR SHOW (I,4042); THE SID CAESAR SHOW (I,4043); THE SID CAESAR SPECIAL (I,4044); TIPTOE THROUGH TV (I,4524); VARIETY: THE WORLD OF SHOW BIZ (I,4704); YOUR SHOW OF SHOWS (I,4957); YOUR SHOW OF SHOWS (II,2875)

CAFFREY, STEPHEN
HARD KNOX (II,1087)

CAGLE, WADE
STRATEGIC AIR COMMAND (I,4252)

CAGNEY, JAMES
THE BOB HOPE SHOW (I,550); NAVY LOG (I,3233); ROBERT MONTGOMERY PRESENTS YOUR LUCKY STRIKE THEATER (I,3809); TOM SNYDER'S CELEBRITY SPOTLIGHT (II,2632)

CAGNEY, JEANNE
QUEEN FOR A DAY (I,3705); STORY THEATER (I,4240); THE UNEXPECTED (I,4674)

CAHILL, BARRY
CENTENNIAL (II,477); THE YOUNG AND THE RESTLESS (II,2862)

CAHILL, KATHY
THE BOBBY DARIN AMUSEMENT COMPANY (I,672); THE SMOTHERS BROTHERS COMEDY HOUR (I,4095)

CAHILL, SALLY
PRISONER: CELL BLOCK H (II,2085)

CAHILL, THERESA
MELODY OF HATE (I,2999)

CAHN, SAMMY
HI, I'M GLEN CAMPBELL (II,1141); PLAY IT AGAIN, UNCLE SAM (II,2052)

CAHOUNE, ANDRE
THE MONTE CARLO SHOW (II,1726)

CAILLOU, ALAN
GOLIATH AWAITS (II,1025); MY THREE SONS (I,3205);

QUARK (II,2095); SHERLOCK HOLMES: THE HOUND OF THE BASKERVILLES (I,4011); TARZAN (I,4361)

CAIN, JEFF
MARGE AND JEFF (I,2908)

CAIN, JERRY
THE BOOK OF LISTS (II,350)

CAIN, LISA
KENNEDY (II,1377)

CAIN, MADELYN
WHIZ KIDS (II,2790)

CAIN, SHIRLEY
COLOR HIM DEAD (II,554)

CAIN, WILLIAM
ALL MY CHILDREN (II,39)

CAINE, HOWARD
CAPTAIN BRASSBOUND'S CONVERSION (I,828); THE GOLDEN AGE OF TELEVISION (II,1016); HOGAN'S HEROES (I,2069); A TIME TO LIVE (I,4510)

CAINE, MICHAEL
RICKLES (II,2170)

CALA, FRANK
THAT'S CAT (II,2576)

CALCOTE, CAROLYN
THE ROWAN AND MARTIN REPORT (II,2221)

CALDER, BOB
JACQUES FRAY'S MUSIC ROOM (I,2329)

CALDER, KING
THE MAN FROM EVERYWHERE (I,2864); MARTIN KANE, PRIVATE EYE (I,2928); READY FOR THE PEOPLE (I,3739)

CALDERT, BILL
ADAMS HOUSE (II,9)

CALFA, DON
ALLISON SIDNEY HARRISON (II,54); THE BEACH GIRLS (II,191); LEGMEN (II,1458); THE NATIONAL SNOOP (II,1797); PARK PLACE (II,1958); RIDING HIGH (II,2173)

CALHOUN, JEFF
DEBBY BOONE. . . ONE STEP CLOSER (II,655)

CALHOUN, RORY
THE BLUE AND THE GRAY (II,273); CAPITOL (II,426); CIRCLE OF FEAR (I,958); CLIMAX! (I,976); THE DESILU REVUE (I,1238); FLATBED ANNIE AND SWEETIEPIE: LADY TRUCKERS (II,876); FORD TELEVISION THEATER (I,1634); LUXURY LINER (I,2796); POTTSVILLE (II,2072); SCREEN DIRECTOR'S PLAYHOUSE (I,3946); THE TEXAN (I,4413); WESTERN STAR THEATER (I,4798); ZANE GREY THEATER (I,4979)

CALI, JOSEPH
FLATBUSH (II,877); TODAY'S FBI (II,2628)

CALIFORNIANS, THE
THE CBS NEWCOMERS (I,879)

CALIRE, DOROTHY
JERRY MAHONEY'S CLUB HOUSE (I,2371)

CALIRI, JON
DOUBLE TROUBLE (II,723); SQUARE PEGS (II,2431)

CALL, ANTHONY
THE EDGE OF NIGHT (II,760); THE GUIDING LIGHT (II,1064); ONE LIFE TO LIVE (II,1907)

CALL, ED
ETHEL IS AN ELEPHANT (II,783); NORMA RAE (II,1860)

CALLAHAN, BILL
JACK CARTER AND COMPANY (I,2305); THE JACK CARTER SHOW (I,2306)

CALLAHAN, JACK
TWIGS (II,2684)

CALLAHAN, JAMES
CONVOY (I,1043); THE GOVERNOR AND J.J. (I,1863); HERCULE POIROT (I,2021); THE JUGGLER OF NOTRE DAME (II,1350); LITTLE HOUSE: LOOK BACK TO YESTERDAY (II,1490); MCNAUGHTON'S DAUGHTER (II,1670); THE RUNAWAYS (II,2231); STUDS LONIGAN (II,2484); TRAVIS LOGAN, D.A. (I,4598); WENDY AND ME (I,4793)

CALLAHAN, MATT
GEORGE WASHINGTON (II,978)

CALLAN, K
CUTTER TO HOUSTON (II,612); THE FOUR OF US (II,914); IKE (II,1223); JAMES AT 15 (II,1271); JOE'S WORLD (II,1305); LEADFOOT (II,1452); MARRIED: THE FIRST YEAR (II,1634); ONE DAY AT A TIME (II,1900); PRIVATE BENJAMIN (II,2087)

CALLAN, MICHAEL
JOURNEY TO THE UNKNOWN (I,2457); KISS ME, KATE (I,2569); OCCASIONAL WIFE (I,3325); SCRUPLES (II,2280); YOUNG HEARTS (II,2865)

CALLARD, KAY
SECRET FILE, U.S.A. (I,3963)

CALLAS, CHARLIE
THE ALAN KING SHOW (I,101); THE ALL-STAR SALUTE TO MOTHER'S DAY (II,51); THE ANDY WILLIAMS SHOW (I,210); BUNGLE ABBEY (II,396); THE CARPENTERS. . . SPACE ENCOUNTERS (II,447); CIRCUS OF THE STARS (II,532); CIRCUS OF THE STARS (II,537); COMEDY IS KING (I,1019); DEAN MARTIN'S CELEBRITY ROAST (II,643); THE DES O'CONNOR SHOW (I,1235); DOM DELUISE AND FRIENDS, PART 2 (II,702); FEMALE INSTINCT (I,1564); THE KLOWNS (I,2574); THE KOPYCATS (I,2580); THE MAD MAD MAD MAD WORLD OF THE SUPER BOWL (II,1586); THE MICKIE FINNS FINALLY PRESENT HOW THE WEST WAS LOST (II,1695); ROOSTER (II,2210); SINATRA—THE FIRST 40 YEARS (II,2359); SWITCH (II,2519); SWITCH (II,2520); TAKE ONE STARRING JONATHAN WINTERS (II,2532)

CALLAWAY, BILL
CAT BALLOU (I,870); THE CATTANOOGA CATS (I,874); THE DRAK PACK (II,733); HELP! IT'S THE HAIR BEAR BUNCH (I,2012); NO SOAP, RADIO (II,1856); SEALAB 2020 (I,3949); THE SMURFS (II,2385); SUPER FRIENDS (II,2497); TROLLKINS (II,2660); THE WORLD'S GREATEST SUPER HEROES (II,2842)

CALLAWAY, CHERYL
TOM, DICK, AND HARRY (I,4536)

CALLINAN, DICK
KEY TORTUGA (II,1383)

CALLOWAY, CAB
THE LITTLEST ANGEL (I,2726); UPTOWN (II,2710)

CALLOWAY, CHRIS
THE DOCTORS (II,694)

CALLOWAY, NORTHERN J
SESAME STREET (I,3982)

CALLOWAY, THOMAS
WORKING STIFFS (II,2834)

CALLOWAY, WILLIAM
GUESS WHO'S COMING TO DINNER? (II,1063); THIS IS KATE BENNETT (II,2597) WOLF ROCK TV (II,2815)

CALNAN, MATILDA
THE CABOT CONNECTION (II,408)

CALVERT, HENRY
HOT L BALTIMORE (II,1187)

CALVIN, DONANN
SEARCH AND RESCUE: THE ALPHA TEAM (II,2283)

CALVIN, HENRY
THE YEOMAN OF THE GUARD (I,4926); ZORRO (I,4982)

CALVIN, JOHN
THE LIVING END (I,2731); THE PAUL LYNDE SHOW (I,3489); SITCOM (II,2367); TALES OF THE GOLD MONKEY (II,2537); WINTER KILL (II,2810); THE WONDERFUL WORLD OF PHILIP MALLEY (II,2831)

CALYPSO KINGS, THE
THE SPIKE JONES SHOW (I,4156)

CAMACHO, CORINNE
GIDGET GETS MARRIED (I,1796); PLANET EARTH (I,3615)

CAMBRIDGE, EDMUND
EVIL ROY SLADE (I,1485)

CAMBRIDGE, GODFREY
THE BEST ON RECORD (I,409); THE FURST FAMILY OF WASHINGTON (I,1718); A LAST LAUGH AT THE 60'S (I,2627); THE MANY FACES OF COMEDY (I,2894)

CAMBRIDGE, PHIL
THE GRAY GHOST (I,1867)

CAMEN, JOEY
THE RIGHTEOUS APPLES (II,2174)

CAMERON, BUD
NEW GIRL IN HIS LIFE (I,3265)

CAMERON, CHRISTINE
ANOTHER WORLD (II,97)

CAMERON, DEAN
SPENCER (II,2420); THINGS ARE LOOKING UP (II,2588)

CAMERON, EARL
ESPIONAGE (I,1464)

CAMERON, HUGH
THE FORTUNE HUNTER (I,1646)

CAMERON, JOANNA
HIGH RISK (II,1146); ISIS (II,1248); NIGHT GAMES (II,1845)

CAMERON, KIMBERLY
B.J. AND THE BEAR (II,127)

CAMERON, KIRK
MICKEY SPILLANE'S MIKE HAMMER: MORE THAN MURDER (II,1693); TWO MARRIAGES (II,2691)

CAMERON, LISA
ANOTHER WORLD (II,97); AS THE WORLD TURNS (II,110)

CAMERON, ROD
CITY DETECTIVE (I,966); CORONADO 9 (I,1053); CROSSROADS (I,1105); HAVE GIRLS—WILL TRAVEL (I,1967); STAR ROUTE (I,4196); STATE TROOPER (I,4214); WHEN THE WEST WAS FUN: A WESTERN REUNION (II,2780)

CAMERON, SHIRLEY
1915 (II,1853)

CAMP, COLLEEN
RICH MAN, POOR MAN—BOOK II (II,2162)

CAMP, HAMILTON
A DOG'S LIFE (II,697); DON'T CALL ME MAMA ANYMORE (I,1350); FAMILY IN BLUE (II,818); HE AND SHE (I,1985); I'LL NEVER FORGET WHAT'S HER NAME (II,1217); JUST OUR LUCK (II,1360); THE NASHVILLE PALACE (II,1792); POTTSVILLE (II,2072); PRIME TIMES (II,2083); THE SMURFS (II,2385); STORY THEATER (I,4241); THREE'S COMPANY (II,2614); TOO CLOSE FOR COMFORT (II,2642); TURN ON (I,4625); WINDOWS, DOORS AND KEYHOLES (II,2806)

CAMP, HELEN PAGE
13 QUEENS BOULEVARD (II,2591); DADDY'S GIRL (I,1124); FATHER O FATHER (II,842); FISHERMAN'S WHARF (II,865); LASSIE: THE NEW BEGINNING (II,1433); LAVERNE AND SHIRLEY (II,1446); THE NEW MAVERICK (II,1822); RICHIE BROCKELMAN, PRIVATE EYE (II,2167); RICHIE BROCKELMAN: MISSING 24 HOURS (II,2168); THIS IS KATE BENNETT (II,2597); THE TONY RANDALL SHOW (II,2640); WENDY HOOPER—U. S. ARMY (II,2764)

CAMP, JOANNE
KENNEDY (II,1377)

CAMPANELLA, FRANK
ARSENIC AND OLD LACE (I,270); COREY: FOR THE PEOPLE (II,575); DECOY (I,1217); GIBBSVILLE (II,988); MATT HELM (II,1652); MURDOCK'S GANG (I,3164); SKAG (II,2374); THE TURNING POINT OF JIM MALLOY (II,2670)

CAMPANELLA, JOSEPH
ESPIONAGE (I,1464); EXPOSE (I,1491); THE LAWYERS (I,2646); MANNIX (II,1624); THE NURSES (I,3318); ONE DAY AT A TIME (II,1900); OWEN MARSHALL: COUNSELOR AT LAW (I,3436); POLICE WOMAN: THE GAMBLE (II,2064); RETURN TO FANTASY ISLAND (II,2144); THIS IS YOUR LIFE (II,2598); THE TROUBLE WITH PEOPLE (I,4611); THE WHOLE WORLD IS WATCHING (I,4851)

CAMPANELLA, ROY
TEXACO COMMAND PERFORMANCE (I,4408)

CAMPBELL, ALAN
COUNTERATTACK: CRIME IN AMERICA (II,581); THREE'S A CROWD (II,2613)

CAMPBELL, ANNA
MASTER OF THE GAME (II,1647)

CAMPBELL, ANNE
MOMENT OF TRUTH (I,3084)

CAMPBELL, ARCHIE
HEE HAW (II,1123)

CAMPBELL, CAROLE ANN
THE HARDY BOYS AND THE MYSTERY OF GHOST FARM (I,1950); THE HARDY BOYS AND THE MYSTERY OF THE APPLEGATE TREASURE (I,1951)

CAMPBELL, CAROLEE
THE DOCTORS (II,694)

CAMPBELL, CARRIE
HI, I'M GLEN CAMPBELL (II,1141)

CAMPBELL, CATHERINE
THE GOSSIP COLUMNIST (II,1045)

CAMPBELL, CHARLES L
THE CHISHOLMS (II,512)

CAMPBELL, CHERYL
PENNIES FROM HEAVEN (II,1994); THE SEVEN DIALS MYSTERY (II,2308)

CAMPBELL, DARREL
ANOTHER LIFE (II,95)

CAMPBELL, DEAN
QUADRANGLE (I,3703)

CAMPBELL, DUANE
ALICE (II,33)

CAMPBELL, FLORA
A DATE WITH JUDY (I,1162); FARAWAY HILL (I,1532); JANE EYRE (I,2337); THE SEEKING HEART (I,3975); THE VALIANT (I,4697); VALIANT LADY (I,4695)

CAMPBELL, GLEN
100 YEARS OF GOLDEN HITS (II,1905); ANNE MURRAY'S WINTER CARNIVAL. . . FROM QUEBEC (II,90); THE BEACH BOYS 20TH ANNIVERSARY (II,188); THE BOB HOPE SHOW (I,628); THE BOB HOPE SHOW (I,627); THE BOB HOPE SHOW (I,655); THE BOB HOPE SHOW (I,661); BOB HOPE SPECIAL: BOB HOPE'S ALL-STAR COMEDY BIRTHDAY PARTY AT WEST POINT (II,298); BOB HOPE SPECIAL: BOB HOPE PRESENTS A CELEBRATION WITH STARS OF COMEDY AND MUSIC (II,340); BOB HOPE SPECIAL: THE BOB HOPE COMEDY SPECIAL (II,332); BOB HOPE SPECIAL: THE BOB HOPE SPECIAL (II,333); BURT REYNOLDS' LATE SHOW (I,777); THE CAPTAIN AND TENNILLE SONGBOOK (II,432); CHRISTMAS IN DISNEYLAND (II,519); A COUNTRY CHRISTMAS (II,584); COUNTRY COMES HOME (II,586); COUNTRY COMES HOME (II,585); COUNTRY STARS OF THE 70S (II,594); THE DIONNE WARWICK SPECIAL (I,1289); DISNEY WORLD—A GALA OPENING—DISNEYLAND EAST (I,1292); DOUG HENNING'S WORLD OF MAGIC II (II,726); FELICIANO—VERY SPECIAL (I,1560); THE FIFTH DIMENSION SPECIAL: AN ODYSSEY IN THE COSMIC UNIVERSE OF PETER MAX (I,1571); FIFTY YEARS OF COUNTRY MUSIC (II,853); GLEN CAMPBELL . . . DOWN HOME—DOWN UNDER (II,1008); GLEN CAMPBELL AND FRIENDS: THE SILVER ANNIVERSARY (II,1005); THE GLEN CAMPBELL GOODTIME HOUR (I,1820); THE GLEN CAMPBELL MUSIC SHOW (II,1006); THE GLEN CAMPBELL MUSIC SHOW (II,1007); THE GLEN CAMPBELL SPECIAL: THE MUSICAL WEST (I,1821); HI, I'M GLEN CAMPBELL (II,1141); JERRY REED AND SPECIAL FRIENDS (II,1285); JOHN DENVER—THANK GOD I'M A COUNTRY BOY (II,1317); JOHNNY CASH: COWBOY HEROES (II,1334); OPRYLAND: NIGHT OF STARS AND FUTURE STARS (II,1921); THE RETURN OF THE SMOTHERS BROTHERS (I,3777); THE RICH LITTLE SHOW (II,2154); SHINDIG (I,4013); SOLID GOLD '79 (II,2396); STAR ROUTE (I,4196); THE SUMMER SMOTHERS BROTHERS SHOW (I,4281); THE VERY FIRST GLEN CAMPBELL SPECIAL (I,4714)

CAMPBELL, GRAEME
THE RETURN OF CHARLIE CHAN (II,2136)

CAMPBELL, J KENNETH
ANOTHER WORLD (II,97); THE COPS AND ROBIN (II,573); GEORGE WASHINGTON (II,978)

CAMPBELL, JAMES
THE TALE OF BEATRIX POTTER (II,2533)

CAMPBELL, JERRY
SIX PACK (II,2370)

CAMPBELL, JIMMY
THE MATT DENNIS SHOW (I,2948)

CAMPBELL, JOHN
THE GIRLS (I,1811); KUDZU (II,1415)

CAMPBELL, JOY
THE BUGALOOS (I,757); LIDSVILLE (I,2679)

CAMPBELL, KAREN
ANOTHER WORLD (II,97)

CAMPBELL, KAY
ALL MY CHILDREN (II,39); THE EDGE OF NIGHT (II,760); THE GUIDING LIGHT (II,1064)

CAMPBELL, KELLY
AS THE WORLD TURNS (II,110)

CAMPBELL, KENNETH
CHIEFS (II,509)

CAMPBELL, LAURA
GENERAL HOSPITAL (II,964)

CAMPBELL, NAOMI
JANE EYRE (I,2337)

CAMPBELL, NICHOLAS
THE HITCHHIKER (II,1157)

CAMPBELL, PAUL

THE THREE MUSKETEERS (I,4480)

CAMPBELL, SHAWN
THE DOCTORS (II,694)

CAMPBELL, TOM
CAMOUFLAGE (II,417)

CAMPBELL, TONI
I REMEMBER MAMA (I,2176)

CAMPBELL, VIRGINIA
HAY FEVER (I,1979)

CAMPBELL, WESLEY
HI, I'M GLEN CAMPBELL (II,1141)

CAMPBELL, WILLIAM
CANNONBALL (I,822); DYNASTY (II,746); MR. AND MRS. COP (II,1748)

CAMPELL, GARY
THE SECRET STORM (I,3969)

CAMPO, TONY
GENERAL HOSPITAL (II,964)

CAMPOS, RAFAEL
THE BLUE MEN (I,492); CENTENNIAL (II,477); THE HAT OF SERGEANT MARTIN (I,1963); RHODA (II,2151); V (II,2713); V: THE FINAL BATTLE (II,2714)

CAMPOS, VICTOR
ARCHER—FUGITIVE FROM THE EMPIRE (II,103); CADE'S COUNTY (I,787); DOCTORS HOSPITAL (II,691); THE IMPOSTER (II,1225); ONE OF OUR OWN (II,1910)

CANADA, RON
GEORGE WASHINGTON (II,978)

CANARY, DAVID
ANOTHER WORLD (II,97); BONANZA (II,347); THE DAIN CURSE (II,613); MELVIN PURVIS: G-MAN (II,1679); PEYTON PLACE (I,3574); SEARCH FOR TOMORROW (II,2284)

CANDY, JOHN
BIG CITY COMEDY (II,233); THE NEW SHOW (II,1828); SECOND CITY TELEVISION (II,2287); WELCOME TO THE FUN ZONE (II,2763)

CANFIELD, MARY GRACE
BEWITCHED (I,418); FAMILY (II,813); GREEN ACRES (I,1884); THE GUARDSMAN (I,1893); THE HATHAWAYS (I,1965); HEAVEN HELP US (I,1995); JOHNNY BELINDA (I,2417); POOR MR. CAMPBELL (I,3647)

CANIFF, MILTON
DATELINE (I,1165)

CANN, STAN
THE NOONDAY SHOW (II,1859)

CANNIBAL AND THE HEADHUNTERS
IT'S WHAT'S HAPPENING BABY! (I,2278)

CANNING, JAMES
100 CENTRE STREET (II,1901); BEST FRIENDS (II,213); THE JORDAN CHANCE (II,1343); KINGSTON: THE POWER PLAY (II,1399)

CANNON, ALFRED
THE D.A.: MURDER ONE (I,1122)

CANNON, ANTHONY
THE DOCTORS (II,694); LAND OF HOPE (II,1424)

CANNON, DYAN
THE BOB HOPE SHOW (I,653); BOB HOPE SPECIAL: THE BOB HOPE SPECIAL (II,334); FULL CIRCLE (I,1702); MASTER OF THE GAME (II,1647)

CANNON, GLENN
THE ISLANDER (II,1249); MAGNUM, P.I. (II,1601)

CANNON, J D
ALIAS SMITH AND JONES (I,118); BANJO HACKETT: ROAMIN' FREE (II,141); BEYOND WITCH MOUNTAIN (II,228); CALL TO GLORY (II,413); CANNON (I,820); THE COURT-MARTIAL OF GENERAL GEORGE ARMSTRONG CUSTER (I,1081); IKE (II,1223); LADY LUCK (I,2604); MCCLOUD (II,1660); THE MISADVENTURES OF SHERIFF LOBO (II,1704); NEITHER ARE WE ENEMIES (I,3247); PROFILES IN COURAGE (I,3678); ROOSTER (II,2210); SAM HILL: WHO KILLED THE MYSTERIOUS MR. FOSTER? (I,3896); TESTIMONY OF TWO MEN (II,2564); U.M.C. (I,4666); WEAPONS MAN (I,4783)

CANNON, JIMMY
THE LES CRANE SHOW (I,2657)

CANNON, KATHERINE
THE BLACK SHEEP SQUADRON (II,262); THE CONTENDER (II,571); FATHER MURPHY (II,841); GABE AND WALKER (II,950); THE SURVIVORS (I,4301)

CANNON, LYLE
TROUBLE IN HIGH TIMBER COUNTRY (II,2661)

CANNON, MAUREEN
BROADWAY OPEN HOUSE (I,736); THE GOODYEAR REVUE (I,1856); THE PATRICIA BOWAN SHOW (I,3475); PAUL WHITEMAN'S SATURDAY NIGHT REVUE (I,3490); PAUL WHITEMAN'S TV TEEN CLUB (I,3491); SCHOOL HOUSE (I,3937)

CANNON, MICHAEL PHILIP
TOO CLOSE FOR COMFORT (II,2642)

CANNON, POPPY
HOME (I,2099)

CANNON, WILLIAM THOMAS
TOO CLOSE FOR COMFORT (II,2642)

CANOVA, DIANA
THE ALL-STAR SALUTE TO MOTHER'S DAY (II,51); BATTLE OF THE NETWORK STARS (II,169); THE CELEBRITY FOOTBALL CLASSIC (II,474); DINAH AND HER NEW BEST FRIENDS (II,676); FOOT IN THE DOOR (II,902); I'M A BIG GIRL NOW (II,1218); THE LOVE BOAT II (II,1533); NIGHT PARTNERS (II,1847); PEKING ENCOUNTER (II,1991); SOAP (II,2392)

CANOVA, JUDY
THE ALL-STAR SALUTE TO MOTHER'S DAY (II,51); CAP'N AHAB (I,824); THE COLGATE COMEDY HOUR (I,997); LI'L ABNER (I,2701)

CANTERBURY, TED
AND HERE'S THE SHOW (I,187)

CANTO, DAL
ANNE MURRAY'S LADIES' NIGHT (II,89)

CANTOR, CHARLIE
MAGGIE (I,2810); WHERE'S RAYMOND? (I,4837)

CANTOR, EDDIE
THE BOB HOPE SHOW (I,509); THE COLGATE COMEDY HOUR (I,997); THE EDDIE CANTOR COMEDY THEATER (I,1407); GEORGE BURNS IN THE BIG TIME (I,1764); THE MORT SAHL SPECIAL (I,3105)

CANTOR, MARILYN
THE SKY'S THE LIMIT (I,4080)

CANTOR, MAX
DINER (II,679)

CANTOR, NAT
THE ROBBINS NEST (I,3806); ROBERT Q'S MATINEE (I,3812)

CANTU-PRIMO, DOLORES
FRED AND BARNEY MEET THE SHMOO (II,918); THE NEW SHMOO (II,1827)

CAPERS, VIRGINIA
FEATHERSTONE'S NEST (II,847); JULIA (I,2476); THE SEEKERS (I,3974); WILLOW B: WOMEN IN PRISON (II,2804)

CAPIKA, CAROL
CRAZY TIMES (II,602)

CAPIZZI, BILL
FARRELL: FOR THE PEOPLE (II,834)

CAPLAN, TWINK
LONDON AND DAVIS IN NEW YORK (II,1512)

CAPOTE, TRUMAN
A CHRISTMAS MEMORY (I,948)

CAPP, AL
ANYONE CAN WIN (I,235); WHAT'S THE STORY? (I,4821)

CAPPELL, PETER
EYE WITNESS (I,1494)

CAPRE, ANGELA
B.J. AND THE BEAR (II,127)

CAPRI, ANNA
ROOM FOR ONE MORE (I,3842)

CAPTAIN AND TENNILLE
BOB HOPE SPECIAL: BOB HOPE'S ALL-STAR COMEDY TRIBUTE TO VAUDEVILLE (II,302); BOB HOPE SPECIAL: BOB HOPE'S BICENTENNIAL STAR SPANGLED SPECTACULAR (II,306); DICK CLARK'S GOOD OL' DAYS: FROM BOBBY SOX TO BIKINIS (II,671); THE MUHAMMED ALI VARIETY SPECIAL (II,1762); PERRY COMO'S CHRISTMAS IN MEXICO (II,2009)

CAPUANO, SAM
OPERATION GREASEPAINT (I,3402)

CARA, IRENE
BOB HOPE SPECIAL: BOB HOPE GOES TO COLLEGE (II,283); IRENE (II,1245); ROOTS: THE NEXT GENERATIONS (II,2213)

CARAFOTES, PAUL

NOW WE'RE COOKIN' (II,1866)

CARBONE, ANTHONY
THE SEEKERS (I,3974)

CARBONETTO, LINDA
THE ICE PALACE (I,2182)

CARD, KATHRYN
THE CHARLIE FARRELL SHOW (I,913); I LOVE LUCY (I,2171); THE RED SKELTON SHOW (I,3755)

CARDI, PAT
HAZEL (I,1982); IT'S ABOUT TIME (I,2263)

CARDILLE, LORI
THE EDGE OF NIGHT (II,760); PAROLE (II,1960); RYAN'S HOPE (II,2234)

CARDINALE, CLAUDIA
CIRCUS OF THE STARS (II,530)

CARDINI
FESTIVAL OF MAGIC (I,1566)

CARDONA, ANNETTE
THE MAN FROM ATLANTIS (II,1615)

CARERA, CHRISTINE
BLUE LIGHT (I,491)

CAREY JR, HARRY
THE FURTHER ADVENTURES OF SPIN AND MARTY (I,1720); KATE BLISS AND THE TICKER TAPE KID (II,1366); THE NEW ADVENTURES OF SPIN AND MARTY (I,3254); SPIN AND MARTY (I,4158); TEXAS JOHN SLAUGHTER (I,4414); WILD TIMES (II,2799)

CAREY, CHRISTOPHER
PLANET EARTH (I,3615)

CAREY, GEOFFREY
MISTRAL'S DAUGHTER (II,1708)

CAREY, GEORGE E
GENERAL HOSPITAL (II,964)

CAREY, JOSIE
THE CHILDREN'S CORNER (I,938)

CAREY, JOYCE
FATHER DEAR FATHER (II,837)

CAREY, MACDONALD
CELANESE THEATER (I,881); CONDOMINIUM (II,566); DAYS OF OUR LIVES (II,629); DOCTOR CHRISTIAN (I,1308); GIDGET GETS MARRIED (I,1796); THE GIRL, THE GOLD WATCH AND EVERYTHING (II,1002); THE GREEN FELT JUNGLE (I,1885); HOLLYWOOD OPENING NIGHT (I,2087); LAST OF THE PRIVATE EYES (I,2628); LOCK UP (I,2737); MARKHAM (I,2916); MIRACLE ON 34TH STREET (I,3056); ON TRIAL (I,3363); PURSUIT (I,3700); THE REBELS (II,2121); ROOTS (II,2211); STRANGER IN OUR HOUSE (II,2477); TOP OF THE HILL (II,2645)

CAREY, MICHELE
DEATH RAY 2000 (II,654); DELTA COUNTY, U.S.A. (II,657); THE LEGEND OF THE GOLDEN GUN (II,1455); A MAN CALLED SLOANE (II,1613); MISSION: IMPOSSIBLE (I,3067); THE NORLISS TAPES (I,3305); ROOSTER (II,2210); SAVAGE (I,3924); THE SIX MILLION DOLLAR MAN (I,4067)

CAREY, OLIVE
MAISIE (I,2833); MR. ADAMS AND EVE (I,3121); THE TEENAGE IDOL (I,4375)

CAREY, PHILIP
BRIGHT PROMISE (I,727); FBI CODE 98 (I,1550); LAREDO (I,2615); THE LEATHERNECKS (I,2647); ONE LIFE TO LIVE (II,1907); PHILIP MARLOWE (I,3584); TALES OF THE 77TH BENGAL LANCERS (I,4348)

CAREY, RICHENDA
LACE (II,1418)

CAREY, ROBERT
MALIBU (II,1607)

CAREY, RON
20TH CENTURY FOLLIES (I,4641); BARNEY MILLER (II,154); THE CORNER BAR (I,1052); THE GARRY MOORE SHOW (I,1740); JOHNNY GARAGE (II,1337); THE MELBA MOORE-CLIFTON DAVIS SHOW (I,2996); THE MONTEFUSCOS (II,1727); PEEPING TIMES (II,1990); PUMPBOYS AND DINETTES ON TELEVISION (II,2090); THE WONDERFUL WORLD OF AGGRAVATION (I,4894)

CAREY, TIMOTHY
EAST OF EDEN (II,752); NIGHTSIDE (II,1851)

CAREY-JONES, SELENA
LOVE IN A COLD CLIMATE (II,1539)

CARGILL, PATRICK
FATHER DEAR FATHER (II,837)

CARIDI, CARMINE
ALICE (II,33); FAME (II,812); GOOD PENNY (II,1036); PETE 'N' TILLIE (II,2027); PHYLLIS (II,2038)

CARL JABLONSKI DANCERS, THE
THE JIM STAFFORD SHOW (II,1291); LIGHTS, CAMERA, MONTY! (II,1475); THE NBC FOLLIES (I,3241)

CARL, ADAM
FAMILY TIES (II,819)

CARL, GEORGE
CLOWNAROUND (I,980)

CARLE, FRANKIE
FRANKIE CARLE TIME (I,1674)

CARLETON, CLAIRE
CIMARRON CITY (I,953); HEY MULLIGAN (I,2040); NEW GIRL IN HIS LIFE (I,3265)

CARLIN, GEORGE
100 YEARS OF GOLDEN HITS (II,1905); AWAY WE GO (I,317); THE FLIP WILSON COMEDY SPECIAL (II,886); GEORGE CARLIN AT CARNEGIE HALL (II,974); KRAFT SUMMER MUSIC HALL (I,2590); MAC DAVIS. . . SOUNDS LIKE HOME (II,1582); THE MAD MAD MAD MAD WORLD OF THE SUPER BOWL (II,1586); THE PERRY COMO SPRINGTIME SHOW (I,3537); PERRY COMO'S HAWAIIAN HOLIDAY (II,2017); STAR SEARCH (II,2437); THAT GIRL (II,2570); TONY ORLANDO AND DAWN (II,2639); A TRIBUTE TO "MR. TELEVISION" MILTON BERLE (II,2658)

CARLIN, JOHN
EGAN (I,1419)

CARLIN, LYNN
BRAVO TWO (II,368); JAMES AT 15 (II,1270); JAMES AT 15 (II,1271); THE LIVES OF JENNY DOLAN (II,1502); NOT UNTIL TODAY (II,1865); STRIKE FORCE (II,2480)

CARLIN, PAUL
RYAN'S HOPE (II,2234)

CARLIN, TOM
TODAY IS OURS (I,4531)

CARLISLE, JOHN
THE OMEGA FACTOR (II,1888)

CARLISLE, KITTY
CELEBRITY TIME (I,887); HOLIDAY (I,2074); THE NASH AIRFLYTE THEATER (I,3227); TO TELL THE TRUTH (II,2625); WHAT'S GOING ON? (I,4814)

CARLISLE, THOMAS
BRIGADOON (I,726)

CARLO, EAST
FARRELL: FOR THE PEOPLE (II,834); THE TARZAN/LONE RANGER/ZORRO ADVENTURE HOUR (II,2543)

CARLO, ISMAEL
SANTA BARBARA (II,2258)

CARLON, FRAN
AS THE WORLD TURNS (II,110); THE HAMPTONS (II,1076); PORTIA FACES LIFE (I,3653)

CARLSEN, ALAN
THE ADAMS CHRONICLES (II,8)

CARLSON, CASEY
IT'S MAGIC, CHARLIE BROWN (I,2271); LIFE IS A CIRCUS, CHARLIE BROWN (I,2683); SHE'S A GOOD SKATE, CHARLIE BROWN (I,4012); YOU'RE THE GREATEST, CHARLIE BROWN (I,4973)

CARLSON, CHARLES
THREE FOR DANGER (I,4475)

CARLSON, GAIL RAE
MICKEY SPILLANE'S MIKE HAMMER: MORE THAN MURDER (II,1693)

CARLSON, KAREN
THE AMERICAN DREAM (II,71); CENTENNIAL (II,477); HERE COME THE BRIDES (I,2024); TWO MARRIAGES (II,2691); THE YELLOW ROSE (II,2847)

CARLSON, KATE
GENERAL HOSPITAL (II,964)

CARLSON, LINDA
KAZ (II,1370); NEWHART (II,1835); PALS (II,1947); SUTTERS BAY (II,2508); WESTSIDE MEDICAL (II,2765)

CARLSON, PHILIP
MANDY'S GRANDMOTHER (II,1618)

CARLSON, RICHARD
CAMEO THEATER (I,808); EYE WITNESS (I,1494); HOLLYWOOD OPENING NIGHT (I,2087); I LED THREE LIVES (I,2169); MACKENZIE'S RAIDERS (I,2803); THE PHILADELPHIA STORY (I,3581); PULITZER PRIZE PLAYHOUSE (I,3692); THRILLER (I,4492)

CARLSON, RUTH
BOZO THE CLOWN (I,702)

CARLSON, STAN
SING ALONG WITH MITCH (I,4057)

CARLSON, STEVE
THE YOUNG AND THE RESTLESS (II,2862)

CARLSSON, CARL
THE JUGGLER OF NOTRE DAME (II,1350)

CARLTON AND COMPANY
LIKE MAGIC (II,1476); MAGIC WITH THE STARS (II,1599)

CARLTON, KYRA
ROWAN AND MARTIN'S LAUGH-IN (I,3856)

CARLTON, MARK
LONDON AND DAVIS IN NEW YORK (II,1512)

CARLTON, SUE
EXPERT WITNESS (I,1488); NAVY LOG (I,3233)

CARLTON, TIMOTHY
CORONATION STREET (I,1054)

CARLYLE, AILEEN
MY FRIEND IRMA (I,3191)

CARLYLE, RICHARD
THE CLOCK (I,978); CRIME PHOTOGRAPHER (I,1097)

CARMEL, ROGER C
BATMAN (I,366); THE EYES OF TEXAS (II,799); FITZ AND BONES (II,867); THE MOTHERS-IN-LAW (I,3110); STAR TREK (II,2440); STUMP THE STARS (I,4274); TERROR AT ALCATRAZ (II,2563)

CARMEN, JULIE
CONDO (II,565); SHE'S IN THE ARMY NOW (II,2320)

CARMEN, MICHAEL
ALL THE RIVERS RUN (II,43)

CARMICHAEL, HOAGY
COUNTRY COMES HOME (II,585); GULF PLAYHOUSE (I,1904); LARAMIE (I,2614); MONSANTO PRESENTS MANCINI (I,3094); THE SATURDAY NIGHT REVUE (I,3921); TIMEX ALL-STAR JAZZ SHOW III (I,4517)

CARMINO, LEONARDO
THE MONEYCHANGERS (II,1724)

CARNE, JUDY
THE BAILEYS OF BALBOA (I,334); FAIR EXCHANGE (I,1512); KRAFT MUSIC HALL PRESENTS SANDLER AND YOUNG (I,2585); LOVE ON A ROOFTOP (I,2775); ROWAN AND MARTIN'S LAUGH-IN (I,3856); ROWAN AND MARTIN'S LAUGH-IN (I,3857); SOMEONE AT THE TOP OF THE STAIRS (I,4114); SUPER COMEDY BOWL 1 (I,4288)

CARNEGIE, ROBERT
THE RETURN OF MARCUS WELBY, M.D (II,2138)

CARNES, KIM
MUSIC CENTRAL (II,1771)

CARNEY, ALAN
COOGAN'S REWARD (I,1044); THE JEAN CARROLL SHOW (I,2353)

CARNEY, ART
ALFRED HITCHCOCK PRESENTS (I,115); AMERICA PAUSES FOR THE MERRY MONTH OF MAY (I,167); ART CARNEY MEETS PETER AND THE WOLF (I,272); ART CARNEY MEETS THE SORCERER'S APPRENTICE (I,273); THE ART CARNEY SHOW (I,274); BATMAN (I,366); THE BEST OF ANYTHING (I,404); THE BOB HOPE CHRYSLER THEATER (I,502); CALL ME BACK (I,798); THE CAMPBELL TELEVISION SOUNDSTAGE (I,812); CAVALCADE OF STARS (I,877); CBS: ON THE AIR (II,460); THE CHEVROLET GOLDEN ANNIVERSARY SHOW (I,924); CHRISTMAS IN DISNEYLAND (II,519); CLIMAX! (I,976); THE CONNIE FRANCIS SHOW (I,1039); DANGER (I,1134); DUPONT SHOW OF THE MONTH (I,1387); THE DUPONT SHOW OF THE WEEK (I,1388); FEMALE INSTINCT (I,1564); FULL MOON OVER BROOKLYN (I,1703); HAPPY ANNIVERSARY AND GOODBYE (II,1082); HAPPY ENDINGS (II,1085); HARVEY (I,1961); HENRY MORGAN'S GREAT TALENT HUNT (I,2017); THE HONEYMOONERS (I,2110); THE HONEYMOONERS (I,2111); THE HONEYMOONERS (II,1175); THE HONEYMOONERS CHRISTMAS SPECIAL (II,1176); THE HONEYMOONERS CHRISTMAS (II,1177); THE HONEYMOONERS SECOND HONEYMOON (II,1178); THE HONEYMOONERS VALENTINE SPECIAL (II,1179); HOORAY FOR LOVE (I,2115); THE JACKIE GLEASON SHOW (I,2323); THE JACKIE GLEASON SHOW (I,2322); THE JACKIE GLEASON SPECIAL (I,2324); THE JACKIE GLEASON SPECIAL (I,2325); THE JANE POWELL SHOW (I,2344); LANIGAN'S RABBI (II,1427); LANIGAN'S RABBI (II,1428); THE LAST LEAF (II,1437); THE LEPRECHAUN'S CHRISTMAS GOLD (II,1461); LOLA (II,1511); A LUCILLE BALL SPECIAL: WHAT NOW CATHERINE CURTIS? (II,1560); THE MAN IN THE DOG SUIT (I,2869); THE MICKEY ROONEY SHOW (I,3027); THE MOREY AMSTERDAM SHOW (I,3101); OUR TOWN (I,3420); PANAMA HATTIE (I,3444); THE PERRY COMO WINTER SHOW (I,3543); PERRY COMO'S WINTER SHOW (I,3546); PERRY COMO'S WINTER SHOW (I,3547); THE RIGHT MAN (I,3790); RINGO (II,2176); THE SNOOP SISTERS (II,2389); SOUND OF THE 60S (I,4134); STAR STAGE (I,4197); STUDIO ONE (I,4268); SUSPENSE (I,4305); THREE IN ONE (I,4479); THE TWILIGHT ZONE (I,4651); VICTORY (I,4729); YOU CAN'T TAKE IT WITH YOU (II,2857)

CARNEY, BARBARA
LANIGAN'S RABBI (II,1427); LANIGAN'S RABBI (II,1428)

CARNEY, DEBORAH
OUR FAMILY BUSINESS (II,1931)

CARNEY, GRACE
ROCKY KING, INSIDE DETECTIVE (I,3824)

CARNEY, HARRY
TIMEX ALL-STAR JAZZ SHOW IV (I,4518)

CARNEY, JOHN
SHOGUN (II,2339)

CARNEY, KIM
THE DAVID LETTERMAN SHOW (II,624)

CARNEY, MARY
RYAN'S HOPE (II,2234)

CAROL, CINDY
THE EVE ARDEN SHOW (I,1471); NEVER TOO YOUNG (I,3248)

CAROL, DIANE
THE KATE SMITH HOUR (I,2508)

CAROL, RONNIE
THE BILLION DOLLAR THREAT (II,246)

CAROLEO, DINA
CAP'N AHAB (I,824); KIBBE HATES FINCH (I,2530); OUTLAW LADY (II,1935); VELVET (II,2725); WONDER GIRL (I,4891)

CAROLINA CLOGGERS, THE
COUNTRY MUSIC CARAVAN (I,1066)

CARON, LESLIE
AN HOUR WITH ROBERT GOULET (I,2137); MASTER OF THE GAME (II,1647)

CARON, MARY
CONCRETE BEAT (II,562)

CARON, RENE
TOMAHAWK (I,4549)

CARPENTER, CARLETON
A DATE WITH DEBBIE (I,1161); LADY IN THE DARK (I,2602); LUKE AND THE TENDERFOOT (I,2792); PARIS IN THE SPRINGTIME (I,3457); PETE AND GLADYS (I,3558); THE SHOWOFF (I,4035)

CARPENTER, GARY
ANOTHER WORLD (II,97)

CARPENTER, KAREN
THE BOB HOPE SHOW (I,656); THE BOB HOPE SHOW (I,663); THE CARPENTERS (II,444); THE CARPENTERS (II,445); THE CARPENTERS AT CHRISTMAS (II,446); THE CARPENTERS. . . SPACE ENCOUNTERS (II,447); THE DOROTHY HAMILL WINTER CARNIVAL SPECIAL (II,720); THE FIFTH DIMENSION TRAVELING SHOW (I,1572); MAKE YOUR OWN KIND OF MUSIC (I,2847); OLIVIA NEWTON-JOHN'S HOLLYWOOD NIGHTS (II,1886); PEGGY FLEMING AT SUN VALLEY (I,3507); THE PERRY COMO CHRISTMAS SHOW (II,2002); ROBERT YOUNG WITH THE YOUNG (I,3814); SPECIAL LONDON BRIDGE SPECIAL (I,4150)

CARPENTER, KEN
THE KRAFT MUSIC HALL (I,2587); THE LUX VIDEO THEATER (I,2795)

CARPENTER, RICHARD
THE BOB HOPE SHOW (I,656); THE BOB HOPE SHOW (I,663); THE CARPENTERS (II,444); THE CARPENTERS (II,445); THE CARPENTERS AT CHRISTMAS (II,446); THE CARPENTERS. . . SPACE ENCOUNTERS (II,447); THE DOROTHY HAMILL WINTER CARNIVAL SPECIAL (II,720); THE FIFTH DIMENSION TRAVELING SHOW (I,1572); MAKE YOUR OWN KIND OF

MUSIC (I,2847); PEGGY FLEMING AT SUN VALLEY (I,3507); THE PERRY COMO CHRISTMAS SHOW (II,2002); ROBERT YOUNG WITH THE YOUNG (I,3814); SPECIAL LONDON BRIDGE SPECIAL (I,4150)

CARPENTER, SUE ANN
THE CITY (I,969)

CARPENTER, THELMA
BAREFOOT IN THE PARK (I,355); CALL HER MOM (I,796)

CARPINELLI, PAUL
THE GUIDING LIGHT (II,1064)

CARR, ANDREW
THE WOMAN IN WHITE (II,2819)

CARR, BETTY ANN
CADE'S COUNTY (I,787); RETURN TO PEYTON PLACE (I,3779); ROWAN AND MARTIN'S LAUGH-IN (I,3856)

CARR, CAMILLA
ANOTHER WORLD (II,97)

CARR, CHRISTOPHER
CINDERELLA (II,526)

CARR, DARLEEN
BATTLE OF THE NETWORK STARS (II,163); BATTLE OF THE NETWORK STARS (II,164); BRET MAVERICK (II,374); THE CHADWICK FAMILY (II,478); CIRCUS OF THE STARS (II,536); DEAN MARTIN PRESENTS THE GOLDDIGGERS (I,1194); THE JOHN FORSYTHE SHOW (I,2410); MISS WINSLOW AND SON (II,1705); ONCE AN EAGLE (II,1897); THE OREGON TRAIL (II,1923); SLEEPWALKER (I,4084); THE SMITH FAMILY (I,4092); THE STREETS OF SAN FRANCISCO (II,2478)

CARR, DIDI
SUGAR TIME (II,2489)

CARR, GERALDINE
I MARRIED JOAN (I,2174); MR. TUTT (I,3153)

CARR, JAMIE
THE KOWBOYS (I,2584)

CARR, JANE
TO SIR, WITH LOVE (II,2624)

CARR, KAREN GLOW
TRAPPER JOHN, M.D. (II,2654)

CARR, MICHAEL
THE BARBARY COAST (II,147); THE FRENCH ATLANTIC AFFAIR (II,924)

CARR, NANCY
THIS IS MUSIC (I,4457)

CARR, PAUL
BUCK ROGERS IN THE 25TH CENTURY (II,384); THE DOCTORS (II,694); OF THIS TIME, OF THAT PLACE (I,3335); SCRUPLES (II,2280); THE WILD WOMEN OF CHASTITY GULCH (II,2801)

CARR, VIKKI
THE BOB HOPE SHOW (I,625); BURT BACHARACH! (I,771); THE DANNY KAYE SHOW (I,1143); GIRL FRIENDS AND NABORS (I,1807); THE ICE PALACE (I,2182); JOHNNY CARSON DISCOVERS CYPRESS GARDENS (I,2419); PERRY COMO'S CHRISTMAS IN MEXICO (II,2009); THE RAY ANTHONY SHOW (I,3729); TEXACO STAR PARADE II (I,4410); A VERY SPECIAL OCCASION (I,4716); THE VIKKI CARR SHOW (I,4732)

CARR-BOSLEY, PATRICIA
THE DRUNKARD (II,737)

CARRABASSE GRANGE HALL TALENT CONTEST WINNING BAND, THE
THE JUD STRUNK SHOW (I,2462)

CARRADINE, DAVID
JOHNNY BELINDA (I,2417); KUNG FU (II,1416); SHANE (I,3995)

CARRADINE, JOHN
THE ADVENTURES OF DR. FU MANCHU (I,58); BRANDED (I,707); CAPTAINS AND THE KINGS (II,435); THE CHEVROLET TELE-THEATER (I,926); CLIMAX! (I,976); DAMON RUNYON THEATER (I,1129); DUPONT SHOW OF THE MONTH (I,1387); FAITH BALDWIN'S THEATER OF ROMANCE (I,1515); FRONT ROW CENTER (I,1694); GOLIATH AWAITS (II,1025); THE HORROR OF IT ALL (II,1181); LIGHT'S OUT (I,2699); MATINEE THEATER (I,2947); MUNSTER, GO HOME! (I,3157); THE MUNSTERS (I,3158); MY FRIEND IRMA (I,3191); NIGHT GALLERY (I,3287); THE NIGHT STRANGLER (I,3293); SCHLITZ PLAYHOUSE OF STARS (I,3936); STUDIO ONE (I,4268); SURE AS FATE (I,4297); SUSPENSE (I,4305); THE TWILIGHT ZONE (I,4651)

CARRADINE, KEITH
CHIEFS (II,509); MAN ON A STRING (I,2877)

CARRADINE, ROBERT
THE COWBOYS (II,598)

CARRAHER, HARLAN
THE GHOST AND MRS. MUIR (I,1786); GIDGET GROWS UP (I,1797)

CARRASCO, CARLOS
THE MADHOUSE BRIGADE (II,1588)

CARREL, ROGER
DON QUIXOTE (I,1339)

CARRELL, LORI
THE FOUR SEASONS (II,915)

CARRER, CHARLIE
NIGHT CLUB (I,3284)

CARRERA, BARBARA
CENTENNIAL (II,477); MASADA (II,1642)

CARREY, JIM
THE DUCK FACTORY (II,738)

CARRI, THOMAS
THE NEW ZOO REVUE (I,3277)

CARRICABURU, CATHY
THE YOUNG AND THE RESTLESS (II,2862)

CARRICART, ROBERT
CAPTAIN BRASSBOUND'S CONVERSION (I,828); CHARLIE ANGELO (I,906); T.H.E. CAT (I,4430)

CARRIER, ALBERT
RAPTURE AT TWO-FORTY (I,3724)

CARRILLO, LEO
THE CISCO KID (I,961); HANDS OF MURDER (I,1929)

CARRINGTON, CARRIE
EVENING AT THE IMPROV (II,786)

CARRINGTON, DEBBIE
THE EARTHLINGS (II,751)

CARRIORT, ROBERT
THE TRAVELS OF JAIMIE MCPHEETERS (I,4596)

CARRIOU, LEN
100 CENTRE STREET (II,1901)

CARROLE, MONA
BELLE STARR (I,391)

CARROLL AND THE ESCORTS, HELEN
TEX AND JINX (I,4406)

CARROLL, ALLISON
TOP OF THE HILL (II,2645)

CARROLL, BEESON
THE BUREAU (II,397); THE DAIN CURSE (II,613); M*A*S*H (II,1569); MARY HARTMAN, MARY HARTMAN (II,1638); P.O.P. (II,1939); PALMERSTOWN U.S.A. (II,1945); STARSTRUCK (II,2446)

CARROLL, BOB
HEAVEN WILL PROTECT THE WORKING GIRL (I,1996); JUDGE FOR YOURSELF (I,2466); SONGS AT TWILIGHT (I,4123)

CARROLL, CLINTON DERRICKS
HIGH FIVE (II,1143)

CARROLL, DAVID-JAMES
BALL FOUR (II,137); HOT W.A.C.S. (II,1191)

CARROLL, DEE
79 PARK AVENUE (II,2310); LOVE NEST (II,1541); MCNAB'S LAB (I,2972)

CARROLL, DIAHANN
20TH CENTURY FOLLIES (I,4641); ABC STAGE '67 (I,6); THE ANTHONY NEWLEY SHOW (I,233); THE BOB GOULET SHOW STARRING ROBERT GOULET (I,501); BOB HOPE SPECIAL: BOB HOPE'S ALL-STAR BIRTHDAY PARTY (II,295); BOB HOPE SPECIAL: BOB HOPE—HOPE, WOMEN AND SONG (II,324); CHRISTMAS IN WASHINGTON (II,520); CRESCENDO (I,1094); THE DIAHANN CARROLL SHOW (I,1252); THE DIAHANN CARROLL SHOW (II,665); DYNASTY (II,746); THE FLIP WILSON SPECIAL (II,887); FRANCIS ALBERT SINATRA DOES HIS THING (I,1658); GEORGE BURNS' HOW TO LIVE TO BE 100 (II,972); HOTEL 90 (I,2130); JACK LEMMON—GET HAPPY (I,2312); JULIA (I,2476); THE MAN IN THE MOON (I,2871); MOVIN' (I,3117); MUSIC U.S.A. (I,3179); ON PARADE (I,3356); ROOTS: THE NEXT GENERATIONS (II,2213); TELLY. . . WHO LOVES YA, BABY? (II,2558)

CARROLL, ED E
THE GIRL, THE GOLD WATCH AND EVERYTHING (II,1002)

CARROLL, EDDY

THE DON KNOTTS SHOW (I,1334)

CARROLL, ERIKA
FAT ALBERT AND THE COSBY KIDS (II,835)

CARROLL, HELENA
BACKSTAIRS AT THE WHITE HOUSE (II,133); THE EDGE OF NIGHT (II,760); THE EYES OF TEXAS II (II,800); GENERAL HOSPITAL (II,964); RYAN'S FOUR (II,2233)

CARROLL, IRENE
THE JULIUS LAROSA SHOW (I,2486)

CARROLL, JAMES
THE RETURN OF LUTHER GILLIS (II,2137)

CARROLL, JANET
I GAVE AT THE OFFICE (II,1213)

CARROLL, JANICE
THE BLUE AND THE GRAY (II,273); THE DESILU REVUE (I,1238); THE HOME FRONT (II,1171); ROUGHNECKS (II,2219)

CARROLL, JEAN
FUN FOR 51 (I,1707); THE JEAN CARROLL SHOW (I,2353)

CARROLL, JILL
SUMMER (II,2491)

CARROLL, JIMMY
MR. AND MRS. CARROLL (I,3122); THE NASH AIRFLYTE THEATER (I,3227)

CARROLL, JON
THE STARLAND VOCAL BAND (II,2441)

CARROLL, JUNE
FIRESIDE THEATER (I,1580)

CARROLL, LARRY
CONDOMINIUM (II,566)

CARROLL, LAURIE
THE JOHNNY CARSON SHOW (I,2422)

CARROLL, LEO G
ALCOA PREMIERE (I,109); CHANNING (I,900); THE GIRL FROM U.N.C.L.E. (I,1808); GOING MY WAY (I,1833); KRAFT MYSTERY THEATER (I,2589); THE MAN FROM U.N.C.L.E. (I,2867); THE MOONGLOW AFFAIR (I,3099); PHILCO TELEVISION PLAYHOUSE (I,3583); STAR TONIGHT (I,4199); TOPPER (I,4582)

CARROLL, LINDA
THE BEST OF FRIENDS (II,216)

CARROLL, MADELEINE
THE BOB HOPE SHOW (I,611)

CARROLL, NANCY
THE ALDRICH FAMILY (I,111); FAITH BALDWIN'S THEATER OF ROMANCE (I,1515)

CARROLL, PAT
BEST FOOT FORWARD (I,401); BUSTING LOOSE (II,402); CINDERELLA (I,956); ENTERTAINMENT —1955 (I,1450); GETTING TOGETHER (I,1784); MAKE ROOM FOR DADDY (I,2843); MAKE ROOM FOR GRANDDADDY (I,2844); THE MAURICE CHEVALIER SHOW (I,2953); OH, NURSE! (I,3343); PANDORA AND FRIEND (I,3445); PRIVATE EYE, PRIVATE EYE (I,3671); THE SATURDAY NIGHT REVUE (I,3921); SECOND CHANCE (I,3958); SPOTLIGHT (I,4162); STEP ON THE GAS (I,4215); STUDIO '57 (I,4267); VARIETY (I,4703); THE WOMEN (I,4890)

CARROLL, RENA
LASSITER (I,2625); LASSITER (II,1434)

CARROLL, RITA
MR. AND MRS. CARROLL (I,3122)

CARROLL, ROBERT
THE STRANGER (I,4250)

CARROLL, TED
CORONATION STREET (I,1054)

CARROLL, VICTORIA
ALICE (II,33); I'VE HAD IT UP TO HERE (II,1221); THE LIFE AND TIMES OF EDDIE ROBERTS (II,1468); NO SOAP, RADIO (II,1856); SMALL AND FRYE (II,2380); W*A*L*T*E*R (II,2731)

CARROLL, VIRGINIA
IT'S A SMALL WORLD (I,2260); THE ROY ROGERS SHOW (I,3859)

CARRONS, GENE
WONDERFUL TOWN (I,4893)

CARROTT, RIC
HAPPY DAYS (II,1084); KING OF THE ROAD (II,1396); THE LOVE BOAT I (II,1532); SPACE ACADEMY (II,2406)

CARRY III, JULIUS J
DRIBBLE (II,736)

CARSON, CHARLES
GLENCANNON (I,1822); MACBETH (I,2801)

CARSON, JACK
ALCOA/GOODYEAR THEATER (I,107); THE ALL-STAR REVUE (I,138); AMOS BURKE: WHO KILLED JULIE GREER? (I,181); ANOTHER DAY, ANOTHER DOLLAR (I,228); ARROYO (I,266); FORD THEATER HOUR (I,1635); HOW'S BUSINESS? (I,2153); THE JACK CARSON SHOW (I,2304); JOHNNY COME LATELY (I,2425); OPERATION ENTERTAINMENT (I,3400); SCREEN DIRECTOR'S PLAYHOUSE (I,3946); STRIKE IT RICH (I,4262); U.S. ROYAL SHOWCASE (I,4688); THE UNITED STATES STEEL HOUR (I,4677)

CARSON, JEAN
THE ANDY GRIFFITH SHOW (I,192); THE BETTY HUTTON SHOW (I,413); EYE WITNESS (I,1494); FORD THEATER HOUR (I,1635); PHILCO TELEVISION PLAYHOUSE (I,3583); THE RED BUTTONS SHOW (I,3750); ROBERT MONTGOMERY PRESENTS YOUR LUCKY STRIKE THEATER (I,3809); THE TRAP (I,4593); THE TWILIGHT ZONE (I,4651)

CARSON, JEANNIE
BERKELEY SQUARE (I,399); BEST FOOT FORWARD (I,401); HEIDI (I,2004); HEY, JEANNIE! (I,2038); JANE WYMAN PRESENTS THE FIRESIDE THEATER (I,2345); LITTLE WOMEN (I,2722); MATINEE THEATER (I,2947); THE MAURICE CHEVALIER SHOW (I,2953); SEARCH FOR TOMORROW (II,2284); SPOTLIGHT (I,4162)

CARSON, JOANNE
ALIVE AND WELL (II,36); JOANNE CARSON'S V.I.P.'S (I,2397); THE NIGHT OF CHRISTMAS (I,3290)

CARSON, JOHN
COME OUT, COME OUT, WHEREVER YOU ARE (I,1014); THE MURDOCKS AND THE MCCLAYS (I,3163); POSSESSION (I,3655)

CARSON, JOHN DAVID
CALL HER MOM (I,796); MITCHELL AND WOODS (II,1709); OF MEN OF WOMEN (I,3332)

CARSON, JOHN E
SPACE GHOST (I,4141)

CARSON, JOHNNY
THE ALL-STAR COMEDY SHOW (I,137); THE ARTHUR MURRAY PARTY FOR BOB HOPE (I,292); THE BOB HOPE SHOW (I,633); BOB HOPE SPECIAL: BOB HOPE FOR PRESIDENT (II,282); BOB HOPE SPECIAL: HAPPY BIRTHDAY, BOB! (II,326); DON RICKLES—ALIVE AND KICKING (I,1340); EARN YOUR VACATION (I,1394); THE FIRST 50 YEARS (II,862); GEORGE BURNS CELEBRATES 80 YEARS IN SHOW BUSINESS (II,967); THE GEORGE BURNS SPECIAL (II,970); GEORGE BURNS' 100TH BIRTHDAY PARTY (II,971); THE JACK BENNY SPECIAL (I,2296); JACK BENNY'S FIRST FAREWELL SHOW (I,2300); JACK BENNY'S SECOND FAREWELL SHOW (I,2302); JOHNNY CARSON DISCOVERS CYPRESS GARDENS (I,2419); JOHNNY CARSON PRESENTS THE SUN CITY SCANDALS (I,2420); JOHNNY CARSON PRESENTS THE SUN CITY SCANDALS '72 (I,2421); THE JOHNNY CARSON SHOW (I,2422); JOHNNY CARSON'S REPERTORY COMPANY IN AN EVENING OF COMEDY (I,2423); JOHNNY COME LATELY (I,2425); A LOVE LETTER TO JACK BENNY (II,1540); LUCY MOVES TO NBC (II,1562); MURDER AT NBC (A BOB HOPE SPECIAL) (I,3159); PURE GOLDIE (I,3696); THE ROYAL FOLLIES OF 1933 (I,3862); SHEENA EASTON, ACT 1 (II,2323); THE STEVE ALLEN SHOW (I,4220); THE TONIGHT SHOW (I,4565); THE TONIGHT SHOW STARRING JOHNNY CARSON (II,2638); A TRIBUTE TO "MR. TELEVISION" MILTON BERLE (II,2658); WHO DO YOU TRUST? (I,4846)

CARSON, KEN
THE GARRY MOORE SHOW (I,1738); THE GARRY MOORE SHOW (I,1739)

CARSON, MARTHA
THE ROCK N' ROLL SHOW (I,3820)

CARSON, MINDY
CLUB EMBASSY (I,983); FORD STAR REVUE (I,1632); MINDY CARSON SINGS (I,3055); THE PERRY COMO SHOW (I,3532); SHOW OF THE YEAR (I,4031); UPBEAT (I,4684)

CARSON, ROBERT
THE PATTY DUKE SHOW (I,3483)

CARSON, SHAWN
REAL KIDS (II,2116)

CARSON, SI MING
WOLF ROCK TV (II,2815)

CARSON, VIOLET
CORONATION STREET (I,1054)

CARSON, WHITEY
TOOTSIE HIPPODROME (I,4575)

CARSTAIRS, ALAN
WHY DIDN'T THEY ASK EVANS? (II,2795)

CARTER FAMILY, THE
COUNTRY MUSIC CARAVAN (I,1066); THE JOHNNY CASH SHOW (I,2424); JOHNNY CASH: THE FIRST 25 YEARS (II,1336)

CARTER SISTERS, THE
COUNTRY MUSIC CARAVAN (I,1066)

CARTER, ANITA
JOHNNY CASH CHRISTMAS 1983 (II,1324)

CARTER, BEVERLY
SONNY BOY (II,2402)

CARTER, BILLY
FLATBED ANNIE AND SWEETIEPIE: LADY TRUCKERS (II,876)

CARTER, CONLAN
COMBAT (I,1011); THE LAW AND MR. JONES (I,2637); WHICH WAY'D THEY GO? (I,4838)

CARTER, DIXIE
DIFF'RENT STROKES (II,674); THE EDGE OF NIGHT (II,760); FILTHY RICH (II,856); ON OUR OWN (II,1892); OUT OF THE BLUE (II,1934)

CARTER, HELEN
JOHNNY CASH CHRISTMAS 1983 (II,1324)

CARTER, JACK
THE ALAN KING SHOW (I,102); AMERICAN MINSTRELS (I,171); THE BOB HOPE SHOW (I,633); CAVALCADE OF STARS (I,877); GEORGE BURNS CELEBRATES 80 YEARS IN SHOW BUSINESS (II,967); HAVE I GOT A CHRISTMAS FOR YOU (II,1105); HUMAN FEELINGS (II,1203); THE HUSTLER OF MUSCLE BEACH (II,1210); JACK CARTER AND COMPANY (I,2305); THE JACK CARTER SHOW (I,2306); THE LAST HURRAH (I,2626); MADHOUSE 90 (I,2807); MURDER AT NBC (A BOB HOPE SPECIAL) (I,3159); PICK AND PAT (I,3590); PLIMPTON! DID YOU HEAR THE ONE ABOUT. . .? (I,3629); THE ROWAN AND MARTIN SPECIAL (I,3855); THE SATURDAY NIGHT REVUE (I,3921); SHOOT-IN AT NBC (A BOB HOPE SPECIAL) (I,4021); TALES OF TOMORROW (I,4352); VAUDEVILLE (II,2722)

CARTER, JANETTE
JOHNNY CASH CHRISTMAS 1983 (II,1324)

CARTER, JANICE
BEN HECHT'S TALES OF THE CITY (I,395); THE ELGIN HOUR (I,1428)

CARTER, JANIS
FEATHER YOUR NEST (I,1555)

CARTER, JOE
JOHNNY CASH CHRISTMAS 1983 (II,1324)

CARTER, JOHN
BARNABY JONES (II,153); DYNASTY (II,747); IF IT'S A MAN, HANG UP (I,2186); THE KILLIN' COUSIN (II,1394); OUR FAMILY BUSINESS (II,1931); PALMERSTOWN U.S.A. (II,1945); ROOTS: THE NEXT GENERATIONS (II,2213); THE SMITH FAMILY (I,4092); THE WINDS OF WAR (II,2807)

CARTER, JOHN ARCH
TEXAS (II,2566)

CARTER, JUDY
THE SHAPE OF THINGS (II,2315); THAT SECOND THING ON ABC (II,2572); THAT THING ON ABC (II,2574)

CARTER, JUNE
COUNTRY MUSIC CARAVAN (I,1066); THE JOHNNY CASH SHOW (I,2424)

CARTER, KIMBERLY
SENSE OF HUMOR (II,2303)

CARTER, LILLIAN
THE LUCILLE BALL SPECIAL (II,1559); SINATRA—THE FIRST 40 YEARS (II,2359)

CARTER, LYNDA
BATTLE OF THE NETWORK STARS (II,163); BOB HOPE SPECIAL: HAPPY BIRTHDAY, BOB! (II,325); CIRCUS OF THE STARS (II,530); CIRCUS OF THE STARS (II,531); LYNDA CARTER'S CELEBRATION (II,1563); LYNDA CARTER'S SPECIAL (II,1564); LYNDA CARTER: BODY AND SOUL (II,1565); LYNDA CARTER: ENCORE (II,1566); LYNDA CARTER: STREET LIGHTS (II,1567); THE OLIVIA NEWTON-JOHN SHOW (II,1885); PARTNERS IN CRIME (II,1961); A SPECIAL OLIVIA NEWTON-JOHN (II,2418); WONDER WOMAN (PILOT 2) (II,2827); WONDER WOMAN (SERIES 1) (II,2828); WONDER WOMAN (SERIES 2) (II,2829)

CARTER, MARILYN
CHIEFS (II,509)

CARTER, MAYBELLE
JOHNNY CASH: THE FIRST 25 YEARS (II,1336)

CARTER, MEL
MAGNUM, P.I. (II,1601)

CARTER, NELL
20TH CENTURY FOLLIES (I,4641); AIN'T MISBEHAVIN' (II,25); BARYSHNIKOV ON BROADWAY (II,158); CHRISTMAS IN WASHINGTON (II,521); THE DEAN MARTIN CELEBRITY ROAST (II,639); GIMME A BREAK (II,995); LOBO (II,1504); MAX (II,1658); RYAN'S HOPE (II,2234)

CARTER, PHIL
CAR 54, WHERE ARE YOU? (I,842)

CARTER, RALPH
GOOD TIMES (II,1038)

CARTER, ROSIE
COUNTRY COMES HOME (II,586)

CARTER, SAM
SING ALONG WITH MITCH (I,4057)

CARTER, SUE
20 MINUTE WORKOUT (II,2680)

CARTER, T K
ADAMS HOUSE (II,9); BORDER PALS (II,352); CAR WASH (II,437); JUST OUR LUCK (II,1360); WILDER AND WILDER (II,2802)

CARTER, TERRY
BATTLESTAR GALACTICA (II,181); THE GREEN PASTURES (I,1887); MCCLOUD (II,1660); MCCLOUD: WHO KILLED MISS U.S.A.? (I,2965)

CARTER, THOMAS
SZYSZNYK (II,2523); THE WHITE SHADOW (II,2788)

CARTIER, WALTER
THE PHIL SILVERS SHOW (I,3580)

CARTWRIGHT, ANGELA
DANNY THOMAS LOOKS AT YESTERDAY, TODAY AND TOMORROW (I,1148); THE DANNY THOMAS TV FAMILY REUNION (I,1154); DICK CLARK'S GOOD OL' DAYS: FROM BOBBY SOX TO BIKINIS (II,671); EVERYTHING HAPPENS TO ME (I,1481); HIGH SCHOOL, U.S.A. (II,1148); LOST IN SPACE (I,2758); MAKE MORE ROOM FOR DADDY (I,2842); MAKE ROOM FOR DADDY (I,2843); MAKE ROOM FOR GRANDDADDY (I,2844); MAKE ROOM FOR GRANDDADDY (I,2845); MR. & MS. AND THE MAGIC STUDIO MYSTERY (II,1747); THE SECRET WORLD OF KIDS (I,3970); U.M.C. (I,4666); WHATEVER BECAME OF. . .? (II,2772)

CARTWRIGHT, NANCY
IN TROUBLE (II,1230); THE RICHIE RICH SHOW (II,2169)

CARTWRIGHT, VERONICA
DANIEL BOONE (I,1142); JOE DANCER: THE BIG BLACK PILL (II,1300); LEAVE IT TO BEAVER (I,2648); WHO HAS SEEN THE WIND? (I,4847)

CARUSO, ANTHONY
BROKEN ARROW (I,743); GREAT BIBLE ADVENTURES (I,1872); THE JOE PISCOPO SPECIAL (II,1304); RUN JACK RUN (I,3872)

CARUSO, CARL
CUT (I,1116); SPIN THE PICTURE (I,4159)

CARUSO, DAVID
CRAZY TIMES (II,602); FOR LOVE AND HONOR (II,903)

CARVALHO, BETTY
INSPECTOR PEREZ (II,1237); SIDNEY SHORR (II,2349); WAIKIKI (II,2734)

CARVEN, MICHAEL
EMERALD POINT, N.A.S. (II,773)

CARVER, MARY
I MARRIED A DOG (I,2173); THE MOONGLOW AFFAIR (I,3099); SIMON AND SIMON (II,2357); THROUGH THE MAGIC PYRAMID (II,2615)

CARVER, PETER
BARRIER REEF (I,362)

CARVER, RANDALL

AMERICA, YOU'RE ON (II,69); THE DAUGHTERS OF JOSHUA CABE RETURN (I,1170); FOREVER FERNWOOD (II,909); THE NEW DAUGHTERS OF JOSHUA CABE (I,3261); THE SIX O'CLOCK FOLLIES (II,2368); TAXI (II,2546)

CARVER, TINA
THE WEB (I,4785)

CARVER, ZEBE
VILLAGE BARN (I,4733)

CARVEY, DANA
ALONE AT LAST (II,60); BLUE THUNDER (II,278); ONE OF THE BOYS (II,1911); WHACKED OUT (II,2767)

CARY, CHRISTOPHER
CAPTAIN AMERICA (II,428); GARRISON'S GORILLAS (I,1735)

CARYL, JEANNE
CAMELOT (II,416)

CASADOS, ELOY PHIL
YOUNG DAN'L BOONE (II,2863)

CASE, ALLEN
AN APARTMENT IN ROME (I,242); ARTHUR GODFREY AND FRIENDS (I,278); THE CHEVROLET GOLDEN ANNIVERSARY SHOW (I,924); THE DEPUTY (I,1234); HOW TO SURVIVE THE 70S AND MAYBE EVEN BUMP INTO HAPPINESS (II,1199); THE LEGEND OF JESSE JAMES (I,2653); THE LIFE AND TIMES OF EDDIE ROBERTS (II,1468); RIDING HIGH (II,2173)

CASE, NELSON
THE ARMSTRONG CIRCLE THEATER (I,260); ARTHUR MURRAY'S DANCE PARTY (I,293); SNEAK PREVIEW (I,4100); TRASH OR TREASURE (I,4594); WHAT'S IT WORTH? (I,4818)

CASEY
YOUR HIT PARADE (I,4952)

CASEY, ANGEL
HAIL THE CHAMP (I,1918)

CASEY, BERNIE
BAY CITY BLUES (II,186); HARRIS AND COMPANY (II,1096); HEAR NO EVIL (II,1116); THE MARTIAN CHRONICLES (II,1635)

CASEY, ED
SAY IT WITH ACTING (I,3928)

CASEY, LAWRENCE
RETURN TO PEYTON PLACE (I,3779); REX HARRISON PRESENTS SHORT STORIES OF LOVE (II,2148)

CASEY, LEE J
CAT BALLOU (I,870)

CASEY, MELINDA
HOT WHEELS (I,2126); SKYHAWKS (I,4079)

CASEY, ROBERT
THE ALDRICH FAMILY (I,111)

CASH, CARLENE
JOHNNY CASH: CHRISTMAS IN SCOTLAND (II,1332)

CASH, JOHN CARTER
JOHNNY CASH: CHRISTMAS IN SCOTLAND (II,1332)

CASH, JOHNNY
100 YEARS OF GOLDEN HITS (II,1905); THE BOB HOPE SHOW (I,640); BOB HOPE SPECIAL: BOB HOPE'S SUPER BIRTHDAY SPECIAL (II,319); COUNTRY COMES HOME (II,585); COUNTRY COMES HOME (II,586); COUNTRY STARS OF THE 70S (II,594); DICK CLARK PRESENTS THE ROCK AND ROLL YEARS (I,1265); FIFTY YEARS OF COUNTRY MUSIC (II,853); GLEN CAMPBELL AND FRIENDS: THE SILVER ANNIVERSARY (II,1005); JOHN DENVER—THANK GOD I'M A COUNTRY BOY (II,1317); JOHNNY CASH AND FRIENDS (II,1322); JOHNNY CASH AND THE COUNTRY GIRLS (II,1323); A JOHNNY CASH CHRISTMAS (II,1327); A JOHNNY CASH CHRISTMAS (II,1328); JOHNNY CASH CHRISTMAS 1983 (II,1324); THE JOHNNY CASH CHRISTMAS SPECIAL (II,1325); THE JOHNNY CASH CHRISTMAS SPECIAL (II,1326); THE JOHNNY CASH SHOW (I,2424); THE JOHNNY CASH SPRING SPECIAL (II,1329); JOHNNY CASH'S AMERICA (II,1330); JOHNNY CASH: CHRISTMAS IN SCOTLAND (II,1332); JOHNNY CASH: CHRISTMAS ON THE ROAD (II,1333); JOHNNY CASH: COWBOY HEROES (II,1334); JOHNNY CASH: SPRING FEVER (II,1335); JOHNNY CASH: THE FIRST 25 YEARS (II,1336); JOHNNY CASH—A MERRY MEMPHIS CHRISTMAS (II,1331); LARRY GATLIN AND THE GATLIN BROTHERS (II,1429); RANCH PARTY (I,3719); SWING OUT, SWEET LAND (II,2515)

CASH, JUNE CARTER
100 YEARS OF GOLDEN HITS (II,1905); BOB HOPE SPECIAL: BOB HOPE'S SUPER BIRTHDAY SPECIAL (II,319); COUNTRY COMES HOME (II,585); COUNTRY COMES HOME (II,586); JOHNNY CASH AND FRIENDS (II,1322); JOHNNY CASH AND THE COUNTRY GIRLS (II,1323); A JOHNNY CASH CHRISTMAS (II,1327); JOHNNY CASH CHRISTMAS 1983 (II,1324); THE JOHNNY CASH CHRISTMAS SPECIAL (II,1325); THE JOHNNY CASH CHRISTMAS SPECIAL (II,1326); THE JOHNNY CASH SPRING SPECIAL (II,1329); JOHNNY CASH'S AMERICA (II,1330); JOHNNY CASH: CHRISTMAS IN SCOTLAND (II,1332); JOHNNY CASH: CHRISTMAS ON THE ROAD (II,1333); JOHNNY CASH: COWBOY HEROES (II,1334); JOHNNY CASH: SPRING FEVER (II,1335); JOHNNY CASH: THE FIRST 25 YEARS (II,1336); JOHNNY CASH—A MERRY MEMPHIS CHRISTMAS (II,1331)

CASH, ROSEANNE
JOHNNY CASH AND THE COUNTRY GIRLS (II,1323); JOHNNY CASH: SPRING FEVER (II,1335); JOHNNY CASH—A MERRY MEMPHIS CHRISTMAS (II,1331)

CASH, TOMMY
MUSIC WORLD (II,1775)

CASHELL, JEANNIE
IT'S A MAN'S WORLD (I,2258)

CASNOFF, PHILIP
GEORGE WASHINGTON (II,978); THE RENEGADES (II,2132)

CASON, BARBARA
AMERICA, YOU'RE ON (II,69); THE BRADY BRIDES (II,361); THE BRADY GIRLS GET MARRIED (II,364); CARTER COUNTRY (II,449); COMEDY TONIGHT (I,1027); MADAME'S PLACE (II,1587); THE NEW TEMPERATURES RISING SHOW (I,3273); TABITHA (II,2526); TONY THE PONY (II,2641)

CASORLA, RICK
TWO GUYS FROM MUCK (II,2690)

CASPER, ROBERT
ALICE IN WONDERLAND (I,120) LITTLE HOUSE: BLESS ALL THE DEAR CHILDREN (II,1489); LITTLE HOUSE: LOOK BACK TO YESTERDAY (II,1490); LITTLE HOUSE: THE LAST FAREWELL (II,1491)

CASPERS, LORNA
ENCHANTED LANDS (I,1440)

CASS, DAVE
ROYCE (II,2224)

CASS, MAURICE
ROCKY JONES, SPACE RANGER (I,3823); WALTER FORTUNE (I,4758)

CASS, PEGGY
THE DOCTORS (II,694); THE HATHAWAYS (I,1965); JACK PAAR TONIGHT (I,2320); LADIES IN BLUE (II,1422); THE LEPRECHAUN'S CHRISTMAS GOLD (II,1461); TO TELL THE TRUTH (II,2625); TO TELL THE TRUTH (II,2626); THE TONIGHT SHOW (I,4564)

CASS, RAYMOND
LOVE OF LIFE (I,2774)

CASSAVETES, JOHN
ALEXANDER THE GREAT (I,113); THE ELGIN HOUR (I,1428); THE FLIERS (I,1596); HEDDA HOPPER'S HOLLYWOOD (I,2002); JOHNNY STACCATO (I,2437); NIGHTSIDE (I,3296); WON'T IT EVER BE MORNING? (I,4902)

CASSEL, BARRY
KID GLOVES (I,2531)

CASSEL, GEORGE
ONE NIGHT BAND (II,1909)

CASSEL, JEAN-PIERRE
LOVE IN A COLD CLIMATE (II,1539)

CASSELL, ARTHUR
ONE MAN'S FAMILY (I,3383)

CASSELL, CINDY
ME AND BENJY (I,2973)

CASSELL, MALCOLM
THE CLAUDETTE COLBERT SHOW (I,971); LIFE WITH FATHER (I,2690)

CASSELL, SEYMOUR
NIGHTSIDE (I,3296)

CASSELL, WALLY
THE MCGONIGLE (I,2968)

CASSERT, VALERIE
KITTY FOYLE (I,2571)

CASSIDY, DAVID
THE BOB HOPE SHOW (I,656); CELEBRITY CHALLENGE OF THE SEXES 2 (II,466); DAVID CASSIDY—MAN UNDERCOVER (II,623); A KNIGHT IN SHINING ARMOUR (I,2575); PARADE OF STARS (II,1954); THE PARTRIDGE FAMILY

(II,1962); THANKSGIVING REUNION WITH THE PARTRIDGE FAMILY AND MY THREE SONS (II,2569)

CASSIDY, JACK
ANNIE, THE WOMAN IN THE LIFE OF A MAN (I,226); THE ARTHUR GODFREY SPECIAL (I,288); ARTHUR GODFREY'S PORTABLE ELECTRIC MEDICINE SHOW (I,290); BENNY AND BARNEY: LAS VEGAS UNDERCOVER (II,206); CIRCUS OF THE STARS (II,530); DEAN'S PLACE (II,650); ED SULLIVAN'S BROADWAY (I,1402); FBI CODE 98 (I,1550); FOOLS, FEMALES AND FUN: I'VE GOTTA BE ME (II,899); GEORGE M! (I,1772); GEORGE M! (II,976); HE AND SHE (I,1985); JACK CASSIDY'S ST. PATRICK'S DAY SPECIAL (I,2307); THE MARY TYLER MOORE SHOW (II,1640); MRS R.—DEATH AMONG FRIENDS (II,1759); OF MEN OF WOMEN (I,3332)

CASSIDY, JOANNA
240-ROBERT (II,2688); BATTLE OF THE NETWORK STARS (II,169); BUFFALO BILL (II,387); THE CELEBRITY FOOTBALL CLASSIC (II,474); FALCON CREST (II,808); THE FAMILY TREE (II,820); THE ROLLERGIRLS (II,2203); SHIELDS AND YARNELL (II,2328); STRIKE FORCE (II,2480)

CASSIDY, JOHN
THE MARTIAN CHRONICLES (II,1635); OPPENHEIMER (II,1919)

CASSIDY, MARTIN
THE DAIN CURSE (II,613)

CASSIDY, MAUREEN
KAY KYSER'S KOLLEGE OF MUSICAL KNOWLEDGE (I,2512)

CASSIDY, MIKE
THE BEST OF FRIENDS (II,216)

CASSIDY, PATRICK
BAY CITY BLUES (II,186); THE SIX OF US (II,2369)

CASSIDY, RYAN
THE FACTS OF LIFE (II,805)

CASSIDY, SHAUN
BREAKING AWAY (II,371); THE GOLDIE HAWN SPECIAL (II,1024); THE HARDY BOYS MYSTERIES (II,1090)

CASSIDY, STEPHEN
BIG JOHN, LITTLE JOHN (II,240)

CASSIDY, TED
THE ADDAMS FAMILY (II,11); THE ADDAMS FAMILY (II,12); THE ADDAMS FAMILY (II,13); BENNY AND BARNEY: LAS VEGAS UNDERCOVER (II,206); BIRDMAN (II,256); BIRDMAN AND THE GALAXY TRIO (I,478); THE FANTASTIC FOUR (II,824); FLASH GORDON—THE GREATEST ADVENTURE OF ALL (II,875); FRANKENSTEIN JR. AND THE IMPOSSIBLES (I,1672); GENESIS II (I,1760); GODZILLA (II,1014); I DREAM OF JEANNIE (I,2167); JACK AND THE BEANSTALK (I,2288); THE NEW ADVENTURES OF HUCKLEBERRY FINN (I,3250); PLANET EARTH (I,3615); THE SIX-MILLION-DOLLAR MAN (II,2372); TARZAN: LORD OF THE JUNGLE (II,2544); THE WORLD'S GREATEST SUPER HEROES (II,2842)

CASSINI, IGOR
THE IGOR CASSINI SHOW (I,2188)

CASSISI, JOHN
FISH (II,864); SHEEHY AND THE SUPREME MACHINE (II,2322)

CASSITY, KRAIG
OPERATION PETTICOAT (II,1916); OPERATION PETTICOAT (II,1917); STEELTOWN (II,2452)

CASSMORE, JUDY
THE DON RICKLES SHOW (I,1342)

CASSON, MEL
DRAW ME A LAUGH (I,1372)

CASSORT, VALERIE
THE FIRST HUNDRED YEARS (I,1581)

CAST, EDWARD
SECRETS OF THE OLD BAILEY (I,3971)

CAST, TRICIA
THE BAD NEWS BEARS (II,134); IT'S A LIVING (II,1254); IT'S YOUR MOVE (II,1259); NIGHT PARTNERS (II,1847)

CASTANG, VERONICA
KENNEDY (II,1377)

CASTEL, NICO
AMAHL AND THE NIGHT VISITORS (I,150)

CASTELLANO, MARGARET E
THE SUPER (I,4293)

CASTELLANO, RICHARD
THE GANGSTER CHRONICLES (II,957); JOE AND SONS (II,1296); JOE AND SONS (II,1297); THE SUPER (I,4293)

CASTELLARI, ENZO
THE WINDS OF WAR (II,2807)

CASTELLO, ANTHONY
SAVAGE: IN THE ORIENT (II,2264)

CASTELLONOS, RAUL
TONIGHT IN HAVANA (I,4557)

CASTELNUOVO, NINO
THREE COINS IN THE FOUNTAIN (I,4473)

CASTILLO, GERALD
I'D RATHER BE CALM (II,1216); THE RENEGADES (II,2132)

CASTLE SISTERS, THE
GLENN MILLER TIME (I,1824)

CASTLE, GENE
BONNIE AND THE FRANKLINS (II,349)

CASTLE, JO ANN
THE LAWRENCE WELK SHOW (I,2643); MEMORIES WITH LAWRENCE WELK (II,1681)

CASTLE, JOAN
A WOMAN TO REMEMBER (I,4888)

CASTLE, JOHN
PENMARRIC (II,1993); THE PRIME OF MISS JEAN BRODIE (II,2082)

CASTLE, KATE
AMAHL AND THE NIGHT VISITORS (I,150)

CASTLE, MARY
LAST STAGECOACH WEST (I,2629); STORIES OF THE CENTURY (I,4235)

CASTLE, NICK
THE CATERINA VALENTE SHOW (I,873)

CASTLE, PAUL
JACK CARTER AND COMPANY (I,2305)

CASTLE, PEGGIE
FOUR STAR PLAYHOUSE (I,1652); LAWMAN (I,2642); ZANE GREY THEATER (I,4979)

CASTLE, ROY
ALICE THROUGH THE LOOKING GLASS (I,122)

CASTLEMAN, BOOMER
THE KOWBOYS (I,2584)

CASTRO SISTERS, THE
THE BOB HOPE SHOW (I,524)

CASTRONOVA, T J
POLICE WOMAN: THE GAMBLE (II,2064); TAXI (II,2546)

CATALLI, LISA
CAMP WILDERNESS (II,419)

CATALLI, PHIL
CAMP WILDERNESS (II,419)

CATANZARO, TONY
ANNIE GET YOUR GUN (I,224)

CATCHING, BILL
THE LONER (I,2744)

CATES, MADELYN
ARCHIE BUNKER'S PLACE (II,105)

CATES, MARCY
THAT GOOD OLD NASHVILLE MUSIC (II,2571)

CATES, MARGIE
THAT GOOD OLD NASHVILLE MUSIC (II,2571)

CATES, PHOEBE
LACE (II,1418)

CATHCART, DICK
PETE KELLY'S BLUES (I,3559)

CATHEY, DALTON
NEVER AGAIN (II,1808)

CATLETT, MARY JO
DIFF'RENT STROKES (II,674); FOUL PLAY (II,910); HANDLE WITH CARE (II,1077); M*A*S*H (II,1569); THE MCLEAN STEVENSON SHOW (II,1664); OVER AND OUT (II,1938); SEMI-TOUGH (II,2300)

CATLETT, WALTER
THE DONALD O'CONNOR SHOW (I,1344)

CATLIN, FAITH
RYAN'S HOPE (II,2234)

CATLIN, JODY
RYAN'S HOPE (II,2234)

CATLIN, MICHAEL
CAPITOL (II,426)

CATON, MICHAEL
THE SULLIVANS (II,2490)

CATTONI, RICO
THE ROYAL FOLLIES OF 1933 (I,3862)

CATTRALL, KIM
THE BASTARD/KENT FAMILY CHRONICLES (II,159); THE GOSSIP COLUMNIST (II,1045); THE NIGHT RIDER

(II,1848); THE REBELS (II,2121); SCRUPLES (II,2280); TUCKER'S WITCH (II,2667)

CATUSI, JIM
THE DAVID FROST REVUE (I,1176); ROXY PAGE (II,2222)

CAUDELL, LANE
ARCHER—FUGITIVE FROM THE EMPIRE (II,103); BATTLES: THE MURDER THAT WOULDN'T DIE (II,180); DAYS OF OUR LIVES (II,629); GOOD OL' BOYS (II,1035)

CAULDWELL, ALAN
IVAN THE TERRIBLE (II,1260)

CAULFIELD, JOAN
THE BOB HOPE SHOW (I,558); THE BOB HOPE SHOW (I,576); THE BOB HOPE SHOW (I,611); THE HIGH CHAPARRAL (I,2051); THE LUX VIDEO THEATER (I,2795); THE MAGICIAN (I,2824); MY FAVORITE HUSBAND (I,3188); PURSUIT (I,3700); SALLY (I,3889); SALLY (I,3890); SCREEN DIRECTOR'S PLAYHOUSE (I,3946)

CAULK, RICHARD
THAT MCMAHON'S HERE AGAIN (I,4419)

CAVA, MARIO DELLA
THE SMALL MIRACLE (I,4090)

CAVADINI, CATHY
KIDD VIDEO (II,1388)

CAVALLERO, GAYLORD
THE FREEWHEELERS (I,1683)

CAVANAUGH, MICHAEL
DOCTOR SCORPION (II,685); HUNTER (II,1206); NIGHT PARTNERS (II,1847); THE TEXAS RANGERS (II,2567)

CAVANAUGH, PAUL
ANDY'S GANG (I,214); SPACE PATROL (I,4144)

CAVANAUGH, VIRGINIA
THE THORNTON SHOW (I,4468)

CAVELL, BUTCH
MR. I. MAGINATION (I,3140)

CAVELL, MARC
DIAGNOSIS: DANGER (I,1250); PISTOLS 'N' PETTICOATS (I,3608)

CAVENDISH, BRENDA
THE NEXT VICTIM (I,3281)

CAVETT, DICK
BOB HOPE SPECIAL: BOB HOPE'S SUPER BIRTHDAY SPECIAL (II,319); THE DICK CAVETT SHOW (I,1264); THE DICK CAVETT SHOW (II,669); DICK CAVETT'S BACKLOT USA (II,670); FEELING GOOD (I,1559); FUNNY GIRL TO FUNNY LADY (II,942); NIGHTSIDE (I,3296); PARADE OF STARS (II,1954); PLIMPTON! DID YOU HEAR THE ONE ABOUT. . .? (I,3629); THIS MORNING (I,4464); TIN PAN ALLEY TODAY (I,4521); YESTERYEAR (II,2851)

CAVONIS, PAUL
THE CITY (II,541); THE LIVING END (I,2731)

CAWDRON, ROBERT
FROM A BIRD'S EYE VIEW (I,1686)

CAZENOVE, CHRISTOPHER
COLOR HIM DEAD (I,1007); COLOR HIM DEAD (II,554)

CEASAR, JOHN
AN ECHO OF THERESA (I,1399)

CEDAR, LARRY
AND THEY ALL LIVED HAPPILY EVER AFTER (II,75); FRIENDS (II,928); W*A*L*T*E*R (II,2731)

CELLIER, PETER
THE LAST DAYS OF POMPEII (II,1436); SOMEONE AT THE TOP OF THE STAIRS (I,4114)

CELLINO, CATERINA
WELCOME BACK, KOTTER (II,2761)

CENEDELLA, ROBERT
JOHNNY GARAGE (II,1337)

CERUSICO, ENZO
MY FRIEND TONY (I,3192)

CERVANTES, DON
THE GREATEST AMERICAN HERO (II,1060)

CERVERA JR, JORGE
THE SAN PEDRO BUMS (II,2251); VIVA VALDEZ (II,2728)

CESANA, RENZO
THE BACHELOR (I,325); THE CONTINENTAL (I,1042)

CESAR, GEORGE
HOLLYWOOD JUNIOR CIRCUS (I,2084)

CEZON, CONNIE
PERRY MASON (II,2025)

CHAD MITCHELL TRIO, THE
76 MEN AND PEGGY LEE (I,3989)

CHADWICK, JOHN
WOMAN'S PAGE (II,2822)

CHADWICK, JUNE
V: THE SERIES (II,2715)

CHADWICK, ROBERT
THE INDIAN (I,2210)

CHADWICK, SUZANNE
WOMAN'S PAGE (II,2822)

CHAFFEE, SUZY
PERRY COMO'S LAKE TAHOE HOLIDAY (II,2018)

CHAGRIN, JULIE
THE GOLDDIGGERS IN LONDON (I,1837)

CHAGRIN, NICHOLAS
LACE (II,1418)

CHAI-LEE
THE BENNY HILL SHOW (II,207)

CHAKIRIS, GEORGE
KISMET (I,2565); RETURN TO FANTASY ISLAND (II,2144); A SALUTE TO TELEVISION'S 25TH ANNIVERSARY (I,3893); THE SAVAGE CURSE (I,3925); YOU'RE THE TOP (I,4974)

CHALLEE, WILLIAM
THE CODE OF JONATHAN WEST (I,992)

CHALMERS, THOMAS
THE FARMER'S DAUGHTER (I,1534)

CHAMBERLAIN, CHARLES
DON MESSER'S JUBILEE (I,1338)

CHAMBERLAIN, JEFF
CAPITOL (II,426)

CHAMBERLAIN, LEE
ALL'S FAIR (II,46); PARIS (II,1957); ROOTS: THE NEXT GENERATIONS (II,2213); WHAT'S HAPPENING!! (II,2769)

CHAMBERLAIN, RICHARD
THE BOB HOPE SHOW (I,602); CENTENNIAL (II,477); DOCTOR KILDARE (I,1315); HAMLET (I,1925); HOLLYWOOD MELODY (I,2085); PERRY COMO'S CHRISTMAS IN THE HOLY LAND (II,2013); SHOGUN (II,2339); THE THORN BIRDS (II,2600); THRILLER (I,4492)

CHAMBERLAIN, SALLY
MAGNAVOX THEATER (I,2827)

CHAMBERLAIN, WILT
THE CANDID CAMERA SPECIAL (II,422); THE CELEBRITY FOOTBALL CLASSIC (II,474); THE SHAPE OF THINGS (I,3999)

CHAMBERS, DAN
RITUALS (II,2182)

CHAMBERS, JOAN
PORTIA FACES LIFE (I,3653)

CHAMBERS, JULIA
FANNY BY GASLIGHT (II,823)

CHAMBERS, PHIL
THE LONG HOT SUMMER (I,2747)

CHAMBERS, RALPH
ABE LINCOLN IN ILLINOIS (I,10)

CHAMBERS, WHEATON
WILLY (I,4868)

CHAMBLISS, WOODY
GUNSMOKE (II,1069); LASSIE: THE NEW BEGINNING (II,1433); THERE'S ALWAYS ROOM (II,2583)

CHAMINO, LEONARDO
RYAN'S HOPE (II,2234)

CHAMO, STUART
THE DAY THE WOMEN GOT EVEN (II,628)

CHAMPION, GOWER
ACCENT ON LOVE (I,16); THE ADMIRAL BROADWAY REVUE (I,37); AMERICA PAUSES FOR SPRINGTIME (I,166); THE MARGE AND GOWER CHAMPION SHOW (I,2907); MARRIAGE—HANDLE WITH CARE (I,2921); STANDARD OIL ANNIVERSARY SHOW (I,4189)

CHAMPION, JEAN
SMILEY'S PEOPLE (II,2383)

CHAMPION, MARGE
ACCENT ON LOVE (I,16); THE ADMIRAL BROADWAY REVUE (I,37); AMERICA PAUSES FOR SPRINGTIME (I,166); THE MARGE AND GOWER CHAMPION SHOW (I,2907); MARRIAGE—HANDLE WITH CARE (I,2921); STANDARD OIL ANNIVERSARY SHOW (I,4189)

CHAMPION, SANDY
PLEASURE COVE (II,2058)

CHAMPLIN, IRENE
FLASH GORDON (I,1593)

CHAN, AGNES
MARCO POLO (II,1626)

CHANCE, LARRY
NORTHWEST PASSAGE (I,3306)

CHANCE, NAOMI
SECRETS OF THE OLD BAILEY (I,3971)

CHANCER, NORMAN
MASTER OF THE GAME (II,1647)

CHANDLER, CHICK
IT'S ALWAYS SUNDAY (I,2265); THE LORETTA YOUNG THEATER (I,2756); NAVY LOG (I,3233); ONE HAPPY FAMILY (I,3375); SOLDIERS OF FORTUNE (I,4111)

CHANDLER, FREDDI
RAGE OF ANGELS (II,2107)

CHANDLER, GEORGE
THE ABBOTT AND COSTELLO SHOW (I,2); THE BASTARD/KENT FAMILY CHRONICLES (II,159); ICHABOD AND ME (I,2183); INDEMNITY (I,2209); MCGHEE (I,2967); TIMMY AND LASSIE (I,4520)

CHANDLER, JAMES
THE MISS AND THE MISSILES (I,3062); THE TRACER (I,4588)

CHANDLER, JOAN
TELLER OF TALES (I,4388)

CHANDLER, JOHN
SAWYER AND FINN (II,2265); THE TRAVELS OF JAIMIE MCPHEETERS (I,4596)

CHANDLER, KEN
FITZ AND BONES (II,867); TERROR AT ALCATRAZ (II,2563)

CHANDLER, ROBIN
MEET YOUR COVER GIRL (I,2992); QUICK ON THE DRAW (I,3711); REVLON MIRROR THEATER (I,3781); VANITY FAIR (I,4702)

CHANDLER, SIMON
LACE (II,1418)

CHANDRA KAY DANCERS, THE
YOUR SHOW OF SHOWS (I,4957)

CHANEL, LORAINE
FALCON'S GOLD (II,809); NEVADA SMITH (II,1807)

CHANEY JR, LON
CAVALCADE OF AMERICA (I,875); COSMOPOLITAN THEATER (I,1057); THE GENERAL ELECTRIC THEATER (I,1753); HAWKEYE AND THE LAST OF THE MOHICANS (I,1975); NUMBER 13 DEMON STREET (I,3316); PISTOLS 'N' PETTICOATS (I,3608); TALES OF TOMORROW (I,4352); TELEPHONE TIME (I,4379)

CHANEY, FRANCES
TALES FROM THE DARKSIDE (II,2534)

CHANEY, JAN
THE ALAN KING SHOW (I,100); ON TRIAL (I,3363)

CHANEY, MELODIE
THE LIFE OF RILEY (I,2686)

CHANG, HARRY
THE MACKENZIES OF PARADISE COVE (II,1585)

CHANNING, CAROL
THE BEST ON RECORD (I,409); THE BIG PARTY FOR REVLON (I,427); BOB HOPE SPECIAL: THE BOB HOPE SPECIAL (II,333); CAROL CHANNING AND 101 MEN (I,849); CAROL CHANNING AND PEARL BAILEY ON BROADWAY (I,850); CAROL CHANNING PROUDLY PRESENTS THE SEVEN DEADLY SINS (I,851); CRESCENDO (I,1094); DANNY THOMAS LOOKS AT YESTERDAY, TODAY AND TOMORROW (I,1148); AN EVENING WITH CAROL CHANNING (I,1473); GEORGE BURNS CELEBRATES 80 YEARS IN SHOW BUSINESS (II,967); I'M A FAN (I,2192); THE LOVE BOAT (II,1535); ONE MORE TIME (I,3386); PARADE OF STARS (II,1954); PRUDENTIAL FAMILY PLAYHOUSE (I,3683); SVENGALI AND THE BLONDE (I,4310); THE WONDERFUL WORLD OF BURLESQUE II (I,4896)

CHANNING, STOCKARD
THE EDDIE RABBITT SPECIAL (II,759); LUCAN (II,1553); STOCKARD CHANNING IN JUST FRIENDS (II,2467); THE STOCKARD CHANNING SHOW (II,2468)

CHAO, ROSALIND
AFTERMASH (II,23); ALMOST AMERICAN (II,55); ANNA AND THE KING (I,221); BATTLE OF THE NETWORK STARS (II,177); DIFF'RENT STROKES (II,674); GILLIGAN'S ISLAND: THE HARLEM GLOBETROTTERS ON GILLIGAN'S ISLAND (II,993); M*A*S*H (II,1569); MOONLIGHT (II,1730); THE ULTIMATE IMPOSTER (II,2700)

CHAPEL, LOYITA
BEHIND THE SCREEN (II,200); THE LIFE AND TIMES OF EDDIE ROBERTS (II,1468); NEVER AGAIN (II,1808); PUNKY BREWSTER (II,2092); TWO THE HARD WAY (II,2695)

CHAPIN, BILLY
TV READERS DIGEST (I,4630); WATERFRONT (I,4774)

CHAPIN, DOUG
SOMERSET (I,4115)

CHAPIN, LAUREN
FATHER KNOWS BEST (II,838); FATHER KNOWS BEST: HOME FOR CHRISTMAS (II,839); FATHER KNOWS BEST: THE FATHER KNOWS BEST REUNION (II,840)

CHAPIN, MILES
BULBA (II,392)

CHAPIN, TOM
EVERYDAY (II,793); MAKE A WISH (I,2838)

CHAPLIN, GERALDINE
THE WORD (II,2833)

CHAPLIN, SYDNEY
WONDERFUL TOWN (I,4893)

CHAPLINE, ROGER
THE CHEAP SHOW (II,491)

CHAPMAN, GRAHAM
THE BIG SHOW (II,243); MONTY PYTHON'S FLYING CIRCUS (II,1729)

CHAPMAN, JOSEPH
COMPUTERCIDE (II,561); JUDGEMENT DAY (II,1348)

CHAPMAN, JUDITH
AS THE WORLD TURNS (II,110); BEYOND WESTWORLD (II,227); THE FALL GUY (II,811); FARRELL: FOR THE PEOPLE (II,834); NICK AND THE DOBERMANS (II,1843); RETURN OF THE MAN FROM U.N.C.L.E.: THE 15 YEARS LATER AFFAIR (II,2140); RYAN'S HOPE (II,2234)

CHAPMAN, LIZA
ANOTHER WORLD (II,97)

CHAPMAN, LONNY
BIG ROSE (II,241); THE DANGEROUS DAYS OF KIOWA JONES (I,1140); FOR THE PEOPLE (I,1626); THE INVESTIGATOR (I,2230); THE LLOYD BRIDGES SHOW (I,2733); ONE STEP BEYOND (I,3388); THE RAINMAKER (II,2109); RIDING HIGH (II,2173); STARLIGHT THEATER (I,4201)

CHAPMAN, MARGOT
THE STARLAND VOCAL BAND (II,2441)

CHAPMAN, MARGUERITE
HOLLYWOOD OPENING NIGHT (I,2087); PEPSI-COLA PLAYHOUSE (I,3523); STUDIO '57 (I,4267); TV READERS DIGEST (I,4630)

CHAPMAN, PATTE
DUFFY'S TAVERN (I,1379); YOU SHOULD MEET MY SISTER (I,4935)

CHAPMAN, PAUL
TEN WHO DARED (II,2559)

CHAPMAN, ROBERT
FATHER KNOWS BEST (II,838)

CHAPMAN, SYDNEY
THE PHIL SILVERS PONTIAC SPECIAL: KEEP IN STEP (I,3579)

CHAPPEL, BETTY
GARROWAY AT LARGE (I,1737)

CHAPPELL, BILLY
THE ICE PALACE (I,2182)

CHAPPELL, ERNEST
VOLUME ONE (I,4742); WORLD'S FAIR BEAUTY SHOW (I,4920)

CHAPPELL, JAN
BLAKE'S SEVEN (II,264)

CHAPPELL, JOHN
AFTERMASH (II,23); GOOBER AND THE TRUCKERS' PARADISE (II,1029); THE REBELS (II,2121)

CHAPTER 5
THE ANITA BRYANT SPECTACULAR (II,82); THE PETER MARSHALL VARIETY SHOW (II,2029)

CHARISSE, CYD
THE BOB HOPE SHOW (I,625); THE BOB HOPE SHOW (I,631); CENTER STAGE: CYD CHARISSE (I,889); THE CYD CHARISSE SPECIAL (I,1118); FOL-DE-ROL (I,1613); GENE KELLY. . . AN AMERICAN IN PASADENA (II,962); MEET CYD CHARISSE (I,2982); THE PERRY COMO SPECIAL (I,3534); SPRING HOLIDAY (I,4168); THE WONDERFUL WORLD OF BURLESQUE III (I,4897)

CHARISSE, ZAN
MARIE (II,1628)

CHARKHAM, DAVID
ABC STAGE '67 (I,6)

CHARLES FALER DANCERS, THE
DON'S MUSICAL PLAYHOUSE (I,1348)

CHARLES, ANNETTE
MUSICAL COMEDY TONIGHT (II,1777)

CHARLES, KEITH
AFTER HOURS: SINGIN', SWINGIN' AND ALL THAT JAZZ (II,22); AS THE WORLD TURNS (II,110); THE EDGE OF NIGHT (II,760); LOVE OF LIFE (I,2774); MASON (II,1643); SEARCH FOR TOMORROW (II,2284); THE SECRET STORM (I,3969); SIDE BY SIDE (II,2347)

CHARLES, LEWIS
THE FEATHER AND FATHER GANG (II,845); THE INDIAN (I,2210)

CHARLES, RAY
ANDY WILLIAMS KALEIDOSCOPE COMPANY (I,201); BARBRA STREISAND AND OTHER MUSICAL INSTRUMENTS (I,353); THE BOB HOPE SHOW (I,631); CELEBRATION: THE AMERICAN SPIRIT (II,461); COTTON CLUB '75 (II,580); COUNTRY COMES HOME (II,585); THE DINAH SHORE SPECIAL (I,1283); DUKE ELLINGTON. . . WE LOVE YOU MADLY (I,1380); FIFTY YEARS OF COUNTRY MUSIC (II,853); IT'S WHAT'S HAPPENING, BABY! (I,2278); JOHNNY CASH: SPRING FEVER (II,1335); THE KENNY ROGERS SPECIAL (II,1382); LYNDA CARTER'S CELEBRATION (II,1563); SALUTE TO LADY LIBERTY (II,2243); THE SECOND BARRY MANILOW SPECIAL (II,2285); A SPECIAL KENNY ROGERS (II,2417)

CHARLES, WALTER
SWEENEY TODD (II,2512)

CHARLES, ZACHARY
ANNIE GET YOUR GUN (I,223); A TIME TO LIVE (I,4510)

CHARLESON, IAN
MASTER OF THE GAME (II,1647)

CHARLESON, LESLIE
ANOTHER APRIL (II,93); THE BLACK SHEEP SQUADRON (II,262); GENERAL HOSPITAL (II,964); GUESS WHO'S COMING TO DINNER? (II,1063); MOST WANTED (II,1738)

CHARLESON, RAY
PRISONERS OF THE LOST UNIVERSE (II,2086)

CHARLIE BYRD TRIO, THE
THE KATE SMITH SHOW (I,2511)

CHARLIE DANIELS BAND, THE
CHERYL LADD. . . LOOKING BACK—SOUVENIRS (II,502); CRYSTAL (II,607); THE ROBERT KLEIN SHOW (II,2185); SALUTE TO LADY LIBERTY (II,2243)

CHARLIE, C D
SUPER CIRCUS (I,4287)

CHARLTON JOHNSON DANCERS, THE
THE SCOEY MITCHLLL SHOW (I,3939)

CHARLTON, ALETHA
SOMEONE AT THE TOP OF THE STAIRS (I,4114)

CHARNEY, JORDAN
THE ANDROS TARGETS (II,76); ANOTHER WORLD (II,97); MARRIAGE IS ALIVE AND WELL (II,1633); ONE LIFE TO LIVE (II,1907); PRIVATE BENJAMIN (II,2087); SOMERSET (I,4115); THREE'S COMPANY (II,2614)

CHARNIN, MARTIN
OPRYLAND: NIGHT OF STARS AND FUTURE STARS (II,1921)

CHARNY, SUZANNE
THE BOB HOPE SHOW (I,639)

CHARO
BOB HOPE SPECIAL: BOB HOPE'S ALL-STAR BIRTHDAY PARTY (II,295); BOB HOPE SPECIAL: BOB HOPE'S ALL-STAR COMEDY SPECIAL FROM AUSTRALIA (II,300); BOB HOPE SPECIAL: BOB HOPE'S ALL-STAR COMEDY SPECTACULAR FROM LAKE TAHOE (II,301); BOB HOPE SPECIAL: HAPPY BIRTHDAY, BOB! (II,326); BOB HOPE SPECIAL: THE BOB HOPE SPECIAL (II,335); CELEBRITY CHALLENGE OF THE SEXES 4 (II,468); THE CELEBRITY FOOTBALL CLASSIC (II,474); CHARO (II,488); CHARO AND THE SERGEANT (II,489); CHICO AND THE MAN (II,508); THE LOVE BOAT (II,1535); THE MAD MAD MAD MAD WORLD OF THE SUPER BOWL (II,1586); PAUL LYNDE GOES M-A-A-A-AD (II,1980); PERRY COMO'S EASTER IN GUADALAJARA (II,2015)

CHARONE, ERVIN
CHICAGOLAND MYSTERY PLAYERS (I,936)

CHARTOFF, MELANIE
BATTLE OF THE NETWORK STARS (II,172); FRIDAYS (II,926); OH MADELINE (II,1880); WHATEVER BECAME OF. . .? (II,2773)

CHASE, BARRIE
ANOTHER EVENING WITH FRED ASTAIRE (I,229); ASTAIRE TIME (I,305); AN EVENING WITH FRED ASTAIRE (I,1474); THE FRED ASTAIRE SHOW (I,1677); THE FRED ASTAIRE SPECIAL (I,1678); THINK PRETTY (I,4448)

CHASE, BUD
THIS IS MUSIC (I,4458)

CHASE, CHAZ
THE TWO RONNIES (II,2694)

CHASE, CHEVY
THE CHEVY CHASE NATIONAL HUMOR TEST (II,504); THE CHEVY CHASE SHOW (II,505); THE GREAT AMERICAN DREAM MACHINE (I,1871); THE PAUL SIMON SPECIAL (II,1983); PLAYBOY'S 25TH ANNIVERSARY CELEBRATION (II,2054); SATURDAY NIGHT LIVE (II,2261); TOM SNYDER'S CELEBRITY SPOTLIGHT (II,2632)

CHASE, CHRISTY
MUSIC CITY NEWS TOP COUNTRY HITS OF THE YEAR (II,1772)

CHASE, ERIC
AMANDA FALLON (I,152); HERE COME THE BRIDES (I,2024); ROBERT YOUNG AND THE FAMILY (I,3813)

CHASE, ILKA
CELEBRITY TIME (I,887); THE CHEVROLET TELE-THEATER (I,926); FASHION MAGIC (I,1536); GLAMOUR-GO-ROUND (I,1818); THE PATTY DUKE SHOW (I,3483); SILVER THEATER (I,4051); TONIGHT AT 8:30 (I,4556); THE TRIALS OF O'BRIEN (I,4605)

CHASE, LAUREN
MR. SMITH (II,1755)

CHASE, SENATOR CLIFFORD
MR. CITIZEN (I,3133)

CHASE, STEPHEN
CONCERNING MISS MARLOWE (I,1032); MEET CORLISS ARCHER (I,2981)

CHASNOFF, PHIL
THE HAMPTONS (II,1076)

CHASON, MYRA
FLAMINGO ROAD (II,872)

CHASSE, BETSY
CHARLES IN CHARGE (II,482); GOLDIE AND KIDS: LISTEN TO ME (II,1021)

CHASTAIN, DON
CALAMITY JANE (I,793); THE DEBBIE REYNOLDS SHOW (I,1214); GENERAL HOSPITAL (II,964); THE GUIDING LIGHT (II,1064); HELLO, LARRY (II,1128); OF MEN OF WOMEN (I,3332); SEARCH FOR TOMORROW (II,2284)

CHATER, GEOFFREY
LOOK BACK IN DARKNESS (I,2752); MOLL FLANDERS (II,1720)

CHATTERTON, RUTH
PRUDENTIAL FAMILY PLAYHOUSE (I,3683)

CHATTIN, SARA
PLANET EARTH (I,3615)

CHATWIN, CLAIR
YOUR HIT PARADE (I,4952)

CHAUVAN, LILYAN
FLYING HIGH (II,895); RYAN'S FOUR (II,2233); THE YOUNG AND THE RESTLESS (II,2862)

CHAYETTE, JEFF
MODESTY BLAISE (II,1719)

CHEATHAM, MARIE
DAYS OF OUR LIVES (II,629); SEARCH FOR TOMORROW (II,2284)

CHECKER, CHUBBY
20TH CENTURY FOLLIES (I,4641); DICK CLARK PRESENTS THE ROCK AND ROLL YEARS (I,1265)

CHECKMATES, THE
PLAYBOY AFTER DARK (I,3620)

CHEE, CURTIS
THE RETURN OF LUTHER GILLIS (II,2137)

CHEEK, MOLLY
CHICAGO STORY (II,506); CHIPS (II,511); THE YEAGERS (II,2843)

CHEERLEADERS, THE
THE GORDON MACRAE SHOW (I,1859)

CHEFFY, MARY
BROTHER RAT (I,746)

CHELSOM, PETER
A WOMAN OF SUBSTANCE (II,2820)

CHER
CHER (II,495); CHER (II,496); CHER AND OTHER FANTASIES (II,497); CHER. . . SPECIAL (II,499); CHER—A CELEBRATION AT CAESAR'S PALACE (II,498); THE FIRST NINE MONTHS ARE THE HARDEST (I,1584); THE FLIP WILSON SPECIAL (II,889); HOW TO HANDLE A WOMAN (I,2146); THE SONNY AND CHER COMEDY HOUR (II,2400); THE SONNY AND CHER SHOW (II,2401); TOM SNYDER'S CELEBRITY SPOTLIGHT (II,2632); WHERE THE GIRLS ARE (I,4830)

CHERNEY, MICKEY
COUNTERATTACK: CRIME IN AMERICA (II,581); KUDZU (II,1415)

CHERRELL, GWEN
BRIEF ENCOUNTER (I,724)

CHERRY, BYRON
BATTLE OF THE NETWORK STARS (II,175); THE DUKES (II,741); THE DUKES OF HAZZARD (II,742)

CHERRY, DON
THE BOB HOPE SHOW (I,518); BYLINE—BETTY FURNESS (I,785); MEET BETTY FURNESS (I,2978); PENTHOUSE PARTY (I,3515); THE PETER LIND HAYES SHOW (I,3564)

CHERRY, FRANK
THE GANGSTER CHRONICLES (II,957)

CHERRY, ROBERT
THE ABBOTT AND COSTELLO SHOW (I,2)

CHESHIRE, ELIZABETH
CAPTAINS AND THE KINGS (II,435); THE FAMILY HOLVAK (II,817); OUR TOWN (I,3421); SUNSHINE (II,2495)

CHESHIRE, HARRY
BUFFALO BILL JR. (I,755); IF YOU KNEW TOMORROW (I,2187); MEET MILLIE (I,2988)

CHESIS, EILEEN
THE TOM EWELL SHOW (I,4538)

CHESNEY, CHARLES
THE FORTUNE HUNTER (I,1646)

CHESNEY, DIANA
FAIR EXCHANGE (I,1512)

CHESTER, COLBY
ABC'S MATINEE TODAY (I,7); FROM HERE TO ETERNITY (II,933); THE MUNSTERS' REVENGE (II,1765); MURDOCK'S GANG (I,3164); THE RANGERS (II,2112); REX HARRISON PRESENTS SHORT STORIES OF LOVE (II,2148); VEGAS (II,2723)

CHESTER, MARY
A HORSEMAN RIDING BY (II,1182)

CHETTERTON, RUTH
HAMLET (I,1924)

CHEUNG, GEORGE KEE
BRING 'EM BACK ALIVE (II,377); JOHNNY BLUE (II,1321); THE SIX O'CLOCK FOLLIES (II,2368)

CHEVALIER, MAURICE
ABC STAGE '67 (I,6); THE BOB HOPE SHOW (I,535); THE GERSHWIN YEARS (I,1779); INVITATION TO PARIS (I,2233); THE MAURICE CHEVALIER SHOW (I,2953); THE MAURICE CHEVALIER SHOW (I,2954); MAURICE CHEVALIER'S PARIS (I,2955); MUSIC BY COLE PORTER (I,3166); THE ROSALIND RUSSELL SHOW (I,3845); THE SAMMY DAVIS JR. SPECIAL (I,3900)

CHEW JR, SAM
THE BANANA COMPANY (II,139); THE BASTARD/KENT FAMILY CHRONICLES (II,159); THE BIONIC WOMAN (II,255); THE NEW OPERATION PETTICOAT (II,1826)

CHEYNE, ANGELA
GENERAL HOSPITAL (II,964)

CHIANESE, DOMINIC
RYAN'S HOPE (II,2234)

CHIANG, GEORGE
GENERAL HOSPITAL (II,964)

CHICO HAMILTON AND HIS QUINTET
TIMEX ALL-STAR JAZZ SHOW III (I,4517)

CHIEFTAINS, THE
THE MOHAWK SHOWROOM (I,3080)

CHILD, JEREMY
FATHER DEAR FATHER (II,837)

CHILD, KRISTY
PRISONER: CELL BLOCK H (II,2085)

CHILDRESS, ALVIN
THE AMOS AND ANDY SHOW (I,179); BANYON (I,345)

CHILDRESS, CAROL
MATINEE AT THE BIJOU (II,1651)

CHILDS, SUZANNE
THAT'S MY LINE (II,2579)

CHILDS, TRACEY
THE PRIME OF MISS JEAN BRODIE (II,2082)

CHILES, LINDEN
COME A RUNNING (I,1013); CONVOY (I,1043); EAST SIDE/WEST SIDE (I,1396); FREEMAN (II,923); JAMES AT 15 (II,1270); JAMES AT 15 (II,1271); THE SECRET STORM (I,3969); WASHINGTON: BEHIND CLOSED DOORS (II,2744)

CHILES, LOIS
DALLAS (II,614)

CHIN, CHAO-LI
FALCON CREST (II,808); INSPECTOR PEREZ (II,1237)

CHIN, JADE
CHARLES IN CHARGE (II,482)

CHING, WILLIAM
OUR MISS BROOKS (I,3415)

CHINH, KIEU
FLY AWAY HOME (II,893)

CHINN, ANTHONY
THE PROTECTORS (I,3681)

CHINOOK
CORKY AND WHITE SHADOW (I,1049)

CHINOWITZ, DEBORAH
I'VE HAD IT UP TO HERE (II,1221)

CHIOUS, NANG SHEEN
MEN OF THE DRAGON (II,1683)

CHIPPENDALE DANCERS, THE
THE SHAPE OF THINGS (II,2315)

CHISHOLM, SUE
NOT THE NINE O'CLOCK NEWS (II,1864)

CHITJIAN, HOWARD
H.M.S. PINAFORE (I,2066)

CHODER, JILL
NUMBER 96 (II,1869)

CHOIRLOY, MITCHELL
FRANKIE LAINE TIME (I,1676)

CHONG, CHEN
PEKING ENCOUNTER (II,1991)

CHONG, RAE DAWN
FRIENDS (II,928)

CHOON, LIM BENG
A TOWN LIKE ALICE (II,2652)

CHORDETTES, THE
ARTHUR GODFREY AND FRIENDS (I,278)

CHOTIN, AL
IT'S A HIT (I,2257)

CHOWN, AVRIL
THE HUDSON BROTHERS RAZZLE DAZZLE COMEDY SHOW (II,1201)

CHOY, KANANI
THE RETURN OF LUTHER GILLIS (II,2137)

CHRIS, MARILYN
ALL MY CHILDREN (II,39); ONE LIFE TO LIVE (II,1907); THE SECRET WAR OF JACKIE'S GIRLS (II,2294)

CHRISTIAN, LEIGH
DELTA COUNTY, U.S.A. (II,657)

CHRISTIAN, LINDA
LAST OF THE PRIVATE EYES (I,2628)

CHRISTIAN, MICHAEL
PEYTON PLACE (I,3574)

CHRISTIAN, ROBERT
ANOTHER WORLD (II,97); PIAF (II,2039); ROLL OF THUNDER, HEAR MY CRY (II,2202)

CHRISTIAN, ROSEANNA
THIS IS KATE BENNETT (II,2597)

CHRISTIANSON, TODD
ON STAGE AMERICA (II,1893)

CHRISTIE, AUDREY
THE ALCOA HOUR (I,108); FAIR EXCHANGE (I,1512); GREAT DAY (II,1053); THE INNER SANCTUM (I,2216); THE JIMMY DURANTE SHOW (I,2389); JOEY FAYE'S FROLICS (I,2406)

CHRISTIE, BOB
MOMENT OF TRUTH (I,3084)

CHRISTIE, DICK
ACE CRAWFORD, PRIVATE EYE (II,6)

CHRISTIE, JULIE
SEPARATE TABLES (II,2304)

CHRISTIE, MADELEINE
THE PRIME OF MISS JEAN BRODIE (II,2082)

CHRISTINE, VIRGINIA
THE TALES OF WELLS FARGO (I,4353); THE UNEXPLAINED (I,4675)

CHRISTMAS, ERIC
THE BLUE KNIGHT (II,277); GIDEON (I,1792); THE SANDY DUNCAN SHOW (I,3910)

CHRISTOPHER, ANTHONY
MR. BLACK (I,3129)

CHRISTOPHER, EUNICE
THE HOME FRONT (II,1171); KOJAK (II,1406)

CHRISTOPHER, JORDAN
SECRETS OF MIDLAND HEIGHTS (II,2296)

CHRISTOPHER, JOSEPH
AS THE WORLD TURNS (II,110)

CHRISTOPHER, JOYCE REEHLING
FIFTH OF JULY (II,851)

CHRISTOPHER, JUDITH
WONDER WOMAN (SERIES 2) (II,2829)

CHRISTOPHER, KAY
DOCTOR I.Q. (I,1314); THE MARSHAL OF GUNSIGHT PASS (I,2925)

CHRISTOPHER, MICHAEL
SANDBURG'S LINCOLN (I,3907)

CHRISTOPHER, MILBOURNE
FESTIVAL OF MAGIC (I,1566)

CHRISTOPHER, STEFANIANNA
HERE COMES THE GRUMP (I,2027)

CHRISTOPHER, THOM
BUCK ROGERS IN THE 25TH CENTURY (II,384); HELLINGER'S LAW (II,1127); S*H*E (II,2235)

CHRISTOPHER, TONY
BOWZER (II,358)

CHRISTOPHER, WILLIAM
AFTERMASH (II,23); GOMER PYLE, U.S.M.C. (I,1843); M*A*S*H (II,1569)

CHRISTY, AL
THE RETURN OF MARCUS WELBY, M.D (II,2138)

CHRISTY, JULIE
FARAWAY HILL (I,1532)

CHRISTY, JUNE
TIMEX ALL-STAR JAZZ SHOW I (I,4515)

CHRISTY, KEN
GETAWAY CAR (I,1783)

CHU, GLORIA
SING ALONG WITH MITCH (I,4057)

CHUCK CASSEY SINGERS, THE
THE JIMMY DEAN SHOW (I,2385); THE NEW CHRISTY MINSTRELS SHOW (I,3259)

CHUCK DAVIS DANCERS, THE
THE RICHARD PRYOR SHOW (II,2163)

CHULEY, BENJAMIN
THE MARY TYLER MOORE SHOW (II,1640)

CHUNG, BYRON
THE BLACK SHEEP SQUADRON (II,262); SEARCH (I,3951)

CHURCH, ELAINE
DOCTORS HOSPITAL (II,691)

CHURCH, SANDRA
THE WOMEN (I,4890)

CHURCHILL, SARAH
CELANESE THEATER (I,881); FAITH BALDWIN'S THEATER OF ROMANCE (I,1515); THE HALLMARK HALL OF FAME (I,1922); HAMLET (I,1924); KING RICHARD II (I,2559); MATINEE THEATER (I,2947); THE SARAH CHURCHILL SHOW (I,3913)

CHURCHILL, STUART
THE FRED WARING SHOW (I,1679)

CIAMPA, CHRISTOPHER
MULLIGAN'S STEW (II,1763); MULLIGAN'S STEW (II,1764)

CIANNELLI, EDUARDO
GREAT BIBLE ADVENTURES (I,1872); JOHNNY STACCATO (I,2437); NO WARNING (I,3301)

CIARROCCHI, BETTY
THE FACTS (II,804)

CILENTO, DIANE
DIAL "M" FOR MURDER (I,1255); ESPIONAGE (I,1464); SPELL OF EVIL (I,4152); TAMING OF THE SHREW (I,4356); VANITY FAIR (I,4700)

CIMINO, LEONARDO
ABC STAGE '67 (I,6); COCAINE AND BLUE EYES (II,548); GIVE US BARABBAS! (I,1816); V (II,2713)

CINDER, ROBERT
CODE NAME: HERACLITUS (I,990); SCALPLOCK (I,3930)

CINTRON, SHARON
BARETTA (II,152)

CIOFFI, CHARLES
ANOTHER WORLD (II,97); ASSIGNMENT: VIENNA (I,304); DOG AND CAT (II,695); FLAMINGO ROAD (II,872); GET CHRISTIE LOVE! (II,981); KATE MCSHANE (II,1369); MODESTY BLAISE (II,1719); SAMURAI (II,2249)

CIOFFI, LOU
YOU ARE THERE (I,4929)

CIOLLI, AUGUSTA
THE GOLDEN AGE OF TELEVISION (II,1016)

CIPRIANI, BARNEY
AQUA VARIETIES (I,248)

CIRILLO, JOE
EISCHIED (II,763)

CISAR, GEORGE
DENNIS THE MENACE (I,1231); STAND BY FOR CRIME (I,4186)

CIVITA, DIANE
HAWAIIAN HEAT (II,1111); V (II,2713); V: THE FINAL BATTLE (II,2714)

CLAIR, DICK
THE EVERLY BROTHERS SHOW (I,1479); THE FUNNY SIDE (I,1714); THE FUNNY SIDE (I,1715); WHAT'S IT ALL ABOUT WORLD? (I,4816)

CLAIR, SYLVIE ST
CAFE DE PARIS (I,791)

CLAIRE, DOROTHY
BROADWAY TO HOLLYWOOD (I,739); HENRY MORGAN'S GREAT TALENT HUNT (I,2017); THE PAUL WINCHELL AND JERRY MAHONEY SHOW (I,3492)

CLAIRE, EDITH
MEETING AT APALACHIN (I,2994)

CLAIRE, RICHARD
DOBIE GILLIS (I,1302)

CLAMPETT, BOB
THE BEANY AND CECIL SHOW (I,374); THE BUFFALO BILLY SHOW (I,756)

CLAMPETT, SODY
THE BEANY AND CECIL SHOW (I,374)

CLANCY BROTHERS AND TOMMY MAKEM, THE
MERV GRIFFIN'S ST. PATRICK'S DAY SPECIAL (I,3018)

CLANCY, DOUG
SAINT PETER (II,2238)

CLANCY, TOM
COMPUTERCIDE (II,561); LITTLE MOON OF ALBAN (I,2714)

CLANTON, RALPH
SOMERSET (I,4115)

CLARA WARD GOSPEL SINGERS
THE BOB GOULET SHOW STARRING ROBERT GOULET (I,501)

CLARA WARD SINGERS, THE
33 1/3 REVOLUTIONS PER MONKEE (I,4451)

CLARE, DIENE
COURT-MARTIAL (I,1080)

CLARE, MARY
THE LILLI PALMER THEATER (I,2704)

CLARIDGE, SHARON
ADAM-12 (I,31)

CLARK BROTHERS, THE
THE BOB HOPE SHOW (I,515); CAVALCADE OF BANDS (I,876)

CLARK SISTERS, THE
ON THE CORNER (I,3359)

CLARK, ALEX
CLAUDIA: THE STORY OF A MARRIAGE (I,972)

CLARK, ALEXANDER
BLITHE SPIRIT (I,483)

CLARK, BOBBY
ALICE IN WONDERLAND (I,120); CASEY JONES (I,865); FOREST RANGER (I,1641); MEET THE GOVERNOR (I,2991); TEXACO STAR THEATER (I,4411)

CLARK, BRETT
NO SOAP, RADIO (II,1856)

CLARK, BRUCE
HERE COME THE DOUBLE DECKERS (I,2025)

CLARK, CANDY
CIRCUS OF THE STARS (II,533); COCAINE AND BLUE EYES (II,548); I GAVE AT THE OFFICE (II,1213); JOHNNY BELINDA (I,2418); LOVE AND LEARN (II,1529)

CLARK, CAROLYN ANN
THE GUIDING LIGHT (II,1064)

CLARK, CHERYL ANN
YOUNG LIVES (II,2866)

CLARK, CLIFF
COMBAT SERGEANT (I,1012)

CLARK, DANE
BEN HECHT'S TALES OF THE CITY (I,395); BOLD VENTURE (I,682); CONDOMINIUM (II,566); FORD THEATER HOUR (I,1635); THE FRENCH ATLANTIC AFFAIR (II,924); GRUEN GUILD PLAYHOUSE (I,1892); JUSTICE (I,2499); MAGNAVOX THEATER (I,2827); THE NASH AIRFLYTE THEATER (I,3227); THE NEW ADVENTURES OF PERRY MASON (I,3252); SHADOW OF SAM PENNY (II,2313); SURE AS FATE (I,4297); WIRE SERVICE (I,4883)

CLARK, DAVE
TV GENERAL STORE (I,4629)

CLARK, DICK
THE $20,000 PYRAMID (II,2681); AMERICAN BANDSTAND (II,70); ANIMALS ARE THE FUNNIEST PEOPLE (II,81); THE DAVID SOUL AND FRIENDS SPECIAL (II,626); DICK CLARK PRESENTS THE ROCK AND ROLL YEARS (I,1265); DICK CLARK PRESENTS THE ROCK AND ROLL YEARS (I,1266); THE DICK CLARK SATURDAY NIGHT BEECHNUT SHOW (I,1267); DICK CLARK'S GOOD OL' DAYS: FROM BOBBY SOX TO BIKINIS (II,671); DICK CLARK'S LIVE WEDNESDAY (II,672); DICK CLARK'S WORLD OF TALENT (I,1268); INSIDE AMERICA (II,1234); KINCAID (I,2543); THE KRYPTON FACTOR (II,1414); MISSING LINKS (I,3066); MR. & MS. AND THE BANDSTAND MURDERS (II,1746); THE NEW $25,000 PYRAMID (II,1811); THE OBJECT IS (I,3324); OLIVIA NEWTON-JOHN'S HOLLYWOOD NIGHTS (II,1886); SALUTE (II,2242); THE SENSATIONAL, SHOCKING, WONDERFUL, WACKY 70S (II,2302); TV'S BLOOPERS AND PRACTICAL JOKES (II,2675); TWILIGHT THEATER II (II,2686); WHERE THE ACTION IS (I,4828)

CLARK, DORAN
EMERALD POINT, N.A.S. (II,773); KING'S CROSSING (II,1397); SECRETS OF MIDLAND HEIGHTS (II,2296)

CLARK, DORT
RUN, BUDDY, RUN (I,3870); THREE STEPS TO HEAVEN (I,4485); A TIME TO LIVE (I,4510); WONDERFUL TOWN (I,4893); THE YOUNG MARRIEDS (I,4942)

CLARK, ERNEST
DOCTOR IN THE HOUSE (I,1313); THE INVISIBLE MAN (I,2232)

CLARK, EUGENE
OUT OF OUR MINDS (II,1933)

CLARK, FRED
ABC STAGE '67 (I,6); THE BEVERLY HILLBILLIES (I,417); THE BIG TIME (I,434); THE DOUBLE LIFE OF HENRY PHYFE (I,1361); EDDIE (I,1405); THE GEORGE BURNS AND GRACIE ALLEN SHOW (I,1763); THE MILTON BERLE SHOW (I,3047); MY DARLING JUDGE (I,3186); PANTOMIME QUIZ (I,3449); WHO'S AFRAID OF MOTHER GOOSE? (I,4852)

CLARK, GAGE
THE HARTMANS (I,1960); MICHAEL SHAYNE, DETECTIVE (I,3022); MR. PEEPERS (I,3147); THREE WISHES (I,4488); TWENTIETH CENTURY (I,4639)

CLARK, HARRY
THE PHIL SILVERS SHOW (I,3580)

CLARK, HOPE
LET'S CELEBRATE (I,2663)

CLARK, JACK
THE CROSS-WITS (II,605); DEALER'S CHOICE (II,635); KODAK REQUEST PERFORMANCE (I,2578); ONE HUNDRED GRAND (I,3377); THE PRICE IS RIGHT (I,3664)

CLARK, JAMES PATRICK
RYAN'S HOPE (II,2234)

CLARK, JEFF
THE KATE SMITH HOUR (I,2508)

CLARK, JOSH
GEORGE WASHINGTON (II,978)

CLARK, JUDY
TV GENERAL STORE (I,4629)

CLARK, KEN
BROCK CALLAHAN (I,741)

CLARK, LIDDY
PRISONER: CELL BLOCK H (II,2085)

CLARK, LYNNE
ROMEO AND JULIET (I,3837)

CLARK, MARLENE
SANFORD AND SON (II,2256)

CLARK, MARSHA
THE GUIDING LIGHT (II,1064)

CLARK, MATT
THE BIG EASY (II,234); DOG AND CAT (II,695); DOG AND CAT (II,696); HIGHWAY HONEYS (II,1151); MELVIN PURVIS: G-MAN (II,1679)

CLARK, MAURICE
PENMARRIC (II,1993)

CLARK, MICHAEL
DOCTOR STRANGE (II,688)

CLARK, NELL
A TIME TO LIVE (I,4510)

CLARK, OLIVER
FAME (I,1517); KAREN (II,1363); OPERATING ROOM (II,1915); THE ORPHAN AND THE DUDE (II,1924); SAWYER AND FINN (II,2265); THREE GIRLS THREE (II,2608); THE TWO OF US (II,2693); WE'VE GOT EACH OTHER (II,2757)

CLARK, PATTY
GLENN MILLER TIME (I,1824)

CLARK, PAULINE
PARADISE BAY (I,3454)

CLARK, PETULA
THE ANDY WILLIAMS SPECIAL (I,213); THE BEST ON RECORD (I,409); THE BOB HOPE SHOW (I,647); THE BRASS ARE COMING (I,708); OPRYLAND U.S.A. (I,3405); PERRY COMO'S HAWAIIAN HOLIDAY (II,2017); PERRY COMO'S OLDE ENGLISH CHRISTMAS (II,2020); PETULA (I,3572); PETULA (I,3573); PORTRAIT OF PETULA (I,3654); RODGERS AND HART TODAY (I,3829); SUNDAY NIGHT AT THE LONDON PALLADIUM (I,4286)

CLARK, PHILIP
ABC'S MATINEE TODAY (I,7); FEEL THE HEAT (II,848); TEXAS (II,2566); THE YOUNG LAWYERS (I,4941)

CLARK, REECE
TV FUNNIES (II,2672)

CLARK, ROBIN
LITTLE HOUSE: BLESS ALL THE DEAR CHILDREN (II,1489)

CLARK, ROY
THE BEVERLY HILLBILLIES (I,417); BING CROSBY AND CAROL BURNETT—TOGETHER AGAIN FOR THE FIRST TIME (I,454); BOB HOPE SPECIAL: THE BOB HOPE COMEDY SPECIAL (II,332); A COUNTRY CHRISTMAS (II,582); A COUNTRY CHRISTMAS (II,583); COUNTRY COMES HOME (II,585); COUNTRY COMES HOME (II,586); COUNTRY GALAXY OF STARS (II,587); COUNTRY MUSIC HIT PARADE (II,588); COUNTRY STARS OF THE 70S (II,594); FIFTY YEARS OF COUNTRY MUSIC (II,853); THE GEORGE HAMILTON IV SHOW (I,1769); HEE HAW (II,1123); THE JOHNNY CASH CHRISTMAS SPECIAL (II,1325); THE JOHNNY CASH CHRISTMAS SPECIAL (II,1326); JUBILEE (II,1347); THE KRAFT 75TH ANNIVERSARY SPECIAL (II,1409); THE MAC DAVIS CHRISTMAS SPECIAL (II,1572); MITZI. . . ZINGS INTO SPRING (II,1713); MOVIN' (I,3117); THE NASHVILLE PALACE (II,1793); THE NASHVILLE PALACE (II,1792); PIONEER SPIRIT (I,3604); THE SENSATIONAL, SHOCKING, WONDERFUL, WACKY 70S (II,2302); STAR SEARCH (II,2437); SWING OUT, SWEET LAND (II,2515); SWINGING COUNTRY (I,4322); USA COUNTRY MUSIC (II,591)

CLARK, SUSAN
MCNAUGHTON'S DAUGHTER (II,1670); SHERLOCK HOLMES (II,2327); WEBSTER (II,2758)

CLARK, TOM
THE WILD WOMEN OF CHASTITY GULCH (II,2801)

CLARKE, BASIL
CASANOVA (II,451)

CLARKE, BLAKE
NATIONAL LAMPOON'S HOT FLASHES (II,1795); REMINGTON STEELE (II,2130)

CLARKE, BRIAN PATRICK
DELTA HOUSE (II,658); EIGHT IS ENOUGH (II,762); GENERAL HOSPITAL (II,964); KING'S CROSSING (II,1397)

TO HOLLYWOOD (II,313); THE EDSEL SHOW (I,1417); THE GENERAL FOODS 25TH ANNIVERSARY SHOW (I,1756); THE JOHNNY JOHNSTON SHOW (I,2428); THE LOSERS (I,2757); THE MOREY AMSTERDAM SHOW (I,3101); ON PARADE (I,3356); THE PAT BOONE AND FAMILY CHRISTMAS SPECIAL (II,1969); ROBERT Q'S MATINEE (I,3812); THE ROSEMARY CLOONEY SHOW (I,3847); THE ROSEMARY CLOONEY SHOW (I,3848); SHOW BIZ (I,4027); TWILIGHT THEATER (II,2685)

CLOUGH, APRIL
CALIFORNIA FEVER (II,411); FLATFOOTS (II,879); T.J. HOOKER (II,2524)

CLOUGH, JOHN SCOTT
NIGHT COURT (II,1844)

CLOWES, ALLEN
ALL THE WAY HOME (I,140)

CLUNES, ALEC
THE BUCCANEERS (I,749)

CLUTE, SIDNEY
CATALINA C-LAB (II,458); INSPECTOR PEREZ (II,1237); LOU GRANT (II,1526); THE MICHELE LEE SHOW (II,1690)

CLUTESI, CHIEF GEORGE
NAKIA (II,1784)

CLYDE, ANDY
CIRCUS BOY (I,959); NO TIME FOR SERGEANTS (I,3300); THE REAL MCCOYS (I,3741); TIMMY AND LASSIE (I,4520)

CLYDE, JEREMY
THE JULIE LONDON SPECIAL (I,2484); MOLL FLANDERS (II,1720)

CLYNE, HAZEL
JANE EYRE (II,1273)

COALE, JOANIE
CONTEST CARNIVAL (I,1040)

COATES, CAROLINE
WHERE THE HEART IS (I,4831)

COATES, PAUL
CONFIDENTIAL FILE (I,1033); TONIGHT! AMERICA AFTER DARK (I,4555)

COATES, PHYLLIS
THE ADVENTURES OF SUPERMAN (I,77); CAVALCADE OF AMERICA (I,875); CROSSROADS (I,1105); CROWN THEATER WITH GLORIA SWANSON (I,1107); THE DESILU PLAYHOUSE (I,1237); THE DUKE (I,1381); FOUR STAR PLAYHOUSE (I,1652); FRONTIER (I,1695); THE JUNE ALLYSON SHOW (I,2488); THE LUX VIDEO THEATER (I,2795); THE PATTY DUKE SHOW (I,3483); PROFESSIONAL FATHER (I,3677); SCIENCE FICTION THEATER (I,3938); STAGE 7 (I,4181); STORY THEATER (I,4240); THIS IS ALICE (I,4453); THOMPSON'S GHOST (I,4467); YOUR JEWELER'S SHOWCASE (I,4953); YOUR SHOW TIME (I,4958)

COATS, ATHOL
POSSESSION (I,3655)

COBB, BUFF
ALL AROUND TOWN (I,128); MIKE AND BUFF (I,3037)

COBB, JULIE
CHARLES IN CHARGE (II,482); THE D.A. (I,1123); FIRST TIME, SECOND TIME (II,863); LADIES' MAN (II,1423); MOBILE MEDICS (II,1716); STATE FAIR (II,2448); A YEAR AT THE TOP (II,2844)

COBB, LEE J
ANNIE, THE WOMAN IN THE LIFE OF A MAN (I,226); DEATH OF A SALESMAN (I,1208); DUPONT SHOW OF THE MONTH (I,1387); THE GENERAL ELECTRIC THEATER (I,1753); HEAT OF ANGER (I,1992); THE LUX VIDEO THEATER (I,2795); MEDIC (I,2976); PLAYHOUSE 90 (I,3623); TALES OF TOMORROW (I,4352); THE VIRGINIAN (I,4738); THE YOUNG LAWYERS (I,4941)

COBB, VINCENT
THE DAY THE WOMEN GOT EVEN (II,628)

COBBS, BILL
THE MEMBER OF THE WEDDING (II,1680); RAGE OF ANGELS (II,2107)

COBERT, MICHAEL
THE ADVENTURES OF BLACK BEAUTY (I,51)

COBERT, PATRICIA
THE JAMES BOYS (II,1272)

COBURN, BRIAN
THE LAST DAYS OF POMPEII (II,1436)

COBURN, CHARLES
BEN HECHT'S TALES OF THE CITY (I,395); THE ETHEL BARRYMORE THEATER (I,1467); FORD TELEVISION THEATER (I,1634); HENRY FONDA PRESENTS THE STAR AND THE STORY (I,2016); PULITZER PRIZE PLAYHOUSE (I,3692); THE ROYAL FAMILY (I,3861)

COBURN, JAMES
ACAPULCO (I,15); ALFRED HITCHCOCK PRESENTS (I,115); BOB HOPE SPECIAL: BOB HOPE'S ALL-STAR BIRTHDAY AT ANNAPOLIS (II,294); THE DAIN CURSE (II,613); DARKROOM (II,619); ESCAPE (II,782); JACQUELINE SUSANN'S VALLEY OF THE DOLLS, 1981 (II,1268); KLONDIKE (I,2573); MALIBU (II,1607); THE MAN FROM GALVESTON (I,2865); SAFARI (I,3882); ZANE GREY THEATER (I,4979)

COCA, IMOGENE
THE ADMIRAL BROADWAY REVUE (I,37); THE BOB HOPE SHOW (I,509); THE BOB HOPE SHOW (I,643); THE BOB HOPE SHOW (I,649); THE BRADY BUNCH (II,362); BUZZY WUZZY (I,782); THE DICK CAVETT SHOW (II,669); GETTING THERE (II,985); GRINDL (I,1889); HOLLYWOOD STARS' SCREEN TESTS (II,1165); THE IMOGENE COCA SHOW (I,2198); IT'S ABOUT TIME (I,2263); PANORAMA (I,3448); THE RETURN OF THE BEVERLY HILLBILLIES (II,2139); RUGGLES OF RED GAP (I,3867); YOUR SHOW OF SHOWS (I,4957); YOUR SHOW OF SHOWS (II,2875)

COCH JR, EDWARD
BEYOND WESTWORLD (II,227)

COCHRAN, PAT
MR. SUCCESS (II,1756)

COCHRANE, MICHAEL
THE CITADEL (II,539); THE FAR PAVILIONS (II,832)

COCHRON, RON
THE ARMSTRONG CIRCLE THEATER (I,260)

COCO, JAMES
CALUCCI'S DEPARTMENT (I,805); THE DUMPLINGS (II,743); THE FRENCH ATLANTIC AFFAIR (II,924); LILY FOR PRESIDENT (II,1478); THE LITTLEST ANGEL (I,2726); MR. SUCCESS (II,1756); RAQUEL (II,2114); THAT WAS THE YEAR THAT WAS (II,2575); THE TROUBLE WITH PEOPLE (I,4611); WHO'S THE BOSS? (II,2792)

CODA, FRANK
THE KILLING GAME (I,2540)

CODE, GRANT
FROM THESE ROOTS (I,1688)

CODINEZ, SALVADOR
FALCON'S GOLD (II,809)

CODY, IRON EYES
HOW THE WEST WAS WON (II,1196); THE QUEST (II,2096); THE SAGA OF ANDY BURNETT (I,3883)

CODY, KATHLEEN
AS THE WORLD TURNS (II,110); THE CHEERLEADERS (II,493); DARK SHADOWS (I,1157); THE EDGE OF NIGHT (II,760); THE SECRET STORM (I,3969)

CODY, MICHAEL
AS THE WORLD TURNS (II,110)

COE, BARRY
FOLLOW THE SUN (I,1617)

COE, DAVID ALLAN
JOHNNY CASH: THE FIRST 25 YEARS (II,1336)

COE, GEORGE
GOODNIGHT, BEANTOWN (II,1041); HOW TO SUCCEED IN BUSINESS WITHOUT REALLY TRYING (II,1197); RAGE OF ANGELS (II,2107)

COFFEY, FRANCINE
COFFEY BREAK (I,994)

COFFEY, SCOTT
KENNEDY (II,1377)

COFFIELD, PETER
ENIGMA (II,778); O'MALLEY (II,1872); W.E.B. (II,2732); WASHINGTON: BEHIND CLOSED DOORS (II,2744)

COFFIN, TRIS
26 MEN (I,4648); THE LONE RANGER (I,2740)

COGAN, SHAYE
FACE THE MUSIC (I,1509); OPEN HOUSE (I,3394); THE VAUGHN MONROE SHOW (I,4707)

COHEN, BESS
DON'T CALL ME MAMA ANYMORE (I,1350)

COHEN, BURT M
VEGAS (II,2724)

COHEN, EVAN
IT'S NOT EASY (II,1256); THE ROPERS (II,2214)

COHEN, JEFF
LITTLE SHOTS (II,1495)

COHEN, JUDITH

SUNDAY FUNNIES (II,2493); A YEAR AT THE TOP (II,2845)

COHEN, MARTY
THE JACKSONS (II,1267); SOLID GOLD (II,2395); SOLID GOLD '79 (II,2396); THE STOCKARD CHANNING SHOW (II,2468)

COHEN, SCOOTER
SCAMPS (II,2267)

COHN, MINDY
THE ACADEMY (II,3); THE ACADEMY II (II,4); THE FACTS OF LIFE (II,805); FACTS OF LIFE: THE FACTS OF LIFE GOES TO PARIS (II,806); THE PARKERS (II,1959)

COHN, SHANNON
LIFE IS A CIRCUS, CHARLIE BROWN (I,2683)

COHOON, PATTI
APPLE'S WAY (II,101); DOCTOR DAN (II,684); HERE COME THE BRIDES (I,2024); THE PARTRIDGE FAMILY (II,1962); THE RUNAWAYS (II,2231)

COLASANTO, NICHOLAS
CHEERS (II,494); LASSITER (I,2625); LASSITER (II,1434); THE OUTSIDE MAN (II,1937); RETURN OF THE WORLD'S GREATEST DETECTIVE (II,2142); TOMA (I,4547)

COLBERT, CLAUDETTE
THE BELLS OF ST. MARY'S (I,392); BLITHE SPIRIT (I,484); THE CLAUDETTE COLBERT SHOW (I,971); CLIMAX! (I,976); THE GENERAL MOTORS 50TH ANNIVERSARY SHOW (I,1758); THE GUARDSMAN (I,1893); PLAYHOUSE 90 (I,3623); THE ROYAL FAMILY (I,3861); TEXACO COMMAND PERFORMANCE (I,4408)

COLBERT, HATTIE
THE GARRY MOORE SHOW (I,1738)

COLBERT, KEITH
CHIEFS (II,509)

COLBERT, PAT
FLAMINGO ROAD (II,872)

COLBERT, ROBERT
ALL THAT GLITTERS (II,41); CITY BENEATH THE SEA (I,965); MAVERICK (II,1657); THE TIME TUNNEL (I,4511); THE YOUNG AND THE RESTLESS (II,2862)

COLBIN, ROD
HARPER VALLEY PTA (II,1095); LITTLE HOUSE: THE LAST FAREWELL (II,1491); RED GOOSE KID'S SPECTACULAR (I,3751) THE ROPERS (II,2214); YOUNG DR. MALONE (I,4938)

COLBOURNE, MAURICE
DAY OF THE TRIFFIDS (II,627)

COLBY, ANITA
PEPSI-COLA PLAYHOUSE (I,3523)

COLBY, BARBARA
PHYLLIS (II,2038)

COLBY, CARROLL
BILLY BOONE AND COUSIN KIBB (I,450)

COLBY, DANNY
SNOOPY'S GETTING MARRIED, CHARLIE BROWN (I,4102)

COLBY, MARION
BROADWAY OPEN HOUSE (I,736); THE DOODLES WEAVER SHOW (I,1351); HENNY AND ROCKY (I,2014); INSIDE U.S.A. WITH CHEVROLET (I,2220); THE MILTON BERLE SPECIAL (I,3052)

COLBY, SKIPPY
THE PHIL SILVERS SHOW (I,3580)

COLE, BARRY
CHALK ONE UP FOR JOHNNY (I,893)

COLE, BUDDY
THE BING CROSBY SHOW (I,462)

COLE, CAROL
GRADY (II,1047)

COLE, CLAY
THE TONY BENNETT SPECIAL (I,4571)

COLE, COZY
TIMEX ALL-STAR JAZZ SHOW I (I,4515)

COLE, DAVID
THE CORN IS GREEN (I,1050)

COLE, DEBBIE
PARADISE BAY (I,3454)

COLE, DENNIS
THE BARBARY COAST (II,147); BEARCATS! (I,375); BIG SHAMUS, LITTLE SHAMUS (II,242); BRACKEN'S WORLD (I,703); THE FELONY SQUAD (I,1563); PARADISE BAY (I,3454); POWDERKEG (I,3657); THE YOUNG AND THE RESTLESS (II,2862)

COLE, EDDIE
BOURBON STREET BEAT (I,699)

COLE, GENE
OUR STREET (I,3418)

COLE, GEORGE
THE SCARECROW OF ROMNEY MARSH (I,3931)

COLE, JACK
THE JACK COLE DANCERS (I,2308)

COLE, JOHN
THE CODE OF JONATHAN WEST (I,992)

COLE, KAY
CARLTON YOUR DOORMAN (II,440)

COLE, MARIA
A SALUTE TO TELEVISION'S 25TH ANNIVERSARY (I,3893)

COLE, MICHAEL
THE MOD SQUAD (I,3078); THE RETURN OF THE MOD SQUAD (I,3776)

COLE, NAT KING
ACADEMY AWARD SONGS (I,13); FIVE STARS IN SPRINGTIME (I,1590); THE NAT KING COLE SHOW (I,3230)

COLE, NATALIE
HERBIE, THE LOVE BUG (II,1131); HI, I'M GLEN CAMPBELL (II,1141); THE NATALIE COLE SPECIAL (II,1794); SINATRA AND FRIENDS (II,2358); UPTOWN (II,2710)

COLE, OLESLEY
THE AWAKENING LAND (II,122)

COLE, OLIVIA
BACKSTAIRS AT THE WHITE HOUSE (II,133); FLY AWAY HOME (II,893); THE GUIDING LIGHT (II,1064); REPORT TO MURPHY (II,2134); ROOTS (II,2211); SZYSZNYK (II,2523); WHEN, JENNY? WHEN (II,2783)

COLE, ROCKY
THE PATTI PAGE SHOW (I,3482)

COLE, TINA
THE JUD STRUNK SHOW (I,2462); THE KING FAMILY CELEBRATE THANKSGIVING (I,2545); THE KING FAMILY CHRISTMAS SPECIAL (I,2546); THE KING FAMILY IN ALASKA (I,2547); THE KING FAMILY IN HAWAII (I,2548); THE KING FAMILY IN WASHINGTON (I,2549); THE KING FAMILY JUNE SPECIAL (I,2550); THE KING FAMILY OCTOBER SPECIAL (I,2551); THE KING FAMILY SPECIAL (I,2553); THE KING FAMILY VALENTINE'S DAY SPECIAL (I,2554); MY THREE SONS (I,3205); THE OSMOND BROTHERS SHOW (I,3409); SUPER COMEDY BOWL 1 (I,4288); THANKSGIVING REUNION WITH THE PARTRIDGE FAMILY AND MY THREE SONS (II,2569)

COLE, TOMMY
THE MICKEY MOUSE CLUB (I,3025); THE MOUSEKETEERS REUNION (II,1742)

COLEMAN, ANN
MITCHELL AND WOODS (II,1709); MOVIN' ON (II,1743)

COLEMAN, BARBARA
HERE'S BARBARA (I,2029)

COLEMAN, BOOTH
DANIEL BOONE (I,1142); RETURN OF THE WORLD'S GREATEST DETECTIVE (II,2142)

COLEMAN, CAROLE
FACE THE MUSIC (I,1509); MAKE MINE MUSIC (I,2840); THE MOREY AMSTERDAM SHOW (I,3100)

COLEMAN, DABNEY
APPLE PIE (II,100); BRIGHT PROMISE (I,727); BUFFALO BILL (II,387); CANNON (II,424); THE COMEDY ZONE (II,559); EGAN (I,1419); FOREVER FERNWOOD (II,909); KISS ME, KILL ME (II,1400); MARY HARTMAN, MARY HARTMAN (II,1638); THAT GIRL (II,2570)

COLEMAN, GARY
ALMOST AMERICAN (II,55); ANSON AND LORRIE (II,98); THE BIG SHOW (II,243); CELEBRITY CHALLENGE OF THE SEXES 3 (II,467); THE DEAN MARTIN CELEBRITY ROAST (II,639); DIFF'RENT STROKES (II,674); THE GARY COLEMAN SHOW (II,958); THE KID WITH THE BROKEN HALO (II,1387); LUCY MOVES TO NBC (II,1562); PAUL LYNDE AT THE MOVIES (II,1976); TOM SNYDER'S CELEBRITY SPOTLIGHT (II,2632)

COLEMAN, HERBERT
THE PHIL SILVERS ARROW SHOW (I,3576)

COLEMAN, JACK

DYNASTY (II,746)

COLEMAN, JAMES
S.W.A.T. (II,2236)

COLEMAN, JOEY
CAMP GRIZZLY (II,418)

COLEMAN, KATHY
LAND OF THE LOST (II,1425)

COLEMAN, MARILYN
GOOD TIMES (II,1038)

COLEMAN, NANCY
THE ADAMS CHRONICLES (II,8); THE KAISER ALUMINUM HOUR (I,2503); SILVER THEATER (I,4051); STAR TONIGHT (I,4199); TALES OF TOMORROW (I,4352); VALIANT LADY (I,4695)

COLEMAN, PETER
ONE LIFE TO LIVE (II,1907)

COLEMAN, RICHARD
KISS KISS, KILL KILL (I,2566)

COLEMAN, TOWNSEND
THE DANCE SHOW (II,616); WE'RE DANCIN' (II,2755)

COLEN, BEATRICE
CHARO (II,488); HAPPY DAYS (II,1084); JERRY (II,1283); WONDER WOMAN (SERIES 1) (II,2828)

COLENBECK, JOHN
AS THE WORLD TURNS (II,110); FROM THESE ROOTS (I,1688)

COLERIDGE, SYLVIA
CORONATION STREET (I,1054)

COLES, ZAIDA
THE DOCTORS (II,694)

COLETTI, RICK
OFF THE RACK (II,1878)

COLICOS, JOHN
THE BASTARD/KENT FAMILY CHRONICLES (II,159); BATTLESTAR GALACTICA (II,181); BERKELEY SQUARE (I,399); CYRANO DE BERGERAC (I,1120); PROFILES IN COURAGE (I,3678); THE THREE MUSKETEERS (I,4481); VANITY FAIR (I,4700)

COLIN, MARGARET
AS THE WORLD TURNS (II,110); THE EDGE OF NIGHT (II,760)

COLL, CHRISTOPHER
CORONATION STREET (I,1054)

COLLA, DICK
DAYS OF OUR LIVES (II,629)

COLLEGE KIDS SQUARE DANCERS
COUNTRY MUSIC CARAVAN (I,1066)

COLLERAN, TERESA
THE BOOK OF LISTS (II,350)

COLLETT, VERNE
WITNESS (I,4885)

COLLEY, DON PEDRO
CASINO (II,452); DANIEL BOONE (I,1142); THE DUKES OF HAZZARD (II,742)

COLLEY, KENNETH
PENNIES FROM HEAVEN (II,1994)

COLLI, MARIO
S*H*E (II,2235)

COLLIER, DICK
CHANCE OF A LIFETIME (I,898)

COLLIER, DON
THE HIGH CHAPARRAL (I,2051); HIGHWAY HONEYS (II,1151); KEY WEST (I,2528); THE OUTLAWS (I,3427); OUTPOST (I,3428)

COLLIER, JOAN
BACKBONE OF AMERICA (I,328)

COLLIER, LESLIE
CINDERELLA (II,526)

COLLIER, LOIS
BLONDIE (I,486); BOSTON BLACKIE (I,695); CROWN THEATER WITH GLORIA SWANSON (I,1107); THE UNEXPECTED (I,4674)

COLLIER, MARIAN
CONGRESSIONAL INVESTIGATOR (I,1037); EGAN (I,1419); MR. NOVAK (I,3145); STAT! (I,4213)

COLLIER, RICHARD
BACHELOR PARTY (I,324); COLONEL STOOPNAGLE'S STOOP (I,1006); GETAWAY CAR (I,1783); MANY HAPPY RETURNS (I,2895)

COLLIN, JOHN
ALL CREATURES GREAT AND SMALL (I,129)

COLLIN, KATHY
THE JERRY LESTER SHOW (I,2360)

COLLINGE, PATRICIA
LOVE STORY (I,2776); STAR TONIGHT (I,4199)

COLLINGS, ANNE
GENERAL HOSPITAL (II,964)

COLLINGS, BENITA
RETURN TO EDEN (II,2143)

COLLINGWOOD, CHARLES
ODYSSEY (I,3329); PERSON TO PERSON (I,3550)

COLLINS III, JOHNNIE
MCDUFF, THE TALKING DOG (II,1662); THE TIM CONWAY SHOW (I,4502)

COLLINS KIDS, THE
STAR ROUTE (I,4196)

COLLINS, AL 'JAZZBO'
TONIGHT! AMERICA AFTER DARK (I,4555)

COLLINS, BETH
ANOTHER WORLD (II,97)

COLLINS, BRENT
AS THE WORLD TURNS (II,110); MURDER INK (II,1769)

COLLINS, DOROTHY
CANDID CAMERA (I,819); THE MOREY AMSTERDAM SHOW (I,3100); SOUNDS OF HOME (I,4136); THIS IS SHOW BUSINESS (I,4459)

COLLINS, FORBES
A HORSEMAN RIDING BY (II,1182)

COLLINS, GARY
BORN FREE (I,691); BORN FREE (II,353); CIRCUS OF THE STARS (II,530); CIRCUS OF THE STARS (II,531); CIRCUS OF THE STARS (II,532); CIRCUS OF THE STARS (II,533); CIRCUS OF THE STARS (II,538); DIAL A DEADLY NUMBER (I,1253); THE DOUBLE KILL (I,1360); HOUR MAGAZINE (II,1193); IRON HORSE (I,2238); JACQUELINE SUSANN'S VALLEY OF THE DOLLS, 1981 (II,1268); ONLY A SCREAM AWAY (I,3393); ROOTS (II,2211); THE SIXTH SENSE (I,4069); SUCCESS: IT CAN BE YOURS (II,2488); THE WACKIEST SHIP IN THE ARMY (I,4744)

COLLINS, GEOFF
PRISONER: CELL BLOCK H (II,2085)

COLLINS, HARRIET
MACKENZIE (II,1584)

COLLINS, JACK
THE BRADY BUNCH (II,362); HIGH BUTTON SHOES (I,2050); OCCASIONAL WIFE (I,3325)

COLLINS, JAMES
THE ADVENTURES OF POLLYANNA (II,14)

COLLINS, JANE
ALL CREATURES GREAT AND SMALL (I,129)

COLLINS, JOAN
BATMAN (I,366); BATTLE OF THE NETWORK STARS (II,174); BLONDES VS. BRUNETTES (II,271); THE BOB HOPE SHOW (I,571); THE BOB HOPE SHOW (I,588); THE BOB HOPE SHOW (I,611); THE DEAN MARTIN CELEBRITY ROAST (II,638); DYNASTY (II,746); THE MAN WHO CAME TO DINNER (I,2881); THE MONEYCHANGERS (II,1724); PAPER DOLLS (II,1951); STEVE MARTIN: COMEDY IS NOT PRETTY (II,2462); WARNING SHOT (I,4770); THE WILD WOMEN OF CHASTITY GULCH (II,2801)

COLLINS, JUDY
THE CRYSTAL GAYLE SPECIAL (II,609); THE TOM JONES CHRISTMAS SPECIAL (I,4539)

COLLINS, LAURA
SOMEONE AT THE TOP OF THE STAIRS (I,4114)

COLLINS, LORRIE
STAR ROUTE (I,4196)

COLLINS, MARY
ABE LINCOLN IN ILLINOIS (I,10)

COLLINS, MICHAEL
FAMILY IN BLUE (II,818)

COLLINS, NANCY
LEGENDS OF THE SCREEN (II,1456)

COLLINS, PAT
I'VE GOT A SECRET (II,1220)

COLLINS, PATRICIA
STRANGE PARADISE (I,4246)

COLLINS, PATRICK
CHECKING IN (II,492); DANNY AND THE MERMAID (II,618); THE JUGGLER OF NOTRE DAME (II,1350); MR. & MRS. & MR. (II,1744); SUPERTRAIN (II,2504); THERE GOES THE NEIGHBORHOOD (II,2582)

COLLINS, PAULINE
NO—HONESTLY (II,1857)

COLLINS, RAY
MIRACLE ON 34TH STREET (I,3056); ON TRIAL (I,3363); PERRY MASON (II,2025); REAL GEORGE (I,3740); SNEAK PREVIEW (I,4100); TWENTIETH CENTURY (I,4639)

COLLINS, RUSSELL
THE BOSTON TERRIER (I,696); THE JOKE AND THE VALLEY (I,2438); MANY HAPPY RETURNS (I,2895); SHIRLEY TEMPLE'S STORYBOOK: THE LEGEND OF SLEEPY HOLLOW (I,4018); SHORT, SHORT DRAMA (I,4023)

COLLINS, SHELDON
MAYBERRY R.F.D. (I,2961)

COLLINS, STEPHEN
BATTLE OF THE NETWORK STARS (II,175); CHIEFS (II,509); THE MICHELE LEE SHOW (II,1690); THE RHINEMANN EXCHANGE (II,2150); SHERLOCK HOLMES (II,2327); SUMMER SOLSTICE (II,2492); TALES OF THE GOLD MONKEY (II,2537)

COLLINS, TED
THE KATE SMITH EVENING HOUR (I,2507); MATINEE IN NEW YORK (I,2946)

COLLINS, TOMMY
BUCK OWENS TV RANCH (I,750)

COLLINS, YUKIO G
MASSARATI AND THE BRAIN (II,1646)

COLLYER, CLAYTON 'BUD'
BEAT THE CLOCK (I,377); BREAK THE BANK (I,712); FEATHER YOUR NEST (I,1555); FIVE STARS IN SPRINGTIME (I,1590); MASQUERADE PARTY (I,2941); THE MISSUS GOES A SHOPPING (I,3069); THE NEW ADVENTURES OF SUPERMAN (I,3255); NOTHING BUT THE TRUTH (I,3313); NUMBER PLEASE (I,3315); ON YOUR WAY (I,3366); QUICK AS A FLASH (I,3709); SAY IT WITH ACTING (I,3928); TALENT JACKPOT (I,4339); TALENT PATROL (I,4340); TO TELL THE TRUTH (I,4527); WINNER TAKE ALL (I,4881)

COLLYER, JUNE
TROUBLE WITH FATHER (I,4610)

COLMAN, ANDREW
FULL CIRCLE (I,1702)

COLMAN, ANN
IN TANDEM (II,1228)

COLMAN, BOOTH
FOOLS, FEMALES AND FUN: WHAT ABOUT THAT ONE? (II,901); GIDEON (I,1792); PLANET OF THE APES (II,2049)

COLMAN, RONALD
FOUR STAR PLAYHOUSE (I,1652); THE HALLS OF IVY (I,1923)

COLMANS, EDWARD
TONIGHT IN HAVANA (I,4557)

COLOCOS, JOHN
TAMING OF THE SHREW (I,4356)

COLODNER, JOEL
TEXAS (II,2566)

COLOMBY, BOBBY
ENTERTAINMENT TONIGHT (II,780)

COLOMBY, SCOTT
ONE DAY AT A TIME (II,1900); SENIOR YEAR (II,2301); SISTER TERRI (II,2366); SONS AND DAUGHTERS (II,2404); SZYSZNYK (II,2523)

COLONNA, JERRY
THE BOB HOPE SHOW (I,628); THE BOB HOPE SHOW (I,570); THE BOB HOPE SHOW (I,510); THE BOB HOPE SHOW (I,529); THE BOB HOPE SHOW (I,536); THE BOB HOPE SHOW (I,552); THE BOB HOPE SHOW (I,566); THE BOB HOPE SHOW (I,574); THE BOB HOPE SHOW (I,581); THE BOB HOPE SHOW (I,587); THE BOB HOPE SHOW (I,593); THE BOB HOPE SHOW (I,598); THE BOB HOPE SHOW (I,609); THE BOB HOPE SHOW (I,611); THE BOB HOPE SHOW (I,633); THE BOB HOPE SHOW (I,642); THE BOB HOPE SHOW (I,659); FIVE STAR COMEDY (I,1588); FRANCES LANGFORD PRESENTS (I,1656); HOLLYWOOD STAR REVUE (I,2093); THE JERRY COLONNA SHOW (I,2359); OPERATION ENTERTAINMENT (I,3400); PINOCCHIO (I,3602); SUPER CIRCUS (I,4287); TIME FOR BEANY (I,4505)

COLONNA, LOUIS
THE JERRY COLONNA SHOW (I,2359)

COLORADO, HORTENSIA
NURSE (II,1870)

COLPITTS, CISSY
THE TED KNIGHT SHOW (II,2550)

COLSON, DAVID
AS THE WORLD TURNS (II,110)

COLT, MARSHALL
COLORADO C.I. (II,555); LOTTERY (II,1525); MCCLAIN'S LAW (II,1659)

COLTER, JESSI
DEAN'S PLACE (II,650); THE GEORGE JONES SPECIAL (II,975); JOHNNY CASH: CHRISTMAS ON THE ROAD (II,1333); JOHNNY CASH: SPRING FEVER (II,1335)

COLTON, JACQUE
THE BURNS AND SCHREIBER COMEDY HOUR (I,768); MASQUERADE (I,2940); PIPER'S PETS (II,2046); WIVES (II,2812)

COLTON, RONALD
SUNDAY IN TOWN (I,4285)

COLUMBA, FRANCO
THE HUSTLER OF MUSCLE BEACH (II,1210)

COLUZZI, FRANCESCA ROMANA
THE LAST DAYS OF POMPEII (II,1436)

COLVIG, VANCE
YOGI BEAR (I,4928)

COLVIN, JACK
BENNY AND BARNEY: LAS VEGAS UNDERCOVER (II,206); THE INCREDIBLE HULK (II,1232)

COMACHO, CORINNE
MEDICAL CENTER (II,1675)

COMDEN, BETTY
THE FABULOUS 50S (I,1501)

COMEGYS, KATHLEEN
JAMIE (I,2334)

COMER, ANJANETTE
BANYON (I,345); NIGHT GAMES (II,1845); THE YOUNG LAWYERS (I,4940)

COMERATE, SHERIDAN
BELLE STARR (I,391)

COMFORT, BOB
WACKO (II,2733)

COMI, PAUL
THE BOSTON TERRIER (I,697); FBI CODE 98 (I,1550); LOLLIPOP LOUIE (I,2738); OF THIS TIME, OF THAT PLACE (I,3335); PURSUE AND DESTROY (I,3699); RAWHIDE (I,3727); TWO FACES WEST (I,4653)

COMMERON, ANNA
PAUL BERNARD—PSYCHIATRIST (I,3485)

COMMODORES, THE
THE MAC DAVIS CHRISTMAS SPECIAL (II,1573)

COMO, PERRY
ANN-MARGRET. . . RHINESTONE COWGIRL (II,87); THE BING CROSBY SHOW (I,463); THE BOB HOPE SHOW (I,551); THE BOB HOPE SHOW (I,578); THE BOB HOPE SHOW (I,641); BOB HOPE SPECIAL: BOB HOPE'S CHRISTMAS SPECIAL (II,308); BOB HOPE SPECIAL: BOB HOPE'S SALUTE TO NASA—25 YEARS OF REACHING FOR THE STARS (II,314); THE CHESTERFIELD SUPPER CLUB (I,923); DATELINE (I,1165); THE DORIS MARY ANNE KAPPELHOFF SPECIAL (I,1356); JULIE ON SESAME STREET (I,2485); THE KRAFT MUSIC HALL (I,2587); THE MANY MOODS OF PERRY COMO (I,2897); THE PERRY COMO CHRISTMAS SHOW (I,3526); THE PERRY COMO CHRISTMAS SHOW (I,3527); THE PERRY COMO CHRISTMAS SHOW (I,3528); THE PERRY COMO CHRISTMAS SHOW (I,3529); THE PERRY COMO CHRISTMAS SHOW (II,2002); PERRY COMO IN LAS VEGAS (II,2003); THE PERRY COMO SHOW (I,3531); THE PERRY COMO SHOW (I,3532); THE PERRY COMO SHOW (I,3533); THE PERRY COMO SPECIAL (I,3534); THE PERRY COMO SPECIAL (I,3535); THE PERRY COMO SPECIAL (I,3536); THE PERRY COMO SPRINGTIME SHOW (I,3537); THE PERRY COMO SPRINGTIME SPECIAL (I,3538); THE PERRY COMO SPRINGTIME SPECIAL (II,2004); THE PERRY COMO SUNSHINE SHOW (II,2005); THE PERRY COMO THANKSGIVING SHOW (I,3539); THE PERRY COMO THANKSGIVING SPECIAL (I,3540); THE PERRY COMO VALENTINE SPECIAL (I,3541); THE PERRY COMO WINTER SHOW (I,3542); THE PERRY COMO WINTER SHOW (I,3543); THE PERRY COMO WINTER SHOW (I,3544); PERRY COMO'S BAHAMA HOLIDAY (II,2006); PERRY COMO'S CHRISTMAS IN AUSTRIA (II,2007); PERRY COMO'S CHRISTMAS IN ENGLAND (II,2008); PERRY COMO'S CHRISTMAS IN MEXICO (II,2009); PERRY COMO'S CHRISTMAS IN NEW MEXICO (II,2010); PERRY COMO'S CHRISTMAS IN NEW YORK (II,2011); PERRY COMO'S CHRISTMAS IN PARIS (II,2012); PERRY COMO'S CHRISTMAS IN THE HOLY LAND (II,2013); PERRY COMO'S EASTER BY THE SEA

(II,2014); PERRY COMO'S EASTER IN GUADALAJARA (II,2015); PERRY COMO'S FRENCH-CANADIAN CHRISTMAS (II,2016); PERRY COMO'S HAWAIIAN HOLIDAY (II,2017); PERRY COMO'S LAKE TAHOE HOLIDAY (II,2018); PERRY COMO'S MUSIC FROM HOLLYWOOD (II,2019); PERRY COMO'S OLDE ENGLISH CHRISTMAS (II,2020); PERRY COMO'S SPRING IN NEW ORLEANS (II,2021); PERRY COMO'S SPRING IN SAN FRANCISCO (II,2022); PERRY COMO'S SPRINGTIME SPECIAL (II,2023); PERRY COMO'S SUMMER SHOW (I,3545); PERRY COMO'S SUMMER OF '74 (II,2024); PERRY COMO'S WINTER SHOW (I,3546); PERRY COMO'S WINTER SHOW (I,3547); PERRY COMO'S WINTER SHOW (I,3548); VARIETY (I,4703)

COMO, ROSELLA
MARCO POLO (II,1626)

COMPTON, FORREST
THE BRIGHTER DAY (I,728); CALL TO DANGER (I,802); THE EDGE OF NIGHT (II,760); GOMER PYLE, U.S.M.C. (I,1843); THAT GIRL (II,2570)

COMPTON, FRANCIS
CAESAR AND CLEOPATRA (I,788)

COMPTON, JOHN
THE D.A.'S MAN (I,1158)

COMPTON, JOYCE
THE ABBOTT AND COSTELLO SHOW (I,2)

COMYN, CHARLES
PRISONERS OF THE LOST UNIVERSE (II,2086)

CONAN, DICK
THE GOOK FAMILY (I,1858)

CONATO, ANN MARIE
OUTLAW LADY (II,1935); WONDER GIRL (I,4891)

CONAWAY, JEFF
BATTLE OF THE NETWORK STARS (II,172); CHERYL LADD. . . LOOKING BACK—SOUVENIRS (II,502); DELTA COUNTY, U.S.A. (II,657); THE JIMMY MCNICHOL SPECIAL (II,1292); TAXI (II,2546); WIZARDS AND WARRIORS (II,2813)

CONDE, RITA
AT EASE (II,116)

CONDEN, EVE
CAMEO THEATER (I,808)

CONDOLI, PETE
FAMILY NIGHT WITH HORACE HEIDT (I,1523)

CONDON, EDDIE
THE EDDIE CONDON SHOW (I,1408); FLOOR SHOW (I,1605)

CONFORTI, GINO
BETWEEN THE LINES (II,225); BUNGLE ABBEY (II,396); HOW TO SURVIVE THE 70S AND MAYBE EVEN BUMP INTO HAPPINESS (II,1199); JOE DANCER: MURDER ONE, DANCER 0 (II,1302); THE KROFFT KOMEDY HOUR (II,1411); A LUCILLE BALL SPECIAL STARRING LUCILLE BALL AND DEAN MARTIN (II,1557); MORE WILD WILD WEST (II,1733); POOR DEVIL (I,3646); THAT GIRL (II,2570); THREE'S COMPANY (II,2614); THROUGH THE MAGIC PYRAMID (II,2615)

CONGDON, JAMES
ANOTHER WORLD (II,97)

CONIFF, FRANK
ARE YOU POSITIVE? (I,255)

CONKLIN, CHESTER
DOC CORKLE (I,1304)

CONKLIN, HAL
CAPTAIN VIDEO AND HIS VIDEO RANGERS (I,838)

CONKLIN, HAROLD
OPERATION: NEPTUNE (I,3404)

CONKLIN, PATRICIA
THE PHOENIX (II,2035); SKEEZER (II,2376)

CONKLIN, PEGGY
COSMOPOLITAN THEATER (I,1057); HANDS OF MURDER (I,1929)

CONLEE, JOHN
CHRISTMAS LEGEND OF NASHVILLE (II,522)

CONLEY, CORINNE
DAYS OF OUR LIVES (II,629); DEAR DETECTIVE (II,652)

CONLEY, DARLENE
THE YOUNG AND THE RESTLESS (II,2862)

CONLEY, JOE
THE LIFE OF RILEY (I,2686); THE WALTONS (II,2740); THE WALTONS: A DAY FOR THANKS ON WALTON'S MOUNTAIN (EPISODE 3) (II,2741); THE WALTONS: MOTHER'S DAY ON WALTON'S MOUNTAIN (EPISODE 2) (II,2742); THE WALTONS: WEDDING ON WALTON'S MOUNTAIN (EPISODE 1) (II,2743)

CONLIN, JIMMY
DUFFY'S TAVERN (I,1379)

CONLON, JUD
FRANKIE LAINE TIME (I,1676)

CONLOW, PETER
THE MAGIC SLATE (I,2820); THE MUSIC OF GERSHWIN (I,3174)

CONN, DIDI
BENSON (II,208); CIRCUS OF THE STARS (II,536); FONZ AND THE HAPPY DAYS GANG (II,898); HANDLE WITH CARE (II,1077); KEEP ON TRUCKIN' (II,1372); THE PRACTICE (II,2076); THE SCOOBY-DOO AND SCRAPPY-DOO SHOW (II,2275)

CONN, KELLY ANN
HIGH SCHOOL, U.S.A. (II,1148); SKYWARD CHRISTMAS (II,2378)

CONNALLY, MERILL
JAKE'S WAY (II,1269)

CONNEAUR, GEORGE
THE THREE MUSKETEERS (I,4480)

CONNELL, JANE
AS CAESAR SEES IT (I,294); BEWITCHED (I,418); THE DUMPLINGS (II,743); FUNNY PAPERS (I,1713); GETTING THERE (II,985); MR. MAYOR (I,3143); THE NUT HOUSE (I,3320); STANLEY (I,4190)

CONNELL, JIM
ARNOLD'S CLOSET REVIEW (I,262); CALL HOLME (II,412); DONNY AND MARIE (II,712); THE PIRATES OF FLOUNDER BAY (I,3607); THE ROWAN AND MARTIN REPORT (II,2221); RUN, BUDDY, RUN (I,3870); THICKER THAN WATER (I,4445)

CONNELL, JOHN
MEET CORLISS ARCHER (I,2981)

CONNELL, ZOE
LOVE OF LIFE (I,2774)

CONNELLY, BILLY
NOT THE NINE O'CLOCK NEWS (II,1864)

CONNELLY, CHRISTOPHER
AIRWOLF (II,27); CHARLIE COBB: NICE NIGHT FOR A HANGING (II,485); THE MARTIAN CHRONICLES (II,1635); PAPER MOON (II,1953); PEYTON PLACE (I,3574); SKYWARD CHRISTMAS (II,2378); STUNT SEVEN (II,2486)

CONNELLY, JOHN
YOUNG DR. MALONE (I,4938)

CONNELLY, MAUREEN
ESPIONAGE (I,1464)

CONNELLY, PEGGY
TAKE A GOOD LOOK (II,2529); TAKE A GOOD LOOK (I,4332); WORDS AND MUSIC (I,4909)

CONNER, BETTY
GIDGET (I,1795)

CONNER, KAYE
THE RAY KNIGHT REVUE (I,3733)

CONNIFF, RAY
THE RAY CONNIFF CHRISTMAS SPECIAL (I,3732)

CONNOLLY, ANN
PETER PAN (I,3566)

CONNOLLY, NORMA
DOWN HOME (II,731); GENERAL HOSPITAL (II,964); HOW'S BUSINESS? (I,2153); MR. AND MRS. COP (II,1748); RANSOM FOR A DEAD MAN (I,3722); THE YOUNG MARRIEDS (I,4942)

CONNOLLY, PAT
NEVER TOO YOUNG (I,3248)

CONNOLLY, SHELIA
WINNER TAKE ALL (I,4881)

CONNOLLY, THOMAS
THE DOCTORS (II,694)

CONNOLLY, VINCENT
THE MARY MARGARET MCBRIDE SHOW (I,2933)

CONNOR, BETTY
THE YOUNG MARRIEDS (I,4942)

CONNOR, ERINS
A WORLD APART (I,4910)

CONNOR, KENNETH
CARRY ON LAUGHING (II,448)

CONNOR, KEVIN
SPACE: 1999 (II,2409)

CONNOR, LYNN
BALANCE YOUR BUDGET (I,336)

CONNOR, WHITFIELD
THE GUIDING LIGHT (II,1064); SCHAEFER CENTURY THEATER (I,3935); WILLY (I,4868)

CONNORS, CHUCK
ARREST AND TRIAL (I,264); BANJO HACKETT: ROAMIN' FREE (II,141); THE BOB HOPE SHOW (I,569); BRANDED (I,707); CAVALCADE OF AMERICA (I,875); CELEBRITY DAREDEVILS (II,473); COWBOY IN AFRICA (I,1085);

FRONTIER (I,1695); THE GREAT MYSTERIES OF HOLLYWOOD (II,1058); THE INDIAN (I,2210); THE JUNE ALLYSON SHOW (I,2488); LONE STAR (II,1513); THE LORETTA YOUNG THEATER (I,2756); MATINEE THEATER (I,2947); NIGHT GALLERY (I,3287); THE POLICE STORY (I,3638); THE RIFLEMAN (I,3789); ROOTS (II,2211); THE SHARPSHOOTER (I,4001); THRILL SEEKERS (I,4491); WESTERN HOUR (I,4797); WHEN THE WEST WAS FUN: A WESTERN REUNION (II,2780); WHICH WAY'D THEY GO? (I,4838); THE YELLOW ROSE (II,2847); ZANE GREY THEATER (I,4979)

CONNORS, MATT
SCAMPS (II,2267)

CONNORS, MIKE
BOB HOPE SPECIAL: BOB HOPE IN "JOYS" (II,284); CASINO (II,452); CHARO (II,488); CIRCUS OF THE STARS (II,535); THE EARTHLINGS (II,751); FRONTIER (I,1695); THE FUNNIEST JOKE I EVER HEARD (II,940); GETAWAY CAR (I,1783); THE KILLER WHO WOULDN'T DIE (II,1393); MANNIX (II,1624); MITZI...THE FIRST TIME (I,3073); SUPER COMEDY BOWL 2 (I,4289); TIGHTROPE (I,4500); TODAY'S FBI (II,2628)

CONOLEY, TERENCE
THE CITADEL (II,539); THE FALL AND RISE OF REGINALD PERRIN (II,810)

CONRAD, CHRISTIAN
HARD KNOX (II,1087)

CONRAD, GEORGE
RAPTURE AT TWO-FORTY (I,3724)

CONRAD, LINDA
THIS BETTER BE IT (II,2595)

CONRAD, MICHAEL
ALL IN THE FAMILY (II,38); BOB HOPE SPECIAL: BOB HOPE'S STAND UP AND CHEER FOR THE NATIONAL FOOTBALL LEAGUE'S 60TH YEAR (II,316); DELVECCHIO (II,659); THE DESPERATE HOURS (I,1239); HILL STREET BLUES (II,1154); HOW THE WEST WAS WON (II,1196); THE RANGERS (II,2112); STARSKY AND HUTCH (II,2445); WORKING STIFFS (II,2834)

CONRAD, NANCY
THE BLACK SHEEP SQUADRON (II,262)

CONRAD, ROBERT
ADAM-12 (I,31); THE ADVENTURES OF NICK CARTER (I,68); ASSIGNMENT: VIENNA (I,304); BATTLE OF THE NETWORK STARS (II,163); BATTLE OF THE NETWORK STARS (II,164); BATTLE OF THE NETWORK STARS (II,165); BATTLE OF THE NETWORK STARS (II,167); BATTLE OF THE NETWORK STARS (II,168); BATTLE OF THE NETWORK STARS (II,169); BATTLE OF THE NETWORK STARS (II,177); THE BLACK SHEEP SQUADRON (II,262); CELEBRITY CHALLENGE OF THE SEXES 1 (II,465); CELEBRITY CHALLENGE OF THE SEXES 2 (II,466); CELEBRITY CHALLENGE OF THE SEXES 5 (II,469); CENTENNIAL (II,477); CIRCUS OF THE STARS (II,531); THE D.A. (I,1123); THE D.A.: CONSPIRACY TO KILL (I,1121); THE D.A.: MURDER ONE (I,1122); THE DUKE (II,740); HARD KNOX (II,1087); HAWAIIAN EYE (I,1973); KRAFT SUSPENSE THEATER (I,2591); A MAN CALLED SLOANE (II,1613); MORE WILD WILD WEST (II,1733); THAT'S TV (II,2581); THE WAY THEY WERE (II,2748); THE WILD WILD WEST (I,4863); THE WILD WILD WEST REVISITED (II,2800)

CONRAD, SHANE
HARD KNOX (II,1087)

CONRAD, SID
NIGHTSIDE (II,1851)

CONRAD, WILLIAM
ALFRED HITCHCOCK PRESENTS (I,115); BATTLES: THE MURDER THAT WOULDN'T DIE (II,180); THE BULLWINKLE SHOW (I,761); CANNON (I,820); CANNON (II,424); CANNON: THE RETURN OF FRANK CANNON (II,425); THE D.A.: CONSPIRACY TO KILL (I,1121); THE DUDLEY DO-RIGHT SHOW (I,1378); ESCAPE (I,1459); THE FLIP WILSON SPECIAL (I,1603); THE FLIP WILSON SPECIAL (II,888); THE FUGITIVE (I,1701); GENERAL ELECTRIC TRUE (I,1754); GEORGE OF THE JUNGLE (I,1773); NERO WOLFE (II,1806); O'HARA, UNITED STATES TREASURY: OPERATION COBRA (I,3347); POLICE SQUAD! (II,2061); ROCKY AND HIS FRIENDS (I,3822); TALES OF THE UNEXPECTED (II,2539); THE TARZAN/LONE RANGER/ZORRO ADVENTURE HOUR (II,2543); TURNOVER SMITH (II,2671); WILD WILD WORLD OF ANIMALS (I,4864)

CONRIED, HANS
ADVENTURES OF HOPPITY HOPPER FROM FOGGY BOGG (I,62); ALLISON SIDNEY HARRISON (II,54); AMERICAN COWBOY (I,169); THE AMERICAN DREAM (II,71); BAREFOOT IN THE PARK (II,151); THE BULLWINKLE SHOW (I,761); CAVALCADE OF AMERICA (I,875); CROWN THEATER WITH GLORIA SWANSON (I,1107); THE DANNY THOMAS TV FAMILY REUNION (I,1154); DAVY CROCKETT (I,1181); THE DRAK PACK (II,733); THE DUDLEY DO-RIGHT SHOW (I,1378); FEATHERTOP (I,1556); FRACTURED FLICKERS (I,1653); THE GENERAL MOTORS 50TH ANNIVERSARY SHOW (I,1758); GEORGE BURNS EARLY, EARLY, EARLY, CHRISTMAS SHOW (II,968); GILLIGAN'S ISLAND (II,990); HANSEL AND GRETEL (I,1934); HIGH TOR (I,2056); KISMET (I,2565); MAKE ROOM FOR DADDY (I,2843); MAKE ROOM FOR GRANDDADDY (I,2844); MAKE ROOM FOR GRANDDADDY (I,2845); MANHATTAN TOWER (I,2889); MIRACLE ON 34TH STREET (I,3056); ONE HOUR IN WONDERLAND (I,3376); PANTOMIME QUIZ (I,3449); PERRY PRESENTS (I,3549); PRIVATE EYE, PRIVATE EYE (I,3671); ROCKY AND HIS FRIENDS (I,3822); THE ROYAL FOLLIES OF 1933 (I,3862); SCHLITZ PLAYHOUSE OF STARS (I,3936); STEP ON THE GAS (I,4215); STUMP THE STARS (I,4273); THROUGH THE MAGIC PYRAMID (II,2615); THE TONY RANDALL SHOW (II,2640); THE WALT DISNEY CHRISTMAS SHOW (I,4755)

CONROY, EDWARD E
FEDERAL AGENT (I,1557)

CONROY, FRANCES
KENNEDY (II,1377)

CONROY, FRANK
LITTLE MOON OF ALBAN (I,2714); THE POWER AND THE GLORY (I,3658); THREE IN ONE (I,4479)

CONROY, KEVIN
ANOTHER WORLD (II,97); A FINE ROMANCE (II,858); KENNEDY (II,1377)

CONSIDINE, BOB
TONIGHT! AMERICA AFTER DARK (I,4555)

CONSIDINE, JOHN
ANOTHER WORLD (II,97); BRIGHT PROMISE (I,727); MICKEY SPILLANE'S MARGIN FOR MURDER (II,1691); TIME FOR ELIZABETH (I,4506)

CONSIDINE, TIM
ANNETTE (I,222); THE FURTHER ADVENTURES OF SPIN AND MARTY (I,1720); THE HARDY BOYS AND THE MYSTERY OF GHOST FARM (I,1950); THE HARDY BOYS AND THE MYSTERY OF THE APPLEGATE TREASURE (I,1951); THE MOUSEKETEERS REUNION (II,1742); MY THREE SONS (I,3205); THE NEW ADVENTURES OF SPIN AND MARTY (I,3254); SPIN AND MARTY (I,4158); THANKSGIVING REUNION WITH THE PARTRIDGE FAMILY AND MY THREE SONS (II,2569)

CONSTANTINE, MICHAEL
79 PARK AVENUE (II,2310); AMANDA'S (II,62); THE BAIT (I,335); BIG ROSE (II,241); CONSPIRACY OF TERROR (II,569); DAUGHTERS (II,620); ELECTRA WOMAN AND DYNA GIRL (II,764); GHOSTBREAKER (I,1789); GREAT ADVENTURE (I,1868); HEY, LANDLORD (I,2039); HIDE AND SEEK (I,2047); PROFILES IN COURAGE (I,3678); REMINGTON STEELE (II,2130); ROOM 222 (I,3843); ROOTS: THE NEXT GENERATIONS (II,2213); SIROTA'S COURT (II,2365); TWIN DETECTIVES (II,2687)

CONTE, JOHN
ANYTHING GOES (I,237); BEST OF THE POST (I,408); A CONNECTICUT YANKEE (I,1038); THE DESERT SONG (I,1236); THE FEMININE TOUCH (I,1565); HOUR OF STARS (I,2135); JOHN CONTE'S LITTLE SHOW (I,2408); THE MARCH OF DIMES FASHION SHOW (I,2901); THE MERRY WIDOW (I,3012); MIDNIGHT MYSTERY (I,3030); NAUGHTY MARIETTA (I,3232); PERSONALITY PUZZLE

(I,3554); TALES OF TOMORROW (I,4352)

CONTE, RICHARD
THE FOUR JUST MEN (I,1650); THE GREEN FELT JUNGLE (I,1885); THE JEAN ARTHUR SHOW (I,2352)

CONTI, AL
KENNEDY (II,1377)

CONTI, SAM
HEAR NO EVIL (II,1116)

CONTI, TOM
IF IT'S A MAN, HANG UP (I,2186)

CONTI, VINCE
KOJAK (II,1406)

CONTINENTALS, THE
FORD STAR REVUE (I,1632)

CONTOURI, CHANTAL
ALL THE RIVERS RUN (II,43)

CONTRERAS, DONALD
ROARING CAMP (I,3804)

CONTRERAS, LOUIS
ALL THAT GLITTERS (II,41)

CONTRERAS, ROBERTO
THE HIGH CHAPARRAL (I,2051)

CONTRERAS, RON
HEAVEN ON EARTH (II,1119)

CONVERSE, FRANK
CIRCLE OF FEAR (I,958); CORONET BLUE (I,1055); D.H.O. (I,1249); THE FAMILY TREE (II,820); GABE AND WALKER (II,950); IN TANDEM (II,1228); MOMMA THE DETECTIVE (II,1721); MOVIN' ON (II,1743); N.Y.P.D. (I,3321); STAT! (I,4213); STEELTOWN (II,2452)

CONVERSE, MELISSA
THE FRENCH ATLANTIC AFFAIR (II,924); THUNDER (II,2617)

CONVY, BERT
BERT CONVY SPECIAL—THERE'S A MEETING HERE TONIGHT (II,210); CELEBRITY DAREDEVILS (II,473); THE CLIFF DWELLERS (I,974); DEAR MOM, DEAR DAD (I,1202); EBONY, IVORY AND JADE (II,755); IT'S NOT EASY (II,1256); JACQUELINE SUSANN'S VALLEY OF THE DOLLS, 1981 (II,1268); KEEP THE FAITH (I,2518); LADY LUCK (I,2604); THE LATE FALL, EARLY SUMMER BERT CONVY SHOW (II,1441); THE LOVE BOAT II (II,1533); LOVE OF LIFE (I,2774); THE PARTRIDGE FAMILY (II,1962); PEOPLE DO THE CRAZIEST THINGS (II,1997); PEOPLE DO THE CRAZIEST THINGS (II,1998); POLICE WOMAN: THE GAMBLE (II,2064); THE SNOOP SISTERS (II,2389); SUPER PASSWORD (II,2499); TATTLETALES (II,2545); TV'S FUNNIEST GAME SHOW MOMENTS (II,2676)

CONWAY, ANTHONY
CINDERELLA (II,526)

CONWAY, BLAKE
THIS IS KATE BENNETT (II,2597)

CONWAY, CURT
BACHELOR AT LAW (I,322); PEYTON PLACE (I,3574)

CONWAY, GARY
BURKE'S LAW (I,762); THE JUDGE AND JAKE WYLER (I,2464); LAND OF THE GIANTS (I,2611)

CONWAY, KEVIN
ALL MY CHILDREN (II,39); THE FIRM (II,860); RAGE OF ANGELS (II,2107); RX FOR THE DEFENSE (I,3878); THE SCARLET LETTER (II,2270)

CONWAY, MICHELE
GENERAL HOSPITAL (II,964)

CONWAY, PAT
THE GREATEST MAN ON EARTH (I,1882); THE KEN MURRAY SHOW (I,2525); SQUADRON (I,4170); TOMBSTONE TERRITORY (I,4550)

CONWAY, RUSS
THE D.A. (I,1123); THE HARDY BOYS AND THE MYSTERY OF GHOST FARM (I,1950); THE HARDY BOYS AND THE MYSTERY OF THE APPLEGATE TREASURE (I,1951); THE JACK BENNY PROGRAM (I,2294); THE MONROES (I,3089); MY THREE SONS (I,3205); NAVY LOG (I,3233); RICHARD DIAMOND, PRIVATE DETECTIVE (I,3785)

CONWAY, SHIRL
CAESAR'S HOUR (I,790); THE NURSES (I,3318)

CONWAY, SUSAN
ADVENTURES IN RAINBOW COUNTRY (I,47); FOREST RANGERS (I,1642)

CONWAY, TIM
ACE CRAWFORD, PRIVATE EYE (II,6); THE BOYS (II,360); CAROL BURNETT AND COMPANY (II,441); THE CAROL BURNETT SHOW (II,443); THE CELEBRITY FOOTBALL CLASSIC (II,474); THE CHEVY CHASE SHOW (II,505); CIRCUS OF THE STARS (II,538); THE COMICS (I,1028); DANNY THOMAS LOOKS AT YESTERDAY, TODAY AND TOMORROW (I,1148); DORIS DAY TODAY (II,716); GREAT DAY (II,1054); HOLLYWOOD'S PRIVATE HOME MOVIES II (II,1168); HOTEL 90 (I,2130); THE JOHN DAVIDSON CHRISTMAS SHOW (II,1308); KEEP U.S. BEAUTIFUL (I,2519); MCHALE'S NAVY (I,2969); THE NEW STEVE ALLEN SHOW (I,3271); THE PAUL LYNDE HALLOWEEN SPECIAL (II,1981); RANGO (I,3721); TELEVISION'S GREATEST COMMERCIALS (II,2553); THE TIM CONWAY COMEDY HOUR (I,4501); THE TIM CONWAY SHOW (I,4502); THE TIM CONWAY SHOW (II,2619); THE TIM CONWAY SPECIAL (I,4503); TURN ON (I,4625); UNCLE TIM WANTS YOU! (II,2704)

CONWAY, TOM
THE BETTY HUTTON SHOW (I,413); MARK SABER (I,2914)

CONWELL, CAROLYN
THE YOUNG AND THE RESTLESS (II,2862)

CONWELL, GORDON
AS CAESAR SEES IT (I,294)

CONWELL, JOHN
SKYFIGHTERS (I,4078)

CONWELL, PATRICIA
COMEDY OF HORRORS (II,557); THE DUKES (II,740); THE EDGE OF NIGHT (II,760)

CONWELL, VIRGINIA
SUMMER IN THE CITY (I,4280)

COO, PETER
ANDY'S GANG (I,214)

COOGAN, JACKIE
THE ADDAMS FAMILY (II,11); THE ADDAMS FAMILY (II,12); THE ADDAMS FAMILY (II,13); THE BENNY RUBIN SHOW (I,398); CLOWN ALLEY (I,979); COWBOY G-MEN (I,1084); DAMON RUNYON THEATER (I,1129); THE DICK POWELL SHOW (I,1269); THE GENERAL ELECTRIC THEATER (I,1753); HAPPY DAYS (II,1084); HER SCHOOL FOR BACHELORS (I,2018); THE HOOFER (I,2113); THE LORETTA YOUNG THEATER (I,2756); A LUCILLE BALL SPECIAL STARRING LUCILLE BALL AND DEAN MARTIN (II,1557); THE MASK OF MARCELLA (I,2937); MCKEEVER AND THE COLONEL (I,2970); PANTOMIME QUIZ (I,3449); THE PARTRIDGE FAMILY (II,1962); TOO MANY SERGEANTS (I,4573); WHEN THE WEST WAS FUN: A WESTERN REUNION (II,2780)

COOGAN, RICHARD
THE CALIFORNIANS (I,795); CAPTAIN VIDEO AND HIS VIDEO RANGERS (I,838); CLEAR HORIZON (I,973); EYE WITNESS (I,1494); LOVE OF LIFE (I,2774)

COOGAN, ROBERT
DANIEL BOONE (I,1142)

COOGAN, SHAYE
SPIN THE PICTURE (I,4159)

COOK, ANCEL
DANNY AND THE MERMAID (II,618)

COOK, BARBARA
BABES IN TOYLAND (I,318); BLOOMER GIRL (I,488); GOLDEN WINDOWS (I,1841); HANSEL AND GRETEL (I,1934); THE YEOMAN OF THE GUARD (I,4926)

COOK, BART
LITTLE WOMEN (I,2724)

COOK, BILL
MINSKY'S FOLLIES (II,1703)

COOK, CAROLE
THE DESILU REVUE (I,1238); GENERAL HOSPITAL (II,964); THE HOOFER (I,2113); KOJAK (II,1406); LADY LUCK (I,2604); RENDEZVOUS HOTEL (II,2131)

COOK, CHRISTOPHER
A CHRISTMAS CAROL (I,947)

COOK, DAVID
LITTLE LORD FAUNTLEROY (II,1492)

COOK, DONALD
ABC ALBUM (I,4); THE DOCTOR (I,1320)

COOK, DORIA
HAPPY ANNIVERSARY AND GOODBYE (II,1082)

COOK, DORRA
SOUTHERN FRIED (I,4139)

COOK, ELISHA
THE JUDGE (I,2468); MAGNUM, P.I. (II,1601); MCNAB'S LAB (I,2972); THE MOTOROLA TELEVISION HOUR (I,3112); NO WARNING (I,3301); SHADOW OF SAM PENNY (II,2313); TERROR AT ALCATRAZ (II,2563); THIS GIRL FOR HIRE (II,2596); THE

TROUBLE WITH GRANDPA (II,2662); TV READERS DIGEST (I,4630)

COOK, FREDRIC
THE BLUE AND THE GRAY (II,273)

COOK, LAWRENCE
ADVENTURING WITH THE CHOPPER (II,15); CHIPS (II,511); DAYS OF OUR LIVES (II,629); FROM HERE TO ETERNITY (II,933); HARRY O (II,1098)

COOK, LINDA
THE EDGE OF NIGHT (II,760); ME AND DUCKY (II,1671)

COOK, MAURICE
THE SKATEBIRDS (II,2375)

COOK, NATHAN
THE ASTRONAUTS (II,114); HOTEL (II,1192); THE WHITE SHADOW (II,2788)

COOK, PETER
THE TWO OF US (II,2693)

COOK, TOMMY
THE BANANA SPLITS ADVENTURE HOUR (I,340); BROKEN ARROW (I,743); THE FUNKY PHANTOM (I,1709); JABBERJAW (II,1261)

COOK, VIRGINIA
FIT FOR A KING (II,866)

COOKE, ALISTAIR
AMERICA (I,165); OMNIBUS (I,3353); UPSTAIRS, DOWNSTAIRS (II,2709)

COOKE, CAROL
NOW WE'RE COOKIN' (II,1866)

COOKE, DONNA
H.M.S. PINAFORE (I,2066)

COOKE, JENNIFER
THE GUIDING LIGHT (II,1064); SUMMER (II,2491); TOM AND JOANN (II,2630); V: THE SERIES (II,2715)

COOKE, RAY
YOUR HIT PARADE (I,4952)

COOKSEY, DANNY
DIFF'RENT STROKES (II,674)

COOKSEY, DARRYL
EMERALD POINT, N.A.S. (II,773)

COOKSON, GARY
DELTA HOUSE (II,658)

COOKSON, PETER
FIRST LOVE (I,1583)

COOLEY, CHARLEY
THE BOB HOPE SHOW (I,552)

COOLIDGE, PHILIP
ABC STAGE '67 (I,6); ACRES AND PAINS (I,19); BACHELOR PARTY (I,324); CONCERNING MISS MARLOWE (I,1032); THE FARMER'S DAUGHTER (I,1533); ICHABOD AND ME (I,2183); TONIGHT AT 8:30 (I,4556)

COOLIDGE, RITA
I BELIEVE IN MUSIC (I,2165); MARLO THOMAS AND FRIENDS IN FREE TO BE. . . YOU AND ME (II,1632); ROCK COMEDY (II,2191)

COONEY, DENNIS
AS THE WORLD TURNS (II,110); LOVE OF LIFE (I,2774); THE MAGNIFICENT YANKEE (I,2829)

COONEY, GERRY
BOB HOPE SPECIAL: BOB HOPE'S ALL-STAR BIRTHDAY AT ANNAPOLIS (II,294)

COONS, JOEY
DANIEL BOONE (I,1142)

COONS, JOHNNY
UNCLE JOHNNY COONS (I,4668); UNCLE MISTLETOE AND HIS ADVENTURES (I,4669)

COOPER, ALICE
ROCK 'N' ROLL: THE FIRST 25 YEARS (II,2189)

COOPER, ANN
BLUE THUNDER (II,278); THE SUNSHINE BOYS (II,2496)

COOPER, BEN
THE FALL GUY (II,811); WON'T IT EVER BE MORNING? (I,4902)

COOPER, BOBBY
CELEBRITY BOWLING (I,883)

COOPER, BURTON
THE GUIDING LIGHT (II,1064)

COOPER, CATHY
SANFORD (II,2255)

COOPER, CHARLES
FATHER MURPHY (II,841); FBI CODE 98 (I,1550); O'CONNOR'S OCEAN (I,3326); STEELTOWN (II,2452)

COOPER, DANNY
MUGGSY (II,1761)

COOPER, DAVID
ADAMS HOUSE (II,9)

COOPER, DEBBIE
THE LAST CONVERTIBLE (II,1435)

COOPER, GARY
HEDDA HOPPER'S HOLLYWOOD (I,2002)

COOPER, GLADYS
MY FAIR LADY (I,3187); ON TRIAL (I,3363); THE ROGUES (I,3832)

COOPER, GORDON
HAPPY BIRTHDAY, AMERICA (II,1083)

COOPER, HAL
HANDS OF MURDER (I,1929)

COOPER, JACKIE
ALCOA/GOODYEAR THEATER (I,107); THE ARMSTRONG CIRCLE THEATER (I,260); DEAN MARTIN'S COMEDY WORLD (II,647); THE DICK POWELL SHOW (I,1269); DOCTOR DAN (II,684); GREAT ADVENTURE (I,1868); HENNESSEY (I,2013); THE INVISIBLE MAN (II,1243); KEEPING AN EYE ON DENISE (I,2520); MOBILE ONE (II,1717); MOBILE TWO (II,1718); THE MOTOROLA TELEVISION HOUR (I,3112); OF MEN OF WOMEN (I,3331); OPERATION PETTICOAT (II,1916); THE PEOPLE'S CHOICE (I,3521); REVLON MIRROR THEATER (I,3781); ROBERT MONTGOMERY PRESENTS YOUR LUCKY STRIKE THEATER (I,3809); SCHLITZ PLAYHOUSE OF STARS (I,3936); STARLIGHT THEATER (I,4201); SUSPENSE (I,4305); TALES OF TOMORROW (I,4352); THE TWILIGHT ZONE (I,4651); WHAT'S UP, AMERICA? (I,4824); WHAT'S UP? (I,4823)

COOPER, JEANNE
BRACKEN'S WORLD (I,703); THE SAN PEDRO BUMS (II,2251); THE YOUNG AND THE RESTLESS (II,2862)

COOPER, JED
STUDS LONIGAN (II,2484)

COOPER, JEFF
DALLAS (II,614); FRANK MERRIWELL (I,1659)

COOPER, JOEL
THE JACKSON FIVE (I,2326)

COOPER, JUNE
MISTER JERICO (I,3070)

COOPER, KEVIN KING
THE CLIFFWOOD AVENUE KIDS (II,543); LITTLE LULU (II,1493)

COOPER, LINDA
GENERAL HOSPITAL (II,964)

COOPER, MAGGIE
THE BOYS IN BLUE (II,359); DEATH RAY 2000 (II,654); SPACE ACADEMY (II,2406); WOMEN IN WHITE (II,2823)

COOPER, MELVILLE
THE CORN IS GREEN (I,1050); CURTAIN CALL THEATER (I,1115); THE DATCHET DIAMONDS (I,1160); EYE WITNESS (I,1494); MR. GLENCANNON TAKES ALL (I,3139); STORY THEATER (I,4240); YOUR SHOW TIME (I,4958)

COOPER, PAT
CELEBRATION: THE AMERICAN SPIRIT (II,461); LOVE AT FIRST SIGHT (II,1530); THE MAD MAD MAD MAD WORLD OF THE SUPER BOWL (II,1586); THE SON-IN-LAW (II,2398)

COOPER, RALPH
AIR SUPPLY IN HAWAII (II,26)

COOPER, RAY
BEACON HILL (II,193)

COOPER, SAM J
FULL HOUSE (II,937)

COOPER, TABI
BRENDA STARR (II,372)

COOPER, TAMARA
BIG BEND COUNTRY (II,230); BLACK BART (II,260); FRONTIER (I,1695); THE LIFE OF RILEY (I,2686)

COOPER, TERRENCE
THE BUCCANEERS (I,749)

COOPER, TOMMY
THE ROBERT GOULET SPECIAL (I,3807)

COOPER, TREVOR
SMILEY'S PEOPLE (II,2383)

COOPER, TYRONE
ON THE FLIP SIDE (I,3360)

COOPER, WILMA LEE
JOHNNY CASH AND THE COUNTRY GIRLS (II,1323)

COOTE, ROBERT
ALICE THROUGH THE LOOKING GLASS (I,122); INSTITUTE FOR REVENGE (II,1239); NERO WOLFE (II,1806); THE ROGUES (I,3832)

COPAGE, MARC
THE BEST OF TIMES (II,219); JULIA (I,2476)

COPE, KENNETH
MY PARTNER THE GHOST (I,3199)

COPE, KEVIN
A HORSEMAN RIDING BY (II,1182)

COPELAND, ALAN
HAPPY DAYS (I,1942)

COPELAND, ERIC
GOLDEN GATE (II,1020); THE JAMES BOYS (II,1272)

COPELAND, JOAN
CAGNEY AND LACEY (II,410); HOW TO SURVIVE A MARRIAGE (II,1198); LOVE OF LIFE (I,2774); ONE LIFE TO LIVE (II,1907); SEARCH FOR TOMORROW (II,2284); THE WEB (I,4784)

COPELAND, JOANNE
VIDEO VILLAGE (I,4731)

COPELAND, MAURICE
THOSE ENDEARING YOUNG CHARMS (I,4469)

COPLEY, PETER
JANE EYRE (I,2339); TYCOON: THE STORY OF A WOMAN (II,2697)

COPLEY, TERI
BATTLE OF THE NETWORK STARS (II,177); FLY AWAY HOME (II,893); WE GOT IT MADE (II,2752)

COPPERFIELD, DAVID
THE ALL-STAR SALUTE TO MOTHER'S DAY (II,51); THE MAGIC OF DAVID COPPERFIELD (II,1593); THE MAGIC OF DAVID COPPERFIELD (II,1594); THE MAGIC OF DAVID COPPERFIELD (II,1595); MAGIC WITH THE STARS (II,1599)

COPPIN, NICHOLAS
THE OMEGA FACTOR (II,1888)

COPPOLA, FRANK
JOE'S WORLD (II,1305)

COPPOLA, NICHOLAS
THE BEST OF TIMES (II,219)

CORAN, JOHN
MICKEY SPILLANE'S MIKE HAMMER: MORE THAN MURDER (II,1693)

CORBERT, ROBERT
SOMEONE AT THE TOP OF THE STAIRS (I,4114)

CORBETT, CAROL
COOL MCCOOL (I,1045); THE GARRY MOORE SHOW (I,1740)

CORBETT, GLENN
THE DOCTORS (II,694); EGAN (I,1419); ESCAPE (I,1461); IT'S A MAN'S WORLD (I,2258); THE LOG OF THE BLACK PEARL (II,1506); THE ROAD WEST (I,3801); ROUTE 66 (I,3852); SILVER THEATER (I,4051); THE STRANGER (I,4251); STUNTS UNLIMITED (II,2487)

CORBETT, GRETCHEN
MANDRAKE (II,1617); ONE DAY AT A TIME (II,1900); THE ROCKFORD FILES (II,2197); THINGS ARE LOOKING UP (II,2588)

CORBETT, LENORE
BLITHE SPIRIT (I,483)

CORBETT, LOIS
THE JACK BENNY PROGRAM (I,2294)

CORBETT, MICHAEL
RYAN'S HOPE (II,2234); SEARCH FOR TOMORROW (II,2284)

CORBETT, RONNIE
THE TWO RONNIES (II,2694)

CORBIN, BARRY
BOONE (II,351); NORMA RAE (II,1860); THE THORN BIRDS (II,2600); TRAVIS MCGEE (II,2656)

CORBIN, CLAYTON
OUR STREET (I,3418)

CORBY, ELLEN
THE ADDAMS FAMILY (II,11); ALIAS MIKE HERCULES (I,117); ALL THE WAY HOME (II,44); FOUR STAR PLAYHOUSE (I,1652); G.E. SUMMMER ORIGINALS (I,1747); THE GRADUATION DRESS (I,1864); HOLLOWAY'S DAUGHTERS (I,2079); THE MARRIAGE BROKER (I,2920); MCKEEVER AND THE COLONEL (I,2970); MEET MCGRAW (I,2984); MR. TERRIFIC (I,3152); ON TRIAL (I,3362); PLEASE DON'T EAT THE DAISIES (I,3628); SAFARI (I,3882); STARS OVER HOLLYWOOD (I,4211); THE WALTONS (II,2740); THE WALTONS: A DAY FOR THANKS ON WALTON'S MOUNTAIN (EPISODE 3) (II,2741); THE WALTONS: MOTHER'S DAY ON WALTON'S MOUNTAIN (EPISODE 2) (II,2742); THE WALTONS: WEDDING ON WALTON'S MOUNTAIN (EPISODE 1) (II,2743); YOUR JEWELER'S SHOWCASE (I,4953)

CORCORAN, BRIAN
TEXAS JOHN SLAUGHTER (I,4414)

CORCORAN, KELLY
THE ROAD WEST (I,3801)

CORCORAN, KEVIN
ADVENTURES IN DAIRYLAND (I,44); DANIEL BOONE (I,1141); THE FURTHER ADVENTURES OF SPIN AND MARTY (I,1720); THE NEW ADVENTURES OF SPIN AND MARTY (I,3254)

CORCORAN, NOREEN
BACHELOR FATHER (I,323); DICK CLARK'S GOOD OL' DAYS: FROM BOBBY SOX TO BIKINIS (II,671); MR. NOVAK (I,3145); NEW GIRL IN HIS LIFE (I,3265)

CORD, ALEX
AIRWOLF (II,27); THE BEST OF FRIENDS (II,216); CASSIE AND COMPANY (II,455); GENESIS II (I,1760); GOLIATH AWAITS (II,1025); HAVE I GOT A CHRISTMAS FOR YOU (II,1105); HUNTER'S MOON (II,1207); W.E.B. (II,2732)

CORD, BILL
PONY EXPRESS (I,3645)

CORD, ERIC
NEVADA SMITH (II,1807)

CORDARO, JACK
POLKA-GO-ROUND (I,3640)

CORDAY, MARA
COMBAT SERGEANT (I,1012)

CORDAY, PAULA
G.E. SUMMMER ORIGINALS (I,1747)

CORDELL, CARL
LUCKY PARTNERS (I,2787)

CORDELL, KATHLEEN
GREAT CATHERINE (I,1873); STAGE 13 (I,4183)

CORDELL, MELINDA
THE YOUNG AND THE RESTLESS (II,2862)

CORDEN, HENRY
ALVIN AND THE CHIPMUNKS (II,61); THE ATOM ANT/SECRET SQUIRREL SHOW (I,311); THE BANANA SPLITS ADVENTURE HOUR (I,340); THE BARKLEYS (I,356); THE BOB HOPE SHOW (I,625); BUFORD AND THE GHOST (II,389); THE C.B. BEARS (II,406); FLINTSTONE FAMILY ADVENTURES (II,882); THE FLINTSTONE FUNNIES (II,883); THE FLINTSTONES (II,885); THE FLYING NUN (I,1611); FRED AND BARNEY MEET THE THING (II,919); FRED FLINTSTONE AND FRIENDS (II,920); GOLDIE GOLD AND ACTION JACK (II,1023); HELLO DERE (I,2010); I DREAM OF JEANNIE (I,2167); INNOCENT JONES (I,2217); KEEP THE FAITH (I,2518); THE MONKEES (I,3088); THE NEW FRED AND BARNEY SHOW (II,1817); RETURN TO THE PLANET OF THE APES (II,2145); THUNDARR THE BARBARIAN (II,2616); YOGI'S GANG (II,2852); YOU SHOULD MEET MY SISTER (I,4935)

CORDEN, RAY
THE KELLY MONTEITH SHOW (II,1376)

CORDERO, MARIA-ELENA
THE JORDAN CHANCE (II,1343); LOOSE CHANGE (II,1518); NAKIA (II,1784)

CORDOVA, FRANCISCO
TEN WHO DARED (II,2559)

CORDOVA, MARGARITA
BENDER (II,204); BRIDGER (II,376); SANTA BARBARA (II,2258)

CORE, NATALIE
LOVE AND LEARN (II,1529)

COREY, IRWIN
DOC (II,682); THE MAD MAD MAD MAD WORLD OF THE SUPER BOWL (II,1586)

COREY, JEFF
BANJO HACKETT: ROAMIN' FREE (II,141); THE FUZZ BROTHERS (I,1722); THE GUN AND THE PULPIT (II,1067); THE OATH: THE SAD AND LONELY SUNDAYS (II,1874); ONE DAY AT A TIME (II,1900); THE RICHARD PRYOR SHOW (II,2163); ROXY PAGE (II,2222); TESTIMONY OF TWO MEN (II,2564)

COREY, JILL
THE DAVE GARROWAY SHOW (I,1173); THE JOHNNY CARSON SHOW (I,2422); MUSIC ON ICE (I,3175); ROBERT Q'S MATINEE (I,3812); TEXACO COMMAND PERFORMANCE (I,4407)

COREY, JOE
FRANK MERRIWELL (I,1659); SKIP TAYLOR (I,4075)

COREY, JOSEF
DEAR PHOEBE (I,1204)

COREY, PROFESSOR IRWIN
THE ANDY WILLIAMS SHOW (I,210); DOROTHY HAMILL'S CORNER OF THE SKY (II,721); WHERE THE GIRLS ARE (I,4830)

COREY, WENDELL
BACKBONE OF AMERICA (I,328); BEN HECHT'S TALES OF THE CITY (I,395); THE

BOB HOPE SHOW (I,571); CELANESE THEATER (I,881); CURTAIN CALL THEATER (I,1115); THE ELEVENTH HOUR (I,1425); GULF PLAYHOUSE (I,1904); HARBOR COMMAND (I,1945); PECK'S BAD GIRL (I,3503); ROBERT MONTGOMERY PRESENTS YOUR LUCKY STRIKE THEATER (I,3809); STUDIO ONE (I,4268); YES YES NANETTE (I,4927)

CORFF, ROBERT
EVERYDAY (II,793)

CORI, LISA
CALIFORNIA FEVER (II,411)

CORIO, ANN
HERE IT IS, BURLESQUE! (II,1133)

CORLAN, ANTHONY
THE ROADS TO FREEDOM (I,3802)

CORLEY, AL
BARE ESSENCE (II,150); DYNASTY (II,746)

CORLEY, MARJORIE
MR. NOVAK (I,3145)

CORLEY, PAT
BAY CITY BLUES (II,186); HILL STREET BLUES (II,1154); THE OUTSIDE MAN (II,1937); USED CARS (II,2712)

CORMACK, GEORGE
THE PRIME OF MISS JEAN BRODIE (II,2082)

CORMAN, ROGER
THE HORROR OF IT ALL (II,1181)

CORNEILSON, MICHAEL
FAMILY IN BLUE (II,818); INSPECTOR PEREZ (II,1237); NIGHTSIDE (II,1851)

CORNELIUS, DON
SOUL TRAIN (I,4131)

CORNELIUS, HELEN
COUNTRY NIGHT OF STARS II (II,593); JOHNNY CASH AND THE COUNTRY GIRLS (II,1323)

CORNELIUS, JULIE
CASANOVA (II,451)

CORNELL, DON
OPERA VS. JAZZ (I,3399); UPBEAT (I,4684)

CORNELL, KATHERINE
THERE SHALL BE NO NIGHT (I,4437)

CORNELL, LILLIAN
WINDOW SHADE REVUE (I,4874)

CORNELL, LYDIA
BATTLE OF THE NETWORK STARS (II,174); FAMILY BUSINESS (II,815); TOO CLOSE FOR COMFORT (II,2642)

CORNELL, SANDI
SKEEZER (II,2376)

CORNELL, TITIA
THE THORNTON SHOW (I,4468)

CORNER, JAMES
BROTHER RAT (I,746); THE MILKY WAY (I,3044)

CORNTHWAITE, ROBERT
THE ADVENTURES OF JIM BOWIE (I,63); DANIEL BOONE (I,1142); GET SMART (II,983)

CORNWELL, CHARLOTTE
THE ROCK FOLLIES (II,2192)

CORREA, DON
PARADE OF STARS (II,1954); PINOCCHIO (II,2044)

CORRELL, CHARLES
CALVIN AND THE COLONEL (I,806); THE FORD 50TH ANNIVERSARY SHOW (I,1630)

CORRELL, RICHARD
LEAVE IT TO BEAVER (I,2648); STILL THE BEAVER (II,2466)

CORRI, ADRIENNE
THE HIGHWAYMAN (I,2059); ONE STEP BEYOND (I,3388); THE SWORD OF FREEDOM (I,4326)

CORRIGAN, LLOYD
ARROYO (I,266); CORKY AND WHITE SHADOW (I,1049); HANK (I,1931); HAPPY (I,1938); HIGH TOR (I,2056); THE LIFE AND LEGEND OF WYATT EARP (I,2681); THE MARRIAGE BROKER (I,2920); THE MILTON BERLE SPECIAL (I,3052); PEPSI-COLA PLAYHOUSE (I,3523); THE RAY MILLAND SHOW (I,3734); THE REAL MCCOYS (I,3741)

CORRIGAN, RAY 'CRASH'
CRASH CORRIGAN'S RANCH (I,1090)

CORSAIR, BILL
SPRAGGUE (II,2430)

CORSAIRS, THE
SONG AND DANCE (I,4120)

CORSAUT, ANETA
THE ANDY GRIFFITH SHOW (I,192); THE BLUE KNIGHT (II,276); FULL HOUSE (II,936); HOUSE CALLS (II,1194); MRS. G. GOES TO COLLEGE (I,3155)

CORSEAU, JOHN
ANOTHER LIFE (II,95)

CORT, BILL
A.E.S. HUDSON STREET (II,17); DUSTY'S TRAIL (I,1390); THE MONTEFUSCOS (II,1727); THE PIRATES OF FLOUNDER BAY (I,3607); THINGS ARE LOOKING UP (II,2588)

CORTES, KEVIN
THE BEST OF TIMES (II,219)

CORTESE, JOSEPH
COMPUTERCIDE (II,561)

CORWIN, CHRISTOPHER
ANOTHER WORLD (II,97)

CORWIN, NORMAN
NORMAN CORWIN PRESENTS (I,3307)

COSBY, BILL
BILL COSBY DOES HIS OWN THING (I,441); THE BILL COSBY SHOW (I,442); THE BILL COSBY SPECIAL (I,444); THE BILL COSBY SPECIAL, OR? (I,443); CAPTAIN KANGAROO (II,434); CELEBRITY CHALLENGE OF THE SEXES 1 (II,465); CELEBRITY CHALLENGE OF THE SEXES 4 (II,468); THE COMICS (I,1028); COS (II,576); COS: THE BILL COSBY COMEDY SPECIAL (II,577); THE COSBY SHOW (II,578); DIANA (I,1258); DICK VAN DYKE MEETS BILL COSBY (I,1274); DOUG HENNING'S WORLD OF MAGIC III (II,727); FAT ALBERT AND THE COSBY KIDS (II,835); FEELING GOOD (I,1559); HOLLYWOOD'S PRIVATE HOME MOVIES (II,1167); I SPY (I,2179); IT'S WHAT'S HAPPENING, BABY! (I,2278); LOLA (II,1509); THE NEW BILL COSBY SHOW (I,3257); PLAYBOY'S 25TH ANNIVERSARY CELEBRATION (II,2054); A SPECIAL BILL COSBY SPECIAL (I,4147); STEVE MARTIN'S THE WINDS OF WHOOPIE (II,2461); TOP SECRET (II,2647); WAKE UP (II,2736); THE WONDERFUL WORLD OF BURLESQUE II (I,4896); A WORLD OF LOVE (I,4914); THE WORLD OF MAGIC (II,2838)

COSELL, GORDON
THE EVERGLADES (I,1478)

COSELL, HOWARD
BATTLE OF THE NETWORK STARS (II,163); BATTLE OF THE NETWORK STARS (II,164); BATTLE OF THE NETWORK STARS (II,165); BATTLE OF THE NETWORK STARS (II,166); BATTLE OF THE NETWORK STARS (II,167); BATTLE OF THE NETWORK STARS (II,168); BATTLE OF THE NETWORK STARS (II,169); BATTLE OF THE NETWORK STARS (II,170); BATTLE OF THE NETWORK STARS (II,171); BATTLE OF THE NETWORK STARS (II,172); BATTLE OF THE NETWORK STARS (II,173); BATTLE OF THE NETWORK STARS (II,174); BATTLE OF THE NETWORK STARS (II,175); BATTLE OF THE NETWORK STARS (II,176); BATTLE OF THE NETWORK STARS (II,177); BATTLE OF THE NETWORK STARS (II,178); BOB HOPE SPECIAL: BOB HOPE'S STAND UP AND CHEER FOR THE NATIONAL FOOTBALL LEAGUE'S 60TH YEAR (II,316); BOB HOPE SPECIAL: THE BOB HOPE SPECIAL (II,335); CELEBRATION: THE AMERICAN SPIRIT (II,461); THE DEAN MARTIN CELEBRITY ROAST (II,639); FOL-DE-ROL (I,1613); THE MUHAMMED ALI VARIETY SPECIAL (II,1762); THE ODD COUPLE (II,1875); SATURDAY NIGHT LIVE WITH HOWARD COSELL (II,2262); SINATRA—THE MAIN EVENT (II,2360)

COSHAM, RALPH
GEORGE WASHINGTON (II,978); KENNEDY (II,1377)

COSLER, LOU
PERRY COMO PRESENTS THE JULIUS LAROSA SHOW (I,3530)

COSSART, ERNEST
LITTLE WOMEN (I,2719)

COSSART, VALERIE
BLITHE SPIRIT (I,483); THE HARTMANS (I,1960); LOVE OF LIFE (I,2774); MY FAIR LADY (I,3187); STAR TONIGHT (I,4199)

COSSELL, BARRY
THE M AND M CANDY CARNIVAL (I,2797)

COSSINS, JAMES
POSSESSION (I,3655); WHY DIDN'T THEY ASK EVANS?

(II,2795)

COSTA, COSIE
CALIFORNIA FEVER (II,411); IN THE BEGINNING (II,1229); THE KIDS FROM C.A.P.E.R. (II,1391)

COSTA, DON
TONY BENNETT IN WAIKIKI (I,4568)

COSTA, MARY
BING CROSBY AND THE SOUNDS OF CHRISTMAS (I,457); CLIMAX! (I,976); THE FRANCES LANGFORD SHOW (I,1657); THE FRANK SINATRA TIMEX SHOW (I,1667); GIRL FRIENDS AND NABORS (I,1807)

COSTANZO, LINDA
SCRUPLES (II,2280)

COSTANZO, ROBERT
CHECKING IN (II,492); JOE AND VALERIE (II,1298); JOE AND VALERIE (II,1299); THE LAST RESORT (II,1440); MAMA MALONE (II,1609); USED CARS (II,2712)

COSTEAU, JACQUES
THOSE AMAZING ANIMALS (II,2601)

COSTELLO, ANTHONY
MR. AND MRS. COP (II,1748); WHEELS (II,2777)

COSTELLO, DANNY
ACCENT ON LOVE (I,16)

COSTELLO, LOU
THE ABBOTT AND COSTELLO SHOW (I,2); THE COLGATE COMEDY HOUR (I,997)

COSTELLO, MARICLARE
CONSPIRACY OF TERROR (II,569); THE FITZPATRICKS (II,868); SARA (II,2260); SKEEZER (II,2376); THE WALTONS (II,2740)

COSTELLO, WARD
THE CITY (II,541); CODE NAME: DIAMOND HEAD (II,549); THE EDGE OF NIGHT (II,760); THE GREATEST GIFT (I,1880); THE SECRET STORM (I,3969)

COSTER, NICHOLAS
ANOTHER WORLD (II,97); BENDER (II,204); THE COURT-MARTIAL OF GENERAL GEORGE ARMSTRONG CUSTER (I,1081); EBONY, IVORY AND JADE (II,755); THE FACTS OF LIFE (II,805); LOBO (II,1504); OUR PRIVATE WORLD (I,3417); RYAN'S FOUR (II,2233); SOMERSET (I,4115); TENSPEED AND BROWN SHOE (II,2562); WHERE'S EVERETT? (I,4835); THE WORD (II,2833)

COSTER, NICK
THE SECRET STORM (I,3969)

COSTI, MARTIN
THE FRENCH ATLANTIC AFFAIR (II,924)

COSTRO, RON
SEARCH (I,3951)

COTE, ELIZABETH
HEAVENS TO BETSY (I,1997)

COTLER, JEFF
EIGHT IS ENOUGH (II,762); GALACTICA 1980 (II,953); ONCE AN EAGLE (II,1897); STATE FAIR (II,2448); STRUCK BY LIGHTNING (II,2482); THE YOUNG PIONEERS (II,2869)

COTLER, KAMI
ME AND THE CHIMP (I,2974); THE WALTONS (II,2740); THE WALTONS: A DAY FOR THANKS ON WALTON'S MOUNTAIN (EPISODE 3) (II,2741); THE WALTONS: MOTHER'S DAY ON WALTON'S MOUNTAIN (EPISODE 2) (II,2742); THE WALTONS: WEDDING ON WALTON'S MOUNTAIN (EPISODE 1) (II,2743)

COTRIGHT, SHARLEEN
THE NEW TEMPERATURES RISING SHOW (I,3273)

COTSWORTH, STAATS
ABE LINCOLN IN ILLINOIS (I,11); BELLE STARR (I,391); THE DOCTORS (II,694); MACBETH (I,2800)

COTTEN, JOSEPH
ALEXANDER THE GREAT (I,113); ALFRED HITCHCOCK PRESENTS (I,115); BROADWAY (I,733); CASINO (II,452); CITY BENEATH THE SEA (I,965); THE DESILU PLAYHOUSE (I,1237); HOLLYWOOD AND THE STARS (I,2081); HOLLYWOOD: THE SELZNICK YEARS (I,2095); THE JUNE ALLYSON SHOW (I,2488); LIGHT'S DIAMOND JUBILEE (I,2698); NOTORIOUS (I,3314); ON TRIAL (I,3362); ON TRIAL (I,3363); RETURN TO FANTASY ISLAND (II,2144); TEXACO COMMAND PERFORMANCE (I,4408)

COTTERILL, CHRISSIE
SEPARATE TABLES (II,2304)

COTTIER, STEPHEN
ADVENTURES IN RAINBOW COUNTRY (I,47)

COTTING, RICHARD
CLUTCH CARGO (I,988)

COTTLE, BOB
THE RUFF AND REDDY SHOW (I,3866)

COTTLE, TOM
REAL LIFE STORIES (II,2117); TOM COTTLE: UP CLOSE (II,2631)

COTTRELL, WILLIAM
NAKED CITY (I,3216)

COUCHER, BRIAN
THE LAST DAYS OF POMPEII (II,1436)

COUFAS, PAUL
BATTLESTAR GALACTICA (II,181)

COUGHLIN, KEVIN
I REMEMBER MAMA (I,2176)

COULOURIS, GEORGE
THE STRANGER (I,4251)

COUNT BASSIE AND HIS ORCHESTRA
SINATRA: THE MAN AND HIS MUSIC (II,2362)

COUNTRY BRIAR HOPPERS, THE
MIDWESTERN HAYRIDE (I,3033)

COUNTRY LADS, THE
THE GEORGE HAMILTON IV SHOW (I,1769); THE JIMMY DEAN SHOW (I,2384)

COUNTRY SQUARE DANCERS, THE
COUNTRY MUSIC CARAVAN (I,1066)

COUPER, BARBARA
VANITY FAIR (I,4701)

COUPLAND, DIANA
BLESS THIS HOUSE (II,268)

COURT, GERALDINE
ANOTHER WORLD (II,97); AS THE WORLD TURNS (II,110); THE DOCTORS (II,694); THE GUIDING LIGHT (II,1064)

COURT, HAZEL
DICK AND THE DUCHESS (I,1263); GIDGET (I,1795); THRILLER (I,4492)

COURT, JENNIFER
THE GUIDING LIGHT (II,1064)

COURTENAY, MARGARET
CAESAR AND CLEOPATRA (I,789); KISS KISS, KILL KILL (I,2566); ROMEO AND JULIET (I,3837)

COURTLAND, JEROME
THE RIFLEMAN (I,3789); THE SAGA OF ANDY BURNETT (I,3883); TALES OF THE VIKINGS (I,4351)

COURTLAND, PAULA
PERRY MASON (II,2025)

COURTLEIGH, BOB
THE ATOM SQUAD (I,312); FIRST LOVE (I,1583)

COURTLEIGH, STEPHEN
ABE LINCOLN IN ILLINOIS (I,10)

COURTNEY, ALEX
SWORD OF JUSTICE (II,2521)

COURTNEY, C C
THE DOCTORS (II,694)

COURTNEY, CHUCK
THE LONE RANGER (I,2740)

COURTNEY, DEBORAH
LOVE OF LIFE (I,2774)

COURTNEY, DIANE
SONG AND DANCE (I,4120)

COURTNEY, FRITZI JANE
SPRAGGUE (II,2430)

COURTNEY, JACQUELINE
ANOTHER WORLD (II,97); ONE LIFE TO LIVE (II,1907); OUR FIVE DAUGHTERS (I,3413)

COURTNEY, NICHOLAS
DOCTOR WHO (II,689); DOCTOR WHO—THE FIVE DOCTORS (II,690)

COVAN, DEFOREST
THAT'S MY MAMA (II,2580)

COVAY, CAB
HEAR NO EVIL (II,1116)

COVER, FRANKLIN
CHANGE AT 125TH STREET (II,480); THE JEFFERSONS (II,1276); A WOMAN CALLED GOLDA (II,2818)

COVINGTON, JULIA
THE ROCK FOLLIES (II,2192)

COVSENTINO, FRANK
WHO'S ON CALL? (II,2791)

COWAN, BERNARD
SPIDER-MAN (I,4154)

COWAN, JEROME
THE ALCOA HOUR (I,108); THE FARMER'S DAUGHTER (I,1534); THE HOOFER (I,2113); THE HOUSE NEXT DOOR (I,2139); MANY HAPPY RETURNS (I,2895); MEET CORLISS ARCHER (I,2981); NOT FOR PUBLICATION (I,3309); THE TAB HUNTER SHOW (I,4328); THE TYCOON (I,4660); VALIANT LADY (I,4695)

COWAN, PHYLLIS
LONDON AND DAVIS IN NEW YORK (II,1512)

COWARD, NOEL
ANDROCLES AND THE LION (I,189); BLITHE SPIRIT (I,484); TOGETHER WITH MUSIC (I,4532)

COWLES, CHANDLER
TEN LITTLE INDIANS (I,4394)

COWLES, MATTHEW
ALL MY CHILDREN (II,39); A WORLD APART (I,4910)

COWLING, SAM
DON MCNEILL'S DINNER CLUB (I,1336); DON MCNEILL'S TV CLUB (I,1337)

COWPER, GERRY
LITTLE LORD FAUNTLEROY (II,1492)

COWSILLS, THE
THE WONDERFUL WORLD OF PIZZAZZ (I,4901)

COX, ALAN
PENMARRIC (II,1993)

COX, BRIAN
THERESE RAQUIN (II,2584)

COX, DOUG
NIGHTSIDE (II,1851); PEN 'N' INC. (II,1992); WACKO (II,2733)

COX, RICHARD
CAMP GRIZZLY (II,418); EXECUTIVE SUITE (II,796); LOVE OF LIFE (I,2774); SEARCH FOR TOMORROW (II,2284)

COX, RONNY
APPLE'S WAY (II,101); COREY: FOR THE PEOPLE (II,575); FIRST TIME, SECOND TIME (II,863); HAVING BABIES I (II,1107); HERNANDEZ, HOUSTON P.D. (I,2034); OUR TOWN (I,3421); RX FOR THE DEFENSE (I,3878); SPENCER (II,2420)

COX, ROSETTA
ROWAN AND MARTIN'S LAUGH-IN (I,3856)

COX, RUTH
BEULAH LAND (II,226); CALIFORNIA FEVER (II,411); THE NANCY DREW MYSTERIES (II,1789); OPERATION: RUNAWAY (II,1918)

COX, WALLY
THE ADVENTURES OF HIRAM HOLIDAY (I,61); ALFRED OF THE AMAZON (I,116); AMERICAN COWBOY (I,169); BABES IN TOYLAND (I,318); THE BOB HOPE SHOW (I,541); THE BOB HOPE SHOW (I,554); THE BOB HOPE SHOW (I,560); THE BOB HOPE SHOW (I,565); THE BOB HOPE SHOW (I,567); THE BOB HOPE SHOW (I,575); THE BOB HOPE SHOW (I,619); THE BOB HOPE SHOW (I,633); THE BOB HOPE SHOW (I,642); THE BOB HOPE SHOW (I,648); FIRESIDE THEATER (I,1580); THE FORD 50TH ANNIVERSARY SHOW (I,1630); HEIDI (I,2004); THE HOLLYWOOD SQUARES (II,1164); IRONSIDE (I,2240); THE LORETTA YOUNG THEATER (I,2756); THE MOUSE FACTORY (I,3113); MR. PEEPERS (I,3147); MURDER AT NBC (A BOB HOPE SPECIAL) (I,3159); THE NIGHT STRANGLER (I,3293); ONCE UPON A MATTRESS (I,3371); SCHOOL HOUSE (I,3937); SHOOT-IN AT NBC (A BOB HOPE SPECIAL) (I,4021); STARLIGHT THEATER (I,4201); UNDERDOG (I,4673); WHAT GAP? (I,4808)

COY, WALTER
FRONTIER (I,1695); A TIME FOR US (I,4507)

COYLE, JAMES
THE MANIONS OF AMERICA (II,1623)

COYLER, CARLTON
OUR FIVE DAUGHTERS (I,3413)

CRABBE, BUSTER
CAPTAIN GALLANT OF THE FOREIGN LEGION (I,831); STAR TONIGHT (I,4199)

CRABBE, CULLEN
CAPTAIN GALLANT OF THE FOREIGN LEGION (I,831)

CRABTREE, MICHAEL WESTON
HIGHWAY HONEYS (II,1151)

CRAGGS, PAUL
LOVE OF LIFE (I,2774)

CRAIG, CAROLYN
GENERAL HOSPITAL (II,964); THE TEENAGE IDOL (I,4375)

CRAIG, COLONEL JOHN D
DANGER IS MY BUSINESS (I,1135); EXPEDITION (I,1487)

CRAIG, DIANE
ALL THE RIVERS RUN (II,43)

CRAIG, HELEN
THE DETECTIVE (II,662); RICH MAN, POOR MAN—BOOK I (II,2161)

CRAIG, IVAN
THE ERROL FLYNN THEATER (I,1458)

CRAIG, JAMES
STUDIO '57 (I,4267); THE WESTERNER (I,4800)

CRAIG, JOHN
THE HELLCATS (I,2009); WHICH WAY'D THEY GO? (I,4838)

CRAIG, L MICHAEL
HOT HERO SANDWICH (II,1186)

CRAIG, MICKEY
CULTURE CLUB IN CONCERT (II,610)

CRAIG, NOEL
THE SECRET STORM (I,3969)

CRAIG, SCOTT
THE NEW MICKEY MOUSE CLUB (II,1823)

CRAIG, SKIP
THE DUDLEY DO-RIGHT SHOW (I,1378); GEORGE OF THE JUNGLE (I,1773)

CRAIG, TONY
THE EDGE OF NIGHT (II,760); THE RAY KNIGHT REVUE (I,3733)

CRAIG, WENDY
BUTTERFLIES (II,404)

CRAIG, YVONNE
THE BARBARA STANWYCK THEATER (I,350); BATMAN (I,366); CHANNING (I,900); DOBIE GILLIS (I,1302); HOORAY FOR LOVE (I,2116); JARRETT (I,2349); PAPA SAID NO (I,3452); THREE COINS IN THE FOUNTAIN (I,4473)

CRAIN, JEANNE
THE BOB HOPE SHOW (I,542); HIS MODEL WIFE (I,2064); LAST OF THE PRIVATE EYES (I,2628); MEET ME IN ST. LOUIS (I,2986)

CRAM, CUSI
ONE LIFE TO LIVE (II,1907)

CRAMER, GRANT
JO'S COUSINS (II,1293)

CRAMER, MARC
VALIANT LADY (I,4695)

CRAMPTON, CYDNEY
THE MISADVENTURES OF SHERIFF LOBO (II,1704)

CRANE, BOB
ARSENIC AND OLD LACE (I,270); THE BOB CRANE SHOW (II,280); CHANNING (I,900); THE DELPHI BUREAU (I,1224); THE DONNA REED SHOW (I,1347); HOGAN'S HEROES (I,2069); MAKE MINE RED, WHITE, AND BLUE (I,2841); MITZI AND A HUNDRED GUYS (II,1710)

CRANE, DAGNE
AS THE WORLD TURNS (II,110)

CRANE, GENE
CONTEST CARNIVAL (I,1040); THE M AND M CANDY CARNIVAL (I,2797)

CRANE, LES
I LOVE A MYSTERY (I,2170); THE LES CRANE SHOW (I,2657)

CRANE, MICHAEL
PETER PAN (II,2030)

CRANE, NORMA
THE FLYING NUN (I,1611); THE INNER SANCTUM (I,2216); MR. PEEPERS (I,3147); NIGHT GALLERY (I,3288); ONE STEP BEYOND (I,3388)

CRANE, RICHARD
COMMANDO CODY (I,1030); MYSTERIES OF CHINATOWN (I,3209); NAVY LOG (I,3233); ROCKY JONES, SPACE RANGER (I,3823); SURFSIDE 6 (I,4299); YOUR SHOW TIME (I,4958)

CRANHAM, KENNETH
THERESE RAQUIN (II,2584)

CRANSHAW, PAT
AFTERMASH (II,23); ALICE (II,33); GREAT DAY (II,1053); MORK AND MINDY (II,1735); ON THE ROCKS (II,1894); RIDING HIGH (II,2173); THICKER THAN WATER (I,4445)

CRANSTON, BRYAN
LOVING (II,1552)

CRANSTON, JOE
THE GALE STORM SHOW (I,1725)

CRANSTON, TOLLER
DOROTHY HAMILL IN ROMEO & JULIET ON ICE (II,718); THE MAGIC PLANET (II,1598)

CRAVAT, NICK
THE COUNT OF MONTE CRISTO (I,1058); DAVY CROCKETT (I,1181)

CRAVEN, FRANK
THE EGG AND I (I,1420)

CRAVEN, GEMMA
PENNIES FROM HEAVEN (II,1994); SONG BY SONG (II,2399)

CRAWFORD JR, ROBERT
LARAMIE (I,2614)

CRAWFORD, BILLY
THE WHITE SHADOW (II,2788)

CRAWFORD, BOBBY
SHIRLEY TEMPLE'S STORYBOOK (I,4017)

CRAWFORD, BRODERICK
THE ADVENTURES OF NICK CARTER (I,68); BRILLIANT BENJAMIN BOGGS (I,730); DAMON RUNYON THEATER (I,1129); FOUR STAR PLAYHOUSE (I,1652); HIGHWAY PATROL (I,2058); HUNTER (II,1204); THE INTERNS (I,2226); KING OF DIAMONDS (I,2558); PARADISE (II,1955)

CRAWFORD, EDWARD
SANFORD AND SON (II,2256)

CRAWFORD, ELLEN
AT YOUR SERVICE (II,118)

CRAWFORD, H MARION
SHERLOCK HOLMES (I,4010)

CRAWFORD, JAN
FAT ALBERT AND THE COSBY KIDS (II,835); MCNAB'S LAB (I,2972)

CRAWFORD, JOAN
BATMAN (I,366); THE BOB HOPE SHOW (I,564); THE BOB HOPE SHOW (I,577); NIGHT GALLERY (I,3288); REVLON MIRROR THEATER (I,3781); THE TALENT SCOUTS PROGRAM (I,4341); TALENT SEARCH (I,4342); ZANE GREY THEATER (I,4979)

CRAWFORD, JOHN
ANOTHER WORLD (II,97); FROM HERE TO ETERNITY (II,933); THE MACAHANS (II,1583); THE POWERS OF MATTHEW STAR (II,2075); SWISS FAMILY ROBINSON (II,2516); THE WALTONS (II,2740)

CRAWFORD, JOHNNY
CAVALCADE OF AMERICA (I,875); DICK CLARK'S GOOD OL' DAYS: FROM BOBBY SOX TO BIKINIS (II,671); THE INDIAN (I,2210); MATINEE THEATER (I,2947); THE MICKEY MOUSE CLUB (I,3025); THE MOUSEKETEERS REUNION (II,1742); THE RIFLEMAN (I,3789); THE SHARPSHOOTER (I,4001); TELEPHONE TIME (I,4379); WHEN THE WEST WAS FUN: A WESTERN REUNION (II,2780); WHICH WAY'D THEY GO? (I,4838)

CRAWFORD, KATHERINE
CAPTAINS AND THE KINGS (II,435); THE GEMINI MAN (II,960); THE GEMINI MAN (II,961); HERE COME THE BRIDES (I,2024); RAPTURE AT TWO-FORTY (I,3724)

CRAWFORD, LEE
KING OF THE ROAD (II,1396); QUICK AND QUIET (II,2099); THE YOUNG AND THE RESTLESS (II,2862)

CRAWFORD, MELISSA
A TOWN LIKE ALICE (II,2652)

CRAWFORD, MICHAEL
THE ADVENTURES OF SIR FRANCIS DRAKE (I,75)

CRAWFORD, ROBERT
THE JUDGE (I,2468)

CRAWFORD, SCOTT
AS THE WORLD TURNS (II,110)

CRAWFORD, TERRY
DARK SHADOWS (I,1157)

CRAWLEY, KATHLEEN
BRAVO DUKE (I,711)

CRAWLEY, MATT
NAKED CITY (I,3216)

CRAZE, PETER
BLAKE'S SEVEN (II,264)

CRAZE, SARAH
LITTLE WOMEN (I,2723)

CREDEL, CURTIS
THE LAST NINJA (II,1438)

CREE, ED
LEGS (II,1459)

CREHAN, JOSEPH
MYSTERY AND MRS. (I,3210)

CRELEY, JACK
THE IMPERIAL GRAND BAND (II,1224); STRANGE PARADISE (I,4246)

CRENNA, RICHARD
ALL'S FAIR (II,46); CBS: ON THE AIR (II,460); CENTENNIAL (II,477); FRONTIER (I,1695); IT TAKES TWO (II,1252); JOSHUA'S WORLD (II,1344); KRAFT SUSPENSE THEATER (I,2591); THE LILY TOMLIN SHOW (I,2706); LONDON AND DAVIS IN NEW YORK (II,1512); LOOK AT US (II,1515); MEDIC (I,2976); MUSICAL COMEDY TONIGHT (II,1777); OUR MISS BROOKS (I,3415); THE REAL MCCOYS (I,3741); SLATTERY'S PEOPLE (I,4082)

CRENSHAW, DR THERESA
THE LOVE REPORT (II,1542)

CRESPI, TODD
THE MAGICIAN (I,2823); THE MAGICIAN (I,2824)

CREWMAN, THE
S. S. TELECRUISE (I,4174)

CRIBBINS, BERNARD
THE VAL DOONICAN SHOW (I,4692)

CRICHTON, CHARLES
SPACE: 1999 (II,2409)

CRICHTON, DON
THE CAROL BURNETT SHOW (II,443); THE JIMMIE RODGERS SHOW (I,2383)

CRIDER, DOROTHY
ANOTHER DAY, ANOTHER DOLLAR (I,228)

CRISCUOLO, LOU
THE EDGE OF NIGHT (II,760); INSTANT FAMILY (II,1238); POPI (II,2068); POPI (II,2069); SEVENTH AVENUE (II,2309); STOCKARD CHANNING IN JUST FRIENDS (II,2467); WILDER AND WILDER (II,2802)

CRISP, MILLICENT
BATTLESTAR GALACTICA (II,181); WACKO (II,2733)

CRISTAL, LINDA
THE BOB HOPE SHOW (I,649); CALL HOLME (II,412); CONDOMINIUM (II,566); THE HIGH CHAPARRAL (I,2051); WHEN THE WEST WAS FUN: A WESTERN REUNION (II,2780)

CRITTENDEN, JAMES
CODE RED (II,553); DELTA COUNTY, U.S.A. (II,657)

CRITTENDEN, JORDAN
MY WIFE NEXT DOOR (II,1780)

CROCKETT, JAN
THE GEORGE HAMILTON IV SHOW (I,1769); JACKIE GLEASON AND HIS AMERICAN SCENE MAGAZINE (I,2321); THE JIMMY DEAN SHOW (I,2384)

CROCKETT, KARLENE
BOONE (II,351)

CROFT, MARTIN
STOP THE MUSIC (I,4232)

CROFT, MARY JANE
THE ADVENTURES OF OZZIE AND HARRIET (I,71); HERE'S LUCY (II,1135); I LOVE LUCY (I,2171); THE LIFE OF RILEY (I,2686); THE LUCY SHOW (I,2791); THE PEOPLE'S CHOICE (I,3521); THE TWO OF US (I,4658)

CROMARTY, ANDEE
THE BENNY HILL SHOW (II,207)

CROMPTON, NANCY
THE BOB HOPE SHOW (I,544)

CROMWELL, GLORIA
RYAN'S HOPE (II,2234)

CROMWELL, JAMES
ALL IN THE FAMILY (II,38); BAREFOOT IN THE PARK (II,151); BORN TO THE WIND (II,354); THE DEADLY GAME (II,633); THE EARTHLINGS (II,751); EDDIE AND HERBERT (II,757); THE GIRL IN THE EMPTY GRAVE (II,997); HOT L BALTIMORE (II,1187); LITTLE HOUSE ON THE PRAIRIE (II,1487); THE NANCY WALKER SHOW (II,1790); POTTSVILLE (II,2072); THE RAINMAKER (II,2109); SNAFU (II,2386); SPRAGGUE (II,2430); STRANDED (II,2475)

CRONEN, LOIS
THE INA RAY HUTTON SHOW (I,2207)

CRONIN, PATRICK J
ALICE (II,33); THE BOYS IN BLUE (II,359); SHE'S WITH ME (II,2321); WINDOWS, DOORS AND KEYHOLES (II,2806)

CRONIN, PAUL
THE SULLIVANS (II,2490)

CRONKITE, KATHY
HIZZONER (II,1159); YOU ASKED FOR IT (II,2855)

CRONKITE, WALTER
20TH CENTURY (I,4640); AIR POWER (I,92); CBS: ON THE AIR (II,460); IT'S NEWS TO ME (I,2272); SINATRA (I,4053); WALTER CRONKITE'S UNIVERSE (II,2739); YOU ARE

THERE (I,4929); YOU ARE THERE (I,4930)

CRONYN, HUME
THE ALCOA HOUR (I,108); BEN HECHT'S TALES OF THE CITY (I,395); A DOLL'S HOUSE (I,1327); THE FOUR POSTER (I,1651); THE GOLDEN AGE OF TELEVISION (II,1017); THE MARRIAGE (I,2922); THE MOON AND SIXPENCE (I,3098); THE MOTOROLA TELEVISION HOUR (I,3112); THE OATH: 33 HOURS IN THE LIFE OF GOD (II,1873)

CRONYN, TANDY
THE FEAR IS SPREADING (I,1553)

CROPPER, ANNA
CROWN MATRIMONIAL (I,1106)

CROSBIE, ANNETTE
THE SIX WIVES OF HENRY VIII (I,4068)

CROSBY BROTHERS, THE
THE BOB HOPE SHOW (I,572)

CROSBY, BARBARA
THE AMATEUR'S GUIDE TO LOVE (I,155)

CROSBY, BILL
MURDER AT NBC (A BOB HOPE SPECIAL) (I,3159)

CROSBY, BING
BING CROSBY AND CAROL BURNETT—TOGETHER AGAIN FOR THE FIRST TIME (I,454); BING CROSBY AND HIS FRIENDS (I,455); BING CROSBY AND HIS FRIENDS (I,456); BING CROSBY AND HIS FRIENDS (II,251); BING CROSBY AND THE SOUNDS OF CHRISTMAS (I,457); THE BING CROSBY CHRISTMAS SHOW (I,458); THE BING CROSBY SHOW (I,462); THE BING CROSBY SHOW (I,463); THE BING CROSBY SHOW (I,464); THE BING CROSBY SHOW (I,465); THE BING CROSBY SHOW (I,466); THE BING CROSBY SHOW (I,467); THE BING CROSBY SHOW (I,468); THE BING CROSBY SHOW (I,469); THE BING CROSBY SPECIAL (I,470); THE BING CROSBY SPECIAL (I,471); THE BING CROSBY SPECIAL (I,472); THE BING CROSBY SPRINGTIME SPECIAL (I,473); BING CROSBY'S CHRISTMAS SHOW (I,474); BING CROSBY'S MERRIE OLDE CHRISTMAS (II,252); BING CROSBY'S SUN VALLEY CHRISTMAS SHOW (I,475); BING CROSBY'S WHITE CHRISTMAS (II,253); BING CROSBY—COOLING IT (I,459); BING CROSBY—COOLING IT (I,460); BING!. . . A 50TH ANNIVERSARY GALA (II,254); THE BOB HOPE SHOW (I,540); THE BOB HOPE SHOW (I,561); THE BOB HOPE SHOW (I,588); THE BOB HOPE SHOW (I,591); THE BOB HOPE SHOW (I,607); THE BOB HOPE SHOW (I,612); THE BOB HOPE SHOW (I,621); THE BOB HOPE SHOW (I,630); THE BOB HOPE SHOW (I,640); THE BOB HOPE SHOW (I,647); BOB HOPE SPECIAL: A QUARTER CENTURY OF BOB HOPE ON TELEVISION (II,281); CHRISTMAS WITH THE BING CROSBYS (I,950); CHRISTMAS WITH THE BING CROSBYS (II,524); THE DANNY THOMAS SPECIAL (I,1150); DANNY THOMAS: THE ROAD TO LEBANON (I,1153); THE DEAN MARTIN SHOW (I,1197); THE EDSEL SHOW (I,1417); THE FRANK SINATRA SHOW (I,1666); HIGH TOR (I,2056); JUBILEE (II,1347); MERRY CHRISTMAS FROM THE CROSBYS (II,1685); MONSANTO PRESENTS MANCINI (I,3094); PHILCO TELEVISION PLAYHOUSE (I,3583); SWING OUT, SWEET LAND (II,2515); VER-R-R-RY INTERESTING (I,4712); YOU'RE THE TOP (I,4974)

CROSBY, BOB
THE BOB CROSBY SHOW (I,497); THE BOB CROSBY SHOW (I,498); THE BOB HOPE SHOW (I,506); THE KRAFT 75TH ANNIVERSARY SPECIAL (II,1409); SOLDIERS IN GREASEPAINT (I,4110); TIMEX ALL-STAR JAZZ SHOW III (I,4517)

CROSBY, CATHY
THE BOB CROSBY SHOW (I,497)

CROSBY, CATHY LEE
BATTLE OF THE NETWORK STARS (II,170); BATTLE OF THE NETWORK STARS (II,171); BATTLE OF THE NETWORK STARS (II,173); BILLY CRYSTAL: A COMIC'S LINE (II,249); BOB HOPE SPECIAL: BOB HOPE'S ALL-STAR COMEDY LOOK AT THE FALL SEASON: IT'S STILL FREE AND WORTH IT! (II,299); BOB HOPE SPECIAL: BOB HOPE'S ALL-STAR LOOK AT TV'S PRIME TIME WARS (II,303); BOB HOPE SPECIAL: BOB HOPE'S FUNNY VALENTINE (II,309); BOB HOPE SPECIAL: BOB HOPE'S USO CHRISTMAS IN BEIRUT (II,320); CELEBRITY CHALLENGE OF THE SEXES 5 (II,469); CELEBRITY DAREDEVILS (II,473); THE CELEBRITY FOOTBALL CLASSIC (II,474); CIRCUS OF THE STARS (II,532); CIRCUS OF THE STARS (II,533); HOTEL (II,1192); LIFE'S MOST EMBARRASSING MOMENTS (II,1470); MAGIC WITH THE STARS (II,1599); ROUGHNECKS (II,2219); THAT'S INCREDIBLE! (II,2578); THE THREE WIVES OF DAVID WHEELER (II,2611); WOMEN WHO RATE A "10" (II,2825); WONDER WOMAN (PILOT 1) (II,2826)

CROSBY, DENNIS
THE BING CROSBY SHOW (I,463); THE BING CROSBY SHOW (I,464); THE BING CROSBY SPECIAL (I,471); SWINGIN' TOGETHER (I,4320)

CROSBY, ELIZABETH
THE EVIL TOUCH (I,1486)

CROSBY, GARY
ADAM-12 (I,31); THE BILL DANA SHOW (I,446); THE BOB HOPE SHOW (I,557); CHASE (I,917); HUNTER (II,1206); MOBILE ONE (II,1717); O'HARA, UNITED STATES TREASURY: OPERATION COBRA (I,3347); SAM (II,2246); SAM (II,2247); TIME OUT FOR GINGER (I,4508)

CROSBY, HARRY
BING CROSBY AND THE SOUNDS OF CHRISTMAS (I,457); BING CROSBY'S CHRISTMAS SHOW (I,474); BING CROSBY'S MERRIE OLDE CHRISTMAS (II,252); BING CROSBY'S SUN VALLEY CHRISTMAS SHOW (I,475); BING CROSBY'S WHITE CHRISTMAS (II,253); BING!. . . A 50TH ANNIVERSARY GALA (II,254); CHRISTMAS WITH THE BING CROSBYS (I,950); RIDING FOR THE PONY EXPRESS (II,2172)

CROSBY, JOAN
FIRST TIME, SECOND TIME (II,863)

CROSBY, JOHN
PLANET EARTH (I,3615); THE SEVEN LIVELY ARTS (I,3987)

CROSBY, KATHRYN
BING CROSBY AND THE SOUNDS OF CHRISTMAS (I,457); THE BING CROSBY SHOW (I,466); BING CROSBY'S CHRISTMAS SHOW (I,474); BING CROSBY'S MERRIE OLDE CHRISTMAS (II,252); BING CROSBY'S SUN VALLEY CHRISTMAS SHOW (I,475); BING CROSBY'S WHITE CHRISTMAS (II,253); BING!. . . A 50TH ANNIVERSARY GALA (II,254); BOB HOPE SPECIAL: HAPPY BIRTHDAY, BOB! (II,325); BOB HOPE SPECIAL: HAPPY BIRTHDAY, BOB! (II,326); BOB HOPE SPECIAL: THE BOB HOPE CHRISTMAS SPECIAL (II,330); CELEBRITY CHALLENGE OF THE SEXES 1 (II,465); CHRISTMAS WITH THE BING CROSBYS (I,950); THE HOUSE NEXT DOOR (I,2139); WOMEN OF RUSSIA (II,2824)

CROSBY, LINDSAY
THE BING CROSBY SHOW (I,463); THE BING CROSBY SHOW (I,464); THE EDSEL SHOW (I,1417); SWINGIN' TOGETHER (I,4320)

CROSBY, LOU
MAYOR OF HOLLYWOOD (I,2962)

CROSBY, MARY
BATTLE OF THE NETWORK STARS (II,168); THE BIG EASY (II,234); BING CROSBY AND THE SOUNDS OF CHRISTMAS (I,457); BING CROSBY'S CHRISTMAS SHOW (I,474); BING CROSBY'S MERRIE OLDE CHRISTMAS (II,252); BING CROSBY'S SUN VALLEY CHRISTMAS SHOW (I,475); BING CROSBY'S WHITE CHRISTMAS (II,253); BING!. . . A 50TH ANNIVERSARY GALA (II,254); BROTHERS AND SISTERS (II,382); CHRISTMAS WITH THE BING CROSBYS (I,950); DALLAS (II,614); GOLDEN GATE (II,1020); PEARL (II,1986)

CROSBY, NATHANIEL
BING CROSBY AND THE SOUNDS OF CHRISTMAS (I,457); BING CROSBY'S CHRISTMAS SHOW (I,474); BING CROSBY'S MERRIE OLDE CHRISTMAS (II,252); BING CROSBY'S SUN VALLEY CHRISTMAS SHOW (I,475); BING CROSBY'S WHITE CHRISTMAS (II,253); BING!. . . A 50TH ANNIVERSARY GALA (II,254); CHRISTMAS WITH THE BING CROSBYS (I,950)

CROSBY, NORM
THE BEAUTIFUL PHYLLIS DILLER SHOW (I,381); THE COMEDY SHOP (II,558); EVERYTHING GOES (I,1480); THE MAD MAD MAD MAD WORLD OF THE SUPER BOWL (II,1586); SUPER COMEDY BOWL 1 (I,4288);

SUPER COMEDY BOWL 2 (I,4289)

CROSBY, PHILIP
THE BING CROSBY SHOW (I,463); THE BING CROSBY SHOW (I,464); THE BING CROSBY SPECIAL (I,471); THE BOB HOPE SHOW (I,598); THE BOB HOPE SHOW (I,620); SWINGIN' TOGETHER (I,4320)

CROSLEY, CHASE
SEARCH FOR TOMORROW (II,2284); TODAY IS OURS (I,4531)

CROSS, ALEXANDER
THE MILKY WAY (I,3044)

CROSS, BEN
THE CITADEL (II,539); THE FAR PAVILIONS (II,832)

CROSS, BILL
NICHOLS AND DYMES (II,1841)

CROSS, BUD
HELLINGER'S LAW (II,1127)

CROSS, CHERYL
THE BENNY HILL SHOW (II,207)

CROSS, DENNIS
THE BLUE ANGELS (I,490)

CROSS, JAMES
WELCOME TO PARADISE (II,2762)

CROSS, JIMMY
ADVENTURES OF A MODEL (I,50); HOW TO MARRY A MILLIONAIRE (I,2147)

CROSS, LARRY
THE KILLING GAME (I,2540)

CROSS, MARY
MANHUNTER (II,1620)

CROSS, MURPHY
PHYL AND MIKHY (II,2037); SCRUPLES (II,2280)

CROSS, RICHARD
AMAHL AND THE NIGHT VISITORS (I,149)

CROSSE, RUPERT
DIAGNOSIS: DANGER (I,1250); THE PARTNERS (I,3462)

CROSSLAND, GORD
DOROTHY HAMILL IN ROMEO & JULIET ON ICE (II,718)

CROSSON, ROBERT
THE RESTLESS GUN (I,3774)

CROTHERS, JOEL
DARK SHADOWS (I,1157); THE EDGE OF NIGHT (II,760); THE SECRET STORM (I,3969); SOMERSET (I,4115)

CROTHERS, SCATMAN
CASABLANCA (II,450); CHICO AND THE MAN (II,508); GILLIGAN'S ISLAND: THE HARLEM GLOBETROTTERS ON GILLIGAN'S ISLAND (II,993); THE HARLEM GLOBETROTTERS (I,1953); HONG KONG PHOOEY (II,1180); JONATHAN WINTERS PRESENTS 200 YEARS OF AMERICAN HUMOR (II,1342); ONE OF THE BOYS (II,1911); REVENGE OF THE GRAY GANG (II,2146); ROOTS (II,2211); SCOOBY'S ALL-STAR LAFF-A-LYMPICS (II,2273); THE SUPER GLOBETROTTERS (II,2498); TIME FOR BEANY (I,4505); VEGAS (II,2723)

CROUCH, ANDRE
THE MAC DAVIS CHRISTMAS SPECIAL (II,1573)

CROUCHER, BRIAN
BLAKE'S SEVEN (II,264)

CROUGH, SUZANNE
GOOBER AND THE GHOST CHASERS (II,1028); MULLIGAN'S STEW (II,1763); MULLIGAN'S STEW (II,1764); THE PARTRIDGE FAMILY (II,1962); PARTRIDGE FAMILY: 2200 A.D. (II,1963); THANKSGIVING REUNION WITH THE PARTRIDGE FAMILY AND MY THREE SONS (II,2569)

CROUSE, LINDSAY
SUMMER SOLSTICE (II,2492)

CROW, CARL
NATIONAL VELVET (I,3231)

CROWDEN, GRAHAM
THE SNOW GOOSE (I,4103)

CROWDEN, SARAH
THE SEVEN DIALS MYSTERY (II,2308)

CROWDER, DEENA
DOWN HOME (II,731); TV FUNNIES (II,2672); WOMEN IN WHITE (II,2823)

CROWE, TONYA
THE HOYT AXTON SHOW (II,1200); JOSHUA'S WORLD (II,1344); KNOTS LANDING (II,1404)

CROWELL, MCLIN
ANOTHER WORLD (II,97)

CROWELL, RODNEY
JOHNNY CASH'S AMERICA (II,1330)

CROWLEY, ANN
TELE-VARIETIES (I,4377)

CROWLEY, ED
LOVE OF LIFE (I,2774)

CROWLEY, JEANANNE
REILLY, ACE OF SPIES (II,2129)

CROWLEY, KATHLEEN
BATMAN (I,366); CROWN THEATER WITH GLORIA SWANSON (I,1107); FBI CODE 98 (I,1550); MARKHAM (I,2916); STAR STAGE (I,4197); THE WESTERNER (I,4800)

CROWLEY, PATRICIA
ALL IN THE FAMILY (I,133); BACHELOR PARTY (I,324); THE CHEVROLET TELE-THEATER (I,926); CLIMAX! (I,976); CROSSROADS (I,1105); A DATE WITH JUDY (I,1162); THE DESILU PLAYHOUSE (I,1237); I REMEMBER CAVIAR (I,2175); JOE FORRESTER (II,1303); THE JUNE ALLYSON SHOW (I,2488); THE LORETTA YOUNG THEATER (I,2756); THE LUX VIDEO THEATER (I,2795); MAGNAVOX THEATER (I,2827); THE MILLIONAIRE (II,1700); PLEASE DON'T EAT THE DAISIES (I,3628); RETURN TO FANTASY ISLAND (II,2144); SCHLITZ PLAYHOUSE OF STARS (I,3936); SUSPENSE (I,4305); THE TWILIGHT ZONE (I,4651); THE TWO OF US (I,4658); THE UNITED STATES STEEL HOUR (I,4677); THE UNTOUCHABLES (I,4681); THE WEB (I,4784); THE WORLD OF ENTERTAINMENT (II,2837); YOU'RE ONLY YOUNG TWICE (I,4971)

CROWTHER, CLAIRE
1915 (II,1853)

CROWTHER, JEANNIE
THE WOMAN IN WHITE (II,2819)

CROWTHER, LIZ
SHOESTRING (II,2338)

CRUICKSHANKS, REID
THE BLUE KNIGHT (II,277)

CRUIKSHANK, ANDREW
THE LILLI PALMER THEATER (I,2704)

CRUIKSHANK, RUFUS
THE ADVENTURES OF ROBIN HOOD (I,74)

CRUMBY, WILLIAM T
CHIEFS (II,509)

CRUPI, TONY
THE OUTLAWS (II,1936)

CRUTCHLEY, ROSALIE
THE SIX WIVES OF HENRY VIII (I,4068)

CRUZ, BRANDON
THE COURTSHIP OF EDDIE'S FATHER (I,1083); JEREMIAH OF JACOB'S NECK (II,1281); THE SENSATIONAL, SHOCKING, WONDERFUL, WACKY 70S (II,2302)

CRYER, DAVID
WHERE THE HEART IS (I,4831)

CRYSTAL, BILLY
THE 36 MOST BEAUTIFUL GIRLS IN TEXAS (II,2594); BATTLE OF THE NETWORK STARS (II,165); BATTLE OF THE NETWORK STARS (II,167); BATTLE OF THE NETWORK STARS (II,168); BATTLE OF THE NETWORK STARS (II,169); THE BILLY CRYSTAL COMEDY HOUR (II,248); BILLY CRYSTAL: A COMIC'S LINE (II,249); THE CELEBRITY FOOTBALL CLASSIC (II,474); DOUG HENNING'S WORLD OF MAGIC V (II,729); HUMAN FEELINGS (II,1203); SATURDAY NIGHT LIVE (II,2261); SOAP (II,2392); THE TV SHOW (II,2673)

CUCCIOLLA, RICCARDO
MARCO POLO (II,1626)

CUDLEFF, RUSTY
WELCOME TO THE FUN ZONE (II,2763)

CUDNEY, ROGER
COMPUTERCIDE (II,561); FALCON'S GOLD (II,809)

CUERVO, ALMA
A.K.A. PABLO (II,28)

CUESTA, HENRY
THE LAWRENCE WELK SHOW (I,2643); THE LAWRENCE WELK SHOW (I,2644); MEMORIES WITH LAWRENCE WELK (II,1681)

CUEVAS, NELSON D
VIVA VALDEZ (II,2728)

CUFF, SIMON
DOCTOR IN THE HOUSE (I,1313)

CUGAT, XAVIER
THE XAVIER CUGAT SHOW (I,4924)

CUKOR, GEORGE

HOLLYWOOD: THE SELZNICK YEARS (I,2095)

CULE, MICHAEL
THE HITCHHIKER'S GUIDE TO THE GALAXY (II,1158)

CULEA, MELINDA
THE A-TEAM (II,119); BATTLE OF THE NETWORK STARS (II,177); DEAR TEACHER (II,653); GLITTER (II,1009); WHACKED OUT (II,2767)

CULLEN, BILL
THE $25,000 PYRAMID (II,2679); BANK ON THE STARS (I,343); THE BILL CULLEN SHOW (I,445); BLANKETY BLANKS (II,266); CHAIN REACTION (II,479); CHILD'S PLAY (II,510); DOWN YOU GO (I,1367); EYE GUESS (I,1493); GIVE AND TAKE (I,1814); HOT POTATO (II,1188); I'VE GOT A SECRET (II,1220); THE JOKER'S WILD (II,1341); THE LOVE EXPERTS (II,1537); MATINEE IN NEW YORK (I,2946); NAME THAT TUNE (I,3218); PASS THE BUCK (II,1964); PASSWORD PLUS (II,1966); PLACE THE FACE (I,3611); THE PRICE IS RIGHT (I,3664); SAY IT WITH ACTING (I,3928); THREE ON A MATCH (I,4482); TO TELL THE TRUTH (II,2625); TV'S FUNNIEST GAME SHOW MOMENTS (II,2676); WHY? (I,4856); WINNER TAKE ALL (I,4881); WINNING STREAK (II,2809); YOU'RE PUTTING ME ON (I,4972)

CULLEN, BRETT
THE CHISHOLMS (II,513); KEY TORTUGA (II,1383); THE THORN BIRDS (II,2600)

CULLEN, KATHLEEN
THE GUIDING LIGHT (II,1064)

CULLEN, PETER
THE BOBBY GOLDSBORO SHOW (I,673); DUNGEONS AND DRAGONS (II,744); THE HUDSON BROTHERS RAZZLE DAZZLE COMEDY SHOW (II,1201); KNIGHT RIDER (II,1402); THE KWICKY KOALA SHOW (II,1417); THE LITTLE RASCALS (II,1494); LOLA (II,1508); LOLA (II,1509); LOLA (II,1510); LOLA (II,1511); PAC-MAN (II,1940); THE PLASTICMAN COMEDY/ADVENTURE SHOW (II,2051); THE SONNY AND CHER COMEDY HOUR (II,2400); THE SONNY AND CHER SHOW (II,2401); THE SONNY COMEDY REVUE (II,2403); THE WOLFMAN JACK SHOW (II,2817)

CULLEN, WILLIAM KIRBY
HOW THE WEST WAS WON (II,1196); THE MACAHANS (II,1583)

CULLETON, JIM
THE MANIONS OF AMERICA (II,1623)

CULLINANE, DAVID
SKEEZER (II,2376)

CULLINGWOOD, PETER
A TOWN LIKE ALICE (II,2652)

CULLUM, JOHN
ANDROCLES AND THE LION (I,189); THE DOCTORS (II,694); ROLL OF THUNDER, HEAR MY CRY (II,2202)

CULLY, ZARA
THE JEFFERSONS (II,1276)

CULP, ROBERT
THE BARBARA STANWYCK THEATER (I,350); THE BOUNTY HUNTER (I,698); CIRCUS OF THE STARS (II,536); GREAT ADVENTURE (I,1868); THE GREATEST AMERICAN HERO (II,1060); I SPY (I,2179); JOSEPHINE LITTLE: ADVENTURES IN HAPPINESS (I,2450); THE JUNE ALLYSON SHOW (I,2488); MARRIED ALIVE (I,2923); THE OUTER LIMITS (I,3426); STAR TONIGHT (I,4199); THOU SHALT NOT KILL (II,2604); TRACKDOWN (I,4589); WOMEN IN WHITE (II,2823); THE WONDERFUL WORLD OF BURLESQUE II (I,4896)

CULPEPPER, STUART
CHIEFS (II,509)

CULVER, HOWARD
THE 25TH MAN (MS) (II,2678); GUNSMOKE (II,1069); HARD CASE (I,1947); THE MAN FROM DENVER (I,2863); THE RETURN OF THE BEVERLY HILLBILLIES (II,2139)

CULVER, LILLIAN
BEN BLUE'S BROTHERS (I,393)

CULVER, MICHAEL
THE DEVIL'S WEB (I,1248); FANNY BY GASLIGHT (II,823); SECRET ARMY (II,2290); SPACE: 1999 (II,2409)

CULVER, RONALD
THE PALLISERS (II,1944)

CUMBUKA, JI-TU
THE BLUE KNIGHT (II,277); THE CRACKER BROTHERS (II,599); DEATH RAY 2000 (II,654); EBONY, IVORY AND JADE (II,755); A MAN CALLED SLOANE (II,1613); MANDRAKE (II,1617); THE NIGHTENGALES (II,1850); ROOTS (II,2211); YOUNG DAN'L BOONE (II,2863)

CUMMINGS, BOB
BEST FOOT FORWARD (I,401); THE BOB CUMMINGS SHOW (I,500); THE BOB HOPE CHRYSLER THEATER (I,502); THE BOB HOPE SHOW (I,508); THE FRANCES LANGFORD SHOW (I,1657); GIDGET GROWS UP (I,1797); GREAT ADVENTURE (I,1868); HEDDA HOPPER'S HOLLYWOOD (I,2002); LAST OF THE PRIVATE EYES (I,2628); THE LOVE BOAT (II,1535); LOVE THAT BOB (I,2779); MY HERO (I,3193); MY LIVING DOLL (I,3195); PARTNERS IN CRIME (I,3461); SCHLITZ PLAYHOUSE OF STARS (I,3936); STUDIO ONE (I,4268)

CUMMINGS, CONSTANCE
LONG DAY'S JOURNEY INTO NIGHT (I,2746)

CUMMINGS, QUINN
BIG EDDIE (II,235); BIG EDDIE (II,236); CELEBRITY CHALLENGE OF THE SEXES 3 (II,467); CIRCUS OF THE STARS (II,533); FAMILY (II,813); JEREMIAH OF JACOB'S NECK (II,1281); REMINGTON STEELE (II,2130)

CUMMINGS, RONN
OUR FIVE DAUGHTERS (I,3413)

CUMMINGS, SUSAN
FOREST RANGER (I,1641); UNION PACIFIC (I,4676)

CUMMINGS, TONY
ANOTHER WORLD (II,97)

CUMMINGS, VICKI
BURLESQUE (I,764); CHRYSLER MEDALLION THEATER (I,951); GOLDEN WINDOWS (I,1841); THE MOTOROLA TELEVISION HOUR (I,3112); SILVER THEATER (I,4051)

CUNNINGHAM, ARLENE
THE ART FORD SHOW (I,275)

CUNNINGHAM, BOB
CRISIS (I,1101); WELCOME TRAVELERS (I,4791)

CUNNINGHAM, HEATHER
THE BIG SHOWDOWN (II,244)

CUNNINGHAM, JOHN
ANOTHER WORLD (II,97); GABE AND WALKER (II,950); HAPPY ENDINGS (II,1085); LOVING (II,1552); NICK AND THE DOBERMANS (II,1843); SEARCH FOR TOMORROW (II,2284)

CUNNINGHAM, LINDA
BONKERS! (II,348); SPELL OF EVIL (I,4152)

CUNNINGHAM, ROBERT
THE POPPY IS ALSO A FLOWER (I,3650)

CUNNINGHAM, SARAH
ANOTHER WORLD (II,97); NERO WOLFE (II,1805); THE TURNING POINT OF JIM MALLOY (II,2670)

CUNNINGHAM, ZANAH
CAVALCADE OF STARS (I,877) MENASHA THE MAGNIFICENT (I,3010)

CUNNINGTON, LLOYD
ALL THE RIVERS RUN (II,43)

CUPITO, SUZANNE
THE LLOYD BRIDGES SHOW (I,2733); THE TWILIGHT ZONE (I,4651)

CUPO, PAT
THE UNCLE FLOYD SHOW (II,2703)

CURAN, WILLIAM BRIAN
OFF THE RACK (II,1878)

CURLY, WILLIAM
SWING INTO SPRING (I,4317)

CURRAN, MARK
HIDDEN FACES (I,2044)

CURRAN, WILLIAM BRIAN
THE BLUE AND THE GRAY (II,273); CATALINA C-LAB (II,458); FARRELL: FOR THE PEOPLE (II,834); NUMBER 96 (II,1869); PAPER DOLLS (II,1951)

CURRERI, LEE
BATTLE OF THE BEAT (II,161); FAME (II,812); THE KIDS FROM FAME (II,1392)

CURRIE, FINDLAY
BRIGADOON (I,726)

CURRIE, MICHAEL
OUR FAMILY BUSINESS (II,1931)

CURRIE, TAD

SIX PACK (II,2370)

CURRIER, HAL
FIRST LOVE (I,1583)

CURRIER, TERRY
GEORGE WASHINGTON (II,978)

CURRIN, BRENDA
TALES FROM THE DARKSIDE (II,2534)

CURRY, JOHN
PERRY COMO'S OLDE ENGLISH CHRISTMAS (II,2020)

CURRY, JULIAN
THE MANIONS OF AMERICA (II,1623)

CURTIN, JANE
BATTLE OF THE NETWORK STARS (II,166); BATTLE OF THE NETWORK STARS (II,168); BEDROOMS (II,198); BOB & RAY & JANE, LARAINE & GILDA (II,279); CANDIDA (II,423); THE COMEDY ZONE (II,559); THE CONEHEADS (II,567); KATE AND ALLIE (II,1365); THE ROBERT KLEIN SHOW (II,2186); SATURDAY NIGHT LIVE (II,2261); THINGS WE DID LAST SUMMER (II,2589)

CURTIN, VALERIE
9 TO 5 (II,1852); THE JIM STAFFORD SHOW (II,1291); THE PRIMARY ENGLISH CLASS (II,2081)

CURTIS, BARRY
THE ADVENTURES OF CHAMPION (I,54); SIX GUNS FOR DONEGAN (I,4065)

CURTIS, CHRISTY
CIRCUS OF THE STARS (II,533)

CURTIS, CRAIG
GENERAL HOSPITAL (II,964); KINCAID (I,2543)

CURTIS, DICK
THE CATTANOOGA CATS (I,874); THE JONATHAN WINTERS SHOW (I,2443); MOTOR MOUSE (I,3111); OFF WE GO (I,3338); QUEEN FOR A DAY (I,3706); SKYHAWKS (I,4079)

CURTIS, DONALD
THE DETECTIVE'S WIFE (I,1244)

CURTIS, HELEN
JACKIE GLEASON AND HIS AMERICAN SCENE MAGAZINE (I,2321)

CURTIS, JACK
THE AMAZING THREE (I,162); MARINE BOY (I,2911); SPEED RACER (I,4151)

CURTIS, JAMIE LEE
THE ALL-STAR SALUTE TO MOTHER'S DAY (II,51); CALLAHAN (II,414); THE CELEBRITY FOOTBALL CLASSIC (II,474); CIRCUS OF THE STARS (II,532); OPERATION PETTICOAT (II,1916); OPERATION PETTICOAT (II,1917); SHE'S IN THE ARMY NOW (II,2320)

CURTIS, JANET LYN
BATTLESTAR GALACTICA (II,181); FITZ AND BONES (II,867); THE MISADVENTURES OF SHERIFF LOBO (II,1704); TERROR AT ALCATRAZ (II,2563)

CURTIS, KEENE
AMANDA'S (II,62); EMPIRE (II,777); THE FANTASTIC FUNNIES (II,825); THE MAGICIAN (I,2823); THE MAGICIAN (I,2824); MODESTY BLAISE (II,1719); ONE IN A MILLION (II,1906); THERE GOES THE NEIGHBORHOOD (II,2582); UNIT 4 (II,2707)

CURTIS, KELLY
THE ALL-STAR SALUTE TO MOTHER'S DAY (II,51)

CURTIS, KEN
GUNSMOKE (II,1069); RIPCORD (I,3793); WHEN THE WEST WAS FUN: A WESTERN REUNION (II,2780); THE YELLOW ROSE (II,2847)

CURTIS, MINNIE JO
THE S. S. HOLIDAY (I,4173)

CURTIS, TODD
CAPITOL (II,426)

CURTIS, TONY
ANNIE AND THE HOODS (II,91); THE BOB HOPE SHOW (I,641); CIRCUS OF THE STARS (II,537); THE GENERAL ELECTRIC THEATER (I,1753); MAGIC WITH THE STARS (II,1599); MCCOY (II,1661); THE PERSUADERS! (I,3556); PLAYBOY'S 25TH ANNIVERSARY CELEBRATION (II,2054); THE RED SKELTON REVUE (I,3754); SCHLITZ PLAYHOUSE OF STARS (I,3936); SUPER COMEDY BOWL 2 (I,4289); VEGAS (II,2723); VEGAS (II,2724)

CURTIS, VIRGINIA
CAESAR'S HOUR (I,790); YOUR SHOW OF SHOWS (I,4957)

CURTIS, WILLA
THE AMOS AND ANDY SHOW (I,179)

CURTIS, WILLIE
TOP TEN (II,2648)

CUSACK, CYRIL
DOUGLAS FAIRBANKS JR. PRESENTS THE RHEINGOLD THEATER (I,1364); THE HANDS OF CORMAC JOYCE (I,1927)

CUSACK, JOAN
ALL TOGETHER NOW (II,45)

CUSACK, SINEAD
THE EYES HAVE IT (I,1496); QUILLER: NIGHT OF THE FATHER (II,2100); QUILLER: PRICE OF VIOLENCE (II,2101)

CUSACK, SORCHA
JANE EYRE (II,1273)

CUSAK, CYRIL
DIAL "M" FOR MURDER (I,1255); THE MOON AND SIXPENCE (I,3098)

CUSHING, PETER
DOCTOR WHO (II,689)

CUST, BOB
BUTCH AND BILLY AND THEIR BANG BANG WESTERN MOVIES (I,780)

CUTELL, LOU
TV FUNNIES (II,2672)

CUTHBERTSON, ALLAN
THE HIGHWAYMAN (I,2059); THE WINDS OF WAR (II,2807)

CUTHBERTSON, IAIN
CAESAR AND CLEOPATRA (I,789)

CUTLER, BARRY
SPIDER-MAN (II,2424)

CUTLER, BRIAN
FRIENDS (II,928); ISIS (II,1248); THE LONG HOT SUMMER (I,2747)

CUTLER, CATHY
IN TROUBLE (II,1230)

CUTLER, JON
AMERICA 2100 (II,66); BROTHERS AND SISTERS (II,382)

CUTLER, WENDY
COMPLETELY OFF THE WALL (II,560); MADAME'S PLACE (II,1587)

CUTRER, T TOMMY
THE COUNTRY MUSIC MURDERS (II,590)

CUTTERS, THE
OH, BOY! (I,3341)

CUTTS, PATRICIA
THE THREE MUSKETEERS (I,4481)

CUTTSHALL, CUTTY
FLOOR SHOW (I,1605)

CYPHER, JON
AS THE WORLD TURNS (II,110); HILL STREET BLUES (II,1154); NIGHT GAMES (II,1845); OUR FIVE DAUGHTERS (I,3413)

CYPHERS, CHARLES
THE BETTY WHITE SHOW (II,224); LITTLE HOUSE: LOOK BACK TO YESTERDAY (II,1490); ROOTS (II,2211)

CYRUS, DONNA
TEXAS (II,2566)

D D
MICKEY SPILLANE'S MIKE HAMMER (II,1692)

D'ABO, MARYAM
MASTER OF THE GAME (II,1647)

D'ALLISO, LUCY
THE YESTERDAY SHOW (II,2850)

D'AMATO, PAUL
MURDER INK (II,1769)

D'AMBOISE, JACQUES
HOLIDAY (I,2074); SUNDAY IN TOWN (I,4285)

D'AMBROSE, STEPHEN
A CHRISTMAS CAROL (II,517)

D'ANDREA, TOM
BOBBY PARKER AND COMPANY (II,344); DANTE (I,1155); THE LIFE OF RILEY (I,2686); THE MCGONIGLE (I,2968); THE SOLDIERS (I,4112)

D'ANGELO, BEVERLY
CAPTAINS AND THE KINGS (II,435); CIRCUS OF THE STARS (II,537)

D'ARCY, JACK
THE OMEGA FACTOR (II,1888)

D'ATRY, JAN
THE WORLD OF PEOPLE (II,2839)

D'ORSAY, FIFI
BAND OF GOLD (I,341); JOHNNY CARSON PRESENTS THE SUN CITY SCANDALS (I,2420); PEPSI-COLA PLAYHOUSE (I,3523); THE RAY PERKINS REVUE (I,3735); ROLL OUT! (I,3833)

DASILVA, HOWARD
ARCHIE BUNKER'S PLACE (II,105); FOR THE PEOPLE (I,1626); KEEP THE FAITH (I,2518); SMILE JENNY, YOU'RE DEAD (II,2382); WALTER FORTUNE (I,4758)

DABNEY, AUGUSTA
GENERAL HOSPITAL (II,964); LOVING (II,1552); MONTGOMERY'S SUMMER STOCK (I,3096); THE TRAP (I,4593); A WORLD APART (I,4910); YOUNG DR. MALONE (I,4938)

DABNEY, SHEILA
RAGE OF ANGELS (II,2107)

DABODOUB, JACK
ALL ABOUT MUSIC (I,127); ANNIE GET YOUR GUN (I,224)

DAHL, ARLENE
ARLENE DAHL'S BEAUTY SPOT (I,257); THE BOB HOPE SHOW (I,527); THE BOB HOPE SHOW (I,611); FORD TELEVISION THEATER (I,1634); ONE LIFE TO LIVE (II,1907); OPENING NIGHT (I,3396); PEPSI-COLA PLAYHOUSE (I,3523)

DAHL, PAT
JACKIE GLEASON AND HIS AMERICAN SCENE MAGAZINE (I,2321)

DAHL, ROALD
TALES OF THE UNEXPECTED (II,2540); WAY OUT (I,4777)

DAILEY, DAN
BURLESQUE (I,765); THE DAUGHTERS OF JOSHUA CABE RETURN (I,1170); FARADAY AND COMPANY (II,833); THE FOUR JUST MEN (I,1650); THE GENERAL MOTORS 50TH ANNIVERSARY SHOW (I,1758); THE GOVERNOR AND J.J. (I,1863); LOW MAN ON THE TOTEM POLE (I,2782); MITZI...THE FIRST TIME (I,3073); PAPA G.I. (I,3451); PARIS IN THE SPRINGTIME (I,3457); TESTIMONY OF TWO MEN (II,2564)

DAILEY, FRANK
MUSIC FROM THE MEADOWBROOK (I,3171)

DAILEY, IRENE
ANOTHER WORLD (II,97); THE EDGE OF NIGHT (II,760)

DAILY, BILL
ALOHA PARADISE (II,58); ALONE AT LAST (II,59); THE BOB NEWHART SHOW (II,342); I DREAM OF JEANNIE (I,2167); INSIDE O.U.T. (II,1235); RENDEZVOUS HOTEL (II,2131); SMALL AND FRYE (II,2380); VALENTINE MAGIC ON LOVE ISLAND (II,2716)

DAILY, ELIZABETH
JOSIE (II,1345); THE RIGHTEOUS APPLES (II,2174)

DAINARD, NEIL
CAGNEY AND LACEY (II,410); STRANGE PARADISE (I,4246)

DAINE, LOIS
THE TALE OF BEATRIX POTTER (II,2533)

DAINTON, JOANNE
MISTER JERICO (I,3070)

DAKER, DAVID
CRY TERROR! (I,1112)

DALBY, LYNN
DOCTOR IN THE HOUSE (I,1313)

DALCE, JOHN
WINDY CITY JAMBOREE (I,4877)

DALE, ALAN
ACTOR'S HOTEL (I,27); THE ALAN DALE SHOW (I,96); OPERA VS. JAZZ (I,3399); PANORAMA (I,3448); SING IT AGAIN (I,4059)

DALE, DICK
THE LAWRENCE WELK SHOW (I,2643); THE LAWRENCE WELK SHOW (I,2644); LAWRENCE WELK'S TOP TUNES AND NEW TALENT (I,2645); MEMORIES WITH LAWRENCE WELK (II,1681)

DALE, ELLIS
TYCOON: THE STORY OF A WOMAN (II,2697)

DALE, ESTHER
DOBIE GILLIS (I,1302)

DALE, JANET
NICHOLAS NICKLEBY (II,1840)

DALE, JIM
CINDERELLA (II,526); SUNDAY NIGHT AT THE LONDON PALLADIUM (I,4286)

DALE, SHIRLEY
ACADEMY THEATER (I,14)

DALE, VINCENT
SALTY (II,2241)

DALEY, CASS
THE BOB HOPE SHOW (I,528); THE BOB HOPE SHOW (I,530); THE BOB HOPE SHOW (I,533); HER SCHOOL FOR BACHELORS (I,2018)

DALEY, JAN
FRANKIE AVALON'S EASTER SPECIAL (I,1673); JACK CASSIDY'S ST. PATRICK'S DAY SPECIAL (I,2307)

DALEY, JANE
JOE DANCER: MURDER ONE, DANCER 0 (II,1302)

DALIO, MARCEL
AT YOUR SERVICE (I,310); CASABLANCA (I,860)

DALLAS COWBOYS CHEERLEADERS, THE
SALUTE TO LADY LIBERTY (II,2243)

DALLIMORE, MAURICE
FAIR EXCHANGE (I,1512)

DALTON, ABBY
ADAMS OF EAGLE LAKE (II,10); ANDERSON AND COMPANY (I,188); BARNEY MILLER (II,154); BATTLE OF THE NETWORK STARS (II,178); BELLE STARR (I,391); FALCON CREST (II,808); HENNESSEY (I,2013); THE JOEY BISHOP SHOW (I,2404); THE JONATHAN WINTERS SHOW (I,2443); THE RIFLEMAN (I,3789); SCHLITZ PLAYHOUSE OF STARS (I,3936)

DALTON, ARLENE
HOWDY DOODY (I,2151); THE KATE SMITH HOUR (I,2508); RED GOOSE KID'S SPECTACULAR (I,3751)

DALTON, AUDREY
ME AND BENJY (I,2973)

DALTON, DANIE WADE
CRASH ISLAND (II,600)

DALTON, LACEY J
MUSIC CITY NEWS TOP COUNTRY HITS OF THE YEAR (II,1772); USA COUNTRY MUSIC (II,591)

DALTON, LEZLIE
THE GUIDING LIGHT (II,1064); MR. & MS. AND THE BANDSTAND MURDERS (II,1746); THE NEW DAUGHTERS OF JOSHUA CABE (I,3261); THE YOUNG AND THE RESTLESS (II,2862)

DALTON, SUSANNA
DELTA HOUSE (II,658); STUNTS UNLIMITED (II,2487); WHACKED OUT (II,2767)

DALTON, TIMOTHY
CENTENNIAL (II,477); MISTRAL'S DAUGHTER (II,1708)

DALTON, WALLY
SNAFU (II,2386); THE THORN BIRDS (II,2600)

DALY, JAMES
THE ALCOA HOUR (I,108); APPOINTMENT WITH ADVENTURE (I,246); EAGLE IN A CAGE (I,1392); THE ELGIN HOUR (I,1428); FOREIGN INTRIGUE: OVERSEAS ADVENTURES (I,1639); FRONT ROW CENTER (I,1694); GIVE US BARABBAS! (I,1816); LADY IN THE DARK (I,2602); THE MAGNIFICENT YANKEE (I,2829); MEDICAL CENTER (II,1675); ST. JOAN (I,4175); U.M.C. (I,4666)

DALY, JANE
CAPITOL (II,426); DID YOU HEAR ABOUT JOSH AND KELLY?! (II,673); I LOVE HER ANYWAY! (II,1214); POTTSVILLE (II,2072)

DALY, JOHN
CELEBRITY TIME (I,887); DATELINE (I,1165); THE FRONT PAGE (I,1691); IT'S NEWS TO ME (I,2272); MARCH OF TIME THROUGH THE YEARS (I,2903); SOUND OF THE 60S (I,4134); VOICE OF FIRESTONE (I,4741); WE TAKE YOUR WORD (I,4781); WHAT'S MY LINE? (I,4819); WHO SAID THAT? (I,4849)

DALY, JONATHAN
C.P.O. SHARKEY (II,407); THE JIMMY STEWART SHOW (I,2391); O.K. CRACKERBY (I,3348)

DALY, KATHRYN
THE GOSSIP COLUMNIST (II,1045)

DALY, RAD
THE BAD NEWS BEARS (II,134); WALKING TALL (II,2738)

DALY, SARAH
GOSSIP (II,1043)

DALY, SUSAN
THE BENNY HILL SHOW (II,207)

DALY, TIMOTHY
RYAN'S FOUR (II,2233)

DALY, TYNE
CAGNEY AND LACEY (II,409); CAGNEY AND LACEY (II,410); HEAT OF ANGER (I,1992); MEDICAL CENTER (II,1675)

DAMAIN, MICHAEL
THE YOUNG AND THE RESTLESS (II,2862)

DAMON, CATHRYN
CALAMITY JANE (I,793); GETTING THERE (II,985); JOHN RITTER: BEING OF SOUND MIND AND BODY (II,1319); SOAP (II,2392); WEBSTER (II,2758)

DAMON, GABRIEL
CALL TO GLORY (II,413)

DAMON, LES
330 INDEPENDENCE S.W. (I,4486); AS THE WORLD TURNS (II,110); THE GUIDING LIGHT (II,1064); KITTY FOYLE (I,2571)

DAMON, MARK
THE MCGONIGLE (I,2968)

DAMON, STUART
THE ADVENTURER (I,43); THE CHAMPIONS (I,896); CINDERELLA (I,956); GENERAL HOSPITAL (II,964); MELODY OF HATE (I,2999)

DAMONE, VIC
THE BOB HOPE SHOW (I,548); THE BOB HOPE SHOW (I,613); BOB HOPE SPECIAL: BOB HOPE'S USO CHRISTMAS IN BEIRUT (II,320); THE DANGEROUS CHRISTMAS OF RED RIDING HOOD (I,1139); THE DANNY THOMAS SPECIAL (I,1151); DEAN MARTIN PRESENTS THE VIC DAMONE SHOW (I,1195); HOLIDAY IN LAS VEGAS (I,2076); THE LIVELY ONES (I,2729); THE MOREY AMSTERDAM SHOW (I,3101); SOUND OF THE 60S (I,4134); THE VIC DAMONE SHOW (I,4717); THE VIC DAMONE SHOW (I,4718)

DANA, BILL
THE BILL DANA SHOW (I,446); THE BOB HOPE SHOW (I,633); FEMALE INSTINCT (I,1564); I'VE HAD IT UP TO HERE (II,1221); THE LAS VEGAS SHOW (I,2619); A LOOK AT THE LIGHT SIDE (I,2751); MAKE ROOM FOR DADDY (I,2843); MITZI AND A HUNDRED GUYS (II,1710); MURDER AT NBC (A BOB HOPE SPECIAL) (I,3159); THE NEW STEVE ALLEN SHOW (I,3271); NO SOAP, RADIO (II,1856); ROSETTI AND RYAN: MEN WHO LOVE WOMEN (II,2217); SHOOT-IN AT NBC (A BOB HOPE SPECIAL) (I,4021); SHOWTIME (I,4036); THE STEVE ALLEN SHOW (I,4219); STEVE ALLEN'S LAUGH-BACK (II,2455); TOO CLOSE FOR COMFORT (II,2642); WINDOWS, DOORS AND KEYHOLES (II,2806); ZORRO AND SON (II,2878)

DANA, CEORA
JACK AND THE BEANSTALK (I,2287)

DANA, DICK
THE DOODLES WEAVER SHOW (I,1351)

DANA, JUSTIN
KNOTS LANDING (II,1404); NUTS AND BOLTS (II,1871); SKYWARD CHRISTMAS (II,2378); UNITED STATES (II,2708)

DANA, LEORA
THE ADAMS CHRONICLES (II,8); ANOTHER WORLD (II,97); CURTAIN CALL THEATER (I,1115); THE MOTOROLA TELEVISION HOUR (I,3112); PHILCO TELEVISION PLAYHOUSE (I,3583); SEVENTH AVENUE (II,2309)

DANA, LINDA
ONE LIFE TO LIVE (II,1907)

DANCERS ZEDAN AND CAROL
STEP THIS WAY (I,4216)

DANCING BLADES, THE
MUSIC ON ICE (I,3175)

DANDRIDGE, DOROTHY
LIGHT'S DIAMOND JUBILEE (I,2698); YOU'RE THE TOP (I,4974)

DANDRIDGE, PAUL
THE LOVE REPORT (II,1543)

DANDRIDGE, RUBY
BEULAH (I,416); FATHER OF THE BRIDE (I,1543)

DANE, ANN
MINSKY'S FOLLIES (II,1703)

DANE, DIANE
DANCE FEVER (II,615)

DANE, FRANK
THE GOOK FAMILY (I,1858); HAWKINS FALLS, POPULATION 6200 (I,1977)

DANE, LAWRENCE
OUR MAN FLINT: DEAD ON TARGET (II,1932)

DANE, PATRICIA
FIRESIDE THEATER (I,1580)

DANELLE, JOHN
ALL MY CHILDREN (II,39)

DANESE, SHERA
ACE CRAWFORD, PRIVATE EYE (II,6); FAME (I,1517); SUZANNE PLESHETTE IS MAGGIE BRIGGS (II,2509)

DANFORD, CHARLES
THE TED STEELE SHOW (I,4373)

DANGCIL, LINDA
THE FLYING NUN (I,1611); HOUSE CALLS (II,1194)

DANGERFIELD, RODNEY
BENNY AND BARNEY: LAS VEGAS UNDERCOVER (II,206); THE DEAN MARTIN SHOW (I,1201); THE MAD MAD MAD MAD WORLD OF THE SUPER BOWL (II,1586); THE ROBERT KLEIN SHOW (II,2186); RODNEY DANGERFIELD SHOW: I CAN'T TAKE IT NO MORE (II,2198); RODNEY DANGERFIELD SPECIAL: IT'S NOT EASY BEIN' ME (II,2199)

DANGLER, ANITA
THE MUNSTERS' REVENGE (II,1765)

DANIEL, CHUCK
RETURN TO PEYTON PLACE (I,3779)

DANIEL, DAN
MUSIC CENTRAL (II,1771)

DANIEL, JENNIFER
SPELL OF EVIL (I,4152)

DANIEL, PIERRE
SHEEHY AND THE SUPREME MACHINE (II,2322)

DANIELL, ALLYSON
MRS. G. GOES TO COLLEGE (I,3155)

DANIELL, HENRY
THE ARMSTRONG CIRCLE THEATER (I,260); MY THREE ANGELS (I,3204); THE SWORD (I,4325); TELEPHONE TIME (I,4379); THRILLER (I,4492)

DANIELS, BILLY
THE BILLY DANIELS SHOW (I,451); COTTON CLUB '75 (II,580); THE SAMMY DAVIS JR. SPECIAL (I,3899)

DANIELS, CAROLYN
JOHNNY BELINDA (I,2417); THAT GIRL (II,2570)

DANIELS, CHARLIE
THE COUNTRY MUSIC MURDERS (II,590); DEAN MARTIN'S CHRISTMAS AT SEA WORLD (II,644)

DANIELS, DANNY
THE MOREY AMSTERDAM SHOW (I,3100)

DANIELS, DAVID MASON
CAPITOL (II,426)

DANIELS, JEFF
CATALINA C-LAB (II,458); FIFTH OF JULY (II,851)

DANIELS, LEROY
SANFORD AND SON (II,2256)

DANIELS, LISA
STAR TONIGHT (I,4199)

DANIELS, MARC
THAT'S OUR SHERMAN (I,4429)

DANIELS, MARV
LEGS (II,1459)

DANIELS, WILLIAM
THE ADAMS CHRONICLES (II,8); ALL THAT GLITTERS (II,41); THE BASTARD/KENT FAMILY CHRONICLES (II,159); BIG BOB JOHNSON AND HIS FANTASTIC SPEED CIRCUS (II,231); BOSTON AND KILBRIDE (II,356); CAPTAIN NICE (I,835); THE COURT-MARTIAL OF GENERAL GEORGE ARMSTRONG CUSTER (I,1081); THE FABULOUS DR. FABLE (I,1500); FREEBIE AND THE BEAN (II,922); HEAVEN ON EARTH (II,1120); INSTANT FAMILY (II,1238); KNIGHT RIDER (II,1402); MURDOCK'S GANG (I,3164); THE NANCY WALKER SHOW (II,1790); NUTS AND BOLTS (II,1871); ONE OF OUR OWN (II,1910); PRIVATE BENJAMIN (II,2087); THE REBELS (II,2121); ROOSTER (II,2210); ST. ELSEWHERE (II,2432); THAT WAS THE YEAR THAT WAS (II,2575); THE WONDERFUL WORLD OF PHILIP MALLEY (II,2831)

DANIELY, LISA
THE INVISIBLE MAN (I,2232); KRAFT MYSTERY THEATER (I,2589)

DANKWORTH, JOHNNY
COTTON CLUB '75 (II,580)

DANN, ROGER
THE BILL GOODWIN SHOW (I,447)

DANNENBAUM, JULIE
CREATIVE COOKING WITH JULIE DANNENBAUM (I,1093)

DANNER, BLYTHE
ADAM'S RIB (I,32); THE COURT-MARTIAL OF GENERAL GEORGE ARMSTRONG CUSTER (I,1081); GEORGE M! (I,1772); GEORGE M! (II,976); SIDEKICKS (II,2348); THAT WAS THE YEAR THAT WAS (II,2575); YOU CAN'T TAKE IT WITH YOU (II,2857)

DANNY DANIELS DANCERS, THE
THE PERRY COMO SPRINGTIME SPECIAL (I,3538); THE PERRY COMO THANKSGIVING SPECIAL (I,3540); THE PERRY COMO WINTER SHOW (I,3542); PERRY COMO'S SUMMER SHOW (I,3545)

DANNY DANIELS SINGERS, THE
WASHINGTON SQUARE (I,4772)

DANO, LINDA
AS THE WORLD TURNS (II,110); THE FESS PARKER SHOW (II,850); THE MONTEFUSCOS (II,1727)

DANO, ROYAL
THE DANGEROUS DAYS OF KIOWA JONES (I,1140); HOW THE WEST WAS WON (II,1196); JOE DANCER: MURDER ONE, DANCER 0 (II,1302)

DANO, SAL
THE PHIL SILVERS SHOW (I,3580)

DANOFF, BILL
THE STARLAND VOCAL BAND (II,2441)

DANOFF, TAFFY
THE STARLAND VOCAL BAND (II,2441)

DANOVA, CESARE
BORDER TOWN (I,689); GARRISON'S GORILLAS (I,1735); THE GINGER ROGERS SHOW (I,1805); POLICE WOMAN: THE GAMBLE (II,2064); TARZAN THE APE MAN (I,4363)

DANSON, RANDY
RYAN'S HOPE (II,2234)

DANSON, TED
ALLISON SIDNEY HARRISON (II,54); BENSON (II,208); CHEERS (II,494); DEAR TEACHER (II,653); ONCE UPON A SPY (II,1898); OUR FAMILY BUSINESS (II,1931)

DANTE, MICHAEL
THE LEGEND OF CUSTER (I,2652)

DANTINE, HELMUT
CALL HOLME (II,412); THE CLOCK (I,978); THE FILE ON DEVLIN (I,1575); SHADOW OF THE CLOAK (I,3992)

DANTINE, NIKI
THE WESTWIND (II,2766)

DANTON, RAY
THE ALASKANS (I,106); BANYON (I,345); FBI CODE 98 (I,1550); OUR MAN FLINT: DEAD ON TARGET (II,1932); A VERY MISSING PERSON (I,4715)

DANUTA
CHICO AND THE MAN (II,508)

DANZA, TONY
MURDER CAN HURT YOU! (II,1768); TAXI (II,2546); WHO'S THE BOSS? (II,2792)

DAPO, PAMELA
MY BOY GOOGIE (I,3185); THOMPSON'S GHOST (I,4467)

DAPO, RONNIE
BABY CRAZY (I,319); HOW'S BUSINESS? (I,2153); THE NEW PHIL SILVERS SHOW (I,3269); ROOM FOR ONE MORE (I,3842)

DARBO, PATRIKA
ONE NIGHT BAND (II,1909)

DARBY, KIM
CIRCLE OF FEAR (I,958); CLOSE TIES (II,545); THE FACTS OF LIFE (II,805); FLATBED ANNIE AND SWEETIEPIE: LADY TRUCKERS (II,876); FLESH AND BLOOD (I,1595); IRONSIDE (I,2240); THE LAST CONVERTIBLE (II,1435); LOVE STORY (I,2778); MIRROR OF DECEPTION (I,3060); MR. NOVAK (I,3145); RICH MAN, POOR MAN—BOOK I (II,2161); THE STREETS OF SAN FRANCISCO (I,4260)

DARCELL, DENISE
CHANCE OF A LIFETIME (I,898); GAMBLE ON LOVE (I,1731); PANTOMIME QUIZ (I,3449)

DARCY, GEORGINE
THE DESILU REVUE (I,1238); HARRIGAN AND SON (I,1955); THE JERRY LEWIS SHOW (I,2362)

DARDEN, SEVERN
BEYOND WESTWORLD (II,227); DISAPPEARANCE OF AIMEE (I,1290); THE FEATHER AND FATHER GANG: NEVER CON A KILLER (II,846); FOREVER FERNWOOD (II,909); HOME ROOM (II,1172); RENDEZVOUS HOTEL (II,2131); SANDBURG'S LINCOLN (I,3907); THE SIX-MILLION-DOLLAR MAN (II,2372); STORY THEATER (I,4241); WONDER WOMAN (PILOT 2) (II,2827)

DARETH, JOAN
SHE'S WITH ME (II,2321)

DARIAS
THE LIBERACE SHOW (I,2676)

DARIN, BOBBY
...AND DEBBIE MAKES SIX (I,185); THE BIG BEAT (I,423); THE BIG PARTY FOR REVLON (I,427); THE BOB HOPE SHOW (I,577); THE BOB HOPE SHOW (I,592); THE BOBBY DARIN AMUSEMENT COMPANY (I,672); AN EVENING WITH JIMMY DURANTE (I,1475); GEORGE BURNS IN THE BIG TIME (I,1764); RODGERS AND HART TODAY (I,3829)

DARIN, ROBERT
BELLE STARR (I,391)

DARK, CHRISTOPHER
FRONTIER (I,1695)

DARK, DANNY
THE NEW SUPER FRIENDS HOUR (II,1830); SUPER FRIENDS (II,2497); THE WORLD'S GREATEST SUPER HEROES (II,2842)

DARK, JOHNNY
THE OSMOND FAMILY SHOW (II,1927)

DARLING, ERICK
THE LORENZO AND HENRIETTA MUSIC SHOW (II,1519)

DARLING, JENNIFER
THE BIONIC WOMAN (II,255); EIGHT IS ENOUGH (II,762); THE GARY COLEMAN SHOW (II,958); THE NEW TEMPERATURES RISING SHOW (I,3273); THE SECRET STORM (I,3969); THE SIX-MILLION-DOLLAR MAN (II,2372); THE SMURFS (II,2385); TROLLKINS (II,2660)

DARLING, JOAN
OWEN MARSHALL: COUNSELOR AT LAW (I,3435); OWEN MARSHALL: COUNSELOR AT LAW (I,3436)

DARLOW, DAVID
THROUGH THE MAGIC PYRAMID (II,2615)

DARNAY, TONI
AS THE WORLD TURNS (II,110); GIVE US BARABBAS! (I,1816)

DARNE, PIERRE
FACTS OF LIFE: THE FACTS OF LIFE GOES TO PARIS (II,806)

DARNELL, LINDA
FORD TELEVISION THEATER (I,1634); PURSUIT (I,3700); SCHLITZ PLAYHOUSE OF STARS (I,3936); SCREEN DIRECTOR'S PLAYHOUSE (I,3946)

DARNELL, MIKE
BIG JOHN, LITTLE JOHN (II,240); NO SOAP, RADIO (II,1856)

DARNELL, NANA
THE MAGIC LAND OF ALLAKAZAM (I,2817); THE MAGIC OF MARK WILSON (II,1596)

DARRELL, STEVE
MEET MCGRAW (I,2984)

DARREN, JAMES
BATTLE OF THE NETWORK STARS (II,178); THE BOB HOPE SHOW (I,583); CITY BENEATH THE SEA (I,965); THE JO STAFFORD SHOW (I,2393); THE LIVES OF JENNY DOLAN (II,1502); PORTRAIT OF A LEGEND (II,2070); ROMP (I,3838); SCRUPLES (II,2281); T.J. HOOKER (II,2524); THE TIME TUNNEL (I,4511); TURNOVER SMITH (II,2671); THE WEB (I,4785)

DARROW, BARBARA
BROCK CALLAHAN (I,741); DOCTORS HOSPITAL (II,691); NEW GIRL IN HIS LIFE (I,3265)

DARROW, HENRY
100 CENTRE STREET (II,1901); THE ADDAMS FAMILY (II,13); BORN TO THE WIND (II,354); BROCK'S LAST CASE (I,742); CENTENNIAL (II,477); HARRY O (II,1099); HERNANDEZ, HOUSTON P.D. (I,2034); THE HIGH CHAPARRAL (I,2051); THE INVISIBLE MAN (II,1242); THE INVISIBLE MAN (II,1243); THE NEW DICK VAN DYKE SHOW (I,3263); NIGHT GAMES (II,1845); ROOSTER (II,2210); THE TARZAN/LONE RANGER/ZORRO ADVENTURE HOUR (II,2543); ZORRO AND SON (II,2878)

DARROW, MIKE
THE $128,000 QUESTION (II,1903); DREAM HOUSE (I,1376); EVERYTHING GOES (I,1480)

DARROW, PAUL
BLAKE'S SEVEN (II,264); MISTER JERICO (I,3070)

DARROW, PETER
OUTLAW LADY (II,1935)

DARROW, SUSAN
BRIGHT PROMISE (I,727)

DARROW, SUSANNAH
HERE COME THE BRIDES (I,2024)

DARTMOR, ELIZABETH
FANTASY ISLAND (II,830)

DARVAS, LILI
REUNION IN VIENNA (I,3780); TWENTY-FOUR HOURS IN A WOMAN'S LIFE (I,4644)

DARVEY, DIANA
THE BENNY HILL SHOW (II,207)

DARWELL, JANE
YOU'RE ONLY YOUNG TWICE (I,4970)

DASH, DARIEN
FATHER MURPHY (II,841); ROOSTER (II,2210); SILVER SPOONS (II,2355)

DASH, STACEY
FARRELL: FOR THE PEOPLE (II,834)

DASTE, HELEN
SMILEY'S PEOPLE (II,2383)

DAUGHTON, JAMES
FUTURE COP (II,946)

DAUPHIN, CLAUDE
THE HAPPY TIME (I,1943); PARIS PRECINCT (I,3458)

DAVALOS, DICK
AMERICANS (I,176)

DAVALOS, ELYSSA
HOW THE WEST WAS WON (II,1196); WILD AND WOOLEY (II,2798)

DAVE BRUBECK QUARTET, THE
THE BIG BAND AND ALL THAT JAZZ (I,422)

DAVE CLARK FIVE, THE
IT'S WHAT'S HAPPENING, BABY! (I,2278); LUCY IN LONDON (I,2789)

DAVENPORT, CLAIRE
CASANOVA (II,451)

DAVENPORT, MAVIS
TOM, DICK, AND HARRY (I,4536)

DAVENPORT, NIGEL
DIAL "M" FOR MURDER (I,1255); MASADA (II,1642); THE PICTURE OF DORIAN GRAY (I,3592)

DAVENPORT, RITA
WOMAN'S PAGE (II,2822)

DAVEY, DICK
THE GARRY MOORE SHOW (I,1740)

DAVEY, GILLIAN
FLAMBARDS (II,869)

DAVEY, JOHN
SHAZAM! (II,2319)

DAVEY, SCOTT
DO NOT DISTURB (I,1300)

DAVI, RICHARD
FROM HERE TO ETERNITY (II,932)

DAVI, ROBERT
THE GANGSTER CHRONICLES (II,957); THE LEGEND OF THE GOLDEN GUN (II,1455); NICK AND THE DOBERMANS (II,1843)

DAVICH, MARTY
DAYS OF OUR LIVES (II,629)

DAVID WINTERS DANCERS
THE BIG SHOW (II,243)

DAVID WINTERS DANCERS, THE
GO! (I,1830); HULLABALOO (I,2156); MONTE CARLO, C'EST LA ROSE (I,3095); MOVIN' WITH NANCY (I,3119); THE STEVE ALLEN COMEDY HOUR (I,4218)

DAVID, BRAD
FIREHOUSE (II,859)

DAVID, CLIFFORD
STAR TONIGHT (I,4199)

DAVID, CLIFTON
THAT'S MY MAMA (II,2580)

DAVID, JEFF
BUCK ROGERS IN THE 25TH CENTURY (II,384); GODZILLA (II,1014)

DAVID, JOANNA
WAR AND PEACE (I,4765)

DAVID, LARRY
FRIDAYS (II,926)

DAVID, MOLLY
KNOTS LANDING (II,1404)

DAVID, NICK
EXO-MAN (II,797)

DAVID, PAUL
TESTIMONY OF TWO MEN (II,2564)

DAVID, THAYER
DARK SHADOWS (I,1157); IN THE DEAD OF NIGHT (I,2202); LAMP AT MIDNIGHT (I,2608); NERO WOLFE (II,1805); ROOTS (II,2211); SPIDER-MAN (II,2424); THE THREE MUSKETEERS (I,4481); WASHINGTON: BEHIND CLOSED DOORS (II,2744)

DAVID, TODD
GENERAL HOSPITAL (II,964)

DAVIDS, SUSAN
AS THE WORLD TURNS (II,110)

DAVIDSON, BEN
BALL FOUR (II,137); CODE R (II,551); LUCAN (II,1553)

DAVIDSON, DOUG
THE YOUNG AND THE RESTLESS (II,2862)

DAVIDSON, EILEEN
THE YOUNG AND THE RESTLESS (II,2862)

DAVIDSON, IAN
MONTY PYTHON'S FLYING CIRCUS (II,1729)

DAVIDSON, JOHN
100 YEARS OF GOLDEN HITS (II,1905); BATTLE OF THE NETWORK STARS (II,171); BATTLE OF THE NETWORK STARS (II,174); THE BOB HOPE SHOW (I,626); THE BOB HOPE SHOW (I,636); THE CARPENTERS (II,445); THE CARPENTERS. . . SPACE ENCOUNTERS (II,447); THE ENTERTAINERS (I,1449); THE FANTASTICKS (I,1531); THE GIRL WITH SOMETHING EXTRA (II,999); THE GOLDDIGGERS (I,1838); THE ICE PALACE (I,2182); THE JOHN DAVIDSON CHRISTMAS SHOW (II,1307); THE JOHN DAVIDSON CHRISTMAS SHOW (II,1308); THE JOHN DAVIDSON SHOW (I,2409); THE JOHN DAVIDSON SHOW (II,1309); THE JOHN DAVIDSON SHOW (II,1310); KING OF THE ROAD (II,1396); KRAFT SUMMER MUSIC HALL (I,2590); THE NBC FOLLIES (I,3242); PERRY COMO'S SUMMER SHOW (I,3545); ROBERTA (I,3816); ROGER AND HARRY (II,2200); THE SANDY DUNCAN SHOW (II,2253); SANDY IN DISNEYLAND (II,2254); STARS AND STRIPES SHOW (II,2442); THAT'S INCREDIBLE! (II,2578)

DAVIDSON, JOYCE
THE JOYCE DAVIDSON SHOW (I,2459)

DAVIDSON, SUZANNE
AS THE WORLD TURNS (II,110); HOW TO SURVIVE A MARRIAGE (II,1198); MIRACLE ON 34TH STREET (I,3058)

DAVIDSON, WILD BILL
FLOOR SHOW (I,1605)

DAVIES, BLAIR
THE BRIGHTER DAY (I,728)

DAVIES, BRIAN
THE DAIN CURSE (II,613)

DAVIES, DEBBIE
VANITY FAIR (I,4701)

DAVIES, DIANA
CORONATION STREET (I,1054)

DAVIES, GEOFFREY
DOCTOR IN THE HOUSE (I,1313)

DAVIES, JACKSON
THE BEST CHRISTMAS PAGEANT EVER (II,212); THE HITCHHIKER (II,1157)

DAVIES, JANET
THE CITADEL (II,539)

DAVIES, JOHN RHYS
THE QUEST (II,2098)

DAVIES, KATHY
THE ADVENTURES OF OZZIE AND HARRIET (I,71)

DAVIES, LANE
DAYS OF OUR LIVES (II,629); SANTA BARBARA (II,2258)

DAVIES, LYNETTE
TYCOON: THE STORY OF A WOMAN (II,2697)

DAVIES, MARJORIE
BROTHER RAT (I,746)

DAVIES, PETER
LOVING (II,1552)

DAVIES, PETRA
KRAFT MYSTERY THEATER (I,2589)

DAVIES, RICHARD
THE CITADEL (II,539)

DAVIES, ROWLAND
WHY DIDN'T THEY ASK EVANS? (II,2795)

DAVIES, RUPERT
THE NEW ADVENTURES OF CHARLIE CHAN (I,3249); SAILOR OF FORTUNE (I,3885); WAR AND PEACE (I,4765)

DAVIES, STEPHEN
MASTER OF THE GAME (II,1647); PIAF (II,2039); VELVET (II,2725)

DAVIES, WINDSOR
MURDER IS A ONE-ACT PLAY (I,3160)

DAVIES-PROWLES, PAUL
GREAT EXPECTATIONS (II,1055)

DAVILLA, DIANE
THE DIARY OF ANNE FRANK (I,1262)

DAVILLE, DAVID
TYCOON: THE STORY OF A WOMAN (II,2697)

DAVION, ALEX
OPPENHEIMER (II,1919)

DAVIS JR, BILLY
THE FIFTH DIMENSION SPECIAL: AN ODYSSEY IN THE COSMIC UNIVERSE OF PETER MAX (I,1571); THE FIFTH DIMENSION TRAVELING SHOW (I,1572); THE MARILYN MCCOO AND BILLY DAVIS JR. SHOW (II,1630)

DAVIS JR, SAMMY
THE BEST ON RECORD (I,409); THE BOB HOPE SHOW (I,648); BOB HOPE SPECIAL: BOB HOPE FOR PRESIDENT (II,282); BOB HOPE SPECIAL: BOB HOPE'S 30TH ANNIVERSARY TV SPECIAL (II,293); BOB HOPE SPECIAL: BOB HOPE'S ALL-STAR COMEDY SPECTACULAR FROM LAKE TAHOE (II,301); BOB HOPE SPECIAL: BOB HOPE'S ALL-STAR TRIBUTE TO THE PALACE THEATER (II,305); BOB HOPE SPECIAL: BOB HOPE'S BICENTENNIAL STAR SPANGLED SPECTACULAR (II,306); BOB HOPE SPECIAL: BOB HOPE PRESENTS A CELEBRATION WITH STARS OF COMEDY AND MUSIC (II,340); BOB HOPE SPECIAL: THE BOB HOPE SPECIAL (II,336); BURT BACHARACH! (I,771); CELEBRITY CHALLENGE OF THE SEXES 3 (II,467); CIRCUS OF THE STARS (II,532); THE DANNY THOMAS SPECIAL (I,1151); DUKE ELLINGTON. . . WE LOVE YOU MADLY (I,1380); THE FLIP WILSON SPECIAL (I,1603); THE FLIP WILSON SPECIAL (II,888); FRANK SINATRA JR. WITH FAMILY AND FRIENDS (I,1663); THE FRANK SINATRA TIMEX SHOW (I,1670); THE GENERAL ELECTRIC THEATER (I,1753); GENERAL HOSPITAL (II,964); HOLIDAY IN LAS VEGAS (I,2076); THE KLOWNS (I,2574); LAS VEGAS PALACE OF STARS (II,1431); MOVIN' WITH NANCY (I,3119); THE NBC FOLLIES (I,3241); THE NBC FOLLIES (I,3242); ONE LIFE TO LIVE (II,1907); THE PIGEON (I,3596); POOR DEVIL (I,3646); ROMP (I,3838); ROWAN AND MARTIN BITE THE HAND THAT FEEDS THEM (I,3853); SAMMY AND COMPANY (II,2248); THE SAMMY DAVIS JR. SHOW (I,3898); THE SAMMY DAVIS JR. SPECIAL (I,3899); THE SAMMY DAVIS JR. SPECIAL (I,3900); SHOW BUSINESS SALUTE TO MILTON BERLE (I,4029); SINATRA—THE FIRST 40 YEARS (II,2359); STEVE AND EYDIE CELEBRATE IRVING BERLIN (II,2456); TEXACO STAR THEATER: OPENING NIGHT (II,2565)

DAVIS, ALANNA
WOMAN'S PAGE (II,2822); YOU ASKED FOR IT (II,2855)

DAVIS, ALFRED
THE HUSTLER OF MUSCLE BEACH (II,1210)

DAVIS, ANN B
THE BRADY BRIDES (II,361); THE BRADY BUNCH (II,362); THE BRADY BUNCH HOUR (II,363); THE BRADY GIRLS GET MARRIED (II,364); THE JOHN FORSYTHE SHOW (I,2410); THE KEEFE BRASSELLE SHOW (I,2514); KELLY'S KIDS (I,2523); LOVE THAT BOB (I,2779); TOO MANY SERGEANTS (I,4573)

DAVIS, BERYL
THE BOB HOPE SHOW (I,597)

DAVIS, BETTE
ALFRED HITCHCOCK PRESENTS (I,115); DISAPPEARANCE OF AIMEE (I,1290); THE ELISABETH MCQUEENEY STORY (I,1429); FORD TELEVISION THEATER (I,1634); HOTEL (II,1192); JOHNNY CARSON PRESENTS THE SUN CITY SCANDALS '72 (I,2421); THE JUDGE AND JAKE WYLER (I,2464); THE JUNE ALLYSON SHOW (I,2488); MADAME SIN (I,2805); THE STAR MAKER (I,4193); SUSPICION (I,4309)

DAVIS, BRAD
CHIEFS (II,509); MCLAREN'S RIDERS (II,1663); ROOTS (II,2211)

DAVIS, CHARLES
THE WILD WILD WEST (I,4863)

DAVIS, CLIFTON
CELEBRATION: THE AMERICAN SPIRIT (II,461); COTTON CLUB '75 (II,580); THE MELBA MOORE-CLIFTON DAVIS SHOW (I,2996); MITZI AND A HUNDRED GUYS (II,1710)

DAVIS, DANIEL
JOHNNY CASH: THE FIRST 25 YEARS (II,1336); SKINFLINT (II,2377); TEXAS (II,2566)

DAVIS, DANNY
I BELIEVE IN MUSIC (I,2165); THE VINCENT LOPEZ SHOW (I,4735)

DAVIS, DAVE
CELEBRITY BOWLING (I,883)

DAVIS, DIANE
GENE KELLY'S WONDERFUL WORLD OF GIRLS (I,1752); THE JONATHAN WINTERS SHOW (I,2443); THE WONDERFUL WORLD OF GIRLS (I,4898)

DAVIS, DON
DON'T CALL US (II,709)

DAVIS, ELMER
PULITZER PRIZE PLAYHOUSE (I,3692)

DAVIS, FRED
BRAINS AND BRAWN (I,706)

DAVIS, GAIL
ANNIE OAKLEY (I,225); THE BOB HOPE SHOW (I,569); THE GENE AUTRY SHOW (I,1748)

DAVIS, GAIL M
HAPPY ANNIVERSARY, CHARLIE BROWN (I,1939); IT'S ARBOR DAY, CHARLIE BROWN (I,2267); YOU'RE A GOOD SPORT, CHARLIE BROWN (I,4963)

DAVIS, GARY
HUSBANDS, WIVES AND LOVERS (II,1209)

DAVIS, GEENA
BUFFALO BILL (II,387); FAMILY TIES (II,819); RIPTIDE (II,2178)

DAVIS, GERALD
ANOTHER WORLD (II,97)

DAVIS, GWEN
CASPER AND FRIENDS (I,867)

DAVIS, HARRY
GIDEON (I,1792)

DAVIS, HENRY
THE VAUGHN MONROE SHOW (I,4707)

DAVIS, HERB
THE EDGE OF NIGHT (II,760)

DAVIS, HUMPHREY
FIRST LOVE (I,1583); OPERATION: NEPTUNE (I,3404)

DAVIS, JACK
ACADEMY THEATER (I,14)

DAVIS, JACKIE
THE NEW HOWDY DOODY SHOW (II,1818)

DAVIS, JAN
THE RED SKELTON SHOW (I,3755)

DAVIS, JANETTE
ARTHUR GODFREY AND FRIENDS (I,278)

DAVIS, JANICE
THE INA RAY HUTTON SHOW (I,2207)

DAVIS, JEFF
PATRICK STONE (I,3476)

DAVIS, JEFFERSON
MY SON THE DOCTOR (I,3203)

DAVIS, JENNIFER
M*A*S*H (II,1569)

DAVIS, JIM
AMANDA FALLON (I,151); CAVALCADE OF AMERICA (I,875); THE COWBOYS (II,598); DALLAS (II,614); ENIGMA (II,778); FIRESIDE THEATER (I,1580); LAST STAGECOACH WEST (I,2629); RESCUE 8 (I,3772); THE RUNAWAY BARGE (II,2230); STORIES OF THE CENTURY (I,4235)

DAVIS, JO
THE JIMMY DEAN SHOW (I,2384)

DAVIS, JOAN
THE BOB HOPE SHOW (I,551); I MARRIED JOAN (I,2174)

DAVIS, JOANNE
333 MONTGOMERY (I,4487); STUDIO '57 (I,4267)

DAVIS, JOE
FOR MEMBERS ONLY (II,905); THE GEORGE HAMILTON IV SHOW (I,1769)

DAVIS, JUDY
A WOMAN CALLED GOLDA (II,2818)

DAVIS, KAREN
MATINEE AT THE BIJOU (II,1651)

DAVIS, KATHY
HARRY AND MAGGIE (II,1097); THE MURDOCKS AND THE MCCLAYS (I,3163)

DAVIS, KENNY
THE HOYT AXTON SHOW (II,1200)

DAVIS, LINDY
CAMP RUNAMUCK (I,811)

DAVIS, LISA
THE BOB HOPE SHOW (I,578); THE GEORGE BURNS SHOW (I,1765); NORTHWEST PASSAGE (I,3306)

DAVIS, MAC
AMERICA (I,164); BOB HOPE SPECIAL: BOB HOPE'S ALL-STAR COMEDY SPECTACULAR FROM LAKE TAHOE (II,301); CHRISTMAS SPECIAL. . . WITH LOVE, MAC DAVIS (II,523); CHRISTMAS WITH THE BING CROSBYS (II,524); DEAN MARTIN PRESENTS MUSIC COUNTRY, U.S.A. (I,1192); I BELIEVE IN MUSIC (I,2165); A JOHNNY CASH CHRISTMAS (II,1327); MAC DAVIS 10TH ANNIVERSARY SPECIAL: I STILL BELIEVE IN MUSIC (II,1571); MAC DAVIS CHRISTMAS SPECIAL. . . WHEN I GROW UP (II,1574); THE MAC DAVIS CHRISTMAS SPECIAL (II,1572); THE MAC DAVIS CHRISTMAS SPECIAL (II,1573); THE MAC DAVIS SHOW (II,1575); THE MAC DAVIS SHOW (II,1576); THE MAC DAVIS SPECIAL (II,1577); MAC DAVIS SPECIAL: THE MUSIC OF CHRISTMAS (II,1578); MAC DAVIS'S CHRISTMAS ODYSSEY: TWO THOUSAND AND TEN (II,1579); MAC DAVIS. . . I BELIEVE IN CHRISTMAS (II,1581); MAC DAVIS. . . SOUNDS LIKE HOME (II,1582); MAC DAVIS—I'LL BE HOME FOR CHRISTMAS (II,1580); TENNESSEE ERNIE FORD'S WHITE CHRISTMAS (I,4402); TOUCH OF GOLD '75 (II,2650)

DAVIS, MICHAEL
THE LOSERS (I,2757); LUXURY LINER (I,2796); THE NEWS IS THE NEWS (II,1838); PARADE OF STARS (II,1954)

DAVIS, MURIEL
EL COYOTE (I,1423)

DAVIS, NICHOLAS
GALACTICA 1980 (II,953)

DAVIS, O C
THE BOOK OF LISTS (II,350)

DAVIS, OSSIE
GREAT ADVENTURE (I,1868); NIGHT GALLERY (I,3288); THE OUTSIDER (I,3430); ROOTS: THE NEXT GENERATIONS (II,2213); TEACHER, TEACHER (I,4366)

DAVIS, PATSY
THE FASHION STORY (I,1537)

DAVIS, PATTI
NIGHT PARTNERS (II,1847); RITUALS (II,2182); TWILIGHT THEATER II (II,2686)

DAVIS, PAUL
MIDNIGHT MYSTERY (I,3030)

DAVIS, PHYLLIS
BATTLE OF THE NETWORK STARS (II,171); THE BOYS (II,360); FANTASY ISLAND (II,829); LADIES IN BLUE (II,1422); MR. MOM (II,1752); OUT OF THE BLUE (I,3422); VEGAS (II,2723); VEGAS (II,2724); THE WILD WOMEN OF CHASTITY GULCH (II,2801); WIVES (II,2812)

DAVIS, RANDIE JEAN
AS THE WORLD TURNS (II,110)

DAVIS, RITA
MONTY PYTHON'S FLYING CIRCUS (II,1729)

DAVIS, ROGER
ALIAS SMITH AND JONES (I,118); ALIAS SMITH AND JONES (I,119); DARK SHADOWS (I,1157); THE GALLANT MEN (I,1728); REDIGO (I,3758)

DAVIS, ROY
ON THE CORNER (I,3359)

DAVIS, RUFE
PETTICOAT JUNCTION (I,3571)

DAVIS, SKEETER
JOHNNY CASH AND THE COUNTRY GIRLS (II,1323)

DAVIS, STEVE
THE HUSTLER OF MUSCLE BEACH (II,1210)

DAVIS, SUSAN
THE C.B. BEARS (II,406); HOT WHEELS (I,2126); SAINT PETER (II,2238); THE SKATEBIRDS (II,2375); WHATEVER HAPPENED TO DOBIE GILLIS? (II,2774)

DAVIS, TERRY
THE EDGE OF NIGHT (II,760)

DAVIS, TODD
GENERAL HOSPITAL (II,964); ONE LIFE TO LIVE (II,1907)

DAVIS, TOM
THE CONEHEADS (II,567); THE NEW SHOW (II,1828)

DAVIS, TOMMY
THE BOB HOPE SHOW (I,597)

DAVIS, VANCE
THIS IS KATE BENNETT (II,2597)

DAVIS, VIVEKA
THE JAMES BOYS (II,1272); MR. SUCCESS (II,1756); V (II,2713); V: THE FINAL BATTLE (II,2714)

DAVIS, WALT
BOSTON AND KILBRIDE (II,356); DOG AND CAT (II,695); HARDCASE (II,1088); THE LEGEND OF THE GOLDEN GUN (II,1455)

DAVISON, BRUCE
THE ASTRONAUTS (II,114); COPS (I,1047); MA AND PA (II,1570); V: THE SERIES (II,2715)

DAVISON, DAVEY
CRISIS (I,1102); GENERAL HOSPITAL (II,964); HAZEL (I,1982); THE ROOKIES (I,3840)

DAVISON, JOEL
DANIEL BOONE (I,1142); O.K. CRACKERBY (I,3348)

DAVISON, PETER
DOCTOR WHO (II,689); DOCTOR WHO—THE FIVE DOCTORS (II,690)

DAWBER, PAM
THE CHEVY CHASE NATIONAL HUMOR TEST (II,504); THE GIRL, THE GOLD WATCH AND EVERYTHING (II,1002); MORK AND MINDY (II,1735); MORK AND MINDY (II,1736); PARADE OF STARS (II,1954); THE PERRY COMO SPRINGTIME SPECIAL (II,2004); SISTER TERRI (II,2366); TEXACO STAR THEATER: OPENING NIGHT (II,2565); TWILIGHT THEATER (II,2685)

DAWN JR, HAZEL
FAIRMEADOWS, U.S.A. (I,1514)

DAWN-PORTER, NYREE
DEATH IN SMALL DOSES (I,1207)

DAWNE, DOREEN
THE BOB HOPE SHOW (I,535)

DAWS, MILLIE
THE GEORGE BURNS AND GRACIE ALLEN SHOW (I,1763)

DAWSON, ANTHONY
DIAL "M" FOR MURDER (I,1256)

DAWSON, CURT
ANOTHER WORLD (II,97); THE GUIDING LIGHT (II,1064)

DAWSON, GREGORY
THE ADVENTURES OF OZZIE AND HARRIET (I,71)

DAWSON, JIM
CAMPO 44 (I,813)

DAWSON, MARK
ALL MY CHILDREN (II,39)

DAWSON, RICHARD
BIZARRE (II,258); CAN YOU TOP THIS? (I,817); CELEBRITY CHALLENGE OF THE SEXES 4 (II,468); FAMILY FEUD (II,816); HOGAN'S HEROES (I,2069); I'VE GOT A SECRET (II,1220); KEEPING AN EYE ON DENISE (I,2520); MASQUERADE PARTY (II,1645); THE MATCH GAME (II,1649); MUNSTER, GO HOME! (I,3157); THE NEW DICK VAN DYKE SHOW (I,3263); ROWAN AND MARTIN'S LAUGH-IN (I,3856); TV'S FUNNIEST GAME SHOW MOMENTS (II,2676)

DAWSON, VICKY
ANOTHER WORLD (II,97); AS THE WORLD TURNS (II,110); THE FOUR OF US (II,914); HOT HERO SANDWICH (II,1186); LOVERS AND FRIENDS (II,1549)

DAY, BOBBY
INCREDIBLE KIDS AND COMPANY (II,1233)

DAY, DENNIS
BABES IN TOYLAND (I,318); THE DANNY THOMAS SPECIAL (I,1151); FROSTY'S WINTER WONDERLAND (II,935); THE JACK BENNY PROGRAM (I,2294); JACK BENNY'S 20TH ANNIVERSARY TV SPECIAL (I,2303); THE MICKEY MOUSE CLUB (I,3025); THE MOUSEKETEERS REUNION (II,1742); THE R.C.A. VICTOR SHOW (I,3737); SHOW BIZ (I,4027)

DAY, DORIS
THE DORIS DAY SHOW (I,1355); DORIS DAY TODAY (II,716); THE DORIS MARY ANNE KAPPELHOFF SPECIAL (I,1356); THE JOHN DENVER SPECIAL (II,1316)

DAY, DOROTHY
THE KATE SMITH HOUR (I,2508)

DAY, GARY
HOMICIDE (I,2101)

DAY, JACK
HAYLOFT HOEDOWN (I,1981)

DAY, LARAINE
BEN HECHT'S TALES OF THE CITY (I,395); DAYDREAMING WITH LARAINE (I,1184); FORD TELEVISION THEATER (I,1634); THE LUX VIDEO THEATER (I,2795); THE NASH AIRFLYTE THEATER (I,3227); PLAYHOUSE 90 (I,3623); PURSUIT (I,3700); SCHLITZ PLAYHOUSE OF STARS (I,3936); SWISS FAMILY ROBINSON (I,4324)

DAY, LORETTA

SHORT, SHORT DRAMA (I,4023)

DAY, LORNA
TESTIMONY OF TWO MEN (II,2564)

DAY, LYNDA *see* **GEORGE, LYNDA DAY**

DAY, MARILYN
54TH STREET REVUE (I,1573); THE FASHION STORY (I,1537)

DAY, NICHOLAS
THE CITADEL (II,539)

DAY, NOLA
THE TED STEELE SHOW (I,4373)

DAY, OTIS
TIGER! TIGER! (I,4499)

DAY, TERRY
THE BENNY HILL SHOW (II,207)

DAY, VENETIA
SMILEY'S PEOPLE (II,2383)

DAYDE, LIANA
THE BOB HOPE SHOW (I,535)

DAYKARHANOVA, TAMARA
REUNION IN VIENNA (I,3780)

DAYMAN, LES
HOMICIDE (I,2101)

DAYTON, DANNY
ADVICE TO THE LOVELORN (II,16); ARCHIE BUNKER'S PLACE (II,105); JOEY FAYE'S FROLICS (I,2406); THE PHIL SILVERS SHOW (I,3580)

DAYTON, JUNE
THE ALDRICH FAMILY (I,111); THE BRIGHTER DAY (I,728); CAPTAIN AMERICA (II,427); CAPTAIN AMERICA (II,428); A DATE WITH LIFE (I,1163); THE DOCTOR (I,1320); HEAVEN ON EARTH (II,1119); THE INNER SANCTUM (I,2216); LITTLE WOMEN (I,2720); LUCAS TANNER (II,1556); MR. O'MALLEY (I,3146); SHORT, SHORT DRAMA (I,4023); WASHINGTON: BEHIND CLOSED DOORS (II,2744)

DE LA CROIX, RAVEN
HEAR NO EVIL (II,1116)

DE SAN JUAN, HECTOR
BROADWAY GOES LATIN (I,734)

DEANDA, PETER
ADVICE TO THE LOVELORN (II,16); BEULAH LAND (II,226); CUTTER (I,1117); ONE LIFE TO LIVE (II,1907)

DEANGELO, CARLO
PAPA CELLINI (I,3450)

DEBAER, JEAN
LOVE, NATALIE (II,1545)

DEBANZIE, LOIS
FAMILY TIES (II,819); NEWHART (II,1835); RETURN OF THE MAN FROM U.N.C.L.E.: THE 15 YEARS LATERAFFAIR (II,2140)

DEBARI, IRENE
SANTA BARBARA (II,2258)

DEBELL, KRISTINE
B.J. AND THE BEAR (II,127); FOR MEMBERS ONLY (II,905)

DEBENNING, BURR
CANNON: THE RETURN OF FRANK CANNON (II,425); CITY BENEATH THE SEA (I,965); CODE RED (II,553); THE EYES OF TEXAS II (II,800); FATHER MURPHY (II,841); THE RETURN OF CAPTAIN NEMO (II,2135)

DEBENNING, JEFF
SANDY (I,3908)

DEBORD, SHARON
GENERAL HOSPITAL (II,964); HOUSE CALLS (II,1194)

DEBROUX, LEE
THE LONG DAYS OF SUMMER (II,1514); ROOTS (II,2211); SALVAGE (II,2244); THE SEAL (II,2282)

DEBURGH, CELIA
ALL THE RIVERS RUN (II,43)

DECAMP, ROSEMARY
B.J. AND THE BEAR (II,126); CALL HOLME (II,412); CHALK ONE UP FOR JOHNNY (I,893); DEATH VALLEY DAYS (I,1211); THE LIFE OF RILEY (I,2685); LOVE THAT BOB (I,2779); THE MISADVENTURES OF SHERIFF LOBO (II,1704); THE PARTRIDGE FAMILY (II,1962); PETTICOAT JUNCTION (I,3571); PHILLIP AND BARBARA (II,2034); THAT GIRL (II,2570); TV READERS DIGEST (I,4630)

DECARLO, YVONNE
BACKBONE OF AMERICA (I,328); THE GREATEST SHOW ON EARTH (I,1883); LIGHT'S OUT (I,2699); MARINELAND CARNIVAL (I,2913); MUNSTER, GO HOME! (I,3157); THE MUNSTERS (I,3158); THE MUNSTERS' REVENGE (II,1765); STAR STAGE (I,4197)

DECAROL, NANCY
THE NEW PEOPLE (I,3268)

DECENZO, JOE
THINGS ARE LOOKING UP (II,2588)

DECLOSS, JAMES
LONGSTREET (I,2750)

DECORDOVA, FRED
THE JACK BENNY PROGRAM (I,2294)

DECORSIA, TED
HUMAN ADVENTURE (I,2157); KNIGHT'S GAMBIT (I,2576); THE NEW ADVENTURES OF HUCKLEBERRY FINN (I,3250); THE SLOWEST GUN IN THE WEST (I,4088); STEVE CANYON (I,4224); WINCHESTER (I,4872)

DECORT, DICK
DAYS OF OUR LIVES (II,629); JACKIE AND DARLENE (II,1265); THE YOUNG AND THE RESTLESS (II,2862)

DECOSTA, DOUG
333 MONTGOMERY (I,4487)

DECOSTA, THORTON
THE AD-LIBBERS (I,29)

DECOSTA, TONY
SEARCH (I,3951)

DECRESPO, HELENA
TWENTY-FOUR HOURS IN A WOMAN'S LIFE (I,4644)

DEFARIA, GAIL
HE'S YOUR DOG, CHARLIE BROWN (I,2037); YOU'RE IN LOVE, CHARLIE BROWN (I,4964)

DEFARIA, KIP
PLAY IT AGAIN, CHARLIE BROWN (I,3617)

DEFINIZIO, BONNIE
BOB HOPE SPECIAL: BOB HOPE'S WICKI-WACKY SPECIAL FROM WAIKIKI (II,321)

DEFORE, DON
THE ADVENTURES OF OZZIE AND HARRIET (I,71); BLACK BEAUTY (II,261); HAZEL (I,1982); HOLLYWOOD OPENING NIGHT (I,2087); MYSTERY AND MRS. (I,3210); PETE SMITH SPECIALTIES (I,3560); THE PHILADELPHIA STORY (I,3582); A PUNT, A PASS, AND A PRAYER (I,3694); SCIENCE FICTION THEATER (I,3938); SILVER THEATER (I,4051)

DEFRANCO, TONY
JACK BENNY'S SECOND FAREWELL SHOW (I,2302)

DEGANON, MATT
I GAVE AT THE OFFICE (II,1213)

DEGORE, JANET
THE LAW AND MR. JONES (I,2637); THE REAL MCCOYS (I,3741)

DEGRAFT, JOSEPH
BORN FREE (I,691)

DEGROOT, MYRA
HERE COME THE BRIDES (I,2024)

DEGRURY, OSCAR
THE NEW BILL COSBY SHOW (I,3257)

DEHART, JEFF
THE BOOK OF LISTS (II,350)

DEHART, JUDITH
GOING PLACES (I,1834)

DEHAVEN, GLORIA
THE ARTHUR MURRAY PARTY FOR BOB HOPE (I,292); AS THE WORLD TURNS (II,110); BANJO HACKETT: ROAMIN' FREE (II,141); THE BOB HOPE SHOW (I,524); THE BOB HOPE SHOW (I,526); THE BOB HOPE SHOW (I,529); THE CABOT CONNECTION (II,408); CALL HER MOM (I,796); DELTA HOUSE (II,658); GENE KELLY. . . AN AMERICAN IN PASADENA (II,962); GIRL TALK (I,1810); THE GLORIA DEHAVEN SHOW (I,1826); THE LLOYD BRIDGES SHOW (I,2733); THE MANY SIDES OF MICKEY ROONEY (I,2899); MARY HARTMAN, MARY HARTMAN (II,1638); THE MICKEY ROONEY SHOW (I,3027); MR. BROADWAY (I,3130); THE MUSIC MART (II,1774); NAKIA (II,1785); THE RIFLEMAN (I,3789); RYAN'S HOPE (II,2234); WEDNESDAY NIGHT OUT (II,2760)

DEHAVILLAND, OLIVIA
ROOTS: THE NEXT GENERATIONS (II,2213)

DEHETRE, KATHERINE
THE RETURN OF MARCUS WELBY, M.D (II,2138)

DEJEAN, WILLIE
DEAR TEACHER (II,653)

DEJONG, HOLLY
THE ASSASSINATION RUN (II,112); PENMARRIC (II,1993)

DEKEYSER, DAVID
SOMEONE AT THE TOP OF THE STAIRS (I,4114); THE STRAUSS FAMILY (I,4253); A

WOMAN CALLED GOLDA (II,2818)

DEKOVA, FRANK
THE ADVENTURES OF RIN TIN TIN (I,73); THE ALASKANS (I,106); CROSSFIRE (II,606); F TROOP (I,1499); THE INDIAN (I,2210); OUTPOST (I,3428); SECOND LOOK (I,3960)

DELAMATER, MARLENE
PINE LAKE LODGE (I,3597)

DELANCEY, KAY
RYAN'S HOPE (II,2234)

DELANCIE, JOHN
THE BASTARD/KENT FAMILY CHRONICLES (II,159); EMERGENCY! (II,776); LITTLE WOMEN (II,1497); NIGHTSIDE (II,1851); TESTIMONY OF TWO MEN (II,2564); THE THORN BIRDS (II,2600)

DELANO, MICHAEL
GENERAL HOSPITAL (II,964); MY WIFE NEXT DOOR (II,1781); RHODA (II,2151); SUPERTRAIN (II,2504)

DELAVALLAD, CARMEN
THE NEW VOICE (II,1833)

DELEON, JACK
CAROUSEL (I,856)

DELISLE, CHRISTINE
BEACH PATROL (II,192); COLORADO C.I. (II,555); THE HOME FRONT (II,1171); WILD AND WOOLEY (II,2798)

DELORENZO, MICHAEL
FAME (II,812)

DELUCA, EVE
HIDDEN TREASURE (I,2045)

DELUCA, RUDY
BEDROOMS (II,198)

DELUGG, MILTON
SEVEN AT ELEVEN (I,3984)

DELUISE, DOM
ANN-MARGRET'S HOLLYWOOD MOVIE GIRLS (II,86); THE ARTHUR GODFREY SPECIAL (I,288); ARTHUR GODFREY'S PORTABLE ELECTRIC MEDICINE SHOW (I,290); BARYSHNIKOV IN HOLLYWOOD (II,157); THE BEST LITTLE SPECIAL IN TEXAS (II,215); DEAN MARTIN AT THE WILD ANIMAL PARK (II,636); THE DEAN MARTIN CELEBRITY ROAST (II,638); THE DEAN MARTIN SHOW (I,1201); DEAN MARTIN'S COMEDY CLASSICS (II,646); DEAN MARTIN'S RED HOT SCANDALS OF 1926 (II,648); DEAN MARTIN'S RED HOT SCANDALS PART 2 (II,649); DOM DELUISE AND FRIENDS (II,701); DOM DELUISE AND FRIENDS, PART 2 (II,702); THE DOM DELUISE SHOW (I,1328); EVIL ROY SLADE (I,1485); THE FUNNIEST JOKE I EVER HEARD (II,939); THE GOLDDIGGERS (I,1838); MAGIC WITH THE STARS (II,1599); THE MUNSTERS (I,3158); THE ROMAN HOLIDAYS (I,3835)

DELUISE, JACK
THE ENTERTAINERS (I,1449)

DELYON, LEO
IT'S A BUSINESS (I,2254); TOP CAT (I,4576)

DEMANN, JEFFREY
SANCTUARY OF FEAR (II,2252)

DEMARCO SISTERS, THE
THE ROBERT Q. LEWIS CHRISTMAS SHOW (I,3811)

DEMARCO, ARLENE
KEEFE BRASSELLE'S VARIETY GARDEN (I,2515)

DEMARNEY, DERRICK
THE LILLI PALMER THEATER (I,2704)

DEMARNEY, TERENCE
JOHNNY RINGO (I,2434)

DEMAURO, NICK
MICKEY SPILLANE'S MIKE HAMMER: MORE THAN MURDER (II,1693)

DEMAVE, JACK
LASSIE (I,2622); SKEEZER (II,2376)

DEMAY, JANET
REMINGTON STEELE (II,2130)

DEMEO, ANGELO
DOBIE GILLIS (I,1302)

DEMILLE, AGNES
MERV GRIFFIN'S SIDEWALKS OF NEW ENGLAND (I,3017)

DEMUNN, JEFFREY
O'MALLEY (II,1872)

DEPEW, ART
THE LAWRENCE WELK SHOW (I,2643); MEMORIES WITH LAWRENCE WELK (II,1681)

DERITA, JOE
THE COMICS (I,1028); THE NEW THREE STOOGES (I,3274)

DEROSA, PATRICIA
THE EDGE OF NIGHT (II,760)

DEROSE, CHRIS
BATTLE OF THE NETWORK STARS (II,165); THE SAN PEDRO BEACH BUMS (II,2250)

DEROY, GEORGE
THE SECRET WAR OF JACKIE'S GIRLS (II,2294)

DESALES, FRANCIS
THE ADVENTURES OF OZZIE AND HARRIET (I,71); DYNASTY (II,747); LEAVE IT TO BEAVER (I,2648); MR. AND MRS. NORTH (I,3125); TWO FACES WEST (I,4653)

DESANTIS, JOE
DOCTOR MIKE (I,1316); FEDERAL AGENT (I,1557); JOHNNY NIGHTHAWK (I,2432)

DESANTIS, JOHN
THE BOYS IN BLUE (II,359)

DESANTIS, STANLEY
THE PAPER CHASE (II,1949)

DESAPLO, FRANCESCA
TOP SECRET (II,2647)

DESHANNON, JACKIE
THE BUDDY GRECO SHOW (I,754); THE CATCHER (I,872); JACK CASSIDY'S ST. PATRICK'S DAY SPECIAL (I,2307)

DESHANNON, JIMMY
JACQUELINE SUSANN'S VALLEY OF THE DOLLS, 1981 (II,1268)

DESHIELDS, ANDRE
AIN'T MISBEHAVIN' (II,25)

DESICA, VITTORIO
THE FOUR JUST MEN (I,1650); THE SMALL MIRACLE (I,4090); SOPHIA! (I,4128); THE WORLD OF SOPHIA LOREN (I,4918)

DESOUZA, EDWARD
JANE EYRE (II,1273); SPELL OF EVIL (I,4152)

DESOUZA, NOEL
THE MAN WITH THE POWER (II,1616)

DEVARONA, DONNA
GAMES PEOPLE PLAY (II,956)

DEVEGH, DIANA
ALL MY CHILDREN (II,39)

DEVERITCH, SEAN
HIGHWAY TO HEAVEN (II,1152); MR. MOM (II,1752)

DEVITO, DANNY
ANN-MARGRET'S HOLLYWOOD MOVIE GIRLS (II,86); TAXI (II,2546)

DEVITO, JULIA
TAXI (II,2546)

DEVOL, FRANK
AMERICA 2-NIGHT (II,65); CAMP RUNAMUCK (I,811); FERNWOOD 2-NIGHT (II,849); I'M DICKENS...HE'S FENSTER (I,2193); THE JEFFERSONS (II,1276); PANTOMIME QUIZ (I,3449); WHERE'S EVERETT? (I,4835)

DEVRIES, JOHN
ANOTHER WORLD (II,97)

DEVRIES, SHARA
THE AT LIBERTY CLUB (I,309)

DEWILDE, BRANDON
JAMIE (I,2333); JAMIE (I,2334); LIGHT'S DIAMOND JUBILEE (I,2698); STANDARD OIL ANNIVERSARY SHOW (I,4189); STAR STAGE (I,4197)

DEWILDE, FREDERICK
BROTHER RAT (I,746); MUSICAL MERRY-GO-ROUND (I,3184)

DEWINDT, SHEILA
B.J. AND THE BEAR (II,126); BATTLESTAR GALACTICA (II,181); CRASH ISLAND (II,600); WOMEN WHO RATE A "10" (II,2825)

DEWINTER, JO
COMEDY OF HORRORS (II,557); GLORIA (II,1010)

DEWINTER, ROZ
AGAINST THE WIND (II,24); THE HANDS OF CORMAC JOYCE (I,1927)

DEWITNER, JOHANA
PLANET EARTH (I,3615)

DEWITT, ALAN
IT'S ABOUT TIME (I,2263)

DEWITT, FAYE
20TH CENTURY FOLLIES (I,4641); FOR THE LOVE OF MIKE (I,1625); HARRIS AGAINST THE WORLD (I,1956); THE NUT HOUSE (I,3320)

DEWITT, GEORGE
ALL IN FUN (I,131); ALL IN ONE (I,132); BE OUR GUEST (I,370); THE JULIUS LAROSA SHOW (I,2487); NAME THAT TUNE (I,3218); SEVEN AT ELEVEN (I,3984)

DEWITT, JACQUELINE
BRIEF ENCOUNTER (I,723); MEET MR. MCNUTLEY (I,2989)

DEWITT, JOYCE
THE B.B. BEEGLE SHOW (II,124); BATTLE OF THE NETWORK STARS (II,167); BATTLE OF THE NETWORK STARS (II,170); CELEBRITY CHALLENGE OF THE SEXES 3 (II,467); THE CELEBRITY FOOTBALL CLASSIC (II,474); CHERYL LADD. . . LOOKING BACK—SOUVENIRS (II,502); JOHN RITTER: BEING OF SOUND MIND AND BODY (II,1319); PERRY COMO'S CHRISTMAS IN NEW MEXICO (II,2010); RISKO (II,2180); STEVE MARTIN: COMEDY IS NOT PRETTY (II,2462); THREE'S COMPANY (II,2614)

DEWOLFE, BILLY
ARSENIC AND OLD LACE (I,270); THE DORIS DAY SHOW (I,1355); FROSTY THE SNOWMAN (II,934); GOOD MORNING WORLD (I,1850); THE IMOGENE COCA SHOW (I,2198); MARLO THOMAS AND FRIENDS IN FREE TO BE. . . YOU AND ME (II,1632); THE PRUITTS OF SOUTHAMPTON (I,3684); THE QUEEN AND I (I,3704); THAT GIRL (II,2570); VER-R-R-RY INTERESTING (I,4712)

DEWOODY, CRYSTAL
TRUE LIFE STORIES (II,2665)

DEYOUNG, CLIFF
CAPTAINS AND THE KINGS (II,435); CENTENNIAL (II,477); HUNTER'S MOON (II,1207); MASTER OF THE GAME (II,1647); THE SECRET STORM (I,3969); SUNSHINE (II,2495); THIS GIRL FOR HIRE (II,2596)

DEZURIK, CAROLYN
IT'S POLKA TIME (I,2273); POLKA TIME (I,3642); POLKA-GO-ROUND (I,3640)

DEACON, BRIAN
SEPARATE TABLES (II,2304)

DEACON, ERIC
PENMARRIC (II,1993)

DEACON, KIM
PRISONER: CELL BLOCK H (II,2085)

DEACON, RICHARD
ANNETTE (I,222); B.J. AND THE BEAR (II,125); THE BEVERLY HILLBILLIES (I,417); THE BOB HOPE SHOW (I,633); BROCK CALLAHAN (I,741); CAROL (I,846); THE CHARLIE FARRELL SHOW (I,913); DATE WITH THE ANGELS (I,1164); THE DICK VAN DYKE SHOW (I,1275); THE GOSSIP COLUMNIST (II,1045); HAVE GIRLS—WILL TRAVEL (I,1967); HONEYMOON SUITE (I,2108); IT'S A SMALL WORLD (I,2260); LEAVE IT TO BEAVER (I,2648); THE MOTHERS-IN-LAW (I,3110); MR. ED (I,3137); MURDER CAN HURT YOU! (II,1768); THE PRUITTS OF SOUTHAMPTON (I,3684); STAGE 7 (I,4181); STEVE MARTIN: COMEDY IS NOT PRETTY (II,2462); STILL THE BEAVER (II,2466)

DEAL, DARVANY
THINGS ARE LOOKING UP (II,2588)

DEAN, BUBBA
SIX PACK (II,2370)

DEAN, DICK
THE RAINBOW GIRL (II,2108)

DEAN, E BRIAN
RAGE OF ANGELS (II,2107)

DEAN, EDDIE
THE MARSHAL OF GUNSIGHT PASS (I,2925)

DEAN, EMILY
DAY OF THE TRIFFIDS (II,627)

DEAN, FABIAN
READY AND WILLING (II,2115); THE TIM CONWAY SHOW (I,4502)

DEAN, FELICITY
THE FAR PAVILIONS (II,832)

DEAN, FLOY
MORNING STAR (I,3104); THE YOUNG MARRIEDS (I,4942)

DEAN, HANNAH
OUT OF THE BLUE (II,1934)

DEAN, IVOR
MY PARTNER THE GHOST (I,3199); THE SAINT (II,2237)

DEAN, JAMES
THE ARMSTRONG CIRCLE THEATER (I,260); THE CAMPBELL TELEVISION SOUNDSTAGE (I,812); DANGER (I,1134); I REMEMBER MAMA (I,2176)

DEAN, JIMMY
THE CITY (II,541); COUNTRY MUSIC HIT PARADE (II,589); COUNTRY NIGHT OF STARS (II,592); DANIEL BOONE (I,1142); JERRY REED AND SPECIAL FRIENDS (II,1285); THE JIMMY DEAN SHOW (I,2384); THE JIMMY DEAN SHOW (I,2385)

DEAN, KATHY
CELEBRITY (II,463)

DEAN, LARRY
THE LAWRENCE WELK SHOW (I,2643); LAWRENCE WELK'S TOP TUNES AND NEW TALENT (I,2645); MEMORIES WITH LAWRENCE WELK (II,1681)

DEAN, MARTIN
MRS. G. GOES TO COLLEGE (I,3155); ONE MAN'S FAMILY (I,3383)

DEAN, PAMELA
TOM, DICK, AND HARRY (I,4536)

DEAN, PHIL
TIGER! TIGER! (I,4499)

DEAN, ROBERT
PAROLE (II,1960)

DEAN, SUZI
HARPER VALLEY (II,1094); HARPER VALLEY PTA (II,1095)

DEANE, PALMER
THE DOCTORS (II,694)

DEARDEN, ROBIN
THE ASPHALT COWBOY (II,111); THE EARTHLINGS (II,751); JOE DANCER: MURDER ONE, DANCER 0 (II,1302); MAGIC MONGO (II,1592); TROUBLE IN HIGH TIMBER COUNTRY (II,2661)

DEARE, MORGAN
MASTER OF THE GAME (II,1647)

DEARING, MARY LEE
HIS HONOR, HOMER BELL (I,2063)

DEAS, JUSTIN
AS THE WORLD TURNS (II,110); RYAN'S HOPE (II,2234)

DEAS, KEN
ALL ABOUT FACES (I,126)

DEAS, YVETTE
MANDY'S GRANDMOTHER (II,1618)

DEAUVILLE, RONNIE
THE FLORIAN ZABACH SHOW (I,1607)

DECHESER, ALLEN
RAGE OF ANGELS (II,2107)

DECHESER, ARTIE
RAGE OF ANGELS (II,2107)

DECKER, ALAN
DAYS OF OUR LIVES (II,629)

DECKER, BILL
BIZARRE (II,258)

DECKER, DIANE
SABER OF LONDON (I,3879)

DEE DEE WOOD DANCERS, THE
PURE GOLDIE (I,3696)

DEE, JOHN J
KING OF KENSINGTON (II,1395)

DEE, RUBY
D.H.O. (I,1249); DEADLOCK (I,1190); THE DUPONT SHOW OF THE WEEK (I,1388); GREAT ADVENTURE (I,1868); THE GUIDING LIGHT (II,1064); PEYTON PLACE (I,3574); ROOTS: THE NEXT GENERATIONS (II,2213)

DEE, SANDRA
THE DAUGHTERS OF JOSHUA CABE (I,1169); FANTASY ISLAND (II,830); GHOST STORY (I,1788); NEEDLES AND PINS (I,3246)

DEEB, GARY
TELEVISION INSIDE AND OUT (II,2552)

DEEBANK, FELIX
DIAL "M" FOR MURDER (I,1256)

DEEI, KAIE
VICTORY (I,4729)

DEEKS, MICHAEL
PETER PAN (II,2030)

DEEL, SONDRA
CAESAR'S HOUR (I,790); FACE THE MUSIC (I,1509); THE MARCH OF DIMES BENEFIT SHOW (I,2900)

DEEMER, ED
DRAGNET (I,1371)

DEEMS, MICKEY
CAR 54, WHERE ARE YOU? (I,842); THE DON KNOTTS SHOW (I,1334); HIZZONER (II,1159); MACK AND MYER FOR HIRE (I,2802); THE MUNSTERS' REVENGE (II,1765); WINDOWS, DOORS AND KEYHOLES (II,2806)

DEEN, NEDRA
ROSENTHAL AND JONES (II,2215)

DEEP RIVER BOYS, THE
THE R.C.A. THANKSGIVING SHOW (I,3736)

DEER, GARY MULE
DINAH AND HER NEW BEST FRIENDS (II,676)

DEERFIELD, LYNN

THE GUIDING LIGHT (II,1064)

DEERING, OLIVE
JANE EYRE (I,2337); STARLIGHT THEATER (I,4201); TALES OF TOMORROW (I,4352)

DEERMAN, YVONNE
THE BENNY HILL SHOW (II,207)

DEES, RICK
SOLID GOLD (II,2395); TOP TEN (II,2648)

DEEZEN, EDDIE
THE ACADEMY II (II,4); HOME ROOM (II,1172); PUNKY BREWSTER (II,2092)

DEFARIA, CHRISTOPHER
A CHARLIE BROWN THANKSGIVING (I,909)

DEFARIA, KIP
THERE'S NO TIME FOR LOVE, CHARLIE BROWN (I,4438)

DEFENDORF, DIANE
ROUGHNECKS (II,2219)

DEGREY, SLIM
WOOBINDA—ANIMAL DOCTOR (II,2832)

DEHM, ADOLPH
THE KATE SMITH HOUR (I,2508)

DEHNER, JOHN
THE ALASKANS (I,106); THE BAILEYS OF BALBOA (I,334); BARE ESSENCE (II,149); BARE ESSENCE (II,150); THE BETTY WHITE SHOW (I,415); BIG HAWAII (II,238); THE BILL COSBY SPECIAL, OR? (I,443); DANGER IN PARADISE (II,617); THE DAVID NIVEN THEATER (I,1178); THE DON KNOTTS SHOW (I,1334); THE DORIS DAY SHOW (I,1355); ENOS (II,779); FOREST RANGER (I,1641); FRONTIER (I,1695); HOW THE WEST WAS WON (II,1196); JANE WYMAN PRESENTS THE FIRESIDE THEATER (I,2345); THE LUCILLE BALL COMEDY HOUR (I,2783); MY WIVES JANE (I,3207); THE NEW DAUGHTERS OF JOSHUA CABE (I,3261); THE NEW TEMPERATURES RISING SHOW (I,3273); THE REBEL (I,3746); THE ROARING TWENTIES (I,3805); THE SOLDIERS (I,4112); TENNESSEE ERNIE FORD MEETS KING ARTHUR (I,4397); THE VIRGINIAN (I,4738); THE WESTERNER (I,4801); THE WINDS OF WAR (II,2807); YOUNG MAVERICK (II,2867)

DEIGNAN, MARTINA
AS THE WORLD TURNS (II,110); CODE RED (II,552); CODE RED (II,553); THE HOME FRONT (II,1171)

DEKKER, ALBERT
CELANESE THEATER (I,881); DEATH OF A SALESMAN (I,1208); FIRESIDE THEATER (I,1580); HOLLYWOOD OPENING NIGHT (I,2087); LIGHT'S OUT (I,2699); PULITZER PRIZE PLAYHOUSE (I,3692)

DEL GRANDE, LOUIS
SEEING THINGS (II,2297); TOM AND JOANN (II,2630)

DEL GROSSO, A
BOZO THE CLOWN (I,702)

DEL REGNO, JOHN
BAKER'S DOZEN (II,136)

DEL RUBIO TRIPLETS, THE
THE BOB HOPE SHOW (I,552)

DEL VANDO, AMAPOLA
HERNANDEZ, HOUSTON P.D. (I,2034)

DELANCEY, JOHN
DAYS OF OUR LIVES (II,629)

DELANEY, BOB
THE AMAZING DUNNINGER (I,158)

DELANEY, DANA
LOVE OF LIFE (I,2774)

DELANEY, DELVENE
THE PAUL HOGAN SHOW (II,1975); YOU ASKED FOR IT (II,2855)

DELANEY, GLORIA
WOMEN IN WHITE (II,2823)

DELANEY, JOAN
THE GOOD GUYS (I,1847)

DELANEY, KIM
ALL MY CHILDREN (II,39)

DELANEY, LAURENCE
THE RANGERS (II,2112)

DELANEY, PAT
CALL HOLME (II,412); COS (II,576)

DELANO, LEE
A LUCILLE BALL SPECIAL STARRING LUCILLE BALL AND DEAN MARTIN (II,1557); PEEPING TIMES (II,1990)

DELANO, MICHAEL
FIREHOUSE (II,859); FLAMINGO ROAD (II,871); FLAMINGO ROAD (II,872); MURDER CAN HURT YOU! (II,1768); STAT! (I,4213)

DELANY, DANA
AS THE WORLD TURNS (II,110); THE STREETS (II,2479)

DELANY, PAT
SWISS FAMILY ROBINSON (II,2516); SWISS FAMILY ROBINSON (II,2517)

DELAPPE, GOMEZ
ALL ABOUT MUSIC (I,127)

DELBAT, GERMAINE
SMILEY'S PEOPLE (II,2383)

DELEGALL, BOB
DOCTOR STRANGE (II,688)

DELEVANTI, CYRIL
THE DICK VAN DYKE SHOW (I,1275); JEFFERSON DRUM (I,2355)

DELFINO, FRANK
THE FEATHER AND FATHER GANG (II,845); THE FEATHER AND FATHER GANG: NEVER CON A KILLER (II,846)

DELGADO, EMILIO
BORN TO THE WIND (II,354); LOU GRANT (II,1526); SESAME STREET (I,3982)

DELGADO, JOHN
SLITHER (II,2379)

DELGADO, ROGER
THE ADVENTURES OF SIR FRANCIS DRAKE (I,75); DOCTOR WHO (II,689)

DELIA, TONY
COMPLETELY OFF THE WALL (II,560)

DELL, CHARLIE
JASON OF STAR COMMAND (II,1274); THAT'S TV (II,2581); TV FUNNIES (II,2672)

DELL, DIANE
THE SECRET STORM (I,3969)

DELL, GABRIEL
THE CORNER BAR (I,1051); CUTTER (I,1117); I DREAM OF JEANNIE (I,2167); THE JERK, TOO (II,1282); RISKO (II,2180); THE STEVE ALLEN SHOW (I,4219); THE STEVE ALLEN SHOW (I,4220); A YEAR AT THE TOP (II,2844)

DELLA FAVE, MICHELLE
AMERICA 2-NIGHT (II,65)

DELLA FAVE, TANYA
AMERICA 2-NIGHT (II,65)

DELLA, JAY
FRANK MERRIWELL (I,1659)

DELLS, DOROTHY
THEY ONLY COME OUT AT NIGHT (II,2586)

DELMAR, KENNY
THE ERNIE KOVACS SHOW (I,1455); THE GO GO GOPHERS (I,1831); HUMAN ADVENTURE (I,2157); KING LEONARDO AND HIS SHORT SUBJECTS (I,2557); SCHOOL HOUSE (I,3937); UNDERDOG (I,4673)

DELMAR, LAUREL
AS THE WORLD TURNS (II,110)

DELMARS, THE
VERSATILE VARIETIES (I,4713)

DELMONTE, MARY LU
ADVENTURES IN DAIRYLAND (I,44)

DELO, KEN
THE LAWRENCE WELK SHOW (I,2643); THE LAWRENCE WELK SHOW (I,2644); MEMORIES WITH LAWRENCE WELK (II,1681)

DELORME, REBECCA
MATINEE AT THE BIJOU (II,1651)

DELOY, GEORGE
9 TO 5 (II,1852); FAMILY BUSINESS (II,815); ONE NIGHT BAND (II,1909); THE SEEKERS (II,2298); STAR OF THE FAMILY (II,2436)

DELVANTI, CYRIL
THE MILTON BERLE SPECIAL (I,3052)

DELVE, DAVID
A HORSEMAN RIDING BY (II,1182)

DELYON, LEO
JACK AND THE BEANSTALK (I,2288)

DEMAREST, WILLIAM
THE DESILU REVUE (I,1238); THE GREATEST SHOW ON EARTH (I,1883); MAKE ROOM FOR DADDY (I,2843); THE MILLIONAIRE (II,1700); MY THREE SONS (I,3205); THE RED SKELTON TIMEX SPECIAL (I,3757); THE TALES OF WELLS FARGO (I,4353); THANKSGIVING REUNION WITH THE PARTRIDGE FAMILY AND MY THREE SONS (II,2569)

DEMAS, ALEX
A MAGIC GARDEN CHRISTMAS (II,1590)

DEMAS, CAROLE
THE MAGIC GARDEN (II,1591); A MAGIC GARDEN CHRISTMAS (II,1590)

DEMERS, TED
CELEBRITY (II,463)

DEMMING, ELLEN
THE GUIDING LIGHT (II,1064)

DEMONGEOT, ROLAND
MARK TWAIN'S TOM SAWYER-HUCKLEBERRY FINN (I,2915)

DEMYAN, LINCOLN
ASSIGNMENT: EARTH (I,299); MAN ON A STRING (I,2877)

DEN, JUDY
PAPA G.I. (I,3451)

DENCE, MAGGIE
THE SULLIVANS (II,2490)

DENCH, JUDI
LOVE IN A COLD CLIMATE (II,1539)

DENDY, LASHANA
DEAR TEACHER (II,653); THE GARY COLEMAN SHOW (II,958); GOLDIE AND KIDS: LISTEN TO ME (II,1021); JOSHUA'S WORLD (II,1344); MAX (II,1658)

DENGEL, RONI
THE DAIN CURSE (II,613); THE DOCTORS (II,694)

DENHAM, MARY
THE DOCTORS (II,694); ON OUR OWN (II,1892)

DENISON, BOB
BATMAN AND THE SUPER SEVEN (II,160)

DENN, MARIE
GOING BANANAS (II,1015)

DENNEHY, BRIAN
BIG SHAMUS, LITTLE SHAMUS (II,242); BUMPERS (II,394); FLY AWAY HOME (II,893); HANDLE WITH CARE (II,1077); PEARL (II,1986); STAR OF THE FAMILY (II,2436)

DENNEN, BARRY
OUT OF THE BLUE (I,3422)

DENNEY, DODO
THE GOOD OLD DAYS (I,1851)

DENNEY, KATHY
YOUNG LIVES (II,2866)

DENNEY, NORA
ETHEL IS AN ELEPHANT (II,783)

DENNEY, REGINALD
ALIAS MIKE HERCULES (I,117); THE DATCHET DIAMONDS (I,1160)

DENNING, DOREEN
FOREIGN INTRIGUE: DATELINE EUROPE (I,1638)

DENNING, GLEN
SPACE PATROL (I,4144)

DENNING, RICHARD
ALICE THROUGH THE LOOKING GLASS (I,122); THE ASPHALT COWBOY (II,111); CAVALCADE OF AMERICA (I,875); CROSSROADS (I,1105); THE FLYING DOCTOR (I,1610); FORD TELEVISION THEATER (I,1634); HAWAII FIVE-O (II,1110); KAREN (I,2505); MICHAEL SHAYNE, PRIVATE DETECTIVE (I,3021); MR. AND MRS. NORTH (I,3125); STUDIO ONE (I,4268)

DENNIS, ALFRED
AFTERMASH (II,23)

DENNIS, BEVERLY
THE RED BUTTONS SHOW (I,3750)

DENNIS, JOHN
THE BARBARY COAST (II,146); THE BLUE KNIGHT (II,277); IN SECURITY (II,1227)

DENNIS, KYLE
THE LAST LEAF (II,1437)

DENNIS, MATT
THE ERNIE KOVACS SHOW (I,1455); THE MATT DENNIS SHOW (I,2948)

DENNIS, NICK
BEN CASEY (I,394); HEAVENS TO BETSY (I,1997)

DENNIS, RES
THE ALVIN SHOW (I,146)

DENNIS, RICHARD
CASANOVA (II,451)

DENNIS, SANDY
A HATFUL OF RAIN (I,1964); THE TROUBLE WITH MOTHER (II,2663); WILSON'S REWARD (II,2805)

DENNISON, RACHEL
9 TO 5 (II,1852); BATTLE OF THE NETWORK STARS (II,176)

DENNY, DODO
ONCE UPON A MATTRESS (I,3371)

DENNY, PAT
NEW GIRL IN HIS LIFE (I,3265)

DENNY, REGINALD
G.E. SUMMMER ORIGINALS (I,1747)

DENON, BOB
THE BOB AND RAY SHOW (I,496)

DENOR, SUZANNE
THE IMPOSTER (II,1225)

DENOS, JOHN
THE YOUNG AND THE RESTLESS (II,2862)

DENSFORD, RONNIE
GALACTICA 1980 (II,953)

DENT, FRANK
MY WIFE NEXT DOOR (II,1781)

DENT, MAGGIE
A TOWN LIKE ALICE (II,2652)

DENTON, CRAHAN
MEMBER OF THE WEDDING (I,3002)

DENTON, LISA
AS THE WORLD TURNS (II,110)

DENVER, BOB
THE ALL-STAR SALUTE TO MOTHER'S DAY (II,51); THE BOB GOULET SHOW STARRING ROBERT GOULET (I,501); DOBIE GILLIS (I,1302); DUSTY'S TRAIL (I,1390); FAR OUT SPACE NUTS (II,831); GILLIGAN'S ISLAND (II,990); GILLIGAN'S ISLAND: RESCUE FROM GILLIGAN'S ISLAND (II,991); GILLIGAN'S ISLAND: THE CASTAWAYS ON GILLIGAN'S ISLAND (II,992); GILLIGAN'S ISLAND: THE HARLEM GLOBETROTTERS ON GILLIGAN'S ISLAND (II,993); GILLIGAN'S PLANET (II,994); THE GOOD GUYS (I,1847); HIGH SCHOOL, U.S.A. (II,1148); THE INVISIBLE WOMAN (II,1244); THE NEW ADVENTURES OF GILLIGAN (II,1813); SCAMPS (II,2267); TWILIGHT THEATER II (II,2686); WHATEVER HAPPENED TO DOBIE GILLIS? (II,2774)

DENVER, DREAMA
GILLIGAN'S ISLAND: THE HARLEM GLOBETROTTERS ON GILLIGAN'S ISLAND (II,993); SCAMPS (II,2267)

DENVER, JOHN
THE BOB HOPE SHOW (I,662); BOB HOPE SPECIAL: BOB HOPE'S SALUTE TO NASA—25 YEARS OF REACHING FOR THE STARS (II,314); THE CARPENTERS (II,444); DORIS DAY TODAY (II,716); JOHN DENVER AND FRIEND (II,1311); JOHN DENVER AND THE LADIES (II,1312); JOHN DENVER AND THE MUPPETS: A CHRISTMAS TOGETHER (II,1313); JOHN DENVER ROCKY MOUNTAIN CHRISTMAS (II,1314); THE JOHN DENVER SPECIAL (II,1315); THE JOHN DENVER SPECIAL (II,1316); JOHN DENVER: MUSIC AND THE MOUNTAINS (II,1318); JOHN DENVER—THANK GOD I'M A COUNTRY BOY (II,1317); SALUTE TO LADY LIBERTY (II,2243); SINATRA AND FRIENDS (II,2358); THE THIRD BARRY MANILOW SPECIAL (II,2590); TWO OF A KIND: GEORGE BURNS AND JOHN DENVER (II,2692)

DENVER, TOM
MADHOUSE 90 (I,2807)

DENZER, GEORJA
OFF THE RACK (II,1878)

DEPAL, WILFRED
WILSON'S REWARD (II,2805)

DER, RICKY
KENTUCKY JONES (I,2526)

DERBY, BRIAN
THE OMEGA FACTOR (II,1888)

DERBYSHIRE, EILEEN
CORONATION STREET (I,1054)

DEREK, BO
TOM SNYDER'S CELEBRITY SPOTLIGHT (II,2632)

DEREK, JOHN
FRONTIER CIRCUS (I,1697)

DERMAN, VERGIE
CINDERELLA (II,526)

DERN, BRUCE
SAM HILL: WHO KILLED THE MYSTERIOUS MR. FOSTER? (I,3896); STONEY BURKE (I,4229)

DERR, RICHARD
FANFARE (I,1527); GALLAGHER (I,1727)

DERRICKS-CARROLL, CLINTON
THE FURTHER ADVENTURES OF WALLY BROWN (II,944); SANFORD (II,2255)

DESCHER, FRAN
STRANGER IN OUR HOUSE (II,2477)

DESCHER, SANDY
THE DONNA REED SHOW (I,1347); THE LORETTA YOUNG SHOW (I,2755); MIRACLE ON 34TH STREET (I,3056); THE NEW PHIL SILVERS SHOW (I,3269)

DESIDERIO, ROBERT
MOONLIGHT (II,1730); RYAN'S HOPE (II,2234)

DESMOND, CONNIE

THE JULIUS LAROSA SHOW (I,2486)

DESMOND, JERRY
THE BOB HOPE SHOW (I,535)

DESMOND, JOHNNY
ALL IN FUN (I,130); BLANSKY'S BEAUTIES (II,267); DON MCNEILL'S TV CLUB (I,1337); FACE THE MUSIC (I,1509); GLENN MILLER TIME (I,1824); THE JACK PAAR SHOW (I,2316); MUSIC ON ICE (I,3175); OPEN HOUSE (I,3394); PANORAMA (I,3448); SALLY (I,3890); TIN PAN ALLEY TV (I,4522)

DESMOND, JULIA
CASANOVA (II,451); THE GOODIES (II,1040)

DESMOND, MARY FRANCES
HAWKINS FALLS, POPULATION 6200 (I,1977); PADDY THE PELICAN (I,3442)

DESMOND, TRUDY
CLASS OF '67 (I,970)

DESPO
IVAN THE TERRIBLE (II,1260)

DESTOR, SAM
SPACE: 1999 (II,2409)

DESTRI, JAMES
BLONDIE (II,272)

DETRICK, BRUCE
ONE LIFE TO LIVE (II,1907)

DEUEL, GEOFFREY
MOVIN' ON (I,3118)

DEUEL, PETER
GIDGET (I,1795); LOVE ON A ROOFTOP (I,2775)

DEUTSCH, BARBARA
TAXI (II,2546)

DEUTSCH, PATTI
BIG DADDY (I,425); THE JOHN BYNER COMEDY HOUR (I,2407); ROWAN AND MARTIN'S LAUGH-IN (I,3856); SLITHER (II,2379)

DEVANE, WILLIAM
THE BAIT (I,335); BATTLE OF THE NETWORK STARS (II,167); BATTLE OF THE NETWORK STARS (II,168); BATTLE OF THE NETWORK STARS (II,170); BATTLE OF THE NETWORK STARS (II,177); BATTLE OF THE NETWORK STARS (II,178); THE BIG EASY (II,234); BLACK BEAUTY (II,261); THE CELEBRITY FOOTBALL CLASSIC (II,474); THE CRIME CLUB (I,1096); FROM HERE TO ETERNITY (II,932); FROM HERE TO ETERNITY (II,933); KNOTS LANDING (II,1404)

DEVENNEY, W SCOTT
HEAR NO EVIL (II,1116); INSPECTOR PEREZ (II,1237); MATINEE AT THE BIJOU (II,1651)

DEVER, TOM
THE RENEGADES (II,2132)

DEVEREAUX, ED
OPPENHEIMER (II,1919); SKIPPY, THE BUSH KANGAROO (I,4076)

DEVERY, ELAINE
AMOS BURKE: WHO KILLED JULIE GREER? (I,181)

DEVIN, MARILYN
TIGER! TIGER! (I,4499)

DEVINE, ANDY
ANDY'S GANG (I,214); FLIPPER (I,1604); THE OVER-THE-HILL GANG RIDES AGAIN (I,3432); WILD BILL HICKOK (I,4861); YOU'RE ONLY YOUNG TWICE (I,4971)

DEVINE, ERICK
THE DRUNKARD (II,737)

DEVINE, KATHLEEN
ONE LIFE TO LIVE (II,1907); SEARCH FOR TOMORROW (II,2284)

DEVION, ALEX
THE MAN WHO NEVER WAS (I,2882)

DEVLIN, DONALD
MR. I. MAGINATION (I,3140); WESLEY (I,4795)

DEVLIN, JOE
DICK TRACY (I,1271)

DEVLIN, JOHN
THE OATH: 33 HOURS IN THE LIFE OF GOD (II,1873); A WORLD APART (I,4910); THE YOUNG AND THE RESTLESS (II,2862)

DEVOE, JOHN
A TIME TO LIVE (I,4510)

DEVON, LAURA
THE RICHARD BOONE SHOW (I,3784)

DEVON, RICHARD
RICHARD DIAMOND, PRIVATE DETECTIVE (I,3785)

DEVORE, CAIN
DREAMS (II,735)

DEWEY, BRIAN
HEC RAMSEY (II,1121)

DEWHURST, COLLEEN
THE BLUE AND THE GRAY (II,273); THE DUPONT SHOW OF THE WEEK (I,1388); THE HANDS OF CORMAC JOYCE (I,1927); THE PRICE (I,3666); STUDS LONIGAN (II,2484); TALENT SEARCH (I,4342)

DEXTER, ALAN
MR. AND MRS. COP (II,1748); THE SEEKERS (I,3974)

DEXTER, JERRY
THE ADVENTURES OF GULLIVER (I,60); AQUAMAN (I,249); THE DRAK PACK (II,733); FANG FACE (II,822); THE FUNKY PHANTOM (I,1709); GOOBER AND THE GHOST CHASERS (II,1028); JOSIE AND THE PUSSYCATS (I,2453); JOSIE AND THE PUSSYCATS IN OUTER SPACE (I,2454); THE PLASTICMAN COMEDY/ADVENTURE SHOW (II,2051); SEALAB 2020 (I,3949); SHAZZAN! (I,4002)

DEXTER, JIMMY
HAPPY DAYS (I,1942)

DEXTER, WILLIAM
SPELL OF EVIL (I,4152); THE STRAUSS FAMILY (I,4253)

DEY, SUSAN
CIRCLE OF FEAR (I,958); EMERALD POINT, N.A.S. (II,773); GOOBER AND THE GHOST CHASERS (II,1028); A KNIGHT IN SHINING ARMOUR (I,2575); LITTLE WOMEN (II,1497); LOVES ME, LOVES ME NOT (II,1551); MALIBU (II,1607); THE PARTRIDGE FAMILY (II,1962); PARTRIDGE FAMILY: 2200 A.D. (II,1963); THANKSGIVING REUNION WITH THE PARTRIDGE FAMILY AND MY THREE SONS (II,2569)

DEZINA, KATE
ALL MY CHILDREN (II,39); THE HAMPTONS (II,1076)

DHIEGH, KHIGH
HAWAII FIVE-O (I,1972); HAWAII FIVE-O (II,1110); JUDGE DEE IN THE MONASTERY MURDERS (I,2465); KHAN! (II,1384)

DHOOGE, DESMOND
HERNANDEZ, HOUSTON P.D. (I,2034)

DIBELLA, FAUSTO
MOSES THE LAWGIVER (II,1737)

DIBENEDETTO, TONY
JOHNNY GARAGE (II,1337)

DICENZO, GEORGE
BLACKSTAR (II,263); THE JORDAN CHANCE (II,1343); MCCLAIN'S LAW (II,1659); MCLAREN'S RIDERS (II,1663); SWISS FAMILY ROBINSON (II,2516); THE TOM SWIFT AND LINDA CRAIG MYSTERY HOUR (II,2633); THE YOUNG SENTINELS (II,2871)

DIGIAMPAOLO, NITA
LITTLE LULU (II,1493); THE NEW MICKEY MOUSE CLUB (II,1823)

DIGIUSEPPE, ENRICO
GOLDEN CHILD (I,1839)

DIMAGGIO, JOE
BOB HOPE SPECIAL: THE BOB HOPE SPECIAL (II,335); THE JOE DIMAGGIO SHOW (I,2399)

DINAPOLI, MARC
MARK TWAIN'S TOM SAWYER-HUCKLEBERRY FINN (I,2915)

DIREDA, JOE
IF I LOVE YOU, AM I TRAPPED FOREVER? (II,1222)

DIAGA, GEORGE
WONDER WOMAN (PILOT 1) (II,2826)

DIAL, BILL
WKRP IN CINCINNATI (II,2814)

DIAMOND, BARRY
COUNTRY MUSIC CARAVAN (I,1066)

DIAMOND, BILL
THE ADVENTURES OF KIT CARSON (I,66)

DIAMOND, BOBBY
DOBIE GILLIS (I,1302); FURY (I,1721); MOBY DICK AND THE MIGHTY MIGHTOR (I,3077); THE PATTY DUKE SHOW (I,3483); YES YES NANETTE (I,4927)

DIAMOND, DON
F TROOP (I,1499); THE FLYING NUN (I,1611); SECOND EDITION (II,2288); THE TARZAN/LONE RANGER/ZORRO ADVENTURE HOUR (II,2543); ZORRO (I,4982)

DIAMOND, EILEEN
THE MICKEY MOUSE CLUB (I,3025); THE MOUSEKETEERS REUNION (II,1742)

DIAMOND, JACK
FAMILY GENIUS (I,1521)

DIAMOND, MARCIA
PAUL BERNARD—PSYCHIATRIST (I,3485)

DIAMOND, NEIL
NEIL DIAMOND SPECIAL: I'M GLAD YOU'RE HERE WITH ME TONIGHT (II,1803); THE NEIL DIAMOND SPECIAL (II,1802)

DIAMOND, SELMA
NIGHT COURT (II,1844); THIS WILL BE THE YEAR THAT WILL BE (I,4466); TOO CLOSE FOR COMFORT (II,2642)

DIANA ROSS AND THE SUPREMES
THE BING COSBY SHOW (I,468); RODGERS AND HART TODAY (I,3829); THE SMOKEY ROBINSON SHOW (I,4093); THE TENNESSEE ERNIE FORD SPECIAL (I,4401)

DIAZ, EDITH
POPI (II,2069); WILLOW B: WOMEN IN PRISON (II,2804)

DIAZ, RUDY
BORN TO THE WIND (II,354); THE MACAHANS (II,1583)

DIAZ, VICTOR
COUNTERTHRUST (I,1061)

DIBBS, KEM
BUCK ROGERS IN THE 25TH CENTURY (I,751); MEN INTO SPACE (I,3007)

DICK LEE AND THE HONEYDREAMERS
TED MACK'S MATINEE (I,4372)

DICK WILLIAMS SINGERS, THE
THE BIG SHOW (II,243); THE JULIE ANDREWS HOUR (I,2480); MUSIC BY COLE PORTER (I,3166)

DICK, GENA
HIGH HOPES (II,1144)

DICK, JAMIE
CAGNEY AND LACEY (II,410)

DICK, JARO
THE IMPERIAL GRAND BAND (II,1224)

DICK, JUDITH
PRISONER: CELL BLOCK H (II,2085)

DICKENS SISTERS, THE
THE EDDY ARNOLD SHOW (I,1411)

DICKENS, JIMMY
THE GRAND OLE OPRY (I,1866)

DICKENSON, CHRISTINE
BRANAGAN AND MAPES (II,367)

DICKENSON, VIC
TIMEX ALL-STAR JAZZ SHOW IV (I,4518)

DICKERSON, BEACH
COCAINE AND BLUE EYES (II,548)

DICKERSON, NATHANIEL
ARTHUR GODFREY AND FRIENDS (I,278)

DICKINS, JIMMY
COUNTRY MUSIC CARAVAN (I,1066)

DICKINSON, ANGIE
60 YEARS OF SEDUCTION (II,2373); ALAN KING'S FINAL WARNING (II,29); THE BOB HOPE CHRYSLER THEATER (I,502); THE BOB HOPE SHOW (I,625); THE BOB HOPE SHOW (I,649); BOB HOPE SPECIAL: BOB HOPE IN "JOYS" (II,284); BOB HOPE SPECIAL: BOB HOPE FOR PRESIDENT (II,282); BOB HOPE SPECIAL: BOB HOPE'S CHRISTMAS PARTY (II,307); BOB HOPE SPECIAL: THE BOB HOPE CHRISTMAS SPECIAL (II,330); BOB HOPE SPECIAL: BOB HOPE'S HILARIOUS UNREHEARSED ANTICS OF THE STARS (II,341); CASSIE AND COMPANY (II,455); THE DEAN MARTIN CELEBRITY ROAST (II,638); DEAN'S PLACE (II,650); THE DICK POWELL SHOW (I,1269); DOM DELUISE AND FRIENDS (II,701); THE FIRST 50 YEARS (II,862); THE GENERAL ELECTRIC THEATER (I,1753); GHOST STORY (I,1788); THE HOMEMADE COMEDY SPECIAL (II,1173); THE MANY FACES OF COMEDY (I,2894); MATINEE THEATER (I,2947); MEN INTO SPACE (I,3007); THE NORLISS TAPES (I,3305); PEARL (II,1986); PERRY COMO'S CHRISTMAS IN PARIS (II,2012); POLICE WOMAN (II,2063); POLICE WOMAN: THE GAMBLE (II,2064); THE POPPY IS ALSO A FLOWER (I,3650); RINGO (II,2176); RODNEY DANGERFIELD SHOW: I CAN'T TAKE IT NO MORE (II,2198); A TRIBUTE TO "MR. TELEVISION" MILTON BERLE (II,2658)

DICKINSON, SAUNDRA
THE HITCHHIKER'S GUIDE TO THE GALAXY (II,1158)

DICKSON, BRENDA
THE YOUNG AND THE RESTLESS (II,2862)

DICKSON, JAMES
MOMMA THE DETECTIVE (II,1721)

DICKSON, PATRICK
A TOWN LIKE ALICE (II,2652)

DIDLEY, BO
LIVE FROM THE LONE STAR (II,1501)

DIEHL, ILKA
CHICAGOLAND MYSTERY PLAYERS (I,936)

DIEHL, JOHN
MIAMI VICE (II,1689)

DIEIKES, JOHN
THE SLOWEST GUN IN THE WEST (I,4088)

DIERKING, NUELLA
OUR FIVE DAUGHTERS (I,3413)

DIERKOP, CHARLES
ALIAS SMITH AND JONES (I,119); CITY BENEATH THE SEA (I,965); LOCK, STOCK, AND BARREL (I,2736); MURDOCK'S GANG (I,3164); THE PIRATES OF FLOUNDER BAY (I,3607); POLICE WOMAN (II,2063); POLICE WOMAN: THE GAMBLE (II,2064); REVENGE OF THE GRAY GANG (II,2146)

DIETRICH, DENA
ADAM'S RIB (I,32); BUT MOTHER! (II,403); GOSSIP (II,1044); KAREN (II,1363); PAUL SAND IN FRIENDS AND LOVERS (II,1982); THE PRACTICE (II,2076); THE ROPERS (II,2214); SCAMPS (II,2267); THE TROUBLE WITH PEOPLE (I,4611)

DIETRICH, MARLENE
MARLENE DIETRICH: I WISH YOU LOVE (I,2918)

DIETZ, EILEEN
GENERAL HOSPITAL (II,964)

DIFFRING, ANTON
ASSIGNMENT: VIENNA (I,304); THE SAVAGE CURSE (I,3925); THE WINDS OF WAR (II,2807)

DIGNAM, BASIL
THE SIX WIVES OF HENRY VIII (I,4068)

DILLARD, DOUG
DEAN MARTIN PRESENTS MUSIC COUNTRY, U.S.A. (I,1192)

DILLARD, MIMI
VALENTINE'S DAY (I,4693)

DILLARDS, THE
THE ANDY GRIFFITH SHOW (I,192)

DILLAWAY, DON
LEAVE IT TO BEAVER (I,2648)

DILLER, PHYLLIS
THE BEAUTIFUL PHYLLIS DILLER SHOW (I,381); BLONDES VS. BRUNETTES (II,271); THE BOB HOPE SHOW (I,604); THE BOB HOPE SHOW (I,610); THE BOB HOPE SHOW (I,611); THE BOB HOPE SHOW (I,613); THE BOB HOPE SHOW (I,615); THE BOB HOPE SHOW (I,642); THE BOB HOPE SHOW (I,643); THE BOB HOPE SHOW (I,649); THE BOB HOPE SHOW (I,657); BOB HOPE SPECIAL: BOB HOPE'S FUNNY VALENTINE (II,309); BOB HOPE SPECIAL: HAPPY BIRTHDAY, BOB! (II,325); BOB HOPE SPECIAL: THE BOB HOPE CHRISTMAS SPECIAL (II,328); BOB HOPE SPECIAL: THE BOB HOPE SPECIAL FROM PALM SPRINGS (II,339); CIRCUS OF THE STARS (II,537); THE COLGATE COMEDY HOUR (I,998); THE DEAN MARTIN CELEBRITY ROAST (II,638); EVENING AT THE IMPROV (II,786); GEORGE BURNS CELEBRATES 80 YEARS IN SHOW BUSINESS (II,967); THE JACK BENNY CHRISTMAS SPECIAL (I,2289); LEAPIN' LIZARDS, IT'S LIBERACE (II,1453); MINSKY'S FOLLIES (II,1703); PLIMPTON! DID YOU HEAR THE ONE ABOUT. . .? (I,3629); THE PRUITTS OF SOUTHAMPTON (I,3684); SHOW STREET (I,4032); SHOWTIME (I,4036); THE SINGERS (I,4060); SWING OUT, SWEET LAND (II,2515); TAKE ONE STARRING JONATHAN WINTERS (II,2532); WHATEVER BECAME OF. . . ? (II,2772)

DILLEY, J C
ONE DAY AT A TIME (II,1900)

DILLEY, R C
ONE DAY AT A TIME (II,1900)

DILLMAN, BRADFORD
THE CASE AGAINST PAUL RYKER (I,861); COURT-MARTIAL (I,1080); DEATH IN DEEP WATER (I,1206); THE DELPHI BUREAU (I,1224); ESPIONAGE (I,1464); THE EYES OF CHARLES SAND (I,1497); FORCE FIVE (II,907); THE GREATEST SHOW ON EARTH (I,1883); HOT PURSUIT (II,1189); KING'S

CROSSING (II,1397); KINGSTON: THE POWER PLAY (II,1399); LONGSTREET (I,2750); LOOK BACK IN DARKNESS (I,2752); THERE SHALL BE NO NIGHT (I,4437); TOURIST (II,2651)

DILLON, BRENDON
ALL IN THE FAMILY (II,38); THE YOUNG PIONEERS (II,2868); THE YOUNG PIONEERS' CHRISTMAS (II,2870)

DILLON, DAN
LONDON AND DAVIS IN NEW YORK (II,1512)

DILLON, DAVEY
ANSON AND LORRIE (II,98)

DILLON, DENNY
HOT HERO SANDWICH (II,1186); SATURDAY NIGHT LIVE (II,2261)

DILLON, MELINDA
ENIGMA (II,778); FREEMAN (II,923); HELLINGER'S LAW (II,1127); THE JUGGLER OF NOTRE DAME (II,1350); MARRIAGE IS ALIVE AND WELL (II,1633); STORY THEATER (I,4241)

DILLON, SARA
ROBERTA (I,3815)

DILLWORTH, GORDON
THE GREAT WALTZ (I,1878)

DILWORTH, GLORIA
COUNTRY STYLE (I,1072)

DIMAS, RAY
ANDERSON AND COMPANY (I,188); BOBBY PARKER AND COMPANY (II,344)

DIMITRI, JAMES
ROAD TO REALITY (I,3800)

DIMITRI, JIM
CHEZ PAREE REVUE (I,932); OH, KAY! (I,3342)

DIMITRI, NICK
THE NORLISS TAPES (I,3305)

DIMITRI, RICHARD
THE 416TH (II,913); HUMAN FEELINGS (II,1203); SEVENTH AVENUE (II,2309); TURNOVER SMITH (II,2671); WHACKED OUT (II,2767); WHEN THINGS WERE ROTTEN (II,2782)

DIMSTER, DENNIS
THE AWAKENING LAND (II,122); THE FRENCH ATLANTIC AFFAIR (II,924)

DINEHART, ALAN
BATTLE OF THE PLANETS (II,179); THE LIFE AND LEGEND OF WYATT EARP (I,2681); THUNDARR THE BARBARIAN (II,2616)

DINEHART, MASON ALAN
SKYFIGHTERS (I,4078)

DINGHAM, BASIL
THE FORSYTE SAGA (I,1645)

DINGLE, CHARLES
ROAD OF LIFE (I,3798)

DINGLE, KAY
HEREAFTER (II,1138); THE MAC DAVIS SHOW (II,1575); MADHOUSE 90 (I,2807); SAMMY AND COMPANY (II,2248); A YEAR AT THE TOP (II,2845)

DINI, RICHARD
SAMMY KAYE'S MUSIC FROM MANHATTAN (I,3902)

DINMAN, DICK
THEY ONLY COME OUT AT NIGHT (II,2586)

DINOME, JERRY
WELCOME TO PARADISE (II,2762)

DINSDALE, SHIRLEY
JUDY SPLINTERS (I,2475)

DINSEN, ALICE
HAWKINS FALLS, POPULATION 6200 (I,1977)

DION, LITTLE
COPS (I,1047)

DIRKSON, DOUGLAS
MALIBU (II,1607); MODESTY BLAISE (II,1719); SHE'S IN THE ARMY NOW (II,2320)

DIRT BAND, THE
ALL COMMERCIALS—A STEVE MARTIN SPECIAL (II,37)

DISHELL, DR WALTER
WOMAN'S PAGE (II,2822)

DISHY, BOB
A.E.S. HUDSON STREET (II,17); ACE (II,5); THE COMEDY ZONE (II,559); DAMN YANKEES (I,1128); THE JONES BOYS (I,2446); PURE GOLDIE (I,3696); STORY THEATER (I,4241); THAT WAS THE WEEK THAT WAS (I,4423)

DISNEY, SHARON
THE WALT DISNEY CHRISTMAS SHOW (I,4755)

DISNEY, WALT
ONE HOUR IN WONDERLAND (I,3376); WALT DISNEY (I,4754); THE WALT DISNEY CHRISTMAS SHOW (I,4755)

DISTEL, SACHA
BRIGITTE BARDOT (I,729); THE MAGICAL MUSIC OF BURT BACHARACH (I,2822); PORTRAIT OF PETULA (I,3654)

DITOSTI, BEN
DAYS OF OUR LIVES (II,629)

DIX, JOHN
MARCO POLO (II,1626)

DIX, RICHARD
ARCHER—FUGITIVE FROM THE EMPIRE (II,103)

DIX, TOMMY
SCHOOL HOUSE (I,3937)

DIXIE PIXIES, THE
THE SPIKE JONES SHOW (I,4156)

DIXIELAND QUARTET, THE
SAMMY KAYE'S MUSIC FROM MANHATTAN (I,3902)

DIXON, BEVERLY
MISSING PIECES (II,1706)

DIXON, BOB
HOLIDAY HOTEL (I,2075)

DIXON, DAVID
THE HITCHHIKER'S GUIDE TO THE GALAXY (II,1158)

DIXON, DONNA
BATTLE OF THE NETWORK STARS (II,171); BATTLE OF THE NETWORK STARS (II,173); BOB HOPE SPECIAL: BOB HOPE'S SPRING FLING OF COMEDY AND GLAMOUR (II,315); BOSOM BUDDIES (II,355); MICKEY SPILLANE'S MARGIN FOR MURDER (II,1691); NO MAN'S LAND (II,1855); RODNEY DANGERFIELD SHOW: I CAN'T TAKE IT NO MORE (II,2198); THE SHAPE OF THINGS (II,2315); WHATEVER BECAME OF. . .? (II,2772); WOMEN WHO RATE A "10" (II,2825)

DIXON, GAIL
ANOTHER WORLD (II,97)

DIXON, HELEN
THE TONIGHT SHOW (I,4561); THE TONIGHT SHOW (I,4562)

DIXON, IVAN
HOGAN'S HEROES (I,2069)

DIXON, JERRY
OPRYLAND: NIGHT OF STARS AND FUTURE STARS (II,1921)

DIXON, MACINTYRE

DIXON, MACINTYRE
COMEDY TONIGHT (I,1027); I'M A FAN (I,2192); POPI (II,2068); WE INTERRUPT THIS SEASON (I,4780); WINDOWS, DOORS AND KEYHOLES (II,2806)

DIXON, N'GAI
A STORM IN SUMMER (I,4237)

DIXON, PAT
ALL MY CHILDREN (II,39)

DIXON, PAUL
MIDWESTERN HAYRIDE (I,3033); THE PAUL DIXON SHOW (I,3487); THE PAUL DIXON SHOW (II,1974)

DIXON, PEG
SPIDER-MAN (I,4154)

DIZON, JESSE
MARCO POLO (II,1626); OPERATION PETTICOAT (II,1917)

DOQUI, ROBERT
THE HARLEM GLOBETROTTERS (I,1953); HUNTER'S MOON (II,1207); LEGMEN (II,1458); MOBILE MEDICS (II,1716); ROGER AND HARRY (II,2200)

DOAN, DOROTHY
THE MARCH OF DIMES FASHION SHOW (I,2902); VANITY FAIR (I,4702)

DOBB, GILLIAN
MAGNUM, P.I. (II,1601)

DOBBINS, BERNIE
THE PIGEON (I,3596)

DOBIE, ALAN
MASTER OF THE GAME (II,1647); WAR AND PEACE (I,4765)

DOBKIN, LAWRENCE
I WAS A BLOODHOUND (I,2180); KRAFT MYSTERY THEATER (I,2589); THE LADY DIED AT MIDNIGHT (I,2601); LAST OF THE PRIVATE EYES (I,2628); MR. ADAMS AND EVE (I,3121); STUDIO '57 (I,4267); THE UNTOUCHABLES (I,4682)

DOBRITCH, SANDY
SUPER CIRCUS (I,4287)

DOBSON, CHARLES
MELODY, HARMONY, RHYTHM (I,2998); WHERE THE HEART IS (I,4831)

DOBSON, JAMES
BACKBONE OF AMERICA (I,328)

DOBSON, KEVIN
BATTLE OF THE NETWORK STARS (II,163); BATTLE OF THE NETWORK STARS

(II,164); BATTLE OF THE NETWORK STARS (II,165); BATTLE OF THE NETWORK STARS (II,166); BATTLE OF THE NETWORK STARS (II,175); KNOTS LANDING (II,1404); KOJAK (II,1406); MICKEY SPILLANE'S MARGIN FOR MURDER (II,1691); PEOPLE DO THE CRAZIEST THINGS (II,1997); SHANNON (II,2314); STRANDED (II,2475)

DOBSON, TAMARA
JASON OF STAR COMMAND (II,1274)

DOBTCHEFF, VERNON
AN ECHO OF THERESA (I,1399); IKE (II,1223); MASADA (II,1642)

DOCKERY, MICHAEL
IS THIS GOODBYE, CHARLIE BROWN? (I,2242); WHAT HAVE WE LEARNED, CHARLIE BROWN? (I,4810)

DODA, CAROL
THE CANDID CAMERA SPECIAL (II,422)

DODD, CAL
CIRCUS OF THE 21ST CENTURY (II,529)

DODD, DICKIE
THE MICKEY MOUSE CLUB (I,3025); THE MOUSEKETEERS REUNION (II,1742)

DODD, GILLIAN
GIDGET'S SUMMER REUNION (I,1799)

DODD, JIMMIE
THE MICKEY MOUSE CLUB (I,3025)

DODD, MOLLY
HAZEL (I,1982); KELLY'S KIDS (I,2523)

DODDS, MICHAEL
MOMENT OF TRUTH (I,3084)

DODSON, ERIC
JANE EYRE (II,1273)

DODSON, JACK
ALL'S FAIR (II,46); THE ANDY GRIFFITH SHOW (I,192); HAPPY DAYS (II,1084); IN THE BEGINNING (II,1229); MAYBERRY R.F.D. (I,2961); PHYL AND MIKHY (II,2037); SNAVELY (II,2387); WALKIN' WALTER (II,2737)

DOERR HUTCHINSON DANCERS, THE
THE JIMMY DEAN SHOW (I,2385); THE NEW CHRISTY MINSTRELS SHOW (I,3259)

DOHERTY, CHARLA
DAYS OF OUR LIVES (II,629); WON'T IT EVER BE MORNING? (I,4902)

DOHERTY, SHANNEN
HIS AND HERS (II,1155); LITTLE HOUSE: A NEW BEGINNING (II,1488); LITTLE HOUSE: BLESS ALL THE DEAR CHILDREN (II,1489); LITTLE HOUSE: LOOK BACK TO YESTERDAY (II,1490); LITTLE HOUSE: THE LAST FAREWELL (II,1491)

DOLAN, DON
ONE NIGHT BAND (II,1909)

DOLAN, ELLEN
THE GUIDING LIGHT (II,1064)

DOLAND, HERBERT
A BELL FOR ADANO (I,388)

DOLEMAN, GUY
ENIGMA (II,778); THE PRISONER (I,3670)

DOLEN, JOSEPH
WHERE THE HEART IS (I,4831)

DOLENZ, GEORGE
THE COUNT OF MONTE CRISTO (I,1058)

DOLENZ, MICKEY
33 1/3 REVOLUTIONS PER MONKEE (I,4451); DEVLIN (II,663); THE FUNKY PHANTOM (I,1709); THE MONKEES (I,3088); THE SKATEBIRDS (II,2375)

DOLIE, JOHN
JAMBOREE (I,2331)

DOLIN, KAREN NOEL
THE MISS AND THE MISSILES (I,3062)

DOLLAR, LYNN
THE $64,000 QUESTION (I,4071)

DOLMAN, NANCY
FAMILY IN BLUE (II,818); LANDON, LANDON & LANDON (II,1426); MAKING A LIVING (II,1604); SOAP (II,2392)

DOMBASLE, ARIELLE
LACE (II,1418)

DOMERGUE, FAITH
THE COUNT OF MONTE CRISTO (I,1058); REVLON MIRROR THEATER (I,3781)

DOMINGO, PLACIDO
BOB HOPE SPECIAL: BOB HOPE'S SUPER BIRTHDAY SPECIAL (II,319); BURNETT "DISCOVERS" DOMINGO (II,398); THE NATIVITY (II,1798); TEXACO STAR THEATER: OPENING NIGHT (II,2565)

DOMINO, FATS
33 1/3 REVOLUTIONS PER MONKEE (I,4451); THE CAPTAIN AND TENNILLE IN NEW ORLEANS (II,431)

DON CRAIG CHORUS, THE
DON'S MUSICAL PLAYHOUSE (I,1348); HOLIDAY HOTEL (I,2075)

DON CRICHTON DANCERS, THE
THE OSMOND FAMILY THANKSGIVING SPECIAL (II,1928); THE TIM CONWAY SHOW (II,2619)

DON JAMES ORCHESTA, THE
100 YEARS OF AMERICA'S POPULAR MUSIC (II,1904)

DON LARGE CHORUS, THE
FAYE EMERSON'S WONDERFUL TOWN (I,1549); THE WAYNE KING SHOW (I,4779)

DON LIBERTI CHORUS, THE
STARTIME (I,4212)

DON LINDLEY AND THE VELVETEERS
MUSIC IN VELVET (I,3172)

DONHOWE, GWYDA
EXECUTIVE SUITE (II,796)

DONAHUE, ELINOR
ALCOA/GOODYEAR THEATER (I,107); THE ANDY GRIFFITH SHOW (I,192); BATTLE OF THE NETWORK STARS (II,165); CONDOMINIUM (II,566); CROSSROADS (I,1105); DENNIS THE MENACE (I,1231); DICK CLARK'S GOOD OL' DAYS: FROM BOBBY SOX TO BIKINIS (II,671); DOCTORS' PRIVATE LIVES (II,692); DOCTORS' PRIVATE LIVES (II,693); FATHER KNOWS BEST (II,838); FATHER KNOWS BEST: HOME FOR CHRISTMAS (II,839); FATHER KNOWS BEST: THE FATHER KNOWS BEST REUNION (II,840); THE FLYING NUN (I,1611); FORD TELEVISION THEATER (I,1634); THE GENERAL ELECTRIC THEATER (I,1753); GIDGET GETS MARRIED (I,1796); THE GRADY NUTT SHOW (II,1048); HIGH SCHOOL, U.S.A. (II,1148); IF I LOVE YOU, AM I TRAPPED FOREVER? (II,1222); MANY HAPPY RETURNS (I,2895); MULLIGAN'S STEW (II,1763); MULLIGAN'S STEW (II,1764); THE ODD COUPLE (II,1875); ONE DAY AT A TIME (II,1900); PLEASE STAND BY (II,2057); THE ROOKIES (II,2208); SIGN-ON (II,2353)

DONAHUE, PATRICIA
FIREBALL FUN FOR ALL (I,1577); MICHAEL SHAYNE, PRIVATE DETECTIVE (I,3021); ONCE THE KILLING STARTS (I,3367); THE THIN MAN (I,4446)

DONAHUE, PHIL
THE PHIL DONAHUE SHOW (II,2032)

DONAHUE, TROY
THE BOB HOPE SHOW (I,575); HAWAIIAN EYE (I,1973); MALIBU (II,1607); THE SECRET STORM (I,3969); SURFSIDE 6 (I,4299)

DONALD MCKAYLE DANCERS, THE
...AND BEAUTIFUL (I,184); FANFARE (I,1528); THE LESLIE UGGAMS SHOW (I,2660); THE NEW BILL COSBY SHOW (I,3257)

DONALD, JAMES
MY FAIR LADY (I,3187); ST. JOAN (I,4175); VICTORIA REGINA (I,4728)

DONALD, PETER
THE AD-LIBBERS (I,29); MASQUERADE PARTY (I,2941); PRIZE PERFORMANCE (I,3673)

DONALDSON, BRUCE
SPRAGGUE (II,2430)

DONALDSON, DAVID
MACKENZIE (II,1584)

DONALDSON, NORMA
FARRELL: FOR THE PEOPLE (II,834); LIFE'S MOST EMBARRASSING MOMENTS IV (II,1473); WILLOW B: WOMEN IN PRISON (II,2804)

DONALES, LOS DIA
WILSON'S REWARD (II,2805)

DONAT, PETER
CAPTAINS AND THE KINGS (II,435); DALLAS (II,614); DELTA COUNTY, U.S.A. (II,657); FLAMINGO ROAD (II,872); GOLDEN GATE (II,1020); MOMENT OF TRUTH (I,3084); RETURN ENGAGEMENT (I,3775); THE RETURN OF CHARLIE CHAN (II,2136); RICH MAN, POOR MAN—BOOK II (II,2162)

DONATH, LUDWIG
GIVE US BARABBAS! (I,1816); GRUEN GUILD PLAYHOUSE (I,1892); LUXURY LINER (I,2796)

DONEGAN, LONNIE
THE TOM JONES SPECIAL (I,4545)

DONLEVY, BRIAN
CHRYSLER MEDALLION THEATER (I,951); DANGEROUS ASSIGNMENT (I,1138); HARD CASE (I,1947)

DONLEY, ROBERT
HEREAFTER (II,1138); THE ROCKFORD FILES (II,2196)

DONN, CARL
RIVKIN: BOUNTY HUNTER (II,2183)

DONNELL, JEFF
COUNTERPOINT (I,1059); CROWN THEATER WITH GLORIA SWANSON (I,1107); THE ETHEL BARRYMORE THEATER (I,1467); GENERAL HOSPITAL (II,964); THE GEORGE GOBEL SHOW (I,1767); GIDGET (I,1795); JULIA (I,2476); MATT HELM (II,1653); RENDEZVOUS HOTEL (II,2131); SPIDER-MAN (II,2424)

DONNELLEY, ALEX
THE YOUNG AND THE RESTLESS (II,2862)

DONNELLY, CAROL
THE STEVE ALLEN COMEDY HOUR (II,2454)

DONNELLY, ELAINE
CASANOVA (II,451); SLEEPWALKER (I,4084)

DONNELLY, JAMIE
FLATBUSH/AVENUE J (II,878)

DONNELLY, RUTH
THE IMOGENE COCA SHOW (I,2198)

DONNELLY, TIM
EMERGENCY! (II,775)

DONNER, ROBERT
MORK AND MINDY (II,1735); NAKIA (II,1784); THE OUTSIDE MAN (II,1937); SCENE OF THE CRIME (II,2271); THE WALTONS (II,2740); THE WALTONS: A DAY FOR THANKS ON WALTON'S MOUNTAIN (EPISODE 3) (II,2741); THE WALTONS: MOTHER'S DAY ON WALTON'S MOUNTAIN (EPISODE 2) (II,2742); THE WALTONS: WEDDING ON WALTON'S MOUNTAIN (EPISODE 1) (II,2743); THE YOUNG PIONEERS (II,2868); THE YOUNG PIONEERS (II,2869); THE YOUNG PIONEERS' CHRISTMAS (II,2870)

DONOHOE, CHRISTOPHER
IT'S MAGIC, CHARLIE BROWN (I,2271); LIFE IS A CIRCUS, CHARLIE BROWN (I,2683)

DONOHUE, BRENDA
TOM AND JOANN (II,2630)

DONOHUE, DOROTHY
EYE WITNESS (I,1494)

DONOHUE, JILL
SAN FRANCISCO INTERNATIONAL AIRPORT (I,3906)

DONOHUE, NANCY
CAR 54, WHERE ARE YOU? (I,842); THE DOCTORS (II,694)

DONOVAN
ANDY'S LOVE CONCERT (I,215)

DONOVAN, KING
FOUR STAR PLAYHOUSE (I,1652); FRONTIER (I,1695); THE GEORGE BURNS AND GRACIE ALLEN SHOW (I,1763); THE GOLDEN AGE OF TELEVISION (II,1018); LOVE THAT BOB (I,2779); PLEASE DON'T EAT THE DAISIES (I,3628); THE RETURN OF THE BEVERLY HILLBILLIES (II,2139)

DONOVAN, LISA
FACE THE MUSIC (II,803)

DONOVAN, MARTIN
THE WOMAN IN WHITE (II,2819)

DOODER, DENNIS
CAMP GRIZZLY (II,418)

DOODLETOWN PIPERS, THE
BING CROSBY'S CHRISTMAS SHOW (I,474); CLASS OF '67 (I,970); DANNY THOMAS GOES COUNTRY AND WESTERN (I,1146); FRANK SINATRA JR. WITH FAMILY AND FRIENDS (I,1663); THE JANE MORGAN SHOW (I,2342); OUR PLACE (I,3416); RODGERS AND HART TODAY (I,3829); THE ROGER MILLER SHOW (I,3830)

DOOHAN, JAMES
ASSIGNMENT: EARTH (I,299); STAR TREK (II,2439); STAR TREK (II,2440)

DOOHAN, JOHN
JASON OF STAR COMMAND (II,1274)

DOOLEY, JIM
AS CAESAR SEES IT (I,294); GREAT ADVENTURE (I,1869)

DOOLEY, PAUL
THE DOM DELUISE SHOW (I,1328); THE FIRM (II,860); LET'S CELEBRATE (I,2663); MOMMA THE DETECTIVE (II,1721)

DOOLING, LUCINDA
THE THORN BIRDS (II,2600); THREE'S COMPANY (II,2614)

DOONICAN, VAL
THE VAL DOONICAN SHOW (I,4692)

DORAN, ANN
ALL THE WAY HOME (II,44); BACKSTAIRS AT THE WHITE HOUSE (II,133); CRAZY TIMES (II,602); FOR THE DEFENSE (I,1624); HEY, LANDLORD (I,2039); LEAVE IT TO BEAVER (I,2648); THE LEGEND OF JESSE JAMES (I,2653); LONGSTREET (I,2749); THE MACAHANS (II,1583); NATIONAL VELVET (I,3231); SHIRLEY (II,2330); WE'VE GOT EACH OTHER (II,2757)

DORAN, BOBBY
ANOTHER WORLD (II,97)

DORAN, BRANDI
MINSKY'S FOLLIES (II,1703)

DORAN, CHRIS
CHARLIE BROWN'S ALL STARS (I,910)

DORAN, JOHNNY
CAPTAINS AND THE KINGS (II,435); MULLIGAN'S STEW (II,1764); SALTY (II,2241)

DORAN, KELLY
THE DAY THE WOMEN GOT EVEN (II,628)

DORAN, ROBERT
BUTTERFLIES (II,405)

DORAN, TAKAYO
LUCY MOVES TO NBC (II,1562)

DORELL, DON
PONY EXPRESS (I,3645)

DOREMUS, DAVID
NANNY AND THE PROFESSOR (I,3225); NANNY AND THE PROFESSOR AND THE PHANTOM OF THE CIRCUS (I,3226); THE WALTONS (II,2740)

DORIAN, ANGELA
THE BILL DANA SHOW (I,446)

DORIN, PHOEBE
ALAN KING'S FINAL WARNING (II,29); AS THE WORLD TURNS (II,110); THE MONTEFUSCOS (II,1727); THE WILD WILD WEST (I,4863)

DORMAN, ELIZABETH
EIGHT IS ENOUGH (II,762)

DORN, DOLORES
RIDDLE AT 24000 (II,2171); THE TURNING POINT OF JIM MALLOY (II,2670)

DORN, MICHAEL
CHIPS (II,511)

DORN-HEFT, DOLORES
THE STRAWBERRY BLONDE (I,4254)

DORNING, STACY
THE ADVENTURES OF BLACK BEAUTY (I,51)

DORR, LESTER
ANOTHER DAY, ANOTHER DOLLAR (I,228)

DORS, DIANA
THE BOB HOPE SHOW (I,544); THE BOB HOPE SHOW (I,550); THE DEVIL'S WEB (I,1248); DOUGLAS FAIRBANKS JR. PRESENTS THE RHEINGOLD THEATER (I,1364); THE PHIL SILVERS PONTIAC SPECIAL: KEEP IN STEP (I,3579); STUMP THE STARS (I,4273)

DORSETT, CHUCK
HEAR NO EVIL (II,1116)

DORSETT, VIVIAN
THE BRIGHTER DAY (I,728)

DORSEY, JIMMY
STAGE SHOW (I,4182)

DORSEY, TOMMY
STAGE SHOW (I,4182)

DORSEY, WRIGHT
100 CENTRE STREET (II,1901)

DOTRICE, ROY
SPACE: 1999 (II,2409)

DOTSON, RHONDA
CELEBRITY (II,463)

DOTTIE LEE SINGERS, THE
THAT GOOD OLD NASHVILLE MUSIC (II,2571)

DOUBET, STEVE
LOOK OUT WORLD (II,1516); MCNAMARA'S BAND (II,1668); MCNAMARA'S BAND (II,1669); TROUBLE IN HIGH TIMBER COUNTRY (II,2661)

DOUBLE DATERS, THE
AND HERE'S THE SHOW (I,187); THE GISELE MACKENZIE SHOW (I,1812); HERE'S THE SHOW (I,2033);

THE JIMMY DEAN SHOW (I,2384)

DOUBLEDAY, KAY
ROAD TO REALITY (I,3800)

DOUCET, CATHERINE
THE MAN WHO CAME TO DINNER (I,2880)

DOUCETTE, JOHN
BIG TOWN (I,436); BRADDOCK (I,704); LOCK UP (I,2737); NAVY LOG (I,3233); THE PARTNERS (I,3462); SABU AND THE MAGIC RING (I,3881); THE SWORD (I,4325)

DOUGHERTY, MICKEY
INCREDIBLE KIDS AND COMPANY (II,1233)

DOUGLAS, ANGELA
ESPIONAGE (I,1464)

DOUGLAS, ANN
SALUTE TO LADY LIBERTY (II,2243)

DOUGLAS, BARBARA
OF MEN OF WOMEN (I,3332)

DOUGLAS, BETH
THREE STEPS TO HEAVEN (I,4485)

DOUGLAS, BILL HENRY
BEST FRIENDS (II,213)

DOUGLAS, BUDDY
H.R. PUFNSTUF (I,2154); LIDSVILLE (I,2679); STARSTRUCK (II,2446); WHITNEY AND THE ROBOT (II,2789)

DOUGLAS, BURT
DAYS OF OUR LIVES (II,629); THE EDGE OF NIGHT (II,760); MORNING STAR (I,3104)

DOUGLAS, CATHY
JOSIE AND THE PUSSYCATS (I,2453); JOSIE AND THE PUSSYCATS IN OUTER SPACE (I,2454)

DOUGLAS, DIANA
BEN HECHT'S TALES OF THE CITY (I,395); CHRYSLER MEDALLION THEATER (I,951); THE COWBOYS (II,598); DAYS OF OUR LIVES (II,629); DEAD MAN ON THE RUN (II,631); THREE STEPS TO HEAVEN (I,4485); THE WEB (I,4784)

DOUGLAS, DONALD
WAR AND PEACE (I,4765)

DOUGLAS, DONNA
BACHELOR FATHER (I,323); THE BEVERLY HILLBILLIES (I,417); GO! (I,1830); MARINELAND CARNIVAL (I,2912); THE RETURN OF THE BEVERLY HILLBILLIES (II,2139)

DOUGLAS, JACK
BOLD JOURNEY (I,680); THE DES O'CONNOR SHOW (I,1235); GOLDEN VOYAGE (I,1840); I SEARCH FOR ADVENTURE (I,2177); KEYHOLE (I,2529); MAN TO WOMAN (I,2879)

DOUGLAS, JAMES
ANOTHER WORLD (II,97); AS THE WORLD TURNS (II,110); THE DOCTORS (II,694); OCTAVIUS AND ME (I,3327); PEYTON PLACE (I,3574)

DOUGLAS, JERRY
COVER GIRLS (II,596); THE SEEKERS (I,3974); THE STRANGER (I,4251); THE YOUNG AND THE RESTLESS (II,2862)

DOUGLAS, KATYA
ZERO ONE (I,4981)

DOUGLAS, KIRK
CELEBRITY DAREDEVILS (II,473); DR JEKYLL AND MR. HYDE (I,1368); THE GENERAL MOTORS 50TH ANNIVERSARY SHOW (I,1758); JOHNNY CASH: THE FIRST 25 YEARS (II,1336); THE LEGEND OF SILENT NIGHT (I,2655); THE MONEYCHANGERS (II,1724); SALUTE TO LADY LIBERTY (II,2243); SHOW BUSINESS SALUTE TO MILTON BERLE (I,4029); SPECIAL LONDON BRIDGE SPECIAL (I,4150); A TRIBUTE TO "MR. TELEVISION" MILTON BERLE (II,2658)

DOUGLAS, LARRY
ARCHER—FUGITIVE FROM THE EMPIRE (II,103); MAKE MINE MUSIC (I,2840); THE MOREY AMSTERDAM SHOW (I,3100); SING IT AGAIN (I,4059)

DOUGLAS, MELVYN
BLIND DATE (I,482); CELANESE THEATER (I,881); FRONTIER JUSTICE (I,1700); GHOST STORY (I,1788); HOLLYWOOD OFF BEAT (I,2086); INHERIT THE WIND (I,2213); LAMP AT MIDNIGHT (I,2608); PHILCO TELEVISION PLAYHOUSE (I,3583)

DOUGLAS, MICHAEL
CELEBRITY CHALLENGE OF THE SEXES 4 (II,468); THE STREETS OF SAN FRANCISCO (I,4260); THE STREETS OF SAN FRANCISCO (II,2478)

DOUGLAS, MIKE
CLUB 60 (I,987); DINAH IN SEARCH OF THE IDEAL MAN (I,1278); KAY KYSER'S KOLLEGE OF MUSICAL KNOWLEDGE (I,2512); MIKE AND PEARL (I,3038); THE MIKE DOUGLAS CHRISTMAS SPECIAL (I,3040); THE MIKE DOUGLAS SHOW (II,1698); THE MUSIC SHOW (I,3178); RINGO (II,2176); THE STARS AND STRIPES SHOW (II,2443)

DOUGLAS, NORMAN
AS CAESAR SEES IT (I,294)

DOUGLAS, PAUL
ADVENTURE THEATER (I,42); THE BOB HOPE SHOW (I,514); BORN YESTERDAY (I,692); JUSTICE (I,2498); THE LADY DIED AT MIDNIGHT (I,2601); THE SID CAESAR SHOW (I,4042)

DOUGLAS, PHYLLIS
SALLY AND SAM (I,3891)

DOUGLAS, ROBERT
THE BARRETTS OF WIMPOLE STREET (I,361); CENTENNIAL (II,477); FLASH GORDON—THE GREATEST ADVENTURE OF ALL (II,875); TARZAN THE APE MAN (I,4363)

DOUGLAS, RONALDA
ISIS (II,1248); THE NEW ODD COUPLE (II,1825)

DOUGLAS, SARAH
BATTLE OF THE NETWORK STARS (II,178); FALCON CREST (II,808); V: THE FINAL BATTLE (II,2714)

DOUGLAS, SUSAN
THE CAMPBELL TELEVISION SOUNDSTAGE (I,812); COSMOPOLITAN THEATER (I,1057); FAITH BALDWIN'S THEATER OF ROMANCE (I,1515); THE GUIDING LIGHT (II,1064); THE KATE SMITH EVENING HOUR (I,2507); STARLIGHT THEATER (I,4201); SUSPENSE (I,4305); TELLER OF TALES (I,4388)

DOUGLASS, AMY
MARKHAM (I,2916)

DOUGLASS, ROBYN
BATTLE OF THE NETWORK STARS (II,170); GALACTICA 1980 (II,953); GOLDEN GATE (II,1020); MICKEY SPILLANE'S MIKE HAMMER: MORE THAN MURDER (II,1693)

DOUGLASS, STEPHEN
GOLDEN CHILD (I,1839)

DOURIF, BRAD
STUDS LONIGAN (II,2484)

DOVE, ANDREW
TYCOON: THE STORY OF A WOMAN (II,2697)

DOVER, ARNOLD
MADHOUSE 90 (I,2807)

DOVER, BEN
MUSIC WORLD (II,1775)

DOW, GRAHAM
ALL THE RIVERS RUN (II,43)

DOW, TONY
HIGH SCHOOL, U.S.A. (II,1148); LEAVE IT TO BEAVER (I,2648); MR. NOVAK (I,3145); NEVER TOO YOUNG (I,3248); STILL THE BEAVER (II,2466); WEEKEND (I,4788)

DOWD, HARRISON
ETHEL AND ALBERT (I,1466); FIRST LOVE (I,1583)

DOWD, M'EL
THE BEST OF EVERYTHING (I,406)

DOWDELL, BOB
CITY BENEATH THE SEA (I,965); VOYAGE TO THE BOTTOM OF THE SEA (I,4743)

DOWDELL, ROBERT
ADAM-12 (I,31); AH!, WILDERNESS (I,91); STONEY BURKE (I,4229)

DOWELL, ANTHONY
CINDERELLA (II,526)

DOWELL, MARY
THE GEORGE ROSS SHOW (I,1775)

DOWLING, CONSTANCE
COSMOPOLITAN THEATER (I,1057)

DOWLING, DORIS
LIGHT'S OUT (I,2699); MY LIVING DOLL (I,3195); ONE STEP BEYOND (I,3388)

DOWLING, EDWARD
ANYWHERE, U.S.A. (I,240)

DOWLING, KATHRYN
CHIEFS (II,509); MARCO POLO (II,1626)

DOWN, ANGELA
LITTLE WOMEN (I,2723); WAR AND PEACE (I,4765)

DOWN, LESLEY-ANNE
THE LAST DAYS OF POMPEII (II,1436); PEEK-A-BOO: THE ONE AND ONLY PHYLLIS DIXEY (II,1989); UPSTAIRS, DOWNSTAIRS (II,2709)

DOWNES, JOHN

THE DOCTORS (II,694)

DOWNEY, L C
MARY HARTMAN, MARY HARTMAN (II,1638)

DOWNEY, MICHELLE
SIGN-ON (II,2353); YOUNG HEARTS (II,2865); YOUNG LIVES (II,2866)

DOWNEY, MORTON
THE MOHAWK SHOWROOM (I,3080); STAR OF THE FAMILY (I,4194)

DOWNIE, PENNY
PRISONER: CELL BLOCK H (II,2085)

DOWNING, DAVID
BACKSTAIRS AT THE WHITE HOUSE (II,133)

DOWNING, FRANK
THE CABOT CONNECTION (II,408)

DOWNING, LARRY
BREAKFAST PARTY (I,716)

DOWNS, ANSON
THE MAN FROM ATLANTIS (II,1615)

DOWNS, CATHY
THE JOE PALOOKA STORY (I,2401)

DOWNS, DERMOTT
FATHER ON TRIAL (I,1544)

DOWNS, FREDERIC
CAMP RUNAMUCK (I,811); DAYS OF OUR LIVES (II,629); FIRST LOVE (I,1583); LITTLE HOUSE ON THE PRAIRIE (II,1487)

DOWNS, HUGH
CONCENTRATION (I,1031); NOT FOR WOMEN ONLY (I,3310); THE TALENT SCOUTS PROGRAM (I,4341); TALENT SEARCH (I,4342); VARIETY (II,2721); YOUR LUNCHEON DATE (I,4955)

DOWNS, JANE
NICHOLAS NICKLEBY (II,1840)

DOWNS, JOHNNY
CAPTAIN BILLY'S MISSISSIPPI MUSIC HALL (I,826); GIRL ABOUT TOWN (I,1806); MANHATTAN SHOWCASE (I,2888)

DOXEE, DIANE
THE DIANE DOXEE-JIMMY BLAINE SHOW (I,1261)

DOYLE, AGNES
THE KING AND MRS. CANDLE (I,2544); THE WOMEN (I,4890)

DOYLE, CLAIRE
AS THE WORLD TURNS (II,110)

DOYLE, DAVID
ACRES AND PAINS (I,19); THE ART CARNEY SHOW (I,274); THE BLUE AND THE GRAY (II,273); BRIDGET LOVES BERNIE (I,722); THE CELEBRITY FOOTBALL CLASSIC (II,474); CHARLIE'S ANGELS (II,486); CHARLIE'S ANGELS (II,487); CIRCUS OF THE STARS (II,530); THE CONFESSIONS OF DICK VAN DYKE (II,568); THE INVISIBLE WOMAN (II,1244); JOHN RITTER: BEING OF SOUND MIND AND BODY (II,1319); JONATHAN WINTERS PRESENTS 200 YEARS OF AMERICAN HUMOR (II,1342); KISS ME, KATE (I,2569); MIRACLE ON 34TH STREET (I,3058); THE NEW DICK VAN DYKE SHOW (I,3262); OF THEE I SING (I,3334); OZZIE'S GIRLS (I,3440); THE PATTY DUKE SHOW (I,3483); THE POLICE STORY (I,3638); THE RIGHT MAN (I,3790); SHAUGHNESSEY (II,2318); TONI'S BOYS (II,2637); WILD AND WOOLEY (II,2798)

DOYLE, KATHLEEN
BACKSTAIRS AT THE WHITE HOUSE (II,133); ME AND DUCKY (II,1671)

DOYLE, LEN
MR. DISTRICT ATTORNEY (I,3136)

DOYLE, PEGGY
DOMINIC'S DREAM (II,704)

DOYLE, ROBERT
LANIGAN'S RABBI (II,1427); LANIGAN'S RABBI (II,1428)

DOYLE, WALLY
GEORGE WASHINGTON (II,978)

DOYLE-MURRAY, BRIAN
THE CHEVY CHASE SHOW (II,505); SATURDAY NIGHT LIVE (II,2261)

DOZIER, DEBBIE
GET CHRISTIE LOVE! (II,982)

DOZIER, WILLIAM
BATMAN (I,366)

DR. DEMENTO
PRIME TIMES (II,2083)

DRAGER, JASON
ALL IN THE FAMILY (II,38)

DRAGO, BILLY
THE CHISHOLMS (II,512); JOHNNY BELINDA (I,2418); THE SHERIFF AND THE ASTRONAUT (II,2326)

DRAGON, DARYL
BATTLE OF THE NETWORK STARS (II,166); THE BEACH BOYS 20TH ANNIVERSARY (II,188); THE CAPTAIN AND TENNILLE (II,429); THE CAPTAIN AND TENNILLE IN HAWAII (II,430); THE CAPTAIN AND TENNILLE IN NEW ORLEANS (II,431); THE CAPTAIN AND TENNILLE SONGBOOK (II,432); PERRY COMO'S BAHAMA HOLIDAY (II,2006); ROCK 'N' ROLL: THE FIRST 25 YEARS (II,2189); THE TONI TENNILLE SHOW (II,2636)

DRAGONI, CARLA
ALL MY CHILDREN (II,39)

DRAKE HOOKS, BEE BEE
BACKSTAIRS AT THE WHITE HOUSE (II,133); SANFORD ARMS (II,2257)

DRAKE, ALFRED
CELANESE THEATER (I,881); KISS ME, KATE (I,2568); MARCO POLO (I,2904); THE MUSIC OF GERSHWIN (I,3174); NAUGHTY MARIETTA (I,3232); TEXACO COMMAND PERFORMANCE (I,4407); THE YEOMAN OF THE GUARD (I,4926)

DRAKE, ALICE
THE LIFE OF RILEY (I,2684)

DRAKE, ALLAN
SANFORD AND SON (II,2256)

DRAKE, CHARLES
MONTGOMERY'S SUMMER STOCK (I,3096); RENDEZVOUS (I,3766); STAGE DOOR (I,4178)

DRAKE, CHRISTIAN
SHEENA, QUEEN OF THE JUNGLE (I,4003)

DRAKE, DEBBIE
THE DEBBIE DRAKE SHOW (I,1212)

DRAKE, DONNA
DEAR ALEX AND ANNIE (II,651)

DRAKE, FABIA
THE PALLISERS (II,1944)

DRAKE, GABE
SATURDAY NIGHT JAMBOREE (I,3920)

DRAKE, GABRIELLE
CRY TERROR! (I,1112); KELLY MONTEITH (II,1375); U.F.O. (I,4662)

DRAKE, GALEN
THE GALEN DRAKE SHOW (I,1726); THIS IS GALEN DRAKE (I,4456)

DRAKE, GEORGIA
POLKA-GO-ROUND (I,3640)

DRAKE, JAMES
THE ASSASSINATION RUN (II,112)

DRAKE, JONATHAN
THE WEB (I,4784)

DRAKE, KEN
NOT FOR HIRE (I,3308)

DRAKE, PAUL
HEAR NO EVIL (II,1116)

DRAKE, PAULINE
HEY MULLIGAN (I,2040)

DRAKE, RONALD
ALL MY CHILDREN (II,39)

DRAKE, TOM
CITY BENEATH THE SEA (I,965)

DRAKELY, ROY
SATINS AND SPURS (I,3917); YOUR SHOW OF SHOWS (I,4957)

DRAPER, JANE
DARK SHADOWS (I,1157)

DRAPER, JOSEF
THE GREATEST GIFT (I,1880)

DRAPER, MARGARET
THE INNER SANCTUM (I,2216); SEARCH FOR TOMORROW (II,2284)

DRAPER, PAUL
HOLLYWOOD STAR REVUE (I,2093)

DRAPER, POLLY
RYAN'S HOPE (II,2234)

DRAPER, RUSTY
SWINGING COUNTRY (I,4322); WASHINGTON SQUARE (I,4772)

DRASIN, RIC
THE INCREDIBLE HULK (II,1232)

DRAYTON, NOEL
THE BARRETTS OF WIMPOLE STREET (I,361)

DRESCHER, FRAN
FAME (II,812); I'D RATHER BE CALM (II,1216); P.O.P. (II,1939)

DRESDEL, SONIA
THE PALLISERS (II,1944); THE STRAUSS FAMILY (I,4253)

DRESSER, PHYLLIS
PEEK-A-BOO: THE ONE AND ONLY PHYLLIS DIXEY (II,1989)

DRESSLER, ERIC
FIRST LOVE (I,1583); GOLDEN WINDOWS (I,1841); SEARCH FOR TOMORROW (II,2284)

DRESSLER, LIEUX
THE BLUE KNIGHT (II,277); GENERAL HOSPITAL (II,964); RETURN OF THE WORLD'S GREATEST DETECTIVE (II,2142); A SHADOW IN THE STREETS (II,2312)

DREW, BARBARA
MIDNIGHT MYSTERY (I,3030)

DREW, DAVID
CINDERELLA (II,526)

DREW, DORIS
THE TENNESSEE ERNIE FORD SHOW (I,4398)

DREW, ELLEN
FORD TELEVISION THEATER (I,1634)

DREW, MICHAEL
MASTER OF THE GAME (II,1647)

DREW, PAULA
FRONT PAGE DETECTIVE (I,1689); THE UNEXPECTED (I,4674)

DREW, ROBERT
ROAD TO REALITY (I,3800)

DREW, WENDY
ANNIE OAKLEY (I,225); AS THE WORLD TURNS (II,110); HANDS OF MURDER (I,1929); JIMMY HUGHES, ROOKIE COP (I,2390)

DREWES, TOM
PEARL (II,1986)

DREXLER, ROSALYN
LILY (I,2705)

DREYFUSS, RICHARD
CATCH 22 (I,871); KAREN (I,2505); LILY (II,1477); ROMP (I,3838); TWO FOR THE MONEY (I,4655)

DRIER, ALEX
HART TO HART (II,1102); MURDOCK'S GANG (I,3164); THE RETURN OF THE SMOTHERS BROTHERS (I,3777); SWEET, SWEET RACHEL (I,4313); THIS WEEK IN NEMTIN (I,4465); WHAT'S IT ALL ABOUT WORLD? (I,4816)

DRIER, MOOSIE
THE BARBARA EDEN SHOW (I,346); THE BOB NEWHART SHOW (II,342); THE CONFESSIONS OF DICK VAN DYKE (II,568); DOC (II,682); EXECUTIVE SUITE (II,796); FATHER ON TRIAL (I,1544); ROWAN AND MARTIN'S LAUGH-IN (I,3856); ROYCE (II,2224); SHEEHY AND THE SUPREME MACHINE (II,2322); TOO GOOD TO BE TRUE (II,2643)

DRIFTERS, THE
IT'S WHAT'S HAPPENING, BABY! (I,2278)

DRISCHELL, RALPH
O'MALLEY (II,1872)

DRISCOLL, BOBBY
ONE HOUR IN WONDERLAND (I,3376); THE WALT DISNEY CHRISTMAS SHOW (I,4755)

DRISCOLL, JOAN
FROM HERE TO ETERNITY (II,933)

DRISCOLL, JULIE
33 1/3 REVOLUTIONS PER MONKEE (I,4451)

DRISCOLL, KATHRYN
THE WALT DISNEY CHRISTMAS SHOW (I,4755)

DRISCOLL, LAWRIE
A CHRISTMAS FOR BOOMER (II,518)

DRISCOLL, PATRICIA
THE ADVENTURES OF ROBIN HOOD (I,74)

DRIVAS, ROBERT
CRISIS (I,1102); OUR PRIVATE WORLD (I,3417) YOUNG DR. MALONE (I,4938)

DRIVER, DON
NAUGHTY MARIETTA (I,3232)

DRIVER, JOHN
THE EDGE OF NIGHT (II,760); SEARCH FOR TOMORROW (II,2284)

DRU, JOANNE
ADVENTURES OF A MODEL (I,50); ALCOA/GOODYEAR THEATER (I,107); FORD TELEVISION THEATER (I,1634); FOUR STAR PLAYHOUSE (I,1652); GUESTWARD HO! (I,1900); HENRY FONDA PRESENTS THE STAR AND THE STORY (I,2016); THE LUX VIDEO THEATER (I,2795); RHEINGOLD THEATER (I,3783)

DRUCE, JEFF
THE SAN PEDRO BUMS (II,2251)

DRUCKER, DAVID
THE LAST CONVERTIBLE (II,1435)

DRUMMOND, ALICE
PARK PLACE (II,1958); SANCTUARY OF FEAR (II,2252); WHERE THE HEART IS (I,4831)

DRURY, JAMES
ALIAS SMITH AND JONES (I,118); ALIAS SMITH AND JONES (I,119); CAMEO THEATER (I,808); FIREHOUSE (II,859); THE MEN FROM SHILOH (I,3005); SHOOT-IN AT NBC (A BOB HOPE SPECIAL) (I,4021); THE VIRGINIAN (I,4737); THE VIRGINIAN (I,4738)

DRUSS, LOU
POLKA-GO-ROUND (I,3640)

DRYDEN, MACK
ALL NIGHT RADIO (II,40)

DRYER, FRED
FORCE SEVEN (II,908); A GIRL'S LIFE (II,1000); HUNTER (II,1206)

DRYER, ROBERT
HEAR NO EVIL (II,1116); JOHNNY BELINDA (I,2418)

DRYER, SALLY
CHARLIE BROWN'S ALL STARS (I,910); IT'S THE GREAT PUMPKIN, CHARLIE BROWN (I,2276); YOU'RE IN LOVE, CHARLIE BROWN (I,4964)

DRYNAN, JEANNIE
PRISONER: CELL BLOCK H (II,2085)

DRYSDALE, DON
THE BOB HOPE SHOW (I,597)

DUBARRY, DENISE
THE BLACK SHEEP SQUADRON (II,262); TOP OF THE HILL (II,2645)

DUBIEF, JACQUELINE
MUSIC ON ICE (I,3175)

DUBOIS, JA'NET
THE BIG EASY (II,234); GOOD EVENING, CAPTAIN (II,1031); GOOD TIMES (II,1038); HELLINGER'S LAW (II,1127); THE PARKERS (II,1959); THE TOM SWIFT AND LINDA CRAIG MYSTERY HOUR (II,2633)

DUBOIS, MARTA
TALES OF THE GOLD MONKEY (II,2537)

DUBOIS, RICHARD
THE MAGIC CLOWN (I,2814)

DUBOSE, WILLIAM
CHIEFS (II,509)

DUBUS, ALMA
HAWKINS FALLS, POPULATION 6200 (I,1977)

DUCORPS, SERVANNE
LACE (II,1418)

DUMONT, MARGARET
MY FRIEND IRMA (I,3191)

DUMONT, PAUL
ALL MY CHILDREN (II,39)

DUPOIS, STARLETTA
UPTOWN SATURDAY NIGHT (II,2711)

DUPONT, PENNY
DAYTIME (II,630)

DUPREE, CAL
THE DANCE SHOW (II,616)

DUSAUTOY, CARMEN
THE CITADEL (II,539)

DUVALL, MARIA
AT YOUR SERVICE (II,118)

DUVALL, SUSAN
THOU SHALT NOT KILL (II,2604)

DUART, LOUISE
THE BAY CITY ROLLERS SHOW (II,187); THE KROFFT SUPERSHOW (II,1412); THE KROFFT SUPERSHOW II (II,1413)

DUBAC, BOB
DIFF'RENT STROKES (II,674)

DUBBINS, DON
BANACEK: DETOUR TO NOWHERE (I,339); STAR TONIGHT (I,4199)

DUBIN, GARY
BRACKEN'S WORLD (I,703); THE LETTERS (I,2670); THE PARTRIDGE FAMILY (II,1962)

DUBOV, PAUL
THE UNTOUCHABLES (I,4681)

DUCEY, CHRIS
THE HAPPENERS (I,1935)

DUCLOS, DANIELLE
AS THE WORLD TURNS (II,110)

DUCOMMUN, RICK
ROCK-N-AMERICA (II,2195); THICKE OF THE NIGHT (II,2587)

DUCROCQ, CAROLYN
FACTS OF LIFE: THE FACTS OF LIFE GOES TO PARIS (II,806)

DUDA, ANDREA
THE JACKIE GLEASON SHOW (I,2323)

DUDA, DOLORES ANN
POLKA PARTY (I,3641)

DUDIKOFF, MICHAEL
THE GRADY NUTT SHOW (II,1048); SAWYER AND FINN (II,2265); STAR OF THE FAMILY (II,2436); YOUNG

HEARTS (II,2865)

DUDLEY, ALAN
THE WOMAN IN WHITE (II,2819)

DUDLEY, DICK
VILLAGE BARN (I,4733)

DUEL, PETER
ALIAS SMITH AND JONES (I,118); ALIAS SMITH AND JONES (I,119); MARCUS WELBY, M.D. (I,2905); THE PSYCHIATRIST: GOD BLESS THE CHILDREN (I,3685)

DUELL, WILLIAM
POLICE SQUAD! (II,2061); SHERLOCK HOLMES (II,2327)

DUERING, CARL
SMILEY'S PEOPLE (II,2383)

DUFF, HOWARD
BATMAN (I,366); BOB HOPE SPECIAL: BOB HOPE'S ALL-STAR LOOK AT TV'S PRIME TIME WARS (II,303); CROSSROADS (I,1105); THE D.A.: MURDER ONE (I,1122); DANTE (I,1155); EAST OF EDEN (II,752); THE FELONY SQUAD (I,1563); FLAMINGO ROAD (II,871); FLAMINGO ROAD (II,872); FRONT ROW CENTER (I,1694); HENRY FONDA PRESENTS THE STAR AND THE STORY (I,2016); KNOTS LANDING (II,1404); LILY FOR PRESIDENT (II,1478); MR. ADAMS AND EVE (I,3121); RHEINGOLD THEATER (I,3783); THE TEENAGE IDOL (I,4375); THIS GIRL FOR HIRE (II,2596); VALENTINE MAGIC ON LOVE ISLAND (II,2716); THE WILD WOMEN OF CHASTITY GULCH (II,2801)

DUFFIN, SHAY
MOTHER, JUGGS AND SPEED (II,1741)

DUFFY, COLIN
LAND OF HOPE (II,1424); THE YEAR WITHOUT A SANTA CLAUS (II,2846)

DUFFY, JACK
THE BOBBY VINTON SHOW (II,345); HALF THE GEORGE KIRBY COMEDY HOUR (I,1920); RED SKELTON'S CHRISTMAS DINNER (II,2123)

DUFFY, JULIA
THE BLUE AND THE GRAY (II,273); THE DOCTORS (II,694); IRENE (II,1245); LOVE OF LIFE (I,2774); NEWHART (II,1835); WIZARDS AND WARRIORS (II,2813)

DUFFY, PATRICK
BATTLE OF THE NETWORK STARS (II,165); BATTLE OF THE NETWORK STARS (II,168); CELEBRITY CHALLENGE OF THE SEXES 3 (II,467); DALLAS (II,614); THE MAN FROM ATLANTIS (II,1614); THE MAN FROM ATLANTIS (II,1615)

DUFFY, SHEILA
THE OMEGA FACTOR (II,1888)

DUFOUR, VAL
ANOTHER WORLD (II,97); CURTAIN CALL THEATER (I,1115); THE DOCTOR (I,1320); THE EDGE OF NIGHT (II,760); FIRST LOVE (I,1583); SEARCH FOR TOMORROW (II,2284)

DUGAN, BOB
CRAZY TIMES (II,602)

DUGAN, DENNIS
ALICE (II,34); BATTLE OF THE NETWORK STARS (II,166); DID YOU HEAR ABOUT JOSH AND KELLY?! (II,673); EMPIRE (II,777); FATHER O FATHER (II,842); FULL HOUSE (II,937); LEADFOOT (II,1452); MAKING A LIVING (II,1604); RICH MAN, POOR MAN—BOOK I (II,2161); RICHIE BROCKELMAN, PRIVATE EYE (II,2167); RICHIE BROCKELMAN: MISSING 24 HOURS (II,2168); THE ROCKFORD FILES (II,2197)

DUGAN, JOHNNY
BREAKFAST IN HOLLYWOOD (I,715); THE JOHNNY DUGAN SHOW (I,2426); LADIES' CHOICE (I,2600)

DUGGAN, ANDREW
BACKSTAIRS AT THE WHITE HOUSE (II,133); BOURBON STREET BEAT (I,699); THE CAT AND THE CANARY (I,868); DOWN HOME (II,731); EXPOSE (I,1491); FARADAY AND COMPANY (II,833); FBI CODE 98 (I,1550); FIREHOUSE (I,1579); GREAT ADVENTURE (I,1868); HAWAII FIVE-O (I,1972); JAKE'S WAY (II,1269); THE KAISER ALUMINUM HOUR (I,2503); KRAFT SUSPENSE THEATER (I,2591); LANCER (I,2610); THE LONG DAYS OF SUMMER (II,1514); M STATION: HAWAII (II,1568); MAN ON THE MOVE (I,2878); MARKHAM (I,2916); MOMMA THE DETECTIVE (II,1721); ONCE AN EAGLE (II,1897); PINE CANYON IS BURNING (II,2040); THE RESTLESS GUN (I,3774); RICH MAN, POOR MAN—BOOK I (II,2161); ROOM FOR ONE MORE (I,3842); THE SAGA OF ANDY BURNETT (I,3883); THE STREETS OF SAN FRANCISCO (I,4260); TWELVE O'CLOCK HIGH (I,4637); A WALK IN THE NIGHT (I,4750); THE WALTONS (II,2740); THE WINDS OF WAR (II,2807); YOU'RE ONLY YOUNG TWICE (I,4971)

DUGGAN, BOB
OF THEE I SING (I,3334); THE RED SKELTON SHOW (I,3755)

DUHIG, MICHAEL
THE SWISS FAMILY ROBINSON (II,2518)

DUKAKIS, JOHN
FAMILY TIES (II,819)

DUKAS, JAMES
KENNEDY (II,1377); O'MALLEY (II,1872)

DUKE, BILL
PALMERSTOWN U.S.A. (II,1945)

DUKE, PATTY
THE ARMSTRONG CIRCLE THEATER (I,260); THE BRIGHTER DAY (I,728); CAPTAINS AND THE KINGS (II,435); CIRCLE OF FEAR (I,958); THE COMEDY ZONE (II,559); GEORGE WASHINGTON (II,978); HAVING BABIES III (II,1109); I LOVE LIBERTY (II,1215); IT TAKES TWO (II,1252); JOURNEY TO THE UNKNOWN (I,2457); KITTY FOYLE (I,2571); MEET ME IN ST. LOUIS (I,2986); ONCE UPON A CHRISTMAS TIME (I,3368); THE PATTY DUKE SHOW (I,3483); PHILLIP AND BARBARA (II,2034); THE POWER AND THE GLORY (I,3658); ROSETTI AND RYAN: MEN WHO LOVE WOMEN (II,2217); SWISS FAMILY ROBINSON (I,4324); WOMEN IN WHITE (II,2823)

DUKE, ROBIN
SATURDAY NIGHT LIVE (II,2261); SECOND CITY TELEVISION (II,2287)

DUKE OF IRON, THE
ALL ABOUT MUSIC (I,127)

DUKES, DAVID
79 PARK AVENUE (II,2310); BEACON HILL (II,193); GEORGE WASHINGTON (II,978); GO WEST YOUNG GIRL (II,1013); HANDLE WITH CARE (II,1077); THE HITCHHIKER (II,1157); THE MANY LOVES OF ARTHUR (II,1625); VALLEY FORGE (I,4698); THE WINDS OF WAR (II,2807)

DULCE, ELISSA
M STATION: HAWAII (II,1568)

DULLAGHAN, JOHN
B.J. AND THE BEAR (II,126)

DULLEA, KEIR
CHANNING (I,900); GIVE US BARABBAS! (I,1816); THE LEGEND OF THE GOLDEN GUN (II,1455); THE STARLOST (I,4203)

DULO, JANE
DUFFY (II,739); GET SMART (II,983); GIMME A BREAK (II,995); HEY, JEANNIE! (I,2038); LEAVE IT TO BEAVER (I,2648); MCHALE'S NAVY (I,2969); THE NOONDAY SHOW (II,1859); THE NORLISS TAPES (I,3305); THE ODD COUPLE (II,1875); REVENGE OF THE GRAY GANG (II,2146); SHA NA NA (II,2311); TWO GIRLS NAMED SMITH (I,4656)

DUMAS, DAVID
HOME ROOM (II,1172)

DUMAS, HELEN
LOVE OF LIFE (I,2774)

DUMBRILLE, DOUGLAS
THE GRAND JURY (I,1865); THE LIFE OF RILEY (I,2686); THE LIFE OF VERNON HATHAWAY (I,2687); PETTICOAT JUNCTION (I,3571)

DUMKE, RALPH
CAPTAIN BILLY'S MISSISSIPPI MUSIC HALL (I,826); MOVIELAND QUIZ (I,3116); ROSALINDA (I,3846); THE SKY'S THE LIMIT (I,4081); SUDDEN SILENCE (I,4275)

DUMONT, SKY
THE WINDS OF WAR (II,2807)

DUNAWAY, FAYE
DISAPPEARANCE OF AIMEE (I,1290); ELLIS ISLAND (II,768); THE SENSATIONAL, SHOCKING, WONDERFUL, WACKY 70S (II,2302)

DUNBAR, OLIVE
BIG JOHN, LITTLE JOHN (II,240); MORNING STAR (I,3104); MY WORLD . . . AND WELCOME TO IT (I,3208); SECOND CHANCE (I,3958); TEMPERATURES RISING (I,4390)

DUNCAN, ALASTAIR
AROUND THE WORLD IN 80 DAYS (I,263)

DUNCAN, ANDREA
COMEDY NEWS (I,1020)

DUNCAN, ANDREW
AS CAESAR SEES IT (I,294); COMEDY NEWS II (I,1021); THE DAY THE WOMEN GOT EVEN (II,628); HOT HERO SANDWICH (II,1186) THE NUT HOUSE (I,3320)

DUNCAN, ANGUS
THE FEATHER AND FATHER GANG: NEVER CON A KILLER (II,846); THE MARY TYLER MOORE SHOW (II,1640); THE PHOENIX (II,2035); WAIKIKI (II,2734)

DUNCAN, ARCHIE
THE ADVENTURES OF ROBIN HOOD (I,74); THE LAWRENCE WELK SHOW (I,2643); SHERLOCK HOLMES (I,4010)

DUNCAN, ARTHUR
THE JERRY COLONNA SHOW (I,2359); THE LAWRENCE WELK SHOW (I,2644); MEMORIES WITH LAWRENCE WELK (II,1681)

DUNCAN, CRAIG
MY FRIEND FLICKA (I,3190)

DUNCAN, DUNCAN
ROBERT YOUNG WITH THE YOUNG (I,3814)

DUNCAN, JAYSON
RETURN TO EDEN (II,2143)

DUNCAN, MARY DAVIS
HIGHWAY HONEYS (II,1151)

DUNCAN, PAMELA
BLONDIE (I,486); SCREEN DIRECTOR'S PLAYHOUSE (I,3946)

DUNCAN, PETER
A FAMILY AFFAIR (II,814)

DUNCAN, SANDY
100 YEARS OF GOLDEN HITS (II,1905); BING CROSBY AND HIS FRIENDS (II,251); BING!. . . A 50TH ANNIVERSARY GALA (II,254); THE BURT BACHARACH SPECIAL (I,776); CELEBRATION: THE AMERICAN SPIRIT (II,461); CHRISTMAS IN DISNEYLAND (II,519); DOUG HENNING'S WORLD OF MAGIC II (II,726); FUNNY FACE (I,1711); THE FUNNY WORLD OF FRED & BUNNI (II,943); HAPPY BIRTHDAY, AMERICA (II,1083); KEEP U.S. BEAUTIFUL (I,2519); MERV GRIFFIN'S ST. PATRICK'S DAY SPECIAL (I,3018); OPRYLAND USA—1975 (II,1920); PARADE OF STARS (II,1954); PERRY COMO'S MUSIC FROM HOLLYWOOD (II,2019); PINOCCHIO (II,2044); THE RICH LITTLE SHOW (II,2154); ROOTS (II,2211); THE SANDY DUNCAN SHOW (I,3910); THE SANDY DUNCAN SHOW (II,2253); SANDY IN DISNEYLAND (II,2254); SEARCH FOR TOMORROW (II,2284); THE SIX-MILLION-DOLLAR MAN (II,2372)

DUNFEE, KATHERINE
THE PAPER CHASE (II,1949)

DUNHAM, JOANNA
POSSESSION (I,3655)

DUNHILLS, THE
TEXACO STAR THEATER (I,4411)

DUNIGAN, TIM
THE A-TEAM (II,119); MISSING PIECES (II,1706); MR. SMITH (II,1755); WIZARDS AND WARRIORS (II,2813)

DUNLAP, AL
RETURN OF THE WORLD'S GREATEST DETECTIVE (II,2142)

DUNLAP, DAWN
ME AND DUCKY (II,1671)

DUNLAP, JACK
LITTLE HOUSE: BLESS ALL THE DEAR CHILDREN (II,1489)

DUNLAP, PAMELA
THREE'S COMPANY (II,2614)

DUNLAP, ROBERT
ADVICE TO THE LOVELORN (II,16); FATHER ON TRIAL (I,1544)

DUNLOP, LESLEY
PENMARRIC (II,1993)

DUNLOP, VIC
THE HALF-HOUR COMEDY HOUR (II,1074); HARPER VALLEY PTA (II,1095); TV FUNNIES (II,2672)

DUNN, BILL
FACE TO FACE (I,1510)

DUNN, BOB
QUICK ON THE DRAW (I,3711); SPIN THE PICTURE (I,4159)

DUNN, CONRAD
LONDON AND DAVIS IN NEW YORK (II,1512)

DUNN, EDDIE
FACE TO FACE (I,1510); LEAVE IT TO THE GIRLS (I,2650); SPIN THE PICTURE (I,4159)

DUNN, ELAINE
THE BING CROSBY SHOW (I,463); THE BOB HOPE SHOW (I,566); THE BOB HOPE SHOW (I,620); BROADWAY OPEN HOUSE (I,736); THE BUDDY GRECO SHOW (I,754)

DUNN, GEORGE
CAMP RUNAMUCK (I,811); FULL SPEED ANYWHERE (I,1704)

DUNN, J MALCOLM
WHEN THE NIGHTINGALE SANG IN BERKELEY SQUARE (I,4827)

DUNN, JAMES
BEN CASEY (I,394); THE CHEVROLET TELE-THEATER (I,926); CURTAIN CALL THEATER (I,1115); ELFEGO BACA (I,1427); HOLLYWOOD OPENING NIGHT (I,2087); IT'S A GREAT LIFE (I,2256); MATINEE THEATER (I,2947); MR. BROADWAY (I,3130); PLAYHOUSE 90 (I,3623); PULITZER PRIZE PLAYHOUSE (I,3692); SCHLITZ PLAYHOUSE OF STARS (I,3936); STUDIO ONE (I,4268); SWINGIN' TOGETHER (I,4320); THE WEB (I,4784)

DUNN, JOHN
LAND OF HOPE (II,1424)

DUNN, KATHY
DAYS OF OUR LIVES (II,629)

DUNN, LARRY
EARTH, WIND & FIRE IN CONCERT (II,750)

DUNN, LIAM
ACE (II,5); CAPTAIN NICE (I,835); DIANA (I,1259); FULL HOUSE (II,936); GENESIS II (I,1760); THE GOOD GUYS (I,1847); THE HEALERS (II,1115); IF I LOVE YOU, AM I TRAPPED FOREVER? (II,1222); KAREN (II,1363); THE MARY TYLER MOORE SHOW (II,1640); THE QUEEN AND I (I,3704); THREE TIMES DALEY (II,2610); TWIGS (II,2683)

DUNN, MICHAEL
THE WILD WILD WEST (I,4863)

DUNN, RALPH
CAPTAIN BILLY'S MISSISSIPPI MUSIC HALL (I,826); KITTY FOYLE (I,2571); NORBY (I,3304)

DUNN, WINIFRED
BURT REYNOLDS' LATE SHOW (I,777)

DUNNAM, STEPHANIE
EMERALD POINT, N.A.S. (II,773); MISTRAL'S DAUGHTER (II,1708)

DUNNCOCK, MILDRED
THE INNER SANCTUM (I,2216)

DUNNE, DOMINIQUE
BREAKING AWAY (II,371); UNIT 4 (II,2707); VALENTINE MAGIC ON LOVE ISLAND (II,2716)

DUNNE, EDITH
LAURA (I,2636)

DUNNE, IRENE
FORD TELEVISION THEATER (I,1634); THE GENERAL ELECTRIC THEATER (I,1753); THE JUNE ALLYSON SHOW (I,2488); PLAYHOUSE OF STARS (I,3625); SCHLITZ PLAYHOUSE OF STARS (I,3936)

DUNNE, LENNIE
THE WOMEN (I,4890)

DUNNE, MURPHY
CASS MALLOY (II,454); THE LORENZO AND HENRIETTA MUSIC SHOW (II,1519); MR. SUCCESS (II,1756); PEEPING TIMES (II,1990); THE PRIMARY ENGLISH CLASS (II,2081); SUNDAY FUNNIES (II,2493)

DUNNE, RICHARD
ANOTHER WORLD (II,97)

DUNNE, STEVE
AMANDA FALLON (I,152); THE BROTHERS BRANNAGAN (I,747); DOUBLE EXPOSURE (I,1359); PROFESSIONAL FATHER (I,3677); YOU'RE ON YOUR OWN (I,4968); YOU'RE ONLY YOUNG TWICE (I,4971)

DUNNINGER, JOSEPH
THE AMAZING DUNNINGER (I,158); DUNNINGER AND WINCHELL (I,1384); THE DUNNINGER SHOW (I,1385)

DUNNOCK, MILDRED
BEN HECHT'S TALES OF THE CITY (I,395); CELANESE THEATER (I,881); DEATH OF A SALESMAN (I,1208); GULF PLAYHOUSE (I,1904); THE POWER AND THE GLORY (I,3658); THE WEB (I,4784)

DUNPHY, JERRY
THE FRENCH ATLANTIC AFFAIR (II,924); JERRY VISITS (I,2373); ROOSTER (II,2210)

DUNSON, SONJA
THE D.A. (I,1123)

DUNSON-FRANKS, SONDRA
CHIEFS (II,509)

DUNUS, RICHARD
THREE EYES (II,2605)

DUPEREY, ANNY
CIRCUS OF THE STARS (II,530)

DURAN, JOHNNY
BEHIND THE SCREEN (II,200)

DURANT, DON
JOHNNY RINGO (I,2434); THE LONER (I,2744); MACREEDY'S WOMAN (I,2804); THE RAY ANTHONY SHOW (I,3728)

DURANTE GIRLS, THE
TEXACO STAR THEATER (I,4412)

DURANTE, JIMMY
ALICE THROUGH THE LOOKING GLASS (I,122); THE ALL-STAR REVUE (I,138); THE BOB HOPE SHOW (I,509); THE BOB HOPE SHOW (I,531); THE BOB HOPE SHOW (I,538); THE BOB HOPE SHOW (I,558); THE BOB HOPE SHOW (I,580); THE BOB HOPE SHOW (I,615); THE BOB HOPE SHOW (I,634); THE COLGATE COMEDY HOUR (I,997); THE DANNY THOMAS SPECIAL (I,1151); ENTERTAINMENT —1955 (I,1450); AN EVENING WITH JIMMY DURANTE (I,1475); THE FRANK SINATRA SHOW (I,1666); FROSTY THE SNOWMAN (II,934); GIVE MY REGARDS TO BROADWAY (I,1815); HOWDY (I,2150); JIMMY DURANTE MEETS THE LIVELY ARTS (I,2386); JIMMY DURANTE PRESENTS THE LENNON SISTERS HOUR (I,2387); THE JIMMY DURANTE SHOW (I,2388); THE JIMMY DURANTE SHOW (I,2389); THE LENNON SISTERS SHOW (I,2656); MONSANTO NIGHT PRESENTS BURL IVES (I,3090); MURDER AT NBC (A BOB HOPE SPECIAL) (I,3159); ROMP (I,3838); A SALUTE TO TELEVISION'S 25TH ANNIVERSARY (I,3893); SPOTLIGHT (I,4162); STANDARD OIL ANNIVERSARY SHOW (I,4189); TEXACO STAR THEATER (I,4412); THE WONDERFUL WORLD OF BURLESQUE I (I,4895)

DURHAM, CHRISTOPHER
CAPITOL (II,426)

DURHAM, MUFFY
BUCKSHOT (II,385); THE HEE HAW HONEYS (II,1124); MURDER CAN HURT YOU! (II,1768)

DURKIN, BETSY
DARK SHADOWS (I,1157)

DURNING, CHARLES
ANOTHER WORLD (II,97); THE BEST LITTLE SPECIAL IN TEXAS (II,215); CAPTAINS AND THE KINGS (II,435); THE COP AND THE KID (II,572); MR. ROBERTS (II,1754); P.O.P. (II,1939); THE RIVALRY (I,3796); RX FOR THE DEFENSE (I,3878); SIDE BY SIDE (II,2346); STUDS LONIGAN (II,2484); SWITCH (II,2520)

DUROCHER, LEO
THE BOB HOPE SHOW (I,548); THE JONATHAN WINTERS SHOW (I,2442); TEXACO COMMAND PERFORMANCE (I,4408)

DUROCK, DICK
BATTLESTAR GALACTICA (II,181); THE BIG EASY (II,234); RETURN OF THE MAN FROM U.N.C.L.E.: THE 15 YEARS LATER AFFAIR (II,2140)

DUROV, YURI
THE BOB HOPE SHOW (I,562)

DURR, JESSICA ANN
KENNEDY (II,1377)

DURRELL, MICHAEL
ALICE (II,33); CHIEFS (II,509); THE GUIDING LIGHT (II,1064); HOW TO SURVIVE THE 70S AND MAYBE EVEN BUMP INTO HAPPINESS (II,1199); I'M A BIG GIRL NOW (II,1218); NOBODY'S PERFECT (II,1858); REMINGTON STEELE (II,2130); SHANNON (II,2314); THE SUNSHINE BOYS (II,2496); V (II,2713); V: THE FINAL BATTLE (II,2714)

DURREN, JOHN
RISKO (II,2180); WHEELS (II,2777)

DURSO, KIM
THE CHADWICK FAMILY (II,478)

DURSTEN, DAVE
CHEZ PAREE REVUE (I,932)

DURSTON, GIGI
THE SONNY KENDIS SHOW (I,4127)

DURY, SUSAN
MIRROR OF DECEPTION (I,3060)

DURYEA, DAN
THE AFFAIRS OF CHINA SMITH (I,87); THE BARBARA STANWYCK THEATER (I,350); CONFIDENTIALLY YOURS (I,1035); THE DAVID NIVEN THEATER (I,1178); THE DESILU PLAYHOUSE (I,1237); HENRY FONDA PRESENTS THE STAR AND THE STORY (I,2016); JUSTICE OF THE PEACE (I,2500); PEYTON PLACE (I,3574); PURSUIT (I,3700); RHEINGOLD THEATER (I,3783); RIVERBOAT (I,3797); SHIRLEY TEMPLE'S STORYBOOK (I,4017)

DUSAY, MARJ
BATTLES: THE MURDER THAT WOULDN'T DIE (II,180); BLONDIE (I,487); BOBBY PARKER AND COMPANY (II,344); BRET MAVERICK (II,374); CAPITOL (II,426); THE FACTS OF LIFE (II,805); IN THE DEAD OF NIGHT (I,2202); MOST WANTED (II,1738); SORORITY '62 (II,2405); SQUARE PEGS (II,2431); STOP SUSAN WILLIAMS (II,2473); WHEELS (II,2777)

DUSENBERRY, ANN
CAPTAINS AND THE KINGS (II,435); CLOSE TIES (II,545); THE FAMILY TREE (II,820); LITTLE WOMEN (II,1497); LITTLE WOMEN (II,1498); THE POSSESSED (II,2071); THE SECRET WAR OF JACKIE'S GIRLS (II,2294); STONESTREET: WHO KILLED THE CENTERFOLD MODEL? (II,2472)

DUSENBERRY, J R
KENNEDY (II,1377)

DUSOME, GAIL
THE HOUSE WITHOUT A CHRISTMAS TREE (I,2143)

DUSS, GOBIND
THE FAR PAVILIONS (II,832)

DUSSAULT, NANCY
ALAN KING LOOKS BACK IN ANGER—A REVIEW OF 1972 (I,99); BURT AND THE GIRLS (I,770); FAMILY BUSINESS (II,815); HOW TO HANDLE A WOMAN (I,2146); THE LILY TOMLIN SHOW (I,2706); THE MANY FACES OF COMEDY (I,2894); MUSICAL COMEDY TONIGHT (II,1777); THE NANCY DUSSAULT SHOW (I,3222); THE NEW DICK VAN DYKE SHOW (I,3262); A PUNT, A PASS, AND A PRAYER (I,3694); THE SHAPE OF THINGS (II,2315); TOO CLOSE FOR COMFORT (II,2642); THE WAY THEY WERE (II,2748)

DUSTINE, JEAN LOUISE
THE MEL TORME SHOW (I,2995)

DUTRA, JOHN
FIRST LOVE (I,1583); GOLDEN WINDOWS (I,1841)

DUTTINE, JOHN
DAY OF THE TRIFFIDS (II,627); A WOMAN OF SUBSTANCE (II,2820)

DUVALL, ROBERT
FAME IS THE NAME OF THE GAME (I,1518); FLESH AND BLOOD (I,1595); GUILTY OR NOT GUILTY (I,1903); IKE (II,1223); THE OUTER LIMITS (I,3426)

DUVALL, SHELLEY
FAERIE TALE THEATER (II,807)

DUVALL, SUSAN
ANGIE (II,80); HOT W.A.C.S. (II,1191); ME AND DUCKY (II,1671); STARTING FRESH (II,2447)

DVORAK, ANN
GRUEN GUILD PLAYHOUSE (I,1892)

DWIGHT-SMITH, MICHAEL
LUCAS TANNER (II,1556)

DWORKIN, PATTI
HARPER VALLEY PTA (II,1095)

DWYER, BILL
YOUNG DR. MALONE (I,4938)

DWYER, JENNY REBECCA
RYAN'S HOPE (II,2234)

DWYER, VIRGINIA
ANOTHER WORLD (II,97); ROAD OF LIFE (I,3798); THE SECRET STORM (I,3969); WONDERFUL JOHN ACTION (I,4892); YOUNG DR. MALONE (I,4938)

DYALL, VALENTINE
THE HITCHHIKER'S GUIDE TO THE GALAXY (II,1158)

DYE, CAMERON
COUNTERATTACK: CRIME IN AMERICA (II,581)

DYER, EDDY C
THE FURST FAMILY OF WASHINGTON (I,1718); INSPECTOR PEREZ (II,1237)

DYER, JEANNE
MAYOR OF HOLLYWOOD (I,2962)

DYKERS, REAR ADMIRAL THOMAS
THE SILENT SERVICE (I,4049)

DYLAN, ELLIE
KIDS ARE PEOPLE TOO (II,1390); THE LOVE REPORT (II,1542)

DYNELEY, PETER
COLOR HIM DEAD (I,1007); COLOR HIM DEAD (II,554); THUNDERBIRDS (I,4497)

DYSART, RICHARD
THE COURT-MARTIAL OF GENERAL GEORGE ARMSTRONG CUSTER (I,1081); THE GEMINI MAN (II,961); GUESS WHO'S COMING TO DINNER? (II,1063); NORMA RAE (II,1860); SANDBURG'S LINCOLN (I,3907); THE SEAL (II,2282)

DYSART, TOMMY
PRISONER: CELL BLOCK H (II,2085)

DYSERT, ALAN
ALL MY CHILDREN (II,39)

DYSON, NOEL
FATHER DEAR FATHER (II,837); THE TWO RONNIES (II,2694)

DZUNDZA, GEORGE
OPEN ALL NIGHT (II,1914)

EAGER, CLAY
MIDWESTERN HAYRIDE (I,3033)

EAGLE, ELEANOR
THE GOOK FAMILY (I,1858)

EAGLE, JACK
THE NEW LORENZO MUSIC SHOW (II,1821)

EAGLES, THE
GOOD VIBRATIONS FROM CENTRAL PARK (I,1855)

EAMES, KATHRYN
TIME FOR ELIZABETH (I,4506)

EARL BROWN DANCERS, THE
THE ANDY WILLIAMS SHOW (I,210); THE LIVELY ONES (I,2729)

EARL BROWN SINGERS, THE
THE DANNY KAYE SHOW (I,1143); THE SONNY AND CHER COMEDY HOUR (II,2400)

EARL TWINS, THE
ASTAIRE TIME (I,305)

EARL OF MT CHARLES
THE MANIONS OF AMERICA (II,1623)

EARL, JOHN
SANFORD ARMS (II,2257)

EARLE, MERIE
ALICE (II,33); GREEN ACRES (I,1884); THE JERRY REED WHEN YOU'RE HOT, YOU'RE HOT HOUR (I,2372); THE WALTONS (II,2740)

EARLE, ROBERT
THE G.E. COLLEGE BOWL (I,1745)

EARLEY, CANDICE
ALL MY CHILDREN (II,39)

EASIER, FRED
BEN BLUE'S BROTHERS (I,393)

EAST, JEFF
STRANGER IN OUR HOUSE (II,2477); WHEN, JENNY? WHEN (II,2783)

EAST, ROBERT
DAVE ALLEN AT LARGE (II,621)

EASTERBROOK, LESLIE
FIRST & TEN (II,861); HIS AND HERS (II,1155); LAVERNE AND SHIRLEY (II,1446)

EASTHAM, RICHARD
ALONG THE BARBARY COAST (I,144); BRIGHT PROMISE (I,727); CONDOMINIUM (II,566); GRANDPA GOES TO WASHINGTON (II,1050); HEIDI (I,2004); SALVAGE (II,2244); TOMBSTONE TERRITORY (I,4550); WONDER WOMAN (SERIES 1) (II,2828)

EASTON, JANE
MIGHTY O (I,3036)

EASTON, JOYCE
MALLORY: CIRCUMSTANTIAL EVIDENCE (II,1608)

EASTON, MYLES
THE GUIDING LIGHT (II,1064)

EASTON, RICHARD
THE ADMIRABLE CRICHTON (I,36)

EASTON, ROBERT
ANNIE OAKLEY (I,225); CENTENNIAL (II,477); STINGRAY (I,4228); SUDDEN SILENCE (I,4275); A VERY MISSING PERSON (I,4715)

EASTON, SHEENA
THE GLEN CAMPBELL MUSIC SHOW (II,1007); KENNY ROGERS IN CONCERT (II,1381); SHEENA EASTON, ACT 1 (II,2323); SHEENA EASTON—LIVE AT THE PALACE (II,2324)

EASTWOOD, CLINT
NAVY LOG (I,3233); RAWHIDE (I,3727); TOM SNYDER'S CELEBRITY SPOTLIGHT (II,2632); THE WEST POINT STORY (I,4796)

EASTWOOD, JAYNE
HIGH HOPES (II,1144); THE WORLD OF DARKNESS (II,2836)

EATON, DOROTHY
ON BORROWED TIME (I,3354)

EATON, SHIRLEY
THE BOB HOPE SHOW (I,535); THE BOB HOPE SHOW (I,614)

EBARA, MISASHI
SHOGUN (II,2339)

EBBLING, PETER
THE PLANT FAMILY (II,2050)

EBELING, GEORGE
THE CHOCOLATE SOLDIER (I,944)

EBEN, AL
HAWAII FIVE-O (II,1110)

EBERG, VICTOR
THE BANANA SPLITS ADVENTURE HOUR (I,340)

EBERHARDT, NORMA
CAPTAIN GALLANT OF THE FOREIGN LEGION (I,831)

EBERLE, RAY
PETER MARSHALL SALUTES THE BIG BANDS (II,2028)

EBERLY, BOB
TV'S TOP TUNES (I,4635)

EBERSOLE, CHRISTINE
RYAN'S HOPE (II,2234); SATURDAY NIGHT LIVE (II,2261)

EBERT, JOYCE
CLOSE TIES (II,545)

EBERT, ROGER
AT THE MOVIES (II,117); SNEAK PREVIEWS (II,2388)

EBSEN, BONNIE
THE KALLIKAKS (II,1361)

EBSEN, BUDDY
THE ANDY WILLIAMS SHOW (I,207); BARNABY JONES (II,153); THE BASTARD/KENT FAMILY CHRONICLES (II,159); THE BEVERLY HILLBILLIES (I,417); THE BING CROSBY SHOW (I,465); CBS: ON THE AIR (II,460); THE CHEVROLET TELE-THEATER (I,926); CORKY AND WHITE SHADOW (I,1049); THE DAUGHTERS OF JOSHUA CABE (I,1169); DAVY CROCKETT (I,1181); THE DESILU PLAYHOUSE (I,1237); THE GENERAL ELECTRIC THEATER (I,1753); THE GRADUATION DRESS (I,1864); GRUEN GUILD PLAYHOUSE (I,1892); THE KILLIN' COUSIN (II,1394); MARINELAND CARNIVAL (I,2912); MATT HOUSTON (II,1654); NORTHWEST PASSAGE (I,3306); PLAYHOUSE 90 (I,3623); THE RETURN OF THE BEVERLY HILLBILLIES (II,2139); A SPECIAL HOUR WITH DINAH SHORE (I,4149); STARS OVER HOLLYWOOD (I,4211)

ECCLES, AMIE
QUINCY, M. E. (II,2102)

ECCLES, EDDY
THE BEVERLY HILLBILLIES (I,417)

ECCLES, ROBIN
ANDERSON AND COMPANY (I,188)

ECCLES, TEDDY
ANDERSON AND COMPANY (I,188); THE BANANA SPLITS ADVENTURE HOUR (I,340); DOCTOR SHRINKER (II,686); THE HERCULOIDS (I,2023); IF I LOVE YOU, AM I TRAPPED FOREVER? (II,1222); THE LITTLE DRUMMER BOY (II,1485); MY BOY GOOGIE (I,3185); SAN FRANCISCO INTERNATIONAL AIRPORT (I,3906)

ECHEVARRIA, ROCKY
FROM HERE TO ETERNITY (II,933)

ECHOS, THE
COKE TIME WITH EDDIE FISHER (I,996)

ECKLES, KATHLEEN
THE DOCTORS (II,694)

ECKSTEIN, BILLY
THE BILL COSBY SPECIAL, OR? (I,443); DUKE ELLINGTON. . . WE LOVE YOU MADLY (I,1380)

EDA YOUNG, BARBARA
ANOTHER WORLD (II,97)

EDDER, TONY
FORCE SEVEN (II,908)

EDDIE KENDRIX SINGERS, THE
DIANA ROSS IN CONCERT (II,668)

EDDIE LITTLE SKY
THE INDIAN (I,2210); ROYCE (II,2224)

EDDINGTON, PAUL
THE ADVENTURES OF ROBIN HOOD (I,74); GOOD NEIGHBORS (II,1034); SPECIAL BRANCH (II,2414); TO SIR, WITH LOVE (II,2624); YES MINISTER (II,2848)

EDDY, NELSON
THE BOB HOPE SHOW (I,519); THE BOB HOPE SHOW (I,529); THE DESERT SONG (I,1236); THE LUX VIDEO THEATER (I,2795)

EDDY, WALLACE
CAMP WILDERNESS (II,419)

EDE, GEORGE
THE EDGE OF NIGHT (II,760)

EDELMAN, HERB
9 TO 5 (II,1852); BANYON (I,345); BIG JOHN, LITTLE JOHN (II,240); THE BILL COSBY SPECIAL, OR? (I,443); THE BOYS (II,360); CROSSFIRE (II,606); FRANKIE AND ANNETTE: THE SECOND TIME AROUND (II,917); THE GOOD GUYS (I,1847); HAVE I GOT A CHRISTMAS FOR YOU (II,1105); HONEST AL'S A-OK USED CAR AND TRAILER RENTAL TIGERS (II,1174); KOSTA AND HIS FAMILY (I,2581); LADIES' MAN (II,1423); OCCASIONAL WIFE (I,3325); OF THEE I SING (I,3334); ONCE UPON A DEAD MAN (I,3369); THE REASON NOBODY HARDLY EVER SEEN A FAT OUTLAW IN THE OLD WEST IS AS FOLLOWS: (I,3744); SHOOTING STARS (II,2341); ST. ELSEWHERE (II,2432); STRIKE FORCE (II,2480); WELCOME BACK, KOTTER (II,2761)

EDELMAN, PETER
THINGS WE DID LAST SUMMER (II,2589)

EDEN, BARBARA
THE BARBARA EDEN SHOW (I,346); THE BIG SHOW (II,243); THE BOB HOPE SHOW (I,621); THE BOB HOPE SHOW (I,641); THE BOB HOPE SHOW (I,651); BOB HOPE SPECIAL: BOB HOPE'S ALL-STAR COMEDY LOOK AT THE FALL SEASON: IT'S STILL FREE AND WORTH IT! (II,299); BOB HOPE SPECIAL: BOB HOPE'S ALL-STAR COMEDY SPECIAL FROM AUSTRALIA (II,300); BOB HOPE SPECIAL: BOB HOPE'S ALL-STAR LOOK AT TV'S PRIME TIME WARS (II,303); CHANGING SCENE (I,899); CONDOMINIUM (II,566); CROSSROADS (I,1105); THE ENGELBERT HUMPERDINCK SPECIAL (I,1446); GENE KELLY'S WONDERFUL WORLD OF GIRLS (I,1752); HARPER VALLEY (II,1094); HARPER VALLEY PTA (II,1095); THE HELLCATS (I,2009); HOW TO MARRY A MILLIONAIRE (I,2147); HOWDY (I,2150); I DREAM OF JEANNIE (I,2167); IT'S ONLY HUMAN (II,1257); KISMET (I,2565); MEN WHO RATE A "10" (II,1684); ROMP (I,3838); STONESTREET: WHO KILLED THE CENTERFOLD MODEL? (II,2472); TELLY. . . WHO LOVES YA, BABY? (II,2558); THE WONDERFUL WORLD OF GIRLS (I,4898)

EDEN, CHANA
THE RELUCTANT SPY (I,3763)

EDGCOMB, JAMES
CELEBRITY (II,463)

EDITH BARSTON DANCERS, THE
FRANKIE CARLE TIME (I,1674)

EDMISTON, WALKER
THE BUFFALO BILLY SHOW (I,756); THE BUGALOOS (I,757); THE GREEN FELT JUNGLE (I,1885); HOME COOKIN' (II,1170); LAND OF THE LOST (II,1425); LIDSVILLE (I,2679); NANNY AND THE PROFESSOR AND THE PHANTOM OF THE CIRCUS (I,3226); THE SCOOBY-DOO AND SCRAPPY-DOO SHOW (II,2275); SCRUPLES (II,2280)

EDMOND JR, LADA
CLASS OF '67 (I,970) THE FLIM-FLAM MAN (I,1599); HULLABALOO (I,2156); ON THE FLIP SIDE (I,3360)

EDMONDS, DON
BROADSIDE (I,732); HOORAY FOR LOVE (I,2116)

EDMONDS, LOUIS
ALL MY CHILDREN (II,39); DARK SHADOWS (I,1157); IN THE DEAD OF NIGHT (I,2202); VICTORIA REGINA (I,4728)

EDMONDS, NOEL
FOUL UPS, BLEEPS AND BLUNDERS (II,911)

EDMUNDS, DAVE
THE TOM JONES SPECIAL (I,4542)

EDMUNDS, WILLIAM
ACTOR'S HOTEL (I,27)

EDNEY, BENJAMIN
BRIEF ENCOUNTER (I,724)

EDNEY, FLORENCE
HAY FEVER (I,1979)

EDWARD, MARION
THE PAUL HOGAN SHOW (II,1975)

EDWARDS, ADAM
RAGE OF ANGELS (II,2107)

EDWARDS, ALLYN
A COUPLE OF JOES (I,1078); MR. CITIZEN (I,3133); ONE MINUTE PLEASE (I,3384)

EDWARDS, ANTHONY
HIGH SCHOOL, U.S.A. (II,1148); IT TAKES TWO (II,1252)

EDWARDS, ARTHUR
FAMILY GENIUS (I,1521)

EDWARDS, BARBARA
TV'S BLOOPERS AND PRACTICAL JOKES (II,2675)

EDWARDS, BILL
GIDGET'S SUMMER REUNION (I,1799); HAWAII FIVE-O (II,1110)

EDWARDS, BURT
WISHMAN (II,2811)

EDWARDS, CLIFF
54TH STREET REVUE (I,1573); THE CLIFF EDWARDS SHOW (I,975); THE MICKEY MOUSE CLUB (I,3025); THE NEW MICKEY MOUSE CLUB (II,1823); UKULELE IKE (I,4664)

EDWARDS, DICK
JAMBOREE (I,2331); WINDY CITY JAMBOREE (I,4877)

EDWARDS, DOUGLAS
THE ARMSTRONG CIRCLE THEATER (I,260); CELEBRITY TIME (I,887); THE EYES HAVE IT (I,1495); MASQUERADE PARTY (I,2941)

EDWARDS, EDWARD
DRIBBLE (II,736); THE DUKES OF HAZZARD (II,742); THE JAMES BOYS (II,1272); SUZANNE PLESHETTE IS MAGGIE BRIGGS (II,2509)

EDWARDS, ELAINE
THE SEVEN LITTLE FOYS (I,3986); SIMON LASH (I,4052)

EDWARDS, ELIZABETH
SCRUPLES (II,2281)

EDWARDS, GAIL
IT'S A LIVING (II,1254); MAKING A LIVING (II,1604)

EDWARDS, GEOFF
THE BOBBY DARIN AMUSEMENT COMPANY (I,672); CHAIN REACTION (II,479); THE HIS AND HER OF IT (I,2062); HOLLYWOOD'S TALKING (I,2097); JACKPOT (II,1266); THE LOVE EXPERTS (II,1537); THE NEW TREASURE HUNT (II,1831); THE OUTLAWS (II,1936); PLAY THE PERCENTAGES (II,2053); SHOOT FOR THE STARS (II,2340); TREASURE HUNT (II,2657)

EDWARDS, GERALD
COWBOY IN AFRICA (I,1085); FAT ALBERT AND THE COSBY KIDS (II,835)

EDWARDS, GLYNN
A PLACE TO DIE (I,3612)

EDWARDS, JACK
ONE MAN'S FAMILY (I,3383)

EDWARDS, JANE
THE NEW ADVENTURES OF GILLIGAN (II,1813)

EDWARDS, JEFF
THE NEW TREASURE HUNT (I,3275)

EDWARDS, JENNIFER B
LITTLE SHOTS (II,1495)

EDWARDS, JOAN
CAVALCADE OF STARS (I,877); THE JOAN EDWARDS SHOW (I,2396)

EDWARDS, LESLIE
CINDERELLA (II,526)

EDWARDS, MARY ANN
YOUNG IN HEART (I,4939)

EDWARDS, NORMAN
ANNIE GET YOUR GUN (I,223)

EDWARDS, PADDI
THE MANY LOVES OF ARTHUR (II,1625); TOO CLOSE FOR COMFORT (II,2642)

EDWARDS, RALPH
THE BOB HOPE SHOW (I,539); ENTERTAINMENT —1955 (I,1450); THIS IS YOUR LIFE (I,4461); THOSE WONDERFUL TV GAME SHOWS (II,2603); TRUTH OR CONSEQUENCES (I,4619)

EDWARDS, RANDALL
AS THE WORLD TURNS (II,110); RYAN'S HOPE (II,2234)

EDWARDS, RICK
HIGH PERFORMANCE (II,1145)

EDWARDS, RONNIE CLAIRE
BOONE (II,351); FUTURE COP (II,946); THE WALTONS (II,2740); THE WALTONS: A DAY FOR THANKS ON WALTON'S MOUNTAIN (EPISODE 3) (II,2741); THE WALTONS: MOTHER'S DAY ON WALTON'S MOUNTAIN (EPISODE 2) (II,2742); THE WALTONS: WEDDING ON WALTON'S MOUNTAIN (EPISODE 1) (II,2743)

EDWARDS, SAM
FULL CIRCLE (I,1702); LITTLE HOUSE: A NEW BEGINNING (II,1488)

EDWARDS, SARAH
THE WORLD OF PEOPLE (II,2839)

EDWARDS, STEPHANIE
COMEDY NEWS II (I,1021); EVERYDAY (II,793); THE GIRL WITH SOMETHING EXTRA (II,999); THE HUDSON BROTHERS SHOW (II,1202); LEAVE IT TO THE WOMEN (II,1454)

EDWARDS, STEVE
ENTERTAINMENT TONIGHT (II,780); ON STAGE AMERICA (II,1893); PERSONAL AND CONFIDENTIAL (II,2026)

EDWARDS, SUSAN
THE SHOW MUST GO ON (II,2344)

EDWARDS, SUZANNE
THE HIS AND HER OF IT (I,2062)

EDWARDS, THEODORE
RAGE OF ANGELS (II,2107)

EDWARDS, VINCE
BEN CASEY (I,394); CIRCUS OF THE STARS (II,533); COVER GIRLS (II,596); DIAL HOT LINE (I,1254); FIREHOUSE (I,1579); MATT LINCOLN (I,2949); THE RHINEMANN EXCHANGE (II,2150); SAGA OF SONORA (I,3884)

EDWARDS, WEBLEY
HAWAII CALLS (I,1970)

EFROM, MARSHALL
THE DICK CAVETT SHOW (II,669); THE GREAT AMERICAN DREAM MACHINE (I,1871); KIDD VIDEO (II,1388); THE KWICKY KOALA SHOW (II,1417); THE SMURFS (II,2385); TROLLKINS (II,2660)

EGAN, EDDIE
CRAZY TIMES (II,602); EISCHIED (II,763); JOE FORRESTER (II,1303); MICKEY SPILLANE'S MIKE HAMMER: MURDER ME, MURDER YOU (II,1694)

EGAN, JENNY
MR. PEEPERS (I,3147)

EGAN, MIKE
IT'S ROCK AND ROLL (II,1258)

EGAN, PETER
A WOMAN OF SUBSTANCE (II,2820)

EGAN, RICHARD
CAPITOL (II,426); EMPIRE (I,1439); HENRY FONDA PRESENTS THE STAR AND THE STORY (I,2016); HOLLYWOOD OPENING NIGHT (I,2087); REDIGO (I,3758)

EGERTON, NANCY
MISTER JERICO (I,3070)

EGGAR, SAMANTHA
ANNA AND THE KING (I,221); FANTASY ISLAND (II,829); THE KILLER WHO WOULDN'T DIE (II,1393); LOVE STORY (I,2778)

EGGERT, NICOLE
DENNIS THE MENACE: MAYDAY FOR MOTHER (II,660); SOMEDAY YOU'LL FIND HER, CHARLIE BROWN (I,4113); T.J. HOOKER (II,2524)

EGI, TOSHIO
SPACE GIANTS (I,4142)

EGO, SANDRA
CADE'S COUNTY (I,787)

EHLERS, BETH
THINGS ARE LOOKING UP (II,2588)

EHRHARDT, BARBARA JEAN
AS THE WORLD TURNS (II,110)

EICHHORN, LISA
FEEL THE HEAT (II,848)

EIKEN, SHERRY
PINK LADY (II,2041)

EIKHARD, SHIRLEY
ANNE MURRAY'S LADIES' NIGHT (II,89)

EILBACHER, CYNTHIA
ANDERSON AND COMPANY (I,188); THE FESS PARKER SHOW (II,850); MY MOTHER THE CAR (I,3197); ROBERT YOUNG AND THE FAMILY (I,3813); THE SENATOR (I,3976); SHIRLEY (II,2330); THE YOUNG AND THE RESTLESS (II,2862)

EILBACHER, LISA
BATTLE OF THE NETWORK STARS (II,176); THE HARDY BOYS MYSTERIES (II,1090); RYAN'S FOUR (II,2233); SPIDER-MAN (II,2424); THE TEXAS WHEELERS (II,2568); WHEELS (II,2777); THE WINDS OF WAR (II,2807)

EILBER, JANET
THIS IS KATE BENNETT (II,2597); TWO MARRIAGES (II,2691)

EILEEN BARTON DANCERS, THE
BROADWAY OPEN HOUSE (I,736)

EILER, BARBARA
STUDIO '57 (I,4267); SWAMP FOX (I,4311); WHERE WERE YOU? (I,4834)

EILER, VIRGINIA
BEN CASEY (I,394); BIRDMAN AND THE GALAXY TRIO (I,478)

EIMEN, JOHNNY
MCKEEVER AND THE COLONEL (I,2970); TOO MANY SERGEANTS (I,4573)

EINSTEIN, BOB
JOEY AND DAD (II,1306); JUST FRIENDS (I,2495); PAT PAULSEN'S HALF A COMEDY HOUR (I,3473); THE RETURN OF THE SMOTHERS BROTHERS (I,3777); THE SMOTHERS BROTHERS COMEDY HOUR (I,4095); THE SMOTHERS BROTHERS SHOW (II,2384); THREE FOR TAHITI (I,4476); VAN DYKE AND COMPANY (II,2719)

EISENMANN, AL
ANOTHER DAY (II,94); GODZILLA (II,1014)

EISENMANN, CHARLES P
THE LITTLEST HOBO (II,1500)

EISENMANN, IKE
BANJO HACKETT: ROAMIN' FREE (II,141); THE BASTARD/KENT FAMILY CHRONICLES (II,159); BLACK BEAUTY (II,261); THE FANTASTIC JOURNEY (II,826); STARTING FRESH (II,2447)

EISENMANN, ROBIN
I DO, I DON'T (II,1212)

EISLEY, ANTHONY
BRIGHT PROMISE (I,727); A DATE WITH LIFE (I,1163); THE FBI (I,1551); HAWAIIAN EYE (I,1973); HOLIDAY (I,2074); PETE KELLY'S BLUES (I,3559)

EKBERG, ANITA
THE BOB HOPE SHOW (I,536); THE BOB HOPE SHOW (I,561); THE BOB HOPE SHOW (I,611); CASABLANCA (I,860); S*H*E (II,2235)

EKELUND, JANE
THIS IS YOUR MUSIC (I,4462)

EKLAND, BRITT
BRITT EKLAND'S JUKE BOX (II,378); CAROL FOR ANOTHER CHRISTMAS (I,852); CIRCUS OF THE STARS (II,535); JACQUELINE SUSANN'S VALLEY OF THE DOLLS, 1981 (II,1268); THE SIX MILLION DOLLAR MAN (I,4067); WOMEN WHO RATE A "10" (II,2825)

ELAM, JACK
BLACK BEAUTY (II,261); BRONCO (I,745); CAT BALLOU (I,869); CHEYENNE (I,931); THE DAKOTAS (I,1126); THE DAUGHTERS OF JOSHUA CABE (I,1169); EIGHT IS ENOUGH (II,762); THE GIRL, THE GOLD WATCH AND DYNAMITE (II,1001); HOW THE WEST WAS WON (II,1196); LACY AND THE MISSISSIPPI QUEEN (II,1419); LEGENDS OF THE WEST: TRUTH AND TALL TALES (II,1457); A MAN CALLED RAGAN (I,2858); THE NEW DAUGHTERS OF JOSHUA CABE (I,3261); PHYLLIS (II,2038); PLIMPTON! SHOWDOWN AT RIO LOBO (I,3630); SAWYER AND FINN (II,2265); SCROOGE'S ROCK 'N' ROLL CHRISTMAS (II,2279); SIDEKICKS (II,2348); SIX GUNS FOR DONEGAN (I,4065); SKYWARD CHRISTMAS (II,2378); THE SLOWEST GUN IN THE WEST (I,4088); STRUCK BY LIGHTNING (II,2482); SUGARFOOT (I,4277); TEMPLE HOUSTON (I,4392); THE TEXAS WHEELERS (II,2568)

ELAN, JOAN
THE BARRETTS OF WIMPOLE STREET (I,361)

ELCAR, DANA
BARETTA (II,152); THE BLACK SHEEP SQUADRON (II,262); CATCH 22 (I,871); CENTENNIAL (II,477); THE D.A.: MURDER ONE (I,1122); DARK SHADOWS (I,1157); DEADLOCK (I,1190); THE GEMINI MAN (II,961); HAWKINS ON MURDER

(I,1978); HERNANDEZ, HOUSTON P.D. (I,2034); INSPECTOR PEREZ (II,1237); LOVE NEST (II,1541); OF MICE AND MEN (I,3333); OUR TOWN (I,3420); THE PATRIOTS (I,3477); SAMURAI (II,2249); SAN FRANCISCO INTERNATIONAL AIRPORT (I,3906); SARGE: THE BADGE OR THE CROSS (I,3915); THE SOUND OF ANGER (I,4133); A TIME TO LIVE (I,4510); WENDY HOOPER—U. S. ARMY (II,2764); THE WHOLE WORLD IS WATCHING (I,4851)

ELDER, ANN
THE LAS VEGAS SHOW (I,2619); ROWAN AND MARTIN'S LAUGH-IN (I,3856); THE SMOTHERS BROTHERS SHOW (I,4096); YOU CAN'T DO THAT ON TELEVISION (I,4933)

ELDER, JUDYANN
MATT HOUSTON (II,1654)

ELDRIDGE, JOHN
MEET CORLISS ARCHER (I,2980)

ELENE, SUESIE
THE PRIMARY ENGLISH CLASS (II,2081)

ELENE, SUSIE
JUDGE DEE IN THE MONASTERY MURDERS (I,2465)

ELERICK, JOHN
COLORADO C.I. (II,555); HAZARD'S PEOPLE (II,1112); MCNAUGHTON'S DAUGHTER (II,1670)

ELES, SANDOR
THE ASSASSINATION RUN (II,112); TYCOON: THE STORY OF A WOMAN (II,2697)

ELG, TAINA
ONE LIFE TO LIVE (II,1907)

ELGAR, AVRIL
THE CITADEL (II,539); GEORGE AND MILDRED (II,965)

ELHARDT, JAMES
FIT FOR A KING (II,866)

ELHARDT, KAYE
LOVE THAT JILL (I,2780); OCTAVIUS AND ME (I,3327)

ELIAS, ALIX
GRADY (II,1047); KAREN (II,1363)

ELIAS, HECTOR
MADAME'S PLACE (II,1587); MODESTY BLAISE (II,1719); MORE WILD WILD WEST (II,1733)

ELIAS, JEANNIE
ST. ELSEWHERE (II,2432)

ELIAS, KELLY
DYNASTY (II,746)

ELIAS, LOUIE
ROLL OUT! (I,3833)

ELIC, JOSEPH
GREAT DAY (II,1053)

ELIC, JOSIP
THE HALLOWEEN THAT ALMOST WASN'T (II,1075)

ELIOT, MARGE
RX FOR THE DEFENSE (I,3878)

ELIZONDO, HECTOR
A.K.A. PABLO (II,28); CASABLANCA (II,450); THE DAIN CURSE (II,613); FEEL THE HEAT (II,848); FREEBIE AND THE BEAN (II,922); POPI (II,2068); POPI (II,2069)

ELKINS, MARY
ENTERTAINMENT TONIGHT (II,780)

ELKINS, RICHARD
THE HARLEM GLOBETROTTERS (I,1953)

ELKINS, ROBERT
THE TROUBLE WITH MOTHER (II,2663)

ELLEN, CLIFF
ALL THE RIVERS RUN (II,43)

ELLENSTEIN, ROBERT
A BELL FOR ADANO (I,388); MR. BROADWAY (I,3130)

ELLERBEE, BOBBY
BIG JOHN (II,239); THE FURTHER ADVENTURES OF WALLY BROWN (II,944); IN THE BEGINNING (II,1229); THE PLASTICMAN COMEDY/ADVENTURE SHOW (II,2051)

ELLERBEE, HARRY
CAPTAIN BRASSBOUND'S CONVERSION (I,828); THE GOOD FAIRY (I,1846); THE WILD WILD WEST (I,4863)

ELLIG, BELLE
THE WINDS OF WAR (II,2807)

ELLINGTON, DUKE
ALL-STAR SWING FESTIVAL (II,53); THE BARBARA MCNAIR AND DUKE ELLINGTON SPECIAL (I,347); THE BIG BAND AND ALL THAT JAZZ (I,422); ON STAGE WITH BARBARA MCNAIR (I,3358); STANDARD OIL ANNIVERSARY SHOW (I,4189); TIMEX ALL-STAR JAZZ SHOW I (I,4515); TIMEX ALL-STAR JAZZ SHOW IV (I,4518)

ELLIOT, CASS
ANDY WILLIAMS KALEIDOSCOPE COMPANY (I,201); DON'T CALL ME MAMA ANYMORE (I,1350); GET IT TOGETHER (I,1780); JACK LEMMON—GET HAPPY (I,2312); SAGA OF SONORA (I,3884)

ELLIOT, EVAN
HARBOURMASTER (I,1946); NORBY (I,3304)

ELLIOT, JANE
ELECTRA WOMAN AND DYNA GIRL (II,764); THE FABULOUS DR. FABLE (I,1500); GENERAL HOSPITAL (II,964); THE GUIDING LIGHT (II,1064); ROSETTI AND RYAN (II,2216); ROSETTI AND RYAN: MEN WHO LOVE WOMEN (II,2217); A TIME FOR US (I,4507)

ELLIOT, LAURA
FRONTIER (I,1695)

ELLIOT, PATSY
THE GARRY MOORE SHOW (I,1740)

ELLIOT, SUSAN
13 QUEENS BOULEVARD (II,2591); BUCKSHOT (II,385); COMPLETELY OFF THE WALL (II,560); THAT'S TV (II,2581); THE UGILY FAMILY (II,2699)

ELLIOTT, BOB
BOB & RAY & JANE, LARAINE & GILDA (II,279); THE BOB AND RAY SHOW (I,496); CLUB EMBASSY (I,982); COMEDY NEWS (I,1020); FROM CLEVELAND (II,931); HAPPY DAYS (I,1942); IT ONLY HURTS WHEN YOU LAUGH (II,1251); THE MIKE DOUGLAS CHRISTMAS SPECIAL (I,3040); THE NAME'S THE SAME (I,3220); ONE MAN SHOW (I,3381)

ELLIOTT, DAVID
THE DOCTORS (II,694); JOE AND VALERIE (II,1298); JOE AND VALERIE (II,1299)

ELLIOTT, DENHOLM
THE FEAR IS SPREADING (I,1553); THE HOLY TERROR (I,2098); THE LARK (I,2616); MADAME SIN (I,2805); MARCO POLO (II,1626); THE MOON AND SIXPENCE (I,3098); THE STRANGE CASE OF DR. JEKYLL AND MR. HYDE (I,4244); VANITY FAIR (I,4700)

ELLIOTT, DICK
THE ANDY GRIFFITH SHOW (I,192); DICK TRACY (I,1271); THE PETRIFIED FOREST (I,3570)

ELLIOTT, LEONARD
A CONNECTICUT YANKEE (I,1038); SATINS AND SPURS (I,3917)

ELLIOTT, LORNA
JANE EYRE (I,2337)

ELLIOTT, PATRICIA
THE ADAMS CHRONICLES (II,8); EMPIRE (II,777); SUMMER SOLSTICE (II,2492)

ELLIOTT, ROSS
THE BLUE ANGELS (I,490); DOCTORS' PRIVATE LIVES (II,693); FBI CODE 98 (I,1550); HARD CASE (I,1947); LEAVE IT TO BEAVER (I,2648); THE LIFE AND LEGEND OF WYATT EARP (I,2681); NAVY LOG (I,3233); TAMMY (I,4357); THE VIRGINIAN (I,4738)

ELLIOTT, SAM
EVEL KNIEVEL (II,785); MISSION: IMPOSSIBLE (I,3067); ONCE AN EAGLE (II,1897); TRAVIS MCGEE (II,2656); WILD TIMES (II,2799); THE YELLOW ROSE (II,2847)

ELLIOTT, SHAWN
THE NEW VOICE (II,1833)

ELLIOTT, STEPHEN
BEACON HILL (II,193); BENDER (II,204); BENSON (II,208); CAPTAIN VIDEO AND HIS VIDEO RANGERS (I,838); THE COURT-MARTIAL OF GENERAL GEORGE ARMSTRONG CUSTER (I,1081); EXECUTIVE SUITE (II,796); FALCON CREST (II,808); HANDS OF MURDER (I,1929); HARDCASE (II,1088); TAXI (II,2546); A WORLD APART (I,4910)

ELLIOTT, TIM
AGAINST THE WIND (II,24); PRISONER: CELL BLOCK H (II,2085)

ELLIOTT, WIN
ON YOUR ACCOUNT (I,3364); WIN WITH A WINNER (I,4870)

ELLIS, ADRIAN
MRS. G. GOES TO COLLEGE (I,3155)

ELLIS, ADRIENNE
330 INDEPENDENCE S.W. (I,4486); MORNING STAR (I,3104)

ELLIS, ANTONI
U.F.O. (I,4662)

ELLIS, BOBBY

THE ALDRICH FAMILY (I,111); THE GEORGE BURNS AND GRACIE ALLEN SHOW (I,1763); THE JACKIE GLEASON SPECIAL (I,2324); LOVE THAT BOB (I,2779); MEET CORLISS ARCHER (I,2979); MEET CORLISS ARCHER (I,2980); SKIP TAYLOR (I,4075)

ELLIS, CAROLINE
THE BUGALOOS (I,757)

ELLIS, CHRIS
CHIEFS (II,509)

ELLIS, DONALD
THE ADAMS CHRONICLES (II,8)

ELLIS, HERB
THE CLAUDETTE COLBERT SHOW (I,971); DRAGNET (I,1369); HENNESSEY (I,2013); MARGIE (I,2909); PETER GUNN (I,3562)

ELLIS, IAN
THE MAGNIFICENT SIX AND A HALF (I,2828)

ELLIS, JUNE
PENMARRIC (II,1993); TYCOON: THE STORY OF A WOMAN (II,2697)

ELLIS, KATHLEEN
MR. NOVAK (I,3145)

ELLIS, LARRY
SAMMY KAYE'S MUSIC FROM MANHATTAN (I,3902)

ELLIS, LOUISE
NIGHTSIDE (II,1851)

ELLIS, MERLE
DINAH'S PLACE (II,678)

ELLIS, MONIE
GIDGET GETS MARRIED (I,1796)

ELLIS, PATTY MARIE
THE AMOS AND ANDY SHOW (I,179)

ELLIS, PEGGY ANNE
COUNTRY STYLE (I,1072); COUNTRY STYLE (I,1073)

ELLIS, RALPH
OUR FIVE DAUGHTERS (I,3413)

ELLIS, REX
GEORGE WASHINGTON (II,978)

ELLIS, RITA
THE GALEN DRAKE SHOW (I,1726); THIS IS GALEN DRAKE (I,4456)

ELLISON, CASEY
PUNKY BREWSTER (II,2092)

ELLSWORTH AND FAIRCHILD
SONG AND DANCE TIME (I,4121)

ELMAN, DAVE
DAVE ELMAN'S CURIOSITY SHOP (I,1172)

ELMAN, MISHA
THE SATURDAY NIGHT REVUE (I,3921)

ELMORE, STEVE
HOW TO SURVIVE A MARRIAGE (II,1198)

ELPHICK, MICHAEL
MASADA (II,1642); PEEK-A-BOO: THE ONE AND ONLY PHYLLIS DIXEY (II,1989); ROGER DOESN'T LIVE HERE ANYMORE (II,2201)

ELRARDT, HORST
THE LONELY WIZARD (I,2743)

ELSOM, ISABEL
DATE WITH THE ANGELS (I,1164); FAITH BALDWIN'S THEATER OF ROMANCE (I,1515); THE TURN OF THE SCREW (I,4624)

ELSOM, JONATHAN
ONLY A SCREAM AWAY (I,3393)

ELSON, ANDREA
WHIZ KIDS (II,2790)

ELSON, STEVE
DAVID BOWIE—SERIOUS MOONLIGHT (II,622)

ELSTON, ROBERT
AS THE WORLD TURNS (II,110); GEORGE WASHINGTON (II,978)

ELY, RICK
THE WHOLE WORLD IS WATCHING (I,4851); THE YOUNG REBELS (I,4944)

ELY, RON
THE AQUANAUTS (I,250); FACE THE MUSIC (II,803); MALIBU RUN (I,2848); THE SEAL (II,2282); TARZAN (I,4361)

EMBERG, BELLA
THE BENNY HILL SHOW (II,207); THE UNEXPURGATED BENNY HILL SHOW (II,2705)

EMBREY, JOAN
THOSE AMAZING ANIMALS (II,2601)

EMERSON, DOROTHY
FUNNY PAPERS (I,1713)

EMERSON, DOUGLAS
HERBIE, THE LOVE BUG (II,1131)

EMERSON, FAYE
15 WITH FAYE (I,1570); THE BOB HOPE SHOW (I,509); THE CHEVROLET TELE-THEATER (I,926); FAYE AND SKITCH (I,1548); FAYE EMERSON'S WONDERFUL TOWN (I,1549); OF ALL THINGS (I,3330); PARIS CAVALCADE OF FASHION (I,3456); QUICK AS A FLASH (I,3709); SILVER THEATER (I,4051); THE UNITED STATES STEEL HOUR (I,4677)

EMERSON, HOPE
THE DENNIS O'KEEFE SHOW (I,1230); DOC CORKLE (I,1304); I MARRIED JOAN (I,2174); KOBB'S CORNER (I,2577); PETER GUNN (I,3562)

EMERSON, KARRIE
DELTA HOUSE (II,658)

EMERY, BOB
THE SMALL FRY CLUB (I,4089)

EMERY, DOROTHY
ABE LINCOLN IN ILLINOIS (I,10)

EMERY, JOHN
ADVENTURES OF A MODEL (I,50); ALWAYS APRIL (I,147)

EMERY, KATHERINE
ART AND MRS. BOTTLE (I,271)

EMERY, RALPH
POP! GOES THE COUNTRY (II,2067)

EMHARDT, ROBERT
ANOTHER WORLD (II,97); INSTITUTE FOR REVENGE (II,1239); KILROY (I,2541); LOCK, STOCK, AND BARREL (I,2736); THE MAGNIFICENT YANKEE (I,2829); NIGHT GAMES (II,1845); SANDBURG'S LINCOLN (I,3907); TENNESSEE ERNIE FORD MEETS KING ARTHUR (I,4397)

EMMET, MICHAEL
MAGGIE (I,2810)

EMMICH, CLIFF
HAZARD'S PEOPLE (II,1112); INSIDE O.U.T. (II,1235); MALLORY: CIRCUMSTANTIAL EVIDENCE (II,1608); SLITHER (II,2379)

EMMONS, BLACK
THE COUNTRY PLACE (I,1071)

ENBERG, DICK
BAFFLE (I,331); THE PERFECT MATCH (I,3524); THREE FOR THE MONEY (II,2606)

ENDERS, RUTH
MR. I. MAGINATION (I,3140)

ENDERSBY, RALPH
FOREST RANGERS (I,1642)

ENDO, HARRY
CODE NAME: DIAMOND HEAD (II,549); HAWAII FIVE-O (II,1110)

ENGEL, GEORGIA
THE BETTY WHITE SHOW (II,224); THE DAY THE WOMEN GOT EVEN (II,628); DEAN MARTIN'S CALIFORNIA CHRISTMAS (II,642); DEAN MARTIN'S CELEBRITY ROAST (II,643); DEAN MARTIN'S RED HOT SCANDALS OF 1926 (II,648); DEAN MARTIN'S RED HOT SCANDALS PART 2 (II,649); DEAN'S PLACE (II,650); GOODTIME GIRLS (II,1042); JENNIFER SLEPT HERE (II,1278); THE LOVE BOAT III (II,1534); THE MARY TYLER MOORE SHOW (II,1640); MORK AND MINDY (II,1735)

ENGLAND, HAL
FATHER KNOWS BEST: HOME FOR CHRISTMAS (II,839); FATHER KNOWS BEST: THE FATHER KNOWS BEST REUNION (II,840)

ENGLANDER, MELBA
MOSES THE LAWGIVER (II,1737)

ENGLE, HOLLAND
CARTUNE-O (I,859)

ENGLE, PAUL
ACTION (I,23); HAZEL (I,1982); NOAH'S ARK (I,3302)

ENGLE, ROY
DATE WITH THE ANGELS (I,1164); MY FAVORITE MARTIAN (I,3189); THE WILD WILD WEST (I,4863)

ENGLISH, ARTHUR
KISS KISS, KILL KILL (I,2566)

ENGLISH, CAMERON
FAME (II,812)

ENGLISH, CHRIS
EIGHT IS ENOUGH (II,762)

ENGLISH, JON
AGAINST THE WIND (II,24)

ENGLISH, PHILIP
THE DOCTORS (II,694)

ENGLISH, RALNA
THE LAWRENCE WELK SHOW (I,2643); MEMORIES WITH LAWRENCE WELK (II,1681)

ENGLUND, PATRICIA
LOVERS AND FRIENDS (II,1549) THAT WAS THE WEEK THAT WAS (I,4423)

ENGLUND, ROBERT
THE MYSTERIOUS TWO (II,1783); V (II,2713); V: THE FINAL BATTLE (II,2714); V: THE SERIES (II,2715)

ENGLUND, SUE
BRACKEN'S WORLD (I,703); LOST IN SPACE (I,2758)

ENGSTROM, JEAN
THE TRAVELS OF JAIMIE MCPHEETERS (I,4596)

ENHARDT, ROBERT
THE CHOPPED LIVER BROTHERS (II,514)

ENOKI, HYOEI
SHOGUN (II,2339)

ENRIQUEZ, RENE
HIGH RISK (II,1146); HILL STREET BLUES (II,1154)

ENSERRO, MICHAEL
ALICE IN WONDERLAND (I,120)

ENSIGN, MICHAEL
THE GREATEST AMERICAN HERO (II,1060); VELVET (II,2725)

ENTEN, BONNIE
COMEDY TONIGHT (I,1027)

EPP, GARY
BEST FRIENDS (II,213); FLATFOOTS (II,879); MARRIED: THE FIRST YEAR (II,1634)

EPSTEIN, PIERRE
PINOCCHIO (I,3603)

ER-KONG, ZHAO
MARCO POLO (II,1626)

ERALI, ELISA
THE SCARLET LETTER (II,2270); SUMMER SOLSTICE (II,2492)

ERANGEY, PAUL
PENMARRIC (II,1993)

ERBY, MORRIS
DRAGNET (I,1371); PETER GUNN (I,3562)

ERCOLI, LINDA
BE MY VALENTINE, CHARLIE BROWN (I,369); IT'S THE EASTER BEAGLE, CHARLIE BROWN (I,2275); YOU'RE NOT ELECTED, CHARLIE BROWN (I,4967)

ERDMAN, RICHARD
EXPERT WITNESS (I,1488); FROM HERE TO ETERNITY (II,933); JOHNNY NIGHTHAWK (I,2432); OUT OF THE BLUE (I,3422); SAINTS AND SINNERS (I,3887); THE TAB HUNTER SHOW (I,4328); THE THIRD COMMANDMENT (I,4449); WHERE'S RAYMOND? (I,4837)

ERIC, ELSPETH
ROAD OF LIFE (I,3798)

ERIC, JAMES
CHIEFS (II,509)

ERICKSON, ALAN
YOU ASKED FOR IT (II,2855)

ERICKSON, CHRISTOPHER
MANDY'S GRANDMOTHER (II,1618)

ERICKSON, LEIF
THE CLAUDETTE COLBERT SHOW (I,971); THE DAUGHTERS OF JOSHUA CABE (I,1169); FORCE FIVE (II,907); GUILTY OR NOT GUILTY (I,1903); HARD CASE (I,1947); THE HIGH CHAPARRAL (I,2051); HUNTER'S MOON (II,1207); THE LORETTA YOUNG SHOW (I,2755); THE NEW HEALERS (I,3266); SAVAGE: IN THE ORIENT (II,2264); THE SHARPSHOOTER (I,4001); STORY THEATER (I,4240); WILD TIMES (II,2799); YOUR SHOW TIME (I,4958)

ERICKSON, PHIL
THE DICK VAN DYKE SPECIAL (I,1276)

ERICSON, DEVON
THE AWAKENING LAND (II,122); THE BUSTERS (II,401); THE CHISHOLMS (II,513); FAMILY (II,813); THE RUNAWAY BARGE (II,2230); STRANDED (II,2475); STUDS LONIGAN (II,2484); TESTIMONY OF TWO MEN (II,2564); THREE'S COMPANY (II,2614); YOUNG DAN'L BOONE (II,2863)

ERICSON, JOHN
CAVALCADE OF AMERICA (I,875); ESCAPE (I,1461); HONEY WEST (I,2105); HONEY WEST: WHO KILLED THE JACKPOT? (I,2106); HUNTER'S MOON (II,1207); KRAFT MYSTERY THEATER (I,2589); SHIRLEY TEMPLE'S STORYBOOK: THE LEGEND OF SLEEPY HOLLOW (I,4018); SYBIL (I,4327); TENAFLY (I,4395)

ERICSON, JUNE
THE BOB NEWHART SHOW (I,666)

ERLET, JANET
ARTHUR GODFREY AND FRIENDS (I,278)

ERNEST FLATT DANCERS, THE
THE CAROL BURNETT SHOW (II,443); THE ENTERTAINERS (I,1449)

ERRICKSON, KRISTA
THE BEST OF TIMES (II,220); HELLO, LARRY (II,1128)

ERSKINE, MARILYN
CHRYSLER MEDALLION THEATER (I,951); DAMON RUNYON THEATER (I,1129); THE LONER (I,2744); THE LUX VIDEO THEATER (I,2795); PEPSI-COLA PLAYHOUSE (I,3523); SCIENCE FICTION THEATER (I,3938); THE TOM EWELL SHOW (I,4538); THE UNITED STATES STEEL HOUR (I,4677)

ERVING, JIM
THE SCOEY MITCHLLL SHOW (I,3939)

ERWIN, BILL
HARD KNOX (II,1087); LONE STAR (II,1513); MACREEDY'S WOMAN (I,2804); MOONLIGHT (II,1730); STRUCK BY LIGHTNING (II,2482)

ERWIN, JOHN
THE BANG-SHANG LALAPALOOZA SHOW (II,140); FRED AND BARNEY MEET THE THING (II,919); RAWHIDE (I,3727)

ERWIN, JUDY
HAZEL (I,1982)

ERWIN, STU
COME A RUNNING (I,1013); FLOYD GIBBONS, REPORTER (I,1608); THE GREATEST SHOW ON EARTH (I,1883); PURSUIT (I,3700); TROUBLE WITH FATHER (I,4610)

ESCORTS, THE
THE JUDY GARLAND SHOW (I,2470)

ESHLEY, NORMAN
GEORGE AND MILDRED (II,965); MAN ABOUT THE HOUSE (II,1611)

ESMOND, JILL
THE ADVENTURES OF ROBIN HOOD (I,74)

ESPINOSA, JOSE ANGEL
THE LOG OF THE BLACK PEARL (II,1506)

ESPINOSA, MARY
THE MICKEY MOUSE CLUB (I,3025); THE MOUSEKETEERS REUNION (II,1742)

ESPINOZA, JAMES
ALEX AND THE DOBERMAN GANG (II,32)

ESPOSITO, GIANCARLO
THE GUIDING LIGHT (II,1064)

ESPOSITO, PHIL
LIFE'S MOST EMBARRASSING MOMENTS (II,1470)

ESPY, WILLIAM GRAY
ANOTHER WORLD (II,97)

ESQUIRE CALENDER GIRLS, THE
THE BOB HOPE SHOW (I,517)

ESSEN, VIOLA
THE DESERT SONG (I,1236)

ESSER, CARL
THE FLINTSTONE COMEDY HOUR (II,881); PEBBLES AND BAMM BAMM (II,1987)

ESSLER, FRED
SHIRLEY TEMPLE'S STORYBOOK: THE LEGEND OF SLEEPY HOLLOW (I,4018)

ESTABLISHMENT, THE
ARTHUR GODFREY'S PORTABLE ELECTRIC MEDICINE SHOW (I,290); THE JONATHAN WINTERS SHOW (I,2443); THE PERRY COMO WINTER SHOW (I,3544); PERRY COMO'S WINTER SHOW (I,3546); PERRY COMO'S WINTER SHOW (I,3548)

ESTABROOK, CHRISTINE
GEORGE WASHINGTON (II,978)

ESTRADA, ANGELINA
CHICO AND THE MAN (II,508)

ESTRADA, ERIK
BOB HOPE SPECIAL: BOB HOPE ON CAMPUS (II,290); BOB HOPE SPECIAL: BOB HOPE'S ALL-STAR LOOK AT TV'S PRIME TIME WARS (II,303); CELEBRITY CHALLENGE OF THE SEXES 3 (II,467); CHIPS (II,511); CIRCUS OF THE STARS (II,533); THE DEAN MARTIN CHRISTMAS SPECIAL (II,640); THE DONNY AND MARIE CHRISTMAS SPECIAL (II,713); DOUG HENNING: MAGIC ON BROADWAY (II,730); FORCE SEVEN (II,908); MAGIC WITH THE STARS (II,1599); THAT'S TV (II,2581); TOM SNYDER'S CELEBRITY SPOTLIGHT

(II,2632); WOMEN WHO RATE A "10" (II,2825)

ESTRIN, PATRICIA
LOVERS AND FRIENDS (II,1549); SEARCH FOR TOMORROW (II,2284)

ETHEL
ETHEL IS AN ELEPHANT (II,783)

ETHERINGTON, COLIN
IF IT'S A MAN, HANG UP (I,2186)

ETLENNE, ROGER
MIMI (I,3054)

ETTLINGER, NORMAN
SCARLETT HILL (I,3933)

EUBANKS, BOB
ALL-STAR SECRETS (II,52); THE DIAMOND HEAD GAME (II,666); DREAM HOUSE (II,734); LIFE'S MOST EMBARRASSING MOMENTS (II,1470); THE NEWLYWED GAME (II,1836); RHYME AND REASON (II,2152); TRIVIA TRAP (II,2659)

EUBANKS, SHANNON
LOVING (II,1552)

EURE, WESLEY
DAYS OF OUR LIVES (II,629); LAND OF THE LOST (II,1425)

EURICH, ROBIN
ALICE (II,33)

EVAN, JESSIE
CORONATION STREET (I,1054)

EVANKE, ED
RYAN'S HOPE (II,2234)

EVANS DANCERS, THE
THE JULIUS LAROSA SHOW (I,2487)

EVANS, ANDREA
ONE LIFE TO LIVE (II,1907)

EVANS, ART
THE ORPHAN AND THE DUDE (II,1924); ROOSEVELT AND TRUMAN (II,2209)

EVANS, BARRY
DOCTOR IN THE HOUSE (I,1313)

EVANS, BOBBY
BORDER COLLIE (I,688)

EVANS, BRUCE
WHEN THE NIGHTINGALE SANG IN BERKELEY SQUARE (I,4827)

EVANS, CLIFFORD
DOUGLAS FAIRBANKS JR. PRESENTS THE RHEINGOLD THEATER (I,1364); THE PRISONER (I,3670); STRYKER OF SCOTLAND YARD (I,4264)

EVANS, COLIN
MOMMA THE DETECTIVE (II,1721)

EVANS, DALE
THE BOB HOPE SHOW (I,537); A COUNTRY HAPPENING (I,1065); HAPPY BIRTHDAY, AMERICA (II,1083); THE NASHVILLE PALACE (II,1792); THE ROY ROGERS AND DALE EVANS SHOW (I,3858); THE ROY ROGERS SHOW (I,3859); SAGA OF SONORA (I,3884); WHEN THE WEST WAS FUN: A WESTERN REUNION (II,2780)

EVANS, DAME JUDITH
PURSUE AND DESTROY (I,3699)

EVANS, DAMON
THE JEFFERSONS (II,1276); ROOTS: THE NEXT GENERATIONS (II,2213)

EVANS, DENNY
ALL NIGHT RADIO (II,40); GIDGET MAKES THE WRONG CONNECTION (I,1798); ME AND MAXX (II,1672); THE PETER MARSHALL VARIETY SHOW (II,2029); THAT SECOND THING ON ABC (II,2572); THAT THING ON ABC (II,2574)

EVANS, DONNY
THE GUINNESS GAME (II,1066)

EVANS, DOUGLAS
MR. TUTT (I,3153)

EVANS, DR BERGEN
DOWN YOU GO (I,1367); GHOST (I,1785); IT'S ABOUT TIME (I,2262); THE LAST WORD (I,2631); TOP DOLLAR (I,4577)

EVANS, ERICK RAY
NANCY ASTOR (II,1788)

EVANS, FRED
PETER PAN (II,2030)

EVANS, GENE
THE BIG VALLEY (I,437); CASINO (II,452); THE CONCRETE COWBOYS (II,564); THE DOG TROOP (I,1325); DRAGNET (I,1370); THE FLIM-FLAM MAN (I,1599); GENERAL ELECTRIC TRUE (I,1754); KATE BLISS AND THE TICKER TAPE KID (II,1366); LASSIE: THE NEW BEGINNING (II,1433); THE MACAHANS (II,1583); MATT HELM (II,1652); MATT HELM (II,1653); MY FRIEND FLICKA (I,3190); SIDEKICKS (II,2348); SPENCER'S PILOTS (II,2421); SPENCER'S PILOTS (II,2422); TRAVIS MCGEE (II,2656); WILD TIMES (II,2799)

EVANS, HARRY
APPLAUSE (I,244)

EVANS, JESSIE
LADY KILLER (I,2603)

EVANS, JUDI
THE GUIDING LIGHT (II,1064)

EVANS, LINDA
BARE ESSENCE (II,149); BATTLE OF THE NETWORK STARS (II,172); THE BIG VALLEY (I,437); BOB HOPE SPECIAL: BOB HOPE'S ALL-STAR COMEDY LOOK AT THE FALL SEASON: IT'S STILL FREE AND WORTH IT! (II,299); BOB HOPE SPECIAL: BOB HOPE'S STAR-STUDDED SPOOF OF THE NEW TV SEASON— G RATED—WITH GLAMOUR, GLITTER & GAGS (II,317); CIRCUS OF THE STARS (II,535); DYNASTY (II,746); GEORGE BURNS AND OTHER SEX SYMBOLS (II,966); HUNTER (II,1204); HUNTER (II,1205); MAGIC WITH THE STARS (II,1599); NAKIA (II,1784)

EVANS, MADGE
THE ALCOA HOUR (I,108)

EVANS, MARY BETH
FOUR EYES (II,912)

EVANS, MAURICE
BATMAN (I,366); BEWITCHED (I,418); THE DEVIL'S DISCIPLE (I,1247); DIAL "M" FOR MURDER (I,1256); THE GIRL, THE GOLD WATCH AND EVERYTHING (II,1002); THE HALLMARK HALL OF FAME (I,1922); HAMLET (I,1924); KING RICHARD II (I,2559); MACBETH (I,2800); MACBETH (I,2801); MAN AND SUPERMAN (I,2854); ST. JOAN (I,4175); TAMING OF THE SHREW (I,4356); THE TEMPEST (I,4391); TWELFTH NIGHT (I,4636); U.M.C. (I,4666)

EVANS, MICHAEL
EYE WITNESS (I,1494); GOLIATH AWAITS (II,1025); NIGHT PARTNERS (II,1847); THE YOUNG AND THE RESTLESS (II,2862)

EVANS, MIKE
ALL IN THE FAMILY (II,38); CALL HER MOM (I,796); THE JEFFERSONS (II,1276); THE PRACTICE (II,2076); RICH MAN, POOR MAN—BOOK I (II,2161); THE RICHARD PRYOR SPECIAL (II,2164); THE RICHARD PRYOR SPECIAL? (II,2165)

EVANS, MITCHELL YOUNG
NUTS AND BOLTS (II,1871)

EVANS, MONICA
THE ODD COUPLE (II,1875)

EVANS, NANCY
DENNIS THE MENACE (I,1231); THE LITTLE REVUE (I,2717); THE WAYNE KING SHOW (I,4779)

EVANS, ORRIN B
THE PEOPLE'S COURT OF SMALL CLAIMS (I,3522)

EVANS, PETER
GEORGE WASHINGTON (II,978)

EVANS, RICHARD
PEYTON PLACE (I,3574)

EVANS, ROBIN
THE EYES OF TEXAS II (II,800); HIGH SCHOOL, U.S.A. (II,1148); RIPTIDE (II,2178); SWEEPSTAKES (II,2514)

EVANS, STEVE
CAR CARE CENTRAL (II,436)

EVANS, TENNIEL
THE CITADEL (II,539)

EVANS-MASSEY, ANDREA
ONE LIFE TO LIVE (II,1907); THE YOUNG AND THE RESTLESS (II,2862)

EVANSON, PAT
A TOWN LIKE ALICE (II,2652)

EVE, MARIE
MARIE AND GEORGE (I,2910)

EVE, TREVOR
LACE (II,1418); SHOESTRING (II,2338)

EVELYN, JUDITH
MATINEE THEATER (I,2947); SURE AS FATE (I,4297); WINDOWS (I,4876)

EVEN DOZEN, THE
MUSIC HALL AMERICA (II,1773)

EVENS, EDITH
TIME REMEMBERED (I,4509)

EVERETT, CHAD
BATTLE OF THE NETWORK STARS (II,170); BATTLE OF THE NETWORK STARS (II,177); CENTENNIAL (II,477); THE DAKOTAS (I,1126); THE FRENCH ATLANTIC AFFAIR (II,924); HAGEN (II,1071); MALIBU (II,1607); A MAN CALLED RAGAN (I,2858); MEDICAL CENTER (II,1675); THE ROUSTERS (II,2220)

EVERETT, ETHEL
AS THE WORLD TURNS (II,110)

EVERETT, KENNY
THE KENNY EVERETT VIDEO SHOW (II,1379)

EVERETT, MARK
GALACTICA 1980 (II,953)

EVERETT, RUPERT
THE FAR PAVILIONS (II,832)

EVERHART, REX
ABC STAGE '67 (I,6); FEELING GOOD (I,1559); FISHERMAN'S WHARF (II,865); STRANDED (II,2475)

EVERLY BROTHERS, THE
THE BIG BEAT (I,423); PETULA (I,3573)

EVERLY, DON
THE EVERLY BROTHERS REUNION CONCERT (II,791); THE EVERLY BROTHERS SHOW (I,1479)

EVERLY, PHIL
THE EVERLY BROTHERS REUNION CONCERT (II,791); THE EVERLY BROTHERS SHOW (I,1479)

EVERS, JASON
CHANNING (I,900); GOLDEN GATE (II,1020); THE GUNS OF WILL SONNETT (I,1907); OF THIS TIME, OF THAT PLACE (I,3335); THREE FOR DANGER (I,4475); WRANGLER (I,4921); THE YOUNG LAWYERS (I,4940)

EVERT-LLOYD, CHRIS
BOB HOPE SPECIAL: THE BOB HOPE SPECIAL (II,337); BURT BACHARACH IN SHANGRI-LA (I,773); LYNDA CARTER'S CELEBRATION (II,1563)

EVERY MOTHER'S SON
JOHNNY CARSON DISCOVERS CYPRESS GARDENS (I,2419)

EVES, GRENVILLE
SECRETS OF THE OLD BAILEY (I,3971)

EVIGAN, GREG
B.J. AND THE BEAR (II,125); B.J. AND THE BEAR (II,126); B.J. AND THE BEAR (II,127); BATTLE OF THE NETWORK STARS (II,168); BATTLE OF THE NETWORK STARS (II,169); BATTLE OF THE NETWORK STARS (II,171); CIRCUS OF THE STARS (II,535); DEBBY BOONE. . . THE SAME OLD BRAND NEW ME (II,656); THE EYES OF TEXAS (II,799); THE EYES OF TEXAS II (II,800); HEREAFTER (II,1138); MASQUERADE (II,1644); MEN WHO RATE A "10" (II,1684); THE OSMOND FAMILY CHRISTMAS SPECIAL (II,1926); SCENE OF THE CRIME (II,2271); A YEAR AT THE TOP (II,2844); A YEAR AT THE TOP (II,2845); THE YELLOW ROSE (II,2847)

EWALD, YVONNE
THE RED SKELTON SHOW (I,3756)

EWELL, TOM
ACTOR'S STUDIO (I,28); ALL IN FUN (I,130); BARETTA (II,152); BEST OF THE WEST (II,218); BROTHER RAT (I,746); COSMOPOLITAN THEATER (I,1057); LIGHT'S OUT (I,2699); SEARCH FOR TOMORROW (II,2284); TERROR AT ALCATRAZ (II,2563); THE TOM EWELL SHOW (I,4538)

EWING, BILL
KORG: 70,000 B.C. (II,1408)

EWING, DIANA
WASHINGTON: BEHIND CLOSED DOORS (II,2744)

EWING, JOHN CHRISTIE
HELLINGER'S LAW (II,1127)

EWING, ROGER
GUNSMOKE (II,1069)

EYER, RICHARD
FATHER KNOWS BEST (II,838); MY FRIEND IRMA (I,3191); STAGECOACH WEST (I,4185)

EYTHE, WILLIAM
MYSTERIES OF CHINATOWN (I,3209)

FABARES, SHELLEY
ANNETTE (I,222); ANNIE OAKLEY (I,225); THE CLAUDETTE COLBERT SHOW (I,971); THE DONNA REED SHOW (I,1347); FOREVER FERNWOOD (II,909); HELLO, LARRY (II,1128); HIGHCLIFFE MANOR (II,1150); HIS AND HERS (II,1155); THE LITTLE PEOPLE (I,2716); MATINEE THEATER (I,2947); MEET ME IN ST. LOUIS (I,2987); MORK AND MINDY (II,1735); MR. NOVAK (I,3145); ONE DAY AT A TIME (II,1900); PLEASURE COVE (II,2058); THE PRACTICE (II,2076); TWO FOR THE MONEY (I,4655); U.M.C. (I,4666)

FABIAN
DICK CLARK PRESENTS THE ROCK AND ROLL YEARS (I,1265)

FABIAN, OLGA
THE GOLDBERGS (I,1835); KRAFT TELEVISION THEATER (I,2592)

FABIANI, JOEL
ABC'S MATINEE TODAY (I,7); BRENDA STARR (II,372); DALLAS (II,614); DARK SHADOWS (I,1157); DEPARTMENT S (I,1232); IRONSIDE (I,2240); THE NEW DAUGHTERS OF JOSHUA CABE (I,3261); RISKO (II,2180); TOM AND JOANN (II,2630)

FABRAY, NANETTE
THE ALCOA HOUR (I,108); ALICE THROUGH THE LOOKING GLASS (I,122); THE BOB HOPE SHOW (I,643); THE BOB HOPE SHOW (I,649); BURT AND THE GIRLS (I,770); CAESAR'S HOUR (I,790); THE CHEVROLET GOLDEN ANNIVERSARY SHOW (I,924); THE CHEVROLET TELE-THEATER (I,926); THE COLGATE COMEDY HOUR (I,998); THE DEAN MARTIN SHOW (I,1199); FAME IS THE NAME OF THE GAME (I,1518); GEORGE M! (I,1772); GEORGE M! (II,976); HAPPY ANNIVERSARY AND GOODBYE (II,1082); HIGH BUTTON SHOES (I,2050); HIGH ROLLERS (II,1147); HOLLYWOOD MELODY (I,2085); HOWDY (I,2150); THE KAISER ALUMINUM HOUR (I,2503); THE MARCH OF DIMES FASHION SHOW (I,2901); THE MARY TYLER MOORE SHOW (II,1640); ONE DAY AT A TIME (II,1900); SO HELP ME, APHRODITE (I,4104); THE WONDERFUL WORLD OF BURLESQUE III (I,4897); YES YES NANETTE (I,4927)

FADDEN, TOM
THE ADVENTURES OF SUPERMAN (I,77); BROKEN ARROW (I,743); BROKEN ARROW (I,744); PETTICOAT JUNCTION (I,3571); THE SLOWEST GUN IN THE WEST (I,4088)

FADIMAN, CLIFTON
ALUMNI FUN (I,145); INFORMATION PLEASE (I,2211); THE QUIZ KIDS (I,3712); THIS IS SHOW BUSINESS (I,4459); WHAT'S IN A WORD? (I,4815)

FAFARA, STANLEY
LEAVE IT TO BEAVER (I,2648)

FAGLER, NICK
CELEBRITY (II,463)

FAHEY, JEFF
ONE LIFE TO LIVE (II,1907)

FAHEY, LEE ANN
BLACK BEAUTY (II,261)

FAHEY, MARY ANN
ALL THE RIVERS RUN (II,43)

FAHEY, MYRNA
FATHER OF THE BRIDE (I,1543); PEYTON PLACE (I,3574); YOUNG IN HEART (I,4939)

FAHY, KATE
ROGER DOESN'T LIVE HERE ANYMORE (II,2201)

FAIGHT, GORDON
CELEBRITY (II,463)

FAIN, SAMMY
JOHNNY CARSON PRESENTS THE SUN CITY SCANDALS '72 (I,2421)

FAIR, JODY
YOUNG IN HEART (I,4939)

FAIRBAIRN, BRUCE
BEHIND THE SCREEN (II,200); BRAVO TWO (II,368); THE ROOKIES (II,2208)

FAIRBANKS JR, DOUGLAS
ABC STAGE '67 (I,6); THE AMAZING YEARS OF CINEMA (II,64); THE BOB HOPE SHOW (I,545); THE BOB HOPE SHOW (I,504); THE BOB HOPE SHOW (I,603); BOB HOPE SPECIAL: BOB HOPE'S 30TH ANNIVERSARY TV SPECIAL (II,293); CIRCUS OF THE STARS (II,533); DOUGLAS FAIRBANKS JR. PRESENTS THE RHEINGOLD THEATER (I,1364); RAQUEL (II,2114); RHEINGOLD THEATER (I,3783)

FAIRBANKS, NOLA
THE MADHOUSE BRIGADE (II,1588)

FAIRBANKS, PEGGY
THE INA RAY HUTTON SHOW (I,2207)

FAIRCHILD, CHARLOTTE
WE INTERRUPT THIS SEASON (I,4780)

FAIRCHILD, IRIS
GENERAL HOSPITAL (II,964)

FAIRCHILD, MORGAN
79 PARK AVENUE (II,2310); BATTLE OF THE NETWORK STARS (II,176); BLONDES VS.

BRUNETTES (II,271); BOB HOPE SPECIAL: BOB HOPE GOES TO COLLEGE (II,283); BOB HOPE SPECIAL: BOB HOPE'S SPRING FLING OF COMEDY AND GLAMOUR (II,315); BOB HOPE SPECIAL: BOB HOPE'S STARS OVER TEXAS (II,318); CIRCUS OF THE STARS (II,536); THE CONCRETE COWBOYS (II,564); THE COUNTRY MUSIC MURDERS (II,590); ESCAPADE (II,781); FLAMINGO ROAD (II,871); FLAMINGO ROAD (II,872); THE GIRL, THE GOLD WATCH AND DYNAMITE (II,1001); THE MAGIC OF DAVID COPPERFIELD (II,1595); MAGIC WITH THE STARS (II,1599); MORK AND MINDY (II,1735); PAPER DOLLS (II,1952); SEARCH FOR TOMORROW (II,2284); THE SHAPE OF THINGS (II,2315); WHATEVER BECAME OF. . .? (II,2773); WOMEN WHO RATE A "10" (II,2825)

FAIRFAX, DIANA
LOVE IN A COLD CLIMATE (II,1539); MOLL FLANDERS (II,1720)

FAIRFAX, JAMES
THE GALE STORM SHOW (I,1725)

FAIRMAN, MICHAEL
FLY AWAY HOME (II,893); LOVE OF LIFE (I,2774); PEEPING TIMES (II,1990); THE POWERS OF MATTHEW STAR (II,2075); RYAN'S HOPE (II,2234)

FAIRMONT, LOUIS
MISTRAL'S DAUGHTER (II,1708)

FAISON, FRANKIE
HOT HERO SANDWICH (II,1186)

FAISON, MATTHEW
STEPHANIE (II,2453)

FAISON, SANDY
MAKING IT (II,1605)

FAITH, ROSEMARY
OPPENHEIMER (II,1919)

FALAN, TANYA
THE LAWRENCE WELK SHOW (I,2643); MEMORIES WITH LAWRENCE WELK (II,1681)

FALANA, AVIE
DINAH AND HER NEW BEST FRIENDS (II,676)

FALANA, LOLA
BEN VEREEN. . . COMIN' AT YA (II,203); THE BOB HOPE SHOW (I,646); THE BOB HOPE SHOW (I,658); BOB HOPE SPECIAL: BOB HOPE'S ALL-STAR SUPER BOWL PARTY (II,304); CELEBRITY CHALLENGE OF THE SEXES 1 (II,465); CELEBRITY CHALLENGE OF THE SEXES 2 (II,466); CELEBRITY CHALLENGE OF THE SEXES 4 (II,468); THE CELEBRITY FOOTBALL CLASSIC (II,474); CIRCUS OF THE STARS (II,531); CIRCUS OF THE STARS (II,533); LOLA (II,1508); LOLA (II,1509); LOLA (II,1510); LOLA (II,1511); THE LOU RAWLS SPECIAL (II,1527); THE NEW BILL COSBY SHOW (I,3257); THE SAMMY DAVIS JR. SPECIAL (I,3899); SIEGFRIED AND ROY (II,2350)

FALIS, TERRI
THE SECRET STORM (I,3969)

FALK, LEONARD
THE WILD WILD WEST (I,4863)

FALK, PETER
ALFRED HITCHCOCK PRESENTS (I,115); THE BARBARA STANWYCK THEATER (I,350); THE BOB HOPE CHRYSLER THEATER (I,502); BRIGADOON (I,726); COLUMBO (II,556); THE DICK POWELL SHOW (I,1269); THE DUPONT SHOW OF THE WEEK (I,1388); A HATFUL OF RAIN (I,1964); JOHNNY CASH: THE FIRST 25 YEARS (II,1336); LOVE OF LIFE (I,2774); THE MILLION DOLLAR INCIDENT (I,3045); PRESCRIPTION: MURDER (I,3660); RANSOM FOR A DEAD MAN (I,3722); THE TRIALS OF O'BRIEN (I,4605); THE TWILIGHT ZONE (I,4651)

FALK, TOM
MR. AND MRS. COP (II,1748)

FALKENBERG, JINX
THE BOB HOPE SHOW (I,509); SHOW OF THE YEAR (I,4031); TEX AND JINX (I,4406); THEATER '62 (I,4433)

FALLON, HANNAH
KENNEDY (II,1377)

FALLON, RONNIE
DOCTOR DOLITTLE (I,1309)

FALZENE, PAUL
ALL MY CHILDREN (II,39)

FAMISON, JANIS
THE YOUNG PIONEERS (II,2868)

FANCHER, HAMPTON
OF MEN OF WOMEN (I,3332)

FANCY, RICHARD
GEORGE WASHINGTON (II,978)

FANN, AL
HOW TO SURVIVE A MARRIAGE (II,1198); THE JERK, TOO (II,1282); THE PLASTICMAN COMEDY/ADVENTURE SHOW (II,2051)

FANNING, BILL
THE JIMMIE RODGERS SHOW (I,2383)

FANNING, GENE
SOMERSET (I,4115); YOUNG DR. MALONE (I,4938)

FANT, LOU
THICKER THAN WATER (I,4445)

FANTASY FACTORY, THE
MAGIC WITH THE STARS (II,1599)

FARACY, STEPHANIE
BUMPERS (II,394); THE FIGHTING NIGHTINGALES (II,855); GOODNIGHT, BEANTOWN (II,1041); THE LAST RESORT (II,1440); PRIVATE BENJAMIN (II,2087); RETURN OF THE WORLD'S GREATEST DETECTIVE (II,2142); STEPHANIE (II,2453); THE THORN BIRDS (II,2600)

FARAKER, LOIS
SUNDAY FUNNIES (II,2493)

FARENTINO, JAMES
BLUE THUNDER (II,278); CELEBRITY CHALLENGE OF THE SEXES 2 (II,466); COOL MILLION (I,1046); CROSSFIRE (II,606); DEATH OF A SALESMAN (I,1208); DYNASTY (II,746); EMILY, EMILY (I,1438); THE FIRST NINE MONTHS ARE THE HARDEST (I,1584); THE LAWYERS (I,2646); LOVE STORY (I,2778); THE MASK OF MARCELLA (I,2937); MITZI AND A HUNDRED GUYS (II,1710); MY WIFE NEXT DOOR (II,1780); THE POSSESSED (II,2071); THE SINGERS (I,4060); THE SOUND OF ANGER (I,4133); THE WHOLE WORLD IS WATCHING (I,4851)

FARGAS, ANTONIO
ADVENTURING WITH THE CHOPPER (II,15); ALL COMMERCIALS—A STEVE MARTIN SPECIAL (II,37); HEREAFTER (II,1138); P.O.P. (II,1939); PAPER DOLLS (II,1951); STARSKY AND HUTCH (II,2444); STARSKY AND HUTCH (II,2445); STEVE MARTIN'S THE WINDS OF WHOOPIE (II,2461)

FARGE, ANNIE
ANGEL (I,216); KRAFT SUSPENSE THEATER (I,2591)

FARGO, DONNA
COUNTRY MUSIC HIT PARADE (I,1069); COUNTRY MUSIC HIT PARADE (II,588); COUNTRY MUSIC HIT PARADE (II,589); DEAN MARTIN PRESENTS MUSIC COUNTRY, U.S.A. (I,1192); THE DONNA FARGO SHOW (II,710)

FARKASH, HANS
1915 (II,1853)

FARLEIGH, LYNN
THE STRAUSS FAMILY (I,4253); THE WORD (II,2833)

FARLEY, DUKE
CAR 54, WHERE ARE YOU? (I,842); OUR FIVE DAUGHTERS (I,3413)

FARLEY, ELIZABETH
THE EDGE OF NIGHT (II,760)

FARLEY, MORGAN
BEANE'S OF BOSTON (II,194); ENIGMA (II,778)

FARMER, LIA
YOUR HIT PARADE (I,4952)

FARMER, LILLIAN
THE KEN MURRAY SHOW (I,2525)

FARMER, SUSAN
DEATH IN DEEP WATER (I,1206)

FARNON, SHANNON
EGAN (I,1419); THE NEW SUPER FRIENDS HOUR (II,1830); NIGHT GALLERY (I,3288); THE PSYCHIATRIST: GOD BLESS THE CHILDREN (I,3685); THE SOUND OF ANGER (I,4133); SUPER FRIENDS (II,2497); VALLEY OF THE DINOSAURS (II,2717); THE WORLD'S GREATEST SUPER HEROES (II,2842)

FARNSWORTH, RICHARD
THE CHEROKEE TRAIL (II,500); THE TEXAS RANGERS (II,2567); TRAVIS MCGEE (II,2656)

FARR, DEREK
STAR MAIDENS (II,2435)

FARR, FELICIA
BONANZA (II,347); THE THREE MUSKETEERS (I,4481)

FARR, JAMIE

AFTERMASH (II,23); BATTLE OF THE NETWORK STARS (II,165); BATTLE OF THE NETWORK STARS (II,168); BATTLE OF THE NETWORK STARS (II,171); CELEBRITY CHALLENGE OF THE SEXES 5 (II,469); CELEBRITY CHARADES (II,470); THE CHICAGO TEDDY BEARS (I,934); CIRCUS OF THE STARS (II,532); CIRCUS OF THE STARS (II,533); CIRCUS OF THE STARS (II,535); CIRCUS OF THE STARS (II,537); CIRCUS OF THE STARS (II,538); DEAR PHOEBE (I,1204); M*A*S*H (II,1569); THE MAD MAD MAD MAD WORLD OF THE SUPER BOWL (II,1586); MURDER CAN HURT YOU! (II,1768); RHYME AND REASON (II,2152)

FARR, LEE
THE DETECTIVES (I,1245)

FARR, STEPHEN
THE 25TH MAN (MS) (II,2678)

FARRAND, JAN
THE EDGE OF NIGHT (II,760)

FARRAR, CATHERINE
THE SIXTH SENSE (I,4069)

FARRAR, VIVIAN
SENSE AND NONSENSE (I,3977)

FARRELL, BILLY
THAT'S OUR SHERMAN (I,4429)

FARRELL, BRIAN
CONDOMINIUM (II,566); LOVE OF LIFE (I,2774)

FARRELL, BRIONI
ALLAN (I,142); TO ROME WITH LOVE (I,4526)

FARRELL, CHARLIE
THE CHARLIE FARRELL SHOW (I,913); MY LITTLE MARGIE (I,3194)

FARRELL, GAIL
THE LAWRENCE WELK SHOW (I,2643); THE LAWRENCE WELK SHOW (I,2644); MEMORIES WITH LAWRENCE WELK (II,1681)

FARRELL, GLENDA
THE ELGIN HOUR (I,1428); FAITH BALDWIN'S THEATER OF ROMANCE (I,1515); FRONT ROW CENTER (I,1694); KRAFT TELEVISION THEATER (I,2592); THE MARRIAGE BROKER (I,2920); SILVER THEATER (I,4051); STARLIGHT THEATER (I,4201); TALES OF TOMORROW (I,4352)

FARRELL, GWEN
M*A*S*H (II,1569)

FARRELL, JUDY
FAME (II,812); M*A*S*H (II,1569)

FARRELL, MARY
9 TO 5 (II,1852); THE LOVE BOAT (II,1535)

FARRELL, MIKE
AMANDA FALLON (I,151); BATTLE OF THE NETWORK STARS (II,164); DAYS OF OUR LIVES (II,629); GOOD EVENING, CAPTAIN (II,1031); THE INTERNS (I,2226); M*A*S*H (II,1569); THE MAN AND THE CITY (I,2856)

FARRELL, RAY
THE BOB HOPE SHOW (I,567)

FARRELL, RICHARD
CASTLE ROCK (II,456)

FARRELL, SHARON
THE EYES OF CHARLES SAND (I,1497); HAWAII FIVE-O (II,1110); KRAFT SUSPENSE THEATER (I,2591); LASSITER (I,2625); LASSITER (II,1434); RITUALS (II,2182); SAINTS AND SINNERS (I,3887)

FARRELL, SHEA
CAPITOL (II,426); HOTEL (II,1192)

FARRELL, SKIP
THE SKIP FARRELL SHOW (I,4074); U.S. TREASURY SALUTES (I,4689)

FARRELL, STEVE
ALL THE RIVERS RUN (II,43)

FARRELL, TERRY
PAPER DOLLS (II,1952)

FARRELL, TIM
ACCUSED (I,18)

FARRELL, TOMMY
THE ADVENTURES OF RIN TIN TIN (I,73); DOBIE GILLIS (I,1302); MANHATTAN TOWER (I,2889); TELE-VARIETIES (I,4377); THIS IS ALICE (I,4453)

FARRINGTON, KENNETH
CORONATION STREET (I,1054)

FARROW, JULIA
CINDERELLA (I,955)

FARROW, MIA
JOHNNY BELINDA (I,2417); PETER PAN (II,2030); PEYTON PLACE (I,3574)

FASCIANO, RICHARD
BEHIND THE SCREEN (II,200)

FASO, LAURIE
MARLO AND THE MAGIC MOVIE MACHINE (II,1631); SOAP (II,2392); THE STEVE LANDESBERG TELEVISION SHOW (II,2458)

FATE, FRED
SNAFU (II,2386)

FATES, GIL
HOLD IT PLEASE (I,2071); THE MARCH OF DIMES FASHION SHOW (I,2901); OPEN HOUSE (I,3394); WHAT'S IT WORTH? (I,4818)

FAULK, HENRY
HEE HAW (II,1123)

FAULK, HUGH
WILD WILD WORLD OF ANIMALS (I,4864)

FAULKNER, EDWARD
ADAM-12 (I,31); THE DOUBLE LIFE OF HENRY PHYFE (I,1361); THE INTERNS (I,2226); THE LOG OF THE BLACK PEARL (II,1506); SKYFIGHTERS (I,4078)

FAULKNER, JAMES
THE MARTIAN CHRONICLES (II,1635)

FAULKNER, STEPHANIE
DYNASTY (II,747); FANTASY ISLAND (II,829)

FAULKONBRIDGE, CLAIRE
LAUGH-IN (II,1444)

FAUN, TOM
SENSE OF HUMOR (II,2303)

FAUNTELLE, DIAN
ROCKY JONES, SPACE RANGER (I,3823); WATERFRONT (I,4774)

FAUSTINO, DAVID
VELVET (II,2725); VENICE MEDICAL (II,2726)

FAUSTINO, RANDY
EXO-MAN (II,797)

FAWCETT, ALLEN
THE EDGE OF NIGHT (II,760); PUTTIN' ON THE HITS (II,2093)

FAWCETT, FARRAH
BATTLE OF THE NETWORK STARS (II,163); CELEBRITY CHALLENGE OF THE SEXES 1 (II,465); CELEBRITY CHALLENGE OF THE SEXES 2 (II,466); CHARLIE'S ANGELS (II,486); CHARLIE'S ANGELS (II,487); HARRY O (II,1099); INSIDE O.U.T. (II,1235); OF MEN OF WOMEN (I,3332); THE SIX-MILLION-DOLLAR MAN (II,2372); THE WAYNE NEWTON SPECIAL (II,2749)

FAWCETT, HUDSON
RETURN TO EDEN (II,2143)

FAWCETT, WILLIAM
ANNIE OAKLEY (I,225); THE CODE OF JONATHAN WEST (I,992); FURY (I,1721); THE MURDOCKS AND THE MCCLAYS (I,3163)

FAX, JESSLYN
THE DANNY KAYE SHOW (I,1144); MAGGIE (I,2810); MANY HAPPY RETURNS (I,2895); THE MONKEES (I,3088); OUR MISS BROOKS (I,3415); THE TIME ELEMENT (I,4504)

FAY, FRANK
TOM AND JERRY (I,4533)

FAYE, ALICE
THE PHIL HARRIS SHOW (I,3575)

FAYE, DORIS
POETRY AND MUSIC (I,3632)

FAYE, HERBIE
ACCIDENTAL FAMILY (I,17); DOC (II,682); EVEL KNIEVEL (II,785); THE MICHELE LEE SHOW (II,1690); THE NEW PHIL SILVERS SHOW (I,3269); THE PHIL SILVERS SHOW (I,3580)

FAYE, JANINA
LITTLE WOMEN (I,2723); MIRROR OF DECEPTION (I,3060)

FAYE, JOEY
54TH STREET REVUE (I,1573); CANDID CAMERA (I,819); DAGMAR'S CANTEEN (I,1125); THE ETHEL WATERS SHOW (I,1468); GUESS AGAIN (I,1894); HIGH BUTTON SHOES (I,2050); THE JAZZ SINGER (I,2351); JOEY FAYE'S FROLICS (I,2406); MACK AND MYER FOR HIRE (I,2802); THE NEW TREASURE HUNT (I,3275); OFF THE RECORD (I,3336); SHOW BUSINESS, INC. (I,4028)

FAYE, MEGAN
THE YESTERDAY SHOW (II,2850)

FAYE, RITA
COUNTRY MUSIC CARAVAN (I,1066)

FAYLEN, CAROL
THE BING CROSBY SHOW (I,467); DO NOT DISTURB (I,1300)

FAYLEN, FRANK

DOBIE GILLIS (I,1302); EDDIE (I,1405); THAT GIRL (II,2570); WHATEVER HAPPENED TO DOBIE GILLIS? (II,2774)

FAYLEN, KAY
DOCTOR CHRISTIAN (I,1308)

FAZEL, RICK
DANNY AND THE MERMAID (II,618)

FAZIO, DINO
CAMPO 44 (I,813)

FEARN, SHEILA
GEORGE AND MILDRED (II,965); SIGN IT DEATH (I,4046)

FEAST, FRED
CORONATION STREET (I,1054)

FEDDERSON, GREGG
FAMILY AFFAIR (I,1519)

FEDDY, DARRELL
CENTENNIAL (II,477)

FEE, MELINDA
THE ALIENS ARE COMING (II,35); DAYS OF OUR LIVES (II,629); THE GUIDING LIGHT (II,1064); THE INVISIBLE MAN (II,1242); THE INVISIBLE MAN (II,1243)

FEEBACK, JENNY
I'M SOOO UGLY (II,1219)

FEENEY, JOE
THE LAWRENCE WELK SHOW (I,2643); THE LAWRENCE WELK SHOW (I,2644); MEMORIES WITH LAWRENCE WELK (II,1681)

FEGAN, JOHN
HOMICIDE (I,2101)

FEIGEL, SYLVIA
STRANGE PARADISE (I,4246)

FEIN, BERNIE
THE PHIL SILVERS SHOW (I,3580)

FEIN, DONNA
HAPPY DAYS (II,1084)

FEINBERG, RON
ACES UP (II,7); THE SCOOBY-DOO/DYNOMUTT HOUR (II,2277)

FEINSTEIN, ALAN
BUNCO (II,395); THE EDGE OF NIGHT (II,760); JIGSAW JOHN (II,1288); MASADA (II,1642); THE RUNAWAYS (II,2231); SUPERCOPS (II,2502)

FELD, FRITZ
THE FREEWHEELERS (I,1683)

FELD, NORMAN
AH!, WILDERNESS (I,91)

FELDER, CLARENCE
CONCRETE BEAT (II,562); THE DAIN CURSE (II,613)

FELDER, SARAH
RYAN'S HOPE (II,2234)

FELDMAN, COREY
ANOTHER MAN'S SHOES (II,96); THE BAD NEWS BEARS (II,134); CASS MALLOY (II,454); THE KID WITH THE BROKEN HALO (II,1387); LOVE, NATALIE (II,1545); MADAME'S PLACE (II,1587); MORK AND MINDY (II,1735); STILL THE BEAVER (II,2466)

FELDMAN, MARTY
CELEBRITY DAREDEVILS (II,473); THE GOLDDIGGERS IN LONDON (I,1837); LIGHTS, CAMERA, MONTY! (II,1475); THE MAN WHO CAME TO DINNER (I,2881); THE MARTY FELDMAN COMEDY MACHINE (I,2929)

FELDMAN, MINDY
THE NEW MICKEY MOUSE CLUB (II,1823)

FELDON, BARBARA
THE ARTHUR GODFREY SPECIAL (I,288); ARTHUR GODFREY'S PORTABLE ELECTRIC MEDICINE SHOW (I,290); DEAN MARTIN'S COMEDY WORLD (II,647); FATHER ON TRIAL (I,1544); THE FOUR OF US (II,914); GET SMART (II,983); LADY KILLER (I,2603); THE MARTY FELDMAN COMEDY MACHINE (I,2929); THE NATURAL LOOK (II,1799); OF MEN OF WOMEN (I,3332); PROFILES IN COURAGE (I,3678); REAL LIFE STORIES (II,2117); ROWAN AND MARTIN'S LAUGH-IN (I,3857); SPECIAL EDITION (II,2416)

FELDSHUH, TOVAH
MURDER INK (II,1769); THE WORLD OF DARKNESS (II,2836)

FELICIANO, JOSE
ANDY'S LOVE CONCERT (I,215); THE BING CROSBY SHOW (I,468); THE BING CROSBY SHOW (I,469); THE ENGELBERT HUMPERDINCK SPECIAL (I,1446); FELICIANO—VERY SPECIAL (I,1560); MONSANTO NIGHT PRESENTS JOSE FELICIANO (I,3091); MONSANTO PRESENTS MANCINI (I,3093); MOVIN' (I,3117); PEGGY FLEMING AT MADISON SQUARE GARDEN (I,3506)

FELIX, OTTO
B.J. AND THE BEAR (II,125); TERROR AT ALCATRAZ (II,2563)

FELL, NORMAN
87TH PRECINCT (I,1422); THE CELEBRITY FOOTBALL CLASSIC (II,474); DAN AUGUST (I,1130); ESCAPE (I,1461); EXECUTIVE SUITE (II,796); GETTING THERE (II,985); GHOSTBREAKER (I,1789); GOING PLACES (I,1834); KRAFT SUSPENSE THEATER (I,2591); THE LLOYD BRIDGES SHOW (I,2733); LOOK WHAT THEY'VE DONE TO MY SONG (II,1517); THE LOVE BOAT (II,1535); MAGIC WITH THE STARS (II,1599); THE MOONGLOW AFFAIR (I,3099); NEEDLES AND PINS (I,3246); THE PAT BOONE AND FAMILY CHRISTMAS SPECIAL (II,1969); RICH MAN, POOR MAN—BOOK I (II,2161); RICHIE BROCKELMAN: MISSING 24 HOURS (II,2168); RISKO (II,2180); ROOTS: THE NEXT GENERATIONS (II,2213); THE ROPERS (II,2214); STUDIO ONE (I,4268); TEACHERS ONLY (II,2547); TEACHERS ONLY (II,2548); THREE'S COMPANY (II,2614)

FELL, STEWART
NOT THE NINE O'CLOCK NEWS (II,1864)

FELLINI, FEDERICO
FELLINI: A DIRECTOR'S NOTEBOOK (I,1562)

FELLOWS, DON
THE CITADEL (II,539)

FELLOWS, EDDIE
THE CLIFF EDWARDS SHOW (I,975)

FELLOWS, EDITH
THE BENNY RUBIN SHOW (I,398)

FELLOWS, SUSANNAH
MASTER OF THE GAME (II,1647); SEPARATE TABLES (II,2304)

FELTON, GREG
BE MY VALENTINE, CHARLIE BROWN (I,369); IT'S ARBOR DAY, CHARLIE BROWN (I,2267)

FELTON, HAPPY
IT'S A HIT (I,2257)

FELTON, VERNA
DECEMBER BRIDE (I,1215); DENNIS THE MENACE (I,1231); HENRY FONDA AND THE FAMILY (I,2015); PETE AND GLADYS (I,3558); THE R.C.A. VICTOR SHOW (I,3737); THE REAL MCCOYS (I,3741); WHERE'S RAYMOND? (I,4837)

FEMIA, JOHN
HELLO, LARRY (II,1128); SQUARE PEGS (II,2431)

FENDER, FREDDY
COUNTRY MUSIC HIT PARADE (II,588); COUNTRY MUSIC HIT PARADE (II,589); COUNTRY NIGHT OF STARS (II,592); COUNTRY STARS OF THE 70S (II,594); DEAN MARTIN'S CALIFORNIA CHRISTMAS (II,642); DEAN'S PLACE (II,650)

FENEMORE, HILDA
THE DOUBLE KILL (I,1360)

FENICHEL, JAY
BLUE JEANS (II,275); CAMP GRIZZLY (II,418)

FENMORE, TANYA
FAMILY TIES (II,819); MAMA'S FAMILY (II,1610); TRAPPER JOHN, M.D. (II,2654); TUCKER'S WITCH (II,2667)

FENN, JEAN
ROSALINDA (I,3846)

FENNELLY, PARKER
GULF PLAYHOUSE (I,1904); HEADMASTER (I,1990); THE NASH AIRFLYTE THEATER (I,3227); TROUBLE WITH RICHARD (I,4612)

FENNEMAN, GEORGE
ANYBODY CAN PLAY (I,234); THE SECRET OF MYSTERY LAKE (I,3967); THOSE WONDERFUL TV GAME SHOWS (II,2603); YOU BET YOUR LIFE (I,4932); YOUR FUNNY FUNNY FILMS (I,4950); YOUR SURPRISE PACKAGE (I,4960)

FENTON, BOB
THE MAGIC LAND OF ALLAKAZAM (I,2817)

FENTON, LUCILLE
ABE LINCOLN IN ILLINOIS (I,10)

FENWICK, EILEEN
CHARADE QUIZ (I,901)

FENWICK, GILLIE
THE STRANGE CASE OF DR. JEKYLL AND MR. HYDE (I,4244)

FENWICK, MYA
RED SKELTON'S CHRISTMAS DINNER (II,2123)

FERDIN, PAMELYN
BABY CRAZY (I,319); BLONDIE (I,487); CURIOSITY SHOP (I,1114); THE DELPHI

BUREAU (I,1224); THE FLYING NUN (I,1611); GUESS WHAT I DID TODAY (I,1897); IT WAS A SHORT SUMMER, CHARLIE BROWN (I,2252); THE JOHN FORSYTHE SHOW (I,2410); LASSIE (I,2624); THE ODD COUPLE (II,1875); THE PAUL LYNDE SHOW (I,3489); PLAY IT AGAIN, CHARLIE BROWN (I,3617); THE ROMAN HOLIDAYS (I,3835); SEALAB 2020 (I,3949); SPACE ACADEMY (II,2406); THESE ARE THE DAYS (II,2585)

FERGUS, WARREN
I'VE HAD IT UP TO HERE (II,1221)

FERGUSON, BIANCA
GENERAL HOSPITAL (II,964)

FERGUSON, BOBBIE
LONDON AND DAVIS IN NEW YORK (II,1512)

FERGUSON, FRANK
EXPERT WITNESS (I,1488); THE MACAHANS (II,1583); MY FRIEND FLICKA (I,3190); NO TIME FOR SERGEANTS (I,3300); PEYTON PLACE (I,3574); THE REAL MCCOYS (I,3741); RETURN TO PEYTON PLACE (I,3779)

FERGUSON, HELEN
ALL ABOUT MUSIC (I,127)

FERGUSON, J DON
THE GREATEST GIFT (II,1061)

FERGUSON, JUNE
THE PRICE IS RIGHT (I,3664)

FERGUSON, STACY
SNOOPY'S GETTING MARRIED, CHARLIE BROWN (I,4102)

FERIA, RITA
THE BOB HOPE SHOW (I,613)

FERMAN, BEN
THE SECRET WAR OF JACKIE'S GIRLS (II,2294)

FERNANDE, GLYN
HUNTER (I,2161)

FERNANDEL
THE BOB HOPE SHOW (I,544)

FERNANDEZ, ABEL
THE SAGA OF ANDY BURNETT (I,3883); STEVE CANYON (I,4224); THE UNTOUCHABLES (I,4681); THE UNTOUCHABLES (I,4682)

FERNANDEZ, MARGARITA
WISHMAN (II,2811)

FERNANDEZ, PETER
MARINE BOY (I,2911)

FERRADAY, LISA
THE GREAT SEBASTIANS (I,1877); PERSONALITY PUZZLE (I,3554); SHORT, SHORT DRAMA (I,4023)

FERRAND, MICHAEL
U.F.O. (I,4662)

FERRAR, CATHERINE
DAYS OF OUR LIVES (II,629); MEDICAL CENTER (II,1675); THE SIX MILLION DOLLAR MAN (I,4067)

FERRARI, MARIO
MOSES THE LAWGIVER (II,1737)

FERRELL, CONCHATA
B.J. AND THE BEAR (II,125); E/R (II,748); HOT L BALTIMORE (II,1187); MCCLAIN'S LAW (II,1659); MIXED NUTS (II,1714); THE RAG BUSINESS (II,2106)

FERRELL, ROY
PECK'S BAD GIRL (I,3503)

FERRELL, TODD
DO NOT DISTURB (I,1300); TIMMY AND LASSIE (I,4520)

FERRER, JOSE
ANOTHER WORLD (II,97); THE ART OF CRIME (II,108); BANYON (I,345); BATTLES: THE MURDER THAT WOULDN'T DIE (II,180); CYRANO DE BERGERAC (I,1119); DEBBY BOONE. . . THE SAME OLD BRAND NEW ME (II,656); EXO-MAN (II,797); FAME (I,1517); THE FRENCH ATLANTIC AFFAIR (II,924); GEORGE WASHINGTON (II,978); KISMET (I,2565); THE LITTLE DRUMMER BOY (II,1485); MARRIAGE—HANDLE WITH CARE (I,2921); NEWHART (II,1835); PHILCO TELEVISION PLAYHOUSE (I,3583); THE RETURN OF CAPTAIN NEMO (II,2135); THE RHINEMANN EXCHANGE (II,2150); RICKLES (II,2170); THIS GIRL FOR HIRE (II,2596); TRUMAN AT POTSDAM (I,4618)

FERRER, MEL
BEHIND THE SCREEN (II,200); BLACK BEAUTY (II,261); DALLAS (II,614); FALCON CREST (II,808); HOW THE WEST WAS WON (II,1196); TENAFLY (I,4395); TOP OF THE HILL (II,2645)

FERRERO, MARTIN
ANGIE (II,80); KANGAROOS IN THE KITCHEN (II,1362); MIAMI VICE (II,1689); STEPHANIE (II,2453); SUNDAY FUNNIES (II,2493)

FERRIGNO, LOU
BATTLE OF THE NETWORK STARS (II,167); BATTLE OF THE NETWORK STARS (II,168); BOB HOPE SPECIAL: BOB HOPE FOR PRESIDENT (II,282); CELEBRITY CHALLENGE OF THE SEXES 3 (II,467); THE INCREDIBLE HULK (II,1232); TRAUMA CENTER (II,2655)

FERRIN, FRANK
ANDY'S GANG (I,214)

FERRIS, BARBARA
THE STRAUSS FAMILY (I,4253)

FERRIS, IRENA
COCAINE AND BLUE EYES (II,548); COVER UP (II,597)

FERRIS, PATRICIA
TOM CORBETT, SPACE CADET (I,4535)

FERRIS, PAUL
THE BARON (I,360)

FERRITER, BILL
ALL MY CHILDREN (II,39)

FERRO, TALYA
RENDEZVOUS HOTEL (II,2131)

FESTER, JOHNNY
CHARADE QUIZ (I,901)

FETCHIT, STEPHAN
CUTTER (I,1117)

FETON, MARY JEAN
ANOTHER LIFE (II,95)

FETTY, DARRELL
FRIENDS (II,928); MR. & MS. AND THE MAGIC STUDIO MYSTERY (II,1747)

FEUER, DEBRA
HARDCASE (II,1088); LACY AND THE MISSISSIPPI QUEEN (II,1419)

FEURY, PEGGY
YOUNG DR. MALONE (I,4938)

FIALA, JERI
THE $100,000 NAME THAT TUNE (II,1902); THE CROSS-WITS (II,605)

FICKETT, MARY
ALL MY CHILDREN (II,39); DREAM GIRL (I,1374); THE EDGE OF NIGHT (II,760); THE NURSES (I,3319)

FIEDLER, ARTHUR
THE BEST ON RECORD (I,409); MARY'S INCREDIBLE DREAM (II,1641)

FIEDLER, JOHN
BUFFALO BILL (II,387); CANNON (I,820); HUMAN FEELINGS (II,1203); JOE DANCER: THE MONKEY MISSION (II,1301); WINNER TAKE ALL (II,2808)

FIELD III, TED
DUNGEONS AND DRAGONS (II,774)

FIELD, BETTY
AH!, WILDERNESS (I,91); COSMOPOLITAN THEATER (I,1057); HAPPY BIRTHDAY (I,1941); LOVE STORY (I,2776); TELLER OF TALES (I,4388)

FIELD, BYRON
THEY'RE OFF (I,4444)

FIELD, CHELSEA
JUMP (II,1356)

FIELD, FLIP
QUINCY, M. E. (II,2102)

FIELD, FRANK
NOT FOR WOMEN ONLY (I,3310)

FIELD, FRITZ
LOST IN SPACE (I,2758)

FIELD, LISABETH
LONGSTREET (I,2750)

FIELD, LOGAN
A DATE WITH LIFE (I,1163)

FIELD, MARGOT
QUILLER: NIGHT OF THE FATHER (II,2100)

FIELD, MARY
TOPPER (I,4582)

FIELD, NORMAN
JOE AND MABEL (I,2398)

FIELD, SALLY
ALIAS SMITH AND JONES (I,118); ALL THE WAY HOME (II,44); BRIDGER (II,376); CALIFORNIA GIRL (I,794); THE FLYING NUN (I,1611); GIDGET (I,1795); THE GIRL WITH SOMETHING EXTRA (II,999); HEY, LANDLORD (I,2039); HITCHED (I,2065); LILY FOR PRESIDENT (II,1478)

FIELD, SYLVIA
ANNETTE (I,222); DENNIS THE MENACE (I,1231); FAITH BALDWIN'S THEATER OF ROMANCE (I,1515); FATHER KNOWS BEST (II,838); THE MAN WHO CAME TO DINNER (I,2880); MR. PEEPERS (I,3147); OUR TOWN (I,3419); THE TIMID SOUL (I,4519)

FIELD, VIRGINIA
MEET THE GIRLS: THE SHAPE, THE FACE, AND THE BRAIN (I,2990); PANTOMIME

QUIZ (I,3449)

FIELD, WILLIE
TEXACO STAR THEATER (I,4411)

FIELD-HOLDEN, REBECCA
PRIVATE BENJAMIN (II,2087)

FIELDER, JOHN
CHEERS (II,494); HITCHED (I,2065); MICKEY AND THE CONTESSA (I,3024); THE NIGHT STALKER (II,1849)

FIELDER, RICHARD
GEORGE WASHINGTON (II,978)

FIELDING, DOROTHY
BIG BEND COUNTRY (II,230); THE DOCTORS (II,694); ONE IN A MILLION (II,1906); ST. ELSEWHERE (II,2432)

FIELDING, JANET
DOCTOR WHO (II,689); DOCTOR WHO—THE FIVE DOCTORS (II,690)

FIELDING, MARJORIE
DOUGLAS FAIRBANKS JR. PRESENTS THE RHEINGOLD THEATER (I,1364)

FIELDING, PAUL
AT YOUR SERVICE (II,118)

FIELDING, TOM
MOMENT OF TRUTH (I,3084)

FIELDMAN, MICHAEL
SMILEY'S PEOPLE (II,2383)

FIELDS, ARTHUR
YOUR SHOW TIME (I,4958)

FIELDS, BONNI LYNN
THE MICKEY MOUSE CLUB (I,3025); THE MOUSEKETEERS REUNION (II,1742)

FIELDS, CHARLIE
SHANNON (II,2314)

FIELDS, CHIP
THE AMAZING SPIDER-MAN (II,63); BROTHERS (II,381); CHANGE AT 125TH STREET (II,480); DAYS OF OUR LIVES (II,629); WHAT'S HAPPENING!! (II,2769)

FIELDS, DARLENE
MEET THE GIRLS: THE SHAPE, THE FACE, AND THE BRAIN (I,2990)

FIELDS, JERE
CURIOSITY SHOP (I,1114); THE KID SUPER POWER HOUR WITH SHAZAM (II,1386); T.L.C. (II,2525)

FIELDS, JIMMY
HAPPY ENDINGS (II,1085)

FIELDS, JOAN
FRONT ROW CENTER (I,1693); THE RAY KNIGHT REVUE (I,3733)

FIELDS, KIM
THE ACADEMY (II,3); BABY, I'M BACK! (II,130); BATTLE OF THE NETWORK STARS (II,178); THE FACTS OF LIFE (II,805); FACTS OF LIFE GOES TO PARIS (II,806); GOOD EVENING, CAPTAIN (II,1031); JO'S COUSINS (II,1293); THE KID WITH THE BROKEN HALO (II,1387); THE PARKERS (II,1959)

FIELDS, MAURIE
A TOWN LIKE ALICE (II,2652)

FIELDS, SIDNEY
THE ABBOTT AND COSTELLO SHOW (I,2); THE FRANK SINATRA SHOW (I,1664); THE JACK BENNY SPECIAL (I,2296); JACKIE GLEASON AND HIS AMERICAN SCENE MAGAZINE (I,2321); SHOW BIZ (I,4027)

FIELDS, THOR
CAMELOT (II,416)

FIELDS, TOTIE
ALAN KING IN LAS VEGAS, PART I (I,97); THE BOB HOPE SHOW (I,643); THE MANY FACES OF COMEDY (I,2894)

FIFIELD, ELAINE
CINDERELLA (I,955)

FIFTH DIMENSION, THE
THE BOBBY SHERMAN SPECIAL (I,676); BURT BACHARACH IN SHANGRI-LA (I,773); THE FIFTH DIMENSION TRAVELING SHOW (I,1572); FRANCIS ALBERT SINATRA DOES HIS THING (I,1658); POLLY BERGEN SPECIAL (I,3644); MAKE MINE RED, WHITE, AND BLUE (I,2841); ONE NIGHT STANDS (I,3387); THE WOODY ALLEN SPECIAL (I,4904)

FILES, GARY
CAPTAIN SCARLET AND THE MYSTERONS (I,837)

FILIPI, CARMEN
FREEBIE AND THE BEAN (II,922); THE LAST CONVERTIBLE (II,1435); ONE NIGHT BAND (II,1909)

FIMPLE, DENNIS
ALIAS SMITH AND JONES (I,118); ALIAS SMITH AND JONES (I,119); B.J. AND THE BEAR (II,127); CENTENNIAL (II,477); GIDGET GETS MARRIED (I,1796); MATT HOUSTON (II,1654); SAM HILL: WHO KILLED THE MYSTERIOUS MR. FOSTER? (I,3896); THE WILD WOMEN OF CHASTITY GULCH (II,2801); THE YOUNG PIONEERS (II,2868)

FINA, JACK
THE SWINGIN' SINGIN' YEARS (I,4319)

FINALY, PETER
ALL THE RIVERS RUN (II,43)

FINCH, BOB
THE HALLOWEEN THAT ALMOST WASN'T (II,1075)

FINCH, JACK
THE MISS AND THE MISSILES (I,3062)

FINCH, ROGER
THE BENNY HILL SHOW (II,207); THE UNEXPURGATED BENNY HILL SHOW (II,2705)

FINE, LARRY
THE COMICS (I,1028); THE NEW THREE STOOGES (I,3274)

FINK, JOHN
HIGH RISK (II,1146); MARY HARTMAN, MARY HARTMAN (II,1638); NANCY (I,3221); RANSOM FOR A DEAD MAN (I,3722); TOPPER RETURNS (I,4583); WHERE'S THE FIRE? (II,2785)

FINLAY, FRANK
CASANOVA (II,451)

FINLEY, EILEEN
THE EDGE OF NIGHT (II,760)

FINLEY, PAT
THE BOB NEWHART SHOW (II,342); FLANNERY AND QUILT (II,873); FROM A BIRD'S EYE VIEW (I,1686); THE FUNNY SIDE (I,1714); THE FUNNY SIDE (I,1715); KEEPING UP WITH THE JONESES (II,1374); OVER AND OUT (II,1938); THE ROCKFORD FILES (II,2197)

FINN, CHRISTINE
THUNDERBIRDS (I,4497)

FINN, FRED
HAPPY TIMES ARE HERE AGAIN (I,1944); JACK CASSIDY'S ST. PATRICK'S DAY SPECIAL (I,2307); THE MICKIE FINN SPECIAL (I,3028); MICKIE FINN'S (I,3029); THE MICKIE FINNS FINALLY PRESENT HOW THE WEST WAS LOST (II,1695)

FINN, JANICE
1915 (II,1853)

FINN, LILA
COUNTERATTACK: CRIME IN AMERICA (II,581)

FINN, MICKIE
HAPPY TIMES ARE HERE AGAIN (I,1944); JACK CASSIDY'S ST. PATRICK'S DAY SPECIAL (I,2307); THE JULIE LONDON SPECIAL (I,2484); THE MICKIE FINN SPECIAL (I,3028); MICKIE FINN'S (I,3029); THE MICKIE FINNS FINALLY PRESENT HOW THE WEST WAS LOST (II,1695)

FINN, SALLY
THE LAWRENCE WELK SHOW (I,2643)

FINNEGAN, JOHN
FITZ AND BONES (II,867); LUCAN (II,1553); TERROR AT ALCATRAZ (II,2563)

FINNERTY, WARREN
NOTORIOUS (I,3314)

FINNEY, MARY
THE GREATEST GIFT (I,1880); HONESTLY, CELESTE! (I,2104)

FIORE AND ELDRIDGE
DEAN MARTIN PRESENTS THE GOLDDIGGERS (I,1194)

FIORE, BILL
THE CORNER BAR (I,1051); THE CORNER BAR (I,1052)

FIRBANK, ANN
BRIEF ENCOUNTER (I,724)

FIRESTONE, EDDIE
THE DICK VAN DYKE SHOW (I,1275); DRAGNET (I,1370); IRONSIDE (I,2240); THE LADY DIED AT MIDNIGHT (I,2601); MIXED DOUBLES (I,3075); THE OATH: THE SAD AND LONELY SUNDAYS (II,1874); THE TROUBLESHOOTERS (I,4614); THE UNTOUCHABLES (I,4681)

FIRESTONE, SCOTT
'TWAS THE NIGHT BEFORE CHRISTMAS (II,2677); ANOTHER WORLD (II,97)

FIRM OF HODGES, THE
ANN-MARGRET—WHEN YOUR SMILING (I,219)

FIRST EDITION, THE
THE MIKE DOUGLAS CHRISTMAS SPECIAL (I,3040)

FIRTH, PETER
HERE COME THE DOUBLE DECKERS (I,2025)

FISCHER, BRUCE
PALMS PRECINCT (II,1946); THE SHERIFF AND THE ASTRONAUT (II,2326)

FISCHER, COREY
MR. & MS. AND THE BANDSTAND MURDERS (II,1746); SUNSHINE (II,2495)

FISH, TED
HENNESSEY (I,2013)

FISHBEIN, BEN
THE O'NEILLS (I,3392)

FISHBURNE, LAURENCE
ONE LIFE TO LIVE (II,1907); THE SIX O'CLOCK FOLLIES (II,2368)

FISHER, AL
ALL IN FUN (I,131)

FISHER, BRUCE
GOOBER AND THE TRUCKERS' PARADISE (II,1029)

FISHER, CARRIE
RINGO (II,2176)

FISHER, CHET
H.M.S. PINAFORE (I,2066)

FISHER, CINDY
HELLINGER'S LAW (II,1127); SWISS FAMILY ROBINSON (II,2516); YOU ARE THE JURY (II,2854)

FISHER, COLIN
FILE IT UNDER FEAR (I,1574)

FISHER, DOUG
MAN ABOUT THE HOUSE (II,1611)

FISHER, EDDIE
THE BOB HOPE SHOW (I,553); THE BOB HOPE SHOW (I,557); COKE TIME WITH EDDIE FISHER (I,996); DATELINE (I,1165); THE EDDIE FISHER SHOW (I,1409); THE EDDIE FISHER SPECIAL (I,1410); THE FORD 50TH ANNIVERSARY SHOW (I,1630); THE GEORGE GOBEL SHOW (I,1768); OPERATION ENTERTAINMENT (I,3400); SHOW OF THE YEAR (I,4031)

FISHER, FRANCES
THE EDGE OF NIGHT (II,760)

FISHER, GAIL
MANNIX (II,1624)

FISHER, GEORGE
BOSTON AND KILBRIDE (II,356); KEY WEST (I,2528); THE MURDOCKS AND THE MCCLAYS (I,3163)

FISHER, HAM
KIDS AND COMPANY (I,2534)

FISHER, JACK
SHOW BIZ (I,4027)

FISHER, JIM
LAUGH TRAX (II,1443); TWILIGHT THEATER (II,2685)

FISHER, LOLA
MISSION MAGIC (I,3068); TEACHER'S PET (I,4367)

FISHER, LOU
ALL IN FUN (I,131)

FISHER, MADELEINE
THE CHOPPED LIVER BROTHERS (II,514); THE KILLIN' COUSIN (II,1394)

FISHER, NANCY
GENERAL HOSPITAL (II,964)

FISHER, NELLIE
AMERICAN SONG (I,172); MELODY TOUR (I,3001); YOUR SHOW OF SHOWS (I,4957); YOUR SHOW OF SHOWS (II,2875)

FISHER, PHIL
LAND OF HOPE (II,1424)

FISHER, SCOTT
THE HUDSON BROTHERS RAZZLE DAZZLE COMEDY SHOW (II,1201)

FISHER, SHUG
THE BEVERLY HILLBILLIES (I,417); JUBILEE U.S.A. (I,2461); THE RETURN OF THE BEVERLY HILLBILLIES (II,2139)

FISHKA, RAIS
THE HILARIOUS HOUSE OF FRIGHTENSTEIN (II,1153)

FISK, MARILYN
F TROOP (I,1499)

FISK, MARTIN
A HORSEMAN RIDING BY (II,1182)

FISKE, ALISON
THE ROADS TO FREEDOM (I,3802)

FIST, FLETCHER
COMBAT (I,1011)

FITCH, LOUISE
MEDICAL CENTER (II,1675)

FITCH, ROBERT
OPERATION GREASEPAINT (I,3402)

FITE, BEVERLY
QUADRANGLE (I,3703); UKULELE IKE (I,4664)

FITHIAN, JEFF
PLEASE DON'T EAT THE DAISIES (I,3628)

FITHIAN, JOE
PLEASE DON'T EAT THE DAISIES (I,3628)

FITTS, RICK
THE BOYS IN BLUE (II,359)

FITZ, ERICA
DARK SHADOWS (I,1157)

FITZELL, ROY
THIS IS YOUR MUSIC (I,4462)

FITZGERALD, BARRY
ALFRED HITCHCOCK PRESENTS (I,115); FORD THEATER HOUR (I,1635); MAGNAVOX THEATER (I,2827)

FITZGERALD, ELLA
ALL-STAR SWING FESTIVAL (II,53); THE BIG BAND AND ALL THAT JAZZ (I,422); THE CAPTAIN AND TENNILLE SONGBOOK (II,432); THE CARPENTERS (II,445); FRANK SINATRA (I,1660); THE FRANK SINATRA TIMEX SHOW (I,1668); FRANK SINATRA: A MAN AND HIS MUSIC (I,1662); MUSIC '55 (I,3168); PERRY COMO'S SUMMER SHOW (I,3545); SWING INTO SPRING (I,4317)

FITZGERALD, FERN
DALLAS (II,614); GOSSIP (II,1043)

FITZGERALD, GERALDINE
THE BARRETTS OF WIMPOLE STREET (I,361); THE BEST OF EVERYTHING (I,406); KENNEDY (II,1377); THE MOON AND SIXPENCE (I,3098); OH MADELINE (II,1880); OUR PRIVATE WORLD (I,3417); STUDIO ONE (I,4268); TELLER OF TALES (I,4388)

FITZGERALD, MIKE
LIVE FROM THE LONE STAR (II,1501)

FITZGERALD, NUALA
DOCTOR SIMON LOCKE (I,1317); HIGH HOPES (II,1144); PAUL BERNARD—PSYCHIATRIST (I,3485)

FITZPATRICK, AILEEN
ANGIE (II,80); SILVER SPOONS (II,2355)

FITZPATRICK, JOHN
ANOTHER WORLD (II,97)

FITZPATRICK, RICHARD
THE WORLD BEYOND (II,2835)

FITZSIMMONS, JEANNIE
THE JORDAN CHANCE (II,1343); RISE AND SHINE (II,2179)

FITZSIMMONS, TOM
THE PAPER CHASE (II,1949); THE PAPER CHASE: THE SECOND YEAR (II,1950)

FITZWILLIAMS, NEAL
YES MINISTER (II,2848)

FIVE KING COUSINS, THE
KRAFT SUMMER MUSIC HALL (I,2590)

FIX, PAUL
THE CITY (II,541); THE INDIAN (I,2210); THE RIFLEMAN (I,3789); SANDBURG'S LINCOLN (I,3907)

FJELD, JULIANNA
HEAR NO EVIL (II,1116)

FLACK, ROBERTA
DUKE ELLINGTON. . . WE LOVE YOU MADLY (I,1380); JOHN LENNON AND YOKO ONO PRESENT THE ONE-TO-ONE CONCERT (I,2414); MARLO THOMAS AND FRIENDS IN FREE TO BE. . . YOU AND ME (II,1632); MONSANTO PRESENTS MANCINI (I,3092); MUSIC CENTRAL (II,1771); ROBERTA FLACK. . . THE FIRST TIME EVER (I,3817)

FLACKS, NIKI
ONE LIFE TO LIVE (II,1907)

FLAGG, FANNIE
THE BOBBIE GENTRY SPECIAL (I,670); CANDID CAMERA (I,819); CANDID CAMERA (II,420); COMEDY NEWS (I,1020); COMEDY NEWS II (I,1021); HARPER VALLEY (II,1094); HARPER VALLEY PTA (II,1095); HOME COOKIN' (II,1170); THE NEW DICK VAN DYKE SHOW (I,3262); WONDER WOMAN (PILOT 2) (II,2827)

FLAGG, LAURA
LITTLE WOMEN (I,2724)

FLAHERTY, JOE
FROM CLEVELAND (II,931); SECOND CITY TELEVISION (II,2287)

FLANAGAN, FIONNUALA
HOW THE WEST WAS WON (II,1196); THE PICTURE OF DORIAN GRAY (I,3592); RICH MAN, POOR MAN—BOOK I (II,2161)

FLANAGAN, JOHN
THE DOUBLE KILL (I,1360)

FLANAGAN, KELLIE
ALL THINGS BRIGHT AND BEAUTIFUL (I,141); THE GHOST AND MRS. MUIR (I,1786)

FLANAGAN, KIT
BRAIN GAMES (II,366)

FLANAGAN, NEIL
CASINO (II,452); GOOD TIME HARRY (II,1037)

FLANAGAN, PAMELA
LITTLE MOON OF ALBAN (I,2714)

FLANAGAN, PAULINE
RAGE OF ANGELS (II,2107)

FLANAGAN, RON
MASTER OF THE GAME (II,1647)

FLANDERS, ED
BACKSTAIRS AT THE WHITE HOUSE (II,133); THE OATH: THE SAD AND LONELY SUNDAYS (II,1874); SANDBURG'S LINCOLN (I,3907); ST. ELSEWHERE (II,2432); TRUMAN AT POTSDAM (I,4618)

FLANEGAN, BRIAN
LET'S TAKE A TRIP (I,2668)

FLANINGAM, LOUISA
GEORGE WASHINGTON (II,978); THE ROCK RAINBOW (II,2194)

FLANNERY, SUSAN
DALLAS (II,614); DAYS OF OUR LIVES (II,629); MELODY OF HATE (I,2999); THE MONEYCHANGERS (II,1724); WOMEN IN WHITE (II,2823)

FLANNIGAN, BILL
GUY LOMBARDO AND HIS ROYAL CANADIANS (I,1910)

FLANNIGAN, DENNIS
THE JOHN BYNER COMEDY HOUR (I,2407)

FLASH, LADY
THE BARRY MANILOW SPECIAL (II,155)

FLATELY, WILLIAM
PEOPLE LIKE US (II,1999)

FLATT, EARL
FLATT AND SCRUGGS (I,1594)

FLATT, LESTER
THE BEVERLY HILLBILLIES (I,417); COUNTRY NIGHT OF STARS II (II,593); FOLK SOUND U.S.A. (I,1614)

FLAVIN, JAMES
FRANK MERRIWELL (I,1659); GALLAGHER (I,1727); THE GEORGE BURNS AND GRACIE ALLEN SHOW (I,1763); MAN WITH A CAMERA (I,2883); MEET THE GOVERNOR (I,2991); THE SOFT TOUCH (I,4107)

FLAX, MARGO
LOVE OF LIFE (I,2774)

FLEER, ALICIA
FRIENDS (II,927)

FLEETWOOD, MICK
FLEETWOOD MAC IN CONCERT (II,880)

FLEISCHER, CHARLES
ALOHA PARADISE (II,58); BLUE JEANS (II,275); CHECKING IN (II,492); KEEP ON TRUCKIN' (II,1372); THE RICHARD PRYOR SHOW (II,2163); SUGAR TIME (II,2489); THICKE OF THE NIGHT (II,2587); WACKO (II,2733); WELCOME BACK, KOTTER (II,2761)

FLEISCHMAN, LARRY
ALL MY CHILDREN (II,39); INSIDE O.U.T. (II,1235)

FLEISS, JANE
THE DOCTORS (II,694)

FLEMING, ART
THE CALIFORNIANS (I,795); DOCTOR I.Q. (I,1314); INTERNATIONAL DETECTIVE (I,2224); JEOPARDY (II,1279); THE MAN AGAINST CRIME (I,2853); NBC ADVENTURE THEATER (I,3235)

FLEMING, ERIC
MAJOR DELL CONWAY OF THE FLYING TIGERS (I,2835); RAWHIDE (I,3727); STUDIO ONE (I,4268)

FLEMING, ERIN
THE NEW BILL COSBY SHOW (I,3257)

FLEMING, KARA
WHERE THE HEART IS (I,4831)

FLEMING, LUCY
PEGGY FLEMING AT SUN VALLEY (I,3507); THE PEGGY FLEMING SHOW (I,3508); PEGGY FLEMING VISITS THE SOVIET UNION (I,3509); SMILEY'S PEOPLE (II,2383); TO EUROPE WITH LOVE (I,4525); WOMEN OF RUSSIA (II,2824)

FLEMING, PEGGY
THE MAC DAVIS CHRISTMAS SPECIAL (II,1572); THE OSMOND FAMILY CHRISTMAS SPECIAL (II,1926); THE PERRY COMO CHRISTMAS SHOW (II,2002)

FLEMING, RHONDA
THE BOB HOPE SHOW (I,558); THE BOB HOPE SHOW (I,611); BOB HOPE SPECIAL: BOB HOPE'S ROAD TO HOLLYWOOD (II,313); THE DICK POWELL SHOW (I,1269); FORD TELEVISION THEATER (I,1634); HAVE GIRLS—WILL TRAVEL (I,1967); LEGENDS OF THE SCREEN (II,1456); THE RED SKELTON CHEVY SPECIAL (I,3753); STAGE DOOR (I,4178)

FLEMING, ROBERT
THE DATCHET DIAMONDS (I,1160); REUNION IN VIENNA (I,3780); THE WOMAN IN WHITE (I,4887)

FLEMING, SUSAN
THE INNER SANCTUM (I,2216)

FLEMING, TONI
SENSE OF HUMOR (II,2303)

FLESER, APRIL
OF MEN OF WOMEN (I,3332)

FLETCHER, AARON
LEWIS AND CLARK (II,1465)

FLETCHER, BARRIE
ONLY A SCREAM AWAY (I,3393)

FLETCHER, BILL
ALIAS SMITH AND JONES (I,119); ENIGMA (II,778); THE MAN WITH THE POWER (II,1616)

FLETCHER, BRAMWELL
THE PARADINE CASE (I,3453)

FLETCHER, DIANE
ROGER DOESN'T LIVE HERE ANYMORE (II,2201)

FLETCHER, JACK
THE ALAN KING SHOW (I,100); ANNIE FLYNN (II,92); THE BOB CRANE SHOW (II,280); CALUCCI'S DEPARTMENT (I,805); GRADY (II,1047); PINOCCHIO (I,3603); PRESENTING SUSAN ANTON (II,2077); TRUE GRIT (II,2664); WONDERFUL TOWN (I,4893)

FLETCHER, JAY
CONCRETE BEAT (II,562)

FLETCHER, LARRY
DECOY (I,1217); PENTAGON U.S.A. (I,3514); THE TRAP (I,4593)

FLETCHER, LESTER
THE ASPHALT COWBOY (II,111)

FLETCHER, PAGE
THE HITCHHIKER (II,1157)

FLETCHER, SKIP
INDEMNITY (I,2209)

FLETCHER, STEVE
ONE LIFE TO LIVE (II,1907)

FLETCHER, WILLIAM
CINDERELLA (II,526)

FLICKER, BARBARA
LANIGAN'S RABBI (II,1427)

FLINT, SAM
BROKEN ARROW (I,744); FATHER KNOWS BEST (II,838)

FLIPETTES, THE
THE FLIP WILSON SHOW (I,1602)

FLIPPEN, JAY C
THE DEAN JONES SHOW (I,1191); ENSIGN O'TOOLE (I,1448); FAME IS THE NAME OF THE GAME (I,1518); GENTRY'S PEOPLE (I,1762); LAST OF THE PRIVATE EYES (I,2628); SHOW BIZ (I,4027); THE SOUND OF ANGER (I,4133); THRILLER (I,4492)

FLIPPEN, LUCY LEE
FLO (II,891); LITTLE HOUSE ON THE PRAIRIE (II,1487)

FLOAT, RAY
DOCTOR WHO—THE FIVE DOCTORS (II,690)

FLOOD, ANN
THE EDGE OF NIGHT (II,760); FROM THESE ROOTS (I,1688)

FLORANCE, SHEILA
PRISONER: CELL BLOCK H (II,2085)

FLOREN, MYRON
THE DANNY THOMAS SPECIAL (I,1151); THE LAWRENCE WELK SHOW (I,2643); THE LAWRENCE WELK SHOW (I,2644); LAWRENCE WELK'S TOP TUNES AND NEW TALENT (I,2645); MEMORIES WITH LAWRENCE WELK (II,1681)

FLORES, JOSE
ARK II (II,107); THE BANG-SHANG LALAPALOOZA SHOW (II,140)

FLOREZ, ANGEL
THE NEW MICKEY MOUSE CLUB (II,1823)

FLORY, MED
THE DEADLY GAME (II,633); KEY TORTUGA (II,1383); LITTLE VIC (II,1496); MIKE AND THE MERMAID (I,3039);

THE RAY ANTHONY SHOW (I,3728)

FLOUNDERS, JACK
THE FANTASTIC FOUR (I,1529)

FLOWER, BELLE
BORN YESTERDAY (I,692)

FLOWER, GILLY
FAWLTY TOWERS (II,843)

FLOWERS, WAYLAND
THE ANDY WILLIAMS SHOW (II,78); THE BEATRICE ARTHUR SPECIAL (II,196); KEEP ON TRUCKIN' (II,1372); LAUGH-IN (II,1444); MADAME'S PLACE (II,1587); MEN WHO RATE A "10" (II,1684); SOLID GOLD (II,2395)

FLUEGEL, DARLANNE
CONCRETE BEAT (II,562)

FLYNN, BERNADINE
THE GOOK FAMILY (I,1858); HAWKINS FALLS, POPULATION 6200 (I,1977)

FLYNN, BILL
PRISONERS OF THE LOST UNIVERSE (II,2086)

FLYNN, ERIC
A KILLER IN EVERY CORNER (I,2538)

FLYNN, ERROL
ALCOA/GOODYEAR THEATER (I,107); THE ERROL FLYNN THEATER (I,1458); SCREEN DIRECTOR'S PLAYHOUSE (I,3946)

FLYNN, JOE
THE ADVENTURES OF OZZIE AND HARRIET (I,71); AN AMATEUR'S GUIDE TO LOVE (I,154); THE BARBARA EDEN SHOW (I,346); THE BOB NEWHART SHOW (I,666); THE JOEY BISHOP SHOW (I,2403); MCHALE'S NAVY (I,2969); THE MOUSE FACTORY (I,3113); READY AND WILLING (II,2115); SKYFIGHTERS (I,4078); THE TIM CONWAY SHOW (I,4502); THE TIM CONWAY SPECIAL (I,4503); VER-R-R-RY INTERESTING (I,4712)

FLYNN, MIRIAM
FULL HOUSE (II,937); MAGGIE (II,1589); MR. SUCCESS (II,1756); SILVER SPOONS (II,2355); THE TIM CONWAY SHOW (II,2619)

FLYNN, SALLY
THE LAWRENCE WELK SHOW (I,2643); MEMORIES WITH LAWRENCE WELK (II,1681)

FOAD, GENE
PENMARRIC (II,1993)

FOCH, NINA
THE CHEVROLET TELE-THEATER (I,926); EBONY, IVORY AND JADE (II,755); FAITH BALDWIN'S THEATER OF ROMANCE (I,1515); GIDGET GROWS UP (I,1797); GULF PLAYHOUSE (I,1904); HERCULE POIROT (I,2021); LIGHT'S OUT (I,2699); THE NASH AIRFLYTE THEATER (I,3227); PLAYWRIGHTS '56 (I,3627); POTTSVILLE (II,2072); PRESCRIPTION: MURDER (I,3660); REBECCA (I,3745); SUSPENSE (I,4305); TEN LITTLE INDIANS (I,4394); THE UNITED STATES STEEL HOUR (I,4677)

FODDY, RALPH
THE CHICAGO STORY (II,507)

FOGEL, JERRY
THE MOTHERS-IN-LAW (I,3110); ROSENTHAL AND JONES (II,2215); SINGLES (I,4062); THE WHITE SHADOW (II,2788)

FOGEL, LEE
THIS IS MUSIC (I,4458)

FOL, BOB
CHERYL LADD: SCENES FROM A SPECIAL (II,503)

FOLEY, BARBARA
A TIME TO LIVE (I,4510)

FOLEY, DANIEL
MASTER OF THE GAME (II,1647)

FOLEY, ELLEN
NIGHT COURT (II,1844); THREE GIRLS THREE (II,2608)

FOLEY, JOHN
PUMPBOYS AND DINETTES ON TELEVISION (II,2090)

FOLEY, JOSEPH
THE ALDRICH FAMILY (I,111); MR. PEEPERS (I,3147)

FOLEY, KARIN
A HORSEMAN RIDING BY (II,1182)

FOLEY, LOUISE
FAMILY (II,813)

FOLEY, RED
THE GRAND OLE OPRY (I,1866); JUBILEE U.S.A. (I,2461); MR. SMITH GOES TO WASHINGTON (I,3151)

FOLK DANCERS, THE
COUNTRY STYLE (I,1072)

FOLLETT, CHARON
THOSE ENDEARING YOUNG CHARMS (I,4469)

FOLLOWS, MEGAN
THE BAXTERS (II,184); DOMESTIC LIFE (II,703); JO'S COUSINS (II,1293)

FOLQUET, DANIELLE
YOU ASKED FOR IT (II,2855)

FONDA, HENRY
THE AMERICAN WEST OF JOHN FORD (I,173); CAPTAINS AND THE KINGS (II,435); CHRYSLER MEDALLION THEATER (I,951); THE DEPUTY (I,1234); THE DICK POWELL SHOW (I,1269); THE FABULOUS 50S (I,1501); FAMILY (II,813); GENERAL ELECTRIC'S ALL-STAR ANNIVERSARY (II,963); HENRY FONDA AND THE FAMILY (I,2015); HENRY FONDA PRESENTS THE STAR AND THE STORY (I,2016); HOLLYWOOD: THE SELZNICK YEARS (I,2095); HOWDY (I,2150); THE MARCH OF DIMES BENEFIT SHOW (I,2900); THE OLDEST LIVING GRADUATE (II,1882); PAT PAULSEN FOR PRESIDENT (I,3472); THE PETRIFIED FOREST (I,3570); RHEINGOLD THEATER (I,3783); ROOTS: THE NEXT GENERATIONS (II,2213); THE SENSATIONAL, SHOCKING, WONDERFUL, WACKY 70S (II,2302); THE SMITH FAMILY (I,4092); STAND UP AND CHEER (I,4188); SUMMER SOLSTICE (II,2492); THAT WAS THE WEEK THAT WAS (I,4422); TRAVELS WITH CHARLEY (I,4597)

FONDA, JANE
9 TO 5 (II,1852); THE HELEN REDDY SPECIAL (II,1126); I LOVE LIBERTY (II,1215); LILY FOR PRESIDENT (II,1478); LILY—SOLD OUT (II,1480); THE SENSATIONAL, SHOCKING, WONDERFUL, WACKY 70S (II,2302); A STRING OF BEADS (I,4263)

FONDA, PETER
CIRCUS OF THE STARS (II,530); CIRCUS OF THE STARS (II,531); THE RETURN OF THE SMOTHERS BROTHERS (I,3777)

FONG, BENSON
BACHELOR FATHER (I,323); FAMILY IN BLUE (II,818); MOONLIGHT (II,1730); SAINT PETER (II,2238)

FONG, BRIAN
CHASE (I,917); THE HARDY BOYS (I,1948)

FONG, FRANCES
BACHELOR FATHER (I,323); BROTHERS (II,381)

FONG, HAROLD
THE GREEN FELT JUNGLE (I,1885); HONG KONG (I,2112)

FONG, KAM
HAWAII FIVE-O (I,1972); HAWAII FIVE-O (II,1110)

FONTAINE SISTERS, THE
ALL IN FUN (I,131)

FONTAINE, BEA
THE CHESTERFIELD SUPPER CLUB (I,923); THE PERRY COMO SHOW (I,3531); THE PERRY COMO SHOW (I,3532)

FONTAINE, EDDIE
THE BARBARY COAST (II,146); THE GALLANT MEN (I,1728); HAPPY DAYS (II,1084)

FONTAINE, FRANK
FRONT ROW CENTER (I,1693); JACKIE GLEASON AND HIS AMERICAN SCENE MAGAZINE (I,2321); THE SCOTT MUSIC HALL (I,3943); THE SWIFT SHOW (I,4316)

FONTAINE, GERI
THE CHESTERFIELD SUPPER CLUB (I,923); THE PERRY COMO SHOW (I,3531); THE PERRY COMO SHOW (I,3532)

FONTAINE, JOAN
THE BOB HOPE SHOW (I,611); FORD TELEVISION THEATER (I,1634); FOUR STAR PLAYHOUSE (I,1652); HOLLYWOOD: THE SELZNICK YEARS (I,2095); ON TRIAL (I,3363); ONE STEP BEYOND (I,3388); PERSPECTIVE ON GREATNESS (I,3555); STAR STAGE (I,4197); THE THREE MUSKETEERS (I,4481)

FONTAINE, LORAINE
GAY NINETIES REVUE (I,1743)

FONTAINE, LYNN
ANASTASIA (I,182); THE GREAT SEBASTIANS (I,1877); THE MAGNIFICENT YANKEE (I,2829)

FONTAINE, MARGIE
THE CHESTERFIELD SUPPER CLUB (I,923); THE PERRY COMO SHOW (I,3531); THE PERRY COMO SHOW (I,3532)

FONTAINE, MICHAEL
SHAPING UP (II,2316)

FONTAINES, THE
CAVALCADE OF STARS (I,877)

FONTANA, D J
ELVIS (I,1435)

FONTANE, CHAR
BANYON (I,344); JOE AND VALERIE (II,1298); JOE AND VALERIE (II,1299); PEARL (II,1986)

FONTE, ALLISON
THE NEW MICKEY MOUSE CLUB (II,1823)

FONTELIEU, STOCKER
THE BIG EASY (II,234)

FONTEYN, MARGOT
CINDERELLA (I,955); SLEEPING BEAUTY (I,4083)

FOODY, RALPH
TWIGS (II,2684)

FOOTE, BRUCE
THIS IS MUSIC (I,4457)

FORAKER, LOIS
AFTERMASH (II,23)

FORAN, DICK
THE BOB HOPE SHOW (I,546); BURLESQUE (I,765); THE CHEVROLET TELE-THEATER (I,926); DAMON RUNYON THEATER (I,1129); FIVE'S A FAMILY (I,1591); FOREST RANGER (I,1641); MIRACLE ON 34TH STREET (I,3056); O.K. CRACKERBY (I,3348); OFF WE GO (I,3338); THE PHILADELPHIA STORY (I,3581); STAGE 7 (I,4181); SWAMP FOX (I,4311)

FORAY, JUNE
THE ALVIN SHOW (I,146); ANDY'S GANG (I,214); ARTHUR GODFREY IN HOLLYWOOD (I,281); THE BUGS BUNNY/ROAD RUNNER SHOW (II,391); THE BULLWINKLE SHOW (I,761); CURIOSITY SHOP (I,1114); DEATH OF A SALESMAN (I,1208); THE DUDLEY DO-RIGHT SHOW (I,1378); FROSTY THE SNOWMAN (II,934); GEORGE OF THE JUNGLE (I,1773); ROCKY AND HIS FRIENDS (I,3822); THE SMURFS (II,2385); THE TOM AND JERRY SHOW (I,4534); THE WOODY WOODPECKER SHOW (I,4906)

FORBES, BRENDA
ANASTASIA (I,182); THE BIG TIME (I,434); BLITHE SPIRIT (I,484); THE MAGNIFICENT YANKEE (I,2829); THE NATURAL LOOK (II,1799)

FORBES, JULIET
BROTHER RAT (I,746)

FORBES, KENNETH
THE GIRLS (I,1811)

FORBES, ROBERTSON
THE FAR PAVILIONS (II,832)

FORBES, SCOTT
THE ADVENTURES OF JIM BOWIE (I,63); FIRESIDE THEATER (I,1580); FOUR STAR PLAYHOUSE (I,1652); FRONTIER (I,1695); G.E. SUMMMER ORIGINALS (I,1747); LAURA (I,2635); THE LONER (I,2744); THE SEEKING HEART (I,3975); SOMEONE AT THE TOP OF THE STAIRS (I,4114)

FORBES, SHEILA
HULLABALOO (I,2156)

FORD, ANITRA
THE NEW PRICE IS RIGHT (I,3270); THE PRICE IS RIGHT (I,3665); THE PRICE IS RIGHT (II,2079); WE'VE GOT EACH OTHER (II,2757); WONDER WOMAN (PILOT 1) (II,2826)

FORD, ART
THE ART FORD SHOW (I,275)

FORD, BARRY
THE BEST OF EVERYTHING (I,406); THE EDGE OF NIGHT (II,760); RIVKIN: BOUNTY HUNTER (II,2183)

FORD, BETTE
NOT IN FRONT OF THE KIDS (II,1861)

FORD, CAROLE ANN
DOCTOR WHO (II,689); DOCTOR WHO—THE FIVE DOCTORS (II,690)

FORD, CONSTANCE
ANOTHER WORLD (II,97); COLONEL HUMPHREY J. FLACK (I,1004); THE GOLDEN AGE OF TELEVISION (II,1018); KRAFT TELEVISION THEATER (I,2592); MATINEE THEATER (I,2947); PHILCO TELEVISION PLAYHOUSE (I,3583); SEARCH FOR TOMORROW (II,2284); STUDIO ONE (I,4268); SUSPENSE (I,4305); THRILLER (I,4492); WOMAN WITH A PAST (I,4889)

FORD, CORKEY
THE RENEGADES (II,2132); YOUNG LIVES (II,2866)

FORD, DAVID
ADAM'S RIB (I,32); DARK SHADOWS (I,1157); SEARCH FOR TOMORROW (II,2284)

FORD, FRANCIS
GRUEN GUILD PLAYHOUSE (I,1892)

FORD, GERALD R
BOB HOPE SPECIAL: BOB HOPE FOR PRESIDENT (II,282); BOB HOPE SPECIAL: BOB HOPE PRESENTS A CELEBRATION WITH STARS OF COMEDY AND MUSIC (II,340)

FORD, GLENN
AMERICA (I,164); CADE'S COUNTY (I,787); THE FAMILY HOLVAK (II,817); GLENN FORD'S SUMMERTIME, U.S.A. (I,1823); THE GOLDDIGGERS (I,1838); THE GREATEST GIFT (II,1061); HOWDY (I,2150); JARRETT (I,2349); ONCE AN EAGLE (II,1897); PUNCH AND JODY (II,2091); THE ROWAN AND MARTIN SPECIAL (I,3855); SINATRA—THE FIRST 40 YEARS (II,2359); WHEN HAVOC STRUCK (II,2778); WHEN THE WEST WAS FUN: A WESTERN REUNION (II,2780)

FORD, HANK
MEET BETTY FURNESS (I,2978)

FORD, HARRISON
DYNASTY (II,747); LEAVE IT TO BEAVER (I,2648); THE POSSESSED (II,2071)

FORD, JANIE
THE ALAN DALE SHOW (I,96)

FORD, JINEANNE
THE CIRCLE FAMILY (II,527); MICKEY SPILLANE'S MIKE HAMMER: MORE THAN MURDER (II,1693)

FORD, JOHN
THE AMERICAN WEST OF JOHN FORD (I,173)

FORD, LYNETTE
THE HANDS OF CORMAC JOYCE (I,1927)

FORD, MARY
FAMILY NIGHT WITH HORACE HEIDT (I,1523); THE GRAND OLE OPRY (I,1866); THE LES PAUL AND MARY FORD SHOW (I,2658)

FORD, MICK
A WOMAN OF SUBSTANCE (II,2820)

FORD, PAUL
AUNT JENNY'S REAL LIFE STORIES (I,314); THE BAILEYS OF BALBOA (I,334); BLOOMER GIRL (I,488); CURTAIN CALL THEATER (I,1115); JUNIOR MISS (I,2493); THE LLOYD BRIDGES SHOW (I,2733); NORBY (I,3304); THE PHIL SILVERS SHOW (I,3580); THE RIGHT MAN (I,3790); THE TEAHOUSE OF THE AUGUST MOON (I,4368)

FORD, PETER
CADE'S COUNTY (I,787); PUNCH AND JODY (II,2091)

FORD, PHIL
MIMI (I,3054); ON PARADE (I,3356)

FORD, ROSS
MEET MILLIE (I,2988); STARS OVER HOLLYWOOD (I,4211)

FORD, RUTH
CAMEO THEATER (I,808)

FORD, SENATOR EDWARD
CAN YOU TOP THIS? (I,816)

FORD, SHERYL
HELLO, LARRY (II,1128)

FORD, STEVEN
CIRCUS OF THE STARS (II,535); THE YOUNG AND THE RESTLESS (II,2862)

FORD, SUSAN
THE JIM NABORS SHOW (II,1290)

FORD, TENNESSEE ERNIE
ANDY GRIFFITH'S UPTOWN-DOWNTOWN SHOW (I,194); CITY VS. COUNTRY (I,968); COUNTRY MUSIC HIT PARADE (I,1069); COUNTRY NIGHT OF STARS (II,592); THE FABULOUS FORDIES (I,1502); FRIENDS AND NABORS (I,1684); H.M.S. PINAFORE (I,2066); KAY KYSER'S KOLLEGE OF MUSICAL KNOWLEDGE (I,2512); LOOKING BACK (I,2754); ONE MORE TIME (I,3385); OPRYLAND U.S.A. (I,3405); THE PEAPICKER IN PICCADILLY (I,3500); THE STARS AND STRIPES SHOW (I,4208); THE STARS AND STRIPES SHOW (II,2443); TENNESSEE ERNIE FORD MEETS KING ARTHUR (I,4397); TENNESSEE ERNIE FORD'S WHITE CHRISTMAS (I,4402); THE TENNESSEE ERNIE FORD SHOW (I,4398); THE TENNESSEE ERNIE FORD SHOW (I,4399); THE TENNESSEE ERNIE FORD SHOW (I,4400); THE TENNESSEE ERNIE FORD SPECIAL (I,4401); TENNESSEE ERNIE'S NASHVILLE-MOSCOW EXPRESS (II,2561); THE WONDERFUL WORLD OF BURLESQUE III (I,4897)

FORD, WALLACE
THE DEPUTY (I,1234); GETAWAY CAR (I,1783); GREAT ADVENTURE (I,1868); HEY TEACHER (I,2041); JANE

WYMAN PRESENTS THE FIRESIDE THEATER (I,2345); THE MOTOROLA TELEVISION HOUR (I,3112); THREE WISHES (I,4488)

FORDE, BRINSLEY
HERE COME THE DOUBLE DECKERS (I,2025); THE MAGNIFICENT SIX AND A HALF (I,2828); TO SIR, WITH LOVE (II,2624)

FOREE, KEN
REPORT TO MURPHY (II,2134)

FOREMAN, DEBORAH
HOT PURSUIT (II,1189)

FOREMAN, GEORGE
THE BOB HOPE SHOW (I,659)

FOREST, JACK V HARRISS
ACTION IN THE AFTERNOON (I,25)

FOREST, MICHAEL
AS THE WORLD TURNS (II,110); THE YOUNG AND THE RESTLESS (II,2862)

FORMAN, BILL
THE WHISTLER (I,4843)

FORMAN, DONNA
IT'S A MYSTERY, CHARLIE BROWN (I,2259)

FORMAN, JOEY
BARNEY AND ME (I,358); FRANKIE AVALON'S EASTER SPECIAL (I,1673); GET SMART (II,983); HEY MULLIGAN (I,2040); THE JOEY BISHOP SHOW (I,2404); A LUCILLE BALL SPECIAL STARRING LUCILLE BALL AND DEAN MARTIN (II,1557); THE MANY SIDES OF MICKEY ROONEY (I,2899); THE NEW STEVE ALLEN SHOW (I,3271); THE SID CAESAR SHOW (I,4043); THE SKY'S THE LIMIT (I,4081); THE STEVE ALLEN COMEDY HOUR (II,2454); STONE (II,2470)

FORMAN, RUTH
THE WORLD'S GREATEST SUPER HEROES (II,2842)

FORMBY, HOMER
FORMBY'S ANTIQUE WORKSHOP (I,1644)

FORMICOLA, FIL
BIG JOHN (II,239); CONCRETE BEAT (II,562); FARRELL: FOR THE PEOPLE (II,834)

FORONGY, RICHARD
HIS AND HERS (II,1155); PEOPLE LIKE US (II,1999)

FORQUET, PHILIPPE
THE YOUNG REBELS (I,4944)

FORREST, GREGG
THE BAD NEWS BEARS (II,134)

FORREST, HAL
SPACE PATROL (I,4144)

FORREST, RAY
GREAT ADVENTURES (I,1870); TELEVISION SCREEN MAGAZINE (I,4383); TV SCREEN MAGAZINE (I,4631); VILLAGE BARN (I,4733)

FORREST, SALLY
COMMAND PERFORMANCE (I,1029); FRONT ROW CENTER (I,1694); SCREEN DIRECTOR'S PLAYHOUSE (I,3946); SUSPENSE (I,4305); THE UNITED STATES STEEL HOUR (I,4677); YOU'RE THE TOP (I,4974)

FORREST, STEVE
THE BARON (I,360); CAPTAIN AMERICA (II,427); CELEBRATION: THE AMERICAN SPIRIT (II,461); CONDOMINIUM (II,566); GHOST STORY (I,1788); HONEY WEST: WHO KILLED THE JACKPOT? (I,2106); MALIBU (II,1607); THE MANIONS OF AMERICA (II,1623); ROUGHNECKS (II,2219); S.W.A.T. (II,2236); TESTIMONY OF TWO MEN (II,2564)

FORREST, WILLIAM
THE ADVENTURES OF RIN TIN TIN (I,73); MEET THE GOVERNOR (I,2991)

FORSBERG, GRANT
OLD FRIENDS (II,1881)

FORSEY, NORMAN
WELCOME TO PARADISE (II,2762)

FORSLUND, CONSTANCE
BIG BOB JOHNSON AND HIS FANTASTIC SPEED CIRCUS (II,231); GILLIGAN'S ISLAND: THE HARLEM GLOBETROTTERS ON GILLIGAN'S ISLAND (II,993); PLEASURE COVE (II,2058)

FORSTER, BRIAN
GOOBER AND THE GHOST CHASERS (II,1028); THE PARTRIDGE FAMILY (II,1962); PARTRIDGE FAMILY: 2200 A.D. (II,1963)

FORSTER, ROBERT
BANYON (I,344); BANYON (I,345); THE CITY (II,541); GOLIATH AWAITS (II,1025); HIGHER AND HIGHER, ATTORNEYS AT LAW (I,2057); NAKIA (II,1784); NAKIA (II,1785); ROYCE (II,2224)

FORSYTH, DAVID
TEXAS (II,2566)

FORSYTH, ROSEMARY
CITY BENEATH THE SEA (I,965); DAYS OF OUR LIVES (II,629); THE DEFENDERS (I,1220); IS THERE A DOCTOR IN THE HOUSE (I,2241); IS THERE A DOCTOR IN THE HOUSE? (II,1247)

FORSYTHE, BROOKE
THE JOHN FORSYTHE SHOW (I,2410)

FORSYTHE, HENDERSON
AS THE WORLD TURNS (II,110); FROM THESE ROOTS (I,1688); HOTEL COSMOPOLITAN (I,2128)

FORSYTHE, JOHN
ALCOA PREMIERE (I,109); BACHELOR FATHER (I,323); A BELL FOR ADANO (I,388); BELLE STARR (I,391); BOB HOPE SPECIAL: BOB HOPE'S MERRY CHRISTMAS SHOW (II,310); CHARLIE'S ANGELS (II,486); CHARLIE'S ANGELS (II,487); CIRCUS OF THE STARS (II,530); CLIMAX! (I,976); COSMOPOLITAN THEATER (I,1057); CURTAIN CALL THEATER (I,1115); DANGER (I,1134); THE DEAN MARTIN CELEBRITY ROAST (II,638); DOM DELUISE AND FRIENDS (II,701); DYNASTY (II,746); THE ELGIN HOUR (I,1428); EMILY, EMILY (I,1438); THE FEATHER AND FATHER GANG: NEVER CON A KILLER (II,846); FORD THEATER HOUR (I,1635); GEORGE BURNS CELEBRATES 80 YEARS IN SHOW BUSINESS (II,967); THE HEALERS (II,1115); THE JOHN FORSYTHE SHOW (I,2410); KRAFT MYSTERY THEATER (I,2589); THE LETTERS (I,2670); THE LIGHT FANTASTIC, OR HOW TO TELL YOUR PAST, PRESENT AND MAYBE YOUR FUTURE THROUGH SOCIAL DANCING (I,2696); LIGHT'S OUT (I,2699); LISA, BRIGHT AND DARK (I,2711); MAGNAVOX THEATER (I,2827); THE MISS AND THE MISSILES (I,3062); THE MUSIC MAKER (I,3173); THE MYSTERIOUS TWO (II,1783); NEW GIRL IN HIS LIFE (I,3265); PHILCO TELEVISION PLAYHOUSE (I,3583); PULITZER PRIZE PLAYHOUSE (I,3692); SCHLITZ PLAYHOUSE OF STARS (I,3936); STAGE DOOR (I,4177); STAR STAGE (I,4197); STARLIGHT THEATER (I,4201); SUSPENSE (I,4305); THE TEAHOUSE OF THE AUGUST MOON (I,4368); TO ROME WITH LOVE (I,4526); TONI'S BOYS (II,2637); ZANE GREY THEATER (I,4979)

FORSYTHE, PAGE
THE JOHN FORSYTHE SHOW (I,2410)

FORTAS, ALAN
ELVIS (I,1435)

FORTE, FABIAN
THE BOB HOPE SHOW (I,589); THE DEAN MARTIN SHOW (I,1199); THE GREATEST SHOW ON EARTH (I,1883)

FORTE, JOE
LIFE WITH LUIGI (I,2692)

FORTIER, BOB
THE BARBARA RUSH SHOW (I,349); THE TROUBLESHOOTERS (I,4614)

FORTIER, MARCELLE
THE LAWYERS (I,2646)

FORTIER, ROBERT
FULL CIRCLE (I,1702); LADY IN THE DARK (I,2602)

FORTINA, CARL
CLUB OASIS (I,985)

FORTUS, DANNY
WHERE'S THE FIRE? (II,2785)

FOSSE, BOB
THE BOB HOPE SHOW (I,626)

FOSTER, AMI
PUNKY BREWSTER (II,2092)

FOSTER, BARRY
ESPIONAGE (I,1464); A FAMILY AFFAIR (II,814); THE SEARCH (I,3950); A WOMAN CALLED GOLDA (II,2818)

FOSTER, BUDDY
THE ANDY GRIFFITH SHOW (I,192); HONDO (I,2103); MAYBERRY R.F.D. (I,2961)

FOSTER, CAROL TRU
THIS IS KATE BENNETT (II,2597)

FOSTER, CASSANDRA
LOU GRANT (II,1526)

FOSTER, CHERIE
BARNEY AND ME (I,358)

FOSTER, CONNIE
CHICAGO STORY (II,506); THE CHICAGO STORY (II,507)

FOSTER, DIANNE
CONFIDENTIALLY YOURS (I,1035); COUNTY GENERAL (I,1076); KRAFT MYSTERY THEATER (I,2589); THE

LLOYD BRIDGES SHOW (I,2733)

FOSTER, DONALD
HAZEL (I,1982)

FOSTER, FRANCES
OUR STREET (I,3418)

FOSTER, GLORIA
TOP SECRET (II,2647)

FOSTER, JEAN RENEE
BEULAH LAND (II,226)

FOSTER, JILL
IS THERE A DOCTOR IN THE HOUSE? (II,1247)

FOSTER, JODIE
THE ADDAMS FAMILY (II,12); THE AMAZING CHAN AND THE CHAN CLAN (I,157); BOB & CAROL & TED & ALICE (I,495); THE COURTSHIP OF EDDIE'S FATHER (I,1083); GHOST STORY (I,1788); LOVE STORY (I,2778); MY SISTER HANK (I,3201); MY THREE SONS (I,3205); PAPER MOON (II,1953); THE PARTRIDGE FAMILY (II,1962); SMILE JENNY, YOU'RE DEAD (II,2382)

FOSTER, JULIA
MOLL FLANDERS (II,1720)

FOSTER, LAURENCE
THE MANIONS OF AMERICA (II,1623)

FOSTER, LIN
YOUNG IN HEART (I,4939)

FOSTER, LINDA
HANK (I,1931); RIDDLE AT 24000 (II,2171); THINK PRETTY (I,4448)

FOSTER, MARGIE
240-ROBERT (II,2689)

FOSTER, MEG
CAGNEY AND LACEY (II,409); THE SCARLET LETTER (II,2270); SUNSHINE (II,2495); WASHINGTON: BEHIND CLOSED DOORS (II,2744)

FOSTER, PATTY
SHORTY (I,4025)

FOSTER, PEGGY
THE LAST CONVERTIBLE (II,1435)

FOSTER, PHIL
THE ALL-STAR REVUE (I,138); ARTHUR GODFREY AND THE SOUNDS OF NEW YORK (I,280); HAVING BABIES III (II,1109); LAVERNE AND SHIRLEY (II,1446); SID CAESAR PRESENTS COMEDY PREVIEW (I,4041); STEVE MARTIN: COMEDY IS NOT PRETTY (II,2462)

FOSTER, PHILIP
THE GREATEST GIFT (I,1880)

FOSTER, PRESTON
CAMEO THEATER (I,808); GUNSLINGER (I,1908); THE MAN FROM GALVESTON (I,2865); WATERFRONT (I,4774)

FOSTER, RON
THE D.A. (I,1123); SEVEN AGAINST THE SEA (I,3983)

FOSTER, RUTH
LITTLE HOUSE: BLESS ALL THE DEAR CHILDREN (II,1489); LITTLE HOUSE: THE LAST FAREWELL (II,1491)

FOSTER, STUART
THE GALEN DRAKE SHOW (I,1726); THIS IS GALEN DRAKE (I,4456)

FOSTER, SUSAN
LOVERS AND FRIENDS (II,1549); OH, NURSE! (I,3343); SIERRA (II,2351)

FOULGER, BYRON
CAPTAIN NICE (I,835); DENNIS THE MENACE (I,1231); FIRESIDE THEATER (I,1580); PETTICOAT JUNCTION (I,3571)

FOULK, ROBERT
FATHER KNOWS BEST (II,838); GREEN ACRES (I,1884); THE HARDY BOYS AND THE MYSTERY OF THE APPLEGATE TREASURE (I,1951); TESTIMONY OF TWO MEN (II,2564); TIMMY AND LASSIE (I,4520); WICHITA TOWN (I,4857)

FOUNTAIN, PETE
THE BING CROSBY SPRINGTIME SPECIAL (I,473); THE BOB HOPE SHOW (I,610); THE BOB HOPE SHOW (I,660); THE LAWRENCE WELK SHOW (I,2643); MEMORIES WITH LAWRENCE WELK (II,1681); THE NASHVILLE SOUND OF BOOTS RANDOLPH (I,3229)

FOUR ACES, THE
ACADEMY AWARD SONGS (I,13)

FOUR BROTHERS, THE
SING ALONG WITH MITCH (I,4056)

FOUR FELLOWS, THE
MITZI'S SECOND SPECIAL (I,3074)

FOUR FRESHMAN, THE
FRANCES LANGFORD PRESENTS (I,1656); THE RAY ANTHONY SHOW (I,3728)

FOUR LADS, THE
THE GUY MITCHELL SHOW (I,1912); PERRY PRESENTS (I,3549); UPBEAT (I,4684)

FOUR STEP BROTHERS, THE
THE BOB HOPE SHOW (I,516); WELCOME ABOARD (I,4789)

FOUR TOPS, THE
IT'S WHAT'S HAPPENING, BABY! (I,2278)

FOUR VOICES, THE
THE PETER LIND HAYES SHOW (I,3564)

FOURNIER, NOEL
NOWHERE TO HIDE (II,1868)

FOUTS, TOM
POLKA-GO-ROUND (I,3640)

FOWKES, CONRAD
AS THE WORLD TURNS (II,110); THE EDGE OF NIGHT (II,760); HIDDEN FACES (I,2044); KITTY FOYLE (I,2571); SEARCH FOR TOMORROW (II,2284); A TIME FOR US (I,4507)

FOWLDS, DEREK
YES MINISTER (II,2848)

FOWLEY, DOUGLAS V
DETECTIVE SCHOOL (II,661); FATHER ON TRIAL (I,1544); THE LIFE AND LEGEND OF WYATT EARP (I,2681); THE MONEYCHANGERS (II,1724); NUTS AND BOLTS (II,1871); THE OREGON TRAIL (II,1922); PISTOLS 'N' PETTICOATS (I,3608)

FOX BROTHERS, THE
MUSIC WORLD (II,1775)

FOX JR, FRED
LEGS (II,1459)

FOX, ANN
THE ROADS TO FREEDOM (I,3802)

FOX, BERNARD
THE ANDY GRIFFITH SHOW (I,192); BEWITCHED (I,418); GENERAL HOSPITAL (II,964); HOGAN'S HEROES (I,2069); INTERTECT (I,2228); MAKE ROOM FOR DADDY (I,2843); SHERLOCK HOLMES: THE HOUND OF THE BASKERVILLES (I,4011); SOAP (II,2392); THE SON-IN-LAW (II,2398); TABITHA (II,2528); THE WIDE OPEN DOOR (I,4859)

FOX, COLIN
HIGH HOPES (II,1144); STRANGE PARADISE (I,4246)

FOX, JAMES
NANCY ASTOR (II,1788)

FOX, JANET
THE WORLD OF MISTER SWEENY (I,4917)

FOX, JOHN J
HEREAFTER (II,1138)

FOX, MARCIA
KISS KISS, KILL KILL (I,2566)

FOX, MICHAEL
MORNING STAR (I,3104)

FOX, MICHAEL J
BATTLE OF THE NETWORK STARS (II,178); FAMILY TIES (II,819); HIGH SCHOOL, U.S.A. (II,1148); THE HOMEMADE COMEDY SPECIAL (II,1173); PALMERSTOWN U.S.A. (II,1945); TROUBLE IN HIGH TIMBER COUNTRY (II,2661)

FOX, NANCY
GENERAL HOSPITAL (II,964); HOME COOKIN' (II,1170); MCCLOUD (II,1660); TEMPERATURES RISING (I,4390)

FOX, PATRICK
THE SECRET STORM (I,3969)

FOX, PETER
DELTA HOUSE (II,658); THE MUNSTERS' REVENGE (II,1765); THE WALTONS (II,2740)

FOX, SAMANTHA
BOSOM BUDDIES (II,355)

FOX, SONNY
THE $64,000 CHALLENGE (I,4070); LET'S TAKE A TRIP (I,2668); THE MOVIE GAME (I,3114); ON YOUR MARK (I,3365); WAY OUT GAMES (II,2747)

FOXWORTH, ROBERT
FALCON CREST (II,808); FRANKENSTEIN (I,1671); MEN AT LAW (I,3004); THE NEW HEALERS (I,3266); SANDBURG'S LINCOLN (I,3907); THE STOREFRONT LAWYERS (I,4233); SUSAN AND SAM (II,2506)

FOXX, ELIZABETH
VELVET (II,2725)

FOXX, REDD
...AND BEAUTIFUL (I,184); ABC STAGE '67 (I,6); THE BOB HOPE SHOW (I,658); THE BOB HOPE SHOW (I,663); BOB HOPE SPECIAL: BOB HOPE'S CHRISTMAS PARTY (II,307); BOB HOPE SPECIAL: HAPPY BIRTHDAY, BOB! (II,326);

CELEBRITY CHALLENGE OF THE SEXES 1 (II,465); CELEBRITY CHALLENGE OF THE SEXES 4 (II,468); COTTON CLUB '75 (II,580); JACK BENNY'S SECOND FAREWELL SHOW (I,2302); KEEP U.S. BEAUTIFUL (I,2519); THE KROFFT KOMEDY HOUR (II,1411); LOLA (II,1510); MY BUDDY (II,1778); THE REDD FOXX COMEDY HOUR (II,2126); THE ROWAN AND MARTIN SPECIAL (I,3855); SANFORD (II,2255); SANFORD AND SON (II,2256); SHOW BUSINESS SALUTE TO MILTON BERLE (I,4029)

FOXX, RHONDA
FEATHERSTONE'S NEST (II,847); SUPERTRAIN (II,2505)

FOY JR, EDDIE
ABC STAGE '67 (I,6); THE CONNIE FRANCIS SHOW (I,1039); THE DEADLY GAME (II,633); FAIR EXCHANGE (I,1512); JOHNNY CARSON PRESENTS THE SUN CITY SCANDALS '72 (I,2421); KING OF THE ROAD (II,1396); MR. BROADWAY (I,3130); REAR GUARD (II,2120); THE ROSALIND RUSSELL SHOW (I,3845); THE SEVEN LITTLE FOYS (I,3986); VAUDEVILLE (II,2722)

FRABOTTA, DON
DAYS OF OUR LIVES (II,629)

FRAKES, JONATHAN
BARE ESSENCE (II,149); BARE ESSENCE (II,150); BEACH PATROL (II,192); BEULAH LAND (II,226); THE DUKES OF HAZZARD (II,742); PAPER DOLLS (II,1952)

FRANCE, DICK
THE LITTLE REVUE (I,2717)

FRANCESCO, SILVIO
THE CATERINA VALENTE SHOW (I,873)

FRANCHI, SERGIO
THE BOB HOPE SHOW (I,599); MUSICAL COMEDY TONIGHT (II,1777); TEXACO STAR PARADE I (I,4409)

FRANCINE, ANNE
THE GREAT SEBASTIANS (I,1877); HARPER VALLEY (II,1094); HARPER VALLEY PTA (II,1095)

FRANCIOSA, TONY
THE CATCHER (I,872); THE CRADLE SONG (I,1088); FAME IS THE NAME OF THE GAME (I,1518); FINDER OF LOST LOVES (II,857); KRAFT TELEVISION THEATER (I,2592); MATT HELM (II,1652); MATT HELM (II,1653); THE NAME OF THE GAME (I,3217); SEARCH (I,3951); VALENTINE'S DAY (I,4693); WHEELS (II,2777)

FRANCIS, ANNE
BANJO HACKETT: ROAMIN' FREE (II,141); CHARLEY'S AUNT (II,483); CLIMAX! (I,976); DALLAS (II,614); THE DAVID NIVEN THEATER (I,1178); GALLAGHER (I,1727); HONEY WEST (I,2105); HONEY WEST: WHO KILLED THE JACKPOT? (I,2106); KRAFT SUSPENSE THEATER (I,2591); MY THREE SONS (I,3205); O'MALLEY (II,1872); THE REBELS (II,2121); RIPTIDE (II,2178); SHADOW OF SAM PENNY (II,2313); STUDIO ONE (I,4268); THE THIRD COMMANDMENT (I,4449); THE TWILIGHT ZONE (I,4651); VERSATILE VARIETIES (I,4713)

FRANCIS, ARLENE
THE ARLENE FRANCIS SHOW (I,258); BLIND DATE (I,482); BY POPULAR DEMAND (I,784); THE CLOCK (I,978); THE COMEBACK STORY (I,1017); FASHION MAGIC (I,1536); HARVEY (I,1962); HOME (I,2099); LAURA (I,2636); LIGHT'S OUT (I,2699); PRIZE PERFORMANCE (I,3673); SOLDIER PARADE (I,4109); SURE AS FATE (I,4297); TALENT PATROL (I,4340); THAT REMINDS ME (I,4420); TONIGHT PREVIEW (I,4560); WHO'S THERE? (I,4854)

FRANCIS, CHERYL
THE GREATEST AMERICAN HERO (II,1060); RYAN'S FOUR (II,2233); THIS IS KATE BENNETT (II,2597)

FRANCIS, CLIVE
CAESAR AND CLEOPATRA (I,789); THE FAR PAVILIONS (II,832); THE GATHERING STORM (I,1741); MASADA (II,1642)

FRANCIS, CONNIE
THE BIG BEAT (I,423); THE BOB HOPE CHRYSLER THEATER (I,502); THE CONNIE FRANCIS SHOW (I,1039); DICK CLARK PRESENTS THE ROCK AND ROLL YEARS (I,1265); THE JIMMIE RODGERS SHOW (I,2382); THE JONATHAN WINTERS SPECIAL (I,2445); MARRIAGE—HANDLE WITH CARE (I,2921); REMEMBER HOW GREAT? (I,3764)

FRANCIS, DEREK
COLOR HIM DEAD (II,554); GREAT EXPECTATIONS (II,1055); MURDER MOTEL (I,3161)

FRANCIS, GENIE
BARE ESSENCE (II,149); BARE ESSENCE (II,150); GENERAL HOSPITAL (II,964); MURDER, SHE WROTE (II,1770)

FRANCIS, IVOR
DUSTY'S TRAIL (I,1390); GENERAL HOSPITAL (II,964); THE HOME FRONT (II,1171); THE NIGHT STRANGLER (I,3293); RETURN OF THE WORLD'S GREATEST DETECTIVE (II,2142); ROOM 222 (I,3843); SNAVELY (II,2387); SPIDER-MAN (II,2424); STATE FAIR (II,2448); THE TURNING POINT OF JIM MALLOY (II,2670); THE WALTONS (II,2740)

FRANCIS, JAN
FILE IT UNDER FEAR (I,1574); SECRET ARMY (II,2290)

FRANCIS, JOAN
JACQUES FRAY'S MUSIC ROOM (I,2329)

FRANCIS, MISSY
JOE'S WORLD (II,1305); LITTLE HOUSE ON THE PRAIRIE (II,1487); MORK AND MINDY (II,1735)

FRANCIS, NICKY
SUPER CIRCUS (I,4287)

FRANCIS, STAN
RUDOLPH THE RED-NOSED REINDEER (II,2227); TUGBOAT ANNIE (I,4622)

FRANCISCUS, JAMES
BAND OF GOLD (I,341); CELEBRITY CHALLENGE OF THE SEXES 2 (II,466); DOC ELLIOT (I,1305); DOC ELLIOT (I,1306); THE GENERAL ELECTRIC THEATER (I,1753); HUNTER (II,1204); HUNTER (II,1205); THE INVESTIGATORS (I,2231); THE JUNE ALLYSON SHOW (I,2488); LONGSTREET (I,2749); LONGSTREET (I,2750); MR. NOVAK (I,3145); NAKED CITY (I,3215); SIX GUNS FOR DONEGAN (I,4065)

FRANCKS, DON
JERICHO (I,2358); THE NUT HOUSE (I,3320); ROYAL CANADIAN MOUNTED POLICE (I,3860)

FRANCKS, LILI
CAGNEY AND LACEY (II,410)

FRANCO, RAMON
VENICE MEDICAL (II,2726)

FRANE, VIOLA
ABE LINCOLN IN ILLINOIS (I,10)

FRANGBLAU, ROSE
HOME (I,2099)

FRANGIONE, NANCY
ALL MY CHILDREN (II,39); ANOTHER WORLD (II,97)

FRANK, ALLAN
CHARADE QUIZ (I,901)

FRANK, BEN
MANHUNTER (II,1620)

FRANK, BOB
THE BARKLEYS (I,356)

FRANK, CARL
THE EDGE OF NIGHT (II,760); I REMEMBER MAMA (I,2176)

FRANK, CHARLES
ALL MY CHILDREN (II,39); ANNIE FLYNN (II,92); BATTLE OF THE NETWORK STARS (II,177); THE CHISHOLMS (II,512); EMERALD POINT, N.A.S. (II,773); FILTHY RICH (II,856); GO WEST YOUNG GIRL (II,1013); THE NEW MAVERICK (II,1822); RIDING HIGH (II,2173); YOUNG MAVERICK (II,2867)

FRANK, DAVID
BACHELOR AT LAW (I,322)

FRANK, ELENA EILEEN
B.J. AND THE BEAR (II,127)

FRANK, GARY
DOROTHY HAMILL'S CORNER OF THE SKY (II,721); FAMILY (II,813); THE LOVE BOAT III (II,1534); NORMA RAE (II,1860); SENIOR YEAR (II,2301); SONS AND DAUGHTERS (II,2404)

FRANK, TONY
CELEBRITY (II,463)

FRANKBONER, SARAH
BANYON (I,344); THE BOBBY DARIN AMUSEMENT COMPANY (I,672)

FRANKEN, STEVE
DOBIE GILLIS (I,1302); FORCE SEVEN (II,908); HIGH SCHOOL, U.S.A. (II,1148); KISS ME, KILL ME (II,1400); THE LIEUTENANT (I,2680); SCARED SILLY (II,2269); THE SHERIFF AND THE ASTRONAUT (II,2326); THE STRANGER (I,4251); THREE FOR TAHITI (I,4476); TOM, DICK, AND MARY (I,4537);

TONY ORLANDO AND DAWN (II,2639)

FRANKENHEIMER, JOHN
THE GOLDEN AGE OF TELEVISION (II,1018)

FRANKFATHER, WILLIAM
THE GREATEST AMERICAN HERO (II,1060)

FRANKLIN, ANNA
BROTHER RAT (I,746)

FRANKLIN, ARETHA
ANDY WILLIAMS MAGIC LANTERN SHOW COMPANY (I,202); BOB HOPE SPECIAL: BOB HOPE ON CAMPUS (II,289); DUKE ELLINGTON. . . WE LOVE YOU MADLY (I,1380); MERV GRIFFIN'S SIDEWALKS OF NEW ENGLAND (I,3017); THE MUHAMMED ALI VARIETY SPECIAL (II,1762); RODNEY DANGERFIELD SPECIAL: IT'S NOT EASY BEIN' ME (II,2199)

FRANKLIN, BONNIE
BOB HOPE SPECIAL: THE BOB HOPE CHRISTMAS SPECIAL (II,330); BONNIE AND THE FRANKLINS (II,349); CBS: ON THE AIR (II,460); A CHRISTMAS CAROL (I,947); THE COMEDY ZONE (II,559); GIDGET (I,1795); THE LOVE BOAT (II,1535); MUSICAL COMEDY TONIGHT (II,1777); ONE DAY AT A TIME (II,1900); PARADE OF STARS (II,1954)

FRANKLIN, CARL
THE A-TEAM (II,119); BATTLE OF THE NETWORK STARS (II,164); CARIBE (II,439); THE FANTASTIC JOURNEY (II,826); JOSHUA'S WORLD (II,1344); THE LEGEND OF THE GOLDEN GUN (II,1455); MCCLAIN'S LAW (II,1659)

FRANKLIN, CASS
FRONT ROW CENTER (I,1693)

FRANKLIN, DIANE
TOO GOOD TO BE TRUE (II,2643)

FRANKLIN, DOUG
BAY CITY BLUES (II,186)

FRANKLIN, GRETCHEN
GEORGE AND MILDRED (II,965)

FRANKLIN, HUGH
ALL MY CHILDREN (II,39); YOUNG DR. MALONE (I,4938)

FRANKLIN, JUDY
A CHRISTMAS CAROL (I,947)

FRANKLIN, NANCY
ONE MAN'S FAMILY (I,3383); A TIME FOR US (I,4507); THE TRAP (I,4593)

FRANKLIN, PAMELA
CROSSFIRE (II,606); EAGLE IN A CAGE (I,1392); GHOST STORY (I,1788); INTERTECT (I,2228); THE LETTERS (I,2670); SCREAMER (I,3945); TERROR FROM WITHIN (I,4404)

FRANKLIN, RICHARD
DOCTOR WHO (II,689); DOCTOR WHO—THE FIVE DOCTORS (II,690)

FRANKLIN, SUSAN
DOCTORS HOSPITAL (II,691)

FRANKLYN, FREDRIC
GOOD PENNY (II,1036)

FRANKLYN, SABINA
PRIDE AND PREJUDICE (II,2080)

FRANKLYN, WILLIAM
TOP SECRET (I,4580)

FRANN, MARY
DAYS OF OUR LIVES (II,629); FIRST TIME, SECOND TIME (II,863); GIDGET'S SUMMER REUNION (I,1799); KING'S CROSSING (II,1397); NEWHART (II,1835); THE TV TV SHOW (II,2674)

FRANZ, ARTHUR
CROSSROADS (I,1105); DOC HOLLIDAY (I,1307); FOUR STAR PLAYHOUSE (I,1652); THE LAST HURRAH (I,2626); THE NURSES (I,3319); PROFILES IN COURAGE (I,3678); TV READERS DIGEST (I,4630); WORLD OF GIANTS (I,4912)

FRANZ, DENNIS
BAY CITY BLUES (II,186); CHICAGO STORY (II,506); THE CHICAGO STORY (II,507); HILL STREET BLUES (II,1154)

FRANZ, EDUARD
BREAKING POINT (I,717); THE JAZZ SINGER (I,2351); PANIC! (I,3447)

FRANZEN, CHARLES
CHIEFS (II,509)

FRAPPIER, PAUL-EMILE
EMILY, EMILY (I,1438)

FRASER, ELISABETH
COUNTERPOINT (I,1059); FIBBER MCGEE AND MOLLY (I,1569); FOUR STAR PLAYHOUSE (I,1652); GRUEN GUILD PLAYHOUSE (I,1892); MCKEEVER AND THE COLONEL (I,2970); OFF WE GO (I,3338); ONE HAPPY FAMILY (I,3375); THE PHIL SILVERS SHOW (I,3580); STORY THEATER (I,4240); YOUR JEWELER'S SHOWCASE (I,4953)

FRASER, JERI LYNNE
MIKE AND THE MERMAID (I,3039)

FRASER, JOHN
THE PATRIOTS (I,3477); THE PICTURE OF DORIAN GRAY (I,3591)

FRASER, JUDE
RETURN OF THE WORLD'S GREATEST DETECTIVE (II,2142)

FRASER, SALLY
ANNIE OAKLEY (I,225); A CHRISTMAS CAROL (I,947); NAVY LOG (I,3233)

FRAWLEY, JAMES
BULBA (II,392)

FRAWLEY, WILLIAM
THE BOB HOPE SHOW (I,550); THE DESILU REVUE (I,1238); I LOVE LUCY (I,2171); THE LUCILLE BALL-DESI ARNAZ SHOW (I,2784); MY THREE SONS (I,3205); SILVER THEATER (I,4051); YOUR SHOW TIME (I,4958)

FRAY, JACQUES
JACQUES FRAY'S MUSIC ROOM (I,2329)

FRAZEE, JANE
BEULAH (I,416)

FRAZEE, TREVA
MY SISTER EILEEN (I,3200)

FRAZER, DAN
KOJAK (II,1406)

FRAZER, KAY
THE BENNY HILL SHOW (II,207)

FRAZER, TONY
ME AND BENJY (I,2973)

FRAZIER, DAN
DRIBBLE (II,736)

FRAZIER, JACK
WORLD'S FAIR BEAUTY SHOW (I,4920)

FRAZIER, SHEILA
FIREHOUSE (I,1579); THE LAZARUS SYNDROME (II,1451); THE NATIONAL SNOOP (II,1797); THE PHOENIX (II,2036)

FREAS, DANIELLE
GOLDIE AND KIDS: LISTEN TO ME (II,1021)

FREBERG, STAN
THE CHEVY SUMMER SHOW (I,929); CHRIS AND THE MAGICAL DRIP (II,516); COMEDY NEWS (I,1020); TIME FOR BEANY (I,4505); WILLIE WONDERFUL (I,4867)

FRECHETTE, PETER
THE ACADEMY (II,3); THE ACADEMY II (II,4)

FRED TOMALSON SINGERS, THE
MONTY PYTHON'S FLYING CIRCUS (II,1729); THE TWO RONNIES (II,2694)

FRED WARING GLEE CLUB, THE
THE FRED WARING SHOW (I,1679)

FREDERICK, HAL
BORN FREE (I,691); BORN FREE (II,353); GIDGET GROWS UP (I,1797); THE INTERNS (I,2226)

FREDERICK, KAREN
A COUPLE OF DONS (I,1077); THREE'S COMPANY (II,2614)

FREDERICK, LYNNE
KEEPING AN EYE ON DENISE (I,2520)

FREDERICKS, DEAN
DANIEL BOONE (I,1141); STEVE CANYON (I,4224)

FREDERICKS, PAULINE
PUREX SPECIAL FOR WOMEN (I,3698)

FREDERICKS, RITA
FIRST LOVE (I,1583)

FREDERICKS, WENDY
INSTANT FAMILY (II,1238)

FREDEY, ALICE
SPRAGGUE (II,2430)

FREDRIC, NORMAN
THE ADVENTURES OF RIN TIN TIN (I,73); JUNGLE JIM (I,2491)

FREED, ALAN
THE BIG BEAT (I,423); THE ROCK N' ROLL SHOW (I,3820)

FREED, BERT
330 INDEPENDENCE S.W. (I,4486); ARSENIC AND OLD LACE (I,267); HEAVEN HELP US (I,1995); THE KAISER ALUMINUM HOUR (I,2503); A PUNT, A PASS, AND A PRAYER (I,3694); SHANE (I,3995); THEN CAME BRONSON (I,4436)

FREED, FRITZ
SUPERTRAIN (II,2504)

FREED, IAN
ST. ELSEWHERE (II,2432)

FREED, SAM
KANGAROOS IN THE KITCHEN (II,1362)

FREEDLEY, VINTON
SHOWTIME, U.S.A. (I,4038); TALENT JACKPOT (I,4339)

FREEDMAN, WINIFRED
JOANIE LOVES CHACHI (II,1295)

FREEMAN JR, AL
HOT L BALTIMORE (II,1187); ONE LIFE TO LIVE (II,1907); ROOTS: THE NEXT GENERATIONS (II,2213)

FREEMAN, ANDY
BEYOND WITCH MOUNTAIN (II,228)

FREEMAN, ARNY
THE GREAT SEBASTIANS (I,1877); THE GREATEST GIFT (I,1880); MR. BROADWAY (I,3130)

FREEMAN, DAMITA JO
HOT W.A.C.S. (II,1191); LOOK WHAT THEY'VE DONE TO MY SONG (II,1517); PRIVATE BENJAMIN (II,2087); THE REDD FOXX COMEDY HOUR (II,2126); THE SCOEY MITCHLLL SHOW (I,3939); SHE'S IN THE ARMY NOW (II,2320)

FREEMAN, DEENA
IN TROUBLE (II,1230); TOO CLOSE FOR COMFORT (II,2642)

FREEMAN, HOWARD
THE DEVIL AND DANIEL WEBSTER (I,1246)

FREEMAN, JILL
TWILIGHT THEATER (II,2685)

FREEMAN, JOAN
BUS STOP (I,778); CODE R (II,551); FOUR EYES (II,912); THE OUTER LIMITS (I,3426)

FREEMAN, JUSTIN
THE DAY THE WOMEN GOT EVEN (II,628)

FREEMAN, KATHLEEN
THE BEVERLY HILLBILLIES (I,417); CALL HER MOM (I,796); THE DAUGHTERS OF JOSHUA CABE RETURN (I,1170); THE DONNA REED SHOW (I,1347); FATHER O FATHER (II,842); FOR LOVE OR $$$ (I,1623); THE FREEWHEELERS (I,1683); FUNNY FACE (I,1711); THE GOOD OLD DAYS (I,1851); HITCHED (I,2065); HOGAN'S HEROES (I,2069); IT'S ABOUT TIME (I,2263); THE JERRY LEWIS SHOW (I,2365); LOVE THAT BOB (I,2779); MAYOR OF THE TOWN (I,2963); THE MISS AND THE MISSILES (I,3062); READY AND WILLING (II,2115); SUTTERS BAY (II,2508); TOPPER (I,4582)

FREEMAN, KAY
THREE'S COMPANY (II,2614)

FREEMAN, LARRY
WOMAN'S PAGE (II,2822)

FREEMAN, LISA
IN TROUBLE (II,1230); THE RENEGADES (II,2132); SUNDAY FUNNIES (II,2493)

FREEMAN, MATT
FATHER MURPHY (II,841)

FREEMAN, MICKEY
THE PHIL SILVERS SHOW (I,3580)

FREEMAN, MONA
DAMON RUNYON THEATER (I,1129); FRONT ROW CENTER (I,1694); THE JUNE ALLYSON SHOW (I,2488); THE LUX VIDEO THEATER (I,2795); PURSUIT (I,3700); SCHLITZ PLAYHOUSE OF STARS (I,3936); ZANE GREY THEATER (I,4979)

FREEMAN, MORGAN
PALMERSTOWN U.S.A. (II,1945); ROLL OF THUNDER, HEAR MY CRY (II,2202)

FREEMAN, MORT
RIVKIN: BOUNTY HUNTER (II,2183)

FREEMAN, PAMELA
THE PRUITTS OF SOUTHAMPTON (I,3684)

FREEMAN, PAUL
FALCON CREST (II,808)

FREEMAN, SANDY
DREAMS (II,735); THE NEW TEMPERATURES RISING SHOW (I,3273)

FREEMAN, STAN
THE GYPSY ROSE LEE SHOW (I,1916); MELODY TOUR (I,3001); THREE'S COMPANY (I,4489)

FREES, PAUL
ADVENTURES OF HOPPITY HOPPER FROM FOGGY BOGG (I,62); THE ATOM ANT/SECRET SQUIRREL SHOW (I,311); THE BANANA SPLITS ADVENTURE HOUR (I,340); THE BEATLES (I,380); THE BULLWINKLE SHOW (I,761); CALVIN AND THE COLONEL (I,806); THE DICK TRACY SHOW (I,1272); THE DUDLEY DO-RIGHT SHOW (I,1378); THE FAMOUS ADVENTURES OF MR. MAGOO (I,1524); THE FANTASTIC FOUR (I,1529); FRANKENSTEIN JR. AND THE IMPOSSIBLES (I,1672); FROSTY THE SNOWMAN (II,934); FROSTY'S WINTER WONDERLAND (II,935); GEORGE OF THE JUNGLE (I,1773); THE HECKLE AND JECKLE SHOW (I,1999); HERE COMES PETER COTTONTAIL (II,1132); JACK FROST (II,1263); THE JACKSON FIVE (I,2326); KING FEATURES TRILOGY (I,2555); THE LITTLE DRUMMER BOY (II,1485); THE MILLIONAIRE (I,3046); THE NEW ADVENTURES OF HUCKLEBERRY FINN (I,3250); THE OSMONDS (I,3410); ROCKY AND HIS FRIENDS (I,3822); RUDOLPH'S SHINY NEW YEAR (II,2228); SANTA CLAUS IS COMIN' TO TOWN (II,2259); SUPER PRESIDENT (I,4291); THE SUPER SIX (I,4292); SUSPENSE (I,4305); THE SWORD (I,4325); THE TOM AND JERRY SHOW (I,4534); THE WOODY WOODPECKER SHOW (I,4906)

FREES, PEGGY
HOUSE CALLS (II,1194)

FREES, SHARI
BIG DADDY (I,425)

FREIMAN, JUSTIN
GEORGE WASHINGTON (II,978)

FREMIN, JOURDAN
AT EASE (II,115)

FRENCH, ARTHUR
MOMMA THE DETECTIVE (II,1721); OUR STREET (I,3418)

FRENCH, BRUCE
DALLAS (II,614); RENDEZVOUS HOTEL (II,2131)

FRENCH, ED
TALES FROM THE DARKSIDE (II,2534)

FRENCH, JACK
THE GEORGE HAMILTON IV SHOW (I,1769)

FRENCH, LEIGH
THE DICK CAVETT SHOW (II,669); GOOBER AND THE TRUCKERS' PARADISE (II,1029); THE LONG DAYS OF SUMMER (II,1514); MAUREEN (II,1656); THE SMOTHERS BROTHERS COMEDY HOUR (I,4095); THE SMOTHERS BROTHERS SHOW (II,2384); WHERE'S THE FIRE? (II,2785)

FRENCH, PEGGY
LIGHT'S OUT (I,2699); TWO GIRLS NAMED SMITH (I,4656)

FRENCH, SUSAN
BARE ESSENCE (II,149); BARE ESSENCE (II,150)

FRENCH, VALERIE
THE EDGE OF NIGHT (II,760); ONE LIFE TO LIVE (II,1907); ONE OF THE BOYS (II,1911); TEN LITTLE INDIANS (I,4394)

FRENCH, VICTOR
BATTLE OF THE NETWORK STARS (II,165); CARTER COUNTRY (II,449); THE CHEROKEE TRAIL (II,500); THE DEAN MARTIN CELEBRITY ROAST (II,637); GET SMART (II,983); THE HERO (I,2035); HIGHWAY TO HEAVEN (II,1152); LITTLE HOUSE ON THE PRAIRIE (II,1487); LITTLE HOUSE: A NEW BEGINNING (II,1488); LITTLE HOUSE: BLESS ALL THE DEAR CHILDREN (II,1489); LITTLE HOUSE: LOOK BACK TO YESTERDAY (II,1490); PLIMPTON! SHOWDOWN AT RIO LOBO (I,3630); RIDING FOR THE PONY EXPRESS (II,2172)

FRERE, DOROTHY
THE FALL AND RISE OF REGINALD PERRIN (II,810)

FRESCO, DAVID
AMANDA FALLON (I,152); THE DICK VAN DYKE SHOW (I,1275); TRUE LIFE STORIES (II,2665); TWILIGHT THEATER II (II,2686)

FRESH, DEBBI
DYNASTY (II,747)

FREY, LEONARD
BEST OF THE WEST (II,218); THE EARTHLINGS (II,751); FIT FOR A KING (II,866); MR. SMITH (II,1755); NEITHER ARE WE ENEMIES (I,3247); PHILLIP AND BARBARA (II,2034); TESTIMONY OF TWO MEN (II,2564)

FREY, NATHANIEL
CAR 54, WHERE ARE YOU? (I,842)

FRICHMAN, DAN
MAKING A LIVING (II,1604)

FRICKIE, JANIE
LOUISE MANDRELL: DIAMONDS, GOLD AND PLATINUM (II,1528)

FRID, JONATHAN
DARK SHADOWS (I,1157)

FRIDELL, SQUIRE
HUMAN FEELINGS (II,1203); ROSETTI AND RYAN (II,2216); ROSETTI AND RYAN: MEN WHO LOVE WOMEN (II,2217)

FRIEBUS, FLORIDA
THE BOB NEWHART SHOW (II,342); DOBIE GILLIS (I,1302); THE GOOD FAIRY (I,1846); LOVE NEST (II,1541); MEET CORLISS ARCHER (I,2981)

FRIED, IAN
TWO MARRIAGES (II,2691)

FRIED, MYRA
FRAGGLE ROCK (II,916)

FRIEDELL, SQUIRE
BETWEEN THE LINES (II,225)

FRIEDKIN, GARY
HAPPY DAYS (II,1084)

FRIEDMAN, ADA
MIXED DOUBLES (I,3075)

FRIEDMAN, BUDD
DON KIRSHNER'S ROCK CONCERT (I,1332); EVENING AT THE IMPROV (II,786)

FRIEDMAN, DAVID
LITTLE HOUSE: A NEW BEGINNING (II,1488); LITTLE HOUSE: BLESS ALL THE DEAR CHILDREN (II,1489); LITTLE HOUSE: LOOK BACK TO YESTERDAY (II,1490); LITTLE HOUSE: THE LAST FAREWELL (II,1491); MAMA'S FAMILY (II,1610); THE WALTONS: A DAY FOR THANKS ON WALTON'S MOUNTAIN (EPISODE 3) (II,2741); THE WALTONS: MOTHER'S DAY ON WALTON'S MOUNTAIN (EPISODE 2) (II,2742); THE WALTONS: WEDDING ON WALTON'S MOUNTAIN (EPISODE 1) (II,2743)

FRIEDMAN, DR SONIA
PEOPLE DO THE CRAZIEST THINGS (II,1997)

FRIEDMAN, NORMAN
BILLY CRYSTAL: A COMIC'S LINE (II,249)

FRIEDMAN, PETER
PIAF (II,2039); RAGE OF ANGELS (II,2107)

FRIEDRICH, JOHN
THE GANGSTER CHRONICLES (II,957); THE THORN BIRDS (II,2600)

FRIESEN, PAUL
SING ALONG WITH MITCH (I,4057)

FRIN, CLAIRE
THE KATE SMITH HOUR (I,2508)

FRISCHMAN, DAN
BLISS (II,269); MAKING THE GRADE (II,1606)

FRITZ, STAN
KOBB'S CORNER (I,2577)

FRIZZELL, DAVID
CHRISTMAS LEGEND OF NASHVILLE (II,522); A COUNTRY CHRISTMAS (II,583)

FRIZZELL, LOU
BONANZA (II,347); CENTENNIAL (II,477); CHOPPER ONE (II,515); COLORADO C.I. (II,555); CROSSFIRE (II,606); FOREVER FERNWOOD (II,909); JACKIE AND DARLENE (II,1265); LETTERS FROM THREE LOVERS (I,2671); LUCAN (II,1553); THE NEW LAND (II,1820)

FROMAN, JANE
THE ALL-STAR REVUE (I,138); JANE FROMAN'S U.S.A. CANTEEN (I,2341)

FROME, DAVID
THE EDGE OF NIGHT (II,760)

FROME, MILTON
THE BEVERLY HILLBILLIES (I,417); THE CAPTAIN AND TENNILLE (II,429); THE DOCTOR WAS A LADY (I,1318); EVIL ROY SLADE (I,1485); THE LARRY STORCH SHOW (I,2618)

FROMMER, BEN
THE ASPHALT COWBOY (II,111); F TROOP (I,1499)

FROMY, BOB
JABBERWOCKY (II,1262)

FRONER, BARRY
THE UNEXPLAINED (I,4675)

FROST, ALICE
THE FARMER'S DAUGHTER (I,1533); HIS MODEL WIFE (I,2064); I REMEMBER MAMA (I,2176); OF THIS TIME, OF THAT PLACE (I,3335); STONE (II,2470)

FROST, DAVID
THE DAVID FROST REVUE (I,1176); THE DAVID FROST SHOW (I,1177); HEADLINERS WITH DAVID FROST (II,1114); PETULA (I,3573); SHOW BUSINESS (II,2343); THAT WAS THE WEEK THAT WAS (I,4423); THAT WAS THE YEAR THAT WAS (I,4424)

FROST, HANK
BYLINE—BETTY FURNESS (I,785)

FROST, MICHAEL
THE OUTSIDE MAN (II,1937)

FRUMKIN, MARLA
SCOOBY-DOO AND SCRAPPY-DOO (II,2274)

FRYE, CHIP
THE POWERS OF MATTHEW STAR (II,2075)

FRYE, DAVID
PLIMPTON! DID YOU HEAR THE ONE ABOUT. . .? (I,3629); THE TOM JONES CHRISTMAS SPECIAL (I,4539)

FRYE, SEAN
THE AWAKENING LAND (II,122)

FRYE, SOLEIL MOON
LITTLE SHOTS (II,1495); PUNKY BREWSTER (II,2092)

FUCCELLO, TOM
DALLAS (II,614); ONE LIFE TO LIVE (II,1907); SHAUGHNESSEY (II,2318)

FUCHS, MATTHIAS
TEN WHO DARED (II,2559)

FUDENNA, MARGENE
VALLEY OF THE DINOSAURS (II,2717)

FUDGE, ALAN
THE CHADWICK FAMILY (II,478); EISCHIED (II,763); EVERY STRAY DOG AND KID (II,792); GOLIATH AWAITS (II,1025); KATE MCSHANE (II,1369); KISS ME, KILL ME (II,1400); THE MAN FROM ATLANTIS (II,1615)

FUENTAS, ERNIE
PALS (II,1947)

FUGATE, JUDITH
LITTLE WOMEN (I,2724)

FUHRAM, BEN
OUR FAMILY BUSINESS (II,1931)

FUJIKAWA, JERRY HATSUO
MR. T. AND TINA (II,1758)

FUJIOKA, JOHN
THE LAST RESORT (II,1440); THE SIX-MILLION-DOLLAR MAN (II,2372); STRANDED (II,2475); TALES OF THE GOLD MONKEY (II,2537)

FULINGTON, MIKE
TOP TEN (II,2648)

FULLER, ANN
ANOTHER WORLD (II,97)

FULLER, BARBARA
GENE KELLY'S WONDERFUL WORLD OF GIRLS (I,1752); MR. TUTT (I,3153)

FULLER, JONATHAN
A CHRISTMAS CAROL (II,517)

FULLER, LORENZO
THE JERRY LESTER SHOW (I,2360)

FULLER, NANCY
FORCE FIVE (II,907)

FULLER, PENNY
ANN IN BLUE (II,83); APPLAUSE (I,244); BARE ESSENCE (II,150); THE WAY THEY WERE (II,2748)

FULLER, ROBERT
EMERGENCY! (II,775); JAKE'S WAY (II,1269); LARAMIE (I,2614); NO WARNING (I,3301); WAGON TRAIN (I,4747)

FULLER, WENDY
BIG BEND COUNTRY (II,230)

FULLERTON, KRISTEN
IT'S MAGIC, CHARLIE BROWN (I,2271); LIFE IS A CIRCUS, CHARLIE BROWN (I,2683)

FULLERTON, MELANIE
THE GUN AND THE PULPIT (II,1067); TO ROME WITH LOVE (I,4526)

FULLERTON, RICHARD
THE RENEGADES (II,2132)

FULLERTON, SCOTT
LAND OF THE LOST (II,1425)

FULLILOVE, DONALD
EMERGENCY PLUS FOUR (II,774); HOME ROOM (II,1172); THE JACKSON FIVE (I,2326)

FULMORE, RAY
HAZEL (I,1982)

FULTON, JESSIE LEE
HIGHWAY HONEYS (II,1151)

FULTON, WENDY
BARE ESSENCE (II,149); BARE ESSENCE (II,150)

FUNAI, HELEN
BEHIND THE SCREEN (II,200); DAYS OF OUR LIVES (II,629); THE RED SKELTON SHOW (I,3755); SOMERSET (I,4115); TERROR AT ALCATRAZ (II,2563)

FUNCCELLO, FRANK
SIDNEY SHORR (II,2349)

FUNIA, HELEN
GIDGET GROWS UP (I,1797)

FUNICELLO, ANNETTE
ADVENTURES IN DAIRYLAND (I,44); ANNETTE (I,222); THE ARTHUR GODFREY SPECIAL (II,109); THE BOB HOPE SHOW (I,602); DICK CLARK'S GOOD OL' DAYS: FROM BOBBY SOX TO BIKINIS (II,671); DISNEYLAND'S 25TH ANNIVERSARY (II,681); EASY DOES IT. . . STARRING FRANKIE AVALON (II,754); FRANKIE AND ANNETTE: THE SECOND TIME AROUND (II,917); THE FURTHER ADVENTURES OF SPIN AND MARTY (I,1720); THE GREATEST SHOW ON EARTH (I,1883); MAKE ROOM FOR DADDY (I,2843); MEN WHO RATE A "10" (II,1684); THE MICKEY MOUSE CLUB (I,3025); THE MOUSE FACTORY (I,3113); THE MOUSEKETEERS REUNION (II,1742); THE NEW ADVENTURES OF SPIN AND MARTY (I,3254); ZORRO (I,4982)

FUNT, ALLEN
CANDID CAMERA (I,818); CANDID CAMERA (I,819); CANDID CAMERA (II,420); CANDID CAMERA LOOKS AT THE DIFFERENCE BETWEEN MEN AND WOMEN (II,421); THE CANDID CAMERA SPECIAL (II,422); IT'S ONLY HUMAN (II,1257); THE JERRY LEWIS SHOW (I,2368)

FUREY, JOHN
THE HOME FRONT (II,1171)

FURLAN, ALAN
THE THREE MUSKETEERS (I,4480)

FURLONG, KIRBY
THE JIMMY STEWART SHOW (I,2391)

FURNESS, BETTY
BYLINE—BETTY FURNESS (I,785); THE DESILU PLAYHOUSE (I,1237); FASHION COMING AND BECOMING (I,1535); HOLLYWOOD SCREEN TEST (I,2089); MEET BETTY FURNESS (I,2978); PENTHOUSE PARTY (I,3515); STUDIO ONE (I,4268)

FURST, STEPHEN
DELTA HOUSE (II,658); FOR MEMBERS ONLY (II,905); NATIONAL LAMPOON'S TWO REELERS (II,1796); REVENGE OF THE GRAY GANG (II,2146); ST. ELSEWHERE (II,2432)

FURTH, GEORGE
BROADSIDE (I,732); CAROL (I,846); CHARLIE COBB: NICE NIGHT FOR A HANGING (II,485); THE DUMPLINGS (II,743); THE GOOD GUYS (I,1847); SAM HILL: WHO KILLED THE MYSTERIOUS MR. FOSTER? (I,3896); TAMMY (I,4357)

FUSTIANO, DAVID
AND THEY ALL LIVED HAPPILY EVER AFTER (II,75)

GABEL, MARTIN
HARVEY (I,1962); HERCULE POIROT (I,2021); THE POWER AND THE GLORY (I,3658); THE RIGHT MAN (I,3790); SMILE JENNY, YOU'RE DEAD (II,2382); TONIGHT IN SAMARKAND (I,4558)

GABET, SHARON
THE EDGE OF NIGHT (II,760)

GABIOLA, SANDY
WONDER WOMAN (PILOT 1) (II,2826)

GABLE, CHRISTOPHER
A WOMAN OF SUBSTANCE (II,2820)

GABLE, FRANCINE
W*A*L*T*E*R (II,2731)

GABLE, JUNE
BARNEY MILLER (II,154); THE BAY CITY AMUSEMENT COMPANY (II,185); LAUGH-IN (II,1444); NEWMAN'S DRUGSTORE (II,1837); SHA NA NA (II,2311)

GABLE, SANDRA
THE SWIFT SHOW (I,4316); TEX AND JINX (I,4406)

GABLER, MUNIA
FARAWAY HILL (I,1532)

GABLER, NEAL
SNEAK PREVIEWS (II,2388)

GABOR, EVA
THE ALL-STAR SALUTE TO MOTHER'S DAY (II,51); ALMOST HEAVEN (II,57); THE BIG PARTY FOR REVLON (I,427); THE BOB HOPE SHOW (I,653); CAROL (I,846); THE EVA GABOR SHOW (I,1469); THE GREAT IMPERSONATION (I,1876); GREEN ACRES (I,1884); HOWDY (I,2150); THE MARCH OF DIMES FASHION SHOW (I,2902); MICKEY AND THE CONTESSA (I,3024); PULITZER PRIZE PLAYHOUSE (I,3692); SILVER THEATER (I,4051); STORY THEATER (I,4240); SUSPENSE (I,4305); TALES OF TOMORROW (I,4352); YOUR SHOW TIME (I,4958)

GABOR, ZSA ZSA
THE ALL-STAR SALUTE TO MOTHER'S DAY (II,51); AS THE WORLD TURNS (II,110); BATMAN (I,366); THE BOB HOPE SHOW (I,625); THE BOB HOPE SHOW (I,528); THE BOB HOPE SHOW (I,533); THE BOB HOPE SHOW (I,573); THE BOB HOPE SHOW (I,581); THE BOB HOPE SHOW (I,643); THE BOB HOPE SHOW (I,649); CLIMAX! (I,976); THE DEAN MARTIN CELEBRITY ROAST (II,638); DOM DELUISE AND FRIENDS, PART 2 (II,702); THE DUPONT SHOW OF THE WEEK (I,1388); MATINEE THEATER (I,2947); MEN WHO RATE A "10" (II,1684); NINOTCHKA (I,3299); SNEAK PREVIEW (I,4100); TEXACO STAR THEATER: OPENING NIGHT (II,2565)

GABRIEL, BEN
PRISONER: CELL BLOCK H (II,2085)

GABRIEL, BOB
SOMERSET (I,4115)

GABRIEL, JOHN
GENERAL HOSPITAL (II,964); LOVE OF LIFE (I,2774); THE MARY TYLER MOORE SHOW (II,1640); RYAN'S HOPE (II,2234)

GABRIEL, ROMAN
THE BOB HOPE SHOW (I,658)

GABRIEL, SANDY
ALL MY CHILDREN (II,39)

GABRIELLE
GABRIELLE (I,1724)

GACKLE, KATHLEEN
HARRY O (II,1098)

GAE, DEBBIE
THE BENNY HILL SHOW (II,207)

GAE, NADINE
THE FRED WARING SHOW (I,1679)

GAFFIN, JENNIFER
SOMEDAY YOU'LL FIND HER, CHARLIE BROWN (I,4113)

GAFFIN, MELANIE
BARBARA MANDRELL AND THE MANDRELL SISTERS (II,143); TAXI (II,2546); WHIZ KIDS (II,2790)

GAGANNIS, BONNIE
THE BILLY CRYSTAL COMEDY HOUR (II,248)

GAGE, PATRICIA
THE RETURN OF CHARLIE CHAN (II,2136)

GAGE, PHIL
THE BOOK OF LISTS (II,350)

GAGNON, ANDRE
PERRY COMO'S FRENCH-CANADIAN CHRISTMAS (II,2016)

GAHLE, SANDRA
THE LANNY ROSS SHOW (I,2612)

GAIGE, RUSSELL
THE ALAN YOUNG SHOW (I,104)

GAIL, MAX
THE ALIENS ARE COMING (II,35); BARNEY MILLER (II,154); BATTLE OF THE NETWORK STARS (II,169); BATTLE OF THE NETWORK STARS (II,171); MR. AND MRS. COP (II,1748); PEARL (II,1986); WHIZ KIDS (II,2790)

GAINER, CAROL
THE NURSES (I,3319)

GAINER, JACK
DANIEL BOONE (I,1142)

GAINES, BOYD
ONE DAY AT A TIME (II,1900)

GAINES, GORDON
THE AT LIBERTY CLUB (I,309)

GAINES, JIMMY
HIS MODEL WIFE (I,2064)

GAINES, RICHARD
BOSS LADY (I,694); MCGARRY AND ME (I,2966); THE PETRIFIED FOREST (I,3570)

GAINES, SONNY JIM
DOWN HOME (II,731)

GAINSBOURG, SERGE
BRIGITTE BARDOT (I,729)

GALATI, DOMENICA
POWERHOUSE (II,2074)

GALE, BILL
FARAWAY HILL (I,1532)

GALE, DAVID
THE EDGE OF NIGHT (II,760); THE SECRET STORM (I,3969)

GALE, JERI
ARTHUR MURRAY'S DANCE PARTY (I,293)

GALE, NANCY
THE RETURN OF THE BEVERLY HILLBILLIES (II,2139)

GALE, PAUL

THE BAY CITY ROLLERS SHOW (II,187); MITCHELL AND WOODS (II,1709); SIGMUND AND THE SEA MONSTERS (II,2352)

GALE, RONA
INTERMEZZO (I,2223)

GALIK, DENISE
THE BEST OF TIMES (II,220); DID YOU HEAR ABOUT JOSH AND KELLY?! (II,673); FLAMINGO ROAD (II,872); MCNAMARA'S BAND (II,1668); V: THE FINAL BATTLE (II,2714)

GALLACHER, FRANK
AGAINST THE WIND (II,24); ALL THE RIVERS RUN (II,43)

GALLAGHER
BOB HOPE SPECIAL: BOB HOPE IN THE STAR-MAKERS (II,285); CELEBRITY CHALLENGE OF THE SEXES 4 (II,468); THE JIM STAFFORD SHOW (II,1291)

GALLAGHER, DON
RENDEZVOUS WITH MUSIC (I,3768)

GALLAGHER, HELEN
HEAVEN WILL PROTECT THE WORKING GIRL (I,1996); MANHATTAN SHOWCASE (I,2888); PARIS IN THE SPRINGTIME (I,3457); RYAN'S HOPE (II,2234); SHANGRI-LA (I,3996); YVES MONTAND ON BROADWAY (I,4978)

GALLAGHER, MEGAN
AT YOUR SERVICE (II,118); DALLAS (II,614); GEORGE WASHINGTON (II,978)

GALLAGHER, PETER
SKAG (II,2374)

GALLAGHER, ROBERT
NAUGHTY MARIETTA (I,3232); PANORAMA (I,3448); STAGE 13 (I,4183)

GALLANTYNE, PAUL
THE GUIDING LIGHT (II,1064)

GALLARDO, SILVANA
BORN TO THE WIND (II,354); FALCON CREST (II,808); THE MANY LOVES OF ARTHUR (II,1625); RYAN'S FOUR (II,2233)

GALLAUDET, JOHN
MY THREE SONS (I,3205)

GALLEGO, GINA
FLAMINGO ROAD (II,872); RITUALS (II,2182)

GALLEGOS, JOSHUA
HOW THE WEST WAS WON (II,1196)

GALLERY, JAMES
BILLY (II,247); THE BRADY GIRLS GET MARRIED (II,364); THE EDGE OF NIGHT (II,760); OPEN ALL NIGHT (II,1914); PALMS PRECINCT (II,1946)

GALLICO, PAUL
CELEBRITY TIME (I,887)

GALLICO, ROBERT
O.S.S. (I,3411)

GALLIGAN, BILL
THE ACADEMY II (II,4)

GALLISON, JOE
ANOTHER WORLD (II,97); DAYS OF OUR LIVES (II,629); ONE LIFE TO LIVE (II,1907); RETURN TO PEYTON PLACE (I,3779)

GALLO, INIOO
SMILEY'S PEOPLE (II,2383)

GALLO, LEW
CHALK ONE UP FOR JOHNNY (I,893)

GALLO, MARIO
DELVECCHIO (II,659)

GALLOP, FRANK
BROADWAY OPEN HOUSE (I,736); BUICK CIRCUS HOUR (I,759); GREAT GHOST TALES (I,1874); HUMAN ADVENTURE (I,2157); KRAFT MYSTERY THEATER (I,2589); LIGHT'S OUT (I,2699)

GALLOW, LEW
TWELVE O'CLOCK HIGH (I,4637)

GALLOWAY, BILL
THE NEW ZOO REVUE (I,3277)

GALLOWAY, CHERYL
DOCTOR HUDSON'S SECRET JOURNAL (I,1312)

GALLOWAY, DON
ARREST AND TRIAL (I,264); CONDOMINIUM (II,566); COVER GIRLS (II,596); THE GUINNESS GAME (II,1066); HIZZONER (II,1159); IRONSIDE (I,2240); THE SECRET STORM (I,3969); ZERO INTELLIGENCE (II,2876)

GALLOWAY, ED
IRONSIDE (II,1246)

GALLOWAY, JACK
MACKENZIE (II,1584); MOLL FLANDERS (II,1720)

GALLOWAY, JENNY
THERESE RAQUIN (II,2584)

GALLOWAY, MIKE
THE BLUE ANGELS (I,490); THE SECRET STORM (I,3969)

GALLOWAY, PAM
COUNTERATTACK: CRIME IN AMERICA (II,581); THIS IS KATE BENNETT (II,2597)

GALLOWAY, PAT
TALES FROM MUPPETLAND (I,4344)

GALLOWAY, TOM
TOM, DICK, AND MARY (I,4537)

GALLUCCI, JEFF
240-ROBERT (II,2689)

GALMAN, PETER
AS THE WORLD TURNS (II,110); THE PEOPLE NEXT DOOR (I,3519)

GALVIN, JAMES
THE DAY THE WOMEN GOT EVEN (II,628)

GALWAY, JAMES
ANDY WILLIAMS' EARLY NEW ENGLAND CHRISTMAS (II,79); JOHN DENVER: MUSIC AND THE MOUNTAINS (II,1318)

GAM, RITA
CAMEO THEATER (I,808); FRONT ROW CENTER (I,1694); HIDDEN FACES (I,2044)

GAMBLE, DUNCAN
CAPITOL (II,426); CONDOMINIUM (II,566)

GAMBLE, RALPH
ANOTHER DAY, ANOTHER DOLLAR (I,228)

GAMMELL, ROBIN
THE BLUE AND THE GRAY (II,273); THE LAST NINJA (II,1438); MISSING PIECES (II,1706); WAIT UNTIL DARK (II,2735); WISHMAN (II,2811)

GAMMON, EVERETT
MR. MERGENTHWIRKER'S LOBBLIES (I,3144)

GAMMON, JAMES
JOE DANCER: THE BIG BLACK PILL (II,1300)

GAMPEL, CHRIS
CAPTAIN BRASSBOUND'S CONVERSION (I,828)

GANAS, MONICA
THE PEE WEE HERMAN SHOW (II,1988)

GANLEY, GIL
THE JERK, TOO (II,1282)

GANNAWAY, AL
HALF-PINT PARTY (I,1919)

GANNON, BARBARA
EVENING AT THE IMPROV (II,786)

GANTNER, CARRILLO
PRISONER: CELL BLOCK H (II,2085)

GANZEL, MARK
MR. MOON'S MAGIC CIRCUS (II,1753)

GANZEL, TERESA
THE DUCK FACTORY (II,738); PUMPBOYS AND DINETTES ON TELEVISION (II,2090); TEACHERS ONLY (II,2548)

GAR, SIR FREDERICK
THE SPIKE JONES SHOW (I,4155)

GARAGIOLA, JOE
DAMN YANKEES (I,1128); THE FIRST 50 YEARS (II,862); HE SAID, SHE SAID (I,1986); THE MEMORY GAME (I,3003); SALE OF THE CENTURY (I,3888); TO TELL THE TRUTH (I,4528)

GARANT, SYLVIE
THE $128,000 QUESTION (II,1903)

GARAS, KAZ
MASSARATI AND THE BRAIN (II,1646); THE STRANGE REPORT (I,4248); WONDER WOMAN (PILOT 1) (II,2826)

GARAY III, JOAQUIN
CODE RED (II,553)

GARBER, GLEN
ADVENTURES IN DAIRYLAND (I,44)

GARBER, TERRI
LONE STAR (II,1513); MR. SMITH (II,1755); NO MAN'S LAND (II,1855); TEXAS (II,2566)

GARBER, VICTOR
CHARLEY'S AUNT (II,483); VALLEY FORGE (I,4698)

GARBUTT, JAMES
THE ONEDIN LINE (II,1912)

GARCIA, JOE
THE CHISHOLMS (II,512)

GARCIA, RAUL
AQUA VARIETIES (I,248)

GARCIA, RICK
VENICE MEDICAL (II,2726)

GARCIA, STELLA
RAPTURE AT TWO-FORTY (I,3724)

GARDE, BETTY
ALL THE WAY HOME (I,140); EASY ACES (I,1398); THE EDGE OF NIGHT (II,760); HANDS OF MURDER (I,1929);

ONE STEP BEYOND (I,3388); STARLIGHT THEATER (I,4201); THE WORLD OF MISTER SWEENY (I,4917)

GARDEN, GRAEME
THE GOODIES (II,1040)

GARDENIA, VINCENT
ALL IN THE FAMILY (II,38); BREAKING AWAY (II,371); CHARLEY'S AUNT (II,483); COPS (I,1047); DEAN'S PLACE (II,650); KENNEDY (II,1377)

GARDINER, JOHN
KID POWER (I,2532)

GARDINER, MARY
THE $100,000 BIG SURPRISE (I,3378)

GARDINER, REGINALD
ALICE IN WONDERLAND (I,120); THE APARTMENT HOUSE (I,241); THE GUARDSMAN (I,1893); THE MAN WHO CAME TO DINNER (I,2880); THE PRUITTS OF SOUTHAMPTON (I,3684); THE TENNESSEE ERNIE FORD SHOW (I,4398); THE WIDE OPEN DOOR (I,4859)

GARDNER, AVA
KNOTS LANDING (II,1404)

GARDNER, CHRISTOPHER
CHRIS AND THE MAGICAL DRIP (II,516); FATHER KNOWS BEST: HOME FOR CHRISTMAS (II,839); FATHER KNOWS BEST: THE FATHER KNOWS BEST REUNION (II,840); THE KILLER WHO WOULDN'T DIE (II,1393)

GARDNER, DEE
DOMINIC'S DREAM (II,704)

GARDNER, DON
THE BUFFALO BILLY SHOW (I,756)

GARDNER, ED
DUFFY'S TAVERN (I,1379)

GARDNER, ELAINE
THE NEW PHIL SILVERS SHOW (I,3269)

GARDNER, GRAIG
CHASE (I,917)

GARDNER, HARVEY
BOBBY PARKER AND COMPANY (II,344)

GARDNER, HUGH
DARK SHADOWS (I,1157)

GARDNER, HY
TONIGHT! AMERICA AFTER DARK (I,4555); WHAT'S GOING ON? (I,4814)

GARDNER, JOAN
THE BUFFALO BILLY SHOW (I,756); HERE COMES PETER COTTONTAIL (II,1132); SANTA CLAUS IS COMIN' TO TOWN (II,2259); SPUNKY AND TADPOLE (I,4169); VALLEY OF THE DINOSAURS (II,2717)

GARDNER, JOE
THE FAMOUS ADVENTURES OF MR. MAGOO (I,1524)

GARDNER, KENNY
GUY LOMBARDO AND HIS ROYAL CANADIANS (I,1910)

GARDNER, LIZ
PLAY YOUR HUNCH (I,3619); YOUNG DR. MALONE (I,4938)

GARDNER, RANDY
BOB HOPE SPECIAL: BOB HOPE'S ALL-STAR COMEDY BIRTHDAY PARTY (II,297)

GARDNER, TERRI
THE NASHVILLE PALACE (II,1792)

GARDUNO, ANNA
FOR LOVERS ONLY (II,904); STEELTOWN (II,2452)

GARFIELD JR, JOHN
WARNING SHOT (I,4770)

GARFIELD, ALLEN
NEVER AGAIN (II,1808); SONNY BOY (II,2402); TAXI (II,2546)

GARFIELD, DAVID
THE ONEDIN LINE (II,1912)

GARFIELD, JULIE
PLAZA SUITE (II,2056)

GARFUNKEL, ART
ANDY WILLIAMS KALEIDOSCOPE COMPANY (I,201); MARLO THOMAS IN ACTS OF LOVE—AND OTHER COMEDIES (I,2919); THE PAUL SIMON SPECIAL (II,1983); SIMON AND GARFUNKEL IN CONCERT (II,2356)

GARGAN, WILLIAM
MARTIN KANE, PRIVATE EYE (I,2928); THE NEW ADVENTURES OF MARTIN KANE (I,3251)

GARLAND, BEVERLY
THE BING CROSBY SHOW (I,467); CLIMAX! (I,976); DAMON RUNYON THEATER (I,1129); DECOY (I,1217); THE DICK POWELL SHOW (I,1269); ELFEGO BACA (I,1427); FLAMINGO ROAD (II,872); FORD TELEVISION THEATER (I,1634); FOUR STAR PLAYHOUSE (I,1652); FRONT ROW CENTER (I,1694); FRONTIER (I,1695); GALLAGHER (I,1727); THE HEALERS (II,1115); JUDGEMENT DAY (II,1348); KRAFT MYSTERY THEATER (I,2589); KRAFT SUSPENSE THEATER (I,2591); THE LUX VIDEO THEATER (I,2795); A MAN CALLED SHENANDOAH (I,2859); MEDIC (I,2976); MY THREE SONS (I,3205); NAVY LOG (I,3233); PEPSI-COLA PLAYHOUSE (I,3523); REMINGTON STEELE (II,2130); SCARECROW AND MRS. KING (II,2268); STAR STAGE (I,4197); STUMP THE STARS (I,4273); TELEPHONE TIME (I,4379); THANKSGIVING REUNION WITH THE PARTRIDGE FAMILY AND MY THREE SONS (II,2569); THIS GIRL FOR HIRE (II,2596); THRILLER (I,4492); THE WEB (I,4785); ZANE GREY THEATER (I,4979)

GARLAND, JUDY
A FUNNY THING HAPPENED ON THE WAY TO HOLLYWOOD (I,1716); THE JACK PAAR SPECIAL (I,2319); JUDY GARLAND AND HER GUESTS, PHIL SILVERS AND ROBERT GOULET (I,2469); THE JUDY GARLAND SHOW (I,2470); THE JUDY GARLAND SHOW (I,2471); THE JUDY GARLAND SHOW (I,2472); THE JUDY GARLAND SHOW (I,2473); THE JUDY GARLAND SHOW (II,1349)

GARLAND, MARGARET
TOM CORBETT, SPACE CADET (I,4535)

GARLAND, RICHARD
JEFF'S COLLIE (I,2356); MEDIC (I,2976)

GARNER, ERROLL
THE ARTHUR GODFREY SHOW (I,286)

GARNER, JACK
BRET MAVERICK (II,374); MEDICAL CENTER (II,1675); THE NEW MAVERICK (II,1822); NO MAN'S LAND (II,1855); NORMA RAE (II,1860); THE ROCKFORD FILES (II,2197); STATE FAIR (II,2448)

GARNER, JAMES
60 YEARS OF SEDUCTION (II,2373); THE BING CROSBY SPECIAL (I,471); THE BOB HOPE SHOW (I,576); THE BOB HOPE SHOW (I,584); THE BOB HOPE SHOW (I,586); THE BOB HOPE SHOW (I,596); THE BOB HOPE SHOW (I,604); BRET MAVERICK (II,374); CONFLICT (I,1036); LILY FOR PRESIDENT (II,1478); MAVERICK (II,1657); THE NEW MAVERICK (II,1822); NICHOLS (I,3283); THE ROCKFORD FILES (II,2196); THE ROCKFORD FILES (II,2197)

GARNER, JAY
BUCK ROGERS IN THE 25TH CENTURY (II,384)

GARNER, LILI
MIMI (I,3054)

GARNER, MARTIN
ALONE AT LAST (II,59); ALONE AT LAST (II,60)

GARNER, MOUSIE
SURFSIDE 6 (I,4299)

GARNER, PAUL
BILLY CRYSTAL: A COMIC'S LINE (II,249)

GARNER, PEGGY ANN
ALCOA PREMIERE (I,109); ALFRED HITCHCOCK PRESENTS (I,115); CLIMAX! (I,976); DANGER (I,1134); DUPONT SHOW OF THE MONTH (I,1387); HOLLYWOOD OPENING NIGHT (I,2087); KRAFT TELEVISION THEATER (I,2592); THE LUX VIDEO THEATER (I,2795); ONE STEP BEYOND (I,3388); THE OUTER LIMITS (I,3426); THE PATRIOTS (I,3477); PRUDENTIAL FAMILY PLAYHOUSE (I,3683); REVLON MIRROR THEATER (I,3781); ROBERT MONTGOMERY PRESENTS YOUR LUCKY STRIKE THEATER (I,3809); THE SEEKERS (I,3973); STAGE 7 (I,4181); STAGE DOOR (I,4178); TWO GIRLS NAMED SMITH (I,4656); THE UNITED STATES STEEL HOUR (I,4677); ZANE GREY THEATER (I,4979)

GARNET, GALE
LOOK WHAT THEY'VE DONE TO MY SONG (II,1517)

GARNETTE, GALE
PAUL BERNARD—PSYCHIATRIST (I,3485)

GARR, TERI
ASSIGNMENT: EARTH (I,299); BANYON (I,344); THE BURNS AND SCHREIBER COMEDY HOUR (I,768); THE KEN BERRY WOW SHOW (I,2524); THE SONNY AND CHER COMEDY HOUR (II,2400); THE SONNY COMEDY REVUE (II,2403)

GARRALAGA, MARTIN
EL COYOTE (I,1423)

GARRETT, BETTY
ALL IN THE FAMILY (II,38); ALL THE WAY HOME (II,44); THE ART CARNEY SHOW (I,274); THE BEST OF ANYTHING (I,404); GENE KELLY. . . AN AMERICAN IN PASADENA (II,962); LAVERNE AND SHIRLEY (II,1446); THE LLOYD BRIDGES SHOW (I,2733); THE LOVE BOAT (II,1535); MR. MERLIN (II,1751)

GARRETT, DONNA
WONDER WOMAN (PILOT 1) (II,2826)

GARRETT, EDDIE
QUINCY, M. E. (II,2102)

GARRETT, ERNIE
THE GOSSIP COLUMNIST (II,1045)

GARRETT, HANK
CAR 54, WHERE ARE YOU? (I,842); DUSTY (II,745); PARIS (II,1957); SPRAGGUE (II,2430)

GARRETT, JIMMY
THE LUCY SHOW (I,2791)

GARRETT, JOY
THE HOYT AXTON SHOW (II,1200); WINDOWS, DOORS AND KEYHOLES (II,2806)

GARRETT, KELLY
CAR CARE CENTRAL (II,436); HEADLINERS WITH DAVID FROST (II,1114); THE REAL TOM KENNEDY SHOW (I,3743); THIS WILL BE THE YEAR THAT WILL BE (I,4466); THE TIM CONWAY SHOW (II,2619); YOUR HIT PARADE (II,2872)

GARRETT, LEIF
BATTLE OF THE NETWORK STARS (II,168); BOB HOPE SPECIAL: THE BOB HOPE SPECIAL (II,338); CELEBRITY CHALLENGE OF THE SEXES 4 (II,468); CELEBRITY CHALLENGE OF THE SEXES 5 (II,469); THE ODD COUPLE (II,1875); THE SENSATIONAL, SHOCKING, WONDERFUL, WACKY 70S (II,2302); THREE FOR THE ROAD (II,2607); THE WOLFMAN JACK RADIO SHOW (II,2816)

GARRETT, MAUREEN
THE GUIDING LIGHT (II,1064); RYAN'S HOPE (II,2234)

GARRETT, MICHAEL
I WAS A BLOODHOUND (I,2180)

GARRETT, PATSY
THE BANANA SPLITS ADVENTURE HOUR (I,340); THE BRIGHTER DAY (I,728); MOBY DICK AND THE MIGHTY MIGHTOR (I,3077); NANNY AND THE PROFESSOR (I,3225); ROOM 222 (I,3843)

GARRETT, STEVE
HERCULES (I,2022)

GARRETT, SUSIE
PUNKY BREWSTER (II,2092)

GARRETT-BONNER, LILLIAN
G.I.'S (II,949)

GARRICK, BEULAH
THE GUIDING LIGHT (II,1064); JOHNNY BELINDA (I,2416)

GARRIPOLI, MARY
T.L.C. (II,2525)

GARRIS, PHIL
THE DONALD O'CONNOR SHOW (I,1344)

GARRISON, DAVID
THE EDGE OF NIGHT (II,760); IT'S YOUR MOVE (II,1259)

GARRISON, IRIS
GENERAL HOSPITAL (II,964)

GARRISON, R CHANDLER
MADAME'S PLACE (II,1587)

GARRISON, ROBERT SCOTT
THE BEST OF TIMES (II,220)

GARRISON, SEAN
THE ADVENTURES OF NICK CARTER (I,68); COVER GIRLS (II,596); DUNDEE AND THE CULHANE (I,1383); THE OUTSIDER (I,3430); RENDEZVOUS HOTEL (II,2131); THE SECRET EMPIRE (II,2291)

GARROWAY, DAVE
BABES IN TOYLAND (I,318); THE CBS NEWCOMERS (I,879); THE DAVE GARROWAY SHOW (I,1173); DAVE'S PLACE (I,1175); GARROWAY (I,1736); GARROWAY AT LARGE (I,1737); HAPPY TIMES ARE HERE AGAIN (I,1944); JACK FROST (II,1263); A SALUTE TO TELEVISION'S 25TH ANNIVERSARY (I,3893); THE TALENT SCOUTS PROGRAM (I,4341); TALENT SEARCH (I,4342); TONIGHT PREVIEW (I,4560)

GARSON, GREER
THE BIG PARTY FOR REVLON (I,427); THE BOB HOPE SHOW (I,546); CAPTAIN BRASSBOUND'S CONVERSION (I,828); CROWN MATRIMONIAL (I,1106); THE LITTLE FOXES (I,2712); LITTLE WOMEN (II,1497); PERRY COMO'S CHRISTMAS IN NEW MEXICO (II,2010); REUNION IN VIENNA (I,3780); STAR STAGE (I,4197); TELEPHONE TIME (I,4379)

GARVER, KATHY
DENNIS THE MENACE: MAYDAY FOR MOTHER (II,660); FAMILY AFFAIR (I,1519); PANIC! (I,3447); THE PATTY DUKE SHOW (I,3483)

GARVEY, CYNDY
THE CELEBRITY FOOTBALL CLASSIC (II,474); GAMES PEOPLE PLAY (II,956); SUCCESS: IT CAN BE YOURS (II,2488)

GARVEY, GERALD
THOSE ENDEARING YOUNG CHARMS (I,4469)

GARVEY, STEVE
CELEBRITY CHALLENGE OF THE SEXES 2 (II,466); THE CELEBRITY FOOTBALL CLASSIC (II,474)

GARVIE, ELIZABETH
PRIDE AND PREJUDICE (II,2080)

GARVIN, JOHN
THE CITADEL (II,539)

GARY LEWIS AND THE PLAYBOYS
IT'S WHAT'S HAPPENING BABY! (I,2296)

GARY PUCKETT AND THE UNION GAP
JACK BENNY'S NEW LOOK (I,2301)

GARY, JOHN
FAMILY NIGHT WITH HORACE HEIDT (I,1523); THE JOHN GARY SHOW (I,2411); THE JOHN GARY SHOW (I,2412)

GARY, LORRAINE
THE CITY (I,969)

GARYA, JOEY
THE CBS NEWCOMERS (I,879)

GASCOINE, JILL
PETER PAN (II,2030)

GATES, JOYCE
FOUR STAR PLAYHOUSE (I,1652)

GATES, LARRY
BACKSTAIRS AT THE WHITE HOUSE (II,133); THE GUIDING LIGHT (II,1064); KATE MCSHANE (II,1369); ON BORROWED TIME (I,3354); SARGE: THE BADGE OR THE CROSS (I,3915)

GATES, NANCY
HONEY WEST: WHO KILLED THE JACKPOT? (I,2106); JANE WYMAN PRESENTS THE FIRESIDE THEATER (I,2345); THE LLOYD BRIDGES SHOW (I,2733); THE LUX VIDEO THEATER (I,2795); PEPSI-COLA PLAYHOUSE (I,3523); STUDIO '57 (I,4267); TOM AND JERRY (I,4533)

GATES, RICK
THE HARDY BOYS (I,1948)

GATES, RUTH
I REMEMBER MAMA (I,2176)

GATESON, MARJORIE
CAMEO THEATER (I,808); ONE MAN'S FAMILY (I,3383); THE SECRET STORM (I,3969)

GATLEY, JIMMY
THE BILL ANDERSON SHOW (I,440)

GATLIN, JERRY
GEORGE WASHINGTON (II,978); NEVADA SMITH (II,1807)

GATLIN, LARRY
BOB HOPE SPECIAL: BOB HOPE'S STARS OVER TEXAS (II,318); BOB HOPE SPECIAL: THE BOB HOPE CHRISTMAS SPECIAL (II,331); COUNTRY COMES HOME (II,585); COUNTRY GALAXY OF STARS (II,587); COUNTRY MUSIC HIT PARADE (II,589); THE COUNTRY MUSIC MURDERS (II,590); ELVIS REMEMBERED: NASHVILLE TO HOLLYWOOD (II,772); GEORGE BURNS CELEBRATES 80 YEARS IN SHOW BUSINESS (II,967); GEORGE BURNS IN NASHVILLE?? (II,969); JOHN SCHNEIDER'S CHRISTMAS HOLIDAY (II,1320); A JOHNNY CASH CHRISTMAS (II,1327); JOHNNY CASH: THE FIRST 25 YEARS (II,1336); LARRY GATLIN AND THE GATLIN BROTHERS (II,1429); MERRY CHRISTMAS FROM THE GRAND OLE OPRY (II,1686); PERRY COMO'S SPRING IN SAN FRANCISCO (II,2022); ROY ACUFF—50 YEARS THE KING OF COUNTRY MUSIC (II,2223); SKINFLINT (II,2377)

GATLIN, RUDY
LARRY GATLIN AND THE GATLIN BROTHERS (II,1429)

GATLIN, STEVE
LARRY GATLIN AND THE GATLIN BROTHERS (II,1429)

GATTEYS, BENNYE
THE BOSTON TERRIER (I,696); THE BRIGHTER DAY (I,728); CAPTAIN KANGAROO (II,434); DAYS OF OUR LIVES (II,629); THE LLOYD BRIDGES SHOW (I,2733)

GATTO, PETER
LOVE OF LIFE (I,2774)

GAUGE, ALEXANDER
THE ADVENTURES OF ROBIN HOOD (I,74)

GAUL, PATRICIA
AFTERMASH (II,23)

GAULI, LORRAINE
THE NEW VOICE (II,1833)

GAUNT, FIONA
A HORSEMAN RIDING BY (II,1182); WAR AND PEACE (I,4765)

GAUNT, WILLIAM
THE CHAMPIONS (I,896); THE FAR PAVILIONS (II,832); TYCOON: THE STORY OF A WOMAN (II,2697)

GAUTIER, DICK
BENNY AND BARNEY: LAS VEGAS UNDERCOVER (II,206); CAN YOU TOP THIS? (I,817); GET SMART (II,983); HERE WE GO AGAIN (I,2028); IT'S YOUR BET (I,2279); THE JONES BOYS (I,2446); THE LIAR'S CLUB (II,1466); MR. TERRIFIC (I,3152); THIS WILL BE THE YEAR THAT WILL BE (I,4466); WHEN THINGS WERE ROTTEN (II,2782)

GAUTREAUX, DAVID
SEARCH FOR TOMORROW (II,2284)

GAVA, CASSANDRA
THE CRACKER BROTHERS (II,599)

GAVALA, YULA
DRIBBLE (II,736)

GAVIN, JAMES
THE BIG VALLEY (I,437); JOHNNY BELINDA (I,2415); THE LIFE OF RILEY (I,2686); TONIGHT IN HAVANA (I,4557)

GAVIN, JOHN
CONVOY (I,1043); DESTRY (I,1241); DOCTORS' PRIVATE LIVES (II,692); DOCTORS' PRIVATE LIVES (II,693); FROM HERE TO ETERNITY (II,933)

GAXTON, WILLIAM
THE NASH AIRFLYTE THEATER (I,3227)

GAY, BARBARA
APARTMENT 3-C (I,243); MR. AND MRS. MYSTERY (I,3123)

GAY, JOHN
APARTMENT 3-C (I,243); MR. AND MRS. MYSTERY (I,3123)

GAYE, LISA
HANK (I,1931); HOW TO MARRY A MILLIONAIRE (I,2147); THE JOHN FORSYTHE SHOW (I,2410); LOVE THAT BOB (I,2779); SCIENCE FICTION THEATER (I,3938); SKYFIGHTERS (I,4078)

GAYE, MARVIN
IT'S WHAT'S HAPPENING, BABY! (I,2278)

GAYE, NORA
MICKEY SPILLANE'S MIKE HAMMER: MORE THAN MURDER (II,1693)

GAYE, PAT
THE FRANK SINATRA SHOW (I,1664)

GAYLE, CRYSTAL
BOB HOPE SPECIAL: BOB HOPE ON THE ROAD TO CHINA (II,291); COUNTRY COMES HOME (II,585); COUNTRY COMES HOME (II,586); COUNTRY MUSIC HIT PARADE (II,589); COUNTRY NIGHT OF STARS II (II,593); THE COUNTRY PLACE (I,1071); CRYSTAL (II,607); CRYSTAL GAYLE IN CONCERT (II,608); THE CRYSTAL GAYLE SPECIAL (II,609); DEAN MARTIN'S CHRISTMAS IN CALIFORNIA (II,645); FIFTY YEARS OF COUNTRY MUSIC (II,853); JOHNNY CASH—A MERRY MEMPHIS CHRISTMAS (II,1331); LORETTA LYNN: THE LADY. . .THE LEGEND (II,1521); THE LOU RAWLS SPECIAL (II,1527); THE OSMOND BROTHERS SPECIAL (II,1925); ROY ACUFF—50 YEARS THE KING OF COUNTRY MUSIC (II,2223); THE SENSATIONAL, SHOCKING, WONDERFUL, WACKY 70S (II,2302)

GAYLE, MONICA
THE WILDS OF TEN THOUSAND ISLANDS (II,2803)

GAYLE, TINA
BATTLE OF THE NETWORK STARS (II,175); CHIPS (II,511)

GAYLOR, ANNA
INVITATION TO PARIS (I,2233)

GAYLOR, CLARISSA
THE DOCTORS (II,694)

GAYNEE, RANDY
THE GOLDEN AGE OF TELEVISION (II,1017)

GAYNES, EDMUND
SLEEPING BEAUTY (I,4083)

GAYNES, GEORGE
THE GIRL IN THE EMPTY GRAVE (II,997); ONE TOUCH OF VENUS (I,3389); PUNKY BREWSTER (II,2092); SCRUPLES (II,2280); SEARCH FOR TOMORROW (II,2284); WASHINGTON: BEHIND CLOSED DOORS (II,2744)

GAYNOR, JACK
THE OUTLAWS (I,3427)

GAYNOR, JANET
CHRYSLER MEDALLION THEATER (I,951); HEDDA HOPPER'S HOLLYWOOD (I,2002); HOLLYWOOD: THE SELZNICK YEARS (I,2095); THE LUX VIDEO THEATER (I,2795)

GAYNOR, MITZI
THE DONALD O'CONNOR SHOW (I,1343); THE FRANK SINATRA SHOW (I,1666); MITZI (I,3071); MITZI AND A HUNDRED GUYS (II,1710); MITZI'S SECOND SPECIAL (I,3074); MITZI. . . ROARIN' IN THE 20S (II,1711); MITZI. . . WHAT'S HOT, WHAT'S NOT (II,1712); MITZI. . . ZINGS INTO SPRING (II,1713); MITZI. . . THE FIRST TIME (I,3073); MITZI: A TRIBUTE TO THE AMERICAN HOUSEWIFE (I,3072); PERRY COMO'S WINTER SHOW (I,3546)

GAZZARA, BEN
ARREST AND TRIAL (I,264); CAROL FOR ANOTHER CHRISTMAS (I,852); DANGER (I,1134); DUPONT SHOW OF THE MONTH (I,1387); RAPTURE AT TWO-FORTY (I,3724); RUN FOR YOUR LIFE (I,3871); THE WEB (I,4784)

GAZZARRI DANCERS, THE
HOLLYWOOD A GO GO (I,2080)

GAZZO, MICHAEL V
BEACH PATROL (II,192)

GEAHLE, CLIVE
MASTER OF THE GAME (II,1647)

GEAR, LUELLA
HAPPY BIRTHDAY (I,1941); JOE AND MABEL (I,2398); LADY IN THE DARK (I,2602); SURE AS FATE (I,4297)

GEARHART, LIVINGSTON
THE FRED WARING SHOW (I,1679)

GEARHART, VIRGINIA
THE FRED WARING SHOW (I,1679)

GEARON, VALERIE
CASANOVA (II,451)

GEARY, ANTHONY
CELEBRITY DAREDEVILS (II,473); THE FUNNIEST JOKE I EVER HEARD (II,939); GENERAL HOSPITAL (II,964); HOLLYWOOD'S PRIVATE HOME MOVIES II (II,1168); I LOVE LIBERTY (II,1215); THE OSMOND FAMILY THANKSGIVING SPECIAL (II,1928)

GEARY, PAUL
THE LONG HOT SUMMER (I,2747); SLATTERY'S PEOPLE (I,4082)

GEARY, TONY
BRIGHT PROMISE (I,727)

GEE, DONALD
THE TALE OF BEATRIX POTTER (II,2533)

GEER, ELLEN
THE JIMMY STEWART SHOW (I,2391); VELVET (II,2725)

GEER, KEVIN
ROUGHNECKS (II,2219); STEELTOWN (II,2452)

GEER, WILL
BROCK'S LAST CASE (I,742); BUNCO (II,395); HARRY O (II,1098); THE JIMMY STEWART SHOW (I,2391); THE OATH: THE SAD AND LONELY SUNDAYS (II,1874); OF MICE AND MEN (I,3333); SAM HILL: WHO KILLED THE MYSTERIOUS MR. FOSTER? (I,3896); THE WALTONS (II,2740)

GEESON, JUDY
MURDER ON THE MIDNIGHT EXPRESS (I,3162); SAM HILL: WHO KILLED THE MYSTERIOUS MR. FOSTER? (I,3896); STAR MAIDENS (II,2435)

GEESON, SALLY
BLESS THIS HOUSE (II,268)

GEFFNER, DEBORAH
JOE DANCER: MURDER ONE, DANCER 0 (II,1302)

GEHRIG, JAMES
THE HOUSE ON HIGH STREET (I,2141)

GEHRIG, PHYLLIS
THE RAY KNIGHT REVUE (I,3733)

GEHRING, TED
ALICE (II,33); B.J. AND THE BEAR (II,127); THE BANANA COMPANY (II,139); THE BAY CITY AMUSEMENT COMPANY (II,185); DALLAS (II,614); THE FAMILY HOLVAK (II,817); THE INVISIBLE MAN (II,1242); THE INVISIBLE MAN (II,1243); LITTLE HOUSE ON THE PRAIRIE (II,1487); PALMERSTOWN U.S.A. (II,1945); SAM HILL: WHO KILLED THE MYSTERIOUS MR. FOSTER? (I,3896); SUPERTRAIN (II,2505)

GELB, ED
LIE DETECTOR (II,1467)

GELBWAKS, JEREMY
THE PARTRIDGE FAMILY (II,1962)

GELDEN, MARY MARGARET
THIS IS YOUR MUSIC (I,4462)

GELDER, IAN
THE NEXT VICTIM (I,3281)

GELLIS, DANNY
KNOTS LANDING (II,1404); NEVER SAY NEVER (II,1809)

GELLMAN, DR MERYLE
THE LOVE REPORT (II,1542)

GELMAN, LARRY
ALMOST HEAVEN (II,57); THE BOB NEWHART SHOW (II,342); FREE COUNTRY (II,921); GIDGET GETS MARRIED (I,1796); MAUDE (II,1655); MORK AND MINDY (II,1735); NEEDLES AND PINS (I,3246); THE ODD COUPLE (II,1875); ONE OF OUR OWN (II,1910); THE RIGHTEOUS APPLES (II,2174); TOPPER (II,2649)

GEMA, THE
THE COUNTRY PLACE (I,1071)

GEMIGNANI, RHODA
CONCRETE BEAT (II,562); FARRELL: FOR THE PEOPLE (II,834)

GENE GENE THE DANCING MACHINE
THE CHUCK BARRIS RAH-RAH SHOW (II,525)

GENEVIEVE
AMERICA PAUSES FOR SPRINGTIME (I,166); SATINS AND SPURS (I,3917); SCRUPLES (II,2280)

GENGE, PAUL
MR. LUCKY (I,3141); SIX GUNS FOR DONEGAN (I,4065)

GENISE, LIVIA
THE FIGHTING NIGHTINGALES (II,855); SCALPELS (II,2266)

GENKY, LES
TALES OF THE GOLD MONKEY (II,2537)

GENN, LEO
ST. JOAN (I,4175)

GENNARO, PETER
THE BING CROSBY SHOW (I,466); THE JUDY GARLAND SHOW (I,2471)

GENTRY BROTHERS, THE
THE DOM DELUISE SHOW (I,1328)

GENTRY, BOBBIE
THE ALL-STAR SALUTE TO MOTHER'S DAY (II,51); THE BOBBIE GENTRY SPECIAL (I,669); THE BOBBIE GENTRY SPECIAL (I,670); BOBBIE GENTRY'S HAPPINESS HOUR (I,671); BOBBIE GENTRY'S HAPPINESS HOUR (II,343); SHOOT-IN AT NBC (A BOB HOPE SPECIAL) (I,4021); THE SOUND AND THE SCENE (I,4132); THE SPECIAL GENTRY ONE (I,4148)

GENTRY, ROBERT
ANOTHER WORLD (II,97); FRANKENSTEIN (I,1671); THE GUIDING LIGHT (II,1064); ONE LIFE TO LIVE (II,1907); A WORLD APART (I,4910)

GEORGE BAINES TRIO, THE
THE SKIP FARRELL SHOW (I,4074)

GEORGE FARSON DANCERS, THE
THE MAGIC PLANET (II,1598)

GEORGE MORGAN SQUARE DANCERS
COUNTRY MUSIC CARAVAN (I,1066)

GEORGE WYLE SINGERS, THE
THE ANDY WILLIAMS SHOW (I,205); THE JERRY LEWIS SHOW (I,2370)

GEORGE, ANTHONY
BROKEN ARROW (I,743); CHECKMATE (I,919); DARK SHADOWS (I,1157); ONE LIFE TO LIVE (II,1907); SANDY (I,3908); SEARCH FOR TOMORROW (II,2284); THE UNTOUCHABLES (I,4682)

GEORGE, BARBARA
BANYON (I,344)

GEORGE, BETTY
THE JERRY LESTER SHOW (I,2360)

GEORGE, CHIEF DAN
CENTENNIAL (II,477)

GEORGE, CHRISTOPHER
ESCAPE (I,1460); THE HOUSE ON GREENAPPLE ROAD (I,2140); THE IMMORTAL (I,2196); THE IMMORTAL (I,2197); MAN ON A STRING (I,2877); MITZI AND A HUNDRED GUYS (II,1710); NOT GUILTY! (I,3311); THE RAT PATROL (I,3726)

GEORGE, EARL
THREE STEPS TO HEAVEN (I,4485)

GEORGE, EBEN
THE CHADWICK FAMILY (II,478)

GEORGE, GLADYS
PEPSI-COLA PLAYHOUSE (I,3523); YOUR JEWELER'S SHOWCASE (I,4953)

GEORGE, HOWARD
HANDLE WITH CARE (II,1077)

GEORGE, JENNIFER
SCAMPS (II,2267)

GEORGE, JOHN
THE ADVENTURES OF DR. FU MANCHU (I,58)

GEORGE, KARL
THE ADVENTURES OF OZZIE AND HARRIET (I,71)

GEORGE, LYNDA DAY
THE BARBARY COAST (II,147); BATTLE OF THE NETWORK STARS (II,164); CANNON (I,820); CASINO (II,452); COME OUT, COME OUT, WHEREVER YOU ARE (I,1014); FANTASY ISLAND (II,829); THE HOUSE ON GREENAPPLE ROAD (I,2140); MISSION: IMPOSSIBLE (I,3067); MRS R.—DEATH AMONG FRIENDS (II,1759); ONCE AN EAGLE (II,1897); QUICK AND QUIET (II,2099); THE RETURN OF CAPTAIN NEMO (II,2135); RICH MAN, POOR MAN—BOOK I (II,2161); ROOTS (II,2211); THE SILENT FORCE (I,4047); THE SOUND OF ANGER (I,4133); TWIN DETECTIVES (II,2687)

GEORGE, PHYLLIS
CANDID CAMERA (II,420); CELEBRITY CHALLENGE OF THE SEXES 1 (II,465); CELEBRITY CHALLENGE OF THE SEXES 2 (II,466); CELEBRITY CHALLENGE OF THE SEXES 4 (II,468); I'VE GOT A SECRET (II,1220); PEOPLE (II,1995); THE STARS AND STRIPES SHOW (I,4206)

GEORGE, SUE
FATHER KNOWS BEST (II,838)

GEORGE, SUSAN
BOB HOPE SPECIAL: THE BOB HOPE SPECIAL (II,338); COMPUTERCIDE (II,561); CROSSROADS (I,1105); DR JEKYLL AND MR. HYDE (I,1368); PAJAMA TOPS (II,1942)

GEORGES, CONNIE
THE BENNY HILL SHOW (II,207)

GEORGIADE, NICHOLAS
MEETING AT APALACHIN (I,2994); RUN, BUDDY, RUN (I,3870); THE UNTOUCHABLES (I,4682)

GEORGOV, WALTER
SUNDAY IN TOWN (I,4285)

GERADO, ANDREW
THE GOLDEN AGE OF TELEVISION (II,1016)

GERARD, GIL
BATTLE OF THE NETWORK STARS (II,169); BATTLE OF THE NETWORK STARS (II,170); BOB HOPE SPECIAL: BOB HOPE'S ALL-STAR LOOK AT TV'S PRIME TIME WARS (II,303); BUCK ROGERS IN THE 25TH CENTURY (II,383); BUCK ROGERS IN THE 25TH CENTURY (II,384); CELEBRITY CHALLENGE OF THE SEXES 5 (II,469); CIRCUS OF THE STARS (II,534); THE DOCTORS (II,694); HEAR NO EVIL (II,1116); JOHNNY BLUE (II,1321); RANSOM FOR ALICE (II,2113)

GERARD, PENNY
MUSICAL MERRY-GO-ROUND (I,3184)

GERBER, BILL
FAY (II,844); LILY (I,2705)

GERBER, JAY
AMERICA, YOU'RE ON (II,69); PEN 'N' INC. (II,1992)

GERBER, JOAN
ARNOLD'S CLOSET REVIEW (I,262); THE BARKLEYS (I,356); THE BEANY AND CECIL SHOW (I,374); THE BUGALOOS (I,757); DOROTHY IN THE LAND OF OZ (II,722); H.R. PUFNSTUF (I,2154); HELP! IT'S THE HAIR BEAR BUNCH (I,2012); LIDSVILLE (I,2679); MITZI (I,3071); NANNY AND THE PROFESSOR AND THE PHANTOM OF THE CIRCUS (I,3226); THE ODDBALL COUPLE (II,1876); THE RICHIE RICH SHOW (II,2169); SKYHAWKS (I,4079); TARZAN: LORD OF THE JUNGLE (II,2544); WAIT TIL YOUR FATHER GETS HOME (I,4748)

GERBER, WILLIAM
MALIBU (II,1607)

GERBERT, GORDON
THE MISS AND THE MISSILES (I,3062)

GERE, RICHARD
D.H.O. (I,1249)

GERET, GEORGES
THE POPPY IS ALSO A FLOWER (I,3650)

GERING, RICHARD
MARGIE (I,2909)

GERKIN, ELLIE
YOUNG LIVES (II,2866)

GERMAIN, STUART
THE DEVIL AND DANIEL WEBSTER (I,1246)

GERRINGER, ROBERT
DARK SHADOWS (I,1157); TEXAS (II,2566)

GERRISH, LEANNE
FAME (II,812)

GERRITSEN, LISA
THE AMAZING CHAN AND THE CHAN CLAN (I,157); IT CAN'T HAPPEN TO ME (II,1250); THE MARY TYLER MOORE SHOW (II,1640); MY WORLD . . . AND WELCOME TO IT (I,3208); PHYLLIS (II,2038)

GERRITY, PATTY ANN
THIS IS ALICE (I,4453)

GERROLL, DANIEL
THE WOMAN IN WHITE (II,2819)

GERRY MULLIGAN QUARTET, THE
TIMEX ALL-STAR JAZZ SHOW II (I,4516)

GERSON, BETTY LOU
MORNING STAR (I,3104)

GERSON, NATASHA
THE OMEGA FACTOR (II,1888)

GERSTAD, JOHN
GIVE US BARABBAS! (I,1816)

GERSTEN, BAILLIE
THIS BETTER BE IT (II,2595)

GERSTLE, FRANK
THE BANANA SPLITS ADVENTURE HOUR (I,340); I AND CLAUDIE (I,2164); THE MCGONIGLE (I,2968); ON TRIAL (I,3362)

GERTZ, JAMI
DREAMS (II,735); THE FACTS OF LIFE (II,805); FOR MEMBERS ONLY (II,905); SQUARE PEGS (II,2431)

GERUSSI, BRUNO
CELEBRITY COOKS (II,472)

GESSLER, TIMOTHY
THE EDGE OF NIGHT (II,760)

GETHER, STEVEN
LOVE OF LIFE (I,2774)

GETTEYS, BENNYE
BRIEF ENCOUNTER (I,723)

GETTY, ESTELLE
NO MAN'S LAND (II,1855)

GETZ, JOHN
ANOTHER WORLD (II,97); CONCRETE BEAT (II,562); LOOSE CHANGE (II,1518); RAFFERTY (II,2105); SUZANNE PLESHETTE IS MAGGIE BRIGGS (II,2509); THREE'S COMPANY (II,2614)

GETZ, PETER MICHAEL
IN TROUBLE (II,1230)

GETZ, STUART
THE PAUL LYNDE SHOW (I,3489); WHO'S AFRAID OF MOTHER GOOSE? (I,4852)

GHANI, TANVEER
THE FAR PAVILIONS (II,832)

GHOSTLEY, ALICE
BEWITCHED (I,418); CAPTAIN NICE (I,835); CAR 54, WHERE ARE YOU? (I,842); THE DATCHET DIAMONDS (I,1160); FREEDOM RINGS (I,1682); THE GOLDDIGGERS (I,1838); HOORAY FOR LOVE (I,2115); JACKIE GLEASON AND HIS AMERICAN SCENE MAGAZINE (I,2321); THE JONATHAN WINTERS SHOW (I,2443); THE JUD STRUNK SHOW (I,2462); THE JULIE ANDREWS HOUR (I,2480); MAMA MALONE (II,1609); MAYBERRY R.F.D. (I,2961); MERRY CHRISTMAS. . . WITH LOVE, JULIE (II,1687); THE NEW TEMPERATURES RISING SHOW (I,3273); THE ODD COUPLE (II,1875); SHANGRI-LA (I,3996); THE SHOWOFF (I,4035); SUTTERS BAY (II,2508); TWELFTH NIGHT (I,4636)

GIAMALVA, JOE
THE SKATEBIRDS (II,2375)

GIAMBALVO, LOUIS
THE DEVLIN CONNECTION (II,664); FAME (II,812); FLY AWAY HOME (II,893); THE GANGSTER CHRONICLES (II,957); OH MADELINE (II,1880); REWARD (II,2147)

GIBB, ANDY
BOB HOPE SPECIAL: BOB HOPE LAUGHS WITH THE MOVIE AWARDS (II,288); BOB HOPE SPECIAL: BOB HOPE'S ALL-STAR CHRISTMAS SHOW (II,296); BOB HOPE SPECIAL: BOB HOPE'S ALL-STAR COMEDY BIRTHDAY PARTY (II,297); CELEBRITY CHALLENGE OF THE SEXES 5 (II,469); THE DEAN MARTIN CHRISTMAS SPECIAL (II,640); GEORGE BURNS' 100TH BIRTHDAY PARTY (II,971); OLIVIA (II,1883); OLIVIA NEWTON-JOHN'S HOLLYWOOD NIGHTS (II,1886); THE OSMOND BROTHERS SPECIAL (II,1925)

GIBB, CYNTHIA
FAME (II,812); SEARCH FOR TOMORROW (II,2284)

GIBB, DON RICHARD
NICHOLS AND DYMES (II,1841)

GIBBERSON, BILL
KENNEDY (II,1377)

GIBBONEY, LINDA
ALL MY CHILDREN (II,39); SEARCH FOR TOMORROW (II,2284)

GIBBONS, ROD
LOVE OF LIFE (I,2774)

GIBBONS, AYLLENE
IRONSIDE (I,2240)

GIBBONS, LEEZA
ENTERTAINMENT TONIGHT (II,780)

GIBBONS, ROD
LOVE OF LIFE (I,2774)

GIBBS, ALAN
ALEX AND THE DOBERMAN GANG (II,32)

GIBBS, GEORGIA
THE ALL-STAR SUMMER REVUE (I,139); GEORGIA GIBBS MILLION RECORD SHOW (I,1777); UPBEAT (I,4684)

GIBBS, JOAN
RAPTURE AT TWO-FORTY (I,3724)

GIBBS, JORDAN
CHECKING IN (II,492)

GIBBS, LYNN
MELODY STREET (I,3000)

GIBBS, MARLA
CHECKING IN (II,492); THE JEFFERSONS (II,1276); PRYOR'S PLACE (II,2089); YOU CAN'T TAKE IT WITH YOU (II,2857)

GIBBS, NORMAN
TRAPPER JOHN, M.D. (II,2654)

GIBBS, TERRY
COUNTRY GALAXY OF STARS (II,587)

GIBBS, TIMOTHY
FATHER MURPHY (II,841); THE ROUSTERS (II,2220)

GIBSON, AMY
LOVE OF LIFE (I,2774)

GIBSON, BEAU
THE RUNAWAY BARGE (II,2230)

GIBSON, BERNADETTE
PRISONER: CELL BLOCK H (II,2085)

GIBSON, BILLY
THE GEORGE HAMILTON IV SHOW (I,1769)

GIBSON, BOB
THE LORENZO AND HENRIETTA MUSIC SHOW (II,1519)

GIBSON, CAL
PARK PLACE (II,1958)

GIBSON, DON
THE BENNETS (I,397); HEE HAW (II,1123); ROY ACUFF—50 YEARS THE KING OF COUNTRY MUSIC (II,2223)

GIBSON, GERRY
CONCRETE BEAT (II,562)

GIBSON, HENRY
THE BUREAU (II,397); DOROTHY HAMILL'S CORNER OF THE SKY (II,721); EVIL ROY SLADE (I,1485); F TROOP (I,1499); THE HALLOWEEN THAT ALMOST WASN'T (II,1075); HIGH SCHOOL, U.S.A. (II,1149); HONEYMOON SUITE (I,2108);

THE KAREN VALENTINE SHOW (I,2506); ROWAN AND MARTIN'S LAUGH-IN (I,3856); ROWAN AND MARTIN'S LAUGH-IN (I,3857); WONDER WOMAN (PILOT 2) (II,2827)

GIBSON, JEFF
EIGHT IS ENOUGH (II,762)

GIBSON, JOHN
CONCERNING MISS MARLOWE (I,1032); CRIME PHOTOGRAPHER (I,1097); A DATE WITH JUDY (I,1162); THE EDGE OF NIGHT (II,760); HUMAN ADVENTURE (I,2157); MIRACLE ON 34TH STREET (I,3057); MONTGOMERY'S SUMMER STOCK (I,3096); THE YOUNG AND THE RESTLESS (II,2862)

GIBSON, JUDY
EVERYDAY (II,793); ONE MORE TRY (II,1908)

GIBSON, JULIE
PARIS CAVALCADE OF FASHION (I,3456)

GIBSON, LAUREN
KNOTS LANDING (II,1404)

GIBSON, LEE
THE BENNY HILL SHOW (II,207)

GIBSON, MARTHA
SEEING THINGS (II,2297)

GIBSON, MARY ANN
TENAFLY (II,2560)

GIBSON, VIRGINIA
CAPTAIN BILLY'S MISSISSIPPI MUSIC HALL (I,826); THE JOHNNY CARSON SHOW (I,2422); RYAN'S HOPE (II,2234); SO THIS IS HOLLYWOOD (I,4105); STUDIO '57 (I,4267); THE WEB (I,4784)

GIBSON, WYNNE
HAPPY BIRTHDAY (I,1941)

GIDDINGS, ALISON
MASTER OF THE GAME (II,1647)

GIDDINGS, MICHELL
MASTER OF THE GAME (II,1647)

GIDEON, BOND
BLANSKY'S BEAUTIES (II,267); OPERATION PETTICOAT (II,1916); OPERATION PETTICOAT (II,1917); THE YOUNG AND THE RESTLESS (II,2862)

GIELGUD, JOHN
BRIDESHEAD REVISITED (II,375); DUPONT SHOW OF THE MONTH (I,1387); THE FAR PAVILIONS (II,832); HAMLET (I,1925); MARCO POLO (II,1626); THE PALLISERS (II,1944); PROBE (I,3674); THE SEVEN DIALS MYSTERY (II,2308); WHY DIDN'T THEY ASK EVANS? (II,2795)

GIEM, GARY
LOVE OF LIFE (I,2774); THE YOUNG AND THE RESTLESS (II,2862)

GIERASCH, STEFAN
A.E.S. HUDSON STREET (II,17); BIG BEND COUNTRY (II,230); DIAGNOSIS: DANGER (I,1250); GABE AND WALKER (II,950); HAZARD'S PEOPLE (II,1112); KATE MCSHANE (II,1369); THE WINDS OF WAR (II,2807)

GIFFORD, ALAN
THE BOB HOPE SHOW (I,559); THE BOB HOPE SHOW (I,567)

GIFFORD, CHRIS
THE GREAT SPACE COASTER (II,1059)

GIFFORD, GLORIA
SHE'S WITH ME (II,2321)

GIFFORD, WENDY
A FAMILY AFFAIR (II,814)

GIFTOS, ELAINE
BANYON (I,344); FISHERMAN'S WHARF (II,865); THE INTERNS (I,2226); NUMBER 96 (II,1869); THE PARTRIDGE FAMILY (II,1962); STONESTREET: WHO KILLED THE CENTERFOLD MODEL? (II,2472); THROUGH THE MAGIC PYRAMID (II,2615); VENICE MEDICAL (II,2726); WILD ABOUT HARRY (II,2797)

GIL, JORGE
KEY TORTUGA (II,1383)

GILBERT, BILLY
ANDY'S GANG (I,214); EL COYOTE (I,1423); EVERYTHING HAPPENS TO ME (I,1481); JACK AND THE BEANSTALK (I,2287); JOHNNY CARSON PRESENTS THE SUN CITY SCANDALS (I,2420)

GILBERT, CAROLYN
THE CAROLYN GILBERT SHOW (I,855); GARROWAY AT LARGE (I,1737); KUKLA, FRAN, AND OLLIE (I,2594); STUD'S PLACE (I,4271)

GILBERT, EDMUND
THE HARDY BOYS MYSTERIES (II,1090); THE NORLISS TAPES (I,3305)

GILBERT, HENRY
THE ADVENTURES OF LONG JOHN SILVER (I,67)

GILBERT, JANICE
BREAK THE BANK (I,712); THE O'NEILLS (I,3392)

GILBERT, JO
THE LIFE OF RILEY (I,2684)

GILBERT, JOANNE
THE BING CROSBY SHOW (I,462); CHALK ONE UP FOR JOHNNY (I,893)

GILBERT, JODY
LIFE WITH LUIGI (I,2692)

GILBERT, JOHNNY
BEAT THE ODDS (I,379); FAST DRAW (I,1539); MUSIC BINGO (I,3165)

GILBERT, JONATHAN
LITTLE HOUSE ON THE PRAIRIE (II,1487); LITTLE HOUSE: A NEW BEGINNING (II,1488); LITTLE HOUSE: LOOK BACK TO YESTERDAY (II,1490)

GILBERT, LAUREN
CONCERNING MISS MARLOWE (I,1032); THE EDGE OF NIGHT (II,760); FAIRMEADOWS, U.S.A. (I,1514); THE KATE SMITH HOUR (I,2508); THE LITTLE FOXES (I,2712); LOVE OF LIFE (I,2774); THREE STEPS TO HEAVEN (I,4485)

GILBERT, LISA
YOUNG LIVES (II,2866)

GILBERT, MARCUS
MASTER OF THE GAME (II,1647)

GILBERT, MELISSA
BATTLE OF THE NETWORK STARS (II,166); BATTLE OF THE NETWORK STARS (II,169); BATTLE OF THE NETWORK STARS (II,172); BATTLE OF THE NETWORK STARS (II,173); BATTLE OF THE NETWORK STARS (II,175); CELEBRITY CHALLENGE OF THE SEXES 5 (II,469); CIRCUS LIONS, TIGERS AND MELISSAS TOO (II,528); THE DEAN MARTIN CELEBRITY ROAST (II,637); LIKE MAGIC (II,1476); LITTLE HOUSE ON THE PRAIRIE (II,1487); LITTLE HOUSE: A NEW BEGINNING (II,1488); LITTLE HOUSE: BLESS ALL THE DEAR CHILDREN (II,1489); LITTLE HOUSE: LOOK BACK TO YESTERDAY (II,1490); LITTLE HOUSE: THE LAST FAREWELL (II,1491)

GILBERT, MICKEY
STUNTS UNLIMITED (II,2487)

GILBERT, NANCY
BUFFALO BILL JR. (I,755)

GILBERT, PATTI
RITUALS (II,2182); THE UGILY FAMILY (II,2699)

GILBERT, PAUL
THE DUKE (I,1381); SHORE LEAVE (I,4022); SHOW BIZ (I,4027)

GILBERT, RUTH
THE MILTON BERLE SHOW (I,3047); PROJECTION ROOM (I,3680); RUTHIE ON THE TELEPHONE (I,3877); TEXACO STAR THEATER (I,4411)

GILCHRIST, CONNIE
THE ADVENTURES OF LONG JOHN SILVER (I,67); LEAVE IT TO BEAVER (I,2648); THE REAL MCCOYS (I,3741)

GILES, SANDRA
CRAZY TIMES (II,602)

GILFILLAN, SUE ANNE
THERE GOES THE NEIGHBORHOOD (II,2582)

GILFORD, FRANK
BATTLE OF THE NETWORK STARS (II,167)

GILFORD, GWYNNE
THE BOYS (II,360); THE CHOPPED LIVER BROTHERS (II,514); A NEW KIND OF FAMILY (II,1819); SHEEHY AND THE SUPREME MACHINE (II,2322); THE WAVERLY WONDERS (II,2746)

GILFORD, JACK
APPLE PIE (II,100); ARSENIC AND OLD LACE (I,270); THE DAVID FROST REVUE (I,1176); DON'T CALL US (II,709); THE DUCK FACTORY (II,738); THE GHOST AND MRS. MUIR (I,1786); HEAVEN ON EARTH (II,1119); HICKEY VS. ANYBODY (II,1142); OF THEE I SING (I,3334); ONCE UPON A MATTRESS (I,3371); PARADE OF STARS (II,1954); PAUL SAND IN FRIENDS AND LOVERS (II,1982); THE PHIL SILVERS ARROW SHOW (I,3576); SEVENTH AVENUE (II,2309); SOAP (II,2392); SUPER COMEDY BOWL 1 (I,4288); TAXI (II,2546); TWIGS (II,2683)

GILGER, GLEN
IT WAS A SHORT SUMMER, CHARLIE BROWN (I,2252)

GILGREEN, JOHN
DOWN HOME (II,731)

GILHAM, CHERYL
THE BENNY HILL SHOW (II,207)

GILL, BEVERLY
WONDER WOMAN (PILOT 1) (II,2826)

GILL, JOHN
SECRETS OF THE OLD BAILEY (I,3971)

GILL, RUSTY
POLKA-GO-ROUND (I,3640)

GILLAM, LORNA
TOO YOUNG TO GO STEADY (I,4574)

GILLARD, STUART
YOUR PLACE OR MINE? (II,2874)

GILLERON, TOM
ALL THAT GLITTERS (II,41)

GILLESPIE, ANN
RYAN'S HOPE (II,2234); THE SHERIFF AND THE ASTRONAUT (II,2326)

GILLESPIE, DARLENE
CORKY AND WHITE SHADOW (I,1049); THE FURTHER ADVENTURES OF SPIN AND MARTY (I,1720); THE MICKEY MOUSE CLUB (I,3025); THE MOUSEKETEERS REUNION (II,1742); THE NEW ADVENTURES OF SPIN AND MARTY (I,3254)

GILLESPIE, DIZZY
ALL-STAR SWING FESTIVAL (II,53); THE BIG BAND AND ALL THAT JAZZ (I,422); THE BILL COSBY SPECIAL, OR? (I,443); TIMEX ALL-STAR JAZZ SHOW IV (I,4518)

GILLESPIE, GINA
THE CODE OF JONATHAN WEST (I,992); KAREN (I,2505); LAW OF THE PLAINSMAN (I,2639); PLAYHOUSE 90 (I,3623); THE SLOWEST GUN IN THE WEST (I,4088); THRILLER (I,4492)

GILLESPIE, JEAN
ANOTHER DAY, ANOTHER DOLLAR (I,228); THE CAMPBELL TELEVISION SOUNDSTAGE (I,812); HOW'S BUSINESS? (I,2153); MAGNAVOX THEATER (I,2827); ST. ELSEWHERE (II,2432); SURE AS FATE (I,4297)

GILLESPIE, LARRAIN
DOBIE GILLIS (I,1302)

GILLESPIE, LARRY
THE CODE OF JONATHAN WEST (I,992)

GILLETTE, ANITA
ALL THAT GLITTERS (II,42); THE BAXTERS (II,183); BOB & CAROL & TED & ALICE (I,495); GEORGE M! (I,1772); GEORGE M! (II,976); ME AND THE CHIMP (I,2974); PINOCCHIO (I,3603); QUINCY, M. E. (II,2102)

GILLETTE, PENELOPE
THE DOCTORS (I,1323)

GILLETTE, PRISCILLA
CURTAIN CALL THEATER (I,1115)

GILLEY, MICKEY
COUNTRY GALAXY OF STARS (II,587); MUSIC CITY NEWS TOP COUNTRY HITS OF THE YEAR (II,1772); THE NASHVILLE PALACE (II,1793); OPRYLAND: NIGHT OF STARS AND FUTURE STARS (II,1921)

GILLIAM, BURTON
BIG BOB JOHNSON AND HIS FANTASTIC SPEED CIRCUS (II,231); THE GEORGIA PEACHES (II,979); THE GIRL, THE GOLD WATCH AND DYNAMITE (II,1001); THE GIRL, THE GOLD WATCH AND EVERYTHING (II,1002); HOME COOKIN' (II,1170)

GILLIAM, BYRON
ROWAN AND MARTIN'S LAUGH-IN (I,3856)

GILLIAM, DAVID
BRAVO TWO (II,368); MASTER OF THE GAME (II,1647)

GILLIAM, STU
DEAN MARTIN PRESENTS THE GOLDDIGGERS (I,1193); FREEMAN (II,923); THE HARLEM GLOBETROTTERS (I,1953); HARRIS AND COMPANY (II,1096); THE HOUNDCATS (I,2132); ROLL OUT! (I,3833); THE SUPER GLOBETROTTERS (II,2498)

GILLIAM, TERRY
MONTY PYTHON'S FLYING CIRCUS (II,1729)

GILLIAN, HUGH
THE FACTS OF LIFE (II,805)

GILLIART, MELVILLE
FARAWAY HILL (I,1532)

GILLIGAN, MAURA
AS THE WORLD TURNS (II,110)

GILLIGAN, RUSTY
CRASH ISLAND (II,600)

GILLILAND, RICHARD
THE BUREAU (II,397); JUST OUR LUCK (II,1360); LITTLE WOMEN (II,1497); LITTLE WOMEN (II,1498); MCMILLAN (II,1666); OPERATION PETTICOAT (II,1916); OPERATION PETTICOAT (II,1917); THE WALTONS: A DAY FOR THANKS ON WALTON'S MOUNTAIN (EPISODE 3) (II,2741); THE WALTONS: MOTHER'S DAY ON WALTON'S MOUNTAIN (EPISODE 2) (II,2742); THE WALTONS: WEDDING ON WALTON'S MOUNTAIN (EPISODE 1) (II,2743)

GILLIN, HUGH
THE BIG EASY (II,234); BORDER PALS (II,352); GABE AND WALKER (II,950); SEMI-TOUGH (II,2299); SEMI-TOUGH (II,2300); TROUBLE IN HIGH TIMBER COUNTRY (II,2661)

GILLIN, LINDA
COMPUTERCIDE (II,561)

GILLING, REBECCA
RETURN TO EDEN (II,2143)

GILLIS, GWYN
THE ANDROS TARGETS (II,76)

GILLMER, CAROLINE
PRISONER: CELL BLOCK H (II,2085)

GILLMORE, LOWELL
THE MILKY WAY (I,3044)

GILLMORE, MARGALO
PETER PAN (I,3566)

GILMAN, KENNETH
AT EASE (II,116); DOROTHY (II,717); FOOT IN THE DOOR (II,902); G.I.'S (II,949); JESSICA NOVAK (II,1286); LOVES ME, LOVES ME NOT (II,1551); MOTHER AND ME, M.D. (II,1740)

GILMAN, LARRY
THE TEXAS RANGERS (II,2567)

GILMAN, SAM
SECOND LOOK (I,3960); SHANE (I,3995)

GILMAN, SID
BROCK CALLAHAN (I,741)

GILMAN, TONI
EVERY STRAY DOG AND KID (II,792); THE GOOD GUYS (I,1847); A TIME TO LIVE (I,4510)

GILMORE, ART
ADAM-12 (I,31); COMEDY SPOT (I,1025); HIGHWAY PATROL (I,2058); MEN OF ANNAPOLIS (I,3008)

GILMORE, LOWELL
THE FORTUNE HUNTER (I,1646); HAY FEVER (I,1979)

GILMORE, PETER
A MAN CALLED INTREPID (II,1612); THE MANIONS OF AMERICA (II,1623); THE ONEDIN LINE (II,1912)

GILMORE, VIOLET
ACCUSED (I,18)

GILMORE, VIRGINIA
SEARCH FOR TOMORROW (II,2284); STARLIGHT THEATER (I,4201)

GILMOUR, IAN
PRISONER: CELL BLOCK H (II,2085)

GILPIN, JACK
KATE AND ALLIE (II,1365)

GILSON, JOHN
THE PHIL SILVERS SHOW (I,3580)

GILSTRAP, SUZANNE
SKYWARD CHRISTMAS (II,2378)

GILYARD JR, CLARENCE
THE DUCK FACTORY (II,738); THINGS ARE LOOKING UP (II,2588)

GIM, H W
BACHELOR FATHER (I,323)

GIMBLE, JOHN
THAT GOOD OLD NASHVILLE MUSIC (II,2571)

GIMIGNANI, RHODA
ROXY PAGE (II,2222)

GIMPEL, ERICA
FAME (II,812); THE KIDS FROM FAME (II,1392)

GING, JACK
DEAR DETECTIVE (II,652); THE ELEVENTH HOUR (I,1425); THE FABULOUS DR. FABLE (I,1500); FOUR EYES (II,912); HOW DO I KILL A THIEF—LET ME COUNT THE WAYS (II,1195); THE IMPOSTER (II,1225); MANNIX (II,1624); MY SISTER HANK (I,3201); O'HARA, UNITED STATES TREASURY: OPERATION COBRA (I,3347); RIPTIDE (II,2178); THE TALES OF WELLS FARGO (I,4353); THE WINDS OF WAR (II,2807)

GINGOLD, HERMIONE

BANYON (I,345); A CRY OF ANGELS (I,1111); THE ELGIN HOUR (I,1428); THE FRANCES LANGFORD SHOW (I,1657); THE FRANK SINATRA TIMEX SHOW (I,1668); MUNSTER, GO HOME! (I,3157); SPECIAL LONDON BRIDGE SPECIAL (I,4150); VICTOR BORGE'S 20TH ANNIVERSARY SHOW (I,4727)

GINSBURG, JENNIFER
THE TROUBLE WITH MOTHER (II,2663)

GINTY, ROBERT
THE BLACK SHEEP SQUADRON (II,262); HAWAIIAN HEAT (II,1111); THE PAPER CHASE (II,1949); THE TURNING POINT OF JIM MALLOY (II,2670)

GIORGIO, BILL
OPERATION GREASEPAINT (I,3402)

GIORGIO, TONY
BOBBY PARKER AND COMPANY (II,344)

GIOVANNI, JIM
LAUGH-IN (II,1444)

GIOVANNINI, CAESAR
KUKLA, FRAN, AND OLLIE (I,2594)

GIOVANNINI, RITA
CRY TERROR! (I,1112)

GIRARD, HENRY
THE ALDRICH FAMILY (I,111)

GIRARDIN, RAY
CONCRETE BEAT (II,562); FULL HOUSE (II,937); GENERAL HOSPITAL (II,964); JOSHUA'S WORLD (II,1344); THUNDER (II,2617)

GIRLING, CYNDI
CASTLE ROCK (II,456)

GIRLS OF THE GOLDEN WEST, THE
MIDWESTERN HAYRIDE (I,3033)

GIROUX, FERNANDE
MOMENT OF TRUTH (I,3084)

GIROUX, LEE
JOHNNY NIGHTHAWK (I,2432); TONIGHT! AMERICA AFTER DARK (I,4555)

GISH, DOROTHY
THE ALCOA HOUR (I,108); THE ELGIN HOUR (I,1428); MAGNAVOX THEATER (I,2827); PHILCO TELEVISION PLAYHOUSE (I,3583); PRUDENTIAL FAMILY PLAYHOUSE (I,3683); PULITZER PRIZE PLAYHOUSE (I,3692); STARLIGHT THEATER (I,4201)

GISH, LILLIAN
THE ALCOA HOUR (I,108); ARSENIC AND OLD LACE (I,270); CELANESE THEATER (I,881); THE DUPONT SHOW OF THE WEEK (I,1388); FORD THEATER HOUR (I,1635); KRAFT TELEVISION THEATER (I,2592); PHILCO TELEVISION PLAYHOUSE (I,3583); PLAYWRIGHTS '56 (I,3627); SPARROW (II,2411); THE SPIRAL STAIRCASE (I,4160); TWIN DETECTIVES (II,2687); WARNING SHOT (I,4770)

GIST, ROBERT
COOGAN'S REWARD (I,1044); HENNESSEY (I,2013)

GIST, ROD
COMPLETELY OFF THE WALL (II,560); THE PETER MARSHALL VARIETY SHOW (II,2029); ROLL OUT! (I,3833)

GITHENS, JAY
LONDON AND DAVIS IN NEW YORK (II,1512)

GITTENS, JEREMY
MASTER OF THE GAME (II,1647)

GITTINGS, RICHARD
BACHELOR AT LAW (I,322)

GIVNEY, JERRY
THE ADVENTURES OF CAPTAIN HARTZ (I,53)

GIVNEY, KATHRYN
MY THREE SONS (I,3205); PRUDENCE AND THE CHIEF (I,3682)

GIVOT, GEORGE
THE FABULOUS SYCAMORES (I,1505); VERSATILE VARIETIES (I,4713)

GLADKE, PETER
THE JACKIE GLEASON SPECIAL (I,2324); MELODY TOUR (I,3001)

GLADSTONE, JASON
ANOTHER WORLD (II,97)

GLADYS KNIGHT AND THE PIPS
MAC DAVIS SPECIAL: THE MUSIC OF CHRISTMAS (II,1578); THE SUZANNE SOMERS SPECIAL (II,2510)

GLAMOUR LOVELIES, THE
THE KEN MURRAY SHOW (I,2525)

GLASER, DAREL
THE CHEERLEADERS (II,493)

GLASER, PAUL MICHAEL
ACES UP (II,7); LOVE OF LIFE (I,2774); STARSKY AND HUTCH (II,2444); STARSKY AND HUTCH (II,2445)

GLASS, KATHY
THE BEST OF EVERYTHING (I,406); THE DOCTORS (II,694); ONE LIFE TO LIVE (II,1907); RETURN TO PEYTON PLACE (I,3779)

GLASS, NED
BANYON (I,345); BRIDGET LOVES BERNIE (I,722); CROSSFIRE (II,606); THE GOOD OLD DAYS (I,1851); THE HOOFER (I,2113); JULIA (I,2476); LITTLE LEATHERNECK (I,2713); MACREEDY'S WOMAN (I,2804); MR. TERRIFIC (I,3152); THE PHIL SILVERS SHOW (I,3580); READY AND WILLING (II,2115); ROSENTHAL AND JONES (II,2215); THREE ON AN ISLAND (I,4483); VEGAS (II,2723); THE WEB (I,4785)

GLASS, RON
BARNEY MILLER (II,154); CHANGE AT 125TH STREET (II,480); THE NEW ODD COUPLE (II,1825)

GLASS, SEAMAN
SLITHER (II,2379); WISHMAN (II,2811)

GLASSMAN, MIKE
MICKEY SPILLANE'S MIKE HAMMER: MORE THAN MURDER (II,1693)

GLAZER, VIC
EASY DOES IT. . . STARRING FRANKIE AVALON (II,754)

GLEA GIRLS, THE
JACKIE GLEASON AND HIS AMERICAN SCENE MAGAZINE (I,2321); THE JACKIE GLEASON SHOW (I,2323)

GLEASON GIRLS, THE
THE JACKIE GLEASON SHOW (I,2322)

GLEASON, JACKIE
THE ARTHUR GODFREY SHOW (I,285); THE BIG SELL (I,431); BOB HOPE SPECIAL: THE BOB HOPE SPECIAL (II,333); CAVALCADE OF STARS (I,877); THE FABULOUS 50S (I,1501); THE HONEYMOONERS (I,2110); THE HONEYMOONERS (I,2111); THE HONEYMOONERS (II,1175); THE HONEYMOONERS CHRISTMAS SPECIAL (II,1176); THE HONEYMOONERS CHRISTMAS (II,1177); THE HONEYMOONERS SECOND HONEYMOON (II,1178); THE HONEYMOONERS VALENTINE SPECIAL (II,1179); JACKIE GLEASON AND HIS AMERICAN SCENE MAGAZINE (I,2321); THE JACKIE GLEASON SHOW (I,2322); THE JACKIE GLEASON SHOW (I,2323); THE JACKIE GLEASON SPECIAL (I,2324); THE JACKIE GLEASON SPECIAL (I,2325); THE LIFE OF RILEY (I,2685); A LUCILLE BALL SPECIAL STARRING LUCILLE BALL AND JACKIE GLEASON (II,1558); THE MILLION DOLLAR INCIDENT (I,3045); PLAYHOUSE 90 (I,3623); SHOW BUSINESS SALUTE TO MILTON BERLE (I,4029); THE SHOWOFF (I,4035); STEP ON THE GAS (I,4215); TIMEX ALL-STAR JAZZ SHOW IV (I,4518); YOU'RE IN THE PICTURE (I,4965)

GLEASON, JAMES
CROWN THEATER WITH GLORIA SWANSON (I,1107); DAMON RUNYON THEATER (I,1129); THE LIFE OF RILEY (I,2686)

GLEASON, JOANNA
CHAIN REACTION (II,479); GREAT DAY (II,1054); HELLO, LARRY (II,1128); STILL THE BEAVER (II,2466); WHY US? (II,2796)

GLEASON, PAUL
ALL MY CHILDREN (II,39); IKE (II,1223)

GLEASON, REGINA
BEN JERROD (I,396); BRIGHT PROMISE (I,727); THE DONALD O'CONNOR SHOW (I,1344); THE LORETTA YOUNG SHOW (I,2755)

GLEESON, REDMOND
MR. AND MRS. COP (II,1748)

GLENISTER, LIONEL
WHEN THE NIGHTINGALE SANG IN BERKELEY SQUARE (I,4827)

GLENN OSSER CHORUS, THE
MUSIC FOR A SPRING NIGHT (I,3169); MUSIC FOR A SUMMER NIGHT (I,3170)

GLENN SR, ROY
THE PIGEON (I,3596)

GLENN, LEO

THE STRANGE CASE OF DR. JEKYLL AND MR. HYDE (I,4244)

GLENN, LOUISE
DON'T CALL ME CHARLIE (I,1349); WHERE THERE'S SMOKEY (I,4832)

GLENNON, TOM
SUSPENSE THEATER ON THE AIR (II,2507)

GLESS, SHARON
CAGNEY AND LACEY (II,409); CENTENNIAL (II,477); CLINIC ON 18TH STREET (II,544); FARADAY AND COMPANY (II,833); HOUSE CALLS (II,1194); THE ISLANDER (II,1249); THE LAST CONVERTIBLE (II,1435); MARCUS WELBY, M.D. (II,1627); MCCLOUD (II,1660); PALMS PRECINCT (II,1946); RICHIE BROCKELMAN: MISSING 24 HOURS (II,2168); SWITCH (II,2519); SWITCH (II,2520); TURNABOUT (II,2669)

GLICK, FRANCIE
KENNEDY (II,1377)

GLICK, PHYLLIS
LOVES ME, LOVES ME NOT (II,1551)

GLICKMAN, MARTY
BRAIN GAMES (II,366)

GLOVER, BETTY
THE BEST OF TIMES (II,219)

GLOVER, BRUCE
HAWK (I,1974); KISS ME, KILL ME (II,1400); MCNAMARA'S BAND (II,1669); YUMA (I,4976)

GLOVER, CRISPIN
THE ACADEMY II (II,4); THE BEST OF TIMES (II,219); HIGH SCHOOL, U.S.A. (II,1148); HIGH SCHOOL, U.S.A. (II,1149)

GLOVER, DANNY
CHIEFS (II,509)

GLOVER, EDWARD
BARETTA (II,152)

GLOVER, JOHN
GEORGE WASHINGTON (II,978); KENNEDY (II,1377); RAGE OF ANGELS (II,2107)

GLOVER, JULIAN
MIRROR OF DECEPTION (I,3060); NANCY ASTOR (II,1788); Q. E. D. (II,2094); QUILLER: NIGHT OF THE FATHER (II,2100)

GLOVER, WILLIAM
MALIBU (II,1607); THERE GOES THE NEIGHBORHOOD (II,2582)

GLOWNA, VADIM
THE MARTIAN CHRONICLES (II,1635)

GLYNN, CARLIN
JOHNNY GARAGE (II,1337)

GLYNN, NORMAN
CASANOVA (II,451)

GMILAMI, AFEMO
RAGE OF ANGELS (II,2107)

GOBEL, GEORGE
'TWAS THE NIGHT BEFORE II,206 BENNY AND BARNEY: LAS VEGAS UNDERCOVER (II,2677); THE APARTMENT HOUSE (I,241); THE BOB HOPE SHOW (I,547); THE BOB HOPE SHOW (I,633); BOB HOPE SPECIAL: BOB HOPE'S STAND UP AND CHEER FOR THE NATIONAL FOOTBALL LEAGUE'S 60TH YEAR (II,316); THE EDDIE FISHER SHOW (I,1409); F TROOP (I,1499); GEORGE GOBEL PRESENTS (I,1766); THE GEORGE GOBEL SHOW (I,1767); THE GEORGE GOBEL SHOW (I,1768); HARPER VALLEY (II,1094); HARPER VALLEY PTA (II,1095); THE HOLLYWOOD SQUARES (II,1164); THE INVISIBLE WOMAN (II,1244); THE JOHN DENVER SPECIAL (II,1316); ONE MORE TIME (I,3386); A PUREX DINAH SHORE SPECIAL (I,3697); SAGA OF SONORA (I,3884)

GOBET, CLAUDE
WOMAN'S PAGE (II,2822)

GOCHMAN, LEN
SOMERSET (I,4115)

GODART, LEE
ALL MY CHILDREN (II,39); THE EDGE OF NIGHT (II,760)

GODDARD, LIZA
SKIPPY, THE BUSH KANGAROO (I,4076)

GODDARD, MARK
THE DETECTIVES (I,1245); THE DOCTORS (II,694); JOHNNY RINGO (I,2434); THE KILLIN' COUSIN (II,1394); LOST IN SPACE (I,2758); MAGGIE BROWN (I,2811); MANY HAPPY RETURNS (I,2895); ONE LIFE TO LIVE (II,1907)

GODDARD, PAULETTE
THE ERROL FLYNN THEATER (I,1458); FEMALE INSTINCT (I,1564); FORD TELEVISION THEATER (I,1634); ON TRIAL (I,3363); THE WOMEN (I,4890)

GODDARD, PHIL
NOT THE NINE O'CLOCK NEWS (II,1864)

GODDARD, WILLOUGHBY
THE ADVENTURES OF WILLIAM TELL (I,85)

GODFREDSON, CHRIS
JAKE'S WAY (II,1269)

GODFREY, ARTHUR
ARTHUR GODFREY AND FRIENDS (I,278); ARTHUR GODFREY AND HIS UKULELE (I,279); ARTHUR GODFREY AND THE SOUNDS OF NEW YORK (I,280); ARTHUR GODFREY IN HOLLYWOOD (I,281); ARTHUR GODFREY LOVES ANIMALS (I,282); THE ARTHUR GODFREY SHOW (I,283); THE ARTHUR GODFREY SHOW (I,284); THE ARTHUR GODFREY SHOW (I,285); THE ARTHUR GODFREY SHOW (I,286); THE ARTHUR GODFREY SHOW (I,287); THE ARTHUR GODFREY SPECIAL (I,288); THE ARTHUR GODFREY SPECIAL (II,109); ARTHUR GODFREY TIME (I,289); ARTHUR GODFREY'S PORTABLE ELECTRIC MEDICINE SHOW (I,290); ARTHUR GODFREY'S TALENT SCOUTS (I,291); CANDID CAMERA (I,819); CBS: ON THE AIR (II,460); FLATBED ANNIE AND SWEETIEPIE: LADY TRUCKERS (II,876); FUN FOR 51 (I,1707); THE REASON NOBODY HARDLY EVER SEEN A FAT OUTLAW IN THE OLD WEST IS AS FOLLOWS: (I,3744); RICKLES (II,2170); THE ROGER MILLER SHOW (I,3830); YOUR ALL AMERICAN COLLEGE SHOW (I,4946)

GODFREY, KATHY
ON YOUR WAY (I,3366)

GODFREY, PATRICK
NICHOLAS NICKLEBY (II,1840); THE SIX WIVES OF HENRY VIII (I,4068)

GODIN, JACQUES
TOMAHAWK (I,4549)

GODSHALL, LIBERTY
B.J. AND THE BEAR (II,125); ETHEL IS AN ELEPHANT (II,783); THE JAMES BOYS (II,1272)

GODWIN, DORAN
SHOESTRING (II,2338)

GOELZ, DAVE
FRAGGLE ROCK (II,916); THE MUPPET SHOW (II,1766)

GOETZ, PETER MICHAEL
AFTERMASH (II,23); ALL TOGETHER NOW (II,45); ONE OF THE BOYS (II,1911)

GOETZ, THEO
THE GUIDING LIGHT (II,1064)

GOFF, NORRIS
LUM AND ABNER (I,2793)

GOH, REX
AIR SUPPLY IN HAWAII (II,26)

GOINS, JESSE D
THE GREATEST AMERICAN HERO (II,1060); PAPER DOLLS (II,1952)

GOITMAN, CHRIS
THE EDGE OF NIGHT (II,760)

GOLD, BRANDY
BABY MAKES FIVE (II,129); V: THE FINAL BATTLE (II,2714)

GOLD, HARRY
BUT MOTHER! (II,403); THE HARVEY KORMAN SHOW (II,1104); OFF THE WALL (II,1879)

GOLD, HEIDI
POTTSVILLE (II,2072)

GOLD, MARTY
THE 25TH MAN (MS) (II,2678); KOBB'S CORNER (I,2577)

GOLD, MISSY
BENSON (II,208); CAPTAINS AND THE KINGS (II,435); CIRCUS OF THE STARS (II,536); CIRCUS OF THE STARS (II,537); TESTIMONY OF TWO MEN (II,2564)

GOLD, TRACEY
BEYOND WITCH MOUNTAIN (II,228); CIRCUS OF THE STARS (II,537); FATHER MURPHY (II,841); GOODNIGHT, BEANTOWN (II,1041); SHIRLEY (II,2330)

GOLDA, MICHAEL
HAWKINS FALLS, POPULATION 6200 (I,1977)

GOLDBERG, ANDY
COMPLETELY OFF THE WALL (II,560)

GOLDBLUM, JEFF
TENSPEED AND BROWN SHOE (II,2562)

GOLDDIGGERS, THE
THE BOB HOPE SHOW (I,629); THE BOB HOPE SHOW (I,638); THE BOB HOPE SHOW (I,639); THE BOB HOPE SHOW (I,646); DEAN MARTIN PRESENTS THE GOLDDIGGERS (I,1193);

DEAN MARTIN PRESENTS THE GOLDDIGGERS (I,1194); THE DEAN MARTIN SHOW (I,1201); DEAN MARTIN'S CHRISTMAS IN CALIFORNIA (II,645); DEAN'S PLACE (II,650); DEAN MARTIN'S RED HOT SCANDALS OF 1926 (II,648); DEAN MARTIN'S RED HOT SCANDALS PART 2 (II,649); MONSANTO NIGHT PRESENTS BURL IVES (I,3090)

GOLDEN WEST COWBOYS, THE
PEE WEE KING SHOW (I,3504)

GOLDEN, BOB
FOOLS, FEMALES AND FUN: I'VE GOTTA BE ME (II,899)

GOLDEN, GINNY
EMERGENCY! (II,775); SEARCH (I,3951)

GOLDENBERG, DEVIN
THE SECRET STORM (I,3969)

GOLDENBERG, HARVEY J
CATALINA C-LAB (II,458)

GOLDFINGER AND DOVE
LIKE MAGIC (II,1476)

GOLDIE, LESLIE
THE BENNY HILL SHOW (II,207)

GOLDMAN, CAROLE
IRENE (II,1245); THE PAPER CHASE (II,1949)

GOLDMAN, DANNY
AT EASE (II,116); BUSTING LOOSE (II,402); DRIBBLE (II,736); THE GOOD LIFE (I,1848); THE GOOD LIFE (II,1033); MICKEY SPILLANE'S MIKE HAMMER (II,1692); MICKEY SPILLANE'S MIKE HAMMER: MORE THAN MURDER (II,1693); THE SMURFS (II,2385)

GOLDMAN, LORRY
G.I.'S (II,949)

GOLDMAN, WENDY
AND THEY ALL LIVED HAPPILY EVER AFTER (II,75); BOSOM BUDDIES (II,355); DELTA HOUSE (II,658); NATIONAL LAMPOON'S HOT FLASHES (II,1795); THE SHOW MUST GO ON (II,2344)

GOLDONI, LELIA
DISAPPEARANCE OF AIMEE (I,1290); NOWHERE TO HIDE (II,1868); SCRUPLES (II,2280)

GOLDSBORO, BOBBY
THE BOBBIE GENTRY SPECIAL (I,670); THE BOBBY GOLDSBORO SHOW (I,673); BURT REYNOLDS' LATE SHOW (I,777); A COUNTRY HAPPENING (I,1065); THE LENNON SISTERS SHOW (I,2656); SCROOGE'S ROCK 'N' ROLL CHRISTMAS (II,2279)

GOLDSMITH, JONATHAN
EISCHIED (II,763)

GOLDSMITH, MERWIN
GOODTIME GIRLS (II,1042); KENNEDY (II,1377)

GOLDSMITH, SYDNEY
STOCKARD CHANNING IN JUST FRIENDS (II,2467); THE STOCKARD CHANNING SHOW (II,2468)

GOLDSTEIN, BOB
FARRELL: FOR THE PEOPLE (II,834)

GOLDSTEIN, JAN
SAWYER AND FINN (II,2265)

GOLDSTEIN, MARK
EVENING AT THE IMPROV (II,786); THE FACTS (II,804)

GOLDWATER, BARRY
I LOVE LIBERTY (II,1215)

GOLIN, CHARLES
THE AWAKENING LAND (II,122)

GOLLAND, JOSEPH
A MAN CALLED INTREPID (II,1612)

GOLONKA, ARLENE
ABC STAGE '67 (I,6); ALAN KING'S FINAL WARNING (II,29); THE ANDY GRIFFITH SHOW (I,192); THE BAIT (I,335); THE BEAN SHOW (I,373); THE BEST OF TIMES (II,220); CALL HOLME (II,412); THE FLYING NUN (I,1611); GIMME A BREAK (II,995); JEREMIAH OF JACOB'S NECK (II,1281); JOE AND VALERIE (II,1299); MAYBERRY R.F.D. (I,2961); SECRETS OF MIDLAND HEIGHTS (II,2296); SPEED BUGGY (II,2419); THAT GIRL (II,2570)

GOMEZ, PANCHITO
BEACH PATROL (II,192); FLATFOOTS (II,879); THE OUTLAWS (II,1936)

GOMEZ, THOMAS
CAESAR AND CLEOPATRA (I,788); LIFE WITH LUIGI (I,2692); MIMI (I,3054); THE RIFLEMAN (I,3789)

GONZALES, CLAUDIO
A.K.A. PABLO (II,28)

GONZALES, MARTHA
A.K.A. PABLO (II,28)

GONZALES, PANCHO
GLENN FORD'S SUMMERTIME, U.S.A. (I,1823)

GONZALEZ, PEARL
ALL ABOUT MUSIC (I,127)

GONZALEZ-GONZALEZ, JOSE
FULL SPEED ANYWHERE (I,1704)

GONZALEZ-GONZALEZ, PEDRO
LOW MAN ON THE TOTEM POLE (I,2782)

GOOD HUMOR COMPANY, THE
THE CBS NEWCOMERS (I,879)

GOOD TIME SINGERS, THE
THE ANDY WILLIAMS NEW YEAR'S EVE SPECIAL (I,203); THE ANDY WILLIAMS SHOW (I,205); THE ANDY WILLIAMS SHOW (I,209)

GOODER, CAROL
FORD THEATER HOUR (I,1635)

GOODEVE, GRANT
BATTLE OF THE NETWORK STARS (II,170); THE CELEBRITY FOOTBALL CLASSIC (II,474); DYNASTY (II,746); EIGHT IS ENOUGH (II,762); PEOPLE LIKE US (II,1999); SOLID GOLD HITS (II,2397)

GOODFELLOW, JOAN
FROM HERE TO ETERNITY (II,933); THE GUN AND THE PULPIT (II,1067); MCLAREN'S RIDERS (II,1663); TROUBLE IN HIGH TIMBER COUNTRY (II,2661)

GOODFRIEND, LYNDA
BLANSKY'S BEAUTIES (II,267); HAPPY DAYS (II,1084); LEGS (II,1459); WHO'S WATCHING THE KIDS? (II,2793)

GOODIER, ROBERT
MOMENT OF TRUTH (I,3084); SAM HILL: WHO KILLED THE MYSTERIOUS MR. FOSTER? (I,3896)

GOODLET, KEN
PRISONER: CELL BLOCK H (II,2085); RETURN TO EDEN (II,2143)

GOODMAN, BENNY
ALL-STAR SWING FESTIVAL (II,53); THE BIG BAND AND ALL THAT JAZZ (I,422); CRESCENDO (I,1094); MUSIC U.S.A. (I,3179); SWING INTO SPRING (I,4317); TWELVE STAR SALUTE (I,4638)

GOODMAN, DODY
DIFF'RENT STROKES (II,674); FOREVER FERNWOOD (II,909); THE LIAR'S CLUB (II,1466); MARY HARTMAN, MARY HARTMAN (II,1638); THE MARY TYLER MOORE COMEDY HOUR (II,1639); TEXAS (II,2566); THE TONIGHT SHOW (I,4564); VALENTINE MAGIC ON LOVE ISLAND (II,2716)

GOODMAN, FRANK
KID GLOVES (I,2531)

GOODMAN, GORDON
THE FRED WARING SHOW (I,1679)

GOODMAN, JOHN
CHIEFS (II,509)

GOODMAN, LEE
DAMN YANKEES (I,1128); DOBIE GILLIS (I,1302); THE GREAT WALTZ (I,1878); THE MAGNIFICENT YANKEE (I,2829)

GOODMAN, MARK
THE ROCK 'N ROLL SHOW (II,2188)

GOODMAN, STEVE
JOHNNY CASH'S AMERICA (II,1330)

GOODNER, CAROL
BLITHE SPIRIT (I,483)

GOODRICH, DEBORAH
ALL MY CHILDREN (II,39); OLD FRIENDS (II,1881)

GOODROW, GARY
NIGHT GALLERY (I,3288); THE OUTLAWS (II,1936); THE TV TV SHOW (II,2674)

GOODROW, MICHAEL
EIGHT IS ENOUGH (II,762)

GOODSON, DIANA
OPRYLAND: NIGHT OF STARS AND FUTURE STARS (II,1921)

GOODSON, MARK
TV'S FUNNIEST GAME SHOW MOMENTS (II,2676)

GOODTIME CHARLIES, THE
COUNTRY CARNIVAL (I,1062)

GOODWIN, ARNOLD
SHOW BIZ (I,4027)

GOODWIN, BILL
THE BILL GOODWIN SHOW (I,447); THE BOB HOPE SHOW (I,516); GERALD MCBOING-BOING (I,1778); IT PAYS TO

BE MARRIED (I,2249); MATINEE IN NEW YORK (I,2946); PENNY TO A MILLION (I,3512)

GOODWIN, HOWARD
CELEBRITY (II,463)

GOODWIN, IAN
GEORGE WASHINGTON (II,978)

GOODWIN, JIM
STAR TREK (II,2440)

GOODWIN, JOSHUA
TOO CLOSE FOR COMFORT (II,2642)

GOODWIN, JULIA
FANNY BY GASLIGHT (II,823)

GOODWIN, LAUREL
CALL TO DANGER (I,803); THE HERO (I,2035); RUN, BUDDY, RUN (I,3870)

GOODWIN, LEE
TONI TWIN TIME (I,4554)

GOODWIN, MICHAEL
ANOTHER WORLD (II,97); BIG JOHN (II,239); THE HAMPTONS (II,1076); MATT HOUSTON (II,1654); STRIKE FORCE (II,2480)

GOODWIN, P J
THE BOYS IN BLUE (II,359)

GOODWIN, TOMMY
RADIO PICTURE SHOW (II,2104)

GOODYEAR, JULIE
CORONATION STREET (I,1054)

GOOFERS, THE
ALL IN FUN (I,131); THE JUDY GARLAND SHOW (I,2470)

GORDEN, MARJORIE
THE YEOMAN OF THE GUARD (I,4926)

GORDENO, PETER
U.F.O. (I,4662)

GORDON, ANITA
THE KEN MURRAY SHOW (I,2525); THE TENNESSEE ERNIE FORD SHOW (I,4400)

GORDON, BARRY
ARCHIE BUNKER'S PLACE (II,105); THE DON RICKLES SHOW (I,1342); FISH (II,864); GOOD TIME HARRY (II,1037); JABBERJAW (II,1261); THE JAZZ SINGER (I,2351); THE JIMMY DURANTE SHOW (I,2389); THE KID SUPER POWER HOUR WITH SHAZAM (II,1386); MR. & MRS. DRACULA (II,1745); THE NEW DICK VAN DYKE SHOW (I,3263); PAC-MAN (II,1940); THE PRACTICE (II,2076); YOU'RE JUST LIKE YOUR FATHER (II,2861)

GORDON, BEN
OUT OF OUR MINDS (II,1933)

GORDON, BRUCE
BEHIND CLOSED DOORS (I,385); CALLING DR. STORM, M.D. (II,415); FISHERMAN'S WHARF (II,865); HOLLOWAY'S DAUGHTERS (I,2079); IF YOU KNEW TOMORROW (I,2187); KING RICHARD II (I,2559); THE LARK (I,2616); A LUCILLE BALL SPECIAL STARRING LUCILLE BALL AND DEAN MARTIN (II,1557); PEYTON PLACE (I,3574); RUN, BUDDY, RUN (I,3870); THE UNTOUCHABLES (I,4681); THE UNTOUCHABLES (I,4682)

GORDON, CHRISTINE
CIRCUS OF THE STARS (II,533); CIRCUS OF THE STARS (II,534)

GORDON, CLARKE
DIAGNOSIS: DANGER (I,1250)

GORDON, COLIN
THE BARON (I,360); THE PRISONER (I,3670)

GORDON, DON
THE BLUE ANGELS (I,490); BROADWAY (I,733); THE CONTENDER (II,571); THE GOLDEN AGE OF TELEVISION (II,1016); LUCAN (II,1554); OF MICE AND MEN (I,3333); THE RETURN OF CHARLIE CHAN (II,2136); SPARROW (II,2410)

GORDON, FIONA
DEATH RAY 2000 (II,654)

GORDON, GALE
THE BROTHERS (I,748); BUNGLE ABBEY (II,396); DENNIS THE MENACE (I,1231); FOR THE LOVE OF MIKE (I,1625); HERE'S LUCY (II,1135); IT TAKES TWO (II,1252); THE LUCILLE BALL COMEDY HOUR (I,2783); THE LUCILLE BALL SPECIAL (II,1559); LUCY MOVES TO NBC (II,1562); THE LUCY SHOW (I,2791); MAKE ROOM FOR DADDY (I,2843); OUR MISS BROOKS (I,3415); PETE AND GLADYS (I,3558); THE ROYAL FOLLIES OF 1933 (I,3862); SALLY (I,3890); WHERE THERE'S SMOKEY (I,4832)

GORDON, GERALD
THE DAY THE WOMEN GOT EVEN (II,628); THE DOCTORS (II,694); FORCE FIVE (II,907); GENERAL HOSPITAL (II,964); HIGHCLIFFE MANOR (II,1150); THE SEEKERS (I,3974)

GORDON, GLEN
THE ADVENTURES OF DR. FU MANCHU (I,58)

GORDON, GLORIA
MY FRIEND IRMA (I,3191)

GORDON, HANNAH
UPSTAIRS, DOWNSTAIRS (II,2709)

GORDON, JEFF
THE GARY COLEMAN SHOW (II,958)

GORDON, JOANN
FROM HERE TO ETERNITY (II,933)

GORDON, JOYCE
THE AD-LIBBERS (I,29); TALES FROM MUPPETLAND (I,4344)

GORDON, LEO V
THE BARBARY COAST (II,147); CIRCUS BOY (I,959); ENOS (II,779); ESCAPE (I,1461); THE HOUSE NEXT DOOR (I,2139); PISTOLS 'N' PETTICOATS (I,3608); THE UNEXPLAINED (I,4675); WHICH WAY'D THEY GO? (I,4838); THE WINDS OF WAR (II,2807)

GORDON, LOUIS
FISHERMAN'S WHARF (II,865)

GORDON, LYNNE
SEEING THINGS (II,2297)

GORDON, MARK
THE MARY TYLER MOORE SHOW (II,1640); THE TED BESSELL SHOW (I,4369); WINNER TAKE ALL (II,2808)

GORDON, MARY ANN
HEE HAW (II,1123)

GORDON, MICHELLE
THE BOB HOPE SHOW (I,575)

GORDON, PAMELA
SUSPENSE (I,4305)

GORDON, PAUL
ARCHIE (II,104); WINDOW SHADE REVUE (I,4874)

GORDON, PHIL
THE BEVERLY HILLBILLIES (I,417); PETE KELLY'S BLUES (I,3559)

GORDON, ROGER
CONSPIRACY OF TERROR (II,569)

GORDON, ROY
THE MILLIONAIRE (I,3046)

GORDON, RUTH
BLITHE SPIRIT (I,485); RHODA (II,2151); THAT WAS THE YEAR THAT WAS (II,2575)

GORDON, SCOTT
THE UNCLE FLOYD SHOW (II,2703)

GORDON, SUSAN
MCKEEVER AND THE COLONEL (I,2970); MIRACLE ON 34TH STREET (I,3057); THE SECRET LIFE OF JOHN MONROE (I,3966)

GORDON, VIRGINIA
OUR MISS BROOKS (I,3415)

GORDON, WILLIAM
CAPTAINS AND THE KINGS (II,435)

GORDONAIRES QUARTET, THE
EDDY ARNOLD TIME (I,1412)

GORE, SANDY
PRISONER: CELL BLOCK H (II,2085)

GORI, KATHI
GIDGET MAKES THE WRONG CONNECTION (I,1798); HONG KONG PHOOEY (II,1180); INCH HIGH, PRIVATE EYE (II,1231)

GORING, MARIUS
THE ADVENTURES OF THE SCARLET PIMPERNEL (I,80); DOUGLAS FAIRBANKS JR. PRESENTS THE RHEINGOLD THEATER (I,1364); THE EXPERT (I,1489)

GORMAN, ANNETTE
DOBIE GILLIS (I,1302); TEXAS JOHN SLAUGHTER (I,4414)

GORMAN, BREON
FOR RICHER, FOR POORER (II,906)

GORMAN, CLIFF
COCAINE AND BLUE EYES (II,548); HAVING BABIES II (II,1108)

GORMAN, LYNNE
MOMENT OF TRUTH (I,3084)

GORMAN, MARI
HARPER VALLEY PTA (II,1095)

GORMAN, MARK
THE FRENCH ATLANTIC AFFAIR (II,924)

GORMAN, PATRICK
G.I.'S (II,949)

GORMAN, REG
THE SULLIVANS (II,2490)

GORMAN, TOM
THE RIGHT MAN (I,3790)

GORME, EYDIE

100 YEARS OF AMERICA'S POPULAR MUSIC (II,1904); THE BOB HOPE SHOW (I,616); THE BOB HOPE SHOW (I,635); BOB HOPE SPECIAL: BOB HOPE'S 30TH ANNIVERSARY TV SPECIAL (II,293); BOB HOPE SPECIAL: BOB HOPE'S ALL-STAR TRIBUTE TO THE PALACE THEATER (II,305); FAVORITE SONGS (I,1545); THE JERRY LEWIS SHOW (I,2363); JUBILEE (II,1347); ON PARADE (I,3356); STEVE AND EYDIE CELEBRATE IRVING BERLIN (II,2456); STEVE AND EYDIE. . . ON STAGE (I,4223); STEVE AND EYDIE: OUR LOVE IS HERE TO STAY (II,2457); STEVE LAWRENCE AND EYDIE GORME FROM THIS MOMENT ON. . . COLE PORTER (II,2459); THE STEVE LAWRENCE AND EYDIE GORME SHOW (I,4226); THE TONIGHT SHOW (I,4562)

GORMLEY, STORY
MICKIE FINN'S (I,3029)

GORNEY, KAREN
ALL MY CHILDREN (II,39)

GORRIAN, LENNIE
THE PAUL DIXON SHOW (I,3487)

GORSHIN, FRANK
. . .AND DEBBIE MAKES SIX (I,185); BATMAN (I,366); CAROL AND COMPANY (I,847); THE EDGE OF NIGHT (II,760); THE FABULOUS FORDIES (I,1502); GOLIATH AWAITS (II,1025); HENNESSEY (I,2013); THE KOPYCATS (I,2580); NEW GIRL IN HIS LIFE (I,3265); RHYME AND REASON (II,2152); RUDOLPH'S SHINY NEW YEAR (II,2228); THE STARS AND STRIPES SHOW (II,2443); THE TOM JONES SPECIAL (I,4540); THE TRINI LOPEZ SHOW (I,4606)

GORTNER, MARJOE
CIRCUS OF THE STARS (II,533); CIRCUS OF THE STARS (II,534); CIRCUS OF THE STARS (II,535); CIRCUS OF THE STARS (II,536); THE GUN AND THE PULPIT (II,1067); SPEAK UP AMERICA (II,2412)

GOSCH, MARTIN
TONIGHT ON BROADWAY (I,4559)

GOSDEN, FREEMAN
CALVIN AND THE COLONEL (I,806); THE FORD 50TH ANNIVERSARY SHOW (I,1630)

GOSE, CARL
ART AND MRS. BOTTLE (I,271)

GOSFIELD, MAURICE
THE PHIL SILVERS SHOW (I,3580); SUMMER IN NEW YORK (I,4279); TOP CAT (I,4576); TROUBLE INC. (I,4609)

GOSLEY, TIM
FRAGGLE ROCK (II,916)

GOSSETT, LOU
BACKSTAIRS AT THE WHITE HOUSE (II,133); BEN VEREEN—HIS ROOTS (II,202); BLACK BART (II,260); CIRCUS OF THE STARS (II,537); THE FUZZ BROTHERS (I,1722); THE HAPPENERS (I,1935); THE LAZARUS SYNDROME (II,1451); THE LIVING END (I,2731); THE POWERS OF MATTHEW STAR (II,2075); ROOTS (II,2211); SIDEKICKS (II,2348); THE YOUNG REBELS (I,4944)

GOTFRIED, GILBERT
THE FURTHER ADVENTURES OF WALLY BROWN (II,944); SATURDAY NIGHT LIVE (II,2261); THICKE OF THE NIGHT (II,2587)

GOTHIE, ROBERT
DOC HOLLIDAY (I,1307); THE GALLANT MEN (I,1728)

GOTTELL, WALTER
THE WORD (II,2833)

GOTTLEIB, CARL
THE KEN BERRY WOW SHOW (I,2524); THE SMOTHERS BROTHERS COMEDY HOUR (I,4095); THE TV TV SHOW (II,2674)

GOTTLIEB, STAN
HOT L BALTIMORE (II,1187)

GOTTLIEB, THEODORE
THE BILLY CRYSTAL COMEDY HOUR (II,248)

GOTTON, SUSAN
BROTHERS AND SISTERS (II,382)

GOTTSCHALK, NORMAN
THE GOOK FAMILY (I,1858); THOSE ENDEARING YOUNG CHARMS (I,4469)

GOUDE, INGRID
LOVE THAT BOB (I,2779); STEVE CANYON (I,4224)

GOUGH, GERALD F
KENNEDY (II,1377)

GOUGH, LLOYD
THE GREEN HORNET (I,1886)

GOUGH, MICHAEL
BRIDESHEAD REVISITED (II,375); DOUGLAS FAIRBANKS JR. PRESENTS THE RHEINGOLD THEATER (I,1364); MISTRAL'S DAUGHTER (II,1708); SEARCH FOR THE NILE (I,3954); SMILEY'S PEOPLE (II,2383)

GOULD, ELLIOTT
CELEBRITY CHALLENGE OF THE SEXES 1 (II,465); CHER AND OTHER FANTASIES (II,497); CIRCUS OF THE STARS (II,535); E/R (II,748); THE HELEN REDDY SPECIAL (II,1126); THE OLIVIA NEWTON-JOHN SHOW (II,1885); RICKLES (II,2170); SPECIAL LONDON BRIDGE SPECIAL (I,4150); A SPECIAL OLIVIA NEWTON-JOHN (II,2418)

GOULD, GORDON
MOMMA THE DETECTIVE (II,1721)

GOULD, GRAYDON
FOREST RANGERS (I,1642)

GOULD, HAROLD
BACHELOR AT LAW (I,322); THE FEATHER AND FATHER GANG (II,845); THE FEATHER AND FATHER GANG: NEVER CON A KILLER (II,846); FLANNERY AND QUILT (II,873); FOOT IN THE DOOR (II,902); HAPPY DAYS (II,1084); HAVE I GOT A CHRISTMAS FOR YOU (II,1105); HE AND SHE (I,1985); THE LONG HOT SUMMER (I,2747); THE MARY TYLER MOORE SHOW (II,1640); MURDOCK'S GANG (I,3164); PARK PLACE (II,1958); READY FOR THE PEOPLE (I,3739); RHODA (II,2151); UNDER THE YUM YUM TREE (I,4671); WASHINGTON: BEHIND CLOSED DOORS (II,2744)

GOULD, MORTON
TWELVE STAR SALUTE (I,4638)

GOULD, SANDRA
BEWITCHED (I,418); I MARRIED JOAN (I,2174); TV FUNNIES (II,2672)

GOULD, SID
CAP'N AHAB (I,824); THE LOVE BOAT (II,1535); MARINELAND CARNIVAL (I,2913); SEVEN AT ELEVEN (I,3984); SID CAESAR PRESENTS COMEDY PREVIEW (I,4041)

GOULDING, RAY
BOB & RAY & JANE, LARAINE & GILDA (II,279); THE BOB AND RAY SHOW (I,496); CLUB EMBASSY (I,982); COMEDY NEWS (I,1020); FROM CLEVELAND (II,931); HAPPY DAYS (I,1942); IT ONLY HURTS WHEN YOU LAUGH (II,1251); THE MIKE DOUGLAS CHRISTMAS SPECIAL (I,3040); THE NAME'S THE SAME (I,3220); ONE MAN SHOW (I,3381)

GOULET, NICOLETTE
RYAN'S HOPE (II,2234); SEARCH FOR TOMORROW (II,2284)

GOULET, ROBERT
BING CROSBY AND THE SOUNDS OF CHRISTMAS (I,457); BLUE LIGHT (I,491); THE BOB GOULET SHOW STARRING ROBERT GOULET (I,501); THE BOB HOPE SHOW (I,594); THE BOB HOPE SHOW (I,651); BRIGADOON (I,726); THE BROADWAY OF LERNER AND LOEWE (I,735); CAROUSEL (I,856); AN HOUR WITH ROBERT GOULET (I,2137); JUDY GARLAND AND HER GUESTS, PHIL SILVERS AND ROBERT GOULET (I,2469); KISS ME, KATE (I,2569); THE MANY SIDES OF DON RICKLES (I,2898); MUSIC BY COLE PORTER (I,3166); THE ROBERT GOULET SPECIAL (I,3807); THE WAYNE NEWTON SPECIAL (II,2749)

GOUTMAN, CHRIS
SEARCH FOR TOMORROW (II,2284); TEXAS (II,2566)

GOUZER, JAMES JOHN
I LOVE LUCY (I,2171)

GOWER, ANDRE
BABY MAKES FIVE (II,129); CIRCUS OF THE STARS (II,537); THE YOUNG AND THE RESTLESS (II,2862)

GOWER, CARLENE
THE CHADWICK FAMILY (II,478)

GOWLING, PENNY
CORONATION STREET (I,1054)

GOYETTE, DESARAE
SUNDAY FUNNIES (II,2493); YOU ASKED FOR IT (II,2855)

GOZ, HARRY
KENNEDY (II,1377)

GOZIER, BERNIE

BOLD VENTURE (I,682)

GRABER, YOSSI
A WOMAN CALLED GOLDA (II,2818)

GRABLE, BETTY
THE BOB HOPE SHOW (I,543); THE BOB HOPE SHOW (I,549); THE BOB HOPE SHOW (I,553); THE BOB HOPE SHOW (I,565); THE FABULOUS FORDIES (I,1502); STAR STAGE (I,4197); TWENTIETH CENTURY (I,4639)

GRABOWSKI, NORMAN
THE JONES BOYS (I,2446); THE MCGONIGLE (I,2968); THE NEW PHIL SILVERS SHOW (I,3269); OH, NURSE! (I,3343)

GRACE, MARY
IT'S ABOUT TIME (I,2263)

GRACE, NICKOLAS
BRIDESHEAD REVISITED (II,375); LACE (II,1418)

GRACE, ROBYN
NEVER TOO YOUNG (I,3248); WENDY AND ME (I,4793)

GRACIE, CHARLES
THE ROCK N' ROLL SHOW (I,3820)

GRACIE, SALLY
THE DOCTORS (II,694); GIFT OF THE MAGI (I,1800); OLD NICKERBOCKER MUSIC HALL (I,3351); ONE LIFE TO LIVE (II,1907); PLAYWRIGHTS '56 (I,3627); THE TIMES SQUARE STORY (I,4514); WHY DIDN'T THEY ASK EVANS? (II,2795)

GRADY, DON
THE MICKEY MOUSE CLUB (I,3025); MY THREE SONS (I,3205); THANKSGIVING REUNION WITH THE PARTRIDGE FAMILY AND MY THREE SONS (II,2569)

GRADY, ED L
CHIEFS (II,509)

GRAF, DAVID
THE RENEGADES (II,2132)

GRAFF, ILENE
13 THIRTEENTH AVENUE (II,2592); BEULAH LAND (II,226); CHARLEY'S AUNT (II,483); THE EARTHLINGS (II,751); HEAVEN ON EARTH (II,1119); LEWIS AND CLARK (II,1465); SUPERTRAIN (II,2505)

GRAFF, TODD
P.O.P. (II,1939)

GRAFF, WILSON
SWAMP FOX (I,4311)

GRAFT, JOSEPH DE
BORN FREE (II,353)

GRAHAM, DAVID
FIREBALL XL-5 (I,1578); SUPERCAR (I,4294); THUNDERBIRDS (I,4497)

GRAHAM, ED
LINUS THE LIONHEARTED (I,2709)

GRAHAM, GARRET
BLACK BART (II,260); THE BOYS IN BLUE (II,359); DYNASTY (II,747); STOCKARD CHANNING IN JUST FRIENDS (II,2467); THE TV TV SHOW (II,2674)

GRAHAM, GARY
SCRUPLES (II,2280); THOU SHALT NOT KILL (II,2604)

GRAHAM, HEATHER
THE SWISS FAMILY ROBINSON (II,2518)

GRAHAM, JOHNNY
EARTH, WIND & FIRE IN CONCERT (II,750)

GRAHAM, JUNE
THE SECRET STORM (I,3969)

GRAHAM, KENNETH
BRIDESHEAD REVISITED (II,375)

GRAHAM, LEE
THE GABBY HAYES SHOW (I,1723)

GRAHAM, MARTHA
THE RAY PERKINS REVUE (I,3735)

GRAHAM, REV BILLY
GEORGE BURNS CELEBRATES 80 YEARS IN SHOW BUSINESS (II,967); THE JOHNNY CASH CHRISTMAS SPECIAL (II,1325); THE WOODY ALLEN SPECIAL (I,4904)

GRAHAM, RONNY
THE BOB CRANE SHOW (II,280); CHICO AND THE MAN (II,508); THE HUDSON BROTHERS SHOW (II,1202); JONATHAN WINTERS PRESENTS 200 YEARS OF AMERICAN HUMOR (II,1342); THE WACKY WORLD OF JONATHAN WINTERS (I,4746)

GRAHAM, SCOTT
THE DOCTORS (II,694); THE YOUNG MARRIEDS (I,4942)

GRAHAM, SHEILA
THE SHEILA GRAHAM SHOW (I,4004)

GRAHAM, STACY
CALL TO DANGER (I,802)

GRAHAM, TIM
THE CODE OF JONATHAN WEST (I,992); EXPERT WITNESS (I,1488); IT'S A SMALL WORLD (I,2260); NATIONAL VELVET (I,3231)

GRAHAM, VIRGINIA
AMERICA ALIVE! (II,67); THE BOB HOPE SHOW (I,643); GIRL TALK (I,1810); THE STRAWHATTERS (I,4257); TEXAS (II,2566); THE VIRGINIA GRAHAM SHOW (I,4736)

GRAHAME, GLORIA
ESCAPE (I,1460); RICH MAN, POOR MAN—BOOK I (II,2161); SEVENTH AVENUE (II,2309)

GRAIMAN, JACK
THE FACTS (II,804)

GRAMMER, KELSEY
CHEERS (II,494); GEORGE WASHINGTON (II,978); KENNEDY (II,1377)

GRANADOS JR, ANGEL
THE RENEGADES (II,2132)

GRAND OLE OPRY SQUARE DANCERS, THE
COUNTRY MUSIC CARAVAN (I,1066); THE GRAND OLE OPRY (I,1866)

GRANDIN, ISABEL
SECOND EDITION (II,2288); THICKE OF THE NIGHT (II,2587)

GRANDPA JONES
HEE HAW (II,1123)

GRANDY, CRAIG
THE JACKSON FIVE (I,2326)

GRANDY, FRED
THE CELEBRITY FOOTBALL CLASSIC (II,474); DUFFY (II,739); FANTASY ISLAND (II,829); THE LOVE BOAT (II,1535); THE LOVE BOAT II (II,1533); THE LOVE BOAT III (II,1534); MAUDE (II,1655); THE MONSTER SQUAD (II,1725)

GRANFIELD, SUZANNE
ANOTHER LIFE (II,95)

GRANGER, FARLEY
BLACK BEAUTY (II,261); CAESAR AND CLEOPATRA (I,788); THE HEIRESS (I,2006); INN OF THE FLYING DRAGON (I,2214); KRAFT TELEVISION THEATER (I,2592); LAURA (I,2636); THE LIVES OF JENNY DOLAN (II,1502); ONE LIFE TO LIVE (II,1907)

GRANGER, GERRI
THE SINGERS (I,4060); WHAT'S IT ALL ABOUT WORLD? (I,4816)

GRANGER, JOHN
JET FIGHTER (I,2374)

GRANGER, STEWART
THE MEN FROM SHILOH (I,3005); SHERLOCK HOLMES: THE HOUND OF THE BASKERVILLES (I,4011)

GRANIK, JOHN
STRANGE PARADISE (I,4246)

GRANLUND, BARBARA
THE JERRY LEWIS SHOW (I,2365)

GRANROTT, BRIAN
PRISONER: CELL BLOCK H (II,2085)

GRANSTEDT, GRETA
DOCTOR MIKE (I,1316)

GRANT, ALBERTA
THE EDGE OF NIGHT (II,760)

GRANT, ALEXANDER
CINDERELLA (I,955)

GRANT, BARRA
THE SUNSHINE BOYS (II,2496)

GRANT, BERNARD
THE GUIDING LIGHT (II,1064); ONE LIFE TO LIVE (II,1907)

GRANT, BETH
THE EYES OF TEXAS (II,799)

GRANT, BILL
THE HUSTLER OF MUSCLE BEACH (II,1210)

GRANT, CARY
SINATRA—THE FIRST 40 YEARS (II,2359)

GRANT, CY
CAPTAIN SCARLET AND THE MYSTERONS (I,837)

GRANT, FAYE
THE GREATEST AMERICAN HERO (II,1060); HOME ROOM (II,1172); V (II,2713); V: THE FINAL BATTLE (II,2714); V: THE SERIES (II,2715)

GRANT, GILLIAN
A CHRISTMAS FOR BOOMER (II,518)

GRANT, GOGI
ACADEMY AWARD SONGS (I,13); SOUND OF THE 60S (I,4134)

GRANT, HANK
GUEST SHOT (I,1899)

GRANT, HARVEY
HAZEL (I,1982); LIFE WITH FATHER (I,2690)

GRANT, KATHRYN
BING CROSBY AND HIS FRIENDS (I,455)

GRANT, KATIE
WORDS AND MUSIC (I,4909)

GRANT, KEITH
THE EDGE OF NIGHT (II,760)

GRANT, KIRBY
SKY KING (I,4077); STORY THEATER (I,4240)

GRANT, LEE
ABC STAGE '67 (I,6); BACKSTAIRS AT THE WHITE HOUSE (II,133); BARE ESSENCE (II,149); DANGER (I,1134); FAY (II,844); JUSTICE (I,2498); KRAFT TELEVISION THEATER (I,2592); ONCE UPON A TIME. . . IS NOW THE STORY OF PRINCESS GRACE (II,1899); PARTNERS IN CRIME (I,3461); PEYTON PLACE (I,3574); PLAYWRIGHTS '56 (I,3627); PLAZA SUITE (II,2056); RANSOM FOR A DEAD MAN (I,3722); ROBERT YOUNG AND THE FAMILY (I,3813); SATURDAY'S CHILDREN (I,3923); SEARCH FOR TOMORROW (II,2284); THE TED BESSELL SHOW (I,4369); THOU SHALT NOT KILL (II,2604); THREE FOR THE GIRLS (I,4477); THE WONDERFUL WORLD OF AGGRAVATION (I,4894)

GRANT, LINDA
GENESIS II (I,1760)

GRANT, MICKI
ANOTHER WORLD (II,97)

GRANT, NORMAN
THE MAGIC SLATE (I,2820)

GRANT, PAUL
MR. ADAMS AND EVE (I,3121)

GRANT, SARINA
JACKIE AND DARLENE (II,1265); MR. DUGAN (II,1749); PALMERSTOWN U.S.A. (II,1945); THE RAG BUSINESS (II,2106); THE SUNSHINE BOYS (II,2496); WILLOW B: WOMEN IN PRISON (II,2804)

GRANT, TED
SATURDAY NIGHT JAMBOREE (I,3920)

GRANVILLE, BONITA
THE ARMSTRONG CIRCLE THEATER (I,260); CLIMAX! (I,976); CROWN THEATER WITH GLORIA SWANSON (I,1107); THE ETHEL BARRYMORE THEATER (I,1467); FORD TELEVISION THEATER (I,1634); GRUEN GUILD PLAYHOUSE (I,1892); THE LUX VIDEO THEATER (I,2795); SCHAEFER CENTURY THEATER (I,3935); STUDIO ONE (I,4268); TELLER OF TALES (I,4388); THE UNEXPECTED (I,4674)

GRANVILLE, CLAIRE
QUADRANGLE (I,3703)

GRASS ROOTS BAND, THE
THE JIMMY DEAN SHOW (I,2385)

GRASSLE, KAREN
BATTLE OF THE NETWORK STARS (II,163); BATTLE OF THE NETWORK STARS (II,164); BATTLE OF THE NETWORK STARS (II,170); EMILY, EMILY (I,1438); LITTLE HOUSE ON THE PRAIRIE (II,1487); LITTLE HOUSE: THE LAST FAREWELL (II,1491)

GRATE, GAIL
CHIEFS (II,509)

GRAUBART, JUDY
COMEDY TONIGHT (I,1027); STORY THEATER (I,4241); YOUR PLACE OR MINE? (II,2874)

GRAUER, BEN
AMERICANA (I,175); THE BIG STORY (I,432); GUESS WHAT HAPPENED? (I,1896); IN TOWN TODAY (I,2206); WHAT HAPPENED? (I,4809)

GRAVES, CHARLES
MARIE (II,1629)

GRAVES, ERNEST
THE GUIDING LIGHT (II,1064); ONE LIFE TO LIVE (II,1907); TODAY IS OURS (I,4531)

GRAVES, JOEL
AFTERMASH (II,23); LITTLE HOUSE: BLESS ALL THE DEAR CHILDREN (II,1489)

GRAVES, JOHNNY
BIG BEND COUNTRY (II,230)

GRAVES, LEONARD
THE BIG PICTURE (I,429); VICTORY AT SEA (I,4730)

GRAVES, LESLIE
CAPITOL (II,426); HERE WE GO AGAIN (I,2028)

GRAVES, PETER
ALFRED HITCHCOCK PRESENTS (I,115); THE BEST OF FRIENDS (II,216); CALL TO DANGER (I,803); CALL TO DANGER (I,804); THE CASE AGAINST PAUL RYKER (I,861); CAVALCADE OF AMERICA (I,875); COURT-MARTIAL (I,1080); DEAD MAN ON THE RUN (II,631); DEAN'S PLACE (II,650); DINAH IN SEARCH OF THE IDEAL MAN (I,1278); FIRESIDE THEATER (I,1580); FURY (I,1721); MISSION: IMPOSSIBLE (I,3067); PEPSI-COLA PLAYHOUSE (I,3523); THE REBELS (II,2121); STEVE MARTIN: COMEDY IS NOT PRETTY (II,2462); STRANGE PLACES (I,4247); STUDIO '57 (I,4267); WHIPLASH (I,4839); THE WINDS OF WAR (II,2807)

GRAVES, TERESA
ARTHUR GODFREY'S PORTABLE ELECTRIC MEDICINE SHOW (I,290); THE BOB HOPE SHOW (I,639); THE BOB HOPE SHOW (I,643); THE BOB HOPE SHOW (I,647); THE FUNNY SIDE (I,1714); THE FUNNY SIDE (I,1715); GET CHRISTIE LOVE! (II,981); GET CHRISTIE LOVE! (II,982); KEEPING UP WITH THE JONESES (II,1374); ROWAN AND MARTIN'S LAUGH-IN (I,3856); SUPER COMEDY BOWL 1 (I,4288); SUPER COMEDY BOWL 2 (I,4289); TURN ON (I,4625)

GRAVITTE, BEAU
AS THE WORLD TURNS (II,110)

GRAY, ALEXANDRA
THIS IS MUSIC (I,4457)

GRAY, ANDREW
MACKENZIE (II,1584)

GRAY, BARRY
WINNER TAKE ALL (I,4881)

GRAY, BILLY
FATHER KNOWS BEST (II,838); FATHER KNOWS BEST: HOME FOR CHRISTMAS (II,839); FATHER KNOWS BEST: THE FATHER KNOWS BEST REUNION (II,840); GRUEN GUILD PLAYHOUSE (I,1892); SCHAEFER CENTURY THEATER (I,3935)

GRAY, BRUCE
HIGH HOPES (II,1144); STRANGE PARADISE (I,4246)

GRAY, CARMINE
ANOTHER WORLD (II,97)

GRAY, CHARLES
ANASTASIA (I,182); CAPTAINS AND THE KINGS (II,435); GUNSLINGER (I,1908); IKE (II,1223); MARIE (II,1629); RAWHIDE (I,3727); ROMEO AND JULIET (I,3837)

GRAY, CINEY
BELLE STARR (I,391)

GRAY, COLEEN
BRIGHT PROMISE (I,727); DAYS OF OUR LIVES (II,629); ELLERY QUEEN: DON'T LOOK BEHIND YOU (I,1434); FAITH BALDWIN'S THEATER OF ROMANCE (I,1515); PANTOMIME QUIZ (I,3449); PINE LAKE LODGE (I,3597); PULITZER PRIZE PLAYHOUSE (I,3692)

GRAY, DEREK
MACKENZIE (II,1584)

GRAY, DOLORES
BUICK CIRCUS HOUR (I,759); THE SID CAESAR SHOW (I,4042); TEXACO COMMAND PERFORMANCE (I,4408); TEXACO STAR THEATER (I,4411); YOU'RE THE TOP (I,4974)

GRAY, DONALD
CAPTAIN SCARLET AND THE MYSTERONS (I,837); SABER OF LONDON (I,3879); THE VISE (I,4739)

GRAY, ERIN
BATTLE OF THE NETWORK STARS (II,169); BATTLE OF THE NETWORK STARS (II,171); BATTLE OF THE NETWORK STARS (II,172); BUCK ROGERS IN THE 25TH CENTURY (II,383); BUCK ROGERS IN THE 25TH CENTURY (II,384); CELEBRITY CHALLENGE OF THE SEXES 5 (II,469); MALIBU U. (I,2849); SILVER SPOONS (II,2355); THE ULTIMATE IMPOSTER (II,2700)

GRAY, FIONA
PENMARRIC (II,1993)

GRAY, GARY
ADVENTURES OF THE SEASPRAY (I,82)

GRAY, GORDON
A TIME FOR US (I,4507)

GRAY, GWYNNE
FLAMBARDS (II,869)

GRAY, HAROLD
RIVKIN: BOUNTY HUNTER (II,2183)

GRAY, JAKE
ART AND MRS. BOTTLE (I,271)

GRAY, JOAN
THE GUIDING LIGHT (II,1064)

GRAY, KEN
MASTER OF THE GAME (II,1647)

GRAY, LINDA
ALL THAT GLITTERS (II,42); BATMAN AND THE SUPER SEVEN (II,160); BEAUTY AND THE BEAST (II,197); BLACKSTAR (II,263); BOB HOPE SPECIAL: BOB HOPE IN THE STAR-MAKERS (II,285); CIRCUS OF THE STARS (II,534); DALLAS (II,614); MAC DAVIS—I'LL BE HOME FOR CHRISTMAS (II,1580); SALUTE TO LADY LIBERTY (II,2243); THE SMURFS (II,2385); TARZAN: LORD OF THE JUNGLE (II,2544)

GRAY, LORRAINE
LANIGAN'S RABBI (II,1427); PARTNERS IN CRIME (I,3461)

GRAY, MICHAEL
BOBBY JO AND THE BIG APPLE GOODTIME BAND (I,674); THE LITTLE PEOPLE (I,2716); SHAZAM! (II,2319)

GRAY, RANDY
JOSHUA'S WORLD (II,1344); MERV GRIFFIN AND THE CHRISTMAS KIDS (I,3013)

GRAY, ROBERT
HARPER VALLEY PTA (II,1095)

GRAY, SAM
THE BENNETS (I,397); THE BRIGHTER DAY (I,728); FROM THESE ROOTS (I,1688); THE HAMPTONS (II,1076); HAWKINS FALLS, POPULATION 6200 (I,1977); NAKED CITY (I,3216); RAGE OF ANGELS (II,2107)

GRAY, VELEKA
HOW TO SURVIVE A MARRIAGE (II,1198); LOVE OF LIFE (I,2774); SOMERSET (I,4115); THE YOUNG AND THE RESTLESS (II,2862)

GRAY, VIVEAN
ALL THE RIVERS RUN (II,43); THE SULLIVANS (II,2490)

GRAYCO, HELEN
CLUB OASIS (I,984); CLUB OASIS (I,985); THE SPIKE JONES SHOW (I,4157); THE SPIKE JONES SHOW (I,4155); THE SPIKE JONES SHOW (I,4156)

GRAYSON, KATHRYN
THE BOB HOPE SHOW (I,547); GENE KELLY. . . AN AMERICAN IN PASADENA (II,962); THE GENERAL ELECTRIC THEATER (I,1753); THE NIGHT OF CHRISTMAS (I,3290)

GRAZIANO, ROCKY
HENNY AND ROCKY (I,2014); THE KEEFE BRASSELLE SHOW (I,2514); THE MARTHA RAYE SHOW (I,2926); MIAMI UNDERCOVER (I,3020); PANTOMIME QUIZ (I,3449)

GREAT SPECKLED BIRD, THE
THE SPECIAL GENTRY ONE (I,4148)

GREAT TOMSON, THE
MAGIC WITH THE STARS (II,1599)

GREAZA, WALTER
MAN AND SUPERMAN (I,2854); TREASURY MEN IN ACTION (I,4604)

GRECO, BUDDY
AWAY WE GO (I,317); THE BOB HOPE SHOW (I,642); BROADWAY OPEN HOUSE (I,736); THE BUDDY GRECO SHOW (I,754); THE ENGELBERT HUMPERDINCK SPECIAL (I,1447); SONGS AT TWILIGHT (I,4123)

GRECO, JENA
MICKEY SPILLANE'S MIKE HAMMER: MORE THAN MURDER (II,1693)

GRECO, JOSE
THE BOB HOPE SHOW (I,534); FRANKIE AVALON'S EASTER SPECIAL (I,1673)

GRED, MITCH
TEXAS (II,2566)

GREEN, BELINDA
THE BOB HOPE SHOW (I,658)

GREEN, BILL
THE LIFE OF RILEY (I,2685); THE MARSHAL OF GUNSIGHT PASS (I,2925)

GREEN, BILLIE
YES, VIRGINIA, THERE IS A SANTA CLAUS (II,2849)

GREEN, BRADLEY
GENERAL HOSPITAL (II,964)

GREEN, DAVID
AIR SUPPLY IN HAWAII (II,26)

GREEN, DOROTHY
BENSON (II,208); FOUR STAR PLAYHOUSE (I,1652); JUSTICE OF THE PEACE (I,2500); MARKHAM (I,2916); TAMMY (I,4357); TV READERS DIGEST (I,4630); THE YOUNG AND THE RESTLESS (II,2862)

GREEN, FREDDY
ONE DEADLY OWNER (I,3373)

GREEN, GILBERT
STARSKY AND HUTCH (II,2445)

GREEN, GRIZZLY
THE DOOLEY BROTHERS (II,715)

GREEN, JANET
THE UNCLE AL SHOW (I,4667)

GREEN, JANET LAINE
SEEING THINGS (II,2297)

GREEN, JENNIFER
YES, VIRGINIA, THERE IS A SANTA CLAUS (II,2849)

GREEN, JOEY
LITTLE VIC (II,1496); RISE AND SHINE (II,2179)

GREEN, JOHN
THE BLACK ROBE (I,479)

GREEN, JOHNNY
MUSIC U.S.A. (I,3179)

GREEN, LEE
RENDEZVOUS WITH ADVENTURE (I,3767)

GREEN, LEIF
THE BEST OF TIMES (II,220)

GREEN, MARGE
CANDID CAMERA (I,819); MARGE AND JEFF (I,2908)

GREEN, MARTYN
ALICE IN WONDERLAND (I,120); PINOCCHIO (I,3602); TONIGHT AT 8:30 (I,4556)

GREEN, MAXINE
TURN ON (I,4625)

GREEN, MELISSA ANN
SPRAGGUE (II,2430)

GREEN, MICHAEL
THE DAKOTAS (I,1126); MICKEY AND THE CONTESSA (I,3024)

GREEN, MITZI
SO THIS IS HOLLYWOOD (I,4105)

GREEN, NIGEL
THE ADVENTURES OF WILLIAM TELL (I,85); HORNBLOWER (I,2121)

GREENAN, DAVID
BATTLESTAR GALACTICA (II,181)

GREENBAUM, ALFRED
DAVE'S PLACE (I,1175)

GREENBERG, RALPH
HAPPY DAYS (II,1084)

GREENBUSH, LINDSAY
LITTLE HOUSE ON THE PRAIRIE (II,1487); SUNSHINE (II,2495)

GREENBUSH, SIDNEY
LITTLE HOUSE ON THE PRAIRIE (II,1487); SUNSHINE (II,2495)

GREENE, ANGELA
DICK TRACY (I,1271); EL COYOTE (I,1423)

GREENE, BENSON
THE MILKY WAY (I,3044)

GREENE, CATHY
ANOTHER WORLD (II,97); HOW TO SURVIVE A MARRIAGE (II,1198)

GREENE, CORKY
WHITNEY AND THE ROBOT (II,2789)

GREENE, DEBORAH
JULIE FARR, M.D. (II,1354); TRAUMA CENTER (II,2655)

GREENE, ELLEN
MIAMI VICE (II,1689); THE ROCK RAINBOW (II,2194); SEVENTH AVENUE (II,2309)

GREENE, ERIC
BORN TO THE WIND (II,354); SPACE ACADEMY (II,2406); THROUGH THE MAGIC PYRAMID (II,2615)

GREENE, JACLYNNE
PANIC! (I,3447)

GREENE, JAMES
THE MANIONS OF AMERICA (II,1623); RAGE OF ANGELS (II,2107); SANDBURG'S LINCOLN (I,3907)

GREENE, JANET LYNN
CASTLE ROCK (II,456)

GREENE, KAREN
THE EVE ARDEN SHOW (I,1470)

GREENE, KELLIE
THE RAY ANTHONY SHOW (I,3729)

GREENE, LAURA
COMEDY TONIGHT (I,1027)

GREENE, LEON
MASADA (II,1642)

GREENE, LINDA MASON
CASTLE ROCK (II,456)

GREENE, LORNE
THE ALCOA HOUR (I,108); THE BASTARD/KENT FAMILY CHRONICLES (II,159); BATTLESTAR GALACTICA (II,181); BONANZA (II,347); CODE RED (II,552); CODE RED (II,553); DANGER (I,1134); THE DEAN MARTIN

CELEBRITY ROAST (II,637); THE ELGIN HOUR (I,1428); GALACTICA 1980 (II,953); THE GREAT AMERICAN MUSIC CELEBRATION (II,1052); GRIFF (I,1888); KRAFT TELEVISION THEATER (I,2592); THE LONDON PALLADIUM (I,2739); LORNE GREENE'S LAST OF THE WILD (II,1522); LORNE GREENE'S NEW WILDERNESS (II,1523); THE MONEYCHANGERS (II,1724); MOVIN' (I,3117); NEVADA SMITH (II,1807); POLICE SQUAD! (II,2061); REX HARRISON PRESENTS SHORT STORIES OF LOVE (II,2148); ROOTS (II,2211); SAILOR OF FORTUNE (I,3885); A SALUTE TO TELEVISION'S 25TH ANNIVERSARY (I,3893); SAM HILL (I,3895); SANDY IN DISNEYLAND (II,2254); STAR STAGE (I,4197); STUDIO ONE (I,4268); SUSPICION (I,4309); SWING OUT, SWEET LAND (II,2515)

GREENE, LYNNIE
ON OUR OWN (II,1892)

GREENE, MATTHEW
SOMERSET (I,4115)

GREENE, MICHAEL
PALMERSTOWN U.S.A. (II,1945)

GREENE, MICHELE
BAY CITY BLUES (II,186); DOROTHY (II,717); EIGHT IS ENOUGH (II,762)

GREENE, RICHARD
THE ADVENTURES OF ROBIN HOOD (I,74); FORD THEATER HOUR (I,1635); THE NASH AIRFLYTE THEATER (I,3227); PRUDENTIAL FAMILY PLAYHOUSE (I,3683); TELLER OF TALES (I,4388)

GREENE, RUBEN
WHERE THE HEART IS (I,4831)

GREENE, SHECKY
THE ALL-STAR SALUTE TO MOTHER'S DAY (II,51); THE BOB HOPE SHOW (I,633); COMBAT (I,1011); THE FUNNIEST JOKE I EVER HEARD (II,940); GEORGE BURNS CELEBRATES 80 YEARS IN SHOW BUSINESS (II,967); HAL LINDEN'S BIG APPLE (II,1072)

GREENE, STANLEY
MEMBER OF THE WEDDING (I,3002)

GREENE, TERRY
MAN AGAINST CRIME (I,2852)

GREENER, DOROTHY
DON'S MUSICAL PLAYHOUSE (I,1348); HOLIDAY HOTEL (I,2075)

GREENHALGH, DAWN
PAUL BERNARD—PSYCHIATRIST (I,3485); STRANGE PARADISE (I,4246)

GREENHOUSE, MARTHA
CAR 54, WHERE ARE YOU? (I,842); OUR TOWN (I,3420)

GREENLAW, VERINA
THE SIX WIVES OF HENRY VIII (I,4068)

GREENLEAF, JIM
ALL THAT GLITTERS (II,42)

GREENLEAF, JOHN
THE KID SUPER POWER HOUR WITH SHAZAM (II,1386)

GREENLEE, DAVID
SPENCER (II,2420)

GREENWALD, KEN
TURN ON (I,4625)

GREENWALD, LELAND
MERV GRIFFIN AND THE CHRISTMAS KIDS (I,3013)

GREENWAY, TOM
333 MONTGOMERY (I,4487); MIDNIGHT MYSTERY (I,3030); SATAN'S WAITIN' (I,3916)

GREENWOOD, BRUCE
THE HITCHHIKER (II,1157); LEGMEN (II,1458)

GREENWOOD, DEBBIE
THE BENNY HILL SHOW (II,207)

GREENWOOD, JOAN
ELLIS ISLAND (II,768); THE KING AND MRS. CANDLE (I,2544); MAN AND SUPERMAN (I,2854)

GREENWOOD, KATY
THE EVE ARDEN SHOW (I,1471)

GREENWOOD, LEE
LOUISE MANDRELL: DIAMONDS, GOLD AND PLATINUM (II,1528)

GREER, BRODIE
CHIPS (II,511); CIRCUS OF THE STARS (II,535); CIRCUS OF THE STARS (II,536)

GREER, DABBS
THE BOYS (I,701); DOBIE GILLIS (I,1302); FIRESIDE THEATER (I,1580); FOREST RANGER (I,1641); THE GHOST AND MRS. MUIR (I,1786); THE GREATEST GIFT (II,1061); GUNSMOKE (II,1069); HANK (I,1931); HENRY FONDA PRESENTS THE STAR AND THE STORY (I,2016); LITTLE HOUSE ON THE PRAIRIE (II,1487); LITTLE HOUSE: A NEW BEGINNING (II,1488); LITTLE HOUSE: LOOK BACK TO YESTERDAY (II,1490); LITTLE HOUSE: THE LAST FAREWELL (II,1491); THE MAN FROM DENVER (I,2863); NAVY LOG (I,3233); THE SECRET LIFE OF JOHN MONROE (I,3966)

GREER, JANE
STAGECOACH WEST (I,4185)

GREER, LEIGHTON
BUFFALO BILL (II,387); SKEEZER (II,2376)

GREER, MICHAEL
BOBBIE GENTRY'S HAPPINESS HOUR (I,671); BOBBIE GENTRY'S HAPPINESS HOUR (II,343)

GREER, ROBIN
RYAN'S HOPE (II,2234)

GREER, WILL
SAVAGE (I,3924)

GREGG, HERBERT
THE ADVENTURES OF ROBIN HOOD (I,74)

GREGG, JULIE
B.J. AND THE BEAR (II,125); BANYON (I,344); MISSION: IMPOSSIBLE (I,3067); MOBILE ONE (II,1717); MOBILE TWO (II,1718); MY SON THE DOCTOR (I,3203); THE SEEKERS (II,2298); SKINFLINT (II,2377); THE WILDS OF TEN THOUSAND ISLANDS (II,2803)

GREGG, TAMMY
OPRYLAND: NIGHT OF STARS AND FUTURE STARS (II,1921)

GREGG, VIRGINIA
THE 25TH MAN (MS) (II,2678); THE BARBARA STANWYCK THEATER (I,350); BIRDMAN (II,256); CALVIN AND THE COLONEL (I,806); THE D.A.: CONSPIRACY TO KILL (I,1121); DRAGNET (I,1370); GENTRY'S PEOPLE (I,1762); GIDGET MAKES THE WRONG CONNECTION (I,1798); THE HERCULOIDS (I,2023); LITTLE WOMEN (II,1497); LITTLE WOMEN (II,1498); SPACE STARS (II,2408); STATE FAIR (II,2448)

GREGORY, CELIA
LACE (II,1418)

GREGORY, CONSTANTINE
THE FAR PAVILIONS (II,832)

GREGORY, DICK
COMEDY NEWS II (I,1021)

GREGORY, DOUGLAS
OUR FIVE DAUGHTERS (I,3413)

GREGORY, FRANK
THE ROBBINS NEST (I,3806)

GREGORY, JACK
CRAZY TIMES (II,602)

GREGORY, JAMES
BARNEY MILLER (II,154); THE BASTARD/KENT FAMILY CHRONICLES (II,159); THE BIG VALLEY (I,437); CALL TO DANGER (I,803); THE CAMPBELL TELEVISION SOUNDSTAGE (I,812); DETECTIVE SCHOOL (II,661); ESCAPE (I,1461); EYE WITNESS (I,1494); F TROOP (I,1499); THE FLIM-FLAM MAN (I,1599); HAPPILY EVER AFTER (I,1937); HAWAII FIVE-O (I,1972); THE INNER SANCTUM (I,2216); THE LAWLESS YEARS (I,2641); LOVE STORY (I,2776); ON TRIAL (I,3363); PURSUIT (I,3700); SHORT, SHORT DRAMA (I,4023); STAR TONIGHT (I,4199); THAT GIRL (II,2570); A VERY MISSING PERSON (I,4715); THE WILD WILD WEST (I,4863)

GREGORY, JAY
SOMERSET (I,4115)

GREGORY, MARY
HIDE AND SEEK (I,2047); JONATHAN WINTERS PRESENTS 200 YEARS OF AMERICAN HUMOR (II,1342); NIGHT PARTNERS (II,1847); THE RIGHTEOUS APPLES (II,2174); THE WACKY WORLD OF JONATHAN WINTERS (I,4746)

GREGORY, MICHAEL
BEACH PATROL (II,192); GENERAL HOSPITAL (II,964); JACQUELINE SUSANN'S VALLEY OF THE DOLLS, 1981 (II,1268)

GREGORY, ROXANNE
LOVE OF LIFE (I,2774); SOMERSET (I,4115)

GREGSON, JOHN
AUCTION AIRE (I,313); ESPIONAGE (I,1464); GIDEON, C.I.D. (I,1793); SHIRLEY'S WORLD (I,4019)

GREIF, STEPHEN

BLAKE'S SEVEN (II,264); THE LAST DAYS OF POMPEII (II,1436); THE SAVAGE CURSE (I,3925)

GREKO, KATHERINE
SWEEPSTAKES (II,2514)

GRENROCK, JOSHUA
ANNIE FLYNN (II,92); THE WAVERLY WONDERS (II,2746)

GRENVILLE, CYNTHIA
THE CITADEL (II,539)

GRERASCH, STEFAN
BORDER TOWN (I,689)

GRESHAM, EDITH
THE ETHEL WATERS SHOW (I,1468)

GRESHAM, HARRY
THE WORLD OF MISTER SWEENY (I,4917)

GREY, BERYL
SLEEPING BEAUTY (I,4083)

GREY, CHARLES
MURDER ON THE MIDNIGHT EXPRESS (I,3162)

GREY, DORI ANN
THE BIG PAYOFF (I,428)

GREY, DUANE
THE FRENCH ATLANTIC AFFAIR (II,924); THE LONER (I,2744)

GREY, GREG
COMMANDO CODY (I,1030)

GREY, JANET
THE GUIDING LIGHT (II,1064)

GREY, JOEL
DON'T CALL ME MAMA ANYMORE (I,1350); GEORGE M! (I,1772); GEORGE M! (II,976); JACK AND THE BEANSTALK (I,2287); JUBILEE (II,1347); LITTLE WOMEN (I,2722); THE MAGICAL MUSIC OF BURT BACHARACH (I,2822); MAN ON A STRING (I,2877); MERRY CHRISTMAS. . . WITH LOVE, JULIE (II,1687); MY LUCKY PENNY (I,3196); PADDINGTON BEAR (II,1941)

GREY, SAM
THE EDGE OF NIGHT (II,760)

GREY, VIRGINIA
THE ETHEL BARRYMORE THEATER (I,1467); GENERAL HOSPITAL (II,964); THE MONEYCHANGERS (II,1724)

GREZA, WALTER
THE EDGE OF NIGHT (II,760)

GRIBBON, ROBERT
ONE LIFE TO LIVE (II,1907)

GRICE, WAYNE
HAWK (I,1974)

GRIEBLING, OTTO
SUPER CIRCUS (I,4287)

GRIEGO, SANDRA
BORN TO THE WIND (II,354); THE CHISHOLMS (II,512)

GRIER, MIKE
CHIEFS (II,509)

GRIER, PAM
ROOTS: THE NEXT GENERATIONS (II,2213)

GRIER, ROOSEVELT
DANIEL BOONE (I,1142); THE SEEKERS (II,2298)

GRIER, ROSEY
BIG DADDY (I,425); THE BOB HOPE SHOW (I,629); CIRCUS OF THE STARS (II,530); COTTON CLUB '75 (II,580); THE GOLDDIGGERS (I,1838); THE MAD MAD MAD MAD WORLD OF THE SUPER BOWL (II,1586); MAKE ROOM FOR GRANDDADDY (I,2845); MARLO THOMAS AND FRIENDS IN FREE TO BE. . . YOU AND ME (II,1632); MONSANTO PRESENTS MANCINI (I,3093); MOVIN' ON (II,1743); THE ROSEY GRIER SHOW (I,3849); SECOND CHANCE (I,3958); SUPER COMEDY BOWL 1 (I,4288)

GRIES, JONATHAN
HIGH SCHOOL, U.S.A. (II,1148); HIGH SCHOOL, U.S.A. (II,1149); THE TEXAS RANGERS (II,2567)

GRIEVE, RUSS
CHARLIE'S ANGELS (II,487)

GRIFFARD, BOB
ANOTHER MAN'S SHOES (II,96)

GRIFFETH, SIMONE
AMANDA'S (II,62); BLACK BEAUTY (II,261); BRET MAVERICK (II,374); THE GREATEST AMERICAN HERO (II,1060); LADIES' MAN (II,1423); MANDRAKE (II,1617); THE SECOND TIME AROUND (II,2289)

GRIFFIN, JULIEANN
ALL COMMERCIALS—A STEVE MARTIN SPECIAL (II,37)

GRIFFIN, LYNN
THE MAGIC OF DAVID COPPERFIELD (II,1595)

GRIFFIN, MERV
AIR TIME '57 (I,93); BIOGRAPHY OF A BOY (I,477); CIRCUS OF THE STARS (II,538); THE HAZEL BISHOP SHOW (I,1983); HIPPODROME (I,2060); HOLLYWOOD TALENT SCOUTS (I,2094); KEEP TALKING (I,2517); LAS VEGAS PALACE OF STARS (II,1431); MERV GRIFFIN AND THE CHRISTMAS KIDS (I,3013); THE MERV GRIFFIN SHOW (I,3014); THE MERV GRIFFIN SHOW (I,3015); THE MERV GRIFFIN SHOW (I,3016); THE MERV GRIFFIN SHOW (II,1688); MERV GRIFFIN'S SIDEWALKS OF NEW ENGLAND (I,3017); MERV GRIFFIN'S ST. PATRICK'S DAY SPECIAL (I,3018); NBC SATURDAY PROM (I,3245); PLAY YOUR HUNCH (I,3619); ROBERT Q'S MATINEE (I,3812); SCENE OF THE CRIME (II,2271); SONG SNAPSHOTS ON A SUMMER HOLIDAY (I,4122); WORD FOR WORD (I,4907)

GRIFFIN, PAM
PUNCH AND JODY (II,2091)

GRIFFIN, ROBERT
JOHNNY RISK (I,2435); MAN OF THE COMSTOCK (I,2873); WALDO (I,4749)

GRIFFIN, SEAN
THE ALIENS ARE COMING (II,35)

GRIFFIN, STEPHANIE
THE GREAT GILDERSLEEVE (I,1875)

GRIFFIN, TOD
OPERATION: NEPTUNE (I,3404)

GRIFFIN, VICTOR
SING ALONG WITH MITCH (I,4057)

GRIFFIS, WILLIAM
ALL MY CHILDREN (II,39); THE MAGNIFICENT YANKEE (I,2829)

GRIFFITH, ANDY
ADAMS OF EAGLE LAKE (II,10); THE ANDY GRIFFITH SHOW (I,192); THE ANDY GRIFFITH SHOW (I,193); ANDY GRIFFITH'S UPTOWN-DOWNTOWN SHOW (I,194); THE ANDY GRIFFITH—DON KNOTTS—JIMNABORS SHOW (I,191); THE ANDY WILLIAMS SHOW (I,209); THE ANDY WILLIAMS SPECIAL (I,211); BEST OF THE WEST (II,218); THE BOB HOPE SHOW (I,597); CELEBRATION: THE AMERICAN SPIRIT (II,461); CENTENNIAL (II,477); THE DEADLY GAME (II,633); DINAH IN SEARCH OF THE IDEAL MAN (I,1278); DON KNOTTS NICE CLEAN, DECENT, WHOLESOME HOUR (I,1333); THE DON KNOTTS SPECIAL (I,1335); FOR LOVERS ONLY (II,904); FRIENDS AND NABORS (I,1684); FROM HERE TO ETERNITY (II,932); FROSTY'S WINTER WONDERLAND (II,935); THE GIRL IN THE EMPTY GRAVE (II,997); HEADMASTER (I,1990); THE JUD STRUNK SHOW (I,2462); LOOKING BACK (I,2754); MITZI AND A HUNDRED GUYS (II,1710); THE NASHVILLE PALACE (II,1793); THE NBC FOLLIES (I,3242); THE NEW ANDY GRIFFITH SHOW (I,3256); ROOTS: THE NEXT GENERATIONS (II,2213); SALVAGE (II,2244); SALVAGE 1 (II,2245); THE STEVE ALLEN SHOW (I,4220); THE TENNESSEE ERNIE FORD SPECIAL (I,4401); THE UNITED STATES STEEL HOUR (I,4677); WASHINGTON: BEHIND CLOSED DOORS (II,2744); WINTER KILL (II,2810); THE YEAGERS (II,2843)

GRIFFITH, ED
OUR FIVE DAUGHTERS (I,3413)

GRIFFITH, HUGH
INN OF THE FLYING DRAGON (I,2214); THE POPPY IS ALSO A FLOWER (I,3650)

GRIFFITH, JAMES
DIAL HOT LINE (I,1254); THE DOG TROOP (I,1325); GENERAL ELECTRIC TRUE (I,1754); U.S. MARSHAL (I,4686)

GRIFFITH, KENNETH
JANE EYRE (I,2339); THE PRISONER (I,3670)

GRIFFITH, KRISTIN
JAKE'S WAY (II,1269)

GRIFFITH, MELANIE
CARTER COUNTRY (II,449); GOLDEN GATE (II,1020); ONCE AN EAGLE (II,1897); SHE'S IN THE ARMY NOW (II,2320)

GRIFFITH, OLWEN
THE CITADEL (II,539)

GRIFFITHS, JAMES
THE SEVEN DIALS MYSTERY (II,2308)

GRIFFITHS, JANE
THE BUCCANEERS (I,749)

GRIFFITHS, NELLIE
MOLL FLANDERS (II,1720)

GRIFFITHS, SANDI
THE LAWRENCE WELK SHOW (I,2644); MEMORIES WITH LAWRENCE WELK (II,1681)

GRIGGS, JOHN
OUR FIVE DAUGHTERS (I,3413)

GRIGOR, NANCY
THE THREE WIVES OF DAVID WHEELER (II,2611)

GRILL, KATHERINE
LITTLE WOMEN (I,2719); MISS SUSAN (I,3065)

GRILLO, JOHN
BRIDESHEAD REVISITED (II,375)

GRIMES, CHESTER
THE JERK, TOO (II,1282)

GRIMES, GARY
ONCE AN EAGLE (II,1897)

GRIMES, JACK
THE AMAZING THREE (I,162); MARINE BOY (I,2911); SPEED RACER (I,4151); TOM CORBETT, SPACE CADET (I,4535)

GRIMES, JACKIE
THE ALDRICH FAMILY (I,111)

GRIMES, TAMMY
'TWAS THE NIGHT BEFORE CHRISTMAS (II,2677); ARCHY AND MEHITABEL (I,254); THE BORROWERS (I,693); THE DATCHET DIAMONDS (I,1160); GIFT OF THE MAGI (I,1800); HEAVEN WILL PROTECT THE WORKING GIRL (I,1996); HOLIDAY (I,2074); HOLLYWOOD SINGS (I,2091); KRAFT TELEVISION THEATER (I,2592); THE TAMMY GRIMES SHOW (I,4358); THAT WAS THE YEAR THAT WAS (II,2575)

GRIMM, MARIA
79 PARK AVENUE (II,2310); DAYS OF OUR LIVES (II,629)

GRIMM, PAUL
THE BIG SHOW (II,243)

GRIMMETT, BILL
GEORGE WASHINGTON (II,978)

GRINNAGE, JACK
THE BOB NEWHART SHOW (I,666); THE NIGHT STALKER (II,1849)

GRISWOLD, CLAIRE
MEMBER OF THE WEDDING (I,3002)

GRIVES, STEVEN
FLAMBARDS (II,869)

GRIZZARD, GEORGE
THE ADAMS CHRONICLES (II,8); THE COUNTRY GIRL (I,1064); ESPIONAGE (I,1464); THE FRONT PAGE (I,1692); MY THREE ANGELS (I,3204); THE NIGHT RIDER (II,1848); NOTORIOUS (I,3314); THE OLDEST LIVING GRADUATE (II,1882); ONE STEP BEYOND (I,3388); PROFILES IN COURAGE (I,3678); A STRING OF BEADS (I,4263); TEACHER, TEACHER (I,4366); THREE WISHES (I,4488); THRILLER (I,4492); TRAVIS LOGAN, D.A. (I,4598); WARNING SHOT (I,4770)

GRODIN, CHARLES
CHARLEY'S AUNT (II,483); LOVE, SEX. . . AND MARRIAGE (II,1546); THE PAUL SIMON SPECIAL (II,1983); THE YOUNG MARRIEDS (I,4942)

GRODIN, CHUCK
MY MOTHER THE CAR (I,3197)

GROENENDALL, CRIS
SWEENEY TODD (II,2512)

GROENER, HARRY
GEORGE WASHINGTON (II,978)

GROGAN, KELLY
LITTLE SHOTS (II,1495)

GROH, DAVID
ANOTHER DAY (II,94); BATTLE OF THE NETWORK STARS (II,164); RHODA (II,2151); THIS IS KATE BENNETT (II,2597); TOURIST (II,2651)

GROMLEY, JIM
CAR 54, WHERE ARE YOU? (I,842)

GRONET, EDDIE
POLKA PARTY (I,3641)

GROOM, MALCOLM
RYAN'S HOPE (II,2234)

GROOM, SAM
ANOTHER WORLD (II,97); DOCTOR SIMON LOCKE (I,1317); INSTITUTE FOR REVENGE (II,1239); MICKEY SPILLANE'S MIKE HAMMER: MORE THAN MURDER (II,1693); OUR PRIVATE WORLD (I,3417); POLICE SURGEON (I,3639); THE TIME TUNNEL (I,4511)

GROSS, MARY
SATURDAY NIGHT LIVE (II,2261)

GROSS, MICHAEL
FAMILY TIES (II,819)

GROSS, MILTON
THINK FAST (I,4447)

GROSSMAN, ALAN
I MARRIED JOAN (I,2174)

GROSSMAN, SUZANNE
THE DIARY OF ANNE FRANK (I,1262)

GROSSO, EDDIE
KOBB'S CORNER (I,2577)

GROSSO, SONNY
MR. INSIDE/MR. OUTSIDE (II,1750)

GROSVENOR, DENNIS
HOMICIDE (I,2101)

GROUP, MITCHELL
MAMA MALONE (II,1609)

GROUT, JAMES
FILE IT UNDER FEAR (I,1574); THE TALE OF BEATRIX POTTER (II,2533); TO SIR, WITH LOVE (II,2624)

GROVE, BETTY ANN
ALL IN FUN (I,130); THE BERT PARKS SHOW (I,400); THE BIG PAYOFF (I,428); CIRCUS TIME (I,960); THE RED BUTTONS SHOW (I,3750); SONG SNAPSHOTS ON A SUMMER HOLIDAY (I,4122); STOP THE MUSIC (I,4232); THE STORK CLUB (I,4236)

GROVER, CINDY
FAMILY (II,813); LOVE OF LIFE (I,2774); MARRIED: THE FIRST YEAR (II,1634)

GROVER, ED
D.H.O. (I,1249); SEARCH FOR TOMORROW (II,2284)

GROVER, EDWARD
LOVE OF LIFE (I,2774)

GROVER, JAMIE
THE SECRET STORM (I,3969)

GROVER, MARY
CAROUSEL (I,856)

GROVER, STANLEY
THE EDGE OF NIGHT (II,760); MARRIED: THE FIRST YEAR (II,1634); SOMERSET (I,4115)

GROVER, STEPHEN
THE ADAMS CHRONICLES (II,8); THE BEST OF EVERYTHING (I,406)

GROVERNOR, LINDA
SECRETS OF MIDLAND HEIGHTS (II,2296); STUNTS UNLIMITED (II,2487)

GROVES, CAROLYN
DARK SHADOWS (I,1157)

GROVES, REGINA
OUR MAN HIGGINS (I,3414)

GROVES, ROBIN
DAUGHTERS (II,620)

GRUBBS, GARY
CONDOMINIUM (II,566); FOR LOVE AND HONOR (II,903)

GUARD, DOMINIC
THE HANDS OF CORMAC JOYCE (I,1927); A WOMAN OF SUBSTANCE (II,2820)

GUARDINO, CHARLES
THE BURNS AND SCHREIBER COMEDY HOUR (I,768)

GUARDINO, HARRY
BENDER (II,204); GET CHRISTIE LOVE! (II,982); HAVING BABIES I (II,1107); THE LLOYD BRIDGES SHOW (I,2733); LOVERS AND OTHER STRANGERS (II,1550); MONTY NASH (I,3097); THE NEW ADVENTURES OF PERRY MASON (I,3252); PARTNERS IN CRIME (I,3461); PLEASURE COVE (II,2058); THE REPORTER (I,3771)

GUARDINO, JEROME
GEORGE BURNS' HOW TO LIVE TO BE 100 (II,972)

GUARDINO, JERRY
CAR 54, WHERE ARE YOU? (I,842)

GUARNIERI, JOHNNY
THE BOB SMITH SHOW (I,668)

GUASTAFERRO, VINCENT
TRUE LIFE STORIES (II,2665)

GUDEGAST, HANS
THE RAT PATROL (I,3726)

GUERICO, JOIE
OZZIE'S GIRLS (I,3440)

GUERRA, BLANCA
FALCON'S GOLD (II,809)

GUERRA, ROSALINDA
RYAN'S HOPE (II,2234)

GUERRASIO, JOHN
HOLLYWOOD HIGH (II,1163)

GUEST, CHRISTOPHER
CLOSE TIES (II,545); THE LILY TOMLIN SPECIAL (II,1479); SATURDAY NIGHT LIVE (II,2261); THE TV SHOW (II,2673)

GUEST, EDGAR
A GUEST IN YOUR HOUSE (I,1898)

GUEST, GEORGE
ON THE CORNER (I,3359)

GUEST, LANCE
ST. ELSEWHERE (II,2432); WHY US? (II,2796)

GUEST, SUSIE
THE LATE FALL, EARLY SUMMER BERT CONVY SHOW (II,1441)

GUEST, VAL
SPACE: 1999 (II,2409)

GUET, DANIEL
THE ASSASSINATION RUN (II,112)

GUETARY, FRANCOIS
LACE (II,1418)

GUIDI, GUIDARINO
THE SMALL MIRACLE (I,4090)

GUILBERT, ANN MORGAN
THE D.A.: CONSPIRACY TO KILL (I,1121); THE DICK VAN DYKE SHOW (I,1275); THE DICK VAN DYKE SPECIAL (I,1276); HEY, LANDLORD (I,2039); THE MANY SIDES OF DON RICKLES (I,2898); THE NEW ANDY GRIFFITH SHOW (I,3256); SECOND CHANCE (I,3958); YOU'RE ONLY YOUNG ONCE (I,4969)

GUILD, LYNN
FATHER KNOWS BEST (II,838); PERRY MASON (II,2025)

GUILLAUME, ROBERT
BENSON (II,208); BOB HOPE SPECIAL: BOB HOPE IN THE STAR-MAKERS (II,285); THE DONNA SUMMER SPECIAL (II,711); HAL LINDEN'S BIG APPLE (II,1072); IT ONLY HURTS WHEN YOU LAUGH (II,1251); JACK LEMMON IN 'S WONDERFUL, 'S MARVELOUS, 'S GERSHWIN (I,2313); THE KID WITH THE BROKEN HALO (II,1387); MAGIC WITH THE STARS (II,1599); RICH LITTLE'S WASHINGTON FOLLIES (II,2159); SOAP (II,2392); TEXACO STAR THEATER: OPENING NIGHT (II,2565); THE WORLD'S FUNNIEST COMMERCIAL GOOFS (II,2841)

GUILLIM, JACK
ROMEO AND JULIET (I,3837)

GUILLVEY, BENNETT
DAYS OF OUR LIVES (II,629)

GUINAN, FRAN
THE YESTERDAY SHOW (II,2850)

GUINESS, LINDIS
GEORGE GOBEL PRESENTS (I,1766)

GUINNESS, ALEC
CAESAR AND CLEOPATRA (I,789); SMILEY'S PEOPLE (II,2383); TINKER, TAILOR, SOLDIER, SPY (II,2621)

GUINNESS, MATTHEW
OPPENHEIMER (II,1919)

GUISEWITE, CATHY
THE FANTASTIC FUNNIES (II,825)

GUITAR, BONNIE
RANCH PARTY (I,3719)

GULAGER, CLU
BLACK BEAUTY (II,261); CALL TO DANGER (I,804); CHARLIE COBB: NICE NIGHT FOR A HANGING (II,485); THE GOLDEN AGE OF TELEVISION (II,1019); THE KILLER WHO WOULDN'T DIE (II,1393); THE MACKENZIES OF PARADISE COVE (II,1585); ONCE AN EAGLE (II,1897); SAN FRANCISCO INTERNATIONAL AIRPORT (I,3905); SAN FRANCISCO INTERNATIONAL AIRPORT (I,3906); SMILE JENNY, YOU'RE DEAD (II,2382); STICKIN' TOGETHER (II,2465); THE TALL MAN (I,4354); THE UNTOUCHABLES (I,4682); THE VIRGINIAN (I,4738)

GUMBEL, BRYAN
GAMES PEOPLE PLAY (II,956)

GUNDERSON, KAREN
THE NEW CHRISTY MINSTRELS SHOW (I,3259)

GUNN, BILL
TACK REYNOLDS (I,4329)

GUNN, GEORGE
THE BIG PICTURE (I,429)

GUNN, JOHN
THE AMAZING CHAN AND THE CHAN CLAN (I,157)

GUNN, MOSES
THE CONTENDER (II,571); THE COWBOYS (II,598); FATHER MURPHY (II,841); GOOD TIMES (II,1038); OF MICE AND MEN (I,3333); ROOTS (II,2211)

GUNN, ROCKY
THE RETURN OF CHARLIE CHAN (II,2136)

GUNTHER, JOHN
HIGH ROAD (I,2054)

GUNTON, BOB
THE COMEDY ZONE (II,559)

GUNTY, MORTY
HEAD OF THE FAMILY (I,1988); THAT GIRL (II,2570)

GUPTA, SNEH
THE FAR PAVILIONS (II,832)

GURNEY, RACHEL
UPSTAIRS, DOWNSTAIRS (II,2709)

GURRY, ERIC
FULL HOUSE (II,937)

GUSS, LOUIS
ANNIE FLYNN (II,92); THE ART OF CRIME (II,108); LOVE THY NEIGHBOR (I,2781); READY FOR THE PEOPLE (I,3739)

GUTHRIE, RICHARD
DAYS OF OUR LIVES (II,629)

GUTTENBERG, STEVE
BILLY (II,247); NO SOAP, RADIO (II,1856)

GUTTERIDGE, LUCY
LOVE IN A COLD CLIMATE (II,1539); NICHOLAS NICKLEBY (II,1840); THE SEVEN DIALS MYSTERY (II,2308)

GUY MITCHELL SINGERS, THE
THE GUY MITCHELL SHOW (I,1913)

GUYSE, SHEILA
THE GREEN PASTURES (I,1887)

GWENN, EDMUND
HENRY FONDA PRESENTS THE STAR AND THE STORY (I,2016); RHEINGOLD THEATER (I,3783)

GWILLIM, DAVID
THE CITADEL (II,539); IF IT'S A MAN, HANG UP (I,2186)

GWILLIM, JACK
VANITY FAIR (I,4700)

GWINN, WILLIAM
ACCUSED (I,18); DAY IN COURT (I,1182); IT COULD HAPPEN TO YOU (I,2246); MORNING COURT (I,3102)

GWYNN, MICHAEL
KISS KISS, KILL KILL (I,2566)

GWYNNE, ANNE
PUBLIC PROSECUTOR (I,3689)

GWYNNE, FRED
ANDERSON AND COMPANY (I,188); ARSENIC AND OLD LACE (I,270); CAR 54, WHERE ARE YOU? (I,842); GUESS WHAT I DID TODAY (I,1897); HARVEY (I,1961); HARVEY (I,1962); IT'S WHAT'S HAPPENING, BABY! (I,2278); THE LITTLEST ANGEL (I,2726); MARINELAND CARNIVAL (I,2913); MUNSTER, GO HOME! (I,3157); THE MUNSTERS (I,3158); THE MUNSTERS' REVENGE (II,1765); A SALUTE TO STAN LAUREL (I,3892); SANCTUARY OF FEAR (II,2252)

GWYNNE, MICHAEL C
THE HEALERS (II,1115); MATT HELM (II,1652); SLITHER (II,2379)

GWYNNE, PETER
AGAINST THE WIND (II,24); THE EVIL TOUCH (I,1486); RETURN TO EDEN (II,2143)

GYNT, GRETA
THE ADVENTURES OF ROBIN HOOD (I,74); DOUGLAS FAIRBANKS JR. PRESENTS THE RHEINGOLD THEATER (I,1364)

GYPSY
FURY (I,1721)

HAAS, PETER
EMERGENCY PLUS FOUR (II,774)

HAASE, HEATHER
GOLDIE AND KIDS: LISTEN TO ME (II,1021)

HABERFIELD, GRAHAM
CORONATION STREET (I,1054)

HACK, RICHARD
FLYING HIGH (II,895)

HACK, SHELLEY
CHARLIE'S ANGELS (II,486); CLOSE TIES (II,545); CUTTER TO HOUSTON (II,612); TONI'S BOYS (II,2637); VANITIES (II,2720)

HACKER, JOSEPH
CALL TO GLORY (II,413); DALLAS (II,614); DRIBBLE (II,736); THE ULTIMATE IMPOSTER (II,2700); WASHINGTON: BEHIND CLOSED DOORS (II,2744); THE WINDS OF WAR (II,2807)

HACKETT, BOBBY
ALL-STAR SWING FESTIVAL (II,53); THE BIG BAND AND ALL THAT JAZZ (I,422); THE GLORIA DEHAVEN SHOW (I,1826); TIMEX ALL-STAR JAZZ SHOW I (I,4515); TIMEX ALL-STAR JAZZ SHOW IV (I,4518)

HACKETT, BUDDY

THE ALAN KING SPECIAL (I,103); THE ALL-STAR COMEDY SHOW (I,137); THE ARTHUR GODFREY SHOW (I,286); THE BOB HOPE SHOW (I,633); BUDDY HACKETT—LIVE AND UNCENSORED (II,386); CELEBRITY CHALLENGE OF THE SEXES 3 (II,467); CELEBRITY SWEEPSTAKES (II,476); CIRCUS OF THE STARS (II,532); DISNEY WORLD —A GALA OPENING—DISNEYLAND EAST (I,1292); ENTERTAINMENT —1955 (I,1450); GEORGE BURNS CELEBRATES 80 YEARS IN SHOW BUSINESS (II,967); THE GOLDDIGGERS (I,1838); JACK FROST (II,1263); THE JACKIE GLEASON SHOW (I,2322); THE LIAR'S CLUB (II,1466); PLIMPTON! DID YOU HEAR THE ONE ABOUT. . .? (I,3629); THE SCOEY MITCHLLL SHOW (I,3939); STANLEY (I,4190); THERE GOES THE NEIGHBORHOOD (II,2582); VARIETY (I,4703); YOU BET YOUR LIFE (II,2856)

HACKETT, JOAN
ANOTHER DAY (II,94); THE BOB HOPE CHRYSLER THEATER (I,502); THE DEFENDERS (I,1220); A GIRL'S LIFE (II,1000); LIGHT'S OUT (I,2700); THE LONG DAYS OF SUMMER (II,1514); PAPER DOLLS (II,1951); PLEASURE COVE (II,2058); THE POSSESSED (II,2071); REBECCA (I,3745); STONESTREET: WHO KILLED THE CENTERFOLD MODEL? (II,2472); YOUNG DR. MALONE (I,4938)

HACKETT, SANDY
AND THEY ALL LIVED HAPPILY EVER AFTER (II,75)

HACKETT, WALTER
SHEPHERD'S FLOCK (I,4007)

HACKMAN, GENE
THE DUPONT SHOW OF THE WEEK (I,1388)

HADDAD, AVA
ONE LIFE TO LIVE (II,1907)

HADDOCK, JULIE ANNE
BOB HOPE SPECIAL: BOB HOPES'S STAR-STUDDED SPOOF OF THE NEW TV SEASON— G RATED—WITH GLAMOUR, GLITTER & GAGS (I,317); BOONE (II,351); THE FACTS OF LIFE (II,805); MULLIGAN'S STEW (II,1763); MULLIGAN'S STEW (II,1764)

HADDON, LAWRENCE
THE ALIENS ARE COMING (II,35); EVERY STRAY DOG AND KID (II,792); THE FLYING NUN (I,1611); GOLIATH AWAITS (II,1025); HAZEL (I,1982); LASSITER (I,2625); LASSITER (II,1434); LOU GRANT (II,1526); MARY HARTMAN, MARY HARTMAN (II,1638); TODAY'S FBI (II,2628)

HADDY, ANNE
PRISONER: CELL BLOCK H (II,2085)

HADELMAN, TIM
HARRY O (II,1098)

HADEN, SARAH
THE ABBOTT AND COSTELLO SHOW (I,2); DOWN HOME (I,1366)

HADGE, MICHAEL
DARK SHADOWS (I,1157)

HADLEY, BRETT
THE YOUNG AND THE RESTLESS (II,2862)

HADLEY, NANCY
THE BROTHERS (I,748); FRONTIER (I,1695); THE JOEY BISHOP SHOW (I,2403); OF THIS TIME, OF THAT PLACE (I,3335)

HADLEY, REED
PUBLIC DEFENDER (I,3687); RACKET SQUAD (I,3714)

HAESEN, N
MARCO POLO (II,1626)

HAFLER, MAX
THE ASSASSINATION RUN (II,112)

HAFT, LINAL
GREAT EXPECTATIONS (II,1055)

HAGAN, ANNA
MOMENT OF TRUTH (I,3084)

HAGAN, GERALDINE
THE ADVENTURES OF ROBIN HOOD (I,74)

HAGARTY, MIKE
THE YESTERDAY SHOW (II,2850)

HAGEDORN, CAROL
ARTHUR GODFREY AND FRIENDS (I,278)

HAGEN, ERICA
WONDER WOMAN (SERIES 1) (II,2828)

HAGEN, JEAN
ALFRED HITCHCOCK PRESENTS (I,115); THE DESILU PLAYHOUSE (I,1237); MAKE ROOM FOR DADDY (I,2843); SIX GUNS FOR DONEGAN (I,4065)

HAGEN, KEVIN
BEULAH LAND (II,226); LAND OF THE GIANTS (I,2611); LITTLE HOUSE ON THE PRAIRIE (II,1487); LITTLE HOUSE: A NEW BEGINNING (II,1488); LITTLE HOUSE: LOOK BACK TO YESTERDAY (II,1490); THE LONELY WIZARD (I,2743); THE MAN FROM GALVESTON (I,2865); THE SAN PEDRO BUMS (II,2251); YANCY DERRINGER (I,4925)

HAGEN, ROSS
CANNON (I,820); DAKTARI (I,1127)

HAGER, JIM
TWIN DETECTIVES (II,2687)

HAGER, JON
TWIN DETECTIVES (II,2687)

HAGERS, THE
HEE HAW (II,1123)

HAGERTHY, RON
HARD CASE (I,1947); SKY KING (I,4077); THEY WENT THATAWAY (I,4443)

HAGERTY, JULIE
THE DAY THE WOMEN GOT EVEN (II,628)

HAGGARD, MERLE
BUCK OWENS TV RANCH (I,750); CENTENNIAL (II,477); ELVIS REMEMBERED: NASHVILLE TO HOLLYWOOD (II,772); FIFTY YEARS OF COUNTRY MUSIC (II,853); JOHNNY CASH CHRISTMAS 1983 (II,1324); LET ME TELL YOU ABOUT A SONG (I,2661); LYNDA CARTER: ENCORE (II,1566); USA COUNTRY MUSIC (II,591)

HAGGERTY, CAPTAIN
MURDER INK (II,1769)

HAGGERTY, DAN
BATTLE OF THE NETWORK STARS (II,164); BATTLE OF THE NETWORK STARS (II,165); BATTLE OF THE NETWORK STARS (II,166); CELEBRITY CHALLENGE OF THE SEXES 1 (II,465); CELEBRITY CHALLENGE OF THE SEXES 3 (II,467); CONDOMINIUM (II,566); THE LIFE AND TIMES OF GRIZZLY ADAMS (II,1469)

HAGGERTY, DON
ALIAS MIKE HERCULES (I,117); THE CASES OF EDDIE DRAKE (I,863); THE FILES OF JEFFERY JONES (I,1576); THE LIFE AND LEGEND OF WYATT EARP (I,2681); STORY THEATER (I,4240)

HAGGERTY, H B
BUCK ROGERS IN THE 25TH CENTURY (II,383)

HAGMAN, HEIDI
ARCHIE BUNKER'S PLACE (II,105); RAGE OF ANGELS (II,2107)

HAGMAN, LARRY
APPLAUSE (I,244); DALLAS (II,614); THE DETECTIVE (II,662); DIANA (II,667); THE DUPONT SHOW OF THE WEEK (I,1388); THE EDGE OF NIGHT (II,760); THE GOOD LIFE (I,1848); THE GOOD LIFE (II,1033); HERE WE GO AGAIN (I,2028); I DREAM OF JEANNIE (I,2167); LOVE STORY (I,2778); RETURN OF THE WORLD'S GREATEST DETECTIVE (II,2142); THE RHINEMANN EXCHANGE (II,2150); SIDEKICKS (II,2348)

HAGON, GARRICK
THE ADVENTURER (I,43); THE CARNATION KILLER (I,845)

HAGON, PETER
FOREST RANGERS (I,1642)

HAGUE, ALBERT
FAME (II,812)

HAGUE, STEPHEN
THE LITTLE PEOPLE (I,2716)

HAHN, ARCHIE
GHOST OF A CHANCE (II,987); LILY (II,1477); THE MANHATTAN TRANSFER (II,1619); MARY HARTMAN, MARY HARTMAN (II,1638); THE STOCKERS (II,2469); TABITHA (II,2526); THE THREE WIVES OF DAVID WHEELER (II,2611)

HAID, CHARLES
THE BASTARD/KENT FAMILY CHRONICLES (II,159); BATTLE OF THE NETWORK STARS (II,177); BATTLE OF THE NETWORK STARS (II,178); DELVECCHIO (II,659); HILL STREET BLUES (II,1154); KATE MCSHANE (II,1368); KATE MCSHANE (II,1369); SCALPELS (II,2266)

HAIG, ALEXANDER
BOB HOPE SPECIAL: BOB HOPE PRESENTS A CELEBRATION WITH STARS OF COMEDY AND MUSIC (II,340)

HAIG, SID
ALIAS SMITH AND JONES (I,119); JASON OF STAR COMMAND (II,1274); MARY

HARTMAN, MARY HARTMAN (II,1638); MCNAMARA'S BAND (II,1668); MCNAMARA'S BAND (II,1669); MISSION: IMPOSSIBLE (I,3067); RETURN OF THE WORLD'S GREATEST DETECTIVE (II,2142); THE ROYAL FOLLIES OF 1933 (I,3862); TWO GUYS FROM MUCK (II,2690)

HAIG, TONY
SHIRLEY TEMPLE'S STORYBOOK (I,4017)

HAIGH, KENNETH
MOLL FLANDERS (II,1720); SEARCH FOR THE NILE (I,3954); TEN LITTLE INDIANS (I,4394)

HAILEY, MARIAN
MA AND PA (II,1570); SEARCH FOR TOMORROW (II,2284)

HAILEY, MARION
HARVEY (I,1962)

HAIMES, ROGER
GIDGET (I,1795)

HAINES, CONNIE
THE BOB HOPE SHOW (I,515); THE BOB HOPE SHOW (I,597); THE FRANKIE LAINE SHOW (I,1675); FRANKIE LAINE TIME (I,1676); OPERATION ENTERTAINMENT (I,3400)

HAINES, LARRY
THE COUNTRY GIRL (I,1064); EYE WITNESS (I,1494); THE FIRST HUNDRED YEARS (I,1581); HOW TO SUCCEED IN BUSINESS WITHOUT REALLY TRYING (II,1197); KING OF THE ROAD (II,1396); ON OUR OWN (II,1892); PHYL AND MIKHY (II,2037); SEARCH FOR TOMORROW (II,2284)

HAIRSON, JESTER
THE MAN IN THE MOON (I,2871); THAT'S MY MAMA (II,2580)

HALE JR, ALAN
BIFF BAKER, U.S.A. (I,420); CASEY JONES (I,865); COME A RUNNING (I,1013); CROSSROADS (I,1105); FIRESIDE THEATER (I,1580); GILLIGAN'S ISLAND (II,990); GILLIGAN'S ISLAND: RESCUE FROM GILLIGAN'S ISLAND (II,991); GILLIGAN'S ISLAND: THE CASTAWAYS ON GILLIGAN'S ISLAND (II,992); GILLIGAN'S ISLAND: THE HARLEM GLOBETROTTERS ON GILLIGAN'S ISLAND (II,993); GILLIGAN'S PLANET (II,994); THE GOOD GUYS (I,1847); JOHNNY RISK (I,2435); THE LORETTA YOUNG THEATER (I,2756); MARCUS WELBY, M.D. (II,1627); MATINEE THEATER (I,2947); MIGHTY O (I,3036); THE NEW ADVENTURES OF GILLIGAN (II,1813); REX HARRISON PRESENTS SHORT STORIES OF LOVE (II,2148); WHEN THE WEST WAS FUN: A WESTERN REUNION (II,2780)

HALE, BARBARA
CROSSROADS (I,1105); FORD TELEVISION THEATER (I,1634); THE LORETTA YOUNG THEATER (I,2756); MEET THE GOVERNOR (I,2991); PERRY MASON (II,2025); PLAYHOUSE 90 (I,3623); SCHLITZ PLAYHOUSE OF STARS (I,3936); SCREEN DIRECTOR'S PLAYHOUSE (I,3946); WHATEVER BECAME OF. . . ? (II,2772)

HALE, BIRDIE
OUR STREET (I,3418)

HALE, CHANIN
DRAGNET (I,1371); GENE KELLY'S WONDERFUL WORLD OF GIRLS (I,1752); HOWDY (I,2150); THE RED SKELTON SHOW (I,3755); THE RED SKELTON SHOW (I,3756); THE WONDERFUL WORLD OF GIRLS (I,4898)

HALE, DOUG
FITZ AND BONES (II,867); IT ONLY HURTS WHEN YOU LAUGH (II,1251); TERROR AT ALCATRAZ (II,2563)

HALE, ELVI
THE SIX WIVES OF HENRY VIII (I,4068)

HALE, GEORGIA
THE STRAUSS FAMILY (I,4253)

HALE, JOHN
WON'T IT EVER BE MORNING? (I,4902)

HALE, JONATHAN
MAN OF THE COMSTOCK (I,2873)

HALE, LEN
THE PHIL SILVERS ARROW SHOW (I,3576)

HALE, NANCY
ANNIE OAKLEY (I,225); CAVALCADE OF AMERICA (I,875); THE WHIRLYBIRDS (I,4841)

HALE, RICHARD
THE MUNSTERS (I,3158); NIGHT GALLERY (I,3288)

HALE, RON
RYAN'S HOPE (II,2234)

HALE, VIRGINIA
NAVY LOG (I,3233)

HALEE, RON
CBS CARTOON THEATER (I,878)

HALELOKE
ARTHUR GODFREY AND FRIENDS (I,278)

HALEY, JACK
FORD STAR REVUE (I,1632); OPERATION ENTERTAINMENT (I,3400)

HALEY, JACKIE EARLE
BREAKING AWAY (II,371); EVERY STRAY DOG AND KID (II,792); THESE ARE THE DAYS (II,2585); VALLEY OF THE DINOSAURS (II,2717); WAIT TIL YOUR FATHER GETS HOME (I,4748)

HALIEN, MARET
MOMMA THE DETECTIVE (II,1721)

HALL, ALBERT
RYAN'S FOUR (II,2233)

HALL, ANDRE
BUTTERFLIES (II,404)

HALL, ARSENIO
THE HALF-HOUR COMEDY HOUR (II,1074)

HALL, BOBBIE
BARNEY AND ME (I,358)

HALL, BRAD
SATURDAY NIGHT LIVE (II,2261)

HALL, BRIAN
FAWLTY TOWERS (II,843)

HALL, BRUCE
ROD BROWN OF THE ROCKET RANGERS (I,3825)

HALL, CLIFF
CRIME PHOTOGRAPHER (I,1097); THE FRONT PAGE (I,1691); JOHNNY JUPITER (I,2429); SCALPLOCK (I,3930); SEARCH FOR TOMORROW (II,2284)

HALL, CYNTHIA
THE FABULOUS DR. FABLE (I,1500)

HALL, DEIDRE
DAYS OF OUR LIVES (II,629); ELECTRA WOMAN AND DYNA GIRL (II,764); EMERGENCY! (II,775); HOT PURSUIT (II,1189); THE YOUNG AND THE RESTLESS (II,2862)

HALL, ED
BABY, I'M BACK! (II,130)

HALL, GEORGE
THE EDGE OF NIGHT (II,760)

HALL, GINGER
FOUR STAR PLAYHOUSE (I,1652)

HALL, GRAYSON
DARK SHADOWS (I,1157); ONE LIFE TO LIVE (II,1907)

HALL, HARRIET
ALL MY CHILDREN (II,39); THE DAY THE WOMEN GOT EVEN (II,628); SOMERSET (I,4115); WHAT'S UP DOC? (II,2771)

HALL, HARRY
THE WAYNE KING SHOW (I,4779)

HALL, HUNTZ
THE CHICAGO TEDDY BEARS (I,934); ESCAPE (I,1460); UNCLE CROC'S BLOCK (II,2702)

HALL, JAMES
NORMA RAE (II,1860); OF MICE AND MEN (I,3333)

HALL, JON
RAMAR OF THE JUNGLE (I,3718)

HALL, JUANITA
CAPTAIN BILLY'S MISSISSIPPI MUSIC HALL (I,826)

HALL, LASAUNDRA
THINGS ARE LOOKING UP (II,2588)

HALL, LOIS
LITTLE HOUSE: LOOK BACK TO YESTERDAY (II,1490); LITTLE WOMEN (I,2720)

HALL, MONTY
BEAT THE CLOCK (II,195); COWBOY THEATER (I,1086); IT'S ANYBODY'S GUESS (II,1255); KEEP TALKING (I,2517); LET'S MAKE A DEAL (II,1462); LET'S MAKE A DEAL (II,1463); LET'S MAKE A DEAL (II,1464); LI'L ABNER (I,2702); LIGHTS, CAMERA, MONTY! (II,1475); MADHOUSE 90 (I,2807); MITZI AND A HUNDRED GUYS (II,1710); MONTY HALL'S VARIETY HOUR (II,1728); STRIKE IT RICH (I,4262); THOSE WONDERFUL TV GAME SHOWS (II,2603); VIDEO VILLAGE (I,4731)

HALL, PHILIP BAKER
RIDING FOR THE PONY EXPRESS (II,2172)

HALL, RICH
NOT NECESSARILY THE NEWS (II,1862); NOT NECESSARILY THE NEWS (II,1863); SATURDAY NIGHT LIVE (II,2261); SMALL WORLD (II,2381)

HALL, RILEY
THE MARSHAL OF GUNSIGHT PASS (I,2925)

HALL, ROD
THE HALF-HOUR COMEDY HOUR (II,1074)

HALL, SEAN TYLER
THE MACKENZIES OF PARADISE COVE (II,1585); STICKIN' TOGETHER (II,2465)

HALL, THURSTON
TOPPER (I,4582); TV READERS DIGEST (I,4630)

HALL, TIM
YOU'RE THE GREATEST, CHARLIE BROWN (I,4973)

HALL, TOM T
COUNTRY NIGHT OF STARS (II,592); DEAN MARTIN PRESENTS MUSIC COUNTRY, U.S.A. (I,1192); HARPER VALLEY, U.S.A. (I,1954); A JOHNNY CASH CHRISTMAS (II,1328); JOHNNY CASH: THE FIRST 25 YEARS (II,1336); LUCY COMES TO NASHVILLE (II,1561); MUSIC CITY NEWS TOP COUNTRY HITS OF THE YEAR (II,1772); SKINFLINT (II,2377)

HALL, TONY
OH, BOY! (I,3341)

HALL, TREVOR
THE YESTERDAY SHOW (II,2850)

HALL, VALERIE
THE INVISIBLE WOMAN (II,1244)

HALL, WILBUR
JOHNNY CARSON PRESENTS THE SUN CITY SCANDALS (I,2420)

HALL, ZOOEY
THE NEW PEOPLE (I,3268)

HALLAHAN, CHARLES
ALLISON SIDNEY HARRISON (II,54); THE CHICAGO STORY (II,507); MICKEY SPILLANE'S MARGIN FOR MURDER (II,1691); THE PAPER CHASE (II,1949)

HALLAM, ROSCOE
MURDER MOTEL (I,3161)

HALLARAN, SUSAN
TALES OF TOMORROW (I,4352)

HALLAREN, JANE
SECOND EDITION (II,2288)

HALLER, MELONIE
WELCOME BACK, KOTTER (II,2761)

HALLETT, JACK
JOHNNY GARAGE (II,1337)

HALLEY, RUDOLPH
CRIME SYNDICATED (I,1099)

HALLICK, TOM
ENTERTAINMENT TONIGHT (II,780); HAWKINS ON MURDER (I,1978); THE RETURN OF CAPTAIN NEMO (II,2135); SEARCH (I,3951); THE YOUNG AND THE RESTLESS (II,2862)

HALLIDAY, ELIZABETH
BOSTON AND KILBRIDE (II,356); JOE DANCER: THE MONKEY MISSION (II,1301); YOUR PLACE OR MINE? (II,2874)

HALLIDAY, HELLEN
PETER PAN (I,3566); SKIN OF OUR TEETH (I,4073)

HALLORAN, SUSAN
OUR FIVE DAUGHTERS (I,3413); YOUNG DR. MALONE (I,4938)

HALLOWAY, STERLING
WILLY (I,4869)

HALLSTROM, HOLLY
THE NEW PRICE IS RIGHT (I,3270); THE PRICE IS RIGHT (I,3665); THE PRICE IS RIGHT (II,2079)

HALOP, BILLY
ALL IN THE FAMILY (II,38); BRACKEN'S WORLD (I,703); PAPA G.I. (I,3451); THE UNEXPECTED (I,4674)

HALOP, FLORENCE
ALLAN (I,142); ALONE AT LAST (II,60); ANGIE (II,80); THE BETTY WHITE SHOW (II,224); HARRY'S BATTLES (II,1100); MEET MILLIE (I,2988); ST. ELSEWHERE (II,2432)

HALPIN, HELEN
DAVE'S PLACE (I,1175)

HALPIN, LUKE
ANNIE GET YOUR GUN (I,223); FLIPPER (I,1604); INTERMEZZO (I,2223); YOUNG DR. MALONE (I,4938)

HALSEY, BRETT
BROCK CALLAHAN (I,741); CHALK ONE UP FOR JOHNNY (I,893); FOLLOW THE SUN (I,1617); GENERAL HOSPITAL (II,964); SCRUPLES (II,2281); SEARCH FOR TOMORROW (II,2284)

HAM, GREG
MEN AT WORK IN CONCERT (II,1682)

HAMAGUCHI, TED
FROM HERE TO ETERNITY (II,933)

HAMBRECK, JOHN
THE CODE OF JONATHAN WEST (I,992)

HAMBRO, LEONID
THE VICTOR BORGE SHOW (I,4720)

HAMEL, AL
THE ANNIVERSARY GAME (I,227); MANTRAP (I,2893); WEDDING PARTY (I,4786); YOU CAN'T DO THAT ON TELEVISION (I,4933)

HAMEL, VERONICA
CITY OF ANGELS (II,540); HILL STREET BLUES (II,1154); THE HUSTLER OF MUSCLE BEACH (II,1210); JACQUELINE SUSANN'S VALLEY OF THE DOLLS, 1981 (II,1268)

HAMER, RUSTY
THE DANNY THOMAS TV FAMILY REUNION (I,1154); EVERYTHING HAPPENS TO ME (I,1481); MAKE MORE ROOM FOR DADDY (I,2842); MAKE ROOM FOR DADDY (I,2843); MAKE ROOM FOR GRANDDADDY (I,2844); MAKE ROOM FOR GRANDDADDY (I,2845); WHATEVER BECAME OF. . . ? (II,2772)

HAMILL, DOROTHY
ANDY WILLIAMS' EARLY NEW ENGLAND CHRISTMAS (II,79); DOROTHY HAMILL IN ROMEO & JULIET ON ICE (II,718); THE DOROTHY HAMILL SPECIAL (II,719); THE DOROTHY HAMILL WINTER CARNIVAL SPECIAL (II,720); DOROTHY HAMILL'S CORNER OF THE SKY (II,721); PERRY COMO'S FRENCH-CANADIAN CHRISTMAS (II,2016)

HAMILL, MARK
BOB HOPE SPECIAL: BOB HOPE'S CHRISTMAS SPECIAL (II,308); THE CITY (II,541); EIGHT IS ENOUGH (II,762); ERIC (I,1451); GENERAL HOSPITAL (II,964); MALLORY: CIRCUMSTANTIAL EVIDENCE (II,1608); ONE DAY AT A TIME (II,1900); THE TEXAS WHEELERS (II,2568)

HAMILTON IV, GEORGE
THE GEORGE HAMILTON IV SHOW (I,1769)

HAMILTON, ALEXA
THE INVISIBLE WOMAN (II,1244)

HAMILTON, ANDY
NOT THE NINE O'CLOCK NEWS (II,1864)

HAMILTON, ANN
THE MORECAMBE AND WISE SHOW (II,1734)

HAMILTON, ANTONY
COVER UP (II,597)

HAMILTON, BARBARA
MAX (II,1658); SHE'S WITH ME (II,2321)

HAMILTON, BARRY
ROBERT YOUNG AND THE FAMILY (I,3813)

HAMILTON, BERNIE
ESCAPE (I,1461); ME AND BENJY (I,2973); STARSKY AND HUTCH (II,2444)

HAMILTON, COLIN
LITTLE HOUSE: BLESS ALL THE DEAR CHILDREN (II,1489)

HAMILTON, DAN
ALL MY CHILDREN (II,39); ANOTHER WORLD (II,97); THE EDGE OF NIGHT (II,760); THE GUIDING LIGHT (II,1064); THE SECRET STORM (I,3969)

HAMILTON, FRANK
CHIEFS (II,509)

HAMILTON, GEORGE
CIRCUS OF THE STARS (II,530); CLASS OF '67 (I,970); THE FANTASTIC MISS PIGGY SHOW (II,827); THE HELLCATS (I,2009); INSTITUTE FOR REVENGE (II,1239); MALIBU (II,1607); MITZI (I,3071); PARIS 7000 (I,3459); POOR RICHARD (II,2065); ROOTS (II,2211); THE SEEKERS (II,2298); THE SURVIVORS (I,4301)

HAMILTON, HAL
THE FORSYTE SAGA (I,1645)

HAMILTON, HENRY
THE ARMSTRONG CIRCLE THEATER (I,260); THE GREATEST GIFT (I,1880)

HAMILTON, JOE
THE DINAH SHORE SHOW (I,1280)

HAMILTON, JOHN

THE ADVENTURES OF SUPERMAN (I,77); MEET THE GOVERNOR (I,2991)

HAMILTON, JULIE
THE MANIONS OF AMERICA (II,1623)

HAMILTON, KIM
BATMAN AND THE SUPER SEVEN (II,160); DOCTORS' PRIVATE LIVES (II,693); EXECUTIVE SUITE (II,796); FUTURE COP (II,945); ME AND BENJY (I,2973); STONE (II,2470)

HAMILTON, KIPP
THE MARRIAGE BROKER (I,2920)

HAMILTON, LEIGH
MISSING PIECES (II,1706)

HAMILTON, LINDA
HILL STREET BLUES (II,1154); KING'S CROSSING (II,1397); SECRETS OF MIDLAND HEIGHTS (II,2296); WISHMAN (II,2811)

HAMILTON, LYNN
THE PSYCHIATRIST: GOD BLESS THE CHILDREN (I,3685); SANFORD AND SON (II,2256); THE WALTONS (II,2740)

HAMILTON, MARGARET
THE ADDAMS FAMILY (II,11); THE ALCOA HOUR (I,108); THE BAT (I,364); THE DEVIL'S DISCIPLE (I,1247); THE ELGIN HOUR (I,1428); ETHEL AND ALBERT (I,1466); FUN FAIR (I,1706); GHOSTBREAKER (I,1789); GULF PLAYHOUSE (I,1904); IS THERE A DOCTOR IN THE HOUSE (I,2241); IS THERE A DOCTOR IN THE HOUSE? (II,1247); LIFE WITH VIRGINIA (I,2695); THE MAN WHO CAME TO DINNER (I,2880); THE NIGHT STRANGLER (I,3293); ON BORROWED TIME (I,3354); ONCE UPON A CHRISTMAS TIME (I,3368); THE PATTY DUKE SHOW (I,3483); THE PAUL LYNDE HALLOWEEN SPECIAL (II,1981); THE PAUL WINCHELL AND JERRY MAHONEY SHOW (I,3492); SIGMUND AND THE SEA MONSTERS (II,2352); SILVER THEATER (I,4051); WHO'S AFRAID OF MOTHER GOOSE? (I,4852)

HAMILTON, MARGUERITE
THE FEMININE TOUCH (I,1565); JOHN CONTE'S LITTLE SHOW (I,2408)

HAMILTON, MURRAY
ALL THE WAY HOME (II,44); B.J. AND THE BEAR (II,126); A BELL FOR ADANO (I,388); THE BOYS IN BLUE (II,359); CANNON (I,820); THE CHEAP DETECTIVE (II,490); CONFLICT (I,1036); THE FARMER'S DAUGHTER (I,1534); INHERIT THE WIND (I,2213); THE INNER SANCTUM (I,2216); THE MAN WHO NEVER WAS (I,2882); MURDOCK'S GANG (I,3164); REVLON MIRROR THEATER (I,3781); RICH MAN, POOR MAN—BOOK I (II,2161)

HAMILTON, NEIL
BATMAN (I,366); THE FRANCES LANGFORD-DON AMECHE SHOW (I,1655); GENERAL HOSPITAL (II,964); HOLLYWOOD SCREEN TEST (I,2089); PANAMA HATTIE (I,3444); STARLIGHT THEATER (I,4201); THAT WONDERFUL GUY (I,4425)

HAMILTON, RANDY
FANTASY (II,828); KIDS ARE PEOPLE TOO (II,1390); TEXAS (II,2566)

HAMILTON, RAY
KING OF DIAMONDS (I,2558)

HAMILTON, RICHARD
BACK TOGETHER (II,131); BRET MAVERICK (II,374); THE GUIDING LIGHT (II,1064)

HAMILTON, RICKI
THE SWIFT SHOW (I,4316)

HAMILTON, TED
THE LOVE BOAT I (II,1532); M STATION: HAWAII (II,1568)

HAMLETT, DILYS
HEDDA GABLER (I,2001)

HAMLIN, HARRY
MASTER OF THE GAME (II,1647); STUDS LONIGAN (II,2484)

HAMMEL, SAYRA
LEGS (II,1459)

HAMMER, ART
RAGE OF ANGELS (II,2107); TEXAS (II,2566)

HAMMER, BEN
ADVICE TO THE LOVELORN (II,16); ANOTHER WORLD (II,97); THE GUIDING LIGHT (II,1064); HOLMES AND YOYO (II,1169); JET FIGHTER (I,2374)

HAMMER, DICK
EMERGENCY! (II,775)

HAMMER, DON
GENERAL HOSPITAL (II,964)

HAMMER, JAY
TEXAS (II,2566)

HAMMER, KEN
YOUNG DR. MALONE (I,4938)

HAMMER, SHIRLEY
THE DAVE GARROWAY SHOW (I,1173)

HAMMERLEE, PAT
BLOOMER GIRL (I,488)

HAMMERLEE, PATRICIA
HEAVEN WILL PROTECT THE WORKING GIRL (I,1996)

HAMMERSTEIN II, OSCAR
TEXACO COMMAND PERFORMANCE (I,4407)

HAMMETT, MIKE
ANOTHER WORLD (II,97)

HAMMIL, MARC
JEANNIE (II,1275)

HAMMINGTON, SAM
ALL THE RIVERS RUN (II,43)

HAMMOND, EARL
THE AD-LIBBERS (I,29); ROCKY KING, INSIDE DETECTIVE (I,3824); TROUBLE INC. (I,4609); VALIANT LADY (I,4695)

HAMMOND, FREEMAN
FROM THESE ROOTS (I,1688)

HAMMOND, GEORGE
THE M AND M CANDY CARNIVAL (I,2797)

HAMMOND, HELEN
THE INA RAY HUTTON SHOW (I,2207)

HAMMOND, JOHN
THE BLUE AND THE GRAY (II,273); RIDING FOR THE PONY EXPRESS (II,2172)

HAMMOND, NICHOLAS
THE ADVENTURES OF POLLYANNA (II,14); THE AMAZING SPIDER-MAN (II,63); THE HOME FRONT (II,1171); THE MANIONS OF AMERICA (II,1623); THE MARTIAN CHRONICLES (II,1635); SPIDER-MAN (II,2424); TWO MARRIAGES (II,2691)

HAMMOND, PATRICK
THE MANIONS OF AMERICA (II,1623)

HAMMOND, PETER
THE ADVENTURES OF ROBIN HOOD (I,74); THE BUCCANEERS (I,749)

HAMMOND, ROGER
CASANOVA (II,451)

HAMMOND, RUTH
YOUNG DR. MALONE (I,4938)

HAMMOND, VIKKI
THE SULLIVANS (II,2490)

HAMNER, CAROL
MALIBU (II,1607)

HAMNER, DON
CODE NAME: HERACLITUS (I,990)

HAMNETT, OLIVIA
RETURN TO EDEN (II,2143)

HAMPSHIRE, SUSAN
BAFFLED (I,332); CRY TERROR! (I,1112); DR JEKYLL AND MR. HYDE (I,1368); THE FORSYTE SAGA (I,1645); THE PALLISERS (II,1944); VANITY FAIR (I,4701)

HAMPTON, ADRIENNE
THE ASSOCIATES (II,113)

HAMPTON, JAMES
BRAVO TWO (II,368); THE DORIS DAY SHOW (I,1355); THE DUKES OF HAZZARD (II,742); F TROOP (I,1499); FORCE FIVE (II,907); KUDZU (II,1415); MAGGIE (II,1589); MARY (II,1637); THROUGH THE MAGIC PYRAMID (II,2615)

HAMPTON, LIONEL
ALL-STAR SWING FESTIVAL (II,53); THE BIG BAND AND ALL THAT JAZZ (I,422); THE BURL IVES THANKSGIVING SPECIAL (I,763); THE JERRY LEWIS SHOW (I,2368); THE RED SKELTON CHEVY SPECIAL (I,3753); THE SATURDAY NIGHT DANCE PARTY (I,3919); SWING INTO SPRING (I,4317); TIMEX ALL-STAR JAZZ SHOW II (I,4516); TIMEX ALL-STAR JAZZ SHOW III (I,4517)

HAMPTON, PAUL
MAN ON A STRING (I,2877); NICHOLS (I,3283); RISKO (II,2180)

HAMPTON, RENEE
TREASURE ISLE (I,4601)

HANAN, PEARL
TOO GOOD TO BE TRUE (II,2643)

HANCOCK, JOHN
THE DUCK FACTORY (II,738); HARDCASTLE AND MCCORMICK (II,1089); MICKEY SPILLANE'S MIKE HAMMER: MORE THAN MURDER (II,1693); PALMERSTOWN U.S.A.

(II,1945); SCRUPLES (II,2280)

HANCOCK, PRENTISS
SPACE: 1999 (II,2409)

HANCOCK, STEPHEN
CORONATION STREET (I,1054)

HANCOCK, WYLEY
THE MAGIC SLATE (I,2820)

HANDL, IRENE
THE SEARCH (I,3950)

HANDLEMAN, STANLEY MYRON
DEAN MARTIN PRESENTS THE GOLDDIGGERS (I,1193); DEAN MARTIN PRESENTS THE GOLDDIGGERS (I,1194); MAKE ROOM FOR GRANDDADDY (I,2845)

HANDLEMAN, TIM
VELVET (II,2725)

HANDY, JOE
COUNTERATTACK: CRIME IN AMERICA (II,581)

HANEFY, JENNY
MOLL FLANDERS (II,1720)

HANEY, ANNE
BIG BEND COUNTRY (II,230); BLISS (II,269); CELEBRITY (II,463); THE INVISIBLE WOMAN (II,1244); P.O.P. (II,1939)

HANEY, CAROL
AMERICA PAUSES FOR THE MERRY MONTH OF MAY (I,167); THE BACHELOR (I,325); THE BOB HOPE SHOW (I,568); PANTOMIME QUIZ (I,3449); PHIL SILVERS IN NEW YORK (I,3577); SUMMER IN NEW YORK (I,4279)

HANEY, FRED
TEXACO COMMAND PERFORMANCE (I,4408)

HANKIN, LARRY
EVIL ROY SLADE (I,1485); FREE COUNTRY (II,921); THE GIRL, THE GOLD WATCH AND EVERYTHING (II,1002); THE MUSIC SCENE (I,3176); THE PLANT FAMILY (II,2050)

HANKS, STEVEN
B.A.D. CATS (II,123); THE WILD WOMEN OF CHASTITY GULCH (II,2801)

HANKS, TOM
BOSOM BUDDIES (II,355); FAMILY TIES (II,819)

HANLEY, ANNE
THE ADVENTURES OF POLLYANNA (II,14)

HANLEY, BRIDGET
BELL, BOOK AND CANDLE (II,201); CIRCLE OF FEAR (I,958); GIDGET (I,1795); GUESS WHAT I DID TODAY (I,1897); HARPER VALLEY (II,1094); HARPER VALLEY PTA (II,1095); HERE COME THE BRIDES (I,2024); MALIBU (II,1607); THE PIRATES OF FLOUNDER BAY (I,3607); REWARD (II,2147)

HANLEY, ELLEN
THE EDDIE ALBERT SHOW (I,1406)

HANLEY, PETE
THE TONIGHT SHOW (I,4563)

HANLEY, PETER
THE JOHNNY CARSON SHOW (I,2422); KOVACS UNLIMITED (I,2583)

HANMER, DON
ALICE IN WONDERLAND (I,120)

HANNA, PHIL
BRIDE AND GROOM (I,721); ONCE UPON A TUNE (I,3372); THE S. S. HOLIDAY (I,4173); SEARCH FOR BEAUTY (I,3953)

HANNA, WALT
THE REDD FOXX COMEDY HOUR (II,2126)

HANNAFORD FAMILY, THE
THE ED WYNN SHOW (I,1403); HOLLYWOOD JUNIOR CIRCUS (I,2084)

HANNAFORD, SUSAN
THE SULLIVANS (II,2490)

HANNAH, BOB
THE CONCRETE COWBOYS (II,564); THE GEORGIA PEACHES (II,979)

HANNAH, DARYL
PAPER DOLLS (II,1951)

HANNAH, ROBERT
OUT OF OUR MINDS (II,1933)

HANNEY, DEL
A WOMAN OF SUBSTANCE (II,2820)

HANNIBAL, MARC
DRAGNET (I,1371)

HANOLD, MARILYN
RUN JACK RUN (I,3872)

HANSEN, DONNA
CONDOMINIUM (II,566)

HANSEN, JANIS
I DREAM OF JEANNIE (I,2167); IN TANDEM (II,1228); MOVIN' ON (II,1743); THE ODD COUPLE (II,1875); THE ROUNDERS (I,3851)

HANSEN, JERRY
MR. MAGOO (I,3142)

HANSEN, PETER
THE ADVENTURES OF JIM BOWIE (I,63); BEN JERROD (I,396); GENERAL HOSPITAL (II,964); LITTLE WOMEN (I,2721); MARKHAM (I,2916); MCKEEVER AND THE COLONEL (I,2970); THE MORECAMBE AND WISE SHOW (II,1734); MR. NOVAK (I,3145); NAVY LOG (I,3233)

HANSEN, WILLIAM
ABE LINCOLN IN ILLINOIS (I,11); IS THERE A DOCTOR IN THE HOUSE? (II,1247); THE TEAHOUSE OF THE AUGUST MOON (I,4368)

HANSON, MARCIE
THE ROLLERGIRLS (II,2203); SORORITY '62 (II,2405)

HANSON, TOM
THE BING CROSBY SPECIAL (I,471)

HANTLEY, KATE
THE CHISHOLMS (II,512)

HAPPY DAYS SINGERS, THE
HAPPY DAYS (I,1942)

HARADA, ERNEST
CODE NAME: DIAMOND HEAD (II,549); PLEASURE COVE (II,2058); THE RETURN OF CHARLIE CHAN (II,2136)

HARBARD, PAUL
JANE EYRE (I,2337)

HARBEN, CHANDELER HILL
LOVE OF LIFE (I,2774)

HARBEN, CHANDLER HILL
THE DOCTORS (II,694); TEXAS (II,2566)

HARBIN, ROBERT
FESTIVAL OF MAGIC (I,1566)

HARBORD, CARL
HAY FEVER (I,1979)

HARBOUR, MICHAEL
BEAUTY AND THE BEAST (I,382)

HARCUM, BOB
THE RIGHTEOUS APPLES (II,2174); WELCOME BACK, KOTTER (II,2761)

HARDAWAY, ROSE
THE JERRY LEWIS SHOW (I,2368)

HARDESTER, CROFTON
THE BIG EASY (II,234)

HARDIMAN, HILARY
POSSESSION (I,3655); THE STRAUSS FAMILY (I,4253)

HARDIN JR, ERNEST
NORMA RAE (II,1860)

HARDIN, JERRY
CELEBRITY (II,463); THE CHISHOLMS (II,512); FILTHY RICH (II,856); KATE BLISS AND THE TICKER TAPE KID (II,1366); SARA (II,2260)

HARDIN, JOSEPH
THE ULTIMATE IMPOSTER (II,2700)

HARDIN, MELORA
CATALINA C-LAB (II,458); THE CLIFFWOOD AVENUE KIDS (II,543); THE FAMILY TREE (II,820); LITTLE HOUSE ON THE PRAIRIE (II,1487); LITTLE HOUSE: LOOK BACK TO YESTERDAY (II,1490); SECRETS OF MIDLAND HEIGHTS (II,2296); THUNDER (II,2617); TWO MARRIAGES (II,2691)

HARDIN, TY
BRONCO (I,745); HUNTER'S MOON (II,1207); RIPTIDE (I,3794); WHEN THE WEST WAS FUN: A WESTERN REUNION (II,2780)

HARDING, ANN
DAMON RUNYON THEATER (I,1129); HOLLYWOOD OPENING NIGHT (I,2087); THE LUX VIDEO THEATER (I,2795); MATINEE THEATER (I,2947); STAGE 7 (I,4181)

HARDING, JEFF
NANCY ASTOR (II,1788)

HARDING, JOHN
HERCULE POIROT (I,2021); NICHOLS (I,3283)

HARDING, JUNE
ARENA (I,256); DIAL HOT LINE (I,1254); MATT LINCOLN (I,2949); THE RICHARD BOONE SHOW (I,3784)

HARDING, PAUL
THE DAIN CURSE (II,613); A MAN CALLED INTREPID (II,1612)

HARDSTARK, MICHAEL
DOBIE GILLIS (I,1302)

HARDT, ELOISE
THE DENNIS O'KEEFE SHOW (I,1230); THE WINDS OF WAR (II,2807)

HARDWICK, MARK

PUMPBOYS AND DINETTES ON TELEVISION (II,2090)

HARDWICKE, EDWARD
NOT GUILTY! (I,3311); OPPENHEIMER (II,1919)

HARDWICKE, SIR CEDRIC
THE ADVENTURES OF DR. FU MANCHU (I,58); THE BARRETTS OF WIMPOLE STREET (I,361); CAESAR AND CLEOPATRA (I,788); CHRYSLER MEDALLION THEATER (I,951); DUPONT SHOW OF THE MONTH (I,1387); THE ELGIN HOUR (I,1428); FOUR STAR PLAYHOUSE (I,1652); MRS. G. GOES TO COLLEGE (I,3155); THE PICTURE OF DORIAN GRAY (I,3591); SKETCHBOOK (I,4072); THE UNKNOWN (I,4680)

HARDY, ALAN
ALL THE RIVERS RUN (II,43)

HARDY, GEORGIANA
MORNING COURT (I,3102)

HARDY, MARK
DOCTOR WHO—THE FIVE DOCTORS (II,690)

HARDY, OLIVER
LAUREL AND HARDY LAUGHTOONS (II,1445)

HARDY, ROBERT
THE FAR PAVILIONS (II,832); THE GATHERING STORM (I,1741); KATE LOVES A MYSTERY (II,1367)

HARDY, SARAH
FROM THESE ROOTS (I,1688)

HARE, ERNEST
ROMEO AND JULIET (I,3837)

HARE, WILL
BENDER (II,204); LITTLE WOMEN (I,2719)

HARENS, DEAN
THE D.A.: MURDER ONE (I,1122); A DATE WITH LIFE (I,1163); FIRST LOVE (I,1583)

HAREWOOD, DORIAN
BEULAH LAND (II,226); GLITTER (II,1009); ROOTS: THE NEXT GENERATIONS (II,2213); STRIKE FORCE (II,2480); TRAUMA CENTER (II,2655)

HARFORD, BETTY
DYNASTY (II,746); THE PAPER CHASE (II,1949); THE PAPER CHASE: THE SECOND YEAR (II,1950)

HARGITAY, MICKEY
HOLIDAY IN LAS VEGAS (I,2076)

HARGRAVE, T J
THE GUIDING LIGHT (II,1064)

HARIMOTO, DALE
YOU ASKED FOR IT (II,2855)

HARING, MARY
LOVE OF LIFE (I,2774)

HARKER, CHARMIENNE
MY LITTLE MARGIE (I,3194)

HARKINS, JOHN
CALLAHAN (II,414); DOC (II,682); MARRIAGE IS ALIVE AND WELL (II,1633); RETURN OF THE MAN FROM U.N.C.L.E.: THE 15 YEARS LATER AFFAIR (II,2140)

HARLAN, JEFF
INSTANT FAMILY (II,1238); LASSIE: THE NEW BEGINNING (II,1433); THE RETURN OF LUTHER GILLIS (II,2137)

HARLAN, ROBIN
THE BIONIC WOMAN (II,255)

HARLAND, MICHAEL
S.W.A.T. (II,2236)

HARLAND, ROBERT
LAW OF THE PLAINSMAN (I,2639); TARGET: THE CORRUPTERS (I,4360)

HARLEM GLOBETROTTERS, THE
THE BURT BACHARACH SPECIAL (I,776); THE HARLEM GLOBETROTTERS POPCORN MACHINE (II,1092)

HARMON, DEBORAH
BUCKSHOT (II,385); COMEDY OF HORRORS (II,557); THE FUN FACTORY (II,938); M*A*S*H (II,1569); THE TED KNIGHT SHOW (II,2550); THE TV SHOW (II,2673); USED CARS (II,2712)

HARMON, DREW
READY AND WILLING (II,2115)

HARMON, JENNIFER
DAYS OF OUR LIVES (II,629); HOW TO SURVIVE A MARRIAGE (II,1198); ONE LIFE TO LIVE (II,1907)

HARMON, JOHN
THE HERO (I,2035); THE RIFLEMAN (I,3789)

HARMON, JOY
20TH CENTURY FOLLIES (I,4641); GIDGET (I,1795)

HARMON, KELLY
BAY CITY BLUES (II,186); THE JAMES BOYS (II,1272); THE LORETTA YOUNG SHOW (I,2755)

HARMON, KRISTEN
THE ADVENTURES OF OZZIE AND HARRIET (I,71)

HARMON, LARRY
WELCOME TO THE FUN ZONE (II,2763)

HARMON, MARK
240-ROBERT (II,2688); BATTLE OF THE NETWORK STARS (II,173); BATTLE OF THE NETWORK STARS (II,174); BATTLE OF THE NETWORK STARS (II,178); CENTENNIAL (II,477); FLAMINGO ROAD (II,871); FLAMINGO ROAD (II,872); GOLIATH AWAITS (II,1025); SAM (II,2246); SAM (II,2247); ST. ELSEWHERE (II,2432)

HARMON, PATTY
TELL IT TO GROUCHO (I,4385)

HARMON, PHIL
THE FACTS (II,804)

HARMON, STEVE
ANOTHER WORLD (II,97); MR. ROBERTS (I,3149)

HARMON, TOM
OZZIE'S GIRLS (I,3440)

HARNELL, JACK
THE SONNY AND CHER SHOW (II,2401)

HARNEY, SUSAN
ANOTHER WORLD (II,97)

HARNICK, SHELDON
THE WAY THEY WERE (II,2748)

HAROUT, MEGDA
RUSSIAN ROULETTE (I,3875)

HARP, KEN
H.M.S. PINAFORE (I,2066)

HARPER, CAROL
EMERGENCY PLUS FOUR (II,774)

HARPER, CONSTANCE
THE ADVENTURES OF OZZIE AND HARRIET (I,71)

HARPER, DAVID S
THE BLUE AND THE GRAY (II,273); THE WALTONS (II,2740); THE WALTONS: A DAY FOR THANKS ON WALTON'S MOUNTAIN (EPISODE 3) (II,2741); THE WALTONS: MOTHER'S DAY ON WALTON'S MOUNTAIN (EPISODE 2) (II,2742); THE WALTONS: WEDDING ON WALTON'S MOUNTAIN (EPISODE 1) (II,2743)

HARPER, DIANE
LOVERS AND FRIENDS (II,1549)

HARPER, GERALD
IF IT'S A MAN, HANG UP (I,2186)

HARPER, JESSICA
LITTLE WOMEN (II,1498); STUDS LONIGAN (II,2484)

HARPER, JOHN
GUNSMOKE (II,1069)

HARPER, KATE
LACE (II,1418); MASTER OF THE GAME (II,1647); OPPENHEIMER (II,1919)

HARPER, MATTHEW
EMERGENCY PLUS FOUR (II,774)

HARPER, ROBERT
13 THIRTEENTH AVENUE (II,2592)

HARPER, RON
87TH PRECINCT (I,1422); GARRISON'S GORILLAS (I,1735); THE JEAN ARTHUR SHOW (I,2352); LAND OF THE LOST (II,1425); LOVE OF LIFE (I,2774); PLANET OF THE APES (II,2049); WENDY AND ME (I,4793); WHERE THE HEART IS (I,4831)

HARPER, SAMANTHA
THE LORENZO AND HENRIETTA MUSIC SHOW (II,1519); MARY HARTMAN, MARY HARTMAN (II,1638); THE STOCKERS (II,2469)

HARPER, TESS
CELEBRITY (II,463); CHIEFS (II,509)

HARPER, TOM
FOR RICHER, FOR POORER (II,906)

HARPER, VALERIE
THE BURNS AND SCHREIBER COMEDY HOUR (I,769); THE CANDID CAMERA SPECIAL (II,422); FARRELL: FOR THE PEOPLE (II,834); I LOVE LIBERTY (II,1215); JOHN DENVER AND THE LADIES (II,1312); JOHN DENVER ROCKY MOUNTAIN CHRISTMAS (II,1314); THE MARY TYLER MOORE SHOW (II,1640); RHODA (II,2151); THE SENSATIONAL, SHOCKING, WONDERFUL, WACKY 70S (II,2302); THE TROUBLE WITH PEOPLE (I,4611)

HARRELL, JAMES

THE CHISHOLMS (II,512)

HARRIETT, JUDY
DOBIE GILLIS (I,1302); THE MICKEY MOUSE CLUB (I,3025); THE MOUSEKETEERS REUNION (II,1742)

HARRIGAN, WILLIAM
HAPPY BIRTHDAY (I,1941)

HARRIMAN, FAWNE
DOCTORS' PRIVATE LIVES (II,693); ONE DAY AT A TIME (II,1900); SOMERSET (I,4115)

HARRINGTON JR, PAT
ANOTHER MAN'S SHOES (II,96); BATTLE OF THE NETWORK STARS (II,163); BENNY AND BARNEY: LAS VEGAS UNDERCOVER (II,206); THE BOB HOPE SHOW (I,623); BOBBY JO AND THE BIG APPLE (I,674); CIRCUS OF THE STARS (II,535); A COUPLE OF JOES (I,1078); THE DEAN MARTIN CELEBRITY ROAST (II,637); FOR LOVE OR $$$ (I,1623); THE FUNNY WORLD OF FRED & BUNNI (II,943); THE GERSHWIN YEARS (I,1779); THE HEALERS (II,1115); JOURNEY TO THE CENTER OF THE EARTH (I,2456); THE LAST CONVERTIBLE (II,1435); THE LOVE BOAT III (II,1534); MAKE ROOM FOR DADDY (I,2843); MR. DEEDS GOES TO TOWN (I,3134); THE NEW STEVE ALLEN SHOW (I,3271); ONE DAY AT A TIME (II,1900); OWEN MARSHALL: COUNSELOR AT LAW (I,3435); RHYME AND REASON (II,2152); SAVAGE (I,3924); SOUND OF THE 60S (I,4134); THE STEVE ALLEN SHOW (I,4219); THE STEVE ALLEN SHOW (I,4220); STEVE ALLEN'S LAUGH-BACK (II,2455); STUMP THE STARS (I,4273); THE TONIGHT SHOW (I,4564); WAIT TIL YOUR FATHER GETS HOME (I,4748); WEDNESDAY NIGHT OUT (II,2760); WONDERFUL JOHN ACTIONGOODTIME BAND (I,4892)

HARRINGTON SR, PAT
THE GOLDBERGS (I,1836)

HARRINGTON, AL
HAWAII FIVE-O (II,1110)

HARRINGTON, BILL
HOLIDAY HOTEL (I,2075)

HARRINGTON, CURTIS
THE HORROR OF IT ALL (II,1181)

HARRINGTON, DELPHI
RYAN'S HOPE (II,2234); WHERE THE HEART IS (I,4831)

HARRINGTON, KATE
THE CRADLE SONG (I,1088); THE GUIDING LIGHT (II,1064)

HARRINGTON, MICHAEL
SUNDAY FUNNIES (II,2493)

HARRINGTON, ROBERT
PETER PAN (I,3566)

HARRIS, ANDREW
THE MANIONS OF AMERICA (II,1623)

HARRIS, ARLENE
STAGE TWO REVUE (I,4184)

HARRIS, BARBARA
CHANNING (I,900)

HARRIS, BAXTER
CONCRETE BEAT (II,562)

HARRIS, BERKELEY
DIAGNOSIS: DANGER (I,1250); HOW TO SURVIVE A MARRIAGE (II,1198); OFF WE GO (I,3338)

HARRIS, BLAKE
SWEEPSTAKES (II,2514)

HARRIS, BOB
STAGE TWO REVUE (I,4184); THE TROUBLESHOOTERS (I,4614)

HARRIS, CAROLINE
JANE EYRE (II,1273); THE ONEDIN LINE (II,1912)

HARRIS, CHARLES
FEEL THE HEAT (II,848)

HARRIS, CHARLOTTE
THE LAWRENCE WELK SHOW (I,2643); THE LAWRENCE WELK SHOW (I,2644); MEMORIES WITH LAWRENCE WELK (II,1681)

HARRIS, CY
THE FIRST HUNDRED YEARS (I,1581)

HARRIS, CYNTHIA
ALLISON SIDNEY HARRISON (II,54); ARCHIE BUNKER'S PLACE (II,105); HUSBANDS AND WIVES (II,1208); HUSBANDS, WIVES AND LOVERS (II,1209); ON THE ROCKS (II,1894); SIROTA'S COURT (II,2365)

HARRIS, DON
MR. I. MAGINATION (I,3140)

HARRIS, DONALD
BONINO (I,685)

HARRIS, EMMYLOU
THE EDDIE RABBITT SPECIAL (II,759); THE GEORGE JONES SPECIAL (II,975); JOHNNY CASH AND THE COUNTRY GIRLS (II,1323); ROY ACUFF—50 YEARS THE KING OF COUNTRY MUSIC (II,2223)

HARRIS, FRANK
MR. MERGENTHWIRKER'S LOBBLIES (I,3144)

HARRIS, GEORGE
PETER PAN (II,2030)

HARRIS, GORDON
ADVENTURES OF CLINT AND MAC (I,55)

HARRIS, HOLLY
ONCE UPON A TUNE (I,3372); THE S. S. HOLIDAY (I,4173)

HARRIS, JENNY
AS THE WORLD TURNS (II,110)

HARRIS, JO ANN
B.J. AND THE BEAR (II,125); CAT BALLOU (I,870); DETECTIVE SCHOOL (II,661); THE DUCK FACTORY (II,738); GOOBER AND THE GHOST CHASERS (II,1028); M STATION: HAWAII (II,1568); MOST WANTED (II,1739); RICH MAN, POOR MAN—BOOK I (II,2161); THE WILD WILD WEST REVISITED (II,2800)

HARRIS, JOHN
THE WOLFMAN JACK SHOW (II,2817)

HARRIS, JONATHAN
THE BANANA SPLITS ADVENTURE HOUR (I,340); BATTLESTAR GALACTICA (II,181); THE BILL DANA SHOW (I,446); LOST IN SPACE (I,2758); LOST IN SPACE (I,2759); MACREEDY'S WOMAN (I,2804); MY FAVORITE MARTIANS (II,1779); MY LUCKY PENNY (I,3196); ONCE UPON A DEAD MAN (I,3369); SPACE ACADEMY (II,2406); THE THIRD MAN (I,4450); UNCLE CROC'S BLOCK (II,2702); THE WEB (I,4784)

HARRIS, JULIE
ANASTASIA (I,182); BACKSTAIRS AT THE WHITE HOUSE (II,133); A DOLL'S HOUSE (I,1327); DUPONT SHOW OF THE MONTH (I,1387); ED SULLIVAN'S BROADWAY (I,1402); THE EVIL TOUCH (I,1486); THE FAMILY HOLVAK (II,817); THE GOLDEN AGE OF TELEVISION (II,1017); THE GOOD FAIRY (I,1846); THE GREATEST GIFT (II,1061); THE HEIRESS (I,2006); THE HOLY TERROR (I,2098); THE HOUSE ON GREENAPPLE ROAD (I,2140); JOHNNY BELINDA (I,2416); KNOTS LANDING (II,1404); KRAFT SUSPENSE THEATER (I,2591); THE LARK (I,2616); LITTLE MOON OF ALBAN (I,2714); LITTLE MOON OF ALBAN (I,2715); MY FAIR LADY (I,3187); THE POWER AND THE GLORY (I,3658); THE PRIME OF MISS JEAN BRODIE (II,2082); STARLIGHT THEATER (I,4201); STUBBY PRINGLE'S CHRISTMAS (II,2483); THICKER THAN WATER (I,4445); THE UNITED STATES STEEL HOUR (I,4677); VICTORIA REGINA (I,4728)

HARRIS, JULIUS
B.J. AND THE BEAR (II,127); BENSON (II,208); THE BLUE AND THE GRAY (II,273); MISSING PIECES (II,1706); RICH MAN, POOR MAN—BOOK I (II,2161); SALTY (II,2241); UPTOWN SATURDAY NIGHT (II,2711)

HARRIS, LEE
THE STREETS OF SAN FRANCISCO (II,2478)

HARRIS, LEONARD
ENTERTAINMENT TONIGHT (II,780)

HARRIS, MARC
TO SIR, WITH LOVE (II,2624)

HARRIS, MARK
CALAMITY JANE (I,793); FAMILY TIES (II,819)

HARRIS, MERCER
MARCUS WELBY, M.D. (I,2905)

HARRIS, MILES
BIG BEND COUNTRY (II,230)

HARRIS, PERCY 'BUD'
BEULAH (I,416)

HARRIS, PHIL
ALAN KING IN LAS VEGAS, PART II (I,98); THE ANDY WILLIAMS SHOW (I,208); THE BIG SELL (I,431); THE BOB HOPE SHOW (I,523); THE BOB HOPE SHOW (I,526); THE BOB HOPE SHOW (I,583); THE BOB HOPE SHOW (I,660); THE DEAN MARTIN SHOW (I,1197); EVERYTHING YOU ALWAYS WANTED TO KNOW ABOUT JACK BENNY AND WERE

AFRAID TO ASK (I,1482); MANHATTAN TOWER (I,2889); MITZI (I,3071); THE PHIL HARRIS SHOW (I,3575); THE WONDERFUL WORLD OF BURLESQUE II (I,4896)

HARRIS, PHILLIPPA
RETURN TO THE PLANET OF THE APES (II,2145)

HARRIS, RICHARD
BURT BACHARACH IN SHANGRI-LA (I,773); CAMELOT (II,416); DUPONT SHOW OF THE MONTH (I,1387); THE PEGGY FLEMING SHOW (I,3508); THE SNOW GOOSE (I,4103); VICTORY (I,4729)

HARRIS, ROBERT H
THE GOLDBERGS (I,1835); THE GOLDBERGS (I,1836); THE INNER SANCTUM (I,2216); WAR IN THE AIR (I,4768)

HARRIS, ROSEMARY
BLITHE SPIRIT (I,485); THE CHISHOLMS (II,512); THE CHISHOLMS (II,513); DIAL "M" FOR MURDER (I,1256); PROFILES IN COURAGE (I,3678); TWELFTH NIGHT (I,4636)

HARRIS, ROSSIE
THE ADVENTURES OF POLLYANNA (II,14); UNITED STATES (II,2708)

HARRIS, STACY
THE D.A.: CONSPIRACY TO KILL (I,1121); DOORWAY TO DANGER (I,1353); THE JERK, TOO (II,1282); THE LIFE AND LEGEND OF WYATT EARP (I,2681); MAGGIE MALONE (I,2812); N.O.P.D. (I,3303); O'HARA, UNITED STATES TREASURY (I,3346); O'HARA, UNITED STATES TREASURY: OPERATION COBRA (I,3347); RETURN TO PEYTON PLACE (I,3779)

HARRIS, VIOLA
DOCTORS' PRIVATE LIVES (II,693)

HARRISON, DENNIS
GOOD OL' BOYS (II,1035)

HARRISON, GEORGE
RINGO (II,2176)

HARRISON, GRACIE
BACK TOGETHER (II,131); THE DOCTORS (II,694)

HARRISON, GREGORY
BATTLE OF THE NETWORK STARS (II,169); BATTLE OF THE NETWORK STARS (II,170); BATTLE OF THE NETWORK STARS (II,171); BATTLE OF THE NETWORK STARS (II,172); CENTENNIAL (II,477); LOGAN'S RUN (II,1507); TRAPPER JOHN, M.D. (II,2654)

HARRISON, HEATHER
ANDERSON AND COMPANY (I,188)

HARRISON, JENILEE
BATTLE OF THE NETWORK STARS (II,172); DALLAS (II,614); MALIBU (II,1607); THREE'S COMPANY (II,2614)

HARRISON, LINDA
BRACKEN'S WORLD (I,703)

HARRISON, MARK
LONDON AND DAVIS IN NEW YORK (II,1512)

HARRISON, NIGEL
BLONDIE (II,272)

HARRISON, NOEL
CALL HOLME (II,412); THE GIRL FROM U.N.C.L.E. (I,1808); GO! (I,1830); LESLIE (I,2659); THE SHIRLEY BASSEY SHOW (I,4015); WHERE THE GIRLS ARE (I,4830)

HARRISON, PATTY
THE KROFFT KOMEDY HOUR (II,1411)

HARRISON, RAY
AMERICAN SONG (I,172)

HARRISON, REX
THE BOB HOPE SHOW (I,509); BURT BACHARACH: CLOSE TO YOU (I,772); THE CHEVROLET TELE-THEATER (I,926); CRESCENDO (I,1094); THE DATCHET DIAMONDS (I,1160); THE FABULOUS 50S (I,1501)

HARRISON, SHIRLEY
THE BIG EASY (II,234)

HARRISON, STAN
DAVID BOWIE—SERIOUS MOONLIGHT (II,622)

HARRISS, PAUL
SOAP FACTORY DISCO (II,2393)

HARROLD, KATHRYN
THE BEST LEGS IN 8TH GRADE (II,214); THE DOCTORS (II,694); THE ROCKFORD FILES (II,2197); WOMEN IN WHITE (II,2823)

HARRON, DONALD
ANYTHING YOU CAN DO (I,239); CYRANO DE BERGERAC (I,1120)

HARROW, LISA
ALL CREATURES GREAT AND SMALL (I,129); NANCY ASTOR (II,1788)

HARROW, LIZ
STAR MAIDENS (II,2435)

HARRY JAMES BAND, THE
THE JERRY LEWIS SHOW (I,2365)

HARRY OSTERWALD SEXTET, THE
CONTINENTAL SHOWCASE (I,1041)

HARRY SIMEONE CHORALE, THE
THE KATE SMITH SHOW (I,2510)

HARRY, DEBORAH
BLONDIE (II,272); MUSIC CENTRAL (II,1771)

HART, BILL
RIDING HIGH (II,2173); STONEY BURKE (I,4229)

HART, BOBBY
THE SOUPY SALES SHOW (I,4138)

HART, BUDDY
LEAVE IT TO BEAVER (I,2648)

HART, CECILIA
PARIS (II,1957)

HART, CHRISTINA
ADDIE AND THE KING OF HEARTS (I,35); THE DAUGHTERS OF JOSHUA CABE RETURN (I,1170); HEAR NO EVIL (II,1116); THE RUNAWAY BARGE (II,2230); SUSAN AND SAM (II,2506)

HART, CLAY
THE LAWRENCE WELK SHOW (I,2643); THE LAWRENCE WELK SHOW (I,2644); MEMORIES WITH LAWRENCE WELK (II,1681)

HART, DOROTHY
PANTOMIME QUIZ (I,3449)

HART, JOHN
THE GOSSIP COLUMNIST (II,1045); HAWKEYE AND THE LAST OF THE MOHICANS (I,1975); THE LONE RANGER (I,2740)

HART, JOHNNIE
THE FANTASTIC FUNNIES (II,825)

HART, MARY
ENTERTAINMENT TONIGHT (II,780); THE REGIS PHILBIN SHOW (II,2128); TRUE LIFE STORIES (II,2665)

HART, MAYBETH
ANN IN BLUE (II,83)

HART, MOSS
ANSWER YES OR NO (I,232)

HART, NINA
AS THE WORLD TURNS (II,110)

HART, RALPH
THE LUCY SHOW (I,2791)

HART, RICHARD
THE ADVENTURES OF ELLERY QUEEN (I,59)

HART, TRISHA
THE BOB CRANE SHOW (II,280); WINDOWS, DOORS AND KEYHOLES (II,2806)

HARTE, JERRY
MASTER OF THE GAME (II,1647); NANCY ASTOR (II,1788)

HARTES, EDWARD
HAPPY DAYS (II,1084)

HARTFORD, DEE
LOST IN SPACE (I,2758)

HARTFORD, JOHN
AMERICA (I,164); THE BOBBIE GENTRY SPECIAL (I,669); THE GLEN CAMPBELL GOODTIME HOUR (I,1820); JUST FRIENDS (I,2495); THE SMOTHERS BROTHERS COMEDY HOUR (I,4095); THE SPECIAL GENTRY ONE (I,4148); THE SUMMER SMOTHERS BROTHERS SHOW (I,4281)

HARTLEY, MARIETTE
THE AFRICAN QUEEN (II,18); THE BIG SHOW (II,243); BLOCKHEADS (II,270); CIRCUS OF THE STARS (II,533); THE COMEDY ZONE (II,559); GENESIS II (I,1760); GHOST STORY (I,1788); GOODNIGHT, BEANTOWN (II,1041); THE HALLOWEEN THAT ALMOST WASN'T (II,1075); THE HERO (I,2035); THE KILLER WHO WOULDN'T DIE (II,1393); THE LAST HURRAH (I,2626); PEYTON PLACE (I,3574); A RAINY DAY (II,2110); THE SECOND TIME AROUND (II,2289); THE SECRET WAR OF JACKIE'S GIRLS (II,2294); SMALL WORLD (II,2381); STONE (II,2470); TELEVISION'S GREATEST COMMERCIALS II (II,2554); TELEVISION'S GREATEST COMMERCIALS III (II,2555)

HARTLEY, MICHAEL CARR
THE ADVENTURES OF A JUNGLE BOY (I,49)

HARTLEY, TED

CHOPPER ONE (II,515); SHIPSHAPE (II,2329)

HARTLINE, MARY
THE MARY HARTLINE SHOW (I,2931); SUPER CIRCUS (I,4287)

HARTLING, KAMME
THE ROOKIES (II,2208)

HARTLY, TED
PEYTON PLACE (I,3574)

HARTMAN, DAVID
CHRISTMAS WITH THE BING CROSBYS (I,950); THE DOCTORS (I,1323); I LOVE A MYSTERY (I,2170); LUCAS TANNER (II,1555); LUCAS TANNER (II,1556); MIRACLE ON 34TH STREET (I,3058); SAN FRANCISCO INTERNATIONAL AIRPORT (I,3906); SUPER COMEDY BOWL 2 (I,4289); THE VIRGINIAN (I,4738)

HARTMAN, DON
THE BOB HOPE SHOW (I,540)

HARTMAN, ELIZABETH
WILLOW B: WOMEN IN PRISON (II,2804)

HARTMAN, ENA
DAN AUGUST (I,1130); FAME IS THE NAME OF THE GAME (I,1518)

HARTMAN, GRACE
THE HARTMANS (I,1960); HOLLYWOOD STAR REVUE (I,2093)

HARTMAN, LISA
HIGH PERFORMANCE (II,1145); JACQUELINE SUSANN'S VALLEY OF THE DOLLS, 1981 (II,1268); KNOTS LANDING (II,1404); THE LISA HARTMAN SHOW (II,1484); SCARED SILLY (II,2269); TABITHA (II,2527); TABITHA (II,2528); VALENTINE MAGIC ON LOVE ISLAND (II,2716); THE WALTONS (II,2740)

HARTMAN, PAUL
ALFRED OF THE AMAZON (I,116); THE ANDY GRIFFITH SHOW (I,192); FAITH BALDWIN'S THEATER OF ROMANCE (I,1515); GULF PLAYHOUSE (I,1904); HANSEL AND GRETEL (I,1934); THE HARTMANS (I,1960); HOLLYWOOD STAR REVUE (I,2093); MAYBERRY R.F.D. (I,2961); OCCASIONAL WIFE (I,3325); OF THEE I SING (I,3334); OUR TOWN (I,3419); THE PETRIFIED FOREST (I,3570); PRIDE OF THE FAMILY (I,3667); TIME REMEMBERED (I,4509)

HARTMAN, PHIL
THE NATIONAL SNOOP (II,1797); THE PEE WEE HERMAN SHOW (II,1988); TOP TEN (II,2648)

HARTNAGEL, TIM
THE FURTHER ADVENTURES OF SPIN AND MARTY (I,1720); THE NEW ADVENTURES OF SPIN AND MARTY (I,3254); SPIN AND MARTY (I,4158)

HARTNELL, WILLIAM
DOCTOR WHO (II,689); DOCTOR WHO—THE FIVE DOCTORS (II,690)

HARTSOUGH, EEVIN
TALES FROM THE DARKSIDE (II,2534)

HARTT, MARY
ANNE MURRAY'S LADIES' NIGHT (II,89)

HARTWIG, GAY
THE FLINTSTONE COMEDY HOUR (II,881); PEBBLES AND BAMM BAMM (II,1987)

HARTY, PATRICIA
BLONDIE (I,487); HERBIE, THE LOVE BUG (II,1131); OCCASIONAL WIFE (I,3325); SEARCH FOR TOMORROW (II,2284); YOUNG HEARTS (II,2865)

HARTZELL, CLARENCE
CACTUS JIM (I,786); THOSE ENDEARING YOUNG CHARMS (I,4469)

HARVEY JR, HARRY
IF YOU KNEW TOMORROW (I,2187); MR. TUTT (I,3153)

HARVEY, ALEX
THE BLUE AND THE GRAY (II,273)

HARVEY, BOB
THE JERRY LEWIS SHOW (I,2370)

HARVEY, DON
THE MISS AND THE MISSILES (I,3062)

HARVEY, GENE
THE LONELY WIZARD (I,2743)

HARVEY, GEORGETTE
THE ETHEL WATERS SHOW (I,1468)

HARVEY, HARRY
IT'S A MAN'S WORLD (I,2258); THE ROY ROGERS SHOW (I,3859)

HARVEY, JAN
A FAMILY AFFAIR (II,814)

HARVEY, JANE
BROADWAY OPEN HOUSE (I,736)

HARVEY, JASON
GABE AND WALKER (II,950)

HARVEY, JERRY
THE BENNETS (I,397)

HARVEY, JOHN
GROWING PAYNES (I,1891)

HARVEY, KEN
SEARCH FOR TOMORROW (II,2284)

HARVEY, LAWRENCE
DIAL "M" FOR MURDER (I,1255); THE MILTON BERLE SPECIAL (I,3050)

HARVEY, MICHAEL
330 INDEPENDENCE S.W. (I,4486); TOM CORBETT, SPACE CADET (I,4535)

HARVEY, PAUL
MEET THE GOVERNOR (I,2991)

HARWOOD, ANDREW
THE PAUL HOGAN SHOW (II,1975)

HARWOOD, DICK
GOODNIGHT, BEANTOWN (II,1041)

HARWOOD, VANESSA
THE MAGIC PLANET (II,1598)

HASEGAWA, HIROSHI
SHOGUN (II,2339)

HASELY, RUTH
SAY WHEN (I,3929)

HASKELL, DAVID
ALL THAT GLITTERS (II,42); MICKEY SPILLANE'S MIKE HAMMER: MORE THAN MURDER (II,1693); MISSING PIECES (II,1706); THIS IS KATE BENNETT (II,2597)

HASKELL, JACK
THE DAVE GARROWAY SHOW (I,1173); GARROWAY AT LARGE (I,1737); THE JACK PAAR SHOW (I,2316); OF ALL THINGS (I,3330); STOP THE MUSIC (I,4232)

HASKELL, PETER
BRACKEN'S WORLD (I,703); THE EYES OF CHARLES SAND (I,1497); THE EYES OF TEXAS (II,799); THE JORDAN CHANCE (II,1343); LADIES IN BLUE (II,1422); THE MAN WHO CAME TO DINNER (I,2881); MANDRAKE (II,1617); RICH MAN, POOR MAN—BOOK II (II,2162); RISKO (II,2180); RYAN'S HOPE (II,2234); SEARCH FOR TOMORROW (II,2284); SHIRLEY (II,2330); STUNT SEVEN (II,2486)

HASKINS, DENNIS
ONE NIGHT BAND (II,1909)

HASLAM, LU ANN
THE NEW ADVENTURES OF HUCKLEBERRY FINN (I,3250)

HASSELHOFF, DAVID
AFTER HOURS: GETTING TO KNOW US (II,21); ALL THAT GLITTERS (II,41); KNIGHT RIDER (II,1402); PLEASURE COVE (II,2058); SCENE OF THE CRIME (II,2271); SEMI-TOUGH (II,2300); STAR SEARCH (II,2437); THE YOUNG AND THE RESTLESS (II,2862)

HASSO, SIGNE
THE BOB HOPE SHOW (I,611); CELANESE THEATER (I,881); CODE NAME: HERACLITUS (I,990); THE MAGICIAN (I,2824); WINNER TAKE ALL (II,2808)

HASTING, BILL
THE JERK, TOO (II,1282)

HASTINGS, BOB
ALL IN THE FAMILY (II,38); THE ATOM SQUAD (I,312); THE C.B. BEARS (II,406); CAPTAIN VIDEO AND HIS VIDEO RANGERS (I,838); THE CLUE CLUB (II,546); DEALER'S CHOICE (II,635); THE FLIM-FLAM MAN (I,1599); GENERAL HOSPITAL (II,964); GIDGET MAKES THE WRONG CONNECTION (I,1798); HOWDY (I,2150); I DREAM OF JEANNIE (I,2167); JEANNIE (II,1275); JERRY (II,1283); KITTY FOYLE (I,2571); MCHALE'S NAVY (I,2969); THE MUNSTERS' REVENGE (II,1765); THE NEW ADVENTURES OF SUPERMAN (I,3255); NO SOAP, RADIO (II,1856); SPIDER-MAN (II,2424); A VERY MISSING PERSON (I,4715); WHEELS (II,2777)

HASTINGS, DON
AFTER HOURS: SINGIN', SWINGIN' AND ALL THAT JAZZ (II,22); AS THE WORLD TURNS (II,110); CAPTAIN VIDEO AND HIS VIDEO RANGERS (I,838); THE EDGE OF NIGHT (II,760)

HATCH, DAVID
CAMBRIDGE CIRCUS (I,807)

HATCH, EDDIE EARL
AS THE WORLD TURNS (II,110)

HATCH, RICHARD

ADDIE AND THE KING OF HEARTS (I,35); ALL MY CHILDREN (II,39); BATTLE OF THE NETWORK STARS (II,163); BATTLE OF THE NETWORK STARS (II,164); BATTLE OF THE NETWORK STARS (II,167); BATTLE OF THE NETWORK STARS (II,168); BATTLESTAR GALACTICA (II,181); THE CELEBRITY FOOTBALL CLASSIC (II,474); CIRCUS OF THE STARS (II,531); CIRCUS OF THE STARS (II,533); CIRCUS OF THE STARS (II,534); FOREVER FERNWOOD (II,909); THE HUSTLER OF MUSCLE BEACH (II,1210); THE NIGHTENGALES (II,1850); PRISONERS OF THE LOST UNIVERSE (II,2086); THE STREETS OF SAN FRANCISCO (II,2478)

HATFIELD, HURD
A CRY OF ANGELS (I,1111); HOLLYWOOD SCREEN TEST (I,2089); LAMP AT MIDNIGHT (I,2608); THE MANIONS OF AMERICA (II,1623); THE NORLISS TAPES (I,3305); THE WORD (II,2833)

HATHAWAY, NOAH
BATTLESTAR GALACTICA (II,181); THE LAST CONVERTIBLE (II,1435); THREE EYES (II,2605)

HATHAWAY, SAMANTHA
BEWITCHED (I,418)

HAUER, BRENT
IT'S MAGIC, CHARLIE BROWN (I,2271); LIFE IS A CIRCUS, CHARLIE BROWN (I,2683)

HAUFRECHT, ALAN
ALICE (II,33); STICK AROUND (II,2464)

HAUNANI, MINN
COCAINE AND BLUE EYES (II,548)

HAUSER, FAYETTE
THE MANHATTAN TRANSFER (II,1619)

HAUSER, GAYELORD
THE GAYELORD HAUSER SHOW (I,1744)

HAUSER, GRETCHEN
ANYTHING GOES (I,237)

HAUSER, TIM
THE MANHATTAN TRANSFER (II,1619)

HAUSER, WINGS
HEAR NO EVIL (II,1116); THE YOUNG AND THE RESTLESS (II,2862)

HAUSNER, JERRY
ARTHUR GODFREY IN HOLLYWOOD (I,281); I LOVE LUCY (I,2171); THE PHIL SILVERS ARROW SHOW (I,3576); VALENTINE'S DAY (I,4693)

HAUTT, WHITEY
THE UNEXPLAINED (I,4675)

HAVENS, BOB
THE LAWRENCE WELK SHOW (I,2643); MEMORIES WITH LAWRENCE WELK (II,1681)

HAVENS, JOHNNY
SATURDAY NIGHT JAMBOREE (I,3920)

HAVENS, RICHIE
THE BOBBIE GENTRY SPECIAL (I,669); THE GIRL, THE GOLD WATCH AND DYNAMITE (II,1001); THE SPECIAL GENTRY ONE (I,4148)

HAVENS, THOMAS
THE KILLIN' COUSIN (II,1394)

HAVERON, JOHN
LOW MAN ON THE TOTEM POLE (I,2782)

HAVERS, NIGEL
A HORSEMAN RIDING BY (II,1182); LOOK BACK IN DARKNESS (I,2752); NANCY ASTOR (II,1788)

HAVOC, JUNE
CELANESE THEATER (I,881); CHRYSLER MEDALLION THEATER (I,951); THE ERROL FLYNN THEATER (I,1458); FIRESIDE THEATER (I,1580); HOLLYWOOD OPENING NIGHT (I,2087); THE JUNE HAVOC SHOW (I,2489); MR. BROADWAY (I,3130); NIGHTSIDE (I,3296); THE OUTER LIMITS (I,3426); PANIC! (I,3447); PULITZER PRIZE PLAYHOUSE (I,3692); TELLER OF TALES (I,4388); WILLY (I,4868); WILLY (I,4869)

HAW, JANE
TYCOON: THE STORY OF A WOMAN (II,2697)

HAWK, JEREMY
LITTLE LORD FAUNTLEROY (II,1492)

HAWKES, LIONEL
THE MAGNIFICENT SIX AND A HALF (I,2828)

HAWKINS FAMILY, THE
GEORGE BURNS EARLY, EARLY, EARLY, CHRISTMAS SHOW (II,968)

HAWKINS, CAROL
CARRY ON LAUGHING (II,448)

HAWKINS, DOLORES
THE GUY MITCHELL SHOW (I,1913)

HAWKINS, JACK
CAESAR AND CLEOPATRA (I,788); THE FOUR JUST MEN (I,1650); JANE EYRE (I,2339); THE POPPY IS ALSO A FLOWER (I,3650)

HAWKINS, JERRY
THE DONNA REED SHOW (I,1347)

HAWKINS, JIMMY
ANNIE OAKLEY (I,225); CAROLYN (I,854); ICHABOD AND ME (I,2183); THE RUGGLES (I,3868)

HAWKINS, MICHAEL
AS THE WORLD TURNS (II,110); RYAN'S HOPE (II,2234)

HAWKINS, SUSAN
CAROLYN (I,854)

HAWKINS, TRICIA PURSLEY
ALL MY CHILDREN (II,39)

HAWKINS, VIRGINIA
ADAM'S RIB (I,32); DYNASTY (II,746); HAWKINS ON MURDER (I,1978); LUCAN (II,1553); MEDICAL CENTER (II,1675)

HAWKINS, YVETTE
CAGNEY AND LACEY (II,410)

HAWLEY, ADELAIDE
BETTY CROCKER STAR MATINEE (I,412); FASHIONS ON PARADE (I,1538)

HAWN, EDWARD RUTLEDGE
PURE GOLDIE (I,3696)

HAWN, GOLDIE
GEORGE BURNS' 100TH BIRTHDAY PARTY (II,971); GOLDIE AND KIDS: LISTEN TO ME (II,1021); GOLDIE AND LIZA TOGETHER (II,1022); THE GOLDIE HAWN SPECIAL (II,1024); GOOD MORNING WORLD (I,1850); PURE GOLDIE (I,3696); ROWAN AND MARTIN'S LAUGH-IN (I,3856)

HAWORTH, SUSAN
PRISONER: CELL BLOCK H (II,2085)

HAWORTH, SUSANNE
ADVENTURES OF THE SEASPRAY (I,82)

HAWTHORNE, JAMES
THE EDGE OF NIGHT (II,760)

HAWTHORNE, JOAN
MURDER IS A ONE-ACT PLAY (I,3160)

HAWTHORNE, NIGEL
A WOMAN CALLED GOLDA (II,2818); YES MINISTER (II,2848)

HAWTREY, KAY
ALL THE WAY HOME (I,140); THE IMPERIAL GRAND BAND (II,1224); PAUL BERNARD—PSYCHIATRIST (I,3485); THE THANKSGIVING TREASURE (I,4416)

HAY, ALEXANDRA
A PLACE TO DIE (I,3612)

HAY, COLIN
MEN AT WORK IN CONCERT (II,1682)

HAY, ROY
CULTURE CLUB IN CONCERT (II,610)

HAYDEN, DON
MY LITTLE MARGIE (I,3194)

HAYDEN, HARRY
MY LITTLE MARGIE (I,3194); TROUBLE WITH FATHER (I,4610)

HAYDEN, KELLY
THE CHICAGO STORY (II,507)

HAYDEN, MARY
THE EDGE OF NIGHT (II,760)

HAYDEN, NORA
77 SUNSET STRIP (I,3988); THE REAL MCCOYS (I,3741)

HAYDEN, RUSSELL
COWBOY G-MEN (I,1084); JUDGE ROY BEAN (I,2467); THE MARSHAL OF GUNSIGHT PASS (I,2925)

HAYDEN, STERLING
THE BLUE AND THE GRAY (II,273); CAROL FOR ANOTHER CHRISTMAS (I,852)

HAYDN, LILI
ADAMS HOUSE (II,9); GOODBYE DOESN'T MEAN FOREVER (II,1039); IT'S A LIVING (II,1254); KATE LOVES A MYSTERY (II,1367); MRS. COLUMBO (II,1760)

HAYDN, RICHARD
THE KING AND MRS. CANDLE (I,2544); THE RETURN OF CHARLIE CHAN (II,2136); THE WIDE OPEN DOOR (I,4859)

HAYEK, JULIE
BOB HOPE SPECIAL: BOB HOPE'S USO CHRISTMAS IN BEIRUT (II,320)

HAYES, ALAN
LOVERS AND OTHER STRANGERS (II,1550)

HAYES, ALLISON
ACAPULCO (I,15); GENERAL HOSPITAL (II,964); KRAFT MYSTERY THEATER (I,2589)

HAYES, ANNELLE
HIS MODEL WIFE (I,2064)

HAYES, BARBARA
THE EDGE OF NIGHT (II,760)

HAYES, BEN
OUR FIVE DAUGHTERS (I,3413)

HAYES, BEVERLY
A TIME FOR US (I,4507)

HAYES, BILL
THE ANDY WILLIAMS SPECIAL (I,213); THE BOB HOPE SHOW (I,505); DAYS OF OUR LIVES (II,629); FIREBALL FUN FOR ALL (I,1577); KISS ME, KATE (I,2568); LITTLE WOMEN (I,2722); OLDSMOBILE MUSIC THEATER (I,3352); PANORAMA (I,3448); VARIETY (I,4703); THE YEOMAN OF THE GUARD (I,4926); YOUR SHOW OF SHOWS (I,4957)

HAYES, BILLIE
THE BAY CITY ROLLERS SHOW (II,187); GENERAL HOSPITAL (II,964); H.R. PUFNSTUF (I,2154); LI'L ABNER (I,2702); THE PAUL LYNDE HALLOWEEN SPECIAL (II,1981)

HAYES, BILLY
LIDSVILLE (I,2679)

HAYES, GEORGE
THE GUIDING LIGHT (II,1064)

HAYES, GEORGE 'GABBY'
THE GABBY HAYES SHOW (I,1723); HOWDY DOODY (I,2151)

HAYES, GLORIA
P.O.P. (II,1939)

HAYES, HELEN
AH!, WILDERNESS (I,91); ARSENIC AND OLD LACE (I,268); ARSENIC AND OLD LACE (I,270); THE BAT (I,364); CHRYSLER MEDALLION THEATER (I,951); FEMALE INSTINCT (I,1564); THE FRONT PAGE (I,1692); THE GENERAL MOTORS 50TH ANNIVERSARY SHOW (I,1758); HARVEY (I,1962); HAWAII FIVE-O (II,1110); LIGHT'S DIAMOND JUBILEE (I,2698); THE MARCH OF DIMES FASHION SHOW (I,2902); THE MONEYCHANGERS (II,1724); PRUDENTIAL FAMILY PLAYHOUSE (I,3683); PULITZER PRIZE PLAYHOUSE (I,3692); ROBERT MONTGOMERY PRESENTS YOUR LUCKY STRIKE THEATER (I,3809); THE ROYAL FAMILY (I,3861); SKIN OF OUR TEETH (I,4073); THE SNOOP SISTERS (II,2389)

HAYES, HERBERT
FOR THE DEFENSE (I,1624)

HAYES, ISAAC
BURT BACHARACH: CLOSE TO YOU (I,772); JACK BENNY'S FIRST FAREWELL SHOW (I,2300); THE OSMONDS SPECIAL (II,1929)

HAYES, JIMMY
IT'S ALWAYS SUNDAY (I,2265); THE LONG HOT SUMMER (I,2747)

HAYES, LINDA
EVERY STRAY DOG AND KID (II,792)

HAYES, MARGARET
ANYTHING FOR MONEY (I,236); BIG TOWN (I,436); THE CAMPBELL TELEVISION SOUNDSTAGE (I,812); THE DESILU PLAYHOUSE (I,1237); THE DOCTOR (I,1320); THE LIFE AND LEGEND OF WYATT EARP (I,2681); LIFE WITH VIRGINIA (I,2695); MONTGOMERY'S SUMMER STOCK (I,3096); PHILCO TELEVISION PLAYHOUSE (I,3583); SHORT, SHORT DRAMA (I,4023); STARLIGHT THEATER (I,4201); SUSPENSE (I,4305); A TIME FOR US (I,4507); TONIGHT AT 8:30 (I,4556); TONIGHT IN SAMARKAND (I,4558)

HAYES, MELVYN
HERE COME THE DOUBLE DECKERS (I,2025)

HAYES, PATRICIA
THE BENNY HILL SHOW (II,207); THE GOODIES (II,1040)

HAYES, PETER LIND
ALUMNI FUN (I,145); INSIDE U.S.A. WITH CHEVROLET (I,2220); THE MARCH OF DIMES FASHION SHOW (I,2902); MIRACLE ON 34TH STREET (I,3057); THE PETER AND MARY SHOW (I,3561); THE PETER LIND HAYES SHOW (I,3564); PETER LOVES MARY (I,3565); STAR OF THE FAMILY (I,4194); THE STORK CLUB (I,4236); WHEN TELEVISION WAS LIVE (II,2779); YOU'RE THE TOP (I,4974)

HAYES, RICHARD
ALL ABOUT FACES (I,126); THE BABY GAME (I,320); BROADWAY OPEN HOUSE (I,736); HIDDEN TREASURE (I,2045); THE JACK PAAR SHOW (I,2316); THE JERRY LESTER SPECIAL (I,2361); PLAY YOUR HUNCH (I,3619); SOLDIER PARADE (I,4109); TALENT PATROL (I,4340); TWO GIRLS NAMED SMITH (I,4656)

HAYES, RITA
WIN WITH A WINNER (I,4870)

HAYES, RON
THE DOG TROOP (I,1325); ESCAPE (I,1461); THE EVERGLADES (I,1478); GALLAGHER (I,1727)

HAYES, STEFAN
CAMP WILDERNESS (II,419)

HAYES, SUSAN SEAFORTH
DAYS OF OUR LIVES (II,629)

HAYLING, PATRICIA
THE DOCTORS (II,694)

HAYMAN, BARTON
WHERE THE HEART IS (I,4831)

HAYMAN, LILLIAN
THE LESLIE UGGAMS SHOW (I,2660); ONE LIFE TO LIVE (II,1907)

HAYMAN, RICHARD
THE MUSIC OF GERSHWIN (I,3174)

HAYMER, JOHNNY
ALVIN AND THE CHIPMUNKS (II,61); M*A*S*H (II,1569); MADAME'S PLACE (II,1587); MR. & MRS. DRACULA (II,1745); NO SOAP, RADIO (II,1856); OPERATION GREASEPAINT (I,3402)

HAYMES, BOB
IT'S A BUSINESS (I,2254)

HAYMES, DICK
CLINIC ON 18TH STREET (II,544); THE FABULOUS FORDIES (I,1502); MR. PORTER OF INDIANA (I,3148)

HAYNES, HILDA
ALL MY CHILDREN (II,39); EXECUTIVE SUITE (II,796); GIMME A BREAK (II,995)

HAYNES, LLOYD
79 PARK AVENUE (II,2310); DYNASTY (II,746); LASSITER (I,2625); LASSITER (II,1434); ROOM 222 (I,3843)

HAYNIE, CINDI
THE ROWAN AND MARTIN REPORT (II,2221); TABITHA (II,2526)

HAYS, BILL
YOUR SHOW OF SHOWS (II,2875)

HAYS, KATHRYN
AFTER HOURS: GETTING TO KNOW US (II,21); AS THE WORLD TURNS (II,110); CIRCLE OF FEAR (I,958); NIGHT GALLERY (I,3287); THE ROAD WEST (I,3801); YOU'RE ONLY YOUNG TWICE (I,4971); YUMA (I,4976)

HAYS, ROBERT
ALMOST HEAVEN (II,57); ANGIE (II,80); BATTLE OF THE NETWORK STARS (II,169); BATTLE OF THE NETWORK STARS (II,170); DELTA COUNTY, U.S.A. (II,657); THE GIRL, THE GOLD WATCH AND EVERYTHING (II,1002); MR. ROBERTS (II,1754); THE YOUNG PIONEERS (II,2868); THE YOUNG PIONEERS (II,2869); THE YOUNG PIONEERS' CHRISTMAS (II,2870)

HAYS, RON
THE ROUNDERS (I,3851)

HAYSBERT, DENNIS
CODE RED (II,553)

HAYSBERT, DICK
THE RETURN OF MARCUS WELBY, M.D (II,2138)

HAYTER, JAMES
THE ONEDIN LINE (II,1912)

HAYWARD, CHARD
THE THORN BIRDS (II,2600)

HAYWARD, CHRIS
NICK AND THE DOBERMANS (II,1843)

HAYWARD, CHUCK
DOC HOLLIDAY (I,1307)

HAYWARD, CYNTHIA
THE FAMILY HOLVAK (II,817); PUNCH AND JODY (II,2091)

HAYWARD, DAVID
THE CHEROKEE TRAIL (II,500); THE CHISHOLMS (II,512); FLYING HIGH (II,895); THE GEORGIA PEACHES (II,979); WAIT TIL YOUR FATHER GETS HOME (I,4748)

HAYWARD, JIM
GETAWAY CAR (I,1783); NORTHWEST PASSAGE (I,3306); SUDDEN SILENCE (I,4275)

HAYWARD, LOUIS
THE HIGHWAYMAN (I,2059); HONEY WEST: WHO KILLED THE JACKPOT? (I,2106); THE LONE WOLF (I,2742); THE PICTURE OF DORIAN GRAY (I,3591); THE SURVIVORS (I,4301)

HAYWARD, SUSAN
HEAT OF ANGER (I,1992)

HAYWARD, THOMAS
ROSALINDA (I,3846)

HAYWOODE, SHARON
THE BENNY HILL SHOW (II,207)

HAYWORTH, RITA
THE POPPY IS ALSO A FLOWER (I,3650)

HAYWORTH, VINTON
ABE LINCOLN IN ILLINOIS (I,10); HANDS OF MURDER (I,1929); I DREAM OF JEANNIE (I,2167); MENASHA THE MAGNIFICENT (I,3010); MR. MERGENTHWIRKER'S LOBBLIES (I,3144); STAGE 13 (I,4183); ZORRO (I,4982)

HAZELDINE, JAMES
THE OMEGA FACTOR (II,1888)

HAZELHURST, NONI
THE SULLIVANS (II,2490)

HAZELWOOD, LEE
MOVIN' WITH NANCY (I,3119)

HAZEN, ETHEL
HARRY O (II,1098)

HEAD, ANTHONY
LOVE IN A COLD CLIMATE (II,1539)

HEAD, GISELA
THE ICE PALACE (I,2182)

HEALEY, JAMES
PENMARRIC (II,1993)

HEALEY, MYRON
FIRESIDE THEATER (I,1580); THE LIFE AND LEGEND OF WYATT EARP (I,2681); SWAMP FOX (I,4311)

HEALY, JACK
CAR 54, WHERE ARE YOU? (I,842); THE PHIL SILVERS SHOW (I,3580)

HEALY, JIM
ALL-STAR ANYTHING GOES (II,50)

HEALY, MARY
INSIDE U.S.A. WITH CHEVROLET (I,2220); MIRACLE ON 34TH STREET (I,3057); THE PETER AND MARY SHOW (I,3561); THE PETER LIND HAYES SHOW (I,3564); PETER LOVES MARY (I,3565); STAR OF THE FAMILY (I,4194); THE STORK CLUB (I,4236); WHEN TELEVISION WAS LIVE (II,2779); YOU'RE THE TOP (I,4974)

HEALY, MICHAEL
INDEMNITY (I,2209)

HEARD, JOHN
THE SCARLET LETTER (II,2270)

HEARN, CHICK
THE CELEBRITY FOOTBALL CLASSIC (II,474); GILLIGAN'S ISLAND: THE HARLEM GLOBETROTTERS ON GILLIGAN'S ISLAND (II,993)

HEARN, CONNIE
A NEW KIND OF FAMILY (II,1819)

HEARN, GEORGE
SANCTUARY OF FEAR (II,2252); SWEENEY TODD (II,2512)

HEASLEY, MARLA
THE A-TEAM (II,119); RIPTIDE (II,2178); STAR SEARCH (II,2438)

HEATH, BOYD
SATURDAY NIGHT JAMBOREE (I,3920); TOOTSIE HIPPODROME (I,4575)

HEATH, DODY
ARSENIC AND OLD LACE (I,269)

HEATH, EIRA
THE BENNY HILL SHOW (II,207)

HEATH, LOUISE
ROUGHNECKS (II,2219)

HEATHCOTTE, THOMAS
THE HIGHWAYMAN (I,2059)

HEATHERLY, MAY
THE MAN FROM U.N.C.L.E. (I,2867)

HEATHERTON, JOEY
ALAN KING IN LAS VEGAS, PART I (I,97); THE BOB HOPE SHOW (I,613); THE BOB HOPE SHOW (I,661); CHANNING (I,900); CIRCUS OF THE STARS (II,530); DEAN MARTIN PRESENTS THE GOLDDIGGERS (I,1193); DOUG HENNING'S WORLD OF MAGIC I (II,725); THE FIFTH DIMENSION SPECIAL: AN ODYSSEY IN THE COSMIC UNIVERSE OF PETER MAX (I,1571); JACK BENNY'S FIRST FAREWELL SHOW (I,2300); JOEY AND DAD (II,1306); MR. NOVAK (I,3145); OF MICE AND MEN (I,3333); THE PERRY COMO SHOW (I,3532); THE PERRY COMO WINTER SHOW (I,3543); PERRY COMO'S WINTER SHOW (I,3547); THE POWDER ROOM (I,3656); TONY BENNETT IN WAIKIKI (I,4568)

HEATHERTON, RAY
JOEY AND DAD (II,1306)

HEATON, ANTHONY
MASTER OF THE GAME (II,1647)

HEBERLE, KAY
THE PLANT FAMILY (II,2050)

HEBERT, CHRIS
BOONE (II,351); FAMILY TIES (II,819)

HEBET, MARC
POLICE SURGEON (I,3639)

HECHT, BEN
BEN HECHT'S TALES OF THE CITY (I,395)

HECHT, GINA
EVENING AT THE IMPROV (II,786); HIZZONER (II,1159); MORK AND MINDY (II,1735)

HECHT, PAUL
THE IMPOSTER (II,1225); KATE AND ALLIE (II,1365)

HECKART, EILEEN
ALICE (II,33); ALL THE WAY HOME (I,140); BACKSTAIRS AT THE WHITE HOUSE (II,133); THE BLUE MEN (I,492); THE CAMPBELL TELEVISION SOUNDSTAGE (I,812); A DOLL'S HOUSE (I,1327); FORD THEATER HOUR (I,1635); THE GOLDEN AGE OF TELEVISION (II,1017); JOE DANCER: THE BIG BLACK PILL (II,1300); THE LITTLE FOXES (I,2712); THE MARY TYLER MOORE SHOW (II,1640); OUT OF THE BLUE (II,1934); PARTNERS IN CRIME (II,1961); PLAYHOUSE 90 (I,3623); SHORT, SHORT DRAMA (I,4023); TRAUMA CENTER (II,2655)

HECTOR, LOUIS
THE ADVENTURE OF THE THREE GARRIDEBS (I,40); TONIGHT AT 8:30 (I,4556)

HEDAYA, DAN
CHEERS (II,494); THE EARTHLINGS (II,751); GOOD TIME HARRY (II,1037); HILL STREET BLUES (II,1154)

HEDISON, DAVID
THE ART OF CRIME (II,108); BENSON (II,208); COLORADO C.I. (II,555); THE CRIME CLUB (I,1096); FIVE FINGERS (I,1586); THE LIVES OF JENNY DOLAN (II,1502); THE POWER WITHIN (II,2073); VOYAGE TO THE BOTTOM OF THE SEA (I,4743)

HEDLEY, JACK
BRIEF ENCOUNTER (I,724); BUSH DOCTOR (II,400); JOURNEY TO THE UNKNOWN (I,2457)

HEDREN, TIPPI
THE COURTSHIP OF EDDIE'S FATHER (I,1083); KRAFT SUSPENSE THEATER (I,2591)

HEEBNER, EMILY
HEAR NO EVIL (II,1116)

HEFFER, RICHARD
THE MARTIAN CHRONICLES (II,1635)

HEFFERNAN, JOHN
LOVERS AND FRIENDS (II,1549); WE INTERRUPT THIS SEASON (I,4780)

HEFFLEY, WAYNE
BACHELOR AT LAW (I,322); JOHNNY BELINDA (I,2418); VOYAGE TO THE BOTTOM OF THE SEA (I,4743)

HEFFNER, KYLE
HERNDON AND ME (II,1139)

HEFLIN, FRANCES
ALL MY CHILDREN (II,39)

HEFLIN, MARTA
THE DOCTORS (II,694)

HEFLIN, NORA
THE MARY TYLER MOORE SHOW (II,1640); STUDS LONIGAN (II,2484); THIS IS KATE BENNETT (II,2597)

HEFLIN, VAN
THE NASH AIRFLYTE THEATER (I,3227); NEITHER ARE WE ENEMIES (I,3247); ROBERT MONTGOMERY PRESENTS YOUR LUCKY STRIKE THEATER (I,3809)

HEFLY, WAYNE
NIGHTSIDE (II,1851)

HEFNER, HUGH
THE CHEVY CHASE NATIONAL HUMOR TEST (II,504); PLAYBOY AFTER DARK (I,3620); PLAYBOY'S 25TH ANNIVERSARY CELEBRATION (II,2054); PLAYBOY'S PENTHOUSE (I,3621); THE SENSATIONAL, SHOCKING, WONDERFUL, WACKY 70S (II,2302)

HEGER, KATHERINE
ONCE UPON A FENCE (I,3370)

HEGIRA, ANNE
WOMAN WITH A PAST (I,4889)

HEGYES, ROBERT
BATTLE OF THE NETWORK STARS (II,163); FOR LOVERS ONLY (II,904); WELCOME BACK, KOTTER (II,2761)

HEHIR, PETER
THE SULLIVANS (II,2490)

HEID, KARL
PERRY MASON (II,2025)

HEIDEN, JANICE
CALL TO GLORY (II,413); CITY OF ANGELS (II,540); GENERAL HOSPITAL (II,964)

HEIDT, HORACE
FAMILY NIGHT WITH HORACE HEIDT (I,1523); THE HORACE HEIDT SHOW (I,2120); SWIFT SHOW WAGON (I,4315)

HEIDT, SALLY JANE
RAGE OF ANGELS (II,2107)

HEIGH, HELENE
ALL IN THE FAMILY (I,133)

HEILBROH, VIVIEN
TARGET (II,2541)

HEILVEIL, ELAYNE
ALL NIGHT RADIO (II,40); FAMILY (II,813); WINTER KILL (II,2810)

HEINE, LEINAALA
THE MACKENZIES OF PARADISE COVE (II,1585); STICKIN' TOGETHER (II,2465)

HEINEMAN, EDA
COLONEL STOOPNAGLE'S STOOP (I,1006); HERB SHRINER TIME (I,2020)

HEINEMANN, LAURIE
ALMOST HEAVEN (II,57); ANOTHER WORLD (II,97); BIG CITY BOYS (II,232); LOOSE CHANGE (II,1518)

HEINER, THOMASINE
MASTER OF THE GAME (II,1647)

HEINKEL, SUSAN
SUSAN'S SHOW (I,4303)

HEINRICH OF MIDVALE
RUN RUN, JOE (II,2229)

HEINZ, TAMMY
THE ADAMS CHRONICLES (II,8)

HEISSER, PETER
FATHER KNOWS BEST (II,838)

HEITMAN, BOB
ALL MY CHILDREN (II,39)

HELBERG, CINDY
OFF THE WALL (II,1879)

HELBERG, FRANK
OFF THE WALL (II,1879)

HELBERG, SANDY
79 PARK AVENUE (II,2310); BETWEEN THE LINES (II,225); FLATBUSH (II,877); HOT W.A.C.S. (II,1191); THE LORENZO AND HENRIETTA MUSIC SHOW (II,1519); THE LOVE BOAT I (II,1532); OFF THE WALL (II,1879); THE PARAGON OF COMEDY (II,1956)

HELD, KARL
GENERAL ELECTRIC TRUE (I,1754); READY FOR THE PEOPLE (I,3739)

HELFER, RALPH
THE PET SET (I,3557)

HELKENBERGER, MEG
RYAN'S HOPE (II,2234)

HELLAND, ERIC
ONE NIGHT BAND (II,1909)

HELLER, BARBARA
BEN BLUE'S BROTHERS (I,393); DEAN MARTIN PRESENTS THE GOLDDIGGERS (I,1193); THE FRANK SINATRA TIMEX SHOW (I,1667); THE GOLDDIGGERS (I,1838); JACKIE GLEASON AND HIS AMERICAN SCENE MAGAZINE (I,2321); THE MILTON BERLE SPECIAL (I,3051); THE WONDERFUL WORLD OF GIRLS (I,4898)

HELLER, RANDEE
HUSBANDS AND WIVES (II,1208); HUSBANDS, WIVES AND LOVERS (II,1209); MAMA MALONE (II,1609); NUMBER 96 (II,1869); SOAP (II,2392)

HELLER, SHAD
THE RETURN OF THE BEVERLY HILLBILLIES (II,2139)

HELLER-GUTWILLIG, ANNIE
CELEBRITY (II,463)

HELLMAN, BONNIE
NURSE (II,1870)

HELLMAN, OCEAN
THE BEST CHRISTMAS PAGEANT EVER (II,212)

HELLMORE, TOM
SECRETS OF THE OLD BAILEY (I,3971)

HELM, ANNE
GENERAL HOSPITAL (II,964); THE LONG HOT SUMMER (I,2747); YOU SHOULD MEET MY SISTER (I,4935)

HELM, CONNIE
READY FOR THE PEOPLE (I,3739)

HELM, LEVON
LIVE FROM THE LONE STAR (II,1501)

HELMOND, KATHERINE
ABC'S MATINEE TODAY (I,7); BENSON (II,208); FIT FOR A KING (II,866); FOR LOVERS ONLY (II,904); NOT IN FRONT OF THE KIDS (II,1861); THE PAT BOONE AND FAMILY EASTER SPECIAL (II,1970); PEARL (II,1986); PEOPLE LIKE US (II,1999); SIDE BY SIDE (II,2346); SOAP (II,2392); WHO'S THE BOSS? (II,2792)

HELMORE, TOM
MISS BISHOP (I,3063); THE PARADINE CASE (I,3453)

HELMS, BOBBY
RANCH PARTY (I,3719)

HELMS-IRONS, ANNE
POWERHOUSE (II,2074)

HELSOP, DUDLEY
OH, BOY! (I,3341)

HELSTROM, GUNNAR
A WALK IN THE NIGHT (I,4750)

HELTON, JO
DOCTOR KILDARE (I,1315); READY FOR THE PEOPLE (I,3739)

HELTON, PERCY
THE BEVERLY HILLBILLIES (I,417); LAST STAGECOACH WEST (I,2629); THE LIFE OF VERNON HATHAWAY (I,2687); THE REASON NOBODY HARDLY EVER SEEN A FAT OUTLAW IN THE OLD WEST IS AS FOLLOWS: (I,3744); SCIENCE FICTION THEATER (I,3938)

HEMMINGS, DAVID
AIRWOLF (II,27)

HEMPHILL, JOHN
OUT OF OUR MINDS (II,1933)

HEMPHILL, ROY
DOBIE GILLIS (I,1302)

HEMPHILL, SHIRLEY
ONE IN A MILLION (II,1906); THE RICHARD PRYOR SPECIAL? (II,2165); WHAT'S HAPPENING!! (II,2769)

HEMSLEY, ESTELLE
THE GREEN PASTURES (I,1887)

HEMSLEY, SHERMAN
ALL IN THE FAMILY (II,38); BATTLE OF THE NETWORK STARS (II,170); DEAN'S PLACE (II,650); THE JEFFERSONS (II,1276); THE SENSATIONAL, SHOCKING, WONDERFUL, WACKY 70S (II,2302)

HENDEL, KENNETH
PRISONERS OF THE LOST UNIVERSE (II,2086)

HENDERSON, AL
CAR 54, WHERE ARE YOU? (I,842); SARA (II,2260)

HENDERSON, BILL
ACE CRAWFORD, PRIVATE EYE (II,6); THE BILL COSBY SPECIAL, OR? (I,443); DREAMS (II,735); A SHADOW IN THE STREETS (II,2312)

HENDERSON, CINDY
THE ADDAMS FAMILY (II,12)

HENDERSON, CURTIS
THE BLUE KNIGHT (II,277)

HENDERSON, DON
A KILLER IN EVERY CORNER (I,2538)

HENDERSON, DOUGLAS
CAPTAIN BRASSBOUND'S CONVERSION (I,828); DIAGNOSIS: DANGER (I,1250); THE WILD WILD WEST (I,4863); WON'T IT EVER BE MORNING? (I,4902)

HENDERSON, FLORENCE
BOB HOPE SPECIAL: BOB HOPE'S ALL-STAR COMEDY SPECIAL FROM AUSTRALIA (II,300); THE BRADY BRIDES (II,361); THE BRADY BUNCH (II,362); THE BRADY BUNCH HOUR (II,363); THE BRADY GIRLS GET MARRIED (II,364); CITY VS. COUNTRY (I,968); THE GENERAL FOODS 25TH ANNIVERSARY SHOW (I,1756); THE GERSHWIN YEARS (I,1779); KELLY'S KIDS (I,2523); LITTLE WOMEN (I,2722); THE LOVE BOAT I (II,1532); OLDSMOBILE MUSIC THEATER (I,3352); THE PAUL LYNDE HALLOWEEN SPECIAL (II,1981); POLICE SQUAD! (II,2061); A SALUTE TO TELEVISION'S 25TH ANNIVERSARY (I,3893); SING ALONG (I,4055); A WORLD OF LOVE (I,4914)

HENDERSON, JAMES BROWN
CHIEFS (II,509)

HENDERSON, JAMI

THE JACKIE GLEASON SHOW (I,2323)

HENDERSON, JO
SUMMER SOLSTICE (II,2492)

HENDERSON, KLEO
26 MEN (I,4648)

HENDERSON, MARCIA
THE ALDRICH FAMILY (I,111); CROSSROADS (I,1105); DEAR PHOEBE (I,1204); FOUR STAR PLAYHOUSE (I,1652); I MARRIED A DOG (I,2173); MATINEE THEATER (I,2947); PULITZER PRIZE PLAYHOUSE (I,3692); SCHLITZ PLAYHOUSE OF STARS (I,3936); TWO GIRLS NAMED SMITH (I,4656)

HENDERSON, MICHAEL
THE RIGHTEOUS APPLES (II,2174)

HENDERSON, PAT
WORDS AND MUSIC (I,4909)

HENDERSON, ROBERT
MASTER OF THE GAME (II,1647)

HENDERSON, SKITCH
FAYE AND SKITCH (I,1548); THE STEVE ALLEN SHOW (I,4219); STEVE ALLEN'S LAUGH-BACK (II,2455)

HENDERSON, TORRENCE
THE BRADY BUNCH (II,362)

HENDERSON, TY
BATMAN AND THE SUPER SEVEN (II,160); BIG SHAMUS, LITTLE SHAMUS (II,242); GOODNIGHT, BEANTOWN (II,1041); MADAME'S PLACE (II,1587); SPACE ACADEMY (II,2406)

HENDL, SUSAN
LITTLE WOMEN (I,2724)

HENDLER, LAURI
GIMME A BREAK (II,995); HIGH SCHOOL, U.S.A. (II,1148); LITTLE LULU (II,1493); MAX (II,1658); A NEW KIND OF FAMILY (II,1819); WHY US? (II,2796)

HENDRA, TONY
THE ENTERTAINERS (I,1449); KISS ME, KATE (I,2569)

HENDREN, RON
ENTERTAINMENT TONIGHT (II,780)

HENDRICKS, MARTHA
MIDWESTERN HAYRIDE (I,3033)

HENDRICKSON, BENJAMIN
TEXAS (II,2566)

HENDRICKSON, STEVEN
ARCHIE BUNKER'S PLACE (II,105)

HENDRIX, RONALD
IT'S YOUR FIRST KISS, CHARLIE BROWN (I,2280)

HENDRIX, WANDA
THE LLOYD BRIDGES SHOW (I,2733)

HENDRY, IAN
THE AVENGERS (II,121); THE INFORMER (I,2212); KILLER WITH TWO FACES (I,2539); ROARING CAMP (I,3804)

HENESY, BOBBY
THE DOCTORS (II,694)

HENESY, DAVID
DARK SHADOWS (I,1157)

HENIE, SONJA
SPOTLIGHT (I,4162)

HENLEY, TREVOR
ALICE (II,33); STEELTOWN (II,2452)

HENNER, MARILU
THE CELEBRITY FOOTBALL CLASSIC (II,474); MR. ROBERTS (II,1754); OFF CAMPUS (II,1877); THE PAPER CHASE (II,1949); TAXI (II,2546)

HENNESSY, MICHAEL
RYAN'S HOPE (II,2234)

HENNESSY, SARA
A CHRISTMAS CAROL (II,517)

HENNING, BUNNY
MRS. G. GOES TO COLLEGE (I,3155)

HENNING, CAROL
LOVE THAT BOB (I,2779)

HENNING, DEBBY
DOUG HENNING'S WORLD OF MAGIC V (II,729); DOUG HENNING: MAGIC ON BROADWAY (II,730)

HENNING, DOUG
THE CRYSTAL GAYLE SPECIAL (II,609); DOUG HENNING'S WORLD OF MAGIC I (II,725); DOUG HENNING'S WORLD OF MAGIC II (II,726); DOUG HENNING'S WORLD OF MAGIC III (II,727); DOUG HENNING'S WORLD OF MAGIC IV (II,728); DOUG HENNING'S WORLD OF MAGIC V (II,729); DOUG HENNING: MAGIC ON BROADWAY (II,730); THE OSMOND FAMILY CHRISTMAS SPECIAL (II,1926); THE WORLD OF MAGIC (II,2838)

HENNING, LINDA KAYE
THE BEVERLY HILLBILLIES (I,417); THE CIRCLE FAMILY (II,527); HIGH ROLLERS (II,1147); KUDZU (II,1415); PETTICOAT JUNCTION (I,3571); THE RETURN OF THE BEVERLY HILLBILLIES (II,2139)

HENNING, PAT
ONCE UPON A CHRISTMAS TIME (I,3368)

HENNING, SUSAN
ELVIS (I,1435)

HENNING, TONY
THE CLAUDETTE COLBERT SHOW (I,971)

HENNINGHAM, STEVEN
1915 (II,1853)

HENREID, MONIKA
STAT! (I,4213)

HENRIED, PAUL
MRS R.—DEATH AMONG FRIENDS (II,1759)

HENRIG, MERRIANA
ESCAPE (I,1460)

HENRY, GREGG
THE BLUE AND THE GRAY (II,273)

HENRY, BUCK
THE GEORGE SEGAL SHOW (II,977); A LAST LAUGH AT THE 60'S (I,2627); THE NEW SHOW (II,1828); PLAYBOY'S 25TH ANNIVERSARY CELEBRATION (II,2054); THAT WAS THE WEEK THAT WAS (I,4423); THAT WAS THE YEAR THAT WAS (II,2575)

HENRY, CAROL ANNE
DYNASTY (II,746)

HENRY, CHUCK
EYE ON HOLLYWOOD (II,798); THE LOVE REPORT (II,1542); THE LOVE REPORT (II,1543)

HENRY, EMMALINE
CAROL (I,846); THE FARMER'S DAUGHTER (I,1533); I DREAM OF JEANNIE (I,2167); I'M DICKENS...HE'S FENSTER (I,2193); MICKEY (I,3023); THE RED SKELTON SHOW (I,3756); THREE'S COMPANY (II,2614); WILD ABOUT HARRY (II,2797)

HENRY, GLORIA
THE ABBOTT AND COSTELLO SHOW (I,2); DENNIS THE MENACE (I,1231); THE FILES OF JEFFERY JONES (I,1576); REAL GEORGE (I,3740); SILVER SPOONS (II,2355); SNEAK PREVIEW (I,4100)

HENRY, GREGG
THE BOYS IN BLUE (II,359); LOOSE CHANGE (II,1518); PEARL (II,1986); RICH MAN, POOR MAN—BOOK II (II,2162); THE YEAGERS (II,2843)

HENRY, KEN
SCOTLAND YARD (I,3941)

HENRY, MIKE
INSIDE O.U.T. (II,1235); M*A*S*H (II,1569)

HENSEL, CHRISTOPHER
THE KID SUPER POWER HOUR WITH SHAZAM (II,1386)

HENSEN, LOIS
SECRET FILE, U.S.A. (I,3963)

HENSLEY, PAMELA
240-ROBERT (II,2689); BATTLE OF THE NETWORK STARS (II,167); BATTLE OF THE NETWORK STARS (II,170); BUCK ROGERS IN THE 25TH CENTURY (II,383); CONDOMINIUM (II,566); KINGSTON: CONFIDENTIAL (II,1398); KINGSTON: THE POWER PLAY (II,1399); MARCUS WELBY, M.D. (II,1627); MATT HOUSTON (II,1654); MRS R.—DEATH AMONG FRIENDS (II,1759); THE REBELS (II,2121); ROOSTER (II,2210)

HENSON, BASIL
AN ECHO OF THERESA (I,1399); WAR AND PEACE (I,4765)

HENSON, JANE
THE MUPPET SHOW (II,1766)

HENSON, JIM
FROG PRINCE (I,1685); THE MUPPET SHOW (II,1766); TALES FROM MUPPETLAND (I,4344)

HENSON, NICKEY
TYCOON: THE STORY OF A WOMAN (II,2697)

HENTELOFF, ALEX
BARNEY MILLER (II,154); THE BASTARD/KENT FAMILY CHRONICLES (II,159); THE BETTY WHITE SHOW (II,224); CENTENNIAL (II,477); CODE NAME: DIAMOND HEAD (II,549); ESCAPADE (II,781); THE INVISIBLE MAN (II,1243); JEREMIAH OF JACOB'S NECK (II,1281); NEEDLES AND PINS (I,3246); PHILLIP AND BARBARA (II,2034); PISTOLS

'N' PETTICOATS (I,3608); THE YOUNG REBELS (I,4944)

HEPBURN, AUDREY
A WORLD OF LOVE (I,4914)

HEPBURN, KATHARINE
HOLLYWOOD: THE SELZNICK YEARS (I,2095)

HEPTON, BERNARD
SECRET ARMY (II,2290); THE SIX WIVES OF HENRY VIII (I,4068); SMILEY'S PEOPLE (II,2383); TINKER, TAILOR, SOLDIER, SPY (II,2621)

HERB ROSE DANCERS, THE
THE MILTON BERLE SHOW (I,3047)

HERBERT, CHARLES
LOVE THAT BOB (I,2779); MEN INTO SPACE (I,3007); THE SECRET LIFE OF JOHN MONROE (I,3966); TOM AND JERRY (I,4533)

HERBERT, DIANA
TONIGHT AT 8:30 (I,4556)

HERBERT, DON
MR. WIZARD (I,3154)

HERBERT, PERCY
CIMARRON STRIP (I,954)

HERBERT, PITT
COREY: FOR THE PEOPLE (II,575); JEREMIAH OF JACOB'S NECK (II,1281); JOHNNY BELINDA (I,2417); MARRIED: THE FIRST YEAR (II,1634)

HERBERT, RACHEL
THE PALLISERS (II,1944)

HERBERT, SYLVIE
LACE (II,1418)

HERBERT, TIM
PAULA STONE'S TOY SHOP (I,3495)

HERBSLEB, JOHN
DUFFY (II,739)

HERD, DAPHNE
TO THE MANOR BORN (II,2627)

HERD, JOHN
KATE AND ALLIE (II,1365)

HERD, RICHARD
DOCTOR SCORPION (II,685); FARRELL: FOR THE PEOPLE (II,834); HAZARD'S PEOPLE (II,1112); IKE (II,1223); T.J. HOOKER (II,2524); V (II,2713); V: THE FINAL BATTLE (II,2714)

HERDNAN, RONALD
A HORSEMAN RIDING BY (II,1182)

HERLIE, EILEEN
ALL MY CHILDREN (II,39)

HERLIHY, ED
THE CHILDREN'S HOUR (I,939); MILESTONES OF THE CENTURY (I,3043); WHAT WILL THEY THINK OF NEXT? (I,4813)

HERLIHY, WALTER
ABC-POWERS CHARM SCHOOL (I,5)

HERMAN'S HERMITS
GO! (I,1830); IT'S WHAT'S HAPPENING, BABY! (I,2278)

HERMAN, AMANDA
NOT IN FRONT OF THE KIDS (II,1861)

HERMAN, GIL
WAIT TIL YOUR FATHER GETS HOME (I,4748)

HERMAN, PEE WEE
LILY FOR PRESIDENT (II,1478)

HERMAN, RANDY
SPARROW (II,2410); SPARROW (II,2411)

HERMAN, WOODY
MUSIC '55 (I,3168); ONE NIGHT STANDS (I,3387); THE SWINGIN' SINGIN' YEARS (I,4319); TIMEX ALL-STAR JAZZ SHOW I (I,4515)

HERMINE MIDGETS, THE
LIDSVILLE (I,2679)

HERNANDEZ, JOHN
MIAMI VICE (II,1689)

HERNANDEZ, JUAN
SAFARI (I,3882); SEVEN AGAINST THE SEA (I,3983)

HERNANDEZ, ROGELIO
TONIGHT IN HAVANA (I,4557)

HERNANDEZ, TINA GAIL
HIGHWAY HONEYS (II,1151)

HERNDON, BILL
THE GUIDING LIGHT (II,1064)

HERRERA, ANTHONY
AS THE WORLD TURNS (II,110); LOVING (II,1552); MANDRAKE (II,1617); THE SECRET STORM (I,3969)

HERRIDGE, ROBERT
THE ROBERT HERRIDGE THEATER (I,3808)

HERRIER, MARK
SPRAGGUE (II,2430)

HERRIN, WILLIAM
DANIEL BOONE (I,1141)

HERRINGTON, TABITHA
THE HITCHHIKER (II,1157)

HERRMANN, EDWARD
BEACON HILL (II,193)

HERRON, CINDY
HIGH FIVE (II,1143); TOO GOOD TO BE TRUE (II,2643)

HERRON, KEVIN
ME AND BENJY (I,2973)

HERSHAW, STEVE
THE TONY RANDALL SHOW (II,2640)

HERSHBERGER, GARY
SUMMER (II,2491)

HERSHEWE, MICHAEL
THE AMERICAN DREAM (II,71)

HERSHEY, BARBARA
FROM HERE TO ETERNITY (II,933); GIDGET (I,1795); HOLLOWAY'S DAUGHTERS (I,2079); A MAN CALLED INTREPID (II,1612); THE MONROES (I,3089); TWILIGHT THEATER II (II,2686)

HERSHOLT, JEAN
DOCTOR CHRISTIAN (I,1308)

HERTELENDY, HANNAH
RYAN'S FOUR (II,2233)

HERTZ, ROSS
THE MILKY WAY (I,3044)

HERVEY, IRENE
DAMON RUNYON THEATER (I,1129); FIRESIDE THEATER (I,1580); THE GEORGE BURNS AND GRACIE ALLEN SHOW (I,1763); GOLIATH AWAITS (II,1025); HONEY WEST (I,2105); LITTLE WOMEN (I,2721); MY THREE SONS (I,3205); O'CONNOR'S OCEAN (I,3326); ROBERTA (I,3816); STAGE 7 (I,4181); STUDIO '57 (I,4267)

HERVEY, JEAN
JANE EYRE (II,1273)

HERZBERG, PAUL
SMILEY'S PEOPLE (II,2383)

HERZOG, JOHN
I'D RATHER BE CALM (II,1216)

HESHIMU
ROOM 222 (I,3843)

HESLOV, GRANT
SPENCER (II,2420)

HESS, BILL
THE BEACHCOMBER (I,371)

HESS, DORIS
HAPPY DAYS (II,1084); MARIE (II,1629); THE WONDERFUL WORLD OF PHILIP MALLEY (II,2831)

HESS, DOROTHY
THE STEVE ALLEN COMEDY HOUR (II,2454)

HESSEMAN, HOWARD
ANOTHER APRIL (II,93); BATTLE OF THE NETWORK STARS (II,169); THE BLUE KNIGHT (II,277); CELEBRITY CHALLENGE OF THE SEXES 4 (II,468); THE FANTASTIC FUNNIES (II,825); JOHN RITTER: BEING OF SOUND MIND AND BODY (II,1319); LORETTA LYNN: THE LADY. . .THE LEGEND (II,1521); MARY HARTMAN, MARY HARTMAN (II,1638); MR. ROBERTS (II,1754); ONE DAY AT A TIME (II,1900); THE TV TV SHOW (II,2674); WKRP IN CINCINNATI (II,2814); WOMEN WHO RATE A "10" (II,2825)

HESSIE, DEWAYNE
THE PLANT FAMILY (II,2050)

HESTON, CHARLTON
ALCOA PREMIERE (I,109); BOB HOPE SPECIAL: BOB HOPE'S ALL-STAR BIRTHDAY AT ANNAPOLIS (II,294); BOGART (I,679); CHIEFS (II,509); CHRYSLER MEDALLION THEATER (I,951); CLIMAX! (I,976); THE CLOCK (I,978); THE DON ADAMS SPECIAL: HOORAY FOR HOLLYWOOD (I,1329); ELIZABETH THE QUEEN (I,1431); F.D.R. (I,1552); THE PATRIOTS (I,3477); ROBERT MONTGOMERY PRESENTS YOUR LUCKY STRIKE THEATER (I,3809); SCHLITZ PLAYHOUSE OF STARS (I,3936); SHIRLEY TEMPLE'S STORYBOOK (I,4017); SUPER COMEDY BOWL 1 (I,4288); SUSPENSE (I,4305); TIPTOE THROUGH TV (I,4524); TWELVE STAR SALUTE (I,4638); THE WAY THEY WERE (II,2748)

HETA, CARMEN
WELCOME TO PARADISE (II,2762)

HETH, JIMMY
ONE IN A MILLION (II,1906)

HETWOOD, BILL
SPENCER'S PILOTS (II,2421)

HEWE, MAYBIN
THE PAUL WINCHELL AND JERRY MAHONEY SHOW (I,3492)

HEWELL, CHRISTOPHER
IVAN THE TERRIBLE (II,1260)

HEWITT, ALAN
MY FAVORITE MARTIAN (I,3189)

HEWITT, CHRISTOPHER
E/R (II,748); FANTASY ISLAND (II,829); LOVE, LIFE, LIBERTY & LUNCH (II,1544); MASSARATI AND THE BRAIN (II,1646)

HEWITT, MARTIN
THE FAMILY TREE (II,820)

HEWITT, SEAN
OUT OF OUR MINDS (II,1933)

HEWITT, VIRGINIA
SPACE PATROL (I,4144)

HEXUM, JON-ERIK
COVER UP (II,597); VOYAGERS (II,2730)

HEYDT, LOUIS JEAN
EXPERT WITNESS (I,1488); MACKENZIE'S RAIDERS (I,2803); TV READERS DIGEST (I,4630)

HEYES JR, DOUGLAS
CAPTAINS AND THE KINGS (II,435)

HEYES, DANIEL
LOVERS AND FRIENDS (II,1549)

HEYES, HERBERT
MIRACLE ON 34TH STREET (I,3056)

HEYMAN, BARTON
ANOTHER WORLD (II,97); KENNEDY (II,1377)

HEYMAN, BURT
DOMINIC'S DREAM (II,704)

HEYWOOD, PAT
TYCOON: THE STORY OF A WOMAN (II,2697)

HI, MAE
FOG (II,897)

HI-LO'S, THE
FRANK SINATRA (I,1660); THE FRANK SINATRA TIMEX SHOW (I,1668); THE ROSEMARY CLOONEY SHOW (I,3847)

HIATT, SHELBY
GENERAL HOSPITAL (II,964)

HICE, FRED
SAM (II,2246)

HICKEY, JIM
WELCOME TO PARADISE (II,2762)

HICKEY, TOM
THE MANIONS OF AMERICA (II,1623)

HICKLAND, CATHERINE
CAPITOL (II,426); THE HEE HAW HONEYS (II,1124); TEXAS (II,2566)

HICKMAN, DARRYL
ALAN KING'S FINAL WARNING (II,29); AMERICANS (I,176); DOBIE GILLIS (I,1302); HEAVE HO HARRIGAN (I,1993); HOORAY FOR LOVE (I,2116); PANIC! (I,3447); PURSUIT (I,3700); SPACE STARS (II,2408)

HICKMAN, DWAYNE
DOBIE GILLIS (I,1302); THE GOOD OLD DAYS (I,1851); HEY TEACHER (I,2041); HIGH SCHOOL, U.S.A. (II,1148); LOVE THAT BOB (I,2779); WE'LL TAKE MANHATTAN (I,4792); WHATEVER HAPPENED TO DOBIE GILLIS? (II,2774); YOU'RE ONLY YOUNG TWICE (I,4971)

HICKMAN, EARL
SPENCER'S PILOTS (II,2421)

HICKMAN, HERMAN
CELEBRITY TIME (I,887)

HICKMONT, JACKI
RIPTIDE (I,3794)

HICKOX, HARRY
THE FABULOUS SYCAMORES (I,1505); NANNY AND THE PROFESSOR (I,3225); NO TIME FOR SERGEANTS (I,3300)

HICKOX, JANE
MOMMA THE DETECTIVE (II,1721)

HICKS, BILL
BULBA (II,392)

HICKS, CATHERINE
THE BAD NEWS BEARS (II,134); JACQUELINE SUSANN'S VALLEY OF THE DOLLS, 1981 (II,1268); RYAN'S HOPE (II,2234); SPARROW (II,2411); TUCKER'S WITCH (II,2667)

HICKS, CHUCK
FORCE SEVEN (II,908)

HICKS, EDWARD
PENMARRIC (II,1993)

HICKS, HILLY
THE BUFFALO SOLDIERS (II,388); GODZILLA (II,1014); LOCAL 306 (II,1505); ROLL OUT! (I,3833); ROOTS (II,2211); SPACE FORCE (II,2407); SPIDER-MAN (II,2424); TURNOVER SMITH (II,2671)

HICKS, RUSSELL
DATE WITH THE ANGELS (I,1164); THE GEORGE BURNS AND GRACIE ALLEN SHOW (I,1763); THE LARRY STORCH SHOW (I,2618); NAVY LOG (I,3233); THE SECRET STORM (I,3969)

HICKSON, JOAN
GREAT EXPECTATIONS (II,1055); WHY DIDN'T THEY ASK EVANS? (II,2795)

HIESTAND, JOHN
IS THERE A DOCTOR IN THE HOUSE? (II,1247)

HIGGINS, ANN
FLYING HIGH (II,895)

HIGGINS, ANTHONY
LACE (II,1418)

HIGGINS, CLAIRE
THE CITADEL (II,539); PRIDE AND PREJUDICE (II,2080)

HIGGINS, DOUG
AS THE WORLD TURNS (II,110)

HIGGINS, JOE
ARREST AND TRIAL (I,264); THE EVERLY BROTHERS SHOW (I,1479); THE GEORGE KIRBY SPECIAL (I,1771); THE RIFLEMAN (I,3789); SIGMUND AND THE SEA MONSTERS (II,2352)

HIGGINS, JOEL
BARE ESSENCE (II,149); BEST OF THE WEST (II,218); CAT BALLOU (I,869); SALVAGE (II,2244); SALVAGE 1 (II,2245); SEARCH FOR TOMORROW (II,2284); SILVER SPOONS (II,2355)

HIGGINS, MICHAEL
ACADEMY THEATER (I,14); THE FIRM (II,860); THE HAMPTONS (II,1076); THE LARK (I,2616); ONE MAN'S FAMILY (I,3383); OUR FIVE DAUGHTERS (I,3413); THE PATRIOTS (I,3477)

HIGGINS, ROSS
AROUND THE WORLD IN 80 DAYS (I,263)

HIGGINSON, JOHN
PRISONER: CELL BLOCK H (II,2085)

HIGGS, RICHARD
LOVE OF LIFE (I,2774)

HIGHTOWER, SALLY
THE FURTHER ADVENTURES OF WALLY BROWN (II,944); OFF THE WALL (II,1879); WELCOME BACK, KOTTER (II,2761)

HIKEN, GERALD
THE BALLAD OF LOUIE THE LOUSE (I,337); THE PARTRIDGE FAMILY (II,1962); THE PHIL SILVERS SHOW (I,3580); SANDBURG'S LINCOLN (I,3907); SPENCER'S PILOTS (II,2421); THERE SHALL BE NO NIGHT (I,4437)

HILBOLDT, LISE
NANCY ASTOR (II,1788)

HILD, CATHY
THE KEN MURRAY SHOW (I,2525)

HILGER, RICHARD
A CHRISTMAS CAROL (II,517)

HILKA, TRISHA
I'D RATHER BE CALM (II,1216)

HILL'S ANGELS
THE UNEXPURGATED BENNY HILL SHOW (II,2705)

HILL, ARTHUR
BORN YESTERDAY (I,692); CANNON: THE RETURN OF FRANK CANNON (II,425); THE DESPERATE HOURS (I,1239); GLITTER (II,1009); HAGEN (II,1071); OWEN MARSHALL: COUNSELOR AT LAW (I,3435); OWEN MARSHALL: COUNSELOR AT LAW (I,3436); THE RIVALRY (I,3796); THE WOMAN IN WHITE (I,4887)

HILL, BENNY
THE BENNY HILL SHOW (II,207); THE UNEXPURGATED BENNY HILL SHOW (II,2705)

HILL, CAROL
HANDS OF MURDER (I,1929); TROUBLE INC. (I,4609)

HILL, CHARLES
THE BIG SHOW (II,243); THE EARTHLINGS (II,751); SOMEONE AT THE TOP OF THE STAIRS (I,4114); TEXAS (II,2566)

HILL, CRAIG
A CHRISTMAS CAROL (I,947); TOP SECRET (II,2647); THE WHIRLYBIRDS (I,4841)

HILL, DANA

BRANAGAN AND MAPES (II,367); THE FALL GUY (II,811); FEATHERSTONE'S NEST (II,847); THE FRENCH ATLANTIC AFFAIR (II,924); THE MEMBER OF THE WEDDING (II,1680); THE PAUL WILLIAMS SHOW (II,1984); THE TWO OF US (II,2693)

HILL, DARRYL F
OUR STREET (I,3418)

HILL, DEAN
THE CHISHOLMS (II,512); HARD KNOX (II,1087)

HILL, FRANKIE
LITTLE SHOTS (II,1495)

HILL, JACQUELINE
DOCTOR WHO (II,689)

HILL, JAMES
OPPENHEIMER (II,1919)

HILL, JANA
BROTHERS AND SISTERS (II,382); CAPTAIN SCARLET AND THE MYSTERONS (I,837)

HILL, JOHNNY
MUSIC IN VELVET (I,3172)

HILL, JULIE
CHICO AND THE MAN (II,508); SECRET ARMY (II,2290)

HILL, MARIANA
HARRY O (II,1098)

HILL, MAURICE
LANDON, LANDON & LANDON (II,1426)

HILL, MAURY
CAP'N AHAB (I,824)

HILL, PHYLLIS
GENERAL HOSPITAL (II,964); MORNING STAR (I,3104); THAT GIRL (II,2570); THREE STEPS TO HEAVEN (I,4485)

HILL, RICHARD
BEACH PATROL (II,192); TODAY'S FBI (II,2628)

HILL, ROGER
HAZARD'S PEOPLE (II,1112)

HILL, ROSE
NICHOLAS NICKLEBY (II,1840)

HILL, SANDY
SHOW BUSINESS (II,2343)

HILL, SEAN TYLER
THE LITTLE PEOPLE (I,2716)

HILL, STEPHANIE
OF THIS TIME, OF THAT PLACE (I,3335)

HILL, STEVEN
ACTOR'S STUDIO (I,28); ESPIONAGE (I,1464); THE GREATEST SHOW ON EARTH (I,1883); MISSION: IMPOSSIBLE (I,3067)

HILL, THOMAS
GOSSIP (II,1043); V: THE FINAL BATTLE (II,2714); WIZARDS AND WARRIORS (II,2813); YOU ARE THE JURY (II,2854)

HILL, TIFFANY
GENERAL HOSPITAL (II,964)

HILLAIRE, MARCEL
ADVENTURES IN PARADISE (I,46); NINOTCHKA (I,3299); PARIS IN THE SPRINGTIME (I,3457); RAPTURE AT TWO-FORTY (I,3724); THREE FOR TAHITI (I,4476)

HILLARD, SUZANNE
MY SISTER HANK (I,3201)

HILLARY, ANN
OUR FIVE DAUGHTERS (I,3413)

HILLEBRAND, FRED
MARTIN KANE, PRIVATE EYE (I,2928)

HILLER, WENDY
THE LILLI PALMER THEATER (I,2704)

HILLERMAN, JOHN
BATTLES: THE MURDER THAT WOULDN'T DIE (II,180); BEANE'S OF BOSTON (II,194); THE BETTY WHITE SHOW (II,224); ELLERY QUEEN (II,766); ELLERY QUEEN: TOO MANY SUSPECTS (II,767); THE FUNNIEST JOKE I EVER HEARD (II,939); GOSSIP (II,1044); INSTITUTE FOR REVENGE (II,1239); MAGNUM, P.I. (II,1601); ONE DAY AT A TIME (II,1900); THE RETURN OF LUTHER GILLIS (II,2137); TENSPEED AND BROWN SHOE (II,2562)

HILLIAS, PEG
THE DOCTOR (I,1320); THE INNER SANCTUM (I,2216); LITTLE WOMEN (I,2720); OUR TOWN (I,3419)

HILLS, BEVERLY
THE JACK BENNY PROGRAM (I,2294)

HILLS, GILLIAN
CASANOVA (II,451)

HILLS, THOMAS
NEWHART (II,1835)

HILLYER, SHARON
THE MAN FROM U.N.C.L.E. (I,2867)

HILTON, BOB
THE NEW TRUTH OR CONSEQUENCES (II,1832)

HILTON, JASMINA
MISTER JERICO (I,3070)

HILTON, JOY
FREEDOM RINGS (I,1682)

HILTON, KITTY
THE CROSS-WITS (II,605)

HILTON, ROBYN
MRS R.—DEATH AMONG FRIENDS (II,1759)

HIMES, JOHN
A TIME TO LIVE (I,4510)

HINCHMAN, BERT
FARRELL: FOR THE PEOPLE (II,834)

HINCKLEY, AL
DECOY (I,1217)

HINCKLEY, PAUL
MAGIC MONGO (II,1592)

HINDLE, ART
BATTLE OF THE NETWORK STARS (II,164); DALLAS (II,614); KINGSTON: CONFIDENTIAL (II,1398); THE POWER WITHIN (II,2073); THE THREE WIVES OF DAVID WHEELER (II,2611); TUCKER'S WITCH (II,2667)

HINDMAN, EARL
KEY WEST (I,2528); RYAN'S HOPE (II,2234)

HINDS, CINDY
TALES OF THE HAUNTED (II,2538)

HINDY, JOSEPH
THE RETURN OF CHARLIE CHAN (II,2136)

HINES, BARBARA
FAMILY NIGHT WITH HORACE HEIDT (I,1523)

HINES, CONNIE
MR. ED (I,3137); PINE LAKE LODGE (I,3597)

HINES, DAMON
SCAMPS (II,2267)

HINES, FRAZER
DOCTOR WHO (II,689); DOCTOR WHO—THE FIVE DOCTORS (II,690)

HINES, GRANGER
THE CIRCLE FAMILY (II,527)

HINES, GREGORY
100 YEARS OF AMERICA'S POPULAR MUSIC (II,1904); I LOVE LIBERTY (II,1215); PARADE OF STARS (II,1954); SHIRLEY MACLAINE: ILLUSIONS (II,2336)

HINES, JANEAR
JULIA (I,2476); SAM (II,2246)

HINES, JEROME
GOLDEN CHILD (I,1839)

HINES, MIMI
MIMI (I,3054); ON PARADE (I,3356); THE UGILY FAMILY (II,2699)

HINES, SHELLEY
FROSTY'S WINTER WONDERLAND (II,935)

HINGERT, MAUREEN
CAPTAIN DAVID GRIEF (I,829)

HINGLE, PAT
ALL THE WAY HOME (I,140); BUS STOP (II,399); CAROL FOR ANOTHER CHRISTMAS (I,852); THE CITY (I,969); GENERAL ELECTRIC'S ALL-STAR ANNIVERSARY (II,963); THE GLASS MENAGERIE (I,1819); GUNSMOKE (II,1069); STONE (II,2470); STONE (II,2471); SUSPENSE (I,4305); SWEET, SWEET RACHEL (I,4313); TWIGS (II,2683); WILD TIMES (II,2799)

HINKER, DAVID
THREE PASSPORTS TO ADVENTURE (I,4484)

HINKER, HAL
THREE PASSPORTS TO ADVENTURE (I,4484)

HINKER, HILLAS
THREE PASSPORTS TO ADVENTURE (I,4484)

HINKLEY, DEL
LITTLE VIC (II,1496); RIDING FOR THE PONY EXPRESS (II,2172)

HINN, MICHAEL
BOOTS AND SADDLES: THE STORY OF THE FIFTH CAVALRY (I,686); EL COYOTE (I,1423); NUMBER 13 DEMON STREET (I,3316)

HINNANT, BILL
FUNNY PAPERS (I,1713); MASQUERADE (I,2940); PETE AND GLADYS (I,3558); YOU'RE A GOOD MAN, CHARLIE BROWN (I,4962)

HINNANT, SKIP
THE PATTY DUKE SHOW (I,3483)

HINTON, ANN
A TOWN LIKE ALICE (II,2652)

HINTON, DARBY
DANIEL BOONE (I,1142)

HINTON, DAVID
SHAUGHNESSEY (II,2318)

HINTON, ED
THE LONELY WIZARD (I,2743)

HINTON, JAMES DAVID
JESSIE (II,1287)

HINTON, SEAN
A TOWN LIKE ALICE (II,2652)

HINZ, TERRY
SNAFU (II,2386)

HIRSCH, DAVID
MASTER OF THE GAME (II,1647)

HIRSCH, JUDD
THE COMEDY ZONE (II,559); DELVECCHIO (II,659); THE HALLOWEEN THAT ALMOST WASN'T (II,1075); I LOVE LIBERTY (II,1215); LORETTA LYNN IN THE BIG APPLE (II,1520); MARRIAGE IS ALIVE AND WELL (II,1633); THE ROBERT KLEIN SHOW (II,2185); TAXI (II,2546)

HIRSCHFELD, ROBERT
HILL STREET BLUES (II,1154)

HIRSH, ADAM
SENSE OF HUMOR (II,2303)

HIRSON, ALICE
KATE BLISS AND THE TICKER TAPE KID (II,1366); LOOSE CHANGE (II,1518); ONE LIFE TO LIVE (II,1907); SITCOM (II,2367); SOMERSET (I,4115); WHEN THE WHISTLE BLOWS (II,2781); YOUR PLACE OR MINE? (II,2874)

HIRST, BETSY
BEAT THE CLOCK (I,378)

HIRT, AL
THE BOB HOPE SHOW (I,660); FAMILY NIGHT WITH HORACE HEIDT (I,1523); FANFARE (I,1528); MAKE YOUR OWN KIND OF MUSIC (I,2847); RAINBOW OF STARS (I,3717)

HISATAKE, KACK
WAIKIKI (II,2734)

HIT PARADE DANCERS, THE
YOUR HIT PARADE (I,4951)

HIT PARADE SINGERS, THE
YOUR HIT PARADE (I,4951)

HITCHCOCK, ALFRED
ALFRED HITCHCOCK PRESENTS (I,115); HOLLYWOOD: THE SELZNICK YEARS (I,2095)

HITCHCOCK, PATRICIA
ALFRED HITCHCOCK PRESENTS (I,115)

HITCHCOCK, RUSSELL
AIR SUPPLY IN HAWAII (II,26)

HITT, RANDALL
CHIEFS (II,509)

HITT, ROBERT
STEPHANIE (II,2453)

HJELM, DANNY
PLAY IT AGAIN, CHARLIE BROWN (I,3617)

HO HO KIDS, THE
THE RED BUTTONS SHOW (I,3750)

HO, DON
THE BOB HOPE SHOW (I,652); CELEBRATION: THE AMERICAN SPIRIT (II,461); THE DON HO SHOW (II,706); PERRY COMO'S HAWAIIAN HOLIDAY (II,2017); THE TOM JONES SPECIAL (I,4545)

HOAG, MITZI
THE ADVENTURES OF POLLYANNA (II,14); THE DEADLY GAME (II,633); DOCTOR DAN (II,684); THE EYES OF TEXAS (II,799); THE FACTS OF LIFE (II,805); THE GIRL IN THE EMPTY GRAVE (II,997); HERE COME THE BRIDES (I,2024); WE'LL GET BY (II,2753); WE'LL GET BY (II,2754)

HOBART, DEBORAH
AS THE WORLD TURNS (II,110)

HOBART, DON
HIGH AND WILD (I,2049); VAGABOND (I,4691)

HOBBIE, DUKE
THE WACKIEST SHIP IN THE ARMY (I,4744)

HOBBS, HEATHER
CAGNEY AND LACEY (II,409); CASS MALLOY (II,454); CRASH ISLAND (II,600); THE HUSTLER OF MUSCLE BEACH (II,1210); VEGAS (II,2724)

HOBBS, PETER
9 TO 5 (II,1852); BEYOND WITCH MOUNTAIN (II,228); BRIGHT PROMISE (I,727); GLORIA (II,1010); MAGNAVOX THEATER (I,2827); THE MARY TYLER MOORE SHOW (II,1640); PEN 'N' INC. (II,1992); SAINT PETER (II,2238); THE SECRET STORM (I,3969); SHORT, SHORT DRAMA (I,4023); YOUR PLACE OR MINE? (II,2874)

HOCHSTRAATE, LUTZ
WOOBINDA—ANIMAL DOCTOR (II,2832)

HOCKING, GLORIA
CELEBRITY (II,463)

HOCKS, LINDA
CARD SHARKS (II,438)

HOCTOR, DANNY
CLUB EMBASSY (I,983)

HODDINOTT, DIANA
YES MINISTER (II,2848)

HODES, STUART
ANNIE GET YOUR GUN (I,223)

HODGE, AL
CAPTAIN VIDEO AND HIS VIDEO RANGERS (I,838)

HODGE, CHARLES
ELVIS (I,1435)

HODGE, MIKE
GEORGE WASHINGTON (II,978)

HODGE, PATRICIA
QUILLER: NIGHT OF THE FATHER (II,2100)

HODGES, EDDIE
GIVE MY REGARDS TO BROADWAY (I,1815); HOLIDAY U.S.A. (I,2078); THE JIMMY DURANTE SHOW (I,2389); THE ROSALIND RUSSELL SHOW (I,3845)

HODGES, JOHNNY
TIMEX ALL-STAR JAZZ SHOW IV (I,4518)

HODGES, RALPH
SKIP TAYLOR (I,4075)

HODGINS, EARL
GUESTWARD HO! (I,1900)

HODGSON, JEFFREY
ALL THE RIVERS RUN (II,43)

HODIAK, KEITH
DOCTOR WHO—THE FIVE DOCTORS (II,690)

HODSON, JIM
ACCUSED (I,18)

HOEBEKE, DANIELLE
THE SCARLET LETTER (II,2270)

HOEDOWN RANCH SQUARE DANCERS, THE
HAYLOFT HOEDOWN (I,1981)

HOEY, DENNIS
COSMOPOLITAN THEATER (I,1057); HAY FEVER (I,1979); JANE EYRE (I,2337); THE SWAN (I,4312)

HOFF, CHRISTIAN
THE RICHIE RICH SHOW (II,2169)

HOFFMAN, ALFRED
CASANOVA (II,451); THE WOMAN IN WHITE (II,2819)

HOFFMAN, BASIL
DRIBBLE (II,736); MY BUDDY (II,1778); SQUARE PEGS (II,2431)

HOFFMAN, BERN
KISMET (I,2565); MAJOR DELL CONWAY OF THE FLYING TIGERS (I,2835)

HOFFMAN, DUSTIN
BETTE MIDLER—OL' RED HAIR IS BACK (II,222); HIGHER AND HIGHER, ATTORNEYS AT LAW (I,2057); MARLO THOMAS AND FRIENDS IN FREE TO BE. . . YOU AND ME (II,1632)

HOFFMAN, ELIZABETH
THE WINDS OF WAR (II,2807)

HOFFMAN, GERTRUDE
MY LITTLE MARGIE (I,3194)

HOFFMAN, JANE
THE HAMPTONS (II,1076); LOVE OF LIFE (I,2774); LOVE, SEX. . . AND MARRIAGE (II,1546); THE SWAN (I,4312)

HOFFMAN, ROBERT
LITTLE HOUSE ON THE PRAIRIE (II,1487); ROBINSON CRUSOE (I,3819)

HOFFMAN, TOBY
THE GUINNESS GAME (II,1066)

HOFFMAN, WENDY
GILLIGAN'S ISLAND: THE HARLEM GLOBETROTTERS ON GILLIGAN'S ISLAND (II,993); HAPPY DAYS (II,1084); MAKIN' IT (II,1603)

HOFMANN, SONIA
WOOBINDA—ANIMAL DOCTOR (II,2832)

HOFT, SYD
SHORTY (I,4025)

HOGAN, BOB
FBI CODE 98 (I,1550)

HOGAN, BRENDA
THE LILLI PALMER THEATER (I,2704)

HOGAN, GARRICK
OPPENHEIMER (II,1919)

HOGAN, JACK
ADAM-12 (I,31); COMBAT (I,1011); MOBILE TWO (II,1718)

HOGAN, JONATHAN
THE DOCTORS (II,694); FIFTH OF JULY (II,851)

HOGAN, PAT
BRAVE EAGLE (I,709); CASEY JONES (I,865); DAVY CROCKETT (I,1181); NORTHWEST PASSAGE (I,3306); TEXAS JOHN SLAUGHTER (I,4414)

HOGAN, PAUL
THE PAUL HOGAN SHOW (II,1975)

HOGAN, ROBERT
AMANDA FALLON (I,151); CALLING DR. STORM, M.D. (II,415); DAYS OF OUR LIVES (II,629); THE DON RICKLES SHOW (I,1342); FOOLS, FEMALES AND FUN: WHAT ABOUT THAT ONE? (II,901); GENERAL HOSPITAL (II,964); LAVERNE AND SHIRLEY (II,1446); MANHUNTER (II,1620); MANHUNTER (II,1621); A NEW KIND OF FAMILY (II,1819); THE NEW OPERATION PETTICOAT (II,1826); ONCE AN EAGLE (II,1897); RANSOM FOR ALICE (II,2113); RICHIE BROCKELMAN, PRIVATE EYE (II,2167); SECRETS OF MIDLAND HEIGHTS (II,2296); SIERRA (II,2351); SUMMER (II,2491); THREE FOR TAHITI (I,4476); A TIME FOR US (I,4507)

HOGESTYN, DRAKE
SEVEN BRIDES FOR SEVEN BROTHERS (II,2307)

HOGLE, FRANCI
CAMP WILDERNESS (II,419)

HOING, HENRY
BETWEEN THE LINES (II,225)

HOINXT, LINDA
DEBBY BOONE. . . ONE STEP CLOSER (II,655)

HOKANSON, MARY ALAN
MARKHAM (I,2916)

HOLBROOK, ANNA
BENJI, ZAX AND THE ALIEN PRINCE (II,205)

HOLBROOK, DAVID
THE LEGEND OF THE GOLDEN GUN (II,1455)

HOLBROOK, HAL
THE AWAKENING LAND (II,122); THE BRIGHTER DAY (I,728); CELEBRITY (II,463); THE CLIFF DWELLERS (I,974); GEORGE WASHINGTON (II,978); THE GLASS MENAGERIE (I,1819); THE LEGEND OF THE GOLDEN GUN (II,1455); THE OATH: 33 HOURS IN THE LIFE OF GOD (II,1873); OMNIBUS (II,1890); OUR TOWN (I,3421); SANDBURG'S LINCOLN (I,3907); THE SENATOR (I,3976); TRAVIS LOGAN, D.A. (I,4598); THE WHOLE WORLD IS WATCHING (I,4851)

HOLCHAK, VICTOR
DAYS OF OUR LIVES (II,629)

HOLCOMB, KATHRYN
HOW THE WEST WAS WON (II,1196); THE MACAHANS (II,1583); SKAG (II,2374)

HOLCOMB, LAWRENCE
THE DAY THE WOMEN GOT EVEN (II,628)

HOLCOMBE, HARRY
THE BACHELOR (I,325); BAREFOOT IN THE PARK (I,355); FIRST LOVE (I,1583); HENNESSEY (I,2013); ROAD OF LIFE (I,3798); SEARCH FOR TOMORROW (II,2284); WONDERFUL JOHN ACTION (I,4892)

HOLCOMBE, WENDY
THE EDDIE RABBITT SPECIAL (II,759); LEWIS AND CLARK (II,1465); MERRY CHRISTMAS FROM THE GRAND OLE OPRY (II,1686); WENDY HOOPER—U. S. ARMY (II,2764)

HOLDEN, GLORIA
YOUR SHOW TIME (I,4958)

HOLDEN, JAMES
ADVENTURES IN PARADISE (I,46); THE CAMPBELL TELEVISION SOUNDSTAGE (I,812)

HOLDEN, JOYCE
THE DONALD O'CONNOR SHOW (I,1344); THE FABULOUS SYCAMORES (I,1505); TV READERS DIGEST (I,4630)

HOLDEN, MARK
TOP TEN (II,2648)

HOLDEN, REBECCA
AS THE WORLD TURNS (II,110); HOT W.A.C.S. (II,1191); JOHNNY BLUE (II,1321); KNIGHT RIDER (II,1402); VENICE MEDICAL (II,2726)

HOLDEN, ROY
PENMARRIC (II,1993)

HOLDEN, WILLIAM
THE BOB HOPE SHOW (I,536); OPERATION ENTERTAINMENT (I,3400)

HOLDER, CHRISTOPHER
THE YOUNG AND THE RESTLESS (II,2862)

HOLDER, GEOFFREY
ANDROCLES AND THE LION (I,189)

HOLDERNESS, SUSAN
IF IT'S A MAN, HANG UP (I,2186)

HOLDING, BONNIE
JOHNNY RISK (I,2435)

HOLDREN, JUDD
COMMANDO CODY (I,1030)

HOLDRIDGE, CHERYL
ANNETTE (I,222); BACHELOR FATHER (I,323); LEAVE IT TO BEAVER (I,2648); THE MICKEY MOUSE CLUB (I,3025); THE MOUSEKETEERS REUNION (II,1742); THE RIFLEMAN (I,3789)

HOLE, JONATHAN
MCNAB'S LAB (I,2972); PIONEER SPIRIT (I,3604)

HOLICKER, HEIDI
YOUNG LIVES (II,2866)

HOLIDAYS, THE
ARTHUR GODFREY'S TALENT SCOUTS (I,291)

HOLLAND, ANTHONY
AND THEY ALL LIVED HAPPILY EVER AFTER (II,75); THE BEST CHRISTMAS PAGEANT EVER (II,212); COMEDY NEWS (I,1020); COMEDY NEWS II (I,1021); P.O.P. (II,1939); THE PARKERS (II,1959); YOUNG HEARTS (II,2865)

HOLLAND, BETTY LOU
THE DEVIL AND DANIEL WEBSTER (I,1246); JOHNNY BELINDA (I,2416); LOVE STORY (I,2776)

HOLLAND, DENISE
CELEBRITY (II,463)

HOLLAND, ERIK
LITTLE HOUSE: LOOK BACK TO YESTERDAY (II,1490); A RAINY DAY (II,2110)

HOLLAND, JEFF
SPARROW (II,2410)

HOLLAND, JOHN
ADAM'S RIB (I,32)

HOLLAND, KRISTINA
THE COURTSHIP OF EDDIE'S FATHER (I,1083); THE FUNKY PHANTOM (I,1709); WAIT TIL YOUR FATHER GETS HOME (I,4748)

HOLLAND, RANDY
THE YOUNG AND THE RESTLESS (II,2862)

HOLLAND, RICHARD
OPERATION: NEPTUNE (I,3404)

HOLLAND, STEVE
FLASH GORDON (I,1593)

HOLLAND, TOM
THE NUT HOUSE (I,3320)

HOLLAND, TOM LEE
THIS IS KATE BENNETT (II,2597)

HOLLANDER, DAVID
CALL TO GLORY (II,413); DEAR TEACHER (II,653); FIRST TIME, SECOND TIME (II,863); LEWIS AND CLARK (II,1465); THE MCLEAN STEVENSON SHOW (II,1665); NERO WOLFE (II,1806); A NEW KIND OF FAMILY (II,1819)

HOLLAR, JANE
THE ALAN YOUNG SHOW (I,104)

HOLLAR, LLOYD
RAGE OF ANGELS (II,2107)

HOLLEN, REBECCA
THE GUIDING LIGHT (II,1064)

HOLLEY, BERNARD
COME OUT, COME OUT, WHEREVER YOU ARE (I,1014)

HOLLIDAY, ART
THE WHITE SHADOW (II,2788)

HOLLIDAY, DAVID
THUNDERBIRDS (I,4497)

HOLLIDAY, FRED
EGAN (I,1419); GALACTICA 1980 (II,953); THE GIRL IN MY LIFE (II,996); HARPER VALLEY PTA (II,1095); MEDICAL CENTER (II,1675); ROGER AND HARRY (II,2200); THIS IS KATE BENNETT (II,2597)

HOLLIDAY, HOPE
BEST FOOT FORWARD (I,401)

HOLLIDAY, JENNIFER
LORETTA LYNN IN THE BIG APPLE (II,1520)

HOLLIDAY, JUDY
ENTERTAINMENT —1955 (I,1450); FANFARE (I,1526); GOOD TIMES (I,1853); KALEIDOSCOPE (I,2504); THE MARCH OF DIMES FASHION SHOW (I,2901); SUNDAY IN TOWN (I,4285)

HOLLIDAY, KENE
BATTLE OF THE NETWORK STARS (II,166); BENSON (II,208); THE BEST OF TIMES (II,220); CARTER COUNTRY (II,449); THE CHICAGO

STORY (II,507); MOMMA THE DETECTIVE (II,1721); THE SHERIFF AND THE ASTRONAUT (II,2326); SOAP (II,2392)

HOLLIDAY, POLLY
ALICE (II,33); ALICE (II,34); ALL THE WAY HOME (II,44); FLO (II,891); PRIVATE BENJAMIN (II,2087); YOU CAN'T TAKE IT WITH YOU (II,2857)

HOLLIMAN, EARL
ALIAS SMITH AND JONES (I,118); ALIAS SMITH AND JONES (I,119); CANNON (I,820); CIRCUS OF THE STARS (II,531); HOTEL DE PAREE (I,2129); THE LADY DIED AT MIDNIGHT (I,2601); POLICE WOMAN (II,2063); SECOND CHANCE (I,3957); THE SIX MILLION DOLLAR MAN (I,4067); THE THORN BIRDS (II,2600); THE TWILIGHT ZONE (I,4651); WIDE COUNTRY (I,4858)

HOLLINGSHEAD, LEE
MISS STEWART, SIR (I,3064)

HOLLINGSWORTH, JOEY
HALF THE GEORGE KIRBY COMEDY HOUR (I,1920)

HOLLIS, JEFF
C.P.O. SHARKEY (II,407); JACKIE AND DARLENE (II,1265)

HOLLISTER, LOUIS
FOLLOW YOUR HEART (I,1618)

HOLLORAN, JACK
FARAWAY HILL (I,1532)

HOLLORAN, SUSAN
NORBY (I,3304)

HOLLOWAY, ANN
FATHER DEAR FATHER (II,837)

HOLLOWAY, ANTHEA
TYCOON: THE STORY OF A WOMAN (II,2697)

HOLLOWAY, FREDA
YOUNG DR. MALONE (I,4938)

HOLLOWAY, JOAN
THE COLLEGE BOWL (I,1001); RUGGLES OF RED GAP (I,3867)

HOLLOWAY, PATRICIA
JOSIE AND THE PUSSYCATS (I,2453); JOSIE AND THE PUSSYCATS IN OUTER SPACE (I,2454)

HOLLOWAY, STANLEY
THE BROADWAY OF LERNER AND LOEWE (I,735); DR JEKYLL AND MR. HYDE (I,1368); THE FANTASTICKS (I,1531); THE JO STAFFORD SHOW (I,2394); LINUS THE LIONHEARTED (I,2709); THE MAURICE CHEVALIER SHOW (I,2954); OUR MAN HIGGINS (I,3414); THE PERRY COMO SPECIAL (I,3535); STORY THEATER (I,4240)

HOLLOWAY, STERLING
THE ADVENTURES OF RIN TIN TIN (I,73); THE BAILEYS OF BALBOA (I,334); THE LIFE OF RILEY (I,2686); SHIRLEY TEMPLE'S STORYBOOK (I,4017); TONY THE PONY (II,2641)

HOLLOWAY, VALERIE
A HORSEMAN RIDING BY (II,1182)

HOLLOWAY, WENDY
A HORSEMAN RIDING BY (II,1182)

HOLLOWELL, TODD
THE JERK, TOO (II,1282)

HOLLY, DENNIS
CELEBRITY (II,463)

HOLLY, ELLEN
ONE LIFE TO LIVE (II,1907)

HOLM, CELESTE
ALCOA PREMIERE (I,109); ARCHIE BUNKER'S PLACE (II,105); BACKSTAIRS AT THE WHITE HOUSE (II,133); CAPTAINS AND THE KINGS (II,435); CAROLYN (I,854); CINDERELLA (I,956); CLIMAX! (I,976); THE DELPHI BUREAU (I,1224); THE DELPHI BUREAU (I,1225); FUN FAIR (I,1706); HOLLYWOOD OPENING NIGHT (I,2087); HONESTLY, CELESTE! (I,2104); JACK AND THE BEANSTALK (I,2287); JESSIE (II,1287); KILROY (I,2541); THE LOVE BOAT II (II,1533); THE LUX VIDEO THEATER (I,2795); THE MAN IN THE DOG SUIT (I,2869); THE MARCH OF DIMES FASHION SHOW (I,2902); MEET ME IN ST. LOUIS (I,2987); NANCY (I,3221); PERSPECTIVE ON GREATNESS (I,3555); THE RIGHT MAN (I,3790); SNEAK PREVIEW (I,4100); SWING OUT, SWEET LAND (II,2515); THIS GIRL FOR HIRE (II,2596); THE UNITED STATES STEEL HOUR (I,4677); THE YEOMAN OF THE GUARD (I,4926); YOUR JEWELER'S SHOWCASE (I,4953); ZANE GREY THEATER (I,4979)

HOLM, JOHN CECIL
JOHNNY BELINDA (I,2416)

HOLM, SHARON
MASTER OF THE GAME (II,1647)

HOLMAN, MICHAEL
GRAFFITI ROCK (II,1049)

HOLMAN, REX
THE WILD WOMEN OF CHASTITY GULCH (II,2801)

HOLMES, BILL
MIDWESTERN HAYRIDE (I,3033)

HOLMES, DENNIS
LARAMIE (I,2614); ROMEO AND JULIET (I,3837)

HOLMES, ED
DECOY (I,1217); GROWING PAYNES (I,1891); ONCE UPON A TUNE (I,3372); THE S. S. HOLIDAY (I,4173)

HOLMES, FRANK
THE SECRET WAR OF JACKIE'S GIRLS (II,2294)

HOLMES, GARY
1915 (II,1853)

HOLMES, JENNIFER
THE ASPHALT COWBOY (II,111); THE FALL GUY (II,811); NEWHART (II,1835)

HOLMES, JOHN
HOTEL COSMOPOLITAN (I,2128)

HOLMES, LARRY
BOB HOPE SPECIAL: BOB HOPE'S ALL-STAR BIRTHDAY AT ANNAPOLIS (II,294)

HOLMES, LOIS
CONCERNING MISS MARLOWE (I,1032)

HOLMES, LYNNE
THE ORPHAN AND THE DUDE (II,1924)

HOLMES, PEGGY
JUMP (II,1356); SUMMER (II,2491)

HOLMES, PHYLLIS
MIDWESTERN HAYRIDE (I,3033)

HOLMES, SCOTT
RYAN'S HOPE (II,2234)

HOLMES, SIMON
MASTER OF THE GAME (II,1647)

HOLMES, TONY
BABY, I'M BACK! (II,130)

HOLNESS, JEAN
CASANOVA (II,451)

HOLT, BOB
THE BARKLEYS (I,356); CURIOSITY SHOP (I,1114); DENNIS THE MENACE: MAYDAY FOR MOTHER (II,660); DOCTOR DOLITTLE (I,1309) FLASH GORDON—THE GREATEST ADVENTURE OF ALL (II,875); THE GREAT GRAPE APE SHOW (II,1056); THE NEW ZOO REVUE (I,3277); THE PRIMARY ENGLISH CLASS (II,2081); SCOOBY'S ALL-STAR LAFF-A-LYMPICS (II,2273); THE SCOOBY-DOO AND SCRAPPY-DOO SHOW (II,2275); THE SMURFS (II,2385)

HOLT, CHARLENE
WONDER WOMAN (PILOT 1) (II,2826)

HOLT, JACQUELINE
STAR TONIGHT (I,4199)

HOLT, JEFF
SCREAMER (I,3945)

HOLT, JENNIFER
UNCLE MISTLETOE AND HIS ADVENTURES (I,4669)

HOLT, PATRICK
FATHER DEAR FATHER (II,837)

HOLTZMAN, GINI
SNOOPY'S GETTING MARRIED, CHARLIE BROWN (I,4102)

HOLZMAN, ROBIN
THE DOCTORS (II,694)

HOME, SALLY
THE TALE OF BEATRIX POTTER (II,2533)

HOMEIER, SKIP
DAN RAVEN (I,1131); ELFEGO BACA (I,1427); THE INTERNS (I,2226); ONE STEP BEYOND (I,3388); SCIENCE FICTION THEATER (I,3938); WASHINGTON: BEHIND CLOSED DOORS (II,2744)

HOMER, JIMMY
THE HUMAN COMEDY (I,2158)

HOMETOWN SINGERS, THE
SWINGING COUNTRY (I,4322)

HOMETOWNERS, THE
MIDWESTERN HAYRIDE (I,3033)

HOMOLKA, OSCAR
THE MOTOROLA TELEVISION HOUR (I,3112); ONE OF OUR OWN (II,1910); SPELLBOUND (I,4153); THE

STRANGE CASE OF DR. JEKYLL AND MR. HYDE (I,4244); VICTORY (I,4729)

HON, JEAN MARIE
ARK II (II,107); DANGER IN PARADISE (II,617); THE MAN FROM ATLANTIS (II,1615)

HONEYDREAMERS, THE
ALL IN FUN (I,130); FIVE STARS IN SPRINGTIME (I,1590); KAY KYSER'S KOLLEGE OF MUSICAL KNOWLEDGE (I,2512); SUMMERTIME U.S.A. (I,4283); U.S. TREASURY SALUTES (I,4689)

HONG, DICK KAY
JOSEPHINE LITTLE: THE MIRACULOUS JOURNEY OFTADPOLE CHAN (I,2452)

HONG, JAMES
BROTHERS (II,381); CANNON: THE RETURN OF FRANK CANNON (II,425); THE HUSTLER OF MUSCLE BEACH (II,1210); INSPECTOR PEREZ (II,1237); JUDGE DEE IN THE MONASTERY MURDERS (I,2465); MARCO POLO (II,1626); MICKEY SPILLANE'S MIKE HAMMER (II,1692); THE NEW ADVENTURES OF CHARLIE CHAN (I,3249); SWITCH (II,2519); WINNER TAKE ALL (II,2808)

HONG, PEARL
THE RETURN OF CHARLIE CHAN (II,2136)

HONIG, HOWARD
MR. MOM (II,1752)

HOOD, DARLA JEAN
THE KEN MURRAY SHOW (I,2525)

HOOD, DON
DEAD MAN ON THE RUN (II,631); THE SHERIFF AND THE ASTRONAUT (II,2326)

HOOD, MORAG
WAR AND PEACE (I,4765)

HOOD, NOEL
FROM A BIRD'S EYE VIEW (I,1686)

HOOKER, JOHN LEE
FOLK SOUND U.S.A. (I,1614)

HOOKER, TRISHA
I'M THE GIRL HE WANTS TO KILL (I,2194)

HOOKS, DAVID
DOCTOR STRANGE (II,688); THE EDGE OF NIGHT (II,760); MANDRAKE (II,1617)

HOOKS, ED
FLAMINGO ROAD (II,872)

HOOKS, ERIC
DOWN HOME (II,731)

HOOKS, JAN
THE HALF-HOUR COMEDY HOUR (II,1074); THE JOE PISCOPO SPECIAL (II,1304); PRIME TIMES (II,2083)

HOOKS, KEVIN
BACKSTAIRS AT THE WHITE HOUSE (II,133); CELEBRITY CHALLENGE OF THE SEXES 5 (II,469); DOWN HOME (II,731); FOR MEMBERS ONLY (II,905); THE WHITE SHADOW (II,2788)

HOOKS, ROBERT
BACKSTAIRS AT THE WHITE HOUSE (II,133); THE CLIFF DWELLERS (I,974); DOWN HOME (II,731); THE FACTS OF LIFE (II,805); FEEL THE HEAT (II,848); THE KILLER WHO WOULDN'T DIE (II,1393); N.Y.P.D. (I,3321); TWO FOR THE MONEY (I,4655)

HOOPAI, JAKE
THE RETURN OF LUTHER GILLIS (II,2137)

HOOPER, LARRY
THE LAWRENCE WELK SHOW (I,2643); THE LAWRENCE WELK SHOW (I,2644); LAWRENCE WELK'S TOP TUNES AND NEW TALENT (I,2645); MEMORIES WITH LAWRENCE WELK (II,1681)

HOOTEN, PETER
DOCTOR STRANGE (II,688); ONE OF OUR OWN (II,1910)

HOOTKINS, WILLIAM
OPPENHEIMER (II,1919)

HOOVER, HAL
DOCTOR MIKE (I,1316)

HOOVER, JOE
HARRY O (II,1098)

HOOVER, ROBERT
ANOTHER WORLD (II,97)

HOPE, BARRY
THE KID WITH THE BROKEN HALO (II,1387); SHE'S WITH ME (II,2321)

HOPE, BOB
. . . AND DEBBIE MAKES SIX (I,185); THE ALL-STAR REVUE (I,138); THE ANITA BRYANT SPECTACULAR (II,82); THE ANN-MARGRET SHOW (I,218); ANN-MARGRET. . . RHINESTONE COWGIRL (II,87); ANN-MARGRET—WHEN YOU'RE SMILING (I,219); THE ARTHUR MURRAY PARTY FOR BOB HOPE (I,292); BING CROSBY AND HIS FRIENDS (I,456); BING CROSBY AND HIS FRIENDS (II,251); THE BING CROSBY SHOW (I,466); THE BING CROSBY SHOW (I,468); THE BING CROSBY SHOW (I,469); THE BING CROSBY SPRINGTIME SPECIAL (I,473); BING!. . . A 50TH ANNIVERSARY GALA (II,254); THE BOB GOULET SHOW STARRING ROBERT GOULET (I,501); THE BOB HOPE CHRYSLER THEATER (I,502); THE BOB HOPE SHOW (I,504); THE BOB HOPE SHOW (I,505); THE BOB HOPE SHOW (I,506); THE BOB HOPE SHOW (I,507); THE BOB HOPE SHOW (I,508); THE BOB HOPE SHOW (I,509); THE BOB HOPE SHOW (I,510); THE BOB HOPE SHOW (I,511); THE BOB HOPE SHOW (I,512); THE BOB HOPE SHOW (I,513); THE BOB HOPE SHOW (I,514); THE BOB HOPE SHOW (I,515); THE BOB HOPE SHOW (I,516); THE BOB HOPE SHOW (I,517); THE BOB HOPE SHOW (I,518); THE BOB HOPE SHOW (I,519); THE BOB HOPE SHOW (I,520); THE BOB HOPE SHOW (I,521); THE BOB HOPE SHOW (I,522); THE BOB HOPE SHOW (I,523); THE BOB HOPE SHOW (I,524); THE BOB HOPE SHOW (I,525); THE BOB HOPE SHOW (I,526); THE BOB HOPE SHOW (I,527); THE BOB HOPE SHOW (I,528); THE BOB HOPE SHOW (I,529); THE BOB HOPE SHOW (I,530); THE BOB HOPE SHOW (I,531); THE BOB HOPE SHOW (I,532); THE BOB HOPE SHOW (I,533); THE BOB HOPE SHOW (I,534); THE BOB HOPE SHOW (I,535); THE BOB HOPE SHOW (I,536); THE BOB HOPE SHOW (I,537); THE BOB HOPE SHOW (I,538); THE BOB HOPE SHOW (I,539); THE BOB HOPE SHOW (I,540); THE BOB HOPE SHOW (I,541); THE BOB HOPE SHOW (I,542); THE BOB HOPE SHOW (I,543); THE BOB HOPE SHOW (I,544); THE BOB HOPE SHOW (I,545); THE BOB HOPE SHOW (I,546); THE BOB HOPE SHOW (I,547); THE BOB HOPE SHOW (I,548); THE BOB HOPE SHOW (I,549); THE BOB HOPE SHOW (I,550); THE BOB HOPE SHOW (I,551); THE BOB HOPE SHOW (I,552); THE BOB HOPE SHOW (I,553); THE BOB HOPE SHOW (I,554); THE BOB HOPE SHOW (I,555); THE BOB HOPE SHOW (I,556); THE BOB HOPE SHOW (I,557); THE BOB HOPE SHOW (I,558); THE BOB HOPE SHOW (I,559); THE BOB HOPE SHOW (I,560); THE BOB HOPE SHOW (I,561); THE BOB HOPE SHOW (I,562); THE BOB HOPE SHOW (I,563); THE BOB HOPE SHOW (I,564); THE BOB HOPE SHOW (I,565); THE BOB HOPE SHOW (I,566); THE BOB HOPE SHOW (I,567); THE BOB HOPE SHOW (I,568); THE BOB HOPE SHOW (I,569); THE BOB HOPE SHOW (I,570); THE BOB HOPE SHOW (I,571); THE BOB HOPE SHOW (I,572); THE BOB HOPE SHOW (I,573); THE BOB HOPE SHOW (I,574); THE BOB HOPE SHOW (I,575); THE BOB HOPE SHOW (I,576); THE BOB HOPE SHOW (I,577); THE BOB HOPE SHOW (I,578); THE BOB HOPE SHOW (I,579); THE BOB HOPE SHOW (I,580); THE BOB HOPE SHOW (I,581); THE BOB HOPE SHOW (I,582); THE BOB HOPE SHOW (I,583); THE BOB HOPE SHOW (I,584); THE BOB HOPE SHOW (I,585); THE BOB HOPE SHOW (I,586); THE BOB HOPE SHOW (I,587); THE BOB HOPE SHOW (I,588); THE BOB HOPE SHOW (I,589); THE BOB HOPE SHOW (I,590); THE BOB HOPE SHOW (I,591); THE BOB HOPE SHOW (I,592); THE BOB HOPE SHOW (I,593); THE BOB HOPE SHOW (I,594); THE BOB HOPE SHOW (I,595); THE BOB HOPE SHOW (I,596); THE BOB HOPE SHOW (I,597); THE BOB HOPE SHOW (I,598); THE BOB HOPE SHOW (I,599); THE BOB HOPE SHOW (I,601); THE BOB HOPE SHOW (I,602); THE BOB HOPE SHOW (I,603); THE BOB HOPE SHOW (I,604); THE BOB HOPE SHOW (I,606); THE BOB HOPE SHOW (I,607); THE BOB HOPE SHOW (I,609); THE BOB HOPE SHOW (I,610); THE BOB HOPE SHOW (I,611); THE BOB HOPE SHOW (I,612); THE BOB HOPE SHOW (I,613); THE BOB HOPE SHOW (I,614); THE BOB HOPE SHOW (I,615); THE BOB HOPE SHOW (I,616); THE BOB HOPE SHOW (I,617); THE BOB HOPE SHOW (I,619); THE BOB HOPE SHOW (I,620); THE BOB HOPE SHOW (I,621); THE BOB HOPE SHOW (I,622); THE BOB HOPE SHOW (I,623); THE BOB HOPE SHOW (I,625); THE BOB HOPE SHOW (I,626); THE BOB HOPE SHOW (I,627); THE BOB HOPE SHOW (I,628); THE BOB HOPE SHOW (I,629); THE BOB HOPE SHOW (I,630); THE BOB HOPE SHOW (I,631); THE BOB HOPE SHOW (I,632); THE BOB HOPE SHOW (I,633); THE BOB HOPE SHOW (I,634); THE BOB HOPE SHOW (I,635); THE BOB HOPE SHOW (I,636); THE BOB HOPE SHOW (I,637); THE BOB HOPE SHOW (I,638); THE BOB HOPE SHOW (I,639); THE BOB HOPE SHOW (I,640); THE BOB HOPE SHOW (I,641); THE BOB HOPE SHOW (I,642); THE BOB HOPE SHOW

(I,643); THE BOB HOPE SHOW (I,644); THE BOB HOPE SHOW (I,645); THE BOB HOPE SHOW (I,646); THE BOB HOPE SHOW (I,647); THE BOB HOPE SHOW (I,648); THE BOB HOPE SHOW (I,649); THE BOB HOPE SHOW (I,650); THE BOB HOPE SHOW (I,651); THE BOB HOPE SHOW (I,652); THE BOB HOPE SHOW (I,653); THE BOB HOPE SHOW (I,654); THE BOB HOPE SHOW (I,655); THE BOB HOPE SHOW (I,656); THE BOB HOPE SHOW (I,657); THE BOB HOPE SHOW (I,658); THE BOB HOPE SHOW (I,659); THE BOB HOPE SHOW (I,660); THE BOB HOPE SHOW (I,661); THE BOB HOPE SHOW (I,662); THE BOB HOPE SHOW (I,663); THE BOB HOPE SHOW (I,664); BOB HOPE SPECIAL: A QUARTER CENTURY OF BOB HOPE ON TELEVISION (II,281); BOB HOPE SPECIAL: BOB HOPE FOR PRESIDENT (II,282); BOB HOPE SPECIAL: BOB HOPE GOES TO COLLEGE (II,283); BOB HOPE SPECIAL: BOB HOPE IN "JOYS" (II,284); BOB HOPE SPECIAL: BOB HOPE IN THE STAR-MAKERS (II,285); BOB HOPE SPECIAL: BOB HOPE IN WHO MAKES THE WORLD LAUGH, PART 2 (II,286); BOB HOPE SPECIAL: BOB HOPE IN WHO MAKES THE WORLD LAUGH? (II,287); BOB HOPE SPECIAL: BOB HOPE LAUGHS WITH THE MOVIE AWARDS (II,288); BOB HOPE SPECIAL: BOB HOPE ON CAMPUS (II,289); BOB HOPE SPECIAL: BOB HOPE ON CAMPUS (II,290); BOB HOPE SPECIAL: BOB HOPE ON THE ROAD TO CHINA (II,291); BOB HOPE SPECIAL: BOB HOPE PRESENTS THE STARS OF TOMORROW (II,292); BOB HOPE SPECIAL: BOB HOPE'S 30TH ANNIVERSARY TV SPECIAL (II,293); BOB HOPE SPECIAL: BOB HOPE'S ALL-STAR BIRTHDAY AT ANNAPOLIS (II,294); BOB HOPE SPECIAL: BOB HOPE'S ALL-STAR BIRTHDAY PARTY (II,295); BOB HOPE SPECIAL: BOB HOPE'S ALL-STAR CHRISTMAS SHOW (II,296); BOB HOPE SPECIAL: BOB HOPE'S ALL-STAR COMEDY BIRTHDAY PARTY (II,297); BOB HOPE SPECIAL: BOB HOPE'S ALL-STAR COMEDY BIRTHDAY PARTY AT WEST POINT (II,298); BOB HOPE SPECIAL: BOB HOPE'S ALL-STAR COMEDY LOOK AT THE FALL SEASON: IT'S STILL FREE AND WORTH IT! (II,299); BOB HOPE SPECIAL: BOB HOPE'S ALL-STAR COMEDY SPECIAL FROM AUSTRALIA (II,300); BOB HOPE SPECIAL: BOB HOPE'S ALL-STAR COMEDY SPECTACULAR FROM LAKE TAHOE (II,301); BOB HOPE SPECIAL: BOB HOPE'S ALL-STAR COMEDY TRIBUTE TO VAUDEVILLE (II,302); BOB HOPE SPECIAL: BOB HOPE'S ALL-STAR LOOK AT TV'S PRIME TIME WARS (II,303); BOB HOPE SPECIAL: BOB HOPE'S ALL-STAR SUPER BOWL PARTY (II,304); BOB HOPE SPECIAL: BOB HOPE'S ALL-STAR TRIBUTE TO THE PALACE THEATER (II,305); BOB HOPE SPECIAL: BOB HOPE'S BICENTENNIAL STAR SPANGLED SPECTACULAR (II,306); BOB HOPE SPECIAL: BOB HOPE'S CHRISTMAS PARTY (II,307); BOB HOPE SPECIAL: BOB HOPE'S CHRISTMAS SPECIAL (II,308); BOB HOPE SPECIAL: BOB HOPE'S FUNNY VALENTINE (II,309); BOB HOPE SPECIAL: BOB HOPE'S MERRY CHRISTMAS SHOW (II,310); BOB HOPE SPECIAL: BOB HOPE'S OVERSEAS CHRISTMAS TOURS (II,311); BOB HOPE SPECIAL: BOB HOPE'S PINK PANTHER THANKSGIVING GALA (II,312); BOB HOPE SPECIAL: BOB HOPE'S ROAD TO HOLLYWOOD (II,313); BOB HOPE SPECIAL: BOB HOPE'S SALUTE TO NASA—25 YEARS OF REACHING FOR THE STARS (II,314); BOB HOPE SPECIAL: BOB HOPE'S SPRING FLING OF COMEDY AND GLAMOUR (II,315); BOB HOPE SPECIAL: BOB HOPE'S STAND UP AND CHEER FOR THE NATIONAL FOOTBALL LEAGUE'S 60TH YEAR (II,316); BOB HOPE SPECIAL: BOB HOPE'S STARS OVER TEXAS (II,318); BOB HOPE SPECIAL: BOB HOPE'S STAR-STUDDED SPOOF OF THE NEW TV SEASON— G RATED—WITH GLAMOUR, GLITTER & GAGS (II,317); BOB HOPE SPECIAL: BOB HOPE'S SUPER BIRTHDAY SPECIAL (II,319); BOB HOPE SPECIAL: BOB HOPE'S USO CHRISTMAS IN BEIRUT (II,320); BOB HOPE SPECIAL: BOB HOPE'S WICKI-WACKY SPECIAL FROM WAIKIKI (II,321); BOB HOPE SPECIAL: BOB HOPE'S WOMEN I LOVE—BEAUTIFUL BUT FUNNY (II,322); BOB HOPE SPECIAL: BOB HOPE'S WORLD OF COMEDY (II,323); BOB HOPE SPECIAL: BOB HOPE—HOPE, WOMEN AND SONG (II,324); BOB HOPE SPECIAL: BOB HOPE PRESENTS A CELEBRATION WITH STARS OF COMEDY AND MUSIC (II,340); BOB HOPE SPECIAL: HAPPY BIRTHDAY, BOB! (II,325); BOB HOPE SPECIAL: HAPPY BIRTHDAY, BOB! (II,326); BOB HOPE SPECIAL: HO HO HOPE'S JOLLY CHRISTMAS HOUR (II,327); BOB HOPE SPECIAL: THE BOB HOPE CHRISTMAS SPECIAL (II,328); BOB HOPE SPECIAL: THE BOB HOPE CHRISTMAS SPECIAL (II,329); BOB HOPE SPECIAL: THE BOB HOPE CHRISTMAS SPECIAL (II,330); BOB HOPE SPECIAL: THE BOB HOPE CHRISTMAS SPECIAL (II,331); BOB HOPE SPECIAL: THE BOB HOPE COMEDY SPECIAL (II,332); BOB HOPE SPECIAL: THE BOB HOPE SPECIAL (II,333); BOB HOPE SPECIAL: THE BOB HOPE SPECIAL (II,334); BOB HOPE SPECIAL: THE BOB HOPE SPECIAL (II,335); BOB HOPE SPECIAL: THE BOB HOPE SPECIAL (II,336); BOB HOPE SPECIAL: THE BOB HOPE SPECIAL (II,337); BOB HOPE SPECIAL: THE BOB HOPE SPECIAL (II,338); BOB HOPE SPECIAL: THE BOB HOPE SPECIAL FROM PALM SPRINGS (II,339); BOB HOPE SPECIAL: BOB HOPE'S HILARIOUS UNREHEARSED ANTICS OF THE STARS (II,341); THE COLGATE COMEDY HOUR (I,997); THE COLGATE COMEDY HOUR (I,998); DANNY THOMAS LOOKS AT YESTERDAY, TODAY AND TOMORROW (I,1148); DATELINE (I,1165); THE DEAN MARTIN CELEBRITY ROAST (II,639); DEBBY BOONE. . . THE SAME OLD BRAND NEW ME (II,656); DISNEY WORLD —A GALA OPENING—DISNEYLAND EAST (I,1292); THE DONNY AND MARIE OSMOND SHOW (II,714); ENTERTAINMENT —1955 (I,1450); THE FIRST 50 YEARS (II,862); FOR LOVE OR $$$ (I,1623); FRANCES LANGFORD PRESENTS (I,1656); GENERAL ELECTRIC'S ALL-STAR ANNIVERSARY (II,963); GEORGE BURNS CELEBRATES 80 YEARS IN SHOW BUSINESS (II,967); GEORGE BURNS EARLY, EARLY, EARLY, CHRISTMAS SHOW (II,968); GEORGE BURNS' 100TH BIRTHDAY PARTY (II,971); GEORGE BURNS' HOW TO LIVE TO BE 100 (II,972); THE GREAT MYSTERIES OF HOLLYWOOD (II,1058); HAVE GIRLS—WILL TRAVEL (I,1967); HER SCHOOL FOR BACHELORS (I,2018); THE HOUSE NEXT DOOR (I,2139); HOW TO HANDLE A WOMAN (I,2146); THE JACK BENNY HOUR (I,2293); THE JACK BENNY SPECIAL (I,2295); JACK BENNY'S 20TH ANNIVERSARY TV SPECIAL (I,2303); JACK BENNY'S FIRST FAREWELL SHOW (I,2300); THE JO STAFFORD SHOW (I,2393); THE KRAFT 75TH ANNIVERSARY SPECIAL (II,1409); A LOVE LETTER TO JACK BENNY (II,1540); THE LUCILLE BALL COMEDY HOUR (I,2783); LUCY MOVES TO NBC (II,1562); MERRY CHRISTMAS FROM THE CROSBYS (II,1685); MITZI AND A HUNDRED GUYS (II,1710); MURDER AT NBC (A BOB HOPE SPECIAL) (I,3159); OPERATION ENTERTAINMENT (I,3400); THE OSMOND BROTHERS SPECIAL (II,1925); PERRY COMO'S LAKE TAHOE HOLIDAY (II,2018); PLAY IT AGAIN, UNCLE SAM (II,2052); PLIMPTON! DID YOU HEAR THE ONE ABOUT. . .? (I,3629); RAQUEL (I,3725); ROBERTA (I,3815); ROBERTA (I,3816); THE ROYAL FOLLIES OF 1933 (I,3862); RUSSIAN ROULETTE (I,3875); A SALUTE TO TELEVISION'S 25TH ANNIVERSARY (I,3893); SCREEN DIRECTOR'S PLAYHOUSE (I,3946); SHOOT-IN AT NBC (A BOB HOPE SPECIAL) (I,4021); SHOW BUSINESS SALUTE TO MILTON BERLE (I,4029); SOUND OFF TIME (I,4135); STARS AND STRIPES SHOW (II,2442); THE STARS AND STRIPES SHOW (I,4206); THE STARS AND STRIPES SHOW (I,4207); THE STARS AND STRIPES SHOW (I,4208); STEVE LAWRENCE AND EYDIE GORME FROM THIS MOMENT ON. . . COLE PORTER (II,2459); SWING OUT, SWEET LAND (II,2515); A TRIBUTE TO "MR. TELEVISION" MILTON BERLE (II,2658)

HOPE, DOLORES

THE BOB HOPE SHOW (I,658); BOB HOPE SPECIAL: BOB HOPE'S ALL-STAR COMEDY BIRTHDAY PARTY AT WEST POINT (II,298); BOB HOPE SPECIAL: BOB HOPE'S SUPER BIRTHDAY SPECIAL (II,319); BOB HOPE SPECIAL: HAPPY BIRTHDAY, BOB!

(II,325)

HOPE, ERICA
THE YOUNG AND THE RESTLESS (II,2862)

HOPE, JACK
THE BOB HOPE SHOW (I,585)

HOPE, LINDA
HER SCHOOL FOR BACHELORS (I,2018)

HOPE, RICHARD
BRIDESHEAD REVISITED (II,375)

HOPE, WILLIAM
MASTER OF THE GAME (II,1647); NANCY ASTOR (II,1788)

HOPES, ISABELLE
WONDERFUL TOWN (I,4893)

HOPKINS, ANTHONY
ALL CREATURES GREAT AND SMALL (I,129); WAR AND PEACE (I,4765)

HOPKINS, BO
THE BUSTERS (II,401); CASINO (II,452); CAT BALLOU (I,869); THE CELEBRITY FOOTBALL CLASSIC (II,474); CHARLIE'S ANGELS (II,487); CRISIS IN SUN VALLEY (II,604); DOC ELLIOT (I,1305); DOC ELLIOT (I,1306); DYNASTY (II,746); THE ROCKFORD FILES (II,2197); THE RUNAWAY BARGE (II,2230); WOMAN ON THE RUN (II,2821)

HOPKINS, HOWARD
BARRIER REEF (I,362)

HOPKINS, KIM
TV'S BLOOPERS AND PRACTICAL JOKES (II,2675)

HOPKINS, LINDA
MITZI. . . ROARIN' IN THE 20S (II,1711)

HOPKINS, MIRIAM
CURTAIN CALL THEATER (I,1115); THE LUX VIDEO THEATER (I,2795)

HOPKINS, SHIRLEY KNIGHT
CIRCLE OF FEAR (I,958); THE COUNTRY GIRL (I,1064)

HOPKINS, TELMA
BATTLE OF THE NETWORK STARS (II,173); BATTLE OF THE NETWORK STARS (II,174); BOSOM BUDDIES (II,355); GIMME A BREAK (II,995); THE KID WITH THE BROKEN HALO (II,1387); MARIE (II,1628); A NEW KIND OF FAMILY (II,1819); THE NEW ODD COUPLE (II,1825); TONY ORLANDO AND DAWN (II,2639)

HOPLIN, MERT
MARLO AND THE MAGIC MOVIE MACHINE (II,1631)

HOPPER, DENNIS
THE KAISER ALUMINUM HOUR (I,2503); THE SHARPSHOOTER (I,4001); SWISS FAMILY ROBINSON (I,4324); WILD TIMES (II,2799)

HOPPER, HAL
THE ADVENTURES OF RIN TIN TIN (I,73)

HOPPER, HEDDA
THE BOB HOPE SHOW (I,536); THE BOB HOPE SHOW (I,552); THE BOB HOPE SHOW (I,559); THE BOB HOPE SHOW (I,566); THE DESILU REVUE (I,1238); HEDDA HOPPER'S HOLLYWOOD (I,2002)

HOPPER, WILLIAM
PERRY MASON (II,2025); THE RESTLESS GUN (I,3774)

HORAN, BARBARA
B.J. AND THE BEAR (II,126); FANTASY ISLAND (II,829); WOMEN WHO RATE A "10" (II,2825)

HORAN, BOBBY
THE HERO (I,2035)

HORAN, HILDY
THE MUNSTERS' REVENGE (II,1765)

HORDEN, MICHAEL
PADDINGTON BEAR (II,1941); SHOGUN (II,2339)

HORDERN, MICHAEL
LAMP AT MIDNIGHT (I,2608); MACBETH (I,2801); THE SCARECROW OF ROMNEY MARSH (I,3931)

HORGAN, PATRICK
CASABLANCA (II,450); THE DOCTORS (II,694); THE EDGE OF NIGHT (II,760); GEORGE WASHINGTON (II,978); RYAN'S HOPE (II,2234)

HORINO, TED
THE OUTLAWS (II,1936)

HORMANN, NICHOLAS
THE RETURN OF MARCUS WELBY, M.D (II,2138)

HORN, LEW
FRIENDS (II,927); G.I.'S (II,949)

HORNE, LENA
THE FRANK SINATRA TIMEX SHOW (I,1667); HARRY AND LENA (I,1957); KEEP U.S. BEAUTIFUL (I,2519); LENA HORNE: THE LADY AND HER MUSIC (II,1460); THE MILTON BERLE SPECIAL (I,3050); MUSIC '55 (I,3168)

HORNE, MARILYN
100 YEARS OF GOLDEN HITS (II,1905); FRIENDS AND NABORS (I,1684)

HORNER JR, JED
NAKIA (II,1784)

HORNER, PENELOPE
THE DOUBLE KILL (I,1360); STAR MAIDENS (II,2435)

HOROWITZ, DAVID
AMERICA ALIVE! (II,67); DINAH'S PLACE (II,678); FIGHT BACK: WITH DAVID HOROWITZ (II,854)

HORSFORD, ANNA MARIA
THE FIRM (II,860); MURDER INK (II,1769)

HORSLEY, JOHN
THE FALL AND RISE OF REGINALD PERRIN (II,810)

HORSLEY, LEE
MATT HOUSTON (II,1654); NERO WOLFE (II,1806); THE WILD WOMEN OF CHASTITY GULCH (II,2801)

HORTON, DANNY
MINDY CARSON SINGS (I,3055)

HORTON, EDWARD EVERETT
ARSENIC AND OLD LACE (I,268); THE BULLWINKLE SHOW (I,761); THE CARA WILLIAMS SHOW (I,843); DENNIS THE MENACE (I,1231); F TROOP (I,1499); FORD THEATER HOUR (I,1635); HOLIDAY HOTEL (I,2075); JOHNNY CARSON PRESENTS THE SUN CITY SCANDALS (I,2420); MAGNAVOX THEATER (I,2827); MANHATTAN TOWER (I,2889); THE MERRY WIDOW (I,3012); ROCKY AND HIS FRIENDS (I,3822); TIME OUT FOR GINGER (I,4508)

HORTON, HELEN
THE BENNY HILL SHOW (II,207); JENNIE: LADY RANDOLPH CHURCHILL (II,1277); MASTER OF THE GAME (II,1647)

HORTON, JOHN
ANOTHER WORLD (II,97); GEORGE WASHINGTON (II,978); MOMENT OF TRUTH (I,3084)

HORTON, LOUISA
THE INNER SANCTUM (I,2216); LIGHT'S OUT (I,2700); LIGHT'S OUT (I,2699); STAGE DOOR (I,4177); A STRING OF BEADS (I,4263)

HORTON, MICHAEL
ALL THE WAY HOME (II,44); ALONE AT LAST (II,59); THE BLUE AND THE GRAY (II,273); THE CHICAGO STORY (II,507); THE EDDIE CAPRA MYSTERIES (II,758); MURDER, SHE WROTE (II,1770); NOT UNTIL TODAY (II,1865); THREE EYES (II,2605); WHIZ KIDS (II,2790)

HORTON, PETER
SAWYER AND FINN (II,2265); SEVEN BRIDES FOR SEVEN BROTHERS (II,2307)

HORTON, ROBERT
THE DANGEROUS DAYS OF KIOWA JONES (I,1140); KING'S ROW (I,2563); A MAN CALLED SHENANDOAH (I,2859); STUDIO '57 (I,4267); WAGON TRAIN (I,4747)

HORVATH, CHARLES
BROKEN ARROW (I,744); THE GOOD OLD DAYS (I,1851)

HORWICH, DR FRANCES
DING DONG SCHOOL (I,1285)

HOSHI, SHIZUKO
THE LAST NINJA (II,1438)

HOSKINS, BOB
CRY TERROR! (I,1112); PENNIES FROM HEAVEN (II,1994)

HOSLEY, PATRICIA
ABC-POWERS CHARM SCHOOL (I,5); THE AD-LIBBERS (I,29); YOUNG MR. BOBBIN (I,4943)

HOSTEN, JENNIFER
THE BOB HOPE SHOW (I,638)

HOSTETTER, JOHN
ONCE UPON A SPY (II,1898); THE SHERIFF AND THE ASTRONAUT (II,2326)

HOT GOSSIP
THE KENNY EVERETT VIDEO SHOW (II,1379)

HOTCHKIS, JOAN
THE 416TH (II,913); FRONTIER (I,1695); THE LIFE AND TIMES OF EDDIE ROBERTS (II,1468); MY WORLD . . . AND WELCOME TO IT (I,3208); THE ODD COUPLE (II,1875); THE SECRET STORM (I,3969)

HOTSY TOTSY BOYS, THE
THE LAWRENCE WELK SHOW (I,2643); MEMORIES

WITH LAWRENCE WELK (II,1681)

HOUGHTON, JAMES
CODE R (II,551); DYNASTY (II,747); KNOTS LANDING (II,1404); THE YOUNG AND THE RESTLESS (II,2862)

HOUGHTON, KATHARINE
THE ADAMS CHRONICLES (II,8)

HOULIHAN, KERI
GIMME A BREAK (II,995); LITTLE SHOTS (II,1495); LOVERS AND OTHER STRANGERS (II,1550); SNOOPY'S GETTING MARRIED, CHARLIE BROWN (I,4102)

HOULTON, JENNIFER
THE DOCTORS (II,694)

HOUSE, BILLY
FOUNTAIN OF YOUTH (I,1647)

HOUSE, DANA
BEANE'S OF BOSTON (II,194); OFF THE WALL (II,1879); WENDY HOOPER—U. S. ARMY (II,2764)

HOUSE, JANE
AS THE WORLD TURNS (II,110)

HOUSEMAN, JOHN
CAPTAINS AND THE KINGS (II,435); THE FRENCH ATLANTIC AFFAIR (II,924); HAZARD'S PEOPLE (II,1112); THE LAST CONVERTIBLE (II,1435); MARCO POLO (II,1626); THE PAPER CHASE (II,1949); THE PAPER CHASE: THE SECOND YEAR (II,1950); SILVER SPOONS (II,2355); TALES OF THE UNEXPECTED (II,2540); TRUMAN AT POTSDAM (I,4618); WASHINGTON: BEHIND CLOSED DOORS (II,2744); THE WINDS OF WAR (II,2807)

HOUSER, JERRY
THE BRADY BRIDES (II,361); THE BRADY GIRLS GET MARRIED (II,364); THE FIGHTING NIGHTINGALES (II,855); THE GARY COLEMAN SHOW (II,958); THE GIRL, THE GOLD WATCH AND DYNAMITE (II,1001); IT TAKES TWO (II,1252); LIVING IN PARADISE (II,1503); THE NEW TEMPERATURES RISING SHOW (I,3273); ONE DAY AT A TIME (II,1900); THREE TIMES DALEY (II,2610); WE'LL GET BY (II,2753); WE'LL GET BY (II,2754)

HOUSTON, GLYN
A HORSEMAN RIDING BY (II,1182)

HOUSTON, MARTIN
DIAGNOSIS: UNKNOWN (I,1251); MY SON JEEP (I,3202)

HOUSTON, PAULA
THE BENNETS (I,397)

HOUSTON, THELMA
THE MARTY FELDMAN COMEDY MACHINE (I,2929)

HOVEN, LOUISE
THE 416TH (II,913); GENERAL HOSPITAL (II,964); THE SAN PEDRO BEACH BUMS (II,2250); THE SAN PEDRO BUMS (II,2251)

HOVER, BOB
ALL MY CHILDREN (II,39); AS THE WORLD TURNS (II,110)

HOVEY, TIM
IVY LEAGUE (I,2286)

HOVIS, GUY
THE LAWRENCE WELK SHOW (I,2643); THE LAWRENCE WELK SHOW (I,2644); MEMORIES WITH LAWRENCE WELK (II,1681)

HOVIS, LARRY
HOGAN'S HEROES (I,2069); HOLMES AND YOYO (II,1169); THE LIAR'S CLUB (II,1466); ROWAN AND MARTIN'S LAUGH-IN (I,3856); ROWAN AND MARTIN'S LAUGH-IN (I,3857)

HOW, JANE
THE CITADEL (II,539)

HOWAR, BARBARA
ENTERTAINMENT TONIGHT (II,780); JOYCE AND BARBARA: FOR ADULTS ONLY (I,2458)

HOWARD BARLOW CHORUS, THE
VOICE OF FIRESTONE (I,4741)

HOWARD ROBERTS SINGERS, THE
THE LESLIE UGGAMS SHOW (I,2660)

HOWARD, ANDREA
HOLMES AND YOYO (II,1169); ROSETTI AND RYAN: MEN WHO LOVE WOMEN (II,2217); SANTA BARBARA (II,2258); WILD ABOUT HARRY (II,2797)

HOWARD, ANNE
ANSON AND LORRIE (II,98); YOUNG HEARTS (II,2865)

HOWARD, BOB
THE BOB HOWARD SHOW (I,665); SING IT AGAIN (I,4059)

HOWARD, CLINT
THE ANDY GRIFFITH SHOW (I,192); THE BAILEYS OF BALBOA (I,334); THE COWBOYS (II,598); GENTLE BEN (I,1761)

HOWARD, CURLY JOE
THE NEW THREE STOOGES (I,3274)

HOWARD, CY
SNEAK PREVIEW (I,4100)

HOWARD, D D
FOUR EYES (II,912); MATT HOUSTON (II,1654); NIGHT COURT (II,1844); VENICE MEDICAL (II,2726)

HOWARD, DENNIS
THE BAY CITY AMUSEMENT COMPANY (II,185); THE FRENCH ATLANTIC AFFAIR (II,924); HAVING BABIES III (II,1109); JULIE FARR, M.D. (II,1354); VENICE MEDICAL (II,2726)

HOWARD, EDDY
SATURDAY NIGHT JAMBOREE (I,3920); THE SWINGIN' SINGIN' YEARS (I,4319)

HOWARD, FRANK
HARD KNOX (II,1087)

HOWARD, HERB
TELE-VARIETIES (I,4377)

HOWARD, JAN
THE BILL ANDERSON SHOW (I,440); COUNTRY MUSIC HIT PARADE (II,589)

HOWARD, JOE
GAY NINETIES REVUE (I,1743)

HOWARD, JOHN
THE ADVENTURES OF THE SEA HAWK (I,81); DAYS OF OUR LIVES (II,629); DOCTOR HUDSON'S SECRET JOURNAL (I,1312); PUBLIC PROSECUTOR (I,3689)

HOWARD, JON
A TOWN LIKE ALICE (II,2652)

HOWARD, KEN
ADAM'S RIB (I,32); THE CELEBRITY FOOTBALL CLASSIC (II,474); THE COURT-MARTIAL OF GENERAL GEORGE ARMSTRONG CUSTER (I,1081); IT'S NOT EASY (II,1256); MANHUNTER (II,1620); MANHUNTER (II,1621); RAGE OF ANGELS (II,2107); THE THORN BIRDS (II,2600); THE WHITE SHADOW (II,2788)

HOWARD, KEVIN
JOE DANCER: THE BIG BLACK PILL (II,1300)

HOWARD, LELAND
MR. BELVEDERE (I,3127)

HOWARD, LISA
THE EDGE OF NIGHT (II,760); THE HUNTER (I,2162)

HOWARD, MOE
THE COMICS (I,1028); THE NEW THREE STOOGES (I,3274)

HOWARD, NORAH
BERKELEY SQUARE (I,399); JANE EYRE (I,2338)

HOWARD, RANCE
COUNTERATTACK: CRIME IN AMERICA (II,581); FLATBED ANNIE AND SWEETIEPIE: LADY TRUCKERS (II,876); GENTLE BEN (I,1761); STATE FAIR (II,2448); THE THORN BIRDS (II,2600)

HOWARD, ROBIN
ROAD TO REALITY (I,3800)

HOWARD, RON
AMANDA FALLON (I,151); THE ANDY GRIFFITH SHOW (I,192); THE ANDY GRIFFITH SHOW (I,193); ANSON AND LORRIE (II,98); BATTLE OF THE NETWORK STARS (II,163); BATTLE OF THE NETWORK STARS (II,164); BOB HOPE SPECIAL: THE BOB HOPE SPECIAL (II,337); DOBIE GILLIS (I,1302); FONZ AND THE HAPPY DAYS GANG (II,898); THE GENERAL ELECTRIC THEATER (I,1753); HAPPY DAYS (II,1084); MR. O'MALLEY (I,3146); THE OLIVIA NEWTON-JOHN SHOW (II,1885); THE SMITH FAMILY (I,4092); A SPECIAL OLIVIA NEWTON-JOHN (II,2418)

HOWARD, RONALD
THE ADVENTURES OF ROBIN HOOD (I,74); COWBOY IN AFRICA (I,1085); SHERLOCK HOLMES (I,4010)

HOWARD, SUSAN
THE BUSTERS (II,401); CELEBRITY CHALLENGE OF THE SEXES 1 (II,465); DALLAS (II,614); THE FANTASTIC JOURNEY (II,826); NIGHT GAMES (II,1845); THE PAPER CHASE (II,1949); PETROCELLI (II,2031); THE POWER WITHIN (II,2073); SAVAGE (I,3924)

HOWARD, TAMA

BIG BEND COUNTRY (II,230)

HOWARD, TOM
IT PAYS TO BE IGNORANT (I,2248)

HOWARD, TREVOR
DEADLY GAME (II,632); EAGLE IN A CAGE (I,1392); GEORGE WASHINGTON (II,978); HEDDA GABLER (I,2001); PLAYHOUSE 90 (I,3623); THE POPPY IS ALSO A FLOWER (I,3650); TONIGHT AT 8:30 (I,4556)

HOWARD, TRUDY
THOMPSON'S GHOST (I,4467)

HOWARD, VANESSA
THE PICTURE OF DORIAN GRAY (I,3592)

HOWARD, VINCE
BARNABY JONES (II,153); HOUSE CALLS (II,1194); MOONLIGHT (II,1730); MR. NOVAK (I,3145)

HOWARD, WENDY SMITH
LEADFOOT (II,1452)

HOWARTH, BETTY JANE
THE PINKY LEE SHOW (I,3601)

HOWARTH, JACK
CORONATION STREET (I,1054)

HOWAT, CLARK
THE ADVENTURES OF DR. FU MANCHU (I,58); DRAGNET (I,1371)

HOWE, DARRELL
THE TEENAGE IDOL (I,4375)

HOWE, LINDA
BRIGADOON (I,726); CAROUSEL (I,856)

HOWE, QUINCY
IT'S NEWS TO ME (I,2272)

HOWEL, JEAN
DOCTOR HUDSON'S SECRET JOURNAL (I,1312)

HOWELL, ARLENE
BOURBON STREET BEAT (I,699)

HOWELL, C THOMAS
BATTLE OF THE NETWORK STARS (II,178); TWO MARRIAGES (II,2691)

HOWELL, ERIK
THE GUIDING LIGHT (II,1064)

HOWELL, HOKE
THE BLUE KNIGHT (II,277); HERE COME THE BRIDES (I,2024); THE SHAMEFUL SECRETS OF HASTINGS CORNERS (I,3994)

HOWELL, JEAN
WASHINGTON: BEHIND CLOSED DOORS (II,2744)

HOWELL, PETER
SCREAMER (I,3945)

HOWELL, WAYNE
THE LEFT OVER REVUE (I,2651)

HOWELLS, URSULA
FATHER DEAR FATHER (II,837); THE FORSYTE SAGA (I,1645); KRAFT MYSTERY THEATER (I,2589)

HOWES, SALLY ANN
BRIGADOON (I,726); GIFT OF THE MAGI (I,1800); JANE EYRE (I,2338); MISSION: IMPOSSIBLE (I,3067); PRUDENCE AND THE CHIEF (I,3682); SHERLOCK HOLMES: THE HOUND OF THE BASKERVILLES (I,4011)

HOWETH, EVIE
THE INA RAY HUTTON SHOW (I,2207)

HOWLAND, BETH
ALICE (II,33); ALICE (II,34); THE TED BESSELL SHOW (I,4369); YOU CAN'T TAKE IT WITH YOU (II,2857)

HOWLIN, OLIN
CIRCUS BOY (I,959)

HOWORTH, NIKI
MONTY PYTHON'S FLYING CIRCUS (II,1729)

HOWSER, HUELL
ENTERTAINMENT TONIGHT (II,780); WEDDING DAY (II,2759)

HOY, BOB
THE BARBARY COAST (II,147); THE HIGH CHAPARRAL (I,2051)

HOY, BRUCE
THE SKATEBIRDS (II,2375)

HOY, ELIZABETH
REAL KIDS (II,2116)

HOY, LINDA
THE RETURN OF MARCUS WELBY, M.D (II,2138)

HOY, ROBERT
STEVE CANYON (I,4224)

HOYE, GLORIA
THE BRIGHTER DAY (I,728); SOMERSET (I,4115)

HOYOS, RODOLFO
CHICO AND THE MAN (II,508); THE FLYING NUN (I,1611); VIVA VALDEZ (II,2728)

HOYT, CLEGG
THE HOUSE NEXT DOOR (I,2139)

HOYT, JOHN
THE ADVENTURES OF RIN TIN TIN (I,73); THE BOSTON TERRIER (I,697); COSMOPOLITAN THEATER (I,1057); CROSSROADS (I,1105); FAME IS THE NAME OF THE GAME (I,1518); FIRESIDE THEATER (I,1580); FOR THE DEFENSE (I,1624); FRONTIER (I,1695); GIMME A BREAK (II,995); GRUEN GUILD PLAYHOUSE (I,1892); MATINEE THEATER (I,2947); PROFILES IN COURAGE (I,3678); RETURN TO PEYTON PLACE (I,3779); THE RHINEMANN EXCHANGE (II,2150); RUN FOR YOUR LIFE (I,3871); THE THIRD COMMANDMENT (I,4449); TOM, DICK, AND MARY (I,4537); THE TURNING POINT OF JIM MALLOY (II,2670)

HQUGEN, PER THEODOR
TEN WHO DARED (II,2559)

HUBBARD, DAVID
THE ACADEMY (II,3); HARRIS AND COMPANY (II,1096); JAMES AT 15 (II,1270); THE SIX O'CLOCK FOLLIES (II,2368); SPACE STARS (II,2408); SQUARE PEGS (II,2431); THE WHITE SHADOW (II,2788)

HUBBARD, EDDIE
VAUDEO VARIETIES (I,4705)

HUBBARD, ELIZABETH
THE DOCTORS (II,694); THE GUIDING LIGHT (II,1064)

HUBBARD, JOHN
DON'T CALL ME CHARLIE (I,1349); FAMILY AFFAIR (I,1519); HEY MULLIGAN (I,2040); OUT OF THE BLUE (I,3422); TALES OF THE 77TH BENGAL LANCERS (I,4348)

HUBBEL, ELAINE DANIELE
PETTICOAT JUNCTION (I,3571)

HUBBELL, CHRIS
ALL MY CHILDREN (II,39)

HUBBOD, IRENE
A DATE WITH LIFE (I,1163)

HUBER, FRANK
NIGHT CLUB (I,3284)

HUBER, HAROLD
I COVER TIMES SQUARE (I,2166)

HUBER, JEAN
NIGHT CLUB (I,3284)

HUBERT, GEORGE
SMILEY'S PEOPLE (II,2383)

HUBLEY, SEASON
BOBBY JO AND THE BIG APPLE GOODTIME BAND (I,674); FAMILY (II,813); THE HEALERS (II,1115); KUNG FU (II,1416); LONDON AND DAVIS IN NEW YORK (II,1512); LOOSE CHANGE (II,1518); THE ROOKIES (II,2208)

HUBLY, SUSAN
LOVE OF LIFE (I,2774)

HUCKABEE, COOPER
THE BLUE AND THE GRAY (II,273); BRAVO TWO (II,368); LITTLE HOUSE: LOOK BACK TO YESTERDAY (II,1490)

HUCKO, PEANUTS
THE LAWRENCE WELK SHOW (I,2643); MEMORIES WITH LAWRENCE WELK (II,1681)

HUDDLE, ELIZABETH
BOONE (II,351)

HUDDLESTON, DAVID
BROCK'S LAST CASE (I,742); COMPUTERCIDE (II,561); THE GUN AND THE PULPIT (II,1067); HAWKINS (I,1976); HAWKINS ON MURDER (I,1978); HIZZONER (II,1159); THE KALLIKAKS (II,1361); KATE BLISS AND THE TICKER TAPE KID (II,1366); ONCE AN EAGLE (II,1897); THE OREGON TRAIL (II,1922); PETROCELLI (II,2031); PLIMPTON! SHOWDOWN AT RIO LOBO (I,3630); SARGE: THE BADGE OR THE CROSS (I,3915); SUPER COMEDY BOWL 2 (I,4289); TENAFLY (I,4395); TENAFLY (II,2560); WINNER TAKE ALL (II,2808)

HUDDLESTON, MICHAEL
A DOG'S LIFE (II,697); LOOK OUT WORLD (II,1516); ZERO INTELLIGENCE (II,2876)

HUDGINS, WAYNE
AS THE WORLD TURNS (II,110)

HUDIS, STEPHEN
SAM HILL: WHO KILLED THE MYSTERIOUS MR. FOSTER? (I,3896)

HUDSON BROTHERS, THE
THE PAT BOONE AND FAMILY CHRISTMAS SPECIAL (II,1969)

HUDSON, BILL
BONKERS! (II,348); THE HUDSON BROTHERS RAZZLE DAZZLE COMEDY SHOW (II,1201); THE HUDSON BROTHERS SHOW (II,1202); THE MILLIONAIRE (II,1700)

HUDSON, BRETT
BONKERS! (II,348); THE HUDSON BROTHERS RAZZLE DAZZLE COMEDY SHOW (II,1201); THE HUDSON BROTHERS SHOW (II,1202); THE MILLIONAIRE (II,1700)

HUDSON, ERNIE
100 CENTRE STREET (II,1901); ALMOST AMERICAN (II,55); CRAZY TIMES (II,602); HIGHCLIFFE MANOR (II,1150)

HUDSON, GARY
AS THE WORLD TURNS (II,110)

HUDSON, JIM
MARIE (II,1629)

HUDSON, JOHN
THE TRAP (I,4593); THE WEB (I,4785)

HUDSON, MARK
BONKERS! (II,348); THE HUDSON BROTHERS RAZZLE DAZZLE COMEDY SHOW (II,1201); THE HUDSON BROTHERS SHOW (II,1202); THE MILLIONAIRE (II,1700)

HUDSON, ROCHELLE
THAT'S MY BOY (I,4427)

HUDSON, ROCK
THE BEATRICE ARTHUR SPECIAL (II,196); THE BIG PARTY FOR REVLON (I,427); CAROL AND COMPANY (I,847); CIRCUS OF THE STARS (II,534); THE DEVLIN CONNECTION (II,664); DYNASTY (II,746); HOLLYWOOD: THE SELZNICK YEARS (I,2095); THE MARTIAN CHRONICLES (II,1635); MCMILLAN (II,1666); MCMILLAN AND WIFE (II,1667); THE OLIVIA NEWTON-JOHN SHOW (II,1885); ONCE UPON A DEAD MAN (I,3369); WHEELS (II,2777)

HUDSON, WILLIAM
FATHER KNOWS BEST (II,838)

HUDSON, YVONNE
SATURDAY NIGHT LIVE (II,2261)

HUEBING, CRAIG
THE DOCTORS (II,694); FROM THESE ROOTS (I,1688); GENERAL HOSPITAL (II,964); PIONEER SPIRIT (I,3604)

HUEY, DUC
FLY AWAY HOME (II,893)

HUFFMAN, DAVID
CAPTAINS AND THE KINGS (II,435); SANDBURG'S LINCOLN (I,3907); SIDNEY SHORR (II,2349); TESTIMONY OF TWO MEN (II,2564)

HUFFMAN, ROSANNA
FREEBIE AND THE BEAN (II,922); TENAFLY (I,4395); TENAFLY (II,2560)

HUFMAN, BASIL
THE MONEYCHANGERS (II,1724)

HUFSEY, BILLY
FAME (II,812)

HUG, JOHN
SPACE: 1999 (II,2409)

HUGH HATTER DANCERS, THE
THE BOB HOPE SHOW (I,507)

HUGH-KELLY, DANIEL
BATTLE OF THE NETWORK STARS (II,177); CHICAGO STORY (II,506); HARDCASTLE AND MCCORMICK (II,1089); MURDER INK (II,1769); RYAN'S HOPE (II,2234)

HUGHES, BARNARD
ANOTHER APRIL (II,93); THE BOB NEWHART SHOW (II,342); THE BORROWERS (I,693); DOC (II,682); DOC (II,683); THE GUIDING LIGHT (II,1064); THE MAGIC OF DAVID COPPERFIELD (II,1594); THE MILLION DOLLAR INCIDENT (I,3045); MR. MERLIN (II,1751); RANSOM FOR ALICE (II,2113); SANCTUARY OF FEAR (II,2252); TALES FROM THE DARKSIDE (II,2534); THE THANKSGIVING TREASURE (I,4416); THE WORLD BEYOND (II,2835)

HUGHES, BARNEY
ALL THE WAY HOME (I,140)

HUGHES, DEL
THE BRIGHTER DAY (I,728)

HUGHES, GEOFFREY
CORONATION STREET (I,1054)

HUGHES, HELEN
TALES OF THE HAUNTED (II,2538)

HUGHES, KATHLEEN
BRACKEN'S WORLD (I,703)

HUGHES, LINDA
THE MICKEY MOUSE CLUB (I,3025); THE MOUSEKETEERS REUNION (II,1742)

HUGHES, MARY BETH
ARTHUR MURRAY'S DANCE PARTY (I,293); FRONT ROW CENTER (I,1694)

HUGHES, MELANIE
GREAT EXPECTATIONS (II,1055)

HUGHES, MICHAEL
MAKE ROOM FOR GRANDDADDY (I,2845)

HUGHES, MILDRED
THE LARRY STORCH SHOW (I,2618)

HUGHES, ROBIN
THE BARRETTS OF WIMPOLE STREET (I,361); THE BROTHERS (I,748)

HUGHES, TRESA
ANOTHER WORLD (II,97)

HUGHES, WENDY
RETURN TO EDEN (II,2143)

HUGHS, BILL
HOLLYWOOD JUNIOR CIRCUS (I,2084)

HUGO, LAWRENCE
THE EDGE OF NIGHT (II,760); SEARCH FOR TOMORROW (II,2284); STAR TONIGHT (I,4199)

HUGO, MAURITZ
COMMANDO CODY (I,1030)

HULCE, THOMAS
EMILY, EMILY (I,1438)

HULETT, OTTO
BORN YESTERDAY (I,692)

HULL, CYNTHIA
THE FLYING NUN (I,1611); HERE COME THE BRIDES (I,2024); IS THERE A DOCTOR IN THE HOUSE? (II,1247)

HULL, DIANNE
THE POLICE STORY (I,3638)

HULL, GOLDIE
THE GRAND OLE OPRY (I,1866)

HULL, HENRY
ALCOA PREMIERE (I,109); ALL IN THE FAMILY (I,133); THE ARMSTRONG CIRCLE THEATER (I,260); THE CAMPBELL TELEVISION SOUNDSTAGE (I,812); FORD THEATER HOUR (I,1635); THE KAISER ALUMINUM HOUR (I,2503); MAGNAVOX THEATER (I,2827); SUSPENSE (I,4305); WINDOWS (I,4876)

HULL, JOSEPHINE
ARSENIC AND OLD LACE (I,267)

HULL, ROD
THE HUDSON BROTHERS RAZZLE DAZZLE COMEDY SHOW (II,1201); THE HUDSON BROTHERS SHOW (II,1202)

HULL, SETH
THE NASHVILLE PALACE (II,1792)

HULL, WARREN
CAVALCADE OF BANDS (I,876); A COUPLE OF JOES (I,1078); PUBLIC PROSECUTOR (I,3689); STRIKE IT RICH (I,4262); THE WARREN HULL SHOW (I,4771)

HULLABALOO DANCERS, THE
HULLABALOO (I,2156)

HULSWIT, MART
THE DESPERATE HOURS (I,1239); THE GUIDING LIGHT (II,1064); HERCULES (I,2022)

HUMBERT, WILLIAM
PENMARRIC (II,1993)

HUMBLE, GWEN
DOCTORS' PRIVATE LIVES (II,692); THE REBELS (II,2121); SKAG (II,2374); TWO GUYS FROM MUCK (II,2690)

HUME, BENITA
THE HALLS OF IVY (I,1923)

HUME, DOUG
NIGHTSIDE (II,1851)

HUME, ROGER
AN ECHO OF THERESA (I,1399)

HUME, VALERIE-JEAN
PAUL BERNARD—PSYCHIATRIST (I,3485)

HUMES, MARY-MARGARET
VELVET (II,2725)

HUMPERDINCK, ENGELBERT
THE BOB HOPE SHOW (I,645); CHANGING SCENE (I,899); THE ENGELBERT HUMPERDINCK SHOW (I,1444); THE ENGELBERT HUMPERDINCK SHOW (I,1445); THE ENGELBERT HUMPERDINCK SPECIAL (I,1446); THE ENGELBERT HUMPERDINCK SPECIAL (I,1447); MAC DAVIS. . . I BELIEVE IN CHRISTMAS (II,1581)

HUMPHREY, CAVADA

DARK SHADOWS (I,1157)

HUMPHREY, DICK
THE JERRY LEWIS SHOW (I,2363)

HUMPHREYS, SUZIE
JAKE'S WAY (II,1269)

HUNDLEY, CRAIG
BABY CRAZY (I,319)

HUNICUTT, ARTHUR
THE DAUGHTERS OF JOSHUA CABE RETURN (I,1170)

HUNLEY, EDDIE
THE BILL COSBY SPECIAL (I,444)

HUNLEY, GARY
SKY KING (I,4077); YOU'RE ONLY YOUNG ONCE (I,4969)

HUNLEY, LEANN
BATTLESTAR GALACTICA (II,181); THE MISADVENTURES OF SHERIFF LOBO (II,1704)

HUNNICUTT, ARTHUR
ELFEGO BACA (I,1427); KILROY (I,2541)

HUNNICUTT, GAYLE
COLOR HIM DEAD (I,1007); COLOR HIM DEAD (II,554); A MAN CALLED INTREPID (II,1612); THE MARTIAN CHRONICLES (II,1635); RETURN OF THE MAN FROM U.N.C.L.E.: THE 15 YEARS LATER AFFAIR (II,2140); SAVAGE: IN THE ORIENT (II,2264); A WOMAN OF SUBSTANCE (II,2820)

HUNSAKER, JOHN
POOR RICHARD (II,2065)

HUNT, ALLAN
VOYAGE TO THE BOTTOM OF THE SEA (I,4743)

HUNT, BILL
SOMERSET (I,4115)

HUNT, GARETH
THE NEW AVENGERS (II,1816)

HUNT, HELEN
AMY PRENTISS (II,74); BATTLE OF THE NETWORK STARS (II,175); FAMILY (II,813); THE FITZPATRICKS (II,868); IT TAKES TWO (II,1252); THE MARY TYLER MOORE SHOW (II,1640); ST. ELSEWHERE (II,2432); SWISS FAMILY ROBINSON (II,2517)

HUNT, JOHN
POLKA-GO-ROUND (I,3640)

HUNT, LINDA
FAME (I,1517)

HUNT, LOIS
ROBERT Q'S MATINEE (I,3812); ROSALINDA (I,3846)

HUNT, MARSHA
ACTION (I,23); COSMOPOLITAN THEATER (I,1057); DANGER (I,1134); THE MAN FROM DENVER (I,2863); MAN ON THE MOVE (I,2878); MY THREE SONS (I,3205); NO WARNING (I,3301); PECK'S BAD GIRL (I,3503); SILVER THEATER (I,4051); SURE AS FATE (I,4297)

HUNT, MAX
SKEEZER (II,2376)

HUNT, MEGAN
THE BEST CHRISTMAS PAGEANT EVER (II,212)

HUNT, MIE
DOMESTIC LIFE (II,703)

HUNT, RICHARD
FRAGGLE ROCK (II,916); THE MUPPET SHOW (II,1766)

HUNT, WILL
HELLO, LARRY (II,1128)

HUNTER, BARBARA JEAN
OF THIS TIME, OF THAT PLACE (I,3335)

HUNTER, BILL
1915 (II,1853); PRISONER: CELL BLOCK H (II,2085)

HUNTER, EVE
THE EVE HUNTER SHOW (I,1472); HOME (I,2099)

HUNTER, IAN
THE ADVENTURES OF ROBIN HOOD (I,74)

HUNTER, IVORY JOE
THE ROCK N' ROLL SHOW (I,3820)

HUNTER, JEFFREY
THE MAN FROM GALVESTON (I,2865); TEMPLE HOUSTON (I,4392)

HUNTER, KIM
ACTOR'S STUDIO (I,28); ALCOA/GOODYEAR THEATER (I,107); APPOINTMENT WITH ADVENTURE (I,246); BACKSTAIRS AT THE WHITE HOUSE (II,133); CELANESE THEATER (I,881); THE EDGE OF NIGHT (II,760); ELLERY QUEEN: TOO MANY SUSPECTS (II,767); FORD THEATER HOUR (I,1635); GIVE US BARABBAS! (I,1816); THE GOLDEN AGE OF TELEVISION (II,1018); THE KAISER ALUMINUM HOUR (I,2503); LAMP AT MIDNIGHT (I,2608); LITTLE WOMEN (I,2719); THE MAGICIAN (I,2824); THE MARCH OF DIMES FASHION SHOW (I,2901); ON TRIAL (I,3363); ONCE AN EAGLE (II,1897); THE PEOPLE NEXT DOOR (I,3519); SCENE OF THE CRIME (II,2271); SCREEN DIRECTOR'S PLAYHOUSE (I,3946); SILVER THEATER (I,4051); STAR TONIGHT (I,4199); STUBBY PRINGLE'S CHRISTMAS (II,2483)

HUNTER, LESLYE
JANE EYRE (I,2338)

HUNTER, ROLAND
CAGNEY AND LACEY (II,410)

HUNTER, RONALD
THE LAZARUS SYNDROME (II,1451); RAGE OF ANGELS (II,2107)

HUNTER, ROSS
THE BOB HOPE SHOW (I,575); MITZI AND A HUNDRED GUYS (II,1710)

HUNTER, RUSSELL
THE SAVAGE CURSE (I,3925)

HUNTER, TAB
CELEBRITY CHALLENGE OF THE SEXES 2 (II,466); CIRCLE OF FEAR (I,958); CIRCUS OF THE STARS (II,535); CONFLICT (I,1036); THE CONNIE FRANCIS SHOW (I,1039); FOREVER FERNWOOD (II,909); HANS BRINKER OR THE SILVER SKATES (I,1933); MEET ME IN ST. LOUIS (I,2986); MERMAN ON BROADWAY (I,3011); PLAYHOUSE 90 (I,3623); SAN FRANCISCO INTERNATIONAL AIRPORT (I,3906); THE TAB HUNTER SHOW (I,4328)

HUNTER, TODD
YOU ARE THERE (I,4929)

HUNTER, TOMMY
THE TOMMY HUNTER SHOW (I,4552)

HUNTER, WALT
A ROCK AND A HARD PLACE (II,2190)

HUNTER, WAYNE
ANNIE GET YOUR GUN (I,224)

HUNTINGTON, JOAN
GENERAL ELECTRIC TRUE (I,1754)

HUNTINGTON, PAM
THE FACTS OF LIFE (II,805)

HURBER, PAUL
HERB SHRINER TIME (I,2020)

HURDY GURDY GIRLS, THE
HURDY GURDY (I,2163)

HURLBUT, GLADYS
THE DONNA REED SHOW (I,1347)

HURLEY, WALTER
DRAW ME A LAUGH (I,1372)

HURSEY, SHERRY
BEST FRIENDS (II,213); THE MARY TYLER MOORE SHOW (II,1640); THE MEMBER OF THE WEDDING (II,1680)

HURST, DAVID
ANASTASIA (I,182); THE DATCHET DIAMONDS (I,1160); THE FLYING NUN (I,1611)

HURST, JAMES
SOUNDS OF HOME (I,4136)

HURST, JUDITH
HELLO, LARRY (II,1128)

HURST, RICHARD
THE BLUE KNIGHT (II,277)

HURST, RICK
AMANDA'S (II,62); BIG BOB JOHNSON AND HIS FANTASTIC SPEED CIRCUS (II,231); THE DUKES OF HAZZARD (II,742); FROM HERE TO ETERNITY (II,932); I'LL NEVER FORGET WHAT'S HER NAME (II,1217); ON THE ROCKS (II,1894); VALENTINE MAGIC ON LOVE ISLAND (II,2716)

HURT, JO
KOBB'S CORNER (I,2577); MEMBER OF THE WEDDING (I,3002)

HURT, MARYBETH
ROYCE (II,2224)

HURT, WILLIAM
ALL THE WAY HOME (II,44); CYRANO DE BERGERAC (I,1120); MACBETH (I,2801)

HURTES, HIDI LYNN
BEYOND WITCH MOUNTAIN (II,228)

HUSKEY, KENNI
BUCK OWENS TV RANCH (I,750)

HUSKY, FERLIN
COUNTRY MUSIC CARAVAN (I,1066); HOWDY (I,2150); THE SOUND AND THE SCENE (I,4132)

HUSMANN, RON
DAYS OF OUR LIVES (II,629); THE GERSHWIN YEARS

(I,1779); ONCE UPON A MATTRESS (I,3371); SEARCH FOR TOMORROW (II,2284); THREE ON AN ISLAND (I,4483); THE WAY THEY WERE (II,2748); YOU BET YOUR LIFE (II,2856)

HUSSEY, OLIVIA
THE 13TH DAY: THE STORY OF ESTHER (II,2593); THE BASTARD/KENT FAMILY CHRONICLES (II,159); THE LAST DAYS OF POMPEII (II,1436)

HUSSEY, P J
SUMMER SOLSTICE (II,2492)

HUSSEY, RUTH
COME A RUNNING (I,1013); JANE WYMAN PRESENTS THE FIRESIDE THEATER (I,2345); TIME OUT FOR GINGER (I,4508); THE WOMEN (I,4890)

HUSSEY, SHERRY
NUMBER 96 (II,1869)

HUSSING, WILL
A TIME TO LIVE (I,4510)

HUSTON, CHRIS
MEDICAL CENTER (II,1675)

HUSTON, DULCIE
MASTER OF THE GAME (II,1647)

HUSTON, GAYE
ANOTHER WORLD (II,97); BONINO (I,685); THE PHILADELPHIA STORY (I,3582)

HUSTON, JOHN
THE LEGEND OF MARILYN MONROE (I,2654); MERV GRIFFIN'S ST. PATRICK'S DAY SPECIAL (I,3018); THE RHINEMANN EXCHANGE (II,2150); THE WORD (II,2833)

HUSTON, MARTIN
THE INNER SANCTUM (I,2216); JUNGLE JIM (I,2491); TOO YOUNG TO GO STEADY (I,4574)

HUSTON, PATRICIA
DAYS OF OUR LIVES (II,629)

HUSTON, PAULA
THE COLLEGE BOWL (I,1001)

HUTCHINS, WILL
BLONDIE (I,487); CONFLICT (I,1036); HEY, LANDLORD (I,2039); THE QUEST (II,2096); SUGARFOOT (I,4277); WHEN THE WEST WAS FUN: A WESTERN REUNION (II,2780)

HUTCHINSON, BILL
MASTER OF THE GAME (II,1647)

HUTCHINSON, DAVID
INSTITUTE FOR REVENGE (II,1239)

HUTCHINSON, JIMMY
POLKA-GO-ROUND (I,3640)

HUTCHINSON, JOSEPHINE
KAREN (I,2505); TRAVIS LOGAN, D.A. (I,4598); THE WALTONS (II,2740)

HUTCHINSON, KEN
MASADA (II,1642)

HUTCHINSON, RAWN
CHIPS (II,511)

HUTTER, CHRISTINA
SKEEZER (II,2376)

HUTTON, BETTY
THE BETTY HUTTON SHOW (I,413); THE BOB HOPE SHOW (I,542); THE PHIL HARRIS SHOW (I,3575); SATINS AND SPURS (I,3917)

HUTTON, GUNILLA
HEE HAW (II,1123); HIGHER AND HIGHER, ATTORNEYS AT LAW (I,2057); MURDER CAN HURT YOU! (II,1768); PETTICOAT JUNCTION (I,3571)

HUTTON, INA RAY
THE INA RAY HUTTON SHOW (I,2207)

HUTTON, JIM
BOB HOPE SPECIAL: BOB HOPE IN "JOYS" (II,284); BUTTERFLIES (II,405); CALL HER MOM (I,796); CALL HOLME (II,412); ELLERY QUEEN (II,766); ELLERY QUEEN: TOO MANY SUSPECTS (II,767); EVERYTHING'S RELATIVE (I,1484); FLYING HIGH (II,895); THEY CALL IT MURDER (I,4441); WEDNESDAY NIGHT OUT (II,2760); YOU'RE ONLY YOUNG ONCE (I,4969)

HUTTON, LAUREN
INSTITUTE FOR REVENGE (II,1239); PAPER DOLLS (II,1952); THE RHINEMANN EXCHANGE (II,2150); STEVE MARTIN'S BEST SHOW EVER (II,2460)

HUTTON, ROBERT
GRUEN GUILD PLAYHOUSE (I,1892); SCHAEFER CENTURY THEATER (I,3935); TV READERS DIGEST (I,4630); YOUR JEWELER'S SHOWCASE (I,4953)

HUTTON, TIMOTHY
THE OLDEST LIVING GRADUATE (II,1882)

HUTTON, WENDY
DEBBY BOONE. . . ONE STEP CLOSER (II,655)

HUXTABLE, VICKY
IF I LOVE YOU, AM I TRAPPED FOREVER? (II,1222); MR. & MS. AND THE BANDSTAND MURDERS (II,1746)

HYAMS, JOE
GUEST SHOT (I,1899)

HYATT, BOBBY
PRIDE OF THE FAMILY (I,3667)

HYDE, BRUCE
FRANK MERRIWELL (I,1659); STAR TREK (II,2440)

HYDE, DICK
THE ERN WESTMORE SHOW (I,1453)

HYDE, JACQUELINE
THE RETURN OF MARCUS WELBY, M.D (II,2138)

HYDE, JONATHAN
LACE (II,1418); MISTRAL'S DAUGHTER (II,1708)

HYDE, KENNETH
THE SWORD OF FREEDOM (I,4326)

HYDE-WHITE, ALEX
BUCK ROGERS IN THE 25TH CENTURY (II,384); THE SEEKERS (II,2298)

HYDE-WHITE, WILFRID
THE ASSOCIATES (II,113); BUCK ROGERS IN THE 25TH CENTURY (II,384); DOUGLAS FAIRBANKS JR. PRESENTS THE RHEINGOLD THEATER (I,1364); LUCY IN LONDON (I,2789); THE REBELS (II,2121); SYBIL (I,4327)

HYER, LEILA
THE FLORIAN ZABACH SHOW (I,1606)

HYER, MARTHA
BROADWAY (I,733); FOUR STAR PLAYHOUSE (I,1652); THE LUX VIDEO THEATER (I,2795); THE WAY THEY WERE (II,2748); YOUR JEWELER'S SHOWCASE (I,4953)

HYLAND, BRIAN
GO! (I,1830)

HYLAND, DIANA
THE DUPONT SHOW OF THE WEEK (I,1388); EIGHT IS ENOUGH (II,762); GUILTY OR NOT GUILTY (I,1903); HERCULES (I,2022); PEYTON PLACE (I,3574); SCALPLOCK (I,3930)

HYLAND, FRANCES
TALES OF THE HAUNTED (II,2538)

HYLAND, PATRICIA
THE NURSES (I,3319)

HYLANDS, SCOTT
ENIGMA (II,778); GEORGE WASHINGTON (II,978); THE WALTONS (II,2740)

HYLTON, JANE
THE ADVENTURES OF SIR LANCELOT (I,76); KRAFT MYSTERY THEATER (I,2589)

HYMAN, EARLE
THE COSBY SHOW (II,578); THE GREEN PASTURES (I,1887)

HYMAN, SID
THE TWO RONNIES (II,2694)

HYMAS, ELAINE
A TIME FOR US (I,4507)

HYNDLEY, CRAIG
STAR TREK (II,2440)

HYNES, KATHERINE
LITTLE MOON OF ALBAN (I,2715)

IANNONE, PATTY
SANDY DREAMS (I,3909)

IANNONE, ROSE MARIE
SANDY DREAMS (I,3909)

ICE ANGELS, THE
THE OSMOND FAMILY SHOW (II,1927)

ICE FOLLIES, THE
PEGGY FLEMING AT MADISON SQUARE GARDEN (I,3506)

ICE VANITIES, THE
DONNY AND MARIE (II,712)

ICHINO, LAURIE
THE DANNY KAYE SHOW (I,1143)

IDELSON, BILL
THE DICK VAN DYKE SHOW (I,1275); FRED AND BARNEY MEET THE SHMOO (II,918); MIXED DOUBLES (I,3075); MY FAVORITE MARTIAN (I,3189); THE NEW SHMOO (II,1827); ONE MAN'S FAMILY (I,3383)

IDEN, MINDI
MICKEY SPILLANE'S MIKE HAMMER: MORE THAN MURDER (II,1693)

IDLE, ERIC
MONTY PYTHON'S FLYING CIRCUS (II,1729)

IEODORESCU, ION

CALLAHAN (II,414)

IGLESIAS, EUGENE
WAR CORRESPONDENT (I,4767)

IGNICIO, JENNIFER
BRANAGAN AND MAPES (II,367)

IGUCHI, YONEO
MIMI (I,3054)

IGUS, DARROW
ANSON AND LORRIE (II,98); FRIDAYS (II,926); ROLL OUT! (I,3833)

IHNAT, STEVE
THE D.A.: CONSPIRACY TO KILL (I,1121); POLICE STORY (I,3636); SWEET, SWEET RACHEL (I,4313)

ILAND, TOBY
SHE'S WITH ME (II,2321)

IMADA, JEFF
MASSARATI AND THE BRAIN (II,1646); THE RENEGADES (II,2132)

IMEL, JACK
THE LAWRENCE WELK SHOW (I,2643); MEMORIES WITH LAWRENCE WELK (II,1681)

IMOFF, GARY
LUCY MOVES TO NBC (II,1562); ME AND DUCKY (II,1671)

IMPERATO, CARLO
ANGIE (II,80); FAME (II,812); THE KIDS FROM FAME (II,1392)

IMPERIALS, THE
THE ANITA BRYANT SPECTACULAR (II,82)

IMPERT, MARGIE
ANOTHER WORLD (II,97); A CHRISTMAS FOR BOOMER (II,518); MAGGIE (II,1589); MAGNUM, P.I. (II,1601); SPENCER'S PILOTS (II,2421); SPENCER'S PILOTS (II,2422)

INDRISANO, JOHN
O.K. CRACKERBY (I,3348)

INESCORT, FRIEDA
MEET CORLISS ARCHER (I,2979); THE TAB HUNTER SHOW (I,4328)

ING, ALVIN
BROTHERS (II,381)

ING, DEBBIE
MARIE (II,1629)

INGBER, MANDY
CHARLES IN CHARGE (II,482)

INGE, JOHN
SILVER SPOONS (II,2355)

INGELS, MARTY
ALWAYS APRIL (I,147); THE CATTANOOGA CATS (I,874); THE DICK VAN DYKE SHOW (I,1275); THE GREAT GRAPE APE SHOW (II,1056); I'M DICKENS...HE'S FENSTER (I,2193); KISS ME, KATE (I,2569); MOTOR MOUSE (I,3111); PAC-MAN (II,1940); THE PRUITTS OF SOUTHAMPTON (I,3684); SUPER COMEDY BOWL 1 (I,4288)

INGERSOL, JAMES
CAPTAIN AMERICA (II,427); LONDON AND DAVIS IN NEW YORK (II,1512)

INGERSOLL, AMY
THE DOCTORS (II,694)

INGERSOLL, RUTH
CAMP WILDERNESS (II,419)

INGLEDEW, ROSALIND
T.L.C. (II,2525)

INGLIS, CHRIS
PLAY IT AGAIN, CHARLIE BROWN (I,3617)

INGRAM, BARRIE
CAMELOT (II,416); CHARLEY'S AUNT (II,483); GEORGE WASHINGTON (II,978); THE JERK, TOO (II,1282); P.O.P. (II,1939); TALES OF THE GOLD MONKEY (II,2537)

INGRAM, MICHAEL
ONE LIFE TO LIVE (II,1907); YOUNG DR. MALONE (I,4938)

INHAT, STEVE
THE WHOLE WORLD IS WATCHING (I,4851)

INNES, GEORGE
ARCHER—FUGITIVE FROM THE EMPIRE (II,103); GOLIATH AWAITS (II,1025); THE KILLING GAME (I,2540); MASADA (II,1642); Q. E. D. (II,2094); SHOGUN (II,2339); UPSTAIRS, DOWNSTAIRS (II,2709)

INNES, JEAN
DOCTOR KILDARE (I,1315); LITTLE LEATHERNECK (I,2713)

INONE, YANCO
THE NEW ZOO REVUE (I,3277)

INSINNIA, ALBERT
CRAZY TIMES (II,602); STOCKARD CHANNING IN JUST FRIENDS (II,2467)

INSPIRATION, THE
HEE HAW (II,1123)

INTERLUDES, THE
THE TONY MARTIN SHOW (I,4572)

INWOOD, STEVE
FARRELL: FOR THE PEOPLE (II,834); JACQUELINE SUSANN'S VALLEY OF THE DOLLS, 1981 (II,1268)

IPALE, AHRON
MOSES THE LAWGIVER (II,1737)

IRELAND, IAN
SHERLOCK HOLMES: THE HOUND OF THE BASKERVILLES (I,4011)

IRELAND, JILL
THE GIRL, THE GOLD WATCH AND EVERYTHING (II,1002); SHANE (I,3995)

IRELAND, JOHN
CASSIE AND COMPANY (II,455); THE CHEATERS (I,918); THE ELGIN HOUR (I,1428); THE MILLIONAIRE (II,1700); PHILCO TELEVISION PLAYHOUSE (I,3583); RAWHIDE (I,3727); TOURIST (II,2651); WHEN THE WEST WAS FUN: A WESTERN REUNION (II,2780)

IRENE, GEORGI
BARBARA MANDRELL AND THE MANDRELL SISTERS (II,143); GALACTICA 1980 (II,953); SILVER SPOONS (II,2355)

IRESON, RICHARD
MASTER OF THE GAME (II,1647)

IRISH ROVERS, THE
THE VIRGINIAN (I,4738)

IRONS, JEREMY
BRIDESHEAD REVISITED (II,375)

IRONSIDE, MICHAEL
V: THE FINAL BATTLE (II,2714); V: THE SERIES (II,2715)

IRVIN, SMITTY
THE GEORGE HAMILTON IV SHOW (I,1769)

IRVING DAVIES DANCERS, THE
THE ENGELBERT HUMPERDINCK SHOW (I,1444); THE LIBERACE SHOW (I,2677); THE MOST IN MUSIC (I,3108)

IRVING, AMY
DYNASTY (II,747); THE FAR PAVILIONS (II,832); ONCE AN EAGLE (II,1897)

IRVING, CHARLES
THE WACKIEST SHIP IN THE ARMY (I,4744)

IRVING, GEORGE S
ALL IN THE FAMILY (II,38); ANASTASIA (I,182); BLUE JEANS (II,275); CAR 54, WHERE ARE YOU? (I,842); THE DAVID FROST REVUE (I,1176); THE DUMPLINGS (II,743); GETTING THERE (II,985); THE GO GO GOPHERS (I,1831); HOLIDAY (I,2074); PINOCCHIO'S CHRISTMAS (II,2045); RYAN'S HOPE (II,2234); THAT WAS THE YEAR THAT WAS (I,4424); UNDERDOG (I,4673); THE YEAR WITHOUT A SANTA CLAUS (II,2846)

IRVING, HOLLIS
BLONDIE (I,486); FIVE'S A FAMILY (I,1591); MARGIE (I,2909); ROAD OF LIFE (I,3798); ROOM 222 (I,3843); WHERE THERE'S SMOKEY (I,4832)

IRVING, JAY
DRAW ME A LAUGH (I,1372)

IRVING, JOHN
THE BARRETTS OF WIMPOLE STREET (I,361); MY FAIR LADY (I,3187)

IRVING, MARGARET
THE PEOPLE'S CHOICE (I,3521)

IRWIN, STAN
THE ABBOTT AND COSTELLO CARTOON SHOW (I,1); MR. SMITH GOES TO WASHINGTON (I,3151)

IRWIN, WYNN
COREY: FOR THE PEOPLE (II,575); FROM HERE TO ETERNITY (II,932); HART TO HART (II,1102); HOME COOKIN' (II,1170); LAVERNE AND SHIRLEY (II,1446); SUGAR TIME (II,2489); THE SUPER (I,4293); TV FUNNIES (II,2672); WHATEVER HAPPENED TO DOBIE GILLIS? (II,2774)

ISACKSEN, PETER
BATTLE OF THE NETWORK STARS (II,165); C.P.O. SHARKEY (II,407); THE HALF-HOUR COMEDY HOUR (II,1074); JESSIE (II,1287); PIPER'S PETS (II,2046)

ISH, KATHRYN
THE LOVE BOAT I (II,1532)

ISHIDA, JUNICHI
MARCO POLO (II,1626)

ISKOWITZ, HOWARD

LOOK WHAT THEY'VE DONE TO MY SONG (II,1517)

ITKIN, PAUL
THE DOCTORS (II,694)

ITKOWITZ, HOWARD
MARIE (II,1629)

ITO, AKJO
JOHNNY SOKKO AND HIS FLYING ROBOT (I,2436)

ITO, ROBERT
THE AMAZING CHAN AND THE CHAN CLAN (I,157); THE BURNS AND SCHREIBER COMEDY HOUR (I,768); THE EYES OF TEXAS II (II,800); MEN OF THE DRAGON (II,1683); QUINCY, M. E. (II,2102)

ITO, YOSHI
MR. ROGERS' NEIGHBORHOOD (I,3150)

ITZIN, GREGORY
BULBA (II,392)

IVAR, STAN
LITTLE HOUSE: A NEW BEGINNING (II,1488); LITTLE HOUSE: LOOK BACK TO YESTERDAY (II,1490); LITTLE HOUSE: THE LAST FAREWELL (II,1491)

IVES, BURL
ALIAS SMITH AND JONES (I,118); ALL THINGS BRIGHT AND BEAUTIFUL (I,141); THE BURL IVES THANKSGIVING SPECIAL (I,763); CAPTAINS AND THE KINGS (II,435); DANIEL BOONE (I,1142); THE GLEN CAMPBELL SPECIAL: THE MUSICAL WEST (I,1821); HOLIDAY U.S.A. (I,2078); THE LAWYERS (I,2646); MERV GRIFFIN'S ST. PATRICK'S DAY SPECIAL (I,3018); MONSANTO NIGHT PRESENTS BURL IVES (I,3090); O.K. CRACKERBY (I,3348); PINOCCHIO (I,3603); THE RED SKELTON CHEVY SPECIAL (I,3753); ROOTS (II,2211); RUDOLPH THE RED-NOSED REINDEER (II,2227); THE SOUND OF ANGER (I,4133); THE WHOLE WORLD IS WATCHING (I,4851); ZANE GREY THEATER (I,4979)

IVES, GEORGE
THE CRACKER BROTHERS (II,599); GET SMART (II,983); MR. ROBERTS (I,3149)

IVEY, LELA
100 CENTRE STREET (II,1901); THE EDGE OF NIGHT (II,760)

IVINS, PERRY
THE BRIGHTER DAY (I,728)

IVO, TOMMY
THE DONNA REED SHOW (I,1347); LEAVE IT TO BEAVER (I,2648); MARGIE (I,2909)

IZAY, CONNIE
M*A*S*H (II,1569)

IZAY, VICTOR
THE D.A. (I,1123); LITTLE HOUSE: LOOK BACK TO YESTERDAY (II,1490)

IZUCHARA, SHANNON
SCAMPS (II,2267)

J C
MR. SMITH (II,1755)

J R
GOING BANANAS (II,1015)

JABARA, PAUL
FLATBUSH/AVENUE J (II,878); THE RUBBER GUN SQUAD (II,2225)

JABER, ZARA
JANE EYRE (II,1273)

JACK ALLISON SINGERS, THE
THE KATE SMITH EVENING HOUR (I,2507); THE KATE SMITH HOUR (I,2508)

JACK COLE DANCERS, THE
THE BOB HOPE SHOW (I,506)

JACK ELLIOTT ORCHESTRA, THE
100 YEARS OF AMERICA'S POPULAR MUSIC (II,1904)

JACK KANE AND HIS MUSICAL MAKERS
THE GISELE MACKENZIE SHOW (I,1813)

JACK REGAS DANCERS, THE
THE BEAUTIFUL PHYLLIS DILLER SHOW (I,381); DANNY THOMAS: AMERICA I LOVE YOU (I,1145); THE JOHN GARY SHOW (I,2411)

JACK AND JILL
FOR YOUR PLEASURE (I,1627)

JACKIE
ME AND THE CHIMP (I,2974)

JACKIE AND GAYLE
KRAFT SUMMER MUSIC HALL (I,2590)

JACKSON FIVE, THE
THE BOB HOPE SHOW (I,662); DIANA (I,1258); HELLZAPOPPIN (I,2011); ONE MORE TIME (I,3386); SANDY IN DISNEYLAND (II,2254)

JACKSON JR, JOE
NIGHT CLUB (I,3284)

JACKSON JR, PAUL M
TENAFLY (I,4395)

JACKSON SISTERS, THE
A COUPLE OF DONS (I,1077)

JACKSON, ANNE
ACADEMY THEATER (I,14); ACRES AND PAINS (I,19); THE ARMSTRONG CIRCLE THEATER (I,260); DANGER (I,1134); THE DOCTOR (I,1320); LOVE OF LIFE (I,2774); ROBERT MONTGOMERY PRESENTS YOUR LUCKY STRIKE THEATER (I,3809); SUSPENSE (I,4305); THE WEB (I,4784); A WOMAN CALLED GOLDA (II,2818)

JACKSON, ANTHONY
BLESS THIS HOUSE (II,268)

JACKSON, ARNOLD
DETECTIVE SCHOOL (II,661)

JACKSON, BARRY
MOLL FLANDERS (II,1720)

JACKSON, BILL
GIGGLESNORT HOTEL (II,989)

JACKSON, CHUCK
IT'S WHAT'S HAPPENING, BABY! (I,2278)

JACKSON, COLETTE
BRADDOCK (I,704)

JACKSON, DAN
THE BENNY HILL SHOW (II,207)

JACKSON, DAVID
BLAKE'S SEVEN (II,264)

JACKSON, EDDIE
THE JIMMY DURANTE SHOW (I,2388); TEXACO STAR THEATER (I,4412)

JACKSON, ELMA
ROOTS (II,2211)

JACKSON, EUGENE
JULIA (I,2476)

JACKSON, GORDON
MADAME SIN (I,2805); A TOWN LIKE ALICE (II,2652); UPSTAIRS, DOWNSTAIRS (II,2709)

JACKSON, HARRY
PAPA SAID NO (I,3452)

JACKSON, JACKIE
THE JACKSON FIVE (I,2326); THE JACKSONS (II,1267)

JACKSON, JAMIE SMITH
HAVING BABIES III (II,1109); LISA, BRIGHT AND DARK (I,2711)

JACKSON, JANET
DIFF'RENT STROKES (II,674); FAME (II,812); GOOD TIMES (II,1038); THE JACKSONS (II,1267); A NEW KIND OF FAMILY (II,1819)

JACKSON, JAY
TIC TAC DOUGH (I,4498); TWENTY QUESTIONS (I,4647)

JACKSON, JOHN
CELEBRITY (II,463)

JACKSON, KATE
CHARLIE'S ANGELS (II,486); CHARLIE'S ANGELS (II,487); DARK SHADOWS (I,1157); JAMES AT 15 (II,1271); THE MAD MAD MAD MAD WORLD OF THE SUPER BOWL (II,1586); MOVIN' ON (I,3118); THE NEW HEALERS (I,3266); THE ROOKIES (II,2208); SCARECROW AND MRS. KING (II,2268); THE SENSATIONAL, SHOCKING, WONDERFUL, WACKY 70S (II,2302); TOPPER (II,2649)

JACKSON, LATOYA
THE JACKSONS (II,1267)

JACKSON, LEONARD
THE JEFFERSONS (II,1276); RAGE OF ANGELS (II,2107)

JACKSON, LIA
HARRIS AND COMPANY (II,1096)

JACKSON, MAHALIA
CRESCENDO (I,1094); MAHALIA JACKSON SINGS (I,2830)

JACKSON, MALLIE
KUDZU (II,1415)

JACKSON, MARILYN
YOUR HIT PARADE (I,4952)

JACKSON, MARLON
THE JACKSON FIVE (I,2326); THE JACKSONS (II,1267)

JACKSON, MARY
HARDCASTLE AND MCCORMICK (II,1089); INSPECTOR PEREZ (II,1237); OPEN ALL NIGHT (II,1914); TWO THE HARD WAY (II,2695); THE WALTONS (II,2740); THE WALTONS: A DAY FOR THANKS ON WALTON'S MOUNTAIN (EPISODE 3) (II,2741); THE WALTONS: MOTHER'S DAY ON WALTON'S MOUNTAIN (EPISODE 2) (II,2742); THE WALTONS: WEDDING ON WALTON'S MOUNTAIN (EPISODE 1) (II,2743)

JACKSON, MAUREEN
THE JACKSONS (II,1267)

JACKSON, MICHAEL
DIANA (II,667); DISNEYLAND'S 25TH ANNIVERSARY (II,681); THE JACKSON FIVE (I,2326); THE JACKSONS (II,1267); MARLO THOMAS AND FRIENDS IN FREE TO BE. . . YOU AND ME (II,1632); THE WOLFMAN JACK RADIO SHOW (II,2816)

JACKSON, NAGLE
WE INTERRUPT THIS SEASON (I,4780)

JACKSON, PAUL
TENAFLY (II,2560)

JACKSON, RANDY
THE JACKSON FIVE (I,2326); THE JACKSONS (II,1267)

JACKSON, REGGIE
BOB HOPE SPECIAL: THE BOB HOPE SPECIAL (II,335); IT'S ONLY HUMAN (II,1257)

JACKSON, RON
MORNING STAR (I,3104)

JACKSON, ROWENA
SLEEPING BEAUTY (I,4083)

JACKSON, SAM
MURDER INK (II,1769)

JACKSON, SAMMY
THE CODE OF JONATHAN WEST (I,992); LI'L ABNER (I,2701); NO TIME FOR SERGEANTS (I,3300)

JACKSON, SHERRY
BRENDA STARR, REPORTER (II,373); CASINO (II,452); COME A RUNNING (I,1013); ENIGMA (II,778); MAKE ROOM FOR DADDY (I,2843); MAKE ROOM FOR GRANDDADDY (I,2845)

JACKSON, STONEY
THE WHITE SHADOW (II,2788)

JACKSON, TITO
THE JACKSON FIVE (I,2326); THE JACKSONS (II,1267)

JACKSON, TYRONE
THE AFRICAN QUEEN (II,18)

JACKSON, VICTORIA
THE HALF-HOUR COMEDY HOUR (II,1074); W*A*L*T*E*R (II,2731)

JACKSON, WANDA
JUBILEE U.S.A. (I,2461)

JACOBI, DEREK
THE PALLISERS (II,1944); THE STRAUSS FAMILY (I,4253)

JACOBI, LOU
ALLAN (I,142); THE DEAN MARTIN SHOW (I,1201); DOUGLAS FAIRBANKS JR. PRESENTS THE RHEINGOLD THEATER (I,1364); ED SULLIVAN'S BROADWAY (I,1402); THE HAPPENERS (I,1935); IVAN THE TERRIBLE (II,1260); THE JUDGE AND JAKE WYLER (I,2464); KIBBE HATES FINCH (I,2530); REAR GUARD (II,2120); SAINT PETER (II,2238)

JACOBS, ANTHONY
WAR AND PEACE (I,4765)

JACOBS, BETH
HOUSE CALLS (II,1194)

JACOBS, CHRISTIAN
GLORIA (II,1010); MAGGIE (II,1589)

JACOBS, LAWRENCE-HILTON
BATTLE OF THE NETWORK STARS (II,164); KOMEDY TONITE (II,1407); ROOTS (II,2211); WELCOME BACK, KOTTER (II,2761)

JACOBS, MICHAEL
NO SOAP, RADIO (II,1856)

JACOBS, PETER
SCAMPS (II,2267)

JACOBS, RACHEL
13 QUEENS BOULEVARD (II,2591); DEAR TEACHER (II,653); IT'S NOT EASY (II,1256); THE KID WITH THE BROKEN HALO (II,1387); MAGGIE (II,1589)

JACOBSON, ART
HOLLYWOOD JUNIOR CIRCUS (I,2084)

JACOBSON, JILL
I GAVE AT THE OFFICE (II,1213)

JACOBSON, JO
BULBA (II,392)

JACOBSON, MAGGIE
THE NEW SHOW (II,1828)

JACOBY, BILLY
THE BAD NEWS BEARS (II,134); CRAZY TIMES (II,602); IT'S NOT EASY (II,1256); MAGGIE (II,1589)

JACOBY, BOBBY
CRAZY TIMES (II,602); KNOTS LANDING (II,1404); TRUE LIFE STORIES (II,2665)

JACOBY, LAURA
JOE DANCER: THE MONKEY MISSION (II,1301); MR. SMITH (II,1755)

JACOBY, SCOTT
79 PARK AVENUE (II,2310)

JACQUET, JEFFREY
GOLDIE AND KIDS: LISTEN TO ME (II,1021); MORK AND MINDY (II,1735); WHIZ KIDS (II,2790)

JADA QUARTET, THE
THE PAT BOONE SHOW (I,3470)

JAECKEL, RICHARD
AT EASE (II,115); BANYON (I,344); CENTENNIAL (II,477); CROSSROADS (I,1105); FIREHOUSE (I,1579); FIREHOUSE (II,859); FRONT ROW CENTER (I,1694); FRONTIER CIRCUS (I,1697); GO WEST YOUNG GIRL (II,1013); HOT W.A.C.S. (II,1191); JANE WYMAN PRESENTS THE FIRESIDE THEATER (I,2345); LUKE AND THE TENDERFOOT (I,2792); MATINEE THEATER (I,2947); MIGHTY O (I,3036); NO WARNING (I,3301); THE OUTER LIMITS (I,3426); PANIC! (I,3447); PARTNERS IN CRIME (I,3461); THE PETRIFIED FOREST (I,3570); PRINCESS (II,2084); REWARD (II,2147); SALVAGE (II,2244); SALVAGE 1 (II,2245)

JAEGER, FREDERICK
LITTLE WOMEN (I,2723); SPECIAL BRANCH (II,2414)

JAFFE, NICHOLE
SCOOBY-DOO, WHERE ARE YOU? (II,2276)

JAFFE, SAM
BEN CASEY (I,394); NIGHT GALLERY (I,3288); THE OATH: THE SAD AND LONELY SUNDAYS (II,1874); SAGA OF SONORA (I,3884); SAM HILL: WHO KILLED THE MYSTERIOUS MR. FOSTER? (I,3896)

JAFFREY, SAEED
THE FAR PAVILIONS (II,832)

JAGGER, DEAN
THE DELPHI BUREAU (I,1224); GULF PLAYHOUSE (I,1904); MATT LINCOLN (I,2949); MR. NOVAK (I,3145); THE STRANGER (I,4251)

JAHN, STEPHEN
THE HARDY BOYS (I,1948)

JAIME ROGERS DANCERS, THE
THE HELEN REDDY SHOW (I,2008); THE HUDSON BROTHERS SHOW (II,1202); THE KEN BERRY WOW SHOW (I,2524); THE SONNY AND CHER COMEDY HOUR (II,2400)

JAIME ROGERS SINGERS, THE
THE ANDY WILLIAMS SHOW (I,210)

JAK, LISA
THE MONROES (I,3089)

JALBERT, PIERRE
COMBAT (I,1011)

JAMAL, IGBAL
THE FAR PAVILIONS (II,832)

JAMES HIBBARD DANCERS, THE
THE BUDDY GRECO SHOW (I,754); FRANKIE AVALON'S EASTER SPECIAL (I,1673)

JAMES STARBUCK DANCERS, THE
THE ANDY WILLIAMS SHOW (I,205); FRANKIE CARLE TIME (I,1674); FRANKIE LAINE TIME (I,1676); JACK PAAR PRESENTS (I,2315)

JAMES STARBUCK SINGERS, THE
SING ALONG WITH MITCH (I,4057)

JAMES AND SMITH
ANN-MARGRET—WHEN YOU'RE SMILING (I,219)

JAMES, ART
BLANK CHECK (II,265); CONCENTRATION (I,1031); FRACTURED PHRASES (I,1654); THE MAGNIFICENT MARBLE MACHINE (II,1600); MATCHES 'N' MATES (I,2945); PAY CARDS (I,3496); SAY WHEN (I,3929); SUPER PAY CARDS (II,2500); TEMPTATION (I,4393); THE WHO, WHAT OR WHERE GAME (I,4850)

JAMES, BARRY
THE SAVAGE CURSE (I,3925)

JAMES, BRION
GOOBER AND THE TRUCKERS' PARADISE (II,1029); HEAR NO EVIL (II,1116); JOSHUA'S WORLD (II,1344); LONE STAR (II,1513); THE MCLEAN STEVENSON SHOW (II,1664); TROUBLE IN HIGH TIMBER COUNTRY (II,2661)

JAMES, CARROLL
GEORGE WASHINGTON (II,978); KENNEDY (II,1377)

JAMES, CLIFTON
CAPTAINS AND THE KINGS (II,435); CITY OF ANGELS (II,540); HART TO HART (II,1101); LEWIS AND CLARK (II,1465); THE RUNAWAY

BARGE (II,2230); TEXAS (II,2566)

JAMES, DENNIS
CASH AND CARRY (I,866); CHANCE OF A LIFETIME (I,898); CLUB 60 (I,987); THE DENNIS JAMES SHOW (I,1228); DENNIS JAMES SPORTS PARADE (I,1229); DENNIS JAMES' CARNIVAL (I,1227); HAGGIS BAGGIS (I,1917); HIGH FINANCE (I,2052); JUDGE FOR YOURSELF (I,2466); NAME THAT TUNE (II,1786); THE NAME'S THE SAME (I,3220); OKAY MOTHER (I,3349); ON YOUR ACCOUNT (I,3364); P.D.Q. (I,3497); PEOPLE WILL TALK (I,3520); THE PRICE IS RIGHT (I,3665); TELEVISION ROOF (I,4382); TURN TO A FRIEND (I,4626); TWO FOR THE MONEY (I,4654); YOUR ALL AMERICAN COLLEGE SHOW (I,4946); YOUR FIRST IMPRESSION (I,4949)

JAMES, FRANCESCA
ALL MY CHILDREN (II,39)

JAMES, GODFREY
LITTLE LORD FAUNTLEROY (II,1492)

JAMES, HARRY
THE BOB HOPE SHOW (I,553); SINATRA—THE FIRST 40 YEARS (II,2359)

JAMES, IRENE
GETAWAY CAR (I,1783)

JAMES, JACQUELINE
THIS IS MUSIC (I,4457)

JAMES, JERI LOU
DOBIE GILLIS (I,1302); IT'S ALWAYS JAN (I,2264); THE R.C.A. VICTOR SHOW (I,3737)

JAMES, JESSICA
DINER (II,679)

JAMES, JOANELLE
THE SKIP FARRELL SHOW (I,4074)

JAMES, JOE
BARRIER REEF (I,362)

JAMES, JOHN
BATTLE OF THE NETWORK STARS (II,174); BATTLE OF THE NETWORK STARS (II,175); BATTLE OF THE NETWORK STARS (II,176); BATTLE OF THE NETWORK STARS (II,178); DYNASTY (II,746); SEARCH FOR TOMORROW (II,2284)

JAMES, KEN
BARRIER REEF (I,362); A MAN CALLED INTREPID (II,1612); SKIPPY, THE BUSH KANGAROO (I,4076)

JAMES, LEN
THE MAGNIFICENT SIX AND A HALF (I,2828)

JAMES, MAGGIE
MACKENZIE (II,1584)

JAMES, OLGA
THE BILL COSBY SHOW (I,442); YOUNG DR. KILDARE (I,4937)

JAMES, OSCAR
NOT THE NINE O'CLOCK NEWS (II,1864); PETER PAN (II,2030)

JAMES, POLLY
SONG BY SONG (II,2399)

JAMES, RALPH
LOST IN SPACE (I,2759); MORK AND MINDY (II,1735); MORK AND MINDY (II,1736)

JAMES, ROBERT
THE ONEDIN LINE (II,1912)

JAMES, SHEILA
BROADSIDE (I,732); DOBIE GILLIS (I,1302); TROUBLE WITH FATHER (I,4610); WHATEVER HAPPENED TO DOBIE GILLIS? (II,2774)

JAMES, SIDNEY
BLESS THIS HOUSE (II,268); CARRY ON LAUGHING (II,448)

JAMES, SONNY
THE BOB HOPE SHOW (I,556); CRESCENDO (I,1094); JOHNNY CASH: THE FIRST 25 YEARS (II,1336)

JAMES, WILLIAM
CAMELOT (II,416)

JAMES-REESE, CINDI
THE RETURN OF MARCUS WELBY, M.D (II,2138)

JAMES-REESE, CYNDI
BRANAGAN AND MAPES (II,367); FLYING HIGH (II,895)

JAMESON, HOUSE
THE ALDRICH FAMILY (I,111); MACBETH (I,2800); MONTGOMERY'S SUMMER STOCK (I,3096)

JAMESON, JOYCE
THE ANDY GRIFFITH SHOW (I,192); THE BOB HOPE CHRYSLER THEATER (I,502); CLUB OASIS (I,984); CLUB OASIS (I,985); HOORAY FOR HOLLYWOOD (I,2114); THE LOVE BOAT I (II,1532); OF MEN OF WOMEN (I,3332); THE SCOOBY-DOO AND SCRAPPY-DOO SHOW (II,2275); THE WILD WILD WEST REVISITED (II,2800)

JAMESON, LOUISE
DOCTOR WHO (II,689); THE OMEGA FACTOR (II,1888)

JAMESON, PAULINE
THE WOMAN IN WHITE (II,2819)

JAMIE ROGERS DANCERS, THE
BURT BACHARACH IN SHANGRI-LA (I,773)

JAMIESON, MALCOLM
THE LAST DAYS OF POMPEII (II,1436)

JAMISON, JEFF
ANOTHER LIFE (II,95)

JAMISON, MIKKI
ADAM-12 (I,31)

JAMISON, RICHARD L
THE CHISHOLMS (II,512)

JAN AND DEAN
IT'S WHAT'S HAPPENING, BABY! (I,2278)

JANE, PAULA
THIS IS MUSIC (I,4458)

JANES, LOREN
EL COYOTE (I,1423)

JANES, SALLIE
THE LATE FALL, EARLY SUMMER BERT CONVY SHOW (II,1441); MARY HARTMAN, MARY HARTMAN (II,1638)

JANIS, CONRAD
BERT CONVY SPECIAL—THERE'S A MEETING HERE TONIGHT (II,210); BOB HOPE SPECIAL: BOB HOPE IN THE STAR-MAKERS (II,285); BONINO (I,685); DANGER (I,1134); DANNY AND THE MERMAID (II,618); THE DOCTOR (I,1320); FULL SPEED ANYWHERE (I,1704); THE GOSSIP COLUMNIST (II,1045); JIMMY HUGHES, ROOKIE COP (I,2390); MIRACLE ON 34TH STREET (I,3058); MORK AND MINDY (II,1735); MORK AND MINDY (II,1736); QUARK (II,2095); REAR GUARD (II,2120); STARLIGHT THEATER (I,4201); SUSPENSE (I,4305); THERE'S ALWAYS ROOM (II,2583)

JANIS, ED
SPUNKY AND TADPOLE (I,4169)

JANIS, PAULA
THE MAGIC GARDEN (II,1591); A MAGIC GARDEN CHRISTMAS (II,1590)

JANN, GERALD
HONG KONG (I,2112)

JANNEY, LEON
ANOTHER WORLD (II,97); HAWK (I,1974); THE PHILADELPHIA STORY (I,3582); STOP ME IF YOU HEARD THIS ONE (I,4231); YOUNG DR. MALONE (I,4938)

JANNIS, VIVI
FATHER KNOWS BEST (II,838)

JANSEN, JIM
HOW TO SUCCEED IN BUSINESS WITHOUT REALLY TRYING (II,1197)

JANSSEN, DAVID
THE BOB HOPE SHOW (I,617); BOB HOPE SPECIAL: BOB HOPE IN "JOYS" (II,284); CENTENNIAL (II,477); CIRCUS OF THE STARS (II,530); THE FUGITIVE (I,1701); THE GENERAL ELECTRIC THEATER (I,1753); HARRY O (II,1098); HARRY O (II,1099); KRAFT MYSTERY THEATER (I,2589); THE LUX VIDEO THEATER (I,2795); O'HARA, UNITED STATES TREASURY (I,3346); O'HARA, UNITED STATES TREASURY: OPERATION COBRA (I,3347); A PUREX DINAH SHORE SPECIAL (I,3697); RICHARD DIAMOND, PRIVATE DETECTIVE (I,3785); SMILE JENNY, YOU'RE DEAD (II,2382); WARNING SHOT (I,4770); THE WORD (II,2833); ZANE GREY THEATER (I,4979)

JANSSEN, EILEEN
THE DONALD O'CONNOR SHOW (I,1344); IT'S ALWAYS SUNDAY (I,2265); MAKE ROOM FOR DADDY (I,2843)

JANUARY, LOIS
LOLA (II,1508); LOLA (II,1509); VAN DYKE AND COMPANY (II,2719)

JARDIN, AL
THE BEACH BOYS 20TH ANNIVERSARY (II,188); THE BEACH BOYS IN CONCERT (II,189)

JARESS, JILL
THE NEW ODD COUPLE (II,1825); THE NEW PEOPLE (I,3268); SHEPHERD'S FLOCK (I,4007)

JARNAGIN JR, PAUL
CRASH ISLAND (II,600)

JARNAGIN, JAMES
STRANGER IN OUR HOUSE (II,2477)

JARNAGIN, NANCY
NORMA RAE (II,1860)

JAROWSKY, ANDREW
ANOTHER WORLD (II,97)

JARREAU, AL
SHEENA EASTON, ACT 1 (II,2323)

JARRELL, ANDY
CATCH 22 (I,871)

JARRETT, CHRIS
THE EDGE OF NIGHT (II,760)

JARRETT, HUGH
CHIEFS (II,509)

JARRETT, RENNE
ABC'S MATINEE TODAY (I,7); THE EDGE OF NIGHT (II,760); NANCY (I,3221); THE NEW DAUGHTERS OF JOSHUA CABE (I,3261); PORTIA FACES LIFE (I,3653); SOMERSET (I,4115)

JARVIS, CAROL
THE BOB HOPE SHOW (I,559)

JARVIS, GRAHAM
BORDER PALS (II,352); BUNGLE ABBEY (II,396); CASS MALLOY (II,454); THE DESPERATE HOURS (I,1239); FOREVER FERNWOOD (II,909); THE GUIDING LIGHT (II,1064); MAKING THE GRADE (II,1606); MARY HARTMAN, MARY HARTMAN (II,1638); THE NEW MAVERICK (II,1822); NUMBER 96 (II,1869); THERE GOES THE NEIGHBORHOOD (II,2582); TWO GUYS FROM MUCK (II,2690)

JARVIS, MARTIN
THE FORSYTE SAGA (I,1645); LITTLE WOMEN (I,2723)

JARVIS, ROBERT
THE 25TH MAN (MS) (II,2678); THIS IS KATE BENNETT (II,2597)

JASON, DAVID
OPEN ALL HOURS (II,1913)

JASON, GEORGE
CAPTAIN BILLY'S MISSISSIPPI MUSIC HALL (I,826)

JASON, HARVEY
ARNOLD'S CLOSET REVIEW (I,262); BRING 'EM BACK ALIVE (II,377); CAPTAINS AND THE KINGS (II,435); THE FRENCH ATLANTIC AFFAIR (II,924); GENESIS II (I,1760); THE PRIMARY ENGLISH CLASS (II,2081); RICH MAN, POOR MAN—BOOK I (II,2161); ROWAN AND MARTIN'S LAUGH-IN (I,3856)

JASON, KATE
PRISONER: CELL BLOCK H (II,2085)

JASON, MARC
HONEST AL'S A-OK USED CAR AND TRAILER RENTAL TIGERS (II,1174)

JASON, PETER
TIGER! TIGER! (I,4499)

JASON, RICK
THE CASE OF THE DANGEROUS ROBIN (I,862); COMBAT (I,1011); DAMON RUNYON THEATER (I,1129); FOUNTAIN OF YOUTH (I,1647); THE MONK (I,3087); PEPSI-COLA PLAYHOUSE (I,3523); PRUDENCE AND THE CHIEF (I,3682)

JASTROM, TERRY
THE PHOENIX (II,2035)

JAY, DAVID
LITTLE DRUMMER BOY, BOOK II (II,1486)

JAY, OREN
LOVE OF LIFE (I,2774)

JAY, RICKY
LIKE MAGIC (II,1476)

JAYNE, JENNIFER
THE ADVENTURES OF WILLIAM TELL (I,85)

JAYNES, MICHAEL
JAKE'S WAY (II,1269)

JAYSTON, MICHAEL
DEATH IN SMALL DOSES (I,1207); JANE EYRE (II,1273); KISS KISS, KILL KILL (I,2566); QUILLER: NIGHT OF THE FATHER (II,2100); QUILLER: PRICE OF VIOLENCE (II,2101); TINKER, TAILOR, SOLDIER, SPY (II,2621)

JEAN, GLORIA
COUNTERPOINT (I,1059)

JEAN, NORMA
JUBILEE U.S.A. (I,2461)

JEANS, ISABEL
VICTORIA REGINA (I,4728)

JEAVONS, COLIN
GREAT EXPECTATIONS (II,1055)

JEETER, JAMES
THE FRENCH ATLANTIC AFFAIR (II,924)

JEFF ALEXANDER CHORUS, THE
THE AMOS AND ANDY SHOW (I,179)

JEFF KUTASH'S DANCIN' MACHINE
THE JIMMY MCNICHOL SPECIAL (II,1292)

JEFF KUTECT DANCERS, THE
HOT CITY DISCO (II,1185)

JEFFERS, ANN
THE GUIDING LIGHT (II,1064)

JEFFERSON JR, HERB
THE BASTARD/KENT FAMILY CHRONICLES (II,159); BATTLESTAR GALACTICA (II,181); CUTTER (I,1117); THE DEVLIN CONNECTION (II,664); GALACTICA 1980 (II,953); POPI (II,2068); RICH MAN, POOR MAN—BOOK I (II,2161); RICH MAN, POOR MAN—BOOK II (II,2162)

JEFFERY, PETER
COME OUT, COME OUT, WHEREVER YOU ARE (I,1014)

JEFFRES, BRAD
JUMP (II,1356)

JEFFREYS, ANNE
BRIGHT PROMISE (I,727); CAVALCADE OF AMERICA (I,875); DEAREST ENEMY (I,1205); THE DELPHI BUREAU (I,1225); FINDER OF LOST LOVES (II,857); GHOSTBREAKER (I,1789); LOVE THAT JILL (I,2780); THE MERRY WIDOW (I,3012); TELEPHONE TIME (I,4379); TOPPER (I,4582)

JEFFRIES, FRAN
THE SMOKEY ROBINSON SHOW (I,4093)

JEFFRIES, GEORGIA
HOUSE CALLS (II,1194)

JEFFRIES, HERB
WHERE'S HUDDLES (I,4836)

JEFFRIES, JIM
ALL IN FUN (I,131)

JEFFRIES, LANG
RESCUE 8 (I,3772)

JELLIFFE, BILL
INSPECTOR PEREZ (II,1237)

JELLISON, BOB
THE BEAUTIFUL PHYLLIS DILLER SHOW (I,381); THE LIFE OF RILEY (I,2685)

JELLISON, ROBERT D
ADVENTURES OF A MODEL (I,50)

JENKINS, ALLEN
THE DUKE (I,1381); FOREST RANGER (I,1641); HEY, JEANNIE! (I,2038); TOP CAT (I,4576)

JENKINS, BO
THE LAST CONVERTIBLE (II,1435)

JENKINS, CAROL MAYO
FAME (II,812)

JENKINS, DAN
GUEST SHOT (I,1899)

JENKINS, HAYES ALAN
SONJA HENIE'S HOLIDAY ON ICE (I,4125)

JENKINS, LARRY FLASH
BAY CITY BLUES (II,186); FINDER OF LOST LOVES (II,857); THE WHITE SHADOW (II,2788)

JENKINS, MARK
THE MAN FROM ATLANTIS (II,1614); YOUNG DR. KILDARE (I,4937)

JENKINS, MEGS
JANE EYRE (II,1273); MACBETH (I,2801); A WOMAN OF SUBSTANCE (II,2820)

JENKINS, PAUL
THE WALTONS (II,2740); THE YOUNG AND THE RESTLESS (II,2862)

JENKINS, RICHARD
PAROLE (II,1960)

JENKS, FRANK
ADVENTURES OF COLONEL FLACK (I,56); THE EDDIE CANTOR COMEDY THEATER (I,1407); FRONT PAGE DETECTIVE (I,1689); MAKE ROOM FOR DADDY (I,2843); STARS OVER HOLLYWOOD (I,4211); WALDO (I,4749)

JENNER, BARRY
ANOTHER WORLD (II,97); FLY AWAY HOME (II,893); KNOTS LANDING (II,1404); SOMERSET (I,4115)

JENNER, BRUCE
AMERICA ALIVE! (II,67); BATTLE OF THE NETWORK STARS (II,166); BOB HOPE SPECIAL: BOB HOPE'S ALL-STAR COMEDY LOOK AT THE FALL SEASON: IT'S STILL FREE AND WORTH IT! (II,299); CELEBRITY CHALLENGE OF THE SEXES 2 (II,466); CHIPS (II,511); DOUG HENNING'S WORLD OF MAGIC V (II,729); JOHN SCHNEIDER'S CHRISTMAS HOLIDAY (II,1320); THE SUNDAY GAMES (II,2494); THAT'S TV (II,2581)

JENNINGS, BRENT

PAROLE (II,1960)

JENNINGS, JACK
SAMMY KAYE'S MUSIC FROM MANHATTAN (I,3902)

JENNINGS, KEN
THE LEPRECHAUN'S CHRISTMAS GOLD (II,1461); SWEENEY TODD (II,2512)

JENNINGS, MARK
ALLAN (I,142)

JENNINGS, WAYLON
THE CHERYL LADD SPECIAL (II,501); THE DUKES OF HAZZARD (II,742); THE JOHNNY CASH SPRING SPECIAL (II,1329); JOHNNY CASH: CHRISTMAS ON THE ROAD (II,1333); JOHNNY CASH: SPRING FEVER (II,1335); JOHNNY CASH: THE FIRST 25 YEARS (II,1336)

JENS, SALOME
ALCOA PREMIERE (I,109); ALL'S FAIR (II,46); BAREFOOT IN ATHENS (I,354); FROM HERE TO ETERNITY (II,932); FROM HERE TO ETERNITY (II,933); MARY HARTMAN, MARY HARTMAN (II,1638); THE MILLION DOLLAR INCIDENT (I,3045)

JENSEN, BOBBY
MICKIE FINN'S (I,3029)

JENSEN, GAIL
MASSARATI AND THE BRAIN (II,1646)

JENSEN, KAREN
BRACKEN'S WORLD (I,703); I LOVE A MYSTERY (I,2170); THE TED BESSELL SHOW (I,4369)

JENSEN, MAREN
BATTLE OF THE NETWORK STARS (II,167); BATTLESTAR GALACTICA (II,181)

JENSEN, SANDI
THE LAWRENCE WELK SHOW (I,2643); MEMORIES WITH LAWRENCE WELK (II,1681)

JENSEN, STAN
FAR OUT SPACE NUTS (II,831)

JENSON, DICK
THE ISLANDER (II,1249)

JENSON, ROY
CALL TO DANGER (I,804); FORCE FIVE (II,907); NIGHTSIDE (II,1851); RICH MAN, POOR MAN—BOOK I (II,2161)

JENTLE, IAN
SHOGUN (II,2339)

JERGENS, ADELE
FORD TELEVISION THEATER (I,1634); STARS OVER HOLLYWOOD (I,4211)

JERGENS, DIANE
COUNTERTHRUST (I,1061); DANIEL BOONE (I,1141); IT'S ALWAYS SUNDAY (I,2265); LITTLE WOMEN (I,2721); LOVE THAT BOB (I,2779); THE LUX VIDEO THEATER (I,2795); MATINEE THEATER (I,2947); THE MCGONIGLE (I,2968); SCHLITZ PLAYHOUSE OF STARS (I,3936); THREE WISHES (I,4488)

JERICHO, PAUL
DOCTOR WHO—THE FIVE DOCTORS (II,690)

JERIS, NANCY
MEN AT LAW (I,3004); SALLY AND SAM (I,3891)

JEROME, ED
LOVE OF LIFE (I,2774)

JEROME, PATTI
PHILLIP AND BARBARA (II,2034)

JERRICK, MIKE
ALIVE AND WELL (II,36)

JERRY MURAD'S HARMONICATS
GEORGE GOBEL PRESENTS (I,1766)

JERRY PACKER SINGERS, THE
THE KRAFT MUSIC HALL (I,2587); KRAFT MUSIC HALL PRESENTS THE DAVE KING SHOW (I,2586); THE PATTI PAGE SHOW (I,3482)

JERVIS, BILL
MEN OF THE DRAGON (II,1683)

JERVIS, HARRY
RETURN TO EDEN (II,2143)

JESSAYE, EVE
KISS ME, KATE (I,2568)

JESSEL, GEORGE
THE ALL-STAR REVUE (I,138); THE BOB HOPE SHOW (I,520); THE BOB HOPE SHOW (I,556); THE COMEBACK STORY (I,1017); GEORGE BURNS IN THE BIG TIME (I,1764); GEORGE BURNS' 100TH BIRTHDAY PARTY (II,971); GEORGE JESSEL'S SHOW BUSINESS (I,1770); HERE COME THE STARS (I,2026); HOLLYWOOD STAR REVUE (I,2093); THE RAY BOLGER SHOW (I,3731)

JESSIE, DEWAYNE
THE FURST FAMILY OF WASHINGTON (I,1718)

JETER, FELICIA
SPEAK UP AMERICA (II,2412)

JETER, JAMES
BENDER (II,204); FATHER MURPHY (II,841)

JETER, MICHAEL
FROM HERE TO ETERNITY (II,933)

JETTNER, CHRISTIAN
THE HEALERS (II,1115)

JEWELL, GERI
THE FACTS OF LIFE (II,805); I LOVE LIBERTY (II,1215)

JEZEK, KENNETH
JUMP (II,1356)

JILLIAN, ANN
BATTLE OF THE NETWORK STARS (II,171); BATTLE OF THE NETWORK STARS (II,172); BATTLE OF THE NETWORK STARS (II,173); BOB HOPE SPECIAL: BOB HOPE LAUGHS WITH THE MOVIE AWARDS (II,288); BOB HOPE SPECIAL: BOB HOPE'S ALL-STAR SUPER BOWL PARTY (II,304); BOB HOPE SPECIAL: BOB HOPE'S USO CHRISTMAS IN BEIRUT (II,320); BOB HOPE SPECIAL: HAPPY BIRTHDAY, BOB! (II,325); CIRCUS OF THE STARS (II,537); THE DEAN JONES SHOW (I,1191); THE DEAN MARTIN CELEBRITY ROAST (II,639); DOUG HENNING'S WORLD OF MAGIC V (II,729); ELLIS ISLAND (II,768); HAZEL (I,1982); IT'S A LIVING (II,1254); JENNIFER SLEPT HERE (II,1278); THE MAGIC PLANET (II,1598); MAKING A LIVING (II,1604); MALIBU (II,1607); OFF WE GO (I,3338); PARADE OF STARS (II,1954); PERRY COMO'S EASTER IN GUADALAJARA (II,2015); THE RAINBOW GIRL (II,2108); SEALAB 2020 (I,3949); TELEVISION'S GREATEST COMMERCIALS IV (II,2556); TEXACO STAR THEATER: OPENING NIGHT (II,2565); WOMEN WHO RATE A "10" (II,2825)

JILLSON, JOYCE
ENTERTAINMENT TONIGHT (II,780); THE JOYCE JILLSON SHOW (II,1346); PEYTON PLACE (I,3574); SAMMY AND COMPANY (II,2248); TOP TEN (II,2648)

JIM HENSON'S MUPPETS
THE ED SULLIVAN SHOW (I,1401); SESAME STREET (I,3982)

JIMENEZ, ALVERNETTE
GIMME A BREAK (II,995); THE SHAPE OF THINGS (II,2315); THICKE OF THE NIGHT (II,2587)

JIMMY JOYCE SINGERS, THE
THE BOBBY DARIN AMUSEMENT COMPANY (I,672); THE JOHN GARY SHOW (I,2411); MAKE MINE RED, WHITE, AND BLUE (I,2841); THE SMOTHERS BROTHERS COMEDY HOUR (I,4095)

JINNETTE, CANDY
THE RENEGADES (II,2132)

JOB, WILLIAM
AN ECHO OF THERESA (I,1399)

JOBIM, ANTONIO CARLOS
FRANK SINATRA: A MAN AND HIS MUSIC (I,1662)

JOCELYN, JUNE
I DREAM OF JEANNIE (I,2167)

JOCHIM, ANTHONY
GETAWAY CAR (I,1783)

JODELSOHN, ANITA
FEATHERSTONE'S NEST (II,847); THE YOUNG AND THE RESTLESS (II,2862)

JOE BASILE'S BAND
INVITATION TO PARIS (I,2233)

JOE, TOKYO
THE DON HO SHOW (II,706)

JOEL, BILLY
BILLY JOEL—A TV FIRST (II,250)

JOEL, DENNIS
THE BETTY HUTTON SHOW (I,413)

JOEL, PAUL
CLUB OASIS (I,985)

JOFFREY BALLET, THE
DIANA (II,667)

JOHN BUTLER BALLET GROUP, THE
THE KATE SMITH HOUR (I,2508)

JOHN BUTLER DANCERS, THE

THE KATE SMITH EVENING HOUR (I,2507)

JOHN WAR EAGLE
THE SAGA OF ANDY BURNETT (I,3883)

JOHN, CAROLINE
DOCTOR WHO (II,689)

JOHN, CAROLYN
DOCTOR WHO—THE FIVE DOCTORS (II,690)

JOHN, ELTON
CHER (II,496); ELTON JOHN IN CENTRAL PARK (II,769); ELTON JOHN: IN CONCERT (II,770); MONSANTO PRESENTS MANCINI (I,3092); OLIVIA NEWTON-JOHN'S HOLLYWOOD NIGHTS (II,1886); THE ROOTS OF ROCK 'N' ROLL (II,2212)

JOHN, MARGARET
A HORSEMAN RIDING BY (II,1182)

JOHNNY MANN SINGERS, THE
THE DANNY KAYE SHOW (I,1143); THE EDDIE FISHER SHOW (I,1409); THE GEORGE GOBEL SHOW (I,1768); STAND UP AND CHEER (I,4188)

JOHNNY AND THE AMERICANS
NBC SATURDAY PROM (I,3245)

JOHNS, GLYNIS
BATMAN (I,366); CHEERS (II,494); THE DICK POWELL SHOW (I,1269); THE DUPONT SHOW OF THE WEEK (I,1388); THE ERROL FLYNN THEATER (I,1458); GLYNIS (I,1829); HIDE AND SEEK (I,2047); KRAFT MYSTERY THEATER (I,2589); THE LLOYD BRIDGES SHOW (I,2733); SAFARI (I,3882); SCHLITZ PLAYHOUSE OF STARS (I,3936); SPRAGGUE (II,2430)

JOHNS, MILTON
A HORSEMAN RIDING BY (II,1182)

JOHNS, STRATFORD
GREAT EXPECTATIONS (II,1055); MASTER OF THE GAME (II,1647)

JOHNSON, ALEXANDRA
CONCRETE BEAT (II,562)

JOHNSON, AMY
SISTER TERRI (II,2366)

JOHNSON, ANDREW
THE REDD FOXX COMEDY HOUR (II,2126)

JOHNSON, ANN-MARIE
DOUBLE TROUBLE (II,724); HIGH SCHOOL, U.S.A. (II,1149)

JOHNSON, ARCH
THE ASPHALT JUNGLE (I,297); CAMP RUNAMUCK (I,811); THE GOLDEN AGE OF TELEVISION (II,1019); THE INVISIBLE MAN (II,1243); IT'S ALWAYS JAN (I,2264); A MAN CALLED RAGAN (I,2858); PETER LOVES MARY (I,3565); THE ROOKIES (I,3840)

JOHNSON, ARTE
ARNOLD'S CLOSET REVIEW (I,262); THE ARTHUR GODFREY SPECIAL (I,288); ARTHUR GODFREY'S PORTABLE ELECTRIC MEDICINE SHOW (I,290); THE B.B. BEEGLE SHOW (II,124); BAGGY PANTS AND THE NITWITS (II,135); BATTLE OF THE NETWORK STARS (II,166); BATTLE OF THE NETWORK STARS (II,171); BEN VEREEN. . . COMIN' AT YA (II,203); BEST FOOT FORWARD (I,401); BEWITCHED (I,418); THE BOBBY VINTON SHOW (II,345); BRIEF ENCOUNTER (I,723); BUNCO (II,395); CALL HOLME (II,412); CONDOMINIUM (II,566); DON'T CALL ME CHARLIE (I,1349); FANTASY ISLAND (II,829); THE FIFTH DIMENSION SPECIAL: AN ODYSSEY IN THE COSMIC UNIVERSE OF PETER MAX (I,1571); FRANK SINATRA JR. WITH FAMILY AND FRIENDS (I,1663); GENERAL ELECTRIC TRUE (I,1754); GEORGE BURNS' HOW TO LIVE TO BE 100 (II,972); GLITTER (II,1009); HENNESSEY (I,2013); THE HOUNDCATS (I,2132); IVY LEAGUE (I,2286); KNOCKOUT (II,1403); THE MICKIE FINN SPECIAL (I,3028); ROBERT YOUNG WITH THE YOUNG (I,3814); ROWAN AND MARTIN'S LAUGH-IN (I,3856); ROWAN AND MARTIN'S LAUGH-IN (I,3857); SALLY (I,3890); SUPER COMEDY BOWL 1 (I,4288); SUPER COMEDY BOWL 2 (I,4289); THE SUPER SIX (I,4292); TALES OF THE APPLE DUMPLING GANG (II,2536); VER-R-R-RY INTERESTING (I,4712)

JOHNSON, AUDRE
THE SECRET STORM (I,3969)

JOHNSON, BEEJ
A DOG'S LIFE (II,697)

JOHNSON, BEN
THE MONROES (I,3089); WILD TIMES (II,2799)

JOHNSON, BESS
SEARCH FOR TOMORROW (II,2284)

JOHNSON, BETTY
EDDY ARNOLD TIME (I,1412); THE TONIGHT SHOW (I,4564)

JOHNSON, BILL
WELCOME TO PARADISE (II,2762)

JOHNSON, BILLY
THE LITTLE REVUE (I,2717)

JOHNSON, BOB
MISSION: IMPOSSIBLE (I,3067)

JOHNSON, BRAD
ANNIE OAKLEY (I,225); WINCHESTER (I,4872)

JOHNSON, BRETT
BRANAGAN AND MAPES (II,367); SNOOPY'S GETTING MARRIED, CHARLIE BROWN (I,4102)

JOHNSON, BYAN
WHAT'S IT ALL ABOUT WORLD? (I,4816)

JOHNSON, CARLTON
ON STAGE WITH BARBARA MCNAIR (I,3358)

JOHNSON, CHARLES
THEY STAND ACCUSED (I,4442)

JOHNSON, CHERIE
PUNKY BREWSTER (II,2092)

JOHNSON, CHIP
BATTLESTAR GALACTICA (II,181); FARRELL: FOR THE PEOPLE (II,834); THE LAST CONVERTIBLE (II,1435)

JOHNSON, DE'LILA
THIS IS KATE BENNETT (II,2597)

JOHNSON, DON
BEULAH LAND (II,226); THE CITY (II,541); COVER GIRLS (II,596); FROM HERE TO ETERNITY (II,933); MIAMI VICE (II,1689); THE REBELS (II,2121); SIX PACK (II,2370)

JOHNSON, GAIL
DAYS OF OUR LIVES (II,629); THIS IS MUSIC (I,4458)

JOHNSON, GEORGANN
ARCHIE BUNKER'S PLACE (II,105); THE BACHELOR (I,325); THE CAMPBELL TELEVISION SOUNDSTAGE (I,812); CUTTER TO HOUSTON (II,612); THE GOLDEN AGE OF TELEVISION (II,1019); THE LARRY STORCH SHOW (I,2618); LIVING IN PARADISE (II,1503); MR. PEEPERS (I,3147); PHIL SILVERS ON BROADWAY (I,3578); REVLON MIRROR THEATER (I,3781); THE SECRET LIFE OF JOHN MONROE (I,3966); SILVER SPOONS (II,2355); SOMERSET (I,4115); SUTTERS BAY (II,2508); TALES OF TOMORROW (I,4352); THREE'S COMPANY (II,2614)

JOHNSON, GERRY
THE FLINTSTONES (II,884)

JOHNSON, HAROLD 'CHICK'
FIREBALL FUN FOR ALL (I,1577)

JOHNSON, JANET LOUISE
B.J. AND THE BEAR (II,125); BATTLESTAR GALACTICA (II,181); THE NANCY DREW MYSTERIES (II,1789)

JOHNSON, JARROD
DUFFY (II,739); FRIENDS (II,927); THE LOST SAUCER (II,1524); SZYSZNYK (II,2523)

JOHNSON, JASON
CRISIS IN SUN VALLEY (II,604); GETAWAY CAR (I,1783)

JOHNSON, JAY
BLOCKHEADS (II,270); CELEBRITY CHALLENGE OF THE SEXES 5 (II,469); CELEBRITY CHARADES (II,470); SO YOU THINK YOU GOT TROUBLES?! (II,2391); SOAP (II,2392); SUTTERS BAY (II,2508)

JOHNSON, JENNIFER
GEORGE WASHINGTON (II,978)

JOHNSON, JERMAIN H
PALMERSTOWN U.S.A. (II,1945)

JOHNSON, JOHN LEE
THE MICKEY MOUSE CLUB (I,3025)

JOHNSON, JOHNNY
HENRY FONDA PRESENTS THE STAR AND THE STORY (I,2016)

JOHNSON, JUDY
HIDDEN TREASURE (I,2045); JUDGE FOR YOURSELF (I,2466); TONIGHT! AMERICA AFTER DARK (I,4555); YOUR SHOW OF SHOWS (I,4957); YOUR SHOW OF SHOWS (II,2875)

JOHNSON, JULIE ANN
THE MAN FROM U.N.C.L.E. (I,2867)

JOHNSON, JUNE
FIREBALL FUN FOR ALL (I,1577)

JOHNSON, KATHIE LEE
THE FUNNIEST JOKE I EVER HEARD (II,940); HEE HAW (II,1123); THE HEE HAW HONEYS (II,1124); THE HEE HAW HONEYS (II,1125)

JOHNSON, KEN
GOOBER AND THE TRUCKERS' PARADISE (II,1029)

JOHNSON, KNOWL
TALES FROM THE DARKSIDE (II,2534)

JOHNSON, LAMONT
THE LILLI PALMER THEATER (I,2704)

JOHNSON, LAURA
DALLAS (II,614); FALCON CREST (II,808); FLY AWAY HOME (II,893)

JOHNSON, LINDA
HEE HAW (II,1123)

JOHNSON, LYN
THE SUPER SIX (I,4292)

JOHNSON, LYNDA BAINES
JUMP (II,1356)

JOHNSON, LYNN-HOLLY
CHIPS (II,511); MICKEY SPILLANE'S MIKE HAMMER: MORE THAN MURDER (II,1693)

JOHNSON, MAGIC
FAMOUS LIVES (II,821); THE JIMMY MCNICHOL SPECIAL (II,1292); SPEAK UP AMERICA (II,2412)

JOHNSON, MARY ANN
LOVE OF LIFE (I,2774)

JOHNSON, MARY JANE
MIDWESTERN HAYRIDE (I,3033)

JOHNSON, MELODIE
A BEDTIME STORY (I,384); ENIGMA (II,778); FAME IS THE NAME OF THE GAME (I,1518); I LOVE A MYSTERY (I,2170); POWDERKEG (I,3657); READY AND WILLING (II,2115)

JOHNSON, MICHAEL
THE HUMAN JUNGLE (I,2159)

JOHNSON, MICHELE
KID POWER (I,2532)

JOHNSON, NOEL
THE EXPERT (I,1489); THE SNOW GOOSE (I,4103)

JOHNSON, PENNY
THE PAPER CHASE: THE SECOND YEAR (II,1950)

JOHNSON, PETER
JABBERWOCKY (II,1262)

JOHNSON, RAFER
POLICE STORY (I,3636)

JOHNSON, RALPH
ANOTHER LIFE (II,95); EARTH, WIND & FIRE IN CONCERT (II,750)

JOHNSON, RICHARD
HAMLET (I,1925)

JOHNSON, ROBERT
COMMAND PERFORMANCE (I,1029)

JOHNSON, RUSSELL
THE BASTARD/KENT FAMILY CHRONICLES (II,159); BLACK SADDLE (I,480); CROSSROADS (I,1105); GENERAL ELECTRIC TRUE (I,1754); GILLIGAN'S ISLAND (II,990); GILLIGAN'S ISLAND: RESCUE FROM GILLIGAN'S ISLAND (II,991); GILLIGAN'S ISLAND: THE CASTAWAYS ON GILLIGAN'S ISLAND (II,992); GILLIGAN'S ISLAND: THE HARLEM GLOBETROTTERS ON GILLIGAN'S ISLAND (II,993); GILLIGAN'S PLANET (II,994); MEDIC (I,2976); THE NEW ADVENTURES OF GILLIGAN (II,1813); NOWHERE TO HIDE (II,1868); OWEN MARSHALL: COUNSELOR AT LAW (I,3435)

JOHNSON, SALLY
DAYS OF OUR LIVES (II,629)

JOHNSON, STEPHEN
THE CURSE OF DRACULA (II,611)

JOHNSON, TANIA
BACKSTAIRS AT THE WHITE HOUSE (II,133)

JOHNSON, TINA
TEXAS (II,2566)

JOHNSON, TOMMY
RENDEZVOUS WITH MUSIC (I,3768)

JOHNSON, VAN
AT YOUR SERVICE (I,310); BATMAN (I,366); BLACK BEAUTY (II,261); CALL HER MOM (I,796); DANNY THOMAS: AMERICA I LOVE YOU (I,1145); GLITTER (II,1009); THE GOLDDIGGERS (I,1838); JOHN SCHNEIDER'S CHRISTMAS HOLIDAY (II,1320); THE JUNE ALLYSON SHOW (I,2488); MAN IN THE MIDDLE (I,2870); ONE DAY AT A TIME (II,1900); THE PIED PIPER OF HAMELIN (I,3595); RICH MAN, POOR MAN—BOOK I (II,2161); RICH MAN, POOR MAN—BOOK II (II,2162); SAN FRANCISCO INTERNATIONAL AIRPORT (I,3906)

JOHNSON, VICTORIA
WOMEN WHO RATE A "10" (II,2825)

JOHNSTON, AMY
BROTHERS AND SISTERS (II,382); BUT MOTHER! (II,403)

JOHNSTON, CHRISTOPHER
GLORIA COMES HOME (II,1011)

JOHNSTON, ERIC
V (II,2713); V: THE FINAL BATTLE (II,2714)

JOHNSTON, JANE
WHEN THINGS WERE ROTTEN (II,2782)

JOHNSTON, JOHN DENNIS
THE BLUE AND THE GRAY (II,273); DEAR DETECTIVE (II,652); THE JORDAN CHANCE (II,1343)

JOHNSTON, JOHNNY
HOME (I,2099); THE JOHNNY JOHNSTON SHOW (I,2428); THE KEN MURRAY SHOW (I,2525); THE STORK CLUB (I,4236)

JOHNSTON, LIONEL
ANOTHER WORLD (II,97); SENIOR YEAR (II,2301); SONS AND DAUGHTERS (II,2404)

JOHNSTON, LYNN MARIE
FLYING HIGH (II,895); THE SKATEBIRDS (II,2375)

JOHNSTONE, WILLIAM
AS THE WORLD TURNS (II,110)

JOLIFFE, DOROTHY
THE RED BUTTONS SHOW (I,3750)

JOLLEY, I STANFORD
DANIEL BOONE (I,1142); SPACE PATROL (I,4144); WON'T IT EVER BE MORNING? (I,4902)

JOLLEY, SCOOTER
WHO'S AFRAID OF MOTHER GOOSE? (I,4852)

JOLLIFFE, DAVID
THE CLUE CLUB (II,546); EMERGENCY PLUS FOUR (II,774); ROOM 222 (I,3843)

JOLLS, MERRILL
JIM AND JUDY IN TELELAND (I,2378)

JON CHARLES ORCHESTRA, THE
100 YEARS OF AMERICA'S POPULAR MUSIC (II,1904)

JONAH JONES QUARTET, THE
ANOTHER EVENING WITH FRED ASTAIRE (I,229); ARTHUR GODFREY AND THE SOUNDS OF NEW YORK (I,280); AN EVENING WITH FRED ASTAIRE (I,1474); THE FRED ASTAIRE SPECIAL (I,1678)

JONES BOYS, THE
THE ROSEMARY CLOONEY SHOW (I,3848)

JONES JR, EDGAR ALLAN
ACCUSED (I,18); DAY IN COURT (I,1182); TRAFFIC COURT (I,4590)

JONES JR, SPIKE
THE COMICS (I,1028)

JONES, AMANDA
AT EASE (II,116)

JONES, ANISSA
FAMILY AFFAIR (I,1519)

JONES, ANTHONY
TEACHER, TEACHER (I,4366)

JONES, ARTHUR
CAPTURE (I,840); WILD CARGO (I,4862)

JONES, BARBARA O
A.E.S. HUDSON STREET (II,17); ENIGMA (II,778)

JONES, BARRY
THE CRADLE SONG (I,1088); HAMLET (I,1924); LITTLE MOON OF ALBAN (I,2714); TEN LITTLE INDIANS (I,4394); TIME REMEMBERED (I,4509)

JONES, BEN
THE DUKES OF HAZZARD (II,742)

JONES, BRENT
DYNASTY (II,747)

JONES, BUSTER
THE SUPER GLOBETROTTERS (II,2498); THE WORLD'S GREATEST SUPER HEROES (II,2842)

JONES, CAROL
THE POSSESSED (II,2071)

JONES, CAROLYN
THE ADDAMS FAMILY (II,11); THE ADDAMS FAMILY (II,13); ALFRED HITCHCOCK PRESENTS (I,115); AMOS BURKE: WHO KILLED JULIE GREER? (I,181); BATMAN (I,366); CAPITOL (II,426); THE DAVID NIVEN THEATER (I,1178); THE DICK POWELL SHOW (I,1269); THE DUPONT SHOW OF THE WEEK (I,1388); FOUR STAR PLAYHOUSE (I,1652); GHOST STORY (I,1788); JANE WYMAN PRESENTS THE FIRESIDE THEATER (I,2345); THE JUNE ALLYSON SHOW (I,2488); THE LLOYD BRIDGES SHOW (I,2733); PANIC! (I,3447); PEPSI-COLA PLAYHOUSE (I,3523); ROOTS (II,2211); SCHLITZ PLAYHOUSE OF STARS (I,3936); STAR STAGE (I,4197); STUDIO '57 (I,4267); WONDER WOMAN (SERIES 1) (II,2828); ZANE GREY THEATER (I,4979)

JONES, CHARLIE
ALMOST ANYTHING GOES (II,56)

JONES, CHERRY
O'MALLEY (II,1872)

JONES, CHRIS
THE LEGEND OF JESSE JAMES (I,2653)

JONES, CHRISTINE
ANOTHER WORLD (II,97); FOR RICHER, FOR POORER (II,906); LOVERS AND FRIENDS (II,1549); NUMBER 96 (II,1869); ONE LIFE TO LIVE (II,1907); RITUALS (II,2182); RYAN'S HOPE (II,2234)

JONES, CHUCK
CURIOSITY SHOP (I,1114)

JONES, CLAUDE EARL
BUFFALO BILL (II,387); DALLAS (II,614); THE DEADLY GAME (II,633); THE GIRL IN THE EMPTY GRAVE (II,997); LONE STAR (II,1513)

JONES, CODY
ROOSTER (II,2210)

JONES, DAVEY MARILYN
EVENING AT THE IMPROV (II,786)

JONES, DAVID
HOT CITY DISCO (II,1185)

JONES, DAVY
33 1/3 REVOLUTIONS PER MONKEE (I,4451); THE MONKEES (I,3088); THE PEAPICKER IN PICCADILLY (I,3500)

JONES, DEAN
AMOS BURKE: WHO KILLED JULIE GREER? (I,181); THE CHICAGO TEDDY BEARS (I,934); THE DEAN JONES SHOW (I,1191); ENSIGN O'TOOLE (I,1448); HARD CASE (I,1947); HERBIE, THE LOVE BUG (II,1131); I LOVE HER ANYWAY! (II,1214); THE ICE PALACE (I,2182); KRAFT SALUTES WALT DISNEY WORLD'S 10TH ANNIVERSARY (II,1410); THE LONG DAYS OF SUMMER (II,1514); MITZI AND A HUNDRED GUYS (II,1710); THE NIGHT OF CHRISTMAS (I,3290); WHAT'S IT ALL ABOUT WORLD? (I,4816)

JONES, DERRICK
TALES OF THE HAUNTED (II,2538)

JONES, DICK
BUFFALO BILL JR. (I,755); THE RANGE RIDER (I,3720); WHEN THE WEST WAS FUN: A WESTERN REUNION (II,2780)

JONES, DUDLEY
MARRIED ALIVE (I,2923)

JONES, EDDIE
TALES FROM THE DARKSIDE (II,2534)

JONES, EMRY
THE LILLI PALMER THEATER (I,2704)

JONES, FREDDIE
APPOINTMENT WITH A KILLER (I,245)

JONES, G STANLEY
EXPERT WITNESS (I,1488)

JONES, GEMMA
DIAL A DEADLY NUMBER (I,1253)

JONES, GEORGE
COUNTRY COMES HOME (II,585); COUNTRY MUSIC HIT PARADE (II,589); COUNTRY NIGHT OF STARS (II,592); THE GEORGE JONES SPECIAL (II,975); THE JOHNNY CASH SPRING SPECIAL (II,1329); JOHNNY CASH: THE FIRST 25 YEARS (II,1336); MUSIC CITY NEWS TOP COUNTRY HITS OF THE YEAR (II,1772); THE NASHVILLE PALACE (II,1792)

JONES, GINGER
BEULAH (I,416)

JONES, GLENDA
ALL ABOUT FACES (I,126)

JONES, GORDON
THE ABBOTT AND COSTELLO SHOW (I,2); DOBIE GILLIS (I,1302); MEET MR. MCNUTLEY (I,2989); SO THIS IS HOLLYWOOD (I,4105)

JONES, GRANDPA
CHRISTMAS LEGEND OF NASHVILLE (II,522); COUNTRY MUSIC CARAVAN (I,1066); JOHNNY CASH: THE FIRST 25 YEARS (II,1336)

JONES, GRIFF RHYS
NOT THE NINE O'CLOCK NEWS (II,1864)

JONES, HARRY
MASTER OF THE GAME (II,1647)

JONES, HENRY
B.J. AND THE BEAR (II,126); CHANNING (I,900); DANGER (I,1134); THE DAUGHTERS OF JOSHUA CABE (I,1169); THE DOCTOR (I,1320); EYE WITNESS (I,1494); THE GIRL WITH SOMETHING EXTRA (II,999); GULF PLAYHOUSE (I,1904); GUN SHY (II,1068); HANDS OF MURDER (I,1929); HITCHED (I,2065); HONESTLY, CELESTE! (I,2104); JANE WYMAN PRESENTS THE FIRESIDE THEATER (I,2345); KATE LOVES A MYSTERY (II,1367); THE LETTERS (I,2670); LETTERS FROM THREE LOVERS (I,2671); LIGHT'S OUT (I,2699); LOST IN SPACE (I,2758); MR. BEVIS (I,3128); MRS. COLUMBO (II,1760); OF THIS TIME, OF THAT PLACE (I,3335); PHYLLIS (II,2038); PLAYWRIGHTS '56 (I,3627); QUICK AND QUIET (II,2099); REVLON MIRROR THEATER (I,3781); SCENE OF THE CRIME (II,2271); SHORT, SHORT DRAMA (I,4023); SINGLES (I,4062); TALES OF THE APPLE DUMPLING GANG (II,2536); THESE ARE THE DAYS (II,2585); WE'LL GET BY (II,2753); THE WEB (I,4784)

JONES, HERBIE
THE JIMMY DEAN SHOW (I,2384)

JONES, J BILL
ANOTHER MAN'S SHOES (II,96)

JONES, JACK
THE BEST ON RECORD (I,409); THE BOB HOPE SHOW (I,601); THE BOB HOPE SHOW (I,606); THE BOB HOPE SHOW (I,609); THE BOB HOPE SHOW (I,615); THE BOB HOPE SHOW (I,617); THE BURT BACHARACH SPECIAL (I,776); CONDOMINIUM (II,566); GUYS 'N' GEISHAS (I,1914); HOWDY (I,2150); THE ICE PALACE (I,2182); THE JACK JONES SPECIAL (I,2310); THE PALACE (II,1943); THE SINGERS (I,4060); TEXACO STAR PARADE II (I,4410); A VERY SPECIAL OCCASION (I,4716)

JONES, JACKIE
THE WAYNE KING SHOW (I,4779)

JONES, JAMES EARL
AS THE WORLD TURNS (II,110); THE GUIDING LIGHT (II,1064); HAPPY ENDINGS (II,1085); PARIS (II,1957); ROOTS: THE NEXT GENERATIONS (II,2213)

JONES, JANET
DANCE FEVER (II,615)

JONES, JANICE
SAMMY KAYE'S MUSIC FROM MANHATTAN (I,3902)

JONES, JEFFREY
A FINE ROMANCE (II,858)

JONES, JO
TIMEX ALL-STAR JAZZ SHOW IV (I,4518)

JONES, JOANELLE
SING-CO-PATION (I,4058)

JONES, JOHN CHRISTOPHER
ON OUR OWN (II,1892)

JONES, JONAH
THE NASHVILLE SOUND OF BOOTS RANDOLPH (I,3229)

JONES, JUDY
LONGSTREET (I,2749); LONGSTREET (I,2750)

JONES, KATHY
CORONATION STREET (I,1054)

JONES, KEN
I'M THE GIRL HE WANTS TO KILL (I,2194)

JONES, L Q
THE BUFFALO SOLDIERS (II,388); CHARLIE WOOSTER—OUTLAW (I,915); COLORADO C.I. (II,555); KLONDIKE (I,2573); RAWHIDE (I,3727); RIDDLE AT 24000 (II,2171); THE VIRGINIAN (I,4738); WILD TIMES (II,2799)

JONES, LEE

STRAWHAT THEATER (I,4256)

JONES, LYNNE
MOLL FLANDERS (II,1720)

JONES, MALLORY
ANOTHER WORLD (II,97); EISCHIED (II,763)

JONES, MARILYN
BUS STOP (II,399); KING'S CROSSING (II,1397); V: THE SERIES (II,2715)

JONES, MICKEY
DOWN HOME (II,731); THE DUKES OF HAZZARD (II,742); FATHER MURPHY (II,841); FLO (II,891); GOOBER AND THE TRUCKERS' PARADISE (II,1029); HEAR NO EVIL (II,1116); JOHNNY BELINDA (I,2418); NORMA RAE (II,1860); ROLLIN' ON THE RIVER (I,3834); V: THE FINAL BATTLE (II,2714); V: THE SERIES (II,2715)

JONES, MILTON
BUTTERFLIES (II,404)

JONES, MORGAN
ADVICE TO THE LOVELORN (II,16); ASSIGNMENT: EARTH (I,299)

JONES, NAT
EXECUTIVE SUITE (II,796)

JONES, NICHOLAS
THE FLAME TREES OF THIKA (II,870); HAMLET (I,1925)

JONES, PAMELA
SEARCH (I,3951)

JONES, PETER
FROM A BIRD'S EYE VIEW (I,1686); THE HITCHHIKER'S GUIDE TO THE GALAXY (II,1158)

JONES, QUINCY
DIANA (II,667); DUKE ELLINGTON. . . WE LOVE YOU MADLY (I,1380)

JONES, RENEE
JESSIE (II,1287)

JONES, REV
1915 (II,1853)

JONES, ROSS
THE ADVENTURES OF CYCLONE MALONE (I,57)

JONES, SAM J
BATTLE OF THE NETWORK STARS (II,173); CODE RED (II,552); CODE RED (II,553); NO MAN'S LAND (II,1855); STUNTS UNLIMITED (II,2487)

JONES, SAMMY
PHYL AND MIKHY (II,2037)

JONES, SHIRLEY
THE ADVENTURES OF POLLYANNA (II,14); THE ALAN KING SHOW (I,102); THE BIG SHOW (II,243); THE BOB HOPE SHOW (I,537); THE BOB HOPE SHOW (I,617); THE BOB HOPE SHOW (I,648); THE BOB HOPE SHOW (I,654); THE BOB HOPE SHOW (I,664); BOB HOPE SPECIAL: BOB HOPE—HOPE, WOMEN AND SONG (II,324); BOB HOPE SPECIAL: HO HO HOPE'S JOLLY CHRISTMAS HOUR (II,327); DUPONT SHOW OF THE MONTH (I,1387); FOR THE LOVE OF MIKE (I,1625); FRIENDS AND NABORS (I,1684); GRUEN GUILD PLAYHOUSE (I,1892); HOLLYWOOD MELODY (I,2085); A KNIGHT IN SHINING ARMOUR (I,2575); THE LIVES OF JENNY DOLAN (II,1502); THE LUX VIDEO THEATER (I,2795); OUT OF THE BLUE (I,3422); THE PARTRIDGE FAMILY (II,1962); THE PERRY COMO SHOW (I,3533); PERRY COMO'S MUSIC FROM HOLLYWOOD (II,2019); THE ROYAL FOLLIES OF 1933 (I,3862); SHIRLEY (II,2330); STEP ON THE GAS (I,4215); THANKSGIVING REUNION WITH THE PARTRIDGE FAMILY AND MY THREE SONS (II,2569); WOMEN OF RUSSIA (II,2824); THE WONDERFUL WORLD OF BURLESQUE I (I,4895); YOU'RE THE TOP (I,4974)

JONES, SIMON
BRIDESHEAD REVISITED (II,375); THE HITCHHIKER'S GUIDE TO THE GALAXY (II,1158); THE NEWS IS THE NEWS (II,1838)

JONES, SPIKE
THE ALL-STAR REVUE (I,138); CLUB OASIS (I,984); CLUB OASIS (I,985); THE SPIKE JONES SHOW (I,4155); THE SPIKE JONES SHOW (I,4156); THE SPIKE JONES SHOW (I,4157)

JONES, STAN
SHERIFF OF COCHISE (I,4008)

JONES, STANLEY
MORK AND MINDY (II,1736); THE RICHIE RICH SHOW (II,2169); THE WORLD'S GREATEST SUPER HEROES (II,2842)

JONES, T C
THE BIG SELL (I,431)

JONES, TERRY
MONTY PYTHON'S FLYING CIRCUS (II,1729)

JONES, TOM
THE BOB HOPE SHOW (I,634); THE BOB HOPE SHOW (I,644); THE BURT BACHARACH SPECIAL (I,775); THE DIAHANN CARROLL SHOW (I,1252); THE ENGELBERT HUMPERDINCK SPECIAL (I,1446); IT'S WHAT'S HAPPENING, BABY! (I,2278); LAS VEGAS PALACE OF STARS (II,1431); LYNDA CARTER: ENCORE (II,1566); MAC DAVIS 10TH ANNIVERSARY SPECIAL: I STILL BELIEVE IN MUSIC (II,1571); MAC DAVIS. . . SOUNDS LIKE HOME (II,1582); PLEASURE COVE (II,2058); RAQUEL (I,3725); SHIRLEY MACLAINE AT THE LIDO (II,2332); SPECIAL LONDON BRIDGE SPECIAL (I,4150); THIS IS TOM JONES (I,4460); THE TOM JONES CHRISTMAS SPECIAL (I,4539); THE TOM JONES SPECIAL (I,4540); THE TOM JONES SPECIAL (I,4541); THE TOM JONES SPECIAL (I,4542); THE TOM JONES SPECIAL (I,4543); THE TOM JONES SPECIAL (I,4544); THE TOM JONES SPECIAL (I,4545)

JONES, TOMMY LEE
CAT ON A HOT TIN ROOF (II,457); CHARLIE'S ANGELS (II,487); ONE LIFE TO LIVE (II,1907); THE RAINMAKER (II,2109)

JONES, TRENT
RYAN'S HOPE (II,2234)

JONES, TYRONE
OUR STREET (I,3418)

JONES, WARNER
THE BLUE ANGELS (I,490); WINDOW ON MAIN STREET (I,4873)

JONES-MORELAND, BETSY
THE OUTER LIMITS (I,3426); PROFILES IN COURAGE (I,3678)

JONNS, BIRL
THE JERK, TOO (II,1282)

JOPLIN, JOHN L
THE RUNAWAY BARGE (II,2230)

JORDAN, BEVERLY
THE LIFE OF VERNON HATHAWAY (I,2687)

JORDAN, BOBBI
THE BARBARY COAST (II,146); THE BARBARY COAST (II,147); BLONDIE (I,487); GENERAL HOSPITAL (II,964); JOE AND SONS (II,1296); THE ROUNDERS (I,3851); TURNABOUT (II,2669)

JORDAN, CHRISTY
THE SEVEN LITTLE FOYS (I,3986)

JORDAN, DULCIE
THE SWIFT SHOW (I,4316); WONDER WOMAN (SERIES 2) (II,2829)

JORDAN, HARLAN
CELEBRITY (II,463); HIGHWAY HONEYS (II,1151)

JORDAN, JAMES CARROLL
THE BLUE AND THE GRAY (II,273); LONDON AND DAVIS IN NEW YORK (II,1512); RICH MAN, POOR MAN—BOOK II (II,2162); SANDBURG'S LINCOLN (I,3907); STEELTOWN (II,2452); WHEELS (II,2777)

JORDAN, JOANNE
COMMANDO CODY (I,1030); HERE'S HOLLYWOOD (I,2031)

JORDAN, KIRK
AMAHL AND THE NIGHT VISITORS (I,148)

JORDAN, RICHARD
CAPTAINS AND THE KINGS (II,435); NIGHTSIDE (I,3296); READY FOR THE PEOPLE (I,3739)

JORDAN, TED
GUNSMOKE (II,1069)

JORDAN, TOM
THE MANIONS OF AMERICA (II,1623)

JORDAN, TONY
THE FAR PAVILIONS (II,832)

JORDAN, WILLIAM
BEYOND WESTWORLD (II,227); CALL TO DANGER (I,804); PROJECT UFO (II,2088)

JORDANAIRES, THE
COUNTRY MUSIC CARAVAN (I,1066); ELVIS REMEMBERED: NASHVILLE TO HOLLYWOOD (II,772)

JORDEN, JAN
M*A*S*H (II,1569); NEVER SAY NEVER (II,1809)

JORDON, SUZANNAH
THE HAPPENERS (I,1935)

JORRIN, MAURICE
COS (II,576)

JORY, VICTOR
THE ALCOA HOUR (I,108); CHRYSLER MEDALLION

THEATER (I,951); CROSSROADS (I,1105); GREAT ADVENTURE (I,1868); JOHNNY BELINDA (I,2416); KING'S ROW (I,2563); LAST STAGECOACH WEST (I,2629); MANHUNT (I,2890); MATINEE THEATER (I,2947); NAKIA (II,1785); PURSUIT (I,3700); SURE AS FATE (I,4297); TV READERS DIGEST (I,4630); WHO HAS SEEN THE WIND? (I,4847)

JOSEPH, ALLEN
DIAGNOSIS: DANGER (I,1250)

JOSEPH, JACKIE
THE ALL-NEW POPEYE HOUR (II,49); BARBARA MANDRELL AND THE MANDRELL SISTERS (II,143); THE BOB NEWHART SHOW (I,666); THE DORIS DAY SHOW (I,1355); THE FIRST NINE MONTHS ARE THE HARDEST (I,1584); JOSIE AND THE PUSSYCATS (I,2453); JOSIE AND THE PUSSYCATS IN OUTER SPACE (I,2454); KELLY'S KIDS (I,2523); THE MAGIC LAND OF ALLAKAZAM (I,2817); THE ROYAL FOLLIES OF 1933 (I,3862); RUN, BUDDY, RUN (I,3870)

JOSEPH, JENNIFER
BATTLESTAR GALACTICA (II,181)

JOSEPH, PHILLIP
GREAT EXPECTATIONS (II,1055)

JOSEPHSON, ERIAND
SCENES FROM A MARRIAGE (II,2272)

JOSEPHSON, JEFFREY
LONE STAR (II,1513)

JOSLIN, JACKIE
THE DINAH SHORE SHOW (I,1280)

JOSLYN, ALLYN
THE ADDAMS FAMILY (II,11); THE EVE ARDEN SHOW (I,1470); HEAVE HO HARRIGAN (I,1993); KILROY (I,2541); MCKEEVER AND THE COLONEL (I,2970); TOO MANY SERGEANTS (I,4573); WHERE'S RAYMOND? (I,4837)

JOSLYN, BETSY
SWEENEY TODD (II,2512)

JOSOL, STAN
THE BEACHCOMBER (I,371)

JOSTYN, JAY
THE DOCTOR (I,1320); MR. DISTRICT ATTORNEY (I,3136); NIGHT COURT (I,3285)

JOURDAN, LOUIS
ACCENT ON LOVE (I,16); CLIMAX! (I,976); THE ELGIN HOUR (I,1428); FORD TELEVISION THEATER (I,1634); THE FRENCH ATLANTIC AFFAIR (II,924); KRAFT SUSPENSE THEATER (I,2591); ONE MORE TIME (I,3385); PARIS PRECINCT (I,3458); ROMANCE THEATER (II,2204); SALUTE TO LADY LIBERTY (II,2243)

JOUZIER, ANNIE
MISTRAL'S DAUGHTER (II,1708)

JOVERS, JOLLY
MINSKY'S FOLLIES (II,1703)

JOY PEOPLE, THE
THE WONDERFUL WORLD OF AGGRAVATION (I,4894)

JOY, CHRISTOPHER
THE NEW ODD COUPLE (II,1825)

JOY, DICK
YOU ARE THERE (I,4929)

JOY, NICHOLAS
BOSS LADY (I,694)

JOYCE, BARBARA
EYE WITNESS (I,1494); THE KEN BERRY WOW SHOW (I,2524); PHILCO TELEVISION PLAYHOUSE (I,3583); STAR TONIGHT (I,4199); TALES OF TOMORROW (I,4352)

JOYCE, BILL
SOMERSET (I,4115)

JOYCE, DENIS
INTERMEZZO (I,2223)

JOYCE, ELAINE
ALLISON SIDNEY HARRISON (II,54); ALONE AT LAST (II,59); BOB HOPE SPECIAL: BOB HOPE IN THE STAR-MAKERS (II,285); THE BOBBY VAN AND ELAINE JOYCE SHOW (I,677); CAROL (I,846); CELEBRITY CHALLENGE OF THE SEXES 4 (II,468); CITY OF ANGELS (II,540); THE DON KNOTTS SHOW (I,1334); THE FASHION STORY (I,1537); I'VE GOT A SECRET (II,1220); LAS VEGAS PALACE OF STARS (II,1431); THE MAGIC OF DAVID COPPERFIELD (II,1594); MR. MERLIN (II,1751); THE UGILY FAMILY (II,2699); WE GOT IT MADE (II,2752)

JOYCE, IRIS
OUR FIVE DAUGHTERS (I,3413)

JOYCE, JIMMY
MERV GRIFFIN'S ST. PATRICK'S DAY SPECIAL (I,3018); WON'T IT EVER BE MORNING? (I,4902)

JOYCE, PADDY
CORONATION STREET (I,1054)

JOYCE, ROSALYN
VALENTINE MAGIC ON LOVE ISLAND (II,2716)

JOYCE, STEPHEN
ANOTHER WORLD (II,97); THE CODE OF JONATHAN WEST (I,992); FOR RICHER, FOR POORER (II,906); HIDDEN FACES (I,2044); LOVERS AND FRIENDS (II,1549); RYAN'S HOPE (II,2234); SEARCH FOR TOMORROW (II,2284); THE VIRGINIAN (I,4737)

JOYCE, SUSAN
THE BEST YEARS (I,410)

JOYCE, WILLIAM
JOHNNY GUITAR (I,2427)

JOYCE, YOOTHA
GEORGE AND MILDRED (II,965); MAN ABOUT THE HOUSE (II,1611)

JOYNT, PAUL
ONE LIFE TO LIVE (II,1907)

JUBILAIRES, THE
FIVE STAR JUBILEE (I,1589); SUGAR HILL TIMES (I,4276)

JUDD, EDWARD
FLAMBARDS (II,869); THE ONEDIN LINE (II,1912); SIGN IT DEATH (I,4046)

JUE, DEBBIE
THE AMAZING CHAN AND THE CHAN CLAN (I,157)

JUE, JAY JAY
THE AMAZING CHAN AND THE CHAN CLAN (I,157)

JULIA, RAUL
ACES UP (II,7); MCCLOUD: WHO KILLED MISS U.S.A.? (I,2965)

JULIAN, CHARLES
THE DOOLEY BROTHERS (II,715)

JULIAN, COLIN
DOCTOR DOLITTLE (I,1309)

JULIAN, JANET
KEY TORTUGA (II,1383)

JULIAN, SALLY
SALE OF THE CENTURY (II,2239)

JULIANO, LENNY
FEEL THE HEAT (II,848)

JUMP, GORDON
ARCHIE (II,104); THE ARCHIE SITUATION COMEDY MUSICAL VARIETY SHOW (II,106); FOR LOVERS ONLY (II,904); GREAT DAY (II,1054); MCDUFF, THE TALKING DOG (II,1662); SECOND EDITION (II,2288); SOAP (II,2392); WKRP IN CINCINNATI (II,2814)

JUNE GRAHAM DANCERS, THE
DON'S MUSICAL PLAYHOUSE (I,1348); HOLIDAY HOTEL (I,2075)

JUNE TAYLOR DANCERS, THE
CAVALCADE OF STARS (I,877); THE DOM DELUISE SHOW (I,1328); THE HONEYMOONERS (I,2111); JACKIE GLEASON AND HIS AMERICAN SCENE MAGAZINE (I,2321); THE JACKIE GLEASON SHOW (I,2322); THE JACKIE GLEASON SHOW (I,2323); THE JACKIE GLEASON SPECIAL (I,2325); THE LARRY STORCH SHOW (I,2618); ONE MORE TIME (I,3386); PHIL SILVERS ON BROADWAY (I,3578); SHOW BUSINESS, INC. (I,4028); STAGE SHOW (I,4182)

JUNE TAYLOR SINGERS, THE
THE LARRY STORCH SHOW (I,2618)

JUNG, CALVIN
FARRELL: FOR THE PEOPLE (II,834)

JUNKIN, JOHN
PENMARRIC (II,1993)

JUPITER, JOEY
MAMA MALONE (II,1609)

JURADO, KATY
A.K.A. PABLO (II,28)

JURASIK, PETER
IN SECURITY (II,1227)

JURGENS, KURT
SMILEY'S PEOPLE (II,2383)

JURGENSON, RANDY
MR. INSIDE/MR. OUTSIDE (II,1750)

JURY, RICHARD
OUT OF THE BLUE (I,3422)

JUSTICE, BARRY
THE PALLISERS (II,1944)

JUSTICE, KATHERINE
BUSH DOCTOR (II,400); CAPTAIN AMERICA (II,428);

DEAD MAN ON THE RUN (II,631); PRESCRIPTION: MURDER (I,3660); THE PSYCHIATRIST: GOD BLESS THE CHILDREN (I,3685); ROARING CAMP (I,3804)

JUSTRICH, TRACY
GALACTICA 1980 (II,953); MARRIED: THE FIRST YEAR (II,1634)

JUTTNER, SHELLY
THE HEALERS (II,1115); THE YOUNG PIONEERS (II,2868)

KISS
100 YEARS OF GOLDEN HITS (II,1905)

KAA, WI KUKI
WELCOME TO PARADISE (II,2762)

KAAHEA, ED
THE ISLANDER (II,1249)

KACZMAREK, JANE
FOR LOVERS ONLY (II,904); HILL STREET BLUES (II,1154); THE LAST LEAF (II,1437); THE PAPER CHASE: THE SECOND YEAR (II,1950); ST. ELSEWHERE (II,2432)

KADA, YOKO
SHOGUN (II,2339)

KADELL, CARLTON
A TIME TO LIVE (I,4510)

KADLER, KAREN
NUMBER 13 DEMON STREET (I,3316)

KAGAN, DIANE
THE ART OF CRIME (II,108)

KAHAN, JUDITH
ALL'S FAIR (II,46); BRANAGAN AND MAPES (II,367); DOC (II,682); FOREVER FERNWOOD (II,909); FREE COUNTRY (II,921); LILY (I,2705); LOVE, NATALIE (II,1545); MARY (II,1637); MO AND JO (II,1715)

KAHANA, KIM
THE FIGHTING NIGHTINGALES (II,855)

KAHLER, WOLF
MISTRAL'S DAUGHTER (II,1708)

KAHN, AL
OUR FAMILY BUSINESS (II,1931)

KAHN, DEE DEE
KEEP ON TRUCKIN' (II,1372)

KAHN, MADELINE
COMEDY TONIGHT (I,1027); EVENING AT THE IMPROV (II,786); THE GEORGE BURNS SPECIAL (II,970); HARVEY (I,1962); KLEIN TIME (II,1401); OH MADELINE (II,1880)

KAI, LANI
ADVENTURES IN PARADISE (I,46)

KALANE, ARMAND
TALES OF THE GOLD MONKEY (II,2537)

KALEM, TONI
ANOTHER WORLD (II,97); DOMINIC'S DREAM (II,704); NOW WE'RE COOKIN' (II,1866)

KALEMBER, PATRICIA
LOVING (II,1552)

KALIBAN, ROBERT
OPERATION GREASEPAINT (I,3402); THE ROBERT KLEIN SHOW (II,2185)

KALINYEA, MICHAEL
SWEENEY TODD (II,2512)

KALK, RICHARD
THE BLUE KNIGHT (II,277)

KALLAN, RANDI
SENIOR YEAR (II,2301); SONS AND DAUGHTERS (II,2404)

KALLEN, KITTY
THE JACK CARSON SHOW (I,2304); JUDGE FOR YOURSELF (I,2466); VARIETY (I,4703)

KALLIS, NICOLE
POLICE WOMAN (II,2063)

KALLMAN, DICK
THE DESILU REVUE (I,1238); HANK (I,1931)

KAMEL, STANLEY
DAYS OF OUR LIVES (II,629); OLD FRIENDS (II,1881); THE PHOENIX (II,2035); THE SHERIFF AND THE ASTRONAUT (II,2326)

KAMEN, MIKE
LOVE THY NEIGHBOR (I,2781)

KAMEN, MILT
FOL-DE-ROL (I,1613); PANTOMIME QUIZ (I,3449); THE PERRY COMO SHOW (I,3532); SID CAESAR INVITES YOU (I,4040)

KAMHI, KATHERINE GENE
ALL MY CHILDREN (II,39)

KAMINSKY, LUCIEN
PIP THE PIPER (I,3606)

KAMMER, NANCY
THE DAY THE WOMEN GOT EVEN (II,628)

KAMPMANN, STEVEN
NEWHART (II,1835); RODNEY DANGERFIELD SPECIAL: IT'S NOT EASY BEIN' ME (II,2199)

KANALY, STEVE
CELEBRITY CHALLENGE OF THE SEXES 5 (II,469); DALLAS (II,614); MELVIN PURVIS: G-MAN (II,1679); SCENE OF THE CRIME (II,2271)

KANE, BARRY
THE NEW CHRISTY MINSTRELS SHOW (I,3259)

KANE, BYRON
THE HARDY BOYS (I,1949)

KANE, CAROL
TAXI (II,2546)

KANE, FRANCES
THE GREAT SPACE COASTER (II,1059)

KANE, LARRY
THE LARRY KANE SHOW (I,2617)

KANE, MICHAEL
ROAD OF LIFE (I,3798)

KANE, SID
ONCE UPON A MATTRESS (I,3371)

KANEKO, MITSUNDBU
JOHNNY SOKKO AND HIS FLYING ROBOT (I,2436)

KANEKO, NOBUO
SHOGUN (II,2339)

KANNE, GRETCHEN
ONCE UPON A DEAD MAN (I,3369)

KANNON, JACKIE
THE SATURDAY NIGHT REVUE (I,3921)

KANTER, ABBE
EXECUTIVE SUITE (II,796)

KANTER, HAL
MILTON BERLE'S MAD MAD MAD WORLD OF COMEDY (II,1701)

KANTER, RICHARD
FINDER OF LOST LOVES (II,857)

KANUI, HANK
SAMMY KAYE'S MUSIC FROM MANHATTAN (I,3902)

KAPELOS, JOHN
THE YESTERDAY SHOW (II,2850)

KAPLAN, BROOKE
RENDEZVOUS HOTEL (II,2131)

KAPLAN, GABRIEL
BATTLE OF THE NETWORK STARS (II,163); BATTLE OF THE NETWORK STARS (II,164); BATTLE OF THE NETWORK STARS (II,165); BATTLE OF THE NETWORK STARS (II,166); BATTLE OF THE NETWORK STARS (II,167); BATTLE OF THE NETWORK STARS (II,173); CELEBRATION: THE AMERICAN SPIRIT (II,461); CELEBRITY CHALLENGE OF THE SEXES 1 (II,465); CELEBRITY CHALLENGE OF THE SEXES 2 (II,466); GABRIEL KAPLAN PRESENTS THE FUTURE STARS (II,951); GABRIEL KAPLAN PRESENTS THE SMALL EVENT (II,952); LEWIS AND CLARK (II,1465); THE LOVE BOAT I (II,1532); THE MUHAMMED ALI VARIETY SPECIAL (II,1762); A TRIBUTE TO "MR. TELEVISION" MILTON BERLE (II,2658); VARIETY (II,2721); WELCOME BACK, KOTTER (II,2761)

KAPLAN, MANDY
DINER (II,679); THE EDGE OF NIGHT (II,760)

KAPLAN, MARVIN
ALICE (II,33); THE C.B. BEARS (II,406); THE CHICAGO TEDDY BEARS (I,934); DOBIE GILLIS (I,1302); HOORAY FOR HOLLYWOOD (I,2114); MAGGIE BROWN (I,2811); MEET MILLIE (I,2988); OUT OF THE BLUE (I,3422); TOM, DICK, AND HARRY (I,4536); TOP CAT (I,4576)

KAPLOWITZ, HERB
STARSTRUCK (II,2446)

KAPOOR, GOYA
THE FAR PAVILIONS (II,832)

KAPP, JOE
CAPTAINS AND THE KINGS (II,435); NAKIA (II,1784)

KAPRALL, BO
THE 416TH (II,913); LAVERNE AND SHIRLEY (II,1446); THE MAC DAVIS SHOW (II,1575); WACKO (II,2733)

KAPTAIN KOOL & THE KONGS
THE KROFFT KOMEDY HOUR (II,141)

KAPU JR, SAM
THE DON HO SHOW (II,706)

KARABATSOS, RON
DREAMS (II,735); HEAR NO EVIL (II,1116); MISSING PIECES (II,1706)

KAREMAN, FRED
THE NEW OPERATION PETTICOAT (II,1826); THE TEAHOUSE OF THE AUGUST MOON (I,4368)

KAREN, JAMES
CHEERS (II,494); EIGHT IS ENOUGH (II,762); INSTITUTE

FOR REVENGE (II,1239); LITTLE HOUSE: THE LAST FAREWELL (II,1491); ME AND DUCKY (II,1671); THE POWERS OF MATTHEW STAR (II,2075); TOPPER (II,2649)

KARGLER, DAVE
ONCE UPON A FENCE (I,3370)

KARL, MARTIN
ARTHUR GODFREY AND FRIENDS (I,278)

KARLAN, RICHARD
THE MAN FROM DENVER (I,2863)

KARLATOS, OLGA
SCRUPLES (II,2281)

KARLEN, BETTY
FAME (II,812)

KARLEN, JOHN
CAGNEY AND LACEY (II,409); COLORADO C.I. (II,555); DARK SHADOWS (I,1157); FRANKENSTEIN (I,1671); THE LONG DAYS OF SUMMER (II,1514); THE MASK OF MARCELLA (I,2937); MELVIN PURVIS: G-MAN (II,1679); THE PATRIOTS (I,3477); THE PICTURE OF DORIAN GRAY (I,3592); THE WINDS OF WAR (II,2807)

KARLOFF, BORIS
ARSENIC AND OLD LACE (I,267); ARSENIC AND OLD LACE (I,268); ARSENIC AND OLD LACE (I,269); THE BORIS KARLOFF MYSTERY PLAYHOUSE (I,690); THE CHEVROLET TELE-THEATER (I,926); CLIMAX! (I,976); COLONEL MARCH OF SCOTLAND YARD (I,1005); A CONNECTICUT YANKEE (I,1038); CURTAIN CALL THEATER (I,1115); DUPONT SHOW OF THE MONTH (I,1387); THE ELGIN HOUR (I,1428); THE GENERAL ELECTRIC THEATER (I,1753); HOLLYWOOD OPENING NIGHT (I,2087); HOLLYWOOD SINGS (I,2091); THE KATE SMITH SHOW (I,2509); THE LARK (I,2616); LIGHT'S OUT (I,2699); THE PARADINE CASE (I,3453); PLAYHOUSE 90 (I,3623); SHIRLEY TEMPLE'S STORYBOOK: THE LEGEND OF SLEEPY HOLLOW (I,4018); SUSPENSE (I,4305); SUSPICION (I,4309); TALES OF TOMORROW (I,4352); TELEPHONE TIME (I,4379); THRILLER (I,4492); THE VEIL (I,4709)

KARMAN, JANICE
ALVIN AND THE CHIPMUNKS (II,61)

KARNES, ROBERT
THE LAWLESS YEARS (I,2641); MIDNIGHT MYSTERY (I,3030); THE NANCY DREW MYSTERIES (II,1789)

KARNILOVA, MARIA
IVAN THE TERRIBLE (II,1260)

KARNS, ROSCOE
HENNESSEY (I,2013); ROCKY KING, INSIDE DETECTIVE (I,3824)

KARNS, TODD
ROCKY KING, INSIDE DETECTIVE (I,3824)

KARR, HARRIET
THE SEEKERS (II,2298)

KARR, JOAN
CRISIS (I,1102)

KARRAS, ALEX
CENTENNIAL (II,477); MULLIGAN'S STEW (II,1763); SUPER COMEDY BOWL 1 (I,4288); WEBSTER (II,2758)

KARRON, RICHARD
THE FURTHER ADVENTURES OF WALLY BROWN (II,944); GOOD TIME HARRY (II,1037); MIXED NUTS (II,1714); PEN 'N' INC. (II,1992); ROOSEVELT AND TRUMAN (II,2209); TEACHERS ONLY (II,2547); YOUNG GUY CHRISTIAN (II,2864)

KARTALIAN, BUCK
THE MONSTER SQUAD (II,1725)

KARVELAS, ROBERT
GET SMART (II,983); THE PARTNERS (I,3462)

KASDORF, LENORE
BIG ROSE (II,241); THE GUIDING LIGHT (II,1064)

KASEM, CASEY
AMERICA'S TOP TEN (II,68); BATTLE OF THE PLANETS (II,179); THE CATTANOOGA CATS (I,874); HERE COMES PETER COTTONTAIL (II,1132); HOT WHEELS (I,2126); JOSIE AND THE PUSSYCATS (I,2453); JOSIE AND THE PUSSYCATS IN OUTER SPACE (I,2454); MR. & MS. AND THE BANDSTAND MURDERS (II,1746); THE NEW SUPER FRIENDS HOUR (II,1830); SCOOBY'S ALL-STAR LAFF-A-LYMPICS (II,2273); SCOOBY-DOO AND SCRAPPY-DOO (II,2274); THE SCOOBY-DOO AND SCRAPPY-DOO SHOW (II,2275); SCOOBY-DOO, WHERE ARE YOU? (II,2276); SKYHAWKS (I,4079); SUPER FRIENDS (II,2497); THE WORLD'S GREATEST SUPER HEROES (II,2842)

KASSUL, ART
BENDER (II,204); HOW DO I KILL A THIEF—LET ME COUNT THE WAYS (II,1195); OPEN ALL NIGHT (II,1914)

KASTNER, PETER
DELTA HOUSE (II,658); THE UGLIEST GIRL IN TOWN (I,4663)

KASZNAR, KURT
CODE NAME: HERACLITUS (I,990); LAND OF THE GIANTS (I,2611); ONCE UPON A DEAD MAN (I,3369); THE ROYAL FOLLIES OF 1933 (I,3862)

KATES, BERNARD
CHARLIE ANGELO (I,906); PAPA G.I. (I,3451)

KATIS, DIANA
MASTER OF THE GAME (II,1647)

KATON, ROSANNE
GRADY (II,1047); THE PARKERS (II,1959)

KATSAU, CLYDE
ALLISON SIDNEY HARRISON (II,54)

KATT, NICKY
GOLDIE AND KIDS: LISTEN TO ME (II,1021); HERBIE, THE LOVE BUG (II,1131); V: THE SERIES (II,2715)

KATT, WILLIAM
THE DAUGHTERS OF JOSHUA CABE (I,1169); THE GREATEST AMERICAN HERO (II,1060); THE RAINMAKER (II,2109)

KATZ, ALLAN
THE NATIONAL SNOOP (II,1797)

KATZ, OMRI
DALLAS (II,614)

KATZ, PHYLLIS
THE BILLY CRYSTAL COMEDY HOUR (II,248); WIZARDS AND WARRIORS (II,2813)

KATZ, TEDDY
RENDEZVOUS WITH MUSIC (I,3768)

KATZIN, LEE H
SPACE: 1999 (II,2409)

KATZMAN, SHERILYN
DALLAS (II,614); THE THREE WIVES OF DAVID WHEELER (II,2611)

KAUFFMAN, JOHN
THE WILDS OF TEN THOUSAND ISLANDS (II,2803)

KAUFMAN, ADAM
THE RED SKELTON SHOW (I,3755)

KAUFMAN, ANDY
CHER AND OTHER FANTASIES (II,497); THE FANTASTIC MISS PIGGY SHOW (II,827); A JOHNNY CASH CHRISTMAS (II,1328); THE LISA HARTMAN SHOW (II,1484); RODNEY DANGERFIELD SHOW: I CAN'T TAKE IT NO MORE (II,2198); STICK AROUND (II,2464); TAXI (II,2546); VAN DYKE AND COMPANY (II,2719)

KAUFMANN, MAURICE
DOUGLAS FAIRBANKS JR. PRESENTS THE RHEINGOLD THEATER (I,1364); THE NEXT VICTIM (I,3281)

KAVA, CAROLINE
IVAN THE TERRIBLE (II,1260)

KAVANAUGH, DORRIE
THE AWAKENING LAND (II,122); DAYS OF OUR LIVES (II,629); ONE LIFE TO LIVE (II,1907); RYAN'S HOPE (II,2234)

KAVANAUGH, JAMES
CRAZY TIMES (II,602)

KAVNER, JULIE
A FINE ROMANCE (II,858); RHODA (II,2151); TAXI (II,2546)

KAVNER, STEVEN
GIDGET'S SUMMER REUNION (I,1799)

KAWAI, LESLIE
THE AMAZING CHAN AND THE CHAN CLAN (I,157); LONDON AND DAVIS IN NEW YORK (II,1512)

KAWAYE, JANICE
LITTLE SHOTS (II,1495)

KAY CHOIR, THE
SO YOU WANT TO LEAD A BAND (I,4106)

KAY, ARTHUR
CBS CARTOON THEATER (I,878)

KAY, BEATRICE
CALVIN AND THE COLONEL (I,806); JOHNNY CARSON PRESENTS THE SUN CITY SCANDALS '72 (I,2421); KEEFE BRASSELLE'S VARIETY GARDEN (I,2515); WHICH

WAY'D THEY GO? (I,4838)

KAY, BRENDA
LAWRENCE WELK'S TOP TUNES AND NEW TALENT (I,2645)

KAY, CHARLES
JENNIE: LADY RANDOLPH CHURCHILL (II,1277)

KAY, DIANNE
CASS MALLOY (II,454); EIGHT IS ENOUGH (II,762); FLAMINGO ROAD (II,871); GLITTER (II,1009); REGGIE (II,2127)

KAY, LEON
SAY IT WITH ACTING (I,3928)

KAY, MARY ELLEN
ANNIE OAKLEY (I,225); THE GEORGE BURNS AND GRACIE ALLEN SHOW (I,1763); SHORT STORY THEATER (I,4024)

KAYDETTES, THE
SAMMY KAYE'S MUSIC FROM MANHATTAN (I,3902); SO YOU WANT TO LEAD A BAND (I,4106)

KAYE, CAREN
BATTLE OF THE NETWORK STARS (II,165); THE BETTY WHITE SHOW (II,224); BLANSKY'S BEAUTIES (II,267); EMPIRE (II,777); THE FUTURE: WHAT'S NEXT (II,947); IT'S YOUR MOVE (II,1259); LEGS (II,1459); THE NATURAL LOOK (II,1799); SIDE BY SIDE (II,2346); WHO'S WATCHING THE KIDS? (II,2793)

KAYE, CELIA
THE JOHN FORSYTHE SHOW (I,2410); THE LORETTA YOUNG SHOW (I,2755)

KAYE, DANNY
BOB HOPE SPECIAL: THE BOB HOPE SPECIAL (II,335); CBS: ON THE AIR (II,460); THE DANNY KAYE SHOW (I,1143); THE DANNY KAYE SHOW (I,1144); DISNEYLAND'S 25TH ANNIVERSARY (II,681); HERE COMES PETER COTTONTAIL (II,1132); AN HOUR WITH DANNY KAYE (I,2136); MUSICAL COMEDY TONIGHT (II,1777); OPERATION ENTERTAINMENT (I,3400); PETER PAN (II,2030); PINOCCHIO (II,2044); A SALUTE TO STAN LAUREL (I,3892); TWELVE STAR SALUTE (I,4638)

KAYE, GLORIA
ANNE MURRAY'S LADIES' NIGHT (II,89)

KAYE, LILA
ELLIS ISLAND (II,768); MAMA MALONE (II,1609); A PLACE TO DIE (I,3612)

KAYE, MANDY
GUESS AGAIN (I,1894); JOEY FAYE'S FROLICS (I,2406)

KAYE, ROMANA
THE PRIME OF MISS JEAN BRODIE (II,2082)

KAYE, SAMMY
SAMMY KAYE'S MUSIC FROM MANHATTAN (I,3902); SO YOU WANT TO LEAD A BAND (I,4106)

KAYE, STUBBY
CRESCENDO (I,1094); ELLIS ISLAND (II,768); FULL SPEED ANYWHERE (I,1704); HANSEL AND GRETEL (I,1934); MINSKY'S FOLLIES (II,1703); MY SISTER EILEEN (I,3200); PANTOMIME QUIZ (I,3449); PINOCCHIO (I,3602); SHENANAGANS (I,4006); SIDE BY SIDE (II,2347); SO HELP ME, APHRODITE (I,4104); THE WONDERFUL WORLD OF PHILIP MALLEY (II,2831)

KAYE, SUZIE
THE JONES BOYS (I,2446)

KAYE, SYLVIA FINE
MUSICAL COMEDY TONIGHT (II,1777)

KAYE, VIRGINIA
THE EDGE OF NIGHT (II,760)

KAYZER, BEAU
AFTER HOURS: SINGIN', SWINGIN' AND ALL THAT JAZZ (II,22); HARDCASE (II,1088); THE YOUNG AND THE RESTLESS (II,2862)

KAZAN, LAINIE
COME WITH ME—LAINIE KAZAN (I,1015); FAMILY BUSINESS (II,815); THE GEORGE KIRBY SPECIAL (I,1771); THE JERK, TOO (II,1282); THE LAINIE KAZAN SHOW (I,2606); THE ROBERT GOULET SPECIAL (I,3807); THE ROWAN AND MARTIN SHOW (I,3854)

KAZANJIAN, BRIAN
YOU'RE NOT ELECTED, CHARLIE BROWN (I,4967)

KAZANN, ZITTO
THE EYES OF TEXAS II (II,800); MOONLIGHT (II,1730); TERROR AT ALCATRAZ (II,2563)

KAZURINSKY, TIM
BIG CITY COMEDY (II,233); SATURDAY NIGHT LIVE (II,2261)

KEACH SR, STACY
FLAMINGO ROAD (II,872); THUNDARR THE BARBARIAN (II,2616)

KEACH, JAMES
BIG BEND COUNTRY (II,230); LACY AND THE MISSISSIPPI QUEEN (II,1419); THOU SHALT NOT KILL (II,2604); WISHMAN (II,2811)

KEACH, STACY
BEAUTY AND THE BEAST (II,197); THE BLUE AND THE GRAY (II,273); CARIBE (II,439); DYNASTY (II,747); GET SMART (II,983); HOW TO MARRY A MILLIONAIRE (I,2147); JOHNNY BELINDA (I,2417); KINGSTON: THE POWER PLAY (II,1399); MICKEY SPILLANE'S MIKE HAMMER (II,1692); MICKEY SPILLANE'S MIKE HAMMER: MORE THAN MURDER (II,1693); MICKEY SPILLANE'S MIKE HAMMER: MURDER ME, MURDER YOU (II,1694); MISTRAL'S DAUGHTER (II,1708); WAIT UNTIL DARK (II,2735)

KEAL, ANITA
SEARCH FOR TOMORROW (II,2284)

KEALE, MOE
BIG HAWAII (II,238); DANGER IN PARADISE (II,617); HAWAII FIVE-O (II,1110); THE ISLANDER (II,1249); THE LITTLE PEOPLE (I,2716); M STATION: HAWAII (II,1568); THE MACKENZIES OF PARADISE COVE (II,1585); PEARL (II,1986); STICKIN' TOGETHER (II,2465)

KEAMS, GERALDINE
BORN TO THE WIND (II,354)

KEAN, BETTY
ALL TOGETHER NOW (II,45); LEAVE IT TO LARRY (I,2649); THE R.C.A. THANKSGIVING SHOW (I,3736); WHERE'S RAYMOND? (I,4837)

KEAN, JANE
FIRESIDE THEATER (I,1580); THE HONEYMOONERS (I,2111); THE HONEYMOONERS CHRISTMAS SPECIAL (II,1176); THE HONEYMOONERS CHRISTMAS (II,1177); THE HONEYMOONERS SECOND HONEYMOON (II,1178); THE HONEYMOONERS VALENTINE SPECIAL (II,1179); THE JACKIE GLEASON SHOW (I,2323); THE JACKIE GLEASON SPECIAL (I,2324); THE JACKIE GLEASON SPECIAL (I,2325); THE R.C.A. THANKSGIVING SHOW (I,3736)

KEANE, CHARLOTTE
THE ADVENTURES OF ELLERY QUEEN (I,59)

KEANE, GEORGE
THE SWAN (I,4312)

KEANE, JAMES
IN SECURITY (II,1227); THE PAPER CHASE (II,1949); THE PAPER CHASE: THE SECOND YEAR (II,1950)

KEANE, JOHN
THE KEANE BROTHERS SHOW (II,1371)

KEANE, LYDIA
PRISONER: CELL BLOCK H (II,2085)

KEANE, NOAH
YOUNG DR. MALONE (I,4938)

KEANE, TERI
THE EDGE OF NIGHT (II,760); HOT PURSUIT (II,1189); LOVING (II,1552); THE YELLOW ROSE (II,2847); YOUNG DR. MALONE (I,4938)

KEANE, TOM
THE KEANE BROTHERS SHOW (II,1371)

KEARNEY, BETH
WELCOME BACK, KOTTER (II,2761)

KEARNEY, CAROLYN
THE LIFE OF RILEY (I,2686); SAVAGE SUNDAY (I,3926); YOUNG IN HEART (I,4939)

KEARNEY, EILEEN
THE DOCTORS (II,694)

KEARNEY, KAY
WELCOME BACK, KOTTER (II,2761)

KEARNEY, MARK
THE BOYS (I,701)

KEARNEY, MICHAEL
THE DESPERATE HOURS (I,1239); THANKSGIVING VISITOR (I,4417)

KEARNS, JOSEPH
THE ADVENTURES OF OZZIE AND HARRIET (I,71); THE ALAN YOUNG SHOW (I,104); DENNIS THE MENACE (I,1231); IT'S A SMALL WORLD (I,2260); OUR MISS BROOKS (I,3415); PETER GUNN (I,3562); PROFESSIONAL FATHER (I,3677)

KEARNS, SANDRA
FLAMINGO ROAD (II,872); MOONLIGHT (II,1730); THE PAUL WILLIAMS SHOW

(II,1984); TALES OF THE APPLE DUMPLING GANG (II,2536)

KEARNS, WILLIAM
PHILIP MARLOWE, PRIVATE EYE (II,2033)

KEAST, PAUL
CASEY JONES (I,865); THE FABULOUS SYCAMORES (I,1505)

KEATING, CHARLES
BRIDESHEAD REVISITED (II,375)

KEATING, FRED
TIME OUT FOR GINGER (I,4508)

KEATING, LARRY
ALL ABOUT BARBARA (I,125); THE GEORGE BURNS AND GRACIE ALLEN SHOW (I,1763); THE GEORGE BURNS SHOW (I,1765); THE HANK MCCUNE SHOW (I,1932); MR. ED (I,3137)

KEATING, MICHAEL
BLAKE'S SEVEN (II,264)

KEATON, BUSTER
DOUGLAS FAIRBANKS JR. PRESENTS THE RHEINGOLD THEATER (I,1364); THE GREATEST SHOW ON EARTH (I,1883); THE MAN WHO CAME TO DINNER (I,2880); A SALUTE TO STAN LAUREL (I,3892); SCREEN DIRECTOR'S PLAYHOUSE (I,3946); SHOW BIZ (I,4027); THE TWILIGHT ZONE (I,4651)

KEATON, DIANE
THE SENSATIONAL, SHOCKING, WONDERFUL, WACKY 70S (II,2302)

KEATON, MICHAEL
ALL'S FAIR (II,46); KLEIN TIME (II,1401); KRAFT SALUTES WALT DISNEY WORLD'S 10TH ANNIVERSARY (II,1410); MARY (II,1637); THE MARY TYLER MOORE COMEDY HOUR (II,1639); REPORT TO MURPHY (II,2134); ROOSEVELT AND TRUMAN (II,2209); WORKING STIFFS (II,2834)

KEATS, STEVEN
THE AWAKENING LAND (II,122); GHOST OF A CHANCE (II,987); SEVENTH AVENUE (II,2309); SUPERCOPS (II,2502); WHERE'S POPPA? (II,2784)

KEDROVA, LILA
THE MASK OF MARCELLA (I,2937)

KEEFE, ADAM
THE NUT HOUSE (I,3320); RUN JACK RUN (I,3872)

KEEFE, JOE
THE YESTERDAY SHOW (II,2850)

KEEFE, JOHN JOHN
THE BENNY HILL SHOW (II,207)

KEEFER, DON
ANGEL (I,216); ASSIGNMENT: EARTH (I,299); A RAINY DAY (II,2110); READY FOR THE PEOPLE (I,3739)

KEEGAN, BARRY
GLENCANNON (I,1822)

KEEGAN, CHRIS
THE BIG TOP (I,435)

KEEGAN, JUNE
PAUL WHITEMAN'S TV TEEN CLUB (I,3491)

KEEGAN, JUNIE
ALL ABOARD (I,124)

KEEGAN, KRIS
ACTION IN THE AFTERNOON (I,25)

KEEL, HOWARD
BING CROSBY AND HIS FRIENDS (I,455); THE BOB HOPE SHOW (I,563); DALLAS (II,614); THE GENERAL MOTORS 50TH ANNIVERSARY SHOW (I,1758); HOLLYWOOD MELODY (I,2085); ROBERTA (I,3815)

KEELER, BRIAN
VENICE MEDICAL (II,2726)

KEELER, DONALD
JEFF'S COLLIE (I,2356)

KEELER, RUBY
HOORAY FOR HOLLYWOOD (I,2114); THE ROWAN AND MARTIN SPECIAL (I,3855)

KEELING, KENNETH
THE WOMAN IN WHITE (II,2819)

KEEN, GEOFFREY
THE SCARECROW OF ROMNEY MARSH (I,3931)

KEEN, MALCOLM
THE GABBY HAYES SHOW (I,1723); MACBETH (I,2801); MAN AND SUPERMAN (I,2854)

KEEN, NOAH
ARREST AND TRIAL (I,264); LOVING (II,1552); THE TURNING POINT OF JIM MALLOY (II,2670)

KEENA, TOM
THE DAY THE WOMEN GOT EVEN (II,628)

KEENAN, MICHAEL
THE CHEAP DETECTIVE (II,490)

KEENAN, PAUL
DAYS OF OUR LIVES (II,629); DYNASTY (II,746)

KEENE, MICHAEL
HARBOURMASTER (I,1946); OUR FIVE DAUGHTERS (I,3413)

KEENER, ELLIOTT
THE BIG EASY (II,234)

KEEP, STEPHEN
THE BILLION DOLLAR THREAT (II,246); FLO (II,891)

KEESHAN, BOB
CAPTAIN KANGAROO (II,434); GOOD EVENING, CAPTAIN (II,1031); HOWDY DOODY (I,2151); MR. MAYOR (I,3143); WAKE UP (II,2736)

KEHOE, JACK
THE CHICAGO STORY (II,507); A HATFUL OF RAIN (I,1964); MOST WANTED (II,1738)

KEIM, BETTY LOU
THE DEPUTY (I,1234); MY SON JEEP (I,3202)

KEITH JR, ROBERT
EYE WITNESS (I,1494)

KEITH, AL
THE ERNIE KOVACS SHOW (I,1455)

KEITH, ALLISON
IT'S HAPPENING (I,2269)

KEITH, BONNIE
DYNASTY (II,746)

KEITH, BRIAN
ARCHER (II,102); CENTENNIAL (II,477); THE CHISHOLMS (II,512); CLIMAX! (I,976); THE COURT-MARTIAL OF GENERAL GEORGE ARMSTRONG CUSTER (I,1081); THE CRUSADER (I,1110); ELFEGO BACA (I,1427); THE ELGIN HOUR (I,1428); FAMILY AFFAIR (I,1519); FORD TELEVISION THEATER (I,1634); GREAT ADVENTURE (I,1868); HARDCASTLE AND MCCORMICK (II,1089); HOW THE WEST WAS WON (II,1196); KRAFT SUSPENSE THEATER (I,2591); THE LITTLE PEOPLE (I,2716); THE LUX VIDEO THEATER (I,2795); PROFILES IN COURAGE (I,3678); THE QUEST (II,2096); ROBERT MONTGOMERY PRESENTS YOUR LUCKY STRIKE THEATER (I,3809); SECOND CHANCE (I,3958); THE SEEKERS (II,2298); STUDIO '57 (I,4267); THE UNITED STATES STEEL HOUR (I,4677); THE WESTERNER (I,4801); WINCHESTER (I,4872); ZANE GREY THEATER (I,4979); THE ZOO GANG (II,2877)

KEITH, BYRON
77 SUNSET STRIP (I,3988); BATMAN (I,366)

KEITH, DAVID
CO-ED FEVER (II,547)

KEITH, IAN
PEPSI-COLA PLAYHOUSE (I,3523); RED SKELTON'S CHRISTMAS DINNER (II,2123); THE STAR MAKER (I,4193)

KEITH, KANDI
POLICE WOMAN (II,2063)

KEITH, LAWRENCE
ALL MY CHILDREN (II,39); ANOTHER WORLD (II,97); THE BAXTERS (II,183); THE FOUR OF US (II,914); KENNEDY (II,1377); KISS ME, KATE (I,2569)

KEITH, MARK
ROLL OF THUNDER, HEAR MY CRY (II,2202)

KEITH, PENELOPE
GOOD NEIGHBORS (II,1034); TO THE MANOR BORN (II,2627)

KEITH, RICHARD
THE ANDY GRIFFITH SHOW (I,192); FIRST LOVE (I,1583); I LOVE LUCY (I,2171); THE LUCILLE BALL-DESI ARNAZ SHOW (I,2784)

KEITH, ROLAND
THE GREAT GILDERSLEEVE (I,1875); LIFE WITH FATHER (I,2690)

KEITH, SUSAN
ANOTHER WORLD (II,97); LOVING (II,1552); ONE LIFE TO LIVE (II,1907)

KELK, JACKIE
THE ALDRICH FAMILY (I,111); YOUNG MR. BOBBIN (I,4943)

KELLAIGH, KATHLEEN
THE GUIDING LIGHT (II,1064)

KELLARD, RICK
WACKO (II,2733)

KELLAWAY, CECIL
CALL HOLME (II,412); THE FABULOUS SYCAMORES (I,1505); THE GREATEST SHOW ON EARTH (I,1883);

KISMET (I,2565); MAGNAVOX THEATER (I,2827)

KELLER, JASON
OUT OF THE BLUE (II,1934)

KELLER, MARY
RYAN'S HOPE (II,2234)

KELLER, NICOLE
STRANGER IN OUR HOUSE (II,2477)

KELLER, SHANE
OUT OF THE BLUE (II,1934)

KELLER, SUSAN
THE SIX-MILLION-DOLLAR MAN (II,2372)

KELLER, WILLIAM
DANIEL BOONE (I,1142)

KELLERMAN, SALLY
CENTENNIAL (II,477); CIRCUS OF THE STARS (II,535); A COUPLE OF DONS (I,1077); DOROTHY HAMILL'S CORNER OF THE SKY (II,721); FOR LOVERS ONLY (II,904); HIGHER AND HIGHER, ATTORNEYS AT LAW (I,2057); KRAFT SUSPENSE THEATER (I,2591); THE OUTER LIMITS (I,3426)

KELLERMANN, SUSAN
AT YOUR SERVICE (II,118); SUTTERS BAY (II,2508); TAXI (II,2546); THE WILD WOMEN OF CHASTITY GULCH (II,2801)

KELLETT, BOB
SPACE: 1999 (II,2409)

KELLEY, DEFOREST
333 MONTGOMERY (I,4487); ABC'S MATINEE TODAY (I,7); ASSIGNMENT: EARTH (I,299); JOHNNY RISK (I,2435); POLICE STORY (I,3636); STAR TREK (II,2439); STAR TREK (II,2440); YOU ARE THERE (I,4929)

KELLEY, WILLIAM
KEY TORTUGA (II,1383)

KELLIN, MIKE
THE ART OF CRIME (II,108); ASSIGNMENT: MUNICH (I,302); BONINO (I,685); THE CATCHER (I,872); FITZ AND BONES (II,867); HONESTLY, CELESTE! (I,2104); NIGHTSIDE (I,3296); ONE STEP BEYOND (I,3388); SEVENTH AVENUE (II,2309); TERROR AT ALCATRAZ (II,2563); THE WACKIEST SHIP IN THE ARMY (I,4744)

KELLINS, JOHN
INDEMNITY (I,2209)

KELLMAN, RICK
WEEKEND (I,4788)

KELLOGG, ABIGAIL
AH!, WILDERNESS (I,91); SEARCH FOR TOMORROW (II,2284)

KELLOGG, GAYLE
NAVY LOG (I,3233)

KELLOGG, JOHN
PEYTON PLACE (I,3574)

KELLOGG, RAY
THE GOOD LIFE (II,1033); THE HOOFER (I,2113); ROWAN AND MARTIN BITE THE HAND THAT FEEDS THEM (I,3853)

KELLY, AL
BACK THE FACT (I,327); CANDID CAMERA (I,819); SONJA HENIE'S HOLIDAY ON ICE (I,4125)

KELLY, APRIL
THE PAUL LYNDE COMEDY HOUR (II,1978)

KELLY, BARBARA
HURDY GURDY (I,2163)

KELLY, BARRY
BIG TOWN (I,436); BURKE'S LAW (I,762); MR. ED (I,3137); MR. ROBERTS (I,3149); PETE AND GLADYS (I,3558); THOMPSON'S GHOST (I,4467)

KELLY, BRIAN
21 BEACON STREET (I,4646); FLIPPER (I,1604); SKYFIGHTERS (I,4078); STRAIGHTAWAY (I,4243)

KELLY, CAROL
TV READERS DIGEST (I,4630)

KELLY, COLEEN
TERROR AT ALCATRAZ (II,2563)

KELLY, CRAIG
COMMANDO CODY (I,1030)

KELLY, DAREN
ALL MY CHILDREN (II,39); LANDON, LANDON & LANDON (II,1426)

KELLY, DAVID
ROBIN'S NEST (II,2187)

KELLY, DON
FRONTIER (I,1695)

KELLY, EMMETT
BETTE MIDLER—OL' RED HAIR IS BACK (II,222)

KELLY, GENE
1968 HOLLYWOOD STARS OF TOMORROW (I,3297); THE BIG SHOW (II,243); DEBBY BOONE. . . THE SAME OLD BRAND NEW ME (II,656); DICK CAVETT'S BACKLOT USA (II,670); DOM DELUISE AND FRIENDS (II,701); THE DOROTHY HAMILL SPECIAL (II,719); THE FIRST 50 YEARS (II,862); THE FUNNY SIDE (I,1714); THE FUNNY SIDE (I,1715); THE GENE KELLY PONTIAC SPECIAL (I,1750); THE GENE KELLY SHOW (I,1751); GENE KELLY'S WONDERFUL WORLD OF GIRLS (I,1752); GENE KELLY. . . AN AMERICAN IN PASADENA (II,962); GOING MY WAY (I,1833); JACK AND THE BEANSTALK (I,2288); THE JULIE ANDREWS SHOW (I,2481); THE JULIE ANDREWS SPECIAL (I,2482); LAS VEGAS PALACE OF STARS (II,1431); LUCY MOVES TO NBC (II,1562); MAGNAVOX PRESENTS FRANK SINATRA (I,2826); OLIVIA NEWTON-JOHN'S HOLLYWOOD NIGHTS (II,1886); OPRYLAND: NIGHT OF STARS AND FUTURE STARS (II,1921); THE PEGGY FLEMING SHOW (I,3508); THE SANDY DUNCAN SHOW (II,2253); SINATRA—THE FIRST 40 YEARS (II,2359); STEVE AND EYDIE: OUR LOVE IS HERE TO STAY (II,2457); A TRIBUTE TO "MR. TELEVISION" MILTON BERLE (II,2658); THE WORLD OF ENTERTAINMENT (II,2837); THE WORLD OF MAGIC (II,2838)

KELLY, GRACE
THE ARMSTRONG CIRCLE THEATER (I,260); DANGER (I,1134); LIGHT'S OUT (I,2699); MONTE CARLO, C'EST LA ROSE (I,3095); THE POPPY IS ALSO A FLOWER (I,3650); PRUDENTIAL FAMILY PLAYHOUSE (I,3683); ROBERT MONTGOMERY PRESENTS YOUR LUCKY STRIKE THEATER (I,3809); THE SWAN (I,4312); THE WEB (I,4784)

KELLY, IRENE
B.J. AND THE BEAR (II,127)

KELLY, JACK
FBI CODE 98 (I,1550); GET CHRISTIE LOVE! (II,981); THE HARDY BOYS MYSTERIES (II,1090); JANE WYMAN PRESENTS THE FIRESIDE THEATER (I,2345); KING'S ROW (I,2563); MAVERICK (II,1657); NBC COMEDY THEATER (I,3238); THE NEW MAVERICK (II,1822); PALMS PRECINCT (II,1946); SALE OF THE CENTURY (I,3888); SECRET AGENT (I,3961); SHOOT-IN AT NBC (A BOB HOPE SPECIAL) (I,4021); VEGAS (II,2723); WHEN THE WEST WAS FUN: A WESTERN REUNION (II,2780)

KELLY, JOE
THE QUIZ KIDS (I,3712)

KELLY, JON
U.F.O. (I,4662)

KELLY, JUDY
THE BOB HOPE SHOW (I,507)

KELLY, KAREN
RITUALS (II,2182)

KELLY, KATHY JO
THE FOUR OF US (II,914)

KELLY, MARGO
FOUR EYES (II,912)

KELLY, MARY
TELETIPS ON LOVELINESS (I,4380)

KELLY, MICHAEL G
STOPWATCH: THIRTY MINUTES OF INVESTIGATIVE TICKING (II,2474)

KELLY, NANCY
FAITH BALDWIN'S THEATER OF ROMANCE (I,1515); THE IMPOSTER (II,1225); THE KAISER ALUMINUM HOUR (I,2503); SILVER THEATER (I,4051); STARLIGHT THEATER (I,4201); SUMMER IN THE CITY (I,4280); SUSPENSE (I,4305); THRILLER (I,4492)

KELLY, PATSY
THE COP AND THE KID (II,572); MY SON THE DOCTOR (I,3203); THE PIGEON (I,3596)

KELLY, PAULA
THE CHEAP DETECTIVE (II,490); CHIEFS (II,509); DUKE ELLINGTON. . . WE LOVE YOU MADLY (I,1380); FEEL THE HEAT (II,848); KOMEDY TONITE (II,1407); NIGHT COURT (II,1844); PETER MARSHALL SALUTES THE BIG BANDS (II,2028); PETER PAN (II,2030); THE RICHARD PRYOR SHOW (II,2163); THE SOUPY SALES SHOW (I,4138)

KELLY, PETER
THE LITTLE FOXES (I,2712)

KELLY, RACHEL
AS THE WORLD TURNS (II,110)

KELLY, RON
SAM (II,2246)

KELLY, ROSEMARY
A TIME TO LIVE (I,4510)

KELLY, ROZ
CELEBRITY CHALLENGE OF THE SEXES 1 (II,465); HAPPY DAYS (II,1084); MURDER CAN HURT YOU! (II,1768); THE NEW LORENZO MUSIC SHOW (II,1821); THE PAUL LYNDE HALLOWEEN SPECIAL (II,1981)

KELLY, SEAN
ANDERSON AND COMPANY (I,188); THE COWBOYS (II,598); HORNBLOWER (I,2121)

KELLY, TOM
VAN DYKE AND COMPANY (II,2719)

KELMAN, DEBBIE
PHYL AND MIKHY (II,2037)

KELMAN, RICKY
THE DENNIS O'KEEFE SHOW (I,1230); I LOVE MY DOCTOR (I,2172); NATIONAL VELVET (I,3231); OUR MAN HIGGINS (I,3414)

KELSEY, LINDA
CAPTAINS AND THE KINGS (II,435); DOC (II,682); LOU GRANT (II,1526); M*A*S*H (II,1569); THE PICTURE OF DORIAN GRAY (I,3592); SPENCER'S PILOTS (II,2421)

KELTON, PERT
THE BALLAD OF LOUIE THE LOUSE (I,337); CAVALCADE OF STARS (I,877); HENRY MORGAN'S GREAT TALENT HUNT (I,2017); KIBBE HATES FINCH (I,2530); OCCASIONAL WIFE (I,3325)

KELTON, RICHARD
GO WEST YOUNG GIRL (II,1013); QUARK (II,2095)

KELTON, ROY
DOCTOR WHO—THE FIVE DOCTORS (II,690)

KEMMER, ED
330 INDEPENDENCE S.W. (I,4486); AS THE WORLD TURNS (II,110); CLEAR HORIZON (I,973); THE EDGE OF NIGHT (II,760); LUXURY LINER (I,2796); SIMON LASH (I,4052); SOMERSET (I,4115); SPACE PATROL (I,4144)

KEMMERLING, WARREN
THE CABOT CONNECTION (II,408); HOW THE WEST WAS WON (II,1196); THE LONG HOT SUMMER (I,2747)

KEMP, BRANDIS
AFTERMASH (II,23); FRIDAYS (II,926); PUMPBOYS AND DINETTES ON TELEVISION (II,2090); SUTTERS BAY (II,2508)

KEMP, ELIXABETH
LOVE OF LIFE (I,2774)

KEMP, JANICE
ADVICE TO THE LOVELORN (II,16)

KEMP, JEREMY
GEORGE WASHINGTON (II,978); THE RHINEMANN EXCHANGE (II,2150); THE WINDS OF WAR (II,2807)

KEMP, ROGER
MASTER OF THE GAME (II,1647)

KEMP, SALLY
THE DOCTORS (I,1323); DYNASTY (II,746); THE HOYT AXTON SHOW (II,1200); PLANET EARTH (I,3615); ROOTS (II,2211)

KEMPER, DORIS
YES YES NANETTE (I,4927)

KEMPNER, BRENDA
JANE EYRE (II,1273)

KEMPSON, RACHEL
JANE EYRE (I,2339); JENNIE: LADY RANDOLPH CHURCHILL (II,1277)

KEN MURRAY CHORUS, THE
THE KEN MURRAY SHOW (I,2525)

KEN MURRAY DANCERS, THE
THE KEN MURRAY SHOW (I,2525)

KENAL SIODMY DANCERS, THE
IT'S POLKA TIME (I,2273)

KENAL SIODMY FOLK DANCERS, THE
POLKA TIME (I,3642)

KENASTON, JACK
UNK AND ANDY (I,4679)

KENDAL, FELICITY
GOOD NEIGHBORS (II,1034)

KENDAL, JENNIFER
THE FAR PAVILIONS (II,832)

KENDALL, ARTHUR
THE STAR MAKER (I,4193)

KENDALL, CAVAN
DIAL A DEADLY NUMBER (I,1253)

KENDALL, CY
ARMCHAIR DETECTIVE (I,259)

KENDALL, JO
CAMBRIDGE CIRCUS (I,807)

KENDIS, SONNY
THE SONNY KENDIS SHOW (I,4127)

KENDIS, WILLIAM
ADVENTURES OF A MODEL (I,50); MICHAEL SHAYNE, DETECTIVE (I,3022)

KENDRICK, RICHARD
PORTIA FACES LIFE (I,3653)

KENIN, ALEXA
CO-ED FEVER (II,547); THE HOUSE WITHOUT A CHRISTMAS TREE (I,2143); OFF CAMPUS (II,1877); THE WORD (II,2833)

KENNAN, SHEILA
SENSE OF HUMOR (II,2303)

KENNEALLY, PAMM
DOCTORS' PRIVATE LIVES (II,693)

KENNEALLY, WILLIAM
YOU ARE THERE (I,4929)

KENNEDY JR, CHARLES
KID POWER (I,2532)

KENNEDY, ADAM
THE CALIFORNIANS (I,795); THE DOCTORS (II,694); FOREST RANGER (I,1641)

KENNEDY, ARTHUR
ALCOA PREMIERE (I,109); THE ETHEL BARRYMORE THEATER (I,1467); F.D.R. (I,1552); NAKIA (II,1784); NAKIA (II,1785); THE THIRD COMMANDMENT (I,4449)

KENNEDY, BETTY
THE CIRCLE FAMILY (II,527); LADIES' MAN (II,1423)

KENNEDY, BOB
SENSE AND NONSENSE (I,3977); WINDOW SHOPPING (I,4875); WINGO (I,4878)

KENNEDY, DENNIS
THE CHISHOLMS (II,512)

KENNEDY, DOUGLAS
THE BIG VALLEY (I,437); NAVY LOG (I,3233); STEVE DONOVAN, WESTERN MARSHAL (I,4225)

KENNEDY, GEORGE
ARCHER—FUGITIVE FROM THE EMPIRE (II,103); BACKSTAIRS AT THE WHITE HOUSE (II,133); BLISS (II,269); THE BLUE KNIGHT (II,276); THE BLUE KNIGHT (II,277); COUNTERATTACK: CRIME IN AMERICA (II,581); NEVER SAY NEVER (II,1809); THE PHIL SILVERS SHOW (I,3580); SARGE (I,3914); SARGE: THE BADGE OR THE CROSS (I,3915); THRILLER (I,4492)

KENNEDY, GERARD
1915 (II,1853); AGAINST THE WIND (II,24); HUNTER (I,2161)

KENNEDY, HAROLD J
THE FRONT PAGE (I,1692)

KENNEDY, JAYNE
THE ALL-STAR SALUTE TO MOTHER'S DAY (II,51); BOB HOPE SPECIAL: BOB HOPE FOR PRESIDENT (II,282); CELEBRITY CHALLENGE OF THE SEXES 3 (II,467); CELEBRITY CHALLENGE OF THE SEXES 5 (II,469); THE CELEBRITY FOOTBALL CLASSIC (II,474); CIRCUS OF THE STARS (II,534); COVER GIRLS (II,596); MEN WHO RATE A "10" (II,1684); MITCHELL AND WOODS (II,1709); SPEAK UP AMERICA (II,2412); ULTRA QUIZ (II,2701)

KENNEDY, JOHN MILTON
ARMCHAIR DETECTIVE (I,259)

KENNEDY, LAURIE
KENNEDY (II,1377); SHERLOCK HOLMES (II,2327)

KENNEDY, LINDSAY
GALACTICA 1980 (II,953); LITTLE HOUSE: A NEW BEGINNING (II,1488); LITTLE HOUSE: BLESS ALL THE DEAR CHILDREN (II,1489); LITTLE HOUSE: LOOK BACK TO YESTERDAY (II,1490); LITTLE HOUSE: THE LAST FAREWELL (II,1491)

KENNEDY, MADGE
LEAVE IT TO BEAVER (I,2648); THE ODD COUPLE (II,1875)

KENNEDY, MARY JO
HERE COME THE BRIDES (I,2024)

KENNEDY, MIMI
BATTLE OF THE NETWORK STARS (II,173); THE BIG SHOW (II,243); I'VE HAD IT UP TO HERE (II,1221); SPENCER (II,2420); STOCKARD CHANNING IN JUST FRIENDS (II,2467); THREE GIRLS THREE (II,2608); THE TWO OF US (II,2693)

KENNEDY, PATRICIA
RETURN TO EDEN (II,2143)

KENNEDY, SARAH
BACHELOR AT LAW (I,322); EMERGENCY PLUS FOUR (II,774); ROWAN AND MARTIN'S LAUGH-IN (I,3856); STARSTRUCK (II,2446);

WILLOW B: WOMEN IN PRISON (II,2804)

KENNEDY, SUE
THE DAY THE WOMEN GOT EVEN (II,628)

KENNEDY, TOM
THE $100,000 NAME THAT TUNE (II,1902); 50 GRAND SLAM (II,852); THE BIG GAME (I,426); BODY LANGUAGE (II,346); BREAK THE BANK (II,369); DOCTOR I.Q. (I,1314); IT'S YOUR BET (I,2279); MITZI AND A HUNDRED GUYS (II,1710); NAME THAT TUNE (II,1787); PASSWORD PLUS (II,1966); THE REAL TOM KENNEDY SHOW (I,3743); SPLIT SECOND (II,2428); TO SAY THE LEAST (II,2623); WHEW! (II,2786); YOU DON'T SAY (I,4934); YOU DON'T SAY (II,2858)

KENNELLY, SHEILA
RETURN TO EDEN (II,2143)

KENNERLY, DIANE
LITTLE HOUSE: THE LAST FAREWELL (II,1491)

KENNERY, MARIE
LOVE OF LIFE (I,2774)

KENNEY, GENERAL GEORGE
FLIGHT (I,1597); SKYFIGHTERS (I,4078)

KENNEY, JUNE
TV READERS DIGEST (I,4630)

KENNY ROGERS AND THE FIRST EDITION
JUST FRIENDS (I,2495); SAGA OF SONORA (I,3884)

KENNY, J ANDREW
B.J. AND THE BEAR (II,127)

KENNY, JAMES
TEN LITTLE INDIANS (I,4394)

KENNY, JUNE
THE INVESTIGATORS (I,2231)

KENNY, SEAN
STAR TREK (II,2440)

KENSIT, PATSY
THE ADVENTURES OF POLLYANNA (II,14); GREAT EXPECTATIONS (II,1055); TYCOON: THE STORY OF A WOMAN (II,2697)

KENT, DON
H.M.S. PINAFORE (I,2066)

KENT, ENID
M*A*S*H (II,1569); NORMA RAE (II,1860)

KENT, JANICE
STILL THE BEAVER (II,2466); THE TED KNIGHT SHOW (II,2550)

KENT, JEAN
THE ADVENTURES OF SIR FRANCIS DRAKE (I,75); COLOR HIM DEAD (I,1007); COLOR HIM DEAD (II,554)

KENT, LANITA
THE JACKIE GLEASON SHOW (I,2323)

KENT, LARRY
JACK LEMMON IN 'S WONDERFUL, 'S MARVELOUS, 'S GERSHWIN (I,2313); PARADE OF STARS (II,1954)

KENT, LILA
THE WONDERFUL WORLD OF PHILIP MALLEY (II,2831)

KENT, PATSY
PENMARRIC (II,1993)

KENT, PAUL
THE INVISIBLE MAN (II,1242)

KENT, SUZANNE
GREAT DAY (II,1054); THE PARAGON OF COMEDY (II,1956)

KENTON, STAN
MUSIC '55 (I,3168); THE SATURDAY NIGHT DANCE PARTY (I,3919)

KENTON, WILLIAM
DOCTOR WHO—THE FIVE DOCTORS (II,690)

KENTUCKY BOYS, THE
MIDWESTERN HAYRIDE (I,3033)

KENYON, NANCY
MELODY TOUR (I,3001); OPERA VS. JAZZ (I,3399)

KENYON, SANDY
CRUNCH AND DES (I,1108); THE DOCTOR (I,1320); LOVE ON A ROOFTOP (I,2775); THE TRAVELS OF JAIMIE MCPHEETERS (I,4596); TRUE LIFE STORIES (II,2665)

KEPLER, SHELL
GENERAL HOSPITAL (II,964); SHIPSHAPE (II,2329)

KERCHEVAL, KEN
DALLAS (II,614); SEARCH FOR TOMORROW (II,2284)

KERLEE, DENNISON
EXPERT WITNESS (I,1488)

KERMOTAN, MICHAEL
ARCHY AND MEHITABEL (I,254); THE SKATEBIRDS (II,2375)

KERN, BONNIE LOU
THE MICKEY MOUSE CLUB (I,3025); THE MOUSEKETEERS REUNION (II,1742)

KERN, IRIS
PALMERSTOWN U.S.A. (II,1945)

KERN, ROGER
THE YOUNG PIONEERS (II,2868); THE YOUNG PIONEERS (II,2869); THE YOUNG PIONEERS' CHRISTMAS (II,2870)

KERNAN, DAVID
SONG BY SONG (II,2399)

KERNS, JOANNA
THE FOUR SEASONS (II,915); THE RETURN OF MARCUS WELBY, M.D (II,2138); THE WALTONS: MOTHER'S DAY ON WALTON'S MOUNTAIN (EPISODE 2) (II,2742); THE WALTONS: WEDDING ON WALTON'S MOUNTAIN (EPISODE 1) (II,2743)

KERNS, SANDRA
DOCTOR SCORPION (II,685)

KERR, BILL
RETURN TO EDEN (II,2143)

KERR, DEBORAH
A WOMAN OF SUBSTANCE (II,2820)

KERR, ELAINE
ANOTHER WORLD (II,97)

KERR, ELIZABETH
THE BETTY WHITE SHOW (II,224); DENNIS THE MENACE: MAYDAY FOR MOTHER (II,660); GIMME A BREAK (II,995); MORK AND MINDY (II,1735); ONE DAY AT A TIME (II,1900); YOUR PLACE OR MINE? (II,2874)

KERR, GRAHAM
THE GALLOPING GOURMET (I,1730)

KERR, JAY
CALIFORNIA FEVER (II,411); A ROCK AND A HARD PLACE (II,2190); WIZARDS AND WARRIORS (II,2813)

KERR, JOHN
ARREST AND TRIAL (I,264); BERKELEY SQUARE (I,399); THE CORN IS GREEN (I,1050); THE LONG HOT SUMMER (I,2747); PEYTON PLACE (I,3574); POLICE WOMAN: THE GAMBLE (II,2064); WASHINGTON: BEHIND CLOSED DOORS (II,2744); YUMA (I,4976)

KERR, SANDRA
THE JUDGE (I,2468)

KERRY, ANNE
MURDER, SHE WROTE (II,1770)

KERRY, JOHN
TERROR AT ALCATRAZ (II,2563)

KERRY, LARRY
A TIME TO LIVE (I,4510)

KERRY, MARGARET
CLUTCH CARGO (I,988); THE RUGGLES (I,3868)

KERSH, KATHY
BRADDOCK (I,704); THE NUT HOUSE (I,3320)

KERSHAW, DOUG
THE CHISHOLMS (II,512); CONSTANTINOPLE (II,570); COUNTRY COMES HOME (II,585); DEAN MARTIN PRESENTS MUSIC COUNTRY, U.S.A. (I,1192); I BELIEVE IN MUSIC (I,2165); MARY'S INCREDIBLE DREAM (II,1641)

KERWIN, BRIAN
B.J. AND THE BEAR (II,125); BATTLE OF THE NETWORK STARS (II,170); BATTLE OF THE NETWORK STARS (II,172); THE BLUE AND THE GRAY (II,273); THE BUSTERS (II,401); THE CHISHOLMS (II,512); THE JAMES BOYS (II,1272); LOBO (II,1504); THE MISADVENTURES OF SHERIFF LOBO (II,1704); THE YOUNG AND THE RESTLESS (II,2862)

KERWIN, LANCE
ADVICE TO THE LOVELORN (II,16); BATTLE OF THE NETWORK STARS (II,165); BATTLE OF THE NETWORK STARS (II,166); THE FAMILY HOLVAK (II,817); THE GREATEST GIFT (II,1061); JAMES AT 15 (II,1270); JAMES AT 15 (II,1271)

KESLESKI, MARGE
CHICAGO STORY (II,506)

KESNER, JILLIAN
CO-ED FEVER (II,547); HECK'S ANGELS (II,1122)

KESSLER TWINS, THE
THE CONTINENTAL SHOWCASE (I,1041)

KESSLER, JILLIAN
MORK AND MINDY (II,1735)

KESSLER, QUINN
MICKEY SPILLANE'S MIKE HAMMER: MURDER ME, MURDER YOU (II,1694)

KESTEN, BRAD

IS THIS GOODBYE, CHARLIE BROWN? (I,2242)

KESTEN, BRIAN
THE CHARLIE BROWN AND SNOOPY SHOW (II,484)

KESTER, MORGAN
FROM HERE TO ETERNITY (II,932)

KESTON, BRAD
WHAT HAVE WE LEARNED, CHARLIE BROWN? (I,4810)

KETCHUM, DAVE
CALL HOLME (II,412); CAMP RUNAMUCK (I,811); GET SMART (II,983); I'M DICKENS...HE'S FENSTER (I,2193); A KNIGHT IN SHINING ARMOUR (I,2575); LEGS (II,1459); NANNY AND THE PROFESSOR AND THE PHANTOM OF THE CIRCUS (I,3226); WHERE'S THE FIRE? (II,2785)

KETCHUM, HANK
THE FANTASTIC FUNNIES (II,825)

KEVIN CARLISLE DANCERS, THE
BATTLE OF THE BEAT (II,161); CHANGING SCENE (I,899)

KEVIN CARLISLE THREE, THE
WHAT'S IT ALL ABOUT WORLD? (I,4816)

KEVIN, JAMES
THE STAR MAKER (I,4193)

KEVOIAN, PETER
THE GIRL, THE GOLD WATCH AND EVERYTHING (II,1002)

KEY, JANET
DEATH IN SMALL DOSES (I,1207)

KEYES, DANIEL
DARK SHADOWS (I,1157); THE JOKE AND THE VALLEY (I,2438); WHERE THE HEART IS (I,4831)

KEYES, DON
HOW TO SURVIVE A MARRIAGE (II,1198)

KEYES, JOE
THE CORNER BAR (I,1051)

KEZER, GLENN
MASTER OF THE GAME (II,1647); ONE TOUCH OF VENUS (I,3389)

KHAMBATTA, PERSIS
THE MAN WITH THE POWER (II,1616)

KHAN, SAJID
MAYA (I,2960)

KHOURY, GEORGE
SABU AND THE MAGIC RING (I,3881)

KHOW, BEULAH
MARCO POLO (II,1626)

KIBBEE, LOIS
THE EDGE OF NIGHT (II,760); GENERAL HOSPITAL (II,964); SOMERSET (I,4115)

KIBBER, JEFFREY
RETURN OF THE WORLD'S GREATEST DETECTIVE (II,2142)

KIBERD, JAMES
LOVING (II,1552)

KIDD, MICHAEL
THE BOB HOPE SHOW (I,505)

KIDD, JONATHAN
THE YOUNG PIONEERS (II,2868)

KIDDER, MARGOT
BUS STOP (II,399); HARRY O (II,1098); NICHOLS (I,3283)

KIDS NEXT DOOR, THE
THE KATE SMITH SHOW (I,2511)

KIEHL, WILLIAM
THE EDGE OF NIGHT (II,760)

KIEL, RICHARD
THE BARBARY COAST (II,146); THE BARBARY COAST (II,147); CIRCUS OF THE STARS (II,533); LAND OF THE LOST (II,1425); LOLA (II,1511); THRILLER (I,4492); VAN DYKE AND COMPANY (II,2718); THE WILD WILD WEST (I,4863)

KIERNAN, JOHN
KIERNAN'S KALEIDOSCOPE (I,2535)

KIERNAN, WALTER
SPARRING PARTNERS (I,4145); WHAT'S THE STORY? (I,4821); WHO SAID THAT? (I,4849); WHO'S THE BOSS? (I,4853)

KIFF, KALEENA
FAMILY TIES (II,819); LOVE, SIDNEY (II,1547); SIDNEY SHORR (II,2349)

KIGER, RANDI
THE MACKENZIES OF PARADISE COVE (II,1585); STICKIN' TOGETHER (II,2465)

KIGER, ROBBIE
ALL THE WAY HOME (II,44); CRAZY LIKE A FOX (II,601); LITTLE SHOTS (II,1495)

KILBRIDE, PERCY
THE FORTUNE HUNTER (I,1646)

KILEY, PHIL 'COO COO'
AGIC MIDWAY (I,2818)

KILEY, RICHARD
ALL THE WAY HOME (I,140); THE ELGIN HOUR (I,1428); GEORGE WASHINGTON (II,978); GOLDEN GATE (II,1020); HOW THE WEST WAS WON (II,1196); INDEMNITY (I,2209); THE MACAHANS (II,1583); MAN ON THE MOVE (I,2878); NIGHT GALLERY (I,3288); PARADE OF STARS (II,1954); REVLON MIRROR THEATER (I,3781); THE THORN BIRDS (II,2600); THOSE FABULOUS CLOWNS (II,2602)

KILGALLEN, ROB
THE DOUBLE LIFE OF HENRY PHYFE (I,1361)

KILGORE, MERLE
THE JOHNNY CASH SPRING SPECIAL (II,1329)

KILIAN, VICTOR
FOREVER FERNWOOD (II,909); MARY HARTMAN, MARY HARTMAN (II,1638)

KILLIAM, PAUL
MOVIE MUSEUM (I,3115)

KILLINGBACK, DEBBIE
MASTER OF THE GAME (II,1647)

KILLMOND, FRANK
THE BOSTON TERRIER (I,697); YOU'RE ONLY YOUNG ONCE (I,4969)

KILLY, JEAN-CLAUDE
PEGGY FLEMING AT SUN VALLEY (I,3507)

KILMAN, PETER
GENERAL HOSPITAL (II,964)

KILPATRICK, DACARLA
RISE AND SHINE (II,2179)

KILPATRICK, ERIC
JESSICA NOVAK (II,1286); THE WHITE SHADOW (II,2788)

KILPATRICK, LINCOLN
DOWN HOME (II,731); THE LESLIE UGGAMS SHOW (I,2660); MATT HOUSTON (II,1654); THE MONEYCHANGERS (II,1724)

KILTY, JACK
MUSICAL MERRY-GO-ROUND (I,3184)

KILTY, JEROME
BERKELEY SQUARE (I,399); OUR TOWN (I,3420); TAMING OF THE SHREW (I,4356)

KIM, EVAN
C.P.O. SHARKEY (II,407); COCAINE AND BLUE EYES (II,548); KHAN! (II,1384); MAKING IT (II,1605); V (II,2713); THE YOUNG SENTINELS (II,2871)

KIM, PETER
THE WONDERFUL WORLD OF PHILIP MALLEY (II,2831)

KIM, ROB
THE LAST NINJA (II,1438)

KIM, SALLY
FROM HERE TO ETERNITY (II,932); FROM HERE TO ETERNITY (II,933); SWEEPSTAKES (II,2514)

KIMBALL, ANNE
ALIAS MIKE HERCULES (I,117); G.E. SUMMMER ORIGINALS (I,1747)

KIMBALL, CHRISTINA KUMI
STRIPPERS (II,2481)

KIMBER, TIM
THE HUSTLER OF MUSCLE BEACH (II,1210)

KIMBROUGH, CLINT
OUR TOWN (I,3420)

KIMLER, KAY
THE YOUNG PIONEERS' CHRISTMAS (II,2870)

KIMMEL, BRUCE
DINAH AND HER NEW BEST FRIENDS (II,676); NEVER SAY NEVER (II,1809); TABITHA (II,2526)

KIMMEL, LESLIE
THE ADVENTURES OF JIM BOWIE (I,63)

KIMMEL, STANLEY
THE GOSSIP COLUMNIST (II,1045)

KIMMELL, DANA
DIFF'RENT STROKES (II,674); HIS AND HERS (II,1155); SUTTERS BAY (II,2508); TEXAS (II,2566)

KIMMENS, KEN
A ROCK AND A HARD PLACE (II,2190)

KINCAID, AARON
BACHELOR FATHER (I,323); THE FIRST HUNDRED YEARS (I,1582); PLANET EARTH (I,3615)

KINCAID, JASON
NIGHTSIDE (II,1851)

KIND, ROSALYN
GHOST OF A CHANCE (II,987)

KINDLER, JAN
OZMOE (I,3437)

KING COUSINS, THE
THE KING FAMILY CELEBRATE THANKSGIVING (I,2545); THE KING FAMILY CHRISTMAS SPECIAL (I,2546); THE KING FAMILY IN ALASKA (I,2547); THE KING FAMILY IN HAWAII (I,2548); THE KING FAMILY IN WASHINGTON (I,2549); THE KING FAMILY JUNE SPECIAL (I,2550); THE KING FAMILY OCTOBER SPECIAL (I,2551); THE KING FAMILY SPECIAL (I,2553); THE KING FAMILY VALENTINE'S DAY SPECIAL (I,2554)

KING FAMILY, THE
THE KING FAMILY CELEBRATE THANKSGIVING (I,2545); THE KING FAMILY CHRISTMAS SPECIAL (I,2546); THE KING FAMILY IN ALASKA (I,2547); THE KING FAMILY IN HAWAII (I,2548); THE KING FAMILY IN WASHINGTON (I,2549); THE KING FAMILY JUNE SPECIAL (I,2550); THE KING FAMILY OCTOBER SPECIAL (I,2551); THE KING FAMILY SPECIAL (I,2553); THE KING FAMILY VALENTINE'S DAY SPECIAL (I,2554)

KING SISTERS, THE
THE KING FAMILY CELEBRATE THANKSGIVING (I,2545); THE KING FAMILY CHRISTMAS SPECIAL (I,2546); THE KING FAMILY IN ALASKA (I,2547); THE KING FAMILY IN HAWAII (I,2548); THE KING FAMILY IN WASHINGTON (I,2549); THE KING FAMILY JUNE SPECIAL (I,2550); THE KING FAMILY OCTOBER SPECIAL (I,2551); THE KING FAMILY SPECIAL (I,2553); THE KING FAMILY VALENTINE'S DAY SPECIAL (I,2554)

KING, ALAN
20TH CENTURY FOLLIES (I,4641); ALAN KING IN LAS VEGAS, PART I (I,97); ALAN KING IN LAS VEGAS, PART II (I,98); ALAN KING LOOKS BACK IN ANGER—A REVIEW OF 1972 (I,99); THE ALAN KING SHOW (I,100); THE ALAN KING SHOW (I,101); THE ALAN KING SHOW (I,102); THE ALAN KING SPECIAL (I,103); ALAN KING'S FINAL WARNING (II,29); ALAN KING'S SECOND ANNUAL FINAL WARNING (II,30); ALAN KING'S THIRD ANNUAL FINAL WARNING!! (II,31); THE ARTHUR MURRAY PARTY FOR BOB HOPE (I,292); COMEDY IS KING (I,1019); HAPPY ENDINGS (II,1085); JULIE ANDREWS: ONE STEP INTO SPRING (II,1353); THE KRAFT 75TH ANNIVERSARY SPECIAL (II,1409); LOVE, LIFE, LIBERTY & LUNCH (II,1544); THE MANY FACES OF COMEDY (I,2894); PINOCCHIO'S CHRISTMAS (II,2045); SEVENTH AVENUE (II,2309); THE WONDERFUL WORLD OF AGGRAVATION (I,4894)

KING, ALDINE
HAGEN (II,1071); KAREN (II,1363); PROJECT UFO (II,2088)

KING, ALEXANDER
JACK PAAR PRESENTS (I,2315); THE TONIGHT SHOW (I,4564)

KING, ANDREA
DO NOT DISTURB (I,1300); GRUEN GUILD PLAYHOUSE (I,1892); MAN OF THE COMSTOCK (I,2873)

KING, B B
THE CAPTAIN AND TENNILLE SONGBOOK (II,432); THE CRYSTAL GAYLE SPECIAL (II,609)

KING, BILLIE JEAN
PERRY COMO'S LAKE TAHOE HOLIDAY (II,2018)

KING, BRETT
DOC HOLLIDAY (I,1307); MACKENZIE'S RAIDERS (I,2803)

KING, BRUCE
ALL ABOUT MUSIC (I,127)

KING, CARL
KING'S CROSSROADS (I,2561); KING'S PARTY LINE (I,2562)

KING, CHRIS
TOP TEN (II,2648)

KING, CHRISTOPHER
OUTPOST (I,3428)

KING, CISSY
THE LAWRENCE WELK SHOW (I,2643); MEMORIES WITH LAWRENCE WELK (II,1681)

KING, DAVE
THE KRAFT MUSIC HALL (I,2587); KRAFT MUSIC HALL PRESENTS THE DAVE KING SHOW (I,2586)

KING, DENNIS
THE ALCOA HOUR (I,108); THE DEVIL'S DISCIPLE (I,1247); GIVE US BARABBAS! (I,1816); TWELFTH NIGHT (I,4636)

KING, DIANA
FATHER DEAR FATHER (II,837)

KING, EDITH
CYRANO DE BERGERAC (I,1119); MAN AND SUPERMAN (I,2854)

KING, FREEMAN
THE BOBBY VINTON SHOW (II,345); DANCE FEVER (II,615); THE HUDSON BROTHERS RAZZLE DAZZLE COMEDY SHOW (II,1201); THE PRIMARY ENGLISH CLASS (II,2081); SEMI-TOUGH (II,2300); THE SONNY AND CHER COMEDY HOUR (II,2400); THE SONNY COMEDY REVUE (II,2403)

KING, JOHN MICHAEL
KENNEDY (II,1377)

KING, JOHN REED
BATTLE OF THE AGES (I,367); CHANCE OF A LIFETIME (I,898); GIVE AND TAKE (I,1814); HAVE A HEART (I,1966); IT'S A GIFT (I,2255); THE MISSUS GOES A SHOPPING (I,3069); ON YOUR WAY (I,3366); THERE'S ONE IN EVERY FAMILY (I,4439); TOOTSIE HIPPODROME (I,4575); WHAT HAVE YOU GOT TO LOSE? (I,4811); WHAT'S YOUR BID? (I,4825); WHERE WAS I? (I,4833); WHY? (I,4856)

KING, KEVIN
MICKEY SPILLANE'S MIKE HAMMER: MORE THAN MURDER (II,1693)

KING, KIP
HELLO, LARRY (II,1128); THE SMURFS (II,2385); TV FUNNIES (II,2672)

KING, LOUISE
MOMENT OF TRUTH (I,3084)

KING, MABEL
THE JERK, TOO (II,1282); WHAT'S HAPPENING!! (II,2769)

KING, MAGGIE
A DOLL'S HOUSE (I,1327); THE GOLDEN AGE OF TELEVISION (II,1017)

KING, MARK
AMERICA 2100 (II,66); CHEERS (II,494); THE CRACKER BROTHERS (II,599); THE INVESTIGATORS (II,1241); NATIONAL LAMPOON'S HOT FLASHES (II,1795)

KING, MARTHA
AMAHL AND THE NIGHT VISITORS (I,149)

KING, MEEGAN
STARSTRUCK (II,2446)

KING, MICKI
GLENN FORD'S SUMMERTIME, U.S.A. (I,1823)

KING, NICK
THE STAR MAKER (I,4193)

KING, PATSY
PRISONER: CELL BLOCK H (II,2085)

KING, PEE WEE
COUNTRY NIGHT OF STARS II (II,593); PEE WEE KING SHOW (I,3504); PEE WEE KING'S FLYING RANCH (I,3505)

KING, PEGGY
THE BOB HOPE SHOW (I,552); THE GEORGE GOBEL SHOW (I,1767); THE GEORGE GOBEL SHOW (I,1768); HANS BRINKER OR THE SILVER SKATES (I,1933); JACK AND THE BEANSTALK (I,2287); THE MEL TORME SHOW (I,2995); THE VIC DAMONE SHOW (I,4717)

KING, PERRY
CAPTAINS AND THE KINGS (II,435); FOUR EYES (II,912); GOLDEN GATE (II,1020); THE LAST CONVERTIBLE (II,1435); THE QUEST (II,2098); RIPTIDE (II,2178)

KING, ROBERT
WATCH ME (II,2745)

KING, RORI
I'M A BIG GIRL NOW (II,1218)

KING, SONNY
JACK CARTER AND COMPANY (I,2305)

KING, TONY
BRONK (II,379)

KING, VICTORIA
THE LIFE OF RILEY (I,2686)

KING, VIVIAN
FARAWAY HILL (I,1532)

KING, WALTER WOOLF
EXPERT WITNESS (I,1488); LIGHTS, CAMERA, ACTION! (I,2697); WE'LL TAKE MANHATTAN (I,4792)

KING, WAYNE
THE WAYNE KING SHOW (I,4779)

KING, WRIGHT
JOHNNY JUPITER (I,2429); JUSTICE OF THE PEACE (I,2500); LOGAN'S RUN (II,1507); THE TRAP (I,4593); VOYAGE TO THE BOTTOM OF THE SEA (I,4743); WANTED: DEAD OR ALIVE (I,4764)

KING, ZALMAN
THE LONG HOT SUMMER (I,2747); SMILE JENNY, YOU'RE DEAD (II,2382); THE YOUNG LAWYERS (I,4940); THE YOUNG LAWYERS (I,4941)

KINGI, HENRY
THE RENEGADES (II,2132)

KINGSTON TRIO, THE
GEORGE BURNS IN THE BIG TIME (I,1764)

KINGSTON, HARRY
BUCK ROGERS IN THE 25TH CENTURY (I,751)

KINGSTON, LENORE
FATHER KNOWS BEST (II,838); GENERAL HOSPITAL (II,964); LEAVE IT TO BEAVER (I,2648)

KINGSTON, SCOTT
THE MACKENZIES OF PARADISE COVE (II,1585)

KINKEAD, MAEVE
ANOTHER WORLD (II,97); THE GUIDING LIGHT (II,1064)

KINNEAR, CHRISTOPHER
LITTLE SHOTS (II,1495)

KINNEAR, ROY
GEORGE AND MILDRED (II,965)

KINNEY, TIFFANY
THE DEVIL'S WEB (I,1248)

KINO, LLOYD
MIMI (I,3054); READY AND WILLING (II,2115); THE RED SKELTON SHOW (I,3755)

KINSELLA, WALTER
GOLDEN WINDOWS (I,1841); MARTIN KANE, PRIVATE EYE (I,2928)

KINSEY, LANCE
THE CHICAGO STORY (II,507); THE YESTERDAY SHOW (II,2850)

KINSKEY, LEONID
ALWAYS APRIL (I,147); THE PEOPLE'S CHOICE (I,3521)

KINSKI, KLAUS
THE HITCHHIKER (II,1157)

KINSMAN, JANET
STRIPPERS (II,2481)

KINSOLVING, LEE
AH!, WILDERNESS (I,91)

KIRBY JR, B
THE SUPER (I,4293)

KIRBY STONE FOUR, THE
THE JIMMIE RODGERS SHOW (I,2382); SID CAESAR INVITES YOU (I,4040)

KIRBY STONE QUINTET, THE
BROADWAY OPEN HOUSE (I,736)

KIRBY, AMANDA
THE PRIME OF MISS JEAN BRODIE (II,2082)

KIRBY, BRUCE
AMERICA, YOU'RE ON (II,69); BOWZER (II,358); BRANAGAN AND MAPES (II,367); CAR 54, WHERE ARE YOU? (I,842); THE DORIS DAY SHOW (I,1355); HOLMES AND YOYO (II,1169); LADIES IN BLUE (II,1422); LOVERS AND OTHER STRANGERS (II,1550); MARIE (II,1628); MCNAMARA'S BAND (II,1668); MCNAMARA'S BAND (II,1669); THE PHIL SILVERS SHOW (I,3580); SHANNON (II,2314); TURNABOUT (II,2669)

KIRBY, CHRIS
BLOCKHEADS (II,270); LIKE MAGIC (II,1476)

KIRBY, DURWARD
CANDID CAMERA (I,819); G.E. GUEST HOUSE (I,1746); THE GARRY MOORE SHOW (I,1738); THE GARRY MOORE SHOW (I,1739); THE GARRY MOORE SHOW (I,1740)

KIRBY, GENE
JOHNNY OLSEN'S RUMPUS ROOM (I,2433)

KIRBY, GEORGE
ABC STAGE '67 (I,6); ALAN KING IN LAS VEGAS, PART II (I,98); BOB HOPE SPECIAL: BOB HOPE'S USO CHRISTMAS IN BEIRUT (II,320); THE DIONNE WARWICK SPECIAL (I,1289); THE GEORGE KIRBY SPECIAL (I,1771); HALF THE GEORGE KIRBY COMEDY HOUR (I,1920); THE KOPYCATS (I,2580); ROSENTHAL AND JONES (II,2215); SAGA OF SONORA (I,3884)

KIRBY, PAT
THE TONIGHT SHOW (I,4561)

KIRBY, PAUL
BEAUTY AND THE BEAST (II,197)

KIRBY, RANDY
THE GIRL FROM U.N.C.L.E. (I,1808); THE KID WITH THE BROKEN HALO (II,1387)

KIRBY, WILLIAM CULLEN
ONE DAY AT A TIME (II,1900)

KIRCHENBAUER, BILL
AMERICA 2-NIGHT (II,65); BARBARA MANDRELL AND THE MANDRELL SISTERS (II,143); THE BOOK OF LISTS (II,350); THE INVESTIGATORS (II,1241); THE LISA HARTMAN SHOW (II,1484); STOPWATCH: THIRTY MINUTES OF INVESTIGATIVE TICKING (II,2474)

KIRCHNER, CLAUDE
MAGIC MIDWAY (I,2818); SUPER CIRCUS (I,4287)

KIRK, JOE
THE ABBOTT AND COSTELLO SHOW (I,2)

KIRK, JULIE
THE BENNY HILL SHOW (II,207)

KIRK, LISA
FRONT ROW CENTER (I,1694); THE MAN IN THE MOON (I,2871); THE MOTOROLA TELEVISION HOUR (I,3112); SUSPICION (I,4309)

KIRK, PHYLLIS
APPOINTMENT WITH ADVENTURE (I,246); THE ERROL FLYNN THEATER (I,1458); FORD TELEVISION THEATER (I,1634); PHILCO TELEVISION PLAYHOUSE (I,3583); PLAYHOUSE 90 (I,3623); PLAYWRIGHTS '56 (I,3627); THE RED BUTTONS SHOW (I,3750); STAR STAGE (I,4197); SUSPENSE (I,4305); TALES OF TOMORROW (I,4352); THE THIN MAN (I,4446); THE TWILIGHT ZONE (I,4651); THE UNITED STATES STEEL HOUR (I,4677); THE WEB (I,4784); THE YOUNG SET (I,4945); ZANE GREY THEATER (I,4979)

KIRK, TOMMY
THE HARDY BOYS AND THE MYSTERY OF GHOST FARM (I,1950); THE HARDY BOYS AND THE MYSTERY OF THE APPLEGATE TREASURE (I,1951)

KIRKBRIDE, ANNE
CORONATION STREET (I,1054)

KIRKCONNELL, CLARE
THE PAPER CHASE: THE SECOND YEAR (II,1950)

KIRKES, SHIRLEY
BLANSKY'S BEAUTIES (II,267); THE CAROL BURNETT SHOW (II,443); THE LATE FALL, EARLY SUMMER BERT CONVY SHOW (II,1441); LEGS (II,1459); OF THEE I SING (I,3334); WHO'S WATCHING THE KIDS? (II,2793)

KIRKLAND, DIANE
VEGAS (II,2723)

KIRKLAND, JEFF
CRASH ISLAND (II,600)

KIRKLAND, MURIEL
THE DOCTORS (II,694); THE HEIRESS (I,2006); THE NURSES (I,3319)

KIRKLAND, PATRICIA
THE EGG AND I (I,1420); FORD THEATER HOUR (I,1635); LITTLE WOMEN (I,2719); PHILCO TELEVISION PLAYHOUSE (I,3583)

KIRKLAND, SALLY
BRONK (II,379); THE GEORGIA PEACHES (II,979); SHAUGHNESSEY (II,2318); STONESTREET: WHO KILLED THE CENTERFOLD MODEL? (II,2472); SUMMER (II,2491); WILLOW B: WOMEN IN PRISON (II,2804)

KIRKPATRICK, JESS
I AND CLAUDIE (I,2164)

KIRKWOOD JR, JOE
THE JOE PALOOKA STORY (I,2401)

KIRKWOOD, JACK
THE BOB HOPE SHOW (I,553); FIBBER MCGEE AND MOLLY (I,1569); THE LIFE OF RILEY (I,2686); ONE HAPPY FAMILY (I,3375)

KIRKWOOD, JAMES
TONI TWIN TIME (I,4554); VALIANT LADY (I,4695)

KIROV BALLET, THE
PEGGY FLEMING VISITS THE SOVIET UNION (I,3509)

KIRSCHNER, RAY
THE ARTHUR GODFREY SHOW (I,287); QUADRANGLE (I,3703)

KIRSHNER, DON
DON KIRSHNER'S ROCK CONCERT (I,1332)

KIRSTEN, DOROTHY
THE CHEVY SUMMER SHOW (I,929); THE CHEVY SUMMER SHOW (I,930)

KIRTLAND, LOUISE
CAR 54, WHERE ARE YOU? (I,842)

KISER, PAM
THE LAST CONVERTIBLE (II,1435)

KISER, TERRY
THE BAY CITY AMUSEMENT COMPANY (II,185); BENNY AND BARNEY: LAS VEGAS UNDERCOVER (II,206); CAPTAINS AND THE KINGS (II,435); CHANGE AT 125TH STREET (II,480); THE COPS AND ROBIN (II,573); THE DOCTORS (II,694); MARCUS WELBY, M.D. (II,1627); NIGHT COURT (II,1844); THE ROLLERGIRLS (II,2203)

KISER, VIRGINIA
KEY WEST (I,2528)

KISSINGER, HENRY
BOB HOPE SPECIAL: BOB HOPE PRESENTS A CELEBRATION WITH STARS OF COMEDY AND MUSIC (II,340)

KISSON, JEFFREY
SPACE: 1999 (II,2409)

KITAEN, TAWNY
MALIBU (II,1607)

KITAGAWA, MIKA
SHOGUN (II,2339)

KITCHELL, ALMA
IN THE KELVINATOR KITCHEN (I,2203)

KITCHEN, MICHAEL
ONCE THE KILLING STARTS (I,3367); SLEEPWALKER (I,4084)

KITLER, LARRY
COUNTERATTACK: CRIME IN AMERICA (II,581)

KITSUDA, YOSHIE
SHOGUN (II,2339)

KITT, EARTHA
BATMAN (I,366); THE EARTHA KITT SHOW (I,1395); SHOW BIZ (I,4027); TWELVE STAR SALUTE (I,4638)

KIVEK, MILOS
MURDER ON THE MIDNIGHT EXPRESS (I,3162)

KJAR, JOAN
HARPER VALLEY PTA (II,1095); SEVEN BRIDES FOR SEVEN BROTHERS (II,2307)

KJELLIN, ALF
THE CABOT CONNECTION (II,408)

KLABOE, DANA
ANOTHER WORLD (II,97)

KLATSCHER, LAURIE
THE DOCTORS (II,694)

KLAVIN, GENE
GOOD MORNING WORLD (I,1850)

KLAVIN, WALT
THE RIGHT MAN (I,3790)

KLEEB, HELEN
DEAR MOM, DEAR DAD (I,1202); THE DETECTIVE (II,662); HARRIGAN AND SON (I,1955); PETE AND GLADYS (I,3558); THE WALTONS (II,2740); THE WALTONS: A DAY FOR THANKS ON WALTON'S MOUNTAIN (EPISODE 3) (II,2741); THE WALTONS: MOTHER'S DAY ON WALTON'S MOUNTAIN (EPISODE 2) (II,2742); THE WALTONS: WEDDING ON WALTON'S MOUNTAIN (EPISODE 1) (II,2743)

KLEIN, ADELAIDE
HANDS OF MURDER (I,1929); TWO GIRLS NAMED SMITH (I,4656)

KLEIN, HARRY
MISTRAL'S DAUGHTER (II,1708)

KLEIN, ROBERT
ALL COMMERCIALS—A STEVE MARTIN SPECIAL (II,37); BOB HOPE SPECIAL: BOB HOPE'S ALL-STAR BIRTHDAY PARTY (II,295); BRAIN GAMES (II,366); COMEDY TONIGHT (I,1027); THE COMEDY ZONE (II,559); KLEIN TIME (II,1401); PAJAMA TOPS (II,1942); THE ROBERT KLEIN SHOW (II,2185); THE ROBERT KLEIN SHOW (II,2186); THAT WAS THE YEAR THAT WAS (II,2575); TV'S BLOOPERS AND PRACTICAL JOKES (II,2675)

KLEMPERER, WERNER
ASSIGNMENT: MUNICH (I,302); HOGAN'S HEROES (I,2069); THE RETURN OF THE BEVERLY HILLBILLIES (II,2139); THE RHINEMANN EXCHANGE (II,2150)

KLENCK, MARGARET
ONE LIFE TO LIVE (II,1907)

KLICK, MARY
THE GEORGE HAMILTON IV SHOW (I,1769); THE JIMMY DEAN SHOW (I,2384)

KLIGMAN, PAUL
SPIDER-MAN (I,4154)

KLINE, KEVIN
SEARCH FOR TOMORROW (II,2284)

KLINE, LARRY
THREE'S COMPANY (II,2614)

KLINE, RICHARD
HIS AND HERS (II,1155); SEVENTH AVENUE (II,2309)

KLINGER, NATALIE
ANOTHER MAN'S SHOES (II,96); BIG JOHN (II,239)

KLOUS, PATRICIA
ALOHA PARADISE (II,58); BATTLE OF THE NETWORK STARS (II,167); FLYING HIGH (II,894); FLYING HIGH (II,895); JOHNNY BLUE (II,1321); THE LOVE BOAT (II,1535)

KLUGE, GEORGE
THESE ARE MY CHILDREN (I,4440)

KLUGH, EARL
COUNTRY COMES HOME (II,586)

KLUGMAN, JACK
THE ALCOA HOUR (I,108); ALFRED HITCHCOCK PRESENTS (I,115); APPOINTMENT WITH ADVENTURE (I,246); THE BURNS AND SCHREIBER COMEDY HOUR (I,769); CAPTAIN VIDEO AND HIS VIDEO RANGERS (I,838); CELEBRITY CHALLENGE OF THE SEXES 2 (II,466); CIRCUS OF THE STARS (II,531); THE DON RICKLES SHOW (II,708); FAME IS THE NAME OF THE GAME (I,1518); FAMOUS LIVES (II,821); THE GREATEST GIFT (I,1880); HARRIS AGAINST THE WORLD (I,1956); KISS ME, KATE (I,2568); KRAFT SUSPENSE THEATER (I,2591); LUCY MOVES TO NBC (II,1562); MAGIC WITH THE STARS (II,1599); THE MILLION DOLLAR INCIDENT (I,3045); THE ODD COUPLE (II,1875); PARADE OF STARS (II,1954); THE PETRIFIED FOREST (I,3570); PLAYHOUSE 90 (I,3623); POOR DEVIL (I,3646); QUINCY, M. E. (II,2102); RICKLES (II,2170); THE SHAPE OF THINGS (I,3999); STUDIO ONE (I,4268); SUPER COMEDY BOWL 2 (I,4289); SUSPICION (I,4309); THE TWILIGHT ZONE (I,4651); THE UNITED STATES STEEL HOUR (I,4677); WON'T IT EVER BE MORNING? (I,4902); THE WONDERFUL WORLD OF AGGRAVATION (I,4894)

KLUNIS, TOM
THE GUIDING LIGHT (II,1064); SEARCH FOR TOMORROW (II,2284)

KLUTE, PAUL
CAGNEY AND LACEY (II,409)

KLUTE, SIDNEY
THE GOOD LIFE (II,1033)

KNAPP, ROBERT
THE BLUE ANGELS (I,490); DAYS OF OUR LIVES (II,629)

KNAUB, JIM
FLATFOOTS (II,879)

KNEALE, PATRICIA
SPELL OF EVIL (I,4152)

KNELL, DAVID
BRET MAVERICK (II,374); GIDGET'S SUMMER REUNION (I,1799); WE GOT IT MADE (II,2752); THE WONDERFUL WORLD OF PHILIP MALLEY (II,2831)

KNICKERBOCKER, WILLIS
KEY TORTUGA (II,1383)

KNIEVEL, EVEL
HAPPY BIRTHDAY, AMERICA (II,1083); THE SENSATIONAL, SHOCKING, WONDERFUL, WACKY 70S (II,2302); VARIETY (II,2721)

KNIGHT, CHARLOTTE
SUDDEN SILENCE (I,4275); SVENGALI AND THE BLONDE (I,4310)

KNIGHT, CHRISTOPHER
ANOTHER WORLD (II,97); THE BRADY BUNCH (II,362); THE BRADY BUNCH HOUR (II,363); THE BRADY GIRLS GET MARRIED (II,364); THE BRADY KIDS (II,365); JOE'S WORLD (II,1305); VALENTINE MAGIC ON LOVE ISLAND (II,2716)

KNIGHT, DAVID
APPLAUSE (I,244)

KNIGHT, DON
THE BUFFALO SOLDIERS (II,388); CODE NAME: DIAMOND HEAD (II,549); THE ICE PALACE (I,2182); THE IMMORTAL (I,2197); MANIMAL (II,1622); MURDOCK'S GANG (I,3164); THE SECRET WAR OF JACKIE'S GIRLS (II,2294)

KNIGHT, DUDLEY
LOBO (II,1504)

KNIGHT, FUZZY

CAPTAIN GALLANT OF THE FOREIGN LEGION (I,831)

KNIGHT, GLADYS
100 YEARS OF GOLDEN HITS (II,1905); GLADYS KNIGHT AND THE PIPS (II,1004); UPTOWN (II,2710)

KNIGHT, JACK
CHEERS (II,494); DEAD MAN ON THE RUN (II,631); EVENING AT THE IMPROV (II,786); PRESENTING SUSAN ANTON (II,2077); RISKO (II,2180)

KNIGHT, MARGOT
PRISONER: CELL BLOCK H (II,2085)

KNIGHT, RANDY
CAPTAIN SAFARI OF THE JUNGLE PATROL (I,836)

KNIGHT, RAY
THE RAY KNIGHT REVUE (I,3733)

KNIGHT, ROSALIND
NANCY ASTOR (II,1788)

KNIGHT, SHIRLEY
THE OUTSIDER (I,3430)

KNIGHT, TED
AQUAMAN (I,249); CELEBRITY CHALLENGE OF THE SEXES 3 (II,467); CIRCUS OF THE STARS (II,538); FAMILY BUSINESS (II,815); FANTASTIC VOYAGE (I,1530); JOURNEY TO THE CENTER OF THE EARTH (I,2456); KRAFT SUSPENSE THEATER (I,2591); LASSIE'S RESCUE RANGERS (II,1432); MAC DAVIS'S CHRISTMAS ODYSSEY: TWO THOUSAND AND TEN (II,1579); THE MARY TYLER MOORE SHOW (II,1640); MITZI AND A HUNDRED GUYS (II,1710); MITZI: A TRIBUTE TO THE AMERICAN HOUSEWIFE (I,3072); THE NEW ADVENTURES OF SUPERMAN (I,3255); OF THEE I SING (I,3334); OLIVIA NEWTON-JOHN'S HOLLYWOOD NIGHTS (II,1886); THE OUTER LIMITS (I,3426); PARADISE (II,1955); THE PAT BOONE AND FAMILY EASTER SPECIAL (II,1970); SANDY IN DISNEYLAND (II,2254); THE SENSATIONAL, SHOCKING, WONDERFUL, WACKY 70S (II,2302); SUPER FRIENDS (II,2497); TED KNIGHT MUSICAL COMEDY VARIETY SPECIAL SPECIAL (II,2549); THE TED KNIGHT SHOW (II,2550); TOO CLOSE FOR COMFORT (II,2642)

KNOOTZ, GREGORY
CRASH ISLAND (II,600)

KNOOTZ, JEFFREY
CRASH ISLAND (II,600)

KNOOTZ, MEL
THE PINKY LEE SHOW (I,3601)

KNOTTS, DON
ALAN KING'S FINAL WARNING (II,29); THE ANDY GRIFFITH SHOW (I,192); ANDY GRIFFITH'S UPTOWN-DOWNTOWN SHOW (I,194); THE ANDY GRIFFITH—DON KNOTTS—JIMNABORS SHOW (I,191); BOB HOPE SPECIAL: BOB HOPE'S ALL-STAR BIRTHDAY PARTY (II,295); BOB HOPE SPECIAL: BOB HOPE'S STAND UP AND CHEER FOR THE NATIONAL FOOTBALL LEAGUE'S 60TH YEAR (II,316); THE CAPTAIN AND TENNILLE IN HAWAII (II,430); THE DEAN MARTIN SHOW (I,1200); DINAH IN SEARCH OF THE IDEAL MAN (I,1278); DON KNOTTS NICE CLEAN, DECENT, WHOLESOME HOUR (I,1333); THE DON KNOTTS SHOW (I,1334); THE DON KNOTTS SPECIAL (I,1335); GLENN FORD'S SUMMERTIME, U.S.A. (I,1823); HARRY AND MAGGIE (II,1097); HOWDY DOODY (I,2151); I LOVE A MYSTERY (I,2170); LAS VEGAS PALACE OF STARS (II,1431); LOOKING BACK (I,2754); THE MAN WHO CAME TO DINNER (I,2881); THE MANY FACES OF COMEDY (I,2894); PIPER'S PETS (II,2046); THE REASON NOBODY HARDLY EVER SEEN A FAT OUTLAW IN THE OLD WEST IS AS FOLLOWS: (I,3744); SEARCH FOR TOMORROW (II,2284); THE STEVE ALLEN SHOW (I,4219); THE STEVE ALLEN SHOW (I,4220); STEVE ALLEN'S LAUGH-BACK (II,2455); THREE'S COMPANY (II,2614)

KNOTTS, KAREN
ONE OF OUR OWN (II,1910)

KNOWLES, PATRIC
THE ETHEL BARRYMORE THEATER (I,1467); PAPA SAID NO (I,3452)

KNOX, ALEXANDER
TINKER, TAILOR, SOLDIER, SPY (II,2621); TRUMAN AT POTSDAM (I,4618)

KNOX, BARBARA
CORONATION STREET (I,1054)

KNOX, MICKEY
THE WINDS OF WAR (II,2807)

KNOX, MONA
MAGGIE (I,2810)

KNOX, TERENCE
ST. ELSEWHERE (II,2432)

KNOX, TOMMY
AND HERE'S THE SHOW (I,187); HERE'S THE SHOW (I,2033)

KNUDSEN, PEGGY
DO NOT DISTURB (I,1300); SO THIS IS HOLLYWOOD (I,4105); YOUR SHOW TIME (I,4958)

KNUDSON, GUNILLA
GIDGET GROWS UP (I,1797)

KNUDSON, KURT
GEORGE WASHINGTON (II,978); SPARROW (II,2411)

KNYVETTE, SALLY
BLAKE'S SEVEN (II,264)

KOBART, RUTH
ANOTHER MAN'S SHOES (II,96)

KOBE, GAIL
BRIGHT PROMISE (I,727); PEYTON PLACE (I,3574)

KOCHHEIM, LORY
MULLIGAN'S STEW (II,1763); MULLIGAN'S STEW (II,1764)

KODI, JIM
A TIME TO LIVE (I,4510)

KODICEK, SUSAN
TINKER, TAILOR, SOLDIER, SPY (II,2621)

KOEHLER, FREDERICK
KATE AND ALLIE (II,1365)

KOENIG, CHARLIE
KOBB'S CORNER (I,2577)

KOENIG, TOMMY
NOT NECESSARILY THE NEWS (II,1862)

KOENIG, WALTER
STAR TREK (II,2440); THE STARLOST (I,4203)

KOENING, KARL
DATE WITH THE ANGELS (I,1164)

KOESUN, ANN
TONI TWIN TIME (I,4554)

KOGAN, MILT
BARNEY MILLER (II,154); BUNCO (II,395); KINGSTON: CONFIDENTIAL (II,1398); KINGSTON: THE POWER PLAY (II,1399); ONE OF OUR OWN (II,1910); THE RETURN OF MARCUS WELBY, M.D (II,2138)

KOHLER, DENNIS
MEMBER OF THE WEDDING (I,3002); ON BORROWED TIME (I,3354); SLEEPING BEAUTY (I,4083); SWISS FAMILY ROBINSON (I,4324)

KOHLER, ESTELLE
THE ADMIRABLE CRICHTON (I,36)

KOHN, MELANIE
BE MY VALENTINE, CHARLIE BROWN (I,369); IT'S A MYSTERY, CHARLIE BROWN (I,2259); IT'S THE EASTER BEAGLE, CHARLIE BROWN (I,2275); YOU'RE A GOOD SPORT, CHARLIE BROWN (I,4963)

KOHN, ROBIN
A CHARLIE BROWN THANKSGIVING (I,909); THERE'S NO TIME FOR LOVE, CHARLIE BROWN (I,4438); YOU'RE NOT ELECTED, CHARLIE BROWN (I,4967)

KOLB, CLARENCE
MY LITTLE MARGIE (I,3194)

KOLB, GIDEON
QUILLER: NIGHT OF THE FATHER (II,2100)

KOLB, MINA
BACK TOGETHER (II,131); THE BEAN SHOW (I,373); PETE AND GLADYS (I,3558); STORY THEATER (I,4241); THREE'S COMPANY (II,2614); THE TV TV SHOW (II,2674)

KOLDEN, SCOTT
ME AND THE CHIMP (I,2974); SIGMUND AND THE SEA MONSTERS (II,2352)

KOLLDEHOFF, RENE
THE WINDS OF WAR (II,2807)

KOLLMAR, RICHARD
GUESS WHAT? (I,1895); SHOW BUSINESS, INC. (I,4028)

KOMACK, JAMES
9 TO 5 (II,1852); BEST FOOT FORWARD (I,401); CHARLIE ANGELO (I,906); THE COURTSHIP OF EDDIE'S FATHER (I,1083); HENNESSEY (I,2013); THE JUNE ALLYSON SHOW (I,2488); WELCOME BACK, KOTTER (II,2761)

KOMANT, CAROLYN
THE ROARING TWENTIES (I,3805)

KOMINSKY, SHERRY
CELEBRITY BOWLING (I,883)

KONA, STEPHANIE

MORK AND MINDY (II,1735)

KONDAZIAN, KAREN
SHANNON (II,2314)

KONRAD, DOROTHY
GREAT DAY (II,1053); THE JIMMY DURANTE SHOW (I,2389); THE LAST RESORT (II,1440)

KONSTANTINE, LEOPOLDINE
THE SWAN (I,4312)

KOOCK, GUICH
CARTER COUNTRY (II,449); THE CHISHOLMS (II,513); THE CIRCLE FAMILY (II,527); LEWIS AND CLARK (II,1465); A ROCK AND A HARD PLACE (II,2190)

KOONIA, IVOR
BILLY CRYSTAL: A COMIC'S LINE (II,249)

KOPELL, BERNIE
BEWITCHED (I,418); DEATH OF A SALESMAN (I,1208); THE DORIS DAY SHOW (I,1355); FLO'S PLACE (II,892); GET SMART (II,983); THE LOVE BOAT (II,1535); THE LOVE BOAT II (II,1533); THE LOVE BOAT III (II,1534); MY FAVORITE MARTIAN (I,3189); NEEDLES AND PINS (I,3246); SALLY AND SAM (I,3891); THAT GIRL (II,2570); WHEN THINGS WERE ROTTEN (II,2782); WILD ABOUT HARRY (II,2797)

KOPINS, KAREN
RIPTIDE (II,2178)

KOPOLAN, HARRY
FALCON'S GOLD (II,809)

KOPPOLA, FRANK
EIGHT IS ENOUGH (II,762)

KORKES, JOHN
THE WORD (II,2833)

KORMAN, HARVEY
BOB HOPE SPECIAL: BOB HOPE FOR PRESIDENT (II,282); THE CAROL BURNETT SHOW (II,443); THE CARPENTERS AT CHRISTMAS (II,446); THE CELEBRITY FOOTBALL CLASSIC (II,474); THE CRACKER BROTHERS (II,599); THE DANNY KAYE SHOW (I,1143); DON RICKLES—ALIVE AND KICKING (I,1340); EUNICE (II,784); EVENING AT THE IMPROV (II,786); THE FLINTSTONES (II,884); GALLAGHER (I,1727); THE HARVEY KORMAN SHOW (II,1103); THE HARVEY KORMAN SHOW (II,1104); HI, I'M GLEN CAMPBELL (II,1141); HOW TO SURVIVE THE 70S AND MAYBE EVEN BUMP INTO HAPPINESS (II,1199); THE INVISIBLE WOMAN (II,1244); THE LOVE BOAT I (II,1532); THE MAD MAD MAD MAD WORLD OF THE SUPER BOWL (II,1586); MAMA'S FAMILY (II,1610); THE MANY SIDES OF DON RICKLES (I,2898); A SALUTE TO STAN LAUREL (I,3892); SNAVELY (II,2387); THE TIM CONWAY SHOW (II,2619); THE TIM CONWAY SPECIAL (I,4503)

KORN KOBBLERS, THE
KOBB'S CORNER (I,2577)

KORN, IRIS
LITTLE HOUSE ON THE PRAIRIE (II,1487)

KORPER, RENE
THE LONELY WIZARD (I,2743)

KORT, DENNIS
DOMINIC'S DREAM (II,704); REAR GUARD (II,2120)

KORVIN, CHARLES
INTERPOL CALLING (I,2227)

KORWIN, ANNA
JANE EYRE (II,1273)

KORWIN, DEVRA
WINTER KILL (II,2810)

KOSARIO, ADRIAN
FAME (II,812)

KOSLECK, MARTIN
LONGSTREET (I,2750)

KOSLO, PAUL
THE DAUGHTERS OF JOSHUA CABE (I,1169); DOWN HOME (II,731); ROOTS: THE NEXT GENERATIONS (II,2213); YOU ARE THE JURY (II,2854)

KOSOFF, JILL
THE DOCTORS (II,694)

KOSS, ALAN
CHEERS (II,494); CONCRETE BEAT (II,562)

KOSSLYN, JACK
HARRY O (II,1098)

KOSSOFF, DAVID
THE LILLI PALMER THEATER (I,2704)

KOSTER, WALLY
ADVENTURES IN RAINBOW COUNTRY (I,47)

KOSTICHEK, CHRIS
DAYS OF OUR LIVES (II,629)

KOSUGI, SHO
THE MASTER (II,1648)

KOTERO, PATTY
ALL THAT GLITTERS (II,41)

KOTTO, YAPHET
FOR LOVE AND HONOR (II,903)

KOUFAX, SANDY
THE BOB HOPE SHOW (I,597)

KOVACK, NANCY
OFF WE GO (I,3338)

KOVACS, BELA
SPACE PATROL (I,4144)

KOVACS, ERNIE
ALCOA/GOODYEAR THEATER (I,107); THE COMEDY OF ERNIE KOVACS (I,1022); ERNIE IN KOVACSLAND (I,1454); THE ERNIE KOVACS SHOW (I,1455); THE ERNIE KOVACS SPECIAL (I,1456); FESTIVAL OF MAGIC (I,1566); GAMBLE ON LOVE (I,1731); I WAS A BLOODHOUND (I,2180); IT'S TIME FOR ERNIE (I,2277); THE JERRY LEWIS SHOW (I,2362); KOVACS ON THE CORNER (I,2582); KOVACS UNLIMITED (I,2583); THE NEW ERNIE KOVACS SHOW (I,3264); PRIVATE EYE, PRIVATE EYE (I,3671); THE ROSALIND RUSSELL SHOW (I,3845); SILENT'S, PLEASE (I,4050); SONJA HENIE'S HOLIDAY ON ICE (I,4125); TAKE A GOOD LOOK (I,4332); TAKE A GOOD LOOK (II,2529); TAKE A GOOD LOOK (I,4332); TIME WILL TELL (I,4512); THE TONIGHT SHOW (I,4563)

KOVACS, JONATHAN HALL
THE FAMILY TREE (II,820); THE SIX OF US (II,2369)

KOVE, MARTIN
CAGNEY AND LACEY (II,409); CODE R (II,551); KINGSTON: THE POWER PLAY (II,1399); TROUBLE IN HIGH TIMBER COUNTRY (II,2661); WE'VE GOT EACH OTHER (II,2757)

KOWANKO, PETER
FOR LOVE AND HONOR (II,903); HOME ROOM (II,1172)

KOZAK, HARLEY
THE GUIDING LIGHT (II,1064); TEXAS (II,2566)

KRAFT, BEATRICE
MARCO POLO (I,2904)

KRAFT, JILL
THE INNER SANCTUM (I,2216)

KRAFT, RANDY
ANOTHER LIFE (II,95); OUR FIVE DAUGHTERS (I,3413)

KRAKOFF, ERIC
GOLDIE AND KIDS: LISTEN TO ME (II,1021)

KRAKOFF, MARCELLO
THE FACTS (II,804)

KRALL, HEIDI
MR. PORTER OF INDIANA (I,3148)

KRAMER, BERT
THE FITZPATRICKS (II,868); SARA (II,2260); TEXAS (II,2566)

KRAMER, JEFFREY
STICK AROUND (II,2464); STRUCK BY LIGHTNING (II,2482); WHAT'S UP DOC? (II,2771)

KRAMER, MANDEL
THE EDGE OF NIGHT (II,760)

KRAMER, STANLEY
BOGART (I,679); SOPHIA! (I,4128)

KRAMER, STEPFANIE
FOUR EYES (II,912); HUNTER (II,1206); MARRIED: THE FIRST YEAR (II,1634); ONE NIGHT BAND (II,1909); WE GOT IT MADE (II,2752)

KRAMRIETHER, TONY
SCARLETT HILL (I,3933)

KRANENDONK, LEONARD
THE FRED WARING SHOW (I,1679)

KRATOCHZIL, TOM
WONDER WOMAN (SERIES 2) (II,2829)

KRAUSS, PHILIP
THE DOCTORS (II,694)

KRAVITS, JASON
POWERHOUSE (II,2074)

KREBS, SUSAN
BETWEEN THE LINES (II,225)

KREINDEL, MITCH
MURDER CAN HURT YOU! (II,1768)

KRELL, MAC
PRUDENCE AND THE CHIEF (I,3682)

KREPPEL, PAUL
13 THIRTEENTH AVENUE (II,2592); IT'S A LIVING (II,1254); MAKING A LIVING (II,1604)

KRESKI, CONNIE
CAPTAINS AND THE KINGS (II,435)

KRESKIN
THE AMAZING WORLD OF KRESKIN (I,163)

KRIEGER, DANNY
BUTCH AND BILLY AND THEIR BANG BANG WESTERN MOVIES (I,780)

KRIEGER, GLORIA
PHIL SILVERS ON BROADWAY (I,3578)

KRIEGER, STEVIE
BUTCH AND BILLY AND THEIR BANG BANG WESTERN MOVIES (I,780)

KRIESA, CHRIS
THE RETURN OF MARCUS WELBY, M.D (II,2138)

KRIGE, ALICE
ELLIS ISLAND (II,768)

KRIMMER, WORTHAM
THE PAPER CHASE: THE SECOND YEAR (II,1950)

KRISTEN, DOROTHY
SEE AMERICA WITH ED SULLIVAN (I,3972)

KRISTEN, ILENE
RYAN'S HOPE (II,2234)

KRISTEN, LIDIA
THE BARBARY COAST (II,147)

KRISTEN, MARTA
LOST IN SPACE (I,2758); MR. NOVAK (I,3145)

KRISTOFFERSON, KRIS
COUNTRY COMES HOME (II,586); DEAN MARTIN PRESENTS MUSIC COUNTRY, U.S.A. (I,1192); GLEN CAMPBELL AND FRIENDS: THE SILVER ANNIVERSARY (II,1005); I BELIEVE IN MUSIC (I,2165); JOHNNY CASH: CHRISTMAS ON THE ROAD (II,1333); JOHNNY CASH: THE FIRST 25 YEARS (II,1336); MARLO THOMAS AND FRIENDS IN FREE TO BE. . . YOU AND ME (II,1632); A SPECIAL ANNE MURRAY CHRISTMAS (II,2413)

KRISTOFFERSON, LISA
JOHNNY CASH: CHRISTMAS ON THE ROAD (II,1333)

KRIZA, JACK
TONI TWIN TIME (I,4554)

KROEGER, GARY
SATURDAY NIGHT LIVE (II,2261)

KROEGER, LEE
MR. & MS. AND THE BANDSTAND MURDERS (II,1746); MR. & MS. AND THE MAGIC STUDIO MYSTERY (II,1747)

KROFFT MARIONETTES, THE
THE DEAN MARTIN SHOW (I,1201)

KROFFT PUPPETS, THE
BARBARA MANDRELL AND THE MANDRELL SISTERS (II,143); FOL-DE-ROL (I,1613)

KROH, ERIC
GENERAL HOSPITAL (II,964)

KRONE, FRED
FOREST RANGER (I,1641)

KRONKITE, CATHY
PERSONAL AND CONFIDENTIAL (II,2026)

KROPER, RENE
THE PIED PIPER OF HAMELIN (I,3595)

KRUGER, CHRISTIANE
STAR MAIDENS (II,2435)

KRUGER, OTTO
THE DESERT SONG (I,1236); THE JUDGE (I,2468); THE LUX VIDEO THEATER (I,2795); PULITZER PRIZE PLAYHOUSE (I,3692); SUSPENSE (I,4305)

KRUGER, PAUL
ON TRIAL (I,3362)

KRUGMAN, LOU
ANDY'S GANG (I,214); THE LIFE OF RILEY (I,2684)

KRUPA, GENE
ALL-STAR SWING FESTIVAL (II,53); THE BIG BAND AND ALL THAT JAZZ (I,422); THE SWINGIN' YEARS (I,4321); TIMEX ALL-STAR JAZZ SHOW I (I,4515); TIMEX ALL-STAR JAZZ SHOW II (I,4516); TIMEX ALL-STAR JAZZ SHOW III (I,4517); TIMEX ALL-STAR JAZZ SHOW IV (I,4518)

KRUSCHEN, JACK
BUSTING LOOSE (II,402); E/R (II,748); EMERGENCY! (II,775); HONG KONG (I,2112); THE LOG OF THE BLACK PEARL (II,1506); NICK AND NORA (II,1842); NO SOAP, RADIO (II,1856); THE RIFLEMAN (I,3789)

KUBCINET, KARYN
MRS. G. GOES TO COLLEGE (I,3155)

KUBIE, BERNIE
THE AWAKENING LAND (II,122)

KUBIK, ALEX
BEYOND WESTWORLD (II,227); HARDCASE (II,1088); RIDING FOR THE PONY EXPRESS (II,2172)

KUHLMAN, RON
THE BRADY BRIDES (II,361); THE BRADY GIRLS GET MARRIED (II,364)

KUHLMANN, ROSEMARY
AMAHL AND THE NIGHT VISITORS (I,148)

KUHLS, TESSY
SMILEY'S PEOPLE (II,2383)

KUHNE, STACEY
WHEN THE WHISTLE BLOWS (II,2781)

KUKAKI, ANOZIA
HAWKINS FALLS, POPULATION 6200 (I,1977)

KUKOFF, BERNIE
JIMMY DURANTE PRESENTS THE LENNON SISTERS HOUR (I,2387)

KULKY, HENRY
HENNESSEY (I,2013); THE LIFE OF RILEY (I,2686); LOVE THAT JILL (I,2780); MAISIE (I,2833); NAVY LOG (I,3233); SHORE LEAVE (I,4022); VOYAGE TO THE BOTTOM OF THE SEA (I,4743)

KULP, NANCY
ADVENTURES OF A MODEL (I,50); THE BEVERLY HILLBILLIES (I,417); FOUNTAIN OF YOUTH (I,1647); THE FREEWHEELERS (I,1683); THE LITTLE PEOPLE (I,2716); LOVE THAT BOB (I,2779); THE MILTON BERLE SPECIAL (I,3052); PINE LAKE LODGE (I,3597); THE RETURN OF THE BEVERLY HILLBILLIES (II,2139); SANFORD AND SON (II,2256); THE SOFT TOUCH (I,4107); SVENGALI AND THE BLONDE (I,4310); VER-R-R-RY INTERESTING (I,4712)

KULUVA, WILL
ARCHIE BUNKER'S PLACE (II,105); CYRANO DE BERGERAC (I,1119); MY THREE ANGELS (I,3204); PRIMUS (I,3668)

KUMAMOTA, LESLIE
THE AMAZING CHAN AND THE CHAN CLAN (I,157)

KUNKEL, ANDY
ALL ABOUT FACES (I,126)

KUNNER, ELOISE
THESE ARE MY CHILDREN (I,4440)

KUPCINET, IRVING
THE BOB HOPE SHOW (I,559); TONIGHT! AMERICA AFTER DARK (I,4555)

KUPPERMAN, ALVIN
MAKING IT (II,1605)

KURING, JUDE
PRISONER: CELL BLOCK H (II,2085)

KURNITZ, JULIE
THE DRUNKARD (II,737)

KURTS, ALWYN
HOMICIDE (I,2101)

KURTY, JEAN
DOCTOR KILDARE (I,1315)

KURTZ, JEFF
CAMP WILDERNESS (II,419)

KURTZ, MARCIA JEAN
THE FOUR OF US (II,914); MR. INSIDE/MR. OUTSIDE (II,1750); MURDER INK (II,1769)

KURTZ, SWOOSIE
AS THE WORLD TURNS (II,110); FIFTH OF JULY (II,851); LOVE, SIDNEY (II,1547); MARRIAGE IS ALIVE AND WELL (II,1633); MARY (II,1637)

KURTZMAN, KATY
ALLISON SIDNEY HARRISON (II,54); DYNASTY (II,746); MULLIGAN'S STEW (II,1763)

KUSATSU, CLYDE
BRING 'EM BACK ALIVE (II,377); DOCTOR STRANGE (II,688); VELVET (II,2725); ZERO INTELLIGENCE (II,2876)

KUSHIDA, BEVERLY
THE AMAZING CHAN AND THE CHAN CLAN (I,157)

KUTASH, JEFF
KICKS (II,1385)

KUTEE
THE RIGHTEOUS APPLES (II,2174)

KUTER, KAY E
GREEN ACRES (I,1884); PETTICOAT JUNCTION (I,3571)

KUYZK, MIMI
HILL STREET BLUES (II,1154)

KWAN, NANCY
THE BOB HOPE SHOW (I,586); HAWAII FIVE-O (I,1972); THE LAST NINJA (II,1438)

KWASMAN, SAM
HOLLYWOOD HIGH (II,1162)

KWONG, PETER
THE RENEGADES (II,2132)

KWONG, RICHARD
THE RENEGADES (II,2132)

KYA-HILL, ROBERT
ANOTHER WORLD (II,97)

KYDD, SAM
THE HIGHWAYMAN (I,2059)

KYLE, BARBARA
HIGH HOPES (II,1144); PAUL BERNARD—PSYCHIATRIST (I,3485)

KYSER, KAY
KAY KYSER'S KOLLEGE OF MUSICAL KNOWLEDGE (I,2512)

LABELL, GENE
FORCE SEVEN (II,908)

LACENTRA, PEG
THE MARGE AND GOWER CHAMPION SHOW (I,2907)

LADUE, JOE
THE YOUNG AND THE RESTLESS (II,2862)

LAFLEUR, ART
WEBSTER (II,2758)

LAFORGE, ELEANOR
GENTLE BEN (I,1761)

LAFORTUNE, FELICITY
RYAN'S HOPE (II,2234)

LAGARE, MICKEY
STORY THEATER (I,4241)

LAGIOIA, JOHN
THE EDGE OF NIGHT (II,760); SEARCH FOR TOMORROW (II,2284)

LALANNE, ELAINE
THE JACK LALANNE SHOW (I,2311)

LALANNE, JACK
THE JACK LALANNE SHOW (I,2311); MORE WILD WILD WEST (II,1733)

LALOGGIA, FRANK
SALT AND PEPE (II,2240); SNAVELY (II,2387)

LALONDE, BUDDY
SONJA HENIE'S HOLIDAY ON ICE (I,4125)

LAMOTHE, HENRY
AQUA VARIETIES (I,248)

LAMOY, OLIVE
POPEYE THE SAILOR (I,3648)

LAMURA, MARK
ALL MY CHILDREN (II,39)

LAPAGE, DUANE
THE FACTS OF LIFE (II,805)

LAPAGE, KIMBERLY
SISTER TERRI (II,2366)

LAPALOMA
THE BIG TOP (I,435)

LAPIERE, GEORGANNE
GENERAL HOSPITAL (II,964); WELCOME BACK, KOTTER (II,2761)

LAPLACA, ALISON
DINER (II,679); SUZANNE PLESHETTE IS MAGGIE BRIGGS (II,2509)

LAPLATT, TED
THE BEST OF EVERYTHING (I,406); WHERE THE HEART IS (I,4831)

LAROCHE, MARY
KAREN (I,2505); MATINEE THEATER (I,2947); PICTURE WINDOW (I,3594); THE TWILIGHT ZONE (I,4651); WASHINGTON: BEHIND CLOSED DOORS (II,2744)

LAROSA, JULIUS
ARTHUR GODFREY AND FRIENDS (I,278); THE JULIUS LAROSA SHOW (I,2486); THE JULIUS LAROSA SHOW (I,2487); LET'S DANCE (I,2664); ON PARADE (I,3356); PERRY COMO PRESENTS THE JULIUS LAROSA SHOW (I,3530); SONJA HENIE'S HOLIDAY ON ICE (I,4125); TV'S TOP TUNES (I,4635)

LARUE, FLORENCE
THE FIFTH DIMENSION SPECIAL: AN ODYSSEY IN THE COSMIC UNIVERSE OF PETER MAX (I,1571)

LARUE, JACK
LIGHT'S OUT (I,2699)

LARUE, LASH
LASH OF THE WEST (I,2620); THE LIFE AND LEGEND OF WYATT EARP (I,2681)

LARUE, WALT
COUNTERATTACK: CRIME IN AMERICA (II,581)

LARUSSA, ADRIANNE
CENTENNIAL (II,477); DAYS OF OUR LIVES (II,629); TERROR AT ALCATRAZ (II,2563)

LASALLE, MARTIN
FALCON'S GOLD (II,809)

LASTARZA, ROLAND
THE GALLANT MEN (I,1728); THE GREEN FELT JUNGLE (I,1885)

LATORRE, TONY
9 TO 5 (II,1852); CAGNEY AND LACEY (II,409)

LATURNE, ROBERT
THE DOCTORS (II,694)

LAAKSO, NELS
KOBB'S CORNER (I,2577)

LAAKSO, PAUL
A CHRISTMAS CAROL (II,517)

LABORTEAUX, MATTHEW
THE ALIENS ARE COMING (II,35); BRAVO TWO (II,368); LEGENDS OF THE WEST: TRUTH AND TALL TALES (II,1457); LITTLE HOUSE ON THE PRAIRIE (II,1487); LITTLE HOUSE: LOOK BACK TO YESTERDAY (II,1490); THE RED HAND GANG (II,2122); WHIZ KIDS (II,2790)

LABORTEAUX, PATRICK
LITTLE HOUSE ON THE PRAIRIE (II,1487); TOO GOOD TO BE TRUE (II,2643)

LABRIOLA, TOMMY
THE KEN MURRAY SHOW (I,2525)

LABROSA, DAVID
POWERHOUSE (II,2074)

LACEY, JOHN
ONE DEADLY OWNER (I,3373)

LACEY, LAARA
HAPPY DAYS (I,1942); MADHOUSE 90 (I,2807)

LACEY, RONALD
THE NEXT VICTIM (I,3281)

LACHER, TAYLOR
240-ROBERT (II,2689); BEULAH LAND (II,226); CADE'S COUNTY (I,787); CANNON: THE RETURN OF FRANK CANNON (II,425); CRISIS IN SUN VALLEY (II,604); THE DEADLY TRIANGLE (II,634); JOE FORRESTER (II,1303); NAKIA (II,1784); NAKIA (II,1785)

LACKLAND, BEN
CAPTAIN VIDEO AND HIS VIDEO RANGERS (I,838)

LACY, JERRY
20TH CENTURY FOLLIES (I,4641); AS THE WORLD TURNS (II,110); COMEDY TONIGHT (I,1027); DARK SHADOWS (I,1157); LOVE OF LIFE (I,2774); THE NIGHTENGALES (II,1850); PLEASURE COVE (II,2058); THE YOUNG AND THE RESTLESS (II,2862)

LADD, ALAN
THE GENERAL ELECTRIC THEATER (I,1753)

LADD, CHERYL
BATTLE OF THE NETWORK STARS (II,165); BEN VEREEN—HIS ROOTS (II,202); CHARLIE'S ANGELS (II,486); THE CHERYL LADD SPECIAL (II,501); CHERYL LADD. . . LOOKING BACK—SOUVENIRS (II,502); CHERYL LADD: SCENES FROM A SPECIAL (II,503); GENERAL ELECTRIC'S ALL-STAR ANNIVERSARY (II,963); HARRY O (II,1098); JOHN DENVER AND THE LADIES (II,1312); JOSIE AND THE PUSSYCATS (I,2453); JOSIE AND THE PUSSYCATS IN OUTER SPACE (I,2454); THE KEN BERRY WOW SHOW (I,2524); PERRY COMO'S SPRING IN SAN FRANCISCO (II,2022); SEARCH (I,3951); THAT THING ON ABC (II,2574); TONI'S BOYS (II,2637)

LADD, DAVID
SHIRLEY TEMPLE'S STORYBOOK (I,4017)

LADD, DIANE
ADDIE AND THE KING OF HEARTS (I,35); ALICE (II,33); BATTLE OF THE NETWORK STARS (II,171); BLACK BEAUTY (II,261); SEARCH FOR TOMORROW (II,2284); THE SECRET STORM (I,3969)

LADD, MARGARET
FALCON CREST (II,808); A TIME FOR US (I,4507)

LADYBIRDS, THE
THE BENNY HILL SHOW (II,207)

LAFFERTY, MARCY
BRONK (II,379); CIRCUS OF THE STARS (II,538); STAT! (I,4213)

LAHR, BERT
ANYTHING GOES (I,238); BURLESQUE (I,764); THE FANTASTICKS (I,1531); THE GREAT WALTZ (I,1878); HEAVEN WILL PROTECT THE WORKING GIRL (I,1996); THE MAN WHO CAME TO DINNER (I,2880); MR. O'MALLEY (I,3146); SHOW BIZ (I,4027); STANDARD OIL ANNIVERSARY SHOW (I,4189); THOMPSON'S GHOST (I,4467)

LAHTI, CHRISTINE
DOCTOR SCORPION (II,685); THE HARVEY KORMAN SHOW (II,1104)

LAHTI, GARY
AS THE WORLD TURNS (II,110)

LAINE, CLEO
COTTON CLUB '75 (II,580)

LAINE, FRANKIE
THE BOB HOPE SHOW (I,525); THE FRANKIE LAINE SHOW (I,1675); FRANKIE LAINE TIME (I,1676); THE JERRY COLONNA SHOW (I,2359); MUSIC '55 (I,3168); ONE MORE TIME (I,3385); THE SINGERS (I,4060)

LAIRD, JENNY
THE FORSYTE SAGA (I,1645)

LAIRD, MICHAEL
AMANDA FALLON (I,151)

LAIRE, JUDSON
I REMEMBER MAMA (I,2176); YOUNG DR. MALONE (I,4938)

LAKE, ARTHUR
BLONDIE (I,486)

LAKE, DON
OUT OF OUR MINDS (II,1933)

LAKE, FLORENCE
THE FESS PARKER SHOW (II,850); JEFF'S COLLIE (I,2356)

LAKE, JANET
THE TYCOON (I,4660)

LAKE, VERONICA
CELANESE THEATER (I,881); THE LUX VIDEO THEATER (I,2795); TALES OF TOMORROW (I,4352); TELLER OF TALES (I,4388)

LALA
BARETTA (II,152)

LAMARR, HEDY
THE BOB HOPE SHOW (I,611); ZANE GREY THEATER (I,4979)

LAMART, RENE
CALIFORNIA FEVER (II,411)

LAMAS, FERNANDO
THE BOB HOPE SHOW (I,623); DINAH AND FRIENDS (II,675); ESTHER WILLIAMS AT CYPRESS GARDENS (I,1465); FOR LOVE OR $$$ (I,1623); JANE WYMAN PRESENTS THE FIRESIDE THEATER (I,2345); POWDERKEG (I,3657); SHIRLEY TEMPLE'S STORYBOOK (I,4017)

LAMAS, LORENZO
BATTLE OF THE NETWORK STARS (II,173); CALIFORNIA FEVER (II,411); FALCON CREST (II,808); SECRETS OF MIDLAND HEIGHTS (II,2296); SHIPSHAPE (II,2329); WHATEVER HAPPENED TO DOBIE GILLIS? (II,2774)

LAMB, BRIAN
DAYTIME (II,630)

LAMB, CLAUDIA
FOREVER FERNWOOD (II,909); MARY HARTMAN, MARY HARTMAN (II,1638)

LAMB, GIL
THE GHOST AND MRS. MUIR (I,1786); THE GIL LAMB SHOW (I,1802); PISTOLS 'N' PETTICOATS (I,3608); SHIRLEY TEMPLE'S STORYBOOK (I,4017); THRU THE CRYSTAL BALL (I,4496)

LAMB, WAYNE
STOP THE MUSIC (I,4232)

LAMBERT DANCERS
THE ED SULLIVAN SHOW (I,1401)

LAMBERT, ANNIE
MASTER OF THE GAME (II,1647)

LAMBERT, DOUG
GENERAL HOSPITAL (II,964)

LAMBERT, GLORIA
SING ALONG WITH MITCH (I,4057)

LAMBERT, JACK
THE REASON NOBODY HARDLY EVER SEEN A FAT OUTLAW IN THE OLD WEST IS AS FOLLOWS: (I,3744); RIVERBOAT (I,3797)

LAMBERT, JANE
ARCHIE (II,104); THE ARCHIE SITUATION COMEDY MUSICAL VARIETY SHOW (II,106); DUFFY (II,739)

LAMBERT, LEE JAY
DANIEL BOONE (I,1142); THE NEW PEOPLE (I,3268)

LAMBERT, MARGARET
FIRESIDE THEATER (I,1580)

LAMBERT, MARTHA
AS THE WORLD TURNS (II,110); BREAKAWAY (II,370)

LAMBERT, PAUL
EXECUTIVE SUITE (II,796)

LAMBIE, JOE
THE EDGE OF NIGHT (II,760); FALCON CREST (II,808)

LAMBRINOS, VASSILI
SOMEWHERE IN ITALY. . . COMPANY B (I,4119)

LAMM, KAREN
HARRY O (II,1098); THE POWER WITHIN (II,2073)

LAMOND, BRITT
THE LIFE AND LEGEND OF WYATT EARP (I,2681)

LAMOND, TONI
THE BOB NEWHART SHOW (II,342)

LAMONT, DUNCAN
THE TEXAN (I,4413)

LAMONT, RENE
YOUNG LIVES (II,2866)

LAMOUR, DOROTHY
THE ARTHUR MURRAY PARTY FOR BOB HOPE (I,292); THE BING CROSBY SHOW (I,469); THE BOB HOPE SHOW (I,532); THE BOB HOPE SHOW (I,549); THE BOB HOPE SHOW (I,590); THE BOB HOPE SHOW (I,611); THE BOB HOPE SHOW (I,645); THE BOB HOPE SHOW (I,655); BOB HOPE SPECIAL: BOB HOPE'S ROAD TO HOLLYWOOD (II,313); BOB HOPE SPECIAL: HAPPY BIRTHDAY, BOB! (II,326); DAMON RUNYON THEATER (I,1129); HOLLYWOOD OPENING NIGHT (I,2087); LEGENDS OF THE SCREEN (II,1456)

LAMPARSKI, RICHARD
WHATEVER BECAME OF. . .? (II,2772); WHATEVER BECAME OF. . .? (II,2773)

LAMPEL, DEBORAH
ONE NIGHT BAND (II,1909)

LAMPERT, ZOHRA
ALFRED HITCHCOCK PRESENTS (I,115); BLACK BEAUTY (II,261); THE COMEDY ZONE (II,559); DOCTORS HOSPITAL (II,691); THE GIRL WITH SOMETHING EXTRA (II,999); THE GIRL, THE GOLD WATCH AND DYNAMITE (II,1001); THE GIRL, THE GOLD WATCH AND EVERYTHING (II,1002); MIXED NUTS (II,1714); ONE OF OUR OWN (II,1910); WHERE THE HEART IS (I,4831)

LAMPKIN, CHARLES
NICK AND NORA (II,1842)

LAN, HSAI HO
MEN OF THE DRAGON (II,1683)

LANCASTER, BURT
I LOVE LIBERTY (II,1215); MARCO POLO (II,1626); MOSES THE LAWGIVER (II,1737); SUPER COMEDY BOWL 2 (I,4289)

LANCASTER, LUCIE
TONIGHT AT 8:30 (I,4556)

LANCASTER, WILL
MOSES THE LAWGIVER (II,1737)

LANCE, ASTRIDE
OUR FIVE DAUGHTERS (I,3413)

LANCER, SUSAN
TONY ORLANDO AND DAWN (II,2639)

LANCHESTER, ELSA
ACADEMY AWARD SONGS (I,13); ALICE IN WONDERLAND (I,120); HEIDI (I,2004); THE JOHN FORSYTHE SHOW (I,2410); NANNY AND THE PROFESSOR (I,3225); STAGE DOOR (I,4178) WHERE'S POPPA? (II,2784)

LAND, JUDY
MEET ME IN ST. LOUIS (I,2987)

LANDA, MIGUEL
RAPTURE AT TWO-FORTY (I,3724)

LANDAU, MARTIN
BUFFALO BILL (II,387); GILLIGAN'S ISLAND: THE HARLEM GLOBETROTTERS ON GILLIGAN'S ISLAND (II,993); MISSION: IMPOSSIBLE (I,3067); THE OUTER LIMITS (I,3426); SPACE: 1999 (II,2409)

LANDAU, MICHAEL
SAVAGE (I,3924)

LANDEN, DINSDALE
AN ECHO OF THERESA (I,1399); NOT GUILTY! (I,3311)

LANDER, DAVID L
GIDGET MAKES THE WRONG CONNECTION (I,1798); LAVERNE AND SHIRLEY (II,1446); WILL THE REAL JERRY LEWIS PLEASE SIT DOWN? (I,4866)

LANDER, DIANE
RENDEZVOUS HOTEL (II,2131)

LANDERS, AUDREY
ARCHIE (II,104); THE ARCHIE SITUATION COMEDY MUSICAL VARIETY SHOW (II,106); BATTLE OF THE NETWORK STARS (II,174); BATTLE OF THE NETWORK STARS (II,176); BOB HOPE SPECIAL: BOB HOPE'S ALL-STAR SUPER BOWL PARTY (II,304); DALLAS (II,614); FIT FOR A KING (II,866); GOOBER AND THE TRUCKERS' PARADISE (II,1029); HIGHCLIFFE MANOR (II,1150); THE HITCHHIKER

(II,1157); THE MAGIC OF DAVID COPPERFIELD (II,1594); THE SECRET STORM (I,3969); SOMERSET (I,4115); WHITE AND RENO (II,2787)

LANDERS, HARRY
BEN CASEY (I,394); POOR MR. CAMPBELL (I,3647)

LANDERS, JUDY
B.J. AND THE BEAR (II,126); BATTLE OF THE NETWORK STARS (II,171); CELEBRITY CHARADES (II,470); CHARLIE'S ANGELS (II,486); CIRCUS OF THE STARS (II,536); CIRCUS OF THE STARS (II,537); DAUGHTERS (II,620); FANTASY ISLAND (II,829); GOSSIP (II,1043); HERE'S RICHARD (II,1136); THE HITCHHIKER (II,1157); MADAME'S PLACE (II,1587); THAT'S TV (II,2581); VEGAS (II,2723); VEGAS (II,2724); WHITE AND RENO (II,2787); WOMEN WHO RATE A "10" (II,2825)

LANDERS, MATT
CAR WASH (II,437); WHO'S ON CALL? (II,2791)

LANDERS, MURIEL
LIFE WITH LUIGI (I,2692); THE NUT HOUSE (I,3320); THE RAY BOLGER SHOW (I,3731); ROWAN AND MARTIN'S LAUGH-IN (I,3856)

LANDESBERG, STEVE
BARNEY MILLER (II,154); BATTLE OF THE NETWORK STARS (II,166); BLACK BART (II,260); THE BOBBY DARIN AMUSEMENT COMPANY (I,672); THE COMEDY ZONE (II,559); THE DON RICKLES SHOW (II,708); HOW TO SURVIVE THE 70S AND MAYBE EVEN BUMP INTO HAPPINESS (II,1199); LET'S CELEBRATE (I,2663); THE MANY FACES OF COMEDY (I,2894); PAUL SAND IN FRIENDS AND LOVERS (II,1982); STEPHANIE (II,2453); THE STEVE LANDESBERG TELEVISION SHOW (II,2458)

LANDEY, CLAYTON
USED CARS (II,2712)

LANDFORD, WALLACE
THE RETURN OF LUTHER GILLIS (II,2137)

LANDI, STEVE
REVENGE OF THE GRAY GANG (II,2146)

LANDIS, HARRY
A WOMAN OF SUBSTANCE (II,2820)

LANDIS, JESSIE ROYCE
THE CHEVROLET TELE-THEATER (I,926); THE DOCTOR (I,1320); IRONSIDE (II,1246); THE MAN IN THE DOG SUIT (I,2869)

LANDIS, MONTE
THE FEATHER AND FATHER GANG (II,845); ROWAN AND MARTIN'S LAUGH-IN (I,3857)

LANDON, JERRY
H.R. PUFNSTUF (I,2154)

LANDON, LESLIE
LITTLE HOUSE: A NEW BEGINNING (II,1488); LITTLE HOUSE: BLESS ALL THE DEAR CHILDREN (II,1489); LITTLE HOUSE: LOOK BACK TO YESTERDAY (II,1490)

LANDON, MICHAEL
AN AMATEUR'S GUIDE TO LOVE (I,154); BELLE STARR (I,391); BING CROSBY'S SUN VALLEY CHRISTMAS SHOW (I,475); BONANZA (II,347); CAVALCADE OF AMERICA (I,875); A COUNTRY HAPPENING (I,1065); THE DEAN MARTIN CELEBRITY ROAST (II,637); DOUG HENNING'S WORLD OF MAGIC I (II,725); GENERAL ELECTRIC'S ALL-STAR ANNIVERSARY (II,963); HIGHWAY TO HEAVEN (II,1152); JOHNNY RISK (I,2435); LITTLE HOUSE ON THE PRAIRIE (II,1487); LITTLE HOUSE: LOOK BACK TO YESTERDAY (II,1490); LITTLE HOUSE: THE LAST FAREWELL (II,1491); LUKE AND THE TENDERFOOT (I,2792); MITZI AND A HUNDRED GUYS (II,1710); MONSANTO PRESENTS MANCINI (I,3094); THE RESTLESS GUN (I,3774); SAM HILL (I,3895); SWING OUT, SWEET LAND (II,2515)

LANDON, SOFIA
THE GUIDING LIGHT (II,1064)

LANDOR, ROSALYN
LOVE IN A COLD CLIMATE (II,1539)

LANDRETH, THERESA
THE AWAKENING LAND (II,122)

LANDRUM, MICHAEL
HOW TO SURVIVE A MARRIAGE (II,1198)

LANDRUM, TERI
KUDZU (II,1415)

LANDRY, GAIL
TURNOVER SMITH (II,2671)

LANDRY, KAREN
ST. ELSEWHERE (II,2432)

LANDSBERG, DAVID
BUFORD AND THE GHOST (II,389); C.P.O. SHARKEY (II,407); HELLO, LARRY (II,1128)

LANDSBURG, VALERIE
FAME (II,812); ME AND DUCKY (II,1671); PHYL AND MIKHY (II,2037)

LANDZAAT, ANDRE
GENERAL HOSPITAL (II,964)

LANE, ABBE
MEN WHO RATE A "10" (II,1684); THE OTHER BROADWAY (II,1930); THE XAVIER CUGAT SHOW (I,4924)

LANE, ALLAN 'ROCKY'
MR. ED (I,3137)

LANE, BURTON
MUSICAL COMEDY TONIGHT (II,1777)

LANE, CHARLES
BEWITCHED (I,418); DEAR PHOEBE (I,1204); DENNIS THE MENACE (I,1231); KAREN (II,1363); LOVE NEST (II,1541); LOVE ON A ROOFTOP (I,2775); THE LUCY SHOW (I,2791); MR. BEVIS (I,3128); THE NEW ADVENTURES OF HUCKLEBERRY FINN (I,3250); PETTICOAT JUNCTION (I,3571); THE REAL MCCOYS (I,3741); THE RETURN OF THE BEVERLY HILLBILLIES (II,2139); SIDE BY SIDE (II,2346); SO THIS IS HOLLYWOOD (I,4105); THE SOFT TOUCH (I,4107); TOM AND JERRY (I,4533)

LANE, DON
THE DON LANE SHOW (II,707)

LANE, DORIS
THE FASHION STORY (I,1537)

LANE, FRANCEY
THE MOREY AMSTERDAM SHOW (I,3101)

LANE, JACKIE
DOCTOR WHO (II,689)

LANE, KEN
THE DEAN MARTIN SHOW (I,1201)

LANE, MICHAEL
THE GYPSY WARRIORS (II,1070); THE MONSTER SQUAD (II,1725)

LANE, MIKE
THE SWORD (I,4325)

LANE, NANCY
ANGIE (II,80); BETWEEN THE LINES (II,225); THE DUCK FACTORY (II,738); RHODA (II,2151)

LANE, NATHAN
ONE OF THE BOYS (II,1911)

LANE, OSCAR
THE GOOD GUYS (I,1847)

LANE, POPPY
CORONATION STREET (I,1054)

LANE, ROCKY
RED RYDER (I,3752)

LANE, RUSTY
CRIME WITH FATHER (I,1100); EYE WITNESS (I,1494); JIMMY HUGHES, ROOKIE COP (I,2390); OPERATION: NEPTUNE (I,3404); REVLON MIRROR THEATER (I,3781)

LANE, SAMANTHA
THE BENNY HILL SHOW (II,207)

LANE, SARA
THE VIRGINIAN (I,4738)

LANE, SCOTT
MCKEEVER AND THE COLONEL (I,2970); TOO MANY SERGEANTS (I,4573)

LANEUVILLE, ERIC
THE COP AND THE KID (II,572); FLO'S PLACE (II,892); THE FURST FAMILY OF WASHINGTON (I,1718); ROOM 222 (I,3843); SANFORD AND SON (II,2256); ST. ELSEWHERE (II,2432)

LANEY, CHARLEY
THE MICKEY MOUSE CLUB (I,3025); THE MOUSEKETEERS REUNION (II,1742)

LANG, BARBARA
LAWMAN (I,2642)

LANG, CHARLEY
BETWEEN THE LINES (II,225); MR. ROBERTS (II,1754)

LANG, DON
OH, BOY! (I,3341)

LANG, DOREEN
AS THE WORLD TURNS (II,110); BLITHE SPIRIT (I,483); GIDGET GROWS UP (I,1797); HAZARD'S PEOPLE (II,1112); THE OATH: THE SAD AND LONELY SUNDAYS (II,1874); STAR TONIGHT (I,4199)

LANG, HAROLD
THE BERT PARKS SHOW (I,400); WINDOW SHADE REVUE (I,4874)

LANG, HOWARD
THE LAST DAYS OF POMPEII (II,1436); THE WINDS OF WAR (II,2807)

LANG, PERRY
BAY CITY BLUES (II,186)

LANG, ROBERT
I'M THE GIRL HE WANTS TO KILL (I,2194)

LANG, ROMONA
GAY NINETIES REVUE (I,1743)

LANG, SHIRLEY
AFTERMASH (II,23)

LANGAN, GLENN
BOSS LADY (I,694); CAPSULE MYSTERIES (I,825)

LANGDON, FREDDY
MIDWESTERN HAYRIDE (I,3033)

LANGDON, SUE ANE
ALL IN THE FAMILY (I,133); THE ANDY GRIFFITH SHOW (I,192); THE APARTMENT HOUSE (I,241); ARNIE (I,261); BACHELOR FATHER (I,323); GRANDPA GOES TO WASHINGTON (II,1050); HAPPY DAYS (II,1084); JACKIE GLEASON AND HIS AMERICAN SCENE MAGAZINE (I,2321); LITTLE LEATHERNECK (I,2713); SUPER COMEDY BOWL 2 (I,4289); THREE'S COMPANY (II,2614); THRILLER (I,4492); WHEN THE WHISTLE BLOWS (II,2781)

LANGE, ANNE
THE COMEDY ZONE (II,559)

LANGE, HOPE
BACK THE FACT (I,327); BEULAH LAND (II,226); CYRANO DE BERGERAC (I,1120); THE GHOST AND MRS. MUIR (I,1786); HAZARD'S PEOPLE (II,1112); HEDDA HOPPER'S HOLLYWOOD (I,2002); KRAFT TELEVISION THEATER (I,2592); THE LOVE BOAT II (II,1533); THE NEW DICK VAN DYKE SHOW (I,3262); THE NEW DICK VAN DYKE SHOW (I,3263); THE RIVALRY (I,3796); THE SKY'S THE LIMIT (I,4080)

LANGE, JEANNE
ANOTHER WORLD (II,97)

LANGE, JESSICA
CAT ON A HOT TIN ROOF (II,457)

LANGE, JIM
BULLSEYE (II,393); THE DATING GAME (I,1168); GIVE-N-TAKE (II,1003); THE HOLLYWOOD CONNECTION (II,1161); THE NEW $100,000 NAME THAT TUNE (II,1810); THE NEW NEWLYWED GAME (II,1824); OPERATION: ENTERTAINMENT (I,3401); SPIN-OFF (II,2427)

LANGE, KELLY
BOB HOPE SPECIAL: BOB HOPE'S ALL-STAR COMEDY LOOK AT THE FALL SEASON: IT'S STILL FREE AND WORTH IT! (II,299); JOE DANCER: MURDER ONE, DANCER 0 (II,1302); TAKE MY ADVICE (II,2531)

LANGE, TED
BATTLE OF THE NETWORK STARS (II,178); THE CELEBRITY FOOTBALL CLASSIC (II,474); CIRCUS OF THE STARS (II,534); GOOD EVENING, CAPTAIN (II,1031); THE LOVE BOAT (II,1535); THE LOVE BOAT II (II,1533); THE LOVE BOAT III (II,1534); MR. T. AND TINA (II,1758); THAT'S MY MAMA (II,2580)

LANGELLA, FRANK
LOVE STORY (I,2778); SHERLOCK HOLMES (II,2327)

LANGERMAN, DEBRA
MACKENZIE (II,1584)

LANGFORD, FRANCES
THE BOB HOPE SHOW (I,517); THE BOB HOPE SHOW (I,574); FRANCES LANGFORD PRESENTS (I,1656); THE FRANCES LANGFORD SHOW (I,1657); THE FRANCES LANGFORD-DON AMECHE SHOW (I,1655); THE PERRY COMO VALENTINE SPECIAL (I,3541); STARTIME (I,4212)

LANGHAM, CHRIS
NOT THE NINE O'CLOCK NEWS (II,1864)

LANGHART, JANET
AMERICA ALIVE! (II,67); GOOD DAY! (II,1030); YOU ASKED FOR IT (II,2855)

LANGLAND, LIANE
MASTER OF THE GAME (II,1647)

LANGRIDGE, LYNN
THE CITADEL (II,539)

LANGRISHE, CAROLYN
MISTRAL'S DAUGHTER (II,1708); Q. E. D. (II,2094)

LANGSTON, MURRAY
CRYSTAL GAYLE IN CONCERT (II,608); EVERYDAY (II,793); THE HUDSON BROTHERS RAZZLE DAZZLE COMEDY SHOW (II,1201); LOLA (II,1509); LOLA (II,1510); LOLA (II,1511); LOLA (II,1508); THE REDD FOXX COMEDY HOUR (II,2126); THE SONNY AND CHER COMEDY HOUR (II,2400); THE SONNY COMEDY REVUE (II,2403); THE WOLFMAN JACK SHOW (II,2817)

LANGTON, BASIL
EAGLE IN A CAGE (I,1392)

LANGTON, DAVID
UPSTAIRS, DOWNSTAIRS (II,2709)

LANGTON, DIANE
CARRY ON LAUGHING (II,448)

LANGTON, PAUL
THE BRIGHTER DAY (I,728); FLOYD GIBBONS, REPORTER (I,1608); PEYTON PLACE (I,3574)

LANGUARD, JANET
THE DONNA REED SHOW (I,1347)

LANIER, SUSAN
THE 416TH (II,913); OVER AND OUT (II,1938); SZYSZNYK (II,2523); WELCOME BACK, KOTTER (II,2761); WILDER AND WILDER (II,2802)

LANIN, JAY
OUTPOST (I,3428)

LANKFORD, KIM
HOLLYWOOD HIGH (II,1162); KNOTS LANDING (II,1404); THREE EYES (II,2605); THE WAVERLY WONDERS (II,2746)

LANNING, JERRY
DAMN YANKEES (I,1128); I AND CLAUDIE (I,2164); SEARCH FOR TOMORROW (II,2284); SOUTHERN FRIED (I,4139); TEXAS (II,2566)

LANNOM, LES
CENTENNIAL (II,477); THE CHISHOLMS (II,513); HARRY O (II,1098); HARRY O (II,1099); JEREMIAH OF JACOB'S NECK (II,1281); LACY AND THE MISSISSIPPI QUEEN (II,1419); PEARL (II,1986)

LANPHIER, JAMES
PETER GUNN (I,3562)

LANSBURG, CONSTANCE
ALL THE RIVERS RUN (II,43)

LANSBURY, ANGELA
CIRCUS OF THE STARS (II,534); CLIMAX! (I,976); FIRESIDE THEATER (I,1580); FORD TELEVISION THEATER (I,1634); FOUR STAR PLAYHOUSE (I,1652); FRONT ROW CENTER (I,1694); HENRY FONDA PRESENTS THE STAR AND THE STORY (I,2016); LACE (II,1418); THE LUX VIDEO THEATER (I,2795); MURDER, SHE WROTE (II,1770); PANTOMIME QUIZ (I,3449); THE PERRY COMO CHRISTMAS SHOW (I,3526); THE PERRY COMO THANKSGIVING SPECIAL (I,3540); PLAYHOUSE 90 (I,3623); REVLON MIRROR THEATER (I,3781); RHEINGOLD THEATER (I,3783); ROBERT MONTGOMERY PRESENTS YOUR LUCKY STRIKE

THEATER (I,3809); SCENE OF THE CRIME (II,2271); SCHLITZ PLAYHOUSE OF STARS (I,3936); STAGE 7 (I,4181); SWEENEY TODD (II,2512)

LANSING, JOHN
THE KIDS FROM C.A.P.E.R. (II,1391)

LANSING, JOI
THE BEVERLY HILLBILLIES (I,417); CAVALCADE OF AMERICA (I,875); FOUNTAIN OF YOUTH (I,1647); FOUR STAR PLAYHOUSE (I,1652); THE JONES BOYS (I,2446); KLONDIKE (I,2573); LOVE THAT BOB (I,2779); THE STAR MAKER (I,4193)

LANSING, MARY
THE ANDY GRIFFITH SHOW (I,192); THE BROTHERS (I,748); MAYBERRY R.F.D. (I,2961)

LANSING, ROBERT
87TH PRECINCT (I,1422); ASSIGNMENT: EARTH (I,299); AUTOMAN (II,120); THE BURNING COURT (I,766); THE DEADLY TRIANGLE (II,634); THE EVIL TOUCH (I,1486); KILLER BY NIGHT (I,2537); THE MAN WHO NEVER WAS (I,2882); S*H*E (II,2235); SHADOW OF SAM PENNY (II,2313); TWELVE O'CLOCK HIGH (I,4637); YOUNG DR. MALONE (I,4938)

LANSING, SHERRY
BANYON (I,344)

LANSON, SNOOKY
DICK CLARK'S GOOD OL' DAYS: FROM BOBBY SOX TO BIKINIS (II,671); FIVE STAR JUBILEE (I,1589); THE GISELE MACKENZIE SHOW (I,1813); A SALUTE TO TELEVISION'S 25TH ANNIVERSARY (I,3893); THE SNOOKY LANSON SHOW (I,4101)

LANTEAU, WILLIAM
BUNGLE ABBEY (II,396); THE LUCILLE BALL COMEDY HOUR (I,2783); MY SON THE DOCTOR (I,3203); NEWHART (II,1835); NO SOAP, RADIO (II,1856); PAPA G.I. (I,3451); SUTTERS BAY (II,2508); THIS GIRL FOR HIRE (II,2596)

LANTZ, GRACE
THE WOODY WOODPECKER SHOW (I,4906)

LANTZ, JIM
A WALK IN THE NIGHT (I,4750)

LANTZ, WALTER
THE WOODY WOODPECKER SHOW (I,4906)

LANYON, ANNABELLE
NANCY ASTOR (II,1788)

LAPOTAIRE, JANE
PIAF (II,2039)

LARCENT, GREAT
LIKE MAGIC (II,1476)

LARCH, JOHN
THE CHADWICK FAMILY (II,478); THE CITY (I,969); CONVOY (I,1043); ELLERY QUEEN: TOO MANY SUSPECTS (II,767); FUTURE COP (II,946); JUDGEMENT DAY (II,1348); THE SEEKERS (I,3973); WINTER KILL (II,2810); YOU ARE THERE (I,4929)

LARKEN, SHEILA
MARCUS WELBY, M.D. (I,2905); MEN AT LAW (I,3004); THE STOREFRONT LAWYERS (I,4233)

LARKIN, AUDREY
DARK SHADOWS (I,1157)

LARKIN, BOB
HOUSE CALLS (II,1194)

LARKIN, DICK
THE LITTLE REVUE (I,2717)

LARKIN, JOHN
ADAMSBURG, U.S.A. (I,33); THE DOG TROOP (I,1325); THE EDGE OF NIGHT (II,760); SAINTS AND SINNERS (I,3887); SAVAGE SUNDAY (I,3926); TWELVE O'CLOCK HIGH (I,4637)

LARKIN, KRISTIN
SHIPSHAPE (II,2329)

LARKIN, MARY
AGAINST THE WIND (II,24); THE CHEROKEE TRAIL (II,500)

LARRABE, LOUISE
CURTAIN CALL THEATER (I,1115)

LARRAIN, MICHAEL
DIAL HOT LINE (I,1254); MATT LINCOLN (I,2949)

LARRIVA, TITO
THE PEE WEE HERMAN SHOW (II,1988)

LARROQUETTE, JOHN
THE 416TH (II,913); BARE ESSENCE (II,149); THE BLACK SHEEP SQUADRON (II,262); DOCTORS HOSPITAL (II,691); THE LAST NINJA (II,1438); NIGHT COURT (II,1844)

LARSEN, DON
THE BOB HOPE SHOW (I,550)

LARSEN, GERALDINE
THE MAGIC LADY (I,2816)

LARSEN, KEITH
BRAVE EAGLE (I,709); THE HUNTER (I,2162); NORTHWEST PASSAGE (I,3306); THE WEB (I,4785)

LARSEN, LARRY
MAGIC MONGO (II,1592); THE MOUSEKETEERS REUNION (II,1742)

LARSEN, LISBY
SEARCH FOR TOMORROW (II,2284)

LARSEN, WILLIAM
HEAVEN ON EARTH (II,1119)

LARSON, ANIA
CASANOVA (II,451)

LARSON, DENNIS
THE BORROWERS (I,693); THE JIMMY STEWART SHOW (I,2391)

LARSON, DORIS
UNCLE MISTLETOE AND HIS ADVENTURES (I,4669)

LARSON, ERIC
GALACTICA 1980 (II,953)

LARSON, JACK
THE ADVENTURES OF SUPERMAN (I,77)

LARSON, JEANETTE
ANOTHER LIFE (II,95)

LARSON, KEITH
THE AQUANAUTS (I,250)

LARSON, LARRY
THE LOST SAUCER (II,1524); THE MICKEY MOUSE CLUB (I,3025); SIGMUND AND THE SEA MONSTERS (II,2352)

LARSON, LISBY
TEXAS (II,2566)

LARSON, MICHELE
GALACTICA 1980 (II,953)

LARSON, PAUL
WE INTERRUPT THIS SEASON (I,4780)

LARSSON, JAN
THE SMALL MIRACLE (I,4090)

LARYEA, WAYNE
THE BUGALOOS (I,757)

LASCOE, HARRY
THE ERNIE KOVACS SHOW (I,1455)

LASCOE, HENRY
MR. BROADWAY (I,3130); NINOTCHKA (I,3299); OLD NICKERBOCKER MUSIC HALL (I,3351); THE TIMES SQUARE STORY (I,4514)

LASHLY, JAMES
P.O.P. (II,1939)

LASKAWAY, HARRIS
MURDER INK (II,1769)

LASKY, KATHY
OUT OF OUR MINDS (II,1933)

LASKY, ZANE
THE LAST RESORT (II,1440); MAKING THE GRADE (II,1606); THE TONY RANDALL SHOW (II,2640)

LASSEN, LEIGH
SEARCH FOR TOMORROW (II,2284)

LASSER, LOUISE
BEDROOMS (II,198); THE DOCTORS (II,694); MAKING A LIVING (II,1604); MARY HARTMAN, MARY HARTMAN (II,1638); MASQUERADE (I,2940); MO AND JO (II,1715); ST. ELSEWHERE (II,2432); TAXI (II,2546)

LASSICK, SYDNEY
ARCHIE BUNKER'S PLACE (II,105)

LASSIE
THE BOB HOPE SHOW (I,539); LASSIE (I,2623)

LATENT, ZIGGY
THE VAUGHN MONROE SHOW (I,4707)

LATESSA, DICK
AND THEY ALL LIVED HAPPILY EVER AFTER (II,75); THE EDGE OF NIGHT (II,760)

LATHAM, CYNTHIA
STARSTRUCK (II,2446); TWENTY-FOUR HOURS IN A WOMAN'S LIFE (I,4644)

LATHAM, LOUISE
THE AWAKENING LAND (II,122); BACKSTAIRS AT THE WHITE HOUSE (II,133); THE CONTENDER (II,571); FAMILY (II,813); JOHNNY BELINDA (I,2417); THE OATH: 33 HOURS IN THE LIFE OF GOD (II,1873); SARA (II,2260); SAVAGE (I,3924); SCRUPLES (II,2280); STONESTREET: WHO KILLED THE CENTERFOLD MODEL? (II,2472); SWEET, SWEET RACHEL (I,4313); WINTER KILL (II,2810)

LATHAM, PHILIP
DOCTOR WHO—THE FIVE DOCTORS (II,690); THE PALLISERS (II,1944)

LATHAN, BOBBI JO
A FINE ROMANCE (II,858)

LATIMER, SHEILA

THE OMEGA FACTOR (II,1888)

LATIMORE, FRANK
RYAN'S HOPE (II,2234)

LATORRE, TONY
THE GANGSTER CHRONICLES (II,957); REVENGE OF THE GRAY GANG (II,2146)

LAU, LAWRENCE
ALL MY CHILDREN (II,39)

LAU, ROSE
THE NATIONAL SNOOP (II,1797)

LAU, WESLEY
PERRY MASON (II,2025); THE TIME TUNNEL (I,4511)

LAUCK, CHESTER
LUM AND ABNER (I,2793)

LAUDIGAN, WILLIAM
LAST OF THE PRIVATE EYES (I,2628)

LAUGHER, BOBBY
THE NEW ERNIE KOVACS SHOW (I,3264)

LAUGHLIN, JOHN
THE WHITE SHADOW (II,2788)

LAUGHTON, CHARLES
THIS IS CHARLES LAUGHTON (I,4455)

LAUHER, BOBBY
THE COMEDY OF ERNIE KOVACS (I,1022); THE ERNIE KOVACS SPECIAL (I,1456); TAKE A GOOD LOOK (II,2529); TAKE A GOOD LOOK (I,4332)

LAUNER, S JOHN
HARRY O (II,1098); RAPTURE AT TWO-FORTY (I,3724)

LAUREL, STAN
LAUREL AND HARDY LAUGHTOONS (II,1445)

LAUREN, MO
BEYOND WESTWORLD (II,227)

LAUREN, TAMMY
ANGIE (II,80); THE KID WITH THE BROKEN HALO (II,1387); LEGS (II,1459); MORK AND MINDY (II,1735); NUTS AND BOLTS (II,1871); OUT OF THE BLUE (II,1934); THINGS ARE LOOKING UP (II,2588); WHO'S WATCHING THE KIDS? (II,2793)

LAURENCE, MARGARET
PRISONER: CELL BLOCK H (II,2085)

LAURENCE, MICHAEL
THIS IS KATE BENNETT (II,2597)

LAURENCE, MITCHELL
NOT NECESSARILY THE NEWS (II,1863)

LAURENCE, SHERISSE
CIRCUS OF THE 21ST CENTURY (II,529)

LAURENS, JOHN
GEORGE WASHINGTON (II,978)

LAURENSON, BARBARA
JENNIE: LADY RANDOLPH CHURCHILL (II,1277)

LAURET, LARYSSA
THE DOCTORS (II,694)

LAURIE, KIMETHA
BRIGHT PROMISE (I,727)

LAURIE, PIPER
THE BOB HOPE SHOW (I,589); BROADWAY (I,733); THE DESILU PLAYHOUSE (I,1237); THE GENERAL ELECTRIC THEATER (I,1753); PLAYHOUSE 90 (I,3623); SKAG (II,2374); STUDIO ONE (I,4268); THE THORN BIRDS (II,2600); TWELFTH NIGHT (I,4636); WINTERSET (I,4882)

LAURIN, VERA
THE GIRL, THE GOLD WATCH AND EVERYTHING (II,1002)

LAURITA, DANA
ESCAPE (I,1461); LEWIS AND CLARK (II,1465)

LAUTER, ED
B.J. AND THE BEAR (II,125); HOW THE WEST WAS WON (II,1196); ROOSTER (II,2210); A SHADOW IN THE STREETS (II,2312)

LAUTER, HARRY
THE ROY ROGERS SHOW (I,3859); TALES OF THE TEXAS RANGERS (I,4349); WATERFRONT (I,4774); WHEN THE WEST WAS FUN: A WESTERN REUNION (II,2780)

LAVERS, PAUL
MOLL FLANDERS (II,1720)

LAVERSEE, LORETTA
HERE COME THE BRIDES (I,2024)

LAVIN, LINDA
THE ALAN KING SHOW (I,101); THE ALAN KING SHOW (I,102); THE ALAN KING SPECIAL (I,103); ALAN KING'S FINAL WARNING (II,29); ALICE (II,33); ALICE (II,34); BARNEY MILLER (II,154); BATTLE OF THE NETWORK STARS (II,164); CBS: ON THE AIR (II,460); COMEDY IS KING (I,1019); DAMN YANKEES (I,1128); DEAN MARTIN'S CHRISTMAS IN CALIFORNIA (II,645); JERRY (II,1283); LILY FOR PRESIDENT (II,1478); LINDA IN WONDERLAND (II,1481); PARADE OF STARS (II,1954)

LAW, DONALD F
CHIEFS (II,509)

LAWFORD, PETER
THE BOB HOPE SHOW (I,568); DEAR PHOEBE (I,1204); THE DORIS DAY SHOW (I,1355); ELLERY QUEEN: DON'T LOOK BEHIND YOU (I,1434); FANTASY ISLAND (II,830); THE FARMER'S DAUGHTER (I,1534); THE FRANK SINATRA TIMEX SHOW (I,1668); THE FRANK SINATRA TIMEX SHOW (I,1670); THE MILTON BERLE SPECIAL (I,3051); PROFILES IN COURAGE (I,3678); RUGGLES OF RED GAP (I,3867); THE SAMMY DAVIS JR. SPECIAL (I,3899); SCHLITZ PLAYHOUSE OF STARS (I,3936); THE THIN MAN (I,4446); TOM AND JERRY (I,4533)

LAWLESS, DEIRDRE
THE MANIONS OF AMERICA (II,1623)

LAWLOR, JOHN
THE BUREAU (II,397); THE FACTS OF LIFE (II,805); PHYLLIS (II,2038); POTTSVILLE (II,2072); WHY US? (II,2796)

LAWRENCE, BARBARA
FOUR STAR PLAYHOUSE (I,1652)

LAWRENCE, BILL
ARTHUR GODFREY AND FRIENDS (I,278)

LAWRENCE, BRUNO
WELCOME TO PARADISE (II,2762)

LAWRENCE, CAROL
THE ARTHUR GODFREY SHOW (I,287); ARTHUR GODFREY'S PORTABLE ELECTRIC MEDICINE SHOW (I,290); THE BING CROSBY SHOW (I,464); THE BOB HOPE SHOW (I,628); THE BOB HOPE SHOW (I,604); THE BOB HOPE SHOW (I,614); THE BOB HOPE SHOW (I,655); BOB HOPE SPECIAL: BOB HOPE'S ALL-STAR TRIBUTE TO THE PALACE THEATER (II,305); BOB HOPE SPECIAL: HAPPY BIRTHDAY, BOB! (II,326); DEAN MARTIN PRESENTS THE VIC DAMONE SHOW (I,1195); THE GENE KELLY SHOW (I,1751); THE GIRL, THE GOLD WATCH AND DYNAMITE (II,1001); JACQUELINE SUSANN'S VALLEY OF THE DOLLS, 1981 (II,1268); KISS ME, KATE (I,2569); MONSANTO NIGHT PRESENTS JOSE FELICIANO (I,3091); MR. & MRS. DRACULA (II,1745); PHIL SILVERS IN NEW YORK (I,3577); RAINBOW OF STARS (I,3717); STRANGER IN OUR HOUSE (II,2477); SUMMER IN NEW YORK (I,4279); THE VIC DAMONE SHOW (I,4718); THE WAY THEY WERE (II,2748); THE WAYNE NEWTON SPECIAL (II,2749); WOMEN OF RUSSIA (II,2824)

LAWRENCE, CHARLES
POPEYE THE SAILOR (I,3648)

LAWRENCE, DAVID
BEWITCHED (I,418)

LAWRENCE, EDDIE
THE GARRY MOORE SHOW (I,1740)

LAWRENCE, ELIZABETH
ALL MY CHILDREN (II,39); THE DOCTORS (II,694); ROAD OF LIFE (I,3798); A WORLD APART (I,4910)

LAWRENCE, ELLIOT
MELODY STREET (I,3000)

LAWRENCE, EMILY
ACADEMY THEATER (I,14)

LAWRENCE, GERTRUDE
GREAT CATHERINE (I,1873); PRUDENTIAL FAMILY PLAYHOUSE (I,3683)

LAWRENCE, GREG
BEWITCHED (I,418)

LAWRENCE, JEREMY
THE CLIFFWOOD AVENUE KIDS (II,543)

LAWRENCE, JOAN
THE CAPTAIN AND TENNILLE (II,429)

LAWRENCE, JOEY
BOB HOPE SPECIAL: HO HO HOPE'S JOLLY CHRISTMAS HOUR (II,327); GIMME A BREAK (II,995); LITTLE SHOTS (II,1495); SCAMPS (II,2267)

LAWRENCE, JOHN

BANYON (I,345); THE GOOD LIFE (II,1033); THE SUPER (I,4293)

LAWRENCE, JULIE
SHORT, SHORT DRAMA (I,4023)

LAWRENCE, LINDA
BOSOM BUDDIES (II,355)

LAWRENCE, MARC
BORDER PALS (II,352); TERROR AT ALCATRAZ (II,2563)

LAWRENCE, MARY
CASEY JONES (I,865); LOVE THAT BOB (I,2779)

LAWRENCE, MORT
THE BIG PAYOFF (I,428)

LAWRENCE, PAMELA
THE WOMEN (I,4890)

LAWRENCE, PAULA
HANSEL AND GRETEL (I,1934)

LAWRENCE, STEVE
100 YEARS OF AMERICA'S POPULAR MUSIC (II,1904); BERT CONVY SPECIAL—THERE'S A MEETING HERE TONIGHT (II,210); THE BEST ON RECORD (I,409); THE BIG SHOW (II,243); THE BOB HOPE SHOW (I,616); THE BOB HOPE SHOW (I,635); BOB HOPE SPECIAL: BOB HOPE'S 30TH ANNIVERSARY TV SPECIAL (II,293); BOB HOPE SPECIAL: BOB HOPE'S ALL-STAR TRIBUTE TO THE PALACE THEATER (II,305); CAROL FOR ANOTHER CHRISTMAS (I,852); FOUL UPS, BLEEPS AND BLUNDERS (II,911); THE GENERAL MOTORS 50TH ANNIVERSARY SHOW (I,1758); HARDCASTLE AND MCCORMICK (II,1089); THE JANE POWELL SHOW (I,2344); JUBILEE (II,1347); LIGHTS, CAMERA, MONTY! (II,1475); NBC FOLLIES OF 1965 (I,3240); ON PARADE (I,3356); STEVE AND EYDIE CELEBRATE IRVING BERLIN (II,2456); STEVE AND EYDIE. . . ON STAGE (I,4223); STEVE AND EYDIE: OUR LOVE IS HERE TO STAY (II,2457); STEVE LAWRENCE AND EYDIE GORME FROM THIS MOMENT ON. . . COLE PORTER (II,2459); THE STEVE LAWRENCE AND EYDIE GORME SHOW (I,4226); THE STEVE LAWRENCE SHOW (I,4227); THE TONIGHT SHOW (I,4561); THE TONIGHT SHOW (I,4562)

LAWRENCE, SUSAN
AMERICA, YOU'RE ON (II,69); BOWZER (II,358); DOCTOR SHRINKER (II,686); THE HARVEY KORMAN SHOW (II,1103); THE RAG BUSINESS (II,2106)

LAWRENCE, VICKI
BATTLE OF THE NETWORK STARS (II,177); BATTLE OF THE NETWORK STARS (II,178); CAROL BURNETT AND COMPANY (II,441); THE CAROL BURNETT SHOW (II,443); EUNICE (II,784); THE FUNNY WORLD OF FRED & BUNNI (II,943); HAVING BABIES I (II,1107); JERRY REED AND SPECIAL FRIENDS (II,1285); THE JIMMIE RODGERS SHOW (I,2383); LAVERNE AND SHIRLEY (II,1446); MAMA'S FAMILY (II,1610); PAUL LYNDE AT THE MOVIES (II,1976); PAUL LYNDE GOES M-A-A-A-AD (II,1980)

LAWS, ELOISE
THE TOM JONES SPECIAL (I,4541)

LAWS, SAM
THE HUSTLER OF MUSCLE BEACH (II,1210); THE PRACTICE (II,2076); THE RICHARD PRYOR SHOW (II,2163); ROLL OUT! (I,3833)

LAWSON, BEN
V (II,2713)

LAWSON, LEE
THE GUIDING LIGHT (II,1064); LACE (II,1418); MASON (II,1643); ONE LIFE TO LIVE (II,1907); WHY DIDN'T THEY ASK EVANS? (II,2795)

LAWSON, LEIGH
MURDER IS A ONE-ACT PLAY (I,3160)

LAWSON, LEN
CONDOMINIUM (II,566); I'D RATHER BE CALM (II,1216)

LAWSON, LINDA
ADVENTURES IN PARADISE (I,46); DON'T CALL ME CHARLIE (I,1349)

LAWSON, MICHAEL
THE O'NEILLS (I,3392)

LAWSON, RICHARD
THE BUFFALO SOLDIERS (II,388); CHICAGO STORY (II,506); LEADFOOT (II,1452)

LAYTON, GEORGE
DOCTOR IN THE HOUSE (I,1313)

LAZAR, AVA
THE BEACH GIRLS (II,191); MICKEY SPILLANE'S MIKE HAMMER: MURDER ME, MURDER YOU (II,1694); SANTA BARBARA (II,2258)

LAZARE, CAROL
PAUL BERNARD—PSYCHIATRIST (I,3485)

LAZARETT, SERGE
1915 (II,1853)

LAZARUS, BILL
CALUCCI'S DEPARTMENT (I,805); LOVE, SEX. . . AND MARRIAGE (II,1546); MARLO THOMAS IN ACTS OF LOVE—AND OTHER COMEDIES (I,2919)

LAZARUS, MELL
THE FANTASTIC FUNNIES (II,825)

LAZENBY, GEORGE
COVER GIRLS (II,596); RETURN OF THE MAN FROM U.N.C.L.E.: THE 15 YEARS LATER AFFAIR (II,2140); RITUALS (II,2182)

LAZER, PETER
MCNAB'S LAB (I,2972); MR. NOVAK (I,3145); TODAY IS OURS (I,4531)

LAZZARD, ADAM
THE LOSERS (I,2757)

LEBEAUF, SABRINA
THE COSBY SHOW (II,578)

LEBLANC, CHRIS
AS THE WORLD TURNS (II,110)

LEBLANC, DIANA
THE SWISS FAMILY ROBINSON (II,2518)

LEBOLT, DAVID
DAVID BOWIE—SERIOUS MOONLIGHT (II,622)

LEBOUVIER, JEAN
KING'S CROSSING (II,1397)

LECLAIR, MICHAEL
HAVING BABIES II (II,1108); NEWMAN'S DRUGSTORE (II,1837)

LECORNEC, BILL
THE NEW HOWDY DOODY SHOW (II,1818)

LEFLEUR, ART
THE BOYS IN BLUE (II,359)

LEGALLIENNE, EVA
ALICE IN WONDERLAND (I,120); THE CORN IS GREEN (I,1050)

LEGAULT, LANCE
THE A-TEAM (II,119); THE BUSTERS (II,401); CAPTAIN AMERICA (II,427); CONSTANTINOPLE (II,570); DYNASTY (II,746)

LEMAT, PAUL
FIREHOUSE (I,1579)

LEMESSENA, WILLIAM
NAUGHTY MARIETTA (I,3232); ON BORROWED TIME (I,3354); ONE MORE TRY (II,1908); THE TEAHOUSE OF THE AUGUST MOON (I,4368); TONIGHT IN SAMARKAND (I,4558)

LEMESURIER, JOHN
FILE IT UNDER FEAR (I,1574)

LEMOLE, LISA
JAKE'S WAY (II,1269)

LENOIRE, ROSETTA
ANOTHER WORLD (II,97); CALUCCI'S DEPARTMENT (I,805); GIMME A BREAK (II,995); THE GREEN PASTURES (I,1887); GUESS WHO'S COMING TO DINNER? (II,1063); THE GUIDING LIGHT (II,1064); MANDY'S GRANDMOTHER (II,1618); RYAN'S HOPE (II,2234); A WORLD APART (I,4910)

LEPLATT, TED
LOVE OF LIFE (I,2774)

LEPORE, RICHARD
THE BLUE MEN (I,492)

LEROY, GLORIA
ALL IN THE FAMILY (II,38); AUTOMAN (II,120); BEST FRIENDS (II,213); BUT MOTHER! (II,403); GOOD PENNY (II,1036); HOT L BALTIMORE (II,1187); KAZ (II,1370); RICHIE BROCKELMAN: MISSING 24 HOURS (II,2168); SCRUPLES (II,2280); TOPPER (II,2649)

LEROY, HAL
THE KATE SMITH HOUR (I,2508)

LEROY, MERWYN
SOPHIA! (I,4128)

LEROY, ZOAUNNE
NEWHART (II,1835)

LETOUZEL, SYLVESTRA
A FAMILY AFFAIR (II,814)

LEVANT, RENE
240-ROBERT (II,2689); THE RIGHTEOUS APPLES (II,2174)

LEA, KAREN
THE VIDEO GAME (II,2727)

LEACH, ALAN
KENNEDY (II,1377)

LEACH, BRITT
AFTERMASH (II,23); COPS (I,1047); SPENCER'S PILOTS (II,2421); SPENCER'S PILOTS (II,2422); THE YOUNG PIONEERS' CHRISTMAS (II,2870)

LEACH, ROBIN
ENTERTAINMENT TONIGHT (II,780); LIFESTYLES OF THE RICH AND FAMOUS (II,1474)

LEACH, ROSEMARY
BRIEF ENCOUNTER (I,724); THE ROADS TO FREEDOM (I,3802); TO SIR, WITH LOVE (II,2624)

LEACHMAN, CLORIS
ACTOR'S STUDIO (I,28); ADVICE TO THE LOVELORN (II,16); ALCOA PREMIERE (I,109); ALFRED HITCHCOCK PRESENTS (I,115); THE ARMSTRONG CIRCLE THEATER (I,260); BACKSTAIRS AT THE WHITE HOUSE (II,133); THE BOB AND RAY SHOW (I,496); THE BOOK OF LISTS (II,350); CHARLIE WILD, PRIVATE DETECTIVE (I,914); CLIMAX! (I,976); THE CLOCK (I,978); THE CRIME CLUB (I,1096); DANGER (I,1134); ERNIE, MADGE, AND ARTIE (I,1457); FORD THEATER HOUR (I,1635); THE GENERAL ELECTRIC THEATER (I,1753); HOLD IT PLEASE (I,2071); KRAFT MYSTERY THEATER (I,2589); THE LIFE OF VERNON HATHAWAY (I,2687); THE LOVE BOAT I (II,1532); THE LUX VIDEO THEATER (I,2795); THE MAN IN THE MOON (I,2871); THE MARY TYLER MOORE SHOW (II,1640); MONTY HALL'S VARIETY HOUR (II,1728); NIGHT GALLERY (I,3287); OF MEN OF WOMEN (I,3331); OF THEE I SING (I,3334); THE OLDEST LIVING GRADUATE (II,1882); THE PAUL LYNDE COMEDY HOUR (II,1978); PETE 'N' TILLIE (II,2027); PHILCO TELEVISION PLAYHOUSE (I,3583); PHYLLIS (II,2038); STAR STAGE (I,4197); SUSPENSE (I,4305); TELLY. . . WHO LOVES YA, BABY? (II,2558); THAT GIRL (II,2570); TIMMY AND LASSIE (I,4520); TWIGS (II,2684); THE WAY THEY WERE (II,2748); THE WEB (I,4784); WONDER WOMAN (PILOT 2) (II,2827); WONDER WOMAN (SERIES 1) (II,2828); ZANE GREY THEATER (I,4979)

LEADER, CAROL
FLAMBARDS (II,869)

LEAF, ALLEN
THE GOLDEN AGE OF TELEVISION (II,1019)

LEAGUE, JANET
OUR STREET (I,3418); WHERE THE HEART IS (I,4831)

LEAHY, FRANK
THE RAY ANTHONY SHOW (I,3728)

LEAK, JENNIFER
ANOTHER WORLD (II,97); THE YOUNG AND THE RESTLESS (II,2862)

LEAKE, BARBARA
VANITY FAIR (I,4701)

LEAKE, CYNTHIA
DAYS OF OUR LIVES (II,629)

LEANE, DANNY
MENASHA THE MAGNIFICENT (I,3010)

LEAO, TAMA
PARADISE (II,1955)

LEAR, JOYCE
TODAY IS OURS (I,4531)

LEAR, NORMAN
BOB HOPE SPECIAL: BOB HOPE'S WORLD OF COMEDY (II,323)

LEARMAN, RICHARD
GEORGE WASHINGTON (II,978)

LEARNED, MICHAEL
DEAN MARTIN'S CALIFORNIA CHRISTMAS (II,642); NURSE (II,1870); THE WALTONS (II,2740); THE WALTONS: MOTHER'S DAY ON WALTON'S MOUNTAIN (EPISODE 2) (II,2742)

LEARY, BRIANNE
THE ASTRONAUTS (II,114); BATTLE OF THE NETWORK STARS (II,167); BATTLE OF THE NETWORK STARS (II,168); THE BLACK SHEEP SQUADRON (II,262); CELEBRITY CHALLENGE OF THE SEXES 3 (II,467); CHIPS (II,511); NO SOAP, RADIO (II,1856); PEN 'N' INC. (II,1992); TURNOVER SMITH (II,2671)

LEARY, DAVID
PIAF (II,2039)

LEARY, DAWN S
SNOOPY'S GETTING MARRIED, CHARLIE BROWN (I,4102)

LEARY, ROBIN
AS THE WORLD TURNS (II,110)

LEBERFELD, ALFRED
SPELLBOUND (I,4153)

LECHNER, BILL
ROCKY JONES, SPACE RANGER (I,3823)

LECORNEC, BILL
HOWDY DOODY (I,2151)

LECOURTOIS, DANIEL
CAPTAIN GALLANT OF THE FOREIGN LEGION (I,831)

LEDDY, DANIEL
SEARCH FOR TOMORROW (II,2284)

LEDEBUR, FRED
NINOTCHKA (I,3299)

LEDERER, SUZANNE
EISCHIED (II,763)

LEE HALE SINGERS, THE
THE ENTERTAINERS (I,1449)

LEE, ANGELA
THE CHARLIE BROWN AND SNOOPY SHOW (II,484); IS THIS GOODBYE, CHARLIE BROWN? (I,2242)

LEE, ANNA
THE CHARLIE FARRELL SHOW (I,913); A DATE WITH JUDY (I,1162); THE EYES OF TEXAS (II,799); FORD THEATER HOUR (I,1635); GENERAL HOSPITAL (II,964); MAGNAVOX THEATER (I,2827); THE NIGHT RIDER (II,1848); PEPSI-COLA PLAYHOUSE (I,3523); PULITZER PRIZE PLAYHOUSE (I,3692); ROBERT MONTGOMERY PRESENTS YOUR LUCKY STRIKE THEATER (I,3809); SCRUPLES (II,2280)

LEE, BARBARA
THE PORTER WAGONER SHOW (I,3652)

LEE, BERNARD
BEAUTY AND THE BEAST (I,382); DOUGLAS FAIRBANKS JR. PRESENTS THE RHEINGOLD THEATER (I,1364); TYCOON: THE STORY OF A WOMAN (II,2697)

LEE, BILL
THE ALVIN SHOW (I,146)

LEE, BONNIE
MAGIC MIDWAY (I,2818)

LEE, BRENDA
THE BOB HOPE SHOW (I,594); AN EVENING WITH THE STATLER BROTHERS (II,790); JERRY REED AND SPECIAL FRIENDS (II,1285); JOHNNY CASH: THE FIRST 25 YEARS (II,1336); NAVY LOG (I,3233); OH, BOY! (I,3341); THE ROAD WEST (I,3801); THE SOUND AND THE SCENE (I,4132)

LEE, BRIAN
MASTER OF THE GAME (II,1647)

LEE, BRUCE
BATMAN (I,366); THE GREEN HORNET (I,1886); LONGSTREET (I,2749)

LEE, CAROL
THE GEORGE BURNS AND GRACIE ALLEN SHOW (I,1763)

LEE, CHERYLANE
THE AMAZING CHAN AND THE CHAN CLAN (I,157); BACHELOR FATHER (I,323); THE GENE KELLY PONTIAC SPECIAL (I,1750); HOORAY FOR LOVE (I,2116); KENTUCKY JONES (I,2526); PAPA G.I. (I,3451)

LEE, CHRISTOPHER
AGGIE (I,90); CAPTAIN AMERICA (II,428); DOUGLAS FAIRBANKS JR. PRESENTS THE RHEINGOLD THEATER (I,1364); THE ERROL FLYNN THEATER (I,1458); THE FAR PAVILIONS (II,832); GOLIATH AWAITS (II,1025); MASSARATI AND THE BRAIN (II,1646); ONCE UPON A SPY (II,1898); ONE STEP BEYOND (I,3388); POOR DEVIL (I,3646); TALES OF THE HAUNTED (II,2538)

LEE, COSETTE
SCARLETT HILL (I,3933); STRANGE PARADISE (I,4246)

LEE, CYNTHIA S
STRIPPERS (II,2481)

LEE, DANA
LOLA (II,1509); THE RENEGADES (II,2132)

LEE, DARCY
MICKEY SPILLANE'S MIKE HAMMER: MORE THAN MURDER (II,1693)

LEE, ERNIE
MIDWESTERN HAYRIDE (I,3033)

LEE, EUGENE
STORY FOR AMERICANS (I,4238)

LEE, GEORGE
CHIEFS (II,509)

LEE, GEORGIA
STAGE TWO REVUE (I,4184)

LEE, GYPSY ROSE
GYPSY (I,1915); THE GYPSY ROSE LEE SHOW (I,1916); THE PRUITTS OF SOUTHAMPTON (I,3684); THINK FAST (I,4447); WHO HAS SEEN THE WIND? (I,4847)

LEE, HELEN
SUNDAY DATE (I,4284)

LEE, IRVING
THE EDGE OF NIGHT (II,760)

LEE, JAMES
ONE MAN'S FAMILY (I,3383)

LEE, JOANNA
SUNDAY FUNNIES (II,2493)

LEE, JOHNNY
THE ADVENTURES OF THE SEA HAWK (I,81); THE AMOS AND ANDY SHOW (I,179); DOBIE GILLIS (I,1302); MUSIC CITY NEWS TOP COUNTRY HITS OF THE YEAR (II,1772); OPRYLAND: NIGHT OF STARS AND FUTURE STARS (II,1921); PRISONER: CELL BLOCK H (II,2085); PRUDENCE AND THE CHIEF (I,3682); RETURN TO EDEN (II,2143); ROOTIE KAZOOTIE (I,3844); A TOWN LIKE ALICE (II,2652); USA COUNTRY MUSIC (II,591)

LEE, KAAREN
BIG JOHN (II,239)

LEE, KAIULANI
ANOTHER WORLD (II,97); CHIEFS (II,509); THE WALTONS (II,2740)

LEE, KATHRYN
STARTIME (I,4212)

LEE, KATIE
JIM HENSON'S MUPPET BABIES (II,1289)

LEE, KAY
FIRESIDE THEATER (I,1580)

LEE, LILA
PANIC! (I,3447)

LEE, MADALINE
THE AMOS AND ANDY SHOW (I,179); OLD NICKERBOCKER MUSIC HALL (I,3351); THE TIMES SQUARE STORY (I,4514)

LEE, MAGGIE
KENNEDY (II,1377)

LEE, MICHELE
ALIAS SMITH AND JONES (I,118); THE ALL-STAR SALUTE TO MOTHER'S DAY (II,51); BATTLE OF THE NETWORK STARS (II,172); THE BOB HOPE SHOW (I,636); BONNIE AND THE FRANKLINS (II,349); CIRCUS OF THE STARS (II,534); CIRCUS OF THE STARS (II,537); THE CONFESSIONS OF DICK VAN DYKE (II,568); ED SULLIVAN'S BROADWAY (I,1402); THE FIRST NINE MONTHS ARE THE HARDEST (I,1584); THE GLEN CAMPBELL SPECIAL: THE MUSICAL WEST (I,1821); I LOVE LIBERTY (II,1215); KNOTS LANDING (II,1404); KRAFT SALUTES WALT DISNEY WORLD'S 10TH ANNIVERSARY (II,1410); LIGHTS, CAMERA, MONTY! (II,1475); THE MAGIC OF DAVID COPPERFIELD (II,1595); MAKE MINE RED, WHITE, AND BLUE (I,2841); THE MICHELE LEE SHOW (II,1690); OF THEE I SING (I,3334); OVER AND OUT (II,1938); PARADE OF STARS (II,1954); PERRY COMO'S CHRISTMAS IN NEW YORK (II,2011); PERRY COMO'S SUMMER OF '74 (II,2024); RICKLES (II,2170); ROBERTA (I,3816); ROMP (I,3838); THE SINGERS (I,4060); SINGLES (I,4062); THE WONDERFUL WORLD OF PIZZAZZ (I,4901)

LEE, PATSY
DON MCNEILL'S TV CLUB (I,1337)

LEE, PATTIE
THE $128,000 QUESTION (II,1903)

LEE, PATTY
SOUNDS OF HOME (I,4136)

LEE, PEGGY
76 MEN AND PEGGY LEE (I,3989); THE ANDY WILLIAMS SHOW (I,207); THE BING CROSBY SPECIAL (I,472); THE BOB HOPE SHOW (I,505); CRESCENDO (I,1094); DUKE ELLINGTON. . . WE LOVE YOU MADLY (I,1380); THE GENERAL ELECTRIC THEATER (I,1753); THE JO STAFFORD SHOW (I,2395); MUSIC '55 (I,3168); THE PEGGY LEE SPECIAL (I,3510); PETULA (I,3573); SEE AMERICA WITH ED SULLIVAN (I,3972); SOMETHING SPECIAL (I,4118); SWING INTO SPRING (I,4317); TV'S TOP TUNES (I,4634)

LEE, PENELOPE
TEN WHO DARED (II,2559)

LEE, PINKY
20TH CENTURY FOLLIES (I,4641); GUMBY (I,1905); HERE IT IS, BURLESQUE! (II,1133); PINKY AND COMPANY (I,3599); THE PINKY LEE SHOW (I,3600); THE PINKY LEE SHOW (I,3601); THOSE TWO (I,4470)

LEE, ROBBIE
SISTER TERRI (II,2366)

LEE, ROBERTA
THE FRANK SINATRA SHOW (I,1664)

LEE, RUDY
ANNETTE (I,222); MAISIE (I,2833)

LEE, RUSTY
THE DOOLEY BROTHERS (II,715)

LEE, RUTA
FIRST & TEN (II,861); FLINTSTONE FAMILY ADVENTURES (II,882); THE FLINTSTONE FUNNIES (II,883); HIGH ROLLERS (II,1147); THE LINE-UP (I,2707); LUCY MOVES TO NBC (II,1562); THE MAN FROM EVERYWHERE (I,2864); POOR MR. CAMPBELL (I,3647); THE ROUSTERS (II,2220); THE ROY ROGERS SHOW (I,3859); SCHLITZ PLAYHOUSE OF STARS (I,3936); STUMP THE STARS (I,4273); THE TWILIGHT ZONE (I,4651)

LEE, RUTH
INDEMNITY (I,2209)

LEE, SONDRA
ARCHY AND MEHITABEL (I,254); HANSEL AND GRETEL (I,1934); ONCE UPON A TUNE (I,3372); PETER PAN (I,3566); PINOCCHIO (I,3602); THE S. S. HOLIDAY (I,4173)

LEE, STEPHEN
SUZANNE PLESHETTE IS MAGGIE BRIGGS (II,2509)

LEE, STUART
THE RED SKELTON SHOW (I,3755)

LEE, SUNSHINE
MULLIGAN'S STEW (II,1763); MULLIGAN'S STEW (II,1764)

LEE, TIM
WELCOME TO PARADISE (II,2762)

LEE, TOMMY
CALL HOLME (II,412)

LEE, TRACY
ANNA AND THE KING (I,221)

LEE, VIRGINIA ANN
THE AMAZING CHAN AND THE CHAN CLAN (I,157); ANNIE OAKLEY (I,225); BANYON (I,344); GENERAL HOSPITAL (II,964); THE RETURN OF CHARLIE CHAN (II,2136)

LEE, WILL
SESAME STREET (I,3982)

LEE, WILLIAM A
DREAM GIRL (I,1374); ON BORROWED TIME (I,3354)

LEE-SUNG, RICHARD
THE JERK, TOO (II,1282)

LEEDS, BARBARA
HAY FEVER (I,1979)

LEEDS, ELISSA
ALL MY CHILDREN (II,39); ANOTHER WORLD (II,97); BLUE JEANS (II,275); DOROTHY (II,717); THE GUIDING LIGHT (II,1064); HOW TO SURVIVE A MARRIAGE (II,1198); THE LEGEND OF THE GOLDEN GUN (II,1455); VEGAS (II,2723)

LEEDS, MAUREEN
THE GINGER ROGERS SHOW (I,1805)

LEEDS, PETER
THE BETTY WHITE SHOW (I,415); THE BOB HOPE SHOW (I,536); THE BOB HOPE SHOW (I,552); THE BOB HOPE SHOW (I,559); THE BOB HOPE SHOW (I,578); THE BOB HOPE SHOW (I,593); THE BOB HOPE SHOW (I,611); THE BOB HOPE SHOW (I,656); LIFE WITH VIRGINIA (I,2695); PETE AND GLADYS (I,3558); SO HELP ME, APHRODITE (I,4104); THE UNTOUCHABLES (I,4681)

LEEDS, PHIL
THE BUREAU (II,397); FEATHERSTONE'S NEST (II,847); FRONT ROW CENTER (I,1693); GOOD TIME HARRY (II,1037); HEREAFTER (II,1138); IVAN THE TERRIBLE (II,1260); WHAT'S UP, AMERICA? (I,4824); WHAT'S UP? (I,4823); A YEAR AT THE TOP (II,2845)

LEEGANT, DAN
ALLISON SIDNEY HARRISON (II,54)

LEEK, TIIU
THAT'S MY LINE (II,2579)

LEEPER, PATRICIA
FRAGGLE ROCK (II,916)

LEES, MICHAEL
LOVE IN A COLD CLIMATE (II,1539)

LEESON, JOHN
DOCTOR WHO (II,689); DOCTOR WHO—THE FIVE DOCTORS (II,690)

LEFEBVER, NED

SPACE ANGEL (I,4140)

LEFF, ELVA
GEORGE WASHINGTON (II,978)

LEGAULT, LANCE
ELVIS (I,1435)

LEGRAND, MICHEL
ANN-MARGRET SMITH (II,85); THE DICK VAN DYKE SPECIAL (I,1277); THE MAURICE CHEVALIER SHOW (I,2954); THE SHIRLEY BASSEY SPECIAL (II,2331)

LEHINE, JOHN
JUSTICE (I,2498)

LEHMAN, LILLIAN
AMANDA FALLON (I,151); EMERGENCY! (II,775); FAY (II,844); TENAFLY (I,4395); TENAFLY (II,2560)

LEHMAN, TED
THE MARY TYLER MOORE SHOW (II,1640)

LEHMAN, TRENT
BARNEY AND ME (I,358); NANNY AND THE PROFESSOR (I,3225); NANNY AND THE PROFESSOR AND THE PHANTOM OF THE CIRCUS (I,3226)

LEHMAN, VAL
PRISONER: CELL BLOCK H (II,2085)

LEHMANN, TED
BLACK BART (II,260)

LEHNE, FRED
IN THE BEGINNING (II,1229); THIS GIRL FOR HIRE (II,2596)

LEHNE, JOHN
CHARLIE'S ANGELS (II,487); DALLAS (II,614); DISAPPEARANCE OF AIMEE (I,1290); SERGEANT T.K. YU (II,2305); WASHINGTON: BEHIND CLOSED DOORS (II,2744)

LEHR, LEW
DETECT AND COLLECT (I,1242)

LEHR, ZELLA
HEE HAW (II,1123)

LEHVE, JOHN
THE EDGE OF NIGHT (II,760)

LEI, LYDIA
CRAZY LIKE A FOX (II,601); INSPECTOR PEREZ (II,1237)

LEIBMAN, RON
THE ART OF CRIME (II,108); KAZ (II,1370); LINDA IN WONDERLAND (II,1481); THE OUTSIDE MAN (II,1937); RIVKIN: BOUNTY HUNTER (II,2183); SIDE BY SIDE (II,2346); STEVE MARTIN'S THE WINDS OF WHOOPIE (II,2461); TWILIGHT THEATER II (II,2686)

LEIGH, BARBARA
SMILE JENNY, YOU'RE DEAD (II,2382)

LEIGH, CAROL
THE RIFLEMAN (I,3789)

LEIGH, JANET
THE ALL-STAR SALUTE TO MOTHER'S DAY (II,51); THE BOB HOPE CHRYSLER THEATER (I,502); THE BOB HOPE SHOW (I,628); THE BOB HOPE SHOW (I,599); THE BOB HOPE SHOW (I,607); THE BOB HOPE SHOW (I,623); CIRCLE OF FEAR (I,958); CIRCUS OF THE STARS (II,530); FOR LOVE OR $$$ (I,1623); GENE KELLY. . . AN AMERICAN IN PASADENA (II,962); THE HOUSE ON GREENAPPLE ROAD (I,2140); LOOKING BACK (I,2754); LOVE STORY (I,2778); THE MONK (I,3087); MURDOCK'S GANG (I,3164); MY WIVES JANE (I,3207)

LEIGH, KATIE
DUNGEONS AND DRAGONS (II,744)

LEIGH-HUNT, BARBARA
SEARCH FOR THE NILE (I,3954)

LEIGH-HUNT, RONALD
THE ADVENTURES OF SIR LANCELOT (I,76); MELODY OF HATE (I,2999)

LEIGHTON, BERNIE
TONY BENNETT IN WAIKIKI (I,4568)

LEIGHTON, CYNTHIA
MEET THE GIRLS: THE SHAPE, THE FACE, AND THE BRAIN (I,2990)

LEIGHTON, MARGARET
HAMLET (I,1925)

LEIGHTON, ROBERTA
ROSETTI AND RYAN: MEN WHO LOVE WOMEN (II,2217); THE YOUNG AND THE RESTLESS (II,2862)

LEIGHTON, SHEILA
THE GREEN HORNET (I,1886)

LEINHARD, OWEN
MICKIE FINN'S (I,3029)

LEISTER, JOHANNA
THE EDGE OF NIGHT (II,760)

LEISTON, FREDDIE
LIFE WITH FATHER (I,2690)

LEISURE, DAVID
WAIT UNTIL DARK (II,2735)

LEITH, VIRGINIA
CONDOMINIUM (II,566)

LEIZMAN, JOAN
THE PARAGON OF COMEDY (II,1956); THE PEE WEE HERMAN SHOW (II,1988)

LELMANN, TED
OFF THE RACK (II,1878)

LEMBECK, ELAINE
WELCOME BACK, KOTTER (II,2761)

LEMBECK, HARVEY
ENSIGN O'TOOLE (I,1448); THE GOLDBERGS (I,1836); THE HATHAWAYS (I,1965); KISS ME, KATE (I,2568); LILY—SOLD OUT (II,1480); THE MILLION DOLLAR INCIDENT (I,3045); MOTHER, JUGGS AND SPEED (II,1741); THE PHIL SILVERS SHOW (I,3580)

LEMBECK, HELAINE
FRANKIE AND ANNETTE: THE SECOND TIME AROUND (II,917); MAGIC MONGO (II,1592)

LEMBECK, MICHAEL
FLANNERY AND QUILT (II,873); THE FUNNY SIDE (I,1714); THE FUNNY SIDE (I,1715); GIDGET GROWS UP (I,1797); GOODBYE DOESN'T MEAN FOREVER (II,1039); HAVING BABIES III (II,1109); THE KROFFT SUPERSHOW (II,1412); THE KROFFT SUPERSHOW II (II,1413); MARY HARTMAN, MARY HARTMAN (II,1638); ONE DAY AT A TIME (II,1900)

LEMMO, JOAN
CRAZY TIMES (II,602)

LEMMON, CHRISTOPHER
BROTHERS AND SISTERS (II,382); THE OUTLAWS (II,1936)

LEMMON, COURTNEY
YES, VIRGINIA, THERE IS A SANTA CLAUS (II,2849)

LEMMON, JACK
THE AD-LIBBERS (I,29); BOB HOPE SPECIAL: BOB HOPE'S STARS OVER TEXAS (II,318); THE BRIGHTER DAY (I,728); THE CAMPBELL TELEVISION SOUNDSTAGE (I,812); CELEBRATION: THE AMERICAN SPIRIT (II,461); CHRYSLER MEDALLION THEATER (I,951); DANGER (I,1134); THE FRANCES LANGFORD-DON AMECHE SHOW (I,1655); THE FUNNIEST JOKE I EVER HEARD (II,939); HEAVEN FOR BETSY (I,1994); I REMEMBER MAMA (I,2176); JACK LEMMON IN 'S WONDERFUL, 'S MARVELOUS, 'S GERSHWIN (I,2313); JACK LEMMON—GET HAPPY (I,2312); MUSICAL COMEDY TONIGHT (II,1777); OLD NICKERBOCKER MUSIC HALL (I,3351); PLAYHOUSE 90 (I,3623); PULITZER PRIZE PLAYHOUSE (I,3692); ROAD OF LIFE (I,3798); SHOW BUSINESS SALUTE TO MILTON BERLE (I,4029); SUPER COMEDY BOWL 1 (I,4288); SUPER COMEDY BOWL 2 (I,4289); THAT WONDERFUL GUY (I,4425); THE TIMES SQUARE STORY (I,4514); TOM SNYDER'S CELEBRITY SPOTLIGHT (II,2632); TONI TWIN TIME (I,4554); TURN OF FATE (I,4623); THE WEB (I,4784); ZANE GREY THEATER (I,4979)

LEMOINE, THIERRY
ROLL OUT! (I,3833)

LEMON, BEN
JAKE'S WAY (II,1269)

LEMON, MEADOWLARK
CRASH ISLAND (II,600); HELLO, LARRY (II,1128); THAT'S TV (II,2581)

LENARD, MARK
ANOTHER WORLD (II,97); HERE COME THE BRIDES (I,2024); PLANET OF THE APES (II,2049); THE SECRET EMPIRE (II,2291); STAR TREK (II,2440); THE THREE MUSKETEERS (I,4481)

LENEHAN, NANCY
EMPIRE (II,777)

LENIHAN, DEIRDRE
EMERGENCY! (II,776); FROM HERE TO ETERNITY (II,933); NEEDLES AND PINS (I,3246); THE WALTONS (II,2740)

LENNICK, BEN
THE TROUBLE WITH TRACY (I,4613)

LENNON SISTERS, THE
THE ANDY WILLIAMS CHRISTMAS SHOW (I,198); THE ICE PALACE (I,2182); LIGHTS, CAMERA, MONTY! (II,1475); THE PERRY COMO THANKSGIVING SHOW

(I,3539)

LENNON, DIANNE
GLENN FORD'S SUMMERTIME, U.S.A. (I,1823); JIMMY DURANTE PRESENTS THE LENNON SISTERS HOUR (I,2387); THE LAWRENCE WELK SHOW (I,2643); LAWRENCE WELK'S TOP TUNES AND NEW TALENT (I,2645); THE LENNON SISTERS SHOW (I,2656); MEMORIES WITH LAWRENCE WELK (II,1681)

LENNON, JANET
GLENN FORD'S SUMMERTIME, U.S.A. (I,1823); JIMMY DURANTE PRESENTS THE LENNON SISTERS HOUR (I,2387); THE LAWRENCE WELK SHOW (I,2643); LAWRENCE WELK'S TOP TUNES AND NEW TALENT (I,2645); THE LENNON SISTERS SHOW (I,2656); MEMORIES WITH LAWRENCE WELK (II,1681)

LENNON, JOHN
JOHN LENNON AND YOKO ONO PRESENT THE ONE-TO-ONE CONCERT (I,2414)

LENNON, KATHY
GLENN FORD'S SUMMERTIME, U.S.A. (I,1823); JIMMY DURANTE PRESENTS THE LENNON SISTERS HOUR (I,2387); THE LAWRENCE WELK SHOW (I,2643); LAWRENCE WELK'S TOP TUNES AND NEW TALENT (I,2645); THE LENNON SISTERS SHOW (I,2656); MEMORIES WITH LAWRENCE WELK (II,1681)

LENNON, PEGGY
GLENN FORD'S SUMMERTIME, U.S.A. (I,1823); JIMMY DURANTE PRESENTS THE LENNON SISTERS HOUR (I,2387); THE LAWRENCE WELK SHOW (I,2643); LAWRENCE WELK'S TOP TUNES AND NEW TALENT (I,2645); THE LENNON SISTERS SHOW (I,2656); MEMORIES WITH LAWRENCE WELK (II,1681)

LENO, JAY
ALMOST HEAVEN (II,57); THE MARILYN MCCOO AND BILLY DAVIS JR. SHOW (II,1630); PLAYBOY'S PLAYMATE PARTY (II,2055); SNAFU (II,2386)

LENOX, DON
THE M AND M CANDY CARNIVAL (I,2797)

LENROW, BERNARD
CRIME PHOTOGRAPHER (I,1097); FIRST LOVE (I,1583)

LENSKA, RULA
THE ROCK FOLLIES (II,2192); THE SEVEN DIALS MYSTERY (II,2308)

LENTINI, SUSAN
THE RENEGADES (II,2132)

LENZ, CAROLYN
ANOTHER LIFE (II,95)

LENZ, KAY
THE FALL GUY (II,811); HOW THE WEST WAS WON (II,1196); THE HUSTLER OF MUSCLE BEACH (II,1210); LISA, BRIGHT AND DARK (I,2711); LOVE STORY (I,2778); PRISONERS OF THE LOST UNIVERSE (II,2086); RICH MAN, POOR MAN—BOOK I (II,2161); RICH MAN, POOR MAN—BOOK II (II,2162); SANCTUARY OF FEAR (II,2252)

LENZ, RICK
ADVICE TO THE LOVELORN (II,16); THE BIONIC WOMAN (II,255); THE CENTURY TURNS (I,890); DOC (I,1303); GREEN ACRES (I,1884); HEC RAMSEY (II,1121)

LEON, DANIEL
THROUGH THE MAGIC PYRAMID (II,2615)

LEON, JOSEPH
HEY, LANDLORD (I,2039); IVAN THE TERRIBLE (II,1260)

LEON, MICHAEL
YOUNG LIVES (II,2866)

LEON, MICHELE
SEMI-TOUGH (II,2300)

LEONARD BLAIR DANCERS, THE
THE JO STAFFORD SHOW (I,2394)

LEONARD, BILL
YOU ARE THERE (I,4929)

LEONARD, DAVID
THE LONE RANGER (I,2740)

LEONARD, JACK E
BABES IN TOYLAND (I,318); THE BOB HOPE SHOW (I,633); BROADWAY OPEN HOUSE (I,736); FRANK SINATRA JR. WITH FAMILY AND FRIENDS (I,1663); THE GUY MITCHELL SHOW (I,1912); PANAMA HATTIE (I,3444); SPRING HOLIDAY (I,4168); THIS IS SHOW BUSINESS (I,4459)

LEONARD, LEE
ENTERTAINMENT TONIGHT (II,780)

LEONARD, QUEENIE
A CHRISTMAS CAROL (I,947)

LEONARD, SHELDON
BIG EDDIE (II,235); BIG EDDIE (II,236); THE BILL COSBY SPECIAL (I,444); THE BOB HOPE SHOW (I,532); DANNY THOMAS: THE ROAD TO LEBANON (I,1153); THE DICK VAN DYKE SHOW (I,1275); THE DUKE (I,1381); THE ISLANDER (II,1249); IT'S ALWAYS SUNDAY (I,2265); LINUS THE LIONHEARTED (I,2709); MAKE ROOM FOR DADDY (I,2843); TOP SECRET (II,2647); THE WONDERFUL WORLD OF BURLESQUE I (I,4895); YOUR JEWELER'S SHOWCASE (I,4953)

LEONARD, SUGAR RAY
BOB HOPE SPECIAL: BOB HOPE'S ALL-STAR COMEDY BIRTHDAY PARTY AT WEST POINT (II,298)

LEONARD-BOYNE, EVA
THE CORN IS GREEN (I,1050)

LEONARDO AND ZOLA
DENNIS JAMES' CARNIVAL (I,1227)

LEONARDO, ANN
CAPTAIN KANGAROO (II,434)

LEONE, MARIANNE
KATE AND ALLIE (II,1365)

LEONG, ALBERT
THE RENEGADES (II,2132)

LEONING, JOHN
KUNG FU (II,1416)

LEOPOLD, THOMAS
I'D RATHER BE CALM (II,1216); THE STEVE ALLEN COMEDY HOUR (II,2454); THE TED KNIGHT SHOW (II,2550); THE TV SHOW (II,2673)

LEPLAT, TED
THE GUIDING LIGHT (II,1064); ONE LIFE TO LIVE (II,1907)

LEPP, NANCY
SEARCH FOR TOMORROW (II,2284)

LERMAN, APRIL
CHARLES IN CHARGE (II,482)

LERNER, DAVID
THE HITCHHIKER'S GUIDE TO THE GALAXY (II,1158)

LERNER, FRED
SAWYER AND FINN (II,2265)

LERNER, MICHAEL
FIREHOUSE (I,1579); GRANDPA MAX (II,1051); HART TO HART (II,1101); HART TO HART (II,1102); I GAVE AT THE OFFICE (II,1213); LOVE STORY (I,2778); STARSKY AND HUTCH (II,2444); STARSKY AND HUTCH (II,2445); VEGAS (II,2723)

LES BLUEBELL GIRLS
SHIRLEY MACLAINE AT THE LIDO (II,2332)

LES BROWN AND HIS ORCHESTRA
TIMEX ALL-STAR JAZZ SHOW III (I,4517)

LES GIRLS
THE MONTE CARLO SHOW (II,1726)

LESCOULIE, JACK
THE BIG SELL (I,431); BRAINS AND BRAWN (I,706); FUN FOR THE MONEY (I,1708); ONE, TWO, THREE—GO! (I,3390); TONIGHT! AMERICA AFTER DARK (I,4555)

LESINAWAI, LEONI
ADVENTURES OF THE SEASPRAY (I,82)

LESLIE, BETHEL
ALCOA PREMIERE (I,109); ARENA (I,256); THE BAT (I,364); COSMOPOLITAN THEATER (I,1057); DANGER (I,1134); THE DOCTORS (II,694); THE GIRLS (I,1811); LIGHT'S OUT (I,2699); THE LLOYD BRIDGES SHOW (I,2733); THE MARCH OF DIMES FASHION SHOW (I,2902); THE NASH AIRFLYTE THEATER (I,3227); ONE STEP BEYOND (I,3388); PEPSI-COLA PLAYHOUSE (I,3523); PLAYHOUSE 90 (I,3623); PRUDENTIAL FAMILY PLAYHOUSE (I,3683); THE RICHARD BOONE SHOW (I,3784); SHORT, SHORT DRAMA (I,4023); STUDIO ONE (I,4268); TALES OF TOMORROW (I,4352); THRILLER (I,4492)

LESLIE, DIANE
PINOCCHIO'S CHRISTMAS (II,2045)

LESLIE, EDITH
THE MANY SIDES OF MICKEY ROONEY (I,2899)

LESLIE, JOAN
FORD TELEVISION THEATER (I,1634); SHADOW OF SAM PENNY (II,2313)

LESLIE, LOUISE
HOMEMAKER'S EXCHANGE (I,2100)

LESLIE, NAN
ANNIE OAKLEY (I,225); THE CALIFORNIANS (I,795); FURY (I,1721); KING'S ROW (I,2563)

LESLIE, NOEL
ALICE IN WONDERLAND (I,120); THE CORN IS GREEN (I,1050)

LESLIE, WILLIAM
THE LINE-UP (I,2707)

LESS, GUNTHER
JOURNEY TO ADVENTURE (I,2455)

LESSER, LEN
CHARLIE ANGELO (I,906)

LESSY, BEN
MAKE ROOM FOR DADDY (I,2843)

LESTER WILSON DANCERS, THE
MARIE (II,1629); THE SAMMY DAVIS JR. SHOW (I,3898)

LESTER, BUDDY
MICKEY SPILLANE'S MIKE HAMMER: MORE THAN MURDER (II,1693); SEVEN AT ELEVEN (I,3984); STARSKY AND HUTCH (II,2445)

LESTER, CHEZ
FATHER MURPHY (II,841); GOLDIE AND KIDS: LISTEN TO ME (II,1021)

LESTER, JACK
THE BENNETS (I,397); MCDUFF, THE TALKING DOG (II,1662); A TIME TO LIVE (I,4510)

LESTER, JEFF
WALKING TALL (II,2738)

LESTER, JERRY
BROADWAY OPEN HOUSE (I,736); CANDID CAMERA (I,818); CAVALCADE OF STARS (I,877); THE JERRY LESTER SHOW (I,2360); THE JERRY LESTER SPECIAL (I,2361); LI'L ABNER (I,2701); THE SATURDAY NIGHT DANCE PARTY (I,3919); SOUND OFF TIME (I,4135)

LESTER, KETTY
ADVENTURING WITH THE CHOPPER (II,15); DAYS OF OUR LIVES (II,629); HUNTER (II,1204); LITTLE HOUSE ON THE PRAIRIE (II,1487); LITTLE HOUSE: A NEW BEGINNING (II,1488); RITUALS (II,2182)

LESTER, LOREN
THE FACTS OF LIFE (II,805)

LESTER, NORA
CAMP WILDERNESS (II,419)

LESTER, ROBIE
SANTA CLAUS IS COMIN' TO TOWN (II,2259)

LESTER, TERRY
ARK II (II,107); THE BARBARY COAST (II,147); CHARACTERS (II,481); DALLAS (II,614); ONCE UPON A SPY (II,1898); THAT'S TV (II,2581); THE YOUNG AND THE RESTLESS (II,2862)

LESTER, TOM
CHARO AND THE SERGEANT (II,489); GREEN ACRES (I,1884)

LETCH, PAT
PRISONER: CELL BLOCK H (II,2085)

LETCHWORTH, EILEEN
ALL MY CHILDREN (II,39); LOVE OF LIFE (I,2774)

LETHIN, LILLIE
GOLIATH AWAITS (II,1025)

LETHIN, LORI
THE DUKES OF HAZZARD (II,742); THE WONDERFUL WORLD OF PHILIP MALLEY (II,2831)

LETNER, KEN
FALCON CREST (II,808)

LETTERI, LOUIS
RED RYDER (I,3752)

LETTERMAN, DAVID
BATTLE OF THE NETWORK STARS (II,167); BOB HOPE SPECIAL: BOB HOPE'S SUPER BIRTHDAY SPECIAL (II,319); THE DAVID LETTERMAN SHOW (II,624); LATE NIGHT WITH DAVID LETTERMAN (II,1442); MARY (II,1637); PEEPING TIMES (II,1990); THE STARLAND VOCAL BAND (II,2441)

LEVANG, NEIL
THE LAWRENCE WELK SHOW (I,2643); MEMORIES WITH LAWRENCE WELK (II,1681)

LEVANT, ANDREW
HOME ROOM (II,1172)

LEVANT, OSCAR
G.E. GUEST HOUSE (I,1746)

LEVEILLEE, CLAUDE
ANNE MURRAY'S WINTER CARNIVAL. . . FROM QUEBEC (II,90)

LEVENE, JOHN
DOCTOR WHO (II,689)

LEVENSON, SAM
CELEBRITY TALENT SCOUTS (I,886); THE SAM LEVENSON SHOW (I,3897); TWO FOR THE MONEY (I,4654)

LEVENTON, ANNABLE
PENMARRIC (II,1993)

LEVERSEE, LORETTA
THE ALDRICH FAMILY (I,111); YOUNG DR. MALONE (I,4938)

LEVIEN, PHILIP
DOCTORS' PRIVATE LIVES (II,692); HARD KNOCKS (II,1086); LOVE AT FIRST SIGHT (II,1530); LOVE AT FIRST SIGHT (II,1531)

LEVIN, CHARLES
ALICE (II,33); BROTHERS (II,381); GOSSIP (II,1043); GOSSIP (II,1044); I LOVE HER ANYWAY! (II,1214); I'D RATHER BE CALM (II,1216); IT TAKES TWO (II,1252)

LEVIN, HENRY
CONTEST CARNIVAL (I,1040)

LEVIN, JOHN
CURIOSITY SHOP (I,1114); RETURN TO PEYTON PLACE (I,3779); SANDBURG'S LINCOLN (I,3907)

LEVIN, MICHAEL
RYAN'S HOPE (II,2234)

LEVIN, PETER
GOODNIGHT, BEANTOWN (II,1041)

LEVIN, WENDY
T.L.C. (II,2525)

LEVINE, ANNA
BEACON HILL (II,193)

LEVINE, DEBORAH
JOE DANCER: THE BIG BLACK PILL (II,1300)

LEVINE, EMILY
EVERYDAY (II,793)

LEVINE, FLOYD
FARRELL: FOR THE PEOPLE (II,834); KAZ (II,1370); MALIBU (II,1607); MICKEY SPILLANE'S MARGIN FOR MURDER (II,1691)

LEVITAN, AMY
MANDY'S GRANDMOTHER (II,1618)

LEVITT, AMY
ONE LIFE TO LIVE (II,1907)

LEVITT, SHIRLEY
SUNDAY DATE (I,4284)

LEVITT, STEVE
MALIBU (II,1607)

LEVY, CAROL
JOHNNY GARAGE (II,1337)

LEVY, DAVID
LITTLE VIC (II,1496); WONDERBUG (II,2830)

LEVY, EUGENE
FROM CLEVELAND (II,931); THE LOVEBIRDS (II,1548); THE MAGIC OF DAVID COPPERFIELD (II,1595); SECOND CITY TELEVISION (II,2287)

LEVY, STEVEN
MOMENT OF TRUTH (I,3084)

LEVY, WEAVER
ADVENTURES IN PARADISE (I,46)

LEW, JAMES
THE RENEGADES (II,2132)

LEWANDOWSKI, BOB
POLKA-GO-ROUND (I,3640)

LEWANDOWSKI, JERRY J
THE WILD WOMEN OF CHASTITY GULCH (II,2801)

LEWIN, JOHN
A CHRISTMAS CAROL (II,517)

LEWIS, ABBY
THE BRIGHTER DAY (I,728); CONCERNING MISS MARLOWE (I,1032); SOMERSET (I,4115); VALIANT LADY (I,4695)

LEWIS, AL
CAR 54, WHERE ARE YOU? (I,842); MARINELAND CARNIVAL (I,2913); MUNSTER, GO HOME! (I,3157); THE MUNSTERS (I,3158); THE MUNSTERS' REVENGE (II,1765); THE UNCLE AL SHOW (I,4667)

LEWIS, ARTHUR
TOPPER RETURNS (I,4583)

LEWIS, BOBO
LI'L ABNER (I,2702); THAT GIRL (II,2570)

LEWIS, BRENDA
GOLDEN CHILD (I,1839)

LEWIS, CATHY
FIBBER MCGEE AND MOLLY (I,1569); HAZEL (I,1982); MY FRIEND IRMA (I,3191)

LEWIS, DAVID
BACHELOR FATHER (I,323); THE BLUE MEN (I,492); BRIGHT PROMISE (I,727); THE FARMER'S DAUGHTER (I,1533); GENERAL HOSPITAL (II,964)

LEWIS, DIANE
SIMON AND SIMON (II,2357)

LEWIS, EMMANUEL
CIRCUS OF THE STARS (II,538); SALUTE TO LADY LIBERTY (II,2243); WEBSTER (II,2758); THE WORLD'S FUNNIEST COMMERCIAL GOOFS (II,2841)

LEWIS, FORREST
CALL TO DANGER (I,803); THE GREAT GILDERSLEEVE (I,1875); ICHABOD AND ME (I,2183); JOHNNY RISK (I,2435); SANDY STRONG (I,3911)

LEWIS, FRANK
EVENING AT THE IMPROV (II,786); THE PATTI PAGE SHOW (I,3480); THE TONY BENNETT SHOW (I,4569)

LEWIS, GARRETT
OF THEE I SING (I,3334)

LEWIS, GARY
THE JERRY LEWIS SHOW (I,2367)

LEWIS, GEOFFREY
CENTENNIAL (II,477); THE DEADLY TRIANGLE (II,634); FLO (II,891); THE GUN AND THE PULPIT (II,1067); GUN SHY (II,1068); THE NEW DAUGHTERS OF JOSHUA CABE (I,3261); POOR RICHARD (II,2065); THE RETURN OF LUTHER GILLIS (II,2137); RETURN OF THE MAN FROM U.N.C.L.E.: THE 15 YEARS LATER AFFAIR (II,2140); SAMURAI (II,2249); SKYWARD CHRISTMAS (II,2378); TRAVIS MCGEE (II,2656)

LEWIS, GREG
EVENING AT THE IMPROV (II,786); IN SECURITY (II,1227); MINSKY'S FOLLIES (II,1703); THE NATIONAL SNOOP (II,1797); PAT BOONE AND FAMILY (II,1968)

LEWIS, HARVEY
ANNIE FLYNN (II,92)

LEWIS, HUGH X
COUNTRY CLUB (I,1063)

LEWIS, J J
THE MADHOUSE BRIGADE (II,1588)

LEWIS, JAMES
ARTHUR GODFREY AND FRIENDS (I,278)

LEWIS, JARMA
SECRET AGENT (I,3961)

LEWIS, JENNIE 'DAGMAR'
BROADWAY OPEN HOUSE (I,736); DAGMAR'S CANTEEN (I,1125); DENNIS JAMES' CARNIVAL (I,1227)

LEWIS, JERRY
CIRCUS OF THE STARS (II,532); THE COLGATE COMEDY HOUR (I,997); THE DUPONT SHOW OF THE WEEK (I,1388); THE FIRST 50 YEARS (II,862); JACK BENNY'S BIRTHDAY SPECIAL (I,2299); THE JAZZ SINGER (I,2351); THE JERRY LEWIS SHOW (I,2362); THE JERRY LEWIS SHOW (I,2363); THE JERRY LEWIS SHOW (I,2364); THE JERRY LEWIS SHOW (I,2365); THE JERRY LEWIS SHOW (I,2366); THE JERRY LEWIS SHOW (I,2367); THE JERRY LEWIS SHOW (I,2368); THE JERRY LEWIS SHOW (I,2369); THE JERRY LEWIS SHOW (I,2370); THE JERRY LEWIS SHOW (II,1284); THE KLOWNS (I,2574); SHOW OF THE YEAR (I,4031); THE WONDERFUL WORLD OF BURLESQUE I (I,4895)

LEWIS, JERRY LEE
33 1/3 REVOLUTIONS PER MONKEE (I,4451); THE BIG BEAT (I,423); THE EDDIE RABBITT SPECIAL (II,759); ELVIS REMEMBERED: NASHVILLE TO HOLLYWOOD (II,772); THE JOHNNY CASH CHRISTMAS SPECIAL (II,1326)

LEWIS, JOAN
ROLL OF THUNDER, HEAR MY CRY (II,2202)

LEWIS, JUDY
GENERAL HOSPITAL (II,964); THE GUIDING LIGHT (II,1064); KITTY FOYLE (I,2571); THE OUTLAWS (I,3427); THE SECRET STORM (I,3969)

LEWIS, LINDA
SONG BY SONG (II,2399)

LEWIS, LOUISE
FRONT ROW CENTER (I,1694)

LEWIS, M K
FARRELL: FOR THE PEOPLE (II,834)

LEWIS, MARCIA
GOODTIME GIRLS (II,1042); LEGS (II,1459); MR. MOON'S MAGIC CIRCUS (II,1753); WHO'S WATCHING THE KIDS? (II,2793)

LEWIS, MARY MARGARET
SKEEZER (II,2376)

LEWIS, MICHAEL
AMAHL AND THE NIGHT VISITORS (I,150)

LEWIS, MONICA
SHOTGUN SLADE (I,4026)

LEWIS, NANCY
PAUL WHITEMAN'S TV TEEN CLUB (I,3491)

LEWIS, NAOMI
ROOTIE KAZOOTIE (I,3844)

LEWIS, NICHOLAS
THE SECRET STORM (I,3969)

LEWIS, RHODA
ESPIONAGE (I,1464); MASTER OF THE GAME (II,1647); SOMEONE AT THE TOP OF THE STAIRS (I,4114)

LEWIS, RICHARD
THE 416TH (II,913); THE FACTS (II,804); MELODY OF HATE (I,2999); THE SONNY AND CHER SHOW (II,2401)

LEWIS, ROBERT Q
GET THE MESSAGE (I,1782); HIDDEN TREASURE (I,2045); MAKE ME LAUGH (I,2839); MASQUERADE PARTY (I,2941); MAURICE WOODRUFF PREDICTS (I,2956); THE NAME'S THE SAME (I,3220); PLAY YOUR HUNCH (I,3619); ROBERT Q'S MATINEE (I,3812); THE ROBERT Q. LEWIS CHRISTMAS SHOW (I,3811); THE SHOW GOES ON (I,4030)

LEWIS, RONNIE
THE JERRY LEWIS SHOW (I,2367)

LEWIS, ROSE
IRENE (II,1245)

LEWIS, SAGAN
ST. ELSEWHERE (II,2432)

LEWIS, SAMANTHA
MR. SUCCESS (II,1756)

LEWIS, SHARI
ARTHUR GODFREY LOVES ANIMALS (I,282); THE ARTHUR GODFREY SHOW (I,287); THE BANANA SPLITS ADVENTURE HOUR (I,340); CAPTAIN KANGAROO (II,434); HI MOM (I,2043); THE SHARI LEWIS SHOW (I,4000); THE SHARI LEWIS SHOW (II,2317); STEP ON THE GAS (I,4215)

LEWIS, STEVE
THE BARKLEYS (I,356)

LEWIS, SYLVIA
THE BIG TIME (I,434); WHERE'S RAYMOND? (I,4837)

LEWIS, VICTORIA ANN
THE RETURN OF MARCUS WELBY, M.D (II,2138)

LEWIS, WANDA
THE PAUL DIXON SHOW (I,3487); THIS IS MUSIC (I,4458); THE UNCLE AL SHOW (I,4667)

LEWIS, WILLIAM
CAESAR'S HOUR (I,790)

LEWIS, CY
PAUL WHITEMAN'S TV TEEN CLUB (I,3491)

LEYDEN, BILL
CALL MY BLUFF (I,800); IT COULD BE YOU (I,2245); MUSICAL CHAIRS (I,3182)

LEYDIG, GREG
THE DAUGHTERS OF JOSHUA CABE RETURN (I,1170)

LEYRAC, MONIQUE
THE PERRY COMO SPECIAL (I,3536)

LEYTON, DEBBIE
SARA (II,2260)

LEYTON, JOHN
JERICHO (I,2358); THE LEGEND OF SILENT NIGHT (I,2655)

LI, PAT
THE CHADWICK FAMILY (II,478)

LIBERACE
BATMAN (I,366); THE DINAH SHORE SPECIAL (I,1283); KEEFE BRASSELLE'S VARIETY GARDEN (I,2515); LEAPIN' LIZARDS, IT'S LIBERACE (II,1453); THE LIBERACE SHOW (I,2674); THE LIBERACE SHOW (I,2675); THE LIBERACE SHOW (I,2676); THE LIBERACE SHOW (I,2677); LILY—SOLD OUT (II,1480); THE RED SKELTON REVUE (I,3754); ROMP (I,3838); SHOWTIME (I,4036); THE TOM JONES SPECIAL (I,4541)

LIBERMAN, RICHARD
LAND OF HOPE (II,1424)

LIBERMAN, ROBERT
EVIL ROY SLADE (I,1485)

LIBERTINI, RICHARD
CALLING DR. STORM, M.D. (II,415); LET'S CELEBRATE (I,2663); THE MELBA MOORE-CLIFTON DAVIS SHOW (I,2996); SOAP (II,2392); STORY THEATER (I,4241)

LICHT, JEREMY
ALL THE WAY HOME (II,44); SKEEZER (II,2376)

LICHTERMAN, MARVIN
MO AND JO (II,1715)

LICHTMAN, PAUL
CARLTON YOUR DOORMAN (II,440)

LIDDEL, BARNEY
THE LAWRENCE WELK SHOW (I,2643); MEMORIES WITH LAWRENCE WELK (II,1681)

LIDDY, G GORDON
TWILIGHT THEATER II (II,2686)

LIDO DE PARIS DANCERS, THE
SIEGFRIED AND ROY (II,2350)

LIDO, BOB
THE LAWRENCE WELK SHOW (I,2643); LAWRENCE WELK'S TOP TUNES AND NEW TALENT (I,2645); MEMORIES WITH LAWRENCE WELK (II,1681)

LIEB, ROBERT
HAZEL (I,1982); MR. MERGENTHWIRKER'S LOBBLIES (I,3144); READY FOR THE PEOPLE (I,3739)

LIEBER, PAUL
BARNEY MILLER (II,154); THE GANGSTER CHRONICLES (II,957)

LIEBERMAN, BRADLEY
CRASH ISLAND (II,600)

LIEBERT, JUDY
QUILLER: PRICE OF VIOLENCE (II,2101)

LIGHT, JUDITH
FAMILY TIES (II,819); ONE LIFE TO LIVE (II,1907); WHO'S THE BOSS? (II,2792); YOU ARE THE JURY (II,2854)

LIGHT, KARL
THE BAT (I,364); THE DOCTORS (II,694)

LIGHTER, KAY
THE PHIL SILVERS SHOW (I,3580)

LIGHTFOOT, LEONARD
SILVER SPOONS (II,2355)

LIGHTHOUSE, THE
THE GEORGE KIRBY SPECIAL (I,1771)

LIGON, TOM
LOVING (II,1552); THE YOUNG AND THE RESTLESS (II,2862)

LILES, LINDA
THE DEVIL'S WEB (I,1248); DIAL A DEADLY NUMBER (I,1253); JENNIE: LADY RANDOLPH CHURCHILL (II,1277)

LILLEY, JACK
LITTLE HOUSE: BLESS ALL THE DEAR CHILDREN (II,1489); LITTLE HOUSE: LOOK BACK TO YESTERDAY (II,1490); LITTLE HOUSE: THE LAST FAREWELL (II,1491)

LILLIE, BEATRICE
THE BOB HOPE SHOW (I,504); THE BOB HOPE SHOW (I,505); THE BOB HOPE SHOW (I,535); THE BOB HOPE SHOW (I,603); TEXACO COMMAND PERFORMANCE (I,4407)

LILLY, SUE
BEAUTYBREAK (I,383)

LIM, KWAN HI
THE KILLER WHO WOULDN'T DIE (II,1393); MAGNUM, P.I. (II,1601)

LIMA, BRIAN
ALL MY CHILDREN (II,39)

LIME, YVONNE
CROSSROADS (I,1105); FATHER KNOWS BEST (II,838); THE GEORGE BURNS AND GRACIE ALLEN SHOW (I,1763); HAPPY (I,1938)

LIMELIGHTERS, THE
ON PARADE (I,3356); SEE AMERICA WITH ED SULLIVAN (I,3972)

LIN, TRACI
RYAN'S HOPE (II,2234)

LINCOLN, JULIE
CINDERELLA (II,526)

LINCOLN, PAMELA
LOVE OF LIFE (I,2774)

LIND, BRIT
HOW THE WEST WAS WON (II,1196); PINE CANYON IS BURNING (II,2040)

LINDAN, JANA
THE RETURN OF LUTHER GILLIS (II,2137)

LINDE, CHERYL
SENIOR YEAR (II,2301); SONS AND DAUGHTERS (II,2404)

LINDEN, EATA
THREE STEPS TO HEAVEN (I,4485)

LINDEN, HAL
THE 36 MOST BEAUTIFUL GIRLS IN TEXAS (II,2594); ALAN KING'S THIRD ANNUAL FINAL WARNING!! (II,31); BARNEY MILLER (II,154); BATTLE OF THE NETWORK STARS (II,163); BATTLE OF THE NETWORK STARS (II,164); THE CAPTAIN AND TENNILLE IN NEW ORLEANS (II,431); THE CELEBRITY FOOTBALL CLASSIC (II,474); CHRISTMAS IN WASHINGTON (II,521); HAL LINDEN'S BIG APPLE (II,1072); I LOVE LIBERTY (II,1215); LOLA (II,1508); THE LOVE BOAT I (II,1532); MR. INSIDE/MR. OUTSIDE (II,1750); PERRY COMO'S MUSIC FROM HOLLYWOOD (II,2019); RUGGLES OF RED GAP (I,3867); SEARCH FOR TOMORROW (II,2284); SECOND EDITION (II,2288); THE SHAMEFUL SECRETS OF HASTINGS CORNERS (I,3994)

LINDEN, JENNIE
MURDER IS A ONE-ACT PLAY (I,3160)

LINDER, CEC
THE DEVIL'S WEB (I,1248); THE EDGE OF NIGHT (II,760); RED SKELTON'S CHRISTMAS DINNER (II,2123); THE SECRET STORM (I,3969); SEEING THINGS (II,2297); THE THANKSGIVING TREASURE (I,4416)

LINDFORS, VIVECA
APPOINTMENT WITH ADVENTURE (I,246); THE DIARY OF ANNE FRANK (I,1262); DUPONT SHOW OF THE MONTH (I,1387); THE PARADINE CASE (I,3453)

LINDGREN, BRUCE
THE ADVENTURES OF CAPTAIN HARTZ (I,53)

LINDGREN, LISA
ANOTHER DAY (II,94); GENERAL HOSPITAL (II,964); HILL STREET BLUES (II,1154)

LINDINE, JACK
CODE RED (II,552); CODE RED (II,553); MURDER CAN HURT YOU! (II,1768)

LINDLEY, AUDRA
ANOTHER WORLD (II,97); BRIDGET LOVES BERNIE (I,722); DOC (II,683); FAY (II,844); FROM THESE ROOTS (I,1688); KRAFT TELEVISION THEATER (I,2592); LOVE STORY (I,2776); THE PAT BOONE AND FAMILY CHRISTMAS SPECIAL (II,1969); PEARL (II,1986); PHILCO TELEVISION PLAYHOUSE (I,3583); ROBERT MONTGOMERY PRESENTS YOUR LUCKY STRIKE THEATER (I,3809); THE ROPERS (II,2214); SKYWARD CHRISTMAS (II,2378); THREE'S COMPANY (II,2614)

LINDLEY, DARLYN ANN
B.J. AND THE BEAR (II,125); THE ROOKIES (II,2208)

LINDLEY, ERICA
THE BENNY HILL SHOW (II,207)

LINDREN, LISA
CRASH ISLAND (II,600)

LINDSAY, CYNTHIA
GENE KELLY'S WONDERFUL WORLD OF GIRLS (I,1752)

LINDSAY, FRANK
PETER PAN (I,3566)

LINDSAY, HELEN
THE PALLISERS (II,1944)

LINDSAY, LAURA
BRADDOCK (I,704)

LINDSAY, LEE
THE COLLEGE BOWL (I,1001)

LINDSAY, MARK
AMERICA (I,164); HAPPENING '68 (I,1936); IT'S HAPPENING (I,2269); MAKE YOUR OWN KIND OF MUSIC (I,2847)

LINDSEY, GENE
DARK SHADOWS (I,1157); THE DOCTORS (II,694)

LINDSEY, GEORGE
THE ANDY GRIFFITH SHOW (I,192); CITY VS. COUNTRY (I,968); GOOBER AND THE TRUCKERS' PARADISE (II,1029); HEE HAW (II,1123); MAYBERRY R.F.D. (I,2961); THE NASHVILLE PALACE (II,1793); TAKE ONE STARRING JONATHAN WINTERS (II,2532); THE TYCOON (I,4660)

LINDSEY, HOWARD
THE FORD 50TH ANNIVERSARY SHOW (I,1630)

LINDSEY, MINNIE S
ROOSEVELT AND TRUMAN (II,2209)

LINDSLY, CHUCK
ALL THAT GLITTERS (II,41)

LINDSTROM, FLORENCE
GENERAL HOSPITAL (II,964)

LINDSTROM, PIA
HOLLYWOOD STARS' SCREEN TESTS (II,1165)

LINDUP, ANNA
THE WOMAN IN WHITE (II,2819)

LINEBACK, RICHARD
JOHNNY BELINDA (I,2418); RIDING FOR THE PONY EXPRESS (II,2172)

LINERO, JEANNIE
BILLY CRYSTAL: A COMIC'S LINE (II,249); HOT L BALTIMORE (II,1187); THE RAG BUSINESS (II,2106)

LINHART, BUZZY
COS (II,576)

LINK, FRANK
THE BURNS AND SCHREIBER COMEDY HOUR (I,768)

LINK, MICHAEL
JULIA (I,2476)

LINK, PETER
AS THE WORLD TURNS (II,110)

LINKE, PAUL
CHIPS (II,511)

LINKER, AMY
LEWIS AND CLARK (II,1465); SQUARE PEGS (II,2431)

LINKLETTER, ART
THE ART LINKLETTER SHOW (I,276); ART LINKLETTER'S HOUSE PARTY (I,277); CBS: ON THE AIR (II,460); HOLLYWOOD TALENT SCOUTS (I,2094); THE LID'S OFF (I,2678); LIFE WITH LINKLETTER (I,2691); PEOPLE ARE FUNNY (I,3518); THE SECRET WORLD OF KIDS (I,3970); SHOW BIZ (I,4027); SONJA HENIE'S HOLIDAY ON ICE (I,4125)

LINKLETTER, JACK
AMERICA ALIVE! (II,67); THE ART LINKLETTER SHOW (I,276); HAGGIS BAGGIS (I,1917); HOOTENANNY (I,2117); ON THE GO (I,3361); THE REBUS GAME (I,3748)

LINN, BAMBI
54TH STREET REVUE (I,1573); BABES IN TOYLAND (I,318); THE BIG TIME (I,434); THE CHOCOLATE SOLDIER (I,944); THE GENERAL MOTORS 50TH ANNIVERSARY SHOW (I,1758); GOOD TIMES (I,1853); HOLIDAY (I,2074); KALEIDOSCOPE (I,2504); LADY IN THE DARK (I,2602); THE MAN IN THE MOON (I,2871); MAX LIEBMAN PRESENTS (I,2958); THE MERRY WIDOW (I,3012); NAUGHTY MARIETTA (I,3232); PANORAMA (I,3448); TEXACO COMMAND PERFORMANCE (I,4407); VICTOR BORGE'S COMEDY IN MUSIC III (I,4725); YOUR SHOW OF SHOWS (II,2875)

LINN, DIANA
CLIMAX! (I,976); JUNIOR MISS (I,2493); SILVER THEATER (I,4051)

LINN-BAKER, MARK
THE COMEDY ZONE (II,559); O'MALLEY (II,1872)

LINTON, JOHN
H.R. PUFNSTUF (I,2154)

LINVILLE, JOANNE
BEHIND THE SCREEN (II,200); THE HOUSE ON GREENAPPLE ROAD (I,2140); THE KAISER ALUMINUM HOUR (I,2503); STUDIO ONE (I,4268); SUSPICION (I,4309)

LINVILLE, LARRY
CALLING DR. STORM, M.D. (II,415); CHECKING IN (II,492); A CHRISTMAS FOR BOOMER (II,518); THE GIRL, THE GOLD WATCH AND DYNAMITE (II,1001); GRANDPA GOES TO WASHINGTON (II,1050); HERBIE, THE LOVE BUG (II,1131); M*A*S*H (II,1569); NIGHT PARTNERS (II,1847); THE NIGHT STALKER (I,3292); PAPER DOLLS (II,1952)

LIONEL BLAIR DANCERS, THE
THE JO STAFFORD SHOW (I,2393); THE JO STAFFORD SHOW (I,2395); ON STAGE: PHIL SILVERS (I,3357); THE ROBERT GOULET SPECIAL (I,3807); SPOTLIGHT (I,4163)

LIOTTA, ANDY
ALAN KING'S THIRD ANNUAL FINAL WARNING!! (II,31)

LIOTTA, RAY
ANOTHER WORLD (II,97); CASABLANCA (II,450); CRAZY TIMES (II,602)

LIPMAN, JENNY
THE ASSASSINATION RUN (II,112); DAY OF THE TRIFFIDS (II,627)

LIPMAN, MAUREEN
FILE IT UNDER FEAR (I,1574); LONG DAY'S JOURNEY INTO NIGHT (I,2746); SMILEY'S PEOPLE (II,2383)

LIPPE, JONATHAN
MEDICAL CENTER (II,1675); THE NEW HEALERS (I,3266)

LIPPIN, RENNE
ANNIE FLYNN (II,92); THE BOB NEWHART SHOW (II,342); FREE COUNTRY (II,921)

LIPSCOMB, DENNIS
FARRELL: FOR THE PEOPLE (II,834)

LIPSON, PAUL
OUR FIVE DAUGHTERS (I,3413)

LIPTON, BILL
LITTLE WOMEN (I,2719); ROAD OF LIFE (I,3798)

LIPTON, JAMES
THE GUIDING LIGHT (II,1064); SILVER THEATER (I,4051)

LIPTON, JOHN
THE ARMSTRONG CIRCLE THEATER (I,260)

LIPTON, LYNNE
COMEDY TONIGHT (I,1027); THE DAVID FROST REVUE (I,1176); FRIENDS AND LOVERS (II,930); THE SINGERS (I,4060); STARSTRUCK (II,2446); VAN DYKE AND COMPANY (II,2718)

LIPTON, MICHAEL
AS THE WORLD TURNS (II,110); BUCKSKIN (I,753); SOMERSET (I,4115)

LIPTON, PEGGY
THE JOHN FORSYTHE SHOW (I,2410); THE MOD SQUAD (I,3078); THE RETURN OF THE MOD SQUAD (I,3776)

LIPTON, ROBERT
AS THE WORLD TURNS (II,110)

LIPTON, SUE
THE BENNY HILL SHOW (II,207)

LISHNER, LEON
AMAHL AND THE NIGHT VISITORS (I,148)

LISI, VIRNI
THE BOB HOPE SHOW (I,635)

LISKA, STEPHEN
THE RENEGADES (II,2132); VOYAGERS (II,2730)

LISSEK, LEON
JOURNEY TO THE UNKNOWN (I,2457); SHOGUN (II,2339); THE SULLIVANS (II,2490)

LIST, EUGENE
THE MUSIC OF GERSHWIN (I,3174)

LISTER, CHEZ
JOSHUA'S WORLD (II,1344)

LISTER, MOIRA
THE BOB HOPE SHOW (I,535)

LITEL, JOHN
HOORAY FOR HOLLYWOOD (I,2114); MY HERO (I,3193)

LITHGOW, JOHN
THE COUNTRY GIRL (I,1064); THE OLDEST LIVING GRADUATE (II,1882)

LITONDO, OLIVER
SEARCH FOR THE NILE (I,3954)

LITROFSKY, MITCH
SEARCH FOR TOMORROW (II,2284)

LITTEL, DON
STOP THE MUSIC (I,4232)

LITTLE ANTHONY AND THE IMPERIALS
IT'S WHAT'S HAPPENING, BABY! (I,2278)

LITTLE DION
...AND BEAUTIFUL (I,184)

LITTLE EGYPT
GIDEON (I,1792)

LITTLE, CLEAVON
THE DAVID FROST REVUE (I,1176); KOMEDY TONITE (II,1407); MR. DUGAN (II,1749); THE NEW TEMPERATURES RISING SHOW (I,3273); NOW WE'RE COOKIN' (II,1866); TEMPERATURES RISING (I,4390); UPTOWN SATURDAY NIGHT (II,2711)

LITTLE, JIMMY
CAR 54, WHERE ARE YOU? (I,842); DECOY (I,1217); HEE HAW (II,1123); THE PHIL SILVERS SHOW (I,3580); THE RED BUTTONS SHOW (I,3750)

LITTLE, JOYCE
PRIVATE BENJAMIN (II,2087)

LITTLE, RICH
BERT CONVY SPECIAL—THERE'S A MEETING HERE TONIGHT (II,210); THE DEAN MARTIN CELEBRITY ROAST (II,637); THE DEAN MARTIN CELEBRITY ROAST (II,638); THE DEAN MARTIN CELEBRITY ROAST (II,639); DEAN MARTIN'S CELEBRITY ROAST (II,643); DORIS DAY TODAY (II,716); THE FLYING NUN (I,1611); THE FUNNIEST JOKE I EVER HEARD (II,939); THE GORDON MACRAE SHOW (I,1860); THE JOHN DAVIDSON SHOW (I,2409); THE JULIE ANDREWS HOUR (I,2480); THE KOPYCATS (I,2580); LOVE ON A ROOFTOP (I,2775); THE MANY FACES OF COMEDY (I,2894); MERRY CHRISTMAS.

. . WITH LOVE, JULIE (II,1687); MITZI AND A HUNDRED GUYS (II,1710); NUTS AND BOLTS (II,1871); ON LOCATION WITH RICH LITTLE (II,1891); THE PERRY COMO CHRISTMAS SHOW (II,2002); PERRY COMO IN LAS VEGAS (II,2003); PIONEER SPIRIT (I,3604); THE RICH LITTLE SHOW (II,2153); THE RICH LITTLE SHOW (II,2154); THE RICH LITTLE SPECIAL (II,2155); THE RICH LITTLE SPECIAL (II,2156); RICH LITTLE'S A CHRISTMAS CAROL (II,2157); RICH LITTLE'S ROBIN HOOD (II,2158); RICH LITTLE'S WASHINGTON FOLLIES (II,2159); RICH LITTLE—COME LAUGH WITH ME (II,2160); SINATRA—THE FIRST 40 YEARS (II,2359); TAKE ONE STARRING JONATHAN WINTERS (II,2532); YOU ASKED FOR IT (II,2855); YOUR ALL AMERICAN COLLEGE SHOW (I,4946)

LITTLEFIELD, LUCIEN
THE ABBOTT AND COSTELLO SHOW (I,2); BLONDIE (I,486); CIRCUS BOY (I,959); CROWN THEATER WITH GLORIA SWANSON (I,1107); THE FABULOUS SYCAMORES (I,1505)

LITTLEFIELD, RALPH
THE LONE RANGER (I,2740)

LITTLER, CRAIG
FOUR EYES (II,912); GENERAL HOSPITAL (II,964); JASON OF STAR COMMAND (II,1274)

LITTLETON, TWYLA
BILLY CRYSTAL: A COMIC'S LINE (II,249)

LITTOMY, LESLIE
THE ETHEL WATERS SHOW (I,1468)

LITVALE, ROBIN
THE RETURN OF MARCUS WELBY, M.D (II,2138)

LITZSINGER, SARA
PERRY COMO'S CHRISTMAS IN NEW YORK (II,2011)

LIU, FRANK MICHAEL
THE CHADWICK FAMILY (II,478); CODE NAME: DIAMOND HEAD (II,549)

LIVELY SET, THE
KRAFT SUMMER MUSIC HALL (I,2590)

LIVELY, ROBYN
KNIGHT RIDER (II,1402); PUNKY BREWSTER (II,2092)

LIVINGSTON, BARRY
20TH CENTURY FOLLIES (I,4641); ALVIN AND THE CHIPMUNKS (II,61); THE DANNY KAYE SHOW (I,1144); HIGH SCHOOL, U.S.A. (II,1148); MY THREE SONS (I,3205); SENIOR YEAR (II,2301); SONS AND DAUGHTERS (II,2404); THANKSGIVING REUNION WITH THE PARTRIDGE FAMILY AND MY THREE SONS (II,2569); THE TWO OF US (I,4658); YOU'RE A GOOD MAN, CHARLIE BROWN (I,4962)

LIVINGSTON, DONALD
THE BILL COSBY SHOW (I,442)

LIVINGSTON, MARY
THE JACK BENNY PROGRAM (I,2294); JACK BENNY'S 20TH ANNIVERSARY TV SPECIAL (I,2303)

LIVINGSTON, MICHELLE
TOMA (II,2634); TOMA (I,4547)

LIVINGSTON, STANLEY
MY THREE SONS (I,3205); THE ROMAN HOLIDAYS (I,3835); SARGE: THE BADGE OR THE CROSS (I,3915); THANKSGIVING REUNION WITH THE PARTRIDGE FAMILY AND MY THREE SONS (II,2569)

LIVOTI, JOE
THE LAWRENCE WELK SHOW (I,2643); MEMORIES WITH LAWRENCE WELK (II,1681)

LLEWELLYN, DOUG
THE PEOPLE'S COURT (II,2001)

LLOYD, BOB
THE NEW SUPER FRIENDS HOUR (II,1830)

LLOYD, CHRISTOPHER
BEST OF THE WEST (II,218); LACY AND THE MISSISSIPPI QUEEN (II,1419); OLD FRIENDS (II,1881); STUNT SEVEN (II,2486); TAXI (II,2546)

LLOYD, DAVID
GOLDEN CHILD (I,1839)

LLOYD, GEORGE
MARIE AND GEORGE (I,2910)

LLOYD, HAROLD
HAROLD LLOYD'S WORLD OF COMEDY (II,1093)

LLOYD, JEREMY
THE EVE ARDEN SHOW (I,1471); ROWAN AND MARTIN'S LAUGH-IN (I,3856)

LLOYD, JOSIE
THE LONG HOT SUMMER (I,2747)

LLOYD, KATHLEEN
CALL TO GLORY (II,413); THE GANGSTER CHRONICLES (II,957); LACY AND THE MISSISSIPPI QUEEN (II,1419); MAGNUM, P.I. (II,1601); OWEN MARSHALL: COUNSELOR AT LAW (I,3436); SHOOTING STARS (II,2341)

LLOYD, NORMAN
SLEZAK AND SON (I,4086); ST. ELSEWHERE (II,2432)

LLOYD, RITA
THE EDGE OF NIGHT (II,760); THE GUIDING LIGHT (II,1064); A TIME FOR US (I,4507)

LLOYD, SIMON
GEORGE AND MILDRED (II,965)

LLOYD, SUE
THE BARON (I,360); JOURNEY TO THE UNKNOWN (I,2457); THE TWO RONNIES (II,2694)

LOBIANCO, TONY
CIRCUS OF THE STARS (II,531); CIRCUS OF THE STARS (II,532); HIDDEN FACES (I,2044); JESSIE (II,1287); MARCO POLO (II,1626); MR. INSIDE/MR. OUTSIDE (II,1750); PALS (II,1947); A SHADOW IN THE STREETS (II,2312)

LOBBIN, PEGGY
CRIME WITH FATHER (I,1100)

LOCANE, AMY
SPENCER (II,2420)

LOCATELL, CAROL
M*A*S*H (II,1569)

LOCKARD, THOMAS
ARTHUR GODFREY AND FRIENDS (I,278)

LOCKE, JON
LAND OF THE LOST (II,1425); ROUGHNECKS (II,2219)

LOCKE, KATHERINE
THE UNEXPECTED (I,4674)

LOCKE, NANCY
FAMILY (II,813)

LOCKE, NED
THE ADVENTURES OF CAPTAIN HARTZ (I,53); A TIME TO LIVE (I,4510)

LOCKE, RALPH
ONE MAN'S FAMILY (I,3383)

LOCKE, ROSANNA
THE BEST OF TIMES (II,220); THE LAST CONVERTIBLE (II,1435)

LOCKE, SONDRA
ABC'S MATINEE TODAY (I,7)

LOCKE, TAMMY
THE MONROES (I,3089)

LOCKERBIE, BETH
SCARLETT HILL (I,3933)

LOCKERMAN, BRADLEY
CAPITOL (II,426)

LOCKHARD, TOM
RED GOOSE KID'S SPECTACULAR (I,3751)

LOCKHART, ANNE
ALL COMMERCIALS—A STEVE MARTIN SPECIAL (II,37); B.J. AND THE BEAR (II,125); BATTLESTAR GALACTICA (II,181); GIDGET'S SUMMER REUNION (I,1799); LISA, BRIGHT AND DARK (I,2711); THE MAGICIAN (I,2824); MAGNUM, P.I. (II,1601); SCENE OF THE CRIME (II,2271)

LOCKHART, GENE
BACKBONE OF AMERICA (I,328); THE BLUE RIBBON CHRISTMAS EVE MUSICAL (I,493); THE ETHEL BARRYMORE THEATER (I,1467); GULF PLAYHOUSE (I,1904); HIS HONOR, HOMER BELL (I,2063); JANE WYMAN PRESENTS THE FIRESIDE THEATER (I,2345); THE LUX VIDEO THEATER (I,2795); PRUDENTIAL FAMILY PLAYHOUSE (I,3683); TALES OF TOMORROW (I,4352)

LOCKHART, JUNE
ABC'S MATINEE TODAY (I,7); THE ALCOA HOUR (I,108); THE BAIT (I,335); CLIMAX! (I,976); THE ELGIN HOUR (I,1428); FORD TELEVISION THEATER (I,1634); FORD THEATER HOUR (I,1635); THE GENERAL ELECTRIC THEATER (I,1753); THE GREATEST AMERICAN HERO (II,1060); HOME (I,2099); THE KAISER ALUMINUM HOUR (I,2503); KNOTS LANDING (II,1404); LIGHT'S OUT (I,2699); LITTLE WOMEN (I,2719); LOOSE CHANGE (II,1518); LOST IN SPACE (I,2758); THE MARCH OF DIMES FASHION SHOW (I,2901); ON TRIAL (I,3363); PEKING ENCOUNTER

(II,1991); PETTICOAT JUNCTION (I,3571); PRUDENTIAL FAMILY PLAYHOUSE (I,3683); ROBERT MONTGOMERY PRESENTS YOUR LUCKY STRIKE THEATER (I,3809); SHIRLEY TEMPLE'S STORYBOOK (I,4017); STAR TONIGHT (I,4199); STUDIO ONE (I,4268); THESE ARE THE DAYS (II,2585); TIMMY AND LASSIE (I,4520); THE UNITED STATES STEEL HOUR (I,4677)

LOCKIN, DANNY
DEAN MARTIN PRESENTS THE GOLDDIGGERS (I,1194)

LOCKLEAR, HEATHER
BATTLE OF THE NETWORK STARS (II,175); BATTLE OF THE NETWORK STARS (II,176); BATTLE OF THE NETWORK STARS (II,177); BATTLE OF THE NETWORK STARS (II,178); DYNASTY (II,746); THE RETURN OF THE BEVERLY HILLBILLIES (II,2139); T.J. HOOKER (II,2524)

LOCKMILLER, RICHARD
THE ALIENS ARE COMING (II,35)

LOCKRIDGE, JACKIE
THE SATURDAY NIGHT REVUE (I,3921)

LOCKS, LAURI
THE $128,000 QUESTION (II,1903)

LOCKWOOD, ALEXANDER
LITTLE WOMEN (I,2721)

LOCKWOOD, GARY
CHALK ONE UP FOR JOHNNY (I,893); FOLLOW THE SUN (I,1617); THE GIRL, THE GOLD WATCH AND DYNAMITE (II,1001); THE LIEUTENANT (I,2680); MANHUNTER (II,1620); SALLY AND SAM (I,3891); TOP OF THE HILL (II,2645)

LOCKWOOD, VERA
CALUCCI'S DEPARTMENT (I,805); LOVE OF LIFE (I,2774); RYAN'S HOPE (II,2234)

LODEN, BARBARA
THE GLASS MENAGERIE (I,1819)

LODER, KATHRYN
A CRY OF ANGELS (I,1111)

LODGE, DAVID
CARRY ON LAUGHING (II,448); KILLER WITH TWO FACES (I,2539)

LODGE, WILLIAM
THE SECRET WAR OF JACKIE'S GIRLS (II,2294)

LOEB, PHILIP
THE ETHEL WATERS SHOW (I,1468); THE GOLDBERGS (I,1835)

LOFTHOUSE, PETE
HURDY GURDY (I,2163)

LOFTIN, CAREY
THE TROUBLESHOOTERS (I,4614)

LOG JAMMERS, THE
STRAWHAT THEATER (I,4256)

LOGAN, AL
TEEN TIME TUNES (I,4374)

LOGAN, BARBARA
THE JOHNNY DUGAN SHOW (I,2426)

LOGAN, BRAD
THE RED SKELTON SHOW (I,3756)

LOGAN, ELLA
MERV GRIFFIN'S ST. PATRICK'S DAY SPECIAL (I,3018); NIGHT CLUB (I,3284)

LOGAN, JOSHUA
MUSICAL COMEDY TONIGHT (II,1777)

LOGAN, MARTHA
THE LANNY ROSS SHOW (I,2612); TEX AND JINX (I,4406)

LOGAN, MICHAEL
THE WINDS OF WAR (II,2807)

LOGAN, PAUL
A WORLD APART (I,4910)

LOGAN, ROBERT
77 SUNSET STRIP (I,3988); DEATH RAY 2000 (II,654); FIRST & TEN (II,861)

LOGAN, TERRENCE
THE CLIFF DWELLERS (I,974); A TIME FOR US (I,4507)

LOGGIA, ROBERT
ELFEGO BACA (I,1427); EMERALD POINT, N.A.S. (II,773); KRAFT SUSPENSE THEATER (I,2591); MALLORY: CIRCUMSTANTIAL EVIDENCE (II,1608); THE MONEYCHANGERS (II,1724); ONE STEP BEYOND (I,3388); THE SECRET STORM (I,3969); STUDIO ONE (I,4268); T.H.E. CAT (I,4430); TONI'S BOYS (II,2637); A WOMAN CALLED GOLDA (II,2818)

LOGGINS, KENNY
FRIDAYS (II,926); KENNY LOGGINS IN CONCERT (II,1380)

LOGUE, SPAIN
SWEENEY TODD (II,2512)

LOHMAN, AL
BEDTIME STORIES (II,199); THE NAMEDROPPERS (I,3219)

LOHMAN, RICK
PHYL AND MIKHY (II,2037); QUICK AND QUIET (II,2099)

LOLLOBRIGIDA, GINA
THE BOB HOPE SHOW (I,566); THE ENGELBERT HUMPERDINCK SHOW (I,1445); FALCON CREST (II,808)

LOM, HERBERT
THE ERROL FLYNN THEATER (I,1458); THE HUMAN JUNGLE (I,2159); LACE (II,1418); MISTER JERICO (I,3070)

LOMAN, HAL
FRONT ROW CENTER (I,1693)

LOMBARD, MICHAEL
FILTHY RICH (II,856); LAND OF HOPE (II,1424); THE MARY TYLER MOORE COMEDY HOUR (II,1639)

LOMBARDO, CARMEN
GUY LOMBARDO AND HIS ROYAL CANADIANS (I,1910); GUY LOMBARDO'S DIAMOND JUBILLE (I,1911)

LOMBARDO, GUY
GUY LOMBARDO AND HIS ROYAL CANADIANS (I,1910); GUY LOMBARDO'S DIAMOND JUBILLE (I,1911); THE SWINGIN' YEARS (I,4321)

LOMBARDO, JOHN
DAYS OF OUR LIVES (II,629)

LOMBARDO, LEBERT
GUY LOMBARDO AND HIS ROYAL CANADIANS (I,1910); GUY LOMBARDO'S DIAMOND JUBILLE (I,1911)

LOMBARDO, VICTOR
GUY LOMBARDO AND HIS ROYAL CANADIANS (I,1910); GUY LOMBARDO'S DIAMOND JUBILLE (I,1911)

LOMOND, BRITT
ZORRO (I,4982)

LON, ALICE
THE LAWRENCE WELK SHOW (I,2643); LAWRENCE WELK'S TOP TUNES AND NEW TALENT (I,2645); MEMORIES WITH LAWRENCE WELK (II,1681)

LONDON LINE DANCERS, THE
SHOWTIME (I,4036)

LONDON, DAMIAN
THE CAPTAIN AND TENNILLE (II,429)

LONDON, DIRK
THE LIFE AND LEGEND OF WYATT EARP (I,2681); STRATEGIC AIR COMMAND (I,4252)

LONDON, JULIE
THE BARBARA STANWYCK THEATER (I,350); THE BOB HOPE SHOW (I,551); THE BOB HOPE SHOW (I,569); THE BOB HOPE SHOW (I,582); THE BOB HOPE SHOW (I,584); THE BOB HOPE SHOW (I,599); THE DAVID NIVEN THEATER (I,1178); EMERGENCY! (II,775); FRANCES LANGFORD PRESENTS (I,1656); THE GERSHWIN YEARS (I,1779); THE JULIE LONDON SPECIAL (I,2484); MAGGIE MALONE (I,2812); SOMETHING SPECIAL (I,4118); YUMMY, YUMMY, YUMMY (I,4977); ZANE GREY THEATER (I,4979)

LONDON, MARC
PLIMPTON! DID YOU HEAR THE ONE ABOUT. . .? (I,3629)

LONDON, MICHAEL J
TERROR AT ALCATRAZ (II,2563)

LONDON, STEVE
THE UNTOUCHABLES (I,4682)

LONG JR, WILLIAM
THE YOUNG AND THE RESTLESS (II,2862)

LONG, AVON
HOTEL BROADWAY (I,2127); ROOTS: THE NEXT GENERATIONS (II,2213)

LONG, BEVERLY
THOSE WHITING GIRLS (I,4471)

LONG, DOREEN
PINE CANYON IS BURNING (II,2040)

LONG, ELLEN
PEE WEE KING SHOW (I,3504); PEE WEE KING'S FLYING RANCH (I,3505)

LONG, JAMES
THE GUIDING LIGHT (II,1064)

LONG, LIONEL
HOMICIDE (I,2101)

LONG, LORETTA
SESAME STREET (I,3982)

LONG, PAM
MAGIC WITH THE STARS (II,1599)

LONG, PAMELA
TEXAS (II,2566)

LONG, RICHARD
77 SUNSET STRIP (I,3988); THE BIG VALLEY (I,437); BOURBON STREET BEAT (I,699); MAVERICK (II,1657); NANNY AND THE PROFESSOR (I,3225); NANNY AND THE PROFESSOR AND THE PHANTOM OF THE CIRCUS (I,3226); THICKER THAN WATER (I,4445); TV READERS DIGEST (I,4630)

LONG, RONALD
ALICE IN WONDERLAND (I,120); THE BARBARA MCNAIR SHOW (I,348); LOVE OF LIFE (I,2774); TAMING OF THE SHREW (I,4356); WONDER WOMAN (PILOT 1) (II,2826)

LONG, SHELLEY
CHEERS (II,494); THE DOOLEY BROTHERS (II,715); GHOST OF A CHANCE (II,987); M*A*S*H (II,1569); THAT SECOND THING ON ABC (II,2572); THAT THING ON ABC (II,2574); YOUNG GUY CHRISTIAN (II,2864)

LONG, STACEY
THE DOCTORS (II,694)

LONG, WAYNE
OPERATION PETTICOAT (II,1916); OPERATION PETTICOAT (II,1917)

LONGAKER, RACHEL
THE WALTONS (II,2740)

LONGDEN, JOHN
THE MAN FROM INTERPOL (I,2866)

LONGDEN, ROBERT
THE SEVEN DIALS MYSTERY (II,2308); WHY DIDN'T THEY ASK EVANS? (II,2795)

LONGDON, TERENCE
HORNBLOWER (I,2121)

LONGERGAN, LENORE
HOLIDAY HOTEL (I,2075)

LONGET, CLAUDINE
THE ANDY WILLIAMS CHRISTMAS SHOW (I,196); THE ANDY WILLIAMS CHRISTMAS SHOW (I,197); THE ANDY WILLIAMS CHRISTMAS SHOW (I,198); THE ANDY WILLIAMS CHRISTMAS SHOW (I,199); THE ANDY WILLIAMS CHRISTMAS SHOW (II,77); ANDY WILLIAMS MAGIC LANTERN SHOW COMPANY (I,202); KRAFT SUSPENSE THEATER (I,2591); MONSANTO PRESENTS MANCINI (I,3093); TENNESSEE ERNIE FORD'S WHITE CHRISTMAS (I,4402)

LONGET, DANIELLE
THE ANDY WILLIAMS CHRISTMAS SHOW (II,77)

LONGFELD, MICHAEL
TEXAS (II,2566)

LONGHURST, JEREMY
SPELL OF EVIL (I,4152)

LONGO, STELLA
MARLO THOMAS IN ACTS OF LOVE—AND OTHER COMEDIES (I,2919)

LONGO, TONY
THE BEST OF TIMES (II,220); FORCE SEVEN (II,908); HERNDON AND ME (II,1139)

LONOW, CLAUDIA
KNOTS LANDING (II,1404)

LONOW, MARK
HUSBANDS AND WIVES (II,1208); HUSBANDS, WIVES AND LOVERS (II,1209)

LONTOC, LEON
BURKE'S LAW (I,762)

LOO, RICHARD
MARCUS WELBY, M.D. (I,2905)

LOOKABILL, LAGENA
CHIEFS (II,509)

LOOKINLAND, MICHAEL
THE BRADY BUNCH (II,362); THE BRADY BUNCH HOUR (II,363); THE BRADY GIRLS GET MARRIED (II,364); THE BRADY KIDS (II,365); TEACHER'S PET (I,4367)

LOOKINLAND, TODD
BEULAH LAND (II,226); EVERY STRAY DOG AND KID (II,792); HOW THE WEST WAS WON (II,1196); KELLY'S KIDS (I,2523); THE NEW LAND (II,1820)

LOOMIS, CHRISTOPHER
SEARCH FOR TOMORROW (II,2284)

LOOMIS, DAYTON
JEFF'S COLLIE (I,2356)

LOOS, ANITA
THE MARCH OF DIMES FASHION SHOW (I,2902)

LOPEZ, J VICTOR
THE MAN FROM ATLANTIS (II,1615)

LOPEZ, JOHN PAUL
MASTER OF THE GAME (II,1647)

LOPEZ, MARCO
A.K.A. PABLO (II,28); EMERGENCY! (II,775)

LOPEZ, PERRY
THE CENTURY TURNS (I,890); DOC ELLIOT (I,1305)

LOPEZ, PRISCILLA
FEELING GOOD (I,1559); IN THE BEGINNING (II,1229); A YEAR AT THE TOP (II,2844)

LOPEZ, TRINI
THE BOB HOPE SHOW (I,602); CELEBRATION: THE AMERICAN SPIRIT (II,461); THE POPPY IS ALSO A FLOWER (I,3650); THE TRINI LOPEZ SHOW (I,4606)

LOPEZ, VINCENT
DINNER DATE (I,1286); DINNER DATE WITH VINCENT LOPEZ (I,1287); THE VINCENT LOPEZ SHOW (I,4735)

LOPINTO, DORIAN
THE DOCTORS (II,694); ONE LIFE TO LIVE (II,1907)

LOPTON, ROBERT
THE SURVIVORS (I,4301)

LOR, DENISE
THE ALAN KING SHOW (I,100); THE BIG PAYOFF (I,428); THE GARRY MOORE SHOW (I,1738); THE GARRY MOORE SHOW (I,1739); SEVEN AT ELEVEN (I,3984)

LORCA, NANA
FRANKIE AVALON'S EASTER SPECIAL (I,1673)

LORD ROCKINGHAM'S XI
OH, BOY! (I,3341)

LORD, BASIL
SLEEPWALKER (I,4084)

LORD, BOBBY
THE BOBBY LORD SHOW (I,675)

LORD, DOROTHEA
GETAWAY CAR (I,1783)

LORD, JACK
APPOINTMENT WITH ADVENTURE (I,246); BORDER TOWN (I,689); CONFLICT (I,1036); THE ELGIN HOUR (I,1428); HAWAII FIVE-O (I,1972); HAWAII FIVE-O (II,1110); KINCAID (I,2543); M STATION: HAWAII (II,1568); ONE STEP BEYOND (I,3388); STONEY BURKE (I,4229); TACK REYNOLDS (I,4329); WEAPONS MAN (I,4783)

LORD, JOAN
A TOWN LIKE ALICE (II,2652)

LORD, JUSTIN
THE 25TH MAN (MS) (II,2678); I'D RATHER BE CALM (II,1216); LOOK OUT WORLD (II,1516); SIDE BY SIDE (II,2346); SKEEZER (II,2376)

LORD, MARJORIE
THE ANDY GRIFFITH SHOW (I,193); CAVALCADE OF AMERICA (I,875); CROWN THEATER WITH GLORIA SWANSON (I,1107); DANNY THOMAS LOOKS AT YESTERDAY, TODAY AND TOMORROW (I,1148); THE DANNY THOMAS TV FAMILY REUNION (I,1154); EVERYTHING HAPPENS TO ME (I,1481); FIRESIDE THEATER (I,1580); FORD TELEVISION THEATER (I,1634); FOUR STAR PLAYHOUSE (I,1652); THE LORETTA YOUNG THEATER (I,2756); MAKE MORE ROOM FOR DADDY (I,2842); MAKE ROOM FOR DADDY (I,2843); MAKE ROOM FOR GRANDDADDY (I,2844); MAKE ROOM FOR GRANDDADDY (I,2845); SCHLITZ PLAYHOUSE OF STARS (I,3936); STORY THEATER (I,4240); YOUR SHOW TIME (I,4958); ZANE GREY THEATER (I,4979)

LORD, PHILIP
THE ADVENTURES OF CAPTAIN HARTZ (I,53); HAWKINS FALLS, POPULATION 6200 (I,1977)

LORD, PHILLIPS H
GANGBUSTERS (I,1733)

LORDE, ATHENA
OUR FIVE DAUGHTERS (I,3413)

LOREN, DONNA
THE MILTON BERLE SHOW (I,3049); SHINDIG (I,4013)

LOREN, SOPHIA
BRIEF ENCOUNTER (I,724); SOPHIA LOREN IN ROME (I,4129); SOPHIA! (I,4128); WITH LOVE, SOPHIA (I,4884); THE WORLD OF SOPHIA LOREN (I,4918)

LORENZ, MATT
WAR CORRESPONDENT (I,4767)

LORI REGAS DANCERS, THE

THE JOHN BYNER COMEDY HOUR (I,2407)

LORIMER, ENID
THE HANDS OF CORMAC JOYCE (I,1927)

LORIMER, LOUISE
PROFILES IN COURAGE (I,3678)

LORING, ANN
LOVE OF LIFE (I,2774)

LORING, ESTELLE
STOP THE MUSIC (I,4232)

LORING, GLORIA
THE BOB HOPE SHOW (I,638); THE BOB HOPE SHOW (I,657); CELEBRITY DAREDEVILS (II,473); CIRCUS OF THE STARS (II,536); DAYS OF OUR LIVES (II,629); HAPPY BIRTHDAY, AMERICA (II,1083); PLAY IT AGAIN, UNCLE SAM (II,2052)

LORING, LISA
THE ADDAMS FAMILY (II,11); THE ADDAMS FAMILY (II,13); ANNIE FLYNN (II,92); AS THE WORLD TURNS (II,110); GABE AND WALKER (II,950); THE PRUITTS OF SOUTHAMPTON (I,3684)

LORING, LYNN
BABY CRAZY (I,319); DOBIE GILLIS (I,1302); FAIR EXCHANGE (I,1512); THE FBI (I,1551); GHOST STORY (I,1788); THE GREATEST SHOW ON EARTH (I,1883); THE JEAN CARROLL SHOW (I,2353); KRAFT TELEVISION THEATER (I,2592); RETURN TO PEYTON PLACE (I,3779); ROBERT MONTGOMERY PRESENTS YOUR LUCKY STRIKE THEATER (I,3809); SEARCH FOR TOMORROW (II,2284)

LORMER, JON
CONSPIRACY OF TERROR (II,569); GALLAGHER (I,1727); THE GOLDBERGS (I,1836); THE GUN AND THE PULPIT (II,1067); THE LIFE AND TIMES OF EDDIE ROBERTS (II,1468); LOOSE CHANGE (II,1518); MISS SUSAN (I,3065)

LORNE, MARION
BEWITCHED (I,418); THE GARRY MOORE SHOW (I,1739); HARVEY (I,1961); MR. PEEPERS (I,3147); SALLY (I,3889); SALLY (I,3890)

LOROS, GEORGE
THE NEW MAVERICK (II,1822); SUPERCOPS (II,2502)

LORRE, PETER
ALFRED HITCHCOCK PRESENTS (I,115); ARSENIC AND OLD LACE (I,268); CLIMAX! (I,976); THE DUPONT SHOW OF THE WEEK (I,1388); HENRY FONDA PRESENTS THE STAR AND THE STORY (I,2016); KRAFT SUSPENSE THEATER (I,2591); PLAYHOUSE 90 (I,3623); REUNION IN VIENNA (I,3780); RHEINGOLD THEATER (I,3783); SCREEN DIRECTOR'S PLAYHOUSE (I,3946); THE UNITED STATES STEEL HOUR (I,4677)

LORRING, JOAN
THE CORN IS GREEN (I,1050); THE DOCTOR (I,1320); LOVE STORY (I,2776); NORBY (I,3304)

LORY, ROBIN
MRS. G. GOES TO COLLEGE (I,3155)

LOTE, ELISSA
OUR FIVE DAUGHTERS (I,3413)

LOTTIMER, ED
NIGHT PARTNERS (II,1847)

LOU REGAS DANCERS, THE
THE JERRY REED WHEN YOU'RE HOT, YOU'RE HOT HOUR (I,2372)

LOU, BONNIE
MIDWESTERN HAYRIDE (I,3033); THE PAUL DIXON SHOW (II,1974)

LOUANNE
THE LONG DAYS OF SUMMER (II,1514); MISSING PIECES (II,1706); TWO MARRIAGES (II,2691)

LOUDEN, JAY
CELEBRITY (II,463)

LOUDEN, VERN
CRUSADER RABBIT (I,1109)

LOUDON, DOROTHY
DOROTHY (II,717); THE GARRY MOORE SHOW (I,1739); IT'S A BUSINESS (I,2254); MA AND PA (II,1570); PARADE OF STARS (II,1954)

LOUGHERY, JACKIE
EARN YOUR VACATION (I,1394); THE GEORGE BURNS AND GRACIE ALLEN SHOW (I,1763); JUDGE ROY BEAN (I,2467); MR. DISTRICT ATTORNEY (I,3136)

LOUGHLIN, LORI
THE EDGE OF NIGHT (II,760); THE TOM SWIFT AND LINDA CRAIG MYSTERY HOUR (II,2633)

LOUIE, BEBE
FROM HERE TO ETERNITY (II,932); FROM HERE TO ETERNITY (II,933)

LOUIE, JOHN
EIGHT IS ENOUGH (II,762)

LOUIS DAPRON DANCERS, THE
BEN VEREEN. . . COMIN' AT YA (II,203); THE MILTON BERLE SHOW (I,3049); PERRY COMO PRESENTS THE JULIUS LAROSA SHOW (I,3530); THE PERRY COMO SHOW (I,3532); PERRY PRESENTS (I,3549); THE SMOTHERS BROTHERS COMEDY HOUR (I,4095)

LOUIS, GLORIA
WAY OF THE WORLD (I,4776)

LOUIS-DREYFUS, JULIA
SATURDAY NIGHT LIVE (II,2261)

LOUISE, ANITA
THE ETHEL BARRYMORE THEATER (I,1467); FIRESIDE THEATER (I,1580); FORD TELEVISION THEATER (I,1634); MY FRIEND FLICKA (I,3190); SPOTLIGHT PLAYHOUSE (I,4167); STARS OVER HOLLYWOOD (I,4211); THEATER TIME (I,4434)

LOUISE, HELLI
THE BENNY HILL SHOW (II,207)

LOUISE, LAURA
YOU ASKED FOR IT (II,2855)

LOUISE, MARIE
HOLLYWOOD JUNIOR CIRCUS (I,2084)

LOUISE, TINA
ADVICE TO THE LOVELORN (II,16); THE BOB HOPE SHOW (I,632); CALL TO DANGER (I,804); DALLAS (II,614); THE DAY THE WOMEN GOT EVEN (II,628); DON'T CALL US (II,709); GILLIGAN'S ISLAND (II,990); HAPPY BIRTHDAY (I,1941); THE JAN MURRAY SHOW (I,2335); KRAFT SUSPENSE THEATER (I,2591); RITUALS (II,2182); A SALUTE TO STAN LAUREL (I,3892)

LOUND, TOM
CAP'N AHAB (I,824)

LOUSIMANO, BUDDY
TOO GOOD TO BE TRUE (II,2643)

LOUVIN, CHARLIE
JOHNNY CASH: THE FIRST 25 YEARS (II,1336)

LOVATO, BETO
A.K.A. PABLO (II,28)

LOVE, KELAND
THE BRADY BRIDES (II,361); PRYOR'S PLACE (II,2089)

LOVE, MIKE
THE BEACH BOYS 20TH ANNIVERSARY (II,188); THE BEACH BOYS IN CONCERT (II,189); SCROOGE'S ROCK 'N' ROLL CHRISTMAS (II,2279)

LOVE, PHYLLIS
COSMOPOLITAN THEATER (I,1057); HIDE AND SEEK (I,2047); PROFILES IN COURAGE (I,3678); THERE SHALL BE NO NIGHT (I,4437)

LOVE, REGINALD
TALES OF THE HAUNTED (II,2538)

LOVEDAY, JOHN
THE MUPPET SHOW (II,1766)

LOVEJOY, FRANK
CHICAGO 212 (I,935); COUNTY GENERAL (I,1076); THE DAVID NIVEN THEATER (I,1178); FORD TELEVISION THEATER (I,1634); HENRY FONDA PRESENTS THE STAR AND THE STORY (I,2016); THE JUNE ALLYSON SHOW (I,2488); THE LORETTA YOUNG THEATER (I,2756); THE MAN AGAINST CRIME (I,2853); MEET MCGRAW (I,2984); MEET MCGRAW (I,2985); RHEINGOLD THEATER (I,3783); STAGE 7 (I,4181)

LOVELADY, JOHN
THE GREAT SPACE COASTER (II,1059)

LOVELL, ANDREA HALL
DAYS OF OUR LIVES (II,629)

LOVELL, DYSON
VANITY FAIR (I,4701)

LOVELL, MARILYN
THE DESILU REVUE (I,1238); THE FARMER'S DAUGHTER (I,1533); HEAD OF THE CLASS (I,1987); THE LIBERACE SHOW (I,2676); THE NUT HOUSE (I,3320)

LOVELL, NIGEL
HUNTER (I,2161)

LOVELL, SUSAN
H.M.S. PINAFORE (I,2066)

LOVELL, SUZANNE
THE FRED WARING SHOW (I,1679)

LOVETT, DOROTHY

THE SEEKING HEART (I,3975)

LOVETT, MARJORIE
THE HAMPTONS (II,1076)

LOVETT, MONICA
FAIRMEADOWS, U.S.A. (I,1514); GOLDEN WINDOWS (I,1841); THE KATE SMITH HOUR (I,2508)

LOVIN' BROTHERS, THE
THE GRAND OLE OPRY (I,1866)

LOVIN' SPOONFUL, THE
THE JUD STRUNK SHOW (I,2462)

LOVSKY, CELIA
DIAGNOSIS: DANGER (I,1250)

LOW, BEN
FARAWAY HILL (I,1532)

LOW, CARL
SEARCH FOR TOMORROW (II,2284)

LOW, CLEM
FROM HERE TO ETERNITY (II,932)

LOW, JAMES
DEBBY BOONE. . . ONE STEP CLOSER (II,655)

LOWE, CHAD
SPENCER (II,2420)

LOWE, CHRIS
SEARCH FOR TOMORROW (II,2284)

LOWE, DOROTHY
THE RED SKELTON SHOW (I,3755)

LOWE, EDMUND
CONFLICT (I,1036); FRONT PAGE DETECTIVE (I,1689)

LOWE, FREDERICK
STRANGER IN OUR HOUSE (II,2477)

LOWE, JIM
SING ALONG (I,4055)

LOWE, LORI
THE ASPHALT COWBOY (II,111); HOW TO SURVIVE A MARRIAGE (II,1198)

LOWE, PHYLLIS
FAMILY GENIUS (I,1521)

LOWE, ROB
A NEW KIND OF FAMILY (II,1819)

LOWE, STANJA
THE LIFE OF RILEY (I,2686)

LOWELL, MUNDELL
THE MATT DENNIS SHOW (I,2948)

LOWELL, TOM
COMBAT (I,1011); KILROY (I,2541)

LOWENS, CURT
THE CABOT CONNECTION (II,408); THE NIGHT RIDER (II,1848); THE SECRET WAR OF JACKIE'S GIRLS (II,2294); V (II,2713)

LOWERY, MARCELLA
HOW TO SUCCEED IN BUSINESS WITHOUT REALLY TRYING (II,1197)

LOWERY, ROBERT
CIRCUS BOY (I,959); PISTOLS 'N' PETTICOATS (I,3608)

LOWLER, JOHN H
ELLERY QUEEN (II,766)

LOWLES, PEGGY
THE DOCTORS (I,1323)

LOWRY, JOE
BACKSTAIRS AT THE WHITE HOUSE (II,133); KENNEDY (II,1377)

LOWRY, JUDITH
CALUCCI'S DEPARTMENT (I,805); PHYLLIS (II,2038)

LOWRY, LYNN
HOW TO SURVIVE A MARRIAGE (II,1198)

LOWRY, ROBERT
THE GUIDING LIGHT (II,1064)

LOWY, OTTO
THE RETURN OF CHARLIE CHAN (II,2136)

LOY, MYRNA
THE JUNE ALLYSON SHOW (I,2488); LEGENDS OF THE SCREEN (II,1456); MEET ME IN ST. LOUIS (I,2986); SCHLITZ PLAYHOUSE OF STARS (I,3936); SUMMER SOLSTICE (II,2492)

LU, LISA
ANNA AND THE KING (I,221); HAVE GUN—WILL TRAVEL (I,1968)

LUCAN, LEE
HOME ROOM (II,1172)

LUCARELLI, JACK
CELEBRITY (II,463)

LUCAS, BETTY
ALL THE RIVERS RUN (II,43); PRISONER: CELL BLOCK H (II,2085)

LUCAS, DIONE
DIONE LUCAS' COOKING SCHOOL (I,1288); THE GOURMET CLUB (I,1861)

LUCAS, ISABELLE
NANCY ASTOR (II,1788)

LUCAS, JONATHAN
ALL IN FUN (I,130); MELODY TOUR (I,3001); THE RAY KNIGHT REVUE (I,3733)

LUCAS, KARL
FAMILY AFFAIR (I,1519)

LUCAS, LEE
THE TROUBLE WITH GRANDPA (II,2662)

LUCAS, LISA
THE ADAMS CHRONICLES (II,8); ADDIE AND THE KING OF HEARTS (I,35); THE EASTER PROMISE (I,1397); THE HOUSE WITHOUT A CHRISTMAS TREE (I,2143); THE THANKSGIVING TREASURE (I,4416)

LUCAS, SHIRLEY
SO THIS IS HOLLYWOOD (I,4105)

LUCAS, WILLIAM
THE ADVENTURES OF BLACK BEAUTY (I,51)

LUCCHESI, VINCENT
HILL STREET BLUES (II,1154)

LUCCI, SUSAN
ALL MY CHILDREN (II,39)

LUCE, CLAIRE
LIGHT'S OUT (I,2699); PHILCO TELEVISION PLAYHOUSE (I,3583)

LUCERO, ENRIQUE
FALCON'S GOLD (II,809)

LUCHSINGER, CHARLES
CARTOON TELETALES (I,857); STOP THE MUSIC (I,4232)

LUCHSINGER, JACK
CARTOON TELETALES (I,857)

LUCIA, CHIP
ONE LIFE TO LIVE (II,1907); PEARL (II,1986)

LUCKEY, SUSAN
ANNIE GET YOUR GUN (I,223)

LUCKHAM, CYRIL
JENNIE: LADY RANDOLPH CHURCHILL (II,1277); THE OMEGA FACTOR (II,1888)

LUCKINBILL, LAURENCE
THE DELPHI BUREAU (I,1224); THE DELPHI BUREAU (I,1225); IKE (II,1223); LIGHT'S OUT (I,2700); MOMMA THE DETECTIVE (II,1721); ONE MORE TRY (II,1908); THE SECRET STORM (I,3969); WHERE THE HEART IS (I,4831)

LUCKING, WILLIAM
THE A-TEAM (II,119); BIG HAWAII (II,238); THE BLUE AND THE GRAY (II,273); COLORADO C.I. (II,555); DANGER IN PARADISE (II,617); FORCE FIVE (II,907); THE FRENCH ATLANTIC AFFAIR (II,924); JESSIE (II,1287); MALLORY: CIRCUMSTANTIAL EVIDENCE (II,1608); THE SAN PEDRO BUMS (II,2251); SHANNON (II,2314)

LUCKY PENNIES, THE
MIDWESTERN HAYRIDE (I,3033)

LUDDEN, ALLEN
ALLEN LUDDEN'S GALLERY (I,143); THE G.E. COLLEGE BOWL (I,1745); IT'S A BIRD, IT'S A PLANE, IT'S SUPERMAN (II,1253); THE LIAR'S CLUB (II,1466); MITZI AND A HUNDRED GUYS (II,1710); PASSWORD (I,3468); PASSWORD (II,1965); PASSWORD PLUS (II,1966); STUMPERS (II,2485); WIN WITH THE STARS (I,4871)

LUDGIN, CHESTER
GOLDEN CHILD (I,1839)

LUDWIG, KAREN
THE DAIN CURSE (II,613)

LUEZ, LAURETTE
THE ADVENTURES OF DR. FU MANCHU (I,58); THE DONALD O'CONNOR SHOW (I,1344)

LUGO, FRANK
POPI (II,2069)

LUISI, JAMES
THE ASPHALT COWBOY (II,111); BEYOND WITCH MOUNTAIN (II,228); THE CABOT CONNECTION (II,408); CAT BALLOU (I,870); DON'T CALL US (II,709); FUTURE COP (II,946); HARRIS AND COMPANY (II,1096); OUR FAMILY BUSINESS (II,1931); THE RENEGADES (II,2132); THE RENEGADES (II,2133); THE ROCKFORD FILES (II,2197)

LUKAS, PAUL
THE KATE SMITH EVENING HOUR (I,2507); THE SLOWEST GUN IN THE WEST (I,4088)

LUKATHER, PAUL
BRIGHT PROMISE (I,727)

LUKE, BENNY
TWILIGHT THEATER (II,2685)

LUKE, JORGE
NEVADA SMITH (II,1807)

LUKE, KEYE
ALVIN AND THE CHIPMUNKS (II,61); THE AMAZING CHAN AND THE CHAN CLAN (I,157); AMERICA PAUSES FOR SPRINGTIME (I,166); ANNA AND THE KING (I,221); BATTLE OF THE PLANETS (II,179); BROTHERS (II,381); COCAINE AND BLUE EYES (II,548); FIRESIDE THEATER (I,1580); FLY AWAY HOME (II,893); HARRY O (II,1099); JUDGE DEE IN THE MONASTERY MURDERS (I,2465); KENTUCKY JONES (I,2526); KUNG FU (II,1416); PANIC! (I,3447); STUDIO '57 (I,4267); TV READERS DIGEST (I,4630); UNIT 4 (II,2707)

LUKENS, LINDA
THE MANY LOVES OF ARTHUR (II,1625)

LUKHAM, CYRIL
THE FORSYTE SAGA (I,1645)

LUKOF, DIDI
MOSES THE LAWGIVER (II,1737)

LUKOYE, PETER
BORN FREE (I,691); BORN FREE (II,353)

LULU
THE TOM JONES SPECIAL (I,4543)

LUM, BEN
THE BOYS IN BLUE (II,359)

LUMB, GEOFFREY
ANOTHER WORLD (II,97); HONESTLY, CELESTE! (I,2104); THE TEMPEST (I,4391); VICTORIA REGINA (I,4728)

LUMBLY, CARL
CAGNEY AND LACEY (II,409); CAGNEY AND LACEY (II,410)

LUMLEY, JOANNA
MISTRAL'S DAUGHTER (II,1708); THE NEW AVENGERS (II,1816)

LUMMIS, DAYTON
LAW OF THE PLAINSMAN (I,2639)

LUMP, FRANK
BROTHER RAT (I,746)

LUNA, BARBARA
ALCOA PREMIERE (I,109); BRENDA STARR (II,372); BUCK ROGERS IN THE 25TH CENTURY (II,384); GENERAL ELECTRIC TRUE (I,1754); HAWAII FIVE-O (II,1110); THE OUTER LIMITS (I,3426); PLEASURE COVE (II,2058); SIDE BY SIDE (II,2347); THEY ONLY COME OUT AT NIGHT (II,2586); WIVES (II,2812)

LUND, ART
CALAMITY JANE (I,793); THE CONTENDER (II,571); THE KEN MURRAY SHOW (I,2525); THE MAN FROM ATLANTIS (II,1614); THE QUEST (II,2096); THE WINDS OF WAR (II,2807)

LUND, DEANNA
I LOVE A MYSTERY (I,2170); LAND OF THE GIANTS (I,2611); STUMP THE STARS (I,4274)

LUND, LEE
BEN VEREEN. . . COMIN' AT YA (II,203)

LUND, TRIG
KOVACS UNLIMITED (I,2583)

LUNDI, SAL
CRAZY TIMES (II,602)

LUNDIGAN, WILLIAM
CLIMAX! (I,976); MEN INTO SPACE (I,3007); RHEINGOLD THEATER (I,3783)

LUNDMARK, BILL
FULL CIRCLE (I,1702)

LUNGHI, CHERIE
ELLIS ISLAND (II,768); MASTER OF THE GAME (II,1647)

LUNGREEN, MARCO
NAKED CITY (I,3216)

LUNT, ALFRED
THE GREAT SEBASTIANS (I,1877); THE MAGNIFICENT YANKEE (I,2829)

LUPINO, IDA
BATMAN (I,366); BOGART (I,679); FOUR STAR PLAYHOUSE (I,1652); THE GENERAL ELECTRIC THEATER (I,1753); I LOVE A MYSTERY (I,2170); THE IDA LUPINO THEATER (I,2185); THE LETTERS (I,2670); MR. ADAMS AND EVE (I,3121); THE TEENAGE IDOL (I,4375)

LUPO, FRANK
THE OUTLAWS (II,1936)

LUPTON, JOHN
ABC'S MATINEE TODAY (I,7); ALFRED HITCHCOCK PRESENTS (I,115); BARE ESSENCE (II,149); BROKEN ARROW (I,743); BROKEN ARROW (I,744); DAYS OF OUR LIVES (II,629); DOCTORS' PRIVATE LIVES (II,693); FOUR STAR PLAYHOUSE (I,1652); ME AND BENJY (I,2973); NEVER TOO YOUNG (I,3248); SIDNEY SHORR (II,2349); STUDIO '57 (I,4267); TROUBLE IN HIGH TIMBER COUNTRY (II,2661)

LUPUS, PETER
MISSION: IMPOSSIBLE (I,3067); POLICE SQUAD! (II,2061)

LURIE, ALAN
PAC-MAN (II,1940); SPACE STARS (II,2408)

LURIE, DAN
THE BIG TOP (I,435)

LURIS, PATRICIA
DELTA HOUSE (II,658)

LUSSIER, ROBERT
HANDLE WITH CARE (II,1077); HOT W.A.C.S. (II,1191); IN SECURITY (II,1227); THE KIDS FROM C.A.P.E.R. (II,1391); THE MAN FROM ATLANTIS (II,1615); NEWMAN'S DRUGSTORE (II,1837)

LUSTGARTEN, EDGAR
SCOTLAND YARD (I,3941)

LUTHER, MICHAEL
YOUNG LIVES (II,2866)

LUTHER, PHIL
SPEED BUGGY (II,2419)

LUTTER, ALFRED
ALICE (II,34)

LUTTRELL, ESTELLE
A TIME TO LIVE (I,4510)

LUTZ, STEPHEN
SKINFLINT (II,2377)

LUXTON, BILL
THE AMAZING WORLD OF KRESKIN (I,163)

LUZIN, VLADIMIR
PEGGY FLEMING VISITS THE SOVIET UNION (I,3509)

LYDEN, ROBERT
ROCKY JONES, SPACE RANGER (I,3823)

LYDER, KAY
OUR FIVE DAUGHTERS (I,3413)

LYDON, JIMMY
ELLERY QUEEN (II,766); ELLERY QUEEN: TOO MANY SUSPECTS (II,767); THE FIRST HUNDRED YEARS (I,1581); THE LIFE OF RILEY (I,2685); LOVE THAT JILL (I,2780); NAVY LOG (I,3233); THE REAL MCCOYS (I,3741); ROLL OUT! (I,3833); SO THIS IS HOLLYWOOD (I,4105)

LYE, PEG
SPELL OF EVIL (I,4152)

LYLES, TRACEE
I'D RATHER BE CALM (II,1216)

LYMAN, DOROTHY
ALL MY CHILDREN (II,39); ANOTHER WORLD (II,97); THE EDGE OF NIGHT (II,760); MAMA'S FAMILY (II,1610)

LYMAN, WILL
ANOTHER WORLD (II,97); GEORGE WASHINGTON (II,978)

LYN, DAWN
DADDY'S GIRL (I,1124); THE FESS PARKER SHOW (II,850); MY THREE SONS (I,3205)

LYN, DIANA
JACK FROST (II,1263); SANTA CLAUS IS COMIN' TO TOWN (II,2259)

LYNCH, EDWARD J
RAGE OF ANGELS (II,2107)

LYNCH, KEN
THE ANDY GRIFFITH SHOW (I,192); THE DORIS DAY SHOW (I,1355); FBI CODE 98 (I,1550); THE GREEN FELT JUNGLE (I,1885); MAN ON THE MOVE (I,2878); MCCLOUD (II,1660); THE PLAINCLOTHESMAN (I,3614); ROOSTER (II,2210); THE WEB (I,4785)

LYNCH, PEG
ETHEL AND ALBERT (I,1466); THE GENERAL MOTORS 50TH ANNIVERSARY SHOW (I,1758); THE KATE SMITH HOUR (I,2508)

LYNCH, RAYMOND
HOW DO I KILL A THIEF—LET ME COUNT THE WAYS (II,1195)

LYNCH, RICHARD
DOG AND CAT (II,695); GALACTICA 1980 (II,953); THE LAST NINJA (II,1438); THE PHOENIX (II,2036); ROGER AND HARRY (II,2200); STARSKY AND HUTCH (II,2445)

LYNCH, SEAN
TEN WHO DARED (II,2559)

LYNDE, JANICE
AFTER HOURS: FROM JANICE, JOHN, MARY AND MICHAEL, WITH LOVE (II,20); ESCAPADE (II,781); NIGHTSIDE (II,1851); ROXY PAGE (II,2222); THE YOUNG AND THE RESTLESS (II,2862)

LYNDE, PAT
THE MANIONS OF AMERICA (II,1623)

LYNDE, PAUL

ARTHUR GODFREY LOVES ANIMALS (I,282); BEWITCHED (I,418); THE BOB HOPE SHOW (I,611); BOB HOPE SPECIAL: BOB HOPE FOR PRESIDENT (II,282); THE CATTANOOGA CATS (I,874); DEAN MARTIN PRESENTS THE GOLDDIGGERS (I,1194); DEAN MARTIN PRESENTS THE GOLDDIGGERS (I,1193); DONNY AND MARIE (II,712); THE DONNY AND MARIE OSMOND SHOW (II,714); THE FLYING NUN (I,1611); GIDGET (I,1795); GIDGET GETS MARRIED (I,1796); GIDGET GROWS UP (I,1797); THE GOOD FAIRY (I,1846); HENRY FONDA AND THE FAMILY (I,2015); THE HOLLYWOOD SQUARES (II,1164); THE JONATHAN WINTERS SHOW (I,2443); KOMEDY TONITE (II,1407); MOTOR MOUSE (I,3111); THE MUNSTERS (I,3158); THE NEW TEMPERATURES RISING SHOW (I,3273); PAUL LYNDE AT THE MOVIES (II,1976); THE PAUL LYNDE COMEDY HOUR (II,1977); THE PAUL LYNDE COMEDY HOUR (II,1978); THE PAUL LYNDE COMEDY HOUR (II,1979); PAUL LYNDE GOES M-A-A-A-AD (II,1980); THE PAUL LYNDE HALLOWEEN SPECIAL (II,1981); THE PAUL LYNDE SHOW (I,3489); THE PERILS OF PENELOPE PITSTOP (I,3525); PERRY COMO'S SUMMER OF '74 (II,2024); THE PRUITTS OF SOUTHAMPTON (I,3684); THE RED BUTTONS SHOW (I,3750); RUGGLES OF RED GAP (I,3867); THE SANDY DUNCAN SHOW (II,2253); STANLEY (I,4190); WHAT'S UP, AMERICA? (I,4824); WHERE'S HUDDLES (I,4836)

LYNDECK, EDMUND
SWEENEY TODD (II,2512)

LYNDHURST, NICHOLAS
BUTTERFLIES (II,404)

LYNDHURST, NICKY
PETER PAN (II,2030)

LYNDON, BARBARA
KENNEDY (II,1377)

LYNLEY, CAROL
THE ALCOA HOUR (I,108); ALCOA PREMIERE (I,109); THE BEST OF FRIENDS (II,216); THE BOB HOPE CHRYSLER THEATER (I,502); CIRCUS OF THE STARS (II,532); THE COPS AND ROBIN (II,573); DUPONT SHOW OF THE MONTH (I,1387); THE EVIL TOUCH (I,1486); FANTASY ISLAND (II,829); FANTASY ISLAND (II,830); THE FLIERS (I,1596); THE GENERAL ELECTRIC THEATER (I,1753); HAVING BABIES II (II,1108); HENRY FONDA AND THE FAMILY (I,2015); IF IT'S A MAN, HANG UP (I,2186); THE IMMORTAL (I,2196); THE IMMORTAL (I,2197); JOURNEY TO THE UNKNOWN (I,2457); JUDGEMENT DAY (II,1348); JUNIOR MISS (I,2493); THE NIGHT STALKER (I,3292); WILLOW B: WOMEN IN PRISON (II,2804)

LYNN DUDDY SINGERS, THE
FIREBALL FUN FOR ALL (I,1577); FRANKIE CARLE TIME (I,1674); FRANKIE LAINE TIME (I,1676)

LYNN, BAMBI
THE BABY SITTER (I,321)

LYNN, BETTY
THE ANDY GRIFFITH SHOW (I,192); FAMILY AFFAIR (I,1519); FIRESIDE THEATER (I,1580); JANE WYMAN PRESENTS THE FIRESIDE THEATER (I,2345); LOVE THAT JILL (I,2780); MATINEE THEATER (I,2947); REVLON MIRROR THEATER (I,3781); TEXAS JOHN SLAUGHTER (I,4414); WHERE'S RAYMOND? (I,4837)

LYNN, BILLY
ANYTHING GOES (I,237)

LYNN, CYNTHIA
HOGAN'S HEROES (I,2069)

LYNN, DEBBY
THE GRADY NUTT SHOW (II,1048)

LYNN, DIANA
CHRYSLER MEDALLION THEATER (I,951); DUPONT SHOW OF THE MONTH (I,1387); FRONT ROW CENTER (I,1694); HOLLYWOOD OPENING NIGHT (I,2087); THE LILLI PALMER THEATER (I,2704); LOW MAN ON THE TOTEM POLE (I,2782); ON TRIAL (I,3363); THE PHILADELPHIA STORY (I,3582); STAGE DOOR (I,4178); THE UNITED STATES STEEL HOUR (I,4677)

LYNN, DICK
THE DOM DELUISE SHOW (I,1328)

LYNN, JANIS
THE CHEERLEADERS (II,493)

LYNN, JEFFREY
FAITH BALDWIN'S THEATER OF ROMANCE (I,1515); MY SON JEEP (I,3202); THE SPIRAL STAIRCASE (I,4160)

LYNN, JONATHAN
CAMBRIDGE CIRCUS (I,807); DOCTOR IN THE HOUSE (I,1313)

LYNN, JUDY
FUN FOR 51 (I,1707); THE HAZEL BISHOP SHOW (I,1983); THE JUDY LYNN SHOW (I,2474); SING IT AGAIN (I,4059); THREE'S COMPANY (I,4489); THE VINCENT LOPEZ SHOW (I,4735)

LYNN, KORY
B.J. AND THE BEAR (II,127)

LYNN, LORETTA
BOB HOPE SPECIAL: HAPPY BIRTHDAY, BOB! (II,325); COS: THE BILL COSBY COMEDY SPECIAL (II,577); A COUNTRY CHRISTMAS (II,582); A COUNTRY CHRISTMAS (II,584); COUNTRY COMES HOME (II,585); COUNTRY COMES HOME (II,586); COUNTRY GALAXY OF STARS (II,587); COUNTRY MUSIC HIT PARADE (I,1069); DEAN MARTIN PRESENTS MUSIC COUNTRY, U.S.A. (I,1192); FIFTY YEARS OF COUNTRY MUSIC (II,853); GEORGE BURNS IN NASHVILLE?? (II,969); LORETTA LYNN IN THE BIG APPLE (II,1520); LORETTA LYNN: THE LADY. . .THE LEGEND (II,1521); SINATRA AND FRIENDS (II,2358); THE WILBURN BROTHERS SHOW (I,4860)

LYNN, MARA
THE NUT HOUSE (I,3320); SHOW BIZ (I,4027)

LYNN, RITA
THE ADVENTURES OF JIM BOWIE (I,63); THE INNER SANCTUM (I,2216); LIGHT'S OUT (I,2699); MEET THE GOVERNOR (I,2991); MR. SMITH GOES TO WASHINGTON (I,3151); SCREEN DIRECTOR'S PLAYHOUSE (I,3946); TELEPHONE TIME (I,4379); THE WEB (I,4784); WINCHESTER (I,4872)

LYNN, ROBERT
SPACE: 1999 (II,2409)

LYNN, SHERRY
A DOG'S LIFE (II,697)

LYNNE, ADA
THE VAUGHN MONROE SHOW (I,4707)

LYNNE, GLORIA
BELAFONTE, NEW YORK (I,386)

LYON, JEFFREY
SNEAK PREVIEWS (II,2388)

LYON, SUE
ARSENIC AND OLD LACE (I,270); THE BOB HOPE SHOW (I,649)

LYONS, ANNABELL
54TH STREET REVUE (I,1573)

LYONS, COLLETTE
BRIEF ENCOUNTER (I,723)

LYONS, GENE
EYE WITNESS (I,1494); IRONSIDE (I,2240); IRONSIDE (II,1246); PENTAGON U.S.A. (I,3514); STAR TREK (II,2440); TACK REYNOLDS (I,4329)

LYONS, JENNY
THE EDGE OF NIGHT (II,760)

LYONS, ROBERT F
THE GANGSTER CHRONICLES (II,957); THE ROOKIES (I,3840); WAIKIKI (II,2734)

LYONS, RUTH
RUTH LYONS 50 CLUB (I,3876)

LYSDAHL, TORI
UNIT 4 (II,2707)

LYSOHIR, BONNIE
ONE OF THE BOYS (II,1911)

LYTELL, BERT
HOLLYWOOD SCREEN TEST (I,2089); THE LAMBS GAMBOL (I,2607); ONE MAN'S FAMILY (I,3383); THE ORCHID AWARD (I,3406); PHILCO TELEVISION PLAYHOUSE (I,3583); THE VALIANT (I,4697)

LYTELL, PAT
ROBERT Q'S MATINEE (I,3812)

LYTTON, DEBBIE
DAYS OF OUR LIVES (II,629); THE NEW LAND (II,1820)

MAVEGA, ALICE
TURN ON (I,4625)

MABLEY, MOMS
ABC STAGE '67 (I,6)

MABRAY, STEWART
BEULAH LAND (II,226)

MACARTHUR, AVIS
ALL MY CHILDREN (II,39)

MACARTHUR, JAMES
BATTLE OF THE NETWORK STARS (II,165); BATTLE OF THE NETWORK STARS

(II,166); HAWAII FIVE-O (II,1110); LASSITER (II,1434)

MACBETH, JACK
RITUALS (II,2182)

MACBRIDE, DONALD
MY FRIEND IRMA (I,3191)

MACCARTHY, KARIN
MOLL FLANDERS (II,1720)

MACCLOSKY, YSABEL
BEWITCHED (I,418)

MACCOLL, CATRIONA
THE LAST DAYS OF POMPEII (II,1436)

MACCORKINDALE, SIMON
FALCON CREST (II,808); FALCON'S GOLD (II,809); MANIMAL (II,1622); THE MANIONS OF AMERICA (II,1623); SCALPELS (II,2266)

MACDONALD, CASEY
STAT! (I,4213)

MACDONALD, CHRISTOPHER
CALL TO GLORY (II,413); FOG (II,897)

MACDONALD, EARL
ABE LINCOLN IN ILLINOIS (I,10)

MACDONALD, FRANCIS
THE ADVENTURES OF CHAMPION (I,54)

MACDONALD, IVAN
ACADEMY THEATER (I,14)

MACDONALD, JEANETTE
THE LUX VIDEO THEATER (I,2795); SCREEN DIRECTOR'S PLAYHOUSE (I,3946)

MACDONALD, JIM
THE MICKEY MOUSE CLUB (I,3025)

MACDONALD, RAY
THE KATE SMITH HOUR (I,2508)

MACDONALD, RYAN
THE BRADY GIRLS GET MARRIED (II,364); DAYS OF OUR LIVES (II,629); THE ODD COUPLE (II,1875); TRUE LIFE STORIES (II,2665)

MACDONALD, SUSAN
THE EDGE OF NIGHT (II,760); SOMERSET (I,4115)

MACDONNELL, KYLE
CELEBRITY TIME (I,887); FOR YOUR PLEASURE (I,1627); GIRL ABOUT TOWN (I,1806); HOLD THAT CAMERA (I,2072)

MACDONNELL, RAY
ALL MY CHILDREN (II,39); THE EDGE OF NIGHT (II,760)

MACDONNELL, SARAH
MUGGSY (II,1761)

MACDUFF, EWEN
THE INVISIBLE MAN (I,2232)

MACDUFF, TYLER
CONFIDENTIALLY YOURS (I,1035); SKYFIGHTERS (I,4078)

MACGEORGE, JIM
THE BEANY AND CECIL SHOW (I,374); FLINTSTONE FAMILY ADVENTURES (II,882); THE FLINTSTONES (II,885); HAPPY DAYS (I,1942); THE KWICKY KOALA SHOW (II,1417); SEMI-TOUGH (II,2300)

MACGIBBON, HARRIET
THE BEVERLY HILLBILLIES (I,417); THE SMOTHERS BROTHERS SHOW (I,4096)

MACGRATH, LEUEEN
THE HOLY TERROR (I,2098)

MACGRAW, ALI
DYNASTY (II,746); THE WINDS OF WAR (II,2807)

MACGREEVY, TOM
RYAN'S HOPE (II,2234)

MACGREGOR, KATHERINE
LITTLE HOUSE ON THE PRAIRIE (II,1487); LITTLE HOUSE: A NEW BEGINNING (II,1488)

MACGREGOR, MARY
SCROOGE'S ROCK 'N' ROLL CHRISTMAS (II,2279)

MACGREGOR, PHIL
THE GUIDING LIGHT (II,1064)

MACGREGOR, SCOTTY
SCRAPBOOK JR. EDITION (I,3944)

MACINTOSH, JAY W
THE ADVENTURES OF POLLYANNA (II,14); CENTENNIAL (II,477); THE HEALERS (II,1115); SENIOR YEAR (II,2301); SONS AND DAUGHTERS (II,2404)

MACKAY, JEFF
THE BLACK SHEEP SQUADRON (II,262); MAGNUM, P.I. (II,1601)

MACKAY, NANCY
LOVE OF LIFE (I,2774)

MACKENSIE, RICHARD
LOVE OF LIFE (I,2774)

MACKENZIE, GISELE
CHEVROLET ON BROADWAY (I,925); AN EVENING WITH JIMMY DURANTE (I,1475); THE GISELE MACKENZIE SHOW (I,1812); THE GISELE MACKENZIE SHOW (I,1813); HOLIDAY U.S.A. (I,2078); KRAFT TELEVISION THEATER (I,2592); THE MISS AND THE MISSILES (I,3062); A SALUTE TO TELEVISION'S 25TH ANNIVERSARY (I,3893); THE SID CAESAR SHOW (I,4043); THE SID CAESAR SPECIAL (I,4044); SOLDIER PARADE (I,4109); STUDIO ONE (I,4268)

MACKENZIE, JULIA
FOR RICHER, FOR POORER (II,906)

MACKENZIE, MICKIE
MR. MERLIN (II,1751)

MACKENZIE, PATCH
CHICO AND THE MAN (II,508); E/R (II,748); HEAVEN ON EARTH (II,1119); NIGHTSIDE (II,1851)

MACKENZIE, PHIL
THE JIM STAFFORD SHOW (II,1291)

MACKENZIE, PHILIP CHARLES
BROTHERS (II,380); CHARACTERS (II,481); MAKING THE GRADE (II,1606); THE SIX O'CLOCK FOLLIES (II,2368)

MACKENZIE, RICHARD
IT TAKES TWO (II,1252); MALIBU (II,1607)

MACKENZIE, STEWART
SHOGUN (II,2339)

MACKENZIE, WILL
ANOTHER APRIL (II,93); THE BOB NEWHART SHOW (II,342); ON THE FLIP SIDE (I,3360); RHODA (II,2151)

MACLACHLAN, JANET
ARCHIE BUNKER'S PLACE (II,105); FRIENDS (II,927); JACQUELINE SUSANN'S VALLEY OF THE DOLLS, 1981 (II,1268); LOVE THY NEIGHBOR (I,2781); ROLL OF THUNDER, HEAR MY CRY (II,2202); SHE'S IN THE ARMY NOW (II,2320)

MACLAINE, SHIRLEY
BARYSHNIKOV IN HOLLYWOOD (II,157); CELEBRATION: THE AMERICAN SPIRIT (II,461); THE SENSATIONAL, SHOCKING, WONDERFUL, WACKY 70S (II,2302); SHIRLEY MACLAINE AT THE LIDO (II,2332); THE SHIRLEY MACLAINE SPECIAL: WHERE DO WE GO FROM HERE? (II,2333); SHIRLEY MACLAINE. . . EVERY LITTLE MOVEMENT (II,2334); SHIRLEY MACLAINE: IF THEY COULD SEE ME NOW (II,2335); SHIRLEY MACLAINE: ILLUSIONS (II,2336); SHIRLEY'S WORLD (I,4019); A WORLD OF LOVE (I,4914)

MACLANE, BARTON
ANYTHING FOR MONEY (I,236); FOUR STAR PLAYHOUSE (I,1652); GIRL ON THE RUN (I,1809); I DREAM OF JEANNIE (I,2167); THE KAISER ALUMINUM HOUR (I,2503); THE OUTLAWS (I,3427); YOUR JEWELER'S SHOWCASE (I,4953)

MACLANE, KERRY
CURIOSITY SHOP (I,1114)

MACLAREN, LYNNE
FOR RICHER, FOR POORER (II,906)

MACLEAN, PETER
THE GUIDING LIGHT (II,1064); THE SECRET STORM (I,3969); WHERE THE HEART IS (I,4831)

MACLEHMAN, EVERETT
SUSPENSE THEATER ON THE AIR (II,2507)

MACLEOD, CATHERINE
HAPPILY EVER AFTER (I,1937)

MACLEOD, GAVIN
ALAN KING'S THIRD ANNUAL FINAL WARNING!! (II,31); BABY CRAZY (I,319); THE BIG SHOW (II,243); CELEBRITY CHALLENGE OF THE SEXES 3 (II,467); THE DEAN MARTIN CELEBRITY ROAST (II,638); THE DUPONT SHOW OF THE WEEK (I,1388); THE LOVE BOAT (II,1535); THE LOVE BOAT III (II,1534); THE MARY TYLER MOORE SHOW (II,1640); MCHALE'S NAVY (I,2969); MITZI AND A HUNDRED GUYS (II,1710); MITZI. . . WHAT'S HOT, WHAT'S NOT (II,1712); MURDER CAN HURT YOU! (II,1768); RANSOM FOR ALICE

(II,2113); SCRUPLES (II,2280); WHATEVER BECAME OF. . .? (II,2773)

MACLEOD, MERCER
EYE WITNESS (I,1494)

MACLEOD, MURRAY
KAREN (I,2505)

MACLIAMMOIR, MICHAEL
GREAT CATHERINE (I,1873)

MACLURE, ELLI
BARRIER REEF (I,362)

MACMAHON, HORACE
THE MOTOROLA TELEVISION HOUR (I,3112)

MACMANUS, GINGER
OUR TOWN (I,3420)

MACMICHAEL, FLORENCE
BACHELOR FATHER (I,323); IVY LEAGUE (I,2286); MR. ED (I,3137)

MACMILLAN, KENNETH
CINDERELLA (I,955)

MACMILLAN, NORMA
DAVY AND GOLIATH (I,1180); UNDERDOG (I,4673)

MACMURRAY, FRED
AMERICAN COWBOY (I,169); THE ANDY WILLIAMS NEW YEAR'S EVE SPECIAL (I,203); THE APARTMENT HOUSE (I,241); BING CROSBY AND HIS FRIENDS (I,455); THE BOB HOPE SHOW (I,513); THE BOB HOPE SHOW (I,515); THE BOB HOPE SHOW (I,527); BOB HOPE SPECIAL: HAPPY BIRTHDAY, BOB! (II,326); THE CHADWICK FAMILY (II,478); MY THREE SONS (I,3205); SCREEN DIRECTOR'S PLAYHOUSE (I,3946); TED KNIGHT MUSICAL COMEDY VARIETY SPECIAL SPECIAL (II,2549); THANKSGIVING REUNION WITH THE PARTRIDGE FAMILY AND MY THREE SONS (II,2569)

MACMURRAY, RICHARD
STAGE 13 (I,4183)

MACNAMARA, BRADY
THE HOUSE WITHOUT A CHRISTMAS TREE (I,2143)

MACNAUGHTON, ROBERT
BIG BEND COUNTRY (II,230)

MACNEIL, KATHY
AS THE WORLD TURNS (II,110)

MACNEIL, ROBERT
TINKER, TAILOR, SOLDIER, SPY (II,2621)

MACNELLIE, TRESS
THE PARAGON OF COMEDY (II,1956)

MACRAE, ELIZABETH
GOMER PYLE, U.S.M.C. (I,1843)

MACRAE, FRANK
USED CARS (II,2712)

MACRAE, GORDON
100 YEARS OF GOLDEN HITS (II,1905); AQUA VARIETIES (I,248); BOB HOPE SPECIAL: BOB HOPE PRESENTS A CELEBRATION WITH STARS OF COMEDY AND MUSIC (II,340); THE COLGATE COMEDY HOUR (I,997); FIVE STARS IN SPRINGTIME (I,1590); THE GENERAL FOODS 25TH ANNIVERSARY SHOW (I,1756); GIFT OF THE MAGI (I,1800); THE GORDON MACRAE SHOW (I,1859); THE GORDON MACRAE SHOW (I,1860); THE LUX VIDEO THEATER (I,2795); VAUDEVILLE (II,2722)

MACRAE, HEATHER
THE FOUR OF US (II,914); THE SHEILA MACRAE SHOW (I,4005); A WORLD APART (I,4910)

MACRAE, MEREDITH
FANTASY (II,828); MY THREE SONS (I,3205); PETTICOAT JUNCTION (I,3571); THE SHEILA MACRAE SHOW (I,4005); STEVE MARTIN: COMEDY IS NOT PRETTY (II,2462); THANKSGIVING REUNION WITH THE PARTRIDGE FAMILY AND MY THREE SONS (II,2569); WILDER AND WILDER (II,2802)

MACRAE, MICHAEL
THE COUNTRY MUSIC MURDERS (II,590); THE NATURAL LOOK (II,1799)

MACRAE, SHEILA
AQUA VARIETIES (I,248); THE BOB HOPE SHOW (I,643); THE HONEYMOONERS (I,2111); THE JACKIE GLEASON SHOW (I,2323); THE JACKIE GLEASON SPECIAL (I,2324); THE JACKIE GLEASON SPECIAL (I,2325); THE LUX VIDEO THEATER (I,2795); THE SECRET WAR OF JACKIE'S GIRLS (II,2294); THE SHEILA MACRAE SHOW (I,4005)

MACRAY, FRANK
HOME COOKIN' (II,1170)

MACAULEY, CHARLES
BRADDOCK (I,704); THE MUNSTERS' REVENGE (II,1765); RETURN OF THE WORLD'S GREATEST DETECTIVE (II,2142)

MACAULEY, JOSEPH
FROM THESE ROOTS (I,1688); PANAMA HATTIE (I,3444)

MACCHIO, RALPH
EIGHT IS ENOUGH (II,762)

MACDONALD, EDMUND
MYSTERIES OF CHINATOWN (I,3209)

MACE, MARY
BABES IN TOYLAND (I,318)

MACHADO, MARIO
GILLIGAN'S ISLAND: RESCUE FROM GILLIGAN'S ISLAND (II,991)

MACHADO, TINA MARIE
HAWAIIAN HEAT (II,1111)

MACHIAVERNA, JOE
SAMMY KAYE'S MUSIC FROM MANHATTAN (I,3902)

MACHON, KAREN
RISKO (II,2180); THE RUNAWAYS (II,2231); THE SEAL (II,2282)

MACHON, KATE
OPERATION: RUNAWAY (II,1918)

MACHT, STEPHEN
THE AMERICAN DREAM (II,71); GEORGE WASHINGTON (II,978); KNOTS LANDING (II,1404); LOOSE CHANGE (II,1518)

MACIAS, ERNESTO
NANCY (I,3221)

MACK TRIPLETS, THE
THE PHIL SILVERS ARROW SHOW (I,3576)

MACK, BILLY
ALL MY CHILDREN (II,39)

MACK, DOTTY
THE DOTTY MACK SHOW (I,1358); MOTHERS DAY (I,3109); THE PAUL DIXON SHOW (I,3487)

MACK, GILBERT
JOHNNY JUPITER (I,2429)

MACK, JOE
SAMMY KAYE'S MUSIC FROM MANHATTAN (I,3902)

MACK, MICHAEL
POWERHOUSE (II,2074)

MACK, TED
TED MACK AND THE ORIGINAL AMATEUR HOUR (I,4370); TED MACK'S FAMILY HOUR (I,4371); TED MACK'S MATINEE (I,4372)

MACKAYE, NORMAN
CONCERNING MISS MARLOWE (I,1032)

MACKLIN, ALBERT
DREAMS (II,735); REMINGTON STEELE (II,2130)

MACKLIN, DAVID
HARRIS AGAINST THE WORLD (I,1956); KINCAID (I,2543); TAMMY (I,4357)

MACLEOD, MARY
THE TRAP (I,4593)

MACLEOD, MURRAY
MOONLIGHT (II,1730)

MACNEE, PATRICK
ALFRED HITCHCOCK PRESENTS (I,115); THE AVENGERS (II,121); BATTLESTAR GALACTICA (II,181); THE BILLION DOLLAR THREAT (II,246); CAESAR AND CLEOPATRA (I,788); COMEDY OF HORRORS (II,557); EMPIRE (II,777); GAVILAN (II,959); KRAFT TELEVISION THEATER (I,2592); MATINEE THEATER (I,2947); MATT HELM (II,1652); MISTER JERICO (I,3070); THE NEW AVENGERS (II,1816); PLAYHOUSE 90 (I,3623); RETURN OF THE MAN FROM U.N.C.L.E.: THE 15 YEARS LATER AFFAIR (II,2140); STUNT SEVEN (II,2486); SUSPICION (I,4309)

MACOMBER, DEBBIE
THE JERRY LEWIS SHOW (I,2370); TURN ON (I,4625)

MACREADY, CAROL
THE FLAME TREES OF THIKA (II,870); THE WOMAN IN WHITE (II,2819)

MACREADY, GEORGE
THE CAT AND THE CANARY (I,868); FAME IS THE NAME OF THE GAME (I,1518); HENRY FONDA PRESENTS THE STAR AND THE STORY (I,2016); THE MILTON BERLE SPECIAL (I,3052); NIGHT GALLERY (I,3288); PANTOMIME QUIZ (I,3449); PEYTON PLACE (I,3574); THE THREE MUSKETEERS (I,4481); THE YOUNG LAWYERS (I,4940)

MACY, BILL
ALL IN THE FAMILY (II,38); BATTLE OF THE NETWORK STARS (II,163); HANGING IN (II,1078); MAUDE (II,1655); STUNT SEVEN (II,2486); TWILIGHT THEATER II (II,2686)

MACY, W H
THE AWAKENING LAND (II,122); SITCOM (II,2367)

MADALONE, DENNIS
FEEL THE HEAT (II,848)

MADARIS, TERESA
WEDNESDAY NIGHT OUT (II,2760)

MADDEN, CIARAN
HAMLET (I,1925)

MADDEN, DAVE
ALICE (II,33); CAMP RUNAMUCK (I,811); MORE WILD WILD WEST (II,1733); THE PARTRIDGE FAMILY (II,1962); PARTRIDGE FAMILY: 2200 A.D. (II,1963); ROWAN AND MARTIN'S LAUGH-IN (I,3856)

MADDEN, DONALD
ANOTHER WORLD (II,97); ONE LIFE TO LIVE (II,1907)

MADDEN, PETER
THE FLYING DOCTOR (I,1610)

MADDEN, SHARON
JOSHUA'S WORLD (II,1344)

MADDERN, VICTOR
FAIR EXCHANGE (I,1512)

MADDOX, DIANA
AQUAMAN (I,249)

MADIGAN, AMY
CRAZY TIMES (II,602); TRAVIS MCGEE (II,2656)

MADISON, GUY
BING CROSBY AND HIS FRIENDS (I,455); LIGHT'S DIAMOND JUBILEE (I,2698); WHEN THE WEST WAS FUN: A WESTERN REUNION (II,2780); WILD BILL HICKOK (I,4861)

MADOC, PHILIP
TARGET (II,2541)

MADSEN, HARRY
KENNEDY (II,1377)

MADSEN, HENRY
THE DAY THE WOMEN GOT EVEN (II,628)

MAFFETT, DEBRA SUE
BOB HOPE SPECIAL: BOB HOPE'S ALL-STAR SUPER BOWL PARTY (II,304)

MAGEE, CLIFFY
PRYOR'S PLACE (II,2089)

MAGEE, DAN
TALES OF THE RED CABOOSE (I,4347)

MAGEE, JACK
ALL MY CHILDREN (II,39)

MAGEE, KEN
LITTLE SHOTS (II,1495)

MAGEE, PATRICK
A KILLER IN EVERY CORNER (I,2538)

MAGGART, BRANDON
BROTHERS (II,380); I'M A FAN (I,2192); JENNIFER SLEPT HERE (II,1278)

MAGNESS, MARILYN
MR. MOON'S MAGIC CIRCUS (II,1753)

MAGNOTTA, VICTOR
MICKEY SPILLANE'S MIKE HAMMER: MORE THAN MURDER (II,1693)

MAGRUDER, JOHN
THE RED SKELTON SHOW (I,3756)

MAGUIRE, GERALD
PRISONER: CELL BLOCK H (II,2085)

MAGUIRE, KATHLEEN
THE CHADWICK FAMILY (II,478); STAR TONIGHT (I,4199); A TIME FOR US (I,4507)

MAGUIRE, MAEVE
ANOTHER WORLD (II,97)

MAGUIRE, MICHAEL
MITCHELL AND WOODS (II,1709)

MAHAFFEY, LORRIE
ANSON AND LORRIE (II,98); THE EYES OF TEXAS (II,799); LEGS (II,1459); WHO'S WATCHING THE KIDS? (II,2793)

MAHAR, CHRISTOPHER
THE LAST NINJA (II,1438)

MAHARIS, GEORGE
THE MONK (I,3087); THE MOST DEADLY GAME (I,3106); MURDER IS A ONE-ACT PLAY (I,3160); OF MEN OF WOMEN (I,3331); RETURN TO FANTASY ISLAND (II,2144); RICH MAN, POOR MAN—BOOK I (II,2161); ROUTE 66 (I,3852); SEARCH FOR TOMORROW (II,2284)

MAHER, JOSEPH
ANOTHER WORLD (II,97); AT YOUR SERVICE (II,118)

MAHI, GERALD
THE RETURN OF LUTHER GILLIS (II,2137)

MAHLER, BRUCE
FRIDAYS (II,926)

MAHON, PEGGY
PAUL BERNARD—PSYCHIATRIST (I,3485)

MAHONE, JUANITA
AS THE WORLD TURNS (II,110)

MAHONEY, JEAN
THE ALAN YOUNG SHOW (I,104); THE R.C.A. VICTOR SHOW (I,3737)

MAHONEY, JERRY
RUNAROUND (I,3873)

MAHONEY, JOCK
B.J. AND THE BEAR (II,126); THE RANGE RIDER (I,3720); SIMON LASH (I,4052); YANCY DERRINGER (I,4925)

MAHONEY, JOHN
CHICAGO STORY (II,506)

MAHONEY, LOUIS
A WOMAN CALLED GOLDA (II,2818)

MAHONEY, MAGGIE
JOHNNY NIGHTHAWK (I,2432)

MAHONEY, TOM
THE ALAN YOUNG SHOW (I,104); ALICE (II,33); THE R.C.A. VICTOR SHOW (I,3737)

MAHONEY, TRISH
THE LETTERS (I,2670)

MAIENCZYK, MATTHEW
ANOTHER WORLD (II,97)

MAIKEN, GILDA
THE DINAH SHORE SHOW (I,1280)

MAILER, NORMAN
BRET MAVERICK (II,374)

MAIN, LAURIE
CAT BALLOU (I,869); THE DATCHET DIAMONDS (I,1160); JANE EYRE (I,2338); VANITY FAIR (I,4700)

MAIR, JIMMY
THE PHOENIX (II,2035); THE YEAGERS (II,2843)

MAISAC, MAURICE
TARZAN AND THE TRAPPERS (I,4362)

MAISNIK, KATHY
BATTLE OF THE NETWORK STARS (II,175); STAR OF THE FAMILY (II,2436)

MAITLAND, ARTHUR
THE VALIANT (I,4697)

MAITLAND, BETH
THE YOUNG AND THE RESTLESS (II,2862)

MAITLAND, JULES
PROBE (I,3674)

MAITLAND, MICHAEL
DARK SHADOWS (I,1157)

MAJORS, LEE
BATTLE OF THE NETWORK STARS (II,173); THE BIG VALLEY (I,437); THE BIONIC WOMAN (II,255); THE DONNY AND MARIE OSMOND SHOW (II,714); THE FALL GUY (II,811); HOW DO I KILL A THIEF—LET ME COUNT THE WAYS (II,1195); THE MEN FROM SHILOH (I,3005); THE OLIVIA NEWTON-JOHN SHOW (II,1885); OWEN MARSHALL: COUNSELOR AT LAW (I,3435); THE SIX MILLION DOLLAR MAN (I,4067); THE SIX-MILLION-DOLLAR MAN (II,2372); A SPECIAL OLIVIA NEWTON-JOHN (II,2418); THE WAYNE NEWTON SPECIAL (II,2749)

MAJORS, MATTIE
YOU ASKED FOR IT (II,2855)

MAK, MARII
THE RENEGADES (II,2132)

MAKO
ALFRED OF THE AMAZON (I,116); HAWAIIAN HEAT (II,1111); JUDGE DEE IN THE MONASTERY MURDERS (I,2465); THE LAST NINJA (II,1438)

MALAGON SISTERS, THE
THE PETER LIND HAYES SHOW (I,3564)

MALAVE, CHU CHU
ZERO INTELLIGENCE (II,2876)

MALCOLM, CHRISTOPHER
TERROR FROM WITHIN (I,4404)

MALDEN, BEVERLY
LAS VEGAS GAMBIT (II,1430)

MALDEN, KARL
LITTLE WOMEN (I,2719); SKAG (II,2374); THE STREETS OF SAN FRANCISCO (I,4260); THE STREETS OF SAN FRANCISCO (II,2478)

MALE, COLIN
THE DOTTY MACK SHOW (I,1358); THIS IS MUSIC (I,4458)

MALET, ARTHUR
CASABLANCA (II,450); THE MAN FROM GALVESTON (I,2865); PALMERSTOWN U.S.A. (II,1945); THE WILD WILD WEST (I,4863)

MALET, PIERRE
MISTRAL'S DAUGHTER (II,1708)

MALIBUTIES, THE
MALIBU U. (I,2849)

MALICK, WENDIE
TRAUMA CENTER (II,2655)

MALIK, ART
THE FAR PAVILIONS (II,832)

MALINDA, JIM
KORG: 70,000 B.C. (II,1408)

MALIS, CLAIRE
THE FACTS OF LIFE (II,805); FROM HERE TO ETERNITY (II,933); ONE LIFE TO LIVE (II,1907)

MALKIN, RACHEL
KATE MCSHANE (II,1368)

MALLABY, GEORGE
HOMICIDE (I,2101); PRISONER: CELL BLOCK H (II,2085)

MALLETT, JANE
ALL THE WAY HOME (I,140)

MALLORY, CAROLE
ROGER AND HARRY (II,2200)

MALLORY, EDWARD
ALL IN THE FAMILY (I,133); DAYS OF OUR LIVES (II,629); MORNING STAR (I,3104); THINK PRETTY (I,4448)

MALLORY, KATHLEEN
COME OUT, COME OUT, WHEREVER YOU ARE (I,1014)

MALLORY, VICTORIA
AFTER HOURS: SINGIN', SWINGIN' AND ALL THAT JAZZ (II,22); THE YOUNG AND THE RESTLESS (II,2862)

MALLOY, LARKIN
THE EDGE OF NIGHT (II,760)

MALNICK, MICHAEL
THE FAR PAVILIONS (II,832)

MALONE, BILL
THE HONEYMOON RACE (I,2107); SUPERMARKET SWEEP (I,4296)

MALONE, CHU CHU
FOR RICHER, FOR POORER (II,906)

MALONE, DALE
LI'L ABNER (I,2702)

MALONE, DEBORAH
THE KROFFT KOMEDY HOUR (II,1411); ME AND MRS. C. (II,1673)

MALONE, DOROTHY
THE BOB HOPE SHOW (I,560); THE BOB HOPE SHOW (I,567); CONDOMINIUM (II,566); THE DOCTOR (I,1320); FLOYD GIBBONS, REPORTER (I,1608); THE GENERAL ELECTRIC THEATER (I,1753); THE GREATEST SHOW ON EARTH (I,1883); HIGH HOPES (II,1144); PEYTON PLACE (I,3574); THE PIGEON (I,3596); REVLON MIRROR THEATER (I,3781); RICH MAN, POOR MAN—BOOK I (II,2161)

MALONE, HOWARD
WINNER TAKE ALL (I,4881)

MALONE, LAURA
ANOTHER WORLD (II,97)

MALONE, MARY
THE ALDRICH FAMILY (I,111); THE GIRLS (I,1811); STARLIGHT THEATER (I,4201)

MALONE, MO
THE MYSTERIOUS TWO (II,1783)

MALONE, NANCY
THE FIRST HUNDRED YEARS (I,1581); FOREVER FERNWOOD (II,909); GREAT ADVENTURE (I,1868); KRAFT SUSPENSE THEATER (I,2591); KRAFT TELEVISION THEATER (I,2592); THE LONG HOT SUMMER (I,2747); MAN IN THE MIDDLE (I,2870); NAKED CITY (I,3216); THE OUTER LIMITS (I,3426); THE OUTSIDER (I,3430); RX FOR THE DEFENSE (I,3878); SAN FRANCISCO INTERNATIONAL AIRPORT (I,3906); STAR TONIGHT (I,4199); THE TWILIGHT ZONE (I,4651)

MALONE, RAY
BROADWAY OPEN HOUSE (I,736); DAGMAR'S CANTEEN (I,1125); THE SATURDAY NIGHT DANCE PARTY (I,3919)

MALONEY, JAMES
21 BEACON STREET (I,4646); MAISIE (I,2833)

MALONEY, LAUREN
ONE DAY AT A TIME (II,1900)

MALONEY, PAIGE
ONE DAY AT A TIME (II,1900)

MALONEY, PATTY
THE BAY CITY ROLLERS SHOW (II,187); CHERYL LADD: SCENES FROM A SPECIAL (II,503); DON'T CALL US (II,709); DONNY AND MARIE (II,712); FAR OUT SPACE NUTS (II,831); THE LITTLE RASCALS (II,1494); PUNCH AND JODY (II,2091)

MALONEY, PETER
CALLAHAN (II,414); THE FOUR OF US (II,914); SANCTUARY OF FEAR (II,2252)

MALOOLY, MAGGIE
LITTLE WOMEN (II,1498); MOBILE MEDICS (II,1716)

MALTIN, LEONARD
ENTERTAINMENT TONIGHT (II,780)

MALZ, JERRY
THE DAY THE WOMEN GOT EVEN (II,628)

MAMAKOS, PETER
THE FRENCH ATLANTIC AFFAIR (II,924); THE NEW OPERATION PETTICOAT (II,1826); THE OATH: 33 HOURS IN THE LIFE OF GOD (II,1873); PETER GUNN (I,3562); SABU AND THE MAGIC RING (I,3881); SECOND LOOK (I,3960)

MAMAS AND THE PAPAS, THE
ROGERS AND HART TODAY (I,3829)

MAN, CHRISTOPHER
KORG: 70,000 B.C. (II,1408)

MANATIS, JANINE
OUR FIVE DAUGHTERS (I,3413)

MANCHESTER, MELISSA
BOB HOPE SPECIAL: BOB HOPE ON CAMPUS (II,290); BOB HOPE SPECIAL: BOB HOPE'S SPRING FLING OF COMEDY AND GLAMOUR (II,315); GOOD VIBRATIONS FROM CENTRAL PARK (I,1855); I LOVE LIBERTY (II,1215); MAC DAVIS—I'LL BE HOME FOR CHRISTMAS (II,1580)

MANCINI, HENRY
100 YEARS OF AMERICA'S POPULAR MUSIC (II,1904); 100 YEARS OF GOLDEN HITS (II,1905); THE MANCINI GENERATION (I,2885); MONSANTO PRESENTS MANCINI (I,3092); MONSANTO PRESENTS MANCINI (I,3093); MONSANTO PRESENTS MANCINI (I,3094); ON PARADE (I,3356); PERRY COMO'S MUSIC FROM HOLLYWOOD (II,2019); PLAY IT AGAIN, UNCLE SAM (II,2052)

MANCINI, RIC
MICKEY SPILLANE'S MIKE HAMMER: MURDER ME, MURDER YOU (II,1694); NEVER SAY NEVER (II,1809)

MANCUSO, NICK
DOCTOR SCORPION (II,685); FEEL THE HEAT (II,848); SCRUPLES (II,2280); UNIT 4 (II,2707)

MANDAN, ROBERT
APPLAUSE (I,244); CARIBE (II,439); CELEBRITY CHARADES (II,470); FOR MEMBERS ONLY (II,905); FROM THESE ROOTS (I,1688); KINGSTON: THE POWER PLAY (II,1399); THE NORLISS TAPES (I,3305); THE OUTLAWS (II,1936); PRIVATE BENJAMIN (II,2087); SEARCH FOR TOMORROW (II,2284); SOAP (II,2392); THREE'S A CROWD (II,2613); YOU CAN'T TAKE IT WITH YOU (II,2857)

MANDEL, HOWIE
JIM HENSON'S MUPPET BABIES (II,1289); LAUGH TRAX (II,1443); ROCK COMEDY (II,2191); THE SHAPE OF THINGS (II,2315); ST. ELSEWHERE (II,2432); WELCOME TO THE FUN ZONE (II,2763)

MANDELL, ALAN
79 PARK AVENUE (II,2310)

MANDELL, DENIS
HAPPY DAYS (II,1084)

MANDRELL, BARBARA
THE ALL-STAR SALUTE TO MOTHER'S DAY (II,51); BARBARA MANDRELL AND THE MANDRELL SISTERS (II,143); BARBARA MANDRELL—THE LADY IS A CHAMP (II,144); BATTLE OF THE NETWORK STARS (II,172); BOB HOPE SPECIAL: BOB HOPE'S ALL-STAR COMEDY BIRTHDAY PARTY (II,297); BOB HOPE SPECIAL: BOB HOPE'S FUNNY VALENTINE (II,309); BOB HOPE SPECIAL: BOB HOPE'S STAND UP AND CHEER FOR THE NATIONAL FOOTBALL LEAGUE'S 60TH YEAR (II,316); BOB HOPE SPECIAL: HAPPY BIRTHDAY, BOB! (II,325); CHRISTMAS IN WASHINGTON (II,520); THE CONCRETE COWBOYS (II,564); COUNTRY GALAXY OF STARS (II,587); THE COUNTRY MUSIC MURDERS (II,590); COUNTRY NIGHT OF STARS (II,592); ELVIS REMEMBERED: NASHVILLE

TO HOLLYWOOD (II,772); AN EVENING WITH THE STATLER BROTHERS (II,790); THE FUNNIEST JOKE I EVER HEARD (II,939); GOOD EVENING, CAPTAIN (II,1031); THE JOHNNY CASH CHRISTMAS SPECIAL (II,1325); LOUISE MANDRELL: DIAMONDS, GOLD AND PLATINUM (II,1528); LUCY COMES TO NASHVILLE (II,1561); MAC DAVIS SPECIAL: THE MUSIC OF CHRISTMAS (II,1578); MERRY CHRISTMAS FROM THE GRAND OLE OPRY (II,1686); THE NASHVILLE PALACE (II,1793); PLAYBOY'S PLAYMATE PARTY (II,2055); ROY ACUFF—50 YEARS THE KING OF COUNTRY MUSIC (II,2223); SKINFLINT (II,2377); THE WAYNE NEWTON SPECIAL (II,2749)

MANDRELL, IRLENE
BARBARA MANDRELL AND THE MANDRELL SISTERS (II,143); COUNTRY GALAXY OF STARS (II,587); LOUISE MANDRELL: DIAMONDS, GOLD AND PLATINUM (II,1528); STATE FAIR USA (II,2449); STATE FAIR USA (II,2450)

MANDRELL, LOUISE
THE ALL-STAR SALUTE TO MOTHER'S DAY (II,51); BARBARA MANDRELL AND THE MANDRELL SISTERS (II,143); BATTLE OF THE NETWORK STARS (II,172); COUNTRY GALAXY OF STARS (II,587); JERRY REED AND SPECIAL FRIENDS (II,1285); LOUISE MANDRELL: DIAMONDS, GOLD AND PLATINUM (II,1528); MERRY CHRISTMAS FROM THE GRAND OLE OPRY (II,1686)

MANDY, MICHAEL
IT'S MAGIC, CHARLIE BROWN (I,2271); LIFE IS A CIRCUS, CHARLIE BROWN (I,2683)

MANERA, CANDICE
ALL MY CHILDREN (II,39)

MANET, JEANNE
PAPA SAID NO (I,3452)

MANETTI, LARRY
BATTLESTAR GALACTICA (II,181); THE BLACK SHEEP SQUADRON (II,262); THE DUKES (II,740); MAGNUM, P.I. (II,1601); THREE EYES (II,2605)

MANHATTAN TRANSFER
CONSTANTINOPLE (II,570); MARY'S INCREDIBLE DREAM (II,1641)

MANI, KARIN
FROM HERE TO ETERNITY (II,932)

MANILOW, BARRY
THE BARRY MANILOW SPECIAL (II,155); BARRY MANILOW—ONE VOICE (II,156); GOLDIE AND KIDS: LISTEN TO ME (II,1021); THE SECOND BARRY MANILOW SPECIAL (II,2285); THE SENSATIONAL, SHOCKING, WONDERFUL, WACKY 70S (II,2302); THE THIRD BARRY MANILOW SPECIAL (II,2590); TOM SNYDER'S CELEBRITY SPOTLIGHT (II,2632)

MANKIEWICZ, JOSEPH L
BOGART (I,679)

MANLEY, BEATRICE
STRANGER IN OUR HOUSE (II,2477)

MANLEY, STEPHEN
DUFFY (II,739); KUNG FU (II,1416); THE LAST CONVERTIBLE (II,1435); SARA (II,2260); SECRETS OF MIDLAND HEIGHTS (II,2296)

MANN, ANITA
THE BEST OF TIMES (II,219)

MANN, CATHERINE
ENTERTAINMENT TONIGHT (II,780)

MANN, CHARLIE
THAT WAS THE WEEK THAT WAS (I,4422)

MANN, CLAUD
HOT (II,1183)

MANN, COLETTE
PRISONER: CELL BLOCK H (II,2085)

MANN, DELBART
THE GOLDEN AGE OF TELEVISION (II,1016)

MANN, DOLORES
THE MUNSTERS' REVENGE (II,1765)

MANN, HOWARD
CRAZY TIMES (II,602); INSPECTOR PEREZ (II,1237); JOHNNY CASH AND FRIENDS (II,1322)

MANN, IRIS
DOBIE GILLIS (I,1302); I REMEMBER MAMA (I,2176); MAGNAVOX THEATER (I,2827)

MANN, JEAN
THE GREATEST GIFT (I,1880)

MANN, JODIE
CONDOMINIUM (II,566)

MANN, JOHNNY
JOHNNY MANN'S STAND UP AND CHEER (I,2430); STAND UP AND CHEER (I,4188)

MANN, LARRY D
ACCIDENTAL FAMILY (I,17); CAROL (I,846); DENNIS THE MENACE: MAYDAY FOR MOTHER (II,660); POLICE SURGEON (I,3639); RUDOLPH THE RED-NOSED REINDEER (II,2227)

MANN, LORIE
OH, BOY! (I,3341)

MANN, MICHAEL
THE ADVENTURES OF BLINKEY (I,52); JOE AND MABEL (I,2398)

MANN, PEANUTS
THE DOODLES WEAVER SHOW (I,1351)

MANN, RHODA
THE YEAR WITHOUT A SANTA CLAUS (II,2846)

MANN, SHELLEY
MUSIC U.S.A. (I,3179)

MANN, STUART
DOLLAR A SECOND (I,1326)

MANNERING, SEAN
TOMA (II,2634)

MANNERS, DAVID
BRIEF PAUSE FOR MURDER (I,725)

MANNERS, MICKEY
THE BOB NEWHART SHOW (I,666); MANY HAPPY RETURNS (I,2895); WHICH WAY'D THEY GO? (I,4838)

MANNERY, ED
MR. MERGENTHWIRKER'S LOBBLIES (I,3144)

MANNING, IRENE
THE KING AND MRS. CANDLE (I,2544)

MANNING, JACK
EMERGENCY! (II,775); MAGNAVOX THEATER (I,2827); THE PAPER CHASE (II,1949)

MANNING, KATY
DOCTOR WHO (II,689)

MANNING, LEE
LAS VEGAS GAMBIT (II,1430)

MANNING, MARTHA
KENNEDY (II,1377)

MANNING, RUTH
ACE (II,5); ALL IN THE FAMILY (II,38); BLUE JEANS (II,275); GOOD TIME HARRY (II,1037); PEN 'N' INC. (II,1992); ROSETTI AND RYAN (II,2216); WILD ABOUT HARRY (II,2797)

MANNING, SEAN
TOMA (I,4547); YES, VIRGINIA, THERE IS A SANTA CLAUS (II,2849)

MANNION, ANNE
THE CITADEL (II,539)

MANNION, MICHAEL
THE SECRET OF CHARLES DICKENS (II,2293)

MANNIX, JULIE
THE BEST OF EVERYTHING (I,406); THE DAUGHTERS OF JOSHUA CABE (I,1169); FLATBED ANNIE AND SWEETIEPIE: LADY TRUCKERS (II,876)

MANOFF, DINAH
CELEBRITY (II,463); THE POSSESSED (II,2071); SOAP (II,2392)

MANON, GLORIA
THE PSYCHIATRIST: GOD BLESS THE CHILDREN (I,3685)

MANSFIELD, JAYNE
THE ARTHUR MURRAY PARTY FOR BOB HOPE (I,292); THE BACHELOR (I,325); THE BOB HOPE SHOW (I,559); THE BOB HOPE SHOW (I,574); THE BOB HOPE SHOW (I,582); THE BOB HOPE SHOW (I,587); HOLIDAY IN LAS VEGAS (I,2076); KRAFT MYSTERY THEATER (I,2589)

MANSFIELD, JOHN
ONE LIFE TO LIVE (II,1907)

MANSFIELD, KATHLEEN
DREAM GIRL (I,1374)

MANSFIELD, SALLY
BACHELOR FATHER (I,323); ROCKY JONES, SPACE RANGER (I,3823)

MANSI, LOUIS
KELLY MONTEITH (II,1375)

MANSON, ALAN
THE EDGE OF NIGHT (II,760); HICKEY VS. ANYBODY (II,1142); SWITCH (II,2520); THREE'S COMPANY (II,2614); WHERE THE HEART IS (I,4831)

MANSON, MAURICE
BEN CASEY (I,394); CONFIDENTIALLY YOURS (I,1035); MR. MERGENTHWIRKER'S

LOBBLIES (I,3144); THE THIN MAN (I,4446)

MANTECA, CARLOS
TURN ON (I,4625)

MANTEE, PAUL
BIG ROSE (II,241); GREAT BIBLE ADVENTURES (I,1872); THIS GIRL FOR HIRE (II,2596)

MANTEGNA, JOE
THE COMEDY ZONE (II,559); NOW WE'RE COOKIN' (II,1866); THE OUTLAWS (II,1936)

MANTELL, JOE
THE GOLDEN AGE OF TELEVISION (II,1016); PETE AND GLADYS (I,3558); THEY ONLY COME OUT AT NIGHT (II,2586); TOM, DICK, AND HARRY (I,4536); THE UNTOUCHABLES (I,4681); THE WEB (I,4785)

MANTEZ, DOLORES
U.F.O. (I,4662)

MANTLE, MICKEY
THE BOB HOPE SHOW (I,552); DICK CLARK'S GOOD OL' DAYS: FROM BOBBY SOX TO BIKINIS (II,671); THE STARS AND STRIPES SHOW (I,4206)

MANTOOTH, DONALD
THE SEEKERS (II,2298)

MANTOOTH, RANDOLPH
DETECTIVE SCHOOL (II,661); EMERGENCY PLUS FOUR (II,774); EMERGENCY! (II,775); EMERGENCY! (II,776); THE NEW OPERATION PETTICOAT (II,1826); THE SEEKERS (II,2298); TESTIMONY OF TWO MEN (II,2564)

MANZ, LINDA
DOROTHY (II,717)

MANZA, RALPH
A.E.S. HUDSON STREET (II,17); BANACEK (II,138); BANACEK: DETOUR TO NOWHERE (I,339); THE D.A.'S MAN (I,1158); GENERAL HOSPITAL (II,964); LOLLIPOP LOUIE (I,2738); MAMA MALONE (II,1609); SAMURAI (II,2249); TOPPER RETURNS (I,4583)

MANZANO, SONIA
B.J. AND THE BEAR (II,125); SESAME STREET (I,3982)

MAPIN, JEFFERSON
CAGNEY AND LACEY (II,410)

MARA, ADELE
CONFLICT (I,1036); COOL MILLION (I,1046); WHEELS (II,2777)

MARACHUK, STEVE
WAIKIKI (II,2734)

MARBLE, HARRY
YOU ARE THERE (I,4929)

MARC BREAUX DANCERS, THE
THE FRED WARING SHOW (I,1679)

MARC, TED
PROFESSIONAL FATHER (I,3677)

MARCANTEL, CHRISTOPHER
ANOTHER WORLD (II,97); THE GUIDING LIGHT (II,1064); LOVING (II,1552); NURSE (II,1870)

MARCEAU, MARCEL
FUNNY FACES (II,941); MEET MARCEL MARCEAU (I,2983)

MARCEL, NINO
ANDY'S GANG (I,214)

MARCELINO, MARIO
FALCON CREST (II,808); FEEL THE HEAT (II,848)

MARCH, ELSPETH
LITTLE MOON OF ALBAN (I,2714)

MARCH, FREDRIC
A CHRISTMAS CAROL (I,947); DUPONT SHOW OF THE MONTH (I,1387); THE NASH AIRFLYTE THEATER (I,3227); THE ROYAL FAMILY (I,3861)

MARCH, HAL
THE $64,000 QUESTION (I,4071); THE BACHELOR (I,325); BACHELOR PARTY (I,324); DREAM GIRL (I,1374); THE GEORGE BURNS AND GRACIE ALLEN SHOW (I,1763); GIDGET (I,1795); HIGH BUTTON SHOES (I,2050); I MARRIED A DOG (I,2173); THE IMOGENE COCA SHOW (I,2198); IT'S YOUR BET (I,2279); LAUGHS FOR SALE (I,2634); MY FRIEND IRMA (I,3191); THE R.C.A. VICTOR SHOW (I,3737); THE SOLDIERS (I,4112); SUMMER IN THE CITY (I,4280); WHAT'S IT FOR? (I,4817)

MARCH, JUDY
DOCTOR DAN (II,684)

MARCH, KENDALL
THE EDGE OF NIGHT (II,760)

MARCH, LINDA
KEEP THE FAITH (I,2518)

MARCH, LORI
ALFRED HITCHCOCK PRESENTS (I,115); THE BRIGHTER DAY (I,728); PLAYWRIGHTS '56 (I,3627); THE SECRET STORM (I,3969); TEXAS (II,2566); THREE STEPS TO HEAVEN (I,4485)

MARCH, LYNDA
WHY DIDN'T THEY ASK EVANS? (II,2795)

MARCH, PEGGY
THE DOM DELUISE SHOW (I,1328)

MARCHAND, NANCY
ANOTHER WORLD (II,97); BEACON HILL (II,193); THE BELLS OF ST. MARY'S (I,392); THE DUPONT SHOW OF THE WEEK (I,1388); THE GOLDEN AGE OF TELEVISION (II,1016); KRAFT TELEVISION THEATER (I,2592); LITTLE WOMEN (I,2720); LOU GRANT (II,1526); LOVE OF LIFE (I,2774); LOVERS AND FRIENDS (II,1549); PHILCO TELEVISION PLAYHOUSE (I,3583); VALLEY FORGE (I,4698)

MARCHENSE, PAULA
THE BOYS IN BLUE (II,359)

MARCIA, DENISE
WALKIN' WALTER (II,2737)

MARCO, PAOLO
THE SMALL MIRACLE (I,4090)

MARCOTTE, SUSAN
THE LAST CONVERTIBLE (II,1435)

MARCOVICCI, ANDREA
THE DEVIL'S WEB (I,1248); HILL STREET BLUES (II,1154); SMILE JENNY, YOU'RE DEAD (II,2382); SPRAGGUE (II,2430); VELVET (II,2725)

MARCUS, BILL
JOHNNY GARAGE (II,1337)

MARCUS, DEE
COMPLETELY OFF THE WALL (II,560)

MARCUS, RICHARD
FIRST TIME, SECOND TIME (II,863)

MARCUS, SPARKY
THE BAD NEWS BEARS (II,134); GOODTIME GIRLS (II,1042); GRANDPA GOES TO WASHINGTON (II,1050); INSTANT FAMILY (II,1238); MARY HARTMAN, MARY HARTMAN (II,1638); THE RICHIE RICH SHOW (II,2169); SIGMUND AND THE SEA MONSTERS (II,2352); SPACE STARS (II,2408); WKRP IN CINCINNATI (II,2814)

MARCUS, VITTINA
LOST IN SPACE (I,2758)

MARDEN, ADRIENNE
MATINEE THEATER (I,2947)

MARDER, JEFF
NATIONAL LAMPOON'S HOT FLASHES (II,1795)

MARDIA, JOHN
SUNDAY IN TOWN (I,4285)

MAREDEN, JOEL
GENERAL HOSPITAL (II,964)

MAREN, JERRY
HIGH SCHOOL, U.S.A. (II,1148); THE MAGIC LADY (I,2816)

MARGETTS, MONTY
MCDUFF, THE TALKING DOG (II,1662); THE TYCOON (I,4660)

MARGOLIN, JANET
LANIGAN'S RABBI (II,1427); LANIGAN'S RABBI (II,1428); THE OUTSIDE MAN (II,1937); PLANET EARTH (I,3615)

MARGOLIN, STUART
BRET MAVERICK (II,374); COPS (I,1047); LANIGAN'S RABBI (II,1427); MR. SMITH (II,1755); NICHOLS (I,3283); OCCASIONAL WIFE (I,3325); THE PARTRIDGE FAMILY (II,1962); THE ROCKFORD FILES (II,2196); THE ROCKFORD FILES (II,2197)

MARGOLIS, BENJAMIN
MISS WINSLOW AND SON (II,1705)

MARGOLIS, MARK
RAGE OF ANGELS (II,2107)

MARGOTTA, MICHAEL
THE BLUE KNIGHT (II,277)

MARIANO, BOBBY
MR. BROADWAY (I,3130)

MARIANO, PATRICIA
MR. BROADWAY (I,3130)

MARIE, JULIENNE
OUR PRIVATE WORLD (I,3417)

MARIE, ROSE
AN AMATEUR'S GUIDE TO LOVE (I,154); THE BOB HOPE SHOW (I,649); THE DICK VAN DYKE SHOW (I,1275); THE DORIS DAY SHOW (I,1355); THE HOLLYWOOD SQUARES (II,1164); HOLLYWOOD'S PRIVATE HOME MOVIES II (II,1168); HONEYMOON SUITE (I,2108); LOVE THAT BOB (I,2779); MY SISTER EILEEN (I,3200); A PUREX DINAH SHORE SPECIAL (I,3697)

MARIHUGH, TAMMY
LOVE THAT BOB (I,2779)

MARILLO, DER
MICKIE FINN'S (I,3029)

MARIN, JERRY
ALEX AND THE DOBERMAN GANG (II,32); ANDY'S GANG (I,214); NO SOAP, RADIO (II,1856); TV FUNNIES (II,2672)

MARIN, RUSS
THE ASPHALT COWBOY (II,111); FROM HERE TO ETERNITY (II,932)

MARINARO, ED
HILL STREET BLUES (II,1154); LAVERNE AND SHIRLEY (II,1446); THREE EYES (II,2605)

MARINE, JOE
THE FRED WARING SHOW (I,1679)

MARINERS, THE
ARTHUR GODFREY AND FRIENDS (I,278); THE JULIUS LAROSA SHOW (I,2487)

MARION, RICHARD
MODESTY BLAISE (II,1719); OPERATION PETTICOAT (II,1916); OPERATION PETTICOAT (II,1917); THE RETURN OF MARCUS WELBY, M.D (II,2138)

MARIONI, RAY
CASANOVA (II,451)

MARK, FLIP
ANOTHER DAY, ANOTHER DOLLAR (I,228); DAYS OF OUR LIVES (II,629); FAIR EXCHANGE (I,1512); GUESTWARD HO! (I,1900); HENRY FONDA AND THE FAMILY (I,2015); HOW'S BUSINESS? (I,2153)

MARK, JACKS
O'MALLEY (II,1872)

MARK, NORMAN
BREAKAWAY (II,370)

MARKELA, HELENA
STRANGER IN OUR HOUSE (II,2477)

MARKEY, ENID
BRINGING UP BUDDY (I,731); HAPPY BIRTHDAY (I,1941); STAGE DOOR (I,4177)

MARKHAM, MONTE
BREAKAWAY (II,370); DALLAS (II,614); ELLERY QUEEN: TOO MANY SUSPECTS (II,767); MITZI AND A HUNDRED GUYS (II,1710); MR. DEEDS GOES TO TOWN (I,3134); THE NEW ADVENTURES OF PERRY MASON (I,3252); RITUALS (II,2182); THE SECOND HUNDRED YEARS (I,3959); THE SIX-MILLION-DOLLAR MAN (II,2372)

MARKHAM, PETRA
A KILLER IN EVERY CORNER (I,2538)

MARKHAM, PIGMEAT
ABC STAGE '67 (I,6); ROWAN AND MARTIN'S LAUGH-IN (I,3856)

MARKIM, AL
TOM CORBETT, SPACE CADET (I,4535)

MARKLAND, TED
FATHER MURPHY (II,841); HARDCASE (II,1088); THE HIGH CHAPARRAL (I,2051)

MARKLE, FLETCHER
FRONT ROW CENTER (I,1694)

MARKO, LARRY
THIS GIRL FOR HIRE (II,2596)

MARKOFF, DIANE
QUINCY, M. E. (II,2102); THE SECRET EMPIRE (II,2291)

MARKOVA, ALICIA
VICTOR BORGE'S COMEDY IN MUSIC IV (I,4726)

MARKS, AL
ALL IN FUN (I,131)

MARKS, GUY
BUNGLE ABBEY (II,396); DEAN'S PLACE (II,650); GREAT DAY (II,1053); THE JOEY BISHOP SHOW (I,2404); THE JOHN FORSYTHE SHOW (I,2410); RANGO (I,3721)

MARKS, JOE E
PETER PAN (I,3566); SUNDAY DATE (I,4284)

MARKS, LOU
ALL IN FUN (I,131)

MARKS, MARIANNE
THE BILLION DOLLAR THREAT (II,246)

MARKSMEN, THE
JUBILEE U.S.A. (I,2461)

MARLEY, BEN
THE ACADEMY II (II,4); HIGH SCHOOL, U.S.A. (II,1149); SQUARE PEGS (II,2431)

MARLEY, JOHN
THE BOSTON TERRIER (I,696); FALCON'S GOLD (II,809); HANDS OF MURDER (I,1929); HAWK (I,1974); THE INNER SANCTUM (I,2216); THE SENATOR (I,3976); THREE STEPS TO HEAVEN (I,4485)

MARLING, JERRY
LIDSVILLE (I,2679)

MARLO, BOB
WORDS AND MUSIC (I,4909)

MARLO, STEVE
THE HOME FRONT (II,1171)

MARLOE, JEAN
GEORGE AND MILDRED (II,965)

MARLOW, JUNE
A HORSEMAN RIDING BY (II,1182)

MARLOWE, CHRISTIAN
HIGHCLIFFE MANOR (II,1150); LOVE OF LIFE (I,2774)

MARLOWE, FAY
THE LAST WAR (I,2630)

MARLOWE, FRANK
COMBAT SERGEANT (I,1012)

MARLOWE, HUGH
THE ADVENTURES OF ELLERY QUEEN (I,59); ANOTHER WORLD (II,97); ON TRIAL (I,3363)

MARLOWE, MARION
ARTHUR GODFREY AND FRIENDS (I,278); FUN FAIR (I,1706)

MARLOWE, NORA
EDDIE (I,1405); FULL HOUSE (II,936); THE GOVERNOR AND J.J. (I,1863); HOT WHEELS (I,2126); LAW OF THE PLAINSMAN (I,2639); MISS STEWART, SIR (I,3064); MY WIVES JANE (I,3207); NATIONAL VELVET (I,3231); THE WALTONS (II,2740)

MARLOWE, SCOTT
EXECUTIVE SUITE (II,796); THOU SHALT NOT KILL (II,2604); TRAVIS LOGAN, D.A. (I,4598); THE UNKNOWN (I,4680)

MARLOWE, WILLIAM
THE EYES HAVE IT (I,1496); THE SNOW GOOSE (I,4103)

MARMY, MAE
ARCHIE (II,104); CAGNEY AND LACEY (II,409); SIX PACK (II,2370)

MARNE, LISA
THE RAY ANTHONY SHOW (I,3729)

MARNER, RICHARD
MACKENZIE (II,1584)

MAROFF, BOB
CONCRETE BEAT (II,562)

MARONEY, KELLI
CELEBRITY (II,463); RYAN'S HOPE (II,2234)

MAROSS, JOE
CODE RED (II,552); CODE RED (II,553); SKYFIGHTERS (I,4078); STAR TONIGHT (I,4199)

MARQUEZ, WILLIAM
FARRELL: FOR THE PEOPLE (II,834)

MARQUIS CHIMPS, THE
THE HATHAWAYS (I,1965); THE JACK BENNY HOUR (I,2293)

MARQUIS, DIXIE
THE PICTURE OF DORIAN GRAY (I,3592); YOUNG DR. KILDARE (I,4937)

MARQUIS, KENNETH
THE ADDAMS FAMILY (II,13)

MARQUIS, KRISTOPHER
CAPTAINS AND THE KINGS (II,435); THE CLIFFWOOD AVENUE KIDS (II,543)

MARR, EDDIE
THE BOB HOPE SHOW (I,625); CIRCUS BOY (I,959)

MARREN, JERRY
QUICK AND QUIET (II,2099)

MARRIOTT, JENNIFER
THE TALE OF BEATRIX POTTER (II,2533)

MARROCCO, GINO
CAGNEY AND LACEY (II,410)

MARROW, PETER
THE DOCTORS (I,1323)

MARS, BRUCE
THEN CAME BRONSON (I,4436)

MARS, JANICE
NORBY (I,3304)

MARS, KENNETH
THE ALAN KING SHOW (I,101); AMERICA 2-NIGHT (II,65); BUNCO (II,395); CAROL BURNETT AND COMPANY (II,441); COMEDY IS KING (I,1019); COMEDY NEWS (I,1020); COMEDY NEWS II (I,1021); THE DON KNOTTS SHOW (I,1334); THE FACTS OF LIFE (II,805); THE FIGHTING NIGHTINGALES (II,855); FLINTSTONE FAMILY ADVENTURES (II,882); FULL HOUSE (II,936); FULL HOUSE (II,937); HE AND SHE (I,1985); IT'S A BIRD, IT'S A PLANE, IT'S SUPERMAN (II,1253); THE KAREN VALENTINE SHOW (I,2506); THE KILLIN' COUSIN (II,1394); LAVERNE AND

SHIRLEY IN THE ARMY (II,1448); LAVERNE AND SHIRLEY WITH THE FONZ (II,1449); SECOND CHANCE (I,3958); SHA NA NA (II,2311); SHEPHERD'S FLOCK (I,4007); VAN DYKE AND COMPANY (II,2718); WONDER WOMAN (PILOT 2) (II,2827); YOU CAN'T TAKE IT WITH YOU (II,2857)

MARS, KENNETH

MARSAC, MAURICE
BARE ESSENCE (II,149); THE JERK, TOO (II,1282)

MARSCA, MAURICE
COMMAND PERFORMANCE (I,1029)

MARSDEN, ROY
VANITY FAIR (I,4701)

MARSH, JEAN
9 TO 5 (II,1852); CIRCUS OF THE STARS (II,536); DOCTOR WHO (II,689); GOLIATH AWAITS (II,1025); THE INFORMER (I,2212); THE INTERNATIONAL ANIMATION FESTIVAL (II,1240); MASTER OF THE GAME (II,1647); MOMMA THE DETECTIVE (II,1721); THE MOON AND SIXPENCE (I,3098); UPSTAIRS, DOWNSTAIRS (II,2709)

MARSH, JULIE
THE CARNATION KILLER (I,845)

MARSH, LINDA
THE D.A.: CONSPIRACY TO KILL (I,1121); HOWDY DOODY (I,2151); THE MONK (I,3087); OF MEN OF WOMEN (I,3332)

MARSH, MICHELE
SANDBURG'S LINCOLN (I,3907)

MARSH, REGINALD
GEORGE AND MILDRED (II,965)

MARSH, SPENCER
ANOTHER LIFE (II,95)

MARSH, TIGER JOE
READY AND WILLING (II,2115)

MARSHAK, TRACY
HAVING BABIES II (II,1108)

MARSHALL FIELD DEPARTMENT STORE MODELS,
he DON MCNEILL'S DINNER CLUB (I,1336)

MARSHALL, ALEXANDER
THE MADHOUSE BRIGADE (II,1588)

MARSHALL, ALEXANDRA
CORONATION STREET (I,1054)

MARSHALL, ANN
MICKEY AND THE CONTESSA (I,3024)

MARSHALL, BARBARA
SONG AND DANCE TIME (I,4121); WORDS AND MUSIC (I,4908)

MARSHALL, BRYAN
A PLACE TO DIE (I,3612); VANITY FAIR (I,4701)

MARSHALL, CONNIE
DOC CORKLE (I,1304)

MARSHALL, DOLORES
SERENADE (I,3979); SING-CO-PATION (I,4058)

MARSHALL, DON
BRADDOCK (I,704); JULIA (I,2476); LAND OF THE GIANTS (I,2611)

MARSHALL, E G
ALFRED HITCHCOCK PRESENTS (I,115); AMERICAN LIFESTYLE (I,170); THE CAMPBELL TELEVISION SOUNDSTAGE (I,812); THE CHEVROLET TELE-THEATER (I,926); THE CITY (I,969); COSMOPOLITAN THEATER (I,1057); DANGER (I,1134); THE DEFENDERS (I,1220); THE DESILU PLAYHOUSE (I,1237); THE DOCTORS (I,1323); DUPONT SHOW OF THE MONTH (I,1387); THE ELGIN HOUR (I,1428); ELLERY QUEEN: DON'T LOOK BEHIND YOU (I,1434); FALCON CREST (II,808); FLESH AND BLOOD (I,1595); THE INNER SANCTUM (I,2216); KENNEDY (II,1377); KRAFT TELEVISION THEATER (I,2592); LIGHT'S OUT (I,2699); THE LITTLE FOXES (I,2712); THE LITTLEST ANGEL (I,2726); PHILCO TELEVISION PLAYHOUSE (I,3583); THE PHOENIX (II,2035); THE PHOENIX (II,2036); PLAYWRIGHTS '56 (I,3627); THE POPPY IS ALSO A FLOWER (I,3650); PURSUIT (I,3700); SHIRLEY TEMPLE'S STORYBOOK (I,4017); SHORT, SHORT DRAMA (I,4023); THE SHOW MUST GO ON (II,2344); SURE AS FATE (I,4297); SUSPENSE (I,4305); SUSPICION (I,4309); THE TRAP (I,4593); THE WEB (I,4784)

MARSHALL, ED
THE GRADY NUTT SHOW (II,1048)

MARSHALL, GARRY
THE GEORGE BURNS AND GRACIE ALLEN SHOW (I,1763); THE UGLIEST GIRL IN TOWN (I,4663)

MARSHALL, GLORIA
FIRESIDE THEATER (I,1580); LOVE THAT BOB (I,2779); THE SECRET OF MYSTERY LAKE (I,3967); TV READERS DIGEST (I,4630)

MARSHALL, GREGORY
THE LIFE OF RILEY (I,2686)

MARSHALL, HERBERT
THE KEN MURRAY SHOW (I,2525); THE PHILADELPHIA STORY (I,3581); TIMES SQUARE PLAYHOUSE (I,4513); THE UNEXPECTED (I,4674)

MARSHALL, JACK
THE KEN MURRAY SHOW (I,2525)

MARSHALL, JANE
THE ROBERT Q. LEWIS CHRISTMAS SHOW (I,3811)

MARSHALL, JIM
LIFE'S MOST EMBARRASSING MOMENTS (II,1470)

MARSHALL, JOAN
BOLD VENTURE (I,682)

MARSHALL, KEN
MARCO POLO (II,1626)

MARSHALL, LINDA
TAMMY (I,4357)

MARSHALL, LIZ
A TOWN LIKE ALICE (II,2652)

MARSHALL, MONA
INSPECTOR GADGET (II,1236)

MARSHALL, MORT
ALICE IN WONDERLAND (I,120); THE DUMPLINGS (II,743); HOLD IT PLEASE (I,2071); ONE TOUCH OF VENUS (I,3389); PANAMA HATTIE (I,3444); PINOCCHIO (I,3603); THE RIGHT MAN (I,3790)

MARSHALL, NANCY
YOUNG IN HEART (I,4939)

MARSHALL, PAT
THE TONIGHT SHOW (I,4562)

MARSHALL, PENNY
THE BARRY MANILOW SPECIAL (II,155); BATTLE OF THE NETWORK STARS (II,163); BATTLE OF THE NETWORK STARS (II,164); BATTLE OF THE NETWORK STARS (II,165); CELEBRITY CHALLENGE OF THE SEXES 1 (II,465); THE CELEBRITY FOOTBALL CLASSIC (II,474); CIRCUS OF THE STARS (II,531); THE COMEDY ZONE (II,559); EVIL ROY SLADE (I,1485); FRIENDS AND LOVERS (II,930); GENERAL ELECTRIC'S ALL-STAR ANNIVERSARY (II,963); HAPPY DAYS (II,1084); LAVERNE AND SHIRLEY (II,1446); LAVERNE AND SHIRLEY IN THE ARMY (II,1448); LILY FOR PRESIDENT (II,1478); THE MARY TYLER MOORE SHOW (II,1640); THE ODD COUPLE (II,1875); PAUL SAND IN FRIENDS AND LOVERS (II,1982); THE SUPER (I,4293); WIVES (II,2812)

MARSHALL, PETER
79 PARK AVENUE (II,2310); AN AMATEUR'S GUIDE TO LOVE (I,154); CIRCUS OF THE STARS (II,530); FANTASY (II,828); HAPPY ANNIVERSARY AND GOODBYE (II,1082); THE HOLLYWOOD SQUARES (II,1164); THE LUCY SHOW (I,2791); MANHATTAN TOWER (I,2889); MITZI AND A HUNDRED GUYS (II,1710); NBC ACTION PLAYHOUSE (I,3234); PETER MARSHALL SALUTES THE BIG BANDS (II,2028); THE PETER MARSHALL VARIETY SHOW (II,2029); THE RICH LITTLE SHOW (II,2154); SHORE LEAVE (I,4022); THE STORYBOOK SQUARES (I,4242); VER-R-R-RY INTERESTING (I,4712)

MARSHALL, REBECCA
A TOWN LIKE ALICE (II,2652)

MARSHALL, REX
THE DOODLES WEAVER SHOW (I,1351)

MARSHALL, RON
THE YEAR WITHOUT A SANTA CLAUS (II,2846)

MARSHALL, SARAH
APPLAUSE (I,244); THE CAT AND THE CANARY (I,868); CYRANO DE BERGERAC (I,1119); THE DUPONT SHOW OF THE WEEK (I,1388); KINCAID (I,2543); MISS WINSLOW AND SON (II,1705); ROBERT MONTGOMERY PRESENTS YOUR LUCKY STRIKE THEATER (I,3809); SCRUPLES (II,2280); THRILLER (I,4492)

MARSHALL, SEAN
THE FITZPATRICKS (II,868); THE MACKENZIES OF PARADISE COVE (II,1585); STICKIN' TOGETHER (II,2465)

MARSHALL, WILLIAM
ROSETTI AND RYAN (II,2216); SABU AND THE MAGIC RING (I,3881)

MARSHALOV, BORIS
JOHNNY BELINDA (I,2416); REUNION IN VIENNA (I,3780)

MARSON, ANIA
MOLL FLANDERS (II,1720); THE STRAUSS FAMILY (I,4253)

MARTA, LYNNE
THE DAVID SOUL AND FRIENDS SPECIAL (II,626); GENESIS II (I,1760); KNIGHT RIDER (II,1402); THE LLOYD THAXTON SHOW (I,2735)

MARTAIN, IAN
WONDERFUL JOHN ACTION (I,4892)

MARTEL, DONNA
G.E. SUMMMER ORIGINALS (I,1747)

MARTEL, K C
BEULAH LAND (II,226); EIGHT IS ENOUGH (II,762); MULLIGAN'S STEW (II,1763); MULLIGAN'S STEW (II,1764); THE MUNSTERS' REVENGE (II,1765); THINGS ARE LOOKING UP (II,2588)

MARTEL, MICHAELINA
YOURS FOR A SONG (I,4975)

MARTELL, ARLENE
CONSPIRACY OF TERROR (II,569)

MARTELL, GREGG
ALIAS MIKE HERCULES (I,117)

MARTER, IAN
DOCTOR WHO (II,689)

MARTH, FRANK
BROADWAY (I,733); CAPTAIN AMERICA (II,427); THE CRIME CLUB (I,1096); THE DELPHI BUREAU (I,1224); FROM THESE ROOTS (I,1688); THE HONEYMOONERS (I,2110); JACKIE GLEASON AND HIS AMERICAN SCENE MAGAZINE (I,2321); THE JACKIE GLEASON SHOW (I,2322); PROFILES IN COURAGE (I,3678); THE YOUNG PIONEERS (II,2868)

MARTHA RAYE DANCERS, THE
THE MARTHA RAYE SHOW (I,2926)

MARTHA AND THE VANDELLAS
IT'S WHAT'S HAPPENING BABY! (I,2278)

MARTIC, SANDY
FARRELL: FOR THE PEOPLE (II,834)

MARTIN BROTHERS, THE
SHOW OF THE YEAR (I,4031)

MARTIN, ALBERTO
FEEL THE HEAT (II,848)

MARTIN, ANDREA
ANSON AND LORRIE (II,98); THE COMEDY ZONE (II,559); FROM CLEVELAND (II,931); THE ROBERT KLEIN SHOW (II,2185); SECOND CITY TELEVISION (II,2287); THAT SECOND THING ON ABC (II,2572); THAT THING ON ABC (II,2574)

MARTIN, ANNE MARIE
DAYS OF OUR LIVES (II,629)

MARTIN, ANTHONY
1915 (II,1853)

MARTIN, BARNEY
A DOG'S LIFE (II,697); JACKIE GLEASON AND HIS AMERICAN SCENE MAGAZINE (I,2321); NUMBER 96 (II,1869); THE TONY RANDALL SHOW (II,2640); ZERO HOUR (I,4980); ZORRO AND SON (II,2878)

MARTIN, BERNARD
DOLLAR A SECOND (I,1326)

MARTIN, BILLY
BOB HOPE SPECIAL: THE BOB HOPE SPECIAL (II,335); THE DEAN MARTIN CELEBRITY ROAST (II,639)

MARTIN, BOBBI
THE BOB HOPE SHOW (I,638); THE BOB HOPE SHOW (I,646); GUESS AGAIN (I,1894)

MARTIN, BRUCE
THE EDGE OF NIGHT (II,760)

MARTIN, CHARLES
PHILIP MORRIS PLAYHOUSE (I,3585)

MARTIN, CHRISTOPHER
THE RENEGADES (II,2132)

MARTIN, CON
SIX PACK (II,2370)

MARTIN, CONNIE
HALF THE GEORGE KIRBY COMEDY HOUR (I,1920)

MARTIN, DEAN
ANN-MARGRET: FROM HOLLYWOOD WITH LOVE (I,217); THE BEST ON RECORD (I,409); THE BIG SHOW (II,243); BING CROSBY AND HIS FRIENDS (I,455); THE BING CROSBY SHOW (I,466); BING CROSBY—COOLING IT (I,459); BING CROSBY—COOLING IT (I,460); THE BOB HOPE SHOW (I,572); THE BOB HOPE SHOW (I,582); THE BOB HOPE SHOW (I,595); THE BOB HOPE SHOW (I,596); BOB HOPE SPECIAL: BOB HOPE'S ALL-STAR COMEDY SPECTACULAR FROM LAKE TAHOE (II,301); BOB HOPE SPECIAL: BOB HOPE'S PINK PANTHER THANKSGIVING GALA (II,312); BOB HOPE SPECIAL: THE BOB HOPE SPECIAL (II,334); THE COLGATE COMEDY HOUR (I,997); DEAN MARTIN AT THE WILD ANIMAL PARK (II,636); THE DEAN MARTIN CELEBRITY ROAST (II,637); THE DEAN MARTIN CELEBRITY ROAST (II,638); THE DEAN MARTIN CELEBRITY ROAST (II,639); THE DEAN MARTIN CHRISTMAS SPECIAL (II,640); DEAN MARTIN IN LONDON (II,641); THE DEAN MARTIN SHOW (I,1196); THE DEAN MARTIN SHOW (I,1197); THE DEAN MARTIN SHOW (I,1198); THE DEAN MARTIN SHOW (I,1199); THE DEAN MARTIN SHOW (I,1200); THE DEAN MARTIN SHOW (I,1201); DEAN MARTIN'S CALIFORNIA CHRISTMAS (II,642); DEAN MARTIN'S CELEBRITY ROAST (II,643); DEAN MARTIN'S CHRISTMAS AT SEA WORLD (II,644); DEAN MARTIN'S CHRISTMAS IN CALIFORNIA (II,645); DEAN MARTIN'S COMEDY CLASSICS (II,646); DEAN MARTIN'S RED HOT SCANDALS OF 1926 (II,648); DEAN MARTIN'S RED HOT SCANDALS PART 2 (II,649); DEAN'S PLACE (II,650); DOM DELUISE AND FRIENDS (II,701); DOM DELUISE AND FRIENDS, PART 2 (II,702); THE DON RICKLES SHOW (II,708); FAVORITE SONGS (I,1545); THE FIRST 50 YEARS (II,862); THE FRANK SINATRA SHOW (I,1666); FROM HOLLYWOOD WITH LOVE: THE ANN-MARGRET SPECIAL (I,1687); THE GENERAL MOTORS 50TH ANNIVERSARY SHOW (I,1758); JACK BENNY'S 20TH ANNIVERSARY TV SPECIAL (I,2303); JACK BENNY'S FIRST FAREWELL SHOW (I,2300); THE JUDY GARLAND SHOW (I,2472); LADIES AND GENTLEMAN. . . BOB NEWHART, PART II (II,1421); A LUCILLE BALL SPECIAL STARRING LUCILLE BALL AND DEAN MARTIN (II,1557); MAC DAVIS 10TH ANNIVERSARY SPECIAL: I STILL BELIEVE IN MUSIC (II,1571); MOVIN' WITH NANCY (I,3119); PETULA (I,3573); THE PHIL HARRIS SHOW (I,3575); THE POWDER ROOM (I,3656); SHIRLEY MACLAINE. . . EVERY LITTLE MOVEMENT (II,2334); SHOW OF THE YEAR (I,4031); SINATRA AND FRIENDS (II,2358); SINATRA—THE FIRST 40 YEARS (II,2359); SWING OUT, SWEET LAND (II,2515); THE WONDERFUL WORLD OF BURLESQUE II (I,4896)

MARTIN, DEAN PAUL
ANN-MARGRET'S HOLLYWOOD MOVIE GIRLS (II,86); THE BOYS IN BLUE (II,359)

MARTIN, DEWEY
DANIEL BOONE (I,1141); DOC HOLLIDAY (I,1307); FRONT ROW CENTER (I,1694)

MARTIN, DICK
AN AMATEUR'S GUIDE TO LOVE (I,154); THE BOB HOPE SHOW (I,553); THE BOB HOPE SHOW (I,615); THE CHEAP SHOW (II,491); THE CHEVY SUMMER SHOW (I,929); THE COLGATE COMEDY HOUR (I,998); THE DINAH SHORE SPECIAL—LIKE HEP (I,1282); HI, I'M GLEN CAMPBELL (II,1141); HOUSE CALLS (II,1194); THE JERRY LEWIS SHOW (I,2363); LADIES AND GENTLEMAN. . . BOB NEWHART, PART II (II,1421); THE LUCY SHOW (I,2791); THE MAD MAD MAD MAD WORLD OF THE SUPER BOWL (II,1586); MINDREADERS (II,1702); MURDER AT NBC (A BOB HOPE SPECIAL) (I,3159); PLAYBOY'S PLAYMATE PARTY (II,2055); ROWAN AND MARTIN BITE THE HAND THAT FEEDS THEM (I,3853); THE ROWAN AND MARTIN REPORT (II,2221); THE ROWAN AND MARTIN SHOW (I,3854); THE ROWAN AND MARTIN SPECIAL (I,3855); ROWAN AND MARTIN'S LAUGH-IN (I,3856); ROWAN AND MARTIN'S LAUGH-IN (I,3857); ROYAL VARIETY

PERFORMANCE (I,3865); SUPER COMEDY BOWL 2 (I,4289); SWING OUT, SWEET LAND (II,2515); TAKE ONE STARRING JONATHAN WINTERS (II,2532); ULTRA QUIZ (II,2701); WHO'S AFRAID OF MOTHER GOOSE? (I,4852)

MARTIN, EILEEN
BREAKFAST PARTY (I,716); THE MEL TORME SHOW (I,2995)

MARTIN, ERIN
THIS IS YOUR MUSIC (I,4462)

MARTIN, EUGENE
JEFFERSON DRUM (I,2355)

MARTIN, FRAN
THE DESILU REVUE (I,1238)

MARTIN, FREDDY
THE HAZEL BISHOP SHOW (I,1983)

MARTIN, GAIL
DEAN MARTIN PRESENTS THE GOLDDIGGERS (I,1194); DEAN MARTIN PRESENTS THE GOLDDIGGERS (I,1193); DEAN MARTIN PRESENTS THE VIC DAMONE SHOW (I,1195); THE VIC DAMONE SHOW (I,4718)

MARTIN, GEORGE
KENNEDY (II,1377)

MARTIN, HELEN
BABY, I'M BACK! (II,130); BENSON (II,208); BIG DADDY (I,425); THE JERK, TOO (II,1282); THAT'S MY MAMA (II,2580)

MARTIN, IAN
ALICE IN WONDERLAND (I,120); THE DEVIL'S DISCIPLE (I,1247); THE EDGE OF NIGHT (II,760); MAN AND SUPERMAN (I,2854); THE O'NEILLS (I,3392)

MARTIN, IRENE A
DIAGNOSIS: DANGER (I,1250)

MARTIN, JARED
BATTLE OF THE NETWORK STARS (II,173); THE BIG EASY (II,234); DALLAS (II,614); THE FANTASTIC JOURNEY (II,826); GRIFF (I,1888); HOW THE WEST WAS WON (II,1196); M STATION: HAWAII (II,1568); MEN OF THE DRAGON (II,1683); WILLOW B: WOMEN IN PRISON (II,2804)

MARTIN, JIM
THE GREAT SPACE COASTER (II,1059)

MARTIN, JOANNA
HARDCASE (II,1088)

MARTIN, JOHN PATRICK
A CHRISTMAS CAROL (II,517)

MARTIN, JOHN SCOTT
DOCTOR WHO—THE FIVE DOCTORS (II,690)

MARTIN, KEIL
THE CATCHER (I,872); HILL STREET BLUES (II,1154); THE LOG OF THE BLACK PEARL (II,1506)

MARTIN, LEILA
GOLDEN WINDOWS (I,1841)

MARTIN, LEWIS
DOCTOR MIKE (I,1316)

MARTIN, LIAM
HAPPY ANNIVERSARY, CHARLIE BROWN (I,1939); IT'S ARBOR DAY, CHARLIE BROWN (I,2267); WHAT A NIGHTMARE, CHARLIE BROWN (I,4806); YOU'RE A GOOD SPORT, CHARLIE BROWN (I,4963)

MARTIN, LORI
NATIONAL VELVET (I,3231); SCRUPLES (II,2280); WEEKEND (I,4788)

MARTIN, LUCY
THE EDGE OF NIGHT (II,760)

MARTIN, MARCELLA
ANOTHER WORLD (II,97)

MARTIN, MARJI
ONE NIGHT BAND (II,1909)

MARTIN, MARY
ANNIE GET YOUR GUN (I,223); THE BING CROSBY CHRISTMAS SHOW (I,458); BOB HOPE SPECIAL: BOB HOPE'S ALL-STAR COMEDY BIRTHDAY PARTY AT WEST POINT (II,298); BORN YESTERDAY (I,692); THE FORD 50TH ANNIVERSARY SHOW (I,1630); THE GENERAL FOODS 25TH ANNIVERSARY SHOW (I,1756); HARDCASTLE AND MCCORMICK (II,1089); MAGIC WITH MARY MARTIN (I,2821); THE MARY MARTIN SHOW (I,2934); MUSIC WITH MARY MARTIN (I,3180); PETER PAN (I,3566); SKIN OF OUR TEETH (I,4073); TOGETHER WITH MUSIC (I,4532)

MARTIN, MEL
BREAKFAST PARTY (I,716); THE PALLISERS (II,1944); STRAWHAT MATINEE (I,4255); STRAWHAT THEATER (I,4256)

MARTIN, MERRY
PETER LOVES MARY (I,3565)

MARTIN, MIA
THE BENNY HILL SHOW (II,207)

MARTIN, MICKEY
DAYS OF OUR LIVES (II,629)

MARTIN, MILLICENT
ESPIONAGE (I,1464); FROM A BIRD'S EYE VIEW (I,1686); PICCADILLY PALACE (I,3589); SONG BY SONG (II,2399); TELEVISION SCREEN MAGAZINE (I,4383); THE TOM JONES CHRISTMAS SPECIAL (I,4539); TV SCREEN MAGAZINE (I,4631)

MARTIN, NAN
DEAR TEACHER (II,653); THE THORN BIRDS (II,2600)

MARTIN, NINA
THE BENNY HILL SHOW (II,207)

MARTIN, OWEN
BROTHER RAT (I,746)

MARTIN, PAMELA SUE
DYNASTY (II,746); THE GUN AND THE PULPIT (II,1067); HUMAN FEELINGS (II,1203); THE NANCY DREW MYSTERIES (II,1789)

MARTIN, PEPPER
THE OUTSIDE MAN (II,1937)

MARTIN, RALPH
OUTLAW LADY (II,1935)

MARTIN, ROBERT DALE
KENNEDY (II,1377)

MARTIN, RON
THE GARRY MOORE SHOW (I,1739); SOMERSET (I,4115)

MARTIN, ROSEMARY
THE TALE OF BEATRIX POTTER (II,2533)

MARTIN, ROSS
THE AD-LIBBERS (I,29); CONCERNING MISS MARLOWE (I,1032); DREAM GIRL OF '67 (I,1375); GODZILLA (II,1014); HAWAII FIVE-O (II,1110); I AND CLAUDIE (I,2164); LIGHT'S OUT (I,2699); MITZI AND A HUNDRED GUYS (II,1710); MITZI'S SECOND SPECIAL (I,3074); MORE WILD WILD WEST (II,1733); MR. LUCKY (I,3141); NIGHT GALLERY (I,3287); ONE STEP BEYOND (I,3388); THE RETURN OF CHARLIE CHAN (II,2136); THE RETURN OF THE MOD SQUAD (I,3776); SEALAB 2020 (I,3949); SEARCH FOR TOMORROW (II,2284); THE SECRET JURY (I,3965); THE SEEKERS (II,2298); THE SKATEBIRDS (II,2375); THE SKY'S THE LIMIT (I,4081); STUMP THE STARS (I,4273); SWING OUT, SWEET LAND (II,2515); TEXAS JOHN SLAUGHTER (I,4414); THE THREE ROBONIC STOOGES (II,2609); THE TWILIGHT ZONE (I,4651); WILD AND WOOLEY (II,2798); THE WILD WILD WEST (I,4863); THE WILD WILD WEST REVISITED (II,2800)

MARTIN, RUCI
THE LISA HARTMAN SHOW (II,1484)

MARTIN, STEVE
ALL COMMERCIALS—A STEVE MARTIN SPECIAL (II,37); BOB HOPE SPECIAL: THE BOB HOPE SPECIAL (II,335); HALF THE GEORGE KIRBY COMEDY HOUR (I,1920); JOHN DENVER ROCKY MOUNTAIN CHRISTMAS (II,1314); JOHNNY CASH AND FRIENDS (II,1322); JOHNNY CASH: THE FIRST 25 YEARS (II,1336); THE KEN BERRY WOW SHOW (I,2524); THE SENSATIONAL, SHOCKING, WONDERFUL, WACKY 70S (II,2302); STEVE MARTIN'S BEST SHOW EVER (II,2460); STEVE MARTIN'S THE WINDS OF WHOOPIE (II,2461); STEVE MARTIN: COMEDY IS NOT PRETTY (II,2462); TWILIGHT THEATER (II,2685)

MARTIN, STROTHER
HAWKINS (I,1976); HAWKINS ON MURDER (I,1978); HOTEL DE PAREE (I,2129); MITZI AND A HUNDRED GUYS (II,1710); ONE OF OUR OWN (II,1910); STEVE MARTIN'S THE WINDS OF WHOOPIE (II,2461); STUBBY PRINGLE'S CHRISTMAS (II,2483)

MARTIN, TODD
ALL THAT GLITTERS (II,41); THE BARBARY COAST (II,147); THE LAWYERS (I,2646)

MARTIN, TONY
THE BOB HOPE SHOW (I,517); THE BOB HOPE SHOW (I,528); THE GENERAL FOODS 25TH ANNIVERSARY SHOW (I,1756); MEET CYD CHARISSE (I,2982); SPRING HOLIDAY (I,4168); THE TONY MARTIN SHOW (I,4572); TWELVE STAR SALUTE (I,4638)

MARTIN, TREVOR
MASTER OF THE GAME (II,1647)

MARTIN, W T
AS THE WORLD TURNS (II,110)

MARTINDALE, WINK
CAN YOU TOP THIS? (I,817); DREAM GIRL OF '67 (I,1375); GAMBIT (II,955); HOW'S YOUR MOTHER-IN-LAW? (I,2148); LAS VEGAS GAMBIT (II,1430); TIC TAC DOUGH (II,2618); WHAT'S THIS SONG? (I,4822); WORDS AND MUSIC (I,4909)

MARTINELLI, GIOVANNI
OPERA CAMEOS (I,3398)

MARTINEZ, A
BORN TO THE WIND (II,354); CASSIE AND COMPANY (II,455); CENTENNIAL (II,477); THE COWBOYS (II,598); EXO-MAN (II,797); MALLORY: CIRCUMSTANTIAL EVIDENCE (II,1608); MRS R.—DEATH AMONG FRIENDS (II,1759); ROUGHNECKS (II,2219); SANTA BARBARA (II,2258); THE STOREFRONT LAWYERS (I,4233); WHIZ KIDS (II,2790)

MARTINEZ, ALMA
SCAMPS (II,2267)

MARTINEZ, CLAUDIO
BORN TO THE WIND (II,354); BRIDGER (II,376); THE NEW VOICE (II,1833); VIVA VALDEZ (II,2728)

MARTINEZ, EDDIE
BLONDIE (II,272)

MARTINEZ, JIMMY
LOLA (II,1508); LOLA (II,1509); PRESENTING SUSAN ANTON (II,2077); THE RICHARD PRYOR SHOW (II,2163); THE STEVE LANDESBERG TELEVISION SHOW (II,2458); TONY ORLANDO AND DAWN (II,2639); WHEN THINGS WERE ROTTEN (II,2782)

MARTINEZ, KENNETH
AT EASE (II,116)

MARTINEZ, MANUEL
IVAN THE TERRIBLE (II,1260); RIVKIN: BOUNTY HUNTER (II,2183)

MARTINEZ, MIKE
THE JACKSON FIVE (I,2326)

MARTINEZ, MINA
REDIGO (I,3758); ROUGHNECKS (II,2219)

MARTINEZ, TONY
THE REAL MCCOYS (I,3741)

MARTINS QUARTET, THE
THE PATRICE MUNSEL SHOW (I,3474)

MARVEL, FRANK
THE YOUNG MARRIEDS (I,4942)

MARVIN, LEE
ALCOA PREMIERE (I,109); THE BARBARA STANWYCK THEATER (I,350); THE BOB HOPE SHOW (I,648); THE BOB HOPE SHOW (I,651); BOB HOPE SPECIAL: BOB HOPE LAUGHS WITH THE MOVIE AWARDS (II,288); BOB HOPE SPECIAL: BOB HOPE'S HILARIOUS UNREHEARSED ANTICS OF THE STARS (II,341); THE CASE AGAINST PAUL RYKER (I,861); CLIMAX! (I,976); COUNTERPOINT (I,1059); THE DICK POWELL SHOW (I,1269); THE DOCTOR (I,1320); THE DUPONT SHOW OF THE WEEK (I,1388); THE GENERAL ELECTRIC THEATER (I,1753); GREAT ADVENTURE (I,1868); KRAFT SUSPENSE THEATER (I,2591); LAWBREAKER (I,2640); THE LOSERS (I,2757); M SQUAD (I,2799); MEDIC (I,2976); THE MOTOROLA TELEVISION HOUR (I,3112); SUSPENSE (I,4305); TV READERS DIGEST (I,4630)

MARX, CARL
HOLLYWOOD JUNIOR CIRCUS (I,2084)

MARX, CHICO
THE COLLEGE BOWL (I,1001); SILVER THEATER (I,4051)

MARX, EDEN
TIME FOR ELIZABETH (I,4506)

MARX, GREGG
DAYS OF OUR LIVES (II,629)

MARX, GROUCHO
THE BOB HOPE SHOW (I,601); THE DUPONT SHOW OF THE WEEK (I,1388); THE GENERAL FOODS 25TH ANNIVERSARY SHOW (I,1756); ONE MAN SHOW (I,3381); SHOW BIZ (I,4027); TELL IT TO GROUCHO (I,4385); TIME FOR ELIZABETH (I,4506); YOU BET YOUR LIFE (I,4932)

MARX, HARPO
DUPONT SHOW OF THE MONTH (I,1387); THE DUPONT SHOW OF THE WEEK (I,1388); THE JUNE ALLYSON SHOW (I,2488)

MARX, MELINDA
SHOW BIZ (I,4027)

MARYE, DONALD
THE KING AND MRS. CANDLE (I,2544)

MARZELLO, VINCENT
IKE (II,1223)

MASAK, RON
THE ALIENS ARE COMING (II,35); THE BANANA COMPANY (II,139); THE FURTHER ADVENTURES OF WALLY BROWN (II,944); THE GOOD GUYS (I,1847); HEAT OF ANGER (I,1992); JEREMIAH OF JACOB'S NECK (II,1281); LOVE THY NEIGHBOR (I,2781); MEDICAL CENTER (II,1675); PLEASURE COVE (II,2058)

MASCARINO, PIERRINO
NATIONAL LAMPOON'S TWO REELERS (II,1796)

MASCOLO, JOSEPH
BRONK (II,379); DOMINIC'S DREAM (II,704); THE GANGSTER CHRONICLES (II,957); SIDE BY SIDE (II,2346); STONESTREET: WHO KILLED THE CENTERFOLD MODEL? (II,2472); WHERE THE HEART IS (I,4831)

MASCORINO, PIERRINO
ANOTHER WORLD (II,97)

MASE, MARINO
JERICHO (I,2358)

MASK, DON
A WALK IN THE NIGHT (I,4750)

MASON, ARTHUR
SCOTLAND YARD (I,3941)

MASON, BILL
THE GREATEST GIFT (I,1880)

MASON, CALVIN
THE GARY COLEMAN SHOW (II,958)

MASON, DON
FOREST RANGERS (I,1642); STINGRAY (I,4228)

MASON, DONNA
CAMP GRIZZLY (II,418)

MASON, ERIC
THE TARZAN/LONE RANGER/ZORRO ADVENTURE HOUR (II,2543)

MASON, GREGG
COLONEL STOOPNAGLE'S STOOP (I,1006)

MASON, INGRID
THE SULLIVANS (II,2490)

MASON, JACKIE
THE BEST OF TIMES (II,219); EVENING AT THE IMPROV (II,786)

MASON, JAMES
ALCOA/GOODYEAR THEATER (I,107); THE BOB HOPE SHOW (I,543); G.E. SUMMMER ORIGINALS (I,1747); GEORGE WASHINGTON (II,978); HOLLYWOOD (II,1160); THE LEGEND OF SILENT NIGHT (I,2655); THE LUX VIDEO THEATER (I,2795); PANIC! (I,3447); PLAYHOUSE 90 (I,3623); REBECCA (I,3745); SCHLITZ PLAYHOUSE OF STARS (I,3936); SEARCH FOR THE NILE (I,3954); TONIGHT IN SAMARKAND (I,4558)

MASON, LOLA
BEYOND WITCH MOUNTAIN (II,228); FISHERMAN'S WHARF (II,865)

MASON, MARGARET
DAYS OF OUR LIVES (II,629); RETURN TO PEYTON PLACE (I,3779); THE YOUNG AND THE RESTLESS (II,2862)

MASON, MARLYN
BRIGADOON (I,726); CAROUSEL (I,856); ESCAPE (I,1460); GHOST STORY (I,1788); HANDLE WITH CARE (II,1077); LONGSTREET (I,2749); OF MEN OF WOMEN (I,3331); A STORM IN SUMMER (I,4237); TWO THE HARD WAY (II,2695)

MASON, MARSHA
LOVE OF LIFE (I,2774); YOUNG DR. KILDARE (I,4937)

MASON, MAX
THE NEXT VICTIM (I,3281)

MASON, MONICA
CINDERELLA (II,526)

MASON, PAMELA
ALAN KING IN LAS VEGAS, PART I (I,97); MY BUDDY (II,1778); PANIC! (I,3447); THE WEAKER SEX (?) (I,4782)

MASON, SYDNEY
MY FRIEND FLICKA (I,3190); STUMP THE AUTHORS (I,4272)

MASON, TOM
THE ALIENS ARE COMING (II,35); FREEBIE AND THE BEAN (II,922); GEORGE WASHINGTON (II,978); GRANDPA GOES TO WASHINGTON (II,1050); NERO WOLFE (II,1805); RETURN OF

THE MAN FROM U.N.C.L.E.: THE 15 YEARS LATER AFFAIR (II,2140); TWO MARRIAGES (II,2691)

MASSAGUER, CONRADO
THE RAY PERKINS REVUE (I,3735)

MASSE, LAUREL
THE MANHATTAN TRANSFER (II,1619)

MASSEY, ANNA
THE PALLISERS (II,1944)

MASSEY, DANIEL
THE ROADS TO FREEDOM (I,3802)

MASSEY, DARIA
THE ISLANDERS (I,2243); SABU AND THE MAGIC RING (I,3881)

MASSEY, ILONA
CAMEO THEATER (I,808); CURTAIN CALL THEATER (I,1115); FAITH BALDWIN'S THEATER OF ROMANCE (I,1515); FORD THEATER HOUR (I,1635); THE ILONA MASSEY SHOW (I,2191); RENDEZVOUS (I,3765)

MASSEY, JUDITH
ALL THE RIVERS RUN (II,43)

MASSEY, RAYMOND
THE CLOCK (I,978); DOCTOR KILDARE (I,1315); I SPY (I,2178); KRAFT TELEVISION THEATER (I,2592); LIGHT'S OUT (I,2699); ST. JOAN (I,4175)

MASSEY, VERA
THE LEFT OVER REVUE (I,2651); POETRY AND MUSIC (I,3632)

MASSEY, WAYNE
ONE LIFE TO LIVE (II,1907)

MASTERS, BEN
ANOTHER WORLD (II,97); CELEBRITY (II,463); LOOSE CHANGE (II,1518); MAKING IT (II,1605); MOBILE MEDICS (II,1716); MUGGSY (II,1761); ONE OF OUR OWN (II,1910)

MASTERS, DR WILLIAM H
AMERICA ALIVE! (II,67)

MASTERS, FRANKIE
LUCKY LETTERS (I,2786)

MASTERS, GARY
SOMEONE AT THE TOP OF THE STAIRS (I,4114)

MASTERS, MARIE
AS THE WORLD TURNS (II,110); HIGHER AND HIGHER, ATTORNEYS AT LAW (I,2057); ONE LIFE TO LIVE (II,1907)

MASTERS, NATALIE
ALICE (II,33); DATE WITH THE ANGELS (I,1164); NOAH'S ARK (I,3302)

MASTERS, RUTH
CAR 54, WHERE ARE YOU? (I,842)

MASTERS, WILLIAM
CONGRESSIONAL INVESTIGATOR (I,1037)

MASTERSON, PETER
DELTA COUNTY, U.S.A. (II,657)

MASTREN, CARMEN
THE MOHAWK SHOWROOM (I,3080)

MASTROIANNI, MARCELLO
THE POPPY IS ALSO A FLOWER (I,3650); SOPHIA LOREN IN ROME (I,4129); WITH LOVE, SOPHIA (I,4884)

MASUR, RICHARD
THE BOUNDER (II,357); BUMPERS (II,394); EAST OF EDEN (II,752); EMPIRE (II,777); HAVING BABIES I (II,1107); HOT L BALTIMORE (II,1187); THE MANY LOVES OF ARTHUR (II,1625); ONE DAY AT A TIME (II,1900); RHODA (II,2151)

MATAKEY, MARIE
THE LAST CONVERTIBLE (II,1435)

MATCHETT, CHRISTINE
OWEN MARSHALL: COUNSELOR AT LAW (I,3435); OWEN MARSHALL: COUNSELOR AT LAW (I,3436)

MATHERS, JERRY
DICK CLARK'S GOOD OL' DAYS: FROM BOBBY SOX TO BIKINIS (II,671); THE GIRL, THE GOLD WATCH AND DYNAMITE (II,1001); HIGH SCHOOL, U.S.A. (II,1149); IT'S A SMALL WORLD (I,2260); LEAVE IT TO BEAVER (I,2648); STILL THE BEAVER (II,2466)

MATHERS, JIMMY
ICHABOD AND ME (I,2183)

MATHESON, DON
BRAVO TWO (II,368); GENERAL HOSPITAL (II,964); LAND OF THE GIANTS (I,2611)

MATHESON, MURRAY
BANACEK (II,138); BANACEK: DETOUR TO NOWHERE (I,339); THE GOOD LIFE (II,1033); HARBOURMASTER (I,1946); KNIGHT'S GAMBIT (I,2576); LOVE STORY (I,2776); MISS STEWART, SIR (I,3064); THE MOON AND SIXPENCE (I,3098); REBECCA (I,3745); TELLER OF TALES (I,4388)

MATHESON, RICHARD
CAPTAINS AND THE KINGS (II,435); TUCKER'S WITCH (II,2667)

MATHESON, RUTH
FAIRMEADOWS, U.S.A. (I,1514)

MATHESON, TIM
BATTLE OF THE NETWORK STARS (II,163); THE BEST LEGS IN 8TH GRADE (II,214); BONANZA (II,347); BUS STOP (II,399); HITCHED (I,2065); HOW THE WEST WAS WON (II,1196); LOCK, STOCK, AND BARREL (I,2736); THE QUEST (II,2096); THE QUEST (II,2097); THE RUNAWAY BARGE (II,2230); THE VIRGINIAN (I,4738)

MATHEWS, CARMEN
CHARLIE COBB: NICE NIGHT FOR A HANGING (II,485); THE CORN IS GREEN (I,1050); THE DOCTOR (I,1320); HANS BRINKER OR THE SILVER SKATES (I,1933); MY THREE ANGELS (I,3204); SAM HILL: WHO KILLED THE MYSTERIOUS MR. FOSTER? (I,3896); THE STREETS OF SAN FRANCISCO (I,4260); THEY CALL IT MURDER (I,4441); TOP OF THE HILL (II,2645)

MATHEWS, CAROL
THE CLOCK (I,978)

MATHEWS, CLARE
MASTER OF THE GAME (II,1647)

MATHEWS, GEORGE
GLYNIS (I,1829)

MATHEWS, SARAH
MASTER OF THE GAME (II,1647)

MATHEWS, WALTER
INSPECTOR PEREZ (II,1237)

MATHEY, HARRIET
JOE DANCER: MURDER ONE, DANCER 0 (II,1302)

MATHIAS, ANNA
BACKSTAIRS AT THE WHITE HOUSE (II,133); PINK LADY (II,2041)

MATHIAS, BOB
THE TROUBLESHOOTERS (I,4614)

MATHIAS, DARIAN
ANOTHER MAN'S SHOES (II,96); HANGING IN (II,1078); LOVE, NATALIE (II,1545); PLEASE STAND BY (II,2057); TWILIGHT THEATER (II,2685)

MATHIEU, MIREILLE
DEAN MARTIN'S CHRISTMAS IN CALIFORNIA (II,645); THE JOHN DAVIDSON SHOW (I,2409)

MATHIS, JOHNNY
THE FRANCES LANGFORD SHOW (I,1657); THE ICE PALACE (I,2182); IT'S WHAT'S HAPPENING, BABY! (I,2278); JACK LEMMON—GET HAPPY (I,2312); JOHNNY MATHIS IN THE CANADIAN ROCKIES (II,1338); MONSANTO PRESENTS MANCINI (I,3092); THE NATALIE COLE SPECIAL (II,1794); SEE AMERICA WITH ED SULLIVAN (I,3972); STAR SEARCH (II,2437)

MATHIS, SHERRY
SEARCH FOR TOMORROW (II,2284)

MATLOCK, MATTY
THE PHIL HARRIS SHOW (I,3575)

MATLOCK, NORMAN
CHIEFS (II,509)

MATLOVSKY, NOELLE
YOU'RE A GOOD MAN, CHARLIE BROWN (I,4962)

MATNEE, PAUL
DEATH RAY 2000 (II,654)

MATRANGA, TONY
THE GIRL, THE GOLD WATCH AND EVERYTHING (II,1002)

MATRICK, GRAHAM
THE PAUL HOGAN SHOW (II,1975)

MATSUBARA, AI
SHOGUN (II,2339)

MATT MATTOX DANCERS, THE
THE PATTI PAGE SHOW (I,3480); THE PATTI PAGE SHOW (I,3481); THE PATTI PAGE SHOW (I,3482); THE TONY BENNETT SHOW (I,4569)

MATT MATTOX SINGERS, THE
A SPECIAL HOUR WITH DINAH SHORE (I,4149)

MATTHAEI, KONRAD
AS THE WORLD TURNS (II,110)

MATTHAU, DAVID
BATTLESTAR GALACTICA (II,181)

MATTHAU, WALTER
ACRES AND PAINS (I,19); THE ALCOA HOUR (I,108); ALFRED HITCHCOCK PRESENTS (I,115); THE ARMSTRONG CIRCLE THEATER (I,260); THE BOB HOPE CHRYSLER THEATER (I,502); THE CAMPBELL TELEVISION SOUNDSTAGE (I,812); CAROL CHANNING AND 101 MEN (I,849); THE DUPONT SHOW OF THE WEEK (I,1388); THE GEORGE BURNS SPECIAL (II,970); I LOVE LIBERTY (II,1215); THE MOTOROLA TELEVISION HOUR (I,3112); PHILCO TELEVISION PLAYHOUSE (I,3583); PROFILES IN COURAGE (I,3678); ROBERT MONTGOMERY PRESENTS YOUR LUCKY STRIKE THEATER (I,3809); SHOW BUSINESS SALUTE TO MILTON BERLE (I,4029); SUPER COMEDY BOWL 2 (I,4289); SUSPENSE (I,4305); TALLAHASSEE 7000 (I,4355); THE UNITED STATES STEEL HOUR (I,4677)

MATTHESON, RUTH
JANE EYRE (I,2337); THE KATE SMITH HOUR (I,2508)

MATTHEWS, AL
NANCY ASTOR (II,1788)

MATTHEWS, BRIAN
THE YOUNG AND THE RESTLESS (II,2862)

MATTHEWS, CAROLE
BROADWAY (I,733); THE CALIFORNIANS (I,795); STARS OVER HOLLYWOOD (I,4211)

MATTHEWS, DOROTHY
ART AND MRS. BOTTLE (I,271)

MATTHEWS, FRANCIS
CAPTAIN SCARLET AND THE MYSTERONS (I,837); IKE (II,1223)

MATTHEWS, GEORGE
GREAT CATHERINE (I,1873); WINTERSET (I,4882)

MATTHEWS, GERRY
PINOCCHIO'S CHRISTMAS (II,2045)

MATTHEWS, GRACE
THE GUIDING LIGHT (II,1064)

MATTHEWS, JON
GEORGE WASHINGTON (II,978)

MATTHEWS, JOYCE
THE JOYCE MATTHEWS SHOW (I,2460)

MATTHEWS, KERWIN
GHOSTBREAKER (I,1789); IN THE DEAD OF NIGHT (I,2202); STRATEGIC AIR COMMAND (I,4252)

MATTHEWS, LARRY
THE DICK VAN DYKE SHOW (I,1275)

MATTHEWS, LESTER
THE ADVENTURES OF DR. FU MANCHU (I,58); FIRESIDE THEATER (I,1580); THE RED SKELTON SHOW (I,3755)

MATTHEWS, PAT
THOSE ENDEARING YOUNG CHARMS (I,4469)

MATTHEWS, RICHARD
DOCTOR WHO—THE FIVE DOCTORS (II,690)

MATTHEWS, SEYMOUR
THE ASSASSINATION RUN (II,112)

MATTHEWS, SHEILA
CITY BENEATH THE SEA (I,965)

MATTHEWS, WALTER
SOMERSET (I,4115)

MATTHEY, PETE
ONE LIFE TO LIVE (II,1907)

MATTHIESON, TIM
THE ADVENTURES OF JONNY QUEST (I,64); THE ALVIN SHOW (I,146); THE HARDY BOYS (I,1948); SAMPSON AND GOLIATH (I,3903); SPACE GHOST (I,4141); THOMPSON'S GHOST (I,4467); WEEKEND (I,4788)

MATTHIUS, GAIL
LAUGH TRAX (II,1443); ROCK COMEDY (II,2191); SATURDAY NIGHT LIVE (II,2261)

MATTICK, PATRICIA
RANSOM FOR A DEAD MAN (I,3722)

MATTINGLY, HEDLEY
CASINO (II,452); DAKTARI (I,1127); GOLIATH AWAITS (II,1025); THE TRAVELS OF JAIMIE MCPHEETERS (I,4596)

MATTOX, MATT
PINOCCHIO (I,3602); SWING INTO SPRING (I,4317)

MATTSON, ROBIN
BATTLES: THE MURDER THAT WOULDN'T DIE (II,180); THE CHEERLEADERS (II,493); DOCTORS' PRIVATE LIVES (II,693); GENERAL HOSPITAL (II,964); THE GUIDING LIGHT (II,1064); RYAN'S HOPE (II,2234)

MAUCERI, PATRICIA
AS THE WORLD TURNS (II,110)

MAUDE-ROXBY, RODDY
ROWAN AND MARTIN'S LAUGH-IN (I,3856); TO SIR, WITH LOVE (II,2624)

MAUDSLEY, BONNIE
TREASURE ISLE (I,4601)

MAUGHAM, MONICA
PRISONER: CELL BLOCK H (II,2085)

MAUGHAM, WILLIAM SOMERSET
TELLER OF TALES (I,4388)

MAULE, BRAD
MALIBU (II,1607); ONE NIGHT BAND (II,1909); THREE'S COMPANY (II,2614)

MAUNDER, WAYNE
CHASE (I,917); LANCER (I,2610); THE LEGEND OF CUSTER (I,2652)

MAUREY, JACQUES
SMILEY'S PEOPLE (II,2383)

MAURO, DAVID
MASADA (II,1642); TOMA (I,4547)

MAURO, RALPH
SHE'S WITH ME (II,2321)

MAURY, DERREL
APPLE PIE (II,100); ARCHIE (II,104); THE ARCHIE SITUATION COMEDY MUSICAL VARIETY SHOW (II,106); JOANIE LOVES CHACHI (II,1295)

MAX, RON
FROM HERE TO ETERNITY (II,932); SEVENTH AVENUE (II,2309)

MAXEY, PAUL
JEFF'S COLLIE (I,2356); THE PEOPLE'S CHOICE (I,3521)

MAXTED, ELLEN
JUST OUR LUCK (II,1360); TEXAS (II,2566)

MAXWELL, BOB
MADE IN AMERICA (I,2806)

MAXWELL, CHARLES
I LED THREE LIVES (I,2169); SCIENCE FICTION THEATER (I,3938)

MAXWELL, DON
USED CARS (II,2712)

MAXWELL, ELSA
DATELINE (I,1165); THE TONIGHT SHOW (I,4564)

MAXWELL, FRANK
ALL IN THE FAMILY (II,38); THE FELONY SQUAD (I,1563); GENERAL HOSPITAL (II,964); THE GREATEST GIFT (I,1880); OUR MAN HIGGINS (I,3414); SANDBURG'S LINCOLN (I,3907); THE SECOND HUNDRED YEARS (I,3959); THE YOUNG MARRIEDS (I,4942)

MAXWELL, JAMES
MIRROR OF DECEPTION (I,3060)

MAXWELL, JEFF
M*A*S*H (II,1569)

MAXWELL, LEN
BATFINK (I,365); THE NUT HOUSE (I,3320)

MAXWELL, LOIS
ADVENTURES IN RAINBOW COUNTRY (I,47); DOUGLAS FAIRBANKS JR. PRESENTS THE RHEINGOLD THEATER (I,1364); STINGRAY (I,4228); U.F.O. (I,4662)

MAXWELL, MARILYN
BEST FOOT FORWARD (I,401); THE BOB HOPE SHOW (I,507); THE BOB HOPE SHOW (I,518); THE BOB HOPE SHOW (I,523); THE BOB HOPE SHOW (I,533); THE BOB HOPE SHOW (I,534); THE BOB HOPE SHOW (I,549); THE BOB HOPE SHOW (I,611); BURLESQUE (I,765); BUS STOP (I,778); COUNTY GENERAL (I,1076); HAVE GIRLS—WILL TRAVEL (I,1967)

MAXWELL, NORMAN
GEORGE WASHINGTON (II,978)

MAXWELL, PAISLEY
PAUL BERNARD—PSYCHIATRIST (I,3485); STRANGE PARADISE (I,4246)

MAXWELL, PAUL
CAPTAIN SCARLET AND THE MYSTERONS (I,837); FIREBALL XL-5 (I,1578); MADAME SIN (I,2805); SUPERCAR (I,4294)

MAXWELL, PHYLLIS
PAUL BERNARD—PSYCHIATRIST (I,3485)

MAXWELL, ROBERT

THE BOB HOPE SHOW (I,508); THE KATE SMITH HOUR (I,2508); THE MUSIC OF GERSHWIN (I,3174)

MAXWELL, ROBERTA
TODAY'S FBI (II,2628)

MAXWELL, STACEY
NO TIME FOR SERGEANTS (I,3300)

MAY, ANGELA
FUTURE COP (II,945)

MAY, BEVERLY
RYAN'S HOPE (II,2234)

MAY, BILLY
THE SATURDAY NIGHT DANCE PARTY (I,3919)

MAY, BOB
LOST IN SPACE (I,2758)

MAY, BUNNY
THE TALE OF BEATRIX POTTER (II,2533)

MAY, CHARMAIN
GOOD NEIGHBORS (II,1034)

MAY, DEBORAH
THE GUIDING LIGHT (II,1064); RAGE OF ANGELS (II,2107)

MAY, DONALD
COLT .45 (I,1010); COUNTY GENERAL (I,1076); THE EDGE OF NIGHT (II,760); THE ROARING TWENTIES (I,3805); TEXAS (II,2566); THE WEST POINT STORY (I,4796)

MAY, ELAINE
ACCENT ON LOVE (I,16); THE FABULOUS 50S (I,1501); JACK PAAR PRESENTS (I,2315); THE JACK PAAR SPECIAL (I,2318); A LAST LAUGH AT THE 60'S (I,2627)

MAY, JACQUELINE
MEDIC (I,2976)

MAY, JULIE
A HORSEMAN RIDING BY (II,1182)

MAY, LENORA
PALMS PRECINCT (II,1946); PRINCESS (II,2084)

MAY, MARIN
V (II,2713); V: THE FINAL BATTLE (II,2714)

MAY, MARTY
FIREBALL FUN FOR ALL (I,1577)

MAY, NATALIE
WAIT UNTIL DARK (II,2735)

MAYAMA, MIKO
AMANDA FALLON (I,152); HEY, LANDLORD (I,2039)

MAYEHOFF, EDDIE
THE BOB HOPE SHOW (I,623); BRILLIANT BENJAMIN BOGGS (I,730); DOC CORKLE (I,1304); FOR LOVE OR $$$ (I,1623); THE GLEN CAMPBELL GOODTIME HOUR (I,1820); HOUR GLASS (I,2133); STANDARD OIL ANNIVERSARY SHOW (I,4189); THAT'S MY BOY (I,4427)

MAYER TWINS, THE
I LOVE LUCY (I,2171)

MAYER, CHIP
OUR FAMILY BUSINESS (II,1931); STUNTS UNLIMITED (II,2487)

MAYER, CHRISTOPHER
THE DUKES (II,741); THE DUKES OF HAZZARD (II,742); GLITTER (II,1009)

MAYER, COLIN
DIVORCE COURT (I,1297)

MAYER, HANS
QUILLER: NIGHT OF THE FATHER (II,2100)

MAYER, JIMMY
TROUBLE IN HIGH TIMBER COUNTRY (II,2661)

MAYER, KEN
SPACE PATROL (I,4144)

MAYER, RONALD
LADY KILLER (I,2603); ONLY A SCREAM AWAY (I,3393)

MAYNARD, EARL
HEAVEN ON EARTH (II,1119)

MAYNARD, MIMI
BEACH PATROL (II,192); CONDOMINIUM (II,566); HART TO HART (II,1101); HART TO HART (II,1102)

MAYNE, BELINDA
GOLIATH AWAITS (II,1025); LACE (II,1418); A MAN CALLED INTREPID (II,1612)

MAYNE, FERDY
A MAN CALLED INTREPID (II,1612); THE WINDS OF WAR (II,2807)

MAYO, DON
CONCERNING MISS MARLOWE (I,1032)

MAYO, RAYMOND
SAM (II,2246)

MAYO, VIRGINIA
THE BOB HOPE SHOW (I,611); BOB HOPE SPECIAL: BOB HOPE'S ROAD TO HOLLYWOOD (II,313); MCGARRY AND ME (I,2966); SANTA BARBARA (II,2258)

MAYO, WHITMAN
THE BEST OF TIMES (II,220); GRADY (II,1047); SANFORD (II,2255); SANFORD AND SON (II,2256); SANFORD ARMS (II,2257)

MAYRON, GAIL
MR. & MRS. DRACULA (II,1745)

MAYRON, MELANIE
LILY—SOLD OUT (II,1480); THE LOVE BOAT III (II,1534)

MAYS, DAWSON
DID YOU HEAR ABOUT JOSH AND KELLY?! (II,673); LEGS (II,1459)

MAZER, ARNIE
DINER (II,679)

MAZER, BILL
CONCENTRATION (I,1031); REACH FOR THE STARS (I,3738)

MAZES, MICHAEL
OPERATION PETTICOAT (II,1916); OPERATION PETTICOAT (II,1917)

MAZURKI, MIKE
THE CHICAGO TEDDY BEARS (I,934); DAVY CROCKETT (I,1181); IT'S ABOUT TIME (I,2263); THE MILTON BERLE SPECIAL (I,3052); REVENGE OF THE GRAY GANG (II,2146)

MAZURSKY, PAUL
CALL TO DANGER (I,802)

MAZZA, JEAN
AS THE WORLD TURNS (II,110)

MAZZO, KAY
LOVE, LIFE, LIBERTY & LUNCH (II,1544)

MAZZOLA, EUGENE
TOMA (I,4547)

MAZZUCA, JOE
ANDY'S GANG (I,214)

MCADAM, HEATHER
THE FACTS OF LIFE (II,805); SALVAGE 1 (II,2245); WALKING TALL (II,2738)

MCADOO, MARY
DO IT YOURSELF (I,1299)

MCAFEE, JOHNNY
SAMMY KAYE'S MUSIC FROM MANHATTAN (I,3902)

MCALLISTER, BOB
KIDS ARE PEOPLE TOO (II,1390)

MCALLISTER, JENNIFER
MARRIED: THE FIRST YEAR (II,1634); WE'VE GOT EACH OTHER (II,2757)

MCANDREW, MARIANNE
MR. AND MRS. COP (II,1748)

MCARDLE, ANDREA
DOUG HENNING: MAGIC ON BROADWAY (II,730); SEARCH FOR TOMORROW (II,2284)

MCARDLE, NICHOLAS
KELLY MONTEITH (II,1375)

MCAVIN, ANDY
CAMELOT (II,416)

MCBAIN, DIANE
DALLAS (II,614); THE DEAN JONES SHOW (I,1191); SURFSIDE 6 (I,4299)

MCBEAN, MIKE
WINKY DINK AND YOU (I,4879)

MCBRIDE, HARLEE
DANNY AND THE MERMAID (II,618); INSTITUTE FOR REVENGE (II,1239)

MCBRIDE, HAZEL
SPACE: 1999 (II,2409)

MCBRIDE, JEAN
LOVE OF LIFE (I,2774)

MCBRIDE, MARY MARGARET
THE MARY MARGARET MCBRIDE SHOW (I,2933); THE TONIGHT SHOW (I,4564)

MCBRIDE, SUSAN
ANOTHER LIFE (II,95)

MCBROOM, AMANDA
THE PAUL WILLIAMS SHOW (II,1984)

MCCABE, MARCIA
SEARCH FOR TOMORROW (II,2284)

MCCAFFERTY, JOHN
TEXAS (II,2566)

MCCAFFERY, JOHN K M
THE DUNNINGER SHOW (I,1385); ONE MINUTE PLEASE (I,3384); TAKE A GUESS (I,4333); TELEVISION SCREEN MAGAZINE (I,4383); TV SCREEN MAGAZINE (I,4631); WE TAKE YOUR WORD (I,4781); WHAT'S THE STORY? (I,4821)

MCCAFFRIE, PAT
BACHELOR FATHER (I,323)

MCCAIN, FRANCES LEE
13 QUEENS BOULEVARD (II,2591); ALL THE WAY HOME (II,44); APPLE'S WAY (II,101); THREE'S COMPANY

(II,2614); WASHINGTON: BEHIND CLOSED DOORS (II,2744)

MCCAIN, MARTHA
THESE ARE MY CHILDREN (I,4440)

MCCALL, MARTY
THE NEW ANDY GRIFFITH SHOW (I,3256)

MCCALL, MITZI
THE FLINTSTONE COMEDY HOUR (II,881); FLINTSTONE FAMILY ADVENTURES (II,882); PEBBLES AND BAMM BAMM (II,1987); ROWAN AND MARTIN'S LAUGH-IN (I,3856); THE SCOOBY-DOO AND SCRAPPY-DOO SHOW (II,2275); SVENGALI AND THE BLONDE (I,4310)

MCCALL, SHALANE
DALLAS (II,614)

MCCALLA, IRISH
SHEENA, QUEEN OF THE JUNGLE (I,4003)

MCCALLEN, KATHY
AS THE WORLD TURNS (II,110)

MCCALLION, JAMES
NATIONAL VELVET (I,3231); REAR GUARD (II,2120); WINNER TAKE ALL (II,2808)

MCCALLUM, CHARLES
RETURN TO EDEN (II,2143)

MCCALLUM, DAVID
AN EVENING WITH CAROL CHANNING (I,1473); THE FILE ON DEVLIN (I,1575); THE INVISIBLE MAN (II,1242); THE INVISIBLE MAN (II,1243); THE MAN FROM U.N.C.L.E. (I,2867); THE MOONGLOW AFFAIR (I,3099); NIGHT GALLERY (I,3287); THE OUTER LIMITS (I,3426); PROFILES IN COURAGE (I,3678); RETURN OF THE MAN FROM U.N.C.L.E.: THE 15 YEARS LATER AFFAIR (II,2140); THE SIX MILLION DOLLAR MAN (I,4067); TEACHER, TEACHER (I,4366); THE UNKNOWN (I,4680)

MCCALLUM, NEIL
SABER OF LONDON (I,3879)

MCCALMAN, MACON
ALLISON SIDNEY HARRISON (II,54); LONDON AND DAVIS IN NEW YORK (II,1512); OFF THE RACK (II,1878); THREE'S COMPANY (II,2614); TOPPER (II,2649); THE ULTIMATE IMPOSTER (II,2700)

MCCAMBRIDGE, MERCEDES
THE CHEVROLET TELE-THEATER (I,926); FRONT ROW CENTER (I,1694); JANE WYMAN PRESENTS THE FIRESIDE THEATER (I,2345); KILLER BY NIGHT (I,2537); THE LORETTA YOUNG THEATER (I,2756); NO WARNING (I,3301); ONE MAN'S FAMILY (I,3383); TWO FOR THE MONEY (I,4655); WIRE SERVICE (I,4883); YOUNG IN HEART (I,4939)

MCCANN, ANDY
RENDEZVOUS WITH MUSIC (I,3768)

MCCANN, CHUCK
ALL THAT GLITTERS (II,42); ARNOLD'S CLOSET REVIEW (I,262); THE C.B. BEARS (II,406); CLOWNAROUND (I,980); COOL MCCOOL (I,1045); THE DRAK PACK (II,733); FAR OUT SPACE NUTS (II,831); FRED AND BARNEY MEET THE SHMOO (II,918); HAPPY DAYS (I,1942); A NEW KIND OF FAMILY (II,1819); THE NEW SHMOO (II,1827); ONE DAY AT A TIME (II,1900); PAC-MAN (II,1940); SEMI-TOUGH (II,2300); TURN ON (I,4625); VAN DYKE AND COMPANY (II,2719)

MCCANN, DONAL
THE PALLISERS (II,1944); SCREAMER (I,3945)

MCCANN, MARIE
TOM AND JOANN (II,2630)

MCCANN, SEAN
THE BAXTERS (II,184)

MCCANNON, DIAN
CASPER AND THE ANGELS (II,453)

MCCARDLE, ANDREA
MO AND JO (II,1715)

MCCAREY, ROD
GENERAL HOSPITAL (II,964)

MCCARLEY, KEVIN
IF I LOVE YOU, AM I TRAPPED FOREVER? (II,1222)

MCCARREN, FRED
AMANDA'S (II,62); FREE COUNTRY (II,921); GIMME A BREAK (II,995); THE LAST CONVERTIBLE (II,1435); MARRIAGE IS ALIVE AND WELL (II,1633); THE RAG BUSINESS (II,2106); SAINT PETER (II,2238); THE SINGLE LIFE (II,2364); STICK AROUND (II,2464); TWO THE HARD WAY (II,2695); USED CARS (II,2712)

MCCARROLL, ESTHER
IN SECURITY (II,1227)

MCCARTHY, ANN
TEXAS (II,2566)

MCCARTHY, ANNETTE
CRAZY TIMES (II,602)

MCCARTHY, CLEM
SHOW OF THE YEAR (I,4031)

MCCARTHY, FRANK
FARRELL: FOR THE PEOPLE (II,834); THE FRENCH ATLANTIC AFFAIR (II,924); TWILIGHT THEATER (II,2685)

MCCARTHY, JOSEPHINE
HI MOM (I,2043)

MCCARTHY, JULIANNA
THE YOUNG AND THE RESTLESS (II,2862)

MCCARTHY, KEVIN
AMANDA'S (II,62); EXO-MAN (II,797); FLAMINGO ROAD (II,871); FLAMINGO ROAD (II,872); FRONT ROW CENTER (I,1694); GHOSTBREAKER (I,1789); THE INNER SANCTUM (I,2216); THE JUNE ALLYSON SHOW (I,2488); MATINEE THEATER (I,2947); THE MOONGLOW AFFAIR (I,3099); PRUDENTIAL FAMILY PLAYHOUSE (I,3683); SATINS AND SPURS (I,3917); THE SURVIVORS (I,4301); U.M.C. (I,4666)

MCCARTHY, LILAH
HEAVEN ON EARTH (II,1119)

MCCARTHY, LIN
MAISIE (I,2833); WAR CORRESPONDENT (I,4767)

MCCARTHY, LINWOOD
THE CONTENDER (II,571); UNIT 4 (II,2707); THE WINDS OF WAR (II,2807)

MCCARTHY, NEIL
NANCY ASTOR (II,1788); SHOGUN (II,2339)

MCCARTHY, TOM
GEORGE WASHINGTON (II,978)

MCCARTNEY, LINDA
JAMES PAUL MCCARTNEY (I,2332)

MCCARTNEY, PAUL
JAMES PAUL MCCARTNEY (I,2332)

MCCARTY, MARY
THE ADMIRAL BROADWAY REVUE (I,37); ARTHUR MURRAY'S DANCE PARTY (I,293); CELEBRITY TIME (I,887); DON'T CALL ME MAMA ANYMORE (I,1350); TRAPPER JOHN, M.D. (II,2654)

MCCARY, ROD
AT EASE (II,116); THE FIGHTING NIGHTINGALES (II,855); HARPER VALLEY (II,1094); HARPER VALLEY PTA (II,1095); HEAVEN ON EARTH (II,1119); JUST OUR LUCK (II,1360); MALIBU (II,1607); MOTHER, JUGGS AND SPEED (II,1741); SUSAN AND SAM (II,2506)

MCCARY, TEX
THE BOB HOPE SHOW (I,509)

MCCASHIN, CONSTANCE
KNOTS LANDING (II,1404); THE MANY LOVES OF ARTHUR (II,1625); MARRIED: THE FIRST YEAR (II,1634)

MCCASLIN, MAYO
HARDCASTLE AND MCCORMICK (II,1089); THE HOME FRONT (II,1171); THE KID SUPER POWER HOUR WITH SHAZAM (II,1386)

MCCAVITT, T J
INSTITUTE FOR REVENGE (II,1239)

MCCAY, PEGGY
THE ARMSTRONG CIRCLE THEATER (I,260); THE CITY (I,969); FBI CODE 98 (I,1550); GENERAL HOSPITAL (II,964); GIBBSVILLE (II,988); GREAT ADVENTURE (I,1868); THE LAZARUS SYNDROME (II,1451); LOOSE CHANGE (II,1518); LOU GRANT (II,1526); LOVE AT FIRST SIGHT (II,1530); LOVE OF LIFE (I,2774); LOVE STORY (I,2776); MIDNIGHT MYSTERY (I,3030); NEWHART (II,1835); PROFILES IN COURAGE (I,3678); ROOM FOR ONE MORE (I,3842); TELLER OF TALES (I,4388); THE TURNING POINT OF JIM MALLOY (II,2670); THE YOUNG MARRIEDS (I,4942)

MCCLAIN, MARCIA
AS THE WORLD TURNS (II,110); THE DRUNKARD (II,737)

MCCLAIN, SAUNDRA
HOT HERO SANDWICH (II,1186)

MCCLAMAN, MALCOLM
BEST OF THE WEST (II,218)

MCCLANAHAN, MACON
LOVE AT FIRST SIGHT (II,1531)

MCCLANAHAN, RUE
ABC'S MATINEE TODAY (I,7); AND THEY ALL LIVED HAPPILY EVER AFTER (II,75); APPLE PIE (II,100); CHARLES IN CHARGE (II,482); CIRCUS OF THE STARS (II,530); GIMME A BREAK (II,995); HAVING BABIES III (II,1109); MAMA'S FAMILY (II,1610); MAUDE (II,1655); MOTHER AND ME, M.D. (II,1740); THE SON-IN-LAW (II,2398); TED KNIGHT MUSICAL COMEDY VARIETY SPECIAL SPECIAL (II,2549); TOPPER (II,2649); WHERE THE HEART IS (I,4831)

MCCLAUGHLIN, RITA
AS THE WORLD TURNS (II,110)

MCCLAY, MARIA
HONG KONG (I,2112)

MCCLINTOCK, JEAN
AS THE WORLD TURNS (II,110)

MCCLORY, SEAN
BRING 'EM BACK ALIVE (II,377); THE CALIFORNIANS (I,795); CAPTAINS AND THE KINGS (II,435); DANIEL BOONE (I,1142); KATE MCSHANE (II,1368); KATE MCSHANE (II,1369); MACREEDY'S WOMAN (I,2804); THE NEW DAUGHTERS OF JOSHUA CABE (I,3261); TALES OF THE 77TH BENGAL LANCERS (I,4348)

MCCLOSKEY, LEIGH
BATTLE OF THE NETWORK STARS (II,172); DOCTORS' PRIVATE LIVES (II,692); DOCTORS' PRIVATE LIVES (II,693); EXECUTIVE SUITE (II,796); MARRIED: THE FIRST YEAR (II,1634); RICH MAN, POOR MAN—BOOK I (II,2161); VELVET (II,2725)

MCCLOSKEY, LILLIAN
HARDCASE (II,1088)

MCCLOSKEY, MITCH
DALLAS (II,614)

MCCLOSKEY, YALE
COREY: FOR THE PEOPLE (II,575)

MCCLOUD, MERCER
THREE STEPS TO HEAVEN (I,4485)

MCCLUNG, SUSAN
KNOTS LANDING (II,1404); T.J. HOOKER (II,2524)

MCCLURE, DOUG
THE BARBARY COAST (II,146); CHECKMATE (I,919); GHOST STORY (I,1788); THE GOLDDIGGERS (I,1838); IVY LEAGUE (I,2286); THE JUDGE AND JAKE WYLER (I,2464); THE MASTER (II,1648); THE MEN FROM SHILOH (I,3005); NIGHTSIDE (II,1851); OVERLAND TRAIL (I,3433); THE REBELS (II,2121); ROOTS (II,2211); SEARCH (I,3951); SHOOT-IN AT NBC (A BOB HOPE SPECIAL) (I,4021); THE SKY'S THE LIMIT (I,4081); THE VIRGINIAN (I,4738); WILD AND WOOLEY (II,2798)

MCCLURE, EDIE
THE HUDSON BROTHERS SHOW (II,1202)

MCCLURE, FRANK CHANDLER
TRAFFIC COURT (I,4590)

MCCLURE, MARC
CALIFORNIA FEVER (II,411); JAMES AT 15 (II,1271)

MCCLURG, BOB
ALICE (II,33)

MCCLURG, EDIE
THE BIG SHOW (II,243); THE CHEVY CHASE SHOW (II,505); THE DAVID LETTERMAN SHOW (II,624); HARPER VALLEY PTA (II,1095); THE KALLIKAKS (II,1361); MADAME'S PLACE (II,1587); NO SOAP, RADIO (II,1856); THE PARAGON OF COMEDY (II,1956); THE PEE WEE HERMAN SHOW (II,1988); POTTSVILLE (II,2072); SECOND EDITION (II,2288); TONY ORLANDO AND DAWN (II,2639); TOP TEN (II,2648); WKRP IN CINCINNATI (II,2814)

MCCOLLUM, JOHN
AMAHL AND THE NIGHT VISITORS (I,149)

MCCONNELL, ED
ANDY'S GANG (I,214)

MCCONNELL, JUDITH
AS THE WORLD TURNS (II,110); GENERAL HOSPITAL (II,964)

MCCONNELL, JUDY
GREEN ACRES (I,1884)

MCCOO, MARILYN
ANNE MURRAY'S LADIES' NIGHT (II,89); THE FIFTH DIMENSION SPECIAL: AN ODYSSEY IN THE COSMIC UNIVERSE OF PETER MAX (I,1571); THE FIFTH DIMENSION TRAVELING SHOW (I,1572); THE MARILYN MCCOO AND BILLY DAVIS JR. SHOW (II,1630); MEN WHO RATE A "10" (II,1684); SOLID GOLD (II,2395)

MCCOOK, JOHN
AFTER HOURS: FROM JANICE, JOHN, MARY AND MICHAEL, WITH LOVE (II,20); AFTER HOURS: SINGIN', SWINGIN' AND ALL THAT JAZZ (II,22); DRAGNET (I,1371); MITZI. . . WHAT'S HOT, WHAT'S NOT (II,1712); THE RAINBOW GIRL (II,2108); REAR GUARD (II,2120); TOURIST (II,2651); THE YOUNG AND THE RESTLESS (II,2862)

MCCORD, KENT
ADAM-12 (I,31); THE ADVENTURES OF OZZIE AND HARRIET (I,71); BATTLE OF THE NETWORK STARS (II,170); THE CELEBRITY FOOTBALL CLASSIC (II,474); GALACTICA 1980 (II,953); THE OUTSIDER (I,3430); PINE CANYON IS BURNING (II,2040); THE ROWAN AND MARTIN SPECIAL (I,3855)

MCCORD, MARY
PHILCO TELEVISION PLAYHOUSE (I,3583)

MCCORD, MEGAN
PINE CANYON IS BURNING (II,2040)

MCCORMACK, PATRICIA
ALCOA/GOODYEAR THEATER (I,107); AS THE WORLD TURNS (II,110); BEN HECHT'S TALES OF THE CITY (I,395); THE BEST OF EVERYTHING (I,406); THE CAMPBELL TELEVISION SOUNDSTAGE (I,812); CAVALCADE OF AMERICA (I,875); I REMEMBER MAMA (I,2176); KRAFT TELEVISION THEATER (I,2592); NIGHT PARTNERS (II,1847); ONE STEP BEYOND (I,3388); PECK'S BAD GIRL (I,3503); PLAYHOUSE 90 (I,3623); REVLON MIRROR THEATER (I,3781); THE ROPERS (II,2214); THE UNITED STATES STEEL HOUR (I,4677); THE WEB (I,4784)

MCCORMICK, FRANKLYN
CHICAGO 212 (I,935)

MCCORMICK, JOHN
THE REAL TOM KENNEDY SHOW (I,3743)

MCCORMICK, LARRY
THE JEFFERSONS (II,1276)

MCCORMICK, MARIANNE
THE BERT PARKS SHOW (I,400)

MCCORMICK, MAUREEN
THE BRADY BRIDES (II,361); THE BRADY BUNCH (II,362); THE BRADY BUNCH HOUR (II,363); THE BRADY GIRLS GET MARRIED (II,364); THE BRADY KIDS (II,365); CAMP RUNAMUCK (I,811); FANTASY ISLAND (II,829); WHEN, JENNY? WHEN (II,2783)

MCCORMICK, MYRON
HEAVE HO HARRIGAN (I,1993); MR. GLENCANNON TAKES ALL (I,3139); THREE IN ONE (I,4479)

MCCORMICK, PARKER
HAPPY BIRTHDAY (I,1941)

MCCORMICK, PAT
THE BAY CITY AMUSEMENT COMPANY (II,185); THE CRACKER BROTHERS (II,599); GUN SHY (II,1068); HOW DO I KILL A THIEF—LET ME COUNT THE WAYS (II,1195); THE JERK, TOO (II,1282); ROOSTER (II,2210); SHAUGHNESSEY (II,2318); WE DARE YOU! (II,2751)

MCCORMICK, PATRICIA
EMERGENCY! (II,776)

MCCOURT, MALACHY
CATALINA C-LAB (II,458); THE DAIN CURSE (II,613); REWARD (II,2147); RYAN'S HOPE (II,2234)

MCCOY, CHARLIE
COUNTRY MUSIC HIT PARADE (I,1069)

MCCOY, JACK
GLAMOUR GIRL (I,1817); LIVE LIKE A MILLIONAIRE (I,2728); MEET THE GOVERNOR (I,2991)

MCCOY, MATT
HOT HERO SANDWICH (II,1186); PEN 'N' INC. (II,1992); WE GOT IT MADE (II,2752)

MCCOY, SID
THE BILL COSBY SHOW (I,442)

MCCRACKEN, JEFF
BAY CITY BLUES (II,186); HAWAIIAN HEAT (II,1111); STRANGER IN OUR HOUSE (II,2477)

MCCRACKEN, JOAN
CLAUDIA: THE STORY OF A MARRIAGE (I,972); GREAT CATHERINE (I,1873); REVLON MIRROR THEATER (I,3781)

MCCRARY, TEX
SHOW OF THE YEAR (I,4031); TEX AND JINX (I,4406)

MCCREA, ANN
THE DONNA REED SHOW (I,1347); SIMON LASH (I,4052)

MCCREA, JODY
THREE ON AN ISLAND (I,4483); WICHITA TOWN (I,4857)

MCCREA, JOEL
WICHITA TOWN (I,4857)

MCCUEN, CAROLYN
THAT TEEN SHOW (II,2573)

MCCULLEM, KATHY
THE BLACK SHEEP SQUADRON (II,262); THE CRACKER BROTHERS (II,599)

MCCULLOCH, IAN
SEARCH FOR THE NILE (I,3954)

MCCULLOUGH, DARRYL
THE SAN PEDRO BEACH BUMS (II,2250); THE SAN PEDRO BUMS (II,2251)

MCCULLOUGH, LINDA
B.J. AND THE BEAR (II,126); THE EYES OF TEXAS II (II,800); THE GUIDING LIGHT (II,1064); WELCOME BACK, KOTTER (II,2761); WOMEN WHO RATE A "10" (II,2825)

MCCULLUM, CHARLES
THE EVIL TOUCH (I,1486)

MCCUNE, HANK
THE HANK MCCUNE SHOW (I,1932)

MCCURDY, VIRGINIA
THE KATE SMITH HOUR (I,2508)

MCCURRY, ANN
BEYOND WESTWORLD (II,227)

MCCUTCHEON, BILL
BALL FOUR (II,137); THE DOM DELUISE SHOW (I,1328); MR. MAYOR (I,3143)

MCDADE, PATRICK F
GEORGE WASHINGTON (II,978)

MCDANIEL, GEORGE
DAYS OF OUR LIVES (II,629); DOWN HOME (II,731); FOOLS, FEMALES AND FUN: IS THERE A DOCTOR IN THE HOUSE? (II,900); NICHOLS AND DYMES (II,1841)

MCDANIEL, HATTIE
BEULAH (I,416)

MCDERMOTT, BRIAN
SECRETS OF THE OLD BAILEY (I,3971)

MCDERMOTT, CHRIS
THE BOOK OF LISTS (II,350); JOANIE LOVES CHACHI (II,1295)

MCDERMOTT, HELEN
A HORSEMAN RIDING BY (II,1182)

MCDERMOTT, HUGH
THE ADVENTURES OF ROBIN HOOD (I,74)

MCDERMOTT, TERRY
HOMICIDE (I,2101)

MCDERMOTT, TOM
CAPTAIN VIDEO AND HIS VIDEO RANGERS (I,838)

MCDEVITT, RUTH
ALL IN THE FAMILY (II,38); ARSENIC AND OLD LACE (I,267); BEN BLUE'S BROTHERS (I,393); BRIGHT PROMISE (I,727); THE CHEERLEADERS (II,493); THE DOCTORS (II,694); THE EVERLY BROTHERS SHOW (I,1479); LITTLE WOMEN (I,2719); MAN IN THE MIDDLE (I,2870); MR. PEEPERS (I,3147); THE NEW ANDY GRIFFITH SHOW (I,3256); THE NIGHT STALKER (II,1849); PISTOLS 'N' PETTICOATS (I,3608); WINTER KILL (II,2810); A WOMAN TO REMEMBER (I,4888); YOUNG DR. MALONE (I,4938)

MCDONALD, AMY
THE JOHN DAVIDSON SHOW (I,2409)

MCDONALD, ANNE MAREE
PRISONER: CELL BLOCK H (II,2085)

MCDONALD, ARCH
RED SKELTON'S CHRISTMAS DINNER (II,2123)

MCDONALD, DAN
STRANGE PARADISE (I,4246)

MCDONALD, GEORGE
THE LIFE OF RILEY (I,2685)

MCDONALD, HARL
TELEVISION SYMPHONY (I,4384)

MCDONALD, KENNETH
THE BOUNTY HUNTER (I,698)

MCDONALD, MARIE
THE BOB HOPE SHOW (I,557)

MCDONALD, MARY ANN
ANNE MURRAY'S LADIES' NIGHT (II,89)

MCDONALD, MICHAEL
NO TIME FOR SERGEANTS (I,3300); ROCK 'N' ROLL: THE FIRST 25 YEARS (II,2189); THE ROOTS OF ROCK 'N' ROLL (II,2212)

MCDONALD, RAY
THE JACK CARSON SHOW (I,2304)

MCDONALD, ROBERTA
MELODY STREET (I,3000)

MCDONALD, SEVEN ANN
THE EDDIE CAPRA MYSTERIES (II,758)

MCDONALD, SUSAN
SOMERSET (I,4115)

MCDONALD, TANNY
KENNEDY (II,1377)

MCDONNELL, JO
THE MUNSTERS' REVENGE (II,1765); ONCE UPON A SPY (II,1898); SWEEPSTAKES (II,2514)

MCDONNELL, MARY
E/R (II,748)

MCDONOUGH, EILEEN
ERIC (I,1451); PEOPLE LIKE US (II,1999); THE WALTONS (II,2740)

MCDONOUGH, KIT
ANGIE (II,80); THE RITA MORENO SHOW (II,2181); TEACHERS ONLY (II,2547)

MCDONOUGH, MARY ELIZABETH
CIRCUS OF THE STARS (II,534); MERV GRIFFIN AND THE CHRISTMAS KIDS (I,3013); THE WALTONS (II,2740); THE WALTONS: A DAY FOR THANKS ON WALTON'S MOUNTAIN (EPISODE 3) (II,2741); THE WALTONS: MOTHER'S DAY ON WALTON'S MOUNTAIN (EPISODE 2) (II,2742); THE WALTONS: WEDDING ON WALTON'S MOUNTAIN (EPISODE 1) (II,2743)

MCDONOUGH, TOM
A TALE OF WELLS FARGO (I,4338)

MCDOWALL, RODDY
BATMAN (I,366); THE BEST OF ANYTHING (I,404); CHRYSLER MEDALLION THEATER (I,951); CIRCUS OF THE STARS (II,536); DUPONT SHOW OF THE MONTH (I,1387); THE ELGIN HOUR (I,1428); FAITH BALDWIN'S THEATER OF ROMANCE (I,1515); THE FANTASTIC JOURNEY (II,826); FANTASY ISLAND (II,829); THE GOOD FAIRY (I,1846); HART TO HART (II,1101); JUDGEMENT DAY (II,1348); THE KAISER ALUMINUM HOUR (I,2503); LONDON AND DAVIS IN NEW YORK (II,1512); THE MARTIAN CHRONICLES (II,1635); MATINEE THEATER (I,2947); MIRACLE ON 34TH STREET (I,3058); NIGHT GALLERY (I,3288); PLANET OF THE APES (II,2049); PLAYHOUSE 90 (I,3623); THE POWER AND THE GLORY (I,3658); THE RHINEMANN EXCHANGE (II,2150); ROBERT MONTGOMERY PRESENTS YOUR LUCKY STRIKE THEATER (I,3809); ST. JOAN (I,4175); SUSPICION (I,4309); TALES OF THE GOLD MONKEY (II,2537); THE TEMPEST (I,4391); THIS GIRL FOR HIRE (II,2596); TOPPER RETURNS (I,4583); TWILIGHT THEATER (II,2685)

MCDOWELL, BETTY
DOUGLAS FAIRBANKS JR. PRESENTS THE RHEINGOLD THEATER (I,1364)

MCDOWELL, PAUL
DAVE ALLEN AT LARGE (II,621)

MCEACHIN, JAMES
ALLISON SIDNEY HARRISON (II,54); BEULAH LAND (II,226); ESCAPE (I,1461); SAMURAI (II,2249); TENAFLY (I,4395); TENAFLY (II,2560)

MCELHONE, ELOISE
QUICK ON THE DRAW (I,3711)

MCELORY, VALERIE
CHICAGOLAND MYSTERY PLAYERS (I,936); THE MAGIC SLATE (I,2820)

MCELROY, HOWARD
KOBB'S CORNER (I,2577)

MCELROY, TAFFY
COUNTRY COMES HOME (II,586)

MCENCROE, ANNIE
REWARD (II,2147)

MCENERY, JOHN

THE WORD (II,2833)

MCENROE, JOHN
PAVAROTTI AND FRIENDS (II,1985)

MCEUEN, JOHN
COUNTRY COMES HOME (II,586)

MCEWAN, GERALDINE
THE PRIME OF MISS JEAN BRODIE (II,2082)

MCFADDEN, BARNEY
THE AWAKENING LAND (II,122); CENTENNIAL (II,477); THE GUIDING LIGHT (II,1064)

MCFADDEN, ROBERT
'TWAS THE NIGHT BEFORE CHRISTMAS (II,2677); THE ASTRONAUT SHOW (I,307); THE CONEHEADS (II,567); COOL MCCOOL (I,1045); COURAGEOUS CAT (I,1079); LITTLE DRUMMER BOY, BOOK II (II,1486); THE MILTON THE MONSTER CARTOON SHOW (I,3053); PINOCCHIO'S CHRISTMAS (II,2045); THE YEAR WITHOUT A SANTA CLAUS (II,2846)

MCFADDEN, TOM
BIG BOB JOHNSON AND HIS FANTASTIC SPEED CIRCUS (II,231); M STATION: HAWAII (II,1568); THE WINDS OF WAR (II,2807)

MCFARLAND, NAN
ABE LINCOLN IN ILLINOIS (I,11); THE MAGNIFICENT YANKEE (I,2829); THE WOMEN (I,4890)

MCFARLANE, ANDREW
1915 (II,1853); THE SULLIVANS (II,2490)

MCFARREN, EVE
THE GUIDING LIGHT (II,1064)

MCGARTH, BRIAN
SOMEONE AT THE TOP OF THE STAIRS (I,4114)

MCGARVIN, DICK
THE WONDERFUL WORLD OF PHILIP MALLEY (II,2831)

MCGAVIN, DARREN
THE ALCOA HOUR (I,108); THE ARMSTRONG CIRCLE THEATER (I,260); BANYON (I,345); THE CAMPBELL TELEVISION SOUNDSTAGE (I,812); CRIME PHOTOGRAPHER (I,1097); FATHER ON TRIAL (I,1544); IKE (II,1223); MAN AGAINST CRIME (I,2852); THE MARTIAN CHRONICLES (II,1635); MIKE HAMMER, DETECTIVE (I,3042); THE NIGHT STALKER (I,3292); THE NIGHT STALKER (II,1849); THE NIGHT STRANGLER (I,3293); NOT UNTIL TODAY (II,1865); OLD NICKERBOCKER MUSIC HALL (I,3351); THE OUTSIDER (I,3430); THE OUTSIDER (I,3431); THE RETURN OF MARCUS WELBY, M.D (II,2138); REVLON MIRROR THEATER (I,3781); RIVERBOAT (I,3797); THE ROOKIES (I,3840); THE ROOKIES (II,2208); SHORT, SHORT DRAMA (I,4023); SMALL AND FRYE (II,2380); SUSPENSE (I,4305); TALES OF TOMORROW (I,4352); THE TIMES SQUARE STORY (I,4514); WAIKIKI (II,2734); THE WEB (I,4784)

MCGEE, BONNIE
TONIGHT WITH BELAFONTE (I,4566)

MCGEE, HENRY
THE BENNY HILL SHOW (II,207); THE UNEXPURGATED BENNY HILL SHOW (II,2705)

MCGEE, VONETTA
THE NORLISS TAPES (I,3305); SCRUPLES (II,2281)

MCGEEHAN, MARY KATE
FALCON CREST (II,808)

MCGHEE, ROBERT A
CRIME AND PUNISHMENT (I,1095)

MCGIBBON, HARRIET
GOLDEN WINDOWS (I,1841)

MCGILL, BRUCE
DELTA HOUSE (II,658); SEMI-TOUGH (II,2300)

MCGILL, EVERETT
THE GUIDING LIGHT (II,1064)

MCGILLIN, HOWARD
NUMBER 96 (II,1869); WHEELS (II,2777); WOMEN IN WHITE (II,2823); THE YOUNG AND THE RESTLESS (II,2862)

MCGINLEY, DAVID
GEORGE WASHINGTON (II,978)

MCGINLEY, TED
HAPPY DAYS (II,1084); HERNDON AND ME (II,1139); THE LOVE BOAT (II,1535)

MCGINN, WALTER
SANDBURG'S LINCOLN (I,3907)

MCGINNIS, SCOTT
ALL NIGHT RADIO (II,40)

MCGIVENEY, MAURA
TURN ON (I,4625)

MCGIVER, JOHN
'TWAS THE NIGHT BEFORE CHRISTMAS (II,2677); THE ALCOA HOUR (I,108); ANNIE, THE WOMAN IN THE LIFE OF A MAN (I,226); THE BOSTON TERRIER (I,696); THE FRONT PAGE (I,1692); HARVEY (I,1962); THE JIMMY STEWART SHOW (I,2391); THE LITTLEST ANGEL (I,2726); LOW MAN ON THE TOTEM POLE (I,2782); MANY HAPPY RETURNS (I,2895); MIRROR, MIRROR, OFF THE WALL (I,3059); THE MISS AND THE MISSILES (I,3062); MR. TERRIFIC (I,3152); THE PATTY DUKE SHOW (I,3483); THE PRUITTS OF SOUTHAMPTON (I,3684); SAM HILL: WHO KILLED THE MYSTERIOUS MR. FOSTER? (I,3896)

MCGOOHAN, PATRICK
DANGER MAN (I,1136); THE PRISONER (I,3670); RAFFERTY (II,2105); THE SCARECROW OF ROMNEY MARSH (I,3931); SECRET AGENT (I,3962)

MCGOVERN, JOHN
THE GOLDEN AGE OF TELEVISION (II,1019)

MCGOVERN, TERRENCE
EVENING AT THE IMPROV (II,786); HIS AND HERS (II,1155); ME AND MRS. C. (II,1673); THE NATIONAL SNOOP (II,1797); NO MAN'S LAND (II,1855); PRESENTING SUSAN ANTON (II,2077); WE DARE YOU! (II,2751)

MCGOWAN, JANE
BEN BLUE'S BROTHERS (I,393)

MCGOWAN, TOM
ANOTHER LIFE (II,95)

MCGOWRAN, JACK
SAILOR OF FORTUNE (I,3885)

MCGRATH, BOB
SESAME STREET (I,3982); SING ALONG WITH MITCH (I,4057)

MCGRATH, DEBRA
OUT OF OUR MINDS (II,1933)

MCGRATH, DEREK
AT YOUR SERVICE (II,118); THE CRACKER BROTHERS (II,599)

MCGRATH, DOUG
LONE STAR (II,1513)

MCGRATH, FRANK
THE ADAM MACKENZIE STORY (I,30); CHARLIE WOOSTER—OUTLAW (I,915); THE ELISABETH MCQUEENEY STORY (I,1429); TAMMY (I,4357); WAGON TRAIN (I,4747)

MCGRATH, GRAHAM
GREAT EXPECTATIONS (II,1055)

MCGRATH, PAUL
BLOOMER GIRL (I,488); FIRST LOVE (I,1583); THE INNER SANCTUM (I,2216); KISS ME, KATE (I,2568); LADY IN THE DARK (I,2602); SPELLBOUND (I,4153)

MCGRAW, BILL
HOLD IT PLEASE (I,2071)

MCGRAW, CHARLES
CASABLANCA (I,860); DIAGNOSIS: DANGER (I,1250); THE FALCON (I,1516); THE NIGHT STALKER (I,3292); O'HARA, UNITED STATES TREASURY: OPERATION COBRA (I,3347); THE SMITH FAMILY (I,4092)

MCGRAW, WALTER
WANTED (I,4763)

MCGREEVY, MICHAEL
DIAL HOT LINE (I,1254); HARRY O (II,1098); RIVERBOAT (I,3797)

MCGREEVY, OLIVER
THE KILLING GAME (I,2540)

MCGREGOR, CHRIS
THE LAST CONVERTIBLE (II,1435)

MCGREGOR, SCOTT
1915 (II,1853); FIRST LOVE (I,1583)

MCGRIFF, STEVE
ALICE (II,33)

MCGUINN, JOE
GETAWAY CAR (I,1783)

MCGUIRE SISTERS, THE
THE ARTHUR GODFREY SHOW (I,286); THE JACK BENNY HOUR (I,2291); JACK BENNY WITH GUEST STARS (I,2298); THE KATE SMITH HOUR (I,2508); THE PAT BOONE SHOW (I,3470); REMEMBER HOW GREAT? (I,3764)

MCGUIRE, BARRY
THE FIRST HUNDRED YEARS (I,1582); THE NEW CHRISTY MINSTRELS SHOW (I,3259)

MCGUIRE, BIFF
GIBBSVILLE (II,988); HERB SHRINER TIME (I,2020); KINGSTON: THE POWER PLAY (II,1399); NERO WOLFE (II,1805); ROGER AND HARRY (II,2200); SEARCH FOR TOMORROW (II,2284); THE TURNING POINT OF JIM MALLOY (II,2670)

MCGUIRE, BRYAN
NO MAN'S LAND (II,1855)

MCGUIRE, CHRISTINE
ARTHUR GODFREY AND FRIENDS (I,278)

MCGUIRE, DOROTHY
ARTHUR GODFREY AND FRIENDS (I,278); CLIMAX! (I,976); HOLLYWOOD: THE SELZNICK YEARS (I,2095); LITTLE WOMEN (II,1497); LITTLE WOMEN (II,1498); THE PHILADELPHIA STORY (I,3581); RICH MAN, POOR MAN—BOOK I (II,2161); THE UNITED STATES STEEL HOUR (I,4677)

MCGUIRE, KATHLEEN
THREE STEPS TO HEAVEN (I,4485); A WORLD APART (I,4910)

MCGUIRE, KERRY
AGAINST THE WIND (II,24)

MCGUIRE, MAEVE
BEACON HILL (II,193); THE EDGE OF NIGHT (II,760)

MCGUIRE, MICHAEL
EMPIRE (II,777); FAMILY IN BLUE (II,818); LIGHT'S OUT (I,2700); THE LONG DAYS OF SUMMER (II,1514); SANCTUARY OF FEAR (II,2252); THE WINDS OF WAR (II,2807)

MCHALE, FRANCIS
THE NEW ERNIE KOVACS SHOW (I,3264)

MCHALE, MIKE
THE STAR MAKER (I,4193)

MCHALE, PHILIP
THE GIRL, THE GOLD WATCH AND DYNAMITE (II,1001); ONE LIFE TO LIVE (II,1907)

MCHATTIE, STEPHEN
CENTENNIAL (II,477); HIGHCLIFFE MANOR (II,1150); ROUGHNECKS (II,2219)

MCHUGH, FRANK
THE BING CROSBY SHOW (I,467); COLONEL HUMPHREY J. FLACK (I,1004); FULL MOON OVER BROOKLYN (I,1703); GULF PLAYHOUSE (I,1904); LOVE STORY (I,2776); THE SPIRAL STAIRCASE (I,4160)

MCHUGH, GARY
THE GAY COED (I,1742)

MCHUGH, JACK
SPACE PATROL (I,4144)

MCINDOE, JOHN
THE BUGALOOS (I,757)

MCINERNEY, BERNIE
THE EDGE OF NIGHT (II,760); ETHEL IS AN ELEPHANT (II,783); THE HAMPTONS (II,1076); O'MALLEY (II,1872); ONE LIFE TO LIVE (II,1907); RYAN'S HOPE (II,2234)

MCINTIRE, BOB
THE LEGEND OF JESSE JAMES (I,2653)

MCINTIRE, JAMES
BIG BEND COUNTRY (II,230)

MCINTIRE, JOHN
THE ADAM MACKENZIE STORY (I,30); ALL THE WAY HOME (II,44); THE AMERICAN DREAM (II,71); AMERICANS (I,176); CAVALCADE OF AMERICA (I,875); CHARLIE WOOSTER—OUTLAW (I,915); CRISIS IN SUN VALLEY (II,604); FRONT ROW CENTER (I,1694); GETAWAY CAR (I,1783); GOLIATH AWAITS (II,1025); THE HEALERS (II,1115); THE JIMMY DURANTE SHOW (I,2389); THE JORDAN CHANCE (II,1343); LASSIE: THE NEW BEGINNING (II,1433); LONE STAR (II,1513); LONGSTREET (I,2750); NAKED CITY (I,3215); THE NEW DAUGHTERS OF JOSHUA CABE (I,3261); NIGHT COURT (II,1844); POWDERKEG (I,3657); SHIRLEY (II,2330); THE VIRGINIAN (I,4738); WAGON TRAIN (I,4747); WHEN THE WEST WAS FUN: A WESTERN REUNION (II,2780)

MCINTIRE, MARILYN
RYAN'S HOPE (II,2234)

MCINTIRE, TIM
KUNG FU (II,1416); MICKEY SPILLANE'S MIKE HAMMER: MORE THAN MURDER (II,1693); RICH MAN, POOR MAN—BOOK I (II,2161); SMILE JENNY, YOU'RE DEAD (II,2382); SOAP (II,2392)

MCINTOSH, STUART
YOUNG DR. MALONE (I,4938)

MCINTOSH, SUE
THE PAUL HOGAN SHOW (II,1975)

MCINTYRE, HAL
JOHNNY OLSEN'S RUMPUS ROOM (I,2433)

MCINTYRE, MARILYN
LOVING (II,1552); SEARCH FOR TOMORROW (II,2284)

MCISAAC, MARIANNE
THE BAXTERS (II,184); HIGH HOPES (II,1144)

MCIVER, WILLIAM
AMAHL AND THE NIGHT VISITORS (I,148)

MCKAY, ALLISON
A COUPLE OF DONS (I,1077); DEAN MARTIN PRESENTS THE GOLDDIGGERS (I,1194); THE PAUL LYNDE SHOW (I,3489)

MCKAY, ANDY
KOVACS UNLIMITED (I,2583)

MCKAY, BRUCE
HEAR NO EVIL (II,1116); SOUNDS OF HOME (I,4136)

MCKAY, DAVID
MR. I. MAGINATION (I,3140); RENDEZVOUS (I,3765)

MCKAY, FAY
HAPPY TIMES ARE HERE AGAIN (I,1944)

MCKAY, GARDNER
ADVENTURES IN PARADISE (I,46); BOOTS AND SADDLES: THE STORY OF THE FIFTH CAVALRY (I,686); THE GINGER ROGERS SHOW (I,1805)

MCKAY, JEFF
DOCTOR SHRINKER (II,686); TALES OF THE GOLD MONKEY (II,2537); THE WILD WILD WEST REVISITED (II,2800)

MCKAY, JIM
THE DOROTHY HAMILL SPECIAL (II,719); GLENN FORD'S SUMMERTIME, U.S.A. (I,1823); MAKE THE CONNECTION (I,2846); THE REAL MCKAY (I,3742); THE VERDICT IS YOURS (I,4711)

MCKAY, LIZABETH
ALL MY CHILDREN (II,39)

MCKAY, SCOTT
THE EDGE OF NIGHT (II,760); HONESTLY, CELESTE! (I,2104); IT HAPPENS IN SPAIN (I,2247); ONE OF OUR OWN (II,1910); STAGE DOOR (I,4179); YOUNG DR. MALONE (I,4938)

MCKAY, SHAWN
HARDCASE (II,1088)

MCKAYLE, DONALD
BILL COSBY DOES HIS OWN THING (I,441)

MCKEAN, MICHAEL
THE BOUNDER (II,357); LAVERNE AND SHIRLEY (II,1446)

MCKEARD, NIGAL
BRIGHT PROMISE (I,727)

MCKECHNIE, DONNA
DARK SHADOWS (I,1157); HOTEL 90 (I,2130); I'M A FAN (I,2192); THE KRAFT 75TH ANNIVERSARY SPECIAL (II,1409)

MCKEE, TODD
SANTA BARBARA (II,2258)

MCKEE, TOM
THE LONELY WIZARD (I,2743); RESCUE 8 (I,3772)

MCKEEVER, JACQUELINE
WONDERFUL TOWN (I,4893)

MCKELLY, ROBERT
NEW GIRL IN HIS LIFE (I,3265)

MCKENNA, CHRISTINE
FLAMBARDS (II,869)

MCKENNA, COLIN
TOM AND JOANN (II,2630)

MCKENNA, KATHERINE
EVERYTHING GOES (I,1480)

MCKENNA, MARGO
AS THE WORLD TURNS (II,110); THE EDGE OF NIGHT (II,760); LOVE OF LIFE (I,2774)

MCKENNA, SIOBHAN
THE CRADLE SONG (I,1088); DUPONT SHOW OF THE MONTH (I,1387); THE LAST DAYS OF POMPEII (II,1436); THE WOMAN IN WHITE (I,4887)

MCKENNA, T P
ALL CREATURES GREAT AND SMALL (I,129); LADY KILLER (I,2603); THE MANIONS OF AMERICA (II,1623); NANCY ASTOR (II,1788)

MCKENNA, VIRGINIA
THE ADMIRABLE CRICHTON (I,36); BEAUTY AND THE BEAST (I,382); THE GATHERING STORM (I,1741); PETER PAN (II,2030)

MCKENNON, DALE
WHICH WAY'D THEY GO? (I,4838)

MCKENNON, DALLAS
THE ARCHIE COMEDY HOUR (I,252); ARCHIE'S TV FUNNIES (I,253); THE BANG-SHANG LALAPALOOZA SHOW (II,140); DANIEL BOONE (I,1142); EVERYTHING'S ARCHIE (I,1483); THE HARDY BOYS (I,1949); THE U.S. OF ARCHIE (I,4687); U.S. OF ARCHIE (II,2698)

MCKENSEY, BEVERLEE
ABC'S MATINEE TODAY (I,7)

MCKENZIE, DUNCAN
CONDOMINIUM (II,566)

MCKENZIE, HONEY
JIM AND JUDY IN TELELAND (I,2378)

MCKENZIE, IDA MOE
THE RED SKELTON SHOW (I,3755)

MCKENZIE, JULIA
THE TWO RONNIES (II,2694)

MCKENZIE, MICKI
THE MUSIC MART (II,1774)

MCKENZIE, RICHARD
ALL IN THE FAMILY (II,38); ARCHIE BUNKER'S PLACE (II,105); DANGER IN PARADISE (II,617); I'M A BIG GIRL NOW (II,1218); ROOTS (II,2211); YOU ARE THE JURY (II,2854)

MCKENZIE, WENDY
THE HANNA-BARBERA HAPPINESS HOUR (II,1079)

MCKEON, DOUG
BIG SHAMUS, LITTLE SHAMUS (II,242); THE EDGE OF NIGHT (II,760)

MCKEON, KRIS
EVERY STRAY DOG AND KID (II,792)

MCKEON, MARY ELLEN
THE WILDS OF TEN THOUSAND ISLANDS (II,2803)

MCKEON, NANCY
THE ACADEMY (II,3); BATTLE OF THE NETWORK STARS (II,174); BATTLE OF THE NETWORK STARS (II,176); DUSTY (II,745); THE FACTS OF LIFE (II,805); FACTS OF LIFE: THE FACTS OF LIFE GOES TO PARIS (II,806); HIGH SCHOOL, U.S.A. (II,1148); JO'S COUSINS (II,1293); THE PARKERS (II,1959); RETURN TO FANTASY ISLAND (II,2144); STONE (II,2470); STONE (II,2471)

MCKEON, PHILIP
ALICE (II,33); LEADFOOT (II,1452)

MCKERN, LEO
THE PRISONER (I,3670); REILLY, ACE OF SPIES (II,2129)

MCKEWIN, VINCE
THE BOYS IN BLUE (II,359)

MCKIE, BENJIE
THE SECRET OF CHARLES DICKENS (II,2293)

MCKIERNAN, DENNIS
ANOTHER WORLD (II,97)

MCKILLAN, KEN
JOE DANCER: MURDER ONE, DANCER 0 (II,1302)

MCKINLEY, ANDREW
AMAHL AND THE NIGHT VISITORS (I,148)

MCKINLEY, EDWARD J
THE BOB HOPE SHOW (I,625); DENNIS THE MENACE (I,1231)

MCKINLEY, JEAN
THE PRIME OF MISS JEAN BRODIE (II,2082)

MCKINLEY, RAY
GLENN MILLER TIME (I,1824)

MCKINNEY, BILL
THE HEALERS (II,1115)

MCKINNEY, JOHN
FANTASY ISLAND (II,830)

MCKINNEY, MYRA
PEPSI-COLA PLAYHOUSE (I,3523)

MCKINNEY, WILLIAM
THE FAMILY HOLVAK (II,817)

MCKINSEY, BEVERLEE
ANOTHER WORLD (II,97); THE CLIFF DWELLERS (I,974); THE GUIDING LIGHT (II,1064); TEXAS (II,2566)

MCKNIGHT, MARIAN
JACK AND THE BEANSTALK (I,2288)

MCKORMICK, LARRY
MURDOCK'S GANG (I,3164)

MCKRELL, JIM
CELEBRITY SWEEPSTAKES (II,476); DALLAS (II,614); THE GAME GAME (I,1732); MITZI AND A HUNDRED GUYS (II,1710); THE QUIZ KIDS (II,2103); SEMI-TOUGH (II,2300); WHO'S ON CALL? (II,2791)

MCKUEN, ROD
THE ROD MCKUEN SPECIAL (I,3826)

MCLAIM, JOHN
THE LEGEND OF THE GOLDEN GUN (II,1455)

MCLAIN, HELYNE
YOUR HIT PARADE (I,4952)

MCLAIN, JAMES
DOCTOR I.Q. (I,1314)

MCLANE, KERRY
THE COWBOYS (II,598)

MCLAREN, CONRAD
KATE AND ALLIE (II,1365)

MCLARTY, RON
MURDER INK (II,1769)

MCLAUGHLEN, PATRICK
THE ADVENTURES OF SIR FRANCIS DRAKE (I,75)

MCLAUGHLIN, ANDREA
PAUL WHITEMAN'S TV TEEN CLUB (I,3491)

MCLAUGHLIN, BRUCE
FEEL THE HEAT (II,848)

MCLAUGHLIN, DON
AS THE WORLD TURNS (II,110); ROAD OF LIFE (I,3798)

MCLAUGHLIN, EMILY
GENERAL HOSPITAL (II,964); YOUNG DR. MALONE (I,4938)

MCLAUGHLIN, HENRY
HOWDY DOODY (I,2151)

MCLAUGHLIN, MICHAEL
MACKENZIE (II,1584)

MCLAUGHLIN, RITA
THE PATTY DUKE SHOW (I,3483); SING ALONG WITH MITCH (I,4057)

MCLEAN, BILL
HARRY O (II,1098); HOME COOKIN' (II,1170)

MCLEAN, DAVID
330 INDEPENDENCE S.W. (I,4486); DAYS OF OUR LIVES (II,629); TATE (I,4364)

MCLEAN, DON
100 YEARS OF GOLDEN HITS (II,1905)

MCLEAN, DOREEN
THE PHIL SILVERS SHOW (I,3580)

MCLEMORE, LAMONTE
THE FIFTH DIMENSION SPECIAL: AN ODYSSEY IN THE COSMIC UNIVERSE OF PETER MAX (I,1571)

MCLENNAN, BILL
LITTLE HOUSE: THE LAST FAREWELL (II,1491)

MCLENNAN, CHUCK
PARTRIDGE FAMILY: 2200 A.D. (II,1963)

MCLENNAN, MARGARET
PRISONER: CELL BLOCK H (II,2085)

MCLENNAN, MARGO
ALL THE RIVERS RUN (II,43)

MCLENNAN, ROD
APPLAUSE (I,244)

MCLEOD, CATHERINE
BACHELOR FATHER (I,323); FRONTIER (I,1695); SUSPENSE (I,4305)

MCLEOD, MARY
BRIDESHEAD REVISITED (II,375)

MCLEOD, MERCER
VANITY FAIR (I,4700)

MCLERIE, ALLYN ANN
AFTER GEORGE (II,19); BENSON (II,208); BEULAH LAND (II,226); DOMESTIC LIFE (II,703); THE DUCK FACTORY (II,738); MR. PORTER OF INDIANA (I,3148); THE OLDEST LIVING GRADUATE (II,1882); RETURN ENGAGEMENT (I,3775); SISTER TERRI (II,2366); THE THORN BIRDS (II,2600); THE TONY RANDALL SHOW (II,2640); WKRP IN CINCINNATI (II,2814)

MCLIAM, JOHN
DELTA COUNTY, U.S.A. (II,657); HIGHWAY TO HEAVEN (II,1152); HITCHED (I,2065); THE INVISIBLE MAN (II,1243); THE TURNING POINT OF JIM MALLOY (II,2670); TWO MARRIAGES (II,2691)

MCLOUGHLIN, TOMMY
NUTS AND BOLTS (II,1871)

MCMAHON, ED
20TH CENTURY FOLLIES (I,4641); THE ALL-STAR SALUTE TO MOTHER'S DAY (II,51); THE BIG TOP (I,435); THE CELEBRITY FOOTBALL CLASSIC (II,474); CONCENTRATION (I,1031); ED MCMAHON AND HIS FRIENDS. . . DISCOVER WET AT CYPRESS GARDENS (I,1400); THE LUCILLE BALL SPECIAL (II,1559); THE MAD MAD MAD MAD WORLD OF THE SUPER BOWL (II,1586);

MISSING LINKS (I,3066); NBC ADVENTURE THEATER (I,3235); SNAP JUDGEMENT (I,4099); STAR SEARCH (II,2438); THE STARS AND STRIPES SHOW (I,4207); THE STARS AND STRIPES SHOW (II,2443); STEVE MARTIN'S THE WINDS OF WHOOPIE (II,2461); TELEVISION'S GREATEST COMMERCIALS (II,2553); TELEVISION'S GREATEST COMMERCIALS II (II,2554); TELEVISION'S GREATEST COMMERCIALS III (II,2555); TELEVISION'S GREATEST COMMERCIALS IV (II,2556); THAT MCMAHON'S HERE AGAIN (I,4419); TV'S BLOOPERS AND PRACTICAL JOKES (II,2675); WHODUNIT? (II,2794)

MCMAHON, HORACE
MAKE ROOM FOR DADDY (I,2843); MARTIN KANE, PRIVATE EYE (I,2928); MR. BEVIS (I,3128); MR. BROADWAY (I,3131); NAKED CITY (I,3215); NAKED CITY (I,3216)

MCMAHON, JENNA
THE EVERLY BROTHERS SHOW (I,1479); THE FUNNY SIDE (I,1714); THE FUNNY SIDE (I,1715); WHAT'S IT ALL ABOUT WORLD? (I,4816)

MCMANUS, GINGER
LET'S TAKE A TRIP (I,2668); MEET ME IN ST. LOUIS (I,2986); PORTIA FACES LIFE (I,3653); THREE STEPS TO HEAVEN (I,4485)

MCMANUS, MICHAEL
ANGIE (II,80); THE BILLY CRYSTAL COMEDY HOUR (II,248); BUMPERS (II,394); CAPTAIN AMERICA (II,427); LEWIS AND CLARK (II,1465); THE MUNSTERS' REVENGE (II,1765); A ROCK AND A HARD PLACE (II,2190); THICKE OF THE NIGHT (II,2587)

MCMARTIN, JOHN
BUTTERFLIES (II,405); HIGHER AND HIGHER, ATTORNEYS AT LAW (I,2057); THE LAST NINJA (II,1438); OUT OF THE BLUE (I,3422); A TIME FOR US (I,4507)

MCMASTER, NILES
THE EDGE OF NIGHT (II,760)

MCMEEL, MICKY
THE BAY CITY ROLLERS SHOW (II,187); THE KROFFT SUPERSHOW (II,1412); THE KROFFT SUPERSHOW II (II,1413)

MCMILLAN, GLORIA
CENTENNIAL (II,477); OUR MISS BROOKS (I,3415)

MCMILLAN, JANET
THE BOOK OF LISTS (II,350)

MCMILLAN, KENNETH
CONCRETE BEAT (II,562); LOVE OF LIFE (I,2774); RHODA (II,2151); THE RUBBER GUN SQUAD (II,2225); SUZANNE PLESHETTE IS MAGGIE BRIGGS (II,2509)

MCMILLAN, NORMA
CASPER AND FRIENDS (I,867); MOBY DICK AND THE MIGHTY MIGHTOR (I,3077)

MCMILLAN, RODDY
KILLER WITH TWO FACES (I,2539)

MCMILLAN, SANDY
GOLIATH AWAITS (II,1025)

MCMILLAN, WILLIAM
ANOTHER WORLD (II,97); THE FOUR OF US (II,914)

MCMULLAN, JIM
BEN CASEY (I,394); BEYOND WESTWORLD (II,227); CHOPPER ONE (II,515); CONVOY (I,1043); FATHER KNOWS BEST: HOME FOR CHRISTMAS (II,839); FATHER KNOWS BEST: THE FATHER KNOWS BEST REUNION (II,840); ROARING CAMP (I,3804); SCRUPLES (II,2281); WHIZ KIDS (II,2790)

MCMURRAY, SAM
BAKER'S DOZEN (II,136); NOT NECESSARILY THE NEWS (II,1862); RYAN'S HOPE (II,2234)

MCMYLER, PAMELA
BANYON (I,344); GIDGET (I,1795)

MCNAIR, BARBARA
THE BARBARA MCNAIR AND DUKE ELLINGTON SPECIAL (I,347); THE BARBARA MCNAIR SHOW (I,348); THE BOB HOPE SHOW (I,620); THE BOB HOPE SHOW (I,627); THE BOB HOPE SHOW (I,634); THE BOB HOPE SHOW (I,649); THE BOB HOPE SHOW (I,654); THE BOOTS RANDOLPH SHOW (I,687); CIRCLE (I,957); GLITTER (II,1009); THE GORDON MACRAE SHOW (I,1860); ON STAGE WITH BARBARA MCNAIR (I,3358); ON STAGE: PHIL SILVERS (I,3357); SOMETHING SPECIAL (I,4118); WHERE THE GIRLS ARE (I,4830)

MCNAIR, HEATHER
AUTOMAN (II,120); SAVAGE: IN THE ORIENT (II,2264)

MCNAIR, RALPH
CELEBRITY TIME (I,887); THE EYES HAVE IT (I,1495)

MCNALLY, MARGE
THE ELEVENTH HOUR (I,1425)

MCNALLY, STEPHEN
BRADDOCK (I,704); CALL TO DANGER (I,804); CHRYSLER MEDALLION THEATER (I,951); CROSSROADS (I,1105); HENRY FONDA PRESENTS THE STAR AND THE STORY (I,2016); THE LIVES OF JENNY DOLAN (II,1502); THE LORETTA YOUNG THEATER (I,2756); MOST WANTED (II,1738); NAKIA (II,1784); THE OUTER LIMITS (I,3426); RHEINGOLD THEATER (I,3783); STAGE 7 (I,4181); TARGET: THE CORRUPTERS (I,4360); TELEPHONE TIME (I,4379); W.E.B. (II,2732); THE WHOLE WORLD IS WATCHING (I,4851)

MCNALLY, TERENCE
CONCRETE BEAT (II,562); COUNTERATTACK: CRIME IN AMERICA (II,581)

MCNAMARA, ED
SCARLETT HILL (I,3933)

MCNAMARA, JOHN
FULL CIRCLE (I,1702)

MCNAMARA, KERRY
RYAN'S HOPE (II,2234)

MCNAMARA, MATHEW
ALL MY CHILDREN (II,39)

MCNAMARA, MAUREEN
ALMOST AMERICAN (II,55)

MCNAMARA, PAT
MR. MOM (II,1752); REVENGE OF THE GRAY GANG (II,2146)

MCNAUGHTON, STEVE
THE EDGE OF NIGHT (II,760); JAKE'S WAY (II,1269); SIGN-ON (II,2353)

MCNEAR, HOWARD
THE ANDY GRIFFITH SHOW (I,192); THE BROTHERS (I,748); THE DONNA REED SHOW (I,1347); THE GEORGE BURNS AND GRACIE ALLEN SHOW (I,1763); THE MANY SIDES OF MICKEY ROONEY (I,2899); TOM, DICK, AND HARRY (I,4536)

MCNEELY, LARRY
THE GLEN CAMPBELL GOODTIME HOUR (I,1820)

MCNEIL, CLAUDIA
PALMERSTOWN U.S.A. (II,1945); ROLL OF THUNDER, HEAR MY CRY (II,2202); ROOTS: THE NEXT GENERATIONS (II,2213)

MCNEILL, DON
THE BOB HOPE SHOW (I,524); DON MCNEILL'S DINNER CLUB (I,1336); DON MCNEILL'S TV CLUB (I,1337); TAKE TWO (I,4337)

MCNELLIS, MAGGI
THE CRYSTAL ROOM (I,1113); LEAVE IT TO THE GIRLS (I,2650); MAGGI'S PRIVATE WIRE (I,2813); SATINS AND SPURS (I,3917); SAY IT WITH ACTING (I,3928)

MCNICHOL, JAMES VINCENT
BATTLE OF THE NETWORK STARS (II,165); CALIFORNIA FEVER (II,411); CIRCUS OF THE STARS (II,531); THE FITZPATRICKS (II,868); HOLLYWOOD TEEN (II,1166); THE JIMMY MCNICHOL SPECIAL (II,1292); STRANDED (II,2475)

MCNICHOL, KRISTY
APPLE'S WAY (II,101); BATTLE OF THE NETWORK STARS (II,164); BATTLE OF THE NETWORK STARS (II,165); BATTLE OF THE NETWORK STARS (II,169); THE CARPENTERS AT CHRISTMAS (II,446); CELEBRITY CHALLENGE OF THE SEXES 1 (II,465); CELEBRITY CHALLENGE OF THE SEXES 2 (II,466); CIRCUS OF THE STARS (II,531); FAMILY (II,813); I LOVE LIBERTY (II,1215); THE JIMMY MCNICHOL SPECIAL (II,1292); THE LOVE BOAT II (II,1533)

MCNULTY, AIDAN
ANOTHER WORLD (II,97)

MCNULTY, PATRICIA
THE TYCOON (I,4660)

MCPARTLAND, JIMMY
JAMBOREE (I,2331); WINDY CITY JAMBOREE (I,4877)

MCPEAK, SANDY
BLUE THUNDER (II,278); LONE STAR (II,1513); SCRUPLES (II,2281); THE SEAL (II,2282)

MCPHAIL, MARNIE
THE EDISON TWINS (II,761)

MCPHEE, CHRIS
FLEETWOOD MAC IN CONCERT (II,880)

MCPHEE, JENNIFER
MATINEE AT THE BIJOU (II,1651)

MCPHEE, JOHN
FLEETWOOD MAC IN CONCERT (II,880)

MCPHERSON, JOHN
MYSTERY CHEF (I,3212)

MCPHERSON, PATRICIA
CIRCUS OF THE STARS (II,538); CONCRETE BEAT (II,562); KNIGHT RIDER (II,1402)

MCQUADE, ARLENE
THE GOLDBERGS (I,1835); THE GOLDBERGS (I,1836)

MCQUADE, JOHN
CHARLIE WILD, PRIVATE DETECTIVE (I,914); CYRANO DE BERGERAC (I,1119); THE LAST WAR (I,2630); MR. MERGENTHWIRKER'S LOBBLIES (I,3144)

MCQUEEN, ARMELIA
AIN'T MISBEHAVIN' (II,25)

MCQUEEN, BUTTERFLY
BEULAH (I,416)

MCQUEEN, STEVE
ALFRED HITCHCOCK PRESENTS (I,115); THE BOB HOPE SHOW (I,574); THE BOB HOPE SHOW (I,579); THE BOUNTY HUNTER (I,698); THE DEFENDER (I,1219); WANTED: DEAD OR ALIVE (I,4764)

MCQUEENY, ROBERT
THE GALLANT MEN (I,1728); MISS SUSAN (I,3065)

MCRAE, CARMEN
TIMEX ALL-STAR JAZZ SHOW I (I,4515)

MCRAE, ELLEN
THE BIG BRAIN (I,424); THE DOCTORS (II,694); IRON HORSE (I,2238)

MCRAE, FRANK
THE ORPHAN AND THE DUDE (II,1924); SHOOTING STARS (II,2341)

MCRANEY, GERALD
THE JORDAN CHANCE (II,1343); THE SEAL (II,2282); SHADOW OF SAM PENNY (II,2313); SIMON AND SIMON (II,2357)

MCROBERTS, BRIONY
PETER PAN (II,2030)

MCSHANE, IAN
BARE ESSENCE (II,150); CODE NAME: DIAMOND HEAD (II,549); MARCO POLO (II,1626); ROOTS (II,2211)

MCSHARRY, CARMEL
LITTLE LORD FAUNTLEROY (II,1492)

MCSWAIN, GINNY
THE C.B. BEARS (II,406)

MCVEAGH, EVE
FARAWAY HILL (I,1532); ROBERTA (I,3816)

MCVEGH, EVE
TESTIMONY OF TWO MEN (II,2564)

MCVEY, PATRICK
BIG TOWN (I,436); BOOTS AND SADDLES: THE STORY OF THE FIFTH CAVALRY (I,686); HAZEL (I,1982); MANHUNT (I,2890); TELEPHONE TIME (I,4379)

MCVEY, TYLER
MEN INTO SPACE (I,3007)

MCWHIRTER, JULIE
ALL COMMERCIALS—A STEVE MARTIN SPECIAL (II,37); ALVIN AND THE CHIPMUNKS (II,61); THE BARKLEYS (I,356); CASPER AND THE ANGELS (II,453); THE DRAK PACK (II,733); THE FLINTSTONES (II,885); THE GARY COLEMAN SHOW (II,958); HAPPY DAYS (I,1942); JABBERJAW (II,1261); JEANNIE (II,1275); JONATHAN WINTERS PRESENTS 200 YEARS OF AMERICAN HUMOR (II,1342); LAVERNE AND SHIRLEY WITH THE FONZ (II,1449); THE LITTLE RASCALS (II,1494); PARTRIDGE FAMILY: 2200 A.D. (II,1963); THE RICH LITTLE SHOW (II,2153); THE RICH LITTLE SPECIAL (II,2155); SCOOBY'S ALL-STAR LAFF-A-LYMPICS (II,2273); THUNDARR THE BARBARIAN (II,2616); TOP TEN (II,2648); WACKO (II,2733)

MCWILLIAMS, CAROLINE
THE ALIENS ARE COMING (II,35); ANOTHER WORLD (II,97); BATTLE OF THE NETWORK STARS (II,170); BENSON (II,208); CASS MALLOY (II,454); THE GUIDING LIGHT (II,1064); THE MANY LOVES OF ARTHUR (II,1625); SOAP (II,2392); ST. ELSEWHERE (II,2432); WHAT'S UP DOC? (II,2771)

MCANN, SHAWN
THE WORLD OF DARKNESS (II,2836)

MCKEAN, MICHAEL
THE TV SHOW (II,2673)

MCKENNON, DALLAS
THE U.S. OF ARCHIE (I,4687)

MEACHAM, ANNE
ANOTHER WORLD (II,97)

MEAD, DIANE
BEAT THE CLOCK (I,378)

MEAD, SARA
THE WOMEN (I,4890)

MEADE, JULIA
CLUB EMBASSY (I,982); THE DENNIS JAMES SHOW (I,1228); THE ED SULLIVAN SHOW (I,1401)

MEADER, GEORGE
MY LITTLE MARGIE (I,3194)

MEADOWS, ARLENE
PAUL BERNARD—PSYCHIATRIST (I,3485)

MEADOWS, AUDREY
THE BOB AND RAY SHOW (I,496); CLOWN ALLEY (I,979); CLUB EMBASSY (I,982); THE GENERAL ELECTRIC THEATER (I,1753); THE HONEYMOONERS (I,2110); THE HONEYMOONERS (II,1175); THE HONEYMOONERS CHRISTMAS SPECIAL (II,1176); THE HONEYMOONERS CHRISTMAS (II,1177); THE HONEYMOONERS SECOND HONEYMOON (II,1178); THE HONEYMOONERS VALENTINE SPECIAL (II,1179); THE JACKIE GLEASON SHOW (I,2322); LILY—SOLD OUT (II,1480); MARRIAGE—HANDLE WITH CARE (I,2921); PULITZER PRIZE PLAYHOUSE (I,3692); A SALUTE TO STAN LAUREL (I,3892); THE SID CAESAR SHOW (I,4042); THE SID CAESAR SPECIAL (I,4044); TIPTOE THROUGH TV (I,4524); TOO CLOSE FOR COMFORT (II,2642); VARIETY: THE WORLD OF SHOW BIZ (I,4704)

MEADOWS, JAYNE
DANGER (I,1134); HAVE I GOT A CHRISTMAS FOR YOU (II,1105); IT'S NOT EASY (II,1256); MAN AND THE CHALLENGE (I,2855); MEDICAL CENTER (II,1675); RISE AND SHINE (II,2179); ROBERT MONTGOMERY PRESENTS YOUR LUCKY STRIKE THEATER (I,3809); THE STEVE ALLEN COMEDY HOUR (I,4218); STEVE ALLEN'S LAUGH-BACK (II,2455); STUDIO ONE (I,4268); THE UNITED STATES STEEL HOUR (I,4677); VAUDEVILLE (II,2722); THE WEB (I,4784); WHAT'S GOING ON? (I,4814)

MEADOWS, JOYCE
THE ART LINKLETTER SHOW (I,276); NEW GIRL IN HIS LIFE (I,3265); TWO FACES WEST (I,4653)

MEADOWS, KRISTEN
A FINE ROMANCE (II,858); GLITTER (II,1009); ONE LIFE TO LIVE (II,1907)

MEADOWS, STEVE
SANTA BARBARA (II,2258)

MEALY, BARBARA
OUR STREET (I,3418); THE RETURN OF MARCUS WELBY, M.D (II,2138)

MEANS, JOHN
GEORGE WASHINGTON (II,978)

MEANS, KEN
THE SKATEBIRDS (II,2375)

MEANY, GEORGE
OPERATION ENTERTAINMENT (I,3400)

MEARA, ANNE
ALAN KING LOOKS BACK IN ANGER—A REVIEW OF 1972 (I,99); ARCHIE BUNKER'S PLACE (II,105); CELEBRATION: THE AMERICAN SPIRIT (II,461); THE CORNER BAR (I,1052); DON RICKLES—ALIVE AND KICKING (I,1340); GLORIA COMES HOME (II,1011); THE GREATEST GIFT (I,1880); THE JONATHAN WINTERS SHOW (I,2442); KATE MCSHANE (II,1368); KATE MCSHANE (II,1369); THE MAD MAD MAD MAD WORLD OF THE SUPER BOWL (II,1586); NINOTCHKA (I,3299); THE PAUL LYNDE SHOW (I,3489); THE PERRY COMO CHRISTMAS SHOW (I,3528); RHODA (II,2151); TAKE FIVE WITH STILLER AND MEARA (II,2530); THIS BETTER BE IT (II,2595)

MEARS, DEANN
BEACON HILL (II,193); TESTIMONY OF TWO MEN (II,2564)

MEDAK, PETER

SPACE: 1999 (II,2409)

MEDALIS, JOE
LONE STAR (II,1513); THE LONG DAYS OF SUMMER (II,1514); MCNAMARA'S BAND (II,1669); SPACE FORCE (II,2407)

MEDARIS, TERESA
THE CHEERLEADERS (II,493); SENIOR YEAR (II,2301); SONS AND DAUGHTERS (II,2404)

MEDERIOS III, ROBERT
THE FLINTSTONES (II,885); THE GALAXY GOOFUPS (II,954); THE HUCKLEBERRY HOUND SHOW (I,2155); THE JETSONS (I,2376); MAGNUM, P.I. (II,1601); THE PETER POTAMUS SHOW (I,3568)

MEDFORD, BRIAN
CORONET BLUE (I,1055)

MEDFORD, KAY
THE DEAN MARTIN SHOW (I,1201); ELKE (I,1432); GENE KELLY'S WONDERFUL WORLD OF GIRLS (I,1752); KITTY FOYLE (I,2571); MY SON THE DOCTOR (I,3203); ON OUR OWN (II,1892); STAR TONIGHT (I,4199); THAT'S LIFE (I,4426); TO ROME WITH LOVE (I,4526); THE WONDERFUL WORLD OF GIRLS (I,4898)

MEDIEROS, MIKE
TEXAS (II,2566)

MEDINA, JULIO
THE TARZAN/LONE RANGER/ZORRO ADVENTURE HOUR (II,2543)

MEDINA, PATRICIA
DOUGLAS FAIRBANKS JR. PRESENTS THE RHEINGOLD THEATER (I,1364)

MEDLEY, BILL
AMERICA (I,164); GOOBER AND THE TRUCKERS' PARADISE (II,1029)

MEDWIN, MICHAEL
SHOESTRING (II,2338)

MEEHAN, DANNY
THE LAS VEGAS SHOW (I,2619)

MEEK, BARBARA
ARCHIE BUNKER'S PLACE (II,105)

MEEKER, KEN
ONE LIFE TO LIVE (II,1907)

MEEKER, RALPH
ALFRED HITCHCOCK PRESENTS (I,115); GENTLE BEN (I,1761); JANE WYMAN PRESENTS THE FIRESIDE THEATER (I,2345); THE LORETTA YOUNG THEATER (I,2756); THE LUX VIDEO THEATER (I,2795); NIGHT GAMES (II,1845); THE NIGHT STALKER (I,3292); NOT FOR HIRE (I,3308); THE POLICE STORY (I,3638); A PUNT, A PASS, AND A PRAYER (I,3694); RIDDLE AT 24000 (II,2171)

MEEKIN, EDWARD
SENSE OF HUMOR (II,2303)

MEEKS, BARBARA
PAROLE (II,1960)

MEEKS, EDWARD
THE ADVENTURES (I,86)

MEERAKKER, WILHELMINA
THE GALLOPING GOURMET (I,1730)

MEFFORD, SCOTT
THE SECRET STORM (I,3969)

MEGNA, JOHN
HOLLYWOOD HIGH (II,1162)

MEGOWAN, DEBBIE
MR. O'MALLEY (I,3146)

MEGOWAN, DON
THE BEACHCOMBER (I,371); COUNTERSPY (I,1060); DAVY CROCKETT (I,1181); MR. CHIPS (I,3132); SECRET AGENT (I,3961)

MEGURO, YUKI
SHOGUN (II,2339)

MEHYA, SHAYUR
THE FAR PAVILIONS (II,832)

MEIGS, WILLIAM
SVENGALI AND THE BLONDE (I,4310)

MEIKLE, PAT
LOVE STORY (I,2777); MAGIC COTTAGE (I,2815)

MEIKLE, RICHARD
BARRIER REEF (I,362)

MEIKLEJOHN, LINDA
M*A*S*H (II,1569)

MEINCH, JOAN
DOUBLE OR NOTHING (I,1362)

MEINERT, MARJORIE
LAWRENCE WELK'S TOP TUNES AND NEW TALENT (I,2645)

MEINRAD, JOSEF
DON QUIXOTE (I,1339)

MEISER, EDITH
REUNION IN VIENNA (I,3780)

MEISNER, GUNTER
THE WINDS OF WAR (II,2807)

MEKKA, EDDIE
BLANSKY'S BEAUTIES (II,267); CELEBRITY DAREDEVILS (II,473); CIRCUS OF THE STARS (II,532); CIRCUS OF THE STARS (II,535); CIRCUS OF THE STARS (II,536); LAVERNE AND SHIRLEY (II,1446)

MEL PAHL CHORUS, THE
PERRY PRESENTS (I,3549)

MELCHOIR, LAURITZ
ARTHUR MURRAY'S DANCE PARTY (I,293)

MELENDEZ, BILL
BE MY VALENTINE, CHARLIE BROWN (I,369); THE CHARLIE BROWN AND SNOOPY SHOW (II,484); A CHARLIE BROWN CHRISTMAS (I,908); A CHARLIE BROWN THANKSGIVING (I,909); CHARLIE BROWN'S ALL STARS (I,910); HAPPY ANNIVERSARY, CHARLIE BROWN (I,1939); HE'S YOUR DOG, CHARLIE BROWN (I,2037); IS THIS GOODBYE, CHARLIE BROWN? (I,2242); IT WAS A SHORT SUMMER, CHARLIE BROWN (I,2252); IT'S A MYSTERY, CHARLIE BROWN (I,2259); IT'S ARBOR DAY, CHARLIE BROWN (I,2267); IT'S MAGIC, CHARLIE BROWN (I,2271); IT'S THE EASTER BEAGLE, CHARLIE BROWN (I,2275); IT'S THE GREAT PUMPKIN, CHARLIE BROWN (I,2276); IT'S YOUR FIRST KISS, CHARLIE BROWN (I,2280); LIFE IS A CIRCUS, CHARLIE BROWN (I,2683); PLAY IT AGAIN, CHARLIE BROWN (I,3617); SHE'S A GOOD SKATE, CHARLIE BROWN (I,4012); SNOOPY'S GETTING MARRIED, CHARLIE BROWN (I,4102); SOMEDAY YOU'LL FIND HER, CHARLIE BROWN (I,4113); THERE'S NO TIME FOR LOVE, CHARLIE BROWN (I,4438); WHAT A NIGHTMARE, CHARLIE BROWN (I,4806); WHAT HAVE WE LEARNED, CHARLIE BROWN? (I,4810); YOU'RE A GOOD SPORT, CHARLIE BROWN (I,4963); YOU'RE IN LOVE, CHARLIE BROWN (I,4964); YOU'RE NOT ELECTED, CHARLIE BROWN (I,4967); YOU'RE THE GREATEST, CHARLIE BROWN (I,4973)

MELENDEZ, TONY
DEATH RAY 2000 (II,654)

MELER, GUSTAVE
TONIGHT IN HAVANA (I,4557)

MELGAR, GABRIEL
CHICO AND THE MAN (II,508)

MELIS, JOSE
THE JACK PAAR SHOW (I,2316)

MELIS, MORGAN K
ONE LIFE TO LIVE (II,1907)

MELL, JOSEPH
THE ELISABETH MCQUEENEY STORY (I,1429); I WAS A BLOODHOUND (I,2180); MCNAMARA'S BAND (II,1668)

MELLEY, ANDREA
THE BENNY HILL SHOW (II,207)

MELLIN, USCHI
STAR MAIDENS (II,2435)

MELLINI, SCOTT
FATHER MURPHY (II,841)

MELLOLARKS, THE
BROADWAY OPEN HOUSE (I,736); CAVALCADE OF BANDS (I,876); CHEVROLET ON BROADWAY (I,925); CHEZ PAREE REVUE (I,932); CLUB 60 (I,987); FORD STAR REVUE (I,1632); THE SNOOKY LANSON SHOW (I,4101)

MELLOLARKS, THE
FRANKIE CARLE TIME (I,1674)

MELMAN, LARRY 'BUD'
LATE NIGHT WITH DAVID LETTERMAN (II,1442)

MELODY, JILL
THE BEST OF EVERYTHING (I,406)

MELROSE, BRIAN
THE YOUNG PIONEERS' CHRISTMAS (II,2870)

MELTON, JAMES
FORD FESTIVAL (I,1629)

MELTON, PAUL
WE INTERRUPT THIS SEASON (I,4780)

MELTON, SID
BACHELOR FATHER (I,323); CAPTAIN MIDNIGHT (I,833); CITY VS. COUNTRY (I,968); DAMON RUNYON THEATER (I,1129); GREEN ACRES (I,1884); IT'S ALWAYS JAN (I,2264); MAKE MORE ROOM FOR DADDY (I,2842); MAKE ROOM FOR DADDY (I,2843); MAKE ROOM FOR GRANDDADDY (I,2844);

MAKE ROOM FOR GRANDDADDY (I,2845); TV READERS DIGEST (I,4630)

MELTON, TROY
PRUDENCE AND THE CHIEF (I,3682); THE ROY ROGERS SHOW (I,3859)

MELVILLE, SAM
FOOLS, FEMALES AND FUN: IS THERE A DOCTOR IN THE HOUSE? (II,900); THE ROOKIES (I,3840); THE ROOKIES (II,2208); ROUGHNECKS (II,2219)

MELVIN, ALLAN
THE ADVENTURES OF GULLIVER (I,60); ALFRED OF THE AMAZON (I,116); ALL IN THE FAMILY (II,38); THE ALL-NEW POPEYE HOUR (II,49); ARCHIE BUNKER'S PLACE (II,105); THE BANANA SPLITS ADVENTURE HOUR (I,340); THE BRADY BUNCH (II,362); THE CATTANOOGA CATS (I,874); THE DICK VAN DYKE SHOW (I,1275); FLASH GORDON (II,874); GLORIA COMES HOME (II,1011); GOMER PYLE, U.S.M.C. (I,1843); KID POWER (I,2532); KING FEATURES TRILOGY (I,2555); THE KWICKY KOALA SHOW (II,1417); THE MAGILLA GORILLA SHOW (I,2825); MAN IN THE MIDDLE (I,2870); MOTOR MOUSE (I,3111); PARTRIDGE FAMILY: 2200 A.D. (II,1963); THE PHIL SILVERS SHOW (I,3580); THE SECRET LIVES OF WALDO KITTY (II,2292); WE'LL TAKE MANHATTAN (I,4792); YOGI'S GANG (II,2852)

MELVIN, DONNIE
A CHRISTMAS MEMORY (I,948); THE SECRET STORM (I,3969)

MELVIN, SUSAN
THE PATTY DUKE SHOW (I,3483)

MEMMOLI, GEORGE
HELLO, LARRY (II,1128)

MEN OF THE DEEP, THE
A SPECIAL ANNE MURRAY CHRISTMAS (II,2413)

MENDELS, ROLANDA
ANOTHER WORLD (II,97)

MENDELSON, GLENN
CHARLIE BROWN'S ALL STARS (I,910)

MENDELSON, LINDA
PLAY IT AGAIN, CHARLIE BROWN (I,3617)

MENDENHALL, DAVID
GENERAL HOSPITAL (II,964); TAXI (II,2546)

MENDENHALL, MARISSA
SCAMPS (II,2267)

MENDICK, CHARLES
THE AD-LIBBERS (I,29)

MENDILLO, STEPHEN
KEY WEST (I,2528)

MENGATTI, JOHN
FOR LOVE AND HONOR (II,903); JO'S COUSINS (II,1293); THE WHITE SHADOW (II,2788)

MENGES, JOYCE
TO ROME WITH LOVE (I,4526)

MENJOU, ADOLPHE
SCIENCE FICTION THEATER (I,3938); TARGET (I,4359); YOUR FAVORITE STORY (I,4948)

MENKIN, SHEPARD
THE ALVIN SHOW (I,146); BARNEY AND ME (I,358); THE FAMOUS ADVENTURES OF MR. MAGOO (I,1524); THE LONE RANGER (I,2741); THUNDARR THE BARBARIAN (II,2616)

MENTEER, GARY
THE DESILU REVUE (I,1238)

MENVILLE, SCOTT
THE LITTLE RASCALS (II,1494)

MENZIES, HEATHER
THE BOB NEWHART SHOW (II,342); CAPTAIN AMERICA (II,427); DOCTOR DAN (II,684); LOGAN'S RUN (II,1507); MAN IN THE MIDDLE (I,2870)

MEPHAM, CHERI
FATHER ON TRIAL (I,1544)

MERANDE, DORO
BRINGING UP BUDDY (I,731); THE FRONT PAGE (I,1692); SATURDAY'S CHILDREN (I,3923); THAT WAS THE WEEK THAT WAS (I,4423); TROUBLE WITH RICHARD (I,4612)

MERCER, FRANCES
DOCTOR HUDSON'S SECRET JOURNAL (I,1312); MY FRIEND IRMA (I,3191)

MERCER, JACK
THE ALL-NEW POPEYE HOUR (II,49); FELIX THE CAT (I,1561); POPEYE THE SAILOR (I,3648); POPEYE THE SAILOR (I,3649)

MERCER, JOHNNY
ALL IN FUN (I,131); THE BING CROSBY SHOW (I,464)

MERCER, LOUIS
THE FREEWHEELERS (I,1683); RAPTURE AT TWO-FORTY (I,3724)

MERCER, MARIAN
CALLING DR. STORM, M.D. (II,415); COMEDY NEWS (I,1020); COMEDY NEWS II (I,1021); THE DEAN MARTIN SHOW (I,1201); DOM DELUISE AND FRIENDS, PART 2 (II,702); THE DOM DELUISE SHOW (I,1328); FOOT IN THE DOOR (II,902); FOREVER FERNWOOD (II,909); IT'S A LIVING (II,1254); JOHNNY CARSON'S REPERTORY COMPANY IN AN EVENING OF COMEDY (I,2423); KING OF THE ROAD (II,1396); LADIES AND GENTLEMAN. . . BOB NEWHART (II,1420); LADIES AND GENTLEMAN. . . BOB NEWHART, PART II (II,1421); MAKING A LIVING (II,1604); MARY HARTMAN, MARY HARTMAN (II,1638); THE SANDY DUNCAN SHOW (I,3910); ST. ELSEWHERE (II,2432); A TOUCH OF GRACE (I,4585); THE WACKY WORLD OF JONATHAN WINTERS (I,4746); WHAT'S UP, AMERICA? (I,4824); WHAT'S UP? (I,4823); ZERO HOUR (I,4980)

MERCER, SALLY
GEORGE WASHINGTON (II,978)

MERCER, TOMMY
TV'S TOP TUNES (I,4635)

MERCERN, TONY
THE ADVENTURES OF CAPTAIN HARTZ (I,53)

MERCOURI, MELINA
MELINA MERCOURI'S GREECE (I,2997)

MERCURIO, GUS
ALL THE RIVERS RUN (II,43)

MEREDITH, BURGESS
AH!, WILDERNESS (I,91); AMERICA PAUSES FOR SPRINGTIME (I,166); AMERICA PAUSES FOR THE MERRY MONTH OF MAY (I,167); BATMAN (I,366); THE BIG STORY (I,432); THE BILL COSBY SPECIAL, OR? (I,443); DEAR MR. GABLE (I,1203); FORD THEATER HOUR (I,1635); GLORIA (II,1010); GLORIA COMES HOME (II,1011); KATE BLISS AND THE TICKER TAPE KID (II,1366); THE LAST HURRAH (I,2626); LIGHT'S OUT (I,2699); LOCK, STOCK, AND BARREL (I,2736); MR. NOVAK (I,3145); THE NEW HEALERS (I,3266); NIGHT GALLERY (I,3287); OF MEN OF WOMEN (I,3331); PHILCO TELEVISION PLAYHOUSE (I,3583); PROBE (I,3674); PROFILES IN COURAGE (I,3678); THE RETURN OF CAPTAIN NEMO (II,2135); SEARCH (I,3951); SILVER THEATER (I,4051); SUSPICION (I,4309); TALES OF TOMORROW (I,4352); THOSE AMAZING ANIMALS (II,2601); THE TWILIGHT ZONE (I,4651)

MEREDITH, CHEERIO
THE AMES BROTHERS SHOW (I,178); THE ANDY GRIFFITH SHOW (I,192); ONE HAPPY FAMILY (I,3375)

MEREDITH, DON
BANJO HACKETT: ROAMIN' FREE (II,141); DINAH AND FRIENDS (II,675); THE FIRST 50 YEARS (II,862); KATE BLISS AND THE TICKER TAPE KID (II,1366); LEGENDS OF THE WEST: TRUTH AND TALL TALES (II,1457); THE SOUND AND THE SCENE (I,4132)

MEREDITH, JUDI
AT YOUR SERVICE (I,310); THE GEORGE BURNS AND GRACIE ALLEN SHOW (I,1763); HOTEL DE PAREE (I,2129)

MEREDITH, JULIE
THE GEORGE BURNS SHOW (I,1765)

MEREDITH, LEE
AS THE WORLD TURNS (II,110); THE CRACKER BROTHERS (II,599); THE GREAT AMERICAN DREAM MACHINE (I,1871); THE JACKIE GLEASON SPECIAL (I,2325); MASON (II,1643); MICKEY SPILLANE'S MIKE HAMMER: MORE THAN MURDER (II,1693); MICKEY SPILLANE'S MIKE HAMMER: MURDER ME, MURDER YOU (II,1694); PARADE OF STARS (II,1954)

MEREDITH, LIZA
ONLY A SCREAM AWAY (I,3393)

MEREDITH, LUCILLE
CARLTON YOUR DOORMAN (II,440); MCCOY (II,1661)

MEREDITH, STEPHEN
DOCTOR WHO—THE FIVE DOCTORS (II,690)

MERIE, ANN
THE BIG EASY (II,234)

MERIN, EDA REISS
FOG (II,897)

MERIWETHER, LEE
BARNABY JONES (II,153); BATTLE OF THE NETWORK STARS (II,163); BATTLE OF THE NETWORK STARS (II,164); CIRCUS OF THE STARS (II,531); CIRCUS OF THE STARS (II,532); CIRCUS OF THE STARS (II,533); CIRCUS OF THE STARS (II,538); CLEAR HORIZON (I,973); DREAM GIRL OF '67 (I,1375); HAVING BABIES II (II,1108); THE KILLIN' COUSIN (II,1394); LOOKING BACK (I,2754); MASQUERADE PARTY (II,1645); MEN WHO RATE A "10" (II,1684); MISSION: IMPOSSIBLE (I,3067); MY SON THE DOCTOR (I,3203); THE NEW ANDY GRIFFITH SHOW (I,3256); THE TIME TUNNEL (I,4511); TOURIST (II,2651); TRUE GRIT (II,2664); THE YOUNG MARRIEDS (I,4942)

MERK, WALLACE
CHIEFS (II,509)

MERKEL, UNA
SNEAK PREVIEW (I,4100)

MERLIN, JAN
KITTY FOYLE (I,2571); THE ROUGH RIDERS (I,3850); TOM CORBETT, SPACE CADET (I,4535)

MERLIN, JOANNA
ANOTHER WORLD (II,97)

MERLIN, JUNE
FESTIVAL OF MAGIC (I,1566)

MERLINI, MARISA
TOP SECRET (II,2647)

MERMAN, ETHEL
100 YEARS OF GOLDEN HITS (II,1905); ANNIE GET YOUR GUN (I,224); ANYTHING GOES (I,238); THE ARTHUR MURRAY PARTY FOR BOB HOPE (I,292); BATMAN (I,366); THE BOB HOPE SHOW (I,589); THE BOB HOPE SHOW (I,592); CRESCENDO (I,1094); ED SULLIVAN'S BROADWAY (I,1402); THE FORD 50TH ANNIVERSARY SHOW (I,1630); THE GERSHWIN YEARS (I,1779); JACK LEMMON IN 'S WONDERFUL, 'S MARVELOUS, 'S GERSHWIN (I,2313); KRAFT SUSPENSE THEATER (I,2591); THE LOVE BOAT (II,1535); MAGGIE BROWN (I,2811); MERMAN ON BROADWAY (I,3011); THE MUSIC OF GERSHWIN (I,3174); PANAMA HATTIE (I,3444); STEVE LAWRENCE AND EYDIE GORME FROM THIS MOMENT ON. . . COLE PORTER (II,2459); THE TALENT SCOUTS PROGRAM (I,4341); TALENT SEARCH (I,4342); TED KNIGHT MUSICAL COMEDY VARIETY SPECIAL SPECIAL (II,2549); TEXACO STAR THEATER: OPENING NIGHT (II,2565); THRU THE CRYSTAL BALL (I,4496); YOU'RE GONNA LOVE IT HERE (II,2860)

MERRETT, THERESA
THAT'S MY MAMA (II,2580)

MERRICK, DAVID
VARIETY (II,2721)

MERRICK, JOHN
GETAWAY CAR (I,1783); A TALE OF WELLS FARGO (I,4338)

MERRIEL DANCERS, THE
THE ED WYNN SHOW (I,1403)

MERRILL, BUDDY
THE LAWRENCE WELK SHOW (I,2643); LAWRENCE WELK'S TOP TUNES AND NEW TALENT (I,2645); MEMORIES WITH LAWRENCE WELK (II,1681)

MERRILL, CAROL
LET'S MAKE A DEAL (II,1462)

MERRILL, DINA
BOB HOPE SPECIAL: BOB HOPE'S ROAD TO HOLLYWOOD (II,313); THE DESILU PLAYHOUSE (I,1237); DUPONT SHOW OF THE MONTH (I,1387); HOT PURSUIT (II,1189); KINGSTON: THE POWER PLAY (II,1399); KRAFT SUSPENSE THEATER (I,2591); THE LETTERS (I,2670); PLAYWRIGHTS '56 (I,3627); ROOTS: THE NEXT GENERATIONS (II,2213)

MERRILL, GARY
ALCOA/GOODYEAR THEATER (I,107); ALFRED HITCHCOCK PRESENTS (I,115); BEN HECHT'S TALES OF THE CITY (I,395); THE DANGEROUS DAYS OF KIOWA JONES (I,1140); THE JACKIE GLEASON SPECIAL (I,2325); JUSTICE (I,2499); THE LADY DIED AT MIDNIGHT (I,2601); THE MASK (I,2938); THE OUTER LIMITS (I,3426); THE REPORTER (I,3771); THE STAR MAKER (I,4193); STAR STAGE (I,4197); THEN CAME BRONSON (I,4436); THE VALIANT YEARS (I,4696); THE WORLD OF DARKNESS (II,2836); YOUNG DR. KILDARE (I,4937)

MERRILL, LARRY
MEET ME IN ST. LOUIS (I,2987); SWINGIN' TOGETHER (I,4320)

MERRILL, ROBERT
AMERICA PAUSES FOR SPRINGTIME (I,166); ANNIE AND THE HOODS (II,91); ANNIE, THE WOMAN IN THE LIFE OF A MAN (I,226); CLOWN ALLEY (I,979); OPERA VS. JAZZ (I,3399); SHOW OF THE YEAR (I,4031); SINATRA AND FRIENDS (II,2358); SINATRA—THE FIRST 40 YEARS (II,2359); YOUR SHOW OF SHOWS (I,4957); YOUR SHOW OF SHOWS (II,2875)

MERRIMAN, RANDY
THE BIG PAYOFF (I,428); THE BOB HOPE SHOW (I,525)

MERRITT, GEORGE
THE PERSUADERS! (I,3556)

MERRITT, THELMA
THE FURST FAMILY OF WASHINGTON (I,1718)

MERROW, JANE
ONCE AN EAGLE (II,1897); UNIT 4 (II,2707)

MERROW, JUNE
SHERLOCK HOLMES: THE HOUND OF THE BASKERVILLES (I,4011)

MERRY TEXACO REPAIRMEN, THE
TEXACO STAR THEATER (I,4411)

MERRYL JAY AND THE CURTIN CALLS
THE BEAUTIFUL PHYLLIS DILLER SHOW (I,381)

MERSKY, KRES
KLEIN TIME (II,1401); STEPHANIE (II,2453); THE TV SHOW (II,2673)

MERTON, ZIENIA
CASANOVA (II,451); SPACE: 1999 (II,2409)

MESKILL, KATHERINE
THE DOCTORS (II,694); SATURDAY'S CHILDREN (I,3923); WHERE THE HEART IS (I,4831)

MESSER, DON
DON MESSER'S JUBILEE (I,1338)

MESSICK, DON
THE ADVENTURES OF GULLIVER (I,60); THE ADVENTURES OF JONNY QUEST (I,64); THE AMAZING CHAN AND THE CHAN CLAN (I,157); THE ATOM ANT/SECRET SQUIRREL SHOW (I,311); THE BANANA SPLITS ADVENTURE HOUR (I,340); THE BANG-SHANG LALAPALOOZA SHOW (II,140); THE BARKLEYS (I,356); THE BEANY AND CECIL SHOW (I,374); BIRDMAN (II,256); BIRDMAN AND THE GALAXY TRIO (I,478); THE BUFFALO BILLY SHOW (I,756); THE C.B. BEARS (II,406); THE CATTANOOGA CATS (I,874); DASTARDLY AND MUTTLEY IN THEIR FLYING MACHINES (I,1159); DOCTOR DOLITTLE (I,1309); THE DRAK PACK (II,733); THE DUCK FACTORY (II,738); THE FLINTSTONE COMEDY HOUR (II,881); FLINTSTONE FAMILY ADVENTURES (II,882); THE FLINTSTONES (II,884); THE FLINTSTONES (II,885); FRANKENSTEIN JR. AND THE IMPOSSIBLES (I,1672); FRED AND BARNEY MEET THE THING (II,919); GIDGET MAKES THE WRONG CONNECTION (I,1798); GODZILLA (II,1014); THE HEATHCLIFF AND MARMADUKE SHOW (II,1118); THE HERCULOIDS (I,2023); HONG KONG PHOOEY (II,1180); THE HUCKLEBERRY HOUND SHOW (I,2155); INCH HIGH, PRIVATE EYE (II,1231); JACK FROST (II,1263); THE JETSONS (I,2376); JOSIE AND THE PUSSYCATS (I,2453); JOSIE AND THE PUSSYCATS IN OUTER SPACE (I,2454); LOST IN SPACE (I,2759); THE MAGILLA GORILLA SHOW (I,2825); MOBY DICK AND THE MIGHTY MIGHTOR (I,3077); THE NEW FRED AND BARNEY SHOW (II,1817); THE PERILS OF PENELOPE PITSTOP (I,3525); THE PETER POTAMUS SHOW (I,3568); THE RUFF AND REDDY SHOW (I,3866); SCOOBY'S ALL-STAR LAFF-A-LYMPICS (II,2273); SCOOBY-DOO AND SCRAPPY-DOO (II,2274); THE SCOOBY-DOO AND SCRAPPY-DOO SHOW (II,2275); SCOOBY-DOO, WHERE ARE YOU? (II,2276); THE SMURFS (II,2385); SPACE KIDDETTES (I,4143); SPACE STARS (II,2408); SPUNKY AND TADPOLE (I,4169); THE WACKY RACES (I,4745); WALLY GATOR (I,4751); THE WORLD'S GREATEST SUPER HEROES (II,2842); YOGI BEAR (I,4928); YOGI'S GANG (II,2852)

MESSNER, JOHNNY
THE VINCENT LOPEZ SHOW (I,4735)

MESZAROS, RON
LITTLE HOUSE: THE LAST FAREWELL (II,1491)

METCALF, C W
TV FUNNIES (II,2672)

METCALF, KEN
THE YOUNG MARRIEDS (I,4942)

METCALF, LAURIE
SATURDAY NIGHT LIVE (II,2261)

METCALF, MARK
TEACHERS ONLY (II,2547)

METCALFE, BURT
FATHER OF THE BRIDE (I,1543); FOR THE LOVE OF MIKE (I,1625)

METRANO, ART
AMY PRENTISS (II,74); BARETTA (II,152); BARNEY MILLER (II,154); CENTENNIAL (II,477); THE CHICAGO TEDDY BEARS (I,934); FLO'S PLACE (II,892); FRED AND BARNEY MEET THE THING (II,919); HUMAN FEELINGS (II,1203); JOANIE LOVES CHACHI (II,1295); LOVES ME, LOVES ME NOT (II,1551); MOVIN' ON (II,1743); RISE AND SHINE (II,2179); THE SOUPY SALES SHOW (I,4138); SUPER COMEDY BOWL 1 (I,4288); SWINGIN' TOGETHER (I,4320); THEN CAME BRONSON (I,4436); THE TIM CONWAY COMEDY HOUR (I,4501)

METTEY, LYNNETTE
AMANDA FALLON (I,151); FITZ AND BONES (II,867); HEAT OF ANGER (I,1992); LOVE STORY (I,2778); M*A*S*H (II,1569); QUINCY, M. E. (II,2102); STARTING FRESH (II,2447); TERROR AT ALCATRAZ (II,2563)

METZGER, MARY LOU
THE LAWRENCE WELK SHOW (I,2643); MEMORIES WITH LAWRENCE WELK (II,1681)

METZINGER, KRAIG
MAUDE (II,1655); SARA (II,2260)

METZLER, JIM
CUTTER TO HOUSTON (II,612)

MEYER, DOROTHY
KING'S CROSSING (II,1397); RYAN'S FOUR (II,2233); SALT AND PEPE (II,2240); WINNER TAKE ALL (II,2808)

MEYER, EMILE
THE HARDY BOYS (I,1948); THE ROARING TWENTIES (I,3805)

MEYER, FREDERICK
FARAWAY HILL (I,1532)

MEYER, MARILYN
EDDIE AND HERBERT (II,757)

MEYER, MICHAEL
DYNASTY (II,746)

MEYER, RONALD
JANE EYRE (II,1273)

MEYER, TARO
ANOTHER WORLD (II,97)

MEYERBERG, ADRIENNE
CHAMPAGNE AND ORCHIDS (I,895)

MEYERINK, VICTORIA
THE DANNY KAYE SHOW (I,1143); GREEN ACRES (I,1884); MY THREE SONS (I,3205)

MEYERS, ARI
FULL HOUSE (II,937); KATE AND ALLIE (II,1365)

MEYERS, DOROTHY
JOEY AND DAD (II,1306)

MIANO, ROBERT
BRIDGER (II,376)

MICHAEL BENNETT DANCERS, THE
THE KRAFT MUSIC HALL (I,2588)

MICHAEL, FRANNY
THE EASTER PROMISE (I,1397)

MICHAEL, GEORGE
BWANA MICHAEL OF AFRICA (I,783)

MICHAEL, MARY
THE WOMEN (I,4890); WONDERFUL JOHN ACTION (I,4892)

MICHAEL, PATRICIA
IKE (II,1223)

MICHAEL, PAUL
MUGGSY (II,1761)

MICHAEL, RALPH
DOCTOR IN THE HOUSE (I,1313); MASTER OF THE GAME (II,1647); NO MAN'S LAND (II,1855); THE QUEST (II,2098)

MICHAEL, RAY
TELE PUN (I,4376)

MICHAELMAN, KEN
THE WHITE SHADOW (II,2788)

MICHAELS, CORINNE
PRINCESS (II,2084)

MICHAELS, ED
DOCTOR I.Q. (I,1314)

MICHAELS, JENNA
DALLAS (II,614); THE LAST CONVERTIBLE (II,1435); THE YOUNG AND THE RESTLESS (II,2862)

MICHAELS, KEVIN
THE NEW PEOPLE (I,3268)

MICHAELS, KRISTY
SENSE OF HUMOR (II,2303)

MICHAELS, LORNE
LILY (I,2705)

MICHAELS, MARILYN
ED SULLIVAN'S BROADWAY (I,1402); THE KOPYCATS (I,2580); THE SINGERS (I,4060)

MICHAELS, MARY
ABE LINCOLN IN ILLINOIS (I,10)

MICHAELS, MICHELLE
MICKEY SPILLANE'S MIKE HAMMER: MORE THAN MURDER (II,1693)

MICHAELS, NICOLA
THE SWORD (I,4325)

MICHAELS, RAY
SAMMY KAYE'S MUSIC FROM MANHATTAN (I,3902)

MICHAELS, SANDRA
ADVENTURES OF CLINT AND MAC (I,55)

MICHAELS, SYLVIA
KAY KYSER'S KOLLEGE OF MUSICAL KNOWLEDGE (I,2512)

MICHAELSEN, KARI
CIRCUS OF THE STARS (II,537); GIMME A BREAK (II,995); MAX (II,1658)

MICHAELSEN, MELISSA
ME AND MAXX (II,1672); NO MAN'S LAND (II,1855)

MICHAS, DMITRI
MR. MOM (II,1752)

MICHEL, FRANNY
APPLE'S WAY (II,101)

MICHELL, KEITH
THE SIX WIVES OF HENRY VIII (I,4068); SOLDIER IN LOVE (I,4108)

MICHELLE, BECKY
LOVE, NATALIE (II,1545)

MICHENAUD, GERALD
TO ROME WITH LOVE (I,4526)

MICKIE FINN WAITRESSES, THE
MICKIE FINN'S (I,3029)

MIDDLECOFF, DR CARY
THE BOB HOPE SHOW (I,556)

MIDDLEMAN, FRANK
WAR AND PEACE (I,4765)

MIDDLEMASS, FRANK
A FAMILY AFFAIR (II,814); LACE (II,1418)

MIDDLETON, BURR
THE KID SUPER POWER HOUR WITH SHAZAM (II,1386)

MIDDLETON, RAY
BORDER PALS (II,352); THE CENTURY TURNS (I,890); A CHRISTMAS CAROL (I,947); DAMN YANKEES (I,1128); PANAMA HATTIE (I,3444); TOO CLOSE FOR COMFORT (II,2642)

MIDDLETON, ROBERT
THE DEAN JONES SHOW (I,1191); THE MONROES (I,3089); ON TRIAL (I,3362); RUN JACK RUN (I,3872)

MIDDLETON, TOM
BRIDGER (II,376); MONTGOMERY'S SUMMER STOCK (I,3096)

MIDGLEY, RICHARD
THE ALDRICH FAMILY (I,111)

MIDLER, BETTE
BETTE MIDLER—ART OR BUST (II,221); BETTE MIDLER—OL' RED HAIR IS BACK (II,222); BING!. . . A 50TH ANNIVERSARY GALA (II,254); BURT BACHARACH—OPUS NO. 3 (I,774); CHER (II,496); NEIL SEDAKA STEPPIN' OUT (II,1804)

MIDWESTERNERS, THE
MIDWESTERN HAYRIDE (I,3033)

MIDWOOD, KENNETH
LITTLE LORD FAUNTLEROY (II,1492)

MIFUNE, TOSHIRO
SHOGUN (II,2339)

MIGHTY CLOUDS OF JOY, THE
JOHNNY CASH—A MERRY MEMPHIS CHRISTMAS (II,1331)

MIGINI, CAROLYN

WAKE UP (II,2736)

MIKALOS, JOE
THE COMEDY OF ERNIE KOVACS (I,1022); THE NEW ERNIE KOVACS SHOW (I,3264)

MIKE CURB CONGREGATION, THE
CHANGING SCENE (I,899); THE GLEN CAMPBELL GOODTIME HOUR (I,1820); MONSANTO NIGHT PRESENTS JOSE FELICIANO (I,3091); TENNESSEE ERNIE FORD'S WHITE CHRISTMAS (I,4402)

MIKE SAMMES SINGERS, THE
THE DES O'CONNOR SHOW (I,1235); THE LONDON PALLADIUM (I,2739); THE MOST IN MUSIC (I,3108); PETULA (I,3573); PICCADILLY PALACE (I,3589); THE ROBERT GOULET SPECIAL (I,3807); SHOWTIME (I,4036); THIS IS TOM JONES (I,4460); THE TOM JONES SPECIAL (I,4540); THE TOM JONES SPECIAL (I,4541); THE TOM JONES SPECIAL (I,4542); THE TOM JONES SPECIAL (I,4543); THE VAL DOONICAN SHOW (I,4692)

MIKLER, MICHAEL
THE YOUNG MARRIEDS (I,4942)

MIKOLAS, JOE
THE ERNIE KOVACS SPECIAL (I,1456)

MILAN, FRANK
TONIGHT IN SAMARKAND (I,4558)

MILAN, LITA
TONIGHT IN HAVANA (I,4557)

MILANO, ALYSSA
WHO'S THE BOSS? (II,2792)

MILANO, FRANK
KING LEONARDO AND HIS SHORT SUBJECTS (I,2557); ROOTIE KAZOOTIE (I,3844); TENNESSEE TUXEDO AND HIS TALES (I,4403)

MILANOV, ZINKA
FESTIVAL OF MUSIC (I,1567)

MILDORD, JOHN
THE SOUND OF ANGER (I,4133)

MILES, BERNARD
WHY DIDN'T THEY ASK EVANS? (II,2795)

MILES, DALLAS
BENJI, ZAX AND THE ALIEN PRINCE (II,205)

MILES, JOANNA
DELTA COUNTY, U.S.A. (II,657); A TIME FOR US (I,4507)

MILES, LIZIE
CRESCENDO (I,1094)

MILES, OGDEN
STORIES IN ONE CAMERA (I,4234)

MILES, PETER
A CHRISTMAS CAROL (I,947)

MILES, RICHARD
THE BETTY HUTTON SHOW (I,413)

MILES, ROBERT
LITTLE HOUSE: LOOK BACK TO YESTERDAY (II,1490)

MILES, ROSALIND
THE TURNING POINT OF JIM MALLOY (II,2670)

MILES, SALLY
COLOR HIM DEAD (II,554)

MILES, SARAH
DYNASTY (II,747)

MILES, SHERRY
PAT PAULSEN'S HALF A COMEDY HOUR (I,3473)

MILES, STEVEN
GIDGET (I,1795)

MILES, SYLVIA
HEAD OF THE FAMILY (I,1988)

MILES, VERA
ALFRED HITCHCOCK PRESENTS (I,115); BAFFLED (I,332); THE BOB HOPE CHRYSLER THEATER (I,502); THE BOB HOPE SHOW (I,611); CANNON (I,820); THE CASE AGAINST PAUL RYKER (I,861); CLIMAX! (I,976); CROWN THEATER WITH GLORIA SWANSON (I,1107); THE GENERAL ELECTRIC THEATER (I,1753); GENTLE BEN (I,1761); HOW THE WEST WAS WON (II,1196); JOURNEY TO THE UNKNOWN (I,2457); THE LUX VIDEO THEATER (I,2795); MAN ON THE MOVE (I,2878); MEDIC (I,2976); MR. TUTT (I,3153); OUR FAMILY BUSINESS (II,1931); THE OUTER LIMITS (I,3426); OWEN MARSHALL: COUNSELOR AT LAW (I,3436); PEPSI-COLA PLAYHOUSE (I,3523); ROUGHNECKS (II,2219); SCREEN DIRECTOR'S PLAYHOUSE (I,3946); STATE FAIR (II,2448); STUMP THE STARS (I,4274); TRAVIS MCGEE (II,2656); THE TWILIGHT ZONE (I,4651); THE UNKNOWN (I,4680)

MILFORD, JOHN
THE ALIENS ARE COMING (II,35); ENOS (II,779); THE LAWYERS (I,2646); THE LEGEND OF JESSE JAMES (I,2653)

MILFORD, PENELOPE
THE OLDEST LIVING GRADUATE (II,1882)

MILINAIRE, NIKOLE
FOREIGN INTRIGUE: OVERSEAS ADVENTURES (I,1639)

MILLAIRE, ALBERT
ADVENTURES IN RAINBOW COUNTRY (I,47)

MILLAN, ANDRA
THE PAPER CHASE: THE SECOND YEAR (II,1950)

MILLAN, ROBYN
THE BEST YEARS (I,410); THE DEAN JONES SHOW (I,1191); MURDER MOTEL (I,3161); THE PATTY DUKE SHOW (I,3483); THE SHAMEFUL SECRETS OF HASTINGS CORNERS (I,3994)

MILLAND, RAY
ALCOA/GOODYEAR THEATER (I,107); THE BOB HOPE SHOW (I,654); THE DUPONT SHOW OF THE WEEK (I,1388); ELLERY QUEEN: TOO MANY SUSPECTS (II,767); FORD TELEVISION THEATER (I,1634); HART TO HART (II,1102); MARKHAM (I,2916); MARKHAM (I,2917); MEET MR. MCNUTLEY (I,2989); OUR FAMILY BUSINESS (II,1931); THE RAY MILLAND SHOW (I,3734); RICH MAN, POOR MAN—BOOK I (II,2161); SEVENTH AVENUE (II,2309); TESTIMONY OF TWO MEN (II,2564); TRAILS WEST (I,4592)

MILLAR, COURT
RAGE OF ANGELS (II,2107)

MILLAR, MARGIE
WHERE'S RAYMOND? (I,4837)

MILLARD, HARVEY
OUR FIVE DAUGHTERS (I,3413)

MILLAY, DIANA
THE BOSTON TERRIER (I,696); DARK SHADOWS (I,1157); DOBIE GILLIS (I,1302); MY THREE ANGELS (I,3204); THRILLER (I,4492)

MILLER, ALLAN
A.E.S. HUDSON STREET (II,17); ARCHIE BUNKER'S PLACE (II,105); BLISS (II,269); DON'T CALL US (II,709); GALACTICA 1980 (II,953); HOW TO SURVIVE A MARRIAGE (II,1198); NERO WOLFE (II,1806); ONE LIFE TO LIVE (II,1907); RIDING HIGH (II,2173); SOAP (II,2392); THREE EYES (II,2605); THE TURNING POINT OF JIM MALLOY (II,2670); WHERE'S POPPA? (II,2784); THE WORD (II,2833)

MILLER, ANN
THE BOB HOPE SHOW (I,557)

MILLER, BARRY
JOE AND SONS (II,1296); JOE AND SONS (II,1297); SZYSZNYK (II,2523)

MILLER, BEN
THE CHEROKEE TRAIL (II,500)

MILLER, BRUCE
SPIDER-WOMAN (II,2426)

MILLER, BUZZ
ALL ABOUT MUSIC (I,127)

MILLER, CAROL
ENTERTAINMENT TONIGHT (II,780); THE ROCK 'N ROLL SHOW (II,2188)

MILLER, CHERYL
BRIGHT PROMISE (I,727); DAKTARI (I,1127); THE GEMINI MAN (II,961); KILROY (I,2541)

MILLER, CLINT
THE GEORGE HAMILTON IV SHOW (I,1769)

MILLER, COREY
ALL IN THE FAMILY (II,38)

MILLER, DAMIAN
TEXAS (II,2566)

MILLER, DARE
THE PET SET (I,3557)

MILLER, DEAN
CHOOSE UP SIDES (I,945); DECEMBER BRIDE (I,1215); HERE'S HOLLYWOOD (I,2031); THERE'S ONE IN EVERY FAMILY (I,4439)

MILLER, DENISE
ARCHIE BUNKER'S PLACE (II,105); BATTLE OF THE NETWORK STARS (II,176); EVERY STRAY DOG AND KID (II,792); FISH (II,864); GLORIA COMES HOME (II,1011); MAKIN' IT (II,1603)

MILLER, DENNY
DOCTOR SCORPION (II,685); KEEPER OF THE WILD (II,1373); THE LIFE OF RILEY (I,2686); MICKEY SPILLANE'S MIKE HAMMER: MORE THAN MURDER (II,1693); MONA

MCCLUSKEY (I,3085); NICK AND NORA (II,1842); THE SEAL (II,2282); SHOOTING STARS (II,2341); TARZAN THE APE MAN (I,4363); V (II,2713); WAGON TRAIN (I,4747)

MILLER, DICK
V: THE FINAL BATTLE (II,2714); W*A*L*T*E*R (II,2731)

MILLER, DONNA
THE RELUCTANT DRAGON AND MR. TOAD (I,3762)

MILLER, DOREEN
WHERE'S EVERETT? (I,4835)

MILLER, DR JULIUS SUMNER
THE MORNING SHOW (I,3103)

MILLER, EVE
FIRESIDE THEATER (I,1580); MAYOR OF THE TOWN (I,2963); YOUR SHOW TIME (I,4958)

MILLER, GARY
THE JOHN BYNER COMEDY HOUR (I,2407); NICHOLS AND DYMES (II,1841)

MILLER, HARRISON
BEST FOOT FORWARD (I,401)

MILLER, HOWARD
CLUB 60 (I,987)

MILLER, J J
THE RED HAND GANG (II,2122)

MILLER, J R
HONEST AL'S A-OK USED CAR AND TRAILER RENTAL TIGERS (II,1174)

MILLER, JASON
THE DAIN CURSE (II,613)

MILLER, JODY
A COUNTRY HAPPENING (I,1065)

MILLER, JOSEPH
LAND OF HOPE (II,1424)

MILLER, JULIE
STRIPPERS (II,2481)

MILLER, KATHLEEN
FRIENDS AND LOVERS (II,930); SIROTA'S COURT (II,2365)

MILLER, KAY
A TIME TO LIVE (I,4510)

MILLER, KEN
BENJI, ZAX AND THE ALIEN PRINCE (II,205)

MILLER, KRISTINE
STORIES OF THE CENTURY (I,4235); YOUR SHOW TIME (I,4958)

MILLER, LARA JILL
GIMME A BREAK (II,995); MAX (II,1658)

MILLER, LAURA
THE BOB HOPE SHOW (I,636); ROBERTA (I,3816)

MILLER, LEE
FRANK MERRIWELL (I,1659); PERRY MASON (II,2025)

MILLER, LINDA
HUSBANDS AND WIVES (II,1208); THE MISSISSIPPI (II,1707)

MILLER, LISA
THE BOB HOPE SHOW (I,630)

MILLER, MARCIE
TV'S TOP TUNES (I,4635)

MILLER, MARK
BRIGHT PROMISE (I,727); GENERAL HOSPITAL (II,964); GUESTWARD HO! (I,1900); THE MAN FROM DENVER (I,2863); THE NAME OF THE GAME (I,3217); PLEASE DON'T EAT THE DAISIES (I,3628); THE SEARCH (I,3950)

MILLER, MARVIN
AQUAMAN (I,249); THE FAMOUS ADVENTURES OF MR. MAGOO (I,1524); FANTASTIC VOYAGE (I,1530); THE FBI (I,1551); GERALD MCBOING-BOING (I,1778); THE LONE RANGER (I,2741); MAYA (I,2960); THE MILLIONAIRE (I,3046); MYSTERIES OF CHINATOWN (I,3209); THE NEW ADVENTURES OF HUCKLEBERRY FINN (I,3250); SPACE PATROL (I,4144); STORIES OF THE CENTURY (I,4235)

MILLER, MINDI
SWITCH (II,2519)

MILLER, MITCH
SING ALONG WITH MITCH (I,4056); SING ALONG WITH MITCH (I,4057); TWELVE STAR SALUTE (I,4638)

MILLER, MRS DOROTHY
THE MERV GRIFFIN SHOW (II,1688)

MILLER, OKAY
HANDLE WITH CARE (II,1077)

MILLER, PEGGY
MY MOTHER THE CAR (I,3197)

MILLER, PROFESSOR JULIUS SUMNER
THE HILARIOUS HOUSE OF FRIGHTENSTEIN (II,1153)

MILLER, R MACK
THE DOCTORS (II,694)

MILLER, ROGER
ANDY WILLIAMS MAGIC LANTERN SHOW COMPANY (I,202); BURT REYNOLDS' LATE SHOW (I,777); COUNTRY NIGHT OF STARS II (II,593); THE ICE PALACE (I,2182); JOHN DENVER—THANK GOD I'M A COUNTRY BOY (II,1317); KING OF THE ROAD (II,1396); LARRY GATLIN AND THE GATLIN BROTHERS (II,1429); MONSANTO NIGHT PRESENTS BURL IVES (I,3090); THE ROGER MILLER SHOW (I,3830)

MILLER, SHERRY
OUT OF OUR MINDS (II,1933)

MILLER, SID
LOST IN SPACE (I,2759); STANDARD OIL ANNIVERSARY SHOW (I,4189)

MILLER, SIDNEY
THE DONALD O'CONNOR SHOW (I,1344); THE DONALD O'CONNOR SHOW (I,1343); DUNGEONS AND DRAGONS (II,744); THE GARY COLEMAN SHOW (II,958); THE MUSIC MART (II,1774); NO SOAP, RADIO (II,1856)

MILLER, SKEDGE
ALICE IN WONDERLAND (I,120)

MILLER, STEVEN
SHOOTING STARS (II,2341)

MILLER, TAYLOR
ALL MY CHILDREN (II,39)

MILLER, WYNNE
OUR FIVE DAUGHTERS (I,3413); SOMERSET (I,4115)

MILLERICK, KERRY
BLONDES VS. BRUNETTES (II,271); THAT'S MY LINE (II,2579)

MILLET, NEFTI
MCCLOUD: WHO KILLED MISS U.S.A.? (I,2965)

MILLHOLLIN, JAMES
THE DANNY KAYE SHOW (I,1144); GRINDL (I,1889)

MILLI, ROBERT
ALL MY CHILDREN (II,39); ANOTHER WORLD (II,97); THE GUIDING LIGHT (II,1064)

MILLICAN, JAMES
THE WESTERNER (I,4800)

MILLIGAN, PATTI
BILLY BOONE AND COUSIN KIBB (I,450)

MILLIGAN, SPENCER
LAND OF THE LOST (II,1425)

MILLIGAN, SPIKE
THE MARTY FELDMAN COMEDY MACHINE (I,2929)

MILLIGAN, STEWART
MASTER OF THE GAME (II,1647)

MILLIGAN, TUCK
THE SHERIFF AND THE ASTRONAUT (II,2326)

MILLINAR, ROD
HUNTER (I,2161)

MILLOT, CHARLES
MASTER OF THE GAME (II,1647)

MILLS BROTHERS, THE
100 YEARS OF GOLDEN HITS (II,1905); BING!. . . A 50TH ANNIVERSARY GALA (II,254); THE GEORGE KIRBY SPECIAL (I,1771); ONE MORE TIME (I,3386); ONE MORE TIME (I,3385)

MILLS, ALAN
ADVENTURES IN RAINBOW COUNTRY (I,47)

MILLS, ALISON
JULIA (I,2476); THE LESLIE UGGAMS SHOW (I,2660)

MILLS, ALLEY
THE ASSOCIATES (II,113); HILL STREET BLUES (II,1154); MAKING THE GRADE (II,1606); POOR RICHARD (II,2065); SECOND EDITION (II,2288)

MILLS, BILLY
THE KATE SMITH HOUR (I,2508)

MILLS, BROOKE
THE BARBARY COAST (II,146)

MILLS, CAROL
THE BENNY HILL SHOW (II,207)

MILLS, CLAIRE
GABE AND WALKER (II,950)

MILLS, DONNA
THE BAIT (I,335); BARE ESSENCE (II,149); BATTLE OF THE NETWORK STARS (II,165); BATTLE OF THE NETWORK STARS (II,171); BATTLE OF THE NETWORK STARS (II,173); BATTLE OF THE NETWORK STARS (II,177); BUNCO (II,395); THE CELEBRITY FOOTBALL CLASSIC (II,474); DOCTORS' PRIVATE LIVES (II,693); THE GOOD LIFE (I,1848); THE GOOD LIFE (II,1033); KILLER

WITH TWO FACES (I,2539); KNOTS LANDING (II,1404); ONE DEADLY OWNER (I,3373); SOMEONE AT THE TOP OF THE STAIRS (I,4114); THE STEVE LAWRENCE SHOW (I,4227); WAIKIKI (II,2734); WOMAN ON THE RUN (II,2821)

MILLS, EDMOND
THE BEACHCOMBER (I,371)

MILLS, EDWIN
RETURN TO THE PLANET OF THE APES (II,2145)

MILLS, GLORIA
THE BURNS AND SCHREIBER COMEDY HOUR (I,768)

MILLS, HAYLEY
THE FLAME TREES OF THIKA (II,870); ONLY A SCREAM AWAY (I,3393)

MILLS, JOHN
DOCTOR STRANGE (II,688); DUNDEE AND THE CULHANE (I,1383); THE DUPONT SHOW OF THE WEEK (I,1388); LITTLE LORD FAUNTLEROY (II,1492); A WOMAN OF SUBSTANCE (II,2820); THE ZOO GANG (II,2877)

MILLS, JULIET
ABC'S MATINEE TODAY (I,7); THE BOB HOPE CHRYSLER THEATER (I,502); LETTERS FROM THREE LOVERS (I,2671); MR. DICKENS OF LONDON (I,3135); NANNY AND THE PROFESSOR (I,3225); NANNY AND THE PROFESSOR AND THE PHANTOM OF THE CIRCUS (I,3226); ONCE AN EAGLE (II,1897); REX HARRISON PRESENTS SHORT STORIES OF LOVE (II,2148)

MILLS, MORT
THE BIG VALLEY (I,437); DAVY CROCKETT (I,1181); MAN WITHOUT A GUN (I,2884)

MILLS, ROBERT
FRAGGLE ROCK (II,916)

MILLS, SALLY
ROWAN AND MARTIN BITE THE HAND THAT FEEDS THEM (I,3853)

MILLS, VICKI
NAME THAT TUNE (I,3218)

MILMORE, JANE
CALIFORNIA FEVER (II,411)

MILNER, JESSAMINE
HICKEY VS. ANYBODY (II,1142)

MILNER, MARTIN
ADAM-12 (I,31); BLACK BEAUTY (II,261); THE DUPONT SHOW OF THE WEEK (I,1388); GIDGET (I,1795); THE GOLDDIGGERS (I,1838); THE LAST CONVERTIBLE (II,1435); THE LIFE OF RILEY (I,2686); NAVY LOG (I,3233); ROUTE 66 (I,3852); THE ROWAN AND MARTIN SPECIAL (I,3855); THE SEEKERS (II,2298); STARR, FIRST BASEMAN (I,4204); SWISS FAMILY ROBINSON (II,2516); SWISS FAMILY ROBINSON (II,2517); TROUBLE WITH FATHER (I,4610); TV READERS DIGEST (I,4630); THE TWILIGHT ZONE (I,4651)

MILNER, ROGER
BRIDESHEAD REVISITED (II,375); PENMARRIC (II,1993)

MILSAP, RONNIE
THE COUNTRY MUSIC MURDERS (II,590); LUCY COMES TO NASHVILLE (II,1561); MAC DAVIS SPECIAL: THE MUSIC OF CHRISTMAS (II,1578); MERRY CHRISTMAS FROM THE GRAND OLE OPRY (II,1686); USA COUNTRY MUSIC (II,591)

MIMIEUX, YVETTE
THE ALL-STAR SALUTE TO MOTHER'S DAY (II,51); BELL, BOOK AND CANDLE (II,201); THE DESPERATE HOURS (I,1239); HOLLYWOOD MELODY (I,2085); THE MOST DEADLY GAME (I,3106); NIGHT PARTNERS (II,1847); ONE STEP BEYOND (I,3388); RANSOM FOR ALICE (II,2113)

MIMMO, HARRY
THE WONDERFUL WORLD OF JACK PAAR (I,4899)

MIMS, WILLIAM
THE ADAM MACKENZIE STORY (I,30); ADAMS OF EAGLE LAKE (II,10); CAPTAIN AMERICA (II,428); THE CIRCLE FAMILY (II,527); THE LONG HOT SUMMER (I,2747)

MINCIOTTI, ESTHER
THE GOLDEN AGE OF TELEVISION (II,1016)

MINEO, SAL
THE BUDDY GRECO SHOW (I,754); THE DANGEROUS DAYS OF KIOWA JONES (I,1140); DUPONT SHOW OF THE MONTH (I,1387); AN EVENING WITH JIMMY DURANTE (I,1475); HARRY O (II,1098); KRAFT TELEVISION THEATER (I,2592); MONA MCCLUSKEY (I,3085); PURSUIT (I,3700); THE ROCK N' ROLL SHOW (I,3820)

MINER, JAN
CAMEO THEATER (I,808); CRIME PHOTOGRAPHER (I,1097); LOVE OF LIFE (I,2774); MONTGOMERY'S SUMMER STOCK (I,3096); PAUL SAND IN FRIENDS AND LOVERS (II,1982); POTTSVILLE (II,2072)

MINES, STEPHEN
AS THE WORLD TURNS (II,110); PARADISE BAY (I,3454)

MINKUS, BARBARA
CURIOSITY SHOP (I,1114); MASQUERADE (I,2940); MOTHER, JUGGS AND SPEED (II,1741); PAC-MAN (II,1940)

MINNELLI, LIZA
THE ALAN KING SHOW (I,101); THE ANTHONY NEWLEY SHOW (I,233); THE ARTHUR GODFREY SHOW (I,287); BARYSHNIKOV ON BROADWAY (II,158); COMEDY IS KING (I,1019); THE DANGEROUS CHRISTMAS OF RED RIDING HOOD (I,1139); THE GENE KELLY PONTIAC SPECIAL (I,1750); GENE KELLY. . . AN AMERICAN IN PASADENA (II,962); GOLDIE AND LIZA TOGETHER (II,1022); JUBILEE (II,1347); LIZA WITH A Z (I,2732); MAC DAVIS 10TH ANNIVERSARY SPECIAL: I STILL BELIEVE IN MUSIC (II,1571); THE MAC DAVIS SPECIAL (II,1577); MOVIN' (I,3117); THE PERRY COMO SPRINGTIME SPECIAL (I,3538); ROYAL VARIETY PERFORMANCE (I,3865); SALUTE TO LADY LIBERTY (II,2243)

MINNER, KATHRYN
HEY, LANDLORD (I,2039)

MINOR, BOB
THE GREATEST AMERICAN HERO (II,1060)

MINOR, MIKE
ALL MY CHILDREN (II,39); THE EDGE OF NIGHT (II,760); PETTICOAT JUNCTION (I,3571)

MINOT, ANNA
AS THE WORLD TURNS (II,110); HANDS OF MURDER (I,1929); THE HUNTER (I,2162); A WORLD APART (I,4910)

MINTER, DON
WORDS AND MUSIC (I,4909)

MINTNER, GORDEN
BRIEF PAUSE FOR MURDER (I,725)

MINTON, SKEETS
ALL ABOARD (I,124)

MINTZ, ELI
THE GOLDBERGS (I,1835); THE GOLDBERGS (I,1836)

MIONI, FABRIZIO
AN APARTMENT IN ROME (I,242)

MIRACLES, THE
THE SMOKEY ROBINSON SHOW (I,4093)

MIRANDA, SUSANNA
CITY BENEATH THE SEA (I,965)

MIRIAM NELSON DANCERS, THE
AWAY WE GO (I,317)

MIRREN, HELEN
KISS KISS, KILL KILL (I,2566)

MISHKIN, PHIL
THE SUPER (I,4293)

MISTRETTA, SAL
SWEENEY TODD (II,2512)

MITA, EDA
THE MISS AND THE MISSILES (I,3062)

MITCHELL BOYS CHOIR, THE
BING CROSBY AND THE SOUNDS OF CHRISTMAS (I,457)

MITCHELL TRIO, THE
AQUA VARIETIES (I,248)

MITCHELL, BILLY J
OPPENHEIMER (II,1919)

MITCHELL, BOBBIE
M*A*S*H (II,1569); RENDEZVOUS HOTEL (II,2131); THE SIX MILLION DOLLAR MAN (I,4067); THE SUNSHINE BOYS (II,2496)

MITCHELL, BRIAN
BATTLE OF THE NETWORK STARS (II,174); CIRCUS OF THE STARS (II,538); ROOTS: THE NEXT GENERATIONS (II,2213); TRAPPER JOHN, M.D. (II,2654)

MITCHELL, CAMERON
THE BASTARD/KENT FAMILY CHRONICLES (II,159); THE BEACHCOMBER (I,371); BLACK BEAUTY (II,261); CUTTER (I,1117); THE DAVID NIVEN THEATER (I,1178); THE DELPHI BUREAU (I,1224); THE DESILU PLAYHOUSE (I,1237); EMPIRE (II,777); GRUEN GUILD PLAYHOUSE (I,1892); THE

HIGH CHAPARRAL (I,2051); HOW THE WEST WAS WON (II,1196); LASSITER (I,2625); LASSITER (II,1434); MEETING AT APALACHIN (I,2994); PARTNERS IN CRIME (II,1961); PHILCO TELEVISION PLAYHOUSE (I,3583); THE QUEST (II,2096); RETURN TO FANTASY ISLAND (II,2144); THE ROOKIES (I,3840); SHOOT-IN AT NBC (A BOB HOPE SPECIAL) (I,4021); STARS OVER HOLLYWOOD (I,4211); THE STRANGER (I,4251); SWISS FAMILY ROBINSON (II,2516); SWISS FAMILY ROBINSON (II,2517); TESTIMONY OF TWO MEN (II,2564); TURNOVER SMITH (II,2671); WILD TIMES (II,2799)

MITCHELL, CHARLOTTE
THE ADVENTURES OF BLACK BEAUTY (I,51)

MITCHELL, CHRISTINE
THE LEPRECHAUN'S CHRISTMAS GOLD (II,1461)

MITCHELL, DON
IRONSIDE (I,2240); IRONSIDE (II,1246)

MITCHELL, EWING
THE ADVENTURES OF CHAMPION (I,54); SKY KING (I,4077)

MITCHELL, FLORENCE
YOUNG DR. MALONE (I,4938)

MITCHELL, GEORGE
THE BEACHCOMBER (I,371); MARCO POLO (I,2904)

MITCHELL, GREG
CLUB CELEBRITY (I,981)

MITCHELL, GUY
ARTHUR GODFREY'S TALENT SCOUTS (I,291); THE BOB HOPE SHOW (I,569); THE GUY MITCHELL SHOW (I,1912); THE GUY MITCHELL SHOW (I,1913); PANDORA AND FRIEND (I,3445); THE ROCK N' ROLL SHOW (I,3820); WHISPERING SMITH (I,4842)

MITCHELL, GWEN
AMY PRENTISS (II,74)

MITCHELL, HERBERT
THE ANSWER MAN (I,231)

MITCHELL, JAKE
THE BLACK SHEEP SQUADRON (II,262); PARIS (II,1957)

MITCHELL, JAMES
ALL MY CHILDREN (II,39); THE MAN IN THE MOON (I,2871); MEET CYD CHARISSE (I,2982); WHERE THE HEART IS (I,4831)

MITCHELL, KATIE
TWILIGHT THEATER II (II,2686)

MITCHELL, KEITH
EIGHT IS ENOUGH (II,762); THE GREAT IMPERSONATION (I,1876); GUN SHY (II,1068); IT'S A LIVING (II,1254); THE KID WITH THE BROKEN HALO (II,1387); THE MACKENZIES OF PARADISE COVE (II,1585); NORMA RAE (II,1860); STICKIN' TOGETHER (II,2465); TALES OF THE APPLE DUMPLING GANG (II,2536); THE WALTONS (II,2740)

MITCHELL, LEX
HOMICIDE (I,2101)

MITCHELL, MARY
LEAVE IT TO BEAVER (I,2648)

MITCHELL, MILLARD
EYE WITNESS (I,1494)

MITCHELL, PAT
AMERICA ALIVE! (II,67); HOUR MAGAZINE (II,1193)

MITCHELL, RED
MUSIC U.S.A. (I,3179)

MITCHELL, SHIRLEY
ALL ABOUT BARBARA (I,125); BACHELOR FATHER (I,323); BEHIND THE SCREEN (II,200); THE DOCTOR WAS A LADY (I,1318); G.E. SUMMMER ORIGINALS (I,1747); THE GOOD LIFE (II,1033); THE GREAT GILDERSLEEVE (I,1875); I WAS A BLOODHOUND (I,2180); MARIE (II,1629); PETE AND GLADYS (I,3558); PETTICOAT JUNCTION (I,3571); PLEASE DON'T EAT THE DAISIES (I,3628); THE REAL MCCOYS (I,3741); THE RITA MORENO SHOW (II,2181); THE ROMAN HOLIDAYS (I,3835)

MITCHELL, SPRING
KAY KYSER'S KOLLEGE OF MUSICAL KNOWLEDGE (I,2512)

MITCHELL, STEVE
MICHAEL SHAYNE, DETECTIVE (I,3022); WONDER WOMAN (PILOT 1) (II,2826)

MITCHELL, THOMAS
BACKBONE OF AMERICA (I,328); DAMON RUNYON THEATER (I,1129); THE DOCTOR (I,1320); GLENCANNON (I,1822); HENRY FONDA PRESENTS THE STAR AND THE STORY (I,2016); THE JOKE AND THE VALLEY (I,2438); LIGHT'S DIAMOND JUBILEE (I,2698); THE LONER (I,2744); MAYOR OF THE TOWN (I,2963); MIRACLE ON 34TH STREET (I,3056); THE O. HENRY PLAYHOUSE (I,3322); PULITZER PRIZE PLAYHOUSE (I,3692); RHEINGOLD THEATER (I,3783); THE RIGHT MAN (I,3790); ROBERT MONTGOMERY PRESENTS YOUR LUCKY STRIKE THEATER (I,3809); TALES OF TOMORROW (I,4352)

MITCHELL, YVONNE
DOUGLAS FAIRBANKS JR. PRESENTS THE RHEINGOLD THEATER (I,1364)

MITCHLLL, SCOEY
BAREFOOT IN THE PARK (I,355); COPS (I,1047); FOG (II,897); A NEW KIND OF FAMILY (II,1819); RHODA (II,2151); THE SCOEY MITCHLLL SHOW (I,3939); WHAT'S IT ALL ABOUT WORLD? (I,4816)

MITCHUM, JOHN
F TROOP (I,1499); RIVERBOAT (I,3797)

MITCHUM, ROBERT
DEAN'S PLACE (II,650); THE DICK POWELL SHOW (I,1269); THE LOSERS (I,2757); THE WINDS OF WAR (II,2807)

MITORE, CARMINE
MOONLIGHT (II,1730)

MITRI, TIBERIA
AS THE WORLD TURNS (II,110)

MIXON, ALLEN
THE ANDROS TARGETS (II,76); THE EDGE OF NIGHT (II,760)

MIYA, JO ANNE
ARREST AND TRIAL (I,264)

MIYAGUCHI, SIJI
SHOGUN (II,2339)

MIYAZAKI, LANI
THE JIMMY DEAN SHOW (I,2384); MR. BROADWAY (I,3131); ONE LIFE TO LIVE (II,1907)

MIYORI, KIM
ST. ELSEWHERE (II,2432)

MOAR, ANDREA
ALL MY CHILDREN (II,39); RITUALS (II,2182)

MOBLEY, CANDY
THE MYSTERIOUS TWO (II,1783)

MOBLEY, MARY ANN
BE OUR GUEST (I,370); CIRCUS OF THE STARS (II,530); CIRCUS OF THE STARS (II,531); CIRCUS OF THE STARS (II,532); CIRCUS OF THE STARS (II,533); ELVIS REMEMBERED: NASHVILLE TO HOLLYWOOD (II,772); FANTASY ISLAND (II,829); THE MOONGLOW AFFAIR (I,3099); WEDDING DAY (II,2759); THE WORLD OF ENTERTAINMENT (II,2837)

MOBLEY, ROGER
FURY (I,1721); GALLAGHER (I,1727); I'M DICKENS...HE'S FENSTER (I,2193); SECOND CHANCE (I,3957)

MOCHIZUKI, KEN
THE NEW VOICE (II,1833)

MOCKUS, TONY
THE AWAKENING LAND (II,122); TWIGS (II,2684)

MODEAN, JAYNE
ME AND DUCKY (II,1671); TRAUMA CENTER (II,2655)

MODELS, TORE
DON MCNEILL'S DINNER CLUB (I,1336)

MODERNAIRES, THE
THE BOB CROSBY SHOW (I,497); PERRY PRESENTS (I,3549); PETER MARSHALL SALUTES THE BIG BANDS (II,2028); SOLDIERS IN GREASEPAINT (I,4110)

MOE, DOUGLAS
PAPA G.I. (I,3451)

MOE, MARCIA
THE DONALD O'CONNOR SHOW (I,1344)

MOE, MARILYN
THE MONROES (I,3089)

MOFFAT, DONALD
THE CHISHOLMS (II,513); EBONY, IVORY AND JADE (II,755); THE FILE ON DEVLIN (I,1575); LOGAN'S RUN (II,1507); THE LONG DAYS OF SUMMER (II,1514); THE NEW LAND (II,1820); OF MICE AND MEN (I,3333); ONE LIFE TO LIVE (II,1907); REX HARRISON PRESENTS SHORT STORIES OF LOVE (II,2148); THE WORD (II,2833)

MOFFAT, JOHN
VANITY FAIR (I,4701)

MOFFAT, KITTY

BOONE (II,351)

MOFFATT, JOHN
LOVE IN A COLD CLIMATE (II,1539)

MOFFATT, SUSAN
THE NEW HEALERS (I,3266)

MOFFO, ANNA
THE PERRY COMO CHRISTMAS SHOW (I,3528)

MOHICA, VIC
BANACEK: DETOUR TO NOWHERE (I,339); ELLERY QUEEN: TOO MANY SUSPECTS (II,767); THE MACAHANS (II,1583); MALLORY: CIRCUMSTANTIAL EVIDENCE (II,1608); STUNTS UNLIMITED (II,2487)

MOHR, GERALD
BRAVO DUKE (I,711); THE FANTASTIC FOUR (I,1529); FOREIGN INTRIGUE: CROSS CURRENT (I,1637); FOUR STAR PLAYHOUSE (I,1652); THE JUNE ALLYSON SHOW (I,2488); MAGNAVOX THEATER (I,2827); STARS OVER HOLLYWOOD (I,4211)

MOHRE, KELLY
GETTING THERE (II,985)

MOIR, RICHARD
1915 (II,1853); PRISONER: CELL BLOCK H (II,2085)

MOKIHANA
ALOHA PARADISE (II,58)

MOLINARE, RICHARD
ALL NIGHT RADIO (II,40)

MOLINARO, AL
ANSON AND LORRIE (II,98); A CHRISTMAS FOR BOOMER (II,518); GREAT DAY (II,1053); HAPPY DAYS (II,1084); JOANIE LOVES CHACHI (II,1295); THE ODD COUPLE (II,1875); ROSETTI AND RYAN: MEN WHO LOVE WOMEN (II,2217); THE UGILY FAMILY (II,2699)

MOLL, RICHARD
NIGHT COURT (II,1844)

MOLLOY, WILLIAM
ERNIE, MADGE, AND ARTIE (I,1457)

MOLNER, ROBERT
CAMELOT (II,416)

MOLTKE, ALEXANDRA
DARK SHADOWS (I,1157)

MOMARY, DOUG
THE NEW ZOO REVUE (I,3277)

MOMBERGER, HILARY
A CHARLIE BROWN THANKSGIVING (I,909); IT WAS A SHORT SUMMER, CHARLIE BROWN (I,2252); PLAY IT AGAIN, CHARLIE BROWN (I,3617); THERE'S NO TIME FOR LOVE, CHARLIE BROWN (I,4438); YOU'RE NOT ELECTED, CHARLIE BROWN (I,4967)

MONACHINO, FRANCIS
AMAHL AND THE NIGHT VISITORS (I,148)

MONACO, RALPH
ETHEL IS AN ELEPHANT (II,783)

MONAGHAN, ED
THE ROWAN AND MARTIN REPORT (II,2221)

MONAGHAN, GREG
NOBODY'S PERFECT (II,1858)

MONAGHAN, KELLY
ANOTHER WORLD (II,97)

MONAHAN, LAURA
THE NEW SHOW (II,1828)

MOND, STEPHEN
DIFF'RENT STROKES (II,674); FRIENDS (II,928); GETTING THERE (II,985)

MONDO, PEGGY
MCHALE'S NAVY (I,2969); TO ROME WITH LOVE (I,4526)

MONES, PAUL
THE RENEGADES (II,2132); THE RENEGADES (II,2133)

MONICA, CORBETT
CALL HER MOM (I,796); THE JOEY BISHOP SHOW (I,2404)

MONK, DEBRA
PUMPBOYS AND DINETTES ON TELEVISION (II,2090)

MONKHOUSE, BOB
BONKERS! (II,348)

MONKS, JAMES
TALES OF THE BLACK CAT (I,4345)

MONROE, DEL
EGAN (I,1419)

MONROE, LUCY
THE LUCY MONROE SHOW (I,2790)

MONROE, MICHAEL
PINE LAKE LODGE (I,3597)

MONROE, VAUGHN
AIR TIME '57 (I,93); THE SWINGIN' SINGIN' YEARS (I,4319); THE VAUGHN MONROE SHOW (I,4707); THE VAUGHN MONROE SHOW (I,4708)

MONSELL, MURIEL
A TIME TO LIVE (I,4510)

MONT, DONALD
BIG CITY COMEDY (II,233)

MONTAGUE, BRUCE
BUTTERFLIES (II,404)

MONTALBAN, CARLOS
CONCERNING MISS MARLOWE (I,1032)

MONTALBAN, RICARDO
ALICE THROUGH THE LOOKING GLASS (I,122); BROKEN ARROW (I,743); CELEBRITIES: WHERE ARE THEY NOW? (II,462); CLIMAX! (I,976); CODE NAME: HERACLITUS (I,990); THE DANNY THOMAS SPECIAL (I,1151); DINAH IN SEARCH OF THE IDEAL MAN (I,1278); EXECUTIVE SUITE (II,796); THE FANTASTICKS (I,1531); FANTASY ISLAND (II,829); FANTASY ISLAND (II,830); FORD TELEVISION THEATER (I,1634); GREAT ADVENTURE (I,1868); THE GREATEST SHOW ON EARTH (I,1883); HARD KNOCKS (II,1086); HOW THE WEST WAS WON (II,1196); THE LLOYD BRIDGES SHOW (I,2733); THE LORETTA YOUNG THEATER (I,2756); THE MAGIC OF DAVID COPPERFIELD (II,1593); MCNAUGHTON'S DAUGHTER (II,1670); THE PIGEON (I,3596); PLAYHOUSE 90 (I,3623); RETURN TO FANTASY ISLAND (II,2144); SARGE: THE BADGE OR THE CROSS (I,3915); THE SINGERS (I,4060); TONIGHT IN HAVANA (I,4557); WONDER WOMAN (PILOT 1) (II,2826)

MONTAND, YVES
YVES MONTAND ON BROADWAY (I,4978)

MONTANNA, JOE
THE WORLD OF ENTERTAINMENT (II,2837)

MONTE, MARY ELAINE
ANN IN BLUE (II,83)

MONTE, VINNIE
THE BENNY RUBIN SHOW (I,398)

MONTEAGO, JOE
SOAP (II,2392)

MONTEALEGRE, FELICIA
GULF PLAYHOUSE (I,1904)

MONTEITH AND RAND
THE SHAPE OF THINGS (II,2315)

MONTEITH, KELLY
DEAN'S PLACE (II,650); KELLY MONTEITH (II,1375); THE KELLY MONTEITH SHOW (II,1376); NO HOLDS BARRED (II,1854)

MONTENARO, TONY
GUESTWARD HO! (I,1900)

MONTERO, PEDRO
FALCON'S GOLD (II,809)

MONTEVECCHI, LILIANE
CIRCUS OF THE STARS (II,530)

MONTGOMERY, BELINDA J
BARE ESSENCE (II,149); BATTLE OF THE NETWORK STARS (II,165); THE COUNTRY MUSIC MURDERS (II,590); THE CRIME CLUB (I,1096); THE D.A.: CONSPIRACY TO KILL (I,1121); DYNASTY (II,746); LETTERS FROM THREE LOVERS (I,2671); LOCK, STOCK, AND BARREL (I,2736); THE MAN FROM ATLANTIS (II,1614); THE MAN FROM ATLANTIS (II,1615); MIAMI VICE (II,1689); MURDER, SHE WROTE (II,1770); TALES FROM MUPPETLAND (I,4344); TROUBLE IN HIGH TIMBER COUNTRY (II,2661); TURNOVER SMITH (II,2671)

MONTGOMERY, BRYAN
CAT BALLOU (I,870)

MONTGOMERY, ELIZABETH
THE ARMSTRONG CIRCLE THEATER (I,260); THE AWAKENING LAND (II,122); BEWITCHED (I,418); THE BOSTON TERRIER (I,697); DUPONT SHOW OF THE MONTH (I,1387); HARVEY (I,1961); KRAFT TELEVISION THEATER (I,2592); THE LORETTA YOUNG THEATER (I,2756); MISSING PIECES (II,1706); MONTGOMERY'S SUMMER STOCK (I,3096); PLAYHOUSE 90 (I,3623); THE SPIRAL STAIRCASE (I,4160); STUDIO ONE (I,4268); SUSPICION (I,4309); THRILLER (I,4492); THE TWILIGHT ZONE (I,4651)

MONTGOMERY, ELLIOTT
JARRETT (I,2349)

MONTGOMERY, GEORGE
CIMARRON CITY (I,953); STAGE 7 (I,4181)

MONTGOMERY, JULIE
ONE LIFE TO LIVE (II,1907)

MONTGOMERY, LEE H
DOCTOR DAN (II,684); ESCAPE (I,1461); JAMES AT 15 (II,1270); THE SIX OF US (II,2369); TRUE GRIT (II,2664); WHAT'S UP, AMERICA? (I,4824); WHAT'S UP? (I,4823)

MONTGOMERY, MARK
YOU'RE A GOOD MAN, CHARLIE BROWN (I,4962)

MONTGOMERY, RAY
RAMAR OF THE JUNGLE (I,3718); STARS OVER HOLLYWOOD (I,4211)

MONTGOMERY, ROBERT
NAVY LOG (I,3233); ROBERT MONTGOMERY PRESENTS YOUR LUCKY STRIKE THEATER (I,3809)

MONTGOMERY, TANNIS G
HAPPY DAYS (II,1084); IF I LOVE YOU, AM I TRAPPED FOREVER? (II,1222)

MONTI, MARY ELAINE
PARK PLACE (II,1958)

MONTI, SCOTT
CAMP WILDERNESS (II,419)

MONTS, SUSAN
SEARCH FOR TOMORROW (II,2284)

MOODY, DAVID
THE ORPHAN AND THE DUDE (II,1924)

MOODY, JIM
RIVKIN: BOUNTY HUNTER (II,2183)

MOODY, JOHN
THE PARAGON OF COMEDY (II,1956); THE PEE WEE HERMAN SHOW (II,1988)

MOODY, KING
GET SMART (II,983)

MOODY, LYNNE
BENSON (II,208); E/R (II,748); FLY AWAY HOME (II,893); NOW WE'RE COOKIN' (II,1866); ROOTS (II,2211); ROOTS: THE NEXT GENERATIONS (II,2213); SOAP (II,2392); THAT'S MY MAMA (II,2580); WILLOW B: WOMEN IN PRISON (II,2804)

MOODY, RALPH
THE MARSHAL OF GUNSIGHT PASS (I,2925); THE RIFLEMAN (I,3789)

MOODY, RON
BING CROSBY'S MERRIE OLDE CHRISTMAS (II,252); THE DAVID SOUL AND FRIENDS SPECIAL (II,626); NOBODY'S PERFECT (II,1858); PORTRAIT OF PETULA (I,3654); TALES OF THE GOLD MONKEY (II,2537); THE WORD (II,2833)

MOON MAIDENS, THE
THE VAUGHN MONROE SHOW (I,4707)

MOON, GEORGANNA
U.F.O. (I,4662)

MOON, GEORGIA
THE LIBERACE SHOW (I,2677)

MOONEY, ART
LET'S DANCE (I,2664)

MOONEY, GAVIN
THE LAST CONVERTIBLE (II,1435)

MOONEY, MAUREEN
SPACE FORCE (II,2407)

MOONEY, RALPH
LET'S DANCE (I,2664)

MOONEY, WILLIAM
ALL MY CHILDREN (II,39)

MOONEY, YVONNE
THE RICHARD PRYOR SPECIAL (II,2164)

MOONMAN-BACOTE
THE MOONMAN CONNECTION (II,1731)

MOOR, BILL
ANOTHER WORLD (II,97)

MOOR, MARY
THE EDGE OF NIGHT (II,760)

MOORE, ALVY
BORDER COLLIE (I,688); GREEN ACRES (I,1884); HOORAY FOR LOVE (I,2116); JOHNNY COME LATELY (I,2425); LACY AND THE MISSISSIPPI QUEEN (II,1419); LITTLE HOUSE: THE LAST FAREWELL (II,1491); MY LITTLE MARGIE (I,3194); NAVY LOG (I,3233); PETE AND GLADYS (I,3558); STEPHANIE (II,2453); WHAT I WANT TO BE (I,4812); THE WONDERFUL WORLD OF PHILIP MALLEY (II,2831)

MOORE, ARCHIE
CUTTER (I,1117)

MOORE, BARBARA
THE MAN FROM U.N.C.L.E. (I,2867)

MOORE, CANDI BETH
OUR FAMILY BUSINESS (II,1931)

MOORE, CANDY
THE DONNA REED SHOW (I,1347); LIFE WITH VIRGINIA (I,2695); THE LORETTA YOUNG THEATER (I,2756); THE LUCY SHOW (I,2791)

MOORE, CHARLES
PAROLE (II,1960)

MOORE, CHARLOTTE
THE NEWS IS THE NEWS (II,1838)

MOORE, CLAYTON
THE LONE RANGER (I,2740); WHEN THE WEST WAS FUN: A WESTERN REUNION (II,2780)

MOORE, CONSTANCE
THE BOB HOPE SHOW (I,520); WINDOW ON MAIN STREET (I,4873); THE YOUNG MARRIEDS (I,4942)

MOORE, DEL
BACHELOR FATHER (I,323); LIFE WITH ELIZABETH (I,2689); VOYAGE TO THE BOTTOM OF THE SEA (I,4743)

MOORE, DEMI
BEDROOMS (II,198); GENERAL HOSPITAL (II,964)

MOORE, DENNIS
THE FURTHER ADVENTURES OF SPIN AND MARTY (I,1720); THE NEW ADVENTURES OF SPIN AND MARTY (I,3254)

MOORE, DICK
CAPTAIN VIDEO AND HIS VIDEO RANGERS (I,838)

MOORE, DUDLEY
BOB HOPE SPECIAL: BOB HOPE'S PINK PANTHER THANKSGIVING GALA (II,312); BOB HOPE SPECIAL: HAPPY BIRTHDAY, BOB! (II,325); THE MUPPETS GO TO THE MOVIES (II,1767)

MOORE, EDWARD
THE EDGE OF NIGHT (II,760); LOVE OF LIFE (I,2774)

MOORE, ERIN O'BRIEN
THE RUGGLES (I,3868)

MOORE, GARRY
FIRESIDE THEATER (I,1580); THE GARRY MOORE SHOW (I,1738); THE GARRY MOORE SHOW (I,1739); THE GARRY MOORE SHOW (I,1740); I'VE GOT A SECRET (I,2283); MR. BROADWAY (I,3130); THE RIGHT MAN (I,3790); SCHAEFER CENTURY THEATER (I,3935); TIMEX ALL-STAR JAZZ SHOW II (I,4516); TO TELL THE TRUTH (I,4528); TO TELL THE TRUTH (II,2625)

MOORE, GEOFFREY
DR JEKYLL AND MR. HYDE (I,1368)

MOORE, HENRIETTA
FIRST LOVE (I,1583)

MOORE, IDA
THE R.C.A. VICTOR SHOW (I,3737)

MOORE, IVANA
COMEDY OF HORRORS (II,557); MR. SUCCESS (II,1756); YOU ASKED FOR IT (II,2855)

MOORE, JOANNA
THE ANDY GRIFFITH SHOW (I,192); THE DOG TROOP (I,1325); HEAVEN HELP US (I,1995); THE MAN FROM GALVESTON (I,2865); SQUADRON (I,4170); SUSPICION (I,4309); THREE COINS IN THE FOUNTAIN (I,4473)

MOORE, JOHN
WHEN THE NIGHTINGALE SANG IN BERKELEY SQUARE (I,4827)

MOORE, JOHN W
LOVE OF LIFE (I,2774)

MOORE, JONATHAN
LOVE OF LIFE (I,2774)

MOORE, JUANITA
THE RICHARD PRYOR SHOW (II,2163); THE WHOLE WORLD IS WATCHING (I,4851)

MOORE, LAURENS
CHIEFS (II,509)

MOORE, LISA
RANSOM FOR A DEAD MAN (I,3722)

MOORE, MARY ALICE
THE EDGE OF NIGHT (II,760); FROM THESE ROOTS (I,1688); STAGE DOOR (I,4177); YOUNG DR. MALONE (I,4938)

MOORE, MARY TYLER
THE ADVENTURES OF OZZIE AND HARRIET (I,71); BACHELOR FATHER (I,323); CBS: ON THE AIR (II,460); DICK VAN DYKE AND THE OTHER WOMAN, MARY TYLER MOORE (I,1273); THE DICK VAN DYKE SHOW (I,1275); HOW TO SURVIVE THE 70S AND MAYBE EVEN BUMP INTO HAPPINESS (II,1199); I LOVE LIBERTY (II,1215); IVY LEAGUE (I,2286); MARY (II,1637); THE MARY

TYLER MOORE COMEDY HOUR (II,1639); THE MARY TYLER MOORE SHOW (II,1640); MARY'S INCREDIBLE DREAM (II,1641); RICHARD DIAMOND, PRIVATE DETECTIVE (I,3785); THRILLER (I,4492); VAN DYKE AND COMPANY (II,2718)

MOORE, MELBA
THE BEATRICE ARTHUR SPECIAL (II,196); BING CROSBY'S CHRISTMAS SHOW (I,474); DOUG HENNING'S WORLD OF MAGIC III (II,727); ELLIS ISLAND (II,768); FLAMINGO ROAD (II,871); THE MELBA MOORE-CLIFTON DAVIS SHOW (I,2996); OPRYLAND U.S.A. (I,3405)

MOORE, MICKI
PAUL BERNARD—PSYCHIATRIST (I,3485)

MOORE, MONICA
FRONT ROW CENTER (I,1693)

MOORE, MONNETTE
THE AMOS AND ANDY SHOW (I,179)

MOORE, NORMA
TEXAS JOHN SLAUGHTER (I,4414)

MOORE, PATRICIA
MAKE ROOM FOR DADDY (I,2843); MAN AND SUPERMAN (I,2854)

MOORE, ROBERT
DIANA (I,1259); THE MARY TYLER MOORE SHOW (II,1640)

MOORE, ROGER
THE ALASKANS (I,106); ALFRED HITCHCOCK PRESENTS (I,115); ANN-MARGRET'S HOLLYWOOD MOVIE GIRLS (II,86); THE BURT BACHARACH SPECIAL (I,776); IVANHOE (I,2282); THE LONDON PALLADIUM (I,2739); MATINEE THEATER (I,2947); MAVERICK (II,1657); THE PERSUADERS! (I,3556); ROYAL VARIETY PERFORMANCE (I,3865); THE SAINT (II,2237)

MOORE, SCOTTY
ELVIS (I,1435)

MOORE, STEPHEN
BRIDESHEAD REVISITED (II,375); THE HITCHHIKER'S GUIDE TO THE GALAXY (II,1158); THE ROCK FOLLIES (II,2192)

MOORE, TEDDE
CASTLE ROCK (II,456)

MOORE, TERRY
CLIMAX! (I,976); EMPIRE (I,1439); OPERATION ENTERTAINMENT (I,3400); PLAYHOUSE 90 (I,3623); STUDIO ONE (I,4268); THE UNITED STATES STEEL HOUR (I,4677)

MOORE, TIM
THE AMOS AND ANDY SHOW (I,179)

MOORE, TOM
HERE'S BOOMER (II,1134); LADIES BE SEATED (I,2598)

MOORE, VERA
ANOTHER WORLD (II,97)

MOORE, VICTOR
MAGNAVOX THEATER (I,2827); STAGE DOOR (I,4178)

MOOREHEAD, AGNES
ALCOA/GOODYEAR THEATER (I,107); ALICE THROUGH THE LOOKING GLASS (I,122); BEWITCHED (I,418); DUPONT SHOW OF THE MONTH (I,1387); MY SISTER EILEEN (I,3200); NIGHT GALLERY (I,3287); POOR MR. CAMPBELL (I,3647); REVLON MIRROR THEATER (I,3781); REX HARRISON PRESENTS SHORT STORIES OF LOVE (II,2148); SHIRLEY TEMPLE'S STORYBOOK (I,4017); SUSPICION (I,4309)

MOOREY, FRANK
A HORSEMAN RIDING BY (II,1182)

MORA, DANNY
PLEASE STAND BY (II,2057); ROOSEVELT AND TRUMAN (II,2209); STAR OF THE FAMILY (II,2436); THE SUNSHINE BOYS (II,2496)

MORALES, SANTOS
POPI (II,2068); STICKIN' TOGETHER (II,2465)

MORAN, EDDIE
ONE LIFE TO LIVE (II,1907)

MORAN, ERIN
DAKTARI (I,1127); THE DON RICKLES SHOW (I,1342); HAPPY DAYS (II,1084); JOANIE LOVES CHACHI (II,1295)

MORAN, JOHN
ANOTHER LIFE (II,95)

MORAN, LOIS
WATERFRONT (I,4774)

MORAN, MARSHA
THE EYES OF TEXAS (II,799)

MORAN, MONICA
THREE ON AN ISLAND (I,4483)

MORAN, STACEY
SEARCH FOR TOMORROW (II,2284)

MORAND, SYLVESTER
WAR AND PEACE (I,4765)

MORANIS, RICK
SECOND CITY TELEVISION (II,2287); TWILIGHT THEATER II (II,2686)

MORAY, DEAN
THE GOOD OLD DAYS (I,1851)

MORAY, JOHN
THE BOOK OF LISTS (II,350)

MORAY, STELLA
THE BENNY HILL SHOW (II,207)

MORCAMBE, ERIC
PICCADILLY PALACE (I,3589)

MORDENTE, LISA
COUSINS (II,595); DOC (II,683); VIVA VALDEZ (II,2728); WELCOME BACK, KOTTER (II,2761)

MORE, KENNETH
FATHER BROWN (II,836); THE FORSYTE SAGA (I,1645)

MORECAMB, ERIC
THE MORECAMBE AND WISE SHOW (II,1734)

MORELAND, ROSALIND
MINSKY'S FOLLIES (II,1703)

MORELL, MARISSA
AS THE WORLD TURNS (II,110)

MORENO, BELITA
COCAINE AND BLUE EYES (II,548)

MORENO, GEORGE
EL COYOTE (I,1423)

MORENO, JUAN
THE FEAR IS SPREADING (I,1553)

MORENO, RITA
9 TO 5 (II,1852); BLONDES VS. BRUNETTES (II,271); BROKEN ARROW (I,743); CLIMAX! (I,976); DOMINIC'S DREAM (II,704); FIRESIDE THEATER (I,1580); I'LL NEVER FORGET WHAT'S HER NAME (II,1217); PLAYHOUSE 90 (I,3623); THE RITA MORENO SHOW (II,2181); THE ROCKFORD FILES (II,2197); SCHLITZ PLAYHOUSE OF STARS (I,3936); THE SHAPE OF THINGS (II,2315); ZANE GREY THEATER (I,4979)

MORENO, RUBEN
BLACK BART (II,260)

MOREY, BILL
THE SHERIFF AND THE ASTRONAUT (II,2326); THE THORN BIRDS (II,2600); TUCKER'S WITCH (II,2667)

MOREY, SEAN
ROCK COMEDY (II,2191)

MORFOGEN, GEORGE
SHERLOCK HOLMES (II,2327); V (II,2713)

MORGAN THE WONDER DOG
A COUPLE OF JOES (I,1078)

MORGAN, AL
THE AL MORGAN SHOW (I,94)

MORGAN, BETTY ANN
SCHOOL HOUSE (I,3937)

MORGAN, CASS
PUMPBOYS AND DINETTES ON TELEVISION (II,2090)

MORGAN, CHARLIE
THE JAYE P. MORGAN SHOW (I,2350)

MORGAN, CINDY
BRING 'EM BACK ALIVE (II,377); MITCHELL AND WOODS (II,1709); THAT'S TV (II,2581); VEGAS (II,2724)

MORGAN, CLARK
THE FIRST HUNDRED YEARS (I,1581)

MORGAN, CLAUDIA
THE CAMPBELL TELEVISION SOUNDSTAGE (I,812); CELANESE THEATER (I,881); KRAFT TELEVISION THEATER (I,2592); THE MILKY WAY (I,3044); OUR FIVE DAUGHTERS (I,3413)

MORGAN, DAN
SPELLBOUND (I,4153); STAGE 13 (I,4183)

MORGAN, DEBBI
ALL MY CHILDREN (II,39); BEHIND THE SCREEN (II,200); ROOTS: THE NEXT GENERATIONS (II,2213)

MORGAN, DENNIS
21 BEACON STREET (I,4646); ADAMSBURG, U.S.A. (I,33); CROSSROADS (I,1105); STAGE DOOR (I,4178); TELEPHONE TIME (I,4379)

MORGAN, DICK
THE JAYE P. MORGAN SHOW (I,2350)

MORGAN, DUKE

THE JAYE P. MORGAN SHOW (I,2350)

MORGAN, FREDDIE
THE SPIKE JONES SHOW (I,4155)

MORGAN, GARFIELD
THE SWEENEY (II,2513)

MORGAN, GARY
FROM THESE ROOTS (I,1688); HEAD OF THE FAMILY (I,1988); PINOCCHIO (II,2044)

MORGAN, GEORGE
M*A*S*H (II,1569)

MORGAN, HALLIE
SARA (II,2260)

MORGAN, HARRY
ARENA (I,256); BACKSTAIRS AT THE WHITE HOUSE (II,133); THE BASTARD/KENT FAMILY CHRONICLES (II,159); CAT BALLOU (I,870); THE CENTURY TURNS (I,890); THE D.A. (I,1123); DECEMBER BRIDE (I,1215); DRAGNET (I,1370); DRAGNET (I,1371); ELLERY QUEEN: DON'T LOOK BEHIND YOU (I,1434); EXO-MAN (II,797); HEC RAMSEY (II,1121); JACK BENNY'S SECOND FAREWELL SHOW (I,2302); KATE BLISS AND THE TICKER TAPE KID (II,1366); KENTUCKY JONES (I,2526); THE MARRIAGE BROKER (I,2920); MCLAREN'S RIDERS (II,1663); MORE WILD WILD WEST (II,1733); OH, THOSE BELLS! (I,3345); THE PAUL LYNDE COMEDY HOUR (II,1979); PETE AND GLADYS (I,3558); THE RICHARD BOONE SHOW (I,3784); RIVKIN: BOUNTY HUNTER (II,2183); ROOTS: THE NEXT GENERATIONS (II,2213); ROUGHNECKS (II,2219); SIDEKICKS (II,2348); THE WILD WILD WEST REVISITED (II,2800); YOU CAN'T TAKE IT WITH YOU (II,2857)

MORGAN, HENRY
AFTERMASH (II,23); THE DOCTOR (I,1320); DRAW TO WIN (I,1373); HENRY MORGAN'S GREAT TALENT HUNT (I,2017); I'VE GOT A SECRET (II,1220); M*A*S*H (II,1569); MY WORLD . . . AND WELCOME TO IT (I,3208); ON THE CORNER (I,3359); SILVER THEATER (I,4051); THAT WAS THE WEEK THAT WAS (I,4422); THAT WAS THE WEEK THAT WAS (I,4423)

MORGAN, JANE
THE JANE MORGAN SHOW (I,2342); ON PARADE (I,3356); OUR MISS BROOKS (I,3415); WENDY AND ME (I,4793)

MORGAN, JAYE P
ACCENT ON LOVE (I,16); THE ADVENTURES OF NICK CARTER (I,68); THE ALL-AMERICAN COLLEGE COMEDY SHOW (II,47); BURT AND THE GIRLS (I,770); CAP'N AHAB (I,824); THE CHUCK BARRIS RAH-RAH SHOW (II,525); COKE TIME WITH EDDIE FISHER (I,996); THE DON RICKLES SHOW (II,708); FUN FAIR (I,1706); THE GENERAL ELECTRIC THEATER (I,1753); THE JAYE P. MORGAN SHOW (I,2350); MUSIC '55 (I,3168); THE PAUL WINCHELL SHOW (I,3493); PERRY PRESENTS (I,3549); RHYME AND REASON (II,2152); ROBERT Q'S MATINEE (I,3812); SONJA HENIE'S HOLIDAY ON ICE (I,4125); STOP THE MUSIC (I,4232); TIMEX ALL-STAR JAZZ SHOW II (I,4516); TIMEX ALL-STAR JAZZ SHOW III (I,4517); TOO CLOSE FOR COMFORT (II,2642)

MORGAN, JENNY
STAR MAIDENS (II,2435)

MORGAN, JOHNNY
THE JACKIE GLEASON SPECIAL (I,2324)

MORGAN, LIZ
CAPTAIN SCARLET AND THE MYSTERONS (I,837)

MORGAN, MARCIA
THE EYES OF TEXAS (II,799)

MORGAN, MICHAEL
ADDIE AND THE KING OF HEARTS (I,35); THE AMAZING CHAN AND THE CHAN CLAN (I,157); JOHNNY GUITAR (I,2427); RICH MAN, POOR MAN—BOOK I (II,2161); SONS AND DAUGHTERS (II,2404)

MORGAN, NANCY
BACKSTAIRS AT THE WHITE HOUSE (II,133); THE SAN PEDRO BEACH BUMS (II,2250)

MORGAN, NORA
HEAVEN ON EARTH (II,1119)

MORGAN, PETER
THE MUNSTERS' REVENGE (II,1765)

MORGAN, PRISCILLA
PRIDE AND PREJUDICE (II,2080)

MORGAN, RANDY
CAMELOT (II,416)

MORGAN, RAY
I'D LIKE TO SEE (I,2184)

MORGAN, READ
ALIAS SMITH AND JONES (I,118); THE BILLION DOLLAR THREAT (II,246); THE DEPUTY (I,1234)

MORGAN, RICHARD
THE SULLIVANS (II,2490)

MORGAN, ROBIN
THE ALCOA HOUR (I,108); I REMEMBER MAMA (I,2176); MEET CORLISS ARCHER (I,2981); MR. I. MAGINATION (I,3140); STAR STAGE (I,4197); SUSPENSE (I,4305); TALES OF TOMORROW (I,4352)

MORGAN, RONALD E
THE INVISIBLE WOMAN (II,1244)

MORGAN, RUSS
IN THE MORGAN MANNER (I,2204); THE RUSS MORGAN SHOW (I,3874)

MORGAN, SEAN
THE ADVENTURES OF OZZIE AND HARRIET (I,71)

MORGAN, SHARON
THE CITADEL (II,539)

MORGAN, TERRENCE
THE ADVENTURES OF SIR FRANCIS DRAKE (I,75)

MORGAN, TRACY
JOHNNY COME LATELY (I,2425)

MORGAN, VERNE
THE BENNY HILL SHOW (II,207)

MORGAN, WESLEY
THE LIFE OF RILEY (I,2686)

MORGANSTERN, ALBERT
STARSKY AND HUTCH (II,2445)

MORICK, DAVE
B.J. AND THE BEAR (II,127); THE CHEAP DETECTIVE (II,490); THE QUEEN AND I (I,3704)

MORITA, PAT
THE BARBARA EDEN SHOW (I,346); BLANSKY'S BEAUTIES (II,267); BROCK'S LAST CASE (I,742); CIRCUS OF THE STARS (II,530); COPS (I,1047); CRASH ISLAND (II,600); EVIL ROY SLADE (I,1485); HAPPY DAYS (II,1084); HUMAN FEELINGS (II,1203); LOLA (II,1508); LOLA (II,1509); LOLA (II,1510); LOLA (II,1511); THE MAD MAD MAD MAD WORLD OF THE SUPER BOWL (II,1586); MR. T. AND TINA (II,1758); PUNCH AND JODY (II,2091); THE QUEEN AND I (I,3704); SANFORD AND SON (II,2256); WIVES (II,2812); YOUNG GUY CHRISTIAN (II,2864)

MORITZ, LOUISA
HAPPY ANNIVERSARY AND GOODBYE (II,1082); THE JOE NAMATH SHOW (I,2400); THE SINGERS (I,4060)

MORLEY, ROBERT
DEADLY GAME (II,632); THE JO STAFFORD SHOW (I,2394)

MORONOSK, MICHAEL
THE LEPRECHAUN'S CHRISTMAS GOLD (II,1461)

MORRELL, ANDRE
THE LILLI PALMER THEATER (I,2704)

MORRICK, DAVE
BARNEY AND ME (I,358); MURDOCK'S GANG (I,3164)

MORRILL, PRISCILLA
BABY MAKES FIVE (II,129); BRET MAVERICK (II,374); DOROTHY (II,717); DREAM GIRL (I,1374); FAMILY (II,813); FAMILY TIES (II,819); IN THE BEGINNING (II,1229); THE MARY TYLER MOORE SHOW (II,1640); MORK AND MINDY (II,1735); NEWHART (II,1835); ONE DAY AT A TIME (II,1900); STICK AROUND (II,2464); A YEAR AT THE TOP (II,2844)

MORRIS, ANITA
CIRCUS OF THE STARS (II,536)

MORRIS, BETH
DIAL A DEADLY NUMBER (I,1253)

MORRIS, BRIAN
MAUDE (II,1655)

MORRIS, CAROL
THE BOB HOPE SHOW (I,552)

MORRIS, CHESTER
CAMEO THEATER (I,808); CAPTURED (I,841); DANGER (I,1134); DIAGNOSIS: UNKNOWN (I,1251); ESPIONAGE (I,1464); THE FLIERS (I,1596); GANGBUSTERS (I,1733); KNIGHT'S GAMBIT (I,2576); LIGHT'S OUT (I,2699); PLAYHOUSE 90 (I,3623); PURSUIT (I,3700); A STRING OF BEADS (I,4263); STUDIO ONE (I,4268)

MORRIS, DEIRDRA
THE WOMAN IN WHITE (II,2819)

MORRIS, EDWARD
MY FAVORITE MARTIANS (II,1779)

MORRIS, GARRETT
AT YOUR SERVICE (II,118); CHANGE AT 125TH STREET (II,480); THE INVISIBLE WOMAN (II,1244); IT'S YOUR MOVE (II,1259); ROLL OUT! (I,3833); SATURDAY NIGHT LIVE (II,2261); THINGS WE DID LAST SUMMER (II,2589)

MORRIS, GREG
KILLER BY NIGHT (I,2537); LADIES IN BLUE (II,1422); MISSION: IMPOSSIBLE (I,3067); MITZI AND A HUNDRED GUYS (II,1710); ROOTS: THE NEXT GENERATIONS (II,2213); SWING OUT, SWEET LAND (II,2515); VEGAS (II,2723); VEGAS (II,2724); WHAT'S HAPPENING!! (II,2769)

MORRIS, HOWARD
ALVIN AND THE CHIPMUNKS (II,61); THE ANDY GRIFFITH SHOW (I,192); THE ARCHIE COMEDY HOUR (I,252); ARCHIE'S TV FUNNIES (I,253); THE ATOM ANT/SECRET SQUIRREL SHOW (I,311); THE BANG-SHANG LALAPALOOZA SHOW (II,140); CAESAR'S HOUR (I,790); EVERYTHING'S ARCHIE (I,1483); THE FAMOUS ADVENTURES OF MR. MAGOO (I,1524); THE JETSONS (I,2376); KING FEATURES TRILOGY (I,2555); THE MAGILLA GORILLA SHOW (I,2825); THE MANY FACES OF COMEDY (I,2894); MISSION MAGIC (I,3068); THE MUNSTERS' REVENGE (II,1765); PANTOMIME QUIZ (I,3449); THE PETER POTAMUS SHOW (I,3568); THE SECRET LIVES OF WALDO KITTY (II,2292); SID CAESAR INVITES YOU (I,4040); TWELFTH NIGHT (I,4636); THE U.S. OF ARCHIE (I,4687); U.S. OF ARCHIE (II,2698); VARIETY: THE WORLD OF SHOW BIZ (I,4704); YOUR SHOW OF SHOWS (I,4957); YOUR SHOW OF SHOWS (II,2875)

MORRIS, JEFF
LASSITER (I,2625); LASSITER (II,1434)

MORRIS, LANA
THE FORSYTE SAGA (I,1645)

MORRIS, LLOYD
ALL THE RIVERS RUN (II,43)

MORRIS, MARY
THE PRISONER (I,3670)

MORRIS, VIRGINIA
THINGS ARE LOOKING UP (II,2588)

MORRIS, WAYNE
DAMON RUNYON THEATER (I,1129); THEY WENT THATAWAY (I,4443)

MORRIS, WOLFE
TYCOON: THE STORY OF A WOMAN (II,2697)

MORRIS-GOWDY, KAREN
RYAN'S HOPE (II,2234)

MORRISH, ANN
THE EXPERT (I,1489)

MORRISON, ANN
GENERAL HOSPITAL (II,964); MEDIC (I,2976)

MORRISON, B J
GEORGE WASHINGTON (II,978)

MORRISON, BARBARA
CAPTAINS AND THE KINGS (II,435); THE JONES BOYS (I,2446)

MORRISON, BRETT
BOBO HOBO AND HIS TRAVELING TROUPE (I,678)

MORRISON, HERBERT
JANE EYRE (I,2337)

MORRISON, PATRICIA
THE CASES OF EDDIE DRAKE (I,863); CAVALCADE OF STARS (I,877); THE GENERAL FOODS 25TH ANNIVERSARY SHOW (I,1756); KISS ME, KATE (I,2568); PULITZER PRIZE PLAYHOUSE (I,3692)

MORRISON, SHELLEY
THE FARMER'S DAUGHTER (I,1533); THE FLYING NUN (I,1611); THE PARTRIDGE FAMILY (II,1962)

MORRISON, TOM
CBS CARTOON THEATER (I,878); THE MIGHTY MOUSE PLAYHOUSE (I,3035)

MORRISS, ANN
THE BROTHERS (I,748)

MORROW, BYRON
DRAGNET (I,1371); EXECUTIVE SUITE (II,796); FOOLS, FEMALES AND FUN: IS THERE A DOCTOR IN THE HOUSE? (II,900); THE GIRL IN THE EMPTY GRAVE (II,997); THE NEW BREED (I,3258); RIDING FOR THE PONY EXPRESS (II,2172); SUPERCOPS (II,2502); THE WINDS OF WAR (II,2807)

MORROW, DON
CAMOUFLAGE (I,810); LET'S PLAY POST OFFICE (I,2667); MARTIN KANE, PRIVATE EYE (I,2928)

MORROW, DORETTA
HOLIDAY (I,2074); MARCO POLO (I,2904); PULITZER PRIZE PLAYHOUSE (I,3692); THE ROBERT Q. LEWIS CHRISTMAS SHOW (I,3811); VICTOR BORGE'S COMEDY IN MUSIC III (I,4725)

MORROW, JEFF
FRONTIER (I,1695); HANDS OF MURDER (I,1929); THE NEW TEMPERATURES RISING SHOW (I,3273); POLICE WOMAN: THE GAMBLE (II,2064); SUDDEN SILENCE (I,4275); UNION PACIFIC (I,4676)

MORROW, KAREN
THE ALAN KING SPECIAL (I,103); FRIENDS (II,927); I'M A FAN (I,2192); THE JIM NABORS HOUR (I,2381); LADIES' MAN (II,1423); MAUREEN (II,1656); THE NANCY DUSSAULT SHOW (I,3222); PAUL SAND IN FRIENDS AND LOVERS (II,1982); TABITHA (II,2527); TABITHA (II,2528)

MORROW, PATRICIA
ANNIE GET YOUR GUN (I,223); CAROLYN (I,854); I LED THREE LIVES (I,2169); MR. NOVAK (I,3145); NEW GIRL IN HIS LIFE (I,3265); PEYTON PLACE (I,3574); RETURN TO PEYTON PLACE (I,3779); WHAT I WANT TO BE (I,4812)

MORROW, SCOTT
DO NOT DISTURB (I,1300); HOW'S BUSINESS? (I,2153)

MORROW, SUSAN
THE LIFE OF VERNON HATHAWAY (I,2687)

MORROW, VIC
B.A.D. CATS (II,123); THE BARBARA STANWYCK THEATER (I,350); CAPTAINS AND THE KINGS (II,435); COMBAT (I,1011); THE LAST CONVERTIBLE (II,1435); LOVE STORY (I,2778); THE MAN WITH THE POWER (II,1616); THE POLICE STORY (I,3638); ROOTS (II,2211); THE SEEKERS (II,2298); STONE (II,2470); TRAVIS LOGAN, D.A. (I,4598); WILD AND WOOLEY (II,2798)

MORSE, BARRY
THE ADVENTURER (I,43); THE FUGITIVE (I,1701); THE HEIRESS (I,2006); HIGHER AND HIGHER, ATTORNEYS AT LAW (I,2057); INN OF THE FLYING DRAGON (I,2214); THE MARTIAN CHRONICLES (II,1635); MASTER OF THE GAME (II,1647); PROFILES IN COURAGE (I,3678); SPACE: 1999 (II,2409); THE THREE MUSKETEERS (I,4481); TRUMAN AT POTSDAM (I,4618); THE WINDS OF WAR (II,2807); A WOMAN OF SUBSTANCE (II,2820); THE ZOO GANG (II,2877)

MORSE, DAVID
OUR FAMILY BUSINESS (II,1931); ST. ELSEWHERE (II,2432)

MORSE, HAYWARD
THE TURN OF THE SCREW (I,4624)

MORSE, RICHARD
THE GUIDING LIGHT (II,1064)

MORSE, ROBERT
JACK FROST (II,1263); LESLIE (I,2659); MARLO THOMAS AND FRIENDS IN FREE TO BE. . . YOU AND ME (II,1632); MIDNIGHT MYSTERY (I,3030); THAT'S LIFE (I,4426); TRAPPER JOHN, M.D. (II,2654)

MORSE, ROBIN
SABU AND THE MAGIC RING (I,3881)

MORSE, RONALD
HUNTER (I,2161)

MORSELL, FRED
FEATHERSTONE'S NEST (II,847); WILSON'S REWARD (II,2805)

MORSHOWER, GLENN
SKEEZER (II,2376)

MORTED, CHARLES
KORG: 70,000 B.C. (II,1408)

MORTENSEN, LYNN
BE MY VALENTINE, CHARLIE BROWN (I,369); HAPPY ANNIVERSARY, CHARLIE BROWN (I,1939); IT'S A MYSTERY, CHARLIE BROWN (I,2259); IT'S THE EASTER BEAGLE, CHARLIE BROWN (I,2275)

MORTIMER, CAROLINE
THE PALLISERS (II,1944)

MORTON, CLIVE
JANE EYRE (I,2339)

MORTON, GREG
SCARECROW AND MRS. KING (II,2268)

MORTON, GREGORY
TWELFTH NIGHT (I,4636)

MORTON, HOWARD
GIMME A BREAK (II,995); MARY HARTMAN, MARY HARTMAN (II,1638); ONE DAY AT A TIME (II,1900); SHE'S WITH ME (II,2321)

MORTON, JOE
ANOTHER WORLD (II,97); FEELING GOOD (I,1559); GRADY (II,1047); SEARCH FOR TOMORROW (II,2284)

MORTON, MICHA
COCAINE AND BLUE EYES (II,548)

MORTON, SKEETS
UNCLE MISTLETOE AND HIS ADVENTURES (I,4669)

MORTON, TOMMY
THE COLLEGE BOWL (I,1001)

MORTON, WAYNE
FRED AND BARNEY MEET THE THING (II,919)

MOSCHITTA JR, JOHN
CALLAHAN (II,414); THE HALF-HOUR COMEDY HOUR (II,1074); MADAME'S PLACE (II,1587); MATT HOUSTON (II,1654); ZORRO AND SON (II,2878)

MOSCOW CIRCUS, THE
PEGGY FLEMING VISITS THE SOVIET UNION (I,3509)

MOSCOW'S CIVIC BALLET
PEGGY FLEMING VISITS THE SOVIET UNION (I,3509)

MOSES, DAVID
THE NEW PEOPLE (I,3268)

MOSES, HARRY
DANGER IN PARADISE (II,617); STATE FAIR (II,2448)

MOSES, MARION
THE BEST YEARS (I,410); HOLLOWAY'S DAUGHTERS (I,2079)

MOSES, RICK
GENERAL HOSPITAL (II,964); YOUNG DAN'L BOONE (II,2863)

MOSES, WILLIAM R
BATTLE OF THE NETWORK STARS (II,176); BATTLE OF THE NETWORK STARS (II,177); BATTLE OF THE NETWORK STARS (II,178); FALCON CREST (II,808)

MOSLEY, BRYAN
CORONATION STREET (I,1054)

MOSLEY, ROGER
THE LIVING END (I,2731)

MOSLEY, ROGER E
MAGNUM, P.I. (II,1601); ROOTS: THE NEXT GENERATIONS (II,2213)

MOSS, ARNOLD
THE CAMPBELL TELEVISION SOUNDSTAGE (I,812); THE CLOCK (I,978); GIDEON (I,1792); LIGHT'S OUT (I,2699); MARCO POLO (I,2904); THE MOTOROLA TELEVISION HOUR (I,3112); SUSPENSE (I,4305); TALES OF TOMORROW (I,4352); YES, VIRGINIA, THERE IS A SANTA CLAUS (II,2849)

MOSS, DAVID
MY LUCKY PENNY (I,3196)

MOSS, JOHN
CULTURE CLUB IN CONCERT (II,610)

MOSS, LARRY
THE DAVID FROST REVUE (I,1176)

MOSS, STEWART
BEYOND WESTWORLD (II,227); CATCH 22 (I,871); CONSPIRACY OF TERROR (II,569); FAY (II,844); OVER AND OUT (II,1938); SARGE (I,3914); SARGE: THE BADGE OR THE CROSS (I,3915)

MOSS, WARWICK
1915 (II,1853)

MOSS, WINSTON
BERKELEY SQUARE (I,399)

MOST, DONNY
DUNGEONS AND DRAGONS (II,744); FONZ AND THE HAPPY DAYS GANG (II,898); HAPPY DAYS (II,1084)

MOSTEL, JOSH
AT EASE (II,115); DELTA HOUSE (II,658); HEREAFTER (II,1138); OFF CAMPUS (II,1877); ZERO HOUR (I,4980)

MOSTEL, ZERO
CAROL + 2 (I,853); FORD THEATER HOUR (I,1635); LITTLE DRUMMER BOY, BOOK II (II,1486); LOVE, LIFE, LIBERTY & LUNCH (II,1544); OFF THE RECORD (I,3336); SAGA OF SONORA (I,3884); ZERO HOUR (I,4980)

MOTTA, BESS
20 MINUTE WORKOUT (II,2680)

MOTTOLA, TONY
MELODY STREET (I,3000); PANAMA HATTIE (I,3444)

MOULDER, WALTER
THE MAGNIFICENT YANKEE (I,2829)

MOULDER-BROWN, JOHN
ELLIS ISLAND (II,768)

MOULTRIE, EMILY
BABY MAKES FIVE (II,129); GOING BANANAS (II,1015)

MOUNTAIN, JOHNNY
EYE ON HOLLYWOOD (II,798); THE LOVE REPORT (II,1542)

MOUNTBATTEN, LADY IRIS
VERSATILE VARIETIES (I,4713)

MOUNTFORD, DIANE
ASSIGNMENT: UNDERWATER (I,303)

MOUTRIE, JANE
HIS HONOR, HOMER BELL (I,2063)

MOWBRAY, ALAN
ADVENTURES OF COLONEL FLACK (I,56); THE BEST IN MYSTERY (I,403); DANTE (I,1155); FOUR STAR PLAYHOUSE (I,1652); GRUEN GUILD PLAYHOUSE (I,1892); HEY MULLIGAN (I,2040); SCHAEFER CENTURY THEATER (I,3935); TENNESSEE ERNIE FORD MEETS KING ARTHUR (I,4397); THE UNEXPECTED (I,4674)

MOWER, PATRICK
MARCO POLO (II,1626); SPECIAL BRANCH (II,2414); TARGET (II,2541)

MOWERY, CARRIE
THE GUIDING LIGHT (II,1064)

MOWRY, JEAN
HAWKINS FALLS, POPULATION 6200 (I,1977); THE SECRET STORM (I,3969)

MOXON, HENRY
MASTER OF THE GAME (II,1647)

MOYER, TAWNY
WELCOME TO THE FUN ZONE (II,2763)

MOYER, TIM
GEORGE WASHINGTON (II,978)

MOYNIHAN, DANIEL
CRIMES OF PASSION (II,603); THE SIX WIVES OF HENRY VIII (I,4068)

MOYSE, DAVID
AIR SUPPLY IN HAWAII (II,26)

MR. T
THE A-TEAM (II,119); BATTLE OF THE NETWORK STARS (II,176); BATTLE OF THE NETWORK STARS (II,177); BOB HOPE SPECIAL: BOB HOPE'S SUPER BIRTHDAY SPECIAL (II,319); BOB HOPE SPECIAL: BOB HOPE'S WICKI-WACKY SPECIAL FROM WAIKIKI (II,321); THE DEAN MARTIN CELEBRITY ROAST (II,639); HOLLYWOOD'S PRIVATE HOME MOVIES II (II,1168); MR. T (II,1757); TWILIGHT THEATER II (II,2686)

MUDD, ROGER
CHRISTMAS IN WASHINGTON (II,521)

MUELLER, RITA
MAKE A FACE (I,2837)

MUELLERLEILE, MARIANNE
I'D RATHER BE CALM (II,1216); P.O.P. (II,1939)

MUENZ, RICHARD
CAMELOT (II,416)

MUGGLETON, AMANDA
PRISONER: CELL BLOCK H (II,2085)

MUGHAN, SHARON
THE FLAME TREES OF THIKA (II,870)

MUGSY
THE UNCLE FLOYD SHOW (II,2703)

MUIR, GAVIN
THE BETTY HUTTON SHOW (I,413)

MUIR, GEORGETTE
HUNTER (II,1204)

MUIR, JEAN
THE ALDRICH FAMILY (I,111)

MULDAUR, DIANA
BLACK BEAUTY (II,261); BORN FREE (I,691); BORN FREE (II,353); CALL TO DANGER (I,804); CANNON: THE RETURN OF FRANK CANNON (II,425); CHARLIE'S ANGELS (II,487); THE DEADLY TRIANGLE (II,634); FITZ AND BONES (II,867); HIZZONER (II,1159); IT CAN'T HAPPEN TO ME (II,1250); MCCLOUD (II,1660); MCCLOUD: WHO KILLED MISS U.S.A.? (I,2965); PINE CANYON IS BURNING (II,2040); PLANET EARTH (I,3615); QUINCY, M. E. (II,2102); THE SECRET STORM (I,3969); THE SURVIVORS (I,4301); TERROR AT

ALCATRAZ (II,2563); THE TONY RANDALL SHOW (II,2640); TOO GOOD TO BE TRUE (II,2643); THE WORD (II,2833)

MULGREW, KATE
KATE LOVES A MYSTERY (II,1367); THE MANIONS OF AMERICA (II,1623); MRS. COLUMBO (II,1760); RYAN'S HOPE (II,2234); THE WORD (II,2833)

MULHALL, JACK
I REMEMBER CAVIAR (I,2175); THE KEN MURRAY SHOW (I,2525)

MULHARE, EDWARD
THE GHOST AND MRS. MUIR (I,1786); GIDGET GROWS UP (I,1797); KNIGHT RIDER (II,1402); STUDIO ONE (I,4268)

MULHERN, SCOTT
BEHIND THE SCREEN (II,200); GENERAL HOSPITAL (II,964)

MULL, MARTIN
AMERICA 2-NIGHT (II,65); THE CHEVY CHASE NATIONAL HUMOR TEST (II,504); DOMESTIC LIFE (II,703); EVENING AT THE IMPROV (II,786); FERNWOOD 2-NIGHT (II,849); THE FUNNIEST JOKE I EVER HEARD (II,940); THE JERK, TOO (II,1282); THE JOHNNY CASH SPRING SPECIAL (II,1329); MAGIC WITH THE STARS (II,1599); MARY HARTMAN, MARY HARTMAN (II,1638); PRIME TIMES (II,2083); THE TV SHOW (II,2673); TWILIGHT THEATER II (II,2686)

MULLANEY, BARBARA
CORONATION STREET (I,1054)

MULLANEY, JACK
ENSIGN O'TOOLE (I,1448); IT'S ABOUT TIME (I,2263); MY LIVING DOLL (I,3195); SHAUGHNESSEY (II,2318)

MULLANEY, KIERAN
FATHER ON TRIAL (I,1544)

MULLANEY, SCOTT
MARIE (II,1629)

MULLAVEY, GREG
BIG EDDIE (II,235); CENTENNIAL (II,477); CRASH ISLAND (II,600); FOREVER FERNWOOD (II,909); HAVING BABIES I (II,1107); MARY HARTMAN, MARY HARTMAN (II,1638); NUMBER 96 (II,1869); RITUALS (II,2182); SHE'S WITH ME (II,2321); SWITCH (II,2520); THIS IS KATE BENNETT (II,2597); WEDNESDAY NIGHT OUT (II,2760); WILDER AND WILDER (II,2802)

MULLAVEY, JACK
THE ANN SOTHERN SHOW (I,220); HIS MODEL WIFE (I,2064)

MULLEN, BARBARA
DOUGLAS FAIRBANKS JR. PRESENTS THE RHEINGOLD THEATER (I,1364)

MULLEN, KATHRYN
FRAGGLE ROCK (II,916)

MULLEN, SUSAN
THE SAN PEDRO BEACH BUMS (II,2250); THE SAN PEDRO BUMS (II,2251)

MULLER, DEBBIE
SHE'S A GOOD SKATE, CHARLIE BROWN (I,4012)

MULLER, HARRISON
THE MUSIC OF GERSHWIN (I,3174)

MULLER, MICHELLE
IT'S ARBOR DAY, CHARLIE BROWN (I,2267); IT'S YOUR FIRST KISS, CHARLIE BROWN (I,2280); YOU'RE THE GREATEST, CHARLIE BROWN (I,4973)

MULLER, THOMAS A
IT'S A MYSTERY, CHARLIE BROWN (I,2259)

MULLIGAN, MOON
COUNTRY MUSIC CARAVAN (I,1066)

MULLIGAN, RICHARD
DIANA (I,1259); GHOST STORY (I,1788); HARVEY (I,1962); HAVING BABIES III (II,1109); THE HERO (I,2035); MALIBU (II,1607); REGGIE (II,2127); SOAP (II,2392)

MULLIGAN, TERRY DAVID
STAR CHART (II,2434)

MULLIKEN, BILL
HARRY'S BUSINESS (I,1958)

MULLIN, LAURENCE
CORONATION STREET (I,1054)

MULLINER, ROD
PRISONER: CELL BLOCK H (II,2085)

MULLINS, MICHAEL
ADVICE TO THE LOVELORN (II,16); THE ASPHALT COWBOY (II,111); DANGER IN PARADISE (II,617); KOJAK (II,1406)

MULLOWNEY, DEBORAH
CAPITOL (II,426)

MULQUEEN, KATHLEEN
DENNIS THE MENACE (I,1231)

MULTRAY, MELISSA
MARIE (II,1629)

MUMY, BILLY
ARCHIE (II,104); LOST IN SPACE (I,2758); THE ROCKFORD FILES (II,2196); SUNSHINE (II,2495); THE TWO OF US (I,4658)

MUNCHIN, JULES
PHIL SILVERS IN NEW YORK (I,3577); SUMMER IN NEW YORK (I,4279)

MUNCHOW, WILLIAM
THE LAST LEAF (II,1437)

MUNCKE, CHRISTOPHER
MASTER OF THE GAME (II,1647); OPPENHEIMER (II,1919)

MUNDY, MEG
THE DOCTORS (II,694); FORD THEATER HOUR (I,1635); LITTLE WOMEN (I,2719); PLAYWRIGHTS '56 (I,3627); SORRY, WRONG NUMBER (I,4130); STARLIGHT THEATER (I,4201); TALES OF TOMORROW (I,4352); THAT'S OUR SHERMAN (I,4429)

MUNI, PAUL
PHILCO TELEVISION PLAYHOUSE (I,3583)

MUNI, SCOTT
THE BEACH BOYS 20TH ANNIVERSARY (II,188)

MUNKER, ARIANE
ANOTHER WORLD (II,97); AS THE WORLD TURNS (II,110); LAND OF HOPE (II,1424); RYAN'S HOPE (II,2234)

MUNNE, MARIA
TONIGHT IN HAVANA (I,4557)

MUNOZ, ANTHONY
LEGMEN (II,1458)

MUNRO, C PETE
ENOS (II,779); SIX PACK (II,2370)

MUNRO, JANET
THE ADMIRABLE CRICHTON (I,36); BERKELEY SQUARE (I,399); TIME REMEMBERED (I,4509)

MUNRO, TIM
GREAT EXPECTATIONS (II,1055)

MUNROE, MAE
FROM THESE ROOTS (I,1688)

MUNSEL, PATRICE
THE GREAT WALTZ (I,1878); NAUGHTY MARIETTA (I,3232); THE PATRICE MUNSEL SHOW (I,3474)

MUNSHIN, JULES
ARCHY AND MEHITABEL (I,254); G.E. SUMMMER ORIGINALS (I,1747); KISS ME, KATE (I,2569); SHIRLEY TEMPLE'S STORYBOOK: THE LEGEND OF SLEEPY HOLLOW (I,4018)

MUNSON, WARREN
FATHER MURPHY (II,841); NO SOAP, RADIO (II,1856)

MUPPETS, THE
JULIE ON SESAME STREET (I,2485); KEEP U.S. BEAUTIFUL (I,2519); PERRY COMO'S WINTER SHOW (I,3547); PURE GOLDIE (I,3696); THE TOM JONES SPECIAL (I,4542)

MURCOTT, DEREK
TESTIMONY OF TWO MEN (II,2564)

MURDOCK, GEORGE
ALL THAT GLITTERS (II,41); BANACEK: DETOUR TO NOWHERE (I,339); BARNEY MILLER (II,154); BATTLESTAR GALACTICA (II,181); THE LAWYERS (I,2646); THE LEATHERNECKS (I,2647); NIGHT GALLERY (I,3288); THE WINDS OF WAR (II,2807)

MURDOCK, JACK
FOG (II,897); OPERATION PETTICOAT (II,1917)

MURDOCK, JAMES
OPERATION PETTICOAT (II,1916); RAWHIDE (I,3727)

MURDOCK, KERMIT
BRIEF PAUSE FOR MURDER (I,725)

MURNEY, CHRISTOPHER
THE COUNTRY GIRL (I,1064); FROM HERE TO ETERNITY (II,932); PEEK-A-BOO: THE ONE AND ONLY PHYLLIS DIXEY (II,1989); POTTSVILLE (II,2072); THE SAN PEDRO BEACH BUMS (II,2250); THE SAN PEDRO BUMS (II,2251)

MURON, EARL
OUR FIVE DAUGHTERS (I,3413)

MURPHEY, MICHAEL
A WALK IN THE NIGHT (I,4750)

MURPHY, ALMA
ALL IN THE FAMILY (I,133); TURN ON (I,4625)

MURPHY, AUDIE
WHISPERING SMITH (I,4842)

MURPHY, BEN
ALIAS SMITH AND JONES (I,118); ALIAS SMITH AND JONES (I,119); BATTLE OF THE NETWORK STARS (II,163); BATTLE OF THE NETWORK STARS (II,177); BRIDGER (II,376); THE CHISHOLMS (II,512); THE CHISHOLMS (II,513); THE GEMINI MAN (II,960); THE GEMINI MAN (II,961); GIDGET'S SUMMER REUNION (I,1799); GRIFF (I,1888); THE LETTERS (I,2670); LOTTERY (II,1525); THE NAME OF THE GAME (I,3217); THE SECRET WAR OF JACKIE'S GIRLS (II,2294); UNIT 4 (II,2707); THE WINDS OF WAR (II,2807)

MURPHY, BRIAN
GEORGE AND MILDRED (II,965); MAN ABOUT THE HOUSE (II,1611)

MURPHY, CHARLES THOMAS
PEN 'N' INC. (II,1992)

MURPHY, DIANE
BEWITCHED (I,418)

MURPHY, EDDIE
THE JOE PISCOPO SPECIAL (II,1304); SATURDAY NIGHT LIVE (II,2261)

MURPHY, ERIN
BEWITCHED (I,418)

MURPHY, GEORGE
THE DESILU REVUE (I,1238); M-G-M PARADE (I,2798); YOU'RE ONLY YOUNG TWICE (I,4970)

MURPHY, HARRY
THE NASHVILLE PALACE (II,1792)

MURPHY, JACKI
EMERGENCY! (II,776)

MURPHY, JOHN
A CHRISTMAS CAROL (I,947)

MURPHY, JOHN CULLEN
THE FANTASTIC FUNNIES (II,825); THE MANIONS OF AMERICA (II,1623)

MURPHY, M P
B.J. AND THE BEAR (II,125)

MURPHY, MARY
THE INVESTIGATORS (I,2231); THE JUDGE (I,2468)

MURPHY, MAURA
ANNIE OAKLEY (I,225)

MURPHY, MAUREEN
THE DEAN MARTIN CELEBRITY ROAST (II,637); THE SHAPE OF THINGS (II,2315); WOMEN WHO RATE A "10" (II,2825)

MURPHY, MELISSA
SEARCH FOR TOMORROW (II,2284)

MURPHY, MICHAEL
BELL, BOOK AND CANDLE (II,201); TWO MARRIAGES (II,2691); WINNER TAKE ALL (II,2808)

MURPHY, MICHAEL MARTIN
THE KOWBOYS (I,2584)

MURPHY, PAMELA
BRIGHT PROMISE (I,727); OUR PRIVATE WORLD (I,3417)

MURPHY, ROSEMARY
ALL MY CHILDREN (II,39); GEORGE WASHINGTON (II,978); KATE AND ALLIE (II,1365); LUCAS TANNER (II,1555); LUCAS TANNER (II,1556); THE SECRET STORM (I,3969)

MURPHY, TERRY
THE LOVE REPORT (II,1543)

MURPHY, TIMOTHY PATRICK
DALLAS (II,614); GLITTER (II,1009); HOTEL (II,1192); SEARCH FOR TOMORROW (II,2284); THE SEEKERS (II,2298); TEACHERS ONLY (II,2547)

MURPHY, TURK
CRESCENDO (I,1094)

MURRAY SISTERS, THE
HAYLOFT HOEDOWN (I,1981)

MURRAY THE K
IT'S WHAT'S HAPPENING, BABY! (I,2278)

MURRAY, ANNE
ANNE MURRAY'S CARIBBEAN CRUISE (II,88); ANNE MURRAY'S LADIES' NIGHT (II,89); ANNE MURRAY'S WINTER CARNIVAL. . . FROM QUEBEC (II,90); CITY VS. COUNTRY (I,968); COUNTRY COMES HOME (II,586); COUNTRY MUSIC HIT PARADE (I,1069); COUNTRY NIGHT OF STARS (II,592); GLEN CAMPBELL AND FRIENDS: THE SILVER ANNIVERSARY (II,1005); I BELIEVE IN MUSIC (I,2165); A JOHNNY CASH CHRISTMAS (II,1328); JOHNNY CASH: THE FIRST 25 YEARS (II,1336); MAC DAVIS 10TH ANNIVERSARY SPECIAL: I STILL BELIEVE IN MUSIC (II,1571); PERRY COMO'S CHRISTMAS IN NEW MEXICO (II,2010); PERRY COMO'S LAKE TAHOE HOLIDAY (II,2018); A SPECIAL ANNE MURRAY CHRISTMAS (II,2413); A SPECIAL EDDIE RABBITT (II,2415)

MURRAY, ARTHUR
ANNIE, THE WOMAN IN THE LIFE OF A MAN (I,226); THE ARTHUR MURRAY PARTY FOR BOB HOPE (I,292); ARTHUR MURRAY'S DANCE PARTY (I,293)

MURRAY, BETTE LOU
WHERE WERE YOU? (I,4834)

MURRAY, BILL
RODNEY DANGERFIELD SPECIAL: IT'S NOT EASY BEIN' ME (II,2199); SATURDAY NIGHT LIVE (II,2261); STEVE MARTIN'S BEST SHOW EVER (II,2460); THINGS WE DID LAST SUMMER (II,2589); TWILIGHT THEATER (II,2685)

MURRAY, BILLY
THE ROCK FOLLIES (II,2192)

MURRAY, CHERYL
CORONATION STREET (I,1054)

MURRAY, DON
BRANAGAN AND MAPES (II,367); HEDDA HOPPER'S HOLLYWOOD (I,2002); HOW THE WEST WAS WON (II,1196); KNOTS LANDING (II,1404); LOVE STORY (I,2778); THE OUTCASTS (I,3425); WINTERSET (I,4882)

MURRAY, JAN
BANJO HACKETT: ROAMIN' FREE (II,141); BLIND DATE (I,482); THE BOB HOPE SHOW (I,659); CHAIN LETTER (I,892); CHARGE ACCOUNT (I,903); DOLLAR A SECOND (I,1326); FUN FOR 51 (I,1707); GO LUCKY (I,1832); THE JAN MURRAY SHOW (I,2335); THE JERRY LEWIS SHOW (I,2362); MEET YOUR MATCH (I,2993); THE PRACTICE (II,2076); SHOW BUSINESS SALUTE TO MILTON BERLE (I,4029); SHOW OF THE YEAR (I,4031); SING IT AGAIN (I,4059); SONGS FOR SALE (I,4124); TREASURE HUNT (I,4600)

MURRAY, JEANNE
THE WOMEN (I,4890)

MURRAY, KATHLEEN
THE DOCTORS (II,694); KITTY FOYLE (I,2571); YOUNG DR. MALONE (I,4938)

MURRAY, KATHRYN
THE ARTHUR MURRAY PARTY FOR BOB HOPE (I,292); ARTHUR MURRAY'S DANCE PARTY (I,293)

MURRAY, KEN
THE BOB HOPE SHOW (I,548); FUN FOR 51 (I,1707); HOLLYWOOD'S PRIVATE HOME MOVIES (II,1167); HOLLYWOOD'S PRIVATE HOME MOVIES II (II,1168); THE KEN MURRAY SHOW (I,2525); WHERE WERE YOU? (I,4834)

MURRAY, MARK
LEAVE IT TO BEAVER (I,2648)

MURRAY, MARY
ONE LIFE TO LIVE (II,1907)

MURRAY, MIKE
BAFFLED (I,332)

MURRAY, PAT
THE GLORIA SWANSON HOUR (I,1827)

MURRAY, PEG
LOVE OF LIFE (I,2774); MURDER INK (II,1769)

MURRILL, CHRISTINA
MATT HOUSTON (II,1654)

MURROW, EDWARD R
THE FORD 50TH ANNIVERSARY SHOW (I,1630); OPEN HOUSE (I,3394); PERSON TO PERSON (I,3550)

MURTAUGH, JAMES
CASINO (II,452); FULL HOUSE (II,937); IN SECURITY (II,1227); NUMBER 96 (II,1869); PLEASURE COVE (II,2058); THE ROLLERGIRLS (II,2203)

MURTAUGH, KATE
IT'S A MAN'S WORLD (I,2258); MCNAMARA'S BAND (II,1669); NICHOLS AND DYMES (II,1841)

MURTON, LIONEL
O.S.S. (I,3411)

MUSANTE, TONY
THE 13TH DAY: THE STORY OF ESTHER (II,2593); NOWHERE TO HIDE (II,1868); TOMA (I,4547); TOMA (II,2634)

MUSBURGER, BRENT
CELEBRITY CHALLENGE OF THE SEXES 2 (II,466)

MUSCAT, ANGELO
THE PRISONER (I,3670)

MUSE, CLARENCE
CASABLANCA (I,860)

MUSHIN, JULES
KISS ME, KATE (I,2569); SHIRLEY TEMPLE'S STORYBOOK: THE LEGEND OF SLEEPY HOLLOW (I,4018)

MUSIC, HENRIETTA
THE LORENZO AND HENRIETTA MUSIC SHOW (II,1519); THE NEW LORENZO MUSIC SHOW (II,1821)

MUSIC, LORENZO
CARLTON YOUR DOORMAN (II,440); THE LORENZO AND HENRIETTA MUSIC SHOW (II,1519); THE NEW LORENZO MUSIC SHOW (II,1821); RHODA (II,2151)

MUSICAL YOUTH
A HOT SUMMER NIGHT WITH DONNA (II,1190)

MUSTIN, BURT
ALL IN THE FAMILY (II,38); THE CODE OF JONATHAN WEST (I,992); DATE WITH THE ANGELS (I,1164); THE FUNNY SIDE (I,1714); THE FUNNY SIDE (I,1715); ICHABOD AND ME (I,2183); LEAVE IT TO BEAVER (I,2648); LOVE NEST (II,1541); THE OVER-THE-HILL GANG RIDES AGAIN (I,3432); PHYLLIS (II,2038); THE PIRATES OF FLOUNDER BAY (I,3607)

MUSZYNSKI, JAN
HIGH HOPES (II,1144)

MUZURKI, MIKE
TOO MANY SERGEANTS (I,4573)

MWENESI, STEVE
THE FLAME TREES OF THIKA (II,870)

MYER RAPPOPORT CHORUS, THE
DON'S MUSICAL PLAYHOUSE (I,1348)

MYERS, BARBARA
FIRST LOVE (I,1583); WOMAN WITH A PAST (I,4889)

MYERS, CARMEL
THE CARMEL MYERS SHOW (I,844); NICK AND NORA (II,1842)

MYERS, CEDRIC
RETURN TO EDEN (II,2143)

MYERS, FRAN
THE GUIDING LIGHT (II,1064)

MYERS, GARRY
U.F.O. (I,4662)

MYERS, JEFF
OPRYLAND: NIGHT OF STARS AND FUTURE STARS (II,1921)

MYERS, MARSHA
LOOK WHAT THEY'VE DONE TO MY SONG (II,1517)

MYERS, NANCY
THE NEW PRICE IS RIGHT (I,3270); THE PRICE IS RIGHT (II,2079); QUEEN FOR A DAY (I,3706)

MYERS, PAMELA
AMERICA, YOU'RE ON (II,69); THE MISADVENTURES OF SHERIFF LOBO (II,1704); SHA NA NA (II,2311)

MYERS, PAULA
THE BILLY CRYSTAL COMEDY HOUR (II,248)

MYERS, PAULINE
DAYS OF OUR LIVES (II,629); EXECUTIVE SUITE (II,796); GOOD TIMES (II,1038)

MYERS, RUSSELL
THE FANTASTIC FUNNIES (II,825)

MYERS, STEPHEN
THE UGILY FAMILY (II,2699)

MYERS, SUSAN
JAMES AT 15 (II,1270); RIDING FOR THE PONY EXPRESS (II,2172)

MYERSON, BESS
THE BIG PAYOFF (I,428); THE BOB HOPE SHOW (I,525); CANDID CAMERA (I,819); JACQUES FRAY'S MUSIC ROOM (I,2329)

MYERSON, JESSICA
THE SMOTHERS BROTHERS COMEDY HOUR (I,4095); THICKER THAN WATER (I,4445)

MYHERS, JOHN
THE DOOLEY BROTHERS (II,715); THE HECTOR HEATHCOTE SHOW (I,2000); HELLO, LARRY (II,1128)

MYLES, ALBERT
THE CONTENDER (II,571)

MYLES, KEN
THE GREAT SPACE COASTER (II,1059)

MYLES, LEIGHT
THE BENNY HILL SHOW (II,207)

MYLES, MEG
THE EDGE OF NIGHT (II,760)

MYLROLE, KATHRYN
ANYTHING GOES (I,237)

MYREN, TANIA
STARSTRUCK (II,2446)

NABORS KIDS, THE
THE JIM NABORS HOUR (I,2381)

NABORS, JIM
. . .AND DEBBIE MAKES SIX (I,185); THE ALL-STAR SALUTE TO MOTHER'S DAY (II,51); THE ANDY GRIFFITH SHOW (I,192); THE ANDY GRIFFITH—DON KNOTTS—JIMNABORS SHOW (I,191); THE BEST LITTLE SPECIAL IN TEXAS (II,215); THE BOB HOPE SHOW (I,652); BURT REYNOLDS' LATE SHOW (I,777); CBS: ON THE AIR (II,460); CELEBRITY DAREDEVILS (II,473); FRIENDS AND NABORS (I,1684); GIRL FRIENDS AND NABORS (I,1807); GOMER PYLE, U.S.M.C. (I,1843); THE JIM NABORS HOUR (I,2381); THE JIM NABORS SHOW (II,1290); THE LOST SAUCER (II,1524)

NADAS, BETSY
MR. ROGERS' NEIGHBORHOOD (I,3150)

NADER, GEORGE
THE FURTHER ADVENTURES OF ELLERY QUEEN (I,1719); MAN AND THE CHALLENGE (I,2855); NAKIA (II,1784); SHANNON (I,3998); STAGE 7 (I,4181)

NADER, MICHAEL
DYNASTY (II,746); GIDGET (I,1795)

NADIR, ROBERT
A CHRISTMAS CAROL (II,517)

NADLER, REGGIS
THE FREEWHEELERS (I,1683)

NAFA, IHAB
BARRIER REEF (I,362)

NAGAZUMI, YASUKO
THE PROTECTORS (I,3681)

NAGEL, CONRAD
BROADWAY TO HOLLYWOOD (I,739); CELEBRITY TIME (I,887); SILVER THEATER (I,4051)

NAGLE, JERI
JACQUES FRAY'S MUSIC ROOM (I,2329)

NAGY, BILL
ADVENTURES OF CLINT AND MAC (I,55)

NAGY, JOEY
DENNIS THE MENACE: MAYDAY FOR MOTHER (II,660)

NAIL, JIMMY
MASTER OF THE GAME (II,1647)

NAIL, JOANNE
MOTHER, JUGGS AND SPEED (II,1741)

NAISH, J CARROL
THE BOB HOPE SHOW (I,623); DOUGLAS FAIRBANKS JR. PRESENTS THE RHEINGOLD THEATER (I,1364); FOR LOVE OR $$$ (I,1623); GUESTWARD HO! (I,1900); LIFE WITH LUIGI (I,2692); THE NEW ADVENTURES OF CHARLIE CHAN (I,3249)

NAISMITH, LAURENCE
THE ADMIRABLE CRICHTON (I,36); THE FILE ON DEVLIN (I,1575); THE PERSUADERS! (I,3556)

NAJEE-ULLAH, MANSOOR
THE FIRM (II,860)

NAKAHARA, KELLYE
M*A*S*H (II,1569)

NAKAHAWA, KERRY
THE RETURN OF MARCUS WELBY, M.D (II,2138)

NAKOPOLOU, ASPA
THE THORN BIRDS (II,2600)

NAMATH, JOE
ALL-AMERICAN PIE (II,48); BOB HOPE SPECIAL: BOB HOPE ON CAMPUS (II,290); BONNIE AND THE FRANKLINS (II,349); THE JOE NAMATH SHOW (I,2400); THE MAD MAD MAD MAD WORLD OF THE SUPER BOWL (II,1586); MARRIAGE IS ALIVE AND WELL (II,1633); THE NASHVILLE PALACE (II,1792); PLIMPTON! SHOWDOWN AT RIO LOBO (I,3630); SUPER COMEDY BOWL 1 (I,4288); TEXACO STAR THEATER: OPENING NIGHT (II,2565); THE WAVERLY WONDERS (II,2746)

NAN SCHWARTZ ORCHESTRA, THE
100 YEARS OF AMERICA'S POPULAR MUSIC (II,1904)

NANASI, ANNA MARIA
DOBIE GILLIS (I,1302); THE GEORGE BURNS AND GRACIE ALLEN SHOW (I,1763)

NANNOS, J CRAIG

GEORGE WASHINGTON (II,978)

NAPIER, ALAN
THE BARBARA RUSH SHOW (I,349); BATMAN (I,366); DON'T CALL ME CHARLIE (I,1349); JOE DANCER: THE MONKEY MISSION (II,1301); PANIC! (I,3447); YOUR SHOW TIME (I,4958)

NAPIER, CHARLES
THE A-TEAM (II,119); B.J. AND THE BEAR (II,125); BIG BOB JOHNSON AND HIS FANTASTIC SPEED CIRCUS (II,231); THE BLUE AND THE GRAY (II,273); THE OREGON TRAIL (II,1923); THE OUTLAWS (II,1936); RANSOM FOR ALICE (II,2113)

NAPIER, FRANCIS
NANCY ASTOR (II,1788)

NAPIER, HUGO
AS THE WORLD TURNS (II,110)

NAPIER, JOHN
BEN JERROD (I,396)

NAPIER, PAUL
DYNASTY (II,746); HART TO HART (II,1101)

NAPIER, RUSSELL
SCOTLAND YARD (I,3941)

NAPOLEON, MARTY
TIMEX ALL-STAR JAZZ SHOW II (I,4516)

NARANJO, IVAN
CENTENNIAL (II,477); HOW THE WEST WAS WON (II,1196); THE TARZAN/LONE RANGER/ZORRO ADVENTURE HOUR (II,2543)

NARDINI, NICOLE
20 MINUTE WORKOUT (II,2680)

NARDINI, TOM
CAT BALLOU (I,869); COWBOY IN AFRICA (I,1085)

NARITA, RICHARD
DON'T CALL US (II,709); EXO-MAN (II,797); FOUL PLAY (II,910)

NARIZZANO, DINO
THE DOCTORS (II,694); HOW TO SURVIVE A MARRIAGE (II,1198); SEARCH FOR TOMORROW (II,2284)

NARZ, JACK
BEAT THE CLOCK (I,378); DOTTO (I,1357); I'LL BET (I,2189); LIFE WITH ELIZABETH (I,2689); NOW YOU SEE IT (II,1867); SEVEN KEYS (I,3985); VIDEO VILLAGE (I,4731)

NASH, ANTHONY
MURDER ON THE MIDNIGHT EXPRESS (I,3162)

NASH, BRIAN
MICKEY (I,3023); PLEASE DON'T EAT THE DAISIES (I,3628)

NASH, CARROLL B
E.S.P. (I,1462)

NASH, CHRIS
THE BEST OF TIMES (II,220)

NASH, JOHNNY
ARTHUR GODFREY AND FRIENDS (I,278); THE ARTHUR GODFREY SHOW (I,283); THE ARTHUR GODFREY SHOW (I,284); THE ARTHUR GODFREY SHOW (I,286)

NASH, JOSEPH
FLASH GORDON (I,1593)

NASH, ROBERT
ANNIE GET YOUR GUN (I,223); I WAS A BLOODHOUND (I,2180)

NASHVILLE ADDITION, THE
HEE HAW (II,1123)

NASHVILLE BRASS, THE
I BELIEVE IN MUSIC (I,2165)

NASHVILLE STRINGS, THE
THE NASHVILLE SOUND OF BOOTS RANDOLPH (I,3229)

NASTAGIA, FRANK
THE MADHOUSE BRIGADE (II,1588); THE SOUPY SALES SHOW (I,4137)

NATALI, DINO
THE HARVEY KORMAN SHOW (II,1103)

NATHAN, MARC
THE UNCLE FLOYD SHOW (II,2703)

NATHAN, STEPHEN
BUSTING LOOSE (II,402); THE CHADWICK FAMILY (II,478); THE WONDERFUL WORLD OF PHILIP MALLEY (II,2831)

NATOLI, SARAH
FISH (II,864)

NATSUKI, YOSUKE
SHOGUN (II,2339)

NATWICK, MILDRED
ADDIE AND THE KING OF HEARTS (I,35); ALICE (II,33); ARSENIC AND OLD LACE (I,269); BLITHE SPIRIT (I,484); CAMEO THEATER (I,808); THE CLOCK (I,978); THE DOCTOR (I,1320); THE EASTER PROMISE (I,1397); FEMALE INSTINCT (I,1564); HARDCASTLE AND MCCORMICK (II,1089); THE HOUSE WITHOUT A CHRISTMAS TREE (I,2143); LITTLE WOMEN (II,1498); THE LORETTA YOUNG THEATER (I,2756); LOVE STORY (I,2776); MAGNAVOX THEATER (I,2827); MCMILLAN AND WIFE (II,1667); PULITZER PRIZE PLAYHOUSE (I,3692); THE SNOOP SISTERS (II,2389); STARLIGHT THEATER (I,4201); THE THANKSGIVING TREASURE (I,4416); YOU CAN'T TAKE IT WITH YOU (II,2857)

NATWICK, MYRON
DAYS OF OUR LIVES (II,629); LITTLE VIC (II,1496)

NAUD, LILLIAN
HAGGIS BAGGIS (I,1917)

NAUD, MELINDA
THE 36 MOST BEAUTIFUL GIRLS IN TEXAS (II,2594); THE CRACKER BROTHERS (II,599); DETECTIVE SCHOOL (II,661); FANTASY ISLAND (II,829); THE NEW OPERATION PETTICOAT (II,1826); NIGHTSIDE (II,1851); OPERATION PETTICOAT (II,1916); OPERATION PETTICOAT (II,1917)

NAUGHTON, DAVID
AT EASE (II,115); MAKIN' IT (II,1603)

NAUGHTON, JAMES
BEULAH LAND (II,226); FARADAY AND COMPANY (II,833); MAKING THE GRADE (II,1606); PAROLE (II,1960); PLANET OF THE APES (II,2049); TRAUMA CENTER (II,2655); WHO'S THE BOSS? (II,2792)

NAVAROO, CHI CHI
BROADWAY GOES LATIN (I,734)

NAVARRO, ANNA
CUTTER (I,1117)

NAVIN JR, JOHN P
THE ACADEMY (II,3); THE ACADEMY II (II,4); JENNIFER SLEPT HERE (II,1278)

NAVIN-SMITH, JENNIFER
RETURN TO EDEN (II,2143)

NAVRATILOVA, MARTINA
CELEBRITY CHALLENGE OF THE SEXES 4 (II,468)

NAYLOR, JAYSON
NEVER SAY NEVER (II,1809); NOT IN FRONT OF THE KIDS (II,1861)

NAYLOR, JERRY
MUSIC CITY U.S.A. (I,3167)

NEAL, DAVID
KENNEDY (II,1377)

NEAL, DONALD
GEORGE WASHINGTON (II,978)

NEAL, ELSIE
MR. ROGERS' NEIGHBORHOOD (I,3150)

NEAL, PATRICIA
THE BASTARD/KENT FAMILY CHRONICLES (II,159); ERIC (I,1451); STUDIO ONE (I,4268); THE WALTONS (II,2740); THE WAY THEY WERE (II,2748)

NEAL, TOM
A TIME TO LIVE (I,4510)

NEALY, MALCOLM
MY LITTLE MARGIE (I,3194)

NEAME, CHRISTOPHER
SECRET ARMY (II,2290)

NEAR, HOLLY
MR. AND MRS. COP (II,1748)

NEATHERLAND, TOM
THE LAWRENCE WELK SHOW (I,2643); MEMORIES WITH LAWRENCE WELK (II,1681)

NECKELS, BRUCE
CONDOMINIUM (II,566)

NEDWELL, ROBIN
DOCTOR IN THE HOUSE (I,1313)

NEEDHAM, CONNIE
CELEBRITY CHALLENGE OF THE SEXES 5 (II,469); EIGHT IS ENOUGH (II,762); FAME (II,812)

NEEDHAM, HAL
EGAN (I,1419); STUNTS UNLIMITED (II,2487)

NEEDLE, KAREN
THE EDGE OF NIGHT (II,760)

NEELEY, TED
MCLAREN'S RIDERS (II,1663)

NEELY, MARK
CENTENNIAL (II,477); THE SECRET WAR OF JACKIE'S GIRLS (II,2294)

NEELY, TED
THE BUDDY GRECO SHOW (I,754)

NEENAN, AUDRIE J
BIG CITY COMEDY (II,233); THE COMEDY ZONE (II,559); NOT NECESSARILY THE NEWS (II,1862); NOT NECESSARILY THE NEWS (II,1863)

NEESON, LIAM
ELLIS ISLAND (II,768); A WOMAN OF SUBSTANCE (II,2820)

NEFF, ELSIE
THE LONELY WIZARD (I,2743)

NEGAZUMI, YASUKO
SPACE: 1999 (II,2409)

NEGIN, LOUIS
RED SKELTON'S CHRISTMAS DINNER (II,2123)

NEGRI, JOE
MR. ROGERS' NEIGHBORHOOD (I,3150)

NEGRON, TAYLOR
DETECTIVE SCHOOL (II,661)

NEHER, SUSAN
GETTING TOGETHER (I,1784); HAPPY DAYS (II,1084); TO ROME WITH LOVE (I,4526)

NEIDHARDT, ELKE
SKIPPY, THE BUSH KANGAROO (I,4076)

NEIL, DIANE THOMPSON
TEXAS (II,2566)

NEIL, GLORIA
THE LIVELY ONES (I,2729)

NEIL, MILT
THE NEW HOWDY DOODY SHOW (II,1818)

NEILL, BOB
THE MAN WITH THE POWER (II,1616); STUDS LONIGAN (II,2484)

NEILL, NOEL
THE ADVENTURES OF SUPERMAN (I,77)

NEILL, SAM
REILLY, ACE OF SPIES (II,2129)

NEILS, SYLVIA
THE NEIGHBORS (II,1801)

NEILSON, INGA
TOO CLOSE FOR COMFORT (II,2642)

NEILSON, JOHN
SOUTHERN FRIED (I,4139)

NEISE, GEORGE
JOHNNY NIGHTHAWK (I,2432); WICHITA TOWN (I,4857)

NELKIN, STACEY
THE ADVENTURES OF POLLYANNA (II,14); THE CHISHOLMS (II,512); THE JERK, TOO (II,1282); THE LAST CONVERTIBLE (II,1435); T.L.C. (II,2525)

NELLE FISHER DANCERS, THE
ALL ABOUT MUSIC (I,127)

NELLIGAN, KATE
THE ONEDIN LINE (II,1912); THERESE RAQUIN (II,2584)

NELLOR, BOB
NIGHT CLUB (I,3284)

NELSON, ANN
FAME (II,812)

NELSON, BARRY
BEN HECHT'S TALES OF THE CITY (I,395); THE CHEVROLET TELE-THEATER (I,926); CIRCLE OF FEAR (I,958); CLIMAX! (I,976); THE DAVID NIVEN THEATER (I,1178); DEATH IN SMALL DOSES (I,1207); THE DUPONT SHOW OF THE WEEK (I,1388); FOOLS, FEMALES AND FUN: IS THERE A DOCTOR IN THE HOUSE? (II,900); HEAVEN HELP US (I,1995); THE HUNTER (I,2162); MASON (II,1643); MY FAVORITE HUSBAND (I,3188); MY WIVES JANE (I,3207); PULITZER PRIZE PLAYHOUSE (I,3692); STARLIGHT THEATER (I,4201); THERE'S ALWAYS ROOM (II,2583); WASHINGTON: BEHIND CLOSED DOORS (II,2744)

NELSON, BEK
LAWMAN (I,2642); PEYTON PLACE (I,3574)

NELSON, CHRIS
SENIOR YEAR (II,2301); SONS AND DAUGHTERS (II,2404); WHERE'S RAYMOND? (I,4837)

NELSON, CHRISTOPHER S
CO-ED FEVER (II,547); GENERAL HOSPITAL (II,964)

NELSON, CONNIE
JOHNNY CASH: CHRISTMAS ON THE ROAD (II,1333)

NELSON, CRAIG RICHARD
BACHELOR AT LAW (I,322); CAROL BURNETT AND COMPANY (II,441); PAUL SAND IN FRIENDS AND LOVERS (II,1982); STICK AROUND (II,2464)

NELSON, CRAIG T
CALL TO GLORY (II,413); CHICAGO STORY (II,506); THE CHICAGO STORY (II,507); PAPER DOLLS (II,1951)

NELSON, DANNY
CHIEFS (II,509)

NELSON, DAVID
THE ADVENTURES OF OZZIE AND HARRIET (I,71); CIRCUS OF THE STARS (II,530); CIRCUS OF THE STARS (II,531); CIRCUS OF THE STARS (II,532); CIRCUS OF THE STARS (II,533); CIRCUS OF THE STARS (II,534); CIRCUS OF THE STARS (II,535); CIRCUS OF THE STARS (II,536); HIGH SCHOOL, U.S.A. (II,1148); SWING OUT, SWEET LAND (II,2515)

NELSON, ED
ALCOA PREMIERE (I,109); BANACEK: DETOUR TO NOWHERE (I,339); CANNON: THE RETURN OF FRANK CANNON (II,425); CAPITOL (II,426); CLINIC ON 18TH STREET (II,544); DOCTORS' PRIVATE LIVES (II,692); DOCTORS' PRIVATE LIVES (II,693); ESCAPE (I,1461); THE GIRL, THE GOLD WATCH AND EVERYTHING (II,1002); THE MAN FROM GALVESTON (I,2865); THE MORNING SHOW (I,3103); PEYTON PLACE (I,3574); THE SILENT FORCE (I,4047); TENAFLY (I,4395); THRILLER (I,4492)

NELSON, FELIX
THE BEST OF TIMES (II,220)

NELSON, FRANK
THE ALL-NEW POPEYE HOUR (II,49); DOROTHY IN THE LAND OF OZ (II,722); I LOVE LUCY (I,2171); THE JACK BENNY PROGRAM (I,2294); JACK BENNY'S 20TH ANNIVERSARY TV SPECIAL (I,2303); THE ODDBALL COUPLE (II,1876)

NELSON, GAYE
OZZIE'S GIRLS (I,3440)

NELSON, GENE
BROADWAY (I,733); THE KAISER ALUMINUM HOUR (I,2503); MARRIED ALIVE (I,2923); SHANGRI-LA (I,3996); TOM, DICK, AND HARRY (I,4536)

NELSON, GUNNAR
SAM (II,2246)

NELSON, HARRIET
THE ADVENTURES OF OZZIE AND HARRIET (I,71); A CHRISTMAS FOR BOOMER (II,518); HIGH SCHOOL, U.S.A. (II,1149); OZZIE'S GIRLS (I,3439); OZZIE'S GIRLS (I,3440)

NELSON, HARRY
HAPPY BIRTHDAY (I,1941)

NELSON, HAYWOOD
GRADY (II,1047); THAT TEEN SHOW (II,2573); WHAT'S HAPPENING!! (II,2769)

NELSON, HELAINE
NAKIA (II,1784)

NELSON, HERBERT
DAYS OF OUR LIVES (II,629); FUTURE COP (II,945); FUTURE COP (II,946); THE GUIDING LIGHT (II,1064); THE PATRIOTS (I,3477)

NELSON, JANE
DEALER'S CHOICE (II,635); THE DIAMOND HEAD GAME (II,666); THE FUN FACTORY (II,938); GIVE-N-TAKE (II,1003); THE NEIGHBORS (II,1801); THE NEW TREASURE HUNT (I,3275); THE NEW TREASURE HUNT (II,1831); THREE FOR THE MONEY (II,2606)

NELSON, JERRY
FRAGGLE ROCK (II,916); THE MUPPET SHOW (II,1766)

NELSON, JESSICA
ARCHIE BUNKER'S PLACE (II,105)

NELSON, JIMMY
BANK ON THE STARS (I,343); THE KATE SMITH HOUR (I,2508); QUICK AS A FLASH (I,3709)

NELSON, JOHN
BRIDE AND GROOM (I,721); LIVE LIKE A MILLIONAIRE (I,2728)

NELSON, JOHN ALLEN
SANTA BARBARA (II,2258)

NELSON, KENNETH
THE ALDRICH FAMILY (I,111); CAPTAIN VIDEO AND HIS VIDEO RANGERS (I,838); HOORAY FOR LOVE (I,2115); LACE (II,1418)

NELSON, KRISTIN
ADAM-12 (I,31); SAM (II,2246)

NELSON, LORI
HOW TO MARRY A MILLIONAIRE (I,2147); THE PIED PIPER OF HAMELIN (I,3595)

NELSON, LOU
WINDOW SHADE REVUE (I,4874)

NELSON, MAGGIE
ELLERY QUEEN (II,766)

NELSON, MATTHEW
SAM (II,2246)

NELSON, NAN-LYNN
HOT HERO SANDWICH (II,1186)

NELSON, NOBLE
DOCTOR DAN (II,684)

NELSON, NOVELLA
CHIEFS (II,509)

NELSON, OZZIE
THE ADVENTURES OF OZZIE AND HARRIET (I,71); JANE WYMAN PRESENTS THE FIRESIDE THEATER (I,2345); OZZIE'S GIRLS (I,3439); OZZIE'S GIRLS (I,3440)

NELSON, PETER
CELEBRITY (II,463); V (II,2713); V: THE FINAL BATTLE (II,2714)

NELSON, PORTIA
THE HARDY BOYS (I,1948); RAGE OF ANGELS (II,2107)

NELSON, RICKY
THE ADVENTURES OF OZZIE AND HARRIET (I,71); THE BOBBIE GENTRY SPECIAL (I,670); FIVE STARS IN SPRINGTIME (I,1590); FOL-DE-ROL (I,1613); HIGH SCHOOL, U.S.A. (II,1149); MALIBU U. (I,2849); MEN WHO RATE A "10" (II,1684); ON THE FLIP SIDE (I,3360); SWING OUT, SWEET LAND (II,2515)

NELSON, TERRY
VANITY FAIR (I,4701)

NELSON, TRACY
THE ADVENTURES OF OZZIE AND HARRIET (I,71); BATTLE OF THE NETWORK STARS (II,176); CIRCUS OF THE STARS (II,538); GLITTER (II,1009); SQUARE PEGS (II,2431)

NELSON, WILLIE
BOB & RAY & JANE, LARAINE & GILDA (II,279); BOB HOPE SPECIAL: BOB HOPE'S PINK PANTHER THANKSGIVING GALA (II,312); GLEN CAMPBELL AND FRIENDS: THE SILVER ANNIVERSARY (II,1005); JOHNNY CASH: CHRISTMAS ON THE ROAD (II,1333)

NEMES, SCOTT
FULL HOUSE (II,937); THE INVISIBLE WOMAN (II,1244)

NEMETZ, LENORE
THE LIFE AND TIMES OF EDDIE ROBERTS (II,1468)

NEPHEW, RICHARD
GEORGE WASHINGTON (II,978)

NERO, FRANCO
THE LAST DAYS OF POMPEII (II,1436)

NERO, PETER
CLASS OF '67 (I,970); THE DINAH SHORE SPECIAL (I,1283); GEORGE GOBEL PRESENTS (I,1766); JACK LEMMON IN 'S WONDERFUL, 'S MARVELOUS, 'S GERSHWIN (I,2313)

NERO, TONI
BILLY CRYSTAL: A COMIC'S LINE (II,249)

NERVIK, LEIG
ANDERSON AND COMPANY (I,188)

NESBITT, BERYL
THE CITADEL (II,539)

NESBITT, CATHLEEN
ABC'S MATINEE TODAY (I,7); THE FARMER'S DAUGHTER (I,1533); FEATHERTOP (I,1556); NOTORIOUS (I,3314); REUNION IN VIENNA (I,3780); UPSTAIRS, DOWNSTAIRS (II,2709); VANITY FAIR (I,4700); THE WOMEN (I,4890)

NESBITT, JOHN
THE PASSING PARADE (I,3466); STAR FOR TODAY (I,4192); TELEPHONE TIME (I,4379)

NESBITT, SUSAN
EASY DOES IT. . . STARRING FRANKIE AVALON (II,754)

NESMITH, MIKE
33 1/3 REVOLUTIONS PER MONKEE (I,4451); THE MONKEES (I,3088)

NETHERTON, RADNA
THE LAWRENCE WELK SHOW (I,2644); MEMORIES WITH LAWRENCE WELK (II,1681)

NETHERY, MIRIAM-BYRD
THE DEADLY GAME (II,633); THE DUKES OF HAZZARD (II,742); LOCAL 306 (II,1505)

NETTLES, JOHN
BERGERAC (II,209)

NETTLETON, JOHN
BRIDESHEAD REVISITED (II,375); THE CITADEL (II,539)

NETTLETON, LOIS
ACCIDENTAL FAMILY (I,17); ALL THAT GLITTERS (II,42); THE ARMSTRONG CIRCLE THEATER (I,260); THE BRIGHTER DAY (I,728); CENTENNIAL (II,477); CHAIN REACTION (II,479); MEET ME IN ST. LOUIS (I,2986); NIGHT GALLERY (I,3287); THE TWILIGHT ZONE (I,4651); WASHINGTON: BEHIND CLOSED DOORS (II,2744); THE WOMAN IN WHITE (I,4887)

NETTO
THE UNCLE FLOYD SHOW (II,2703)

NEUCHATEAU, CORINNE
LOVE OF LIFE (I,2774)

NEUFORD, BETH
CIRCUS OF THE STARS (II,534)

NEUHAUS, LACEY
FROM HERE TO ETERNITY (II,933)

NEUMAN, DOROTHY
HANK (I,1931); MR. BEVIS (I,3128)

NEUMANN, DOROTHY
FIBBER MCGEE AND MOLLY (I,1569); MONA MCCLUSKEY (I,3085)

NEUMANN, JENNY
EMPIRE (II,777); KING OF THE ROAD (II,1396)

NEUN, MIKE
DINAH AND HER NEW BEST FRIENDS (II,676)

NEVARD, BILLY
WESLEY (I,4795)

NEVE, SUZANNE
THE FORSYTE SAGA (I,1645); NOT GUILTY! (I,3311); TERROR FROM WITHIN (I,4404)

NEVIL, STEVE
THE BLUE AND THE GRAY (II,273); THE MCLEAN STEVENSON SHOW (II,1665)

NEVILLE, JOHN
ROMEO AND JULIET (I,3837)

NEVILLE, TAMZIN
LOVE IN A COLD CLIMATE (II,1539)

NEVINS, APRIL
THE CAROL BURNETT SHOW (II,443)

NEVINS, CLAUDETTE
ELECTRA WOMAN AND DYNA GIRL (II,764); FAMILY (II,813); HEADMASTER (I,1990); HUSBANDS AND WIVES (II,1208); HUSBANDS, WIVES AND LOVERS (II,1209); KNOTS LANDING (II,1404); MARRIED: THE FIRST YEAR (II,1634); ONE DAY AT A TIME (II,1900); THE POSSESSED (II,2071); RETURN TO THE PLANET OF THE APES (II,2145); THREE'S COMPANY (II,2614)

NEVINS, NATALIE
THE LAWRENCE WELK SHOW (I,2643); MEMORIES WITH LAWRENCE WELK (II,1681); THAT MCMAHON'S HERE AGAIN (I,4419)

NEVINS, SHEILA
BRAIN GAMES (II,366)

NEW CHRISTY MINSTRELS, THE
THE ANDY WILLIAMS SHOW (I,205); THE JULIE ANDREWS SHOW (I,2481); THE JULIE ANDREWS SPECIAL (I,2482); MARINELAND CARNIVAL (I,2913); SOMETHING SPECIAL (I,4118); THE STARS AND STRIPES SHOW (I,4206); THE VIKKI CARR SHOW (I,4732)

NEW DOODLETOWN PIPERS, THE
MAKE YOUR OWN KIND OF MUSIC (I,2847)

NEW FACES, THE
THE DES O'CONNOR SHOW (I,1235)

NEW LIFE SINGERS
BARBARA MANDRELL—THE LADY IS A CHAMP (II,144)

NEW, NANCY
STICK AROUND (II,2464)

NEWAY, PATRICIA
CAROUSEL (I,856); GOLDEN CHILD (I,1839)

NEWCOMBE, HERB
ADVENTURES IN DAIRYLAND (I,44)

NEWELL, JAMES
RENFREW OF THE ROYAL MOUNTED (I,3769)

NEWELL, KAREN
NIGHTSIDE (II,1851)

NEWELL, PATRICK
THE AVENGERS (II,121); THE BENNY HILL SHOW (II,207); CASANOVA (II,451); MOLL FLANDERS (II,1720)

NEWHAM, MARY
JANE EYRE (I,2337)

NEWHART, BOB
THE BOB NEWHART SHOW (I,666); THE BOB NEWHART SHOW (II,342); CIRCUS OF THE STARS (II,535); CIRCUS OF THE STARS (II,536); THE COLGATE COMEDY HOUR (I,998); DEAN MARTIN'S COMEDY CLASSICS (II,646); THE DON RICKLES SHOW

(II,708); DON RICKLES—ALIVE AND KICKING (I,1340); ED MCMAHON AND HIS FRIENDS. . . DISCOVER WET AT CYPRESS GARDENS (I,1400); THE ENTERTAINERS (I,1449); A FUNNY THING HAPPENED ON THE WAY TO HOLLYWOOD (I,1716); THE JACK PAAR SPECIAL (I,2319); LADIES AND GENTLEMAN. . . BOB NEWHART (II,1420); LADIES AND GENTLEMAN. . . BOB NEWHART, PART II (II,1421); A LAST LAUGH AT THE 60'S (I,2627); NEWHART (II,1835); THE PERRY COMO CHRISTMAS SHOW (I,3526); THE PERRY COMO THANKSGIVING SPECIAL (I,3540); PERRY COMO'S SPRINGTIME SPECIAL (II,2023); THE ROWAN AND MARTIN SPECIAL (I,3855); A SALUTE TO STAN LAUREL (I,3892)

NEWKIRK, DEREK
MASADA (II,1642)

NEWLAN, PAUL
DAVY CROCKETT (I,1181); M SQUAD (I,2799)

NEWLAND, JOHN
THE ARMSTRONG CIRCLE THEATER (I,260); THE CLOCK (I,978); EYE WITNESS (I,1494); THE INNER SANCTUM (I,2216); LIGHT'S OUT (I,2699); THE LORETTA YOUNG THEATER (I,2756); MONTGOMERY'S SUMMER STOCK (I,3096); THE NEXT STEP BEYOND (II,1839); ONE MAN'S FAMILY (I,3383); ONE STEP BEYOND (I,3388); PHILCO TELEVISION PLAYHOUSE (I,3583); SURE AS FATE (I,4297); TALES OF TOMORROW (I,4352); THE WEB (I,4784)

NEWLANDS, ANTHONY
THE ADVENTURES OF THE SCARLET PIMPERNEL (I,80); CRIMES OF PASSION (II,603); MURDER IS A ONE-ACT PLAY (I,3160)

NEWLEY, ANTHONY
THE ANTHONY NEWLEY SHOW (I,233); THE BOB HOPE SHOW (I,637); BURT BACHARACH! (I,771); CIRCUS OF THE STARS (II,532); LINDA IN WONDERLAND (II,1481); LUCY IN LONDON (I,2789); MALIBU (II,1607)

NEWMAN, BARRY
THE EDGE OF NIGHT (II,760); NIGHT GAMES (II,1845); PETROCELLI (II,2031)

NEWMAN, EDWIN
THE DAVID LETTERMAN SHOW (II,624); THE EARTHLINGS (II,751); LILY FOR PRESIDENT (II,1478)

NEWMAN, ELMER
HAYLOFT HOEDOWN (I,1981)

NEWMAN, JOLIE
THE RED HAND GANG (II,2122)

NEWMAN, LARAINE
BOB & RAY & JANE, LARAINE & GILDA (II,279); THE CONEHEADS (II,567); THE MANHATTAN TRANSFER (II,1619); PRIME TIMES (II,2083); SATURDAY NIGHT LIVE (II,2261); STEVE MARTIN'S BEST SHOW EVER (II,2460); THINGS WE DID LAST SUMMER (II,2589)

NEWMAN, LIONEL
THE JERRY LEWIS SHOW (I,2367)

NEWMAN, MARTIN
KITTY FOYLE (I,2571)

NEWMAN, PAMELA
HIS AND HERS (II,1155)

NEWMAN, PAUL
APPOINTMENT WITH ADVENTURE (I,246); DANGER (I,1134); THE GOLDEN AGE OF TELEVISION (II,1019); I REMEMBER MAMA (I,2176); THE KAISER ALUMINUM HOUR (I,2503); OUR TOWN (I,3419); PLAYHOUSE 90 (I,3623); PLAYWRIGHTS '56 (I,3627); THE SENSATIONAL, SHOCKING, WONDERFUL, WACKY 70S (II,2302); SUPER COMEDY BOWL 2 (I,4289)

NEWMAN, PHYLLIS
ABC STAGE '67 (I,6); THE BOB HOPE SHOW (I,649); DIAGNOSIS: UNKNOWN (I,1251); THE PEOPLE NEXT DOOR (I,3519); STAR STAGE (I,4197); THAT WAS THE WEEK THAT WAS (I,4423)

NEWMAN, RANDY
RANDY NEWMAN AT THE ODEON (II,2111)

NEWMAN, ROBERT
THE GUIDING LIGHT (II,1064)

NEWMAN, ROGER
BRING 'EM BACK ALIVE (II,377); THE GUIDING LIGHT (II,1064); OPERATION GREASEPAINT (I,3402)

NEWMAN, RUSSELL
1915 (II,1853)

NEWMAN, SHAWN
DEBBY BOONE. . . ONE STEP CLOSER (II,655)

NEWMAN, STEPHEN
TEXAS (II,2566)

NEWMAN, TOM
B.J. AND THE BEAR (II,127); THE MUNSTERS' REVENGE (II,1765)

NEWMAN, WILLIAM
CHIEFS (II,509)

NEWMAR, JULIE
BATMAN (I,366); FOOLS, FEMALES AND FUN: WHAT ABOUT THAT ONE? (II,901); HIGH SCHOOL, U.S.A. (II,1149); MCCLOUD: WHO KILLED MISS U.S.A.? (I,2965); MY LIVING DOLL (I,3195); THE PHIL SILVERS SHOW (I,3580); THREE ON AN ISLAND (I,4483); THE TWILIGHT ZONE (I,4651); A VERY MISSING PERSON (I,4715); THE WORLD OF MAGIC (II,2838)

NEWMARA, TOMMY
NUMBER 13 DEMON STREET (I,3316)

NEWPORT YOUTH BAND, THE
76 MEN AND PEGGY LEE (I,3989)

NEWTH, JONATHAN
DAY OF THE TRIFFIDS (II,627)

NEWTON
OPRYLAND U.S.A. (I,3405)

NEWTON, ADELE
THE DOCTOR (I,1320)

NEWTON, JIMMY C
COUNTRY COMES HOME (II,586)

NEWTON, JOHN
DECOY (I,1217); THE DOCTORS (II,694); HARD CASE (I,1947); HAZEL (I,1982)

NEWTON, RICHARD
ACADEMY THEATER (I,14)

NEWTON, ROBERT
THE ADVENTURES OF LONG JOHN SILVER (I,67); MR. GLENCANNON TAKES ALL (I,3139)

NEWTON, SANDIE
GOOBER AND THE TRUCKERS' PARADISE (II,1029)

NEWTON, THEODORE
MIDNIGHT MYSTERY (I,3030)

NEWTON, WAYNE
ONE MORE TIME (I,3385); THE WAYNE NEWTON SPECIAL (II,2749); THE WAYNE NEWTON SPECIAL (II,2750); THE WONDERFUL WORLD OF BURLESQUE II (I,4896)

NEWTON-JOHN, OLIVIA
BOB HOPE SPECIAL: BOB HOPE'S CHRISTMAS SPECIAL (II,308); BOB HOPE SPECIAL: BOB HOPE'S SALUTE TO NASA—25 YEARS OF REACHING FOR THE STARS (II,314); BOB HOPE SPECIAL: BOB HOPE'S STAND UP AND CHEER FOR THE NATIONAL FOOTBALL LEAGUE'S 60TH YEAR (II,316); BOB HOPE SPECIAL: THE BOB HOPE CHRISTMAS SPECIAL (II,328); BOB HOPE SPECIAL: THE BOB HOPE SPECIAL (II,334); GLEN CAMPBELL . . . DOWN HOME—DOWN UNDER (II,1008); JOHN DENVER ROCKY MOUNTAIN CHRISTMAS (II,1314); MAC DAVIS 10TH ANNIVERSARY SPECIAL: I STILL BELIEVE IN MUSIC (II,1571); MUSIC CENTRAL (II,1771); OLIVIA (II,1883); OLIVIA NEWTON-JOHN IN CONCERT (II,1884); THE OLIVIA NEWTON-JOHN SHOW (II,1885); OLIVIA NEWTON-JOHN'S HOLLYWOOD NIGHTS (II,1886); OLIVIA NEWTON-JOHN: LET'S BE PHYSICAL (II,1887); PERRY COMO'S SPRINGTIME SPECIAL (II,2023); THE SENSATIONAL, SHOCKING, WONDERFUL, WACKY 70S (II,2302); A SPECIAL OLIVIA NEWTON-JOHN (II,2418)

NEY, RICHARD
CONFIDENTIALLY YOURS (I,1035); GHOSTBREAKER (I,1789)

NICASSIO, JOEAL
240-ROBERT (II,2688)

NICASTO, MICHELE
SUZANNE PLESHETTE IS MAGGIE BRIGGS (II,2509)

NICHOLAS BROTHERS, THE
THE BOB HOPE SHOW (I,609)

NICHOLAS, DENISE
BABY, I'M BACK! (II,130); BATTLE OF THE NETWORK STARS (II,166); JACQUELINE SUSANN'S VALLEY OF THE DOLLS, 1981 (II,1268); ROOM 222 (I,3843)

NICHOLAS, JEREMY
THE TALE OF BEATRIX POTTER (II,2533)

NICHOLAS, NICK
JOEY AND DAD (II,1306)

NICHOLAS, NOAH
CAGNEY AND LACEY (II,410)

NICHOLLS, ANTHONY
THE CHAMPIONS (I,896)

NICHOLLS, PHOEBE
BRIDESHEAD REVISITED (II,375)

NICHOLLS, SUE
THE FALL AND RISE OF REGINALD PERRIN (II,810)

NICHOLS, BARBARA
ALL ABOUT BARBARA (I,125); THE BEVERLY HILLBILLIES (I,417); BROADWAY OPEN HOUSE (I,736); DANGER (I,1134); THE DESILU PLAYHOUSE (I,1237); THE DICK POWELL SHOW (I,1269); THE GENERAL ELECTRIC THEATER (I,1753); THE JACK BENNY PROGRAM (I,2294); KRAFT SUSPENSE THEATER (I,2591); LOVE THAT JILL (I,2780); THE MUSIC MAKER (I,3173); SID CAESAR PRESENTS COMEDY PREVIEW (I,4041); THE TWILIGHT ZONE (I,4651); THE UNITED STATES STEEL HOUR (I,4677); THE UNTOUCHABLES (I,4681)

NICHOLS, BOB
SECOND CHANCE (I,3958)

NICHOLS, DAVE
CASTLE ROCK (II,456)

NICHOLS, JOSEPHINE
TEXAS (II,2566); A TIME FOR US (I,4507)

NICHOLS, MIKE
ACCENT ON LOVE (I,16); THE FABULOUS 50S (I,1501); JACK PAAR PRESENTS (I,2315); THE JACK PAAR SPECIAL (I,2318); A LAST LAUGH AT THE 60'S (I,2627)

NICHOLS, NICHELLE
ASSIGNMENT: EARTH (I,299); STAR TREK (II,2439); STAR TREK (II,2440)

NICHOLS, NICK
RED SKELTON'S CHRISTMAS DINNER (II,2123)

NICHOLS, RED
FAMILY NIGHT WITH HORACE HEIDT (I,1523); HOLIDAY U.S.A. (I,2078); THE PHIL HARRIS SHOW (I,3575)

NICHOLS, ROBERT
WHERE THERE'S SMOKEY (I,4832)

NICHOLSEN, CAROL
ROOM FOR ONE MORE (I,3842)

NICHOLSON, BOB
GUMBY (I,1905); HOWDY DOODY (I,2151)

NICHOLSON, NICK
THE NEW HOWDY DOODY SHOW (II,1818)

NICK CASTLE DANCERS, THE
THE ANDY WILLIAMS NEW YEAR'S EVE SPECIAL (I,203); THE ANDY WILLIAMS SHOW (I,205); THE ANDY WILLIAMS SHOW (I,209); FAVORITE SONGS (I,1545); THE JERRY LEWIS SHOW (I,2370)

NICKERSON, DAWN
HARRY'S GIRLS (I,1959)

NICKERSON, DENISE
DARK SHADOWS (I,1157); THE DOCTORS (II,694); IF I LOVE YOU, AM I TRAPPED FOREVER? (II,1222); SEARCH FOR TOMORROW (II,2284)

NICKERSON, GINGER
OPRYLAND: NIGHT OF STARS AND FUTURE STARS (II,1921)

NICKERSON, HELENE
THE MIGHTY HERCULES (I,3034)

NICKERSON, SHANE
ALL THE WAY HOME (I,140); THE GUIDING LIGHT (II,1064)

NICKLAUS, JACK
THE BOB HOPE SHOW (I,632)

NICKS, STEVIE
FLEETWOOD MAC IN CONCERT (II,880); STEVIE NICKS IN CONCERT (II,2463)

NICOL, ALEX
RETURN TO PEYTON PLACE (I,3779); TV READERS DIGEST (I,4630)

NICOLLE, JAIME
BARBARA MANDRELL AND THE MANDRELL SISTERS (II,143)

NICOLS, ROSEMARY
DEPARTMENT S (I,1232)

NIELSEN, CLAIRE
THE FEAR IS SPREADING (I,1553)

NIELSEN, LESLIE
AMANDA FALLON (I,152); THE ARMSTRONG CIRCLE THEATER (I,260); BACKSTAIRS AT THE WHITE HOUSE (II,133); BRACKEN'S WORLD (I,703); CHANNING (I,900); CHRYSLER MEDALLION THEATER (I,951); THE CLOCK (I,978); CODE NAME: HERACLITUS (I,990); DANGER (I,1134); DEADLOCK (I,1190); THE EXPLORERS (I,1490); FORD THEATER HOUR (I,1635); THE GREEN FELT JUNGLE (I,1885); GUILTY OR NOT GUILTY (I,1903); HAWAII FIVE-O (I,1972); INSTITUTE FOR REVENGE (II,1239); KRAFT SUSPENSE THEATER (I,2591); THE LAW ENFORCERS (I,2638); THE LETTERS (I,2670); LOVE STORY (I,2776); MAGNAVOX THEATER (I,2827); THE NEW BREED (I,3258); NIGHT GALLERY (I,3287); PEYTON PLACE (I,3574); PHILCO TELEVISION PLAYHOUSE (I,3583); POLICE SQUAD! (II,2061); PRIME TIMES (II,2083); THE RETURN OF CHARLIE CHAN (II,2136); ROBERT MONTGOMERY PRESENTS YOUR LUCKY STRIKE THEATER (I,3809); SHAPING UP (II,2316); SHORT, SHORT DRAMA (I,4023); STAGE 13 (I,4183); STARLIGHT THEATER (I,4201); STUDIO ONE (I,4268); SURE AS FATE (I,4297); SUSPENSE (I,4305); SWAMP FOX (I,4311); TALES OF TOMORROW (I,4352); THEY CALL IT MURDER (I,4441); THRILLER (I,4492); THE TRAP (I,4593); TWILIGHT THEATER II (II,2686); THE WEB (I,4784)

NIELSON, CHRISTINE
THE APARTMENT HOUSE (I,241)

NIELSON, INGA
THE DEAN MARTIN SHOW (I,1201)

NIELSON, NORM
MAGIC WITH THE STARS (II,1599)

NIELSON, TOM
THE GUIDING LIGHT (II,1064)

NIEMELA, AINA
AH!, WILDERNESS (I,91)

NIESE, GEORGE
THE BIG BRAIN (I,424)

NIGH, JANE
BACHELOR FATHER (I,323); BIG TOWN (I,436); STORY THEATER (I,4240); THE UNEXPECTED (I,4674)

NIGHMAN, BRET
THE WONDERFUL WORLD OF PHILIP MALLEY (II,2831)

NIGRA, CHRISTINA
GOLIATH AWAITS (II,1025)

NILES AND FOSSE
YOUR HIT PARADE (I,4951)

NILES, WENDELL
THE R.C.A. THANKSGIVING SHOW (I,3736)

NILL, LANI
THE ARTHUR GODFREY SHOW (I,283)

NIMMO, BILL
FOR LOVE OR MONEY (I,1622); KEEP IT IN THE FAMILY (I,2516)

NIMOY, LEONARD
ASSIGNMENT: EARTH (I,299); BAFFLED (I,332); THE CORAL JUNGLE (II,574); IN SEARCH OF. . . (II,1226); KRAFT SUSPENSE THEATER (I,2591); MARCO POLO (II,1626); MISSION: IMPOSSIBLE (I,3067); MITZI AND A HUNDRED GUYS (II,1710); NAVY LOG (I,3233); NIGHT GALLERY (I,3287); THE OUTER LIMITS (I,3426); REX HARRISON PRESENTS SHORT STORIES OF LOVE (II,2148); STAR TREK (II,2439); STAR TREK (II,2440); A WOMAN CALLED GOLDA (II,2818)

NINOMIYA, HIDEKI
SPACE GIANTS (I,4142)

NISBET, STUART
THE NIGHT RIDER (II,1848)

NITTIES, AARON
1915 (II,1853)

NITWITS COMEDY ACT, THE
THE JERRY LEWIS SHOW (I,2368)

NIVEN, DAVID
THE BOB HOPE SHOW (I,530); THE BOB HOPE SHOW (I,534); THE BOB HOPE SHOW (I,538); CELANESE THEATER (I,881); THE DAVID NIVEN THEATER (I,1178); DAVID NIVEN'S WORLD (II,625); FOUR STAR PLAYHOUSE (I,1652); HOLLYWOOD OPENING NIGHT (I,2087); LIGHT'S DIAMOND JUBILEE (I,2698); A MAN CALLED INTREPID (II,1612); THE NASH AIRFLYTE THEATER (I,3227); THE ROGUES (I,3832); TEXACO COMMAND PERFORMANCE (I,4408); TURN OF FATE (I,4623)

NIVEN, KIP
COMEDY OF HORRORS (II,557); ESCAPE (I,1461); GOLIATH AWAITS (II,1025); ONCE AN EAGLE (II,1897); SNAFU (II,2386); THE WALTONS (II,2740); THE WALTONS: A DAY FOR

THANKS ON WALTON'S MOUNTAIN (EPISODE 3) (II,2741); THE WALTONS: MOTHER'S DAY ON WALTON'S MOUNTAIN (EPISODE 2) (II,2742); THE WALTONS: WEDDING ON WALTON'S MOUNTAIN (EPISODE 1) (II,2743)

NIVEN, SUSAN
REVENGE OF THE GRAY GANG (II,2146)

NIX, MARTHA
THE WALTONS (II,2740)

NIXON, CYNTHIA
FIFTH OF JULY (II,851)

NIXON, MARNI
JACK AND THE BEANSTALK (I,2288)

NIXON, RICHARD
THE SECRET WORLD OF KIDS (I,3970)

NOBEL, TRISHA
THE WILD WILD WEST REVISITED (II,2800)

NOBLE, JAMES
BENSON (II,208); THE DOCTORS (II,694); HART TO HART (II,1101); THIS IS KATE BENNETT (II,2597); A WORLD APART (I,4910)

NOBLE, KENNY
TOP 40 VIDEOS (II,2644)

NOBLE, ROBERT
GEORGE WASHINGTON (II,978)

NOBLE, TRISHA
EXECUTIVE SUITE (II,796); FLAMINGO ROAD (II,872); I'M A FAN (I,2192); ONE OF OUR OWN (II,1910); THE RHINEMANN EXCHANGE (II,2150); STRIKE FORCE (II,2480); TESTIMONY OF TWO MEN (II,2564); WILLOW B: WOMEN IN PRISON (II,2804)

NOBLES, JOAN
KOBB'S CORNER (I,2577)

NODELL, MELVIN
THE GAY COED (I,1742)

NOEL, CHRIS
FLY AWAY HOME (II,893); THE GOOD OLD DAYS (I,1851); THE LIEUTENANT (I,2680); WILD TIMES (II,2799)

NOEL, DICK
RUTH LYONS 50 CLUB (I,3876); THE TENNESSEE ERNIE FORD SHOW (I,4400)

NOEL, HENRI
THE MUSIC SHOW (I,3178)

NOEL, HUBERT
THE GYPSY WARRIORS (II,1070)

NOEL, KARISSA
REAL KIDS (II,2116)

NOLAN, BRIAN
THE BENNY HILL SHOW (II,207)

NOLAN, DANI SUE
NAVY LOG (I,3233)

NOLAN, DINA
THE ADVENTURES OF SUPERMAN (I,77)

NOLAN, JAMES
THE RESTLESS GUN (I,3774)

NOLAN, JEANETTE
ALIAS SMITH AND JONES (I,119); ALL THE WAY HOME (II,44); THE AWAKENING LAND (II,122); CAROLYN (I,854); CHARLIE WOOSTER—OUTLAW (I,915); CROSSROADS (I,1105); DIRTY SALLY (II,680); F TROOP (I,1499); THE FARMER'S DAUGHTER (I,1533); GENTRY'S PEOPLE (I,1762); GOLIATH AWAITS (II,1025); HOTEL DE PAREE (I,2129); THE HUSTLER OF MUSCLE BEACH (II,1210); LASSIE: THE NEW BEGINNING (II,1433); LONGSTREET (I,2750); THE NEW DAUGHTERS OF JOSHUA CABE (I,3261); NIGHT COURT (II,1844); SNEAK PREVIEW (I,4100); TAMMY (I,4357); THE VIRGINIAN (I,4737); THE VIRGINIAN (I,4738); WHEN THE WEST WAS FUN: A WESTERN REUNION (II,2780); THE WILD WOMEN OF CHASTITY GULCH (II,2801)

NOLAN, JOHN
IN THE STEPS OF A DEAD MAN (I,2205)

NOLAN, JOHNNY
QUINCY, M. E. (II,2102)

NOLAN, KATHLEEN
AMANDA FALLON (I,152); BROADSIDE (I,732); CONFLICT (I,1036); THE ELGIN HOUR (I,1428); JACQUELINE SUSANN'S VALLEY OF THE DOLLS, 1981 (II,1268); JAMIE (I,2334); THE LLOYD BRIDGES SHOW (I,2733); PETER PAN (I,3566); THE REAL MCCOYS (I,3741); TELEPHONE TIME (I,4379); TESTIMONY OF TWO MEN (II,2564); THOSE WHITING GIRLS (I,4471); WEDNESDAY NIGHT OUT (II,2760)

NOLAN, LLOYD
ADAMS HOUSE (II,9); AH!, WILDERNESS (I,91); THE BARBARA STANWYCK THEATER (I,350); THE BOB HOPE SHOW (I,539); CALL TO DANGER (I,802); THE CASE AGAINST PAUL RYKER (I,861); CLIMAX! (I,976); THE DESILU PLAYHOUSE (I,1237); THE DICK POWELL SHOW (I,1269); GREAT ADVENTURE (I,1868); JULIA (I,2476); MARTIN KANE, PRIVATE EYE (I,2928); SANDBURG'S LINCOLN (I,3907); SIX GUNS FOR DONEGAN (I,4065); SPECIAL AGENT 7 (I,4146); THE UNTOUCHABLES (I,4682)

NOLAN, SCOTTY
RYAN'S HOPE (II,2234)

NOLAN, TOMMY
BUCKSKIN (I,753); THE GRADUATION DRESS (I,1864); JESSIE (II,1287)

NOLTE, NICK
ADAMS OF EAGLE LAKE (II,10); RICH MAN, POOR MAN—BOOK I (II,2161); THE RUNAWAY BARGE (II,2230); WINTER KILL (II,2810)

NOMLEENA, KEENA
BRAVE EAGLE (I,709)

NOONAN, CHRISTINE
CASANOVA (II,451)

NOONAN, STAN
ARTHUR GODFREY AND FRIENDS (I,278)

NOONAN, TOMMY
THE FREEWHEELERS (I,1683)

NOONE, KATHLEEN
ALL MY CHILDREN (II,39)

NOONE, PETER
BATTLE OF THE BEAT (II,161); GO! (I,1830); PINOCCHIO (I,3603)

NORBERG, GARY
TWILIGHT THEATER II (II,2686)

NORBERG, GREG
MARIE (II,1629)

NORDE, EVA
THE REAL MCCOYS (I,3741)

NORDINE, KEN
ANOTHER EVENING WITH FRED ASTAIRE (I,229); AN EVENING WITH FRED ASTAIRE (I,1474)

NORDON, TOMMY
FLIPPER (I,1604); SEARCH FOR TOMORROW (II,2284); SING ALONG WITH MITCH (I,4057)

NORELL, HENRY
DENNIS THE MENACE (I,1231)

NORELL, MICHAEL
EMERGENCY! (II,775)

NORIEGA, RICHARD
COMPUTERCIDE (II,561)

NORMAN LUBOFF CHOIR, THE
THE EDSEL SHOW (I,1417); THE JERRY LEWIS SHOW (I,2362)

NORMAN MAEN DANCERS, THE
THE KOPYCATS (I,2580); THIS IS TOM JONES (I,4460); THE TOM JONES SPECIAL (I,4540); THE TOM JONES SPECIAL (I,4541); THE TOM JONES SPECIAL (I,4542); THE TOM JONES SPECIAL (I,4543); THE VAL DOONICAN SHOW (I,4692)

NORMAN PARIS CHORUS, THE
THE MARTHA WRIGHT SHOW (I,2927)

NORMAN PARIS TRIO, THE
HOME (I,2099)

NORMAN AND DEAN
THE MORT SAHL SPECIAL (I,3105)

NORMAN, B G
THE FURTHER ADVENTURES OF SPIN AND MARTY (I,1720); LIFE WITH FATHER (I,2690); THE NEW ADVENTURES OF SPIN AND MARTY (I,3254); SPIN AND MARTY (I,4158)

NORMAN, JANE
MAINTENANCE MS. (I,2832); PIXANNE (II,2048)

NORMAN, MAIDIE
BARE ESSENCE (II,149); NEVER SAY NEVER (II,1809)

NORMAN, TEDDY
THE VINCENT LOPEZ SHOW (I,4735)

NORRIS, CHRISTOPHER
BATTLE OF THE NETWORK STARS (II,175); SENIOR YEAR (II,2301); TRAPPER JOHN, M.D. (II,2654)

NORRIS, JAN
IT'S A MAN'S WORLD (I,2258)

NORRIS, KAREN
U.M.C. (I,4666)

NORRIS, KATHI
SPIN THE PICTURE (I,4159); TRUE STORY (I,4617); TV SHOPPER (I,4632)

NORTH CAROLINA SQUARE DANCERS, THE
COUNTRY MUSIC CARAVAN (I,1066)

NORTH, ALAN
FOR LOVERS ONLY (II,904); LOVE, SIDNEY (II,1547)

NORTH, ALEX
POLICE SQUAD! (II,2061)

NORTH, HEATHER
DAYS OF OUR LIVES (II,629); GIDGET (I,1795); PARADISE BAY (I,3454); SCOOBY-DOO AND SCRAPPY-DOO (II,2274); SCOOBY-DOO, WHERE ARE YOU? (II,2276)

NORTH, JAY
THE BANANA SPLITS ADVENTURE HOUR (I,340); DENNIS THE MENACE (I,1231); THE FLINTSTONE COMEDY HOUR (II,881); HERE COMES THE GRUMP (I,2027); MAYA (I,2960); PEBBLES AND BAMM BAMM (II,1987)

NORTH, LESLEY
TERROR FROM WITHIN (I,4404)

NORTH, NOELLE
FRED AND BARNEY MEET THE THING (II,919); WOLF ROCK TV (II,2815)

NORTH, SHEREE
ANYTHING GOES (I,238); ARCHIE BUNKER'S PLACE (II,105); BAY CITY BLUES (II,186); BIG EDDIE (II,235); BIG EDDIE (II,236); THE BING CROSBY SPECIAL (I,470); THE BOB HOPE SHOW (I,558); A CHRISTMAS FOR BOOMER (II,518); CODE NAME: HERACLITUS (I,990); ELVIS REMEMBERED: NASHVILLE TO HOLLYWOOD (II,772); GREAT ADVENTURE (I,1868); HAVE I GOT A CHRISTMAS FOR YOU (II,1105); I'M A BIG GIRL NOW (II,1218); KEY WEST (I,2528); THE MARY TYLER MOORE SHOW (II,1640); MOST WANTED (II,1738); THE NIGHTENGALES (II,1850); THE RETURN OF LUTHER GILLIS (II,2137); THE SEEKERS (I,3974); A SHADOW IN THE STREETS (II,2312); THEN CAME BRONSON (I,4436); WINTER KILL (II,2810); WOMEN IN WHITE (II,2823)

NORTH, ZEME
THE DOUBLE LIFE OF HENRY PHYFE (I,1361)

NORTHCUTTE, SHAWNTE
THE NEW MICKEY MOUSE CLUB (II,1823)

NORTHROP, WAYNE
DAYS OF OUR LIVES (II,629); DYNASTY (II,746)

NORTHUP, HARRY
KNOTS LANDING (II,1404)

NORTON, CLIFF
CAESAR'S HOUR (I,790); CAROL (I,846); THE DAVE GARROWAY SHOW (I,1173); DID YOU HEAR ABOUT JOSH AND KELLY?! (II,673); THE FEATHER AND FATHER GANG: NEVER CON A KILLER (II,846); THE FUNNY MANNS (I,1712); GARROWAY AT LARGE (I,1737); GEORGE GOBEL PRESENTS (I,1766); THE HOOFER (I,2113); IT'S ABOUT TIME (I,2263); JACK AND THE BEANSTALK (I,2288); THE JONES BOYS (I,2446); LOW MAN ON THE TOTEM POLE (I,2782); MARINELAND CARNIVAL (I,2912); MITZI: A TRIBUTE TO THE AMERICAN HOUSEWIFE (I,3072); MY SON THE DOCTOR (I,3203); PETE AND GLADYS (I,3558); THE PUBLIC LIFE OF CLIFF NORTON (I,3688); REAR GUARD (II,2120); SID CAESAR PRESENTS COMEDY PREVIEW (I,4041); STICK AROUND (II,2464); WHERE'S HUDDLES (I,4836); YOUR SHOW OF SHOWS (I,4957); YOUR SHOW OF SHOWS (II,2875)

NORTON, COE
MANDRAKE THE MAGICIAN (I,2886)

NORTON, JACK
ARTHUR MURRAY'S DANCE PARTY (I,293)

NORTON, JIM
LITTLE LORD FAUNTLEROY (II,1492); QUILLER: NIGHT OF THE FATHER (II,2100)

NORTON, KEN
CIRCUS OF THE STARS (II,532)

NORTON, LEONIE
ANOTHER WORLD (II,97); THE NURSES (I,3319); ONE LIFE TO LIVE (II,1907)

NORTON, WAYNE
WHIZ KIDS (II,2790)

NORTON-TAYLOR, JUDY
BATTLE OF THE NETWORK STARS (II,169); BATTLE OF THE NETWORK STARS (II,171); BATTLE OF THE NETWORK STARS (II,172); CELEBRITY CHALLENGE OF THE SEXES 5 (II,469); CELEBRITY CHARADES (II,470); CIRCUS OF THE STARS (II,537); CIRCUS OF THE STARS (II,538); THE WALTONS (II,2740); THE WALTONS: A DAY FOR THANKS ON WALTON'S MOUNTAIN (EPISODE 3) (II,2741); THE WALTONS: MOTHER'S DAY ON WALTON'S MOUNTAIN (EPISODE 2) (II,2742); THE WALTONS: WEDDING ON WALTON'S MOUNTAIN (EPISODE 1) (II,2743)

NORVO, RED
FRANK SINATRA (I,1660)

NORWINE, JACK
IT'S A HIT (I,2257)

NOSSEM, BRAM
THE ATOM SQUAD (I,312)

NOTABLES, THE
THE DINAH SHORE CHEVY SHOW (I,1279)

NOTEWORTHIES, THE
HEAD OF THE CLASS (I,1987); THE JIMMY DEAN SHOW (I,2384)

NOTRE DAME BOYS CHOIR, THE
PERRY COMO'S CHRISTMAS IN PARIS (II,2012)

NOURI, MICHAEL
AFTER HOURS: FROM JANICE, JOHN, MARY AND MICHAEL, WITH LOVE (II,20); BAY CITY BLUES (II,186); THE CURSE OF DRACULA (II,611); THE GANGSTER CHRONICLES (II,957); THE LAST CONVERTIBLE (II,1435); NICK AND THE DOBERMANS (II,1843); SEARCH FOR TOMORROW (II,2284); SPRAGGUE (II,2430)

NOURSE, ALLEN
THE EDGE OF NIGHT (II,760); OUR FIVE DAUGHTERS (I,3413)

NOVA, LUDMILLA
THE SNOW GOOSE (I,4103)

NOVACK, SHELLY
THE FBI (I,1551); MCCLOUD: WHO KILLED MISS U.S.A.? (I,2965); MOST WANTED (II,1738); MOST WANTED (II,1739); A PUNT, A PASS, AND A PRAYER (I,3694)

NOVAK, KIM
THE BOB HOPE SHOW (I,548); LIGHT'S DIAMOND JUBILEE (I,2698); MALIBU (II,1607)

NOVARRO, RAMON
ELFEGO BACA (I,1427); HEDDA HOPPER'S HOLLYWOOD (I,2002)

NOVELETTES, THE
THE JAN MURRAY SHOW (I,2335)

NOVELLO, DON
BLONDES VS. BRUNETTES (II,271); THE SMOTHERS BROTHERS SHOW (II,2384); TWILIGHT THEATER II (II,2686)

NOVELLO, JAY
BRAVO DUKE (I,711); FIRESIDE THEATER (I,1580); I AND CLAUDIE (I,2164); THE LIFE OF VERNON HATHAWAY (I,2687); MCHALE'S NAVY (I,2969)

NOVICE, ALICE
THE NEW ERNIE KOVACS SHOW (I,3264)

NOVOTNA, JARMILA
THE GREAT WALTZ (I,1878); HANS BRINKER OR THE SILVER SKATES (I,1933)

NOYERS, TIM
THE ICE PALACE (I,2182)

NUCKOLS, WILLIAM
SUPERTRAIN (II,2504)

NUDELL, SAM
LEADFOOT (II,1452)

NUENZ, RICHARD
RYAN'S HOPE (II,2234)

NUGENT, JUDY
ANNETTE (I,222); ANNIE OAKLEY (I,225); FORD TELEVISION THEATER (I,1634); MAKE ROOM FOR DADDY (I,2843); THE RUGGLES (I,3868)

NUGENT, KELBE
WELCOME TO PARADISE (II,2762)

NUGGETT
HEAR NO EVIL (II,1116)

NUNN, ALICE
CAMP RUNAMUCK (I,811); GREAT DAY (II,1053); HAPPY DAYS (II,1084); LOVE NEST (II,1541); THE STOCKERS (II,2469); WESTSIDE MEDICAL (II,2765)

NUNN, JUDY
PRISONER: CELL BLOCK H (II,2085)

NUNN, TERRI
THE GEORGIA PEACHES (II,979)

NUNN, WILLIAM
THE RETURN OF CHARLIE CHAN (II,2136)

NUREYEV, RUDOLF
THE BURT BACHARACH SPECIAL (I,775); JIMMY DURANTE MEETS THE LIVELY ARTS (I,2386); SPECIAL LONDON BRIDGE SPECIAL (I,4150); SUNDAY NIGHT AT THE LONDON PALLADIUM (I,4286)

NUSSER, JAMES
GUNSMOKE (II,1069)

NUTT, GRADY
THE GRADY NUTT SHOW (II,1048); HEE HAW (II,1123)

NUTTER, MAYFI
BUCK OWENS TV RANCH (I,750)

NUYEN, FRANCE
CODE NAME: DIAMOND HEAD (II,549); RETURN TO FANTASY ISLAND (II,2144)

NYBERG, PETER
DEATH RAY 2000 (II,654)

NYE, CARRIE
THE ADMIRABLE CRICHTON (I,36); THE PICTURE OF DORIAN GRAY (I,3591)

NYE, LOUIS
ALL COMMERCIALS—A STEVE MARTIN SPECIAL (II,37); THE ANN SOTHERN SHOW (I,220); THE BEVERLY HILLBILLIES (I,417); HAPPY DAYS (I,1942); HER SCHOOL FOR BACHELORS (I,2018); I'VE HAD IT UP TO HERE (II,1221); THE JUD STRUNK SHOW (I,2462); MITZI AND A HUNDRED GUYS (II,1710); NEEDLES AND PINS (I,3246); THE NEW STEVE ALLEN SHOW (I,3271); THE RITA MORENO SHOW (II,2181); THE STEVE ALLEN COMEDY HOUR (I,4218); THE STEVE ALLEN SHOW (I,4219); THE STEVE ALLEN SHOW (I,4220); STEVE ALLEN'S LAUGH-BACK (II,2455); STEVE MARTIN: COMEDY IS NOT PRETTY (II,2462); THINK PRETTY (I,4448)

NYE, PAT
LITTLE WOMEN (I,2723)

NYE, WILL
FARRELL: FOR THE PEOPLE (II,834); GIDGET'S SUMMER REUNION (I,1799)

NYGH, ANNA
MASTER OF THE GAME (II,1647)

NYLAND, DIANE
THE TROUBLE WITH TRACY (I,4613)

NYPE, RUSSELL
DOROTHY (II,717); HEAVENS TO BETSY (I,1997); KISS ME, KATE (I,2569); ONE TOUCH OF VENUS (I,3389)

O'BRADOVICH, ED
THE DUKE (II,740)

O'BRIAN, HUGH
BENNY AND BARNEY: LAS VEGAS UNDERCOVER (II,206); BUSH DOCTOR (II,400); DAMON RUNYON THEATER (I,1129); THE DESILU REVUE (I,1238); DIAL "M" FOR MURDER (I,1255); FANTASY ISLAND (II,830); FEATHERTOP (I,1556); FIRESIDE THEATER (I,1580); FRANCES LANGFORD PRESENTS (I,1656); THE GOLDDIGGERS (I,1838); THE GRADUATION DRESS (I,1864); GREAT BIBLE ADVENTURES (I,1872); THE LIFE AND LEGEND OF WYATT EARP (I,2681); THE LONDON PALLADIUM (I,2739); PROBE (I,3674); A PUNT, A PASS, AND A PRAYER (I,3694); SEARCH (I,3951); THE SEEKERS (II,2298); A SPECIAL HOUR WITH DINAH SHORE (I,4149); SPELLBOUND (I,4153); STAGE 7 (I,4181); STUDIO '57 (I,4267); SWING OUT, SWEET LAND (II,2515)

O'BRIEN, CARL 'CUBBY'
THE MICKEY MOUSE CLUB (I,3025); THE MOUSEKETEERS REUNION (II,1742)

O'BRIEN, DAN
THE FRENCH ATLANTIC AFFAIR (II,924)

O'BRIEN, DAVE
PETE SMITH SPECIALTIES (I,3560)

O'BRIEN, DAVID
THE DOCTORS (II,694); THE HEIRESS (I,2006); RENFREW OF THE ROYAL MOUNTED (I,3769); SEARCH FOR TOMORROW (II,2284)

O'BRIEN, EDMOND
ACTION (I,23); THE BLUE MEN (I,492); THE BOB HOPE SHOW (I,532); CLIMAX! (I,976); FLESH AND BLOOD (I,1595); GALLAGHER (I,1727); THE GOLDEN AGE OF TELEVISION (II,1018); HENRY FONDA PRESENTS THE STAR AND THE STORY (I,2016); JOHNNY MIDNIGHT (I,2431); THE LONG HOT SUMMER (I,2747); MAN ON THE MOVE (I,2878); MEN IN CRISIS (I,3006); THE OUTSIDER (I,3430); PLAYWRIGHTS '56 (I,3627); SAM BENEDICT (I,3894)

O'BRIEN, ERIN
THE EDDIE FISHER SHOW (I,1409); THE FRANK SINATRA SHOW (I,1664); GIRL ON THE RUN (I,1809); THE LIBERACE SHOW (I,2676)

O'BRIEN, FRANK
BORDER PALS (II,352); OFF THE WALL (II,1879)

O'BRIEN, GEORGE
PANTOMIME QUIZ (I,3449)

O'BRIEN, JOAN
THE BOB CROSBY SHOW (I,497)

O'BRIEN, KEVIN
STUDS LONIGAN (II,2484)

O'BRIEN, KIM
THE BEACH GIRLS (II,191); IS THERE A DOCTOR IN THE HOUSE? (II,1247)

O'BRIEN, LARRY
SAMMY KAYE'S MUSIC FROM MANHATTAN (I,3902)

O'BRIEN, LAURIE
JIM HENSON'S MUPPET BABIES (II,1289)

O'BRIEN, LEO
CHIEFS (II,509)

O'BRIEN, LOUISE
THE PAT BOONE SHOW (I,3470); SING ALONG WITH MITCH (I,4057)

O'BRIEN, MARGARET
FORD TELEVISION THEATER (I,1634); FRONT ROW CENTER (I,1694); THE JUNE ALLYSON SHOW (I,2488); LITTLE WOMEN (I,2722); THE LUX VIDEO THEATER (I,2795); MAGGIE (I,2810); PLAYHOUSE 90 (I,3623); PURSUIT (I,3700); ROBERT MONTGOMERY PRESENTS YOUR LUCKY STRIKE THEATER (I,3809); STUDIO ONE (I,4268); TESTIMONY OF TWO MEN (II,2564)

O'BRIEN, MARIA
CHIPS (II,511); THE LIFE AND TIMES OF EDDIE ROBERTS (II,1468); NUMBER 96 (II,1869); THE PRIMARY ENGLISH CLASS (II,2081); TABITHA (II,2526)

O'BRIEN, MARY
GENERAL HOSPITAL (II,964)

O'BRIEN, MAUREEN
DOCTOR WHO (II,689)

O'BRIEN, MELODY
RIVAK, THE BARBARIAN (I,3795)

O'BRIEN, PAT
ABC'S MATINEE TODAY (I,7); THE ADVENTURES OF NICK CARTER (I,68); AMANDA FALLON (I,152); THE BOB HOPE CHRYSLER THEATER (I,502); CLIMAX! (I,976); CROSSROADS (I,1105); FORD TELEVISION THEATER (I,1634); FRONT ROW CENTER (I,1694); HAPPY DAYS (II,1084); HARRIGAN AND SON (I,1955); HENRY FONDA PRESENTS THE STAR AND THE STORY (I,2016); KISS ME, KILL ME (II,1400); KRAFT SUSPENSE THEATER (I,2591); KRAFT TELEVISION THEATER (I,2592); LIFE'S MOST EMBARRASSING MOMENTS (II,1470); THE LUX VIDEO THEATER (I,2795); OPERATION ENTERTAINMENT (I,3400); PERSPECTIVE ON GREATNESS (I,3555); RHEINGOLD THEATER (I,3783); SUPER COMEDY BOWL 1 (I,4288); THE UNITED STATES STEEL HOUR (I,4677)

O'BRIEN, RICHARD
LANDON, LANDON & LANDON (II,1426); TOP OF THE HILL (II,2645)

O'BRIEN, RORY
THE FARMER'S DAUGHTER (I,1533)

O'BRIEN, SHAWN
SUNDAY IN TOWN (I,4285)

O'BRIEN, STACY
THE BIONIC WOMAN (II,255)

O'BRIEN, STEVE
THE $20,000 PYRAMID (II,2681)

O'BRIEN, THOMAS
CALL TO GLORY (II,413)

O'BRIEN, VINCE
DARK SHADOWS (I,1157); SEARCH FOR TOMORROW (II,2284)

O'BRYAN, LAUREN
RYAN'S HOPE (II,2234)

O'BRYNE, BRYAN
BLONDIE (I,487); OCCASIONAL WIFE (I,3325)

O'BYRNE, BRYAN
SUTTERS BAY (II,2508)

O'CALLAGHAN, RICHARD

FILE IT UNDER FEAR (I,1574)

O'CONNELL, ARTHUR
THE HUMAN COMEDY (I,2158); LOVE STORY (I,2776); THE SECOND HUNDRED YEARS (I,3959); THE SECRET LIFE OF JOHN MONROE (I,3966)

O'CONNELL, BOB
DARK SHADOWS (I,1157)

O'CONNELL, HELEN
THE EDDIE CANTOR COMEDY THEATER (I,1407); THE HELEN O'CONNELL SHOW (I,2007); MANHATTAN TOWER (I,2889); PETER MARSHALL SALUTES THE BIG BANDS (II,2028); THE RUSS MORGAN SHOW (I,3874); THE SWINGIN' YEARS (I,4321); TV'S TOP TUNES (I,4635)

O'CONNELL, JAMES
CAPTAINS AND THE KINGS (II,435)

O'CONNELL, KEVIN
GO (II,1012)

O'CONNELL, PEGGY
A CHRISTMAS CAROL (II,517)

O'CONNELL, TAAFFE
BLANSKY'S BEAUTIES (II,267)

O'CONNELL, WILLIAM
PIONEER SPIRIT (I,3604)

O'CONNER, EDDIE
THE VINCENT LOPEZ SHOW (I,4735)

O'CONNOR, CARROLL
ALCOA PREMIERE (I,109); ALL IN THE FAMILY (II,38); ARCHIE BUNKER'S PLACE (II,105); CBS: ON THE AIR (II,460); THE DICK POWELL SHOW (I,1269); DON RICKLES—ALIVE AND KICKING (I,1340); THE DUPONT SHOW OF THE WEEK (I,1388); FUNNY PAPERS (I,1713); GLORIA COMES HOME (II,1011); GREAT ADVENTURE (I,1868); KEEP U.S. BEAUTIFUL (I,2519); THE LAST HURRAH (I,2626); LUXURY LINER (I,2796); OF THEE I SING (I,3334); THE OUTER LIMITS (I,3426); PROFILES IN COURAGE (I,3678); SHOW BUSINESS SALUTE TO MILTON BERLE (I,4029); THREE FOR THE GIRLS (I,4477); TOM SNYDER'S CELEBRITY SPOTLIGHT (II,2632); A WALK IN THE NIGHT (I,4750)

O'CONNOR, DARREN
HOLLYWOOD HIGH (II,1163)

O'CONNOR, DES
THE DES O'CONNOR SHOW (I,1235); ROYAL VARIETY PERFORMANCE (I,3865)

O'CONNOR, DI
ALL THE RIVERS RUN (II,43)

O'CONNOR, DONALD
...AND DEBBIE MAKES SIX (I,185); ABC STAGE '67 (I,6); THE ALL-STAR SALUTE TO MOTHER'S DAY (II,51); BING!. . . A 50TH ANNIVERSARY GALA (II,254); THE BOB HOPE SHOW (I,602); THE BOB HOPE SHOW (I,634); BRILLIANT BENJAMIN BOGGS (I,730); THE COLGATE COMEDY HOUR (I,997); THE DONALD O'CONNOR SHOW (I,1343); THE DONALD O'CONNOR SHOW (I,1344); THE DONALD O'CONNOR SHOW (I,1345); DONALD'S DUBLIN (I,1346); DUPONT SHOW OF THE MONTH (I,1387); THE GENE KELLY SHOW (I,1751); HOLLYWOOD MELODY (I,2085); THE HOOFER (I,2113); LI'L ABNER (I,2702); THE MUSIC MART (II,1774); STANDARD OIL ANNIVERSARY SHOW (I,4189); TEXACO STAR THEATER: OPENING NIGHT (II,2565)

O'CONNOR, GLYNNIS
BLACK BEAUTY (II,261); THE CHISHOLMS (II,512); OUR TOWN (I,3421); SENIOR YEAR (II,2301); SONS AND DAUGHTERS (II,2404)

O'CONNOR, JIM BOYD
SCRAPBOOK JR. EDITION (I,3944)

O'CONNOR, JOSEPH
THE FORSYTE SAGA (I,1645)

O'CONNOR, MAUREEN
MATT HOUSTON (II,1654)

O'CONNOR, PEGGY
EXPERT WITNESS (I,1488); SCIENCE FICTION THEATER (I,3938)

O'CONNOR, T P
THE NEXT VICTIM (I,3281)

O'CONNOR, TERRENCE
THE NEW ODD COUPLE (II,1825)

O'CONNOR, TIM
ALL IN THE FAMILY (II,38); BUCK ROGERS IN THE 25TH CENTURY (II,383); THE DEVIL AND DANIEL WEBSTER (I,1246); DYNASTY (II,746); ELLERY QUEEN: TOO MANY SUSPECTS (II,767); THE HOUSE ON GREENAPPLE ROAD (I,2140); THE MAN WITH THE POWER (II,1616); MANHUNTER (II,1620); PEYTON PLACE (I,3574); PROFILES IN COURAGE (I,3678); RX FOR THE DEFENSE (I,3878); SPELLBOUND (I,4153); STATE FAIR (II,2448); THE STRANGER (I,4251); THEY ONLY COME OUT AT NIGHT (II,2586); THE THREE MUSKETEERS (I,4481); WHEELS (II,2777); WINTER KILL (II,2810)

O'CONNOR, UNA
LITTLE WOMEN (I,2720)

O'DANIELS, MARK
NOTORIOUS (I,3314)

O'DAY, ANITA
THE SWINGIN' YEARS (I,4321); TIMEX ALL-STAR JAZZ SHOW III (I,4517)

O'DEA, DENNIS
DOUGLAS FAIRBANKS JR. PRESENTS THE RHEINGOLD THEATER (I,1364)

O'DELL, BRYAN
WHAT'S HAPPENING!! (II,2769)

O'DELL, DELL
THE DELL O'DELL SHOW (I,1221)

O'DELL, EDNA
JIMMY DURANTE PRESENTS THE LENNON SISTERS HOUR (I,2387)

O'DONNELL, ANNE
THE LIFE AND TIMES OF EDDIE ROBERTS (II,1468)

O'DONNELL, CATHY
THE SHOWOFF (I,4035)

O'DONNELL, ERIN
WENDY AND ME (I,4793)

O'DONNELL, GENE
BARNEY BLAKE, POLICE REPORTER (I,359); A DATE WITH JUDY (I,1162)

O'DONNELL, JACKLYN
THE ED WYNN SHOW (I,1404); YES YES NANETTE (I,4927)

O'FARRELL, BERNADETTE
THE ADVENTURES OF ROBIN HOOD (I,74)

O'FARRELL, PETER
PETER PAN (II,2030); PRISONERS OF THE LOST UNIVERSE (II,2086)

O'FLYNN, DAMIAN
THE LIFE AND LEGEND OF WYATT EARP (I,2681)

O'GRADY, KAY
AND HERE'S THE SHOW (I,187); HERE'S THE SHOW (I,2033)

O'GRADY, LANI
CELEBRITY CHALLENGE OF THE SEXES 5 (II,469); CELEBRITY CHARADES (II,470); EIGHT IS ENOUGH (II,762); HEADMASTER (I,1990); THE KID WITH THE BROKEN HALO (II,1387)

O'HAGEN, TIMOTHY
THE GOSSIP COLUMNIST (II,1045); THE LAST CONVERTIBLE (II,1435)

O'HALLORAN, MICHAEL
LIVE LIKE A MILLIONAIRE (I,2728)

O'HANLON JR, GEORGE
BEANE'S OF BOSTON (II,194); THE NANCY DREW MYSTERIES (II,1789); POTTSVILLE (II,2072)

O'HANLON, GEORGE
THE JETSONS (I,2376); THE LIFE OF RILEY (I,2686); OCTAVIUS AND ME (I,3327); REAL GEORGE (I,3740); THE REPORTER (I,3771); SNEAK PREVIEW (I,4100)

O'HANLON, PHILIP
YOU'RE ONLY YOUNG ONCE (I,4969)

O'HANLON, SUSAN
GENERAL HOSPITAL (II,964); JASON OF STAR COMMAND (II,1274)

O'HARA, CATHERINE
FROM CLEVELAND (II,931); SECOND CITY TELEVISION (II,2287); THE STEVE ALLEN COMEDY HOUR (II,2454)

O'HARA, JENNY
BLACK BEAUTY (II,261); THE FACTS OF LIFE (II,805); FAMILY (II,813); HIGHCLIFFE MANOR (II,1150); RETURN OF THE WORLD'S GREATEST DETECTIVE (II,2142); SECRETS OF MIDLAND HEIGHTS (II,2296); V: THE FINAL BATTLE (II,2714)

O'HARA, JILL
A TIME FOR US (I,4507)

O'HARA, LESLIE
THE GUIDING LIGHT (II,1064)

O'HARA, MAUREEN

THE ANDY WILLIAMS SHOW (I,209); THE BOB HOPE SHOW (I,568); A CRY OF ANGELS (I,1111); THE FABULOUS FORDIES (I,1502); SPELLBOUND (I,4153); THE TALENT SCOUTS PROGRAM (I,4341); TALENT SEARCH (I,4342); WHO'S AFRAID OF MOTHER GOOSE? (I,4852)

O'HARA, QUINN
THE LIVELY ONES (I,2729)

O'HARA, SHIRLEY
THE DETECTIVE (II,662); FIRST LOVE (I,1583); FUTURE COP (II,946); MANHUNTER (II,1620)

O'HARE, JOHN
LOVE OF LIFE (I,2774)

O'HEANEY, CAITLIN
APPLE PIE (II,100); TALES OF THE GOLD MONKEY (II,2537)

O'HERLIHY, DAN
BANJO HACKETT: ROAMIN' FREE (II,141); CAVALCADE OF AMERICA (I,875); THE DEADLY GAME (II,633); DEATH RAY 2000 (II,654); HUNTER'S MOON (II,1207); JANE WYMAN PRESENTS THE FIRESIDE THEATER (I,2345); JENNIE: LADY RANDOLPH CHURCHILL (II,1277); THE JUNE ALLYSON SHOW (I,2488); KRAFT TELEVISION THEATER (I,2592); THE LONG HOT SUMMER (I,2747); A MAN CALLED SLOANE (II,1613); NANCY ASTOR (II,1788); ON TRIAL (I,3363); SCHLITZ PLAYHOUSE OF STARS (I,3936); SCREEN DIRECTOR'S PLAYHOUSE (I,3946); STAGE 7 (I,4181); THE TRAVELS OF JAIMIE MCPHEETERS (I,4596); THE UNITED STATES STEEL HOUR (I,4677); WHIZ KIDS (II,2790); WOMAN ON THE RUN (II,2821); YOUR SHOW TIME (I,4958)

O'HERLIHY, GAVAN
HAPPY DAYS (II,1084); RICH MAN, POOR MAN—BOOK I (II,2161)

O'KEEFE, DENNIS
BING CROSBY AND HIS FRIENDS (I,455); THE DENNIS O'KEEFE SHOW (I,1230); GULF PLAYHOUSE (I,1904); IT'S ALWAYS SUNDAY (I,2265); NIGHT PROWL (I,3291); SUSPICION (I,4309)

O'KEEFE, MICHAEL
THE OATH: 33 HOURS IN THE LIFE OF GOD (II,1873)

O'KEEFE, PAUL
AS THE WORLD TURNS (II,110); HOT HERO SANDWICH (II,1186); THE PATTY DUKE SHOW (I,3483)

O'KEEFE, WALTER
MAYOR OF HOLLYWOOD (I,2962); TWO FOR THE MONEY (I,4654)

O'KELLY, TIM
HAWAII FIVE-O (I,1972); THE MONROES (I,3089); U.M.C. (I,4666)

O'LAUGHLIN, MAURICE
RETURN TO EDEN (II,2143)

O'LEARY, CAROL
SCRUPLES (II,2280)

O'LEARY, JACK
THE JERK, TOO (II,1282); MARY (II,1637); REPORT TO MURPHY (II,2134)

O'LEARY, JOHN
CATALINA C-LAB (II,458); GHOST OF A CHANCE (II,987); THE GIRL, THE GOLD WATCH AND EVERYTHING (II,1002)

O'LEARY, MICHAEL
ADVICE TO THE LOVELORN (II,16)

O'LOUGHLIN, GERALD S
AUTOMAN (II,120); THE BLUE AND THE GRAY (II,273); THE D.A.: MURDER ONE (I,1122); THE DOCTORS (II,694); DUSTY (II,745); FAME (II,812); LASSITER (I,2625); LASSITER (II,1434); LONDON AND DAVIS IN NEW YORK (II,1512); MCCLAIN'S LAW (II,1659); MEN AT LAW (I,3004); THE POWERS OF MATTHEW STAR (II,2075); THE ROOKIES (II,2208); SPARROW (II,2411); THE STOREFRONT LAWYERS (I,4233); WHEELS (II,2777); WILSON'S REWARD (II,2805); WOMEN IN WHITE (II,2823)

O'MAHONEY, NORA
LITTLE MOON OF ALBAN (I,2714)

O'MAHONEY, PRINCESS
BEACH PATROL (II,192); TV FUNNIES (II,2672)

O'MALLEY, J PAT
ALARM (I,105); ALICE IN WONDERLAND (I,120); BAND OF GOLD (I,341); CONFIDENTIALLY YOURS (I,1035); THE DICK VAN DYKE SHOW (I,1275); DOC (I,1303); THE ELGIN HOUR (I,1428); GAY NINETIES REVUE (I,1743); KLONDIKE (I,2573); LIGHT'S OUT (I,2699); MAUDE (II,1655); MY FAVORITE MARTIAN (I,3189); PLEASE DON'T EAT THE DAISIES (I,3628); THE ROUNDERS (I,3851); SPIN AND MARTY (I,4158); STAGE 13 (I,4183); TIGER! TIGER! (I,4499); A TOUCH OF GRACE (I,4585); WENDY AND ME (I,4793); WHERE'S THE FIRE? (II,2785)

O'MALLEY, KATHLEEN
BACHELOR AT LAW (I,322)

O'MALLEY, MURIAL
THE YEOMAN OF THE GUARD (I,4926)

O'MALLEY, NEAL
HUMAN ADVENTURE (I,2157)

O'MALLEY, TOM
CANDID CAMERA (I,819); EVERYTHING GOES (I,1480)

O'MARA, TERI
THE LOVE BOAT I (II,1532)

O'MARA, TERRY
LET'S CELEBRATE (I,2663)

O'MORRISON, KEVIN
CHARLIE WILD, PRIVATE DETECTIVE (I,914); GOLDEN WINDOWS (I,1841)

O'NEAL, ANN
PROFESSIONAL FATHER (I,3677)

O'NEAL, ED
FARRELL: FOR THE PEOPLE (II,834)

O'NEAL, FREDERICK
THE GREEN PASTURES (I,1887); THE PATRIOTS (I,3477)

O'NEAL, JIMMY
SHINDIG (I,4013)

O'NEAL, JOYCE
THE JERK, TOO (II,1282)

O'NEAL, KEVIN
THE JIMMY DURANTE SHOW (I,2389); THE NEW PEOPLE (I,3268)

O'NEAL, PATRICK
CROSSFIRE (II,606); DIAGNOSIS: UNKNOWN (I,1251); DICK AND THE DUCHESS (I,1263); THE DORIS DAY SHOW (I,1355); EISCHIED (II,763); EMERALD POINT, N.A.S. (II,773); GRUEN GUILD PLAYHOUSE (I,1892); KAZ (II,1370); THE KILLER WHO WOULDN'T DIE (II,1393); THE LAST HURRAH (I,2626); THE MASK OF MARCELLA (I,2937); ONCE THE KILLING STARTS (I,3367); PEPSI-COLA PLAYHOUSE (I,3523); PORTIA FACES LIFE (I,3653); SPRAGGUE (II,2430); TODAY IS OURS (I,4531); TWIN DETECTIVES (II,2687); WILSON'S REWARD (II,2805)

O'NEAL, RON
BRING 'EM BACK ALIVE (II,377)

O'NEAL, RYAN
EMPIRE (I,1439); GO! (I,1830); PEYTON PLACE (I,3574); ROMP (I,3838); THE SEARCH (I,3950); UNDER THE YUM YUM TREE (I,4671)

O'NEAL, THOMAS 'TIP'
BOB HOPE SPECIAL: BOB HOPE PRESENTS A CELEBRATION WITH STARS OF COMEDY AND MUSIC (II,340)

O'NEAL, TRISHA
JACQUELINE SUSANN'S VALLEY OF THE DOLLS, 1981 (II,1268)

O'NEAL, WILLIAM
ANNIE GET YOUR GUN (I,223)

O'NEIL, BETTY
PANAMA HATTIE (I,3444)

O'NEIL, CATHRYN
CO-ED FEVER (II,547)

O'NEIL, COLETTE
COME OUT, COME OUT, WHEREVER YOU ARE (I,1014)

O'NEIL, F J
KENNEDY (II,1377)

O'NEIL, JAMES
ALL THE WAY HOME (I,140)

O'NEIL, KEVIN
NO TIME FOR SERGEANTS (I,3300)

O'NEIL, RICHARD
DOROTHY HAMILL IN ROMEO & JULIET ON ICE (II,718); THE MAGIC PLANET (II,1598)

O'NEIL, SHANA
THE JERK, TOO (II,1282)

O'NEIL, TRICIA
CHARLIE COBB: NICE NIGHT FOR A HANGING (II,485); HOW TO SURVIVE A MARRIAGE (II,1198); PALMS PRECINCT (II,1946); THE POWERS OF MATTHEW STAR (II,2075)

O'NEILL, AMY
MAMA'S FAMILY (II,1610)

O'NEILL, ANNIE
FOG (II,897); IT TAKES TWO (II,1252)

O'NEILL, BARBARA

YOUNG DR. MALONE (I,4938)

O'NEILL, DANNY
JAMBOREE (I,2331); ONE IN A MILLION (I,3379); WINDY CITY JAMBOREE (I,4877)

O'NEILL, DICK
CAGNEY AND LACEY (II,409); CHARLES IN CHARGE (II,482); EMPIRE (II,777); THE FACTS OF LIFE (II,805); FAMILY IN BLUE (II,818); THE FEATHER AND FATHER GANG (II,845); FOG (II,897); JOE AND SONS (II,1297); KAZ (II,1370); THE LIVING END (I,2731); A MAN CALLED INTREPID (II,1612); THE RAG BUSINESS (II,2106); ROSETTI AND RYAN (II,2216); ROSETTI AND RYAN: MEN WHO LOVE WOMEN (II,2217); SUPERCOPS (II,2502)

O'NEILL, ED
THE DAY THE WOMEN GOT EVEN (II,628)

O'NEILL, EILEEN
BURKE'S LAW (I,762)

O'NEILL, JENNIFER
BARE ESSENCE (II,150); COVER UP (II,597); SPECIAL LONDON BRIDGE SPECIAL (I,4150)

O'NEILL, JOAN
DAYTIME (II,630)

O'NEILL, JOHNNY
THE DESILU REVUE (I,1238)

O'NEILL, MICKY
THE SWISS FAMILY ROBINSON (II,2518)

O'NEILL, PAMELA
THE FASHION STORY (I,1537)

O'NEILL, RICHARD
AT EASE (II,116)

O'NEILL, SANDY
SANDY (I,3908)

O'NEILL, TRACY
AS THE WORLD TURNS (II,110)

O'NEILL, TRICIA
THE FALL GUY (II,811)

O'QUINN, TERRY
THE DOCTORS (II,694)

O'REILLY, CYRIL
THE GANGSTER CHRONICLES (II,957); INSPECTOR PEREZ (II,1237)

O'REILLY, JAMES
THE AWAKENING LAND (II,122); THE CHISHOLMS (II,512)

O'REILLY, ROBERT
THE TOM SWIFT AND LINDA CRAIG MYSTERY HOUR (II,2633)

O'ROURKE, HEATHER
HAPPY DAYS (II,1084); MASSARATI AND THE BRAIN (II,1646)

O'ROURKE, THOMAS
THE GUIDING LIGHT (II,1064)

O'SHEA, MICHAEL
IT'S A GREAT LIFE (I,2256); MCGARRY AND ME (I,2966)

O'SHEA, MILO
ELLIS ISLAND (II,768)

O'SHEA, TESSA
THE WORD (II,2833)

O'SHEA, TESSIE
THE ENTERTAINERS (I,1449); THE STRANGE CASE OF DR. JEKYLL AND MR. HYDE (I,4244)

O'SULLIVAN, COLLEEN
MR. BEVIS (I,3128)

O'SULLIVAN, GILBERT
BURT BACHARACH—OPUS NO. 3 (I,774)

O'SULLIVAN, JAMES
ALL MY CHILDREN (II,39); BROTHERS (II,381); OUR FAMILY BUSINESS (II,1931); SEARCH FOR TOMORROW (II,2284); SOMERSET (I,4115)

O'SULLIVAN, MAUREEN
ALL MY CHILDREN (II,39); CAVALCADE OF AMERICA (I,875); THE CHILDREN'S HOUR (I,940); CROSSROADS (I,1105); FIRESIDE THEATER (I,1580); FORD TELEVISION THEATER (I,1634); HOLLYWOOD OPENING NIGHT (I,2087); THE LUX VIDEO THEATER (I,2795); MANDY'S GRANDMOTHER (II,1618); MATINEE THEATER (I,2947); STAR STAGE (I,4197)

O'SULLIVAN, RICHARD
THE HIGHWAYMAN (I,2059); MAN ABOUT THE HOUSE (II,1611); ROBIN'S NEST (II,2187)

O'SULLIVAN, TERRY
THE BEST OF EVERYTHING (I,406); DAYS OF OUR LIVES (II,629); SEARCH FOR TOMORROW (II,2284)

O'TOOLE, ANNETTE
THE BEST LEGS IN 8TH GRADE (II,214); BRIGHT PROMISE (I,727); VANITIES (II,2720)

O'TOOLE, MICHAEL
QUEST FOR ADVENTURE (I,3708)

O'TOOLE, PETER
MASADA (II,1642)

OAK RIDGE BOYS, THE
A COUNTRY CHRISTMAS (II,584); COUNTRY COMES HOME (II,586); COUNTRY GALAXY OF STARS (II,587); JOHNNY CASH: COWBOY HEROES (II,1334); JOHNNY CASH: THE FIRST 25 YEARS (II,1336); LORETTA LYNN: THE LADY. . .THE LEGEND (II,1521); THE NASHVILLE PALACE (II,1792); THE NASHVILLE PALACE (II,1793); A SPECIAL KENNY ROGERS (II,2417)

OAKES, BETTY
VALIANT LADY (I,4695)

OAKES, GARY
PLEASE STAND BY (II,2057)

OAKES, RANDI
BATTLE OF THE NETWORK STARS (II,169); BATTLE OF THE NETWORK STARS (II,170); BATTLE OF THE NETWORK STARS (II,172); BATTLE OF THE NETWORK STARS (II,173); BATTLE OF THE NETWORK STARS (II,174); BATTLE OF THE NETWORK STARS (II,176); BATTLESTAR GALACTICA (II,181); CHIPS (II,511); CIRCUS OF THE STARS (II,534); CIRCUS OF THE STARS (II,535); CIRCUS OF THE STARS (II,538); LOVERS AND OTHER STRANGERS (II,1550); MEN WHO RATE A "10" (II,1684); ON STAGE AMERICA (II,1893); ROSETTI AND RYAN (II,2216); ROSETTI AND RYAN: MEN WHO LOVE WOMEN (II,2217); TWIN DETECTIVES (II,2687)

OAKES, SUSAN
THE SECRET STORM (I,3969)

OAKES, WENDY
FROM HERE TO ETERNITY (II,933)

OAKIE, JACK
BURLESQUE (I,765); THE REAL MCCOYS (I,3741)

OAKLAND, SIMON
ALEXANDER THE GREAT (I,113); THE BLACK SHEEP SQUADRON (II,262); DAVID CASSIDY—MAN UNDERCOVER (II,623); DECOY (I,1217); THE GREAT SEBASTIANS (I,1877); KEY WEST (I,2528); THE NIGHT STALKER (I,3292); THE NIGHT STALKER (II,1849); THE NIGHT STRANGLER (I,3293); THE OUTER LIMITS (I,3426); PROFILES IN COURAGE (I,3678); READY FOR THE PEOPLE (I,3739); THE THIRD COMMANDMENT (I,4449); TOMA (I,4547); TOMA (II,2634); WHO HAS SEEN THE WIND? (I,4847)

OAKLAND, SUE
THE $100,000 BIG SURPRISE (I,3378); WHAT'S GOING ON? (I,4814)

OAKLEY, PATTY
BIG CITY COMEDY (II,233)

OANIARD, NEIL
THE RETURN OF CHARLIE CHAN (II,2136)

OATES, WARREN
THE AFRICAN QUEEN (II,18); BLACK BEAUTY (II,261); THE BLUE AND THE GRAY (II,273); EAST OF EDEN (II,752); STONEY BURKE (I,4229); THRILLER (I,4492); TRUE GRIT (II,2664); WANTED: DEAD OR ALIVE (I,4764)

OATMAN, DONEY
DELTA COUNTY, U.S.A. (II,657); FULL HOUSE (II,936); LITTLE VIC (II,1496); THE ODD COUPLE (II,1875)

OBER, PHILIP
FAMILY AFFAIR (I,1519); HERCULE POIROT (I,2021); SNEAK PREVIEW (I,4100)

OBERDIEAR, KAREN
THE TEXAS WHEELERS (II,2568)

OBERLIN, BILL
PISTOLS 'N' PETTICOATS (I,3608)

OBERMAN, MILT
THE RETURN OF MARCUS WELBY, M.D (II,2138); SUNDAY FUNNIES (II,2493)

OBERON, MERLE
ASSIGNMENT: FOREIGN LEGION (I,300); FORD TELEVISION THEATER (I,1634); THE MAN WHO CAME TO DINNER (I,2880)

OBESTEIN, MAX
AROUND THE WORLD IN 80 DAYS (I,263)

OBLER, PHILIP
GALLAGHER (I,1727)

OBLONG, HAROLD
MR. & MS. AND THE BANDSTAND MURDERS (II,1746)

OBRAZSOVA THEATER, THE
PEGGY FLEMING VISITS THE SOVIET UNION (I,3509)

OCASIO, JOSE
6 RMS RIV VU (II,2371)

ODDIE, BILL
CAMBRIDGE CIRCUS (I,807); THE GOODIES (II,1040)

ODEN, SUSAN
ALFRED OF THE AMAZON (I,116)

ODETTA
TONIGHT WITH BELAFONTE (I,4566)

ODLEY, ELEANOR
WALDO (I,4749)

ODNEY, DOUGLAS
STRATEGIC AIR COMMAND (I,4252)

OEHLER, GRETCHEN
ANOTHER WORLD (II,97); TEXAS (II,2566)

OGG, SAMMY
ADVENTURES IN DAIRYLAND (I,44); BACKBONE OF AMERICA (I,328); THE FURTHER ADVENTURES OF SPIN AND MARTY (I,1720); THE LONELY WIZARD (I,2743); THE NEW ADVENTURES OF SPIN AND MARTY (I,3254); PROFESSIONAL FATHER (I,3677); SPIN AND MARTY (I,4158)

OGILVY, IAN
THE GATHERING STORM (I,1741); MOLL FLANDERS (II,1720); RETURN OF THE SAINT (II,2141)

OGLE, BOB
THE SHIRT TALES (II,2337)

OGLE, NATALIE
PRIDE AND PREJUDICE (II,2080)

OGLE, ROBERT ALLEN
THE KWICKY KOALA SHOW (II,1417)

OH, SOON-TECK
CHARLIE'S ANGELS (II,486); EAST OF EDEN (II,752); ENIGMA (II,778); JUDGE DEE IN THE MONASTERY MURDERS (I,2465); MAGNUM, P.I. (II,1601); MARCO POLO (II,1626); THE RETURN OF CHARLIE CHAN (II,2136); REX HARRISON PRESENTS SHORT STORIES OF LOVE (II,2148); STUNT SEVEN (II,2486)

OHBAYASHI, TAKESHI
SHOGUN (II,2339)

OHMART, CAROL
THE AD-LIBBERS (I,29); LIGHT'S OUT (I,2699); THE LUX VIDEO THEATER (I,2795); MATINEE THEATER (I,2947)

OHRMAN, WARD W
TWIGS (II,2684)

OISTRAKH, DAVID
THE BOB HOPE SHOW (I,562)

OKADA, MASUMI
SPACE GIANTS (I,4142)

OKON, PHIL
SING ALONG WITH MITCH (I,4057)

OKON, TIM
TOM AND JOANN (II,2630)

OLAF, PIERRE
LACE (II,1418)

OLANDT, KEN
CIRCUS OF THE STARS (II,538); RIPTIDE (II,2178); SUMMER (II,2491)

OLAY, RUTH
TIMEX ALL-STAR JAZZ SHOW IV (I,4518)

OLBERDING, ROSEMARY
STRAWHAT MATINEE (I,4255); STRAWHAT THEATER (I,4256)

OLDFIELD, RICHARD
DEATH IN SMALL DOSES (I,1207); THE MARTIAN CHRONICLES (II,1635); MASTER OF THE GAME (II,1647)

OLEK, HENRY
THE PARTRIDGE FAMILY (II,1962); THE SECRET WAR OF JACKIE'S GIRLS (II,2294)

OLFSON, KEN
FLYING HIGH (II,894); THE NANCY WALKER SHOW (II,1790); ROXY PAGE (II,2222)

OLIDI, BOB
THE YEAGERS (II,2843)

OLIN, JESSICA RUTH
THE SCARLET LETTER (II,2270)

OLIN, KEN
BAY CITY BLUES (II,186); HILL STREET BLUES (II,1154)

OLINEY, ALAN
MISSING PIECES (II,1706); THE RENEGADES (II,2132)

OLINEY, RONALD
THE RENEGADES (II,2132)

OLIPHANT, PETER
THE DICK VAN DYKE SHOW (I,1275); PICTURE WINDOW (I,3594)

OLIVER, BARRETT
THE CIRCLE FAMILY (II,527)

OLIVER, HARRY
BORN YESTERDAY (I,692)

OLIVER, JOSIE
BIG HAWAII (II,238)

OLIVER, KYLE
SKEEZER (II,2376)

OLIVER, STEPHEN
BRACKEN'S WORLD (I,703)

OLIVER, SUSAN
AN APARTMENT IN ROME (I,242); THE BARBARA STANWYCK THEATER (I,350); CIRCLE OF FEAR (I,958); DAYS OF OUR LIVES (II,629); THE JUNE ALLYSON SHOW (I,2488); LOVE STORY (I,2778); MATINEE THEATER (I,2947); PEYTON PLACE (I,3574); THE PICTURE OF DORIAN GRAY (I,3591); PLAYHOUSE 90 (I,3623); SUSPICION (I,4309); THRILLER (I,4492); THE TWILIGHT ZONE (I,4651); ZANE GREY THEATER (I,4979)

OLIVIER, FRANCINE
MISTRAL'S DAUGHTER (II,1708)

OLIVIER, LAURENCE
BRIDESHEAD REVISITED (II,375); THE LAST DAYS OF POMPEII (II,1436); LONG DAY'S JOURNEY INTO NIGHT (I,2746); THE MOON AND SIXPENCE (I,3098); THE POWER AND THE GLORY (I,3658)

OLIVIERI, DENNIS
THE NEW PEOPLE (I,3268)

OLKEWICZ, WALTER
THE BLUE AND THE GRAY (II,273); COMEDY OF HORRORS (II,557); THE DUCK FACTORY (II,738); FAMILY TIES (II,819); THE LAST RESORT (II,1440); PARTNERS IN CRIME (II,1961); TAXI (II,2546); TRAVIS MCGEE (II,2656); WIZARDS AND WARRIORS (II,2813)

OLLERENSHAW, MAGGIE
OPEN ALL HOURS (II,1913)

OLMSTEAD, NELSON
THE AD-LIBBERS (I,29); ETHEL AND ALBERT (I,1466); THE PHIL SILVERS SHOW (I,3580); TODAY IS OURS (I,4531)

OLRICH, APRIL
MACBETH (I,2801)

OLSEN AND JOHNSON
THE ALL-STAR REVUE (I,138); FIVE STAR COMEDY (I,1588)

OLSEN, GARY
DAY OF THE TRIFFIDS (II,627)

OLSEN, J C
FIREBALL FUN FOR ALL (I,1577)

OLSEN, JOHN 'OLE'
FIREBALL FUN FOR ALL (I,1577)

OLSEN, JOHNNY
DOORWAY TO FAME (I,1354); FUN FOR THE MONEY (I,1708); JOHNNY OLSEN'S RUMPUS ROOM (I,2433); KIDS AND COMPANY (I,2534); RED GOOSE KID'S SPECTACULAR (I,3751); THE STRAWHATTERS (I,4257)

OLSEN, MERLIN
BOB HOPE SPECIAL: BOB HOPE'S ALL-STAR COMEDY LOOK AT THE FALL SEASON: IT'S STILL FREE AND WORTH IT! (II,299); BOB HOPE SPECIAL: BOB HOPE'S ALL-STAR SUPER BOWL PARTY (II,304); THE DEAN MARTIN CELEBRITY ROAST (II,637); FATHER MURPHY (II,841); THE JUGGLER OF NOTRE DAME (II,1350); LITTLE HOUSE ON THE PRAIRIE (II,1487)

OLSEN, SUSAN
THE BRADY BUNCH (II,362); THE BRADY BUNCH HOUR (II,363); THE BRADY GIRLS GET MARRIED (II,364); THE BRADY KIDS (II,365); TEACHER'S PET (I,4367)

OLSEN, TODD
LUCAN (II,1553)

OLSEN, TRACY
MRS. G. GOES TO COLLEGE (I,3155)

OLSON, ERIC
SWISS FAMILY ROBINSON (II,2516); SWISS FAMILY ROBINSON (II,2517)

OLSON, JAMES
THE COURT-MARTIAL OF GENERAL GEORGE ARMSTRONG CUSTER (I,1081); MANHUNTER (II,1620)

OLSON, JULIE MARIE
HAWAIIAN HEAT (II,1111)

OLSON, LEILA HEE
HAWAIIAN HEAT (II,1111)

OLSON, NANCY
CHANNING (I,900); HIGH TOR (I,2056); KINGSTON: CONFIDENTIAL (II,1398); PAPER DOLLS (II,1952); THE ROYAL FAMILY (I,3861); THE WOMEN (I,4890)

OLSTEAD, NORMAN
SKY KING (I,4077)

OMS, ALBA
FARRELL: FOR THE PEOPLE (II,834)

ONO, YOKO
JOHN LENNON AND YOKO ONO PRESENT THE ONE-TO-ONE CONCERT (I,2414)

ONTKEAN, MICHAEL
THE ROOKIES (I,3840); THE ROOKIES (II,2208)

ONYX, NARDA
BOLD VENTURE (I,682)

OPATOSHU, DAVID
BONINO (I,685); CONSPIRACY OF TERROR (II,569); THE D.A.: MURDER ONE (I,1122); FLASH GORDON—THE GREATEST ADVENTURE OF ALL (II,875); MAN AGAINST CRIME (I,2852); MASADA (II,1642); WOMAN ON THE RUN (II,2821)

OPERATION ENTERTAINMENT GIRLS, THE
OPERATION: ENTERTAINMENT (I,3401)

OPHIR, SHAI K
HANSEL AND GRETEL (I,1934)

OPPENHEIMER, ALAN
BIG EDDIE (II,236); BLACKSTAR (II,263); THE C.B. BEARS (II,406); CLOSE TIES (II,545); DADDY'S GIRL (I,1124); THE DRAK PACK (II,733); FLASH GORDON (II,874); HAPPY DAYS (II,1084); HE AND SHE (I,1985); HOW TO SURVIVE THE 70S AND MAYBE EVEN BUMP INTO HAPPINESS (II,1199); INSIDE O.U.T. (II,1235); THE KID SUPER POWER HOUR WITH SHAZAM (II,1386); THE LIVES OF JENNY DOLAN (II,1502); LIVING IN PARADISE (II,1503); MAUREEN (II,1656); THE NEW ADVENTURES OF MIGHTY MOUSE AND HECKLE AND JECKLE (II,1814); OF MEN OF WOMEN (I,3332); PEEPING TIMES (II,1990); THE SHAMEFUL SECRETS OF HASTINGS CORNERS (I,3994); THE SIX MILLION DOLLAR MAN (I,4067); THE SIX-MILLION-DOLLAR MAN (II,2372); THE SMURFS (II,2385); STOPWATCH: THIRTY MINUTES OF INVESTIGATIVE TICKING (II,2474); SUSAN AND SAM (II,2506); TARZAN: LORD OF THE JUNGLE (II,2544); THREE FOR TAHITI (I,4476); THUNDARR THE BARBARIAN (II,2616); TROLLKINS (II,2660); VALLEY OF THE DINOSAURS (II,2717); WASHINGTON: BEHIND CLOSED DOORS (II,2744)

OPPENHEIMER, JOHN
EISCHIED (II,763)

OQUENDO, LUIS
TONIGHT IN HAVANA (I,4557)

ORBACH, JERRY
ANNIE GET YOUR GUN (I,224); MITZI: A TRIBUTE TO THE AMERICAN HOUSEWIFE (I,3072); PLAZA SUITE (II,2056); THE STREETS (II,2479); TWENTY-FOUR HOURS IN A WOMAN'S LIFE (I,4644); THE WAY THEY WERE (II,2748)

ORBISON, ROY
ELVIS REMEMBERED: NASHVILLE TO HOLLYWOOD (II,772); THE JOHNNY CASH CHRISTMAS SPECIAL (II,1326)

ORCHARD, JOHN
GALLAGHER (I,1727); M*A*S*H (II,1569)

ORENSTEIN, JANET
RUDOLPH THE RED-NOSED REINDEER (II,2227)

ORKIN, DICK
AMERICA ALIVE! (II,67); THE TIM CONWAY SHOW (II,2619)

ORLANDO, DON
AN APARTMENT IN ROME (I,242)

ORLANDO, TONY
BOB HOPE SPECIAL: BOB HOPE PRESENTS A CELEBRATION WITH STARS OF COMEDY AND MUSIC (II,340); THE JOHNNY CASH CHRISTMAS SPECIAL (II,1325); LOUISE MANDRELL: DIAMONDS, GOLD AND PLATINUM (II,1528); LYNDA CARTER: STREET LIGHTS (II,1567); SALUTE TO LADY LIBERTY (II,2243); TONY ORLANDO AND DAWN (II,2639)

ORMAN, ROSCOE
SESAME STREET (I,3982)

ORMES, YVONNE
THE BOB HOPE SHOW (I,638)

ORMSBY-KNOX, IAIN
GREAT EXPECTATIONS (II,1055)

ORR, CORINNE
THE AMAZING THREE (I,162); MARINE BOY (I,2911); SPEED RACER (I,4151)

ORR, MARY
AUNT JENNY'S REAL LIFE STORIES (I,314)

ORRICK, DAVID
THE BROTHERS (I,748)

ORRINSON, JACK
DOWN HOME (I,1366); THE PLAINCLOTHESMAN (I,3614)

ORSEN, JACK
CAPTAIN VIDEO AND HIS VIDEO RANGERS (I,838)

ORTEGA, SANTOS
AS THE WORLD TURNS (II,110); THE BRIGHTER DAY (I,728)

ORTH, FRANK
BOSTON BLACKIE (I,695)

ORTIZ, ANITA
RYAN'S HOPE (II,2234)

OSBORN, JIMMY
ARTHUR GODFREY AND FRIENDS (I,278)

OSBORN, LYN
SPACE PATROL (I,4144)

OSBORN, SCOTT
ENTERTAINMENT TONIGHT (II,780)

OSBORN, STEVE
CAMELOT (II,416)

OSBORNE, BOB
THE DESILU REVUE (I,1238)

OSBORNE, BRUCE
THE PORTER WAGONER SHOW (I,3652)

OSBORNE, FRANCES
ONE OF OUR OWN (II,1910)

OSBORNE, MICHAEL
THE SIX WIVES OF HENRY VIII (I,4068)

OSBURNE, MARY
DON MESSER'S JUBILEE (I,1338)

OSCAR THE WONDER RODENT
THE CHEAP SHOW (II,491)

OSCARD, MIKO
THE HAPPY TIME (I,1943)

OSGOOD, CHARLES
EVERYTHING YOU EVER WANTED TO KNOW ABOUT MONSTERS . . . BUT WERE AFRAID! (II,794)

OSLER, WILLIAM
THE STARLOST (I,4203)

OSMOND BROTHERS, THE
THE ANDY WILLIAMS CHRISTMAS SHOW (I,196); THE ANDY WILLIAMS CHRISTMAS SHOW (I,197); THE ANDY WILLIAMS SHOW (I,205); THE ANDY WILLIAMS SHOW (I,208); THE ANDY WILLIAMS SPECIAL (I,212); THE BOB HOPE SHOW (I,650); CHANGING SCENE (I,899); CHRISTMAS IN WASHINGTON (II,521); DONNY AND MARIE (II,712); THE JERRY LEWIS SHOW (I,2370); MOVIN' WITH NANCY ON STAGE (I,3120); THE PAUL LYNDE COMEDY HOUR (II,1977); THE TRAVELS OF JAIMIE MCPHEETERS (I,4596)

OSMOND FAMILY, THE
THE OSMOND FAMILY CHRISTMAS SPECIAL (II,1926); THE OSMOND FAMILY SHOW (II,1927); THE OSMOND FAMILY THANKSGIVING SPECIAL (II,1928)

OSMOND, ALAN
THE OSMOND BROTHERS SHOW (I,3409); THE OSMOND BROTHERS SPECIAL (II,1925); THE OSMONDS (I,3410); THE OSMONDS SPECIAL (II,1929); THE SEVEN LITTLE FOYS (I,3986)

OSMOND, CLIFF
ALEXANDER THE GREAT (I,113); THE ART OF CRIME (II,108); BEST FRIENDS (II,213)

OSMOND, DONNY
BOB HOPE SPECIAL: BOB HOPE'S BICENTENNIAL STAR SPANGLED SPECTACULAR (II,306); BOB HOPE SPECIAL: BOB HOPE'S CHRISTMAS PARTY (II,307); BOB HOPE SPECIAL: HAPPY BIRTHDAY, BOB! (II,326); CHRISTMAS IN WASHINGTON (II,521); DISNEYLAND'S 25TH ANNIVERSARY (II,681); DONNY AND MARIE (II,712); THE DONNY AND MARIE CHRISTMAS SPECIAL (II,713); THE DONNY AND MARIE OSMOND SHOW (II,714); GENERAL ELECTRIC'S ALL-STAR ANNIVERSARY (II,963); THE OSMOND BROTHERS

SHOW (I,3409); THE OSMOND FAMILY CHRISTMAS SPECIAL (II,1926); THE OSMOND FAMILY SHOW (II,1927); THE OSMOND FAMILY THANKSGIVING SPECIAL (II,1928); THE OSMONDS (I,3410); THE OSMONDS SPECIAL (II,1929); THE SEVEN LITTLE FOYS (I,3986); A TRIBUTE TO "MR. TELEVISION" MILTON BERLE (II,2658); THE WILD WOMEN OF CHASTITY GULCH (II,2801)

OSMOND, ERIC
STILL THE BEAVER (II,2466)

OSMOND, JAY
THE OSMOND BROTHERS SHOW (I,3409); THE OSMOND BROTHERS SPECIAL (II,1925); THE OSMONDS (I,3410); THE OSMONDS SPECIAL (II,1929); THE SEVEN LITTLE FOYS (I,3986)

OSMOND, JIMMY
DONNY AND MARIE (II,712); THE OSMONDS (I,3410); THE OSMONDS SPECIAL (II,1929)

OSMOND, KEN
HAPPY DAYS (II,1084); HIGH SCHOOL, U.S.A. (II,1148); HIGH SCHOOL, U.S.A. (II,1149); LEAVE IT TO BEAVER (I,2648); STILL THE BEAVER (II,2466)

OSMOND, MARIE
THE BIG SHOW (II,243); THE BOB HOPE SHOW (I,664); BOB HOPE SPECIAL: BOB HOPE'S 30TH ANNIVERSARY TV SPECIAL (II,293); BOB HOPE SPECIAL: BOB HOPE'S ALL-STAR COMEDY BIRTHDAY PARTY AT WEST POINT (II,298); BOB HOPE SPECIAL: BOB HOPE'S BICENTENNIAL STAR SPANGLED SPECTACULAR (II,306); BOB HOPE SPECIAL: BOB HOPE'S CHRISTMAS PARTY (II,307); BOB HOPE SPECIAL: BOB HOPE'S SALUTE TO NASA—25 YEARS OF REACHING FOR THE STARS (II,314); BOB HOPE SPECIAL: HAPPY BIRTHDAY, BOB! (II,326); CHRISTMAS IN WASHINGTON (II,521); DONNY AND MARIE (II,712); THE DONNY AND MARIE CHRISTMAS SPECIAL (II,713); THE DONNY AND MARIE OSMOND SHOW (II,714); DOUG HENNING'S WORLD OF MAGIC IV (II,728); GENERAL ELECTRIC'S ALL-STAR ANNIVERSARY (II,963); HOLLYWOOD'S PRIVATE HOME MOVIES II (II,1168); MARIE (II,1628); MARIE (II,1629); THE OSMOND FAMILY CHRISTMAS SPECIAL (II,1926); THE OSMOND FAMILY SHOW (II,1927); THE OSMOND FAMILY THANKSGIVING SPECIAL (II,1928); THE OSMONDS SPECIAL (II,1929); PAUL LYNDE GOES M-A-A-A-AD (II,1980); THE PERRY COMO SUNSHINE SHOW (II,2005); ROOSTER (II,2210); SALUTE TO LADY LIBERTY (II,2243); THE SUZANNE SOMERS SPECIAL (II,2510); A TRIBUTE TO "MR. TELEVISION" MILTON BERLE (II,2658)

OSMOND, MERRILL
THE OSMOND BROTHERS SHOW (I,3409); THE OSMOND BROTHERS SPECIAL (II,1925); THE OSMONDS (I,3410); THE OSMONDS SPECIAL (II,1929); THE SEVEN LITTLE FOYS (I,3986)

OSMOND, WAYNE
THE OSMOND BROTHERS SHOW (I,3409); THE OSMOND BROTHERS SPECIAL (II,1925); THE OSMONDS (I,3410); THE OSMONDS SPECIAL (II,1929); THE SEVEN LITTLE FOYS (I,3986)

OSMONDS, THE
ANN-MARGRET OLSSON (II,84); THE GEORGE BURNS SPECIAL (II,970); ROYAL VARIETY PERFORMANCE (I,3865)

OSSER, GLENN
MUSIC FOR A SPRING NIGHT (I,3169); MUSIC FOR A SUMMER NIGHT (I,3170)

OSTELOH, ROBERT
FOR THE DEFENSE (I,1624)

OSTERHAGE, JEFF
THE LEGEND OF THE GOLDEN GUN (II,1455); THE TEXAS RANGERS (II,2567); TRUE GRIT (II,2664)

OSTERWALD, BIBI
THE ARMSTRONG CIRCLE THEATER (I,260); BEULAH LAND (II,226); BRIDGET LOVES BERNIE (I,722); CAPTAIN BILLY'S MISSISSIPPI MUSIC HALL (I,826); FRONT ROW CENTER (I,1693); GIFT OF THE MAGI (I,1800); THE IMOGENE COCA SHOW (I,2198); OUR TOWN (I,3420); ROARING CAMP (I,3804); THE S. S. HOLIDAY (I,4173); WHERE THE HEART IS (I,4831); WINDOW SHADE REVUE (I,4874); THE WOMEN (I,4890); THE WONDERFUL WORLD OF PHILIP MALLEY (II,2831)

OSTRANDER, RUTH
STOP THE MUSIC (I,4232)

OSTRUS, SHERRY
THE JULIUS LAROSA SHOW (I,2486)

OSWIN, CINDY
BUTTERFLIES (II,404)

OTEL, DUFFY
PAUL WHITEMAN'S SATURDAY NIGHT REVUE (I,3490)

OTIS, GENE
CENTENNIAL (II,477)

OTTENHEIMER, ALBERT
HOW TO SURVIVE A MARRIAGE (II,1198)

OTWELL TWINS, THE
THE LAWRENCE WELK SHOW (I,2644); MEMORIES WITH LAWRENCE WELK (II,1681)

OUSLEY, DINA
BRONK (II,379); STATE FAIR (II,2448)

OVERLAND, BILL
HAPPY DAYS (I,1942)

OVERMIRE, LAURENCE
A CHRISTMAS CAROL (II,517)

OVERSTREET, DENNIS
WOMAN'S PAGE (II,2822)

OVERTON, BILL
BACKSTAIRS AT THE WHITE HOUSE (II,133); FIREHOUSE (II,859); GUESS WHO'S COMING TO DINNER? (II,1063)

OVERTON, FRANK
COLOSSUS (I,1009); EXPOSE (I,1491); ONE STEP BEYOND (I,3388); TWELVE O'CLOCK HIGH (I,4637)

OVERTON, NANCY
THE BERT PARKS SHOW (I,400)

OVERTON, RICK
THE SHOW MUST GO ON (II,2344)

OWEN, BEVERLEY
ALL MY CHILDREN (II,39); ANOTHER WORLD (II,97); KRAFT MYSTERY THEATER (I,2589); THE MUNSTERS (I,3158)

OWEN, BILL
LAST OF THE SUMMER WINE (II,1439)

OWEN, DEIRDRE
STAR TONIGHT (I,4199)

OWEN, ETHEL
THE LARRY STORCH SHOW (I,2618)

OWEN, JACK
DON MCNEILL'S TV CLUB (I,1337)

OWEN, JAY
DOCTOR I.Q. (I,1314)

OWEN, KIM
THE COUNTRY MUSIC MURDERS (II,590)

OWEN, MEG WYNN
UPSTAIRS, DOWNSTAIRS (II,2709); A WOMAN OF SUBSTANCE (II,2820)

OWEN, REGINALD
THE QUEEN AND I (I,3704); TOPPER RETURNS (I,4583)

OWEN, TUDOR
JOHNNY BELINDA (I,2415); MAYOR OF THE TOWN (I,2963); MY FRIEND FLICKA (I,3190)

OWENS, BONNIE
LET ME TELL YOU ABOUT A SONG (I,2661)

OWENS, BUCK
BUCK OWENS TV RANCH (I,750); DEAN MARTIN'S CHRISTMAS AT SEA WORLD (II,644); HEE HAW (II,1123); MURDER CAN HURT YOU! (II,1768)

OWENS, DEIRDRE
THE CRADLE SONG (I,1088)

OWENS, DICK
THE REDD FOXX COMEDY HOUR (II,2126)

OWENS, GARY
BOB HOPE SPECIAL: BOB HOPE FOR PRESIDENT (II,282); THE GONG SHOW (II,1027); THE GREEN HORNET (I,1886); THE HUDSON BROTHERS SHOW (II,1202); LETTERS TO LAUGH-IN (I,2672); MCNAB'S LAB (I,2972); NO SOAP, RADIO (II,1856); THE PERILS OF PENELOPE PITSTOP (I,3525); ROGER RAMJET (I,3831); SCOOBY'S ALL-STAR LAFF-A-LYMPICS (II,2273); THE SCOOBY-DOO/DYNOMUTT HOUR (II,2277); SPACE GHOST (I,4141); SPACE STARS (II,2408); YOGI'S SPACE RACE (II,2853)

OWENS, GRANT
CHARLIE'S ANGELS (II,487)

OWENS, JILL

BLANSKY'S BEAUTIES (II,267)

OWENS, JIM
CHRISTMAS LEGEND OF NASHVILLE (II,522)

OWENS, MARJORIE
DOC HOLLIDAY (I,1307)

OWENS, RAY
ANOTHER LIFE (II,95); ASTRO BOY (I,306); LITTLE DRUMMER BOY, BOOK II (II,1486); PINOCCHIO'S CHRISTMAS (II,2045)

OXENBERG, CATHERINE
DYNASTY (II,746)

OZ, FRANK
THE FANTASTIC MISS PIGGY SHOW (II,827); THE MUPPET SHOW (II,1766)

OZARK GIRLS, THE
A COUNTRY HAPPENING (I,1065)

OZORIO, CECIL
TURN ON (I,4625)

PAAR, JACK
BANK ON THE STARS (I,343); THE BOB HOPE SHOW (I,588); A FUNNY THING HAPPENED ON THE WAY TO HOLLYWOOD (I,1716); JACK PAAR AND A FUNNY THING HAPPENED EVERYWHERE (I,2314); JACK PAAR PRESENTS (I,2315); THE JACK PAAR SHOW (I,2316); THE JACK PAAR SHOW (I,2317); THE JACK PAAR SPECIAL (I,2318); THE JACK PAAR SPECIAL (I,2319); JACK PAAR TONIGHT (I,2320); THE JERRY LEWIS SHOW (I,2365); PLACE THE FACE (I,3611); THE ROSALIND RUSSELL SHOW (I,3845); TAKE ONE STARRING JONATHAN WINTERS (II,2532); THE TONIGHT SHOW (I,4564); UP TO PAAR (I,4683); THE WONDERFUL WORLD OF JACK PAAR (I,4899)

PABLO, ANGEL
THE DON HO SHOW (II,706)

PACAL, FRANCOIS
CORONATION STREET (I,1054)

PACE, JUDY
OH, NURSE! (I,3343); THE ROWAN AND MARTIN REPORT (II,2221); THE YOUNG LAWYERS (I,4940); THE YOUNG LAWYERS (I,4941)

PACE, RALPH
CHIEFS (II,509); SIX PACK (II,2370)

PACE, TOM
GENESIS II (I,1760)

PACEY, STEVEN
BLAKE'S SEVEN (II,264)

PACKER, DAVID
V (II,2713); V: THE FINAL BATTLE (II,2714)

PACKER, DORIS
DOBIE GILLIS (I,1302); HAPPY (I,1938); LEAVE IT TO BEAVER (I,2648); THE THIN MAN (I,4446)

PACKWOOD, HARRY
THE DOCTORS (II,694)

PADBURY, WENDY
DOCTOR WHO (II,689); DOCTOR WHO—THE FIVE DOCTORS (II,690)

PADDY STONE DANCERS, THE
BURT BACHARACH! (I,771); THE DES O'CONNOR SHOW (I,1235); JULIE ON SESAME STREET (I,2485); KRAFT MUSIC HALL PRESENTS SANDLER AND YOUNG (I,2585); THE LONDON PALLADIUM (I,2739); PETULA (I,3573); PICCADILLY PALACE (I,3589)

PADGETT, GLENDA
DOBIE GILLIS (I,1302)

PADILLA JR, MANUEL
THE FLYING NUN (I,1611); TARZAN (I,4361)

PAGAN, ANNA L
JACKIE AND DARLENE (II,1265); THE RAG BUSINESS (II,2106)

PAGAN, PETER
ACADEMY THEATER (I,14)

PAGANO, GIULIA
MASADA (II,1642)

PAGE FIVE SINGERS, THE
THE PATTI PAGE SHOW (I,3479)

PAGE, ANNETTE
CINDERELLA (I,955)

PAGE, ARTHUR
THE CARNATION KILLER (I,845)

PAGE, GERALDINE
BAREFOOT IN ATHENS (I,354); THE BLUE AND THE GRAY (II,273); A CHRISTMAS MEMORY (I,948); MATINEE THEATER (I,2947); THANKSGIVING VISITOR (I,4417); WINDOWS (I,4876)

PAGE, HARRISON
ADVENTURING WITH THE CHOPPER (II,15); BACKSTAIRS AT THE WHITE HOUSE (II,133); C.P.O. SHARKEY (II,407); HIGH FIVE (II,1143); LOVE THY NEIGHBOR (I,2781); SUPERTRAIN (II,2504); SUPERTRAIN (II,2505)

PAGE, KEN
AIN'T MISBEHAVIN' (II,25)

PAGE, LAWANDA
B.A.D. CATS (II,123); BROTHERS AND SISTERS (II,382); DEAN MARTIN'S CELEBRITY ROAST (II,643); DETECTIVE SCHOOL (II,661); GOOD EVENING, CAPTAIN (II,1031); LADIES AND GENTLEMAN. . . BOB NEWHART (II,1420); THE REDD FOXX COMEDY HOUR (II,2126); THE RICHARD PRYOR SPECIAL? (II,2165); SANFORD (II,2255); SANFORD AND SON (II,2256); SANFORD ARMS (II,2257)

PAGE, LAUREL
CAPTAIN CAVEMAN AND THE TEEN ANGELS (II,433); CASPER AND THE ANGELS (II,453); SCOOBY'S ALL-STAR LAFF-A-LYMPICS (II,2273)

PAGE, PATTI
THE BIG RECORD (I,430); THE BOB HOPE SHOW (I,573); THE BOB HOPE SHOW (I,576); THE BOB HOPE SHOW (I,577); THE BOB HOPE SHOW (I,583); THE BOB HOPE SHOW (I,632); COUNTRY NIGHT OF STARS II (II,593); FIVE STARS IN SPRINGTIME (I,1590); I BELIEVE IN MUSIC (I,2165); THE MIKE DOUGLAS CHRISTMAS SPECIAL (I,3040); THE PATTI PAGE SHOW (I,3479); THE PATTI PAGE SHOW (I,3480); THE PATTI PAGE SHOW (I,3481); THE PATTI PAGE SHOW (I,3482); THE SCOTT MUSIC HALL (I,3943); SOMETHING SPECIAL (I,4118)

PAGET, DEBRA
THE JUNE ALLYSON SHOW (I,2488)

PAGETT, GARY
LITTLE HOUSE: THE LAST FAREWELL (II,1491); LOU GRANT (II,1526)

PAGETT, NICOLA
A WOMAN OF SUBSTANCE (II,2820)

PAIGE, JANIS
BABY MAKES FIVE (II,129); THE BETTY WHITE SHOW (II,224); THE BOB HOPE SHOW (I,509); THE BOB HOPE SHOW (I,527); THE BOB HOPE SHOW (I,530); THE BOB HOPE SHOW (I,541); THE BOB HOPE SHOW (I,555); THE BOB HOPE SHOW (I,563); THE BOB HOPE SHOW (I,581); THE BOB HOPE SHOW (I,590); THE BOB HOPE SHOW (I,593); THE BOB HOPE SHOW (I,611); THE BOB HOPE SHOW (I,636); EIGHT IS ENOUGH (II,762); GUN SHY (II,1068); HOORAY FOR LOVE (I,2115); IT'S ALWAYS JAN (I,2264); LANIGAN'S RABBI (II,1427); LANIGAN'S RABBI (II,1428); LAST OF THE PRIVATE EYES (I,2628); MAISIE (I,2833); THE MILTON BERLE SPECIAL (I,3050); NO MAN'S LAND (II,1855); ROBERTA (I,3815); ROBERTA (I,3816); SCHLITZ PLAYHOUSE OF STARS (I,3936); SHOW OF THE YEAR (I,4031); THE TURNING POINT OF JIM MALLOY (II,2670); VALENTINE MAGIC ON LOVE ISLAND (II,2716); VERSATILE VARIETIES (I,4713)

PAIGE, MABEL
THE ALAN YOUNG SHOW (I,104)

PAIGE, ROBERT
THE BIG PAYOFF (I,428); BRIDE AND GROOM (I,721); THE COLGATE COMEDY HOUR (I,997); SCHAEFER CENTURY THEATER (I,3935)

PAINE, DEBBIE
MR. BELVEDERE (I,3127)

PAISNER, DINA
JANE EYRE (I,2338)

PAIVA, NESTOR
COMMAND PERFORMANCE (I,1029); THE DONALD O'CONNOR SHOW (I,1344); ELFEGO BACA (I,1427); FOUR STAR PLAYHOUSE (I,1652); THE GEORGE BURNS AND GRACIE ALLEN SHOW (I,1763)

PAJD, LOUISE
PRISONER: CELL BLOCK H (II,2085); U.F.O. (I,4662)

PAK, JACQUELYN
ANNIE OAKLEY (I,225)

PALANCE, HOLLY
LOOK BACK IN DARKNESS (I,2752); RIPLEY'S BELIEVE IT OR NOT (II,2177); THE THORN BIRDS (II,2600)

PALANCE, JACK
ALICE THROUGH THE LOOKING GLASS (I,122); BRONK (II,379); CURTAIN CALL THEATER (I,1115); DANGER (I,1134); THE GREATEST SHOW ON EARTH (I,1883); GULF PLAYHOUSE

(I,1904); HANDS OF MURDER (I,1929); LIGHT'S OUT (I,2699); THE MOTOROLA TELEVISION HOUR (I,3112); RICKLES (II,2170); RIPLEY'S BELIEVE IT OR NOT (II,2177); RIVAK, THE BARBARIAN (I,3795); THE STRANGE CASE OF DR. JEKYLL AND MR. HYDE (I,4244); SUSPENSE (I,4305); TALES OF THE HAUNTED (II,2538); TEXACO COMMAND PERFORMANCE (I,4407); THE WEB (I,4784); ZANE GREY THEATER (I,4979)

PALAZZO, DONNA
OUTLAW LADY (II,1935); WONDER GIRL (I,4891)

PALER, JANE
HI MOM (I,2043)

PALEY, PATRICIA
THE DOCTORS (II,694)

PALEY, PETRONIA
ANOTHER WORLD (II,97)

PALEY, PHILIP
LAND OF THE LOST (II,1425)

PALEY, WILLIAM
TELEVISION SYMPHONY (I,4384)

PALFREY, YOLANDE
LOVE IN A COLD CLIMATE (II,1539)

PALILLO, RON
THE INVISIBLE WOMAN (II,1244); LAVERNE AND SHIRLEY IN THE ARMY (II,1448); LAVERNE AND SHIRLEY WITH THE FONZ (II,1449); RUBIK, THE AMAZING CUBE (II,2226); WELCOME BACK, KOTTER (II,2761)

PALIN, MICHAEL
MONTY PYTHON'S FLYING CIRCUS (II,1729); THE NEWS IS THE NEWS (II,1838)

PALL, GLORIA
COMMANDO CODY (I,1030); LOW MAN ON THE TOTEM POLE (I,2782)

PALLADINI, JEFFREY
THE MONTEFUSCOS (II,1727)

PALMER, AMY
THE TROUBLE WITH MOTHER (II,2663)

PALMER, ANTHONY
LACY AND THE MISSISSIPPI QUEEN (II,1419)

PALMER, ARNOLD
THE BOB HOPE SHOW (I,622)

PALMER, BETSY
APPOINTMENT WITH ADVENTURE (I,246); AS THE WORLD TURNS (II,110); THE BALLAD OF LOUIE THE LOUSE (I,337); THE CAMPBELL TELEVISION SOUNDSTAGE (I,812); CANDID CAMERA (II,420); CANDID CAMERA (I,819); DANGER (I,1134); GIRL TALK (I,1810); THE GOLDEN AGE OF TELEVISION (II,1016); THE INNER SANCTUM (I,2216); KRAFT TELEVISION THEATER (I,2592); NUMBER 96 (II,1869); A PUNT, A PASS, AND A PRAYER (I,3694); STAR STAGE (I,4197)

PALMER, BOB
THE LINE-UP (I,2707)

PALMER, BURT
ALL CREATURES GREAT AND SMALL (I,129)

PALMER, BYRON
BRIDE AND GROOM (I,721); THE ROBERT Q. LEWIS CHRISTMAS SHOW (I,3811); THIS IS YOUR MUSIC (I,4462)

PALMER, CARLA
HIGHWAY HONEYS (II,1151)

PALMER, DAVID
BOSTON AND KILBRIDE (II,356)

PALMER, DAWSON
LOST IN SPACE (I,2758)

PALMER, GEOFFREY
BUTTERFLIES (II,404); THE FALL AND RISE OF REGINALD PERRIN (II,810)

PALMER, GREGG
THE BLUE AND THE GRAY (II,273); RUN, BUDDY, RUN (I,3870)

PALMER, GRIFF
GO WEST YOUNG GIRL (II,1013)

PALMER, JONI
FAME (II,812)

PALMER, LELAND
DINAH AND HER NEW BEST FRIENDS (II,676); DON'T CALL US (II,709); THE MANHATTAN TRANSFER (II,1619); THERE'S ALWAYS ROOM (II,2583)

PALMER, LILLI
THE BOB HOPE SHOW (I,509); THE DIARY OF ANNE FRANK (I,1262); THE LILLI PALMER SHOW (I,2703); THE LILLI PALMER THEATER (I,2704); PHILCO TELEVISION PLAYHOUSE (I,3583); SUSPENSE (I,4305); TAMING OF THE SHREW (I,4356); THE ZOO GANG (II,2877)

PALMER, LIZA
CHANCE OF A LIFETIME (I,898); KAY KYSER'S KOLLEGE OF MUSICAL KNOWLEDGE (I,2512)

PALMER, MARIA
FOUR STAR PLAYHOUSE (I,1652); ONE STEP BEYOND (I,3388); STARS OVER HOLLYWOOD (I,4211); STORY THEATER (I,4240); THE YOUNG MARRIEDS (I,4942)

PALMER, MAUREEN
STOP THE MUSIC (I,4232)

PALMER, NORMAN
THE HERO (I,2035)

PALMER, PETER
BUNGLE ABBEY (II,396); FUN FAIR (I,1706); THE KALLIKAKS (II,1361); THE LEGEND OF CUSTER (I,2652); WINDOWS, DOORS AND KEYHOLES (II,2806)

PALMER, SCOTT
DAYS OF OUR LIVES (II,629); THE YOUNG AND THE RESTLESS (II,2862)

PALMER, SUE
COMPUTERCIDE (II,561)

PALMER, TOM
JEREMIAH OF JACOB'S NECK (II,1281)

PALMER, TONY
FOREVER FERNWOOD (II,909)

PALMERTON, JEAN
THE BOB NEWHART SHOW (II,342)

PALUZZI, LUCIANA
DOUGLAS FAIRBANKS JR. PRESENTS THE RHEINGOLD THEATER (I,1364); FIVE FINGERS (I,1586); POWDERKEG (I,3657)

PALZIS, KELLY
CAPITOL (II,426); LONE STAR (II,1513)

PANG, JOANNA
ISIS (II,1248)

PANGBORN, FRANKLIN
SVENGALI AND THE BLONDE (I,4310)

PANICH, DAVID
PLIMPTON! DID YOU HEAR THE ONE ABOUT. . .? (I,3629)

PANKHURST, GARRY
SKIPPY, THE BUSH KANGAROO (I,4076)

PANKIN, STUART
CAR WASH (II,437); THE EYES OF TEXAS II (II,800); NO SOAP, RADIO (II,1856); NOT NECESSARILY THE NEWS (II,1863); THE SAN PEDRO BEACH BUMS (II,2250); THE SAN PEDRO BUMS (II,2251); VALENTINE MAGIC ON LOVE ISLAND (II,2716); THE WONDERFUL WORLD OF PHILIP MALLEY (II,2831)

PANNECK, ANTOINETTE
THE DOCTORS (II,694)

PANTER, NICOLE
THE PEE WEE HERMAN SHOW (II,1988)

PANTOLIANO, JOE
FREE COUNTRY (II,921); FROM HERE TO ETERNITY (II,932); MCNAMARA'S BAND (II,1669); MR. ROBERTS (II,1754)

PAOLI, CECILE
BERGERAC (II,209)

PAOLONE, CATHERINE
LONDON AND DAVIS IN NEW YORK (II,1512)

PAPAS, IRENE
MOSES THE LAWGIVER (II,1737)

PAPENFUSS, TONY
NEWHART (II,1835)

PAQUIN, ROBERT
DEAR MOM, DEAR DAD (I,1202)

PARADAY, RON
CANDIDA (II,423)

PARADISE, LYNETTE
KIDS 2 KIDS (II,1389)

PARADY, HERSHA
LITTLE HOUSE ON THE PRAIRIE (II,1487); THE PHOENIX (II,2035)

PARADY, RON
HILL STREET BLUES (II,1154)

PARAGON, JOHN
ELVIRA'S MOVIE MACABRE (II,771); THE FACTS (II,804); THE HALF-HOUR COMEDY HOUR (II,1074); THE PARAGON OF COMEDY (II,1956); THE PEE WEE HERMAN SHOW (II,1988); WELCOME TO THE FUN ZONE (II,2763)

PARE, MICHAEL
CRAZY TIMES (II,602); THE GREATEST AMERICAN HERO (II,1060)

PARFREY, WOODROW
THE APARTMENT HOUSE (I,241); B.J. AND THE BEAR (II,127); DOWN HOME (II,731);

THE FESS PARKER SHOW (II,850); MELVIN PURVIS: G-MAN (II,1679); THE MONEYCHANGERS (II,1724); THE NEW MAVERICK (II,1822); PALS (II,1947); PLANET OF THE APES (II,2049); RETURN OF THE WORLD'S GREATEST DETECTIVE (II,2142); SAM HILL: WHO KILLED THE MYSTERIOUS MR. FOSTER? (I,3896); THE SHAMEFUL SECRETS OF HASTINGS CORNERS (I,3994); TIME EXPRESS (II,2620); VALLEY FORGE (I,4698); A VERY MISSING PERSON (I,4715)

PARHAM, ALFREDINE
OUR STREET (I,3418)

PARIOT, BARBARA
JOSIE AND THE PUSSYCATS (I,2453); JOSIE AND THE PUSSYCATS IN OUTER SPACE (I,2454)

PARIS, BOB
MINSKY'S FOLLIES (II,1703)

PARIS, CHERYL
GOLDEN GATE (II,1020); THE RENEGADES (II,2132)

PARIS, JERRY
THE DICK VAN DYKE SHOW (I,1275); EVIL ROY SLADE (I,1485); MICHAEL SHAYNE, PRIVATE DETECTIVE (I,3021); NAVY LOG (I,3233); STEVE CANYON (I,4224); THOSE WHITING GIRLS (I,4471); TV READERS DIGEST (I,4630); THE UNTOUCHABLES (I,4682); THE WEB (I,4785)

PARIS, ROBBY
THE MONTEFUSCOS (II,1727)

PARIS, ROBIN MARY
O'MALLEY (II,1872)

PARK, MERLE
CINDERELLA (I,955); CINDERELLA (II,526); SUNDAY NIGHT AT THE LONDON PALLADIUM (I,4286)

PARK, PEYTON
CELEBRITY (II,463)

PARKER, ALEXANDER
ANOTHER WORLD (II,97)

PARKER, BRET
SECOND CHANCE (I,3958)

PARKER, DEAN
YOUR HIT PARADE (I,4952)

PARKER, DEE
REHEARSAL CALL (I,3761)

PARKER, DENNIS
THE EDGE OF NIGHT (II,760)

PARKER, EILEEN
DON MCNEILL'S TV CLUB (I,1337)

PARKER, ELEANOR
THE BASTARD/KENT FAMILY CHRONICLES (II,159); THE BOB HOPE CHRYSLER THEATER (I,502); BRACKEN'S WORLD (I,703); FANTASY ISLAND (II,830); GHOST STORY (I,1788); GUESS WHO'S COMING TO DINNER? (II,1063); KNIGHT'S GAMBIT (I,2576); KRAFT SUSPENSE THEATER (I,2591); ONCE UPON A SPY (II,1898); WARNING SHOT (I,4770)

PARKER, ELLEN
CAESAR'S HOUR (I,790); FOODINI THE GREAT (I,1619); KENNEDY (II,1377)

PARKER, F WILLIAM
HELLO, LARRY (II,1128); HERNDON AND ME (II,1139); MITCHELL AND WOODS (II,1709)

PARKER, FESS
ANNIE OAKLEY (I,225); THE BOB HOPE SHOW (I,569); THE CODE OF JONATHAN WEST (I,992); DANIEL BOONE (I,1142); DAVY CROCKETT (I,1181); THE FESS PARKER SHOW (II,850); MERMAN ON BROADWAY (I,3011); MR. SMITH GOES TO WASHINGTON (I,3151); MY LITTLE MARGIE (I,3194)

PARKER, FRANK
ARTHUR GODFREY AND FRIENDS (I,278); BRIDE AND GROOM (I,721)

PARKER, GINNY
THE DUKES OF HAZZARD (II,742)

PARKER, GLORIA
THE CONTINENTAL (I,1042)

PARKER, HELEN
CAR 54, WHERE ARE YOU? (I,842)

PARKER, JAMESON
BATTLE OF THE NETWORK STARS (II,175); ONE LIFE TO LIVE (II,1907); SHADOW OF SAM PENNY (II,2313); SIMON AND SIMON (II,2357)

PARKER, JANET
TIME OUT FOR GINGER (I,4508)

PARKER, JEFF
PRINCESS (II,2084)

PARKER, KATHY
A WORLD APART (I,4910)

PARKER, LARA
THE CHADWICK FAMILY (II,478); DARK SHADOWS (I,1157); JESSICA NOVAK (II,1286); ROOSTER (II,2210); STRANDED (II,2475); WASHINGTON: BEHIND CLOSED DOORS (II,2744)

PARKER, LEW
STARTIME (I,4212); THAT GIRL (II,2570); YOUR SURPRISE STORE (I,4961)

PARKER, MAGGI
HAWAII FIVE-O (II,1110)

PARKER, MONICA
ONE NIGHT BAND (II,1909); WHAT HAVE WE LEARNED, CHARLIE BROWN? (I,4810)

PARKER, NOELLE
RYAN'S HOPE (II,2234)

PARKER, NORMAN
FAMILY TIES (II,819)

PARKER, PAMELA
THE NEW PRICE IS RIGHT (I,3270); THE PRICE IS RIGHT (II,2079)

PARKER, PENNY
MAKE ROOM FOR DADDY (I,2843); MARGIE (I,2909)

PARKER, RHONDA
THE AVENGERS (II,121)

PARKER, SANDY
THE ICE PALACE (I,2182)

PARKER, SARAH JESSICA
SQUARE PEGS (II,2431)

PARKER, STEVE
THE SONNY AND CHER COMEDY HOUR (II,2400)

PARKER, SUZY
THE FABULOUS 50S (I,1501); PLAYHOUSE 90 (I,3623); SYBIL (I,4327); THE TWILIGHT ZONE (I,4651)

PARKER, TODD
PINOCCHIO'S CHRISTMAS (II,2045)

PARKER, WARREN
THE GINGER ROGERS SHOW (I,1805)

PARKER, WES
ALL THAT GLITTERS (II,42); PLEASURE COVE (II,2058)

PARKER, WILLARD
TALES OF THE TEXAS RANGERS (I,4349)

PARKER, WOODY
THE GREATEST GIFT (I,1880)

PARKES, GERALD
CAGNEY AND LACEY (II,410); FRAGGLE ROCK (II,916)

PARKINS, BARBARA
BATTLE OF THE NETWORK STARS (II,163); CAPTAINS AND THE KINGS (II,435); JENNIE: LADY RANDOLPH CHURCHILL (II,1277); THE MANIONS OF AMERICA (II,1623); PEYTON PLACE (I,3574); TESTIMONY OF TWO MEN (II,2564)

PARKINSON, DIAN
JERRY (II,1283); THE NEW PRICE IS RIGHT (I,3270); THE PRICE IS RIGHT (I,3665); THE PRICE IS RIGHT (II,2079); VEGAS (II,2723)

PARKINSON, NANCEE
LI'L ABNER (I,2702)

PARKOW, JOHN
THE DOCTORS (II,694)

PARKS, ANDREW
THESE ARE THE DAYS (II,2585)

PARKS, BERNICE
ONCE UPON A TUNE (I,3372)

PARKS, BERT
BALANCE YOUR BUDGET (I,336); BANDSTAND (I,342); THE BERT PARKS SHOW (I,400); BID N' BUY (I,419); THE BIG PAYOFF (I,428); BREAK THE $250,000 BANK (I,713); BREAK THE BANK (I,712); COUNTY FAIR (I,1075); DOUBLE OR NOTHING (I,1362); THE GIANT STEP (I,1791); HAGGIS BAGGIS (I,1917); HOLD THAT NOTE (I,2073); LIFE'S MOST EMBARRASSING MOMENTS III (II,1472); MASQUERADE PARTY (I,2941); PARTY LINE (I,3464); STOP THE MUSIC (I,4232); THAT SECOND THING ON ABC (II,2572); TWO IN LOVE (I,4657); WKRP IN CINCINNATI (II,2814); YOURS FOR A SONG (I,4975)

PARKS, CATHERINE
BEHIND THE SCREEN (II,200); ZORRO AND SON (II,2878)

PARKS, CHARLES
THE LAST CONVERTIBLE (II,1435); TUCKER'S WITCH (II,2667)

PARKS, DARLENE
FOR RICHER, FOR POORER (II,906)

PARKS, HILDY
LOVE OF LIFE (I,2774); STAR TONIGHT (I,4199)

PARKS, JOEL

NEWMAN'S DRUGSTORE (II,1837); TESTIMONY OF TWO MEN (II,2564)

PARKS, MICHAEL
DIAGNOSIS: DANGER (I,1250); A HATFUL OF RAIN (I,1964); REWARD (II,2147); ROYCE (II,2224); TACK REYNOLDS (I,4329); THEN CAME BRONSON (I,4435); THEN CAME BRONSON (I,4436); TURNOVER SMITH (II,2671); THE YOUNG LAWYERS (I,4940)

PARKS, VAN DYKE
BONINO (I,685); WINDOWS (I,4876)

PARLATO, CHARLIE
THE LAWRENCE WELK SHOW (I,2643); MEMORIES WITH LAWRENCE WELK (II,1681)

PARMENTIER, RICHARD
MASTER OF THE GAME (II,1647)

PARNABY, ALAN
FLAMBARDS (II,869)

PARNELL, DENNIS
LOVE OF LIFE (I,2774)

PARNELL, EMORY
THE LIFE OF RILEY (I,2685); THE LIFE OF RILEY (I,2686); WALDO (I,4749)

PARNELL, MIKE
THE RETURN OF MARCUS WELBY, M.D (II,2138)

PARR, RUSS
ROCK-N-AMERICA (II,2195)

PARR, SALLY
STARS OVER HOLLYWOOD (I,4211)

PARR, STEPHEN
ALL MY CHILDREN (II,39); CALLING DR. STORM, M.D. (II,415); THE LIFE AND TIMES OF EDDIE ROBERTS (II,1468); THE SKATEBIRDS (II,2375)

PARRIS, PAT
BUFORD AND THE GHOST (II,389); FLINTSTONE FAMILY ADVENTURES (II,882); THE FLINTSTONE FUNNIES (II,883); THE FLINTSTONES (II,885); JABBERJAW (II,1261); THE SHIRT TALES (II,2337); YOGI'S SPACE RACE (II,2853)

PARRISH, CLIFFORD
SOMEONE AT THE TOP OF THE STAIRS (I,4114)

PARRISH, ELIZABETH
THE EDGE OF NIGHT (II,760)

PARRISH, FRANNY
GOODNIGHT, BEANTOWN (II,1041)

PARRISH, HELEN
HOUR GLASS (I,2133); LEAVE IT TO BEAVER (I,2648); STARS OVER HOLLYWOOD (I,4211)

PARRISH, JUDY
BARNEY BLAKE, POLICE REPORTER (I,359); CAMEO THEATER (I,808); THE CHEVROLET TELE-THEATER (I,926); GROWING PAYNES (I,1891); MONTGOMERY'S SUMMER STOCK (I,3096); STARLIGHT THEATER (I,4201); THE WEB (I,4784)

PARRISH, JULIE
CAPITOL (II,426); GOOD MORNING WORLD (I,1850); RETURN TO PEYTON PLACE (I,3779)

PARRISH, LESLIE
BANYON (I,345); THE D.A.: CONSPIRACY TO KILL (I,1121)

PARRY, WILLIAM
CAMELOT (II,416)

PARRY, ZALE
MARINELAND CARNIVAL (I,2913)

PARSLOW, FRED
AGAINST THE WIND (II,24); THE SULLIVANS (II,2490)

PARSONS, CANDI
BEST FOOT FORWARD (I,401)

PARSONS, ESTELLE
ALL IN THE FAMILY (II,38); BACKSTAIRS AT THE WHITE HOUSE (II,133); THE DUPONT SHOW OF THE WEEK (I,1388); THE FRONT PAGE (I,1692); THE GUN AND THE PULPIT (II,1067); HOME (I,2099); SENSE OF HUMOR (II,2303); THAT WAS THE YEAR THAT WAS (II,2575)

PARSONS, FERN
ADVENTURES IN DAIRYLAND (I,44); THOSE ENDEARING YOUNG CHARMS (I,4469)

PARSONS, JULIE
CRISIS IN SUN VALLEY (II,604)

PARSONS, KELLY
THE NEW MICKEY MOUSE CLUB (II,1823)

PARSONS, MILTON
BIG EDDIE (II,235); BIG EDDIE (II,236); LOCAL 306 (II,1505)

PARSONS, NICHOLAS
THE BENNY HILL SHOW (II,207); THE EVE ARDEN SHOW (I,1471); THE UGLIEST GIRL IN TOWN (I,4663)

PART, BRIAN
LITTLE HOUSE ON THE PRAIRIE (II,1487)

PARTON, DOLLY
THE BEST LITTLE SPECIAL IN TEXAS (II,215); BURT REYNOLDS' LATE SHOW (I,777); CHER. . . SPECIAL (II,499); CHRISTMAS SPECIAL. . . WITH LOVE, MAC DAVIS (II,523); COUNTRY MUSIC HIT PARADE (II,588); COUNTRY STARS OF THE 70S (II,594); DOLLY (II,699); DOLLY IN CONCERT (II,700); FIFTY YEARS OF COUNTRY MUSIC (II,853); JOHNNY CASH: THE FIRST 25 YEARS (II,1336); KENNY & DOLLY: A CHRISTMAS TO REMEMBER (II,1378); LILY—SOLD OUT (II,1480); MAC DAVIS 10TH ANNIVERSARY SPECIAL: I STILL BELIEVE IN MUSIC (II,1571); MAC DAVIS. . . SOUNDS LIKE HOME (II,1582); THE PORTER WAGONER SHOW (I,3652); THE ROWAN AND MARTIN SPECIAL (I,3855); ROY ACUFF—50 YEARS THE KING OF COUNTRY MUSIC (II,2223)

PARTON, REG
JOHNNY GUITAR (I,2427)

PARTON, STELLA
MUSIC WORLD (II,1775)

PASCOE, DON
WOOBINDA—ANIMAL DOCTOR (II,2832)

PASKEY, EDDIE
STAR TREK (II,2440)

PASQUALE, NICHOLAS
WONDER GIRL (I,4891)

PASSARELLA, ART
MCNAB'S LAB (I,2972); THE STREETS OF SAN FRANCISCO (II,2478)

PASSELTINER, BERNIE
THE HAMPTONS (II,1076)

PASTARINI, DAN
BATTLE OF THE LAS VEGAS SHOW GIRLS (II,162)

PASTELS, THE
ARTHUR MURRAY'S DANCE PARTY (I,293); THE PATRICIA BOWAN SHOW (I,3475)

PASTENE, ROBERT
BUCK ROGERS IN THE 25TH CENTURY (I,751); SHORT, SHORT DRAMA (I,4023); THE TRAP (I,4593)

PASTERNAK, MICHAEL
CO-ED FEVER (II,547)

PASTORELLI, ROBERT
DINER (II,679)

PATAKI, MICHAEL
THE AMAZING SPIDER-MAN (II,63); BENNY AND BARNEY: LAS VEGAS UNDERCOVER (II,206); THE CHOPPED LIVER BROTHERS (II,514); THE EYES OF TEXAS II (II,800); THE FLYING NUN (I,1611); FRIENDS AND LOVERS (II,930); GET CHRISTIE LOVE! (II,981); PAUL SAND IN FRIENDS AND LOVERS (II,1982); PHYL AND MIKHY (II,2037); SAMURAI (II,2249); SPIDER-MAN (II,2424); TERROR AT ALCATRAZ (II,2563); WENDY HOOPER—U. S. ARMY (II,2764)

PATAWABANO, BUCKLEY
ADVENTURES IN RAINBOW COUNTRY (I,47)

PATCH, MARILYN
THE NEW HOWDY DOODY SHOW (II,1818)

PATCHETT, TOM
THE ARTHUR GODFREY SPECIAL (I,288); ARTHUR GODFREY'S PORTABLE ELECTRIC MEDICINE SHOW (I,290); THE CHOPPED LIVER BROTHERS (II,514); MAKE YOUR OWN KIND OF MUSIC (I,2847)

PATE, MICHAEL
THE GREEN FELT JUNGLE (I,1885); HONDO (I,2103); WINCHESTER (I,4872)

PATERSON, VINCENT
JUMP (II,1356)

PATERSON, WILLIAM
HEAR NO EVIL (II,1116)

PATINKIN, MANDY
THAT SECOND THING ON ABC (II,2572); THAT THING ON ABC (II,2574)

PATRICK, BONNIE
GEORGE WASHINGTON (II,978)

PATRICK, BUTCH
GENERAL HOSPITAL (II,964); LIDSVILLE (I,2679); MARINELAND CARNIVAL

(I,2913); MUNSTER, GO HOME! (I,3157); THE MUNSTERS (I,3158); THE REAL MCCOYS (I,3741)

PATRICK, DENNIS
ADVICE TO THE LOVELORN (II,16); BERT D'ANGELO/SUPERSTAR (II,211); THE CLIFFWOOD AVENUE KIDS (II,543); CODE NAME: DIAMOND HEAD (II,549); DALLAS (II,614); DARK SHADOWS (I,1157); THE GEORGIA PEACHES (II,979); NIGHT GAMES (II,1845); RITUALS (II,2182)

PATRICK, JOAN
DEATH OF A SALESMAN (I,1208); DOCTOR KILDARE (I,1315)

PATRICK, JOHN
PENMARRIC (II,1993)

PATRICK, JULIAN
AMAHL AND THE NIGHT VISITORS (I,149)

PATRICK, LEE
THE ALVIN SHOW (I,146); BACKBONE OF AMERICA (I,328); BOSS LADY (I,694); THE MARRIAGE BROKER (I,2920); MIMI (I,3054); TOPPER (I,4582); YOUR SHOW TIME (I,4958)

PATRICK, LORY
THE TALES OF WELLS FARGO (I,4353)

PATRICK, NIGEL
ZERO ONE (I,4981)

PATTEN, ROBERT
ADAM-12 (I,31); GIDGET'S SUMMER REUNION (I,1799); IT CAN'T HAPPEN TO ME (II,1250); JAMES AT 15 (II,1271); LOST IN SPACE (I,2759); THREE FOR THE ROAD (II,2607)

PATTERSON, BILL
SMILEY'S PEOPLE (II,2383)

PATTERSON, CJON DAMITRI
CRASH ISLAND (II,600)

PATTERSON, DICK
DON'T CALL ME MAMA ANYMORE (I,1350); MY SON THE DOCTOR (I,3203); STUMP THE STARS (I,4274)

PATTERSON, ELIZABETH
FRONT ROW CENTER (I,1694); I LOVE LUCY (I,2171); LITTLE WOMEN (I,2720); STAGE 7 (I,4181)

PATTERSON, HANK
GREEN ACRES (I,1884); GUNSMOKE (II,1069); HARD CASE (I,1947); PETTICOAT JUNCTION (I,3571); THE ROY ROGERS SHOW (I,3859)

PATTERSON, HERBERT
GOLDEN WINDOWS (I,1841)

PATTERSON, KARI ANN
IN SECURITY (II,1227)

PATTERSON, KEN
BLACK SADDLE (I,480)

PATTERSON, LEE
DOUGLAS FAIRBANKS JR. PRESENTS THE RHEINGOLD THEATER (I,1364); ONE LIFE TO LIVE (II,1907); SURFSIDE 6 (I,4299); TEXAS (II,2566)

PATTERSON, LORNA
BEANE'S OF BOSTON (II,194); THE BOOK OF LISTS (II,350); GOODTIME GIRLS (II,1042); THE LOVEBIRDS (II,1548); THE OSMOND FAMILY THANKSGIVING SPECIAL (II,1928); PRIVATE BENJAMIN (II,2087); SIDNEY SHORR (II,2349); WORKING STIFFS (II,2834)

PATTERSON, MELODY
F TROOP (I,1499)

PATTERSON, NEVA
CIRCLE OF FEAR (I,958); DOC ELLIOT (I,1306); FAITH BALDWIN'S THEATER OF ROMANCE (I,1515); THE GOVERNOR AND J.J. (I,1863); HANDS OF MURDER (I,1929); JOE DANCER: THE BIG BLACK PILL (II,1300); THE MAN IN THE DOG SUIT (I,2869); MANHATTAN HONEYMOON (I,2887); NICHOLS (I,3283); THE PHILADELPHIA STORY (I,3581); SATINS AND SPURS (I,3917); SHORT, SHORT DRAMA (I,4023); SLEZAK AND SON (I,4086); STAR TONIGHT (I,4199); V (II,2713); V: THE FINAL BATTLE (II,2714); WHAT'S UP DOC? (II,2771)

PATTERSON, WILLIS
AMAHL AND THE NIGHT VISITORS (I,149)

PATTI AND THE BLUEBELLS
IT'S WHAT'S HAPPENING BABY! (I,2278)

PATTISON, JOHN
FRAGGLE ROCK (II,916)

PATTON, BARDY
SUPER CIRCUS (I,4287)

PATTON, BART
OF THIS TIME, OF THAT PLACE (I,3335); RIVERBOAT (I,3797)

PATTON, LUCILLE
JACKIE GLEASON AND HIS AMERICAN SCENE MAGAZINE (I,2321)

PATTON, MARY
BRIEF PAUSE FOR MURDER (I,725); SEARCH FOR TOMORROW (II,2284); SURE AS FATE (I,4297)

PATTON, PHIL
LADIES BE SEATED (I,2598)

PATTON, WILL
RYAN'S HOPE (II,2234)

PATTS, PATRICIA ANN
THE FANTASTIC FUNNIES (II,825); SHE'S A GOOD SKATE, CHARLIE BROWN (I,4012); YOU'RE THE GREATEST, CHARLIE BROWN (I,4973)

PAUL REVERE AND THE RAIDERS
THE JACK BENNY SPECIAL (I,2296); MERV GRIFFIN'S SIDEWALKS OF NEW ENGLAND (I,3017); SCROOGE'S ROCK 'N' ROLL CHRISTMAS (II,2279); WHERE THE ACTION IS (I,4828)

PAUL STEFFIN DANCERS, THE
THE ALL-STAR SUMMER REVUE (I,139)

PAUL, ALAN
THE MANHATTAN TRANSFER (II,1619)

PAUL, ALEXANDRA
PAPER DOLLS (II,1951)

PAUL, EUGENIA
ZORRO (I,4982)

PAUL, FRAVEEN
THE FAR PAVILIONS (II,832)

PAUL, GEORGIA
HARPER VALLEY (II,1094); THE MYSTERIOUS TWO (II,1783)

PAUL, LEE
CONDOMINIUM (II,566); GET CHRISTIE LOVE! (II,982)

PAUL, LES
THE GRAND OLE OPRY (I,1866); THE LES PAUL AND MARY FORD SHOW (I,2658)

PAUL, P R
FAME (II,812)

PAUL, RALPH
SENSE AND NONSENSE (I,3977)

PAUL, RICHARD
CARTER COUNTRY (II,449); HAPPY DAYS (II,1084); HERBIE, THE LOVE BUG (II,1131); ONE IN A MILLION (II,1906); SPACE FORCE (II,2407)

PAUL, ROBERT
THE PEGGY FLEMING SHOW (I,3508)

PAUL, STEVEN
WHATEVER HAPPENED TO DOBIE GILLIS? (II,2774)

PAUL, YVONNE
THE BENNY HILL SHOW (II,207)

PAULA KELLY AND THE MODERNAIRES
THE ROSEMARY CLOONEY SHOW (I,3848)

PAULEY, ALEXANDRA
JOSHUA'S WORLD (II,1344)

PAULIN, SCOTT
THE SHERIFF AND THE ASTRONAUT (II,2326)

PAULL, MORGAN
STUNT SEVEN (II,2486)

PAULL, RICKY
HARD KNOX (II,1087)

PAULSEN, ALBERT
CALL TO DANGER (I,803); DOCTORS HOSPITAL (II,691); GENERAL ELECTRIC TRUE (I,1754); THE GYPSY WARRIORS (II,1070); MCNAMARA'S BAND (II,1669); ONE OF OUR OWN (II,1910); STOP SUSAN WILLIAMS (II,2473); A WALK IN THE NIGHT (I,4750)

PAULSEN, PAT
THE BOB HOPE SHOW (I,633); THE GLEN CAMPBELL GOODTIME HOUR (I,1820); JOEY AND DAD (II,1306); JUST FRIENDS (I,2495); A LAST LAUGH AT THE 60'S (I,2627); THE MOUSE FACTORY (I,3113); PAT PAULSEN FOR PRESIDENT (I,3472); PAT PAULSEN'S HALF A COMEDY HOUR (I,3473); THE SMOTHERS BROTHERS COMEDY HOUR (I,4095); THE SMOTHERS BROTHERS SHOW (II,2384); THE SUMMER SMOTHERS BROTHERS SHOW (I,4281); TOO CLOSE FOR COMFORT (II,2642); THE WILD WILD WEST (I,4863); THE WONDERFUL WORLD OF PIZZAZZ (I,4901)

PAULSON, ALBERT
A WORLD APART (I,4910)

PAVAN, MARISA
THE DIARY OF ANNE FRANK (I,1262); THE KAISER ALUMINUM HOUR (I,2503); THE MONEYCHANGERS (II,1724); SHANGRI-LA (I,3996)

PAVAROTTI, FERNANDO
PAVAROTTI AND FRIENDS (II,1985)

PAVAROTTI, LUCIANO
PAVAROTTI AND FRIENDS (II,1985)

PAVIA, NESTOR
HELLO DERE (I,2010)

PAWLUK, MIRA
MOMENT OF TRUTH (I,3084)

PAXTON, BILL
GREAT DAY (II,1054)

PAXTON, TOM
THE TOM JONES SPECIAL (I,4543)

PAYAN, ILKA
ROOSEVELT AND TRUMAN (II,2209)

PAYCHECK, JOHNNY
COUNTRY NIGHT OF STARS (II,592); JOHNNY CASH: THE FIRST 25 YEARS (II,1336); LIVE FROM THE LONE STAR (II,1501)

PAYMER, DAVID
SHE'S WITH ME (II,2321)

PAYNE, FREDA
THE WAYNE NEWTON SPECIAL (II,2749)

PAYNE, JOHN
CALL OF THE WEST (I,801); THE NASH AIRFLYTE THEATER (I,3227); O'CONNOR'S OCEAN (I,3326); THE PHILADELPHIA STORY (I,3581); THE RESTLESS GUN (I,3773); THE RESTLESS GUN (I,3774)

PAYNE, JULIE
THE DUCK FACTORY (II,738); FULL HOUSE (II,937); HOT W.A.C.S. (II,1191); MISS BISHOP (I,3063); PRIME TIMES (II,2083); WIZARDS AND WARRIORS (II,2813)

PAYNE, TERI
CHICO AND THE MAN (II,508)

PAYNE, THERESA
GEORGE WASHINGTON (II,978)

PAYTON, JO MARIE
THE PLANT FAMILY (II,2050)

PAYTON, PEYTON
SKYWARD CHRISTMAS (II,2378)

PAYTON-WRIGHT, PAMELA
THE ADAMS CHRONICLES (II,8)

PEABODY, DICK
COMBAT (I,1011); SIDEKICKS (II,2348)

PEACH, FRANCES
LAVERNE AND SHIRLEY (II,1446)

PEACH, JOAN
PUNCH AND JODY (II,2091)

PEACH, MARY
THE FAR PAVILIONS (II,832)

PEACOCK DANCERS, THE
PINK LADY (II,2041)

PEAKE-JONES, TESSA
PRIDE AND PREJUDICE (II,2080)

PEAKER, E J
THE GREATEST AMERICAN HERO (II,1060); MADAME'S PLACE (II,1587); NIGHT GALLERY (I,3287); THAT'S LIFE (I,4426); TOP OF THE MONTH (I,4578)

PEARCE SISTERS, THE
THE POLLY BERGEN SPECIAL (I,3644)

PEARCE, AL
THE AL PEARCE SHOW (I,95)

PEARCE, ALAN
SCARLETT HILL (I,3933)

PEARCE, ALICE
ACRES AND PAINS (I,19); ALICE IN WONDERLAND (I,120); THE ALICE PEARCE SHOW (I,121); BEWITCHED (I,418); DENNIS THE MENACE (I,1231); JAMIE (I,2334); THE JEAN CARROLL SHOW (I,2353); MY BOY GOOGIE (I,3185); THE WONDERFUL WORLD OF JACK PAAR (I,4899)

PEARCE, GAVIN
GEORGE WASHINGTON (II,978)

PEARCE, JACQUELINE
BLAKE'S SEVEN (II,264)

PEARCY, PATRICIA
LITTLE HOUSE: BLESS ALL THE DEAR CHILDREN (II,1489); ONE LIFE TO LIVE (II,1907)

PEARDON, PATRICIA
JOHNNY JUPITER (I,2429); OUR FIVE DAUGHTERS (I,3413)

PEARL, BARRY
BEST FRIENDS (II,213); C.P.O. SHARKEY (II,407); THE MUNSTERS' REVENGE (II,1765)

PEARL, MINNIE
ANN-MARGRET. . . RHINESTONE COWGIRL (II,87); THE BOB HOPE SHOW (I,643); BURT REYNOLDS' LATE SHOW (I,777); CHRISTMAS LEGEND OF NASHVILLE (II,522); A COUNTRY CHRISTMAS (II,582); A COUNTRY CHRISTMAS (II,584); A COUNTRY CHRISTMAS (II,583); COUNTRY COMES HOME (II,585); COUNTRY MUSIC CARAVAN (I,1066); GEORGE BURNS IN NASHVILLE?? (II,969); THE GRAND OLE OPRY (I,1866); HEE HAW (II,1123); JOHNNY CASH AND THE COUNTRY GIRLS (II,1323); JOHNNY CASH: THE FIRST 25 YEARS (II,1336); ON STAGE AMERICA (II,1893); ROY ACUFF—50 YEARS THE KING OF COUNTRY MUSIC (II,2223); SWINGING COUNTRY (I,4322)

PEARL, RENEE
THE DOCTORS (II,694)

PEARLMAN, MICHAEL
CHARLES IN CHARGE (II,482)

PEARLMAN, RODNEY
ADVENTURES OF THE SEASPRAY (I,82)

PEARLMAN, STEPHEN
CELEBRITY (II,463); COREY: FOR THE PEOPLE (II,575); ETHEL IS AN ELEPHANT (II,783); FUTURE COP (II,946)

PEARN, NAT
A HORSEMAN RIDING BY (II,1182)

PEARSON, ANN
SEARCH FOR TOMORROW (II,2284); VALIANT LADY (I,4695)

PEARSON, CAROL ANNE
THE PARTRIDGE FAMILY (II,1962)

PEARSON, GARRET
THIS IS KATE BENNETT (II,2597)

PEARSON, GE GE
CRUSADER RABBIT (I,1109)

PEARSON, KAREN
THE BORROWERS (I,693)

PEARSON, RED
THE SOLDIERS (I,4112)

PEARSON, RICHARD
THERESE RAQUIN (II,2584)

PEARTHREE, PIPPA
BUFFALO BILL (II,387)

PEARY, HAL
BLONDIE (I,486); BUFORD AND THE GHOST (II,389); FIBBER MCGEE AND MOLLY (I,1569); THE KRAFT 75TH ANNIVERSARY SPECIAL (II,1409); THE PIRATES OF FLOUNDER BAY (I,3607); THE ROMAN HOLIDAYS (I,3835); RUDOLPH'S SHINY NEW YEAR (II,2228); STARS OVER HOLLYWOOD (I,4211); WILLY (I,4869)

PEASE, PATSY
MR. SUCCESS (II,1756); SEARCH FOR TOMORROW (II,2284)

PECK, BOB
NICHOLAS NICKLEBY (II,1840)

PECK, ED
CALL TO DANGER (I,802); HAPPY DAYS (II,1084); MAJOR DELL CONWAY OF THE FLYING TIGERS (I,2835); SEMI-TOUGH (II,2300); THE SUPER (I,4293); TOPPER RETURNS (I,4583)

PECK, FLETCHER
THE JAN MURRAY SHOW (I,2335)

PECK, GREGORY
THE BLUE AND THE GRAY (II,273); THE FIRST 50 YEARS (II,862); GEORGE BURNS' 100TH BIRTHDAY PARTY (II,971); HOLLYWOOD: THE SELZNICK YEARS (I,2095); JACK BENNY'S NEW LOOK (I,2301); A SALUTE TO STAN LAUREL (I,3892); A TRIBUTE TO "MR. TELEVISION" MILTON BERLE (II,2658)

PECK, HARRIS B
THE HOUSE ON HIGH STREET (I,2141)

PECK, JIM
THE BIG SHOWDOWN (II,244); HOT SEAT (I,2125); THE JOKER'S WILD (II,1341); SECOND CHANCE (II,2286); SIX PACK (II,2370); THREE'S A CROWD (II,2612); YOU DON'T SAY (II,2859)

PECK, STANLEY
MAN AGAINST CRIME (I,2852)

PECK, STEVE
333 MONTGOMERY (I,4487)

PECK, THEODORA
THE FORTUNE HUNTER (I,1646)

PEDDLERS, THE
THE ROBERT GOULET SPECIAL (I,3807)

PEDEN, EMILY
THE NEW ZOO REVUE (I,3277)

PEDI, TOM
ARNIE (I,261); DOMINIC'S DREAM (II,704)

PEDLEY, ANTHONY
APPOINTMENT WITH A KILLER (I,245)

PEDROSKI, LEFTY
CAR WASH (II,437)

PEDROZA, INEZ
CONDOMINIUM (II,566); THE GREEN FELT JUNGLE (I,1885)

PEEPLES, NIA
FAME (II,812)

PEERCE, JAN
FESTIVAL OF MUSIC (I,1567); OPERA VS. JAZZ (I,3399); THE SPIKE JONES SHOW (I,4155); TWELVE STAR SALUTE (I,4638)

PEGGINS, DON
THE MAGIC SLATE (I,2820)

PEIGLEBECK, WILL
HOME (I,2099)

PEINE, JOSH
DON'T CALL ME CHARLIE (I,1349)

PEINE, VIRGINIA
THE STORK CLUB (I,4236)

PEIRCE, ROBERT
THE DOOLEY BROTHERS (II,715); JOANIE LOVES CHACHI (II,1295); SHE'S IN THE ARMY NOW (II,2320)

PELAG, ALEX
MASADA (II,1642)

PELHAM, GEORGE
MYSTERY AND MRS. (I,3210)

PELIKAN, LISA
BEACON HILL (II,193); THE COUNTRY GIRL (I,1064); THE LAST CONVERTIBLE (II,1435); STUDS LONIGAN (II,2484); TRUE GRIT (II,2664)

PELISH, THELMA
CALL HER MOM (I,796)

PELLEGRINI, GENE
LOVE OF LIFE (I,2774)

PELLETIER, GILLES
ROYAL CANADIAN MOUNTED POLICE (I,3860)

PELLETTI, JOHN
PANAMA HATTIE (I,3444)

PELLEY, FARRELL
OLD NICKERBOCKER MUSIC HALL (I,3351)

PELLIZER, DARRYL
ALL THE RIVERS RUN (II,43)

PELLMIGE, THELMA
CANDID CAMERA (I,819)

PELLOW, CLIFF
BIG EDDIE (II,235); BIG EDDIE (II,236); THE BOYS (II,360); THE CONFESSIONS OF DICK VAN DYKE (II,568); MARIE (II,1628); SCAMPS (II,2267)

PELOQUIN, JEAN
THE VIRGINIAN (I,4738)

PELT, TIM
ROOSEVELT AND TRUMAN (II,2209)

PELUCE, MEENO
THE BAD NEWS BEARS (II,134); BEST OF THE WEST (II,218); KANGAROOS IN THE KITCHEN (II,1362); VOYAGERS (II,2730); W*A*L*T*E*R (II,2731)

PELUSO, LISA
SEARCH FOR TOMORROW (II,2284)

PEMBERTON, BONNIE
CELEBRITY (II,463)

PENAU, IVES
SMILEY'S PEOPLE (II,2383)

PENBERTHY, BEVERLY
ANOTHER WORLD (II,97)

PENCE, DENICE
THE GUIDING LIGHT (II,1064)

PENDERGRASS, TEDDY
MEN WHO RATE A "10" (II,1684); TEDDY PENDERGRASS IN CONCERT (II,2551)

PENDLETON, AUSTIN
BIG CITY BOYS (II,232); LET'S CELEBRATE (I,2663); YOU'RE GONNA LOVE IT HERE (II,2860)

PENDLETON, DAVID
THE GUIDING LIGHT (II,1064)

PENDLETON, KAREN
THE MICKEY MOUSE CLUB (I,3025); THE MOUSEKETEERS REUNION (II,1742)

PENDLETON, OLIVE
FLAMBARDS (II,869)

PENDLETON, STEVE
THE LIFE OF RILEY (I,2686)

PENDLETON, WAYNE
THE EDGE OF NIGHT (II,760)

PENDLETON, WYMAN
MURDER INK (II,1769)

PENDREY, RICHARD
FILE IT UNDER FEAR (I,1574)

PENGHLIS, THAAO
DAYS OF OUR LIVES (II,629)

PENHALL, BRUCE
CHIPS (II,511); CIRCUS OF THE STARS (II,537)

PENLOW, ARTHUR
CORONATION STREET (I,1054)

PENN, DALLIA
THE FORSYTE SAGA (I,1645)

PENN, LEO
BEN CASEY (I,394); MRS. G. GOES TO COLLEGE (I,3155)

PENNARIO, LEONARD
THE JUDY GARLAND SHOW (I,2471)

PENNELL, LARRY
THE BEVERLY HILLBILLIES (I,417); CITY BENEATH THE SEA (I,965); THE GREEN FELT JUNGLE (I,1885); HUNTER (II,1204); LASSIE (I,2624); RIPCORD (I,3793); THREE FOR DANGER (I,4475)

PENNELL, NICHOLAS
THE FORSYTE SAGA (I,1645)

PENNEY, HANNIBAL
RYAN'S HOPE (II,2234)

PENNEY, JENNIFER
CINDERELLA (II,526)

PENNINGTON, JANICE
THE NEW PRICE IS RIGHT (I,3270); THE PRICE IS RIGHT (I,3665); THE PRICE IS RIGHT (II,2079)

PENNINGTON, MARLA
THE BUFFALO SOLDIERS (II,388); HERNDON AND ME (II,1139); SOAP (II,2392)

PENNOCK, CHRIS
DARK SHADOWS (I,1157); GENERAL HOSPITAL (II,964); SOMERSET (I,4115)

PENNSYLVANIANS, THE
THE FRED WARING SHOW (I,1679)

PENNY, DON
THE BRIGHTER DAY (I,728); THE LIEUTENANT (I,2680); THAT GIRL (II,2570); THE WACKIEST SHIP IN THE ARMY (I,4744)

PENNY, JENNIFER
HEAT OF ANGER (I,1992)

PENNY, JOE
DELTA COUNTY, U.S.A. (II,657); FOREVER FERNWOOD (II,909); THE GANGSTER CHRONICLES (II,957); THE HOME FRONT (II,1171); MOTHER, JUGGS AND SPEED (II,1741); RIPTIDE (II,2178); SAMURAI (II,2249); SAVAGE: IN THE ORIENT (II,2264)

PENNY, SYDNEY
THE CIRCLE FAMILY (II,527); DEAR TEACHER (II,653); IT'S MAGIC, CHARLIE BROWN (I,2271); THE LAST LEAF (II,1437); THE NIGHT RIDER (II,1848); THE THORN BIRDS (II,2600); THROUGH THE MAGIC PYRAMID (II,2615)

PENROD, HANNA
BE FEMININE WITH HANNA (I,368)

PENROD, JUDITH
CODE RED (II,553)

PENTECOST, GEORGE
AFTER GEORGE (II,19); BLANSKY'S BEAUTIES (II,267); FOG (II,897); SNAVELY (II,2387); YOUR PLACE OR MINE? (II,2874)

PENTELOW, ARTHUR
CASANOVA (II,451)

PENYA, ANTHONY
CODE RED (II,553)

PEPPARD, GEORGE
THE A-TEAM (II,119); BANACEK (II,138); BANACEK: DETOUR TO NOWHERE (I,339); THE DEAN MARTIN CELEBRITY ROAST (II,639); DOCTORS HOSPITAL (II,691); EVENING AT THE MOULIN ROUGE (II,787); THE GOLDEN AGE OF TELEVISION (II,1019); HOLLYWOOD STARS' SCREEN TESTS (II,1165); LITTLE MOON OF ALBAN (I,2714); MATINEE THEATER (I,2947); ONE OF OUR OWN (II,1910); SUSPICION (I,4309); TWILIGHT THEATER (II,2685)

PEPPER, BARBARA
THE GREATEST SHOW ON EARTH (I,1883); GREEN ACRES (I,1884); THE JACK BENNY PROGRAM (I,2294); POOR MR. CAMPBELL (I,3647)

PEPPER, COREY
STUDS LONIGAN (II,2484)

PEPPER, CYNTHIA
MARGIE (I,2909); SALLY AND SAM (I,3891); THREE COINS IN THE FOUNTAIN (I,4473)

PEPPERCORN PLAYERS PUPPETS
PERRY COMO'S CHRISTMAS IN NEW YORK (II,2011)

PERA, LISA
HELLO DERE (I,2010)

PERA, RADAMAS
GIDGET GETS MARRIED (I,1796); KUNG FU (II,1416); LITTLE HOUSE ON THE PRAIRIE (II,1487)

PERAK, JOHN
THE DEADLY GAME (II,633)

PERAULT, ROBERT
ALL MY CHILDREN (II,39)

PERCEVAL, ROBERT
THE BUCCANEERS (I,749)

PERCIVAL, LONIE
THE BEATLES (I,380)

PERCY, PAT
UNCLE MISTLETOE AND HIS ADVENTURES (I,4669)

PERETZ, SUSAN
A.E.S. HUDSON STREET (II,17)

PEREZ, ANTHONY
POPI (II,2068); POPI (II,2069)

PEREZ, JOSE
ACES UP (II,7); CALUCCI'S DEPARTMENT (I,805); I'LL NEVER FORGET WHAT'S HER NAME (II,1217); INSPECTOR PEREZ (II,1237); ON THE ROCKS (II,1894); STEAMBATH (II,2451)

PEREZ, LOUIS
LITTLE SHOTS (II,1495)

PEREZ, PAUL
THE CBS NEWCOMERS (I,879)

PERGAMENT, ASHER
THE ADAMS CHRONICLES (II,8)

PERITO, JENNIFER
MAKIN' IT (II,1603)

PERKINS, ANTHONY
HEDDA HOPPER'S HOLLYWOOD (I,2002); WINDOWS (I,4876); THE WORLD OF SOPHIA LOREN (I,4918)

PERKINS, CARL
ELVIS REMEMBERED: NASHVILLE TO HOLLYWOOD (II,772); THE JOHNNY CASH CHRISTMAS SPECIAL (II,1326); THE JOHNNY CASH SHOW (I,2424)

PERKINS, DARIUS
ALL THE RIVERS RUN (II,43)

PERKINS, GAY
THE CBS NEWCOMERS (I,879)

PERKINS, JACK
THE CIRCLE FAMILY (II,527); THE HERO (I,2035)

PERKINS, JOHN
ROYAL CANADIAN MOUNTED POLICE (I,3860)

PERKINS, JOSEPH
WINDOWS (I,4876)

PERKINS, JUDY
MIDWESTERN HAYRIDE (I,3033)

PERKINS, KENT
ALICE (II,33); STEPHANIE (II,2453)

PERKINS, LESLIE
WE'LL TAKE MANHATTAN (I,4792)

PERKINS, MILLIE
THE BOB HOPE SHOW (I,575); THE TROUBLE WITH GRANDPA (II,2662)

PERKINS, RAY
THE RAY PERKINS REVUE (I,3735)

PERKINS, VOLTAIRE
DIVORCE COURT (I,1297)

PERLE, BECKY
THE KID SUPER POWER HOUR WITH SHAZAM (II,1386)

PERLMAN, CLIFF
ALAN KING IN LAS VEGAS, PART II (I,98)

PERLMAN, ITZHAK
JOHN DENVER: MUSIC AND THE MOUNTAINS (II,1318)

PERLMAN, RHEA
CHEERS (II,494); TAXI (II,2546)

PERPICH, JONATHAN
THE FAMILY TREE (II,820); YOUNG LIVES (II,2866)

PERREAU, GIGI
THE BETTY HUTTON SHOW (I,413); CHALK ONE UP FOR JOHNNY (I,893); CROWN THEATER WITH GLORIA SWANSON (I,1107); FOLLOW THE SUN (I,1617); FORD TELEVISION THEATER (I,1634)

PERREAU, JANINE
A CHRISTMAS CAROL (I,947)

PERREAULT, JACQUES
AS THE WORLD TURNS (II,110)

PERRIN, MAC
THE REAL MCKAY (I,3742)

PERRIN, VIC
DRAGNET (I,1370); FLASH GORDON—THE GREATEST ADVENTURE OF ALL (II,875); FOR THE DEFENSE (I,1624); FRONTIER (I,1695); THE NEW ADVENTURES OF HUCKLEBERRY FINN (I,3250); THE OUTER LIMITS (I,3426); THE WORLD'S GREATEST SUPER HEROES (II,2842)

PERRINE, VALERIE
CELEBRITY CHALLENGE OF THE SEXES 2 (II,466); CIRCUS OF THE STARS (II,530); CIRCUS OF THE STARS (II,531); CIRCUS OF THE STARS (II,532); CIRCUS OF THE STARS (II,534); LADY LUCK (I,2604); LOVE STORY (I,2778); MALIBU (II,1607); RODNEY DANGERFIELD SPECIAL: IT'S NOT EASY BEIN' ME (II,2199)

PERRY, BARBARA
THE DICK VAN DYKE SHOW (I,1275); THE MARGE AND GOWER CHAMPION SHOW (I,2907)

PERRY, CELIA
FARRELL: FOR THE PEOPLE (II,834)

PERRY, ELIZABETH
MORNING STAR (I,3104)

PERRY, FELTON
THE CITY (II,541); DIAL HOT LINE (I,1254); THE FUZZ BROTHERS (I,1722); MATT LINCOLN (I,2949)

PERRY, JIM
CARD SHARKS (II,438); IT'S YOUR MOVE (I,2281); SALE OF THE CENTURY (II,2239)

PERRY, JOE
FOUR EYES (II,912); THE WHIRLYBIRDS (I,4841)

PERRY, JOHN BENNETT
240-ROBERT (II,2688); 240-ROBERT (II,2689); BOBBY JO AND THE BIG APPLE GOODTIME BAND (I,674); EVERYDAY (II,793); PAPER DOLLS (II,1952); TALES OF THE APPLE DUMPLING GANG (II,2536)

PERRY, JOSEPH
I LOVE A MYSTERY (I,2170); M*A*S*H (II,1569)

PERRY, MARY
WINDOWS (I,4876)

PERRY, POSY
THE THORNTON SHOW (I,4468)

PERRY, ROD
S.W.A.T. (II,2236)

PERRY, ROGER
ARREST AND TRIAL (I,264); THE BARBARA EDEN SHOW (I,346); CONSPIRACY OF TERROR (II,569); CRISIS (I,1102); THE D.A.: CONSPIRACY TO KILL (I,1121); THE DESILU REVUE (I,1238); THE FACTS OF LIFE (II,805); THE FIRST HUNDRED YEARS (I,1582); GIDGET GETS MARRIED (I,1796); HARRIGAN AND SON (I,1955); THE HAT OF SERGEANT MARTIN (I,1963); THE MAN WITH THE POWER (II,1616); MOST WANTED (II,1738); THINK PRETTY (I,4448); YOU'RE ONLY YOUNG TWICE (I,4970)

PERRY, STEVE
ON THE FLIP SIDE (I,3360)

PERRY, VICTORIA
BANYON (I,344); FROM HERE TO ETERNITY (II,932)

PERRY, WOLFE
THE WHITE SHADOW (II,2788)

PERRYMAN, CLARA
VENICE MEDICAL (II,2726)

PERSCHY, MARIA
GENERAL HOSPITAL (II,964)

PERSHING, DIANE
BATMAN AND THE SUPER SEVEN (II,160); FLASH GORDON (II,874); FLASH GORDON—THE GREATEST ADVENTURE OF ALL (II,875); THE NEW ADVENTURES OF MIGHTY MOUSE AND HECKLE AND JECKLE (II,1814)

PERSKY, LISA JANE
BACK TOGETHER (II,131)

PERSOFF, NEHEMIAH
THE 13TH DAY: THE STORY OF ESTHER (II,2593); CONDOMINIUM (II,566); THE DANGEROUS DAYS OF KIOWA JONES (I,1140); ERIC (I,1451); THE GOLDEN AGE OF TELEVISION (II,1016); THE HELLCATS (I,2009); HIGH HOPES (II,1144); THE PEOPLE NEXT DOOR (I,3519); PLAYWRIGHTS '56 (I,3627); THE REBELS (II,2121); REUNION IN VIENNA (I,3780); RICH MAN, POOR MAN—BOOK II (II,2162);

PETERSON, OSCAR
THE PERRY COMO SPECIAL (I,3536); STEVE AND EYDIE CELEBRATE IRVING BERLIN (II,2456)

PETERSON, RENNY
MAKE ME LAUGH (I,2839); NO SOAP, RADIO (II,1856)

PETERSON, SANDRA
THE EYES OF TEXAS II (II,800)

PETERSON, TOMMY
THE CHEROKEE TRAIL (II,500); V (II,2713)

PETERSON, VIDAL
MORK AND MINDY (II,1735); THE THORN BIRDS (II,2600)

PETHERBRIDGE, EDWARD
NICHOLAS NICKLEBY (II,1840)

PETINA, IRA
VICTORY (I,4729)

PETIT, MICHAEL
FIVE'S A FAMILY (I,1591)

PETLOCK, JOHN
BRING 'EM BACK ALIVE (II,377)

PETRA, HORTENSE
DALLAS (II,614)

PETRI, SUZANNE
A CHRISTMAS CAROL (II,517)

PETRIE, DANIEL
THE GOLDEN AGE OF TELEVISION (II,1019)

PETRIE, DONALD
THE 416TH (II,913)

PETRIE, DORIS
HIGH HOPES (II,1144)

PETRIE, GEORGE O
THE BLUE AND THE GRAY (II,273); DALLAS (II,614); THE EDGE OF NIGHT (II,760); THE HONEYMOONERS (I,2110); THE JACKIE GLEASON SHOW (I,2322); LEAVE IT TO BEAVER (I,2648); OPERATION PETTICOAT (II,1916); SEARCH FOR TOMORROW (II,2284)

PETRILLO, TONY
THE UNCLE FLOYD SHOW (II,2703)

PETROVITCH, MICHAEL
THE ZOO GANG (II,2877)

PETROVNA, SONIA
THE EDGE OF NIGHT (II,760); SEARCH FOR TOMORROW (II,2284)

PETTERSON, ORVILLE
THE FIRST HUNDRED YEARS (I,1582)

PETTET, JOANNA
ALL THAT GLITTERS (II,41); APPOINTMENT WITH A KILLER (I,245); BATTLE OF THE NETWORK STARS (II,163); CANNON: THE RETURN OF FRANK CANNON (II,425); CAPTAINS AND THE KINGS (II,435); THE DELPHI BUREAU (I,1224); THE DOCTORS (II,694); A KILLER IN EVERY CORNER (I,2538); KNOTS LANDING (II,1404); MISS STEWART, SIR (I,3064); NIGHT GALLERY (I,3287); THREE FOR DANGER (I,4475); WINNER TAKE ALL (II,2808)

PETTEWAY, JAMES
ONE NIGHT BAND (II,1909)

PETTINGER, GARY
HEAR NO EVIL (II,1116)

PETWAY, ALVA
BENSON (II,208)

PEYSER, PENNY
B.J. AND THE BEAR (II,125); B.J. AND THE BEAR (II,127); BATTLE OF THE NETWORK STARS (II,175); THE BLUE AND THE GRAY (II,273); CRAZY LIKE A FOX (II,601); NATIONAL LAMPOON'S TWO REELERS (II,1796); RICH MAN, POOR MAN—BOOK II (II,2162); THE TONY RANDALL SHOW (II,2640); WILD TIMES (II,2799)

PFEIFFER, CONSTANCE
THEY ONLY COME OUT AT NIGHT (II,2586)

PFEIFFER, MICHELLE
B.A.D. CATS (II,123); DELTA HOUSE (II,658)

PFENNING, WESLEY
ANOTHER WORLD (II,97); THE KID WITH THE BROKEN HALO (II,1387)

PFLUG, JO ANN
BURT AND THE GIRLS (I,770); CANDID CAMERA (II,420); THE DAY THE WOMEN GOT EVEN (II,628); THE FALL GUY (II,811); THE FANTASTIC FOUR (I,1529); THE NEW OPERATION PETTICOAT (II,1826); NICK AND NORA (II,1842); THE NIGHT STRANGLER (I,3293); NUTS AND BOLTS (II,1871); RITUALS (II,2182); THEY CALL IT MURDER (I,4441)

PHALEN, ROBERT
BENDER (II,204); CENTENNIAL (II,477); HELLINGER'S LAW (II,1127)

PHELAN, JOSEPH
THE INVISIBLE WOMAN (II,1244)

PHELPS, CLAUDE
HEE HAW (II,1123)

PHELPS, ELEANOR
THE SECRET STORM (I,3969); SOMERSET (I,4115)

PHILBIN, REGIS
BATTLE OF THE LAS VEGAS SHOW GIRLS (II,162); THE KAREN VALENTINE SHOW (I,2506); LILY FOR PRESIDENT (II,1478); THE NEIGHBORS (II,1801); THE REGIS PHILBIN SHOW (I,3760); THE REGIS PHILBIN SHOW (II,2128); TRUE LIFE STORIES (II,2665)

PHILBROOK, JAMES
THE INVESTIGATORS (I,2231); THE ISLANDERS (I,2243); THE LORETTA YOUNG SHOW (I,2755)

PHILIP, STERLING
WHERE THE HEART IS (I,4831)

PHILIPP, KAREN
ADVICE TO THE LOVELORN (II,16); EVEL KNIEVEL (II,785); LOVERS AND FRIENDS (II,1549); M*A*S*H (II,1569); QUINCY, M. E. (II,2102); TWILIGHT THEATER II (II,2686)

PHILIPS, EDWIN
SATINS AND SPURS (I,3917)

PHILIPS, LEE
THE FURTHER ADVENTURES OF ELLERY QUEEN (I,1719); THE GOLDEN AGE OF TELEVISION (II,1016); SATAN'S WAITIN' (I,3916)

PHILLIPPE, ANDRE
MR. NOVAK (I,3145)

PHILLIPS, ANTON
SPACE: 1999 (II,2409)

PHILLIPS, ARLENE
THE KENNY EVERETT VIDEO SHOW (II,1379)

PHILLIPS, BARNEY
THE BANANA SPLITS ADVENTURE HOUR (I,340); THE BETTY WHITE SHOW (II,224); DRAGNET (I,1369); THE FELONY SQUAD (I,1563); INSIDE O.U.T. (II,1235); JOHNNY MIDNIGHT (I,2431); LONGSTREET (I,2750); SHAZZAN! (I,4002); TWELVE O'CLOCK HIGH (I,4637)

PHILLIPS, BILL
THE KITTY WELLS/JOHNNY WRIGHT FAMILY SHOW (I,2572)

PHILLIPS, BRYN
SUNDAY NIGHT AT THE LONDON PALLADIUM (I,4286)

PHILLIPS, CARMEN
THE LIEUTENANT (I,2680)

PHILLIPS, CONRAD
THE ADVENTURES OF WILLIAM TELL (I,85)

PHILLIPS, DEMETRE
CAMP GRIZZLY (II,418); FOG (II,897); SHIPSHAPE (II,2329)

PHILLIPS, EDWIN
BROTHER RAT (I,746)

PHILLIPS, ETHAN
BENSON (II,208); CIRCUS OF THE STARS (II,536)

PHILLIPS, JASON
EMERGENCY PLUS FOUR (II,774)

PHILLIPS, JOE
THE LAST CONVERTIBLE (II,1435)

PHILLIPS, JOHN
THE FORSYTE SAGA (I,1645); JANE EYRE (II,1273); LYNDA CARTER: ENCORE (II,1566); THE ONEDIN LINE (II,1912)

PHILLIPS, JULIE
ANOTHER WORLD (II,97)

PHILLIPS, KATHY
THE R.C.A. VICTOR SHOW (I,3737)

PHILLIPS, LESLIE
THE ERROL FLYNN THEATER (I,1458)

PHILLIPS, MACKENZIE
BATTLE OF THE NETWORK STARS (II,163); BATTLE OF THE NETWORK STARS (II,166); CIRCUS OF THE STARS (II,531); ONE DAY AT A TIME (II,1900); THAT'S TV (II,2581); THINGS WE DID LAST SUMMER (II,2589)

PHILLIPS, MARGARET
CASTLE ROCK (II,456); THE PICTURE OF DORIAN GRAY (I,3591)

PHILLIPS, MICHELLE
BATTLE OF THE NETWORK STARS (II,165); CIRCUS OF THE STARS (II,532); DON'T CALL ME MAMA ANYMORE (I,1350); FANTASY ISLAND (II,829); THE FRENCH ATLANTIC AFFAIR (II,924); LADIES IN BLUE (II,1422); MICKEY SPILLANE'S MIKE

HAMMER: MURDER ME, MURDER YOU (II,1694); MOONLIGHT (II,1730)

PHILLIPS, NANCY
ROWAN AND MARTIN'S LAUGH-IN (I,3856)

PHILLIPS, NEVILLE
AN ECHO OF THERESA (I,1399)

PHILLIPS, PAUL
THE LAST CONVERTIBLE (II,1435)

PHILLIPS, PENNY
KING FEATURES TRILOGY (I,2555)

PHILLIPS, PHIL
THE LONELY WIZARD (I,2743)

PHILLIPS, RANDY
THE EDGE OF NIGHT (II,760)

PHILLIPS, ROBERT
FROM HERE TO ETERNITY (II,933); THE GUN AND THE PULPIT (II,1067); JOE DANCER: THE BIG BLACK PILL (II,1300); TOMA (I,4547); THE ULTIMATE IMPOSTER (II,2700); YUMA (I,4976)

PHILLIPS, SIAN
SMILEY'S PEOPLE (II,2383); TINKER, TAILOR, SOLDIER, SPY (II,2621)

PHILLIPS, TRACEY
CO-ED FEVER (II,547)

PHILLIPS, VALERIE
A HORSEMAN RIDING BY (II,1182)

PHILLIPS, WENDELL
ABE LINCOLN IN ILLINOIS (I,10)

PHILLIPS, WENDY
THE EDDIE CAPRA MYSTERIES (II,758); EXECUTIVE SUITE (II,796); ONE OF OUR OWN (II,1910); RIDING HIGH (II,2173)

PHILLIPS, WILLIAM
BEN JERROD (I,396)

PHILLIS MYLES AND THE WEST TWINS
LUCKY LETTERS (I,2786)

PHILLIS, JOHN
CRIMES OF PASSION (II,603)

PHILPOT, IVAN
VALENTINE MAGIC ON LOVE ISLAND (II,2716)

PHILPOT, JIM
MIDWESTERN HAYRIDE (I,3033)

PHILPOTS, AMBROSINE
THE EVE ARDEN SHOW (I,1471)

PHILPOTT, JOHN
THE BUGALOOS (I,757)

PHIPPS, WILLIAM
THE CODE OF JONATHAN WEST (I,992); THE LIFE AND LEGEND OF WYATT EARP (I,2681); THE WESTERNER (I,4800)

PHIPPS, WILLIAM EDWARD
BOONE (II,351); THE GREEN FELT JUNGLE (I,1885); SARA (II,2260); SPACE FORCE (II,2407); TIME EXPRESS (II,2620)

PHOENIX, LEAF
SIX PACK (II,2370)

PHOENIX, PATRICIA
CORONATION STREET (I,1054)

PHOENIX, RIVER
CELEBRITY (II,463); SEVEN BRIDES FOR SEVEN BROTHERS (II,2307)

PIATIGORSKY, GREGOR
FESTIVAL OF MUSIC (I,1567)

PIAZZA, BEN
BATTLES: THE MURDER THAT WOULDN'T DIE (II,180); BEN CASEY (I,394); BENDER (II,204); THE CIRCLE FAMILY (II,527); DALLAS (II,614); FOREVER FERNWOOD (II,909); THE WAVERLY WONDERS (II,2746); THE WINDS OF WAR (II,2807)

PIAZZA, MARGUERITE
THE NASH AIRFLYTE THEATER (I,3227); YOUR SHOW OF SHOWS (I,4957); YOUR SHOW OF SHOWS (II,2875)

PICARDO, ROBERT
STEAMBATH (II,2451)

PICARDO, ROBERTA
FULL HOUSE (II,937)

PICAROLL, JOHN
HIGH TOR (I,2056)

PICERNI, CHARLES
STUNTS UNLIMITED (II,2487)

PICERNI, PAUL
BIG ROSE (II,241); A LUCILLE BALL SPECIAL STARRING LUCILLE BALL AND DEAN MARTIN (II,1557); MATT HELM (II,1652); THE MCGONIGLE (I,2968); MEET MCGRAW (I,2984); O'HARA, UNITED STATES TREASURY (I,3346); THE UNTOUCHABLES (I,4681); THE UNTOUCHABLES (I,4682); THE YOUNG MARRIEDS (I,4942)

PICK AND PAT
AMERICAN MINSTRELS (I,171) PICK AND PAT (I,3590)

PICKARD, FRED
THE BILL COSBY SHOW (I,442)

PICKARD, JACK
BOOTS AND SADDLES: THE STORY OF THE FIFTH CAVALRY (I,686)

PICKARD, JOHN
GUNSLINGER (I,1908); MOBILE MEDICS (II,1716)

PICKENS, JANE
THE JANE PICKENS SHOW (I,2343); THE R.C.A. THANKSGIVING SHOW (I,3736); SHOW OF THE YEAR (I,4031)

PICKENS, JUNE
STRAWHAT THEATER (I,4256)

PICKENS, SLIM
B.J. AND THE BEAR (II,125); BANJO HACKETT: ROAMIN' FREE (II,141); THE BUSTERS (II,401); FILTHY RICH (II,856); THE GUN AND THE PULPIT (II,1067); HEE HAW (II,1123); HITCHED (I,2065); HOW THE WEST WAS WON (II,1196); JAKE'S WAY (II,1269); THE LEGEND OF CUSTER (I,2652); THE MARY TYLER MOORE SHOW (II,1640); THE NASHVILLE PALACE (II,1792); THE OUTLAWS (I,3427); THE SAGA OF ANDY BURNETT (I,3883); SAM HILL: WHO KILLED THE MYSTERIOUS MR. FOSTER? (I,3896); SAWYER AND FINN (II,2265); WHEN THE WEST WAS FUN: A WESTERN REUNION (II,2780)

PICKERING, BOB
THE GUIDING LIGHT (II,1064)

PICKERING, PATRICE
CAMELOT (II,416)

PICKETT, CINDY
CALL TO GLORY (II,413); THE CHEROKEE TRAIL (II,500); COCAINE AND BLUE EYES (II,548); FAMILY IN BLUE (II,818); THE GUIDING LIGHT (II,1064); MICKEY SPILLANE'S MARGIN FOR MURDER (II,1691)

PICKETT, LENNY
DAVID BOWIE—SERIOUS MOONLIGHT (II,622)

PICKETT, WILSON
...AND BEAUTIFUL (I,184)

PICKETTS, PAT
OUR STREET (I,3418)

PICKLES, CHRISTINA
ANOTHER WORLD (II,97); THE GUIDING LIGHT (II,1064); ST. ELSEWHERE (II,2432); VANITY FAIR (I,4700)

PICKLES, VIVIAN
LOVE IN A COLD CLIMATE (II,1539)

PICKNER, MARGIE
THE GEORGE BURNS AND GRACIE ALLEN SHOW (I,1763)

PICKUP, RONALD
JENNIE: LADY RANDOLPH CHURCHILL (II,1277); LONG DAY'S JOURNEY INTO NIGHT (I,2746)

PICON, MOLLY
CAR 54, WHERE ARE YOU? (I,842); THE JAZZ SINGER (I,2351); THE MOLLY PICON SHOW (I,3081)

PIDGEON, WALTER
CINDERELLA (I,956); THE HOUSE ON GREENAPPLE ROAD (I,2140); M-G-M PARADE (I,2798); MEET ME IN ST. LOUIS (I,2986); SWISS FAMILY ROBINSON (I,4324); WARNING SHOT (I,4770); ZANE GREY THEATER (I,4979)

PIEKARSKI, JULIE
THE BEST OF TIMES (II,219); THE FACTS OF LIFE (II,805); THE NEW MICKEY MOUSE CLUB (II,1823)

PIERCE, BARBARA
FOREST RANGERS (I,1642); MOMENT OF TRUTH (I,3084)

PIERCE, JUDY
THE LATE FALL, EARLY SUMMER BERT CONVY SHOW (II,1441)

PIERCE, MAGGIE
THE ELISABETH MCQUEENEY STORY (I,1429); MY MOTHER THE CAR (I,3197)

PIERCE, ROBERT
THE SECRET WAR OF JACKIE'S GIRLS (II,2294)

PIERCE, STACK
INSIDE O.U.T. (II,1235); PALMERSTOWN U.S.A. (II,1945); PEOPLE LIKE US (II,1999); SAWYER AND FINN (II,2265); V: THE FINAL BATTLE (II,2714)

PIERCE, TIANA
THE BEST OF TIMES (II,219)

PIERCE, VERNA
SEARCH FOR TOMORROW (II,2284)

PIERCE, WEBB
THE GREEN FELT JUNGLE (I,1885)

PIERONI, LEONARDO
MISTER JERICO (I,3070)

PIERPOINT, ERIC
HOT PURSUIT (II,1189)

PIERPONT, LAURA
BORN YESTERDAY (I,692)

PIERRE, OLIVIER
MASTER OF THE GAME (II,1647)

PIERSON, GEOFFREY
RYAN'S HOPE (II,2234)

PIERSON, RICHARD
MASADA (II,1642)

PIERSON, WILLIAM
THREE'S COMPANY (II,2614)

PIETRAGALLO, GENE
SEARCH FOR TOMORROW (II,2284)

PIGEON, CORKY
GREAT DAY (II,1054); SILVER SPOONS (II,2355)

PIGUS, LILA
GENERAL HOSPITAL (II,964)

PIKE, DON
FLATBED ANNIE AND SWEETIEPIE: LADY TRUCKERS (II,876)

PIKE, JOHN
IVANHOE (I,2282)

PILARRE, SUSAN
LITTLE WOMEN (I,2724)

PILLAR, GARRY
ANOTHER WORLD (II,97); BRIGHT PROMISE (I,727); THE GUIDING LIGHT (II,1064)

PILON, DANIEL
THE HAMPTONS (II,1076); MASSARATI AND THE BRAIN (II,1646); MISSING PIECES (II,1706); RYAN'S HOPE (II,2234)

PINA, CHRIS
HOLLYWOOD HIGH (II,1162)

PINASSI, DOMINIQUE
THE MONTEFUSCOS (II,1727); NEWMAN'S DRUGSTORE (II,1837)

PINCARD, RON
EMERGENCY! (II,775); SEALAB 2020 (I,3949)

PINE MOUNTAIN BOYS, THE
STRAWHAT THEATER (I,4256)

PINE, CATHY
BATTLESTAR GALACTICA (II,181)

PINE, PHILIP
THE BLUE KNIGHT (II,276); GOLDEN WINDOWS (I,1841)

PINE, ROBERT
BERT D'ANGELO/SUPERSTAR (II,211); CHIPS (II,511); MAGNUM, P.I. (II,1601); MUNSTER, GO HOME! (I,3157); THE MYSTERIOUS TWO (II,1783)

PINI, KARIN
THE PAUL HOGAN SHOW (II,1975)

PINK LADY
PINK LADY (II,2041)

PINKARD, FRED
ARCHER—FUGITIVE FROM THE EMPIRE (II,103); PALMERSTOWN U.S.A. (II,1945)

PINKARD, RON
DOCTOR DAN (II,684)

PINKERTON, NANCY
ONE LIFE TO LIVE (II,1907); SOMERSET (I,4115)

PINKUS, LULU
PRISONER: CELL BLOCK H (II,2085)

PINNEY, PAT
BLACKSTAR (II,263)

PINSENT, GORDON
FOREST RANGERS (I,1642); QUENTIN DERGENS, M.P. (I,3707); SCARLETT HILL (I,3933)

PINSON, ALLEN
DOC HOLLIDAY (I,1307)

PINTAURO, DANNY
WHO'S THE BOSS? (II,2792)

PINTER, MARK
BEHIND THE SCREEN (II,200); THE GUIDING LIGHT (II,1064); LOVE OF LIFE (I,2774); SECRETS OF MIDLAND HEIGHTS (II,2296)

PINZA, EZIO
THE ALL-STAR REVUE (I,138); BONINO (I,685); THE GENERAL FOODS 25TH ANNIVERSARY SHOW (I,1756); THE MARCH OF DIMES FASHION SHOW (I,2902); THE R.C.A. VICTOR SHOW (I,3737)

PIOLI, JUDY
SHAPING UP (II,2316); STAR OF THE FAMILY (II,2436)

PIPER, JACK
MURDER ON THE MIDNIGHT EXPRESS (I,3162)

PIPER, JACKIE
THE TWO RONNIES (II,2694)

PIPER, RICHARD
MASTER OF THE GAME (II,1647)

PIPS, THE
GLADYS KNIGHT AND THE PIPS (II,1004)

PIRCENI, CHARLES
MICKEY SPILLANE'S MARGIN FOR MURDER (II,1691)

PIRRONE, GEORGE
BACHELOR PARTY (I,324)

PISANI, REMO
THE REPORTER (I,3771)

PISCOPO, JOE
BATTLE OF THE NETWORK STARS (II,174); THE JOE PISCOPO SPECIAL (II,1304); THE MADHOUSE BRIGADE (II,1588); SATURDAY NIGHT LIVE (II,2261)

PISIER, MARIE-FRANCE
THE FRENCH ATLANTIC AFFAIR (II,924); SCRUPLES (II,2280)

PITCHESS, PETER J
240-ROBERT (II,2688); 240-ROBERT (II,2689)

PITHEY, WENSLEY
IKE (II,1223)

PITLIK, NOAM
THE BAIT (I,335); THE BOB NEWHART SHOW (II,342); BOBBY PARKER AND COMPANY (II,344); CHARO AND THE SERGEANT (II,489); PAPA G.I. (I,3451); SANFORD AND SON (II,2256)

PITMAN, ROBERT JOHN
DENNIS THE MENACE (I,1231)

PITONIAK, ANNE
AFTERMASH (II,23)

PITT, INGRID
THE KILLING GAME (I,2540)

PITTMAN, TOM
THE STAR MAKER (I,4193)

PITTS, INGRID
SMILEY'S PEOPLE (II,2383)

PITTS, ZASU
THE GALE STORM SHOW (I,1725); THE MAN WHO CAME TO DINNER (I,2880)

PIVEN, BYRNE
ADAMS HOUSE (II,9)

PIXIEKIN PUPPETS, THE
THE RAY CONNIFF CHRISTMAS SPECIAL (I,3732)

PIZER, J BRIAN
THE LEGEND OF THE GOLDEN GUN (II,1455)

PLACE, MARY KAY
THE CHEERLEADERS (II,493); FOREVER FERNWOOD (II,909); JOHN DENVER—THANK GOD I'M A COUNTRY BOY (II,1317); MARY HARTMAN, MARY HARTMAN (II,1638)

PLANA, TONY
CALLAHAN (II,414)

PLANK, MELINDA
THE NURSES (I,3319); OUR FIVE DAUGHTERS (I,3413); SEARCH FOR TOMORROW (II,2284); SOMERSET (I,4115)

PLANTE, LOUIS
THE AWAKENING LAND (II,122)

PLANTING, LAURA
IT'S YOUR FIRST KISS, CHARLIE BROWN (I,2280)

PLANTT-WINSTON, SUSAN
ALL MY CHILDREN (II,39)

PLATO, DANA
ALMOST AMERICAN (II,55); CELEBRITY DAREDEVILS (II,473); CIRCUS OF THE STARS (II,534); CIRCUS OF THE STARS (II,535); DIFF'RENT STROKES (II,674); FAMILY (II,813); HIGH SCHOOL, U.S.A. (II,1148); KRAFT SALUTES WALT DISNEY WORLD'S 10TH ANNIVERSARY (II,1410)

PLATT, EDWARD
AMOS BURKE: WHO KILLED JULIE GREER? (I,181); FEMALE INSTINCT (I,1564); GENERAL HOSPITAL (II,964); GET SMART (II,983); THE OUTER LIMITS (I,3426)

PLATT, HOWARD
ALONE AT LAST (II,59); EMPIRE (II,777); FLYING HIGH (II,894); FLYING HIGH (II,895); MR. AND MRS. COP (II,1748); SANFORD (II,2255); SANFORD AND SON (II,2256)

PLATT, VICTOR

ESPIONAGE (I,1464)

PLATTERS, THE
THE JONATHAN WINTERS SHOW (I,2440)

PLATTO, ALMA
THESE ARE MY CHILDREN (I,4440)

PLAYBOY BUNNIES, THE
PLAYBOY AFTER DARK (I,3620); PLAYBOY'S PENTHOUSE (I,3621)

PLAYBOY PLAYMATES, THE
GEORGE BURNS EARLY, EARLY, EARLY, CHRISTMAS SHOW (II,968)

PLAYDON, PAUL
THE ADVENTURES OF JIM BOWIE (I,63)

PLAYTEN, ALICE
THE LOST SAUCER (II,1524); MASQUERADE (I,2940); OVER AND OUT (II,1938); THAT'S CAT (II,2576)

PLAZ, TOM
THE HUSTLER OF MUSCLE BEACH (II,1210)

PLEASENCE, ANGELA
THE SIX WIVES OF HENRY VIII (I,4068)

PLEASENCE, DONALD
THE ADVENTURES OF ROBIN HOOD (I,74); THE BASTARD/KENT FAMILY CHRONICLES (II,159); CENTENNIAL (II,477); COMPUTERCIDE (II,561); THE DIARY OF ANNE FRANK (I,1262); DR JEKYLL AND MR. HYDE (I,1368); THE FRENCH ATLANTIC AFFAIR (II,924); MASTER OF THE GAME (II,1647); ONE STEP BEYOND (I,3388); THE OUTER LIMITS (I,3426)

PLEASENT VALLEY BOYS, THE
MIDWESTERN HAYRIDE (I,3033)

PLEDGER, COURTNEY
WALKING TALL (II,2738)

PLEETS, TOM
SKEEZER (II,2376)

PLENER, SHIMMY
THE IMPERIAL GRAND BAND (II,1224)

PLESHETTE, JOHN
BELL, BOOK AND CANDLE (II,201); THE KID WITH THE BROKEN HALO (II,1387); KNOTS LANDING (II,1404); SEVENTH AVENUE (II,2309); ZERO HOUR (I,4980)

PLESHETTE, SUZANNE
ALFRED HITCHCOCK PRESENTS (I,115); BAND OF GOLD (I,341); THE BOB NEWHART SHOW (II,342); CHANNING (I,900); THE DICK POWELL SHOW (I,1269); EXPOSE (I,1491); FLESH AND BLOOD (I,1595); THE GENERAL ELECTRIC THEATER (I,1753); KATE BLISS AND THE TICKER TAPE KID (II,1366); MITZI: A TRIBUTE TO THE AMERICAN HOUSEWIFE (I,3072); RICHIE BROCKELMAN: MISSING 24 HOURS (II,2168); SUZANNE PLESHETTE IS MAGGIE BRIGGS (II,2509)

PLIMPTON, GEORGE
THE FUTURE: WHAT'S NEXT (II,947); PLIMPTON! DID YOU HEAR THE ONE ABOUT. . .? (I,3629); PLIMPTON! SHOWDOWN AT RIO LOBO (I,3630); PLIMPTON! THE MAN ON THE FLYING TRAPEZE (I,3631)

PLOWMAN, MELINDA
THE ADVENTURES OF OZZIE AND HARRIET (I,71); MATINEE THEATER (I,2947); MY DARLING JUDGE (I,3186); THE OUTER LIMITS (I,3426); PLEASE DON'T EAT THE DAISIES (I,3628)

PLUMB, EVE
THE BRADY BRIDES (II,361); THE BRADY BUNCH (II,362); THE BRADY GIRLS GET MARRIED (II,364); THE BRADY KIDS (II,365); THE HOUSE ON GREENAPPLE ROAD (I,2140); KELLY'S KIDS (I,2523); LITTLE WOMEN (II,1497); LITTLE WOMEN (II,1498); TEACHER'S PET (I,4367)

PLUMB, FLORA
FOR RICHER, FOR POORER (II,906); LOVERS AND FRIENDS (II,1549)

PLUMB, SUSAN
NIGHTSIDE (II,1851)

PLUMLEY, DOM
THE RUNAWAY BARGE (II,2230)

PLUMMER, CHRISTOPHER
CAPTAIN BRASSBOUND'S CONVERSION (I,828); CYRANO DE BERGERAC (I,1119); CYRANO DE BERGERAC (I,1120); A DOLL'S HOUSE (I,1327); DUPONT SHOW OF THE MONTH (I,1387); THE GOLDEN AGE OF TELEVISION (II,1017); JOHNNY BELINDA (I,2416); LITTLE MOON OF ALBAN (I,2714); THE MONEYCHANGERS (II,1724); PARADE OF STARS (II,1954); THE PHILADELPHIA STORY (I,3582); THE THORN BIRDS (II,2600); TIME REMEMBERED (I,4509)

PLUMMER, PENNY
THE BOB HOPE SHOW (I,629)

PLUMMER, SCOTTY
THE MUSIC MART (II,1774)

PNIEWSKI, MICHAEL
GIDGET'S SUMMER REUNION (I,1799)

PO' BOYS, THE
THE BILL ANDERSON SHOW (I,440)

POCKOR, BRIAN
DOROTHY HAMILL IN ROMEO & JULIET ON ICE (II,718); THE MAGIC PLANET (II,1598)

PODELL, RICK
THE CHOPPED LIVER BROTHERS (II,514); NEVER SAY NEVER (II,1809); THE PAUL WILLIAMS SHOW (II,1984); SNAFU (II,2386)

PODMORE, WILLIAM
THE ADVENTURE OF THE THREE GARRIDEBS (I,40)

POE, SANDRA
CHARADE QUIZ (I,901)

POHLMANN, ERIC
COLONEL MARCH OF SCOTLAND YARD (I,1005)

POINDEXTER
BLACK BART (II,260); THE CLIFFWOOD AVENUE KIDS (II,543); TONY THE PONY (II,2641)

POINTER SISTERS, THE
THE FLIP WILSON COMEDY SPECIAL (II,886); THE MAC DAVIS CHRISTMAS SPECIAL (II,1573); THE SUZANNE SOMERS SPECIAL (II,2511)

POINTER, ANITA
THE POINTER SISTERS (II,2059)

POINTER, JUNE
THE POINTER SISTERS (II,2059)

POINTER, PRISCILLA
DALLAS (II,614); FROM HERE TO ETERNITY (II,933); JUDGEMENT DAY (II,1348); THE MYSTERIOUS TWO (II,1783); ST. ELSEWHERE (II,2432); TOO CLOSE FOR COMFORT (II,2642); WHERE THE HEART IS (I,4831)

POINTER, RUTH
THE POINTER SISTERS (II,2059)

POITIER, SIDNEY
ABC STAGE '67 (I,6)

POLA, JAIME
MULLIGAN'S STEW (II,1763)

POLAN, LINDA
THE SECRET OF CHARLES DICKENS (II,2293)

POLANSKY, ALAN
MASTER OF THE GAME (II,1647)

POLEN, NAT
AS THE WORLD TURNS (II,110); CAPTAIN VIDEO AND HIS VIDEO RANGERS (I,838); HIDDEN FACES (I,2044); THE NURSES (I,3319)

POLGAR, FRANZ J
THE AMAZING POLGAR (I,160)

POLIC II, HENRY
HECK'S ANGELS (II,1122); HOW TO SURVIVE THE 70S AND MAYBE EVEN BUMP INTO HAPPINESS (II,1199); THE LATE FALL, EARLY SUMMER BERT CONVY SHOW (II,1441); MCNAMARA'S BAND (II,1668); THE MONSTER SQUAD (II,1725); SCRUPLES (II,2281); WEBSTER (II,2578); WHEN THINGS WERE ROTTEN (II,2782)

POLITO, JON
THE GANGSTER CHRONICLES (II,957)

POLK, GORDON
THE JERRY COLONNA SHOW (I,2359)

POLKA CHIPS, THE
IT'S POLKA TIME (I,2273); POLKA TIME (I,3642)

POLKA DOTS, THE
THE JO STAFFORD SHOW (I,2393); THE JO STAFFORD SHOW (I,2394); THE SPIKE JONES SHOW (I,4156)

POLKA ROUNDERS, THE
POLKA-GO-ROUND (I,3640)

POLLACK, DEE
PURSUE AND DESTROY (I,3699)

POLLACK, NAOMI
KORG: 70,000 B.C. (II,1408)

POLLAK, KEVIN
NATIONAL LAMPOON'S HOT FLASHES (II,1795)

POLLARD, MICHAEL J
DOBIE GILLIS (I,1302); HENRY FONDA AND THE FAMILY (I,2015); THE LUCY SHOW (I,2791); THE MISS AND THE MISSILES (I,3062)

POLLEY, DIANE
PAUL BERNARD—PSYCHIATRIST (I,3485)

POLLICK, STEVE
CAMP GRIZZLY (II,418)

POLLOACK, DAVID
THIS BETTER BE IT (II,2595)

POLLOCK, BEN
MARKHAM (I,2916)

POLLOCK, DEE
GUNSLINGER (I,1908)

POLLOCK, NANCY
FIRST LOVE (I,1583); THE GOLDBERGS (I,1836)

POLLOCK, TENO
MR. NOVAK (I,3145)

POLSEN, ALFRED
THE AFRICAN QUEEN (II,18)

POLSON, CECILY
A TOWN LIKE ALICE (II,2652)

POM, MARY
THE WIZARD OF ODDS (I,4886)

POMERANTZ, EARL
BOBBIE GENTRY'S HAPPINESS HOUR (I,671); BOBBIE GENTRY'S HAPPINESS HOUR (II,343)

POMERANTZ, JEFFREY
ONE LIFE TO LIVE (II,1907); THE ROOKIES (I,3840); SEARCH FOR TOMORROW (II,2284); THE SECRET STORM (I,3969); UNIT 4 (II,2707)

POMERHN, MARGUERITE
FAME (II,812)

PONAZECKI, JOE
ANOTHER WORLD (II,97); CONCRETE BEAT (II,562); THE SECRET STORM (I,3969); TALES FROM THE DARKSIDE (II,2534)

PONCE, DANNY
FULL HOUSE (II,937); HAPPY DAYS (II,1084); KNOTS LANDING (II,1404)

PONCE, PONCIE
HAWAIIAN EYE (I,1973)

PONS, BEATRICE
CAR 54, WHERE ARE YOU? (I,842); THE PHIL SILVERS SHOW (I,3580)

PONS, LILY
THE BOB HOPE SHOW (I,508)

PONTEROTTO, DONNA
THE HARVEY KORMAN SHOW (II,1103); HEAVEN ON EARTH (II,1120); JOE AND VALERIE (II,1298); JOE AND VALERIE (II,1299); THE LATE FALL, EARLY SUMMER BERT CONVY SHOW (II,1441); REPORT TO MURPHY (II,2134)

PONTI, SAL
TAMMY (I,4357)

PONTING, SALLY
MAGNUM, P.I. (II,1601)

PONZINI, ANTHONY
ANOTHER WORLD (II,97); FLATBUSH (II,877); MOONLIGHT (II,1730); ONE LIFE TO LIVE (II,1907)

POOLE, GORDON
1915 (II,1853)

POOLE, ROY
ABE LINCOLN IN ILLINOIS (I,11); AH!, WILDERNESS (I,91); THE ANDROS TARGETS (II,76); HELLINGER'S LAW (II,1127); LAND OF HOPE (II,1424); ROLL OF THUNDER, HEAR MY CRY (II,2202); RYAN'S HOPE (II,2234); SANDBURG'S LINCOLN (I,3907); SOLDIER IN LOVE (I,4108); A TIME FOR US (I,4507); THE WINDS OF WAR (II,2807)

POPE, PATTY
A DATE WITH JUDY (I,1162)

POPE, PEGGY
9 TO 5 (II,1852); BETWEEN THE LINES (II,225); BILLY (II,247); CALUCCI'S DEPARTMENT (I,805); FOR MEMBERS ONLY (II,905); THE GRADY NUTT SHOW (II,1048); KANGAROOS IN THE KITCHEN (II,1362); MARLO THOMAS IN ACTS OF LOVE—AND OTHER COMEDIES (I,2919); THE RAG BUSINESS (II,2106); REVENGE OF THE GRAY GANG (II,2146); SIDE BY SIDE (II,2347); SOAP (II,2392)

POPENOE, DR PAUL
DIVORCE HEARING (I,1298)

POPOV
THE BOB HOPE SHOW (I,562)

POPPEL, MARC
RITUALS (II,2182)

POPPEN, DET
POPEYE THE SAILOR (I,3648)

POPPY FAMILY, THE
THE GEORGE KIRBY SPECIAL (I,1771)

POPWELL, ALBERT
SEARCH (I,3951)

PORCELLI, FRED
ALL MY CHILDREN (II,39)

PORRETT, SUSAN
THE CITADEL (II,539)

PORSKY, BOB
JABBERWOCKY (II,1262)

PORTER, ARTHUR GOULD
ANYTHING GOES (I,238); THE BEVERLY HILLBILLIES (I,417)

PORTER, BOBBY
QUARK (II,2095)

PORTER, COLE
MR. PORTER OF INDIANA (I,3148); YOU'RE THE TOP (I,4974)

PORTER, DON
ALL TOGETHER NOW (II,45); THE ANN SOTHERN SHOW (I,220); BATTLES: THE MURDER THAT WOULDN'T DIE (II,180); CAP'N AHAB (I,824); FRANKIE AND ANNETTE: THE SECOND TIME AROUND (II,917); THE FUZZ BROTHERS (I,1722); GIDGET (I,1795); HAPPY ANNIVERSARY AND GOODBYE (II,1082); I LOVE MY DOCTOR (I,2172); THE NORLISS TAPES (I,3305); PANDORA AND FRIEND (I,3445); PRIVATE SECRETARY (I,3672); WHAT'S UP DOC? (II,2771)

PORTER, ERIC
THE FORSYTE SAGA (I,1645); WHY DIDN'T THEY ASK EVANS? (II,2795)

PORTER, NYREE DAWN
THE FORSYTE SAGA (I,1645); JANE EYRE (I,2339); THE MARTIAN CHRONICLES (II,1635); THE PROTECTORS (I,3681)

PORTER, RICK
ANOTHER WORLD (II,97)

PORTER, ROBBIE
MALIBU U. (I,2849)

PORTER, ROBERT
WONDER WOMAN (PILOT 1) (II,2826)

PORTER, SARAH
A HORSEMAN RIDING BY (II,1182)

PORTER, SCOTT
THE ROCK RAINBOW (II,2194)

PORTER, TODD
WHIZ KIDS (II,2790)

PORTER, WILL
THAT THING ON ABC (II,2574)

PORTILLO, ROSE
BORN TO THE WIND (II,354)

PORTMAN, ERIC
THE LILLI PALMER THEATER (I,2704); VICTORY (I,4729)

POST JR, WILLIAM
BEULAH (I,416); CLAUDIA: THE STORY OF A MARRIAGE (I,972); THE EDGE OF NIGHT (II,760); MIRACLE ON 34TH STREET (I,3057); WHERE THE HEART IS (I,4831)

POST, BILL
THE BRIGHTER DAY (I,728)

POST, CLAYTON
DOC HOLLIDAY (I,1307)

POST, MARKIE
THE FALL GUY (II,811); THE GANGSTER CHRONICLES (II,957); HOW DO I KILL A THIEF—LET ME COUNT THE WAYS (II,1195); MASSARATI AND THE BRAIN (II,1646); SCENE OF THE CRIME (II,2271); SEMI-TOUGH (II,2300); SIX PACK (II,2370)

POST, MARTE
YOUNG LIVES (II,2866)

POST, TOM
ON THE ROCKS (II,1894)

POST, WILLIAM
YOUNG DR. MALONE (I,4938)

POSTON, FRANCESCA
ALL MY CHILDREN (II,39); ANOTHER WORLD (II,97)

POSTON, TOM
BEANE'S OF BOSTON (II,194); THE BOB NEWHART SHOW (II,342); BOBBY PARKER AND COMPANY (II,344); THE GIRL, THE GOLD WATCH AND DYNAMITE (II,1001); HARRY AND MAGGIE (II,1097); MERMAN ON BROADWAY (I,3011); MORK AND MINDY (II,1735); NEWHART (II,1835); PANTOMIME QUIZ (I,3449); SPLIT PERSONALITY (I,4161); THE STEVE ALLEN SHOW (I,4219); THE TEMPEST (I,4391); WE'VE GOT EACH OTHER (II,2757)

POTASH, PAUL
THE NEW CHRISTY MINSTRELS SHOW (I,3259)

POTTER, CAROL
ANNIE FLYNN (II,92); TODAY'S FBI (II,2628)

POTTER, CLIFF
THE NAME OF THE GAME (I,3217)

POTTER, JERRY
ALICE (II,33)

POTTER, PAUL
LOVE OF LIFE (I,2774); THE MILKY WAY (I,3044)

POTTER, PETER
PETER POTTER'S JUKE BOX JURY (I,3569)

POTTS, ANNIE
FLATBED ANNIE AND SWEETIEPIE: LADY TRUCKERS (II,876); GOODTIME GIRLS (II,1042); HOLLYWOOD HIGH (II,1162); HOLLYWOOD HIGH (II,1163); IN SECURITY (II,1227)

POTTS, CLIFF
BIG HAWAII (II,238); DANGER IN PARADISE (II,617); FOR LOVE AND HONOR (II,903); LITTLE WOMEN (II,1497); LITTLE WOMEN (II,1498); LOU GRANT (II,1526); NEVADA SMITH (II,1807); ONCE AN EAGLE (II,1897); SAN FRANCISCO INTERNATIONAL AIRPORT (I,3906)

POTTS, MARIA
CENTENNIAL (II,477)

POULTER, LOU
BEHIND THE SCREEN (II,200)

POUSETTE, LENA
BARE ESSENCE (II,149)

POWELL, ADDISON
DARK SHADOWS (I,1157)

POWELL, ANGELA
THE NEW DICK VAN DYKE SHOW (I,3262); THE NEW DICK VAN DYKE SHOW (I,3263)

POWELL, BUDDY
I'D RATHER BE CALM (II,1216); PRESENTING SUSAN ANTON (II,2077)

POWELL, DICK
AMOS BURKE: WHO KILLED JULIE GREER? (I,181); THE BEST IN MYSTERY (I,403); THE DICK POWELL SHOW (I,1269); FOUR STAR PLAYHOUSE (I,1652); THE JUNE ALLYSON SHOW (I,2488); SQUADRON (I,4170); ZANE GREY THEATER (I,4979)

POWELL, FLONISHA
SKEEZER (II,2376)

POWELL, J C
THE ADAMS CHRONICLES (II,8)

POWELL, JACK
BABES IN TOYLAND (I,318)

POWELL, JANE
THE DANNY THOMAS SPECIAL (I,1151); THE DICK POWELL SHOW (I,1269); FEATHERTOP (I,1556); GIVE MY REGARDS TO BROADWAY (I,1815); HOORAY FOR LOVE (I,2115); THE JANE POWELL SHOW (I,2344); THE JUNE ALLYSON SHOW (I,2488); THE LETTERS (I,2670); MEET ME IN ST. LOUIS (I,2986); RUGGLES OF RED GAP (I,3867); STANDARD OIL ANNIVERSARY SHOW (I,4189); TURN OF FATE (I,4623); THE VICTOR BORGE SHOW (I,4719); THE VICTOR BORGE SPECIAL (I,4722)

POWELL, JEAN
M*A*S*H (II,1569)

POWELL, RANDOLPH
COLORADO C.I. (II,555); THE CONCRETE COWBOYS (II,564); DALLAS (II,614); DOCTORS' PRIVATE LIVES (II,692); DOCTORS' PRIVATE LIVES (II,693); LOGAN'S RUN (II,1507)

POWELL, ROBERT
LADY KILLER (I,2603); MURDER IS A ONE-ACT PLAY (I,3160)

POWELL, SUSAN
BOB HOPE SPECIAL: BOB HOPE FOR PRESIDENT (II,282)

POWER, EDWARD
ANOTHER WORLD (II,97); DELTA COUNTY, U.S.A. (II,657); PAPER DOLLS (II,1952)

POWER, TYRONE
OPERATION ENTERTAINMENT (I,3400); STANDARD OIL ANNIVERSARY SHOW (I,4189)

POWER, UDANA
BARNEY AND ME (I,358); THE HANNA-BARBERA HAPPINESS HOUR (II,1079); THE LIFE AND TIMES OF EDDIE ROBERTS (II,1468); LOVES ME, LOVES ME NOT (II,1551); MR. & MS. AND THE MAGIC STUDIO MYSTERY (II,1747); SOAP (II,2392)

POWERS, BEN
GIMME A BREAK (II,995); GOOD TIMES (II,1038); LAUGH-IN (II,1444); MICKEY SPILLANE'S MIKE HAMMER (II,1692)

POWERS, BEVERLY
THE RED SKELTON SHOW (I,3755)

POWERS, HUNT
GENERAL HOSPITAL (II,964)

POWERS, JOHN ROBERT
ABC-POWERS CHARM SCHOOL (I,5)

POWERS, LEONA
THE ALDRICH FAMILY (I,111); THE FRONT PAGE (I,1691); MY SON JEEP (I,3202)

POWERS, MALA
CROSSROADS (I,1105); DANIEL BOONE (I,1141); FORD TELEVISION THEATER (I,1634); HAZEL (I,1982); KRAFT SUSPENSE THEATER (I,2591); THE MAN AND THE CITY (I,2856); ON TRIAL (I,3363)

POWERS, NICK
THE NBC FOLLIES (I,3242)

POWERS, ROY
THE ICE PALACE (I,2182)

POWERS, STEFANIE
BOB HOPE SPECIAL: BOB HOPE FOR PRESIDENT (II,282); CELEBRITY CHALLENGE OF THE SEXES 1 (II,465); ELLERY QUEEN: DON'T LOOK BEHIND YOU (I,1434); FAMOUS LIVES (II,821); THE FEATHER AND FATHER GANG (II,845); THE FEATHER AND FATHER GANG: NEVER CON A KILLER (II,846); THE GIRL FROM U.N.C.L.E. (I,1808); HART TO HART (II,1101); HART TO HART (II,1102); MANHUNTER (II,1620); MISTRAL'S DAUGHTER (II,1708); NIGHT GAMES (II,1845); THE SIX-MILLION-DOLLAR MAN (II,2372); SWEET, SWEET RACHEL (I,4313); SWINGIN' TOGETHER (I,4320); TOPPER RETURNS (I,4583); WARNING SHOT (I,4770); WASHINGTON: BEHIND CLOSED DOORS (II,2744)

POYNER, JIM
ANOTHER WORLD (II,97); TEXAS (II,2566)

PRAETOR, WALLY
MIDWESTERN HAYRIDE (I,3033)

PRAF, EDITH
THE ED WYNN SHOW (I,1403)

PRAGER, EMILY
THE EDGE OF NIGHT (II,760); SATURDAY NIGHT LIVE (II,2261)

PRAGER, SALLY
SKAG (II,2374)

PRAGER, STANLEY
THE COLLEGE BOWL (I,1001)

PRANGE, LAURIE
AMANDA FALLON (I,152); BEULAH LAND (II,226); THE MISSISSIPPI (II,1707); NIGHT GALLERY (I,3287); RANSOM FOR ALICE (II,2113); TESTIMONY OF TWO MEN (II,2564)

PRANSKY, JANELLE
KORG: 70,000 B.C. (II,1408)

PRATHER, JOAN
BATTLE OF THE NETWORK STARS (II,170); EIGHT IS ENOUGH (II,762); EXECUTIVE SUITE (II,796)

PRATT, DEBORAH
AIRWOLF (II,27); HAPPY DAYS (II,1084); MAGNUM, P.I. (II,1601); PHYL AND MIKHY (II,2037); PRIME TIMES (II,2083); THE REDD FOXX COMEDY HOUR (II,2126); SHE'S WITH ME (II,2321)

PRATT, JUDD
UNION PACIFIC (I,4676)

PRATT, JUDSON
GALLAGHER (I,1727)

PRATT, MIKE
MY PARTNER THE GHOST (I,3199)

PRATT, PETER
DOCTOR WHO (II,689)

PRATT, ROBERT
THE CENTURY TURNS (I,890); DIAL HOT LINE (I,1254)

PRATT, ROSE
WHY DIDN'T THEY ASK EVANS? (II,2795)

PRATT, SUSAN
GENERAL HOSPITAL (II,964); THE GUIDING LIGHT (II,1064); WELCOME BACK, KOTTER (II,2761)

PRAVADA, GEORGE
SCREAMER (I,3945)

PREBBLE, SIMON
THE SIX WIVES OF HENRY VIII (I,4068)

PRECHEUR, SARAH
STRANGER IN OUR HOUSE (II,2477)

PREECE, TIM
THE FALL AND RISE OF REGINALD PERRIN (II,810)

PREISS, WOLFGANG
IKE (II,1223)

PREMICE, JOSEPHINE
THE JEFFERSONS (II,1276)

PREMINGER, MICHAEL
DINAH AND HER NEW BEST FRIENDS (II,676)

PREMINGER, OTTO
BATMAN (I,366); RICKLES (II,2170)

PRENDERGAST, GARY
CATALINA C-LAB (II,458); MAKIN' IT (II,1603)

PRENDERGAST, GERALD
BORDER PALS (II,352); SUMMER (II,2491)

PRENEY, PAULETTE
A HORSEMAN RIDING BY (II,1182)

PRENTICE, JESSICA
POWERHOUSE (II,2074)

PRENTICE, KEITH
DARK SHADOWS (I,1157)

PRENTIS, LOU
BUCK ROGERS IN THE 25TH CENTURY (I,751); FEATHER YOUR NEST (I,1555); FOODINI THE GREAT (I,1619)

PRENTISS, ANN
CAPTAIN NICE (I,835); PHILLIP AND BARBARA (II,2034); SEARCH (I,3951); WHAT GAP? (I,4808)

PRENTISS, ED
ACTION AUTOGRAPHS (I,24); MAJORITY RULES (I,2836); MORNING STAR (I,3104)

PRENTISS, PAULA
HAVING BABIES II (II,1108); HE AND SHE (I,1985); TOP OF THE HILL (II,2645); THE WAY THEY WERE (II,2748)

PRESBY, ARCH
THE JOHNNY DUGAN SHOW (I,2426)

PRESBY, SHANNON
CRAZY TIMES (II,602)

PRESCOTT, ALLAN
QUIZZING THE NEWS (I,3713); THE S. S. HOLIDAY (I,4173)

PRESCOTT, ROBERT T
ALL TOGETHER NOW (II,45)

PRESLEY, ELVIS
ELVIS (I,1435); THE FRANK SINATRA TIMEX SHOW (I,1670)

PRESLEY, PRISCILLA
DALLAS (II,614); THOSE AMAZING ANIMALS (II,2601); TOM SNYDER'S CELEBRITY SPOTLIGHT (II,2632)

PRESNELL, HARVE
THE SINGERS (I,4060)

PRESS, BARBARA
ALLAN (I,142)

PRESSMAN, LAWRENCE
6 RMS RIV VU (II,2371); A BEDTIME STORY (I,384); CANNON (I,820); FEMALE INSTINCT (I,1564); LADIES' MAN (II,1423); LOVE AND LEARN (II,1529); THE MAN FROM ATLANTIS (II,1614); THE MARY TYLER MOORE SHOW (II,1640); MULLIGAN'S STEW (II,1763); MULLIGAN'S STEW (II,1764); THE NANCY DUSSAULT SHOW (I,3222); RICH MAN, POOR MAN—BOOK I (II,2161); STOCKARD CHANNING IN JUST FRIENDS (II,2467); THE WINDS OF WAR (II,2807); WINTER KILL (II,2810)

PRESSON, JASON
WISHMAN (II,2811)

PRESTIA, SHIRLEY
ADAMS HOUSE (II,9); E/R (II,748)

PRESTIA, VINCENT
CAMELOT (II,416)

PRESTIDGE, MEL
HAWAIIAN EYE (I,1973)

PRESTON, BILLY
ROCK COMEDY (II,2191)

PRESTON, DUNCAN
MURDER ON THE MIDNIGHT EXPRESS (I,3162)

PRESTON, J A
100 CENTRE STREET (II,1901); ALL'S FAIR (II,46); HILL STREET BLUES (II,1154)

PRESTON, JAMES
ANOTHER WORLD (II,97)

PRESTON, KELLY
BLUE THUNDER (II,278); FOR LOVE AND HONOR (II,903)

PRESTON, MIKE
HOMICIDE (I,2101); HOT PURSUIT (II,1189); OH, BOY! (I,3341)

PRESTON, ROBERT
ANYWHERE, U.S.A. (I,240); THE BELLS OF ST. MARY'S (I,392); BOB HOPE SPECIAL: BOB HOPE'S PINK PANTHER THANKSGIVING GALA (II,312); THE CHISHOLMS (II,512); THE CHISHOLMS (II,513); CIRCUS OF THE STARS (II,537); CLIMAX! (I,976); CURTAIN CALL THEATER (I,1115); HAPPY ENDINGS (II,1085); THE LUX VIDEO THEATER (I,2795); THE MAN AGAINST CRIME (I,2853); THE UNITED STATES STEEL HOUR (I,4677)

PRESTON, WADE
COLT .45 (I,1010)

PRESTON, WILLIAM
THE COP AND THE KID (II,572); O'MALLEY (II,1872)

PREVILLE, ANN
FOREIGN INTRIGUE: OVERSEAS ADVENTURES (I,1639)

PREVIN, ANDRE
THE BING CROSBY CHRISTMAS SHOW (I,458); THE BING CROSBY SHOW (I,465); THE DEAN MARTIN SHOW (I,1199); THE DONALD O'CONNOR SHOW (I,1343); MUSIC U.S.A. (I,3179); SOUND OF THE 60S (I,4134); SWING INTO SPRING (I,4317)

PRICE, ALLEN
THE PRACTICE (II,2076)

PRICE, ANNABELLE
MAKING THE GRADE (II,1606)

PRICE, BECKY
CARD SHARKS (II,438)

PRICE, BRENDAN
TARGET (II,2541)

PRICE, EMLYN
THE ROCK FOLLIES (II,2192)

PRICE, HUGH
ERNIE IN KOVACSLAND (I,1454); IT'S TIME FOR ERNIE (I,2277)

PRICE, IVY
CELEBRITY (II,463)

PRICE, JANELLE
THE CHEAP SHOW (II,491)

PRICE, KAREN
THE JUGGLER OF NOTRE DAME (II,1350)

PRICE, KENNY
HEE HAW (II,1123); THE HEE HAW HONEYS (II,1124); THE HEE HAW HONEYS (II,1125)

PRICE, LEONTYNE
ENTERTAINMENT —1955 (I,1450)

PRICE, MARC
CONDO (II,565); FAMILY TIES (II,819)

PRICE, PAUL B
BUSTING LOOSE (II,402)

PRICE, RAY
COUNTRY MUSIC CARAVAN (I,1066); COUNTRY NIGHT OF STARS II (II,593); RANCH PARTY (I,3719)

PRICE, RED
OH, BOY! (I,3341)

PRICE, ROGER
DROODLES (I,1377); HOW TO (I,2145); IN TROUBLE (II,1230)

PRICE, RON
WHAT'S IT ALL ABOUT WORLD? (I,4816)

PRICE, SARI
ALEX AND THE DOBERMAN GANG (II,32); DYNASTY (II,747)

PRICE, SHERWOOD
COUNTY GENERAL (I,1076); THE MAN FROM GALVESTON (I,2865); MICHAEL SHAYNE, DETECTIVE (I,3022)

PRICE, VINCENT
BATMAN (I,366); THE CHEVY MYSTERY SHOW (I,928); CHRIS AND THE MAGICAL DRIP (II,516); CIRCUS OF THE STARS (II,536); CLOWN ALLEY (I,979); CROSSROADS (I,1105); E.S.P. (I,1462); E.S.P. (I,1463); THE GENERAL ELECTRIC THEATER (I,1753); GRUEN GUILD PLAYHOUSE (I,1892); HERE COMES PETER COTTONTAIL (II,1132); THE HILARIOUS HOUSE OF FRIGHTENSTEIN (II,1153); JANE WYMAN PRESENTS THE FIRESIDE THEATER (I,2345); JOHN RITTER: BEING OF SOUND MIND AND BODY (II,1319); LIGHT'S OUT (I,2699); LINDSAY WAGNER—ANOTHER SIDE OF ME (II,1483); THE LUX VIDEO THEATER (I,2795); MAGIC WITH THE STARS (II,1599); NIGHT GALLERY (I,3287); PANTOMIME QUIZ (I,3449); PLAYHOUSE 90 (I,3623); PULITZER PRIZE PLAYHOUSE (I,3692); RED SKELTON'S CHRISTMAS DINNER (II,2123); RINGO (II,2176); ROBERT MONTGOMERY PRESENTS YOUR LUCKY STRIKE THEATER (I,3809); SCHLITZ PLAYHOUSE OF STARS (I,3936); THE SECRET WORLD OF KIDS (I,3970); TENNESSEE ERNIE FORD MEETS KING ARTHUR (I,4397); THE THREE MUSKETEERS (I,4481); TIME EXPRESS (II,2620); THE WEB (I,4784)

PRICHETT, LIZABETH
THE JACKIE GLEASON SPECIAL (I,2325)

PRICKETT, MAUDIE
BELLE STARR (I,391); THE CLAUDETTE COLBERT SHOW (I,971); DATE WITH THE ANGELS (I,1164); THE DOCTOR WAS A LADY (I,1318); FURY (I,1721); GALLAGHER (I,1727); HAZEL (I,1982); THE JACK BENNY PROGRAM (I,2294); JOHNNY BELINDA (I,2415); THE TAMMY GRIMES SHOW (I,4358)

PRIDDY, NANCY
MEDICAL CENTER (II,1675)

PRIDE, CHARLEY
THE BOB HOPE SHOW (I,652); CELEBRITY CHALLENGE OF THE SEXES 4 (II,468); A COUNTRY CHRISTMAS (II,583); COUNTRY COMES HOME (II,585); COUNTRY NIGHT OF STARS (II,592); GENERAL ELECTRIC'S ALL-STAR ANNIVERSARY (II,963); NASHVILLE 99 (II,1791); THE SOUND AND THE SCENE (I,4132); STARS AND STRIPES SHOW (II,2442); THE TOM JONES SPECIAL (I,4544)

PRIDE, DICKIE
OH, BOY! (I,3341)

PRIEST, NATALIE
MISS SUSAN (I,3065)

PRIEST, PAT
MARINELAND CARNIVAL (I,2913); THE MARY TYLER MOORE SHOW (II,1640); THE MUNSTERS (I,3158)

PRIMA, LOUIS
DANNY THOMAS: AMERICA I LOVE YOU (I,1145); THE SATURDAY NIGHT DANCE PARTY (I,3919)

PRIMBY, KATHY
MARIE (II,1629)

PRIMUS, BARRY
BIG ROSE (II,241); CAGNEY AND LACEY (II,409); THE DESPERATE HOURS (I,1239); PAPER DOLLS (II,1951); ROGER AND HARRY (II,2200); WASHINGTON: BEHIND CLOSED DOORS (II,2744)

PRINCE, AKILI
PRYOR'S PLACE (II,2089)

PRINCE, CHRISTOPHER
SUNDAY FUNNIES (II,2493)

PRINCE, JONATHAN
MR. MERLIN (II,1751)

PRINCE, WILLIAM
THE AMERICAN GIRLS (II,72); ANOTHER WORLD (II,97); ARSENIC AND OLD LACE (I,267); THE CHEVROLET TELE-THEATER (I,926); GEORGE WASHINGTON (II,978); JOE DANCER: MURDER ONE, DANCER 0 (II,1302); JUSTICE (I,2499); KEY WEST (I,2528); THE MAN WHO CAME TO DINNER (I,2880); THE MASK (I,2938); MOONLIGHT (II,1730); NIGHT GAMES (II,1845); THE RHINEMANN EXCHANGE (II,2150); STARLIGHT THEATER (I,4201); WHERE THE HEART IS (I,4831); A WORLD APART (I,4910); YOUNG DR. MALONE (I,4938)

PRINCESS GRACE OF MONACO
THE NATIVITY (II,1798)

PRINCI, ELAINE
AS THE WORLD TURNS (II,110); DAYS OF OUR LIVES (II,629)

PRINCIPAL, VICTORIA
60 YEARS OF SEDUCTION (II,2373); BATTLE OF THE NETWORK STARS (II,166); BATTLE OF THE NETWORK STARS (II,168); CELEBRITY CHALLENGE OF THE SEXES 5 (II,469); DALLAS (II,614); FANTASY ISLAND (II,830); LOVE STORY (I,2778); WOMEN WHO RATE A "10" (II,2825)

PRINE, ANDREW
COME A RUNNING (I,1013); M STATION: HAWAII (II,1568); PROFILES IN COURAGE (I,3678); THE ROAD WEST (I,3801); SECOND CHANCE (I,3957); V (II,2713); V: THE FINAL BATTLE (II,2714); W.E.B. (II,2732); WIDE COUNTRY (I,4858); WONDER WOMAN (PILOT 1) (II,2826)

PRINE, JACK
MIGHTY O (I,3036)

PRINE, JOHN
JOHNNY CASH'S AMERICA (II,1330)

PRINGLE, JANE
IRONSIDE (II,1246)

PRINGLE, JOAN
COREY: FOR THE PEOPLE (II,575); RAFFERTY (II,2105); SHANNON (II,2314); THAT'S MY MAMA (II,2580); THE WHITE SHADOW (II,2788)

PRINZ, ROSEMARY
ALL MY CHILDREN (II,39); AS THE WORLD TURNS (II,110); FIRST LOVE (I,1583); HOW TO SURVIVE A MARRIAGE (II,1198); ONE OF THE BOYS (II,1911)

PRINZE, FREDDIE
CHICO AND THE MAN (II,508); THE FIRST 50 YEARS (II,862); THE FLIP WILSON SPECIAL (II,887)

PRITCHARD, DAVID
QUICK AND QUIET (II,2099)

PRITCHETT, FLORENCE
TELEVISION FASHION FAIR (I,4381)

PRITCHETT, JAMES
AS THE WORLD TURNS (II,110); THE DOCTORS (II,694)

PRITKIN, CAROL
MICKEY SPILLANE'S MIKE HAMMER: MORE THAN MURDER (II,1693)

PRIVNEY, CHUCK
OPERATION GREASEPAINT (I,3402)

PROBERT, SION
THE CITADEL (II,539)

PROCTOR, CATHERINE
THE WOMAN IN WHITE (I,4887)

PROCTOR, PHILIP
ME AND MRS. C. (II,1673)

PROFANATA, GENE
MR. T. AND TINA (II,1758)

PROFT, PAT
ALL COMMERCIALS—A STEVE MARTIN SPECIAL (II,37); THE BURNS AND SCHREIBER COMEDY HOUR (I,768); DETECTIVE SCHOOL (II,661); JOEY AND DAD (II,1306); MADHOUSE 90 (I,2807); TWILIGHT THEATER (II,2685); VAN DYKE AND COMPANY (II,2719)

PROHASKA, JANOS
THE ANDY WILLIAMS SHOW (I,210); BARNEY AND ME (I,358)

PROKOP, PAUL
THE GUIDING LIGHT (II,1064)

PROMENADERS, THE
JUBILEE U.S.A. (I,2461)

PROMENADES, THE
FIVE STAR JUBILEE (I,1589)

PROSKY, ROBERT
HILL STREET BLUES (II,1154)

PROUT, LOU
POLKA-GO-ROUND (I,3640)

PROVENZA, PAUL
BACK TOGETHER (II,131); BLUE JEANS (II,275); THE FACTS (II,804)

PROVINE, DOROTHY
THE ALASKANS (I,106); THE BOB HOPE SHOW (I,587); THE DEAN MARTIN SHOW (I,1200); THE ROARING TWENTIES (I,3805); THE SOUND OF ANGER (I,4133); YOU'RE ONLY YOUNG ONCE (I,4969)

PROVOST, JON
JEFF'S COLLIE (I,2356); THE SECRET WORLD OF KIDS (I,3970); TIMMY AND LASSIE (I,4520)

PROWSE, JULIET
BING CROSBY AND CAROL BURNETT—TOGETHER AGAIN FOR THE FIRST TIME (I,454); THE BOB HOPE SHOW (I,584); THE BOB HOPE SHOW (I,591); THE BOB HOPE SHOW (I,627); DANNY THOMAS LOOKS AT YESTERDAY, TODAY AND TOMORROW (I,1148); DON KNOTTS NICE CLEAN, DECENT, WHOLESOME HOUR (I,1333); THE DON KNOTTS SPECIAL (I,1335); DON RICKLES—ALIVE AND KICKING (I,1340); FAVORITE SONGS (I,1545); FRANK SINATRA (I,1660); THE FRANK SINATRA TIMEX SHOW (I,1668); THE FRANK SINATRA TIMEX SHOW (I,1667); HOLLYWOOD MELODY (I,2085); THE KLOWNS (I,2574); THE LOVE BOAT (II,1535); MONA MCCLUSKEY (I,3085); MUSICAL COMEDY TONIGHT (II,1777); ON PARADE (I,3356); THE PAUL LYNDE COMEDY HOUR (II,1979); REMEMBER HOW GREAT? (I,3764); SECOND CHANCE (I,3958); THE SHAPE OF THINGS (II,2315); SHOWTIME (I,4036); STARS AND STRIPES SHOW (II,2442); WOMEN OF RUSSIA (II,2824)

PRUD'HOMME, CAMERON
THE BLUE MEN (I,492)

PRUDDEN, SUZY
YOUR NEW DAY (II,2873)

PRYCE, JONATHAN
ROGER DOESN'T LIVE HERE ANYMORE (II,2201)

PRYOR, AINSLEE
THE ADVENTURES OF HIRAM HOLIDAY (I,61)

PRYOR, MAUREEN
MOLL FLANDERS (II,1720)

PRYOR, NICHOLAS
THE ADAMS CHRONICLES (II,8); AH!, WILDERNESS (I,91); ALL MY CHILDREN (II,39); ANOTHER WORLD (II,97); THE BIG EASY (II,234); EAST OF EDEN (II,752); EIGHT IS ENOUGH (II,762); FORCE FIVE (II,907); HAVING BABIES II (II,1108); HOME ROOM (II,1172); LITTLE HOUSE: A NEW BEGINNING (II,1488); MARRIAGE IS ALIVE AND WELL (II,1633); THE NURSES (I,3319); THE OUTSIDE MAN (II,1937); REVENGE OF THE GRAY GANG (II,2146); WASHINGTON: BEHIND CLOSED DOORS (II,2744); YOUNG DR. MALONE (I,4938)

PRYOR, RICHARD
ABC STAGE '67 (I,6); COMEDY NEWS (I,1020); THE FLIP WILSON SPECIAL (II,889); FLIP WILSON. . . OF COURSE (II,890); KRAFT SUMMER MUSIC HALL (I,2590); A LAST LAUGH AT THE 60'S (I,2627); LILY (I,2705); THE LILY TOMLIN SHOW (I,2706); PRYOR'S PLACE (II,2089); THE RICHARD PRYOR SHOW (II,2163); THE RICHARD PRYOR SPECIAL (II,2164); THE RICHARD PRYOR SPECIAL? (II,2165); THE YOUNG LAWYERS (I,4940)

PUCKETT, GARY
AMERICA (I,164)

PUDDEN, SARAH
THE LIFE OF RILEY (I,2686)

PUGLIA, FRANK
THAT GIRL (II,2570)

PULFORD, DON
FROM HERE TO ETERNITY (II,932)

PULLIAM, KESHIA KNIGHT
THE COSBY SHOW (II,578)

PULLMAN, DALE
REAL KIDS (II,2116)

PULLMAN, DULCIE
THE ROPERS (II,2214)

PULLMAN, MARCY
HELLINGER'S LAW (II,1127)

PUNN, TANYA
BEULAH LAND (II,226)

PUNT, SHANE
THE BEST CHRISTMAS PAGEANT EVER (II,212)

PUNZLOV, FREDERICK
CELEBRITY (II,463)

PUPA, PICCOLA
MAKE ROOM FOR DADDY (I,2843)

PURCELL, LEE
THE COUNTRY MUSIC MURDERS (II,590); THE GIRL, THE GOLD WATCH AND DYNAMITE (II,1001); MY WIFE NEXT DOOR (II,1781); THE SECRET WAR OF JACKIE'S GIRLS (II,2294); STRANGER IN OUR HOUSE (II,2477)

PURCELL, SARAH
BATTLE OF THE NETWORK STARS (II,169); BATTLE OF THE NETWORK STARS (II,170); BATTLE OF THE NETWORK STARS (II,171); THE BETTER SEX (II,223); THE BIG SHOW (II,243); CELEBRITY CHALLENGE OF THE SEXES 5 (II,469); REAL PEOPLE (II,2118); THE SHAPE OF THINGS (II,2315)

PURCHASE, BRUCE
A HORSEMAN RIDING BY (II,1182)

PURCILL, KAREN
A MAN CALLED SLOANE (II,1613)

PURDHAM, DAVID
PIAF (II,2039); RYAN'S HOPE (II,2234)

PURDOM, EDMUND
THE SWORD OF FREEDOM (I,4326); THE WINDS OF WAR (II,2807)

PURDUM, RALPH
THE ADMIRABLE CRICHTON (I,36)

PURDY, RICHARD
KING RICHARD II (I,2559); LITTLE WOMEN (I,2720)

PURL, LINDA
BEACON HILL (II,193); HAPPY DAYS (II,1084); HAVING BABIES I (II,1107); I DO, I DON'T (II,1212); THE LAST DAYS OF POMPEII (II,1436); LIFE'S MOST EMBARRASSING MOMENTS IV (II,1473); THE MANIONS OF AMERICA (II,1623); THE OREGON TRAIL (II,1922); STATE FAIR (II,2448); TESTIMONY OF TWO MEN (II,2564); THE YOUNG PIONEERS (II,2868); THE YOUNG PIONEERS (II,2869); THE YOUNG PIONEERS' CHRISTMAS (II,2870)

PURVEY, BOB
LOVERS AND FRIENDS (II,1549)

PUSSYCAT DANCERS, THE
THE STEVE LAWRENCE SHOW (I,4227)

PUTCH, JOHN
THE ADVENTURES OF POLLYANNA (II,14); ONE DAY AT A TIME (II,1900)

PUTCH, WILLIAM
FAMILY (II,813)

PUTNAM, GEORGE
BROADWAY TO HOLLYWOOD (I,739); TELEVISION SCREEN MAGAZINE (I,4383)

PUTNAM, LORI
THE FIRM (II,860)

PUZOFF, GARY
TOO GOOD TO BE TRUE (II,2643)

PYLE, DENVER
THE ANDY GRIFFITH SHOW (I,192); CODE 3 (I,993); CROSSROADS (I,1105); CROWN THEATER WITH GLORIA SWANSON (I,1107); THE DORIS DAY SHOW (I,1355); THE DUKES (II,741); THE DUKES OF HAZZARD (II,742); HERE COME THE BRIDES (I,2024); HITCHED (I,2065); KAREN (II,1363); KRAFT MYSTERY THEATER (I,2589); THE LIFE AND LEGEND OF WYATT EARP (I,2681); THE LIFE AND TIMES OF GRIZZLY ADAMS (II,1469); MEDIC (I,2976); MRS R.—DEATH AMONG FRIENDS (II,1759); THE ROY ROGERS SHOW (I,3859); SIDEKICKS (II,2348); TAMMY (I,4357); THRILLER (I,4492); WHEN THE WEST WAS FUN: A WESTERN REUNION (II,2780)

PYNE, JOE
THE JOE PYNE SHOW (I,2402); SHOWDOWN (I,4034)

PYNE, NATASHA
FATHER DEAR FATHER (II,837)

PYNER, NICOLE
RETURN TO EDEN (II,2143)

QUADE, ANTHONY
THE STRANGE REPORT (I,4248)

QUADE, JOHN
BIG BEND COUNTRY (II,230); THE CHEAP DETECTIVE (II,490); HUNTER'S MOON (II,1207); NO MAN'S LAND (II,1855); OF MEN OF WOMEN (I,3332); TROUBLE IN HIGH TIMBER COUNTRY (II,2661); VEGAS (II,2723); THE YEAGERS (II,2843)

QUADFLIEG, CHRISTIAN
STAR MAIDENS (II,2435)

QUAID, DENNIS
JOHNNY BELINDA (I,2418)

QUALEN, JOHN
DOC (I,1303); MAKE ROOM FOR DADDY (I,2843); PINE LAKE LODGE (I,3597); THE REAL MCCOYS (I,3741); SALLY AND SAM (I,3891); SCHAEFER CENTURY THEATER (I,3935)

QUARRY, ROBERT
HOLLYWOOD SCREEN TEST (I,2089); THE MILLIONAIRE (II,1700)

QUATRO, SUZY
HAPPY DAYS (II,1084)

QUAYLE, ANTHONY
BAREFOOT IN ATHENS (I,354); THE EVIL TOUCH (I,1486); JARRETT (I,2349); LACE (II,1418); THE LAST DAYS OF POMPEII (II,1436); THE MANIONS OF AMERICA (II,1623); MASADA (II,1642); MOSES THE LAWGIVER (II,1737); THE POPPY IS ALSO A FLOWER (I,3650); THE SIX WIVES OF HENRY VIII (I,4068)

QUEALY, GERIT
RYAN'S HOPE (II,2234)

QUEANT, GILLES
CAPTAIN GALLANT OF THE FOREIGN LEGION (I,831)

QUEENSBERRY, ANN
THE WOMAN IN WHITE (II,2819)

QUENTIN, JOHN
SEARCH FOR THE NILE (I,3954)

QUERISHI, RAHMANN
LACE (II,1418)

QUESTEL, MAE
THE BETTY BOOP SHOW (I,411); CASPER AND FRIENDS (I,867); THE CORNER BAR (I,1052); POPEYE THE SAILOR (I,3648); POPEYE THE SAILOR (I,3649); WINKY DINK AND YOU (I,4880)

QUICK, DIANA
BRIDESHEAD REVISITED (II,375); THE WOMAN IN WHITE (II,2819)

QUICK, ELDON

ADAMS OF EAGLE LAKE (II,10); BRET MAVERICK (II,374)

QUICK, SUSAN
THE CHEERLEADERS (II,493)

QUILL, MICHAEL
THE LAST DAYS OF POMPEII (II,1436)

QUILLAN, EDDIE
THE ADDAMS FAMILY (II,11); THE BANANA COMPANY (II,139); CAP'N AHAB (I,824); FLATFOOTS (II,879); GEORGE GOBEL PRESENTS (I,1766); HARRY AND MAGGIE (II,1097); I AND CLAUDIE (I,2164); THE RIFLEMAN (I,3789); VALENTINE'S DAY (I,4693)

QUILLEY, DENNIS
LONG DAY'S JOURNEY INTO NIGHT (I,2746); MASADA (II,1642)

QUINE, DON
PEYTON PLACE (I,3574); THE VIRGINIAN (I,4738)

QUINLAN, KATHLEEN
LUCAS TANNER (II,1555); SHE'S IN THE ARMY NOW (II,2320); THE TURNING POINT OF JIM MALLOY (II,2670)

QUINLAN, ROBERTA
THE MOHAWK SHOWROOM (I,3080); THE ROBERTA QUINLAN SHOW (I,3818); SONG AND DANCE TIME (I,4121)

QUINLIVAN, CHARLES
MR. GARLUND (I,3138)

QUINLON, CHUCK
MR. MOON'S MAGIC CIRCUS (II,1753)

QUINN, AILEEN
ANDY WILLIAMS' EARLY NEW ENGLAND CHRISTMAS (II,79)

QUINN, ANTHONY
THE CITY (I,969); DANGER (I,1134); FORD THEATER HOUR (I,1635); THE MAN AND THE CITY (I,2856); PHILCO TELEVISION PLAYHOUSE (I,3583); SALUTE TO LADY LIBERTY (II,2243); TEN WHO DARED (II,2559)

QUINN, BILL
ARCHIE BUNKER'S PLACE (II,105); COREY: FOR THE PEOPLE (II,575); DECOY (I,1217); FOOLS, FEMALES AND FUN: WHAT ABOUT THAT ONE? (II,901); GLORIA COMES HOME (II,1011); THE INDIAN (I,2210); THE MARY TYLER MOORE SHOW (II,1640); MCHALE'S NAVY (I,2969); MCMILLAN (II,1666); MCMILLAN AND WIFE (II,1667); OWEN MARSHALL: COUNSELOR AT LAW (I,3435); PLEASE DON'T EAT THE DAISIES (I,3628); PRISONER: CELL BLOCK H (II,2085); THE RIFLEMAN (I,3789); SCRUPLES (II,2280); VELVET (II,2725)

QUINN, CARMEL
ARTHUR GODFREY AND FRIENDS (I,278)

QUINN, DON
PEYTON PLACE (I,3574)

QUINN, J C
O'MALLEY (II,1872)

QUINN, LOUIS
77 SUNSET STRIP (I,3988); SLITHER (II,2379)

QUINN, MICHAEL
THE GHOST SQUAD (I,1787); THOU SHALT NOT KILL (II,2604)

QUINN, PATRICIA
BEAUTY AND THE BEAST (I,382)

QUINN, PATRICK
BOSOM BUDDIES (II,355)

QUINN, SPENCER
THE JERRY REED WHEN YOU'RE HOT, YOU'RE HOT HOUR (I,2372); THE SMOTHERS BROTHERS SHOW (I,4097)

QUINN, TEDDY
ACCIDENTAL FAMILY (I,17); PRUDENCE AND THE CHIEF (I,3682)

QUINN, TOM
BAKER'S DOZEN (II,136); FISHERMAN'S WHARF (II,865); POPI (II,2068)

QUINONES, ADOLFO
THE BIG SHOW (II,243)

QUO, BEULAH
BACHELOR FATHER (I,323); GENESIS II (I,1760); INSPECTOR PEREZ (II,1237); MARCO POLO (II,1626); SAINT PETER (II,2238); SAMURAI (II,2249)

RABAT, KEN
ALL MY CHILDREN (II,39)

RABB, ELLIE
THE DAIN CURSE (II,613)

RABBITT, EDDIE
ANNE MURRAY'S CARIBBEAN CRUISE (II,88); ANSON AND LORRIE (II,98); BOB HOPE SPECIAL: BOB HOPE GOES TO COLLEGE (II,283); COUNTRY GALAXY OF STARS (II,587); COUNTRY NIGHT OF STARS II (II,593); COUNTRY STARS OF THE 70S (II,594); CRYSTAL (II,607); THE EDDIE RABBITT SPECIAL (II,759); JOHNNY CASH—A MERRY MEMPHIS CHRISTMAS (II,1331); LYNDA CARTER: BODY AND SOUL (II,1565); MUSIC CITY NEWS TOP COUNTRY HITS OF THE YEAR (II,1772); A SPECIAL EDDIE RABBITT (II,2415)

RABY, JOHN
CONCERNING MISS MARLOWE (I,1032); A WOMAN TO REMEMBER (I,4888)

RACE, CLARK
THE PARENT GAME (I,3455)

RACHE, DAVID
SANCTUARY OF FEAR (II,2252)

RACIMO, VICTORIA
THE CHISHOLMS (II,513)

RADD, RONALD
THE SHARI LEWIS SHOW (I,4000); THE TEMPEST (I,4391)

RADILAC, CHARLES
THE FILE ON DEVLIN (I,1575)

RADIO CITY MUSIC HALL ROCKETTES, THE
THE MARY MARTIN SHOW (I,2934)

RADNER, GILDA
BOB & RAY & JANE, LARAINE & GILDA (II,279); SATURDAY NIGHT LIVE (II,2261); STEVE MARTIN'S THE WINDS OF WHOOPIE (II,2461); THINGS WE DID LAST SUMMER (II,2589)

RADO, IVAN J
ARCHER—FUGITIVE FROM THE EMPIRE (II,103)

RAE, CHARLOTTE
THE ACADEMY (II,3); THE ACADEMY II (II,4); BEANE'S OF BOSTON (II,194); BOB HOPE SPECIAL: BOB HOPE'S ALL-STAR LOOK AT TV'S PRIME TIME WARS (II,303); CAR 54, WHERE ARE YOU? (I,842); DIFF'RENT STROKES (II,674); THE FACTS OF LIFE (II,805); FACTS OF LIFE: THE FACTS OF LIFE GOES TO PARIS (II,806); HARVEY (I,1961); HOT L BALTIMORE (II,1187); JO'S COUSINS (II,1293); OUR TOWN (I,3421); THE PARKERS (II,1959); PINOCCHIO (I,3603); THE RICH LITTLE SHOW (II,2153); SESAME STREET (I,3982); THE SID CAESAR SHOW (I,4043); THE WAY THEY WERE (II,2748)

RAE, CLAUDE
THE RELUCTANT DRAGON AND MR. TOAD (I,3762)

RAE, GAIL
GENERAL HOSPITAL (II,964)

RAE, GRETA
THE DOCTORS (II,694)

RAE, MAURINE
BACHELOR PARTY (I,324)

RAE, MELBA
HANDS OF MURDER (I,1929); LIGHT'S OUT (I,2699)

RAEBECK, WENDY
NANCY ASTOR (II,1788); SWEEPSTAKES (II,2514)

RAFFERTY, BILL
LAUGH-IN (II,1444); REAL PEOPLE (II,2118)

RAFFERTY, FRANCES
DECEMBER BRIDE (I,1215); G.E. SUMMMER ORIGINALS (I,1747); PEPSI-COLA PLAYHOUSE (I,3523); PETE AND GLADYS (I,3558); STARS OVER HOLLYWOOD (I,4211)

RAFFIN, DEBORAH
CIRCUS OF THE STARS (II,530); CIRCUS OF THE STARS (II,531); FOR LOVERS ONLY (II,904); FOUL PLAY (II,910); THE LAST CONVERTIBLE (II,1435); OF MEN OF WOMEN (I,3332)

RAFILL, STEWART
TARZAN (I,4361)

RAFT, GEORGE
I'M THE LAW (I,2195); THE RED SKELTON TIMEX SPECIAL (I,3757)

RAGGIO, LISA
PRIVATE BENJAMIN (II,2087)

RAGHEB, OSMAN
THE WINDS OF WAR (II,2807)

RAGIN, JOHN S
THE EYES OF TEXAS II (II,800); THE ISLANDER (II,1249); QUINCY, M. E. (II,2102); SENIOR YEAR (II,2301); SONS AND DAUGHTERS (II,2404)

RAGLAND, LARRY
KEEP ON TRUCKIN' (II,1372)

RAHT, KATHERINE
CONCERNING MISS MARLOWE (I,1032); THE CRADLE SONG (I,1088); HARVEY (I,1961)

RAIDERS, THE
HAPPENING '68 (I,1936)

RAIKIN, DAVID
THE BOB HOPE SHOW (I,562)

RAILSBACH, STEVE
FROM HERE TO ETERNITY (II,932)

RAIN, DOUGLAS
A CRY OF ANGELS (I,1111)

RAINBOW, LYNN
AGAINST THE WIND (II,24)

RAINER, IRIS
HERE COMES PETER COTTONTAIL (II,1132); SKYHAWKS (I,4079)

RAINER, JOE
BENJI, ZAX AND THE ALIEN PRINCE (II,205)

RAINER, LUISE
FAITH BALDWIN'S THEATER OF ROMANCE (I,1515); SUSPENSE (I,4305)

RAINES, CRISTINA
BATTLE OF THE NETWORK STARS (II,171); BATTLE OF THE NETWORK STARS (II,173); BATTLE OF THE NETWORK STARS (II,174); CENTENNIAL (II,477); THE FAMILY HOLVAK (II,817); FLAMINGO ROAD (II,871); FLAMINGO ROAD (II,872); LOOSE CHANGE (II,1518); THE RETURN OF MARCUS WELBY, M.D (II,2138); SUNSHINE (II,2495)

RAINES, ELLA
DOUGLAS FAIRBANKS JR. PRESENTS THE RHEINGOLD THEATER (I,1364); JANET DEAN, REGISTERED NURSE (I,2348); PULITZER PRIZE PLAYHOUSE (I,3692); ROBERT MONTGOMERY PRESENTS YOUR LUCKY STRIKE THEATER (I,3809)

RAINES, JENNIFER
FOUR STAR PLAYHOUSE (I,1652)

RAINES, STEVE
RAWHIDE (I,3727)

RAINEY, ETHEL
AS THE WORLD TURNS (II,110)

RAINEY, FORD
ARENA (I,256); THE BIONIC WOMAN (II,255); CAPTAINS AND THE KINGS (II,435); THE D.A.: MURDER ONE (I,1122); KEY WEST (I,2528); MANHUNTER (II,1620); MANHUNTER (II,1621); THE NEW DAUGHTERS OF JOSHUA CABE (I,3261); THE RICHARD BOONE SHOW (I,3784); SAM HILL (I,3895); SEARCH (I,3951); TACK REYNOLDS (I,4329); TENAFLY (II,2560); WINDOW ON MAIN STREET (I,4873)

RAINS, CLAUDE
ALFRED HITCHCOCK PRESENTS (I,115); CHRYSLER MEDALLION THEATER (I,951); THE DUPONT SHOW OF THE WEEK (I,1388); THE KAISER ALUMINUM HOUR (I,2503); ON BORROWED TIME (I,3354); ONCE UPON A CHRISTMAS TIME (I,3368); THE PIED PIPER OF HAMELIN (I,3595); PLAYHOUSE 90 (I,3623); SHANGRI-LA (I,3996)

RAINWATER, MARVIN
JUBILEE U.S.A. (I,2461)

RAISCH, BILL
THE FUGITIVE (I,1701)

RAITT, JOHN
ANNIE GET YOUR GUN (I,223); THE CHEVY SUMMER SHOW (I,929); THE CHEVY SUMMER SHOW (I,930); THE GENERAL FOODS 25TH ANNIVERSARY SHOW (I,1756); THE MOTOROLA TELEVISION HOUR (I,3112); PULITZER PRIZE PLAYHOUSE (I,3692)

RALLY, PAUL
THE DAVID LETTERMAN SHOW (II,624)

RALPH, SHERYL LEE
THE KROFFT KOMEDY HOUR (II,1411)

RALSTON, BOB
THE LAWRENCE WELK SHOW (I,2643); THE LAWRENCE WELK SHOW (I,2644); MEMORIES WITH LAWRENCE WELK (II,1681)

RALSTON, ESTHER
OUR FIVE DAUGHTERS (I,3413)

RAMBEAU, MARC
DIAGNOSIS: DANGER (I,1250)

RAMBO, DACK
ALL MY CHILDREN (II,39); DIRTY SALLY (II,680); THE GUNS OF WILL SONNETT (I,1907); THE LORETTA YOUNG SHOW (I,2755); NEVER TOO YOUNG (I,3248); NO MAN'S LAND (II,1855); PAPER DOLLS (II,1952); SWORD OF JUSTICE (II,2521); WAIKIKI (II,2734)

RAMBO, DAVID
THE BEST OF TIMES (II,219)

RAMBO, DIRK
THE LORETTA YOUNG SHOW (I,2755)

RAMEY, DIANE
YOUNG DR. MALONE (I,4938)

RAMIERZ, TONY
MARIE (II,1628)

RAMIREZ, FRANK
PARIS (II,1957)

RAMIREZ, MONIKA
BIG FOOT AND WILD BOY (II,237)

RAMIREZ, RAMIRIO RAY
O'MALLEY (II,1872)

RAMIS, HAROLD
RODNEY DANGERFIELD SHOW: I CAN'T TAKE IT NO MORE (II,2198); SECOND CITY TELEVISION (II,2287)

RAMITI, DIDI
WAR CORRESPONDENT (I,4767)

RAMON, ROBERTO
THE CIRCLE FAMILY (II,527)

RAMOS, RUBY
THE HIGH CHAPARRAL (I,2051)

RAMSAY, REMAK
FUNNY PAPERS (I,1713)

RAMSEL, GENA
THE JERK, TOO (II,1282)

RAMSEN, BOBBY
BROTHERS (II,381); DOUBLE TROUBLE (II,724); THE MARY TYLER MOORE COMEDY HOUR (II,1639); P.O.P. (II,1939); W*A*L*T*E*R (II,2731); WHITE AND RENO (II,2787)

RAMSEY, ANNE
CASSIE AND COMPANY (II,455); HERNDON AND ME (II,1139)

RAMSEY, DAVID
LOVERS AND FRIENDS (II,1549)

RAMSEY, LOGAN
CONSPIRACY OF TERROR (II,569); THE DEVIL'S DISCIPLE (I,1247); JOE DANCER: THE MONKEY MISSION (II,1301); THE JOKE AND THE VALLEY (I,2438); LASSIE: THE NEW BEGINNING (II,1433); LETTERS FROM THREE LOVERS (I,2671); LITTLE WOMEN (II,1497); ON THE ROCKS (II,1894); THE ROOKIES (I,3840); THE WINDS OF WAR (II,2807)

RAMSEY, MARION
COS (II,576); KEEP ON TRUCKIN' (II,1372); KOMEDY TONITE (II,1407)

RAMSEY, MORGAN
HUNTER'S MOON (II,1207)

RAMSEY, PENNY
PRISONER: CELL BLOCK H (II,2085)

RAMUS, NICK
BORN TO THE WIND (II,354); CENTENNIAL (II,477); FALCON CREST (II,808)

RANCE, TIA
DOWN HOME (II,731)

RANDALL, ANN
BANYON (I,345); HEE HAW (II,1123)

RANDALL, BOB
ON OUR OWN (II,1892)

RANDALL, CHARLOTTE 'REBEL'
AUCTION AIRE (I,313)

RANDALL, FRANKIE
THE ROWAN AND MARTIN SHOW (I,3854)

RANDALL, GRETA
THE JACKIE GLEASON SPECIAL (I,2324)

RANDALL, JOSH
OUTLAW LADY (II,1935)

RANDALL, LYNNE
THE GOSSIP COLUMNIST (II,1045)

RANDALL, MARIAN
MEET CORLISS ARCHER (I,2981)

RANDALL, STUART
THE ADVENTURES OF SUPERMAN (I,77); CIMARRON CITY (I,953); FRONTIER (I,1695); LARAMIE (I,2614); A TALE OF WELLS FARGO (I,4338)

RANDALL, SUE
HEAVEN HELP US (I,1995); LEAVE IT TO BEAVER (I,2648); PETE AND GLADYS (I,3558); PROFILES IN COURAGE (I,3678); SATAN'S WAITIN' (I,3916); STAR TONIGHT (I,4199); YOU'RE ONLY YOUNG TWICE (I,4970)

RANDALL, TONY
THE ALAN KING SHOW (I,102); ALCOA/GOODYEAR THEATER (I,107); APPOINTMENT WITH ADVENTURE (I,246); ARSENIC AND OLD LACE (I,269); BATTLE OF THE NETWORK STARS (II,166); THE BIG SHOW (II,243); THE

BOB HOPE SHOW (I,601); THE BOB HOPE SHOW (I,606); THE BOB HOPE SHOW (I,659); BOB HOPE SPECIAL: BOB HOPE FOR PRESIDENT (II,282); BOB HOPE SPECIAL: BOB HOPE ON CAMPUS (II,290); CAPTAIN VIDEO AND HIS VIDEO RANGERS (I,838); CELEBRITY CHALLENGE OF THE SEXES 1 (II,465); THE CHEVROLET GOLDEN ANNIVERSARY SHOW (I,924); CIRCUS OF THE STARS (II,538); COOGAN'S REWARD (I,1044); COS: THE BILL COSBY COMEDY SPECIAL (II,577); DOUG HENNING: MAGIC ON BROADWAY (II,730); HEAVEN WILL PROTECT THE WORKING GIRL (I,1996); HIPPODROME (I,2060); HOLIDAY IN LAS VEGAS (I,2076); HOORAY FOR LOVE (I,2115); KATE BLISS AND THE TICKER TAPE KID (II,1366); THE LITTLEST ANGEL (I,2726); LOVE, SIDNEY (II,1547); THE MAN IN THE MOON (I,2871); MR. PEEPERS (I,3147); THE ODD COUPLE (II,1875); ONE MAN'S FAMILY (I,3383); PANORAMA (I,3448); PARADE OF STARS (II,1954); THE PAUL LYNDE COMEDY HOUR (II,1978); PEPSI-COLA PLAYHOUSE (I,3523); SHORT, SHORT DRAMA (I,4023); THE SID CAESAR SPECIAL (I,4044); SIDNEY SHORR (II,2349); SO HELP ME, APHRODITE (I,4104); SOUND OF THE 60S (I,4134); STUDIO ONE (I,4268); THE TONY RANDALL SHOW (II,2640); TOP OF THE MONTH (I,4578); THE WIDE OPEN DOOR (I,4859); THE WONDERFUL WORLD OF AGGRAVATION (I,4894)

RANDALL, TRIP
LOVE OF LIFE (I,2774)

RANDELL, RON
THE VISE (I,4739)

RANDI, DON
MOVIN' WITH NANCY ON STAGE (I,3120)

RANDI, JUSTIN
STEELTOWN (II,2452); THUNDER (II,2617)

RANDO, BOB
IRON HORSE (I,2238)

RANDOLPH, AMANDA
THE AMOS AND ANDY SHOW (I,179); THE DANNY THOMAS TV FAMILY REUNION (I,1154); THE LORETTA YOUNG THEATER (I,2756); MAKE MORE ROOM FOR DADDY (I,2842); MAKE ROOM FOR DADDY (I,2843)

RANDOLPH, BILL
THE COMEDY ZONE (II,559); TRAUMA CENTER (II,2655)

RANDOLPH, BOOTS
THE BOOTS RANDOLPH SHOW (I,687); THE COUNTRY MUSIC MURDERS (II,590); THE NASHVILLE SOUND OF BOOTS RANDOLPH (I,3229)

RANDOLPH, DON
THE YOUNG MARRIEDS (I,4942)

RANDOLPH, ISABEL
DECEMBER BRIDE (I,1215); THE DICK VAN DYKE SHOW (I,1275); THE JERRY COLONNA SHOW (I,2359); MEET MILLIE (I,2988); OUR MISS BROOKS (I,3415)

RANDOLPH, JOHN
THE ADVENTURES OF POLLYANNA (II,14); ANGIE (II,80); THE BOB NEWHART SHOW (II,342); DOCTORS' PRIVATE LIVES (II,693); EMERALD POINT, N.A.S. (II,773); EXECUTIVE SUITE (II,796); THE FACTS OF LIFE (II,805); FAMILY TIES (II,819); IN SECURITY (II,1227); INHERIT THE WIND (I,2213); THE JUDGE AND JAKE WYLER (I,2464); LUCAN (II,1553); LUCAN (II,1554); LUCAS TANNER (II,1556); NERO WOLFE (II,1805); OF MICE AND MEN (I,3333); OLD FRIENDS (II,1881); PARTNERS IN CRIME (I,3461); RICHIE BROCKELMAN, PRIVATE EYE (II,2167); SANDBURG'S LINCOLN (I,3907); THE SENATOR (I,3976); THE SHERIFF AND THE ASTRONAUT (II,2326); SHOOTING STARS (II,2341); TOPPER RETURNS (I,4583); WASHINGTON: BEHIND CLOSED DOORS (II,2744); WONDER WOMAN (PILOT 2) (II,2827)

RANDOLPH, JOYCE
THE HONEYMOONERS (I,2110); THE HONEYMOONERS (II,1175); THE JACKIE GLEASON SHOW (I,2322)

RANDOLPH, LILLIAN
THE AMOS AND ANDY SHOW (I,179); THE BILL COSBY SHOW (I,442); THE GREAT GILDERSLEEVE (I,1875); ROOTS (II,2211); TENAFLY (I,4395)

RANDOM, ROBERT
SCALPLOCK (I,3930)

RANDY SPARKS & THE NEW CHRISTY MINSTRELS
THE ANDY WILLIAMS SHOW (I,205)

RANDY SPARKS AND THE BACKPORCH MAJORITY
ALL THINGS BRIGHT AND BEAUTIFUL (I,141); JACK CASSIDY'S ST. PATRICK'S DAY SPECIAL (I,2307)

RANDY VAN HORNE SINGERS
THE NAT KING COLE SHOW (I,3230)

RANEY, JANET
HEAR NO EVIL (II,1116)

RANEY, WALTER
WHAT'S THE STORY? (I,4821)

RANIER, JEANNE
MY BOY GOOGIE (I,3185)

RANKIN, GIL
TOMBSTONE TERRITORY (I,4550)

RANKIN, STEVE
ALL MY CHILDREN (II,39)

RANSOME, NORMA
THAT'S O'TOOLE (I,4428)

RANSOME, PRUNELLA
A HORSEMAN RIDING BY (II,1182)

RAPELYE, MARY LINDA
AS THE WORLD TURNS (II,110)

RAPHAEL, GERRAINE
CAPTAIN VIDEO AND HIS VIDEO RANGERS (I,838)

RAPLEY, JOHN
THE DEVIL'S WEB (I,1248)

RAPPAPORT, MICHAEL
THE NEW ODD COUPLE (II,1825)

RASCHE, DAVID
RYAN'S HOPE (II,2234)

RASEY, JEAN
THE NANCY DREW MYSTERIES (II,1789)

RASKIN, DAMON
LOOK OUT WORLD (II,1516); THE MONTEFUSCOS (II,1727); THE PRACTICE (II,2076)

RASSULO, JOE
MISS WINSLOW AND SON (II,1705)

RASTATTER, WENDY
BATTLE OF THE NETWORK STARS (II,167); DAVID CASSIDY—MAN UNDERCOVER (II,623); THE PAPER CHASE (II,1949); STEELTOWN (II,2452); WELCOME BACK, KOTTER (II,2761)

RASULALA, THALMUS
THE BAIT (I,335); THE JERK, TOO (II,1282); ROOTS (II,2211); WHAT'S HAPPENING!! (II,2769)

RASUMMY, JAY
ONE NIGHT BAND (II,1909)

RATHBONE, BASIL
THE CHEVROLET TELE-THEATER (I,926); A CHRISTMAS CAROL (I,947); DUPONT SHOW OF THE MONTH (I,1387); HANS BRINKER OR THE SILVER SKATES (I,1933); KRAFT TELEVISION THEATER (I,2592); THE LARK (I,2616); LIGHT'S OUT (I,2699); LOVE STORY (I,2776); THE MARCH OF DIMES FASHION SHOW (I,2901); THE MOTOROLA TELEVISION HOUR (I,3112); THE NASH AIRFLYTE THEATER (I,3227); THE PIRATES OF FLOUNDER BAY (I,3607); SOLDIER IN LOVE (I,4108); SUSPENSE (I,4305); SVENGALI AND THE BLONDE (I,4310); VICTORIA REGINA (I,4728); YOUR LUCKY CLUE (I,4954)

RATHBURN, ROGER
SOMERSET (I,4115)

RATRAY, PETER
ANOTHER WORLD (II,97); BRIGHT PROMISE (I,727); THE NEW PEOPLE (I,3268)

RATZENBERGER, JOHN
CHEERS (II,494); GOLIATH AWAITS (II,1025)

RAVAZZA, CARL
WINDOW SHADE REVUE (I,4874)

RAVEL, MARGIE
BROADWAY GOES LATIN (I,734)

RAVENS, RUPERT
SANTA BARBARA (II,2258)

RAVENSCROFT, THURL
NANNY AND THE PROFESSOR AND THE PHANTOM OF THE CIRCUS (I,3226)

RAWLINGS, ALICE

THE PATTY DUKE SHOW (I,3483)

RAWLINGS, MARGARET
THE TRAP (I,4593)

RAWLINS, LESTER
THE EDGE OF NIGHT (II,760); NICK AND NORA (II,1842); RYAN'S HOPE (II,2234)

RAWLINSON, BRIAN
THE BUCCANEERS (I,749); THE ONEDIN LINE (II,1912)

RAWLINSON, HOWARD
LADY KILLER (I,2603)

RAWLS, EUGENIA
THE GREAT SEBASTIANS (I,1877); ROAD TO REALITY (I,3800)

RAWLS, LOU
THE BOB HOPE SHOW (I,622); DEAN MARTIN PRESENTS THE GOLDDIGGERS (I,1194); THE ENGELBERT HUMPERDINCK SHOW (I,1445); THE LOU RAWLS SPECIAL (II,1527); THE STARS AND STRIPES SHOW (I,4208); TENNESSEE ERNIE FORD'S WHITE CHRISTMAS (I,4402); UPTOWN (II,2710)

RAY ANTHONY CHORUS, THE
TV'S TOP TUNES (I,4635)

RAY CHARLES CHOIR, THE
JACK AND THE BEANSTALK (I,2287)

RAY CHARLES SINGERS, THE
ACCENT ON LOVE (I,16); THE DANNY THOMAS TV FAMILY REUNION (I,1154); THE HOLLYWOOD PALACE (I,2088); THE PERRY COMO SHOW (I,3531); THE PERRY COMO SHOW (I,3532); THE PERRY COMO SHOW (I,3533); THE PERRY COMO SPECIAL (I,3534); THE PERRY COMO SPECIAL (I,3535); THE PERRY COMO SPRINGTIME SHOW (I,3537); THE PERRY COMO VALENTINE SPECIAL (I,3541); THE PERRY COMO WINTER SHOW (I,3542); THE PERRY COMO WINTER SHOW (I,3543); PERRY COMO'S WINTER SHOW (I,3547)

RAY PORTER CHORUS, THE
THE GOODYEAR REVUE (I,1856)

RAY PORTER SINGERS, THE
PAUL WHITEMAN'S TV TEEN CLUB (I,3491)

RAY, ALDO
DEADLOCK (I,1190); HAVE GIRLS—WILL TRAVEL (I,1967); THE HOUNDCATS (I,2132); LOLLIPOP LOUIE (I,2738); WOMEN IN WHITE (II,2823)

RAY, ALLAN
LEAVE IT TO BEAVER (I,2648)

RAY, ANDREW
CARRY ON LAUGHING (II,448); CROWN MATRIMONIAL (I,1106)

RAY, DON
THE KEN BERRY WOW SHOW (I,2524)

RAY, GENE ANTHONY
FAME (II,812); THE KIDS FROM FAME (II,1392)

RAY, GERALD
MR. AND MRS. COP (II,1748)

RAY, HELEN
ETHEL AND ALBERT (I,1466)

RAY, JAMES
THE EDGE OF NIGHT (II,760); OPERATION PETTICOAT (II,1917); RETURN ENGAGEMENT (I,3775); WHEELS (II,2777); THE WINDS OF WAR (II,2807)

RAY, JOANNA
ROWAN AND MARTIN BITE THE HAND THAT FEEDS THEM (I,3853)

RAY, JOHNNIE
20TH CENTURY FOLLIES (I,4641); 100 YEARS OF GOLDEN HITS (II,1905); ONE MORE TIME (I,3385)

RAY, LESLIE ANN
THE DOCTORS (II,694); THE EDGE OF NIGHT (II,760); SEARCH FOR TOMORROW (II,2284)

RAY, MARGUERITE
SANFORD (II,2255); THE YOUNG AND THE RESTLESS (II,2862)

RAYBOULD, HARRY
GENESIS II (I,1760)

RAYBURN, GENE
THE AMATEUR'S GUIDE TO LOVE (I,155); CHOOSE UP SIDES (I,945); DOUGH RE MI (I,1363); HEAD OF THE CLASS (I,1987); MAKE THE CONNECTION (I,2846); THE MATCH GAME (I,2944); THE MATCH GAME (II,1649); THE MATCH GAME—HOLLYWOOD SQUARES HOUR (II,1650); PLAY YOUR HUNCH (I,3619); THE SKY'S THE LIMIT (I,4080); SNAP JUDGEMENT (I,4099); TIC TAC DOUGH (I,4498); TV'S FUNNIEST GAME SHOW MOMENTS (II,2676)

RAYE, FRANK
SING ALONG WITH MITCH (I,4057)

RAYE, MARTHA
ALICE (II,33); THE ALL-STAR REVUE (I,138); ANYTHING GOES (I,237); THE BIG TIME (I,434); BING!. . . A 50TH ANNIVERSARY GALA (II,254); THE BOB HOPE SHOW (I,595); THE BOB HOPE SHOW (I,597); THE BOB HOPE SHOW (I,601); THE BOB HOPE SHOW (I,606); THE BOB HOPE SHOW (I,630); THE BOB HOPE SHOW (I,643); THE BOB HOPE SHOW (I,649); BOB HOPE SPECIAL: BOB HOPE'S ROAD TO HOLLYWOOD (II,313); THE BUGALOOS (I,757); CIRCUS OF THE STARS (II,536); CLOWN ALLEY (I,979); THE COMICS (I,1028); DATELINE (I,1165); THE GOSSIP COLUMNIST (II,1045); THE MARTHA RAYE SHOW (I,2926); MCMILLAN (II,1666); SKINFLINT (II,2377); THE STEVE ALLEN SHOW (I,4220); STEVE ALLEN'S LAUGH-BACK (II,2455)

RAYE, MARY
HOTPOINT HOLIDAY (I,2131); THE MOREY AMSTERDAM SHOW (I,3101)

RAYE, MELBA
SEARCH FOR TOMORROW (II,2284)

RAYE, PAULA
JAMBOREE (I,2331); WINDY CITY JAMBOREE (I,4877)

RAYE, PHILIP
LITTLE WOMEN (I,2723)

RAYE, SUSAN
BUCK OWENS TV RANCH (I,750); HEE HAW (II,1123); SUSAN RAYE TIME (I,4302)

RAYE, TISCH
DELTA COUNTY, U.S.A. (II,657); DYNASTY (II,746); W.E.B. (II,2732)

RAYLOR, RIP
CIRCUS OF THE STARS (II,533)

RAYMOND, GARY
THE RAT PATROL (I,3726)

RAYMOND, GENE
ADAMSBURG, U.S.A. (I,33); FIRESIDE THEATER (I,1580); G.E. SUMMMER ORIGINALS (I,1747); ICHABOD AND ME (I,2183); PARIS 7000 (I,3459); TV READERS DIGEST (I,4630); WHAT'S GOING ON? (I,4814)

RAYMOND, GUY
AMERICA, YOU'RE ON (II,69); DOC (I,1303); DYNASTY (II,747); THE FLIM-FLAM MAN (I,1599); THE GHOST AND MRS. MUIR (I,1786); HARRIS AGAINST THE WORLD (I,1956); KAREN (I,2505); SATINS AND SPURS (I,3917); STARSTRUCK (II,2446); TOM, DICK, AND MARY (I,4537)

RAYMOND, HELEN
ANYTHING GOES (I,237); THE KING AND MRS. CANDLE (I,2544); THE WOMEN (I,4890)

RAYMOND, JOHN
THE FANTASTIC FUNNIES (II,825)

RAYMOND, LIAM
CAPTAIN BRASSBOUND'S CONVERSION (I,828)

RAYMOND, LINA
THE GYPSY WARRIORS (II,1070); THE JERK, TOO (II,1282)

RAYMOND, PAULA
FOUR STAR PLAYHOUSE (I,1652); THE UNEXPECTED (I,4674); YOUR SHOW TIME (I,4958)

RAYMOND, ROBIN
BEN BLUE'S BROTHERS (I,393)

RAYMOND, SID
CASPER AND FRIENDS (I,867); JERRY MAHONEY'S CLUB HOUSE (I,2371); THE PAUL WINCHELL AND JERRY MAHONEY SHOW (I,3492); THE SLIGHTLY FALLEN ANGEL (I,4087)

RAYMOND, TONY
THE GANGSTER CHRONICLES (II,957)

RAYMOND, VERNA
SHOW OF THE YEAR (I,4031)

RAYNR, DAVID
THE ACADEMY II (II,4); THE BOYS IN BLUE (II,359)

RAZ, KAVI
ST. ELSEWHERE (II,2432)

REA, PEGGY
BRONK (II,379); THE DUKES OF HAZZARD (II,742); HAPPY DAYS (II,1084); THE RED SKELTON SHOW (I,3755); THE RED SKELTON SHOW (I,3756); STAT! (I,4213); SUPERCOPS (II,2502); THE WALTONS

(II,2740)

REA, STEPHEN
COLOR HIM DEAD (I,1007); COLOR HIM DEAD (II,554)

READE, PHILIP
SHIELDS AND YARNELL (II,2328)

READING, DONNA
MONTY PYTHON'S FLYING CIRCUS (II,1729)

READY, LYNN
THE MICKEY MOUSE CLUB (I,3025); THE MOUSEKETEERS REUNION (II,1742)

REAGAN, MAUREEN
ENTERTAINMENT TONIGHT (II,780)

REAGAN, NANCY
BOB HOPE SPECIAL: BOB HOPE PRESENTS A CELEBRATION WITH STARS OF COMEDY AND MUSIC (II,340); BOB HOPE SPECIAL: HAPPY BIRTHDAY, BOB! (II,325); CHRISTMAS IN WASHINGTON (II,520); CHRISTMAS IN WASHINGTON (II,521); DEAN'S PLACE (II,650)

REAGAN, RONALD
AMOS BURKE: WHO KILLED JULIE GREER? (I,181); THE BOB HOPE SHOW (I,582); BOB HOPE SPECIAL: BOB HOPE PRESENTS A CELEBRATION WITH STARS OF COMEDY AND MUSIC (II,340); BOB HOPE SPECIAL: HAPPY BIRTHDAY, BOB! (II,325); CALL TO DANGER (I,802); CHRISTMAS IN WASHINGTON (II,520); CHRISTMAS IN WASHINGTON (II,521); CHRYSLER MEDALLION THEATER (I,951); DEAN'S PLACE (II,650); DEATH VALLEY DAYS (I,1211); THE GENERAL ELECTRIC THEATER (I,1753); HOLLYWOOD OPENING NIGHT (I,2087); THE JUDY GARLAND SHOW (I,2471); THE JUNE ALLYSON SHOW (I,2488); KRAFT SUSPENSE THEATER (I,2591); THE LUX VIDEO THEATER (I,2795); THE NASH AIRFLYTE THEATER (I,3227); OPERATION ENTERTAINMENT (I,3400); THE ORCHID AWARD (I,3406); REVLON MIRROR THEATER (I,3781); ROY ACUFF—50 YEARS THE KING OF COUNTRY MUSIC (II,2223); THE SWINGIN' SINGIN' YEARS (I,4319); THE SWINGIN' YEARS (I,4321)

REAMS, LEE ROY
BONNIE AND THE FRANKLINS (II,349)

REASON, REX
CONFLICT (I,1036); MAN WITHOUT A GUN (I,2884); THE ROARING TWENTIES (I,3805); THE WEB (I,4785)

REASON, RHODES
BUS STOP (I,778); COUNTY GENERAL (I,1076); HAPPY ANNIVERSARY AND GOODBYE (II,1082); PRUDENCE AND THE CHIEF (I,3682); THREE ON AN ISLAND (I,4483); TV READERS DIGEST (I,4630); WHITE HUNTER (I,4845)

REASONER, HARRY
A SALUTE TO TELEVISION'S 25TH ANNIVERSARY (I,3893)

REBAR, ALEX
THE YOUNG AND THE RESTLESS (II,2862)

REBHORN, JAMES
THE GUIDING LIGHT (II,1064); TEXAS (II,2566)

RECKELL, PETER
AS THE WORLD TURNS (II,110)

RED NORVO COMBO, THE
THE FRANK SINATRA TIMEX SHOW (I,1668)

REDD, VERONICA
THE BLUE AND THE GRAY (II,273); EVERY STRAY DOG AND KID (II,792)

REDDICK, GERRI
BLANSKY'S BEAUTIES (II,267)

REDDIN, ELIZABETH
DAVID CASSIDY—MAN UNDERCOVER (II,623)

REDDING, CHAD
THE GANGSTER CHRONICLES (II,957)

REDDING, LYNN
JOE FORRESTER (II,1303)

REDDY, HELEN
CELEBRATION: THE AMERICAN SPIRIT (II,461); THE DON RICKLES SHOW (II,708); THE FLIP WILSON SPECIAL (I,1603); THE FLIP WILSON SPECIAL (II,888); GEORGE BURNS' 100TH BIRTHDAY PARTY (II,971); THE HELEN REDDY SHOW (I,2008); THE HELEN REDDY SPECIAL (II,1126); I LOVE LIBERTY (II,1215); THE MIDNIGHT SPECIAL (II,1696)

REDDY, TOM
DOCTOR I.Q. (I,1314); HOW DO YOU RATE? (I,2144)

REDEKER, QUINN
DAN RAVEN (I,1131); DAYS OF OUR LIVES (II,629); THE GEMINI MAN (II,961); THE LOVE BOAT II (II,1533)

REDETTES, THE
THE RED SKELTON REVUE (I,3754)

REDFEARN, LINDA
BORN TO THE WIND (II,354)

REDFIELD, ADAM
FIT FOR A KING (II,866)

REDFIELD, BILLY
A DATE WITH LIFE (I,1163); JIMMY HUGHES, ROOKIE COP (I,2390); KITTY FOYLE (I,2571)

REDFIELD, DENNIS
DALLAS (II,614); FRIENDS (II,927); THE NIGHTENGALES (II,1850); STATE FAIR (II,2448)

REDFIELD, WILLIAM
ADVENTURES OF A MODEL (I,50); ANDROCLES AND THE LION (I,189); AS THE WORLD TURNS (II,110); THE BOB NEWHART SHOW (II,342); HANDS OF MURDER (I,1929); THE MARRIAGE (I,2922); THE MILLION DOLLAR INCIDENT (I,3045); MR. O'MALLEY (I,3146); THE ODD COUPLE (II,1875); PHIL SILVERS ON BROADWAY (I,3578); PHILCO TELEVISION PLAYHOUSE (I,3583)

REDFORD, IAN
SLEEPWALKER (I,4084)

REDFORD, JEFF
RENDEZVOUS HOTEL (II,2131); THE WILD WILD WEST REVISITED (II,2800)

REDFORD, LINDA
WINDOWS, DOORS AND KEYHOLES (II,2806)

REDFORD, ROBERT
PLAYHOUSE 90 (I,3623)

REDGRAVE, LYNN
BATTLE OF THE NETWORK STARS (II,174); CENTENNIAL (II,477); CIRCUS OF THE STARS (II,537); THE EDGE OF NIGHT (II,760); HELLZAPOPPIN (I,2011); HOUSE CALLS (II,1194); LINDA IN WONDERLAND (II,1481); MUSICAL COMEDY TONIGHT (II,1777); NOT FOR WOMEN ONLY (I,3310); THE SHAPE OF THINGS (II,2315); STEVE MARTIN'S BEST SHOW EVER (II,2460); TEACHERS ONLY (II,2547); TEACHERS ONLY (II,2548)

REDGRAVE, MICHAEL
ABC STAGE '67 (I,6); DR JEKYLL AND MR. HYDE (I,1368); THE GREAT WAR (I,1879); HAMLET (I,1925); HEDDA GABLER (I,2001); MR. DICKENS OF LONDON (I,3135); RUGGLES OF RED GAP (I,3867)

REDI, TOM
STONE (II,2470)

REDINA, VICTOR
THE TOM SWIFT AND LINDA CRAIG MYSTERY HOUR (II,2633)

REDMAN, JOYCE
THE SEVEN DIALS MYSTERY (II,2308)

REDMOND, LIAM
LITTLE MOON OF ALBAN (I,2715); THE TEMPEST (I,4391)

REDMOND, MARGE
ABC'S MATINEE TODAY (I,7); DEATH OF A SALESMAN (I,1208); THE DOUBLE LIFE OF HENRY PHYFE (I,1361); THE FLYING NUN (I,1611)

REDMOND, MOIRA
SIGN IT DEATH (I,4046)

REDWING, KIKO MCKEE
THE NEW VOICE (II,1833)

REDWOOD, MANNING
OPPENHEIMER (II,1919)

REECE, BRIAN
THE NEW ADVENTURES OF MARTIN KANE (I,3251)

REED, ALAINA
SESAME STREET (I,3982)

REED, ALAN
330 INDEPENDENCE S.W. (I,4486); ANDY'S GANG (I,214); DUFFY'S TAVERN (I,1379); THE FLINTSTONE COMEDY HOUR (II,881); THE FLINTSTONES (II,884); THE JAZZ SINGER (I,2351); LIFE WITH LUIGI (I,2692); MICKEY (I,3023); MR. ADAMS AND EVE (I,3121); PETER LOVES MARY (I,3565); TOUCHE TURTLE (I,4586); WHERE'S HUDDLES (I,4836)

REED, ALBERT
CHASE (I,917); ISIS (II,1248)

REED, CAROL
MELODY, HARMONY, RHYTHM (I,2998); RENDEZVOUS WITH MUSIC (I,3768)

REED, DONNA
DALLAS (II,614); THE DONNA REED SHOW (I,1347); SUSPICION (I,4309)

REED, ELLIOTT
ALAN KING'S FINAL WARNING (II,29)

REED, FLORENCE
SKIN OF OUR TEETH (I,4073)

REED, HOWARD
TOTAL ECLIPSE (I,4584)

REED, JAMES
REMINGTON STEELE (II,2130)

REED, JANET
THE BOB HOPE SHOW (I,505); SUNDAY IN TOWN (I,4285)

REED, JERRY
THE BEST LITTLE SPECIAL IN TEXAS (II,215); CONCRETE COWBOYS (II,563); THE CONCRETE COWBOYS (II,564); A COUNTRY CHRISTMAS (II,583); DEAN MARTIN AT THE WILD ANIMAL PARK (II,636); DEAN MARTIN PRESENTS MUSIC COUNTRY, U.S.A. (I,1192); THE GLEN CAMPBELL GOODTIME HOUR (I,1820); GOOD OL' BOYS (II,1035); HARPER VALLEY, U.S.A. (I,1954); JERRY REED AND SPECIAL FRIENDS (II,1285); THE JERRY REED WHEN YOU'RE HOT, YOU'RE HOT HOUR (I,2372); LOUISE MANDRELL: DIAMONDS, GOLD AND PLATINUM (II,1528); LYNDA CARTER'S CELEBRATION (II,1563); MAMA'S FAMILY (II,1610); NASHVILLE 99 (II,1791)

REED, KATHY
DEAR MOM, DEAR DAD (I,1202)

REED, LUCILLE
THIS IS MUSIC (I,4457)

REED, LYDIA
THE CAMPBELL TELEVISION SOUNDSTAGE (I,812); THE DOCTOR (I,1320); THE REAL MCCOYS (I,3741); ROBERT MONTGOMERY PRESENTS YOUR LUCKY STRIKE THEATER (I,3809); STAGE 7 (I,4181); THE WORLD OF MISTER SWEENY (I,4917)

REED, MARSHALL
THE LINE-UP (I,2707); THE WALTONS (II,2740)

REED, MARY-ROBIN
THE GIRL IN THE EMPTY GRAVE (II,997); THE REASON NOBODY HARDLY EVER SEEN A FAT OUTLAW IN THE OLD WEST IS AS FOLLOWS: (I,3744)

REED, MAXWELL
CAPTAIN DAVID GRIEF (I,829)

REED, MICHAEL
THE WALTONS (II,2740)

REED, PAMELA
THE ANDROS TARGETS (II,76)

REED, PAUL
CAR 54, WHERE ARE YOU? (I,842); THE CARA WILLIAMS SHOW (I,843); SID CAESAR INVITES YOU (I,4040); THE SLIGHTLY FALLEN ANGEL (I,4087); TIPTOE THROUGH TV (I,4524)

REED, PHILIP
RUTHIE ON THE TELEPHONE (I,3877)

REED, RALPH
BROKEN ARROW (I,743); LIFE WITH FATHER (I,2690)

REED, REX
HELLZAPOPPIN (I,2011); INSIDE AMERICA (II,1234); REX REED'S MOVIE GUIDE (II,2149); THAT WAS THE YEAR THAT WAS (II,2575)

REED, ROBERT
ASSIGNMENT: MUNICH (I,302); THE BRADY BUNCH (II,362); THE BRADY BUNCH HOUR (II,363); THE BRADY GIRLS GET MARRIED (II,364); CASINO (II,452); THE CITY (I,969); THE DEFENDERS (I,1220); GALACTICA 1980 (II,953); INTERTECT (I,2228); KELLY'S KIDS (I,2523); LANIGAN'S RABBI (II,1427); LI'L ABNER (I,2701); THE LOVE BOAT II (II,1533); MANNIX (II,1624); NURSE (II,1870); OPERATION: RUNAWAY (II,1918); RICH MAN, POOR MAN—BOOK I (II,2161); ROOTS (II,2211); SCRUPLES (II,2280); THE SEEKERS (II,2298); SOMEWHERE IN ITALY. . . COMPANY B (I,4119); THE WAY THEY WERE (II,2748)

REED, ROBIN
A CHARLIE BROWN THANKSGIVING (I,909)

REED, RUSS
HAWKINS FALLS, POPULATION 6200 (I,1977); A TIME TO LIVE (I,4510)

REED, SEIDINA
THE CONCRETE COWBOYS (II,564); JERRY REED AND SPECIAL FRIENDS (II,1285)

REED, SHANNA
FOR LOVE AND HONOR (II,903); TEXAS (II,2566)

REED, SUSAN
ALL ABOUT MUSIC (I,127)

REED, SUZANNE
CODE R (II,551)

REED, TOBY
TOP DOLLAR (I,4577)

REED, TRACY
BAREFOOT IN THE PARK (I,355); COCAINE AND BLUE EYES (II,548); FUTURE COP (II,945); ME AND BENJY (I,2973); TOP SECRET (II,2647); TURNOVER SMITH (II,2671); WOMEN IN WHITE (II,2823)

REEHLING, JOYCE
THE SINGLE LIFE (II,2364)

REES, ANGHARAD
BAFFLED (I,332); MASTER OF THE GAME (II,1647); ONCE THE KILLING STARTS (I,3367)

REES, BETTY ANN
GENERAL HOSPITAL (II,964)

REES, JOHN
MEN AT WORK IN CONCERT (II,1682)

REES, JUDSON
A DATE WITH JUDY (I,1162)

REES, LANNY
THE LIFE OF RILEY (I,2685)

REES, ROGER
NICHOLAS NICKLEBY (II,1840)

REES, TANYA
SMILEY'S PEOPLE (II,2383)

REESE, DELLA
. . .AND BEAUTIFUL (I,184); BURT AND THE GIRLS (I,770); CHICO AND THE MAN (II,508); DADDY'S GIRL (I,1124); DELLA (I,1222); FLO'S PLACE (II,892); IT TAKES TWO (II,1252); WELCOME BACK, KOTTER (II,2761)

REESE, JOY
HANDS OF MURDER (I,1929); WESLEY (I,4795)

REESE, MASON
MASON (II,1643)

REESE, ROXANNE
ST. ELSEWHERE (II,2432)

REESE, ROY
MR. ROBERTS (I,3149)

REESE, TOM
BRADDOCK (I,704); ELLERY QUEEN (II,766); ELLERY QUEEN: TOO MANY SUSPECTS (II,767); THE SEEKERS (I,3974); YOU SHOULD MEET MY SISTER (I,4935)

REEVE, ADA
THE LILLI PALMER THEATER (I,2704)

REEVE, CHRISTOPHER
CELEBRITY DAREDEVILS (II,473); I LOVE LIBERTY (II,1215); LOVE OF LIFE (I,2774)

REEVES, CRYSTAL
MY LITTLE MARGIE (I,3194)

REEVES, DAVE
THE SCOEY MITCHLLL SHOW (I,3939)

REEVES, DEL
COUNTRY CARNIVAL (I,1062)

REEVES, DICK
JOHNNY COME LATELY (I,2425)

REEVES, GEORGE
ACTOR'S STUDIO (I,28); THE ADVENTURES OF SUPERMAN (I,77); FIRESIDE THEATER (I,1580); HANDS OF MURDER (I,1929); LIGHT'S OUT (I,2699); SILVER THEATER (I,4051); STARLIGHT THEATER (I,4201); SUSPENSE (I,4305); THE TRAP (I,4593)

REEVES, GLENN
THE GLENN REEVES SHOW (I,1825)

REEVES, LARRY
ANNIE OAKLEY (I,225)

REEVES, LISA
THE SAN PEDRO BEACH BUMS (II,2250); THE SAN PEDRO BUMS (II,2251)

REEVES, MARTHA
FLIP WILSON. . . OF COURSE (II,890)

REEVES, MARY ANN
SCHOOL HOUSE (I,3937)

REEVES, RICHARD
DATE WITH THE ANGELS (I,1164)

REGALADO, YVONNE
BIG FOOT AND WILD BOY (II,237)

REGALBUTO, JOSEPH
ACE CRAWFORD, PRIVATE EYE (II,6); THE ASSOCIATES (II,113); HARRY'S BATTLES (II,1100); MORK AND MINDY (II,1735); YOU ARE THE JURY (II,2854)

REGAN, ELLEN
THE 25TH MAN (MS) (II,2678); ALMOST HEAVEN (II,57); HOT

W.A.C.S. (II,1191); THE LOVEBIRDS (II,1548); MAKING THE GRADE (II,1606); TAXI (II,2546); THE TED KNIGHT SHOW (II,2550)

REGAN, MARGIE
MICHAEL SHAYNE, PRIVATE DETECTIVE (I,3021)

REGAN, MICHAEL
DEBBY BOONE. . . ONE STEP CLOSER (II,655)

REGAN, PATTY
F TROOP (I,1499)

REGAN, PHIL
HOLLYWOOD STAR REVUE (I,2093); STARTIME (I,4212)

REGAS, JACK
THE GISELE MACKENZIE SHOW (I,1813)

REGAS, PEDRO
PAT PAULSEN'S HALF A COMEDY HOUR (I,3473)

REGEHR, DUNCAN
THE BLUE AND THE GRAY (II,273); GOLIATH AWAITS (II,1025); THE LAST DAYS OF POMPEII (II,1436); V: THE SERIES (II,2715); WIZARDS AND WARRIORS (II,2813)

REGGIE BEANE DANCERS, THE
THE S. S. HOLIDAY (I,4173)

REGINA, PAUL
BROTHERS (II,380); THE GANGSTER CHRONICLES (II,957); JOE AND VALERIE (II,1298); JOE AND VALERIE (II,1299); THE RENEGADES (II,2132); THE SINGLE LIFE (II,2364); ZORRO AND SON (II,2878)

REICHMANN, WOLFGANG
THE MARTIAN CHRONICLES (II,1635)

REID, ADRIAN
CINDERELLA (II,526)

REID, BERYL
SMILEY'S PEOPLE (II,2383); TINKER, TAILOR, SOLDIER, SPY (II,2621)

REID, CARL BENTON
AMOS BURKE, SECRET AGENT (I,180); THE DICK VAN DYKE SHOW (I,1275); WON'T IT EVER BE MORNING? (I,4902)

REID, ELLIOTT
BACHELOR PARTY (I,324); THE CAMPBELL TELEVISION SOUNDSTAGE (I,812); I REMEMBER CAVIAR (I,2175); MISS WINSLOW AND SON (II,1705); THAT WAS THE WEEK THAT WAS (I,4423); THE VICTOR BORGE SHOW (I,4719); THE WONDERFUL WORLD OF PHILIP MALLEY (II,2831)

REID, FIONA
KING OF KENSINGTON (II,1395)

REID, FRANCES
BERKELEY SQUARE (I,399); THE CHEVROLET TELE-THEATER (I,926); DAYS OF OUR LIVES (II,629); HANDS OF MURDER (I,1929); PHILCO TELEVISION PLAYHOUSE (I,3583); PORTIA FACES LIFE (I,3653)

REID, KATE
ABE LINCOLN IN ILLINOIS (I,11); GAVILAN (II,959); THE GOOD LIFE (I,1848); THE GOOD LIFE (II,1033); HAWKINS ON MURDER (I,1978); THE HOLY TERROR (I,2098); LOOSE CHANGE (II,1518); MRS R.—DEATH AMONG FRIENDS (II,1759); NEITHER ARE WE ENEMIES (I,3247)

REID, MILTON
THE GOODIES (II,1040)

REID, SHEILA
MOLL FLANDERS (II,1720)

REID, TIM
BATTLE OF THE NETWORK STARS (II,167); BATTLE OF THE NETWORK STARS (II,172); BATTLE OF THE NETWORK STARS (II,173); BUMPERS (II,394); EASY DOES IT. . . STARRING FRANKIE AVALON (II,754); LITTLE LULU (II,1493); THE MARILYN MCCOO AND BILLY DAVIS JR. SHOW (II,1630); SIMON AND SIMON (II,2357); SOLID GOLD '79 (II,2396); TEACHERS ONLY (II,2548); WKRP IN CINCINNATI (II,2814); YOU CAN'T TAKE IT WITH YOU (II,2857)

REIGERT, PAUL
OFF CAMPUS (II,1877)

REIGHTON, LINDA
ONE MAN'S FAMILY (I,3383)

REILLY, CHARLES NELSON
THE 36 MOST BEAUTIFUL GIRLS IN TEXAS (II,2594); ARNIE (I,261); BARYSHNIKOV IN HOLLYWOOD (II,157); CALL HER MOM (I,796); DINAH AND FRIENDS (II,675); FLINTSTONE FAMILY ADVENTURES (II,882); THE FLINTSTONE FUNNIES (II,883); THE GHOST AND MRS. MUIR (I,1786); THE GOLDDIGGERS (I,1838); THE GOLDDIGGERS IN LONDON (I,1837); THE KAREN VALENTINE SHOW (I,2506); LIDSVILLE (I,2679); THE MATCH GAME (II,1649); THE SINGERS (I,4060); STAR SEARCH (II,2437); THE STEVE LAWRENCE SHOW (I,4227); SUPER COMEDY BOWL 1 (I,4288); SUPER COMEDY BOWL 2 (I,4289); TEXACO STAR THEATER: OPENING NIGHT (II,2565); UNCLE CROC'S BLOCK (II,2702)

REILLY, DAVID
ONE LIFE TO LIVE (II,1907)

REILLY, EARL
LIFE IS A CIRCUS, CHARLIE BROWN (I,2683)

REILLY, ED
SOMEDAY YOU'LL FIND HER, CHARLIE BROWN (I,4113)

REILLY, JENNIFER
THE DOCTORS (II,694)

REILLY, JOHN
AFTER GEORGE (II,19); AS THE WORLD TURNS (II,110); THE BANANA COMPANY (II,139); DALLAS (II,614); FLATFOOTS (II,879); THE HAMPTONS (II,1076); HOW THE WEST WAS WON (II,1196); LASSIE: THE NEW BEGINNING (II,1433); MADAME'S PLACE (II,1587); MISSING PIECES (II,1706); NUMBER 96 (II,1869); THE SECRET WAR OF JACKIE'S GIRLS (II,2294); WISHMAN (II,2811)

REILLY, TOM
CHIPS (II,511); FORCE SEVEN (II,908)

REIMBOLD, BILL
MASTER OF THE GAME (II,1647)

REINDEL, CARL
THE NEW PEOPLE (I,3268)

REINER, CARL
THE 2,000 YEAR OLD MAN (II,2696); 54TH STREET REVUE (I,1573); THE ALL-STAR COMEDY SHOW (I,137); ANNIE AND THE HOODS (II,91); THE ART LINKLETTER SHOW (I,276); CAESAR'S HOUR (I,790); THE CAMPBELL TELEVISION SOUNDSTAGE (I,812); CELEBRITY CHALLENGE OF THE SEXES 3 (II,467); THE CELEBRITY GAME (I,884); THE COLGATE COMEDY HOUR (I,998); A DATE WITH DEBBIE (I,1161); THE DICK VAN DYKE SHOW (I,1275); THE DICK VAN DYKE SPECIAL (I,1277); THE EDDIE CONDON SHOW (I,1408); THE FABULOUS FUNNIES (II,801); THE FASHION STORY (I,1537); FLOOR SHOW (I,1605); THE GOLDEN AGE OF TELEVISION (II,1018); GOOD HEAVENS (II,1032); HAPPY ANNIVERSARY, CHARLIE BROWN (I,1939); HEAD OF THE FAMILY (I,1988); JULIE AND DICK IN COVENT GARDEN (II,1352); KEEP TALKING (I,2517); LINUS THE LIONHEARTED (I,2709); MITZI. . . ROARIN' IN THE 20S (II,1711); SID CAESAR INVITES YOU (I,4040); STEVE MARTIN: COMEDY IS NOT PRETTY (II,2462); TAKE A GOOD LOOK (II,2529); THIS WEEK IN NEMTIN (I,4465); THOSE WONDERFUL TV GAME SHOWS (II,2603); A TRIBUTE TO "MR. TELEVISION" MILTON BERLE (II,2658); TWILIGHT THEATER (II,2685); VAN DYKE AND COMPANY (II,2718); THE WONDERFUL WORLD OF PIZZAZZ (I,4901); YOUR SHOW OF SHOWS (I,4957); YOUR SHOW OF SHOWS (II,2875)

REINER, ROB
ALL IN THE FAMILY (II,38); BATTLE OF THE NETWORK STARS (II,164); CELEBRITY CHALLENGE OF THE SEXES 1 (II,465); FREE COUNTRY (II,921); THE MICKIE FINNS FINALLY PRESENT HOW THE WEST WAS LOST (II,1695); THE PARTRIDGE FAMILY (II,1962); THE TV SHOW (II,2673)

REINHEART, ALICE
STAGE 13 (I,4183)

REINHOLD, JUDGE
NEVER AGAIN (II,1808)

REINHOLT, GEORGE
ANOTHER WORLD (II,97); ONE LIFE TO LIVE (II,1907)

REINKING, ANN
DOUG HENNING: MAGIC ON BROADWAY (II,730); PARADE OF STARS (II,1954)

REIS, VIVIAN
KING OF KENSINGTON (II,1395); PAUL BERNARD—PSYCHIATRIST (I,3485)

REISCH, STEVE
B.J. AND THE BEAR (II,126)

REISCHL, GERI
THE BRADY BUNCH HOUR (II,363)

REISER, PAUL
DINER (II,679)

REISER, ROBERT
DOCTOR DAN (II,684)

REISMAN, NAOMI
ANOTHER LIFE (II,95)

REISS, AMANDA
CROWN MATRIMONIAL (I,1106)

REISSER, DORA
ESPIONAGE (I,1464)

REITZEN, JACK
TERRY AND THE PIRATES (I,4405)

RELPH, EMMA
DAY OF THE TRIFFIDS (II,627)

RELTON, WILLIAM
BEAUTY AND THE BEAST (I,382); PENMARRIC (II,1993)

REMES, JANE
MELODY TOUR (I,3001)

REMICK, LEE
THE ANDY WILLIAMS SPECIAL (I,212); DAMN YANKEES (I,1128); THE FARMER'S DAUGHTER (I,1534); IKE (II,1223); JENNIE: LADY RANDOLPH CHURCHILL (II,1277); KRAFT TELEVISION THEATER (I,2592); THE MAN WHO CAME TO DINNER (I,2881); MISTRAL'S DAUGHTER (II,1708); OF MEN OF WOMEN (I,3331); THE TEMPEST (I,4391); WHEELS (II,2777)

REMIS, NADINE
YOUNG LIVES (II,2866)

REMSBERG, CALVIN
SWEENEY TODD (II,2512)

REMSEN, BERT
THE AWAKENING LAND (II,122); CRAZY TIMES (II,602); GIBBSVILLE (II,988); THE GOLDEN AGE OF TELEVISION (II,1019); IT'S A LIVING (II,1254); THE OATH: THE SAD AND LONELY SUNDAYS (II,1874)

REMSEN, GUY
ETHEL IS AN ELEPHANT (II,783)

REMY, ETHEL
AS THE WORLD TURNS (II,110)

RENADY, PETER
KIDD VIDEO (II,1388)

RENALDO, DUNCAN
THE CISCO KID (I,961)

RENARD, KAY
ABE LINCOLN IN ILLINOIS (I,10)

RENARD, KEN
THE GREATEST GIFT (II,1061)

RENAUD, LINE
THE BOB HOPE SHOW (I,537); THE BOB HOPE SHOW (I,539); THE BOB HOPE SHOW (I,545); PERRY COMO'S CHRISTMAS IN PARIS (II,2012)

RENCHER, DEREK
CINDERELLA (II,526)

RENELLA, PAT
GENERAL HOSPITAL (II,964); THE NEW PHIL SILVERS SHOW (I,3269); THE ROCKFORD FILES (II,2196)

RENIER, YVES
THE ADVENTURES (I,86)

RENNEAU, JERRY
THE JONATHAN WINTERS SHOW (I,2443)

RENNICK, NANCY
THE FARMER'S DAUGHTER (I,1533); KENTUCKY JONES (I,2526); THE OUTER LIMITS (I,3426); RESCUE 8 (I,3772)

RENNIE, MICHAEL
THE BARBARA STANWYCK THEATER (I,350); BATMAN (I,366); PURSUIT (I,3700); RAPTURE AT TWO-FORTY (I,3724); THE SEARCH (I,3950); SHIRLEY TEMPLE'S STORYBOOK (I,4017); THE THIRD MAN (I,4450)

RENTERIA, JOE
NAKIA (II,1784)

RENWICK, DAVID
NOT THE NINE O'CLOCK NEWS (II,1864)

RENZI, EVA
PRIMUS (I,3668)

REPP, STAFFORD
BATMAN (I,366); THE GOOD LIFE (II,1033); THE NEW PHIL SILVERS SHOW (I,3269); TEXAS JOHN SLAUGHTER (I,4414); THE THIN MAN (I,4446)

RESCHER, DEE DEE
COUSINS (II,595); DINAH AND HER NEW BEST FRIENDS (II,676); EMPIRE (II,777); IRENE (II,1245); WINDOWS, DOORS AND KEYHOLES (II,2806)

RESER, HARRY
SAMMY KAYE'S MUSIC FROM MANHATTAN (I,3902)

RESIN, ALAN
HOW TO SUCCEED IN BUSINESS WITHOUT REALLY TRYING (II,1197)

RESIN, DAN
THE MADHOUSE BRIGADE (II,1588); ON OUR OWN (II,1892)

RESNICK, AMY
PAPER DOLLS (II,1952)

RESNICK, BOBBY
MOBY DICK AND THE MIGHTY MIGHTOR (I,3077)

RESTIN, STEVE
TROUBLE IN HIGH TIMBER COUNTRY (II,2661)

RETTIG, TOMMY
JEFF'S COLLIE (I,2356); NEVER TOO YOUNG (I,3248); TIMMY AND LASSIE (I,4520)

RETTON, MARY LOU
BOB HOPE SPECIAL: HO HO HOPE'S JOLLY CHRISTMAS HOUR (II,327)

REUBENS, PAUL
ALL COMMERCIALS—A STEVE MARTIN SPECIAL (II,37); BUCKSHOT (II,385); FLINTSTONE FAMILY ADVENTURES (II,882); THE FLINTSTONE FUNNIES (II,883); THE PARAGON OF COMEDY (II,1956); THE PEE WEE HERMAN SHOW (II,1988)

REVERE, ANN
ART AND MRS. BOTTLE (I,271); SEARCH FOR TOMORROW (II,2284); SESAME STREET (I,3982); TWO FOR THE MONEY (I,4655)

REVILL, CLIVE
13 THIRTEENTH AVENUE (II,2592); CENTENNIAL (II,477); DEATH RAY 2000 (II,654); GEORGE WASHINGTON (II,978); JOE DANCER: THE MONKEY MISSION (II,1301); PINOCCHIO (II,2044); WINNER TAKE ALL (II,2808); WIZARDS AND WARRIORS (II,2813)

REY, ALEJANDRO
DAYS OF OUR LIVES (II,629); THE FLYING NUN (I,1611); THE GREATEST SHOW ON EARTH (I,1883); THE LLOYD BRIDGES SHOW (I,2733); NIGHT GALLERY (I,3287); SLATTERY'S PEOPLE (I,4082); STUNTS UNLIMITED (II,2487); THREE FOR DANGER (I,4475)

REY, ANTONIA
ONE MORE TRY (II,1908)

REYES, EVA
THE RAY PERKINS REVUE (I,3735)

REYES, PAUL
THE RAY PERKINS REVUE (I,3735)

REYNALDO, JORGE
FALCON'S GOLD (II,809)

REYNE, JAMES
RETURN TO EDEN (II,2143)

REYNOLDS, BOB
KEY TORTUGA (II,1383)

REYNOLDS, BURT
THE BEST LITTLE SPECIAL IN TEXAS (II,215); BURT AND THE GIRLS (I,770); BURT REYNOLDS' LATE SHOW (I,777); CELEBRITY DAREDEVILS (II,473); THE CELEBRITY FOOTBALL CLASSIC (II,474); DAN AUGUST (I,1130); DINAH IN SEARCH OF THE IDEAL MAN (I,1278); DOM DELUISE AND FRIENDS (II,701); GUNSMOKE (II,1069); HAWK (I,1974); HOW TO HANDLE A WOMAN (I,2146); JERRY REED AND SPECIAL FRIENDS (II,1285); LASSITER (I,2625); LASSITER (II,1434); THE MAN FROM EVERYWHERE (I,2864); PLAYHOUSE 90 (I,3623); RIVERBOAT (I,3797); STEVE MARTIN'S THE WINDS OF WHOOPIE (II,2461); SUPER COMEDY BOWL 2 (I,4289); THE TWILIGHT ZONE (I,4651); THE VERY FIRST GLEN CAMPBELL SPECIAL (I,4714); THE WAYNE NEWTON SPECIAL (II,2749); ZANE GREY THEATER (I,4979)

REYNOLDS, CINDY
THE $128,000 QUESTION (II,1903); MOMMA THE DETECTIVE (II,1721)

REYNOLDS, DALE
THE MYSTERIOUS TWO (II,1783)

REYNOLDS, DEBBIE
...AND DEBBIE MAKES SIX (I,185); THE ALL-STAR SALUTE TO MOTHER'S DAY (II,51); ALOHA PARADISE (II,58); BING!. . . A 50TH ANNIVERSARY GALA (II,254); THE BOB HOPE SHOW (I,616); THE BOB HOPE SHOW (I,650); BOB HOPE SPECIAL: BOB HOPE'S BICENTENNIAL STAR SPANGLED SPECTACULAR (II,306); BOB HOPE SPECIAL: THE BOB HOPE SPECIAL (II,336); CIRCUS OF THE STARS (II,536); A DATE WITH DEBBIE (I,1161); DEBBIE REYNOLDS AND THE SOUND

OF CHILDREN (I,1213); THE DEBBIE REYNOLDS SHOW (I,1214); THE EDDIE FISHER SHOW (I,1409); GIRL FRIENDS AND NABORS (I,1807); HEDDA HOPPER'S HOLLYWOOD (I,2002); JENNIFER SLEPT HERE (II,1278); LEAPIN' LIZARDS, IT'S LIBERACE (II,1453); OPERATION ENTERTAINMENT (I,3400); THE PERRY COMO SUNSHINE SHOW (II,2005)

REYNOLDS, GEORGE
LUCAN (II,1553); THIS GIRL FOR HIRE (II,2596)

REYNOLDS, JAMES
DAYS OF OUR LIVES (II,629); THE GALLANT MEN (I,1728); KEEPER OF THE WILD (II,1373); TIME EXPRESS (II,2620)

REYNOLDS, JAY
SCENE 70 (I,3934)

REYNOLDS, JEFF
DEBBY BOONE. . . ONE STEP CLOSER (II,655)

REYNOLDS, KATHRYN
CANNON (II,424); MOONLIGHT (II,1730); SOAP (II,2392)

REYNOLDS, KAY
THE SECOND HUNDRED YEARS (I,3959)

REYNOLDS, LARRY
A MAN CALLED INTREPID (II,1612); THE THANKSGIVING TREASURE (I,4416)

REYNOLDS, MARJORIE
GRUEN GUILD PLAYHOUSE (I,1892); HANDS OF MURDER (I,1929); LEAVE IT TO BEAVER (I,2648); THE LIFE OF RILEY (I,2686); PEARL (II,1986); THE UNEXPECTED (I,4674)

REYNOLDS, MARY
TOOTSIE HIPPODROME (I,4575)

REYNOLDS, MICHAEL J
SEARCH AND RESCUE: THE ALPHA TEAM (II,2283)

REYNOLDS, PETER
THE ERROL FLYNN THEATER (I,1458)

REYNOLDS, REBECCA
THE EYES OF TEXAS (II,799); THE EYES OF TEXAS II (II,800); THE FACTS (II,804)

REYNOLDS, TOM
THE FBI (I,1551)

REYNOLDS, WILLIAM
FBI CODE 98 (I,1550); THE ISLANDERS (I,2243); PETE KELLY'S BLUES (I,3559)

RHEIN, JED
BRANAGAN AND MAPES (II,367)

RHOADES, BARBARA
THE BLUE KNIGHT (II,276); THE BUREAU (II,397); BUSTING LOOSE (II,402); CELEBRITY CHALLENGE OF THE SEXES (II,464); CONSPIRACY OF TERROR (II,569); THE DAY THE WOMEN GOT EVEN (II,628); HANGING IN (II,1078); THE JUDGE AND JAKE WYLER (I,2464); MR. DUGAN (II,1749); THE PAUL LYNDE COMEDY HOUR (II,1977); THE POLICE STORY (I,3638); PUNCH AND JODY (II,2091); SOAP (II,2392); STONE (II,2471); TABITHA (II,2526); TWIN DETECTIVES (II,2687); VENICE MEDICAL (II,2726)

RHOADES, HARRY
BACKSTAIRS AT THE WHITE HOUSE (II,133)

RHOADES, KENNY
THE MUNSTERS' REVENGE (II,1765)

RHODES, ANGELA
THE WALTONS: A DAY FOR THANKS ON WALTON'S MOUNTAIN (EPISODE 3) (II,2741); THE WALTONS: MOTHER'S DAY ON WALTON'S MOUNTAIN (EPISODE 2) (II,2742); THE WALTONS: WEDDING ON WALTON'S MOUNTAIN (EPISODE 1) (II,2743)

RHODES, CARRIE
VELVET (II,2725)

RHODES, CHRISTOPHER
DOUGLAS FAIRBANKS JR. PRESENTS THE RHEINGOLD THEATER (I,1364)

RHODES, DONNELLY
DOUBLE TROUBLE (II,723); DOUBLE TROUBLE (II,724); JO'S COUSINS (II,1293); MAKING A LIVING (II,1604); OUR MAN FLINT: DEAD ON TARGET (II,1932); REPORT TO MURPHY (II,2134); SOAP (II,2392); TAXI (II,2546); THE YOUNG AND THE RESTLESS (II,2862)

RHODES, ELSIE
BLOOMER GIRL (I,488); TED MACK'S MATINEE (I,4372)

RHODES, ERIK
GREAT CATHERINE (I,1873)

RHODES, GRANDON
ABE LINCOLN IN ILLINOIS (I,10); THE GEORGE BURNS AND GRACIE ALLEN SHOW (I,1763)

RHODES, HARI
DEADLOCK (I,1190); JUDGEMENT DAY (II,1348); THE LAW ENFORCERS (I,2638); MATT HELM (II,1652); MOST WANTED (II,1739); ROOTS (II,2211); TO SIR, WITH LOVE (II,2624)

RHODES, JOAN
THE BOB HOPE SHOW (I,543)

RHODES, JORDAN
THE BLUE AND THE GRAY (II,273); IS THERE A DOCTOR IN THE HOUSE? (II,1247); TESTIMONY OF TWO MEN (II,2564); WONDER WOMAN (PILOT 1) (II,2826)

RHODES, RICHARD
GEORGE WASHINGTON (II,978)

RHUE, MADLYN
BRACKEN'S WORLD (I,703); DYNASTY (II,746); ESPIONAGE (I,1464); EXECUTIVE SUITE (II,796); FAME (II,812); POOR DEVIL (I,3646)

RHUMBA BAND, THE
THE RAY PERKINS REVUE (I,3735)

RHYAL, MARY LOU
SING ALONG WITH MITCH (I,4057)

RHYS-DAVIES, JOHN
NO MAN'S LAND (II,1855); SHOGUN (II,2339)

RHYTHM PALS, THE
THE TOMMY HUNTER SHOW (I,4552)

RHYTHMAIRES, THE
FRANKIE LAINE TIME (I,1676)

RIANO, RENIE
THE ABBOTT AND COSTELLO SHOW (I,2); ANDERSON AND COMPANY (I,188)

RIATT, JOHN
BUICK CIRCUS HOUR (I,759); YVES MONTAND ON BROADWAY (I,4978)

RIBACK, BILL
THE FACTS (II,804)

RIBEIRO, ALFONSO
SILVER SPOONS (II,2355)

RICARD, ADRIAN
DOCTORS HOSPITAL (II,691)

RICARDO, DIANA
MASTER OF THE GAME (II,1647)

RICCI, MARLENE
LAS VEGAS PALACE OF STARS (II,1431)

RICE, ADINA
ONE TOUCH OF VENUS (I,3389)

RICE, DON
THE GOLDDIGGERS (I,1838); THE PERRY COMO SPECIAL (I,3536)

RICE, GREG
FOUL PLAY (II,910)

RICE, HOWARD
ROOM 222 (I,3843)

RICE, JOHN
FOUL PLAY (II,910)

RICE, ROSEMARY
I REMEMBER MAMA (I,2176)

RICE, TIMOTHY
GEORGE WASHINGTON (II,978)

RICH, ADAM
BOB HOPE SPECIAL: THE BOB HOPE CHRISTMAS SPECIAL (II,330); CIRCUS OF THE STARS (II,536); CODE RED (II,552); DISNEYLAND'S 25TH ANNIVERSARY (II,681); THE DONNY AND MARIE CHRISTMAS SPECIAL (II,713); DUNGEONS AND DRAGONS (II,744); EIGHT IS ENOUGH (II,762); GUN SHY (II,1068); WHATEVER BECAME OF. . . ? (II,2772)

RICH, ALLAN
ARCHER—FUGITIVE FROM THE EMPIRE (II,103); BUT MOTHER! (II,403); THE CHICAGO STORY (II,507); MALLORY: CIRCUMSTANTIAL EVIDENCE (II,1608); NEWMAN'S DRUGSTORE (II,1837); THE SEEKERS (II,2298); WHEELS (II,2777)

RICH, BERNARD
SABU AND THE MAGIC RING (I,3881)

RICH, BUDDY
COTTON CLUB '75 (II,580); THE MARGE AND GOWER CHAMPION SHOW (I,2907); THE SHANI WALLIS SHOW (I,3997)

RICH, CHARLIE
BURT REYNOLDS' LATE SHOW (I,777); COUNTRY MUSIC HIT PARADE (II,588); COUNTRY STARS OF THE 70S (II,594); ELVIS REMEMBERED: NASHVILLE TO HOLLYWOOD (II,772); I BELIEVE IN MUSIC (I,2165); USA COUNTRY MUSIC (II,591)

RICH, CHRISTOPHER
ANOTHER WORLD (II,97)

RICH, DON
HEE HAW (II,1123)

RICH, DORIS
THE EGG AND I (I,1420); THE GREATEST GIFT (I,1880); THREE STEPS TO HEAVEN (I,4485)

RICH, JENNIE LEE
THE BENNY HILL SHOW (II,207)

RICH, JUDY
BLANK CHECK (II,265); GIVE-N-TAKE (II,1003)

RICH, KAREN
THE ALAN DALE SHOW (I,96)

RICH, MICHAEL
THE TED STEELE SHOW (I,4373)

RICH, ROYCE
THE ROBERT KLEIN SHOW (II,2185)

RICH, VERNON
TOM AND JERRY (I,4533)

RICHARD, BERYL
A COUPLE OF JOES (I,1078)

RICHARD, CAROL
THE R.C.A. VICTOR SHOW (I,3737)

RICHARD, CULLY
DON'T CALL ME CHARLIE (I,1349)

RICHARD, DARYL
THE DONNA REED SHOW (I,1347); THE JIMMY DURANTE SHOW (I,2389)

RICHARD, ERIC
SHOGUN (II,2339)

RICHARD, LITTLE
33 1/3 REVOLUTIONS PER MONKEE (I,4451)

RICHARD, PAUL
DARK SHADOWS (I,1157)

RICHARDS JR, DANNY
WILLY (I,4868)

RICHARDS, ADDISON
BEN JERROD (I,396); CIMARRON CITY (I,953); DOBIE GILLIS (I,1302); FIBBER MCGEE AND MOLLY (I,1569); PENTAGON U.S.A. (I,3514); TENNESSEE ERNIE FORD MEETS KING ARTHUR (I,4397)

RICHARDS, ANGELA
SECRET ARMY (II,2290)

RICHARDS, BEAH
BENSON (II,208); THE BILL COSBY SHOW (I,442); DOWN HOME (II,731); ROOTS: THE NEXT GENERATIONS (II,2213); TOO GOOD TO BE TRUE (II,2643)

RICHARDS, BILLIE
RUDOLPH THE RED-NOSED REINDEER (II,2227); RUDOLPH'S SHINY NEW YEAR (II,2228)

RICHARDS, CHRISTINE
CAMP GRIZZLY (II,418)

RICHARDS, CULLY
THE LAS VEGAS SHOW (I,2619)

RICHARDS, DEAN
MIDWESTERN HAYRIDE (I,3033)

RICHARDS, DONALD
JACK CARTER AND COMPANY (I,2305); THE JACK CARTER SHOW (I,2304); THE JACK CARTER SHOW (I,2306); THE SATURDAY NIGHT REVUE (I,3921)

RICHARDS, EMILY
NICHOLAS NICKLEBY (II,1840)

RICHARDS, EVAN
MAMA MALONE (II,1609); MOONLIGHT (II,1730)

RICHARDS, FRANK
DAVY CROCKETT (I,1181)

RICHARDS, GEORGE
THE PHIL SILVERS SHOW (I,3580)

RICHARDS, GLENN
FLAMINGO ROAD (II,871); FLAMINGO ROAD (II,872)

RICHARDS, GRANT
THE DOOR WITH NO NAME (I,1352); FOLLOW YOUR HEART (I,1618); THE SEEKERS (I,3974)

RICHARDS, JEFF
JEFFERSON DRUM (I,2355); THE WEB (I,4785)

RICHARDS, JILL
CLUB CELEBRITY (I,981)

RICHARDS, KATHY
HAPPY DAYS (II,1084)

RICHARDS, KIM
HELLO, LARRY (II,1128); HERE WE GO AGAIN (I,2028); JAMES AT 15 (II,1270); JAMES AT 15 (II,1271); MERV GRIFFIN AND THE CHRISTMAS KIDS (I,3013); NANNY AND THE PROFESSOR (I,3225); NANNY AND THE PROFESSOR AND THE PHANTOM OF THE CIRCUS (I,3226); WHY US? (II,2796)

RICHARDS, KYLE
BEULAH LAND (II,226); FATHER KNOWS BEST: HOME FOR CHRISTMAS (II,839); FATHER KNOWS BEST: THE FATHER KNOWS BEST REUNION (II,840); GOOD TIME HARRY (II,1037); HELLINGER'S LAW (II,1127); LITTLE HOUSE ON THE PRAIRIE (II,1487); THE SEAL (II,2282); THIS IS KATE BENNETT (II,2597)

RICHARDS, LISA BLAKE
DARK SHADOWS (I,1157); HELLINGER'S LAW (II,1127); ONE LIFE TO LIVE (II,1907); TRUE LIFE STORIES (II,2665); WHERE THE HEART IS (I,4831)

RICHARDS, LLOYDE G
THE LITTLE FOXES (I,2712)

RICHARDS, LOU
CASS MALLOY (II,454); GLORIA (II,1010); SECOND EDITION (II,2288); ST. ELSEWHERE (II,2432)

RICHARDS, MICHAEL
AT YOUR SERVICE (II,118); FRIDAYS (II,926); HERNDON AND ME (II,1139)

RICHARDS, PAUL
BREAKING POINT (I,717); EL COYOTE (I,1423); THE LLOYD BRIDGES SHOW (I,2733); NAVY LOG (I,3233); THE OVER-THE-HILL GANG RIDES AGAIN (I,3432); SAVAGE (I,3924); THE WEB (I,4785); WHO HAS SEEN THE WIND? (I,4847)

RICHARDS, RICK
OUT OF THE BLUE (I,3422)

RICHARDS, RON
NO SOAP, RADIO (II,1856)

RICHARDS, RUSTY
BURT REYNOLDS' LATE SHOW (I,777)

RICHARDS, SANDRA
MONTY PYTHON'S FLYING CIRCUS (II,1729)

RICHARDS, STAN
SURF'S UP (I,4298)

RICHARDS, TED
AN ECHO OF THERESA (I,1399)

RICHARDS, TESSA
ANGIE (II,80)

RICHARDSON, DAVID
LITTLE WOMEN (I,2724)

RICHARDSON, DUNCAN
WHAT I WANT TO BE (I,4812)

RICHARDSON, IAN
IKE (II,1223); MISTRAL'S DAUGHTER (II,1708); THE WOMAN IN WHITE (II,2819)

RICHARDSON, JAMES G
THE RANGERS (II,2112); SIERRA (II,2351)

RICHARDSON, JUDITH
ONE MAN'S FAMILY (I,3383)

RICHARDSON, LEE
THE GUIDING LIGHT (II,1064)

RICHARDSON, MICHAEL
CHASE (I,917)

RICHARDSON, MIRANDA
A WOMAN OF SUBSTANCE (II,2820)

RICHARDSON, PATRICIA
DOUBLE TROUBLE (II,723)

RICHARDSON, RALPH
HEDDA GABLER (I,2001)

RICHARDSON, SUSAN
BATTLE OF THE NETWORK STARS (II,168); BATTLE OF THE NETWORK STARS (II,171); CELEBRITY CHALLENGE OF THE SEXES 3 (II,467); CIRCUS OF THE STARS (II,534); EIGHT IS ENOUGH (II,762); ONE DAY AT A TIME (II,1900)

RICHENS, ADAM
PETER PAN (II,2030)

RICHFIELD, EDWIN
INTERPOL CALLING (I,2227)

RICHIARDI
MAGIC WITH THE STARS (II,1599)

RICHMAN, CARYN
GIDGET'S SUMMER REUNION (I,1799); TEXAS (II,2566)

RICHMAN, JEFFREY

RILEY, SKIP
THE SEEKERS (II,2298)

RILEY, TRACY JO
PRISONER: CELL BLOCK H (II,2085)

RILO
KEEP ON TRUCKIN' (II,1372)

RIMMER, SHANE
LACE (II,1418); A MAN CALLED INTREPID (II,1612); MASTER OF THE GAME (II,1647); THUNDERBIRDS (I,4497)

RINGA, CHARLES
HOW DO I KILL A THIEF—LET ME COUNT THE WAYS (II,1195)

RINGWALD, ELIZABETH
CRASH ISLAND (II,600)

RINGWALD, MOLLY
THE FACTS OF LIFE (II,805)

RINGWALD, MONICA
THE BENNY HILL SHOW (II,207)

RINKER, MARGARET
THE INA RAY HUTTON SHOW (I,2207)

RINTOUL, DAVID
PRIDE AND PREJUDICE (II,2080)

RIOVANELLI, ANTONIO
MOSES THE LAWGIVER (II,1737)

RIPERTON, MINNIE
MONTY HALL'S VARIETY HOUR (II,1728)

RIPLEY, ROBERT
BELIEVE IT OR NOT (I,387)

RIPPER, MICHAEL
BUTTERFLIES (II,404)

RIPPY, LEON
CHIEFS (II,509)

RIPPY, RODNEY ALLEN
THE HARLEM GLOBETROTTERS POPCORN MACHINE (II,1092); MERV GRIFFIN AND THE CHRISTMAS KIDS (I,3013); VARIETY (II,2721)

RISK, LINDA SUE
THE RED SKELTON SHOW (I,3755)

RISLEY, ANN
OFF CAMPUS (II,1877); SATURDAY NIGHT LIVE (II,2261)

RIST, ROBBIE
BIG JOHN, LITTLE JOHN (II,240); THE BIONIC WOMAN (II,255); THE BRADY BUNCH (II,362); GALACTICA 1980 (II,953); GOSSIP (II,1044); HAVING BABIES II (II,1108); INSTANT FAMILY (II,1238); KIDD VIDEO (II,1388); LITTLE LULU (II,1493); LUCAS TANNER (II,1555); LUCAS TANNER (II,1556); THE MARY TYLER MOORE SHOW (II,1640); THE PAUL LYNDE COMEDY HOUR (II,1977); THE ROWAN AND MARTIN REPORT (II,2221); THROUGH THE MAGIC PYRAMID (II,2615)

RISTIVO, JOE
LOOK WHAT THEY'VE DONE TO MY SONG (II,1517)

RITCH, STEVE
THE PETRIFIED FOREST (I,3570)

RITCHARD, CYRIL
CAESAR AND CLEOPATRA (I,788); THE DANGEROUS CHRISTMAS OF RED RIDING HOOD (I,1139); DEAREST ENEMY (I,1205); THE GENERAL MOTORS 50TH ANNIVERSARY SHOW (I,1758); THE GOOD FAIRY (I,1846); JACK AND THE BEANSTALK (I,2287); THE KING AND MRS. CANDLE (I,2544); LOVE, LIFE, LIBERTY & LUNCH (II,1544); PETER PAN (I,3566); PLAYWRIGHTS '56 (I,3627); PRUDENTIAL FAMILY PLAYHOUSE (I,3683); ROSALINDA (I,3846)

RITCHIE, CLINT
THE BASTARD/KENT FAMILY CHRONICLES (II,159); ONE LIFE TO LIVE (II,1907); THUNDER (II,2617)

RITCHIE, JEAN
AMERICA PAUSES FOR SPRINGTIME (I,166)

RITCHIE, ROBERT
STUNT SEVEN (II,2486)

RITTER, JOHN
BACHELOR AT LAW (I,322); BOB HOPE SPECIAL: BOB HOPE'S SUPER BIRTHDAY SPECIAL (II,319); THE CELEBRITY FOOTBALL CLASSIC (II,474); COMPLETELY OFF THE WALL (II,560); ECHOES OF THE SIXTIES (II,756); EVIL ROY SLADE (I,1485); THE FANTASTIC MISS PIGGY SHOW (II,827); GENERAL ELECTRIC'S ALL-STAR ANNIVERSARY (II,963); THE GOLDIE HAWN SPECIAL (II,1024); HOW TO SURVIVE THE 70S AND MAYBE EVEN BUMP INTO HAPPINESS (II,1199); JOHN RITTER: BEING OF SOUND MIND AND BODY (II,1319); LIFE'S MOST EMBARRASSING MOMENTS (II,1470); RINGO (II,2176); THE SINGING COWBOYS RIDE AGAIN (II,2363); THAT THING ON ABC (II,2574); THREE'S A CROWD (II,2613); THREE'S COMPANY (II,2614); THE WALTONS (II,2740); WHAT'S UP, AMERICA? (I,4824); WHAT'S UP? (I,4823)

RITTER, TEX
ACADEMY AWARD SONGS (I,13); FIVE STAR JUBILEE (I,1589); RANCH PARTY (I,3719)

RITTER, THELMA
THE SHOWOFF (I,4035)

RITTS, MARY
HI MOM (I,2043)

RITTS, PAUL
HI MOM (I,2043)

RITZ BROTHERS, THE
THE ALL-STAR REVUE (I,138); THE GINGER ROGERS SHOW (I,1804); GOOD TIMES (I,1853)

RITZKE, KRISTINE
RETURN TO FANTASY ISLAND (II,2144)

RIVA, MARIA
STAGE 7 (I,4181); SURE AS FATE (I,4297)

RIVAS, CARLOS
SECOND LOOK (I,3960); THE TARZAN/LONE RANGER/ZORRO ADVENTURE HOUR (II,2543)

RIVERA, CHITA
ARTHUR GODFREY AND THE SOUNDS OF NEW YORK (I,280); THE GENERAL MOTORS 50TH ANNIVERSARY SHOW (I,1758); THE GEORGE BURNS SPECIAL (II,970); THE MAURICE CHEVALIER SHOW (I,2954); THE NEW DICK VAN DYKE SHOW (I,3263); THE STARS AND STRIPES SHOW (II,2443); TIPTOE THROUGH TV (I,4524); VARIETY: THE WORLD OF SHOW BIZ (I,4704)

RIVERS, JOAN
CELEBRITY CHALLENGE OF THE SEXES 4 (II,468); CIRCUS OF THE STARS (II,534); COMEDY NEWS II (I,1021); LILY—SOLD OUT (II,1480); THAT SHOW STARRING JOAN RIVERS (I,4421); THE TONIGHT SHOW (I,4565)

RIVERS, JOHNNY
IT'S WHAT'S HAPPENING, BABY! (I,2278); LOUISE MANDRELL: DIAMONDS, GOLD AND PLATINUM (II,1528); ONE NIGHT STANDS (I,3387)

RIVERS, LUCILLE
THE LUCILLE RIVERS SHOW (I,2785)

RIVERS, PAMELA
THE CLOCK (I,978)

RIZZO, CARMINE
ANOTHER WORLD (II,97)

RIZZO, RICHARD
LYNDA CARTER'S SPECIAL (II,1564)

ROACH, DARYL
THE LIFE AND TIMES OF EDDIE ROBERTS (II,1468)

ROACH, JAMES
THE RETURN OF LUTHER GILLIS (II,2137)

ROACHE, WILLIAM
CORONATION STREET (I,1054)

ROAD, MIKE
THE ADVENTURES OF JONNY QUEST (I,64); ALIAS SMITH AND JONES (I,118); BUCKSKIN (I,753); THE FANTASTIC FOUR (II,824); GIDGET MAKES THE WRONG CONNECTION (I,1798); THE HERCULOIDS (I,2023); O'HARA, UNITED STATES TREASURY: OPERATION COBRA (I,3347); THE ROARING TWENTIES (I,3805); SPACE GHOST (I,4141); SPACE STARS (II,2408); VALLEY OF THE DINOSAURS (II,2717)

ROARKE, HAYDEN
THE SOFT TOUCH (I,4107)

ROARKE, JIM
THE MAN FROM DENVER (I,2863)

ROARKE, JOHN
FRIDAYS (II,926)

ROAT, RICHARD
ALMOST HEAVEN (II,57); BACKSTAIRS AT THE WHITE HOUSE (II,133); COMEDY OF HORRORS (II,557); FROM HERE TO ETERNITY (II,933); NEWHART (II,1835); ST. ELSEWHERE (II,2432)

ROB ISCOVE DANCERS, THE
ANN-MARGRET OLSSON (II,84); BURT BACHARACH: CLOSE TO YOU (I,772)

ROBA, MELANIE
BANYON (I,344)

ROBARDS JR, JASON
ABE LINCOLN IN ILLINOIS (I,11); THE BAT (I,364); STAR TONIGHT (I,4199); WINDOWS (I,4876)

ROBARDS, GLENN
RAGE OF ANGELS (II,2107)

ROBARDS, JASON
ADDIE AND THE KING OF HEARTS (I,35); THE BELLE OF 14TH STREET (I,390); A DOLL'S HOUSE (I,1327); THE EASTER PROMISE (I,1397); GHOST STORY (I,1788); THE GOLDEN AGE OF TELEVISION (II,1017); THE HOUSE WITHOUT A CHRISTMAS TREE (I,2143); THE MAGIC OF DAVID COPPERFIELD (II,1594); THE THANKSGIVING TREASURE (I,4416); WASHINGTON: BEHIND CLOSED DOORS (II,2744)

ROBB, DAVID
FANNY BY GASLIGHT (II,823); THE FLAME TREES OF THIKA (II,870); THE LAST DAYS OF POMPEII (II,1436)

ROBBINS, A
CAPTAIN KANGAROO (II,434)

ROBBINS, BARBARA
THE ALDRICH FAMILY (I,111); JANE EYRE (I,2338)

ROBBINS, BRIAN
THREE'S COMPANY (II,2614)

ROBBINS, CINDY
MCHALE'S NAVY (I,2969); THE TOM EWELL SHOW (I,4538)

ROBBINS, DEANN
THE WALTONS: A DAY FOR THANKS ON WALTON'S MOUNTAIN (EPISODE 3) (II,2741); THE WALTONS: MOTHER'S DAY ON WALTON'S MOUNTAIN (EPISODE 2) (II,2742); THE WALTONS: WEDDING ON WALTON'S MOUNTAIN (EPISODE 1) (II,2743)

ROBBINS, FRED
ADVENTURES IN JAZZ (I,45); CAVALCADE OF BANDS (I,876); COKE TIME WITH EDDIE FISHER (I,996); HAGGIS BAGGIS (I,1917); THE ROBBINS NEST (I,3806)

ROBBINS, GALE
THE BOB HOPE SHOW (I,513); DAMON RUNYON THEATER (I,1129); FORD TELEVISION THEATER (I,1634); MEET THE GIRLS: THE SHAPE, THE FACE, AND THE BRAIN (I,2990)

ROBBINS, JANE MARLA
A.E.S. HUDSON STREET (II,17)

ROBBINS, JONI
THE NEW ZOO REVUE (I,3277)

ROBBINS, LOIS
NATIONAL LAMPOON'S HOT FLASHES (II,1795)

ROBBINS, MARTY
COUNTRY MUSIC CARAVAN (I,1066); DEAN MARTIN PRESENTS MUSIC COUNTRY, U.S.A. (I,1192); THE MARTY ROBBINS SPOTLITE (II,1636)

ROBBINS, PATRICIA
ONE MAN'S FAMILY (I,3383)

ROBBINS, PETER
BLONDIE (I,487); A CHARLIE BROWN CHRISTMAS (I,908); CHARLIE BROWN'S ALL STARS (I,910); HE'S YOUR DOG, CHARLIE BROWN (I,2037); IT WAS A SHORT SUMMER, CHARLIE BROWN (I,2252); IT'S THE GREAT PUMPKIN, CHARLIE BROWN (I,2276); YOU'RE IN LOVE, CHARLIE BROWN (I,4964)

ROBBINS, REX
THE DAY THE WOMEN GOT EVEN (II,628); KENNEDY (II,1377)

ROBERSON, CLIFF
ALCOA PREMIERE (I,109); ALCOA/GOODYEAR THEATER (I,107)

ROBERT HERGET DANCERS, THE
THE WOODY ALLEN SPECIAL (I,4904)

ROBERT MITCHELL BOYS CHOIR, THE
MERV GRIFFIN AND THE CHRISTMAS KIDS (I,3013)

ROBERT SIDNEY DANCERS, THE
THE PEARL BAILEY SHOW (I,3501); PERRY COMO'S WINTER SHOW (I,3547)

ROBERTS, ALAN
SOMEONE AT THE TOP OF THE STAIRS (I,4114)

ROBERTS, ALLENE
PEPSI-COLA PLAYHOUSE (I,3523)

ROBERTS, ANDY
BROADWAY OPEN HOUSE (I,736); TED MACK'S FAMILY HOUR (I,4371)

ROBERTS, ART
BRENDA STARR (II,372)

ROBERTS, ARTHUR
GENERAL HOSPITAL (II,964)

ROBERTS, BEVERLY
THE CLOCK (I,978)

ROBERTS, CLETE
CRIME AND PUNISHMENT (I,1095); GOLIATH AWAITS (II,1025); M*A*S*H (II,1569); SHADOW OF SAM PENNY (II,2313); V (II,2713); V: THE FINAL BATTLE (II,2714); W*A*L*T*E*R (II,2731); YOU ARE THERE (I,4929)

ROBERTS, CONRAD
CASINO (II,452); THE DOCTORS (II,694)

ROBERTS, CRIS
MARKHAM (I,2916)

ROBERTS, DAVIS
BOONE (II,351); FILTHY RICH (II,856); PALMERSTOWN U.S.A. (II,1945); ROOTS (II,2211)

ROBERTS, DORIS
ALICE (II,33); ANGIE (II,80); BELL, BOOK AND CANDLE (II,201); IN TROUBLE (II,1230); THE LILY TOMLIN SPECIAL (II,1479); MAGGIE (II,1589); THE MARY TYLER MOORE COMEDY HOUR (II,1639); ME AND MRS. C. (II,1673); THE OATH: 33 HOURS IN THE LIFE OF GOD (II,1873); REMINGTON STEELE (II,2130); SOAP (II,2392); THE TROUBLE WITH PEOPLE (I,4611)

ROBERTS, DOROTHY
HART TO HART (II,1101)

ROBERTS, ERIC
ANOTHER WORLD (II,97)

ROBERTS, EWAN
THE ADVENTURES OF SIR FRANCIS DRAKE (I,75); BAFFLED (I,332); COLONEL MARCH OF SCOTLAND YARD (I,1005)

ROBERTS, FRANCESCA
PRIVATE BENJAMIN (II,2087)

ROBERTS, HOLLY
THE HAMPTONS (II,1076)

ROBERTS, HOWARD
HAIL THE CHAMP (I,1918)

ROBERTS, IVOR
LADY KILLER (I,2603)

ROBERTS, JASON
THE COUNTRY GIRL (I,1064)

ROBERTS, JIM
THE LAWRENCE WELK SHOW (I,2643); THE LAWRENCE WELK SHOW (I,2644); MEMORIES WITH LAWRENCE WELK (II,1681)

ROBERTS, JOAN
COLORADO C.I. (II,555); PRIVATE BENJAMIN (II,2087); QUICK AND QUIET (II,2099); A ROCK AND A HARD PLACE (II,2190)

ROBERTS, JUDITH
ALL MY CHILDREN (II,39)

ROBERTS, KEN
GLENN FORD'S SUMMERTIME, U.S.A. (I,1823); LADIES BEFORE GENTLEMEN (I,2599); WHERE WAS I? (I,4833)

ROBERTS, KENNY
MIDWESTERN HAYRIDE (I,3033)

ROBERTS, LOIS
THE ADAM MACKENZIE STORY (I,30); BROADSIDE (I,732)

ROBERTS, LYNN
FIRESIDE THEATER (I,1580); SAMMY KAYE'S MUSIC FROM MANHATTAN (I,3902); SCHAEFER CENTURY THEATER (I,3935)

ROBERTS, LYNNE
STARS OVER HOLLYWOOD (I,4211)

ROBERTS, MARK
ACADEMY THEATER (I,14); THE BROTHERS BRANNAGAN (I,747); A DATE WITH LIFE (I,1163); THE FRONT PAGE (I,1691); GENERAL HOSPITAL (II,964); LOBO (II,1504); THREE STEPS TO HEAVEN (I,4485); THE YOUNG AND THE RESTLESS (II,2862)

ROBERTS, MICHAEL D
BARETTA (II,152); DOUBLE TROUBLE (II,724); THE EARTHLINGS (II,751); JESSICA NOVAK (II,1286); MANIMAL (II,1622); NIGHTSIDE (II,1851)

ROBERTS, NANCY
THE CRACKER BROTHERS (II,599); I'D RATHER BE CALM (II,1216); MASTER OF THE GAME (II,1647)

ROBERTS, NATHAN
THE OUTLAWS (II,1936)

ROBERTS, PERNELL
THE ADVENTURES OF NICK CARTER (I,68); ASSIGNMENT: MUNICH (I,302); BATTLE OF

THE NETWORK STARS (II,173); BATTLE OF THE NETWORK STARS (II,174); BONANZA (II,347); CAPTAINS AND THE KINGS (II,435); CAROUSEL (I,856); CENTENNIAL (II,477); CHARLIE COBB: NICE NIGHT FOR A HANGING (II,485); DEAD MAN ON THE RUN (II,631); THE LIVES OF JENNY DOLAN (II,1502); THE NIGHT RIDER (II,1848); SAM HILL (I,3895); SAN FRANCISCO INTERNATIONAL AIRPORT (I,3905); SAN FRANCISCO INTERNATIONAL AIRPORT (I,3906); TRAPPER JOHN, M.D. (II,2654)

ROBERTS, RACHEL
BAFFLED (I,332); BLITHE SPIRIT (I,485); THE TONY RANDALL SHOW (II,2640); THE WILDS OF TEN THOUSAND ISLANDS (II,2803)

ROBERTS, RALPH
HANS BRINKER OR THE SILVER SKATES (I,1933)

ROBERTS, RANDOLPH
HAPPY DAYS (II,1084)

ROBERTS, RENEE
FAWLTY TOWERS (II,843)

ROBERTS, ROY
ALIAS MIKE HERCULES (I,117); BEWITCHED (I,418); THE GALE STORM SHOW (I,1725); GUNSMOKE (II,1069); THE LUCY SHOW (I,2791); MAGGIE BROWN (I,2811); THE MARRIAGE BROKER (I,2920); MCHALE'S NAVY (I,2969); PETTICOAT JUNCTION (I,3571); TV READERS DIGEST (I,4630)

ROBERTS, STEPHEN
CONGRESSIONAL INVESTIGATOR (I,1037); IKE (II,1223); LITTLE HOUSE: BLESS ALL THE DEAR CHILDREN (II,1489); MR. NOVAK (I,3145)

ROBERTS, STEVE
HENNESSEY (I,2013)

ROBERTS, TANYA
CHARLIE'S ANGELS (II,486); LADIES IN BLUE (II,1422); MICKEY SPILLANE'S MIKE HAMMER: MURDER ME, MURDER YOU (II,1694); PLEASURE COVE (II,2058); WAIKIKI (II,2734)

ROBERTS, TONY
THE EDGE OF NIGHT (II,760); THE FOUR SEASONS (II,915); LET'S CELEBRATE (I,2663); ROSETTI AND RYAN (II,2216); ROSETTI AND RYAN: MEN WHO LOVE WOMEN (II,2217); SNAFU (II,2386); THE WAY THEY WERE (II,2748)

ROBERTS, TRACY
GET CHRISTIE LOVE! (II,982)

ROBERTSON, AL
SKEEZER (II,2376)

ROBERTSON, CHUCK
FOREST RANGER (I,1641)

ROBERTSON, CLIFF
BATMAN (I,366); THE BOB HOPE CHRYSLER THEATER (I,502); FALCON CREST (II,808); FORD TELEVISION THEATER (I,1634); THE GOLDEN AGE OF TELEVISION (II,1019); MONTGOMERY'S SUMMER STOCK (I,3096); THE OUTER LIMITS (I,3426); PLAYHOUSE 90 (I,3623); ROD BROWN OF THE ROCKET RANGERS (I,3825); SATURDAY'S CHILDREN (I,3923); SECOND CHANCE (I,3957); SHORT, SHORT DRAMA (I,4023); THE SINGERS (I,4060); THE TWILIGHT ZONE (I,4651); WASHINGTON: BEHIND CLOSED DOORS (II,2744)

ROBERTSON, DALE
BIG JOHN (II,239); DEATH VALLEY DAYS (I,1211); DYNASTY (II,746); FORD TELEVISION THEATER (I,1634); FRONTIER ADVENTURES (I,1696); IRON HORSE (I,2238); MELVIN PURVIS: G-MAN (II,1679); SCALPLOCK (I,3930); SHOOT-IN AT NBC (A BOB HOPE SPECIAL) (I,4021); THE STARS AND STRIPES SHOW (I,4206); A TALE OF WELLS FARGO (I,4338); THE TALES OF WELLS FARGO (I,4353)

ROBERTSON, DENNIS
LITTLE HOUSE: THE LAST FAREWELL (II,1491); TAMMY (I,4357); THE TIM CONWAY SHOW (I,4502)

ROBERTSON, ELIZABETH
BEULAH LAND (II,226)

ROBERTSON, GORDON
FRAGGLE ROCK (II,916)

ROBERTSON, MYLES
PRISONERS OF THE LOST UNIVERSE (II,2086)

ROBERTSON, PAUL
HANS BRINKER OR THE SILVER SKATES (I,1933)

ROBERTSON, ROLLA
THE DAY THE WOMEN GOT EVEN (II,628)

ROBERTSON, RONNIE
ONCE UPON A CHRISTMAS TIME (I,3368)

ROBERTSON, WILLIAM
THE RUBBER GUN SQUAD (II,2225)

ROBIERO, JOY
ROWAN AND MARTIN'S LAUGH-IN (I,3856)

ROBIN, ANDREW
ROUGHNECKS (II,2219)

ROBIN, DIANE
ANGIE (II,80); MAKIN' IT (II,1603); WHO'S THE BOSS? (II,2792)

ROBIN, JANET
THE BEST OF TIMES (II,219)

ROBIN, MARK
THE LAST CONVERTIBLE (II,1435)

ROBIN, TINA
SING ALONG (I,4055)

ROBINETTE, DALE
THE BILLION DOLLAR THREAT (II,246); CRISIS IN SUN VALLEY (II,604); THE DEADLY TRIANGLE (II,634); THE DOCTORS (II,694); DOG AND CAT (II,695); SEARCH FOR TOMORROW (II,2284); SHE'S IN THE ARMY NOW (II,2320)

ROBINS, BARBARA
THE HEIRESS (I,2006)

ROBINSON, AL
ALKALI IKE (I,123)

ROBINSON, ANDREW
BIG BEND COUNTRY (II,230); THE CATCHER (I,872); FROM HERE TO ETERNITY (II,932); LADIES IN BLUE (II,1422); LANIGAN'S RABBI (II,1427); ONCE AN EAGLE (II,1897); REWARD (II,2147)

ROBINSON, ANN
FURY (I,1721)

ROBINSON, ANNETTE
WHY DIDN'T THEY ASK EVANS? (II,2795)

ROBINSON, ARNOLD
FEDERAL AGENT (I,1557); KITTY FOYLE (I,2571)

ROBINSON, BARTLETT
IVY LEAGUE (I,2286); MONA MCCLUSKEY (I,3085); OF THIS TIME, OF THAT PLACE (I,3335); READY FOR THE PEOPLE (I,3739); THE TIME ELEMENT (I,4504); WENDY AND ME (I,4793)

ROBINSON, BILL
THE R.C.A. THANKSGIVING SHOW (I,3736)

ROBINSON, BUD
LET'S DANCE (I,2664)

ROBINSON, BUMPER
COCAINE AND BLUE EYES (II,548)

ROBINSON, CAROL
ARNOLD'S CLOSET REVIEW (I,262)

ROBINSON, CECE
LET'S DANCE (I,2664)

ROBINSON, CHARLES KNOX
BUFFALO BILL (II,387); THE BUFFALO SOLDIERS (II,388); NIGHT COURT (II,1844); NOWHERE TO HIDE (II,1868)

ROBINSON, CHRIS
THE DOG TROOP (I,1325); TRAVIS LOGAN, D.A. (I,4598); TWELVE O'CLOCK HIGH (I,4637); THE WILDS OF TEN THOUSAND ISLANDS (II,2803)

ROBINSON, DENNIS
YOUNG DR. KILDARE (I,4937)

ROBINSON, DOUG
ALICE (II,33)

ROBINSON, EARTHA
FAME (II,812)

ROBINSON, EDWARD G
BRACKEN'S WORLD (I,703); CITY BENEATH THE SEA (I,965); THE DEVIL AND DANIEL WEBSTER (I,1246); FOR THE DEFENSE (I,1624); THE FORD TELEVISION THEATER (I,1634); THE LUX VIDEO THEATER (I,2795); NIGHT GALLERY (I,3287); OPERATION ENTERTAINMENT (I,3400); PLAYHOUSE 90 (I,3623); THE RIGHT MAN (I,3790); THE SINGERS (I,4060); TWELVE STAR SALUTE (I,4638); U.M.C. (I,4666); WHO HAS SEEN THE WIND? (I,4847) ZANE GRAY THEATER (I,4979)

ROBINSON, FRANCES
HIS MODEL WIFE (I,2064); A STORM IN SUMMER (I,4237)

ROBINSON, FRANK
RYAN'S HOPE (II,2234)

ROBINSON, GLADIS
THE WOMAN IN WHITE (II,2819)

ROBINSON, JANICE
THE BILL COSBY SPECIAL (I,444)

ROBINSON, JAY
DOCTOR SHRINKER (II,686); THE SECRET EMPIRE (II,2291)

ROBINSON, JOHN MARK
THE SAN PEDRO BEACH BUMS (II,2250); THE SAN PEDRO BUMS (II,2251)

ROBINSON, LANY
KITTY FOYLE (I,2571)

ROBINSON, LARRY
THE GOLDBERGS (I,1835)

ROBINSON, MICHELE
DEVLIN (II,663)

ROBINSON, MURPHY
BACKSTAIRS AT THE WHITE HOUSE (II,133)

ROBINSON, PHILIP
MR. MERGENTHWIRKER'S LOBBLIES (I,3144)

ROBINSON, ROGER
EISCHIED (II,763); FRIENDS (II,927); MALLORY: CIRCUMSTANTIAL EVIDENCE (II,1608)

ROBINSON, RUSSELL PHILIP
THE WHITE SHADOW (II,2788)

ROBINSON, RUTH
BEULAH (I,416)

ROBINSON, RUTHIE
THE DOCTOR WAS A LADY (I,1318); SHIRLEY TEMPLE'S STORYBOOK (I,4017)

ROBINSON, SANDY
SEARCH FOR TOMORROW (II,2284)

ROBINSON, SHEILA
OPRYLAND: NIGHT OF STARS AND FUTURE STARS (II,1921)

ROBINSON, SMOKEY
ROCK 'N' ROLL: THE FIRST 25 YEARS (II,2189); THE SMOKEY ROBINSON SHOW (I,4093)

ROBINSON, SUE
THE SAVAGE CURSE (I,3925)

ROBINSON, SUGAR RAY
THE BOB HOPE SHOW (I,655); CITY BENEATH THE SEA (I,965)

ROBINSON, VIRGINIA
LOVE OF LIFE (I,2774)

ROBLE, CHET
STUD'S PLACE (I,4271)

ROBLES, WALTER
HERNDON AND ME (II,1139)

ROBSON, FLORA
A MAN CALLED INTREPID (II,1612)

ROBSON, ZULEIKA
MARRIED ALIVE (I,2923)

ROCCO, ALEX
79 PARK AVENUE (II,2310); THE BEST OF TIMES (II,220); THE BLUE KNIGHT (II,277); THE FACTS OF LIFE (II,805); HUSBANDS AND WIVES (II,1208); HUSBANDS, WIVES AND LOVERS (II,1209); LILY—SOLD OUT (II,1480); THIS BETTER BE IT (II,2595); THREE FOR THE ROAD (II,2607); TWIGS (II,2683)

ROCHE, CHAD
BENDER (II,204)

ROCHE, CON
AS THE WORLD TURNS (II,110)

ROCHE, EUGENE
ALONE AT LAST (II,60); THE ART OF CRIME (II,108); COCAINE AND BLUE EYES (II,548); COREY: FOR THE PEOPLE (II,575); THE CORNER BAR (I,1052); EGAN (I,1419); FARRELL: FOR THE PEOPLE (II,834); THE FEATHER AND FATHER GANG: NEVER CON A KILLER (II,846); GOOD TIME HARRY (II,1037); HART TO HART (II,1101); HART TO HART (II,1102); HIGHER AND HIGHER, ATTORNEYS AT LAW (I,2057); JOHNNY BLUE (II,1321); THE JUGGLER OF NOTRE DAME (II,1350); LIFE'S MOST EMBARASSING MOMENTS (I,1470); LOCAL 306 (II,1505); MAGNUM, P.I. (II,1601); MALLORY: CIRCUMSTANTIAL EVIDENCE (II,1608); THE NEW MAVERICK (II,1822); PEOPLE LIKE US (II,1999); THE POSSESSED (II,2071); PRINCESS (II,2084); THE RETURN OF LUTHER GILLIS (II,2137); SOAP (II,2392); TWO THE HARD WAY (II,2695); WEBSTER (II,2758); WINTER KILL (II,2810); YOU CAN'T TAKE IT WITH YOU (II,2857)

ROCHE, SEAN
BENDER (II,204); OFF THE WALL (II,1879); STICKIN' TOGETHER (II,2465); THE WALTONS (II,2740)

ROCHE, TAMI
HERE IT IS, BURLESQUE! (II,1133)

ROCHELLE, LISA
HAPPY ENDINGS (II,1085)

ROCK, BLOSSOM
THE ADDAMS FAMILY (II,11)

ROCK, FILIPPA
THE BARRETTS OF WIMPOLE STREET (I,361)

ROCKET, CHARLES
I DO, I DON'T (II,1212); THE INVESTIGATORS (II,1241); THE OUTLAWS (II,1936); SATURDAY NIGHT LIVE (II,2261)

ROCKWELL, JOHN
LIFE WITH VIRGINIA (I,2695)

ROCKWELL, ROBERT
ADAM-12 (I,31); THE ADVENTURES OF SUPERMAN (I,77); THE BILL COSBY SHOW (I,442); LOVE AT FIRST SIGHT (II,1530); THE MAN FROM BLACKHAWK (I,2861); MAN WITHOUT A GUN (I,2884); OUR MISS BROOKS (I,3415); SEARCH FOR TOMORROW (II,2284); THOMPSON'S GHOST (I,4467)

ROCUZZO, MARIO
THE FUZZ BROTHERS (I,1722)

RODAN, ROBERT
DARK SHADOWS (I,1157)

RODANN, ZIVA
ALEXANDER THE GREAT (I,113); GIVE MY REGARDS TO BROADWAY (I,1815); MIRROR, MIRROR, OFF THE WALL (I,3059)

RODD, MARCIA
13 QUEENS BOULEVARD (II,2591); ALL IN THE FAMILY (II,38); THE DUMPLINGS (II,743); FLAMINGO ROAD (II,872); THE FOUR SEASONS (II,915); PIONEER SPIRIT (I,3604); TRAPPER JOHN, M.D. (II,2654)

RODELL, BARBARA
ANOTHER WORLD (II,97); AS THE WORLD TURNS (II,110); THE GUIDING LIGHT (II,1064); THE SECRET STORM (I,3969)

RODENSKY, SHMUEL
MOSES THE LAWGIVER (II,1737)

RODGER, STRUAN
THE MANIONS OF AMERICA (II,1623)

RODGERS, DAVE
BEST OF THE WEST (II,218)

RODGERS, ILONA
1915 (II,1853)

RODGERS, JIMMIE
THE JIMMIE RODGERS SHOW (I,2382); THE JIMMIE RODGERS SHOW (I,2383)

RODGERS, PAMELA
HEY, LANDLORD (I,2039); MAN ON THE MOVE (I,2878)

RODGERS, RICHARD
DATELINE (I,1165); THE GERSHWIN YEARS (I,1779); TALENT SEARCH (I,4342)

RODGERS, TIMMIE
THE WONDERFUL WORLD OF AGGRAVATION (I,4894)

RODMAN, VIC
NOAH'S ARK (I,3302)

RODNEY, JUNE
DOUGLAS FAIRBANKS JR. PRESENTS THE RHEINGOLD THEATER (I,1364)

RODRIGUES, PERCY
CAROL FOR ANOTHER CHRISTMAS (I,852); CUTTER TO HOUSTON (II,612); ENIGMA (II,778); EXECUTIVE SUITE (II,796); GENESIS II (I,1760); GOOD TIMES (II,1038); THE LIVES OF JENNY DOLAN (II,1502); THE MONEYCHANGERS (II,1724); MOST WANTED (II,1738); THE NIGHT RIDER (II,1848); PEYTON PLACE (I,3574); ROOTS: THE NEXT GENERATIONS (II,2213); SANFORD (II,2255); THE SILENT FORCE (I,4047); THIS GIRL FOR HIRE (II,2596)

RODRIQUEZ, JOHNNY
COUNTRY MUSIC HIT PARADE (II,588); THE RUNAWAY BARGE (II,2230)

RODRIQUEZ, JOSE LUIS
ANNE MURRAY'S CARIBBEAN CRUISE (II,88)

RODRIQUEZ, MARCO
BAY CITY BLUES (II,186)

RODRIQUEZ, MIGUEL
HIGHWAY HONEYS (II,1151)

RODRIQUEZ, PAUL
A.K.A. PABLO (II,28)

RODWAY, NORMAN
APPOINTMENT WITH A KILLER (I,245); REILLY, ACE OF SPIES (II,2129)

ROE, PATRICIA
ONE LIFE TO LIVE (II,1907)

ROEBLING, PAUL
ANASTASIA (I,182)

ROELAND, AUGUSTA

FIRST LOVE (I,1583)

ROERICK, WILLIAM
THE GUIDING LIGHT (II,1064)

ROESCH, TIMOTHY
CAMP GRIZZLY (II,418)

ROESSLER, ELMIRA
UNCLE MISTLETOE AND HIS ADVENTURES (I,4669)

ROGERS JR, WILL
THE PIONEERS (I,3605)

ROGERS, ANNE
ELIZABETH THE QUEEN (I,1431)

ROGERS, BROOKS
THE EDGE OF NIGHT (II,760)

ROGERS, BUDDY
CAVALCADE OF BANDS (I,876)

ROGERS, GABY
THE KAISER ALUMINUM HOUR

ROGERS, EILEEN
THE CHEVROLET GOLDEN ANNIVERSARY SHOW (I,924)

ROGERS, ELIZABETH
DRAGNET (I,1370); LACY AND THE MISSISSIPPI QUEEN (II,1419); STAR TREK (II,2440)

ROGERS, EMMA
PENMARRIC (II,1993)

ROGERS, FRED
THE CHILDREN'S CORNER (I,938); MR. ROGERS' NEIGHBORHOOD (I,3150)

ROGERS, GIL
ALL MY CHILDREN (II,39); DO NOT DISTURB (I,1300); WHERE THE HEART IS (I,4831)

ROGERS, GINGER
ACCENT ON LOVE (I,16); THE ALL-STAR SALUTE TO MOTHER'S DAY (II,51); THE BOB HOPE SHOW (I,552); THE BOB HOPE SHOW (I,570); THE BOB HOPE SHOW (I,575); THE BOB HOPE SHOW (I,578); THE BOB HOPE SHOW (I,582); CINDERELLA (I,956); THE GINGER ROGERS SHOW (I,1804); THE GINGER ROGERS SHOW (I,1805); THE JUNE ALLYSON SHOW (I,2488); LEGENDS OF THE SCREEN (II,1456); TONIGHT AT 8:30 (I,4556)

ROGERS, JAIME
ELVIS (I,1435);

ROGERS, JEFFREY
YOUNG HEARTS (II,2865)

ROGERS, JIMMY
THE ISLANDER (II,1249)

ROGERS, JULIE
THE KILLIN' COUSIN (II,1394)

ROGERS, KASEY
BEWITCHED (I,418); PEYTON PLACE (I,3574)

ROGERS, KENNY
THE CAPTAIN AND TENNILLE IN HAWAII (II,430); KENNY & DOLLY: A CHRISTMAS TO REMEMBER (II,1378); KENNY ROGERS IN CONCERT (II,1381); THE KENNY ROGERS SPECIAL (II,1382); ROLLIN' ON THE RIVER (I,3834); A SPECIAL KENNY ROGERS (II,2417)

ROGERS, LYNN
A TIME FOR US (I,4507)

ROGERS, LYNNE
THE GUIDING LIGHT (II,1064)

ROGERS, MELODY
NOBODY'S PERFECT (II,1858)

ROGERS, MIMI
HEAR NO EVIL (II,1116); PAPER DOLLS (II,1952); THE ROUSTERS (II,2220)

ROGERS, MITZIE
WHY DIDN'T THEY ASK EVANS? (II,2795)

ROGERS, PAUL
ROMEO AND JULIET (I,3837)

ROGERS, RICHARD
THE ADVENTURES OF WILLIAM TELL (I,85)

ROGERS, ROY
A COUNTRY HAPPENING (I,1065); THE GREAT MOVIE COWBOYS (II,1057); THE ROY ROGERS AND DALE EVANS SHOW (I,3858); THE ROY ROGERS SHOW (I,3859); SAGA OF SONORA (I,3884)

ROGERS, STEPHEN
CHIEFS (II,509); COMBAT (I,1011)

ROGERS, SUZANNE
DAYS OF OUR LIVES (II,629)

ROGERS, TERRY
BLOCKHEADS (II,270)

ROGERS, TIMMIE
CITY OF ANGELS (II,540)

ROGERS, TRISTAN
GENERAL HOSPITAL (II,964)

ROGERS, WAYNE
THE CELEBRITY FOOTBALL CLASSIC (II,474); CHIEFS (II,509); CIRCUS OF THE STARS (II,530); CIRCUS OF THE STARS (II,534); CITY OF ANGELS (II,540); FAMOUS LIVES (II,821); GREAT ADVENTURE (I,1868); HAVING BABIES II (II,1108); HOUSE CALLS (II,1194); THE LONG HOT SUMMER (I,2747); M*A*S*H (II,1569); STAGECOACH WEST (I,4185); TOP OF THE HILL (II,2645)

ROIDNANSKY, SERGE
FAME (II,812)

ROJAS, CARMINE
DAVID BOWIE—SERIOUS MOONLIGHT (II,622)

ROJAS, MANUEL
ELFEGO BACA (I,1426)

ROJO, GUSTAVO
THREE WISHES (I,4488)

ROKER, RENNY
MAKING IT (II,1605); THE MYSTERIOUS TWO (II,1783); NOBODY'S PERFECT (II,1858)

ROKER, ROXIE
CHANGE AT 125TH STREET (II,480); THE JEFFERSONS (II,1276); ROOTS (II,2211)

ROLAND, ANGELA
STRANGE PARADISE (I,4246)

ROLAND, CHRISTOPHER
ANOTHER LIFE (II,95)

ROLAND, GILBERT
BEWITCHED (I,418); THE DESILU PLAYHOUSE (I,1237); THE GREATEST SHOW ON EARTH (I,1883); JANE WYMAN PRESENTS THE FIRESIDE THEATER (I,2345); THE POPPY IS ALSO A FLOWER (I,3650)

ROLAND, NORMA
U.F.O. (I,4662)

ROLAND, STEVE
HOW TO SUCCEED IN BUSINESS WITHOUT REALLY TRYING (II,1197)

ROLF, FREDERICK
VANITY FAIR (I,4700)

ROLFE, GUY
RIVAK, THE BARBARIAN (I,3795)

ROLIKE, HANK
ROOSEVELT AND TRUMAN (II,2209)

ROLIN, JUDI
ALICE THROUGH THE LOOKING GLASS (I,122)

ROLLASON, JON
CORONATION STREET (I,1054)

ROLLE, ESTHER
GOOD TIMES (II,1038); MAUDE (II,1655); MOMMA THE DETECTIVE (II,1721)

ROLLINS JR, HOWARD E
THE MEMBER OF THE WEDDING (II,1680)

ROLLINS, BARBARA
THE NASH AIRFLYTE THEATER

ROLLINS, HOWARD
OUR STREET (I,3418)

ROLSTON, MARK
MASTER OF THE GAME (II,1647)

ROMAN, GREG
THE NEW BREED (I,3258)

ROMAN, JOSEPH S
QUINCY, M. E. (II,2102)

ROMAN, LULU
THE HEE HAW HONEYS (II,1125)

ROMAN, MURRAY
ON THE FLIP SIDE (I,3360)

ROMAN, NINA
ELLERY QUEEN (II,766)

ROMAN, RUTH
ALFRED HITCHCOCK PRESENTS (I,115); THE BOB HOPE CHRYSLER THEATER (I,502); COPS (I,1047); CRISIS (I,1102); THE FORD TELEVISION THEATER (I,1634); GENERAL ELECTRIC THEATER (I,1753); THE GREATEST SHOW ON EARTH (I,1883); JANE WYMAN PRESENTS THE FIRESIDE THEATER (I,2345); THE LONG HOT SUMMER (I,2747); THE PHILADELPHIA STORY (I,3582); PUNCH AND JODY (II,2091); WILLOW B: WOMEN IN PRISON (II,2804)

ROMANO, ANDY
79 PARK AVENUE (II,2310); CANNON (II,424); FRIENDS (II,927); GET CHRISTIE LOVE! (II,981); GET CHRISTIE LOVE! (II,982); THE RUBBER GUN SQUAD (II,2225)

ROMANS, PIA
THE AWAKENING LAND (II,122)

ROMANUS, RICHARD
FOUL PLAY (II,910); MICKEY SPILLANE'S MIKE HAMMER: MORE THAN MURDER (II,1693); STRIKE FORCE (II,2480); TENSPEED AND BROWN SHOE (II,2562)

ROMANUS, ROBERT
THE BEST OF TIMES (II,220)

ROMAY, LINA
MAYOR OF HOLLYWOOD (I,2962)

ROMBOLA, ED
THE BIG EASY (II,234)

ROMER, DENNIS
AS THE WORLD TURNS (II,110); FOR RICHER, FOR POORER (II,906)

ROMER, TONI
THE PHIL SILVERS SHOW (I,3580)

ROMERO, CARLOS
ADAM-12 (I,31); THE D.A.: MURDER ONE (I,1122); FALCON CREST (II,808); THE HAT OF SERGEANT MARTIN (I,1963); THE TOM SWIFT AND LINDA CRAIG MYSTERY HOUR (II,2633); WICHITA TOWN (I,4857)

ROMERO, CESAR
BATMAN (I,366); CESAR'S WORLD (I,891); CHICO AND THE MAN (II,508); CLIMAX! (I,976); FORD TELEVISION THEATER (I,1634); GENERAL ELECTRIC TRUE (I,1754); GRUEN GUILD PLAYHOUSE (I,1892); THE MAN FROM EVERYWHERE (I,2864); MANHATTAN TOWER (I,2889); PASSPORT TO DANGER (I,3467); PEPSI-COAL PLAYHOUSE (I,3523); THE RAINBOW GIRL (II,2108); TAKE A GOOD LOOK (II,2529); YOUR CHEVROLET SHOWROOM (I,4947)

ROMERO, GIBO
MAGNUM, P.I. (II,1601)

ROMERO, JANELLE
ROUGHNECKS (II,2219)

ROMERO, NED
BORN TO THE WIND (II,354); THE D.A. (I,1123); DAN AUGUST (I,1130); GEORGE WASHINGTON (II,978)

ROMOS, LARRY
THE NEW CHRISTY MINSTRELS SHOW (I,3259)

RONANE, JOHN
THE SIX WIVES OF HENRY VIII (I,4068)

RONDELL, R A
SAM (II,2246)

RONDELL, RON
DOUGLAS FAIRBANKS JR. PRESENTS THE RHEINGOLD THEATER (I,1364); LOVERS AND FRIENDS (II,1549); O.S.S. (I,3411)

RONNIE, JULIE
SANTA BARBARA (II,2258)

RONSTADT, LINDA
HEADMASTER (I,1990); LINDA RONSTADT IN CONCERT (II,1482)

ROONEY JR, MICKEY
THE MOUSEKETEERS REUNION (II,1742)

ROONEY, MICHAEL
FAME (II,812)

ROONEY, MICKEY
AMOS BURKE: WHO KILLED JULIE GREER? (I,181); THE BOB HOPE CHRYSLER THEATER (I,502); CIRCUS OF THE STARS (II,536); THE DICK POWELL SHOW (I,1269); EVIL ROY SLADE (I,1485); THE GOLDEN AGE OF TELEVISION (II,1018); HEY MULLIGAN (I,2040); THE MANY SIDES OF MICKEY ROONEY (I,2899); MICKEY (I,3023); THE MICKEY ROONEY SHOW (I,3027); MR. BROADWAY (I,3130); THE MUSIC MAKER (I,3173); NIGHT GALLERY (I,3287); O'MALLEY (II,1872); ONE OF THE BOYS (II,1911); PINOCCHIO (I,3602); PLAYHOUSE 90 (I,3623); RAQUEL (II,2114); SANTA CLAUS IS COMIN' TO TOWN (II,2259); SCHLITZ PLAYHOUSE OF THE STARS (I,3936); THE SEVEN LITTLE FOYS (I,3986); THE TWILIGHT ZONE (I,4651); A YEAR AT THE TOP (II,2844); A YEAR AT THE TOP (II,2845); THE YEAR WITHOUT A SANTA CLAUS (II,2846)

ROONEY, SHERRY
LOVE OF LIFE (I,2774); SEARCH FOR TOMORROW (II,2284)

ROONEY, TEDDY
SHIRLEY TEMPLE'S STORYBOOK (I,4017)

ROONEY, TIM
GIDGET (I,1795); THE HUMAN COMEDY (I,2158); MICKEY (I,3023); THE MOUSEKETEERS REUNION (II,1742); ROOM FOR ONE MORE (I,3842)

ROOPE, FAY
THE LONELY WIZARD (I,2743)

ROOS, JOANNA
INHERIT THE WIND (I,2213); JOHNNY BELINDA (I,2416); LOVE OF LIFE (I,2774); TODAY IS OURS (I,4531)

RORKE, HAYDEN
COUNTERPOINT (I,1059); I DREAM OF JEANNIE (I,2167); THE LIVES OF JENNY DOLAN (II,1502); THE MONEYCHANGERS (II,1724); MR. ADAMS AND EVE (I,3121); NO TIME FOR SERGEANTS (I,3300)

RORY, RENEE
LOVE OF LIFE (I,2774)

RORY, ROSANNA THE ERROL FLYNN THEATER (I,1458)

ROSARIO, BERT
A.K.A. PABLO (II,28); B.J. AND THE BEAR (II,125); CONCRETE BEAT (II,562); MICKEY SPILLANE'S MIKE HAMMER: MURDER ME, MURDER YOU (II,1694); P.O.P. (II,1939); THE RITA MORENO SHOW (II,2181); ROOSEVELT AND TRUMAN (II,2209); SWORD OF JUSTICE (II,2521)

ROSARIO, JOE
ARCHIE BUNKER'S PLACE (II,105)

ROSATO, TONY
AMANDA'S (II,62); SATURDAY NIGHT LIVE (II,2261)

ROSE, BILLY
BILLY ROSE'S PLAYBILL (I,452)

ROSE, CLIFFORD
THE PALLISERS (II,1944); THE ROADS TO FREEDOM (I,3802); SECRET ARMY (II,2290)

ROSE, GEORGE
BEACON HILL (II,193); CAPTAIN BRASSBOUND'S CONVERSION (I,828); CYRANO DE BERGERAC (I,1120); EAGLE IN A CAGE (I,1392); THE LITTLEST ANGEL (I,2726); MY FAIR LADY (I,3187); ST. JOAN (I,4175)

ROSE, HELEN
A TOWN LIKE ALICE (II,2652)

ROSE, JAMIE
FALCON CREST (II,808); NEVER AGAIN (II,1808)

ROSE, JANE
THE ADDAMS FAMILY (II,13); ALL IN THE FAMILY (II,38); CO-ED FEVER (II,547); LOVE OF LIFE (I,2774); PHYLLIS (II,2038); THE SECRET STORM (I,3969); SOMERSET (I,4115); WONDERFUL JOHN ACTION (I,4892)

ROSE, LOIS
VISUAL GIRL (I,4740)

ROSE, MARGOT
NEVER AGAIN (II,1808); REPORT TO MURPHY (II,2134)

ROSE, NORMAN
ALL MY CHILDREN (II,39); THE BIG STORY (I,432); MAN BEHIND THE BADGE (I,2857); ONE LIFE TO LIVE (II,1907); POLICE STORY (I,3635)

ROSE, PATRICIA
THE GUIDING LIGHT (II,1064)

ROSE, REVA
LANIGAN'S RABBI (II,1428); MARY HARTMAN, MARY HARTMAN (II,1638); TEMPERATURES RISING (I,4390); THAT GIRL (II,2570); WILD ABOUT HARRY (II,2797)

ROSE, ROBIN
LAND OF HOPE (II,1424); THE WHITE SHADOW (II,2788)

ROSEBORO, JOHN
DRAGNET (I,1370)

ROSEDALE, DOLORES (ROXANNE)
BEAT THE CLOCK (I,377)

ROSELIUS, JOHN
THE GIRL, THE GOLD WATCH AND EVERYTHING (II,1002)

ROSELLE, NICOLE
GIMME A BREAK (II,995)

ROSEN, LEONARD
WHAT'S YOUR BID? (I,4825)

ROSEN, STU
DUSTY'S TREEHOUSE (I,1391)

ROSENBERG, ARTHUR
HUNTER (II,1206); THE JAMES BOYS (II,1272)

ROSENBERG, LAWRENCE
THE FIGHTING NIGHTINGALES (II,855)

ROSENBERG, SANDY
MR. MERLIN (II,1751)

ROSENBLATT, MARTIN
O'MALLEY (II,1872)

ROSENBLOOM, MAXIE
THE JOE PALOOKA STORY (I,2401)

ROSENFIELD, JOE
A COUPLE OF JOES (I,1078)

ROSENTHAL, CINDY
THE DRUNKARD (II,737)

ROSESKI, BOB
WOMAN'S PAGE (II,2822)

ROSETTI, ANGELO
BARETTA (II,152); H.R. PUFNSTUF (I,2154); LIDSVILLE (I,2679)

ROSIN, ISABELLE
JANE EYRE (II,1273)

ROSIN, JAMES

MICKEY SPILLANE'S MIKE HAMMER: MORE THAN MURDER (II,1693)

ROSQUI, TOM
ALL MY CHILDREN (II,39); DEAD MAN ON THE RUN (II,631); ZERO INTELLIGENCE (II,2876)

ROSS, ANTHONY
THE INNER SANCTUM (I,2216); THE TELLTALE CLUE (I,4389); THE WEB (I,4784)

ROSS, CAROLINE
ESCAPE (I,1460)

ROSS, CHELCHIE
THE LAST LEAF (II,1437)

ROSS, DAVID
BYLINE—BETTY FURNESS (I,785)

ROSS, DIANA
DIANA (I,1258); DIANA (II,667); DIANA ROSS AND THE SUPREMES AND THE TEMPTATIONS ON BROADWAY (I,1260); DIANA ROSS IN CONCERT (II,668); AN EVENING WITH DIANA ROSS (II,788)

ROSS, DON
DRAGNET (I,1371); SAM (II,2246)

ROSS, DUNCAN
CHEERS (II,494); TUCKER'S WITCH (II,2667)

ROSS, EARLE
THE GREAT GILDERSLEEVE (I,1875); MEET MILLIE (I,2988)

ROSS, EDMUNDO
BROADWAY GOES LATIN (I,734)

ROSS, ELIZABETH
THE CLOCK (I,978); THE HANDS OF MURDER (I,1929)

ROSS, FREDRIC
EXPERT WITNESS (I,1488)

ROSS, GEORGE
THE GEORGE ROSS SHOW (I,1775); MACBETH (I,2801)

ROSS, JARROD
THE GUIDING LIGHT (II,1064)

ROSS, JAMIE
WE INTERRUPT THIS SEASON (I,4780)

ROSS, JEREMY
NO MAN'S LAND (II,1855)

ROSS, JOANNA
THE GREATEST GIFT (I,1880)

ROSS, JOE E
THE C.B. BEARS (II,406); CAR 54, WHERE ARE YOU? (I,842); HELP! IT'S THE HAIR BEAR BUNCH (I,2012); HONG KONG PHOOEY (II,1180); IT'S ABOUT TIME (I,2263); THE PHIL SILVERS SHOW (I,3580)

ROSS, KATHARINE
GREAT BIBLE ADVENTURES (I,1872); KRAFT SUSPENSE THEATER (I,2591); TRAVIS MCGEE (II,2656); WAIT UNTIL DARK (II,2735)

ROSS, KIMBERLY
CAPITOL (II,426)

ROSS, LANNY
THE LANNY ROSS SHOW (I,2612); THE SWIFT SHOW (I,4316)

ROSS, LEONARD
BARNEY AND ME (I,358); THE EYES OF TEXAS II (II,800)

ROSS, LISA HOPE
THE BEST OF TIMES (II,219)

ROSS, LYNNE
THE SEVEN DIALS MYSTERY (II,2308)

ROSS, MARION
ALIVE AND WELL (II,36); BLITHE SPIRIT (I,484); CHANNING (I,900); ESCAPE (I,1461); GREAT ADVENTURE (I,1868); HAPPY DAYS (II,1084); LIFE WITH FATHER (I,2690); MRS. G. GOES TO COLLEGE (I,3155); THE OUTER LIMITS (I,3426); PARADISE BAY (I,3454); PEARL (II,1986); THE PSYCHIATRIST: GOD BLESS THE CHILDREN (I,3685); THE SLOWEST GUN IN THE WEST (I,4088); TRUE LIFE STORIES (II,2665)

ROSS, MERIE LYNN
GENERAL HOSPITAL (II,964); THE TEXAS RANGERS (II,2567)

ROSS, MICHAEL
FULL CIRCLE (I,1702); TAMMY (I,4357)

ROSS, NEILSON
PAC-MAN (II,1940)

ROSS, RICCO
THE RENEGADES (II,2132)

ROSS, SHAVAR
DIFF'RENT STROKES (II,674); THE LITTLE RASCALS (II,1494); MORK AND MINDY (II,1736)

ROSS, STANLEY RALPH
ALLISON SIDNEY HARRISON (II,54); TRUE LIFE STORIES (II,2665); THE WORLD'S GREATEST SUPER HEROES (II,2842)

ROSS, TED
HIGH FIVE (II,1143); PAROLE (II,1960); SIROTA'S COURT (II,2365)

ROSS, TONY
ROLL OF THUNDER, HEAR MY CRY (II,2202)

ROSS, WINSTON
THE GREATEST GIFT (I,1880)

ROSS-LEMING, EUGENIE
HIGHCLIFFE MANOR (II,1150)

ROSSEN, CAROL
THE CLIFF DWELLERS (I,974); COREY: FOR THE PEOPLE (II,575); THE OATH: 33 HOURS IN THE LIFE OF GOD (II,1873)

ROSSEN, PAUL
THE YOUNG MARRIEDS (I,4942)

ROSSI, LEO
PARTNERS IN CRIME (II,1961)

ROSSI, STEVE
HELLO DERE (I,2010)

ROSSINGTON, NORMAN
CASANOVA (II,451); SEARCH FOR THE NILE (I,3954)

ROSSITER, LEONARD
THE FALL AND RISE OF REGINALD PERRIN (II,810)

ROSSLER, ELMIRA
HAWKINS FALLS, POPULATION 6200 (I,1977)

ROSSO, SPRAY
B.J. AND THE BEAR (II,125)

ROSSON, PAT
COUNTY GENERAL (I,1076)

ROSSOVICH, TIM
WHEN THE WHISTLE BLOWS (II,2781)

ROSSULO, JOE
THE POWER WITHIN (II,2073)

ROST, ELAINE
CONCERNING MISS MARLOWE (I,1032)

ROSWELL, MAGGIE
AND THEY ALL LIVED HAPPILY EVER AFTER (II,75); CHARACTERS (II,481); PIPER'S PETS (II,2046)

ROTER, DIANE
THE VIRGINIAN (I,4738)

ROTH, ALLISON
TOO GOOD TO BE TRUE (II,2643)

ROTH, AUDREY
MR. ROGERS' NEIGHBORHOOD (I,3150)

ROTH, PHIL
THE CHOPPED LIVER BROTHERS (II,514)

ROTH, WOLF
THE AFRICAN QUEEN (II,18); HIGH RISK (II,1146)

ROTHSCHILD, SIGMUND
TRASH OR TREASURE (I,4594); WHAT'S IT WORTH? (I,4818)

ROTHWELL, MICHAEL
VANITY FAIR (I,4701)

ROUNDS, DAVID
ALICE (II,33); BEACON HILL (II,193); THE BLUE AND THE GRAY (II,273)

ROUNDTREE, RICHARD
CIRCUS OF THE STARS (II,531); FIREHOUSE (I,1579); ROOTS (II,2211); SHAFT (I,3993)

ROUNTRER, ROSE MARI
CELEBRITY (II,463)

ROURKE, MICKEY
HARDCASE (II,1088)

ROUSE, GRAHAM
1915 (II,1853)

ROUSE, SIMON
THE MANIONS OF AMERICA (II,1623)

ROUSSEAU, DIANE
LOVE OF LIFE (I,2774)

ROUSSEL, ELVERA
THE GUIDING LIGHT (II,1064)

ROUTBARD, EVAN
CAGNEY AND LACEY (II,410)

ROUX, CAROL
ANOTHER WORLD (II,97); SOMERSET (I,4115)

ROWAN, DAN
THE ROWAN AND MARTIN (II,2221); THE ROWAN AND MARTIN (I,3854); THE ROWAN AND MARTIN (I,3855); THE ROWAN AND MARTIN REPORT (II,2221); THE ROWAN AND MARTIN SHOW (I,3854); THE ROWAN AND MARTIN SPECIAL (I,3855); ROWAN AND MARTIN BITE THE HAND THAT FEEDS THEM (I,3853); ROWAN AND MARTIN'S (I,3856); ROWAN AND MARTIN'S (I,3857); ROWAN AND MARTIN'S LAUGH-IN (I,3856); ROWAN AND MARTIN'S LAUGH-IN (I,3857);

ROYAL VARIETY PERFORMANCE (I,3865); ULTRA QUIZ (II,2701); WHO'S AFRAID OF MOTHER GOOSE? (I,4852)

ROWAN, GAYE
OUR MAN FLINT: DEAD ON TARGET (II,1932); SCRUPLES (II,2281); THE STARLOST (I,4203)

ROWBOTTOM, JO
LITTLE WOMEN (I,2723)

ROWE, CHARLES
TERROR AT ALCATRAZ (II,2563)

ROWE, DOUG
FRANKIE AND ANNETTE: THE SECOND TIME AROUND (II,917)

ROWE, FANNY
NO—HONESTLY (II,1857); THE TALE OF BEATRIX POTTER (II,2533)

ROWE, GRAHAM
MASTER OF THE GAME (II,1647)

ROWE, HANSFORD
LOVE, SIDNEY (II,1547); MALIBU (II,1607); SITCOM (II,2367); THANKSGIVING VISITOR (I,4417); V (II,2713); V: THE FINAL BATTLE (II,2714); WHACKED OUT (II,2767)

ROWE, MISTY
HAPPY DAYS (II,1084); THE HEE HAW HONEYS (II,1125); JOE'S WORLD (II,1305); WHEN THINGS WERE ROTTEN (II,2782)

ROWE, RED
FACE THE FACTS (I,1508); TELL IT TO THE CAMERA (I,4386)

ROWLAND, HENRY
JOHNNY GUITAR (I,2427)

ROWLAND, JADA
AS THE WORLD TURNS (II,110); THE DEVIL'S DISCIPLE (I,1247); THE DOCTORS (II,694); THE HAMPTONS (II,1076); THE SECRET STORM (I,3969); SLEEPING BEAUTY (I,4083); STAR TONIGHT (I,4199); THE WOMEN (I,4890)

ROWLAND, JEFFREY
THE SLIGHTLY FALLEN ANGEL (I,4087)

ROWLAND, STEVE
THE LIFE AND LEGEND OF WYATT EARP (I,2681)

ROWLANDS, GENA
87TH PRECINCT (I,1422); GHOST STORY (I,1788); THE LLOYD BRIDGES SHOW (I,2733); PEYTON PLACE (I,3574); STUDIO ONE (I,4268); TOP SECRET U.S.A. (I,4581); WON'T IT EVER BE MORNING? (I,4902)

ROWLANDS, PATSY
BLESS THIS HOUSE (II,268)

ROWLES, POLLY
THE ALCOA HOUR (I,108); THE DEFENDERS (I,1220); THE ELGIN HOUR (I,1428); HOW TO SUCCEED IN BUSINESS WITHOUT REALLY TRYING (II,1197); JAMIE (I,2333); JAMIE (I,2334); LOVE OF LIFE (I,2774); MEET CORLISS ARCHER (I,2981); THE NURSES (I,3319); PULITZER PRIZE PLAYHOUSE (I,3692); SOMERSET (I,4115); THE THREE MUSKETEERS (I,4481)

ROY, CECIL
CASPER AND FRIENDS (I,867)

ROY, RAYMOND
CAMP WILDERNESS (II,419)

ROY, RONALD
THE CHEERLEADERS (II,493); MAUREEN (II,1656)

ROYCE, ROBERTA
THE SECRET STORM (I,3969)

ROYLANCE, PAMELA
LITTLE HOUSE: A NEW BEGINNING (II,1488); LITTLE HOUSE: BLESS ALL THE DEAR CHILDREN (II,1489); LITTLE HOUSE: LOOK BACK TO YESTERDAY (II,1490); LITTLE HOUSE: THE LAST FAREWELL (II,1491)

ROYLE, SELENA
CROWN THEATER WITH GLORIA SWANSON (I,1107); THE ETHEL BARRYMORE THEATER (I,1467); YOUR SHOW TIME (I,4958)

RUBEN, MARTIN
A CHRISTMAS CAROL (II,517)

RUBEN, TOM
C.P.O. SHARKEY (II,407)

RUBENSTEIN, PHIL
GIMME A BREAK (II,995); A GIRL'S LIFE (II,1000); I LOVE HER ANYWAY! (II,1214); MY WIFE NEXT DOOR (II,1781); WORKING STIFFS (II,2834)

RUBIN, ANDREW
JESSICA NOVAK (II,1286)

RUBIN, BENNY
THE BENNY RUBIN SHOW (I,398); THE BRIGHTER DAY (I,728); THE DICK TRACY SHOW (I,1272); THE JACK BENNY PROGRAM (I,2294); RETURN OF THE WORLD'S GREATEST DETECTIVE (II,2142)

RUBIN, DR THEODORE
CONFIDENTIAL FOR WOMEN (I,1034)

RUBINEK, SAUL
DUSTY (II,745)

RUBINO, EDIE MARIE
A.K.A. PABLO (II,28)

RUBINSTEIN, DEIRDRE
PRISONER: CELL BLOCK H (II,2085)

RUBINSTEIN, JOHN
COREY: FOR THE PEOPLE (II,575); CRAZY LIKE A FOX (II,601); FAMILY (II,813); THE FRENCH ATLANTIC AFFAIR (II,924); MR. & MS. AND THE BANDSTAND MURDERS (II,1746); MR. & MS. AND THE MAGIC STUDIO MYSTERY (II,1747); THE PSYCHIATRIST: GOD BLESS THE CHILDREN (I,3685); THE STREETS OF SAN FRANCISCO (I,4260)

RUBINSTEIN, ZELDA
FLINTSTONE FAMILY ADVENTURES (II,882); THE FLINTSTONE FUNNIES (II,883); I GAVE AT THE OFFICE (II,1213)

RUBY, BERNIE
BIG EDDIE (II,235)

RUBY, HARRY
MAKE ROOM FOR DADDY (I,2843)

RUBY, MARTIN
AS THE WORLD TURNS (II,110)

RUCK, ALAN
HARD KNOX (II,1087)

RUCKER, BARBARA
AS THE WORLD TURNS (II,110); THE NEW TEMPERATURES RISING SHOW (I,3273); TEXAS (II,2566)

RUCKER, DENNIS
THE CENTURY TURNS (I,890); ESCAPADE (II,781); ESCAPE (I,1461); FOOLS, FEMALES AND FUN: I'VE GOTTA BE ME (II,899); GET CHRISTIE LOVE! (II,981); JAMES AT 15 (II,1271); THE SIX MILLION DOLLAR MAN (I,4067); A VERY MISSING PERSON (I,4715)

RUDD, PAUL
BEACON HILL (II,193); BEULAH LAND (II,226)

RUDD, RODDY
A HORSEMAN RIDING BY (II,1182)

RUDLEY, HERBERT
THE CALIFORNIANS (I,795); CALL HER MOM (I,796); DALLAS (II,614); THE DEBBIE REYNOLDS SHOW (I,1214); HANDS OF MURDER (I,1929); THE INDIAN (I,2210); MICHAEL SHAYNE, PRIVATE DETECTIVE (I,3021); MONA MCCLUSKEY (I,3085); THE MOTHERS-IN-LAW (I,3110)

RUDLING, JOHN
TO THE MANOR BORN (II,2627)

RUDY, MARTIN
MOONLIGHT (II,1730)

RUFFIN, LARK
ROLL OF THUNDER, HEAR MY CRY (II,2202)

RUFFINO, CLAUDIO
S*H*E (II,2235)

RUGGERI, ADA
PAPA CELLINI (I,3450)

RUGGLES, CHARLIE
ADAMSBURG, U.S.A. (I,33); THE BELLS OF ST. MARY'S (I,392); THE BEVERLY HILLBILLIES (I,417); THE BULLWINKLE SHOW (I,761); CAROUSEL (I,856); CHRYSLER MEDALLION THEATER (I,951); CONFLICT (I,1036); THE GINGER ROGERS SHOW (I,1805); MCKEEVER AND THE COLONEL (I,2970); THE MOTOROLA TELEVISION HOUR (I,3112); ONCE UPON A CHRISTMAS TIME (I,3368); ROCKY AND HIS FRIENDS (I,3822); THE RUGGLES (I,3868); THE SOFT TOUCH (I,4107); THE WORLD OF MISTER SWEENY (I,4917)

RUICK, BARBARA
CINDERELLA (I,956); THE R.C.A. VICTOR SHOW (I,3737)

RUICK, MELVILLE
CITY HOSPITAL (I,967); THE DOOR WITH NO NAME (I,1352); FIRST LOVE (I,1583)

RUISINGER, THOMAS
SUMMER SOLSTICE (II,2492)

RUIZ, ISAAC
CHICO AND THE MAN (II,508)

RULE, JANICE
THE BOB HOPE CHRYSLER THEATER (I,502); THE CAMPBELL TELEVISION SOUNDSTAGE (I,812);

PLAYWRIGHTS '56 (I,3627); PROFILES IN COURAGE (I,3678); TONIGHT IN SAMARKAND (I,4558); THE TWILIGHT ZONE (I,4651); THE WORD (II,2833)

RUMIN, SIG
LIFE WITH LUIGI (I,2692)

RUNDLE, CIS
MATT HOUSTON (II,1654); THE TEXAS RANGERS (II,2567); VELVET (II,2725)

RUNDLE, ROBBIE
CODE R (II,551)

RUNYEON, FRANK
AS THE WORLD TURNS (II,110)

RUNYON, JENNIFER
ANOTHER WORLD (II,97); CHARLES IN CHARGE (II,482); SIX PACK (II,2370)

RUOCHENG, YING
MARCO POLO (II,1626)

RUPPERT, TAIT
CANDIDA (II,423)

RUPRECHT, DAVID
GILLIGAN'S ISLAND: THE HARLEM GLOBETROTTERS ON GILLIGAN'S ISLAND (II,993)

RUSCIO, AL
DIAGNOSIS: DANGER (I,1250); THE OUTSIDE MAN (II,1937); SHANNON (II,2314); STEAMBATH (II,2451)

RUSH, BARBARA
AT YOUR SERVICE (II,118); THE BARBARA RUSH SHOW (I,349); BATMAN (I,366); CLIMAX! (I,976); THE CRIME CLUB (I,1096); CUTTER (I,1117); THE EYES OF CHARLES SAND (I,1497); FLAMINGO ROAD (II,871); FLAMINGO ROAD (II,872); FOOLS, FEMALES AND FUN: IS THERE A DOCTOR IN THE HOUSE? (II,900); THE FUGITIVE (I,1701); THE GENERAL ELECTRIC THEATER (I,1753); THE LUX VIDEO THEATER (I,2795); THE NEW DICK VAN DYKE SHOW (I,3263); NIGHT GALLERY (I,3287); NOTORIOUS (I,3314); OF MEN OF WOMEN (I,3332); THE OUTER LIMITS (I,3426); PEYTON PLACE (I,3574); SAINTS AND SINNERS (I,3887); THE SEEKERS (II,2298); THE UNKNOWN (I,4680)

RUSH, DEBORAH
RYAN'S HOPE (II,2234)

RUSH, DENNIS
BABY CRAZY (I,319); DO NOT DISTURB (I,1300)

RUSH, LIZABETH
THE JUDGE AND JAKE WYLER (I,2464); NOT FOR HIRE (I,3308)

RUSH, SARAH
BATTLESTAR GALACTICA (II,181); MODESTY BLAISE (II,1719); ROUGHNECKS (II,2219); THE SEEKERS (II,2298)

RUSHMORE, KAREN
13 QUEENS BOULEVARD (II,2591); FLYING HIGH (II,895)

RUSKIN, JEANNE
THE EDGE OF NIGHT (II,760); WHERE THE HEART IS (I,4831)

RUSKIN, JOSEPH
CAPTAIN AMERICA (II,427); DOCTOR SCORPION (II,685); THE GYPSY WARRIORS (II,1070); THE MUNSTERS' REVENGE (II,1765)

RUSKIN, SHEILA
MACKENZIE (II,1584)

RUSKIN, SHIMEN
THE CORNER BAR (I,1051); THE CORNER BAR (I,1052); GRANDPA MAX (II,1051)

RUSS, DEBBIE
HERE COME THE DOUBLE DECKERS (I,2025)

RUSS, GENE
THE CHEROKEE TRAIL (II,500)

RUSS, WILLIAM
ANOTHER WORLD (II,97)

RUSSEL, SUZANNE
DOROTHY HAMILL IN ROMEO & JULIET ON ICE (II,718)

RUSSELL, ALAN
RETURN TO EDEN (II,2143)

RUSSELL, ANDY
THE ANDY AND DELLA RUSSELL SHOW (I,190)

RUSSELL, BING
BONANZA (II,347); THE MONEYCHANGERS (II,1724)

RUSSELL, BOB
STAND UP AND BE COUNTED (I,4187); VERSATILE VARIETIES (I,4713)

RUSSELL, BRYAN
JANE EYRE (I,2337); KILROY (I,2541)

RUSSELL, CONNIE
CLUB EMBASSY (I,983)

RUSSELL, DEL
ARNIE (I,261)

RUSSELL, DELLA
THE ANDY AND DELLA RUSSELL SHOW (I,190)

RUSSELL, DON
GUIDE RIGHT (I,1901); STARS ON PARADE (I,4210)

RUSSELL, ELINOR
OZMOE (I,3437)

RUSSELL, EVELYN
THE LITTLEST ANGEL (I,2726)

RUSSELL, FORBESY
WHO'S ON CALL? (II,2791)

RUSSELL, FRANY
OUR MAN FLINT: DEAD ON TARGET (II,1932)

RUSSELL, FRANZ
THE TROUBLE WITH TRACY (I,4613)

RUSSELL, GRAHAM
AIR SUPPLY IN HAWAII (II,26)

RUSSELL, IRIS
SPELL OF EVIL (I,4152)

RUSSELL, JACK
BLOOMER GIRL (I,488); THE MERRY WIDOW (I,3012)

RUSSELL, JACKIE
THE JOEY BISHOP SHOW (I,2403)

RUSSELL, JANE
THE DESILU PLAYHOUSE (I,1237); MACREEDY'S WOMAN (I,2804); THE YELLOW ROSE (II,2847)

RUSSELL, JEANNIE
DENNIS THE MENACE (I,1231)

RUSSELL, JOHN
ALIAS SMITH AND JONES (I,119); IT TAKES A THIEF (I,2250); JASON OF STAR COMMAND (II,1274); LAWMAN (I,2642); SOLDIERS OF FORTUNE (I,4111)

RUSSELL, KURT
BATTLE OF THE NETWORK STARS (II,164); THE NEW LAND (II,1820); THE QUEST (II,2096); THE QUEST (II,2097); THE TRAVELS OF JAIMIE MCPHEETERS (I,4596)

RUSSELL, MARION
THE GREATEST GIFT (I,1880)

RUSSELL, MARK
KOJAK (II,1406); REAL PEOPLE (II,2118)

RUSSELL, NEIL
YUMA (I,4976)

RUSSELL, NIPSEY
BAREFOOT IN THE PARK (I,355); CAR 54, WHERE ARE YOU? (I,842); CHAIN REACTION (II,479); DEAN MARTIN'S COMEDY WORLD (II,647); FAME (I,1517); MASQUERADE PARTY (II,1645)

RUSSELL, RON
RETURN TO PEYTON PLACE (I,3779); VISUAL GIRL (I,4740)

RUSSELL, ROSALIND
THE ROSALIND RUSSELL SHOW (I,3845); THE LORETTA YOUNG THEATER (I,2756); WONDERFUL TOWN (I,4893)

RUSSELL, TODD
PUD'S PRIZE PARTY (I,3691); ROOTIE KAZOOTIE (I,3844); WHEEL OF FORTUNE (I,4826)

RUSSELL, TOMMY
IT'S A SMALL WORLD (I,2260)

RUSSELL, WILLIAM
THE ADVENTURES OF SIR LANCELOT (I,76); DOCTOR WHO (II,689)

RUSSO, BARRY
LONGSTREET (I,2750); THE YOUNG MARRIEDS (I,4942)

RUSSO, CAROLE
KEY TORTUGA (II,1383)

RUSSO, GIANNI
RIPTIDE (II,2178)

RUSSO, J DUKE
MAN ON A STRING (I,2877)

RUSSO, MATT
MURDER INK (II,1769); POPI (II,2068)

RUSSOM, LEON
ANOTHER WORLD (II,97)

RUST, RICHARD
SAM BENEDICT (I,3894)

RUTH, KITTY
B.J. AND THE BEAR (II,125)

RUTHERFORD, ANGELO
GENTLE BEN (I,1761)

RUTHERFORD, ANN
BEN HECHT'S TALES OF THE CITY (I,395); THE BOB NEWHART SHOW (II,342); GRUEN GUILD PLAYHOUSE (I,1892); THE NASH AIRFLYTE THEATER (I,3227); NO WARNING (I,3301)

RUTHERFORD, GENE
THE COPS AND ROBIN (II,573)

RUTHERFORD, LORI ANN

THE EASTER PROMISE (I,1397); THE NEW ANDY GRIFFITH SHOW (I,3256)

RUTLAND, LOIS
CINDERELLA (II,526)

RUTLEDGE, PATRICIA
ANDROCLES AND THE LION (I,189)

RUTTAN, SUSAN
AFTER GEORGE (II,19); BUFFALO BILL (II,387); EMPIRE (II,777)

RUTTER, ANNE
MURDER MOTEL (I,3161)

RUYMAN, AYN
THE MCLEAN STEVENSON SHOW (II,1665); OUR FAMILY BUSINESS (II,1931); THREE TIMES DALEY (II,2610)

RUYSDAEL, BASIL
DAVY CROCKETT (I,1181)

RYAL, DICK
THE WORLD'S GREATEST SUPER HEROES (II,2842)

RYAN, BAYA KENIN
THE HOUSE WITHOUT A CHRISTMAS TREE (I,2143)

RYAN, DEBORAH
B.J. AND THE BEAR (II,125); COREY: FOR THE PEOPLE (II,575); THE EYES OF TEXAS II (II,800); SHIPSHAPE (II,2329)

RYAN, DICK
IF YOU KNEW TOMORROW (I,2187)

RYAN, EDMOND
DREAM GIRL (I,1374); YOUNG DR. MALONE (I,4938)

RYAN, FRAN
DAYS OF OUR LIVES (II,629); THE DEADLY GAME (II,633); THE DOCTORS (I,1323); THE DORIS DAY SHOW (I,1355); GREEN ACRES (I,1884); GUNSMOKE (II,1069); JOHNNY BELINDA (I,2418); THE NEW ZOO REVUE (I,3277); NO SOAP, RADIO (II,1856); THE ODD COUPLE (II,1875); THE RETURN OF MARCUS WELBY, M.D (II,2138); SIGMUND AND THE SEA MONSTERS (II,2352); WHO'S ON CALL? (II,2791)

RYAN, IRENE
THE BEVERLY HILLBILLIES (I,417); MARINELAND CARNIVAL (I,2912); TOM, DICK, AND HARRY (I,4536)

RYAN, JOHN
A HATFUL OF RAIN (I,1964)

RYAN, JOHN P
ARCHER (II,102); SHOOTING STARS (II,2341); WILLOW B: WOMEN IN PRISON (II,2804)

RYAN, LINDA
HAWAII FIVE-O (II,1110); WOMEN IN WHITE (II,2823)

RYAN, MADGE
A HORSEMAN RIDING BY (II,1182); MOLL FLANDERS (II,1720)

RYAN, MARLO
PINE LAKE LODGE (I,3597)

RYAN, MEG
AS THE WORLD TURNS (II,110); CHARLES IN CHARGE (II,482)

RYAN, MICHAEL
ANOTHER WORLD (II,97); BEN JERROD (I,396)

RYAN, MITCHELL
CHASE (I,917); THE CHISHOLMS (II,513); DARK SHADOWS (I,1157); EXECUTIVE SUITE (II,796); THE FUZZ BROTHERS (I,1722); HAVING BABIES III (II,1109); HIGH PERFORMANCE (II,1145); HOT PURSUIT (II,1189); JOE DANCER: THE MONKEY MISSION (II,1301); JULIE FARR, M.D. (II,1354); KING'S CROSSING (II,1397)

RYAN, NATASHA
THE COPS AND ROBIN (II,573); DAYS OF OUR LIVES (II,629); LADIES' MAN (II,1423)

RYAN, PATSY
GENERAL HOSPITAL (II,964)

RYAN, PEGGY
HAWAII FIVE-O (II,1110)

RYAN, ROBERT
THE FRONT PAGE (I,1692); GUILTY OR NOT GUILTY (I,1903); SCREEN DIRECTOR'S PLAYHOUSE (I,3946); WORLD WAR I (I,4919); ZANE GREY THEATER (I,4979)

RYAN, STEVE
OLD FRIENDS (II,1881); PALMS PRECINCT (II,1946); TEACHERS ONLY (II,2548)

RYDBECK, WHITNEY
THE CHEROKEE TRAIL (II,500); GILLIGAN'S ISLAND: THE HARLEM GLOBETROTTERS ON GILLIGAN'S ISLAND (II,993); WHITNEY AND THE ROBOT (II,2789)

RYDELL, BOBBY
SWINGIN' TOGETHER (I,4320)

RYDELL, MARK
AS THE WORLD TURNS (II,110)

RYDER, ALFRED
THE FLIERS (I,1596); LIGHT'S OUT (I,2699); PHILCO TELEVISION PLAYHOUSE (I,3583); THE SWAN (I,4312); TELLER OF TALES (I,4388); U.M.C. (I,4666)

RYDER, EDDIE
THE DENNIS O'KEEFE SHOW (I,1230); DOCTOR KILDARE (I,1315); FBI CODE 98 (I,1550); GENERAL HOSPITAL (II,964); OUR MISS BROOKS (I,3415); THINK PRETTY (I,4448)

RYE, ISABELLA
S*H*E (II,2235)

RYE, MICHAEL
HOT WHEELS (I,2126); THE LONE RANGER (I,2741); SKYHAWKS (I,4079); THE SUPER GLOBETROTTERS (II,2498); THE WORLD'S GREATEST SUPER HEROES (II,2842)

RYERSON, ANN
PRIVATE BENJAMIN (II,2087); THE YESTERDAY SHOW (II,2850)

RYERTON, NAOMI
GOLDEN WINDOWS (I,1841)

RYLAND, JACK
ONE LIFE TO LIVE (II,1907); THE SECRET STORM (I,3969)

RYON, REX
T.L.C. (II,2525)

SAAVEDRA, MARIA
DON QUIXOTE (I,1339)

SABA, SHARI
M*A*S*H (II,1569); THE ROLLERGIRLS (II,2203)

SABATINO, MICHAEL
BEHIND THE SCREEN (II,200); KNOTS LANDING (II,1404)

SABELLA, ERNIE
100 CENTRE STREET (II,1901); 13 THIRTEENTH AVENUE (II,2592); IT'S YOUR MOVE (II,1259); PUNKY BREWSTER (II,2092)

SABER, DAVID
MAYOR OF THE TOWN (I,2963)

SABIN, ALEC
TINKER, TAILOR, SOLDIER, SPY (II,2621)

SABIN, DAVID
WHEN THINGS WERE ROTTEN (II,2782)

SABISTON, ANDREW
THE EDISON TWINS (II,761)

SABRE, JULIA
FROM HERE TO ETERNITY (II,932)

SABU
SABU AND THE MAGIC RING (I,3881)

SACCHI, ROBERT
HARD KNOCKS (II,1086)

SACHS, ANDREW
FAWLTY TOWERS (II,843)

SACHS, DAVID
THE INTERNS (I,2226)

SACINO, ANDREA
MANY HAPPY RETURNS (I,2895)

SACKS, MICHAEL
BUNCO (II,395)

SADLER, KARL
OPERATION GREASEPAINT (I,3402)

SADOFF, FRED
MURDOCK'S GANG (I,3164)

SADOWSKY, ADAM
IT'S YOUR MOVE (II,1259)

SAFRAN, JUDY
THE SECRET STORM (I,3969)

SAGAL, JEAN
DOUBLE TROUBLE (II,723); DOUBLE TROUBLE (II,724)

SAGAL, LIZ
DOUBLE TROUBLE (II,723); DOUBLE TROUBLE (II,724)

SAGAN, CARL
COSMOS (II,579)

SAGAR, CAROL
HERE WE GO AGAIN (I,2028)

SAHL, MORT
COMEDY NEWS II (I,1021); THE MORT SAHL SPECIAL (I,3105)

SAINT, EVA MARIE
CAROL FOR ANOTHER CHRISTMAS (I,852); THE GOLDEN AGE OF TELEVISION (II,1016); HER SCHOOL FOR BACHELORS (I,2018); HOW THE WEST WAS WON (II,1196); JAMIE (I,2333); THE MACAHANS (II,1583); MALIBU (II,1607); ONE MAN'S FAMILY (I,3383); OUR TOWN (I,3419); PHILCO TELEVISION PLAYHOUSE (I,3583); REVLON MIRROR THEATER (I,3781); TAXI (I,4365); THE UNITED STATES STEEL HOUR (I,4677); THE WEB (I,4784)

SAINT, JAN

INDEMNITY (I,2209)

SAINTE MARIE, BUFFY
PERRY COMO'S CHRISTMAS IN NEW MEXICO (II,2010)

SAIRE, REBECCA
LOVE IN A COLD CLIMATE (II,1539)

SAITO, JAMES
THE RENEGADES (II,2132)

SAJAK, PAT
WHEEL OF FORTUNE (II,2775)

SAJAL, JOEY
MASADA (II,1642)

SAKAI, FRANKIE
SHOGUN (II,2339)

SAKAKEENY, WENDY
KENNEDY (II,1377)

SAKATA, HAROLD
HIGHCLIFFE MANOR (II,1150); THE POPPY IS ALSO A FLOWER (I,3650); SARGE (I,3914); SARGE: THE BADGE OR THE CROSS (I,3915)

SAKS, GENE
LOVE, SEX. . . AND MARRIAGE (II,1546); REUNION IN VIENNA (I,3780)

SALAMAN, CHLOE
FANNY BY GASLIGHT (II,823)

SALAMAN, TOBY
MACKENZIE (II,1584)

SALANTA, GREGORY
KATE AND ALLIE (II,1365)

SALDANA, THERESA
THE GANGSTER CHRONICLES (II,957)

SALEIDO, MICHAEL A
DAVID CASSIDY—MAN UNDERCOVER (II,623)

SALEM, KARIO
THE 13TH DAY: THE STORY OF ESTHER (II,2593); CENTENNIAL (II,477); ONCE AN EAGLE (II,1897); TESTIMONY OF TWO MEN (II,2564); THROUGH THE MAGIC PYRAMID (II,2615)

SALERNO, CHARLENE
THE ADVENTURES OF OZZIE AND HARRIET (I,71)

SALES, CLIFFORD
BEULAH (I,416)

SALES, SOUPY
BARNEY AND ME (I,358); THE BEVERLY HILLBILLIES (I,417); CHAIN REACTION (II,479); THE HOOFER (I,2113); JUNIOR ALMOST ANYTHING GOES (II,1357); THE NEW SOUPY SALES SHOW (II,1829); THE SOUPY SALES SHOW (I,4137); THE SOUPY SALES SHOW (I,4138); TO TELL THE TRUTH (II,2626); WHERE THERE'S SMOKEY (I,4832)

SALESBERG, GARY
CAGNEY AND LACEY (II,410)

SALIK, RACHEL
LACE (II,1418)

SALISBURY, COLGATE
ANOTHER WORLD (II,97)

SALISBURY, LYNN
LIGHT'S OUT (I,2699)

SALKILLD, GORDON
THE DOUBLE KILL (I,1360)

SALLIS, PETER
LAST OF THE SUMMER WINE (II,1439)

SALMI, ALBERT
79 PARK AVENUE (II,2310); DANIEL BOONE (I,1142); THE GOLDEN AGE OF TELEVISION (II,1019); THE KAISER ALUMINUM HOUR (I,2503); NIGHT GAMES (II,1845); ONCE AN EAGLE (II,1897); ONE STEP BEYOND (I,3388); PETROCELLI (II,2031); PROFILES IN COURAGE (I,3678); THOU SHALT NOT KILL (II,2604); THE TWILIGHT ZONE (I,4651)

SALSBERG, GERRY
HIGH HOPES (II,1144)

SALT, JENNIFER
OLD FRIENDS (II,1881); SOAP (II,2392)

SALUGA, BILL
GOING BANANAS (II,1015); THE JERK, TOO (II,1282); TOP TEN (II,2648)

SAM
B.J. AND THE BEAR (II,125); B.J. AND THE BEAR (II,126); B.J. AND THE BEAR (II,127); SAM (II,2247)

SAMMS, EMMA
DYNASTY (II,746); ELLIS ISLAND (II,768); GENERAL HOSPITAL (II,964); GOLIATH AWAITS (II,1025); MORE WILD WILD WEST (II,1733)

SAMPLER, PHILECE
DAYS OF OUR LIVES (II,629); RITUALS (II,2182)

SAMPSON, ROBERT
BRADDOCK (I,704); BRIDGET LOVES BERNIE (I,722); JACQUELINE SUSANN'S VALLEY OF THE DOLLS, 1981 (II,1268); THE JERK, TOO (II,1282); KINGSTON: THE POWER PLAY (II,1399); WEDNESDAY NIGHT OUT (II,2760)

SAMPSON, WILL
BORN TO THE WIND (II,354); FROM HERE TO ETERNITY (II,932); FROM HERE TO ETERNITY (II,933); VEGAS (II,2723); VEGAS (II,2724); THE YELLOW ROSE (II,2847)

SAMUELS, JIMMY
DID YOU HEAR ABOUT JOSH AND KELLY?! (II,673); POTTSVILLE (II,2072)

SAN JUAN, GUILLERMO
BORN TO THE WIND (II,354)

SAN MARCO, ROSANNA
THE GOLDEN AGE OF TELEVISION (II,1016)

SAND, PAUL
FRIENDS AND LOVERS (II,930); LADY LUCK (I,2604); PAUL SAND IN FRIENDS AND LOVERS (II,1982); ST. ELSEWHERE (II,2432); YOU CAN'T TAKE IT WITH YOU (II,2857)

SAND, REBECCA
STAR TONIGHT (I,4199)

SANDE, WALTER
THE FARMER'S DAUGHTER (I,1533); PUBLIC PROSECUTOR (I,3689); SIMON LASH (I,4052); TUGBOAT ANNIE (I,4622); WATERFRONT (I,4774); THE WILD WILD WEST (I,4863)

SANDER, IAN
EGAN (I,1419)

SANDERS, BYRON
LOVE OF LIFE (I,2774)

SANDERS, BEVERLY
FREEMAN (II,923); HICKEY VS. ANYBODY (II,1142); ONE DAY AT A TIME (II,1900); SPARROW (II,2410); TWO GUYS FROM MUCK (II,2690)

SANDERS, BRAD
MICKEY SPILLANE'S MIKE HAMMER: MORE THAN MURDER (II,1693)

SANDERS, BYRON
CONCERNING MISS MARLOWE (I,1032); THE DOCTORS (II,694); SEARCH FOR TOMORROW (II,2284)

SANDERS, DAVID
FROM THESE ROOTS (I,1688)

SANDERS, GEORGE
BATMAN (I,366); THE GEORGE SANDERS MYSTERY THEATER (I,1776); LAURA (I,2635); LAURA (I,2636); WARNING SHOT (I,4770)

SANDERS, HENRY
CONCRETE BEAT (II,562)

SANDERS, HUGH
LIFE WITH VIRGINIA (I,2695); THE LONELY WIZARD (I,2743); MY FRIEND FLICKA (I,3190)

SANDERS, JAY O
AFTERMASH (II,23)

SANDERS, KELLY
HAPPY DAYS (II,1084)

SANDERS, LEW
CHIPS (II,511)

SANDERS, LUGENE
THE LIFE OF RILEY (I,2686); MEET CORLISS ARCHER (I,2979)

SANDERS, RICHARD
THE INVISIBLE WOMAN (II,1244); SPENCER (II,2420); TROUBLE IN HIGH TIMBER COUNTRY (II,2661); WKRP IN CINCINNATI (II,2814)

SANDERS, SANDY
THE WESTERNER (I,4800)

SANDERS, STACIE
CHIEFS (II,509)

SANDERSON, WILLIAM
NEWHART (II,1835)

SANDLER, BOBBY
ON THE ROCKS (II,1894)

SANDLER, TONY
KRAFT MUSIC HALL PRESENTS SANDLER AND YOUNG (I,2585)

SANDOR, ALFRED
OUR FIVE DAUGHTERS (I,3413)

SANDOR, STEVE
AMY PRENTISS (II,74); THE YELLOW ROSE (II,2847)

SANDS, BILLY
BIG EDDIE (II,235); BIG EDDIE (II,236); MCHALE'S NAVY (I,2969); THE MUNSTERS' REVENGE (II,1765); THE PHIL SILVERS SHOW (I,3580); WIVES (II,2812)

SANDS, DIANA
THE LIVING END (I,2731); THE OUTER LIMITS (I,3426); TWO'S COMPANY (I,4659)

SANDS, DOROTHY
MY FAIR LADY (I,3187); ROAD OF LIFE (I,3798); THE TRAP (I,4593)

SANDS, LESLIE
THE SEVEN DIALS MYSTERY (II,2308)

SANDY, GARY
ALL THAT GLITTERS (II,42); AS THE WORLD TURNS (II,110); BATTLE OF THE NETWORK STARS (II,168); BATTLE OF THE NETWORK STARS (II,170); FOR LOVERS ONLY (II,904); SOMERSET (I,4115); WKRP IN CINCINNATI (II,2814)

SANFORD, ISABEL
ALL IN THE FAMILY (II,38); CIRCUS OF THE STARS (II,536); THE JEFFERSONS (II,1276); THE LOVE BOAT (II,1535); THE SHAPE OF THINGS (II,2315)

SANFORD, JUDY
GOLDEN CHILD (I,1839)

SANFORD, MARJORIE M
THE CITY (I,969)

SANFORD, RALPH
THE LIFE AND LEGEND OF WYATT EARP (I,2681)

SANFORD, WILLIAM
THE ALVIN SHOW (I,146)

SANIN, JAY
FOLLOW THE SUN (I,1617)

SANO, ATSUKO
SHOGUN (II,2339)

SANSBERRY, HOPE
THE PHIL SILVERS SHOW (I,3580)

SANSOM, KEN
CHARLIE'S ANGELS (II,487); THE INVISIBLE WOMAN (II,1244)

SANTANA, ARNALDO
A.K.A. PABLO (II,28); RAGE OF ANGELS (II,2107)

SANTANGELO, MELODY
MR. INSIDE/MR. OUTSIDE (II,1750)

SANTIAGO, MILLIE
THE NEW VOICE (II,1833)

SANTIAGO, SAUNDRA
MIAMI VICE (II,1689)

SANTON, PENNY
AFTERMASH (II,23); B.A.D. CATS (II,123); DON'T CALL ME CHARLIE (I,1349); MATT HOUSTON (II,1654); MOONLIGHT (II,1730); PHYLLIS (II,2038); ROLL OUT! (I,3833); ROSETTI AND RYAN (II,2216); ROSETTI AND RYAN: MEN WHO LOVE WOMEN (II,2217); A STORM IN SUMMER (I,4237)

SANTONI, RENI
LOVE AT FIRST SIGHT (II,1531); MANIMAL (II,1622); OWEN MARSHALL: COUNSELOR AT LAW (I,3435); POWDERKEG (I,3657)

SANTORO, DEAN
AS THE WORLD TURNS (II,110); CONCRETE BEAT (II,562); THE MAN FROM ATLANTIS (II,1614); OVER AND OUT (II,1938)

SANTOS, JOE
A.K.A. PABLO (II,28); THE HUSTLER OF MUSCLE BEACH (II,1210); ME AND MAXX (II,1672); NIGHTSIDE (I,3296); THE ROCKFORD FILES (II,2196); THE ROCKFORD FILES (II,2197)

SANTOYO, JORGE
FALCON'S GOLD (II,809)

SAPIEN, BUZZ
CASS MALLOY (II,454)

SAPOLSKY, ROBERT
AMAHL AND THE NIGHT VISITORS (I,150)

SAPRU, TAJ
THE FAR PAVILIONS (II,832)

SARANDON, CHRIS
BLISS (II,269)

SARANDON, SUSAN
A WORLD APART (I,4910)

SARANTOS, JOHN
MR. MOON'S MAGIC CIRCUS (II,1753)

SARDOFF, FRED
BRACKEN'S WORLD (I,703); DEATH RAY 2000 (II,654)

SARGENT, ANNE
THE FIRST HUNDRED YEARS (I,1581); MY SON JEEP (I,3202)

SARGENT, BILL
THE SEEKERS (I,3973)

SARGENT, DICK
BEWITCHED (I,418); BRILLIANT BENJAMIN BOGGS (I,730); BROADSIDE (I,732); THE DUKES OF HAZZARD (II,742); FANTASY ISLAND (II,830); FOOLS, FEMALES AND FUN: WHAT ABOUT THAT ONE? (II,901); THE GOSSIP COLUMNIST (II,1045); HAZEL (I,1982); MELVIN PURVIS: G-MAN (II,1679); NOT UNTIL TODAY (II,1865); ONE HAPPY FAMILY (I,3375); THE POWER WITHIN (II,2073); THE TAMMY GRIMES SHOW (I,4358)

SARGENT, JOSEPH
GETAWAY CAR (I,1783)

SARGENT, WILLIAM
THE IMMORTAL (I,2196)

SARIO, IVAN
THE LAST NINJA (II,1438)

SARIS, MARILYN
THE ADVENTURES OF JIM BOWIE (I,63)

SAROYAN, HANK
TROLLKINS (II,2660)

SAROYAN, LUCY
BRAVO TWO (II,368); EIGHT IS ENOUGH (II,762)

SARRACINO, ERNEST
CONDOMINIUM (II,566); IN TANDEM (II,1228); JOHNNY COME LATELY (I,2425); SECRET AGENT (I,3961); THE SWORD (I,4325)

SARRADET, RICHARD
GENERAL HOSPITAL (II,964)

SARRAZIN, MICHAEL
BEULAH LAND (II,226)

SARSER, PIP
AS THE WORLD TURNS (II,110)

SARTAIN, GAILARD
THE BANANA COMPANY (II,139); BULBA (II,392); THE HEE HAW HONEYS (II,1125)

SARTORI, MARY LOU
THE MOUSEKETEERS REUNION (II,1742)

SASSOON, VIDAL
MAURICE WOODRUFF PREDICTS (I,2956); YOUR NEW DAY (II,2873)

SATOW, MARK
THE BAT (I,364)

SATTIN, LONNIE
CIRCLE (I,957)

SATTLES, BARRY
SEARCH FOR TOMORROW (II,2284)

SAULSBERRY, RODNEY
CAPITOL (II,426)

SAUNDERS, BEVERLY
HOLLYWOOD HIGH (II,1163)

SAUNDERS, DAVID
HARD KNOX (II,1087)

SAUNDERS, GLORIA
COUNTERPOINT (I,1059); FRONTIER (I,1695); MYSTERIES OF CHINATOWN (I,3209); SCHLITZ PLAYHOUSE OF STARS (I,3936); STARS OVER HOLLYWOOD (I,4211); TERRY AND THE PIRATES (I,4405)

SAUNDERS, J JAY
DYNASTY (II,747); GOOD TIMES (II,1038); LITTLE VIC (II,1496); SALVAGE (II,2244); SALVAGE 1 (II,2245)

SAUNDERS, JUSTINE
AGAINST THE WIND (II,24)

SAUNDERS, LANNA
DAYS OF OUR LIVES (II,629)

SAUNDERS, LEE
THE GREATEST GIFT (I,1880)

SAUNDERS, LEW
240-ROBERT (II,2688)

SAUNDERS, LORI
DUSTY'S TRAIL (I,1390); OH, NURSE! (I,3343); PETTICOAT JUNCTION (I,3571); THE YOUNG AND THE RESTLESS (II,2862)

SAUNDERS, MARY JANE
THE TALES OF WELLS FARGO (I,4353)

SAUNDERS, NICHOLAS
MARTIN KANE, PRIVATE EYE (I,2928); THE PHIL SILVERS SHOW (I,3580)

SAVAGE, ANN
FIRESIDE THEATER (I,1580)

SAVAGE, BRAD
ANOTHER APRIL (II,93); BOB & CAROL & TED & ALICE (I,495); CHRISTMAS IN DISNEYLAND (II,519); GRANDPA MAX (II,1051); SECOND CHANCE (I,3958); THE SECOND TIME AROUND (II,2289); THE TONY RANDALL SHOW (II,2640)

SAVAGE, DENNIS
LITTLE LORD FAUNTLEROY (II,1492)

SAVAGE, ELIZABETH
13 THIRTEENTH AVENUE (II,2592); THE RENEGADES (II,2132)

SAVAGE, JOHN
ERIC (I,1451); GIBBSVILLE (II,988); THE TURNING POINT OF JIM MALLOY (II,2670)

SAVAGE, TRACIE
FLATFOOTS (II,879); LITTLE HOUSE ON THE PRAIRIE (II,1487)

SAVALAS, GEORGE
KOJAK (II,1406)

SAVALAS, TELLY
ACAPULCO (I,15); ALCOA PREMIERE (I,109); BATTLE OF THE NETWORK STARS

(II,163); BATTLE OF THE NETWORK STARS (II,164); BATTLE OF THE NETWORK STARS (II,165); BOB HOPE SPECIAL: BOB HOPE IN "JOYS" (II,284); THE CAT AND THE CANARY (I,868); CBS: ON THE AIR (II,460); CIRCUS OF THE STARS (II,531); THE DICK POWELL SHOW (I,1269); DINAH IN SEARCH OF THE IDEAL MAN (I,1278); THE FRENCH ATLANTIC AFFAIR (II,924); HELLINGER'S LAW (II,1127); KOJAK (II,1406); KRAFT SUSPENSE THEATER (I,2591); SECOND LOOK (I,3960); TELLY. . . WHO LOVES YA, BABY? (II,2558); WINDOWS, DOORS AND KEYHOLES (II,2806)

SAVIDGE, JENNIFER
MR. SUCCESS (II,1756); ST. ELSEWHERE (II,2432)

SAVILE, DAVID
DOCTOR WHO—THE FIVE DOCTORS (II,690)

SAVIOR, PAUL
GENERAL HOSPITAL (II,964)

SAVO, JIMMY
THROUGH THE CRYSTAL BALL (I,4494)

SAVOIA, JOHN
SPRAGGUE (II,2430)

SAWAYA, GEORGE
THE BEST OF FRIENDS (II,216)

SAWYER, CONNIE
ALLAN (I,142); DOC ELLIOT (I,1305); EVIL ROY SLADE (I,1485); JUST OUR LUCK (II,1360); LEGMEN (II,1458); MCGHEE (I,2967)

SAWYER, HAL
SAWYER VIEWS HOLLYWOOD (I,3927)

SAWYER, JOE
THE ADVENTURES OF RIN TIN TIN (I,73); MAISIE (I,2833)

SAWYER, MARK
NEVER AGAIN (II,1808)

SAXON, JOHN
79 PARK AVENUE (II,2310); CROSSFIRE (II,606); THE DOCTORS (I,1323); DYNASTY (II,746); GOLDEN GATE (II,1020); ONCE AN EAGLE (II,1897); PLANET EARTH (I,3615); PRISONERS OF THE LOST UNIVERSE (II,2086); ROOSTER (II,2210); SAVAGE: IN THE ORIENT (II,2264); SCARECROW AND MRS. KING (II,2268)

SAYER, DIANE
HERE COME THE BRIDES (I,2024)

SAYERS, DORA
CLAUDIA: THE STORY OF A MARRIAGE (I,972)

SAYLOR, KATIE
THE FANTASTIC JOURNEY (II,826); MEN OF THE DRAGON (II,1683)

SCALA, GIA
ALFRED HITCHCOCK PRESENTS (I,115)

SCALES, PRUNELLA
FAWLTY TOWERS (II,843)

SCALIA, JACK
THE DEVLIN CONNECTION (II,664); HIGH PERFORMANCE (II,1145)

SCALLON, BRENDA
RETURN TO EDEN (II,2143)

SCANLAN, MARTIN
THE AWAKENING LAND (II,122)

SCANLON, JOHN G
HEAR NO EVIL (II,1116)

SCANNELL, KEVIN
THE CRACKER BROTHERS (II,599); HARPER VALLEY PTA (II,1095)

SCANNELL, SUSAN
DYNASTY (II,746); SEARCH FOR TOMORROW (II,2284)

SCARBER, SAM
CONCRETE BEAT (II,562); MR. MOM (II,1752); THE OUTLAWS (II,1936)

SCARBURY, JOEY
THE GREATEST AMERICAN HERO (II,1060)

SCARDINO, DON
AS THE WORLD TURNS (II,110); THE FLIM-FLAM MAN (I,1599); THE GUIDING LIGHT (II,1064); HEREAFTER (II,1138); THE PEOPLE NEXT DOOR (I,3519); THE RUBBER GUN SQUAD (II,2225); RYAN'S HOPE (II,2234); SIDE BY SIDE (II,2347)

SCARPELLI, GLENN
FANTASY (II,828); FISHERMAN'S WHARF (II,865); JENNIFER SLEPT HERE (II,1278); ONE DAY AT A TIME (II,1900); RIVKIN: BOUNTY HUNTER (II,2183)

SCARWID, DIANA
THE POSSESSED (II,2071); STUDS LONIGAN (II,2484); THOU SHALT NOT KILL (II,2604)

SCHAAL, RICHARD
AFTER GEORGE (II,19); BACHELOR AT LAW (I,322); IT'S A LIVING (II,1254); JUST OUR LUCK (II,1360); THE MARY TYLER MOORE SHOW (II,1640); PHYLLIS (II,2038); PLEASE STAND BY (II,2057)

SCHAAL, WENDY
AFTERMASH (II,23); FANTASY ISLAND (II,829); IT'S A LIVING (II,1254); THE LIFE AND TIMES OF EDDIE ROBERTS (II,1468)

SCHAAP, DICK
THE JOE NAMATH SHOW (I,2400)

SCHACHT, SAM
THE FOUR OF US (II,914)

SCHACHTER, FELICE
THE FACTS OF LIFE (II,805)

SCHACTER, SIMONE
AS THE WORLD TURNS (II,110)

SCHAEFER, GEORGE
THE GOLDEN AGE OF TELEVISION (II,1017)

SCHAEFFER, LYDIA
LEAVE IT TO LARRY (I,2649)

SCHAFER, NATALIE
ABC STAGE '67 (I,6); THE CHEVROLET TELE-THEATER (I,926); GILLIGAN'S ISLAND (II,990); GILLIGAN'S ISLAND: RESCUE FROM GILLIGAN'S ISLAND (II,991); GILLIGAN'S ISLAND: THE CASTAWAYS ON GILLIGAN'S ISLAND (II,992); GILLIGAN'S ISLAND: THE HARLEM GLOBETROTTERS ON GILLIGAN'S ISLAND (II,993); GILLIGAN'S PLANET (II,994); GUESTWARD HO! (I,1900); LADIES IN BLUE (II,1422); THE NEW ADVENTURES OF GILLIGAN (II,1813); THE PETRIFIED FOREST (I,3570)

SCHAFF, LILLIAN
ONE MAN'S FAMILY (I,3383)

SCHAFFER, GREGORY MARC
ONE LIFE TO LIVE (II,1907)

SCHALLERT, WILLIAM
THE ADVENTURES OF JIM BOWIE (I,63); ARROYO (I,266); THE BARRETTS OF WIMPOLE STREET (I,361); THE BOYS (I,701); COMMANDO CODY (I,1030); DOBIE GILLIS (I,1302); THE DUCK FACTORY (II,738); ESCAPE (I,1460); GET SMART (II,983); GIDGET'S SUMMER REUNION (I,1799); HERE COME THE BRIDES (I,2024); IKE (II,1223); LEAVE IT TO BEAVER (I,2648); LITTLE WOMEN (II,1497); LITTLE WOMEN (II,1498); LOBO (II,1504); THE MAN FROM DENVER (I,2863); MAN ON A STRING (I,2877); THE MISADVENTURES OF SHERIFF LOBO (II,1704); MR. BEVIS (I,3128); THE NANCY DREW MYSTERIES (II,1789); THE PATTY DUKE SHOW (I,3483); TELEPHONE TIME (I,4379); THE WALTONS (II,2740); THE WILD WILD WEST (I,4863)

SCHANKMAN, DANA
THE YOUNG AND THE RESTLESS (II,2862)

SCHANKMAN, LAUREN
THE YOUNG AND THE RESTLESS (II,2862)

SCHANLEY, TOM
THE YELLOW ROSE (II,2847)

SCHANNEL, SUE
ANOTHER LIFE (II,95)

SCHATZBERG, STEVE
THE GARY COLEMAN SHOW (II,958); THE SHIRT TALES (II,2337)

SCHECTMAN, ALBERT
HOW TO SURVIVE A MARRIAGE (II,1198)

SCHEDEEN, ANNE
ALMOST HEAVEN (II,57); E/R (II,748); EXO-MAN (II,797); MARCUS WELBY, M.D. (II,1627); NEVER SAY NEVER (II,1809); PAPER DOLLS (II,1952); THREE'S COMPANY (II,2614)

SCHEIDER, ROY
ASSIGNMENT: MUNICH (I,302); HIDDEN FACES (I,2044); LOVE OF LIFE (I,2774); SEARCH FOR TOMORROW (II,2284); THE SECRET STORM (I,3969)

SCHEIMER, ERICA
THE KID SUPER POWER HOUR WITH SHAZAM (II,1386); LASSIE'S RESCUE RANGERS (II,1432); MISSION MAGIC (I,3068); TEACHER'S PET (I,4367)

SCHEIMER, LANE
LASSIE'S RESCUE RANGERS (II,1432); MISSION MAGIC (I,3068); MY FAVORITE MARTIANS (II,1779); SPORT BILLY (II,2429)

SCHELER, DAMION
TEXAS (II,2566)

SCHELL, CATHERINE
THE ADVENTURER (I,43); LOOK BACK IN DARKNESS (I,2752); MADAME SIN (I,2805); SEARCH FOR THE NILE (I,3954); SPACE: 1999 (II,2409)

SCHELL, MARIA
THE MARTIAN CHRONICLES (II,1635); NINOTCHKA (I,3299); WHO HAS SEEN THE WIND? (I,4847)

SCHELL, MAXIMILIAN
THE THREE MUSKETEERS (I,4481)

SCHELL, RONNIE
BATTLE OF THE PLANETS (II,179); FOREVER FERNWOOD (II,909); GOMER PYLE, U.S.M.C. (I,1843); GOOBER AND THE GHOST CHASERS (II,1028); GOOD MORNING WORLD (I,1850); THE SHIRT TALES (II,2337); THAT GIRL (II,2570)

SCHENKKAN, ROBERT
GEORGE WASHINGTON (II,978); SANCTUARY OF FEAR (II,2252)

SCHEPPS, SHAWN
THE TOM SWIFT AND LINDA CRAIG MYSTERY HOUR (II,2633)

SCHERECK, VICKI
BIG BEND COUNTRY (II,230)

SCHERER, GENE
FROM HERE TO ETERNITY (II,932)

SCHERER, GLENN
FEEL THE HEAT (II,848)

SCHIAVELLI, VINCENT
COMEDY OF HORRORS (II,557); THE CORNER BAR (I,1051); GILLIGAN'S ISLAND: RESCUE FROM GILLIGAN'S ISLAND (II,991); LITTLE SHOTS (II,1495); TAXI (II,2546)

SCHIEGL, KURT
STRANGE PARADISE (I,4246)

SCHIFF, ERIC
RISE AND SHINE (II,2179)

SCHILDKRAUT, JOSEPH
COSMOPOLITAN THEATER (I,1057); FAITH BALDWIN'S THEATER OF ROMANCE (I,1515); HAMLET (I,1924); JOSEPH SCHILDKRAUT PRESENTS (I,2449)

SCHILLING, WILLIAM
THE BEST OF TIMES (II,220); E/R (II,748)

SCHIMMEL, JOHN
PUMPBOYS AND DINETTES ON TELEVISION (II,2090)

SCHMEED TRIO, THE
HEIDI (I,2004)

SCHMIDT, GEORGIA
ADAMS HOUSE (II,9); DOMINIC'S DREAM (II,704); FLO (II,891); TERROR AT ALCATRAZ (II,2563)

SCHMIDTKE, NED
TEXAS (II,2566)

SCHMITTER, MORGAN
CHARGE ACCOUNT (I,903)

SCHMOCK, JONATHAN
DOUBLE TROUBLE (II,724); T.L.C. (II,2525)

SCHNABEL, STEFAN
CAPTAIN VIDEO AND HIS VIDEO RANGERS (I,838); THE GUIDING LIGHT (II,1064); MR. INSIDE/MR. OUTSIDE (II,1750); OF THIS TIME, OF THAT PLACE (I,3335); TALES OF THE VIKINGS (I,4351)

SCHNEIDER, CHRISTINE
TESTIMONY OF TWO MEN (II,2564)

SCHNEIDER, JOHN
CIRCUS OF THE STARS (II,533); THE DUKES (II,741); THE DUKES OF HAZZARD (II,742); JOHN SCHNEIDER'S CHRISTMAS HOLIDAY (II,1320); STAR SEARCH (II,2437)

SCHNEIDER, MARK
THE NIGHTENGALES (II,1850); PAPER DOLLS (II,1952)

SCHNEIDER, TAWNY
EYE ON HOLLYWOOD (II,798); THE LOVE REPORT (II,1542); THE LOVE REPORT (II,1543)

SCHNETZER, STEPHEN
ANOTHER WORLD (II,97); DAYS OF OUR LIVES (II,629); ONE LIFE TO LIVE (II,1907); RAGE OF ANGELS (II,2107)

SCHODITSCH, PEGGY
A CHRISTMAS CAROL (II,517)

SCHOELEN, JILL
THE BEST OF TIMES (II,219); GREAT DAY (II,1054)

SCHOENBERG, JEREMY
IS THIS GOODBYE, CHARLIE BROWN? (I,2242); SNOOPY'S GETTING MARRIED, CHARLIE BROWN (I,4102); WHAT HAVE WE LEARNED, CHARLIE BROWN? (I,4810)

SCHOEPPERLE, JOHN
ROOTIE KAZOOTIE (I,3844)

SCHOFIELD, CHRISTINE
TYCOON: THE STORY OF A WOMAN (II,2697)

SCHOFIELD, FRANK
OUR FIVE DAUGHTERS (I,3413); THE PATRIOTS (I,3477); A TIME FOR US (I,4507)

SCHOFIELD, LESLIE
THE EYES HAVE IT (I,1496)

SCHOFIELD, LESTER
MASTER OF THE GAME (II,1647)

SCHOFIELD, NELL
1915 (II,1853)

SCHOFIELD, PETER
DIAL A DEADLY NUMBER (I,1253)

SCHOLL, DANNY
BLOOMER GIRL (I,488); LUXURY LINER (I,2796)

SCHOLZ, WILL
THE CHOCOLATE SOLDIER (I,944)

SCHORR, MITCHELL
THE GANGSTER CHRONICLES (II,957)

SCHRAMM, DAVID
KENNEDY (II,1377)

SCHRECK, VICKI
ADDIE AND THE KING OF HEARTS (I,35); THE EASTER PROMISE (I,1397); HOW THE WEST WAS WON (II,1196); THE MACAHANS (II,1583)

SCHREFFLER, MARILYN
THE ALL-NEW POPEYE HOUR (II,49); BUFORD AND THE GHOST (II,389); CAPTAIN CAVEMAN AND THE TEEN ANGELS (II,433); FRED AND BARNEY MEET THE THING (II,919); THE HEATHCLIFF AND MARMADUKE SHOW (II,1118); SCOOBY'S ALL-STAR LAFF-A-LYMPICS (II,2273); THUNDARR THE BARBARIAN (II,2616); YOGI'S SPACE RACE (II,2853)

SCHREIBER, AVERY
THE BURNS AND SCHREIBER COMEDY HOUR (I,768); THE BURNS AND SCHREIBER COMEDY HOUR (I,769); ESCAPE (I,1460); FLATBED ANNIE AND SWEETIEPIE: LADY TRUCKERS (II,876); MASQUERADE (I,2940); MORE WILD WILD WEST (II,1733); MY MOTHER THE CAR (I,3197); OPERATION GREASEPAINT (I,3402); OUR PLACE (I,3416); SECOND CHANCE (I,3958); THAT WAS THE YEAR THAT WAS (I,4424); THIS WILL BE THE YEAR THAT WILL BE (I,4466)

SCHREIBMAN, PHILIP
SEEING THINGS (II,2297)

SCHREINER, BILL
GENERAL HOSPITAL (II,964)

SCHRODER, RICKY
BATTLE OF THE NETWORK STARS (II,175); BOB HOPE SPECIAL: BOB HOPE'S STAR-STUDDED SPOOF OF THE NEW TV SEASON— G RATED—WITH GLAMOUR, GLITTER & GAGS (II,317); CIRCUS OF THE STARS (II,536); CIRCUS OF THE STARS (II,538); KRAFT SALUTES WALT DISNEY WORLD'S 10TH ANNIVERSARY (II,1410); SILVER SPOONS (II,2355)

SCHRUM, ED
GIMME A BREAK (II,995)

SCHRUM, PETER
THE DOOLEY BROTHERS (II,715); THE FRENCH ATLANTIC AFFAIR (II,924); THE JERK, TOO (II,1282); NATIONAL LAMPOON'S TWO REELERS (II,1796)

SCHUBB, MARK
CONDO (II,565); THE SHERIFF AND THE ASTRONAUT (II,2326)

SCHUBERT, ARLUND
DAYS OF OUR LIVES (II,629)

SCHUCK, CHARLES
MCMILLAN (II,1666)

SCHUCK, JOHN
BATTLE OF THE NETWORK STARS (II,163); THE HALLOWEEN THAT ALMOST WASN'T (II,1075); HOLMES AND YOYO (II,1169); MCMILLAN AND WIFE (II,1667); THE NEW ODD COUPLE (II,1825); ONCE UPON A DEAD MAN (I,3369); ROOTS (II,2211); TURNABOUT (II,2669)

SCHUCK, PETER
OPERATION PETTICOAT (II,1917)

SCHULL, CHARLES
ABC'S MATINEE TODAY (I,7); SCRUPLES (II,2281)

SCHULLER, FRANK

THE AFRICAN QUEEN (II,18); LONE STAR (II,1513)

SCHULMAN, BILLY
TEACHER, TEACHER (I,4366)

SCHULMAN, JOSH
FATHER MURPHY (II,841)

SCHULTZ, DWIGHT
THE A-TEAM (II,119); SHERLOCK HOLMES (II,2327)

SCHULTZ, JACQUELINE
AS THE WORLD TURNS (II,110)

SCHULTZ, KEITH
THE MONROES (I,3089)

SCHULTZ, KEVIN
THE MONROES (I,3089); THE NEW ADVENTURES OF HUCKLEBERRY FINN (I,3250)

SCHULTZ, LENNIE
BALL FOUR (II,137); THE PINK PANTHER (II,2042)

SCHULTZ, TONY
RYAN'S HOPE (II,2234)

SCHUMANN, ROY
AS THE WORLD TURNS (II,110)

SCHURANBERD, DIANE
I'M SOOO UGLY (II,1219)

SCHUTZMAN, SCOTT
I DO, I DON'T (II,1212); RISE AND SHINE (II,2179)

SCHUYLER, ANN
IT'S A MAN'S WORLD (I,2258)

SCHWARTZ, BRUCE D
BLOCKHEADS (II,270)

SCHWARTZ, NEIL J
HAPPY DAYS (II,1084); STEAMBATH (II,2451)

SCHWARTZ, SAM
THE GREAT WALTZ (I,1878)

SCHWARTZ, STEPHEN M
PLEASE STAND BY (II,2057)

SCHWARZENEGGER, ARNOLD
HAPPY ANNIVERSARY AND GOODBYE (II,1082)

SCHWEID, CAROLE
FITZ AND BONES (II,867); TERROR AT ALCATRAZ (II,2563)

SCOFIELD, DINO
ARCHIE BUNKER'S PLACE (II,105)

SCOFIELD, FRANK
SOMERSET (I,4115)

SCOGGINS, JERRY
THE SINGING COWBOYS RIDE AGAIN (II,2363)

SCOGGINS, TRACY
BIG JOHN (II,239); CIRCUS OF THE STARS (II,537); HAWAIIAN HEAT (II,1111); THE RENEGADES (II,2133)

SCOLARI, PETER
BABY MAKES FIVE (II,129); BOSOM BUDDIES (II,355); CIRCUS OF THE STARS (II,536); CIRCUS OF THE STARS (II,537); CIRCUS OF THE STARS (II,538); THE FURTHER ADVENTURES OF WALLY BROWN (II,944); GOODTIME GIRLS (II,1042); NEWHART (II,1835)

SCOLLAY, FRED J
ANOTHER WORLD (II,97); THE EDGE OF NIGHT (II,760); NOTORIOUS (I,3314); YOUNG DR. MALONE (I,4938)

SCOOLER, ZVEE
INTERMEZZO (I,2223)

SCOTT, BILL
ADVENTURES OF HOPPITY HOPPER FROM FOGGY BOGG (I,62); THE BULLWINKLE SHOW (I,761); THE DUDLEY DO-RIGHT SHOW (I,1378); GEORGE OF THE JUNGLE (I,1773); ROCKY AND HIS FRIENDS (I,3822)

SCOTT, BONNIE
THAT GIRL (II,2570); YOU CAN'T DO THAT ON TELEVISION (I,4933)

SCOTT, BRENDA
HAZEL (I,1982); MR. NOVAK (I,3145); SWEET, SWEET RACHEL (I,4313)

SCOTT, BRYAN
CAMP GRIZZLY (II,418); KIDD VIDEO (II,1388); PLEASE STAND BY (II,2057)

SCOTT, BYRON
SPIDER-WOMAN (II,2426)

SCOTT, DEBRALEE
ANGIE (II,80); CHAIN REACTION (II,479); FOREVER FERNWOOD (II,909); LISA, BRIGHT AND DARK (I,2711); LIVING IN PARADISE (II,1503); MARY HARTMAN, MARY HARTMAN (II,1638); SENIOR YEAR (II,2301); SONS AND DAUGHTERS (II,2404); WELCOME BACK, KOTTER (II,2761)

SCOTT, DEVON
THE TONY RANDALL SHOW (II,2640); WE'LL GET BY (II,2753); WE'LL GET BY (II,2754)

SCOTT, DONOVAN
SCARED SILLY (II,2269)

SCOTT, ERIC
THE WALTONS (II,2740); THE WALTONS: A DAY FOR THANKS ON WALTON'S MOUNTAIN (EPISODE 3) (II,2741); THE WALTONS: MOTHER'S DAY ON WALTON'S MOUNTAIN (EPISODE 2) (II,2742); THE WALTONS: WEDDING ON WALTON'S MOUNTAIN (EPISODE 1) (II,2743)

SCOTT, EVELYN
BACHELOR FATHER (I,323); PEYTON PLACE (I,3574); RETURN TO PEYTON PLACE (I,3779)

SCOTT, FARNHAM
GEORGE WASHINGTON (II,978)

SCOTT, FOUR
THE HOYT AXTON SHOW (II,1200)

SCOTT, FRED
CAPTAIN VIDEO AND HIS VIDEO RANGERS (I,838)

SCOTT, FREDRICK
FARAWAY HILL (I,1532)

SCOTT, GEOFFREY
BATTLE OF THE NETWORK STARS (II,177); CONCRETE COWBOYS (II,563); DOG AND CAT (II,695); DYNASTY (II,746); THE SECRET EMPIRE (II,2291)

SCOTT, GEORGE C
BEAUTY AND THE BEAST (I,382); THE BURNING COURT (I,766); EAST SIDE/WEST SIDE (I,1396); THE FIRST 50 YEARS (II,862); JANE EYRE (I,2339); MIRROR, MIRROR, OFF THE WALL (I,3059); THE PICTURE OF DORIAN GRAY (I,3591); PLAYHOUSE 90 (I,3623); THE POWER AND THE GLORY (I,3658); THE PRICE (I,3666); THE TROUBLE WITH PEOPLE (I,4611); WINTERSET (I,4882)

SCOTT, GORDON
HERCULES (I,2022); TARZAN AND THE TRAPPERS (I,4362)

SCOTT, HAZEL
THE HAZEL SCOTT SHOW (I,1984)

SCOTT, HENRY
THE NEW PHIL SILVERS SHOW (I,3269)

SCOTT, JACQUELINE
THE FUGITIVE (I,1701); GENERAL ELECTRIC TRUE (I,1754); THE KAISER ALUMINUM HOUR (I,2503); SALVAGE (II,2244); SECOND CHANCE (I,3957)

SCOTT, JACQUES
THE ANN SOTHERN SHOW (I,220)

SCOTT, JEAN BRUCE
AIRWOLF (II,27); DAYS OF OUR LIVES (II,629); MAGNUM P.I. (II,1601); ST. ELSEWHERE (II,2432); WISHMAN (II,2811)

SCOTT, JEFFREY
BATTLE OF THE NETWORK STARS (II,176)

SCOTT, JIM
THE BAY CITY AMUSEMENT COMPANY (II,185)

SCOTT, JOEY
LEAVE IT TO BEAVER (I,2648); MR. LUCKY (I,3141); NATIONAL VELVET (I,3231)

SCOTT, JUDSON
THE PHOENIX (II,2035); THE PHOENIX (II,2036); V: THE SERIES (II,2715); VELVET (II,2725)

SCOTT, KATHRYN LEIGH
BATTLE OF THE NETWORK STARS (II,169); BIG SHAMUS, LITTLE SHAMUS (II,242); BOSTON AND KILBRIDE (II,356); DARK SHADOWS (I,1157); THE GYPSY WARRIORS (II,1070); PHILIP MARLOWE, PRIVATE EYE (II,2033); THE RENEGADES (II,2133); THE SCOOBY-DOO AND SCRAPPY-DOO SHOW (II,2275)

SCOTT, KEN
BEN JERROD (I,396)

SCOTT, LARRY B
THE JERK, TOO (II,1282); KUDZU (II,1415); THE NEW VOICE (II,1833); ROLL OF THUNDER, HEAR MY CRY (II,2202)

SCOTT, LORENE
FARAWAY HILL (I,1532)

SCOTT, LUCIEN
THE BOB NEWHART SHOW (II,342)

SCOTT, MARILYN
THE SMOTHERS BROTHERS SHOW (I,4096)

SCOTT, MARTHA
BEN HECHT'S TALES OF THE BIG CITY (I,395); BEULAH LAND (II,226); THE BIONIC WOMAN (II,255); THE BOB NEWHART SHOW (II,342);

DALLAS (II,614); LIGHT'S OUT (I,2699); MARRIED: THE FIRST YEAR (II,1634); MODERN ROMANCES (I,3079); MY WIFE NEXT DOOR (II,1780); REVLON MIRROR THEATER (I,3871); SECRETS OF MIDLAND HEIGHTS (II,2296); TELLER OF TALES (I,4388); YOU'RE ONLY YOUNG TWICE (I,4970); THE WEB (I,4784); THE WORD (II,2833)

SCOTT, MARY
HAZEL (I,1982)

SCOTT, PIPPA
THE GENERAL ELECTRIC THEATER (I,1753); GUILTY OR NOT GUILTY (I,1903); JIGSAW JOHN (II,1288); THE JUNE ALLYSON SHOW (I,2488); KRAFT SUSPENSE THEATER (I,2591); MR. LUCKY (I,3141); MY SISTER HANK (I,3201); THRILLER (I,4492); THE VIRGINIAN (I,4738)

SCOTT, RUSS
THE ROY ROGERS SHOW (I,3859)

SCOTT, SAMANTHA
BEWITCHED (I,418)

SCOTT, SANDRA
MOMENT OF TRUTH (I,3084); OUR PRIVATE WORLD (I,3417); THE TROUBLE WITH TRACY (I,4613)

SCOTT, SIMON
THE BARBARY COAST (II,147); MARKHAM (I,2917); THE MOD SQUAD (I,3078); READY FOR THE PEOPLE (I,3739); THE RELUCTANT SPY (I,3763); THE RETURN OF THE MOD SQUAD (I,3776); THE SIX MILLION DOLLAR MAN (I,4067); THE STAR MAKER (I,4193); TRAPPER JOHN, M.D. (II,2654)

SCOTT, SYDNA
BRIGHT PROMISE (I,727); DANGER (I,1134); FOREIGN INTRIGUE: DATELINE EUROPE (I,1638)

SCOTT, TIMOTHY
THE CHEROKEE TRAIL (II,500); DOWN HOME (II,731); ROUGHNECKS (II,2219); WILD TIMES (II,2799)

SCOTT, TOM
BACKSTAIRS AT THE WHITE HOUSE (II,133); SAM (II,2246)

SCOTT, VERNON
GUEST SHOT (I,1899); TONIGHT! AMERICA AFTER DARK (I,4555)

SCOTT, ZACHARY
THE DUPONT SHOW OF THE WEEK (I,1388); G.E. SUMMER ORIGINALS (I,1747); HENRY FONDA PRESENTS THE STAR AND THE STORY (I,2016); JANE EYRE (I,2338); RHEINGOLD THEATER (I,3783)

SCOTTI, VITO
ANDY'S GANG (I,214); BAREFOOT IN THE PARK (I,355); CAMPO 44 (I,813); THE EYES OF TEXAS II (II,800); THE FLYING NUN (I,1611); GILLIGAN'S ISLAND (II,990); KNIGHT'S GAMBIT (I,2576); LIFE WITH LUIGI (I,2692); TO ROME WITH LOVE (I,4526); WHICH WAY'D THEY GO? (I,4838)

SCOULAR, ANGELA
PENMARRIC (II,1993)

SCOULAR, CHRISTOPHER
THE SEVEN DIALS MYSTERY (II,2308)

SCOURBY, ALEXANDER
THE KAISER ALUMINUM HOUR (I,2503); THE SECRET STORM (I,3969); STRANGE TRUE STORIES (II,2476); THE WEB (I,4785); THE WORLD OF JAMES BOND (I,4913); THE WORLD OF MAURICE CHEVALIER (I,4916)

SCRABER, SAM
W*A*L*T*E*R (II,2731)

SCRANTON, PETER
SKEEZER (II,2376)

SCRIBNER, JIMMY
SLEEPY JOE (I,4085)

SCRIBNER, RONNIE
GILLIGAN'S ISLAND: THE CASTAWAYS ON GILLIGAN'S ISLAND (II,992); THE LONG DAYS OF SUMMER (II,1514)

SCROGGINS, ROBERT
THE ADVENTURES OF SIR LANCELOT (I,76)

SCRUGGS, EARL C
THE BEVERLY HILLBILLIES (I,417); THE RETURN OF THE BEVERLY HILLBILLIES (II,2139)

SCRUGGS, HAROLD
SKYWARD CHRISTMAS (II,2378)

SCRUGGS, LANGHORNE
THE CORNER BAR (I,1051)

SCRUGGS, LESTER
FLATT AND SCRUGGS (I,1594)

SCRUGGS, LINDA
WHIZ KIDS (II,2790)

SCULLY, HELEN
FAMILY HONOR (I,1522)

SCULLY, SEAN
THE SCARECROW OF ROMNEY MARSH (I,3931)

SCULLY, TERRY
THE FORSYTE SAGA (I,1645)

SCULLY, VIN
CELEBRITY CHALLENGE OF THE SEXES 1 (II,465); IT TAKES TWO (I,2251); THE VIN SCULLY SHOW (I,4734)

SEAFORTH, SUSAN
COME A RUNNING (I,1013); THE LORETTA YOUNG THEATER (I,2756); THE YOUNG MARRIEDS (I,4942)

SEAGRAM, LISA
THE BEVERLY HILLBILLIES (I,417)

SEAGREN, BOB
CIRCUS OF THE STARS (II,532); CIRCUS OF THE STARS (II,533); CIRCUS OF THE STARS (II,534); SOAP (II,2392); STUNT SEVEN (II,2486); TONI'S BOYS (II,2637); VALENTINE MAGIC ON LOVE ISLAND (II,2716)

SEAGROVE, JENNY
THE WOMAN IN WHITE (II,2819); A WOMAN OF SUBSTANCE (II,2820)

SEALES, FRANKLYN
BEULAH LAND (II,226); HIGH FIVE (II,1143); HILL STREET BLUES (II,1154); SILVER SPOONS (II,2355)

SEARLE, JUDITH
CONCRETE BEAT (II,562)

SEARS, BILL
KID GLOVES (I,2531)

SEARS, HEATHER
THE INFORMER (I,2212)

SEAWELL, D C
LITTLE SHOTS (II,1495)

SEAY, JAMES
FURY (I,1721); THE LIFE AND LEGEND OF WYATT EARP (I,2681)

SEBASTIAN, JOHN B
GOOD VIBRATIONS FROM CENTRAL PARK (I,1855); THE JERK, TOO (II,1282); WELCOME BACK KOTTER (II,2761)

SECREST, JIM
ANOTHER WORLD (II,97)

SEDAKA, NEIL
NEIL SEDAKA STEPPIN' OUT (II,1804); THE ROOTS OF ROCK 'N' ROLL (II,2212)

SEDAN, ROLFE
THE ADDAMS FAMILY (II,11); THE GALE STORM SHOW (I,1725); THE GEORGE BURNS AND GRACIE ALLEN SHOW (I,1763)

SEDERHOLM, DAVID
RYAN'S HOPE (II,2234)

SEDGLEY, KAREN
SPARROW (II,2410)

SEDGWICK, KYRA
ANOTHER WORLD (II,97)

SEEAR, JETTE
THE LOVE BOAT I (II,1532)

SEEGER, SARA
DENNIS THE MENACE (I,1231); OCCASIONAL WIFE (I,3325); ROOM FOR ONE MORE (I,3842)

SEEL, CHARLES
DENNIS THE MENACE (I,1231); GUNSMOKE (II,1069); THE ROAD WEST (I,3801)

SEER, RICHARD
DELTA HOUSE (II,658)

SEFF, BRIAN
THE PEE WEE HERMAN SHOW (II,1988)

SEFF, RICHARD
RAGE OF ANGELS (II,2107)

SEFLINGER, CAROL ANNE
LITTLE VIC (II,1496); WONDERBUG (II,2830)

SEGAL, FRANCINE
HARDCASE (II,1088)

SEGAL, GEORGE
CIRCUS OF THE STARS (II,538); DEADLY GAME (II,632); DEATH OF A SALESMAN (I,1208); THE DESPERATE HOURS (I,1239); THE GEORGE SEGAL SHOW (II,977); OF MICE AND MEN (I,3333)

SEGAL, JONATHAN
EXO-MAN (II,797); THE PAPER CHASE (II,1949)

SEGAL, MICHAEL
FATHER DEAR FATHER (II,837); MAN ABOUT THE HOUSE (II,1611)

SEGAL, RICKY
THE PARTRIDGE FAMILY (II,1962)

SEGALL, PAMELA
E/R (II,748); THE FACTS OF LIFE (II,805)

SEIFERS, JOEY
YOUNG LIVES (II,2866)

SEIGEL, DONNA
ALL NIGHT RADIO (II,40)

SEIGEL, JANIS
THE MANHATTAN TRANSFER (II,1619)

SEIPOLD, MANFRED
THE LEGEND OF SILENT NIGHT (I,2655)

SEKKA, JOHNNY
THE AFRICAN QUEEN (II,18); MASTER OF THE GAME (II,1647)

SELBY, DAVID
DARK SHADOWS (I,1157); FALCON CREST (II,808); FLAMINGO ROAD (II,872); THE NIGHT RIDER (II,1848); WASHINGTON: BEHIND CLOSED DOORS (II,2744)

SELBY, SARAH
FATHER KNOWS BEST (II,838); THE GEORGE BURNS AND GRACIE ALLEN SHOW (I,1763); GUNSMOKE (II,1069); THE HALLS OF IVY (I,1923); THE HARDY BOYS AND THE MYSTERY OF GHOST FARM (I,1950); THE HARDY BOYS AND THE MYSTERY OF THE APPLEGATE TREASURE (I,1951); THE RIFLEMAN (I,3789)

SELL, JANIE
SIDE BY SIDE (II,2347); STARTING FRESH (II,2447); WIVES (II,2812)

SELLECCA, CONNIE
BEYOND WESTWORLD (II,227); CAPTAIN AMERICA (II,428); CELEBRITY CHALLENGE OF THE SEXES 5 (II,469); THE CELEBRITY FOOTBALL CLASSIC (II,474); CIRCUS OF THE STARS (II,534); FLYING HIGH (II,894); FLYING HIGH (II,895); THE GREATEST AMERICAN HERO (II,1060); HOTEL (II,1192)

SELLECK, TOM
BATTLE OF THE NETWORK STARS (II,171); BATTLE OF THE NETWORK STARS (II,172); BOB HOPE SPECIAL: BOB HOPE'S STAR-STUDDED SPOOF OF THE NEW TV SEASON— G RATED—WITH GLAMOUR, GLITTER & GAGS (II,317); BOSTON AND KILBRIDE (II,356); BUNCO (II,395); THE CONCRETE COWBOYS (II,564); THE GYPSY WARRIORS (II,1070); MAGNUM, P.I. (II,1601); MOST WANTED (II,1738); THE RETURN OF LUTHER GILLIS (II,2137); THE ROCKFORD FILES (II,2197); THE YOUNG AND THE RESTLESS (II,2862)

SELLERS, ELIZABETH
DOUGLAS FAIRBANKS JR PRESENTS THE RHEINGOLD THEATER (I,1364); ONE STEP BEYOND (I,3388)

SELLERS, PETER
CAROL FOR ANOTHER CHRISTMAS (I,852)

SELZER, MILTON
THE FARMER'S DAUGHTER (I,1534); THE HARVEY KORMAN SHOW (II,1103); THE HARVEY KORMAN SHOW (II,1104); KEEP THE FAITH (I,2518); MR. & MS. AND THE BANDSTAND MURDERS (II,1746); MR. & MS. AND THE MAGIC STUDIO MYSTERY (II,1747); MURDOCK'S GANG (I,3164); NEEDLES AND PINS (I,3246); RX FOR THE DEFENSE (I,3878); SAM HILL: WHO KILLED THE MYSTERIOUS MR. FOSTER? (I,3896)

SELZER, WILL
HIZZONER (II,1159); KAREN (II,1363); MARY HARTMAN, MARY HARTMAN (II,1638)

SEMON, MAXINE
THE LIFE OF RILEY (I,2685)

SEMPLER, PHILECA
FANTASY ISLAND (II,829)

SEN, EN HE
MARCO POLO (II,1626)

SEN YUNG, VICTOR
ALIAS MIKE HERCULES (I,117); BACHELOR FATHER (I,323); BONANZA (II,347)

SENECA, JOSE
O'MALLEY (II,1872)

SENNETT, SUSAN
ERNIE, MADGE, AND ARTIE (I,1457); LOCAL 306 (II,1505); OZZIE'S GIRLS (I,3439); OZZIE'S GIRLS (I,3440)

SENNO, HIROMI
SHOGUN (II,2339)

SERBER, ROBERT
OPPENHEIMER (II,1919)

SERGEVA, KATHERINE
GREAT CATHERINE (I,1873)

SERINUS, JASON
SHE'S A GOOD SKATE, CHARLIE BROWN (I,4012)

SERLING, ROD
THE LIAR'S CLUB (I,2673); NIGHT GALLERY (I,3287); NIGHT GALLERY (I,3288); THE TWILIGHT ZONE (I,4651)

SERNA, PEPE
THE 25TH MAN (MS) (II,2678); CAR WASH (II,437); FEEL THE HEAT (II,848); JOE DANCER: THE MONKEY MISSION (II,1301)

SERNAS, JACQUES
THE SWORD (I,4325)

SEROFF, MUNI
GIVE US BARABBAS! (I,1816)

SERRA, RAY
THE EDGE OF NIGHT (II,760)

SERRET, JOHN
MASTER OF THE GAME (II,1647)

SERRIS, IRENE
ONCE UPON A SPY (II,1898)

SERVER, ERIC
B.J. AND THE BEAR (II,126); BUCK ROGERS IN THE 25TH CENTURY (II,383); CENTENNIAL (II,477); GENERAL HOSPITAL (II,964); HAZARD'S PEOPLE (II,1112); TABITHA (II,2527)

SESSIONS, ELMIRA
ALWAYS APRIL (I,147)

SETHI, JONIE
THE FAR PAVILIONS (II,832)

SETON, BRUCE
THE ADVENTURES OF SIR LANCELOT (I,76); INSPECTOR FABIAN OF SCOTLAND YARD (I,2221); IVANHOE (I,2282)

SEURAT, PILAR
WEAPONS MAN (I,4783)

SEUSS, BARBARA
MATINEE AT THE BIJOU (II,1651)

SEVAREID, ERIC
THE FABULOUS 50S (I,1501)

SEVERINSEN, DOC
ALL-STAR SWING FESTIVAL (II,53); THE BIG BAND AND ALL THAT JAZZ (I,422)

SEVERN, MAIDA
SENIOR YEAR (II,2301)

SEWARDS, TERENCE
LOOK BACK IN DARKNESS (I,2752)

SEWELL, GEORGE
RUNNING BLIND (II,2232); SPECIAL BRANCH (II,2414); U.F.O. (I,4662)

SEWELL, GRANT
YOU ARE THERE (I,4929)

SEYMOUR, ANNE
ABC'S MATINEE TODAY (I,7); ADAM'S RIB (I,32); THE BURNING COURT (I,766); CLOSE TIES (II,545); DOWN HOME (II,731); EMPIRE (I,1439); FAMILY TIES (II,819); FOLLOW YOUR HEART (I,1618); MR. NOVAK (I,3145); RETURN TO PEYTON PLACE (I,3779); TENAFLY (I,4395); THE TIM CONWAY SHOW (I,4502); THE WEB (I,4784)

SEYMOUR, CAROLYN
THE LAST NINJA (II,1438); MODESTY BLAISE (II,1719); RETURN OF THE MAN FROM U.N.C.L.E.: THE 15 YEARS LATER AFFAIR (II,2140)

SEYMOUR, DAN
SING IT AGAIN (I,4059); WHERE WAS I? (I,4833)

SEYMOUR, JANE
THE AWAKENING LAND (II,122); BATTLE OF THE NETWORK STARS (II,164); BENNY AND BARNEY: LAS VEGAS UNDERCOVER (II,206); CAPTAINS AND THE KINGS (II,435); CONCERNING MISS MARLOWE (I,1032); EAST OF EDEN (II,752); FORD THEATER HOUR (I,1635); THE INNER SANCTUM (I,2216); KRAFT TELEVISION THEATER (I,2592); LIGHT'S OUT (I,2699); THE ONEDIN LINE (II,1912); PHILCO TELEVISION PLAYHOUSE (I,3583); SEVENTH AVENUE (II,2309); STAR TONIGHT (I,4199); STUDIO ONE (I,4268); SUSPENSE (I,4305); YOUNG MR. BOBBIN (I,4943); THE WEB (I,4784)

SEYMOUR, JEFF
HILL STREET BLUES (II,1154)

SEYMOUR, RALPH
MAKIN' IT (II,1603)

SEYMOUR, SHAUGHAN
PENMARRIC (II,1993)

SEYMOURE, CAROLYN

SHA NA NA
SHA NA NA (II,2311)

SHACKELFORD, DAVID
FAMILY (II,813)

SHACKELFORD, MICHAEL
FAMILY (II,813)

SHACKELFORD, TED
ANOTHER WORLD (II,97); BATTLE OF THE NETWORK STARS (II,176); DALLAS (II,614); THE JORDAN CHANCE (II,1343); KNOTS LANDING (II,1404)

SHADROCK
THE COP AND THE KID (II,572)

SHAFFER, LOUISE
ALL THAT GLITTERS (II,42); THE EDGE OF NIGHT (II,760); HIDDEN FACES (I,2044); RYAN'S HOPE (II,2234); WHERE THE HEART IS (I,4831)

SHAFFER, PAUL
HEREAFTER (II,1138); A YEAR AT THE TOP (II,2844); A YEAR AT THE TOP (II,2845)

SHAFFO, ROBERT
SABU AND THE MAGIC RING (I,3881)

SHAGINYAN, ANATOLY
THE WINDS OF WAR (II,2807)

SHAHAN, ROCKY
RAWHIDE (I,3727)

SHAHEED, HASSAN
DAYS OF OUR LIVES (II,629)

SHAKELFORD, LYNN
ALMOST ANYTHING GOES (II,56)

SHAKER, MARTIN
DAYS OF OUR LIVES (II,629)

SHALET, DIANE
BACKSTAIRS AT THE WHITE HOUSE (II,133)

SHANE, JIM
THE NEW PHIL SILVERS SHOW (I,3269)

SHANKS, DON
THE CHISHOLMS (II,512); THE LIFE AND TIMES OF GRIZZLY ADAMS (II,1469)

SHANNON, EFFIE
JANE EYRE (I,2337)

SHANNON, JAMES
HOW TO SURVIVE A MARRIAGE (II,1198)

SHANNON, MICHAEL
ALL MY CHILDREN (II,39); THE COPS AND ROBIN (II,573); FUTURE COP (II,945); FUTURE COP (II,946); THE PAPER CHASE: THE SECOND YEAR (II,1950); RIKER (II,2175); SEARCH FOR TOMORROW (II,2284)

SHANNON, RICHARD
333 MONTGOMERY (I,4487); BROCK CALLAHAN (I,741)

SHAPIRO, GARY
KID POWER (I,2532); SEALAB 2020 (I,3949)

SHAPS, CYRIL
THE ONEDIN LINE (II,1912)

SHARIF, OMAR
THE FAR PAVILIONS (II,832); THE POPPY IS ALSO A FLOWER (I,3650); S*H*E (II,2235)

SHARKEY, JACK
CAMELOT (II,416)

SHARKEY, JOE
O'MALLEY (II,1872)

SHARKEY, RAY
BEST FRIENDS (II,213)

SHARMA, BARBARA
FATHER O FATHER (II,842); GLITTER (II,1009); MA AND PA (II,1570); MASQUERADE (I,2940); RHODA (II,2151); RISKO (II,2180); THREE FOR THE GIRLS (I,4477)

SHARON, FRAN
ANOTHER WORLD (II,97); THE EDGE OF NIGHT (II,760); SEARCH FOR TOMORROW (II,2284)

SHARP, ALEX
LITTLE HOUSE: THE LAST FAREWELL (II,1491)

SHARP, ANITA
MACBETH (I,2801)

SHARP, ANTHONY
THE FAR PAVILIONS (II,832)

SHARP, HENRY
THE GREAT WALTZ (I,1878)

SHARP, SAUNDRA
OUR STREET (I,3418); WONDER WOMAN (SERIES 2) (II,2829)

SHARP, TOM
THE HALF-HOUR COMEDY HOUR (II,1074)

SHARPE, ALEX
FORCE SEVEN (II,908)

SHARPE, CORNELIA
COVER GIRLS (II,596); S*H*E (II,2235)

SHARPE, JOHN
THE PRISONER (I,3670); THE SAVAGE CURSE (I,3925)

SHARPE, KAREN
CAMEO THEATER (I,808); THE DESILU PLAYHOUSE (I,1237); JOHNNY RINGO (I,2434) THE LUX VIDEO THEATER (I,2795); MATINEE THEATER (I,2947)

SHARPE, SANDRA
WILLOW B: WOMEN IN PRISON (II,2804)

SHARPLES, WINSTON
ROCKET ROBIN HOOD (I,3821)

SHARRETT, MICHAEL
JOE'S WORLD (II,1305); THE NIGHT RIDER (II,1848)

SHARROCK, IAN
PETER PAN (II,2030)

SHARRON, BO
HAPPY DAYS (II,1084)

SHATNER, WILLIAM
ALEXANDER THE GREAT (I,113); ALFRED HITCHCOCK PRESENTS (I,115); ASSIGNMENT: EARTH (I,299); THE BARBARY COAST (II,146); THE BARBARY COAST (II,147); THE BASTARD/KENT FAMILY CHRONICLES (II,159); BATTLE OF THE NETWORK STARS (II,167); BATTLE OF THE NETWORK STARS (II,174); BATTLE OF THE NETWORK STARS (II,175); BATTLE OF THE NETWORK STARS (II,177); CELEBRITY DAREDEVILS (II,473); CIRCUS OF THE STARS (II,535); CIRCUS OF THE STARS (II,538); CLIMAX! (I,976); COLOSSUS (I,1009); THE DEFENDER (I,1219); FOR THE PEOPLE (I,1626); HEROES AND SIDEKICKS—INDIANA JONES AND THE TEMPLE OF DOOM (II,1140); HOW THE WEST WAS WON (II,1196); KRAFT TELEVISION THEATER (I,2592); LITTLE WOMEN (II,1497); THE OUTER LIMITS (I,3426); OWEN MARSHALL: COUNSELOR AT LAW (I,3436); POLICE SQUAD! (II,2061); SHERLOCK HOLMES: THE HOUND OF THE BASKERVILLES (I,4011); STAR TREK (II,2439); STAR TREK (II,2440); STUDIO ONE (I,4268); SUSPICION (I,4309); THRILLER (I,4492); T.J. HOOKER (II,2524); TESTIMONY OF TWO MEN (II,2564); THIS WAS AMERICA (II,2599); TV'S FUNNIEST GAME SHOW MOMENTS (II,2676)

SHAUGHNESSY, MICKEY
THE CHICAGO TEDDY BEARS (I,934); THE JONES BOYS (I,2446); THE MCGONIGLE (I,2968); MICKEY AND THE CONTESSA (I,3024)

SHAVER, HELEN
HILL STREET BLUES (II,1154); JESSICA NOVAK (II,1286); SEARCH AND RESCUE: THE ALPHA TEAM (II,2283); UNITED STATES (II,2708)

SHAW, BETTINA
TRAVEL WITH DON AND BETTINA (I,4595)

SHAW, BRIAN
CINDERELLA (II,526); SLEEPING BEAUTY (I,4083)

SHAW, DIANNE B
HARD KNOX (II,1087)

SHAW, DON
TRAVEL WITH DON AND BETTINA (I,4595)

SHAW, DOROTHY
THE BRADY BUNCH (II,362)

SHAW, JAN
THE FRONT PAGE (I,1691)

SHAW, JENNIFER
JAMES AT 15 (II,1271); MAN ON THE MOVE (I,2878)

SHAW, MARGUERITE
THE BALLAD OF LOUIE THE LOUSE (I,337)

SHAW, MARK
THE WORLD OF PEOPLE (II,2839)

SHAW, MARTIN
DOCTOR IN THE HOUSE (I,1313); HAMLET (I,1925); TEN WHO DARED (II,2559)

SHAW, PAULA
THREE'S COMPANY (II,2614)

SHAW, RETA
THE ANN SOTHERN SHOW (I,220); ANNIE GET YOUR GUN (I,223); BEWITCHED (I,418); CALL HOLME (II,412); THE CARA WILLIAMS SHOW (I,843); COME A RUNNING (I,1013); FIBBER MCGEE AND MOLLY (I,1569); THE GHOST AND MRS. MUIR (I,1786); GUESS WHAT I DID TODAY (I,1897); HEY TEACHER (I,2041); ICHABOD AND ME (I,2183); THE LIFE OF RILEY (I,2686); MEET ME IN ST. LOUIS (I,2986); MEET ME IN ST. LOUIS (I,2987); OH, THOSE BELLS! (I,3345); THE TAB HUNTER SHOW (I,4328); THINK PRETTY (I,4448)

SHAW, ROBERT
THE BUCCANEERS (I,749); CAROL FOR ANOTHER CHRISTMAS (I,852)

SHAW, RODERICK
THE ADVENTURES OF BLACK BEAUTY (I,51)

SHAW, STAN
THE BUFFALO SOLDIERS (II,388); THE MISSISSIPPI (II,1707); THE MONEYCHANGERS (II,1724); ROOTS: THE NEXT GENERATIONS (II,2213); VENICE MEDICAL (II,2726)

SHAW, STEVE
FLYING HIGH (II,895); KNOTS LANDING (II,1404)

SHAW, SUSAN
BENDER (II,204); ONE MAN'S FAMILY (I,3383)

SHAW, VICTORIA
GALLAGHER (I,1727); GENERAL HOSPITAL (II,964); SECOND LOOK (I,3960)

SHAWLEE, JOAN
AGGIE (I,90); THE BETTY HUTTON SHOW (I,413); THE CIRCLE FAMILY (II,527); THE DICK VAN DYKE SHOW (I,1275); THE FEATHER AND FATHER (II,846); THE FEATHER AND FATHER GANG (II,845); THE FEATHER AND FATHER GANG: NEVER CON A KILLER (II,846); JOE'S WORLD (II,1305); MATT HELM (II,1652)

SHAWLEY, ROBERT
THE HARTMANS (I,1960); TONIGHT AT 8:30 (I,4556)

SHAWN, DICK
ABC STAGE '67 (I,6); EVIL ROY SLADE (I,1485); GOOD TIMES (I,1853); LOVE, LIFE, LIBERTY & LUNCH (II,1544); MR. & MRS. DRACULA (II,1745); SHERIFF WHO? (I,4009); SUNDAY IN TOWN (I,4285); THREE'S COMPANY (II,2614); WHO'S AFRAID OF MOTHER GOOSE? (I,4852); THE YEAR WITHOUT A SANTA CLAUS (II,2846); YOU'RE JUST LIKE YOUR FATHER (II,2861)

SHAWN, LYNN
INSIDE AMERICA (II,1234)

SHAY, DOROTHY
THE WALTONS (II,2740)

SHAY, JANET
HARDCASE (II,1088)

SHAY, MICHELLE
ANOTHER WORLD (II,97)

SHAY, PATRICIA
MR. MERGENTHWIRKER'S LOBBLIES (I,3144)

SHAYNE, ROBERT
THE ADVENTURES OF SUPERMAN (I,77); MYSTERY AND MRS. (I,3210); NAVY LOG (I,3233)

SHEA, CHRISTOPHER
A CHARLIE BROWN CHRISTMAS (I,908); CHARLIE BROWN'S ALL STARS (I,910); HE'S YOUR DOG, CHARLIE BROWN (I,2037); IT'S THE GREAT PUMPKIN, CHARLIE BROWN (I,2276); SHANE (I,3995); YOU'RE IN LOVE, CHARLIE BROWN (I,4964)

SHEA, ERIC
ANNA AND THE KING (I,221); BOBBY PARKER AND COMPANY (II,344); MARY HARTMAN, MARY HARTMAN (II,1638)

SHEA, JOHN
KENNEDY (II,1377); THE LAST CONVERTIBLE (II,1435); THE ROCK RAINBOW (II,2194)

SHEA, KATHY
THE ASPHALT COWBOY (II,111); WENDY HOOPER—U. S. ARMY (II,2764)

SHEA, MICHAEL
ANOTHER MAN'S SHOES (II,96); THE NEW ADVENTURES OF HUCKLEBERRY FINN (I,3250); THE NEW DICK VAN DYKE SHOW (I,3262)

SHEA, STEPHEN
BE MY VALENTINE, CHARLIE BROWN (I,369); A CHARLIE BROWN THANKSGIVING (I,909); IT'S A MYSTERY, CHARLIE BROWN (I,2259); IT'S THE EASTER BEAGLE, CHARLIE BROWN (I,2275); PLAY IT AGAIN, CHARLIE BROWN (I,3617); THERE'S NO TIME FOR LOVE, CHARLIE BROWN (I,4438); YOU'RE NOT ELECTED, CHARLIE BROWN (I,4967)

SHEA, WILLIAM
FAMILY HONOR (I,1522)

SHEAN, LISABETH
THE DOCTORS (II,694)

SHEAR, PEARL
THE BUREAU (II,397); CATALINA C-LAB (II,458); MODESTY BLAISE (II,1719); SAINT PETER (II,2238); THE WALTONS (II,2740)

SHEARER, HARRY
IT'S A SMALL WORLD (I,2260)

SHEARER, HARRY J
SATURDAY NIGHT LIVE (II,2261)

SHEARIN, JOHN
BRET MAVERICK (II,374); THE DOCTORS (II,694); FLAMINGO ROAD (II,872); GUILTY OR INNOCENT (II,1065); LOVING (II,1552)

SHEARING, DINAH
ALL THE RIVERS RUN (II,43)

SHEARMAN, ALAN
MASTER OF THE GAME (II,1647)

SHEBBEARE, NORMA
THE WOMAN IN WHITE (II,2819)

SHEEDY, ALLY
HOME ROOM (II,1172)

SHEEHAN, DAVID
THE GOSSIP COLUMNIST (II,1045)

SHEEHAN, DOUGLAS
BATTLE OF THE NETWORK STARS (II,178); GENERAL HOSPITAL (II,964); HEAVEN ON EARTH (II,1119); KNOTS LANDING (II,1404)

SHEEHAN, MICHAEL
THE EARTHLINGS (II,751); FLINTSTONE FAMILY ADVENTURES (II,882); THE FLINTSTONE FUNNIES (II,883)

SHEEN, MARTIN
AS THE WORLD TURNS (II,110); CIRCLE OF FEAR (I,958); CIRCUS OF THE STARS (II,532); THE CRIME CLUB (I,1096); HARRY O (II,1098); KENNEDY (II,1377); LETTERS FROM THREE LOVERS (I,2671); TAXI (I,4365); THEN CAME BRONSON (I,4436)

SHEFFER, CRAIG
THE HAMPTONS (II,1076)

SHEFFIELD, JAY
TAMMY (I,4357)

SHEINER, DAVID
DIANA (I,1259); IRONSIDE (I,2240); MR. NOVAK (I,3145); REX HARRISON PRESENTS SHORT STORIES OF LOVE (II,2148); SCALPLOCK (I,3930)

SHEINER, DAVID S
HOW DO I KILL A THIEF—LET ME COUNT THE WAYS (II,1195); LANIGAN'S RABBI (II,1427)

SHELBY, LAUREL
ONE TOUCH OF VENUS (I,3389)

SHELDEN, JANA
OPPENHEIMER (II,1919)

SHELDON, GAIL
BEAT THE CLOCK (I,378)

SHELDON, GENE
THE AMAZING TALES OF HANS CHRISTIAN ANDERSON (I,161); ONE MORE TIME (I,3386); ZORRO (I,4982)

SHELDON, JACK
THE CARA WILLIAMS SHOW (I,843); THE GIRL WITH SOMETHING EXTRA (II,999); THE MERV GRIFFIN SHOW (II,1688); THE NUT HOUSE (I,3320); RUN, BUDDY, RUN (I,3870); UNDER THE YUM YUM TREE (I,4671)

SHELDON, JEROME
ARMCHAIR DETECTIVE (I,259)

SHELDON, JOHN
DUFFY (II,739)

SHELLE, LORIE
THE GUIDING LIGHT (II,1064)

SHELLEY, BARBARA
KRAFT MYSTERY THEATER (I,2589); THE LLOYD BRIDGES SHOW (I,2733); SOMEWHERE IN ITALY. . . COMPANY B (I,4119)

SHELLEY, CAROLE
THE ODD COUPLE (II,1875)

SHELLEY, DAVE
THE GREATEST AMERICAN HERO (II,1060)

SHELLEY, JOSHUA
B.J. AND THE BEAR (II,125); CASSIE AND COMPANY (II,455); NEEDLES AND PINS (I,3246); SONNY BOY (II,2402); STARLIGHT THEATER (I,4201)

SHELLY, AL
SAM (II,2246)

SHELLY, CINDY
SMILEY'S PEOPLE (II,2383)

SHELLY, NORMAN
PETER PAN (I,3566)

SHELTON, DEBORAH
THE YEAGERS (II,2843); THE YELLOW ROSE (II,2847)

SHELTON, DON
MR. TUTT (I,3153)

SHELTON, LAURA
FBI CODE 98 (I,1550)

SHELTON, REID
FIRST & TEN (II,861); TOO GOOD TO BE TRUE (II,2643); WHIZ KIDS (II,2790)

SHELYNE, CAROLE
HERE COME THE BRIDES (I,2024)

SHENAR, PAUL
BEULAH LAND (II,226); THE GEMINI MAN (II,961); ROOTS (II,2211); THREE EYES (II,2605)

SHEPARD, JAN

CAPTAIN MIDNIGHT (I,833); DOCTOR CHRISTIAN (I,1308)

SHEPHERD, CYBILL
THE YELLOW ROSE (II,2847)

SHEPHERD, JEAN
JEAN SHEPHERD'S AMERICA (I,2354)

SHEPHERD, JEANNE
THE INNER SANCTUM (I,2216)

SHEPHERD, JOHN
CLOSE TIES (II,545); HIS AND HERS (II,1155); THE HOYT AXTON SHOW (II,1200)

SHEPLEY, MICHAEL
DICK AND THE DUCHESS (I,1263)

SHEPODD, JON
TIMMY AND LASSIE (I,4520)

SHEPPARD, ANTHONY
SECRETS OF THE OLD BAILEY (I,3971)

SHEPPARD, MORGAN
SHOGUN (II,2339)

SHEPPARD, STEPHEN
THE ONEDIN LINE (II,1912)

SHERA, MARK
ADAMS HOUSE (II,9); BARNABY JONES (II,153); THE KILLIN' COUSIN (II,1394); S.W.A.T. (II,2236)

SHERALEE
YOUR HIT PARADE (II,2872)

SHEREN, JOHN
THE FRENCH ATLANTIC AFFAIR (II,924)

SHERIDAN, ANN
ANOTHER WORLD (II,97); FORD TELEVISION THEATER (I,1634); HENRY FONDA PRESENTS THE STAR AND THE STORY (I,2016); THE LUX VIDEO THEATER (I,2795); PISTOLS 'N' PETTICOATS (I,3608); PURSUIT (I,3700); SNEAK PREVIEW (I,4100)

SHERIDAN, ARDELL
THE SUPER (I,4293)

SHERIDAN, DAN
LAWMAN (I,2642)

SHERIDAN, DINAH
DOCTOR WHO—THE FIVE DOCTORS (II,690)

SHERIDAN, NANCY
FOLLOW YOUR HEART (I,1618); HAY FEVER (I,1979); STUDIO ONE (I,4268); TODAY IS OURS (I,4531)

SHERIDAN, NICOLETTE
PAPER DOLLS (II,1952)

SHERIDAN, PHIL
CONTEST CARNIVAL (I,1040)

SHERMAN, BOB
APPLAUSE (I,244)

SHERMAN, BOBBY
THE BOBBY SHERMAN SPECIAL (I,676); GETTING TOGETHER (I,1784); THE GOSSIP COLUMNIST (II,1045); HERE COME THE BRIDES (I,2024); A KNIGHT IN SHINING ARMOUR (I,2575)

SHERMAN, BRUCE
CAMELOT (II,416)

SHERMAN, COURTNEY
THE GUIDING LIGHT (II,1064); SEARCH FOR TOMORROW (II,2284)

SHERMAN, DON
THE LIVING END (I,2731); MISSING PIECES (II,1706)

SHERMAN, ELLEN
KOSTA AND HIS FAMILY (I,2581); MISS WINSLOW AND SON (II,1705)

SHERMAN, FRED
CIMARRON CITY (I,953); HEY MULLIGAN (I,2040)

SHERMAN, HIRAM
ABE LINCOLN IN ILLINOIS (I,11); ALICE IN WONDERLAND (I,120); THE MAN IN THE DOG SUIT (I,2869); MIRACLE ON 34TH STREET (I,3057); THE TAMMY GRIMES SHOW (I,4358); THAT'S OUR SHERMAN (I,4429)

SHERMAN, HOWARD
CELEBRITY (II,463)

SHERMAN, JENNY
CRASH ISLAND (II,600); DALLAS (II,614); DAYS OF OUR LIVES (II,629); GENERAL HOSPITAL (II,964)

SHERMAN, KEN
THE ADVENTURES OF JIM BOWIE (I,63)

SHERMAN, KERRY
ALONE AT LAST (II,59)

SHERMAN, MELISSA
ALONE AT LAST (II,59); JOE'S WORLD (II,1305)

SHERMAN, RANSOM
AND HERE'S THE SHOW (I,187); FATHER OF THE BRIDE (I,1543); HERE'S THE SHOW (I,2033); THE RANSOM SHERMAN SHOW (I,3723)

SHERMIT, HAZEL
THE AMAZING CHAN AND THE CHAN CLAN (I,157); BLONDIE (I,486); THE NEW ZOO REVUE (I,3277); TOM, DICK, AND HARRY (I,4536)

SHERRIN, NED
SONG BY SONG (II,2399)

SHERRY, BILL
THE LITTLE REVUE (I,2717)

SHERRY, DIANE
THE BING CROSBY SHOW (I,467)

SHERWOOD, BOBBY
QUICK AS A FLASH (I,3709); SID CAESAR PRESENTS COMEDY PREVIEW (I,4041); STARS ON PARADE (I,4210); STEP THIS WAY (I,4216)

SHERWOOD, BRUCE
THE SECRET STORM (I,3969)

SHERWOOD, DON
CLUB 60 (I,987)

SHERWOOD, GALE
A CONNECTICUT YANKEE (I,1038); THE DESERT SONG (I,1236); NAUGHTY MARIETTA (I,3232); PARIS IN THE SPRINGTIME (I,3457)

SHERWOOD, MADELEINE
DOCTOR DAN (II,684); THE DOCTORS (II,694); THE FLYING NUN (I,1611); THE GUIDING LIGHT (II,1064); THEY ONLY COME OUT AT NIGHT (II,2586)

SHERWOOD, ROBERTA
MR. BROADWAY (I,3130)

SHEYBAL, VLADEK
SHOGUN (II,2339); SMILEY'S PEOPLE (II,2383); U.F.O. (I,4662)

SHIELDS, ARTHUR
THE LIFE OF RILEY (I,2686); STORY THEATER (I,4240)

SHIELDS, BROOKE
BOB HOPE SPECIAL: BOB HOPE'S STAR-STUDDED SPOOF OF THE NEW TV SEASON— G RATED—WITH GLAMOUR, GLITTER & GAGS (II,317); CELEBRITY CHALLENGE OF THE SEXES 5 (II,469); CIRCUS OF THE STARS (II,533); CIRCUS OF THE STARS (II,534); CIRCUS OF THE STARS (II,535); CIRCUS OF THE STARS (II,536); CIRCUS OF THE STARS (II,537); CIRCUS OF THE STARS (II,538); MEN WHO RATE A "10" (II,1684)

SHIELDS, FRED
MEET CORLISS ARCHER (I,2979)

SHIELDS, HELEN
CONCERNING MISS MARLOWE (I,1032); FROM THESE ROOTS (I,1688)

SHIELDS, HERB
FROM THESE ROOTS (I,1688)

SHIELDS, ROBERT
SHIELDS AND YARNELL (II,2328); THE WILD WILD WEST REVISITED (II,2800)

SHIELS, DAVID
HIS AND HERS (II,1155)

SHIGETA, JAMES
CAROL FOR ANOTHER CHRISTMAS (I,852); THE HARDY BOYS (I,1948); THE KILLER WHO WOULDN'T DIE (II,1393); MATT HELM (II,1652); THE MONEYCHANGERS (II,1724); ONCE AN EAGLE (II,1897); THE RENEGADES (II,2132); SAMURAI (II,2249)

SHILLO, MICHAEL
MASADA (II,1642)

SHILOAH, YOUSEG
MOSES THE LAWGIVER (II,1737)

SHIMADA, YOKO
SHOGUN (II,2339)

SHIMERMAN, ARMEN
BULBA (II,392)

SHIMODA, YUKI
JOHNNY MIDNIGHT (I,2431); JUDGE DEE IN THE MONASTERY MURDERS (I,2465); THE TEAHOUSE OF THE AUGUST MOON (I,4368); A TOWN LIKE ALICE (II,2652)

SHIMONO, SAB
MAGIC MONGO (II,1592); MODESTY BLAISE (II,1719)

SHIN-LIAN
CAPTAIN SCARLET AND THE MYSTERONS (I,837)

SHINNIMAN, MICHAEL
DOROTHY HAMILL IN ROMEO & JULIET ON ICE (II,718)

SHIPMAN, NINA
THE ANDY GRIFFITH SHOW (I,192); HELLO DERE (I,2010)

SHIPP, JOHN WESLEY
THE GUIDING LIGHT (II,1064)

SHIPP, MARY
CROWN THEATER WITH GLORIA SWANSON (I,1107); LIFE WITH LUIGI (I,2692); MY

FRIEND IRMA (I,3191)

SHIRE, TALIA
RICH MAN, POOR MAN—BOOK I (II,2161)

SHIRLEY, PEG
AT EASE (II,116)

SHIRLEY, TOM
FROM THESE ROOTS (I,1688); HOTEL COSMOPOLITAN (I,2128); THEY'RE OFF (I,4444)

SHIRRIFF, CATHERINE
THE CABOT CONNECTION (II,408); RIPLEY'S BELIEVE IT OR NOT (II,2177); SHAPING UP (II,2316)

SHIVELY, PAUL
NANNY AND THE PROFESSOR AND THE PHANTOM OF THE CIRCUS (I,3226)

SHOBERG, DICK
SOMERSET (I,4115)

SHOBERG, RICHARD
ALL MY CHILDREN (II,39); THE EDGE OF NIGHT (II,760)

SHOCKLEY, SALLIE
LADY LUCK (I,2604); SARGE (I,3914); SARGE: THE BADGE OR THE CROSS (I,3915)

SHOEMAKER, ANN
THE KAISER ALUMINUM HOUR (I,2503); ROBERTA (I,3816)

SHOEMAKER, DICK
ENTERTAINMENT TONIGHT (II,780)

SHOLDAR, MICKEY
THE FARMER'S DAUGHTER (I,1533)

SHOLTO, PAMELA
A HORSEMAN RIDING BY (II,1182)

SHOOP, PAMELA SUSAN
79 PARK AVENUE (II,2310); GALACTICA 1980 (II,953); KEEPER OF THE WILD (II,1373); MITCHELL AND WOODS (II,1709); WONDER WOMAN (SERIES 1) (II,2828)

SHOR, DAN
THE BLUE AND THE GRAY (II,273); STUDS LONIGAN (II,2484)

SHORE, DINAH
BRIEF ENCOUNTER (I,723); DINAH AND FRIENDS (II,675); DINAH AND HER NEW BEST FRIENDS (II,676); DINAH IN SEARCH OF THE IDEAL MAN (I,1278); THE DINAH SHORE SPECIAL—LIKE HEP (I,1282); THE DINAH SHORE CHEVY SHOW (I,1279); THE DINAH SHORE SHOW (I,1280); THE DINAH SHORE SHOW (I,1281); THE DINAH SHORE SPECIAL (I,1283); THE DINAH SHORE SPECIAL (I,1284); DINAH! (II,677); DINAH'S PLACE (II,678); HOW TO HANDLE A WOMAN (I,2146); A PUREX DINAH SHORE SPECIAL (I,3697); A SPECIAL HOUR WITH DINAH SHORE (I,4149)

SHORE, ELAINE
ARNIE (I,261); THE TROUBLE WITH PEOPLE (I,4611)

SHORE, ROBERTA
ALCOA/GOODYEAR HOUR (I,107); ANNETTE (I,222); THE BOB CUMMINGS SHOW (I,500); LIFE WITH VIRGINIA (I,2695); THE VIRGINIAN (I,4738)

SHORR, LONNIE
BIG EDDIE (II,235); BIG EDDIE (II,236); WILDER AND WILDER (II,2802)

SHORT, DANA
ST. ELSEWHERE (II,2432)

SHORT, MARTIN
THE ASSOCIATES (II,113); I'M A BIG GIRL NOW (II,1218); THE IMPERIAL GRAND BAND (II,1224); SATURDAY NIGHT LIVE (II,2261); WHITE AND RENO (II,2787)

SHORT, ROBERT
STARSTRUCK (II,2446)

SHORT, SYLVIA
THE FIRM (II,860); MAN AND SUPERMAN (I,2854)

SHORTER, BETH
THE FIRM (II,860)

SHORTRIDGE, STEPHEN
ALOHA PARADISE (II,58); MARIE (II,1628); TONI'S BOYS (II,2637); WELCOME BACK, KOTTER (II,2761)

SHOWALTER, MAX
HOW TO SUCCEED IN BUSINESS WITHOUT REALLY TRYING (II,1197); IT'S A SMALL WORLD (I,2260); THE LUCILLE BALL COMEDY HOUR (I,2783); THE STOCKARD CHANNING SHOW (II,2468)

SHRAPNEL, JOHN
THE WOMAN IN WHITE (II,2819)

SHRINER, HERB
HERB SHRINER TIME (I,2020); MEET THE GOVERNOR (I,2991); SCREEN DIRECTOR'S PLAYHOUSE (I,3946); TWO FOR THE MONEY (I,4654)

SHRINER, KIN
GENERAL HOSPITAL (II,964); RITUALS (II,2182); TEXAS (II,2566)

SHRONG, MAURICE
ONE MORE TRY (II,1908)

SHROYER, SONNY
THE DUKES OF HAZZARD (II,742); ENOS (II,779)

SHRYER, BRET
DEAR TEACHER (II,653); SHIRLEY (II,2330)

SHUE, ELISABETH
CALL TO GLORY (II,413)

SHULL, JOHN
HOW DO I KILL A THIEF—LET ME COUNT THE WAYS (II,1195)

SHULL, RICHARD B
DIANA (I,1259); HART TO HART (II,1102); HOLMES AND YOYO (II,1169); SUTTERS BAY (II,2508)

SHUMAN, ROY
SEARCH FOR TOMORROW (II,2284); A WORLD APART (I,4910)

SHUSTER, FRANK
HOLIDAY LODGE (I,2077); WAYNE AND SHUSTER TAKE AN AFFECTIONATE LOOK AT. . . (I,4778)

SHUTAN, JAN
MOTHER, JUGGS AND SPEED (II,1741); SENIOR YEAR (II,2301); SONS AND DAUGHTERS (II,2404)

SHUTTA, ETHEL
FEELING GOOD (I,1559)

SHYDNER, RICH
T.L.C. (II,2525)

SHYST, WILLIAM
OUR FIVE DAUGHTERS (I,3413)

SIBBALD, LAURIE
F TROOP (I,1499); NO TIME FOR SERGEANTS (I,3300)

SIBLEY, DAVID
A FAMILY AFFAIR (II,814)

SICARI, JOSEPH
GUESS WHO'S COMING TO DINNER? (II,1063); THE LOVE BOAT I (II,1532); MCNAMARA'S BAND (II,1668); THE MONEYCHANGERS (II,1724)

SIDDALL, TEDDI
NICHOLS AND DYMES (II,1841)

SIDNEY, GEORGE
HOORAY FOR HOLLYWOOD (I,2114)

SIDNEY, P J
THE PHIL SILVERS SHOW (I,3580)

SIDNEY, SUSANNE
THE GRADUATION DRESS (I,1864); JUNIOR MISS (I,2493)

SIDNEY, SYLVIA
THE GOSSIP COLUMNIST (II,1045); THE JUNE ALLYSON SHOW (I,2488); KRAFT TELEVISION THEATER (I,2592); MAUREEN (II,1656); PLAYWRIGHTS '56 (I,3627); SENSE OF HUMOR (II,2303); WKRP IN CINCINNATI (II,2814)

SIEBERT, CHARLES
THE ADAMS CHRONICLES (II,8); ANOTHER WORLD (II,97); AS THE WORLD TURNS (II,110); HUSBANDS AND WIVES (II,1208); HUSBANDS, WIVES AND LOVERS (II,1209); ONE DAY AT A TIME (II,1900); SEARCH FOR TOMORROW (II,2284); TOPPER (II,2649); TRAPPER JOHN, M.D. (II,2654); WILD AND WOOLEY (II,2798)

SIEGEL, DONNA
THE NASHVILLE PALACE (II,1792)

SIEGEL, LAURA
SONS AND DAUGHTERS (II,2404)

SIEGEL, SAM
THE BENNETS (I,397)

SIEGEL, STANLEY
THE STANLEY SIEGEL SHOW (II,2433)

SIEMASZKO, CASEY
HARD KNOX (II,1087)

SIERRA, GREGORY
A.E.S. HUDSON STREET (II,17); BARNEY MILLER (II,154); FARRELL: FOR THE PEOPLE (II,834); MIAMI VICE (II,1689); SANFORD AND SON (II,2256); SOAP (II,2392); WHERE'S THE FIRE? (II,2785); ZORRO AND SON (II,2878)

SIERRA, MARGARITA
SURFSIDE 6 (I,4299)

SIGEL, BARBARA
NAKIA (II,1784); SAN FRANCISCO INTERNATIONAL AIRPORT (I,3905)

SIGGINS, JEFF

ON THE FLIP SIDE (I,3360); THE PATTY DUKE SHOW (I,3483)

SIGISMONDI, ARISTIDE
PAPA CELLINI (I,3450)

SIGNORELLI, TOM
THE RUBBER GUN SQUAD (II,2225)

SIGNORET, SIMONE
THE GENERAL ELECTRIC THEATER (I,1753)

SIHOL, CAROLINE
SMILEY'S PEOPLE (II,2383)

SIKES, CYNTHIA
BATTLE OF THE NETWORK STARS (II,176); BATTLE OF THE NETWORK STARS (II,177); BIG SHAMUS, LITTLE SHAMUS (II,242); CAPTAINS AND THE KINGS (II,435); FLAMINGO ROAD (II,872); POOR RICHARD (II,2065); ST. ELSEWHERE (II,2432)

SIKKING, JAMES B
CALLING DR. STORM, M.D. (II,415); GENERAL HOSPITAL (II,964); HILL STREET BLUES (II,1154); INSIDE O.U.T. (II,1235); TROUBLE IN HIGH TIMBER COUNTRY (II,2661); TURNABOUT (II,2669)

SILBAR, STUART
BLACK BEAUTY (II,261)

SILBERG, TUSSE
SMILEY'S PEOPLE (II,2383)

SILBERSCHER, MARVIN
THE IMMORTAL (I,2196)

SILETTI, MARIO
THE MAN FROM U.N.C.L.E. (I,2867)

SILIC, IGOR
CASANOVA (II,451)

SILLA, FELIX
THE ADDAMS FAMILY (II,11); THE ADDAMS FAMILY (II,13); BUCK ROGERS IN THE 25TH CENTURY (II,383); BUCK ROGERS IN THE 25TH CENTURY (II,384); H.R. PUFNSTUF (I,2154); LIDSVILLE (I,2679)

SILLIMAN, MAUREEN
6 RMS RIV VU (II,2371); THE GUIDING LIGHT (II,1064); SANCTUARY OF FEAR (II,2252)

SILLS, BEVERLY
SILLS AND BURNETT AT THE MET (II,2354)

SILLS, PAUL
STORY THEATER (I,4241)

SILO, SUSAN
ALWAYS APRIL (I,147); BRILLIANT BENJAMIN BOGGS (I,730); THE C.B. BEARS (II,406); HARRY'S GIRLS (I,1959); HECK'S ANGELS (II,1122); PAC-MAN (II,1940); THE SCOOBY-DOO AND SCRAPPY-DOO SHOW (II,2275); YES, VIRGINIA, THERE IS A SANTA CLAUS (II,2849)

SILVA, GENO
THE CHISHOLMS (II,512); MCLAREN'S RIDERS (II,1663); WILD TIMES (II,2799)

SILVA, HENRY
BUCK ROGERS IN THE 25TH CENTURY (II,383); WEAPONS MAN (I,4783)

SILVA, TRINIDAD
HILL STREET BLUES (II,1154)

SILVER, BORAH
KOJAK (II,1406)

SILVER, DAVID
OPPENHEIMER (II,1919)

SILVER, JEFF
THE CHARLIE FARRELL SHOW (I,913); LOVE THAT BOB (I,2779)

SILVER, JOE
CORONET BLUE (I,1055); FAY (II,844); RYAN'S HOPE (II,2234); STARSTRUCK (II,2446)

SILVER, JOHNNY
FLO'S PLACE (II,892); HELLO, LARRY (II,1128); NEW GIRL IN HIS LIFE (I,3265); PETE KELLY'S BLUES (I,3559); UNCLE CROC'S BLOCK (II,2702)

SILVER, MARK
COCAINE AND BLUE EYES (II,548)

SILVER, RON
BAKER'S DOZEN (II,136); DEAR DETECTIVE (II,652); RETURN OF THE WORLD'S GREATEST DETECTIVE (II,2142); RHODA (II,2151); THE STOCKARD CHANNING SHOW (II,2468)

SILVERA, FRANK
GREAT ADVENTURE (I,1868); THE HIGH CHAPARRAL (I,2051); SKIN OF OUR TEETH (I,4073)

SILVERHEELS JR, JAY
KID POWER (I,2532)

SILVERHEELS, JAY
CAT BALLOU (I,870); THE LONE RANGER (I,2740); PISTOLS 'N' PETTICOATS (I,3608)

SILVERMAN, JONATHAN
E/R (II,748)

SILVERS, CATHY
HAPPY DAYS (II,1084); HIGH SCHOOL, U.S.A. (II,1148); T.L.C. (II,2525)

SILVERS, CHUBBY
SO YOU WANT TO LEAD A BAND (I,4106)

SILVERS, EDMUND
THE JACKSON FIVE (I,2326)

SILVERS, FRANK
WINCHESTER (I,4872)

SILVERS, PHIL
THE BALLAD OF LOUIE THE LOUSE (I,337); THE BEVERLY HILLBILLIES (I,417); DAMN YANKEES (I,1128); EDDIE (I,1405); HAPPY DAYS (II,1084); JUST POLLY AND ME (I,2497); THE LOVE BOAT (II,1535); THE LOVE BOAT III (II,1534); THE NEW PHIL SILVERS SHOW (I,3269); ON STAGE: PHIL SILVERS (I,3357); THE PHIL SILVERS ARROW SHOW (I,3576); PHIL SILVERS IN NEW YORK (I,3577); PHIL SILVERS ON BROADWAY (I,3578); THE PHIL SILVERS PONTIAC SPECIAL: KEEP IN STEP (I,3579); THE PHIL SILVERS SHOW (I,3580); THE SLOWEST GUN IN THE WEST (I,4088); SUMMER IN NEW YORK (I,4279)

SILVERSTONE, LILLIAN
MASTER OF THE GAME (II,1647); NANCY ASTOR (II,1788)

SILVESTRE, ARMANDO
DANIEL BOONE (I,1142)

SILVEY, SUSIE
SMILEY'S PEOPLE (II,2383)

SIM, GERALD
ONCE THE KILLING STARTS (I,3367); TYCOON: THE STORY OF A WOMAN (II,2697)

SIMCOX, TOM
BRADDOCK (I,704); CODE R (II,551); THE FLIERS (I,1596)

SIMMONDS, DOUGLAS
HERE COME THE DOUBLE DECKERS (I,2025)

SIMMONDS, NIKOLAS
THE STRAUSS FAMILY (I,4253)

SIMMONS, GRACE
THE JERK, TOO (II,1282)

SIMMONS, JEAN
THE DAIN CURSE (II,613); THE EASTER PROMISE (I,1397); GOLDEN GATE (II,1020); THE HOME FRONT (II,1171); JACQUELINE SUSANN'S VALLEY OF THE DOLLS, 1981 (II,1268); SOLDIER IN LOVE (I,4108); THE THORN BIRDS (II,2600)

SIMMONS, MICHAEL
SUMMER SOLSTICE (II,2492)

SIMMONS, PATRICK
ROCK 'N' ROLL: THE FIRST 25 YEARS (II,2189); THE ROOTS OF ROCK 'N' ROLL (II,2212)

SIMMONS, RICHARD
GENERAL HOSPITAL (II,964); HERE'S RICHARD (II,1136); HERE'S RICHARD (II,1137); THE RICHARD SIMMONS SHOW (II,2166); ULTRA QUIZ (II,2701)

SIMMONS, RICHARD (Not the same as the exercise king)
SERGEANT PRESTON OF THE YUKON (I,3980)

SIMMONS, RICHARD LEE
I LOVE LUCY (I,2171)

SIMMS, FRANK
DAVID BOWIE—SERIOUS MOONLIGHT (II,622)

SIMMS, GEORGE
DAVID BOWIE—SERIOUS MOONLIGHT (II,622)

SIMMS, GINNY
CLUB CELEBRITY (I,981)

SIMMS, HILDA
THE NURSES (I,3318)

SIMON, ANN
ENTERTAINMENT TONIGHT (II,780)

SIMON, CARLY
GOOD VIBRATIONS FROM CENTRAL PARK (I,1854)

SIMON, JOANNE
ENTERTAINMENT TONIGHT (II,780)

SIMON, JOEL
ANOTHER WORLD (II,97)

SIMON, JOSETTE
BLAKE'S SEVEN (II,264)

SIMON, LEONARD
JUST OUR LUCK (II,1360)

SIMON, PAUL

THE PAUL SIMON SPECIAL (II,1983); SIMON AND GARFUNKEL IN CONCERT (II,2356)

SIMON, PETER
ALL MY CHILDREN (II,39); AS THE WORLD TURNS (II,110); THE GUIDING LIGHT (II,1064); SEARCH FOR TOMORROW (II,2284)

SIMON, ROBERT F
THE AMAZING SPIDER-MAN (II,63); BEWITCHED (I,418); BROKEN ARROW (I,743); ELFEGO BACA (I,1427); THE GIRL IN THE EMPTY GRAVE (II,997); THE LEGEND OF CUSTER (I,2652); NANCY (I,3221); SAINTS AND SINNERS (I,3887); THE SIX MILLION DOLLAR MAN (I,4067); THE WEB (I,4785)

SIMONE, LISA
AN APARTMENT IN ROME (I,242)

SIMONS, ALEX
COCAINE AND BLUE EYES (II,548)

SIMPSON, FRANK
THE RUBBER GUN SQUAD (II,2225)

SIMPSON, MICKEY
THE INDIAN (I,2210); SAM HILL (I,3895)

SIMPSON, O J
CELEBRITY CHALLENGE OF THE SEXES 1 (II,465); COCAINE AND BLUE EYES (II,548); ROOTS (II,2211)

SIMPSON, PAMELA
THE CRADLE SONG (I,1088)

SIMPSON, SANDY
MAKING A LIVING (II,1604); V: THE FINAL BATTLE (II,2714)

SIMS, MARLEY
HARRY'S BATTLES (II,1100)

SIMS, PHILIP
C.P.O. SHARKEY (II,407)

SIMS, RAY
HEAT OF ANGER (I,1992)

SIMS, WARWICK
AGAINST THE WIND (II,24); THE BLUE AND THE GRAY (II,273); GOLIATH AWAITS (II,1025)

SINATRA JR, FRANK
CLINIC ON 18TH STREET (II,544); DEAN MARTIN PRESENTS THE GOLDDIGGERS (I,1193); FRANK SINATRA JR. WITH FAMILY AND FRIENDS (I,1663)

SINATRA, CHRISTINA
ADAM-12 (I,31); FANTASY ISLAND (II,830)

SINATRA, FRANK
ANYTHING GOES (I,238); FRANCIS ALBERT SINATRA DOES HIS THING (I,1658); FRANK SINATRA (I,1660); THE FRANK SINATRA SHOW (I,1664); THE FRANK SINATRA SHOW (I,1665); THE FRANK SINATRA SHOW (I,1666); THE FRANK SINATRA SHOW (I,1667); THE FRANK SINATRA TIMEX SHOW (I,1668); THE FRANK SINATRA TIMEX SHOW (I,1667); THE FRANK SINATRA TIMEX SHOW (I,1670); FRANK SINATRA: A MAN AND HIS MUSIC (I,1662); FRANK SINATRA—A MAN AND HIS MUSIC (I,1661); MAGNAVOX PRESENTS FRANK SINATRA (I,2826); THE MATT DENNIS SHOW (I,2948); OUR TOWN (I,3419); SINATRA (I,4054); SINATRA AND FRIENDS (II,2358); SINATRA: CONCERT FOR THE AMERICAS (II,2361); SINATRA: THE MAN AND HIS MUSIC (II,2362); SINATRA—THE FIRST 40 YEARS (II,2359); SINATRA—THE MAIN EVENT (II,2360)

SINATRA, NANCY
MOVIN' WITH NANCY (I,3119); MOVIN' WITH NANCY ON STAGE (I,3120); WHO'S AFRAID OF MOTHER GOOSE? (I,4852)

SINATRA, RICHARD
MR. ROBERTS (I,3149)

SINCLAIR, BETTY
YOUNG DR. MALONE (I,4938)

SINCLAIR, CAROL
PAPA CELLINI (I,3450)

SINCLAIR, DONALD
VANITY FAIR (I,4701)

SINCLAIR, MADGE
DOWN HOME (II,731); GRANDPA GOES TO WASHINGTON (II,1050); GUESS WHO'S COMING TO DINNER? (II,1063); ROOTS (II,2211); THREE EYES (II,2605); TRAPPER JOHN, M.D. (II,2654); WALKIN' WALTER (II,2737)

SINCLAIR, MARY
FIRESIDE THEATER (I,1580); LITTLE WOMEN (I,2720)

SINCLAIR, MILLIE
EARN YOUR VACATION (I,1394)

SINCLAIR, PEGGY
THE SAVAGE CURSE (I,3925)

SINDEN, DONALD
FATHER DEAR FATHER (II,837)

SINDEN, JEREMY
BRIDESHEAD REVISITED (II,375); THE FAR PAVILIONS (II,832)

SINDEN, LEON
THE ASSASSINATION RUN (II,112)

SING, MAI TAI
HONG KONG (I,2112)

SINGER, ELLEN
BEAT THE CLOCK (I,378)

SINGER, LORI
FAME (II,812); THE KIDS FROM FAME (II,1392)

SINGER, MARC
79 PARK AVENUE (II,2310); THE CONTENDER (II,571); THE FEATHER AND FATHER GANG: NEVER CON A KILLER (II,846); PAPER DOLLS (II,1951); ROOTS: THE NEXT GENERATIONS (II,2213); V (II,2713); V: THE FINAL BATTLE (II,2714); V: THE SERIES (II,2715)

SINGER, RAYMOND
CHARLES IN CHARGE (II,482); FLATFOOTS (II,879); GOSSIP (II,1043); LOCAL 306 (II,1505); MAMA MALONE (II,1609); OPERATION PETTICOAT (II,1916); OPERATION PETTICOAT (II,1917)

SINGER, STUFFY
BEULAH (I,416); BLONDIE (I,486); SANDY DREAMS (I,3909)

SINGLETON, DORIS
ANGEL (I,216); FAMILY IN BLUE (II,818); THE GREAT GILDERSLEEVE (I,1875); THE HOUSE NEXT DOOR (I,2139); LUCY MOVES TO NBC (II,1562); MCKEEVER AND THE COLONEL (I,2970); MY THREE SONS (I,3205)

SINGLETON, EDDIE
HARRIS AND COMPANY (II,1096); LITTLE LULU (II,1493)

SINGLETON, PENNY
THE JETSONS (I,2376)

SINUTKO, SHANE
FEATHERSTONE'S NEST (II,847); LASSIE: THE NEW BEGINNING (II,1433); PINE CANYON IS BURNING (II,2040); RHODA (II,2151); SAMURAI (II,2249)

SIRGO, LOU
N.O.P.D. (I,3303)

SIRIANNI, E A
HAZARD'S PEOPLE (II,1112); NICHOLS AND DYMES (II,1841)

SIRICO, ANTHONY
THE PLANT FAMILY (II,2050)

SIROLA, JOE
THE BRIGHTER DAY (I,728)

SIROLA, JOSEPH
HIGH RISK (II,1146); THE MAGICIAN (I,2823); THE MONTEFUSCOS (II,1727); WASHINGTON: BEHIND CLOSED DOORS (II,2744)

SISKEL, GENE
AT THE MOVIES (II,117); SNEAK PREVIEWS (II,2388)

SISLER, LIZZ
MY LITTLE MARGIE (I,3194)

SISSON, DAVID
THE WORLD OF PEOPLE (II,2839)

SKAFF, GEORGE
HIGH RISK (II,1146)

SKAGGS, WOODY
CAPTAINS AND THE KINGS (II,435)

SKAGS, JIM
THE BEST OF FRIENDS (II,216)

SKALA, LILIA
GREEN ACRES (I,1884); IRONSIDE (I,2240); PROBE (I,3674); WHO HAS SEEN THE WIND? (I,4847)

SKELLEY, ARRIN
IT'S YOUR FIRST KISS, CHARLIE BROWN (I,2280); SHE'S A GOOD SKATE, CHARLIE BROWN (I,4012); YOU'RE THE GREATEST, CHARLIE BROWN (I,4973)

SKELTON, PATRICK
ALL MY CHILDREN (II,39)

SKELTON, RED
CLOWN ALLEY (I,979); FUNNY FACES (II,941); THE RED SKELTON CHEVY SPECIAL (I,3753); THE RED SKELTON REVUE (I,3754); THE RED SKELTON SHOW (I,3755); THE RED SKELTON SHOW (I,3756); THE RED SKELTON TIMEX SPECIAL (I,3757); RED SKELTON'S CHRISTMAS DINNER (II,2123); RED SKELTON'S FUNNY FACES (II,2124); RED SKELTON: A ROYAL PERFORMANCE (II,2125); RUDOLPH'S SHINY NEW YEAR (II,2228)

SKERRITT, TOM
A BELL FOR ADANO (I,388); GALLAGHER (I,1727); RYAN'S FOUR (II,2233)

SKIDMORE, JAN
ANNIE GET YOUR GUN (I,223)

SKINNER, CORNELIA OTIS
THE ALCOA HOUR (I,108); DEAREST ENEMY (I,1205); THE FARMER'S DAUGHTER (I,1534)

SKINNER, EDITH
CHILDREN'S SKETCH BOOK (I,941)

SKINNER, EDNA
MR. ED (I,3137); TOPPER (I,4582)

SKULNIK, MENASHA
MENASHA THE MAGNIFICENT (I,3010)

SKYLARKS, THE
JUDGE FOR YOURSELF (I,2466); TV'S TOP TUNES (I,4634)

SKYLER, TRISTINE
THE HAMPTONS (II,1076)

SLACK, BEN
ARCHIE BUNKER'S PLACE (II,105); FEELING GOOD (I,1559)

SLADE, BETSY
BEULAH LAND (II,226); HANDLE WITH CARE (II,1077)

SLADE, MARK
THE HIGH CHAPARRAL (I,2051); THE RETURN OF THE MOD SQUAD (I,3776); SALTY (II,2241); VOYAGE TO THE BOTTOM OF THE SEA (I,4743); THE WACKIEST SHIP IN THE ARMY (I,4744); WAIKIKI (II,2734)

SLADEN, ELISABETH
DOCTOR WHO (II,689); DOCTOR WHO—THE FIVE DOCTORS (II,690)

SLATE, HENRY
ADVENTURES IN PARADISE (I,46)

SLATE, JEREMY
ACCIDENTAL FAMILY (I,17); THE AQUANAUTS (I,250); MALIBU RUN (I,2848); MCGHEE (I,2967); ONE LIFE TO LIVE (II,1907); STRANGER IN OUR HOUSE (II,2477)

SLATEN, TROY
CAGNEY AND LACEY (II,409)

SLATER, BILL
BROADWAY TO HOLLYWOOD (I,739); CHARADE QUIZ (I,901); HAPPY BIRTHDAY (I,1940); MESSING PRIZE PARTY (I,3019); TWENTY QUESTIONS (I,4647)

SLATER, JUNE
THE EVIL TOUCH (I,1486)

SLATER, SHIRLEY
TEXAS (II,2566)

SLATER, TOM
ABE LINCOLN IN ILLINOIS (I,11)

SLATTERY, KERRY
CONCRETE BEAT (II,562)

SLATTERY, RICHARD X
C.P.O. SHARKEY (II,407); THE GALLANT MEN (I,1728); MR. ROBERTS (I,3149); OF MEN OF WOMEN (I,3332); RICH MAN, POOR MAN—BOOK II (II,2162); SWITCH (II,2519); WONDER WOMAN (PILOT 1) (II,2826)

SLAVENSKA, MIA
THE GREAT WALTZ (I,1878)

SLAVIN, MILLIE
6 RMS RIV VU (II,2371); BLACK BART (II,260); FARRELL: FOR THE PEOPLE (II,834); LANDON, LANDON & LANDON (II,1426); QUICK AND QUIET (II,2099); RAFFERTY (II,2105); STRUCK BY LIGHTNING (II,2482)

SLEEP, WAYNE
CINDERELLA (II,526)

SLEZAK, ERICA
ONE LIFE TO LIVE (II,1907)

SLEZAK, LEO
SLEZAK AND SON (I,4086)

SLEZAK, WALTER
BATMAN (I,366); THE CHEVY MYSTERY SHOW (I,928); A CRY OF ANGELS (I,1111); THE GOOD FAIRY (I,1846); MY THREE ANGELS (I,3204); PINOCCHIO (I,3602); SLEZAK AND SON (I,4086); THE SLIGHTLY FALLEN ANGEL (I,4087); STUDIO ONE (I,4268); TELEPHONE TIME (I,4379); THE WOMAN IN WHITE (I,4887)

SLEZER, MILTON
SCRUPLES (II,2280)

SLICK, EARL
DAVID BOWIE—SERIOUS MOONLIGHT (II,622)

SLIFER, ELIZABETH
RED RYDER (I,3752)

SLIFER, LIZ
THE STAR MAKER (I,4193)

SLOAN, LISA
THE EDGE OF NIGHT (II,760)

SLOAN, TINA
ANOTHER WORLD (II,97); THE GUIDING LIGHT (II,1064); SEARCH FOR TOMORROW (II,2284); SOMERSET (I,4115)

SLOANE, EVERETT
ALCOA/GOODYEAR THEATER (I,107); THE ARMSTRONG CIRCLE THEATER (I,260); THE DICK TRACY SHOW (I,1272); EXPERT WITNESS (I,1488); FRONT ROW CENTER (I,1694); HIGH TOR (I,2056); THE INNER SANCTUM (I,2216); JANE WYMAN PRESENTS THE FIRESIDE THEATER (I,2345); KRAFT MYSTERY THEATER (I,2589); THE MILLION DOLLAR INCIDENT (I,3045); NO WARNING (I,3301); OFFICIAL DETECTIVE (I,3339); ON TRIAL (I,3363); READY FOR THE PEOPLE (I,3739); SUSPENSE (I,4305); THRILLER (I,4492)

SLOANE, JACK
MR. MERGENTHWIRKER'S LOBBLIES (I,3144)

SLOANE, RICKEY
THE ALAN KING SHOW (I,100)

SLOANE, TIM
THAT TEEN SHOW (II,2573)

SLOANE, WILLIAM
THE BIG STORY (I,432)

SLOCUM, CRAIG
DARK SHADOWS (I,1157)

SLOCUM, SOPHIA
LOVERS AND FRIENDS (II,1549)

SLOMAN, ROGER
THE SEVEN DIALS MYSTERY (II,2308)

SLOUT, MARTE BOYLE
ALL THAT GLITTERS (II,42); FLAMINGO ROAD (II,872)

SLOWE, GEORGIA
MARCO POLO (II,1626)

SLOYAN, JAMES
THE ASPHALT COWBOY (II,111); CENTENNIAL (II,477); DISAPPEARANCE OF AIMEE (I,1290); OH MADELINE (II,1880); RYAN'S HOPE (II,2234); TENSPEED AND BROWN SHOE (II,2562); TROUBLE IN HIGH TIMBER COUNTRY (II,2661); WESTSIDE MEDICAL (II,2765)

SLY AND THE FAMILY STONE
GOOD VIBRATIONS FROM CENTRAL PARK (I,1855)

SMALL, MARYA
CHARLEY'S AUNT (II,483); SORORITY '62 (II,2405)

SMALLWOOD, TUCKER
SEARCH FOR TOMORROW (II,2284)

SMART, JEAN
PIAF (II,2039); REGGIE (II,2127); TEACHERS ONLY (II,2548)

SMART, PATSY
UPSTAIRS, DOWNSTAIRS (II,2709)

SMEDBERG, KEN
THE NEW TEMPERATURES RISING SHOW (I,3273)

SMERCZAK, RON
PRISONERS OF THE LOST UNIVERSE (II,2086)

SMETS, ALEXANDRA
THE SCARLET LETTER (II,2270)

SMIFF, KNUCKLEHEAD
RUNAROUND (I,3873)

SMIGHT, JOYCE
THE RESTLESS GUN (I,3774)

SMIKA, GINA MARIA
THE OREGON TRAIL (II,1922); THE OREGON TRAIL (II,1923)

SMILDSIN, KURT
NIGHT PARTNERS (II,1847)

SMILEY, JOSEPH
THE VALIANT (I,4697)

SMILLIE, JAMES
MURDER ON THE MIDNIGHT EXPRESS (I,3162); PRISONER: CELL BLOCK H (II,2085); RETURN TO EDEN (II,2143)

SMINKLEY, TOM
THE DOCTORS (II,694)

SMITH, ALBERT
THE GREATEST GIFT (II,1061)

SMITH, ALEXIS
DALLAS (II,614); HENRY FONDA PRESENTS THE ATSR AND THE STORY (I,2016); NIGHTSIDE (I,3296); ON TRIAL (I,3361); RHEINGOLD THEATER (I,3783)

SMITH, ALLISON
KATE AND ALLIE (II,1365)

SMITH, ANDREA

THE AMERICAN DREAM (II,71)

SMITH, ARCHIE
CRIME PHOTOGRAPHER (I,1097)

SMITH, BERNICE
THE YOUNG PIONEERS (II,2868)

SMITH, BILL
THE ASPHALT JUNGLE (I,297); BORDER TOWN (I,689); THE FUZZ BROTHERS (I,1722); ZERO ONE (I,4981)

SMITH, BOB
THE BOB SMITH SHOW (I,668); CHICAGOLAND MYSTERY PLAYERS (I,936); HOWDY DOODY (I,2151); HOWDY DOODY AND FRIENDS (I,2152); THE NEW HOWDY DOODY SHOW (II,1818)

SMITH, BRETT
PAROLE (II,1960)

SMITH, BUBBA
BLUE THUNDER (II,278); JOE DANCER: THE BIG BLACK PILL (II,1300); OPEN ALL NIGHT (II,1914); SEMI-TOUGH (II,2299); SEMI-TOUGH (II,2300)

SMITH, CAMERON
RITUALS (II,2182)

SMITH, CARL
COUNTRY MUSIC CARAVAN (I,1066); COUNTRY MUSIC HALL (I,1068); FIVE STAR JUBILEE (I,1589)

SMITH, CAROLINE
THE LAST CONVERTIBLE (II,1435); LITTLE LORD FAUNTLEROY (II,1492); THE SECRET WAR OF JACKIE'S GIRLS (II,2294); WOMEN IN WHITE (II,2823)

SMITH, CHARLES
BOSS LADY (I,694); FOREST RANGER (I,1641); HOLIDAY LODGE (I,2077); THE SUPER SIX (I,4292); WHERE'S RAYMOND? (I,4837)

SMITH, CHARLES MARTIN
A DOG'S LIFE (II,697); GABE AND WALKER (II,950)

SMITH, CINDY
OPRYLAND: NIGHT OF STARS AND FUTURE STARS (II,1921)

SMITH, COOPER
ST. ELSEWHERE (II,2432)

SMITH, COTTER
MISTRAL'S DAUGHTER (II,1708)

SMITH, CRAIG
THE HAPPENERS (I,1935)

SMITH, CYRIL
THE ADVENTURES OF SIR LANCELOT (I,76)

SMITH, DAN
CAMP WILDERNESS (II,419)

SMITH, DAVID
TEACHER'S PET (I,4367)

SMITH, DEIDRE
BIG DADDY (I,425)

SMITH, DEREK
THE CARNATION KILLER (I,845); SCREAMER (I,3945)

SMITH, DINNIE
HOTEL COSMOPOLITAN (I,2128)

SMITH, DONEGAN
FLO (II,891)

SMITH, DOROTHY
ADVENTURES OF CLINT AND MAC (I,55)

SMITH, DWAN
ELLERY QUEEN: TOO MANY SUSPECTS (II,767); JOE FORRESTER (II,1303)

SMITH, EBONIE
THE JEFFERSONS (II,1276)

SMITH, ELIZABETH
BIG HAWAII (II,238); DANGER IN PARADISE (II,617)

SMITH, EMANUEL
THE CITY (I,969)

SMITH, FRANK
AIR SUPPLY IN HAWAII (II,26); TWIGS (II,2684)

SMITH, FRAZER
ROCK-N-AMERICA (II,2195)

SMITH, G ALBERT
HAPPY BIRTHDAY (I,1941)

SMITH, GARNETT
NUTS AND BOLTS (II,1871)

SMITH, GEORGE
THE DOCTORS (II,694)

SMITH, GIL
DENNIS THE MENACE (I,1231); PETER LOVES MARY (I,3565)

SMITH, H ALLEN
ARMCHAIR DETECTIVE (I,259)

SMITH, HAL
THE ANDY GRIFFITH SHOW (I,192); DAVY AND GOLIATH (I,1180); DOCTOR DOLITTLE (I,1309); FRANKENSTEIN JR. AND THE IMPOSSIBLES (I,1672); HELP! IT'S THE HAIR BEAR BUNCH (I,2012); I MARRIED JOAN (I,2174); JEFFERSON DRUM (I,2355); THE PETER POTAMUS SHOW (I,3568); THE ROMAN HOLIDAYS (I,3835); SHERIFF WHO? (I,4009)

SMITH, HOWARD
THE ALDRICH FAMILY (I,111); CURTAIN CALL THEATER (I,1115); THE EVE ARDEN SHOW (I,1471); FIRST LOVE (I,1583); HAZEL (I,1982); TROUBLE WITH RICHARD (I,4612)

SMITH, HOWARD K
V (II,2713); V: THE SERIES (II,2715)

SMITH, J BRENNAN
ADAMS HOUSE (II,9); THE BAD NEWS BEARS (II,134); THE CLIFFWOOD AVENUE KIDS (II,543)

SMITH, JACK
THE AMERICAN WEST (I,174); LOVE STORY (I,2777); PLACE THE FACE (I,3611); TRAILS TO ADVENTURE (I,4591); YOU ASKED FOR IT (I,4931); YOU ASKED FOR IT (II,2855)

SMITH, JACLYN
BATTLE OF THE NETWORK STARS (II,164); CHARLIE'S ANGELS (II,486); CHARLIE'S ANGELS (II,487); FOOLS, FEMALES AND FUN: IS THERE A DOCTOR IN THE HOUSE? (II,900); GEORGE WASHINGTON (II,978); THE MAD MAD MAD MAD WORLD OF THE SUPER BOWL (II,1586); MAGIC WITH THE STARS (II,1599); RAGE OF ANGELS (II,2107); SWITCH (II,2520); TONI'S BOYS (II,2637)

SMITH, JAMES S
SAM (II,2246)

SMITH, JAMIE
GOLDEN WINDOWS (I,1841)

SMITH, JIM B
THE RANGERS (II,2112)

SMITH, JOHN
CIMARRON CITY (I,953); FRONTIER (I,1695); LARAMIE (I,2614)

SMITH, JUSTIN
POLICE STORY (I,3636)

SMITH, K L
BELLE STARR (I,391)

SMITH, K T
HARD CASE (I,1947)

SMITH, KATE
THE KATE SMITH EVENING HOUR (I,2507); THE KATE SMITH HOUR (I,2508); THE KATE SMITH SHOW (I,2509); THE KATE SMITH SHOW (I,2510); THE KATE SMITH SHOW (I,2511); THE LONDON PALLADIUM (I,2739); ONCE UPON A CHRISTMAS TIME (I,3368)

SMITH, KEITH
MOLL FLANDERS (II,1720)

SMITH, KENT
THE INVADERS (I,2229); THE JUDGE AND JAKE WYLER (I,2464); KING RICHARD II (I,2559); LITTLE WOMEN (I,2720); THE NIGHT STALKER (I,3292); PEYTON PLACE (I,3574); PHILIP MORRIS PLAYHOUSE (I,3585); PROBE (I,3674); SECRETS OF THE OLD BAILEY (I,3971)

SMITH, KIMRY
JUMP (II,1356)

SMITH, KURTWOOD
THE RENEGADES (II,2132); THE RENEGADES (II,2133)

SMITH, LANE
THE BIG EASY (II,234); CHIEFS (II,509); THE GEORGIA PEACHES (II,979); THE MEMBER OF THE WEDDING (II,1680); THOU SHALT NOT KILL (II,2604); V: THE SERIES (II,2715)

SMITH, LARRY
THE UNCLE AL SHOW (I,4667)

SMITH, LEONARD
OUR MISS BROOKS (I,3415)

SMITH, LEWIS
LONE STAR (II,1513)

SMITH, LIONEL
PARK PLACE (II,1958)

SMITH, LIZ
SEPARATE TABLES (II,2304)

SMITH, LOIS
THE DOCTORS (II,694); LOVE OF LIFE (I,2774); RAGE OF ANGELS (II,2107); SOMERSET (I,4115); STAR TONIGHT (I,4199); VICTORY (I,4729)

SMITH, LORING
CAPTAIN BRASSBOUND'S CONVERSION (I,828); THE HARTMANS (I,1960); HARVEY (I,1961); MIRACLE ON 34TH STREET (I,3057); THE RIGHT MAN (I,3790)

SMITH, MADELINE
WHY DIDN'T THEY ASK EVANS? (II,2795)

SMITH, MARIELLEN

BAND OF GOLD (I,341)

SMITH, MARK
WAY OUT GAMES (II,2747)

SMITH, MARTHA
ALEX AND THE DOBERMAN GANG (II,32); BATTLE OF THE NETWORK STARS (II,177); EBONY, IVORY AND JADE (II,755); HARD KNOCKS (II,1086); SCARECROW AND MRS. KING (II,2268)

SMITH, MATTHEW
WHEN THE NIGHINGALE SANG IN BERKELEY SQUARE (I,4827)

SMITH, MATTHEW DAVID
THE JOEY BISHOP SHOW (I,2404)

SMITH, MEL
NOT THE NINE O'CLOCK NEWS (II,1864)

SMITH, MICHAEL
THE MOUSEKETEERS REUNION (II,1742)

SMITH, MICHELLE
A.K.A. PABLO (II,28)

SMITH, NICHOLAS
DR JEKYLL AND MR. HYDE (I,1368)

SMITH, O C
BIG DADDY (I,425)

SMITH, OLIVER
COLOR HIM DEAD (II,554)

SMITH, PATRICIA
THE BOB NEWHART SHOW (II,342); THE DEBBIE REYNOLDS SHOW (I,1214); PLANET EARTH (I,3615); RIDDLE AT 24000 (II,2171); WHERE'S EVERETT? (I,4835)

SMITH, PAUL
THE DORIS DAY SHOW (I,1355); FIBBER MCGEE AND MOLLY (I,1569); INSIDE O.U.T. (II,1235); MASADA (II,1642); MCNAB'S LAB (I,2972); THE MONSTER SQUAD (II,1725); MR. TERRIFIC (I,3152); MRS. G. GOES TO COLLEGE (I,3155); NO TIME FOR SERGEANTS (I,3300)

SMITH, PETE
M-G-M PARADE (I,2798); PETE SMITH SPECIALTIES (I,3560)

SMITH, QUEENIE
THE FUNNY SIDE (I,1714); THE FUNNY SIDE (I,1715)

SMITH, RAY
THE CITADEL (II,539); LOOK BACK IN DARKNESS (I,2752); MASADA (II,1642)

SMITH, REID
BORDER PALS (II,352); THE CHISHOLMS (II,513); THE HOYT AXTON SHOW (II,1200); MALIBU (II,1607)

SMITH, REX
SOLID GOLD (II,2395)

SMITH, ROGER
77 SUNSET STRIP (I,3988); FORD TELEVISION THEATER (I,1634); KNIGHT'S GAMBIT (I,2576); MR. ROBERTS (I,3149)

SMITH, ROY
PINOCCHIO (II,2044)

SMITH, RUFUS
ANNIE GET YOUR GUN (I,224)

SMITH, SAM
HOW TO SUCCEED IN BUSINESS WITHOUT REALLY TRYING (II,1197)

SMITH, SAMANTHA
CHARLES IN CHARGE (II,482)

SMITH, SAMIE
MOLL FLANDERS (II,1720)

SMITH, SAMMY
THE PATTY DUKE SHOW (I,3483)

SMITH, SANDRA
THE INTERNS (I,2226); SCALPLOCK (I,3930)

SMITH, SANDY
OUR PRIVATE WORLD (I,3417)

SMITH, SCOTT
FIREHOUSE (II,859)

SMITH, SHELLEY
THE ASSOCIATES (II,113); BATTLE OF THE NETWORK STARS (II,169); THE CELEBRITY FOOTBALL CLASSIC (II,474); FOR LOVE AND HONOR (II,903); THE PHOENIX (II,2035); SCRUPLES (II,2281)

SMITH, SYDNEY
HOME (I,2099)

SMITH, TED
THE NEW ADVENTURES OF HUCKLEBERRY FINN (I,3250)

SMITH, TONI GAYLE
DUNGEONS AND DRAGONS (II,744)

SMITH, TRUMAN
AH!, WILDERNESS (I,91); FAMOUS JURY TRIALS (I,1525); HANDS OF MURDER (I,1929)

SMITH, WILLIAM
THE BRIGHTER DAY (I,728); HAWAII FIVE-O (II,1110); THE JERK, TOO (II,1282); LAREDO (I,2615); MRS R.—DEATH AMONG FRIENDS (II,1759); THE REBELS (II,2121); RICH MAN, POOR (II,2162); RICH MAN, POOR MAN—BOOK I (II,2161); RICH MAN, POOR MAN—BOOK II (II,2162); THE ROCKFORD FILES (II,2196); TALES OF THE APPLE DUMPLING GANG (II,2536)

SMITHERS, JAN
BATTLE OF THE NETWORK STARS (II,169); LOVE STORY (I,2778); WKRP IN CINCINNATI (II,2814)

SMITHERS, WILLIAM
CALL TO DANGER (I,803); CANNON: THE RETURN OF FRANK CANNON (II,425); DALLAS (II,614); DOCTORS' PRIVATE LIVES (II,692); EAGLE IN A CAGE (I,1392); ESPIONAGE (I,1464); EXECUTIVE SUITE (II,796); THE GUIDING LIGHT (II,1064); THE MONK (I,3087); PEYTON PLACE (I,3574)

SMITROVICH, BILL
JOHNNY GARAGE (II,1337)

SMOLKA, KEN
GENERAL HOSPITAL (II,964)

SMOOT, FRED
CAMPO 44 (I,813); THE WACKIEST SHIP IN THE ARMY (I,4744)

SMOTHERS, DICK
ALICE THROUGH THE LOOKING GLASS (I,122); FITZ AND BONES (II,867); MARINELAND CARNIVAL (I,2912); THE RETURN OF THE SMOTHERS BROTHERS (I,3777); THE SMOTHERS BROTHERS (I,4095); THE SMOTHERS BROTHERS (I,4096); THE SMOTHERS BROTHERS (I,4097); THE SMOTHERS BROTHERS (II,2384); THE SMOTHERS BROTHERS COMEDY HOUR (I,4095); THE SMOTHERS BROTHERS SHOW (I,4096); THE SMOTHERS BROTHERS SHOW (I,4097); THE SMOTHERS BROTHERS SHOW (II,2384); THE SMOTHERS ORGANIC PRIME TIME SPACE RIDE (I,4098); TERROR AT ALCATRAZ (II,2563)

SMOTHERS, TOM
ALICE THROUGH THE LOOKING GLASS (I,122); FITZ AND BONES (II,867); MARINELAND CARNIVAL (I,2912); THE RETURN OF THE SMOTHERS BROTHERS (I,3777); THE SMOTHERS BROTHERS (I,4095); THE SMOTHERS BROTHERS (I,4096); THE SMOTHERS BROTHERS (I,4097); THE SMOTHERS BROTHERS (II,2384); THE SMOTHERS BROTHERS COMEDY HOUR (I,4095); THE SMOTHERS BROTHERS SHOW (I,4096); THE SMOTHERS BROTHERS SHOW (I,4097); THE SMOTHERS BROTHERS SHOW (II,2384); THE SMOTHERS ORGANIC PRIME TIME SPACE RIDE (I,4098); TERROR AT ALCATRAZ (II,2563)

SMYTH, SHARON
DARK SHADOWS (I,1157); SEARCH FOR TOMORROW (II,2284)

SMYTHE, MARCUS
THE GUIDING LIGHT (II,1064); MOTHER, JUGGS AND SPEED (II,1741); SEARCH FOR TOMORROW (II,2284); SUMMER SOLSTICE (II,2492)

SNEE, JOHN
STILL THE BEAVER (II,2466)

SNEED, ALLYSIA
FAME (II,812)

SNEED, MAURICE
SUSAN AND SAM (II,2506)

SNELL, JAMES
THE FAR PAVILIONS (II,832)

SNIDER, BARRY
RAGE OF ANGELS (II,2107); VALLEY FORGE (I,4698)

SNIVELY, ROBERT
RETURN OF THE WORLD'S GREATEST DETECTIVE (II,2142)

SNODGRESS, CARRIE
THE WHOLE WORLD IS WATCHING (I,4851)

SNOWDEN, VAN
THE BUGALOOS (I,757); LIDSVILLE (I,2679); SIGMUND AND THE SEA MONSTERS (II,2352)

SNYDER, ARLEN DEAN
DEAR DETECTIVE (II,652); HELLINGER'S LAW (II,1127); NIGHT PARTNERS (II,1847); THE TEXAS RANGERS (II,2567); TRAUMA CENTER (II,2655)

SNYDER, BARRY
TOP OF THE HILL (II,2645)

SNYDER, DREW
ANOTHER WORLD (II,97); EMERALD POINT, N.A.S. (II,773); LOVE OF LIFE (I,2774)

SNYDER, NANCY
FOR RICHER, FOR POORER (II,906); ONE LIFE TO LIVE (II,1907)

SNYDER, TOM
TOM SNYDER'S CELEBRITY SPOTLIGHT (II,2632); THE TOMORROW SHOW (II,2635)

SNYDERS, SAMMY
THE BAXTERS (II,184)

SOARES, SILVIA
SARA (II,2260)

SOBALOFF, ABRAHAM
MASQUERADE (I,2940)

SOBIER, CLIFF
SUPER CIRCUS (I,4287)

SOBLE, RON
THE DAUGHTERS OF JOSHUA CABE (I,1169); MCNAMARA'S BAND (II,1669); THE MONROES (I,3089); TOMA (I,4547)

SOBOLOFF, ABRAHAM
HECK'S ANGELS (II,1122)

SOBOLOFF, ARNOLD
LOOK OUT WORLD (II,1516); TABITHA (II,2526); THE YOUNG PIONEERS (II,2868); THE YOUNG PIONEERS' CHRISTMAS (II,2870)

SOCHOR, JUDY
YOUR SHOW TIME (I,4958)

SODSAI, SONDI
ADVENTURES IN PARADISE (I,46)

SOFAER, ABRAHAM
I DREAM OF JEANNIE (I,2167)

SOKOL, MARILYN
SCALPELS (II,2266)

SOKOLOFT, VLADMIR
SABU AND THE MAGIC RING (I,3881)

SOLARI, RUDY
GARRISON'S GORILLAS (I,1735); REDIGO (I,3758); RETURN TO PEYTON PLACE (I,3779); THE WACKIEST SHIP IN THE ARMY (I,4744)

SOLBERG, MIM
A CHRISTMAS CAROL (II,517)

SOLES, P J
CALLING DR. STORM, M.D. (II,415); THE POSSESSED (II,2071); SAWYER AND FINN (II,2265)

SOLES, PAUL
THE RELUCTANT DRAGON AND MR. TOAD (I,3762); RUDOLPH THE RED-NOSED REINDEER (II,2227)

SOLES, PETER
SPIDER-MAN (I,4154)

SOLIANO, JOHN
WILSON'S REWARD (II,2805)

SOLOMAN, DEBORAH
ALL MY CHILDREN (II,39)

SOLOMON, BRUCE
LANIGAN'S RABBI (II,1428); MARY HARTMAN, MARY HARTMAN (II,1638)

SOLOMON, GEORGE
I GAVE AT THE OFFICE (II,1213)

SOLOMON, MARCIA
KNOTS LANDING (II,1404)

SOLOW, PAMELA
THE YOUNG AND THE RESTLESS (II,2862)

SOMACK, JACK
BALL FOUR (II,137); CARLTON YOUR DOORMAN (II,440); RETURN OF THE MAN FROM U.N.C.L.E.: THE 15 YEARS LATER AFFAIR (II,2140); THE STOCKARD CHANNING SHOW (II,2468)

SOMER, LINDA
BEAT THE CLOCK (I,378)

SOMERS, BRETT
GETTING THERE (II,985); THE MARY TYLER MOORE SHOW (II,1640); THE NEW ADVENTURES OF PERRY MASON (I,3252); THE ODD COUPLE (II,1875); THE WOMEN (I,4890)

SOMERS, DON
ALICE IN WONDERLAND (I,120)

SOMERS, JOCELYN
DAYS OF OUR LIVES (II,629); THE DOCTORS (II,694)

SOMERS, JULIAN
CORONATION STREET (I,1054)

SOMERS, SUZANNE
BATTLE OF THE NETWORK STARS (II,165); BATTLE OF THE NETWORK STARS (II,166); CELEBRITY CHALLENGE OF THE SEXES 2 (II,466); CELEBRITY CHALLENGE OF THE SEXES 3 (II,467); ECHOES OF THE SIXTIES (II,756); HIGH ROLLERS (II,1147); THE SUZANNE SOMERS (II,2511); THE SUZANNE SOMERS SPECIAL (II,2510); THE SUZANNE SOMERS SPECIAL (II,2511); THREE'S COMPANY (II,2614)

SOMES, MICHAEL
CINDERELLA (I,955); SLEEPING BEAUTY (I,4083)

SOMMARS, JULIE
CENTENNIAL (II,477); FOOLS, FEMALES AND FUN: I'VE GOTTA BE ME (II,899); THE GOVERNOR AND J.J. (I,1863); GREAT ADVENTURE (I,1868); I'M THE GIRL HE WANTS TO KILL (I,2194); MY WIFE NEXT DOOR (II,1780); REX HARRISON PRESENTS SHORT STORIES OF LOVE (II,2148); RITUALS (II,2182); THE SEARCH (I,3950)

SOMMER, ELKE
CELEBRITY CHALLENGE OF THE SEXES 2 (II,466); CIRCUS OF THE STARS (II,533); ELKE (I,1432); ELKE SOMMER'S WORLD OF SPEED AND BEAUTY (II,765); FOREVER FERNWOOD (II,909); PROBE (I,3674); STUNT SEVEN (II,2486); TOP OF THE HILL (II,2645)

SOMMER, JOSEF
THE SCARLET LETTER (II,2270); VALLEY FORGE (I,4698)

SOMMERFIELD, DIANE
DAYS OF OUR LIVES (II,629); MALIBU (II,1607); SALT AND PEPE (II,2240)

SOMMERS, ELLIE
HANS BRINKER OR THE SILVER SKATES (I,1933)

SOMMERS, JIMMIE
A DATE WITH JUDY (I,1162)

SOMMERS, JOANIE
MR. & MRS. & MR. (II,1744); ON THE FLIP SIDE (I,3360); WHO'S AFRAID OF MOTHER GOOSE? (I,4852)

SOMMERS, SHELLEY
PAUL BERNARD—PSYCHIATRIST (I,3485)

SOMMERS, YALE
DAKTARI (I,1127)

SOMMERVILLE, KAREN
THE FANTASTIC JOURNEY (II,826)

SONDERGAARD, GALE
THE BEST OF EVERYTHING (I,406); CENTENNIAL (II,477); NIGHT GALLERY (I,3287)

SONDOR, VICTOR
WALKING TALL (II,2738)

SONKILLA, PAUL
RETURN TO EDEN (II,2143)

SOO, JACK
BARNEY MILLER (II,154); THE JIMMY STEWART SHOW (I,2391); THE MONK (I,3087); THE PIRATES OF FLOUNDER BAY (I,3607); VALENTINE'S DAY (I,4693)

SOODIK, TRISH
LUCAS TANNER (II,1556)

SOPER, ANDY
PAROLE (II,1960)

SOPER, MARK
WILSON'S REWARD (II,2805)

SORBACK, ROGER
O'MALLEY (II,1872)

SOREL, LOUISE
BEWITCHED (I,418); CIRCLE OF FEAR (I,958); THE CURSE OF DRACULA (II,611); THE DON RICKLES SHOW (I,1342); GET CHRISTIE LOVE! (II,982); KNOTS LANDING (II,1404); LADIES' MAN (II,1423); NIGHT GALLERY (I,3287); ONE IN A MILLION (II,1906); ONE OF OUR OWN (II,1910); THE RETURN OF CHARLIE CHAN (II,2136); SANTA BARBARA (II,2258); THE SURVIVORS (I,4301)

SORELLO, WALTER
THE WILD WILD WEST (I,4863)

SORENSEN, LINDA
OUR MAN FLINT: DEAD ON TARGET (II,1932)

SORENSEN, RICKIE
FATHER OF THE BRIDE (I,1543); GIDGET (I,1795); TARZAN AND THE TRAPPERS (I,4362)

SORENSON, LELANI
MAKE ROOM FOR DADDY (I,2843)

SORIN, LOU
THE GOLDBERGS (I,1836)

SORRELLS, ROBERT
ENSIGN O'TOOLE (I,1448); PAPA G.I. (I,3451)

SORVINO, PAUL
BERT D'ANGELO/SUPERSTAR (II,211); CHIEFS (II,509); SEVENTH AVENUE (II,2309); WE'LL GET BY (II,2753); WE'LL GET BY (II,2754)

SOTHERN, ANN
ALIAS SMITH AND JONES (I,118); ALWAYS APRIL (I,147); THE ANN SOTHERN SHOW (I,220); CAPTAINS AND THE KINGS (II,435); FOL-DE-ROL (I,1613); HOLIDAY IN LAS VEGAS (I,2076); HOLLYWOOD OPENING NIGHT (I,2087); THE

JUNE ALLYSON SHOW (I,2488); LADY IN THE DARK (I,2602); MY MOTHER THE CAR (I,3197); THE OUTSIDER (I,3430); PANDORA AND FRIEND (I,3445); PRIVATE SECRETARY (I,3672); SCHLITZ PLAYHOUSE OF STARS (I,3936)

SOTO, ROSANA
A.E.S. HUDSON STREET (II,17)

SOTOS, GEORGE
SIT OR MISS (I,4064)

SOTTSCHALK, NORM
HAWKINS FALLS, POPULATION 6200 (I,1977)

SOUBIER, CLIFF
THE GOOK FAMILY (I,1858)

SOUL, DAVID
CASABLANCA (II,450); CIRCLE OF FEAR (I,958); A COUNTRY CHRISTMAS (II,582); THE DAVID SOUL AND FRIENDS SPECIAL (II,626); HERE COME THE BRIDES (I,2024); INTERTECT (I,2228); THE MANIONS OF AMERICA (II,1623); MOVIN' ON (I,3118); OWEN MARSHALL: COUNSELOR AT LAW (I,3435); STARSKY AND HUTCH (II,2444); STARSKY AND HUTCH (II,2445); THE YELLOW ROSE (II,2847)

SOULE, OLAN
THE 25TH MAN (MS) (II,2678); ARNIE (I,261); CAPTAIN MIDNIGHT (I,833); CHARLIE COBB: NICE NIGHT FOR A HANGING (II,485); THE D.A.: CONSPIRACY TO KILL (I,1121); GETAWAY CAR (I,1783); HAVE GUN—WILL TRAVEL (I,1968); THE JERK, TOO (II,1282); THE NEW SUPER FRIENDS HOUR (II,1830); SUPER FRIENDS (II,2497); THE WORLD'S GREATEST SUPER HEROES (II,2842)

SOUTHERN, HARRY
BUCK ROGERS IN THE 25TH CENTURY (I,751)

SPACE, ARTHUR
ALICE (II,34); JEFF'S COLLIE (I,2356); MA AND PA (II,1570); NATIONAL VELVET (I,3231); TIMMY AND LASSIE (I,4520)

SPADER, JAMES
DINER (II,679); THE FAMILY TREE (II,820)

SPAIN, FAY
JOHNNY GUITAR (I,2427); PURSUIT (I,3700); THE THIRD COMMANDMENT (I,4449); THE WEB (I,4785)

SPANG, LAURETTE
B.J. AND THE BEAR (II,125); BATTLESTAR GALACTICA (II,181); THE CELEBRITY FOOTBALL CLASSIC (II,474); COLORADO C.I. (II,555); THE LOVE BOAT I (II,1532); MCNAMARA'S BAND (II,1669); TOURIST (II,2651)

SPANO, JOE
HILL STREET BLUES (II,1154)

SPARDLIN, C D
SAM HILL: WHO KILLED THE MYSTERIOUS MR. FOSTER? (I,3896)

SPARER, PAUL
SOMERSET (I,4115)

SPARKS, DANA
COVER UP (II,597)

SPARKS, DON
THE HARVEY KORMAN SHOW (II,1103); MADAME'S PLACE (II,1587); THE NEW OPERATION PETTICOAT (II,1826)

SPARKS, TONY
THE BEST OF FRIENDS (II,216)

SPARV, CAMILLA
THE FEATHER AND FATHER GANG: NEVER CON A KILLER (II,846); JACQUELINE SUSANN'S VALLEY OF THE DOLLS, 1981 (II,1268); MASSARATI AND THE BRAIN (II,1646)

SPAULL, GUY
EAGLE IN A CAGE (I,1392)

SPEAR, JACK
PIP THE PIPER (I,3606)

SPEAR, MARTIN
FARRELL: FOR THE PEOPLE (II,834)

SPEAR, PHYLLIS
PIP THE PIPER (I,3606)

SPEARS, LES
HAWKINS FALLS, POPULATION 6200 (I,1977)

SPEARS, STEVE
SPACE STARS (II,2408); TROLLKINS (II,2660)

SPECHT, BETH
THE BILLION DOLLAR THREAT (II,246); KING OF THE ROAD (II,1396)

SPECHT, KAREN
THE BILLION DOLLAR THREAT (II,246); KING OF THE ROAD (II,1396)

SPEED, DORIS
CORONATION STREET (I,1054)

SPEER, MARTIN
EXO-MAN (II,797); FLYING HIGH (II,895)

SPEIDEL, TOM
ART AND MRS. BOTTLE (I,271)

SPEILBERG, DAVID
FROM HERE TO ETERNITY (II,933)

SPEIRS, ROSALIND
PRISONER: CELL BLOCK H (II,2085)

SPEISER, JERRY
MEN AT WORK IN CONCERT (II,1682)

SPEISS, FANNIE
THE DOCTORS (II,694)

SPELLING, DANIEL
MR. AND MRS. COP (II,1748)

SPELLING, VICTORIA
SHOOTING STARS (II,2341)

SPELMAN, SHARON
ANGIE (II,80); CALLING DR. STORM, M.D. (II,415); THE COP AND THE KID (II,572); THE DEADLY GAME (II,633); GIMME A BREAK (II,995); THE GIRL IN THE EMPTY GRAVE (II,997); NUMBER 96 (II,1869); PAUL SAND IN FRIENDS AND LOVERS (II,1982); PEEPING TIMES (II,1990); SEARCH FOR TOMORROW (II,2284); SECOND EDITION (II,2288)

SPENCE, FIONA
PRISONER: CELL BLOCK H (II,2085)

SPENCE, SANDRA
TERRY AND THE PIRATES (I,4405); THE WHIRLYBIRDS (I,4841)

SPENCER, DANIELLE
WHAT'S HAPPENING!! (II,2769)

SPENCER, ELLEN
SEARCH FOR TOMORROW (II,2284)

SPENCER, EVE
AUNT JENNY'S REAL LIFE STORIES (I,314)

SPENCER, GILLIAN
ALL MY CHILDREN (II,39); AS THE WORLD TURNS (II,110); THE GUIDING LIGHT (II,1064); ONE LIFE TO LIVE (II,1907)

SPENCER, GORDON
CAROL FOR ANOTHER CHRISTMAS (I,852)

SPENCER, JOHN
COCAINE AND BLUE EYES (II,548)

SPENCER, LARRY
THE $1.98 BEAUTY SHOW (II,698)

SPENCER, NEVADA
DOCTOR DAN (II,684)

SPENCER, PENNY
U.F.O. (I,4662)

SPENCER, SALLY-JANE
THE FALL AND RISE OF REGINALD PERRIN (II,810)

SPENCER, STEVEN
EVERY STRAY DOG AND KID (II,792)

SPERBER, MILO
THE WOMAN IN WHITE (II,2819)

SPERBER, WENDIE JO
BOSOM BUDDIES (II,355); PRIVATE BENJAMIN (II,2087); THE STOCKARD CHANNING SHOW (II,2468)

SPERDAKOS, GEORGE
EXO-MAN (II,797); WILD ABOUT HARRY (II,2797)

SPIEGEL, RIC
BENJI, ZAX AND THE ALIEN PRINCE (II,205)

SPIELBERG, DAVID
THE AMERICAN GIRLS (II,72); BOB & CAROL & TED & ALICE (I,495); FORCE FIVE (II,907); FROM HERE TO ETERNITY (II,932); JESSICA NOVAK (II,1286); OPERATING ROOM (II,1915); THE PRACTICE (II,2076); STONE (II,2470); WHEELS (II,2777)

SPIES, ADRIAN
CRISIS (I,1101)

SPIES, CARSON
BACKSTAIRS AT THE WHITE HOUSE (II,133)

SPILKER, SUSAN
I'D RATHER BE CALM (II,1216); SIGN-ON (II,2353)

SPILLMAN, SANDY
MR. & MS. AND THE MAGIC STUDIO MYSTERY (II,1747); NIGHT COURT (I,3285)

SPINELL, JOE
NIGHTSIDE (II,1851)

SPINER, BRENT
THE DAIN CURSE (II,613)

SPINETTI, VICTOR
MISTRAL'S DAUGHTER (II,1708)

SPINKS, JAMES
THE NIGHTENGALES (II,1850)

SPINNEY, CARROLL
SESAME STREET (I,3982)

SPIRIDAKIS, TONY
BAY CITY BLUES (II,186)

SPO-DE-ODEE
FATHER O FATHER (II,842); GREAT DAY (II,1053); WALKIN' WALTER (II,2737)

SPOAN, PATRICK
STEAMBATH (II,2451)

SPOUND, MICHAEL
HOME ROOM (II,1172); HOTEL (II,1192)

SPRADLIN, G D
CALL TO GLORY (II,413); DIAL HOT LINE (I,1254); HERNANDEZ, HOUSTON P.D. (I,2034); THE OREGON TRAIL (II,1922); RICH MAN, POOR MAN—BOOK II (II,2162)

SPRIGGS, ELIZABETH
THE SECRET OF CHARLES DICKENS (II,2293)

SPRINGFIELD, RICK
BATTLESTAR GALACTICA (II,181); MISSION MAGIC (I,3068); THE NANCY DREW MYSTERIES (II,1789)

SPRUANCE, DON
BEN CASEY (I,394)

SPRUILL, STEPHANIE
OLIVIA NEWTON-JOHN IN CONCERT (II,1884)

SPRUNG, SANDY
THE NATURAL LOOK (II,1799)

SPURRIER, PAUL
PENMARRIC (II,1993)

SQUEAKY
ANDY'S GANG (I,214)

SQUIRE, KATHERINE
THE DOCTORS (II,694); ONE LIFE TO LIVE (II,1907)

ST CLEMENT, PAM
A HORSEMAN RIDING BY (II,1182)

ST JACQUES, RAYMOND
THE 416TH (II,913); THE EYES OF TEXAS (II,799); THE MONK (I,3087); RAWHIDE (I,3727); ROOTS (II,2211)

ST JOHN, AL (FUZZY)
LASH OF THE WEST (I,2620)

ST JOHN, BILL
MICKEY AND THE CONTESSA (I,3024)

ST JOHN, CHRISTOFF
THE BAD NEWS BEARS (II,134); BEULAH LAND (II,226); BIG JOHN, LITTLE JOHN (II,240); ROOTS: THE NEXT GENERATIONS (II,2213); THE SAN PEDRO BEACH BUMS (II,2250); WALKIN' WALTER (II,2737)

ST JOHN, ELWOOD
THE YOUNG AND THE RESTLESS (II,2862)

ST JOHN, GREER
THE DAY THE WOMEN GOT EVEN (II,628); YOGI'S GANG (II,2852); YOGI'S SPACE RACE (II,2853)

/bf ST JOHN, HOWARD
AN APARTMENT IN ROME (I,242); BEST FOOT FORWARD (I,401); BRIEF ENCOUNTER (I,723); FAIRMEADOWS, U.S.A. (I,1514); GIFT OF THE MAGI (I,1800); HANK (I,1931); THE INVESTIGATOR (I,2230); THE MAN WHO CAME TO DINNER (I,2880); MEET CORLISS ARCHER (I,2981); THE PATRIOTS (I,3477)

ST JOHN, JANIS
PALMERSTOWN U.S.A. (II,1945)

ST JOHN, JILL
THE BOB HOPE SHOW (I,625); BRENDA STARR (II,372); CELEBRITY DAREDEVILS (II,473); EMERALD POINT, N.A.S. (II,773); FAME IS THE NAME OF THE GAME (I,1518); HART TO HART (II,1101); HAVE GIRLS—WILL TRAVEL (I,1967); THE HOUSE NEXT DOOR (I,2139); JUNIOR MISS (I,2493); ROOSTER (II,2210); RUSSIAN ROULETTE (I,3875); SAGA OF SONORA (I,3884); TWO GUYS FROM MUCK (II,2690)

ST JOHN, MARCO
BALL FOUR (II,137); HARDCASE (II,1088); SEARCH FOR TOMORROW (II,2284); THE SIX OF US (II,2369)

ST JOHN, ROBERT
BELIEVE IT OR NOT (I,387)

ST LAMONT, DEA
FROM HERE TO ETERNITY (II,933)

ST LOUIS, MARC
TOO GOOD TO BE TRUE (II,2643)

ST PIERRE, MONIQUE
MALIBU (II,1067)

STAAHL, JIM
GOODNIGHT, BEANTOWN (II,1041); LAUGH TRAX (II,1443); MORK AND MINDY (II,1735); ROCK COMEDY (II,2191); THE SECOND TIME AROUND (II,2289); THE YESTERDAY SHOW (II,2850)

STAAHL, RICHARD
RHODA (II,2151)

STACEY, OLIVE
THE FIRST HUNDRED YEARS (I,1581)

STACK, ELIZABETH
LOVE OF LIFE (I,2774)

STACK, PATRICK
RAGE OF ANGELS (II,2107)

STACK, ROBERT
CELANESE THEATER (I,881); FLOYD GIBBONS, REPORTER (I,1608); GEORGE WASHINGTON (II,978); HOLLYWOOD OPENING NIGHT (I,2087); LAURA (I,2635); LAURA (I,2636); LIGHT'S OUT (I,2699); MOST WANTED (II,1738); MOST WANTED (II,1739); THE NAME OF THE GAME (I,3217); POLICE SQUAD! (II,2061); PULITZER PRIZE PLAYHOUSE (I,3692); THE SEEKERS (I,3973); THE SEEKERS (I,3974); STRIKE FORCE (II,2480); THE UNTOUCHABLES (I,4681); THE UNTOUCHABLES (I,4682)

STACK, TIMOTHY
MICKEY SPILLANE'S MIKE HAMMER: MURDER ME, MURDER YOU (II,1694); REGGIE (II,2127)

STACY, JAMES
THE ADVENTURES OF OZZIE AND HARRIET (I,71); BABY CRAZY (I,319); HEAT OF ANGER (I,1992); LANCER (I,2610)

STACY, MICHELLE
ARCHIE (II,104); THE AWAKENING LAND (II,122); DYNASTY (II,747); THE FESS PARKER SHOW (II,850); MCDUFF, THE TALKING DOG (II,1662); THE YOUNG PIONEERS (II,2869)

STACY, NEIL
WAR AND PEACE (I,4765)

STADER, PETER
GOLIATH AWAITS (II,1025)

STADLEN, LEWIS J
BENSON (II,208); GEORGE M! (I,1772); GEORGE M! (II,976)

STAFF, KATHY
CORONATION STREET (I,1054); LAST OF THE SUMMER WINE (II,1439); SEPARATE TABLES (II,2304)

STAFFORD, ADAM
PETER PAN (II,2030)

STAFFORD, HANLEY
THE HANK MCCUNE SHOW (I,1932)

STAFFORD, JIM
THE JIM STAFFORD SHOW (II,1291); MUSIC CITY NEWS TOP COUNTRY HITS OF THE YEAR (II,1772); THOSE AMAZING ANIMALS (II,2601)

STAFFORD, JO
THE JO STAFFORD SHOW (I,2392); THE JO STAFFORD SHOW (I,2393); THE JO STAFFORD SHOW (I,2394); THE JO STAFFORD SHOW (I,2395)

STAFFORD, JOSEPH
PETER PAN (I,3566)

STAFFORD, MARIAN
TREASURE HUNT (I,4600)

STAFFORD, NANCY
THE DOCTORS (II,694); LONE STAR (II,1513); POOR RICHARD (II,2065); ST. ELSEWHERE (II,2432)

STAFFORD, SUSAN
WHEEL OF FORTUNE (II,2775)

STAFFORD, TRACY
THE ADVENTURES OF OZZIE AND HARRIET (I,71); A CHARLIE BROWN CHRISTMAS (I,908)

STAHL, RICHARD
THE BOYS (II,360); HOUSE CALLS (II,1194); THE LOVE BOAT I (II,1532); ROSETTI AND RYAN: MEN WHO LOVE WOMEN (II,2217); SHERIFF WHO? (I,4009); STRUCK BY LIGHTNING (II,2482); TURNABOUT (II,2669)

STAINES, ANDREW
THE WOMAN IN WHITE (II,2819)

STALEY, JAMES
AND THEY ALL LIVED HAPPILY EVER AFTER (II,75)

STALEY, JOAN
77 SUNSET STRIP (I,3988); AMOS BURKE: WHO KILLED JULIE GREER? (I,181); BROADSIDE (I,732); COLOSSUS (I,1009)

STALLYBRASS, ANNE
THE ONEDIN LINE (II,1912); THE SIX WIVES OF HENRY VIII (I,4068); THE STRAUSS FAMILY (I,4253)

STAMBAUGH, DAVID
LOVE OF LIFE (I,2774)

STAMBERGER, JACK
LOVE OF LIFE (I,2774)

STAMOS, JOHN

DREAMS (II,735); GENERAL HOSPITAL (II,964)

STANDER, LIONEL
HART TO HART (II,1101); HART TO HART (II,1102); THE NASH AIRFLYTE THEATER (I,3227); THE SUNSHINE BOYS (II,2496)

STANDING, JOHN
TINKER, TAILOR, SOLDIER, SPY (II,2621)

STANDLEE, SHIRLEY
ROD BROWN OF THE ROCKET RANGERS (I,3825)

STANG, ARNOLD
THE BILLY BEAN SHOW (I,449); BROADSIDE (I,732); THE BUREAU (II,397); CASPER AND FRIENDS (I,867); DECEMBER BRIDE (I,1215); DOC CORKLE (I,1304); JACK AND THE BEANSTALK (I,2287); LAST OF THE PRIVATE EYES (I,2628); POPEYE THE SAILOR (I,3648); STARLIGHT THEATER (I,4201); TOP CAT (I,4576)

STANGER, BARBARA
DAYS OF OUR LIVES (II,629)

STANGIN, MIMI
ROAR OF THE RAILS (I,3803)

STANIS, BERNNADETTE
GOOD TIMES (II,1038)

STANLEY, FLORENCE
BARNEY MILLER (II,154); FISH (II,864); JOE AND SONS (II,1296); SONNY BOY (II,2402)

STANLEY, HELEN
BURLESQUE (I,765); DAVY CROCKETT (I,1181)

STANLEY, JAMES
FLATFOOTS (II,879)

STANLEY, KIM
CAT ON A HOT TIN ROOF (II,457); THE ELGIN HOUR (I,1428); FLESH AND BLOOD (I,1595); MAGNAVOX THEATER (I,2827); PLAYWRIGHTS '56 (I,3627); STUDIO ONE (I,4268); SURE AS FATE (I,4297); THE TRAP (I,4593); U.M.C. (I,4666)

STANLEY, LOUISE
RENFREW OF THE ROYAL MOUNTED (I,3769)

STANLEY, PAT
THE EDGE OF NIGHT (II,760)

STANLEY, RICHARD
LOOSE CHANGE (II,1518)

STANTON, HARRY DEAN
THE DANGEROUS DAYS OF KIOWA JONES (I,1140); FLATBED ANNIE AND SWEETIEPIE: LADY TRUCKERS (II,876); FRANK MERRIWELL (I,1659); THE OLDEST LIVING GRADUATE (II,1882); WAR CORRESPONDENT (I,4767)

STANTON, HENRY
FIRST LOVE (I,1583)

STANTON, JOHN
HOMICIDE (I,2101)

STANTON, PENNY
MARCUS WELBY, M.D. (I,2905)

STANWYCK, BARBARA
ALONG THE BARBARY COAST (I,144); THE BARBARA STANWYCK THEATER (I,350); THE BIG VALLEY (I,437); THE DICK POWELL SHOW 9(I,1269); FORD TELEVISION THEATER (I,1634); THE GENERAL ELECTRIC THEATER (I,1753); JOSEPHINE LITTLE: (I,2451); JOSEPHINE LITTLE: ADVENTURES IN HAPPINESS (I,2450); JOSEPHINE LITTLE: DRAGON BY THE TAIL (I,2451); JOSEPHINE LITTLE: THE MIRACULOUS JOURNEY OFTADPOLE CHAN (I,2452); THE LETTERS (I,2670); THE SEEKERS (I,3973); THE SEEKERS (I,3974); SUDDEN SILENCE (I,4275); THE THORN BIRDS (II,2600); TONI'S BOYS (II,2637); WAGON TRAIN (I,4747); ZANE GREY THEATER (I,4979)

STAPANETS, KAREL
THE LILLI PALMER THEATER

STAPLE, WILL
A HORSEMAN RIDING BY (II,1182)

STAPLETON, ALAN
OZMOE (I,3437)

STAPLETON, JEAN
ALL IN THE FAMILY (II,38); CBS: ON THE AIR (II,460); CIRCUS OF THE STARS (II,530); SCARECROW AND MRS. KING (II,2268); WOMAN WITH A PAST (I,4889); YOU CAN'T TAKE IT WITH YOU (II,2857)

STAPLETON, MAUREEN
CURTAIN CALL THEATER (I,1115); MIRROR, MIRROR, OFF THE WALL (I,3059); OLD NICKERBOCKER MUSIC HALL (I,3351); STAR TONIGHT (I,4199); THERE'S ALWAYS ROOM (II,2583); THE TIMES SQUARE STORY (I,4514)

STAR-SHEMAH
MUGGSY (II,1761)

STARK, MICHAEL J
THE DOCTORS (II,694); THE GUIDING LIGHT (II,1064)

STARK, RICHARD
DANGER (I,1134)

STARK, SALLY
DAYS OF OUR LIVES (II,629); LOVE OF LIFE (I,2774)

STARKE, TODD
THE DORIS DAY SHOW (I,1355); FIRST TIME, SECOND TIME (II,863)

STARLIGHTERS, THE
THE JO STAFFORD SHOW (I,2392)

STARR, DON
ALEX AND THE DOBERMAN GANG (II,32); DALLAS (II,614)

STARR, FRANCES
LITTLE WOMEN (I,2719)

STARR, KAY
THE KAY STARR SHOW (I,2513); THE PIED PIPER OF HAMELIN (I,3595)

STARR, MICHAEL
O'MALLEY (II,1872); TALES OF THE HAUNTED (II,2538)

STARR, RINGO
RINGO (II,2176)

STARR, RONALD
MR. ROBERTS (I,3149)

STARR, STEVE
MATINEE AT THE BIJOU (II,1651)

STARRETT, CHARLES
FAITH BALDWIN'S THEATER OF ROMANCE (I,1515)

STARRETT, VALERIE
GENERAL HOSPITAL (II,964)

STATES, CHAD
RUN RUN, JOE (II,2229)

STATFORD, TRACY
THE LORETTA YOUNG SHOW (I,2755)

STATLER BROTHERS, THE
AN EVENING WITH THE STATLER BROTHERS (II,790)

STAUFFER, JACK
ALEX AND THE DOBERMAN GANG (II,32); ALL MY CHILDREN (II,39); THE GUIDING LIGHT (II,1064); MOBILE MEDICS (II,1716); THE YOUNG AND THE RESTLESS (II,2862)

STAUFFORD, JACK
ABC'S MATINEE TODAY (I,7)

STEADMAN, IAN
PRISONERS OF THE LOST UNIVERSE (II,2086)

STEADMAN, JOHN
THE BLUE KNIGHT (II,276); THE BLUE KNIGHT (II,277); ONE NIGHT BAND (II,1909); THE OUTLAWS (II,1936)

STEARNS, CHRISTOPHER
MARY KAY AND JOHNNY (I,2932)

STEARNS, JOHNNY
MARY KAY AND JOHNNY (I,2932)

STEARNS, MARY KAY
MARY KAY AND JOHNNY (I,2932)

STECK, JACK
S. S. TELECRUISE (I,4174)

STEDMAN, JOEL
STATE FAIR (II,2448)

STEEDMAN, SHIRLEY
A MAN CALLED INTREPID (II,1612); PENMARRIC (II,1993)

STEEDMAN, TONY
KISS KISS, KILL KILL (I,2566)

STEEL, ANTHONY
I'M THE GIRL HE WANTS TO KILL (I,2194)

STEEL, SUSAN
THE GOLDBERGS (I,1836)

STEELE, ALISON
BRAIN GAMES (II,366)

STEELE, AMY
FOR LOVE AND HONOR (II,903); THE GUIDING LIGHT (II,1064); THE POWERS OF MATTHEW STAR (II,2075)

STEELE, BARBARA
THE WINDS OF WAR (II,2807)

STEELE, BOB
ARROYO (I,266); DOC (I,1303); F TROOP (I,1499)

STEELE, GORDON
CENTENNIAL (II,477)

STEELE, JAN
B.J. AND THE BEAR (II,127)

STEELE, JARDIEH
RYAN'S HOPE (II,2234)

STEELE, KAREN
BRADDOCK (I,704); CONFLICT (I,1036); DEATH OF A SALESMAN (I,1208); MCGHEE (I,2967); STUDIO '57

(I,4267)

STEELE, KATHRYN
H.M.S. PINAFORE (I,2066)

STEELE, MIKE
THE ROCKFORD FILES (II,2196)

STEELE, RICHARD
MAYBERRY R.F.D. (I,2961)

STEELE, STEPHANIE
ARNIE (I,261); THE NEW LAND (II,1820)

STEELE, TED
CAVALCADE OF BANDS (I,876); THE TED STEELE SHOW (I,4373)

STEFAN, VIRGINIA
I LED THREE LIVES (I,2169)

STEFANELLI, SIMONETTA
MOSES THE LAWGIVER (II,1737)

STEFANI, MICHAEL
THE YOUNG MARRIEDS (I,4942)

STEFANS, ANN
THE GEORGE BURNS AND GRACIE ALLEN SHOW (I,1763)

STEFFIN, SHIRRY
THE BEVERLY HILLBILLIES (I,417)

STEFFINS, ROY
CAPTAIN Z-RO (I,839)

STEIGER, ROD
THE BOB HOPE CHRYLSER THEATER (I,502); THE GOLDEN AGE OF TELEVISION (II,1016); THE LONELY WIZARD (I,2743)

STEIN, RON
THE JERK, TOO (II,1282)

STEINBERG, CATHY
A CHARLIE BROWN CHRISTMAS (I,908); IT'S THE GREAT PUMPKIN, CHARLIE BROWN (I,2276); YOU'RE IN LOVE, CHARLIE BROWN (I,4964)

STEINBERG, DAVID
BILLY CRYSTAL: A COMIC'S LINE (II,249); THE DAVID STEINBERG SHOW (I,1179); THE MUSIC SCENE (I,3176); THE NOONDAY SHOW (II,1859); OUT OF OUR MINDS (II,1933)

STEINBERG, MELISSA
I'D RATHER BE CALM (II,1216); WHO'S ON CALL? (II,2791)

STEINBERG, RON
ACES UP (II,7)

STEINDLER, MAUREEN
THE CHISHOLMS (II,512)

STEINDORSDOTTIR, HEDDA
RUNNING BLIND (II,2232)

STEINER, RONNIE
THE MOUSEKETEERS REUNION (II,1742)

STEINER, SHERRY
PIAF (II,2039)

STEINFELD, JAKE
THE INVISIBLE WOMAN (II,1244); MR. SUCCESS (II,1756); SHAPING UP (II,2316)

STELL, ANN
FARAWAY HILL (I,1532)

STELLA, ANTOINETTE
B.J. AND THE BEAR (II,127); THE CURSE OF DRACULA (II,611)

STELLAR, CLARICE
SAINT PETER (II,2238)

STEMFELD, MICHAEL
TOO GOOD TO BE TRUE (II,2643)

STEMPEL, KYRA
HONEST AL'S A-OK USED CAR AND TRAILER RENTAL TIGERS (II,1174)

STENBORG, HELEN
ANOTHER WORLD (II,97); FIFTH OF JULY (II,851)

STENGAARD, YVETTE
DOCTOR IN THE HOUSE (I,1313)

STEPANEK, KAREL
THERE SHALL BE NO NIGHT (I,4437)

STEPHENS, GARN
PHYLLIS (II,2038)

STEPHENS, HARVEY
STAGE DOOR (I,4177)

STEPHENS, JAMES
THE MYSTERIOUS TWO (II,1783); THE PAPER CHASE (II,1949); THE PAPER CHASE: THE SECOND YEAR (II,1950); TOURIST (II,2651); TRUE GRIT (II,2664)

STEPHENS, LARAINE
THE ADVENTURES OF NICK CARTER (I,68); BRACKEN'S WORLD (I,703); EISCHIED (II,763); JARRETT (I,2349); MATT HELM (II,1652); MATT HELM (II,1653); O.K. CRACKERBY (I,3348); THE RANGERS (II,2112); RICH MAN, POOR MAN—BOOK II (II,2162); RISKO (II,2180); SCRUPLES (II,2281); WOMEN IN WHITE (II,2823)

STEPHENS, MARVIN
THE GREATEST GIFT (I,1880)

STEPHENS, PERRY
LOVING (II,1552)

STEPHENS, RAY
THE GREAT SPACE COASTER (II,1059)

STEPHENS, SALLY
A PLACE TO DIE (I,3612)

STEPHENSON, BARBARA
B.J. AND THE BEAR (II,127)

STEPHENSON, JOHN
THE ADVENTURES OF GULLIVER (I,60); THE ADVENTURES OF JONNY QUEST (I,64); THE ATOM ANT/SECRET SQUIRREL SHOW (I,311); THE BANANA SPLITS ADVENTURE HOUR (I,340); CASPER AND THE ANGELS (II,453); THE CLUE CLUB (II,546); THE FLINTSTONE COMEDY HOUR (II,881); FLINTSTONE FAMILY ADVENTURES (II,882); THE FLINTSTONE FUNNIES (II,883); THE FLINTSTONES (II,884); THE FLINTSTONES (II,885); FRANKENSTEIN JR. AND THE IMPOSSIBLES (I,1672); FRED AND BARNEY MEET THE THING (II,919); THE GALAXY GOOFUPS (II,954); HELP! IT'S THE HAIR BEAR BUNCH (I,2012); INCH HIGH, PRIVATE EYE (II,1231); THE KWICKY KOALA SHOW (II,1417); MOBY DICK AND THE MIGHTY MIGHTOR (I,3077); MORNING STAR (I,3104); THE NEW FRED AND BARNEY SHOW (II,1817); THE PETER POTAMUS SHOW (I,3568); SCOOBY'S ALL-STAR LAFF-A-LYMPICS (II,2273); THE SCOOBY-DOO AND SCRAPPY-DOO SHOW (II,2275); SEALAB 2020 (I,3949); TOP CAT (I,4576); THE WACKY RACES (I,4745); YOGI'S GANG (II,2852)

STEPHENSON, PAMELA
NOT THE NINE O'CLOCK NEWS (II,1864); SATURDAY NIGHT LIVE (II,2261)

STEPHENSON, SKIP
BATTLE OF THE NETWORK STARS (II,172); REAL PEOPLE (II,2118)

STEPPHERD, MORGAN
THE FLAME TREES OF THIKA (II,870)

STERE, MORGAN
ELLERY QUEEN: DON'T LOOK BEHIND YOU (I,1434)

STERKE, JEANETTE
THE GREAT IMPERSONATION (I,1876); VANITY FAIR (I,4700)

STERLING, ALEX
HARD KNOX (II,1087)

STERLING, JACK
54TH STREET REVUE (I,1573); THE BIG TOP (I,435)

STERLING, JAN
AT YOUR SERVICE (I,310); BACKSTAIRS AT THE WHITE HOUSE (II,133); LUXURY LINER (I,2796); PUBLICITY GIRL (I,3690)

STERLING, PHILIP
BLISS (II,269); CHICAGO STORY (II,506); CITY OF ANGELS (II,540); DOCTOR SCORPION (II,685); DOCTOR STRANGE (II,688); SOMERSET (I,4115)

STERLING, ROBERT
DEAREST ENEMY (I,1205); ICHABOD AND ME (I,2183); LETTERS FROM THREE LOVERS (I,2671); LOVE THAT JILL (I,2780); TOPPER (I,4582)

STERLING, TISHA
FRANK MERRIWELL (I,1659); KISS ME, KILL ME (II,1400); THE LONG HOT SUMMER (I,2747); POWDERKEG (I,3657)

STERN, BILL
ARE YOU POSITIVE? (I,255); BYLINE—BETTY FURNESS (I,785)

STERN, CHRIS
BLONDIE (II,272)

STERN, ERIC
FROSTY'S WINTER WONDERLAND (II,935)

STERN, JOE
BATMAN AND THE SUPER SEVEN (II,160)

STERN, LILIBET
ALONE AT LAST (II,60); PAPER DOLLS (II,1951); THE YOUNG AND THE RESTLESS (II,2862)

STERN, TOM
FITZ AND BONES (II,867); TERROR AT ALCATRAZ (II,2563)

STERN, WES
GETTING TOGETHER (I,1784); A KNIGHT IN SHINING ARMOUR (I,2575); RHODA (II,2151)

STERNE, MORGAN
THE DOCTORS (II,694); A TIME FOR US (I,4507)

STERNHAGEN, FRANCES
ANOTHER WORLD (II,97); THREE IN ONE (I,4479)

STERRETT, GILLIAN
THE ADVENTURES OF ROBIN HOOD (I,74)

STETROP, MARTY
SCARLETT HILL (I,3933)

STETSON, LEE
CODE NAME: DIAMOND HEAD (II,549)

STEVEN, CARL
SNOOPY'S GETTING MARRIED, CHARLIE BROWN (I,4102)

STEVEN, JOHNNY
MEETING AT APALACHIN (I,2994)

STEVENS, ALEX
DARK SHADOWS (I,1157)

STEVENS, ANDREW
THE BASTARD/KENT FAMILY CHRONICLES (II,159); BATTLE OF THE NETWORK STARS (II,173); BATTLE OF THE NETWORK STARS (II,177); CIRCUS OF THE STARS (II,535); CODE RED (II,552); CODE RED (II,553); EMERALD POINT, N.A.S. (II,773); ONCE AN EAGLE (II,1897); THE OREGON TRAIL (II,1922); THE OREGON TRAIL (II,1923); THE REBELS (II,2121); TOPPER (II,2649)

STEVENS, ANGELA
MY LITTLE MARGIE (I,3194)

STEVENS, CONNIE
CALL HER MOM (I,796); CELEBRITY CHALLENGE OF THE SEXES 1 (II,465); CELEBRITY CHALLENGE OF THE SEXES 3 (II,467); HARRY'S BATTLES (II,1100); HAWAIIAN EYE (I,1973); THE LITTLEST ANGEL (I,2726); MISTER JERICO (I,3070); MURDER CAN HURT YOU! (II,1768); SCRUPLES (II,2280); WENDY AND ME (I,4793)

STEVENS, CRAIG
THE BEST YEARS (I,410); THE CABOT CONNECTION (II,408); DALLAS (II,614); THE EYES OF TEXAS II (II,800); FEMALE INSTINCT (I,1564); FIRESIDE THEATER (I,1580); FORD TELEVISION THEATER (I,1634); GHOST STORY (I,1788); GRUEN GUILD PLAYHOUSE (I,1892); HAPPY DAYS (II,1084); THE HOME FRONT (II,1171); THE INVISIBLE MAN (II,1242); JANE WYMAN PRESENTS THE FIRESIDE THEATER (I,2345); THE LORETTA YOUNG THEATER (I,2756); THE LOVE BOAT II (II,1533); MAN OF THE WORLD (I,2875); MCCLOUD: WHO KILLED MISS U.S.A.? (I,2965); MATINEE THEATER (I,2947); MIGHTY O (I,3036); MR. BROADWAY (I,3131); NICK AND NORA (II,1842); ON TRIAL (I,3363); PEPSI-COLA PLAYHOUSE (I,3523); PETER GUNN (I,3562); RICH MAN, POOR MAN—BOOK I (II,2161) SCHLITZ PLAYHOUSE OF STARS (I,3936); STUDIO '57 (I,4267);

STEVENS, ERICA
OPPENHEIMER (II,1919)

STEVENS, INGER
ALFRED HITCHCOCK PRESENTS (I,115); DUPONT SHOW OF THE MONTH (I,1387); THE FARMER'S DAUGHTER (I,1533); SATURDAY'S CHILDREN (I,3923); THE TWILIGHT ZONE (I,4651); ZANE GREY THEATER (I,4979);

STEVENS, JULIE
BIG TOWN (I,436)

STEVENS, K T
DAYS OF OUR LIVES (II,629); THE ETHEL BARRYMORE THEATER (I,1467); GENERAL HOSPITAL (II,964); STUDIO '57 (I,4267); THE YOUNG AND THE RESTLESS (II,2862)

STEVENS, KAYE
DAYS OF OUR LIVES (II,629); JUST LIKE A WOMAN (I,2496)

STEVENS, LENORE
OZZIE'S GIRLS (I,3440)

STEVENS, LEON B
LOVE OF LIFE (I,2774); NOTORIOUS (I,3314)

STEVENS, LESTER
THE ADVENTURES OF DR. FU MANCHU (I,58)

STEVENS, LIBBY
THE IMPERIAL GRAND BAND (II,1224)

STEVENS, MARK
BIG TOWN (I,436); HOLLYWOOD OPENING NIGHT (I,2087); MICHAEL SHAYNE, DETECTIVE (I,3022); THE NEW ADVENTURES OF MARTIN KANE (I,3251)

STEVENS, MEL
THE MACAHANS (II,1583)

STEVENS, MICKEY
THE FLINTSTONE COMEDY HOUR (II,881)

STEVENS, MORGAN
BARE ESSENCE (II,149); CIRCUS OF THE STARS (II,537); FAME (II,812); HELLINGER'S LAW (II,1127); HUNTER'S MOON (II,1207); THE RETURN OF MARCUS WELBY, M.D (II,2138); THE WALTONS: A DAY FOR THANKS ON WALTON'S MOUNTAIN (EPISODE 3) (II,2741); THE WALTONS: MOTHER'S DAY ON WALTON'S MOUNTAIN (EPISODE 2) (II,2742); THE WALTONS: WEDDING ON WALTON'S MOUNTAIN (EPISODE 1) (II,2743)

STEVENS, MORTON
ABE LINCOLN IN ILLINOIS (I,10); THE BILLY BEAN SHOW (I,449); THE DEVIL'S DISCIPLE (I,1247)

STEVENS, NANCY
BRIGHT PROMISE (I,727)

STEVENS, NAOMI
ABC'S MATINEE TODAY (I,7); THE DORIS DAY SHOW (I,1355); THE FLYING NUN (I,1611); HEY, LANDLORD (I,2039); THE MONTEFUSCOS (II,1727); SARGE: THE BADGE OR THE CROSS (I,3915); THE SEVEN LITTLE FOYS (I,3986); THE STREETS OF SAN FRANCISCO (I,4260); VEGAS (II,2723); VEGAS (II,2724)

STEVENS, ONSLOW
SCHAEFER CENTURY THEATER (I,3935); TEXAS JOHN SLAUGHTER (I,4414)

STEVENS, PATRICIA
M*A*S*H (II,1569); SCOOBY-DOO AND SCRAPPY-DOO (II,2274)

STEVENS, PAUL
ANOTHER WORLD (II,97); GET CHRISTIE LOVE! (II,982); HERCULES (I,2022); THE NURSES (I,3319); SHIRLEY TEMPLE'S STORYBOOK (I,4017); THE TURN OF THE SCREW (I,4624); THE YOUNG AND THE RESTLESS (II,2862)

STEVENS, PERRY
OPRYLAND: NIGHT OF STARS AND FUTURE STARS (II,1921)

STEVENS, RAY
THE CONCRETE COWBOYS (II,564); THE COUNTRY MUSIC MURDERS (II,590)

STEVENS, RISE
THE CHOCOLATE SOLDIER (I,944); FESTIVAL OF MUSIC (I,1567); HANSEL AND GRETEL (I,1934); LITTLE WOMEN (I,2722)

STEVENS, RORY
HOLLYWOOD HIGH (II,1162); LEAVE IT TO BEAVER (I,2648)

STEVENS, RUSTY
LEAVE IT TO BEAVER (I,2648); STILL THE BEAVER (II,2466)

STEVENS, SHADOE
HOT CITY DISCO (II,1185)

STEVENS, SHAWN
DAYS OF OUR LIVES (II,629); THE LAST CONVERTIBLE (II,1435); THE MACKENZIES OF PARADISE COVE (II,1585)

STEVENS, STELLA
BEN CASEY (I,394); THE BOB HOPE SHOW (I,628); CHARLIE COBB: NICE NIGHT FOR A HANGING (II,485); FLAMINGO ROAD (II,871); FLAMINGO ROAD (II,872); THE FRENCH ATLANTIC AFFAIR (II,924); GENERAL ELECTRIC THEATER (I,1753); THE GRADUATION DRESS (I,1864); HART TO HART (II,1101); THE JORDAN CHANCE (II,1343); KISS ME, KILL ME (II,1400); THE LOVE BOAT III (II,1534); NO MAN'S LAND (II,1855); WONDER WOMAN (PILOT 2) (II,2827)

STEVENS, WARREN
ARENA (I,256); BEHIND THE SCREEN (II,200); MOBILE ONE (II,1717); ONE STEP BEYOND (I,3388); THE REBELS (II,2121); THE RETURN OF CAPTAIN NEMO (II,2135); RETURN TO PEYTON PLACE (I,3779); REVLON MIRROR THEATER (I,3781); SCIENCE FICTION THEATER (I,3938); SIMON LASH (I,4052); STARLIGHT THEATER (I,4201); TALES OF THE 77TH BENGAL LANCERS (I,4348) THE TRAP (I,4593)

STEVENS, WILLIAM
ADAM-12 (I,31)

STEVENSON, ALAN
FIRST LOVE (I,1583)

STEVENSON, BOB
HIGH ROAD TO ADVENTURE (I,2055); MICHAEL SHAYNE, DETECTIVE (I,3022)

STEVENSON, DOUGLAS
SEARCH FOR TOMORROW (II,2284)

STEVENSON, JOHN

BOLD JOURNEY (I,680)

STEVENSON, MCLEAN
THE ASTRONAUTS (II,114); BATTLE OF THE NETWORK STARS (II,167); CELEBRITY CHALLENGE OF THE SEXES (II,464); CELEBRITY CHALLENGE OF THE SEXES 1 (II,465); CELEBRITY CHALLENGE OF THE SEXES 2 (II,466); CONDO (II,565); DINAH IN SEARCH OF THE IDEAL MAN (I,1278); THE DORIS DAY SHOW (I,1355); HELLO, LARRY (II,1128); IN THE BEGINNING (II,1229); M*A*S*H (II,1569); THE MCLEAN STEVENSON (II,1665); THE MCLEAN STEVENSON SHOW (II,1664); THE MCLEAN STEVENSON SHOW (II,1665); MY WIVES JANE (I,3207) YOU CAN'T DO THAT ON TELEVISION (I,4933)

STEVENSON, PARKER
BATTLE OF THE NETWORK STARS (II,165); BATTLE OF THE NETWORK STARS (II,166); FALCON CREST (II,808); THE HARDY BOYS MYSTERIES (II,1090); SHOOTING STARS (II,2341)

STEVENSON, RACHEL
A GUEST IN YOUR HOUSE (I,1898)

STEVENSON, ROBERT J
DAYS OF OUR LIVES (II,629); JEFFERSON DRUM (I,2355); THE KINGDOM OF THE SEA (I,2560)

STEVENSON, SCOTT
TEXAS (II,2566)

STEVENSON, VALERIE
DREAMS (II,735)

STEWART, ALEXANDRA
MISTRAL'S DAUGHTER (II,1708)

STEWART, ANN RANDALL
ROGER AND HARRY (II,2200)

STEWART, BYRON
THE WHITE SHADOW (II,2788)

STEWART, CATHERINE MARY
DAYS OF OUR LIVES (II,629)

STEWART, CHARLES
LITTLE VIC (II,1496); PICTURE WINDOW (I,3594)

STEWART, CHARLOTTE
THE FIRST HUNDRED YEARS (I,1582); LITTLE HOUSE ON THE PRAIRIE (II,1487); MOTHER, JUGGS AND SPEED (II,1741)

STEWART, CURT
OUR STREET (I,3418)

STEWART, DAVID
THE DEFENDER (I,1219); YOUNG DR. MALONE (I,4938)

STEWART, DICK
DREAM GIRL OF '67 (I,1375); THE ROWAN AND MARTIN REPORT (II,2221)

STEWART, DON
AFTER HOURS: GETTING TO KNOW US (II,21); THE GUIDING LIGHT (II,1064)

STEWART, ELAINE
GAMBIT (II,955); HIGH ROLLERS (II,1147)

STEWART, FRED
THE DOCTORS (II,694); INTERMEZZO (I,2223); THE MILKY WAY (I,3044); THE NURSES (I,3318)

STEWART, HOMMY
H.R. PUFNSTUF (I,2154); LIDSVILLE (I,2679)

STEWART, HORACE
THE AMOS AND ANDY SHOW (I,179)

STEWART, JAMES
THE AMERICAN WEST OF JOHN FORD (I,173); THE DICK POWELL SHOW (I,1269); THE GENERAL ELECTRIC THEATER (I,1753); HARVEY (I,1962); HAWKINS (I,1976); HAWKINS ON MURDER (I,1978); THE JIMMY STEWART SHOW (I,2391); LUXURY LINER (I,2796)

STEWART, JAY
IT'S ANYBODY'S GUESS (II,1255); LET'S MAKE A DEAL (II,1462)

STEWART, JOB
ROMEO AND JULIET (I,3837)

STEWART, JOHN
FLY AWAY HOME (II,893); FROM THESE ROOTS (I,1688); WESLEY (I,4795)

STEWART, KAY
DOBIE GILLIS (I,1302); YOUNG IN HEART (I,4939)

STEWART, LYNNE
HUSBANDS, WIVES AND LOVERS (II,1209); LAVERNE AND SHIRLEY WITH THE FONZ (II,1449); M*A*S*H (II,1569); THE PEE WEE HERMAN SHOW (II,1988)

STEWART, MARGARET
LEAVE IT TO BEAVER (I,2648); OPERATION: NEPTUNE (I,3404)

STEWART, MARK
MANNIX (II,1624)

STEWART, MARY
EYE WITNESS (I,1494)

STEWART, MEL
ALL IN THE FAMILY (II,38); DEADLOCK (I,1190); FREEBIE AND THE BEAN (II,922); GOOD OL' BOYS (II,1035); HARRY O (II,1098); THE INVISIBLE WOMAN (II,1244); THE LOVE BOAT (II,1535); MARRIAGE IS ALIVE AND WELL (II,1633); ON THE ROCKS (II,1894); ONE IN A MILLION (II,1906); THE OUTLAWS (II,1936); PUNCH AND JODY (II,2091); ROLL OUT! (I,3833); SALT AND PEPE (II,2240); SCARECROW AND MRS. KING (II,2268); SOAP (II,2392); STONE (II,2470); TABITHA (II,2527); TABITHA (II,2528)

STEWART, MICHAEL G
BEULAH LAND (II,226)

STEWART, NAN
ALL THE WAY HOME (I,140)

STEWART, NICK
BIG DADDY (I,425)

STEWART, PATRICK
SMILEY'S PEOPLE (II,2383)

STEWART, PAUL
CITY BENEATH THE SEA (I,965); THE DAIN CURSE (II,613); DEADLINE (I,1188); FRONT PAGE STORY (I,1690); THE INNER SANCTUM (I,2216); THE MAN WHO NEVER WAS (I,2882); MOBY DICK AND THE MIGHTY MIGHTOR (I,3077); NO WARNING (I,3301); THE SUPER SIX (I,4292); TOP SECRET U.S.A. (I,4581)

STEWART, PAULA
BLOOMER GIRL (I,488)

STEWART, PENNY
PRISONER: CELL BLOCK H (II,2085)

STEWART, RANDY
BROCK CALLAHAN (I,741)

STEWART, RAY
A.E.S. HUDSON STREET (II,17)

STEWART, ROBERT W
OUR FIVE DAUGHTERS (I,3413)

STEWART, ROBIN
BLESS THIS HOUSE (II,268)

STEWART, ROD
THE ROOTS OF ROCK 'N' ROLL (II,2212)

STEWART, ROY
LOCAL 306 (II,1505)

STEWART, THOMAS A
THE ADAMS CHRONICLES (II,8)

STEWART, TRISH
CIRCUS OF THE STARS (II,533); SALVAGE (II,2244); SALVAGE 1 (II,2245); WILD TIMES (II,2799); THE YOUNG AND THE RESTLESS (II,2862)

STICH, PATRICIA
THE CLUE CLUB (II,546); GRIFF (I,1888)

STICKNEY, DOROTHY
ARSENIC AND OLD LACE (I,269); THE WALTONS (II,2740)

STIERS, DAVID OGDEN
CHARLIE'S ANGELS (II,487); COUSINS (II,595); DOC (II,683); M*A*S*H (II,1569); THE OLDEST LIVING GRADUATE (II,1882)

STILES, ROBERT
SLITHER (II,2379)

STILLER, JERRY
ACRES AND PAINS (I,19); ARCHIE BUNKER'S PLACE (II,105); JOE AND SONS (II,1296); THE PAUL LYNDE SHOW (I,3489); TAKE FIVE WITH STILLER AND MEARA (II,2530)

STILLWELL, JACK
ACROBAT RANCH (I,20)

STILWELL, DIANE
BLISS (II,269); THE DUCK FACTORY (II,738); I LOVE HER ANYWAY! (II,1214); THE LOVE BOAT II (II,1533); A MAN CALLED SLOANE (II,1613); SIDE BY SIDE (II,2347)

STINETTE, DOROTHY
THE EDGE OF NIGHT (II,760); SOMERSET (I,4115)

STIRLING, PETER
MALLORY: CIRCUMSTANTIAL EVIDENCE (II,1608)

STIRLING, PHILIP
CALUCCI'S DEPARTMENT (I,805)

STIRNER, BRIAN
ALL CREATURES GREAT AND SMALL (I,129)

STOCK, ALAN
THE CONTENDER (II,571); PALMS PRECINCT (II,1946)

STOCK, BARBARA
THE ACADEMY II (II,4)

STOCK, NIGEL
A MAN CALLED INTREPID (II,1612); TINKER, TAILOR, SOLDIER, SPY (II,2621)

STOCKWELL, DEAN
THE ADVENTURES OF NICK CARTER (I,68); THE JOKE AND THE VALLEY (I,2438)

STOCKWELL, GUY
ADVENTURES IN PARADISE (I,46); RETURN TO PEYTON PLACE (I,3779); THE SOUND OF ANGER (I,4133); THRILLER (I,4492)

STOCKWELL, JOHN
TOO GOOD TO BE TRUE (II,2643)

STODDARD, ALEXANDRA
NOT UNTIL TODAY (II,1865); SPACE STARS (II,2408)

STODDARD, HAILA
PRUDENTIAL FAMILY PLAYHOUSE (I,3683); THE SECRET STORM (I,3969); SURE AS FATE (I,4297)

STODDARD, MALCOLM
THE ASSASSINATION RUN (II,112)

STOHL, HANK
THE AMAZING DUNNINGER (I,158)

STOKER, AUSTIN
RETURN TO THE PLANET OF THE APES (II,2145); ROOTS (II,2211); YOUNG LIVES (II,2866)

STOKER, MIKE
EMERGENCY! (II,775)

STOKES, BARRY
THE LAST DAYS OF POMPEII (II,1436)

STOKEY, MIKE
PANTOMIME QUIZ (I,3449); STUMP THE STARS (I,4273); STUMP THE STARS (I,4274)

STOLARSKY, PAUL
I'D RATHER BE CALM (II,1216)

STOLLERY, DAVID
ANNETTE (I,222); THE FURTHER ADVENTURES OF SPIN AND MARTY (I,1720); THE NEW ADVENTURES OF SPIN AND MARTY (I,3254); SPIN AND MARTY (I,4158)

STOLTZ, ERIC
PAPER DOLLS (II,1951); THINGS ARE LOOKING UP (II,2588)

STONE, BEN
KING LEONARDO AND HIS SHORT SUBJECTS (I,2557); UNDERDOG (I,4673)

STONE, CAROLE
THE LIFE AND LEGEND OF WYATT EARP (I,2681)

STONE, CHRISTOPHER
THE BIONIC WOMAN (II,255); BLACK BEAUTY (II,261); THE BLUE AND THE GRAY (II,273); HARPER VALLEY (II,1094); THE INTERNS (I,2226); PRUDENTIAL FAMILY PLAYHOUSE (I,3683); SPENCER'S PILOTS (II,2421); SPENCER'S PILOTS (II,2422)

STONE, CYNTHIA
THE FRANCES LANGFORD-DON AMECHE SHOW (I,1655); HEAVEN FOR BETSY (I,1994); MEDIC (I,2976); THAT WONDERFUL GUY (I,4425)

STONE, DANTON
FIFTH OF JULY (II,851)

STONE, EZRA
THE MUNSTERS' REVENGE (II,1765)

STONE, GAIL
THE EVE ARDEN SHOW (I,1470)

STONE, HAROLD J
BRIDGET LOVES BERNIE (I,722); THE GOLDBERGS (I,1835); THE GRAND JURY (I,1865); THE HOUSE NEXT DOOR (I,2139); JOE AND SONS (II,1296); MY WORLD . . . AND WELCOME TO IT (I,3208); RUSSIAN ROULETTE (I,3875); SOMEWHERE IN ITALY. . . COMPANY B (I,4119)

STONE, JEFFREY
THE THREE MUSKETEERS (I,4480)

STONE, JOSEPH
KAREN (II,1363)

STONE, LEONARD J
CAMP RUNAMUCK (I,811); CIMARRON STRIP (I,954); FOR LOVE OR $$$ (I,1623); THE GIRL IN THE EMPTY GRAVE (II,997); THE JEAN ARTHUR SHOW (I,2352); LOST IN SPACE (I,2758); ONCE UPON A SPY (II,1898); THE TOM SWIFT AND LINDA CRAIG MYSTERY HOUR (II,2633)

STONE, MILBURN
FRONT ROW CENTER (I,1694); GUNSMOKE (II,1069); TV READERS DIGEST (I,4630)

STONE, PAULA
PAULA STONE'S TOY SHOP (I,3495)

STONE, PHILIP
DEATH IN DEEP WATER (I,1206); LACE (II,1418)

STONE, ROBINSON
ANASTASIA (I,182); GIVE US BARABBAS! (I,1816); TAMING OF THE SHREW (I,4356)

STONE, SHARON
BAY CITY BLUES (II,186)

STONE, SID
TV AUCTION (I,4628)

STONE, SUZANNE
9 TO 5 (II,1852); CAGNEY AND LACEY (II,409); SANFORD (II,2255); TEXAS (II,2566)

STONEBURNER, SAM
RAGE OF ANGELS (II,2107)

STONEMAN, HEATHER
SNOOPY'S GETTING MARRIED, CHARLIE BROWN (I,4102)

STONEMAN, ROBERT
BIG BOB JOHNSON AND HIS FANTASTIC SPEED CIRCUS (II,231); MARY HARTMAN, MARY HARTMAN (II,1638)

STONER, SHERRI
LITTLE HOUSE: LOOK BACK TO YESTERDAY (II,1490); LITTLE HOUSE: THE LAST FAREWELL (II,1491)

STOORS, SUZANNE
THE HEIRESS (I,2006); NAKED CITY (I,3215)

STORCH, ARTHUR
THE DEFENDER (I,1219); FIRST LOVE (I,1583)

STORCH, LARRY
THE BRADY KIDS (II,365); CAR 54, WHERE ARE YOU? (I,842); CAVALCADE OF STARS (I,877); CHARLIE ANGELO (I,906); THE DORIS DAY SHOW (I,1355); F TROOP (I,1499); THE GHOST BUSTERS (II,986); JACK FROST (II,1263); THE LARRY STORCH SHOW (I,2618); MY LUCKY PENNY (I,3196); OUT OF THE INKWELL (I,3423); THE QUEEN AND I (I,3704); TEACHER'S PET (I,4367); TENNESSEE TUXEDO AND HIS TALES (I,4403); TWO GUYS FROM MUCK (II,2690)

STORER, LYNN
THE CABOT CONNECTION (II,408)

STOREY, MARGENE
THE MOUSEKETEERS REUNION (II,1742)

STOREY, RALPH
THE REVLON REVUE (I,3782)

STORM, DEBI
ELKE (I,1432)

STORM, GALE
FORD TELEVISION THEATER (I,1634); THE GALE STORM SHOW (I,1725); MY LITTLE MARGIE (I,3194); MYSTERY AND MRS. (I,3210); NBC COMEDY HOUR (I,3237); THE UNEXPECTED (I,4674)

STORM, HOWARD
DESILU REVUE (I,1238); FLANNERY AND QUILT (II,873)

STORM, JAMES
DARK SHADOWS (I,1157); THE DOCTORS (II,694); THE YOUNG AND THE RESTLESS (II,2862)

STORM, MICHAEL
ONE LIFE TO LIVE (II,1907)

STORM, ROBERTA
CAPTAINS AND THE KINGS (II,435)

STORM, WAYNE
THE PHOENIX (II,2035)

STORY, RALPH
THE $64,000 CHALLENGE (I,4070); IN COMMON (I,2200); WHAT DO YOU HAVE IN COMMON? (I,4807)

STOUT, BILL
THE VERDICT IS YOURS (I,4711)

STOUT, DON
KEY TORTUGA (II,1383)

STOUT, PAUL
SCARECROW AND MRS. KING (II,2268)

STOVALL, COUNT
THE DOCTORS (II,694)

STOVER, MAUDE REAM
THE FORTUNE HUNTER (I,1646)

STOWE, MADELINE
BEULAH LAND (II,226); THE GANGSTER CHRONICLES (II,957)

STRACKLE, WIN

HAWKINS FALLS, POPULATION 6200 (I,1977); STUD'S PLACE (I,4271)

STRAFFORD, TRACY
THE JOHN FORSYTHE SHOW (I,2410)

STRAIGHT, BEATRICE
BEACON HILL (II,193); THE BORROWERS (I,693); COSMOPOLITAN THEATER (I,1057); THE DAIN CURSE (II,613); THE INNER SANCTUM (I,2216); KING'S CROSSING (II,1397); LOVE OF LIFE (I,2774); LOVE STORY (I,2776); WONDER WOMAN (SERIES 2) (II,2829); THE WORLD OF DARKNESS (II,2836)

STRAIGHT, CLARENCE
THE ED WYNN SHOW (I,1404)

STRALSER, LUCIA
BIG HAWAII (II,238); DANGER IN PARADISE (II,617)

STRAND, ROBIN
BEACH PATROL (II,192); NICHOLS AND DYMES (II,1841); STARSTRUCK (II,2446); THREE EYES (II,2605)

STRANGE, GLENN
GUNSMOKE (II,1069); LAST STAGECOACH WEST (I,2629); THE LONE RANGER (I,2740); MAN OF THE COMSTOCK (I,2873)

STRANGIS, JUDY
ELECTRA WOMAN AND DYNA GIRL (II,764); LOOSE CHANGE (II,1518); THE ROMAN HOLIDAYS (I,3835); ROOM 222 (I,3843); WHEELIE AND THE CHOPPER BUNCH (II,2776)

STRASBERG, SUSAN
BOB HOPE CHRYSLER THEATER (I,502); CRISIS (I,1102); THE EVIL TOUCH (I,1486); FRANKENSTEIN (I,1671); MARCUS WELBY, M.D. (I,2905); THE MARRIAGE (I,2922); NIGHT GALLERY (I,3287); TOMA (I,4547); TOMA (II,2634)

STRASSER, ROBIN
ALL MY CHILDREN (II,39); ANOTHER WORLD (II,97); ONE LIFE TO LIVE (II,1907); THE SECRET STORM (I,3969)

STRASSMAN, MARCIA
BRENDA STARR (II,372); E/R (II,748); GOOD TIME HARRY (II,1037); THE LOVE BOAT II (II,1533); M*A*S*H (II,1569); THE NIGHTENGALES (II,1850); WEDNESDAY NIGHT OUT (II,2760); WELCOME BACK, KOTTER (II,2761)

STRATAS, TERESA
AMAHL AND THE NIGHT VISITORS (I,150)

STRATTON JR, GIL
THAT'S MY BOY (I,4427)

STRATTON, ALBERT
FOR RICHER, FOR POORER (II,906); KENNEDY (II,1377); THE NEW ADVENTURES OF PERRY MASON (I,3252); SARA (II,2260)

STRATTON, CHESTER
SURE AS FATE (I,4297); TELLER OF TALES (I,4388)

STRATTON, CHET
THE LONELY WIZARD (I,2743); MAN AND SUPERMAN (I,2854); THE MUNSTERS (I,3158)

STRATTON, DEE
JACK FROST (II,1263)

STRATTON, GIL
WALDO (I,4749)

STRATTON, JOHN
GREAT EXPECTATIONS (II,1055); THE TALE OF BEATRIX POTTER (II,2533)

STRATTON, RON
CAMELOT (II,416)

STRATTON, STAN
LOGAN'S RUN (II,1507)

STRATTON, W K
BATTLE OF THE NETWORK STARS (II,164); THE BLACK SHEEP SQUADRON (II,262)

STRATTON, WILLIAM
THE YEAGERS (II,2843)

STRAUB, JOHN
GIVE US BARABBAS! (I,1816)

STRAUSS, BONNIE
HOUR MAGAZINE (II,1193)

STRAUSS, PETER
MASADA (II,1642); RICH MAN, POOR (II,2162); RICH MAN, POOR MAN—BOOK I (II,2161); RICH MAN, POOR MAN—BOOK II (II,2162)

STRAUSS, ROBERT
THE ELISABETH MCQUEENEY STORY (I,1429); MONA MCCLUSKEY (I,3085); PEPSI-COLA PLAYHOUSE (I,3523); SO HELP ME, APHRODITE (I,4104)

STRAW, JACK
MOMMA THE DETECTIVE (II,1721); THE SOFT TOUCH (I,4107)

STRAWMEYER, MELISSA
SOMEDAY YOU'LL FIND HER, CHARLIE BROWN (I,4113)

STREATER, BILLY
NO MAN'S LAND (II,1855)

STREET, ELLIOTT
CHIEFS (II,509); DIAL HOT LINE (I,1254); JEREMIAH OF JACOB'S NECK (II,1281); MELVIN PURVIS: G-MAN (II,1679)

STREISAND, BARBRA
BARBRA STREISAND AND OTHER MUSICAL INSTRUMENTS (I,353); BARBRA STREISAND: A HAPPENING IN CENTRAL PARK (I,352); THE BELLE OF 14TH STREET (I,390); FUNNY GIRL TO FUNNY LADY (II,942); MY NAME IS BARBRA (I,3198)

STRIBLING, HUGH
SECRETS OF THE OLD BAILEY (I,3971)

STRICKLAND, AMZIE
ARCHIE (II,104); THE BILL DANA SHOW (I,446); CARTER COUNTRY (II,449); FLO (II,891); FULL CIRCLE (I,1702); JEREMIAH OF JACOB'S NECK (II,1281); THE PATTY DUKE SHOW (I,3483); TOPPER RETURNS (I,4583)

STRICKLAND, GAIL
9 TO 5 (II,1852); ELLERY QUEEN: TOO MANY SUSPECTS (II,767); NIGHT COURT (II,1844); THE SIX OF US (II,2369)

STRICKSON, MARK
DOCTOR WHO—THE FIVE DOCTORS (II,690)

STRIDE, VIRGINA
THE EXPERT (I,1489)

STRIMPELL, STEPHEN
MR. TERRIFIC (I,3152)

STRITCH, ELAINE
FULL MOON OVER BROOKLYN (I,1703); GROWING PAYNES (I,1891); THE MOTOROLA TELEVISION HOUR (I,3112); MY SISTER EILEEN (I,3200); THE POWDER ROOM (I,3656); THREE IN ONE (I,4479); THE TRIALS OF O'BRIEN (I,4605); YOU SHOULD MEET MY SISTER (I,4935)

STRITCH, KATHY
MCCLOUD: WHO KILLED MISS U.S.A.? (I,2965)

STRIVELLI, JERRY
THE STREETS (II,2479)

STRODE, WOODY
HOW THE WEST WAS WON (II,1196); KEY WEST (I,2528); MANDRAKE THE MAGICIAN (I,2886); THE OUTSIDE MAN (II,1937)

STROKA, MICHAEL
DARK SHADOWS (I,1157); THE EDGE OF NIGHT (II,760); THE HOME FRONT (II,1171)

STROLL, EDSON
MCHALE'S NAVY (I,2969)

STROLL, EDWARD
CONGRESSIONAL INVESTIGATOR (I,1037)

STROMSOE, FRED
ADAM-12 (I,31)

STROMSTEDT, ULLA
FLIPPER (I,1604)

STRONG, DENNIS
CAGNEY AND LACEY (II,410)

STRONG, JAY
SALLY AND SAM (I,3891)

STRONG, LEONARD
ELFEGO BACA (I,1427)

STRONG, MICHAEL
THE D.A.: CONSPIRACY TO KILL (I,1121); THE IMMORTAL (I,2197); TRAVIS LOGAN, D.A. (I,4598)

STRONG, PAT
MR. & MS. AND THE MAGIC STUDIO MYSTERY (II,1747)

STRONGIN, MIMI
FAIRMEADOWS, U.S.A. (I,1514)

STROOCK, GLORIA
FEMALE INSTINCT (I,1564); THE GIRLS (I,1811); MCMILLAN (II,1666); MCMILLAN AND WIFE (II,1667); THAT'S OUR SHERMAN (I,4429); WILD ABOUT HARRY (II,2797)

STROUD, CLAUDE
THE DUKE (I,1381); THE MAN FROM GALVESTON (I,2865); THE MARY TYLER MOORE SHOW (II,1640); THE PETER AND MARY SHOW (I,3561); THE SOFT TOUCH (I,4107)

STROUD, DON
THE D.A.: CONSPIRACY TO KILL (I,1121); THE DAUGHTERS OF JOSHUA CABE (I,1169); GIDGET'S SUMMER REUNION (I,1799); A HATFUL OF RAIN (I,1964); HIGH RISK (II,1146); KATE LOVES A MYSTERY (II,1367); MICKEY SPILLANE'S MIKE HAMMER (II,1692); MICKEY

SPILLANE'S MIKE HAMMER: MORE THAN MURDER (II,1693); MICKEY SPILLANE'S MIKE HAMMER: MURDER ME, MURDER YOU (II,1694)

STROUD, DUKE
THE OUTLAWS (II,1936)

STROUD, GREG
PRISONER: CELL BLOCK H (II,2085)

STRUDWICK, SHEPPARD
ANOTHER WORLD (II,97); THE BAT (I,364); CURTAIN CALL THEATER (I,1115); GIRL ON THE RUN (I,1809); THE INNER SANCTUM (I,2216); LADY IN THE DARK (I,2602); LOVE OF LIFE (I,2774); ONE LIFE TO LIVE (II,1907); SHORT, SHORT DRAMA (I,4023)

STRUNK, JUD
THE JUD STRUNK SHOW (I,2462)

STRUTHERS, SALLY
ALL IN THE FAMILY (II,38); GLORIA (II,1010); GLORIA COMES HOME (II,1011); HOTEL 90 (I,2130); PEBBLES AND BAMM BAMM (II,1987)

STRYKER, AMY
DRIBBLE (II,736)

STRYKERT, RON
MEN AT WORK IN CONCERT (II,1682)

STUART, ANNA
THE DOCTORS (II,694)

STUART, BARBARA
CIRCLE OF FEAR (I,958); DOCTOR DAN (II,684); THE GEORGE BURNS AND GRACIE ALLEN SHOW (I,1763); THE GEORGE BURNS SHOW (I,1765); GOMER PYLE, U.S.M.C. (I,1843); THE GREAT GILDERSLEEVE (I,1875); I AND CLAUDIE (I,2164); THE JONES BOYS (I,2446); MASON (II,1643); THE MCLEAN STEVENSON SHOW (II,1665); PETE AND GLADYS (I,3558); THE QUEEN AND I (I,3704)

STUART, LINDA P
THIS IS KATE BENNETT (II,2597)

STUART, LYNDEL
CONDOMINIUM (II,566)

STUART, MARTHA
THOSE TWO (I,4470)

STUART, MARY
AFTER HOURS: FROM JANICE, JOHN, MARY AND MICHAEL, WITH LOVE (II,20); SEARCH FOR TOMORROW (II,2284); STARS OVER HOLLYWOOD (I,4211)

STUART, MAXINE
THE EDGE OF NIGHT (II,760); FIT FOR A KING (II,866); FOLLOW YOUR HEART (I,1618); FOOLS, FEMALES AND FUN: IS THERE A DOCTOR IN THE HOUSE? (II,900); MARGIE (I,2909); NORBY (I,3304); ONE OF OUR OWN (II,1910); REVENGE OF THE GRAY GANG (II,2146); ROOM FOR ONE MORE (I,3842); THE ROUSTERS (II,2220); SLATTERY'S PEOPLE (I,4082); THE YOUNG MARRIEDS (I,4942)

STUART, PATRICK
GALACTICA 1980 (II,953)

STUART, RANDY
BIFF BAKER, U.S.A. (I,420); DRAGNET (I,1371); GRUEN GUILD PLAYHOUSE (I,1892); THE LIFE AND LEGEND OF WYATT EARP (I,2681); THE RELUCTANT SPY (I,3763); SCHAEFER CENTURY THEATER (I,3935); STAGE 7 (I,4181)

STUART, ROY
GOMER PYLE, U.S.M.C. (I,1843)

STUDER, HAL
THE EDGE OF NIGHT (II,760); FARAWAY HILL (I,1532)

STUDSTILL, PAT
LANDON, LANDON & LANDON (II,1426)

STUMPF, RANDY
THE FIGHTING NIGHTINGALES (II,855); FLATBUSH (II,877)

STURDEVANT, DENNIS
THE ALAN KING SHOW (I,100)

STURGEON, WINA
EXO-MAN (II,797)

STURGES, ALLISUN
CHARLES IN CHARGE (II,482)

STURGES, PATRICIA
M*A*S*H (II,1569)

STURGESS, DOROTHY
PRISONER: CELL BLOCK H (II,2085)

STURGESS, OLIVE
HAPPILY EVER AFTER (I,1937); STUDIO '57 (I,4267); TWENTIETH CENTURY (I,4639)

STUTHMAN, FRED
THE BLUE AND THE GRAY (II,273); HELLO, LARRY (II,1128); KANGAROOS IN THE KITCHEN (II,1362); THE OUTSIDE MAN (II,1937); SCARED SILLY (II,2269); WKRP IN CINCINNATI (II,2814)

STUTT, BOB
FRAGGLE ROCK (II,916)

STUYCK, JURIS
A WOMAN OF SUBSTANCE (II,2820)

STYLES, SUSAN
CUTTER TO HOUSTON (II,612)

SUBER, RAY
PADDY THE PELICAN (I,3442); SANDY STRONG (I,3911)

SUCHET, DAVID
MASTER OF THE GAME (II,1647); OPPENHEIMER (II,1919); REILLY, ACE OF SPIES (II,2129)

SUDERT, SUSAN
THE SECRET STORM (I,3969)

SUDROW, LYLE
THE GUIDING LIGHT (II,1064)

SUDROW, PENELOPE
CRASH ISLAND (II,600)

SUES, ALAN
BEANE'S OF BOSTON (II,194); THE NUT HOUSE (I,3320); OFF WE GO (I,3338)

SUFFIN, JORDAN
EMERGENCY! (II,776)

SUITS, WENDY
THE NASHVILLE PALACE (II,1792)

SULLIVAN, ANN
GROWING PAYNES (I,1891)

SULLIVAN, BARRY
CANNON (I,820); CASINO (II,452); HARBOURMASTER (I,1946); THE HOUSE ON GREENAPPLE ROAD (I,2140); THE IMMORTAL (I,2196); THE IMMORTAL (I,2197); JOHNNY BELINDA (I,2417); JUDGEMENT DAY (II,1348); LETTERS FROM THREE LOVERS (I,2671); THE MAGICIAN (I,2824); THE MAN CALLED X (I,2860); THE MERRY WIDOW (I,3012); MY THREE ANGELS (I,3204); NIGHT GALLERY (I,3288); ONCE AN EAGLE (II,1897); THE PRICE (I,3666); PURSUIT (I,3700); RICH MAN, POOR MAN—BOOK II (II,2162); THE ROAD WEST (I,3801); SAVAGE (I,3924); THE TALL MAN (I,4354); YUMA (I,4976)

SULLIVAN, BRAD
BEST OF THE WEST (II,218); GEORGE WASHINGTON (II,978)

SULLIVAN, ED
CLOWNAROUND (I,980); THE ED SULLIVAN SHOW (I,1401); ED SULLIVAN'S BROADWAY (I,1402); FUN FOR 51 (I,1707); SEE AMERICA WITH ED SULLIVAN (I,3972)

SULLIVAN, ELLIOTT
THE TRAP (I,4593)

SULLIVAN, FRANCIS L
DESTINY (I,1240); SURE AS FATE (I,4297)

SULLIVAN, GRANT
CAPTAIN VIDEO AND HIS VIDEO RANGERS (I,838); GOLDEN WINDOWS (I,1841); PONY EXPRESS (I,3645)

SULLIVAN, J T
THREE FOR THE GIRLS (I,4477)

SULLIVAN, JEAN
THE DOCTORS (II,694)

SULLIVAN, JENNY
CAPTAINS AND THE KINGS (II,435); ME AND MAXX (II,1672); THE SEAL (II,2282); V (II,2713); V: THE FINAL BATTLE (II,2714)

SULLIVAN, JOE
BRENNER (I,719); MAN AGAINST CRIME (I,2852); MEETING AT APALACHIN (I,2994)

SULLIVAN, KEVIN
HAPPY DAYS (II,1084)

SULLIVAN, KITTY
THE PATTY DUKE SHOW (I,3483)

SULLIVAN, LIAM
COMPUTERCIDE (II,561); GALLAGHER (I,1727); THE MONROES (I,3089); THE SECRET STORM (I,3969)

SULLIVAN, PAUL
IT'S A SMALL WORLD (I,2260)

SULLIVAN, ROGER
THREE STEPS TO HEAVEN (I,4485)

SULLIVAN, SUSAN
ANOTHER WORLD (II,97); BELL, BOOK AND CANDLE (II,201); THE CITY (II,541); FALCON CREST (II,808); HAVING BABIES II (II,1108); HAVING BABIES III (II,1109); THE INCREDIBLE HULK (II,1232); IT'S A LIVING (II,1254); JULIE FARR, M.D. (II,1354); MARRIAGE IS ALIVE AND WELL (II,1633); THE

NEW MAVERICK (II,1822); OUR MAN FLINT: DEAD ON TARGET (II,1932); RICH MAN, POOR MAN—BOOK II (II,2162); ROGER AND HARRY (II,2200); A WORLD APART (I,4910)

SULLIVAN, TOM
CIRCUS OF THE STARS (II,531); MORK AND MINDY (II,1735)

SULLY, FRANK
ALIAS MIKE HERCULES (I,117)

SULLY, PATRICIA
A TIME TO LIVE (I,4510)

SUMANT
FIRST TIME, SECOND TIME (II,863)

SUMMER, DONNA
BILLBOARD'S DISCO PARTY (II,245); THE DONNA SUMMER SPECIAL (II,711); A HOT SUMMER NIGHT WITH DONNA (II,1190)

SUMMERS, ANN
THE DOCTOR (I,1320)

SUMMERS, BUNNY
SKEEZER (II,2376)

SUMMERS, HOPE
THE ANDY GRIFFITH SHOW (I,192); ANOTHER DAY (II,94); THE DETECTIVE (II,662); THE FLIM-FLAM MAN (I,1599); THE GOOK FAMILY (I,1858); HAWKINS FALLS, POPULATION 6200 (I,1977); LOVE ON A ROOFTOP (I,2775); THE RIFLEMAN (I,3789)

SUMMERS, JEREMY
HARDCASE (II,1088)

SUMMERS, JERRY
THE BEACHCOMBER (I,371); THE HIGH CHAPARRAL (I,2051); ROLL OUT! (I,3833)

SUMMERS, YALE
330 INDEPENDENCE S.W. (I,4486); BIG ROSE (II,241); RETURN TO PEYTON PLACE (I,3779)

SUMNER, CAROL
LITTLE WOMEN (I,2724)

SUN, IRENE YAH-LING
AS THE WORLD TURNS (II,110); KHAN! (II,1384); LOVE OF LIFE (I,2774); MAGNUM, P.I. (II,1601)

SUNDQUIST, GERRY
GREAT EXPECTATIONS (II,1055); THE LAST DAYS OF POMPEII (II,1436)

SUNDSTROM, FLORENCE
I DREAM OF JEANNIE (I,2167); THE LIFE OF RILEY (I,2686)

SUNG, LELAND
M*A*S*H (II,1569)

SUPIRAN, JERRY
GALACTICA 1980 (II,953); LITTLE LORD FAUNTLEROY (II,1492); YOUNG HEARTS (II,2865)

SUPIRAN, RICKY
MASSARATI AND THE BRAIN (II,1646)

SUROVY, NICHOLAS
GEORGE WASHINGTON (II,978); RYAN'S HOPE (II,2234); A WORLD APART (I,4910)

SURPRENANT, JENNIFER
THE ADDAMS FAMILY (II,13)

SUSANN, JACQUELINE
DANGER (I,1134); THE JACQUELINE SUSANN SHOW (I,2327); THE MOREY AMSTERDAM SHOW (I,3101); SUSPENSE (I,4305); YOUR SURPRISE STORE (I,4961)

SUSI, CAROL ANN
THE NIGHT STALKER (II,1849); WENDY HOOPER—U. S. ARMY (II,2764)

SUSIE
FLIPPER (I,1604)

SUSMAN, TODD
THE BOB CRANE SHOW (II,280); ETHEL IS AN ELEPHANT (II,783); FLATFOOTS (II,879); GETTING THERE (II,985); GOING PLACES (I,1834); NUMBER 96 (II,1869); OFF THE WALL (II,1879); SPENCER'S PILOTS (II,2421); SPENCER'S PILOTS (II,2422); STAR OF THE FAMILY (II,2436)

SUSSKIND, JOYCE
JOYCE AND BARBARA: FOR ADULTS ONLY (I,2458)

SUTCLIFFE, CLAIRE
DEATH IN SMALL DOSES (I,1207)

SUTCLIFFE, IRENE
CORONATION STREET (I,1054)

SUTER, ERIC
FAT ALBERT AND THE COSBY KIDS (II,835)

SUTHERLAND, ESTHER
THE BOYS (II,360); NICHOLS AND DYMES (II,1841)

SUTHERLAND, JANE
THE PHILADELPHIA STORY (I,3581)

SUTHERLAND, NANCY ELLEN
TALES FROM THE DARKSIDE (II,2534)

SUTORIUS, JAMES
THE ANDROS TARGETS (II,76); THE BOB CRANE SHOW (II,280); HELLINGER'S LAW (II,1127); OPERATING ROOM (II,1915)

SUTTON, DOLORES
A DATE WITH LIFE (I,1163); THE DEFENDER (I,1219); VALIANT LADY (I,4695)

SUTTON, DUDLEY
CRY TERROR! (I,1112); MADAME SIN (I,2805); SMILEY'S PEOPLE (II,2383)

SUTTON, FRANK
ERNIE, MADGE, AND ARTIE (I,1457); GOMER PYLE, U.S.M.C. (I,1843)

SUTTON, GRADY
THE EGG AND I (I,1420); THE PRUITTS OF SOUTHAMPTON (I,3684)

SUTTON, JOHN
THE COUNT OF MONTE CRISTO (I,1058); SCHAEFER CENTURY THEATER (I,3935); SWAMP FOX (I,4311)

SUTTON, LISA
HILL STREET BLUES (II,1154)

SUTTON, SARAH
DOCTOR WHO (II,689)

SUZUKI, PAT
MR. T. AND TINA (II,1758)

SVENSON, BO
BATTLE OF THE NETWORK STARS (II,166); FRANKENSTEIN (I,1671); HERE COME THE BRIDES (I,2024); I DO, I DON'T (II,1212); WALKING TALL (II,2738)

SVENSON, MONIKA
THE ROOKIES (I,3840)

SVENSON, SUNJA
EISCHIED (II,763)

SWACKHAMER, TENEYCK
PEKING ENCOUNTER (II,1991)

SWAIM, CASKEY
BATTLE OF THE NETWORK STARS (II,167); PROJECT UFO (II,2088)

SWAIN, JOHN HOWARD
NIGHT PARTNERS (II,1847)

SWAN, BILLY
ALLISON SIDNEY HARRISON (II,54)

SWAN, MICHAEL
STOP SUSAN WILLIAMS (II,2473)

SWANSON, AUDREY
DIAGNOSIS: DANGER (I,1250)

SWANSON, GARY
FROM HERE TO ETERNITY (II,932); LOOSE CHANGE (II,1518); SOMERSET (I,4115)

SWANSON, GLORIA
CROWN THEATER WITH GLORIA SWANSON (I,1107); THE GLORIA SWANSON HOUR (I,1827); HOLLYWOOD OPENING NIGHT (I,2087); KRAFT SUSPENSE THEATER (I,2591); MEN WHO RATE A "10" (II,1684)

SWANSON, PAUL
THE AWAKENING LAND (II,122)

SWARBRICK, CAROL
SKINFLINT (II,2377)

SWARD, ANNE
AS THE WORLD TURNS (II,110)

SWARTZ, TONY
ARCHER—FUGITIVE FROM THE EMPIRE (II,103); BATTLESTAR GALACTICA (II,181); DYNASTY (II,747); NO MAN'S LAND (II,1855)

SWASEY, NIKKI
DIFF'RENT STROKES (II,674)

SWAYZE, JOHN CAMERON
A CHANCE FOR ROMANCE (I,897); GUESS WHAT HAPPENED? (I,1896); IT'S A WONDERFUL WORLD (I,2261); NOTHING BUT THE TRUTH (I,3313)

SWAYZE, PATRICK
THE RENEGADES (II,2132); THE RENEGADES (II,2133)

SWEENEY, BOB
THE BROTHERS (I,748); A CHRISTMAS CAROL (I,947); FIBBER MCGEE AND MOLLY (I,1569); THE GEORGE BURNS AND GRACIE ALLEN SHOW (I,1763); MY FAVORITE HUSBAND (I,3188); OUR MISS

BROOKS (I,3415); SUMMER IN THE CITY (I,4280)

SWEENEY, JOAN
HARD KNOX (II,1087); THE OUTLAWS (II,1936)

SWEENEY, JOSEPH
THE PETRIFIED FOREST (I,3570); WESLEY (I,4795)

SWEENEY, RENEE
GETTING READY (II,984)

SWEENEY, ROBERT
YOUR JEWELERS SHOWCASE (I,4953)

COREY: FOR THE PEOPLE (II,575)

SWEET, BLANCHE
THE THIN MAN (I,4446)

SWEET, DOLPH
ANOTHER WORLD (II,97); THE DESPERATE HOURS (I,1239); THE EDGE OF NIGHT (II,760); GIMME A BREAK (II,995); MAX (II,1658); RENDEZVOUS HOTEL (II,2131); SPARROW (II,2411); STUDS LONIGAN (II,2484); THE TRIALS OF O'BRIEN (I,4605); WHEN THE WHISTLE BLOWS (II,2781)

SWEET, KATIE
HANK (I,1931)

SWEET, REBECCA
BRANAGAN AND MAPES (II,367)

SWENSON, INGA
ANDROCLES AND THE LION (I,189); BENSON (II,208); BONANZA (II,347); DUPONT SHOW OF THE MONTH (I,1387); TESTIMONY OF TWO MEN (II,2564)

SWENSON, INGRID
SOAP (II,2392)

SWENSON, KARL
ALICE IN WONDERLAND (I,120); BACHELOR FATHER (I,323); THE GUN AND THE PULPIT (II,1067); KRAFT MYSTERY THEATER (I,2589); LEAVE IT TO BEAVER (I,2648); LIFE WITH VIRGINIA (I,2695); LITTLE HOUSE ON THE PRAIRIE (II,1487); THE LONELY WIZARD (I,2743); THE MAN FROM GALVESTON (I,2865); THE NEW HEALERS (I,3266); PORTIA FACES LIFE (I,3653); A WALK IN THE NIGHT (I,4750)

SWIFT, ALLEN
'TWAS THE NIGHT BEFORE CHRISTMAS (II,2677); BARBAPAPA (II,142); HOWDY DOODY (I,2151); KING LEONARDO AND HIS SHORT SUBJECTS (I,2557); LITTLE DRUMMER BOY, BOOK II (II,1486); PINOCCHIO'S CHRISTMAS (II,2045); TENNESSEE TUXEDO AND HIS TALES (I,4403); THE TOM AND JERRY SHOW (I,4534)

SWIFT, DAVID
DAY OF THE TRIFFIDS (II,627); WAR AND PEACE (I,4765)

SWIFT, JAMES
THE FORTUNE HUNTER (I,1646)

SWIFT, SUSAN
THE CHISHOLMS (II,512); THE CHISHOLMS (II,513); FEATHERSTONE'S NEST (II,847); SEVEN BRIDES FOR SEVEN BROTHERS (II,2307); THE SIX OF US (II,2369)

SWIMBURNE, NORA
DOUGLAS FAIRBANKS JR. PRESENTS THE RHEINGOLD THEATER (I,1364)

SWIRE, SYDNEY
BEACON HILL (II,193)

SWIT, LORETTA
ANIMALS ARE THE FUNNIEST PEOPLE (II,81); BATTLE OF THE NETWORK STARS (II,163); BATTLE OF THE NETWORK STARS (II,164); BATTLE OF THE NETWORK STARS (II,165); THE BEST CHRISTMAS PAGEANT EVER (II,212); CAGNEY AND LACEY (II,410); CIRCUS OF THE STARS (II,533); IT'S A BIRD, IT'S A PLANE, IT'S SUPERMAN (II,1253); M*A*S*H (II,1569)

SWOFFORD, KEN
CRISIS IN SUN VALLEY (II,604); THE EDDIE CAPRA MYSTERIES (II,758); ELLERY QUEEN (II,766); FAME (II,812); THE OREGON TRAIL (II,1923); RICH MAN, POOR MAN—BOOK II (II,2162); THE ROUSTERS (II,2220); SWITCH (II,2519); SWITCH (II,2520)

SWOGER, HARRY
THE BIG VALLEY (I,437); READY AND WILLING (II,2115)

SWOPE, TRACY BROOKS
ANOTHER WORLD (II,97); CRISIS IN SUN VALLEY (II,604); THE LAST CONVERTIBLE (II,1435); THE LOVE BOAT II (II,1533); THE SECRET WAR OF JACKIE'S GIRLS (II,2294); THE ULTIMATE IMPOSTER (II,2700); VOYAGERS (II,2730); WHERE THE HEART IS (I,4831)

SYDES, CAROL
THE LORETTA YOUNG SHOW (I,2755); MEDIC (I,2976)

SYE, CYNTHIA
ROOTS: THE NEXT GENERATIONS (II,2213)

SYKES, BRENDA
EXECUTIVE SUITE (II,796); THE NEW PEOPLE (I,3268); OZZIE'S GIRLS (I,3439); OZZIE'S GIRLS (I,3440)

SYLTER, FRED
ADAM'S RIB (I,32)

SYLVAN, PAUL
BUSTING LOOSE (II,402); EMERGENCY! (II,776); FLATBUSH/AVENUE J (II,878); LAVERNE AND SHIRLEY (II,1446)

SYLVESTER, HAROLD
RICHIE BROCKELMAN: MISSING 24 HOURS (II,2168); TODAY'S FBI (II,2628); WALKING TALL (II,2738); WHEELS (II,2777)

SYLVESTER, JOHN
SEARCH FOR TOMORROW (II,2284)

SYLVESTER, WILLIAM
THE GEMINI MAN (II,960)

SYLVIA
COUNTRY GALAXY OF STARS (II,587)

SYMES, BILL
TREASURE UNLIMITED (I,4603)

SYMES, BUDD
YOUNG LIVES (II,2866)

SYMINGTON, DAVID
THE MAGNIFICENT YANKEE (I,2829)

SYMINGTON, DONALD
DREAM GIRL (I,1374); LOVE OF LIFE (I,2774); SANCTUARY OF FEAR (II,2252)

SYMONDS, ROBERT
THE BLUE AND THE GRAY (II,273); LOOSE CHANGE (II,1518); NICK AND NORA (II,1842); SEVENTH AVENUE (II,2309); WHERE THE HEART IS (I,4831)

SYMPSON, TONY
PETER PAN (II,2030)

SYMS, SYLVIA
NANCY ASTOR (II,1788)

SZIEGETI, CYNTHIA
FOG (II,897)

TAFEL, SUSANNA
THE HUNTER (I,2162)

TABBERT, BILL
OUR FIVE DAUGHTERS (I,3413)

TABER, RICHARD
BORN YESTERDAY (I,692)

TABLIAN, VIC
THE ASSASSINATION RUN (II,112)

TABOR, JOAN
THE JACK BENNY PROGRAM (I,2294)

TABORI, KRISTOFFER
BETWEEN THE LINES (II,225); BLACK BEAUTY (II,261); CHICAGO STORY (II,506); THE CHICAGO STORY (II,507); NEITHER ARE WE ENEMIES (I,3247); SEVENTH AVENUE (II,2309)

TACKER, FRANCINE
THE ASSOCIATES (II,113); THE BOUNDER (II,357); DALLAS (II,614); EMPIRE (II,777); GOODTIME GIRLS (II,1042); OH MADELINE (II,1880); THE PAPER CHASE (II,1949)

TAEGER, RALPH
ACAPULCO (I,15); HONDO (I,2103); KLONDIKE (I,2573)

TAFLER, SYDNEY
MELODY OF HATE (I,2999)

TAFT, JERRY
MICKEY SPILLANE'S MIKE HAMMER: MORE THAN MURDER (II,1693)

TAGGART, HAL
I DREAM OF JEANNIE (I,2167)

TAGGART, MILLEE
THE EDGE OF NIGHT (II,760); ONE LIFE TO LIVE (II,1907); SEARCH FOR TOMORROW (II,2284)

TAGGART, RITA
EVERY STRAY DOG AND KID (II,792); HELLO, LARRY (II,1128); STEAMBATH (II,2451)

TAIT, MICHAEL
HIGH HOPES (II,1144)

TAKA, MIIKO
JUDGE DEE IN THE MONASTERY MURDERS (I,2465); SHOGUN (II,2339)

TAKAMATSU, HIDEO
SHOGUN (II,2339)

TAKAMOTO, MICHAEL
THE AMAZING CHAN AND THE CHAN CLAN (I,157)

TAKEI, GEORGE

STAR TREK (II,2439); STAR TREK (II,2440)

TAKEI, MIDORI
SHOGUN (II,2339)

TALBERT, JANE
GOLDEN WINDOWS (I,1841)

TALBOT, GLORIA
ANNIE OAKLEY (I,225); BACKBONE OF AMERICA (I,328); CAVALCADE OF AMERICA (I,875); CONFLICT (I,1036); CROSSROADS (I,1105); THE GENERAL ELECTRIC THEATER (I,1753); THE LIFE AND LEGEND OF WYATT EARP (I,2681); THE LUX VIDEO THEATER (I,2795); MATINEE THEATER (I,2947); MY LITTLE MARGIE (I,3194); THE ROY ROGERS SHOW (I,3859); WANTED: DEAD OR ALIVE (I,4764)

TALBOT, LYLE
THE ADVENTURES OF OZZIE AND HARRIET (I,71); BEN JERROD (I,396); COMMANDO CODY (I,1030); LEAVE IT TO BEAVER (I,2648); THE LONE RANGER (I,2740); LOVE THAT BOB (I,2779); PURSUIT (I,3700); REVLON MIRROR THEATER (I,3781)

TALBOT, NITA
BOURBON STREET BEAT (I,699); EYE WITNESS (I,1494); FUNNY FACE (I,1711); HERE WE GO AGAIN (I,2028); HOGAN'S HEROES (I,2069); THE INNER SANCTUM (I,2216); JANE WYMAN PRESENTS THE FIRESIDE THEATER (I,2345); THE JIM BACKUS SHOW—HOT OFF THE WIRE (I,2379); JOE AND MABEL (I,2398); THE PARTRIDGE FAMILY (II,1962); THE ROCKFORD FILES (II,2196); SEARCH FOR TOMORROW (II,2284); SHAUGHNESSEY (II,2318); SOAP (II,2392); STAGE DOOR (I,4178); STUDIO ONE (I,4268); SUPERTRAIN (II,2504); THEY CALL IT MURDER (I,4441); THE THIN MAN (I,4446); TURNOVER SMITH (II,2671); UNDER THE YUM YUM TREE (I,4671); THE WOMEN (I,4890); YOU ARE THE JURY (II,2854)

TALBOT, STEPHEN
LEAVE IT TO BEAVER (I,2648)

TALBOTT, MICHAEL
IF I LOVE YOU, AM I TRAPPED FOREVER? (II,1222); MIAMI VICE (II,1689); THIS IS KATE BENNETT (II,2597); USED CARS (II,2712)

TALBOY, TARA
THE CLIFFWOOD AVENUE KIDS (II,543); THE CLUE CLUB (II,546)

TALKINGTON, BRUCE
BILLY (II,247)

TALLENTS, JOHN
DOCTOR WHO—THE FIVE DOCTORS (II,690)

TALLMADGE, KIM
THE MAGNIFICENT SIX AND A HALF (I,2828)

TALMAN, WILLIAM
PERRY MASON (II,2025)

TALTON, ALEXANDRA
MY FAVORITE HUSBAND (I,3188)

TAMBA, TETSURO
MARCO POLO (II,1626)

TAMBOR, JEFFREY
9 TO 5 (II,1852); EDDIE AND HERBERT (II,757); HILL STREET BLUES (II,1154); PALS (II,1947); THE ROPERS (II,2214)

TAMBURO, CHUCK
MASSARATI AND THE BRAIN (II,1646)

TAMIROFF, AKIM
BROADWAY (I,733); THE CHOCOLATE SOLDIER (I,944); THE ETHEL BARRYMORE THEATER (I,1467); THE GREAT SEBASTIANS (I,1877); THEN CAME BRONSON (I,4436)

TAMM, DANIEL
BLACK BEAUTY (II,261)

TAMM, MARY
THE ASSASSINATION RUN (II,112); DOCTOR WHO (II,689)

TAMMI, TOM
AS THE WORLD TURNS (II,110)

TANDY, JESSICA
THE FOUR POSTER (I,1651); THE MARRIAGE (I,2922); THE MOON AND SIXPENCE (I,3098); TELLER OF TALES (I,4388)

TANDY, STEVEN
ALL THE RIVERS RUN (II,43); THE SULLIVANS (II,2490)

TANISHA, TA
ROOM 222 (I,3843)

TANNEN, CHARLES
WHERE THERE'S SMOKEY (I,4832)

TANNEN, WILLIAM
THE LIFE AND LEGEND OF WYATT EARP (I,2681)

TANNER, CLAY
BIG BOB JOHNSON AND HIS FANTASTIC SPEED CIRCUS (II,231)

TANNER, GORDON
SABER OF LONDON (I,3879)

TANNER, MARK
I'M THE GIRL HE WANTS TO KILL (I,2194)

TANNER, RICHARD
LITTLE WOMEN (I,2724)

TANNER, STEVE
240-ROBERT (II,2688); 240-ROBERT (II,2689)

TANZINI, PHILIP
GENERAL HOSPITAL (II,964); THE SUNSHINE BOYS (II,2496)

TAPLEY, COLIN
SABER OF LONDON (I,3879)

TAPP, JIMMY
THE MIGHTY HERCULES (I,3034)

TAPSCOTT, MARK
DAYS OF OUR LIVES (II,629); ESCAPE (I,1460); SWEET, SWEET RACHEL (I,4313)

TARBUCK, BARBARA
CAGNEY AND LACEY (II,409); CALIFORNIA FEVER (II,411)

TARKENTON, FRAN
THAT'S INCREDIBLE! (II,2578)

TARKINGTON, ROCKNE
THE BANANA SPLITS ADVENTURE HOUR (I,340); MATT HOUSTON (II,1654); ROLL OF THUNDER, HEAR MY CRY (II,2202); ROUGHNECKS (II,2219); TARZAN (I,4361)

TARNOW, TOBY
MOMENT OF TRUTH (I,3084)

TARR, JUSTIN
THE RAT PATROL (I,3726)

TARSES, JAY
THE CHOPPED LIVER BROTHERS (II,514); THE DUCK FACTORY (II,738); OPEN ALL NIGHT (II,1914)

TARTER, DALE
JOE FORRESTER (II,1303)

TASCO, RAI
PLANET EARTH (I,3615)

TASLITZ, ERIC
GALACTICA 1980 (II,953)

TATA, JOE E
DEAD MAN ON THE RUN (II,631); JACQUELINE SUSANN'S VALLEY OF THE DOLLS, 1981 (II,1268); MOBILE TWO (II,1718); NO TIME FOR SERGEANTS (I,3300)

TATE, DAVID
THE HITCHHIKER'S GUIDE TO THE GALAXY (II,1158)

TATE, JACQUES
THE INVISIBLE WOMAN (II,1244)

TATE, KEVIN
DANIEL BOONE (I,1142)

TATE, SHARON
THE BEVERLY HILLBILLIES (I,417)

TATUM, JEANNE
MAGGIE (I,2810)

TATUM, LEE
ANOTHER LIFE (II,95)

TAWNY, LITTLE
TENSPEED AND BROWN SHOE (II,2562)

TAYBACK, VIC
ALICE (II,33); ALICE (II,34); CALL HOLME (II,412); CELEBRITY CHALLENGE OF THE SEXES 5 (II,469); CIRCUS OF THE STARS (II,533); COPS (I,1047); GRIFF (I,1888); KHAN! (II,1384); MORNING STAR (I,3104); THE MYSTERIOUS TWO (II,1783); THE STREETS OF SAN FRANCISCO (II,2478); THE SUPER (I,4293); THROUGH THE MAGIC PYRAMID (II,2615); TWO'S COMPANY (I,4659)

TAYLERSON, MARILYN
VANITY FAIR (I,4701)

TAYLOR, BENEDICT
THE FAR PAVILIONS (II,832); THE LAST DAYS OF POMPEII (II,1436)

TAYLOR, BENJAMIN
ROGER DOESN'T LIVE HERE ANYMORE (II,2201)

TAYLOR, BILL
THE HUSTLER OF MUSCLE BEACH (II,1210)

TAYLOR, BRIAN
CAPITOL (II,426)

TAYLOR, BROOKE
THE DOCTORS (II,694)

TAYLOR, BUCK
THE BUSTERS (II,401); THE CHEROKEE TRAIL (II,500); GUNSMOKE (II,1069); KATE BLISS AND THE TICKER TAPE KID (II,1366); THE MONROES (I,3089); NO MAN'S LAND (II,1855); WILD TIMES (II,2799)

TAYLOR, BUDGE
THE LEGEND OF THE GOLDEN GUN (II,1455)

TAYLOR, BURT
GOOD MORNING WORLD (I,1850)

TAYLOR, CAROLE
CELEBRITY REVUE (II,475)

TAYLOR, CHARLES
ALL THAT GLITTERS (II,41); PORTIA FACES LIFE (I,3653)

TAYLOR, CLARICE
BEULAH LAND (II,226); THE COSBY SHOW (II,578); HIGH FIVE (II,1143); SALT AND PEPE (II,2240)

TAYLOR, CLYDE PHILLIP
I'D RATHER BE CALM (II,1216)

TAYLOR, DON
COMMAND PERFORMANCE (I,1029)

TAYLOR, DOUGLAS
CONCERNING MISS MARLOWE (I,1032)

TAYLOR, DUB
THE ANDY GRIFFITH SHOW (I,192); CASEY JONES (I,865); THE DAUGHTERS OF JOSHUA CABE RETURN (I,1170); THE DOOLEY BROTHERS (II,715); THE FLIM-FLAM MAN (I,1599); GETTING THERE (II,985); GREAT DAY (II,1053); HAZEL (I,1982); THE MURDOCKS AND THE MCCLAYS (I,3163); OCTAVIUS AND ME (I,3327); THE OUTLAWS (II,1936); PLEASE DON'T EAT THE DAISIES (I,3628); PUMPBOYS AND DINETTES ON TELEVISION (II,2090); THE ROY ROGERS SHOW (I,3859); SAM HILL: WHO KILLED THE MYSTERIOUS MR. FOSTER? (I,3896)

TAYLOR, ELIZABETH
BOB HOPE SPECIAL: BOB HOPE'S STAR-STUDDED SPOOF OF THE NEW TV SEASON— G RATED—WITH GLAMOUR, GLITTER & GAGS (II,317); ELIZABETH TAYLOR IN LONDON (I,1430); GENERAL HOSPITAL (II,964); RETURN ENGAGEMENT (I,3775)

TAYLOR, F CHASE
COLONEL STOOPNAGLE'S STOOP (I,1006)

TAYLOR, FRANK
JESSICA NOVAK (II,1286)

TAYLOR, GRANT
THE ADVENTURES OF LONG JOHN SILVER (I,67); U.F.O. (I,4662)

TAYLOR, HOLLAND
BOSOM BUDDIES (II,355); SILVER SPOONS (II,2355)

TAYLOR, IRVING
THREE STEPS TO HEAVEN (I,4485)

TAYLOR, JANA
GENERAL HOSPITAL (II,964); MAKE MORE ROOM FOR DADDY (I,2842); MAKE ROOM FOR GRANDDADDY (I,2844); MAKE ROOM FOR GRANDDADDY (I,2845)

TAYLOR, JOAN
THE RIFLEMAN (I,3789)

TAYLOR, JOSEPH
THE YOUNG AND THE RESTLESS (II,2862)

TAYLOR, JOSH
ONE LIFE TO LIVE (II,1907); RIKER (II,2175); SEMI-TOUGH (II,2299)

TAYLOR, JOYCE
BOLD VENTURE (I,682); MEN INTO SPACE (I,3007)

TAYLOR, JUD
DOCTOR KILDARE (I,1315)

TAYLOR, JUNE
BOSTON AND KILBRIDE (II,356); CAPTAINS AND THE KINGS (II,435); GILLIGAN'S ISLAND: RESCUE FROM GILLIGAN'S ISLAND (II,991); WOMEN IN WHITE (II,2823)

TAYLOR, KATE
GOOD VIBRATIONS FROM CENTRAL PARK (I,1854)

TAYLOR, KEITH
LEAVE IT TO BEAVER (I,2648); MCKEEVER AND THE COLONEL (I,2970); TOO MANY SERGEANTS (I,4573)

TAYLOR, KENT
BOSTON BLACKIE (I,695); PANIC! (I,3447); THE ROUGH RIDERS (I,3850)

TAYLOR, KIT
THE ADVENTURES OF LONG JOHN SILVER (I,67)

TAYLOR, LARRY
AN ECHO OF THERESA (I,1399); PRISONERS OF THE LOST UNIVERSE (II,2086)

TAYLOR, LAURA LEIGH
ANOTHER LIFE (II,95)

TAYLOR, LAUREN-MARIE
LOVING (II,1552)

TAYLOR, MARC SCOTT
QUINCY, M. E. (II,2102)

TAYLOR, MARK L
HOUSE CALLS (II,1194); MORK AND MINDY (II,1736); V: THE FINAL BATTLE (II,2714)

TAYLOR, MARY LOU
MR. AND MRS. NORTH (I,3124)

TAYLOR, MESHACH
BUFFALO BILL (II,387); I'D RATHER BE CALM (II,1216)

TAYLOR, NATHANIEL
SANFORD (II,2255); SANFORD AND SON (II,2256); WHAT'S HAPPENING!! (II,2769)

TAYLOR, RENEE
BEDROOMS (II,198); FOREVER FERNWOOD (II,909); GOOD PENNY (II,1036); LOVE, SEX. . . AND MARRIAGE (II,1546); PARADISE (II,1955); THE TROUBLE WITH PEOPLE (I,4611)

TAYLOR, RIP
THE $1.98 BEAUTY SHOW (II,698); HERE COMES THE GRUMP (I,2027); MINSKY'S FOLLIES (II,1703); SIGMUND AND THE SEA MONSTERS (II,2352)

TAYLOR, ROBERT
COLOSSUS (I,1009); DEATH VALLEY DAYS (I,1211); THE DETECTIVES (I,1245); THE DICK POWELL SHOW (I,1269)

TAYLOR, ROD
THE ADVENTURES OF LONG JOHN SILVER (I,67); BEARCATS! (I,375); THE DESILU PLAYHOUSE (I,1237); HELLINGER'S LAW (II,1127); HONG KONG (I,2112); THE LUX VIDEO THEATER (I,2795); MASQUERADE (II,1644); THE OREGON TRAIL (II,1922); THE OREGON TRAIL (II,1923); POWDERKEG (I,3657); STUDIO '57 (I,4267)

TAYLOR, ROMMEY
FOOLS, FEMALES AND FUN: WHAT ABOUT THAT ONE? (II,901)

TAYLOR, RUSSI
FLINTSTONE FAMILY ADVENTURES (II,882); THE FLINTSTONE FUNNIES (II,883); THE HEATHCLIFF AND MARMADUKE SHOW (II,1118); JIM HENSON'S MUPPET BABIES (II,1289); PAC-MAN (II,1940)

TAYLOR, TAMMY
DAYS OF OUR LIVES (II,629)

TAYLOR, TERRI
GOLIATH AWAITS (II,1025)

TAYLOR, TOM
THE CHISHOLMS (II,512); FAIRMEADOWS, U.S.A. (I,1514); THE GOLDBERGS (I,1835); THE GOLDBERGS (I,1836)

TAYLOR, VALERIE
BAFFLED (I,332); MACBETH (I,2801)

TAYLOR, VAUGHN
EYE WITNESS (I,1494); FBI CODE 98 (I,1550); FIRESIDE THEATER (I,1580); GALLAGHER (I,1727); THE INNER SANCTUM (I,2216); JOHNNY JUPITER (I,2429); THE LAST WAR (I,2630); MR. MERGENTHWIRKER'S LOBBLIES (I,3144) SHORT, SHORT DRAMA (I,4023); STAR TONIGHT (I,4199)

TAYLOR, WALLY
THE NEW ODD COUPLE (II,1825); ROOTS (II,2211)

TAYLOR, WILLIAM
S*H*E (II,2235)

TAYLOR, ZANDOR
MULLIGAN'S STEW (II,1763)

TAYLOR-YOUNG, LEIGH
BATTLE OF THE NETWORK STARS (II,175); THE DEVLIN CONNECTION (II,664); PEYTON PLACE (I,3574); UNDER THE YUM YUM TREE (I,4671)

TEAD, PHILLIPS
THE ADVENTURES OF SUPERMAN (I,77); THE BILLY BEAN SHOW (I,449)

TEAGUE, GUY
FURY (I,1721)

TEAGUE, MARSHALL
TOPPER (II,2649); TRAVIS MCGEE (II,2656)

TEAL, RAY
BONANZA (II,347); GALLAGHER (I,1727); GIRL ON THE RUN (I,1809)

TEALE, LEONARD
HOMICIDE (I,2101)

TEBALDI, RENATA
FESTIVAL OF MUSIC (I,1567)

TEDROW, IRENE
THE AMAZING SPIDER-MAN (II,63); CENTENNIAL (II,477); DENNIS THE MENACE (I,1231); DIFF'RENT STROKES (II,674); DOROTHY (II,717);

THE LAST NINJA (II,1438); LOW MAN ON THE TOTEM POLE (I,2782); MEET CORLISS ARCHER (I,2979); NEVER SAY NEVER (II,1809); PEOPLE LIKE US (II,1999); THE RUGGLES (I,3868); THREE'S COMPANY (II,2614); THE YOUNG MARRIEDS (I,4942)

TEETER, LARA
DEBBY BOONE. . . ONE STEP CLOSER (II,655)

TEFKIN, BLAIR
V (II,2713); V: THE FINAL BATTLE (II,2714); V: THE SERIES (II,2715)

TEICHER, ROY
BROTHERS AND SISTERS (II,382)

TEIGH, LILA
GOOD PENNY (II,1036)

TEITEL, CAROL
THE EDGE OF NIGHT (II,760); THE GUIDING LIGHT (II,1064); LOVERS AND OTHER STRANGERS (II,1550)

TELFER, FRANK
AS THE WORLD TURNS (II,110); THE DOCTORS (II,694)

TELFORD, ROBERT
BIG BEND COUNTRY (II,230)

TEMPLE, RENNY
THE LIFE AND TIMES OF EDDIE ROBERTS (II,1468)

TEMPLE, SHIRLEY
SHIRLEY TEMPLE'S STORYBOOK (I,4017); SHIRLEY TEMPLE'S STORYBOOK: THE LEGEND OF SLEEPY HOLLOW (I,4018)

TEMPLETON, ALEC
ALEC TEMPLETON TIME (I,112)

TEMPLETON, CHRISTOPHER
THE YOUNG AND THE RESTLESS (II,2862)

TEMPLIN, DEBORAH JEAN
MICKEY SPILLANE'S MIKE HAMMER: MORE THAN MURDER (II,1693)

TEMPTATIONS, THE
GOOD VIBRATIONS FROM CENTRAL PARK (I,1855)

TENNANT, DON
HOLD 'ER NEWT (I,2070)

TENNANT, DOROTHY
THE NEW ADVENTURES OF HUCKLEBERRY FINN (I,3250)

TENNANT, VICTORIA
CHIEFS (II,509); THE WINDS OF WAR (II,2807)

TENNEY, CHRISTOPHER
THE DEADLY GAME (II,633)

TENNILLE, TONI
BATTLE OF THE NETWORK STARS (II,166); BATTLE OF THE NETWORK STARS (II,168); THE CAPTAIN AND TENNILLE (II,429); THE CAPTAIN AND TENNILLE IN HAWAII (II,430); THE CAPTAIN AND TENNILLE IN NEW ORLEANS (II,431); THE CAPTAIN AND TENNILLE SONGBOOK (II,432); CELEBRITY CHALLENGE OF THE SEXES 4 (II,468); FANTASY ISLAND (II,829); ROCK 'N' ROLL: THE FIRST 25 YEARS (II,2189); TMT (II,2622); THE TONI TENNILLE SHOW (II,2636)

TENNY, CHARLES
THE ADAMS CHRONICLES (II,8)

TENNYSA, SCOTT
TWO GIRLS NAMED SMITH (I,4656)

TENORIO JR, JOHN
NAKIA (II,1785)

TENOWICH, AMY
MORK AND MINDY (II,1735)

TEPPER, WILLIAM
CALL HER MOM (I,796)

TERHUNE, SHANNON
RITUALS (II,2182)

TERLESKY, J T
LEGMEN (II,1458)

TERRELL, STEVE
LIFE WITH FATHER (I,2690); TWENTIETH CENTURY (I,4639)

TERRIO, DENEY
DANCE FEVER (II,615)

TERRY, HELEN
CULTURE CLUB IN CONCERT (II,610)

TERRY, JOE
ADVICE TO THE LOVELORN (II,16); THE GOSSIP COLUMNIST (II,1045)

TERRY-THOMAS
I LOVE A MYSTERY (I,2170); MUNSTER, GO HOME! (I,3157)

TERZIEFF, LAURENT
MOSES THE LAWGIVER (II,1737)

TESREAU, KRISTA
THE GUIDING LIGHT (II,1064)

TESSEL, KENNY
THE ACADEMY II (II,4)

TESSIER, MICHAEL
BARNEY MILLER (II,154)

TESSIER, ROBERT
THE BILLION DOLLAR THREAT (II,246); CENTENNIAL (II,477); SAWYER AND FINN (II,2265); THE STOCKERS (II,2469)

TESTI, FABIO
S*H*E (II,2235)

TETLEY, WALTER
THE BULLWINKLE SHOW (I,761); THE DUDLEY DO-RIGHT SHOW (I,1378); GEORGE OF THE JUNGLE (I,1773); THE WOODY WOODPECKER SHOW (I,4906)

TETZEL, JOAN
DOUGLAS FAIRBANKS JR. PRESENTS THE RHEINGOLD THEATER (I,1364); THE THREE MUSKETEERS (I,4481)

TEWES, LAUREN
THE CELEBRITY FOOTBALL CLASSIC (II,474); THE LOVE BOAT (II,1535); THE LOVE BOAT III (II,1534)

TEWKESBURY, JOAN
PETER PAN (I,3566)

TEWSON, JOSEPHINE
KISS KISS, KILL KILL (I,2566)

TEXAS, TEMPLE
THE GOOD FAIRY (I,1846)

THACKITT, WESLEY
MARGIE (I,2909)

THALER, ROBERT
THE RENEGADES (II,2133)

THALL, BILL
MIDWESTERN HAYRIDE (I,3033)

THALL, WILLIE
MIDWESTERN HAYRIDE (I,3033)

THAMES, BYRON
FATHER MURPHY (II,841)

THATCHER, LEORA
CONCERNING MISS MARLOWE (I,1032); INHERIT THE WIND (I,2213); THE JOKE AND THE VALLEY (I,2438); MAGNAVOX THEATER (I,2827); YOUNG DR. MALONE (I,4938)

THATCHER, TORIN
BRENDA STARR (II,372); COSMOPOLITAN THEATER (I,1057); GREAT BIBLE ADVENTURES (I,1872); THE HOLY TERROR (I,2098); THE NASH AIRFLYTE THEATER (I,3227); THE STRANGE CASE OF DR. JEKYLL AND MR. HYDE (I,4244); THE TRAP (I,4593)

THAW, JOHN
THE SWEENEY (II,2513)

THAXTER, PHYLLIS
THE LORETTA YOUNG THEATER (I,2756); THE MOTOROLA TELEVISION HOUR (I,3112); STAGE 7 (I,4181)

THAXTON, LLOYD
FUNNY YOU SHOULD ASK (I,1717); THE LLOYD THAXTON SHOW (I,2735); SHOWCASE '68 (I,4033)

THAYER, BRYNN
ONE LIFE TO LIVE (II,1907)

THAYER, MAX
SKEEZER (II,2376)

THEBOM, BLANCHE
FESTIVAL OF MUSIC (I,1567)

THEODORE, DONNA
SEARCH FOR TOMORROW (II,2284)

THEODORE, PARIS
PETER PAN (I,3566)

THICKE, ALAN
THICKE OF THE NIGHT (II,2587)

THIELMAN, CONSTANTINO
WILSON'S REWARD (II,2805)

THIESS, URSULA
THE DETECTIVES (I,1245)

THIGPEN, LYNNE
GIMME A BREAK (II,995); LOVE, SIDNEY (II,1547); THE NEWS IS THE NEWS (II,1838); POTTSVILLE (II,2072)

THING
THE ADDAMS FAMILY (II,11)

THINNES, ROY
CHICAGO 212 (I,935); CODE NAME: DIAMOND HEAD (II,549); FALCON CREST (II,808); FROM HERE TO ETERNITY (II,932); FROM HERE TO ETERNITY (II,933); GENERAL HOSPITAL (II,964); THE INVADERS (I,2229); THE LONG HOT SUMMER (I,2747); THE NORLISS TAPES (I,3305); THE PSYCHIATRIST (I,3686); THE PSYCHIATRIST: GOD BLESS THE CHILDREN (I,3685); THE RETURN OF THE MOD SQUAD (I,3776); SCRUPLES (II,2281); STONE (II,2470)

ARCHIE (II,104); THE ARCHIE SITUATION COMEDY MUSICAL VARIETY SHOW (II,106); CAMP GRIZZLY (II,418); CHICO AND THE MAN (II,508); MANHUNTER (II,1621); MCLAREN'S RIDERS (II,1663); THE NEW OPERATION PETTICOAT (II,1826); NUMBER 96 (II,1869); RISKO (II,2180); THE YOUNG REBELS (I,4944)

THOMPSON, J D
NEW GIRL IN HIS LIFE (I,3265)

THOMPSON, JACK
A WOMAN CALLED GOLDA (II,2818)

THOMPSON, JENN
HARPER VALLEY (II,1094); HARPER VALLEY PTA (II,1095)

THOMPSON, JOANNE
M*A*S*H (II,1569)

THOMPSON, JOHNNY
CLUB 7 (I,986)

THOMPSON, KAY
AMOS BURKE: WHO KILLED JULIE GREER? (I,181)

THOMPSON, LINDA
GOOD OL' BOYS (II,1035)

THOMPSON, MARSHALL
ANGEL (I,216); DAKTARI (I,1127); THE ETHEL BARRYMORE THEATER (I,1467); JAMBO (I,2330); TV READERS DIGEST (I,4630); WORLD OF GIANTS (I,4912)

THOMPSON, NEIL
WORKING STIFFS (II,2834)

THOMPSON, REX
THE DOCTORS (II,694)

THOMPSON, RUDY
CHIEFS (II,509)

THOMPSON, SADA
FAMILY (II,813); MARCO POLO (II,1626); OUR TOWN (I,3421); SANDBURG'S LINCOLN (I,3907)

THOMPSON, SCOTT
JESSICA NOVAK (II,1286)

THOMPSON, TINA
FATHER KNOWS BEST (II,838)

THOMPSON, TONY
DAVID BOWIE—SERIOUS MOONLIGHT (II,622)

THOMPSON, VICTORIA
ANOTHER WORLD (II,97); THE YOUNG AND THE RESTLESS (II,2862)

THOMSETT, SALLY
MAN ABOUT THE HOUSE (II,1611)

THOMSON, GORDON
DYNASTY (II,746); RYAN'S HOPE (II,2234)

THOR, CAMERON
SUMMER (II,2491)

THOR, JEROME
ALONG THE BARBARY COAST (I,144); FOREIGN INTRIGUE: DATELINE EUROPE (I,1638); HANDS OF MURDER (I, 1929); O'HARA, UNITED STATES TREASURY: OPERATION COBRA (I,3347); SURE AS FATE (I,4297)

THORBURN, JUNE
TALES OF THE VIKINGS (I,4351)

THORN, BERT
CONCERNING MISS MARLOWE (I,1032)

THORNE, ANGELA
MISTRAL'S DAUGHTER (II,1708); TO THE MANOR BORN (II,2627)

THORNE, JUNE
GEORGE WASHINGTON (II,978)

THORNE, TERESA
SABER OF LONDON (I,3879)

THORNTON, ANGELA
JANE EYRE (I,2338)

THORNTON, JUDY
THE THORNTON SHOW (I,4468)

THORNTON, SIGRID
1915 (II,1853); ALL THE RIVERS RUN (II,43); PRISONER: CELL BLOCK H (II,2085)

THORPE, DANIEL
THE CHEAP DETECTIVE (II,490)

THORPE, DAVID
PURSUE AND DESTROY (I,3699)

THORSEN, LESLIE
THE GREATEST GIFT (II,1061)

THORSON, LINDA
THE AVENGERS (II,121); LADY KILLER (I,2603)

THORSON, PAUL
ONE MAN'S FAMILY (I,3383)

THORSON, RUSSELL
THE DETECTIVES (I,1245); IT'S A SMALL WORLD (I,2260); SAVAGE SUNDAY (I,3926)

THOUGHTON, PATRICK
THE DEVIL'S WEB (I,1248)

THRASER, SUNNY
THE EDISON TWINS (II,761)

THRASHER, MARGE
FORMBY'S ANTIQUE WORKSHOP (I,1644)

THRELFALL, DAVID
NICHOLAS NICKLEBY (II,1840)

THRONE, MALACHI
BATMAN (I,366); CODE NAME: HERACLITUS (I,990); ELECTRA WOMAN AND DYNA GIRL (II,764); GENERAL ELECTRIC TRUE (I,1754); IT TAKES A THIEF (I,2250); POLICE STORY (I,3636)

THULIN, INGRID
ESPIONAGE (I,1464); INTERMEZZO (I,2223); MOSES THE LAWGIVER (II,1737)

THURLEY, MARTIN C
PENMARRIC (II,1993)

THURMAN, JIM
SIGN-ON (II,2353)

THURSHY, DAVID
STARR, FIRST BASEMAN (I,4204)

THURSTON, CAROL
FRONTIER (I,1695); THE LIFE AND LEGEND OF WYATT EARP (I,2681)

THURSTON, TOMI
CHICAGO 212 (I,935)

THYHURST, TYLER
MR. ROBERTS (II,1754)

TIANO, LOU
SUPERCOPS (II,2502)

TICOTIN, RACHEL
FOR LOVE AND HONOR (II,903)

TIEGS, CHERYL
BATTLE OF THE NETWORK STARS (II,166)

TIERNEY, GENE
SCRUPLES (II,2280)

TIFFIN, PAMELA
THE SURVIVORS (I,4301); THREE ON AN ISLAND (I,4483)

TIGAR, KENNETH
THE GANGSTER CHRONICLES (II,957); GREAT DAY (II,1054); THE GYPSY WARRIORS (II,1070); LOVE, NATALIE (II,1545); THE MAN FROM ATLANTIS (II,1615); MISSING PIECES (II,1706); THE ROCK RAINBOW (II,2194)

TIGHE, KEVIN
BATTLE OF THE NETWORK STARS (II,163); EMERGENCY PLUS FOUR (II,774); EMERGENCY! (II,775); EMERGENCY! (II,776); THE REBELS (II,2121)

TIL, ROGER
FACTS OF LIFE: THE FACTS OF LIFE GOES TO PARIS (II,806); SCRUPLES (II,2281)

TILLET, JAMES
MASTER OF THE GAME (II,1647)

TILLINGER, JOHN
ANOTHER WORLD (II,97); SHERLOCK HOLMES (II,2327)

TILLIS, MEL
COUNTRY GALAXY OF STARS (II,587); THE COUNTRY MUSIC MURDERS (II,590); MEL AND SUSAN TOGETHER (II,1677); SKINFLINT (II,2377); THE STOCKERS (II,2469)

TILLOTSON, JOHNNY
GIDGET (I,1795)

TILLSTROM, BURR
ALICE IN WONDERLAND (I,120); KUKLA, FRAN, AND OLLIE (I,2594); THE KUKLAPOLITAN EASTER SHOW (I,2595)

TILLY, JENNIFER
SHAPING UP (II,2316)

TILLY, MEG
THE TROUBLE WITH GRANDPA (II,2662)

TILTON, CHARLENE
BATTLE OF THE NETWORK STARS (II,167); BATTLE OF THE NETWORK STARS (II,170); BATTLE OF THE NETWORK STARS (II,172); BATTLE OF THE NETWORK STARS (II,173); BATTLE OF THE NETWORK STARS (II,178); CIRCUS OF THE STARS (II,533); DALLAS (II,614)

TIMBERLAKE, DEE
THE PLASTICMAN COMEDY/ADVENTURE SHOW (II,2051); THE YOUNG SENTINELS (II,2871)

TIMBERLAKE, LUCILLE
ROSENTHAL AND JONES (II,2215)

TIMKO, JOHNNY
THE AWAKENING LAND (II,122); MULLIGAN'S STEW (II,1763); SUMMER (II,2491)

TIMMINS, CALI
RYAN'S HOPE (II,2234)

TINER, PAUL
ANOTHER WORLD (II,97)

TINGWELL, CHARLES
ALL THE RIVERS RUN (II,43); CAPTAIN SCARLET AND THE MYSTERONS (I,837); HOMICIDE (I,2101)

TINLING, TED
THE EDGE OF NIGHT (II,760)

TINY TIM
ONE MORE TIME (I,3386)

TIPPERT, WAYNE
THE SECRET STORM (I,3969)

TIPPIT, WAYNE
THE DOCTORS (II,694); RAGE OF ANGELS (II,2107); SEARCH FOR TOMORROW (II,2284)

TIRELLI, JAIME
BALL FOUR (II,137); MOBILE MEDICS (II,1716)

TIVELL, AMY
THE MONEYCHANGERS (II,1724)

TJAN, WILLIE
THE PLANT FAMILY (II,2050)

TOBEY, JEANNE
THE CRADLE SONG (I,1088)

TOBEY, KENNETH
ADAMSBURG, U.S.A. (I,33); BATTLES: THE MURDER THAT WOULDN'T DIE (II,180); CLINIC ON 18TH STREET (II,544); CONFLICT (I,1036); DAVY CROCKETT (I,1181); FIRESIDE THEATER (I,1580); FRONTIER (I,1695); OUR PRIVATE WORLD (I,3417); PANIC! (I,3447); THE WHIRLYBIRDS (I,4841)

TOBIAS, GEORGE
ADVENTURES IN PARADISE (I,46); BEWITCHED (I,418); THE SEVEN LITTLE FOYS (I,3986); TELEPHONE TIME (I,4379); THE WALTONS (II,2740)

TOBIAS, OLIVER
TERROR FROM WITHIN (I,4404)

TOBIN, DAN
THE FIRST HUNDRED YEARS (I,1581); I MARRIED JOAN (I,2174); MIMI (I,3054); THREE WISHES (I,4488)

TOBIN, MARK
THE DESILU REVUE (I,1238)

TOBIN, MATTHEW
THE BLUE AND THE GRAY (II,273)

TOBIN, MICHELE
CALIFORNIA FEVER (II,411); FATHER ON TRIAL (I,1544); THE FITZPATRICKS (II,868); GRANDPA GOES TO WASHINGTON (II,1050)

TOCHI, BRIAN
THE AMAZING CHAN AND THE CHAN CLAN (I,157); ANNA AND THE KING (I,221); THE RENEGADES (II,2133); SPACE ACADEMY (II,2406)

TOCHI, WENDY
ANNA AND THE KING (I,221)

TODD, ANN
THRILLER (I,4492); TROUBLE WITH FATHER (I,4610)

TODD, BEVERLY
DEADLOCK (I,1190); HAVING BABIES III (II,1109); JULIE FARR, M.D. (II,1354)

TODD, CHUCK
SCRUPLES (II,2280)

TODD, DANIEL
MY THREE SONS (I,3205)

TODD, HALLIE
THE BEST OF TIMES (II,220); BROTHERS (II,380)

TODD, JOSEPH
MY THREE SONS (I,3205)

TODD, MICHAEL
MY THREE SONS (I,3205)

TODD, RICHARD
NOT GUILTY! (I,3311)

TOGNI, SUZANNE
THE MAGNIFICENT SIX AND A HALF (I,2828)

TOKUDA, MARILYN
BROTHERS AND SISTERS (II,382); THE ROLLERGIRLS (II,2203)

TOLAN, ANDREW
LOVE OF LIFE (I,2774)

TOLAN, MICHAEL
HAZARD'S PEOPLE (II,1112); LOOSE CHANGE (II,1518); THE MARY TYLER MOORE SHOW (II,1640); MO AND JO (II,1715); THE NURSES (I,3318); THE SENATOR (I,3976); VALLEY FORGE (I,4698)

TOLBERT, BERLINDA
BATTLE OF THE NETWORK STARS (II,173); THE JEFFERSONS (II,1276)

TOLBERT, R C
THE LEGEND OF THE GOLDEN GUN (II,1455)

TOLETINO, JOAN
STOCKARD CHANNING IN JUST FRIENDS (II,2467)

TOLIN, MIKE
GOLDEN WINDOWS (I,1841)

TOLKIN, STACY HEATHER
IS THIS GOODBYE, CHARLIE BROWN? (I,2242); WHAT HAVE WE LEARNED, CHARLIE BROWN? (I,4810); WKRP IN CINCINNATI (II,2814)

TOLKSDORF, BRIGITTA
LOVE OF LIFE (I,2774)

TOLL, PAMELA
THE DOCTORS (II,694); THE MILLIONAIRE (II,1700); SOMERSET (I,4115)

TOLLAND, KRISTINA
HERE COME THE BRIDES (I,2024)

TOLLEFSON, BUD
ANDY'S GANG (I,214)

TOLO, MARILU
THE LAST DAYS OF POMPEII (II,1436)

TOLSKY, SUSAN
HERE COME THE BRIDES (I,2024); MADAME'S PLACE (II,1587); THE THREE WIVES OF DAVID WHEELER (II,2611)

TOM, LAUREN
THE FACTS OF LIFE (II,805)

TOM SCOTT ORCHESTRA, THE
OLIVIA NEWTON-JOHN IN CONCERT (II,1884)

TOMA, DAVID
EISCHIED (II,763); TOMA (I,4547); TOMA (II,2634)

TOMA, ROBIN
THE AMAZING CHAN AND THE CHAN CLAN (I,157)

TOMACK, SID
THE LIFE OF RILEY (I,2685); MY FRIEND IRMA (I,3191)

TOMARKEN, PETER
HIT MAN (II,1156); PRESS YOUR LUCK (II,2078)

TOMASIN, JENNY
UPSTAIRS, DOWNSTAIRS (II,2709)

TOMEI, MARISA
AS THE WORLD TURNS (II,110)

TOMKINS, JOAN
THE AWAKENING LAND (II,122)

TOMLIN, GARY
ANOTHER WORLD (II,97); SEARCH FOR TOMORROW (II,2284)

TOMLIN, LILY
LILY (I,2705); LILY (II,1477); LILY FOR PRESIDENT (II,1478); THE LILY TOMLIN SHOW (I,2706); THE LILY TOMLIN SPECIAL (II,1479); LILY—SOLD OUT (II,1480)

TOMLIN, PINKY
WATERFRONT (I,4774)

TOMLINSON, DAVID
THE LILLI PALMER THEATER (I,2704)

TOMME, RON
DALLAS (II,614); LOVE OF LIFE (I,2774)

TOMMI, TOM V V
THE ADAMS CHRONICLES (II,8)

TOMPKINS, ANGEL
THE BUFFALO SOLDIERS (II,388); PROBE (I,3674); SEARCH (I,3951)

TOMPKINS, JOAN
FATHER ON TRIAL (I,1544); GENERAL HOSPITAL (II,964); MY THREE SONS (I,3205); RICH MAN, POOR MAN—BOOK II (II,2162); SAM BENEDICT (I,3894)

TONE, BEN
ALICE IN WONDERLAND (I,120)

TONE, FRANCHOT
BEN CASEY (I,394); DUPONT SHOW OF THE MONTH (I,1387); THE ELGIN HOUR (I,1428); THE GUARDSMAN (I,1893); THE KAISER ALUMINUM HOUR (I,2503); LIGHT'S OUT (I,2699); THE LITTLE FOXES (I,2712); PLAYWRIGHTS '56 (I,3627); PURSUIT (I,3700); TALES OF TOMORROW (I,4352); THE UNITED STATES STEEL HOUR (I,4677)

TONES, TOM
KENNEDY (II,1377)

TONG, JACQUELINE
PEEK-A-BOO: THE ONE AND ONLY PHYLLIS DIXEY (II,1989); UPSTAIRS, DOWNSTAIRS (II,2709)

TONG, KAM
HAVE GUN—WILL TRAVEL (I,1968); MR. GARLUND (I,3138)

TONG, SAMMEE
BACHELOR FATHER (I,323); MICKEY (I,3023); NEW GIRL IN HIS LIFE (I,3265); SPIN AND

MARTY (I,4158)

TONGE, PHILIP
BLITHE SPIRIT (I,483); BLITHE SPIRIT (I,484); JANE EYRE (I,2337); NORTHWEST PASSAGE (I,3306)

TONGE, WILLIAM
JANE EYRE (I,2337)

TONY MOTTOLA TRIO, THE
CRIME PHOTOGRAPHER (I,1097)

TONY URBANO AND COMPANY
DUSTY'S TREEHOUSE (I,1391)

TOOMEY, MARILYN
WIN WITH A WINNER (I,4870)

TOOMEY, REGIS
BACKBONE OF AMERICA (I,328); THE BEST IN MYSTERY (I,403); BURKE'S LAW (I,762); HEY MULLIGAN (I,2040); THE LORETTA YOUNG THEATER (1,2756); MAGGIE MALONE (I,2812); NAVY LOG (I,3233); PETTICOAT JUNCTION (I,3571); RICHARD DIAMOND, PRIVATE DETECTIVE (I,3785); SHANNON (I,3998); THE THIRD COMMANDMENT (I,4449)

TOPE, JOANNA
THE OMEGA FACTOR (II,1888)

TOPOL
THE WINDS OF WAR (II,2807)

TOPOLOWSKY, STEVEN
COCAINE AND BLUE EYES (II,548)

TOPPANO, PEITA
PRISONER: CELL BLOCK H (II,2085)

TOPPER, TIM
GOING BANANAS (II,1015); SEVEN BRIDES FOR SEVEN BROTHERS (II,2307)

TOPPING, LYNNE
THE EDDIE CAPRA MYSTERIES (II,758); THE YOUNG AND THE RESTLESS (II,2862)

TORGOV, SARAH
THE CIRCLE FAMILY (II,527)

TORK, PETER
33 1/3 REVOLUTIONS PER MONKEE (I,4451); THE MONKEES (I,3088)

TORME, MEL
THE GOLDEN AGE OF TELEVISION (II,1018); IT WAS A VERY GOOD YEAR (I,2253); THE MEL TORME SHOW (I,2995); SUMMERTIME U.S.A. (I,4283); TV'S TOP TUNES (I,4634)

TORN, RIP
THE ALCOA HOUR (I,108); THE BLUE AND THE GRAY (II,273); CAT ON A HOT TIN ROOF (II,457); JOHNNY BELINDA (I,2416); TWENTY-FOUR HOURS IN A WOMAN'S LIFE (I,4644)

TORP, JOHN
SORORITY '62 (II,2405)

TORRES, ANTONIO
A.K.A. PABLO (II,28); AT YOUR SERVICE (II,118)

TORRES, LIZ
ALL IN THE FAMILY (II,38); CHECKING IN (II,492); MORE WILD WILD WEST (II,1733); MURDER CAN HURT YOU! (II,1768); THE NEW ODD COUPLE (II,1825); NOT IN FRONT OF THE KIDS (II,1861); PHYLLIS (II,2038); PINOCCHIO (II,2044); POPI (II,2068); STOCKARD CHANNING IN JUST FRIENDS (II,2467); WILLOW B: WOMEN IN PRISON (II,2804)

TORREY, ROGER
THE BEVERLY HILLBILLIES (I,417); IRON HORSE (I,2238); LITTLE HOUSE: THE LAST FAREWELL (II,1491); SCALPLOCK (I,3930)

TORTELL, VINCE
THE LAST CONVERTIBLE (II,1435)

TORTY, BOB
WHEN, JENNY? WHEN (II,2783)

TOTTEN, HEATHER
FAMILY (II,813)

TOTTER, AUDREY
FOUR STAR PLAYHOUSE (I,1652); CIMARRON CITY (I,953); CONFIDENTIALLY YOURS (I,1035); MEDICAL CENTER (II,1675); MEET MCGRAW (I,2984); MY DARLING JUDGE (I,3186); OUR MAN HIGGINS (I,3414); THE OUTSIDER (I,3430); U.M.C. (I,4666)

TOUGH, MICHAEL
SEARCH AND RESCUE: THE ALPHA TEAM (II,2283)

TOULIATOS, GEORGE
FALCON'S GOLD (II,809)

TOWAB, HARRY
THE MANIONS OF AMERICA (II,1623)

TOWBIN, BERYL
CALAMITY JANE (I,793)

TOWERS, BARBARA
DOCTOR DOLITTLE (I,1309)

TOWERS, CONSTANCE
CAPITOL (II,426)

TOWERS, ROBERT
GOOBER AND THE TRUCKERS' PARADISE (II,1029); KIDD VIDEO (II,1388)

TOWNE, ALINE
THE ADVENTURES OF SUPERMAN (I,77); COMMANDO CODY (I,1030)

TOWNE, CYNTHIA
RHODA (II,2151)

TOWNER, BOB
THE MAGIC LAND OF ALLAKAZAM (I,2817)

TOWNER, LESLIE
THE ASSASSINATION RUN (II,112)

TOWNES, HARRY
B.J. AND THE BEAR (II,127); CASINO (II,452); CONDOMINIUM (II,566); ELIZABETH THE QUEEN (I,1431); MEDIC (I,2976); SIX GUNS FOR DONEGAN (I,4065)

TOWNSEND, BARBARA
AFTERMASH (II,23); CONCERNING MISS MARLOWE (I,1032)

TOWNSEND, ERNIE
THE EDGE OF NIGHT (II,760)

TOWNSEND, JILL
CIMARRON STRIP (I,954)

TOWNSEND, NED
BUCKSHOT (II,385)

TOWNSEND, PATTI
LIVING IN PARADISE (II,1503)

TOWNSON, RON
THE FIFTH DIMENSION SPECIAL: AN ODYSSEY IN THE COSMIC UNIVERSE OF PETER MAX (I,1571)

TOYOSHIMA, TIFFANY
TWO MARRIAGES (II,2691)

TOZERE, FREDERIC
OUR FIVE DAUGHTERS (I,3413); STANLEY (I,4190)

TOZZI, GIORGIO
AMAHL AND THE NIGHT VISITORS (I,150)

TRACEY, RAY
CENTENNIAL (II,477); HOW THE WEST WAS WON (II,1196)

TRACY, ARTHUR
SUSPENSE THEATER ON THE AIR (II,2507)

TRACY, BARBARA
MASQUERADE (I,2940)

TRACY, DOREEN
ANNETTE (I,222); THE MOUSEKETEERS REUNION (II,1742)

TRACY, LEE
THE AMAZING MR. MALONE (I,159); CHALK ONE UP FOR JOHNNY (I,893); THE CHEVROLET TELE-THEATER (I,926); COSMOPOLITAN THEATER (I,1057); DANGER (I,1134); FORD THEATER HOUR (I,1635); MAGNAVOX THEATER (I,2827); MARTIN KANE, PRIVATE EYE (I,2928); NEW YORK CONFIDENTIAL (I,3276); PROFILES IN COURAGE (I,3678); PULITZER PRIZE PLAYHOUSE (I,3692)

TRACY, MARLENE
LOCK, STOCK, AND BARREL (I,2736)

TRACY, STEVE
LITTLE HOUSE ON THE PRAIRIE (II,1487)

TRACY, WALTER
TERRY AND THE PIRATES (I,4405)

TRAED, MICHAEL
DYNASTY (II,746)

TRANUM, CHUCK
THE GAY COED (I,1742)

TRARES, MILDRED
ABE LINCOLN IN ILLINOIS (I,11); BERKELEY SQUARE (I,399); THE CRADLE SONG (I,1088); GIFT OF THE MAGI (I,1800); THE JOKE AND THE VALLEY (I,2438); THE LITTLE FOXES (I,2712); LITTLE MOON OF ALBAN (I,2714); MY FAIR LADY (I,3187); ON BORROWED TIME (I,3354); ONE TOUCH OF VENUS (I,3389)

TRASK, DIANA
KRAFT MYSTERY THEATER (I,2589)

TRAUBEL, HELEN
VALENTINE'S DAY (I,4693)

TRAVALENA, FRED
ANYTHING FOR MONEY (II,99); THE FUNNY WORLD OF FRED & BUNNI (II,943); THE KOPYCATS (I,2580); THE SHIRT TALES (II,2337); SHOOTING STARS (II,2341); THE SIX O'CLOCK FOLLIES (II,2368)

TRAVANTI, DANIEL J
BATTLE OF THE NETWORK STARS (II,174); BATTLE OF THE NETWORK STARS (II,175); CALL TO DANGER (I,803); GENERAL HOSPITAL (II,964); HILL STREET BLUES (II,1154)

TRAVELLI, DEVORAH
DALLAS (II,614)

TRAVERS, BILL
THE ADMIRABLE CRICHTON (I,36)

TRAVERS, TOM
ALL THE RIVERS RUN (II,43)

TRAVIS, BOB
TMT (II,2622)

TRAVIS, RICHARD
CODE 3 (I,993); THE GRAND JURY (I,1865); YOUR SHOW TIME (I,4958)

TRAVOLTA, ELLEN
ALLISON SIDNEY HARRISON (II,54); COVER GIRLS (II,596); HAPPY DAYS (II,1084); JOANIE LOVES CHACHI (II,1295); MAKIN' IT (II,1603); MARIE (II,1628); NUMBER 96 (II,1869); WELCOME BACK, KOTTER (II,2761)

TRAVOLTA, JOEY
BIG JOHN (II,239)

TRAVOLTA, JOHN
WELCOME BACK, KOTTER (II,2761)

TRAYLOR, WILLIAM
LITTLE WOMEN (I,2721)

TREACHER, ARTHUR
THE MERV GRIFFIN SHOW (I,3015); THE MERV GRIFFIN SHOW (I,3016)

TREAS, TERRI
SEVEN BRIDES FOR SEVEN BROTHERS (II,2307)

TREAT, CLARENCE
THE NEW CHRISTY MINSTRELS SHOW (I,3259)

TREBEK, ALEX
THE $128,000 QUESTION (II,1903); BATTLESTARS (II,182); HIGH ROLLERS (II,1147); JEOPARDY (II,1280); PITFALL (II,2047); THE WIZARD OF ODDS (I,4886)

TREEN, MARY
THE ANDY GRIFFITH SHOW (I,192); DOC (I,1303); THE JOEY BISHOP SHOW (I,2404); LAST STAGECOACH WEST (I,2629); THE LIFE OF RILEY (I,2685); WILLY (I,4868)

TREMAYNE, LES
THE DUKES OF HAZZARD (II,742); THE FURTHER ADVENTURES OF ELLERY QUEEN (I,1719); MCGARRY AND ME (I,2966); NAVY LOG (I,3233); ONE MAN'S FAMILY (I,3383); SHAZAM! (II,2319)

TRENDLER, ROBERT
THE MUSIC SHOW (I,3178)

TRENT, JOE
ANOTHER WORLD (II,97)

TRENT, KAREN SUE
LEAVE IT TO BEAVER (I,2648)

TRENT, PETER
THE THREE MUSKETEERS (I,4480)

TRENTHAM, BARBARA
SHEILA (II,2325)

TRESS, DAVID
THE GEORGIA PEACHES (II,979)

TREVIELLE, ROGER
CAPTAIN GALLANT OF THE FOREIGN LEGION (I,831)

TREVIGLIO, LEONARD
TOP SECRET (II,2647)

TREVINO, GEORGE
ELFEGO BACA (I,1426)

TREVIS, MARK
THE DESILU REVUE (I,1238)

TREVOR, CLAIRE
ALFRED HITCHCOCK PRESENTS (I,115); THE DESILU PLAYHOUSE (I,1237); FORD TELEVISION THEATER (I,1634); THE LUX VIDEO THEATER (I,2795); STAGE 7 (I,4181)

TRIANA, PATRICIA
RYAN'S HOPE (II,2234)

TRIBBEY, VANA
ANOTHER WORLD (II,97)

TRIBBLE, DARRYL
FAME (II,812)

TRIKONIS, GUS
DIAGNOSIS: DANGER (I,1250)

TRINKA, PAUL
VOYAGE TO THE BOTTOM OF THE SEA (I,4743)

TRIPP, JOHN
AS THE WORLD TURNS (II,110)

TRIPP, PAUL
DOBIE GILLIS (I,1302); IT'S MAGIC (I,2270); MR. I. MAGINATION (I,3140); TALES OF TOMORROW (I,4352)

TRIPP, SUZANNE
OUR FIVE DAUGHTERS (I,3413)

TRISKA, JAN
RETURN OF THE MAN FROM U.N.C.L.E.: THE 15 YEARS LATER AFFAIR (II,2140)

TRISTAN, DOROTHY
THE OATH: THE SAD AND LONELY SUNDAYS (II,1874); THE WALTONS (II,2740)

TRONTO, RUDY
ALL ABOUT MUSIC (I,127)

TROOBNICK, GENE
DAMN YANKEES (I,1128)

TROSTER, GAVIN
THE CHISHOLMS (II,512)

TROUGHTON, PATRICK
THE ADVENTURES OF THE SCARLET PIMPERNEL (I,80); DOCTOR WHO (II,689); DOCTOR WHO—THE FIVE DOCTORS (II,690); LITTLE WOMEN (I,2723); THE SIX WIVES OF HENRY VIII (I,4068)

TROUP, BOBBY
THE 25TH MAN (MS) (II,2678); ACAPULCO (I,15); BATTLE OF THE NETWORK STARS (II,163); BENNY AND BARNEY: LAS VEGAS UNDERCOVER (II,206); DRAGNET (I,1370); EMERGENCY! (II,775); HOORAY FOR HOLLYWOOD (I,2114)

TROUP, RONNE
THE BANANA SPLITS ADVENTURE HOUR (I,340); THE DAUGHTERS OF JOSHUA CABE RETURN (I,1170); HIGH RISK (II,1146); MY THREE SONS (I,3205); THE PARTRIDGE FAMILY (II,1962); THANKSGIVING REUNION WITH THE PARTRIDGE FAMILY AND MY THREE SONS (II,2569)

TROUT, ROBERT
WHO SAID THAT? (I,4849)

TROW, ROBERT
MR. ROGERS' NEIGHBORHOOD (I,3150)

TROWE, JOSE CHAVEZ
FALCON'S GOLD (II,809)

TROY, LOUISE
MR. BELVEDERE (I,3127)

TRUE GRIT
THE BLACK SHEEP SQUADRON (II,262)

TRUEMAN, PAULA
BILLY (II,247); MURDER INK (II,1769); O'MALLEY (II,1872)

TRUEX, ERNEST
THE ANN SOTHERN SHOW (I,220); CAMEO THEATER (I,808); THE DOCTOR (I,1320); FATHER KNOWS BEST (II,838); FORD THEATER HOUR (I,1635); JAMIE (I,2333); JAMIE (I,2334); MR. PEEPERS (I,3147); OUR TOWN (I,3419); THE TIMID SOUL (I,4519)

TRUMAN, PAMELA
MR. BELVEDERE (I,3127)

TRUMAN, PAULA
STAR STAGE (I,4197); WHERE THE HEART IS (I,4831)

TRUMBULL, BOBBY
CAPTAIN Z-RO (I,839)

TRUMBULL, BRAD
ALMOST AMERICAN (II,55); CALL TO DANGER (I,803); THE HARVEY KORMAN SHOW (II,1103); THE MARY TYLER MOORE SHOW (II,1640)

TRUMBULL, ROBERT
DEAR MOM, DEAR DAD (I,1202)

TRUMP, GERALD
KINCAID (I,2543)

TRUSEL, LISA
FATHER MURPHY (II,841)

TRUSTMAN, SUSAN
ANOTHER WORLD (II,97)

TRYON, TOM
TEXAS JOHN SLAUGHTER (I,4414)

TSANG, WILLIE
MY DARLING JUDGE (I,3186)

TSE, MARIKO
FROM HERE TO ETERNITY (II,932)

TSU, IRENE
FUTURE COP (II,945); JUDGE DEE IN THE MONASTERY MURDERS (I,2465); RYAN'S FOUR (II,2233)

TUBB, ERNEST
COUNTRY MUSIC CARAVAN (I,1066)

TUCCI, MARIA
LAND OF HOPE (II,1424)

TUCCI, MICHAEL
FRIENDS (II,928); ON OUR OWN (II,1892); THE PAPER CHASE: THE SECOND YEAR (II,1950); THE RAINBOW GIRL (II,2108)

TUCKER, BURNELL
THE MARTIAN CHRONICLES (II,1635); MASTER OF THE GAME (II,1647)

TUCKER, CYNTHIA
MASTER OF THE GAME (II,1647)

TUCKER, DUANE
THE SECRET WAR OF JACKIE'S GIRLS (II,2294)

TUCKER, DYLAN C
FATHER MURPHY (II,841)

TUCKER, FORREST
ALIAS SMITH AND JONES (I,119); ALICE (II,33); BLACK BEAUTY (II,261); BOBBY JO AND THE BIG APPLE GOODTIME BAND (I,674); CAT BALLOU (I,870); CHANNING (I,900); CRUNCH AND DES (I,1108); DOC (I,1303); DUSTY'S TRAIL (I,1390); F TROOP (I,1499); FILTHY RICH (II,856); THE FLIM-FLAM MAN (I,1599); G.E. SUMMER ORIGINALS (I,1747); THE GHOST BUSTERS (II,986); JARRETT (I,2349); THE KAISER ALUMINUM HOUR (I,2503); THE LUX VIDEO THEATER (I,2795); ONCE AN EAGLE (II,1897); POTTSVILLE (II,2072); THE REBELS (II,2121); SCHLITZ PLAYHOUSE OF STARS (I,3936)

TUCKER, IAN
PETER PAN (I,3566)

TUCKER, JOHN BARTHOLOMEW
CANDID CAMERA (II,420); TREASURE ISLE (I,4601)

TUCKER, MADGE
THE LADY NEXT DOOR (I,2605)

TUCKER, MICHAEL
LOVE, SEX. . . AND MARRIAGE (II,1546)

TUCKER, NAN
IVAN THE TERRIBLE (II,1260); RYAN'S HOPE (II,2234)

TUCKER, TANYA
CELEBRITY CHALLENGE OF THE SEXES 3 (II,467); THE GEORGIA PEACHES (II,979); MUSIC CITY NEWS TOP COUNTRY HITS OF THE YEAR (II,1772); PUMPBOYS AND DINETTES ON TELEVISION (II,2090); THE REBELS (II,2121)

TUDDENHAM, PETER
BLAKE'S SEVEN (II,264); QUILLER: PRICE OF VIOLENCE (II,2101)

TUERPE, PAUL
LANDON, LANDON & LANDON (II,1426)

TUFANO, DENNIS
OLIVIA NEWTON-JOHN IN CONCERT (II,1884)

TUFELD, DICK
LOST IN SPACE (I,2758)

TUFELD, LYNN
TESTIMONY OF TWO MEN (II,2564)

TUFTS, SONNY
DAMON RUNYON THEATER (I,1129); HAVE GIRLS—WILL TRAVEL (I,1967)

TUFTS, WARREN
CAPTAIN FATHOM (I,830)

TUGGLE, BRETT
HOME ROOM (II,1172)

TULL, PATRICK
MURDER MOTEL (I,3161)

TULLEY, PAUL
ANOTHER WORLD (II,97)

TULLY, MARGO
KENNEDY (II,1377)

TULLY, PAUL
HARRY O (II,1099)

TULLY, PETER
FOREST RANGERS (I,1642)

TULLY, TOM
THE DICK VAN DYKE SHOW (I,1275); HEY, LANDLORD (I,2039); THE LINE-UP (I,2707); SHANE (I,3995); TELEPHONE TIME (I,4379)

TUNNELL, JANIS
FOR MEMBERS ONLY (II,905)

TUNNEY, JIM
CELEBRITY CHALLENGE OF THE SEXES (II,464)

TUPOU, MANU
BORN TO THE WIND (II,354)

TURCO, PAOLO
TOP SECRET (II,2647)

TURCO, TONY
THE FIRM (II,860); RAGE OF ANGELS (II,2107)

TURENNE, LOUIS
THE EDGE OF NIGHT (II,760); GEORGE WASHINGTON (II,978)

TURGEON, PETER
DARK SHADOWS (I,1157)

TURICH, FELIPE
VENICE MEDICAL (II,2726)

TURKEL, ANN
DEATH RAY 2000 (II,654); MASSARATI AND THE BRAIN (II,1646); MATT HELM (II,1652); MODESTY BLAISE (II,1719)

TURKEL, STUDS
STUD'S PLACE (I,4271)

TURKUS, BURTON
MR. ARSENIC (I,3126)

TURMAN, GLYNN
THE BLUE KNIGHT (II,277); CASS MALLOY (II,454); CENTENNIAL (II,477); MANIMAL (II,1622); PEYTON PLACE (I,3574); POOR RICHARD (II,2065)

TURNBAUGH, BRENDA
LITTLE HOUSE ON THE PRAIRIE (II,1487)

TURNBAUGH, WENDI
LITTLE HOUSE ON THE PRAIRIE (II,1487)

TURNBULL, GLENN
FULL SPEED ANYWHERE (I,1704)

TURNER, ARNOLD
THE SEAL (II,2282)

TURNER, BARBARA
LOLLIPOP LOUIE (I,2738)

TURNER, DAIN
HARRIS AND COMPANY (II,1096)

TURNER, DAVE
THE BARBARY COAST (II,146)

TURNER, E FRANTZ
NICHOLS AND DYMES (II,1841)

TURNER, GEORGE
THE DATCHET DIAMONDS (I,1160); TEN LITTLE INDIANS (I,4394)

TURNER, IKE
GOOD VIBRATIONS FROM CENTRAL PARK (I,1854)

TURNER, JACQUELINE
THE AT LIBERTY CLUB (I,309)

TURNER, JAKE
LOVE OF LIFE (I,2774)

TURNER, JANINE
BEHIND THE SCREEN (II,200)

TURNER, JOHN
A PLACE TO DIE (I,3612)

TURNER, KATE
PRISONER: CELL BLOCK H (II,2085)

TURNER, KATHLEEN
THE DOCTORS (II,694)

TURNER, LANA
FALCON CREST (II,808); THE SURVIVORS (I,4301)

TURNER, LYNN
CASANOVA (II,451)

TURNER, MARYANN
A HORSEMAN RIDING BY (II,1182)

TURNER, PATTY
LEAVE IT TO BEAVER (I,2648)

TURNER, SKIP
WONDER GIRL (I,4891)

TURNER, STEPHEN
LITTLE WOMEN (I,2723)

TURNER, TIERRE
THE COP AND THE KID (II,572); THE WAVERLY WONDERS (II,2746)

TURNER, TIM
WHITE HUNTER (I,4845)

TURNER, TINA
GOOD VIBRATIONS FROM CENTRAL PARK (I,1854); ROCK 'N' ROLL: THE FIRST 25 YEARS (II,2189)

TURNER, VICTORIA
PENMARRIC (II,1993)

TURQUAND, TODD
THE NEW MICKEY MOUSE CLUB (II,1823); RHODA (II,2151)

TUSEO, TONY
RIVKIN: BOUNTY HUNTER (II,2183)

TUTIN, DOROTHY
THE SIX WIVES OF HENRY VIII (I,4068)

TUTTLE, LURENE
ALL IN THE FAMILY (I,133); AN APARTMENT IN ROME (I,242); CARLTON YOUR DOORMAN (II,440); DOROTHY IN THE LAND OF OZ (II,722); FATHER OF THE BRIDE (I,1543); I DREAM OF JEANNIE (I,2167); I REMEMBER CAVIAR (I,2175); JULIA (I,2476); THE KILLIN' COUSIN (II,1394); LIFE WITH FATHER (I,2690); THE REAL MCCOYS (I,3741); THE RETURN OF THE BEVERLY HILLBILLIES (II,2139); SHOOTING STARS (II,2341)

TUTTLE, MICHELLE
YOUNG DR. MALONE (I,4938)

TUZZI, GEORGE MICHAEL

KENNEDY (II,1377)

TWEDELL, FRANK
THE EGG AND I (I,1420); THE JOKE AND THE VALLEY (I,2438); THE MAN WHO CAME TO DINNER (I,2880); THREE STEPS TO HEAVEN (I,4485)

TWEED, BABETTE
SHERLOCK HOLMES (II,2327)

TWEED, SHANNON
FALCON CREST (II,808)

TWEED, TERRY
THE BAXTERS (II,184)

TWIGG, SIMON
SATAN'S WAITIN' (I,3916)

TWIGGY
ECHOES OF THE SIXTIES (II,756); TWIGGY IN NEW YORK (I,4649); TWIGGY'S JUKE BOX (II,2682)

TWITCHELL, TIFNI
ARCHIE (II,104)

TWOHEY, ROZ
HAWKINS FALLS, POPULATION 6200 (I,1977)

TWOMEY, ANNE
THE FIRM (II,860)

TYLER, BEVERLY
BIG TOWN (I,436)

TYLER, GINNY
THE ADVENTURES OF GULLIVER (I,60); THE FANTASTIC FOUR (II,824); SPACE GHOST (I,4141)

TYLER, JANET
KUDA BUX, HINDU MYSTIC (I,2593)

TYLER, JUDY
HOWDY DOODY (I,2151)

TYLER, KIM
PLEASE DON'T EAT THE DAISIES (I,3628)

TYLER, RICHARD
THE ALDRICH FAMILY (I,111); MAKE ROOM FOR DADDY (I,2843)

TYLER, ROBIN
THE KROFFT KOMEDY HOUR (II,1411)

TYLER, WILLIAM
BRACKEN'S WORLD (I,703)

TYLER, WILLIE
BLOCKHEADS (II,270)

TYLO, MICHAEL
THE GUIDING LIGHT (II,1064)

TYLO, MIKE
THE GUIDING LIGHT (II,1064)

TYLONG, MYFANNY
THE CITADEL (II,539)

TYNAN, WILLIAM
A WORLD APART (I,4910)

TYNER, CHARLES
THE AWAKENING LAND (II,122); FATHER MURPHY (II,841); THE GREATEST GIFT (II,1061); LASSIE: THE NEW BEGINNING (II,1433); WINTER KILL (II,2810); THE YOUNG PIONEERS (II,2868)

TYRON, TOM
FRONTIER (I,1695)

TYRRELL, ANN
THE ANN SOTHERN SHOW (I,220); PRIVATE SECRETARY (I,3672); THE UNEXPECTED (I,4674)

TYRRELL, SUSAN
OPEN ALL NIGHT (II,1914); WILLOW B: WOMEN IN PRISON (II,2804)

TYSON, CICELY
CBS: ON THE AIR (II,460); EAST SIDE/WEST SIDE (I,1396); THE GUIDING LIGHT (II,1064); ROOTS (II,2211); WEDNESDAY NIGHT OUT (II,2760)

TYSON, IAN
THE IAN TYSON SHOW (I,2181)

TYSSEN, GRETA
TREASURE HUNT (I,4600)

TYZACK, MARGARET
THE FORSYTE SAGA (I,1645)

TZUDIKER, BOB
MICKEY SPILLANE'S MIKE HAMMER: MORE THAN MURDER (II,1693)

UBELS, STEVE
SHOGUN (II,2339)

UGGAMS, LESLIE
BACKSTAIRS AT THE WHITE HOUSE (II,133); THE BOOK OF LISTS (II,350); CELEBRITY CHALLENGE OF THE SEXES 2 (II,466); FANTASY (II,828); HIGH ROLLERS (II,1147); THE KRAFT 75TH ANNIVERSARY SPECIAL (II,1409); LESLIE (I,2659); THE LESLIE UGGAMS SHOW (I,2660); ROOTS (II,2211); SMALL WORLD (II,2381)

UKENA, PAUL
MARCO POLO (I,2904)

ULLMAN, TRACY
MACKENZIE (II,1584)

ULLMANN, LIV
SCENES FROM A MARRIAGE (II,2272)

ULLRICK, SHARON
THE BLACK SHEEP SQUADRON (II,262)

ULMER, ADRIENNE
LITTLE WOMEN (I,2721)

ULRICH, KIM
AS THE WORLD TURNS (II,110)

UMEKI, MIYOSHI
THE COURTSHIP OF EDDIE'S FATHER (I,1083)

UNDERHILL, DON
THE MOUSEKETEERS REUNION (II,1742)

UNDERWOOD, RAY
THE LAST RESORT (II,1440)

UPJOHN, WILLIAM
ALL THE RIVERS RUN (II,43)

URBANT, JACK
OZMOE (I,3437)

URECAL, MINERVA
THE ADVENTURES OF JIM BOWIE (I,63); THE HUMAN COMEDY (I,2158); MEET MR. MCNUTLEY (I,2989); PETER GUNN (I,3562); THE R.C.A. VICTOR SHOW (I,3737); THE ROY ROGERS SHOW (I,3859); TOM AND JERRY (I,4533); TUGBOAT ANNIE (I,4622)

URICH, ROBERT
60 YEARS OF SEDUCTION (II,2373); BATTLE OF THE NETWORK STARS (II,167); BATTLE OF THE NETWORK STARS (II,168); BATTLE OF THE NETWORK STARS (II,172); BOB & CAROL & TED & ALICE (I,495); BOB HOPE SPECIAL: BOB HOPE IN THE STAR-MAKERS (II,285); BUNCO (II,395); THE CELEBRITY FOOTBALL CLASSIC (II,474); THE FUNNIEST JOKE I EVER HEARD (II,939); THE FUNNIEST JOKE I EVER HEARD (II,940); GAVILAN (II,959); LADIES IN BLUE (II,1422); MERRY CHRISTMAS FROM THE GRAND OLE OPRY (II,1686); MISTRAL'S DAUGHTER (II,1708); S.W.A.T. (II,2236); SOAP (II,2392); TABITHA (II,2527); TABITHA (II,2528); VEGAS (II,2723); VEGAS (II,2724)

URICH, TOM
ANOTHER LIFE (II,95)

URQUHART, GORDON
CHICAGOLAND MYSTERY PLAYERS (I,936)

URSETH, BONNIE
WE GOT IT MADE (II,2752)

USTINOV, PETER
BAREFOOT IN ATHENS (I,354); DOCTOR SNUGGLES (II,687); GIDEON (I,1792); LOVE, LIFE, LIBERTY & LUNCH (II,1544); NICHOLAS NICKLEBY (II,1840); OMNI: THE NEW FRONTIER (II,1889); THE SEVEN DIALS MYSTERY (II,2308); A STORM IN SUMMER (I,4237); WHY DIDN'T THEY ASK EVANS? (II,2795)

UTMAN, BRIAN
HERBIE, THE LOVE BUG (II,1131); SEVEN BRIDES FOR SEVEN BROTHERS (II,2307)

UTTAL, FRED
Q.E.D. (I,3701)

VACCARO, BRENDA
DEAR DETECTIVE (II,652); THE GREATEST SHOW ON EARTH (I,1883); MY LUCKY PENNY (I,3196); PAPER DOLLS (II,1952); SARA (II,2260); THAT WAS THE YEAR THAT WAS (II,2575); TRAVIS LOGAN, D.A. (I,4598)

VACIO, NATIVIDAD
FATHER KNOWS BEST (II,838)

VAGUE, VERA
FOLLOW THE LEADER (I,1616)

VAHANIAN, MARC
THE REBELS (II,2121)

VAHEY, ROBERT
A HORSEMAN RIDING BY (II,1182)

VAHL, WILLIAM
PENMARRIC (II,1993)

VALDIS, SIGRID
HOGAN'S HEROES (I,2069)

VALE, RITA
THE WOMAN IN WHITE (I,4887)

VALENTE, CATERINA
THE CATERINA VALENTE SHOW (I,873); THE ENTERTAINERS (I,1449)

VALENTI, LILLI
ROBERTA (I,3815)

VALENTINE, ANTHONY
THE FEAR IS SPREADING (I,1553); MASADA (II,1642)

VALENTINE, JAMES
CAMELOT (II,416)

VALENTINE, KAREN
ADAMS HOUSE (II,9); AMERICA 2100 (II,66); THE DAUGHTERS OF JOSHUA CABE (I,1169); GIDGET

GROWS UP (I,1797); A GIRL'S LIFE (II,1000); GO WEST YOUNG GIRL (II,1013); GOODBYE DOESN'T MEAN FOREVER (II,1039); HAVING BABIES I (II,1107); KAREN (II,1363); THE KAREN VALENTINE SHOW (I,2506); THE LOVE BOAT I (II,1532); RETURN TO FANTASY ISLAND (II,2144); ROOM 222 (I,3843); SKEEZER (II,2376)

VALENTINE, KATE
LOVE IN A COLD CLIMATE (II,1539)

VALENTINE, PAUL
TONIGHT IN SAMARKAND (I,4558)

VALENTINO, BARRY
VALENTINO (I,4694)

VALENTY, LILI
GIMME A BREAK (II,995); THEY ONLY COME OUT AT NIGHT (II,2586)

VALLANCE, LOUISE
FALCON'S GOLD (II,809); THE ROPERS (II,2214)

VALLATINE, JAMES
SHANGRI-LA (I,3996)

VALLEE, RUDY
ALIAS SMITH AND JONES (I,118); BATMAN (I,366); HANSEL AND GRETEL (I,1934); HOTPOINT HOLIDAY (I,2131); KRAFT TELEVISION THEATER (I,2592); MATINEE AT THE BIJOU (II,1651); NIGHT GALLERY (I,3287); ON BROADWAY TONIGHT (I,3355)

VALLELY, JAMES
DOUBLE TROUBLE (II,724); T.L.C. (II,2525)

VALLEN, LESLIE
KUDZU (II,1415)

VALLEY, BEATRICE
DICK AND THE DUCHESS (I,1263)

VALLI, FRANKIE
EBONY, IVORY AND JADE (II,755)

VALLI, JUNE
THE ANDY WILLIAMS-JUNE VALLI SHOW (I,200); LET'S DANCE (I,2664)

VALLIER, HELENE
MISTRAL'S DAUGHTER (II,1708)

VALLO, RAMON
WATERFRONT (I,4774)

VALLONE, RAF
FAME (I,1517); THE SMALL MIRACLE (I,4090)

VAN ARK, JOAN
BATMAN AND THE SUPER SEVEN (II,160); BATTLE OF THE NETWORK STARS (II,170); BATTLE OF THE NETWORK STARS (II,171); BATTLE OF THE NETWORK STARS (II,174); BATTLE OF THE NETWORK STARS (II,175); BATTLE OF THE NETWORK STARS (II,177); BIG ROSE (II,241); CELEBRITY CHALLENGE OF THE SEXES 5 (II,469); DALLAS (II,614); THE JUDGE AND JAKE WYLER (I,2464); KNOTS LANDING (II,1404); NIGHT GALLERY (I,3287); RHODA (II,2151); SPIDER-WOMAN (II,2426); TEMPERATURES RISING (I,4390); TESTIMONY OF TWO MEN (II,2564); THUNDARR THE BARBARIAN (II,2616); WE'VE GOT EACH OTHER (II,2757)

VAN BEERS, STANLEY
THE ADVENTURES OF THE SCARLET PIMPERNEL (I,80)

VAN BERGEN, LEWIS
MODESTY BLAISE (II,1719)

VAN CLEEF, LEE
LAST STAGECOACH WEST (I,2629); THE MASTER (II,1648); NOWHERE TO HIDE (II,1868); THE SLOWEST GUN IN THE WEST (I,4088)

VAN DER BYL, PHILIP
PRISONERS OF THE LOST UNIVERSE (II,2086)

VAN DER VLIS, DIANA
GHOSTBREAKER (I,1789); RYAN'S HOPE (II,2234); WHERE THE HEART IS (I,4831)

VAN DEVORE, TRISH
BEAUTY AND THE BEAST (I,382); ONE LIFE TO LIVE (II,1907); SEARCH FOR TOMORROW (II,2284)

VAN DISSEL, PETER
THE ASSASSINATION RUN (II,112)

VAN DOREN, CHARLES
THE THIRD COMMANDMENT (I,4449)
.sp .51

VAN DOREN, MAMIE
MEET THE GIRLS: THE SHAPE, THE FACE, AND THE BRAIN (I,2990)

VAN DREELAND, JOHN
THE RHINEMANN EXCHANGE (II,2150); THREE FOR DANGER (I,4475); THE WORD (II,2833)

VAN DUSEN, GRANVILLE
ALLISON SIDNEY HARRISON (II,54); THE ASTRONAUTS (II,114); DOCTOR SCORPION (II,685); DYNASTY (II,747); ESCAPADE (II,781); MY WIFE NEXT DOOR (II,1781); SUTTERS BAY (II,2508); THIS IS KATE BENNETT (II,2597); THE WORLD BEYOND (II,2835); THE WORLD OF DARKNESS (II,2836)

VAN DUSEN, LINDA
MATINEE AT THE BIJOU (II,1651)

VAN DYKE, BARRY
CASINO (II,452); GALACTICA 1980 (II,953); GHOST OF A CHANCE (II,987); GUN SHY (II,1068); THE HARVEY KORMAN SHOW (II,1103); THE HARVEY KORMAN SHOW (II,1104); MR. MOM (II,1752); THE POWERS OF MATTHEW STAR (II,2075); TABITHA (II,2527); WHAT'S UP DOC? (II,2771)

VAN DYKE, BONNIE
SENIOR YEAR (II,2301); SONS AND DAUGHTERS (II,2404)

/bf VAN DYKE, DICK
CBS CARTOON THEATER (I,878); CBS: ON THE AIR (II,460); THE CONFESSIONS OF DICK VAN DYKE (II,568); DICK VAN DYKE AND THE OTHER WOMAN, MARY TYLER MOORE (I,1273); DICK VAN DYKE MEETS BILL COSBY (I,1274); THE DICK VAN DYKE SHOW (I,1275); THE DICK VAN DYKE SPECIAL (I,1276); THE DICK VAN DYKE SPECIAL (I,1277); THE FIRST NINE MONTHS ARE THE HARDEST (I,1584); HARRY'S BATTLES (II,1100); I'M A FAN (I,2192); JULIE AND DICK IN COVENT GARDEN (II,1352); LAUGH LINE (I,2633); MOTHERS DAY (I,3109); THE NEW DICK VAN DYKE SHOW (I,3262); THE NEW DICK VAN DYKE SHOW (I,3263); A SALUTE TO STAN LAUREL (I,3892); TROUBLE WITH RICHARD (I,4612); TRUE LIFE STORIES (II,2665); VAN DYKE AND COMPANY (II,2718); VAN DYKE AND COMPANY (II,2719)

VAN DYKE, JERRY
13 QUEENS BOULEVARD (II,2591); ACCIDENTAL FAMILY (I,17); THE DICK VAN DYKE SHOW (I,1275); HEADMASTER (I,1990); MY BOY GOOGIE (I,3185); MY MOTHER THE CAR (I,3197); PICTURE THIS (I,3593); YOU'RE ONLY YOUNG TWICE (I,4971)

VAN ELTZ, THEODORE
ONE MAN'S FAMILY (I,3383)

VAN EVERA, JEAN
THE WORLD BEYOND (II,2835)

VAN EYCK, ANTHONY
RITUALS (II,2182)

VAN EYCK, PETER
CASABLANCA (I,860)

VAN FLEET, JO
CINDERELLA (I,956); HEIDI (I,2004); SATAN'S WAITIN' (I,3916)

VAN FLEET, MARY
THE HEIRESS (I,2006); OUR TOWN (I,3420)

VAN HOFFMAN, BRANT
240-ROBERT (II,2689)

VAN HORN, BUDDY
THE FLINTSTONE COMEDY HOUR (II,881); PORKY PIG AND FRIENDS (I,3651); THE RETURN OF THE BEVERLY HILLBILLIES (II,2139)

VAN KRUAN, KIMBERLY
MASTER OF THE GAME (II,1647)

VAN NESS, JON
NORMA RAE (II,1860)

VAN NIEL, JAN
SEMI-TOUGH (II,2299)

VAN NORDEN, PETER
MALIBU (II,1607); OLD FRIENDS (II,1881)

VAN NUYS, ED
KENNEDY (II,1377); MR. INSIDE/MR. OUTSIDE (II,1750); O'MALLEY (II,1872)

VAN PATTEN, DICK
ANDY WILLIAMS' EARLY NEW ENGLAND CHRISTMAS (II,79); BATTLE OF THE NETWORK STARS (II,168); BATTLE OF THE NETWORK STARS (II,169); CELEBRITY CHALLENGE OF THE SEXES 2 (II,466); THE CENTURY TURNS (I,890); CHARO AND THE SERGEANT (II,489); EIGHT IS ENOUGH (II,762); ERNIE, MADGE, AND ARTIE (I,1457); FIT FOR A KING (II,866); GRANDPA MAX (II,1051); I LOVE LIBERTY (II,1215); I REMEMBER MAMA (I,2176); THE LOVE BOAT I (II,1532); THE NEW DICK VAN DYKE SHOW (I,3263); THE

NURSES (I,3319); THE PARTNERS (I,3462); PAT BOONE AND FAMILY (II,1968); RICH LITTLE'S WASHINGTON FOLLIES (II,2159); STATE FAIR USA (II,2449); STATE FAIR USA (II,2450); TAKE ONE STARRING JONATHAN WINTERS (II,2532); THAT SECOND THING ON ABC (II,2572); WHATEVER BECAME OF. . . ? (II,2772); WHEN THINGS WERE ROTTEN (II,2782); YOUNG DR. MALONE (I,4938)

VAN PATTEN, JAMES
THE CHISHOLMS (II,512); THE CHISHOLMS (II,513); JOSIE (II,1345); LOVE AND LEARN (II,1529)

VAN PATTEN, JOYCE
ACCIDENTAL FAMILY (I,17); BULBA (II,392); BURT AND THE GIRLS (I,770); BUS STOP (II,399); A CHRISTMAS FOR BOOMER (II,518); DOBIE GILLIS (I,1302); THE DON RICKLES SHOW (I,1342); GEORGE GOBEL PRESENTS (I,1766); THE GOOD GUYS (I,1847); THE MARTIAN CHRONICLES (II,1635); THE MARY TYLER MOORE COMEDY HOUR (II,1639); MAUREEN (II,1656); THE PLANT FAMILY (II,2050); WINTER KILL (II,2810); YOU CAN'T TAKE IT WITH YOU (II,2857); YOUNG DR. MALONE (I,4938)

VAN PATTEN, TIMOTHY
JOHNNY GARAGE (II,1337); THE MASTER (II,1648); THE WHITE SHADOW (II,2788)

VAN PATTEN, VINCENT
APPLE'S WAY (II,101)

VAN REENEN, JAN
MODESTY BLAISE (II,1719)

VAN ROOTEN, LUIS
THE DOCTORS (II,694); THE JOE PALOOKA STORY (I,2401); MAJOR DELL CONWAY OF THE FLYING TIGERS (I,2835); THE RIGHT MAN (I,3790)

VAN VALKENBURGH, DEBORAH
FAMILY BUSINESS (II,815); TOO CLOSE FOR COMFORT (II,2642)

VAN VLEET, RICHARD
ALL MY CHILDREN (II,39)

VAN VOOREN, MONIQUE
THE DUPONT SHOW OF THE WEEK (I,1388)

VAN VOORHIS, WESTBROOK
MARCH OF TIME THROUGH THE YEARS (I,2903); PANIC! (I,3447)

VAN, BILLY
THE HILARIOUS HOUSE OF FRIGHTENSTEIN (II,1153); YOU CAN'T DO THAT ON TELEVISION (I,4933)

VAN, BOBBY
THE BOBBY VAN AND ELAINE JOYCE SHOW (I,677); BUNCO (II,395); CIRCUS OF THE STARS (II,530); THE FUN FACTORY (II,938); THE HUSTLER OF MUSCLE BEACH (II,1210); THE LOVEBIRDS (II,1548); MAKE ME LAUGH (II,1602); SHOWOFFS (II,2345)

VAN, GLORIA
JAMBOREE (I,2331)

VAN, GUS
GAY NINETIES REVUE (I,1743)

VANCE, CHRISTIAN
PEARL (II,1986)

VANCE, DANA
HOT W.A.C.S. (II,1191)

VANCE, VIVIAN
THE FRONT PAGE (I,1692); I LOVE LUCY (I,2171); THE LUCILLE BALL-DESI ARNAZ SHOW (I,2784); THE LUCY SHOW (I,2791)

VANDENBOSCH, TERRI
AS THE WORLD TURNS (II,110)

VANDENBERG, TOM
FRAGGLE ROCK (II,916)

VANDERBILT, GLORIA
KRAFT TELEVISION THEATER (I,2592)

VANDERPYL, JEAN
THE ATOM ANT/SECRET SQUIRREL SHOW (I,311); THE BANANA SPLITS ADVENTURE HOUR (I,340); THE FLINTSTONE COMEDY HOUR (II,881); FLINTSTONE FAMILY ADVENTURES (II,882); THE FLINTSTONE FUNNIES (II,883); THE FLINTSTONES (II,884); THE FLINTSTONES (II,885); FRED AND BARNEY MEET THE THING (II,919); INCH HIGH, PRIVATE EYE (II,1231); THE JETSONS (I,2376); THE MAGILLA GORILLA SHOW (I,2825); THE NEW FRED AND BARNEY SHOW (II,1817); PLEASE DON'T EAT THE DAISIES (I,3628); TOP CAT (I,4576); WHERE'S HUDDLES (I,4836)

VANDERKLOOT, VICTORIA
LONDON AND DAVIS IN NEW YORK (II,1512)

VANDERS, WARREN
DANIEL BOONE (I,1142); DOC ELLIOT (I,1305); EMPIRE (I,1439); NEVADA SMITH (II,1807)

VANDERVEEN, JOYCE
THE ADVENTURES OF JIM BOWIE (I,63)

VANDIS, TITOS
GENESIS II (I,1760); IN TANDEM (II,1228); ROGER AND HARRY (II,2200); THE SAN PEDRO BUMS (II,2251); TERROR AT ALCATRAZ (II,2563)

VANDROSS, LUTHER
FINDER OF LOST LOVES (II,857)

VANNI, RENATA
THAT GIRL (II,2570)

VARDEN, EVELYN
THE CRADLE SONG (I,1088); DREAM GIRL (I,1374); EYE WITNESS (I,1494)

VARDEN, NORMA
DOC (I,1303); HAZEL (I,1982); I MARRIED JOAN (I,2174)

VARELLA, JAY
CONDOMINIUM (II,566); THE GANGSTER CHRONICLES (II,957); MISSING PIECES (II,1706); NAKIA (II,1784)

VARGAS, JOHN
AT EASE (II,115)

VARGAS, VICTORIA
IS THIS GOODBYE, CHARLIE BROWN? (I,2242); WHAT HAVE WE LEARNED, CHARLIE BROWN? (I,4810)

VARLEY, BEATRICE
HEDDA GABLER (I,2001)

VARLEY, SARAH-JANE
GREAT EXPECTATIONS (II,1055)

VARNEY, JIM
AMERICA 2-NIGHT (II,65); THE NEW OPERATION PETTICOAT (II,1826); OPERATION PETTICOAT (II,1916); OPERATION PETTICOAT (II,1917); THE ROUSTERS (II,2220)

VASIL, ART
RAGE OF ANGELS (II,2107)

VASQUEZ, DENNIS
POPI (II,2068); POPI (II,2069)

VAUGHAN, PETER
THE EYES HAVE IT (I,1496)

VAUGHN, HEIDI
THE EDGE OF NIGHT (II,760); FRANKENSTEIN (I,1671)

VAUGHN, MARTIN
1915 (II,1853)

VAUGHN, PAUL
CHEERS (II,494)

VAUGHN, ROBERT
ACTION (I,23); BACKSTAIRS AT THE WHITE HOUSE (II,133); THE BLUE AND THE GRAY (II,273); THE BOSTON TERRIER (I,696); THE BOSTON TERRIER (I,697); CAPTAINS AND THE KINGS (II,435); CENTENNIAL (II,477); EMERALD POINT, N.A.S. (II,773); GENERAL ELECTRIC TRUE (I,1754); THE GOSSIP COLUMNIST (II,1045); THE ISLANDER (II,1249); KISS ME, KILL ME (II,1400); KRAFT MYSTERY THEATER (I,2589); THE LIEUTENANT (I,2680); THE MAN FROM U.N.C.L.E. (I,2867); THE MOONGLOW AFFAIR (I,3099); THE PROTECTORS (I,3681); THE REBELS (II,2121); RETURN OF THE MAN FROM U.N.C.L.E.: THE 15 YEARS LATER AFFAIR (II,2140); TELEPHONE TIME (I,4379); WASHINGTON: BEHIND CLOSED DOORS (II,2744); ZANE GREY THEATER (I,4979)

VAUGHN, SHARON
BEWITCHED (I,418)

VAUGHN, SKEETER
BRIDGER (II,376); THE CHEROKEE TRAIL (II,500)

VAUGHN, WILLIAM
THE LONER (I,2744)

VAUX, LANE
FAT ALBERT AND THE COSBY KIDS (II,835)

VESOTA, BRUNO
MY MOTHER THE CAR (I,3197)

VEAGUE, VERA
THE GREATEST MAN ON EARTH (I,1882)

VEAZIE, CAROL
THE DICK VAN DYKE SHOW (I,1275); FIRST LOVE (I,1583); NORBY (I,3304)

VEGA, ERNESTO
THE RIGHTEOUS APPLES (II,2174)

VEGA, ROBERT
WOLF ROCK TV (II,2815)

VELA, PACO
DALLAS (II,614)

VELEZ, EDDIE
FOR LOVE AND HONOR (II,903)

VELEZ, MARTHA
A.K.A. PABLO (II,28)

VENDIG, LAURIE
THREE STEPS TO HEAVEN (I,4485)

VENDRELL, MICHAEL
GOLIATH AWAITS (II,1025)

VENEY, GARY
SHIPSHAPE (II,2329)

VENEZIANI, ROSETTA
NAKED CITY (I,3216)

VENNERA, CHICK
G.I.'S (II,949); VEGAS (II,2723)

VENOCOUR, JOHN
THE KID SUPER POWER HOUR WITH SHAZAM (II,1386)

VENORA, DIANE
GETTING THERE (II,985)

VENTHAM, WANDA
U.F.O. (I,4662)

VENTON, HARLEY
THE GUIDING LIGHT (II,1064)

VENTURE, RICHARD
THE CHICAGO STORY (II,507); FROM HERE TO ETERNITY (II,932); GENERAL HOSPITAL (II,964); THE SECRET STORM (I,3969); THE THORN BIRDS (II,2600)

VENUTA, BENAY
ANNIE GET YOUR GUN (I,224)

VENZIE, CAROL
OUR TOWN (I,3419)

VER DORN, JERRY
THE GUIDING LIGHT (II,1064)

VERA, RICKY
OUR MISS BROOKS (I,3415)

VERA, TONY
MURDER INK (II,1769)

VERBIT, HELEN
FLATBUSH (II,877); LOVERS AND OTHER STRANGERS (II,1550); RETURN OF THE WORLD'S GREATEST DETECTIVE (II,2142); THE RIGHTEOUS APPLES (II,2174)

VERDIER, PAUL
THE FRENCH ATLANTIC AFFAIR (II,924)

VERDON, GWEN
THE JERK, TOO (II,1282)

VERDUGO, ELENA
THE FLIM-FLAM MAN (I,1599); HARRY'S BUSINESS (I,1958); MANY HAPPY RETURNS (I,2895); MARCUS WELBY, M.D. (II,1627); MEET MILLIE (I,2988); MONA MCCLUSKEY (I,3085); THE NEW PHIL SILVERS SHOW (I,3269); REDIGO (I,3758); THE RETURN OF MARCUS WELBY, M.D (II,2138)

VEREEN, BEN
BEN VEREEN. . . COMIN' AT YA (II,203); BEN VEREEN—HIS ROOTS (II,202); ELLIS ISLAND (II,768); MARY'S INCREDIBLE DREAM (II,1641); ROOTS (II,2211); TENSPEED AND BROWN SHOE (II,2562); WEBSTER (II,2758)

VERMILYEN, HAROLD
HAPPY BIRTHDAY (I,1941); MARCO POLO (I,2904)

VERNAN, RON
THE RITA MORENO SHOW (II,2181)

VERNER, LILLIAN
TYCOON: THE STORY OF A WOMAN (II,2697)

VERNON, DAI
MR. & MS. AND THE MAGIC STUDIO MYSTERY (II,1747)

VERNON, GLENN
FOR THE DEFENSE (I,1624)

VERNON, HARVEY
CARTER COUNTRY (II,449); LITTLE HOUSE: BLESS ALL THE DEAR CHILDREN (II,1489)

VERNON, IRENE
BEWITCHED (I,418); DO NOT DISTURB (I,1300); FIRESIDE THEATER (I,1580)

VERNON, JACKIE
FROSTY THE SNOWMAN (II,934); FROSTY'S WINTER WONDERLAND (II,935)

VERNON, JOHN
THE BARBARY COAST (II,147); THE BLUE AND THE GRAY (II,273); DELTA HOUSE (II,658); THE IMPOSTER (II,1225); THE MASK OF MARCELLA (I,2937); MATT HELM (II,1652); SWISS FAMILY ROBINSON (II,2516)

VERNON, RICHARD
IN THE STEPS OF A DEAD MAN (I,2205)

VERO, DENNIS J
SENSE OF HUMOR (II,2303)

VERSATONES, THE
ALL ABOUT MUSIC (I,127)

VESTOFF, VIRGINIA
ALONE AT LAST (II,59); DARK SHADOWS (I,1157); THE DOCTORS (II,694); WE INTERRUPT THIS SEASON (I,4780)

VETRI, VICTORIA
THE PIGEON (I,3596)

VICARY, JAMES
THE SECRET STORM (I,3969)

VICENZIO, SAM
COCAINE AND BLUE EYES (II,548)

VICKERS, JOHN
YOUNG DR. MALONE (I,4938)

VICKERS, MARTHA
THE UNEXPECTED (I,4674)

VICKERS, YVETTE
BEN BLUE'S BROTHERS (I,393)

VICKERY, JAMES
FAIRMEADOWS, U.S.A. (I,1514)

VICTOR, JAMES
CONDO (II,565); MIXED NUTS (II,1714); TWIN DETECTIVES (II,2687); VIVA VALDEZ (II,2728)

VICTOR, PAULA
RHODA (II,2151)

VIDAL, GORE
HOT LINE (I,2123)

VIGARD, KRISTEN
THE GUIDING LIGHT (II,1064)

VIGODA, ABE
BARNEY MILLER (II,154); BOB HOPE SPECIAL: BOB HOPE IN "JOYS" (II,284); CIRCUS OF THE STARS (II,530); CIRCUS OF THE STARS (II,531); DARK SHADOWS (I,1157); FISH (II,864); HAVING BABIES I (II,1107); TOMA (I,4547)

VIGRAN, HERB
THE ADVENTURES OF (I,77); THE ADVENTURES OF GULLIVER (I,60); THE ADVENTURES OF SUPERMAN (I,77); THE BEST IN MYSTERY (I,403); THE DICK VAN DYKE SHOW (I,1275); THE ED WYNN SHOW (I,1404); IF YOU KNEW TOMORROW (I,2187); INSIDE O.U.T. (II,1235); THE JACK BENNY PROGRAM (I,2294); THE LIFE OF RILEY (I,2684); THE LIFE OF VERNON HATHAWAY (I,2687); MIRACLE ON 34TH STREET (I,3056); THE SHIRT TALES (II,2337); TESTIMONY OF TWO MEN (II,2564)

VIHARO, ROBERT
THE SURVIVORS (I,4301)

VILLARD, TOM
HIGH SCHOOL, U.S.A. (II,1148); REVENGE OF THE GRAY GANG (II,2146); SIDNEY SHORR (II,2349); WE GOT IT MADE (II,2752); WHACKED OUT (II,2767)

VILLECHAIZE, HERVE
CIRCUS OF THE STARS (II,537); FANTASY ISLAND (II,829); FANTASY ISLAND (II,830); HARD KNOCKS (II,1086); RETURN TO FANTASY ISLAND (II,2144)

VILLELLA, EDWARD
BRIGADOON (I,726); LITTLE WOMEN (I,2724)

VILLIERS, JAMES
THE DOUBLE KILL (I,1360); MARRIED ALIVE (I,2923)

VINCENT, JAN-MICHAEL
AIRWOLF (II,27); THE BANANA SPLITS ADVENTURE HOUR (I,340); THE CATCHER (I,872); THE SURVIVORS (I,4301); THE WINDS OF WAR (II,2807)

VINCENT, JUNE
BRIGHT PROMISE (I,727); THE DELPHI BUREAU (I,1224); THE HAPPY TIME (I,1943); THE STREETS OF SAN FRANCISCO (I,4260)

VINCENT, LARRY
SEYMOUR PRESENTS (I,3990)

VINCENT, RALPH
YOU'RE INVITED (I,4966)

VINCENT, VIRGINIA
ABC'S MATINEE TODAY (I,7); CLINIC ON 18TH STREET (II,544); EIGHT IS ENOUGH (II,762); THE JOEY BISHOP SHOW (I,2403); THE PSYCHIATRIST: GOD BLESS THE CHILDREN (I,3685); STAGE DOOR (I,4178); THE SUPER (I,4293); THE WEB (I,4784)

VINCI, BOB
KING OF KENSINGTON (II,1395)

VINE, BILLY
54TH STREET REVUE (I,1573)

VINE, JOHN

THE SEVEN DIALS MYSTERY (II,2308)

VINES, LEE
THE MARY TYLER MOORE SHOW (II,1640)

VINOVICH, STEVE
CALL HER MOM (I,796); CATALINA C-LAB (II,458)

VINSON, GARY
MCHALE'S NAVY (I,2969); MISS STEWART, SIR (I,3064); THE MUNSTERS' REVENGE (II,1765); PISTOLS 'N' PETTICOATS (I,3608); THE ROARING TWENTIES (I,3805)

VINSON, JUNE
YOUR JEWELER'S SHOWCASE (I,4953)

VINT, BILL
BANACEK: DETOUR TO NOWHERE (I,339)

VINT, JESSE
CENTENNIAL (II,477)

VINTON, BOBBY
THE BOBBY VINTON SHOW (II,345); THE GOSSIP COLUMNIST (II,1045)

VIRGIN, NERENE
POLICE SURGEON (I,3639)

VIROLO, TITO
PAPA CELLINI (I,3450)

VIS, ROSEMARIE
WILSON'S REWARD (II,2805)

VISCUSO, SAL
M*A*S*H (II,1569); THE MONTEFUSCOS (II,1727); SOAP (II,2392)

VITALE, MILLY
RIVAK, THE BARBARIAN (I,3795)

VITALIE, JOE
FISHERMAN'S WHARF (II,865)

VITTE, RAY
AMERICA, YOU'RE ON (II,69); DAVID CASSIDY—MAN UNDERCOVER (II,623); DOC (II,683); MOTHER, JUGGS AND SPEED (II,1741); THE QUEST (II,2098)

VIVINO, FLOYD
THE UNCLE FLOYD SHOW (II,2703)

VIVYAN, JOHN
HIS MODEL WIFE (I,2064); MR. LUCKY (I,3141)

VOGAN, EMMETT
MAN OF THE COMSTOCK (I,2873)

VOGEL, CAROL
JOSHUA'S WORLD (II,1344)

VOGEL, MITCH
BONANZA (II,347); THE BOYS (I,701); STATE FAIR (II,2448)

VOGEL, TONY
MARCO POLO (II,1626)

VOGES, DANIE
PRISONERS OF THE LOST UNIVERSE (II,2086)

VOHS, JOAN
BACHELOR FATHER (I,323); FAMILY AFFAIR (I,1519); FRONTIER (I,1695)

VOIGHT, JILL
ONE LIFE TO LIVE (II,1907)

VOLA, VICKI
MR. DISTRICT ATTORNEY (I,3136); SEARCH FOR TOMORROW (II,2284)

VOLAND, HERBERT
ARNIE (I,261); THE GOOD LIFE (II,1033); HOT W.A.C.S. (II,1191); LOVE ON A ROOFTOP (I,2775); M*A*S*H (II,1569); THE MOTHERS-IN-LAW (I,3110); MR. DEEDS GOES TO TOWN (I,3134); THE MUNSTERS' REVENGE (II,1765); THE PAUL LYNDE SHOW (I,3489)

VOLSKA, ANNA
A TOWN LIKE ALICE (II,2652)

VOLSTEDT, JOHN
NEWHART (II,1835)

VOLZ, NEDRA
ALL IN THE FAMILY (II,38); CONDOMINIUM (II,566); DIFF'RENT STROKES (II,674); FILTHY RICH (II,856); HANGING IN (II,1078); KING OF THE ROAD (II,1396); MR. DUGAN (II,1749); PALS (II,1947); THEY ONLY COME OUT AT NIGHT (II,2586); A YEAR AT THE TOP (II,2844)

VON FURSTENBERG, BETSY
AS THE WORLD TURNS (II,110); BACHELOR AT LAW (I,322); TONIGHT IN SAMARKAND (I,4558)

VON LEER, HUNTER
THE DEADLY GAME (II,633); GENERAL HOSPITAL (II,964); THE GIRL IN THE EMPTY GRAVE (II,997); JOSHUA'S WORLD (II,1344)

VON SCHERLER, SASHA
ON OUR OWN (II,1892)

VON SYDOW, MAX
THE DIARY OF ANNE FRANK (I,1262)

VON TOY, MARY
ONE OF OUR OWN (II,1910)

VON ZELL, HARRY
BACHELOR FATHER (I,323); ELLERY QUEEN: TOO MANY SUSPECTS (II,767); THE GEORGE BURNS AND GRACIE ALLEN SHOW (I,763); THE GEORGE BURNS SHOW (I,1765)

VON ZERNECK, DANIELLE
GENERAL HOSPITAL (II,964)

VON ZERNECK, PETER
THE BLUE AND THE GRAY (II,273); GOLIATH AWAITS (II,1025); NOTORIOUS (I,3314)

VORGAN, GEE GEE
MARRIAGE IS ALIVE AND WELL (II,1633)

VORGAN, GIGI
MARRIED: THE FIRST YEAR (II,1634)

VOSKOVEC, GEORGE
ARSENIC AND OLD LACE (I,269); INTERMEZZO (I,2223); LAMP AT MIDNIGHT (I,2608); NERO WOLFE (II,1806); REUNION IN VIENNA (I,3780); SKAG (II,2374)

VOTRIAN, PETER
THE UNEXPLAINED (I,4675)

VOUK, LARRY
THE SKATEBIRDS (II,2375)

VOZOFF, LORRINE
RITUALS (II,2182)

VYE, MURVYN
THE BOB CUMMINGS SHOW (I,500); FRANK MERRIWELL (I,1659); SIMON LASH (I,4052)

WADE, ADAM
THE SUPER GLOBETROTTERS (II,2498); UPTOWN SATURDAY NIGHT (II,2711)

WADE, DAVID
THE GOURMET (I,1862)

WADE, ERNESTINE
THE AMOS AND ANDY SHOW (I,179)

WAGGENHEIM, CHARLES
ADVENTURES OF A MODEL (I,50); GUNSMOKE (II,1069)

WAGGONER, LYLE
THE BARBARA EDEN SHOW (I,346); BATTLE OF THE NETWORK STARS (II,165); BULBA (II,392); THE GOSSIP COLUMNIST (II,1045); IT'S YOUR BET (I,2279); LETTERS FROM THREE LOVERS (I,2671); THE LOVE BOAT II (II,1533); TWO THE HARD WAY (II,2695); THE UGILY FAMILY (II,2699); WONDER WOMAN (PILOT 2) (II,2827); WONDER WOMAN (SERIES 1) (II,2828); WONDER WOMAN (SERIES 2) (II,2829)

WAGNER, ALEXANDRA
THE TURN OF THE SCREW (I,4624)

WAGNER, ARTHUR
WILD TIMES (II,2799)

WAGNER, CHERYL
FRAGGLE ROCK (II,916)

WAGNER, CHUCK
AUTOMAN (II,120)

WAGNER, HELEN
AS THE WORLD TURNS (II,110); THE WORLD OF MISTER SWEENY (I,4917)

WAGNER, JACK
THE ADVENTURES OF OZZIE AND HARRIET (I,71)

WAGNER, LINDSAY
THE BIONIC WOMAN (II,255); JESSIE (II,1287); LINDSAY WAGNER—ANOTHER SIDE OF ME (II,1483); THE ROCKFORD FILES (II,2196); SCRUPLES (II,2280); THE SIX-MILLION-DOLLAR MAN (II,2372); WINDOWS, DOORS AND KEYHOLES (II,2806)

WAGNER, LOU
CHIPS (II,511)

WAGNER, MICHAEL
THE JORDAN CHANCE (II,1343); RYAN'S HOPE (II,2234)

WAGNER, MICHELLE
FALCON'S GOLD (II,809)

WAGNER, MIKE
CAMP RUNAMUCK (I,811); OZZIE'S GIRLS (I,3440); ROARING CAMP (I,3804); ROUGHNECKS (II,2219)

WAGNER, PAULA
LOOSE CHANGE (II,1518)

WAGNER, ROBERT
CITY BENEATH THE SEA (I,965); HART TO HART (II,1101); HART TO HART (II,1102); IT TAKES A THIEF (I,2250); KILLER BY NIGHT (I,2537); MADAME SIN (I,2805); PEARL (II,1986); THE STREETS OF SAN FRANCISCO (I,4260); SWITCH (II,2519); SWITCH (II,2520); YOU CAN'T DO THAT ON TELEVISION (I,4933)

WAGNER, SHERRI
THE YOUNG PIONEERS' CHRISTMAS (II,2870)

WAGNER, WENDE
THE GREEN HORNET (I,1886)

WAGONER, PORTER
THE PORTER WAGONER SHOW (I,3652)

WAHAMA
MY FRIEND FLICKA (I,3190)

WAINRIGHT, JAMES
BEYOND WESTWORLD (II,227); BRIDGER (II,376); DANIEL BOONE (I,1142); JIGSAW (I,2377); MAN ON THE MOVE (I,2878); ONCE UPON A DEAD MAN (I,3369)

WAIRD, ANDREW
CINDERELLA (II,526)

WAITE, JACQUELINE
FARAWAY HILL (I,1532)

WAITE, JOHN
PAPER DOLLS (II,1952)

WAITE, RALPH
THE DESPERATE HOURS (I,1239); THE MISSISSIPPI (II,1707); ROOTS (II,2211); THE WALTONS (II,2740); THE WALTONS: A DAY FOR THANKS ON WALTON'S MOUNTAIN (EPISODE 3) (II,2741); THE WALTONS: MOTHER'S DAY ON WALTON'S MOUNTAIN (EPISODE 2) (II,2742); THE WALTONS: WEDDING ON WALTON'S MOUNTAIN (EPISODE 1) (II,2743)

WAITES, TOM
O'MALLEY (II,1872)

WAITRESSES, THE
SQUARE PEGS (II,2431)

WAKEFIELD, JACK
HEAD OF THE FAMILY (I,1988)

WAKEFIELD, OLIVER
THE ALL-STAR SUMMER REVUE (I,139)

WAKELEY, JIMMY
FIVE STAR JUBILEE (I,1589)

WAKELEY, MARGED
BACKSTAIRS AT THE WHITE HOUSE (II,133); EXECUTIVE SUITE (II,796)

WALAS, CHRIS
STARSTRUCK (II,2446)

WALBERG, GARRY
THE ODD COUPLE (II,1875); QUINCY, M. E. (II,2102)

WALCOTT, GREGORY
87TH PRECINCT (I,1422); ALICE (II,33); ESCAPADE (II,781); THE QUEST (II,2096); SIMON LASH (I,4052)

WALCUTT, JOHN
MR. ROBERTS (II,1754)

WALD, KARI
THE LAST CONVERTIBLE (II,1435)

WALDEN, LOIS
HARRIS AND COMPANY (II,1096)

WALDEN, ROBERT
BATTLE OF THE NETWORK STARS (II,171); BATTLE OF THE NETWORK STARS (II,174); BOBBY JO AND THE BIG APPLE GOODTIME BAND (I,674); BROTHERS (II,380); CENTENNIAL (II,477); THE DOCTORS (I,1323); JERRY (II,1283); LOU GRANT (II,1526); MEDICAL CENTER (II,1675); A RAINY DAY (II,2110); THE TED BESSELL SHOW (I,4369)

WALDEN, SUSAN
THE CONTENDER (II,571); LITTLE WOMEN (II,1498); THE SAN PEDRO BUMS (II,2251)

WALDER, LYVONNE
THE DOCTORS (I,1323)

WALDIS, OTTO
COMMAND PERFORMANCE (I,1029)

WALDO, JANET
THE ADDAMS FAMILY (II,12); THE AMAZING CHAN AND THE CHAN CLAN (I,157); THE ATOM ANT/SECRET SQUIRREL SHOW (I,311); BATTLE OF THE PLANETS (II,179); BEAUTY AND THE BEAST (II,197); THE CATTANOOGA CATS (I,874); HELP! IT'S THE HAIR BEAR BUNCH (I,2012); THE JETSONS (I,2376); JOSIE AND THE PUSSYCATS (I,2453); JOSIE AND THE PUSSYCATS IN OUTER SPACE (I,2454); THE LUCY SHOW (I,2791); THE NEW FRED AND BARNEY SHOW (II,1817); THE PERILS OF PENELOPE PITSTOP (I,3525); THE ROMAN HOLIDAYS (I,3835); SHAZZAN! (I,4002); SPACE KIDDETTES (I,4143); VALENTINE'S DAY (I,4693); THE WACKY RACES (I,4745)

WALDRIP, TIM
THE AMERICAN DREAM (II,71)

WALDRON, JOHN
CAPTAIN AMERICA (II,428)

WALEY, JULIET
JANE EYRE (II,1273)

WALKEN, GLENN
AH!, WILDERNESS (I,91)

WALKER JR, ROBERT
BEULAH LAND (II,226); THE PICTURE OF DORIAN GRAY (I,3591)

WALKER, AMANDA
THE STRAUSS FAMILY (I,4253)

WALKER, BETTY
THE PHIL SILVERS SHOW (I,3580)

WALKER, BILL
INSPECTOR PEREZ (II,1237); TENAFLY (I,4395)

WALKER, BILLY
COUNTRY CARNIVAL (I,1062)

WALKER, CHARLES
ADVICE TO THE LOVELORN (II,16)

WALKER, CHET
THE JERK, TOO (II,1282)

WALKER, CLINT
CENTENNIAL (II,477); CHEYENNE (I,931); KODIAK (II,1405); YUMA (I,4976)

WALKER, DAYTON
SHOW BUSINESS, INC. (I,4028)

WALKER, DIANA
DARK SHADOWS (I,1157); WHERE THE HEART IS (I,4831)

WALKER, ELIZABETH
PEYTON PLACE (I,3574)

WALKER, ERIC
THE CIRCLE FAMILY (II,527)

WALKER, GENE
THE WORLD OF MISTER SWEENY (I,4917)

WALKER, GLENN
THE GUIDING LIGHT (II,1064)

WALKER, HARRY
SCREAMER (I,3945)

WALKER, HELEN
THE MARRIAGE BROKER (I,2920)

WALKER, ISABELLE
SCENE OF THE CRIME (II,2271)

WALKER, JACK DAVID
KOSTA AND HIS FAMILY (I,2581)

WALKER, JANE
FIRESIDE THEATER (I,1580)

WALKER, JIMMIE
AT EASE (II,115); B.A.D. CATS (II,123); BATTLE OF THE NETWORK STARS (II,163); BATTLE OF THE NETWORK STARS (II,165); BATTLE OF THE NETWORK STARS (II,166); COTTON CLUB '75 (II,580); GOOD TIMES (II,1038); THE JERK, TOO (II,1282); MURDER CAN HURT YOU! (II,1768)

WALKER, JUDY
THE GEORGE BURNS AND GRACIE ALLEN SHOW (I,1763)

WALKER, KATHRYN
THE ADAMS CHRONICLES (II,8); ANOTHER WORLD (II,97); BEACON HILL (II,193); THE HOUSE WITHOUT A CHRISTMAS TREE (I,2143); LIGHT'S OUT (I,2700); MANDY'S GRANDMOTHER (II,1618); RX FOR THE DEFENSE (I,3878); SEARCH FOR TOMORROW (II,2284); THE THANKSGIVING TREASURE (I,4416)

WALKER, KAY
CAROLYN (I,854)

WALKER, LOU
CHIEFS (II,509)

WALKER, MARCY
ALL MY CHILDREN (II,39)

WALKER, MICHAEL
MR. NOVAK (I,3145)

WALKER, NANCY
BLANSKY'S BEAUTIES (II,267); BRIDGET LOVES BERNIE (I,722); FAMILY AFFAIR (I,1519); HUMAN FEELINGS (II,1203); KEEP THE FAITH (I,2518); THE MARY TYLER MOORE SHOW (II,1640); MCMILLAN AND WIFE (II,1667); THE NANCY WALKER SHOW (II,1790); RHODA (II,2151); THREE FOR THE GIRLS (I,4477)

WALKER, NICHOLAS
CAPITOL (II,426)

WALKER, PEGGY
THE LAZARUS SYNDROME (II,1451)

WALKER, PETER
AMAHL AND THE NIGHT VISITORS (I,150)

WALKER, RONNIE
WONDERFUL JOHN ACTION (I,4892)

WALKER, ROY
THE ABBOTT AND COSTELLO SHOW (I,2); THE RESTLESS GUN (I,3774)

WALKER, SCOTT
ESCAPE (I,1461)

WALKER, SIDNEY
AS THE WORLD TURNS (II,110); THE GUIDING LIGHT (II,1064); THE SECRET STORM (I,3969)

WALKER, TIPPY
ABC STAGE '67 (I,6)

WALKER, TONJA
CAPITOL (II,426); THE HOYT AXTON SHOW (II,1200)

WALKER II, WILLIAM
MOMMA THE DETECTIVE (II,1721)

WALKER, ZENA
THE ADVENTURES OF SIR LANCELOT (I,76); JOURNEY TO THE UNKNOWN (I,2457); KRAFT MYSTERY THEATER (I,2589)

WALKIN, GLENN
LEAVE IT TO LARRY (I,2649)

WALL, GERALDINE
THE DICK VAN DYKE SHOW (I,1275)

WALL, JAMES E
CAPTAIN KANGAROO (II,434)

WALL, LUCILLE
GENERAL HOSPITAL (II,964)

WALL, MAX
A KILLER IN EVERY CORNER (I,2538)

WALLACE, CHRIS
ALL MY CHILDREN (II,39)

WALLACE, DEE
100 CENTRE STREET (II,1901); DRIBBLE (II,736); THE SECRET WAR OF JACKIE'S GIRLS (II,2294); SKEEZER (II,2376)

WALLACE, ELIZABETH
COMPUTERCIDE (II,561)

WALLACE, GEORGE
HIGHER AND HIGHER, ATTORNEYS AT LAW (I,2057)

WALLACE, GEORGE D
BLISS (II,269); THE CHEERLEADERS (II,493); COLORADO C.I. (II,555)

WALLACE, HELEN
THE BOSTON TERRIER (I,697)

WALLACE, JACK
SPARROW (II,2410)

WALLACE, JENNIFER
GAVILAN (II,959)

WALLACE, MARCIA
BATTLE OF THE NETWORK STARS (II,164); BEWITCHED (I,418); THE BOB NEWHART SHOW (II,342); CHARACTERS (II,481); FLYING HIGH (II,895); GILLIGAN'S ISLAND: THE CASTAWAYS ON GILLIGAN'S ISLAND (II,992)

WALLACE, MARIE
DARK SHADOWS (I,1157); SOMERSET (I,4115)

WALLACE, MARJORIE
ENTERTAINMENT TONIGHT (II,780)

WALLACE, MIKE
THE $100,000 BIG SURPRISE (I,3378); ALL AROUND TOWN (I,128); THE BIG SURPRISE (I,433); BIOGRAPHY (I,476); GUESS AGAIN (I,1894); I'LL BUY THAT (I,2190); MIKE AND BUFF (I,3037); THE MIKE WALLACE PROFILES (II,1699); WHAT'S IN A WORD? (I,4815); WHO PAYS? (I,4848)

WALLACE, MYRON
STAND BY FOR CRIME (I,4186)

WALLACE, PAUL
FATHER KNOWS BEST (II,838)

WALLACE, ROBERTA
FATHER KNOWS BEST (II,838)

WALLACE, ROWENA
BARRIER REEF (I,362)

WALLACE, ROYCE
COPS (I,1047); FAME (II,812); SOAP (II,2392); THE STOREFRONT LAWYERS (I,4233); THIS IS KATE BENNETT (II,2597); WEDNESDAY NIGHT OUT (II,2760)

WALLACE, TONI
THE PRICE IS RIGHT (I,3664)

WALLACH, ELI
ALCOA/GOODYEAR THEATER (I,107); BATMAN (I,366); DUPONT SHOW OF THE MONTH (I,1387); THE KAISER ALUMINUM HOUR (I,2503); THE LARK (I,2616); LIGHT'S OUT (I,2699); PHILCO TELEVISION PLAYHOUSE (I,3583); THE POPPY IS ALSO A FLOWER (I,3650); SEVENTH AVENUE (II,2309); SUSPICION (I,4309)

WALLACH, ROBERTA
HOLLYWOOD HIGH (II,1163); LAND OF HOPE (II,1424)

WALLENBERG, LINDA
A WALK IN THE NIGHT (I,4750)

WALLER, EDDY
CASEY JONES (I,865); DANIEL BOONE (I,1141); STEVE DONOVAN, WESTERN MARSHAL (I,4225)

WALLEY, DEBORAH
THE MOTHERS-IN-LAW (I,3110)

WALLIS, SHANI
THE SHANI WALLIS SHOW (I,3997)

WALMSLEY, JON
THE WALTONS (II,2740); THE WALTONS: A DAY FOR THANKS ON WALTON'S MOUNTAIN (EPISODE 3) (II,2741); THE WALTONS: MOTHER'S DAY ON WALTON'S MOUNTAIN (EPISODE 2) (II,2742); THE WALTONS: WEDDING ON WALTON'S MOUNTAIN (EPISODE 1) (II,2743)

WALSH, ARTHUR
TWO GIRLS NAMED SMITH (I,4656)

WALSH, BILL
DIVORCE COURT (I,1297)

WALSH, BROCK
PAPER DOLLS (II,1951)

WALSH, CHARLIE
RUSSIAN ROULETTE (I,3875)

WALSH, DERMOT
RICHARD THE LION HEART (I,3786)

WALSH, JOHN
CELEBRITY (II,463)

WALSH, KAY
JOURNEY TO THE UNKNOWN (I,2457)

WALSH, KELIHER
A CHRISTMAS CAROL (II,517)

WALSH, LORY
THE ASPHALT COWBOY (II,111); THE KILLIN' COUSIN (II,1394); THE MACKENZIES OF PARADISE COVE (II,1585); STICKIN' TOGETHER (II,2465); THE UGILY FAMILY (II,2699)

WALSH, M EMMET
AMY PRENTISS (II,74); DEAR DETECTIVE (II,652); DOCTOR DAN (II,684); THE DON RICKLES SHOW (I,1342); EAST OF EDEN (II,752); NICHOLS (I,3283); NIGHT PARTNERS (II,1847); THE OUTLAWS (II,1936); THE SANDY DUNCAN SHOW (I,3910); YOU ARE THE JURY (II,2854)

WALSTON, RAY
DANNY AND THE MERMAID (II,618); HARRY'S BUSINESS (I,1958); INSTITUTE FOR REVENGE (II,1239); THE JERK, TOO (II,1282); THE KID WITH THE BROKEN HALO (II,1387); MY FAVORITE MARTIAN (I,3189); OH MADELINE (II,1880); PLAYHOUSE 90 (I,3623); SATAN'S WAITIN' (I,3916); STOP SUSAN WILLIAMS (II,2473); SUSPENSE (I,4305); THERE SHALL BE NO NIGHT (I,4437); THIS GIRL FOR HIRE (II,2596)

WALTER, CY
THREE'S COMPANY (I,4489)

WALTER, GRETCHEN
HIDDEN FACES (I,2044)

WALTER, JESSICA
ALL THAT GLITTERS (II,42); AMY PRENTISS (II,74); BARE ESSENCE (II,150); DOCTOR STRANGE (II,688); FOR THE PEOPLE (I,1626); HAVING BABIES I (II,1107); THE IMMORTAL (I,2196); THE IMMORTAL (I,2197); KISS ME, KATE (I,2569); LOVE OF LIFE (I,2774); MISSION: IMPOSSIBLE (I,3067); PURSUE AND DESTROY (I,3699); THE RETURN OF MARCUS WELBY, M.D (II,2138); SCRUPLES (II,2281); T.L.C. (II,2525); THEY CALL IT MURDER (I,4441); THREE'S A CROWD (II,2613); TRAPPER JOHN, M.D. (II,2654); WHEELS (II,2777); WILD AND WOOLEY (II,2798)

WALTER, RITA
AS THE WORLD TURNS (II,110)

WALTER, TRACEY
BEST OF THE WEST (II,218)

WALTERHOUSE, ANNRAE
ALICE (II,33); LITTLE LULU (II,1493)

WALTERS, BARBARA
THE BARBARA WALTERS SPECIAL (II,145); NOT FOR WOMEN ONLY (I,3310)

WALTERS, FURMAN
THE GREATEST GIFT (II,1061)

WALTERS, JACK
THE CITADEL (II,539)

WALTERS, LAURIE
CELEBRITY CHALLENGE OF THE SEXES 5 (II,469); EIGHT IS ENOUGH (II,762)

WALTERS, MIMI
DOCTOR I.Q. (I,1314)

WALTERS, NICK
ALL THE RIVERS RUN (II,43)

WALTERS, SUSAN
LOVING (II,1552)

WALTERS, THORLEY
DEATH IN SMALL DOSES (I,1207); JENNIE: LADY RANDOLPH CHURCHILL (II,1277)

WALTHER, GRETCHEN
SEARCH FOR TOMORROW (II,2284)

WALTON, JESS
REX HARRISON PRESENTS SHORT STORIES OF LOVE (II,2148)

WALTON, JUDY
T.L.C. (II,2525)

WALTON, PEGGY
BIG ROSE (II,241)

WALTON, SUNNI
TOO CLOSE FOR COMFORT (II,2642)

WAMBAUGH, JOSEPH
THE BLUE KNIGHT (II,277)

WANAMAKER, SAM
CAMEO THEATER (I,808); THE LILLI PALMER THEATER (I,2704); OUR FAMILY BUSINESS (II,1931); WARNING SHOT (I,4770)

WANAMAKER, ZOE
PIAF (II,2039)

WANDERONE JR, RUDOLPH
CELEBRITY BILLIARDS (I,882)

WANDREY, DONNA
AS THE WORLD TURNS (II,110); DARK SHADOWS (I,1157)

WANN, JIM
PUMPBOYS AND DINETTES ON TELEVISION (II,2090)

WAPNER, JOSEPH A
THE PEOPLE'S COURT (II,2001)

WAR, LEE TIT
MEN OF THE DRAGON (II,1683)

WARBECK, DAVID
ONLY A SCREAM AWAY (I,3393); U.F.O. (I,4662)

WARD, B J
GODZILLA (II,1014); THE LITTLE RASCALS (II,1494); SPACE STARS (II,2408)

WARD, BILL
DOC HOLLIDAY (I,1307)

WARD, BURT
BATMAN (I,366); BATMAN AND THE SUPER SEVEN (II,160); HIGH SCHOOL, U.S.A. (II,1149); THE NEW ADVENTURES OF BATMAN (II,1812)

WARD, ELIZABETH
THE DAY THE WOMEN GOT EVEN (II,628)

WARD, EVELYN
THE BIG BRAIN (I,424)

WARD, JANET
ANOTHER WORLD (II,97); THE GREATEST GIFT (I,1880); HIGH BUTTON SHOES (I,2050)

WARD, JANIS
STEAMBATH (II,2451)

WARD, JOHN
BARETTA (II,152)

WARD, JONATHAN
CHARLES IN CHARGE (II,482); A CHRISTMAS FOR BOOMER (II,518)

WARD, KALEY
HIGH SCHOOL, U.S.A. (II,1148)

WARD, LALLA
DOCTOR WHO (II,689); DOCTOR WHO—THE FIVE DOCTORS (II,690)

WARD, LARRY
THE BRIGHTER DAY (I,728); THE DAKOTAS (I,1126); THE GUN AND THE PULPIT (II,1067); A MAN CALLED RAGAN (I,2858)

WARD, LYMAN
A GIRL'S LIFE (II,1000); HART TO HART (II,1102); NOW WE'RE COOKIN' (II,1866); THE PHOENIX (II,2035); W*A*L*T*E*R (II,2731)

WARD, MARY
PRISONER: CELL BLOCK H (II,2085)

WARD, RACHEL
THE THORN BIRDS (II,2600)

WARD, RICHARD
BEACON HILL (II,193); GOOD TIMES (II,1038); MARY HARTMAN, MARY HARTMAN (II,1638); ROLL OUT! (I,3833); STARSKY AND HUTCH (II,2445)

WARD, ROBIN
THE STARLOST (I,4203); TALES FROM MUPPETLAND (I,4344); TO TELL THE TRUTH (II,2626)

WARD, SANDY
DALLAS (II,614); I DO, I DON'T (II,1212)

WARD, SELA
EMERALD POINT, N.A.S. (II,773)

WARD, SIMON
ALL CREATURES GREAT AND SMALL (I,129); VALLEY FORGE (I,4698)

WARD, SKIP
HEAVEN HELP US (I,1995); THE LINE-UP (I,2707); MRS. G. GOES TO COLLEGE (I,3155)

WARD, TONY
HUNTER (I,2161)

WARD, VICTORIA
CURTAIN CALL THEATER (I,1115)

WARDE, HARLAN
RANSOM FOR A DEAD MAN (I,3722); THE RIFLEMAN (I,3789); STRATEGIC AIR COMMAND (I,4252); THE VIRGINIAN (I,4738)

WARDEN, JACK
THE ARMSTRONG CIRCLE THEATER (I,260); THE ASPHALT JUNGLE (I,297); THE BAD NEWS BEARS (II,134); THE BLUE MEN (I,492); THE CAMPBELL TELEVISION SOUNDSTAGE (I,812); CRAZY LIKE A FOX (II,601); DANGER (I,1134); GALLAGHER (I,1727); GREAT ADVENTURE (I,1868); JIGSAW JOHN (II,1288); THE LARK (I,2616); MAN ON A STRING (I,2877); N.Y.P.D. (I,3321); NORBY (I,3304); THE PETRIFIED FOREST (I,3570); SECOND LOOK (I,3960); THEY ONLY COME OUT AT NIGHT (II,2586); TOPPER (II,2649); THE TWILIGHT ZONE (I,4651); THE WACKIEST SHIP IN THE ARMY (I,4744)

WARDER, CARI ANNE
BIG JOHN, LITTLE JOHN (II,240); FATHER KNOWS BEST: HOME FOR CHRISTMAS (II,839); FATHER KNOWS BEST: THE FATHER KNOWS BEST REUNION (II,840); THE GREATEST GIFT (II,1061)

WARE, MIDGE
333 MONTGOMERY (I,4487); THE BEVERLY HILLBILLIES (I,417); GUNSLINGER (I,1908)

WARFIELD, CHRIS
INNOCENT JONES (I,2217)

WARFIELD, DONALD
LAND OF HOPE (II,1424)

WARFIELD, MARLENE
CUTTER (I,1117); MAUDE (II,1655)

WARFIELD, MARSHA
RIPTIDE (II,2178)

WARFIELD, WILLIAM
THE GREEN PASTURES (I,1887)

WARGA, ROBERT
JUSTICE OF THE PEACE (I,2500)

WARICK, LEE
GENERAL HOSPITAL (II,964)

WARING, FRED
CHEVROLET ON BROADWAY (I,925); THE FRED WARING SHOW (I,1679); FRED WARING: WAY BACK HOME (I,1680)

WARING, RICHARD
EAGLE IN A CAGE (I,1392); MACBETH (I,2800)

WARING, TODD
FIT FOR A KING (II,866)

WARLOCK, BILLY
HAPPY DAYS (II,1084); SIX PACK (II,2370)

WARNER, DAVID
MARCO POLO (II,1626); MASADA (II,1642); NANCY ASTOR (II,1788)

WARNER, JACKIE
THE SHARI LEWIS SHOW (I,4000)

WARNER, JODY
DOBIE GILLIS (I,1302)

WARNER, JUDY
ONE HAPPY FAMILY (I,3375)

WARNER, LUCY
MOMENT OF TRUTH (I,3084); SCARLETT HILL (I,3933); STRANGE PARADISE (I,4246)

WARNER, MALCOLM-JAMAL
THE COSBY SHOW (II,578)

WARNER, MARSHA
THE INVISIBLE WOMAN (II,1244)

WARNER, RICHARD
THE ADVENTURES OF SIR FRANCIS DRAKE (I,75)

WARNER, SANDRA
HEAVEN HELP US (I,1995); MR. SMITH GOES TO WASHINGTON (I,3151)

WARREN, BIFF
AS THE WORLD TURNS (II,110); THE KIDS FROM C.A.P.E.R. (II,1391)

WARREN, HELEN
THE GREATEST GIFT (I,1880)

WARREN, LEONARD
FESTIVAL OF MUSIC (I,1567)

WARREN, JENNIFER
BANJO HACKETT: ROAMIN' FREE (II,141); BUTTERFLIES (II,405); CELEBRITY (II,463); JUST FRIENDS (I,2495); PAPER DOLLS (II,1951); PAPER DOLLS (II,1952)

WARREN, JOSEPH
CAR 54, WHERE ARE YOU? (I,842); FIRST LOVE (I,1583); KENNEDY (II,1377); RAGE OF ANGELS (II,2107)

WARREN, KATHERINE
FATHER KNOWS BEST (II,838); LEAVE IT TO BEAVER (I,2648)

WARREN, KATHLEEN
IVY LEAGUE (I,2286)

WARREN, KENNETH J
COURT-MARTIAL (I,1080)

WARREN, LESLEY ANN
79 PARK AVENUE (II,2310); ASSIGNMENT: MUNICH (I,302); BEULAH LAND (II,226); CAT BALLOU (I,869); CINDERELLA (I,956); THE DAUGHTERS OF JOSHUA CABE (I,1169); IT'S A BIRD, IT'S A PLANE, IT'S SUPERMAN (II,1253); THE LETTERS (I,2670); MISSION: IMPOSSIBLE (I,3067); NIGHT GALLERY (I,3287); PEARL (II,1986); SAGA OF SONORA (I,3884)

WARREN, MICHAEL
BATTLE OF THE NETWORK STARS (II,172); BATTLE OF THE NETWORK STARS (II,175); HILL STREET BLUES (II,1154); JOE FORRESTER (II,1303); PARIS (II,1957); SIERRA (II,2351)

WARREN, STEFANI
HERE COME THE BRIDES (I,2024); THE SHAMEFUL SECRETS OF HASTINGS CORNERS (I,3994)

WARRICK, LEE
ONE LIFE TO LIVE (II,1907)

WARRICK, ROBERT
FAITH BALDWIN'S THEATER OF ROMANCE (I,1515)

WARRICK, RUTH
ALL MY CHILDREN (II,39); AS THE WORLD TURNS (II,110); FATHER OF THE BRIDE (I,1543); PEYTON PLACE (I,3574); SHORT, SHORT DRAMA (I,4023); YOUR JEWELER'S SHOWCASE (I,4953)

WARRINER, FREDERIC
EAGLE IN A CAGE (I,1392)

WARSHAW, BARBARA
ANOTHER LIFE (II,95)

WARWICK, DAVID
THE FALL AND RISE OF REGINALD PERRIN (II,810)

WARWICK, DIONNE
THE DIONNE WARWICK SPECIAL (I,1289); FINDER OF LOST LOVES (II,857); JACQUELINE SUSANN'S VALLEY OF THE DOLLS, 1981 (II,1268); SOLID GOLD (II,2395); SOLID GOLD '79 (II,2396)

WARWICK, JAMES
THE SEVEN DIALS MYSTERY (II,2308); WHY DIDN'T THEY ASK EVANS? (II,2795)

WARWICK, JOHN
ADVENTURES OF CLINT AND MAC (I,55); YOUR SHOW TIME (I,4958)

WARWICK, ROBERT
BROKEN ARROW (I,743); ON TRIAL (I,3362)

WARWICK, RUTH
HENRY FONDA PRESENTS THE STAR AND THE STORY (I,2016)

WASHAM, IAN MILLER
ALL MY CHILDREN (II,39)

WASHBOURNE, MONA
BRIDESHEAD REVISITED (II,375); THE LILLI PALMER THEATER (I,2704); THERESE RAQUIN (II,2584)

WASHBROOK, DON
HANK (I,1931); PETTICOAT JUNCTION (I,3571)

WASHBROOK, JOHNNY
HAZEL (I,1982); MY FRIEND FLICKA (I,3190)

WASHBURN, BEVERLY
FOUR STAR PLAYHOUSE (I,1652); GIDGET (I,1795); JANE WYMAN PRESENTS THE FIRESIDE THEATER (I,2345); THE LORETTA YOUNG SHOW (I,2755); ONE STEP BEYOND (I,3388); PROFESSIONAL FATHER (I,3677); STAR STAGE (I,4197); ZANE GREY THEATER (I,4979)

WASHINGTON V, JOHN A
GEORGE WASHINGTON (II,978)

WASHINGTON, DENZEL
ST. ELSEWHERE (II,2432)

WASHINGTON, KENNETH
HOGAN'S HEROES (I,2069); OUR FAMILY BUSINESS (II,1931)

WASHINGTON, VERNON
CHANGE AT 125TH STREET (II,480)

WASS, TED
THE 13TH DAY: THE STORY OF ESTHER (II,2593); HANDLE WITH CARE (II,1077); SOAP (II,2392)

WASSON, CRAIG
BUTTERFLIES (II,405); PHYLLIS (II,2038); SKAG (II,2374)

WATERHOUSE, MATTHEW
DOCTOR WHO (II,689)

WATERMAN, DENNIS
THE EYES HAVE IT (I,1496); FAIR EXCHANGE (I,1512); JOURNEY TO THE UNKNOWN (I,2457); THE SWEENEY (II,2513)

WATERMAN, TEXIE
OPRYLAND: NIGHT OF STARS AND FUTURE STARS (II,1921)

WATERMAN, WILLARD
ALL ABOUT BARBARA (I,125); DENNIS THE MENACE (I,1231); THE DICK VAN DYKE SHOW (I,1275); THE GREAT GILDERSLEEVE (I,1875); HOW TO MARRY A MILLIONAIRE (I,2147); THE REAL MCCOYS (I,3741); THOMPSON'S GHOST (I,4467)

WATERS, ETHEL
BEULAH (I,416); THE ETHEL WATERS SHOW (I,1468); GREAT ADVENTURE (I,1868); MANHATTAN TOWER (I,2889); PLAYWRIGHTS '56 (I,3627); YOU'RE ONLY YOUNG TWICE (I,4971)

WATERS, HELEN
INTERMEZZO (I,2223)

WATERS, JOHN
ALL THE RIVERS RUN (II,43)

WATERS, REBA
CAVALCADE OF AMERICA (I,875); FATHER KNOWS BEST (II,838); ONE STEP BEYOND (I,3388); PECK'S BAD GIRL (I,3503); WE'VE GOT EACH OTHER (II,2757)

WATERSON, SAM
OPPENHEIMER (II,1919); Q. E. D. (II,2094)

WATKINS, CARLENE
B.J. AND THE BEAR (II,125); BEST OF THE WEST (II,218); CONDOMINIUM (II,566); IT'S NOT EASY (II,1256); LITTLE WOMEN (II,1497); THE SECRET EMPIRE (II,2291)

WATKINS, JIM
THE MAGICIAN (I,2823); THE MAGICIAN (I,2824)

WATKINS, LINDA
ONCE UPON A DEAD MAN (I,3369); ONE LIFE TO LIVE (II,1907)

WATLING, DEBORAH
DOCTOR WHO (II,689); THE INVISIBLE MAN (I,2232)

WATROBA, SHARON
KENNEDY (II,1377)

WATSON JR, JAMES
BACKSTAIRS AT THE WHITE HOUSE (II,133)

WATSON, BETTY JANE
THE EARL WRIGHTSON SHOW (I,1393)

WATSON, BOBS
THE JIM BACKUS SHOW—HOT OFF THE WIRE (I,2379)

WATSON, BRUCE
THE BANANA SPLITS ADVENTURE HOUR (I,340); THE CATTANOOGA CATS (I,874); DRAGNET (I,1370)

WATSON, DAVID
NEVER TOO YOUNG (I,3248); RAWHIDE (I,3727)

WATSON, DEBBIE
KAREN (I,2505); MUNSTER, GO HOME! (I,3157); TAMMY (I,4357)

WATSON, DOUGLAS
ABE LINCOLN IN ILLINOIS (I,11); ANOTHER WORLD (II,97); MAN AND SUPERMAN (I,2854); MOMENT OF TRUTH (I,3084); TAMING OF THE SHREW (I,4356)

WATSON, DUNCAN
BE MY VALENTINE, CHARLIE BROWN (I,369); HAPPY ANNIVERSARY, CHARLIE BROWN (I,1939); YOU'RE A GOOD SPORT, CHARLIE BROWN (I,4963)

WATSON, ELLEN
ANOTHER WORLD (II,97)

WATSON, GARY
ONCE THE KILLING STARTS (I,3367); THE PALLISERS (II,1944); WAR AND PEACE (I,4765)

WATSON, IRWIN C
A.E.S. HUDSON STREET (II,17); MY BUDDY (II,1778)

WATSON, JACK
MARCO POLO (II,1626)

WATSON, JUNE
MURDER MOTEL (I,3161)

WATSON, JUSTICE
HOLIDAY LODGE (I,2077)

WATSON, LARRY
READY AND WILLING (II,2115); TESTIMONY OF TWO MEN (II,2564)

WATSON, LUCILLE
PULITZER PRIZE PLAYHOUSE (I,3692)

WATSON, MARSHALL
AS THE WORLD TURNS (II,110)

WATSON, MELANIE
DIFF'RENT STROKES (II,674)

WATSON, MILLS
B.J. AND THE BEAR (II,125); B.J. AND THE BEAR (II,127); THE CRIME CLUB (I,1096); DEAD MAN ON THE RUN (II,631); HARPER VALLEY (II,1094); HEAT OF ANGER (I,1992); LOBO (II,1504); LOCK, STOCK, AND BARREL (I,2736); THE MISADVENTURES OF SHERIFF LOBO (II,1704); RANSOM FOR ALICE (II,2113)

WATSON, MORAY
PRIDE AND PREJUDICE (II,2080); QUILLER: NIGHT OF THE FATHER (II,2100); QUILLER: PRICE OF VIOLENCE (II,2101)

WATSON, SUSAN
THE FANTASTICKS (I,1531); THE FRONT PAGE (I,1692); MAGGIE BROWN (I,2811)

WATSON, VERNEE
BENSON (II,208); CAPTAIN CAVEMAN AND THE TEEN ANGELS (II,433); CARTER COUNTRY (II,449); DRIBBLE (II,736); EIGHT IS ENOUGH (II,762); LONDON AND DAVIS IN NEW YORK (II,1512); SCOOBY'S ALL-STAR LAFF-A-LYMPICS (II,2273); WELCOME BACK, KOTTER (II,2761)

WATSON, WILLIAM
THE CONTENDER (II,571); DOWN HOME (II,731); ROOTS (II,2211)

WATT, BILLIE LOU
ASTRO BOY (I,306); THE BILLY BEAN SHOW (I,449); THE EDGE OF NIGHT (II,760); FROM THESE ROOTS (I,1688); KIMBA, THE WHITE LION (I,2542); SEARCH FOR TOMORROW (II,2284)

WATT, CHARLES
YOUNG IN HEART (I,4939)

WATT, STAN
LOVE OF LIFE (I,2774)

WATTIS, RICHARD
DICK AND THE DUCHESS (I,1263)

WATTS, CHARLES
DENNIS THE MENACE (I,1231)

WATTS, ELIZABETH
THE SLIGHTLY FALLEN ANGEL (I,4087)

WATTS, JEROME
PETER PAN (II,2030)

WAUGH, FRED
THE AMAZING SPIDER-MAN (II,63)

WAXMAN, AL
CAGNEY AND LACEY (II,409); CAGNEY AND LACEY (II,410); KING OF KENSINGTON (II,1395)

WAXMAN, STANLEY
LASSITER (I,2625); LASSITER (II,1434); YOUR SHOW TIME (I,4958)

WAYANS, KEENAN IVORY
FOR LOVE AND HONOR (II,903); IRENE (II,1245)

WAYLAND, LEN
THE BLUE AND THE GRAY (II,273); DRAGNET (I,1371); THE GEMINI MAN (II,961); GENERAL HOSPITAL (II,964); SAM (II,2247); A TIME TO LIVE (I,4510)

WAYLEN, MICHAEL
THE GEORGE BURNS AND GRACIE ALLEN SHOW (I,1763)

WAYLORS, THE
THE DUKES OF HAZZARD (II,742)

WAYNE, CAROL
CELEBRITY CHALLENGE OF THE SEXES 3 (II,467); HEAVEN ON EARTH (II,1120); JOE DANCER: THE BIG BLACK PILL (II,1300); THE TONIGHT SHOW (I,4565)

WAYNE, DAVID
THE ALCOA HOUR (I,108); ALCOA PREMIERE (I,109); ARSENIC AND OLD LACE (I,270); BATMAN (I,366); THE CATCHER (I,872); DALLAS (II,614); THE DEVIL AND DANIEL WEBSTER (I,1246); DO NOT DISTURB (I,1300); ELLERY QUEEN (II,766); ELLERY QUEEN: TOO MANY SUSPECTS (II,767); FAMILY (II,813); THE GOOD LIFE (I,1848); THE GOOD LIFE (II,1033); GREAT CATHERINE (I,1873); HOLLOWAY'S DAUGHTERS (I,2079); HOUSE CALLS (II,1194); IT'S A BIRD, IT'S A PLANE, IT'S SUPERMAN (II,1253); THE JUDY GARLAND SHOW (I,2470); JUNIOR MISS (I,2493); LAMP AT MIDNIGHT (I,2608); LASSIE: THE NEW BEGINNING (II,1433); LOOSE CHANGE (II,1518); MATT HOUSTON (II,1654); NORBY (I,3304); RUGGLES OF RED GAP (I,3867); SLEEPING BEAUTY (I,4083); THE STRAWBERRY BLONDE (I,4254); THE TEAHOUSE OF THE AUGUST MOON (I,4368)

WAYNE, FRED
ANYTHING GOES (I,237); THE SWORD (I,4325)

WAYNE, JESSE
NICHOLS (I,3283)

WAYNE, JOHN
THE AMERICAN WEST OF JOHN FORD (I,173); GENERAL ELECTRIC'S ALL-STAR ANNIVERSARY (II,963); SCREEN DIRECTOR'S PLAYHOUSE (I,3946); SWING OUT, SWEET LAND (II,2515)

WAYNE, JOHNNY
HOLIDAY LODGE (I,2077); WAYNE AND SHUSTER TAKE AN AFFECTIONATE LOOK AT. . . (I,4778)

WAYNE, MARTHA
HOLLYWOOD SCREEN TEST (I,2089)

WAYNE, NINA
CAMP RUNAMUCK (I,811); THE NIGHT STRANGLER (I,3293)

WAYNE, PATRICK
BATTLE OF THE NETWORK STARS (II,169); THE MONTE CARLO SHOW (II,1726); MOVIN' ON (I,3118); MR. ADAMS AND EVE (I,3121); THE ROUNDERS (I,3851); SHIRLEY (II,2330); THE TEENAGE IDOL (I,4375)

WAYNNS, KEENAN
HIGHWAY HONEYS (II,1151)

WEARY, A C
13 THIRTEENTH AVENUE (II,2592); Q. E. D. (II,2094); THE SIX O'CLOCK FOLLIES (II,2368)

WEATHERLY, SHAWN
BATTLE OF THE NETWORK STARS (II,178); INSIDE AMERICA (II,1234); SHAPING UP (II,2316)

WEATHERWAX, KEN
THE ADDAMS FAMILY (II,11); THE ADDAMS FAMILY (II,13)

WEAVER, BUCK
BUCKAROO 500 (I,752)

WEAVER, DENNIS
CENTENNIAL (II,477); EMERALD POINT, N.A.S. (II,773); GENTLE BEN (I,1761); GUNSMOKE (II,1069); THE ISLANDER (II,1249); KENTUCKY JONES (I,2526); MCCLOUD (II,1660); MCCLOUD: WHO KILLED MISS U.S.A.? (I,2965); OPRYLAND USA—1975 (II,1920); PEARL (II,1986); STONE (II,2470); STONE (II,2471)

WEAVER, DOODLES
THE DAUGHTERS OF JOSHUA CABE (I,1169); A DAY WITH DOODLES (I,1183); THE DOODLES WEAVER SHOW (I,1351); THE PIED PIPER OF HAMELIN (I,3595)

WEAVER, FRITZ
FEMALE INSTINCT (I,1564); GREAT ADVENTURE (I,1868); HEAT OF ANGER (I,1992); JANE EYRE (I,2338); THE MARTIAN CHRONICLES (II,1635); MOMMA THE DETECTIVE (II,1721); THE PEOPLE NEXT DOOR (I,3519); THE POWER AND THE GLORY (I,3658); RX FOR THE DEFENSE (I,3878)

WEAVER, LEE
THE BILL COSBY SHOW (I,442); GUESS WHO'S COMING TO DINNER? (II,1063)

WEAVER, NED
MARKHAM (I,2916)

WEAVER, PATTY
DAYS OF OUR LIVES (II,629); THE YOUNG AND THE RESTLESS (II,2862)

WEAVER, ROBBY
CENTENNIAL (II,477); THE GREATEST AMERICAN HERO (II,1060); STONE (II,2471)

WEAVER, SYLVESTER 'PAT'
TONIGHT PREVIEW (I,4560)

WEBB, ALAN
ELIZABETH THE QUEEN (I,1431); THE HOLY TERROR

(I,2098); LITTLE MOON OF ALBAN (I,2715); THE PHILADELPHIA STORY (I,3582)

WEBB, ANTHONY
JANE EYRE (II,1273)

WEBB, GREG
BOONE (II,351)

WEBB, JACK
DRAGNET (I,1369); DRAGNET (I,1370); DRAGNET (I,1371); GENERAL ELECTRIC TRUE (I,1754)

WEBB, JANE
THE ARCHIE COMEDY HOUR (I,252); ARCHIE'S TV FUNNIES (I,253); THE BANG-SHANG LALAPALOOZA SHOW (II,140); THE BRADY KIDS (II,365); EVERYTHING'S ARCHIE (I,1483); FANTASTIC VOYAGE (I,1530); THE HARDY BOYS (I,1949); JOURNEY TO THE CENTER OF THE EARTH (I,2456); LASSIE'S RESCUE RANGERS (II,1432); MY FAVORITE MARTIANS (II,1779); THE NEW ADVENTURES OF GILLIGAN (II,1813); SABRINA, THE TEENAGE WITCH (I,3880); THE SECRET LIVES OF WALDO KITTY (II,2292); SUPER WITCH (II,2501); TARZAN: LORD OF THE JUNGLE (II,2544); THE U.S. OF ARCHIE (I,4687); U.S. OF ARCHIE (II,2698)

WEBB, LUCY
PRIVATE BENJAMIN (II,2087)

WEBB, MARTI
TELL ME ON A SUNDAY (II,2557)

WEBB, RICHARD
ANYTHING FOR MONEY (I,236); CAPTAIN MIDNIGHT (I,833); DANIEL BOONE (I,1142); U.S. BORDER PATROL (I,4685); THE WEB (I,4784)

WEBBER, CHAD
THERE'S NO TIME FOR LOVE, CHARLIE BROWN (I,4438); YOU'RE NOT ELECTED, CHARLIE BROWN (I,4967)

WEBBER, DEAN
THE ANNIVERSARY GAME (I,227)

WEBBER, KARL
KITTY FOYLE (I,2571)

WEBBER, LAURA
YOUNG MR. BOBBIN (I,4943)

WEBBER, PEGGY
CAVALCADE OF AMERICA (I,875); FRONT ROW CENTER (I,1694); KING'S ROW (I,2563); MEDIC (I,2976); THE NEW ADVENTURES OF HUCKLEBERRY FINN (I,3250)

WEBBER, ROBERT
79 PARK AVENUE (II,2310); CUTTER (I,1117); THE EDGE OF NIGHT (II,760); ESPIONAGE (I,1464); EYE WITNESS (I,1494); FULL MOON OVER BROOKLYN (I,1703); HAWKINS ON MURDER (I,1978); JUDGEMENT DAY (II,1348); NO MAN'S LAND (II,1855); THE PARADINE CASE (I,3453); SHOOTING STARS (II,2341)

WEBER, KARL
HANDS OF MURDER (I,1929)

WEBER, LARRY
THE DOCTORS (II,694); LOVE OF LIFE (I,2774); TEXAS (II,2566)

WEBER, LAURA
CAPTAIN VIDEO AND HIS VIDEO RANGERS (I,838); THE CLOCK (I,978)

WEBER, LAWRENCE
THE MAN IN THE DOG SUIT (I,2869); MIRACLE ON 34TH STREET (I,3057)

WEBER, RICHARD K
LOVE OF LIFE (I,2774)

WEBLES, GEORGE
HOW TO SURVIVE A MARRIAGE (II,1198)

WEBSTER, AL
THE VALIANT (I,4697)

WEBSTER, BYRON
ARCHIE (II,104); THE ARCHIE SITUATION COMEDY MUSICAL VARIETY SHOW (II,106); THE BARBARY COAST (II,147); THE SIX O'CLOCK FOLLIES (II,2368); SKINFLINT (II,2377)

WEBSTER, CHUCK
INDEMNITY (I,2209); PHOTOCRIME (I,3587)

WEBSTER, DIANA
REX HARRISON PRESENTS SHORT STORIES OF LOVE (II,2148)

WEBSTER, DIANE
DEATH RAY 2000 (II,654); DOCTOR STRANGE (II,688); MASADA (II,1642)

WEBSTER, MARY
THE ONEDIN LINE (II,1912); PANIC! (I,3447)

WEDDLE, VERNON
EAST OF EDEN (II,752); HANDLE WITH CARE (II,1077); PALMERSTOWN U.S.A. (II,1945); SECOND CHANCE (I,3958)

WEDEMEYER, HERMAN
HAWAII FIVE-O (II,1110)

WEDGEWORTH, ANN
ANOTHER WORLD (II,97); FILTHY RICH (II,856); SOMERSET (I,4115); THREE'S COMPANY (II,2614)

WEED, GENE
SHIVAREE (I,4020)

WEEDEN, KAREN
THE EDGE OF NIGHT (II,760)

WEEKS, ALAN
BAKER'S DOZEN (II,136); FAME (II,812); THE RUBBER GUN SQUAD (II,2225)

WEEKS, JIMMIE RAE
KENNEDY (II,1377)

WEEMS, DEBBIE
CAPTAIN KANGAROO (II,434)

WEEMS, PRISCILLA
SCENE OF THE CRIME (II,2271)

WEENICK, ANNABELLE
CELEBRITY (II,463)

WEHR, GRANT
SOMEDAY YOU'LL FIND HER, CHARLIE BROWN (I,4113)

WEIDMAN, CHARLES
HOLIDAY (I,2074)

WEIGEL, JOHN
BON VOYAGE (I,683)

WEIL, JERI
LEAVE IT TO BEAVER (I,2648)

WEIL, LISI
CHILDREN'S SKETCH BOOK (I,941)

WEINBERGER, ED
MR. SMITH (II,1755)

WEINRIB, LENNIE
THE ADDAMS FAMILY (II,12); THE AMAZING CHAN AND THE CHAN CLAN (I,157); BATMAN AND THE SUPER SEVEN (II,160); BIG DADDY (I,425); THE C.B. BEARS (II,406); DOCTOR DOLITTLE (I,1309); THE FLINTSTONE COMEDY HOUR (II,881); FLINTSTONE FAMILY ADVENTURES (II,882); HELP! IT'S THE HAIR BEAR BUNCH (I,2012); INCH HIGH, PRIVATE EYE (II,1231); LIDSVILLE (I,2679); MAGIC MONGO (II,1592); THE NATIONAL SNOOP (II,1797); THE NEW ADVENTURES OF BATMAN (II,1812); PEBBLES AND BAMM BAMM (II,1987); SCOOBY-DOO AND SCRAPPY-DOO (II,2274); SPACE STARS (II,2408); WHEELIE AND THE CHOPPER BUNCH (II,2776)

WEINTRAUB, CARL
EXECUTIVE SUITE (II,796)

WEINTRAUB, CINDY
BAKER'S DOZEN (II,136)

WEIRE BROTHERS, THE
ROSALINDA (I,3846)

WEIS, LESA
CELEBRITY (II,463)

WEISE, GEORGE
MR. TUTT (I,3153)

WEISGIL, FRED
A WOMAN CALLED GOLDA (II,2818)

WEISMAN, JOSEPH
RAGE OF ANGELS (II,2107)

WEISMAN, SAM
THE LAST CONVERTIBLE (II,1435); STUDS LONIGAN (II,2484); WISHMAN (II,2811)

WEISMAN, SID
MARY HARTMAN, MARY HARTMAN (II,1638)

WEISSMULLER, JOHNNY
JUNGLE JIM (I,2491)

WEITZ, BRUCE
BATTLE OF THE NETWORK STARS (II,174); BATTLE OF THE NETWORK STARS (II,176); CATALINA C-LAB (II,458); EVERY STRAY DOG AND KID (II,792); HILL STREET BLUES (II,1154)

WELCH, JOSEPH N
THE DOW HOUR OF GREAT MYSTERIES (I,1365); THE WOMAN IN WHITE (I,4887)

WELCH, LOUIS
THE GANGSTER CHRONICLES (II,957); THE LOVEBIRDS (II,1548); THE WALTONS: WEDDING ON WALTON'S MOUNTAIN (EPISODE 1) (II,2743)

WELCH, RAQUEL
FUNNY PAPERS (I,1713); THE HOLLYWOOD PALACE (I,2088); RAQUEL (I,3725); RAQUEL (II,2114); REALLY RAQUEL (II,2119)

WELD, TUESDAY

THE DICK POWELL SHOW (I,1269); DOBIE GILLIS (I,1302); THE DUPONT SHOW OF THE WEEK (I,1388); THE GREATEST SHOW ON EARTH (I,1883); THE RAINMAKER (II,2109)

WELDON, ANN
9 TO 5 (II,1852); ONE IN A MILLION (II,1906); SIDNEY SHORR (II,2349)

WELDON, BEN
THE LONE RANGER (I,2740)

WELDON, CHARLES
DYNASTY (II,747); KISS ME, KILL ME (II,1400)

WELDON, JIMMY
FUNNY BONERS (I,1710); THE WORLD'S GREATEST SUPER HEROES (II,2842); YOGI BEAR (I,4928)

WELK, LAWRENCE
THE LAWRENCE WELK CHRISTMAS SPECIAL (II,1450); THE LAWRENCE WELK SHOW (I,2643); THE LAWRENCE WELK SHOW (I,2644); LAWRENCE WELK'S TOP TUNES AND NEW TALENT (I,2645); MEMORIES WITH LAWRENCE WELK (II,1681)

WELKER, FRANK
THE ALL-NEW POPEYE HOUR (II,49); ALVIN AND THE CHIPMUNKS (II,61); THE BARKLEYS (I,356); BUFORD AND THE GHOST (II,389); CATCH 22 (I,871); FANG FACE (II,822); THE FANTASTIC FOUR (II,824); FLINTSTONE FAMILY ADVENTURES (II,882); THE FLINTSTONE FUNNIES (II,883); THE FLINTSTONES (II,885); FONZ AND THE HAPPY DAYS GANG (II,898); FRED AND BARNEY MEET THE SHMOO (II,918); INSPECTOR GADGET (II,1236); JABBERJAW (II,1261); JIM HENSON'S MUPPET BABIES (II,1289); THE KOWBOYS (I,2584); THE KWICKY KOALA SHOW (II,1417); LAVERNE AND SHIRLEY WITH THE FONZ (II,1449); MORK AND MINDY (II,1736); THE NEW ADVENTURES OF MIGHTY MOUSE AND HECKLE AND JECKLE (II,1814); THE NEW SHMOO (II,1827); PAC-MAN (II,1940); THE PLASTICMAN COMEDY/ADVENTURE SHOW (II,2051); THE RICHIE RICH SHOW (II,2169); SCOOBY'S ALL-STAR LAFF-A-LYMPICS (II,2273); SCOOBY-DOO AND SCRAPPY-DOO (II,2274); SCOOBY-DOO, WHERE ARE YOU? (II,2276); THE SCOOBY-DOO/DYNOMUTT HOUR (II,2277); THE SKATEBIRDS (II,2375); THE SMURFS (II,2385); SPACE STARS (II,2408); SPORT BILLY (II,2429); SUPER FRIENDS (II,2497); THE SUPER GLOBETROTTERS (II,2498); THE THREE ROBONIC STOOGES (II,2609); TOO CLOSE FOR COMFORT (II,2642); TROLLKINS (II,2660); WHAT'S NEW MR. MAGOO? (II,2770); WHEELIE AND THE CHOPPER BUNCH (II,2776); WONDERBUG (II,2830); THE WORLD'S GREATEST SUPER HEROES (II,2842); YOGI'S SPACE RACE (II,2853)

WELLER, MARY LOUISE
CALLING DR. STORM, M.D. (II,415); ONCE UPON A SPY (II,1898); ONE DAY AT A TIME (II,1900); SEMI-TOUGH (II,2299); VALENTINE MAGIC ON LOVE ISLAND (II,2716)

WELLER, ROBB
ENTERTAINMENT TONIGHT (II,780)

WELLES, CAROLE
MEDIC (I,2976)

WELLES, JESSE
GOOD TIME HARRY (II,1037); HAZARD'S PEOPLE (II,1112); HUSBANDS, WIVES AND LOVERS (II,1209); OH MADELINE (II,1880); A SHADOW IN THE STREETS (II,2312)

WELLES, JOAN
I'M A BIG GIRL NOW (II,1218)

WELLES, ORSON
FOUNTAIN OF YOUTH (I,1647); MAGIC WITH THE STARS (II,1599); MAGNUM, P.I. (II,1601); THE MAN WHO CAME TO DINNER (I,2881); ORSON WELLES' GREAT MYSTERIES (I,3408); SCENE OF THE CRIME (II,2271); TWENTIETH CENTURY (I,4639)

WELLES, REBECCA
ONE STEP BEYOND (I,3388); THE WEB (I,4785)

WELLINGTONS, THE
GILLIGAN'S ISLAND (II,990)

WELLMAN JR, WILLIAM
THE BLUE AND THE GRAY (II,273); YOUNG LIVES (II,2866)

WELLMAN, JAMES
STAR TREK (II,2440)

WELLMAN, MAGGIE
THE BLUE AND THE GRAY (II,273); THE DEADLY TRIANGLE (II,634); THE PARTRIDGE FAMILY (II,1962)

WELLS, AARIKA
HART TO HART (II,1102); MICKEY SPILLANE'S MARGIN FOR MURDER (II,1691); THE SIX O'CLOCK FOLLIES (II,2368); SUPERTRAIN (II,2504)

WELLS, ANNEKE
DOCTOR WHO (II,689); THE STRANGE REPORT (I,4248)

WELLS, CAROLE
NATIONAL VELVET (I,3231); PISTOLS 'N' PETTICOATS (I,3608); TIME FOR ELIZABETH (I,4506)

WELLS, CHRISTOPHER
FOR LOVERS ONLY (II,904)

WELLS, CLAUDETTE
SQUARE PEGS (II,2431)

WELLS, CLAUDIA
HERBIE, THE LOVE BUG (II,1131); LOVERS AND OTHER STRANGERS (II,1550); OFF THE RACK (II,1878); RISE AND SHINE (II,2179)

WELLS, DANNY
FLO'S PLACE (II,892); FOUR EYES (II,912); GREAT DAY (II,1054); HARRY'S BATTLES (II,1100); NIGHTSIDE (II,1851); NOBODY'S PERFECT (II,1858); VELVET (II,2725)

WELLS, DARRYL
THE NURSES (I,3319)

WELLS, DAVID
STRANGE PARADISE (I,4246)

WELLS, DAWN
CHANNING (I,900); GILLIGAN'S ISLAND (II,990); GILLIGAN'S ISLAND: RESCUE FROM GILLIGAN'S ISLAND (II,991); GILLIGAN'S ISLAND: THE CASTAWAYS ON GILLIGAN'S ISLAND (II,992); GILLIGAN'S ISLAND: THE HARLEM GLOBETROTTERS ON GILLIGAN'S ISLAND (II,993); GILLIGAN'S PLANET (II,994); HIGH ROLLERS (II,1147); HIGH SCHOOL, U.S.A. (II,1148)

WELLS, DEREK
THE FITZPATRICKS (II,868)

WELLS, ELAINE
WHY DIDN'T THEY ASK EVANS? (II,2795)

WELLS, J C
STARSTRUCK (II,2446)

WELLS, JESSE
SOAP (II,2392)

WELLS, JOAN
DIFF'RENT STROKES (II,674)

WELLS, JULIENNE
A.E.S. HUDSON STREET (II,17); ADVICE TO THE LOVELORN (II,16)

WELLS, KIM
STRANGER IN OUR HOUSE (II,2477)

WELLS, KITTY
THE KITTY WELLS/JOHNNY WRIGHT FAMILY SHOW (I,2572)

WELLS, MARY K
BIG TOWN (I,436); THE BRIGHTER DAY (I,728); CAMEO THEATER (I,808); THE EDGE OF NIGHT (II,760); THE GREATEST GIFT (I,1880); RETURN TO PEYTON PLACE (I,3779); THE SECRET STORM (I,3969)

WELLS, REBECCA
THE BROTHERS BRANNAGAN (I,747)

WELLS, TERRY
LOVE, NATALIE (II,1545)

WELLS, TICO
SUMMER (II,2491)

WELLS, VIRGINIA
WALDO (I,4749)

WELSH, CHARLIE
THE RAINBOW GIRL (II,2108)

WELSH, JOHN
BRIDESHEAD REVISITED (II,375); THE CITADEL (II,539); DEAR TEACHER (II,653); LITTLE WOMEN (I,2723); THE TALE OF BEATRIX POTTER (II,2533)

WELSH, KENNETH
PIAF (II,2039)

WENCES, SENOR
FIVE STAR COMEDY (I,1588)

WENDELL, BILL
STAGE A NUMBER (I,4176); TIC TAC DOUGH (I,4498); WHAT'S IT WORTH? (I,4818)

WENDELL, HOWARD
DIAGNOSIS: DANGER (I,1250); LEAVE IT TO BEAVER (I,2648); MARKHAM (I,2916); SAM HILL (I,3895)

WENDORF, REUBEN
THE GOLDBERGS (I,1836)

WENDT, GEORGE

CHEERS (II,494)

WENTWORTH, MARK
THE ADAMS CHRONICLES (II,8)

WERLE, BARBARA
SAN FRANCISCO INTERNATIONAL AIRPORT (I,3905)

WERNER, KAREN
THE MYSTERIOUS TWO (II,1783); PHYL AND MIKHY (II,2037)

WERRIS, SNAG
THE GOOD LIFE (II,1033)

WERTIMER, NED
THE JEFFERSONS (II,1276); PINOCCHIO (I,3603); SECOND CHANCE (I,3958)

WESSELL, DICK
RIVERBOAT (I,3797)

WESSLER, RICHARD
H.M.S. PINAFORE (I,2066)

WESSON, DICK
DOG AND CAT (II,695); FRIENDS AND LOVERS (II,930); LOVE THAT BOB (I,2779); PAUL SAND IN FRIENDS AND LOVERS (II,1982); THE PEOPLE'S CHOICE (I,3521)

WESSON, EILEEN
THE WHOLE WORLD IS WATCHING (I,4851)

WEST, ADAM
ALEXANDER THE GREAT (I,113); ALL IN THE FAMILY (I,133); BATMAN (I,366); BATMAN AND THE SUPER SEVEN (II,160); THE DETECTIVES (I,1245); THE EYES OF CHARLES SAND (I,1497); GUESTWARD HO! (I,1900); NEVADA SMITH (II,1807); THE NEW ADVENTURES OF BATMAN (II,1812); NIGHT GALLERY (I,3287); POOR DEVIL (I,3646)

WEST, BERNIE
CALAMITY JANE (I,793)

WEST, BROOKS
MY FRIEND IRMA (I,3191)

WEST, DOTTIE
CHRISTMAS LEGEND OF NASHVILLE (II,522); CIRCUS OF THE STARS (II,537); SKINFLINT (II,2377)

WEST, JANE
THE O'NEILLS (I,3392)

WEST, JENNIFER
A TOWN LIKE ALICE (II,2652)

WEST, JOHN STUART
THIS IS KATE BENNETT (II,2597)

WEST, MADGE
THE MCLEAN STEVENSON SHOW (II,1665)

WEST, MARTIN
AS THE WORLD TURNS (II,110); GENERAL HOSPITAL (II,964); THE MAN FROM GALVESTON (I,2865); MICKEY SPILLANE'S MIKE HAMMER: MORE THAN MURDER (II,1693)

WEST, NORMA
SMILEY'S PEOPLE (II,2383)

WEST, PAMELA
FITZ AND BONES (II,867)

WEST, RED
THE BLACK SHEEP SQUADRON (II,262); THE CONCRETE COWBOYS (II,564); THE DUKES (II,740); HARD KNOX (II,1087)

WEST, ROY
CARLTON YOUR DOORMAN (II,440)

WEST, SONDRA
EIGHT IS ENOUGH (II,762)

WEST, TEGAN
HIGH SCHOOL, U.S.A. (II,1149)

WEST, TIMOTHY
MASADA (II,1642)

WEST, WALLY
THE ROY ROGERS SHOW (I,3859)

WEST, WILL
TONIGHT AT 8:30 (I,4556)

WESTCOTT, HELEN
DIAGNOSIS: DANGER (I,1250); A TALE OF WELLS FARGO (I,4338)

WESTERFIELD, JAMES
THE BLUE MEN (I,492); ELFEGO BACA (I,1426); EXPERT WITNESS (I,1488); GALLAGHER (I,1727); THE MURDOCKS AND THE MCCLAYS (I,3163); SCALPLOCK (I,3930); THEY WENT THATAWAY (I,4443); THE TRAVELS OF JAIMIE MCPHEETERS (I,4596)

WESTERMAN, CHANTAL
SOAP WORLD (II,2394)

WESTFALL, KAY
THE BENNETS (I,397); OH, KAY! (I,3342); SIT OR MISS (I,4064)

WESTFALL, ROY
THE BENNETS (I,397)

WESTMAN, NYDIA
BLOOMER GIRL (I,488); THE FABULOUS SYCAMORES (I,1505); GOING MY WAY (I,1833); MARY KAY AND JOHNNY (I,2932); THE MURDOCKS AND THE MCCLAYS (I,3163); OUT OF THE BLUE (I,3422); YOUNG MR. BOBBIN (I,4943)

WESTMORE, BETTY
SEARCH FOR BEAUTY (I,3953)

WESTMORE, ERN
THE ERN WESTMORE SHOW (I,1453); SEARCH FOR BEAUTY (I,3953)

WESTMORELAND, JAMES
THE MONROES (I,3089)

WESTON, CELIA
ALICE (II,33); BATTLE OF THE NETWORK STARS (II,178); A ROCK AND A HARD PLACE (II,2190); THE SINGLE LIFE (II,2364)

WESTON, DIANA
DEATH IN DEEP WATER (I,1206)

WESTON, ELLEN
ANOTHER APRIL (II,93); GET SMART (II,983); THE HEALERS (II,1115); LETTERS FROM THREE LOVERS (I,2671); MOBILE MEDICS (II,1716); S.W.A.T. (II,2236); SMILE JENNY, YOU'RE DEAD (II,2382); THE YOUNG AND THE RESTLESS (II,2862)

WESTON, JACK
79 PARK AVENUE (II,2310); BAND OF GOLD (I,341); BEWITCHED (I,418); CODE NAME: HERACLITUS (I,990); D.H.O. (I,1249); FAME IS THE NAME OF THE GAME (I,1518); FOR THE LOVE OF MIKE (I,1625); THE FOUR SEASONS (II,915); HARVEY (I,1961); THE HATHAWAYS (I,1965); HICKEY VS. ANYBODY (II,1142); I LOVE A MYSTERY (I,2170); THE LUCILLE BALL COMEDY HOUR (I,2783); MY SISTER EILEEN (I,3200); READY AND WILLING (II,2115); ROD BROWN OF THE ROCKET RANGERS (I,3825); STAGE DOOR (I,4178); THE THIRD COMMANDMENT (I,4449); THE TROUBLE WITH PEOPLE (I,4611); WHERE THERE'S SMOKEY (I,4832)

WESTON, JIM
ME AND MAXX (II,1672)

WESTON, ROGER
OUTLAW LADY (II,1935)

WESTON, STEVE
THE TROUBLE WITH TRACY (I,4613)

WESTWOOD, PATRICK
CAPTAIN BRASSBOUND'S CONVERSION (I,828)

WETMORE, JOAN
THE NURSES (I,3319); YOUNG DR. MALONE (I,4938)

WETMORE, MICHELE
WOMAN'S PAGE (II,2822)

WEXLER, BERNARD
CALUCCI'S DEPARTMENT (I,805)

WEYAND, RICHARD
VICTORY (I,4729)

WEYAND, RON
THE DAIN CURSE (II,613)

WHALEY, ROY
CRUSADER RABBIT (I,1109)

WHATLEY, DIXIE
ENTERTAINMENT TONIGHT (II,780)

WHEATLEY, ALAN
THE ADVENTURES OF ROBIN HOOD (I,74)

WHEATLEY, WILLIAM
THE GYPSY WARRIORS (II,1070); LANIGAN'S RABBI (II,1427)

WHEATON, FRANK
COUNTERATTACK: CRIME IN AMERICA (II,581)

WHEEL, PATRICIA
CYRANO DE BERGERAC (I,1119); FIRST LOVE (I,1583); A WOMAN TO REMEMBER (I,4888)

WHEELER, BERT
BRAVE EAGLE (I,709)

WHEELER, CAROL
THE INNER SANCTUM (I,2216)

WHEELER, JOHN
GOLDEN CHILD (I,1839)

WHEELER, JOSH
SATINS AND SPURS (I,3917)

WHEELER, LOIS
SHORT, SHORT DRAMA (I,4023)

WHEELER, MARGARET
IN SECURITY (II,1227)

WHEELER, MARK
MOBILE ONE (II,1717); MOBILE TWO (II,1718)

WHEELER, SANDRA
THE AWAKENING LAND (II,122)

WHEELS, JACK
TELL IT TO GROUCHO (I,4385)

WHELAN, JILL
BATTLE OF THE NETWORK STARS (II,177); FANTASY (II,828); FRIENDS (II,927); THE LOVE BOAT (II,1535)

WHELCHEL, LISA
THE ACADEMY (II,3); BATTLE OF THE NETWORK STARS (II,178); THE FACTS OF LIFE (II,805); FACTS OF LIFE: THE FACTS OF LIFE GOES TO PARIS (II,806); JO'S COUSINS (II,1293); THE NEW MICKEY MOUSE CLUB (II,1823); THE PARKERS (II,1959); THE WILD WOMEN OF CHASTITY GULCH (II,2801)

WHETON, WIL
13 THIRTEENTH AVENUE (II,2592)

WHILTON, PEGGY
THE DOCTORS (II,694)

WHINNERY, BARBARA
ST. ELSEWHERE (II,2432)

WHIP, JOSEPH
ADVICE TO THE LOVELORN (II,16)

WHIPPLE, RANDY
MY MOTHER THE CAR (I,3197)

WHIPPLE, SAM
BUFFALO BILL (II,387); OPEN ALL NIGHT (II,1914)

WHIPPLE, SHONDA
LITTLE HOUSE: LOOK BACK TO YESTERDAY (II,1490)

WHITAKER, DORI
BELL, BOOK AND CANDLE (II,201)

WHITAKER, JACK
THE FACE IS FAMILIAR (I,1506)

WHITAKER, JOHNNIE
BABY CRAZY (I,319); FAMILY AFFAIR (I,1519); THE LITTLEST ANGEL (I,2726)

WHITAKER, JOHNNY
MULLIGAN'S STEW (II,1763); SIGMUND AND THE SEA MONSTERS (II,2352)

WHITE, AL C
THE MUNSTERS' REVENGE (II,1765); WHEELS (II,2777)

WHITE, ALAN
THE FLYING DOCTOR (I,1610)

WHITE, ANN
THE NEW CHRISTY MINSTRELS SHOW (I,3259)

WHITE, ARTHUR
COLOR HIM DEAD (II,554)

WHITE, BETTY
THE BETTY WHITE SHOW (I,414); THE BETTY WHITE SHOW (I,415); THE BETTY WHITE SHOW (II,224); CIRCUS OF THE STARS (II,531); CIRCUS OF THE STARS (II,532); DATE WITH THE ANGELS (I,1164); EUNICE (II,784); JUST MEN (II,1359); THE LIAR'S CLUB (I,2673); LIFE WITH ELIZABETH (I,2689); THE LOVE BOAT (II,1535); MAMA'S FAMILY (II,1610); THE MARY TYLER MOORE SHOW (II,1640); THE PET SET (I,3557); THE SHAPE OF THINGS (II,2315); SNAVELY (II,2387); STEPHANIE (II,2453)

WHITE, CALLAN
SPRAGGUE (II,2430)

WHITE, CAROL ITA
LAVERNE AND SHIRLEY (II,1446); STARSKY AND HUTCH (II,2445)

WHITE, CHARLES
THE DAY THE WOMEN GOT EVEN (II,628); FOR LOVERS ONLY (II,904); THE FRONT PAGE (I,1692); LOVE OF LIFE (I,2774); THE PATTY DUKE SHOW (I,3483)

WHITE, CHRISTINE
CONCERNING MISS MARLOWE (I,1032); THE DEAN JONES SHOW (I,1191); ICHABOD AND ME (I,2183)

WHITE, DALE
THE JACK BENNY PROGRAM (I,2294)

WHITE, DAN
FROM THESE ROOTS (I,1688)

WHITE, DANA
STAR TONIGHT (I,4199)

WHITE, DAVID
THE BETTY HUTTON SHOW (I,413); BEWITCHED (I,418); THE SLIGHTLY FALLEN ANGEL (I,4087); SPIDER-MAN (II,2424); TWIN DETECTIVES (II,2687)

WHITE, DEBORAH
FAMILY (II,813); STICKIN' TOGETHER (II,2465)

WHITE, DEVEREAUX
BEULAH LAND (II,226)

WHITE, FRANCES
SCREAMER (I,3945)

WHITE, FRED
EARTH, WIND & FIRE IN CONCERT (II,750)

WHITE, JANE
ONCE UPON A MATTRESS (I,3371)

WHITE, JESSE
THE ANN SOTHERN SHOW (I,220); ANOTHER DAY, ANOTHER DOLLAR (I,228); THE BOB HOPE SHOW (I,625); GEORGE M! (I,1772); GEORGE M! (II,976); HARVEY (I,1962); THE HOUSE NEXT DOOR (I,2139); ICHABOD AND ME (I,2183); INSPECTOR GADGET (II,1236); MAKE ROOM FOR DADDY (I,2843); THE MARRIAGE BROKER (I,2920); OF THEE I SING (I,3334); THE PLANT FAMILY (II,2050); PRIVATE SECRETARY (I,3672); THE TIME ELEMENT (I,4504)

WHITE, JOHN SYLVESTER
A SHADOW IN THE STREETS (II,2312); WELCOME BACK, KOTTER (II,2761)

WHITE, JOHNNY O
THE RUNAWAY BARGE (II,2230)

WHITE, KENNETH
THE FRENCH ATLANTIC AFFAIR (II,924); FROM HERE TO ETERNITY (II,932); PALMERSTOWN U.S.A. (II,1945)

WHITE, LAUREN
THE DOCTORS (II,694); HOW TO SURVIVE A MARRIAGE (II,1198); ONE OF THE BOYS (II,1911)

WHITE, MAURICE
EARTH, WIND & FIRE IN CONCERT (II,750)

WHITE, MORGAN
HAWAII FIVE-O (II,1110)

WHITE, NONI
JOHNNY BELINDA (I,2418)

WHITE, PAMELA
THE BASTARD/KENT FAMILY CHRONICLES (II,159)

WHITE, PATRICIA
DOLLAR A SECOND (I,1326)

WHITE, PETER
ALL MY CHILDREN (II,39)

WHITE, RUTH
HANDS OF MURDER (I,1929); HIGHER AND HIGHER, ATTORNEYS AT LAW (I,2057); JOHNNY BELINDA (I,2417); LITTLE MOON OF ALBAN (I,2715); VICTORY (I,4729)

WHITE, SLAPPY
MY BUDDY (II,1778); SANFORD AND SON (II,2256); WHITE AND RENO (II,2787)

WHITE, STORY
FLYING HIGH (II,895)

WHITE, TED
FORCE SEVEN (II,908)

WHITE, VANNA
WHEEL OF FORTUNE (II,2775)

WHITE, VERDONE
EARTH, WIND & FIRE IN CONCERT (II,750)

WHITE, WILL J
MIDNIGHT MYSTERY (I,3030)

WHITE, WILLARD
AMAHL AND THE NIGHT VISITORS (I,150)

WHITE, YVONNE
THE ADVENTURES OF SUPERMAN (I,77); I WAS A BLOODHOUND (I,2180)

WHITEFIELD, ANNE
MY DARLING JUDGE (I,3186)

WHITEFIELD, PETRONELLA
MASTER OF THE GAME (II,1647)

WHITEHEAD, GEOFFREY
TYCOON: THE STORY OF A WOMAN (II,2697)

WHITEHOUSE, ANTHONY
O'MALLEY (II,1872)

WHITELAW, BILLIE
THE STRANGE CASE OF DR. JEKYLL AND MR. HYDE (I,4244)

WHITELEY, ARKIE
A TOWN LIKE ALICE (II,2652)

WHITELEY, THELMA
NICHOLAS NICKLEBY (II,1840)

WHITEMAN, FRANK
THE FIGHTING NIGHTINGALES (II,855)

WHITEMAN, GEORGE
THE FIGHTING NIGHTINGALES (II,855)

WHITEMAN, MARGO
PAUL WHITEMAN'S TV TEEN CLUB (I,3491)

WHITEMAN, PAUL
AMERICA'S GREATEST BANDS (I,177); PAUL WHITEMAN'S SATURDAY NIGHT REVUE (I,3490); PAUL

WHITEMAN'S TV TEEN CLUB (I,3491)

WHITESIDE, ANN
THE DOCTORS (II,694)

WHITESIDE, RAY
DAVY CROCKETT (I,1181)

WHITFIELD, ANNE
HOLLYWOOD OPENING NIGHT (I,2087); ONE MAN'S FAMILY (I,3383)

WHITFIELD, DAVID
THE INVISIBLE WOMAN (II,1244)

WHITING, ARCH
RUN RUN, JOE (II,2229); THE STRANGER (I,4251); VOYAGE TO THE BOTTOM OF THE SEA (I,4743)

WHITING, BARBARA
THOSE WHITING GIRLS (I,4471); YOUR JEWELER'S SHOWCASE (I,4953)

WHITING, JACK
THE MARGE AND GOWER CHAMPION SHOW (I,2907); PARIS IN THE SPRINGTIME (I,3457)

WHITING, MARGARET
THE STRAUSS FAMILY (I,4253); THOSE WHITING GIRLS (I,4471)

WHITING, NAPOLEON
THE BIG VALLEY (I,437)

WHITING, RICHARD
REVENGE OF THE GRAY GANG (II,2146)

WHITLEY, JUNE
THE FLYING NUN (I,1611)

WHITMAN, ERNEST
BEULAH (I,416)

WHITMAN, PARKER
HEAR NO EVIL (II,1116)

WHITMAN, STUART
THE BOB HOPE CHRYSLER THEATER (I,502); CIMARRON STRIP (I,954); CITY BENEATH THE SEA (I,965); CONDOMINIUM (II,566); FOUR STAR PLAYHOUSE (I,1652); GHOST STORY (I,1788); GO WEST YOUNG GIRL (II,1013); INTERTECT (I,2228); THE LAST CONVERTIBLE (II,1435); ROUGHCUTS (II,2218); THE SEEKERS (II,2298); STARR, FIRST BASEMAN (I,4204); WOMEN IN WHITE (II,2823)

WHITMIRE, STEVE
FRAGGLE ROCK (II,916)

WHITMORE JR, JAMES
THE BASTARD/KENT FAMILY CHRONICLES (II,159); THE BLACK SHEEP SQUADRON (II,262); BOSTON AND KILBRIDE (II,356); THE GYPSY WARRIORS (II,1070); HAZARD'S PEOPLE (II,1112); HUMAN FEELINGS (II,1203); HUNTER (II,1206); THE YEAGERS (II,2843)

WHITMORE, JAMES
ALCOA PREMIERE (I,109); CELEBRITY (II,463); COMEBACK (I,1016); THE DESILU PLAYHOUSE (I,1237); KRAFT TELEVISION THEATER (I,2592); THE LAW AND MR. JONES (I,2637); THE MAN FROM DENVER (I,2863); MY FRIEND TONY (I,3192); PANIC! (I,3447); SURVIVAL (I,4300); TEMPERATURES RISING (I,4390); THE TWILIGHT ZONE (I,4651); WILL ROGERS' U.S.A. (I,4865); THE WORD (II,2833); ZANE GREY THEATER (I,4979)

WHITMORE, STEVE
THE LAST CONVERTIBLE (II,1435)

WHITNEY, BEVERLY
EYE WITNESS (I,1494); HANDS OF MURDER (I,1929)

WHITNEY, EVE
MYSTERY AND MRS. (I,3210)

WHITNEY, GRACE LEE
CALL TO DANGER (I,802); THE MAN FROM GALVESTON (I,2865); POLICE STORY (I,3636); SKYFIGHTERS (I,4078); STAR TREK (II,2440)

WHITNEY, JASON
THE MCLEAN STEVENSON SHOW (II,1665)

WHITNEY, PETER
THE MAN FROM EVERYWHERE (I,2864); MEET MCGRAW (I,2984); NAVY LOG (I,3233); THE ROUGH RIDERS (I,3850); WHICH WAY'D THEY GO? (I,4838)

WHITNEY, RUSSELL
NOAH'S ARK (I,3302)

WHITSON, SAMUEL
TRAFFIC COURT (I,4590)

WHITSUN, PAUL
IVANHOE (I,2282)

WHITTLE, NICHOLAS
CINDERELLA (II,526)

WHITWORTH, DEAN
CHIEFS (II,509)

WHOLEY, DENNIS
THE GENERATION GAP (I,1759)

WHYTE, PAT
TALES OF THE 77TH BENGAL LANCERS (I,4348)

WIBLE, NANCY
DAVY AND GOLIATH (I,1180)

WICKER, IREENE
THE IREENE WICKER SHOW (I,2234); IREENE WICKER SINGS (I,2236); THE SINGING LADY (I,4061)

WICKERT, ANTHONY
WHIPLASH (I,4839)

WICKES, MARY
ANNETTE (I,222); BONINO (I,685); DENNIS THE MENACE (I,1231); DOC (II,682); THE HALLS OF IVY (I,1923); THE JIMMY STEWART SHOW (I,2391); JULIA (I,2476); MA AND PA (II,1570); MAKE ROOM FOR DADDY (I,2843); THE MAN WHO CAME TO DINNER (I,2881); THE MONK (I,3087); MRS. G. GOES TO COLLEGE (I,3155); THE PETER AND MARY SHOW (I,3561); SIGMUND AND THE SEA MONSTERS (II,2352); TIME OUT FOR GINGER (I,4508)

WICKWIRE, NANCY
ANOTHER WORLD (II,97); AS THE WORLD TURNS (II,110); DAYS OF OUR LIVES (II,629); ESPIONAGE (I,1464)

WIDDOES, JAMES
AFTERMASH (II,23); BACK TOGETHER (II,131); CHARLES IN CHARGE (II,482); CHARLEY'S AUNT (II,483); DELTA HOUSE (II,658); PARK PLACE (II,1958)

WIDDOES, KATHLEEN
ANOTHER WORLD (II,97); A BELL FOR ADANO (I,388); OUR TOWN (I,3420); PUNCH AND JODY (II,2091); THE RETURN OF CHARLIE CHAN (II,2136); RYAN'S HOPE (II,2234); YOUNG DR. MALONE (I,4938)

WIDDOWSON-REYNOLDS, ROSINE
NUMBER 96 (II,1869)

WIDMARK, RICHARD
BROCK'S LAST CASE (I,742); MADIGAN (I,2808)

WIERE, HARRY
OH, THOSE BELLS! (I,3345)

WIERE, HERBERT
OH, THOSE BELLS! (I,3345)

WIERE, SYLVESTER
OH, THOSE BELLS! (I,3345)

WIGGINS, CHRIS
A MAN CALLED INTREPID (II,1612); MOMENT OF TRUTH (I,3084); PAUL BERNARD—PSYCHIATRIST (I,3485); THE SWISS FAMILY ROBINSON (II,2518)

WIGGINS, RUSSELL C
BANACEK: DETOUR TO NOWHERE (I,339)

WIGGINS, TOM
ANOTHER WORLD (II,97); BREAKING AWAY (II,371)

WIGGINS, TUDI
AFTER HOURS: GETTING TO KNOW US (II,21); ALL MY CHILDREN (II,39); THE GUIDING LIGHT (II,1064); LOVE OF LIFE (I,2774); PAUL BERNARD—PSYCHIATRIST (I,3485); RED SKELTON'S CHRISTMAS DINNER (II,2123); STRANGE PARADISE (I,4246)

WIGGINTON, ROBERT
ONE MAN'S FAMILY (I,3383)

WIGHTMAN, ROBERT
THE WALTONS (II,2740); THE WALTONS: A DAY FOR THANKS ON WALTON'S MOUNTAIN (EPISODE 3) (II,2741)

WIKES, MICHAEL
POWERHOUSE (II,2074)

WILBUR, GEORGE
THE BEST OF FRIENDS (II,216); HARDCASE (II,1088)

WILBURN, DOYLE
THE WILBURN BROTHERS SHOW (I,4860)

WILBURN, GEORGE
THE GOOD GUYS (I,1847)

WILBURN, TED
THE WILBURN BROTHERS SHOW (I,4860)

WILCOX, CLAIRE
DANIEL BOONE (I,1142); HARRIS AGAINST THE WORLD (I,1956); THE PARTRIDGE FAMILY (II,1962)

WILCOX, COLLIN
THE CAT AND THE CANARY (I,868); GREAT ADVENTURE (I,1868); KRAFT TELEVISION THEATER (I,2592); MEMBER OF THE WEDDING (I,3002); A RAINY DAY (II,2110); THE SOUND OF ANGER (I,4133)

WILCOX, FRANK
THE BEVERLY HILLBILLIES (I,417); THE DOCTOR WAS A LADY (I,1318); THE GEORGE BURNS AND GRACIE ALLEN SHOW (I,1763); IT'S ABOUT TIME (I,2263); LEAVE IT TO BEAVER (I,2648)

WILCOX, HARLOW
YOU ARE THERE (I,4929)

WILCOX, LARRY
BATTLE OF THE NETWORK STARS (II,165); BATTLE OF THE NETWORK STARS (II,166); BATTLE OF THE NETWORK STARS (II,168); BATTLE OF THE NETWORK STARS (II,170); CHIPS (II,511); FORCE SEVEN (II,908); LASSIE (I,2624); MEMBER OF THE WEDDING (I,3002)

WILCOX, NINA
HARBOURMASTER (I,1946); JESSICA NOVAK (II,1286); MO AND JO (II,1715)

WILCOX, PAULA
MAN ABOUT THE HOUSE (II,1611)

WILCOX, RALPH
BIG EDDIE (II,236); THE BUFFALO SOLDIERS (II,388); BUSTING LOOSE (II,402); ONE IN A MILLION (II,1906); SHAUGHNESSEY (II,2318)

WILCOX, SHANNON
CAGNEY AND LACEY (II,409)

WILCOXON, HENRY
PURSUE AND DESTROY (I,3699); SARGE (I,3914); SARGE: THE BADGE OR THE CROSS (I,3915)

WILD, JACK
H.R. PUFNSTUF (I,2154)

WILDE, BRIAN
LAST OF THE SUMMER WINE (II,1439); THE LILLI PALMER THEATER (I,2704); THE SEVEN DIALS MYSTERY (II,2308)

WILDE, COLLETTE
KRAFT MYSTERY THEATER (I,2589)

WILDE, CORNEL
ALCOA/GOODYEAR THEATER (I,107); KRAFT SUSPENSE THEATER (I,2591)

WILDE, HEATHER
YOUR SHOW TIME (I,4958)

WILDE, ROBERT
RETURN TO EDEN (II,2143)

WILDER, GENE
THE TROUBLE WITH PEOPLE (I,4611)

WILDER, GLENN
THE CABOT CONNECTION (II,408); FORCE SEVEN (II,908)

WILDER, JOHN
THE ELISABETH MCQUEENEY STORY (I,1429); MAGGIE MALONE (I,2812)

WILDER, SCOTT
ARCHER—FUGITIVE FROM THE EMPIRE (II,103)

WILDER, YVONNE
AMERICA, YOU'RE ON (II,69); ARCHIE BUNKER'S PLACE (II,105); CONDO (II,565); I'LL NEVER FORGET WHAT'S HER NAME (II,1217); OPERATION PETTICOAT (II,1916); OPERATION PETTICOAT (II,1917); THE PARTRIDGE FAMILY (II,1962); THE RETURN OF MARCUS WELBY, M.D (II,2138); SONNY BOY (II,2402)

WILFRED, HELEN HAYES
THE ALCOA HOUR (I,108)

WILHITE, CHARLES
CAMP GRIZZLY (II,418)

WILKE, ROBERT J
THE BOSTON TERRIER (I,696); THE LEGEND OF JESSE JAMES (I,2653); THE SAGA OF ANDY BURNETT (I,3883); THE SLOWEST GUN IN THE WEST (I,4088); THE TEXAS RANGERS (II,2567); THEY CALL IT MURDER (I,4441)

WILKES, DONNA
HELLO, LARRY (II,1128)

WILKIN, BARBARA
GENERAL ELECTRIC TRUE (I,1754)

WILKIN, BRAD
INSTANT FAMILY (II,1238)

WILKINS, JEREMY
CAPTAIN SCARLET AND THE MYSTERONS (I,837); U.F.O. (I,4662)

WILKINSON, KATE
THE GUIDING LIGHT (II,1064)

WILKINSON, LISA
ALL MY CHILDREN (II,39)

WILKINSON, ROBERT
MAGIC COTTAGE (I,2815)

WILKINSON, WALLACE
CHIEFS (II,509); SIX PACK (II,2370)

WILKOF, LEE
HART TO HART (II,1102); NEWHART (II,1835); W.E.B. (II,2732)

WILL, SANDRA
DANGER IN PARADISE (II,617); GO WEST YOUNG GIRL (II,1013)

WILLARD, CAROL
A WORLD APART (I,4910)

WILLARD, FRED
AMERICA 2-NIGHT (II,65); BATTLE OF THE NETWORK STARS (II,173); FERNWOOD 2-NIGHT (II,849); FLATBED ANNIE AND SWEETIEPIE: LADY TRUCKERS (II,876); FOREVER FERNWOOD (II,909); OPERATION GREASEPAINT (I,3402); PEN 'N' INC. (II,1992); REAL PEOPLE (II,2118); SIROTA'S COURT (II,2365); SPACE FORCE (II,2407)

WILLES, JEAN
THE BOUNTY HUNTER (I,698); THE SLOWEST GUN IN THE WEST (I,4088)

WILLETT, JOANN
HIGH SCHOOL, U.S.A. (II,1149)

WILLIAM, EARL
THE DESERT SONG (I,1236)

WILLIAMS JR, HANK
THE DANGEROUS DAYS OF KIOWA JONES (I,1140)

WILLIAMS JR, SPENCER
THE AMOS AND ANDY SHOW (I,179)

WILLIAMS, ADAM
JUSTICE OF THE PEACE (I,2500)

WILLIAMS, ALFIE
THE MOONMAN CONNECTION (II,1731)

WILLIAMS, ALISTER
MURDER ON THE MIDNIGHT EXPRESS (I,3162)

WILLIAMS, ALLEN
BATTLE OF THE NETWORK STARS (II,169); THE FEATHER AND FATHER GANG (II,845); LOU GRANT (II,1526)

WILLIAMS, AMY
ANOTHER LIFE (II,95)

WILLIAMS, ANDY
THE ANDY WILLIAMS CHRISTMAS SHOW (I,196); THE ANDY WILLIAMS CHRISTMAS SHOW (I,197); THE ANDY WILLIAMS CHRISTMAS SHOW (I,198); THE ANDY WILLIAMS CHRISTMAS SHOW (I,199); THE ANDY WILLIAMS CHRISTMAS SHOW (II,77); ANDY WILLIAMS KALEIDOSCOPE COMPANY (I,201); THE ANDY WILLIAMS SHOW (I,204); THE ANDY WILLIAMS SHOW (I,205); THE ANDY WILLIAMS SHOW (I,206); THE ANDY WILLIAMS SHOW (I,207); THE ANDY WILLIAMS SHOW (I,208); THE ANDY WILLIAMS SHOW (I,209); THE ANDY WILLIAMS SHOW (I,210); THE ANDY WILLIAMS SHOW (II,78); THE ANDY WILLIAMS SPECIAL (I,211); THE ANDY WILLIAMS SPECIAL (I,212); THE ANDY WILLIAMS SPECIAL (I,213); ANDY WILLIAMS MAGIC LANTERN SHOW COMPANY (I,202); THE ANDY WILLIAMS NEW YEAR'S EVE SPECIAL (I,203); ANDY WILLIAMS' EARLY NEW ENGLAND CHRISTMAS (II,79); THE ANDY WILLIAMS-JUNE VALLI SHOW (I,200); ANDY'S LOVE CONCERT (I,215); THE BIG BEAT (I,423); THE MAN IN THE MOON (I,2871)

WILLIAMS, ANN
THE DOCTORS (II,694); THE EDGE OF NIGHT (II,760); LOVING (II,1552); SEARCH FOR TOMORROW (II,2284)

WILLIAMS, ANSON
ANSON AND LORRIE (II,98); HAPPY DAYS (II,1084); INSIDE AMERICA (II,1234); LISA, BRIGHT AND DARK (I,2711); THE PAUL LYNDE SHOW (I,3489)

WILLIAMS, BARRY
THE BRADY BUNCH (II,362); THE BRADY BUNCH HOUR (II,363); THE BRADY GIRLS GET MARRIED (II,364); THE BRADY KIDS (II,365); KELLY'S KIDS (I,2523); THE SHAMEFUL SECRETS OF HASTINGS CORNERS (I,3994)

WILLIAMS, BETTY
STORIES IN ONE CAMERA (I,4234)

WILLIAMS, BIFF
GALLAGHER (I,1727)

WILLIAMS, BILL
ADVENTURES IN JAZZ (I,45); THE ADVENTURES OF KIT CARSON (I,66); ASSIGNMENT: UNDERWATER (I,303); DATE WITH THE ANGELS (I,1164); MUSIC FROM THE MEADOWBROOK (I,3171); THE S. S. HOLIDAY (I,4173); THE UNTOUCHABLES (I,4681)

WILLIAMS, BILLY DEE
ANOTHER WORLD (II,97); CHIEFS (II,509); CRISIS (I,1102); DYNASTY (II,746); HIGHER AND HIGHER, ATTORNEYS AT LAW (I,2057); SHOOTING STARS (II,2341)

WILLIAMS, BINDI
WOOBINDA—ANIMAL DOCTOR (II,2832)

WILLIAMS, BRANDON
THE GREATEST AMERICAN HERO (II,1060)

WILLIAMS, CARA
ALFRED HITCHCOCK PRESENTS (I,115); THE

WILLIAMSON, KATE
LITTLE HOUSE: BLESS ALL THE DEAR CHILDREN (II,1489)

WILLIAMSON, MYKEL T
BAY CITY BLUES (II,186); THE RIGHTEOUS APPLES (II,2174)

WILLIAMSON, NICOL
OF MICE AND MEN (I,3333); THE WORD (II,2833)

WILLIE
JOE DANCER: THE MONKEY MISSION (II,1301)

WILLINGHAM, CARMEN
DEBBY BOONE. . . ONE STEP CLOSER (II,655)

WILLINGHAM, NOBLE
BLACK BART (II,260); THE BLUE AND THE GRAY (II,273); CUTTER TO HOUSTON (II,612); EVEL KNIEVEL (II,785); THE GEORGIA PEACHES (II,979); THE TEXAS WHEELERS (II,2568); W*A*L*T*E*R (II,2731); WHEN THE WHISTLE BLOWS (II,2781)

WILLIS, ANN
THE SHAMEFUL SECRETS OF HASTINGS CORNERS (I,3994)

WILLIS, AUSTIN
CASINO (II,452); RIVAK, THE BARBARIAN (I,3795); SEAWAY (I,3956)

WILLIS, BEVERLY
HOORAY FOR LOVE (I,2116)

WILLIS, CURTIZ
THE COP AND THE KID (II,572)

WILLIS, HOPE ALEXANDER
ADDIE AND THE KING OF HEARTS (I,35)

WILLIS, JOHN
GOOD DAY! (II,1030); HOLLYWOOD BACKSTAGE (I,2082); THE STORY OF—(I,4239)

WILLIS, MARLENE
DOBIE GILLIS (I,1302)

WILLMAN, NOEL
CAESAR AND CLEOPATRA (I,789)

WILLOCK, DAVE
BOOTS AND SADDLES: THE STORY OF THE FIFTH CAVALRY (I,686); DAVE AND CHARLEY (I,1171); DO IT YOURSELF (I,1299); MARGIE (I,2909); MY SON THE DOCTOR (I,3203); OFF WE GO (I,3338); THE QUEEN AND I (I,3704); THE ROMAN HOLIDAYS (I,3835); SAWYER AND FINN (II,2265); YOUR JEWELER'S SHOWCASE (I,4953)

WILLOCK, MARGARET
FAY (II,844); PALS (II,1947)

WILLRICH, RUDY
ALL MY CHILDREN (II,39)

WILLS, BEVERLY
I MARRIED JOAN (I,2174); TOO MANY SERGEANTS (I,4573)

WILLS, CHILL
ALFRED HITCHCOCK PRESENTS (I,115); FRONTIER CIRCUS (I,1697); THE OVER-THE-HILL GANG RIDES AGAIN (I,3432); THE ROUNDERS (I,3851); STUBBY PRINGLE'S CHRISTMAS (II,2483)

WILLS, HENRY
ROUGHNECKS (II,2219)

WILLS, TERRY
FLO (II,891)

WILLSON, PAUL
EMPIRE (II,777); SUNDAY FUNNIES (II,2493)

WILLSON, WALTER
MCDUFF, THE TALKING DOG (II,1662)

WILSON, ALEX
THE BOYS (II,360)

WILSON, BARBARA
BOLD VENTURE (I,682)

WILSON, BEN
THE MACAHANS (II,1583)

WILSON, BILL
CAPTURE (I,840)

WILSON, BRIAN
THE BEACH BOYS 20TH ANNIVERSARY (II,188); THE BEACH BOYS IN CONCERT (II,189)

WILSON, BRIAN G
PALMERSTOWN U.S.A. (II,1945)

WILSON, CARL
THE BEACH BOYS 20TH ANNIVERSARY (II,188); THE BEACH BOYS IN CONCERT (II,189)

WILSON, CLIVE
THE BIG EASY (II,234)

WILSON, DAVID
THE GANGSTER CHRONICLES (II,957); IT'S A BIRD, IT'S A PLANE, IT'S SUPERMAN (II,1253); STUDS LONIGAN (II,2484)

WILSON, DEMOND
BABY, I'M BACK! (II,130); BATTLE OF THE NETWORK STARS (II,163); BATTLE OF THE NETWORK STARS (II,175); THE NEW ODD COUPLE (II,1825); SANFORD AND SON (II,2256)

WILSON, DENNIS
THE BEACH BOYS 20TH ANNIVERSARY (II,188); THE BEACH BOYS IN CONCERT (II,189)

WILSON, DICK
BARNEY AND ME (I,358); BEWITCHED (I,418); MCHALE'S NAVY (I,2969); SMALL AND FRYE (II,2380)

WILSON, DON
THE JACK BENNY PROGRAM (I,2294)

WILSON, DOOLEY
BEULAH (I,416)

WILSON, DORO
OPRYLAND: NIGHT OF STARS AND FUTURE STARS (II,1921)

WILSON, EARL
GUEST SHOT (I,1899); STAGE ENTRANCE (I,4180); TONIGHT! AMERICA AFTER DARK (I,4555)

WILSON, ELIZABETH
ALL IN THE FAMILY (II,38); ANOTHER APRIL (II,93); DOC (II,682); EAST SIDE/WEST SIDE (I,1396); THE EASTER PROMISE (I,1397); HAPPY ENDINGS (II,1085); SANCTUARY OF FEAR (II,2252)

WILSON, ETHEL
THE ALDRICH FAMILY (I,111)

WILSON, FLIP
BATTLE OF THE NETWORK STARS (II,178); CELEBRITY CHALLENGE OF THE SEXES 2 (II,466); THE CHEAP DETECTIVE (II,490); THE FLIP WILSON COMEDY SPECIAL (II,886); THE FLIP WILSON SHOW (I,1602); THE FLIP WILSON SPECIAL (I,1603); THE FLIP WILSON SPECIAL (II,887); THE FLIP WILSON SPECIAL (II,888); THE FLIP WILSON SPECIAL (II,889); FLIP WILSON. . . OF COURSE (II,890); PEOPLE ARE FUNNY (II,1996); PINOCCHIO (II,2044)

WILSON, GRANT
ALICE (II,33)

WILSON, JEANNIE
HANDLE WITH CARE (II,1077); MARRIAGE IS ALIVE AND WELL (II,1633); SIMON AND SIMON (II,2357); A YEAR AT THE TOP (II,2845)

WILSON, JOYCE VINCENT
TONY ORLANDO AND DAWN (II,2639)

WILSON, JULIE
THE BACHELOR (I,325); KISS ME, KATE (I,2568)

WILSON, KARA
MACKENZIE (II,1584)

WILSON, LIONEL
THE ASTRONAUT SHOW (I,307); BRAIN GAMES (II,366); TOM TERRIFIC (I,4546)

WILSON, LOIS
THE ALDRICH FAMILY (I,111); MAGNAVOX THEATER (I,2827)

WILSON, MARIE
MY FRIEND IRMA (I,3191); THE SOFT TOUCH (I,4107); WHERE'S HUDDLES (I,4836)

WILSON, MARK
THE MAGIC LAND OF ALLAKAZAM (I,2817); THE MAGIC OF MARK WILSON (II,1596)

WILSON, MEREDITH
TEXACO STAR PARADE I (I,4409); TEXACO STAR PARADE II (I,4410)

WILSON, MIKE
THE MAGIC LAND OF ALLAKAZAM (I,2817)

WILSON, NANCY
WHO HAS SEEN THE WIND? (I,4847)

WILSON, NED
RICHIE BROCKELMAN: MISSING 24 HOURS (II,2168)

WILSON, PATRICIA
STRANGER IN OUR HOUSE (II,2477)

WILSON, RICHARD
THE SECRET OF CHARLES DICKENS (II,2293)

WILSON, RINI
TEXACO STAR PARADE I (I,4409); TEXACO STAR PARADE II (I,4410)

WILSON, RITA
THE BEACH GIRLS (II,191); THE CHEERLEADERS (II,493); FLYING HIGH (II,895); M*A*S*H (II,1569)

WILSON, ROBIN
HOT L BALTIMORE (II,1187)

WILSON, ROGER

SEVEN BRIDES FOR SEVEN BROTHERS (II,2307)

WILSON, SERETTA
THE ZOO GANG (II,2877)

WILSON, SHEREE
COVER UP (II,597); VELVET (II,2725)

WILSON, SLIM
COUNTRY MUSIC CAROUSEL (I,1067); TALENT VARIETIES (I,4343)

WILSON, STUART
THE DOUBLE KILL (I,1360); RUNNING BLIND (II,2232); THE STRAUSS FAMILY (I,4253)

WILSON, TERRY
THE DAUGHTERS OF JOSHUA CABE RETURN (I,1170); THE ELISABETH MCQUEENEY STORY (I,1429); WAGON TRAIN (I,4747)

WILSON, THEODORE
THE FURST FAMILY OF WASHINGTON (I,1718); IRENE (II,1245); THE LOVE BOAT I (II,1532); POLICE WOMAN: THE GAMBLE (II,2064); RENDEZVOUS HOTEL (II,2131); ROLL OUT! (I,3833); SANFORD ARMS (II,2257); SURE AS FATE (I,4297); THAT'S MY MAMA (II,2580); WALKIN' WALTER (II,2737)

WILSON, TREY
ALL MY CHILDREN (II,39); KENNEDY (II,1377); THE NEWS IS THE NEWS (II,1838)

WILSON, WARD
CAN YOU TOP THIS? (I,816)

WILSON, ZOEY
HONEST AL'S A-OK USED CAR AND TRAILER RENTAL TIGERS (II,1174)

WILTON, PENELOPE
THE TALE OF BEATRIX POTTER (II,2533)

WILZAK, CRISSY
MORK AND MINDY (II,1735)

WINBURN, RANDY
JOE AND VALERIE (II,1298)

WINCHELL, PAUL
ADAMS OF EAGLE LAKE (II,10); THE BANANA SPLITS ADVENTURE HOUR (I,340); THE BEVERLY HILLBILLIES (I,417); THE C.B. BEARS (II,406); CARTOONSVILLE (I,858); CIRCUS TIME (I,960); THE CLUE CLUB (II,546); DASTARDLY AND MUTTLEY IN THEIR FLYING MACHINES (I,1159); DUNNINGER AND WINCHELL (I,1384); FIVE STAR COMEDY (I,1588); GOOBER AND THE GHOST CHASERS (II,1028); THE HEATHCLIFF AND MARMADUKE SHOW (II,1118); HELP! IT'S THE HAIR BEAR BUNCH (I,2012); JERRY MAHONEY'S CLUB HOUSE (I,2371); THE ODDBALL COUPLE (II,1876); THE PAUL WINCHELL AND JERRY MAHONEY SHOW (I,3492); THE PAUL WINCHELL SHOW (I,3493); THE PAUL WINCHELL SHOW (I,3494); THE PERILS OF PENELOPE PITSTOP (I,3525); RUNAROUND (I,3873); THE SKATEBIRDS (II,2375); THE SMURFS (II,2385); THE THREE ROBONIC STOOGES (II,2609); TOYLAND EXPRESS (I,4587); TROLLKINS (II,2660); THE WACKY RACES (I,4745); WHEELIE AND THE CHOPPER BUNCH (II,2776)

WINCHELL, WALTER
TELEPHONE TIME (I,4379); THE WALTER WINCHELL FILE (I,4759); THE WALTER WINCHELL SHOW (I,4760)

WINCHESTER, ARNA MARIA
1915 (II,1853)

WINCOTT, GEOFFREY
CASANOVA (II,451)

WINDING, VICTOR
DEATH IN SMALL DOSES (I,1207); THE EXPERT (I,1489)

WINDOM, WILLIAM
100 YEARS OF GOLDEN HITS (II,1905); BRIDGER (II,376); BROTHERS AND SISTERS (II,382); DIAL "M" FOR MURDER (I,1256); ESCAPE (I,1460); THE FARMER'S DAUGHTER (I,1533); FLAMINGO ROAD (II,872); THE GIRL WITH SOMETHING EXTRA (II,999); HECK'S ANGELS (II,1122); THE HOUSE ON GREENAPPLE ROAD (I,2140); IS THERE A DOCTOR IN THE HOUSE (I,2241); IS THERE A DOCTOR IN THE HOUSE? (II,1247); LANDON, LANDON & LANDON (II,1426); MY WORLD . . . AND WELCOME TO IT (I,3208); THE NEW HEALERS (I,3266); NIGHT GALLERY (I,3287); ONCE AN EAGLE (II,1897); PRESCRIPTION: MURDER (I,3660); QUICK AND QUIET (II,2099); RICHIE BROCKELMAN: MISSING 24 HOURS (II,2168); SECOND CHANCE (I,3958); SEVENTH AVENUE (II,2309); ST. ELSEWHERE (II,2432); THE TOM SWIFT AND LINDA CRAIG MYSTERY HOUR (II,2633); THE TWILIGHT ZONE (I,4651); U.M.C. (I,4666); VELVET (II,2725)

WINDSOR, BERNADETTE
A HORSEMAN RIDING BY (II,1182)

WINDSOR, FRANK
JACK THE RIPPER (II,1264)

WINDSOR, JOYCE
BUTTERFLIES (II,404); ROGER DOESN'T LIVE HERE ANYMORE (II,2201)

WINDSOR, MARIE
ALIAS MIKE HERCULES (I,117); AN APARTMENT IN ROME (I,242); BRACKEN'S WORLD (I,703); THE CHARLIE FARRELL SHOW (I,913); GENERAL HOSPITAL (II,964); JOHNNY COME LATELY (I,2425); THE LLOYD BRIDGES SHOW (I,2733); MANHUNTER (II,1620); PEPSI-COLA PLAYHOUSE (I,3523); PURSUIT (I,3700); SCIENCE FICTION THEATER (I,3938); SHOTGUN SLADE (I,4026); STRANDED (II,2475); TOM AND JERRY (I,4533); THE UNEXPECTED (I,4674)

WINDSOR, SHELLEY
ANNIE GET YOUR GUN (I,223)

WINDUST, PENELOPE
DEATH RAY 2000 (II,654); THE HARVEY KORMAN SHOW (II,1103); JOHNNY BELINDA (I,2418); V (II,2713)

WINES, CHRISTOPHER
ALL MY CHILDREN (II,39)

WINET, BRUCE
FIT FOR A KING (II,866)

WINFIELD, GREG
THE VIDEO GAME (II,2727)

WINFIELD, PAUL
BACKSTAIRS AT THE WHITE HOUSE (II,133); THE BLUE AND THE GRAY (II,273); HOTEL (II,1192); JULIA (I,2476); KEY TORTUGA (II,1383)

WINFILD, CHRIS
GOLIATH AWAITS (II,1025)

WING, ANNA
PENMARRIC (II,1993); THE WOMAN IN WHITE (II,2819)

WING, DOUGLAS
GEORGE WASHINGTON (II,978)

WING-DAVEY, MARK
THE HITCHHIKER'S GUIDE TO THE GALAXY (II,1158)

WINGER, DEBRA
WONDER WOMAN (SERIES 1) (II,2828)

WINGREEN, JASON
ALL IN THE FAMILY (II,38); ARCHIE BUNKER'S PLACE (II,105); BANYON (I,345); THE BARBARY COAST (II,146); CAPTAIN AMERICA (II,427); GLORIA COMES HOME (II,1011); LANDON, LANDON & LANDON (II,1426); THE LONG HOT SUMMER (I,2747); THE ROUNDERS (I,3851); SAN FRANCISCO INTERNATIONAL AIRPORT (I,3906); VEGAS (II,2723)

WINKLEMAN, MICHAEL
THE REAL MCCOYS (I,3741)

WINKLEMAN, WENDY
THE DOCTOR WAS A LADY (I,1318); LEAVE IT TO BEAVER (I,2648)

WINKLER, HENRY
FONZ AND THE HAPPY DAYS GANG (II,898); HAPPY DAYS (II,1084); HENRY WINKLER MEETS WILLIAM SHAKESPEARE (II,1129); LAVERNE AND SHIRLEY WITH THE FONZ (II,1449)

WINKLER, K C
RIPTIDE (II,2178)

WINKLER, MEL
THE DOCTORS (II,694)

WINKWORTH, MARK
ARCHIE (II,104); THE ARCHIE SITUATION COMEDY MUSICAL VARIETY SHOW (II,106); SUGAR TIME (II,2489)

WINN, DAVID
THE YOUNG AND THE RESTLESS (II,2862)

WINN, KITTY
MAN ON A STRING (I,2877); MOST WANTED (II,1738)

WINNINGER, CHARLES
THE CHARLIE FARRELL SHOW (I,913); THE PHILADELPHIA STORY (I,3581)

WINNINGHAM, MARE
STEELTOWN (II,2452); THE THORN BIRDS (II,2600); THE YOUNG PIONEERS (II,2869)

WINONA, KIM
BRAVE EAGLE (I,709)

WINSLOW, BRIGIT
MR. INSIDE/MR. OUTSIDE (II,1750)

WINSLOW, DICK
THE MISADVENTURES OF SHERIFF LOBO (II,1704)

WINSLOW, GEORGE
BLONDIE (I,486)

WINSLOW, MICHAEL
IRENE (II,1245); NIGHTSIDE (II,1851)

WINSLOW, MIKE
SPACE STARS (II,2408)

WINSLOW, PAULA
MY MOTHER THE CAR (I,3197); OUR MISS BROOKS (I,3415)

WINSTON, HATTIE
ANN IN BLUE (II,83); THE DAIN CURSE (II,613); NURSE (II,1870)

WINSTON, HELENE
FRIENDS AND LOVERS (II,930); KING OF KENSINGTON (II,1395); WHACKED OUT (II,2767)

WINSTON, JOHN
STAR TREK (II,2440)

WINSTON, LESLIE
B.J. AND THE BEAR (II,127); CENTENNIAL (II,477); MAKIN' IT (II,1603); THE WALTONS (II,2740); THE WALTONS: A DAY FOR THANKS ON WALTON'S MOUNTAIN (EPISODE 3) (II,2741); THE WALTONS: MOTHER'S DAY ON WALTON'S MOUNTAIN (EPISODE 2) (II,2742); THE WALTONS: WEDDING ON WALTON'S MOUNTAIN (EPISODE 1) (II,2743)

WINTER, CHARLES
'TWAS THE NIGHT BEFORE CHRISTMAS (II,2677)

WINTER, CHRISTINE
THE YEAR WITHOUT A SANTA CLAUS (II,2846)

WINTER, EDWARD
THE 25TH MAN (MS) (II,2678); ADAM'S RIB (I,32); THE ADVENTURES OF POLLYANNA (II,14); BIG DADDY (I,425); EMPIRE (II,777); FALCON CREST (II,808); FAMILY IN BLUE (II,818); THE FEATHER AND FATHER (II,846); THE FEATHER AND FATHER GANG (II,845); THE FEATHER AND FATHER GANG: NEVER CON A KILLER (II,846); FLY AWAY HOME (II,893); THE GIRL IN THE EMPTY GRAVE (II,997); JOE DANCER: THE BIG BLACK PILL (II,1300); KAREN (II,1363); M*A*S*H (II,1569); PROJECT UFO (II,2088); RENDEZVOUS HOTEL (II,2131); SAM (II,2246); THE SECOND TIME AROUND (II,2289); SOAP (II,2392); SOMERSET (I,4115); WAIT UNTIL DARK (II,2735); WOMAN ON THE RUN (II,2821)

WINTER, JANET
WOMEN IN WHITE (II,2823)

WINTER, LYNETTE
GIDGET (I,1795); PETTICOAT JUNCTION (I,3571)

WINTER, MEG
SOMERSET (I,4115)

WINTERS, BERNADETTE
BACHELOR FATHER (I,323); KAREN (I,2505)

WINTERS, DAVID
KINCAID (I,2543)

WINTERS, DEBORAH
THE PEOPLE NEXT DOOR (I,3519); THE WINDS OF WAR (II,2807)

WINTERS, GLORIA
THE LIFE OF RILEY (I,2685); SKY KING (I,4077); WHERE'S RAYMOND? (I,4837)

WINTERS, JANA
HEAR NO EVIL (II,1116)

WINTERS, JONATHAN
GUYS 'N' GEISHAS (I,1914); HERE'S THE SHOW (I,2033); THE JONATHAN WINTERS (I,2440); THE JONATHAN WINTERS (I,2441); THE JONATHAN WINTERS (I,2442); THE JONATHAN WINTERS (I,2443); THE JONATHAN WINTERS (I,2444); THE JONATHAN WINTERS (I,2445); JONATHAN WINTERS PRESENTS 200 YEARS OF AMERICAN HUMOR (II,1342); THE JONATHAN WINTERS SHOW (I,2440); THE JONATHAN WINTERS SHOW (I,2441); THE JONATHAN WINTERS SHOW (I,2442); THE JONATHAN WINTERS SHOW (I,2443); THE JONATHAN WINTERS SHOW (I,2444); THE JONATHAN WINTERS SPECIAL (I,2445); LINUS THE LIONHEARTED (I,2709); THE LONDON PALLADIUM (I,2739); MORE WILD WILD WEST (II,1733); MORK AND MINDY (II,1735); SHIRLEY TEMPLE'S STORYBOOK (I,4017); TAKE ONE STARRING JONATHAN WINTERS (II,2532); THE TWILIGHT ZONE (I,4651); THE WACKY WORLD OF JONATHAN WINTERS (I,4746); THE WONDERFUL WORLD OF JONATHAN WINTERS (I,4900)

WINTERS, LAWRENCE
ALL ABOUT MUSIC (I,127)

WINTERS, MARIAN
LAND OF HOPE (II,1424); STAR TONIGHT (I,4199)

WINTERS, ROLAND
ADVENTURES OF A MODEL (I,50); THE DAIN CURSE (II,613); DOC (I,1303); DOORWAY TO DANGER (I,1353); HELLO DERE (I,2010); THE KAISER ALUMINUM HOUR (I,2503); KRAFT TELEVISION THEATER (I,2592); LITTLE WOMEN (I,2722); MEET MILLIE (I,2988); MIRACLE ON 34TH STREET (I,3058); THE SMOTHERS BROTHERS SHOW (I,4096); TIME FOR ELIZABETH (I,4506)

WINTERS, SHELLEY
THE ADVENTURES OF NICK CARTER (I,68); ALCOA PREMIERE (I,109); BATMAN (I,366); BIG ROSE (II,241); THE BOB HOPE CHRYSLER THEATER (I,502); CLIMAX! (I,976); THE FRENCH ATLANTIC AFFAIR (II,924); FROSTY'S WINTER WONDERLAND (II,935); HAWAIIAN HEAT (II,1111); KOJAK (II,1406); KRAFT TELEVISION THEATER (I,2592); SCHLITZ PLAYHOUSE OF STARS (I,3936); THE WOMEN (I,4890)

WINTERSOLE, WILLIAM
THE LIFE AND TIMES OF EDDIE ROBERTS (II,1468); THE OUTSIDE MAN (II,1937); SARA (II,2260)

WINWOOD, ESTELLE
BEWITCHED (I,418); BLITHE SPIRIT (I,483); THE MOTOROLA TELEVISION HOUR (I,3112); TONIGHT AT 8:30 (I,4556)

WIPF, ALEX
ANOTHER WORLD (II,97)

WIRTH, SANDY
SUPER CIRCUS (I,4287)

WISBAR, FRANK
FIRESIDE THEATER (I,1580)

WISDOM, NORMAN
ANDROCLES AND THE LION (I,189)

WISE, ALFIE
CALL HER MOM (I,796); THE SANDY DUNCAN SHOW (I,3910); TRAUMA CENTER (II,2655); UNCLE CROC'S BLOCK (II,2702)

WISE, ERNIE
THE MORECAMBE AND WISE SHOW (II,1734)

WISE, RAY
LOVE OF LIFE (I,2774)

WISE, ROBERT
THE FURTHER ADVENTURES OF WALLY BROWN (II,944)

WISEMAN, JOSEPH
GREAT BIBLE ADVENTURES (I,1872); MASADA (II,1642); MEN OF THE DRAGON (II,1683); NIGHTSIDE (I,3296); THE OUTSIDER (I,3430)

WISHNER, SUZANNE
SORORITY '62 (II,2405)

WITHE, RUTH
HARVEY (I,1961)

WITHERS, GRANT
IT'S ALWAYS SUNDAY (I,2265); LAST STAGECOACH WEST (I,2629)

WITHERS, ISABEL
THE LIFE OF RILEY (I,2686)

WITHERS, JANE
THE APARTMENT HOUSE (I,241)

WITHERS, MARK
DYNASTY (II,746); KAZ (II,1370)

WITHERSPOON, DANE
THE HOME FRONT (II,1171); SANTA BARBARA (II,2258)

WITHROW, GLENN
RIDING FOR THE PONY EXPRESS (II,2172)

WITNEY, MICHAEL
THE CATCHER (I,872); MISS STEWART, SIR (I,3064); THE TRAVELS OF JAIMIE MCPHEETERS (I,4596)

WITT, DAVID
ANOTHER LIFE (II,95)

WITT, HOWARD
ALICE (II,33); I LOVE HER ANYWAY! (II,1214); MOTHER AND ME, M.D. (II,1740); THE SIX O'CLOCK FOLLIES (II,2368); W.E.B. (II,2732); WHACKED OUT (II,2767)

WITT, KATHRYN
FLYING HIGH (II,894); FLYING HIGH (II,895); MASSARATI AND THE BRAIN (II,1646); RENDEZVOUS HOTEL (II,2131)

WITT, ROZ
NO MAN'S LAND (II,1855)

WIXTED, MICHAEL JAMES

SANDBURG'S LINCOLN (I,3907); THE SMITH FAMILY (I,4092); SWISS FAMILY ROBINSON (II,2516)

WODD, BEVERLY
ROOTS (II,2211)

WODELL, VERNON
FILTHY RICH (II,856)

WOLCOTT, ROBBIE
THE BIONIC WOMAN (II,255)

WOLFE, IAN
AMERICA, YOU'RE ON (II,69); DANIEL BOONE (I,1142); THE DEFENDER (I,1219); IS THERE A DOCTOR IN THE HOUSE? (II,1247); MR. TUTT (I,3153); WIZARDS AND WARRIORS (II,2813)

WOLFE, ISOBEL
WILLOW B: WOMEN IN PRISON (II,2804)

WOLFE, JANICE
BREAK THE BANK (I,712)

WOLFE, KAREN
ANOTHER WORLD (II,97); BABES IN TOYLAND (I,318); THE COURTSHIP OF EDDIE'S FATHER (I,1083); DAYS OF OUR LIVES (II,629); PANAMA HATTIE (I,3444)

WOLFE, NEIL
ADVENTURES OF CLINT AND MAC (I,55)

WOLFE, WENDY
KATE AND ALLIE (II,1365); TUCKER'S WITCH (II,2667)

WOLFF, FRANK
FELLINI: A DIRECTOR'S NOTEBOOK (I,1562)

WOLFINGTON, IGGIE
ADAMS OF EAGLE LAKE (II,10); FATHER O FATHER (II,842); ONE TOUCH OF VENUS (I,3389); THE STRAWBERRY BLONDE (I,4254)

WOLFIT, SIR DONALD
THE GHOST SQUAD (I,1787)

WOLFMAN JACK
FONZ AND THE HAPPY DAYS GANG (II,898); WOLF ROCK TV (II,2815); THE WOLFMAN JACK RADIO SHOW (II,2816); THE WOLFMAN JACK SHOW (II,2817)

WOLPE, LENNY
THE DRUNKARD (II,737)

WOLTER, SHERILYN
B.J. AND THE BEAR (II,126); GENERAL HOSPITAL (II,964)

WOMACK, GEORGE
TWIGS (II,2684)

WONG, ANNA MAY
THE BARBARA STANWYCK THEATER (I,350); THE GALLERY OF MADAME LIU TSONG (I,1729); JOSEPHINE LITTLE: DRAGON BY THE TAIL (I,2451)

WONG, ARTHUR
KENTUCKY JONES (I,2526)

WONG, BEAL
BACHELOR FATHER (I,323)

WONG, CAREY
KELLY'S KIDS (I,2523); KID POWER (I,2532)

WONG, CHARLENE
ESCAPE (I,1461)

WONG, CHRISTOPHER
YES, VIRGINIA, THERE IS A SANTA CLAUS (II,2849)

WONG, CURTIS
THE NEW MICKEY MOUSE CLUB (II,1823)

WONG, INGRID
FANTASY ISLAND (II,829)

WONG, JOE
THE FURTHER ADVENTURES OF SPIN AND MARTY (I,1720); THE NEW ADVENTURES OF SPIN AND MARTY (I,3254); SPIN AND MARTY (I,4158)

WONG, STANLEY
THE AMAZING CHAN AND THE CHAN CLAN (I,157)

WOOD, ALAN G
FROM HERE TO ETERNITY (II,932); MR. ADAMS AND EVE (I,3121)

WOOD, ALLEN
THE DOOLEY BROTHERS (II,715)

WOOD, ANDY
JOE DANCER: THE MONKEY MISSION (II,1301)

WOOD, BARRY
BACKSTAGE WITH BARRY WOOD (I,329); PLACES PLEASE (I,3613)

WOOD, BERT
SCRUPLES (II,2280)

WOOD, CHUCK
JULIA (I,2476)

WOOD, G
HOLMES AND YOYO (II,1169); M*A*S*H (II,1569); ON BORROWED TIME (I,3354)

WOOD, GENE
ANYTHING YOU CAN DO (I,239); BEAT THE CLOCK (I,378)

WOOD, JANET
FROM HERE TO ETERNITY (II,933); HOLLYWOOD HIGH (II,1163); TOPPER (II,2649); THE YOUNG AND THE RESTLESS (II,2862)

WOOD, JEANE
ANOTHER DAY, ANOTHER DOLLAR (I,228); DOBIE GILLIS (I,1302); PANDORA AND FRIEND (I,3445)

WOOD, KELLY
AS THE WORLD TURNS (II,110); THE PATTY DUKE SHOW (I,3483); SEARCH FOR TOMORROW (II,2284)

WOOD, KRISTI JILL
GETTING THERE (II,985)

WOOD, LANA
CAPITOL (II,426); CAPTAIN AMERICA (II,428); COREY:(FOR THE PEOPLE (II,575); THE LONG HOT SUMMER (I,2747); O'HARA, UNITED STATES TREASURY: OPERATION COBRA (I,3347); THE OVER-THE-HILL GANG RIDES AGAIN (I,3432)

WOOD, LEN
THE AMAZING CHAN AND THE CHAN CLAN (I,157)

WOOD, LESLIE
DOG AND CAT (II,695); GENERAL HOSPITAL (II,964)

WOOD, LYNN
DUSTY'S TRAIL (I,1390); STONE (II,2470); THE YOUNG AND THE RESTLESS (II,2862)

WOOD, NAPUNA
PAPA G.I. (I,3451)

WOOD, NATALIE
CONFLICT (I,1036); FORD TELEVISION THEATER (I,1634); FROM HERE TO ETERNITY (II,932); HART TO HART (II,1101); HEIDI (I,2004); THE KAISER ALUMINUM HOUR (I,2503); PRIDE OF THE FAMILY (I,3667); SCHAEFER CENTRUY THEATER (I,3935); STUDIO '57 (I,4267)

WOOD, PEGGY
I REMEMBER MAMA (I,2176); PHILCO TELEVISION PLAYHOUSE (I,3583); PULITZER PRIZE PLAYHOUSE (I,3692)

WOOD, RICHARD
DARK SHADOWS (I,1157)

WOOD, SEAN
SOMERSET (I,4115)

WOOD, TERRI LYNN
THE BAXTERS (II,183); CHRISTMAS IN DISNEYLAND (II,519); ROYCE (II,2224)

WOOD, WARD
MANNIX (II,1624); PURSUE AND DESTROY (I,3699)

WOOD, WILLIAM
WINDOW SHOPPING (I,4875)

WOODARD, ALFRE
TUCKER'S WITCH (II,2667)

WOODBARY, FRANCES
THE WOMEN (I,4890)

WOODBRIDGE, GEORGE
ADVENTURES OF CLINT AND MAC (I,55); THE FORSYTE SAGA (I,1645)

WOODBURNE, JACKIE
1915 (II,1853)

WOODBURY, WOODY
WHO DO YOU TRUST? (I,4846); THE WOODY WOODBURY SHOW (I,4905)

WOODELL, PAT
PETTICOAT JUNCTION (I,3571)

WOODESON, NICHOLAS
PIAF (II,2039)

WOODNER, RUTH
DAYDREAMING WITH LARAINE (I,1184)

WOODRUFF, MAURICE
MAURICE WOODRUFF PREDICTS (I,2956)

WOODS, DONALD
CRAIG KENNEDY, CRIMINOLOGIST (I,1089); DAMON RUNYON THEATER (I,1129); HOTEL COSMOPOLITAN (I,2128); THE ORCHID AWARD (I,3406); PORTIA FACES LIFE (I,3653); STAR STAGE (I,4197); TAMMY (I,4357)

WOODS, GRANT
THE LEGEND OF CUSTER (I,2652); STAR TREK (II,2440)

WOODS, JAMES
ALL THE WAY HOME (I,140); DISAPPEARANCE OF AIMEE (I,1290)

WOODS, LESLEY
BRIGHT PROMISE (I,727); DEAR DETECTIVE (II,652); YOUNG DR. MALONE (I,4938)

WOODS, LINDA
OUR MAN FLINT: DEAD ON TARGET (II,1932)

WOODS, MICHAEL
BARE ESSENCE (II,150); TEXAS (II,2566)

WOODS, REN

IT CAN'T HAPPEN TO ME (II,1250); ROOTS (II,2211); WE'VE GOT EACH OTHER (II,2757)

WOODS, RICHARD
RAGE OF ANGELS (II,2107); SHERLOCK HOLMES (II,2327)

WOODS, ROBERT
ONE LIFE TO LIVE (II,1907)

WOODS, SARA
SWEENEY TODD (II,2512)

WOODSON, WILLIAM
THE C.B. BEARS (II,406); F TROOP (I,1499); THE GREATEST SHOW ON EARTH (I,1883); THE NEW SUPER FRIENDS HOUR (II,1830); THE SHIRT TALES (II,2337); SUPER FRIENDS (II,2497)

WOODVILLE, KATE
DAYS OF OUR LIVES (II,629); EIGHT IS ENOUGH (II,762); THE HEALERS (II,1115); THE RHINEMANN EXCHANGE (II,2150)

WOODVINE, JOHN
NICHOLAS NICKLEBY (II,1840); THE TALE OF BEATRIX POTTER (II,2533)

WOODWARD, ALAN
A CHRISTMAS CAROL (II,517)

WOODWARD, BOB
ANNIE OAKLEY (I,225)

WOODWARD, CHARLAINE
AIN'T MISBEHAVIN' (II,25)

WOODWARD, CHRISTINE
CINDERELLA (II,526)

WOODWARD, JOANNE
THE ALCOA HOUR (I,108); ALFRED HITCHCOCK PRESENTS (I,115); ALL THE WAY HOME (I,140); THE ARMSTRONG CIRCLE THEATER (I,260); CANDIDA (II,423); FORD TELEVISION THEATER (I,1634); FOUR STAR PLAYHOUSE (I,1652); LITTLE WOMEN (I,2724); KRAFT TELEVISION THEATER (I,2592); PLAYHOUSE 90 (I,3623) STAR TONIGHT (I,4199); STUDIO ONE (I,4268)

WOODWARD, JONATHAN B
FROM HERE TO ETERNITY (II,932)

WOODWARD, KIMBERLY
LOVE, NATALIE (II,1545)

WOODWARD, LENORE
THE EYES OF TEXAS (II,799)

WOODWARD, MORGAN
CENTENNIAL (II,477); CHARLIE WOOSTER—OUTLAW (I,915); THE CIRCLE FAMILY (II,527); DALLAS (II,614); DOC ELLIOT (I,1305); HILL STREET BLUES (II,1154); HOW THE WEST WAS WON (II,1196); THE LIFE AND LEGEND OF WYATT EARP (I,2681); LOGAN'S RUN (II,1507); PISTOLS 'N' PETTICOATS (I,3608); THE QUEST (II,2096); YUMA (I,4976)

WOODWARD, PETER
FANNY BY GASLIGHT (II,823)

WOODWORTH, LEONA
TELETIPS ON LOVELINESS (I,4380)

WOOLDRIDGE, IAN
GAMES PEOPLE PLAY (II,956)

WOOLERY, CHUCK
THE LOVE CONNECTION (II,1536); THE NEW ZOO REVUE (I,3277); SCRABBLE (II,2278); WHEEL OF FORTUNE (II,2775); YOUR HIT PARADE (II,2872)

WOOLEY, SHEB
MUSIC U.S.A. (I3179); RAWHIDE (I,3727)

WOOLF, CHARLES
CARLTON YOUR DOORMAN (II,440); DOROTHY IN THE LAND OF OZ (II,722)

WOOLFE, BETTY
AN ECHO OF THERESA (I,1399)

WOOLFE, ERIC
THE STRAUSS FAMILY (I,4253)

WOOLFOLK, ANDREW
EARTH, WIND & FIRE IN CONCERT (II,750)

WOOLLEN, SUSAN
B.J. AND THE BEAR (II,126); RETURN OF THE MAN FROM U.N.C.L.E.: THE 15 YEARS LATER AFFAIR (II,2140)

WOOLLEY, MONTY
THE MAN WHO CAME TO DINNER (I,2880)

WOOLTON, STEPHEN
DO IT YOURSELF (I,1299); THIS IS ALICE (I,4453)

WOOTTON, STEPHEN
LEAVE IT TO BEAVER (I,2648)

WOPAT, TOM
BATTLE OF THE NETWORK STARS (II,174); BATTLE OF THE NETWORK STARS (II,176); CELEBRITY CHALLENGE OF THE SEXES 5 (II,469); THE DUKES (II,741); THE DUKES OF HAZZARD (II,742)

WORDSWORTH, RICHARD
ROMEO AND JULIET (I,3837)

WORKMAN, LINDSAY
HERE COME THE BRIDES (I,2024); OWEN MARSHALL: COUNSELOR AT LAW (I,3435)

WORLEY, JO ANNE
CIRCUS OF THE STARS (II,530); DOBIE GILLIS (I,1302); THROUGH THE MAGIC PYRAMID (II,2615)

WORLOCK, FREDERIC
HONESTLY, CELESTE! (I,2104); KING RICHARD II (I,2559); REUNION IN VIENNA (I,3780)

WORSHAM, MARIE
KITTY FOYLE (I,2571)

WORTH, BRIAN
THE ADVENTURES OF ROBIN HOOD (I,74)

WORTH, HARRY
THE GUARDSMAN (I,1893)

WORTH, HELEN
CORONATION STREET (I,1054)

WORTH, IRENE
SEPARATE TABLES (II,2304)

WORTHINGTON, CAROL
THE DORIS DAY SHOW (I,1355); ROOM 222 (I,3843)

WORTHINGTON, CATHY
FARRELL: FOR THE PEOPLE (II,834)

WOVEN, DAN
THE INVISIBLE WOMAN (II,1244)

WRAY, FAY
ALFRED HITCHCOCK PRESENTS (I,115); DAMON RUNYON THEATER (I,1129); THE DAVID NIVEN THATER (I,1178); THE GENERAL ELECTRIC THEATER (I,1753); IT'S ALWAYS SUNDAY (I,2265); JANE WYMAN PRESENTS THE FIRESIDE THEATER (I,2345); KRAFT TELEVISION THEATER (I,2592); PLAYHOUSE 90 (I,3623); PRIDE OF THE FAMILY (I,3667)

WREN, SAM
WREN'S NEST (I,4922)

WREN, VIRGINIA
WREN'S NEST (I,4922)

WRIGHT JR, COBINA
COBINA (I,989)

WRIGHT, ADRIAN
ALL THE RIVERS RUN (II,43)

WRIGHT, AMY
A FINE ROMANCE (II,858)

WRIGHT, BEN
79 PARK AVENUE (II,2310); BORDER TOWN (I,689); CHARLIE ANGELO (I,906); MCNAMARA'S BAND (II,1668); PROBE (I,3674); THE RHINEMANN EXCHANGE (II,2150); TURNOVER SMITH (II,2671)

WRIGHT, BOBBY
SEVEN AGAINST THE SEA (I,3983)

WRIGHT, BRUCE
BATTLESTAR GALACTICA (II,181); MR. ROBERTS (II,1754)

WRIGHT, ED
STRANGER IN OUR HOUSE (II,2477)

WRIGHT, HOWARD
HARD CASE (I,1947)

WRIGHT, J J
PAROLE (II,1960)

WRIGHT, JANET
THE BEST CHRISTMAS PAGEANT EVER (II,212)

WRIGHT, JOHN
MCHALE'S NAVY (I,2969)

WRIGHT, JOHNNY
THE KITTY WELLS/JOHNNY WRIGHT FAMILY SHOW (I,2572)

WRIGHT, MAGGIE
THE MARTIAN CHRONICLES (II,1635)

WRIGHT, MARTHA
LET'S DANCE (I,2664); THE MARTHA WRIGHT SHOW (I,2927); SOLDIER PARADE (I,4109)

WRIGHT, MARY CATHERINE
GOSSIP (II,1044)

WRIGHT, MAX
BUFFALO BILL (II,387); I GAVE AT THE OFFICE (II,1213); TALES FROM THE DARKSIDE (II,2534)

WRIGHT, MICHAEL
V (II,2713); V: THE FINAL BATTLE (II,2714); V: THE SERIES (II,2715)

WRIGHT, NANCY
YOUR LUNCHEON DATE (I,4955)

WRIGHT, PAULA
DOUGLAS FAIRBANKS JR. PRESENTS THE RHEINGOLD THEATER (I,1364)

WRIGHT, ROBERT
ROSALINDA (I,3846); THE YEOMAN OF THE GUARD (I,4926)

WRIGHT, ROBIN
SANTA BARBARA (II,2258)

WRIGHT, ROY
THE ISLANDERS (I,2243)

WRIGHT, SAM
BALL FOUR (II,137); ENOS (II,779)

WRIGHT, TERESA
THE DESPERATE HOURS (I,1239); THE DEVIL'S DISCIPLE (I,1247); CLIMAX! (I,976); THE ELGIN HOUR (I,1428); FORD TELEVISION THEATER (I,1634); FOUR STAR PLAYHOUSE (I,1652); HENRY FONDA PRESENTS THE STAR AND THE STORY (I,2016); HOLLYWOOD OPENING NIGHT (I,2087); INTERMEZZO (I,2223); MIRACLE ON 34TH STREET (I,3056); RHEINGOLD THEATER (I,3783); ROBERT MONTGOMERY PRESENTS YOUR LUCKY STRIKE THEATER (I,3809); SCREEN DIRECTOR'S PLAYHOUSE (I,3946); THE UNITED STATES STEEL HOUR (I,4677)

WRIGHT, WARREN
JUNIOR HIGH JINKS (I,2492)

WRIGHT, WILL
DENNIS THE MENACE (I,1231); THE DICK VAN DYKE SHOW (I,1275); PINE LAKE LODGE (I,3597)

WRIGHTSON, EARL
THE AT HOME SHOW (I,308); THE EARL WRIGHTSON SHOW (I,1393); THE GOODYEAR REVUE (I,1856); MASLAND AT HOME (I,2939)

WROE, TRUDY
BIG TOWN (I,436)

WU, CHRISTIANA
BROTHERS (II,381)

WULFF, KAI
CASABLANCA (II,450)

WYATT, JANE
ALCOA PREMIERE (I,109); THE BOB HOPE CHRYSLER THEATER (I,502); CONFIDENTIAL FOR WOMEN (I,1034); FATHER KNOWS BEST (II,838); FATHER KNOWS BEST: HOME FOR CHRISTMAS (II,839); FATHER KNOWS BEST: THE FATHER KNOWS BEST REUNION (II,840); FIRESIDE THEATER (I,1580); THE GENERAL ELECTRIC THEATER (I,1753); LIGHT'S OUT (I,2699); THE MILLIONAIRE (II,1700); THE MOTOROLA TELEVISION HOUR (I,3112); THE NASH AIRFLYTE THEATER (I,3227); PLAYWRIGHTS '56 (I,3627); ROBERT MONTGOMERY PRESENTS YOUR LUCKY STRIKE THEATER (I,3809); ST. ELSEWHERE (II,2432); STAR TREK (II,2440); STUDIO ONE (I,4268)

WYATT, SHARON
GENERAL HOSPITAL (II,964)

WYATT, TESSA
ROBIN'S NEST (II,2187)

WYCHERLY, MARGARET
CLAUDIA: THE STORY OF A MARRIAGE (I,972)

WYENN, THANN
PETE KELLY'S BLUES (I,3559); UNIT 4 (II,2707); WON'T IT EVER BE MORNING? (I,4902)

WYLAND, LEN
FIRST LOVE (I,1583)

WYLDE, PIPPA
CINDERELLA (II,526)

WYLDECK, MARTIN
SPELL OF EVIL (I,4152)

WYLE, GEORGE
MADAME'S PLACE (II,1587)

WYLE, MICHAEL
ALL NIGHT RADIO (II,40)

WYLER, GRETCHEN
THE ADVENTURES OF POLLYANNA (II,14); DALLAS (II,614); FOR MEMBERS ONLY (II,905); GHOST OF A CHANCE (II,987); ON OUR OWN (II,1892); SOMERSET (I,4115); STEP THIS WAY (I,4217)

WYLER, JOHN
THE SINGLE LIFE (II,2364)

WYLER, RICHARD
THE MAN FROM INTERPOL (I,2866); RIVAK, THE BARBARIAN (I,3795)

WYLEY, DAVID
USED CARS (II,2712)

WYLIE, FRANK
NOT GUILTY! (I,3311)

WYLIE, ROWAN
CASANOVA (II,451)

WYLLIE, MEG
DAYS OF OUR LIVES (II,629); THE ELISABETH MCQUEENEY STORY (I,1429); HOLLOWAY'S DAUGHTERS (I,2079); MOVIN' ON (I,3118); PROFILES IN COURAGE (I,3678); SUTTERS BAY (II,2508); THE THORN BIRDS (II,2600); THE TRAVELS OF JAIMIE MCPHEETERS (I,4596)

WYMAN, JANE
ACADEMY AWARD SONGS (I,13); AMANDA FALLON (I,151); AMANDA FALLON (I,152); FALCON CREST (II,808); JANE WYMAN PRESENTS THE FIRESIDE THEATER (I,2345); THE JANE WYMAN THEATER (I,2346); JANE WYMAN'S SUMMER PLAYHOUSE (I,2347); THE LUX VIDEO THEATER (I,2795)

WYMARK, PATRICK
THE SCARECROW OF ROMNEY MARSH (I,3931)

WYMORE, PATRICE
THE ERROL FLYNN THEATER (I,1458); NEVER TOO YOUNG (I,3248)

WYNANT, H M
THE GEMINI MAN (II,961); KNIGHT'S GAMBIT (I,2576)

WYNDHAM, ANNE
BARNEY MILLER (II,154); GENERAL HOSPITAL (II,964); SEARCH FOR TOMORROW (II,2284)

WYNDHAM, VICTORIA
ANOTHER WORLD (II,97); FOR RICHER, FOR POORER (II,906); THE GUIDING LIGHT (II,1064)

WYNER, GEORGE
AT EASE (II,115); BIG SHAMUS, LITTLE SHAMUS (II,242); CASS MALLOY (II,454); DELVECCHIO (II,659); DUFFY (II,739); FRIENDS (II,928); HILL STREET BLUES (II,1154); THE ISLANDER (II,1249); KAZ (II,1370); LUCAN (II,1553); MATT HOUSTON (II,1654); NERO WOLFE (II,1806); SHEILA (II,2325); THE SUNSHINE BOYS (II,2496)

WYNGARDE, PETER
DEPARTMENT S (I,1232); GALLAGHER (I,1727); ONE STEP BEYOND (I,3388); THE PRISONER (I,3670)

WYNN, ED
THE ED WYNN SHOW (I,1403); THE ED WYNN SHOW (I,1404); THE GENERAL ELECTRIC THEATER (I,1753); MEET ME IN ST. LOUIS (I,2986); MIRACLE ON 34TH STREET (I,3057); ON BORROWED TIME (I,3354); THE TWILIGHT ZONE (I,4651); YOU'RE ONLY YOUNG TWICE (I,4971)

WYNN, FRED
CODE 3 (I,993)

WYNN, HELEN
ART AND MRS. BOTTLE (I,271)

WYNN, KEENAN
ALCOA/GOODYEAR THEATER (I,107); ASSIGNMENT: MUNICH (I,302); THE BASTARD/KENT FAMILY CHRONICLES (II,159); THE BILLION DOLLAR THREAT (II,246); BROADWAY (I,733); CALL TO GLORY (II,413); CANNON (I,820); DALLAS (II,614); THE DESILU PLAYHOUSE (I,1237); THE DICK POWELL SHOW (I,1269); FORD TELEVISION THEATER (I,1634); THE HOUSE ON GREENAPPLE ROAD (I,2140); JEREMIAH OF JACOB'S NECK (II,1281); JOE DANCER: THE MONKEY MISSION (II,1301); THE JOKE AND THE VALLEY (I,2438); THE JUNE ALLYSON SHOW (I,2488); LAST OF THE PRIVATE EYES (I,2628); THE LOSERS (I,2757); ON TRIAL (I,3363); PATRICK STONE (I,3476); THE PIRATES OF FLOUNDER BAY (I,3607); THE POWER AND THE GLORY (I,3658); THE QUEST (II,2096); RETURN OF THE MAN FROM U.N.C.L.E.: THE 15 YEARS LATER AFFAIR (II,2140); SANTA CLAUS IS COMIN' TO TOWN (II,2259); SCHLITZ PLAYHOUSE OF STARS (I,3963); SCREEN DIRECTOR'S PLAYHOUSE (I,3946); TEXACO COMMAND PERFORMANCE (I,4407); THE TROUBLESHOOTERS (I,4614); TWENTIETH CENTURY (I,4639); THE TWILIGHT ZONE (I,4651); THE UNITED STATES STEEL HOUR (I,4677); WARNING SHOT (I,4770); THE WESTERNERS (I,4802); THE YOUNG LAWYERS (I,4940)

WYNN, KITTY
BEACON HILL (II,193)

WYNN, MAY
NOAH'S ARK (I,3302)

WYNN, PETER
ACADEMY THEATER (I,14)

WYNN, TERRI
COVER UP (II,597)

WYNTER, DANA

THE BOB HOPE CHRYSLER THEATER (I,502); LAURA (I,2635); THE LIVES OF JENNY DOLAN (II,1502); M STATION: HAWAII (II,1568); THE MAN WHO NEVER WAS (I,2882); OWEN MARSHALL: COUNSELOR AT LAW (I,3436); PLAYHOUSE 90 (I,3623); TENSPEED AND BROWN SHOE (II,2562)

WYSS, AMANDA
CASS MALLOY (II,454); LONE STAR (II,1513); STAR OF THE FAMILY (II,2436); TEACHERS ONLY (II,2547); THE TOM SWIFT AND LINDA CRAIG MYSTERY HOUR (II,2633)

YAGHER, JEFF
V: THE SERIES (II,2715)

YAGHIIJIAN, KURT
AMAHL AND THE NIGHT VISITORS (I,149)

YAGI, JAMES
DOBIE GILLIS (I,1302)

YAH LING SUN, IRENE
INSPECTOR PEREZ (II,1237); THE QUEST (II,2096); SAVAGE: IN THE ORIENT (II,2264)

YAHEE
THE EDGE OF NIGHT (II,760)

YAMA, MICHAEL
HOTEL (II,1192)

YAMADA, ROMI
GUYS 'N' GEISHAS (I,1914)

YAMAHA, MICHAEL
THE LEGEND OF THE GOLDEN GUN (II,1455)

YAMAMOTO, RINICHI
SHOGUN (II,2339)

YANCY, EMILY
POOR DEVIL (I,3646); SECOND CHANCE (I,3958)

YANEZ, DAVID
BIG CITY BOYS (II,232); FISH (II,864); WALKIN' WALTER (II,2737)

YANGHA, THOMAS
TARZAN THE APE MAN (I,4363)

YAPP, PETER
MISTER JERICO (I,3070)

YARBOROUGH, BARTON
DRAGNET (I,1369)

YARDLEY, STEPHEN
CRY TERROR! (I,1112); FANNY BY GASLIGHT (II,823)

YARDUM, JET
DEAR DETECTIVE (II,652)

YARLETT, CLAIRE
RITUALS (II,2182)

YARMY, DICK
THE BUREAU (II,397); DUFFY (II,739); HANDLE WITH CARE (II,1077); WILD ABOUT HARRY (II,2797)

YARNELL, BRUCE
ANNIE GET YOUR GUN (I,224); THE GOOD OLD DAYS (I,1851); THE OUTLAWS (I,3427); OUTPOST (I,3428)

YARNELL, CELESTE
RANSOM FOR A DEAD MAN (I,3722)

YARNELL, LORENE
SHIELDS AND YARNELL (II,2328); THE WILD WILD WEST REVISITED (II,2800)

YARWOOD, RICHARD
CAGNEY AND LACEY (II,410)

YASHIMA, MOMO
BEHIND THE SCREEN (II,200); FITZ AND BONES (II,867); THE RETURN OF MARCUS WELBY, M.D (II,2138)

YASHIRO, MAYAKO
SPACE GIANTS (I,4142)

YATES, CASSIE
THE FAMILY TREE (II,820); HAVING BABIES II (II,1108); KNOTS LANDING (II,1404); NOBODY'S PERFECT (II,1858); NORMA RAE (II,1860); RICH MAN, POOR MAN—BOOK II (II,2162)

YATES, CURTIS
SAINT PETER (II,2238)

YATES, JOHN ROBERT
CAMP GRIZZLY (II,418)

YATES, PAULINE
THE FALL AND RISE OF REGINALD PERRIN (II,810)

YATES, STEPHEN
ANOTHER WORLD (II,97); THE GUIDING LIGHT (II,1064)

YEAGER, BILL
ANOTHER MAN'S SHOES (II,96)

YEE, LINDA
HEAR NO EVIL (II,1116)

YEMM, NORMAN
HOMICIDE (I,2101); THE SULLIVANS (II,2490)

YIP, WILLIAM
JOHNNY GUITAR (I,2427)

YLADMAN, LYNN
CASANOVA (II,451)

YNFANTE, CHARLES
THEY ONLY COME OUT AT NIGHT (II,2586)

YNIQUEZ, RICHARD
ALMOST AMERICAN (II,55); MAMA MALONE (II,1609)

YODA, YOSHIO
MCHALE'S NAVY (I,2969)

YOHN, ERICA
THE 13TH DAY: THE STORY OF ESTHER (II,2593); THE FIGHTING NIGHTINGALES (II,855); MARRIAGE IS ALIVE AND WELL (II,1633)

YONG, SHI
PEKING ENCOUNTER (II,1991)

YORDANOFF, WLADIMIR
MASTER OF THE GAME (II,1647)

YORK, DICK
BEWITCHED (I,418); GOING MY WAY (I,1833); HIGH SCHOOL, U.S.A. (II,1149); THRILLER (I,4492); THE TWILIGHT ZONE (I,4651)

YORK, ELIZABETH
PORTIA FACES LIFE (I,3653)

YORK, FRANCINE
THE BARBARY COAST (II,146); CAP'N AHAB (I,824); THE COURTSHIP OF EDDIE'S FATHER (I,1083); I LOVE A MYSTERY (I,2170); PIONEER SPIRIT (I,3604); SLATTERY'S PEOPLE (I,4082)

YORK, HELEN
PADDY THE PELICAN (I,3442); UNCLE MISTLETOE AND HIS ADVENTURES (I,4669)

YORK, JAMES
THE COPS AND ROBIN (II,573)

YORK, JEFF
THE ALASKANS (I,106); DAVY CROCKETT (I,1181); HENRY FONDA PRESENTS THE STAR AND THE STORY (I,2016); THE SAGA OF ANDY BURNETT (I,3883); TAMMY (I,4357)

YORK, KATHLEEN
I GAVE AT THE OFFICE (II,1213)

YORK, MICHAEL
CIRCUS OF THE STARS (II,531); A MAN CALLED INTREPID (II,1612); PARADE OF STARS (II,1954); TWILIGHT THEATER (II,2685)

YORK, REBECCA
DEAR TEACHER (II,653); MR. MOM (II,1752); NEWHART (II,1835)

YORK, SUSANNAH
JANE EYRE (I,2339)

YORKE, ALAN
THE NURSES (I,3319)

YORKE, RUTH
THE GOLDBERGS (I,1836); HUMAN ADVENTURE (I,2157)

YORTY, SAM
FORCE SEVEN (II,908)

YOSHIOKA, ADELE
THE RETURN OF CHARLIE CHAN (II,2136); ROWAN AND MARTIN'S LAUGH-IN (I,3856); THEY ONLY COME OUT AT NIGHT (II,2586)

YOTHERS, BUMPER
THE UGILY FAMILY (II,2699)

YOTHERS, POINDEXTER
FULL HOUSE (II,936)

YOTHERS, TINA
THE CHEROKEE TRAIL (II,500); DOMESTIC LIFE (II,703); FAMILY TIES (II,819); FATHER MURPHY (II,841)

YOUENS, BERNARD
CORONATION STREET (I,1054)

YOUNG AMERICANS, THE
THE BING CROSBY SHOW (I,465); THE PERRY COMO CHRISTMAS SHOW (I,3529); SOMETHING SPECIAL (I,4118); TEXACO STAR PARADE I (I,4409)

YOUNG RASCALS, THE
LESLIE (I,2659)

YOUNG, AGNES
THE GIRLS (I,1811)

YOUNG, ALAN
THE ALAN YOUNG SHOW (I,104); ALVIN AND THE CHIPMUNKS (II,61); BATTLE OF THE PLANETS (II,179); BEAUTY AND THE BEAST (II,197); FROG PRINCE (I,1685); THE LIFE OF VERNON HATHAWAY (I,2687); MATINEE THEATER (I,2947); MR. ED (I,3137); PINE LAKE LODGE (I,3597); THE RAY CONNIFF CHRISTMAS SPECIAL (I,3732); THE SATURDAY NIGHT REVUE (I,3921); SITCOM (II,2367); THE SMURFS (II,2385); TENNESSEE ERNIE FORD MEETS KING ARTHUR (I,4397)

YOUNG, BRUCE A
E/R (II,748)

YOUNG, BUCK
GIDGET'S SUMMER REUNION (I,1799); THE STRANGER (I,4251)

YOUNG, BURT
MURDER CAN HURT YOU! (II,1768)

YOUNG, CARLETON G
LEAVE IT TO BEAVER (I,2648)

YOUNG, CARMEN
FOG (II,897)

YOUNG, DAVID
ALL THE RIVERS RUN (II,43)

YOUNG, DEAN
THE FANTASTIC FUNNIES (II,825)

YOUNG, DEE DEE
THE BILL COSBY SHOW (I,442)

YOUNG, DENNIS
GENESIS II (I,1760)

YOUNG, DEY
SITCOM (II,2367)

YOUNG, DONALD
LYNDA CARTER: ENCORE (II,1566)

YOUNG, DONNA JEAN
ON THE FLIP SIDE (I,3360); PIONEER SPIRIT (I,3604); ROWAN AND MARTIN'S LAUGH-IN (I,3856); VAUDEVILLE (II,2722)

YOUNG, DOUG
THE HUCKLEBERRY HOUND SHOW (I,2155); THE QUICK DRAW MCGRAW SHOW (I,3710)

YOUNG, EVE
THE BOB SMITH SHOW (I,668); MUSICAL MERRY-GO-ROUND (I,3184); THE R.C.A. THANKSGIVING SHOW (I,3736)

YOUNG, FARON
COUNTRY COMES HOME (II,585); COUNTRY MUSIC CARAVAN (I,1066); JERRY REED AND SPECIAL FRIENDS (II,1285)

YOUNG, FELICITY
KRAFT MYSTERY THEATER (I,2589)

YOUNG, GIG
ALCOA/GOODYEAR THEATER (I,107); CASABLANCA (I,860); CONFLICT (I,1036); GIBBSVILLE (II,988); KING'S ROW (I,2563); KRAFT SUSPENSE THEATER (I,2591); NINOTCHKA (I,3299); THE PHILADELPHIA STORY (I,3582); ROBERT MONTGOMERY PRESENTS YOUR LUCKY STRIKE THEATER (I,3809); THE ROGUES (I,3832); SILVER THEATER (I,4051); THE SPIRAL STAIRCASE (I,4160); TONIGHT AT 8:30 (I,4556); THE TURNING POINT OF JIM MALLOY (II,2670); WARNER BROTHERS PRESENTS (I,4769)

YOUNG, HEATHER
LAND OF THE GIANTS (I,2611); OH, NURSE! (I,3343)

YOUNG, J S
LITTLE HOUSE: BLESS ALL THE DEAR CHILDREN (II,1489)

YOUNG, JANIS
ANOTHER WORLD (II,97); OUR FIVE DAUGHTERS (I,3413)

YOUNG, JOE
PAUL WHITEMAN'S SATURDAY NIGHT REVUE (I,3490)

YOUNG, JUDY
WELCOME BACK, KOTTER (II,2761)

YOUNG, KEONE
JERRY (II,1283)

YOUNG, LEIGH TAYLOR
THE HAMPTONS (II,1076)

YOUNG, LORETTA
A LETTER TO LORETTA (I,2669); THE LORETTA YOUNG SHOW (I,2755); THE LORETTA YOUNG THEATER (I,2756)

YOUNG, MAURICE
THE LONER (I,2744)

YOUNG, MICHAEL
KIDS ARE PEOPLE TOO (II,1390); SOAP WORLD (II,2394)

YOUNG, NORMA
CELEBRITY (II,463)

YOUNG, OTIS
THE OUTCASTS (I,3425)

YOUNG, RALPH
KRAFT MUSIC HALL PRESENTS SANDLER AND YOUNG (I,2585)

YOUNG, RAY
BIG FOOT AND WILD BOY (II,237); IS THERE A DOCTOR IN THE HOUSE? (II,1247); LI'L ABNER (I,2702); THE RETURN OF THE BEVERLY HILLBILLIES (II,2139)

YOUNG, ROBERT
CELEBRATION: THE AMERICAN SPIRIT (II,461); CLIMAX! (I,976); FATHER KNOWS BEST (II,838); FATHER KNOWS BEST: HOME FOR CHRISTMAS (II,839); FATHER KNOWS BEST: THE FATHER KNOWS BEST REUNION (II,840); HOLLOWAY'S DAUGHTERS (I,2079); LITTLE WOMEN (II,1497); LITTLE WOMEN (II,1498); MARCUS WELBY, M.D. (I,2905); MARCUS WELBY, M.D. (II,1627); THE RETURN OF MARCUS WELBY, M.D (II,2138); ROBERT YOUNG AND THE FAMILY (I,3813); ROBERT YOUNG WITH THE YOUNG (I,3814); A SALUTE TO TELEVISION'S 25TH ANNIVERSARY (I,3893); WINDOW ON MAIN STREET (I,4873)

YOUNG, ROLAND
THE CHEVROLET TELE-THEATER (I,926); THE DOCTOR (I,1320); HOLLYWOOD OPENING NIGHT (I,2087); PULITZER PRIZE PLAYHOUSE (I,3692)

YOUNG, RONNIE
THE MICKEY MOUSE CLUB (I,3025)

YOUNG, SKIP
THE ADVENTURES OF OZZIE AND HARRIET (I,71)

YOUNG, STEPHEN
JUDD, FOR THE DEFENSE (I,2463); LEADFOOT (II,1452); OH, NURSE! (I,3343); SEAWAY (I,3956)

YOUNG, SUE
THE BENNY HILL SHOW (II,207)

YOUNG, TONY
GUNSLINGER (I,1908); WHEN THE WEST WAS FUN: A WESTERN REUNION (II,2780)

YOUNG, TRUDY
STRANGE PARADISE (I,4246)

YOUNG, VICTORIA
THE LITTLE PEOPLE (I,2716)

YOUNG, WILLIAM ALLEN
THE BOYS IN BLUE (II,359); UNIT 4 (II,2707); WHITE AND RENO (II,2787)

YOUNGER, BEVERLY
THE BENNETS (I,397); THE GOOK FAMILY (I,1858); STUD'S PLACE (I,4271); A TIME TO LIVE (I,4510)

YOUNGFELLOW, BARRIE
A.E.S. HUDSON STREET (II,17); ANNIE FLYNN (II,92); THE BAY CITY AMUSEMENT COMPANY (II,185); IT'S A LIVING (II,1254); LOCAL 306 (II,1505); MAKING A LIVING (II,1604); THE SINGLE LIFE (II,2364)

YOUNGMAN, HENNY
AS THE WORLD TURNS (II,110); HENNY AND ROCKY (I,2014); JOEY AND DAD (II,1306); THE MAD MAD MAD MAD WORLD OF THE SUPER BOWL (II,1586); THE MANY FACES OF COMEDY (I,2894); SHOW BUSINESS SALUTE TO MILTON BERLE (I,4029); SHOW OF THE YEAR (I,4031); THAT THING ON ABC (II,2574)

YOUNGQUIST, ARTHUR
DOCTOR FIX-UM (I,1310)

YOUNGS, GAIL
THE CHICAGO STORY (II,507)

YOUNGS, JIM
SECRETS OF MIDLAND HEIGHTS (II,2296)

YOURMAN, RICHARD
THE GUIDING LIGHT (II,1064)

YUE, MARION
CONCRETE BEAT (II,562); TRAPPER JOHN, M.D. (II,2654)

YUKIMURA, IZUMI
THE VICTOR BORGE SHOW (I,4720)

YULIN, HARRIS
THE 13TH DAY: THE STORY OF ESTHER (II,2593); ABC'S MATINEE TODAY (I,7); DYNASTY (II,747); THE GREATEST GIFT (II,1061); HOW THE WEST WAS WON (II,1196); MELVIN PURVIS: G-MAN (II,1679); THE NIGHT RIDER (II,1848); RANSOM FOR ALICE (II,2113); ROGER AND HARRY (II,2200)

YUME, MARION
TRAUMA CENTER (II,2655)

YUN, LANG
BROTHERS (II,381)

YUNE, JOHNNY
BILLY CRYSTAL: A COMIC'S LINE (II,249); GIDGET'S SUMMER REUNION (I,1799); SERGEANT T.K. YU (II,2305); THE SON-IN-LAW (II,2398)

YUNIPINGLI, CHARLES LATHALU

RETURN TO EDEN (II,2143)

YUNIPINGLI, STEVE DJATI
RETURN TO EDEN (II,2143)

YURO, ROBERT
TOMA (I,4547)

YURS, BOB
OUR FIVE DAUGHTERS (I,3413)

YUSEN, SUSAN
THE EDGE OF NIGHT (II,760)

YUSKIS, ANTOINETTE
BLANSKY'S BEAUTIES (II,267)

ZABACH, FLORIAN
CLUB EMBASSY (I,982); THE FLORIAN ZABACH SHOW (I,1606); THE FLORIAN ZABACH SHOW (I,1607); MINDY CARSON SINGS (I,3055)

ZABRISKIE, GRACE
EAST OF EDEN (II,752)

ZACHA, W T
BRIDGER (II,376); CODE R (II,551); STATE FAIR (II,2448); THE ULTIMATE IMPOSTER (II,2700)

ZACHARIAS, STEFAN
THE DEADLY GAME (II,633); THE GIRL, THE GOLD WATCH AND EVERYTHING (II,1002); LANIGAN'S RABBI (II,1427)

ZACHERLE, JOHN
ACTION IN THE AFTERNOON (I,25)

ZADORA, PIA
CIRCUS OF THE STARS (II,537); PAJAMA TOPS (II,1942)

ZAFRIOS, LOS
THE BENNY HILL SHOW (II,207)

ZAGON, MARTIN
HIGHCLIFFE MANOR (II,1150); STEPHANIE (II,2453)

ZAHL, EDA
MR. & MRS. & MR. (II,1744)

ZAL, ROXANA
THE ADVENTURES OF POLLYANNA (II,14)

ZALE, ALEXANDER
MOONLIGHT (II,1730)

ZANE, FRANK
THE HUSTLER OF MUSCLE BEACH (II,1210)

ZANIN, BRUNO
MARCO POLO (II,1626)

ZAPATA, CARMEN
FLYING HIGH (II,895); HAGEN (II,1071); THE MAN AND THE CITY (I,2856); THE TOM SWIFT AND LINDA CRAIG MYSTERY HOUR (II,2633); VIVA VALDEZ (II,2728)

ZAPPA, MOON UNIT
TWILIGHT THEATER II (II,2686)

ZAREMBA, JOHN
BEN CASEY (I,394); OWEN MARSHALL: COUNSELOR AT LAW (I,3435); READY AND WILLING (II,2115); THE TIME TUNNEL (I,4511)

ZARISH, JANET
AS THE WORLD TURNS (II,110)

ZARIT, PAMELA
THE SANDY DUNCAN SHOW (I,3910)

ZASLOW, MICHAEL
THE GUIDING LIGHT (II,1064); KING'S CROSSING (II,1397); THE LONG HOT SUMMER (I,2747); ONE LIFE TO LIVE (II,1907); UNIT 4 (II,2707)

ZATESLO, GEORGE
GIDGET'S SUMMER REUNION (I,1799)

ZAVAGLIA, RICHARD
ONE MORE TRY (II,1908)

ZEE, ELEANOR
ONE OF OUR OWN (II,1910)

ZEE, JOHN
BRING 'EM BACK ALIVE (II,377)

ZEIGLER, HEIDI
MR. MOM (II,1752)

ZEIGLER, TED
THE HUDSON BROTHERS RAZZLE DAZZLE COMEDY SHOW (II,1201); OF THEE I SING (I,3334); SHIELDS AND YARNELL (II,2328); THE SONNY AND CHER COMEDY HOUR (II,2400); THE SONNY AND CHER SHOW (II,2401); THE SONNY COMEDY REVUE (II,2403); TOP TEN (II,2648)

ZEIN, CHIP
REGGIE (II,2127)

ZELTZER, ARI
JOE'S WORLD (II,1305)

ZEM, KATHY
THE NBC FOLLIES (I,3242)

ZEMAN, JACKIE
GENERAL HOSPITAL (II,964); ONE LIFE TO LIVE (II,1907)

ZENK, COLLEEN
AS THE WORLD TURNS (II,110)

ZENON, MICHAEL
FOREST RANGERS (I,1642)

ZENOR, SUZANNE
CATCH 22 (I,871); DAYS OF OUR LIVES (II,629); HUSBANDS AND WIVES (II,1208); THE YOUNG AND THE RESTLESS (II,2862)

ZENTALL, KATE
FEATHERSTONE'S NEST (II,847)

ZERBE, ANTHONY
CENTENNIAL (II,477); THE CHISHOLMS (II,512); GEORGE WASHINGTON (II,978); HARRY O (II,1099); THE HEALERS (II,1115); HOW THE WEST WAS WON (II,1196); ONCE AN EAGLE (II,1897); RETURN OF THE MAN FROM U.N.C.L.E.: THE 15 YEARS LATER AFFAIR (II,2140); SHERLOCK HOLMES: THE HOUND OF THE BASKERVILLES (I,4011)

ZERI, WALTER G
ONE NIGHT BAND (II,1909)

ZETTERLING, MIA
THE ERROL FLYNN THEATER (I,1458)

ZEVNIK, NEIL
B.J. AND THE BEAR (II,126)

ZEWE, JOHN
MATINEE AT THE BIJOU (II,1651)

ZEWE, MARK
MATINEE AT THE BIJOU (II,1651)

ZIELINSKI, BRUNO 'JUNIOR'
IT'S POLKA TIME (I,2273); POLKA TIME (I,3642)

ZIEN, CHIP
HECK'S ANGELS (II,1122); LOVE, SIDNEY (II,1547); OFF CAMPUS (II,1877)

ZIKA, CHRISTIAN
NUMBER 96 (II,1869)

ZIMBALIST JR, EFREM
77 SUNSET STRIP (I,3988); THE ANITA BRYANT SPECTACULAR (II,82); ANYTHING FOR MONEY (I,236); BEYOND WITCH MOUNTAIN (II,228); CHARLEY'S AUNT (II,483); CONCERNING MISS MARLOWE (I,1032); CONFLICT (I,1036); FAMILY IN BLUE (II,818); THE FBI (I,1551); GIRL ON THE RUN (I,1809); MAVERICK (II,1657); THE RELUCTANT SPY (I,3763); REMINGTON STEELE (II,2130); A SALUTE TO TELEVISION'S 25TH ANNIVERSARY (I,3893); SCRUPLES (II,2280); SHOOTING STARS (II,2341); STAR TONIGHT (I,4199); WILD ABOUT HARRY (II,2797); YOU ARE THE JURY (II,2854)

ZIMBALIST, STEPHANIE
CANDID CAMERA LOOKS AT THE DIFFERENCE BETWEEN MEN AND WOMEN (II,421); CENTENNIAL (II,477); REMINGTON STEELE (II,2130); WILD ABOUT HARRY (II,2797)

ZIMMER, KIM
THE DOCTORS (II,694); THE GUIDING LIGHT (II,1064); ONE LIFE TO LIVE (II,1907)

ZIMMER, NORMA
THE LAWRENCE WELK SHOW (I,2643); THE LAWRENCE WELK SHOW (I,2644); MEMORIES WITH LAWRENCE WELK (II,1681)

ZIMMERMAN, ED
THE GUIDING LIGHT (II,1064)

ZIMMERMAN, MATT
THUNDERBIRDS (I,4497)

ZIPP, DEBBIE
THE CHEERLEADERS (II,493); SMALL AND FRYE (II,2380); THERE'S ALWAYS ROOM (II,2583)

ZIPPI, DANIEL
KING'S CROSSING (II,1397); SECRETS OF MIDLAND HEIGHTS (II,2296)

ZIPPY THE CHIMP
CAPTAIN SAFARI OF THE JUNGLE PATROL (I,836)

ZISKIE, DAN
THE EDGE OF NIGHT (II,760); THE FIRM (II,860)

ZISKIND, TRUDI
DOBIE GILLIS (I,1302)

ZITO, LOUIS
FLYING HIGH (II,895)

ZIVERING, DARRELL
KOJAK (II,1406)

ZIZKIND, TRUDI
HARD CASE (I,1947)

ZMED, ADRIAN
BATTLE OF THE NETWORK STARS (II,176); FLATBUSH (II,877); GOODTIME GIRLS (II,1042); REVENGE OF THE GRAY GANG (II,2146); T.J. HOOKER (II,2524)

ZOBLE, VIC
AQUA VARIETIES (I,248)

ZON, DEBORAH
SNAVELY (II,2387); THAT SECOND THING ON ABC (II,2572); THAT THING ON ABC (II,2574)

ZORBAUGH, DR HENRY
PLAY THE GAME (I,3618)

ZOREK, MICHAEL
HIGH SCHOOL, U.S.A. (II,1148); HIGH SCHOOL, U.S.A. (II,1149); YOUNG HEARTS (II,2865)

ZORICH, LOUIS
THE KAREN VALENTINE SHOW (I,2506); MASTER OF THE GAME (II,1647); RYAN'S HOPE (II,2234)

ZOULR, OLAN
JOHNNY GUITAR (I,2427)

ZOVELLA
THE MAGIC CLOWN (I,2814)

ZUBER, MARC
QUILLER: PRICE OF VIOLENCE (II,2101)

ZUCKER, CHARLES
WELCOME TO THE FUN ZONE (II,2763); YOUNG HEARTS (II,2865)

ZUCKERT, BILL
ADAM'S RIB (I,32); ANOTHER MAN'S SHOES (II,96); BACHELOR AT LAW (I,322); CAPTAIN NICE (I,835); THE CHEVY CHASE SHOW (II,505); COLUMBO (II,556); CONDOMINIUM (II,566); GREEN ACRES (I,1884); THE LILY TOMLIN SPECIAL (II,1479); RHODA (II,2151); THE WACKIEST SHIP IN THE ARMY (I,4744); WHAT'S UP, AMERICA? (I,4824); WHAT'S UP? (I,4823)

ZUCKERT, FRED
MAUDE (II,1655)

ZUKOR, ADOLPH
ENTERTAINMENT /emdash 1955 (I,1450)

ZULU
CODE NAME: DIAMOND HEAD (II,549); HAWAII FIVE-O (I,1972); HAWAII FIVE-O (II,1110)

ZUNGALO III, ALBERT
THE GUIDING LIGHT (II,1064)

ZWERLING, DARRELL
NEWMAN'S DRUGSTORE (I,1837)

Producers

AARDVARK, ED
BACKSTAGE PASS (II,132)

AARON, JOHN
ALUMNI FUN (I,145); THE DUPONT SHOW OF THE WEEK (I,1388); PERSON TO PERSON (I,3550)

AARON, ROBERT
ANOTHER LIFE (II,95)

ABATEMARCO, FRANK
MICKEY SPILLANE'S MIKE HAMMER (II,1692)

ABBEY, PETER
TEDDY PENDERGRASS IN CONCERT (II,2551)

ABBI, CATHY
BEHIND THE SCREEN (II,200); LOVE OF LIFE (I,2774); THE YOUNG AND THE RESTLESS (II,2862)

ABBOTT, GEORGE
U.S. ROYAL SHOWCASE (I,4688)

ABBOTT, MICHAEL
HEDDA HOPPER'S HOLLYWOOD (I,2002)

ABBOTT, NORMAN
DANNY THOMAS LOOKS AT YESTERDAY, TODAY AND TOMORROW (I,1148); EVERYTHING YOU ALWAYS WANTED TO KNOW ABOUT JACK BENNY AND WERE AFRAID TO ASK (I,1482); THE GHOST BUSTERS (II,986); THE JACK BENNY PROGRAM (I,2294); JACK BENNY'S FIRST FAREWELL SHOW (I,2300); JACK BENNY'S NEW LOOK (I,2301); JACK BENNY'S SECOND FAREWELL SHOW (I,2302); A LOVE LETTER TO JACK BENNY (II,1540)

ABDO, NICK
BLANSKY'S BEAUTIES (II,267); BROTHERS AND SISTERS (II,382); STILL THE BEAVER (II,2466); WORKING STIFFS (II,2834)

ABEL, ROBERT
SOPHIA! (I,4128)

ABEL, RUDY E
CANNONBALL (I,822); FATHER OF THE BRIDE (I,1543); THE HUMAN COMEDY (I,2158); TIMMY AND LASSIE (I,4520); WALDO (I,4749)

ABELL, JIMMY
TOUCH OF GOLD '75 (II,2650)

ABEYTA, PAUL
DANCE FEVER (II,615)

ABRAHAMS, JIM
POLICE SQUAD! (II,2061)

ABRAHAMS, MORT
DEADLY GAME (II,632); THE GENERAL ELECTRIC THEATER (I,1753); SEPARATE TABLES (II,2304); SUSPICION (I,4309); TALES OF TOMORROW (I,4352); WINDOWS (I,4876)

ABRAMS, GERALD W
THE BUREAU (II,397); CUTTER TO HOUSTON (II,612); HAVING BABIES I (II,1107); HAVING BABIES II (II,1108); HAVING BABIES III (II,1109); HOLLYWOOD HIGH (II,1162); HOLLYWOOD HIGH (II,1163); JULIE FARR, M.D. (II,1354); STEELTOWN (II,2452)

ACE, GOODMAN
EASY ACES (I,1398); RUTHIE ON THE TELEPHONE (I,3877)

ACKERMAN, HARRY
ALL IN THE FAMILY (I,133); BACHELOR FATHER (I,323); BEWITCHED (I,418); DENNIS THE MENACE (I,1231); THE EVE ARDEN SHOW (I,1471); THE FARMER'S DAUGHTER (I,1533); THE FLYING NUN (I,1611); GIDGET (I,1795); GIDGET GETS MARRIED (I,1796); GIDGET'S SUMMER REUNION (I,1799); GRINDL (I,1889); HAZEL (I,1982); INSIDE O.U.T. (II,1235); IS THERE A DOCTOR IN THE HOUSE? (II,1247); KEEPING AN EYE ON DENISE (I,2520); LEAVE IT TO BEAVER (I,2648); LOVE ON A ROOFTOP (I,2775); MR. DEEDS GOES TO TOWN (I,3134); MY SISTER EILEEN (I,3200); OCCASIONAL WIFE (I,3325); THE PAUL LYNDE SHOW (I,3489); THE SECOND HUNDRED YEARS (I,3959); THE SHAMEFUL SECRETS OF HASTINGS CORNERS (I,3994); TALES OF THE TEXAS RANGERS (I,4349); TEMPERATURES RISING (I,4390); THE UGLIEST GIRL IN TOWN (I,4663); UNDER THE YUM YUM TREE (I,4671)

ACKERMAN, LEONARD
ELLERY QUEEN: DON'T LOOK BEHIND YOU (I,1434); TARGET: THE CORRUPTERS (I,4360)

ACOMBA, DAVID
THE MAGIC PLANET (II,1598)

ADAMS, DON
THE DON ADAMS SCREEN TEST (II,705); THE DON ADAMS SPECIAL: HOORAY FOR HOLLYWOOD (I,1329); THE PARTNERS (I,3462)

ADAMS, ROBERT
THE CHAMBER MUSIC SOCIETY OF LOWER BASIN STREET (I,894); GAMBLE ON LOVE (I,1731)

ADAMS, ROGER
THE GLEN CAMPBELL MUSIC SHOW (II,1006)

ADAMSON, ED
BANYON (I,344); BANYON (I,345); THE ROUNDERS (I,3851); WANTED: DEAD OR ALIVE (I,4764)

ADAMSON, RICHARD
WACKO (II,2733)

ADEE, PETER
THE LOVE REPORT (II,1542)

ADELMAN, BARRY
THE HEE HAW HONEYS (II,1124); THE HEE HAW HONEYS (II,1125); PEOPLE ARE FUNNY (II,1996)

ADELSON, GARY
CASS MALLOY (II,454); EIGHT IS ENOUGH (II,762); FLATBUSH (II,877); LACE (II,1418); OUR FAMILY BUSINESS (II,1931); TOO GOOD TO BE TRUE (II,2643)

ADLER, BILL
KID TALK (I,2533)

ADLER, DIANE
SHE'S IN THE ARMY NOW (II,2320)

ADLER, EDWARD
NURSE (II,1870)

ADLER, JERRY
REWARD (II,2147)

ADMINA, A D
JESSIE (II,1287)

ADREON, FRANKLIN
COMMANDO CODY (I,1030)

AGHAYAN, RAY
THE DIAHANN CARROLL SHOW (II,665)

AHERNS, LYN
THE ABC AFTERSCHOOL SPECIAL (II,1); DEAR ALEX AND ANNIE (II,651)

AHLERS, DAVID
TONIGHT IN HAVANA (I,4557)

AILES, ROGER E
ALLEN LUDDEN'S GALLERY (I,143); THE REAL TOM KENNEDY SHOW (I,3743); STEVE ALLEN'S LAUGH-BACK (II,2455)

AIMER, AL
MYSTERY AND MRS. (I,3210)

ALBERG, MILDRED FREED
THE DEVIL'S DISCIPLE (I,1247); KISS ME, KATE (I,2568)

ALBERT, LOU
MR. CHIPS (I,3132); SUPER PAY CARDS (II,2500)

ALBRECHT, JOIE
TELEVISION'S GREATEST COMMERCIALS II (II,2554); TELEVISION'S GREATEST COMMERCIALS III (II,2555); TELEVISION'S GREATEST COMMERCIALS IV (II,2556)

ALBRECHT, RICHARD
THEY STAND ACCUSED (I,4442)

ALBRIGHT, RICHIE
THE DUKES OF HAZZARD (II,742)

ALCHIN, RAY
1915 (II,1853)

ALCORN, R W
CITIZEN SOLDIER (I,963)

ALDA, ALAN
THE FOUR SEASONS (II,915); SUSAN AND SAM (II,2506); WE'LL GET BY (II,2753)

ALDWORTH, JACK
THE LUCILLE BALL-DESI ARNAZ SHOW (I,2784)

ALEXANDER, ANDREW
SECOND CITY TELEVISION (II,2287)

ALEXANDER, BEN
PARTY TIME AT CLUB ROMA (I,3465)

ALEXANDER, DAVID
KRAFT TELEVISION THEATER (I,2592); THE UNITED STATES STEEL HOUR (I,4677)

ALEXANDER, ELLIOT
IT'S A BIRD, IT'S A PLANE, IT'S SUPERMAN (II,1253)

ALEXANDER, LES
NO SOAP, RADIO (II,1856)

ALEXANDER, M
Q.T. HUSH (I,3702)

ALEXANDER, RONALD
TOO YOUNG TO GO STEADY (I,4574)

ALEXANDER, STEVE
CELEBRITY TIME (I,887); THE EYES HAVE IT (I,1495)

ALEY, ALBERT
IRONSIDE (II,1246); THE PAPER CHASE (II,1949)

ALISI, ART
THE NAMEDROPPERS (I,3219)

ALISI, ROBERT
LAS VEGAS GAMBIT (II,1430)

ALKON, SELIG
THE ARMSTRONG CIRCLE THEATER (I,260); LET'S TAKE A TRIP (I,2668); STRANGE PARADISE (I,4246)

ALLAND, WILLIAM
WORLD OF GIANTS (I,4912)

ALLEN, CHERYL
THE LAST LEAF (II,1437)

ALLEN, CRAIG
GIRL ABOUT TOWN (I,1806); STARS IN KHAKI AND BLUE (I,4209)

ALLEN, HERB
THE BOB CROSBY SHOW (I,497); HAIL THE CHAMP (I,1918)

ALLEN, HOYT
THE FIRST HUNDRED YEARS (I,1581)

ALLEN, IRWIN
CITY BENEATH THE SEA (I,965); CODE RED (II,552); CODE RED (II,553); LAND OF THE GIANTS (I,2611); LOST IN SPACE (I,2758); MATT HELM (II,1652); THE RETURN OF CAPTAIN NEMO (II,2135); SWISS FAMILY ROBINSON (II,2516); SWISS FAMILY ROBINSON (II,2517); THE TIME TUNNEL (I,4511); VOYAGE TO THE BOTTOM OF THE SEA (I,4743)

ALLEN, JAMI
20 MINUTE WORKOUT (II,2680)

ALLEN, RANDY
THE MOONMAN CONNECTION (II,1731)

ALLEN, RAY
ALICE (II,33); BIG JOHN, LITTLE JOHN (II,240); THE LOVE BOAT (II,1535); MAN IN THE MIDDLE (I,2870); MCDUFF, THE TALKING DOG (II,1662); THE MONSTER SQUAD (II,1725); THE RED HAND GANG (II,2122)

ALLEN, ROBERT
SEEING THINGS (II,2297)

ALLEN, ROGER
MIKE AND PEARL (I,3038)

ALLEY, ELMER
THAT GOOD OLD NASHVILLE MUSIC (II,2571)

ALLISON, JUDITH
DOUBLE TROUBLE (II,724); PRIVATE BENJAMIN (II,2087); A ROCK AND A HARD PLACE (II,2190); SUMMER (II,2491); WHITE AND RENO (II,2787); WIZARDS AND WARRIORS (II,2813)

ALLYN, SANDRA
ECHOES OF THE SIXTIES (II,756); ONCE UPON A TIME. . . IS NOW THE STORY OF PRINCESS GRACE (II,1899)

ALLYN, WILLIAM
ECHOES OF THE SIXTIES (II,756); ONCE UPON A TIME. . . IS NOW THE STORY OF PRINCESS GRACE (II,1899)

ALSBERG, ARTHUR
BRIDGET LOVES BERNIE (I,722); CRASH ISLAND (II,600); FRANKIE AND ANNETTE: THE SECOND TIME AROUND (II,917); THE MUNSTERS' REVENGE (II,1765)

ALSTON, HOWARD
CENTENNIAL (II,477); THE RETURN OF MARCUS WELBY, M.D (II,2138)

ALTER, PAUL
BEAT THE CLOCK (II,195); CARD SHARKS (II,438); TATTLETALES (II,2545); TREASURE ISLE (I,4601); TV'S FUNNIEST GAME SHOW MOMENTS (II,2676)

ALTMAN, ROBERT
COMBAT (I,1011); A WALK IN THE NIGHT (I,4750)

ALTON, ROBERT
YOU'RE THE TOP (I,4974)

AMATEAU, ROD
THE BOB HOPE CHRYSLER THEATER (I,502); BORDER PALS (II,352); THE DUKES OF HAZZARD (II,742); ENOS (II,779); THE GEORGE BURNS AND GRACIE ALLEN SHOW (I,1763); THE GEORGE BURNS SHOW (I,1765); HIGHWAY HONEYS (II,1151); MY MOTHER THE CAR (I,3197); THE NEW PHIL SILVERS SHOW (I,3269); O.K. CRACKERBY (I,3348); SIX PACK (II,2370); SUPERTRAIN (II,2504)

AMBLER, ERIC
ALCOA PREMIERE (I,109)

AMGOTT, MADELINE
NOT FOR WOMEN ONLY (I,3310)

AMMONDS, JOHN
THE MORECAMBE AND WISE SHOW (II,1734)

AMSTERDAM, MOREY
CAN YOU TOP THIS? (I,817); THE MOREY AMSTERDAM SHOW (I,3101)

ANDERSON, ALEXANDER
CRUSADER RABBIT (I,1109)

ANDERSON, BILL
DANIEL BOONE (I,1141); SWAMP FOX (I,4311); WEDDING DAY (II,2759)

ANDERSON, BRUCE
TALES OF THE BLACK CAT (I,4345)

ANDERSON, GEORGE
THE MAGIC RANCH (I,2819); SANDY STRONG (I,3911)

ANDERSON, GERRY
CAPTAIN SCARLET AND THE MYSTERONS (I,837); FIREBALL XL-5 (I,1578); THE PROTECTORS (I,3681); SPACE: 1999 (II,2409); STINGRAY (I,4228); SUPERCAR (I,4294); THUNDERBIRDS (I,4497); U.F.O. (I,4662)

ANDERSON, JOHN MAXWELL
ALMOST AMERICAN (II,55); DIFF'RENT STROKES (II,674); E/R (II,748); THE FACTS OF LIFE (II,805); IT'S YOUR MOVE (II,1259)

ANDERSON, JON C
MICKEY SPILLANE'S MIKE HAMMER (II,1692); MICKEY SPILLANE'S MIKE HAMMER: MORE THAN MURDER (II,1693)

ANDERSON, MACK
EYE ON HOLLYWOOD (II,798); THE LOVE REPORT (II,1542); THE LOVE REPORT (II,1543)

ANDERSON, NICK
POPI (II,2069)

ANDERSON, R H
OUR MAN FLINT: DEAD ON TARGET (II,1932)

ANDERSON, SYLVIA
SPACE: 1999 (II,2409); STINGRAY (I,4228); SUPERCAR (I,4294)

ANDERSON, WILLIAM H
ZORRO (I,4982)

ANDOR, GREGORY
COSMOS (II,579)

ANDRE, JACQUES
DOROTHY HAMILL'S CORNER OF THE SKY (II,721); FANFARE (I,1528)

ANDREOLI, FRANK
CELEBRITY COOKS (II,472)

ANDREWS, CHARLES
THE ARTHUR GODFREY SHOW (I,284); THE ARTHUR GODFREY SHOW (I,285); THE ARTHUR GODFREY SHOW (I,286); THE DAVE GARROWAY SHOW (I,1173); MARINELAND CARNIVAL (I,2913); THAT WAS THE YEAR THAT WAS (I,4424); THIS WILL BE THE YEAR THAT WILL BE (I,4466)

ANDREWS, PETER
HOW TO SURVIVE A MARRIAGE (II,1198)

ANDREWS, RALPH
50 GRAND SLAM (II,852); CELEBRITY SWEEPSTAKES (II,476); I'LL BET (I,2189); THE LIAR'S CLUB (II,1466); LIE DETECTOR (II,1467); YOU DON'T SAY (I,4934)

ANDROSKY, JEFF
THE LOVE REPORT (II,1542); THE LOVE REPORT (II,1543)

ANGELO, LEANA
PEOPLE TO PEOPLE (II,2000)

ANGELOS, BILL
THE ARTHUR GODFREY SPECIAL (I,288); ARTHUR GODFREY'S PORTABLE ELECTRIC MEDICINE SHOW (I,290); BING CROSBY AND THE SOUNDS OF CHRISTMAS (I,457); BING CROSBY'S CHRISTMAS SHOW (I,474); BING CROSBY'S SUN VALLEY CHRISTMAS SHOW (I,475); BING CROSBY—COOLING IT (I,459); BING CROSBY—COOLING IT (I,460); CHRISTMAS WITH THE BING CROSBYS (I,950); CHRISTMAS WITH THE BING CROSBYS (II,524); LOOKING BACK (I,2754); OUR PLACE (I,3416); PERRY COMO'S WINTER SHOW (I,3546)

ANGUIST, TOBY
HOPALONG CASSIDY (I,2118)

ANGUS, ROBERT
THE ADVENTURES OF OZZIE AND HARRIET (I,71)

ANHALT, EDWARD
NOWHERE TO HIDE (II,1868)

ANNAKIN, KEN
HUNTER'S MOON (II,1207)

ANSARA, MICHAEL
THE BARBARA EDEN SHOW (I,346)

ANSPAUGH, DAVID
HILL STREET BLUES (II,1154)

ANTONACCI, GREG
BROTHERS (II,380); IT TAKES TWO (II,1252); IT'S A LIVING (II,1254); MAKING A LIVING (II,1604)

APPEL, DON
THE BLUE ANGEL (I,489); THE IMOGENE COCA SHOW (I,2198); THE RED BUTTONS SHOW (I,3750); THIS IS GALEN DRAKE (I,4456); THE VAUGHN MONROE SHOW (I,4707); THE VIC DAMONE SHOW (I,4717)

APPLEBAUM, LAWRENCE
EGAN (I,1419)

ARANGO, DOUGLAS
CARTER COUNTRY (II,449); JENNIFER SLEPT HERE (II,1278); TOO CLOSE FOR COMFORT (II,2642)

ARDEN, CHARLES
THE EARTHA KITT SHOW (I,1395)

ARDEN, DON
AIR SUPPLY IN HAWAII (II,26)

ARDEN, ROBERT
A HATFUL OF RAIN (I,1964)

ARGENT, DOUGLAS
FAWLTY TOWERS (II,843)

ARLEDGE, ROONE
BATTLE OF THE NETWORK STARS (II,163); BATTLE OF THE NETWORK STARS (II,164); BATTLE OF THE NETWORK STARS (II,165); BATTLE OF THE NETWORK STARS (II,166); BATTLE OF THE NETWORK STARS (II,167); BATTLE OF THE NETWORK STARS (II,168); BATTLE OF THE NETWORK STARS (II,169); BATTLE OF THE NETWORK STARS (II,170); BATTLE OF THE NETWORK STARS (II,171); BATTLE OF THE NETWORK STARS (II,172); BATTLE OF THE NETWORK STARS (II,173); BATTLE OF THE NETWORK STARS (II,174); BATTLE OF THE NETWORK STARS (II,175); BATTLE OF THE NETWORK STARS (II,176); BATTLE OF THE NETWORK STARS (II,177); SATURDAY NIGHT LIVE WITH HOWARD COSELL (II,2262); SINATRA—THE MAIN EVENT (II,2360)

ARLEN, DEK
SHEENA EASTON, ACT 1 (II,2323); SONG BY SONG (II,2399)

ARLEN, JILL
SHEENA EASTON, ACT 1 (II,2323)

ARLETT, RICHARD
COMEBACK (I,1016)

ARLEY, JEAN
LOVE OF LIFE (I,2774)

ARMER, ALAN A
CANNON (II,424); THE INVADERS (I,2229); LANCER (I,2610); MAN WITHOUT A GUN (I,2884); MY FRIEND FLICKA (I,3190); THE STRANGER (I,4251); THE UNTOUCHABLES (I,4682); WESTSIDE MEDICAL (II,2765)

ARMSTRONG, BILL
PETER MARSHALL SALUTES THE BIG BANDS (II,2028); PITFALL (II,2047)

ARMUS, BURTON
AIRWOLF (II,27); CASSIE AND COMPANY (II,455); PARIS (II,1957)

ARNAZ, DESI
THE ANN SOTHERN SHOW (I,220); THE DESILU PLAYHOUSE (I,1237); I LOVE LUCY (I,2171); THE LUCILLE BALL-DESI ARNAZ SHOW (I,2784); THE LUCY SHOW (I,2791); THE MOTHERS-IN-LAW (I,3110); YOU'RE ONLY YOUNG TWICE (I,4970)

ARNELL, PETER
BALANCE YOUR BUDGET (I,336); CELEBRITY TALENT SCOUTS (I,886); DICK CLARK'S WORLD OF TALENT (I,1268); FACE THE FACTS (I,1508); HIGH FINANCE (I,2052); I'LL BUY THAT (I,2190); TAKE A GOOD LOOK (I,4332); TAKE A GOOD LOOK (II,2529); TAKE A GUESS (I,4333); TALENT SEARCH (I,4342); WHAT'S IN A WORD? (I,4815)

ARNOLD, DANNY
100 YEARS OF AMERICA'S POPULAR MUSIC (II,1904); A.E.S. HUDSON STREET (II,17); ACES UP (II,7); ANN IN BLUE (II,83); BARNEY MILLER (II,154); BEWITCHED (I,418); FISH (II,864); MY WORLD . . . AND WELCOME TO IT (I,3208); THE REAL MCCOYS (I,3741); SOMEWHERE IN ITALY. . . COMPANY B (I,4119)

ARNOLD, JACK
GILLIGAN'S ISLAND (II,990); THE HOUSE NEXT DOOR (I,2139); IT TAKES A THIEF (I,2250); MR. LUCKY (I,3141); THE SID CAESAR, IMOGENE COCA, CARL REINER, HOWARD MORRIS SPECIAL (I,4039)

ARNOLD, JAY GORDON
EVENING AT THE IMPROV (II,786)

ARNOLD, KAY
MS. FIXER UPPER (I,3156)

ARNOLD, NICK
BAKER'S DOZEN (II,136); CALUCCI'S DEPARTMENT (I,805); LOVE AT FIRST SIGHT (II,1530); LOVE AT FIRST SIGHT (II,1531); A NEW KIND OF FAMILY (II,1819); PRIVATE BENJAMIN (II,2087); SMALL AND FRYE (II,2380); STOCKARD CHANNING IN JUST FRIENDS (II,2467); WELCOME BACK, KOTTER (II,2761); WHACKED OUT (II,2767)

ARNOTT, BOB
THE HUDSON BROTHERS RAZZLE DAZZLE COMEDY SHOW (II,1201); THE HUDSON BROTHERS SHOW (II,1202); JOEY AND DAD (II,1306)

ARONS, RICHARD
THE JACKSONS (II,1267)

ARQUETTE, CLIFF
DAVE AND CHARLEY (I,1171)

ARQUETTE, LEWIS
THE LORENZO AND HENRIETTA MUSIC SHOW (II,1519)

ARRICK, LARRY
EAST SIDE/WEST SIDE (I,1396); MR. BROADWAY (I,3131)

ARTHUR, KAREN
REMINGTON STEELE (II,2130)

ARTHUR, MAVIS
THE VIDEO GAME (II,2727)

ARTHUR, ROBERT
THE BEST OF SULLIVAN (II,217); THE DAVID SOUL AND FRIENDS SPECIAL (II,626); EASY DOES IT. . .

STARRING FRANKIE AVALON (II,754); THE NATALIE COLE SPECIAL (II,1794)

ASHER, WILLIAM
BEWITCHED (I,418); FIBBER MCGEE AND MOLLY (I,1569); JANE WYMAN PRESENTS THE FIRESIDE THEATER (I,2345); THE PATTY DUKE SHOW (I,3483); THE PAUL LYNDE SHOW (I,3489); SHIRLEY TEMPLE'S STORYBOOK (I,4017); TABITHA (II,2526); TEMPERATURES RISING (I,4390)

ASHLEY, JOHN
THE A-TEAM (II,119); THE QUEST (II,2098)

ASHLEY, TED
MISS SUSAN (I,3065)

ASHTON, BARRY
MINSKY'S FOLLIES (II,1703)

ASKEY, DAVID
DOCTOR IN THE HOUSE (I,1313)

ASSELIN, DIANE
THE ABC AFTERSCHOOL SPECIAL (II,1); SANTA BARBARA (II,2258)

ASSELIN, PAUL
THE ABC AFTERSCHOOL SPECIAL (II,1)

ASSEYER, TAMARA
NORMA RAE (II,1860)

ASTAIRE, FRED
ANOTHER EVENING WITH FRED ASTAIRE (I,229); ASTAIRE TIME (I,305); THE FRED ASTAIRE SHOW (I,1677)

ATKINS, IRVIN S
BID N' BUY (I,419); HOLLYWOOD TALENT SCOUTS (I,2094); ON THE GO (I,3361)

ATLASS, FRANK
SUSAN'S SHOW (I,4303)

AUBREY, JAMES
SHANNON (II,2314); TOP TEN (II,2648)

AUCHINCLOSS, GORDON
CHEVROLET ON BROADWAY (I,925); THE MATT DENNIS SHOW (I,2948)

AUER, JOHN H
U.S. MARSHAL (I,4686); THE WHIRLYBIRDS (I,4841)

AUERBACH, DAVID
THE WORLD'S FUNNIEST COMMERCIAL GOOFS (II,2841)

AURITT, JOAN
AMERICA ALIVE! (II,67)

AURTHUR, ROBERT ALAN
CALL ME BACK (I,798); CLOWNAROUND (I,980); ED SULLIVAN'S BROADWAY (I,1402); FANFARE (I,1528); THREE IN ONE (I,4479)

AUSTIN, BOB
SKIPPY, THE BUSH KANGAROO (I,4076)

AUSTIN, BUD
KLEIN TIME (II,1401); THE PLANT FAMILY (II,2050); STILL THE BEAVER (II,2466); WHEW! (II,2786)

AUSTIN, RAY
SHIRLEY'S WORLD (I,4019)

AUSTIN, RONALD
BEACH PATROL (II,192); CHARLIE'S ANGELS (II,486); CHOPPER ONE (II,515); FIREHOUSE (II,859); JIGSAW JOHN (II,1288); Q. E. D. (II,2094); SUPERCOPS (II,2502)

AUTRY, GENE
THE ADVENTURES OF CHAMPION (I,54); BUFFALO BILL JR. (I,755); GENE AUTRY'S MELODY RANCH (I,1749)

AVANT, CLARANCE
THE MUHAMMED ALI VARIETY SPECIAL (II,1762)

AVENT, JON
CALL TO GLORY (II,413)

AVERBACK, HY
AT EASE (II,115); BIG EDDIE (II,236); BROTHERS AND SISTERS (II,382); THE DON RICKLES SHOW (I,1342); THE DUKES OF HAZZARD (II,742); ENSIGN O'TOOLE (I,1448); F TROOP (I,1499); FREEBIE AND THE BEAN (II,922); THE JIMMY DURANTE SHOW (I,2389); KEEPING AN EYE ON DENISE (I,2520); MEET CORLISS ARCHER (I,2980); MRS. G. GOES TO COLLEGE (I,3155); NEEDLES AND PINS (I,3246); NEWMAN'S DRUGSTORE (II,1837); SYBIL (I,4327); THE TOM EWELL SHOW (I,4538); TWO'S COMPANY (I,4659)

AYLESWORTH, JOHN
THE BRASS ARE COMING (I,708); BUCKSHOT (II,385); DON KNOTTS NICE CLEAN, DECENT, WHOLESOME HOUR (I,1333); EVERYTHING GOES (I,1480); THE FUNNY WORLD OF FRED & BUNNI (II,943); THE GRADY NUTT SHOW (II,1048); THE HARLEM GLOBETROTTERS POPCORN MACHINE (II,1092); HECK'S ANGELS (II,1122); HEE HAW (II,1123); THE JUD STRUNK SHOW (I,2462); KEEP ON TRUCKIN' (II,1372); THE NASHVILLE PALACE (II,1792); THE OSMOND BROTHERS SHOW (I,3409); SHIELDS AND YARNELL (II,2328)

AZAROS, DON
THE SUNDAY GAMES (II,2494)

AZOFF, IRVING
STEVIE NICKS IN CONCERT (II,2463)

B B & R PRODUCTIONS
THE MOONMAN CONNECTION (II,1731)

BBC
WAR IN THE AIR (I,4768)

BABBIN, JACQUELINE
ALL MY CHILDREN (II,39); THE ARMSTRONG CIRCLE THEATER (I,260); BEACON HILL (II,193); THE DUPONT SHOW OF THE WEEK (I,1388); THE GLASS MENAGERIE (I,1819); HIGHER AND HIGHER, ATTORNEYS AT LAW (I,2057); NEITHER ARE WE ENEMIES (I,3247); OLDSMOBILE MUSIC THEATER (I,3352); THE PICTURE OF DORIAN GRAY (I,3591); THE THREE MUSKETEERS (I,4481); WAY OUT (I,4777)

BABBISH, ANDREW
SOAP FACTORY DISCO (II,2393)

BABBITT, STACY
HIT MAN (II,1156)

BACAL, JOE
G.I. JOE: A REAL AMERICAN HERO (II,948); THE TRANSFORMERS (II,2653)

BACALL, DAVE
THE GREAT SPACE COASTER (II,1059)

BACH, BOB
THE MEL TORME SHOW (I,2995); MUSIC '55 (I,3168); WHAT'S MY LINE? (I,4820)

BACHANAN, ROBERT
COLONEL BLEEP (I,1002)

BACHARACH, DORO
THE ABC AFTERSCHOOL SPECIAL (II,1)

BACHER, WILLIAM
WASHINGTON SQUARE (I,4772)

BACHMAN, KAY
THE QUIZ KIDS (II,2103)

BACKE, JOHN D
KUDZU (II,1415)

BACON, JAMES
THE GREAT MYSTERIES OF HOLLYWOOD (II,1058)

BACOTE, WILLIE
THE MOONMAN CONNECTION (II,1731)

BACRACH, JAN
AN EVENING WITH ROBIN WILLIAMS (II,789)

BADAMI II, FRANK
OFF THE RACK (II,1878)

BADAMI, FRANK
HAPPY ENDINGS (II,1085); JOE AND VALERIE (II,1299); LOVE, LIFE, LIBERTY & LUNCH (II,1544); THAT WAS THE YEAR THAT WAS (II,2575)

BADDISH, ANDY
LIVE FROM THE LONE STAR (II,1501)

BADHAM, JOHN
DIAL HOT LINE (I,1254)

BADIYI, REZA S
HE AND SHE (I,1985)

BAER, ART
THE COP AND THE KID (II,572); GLITTER (II,1009); IT'S YOUR MOVE (I,2281); THE LOVE BOAT (II,1535); MADHOUSE 90 (I,2807); MAKE A FACE (I,2837); PICTURE THIS (I,3593)

BAER, DONALD A
BLUE THUNDER (II,278); TALES OF THE GOLD MONKEY (II,2537)

BAER, NORMAN
EVERYTHING GOES (I,1480); THE GREAT MYSTERIES OF HOLLYWOOD (II,1058); THE HARLEM GLOBETROTTERS POPCORN MACHINE (II,1092); MELINA MERCOURI'S GREECE (I,2997); SOPHIA LOREN IN ROME (I,4129)

BAFFICO, JAMES
ANOTHER WORLD (II,97)

BAGDASARIAN, ROSS
ALVIN AND THE CHIPMUNKS (II,61)

BAGNI, GWEN
SHIRLEY (II,2330)

BAILEY, DOROTHY J
THE SECRET WAR OF JACKIE'S GIRLS (II,2294)

BAILEY, ED

THE CROSS-WITS (II,605); TRUTH OR CONSEQUENCES (I,4619); TRUTH OR CONSEQUENCES (I,4620)

BAIRD, BIL
THE WHISTLING WIZARD (I,4844)

BAISER, JAMES D
26 MEN (I,4648)

BAKER, CECIL
THE LINE-UP (I,2707)

BAKER, DEE
THE ARTHUR GODFREY SPECIAL (II,109); DOROTHY HAMILL'S CORNER OF THE SKY (II,721); ULTRA QUIZ (II,2701)

BAKER, DIANE
THE ABC AFTERSCHOOL SPECIAL (II,1); A WOMAN OF SUBSTANCE (II,2820)

BAKER, GAIL TREEL
GRAFFITI ROCK (II,1049)

BAKER, MARTIN G
FRAGGLE ROCK (II,916)

BAKER, ROBERT S
THE PERSUADERS! (I,3556); RETURN OF THE SAINT (II,2141); THE SAINT (II,2237)

BAKER, WALT
ELVIRA'S MOVIE MACABRE (II,771)

BAKSHI, ROBERT
SPIDER-MAN (I,4154)

BALDWIN, GERALD
THE SMURFS (II,2385); SNORKS (II,2390)

BALL, JOHN FLEMING
TALES OF THE UNEXPECTED (II,2540)

BALL, LUCILLE
BUNGLE ABBEY (II,396); THE DESILU REVUE (I,1238); A LUCILLE BALL SPECIAL STARRING LUCILLE BALL AND DEAN MARTIN (II,1557); A LUCILLE BALL SPECIAL STARRING LUCILLE BALL AND JACKIE GLEASON (II,1558); THE LUCILLE BALL SPECIAL (II,1559); A LUCILLE BALL SPECIAL: WHAT NOW CATHERINE CURTIS? (II,1560); LUCY IN LONDON (I,2789); LUCY MOVES TO NBC (II,1562); THE LUCY SHOW (I,2791); THE MUSIC MART (II,1774)

BALLARD, JACK
HIDE AND SEEK (I,2047)

BALLEW, JERRY
THE SCARLET LETTER (II,2270)

BALLOU, FRANK
RIVKIN: BOUNTY HUNTER (II,2183)

BALTER, ALLAN
CAPTAIN AMERICA (II,427); CAPTAIN AMERICA (II,428); THE MAN WITH THE POWER (II,1616); SAMURAI (II,2249); SAN FRANCISCO INTERNATIONAL AIRPORT (I,3905); SAN FRANCISCO INTERNATIONAL AIRPORT (I,3906); SHAFT (I,3993); THE SIX-MILLION-DOLLAR MAN (II,2372)

BALTIMORE, LYNN D
PEOPLE ARE FUNNY (II,1996)

BANATTA, DON
POPI (II,2069)

BANKS, GENE
THE $1.98 BEAUTY SHOW (II,698); THE CHUCK BARRIS RAH-RAH SHOW (II,525); DAY IN COURT (I,1182); DREAM GIRL OF '67 (I,1375); GENERAL HOSPITAL (II,964); THE GONG SHOW (II,1026)

BANKS, HENRY
OZMOE (I,3437)

BANKS, TOMMY
CELEBRITY CONCERTS (II,471); CELEBRITY REVUE (II,475)

BANNER, BOB
THE ALAN KING SHOW (I,100); ALL-STAR ANYTHING GOES (II,50); ALMOST ANYTHING GOES (II,56); ANDY WILLIAMS' EARLY NEW ENGLAND CHRISTMAS (II,79); BATTLE OF THE LAS VEGAS SHOW GIRLS (II,162); CALAMITY JANE (I,793); CANDID CAMERA (I,819); CAROL + 2 (I,853); CAROL AND COMPANY (I,847); THE DINAH SHORE SHOW (I,1280); THE DINAH SHORE SHOW (I,1281); THE DON HO SHOW (II,706); THE FRED WARING SHOW (I,1679); THE GARRY MOORE SHOW (I,1739); THE GARRY MOORE SHOW (I,1740); THE GINGER ROGERS SHOW (I,1804); HOT (II,1183); HOT (II,1184); THE JIMMY DEAN SHOW (I,2385); JULIE AND CAROL AT CARNEGIE HALL (I,2478); JULIE ANDREWS: ONE STEP INTO SPRING (II,1353); JUNIOR ALMOST ANYTHING GOES (II,1357); LEAPIN' LIZARDS, IT'S LIBERACE (II,1453); LISA, BRIGHT AND DARK (I,2711); PEGGY FLEMING AT MADISON SQUARE GARDEN (I,3506); PEGGY FLEMING AT SUN VALLEY (I,3507); PEGGY FLEMING VISITS THE SOVIET UNION (I,3509); PERRY COMO IN LAS VEGAS (II,2003); THE PERRY COMO SPRINGTIME SPECIAL (II,2004); PERRY COMO'S BAHAMA HOLIDAY (II,2006); PERRY COMO'S CHRISTMAS IN AUSTRIA (II,2007); PERRY COMO'S CHRISTMAS IN MEXICO (II,2009); PERRY COMO'S CHRISTMAS IN NEW MEXICO (II,2010); PERRY COMO'S CHRISTMAS IN PARIS (II,2012); PERRY COMO'S CHRISTMAS IN THE HOLY LAND (II,2013); PERRY COMO'S EASTER BY THE SEA (II,2014); PERRY COMO'S EASTER IN GUADALAJARA (II,2015); PERRY COMO'S FRENCH-CANADIAN CHRISTMAS (II,2016); PERRY COMO'S HAWAIIAN HOLIDAY (II,2017); PERRY COMO'S LAKE TAHOE HOLIDAY (II,2018); PERRY COMO'S SPRING IN NEW ORLEANS (II,2021); PERRY COMO'S SPRING IN SAN FRANCISCO (II,2022); PLEASE STAND BY (II,2057); ROSALINDA (I,3846); SOLID GOLD (II,2395); SOLID GOLD '79 (II,2396); STAR SEARCH (II,2437); STAR SEARCH (II,2438); THICKER THAN WATER (I,4445); TO EUROPE WITH LOVE (I,4525); THE WAY THEY WERE (II,2748)

BANSKA, HERBERT
NO HOLDS BARRED (II,1854)

BANTA, GLORIA
ANGIE (II,80); COUSINS (II,595); IT'S A LIVING (II,1254); MAKING A LIVING (II,1604)

BANTMAN, JOSEPH
THE GENERAL ELECTRIC THEATER (I,1753)

BARASCH, NORMAN
BENSON (II,208); DOC (II,683); FISH (II,864); THE NEW ODD COUPLE (II,1825); NOBODY'S PERFECT (II,1858); ONE OF THE BOYS (II,1911)

BARBASH, BOB
KINCAID (I,2543)

BARBERA, JOSEPH
THE ABBOTT AND COSTELLO CARTOON SHOW (I,1); THE ABC AFTERSCHOOL SPECIAL (II,1); THE ADDAMS FAMILY (II,12); THE ADVENTURES OF GULLIVER (I,60); THE ADVENTURES OF JONNY QUEST (I,64); THE ALL-NEW POPEYE HOUR (II,49); THE AMAZING CHAN AND THE CHAN CLAN (I,157); THE ATOM ANT/SECRET SQUIRREL SHOW (I,311); THE B.B. BEEGLE SHOW (II,124); THE BANANA SPLITS ADVENTURE HOUR (I,340); THE BEACH GIRLS (II,191); BENJI, ZAX AND THE ALIEN PRINCE (II,205); BIRDMAN (II,256); BIRDMAN AND THE GALAXY TRIO (I,478); THE BISKITTS (II,257); BUFORD AND THE GHOST (II,389); BUTCH CASSIDY AND THE SUNDANCE KIDS (I,781); THE C.B. BEARS (II,406); CAPTAIN CAVEMAN AND THE TEEN ANGELS (II,433); CASPER AND THE ANGELS (II,453); THE CATTANOOGA CATS (I,874); THE CLUE CLUB (II,546); DASTARDLY AND MUTTLEY IN THEIR FLYING MACHINES (I,1159); DEVLIN (II,663); THE DRAK PACK (II,733); THE DUKES (II,741); THE FANTASTIC FOUR (I,1529); THE FLINTSTONE COMEDY HOUR (II,881); FLINTSTONE FAMILY ADVENTURES (II,882); THE FLINTSTONE FUNNIES (II,883); THE FLINTSTONES (II,884); THE FLINTSTONES (II,885); FONZ AND THE HAPPY DAYS GANG (II,898); FRANKENSTEIN JR. AND THE IMPOSSIBLES (I,1672); FRED AND BARNEY MEET THE SHMOO (II,918); FRED AND BARNEY MEET THE THING (II,919); FRED FLINTSTONE AND FRIENDS (II,920); THE FUNKY PHANTOM (I,1709); THE FUNNY WORLD OF FRED & BUNNI (II,943); THE GALAXY GOOFUPS (II,954); THE GARY COLEMAN SHOW (II,958); GIDGET MAKES THE WRONG CONNECTION (I,1798); GODZILLA (II,1014); GOOBER AND THE GHOST CHASERS (II,1028); THE GREAT GRAPE APE SHOW (II,1056); THE HANNA-BARBERA HAPPINESS HOUR (II,1079); THE HARLEM GLOBETROTTERS (I,1953); HELP! IT'S THE HAIR BEAR BUNCH (I,2012); THE HERCULOIDS (I,2023); HERE COME THE STARS (I,2026); HONG KONG PHOOEY (II,1180); THE HUCKLEBERRY HOUND SHOW (I,2155); INCH HIGH, PRIVATE EYE (II,1231); JABBERJAW (II,1261); JACK AND THE BEANSTALK (I,2288); JEANNIE (II,1275); THE JETSONS (I,2376); JOKEBOOK (II,1339); JOSIE AND THE PUSSYCATS (I,2453); JOSIE AND THE

PUSSYCATS IN OUTER SPACE (I,2454); KORG: 70,000 B.C. (II,1408); THE KWICKY KOALA SHOW (II,1417); LAVERNE AND SHIRLEY IN THE ARMY (II,1448); LAVERNE AND SHIRLEY WITH THE FONZ (II,1449); LIPPY THE LION (I,2710); THE LITTLE RASCALS (II,1494); LOST IN SPACE (I,2759); THE MAGILLA GORILLA SHOW (I,2825); MOBY DICK AND THE MIGHTY MIGHTOR (I,3077); MONCHHICHIS (II,1722); MORK AND MINDY (II,1736); MOTOR MOUSE (I,3111); THE NEW ADVENTURES OF HUCKLEBERRY FINN (I,3250); THE NEW FRED AND BARNEY SHOW (II,1817); THE NEW SHMOO (II,1827); THE NEW SUPER FRIENDS HOUR (II,1830); PAC-MAN (II,1940); PARTRIDGE FAMILY: 2200 A.D. (II,1963); PEBBLES AND BAMM BAMM (II,1987); THE PERILS OF PENELOPE PITSTOP (I,3525); THE PETER POTAMUS SHOW (I,3568); PINK PANTHER AND SONS (II,2043); THE QUICK DRAW MCGRAW SHOW (I,3710); THE RICHIE RICH SHOW (II,2169); THE ROMAN HOLIDAYS (I,3835); THE RUFF AND REDDY SHOW (I,3866); SAMPSON AND GOLIATH (I,3903); SCOOBY'S ALL-STAR LAFF-A-LYMPICS (II,2273); SCOOBY-DOO AND SCRAPPY-DOO (II,2274); THE SCOOBY-DOO AND SCRAPPY-DOO SHOW (II,2275); SCOOBY-DOO, WHERE ARE YOU? (II,2276); THE SCOOBY-DOO/DYNOMUTT HOUR (II,2277); SEALAB 2020 (I,3949); SERGEANT T.K. YU (II,2305); SHAZZAN! (I,4002); THE SHIRT TALES (II,2337); THE SKATEBIRDS (II,2375); THE SMURFS (II,2385); SNORKS (II,2390); SPACE GHOST (I,4141); SPACE KIDDETTES (I,4143); SPACE STARS (II,2408); SPEED BUGGY (II,2419); SUPER FRIENDS (II,2497); THE SUPER GLOBETROTTERS (II,2498); THESE ARE THE DAYS (II,2585); THE THREE ROBONIC STOOGES (II,2609); THE TOM AND JERRY SHOW (I,4534); TOP CAT (I,4576); TOUCHE TURTLE (I,4586); TROLLKINS (II,2660); VALLEY OF THE DINOSAURS (II,2717); THE WACKY RACES (I,4745); WAIT TIL YOUR FATHER GETS HOME (I,4748); WALLY GATOR (I,4751); WHEELIE AND THE CHOPPER BUNCH (II,2776); WHERE'S HUDDLES (I,4836); THE WORLD'S GREATEST SUPER HEROES (II,2842); YOGI BEAR (I,4928); YOGI'S SPACE RACE (II,2853)

BARBOUR, JOHN
REAL PEOPLE (II,2118); SPEAK UP AMERICA (II,2412)

BARBOUR, OLIVE
THOSE TWO (I,4470)

BARCLAY, EDWIN
PLACE THE FACE (I,3611)

BARCLAY, HUMPHREY
DOCTOR IN THE HOUSE (I,1313); NO—HONESTLY (II,1857)

BARCLAY, RICHARD
THE HALLOWEEN THAT ALMOST WASN'T (II,1075); STRIPPERS (II,2481); THOSE FABULOUS CLOWNS (II,2602)

BARDNER, GERALD
WHAT'S HAPPENING!! (II,2769)

BARE, RICHARD L
GANGBUSTERS (I,1733); THE ISLANDERS (I,2243); MAN AGAINST CRIME (I,2852); YOU'RE ONLY YOUNG ONCE (I,4969)

BARIMO, GEORGE
THE DOCTORS (II,694); A FINE ROMANCE (II,858); KATE AND ALLIE (II,1365)

BARKER, CECIL
BING CROSBY AND HIS FRIENDS (I,455); BRANDED (I,707); HOLIDAY LODGE (I,2077); THE RED SKELTON CHEVY SPECIAL (I,3753); THE RED SKELTON SHOW (I,3755); THE RED SKELTON TIMEX SPECIAL (I,3757)

BARKLEY, DEANNE
WEDDING DAY (II,2759)

BARLOW, DAVID
SEEING THINGS (II,2297)

BARMAK, IRA
THE GAME GAME (I,1732); MASON (II,1643); ROSENTHAL AND JONES (II,2215)

BARNES, HOWARD G
ARTHUR MURRAY'S DANCE PARTY (I,293)

BARNETT, JACKIE
THE ANTHONY NEWLEY SHOW (I,233); THE BARBARA MCNAIR AND DUKE ELLINGTON SPECIAL (I,347); THE BLUE JEAN NETWORK (II,274); THE CATERINA VALENTE SHOW (I,873); CENTER STAGE: CYD CHARISSE (I,889); THE CYD CHARISSE SPECIAL (I,1118); THE EDDIE FISHER SPECIAL (I,1410); THE GEORGE KIRBY SPECIAL (I,1771); THE JANE MORGAN SHOW (I,2342); THE JULIE LONDON SPECIAL (I,2484); THE KATE SMITH SHOW (I,2511); MUSIC CENTRAL (II,1771); ON STAGE WITH BARBARA MCNAIR (I,3358); ON STAGE: PHIL SILVERS (I,3357); THE PEGGY LEE SPECIAL (I,3510); THE POLLY BERGEN SPECIAL (I,3644); THE ROBERT GOULET SPECIAL (I,3807); THE SHANI WALLIS SHOW (I,3997); THE SHIRLEY BASSEY SHOW (I,4014); THE SHIRLEY BASSEY SHOW (I,4015); THE SMOKEY ROBINSON SHOW (I,4093); THE TRINI LOPEZ SHOW (I,4606); THE VIKKI CARR SHOW (I,4732)

BARNETT, JOAN
KENNEDY (II,1377)

BARNUM, PETE
THE ALL-STAR REVUE (I,138); THE ALL-STAR SUMMER REVUE (I,139)

BARON, ALLEN
HOUSE CALLS (II,1194)

BARR, ANTHONY
CLIMAX! (I,976)

BARRETT, EARL
FAMILY BUSINESS (II,815); POOR DEVIL (I,3646); THE SAN PEDRO BEACH BUMS (II,2250); THE SANDY DUNCAN SHOW (I,3910); TOO CLOSE FOR COMFORT (II,2642)

BARRETT, ELLEN
RYAN'S HOPE (II,2234); SEARCH FOR TOMORROW (II,2284)

BARRETT, JAMES LEE
BIG JOHN (II,239)

BARRETT, JOHN
THE SMOTHERS ORGANIC PRIME TIME SPACE RIDE (I,4098)

BARRETT, TONY
DAN RAVEN (I,1131); THE MOD SQUAD (I,3078); THE MONK (I,3087); THE SNOOP SISTERS (II,2389)

BARRETT, WALTER
MUSIC CITY NEWS TOP COUNTRY HITS OF THE YEAR (II,1772)

BARRIS, CHUCK
THE $1.98 BEAUTY SHOW (II,698); CAMOUFLAGE (II,417); THE CHUCK BARRIS RAH-RAH SHOW (II,525); THE DATING GAME (I,1168); DREAM GIRL OF '67 (I,1375); THE GAME GAME (I,1732); THE GONG SHOW (II,1026); THE GONG SHOW (II,1027); HOW'S YOUR MOTHER-IN-LAW? (I,2148); LEAVE IT TO THE WOMEN (II,1454); THE NEW NEWLYWED GAME (II,1824); THE NEW TREASURE HUNT (I,3275); THE NEW TREASURE HUNT (II,1831); THE NEWLYWED GAME (II,1836); OPERATION: ENTERTAINMENT (I,3401); THE PARENT GAME (I,3455); THREE'S A CROWD (II,2612); TREASURE HUNT (II,2657); YOUR HIT PARADE (II,2872)

BARRON, ARTHUR
THE ABC AFTERSCHOOL SPECIAL (II,1)

BARRON, EVELYN
THE ABC AFTERSCHOOL SPECIAL (II,1)

BARRON, FRED
THE BOUNDER (II,357)

BARRON, JEFFREY
TOP TEN (II,2648)

BARRY, JACK
BACK THE FACT (I,327); BERT CONVY SPECIAL—THERE'S A MEETING HERE TONIGHT (II,210); BLANK CHECK (II,265); BREAK THE BANK (II,369); BULLSEYE (II,393); CONFLICT (I,1036); FAITH BALDWIN'S THEATER OF ROMANCE (I,1515); THE HOLLYWOOD CONNECTION (II,1161); HOT POTATO (II,1188); THE JOE DIMAGGIO SHOW (I,2399); JOKER! JOKER!! JOKER!!! (II,1340); THE JOKER'S WILD (II,1341); JUVENILE JURY (I,2501); LIFE BEGINS AT EIGHTY (I,2682); OH, BABY! (I,3340); PLAY THE PERCENTAGES (II,2053); TALES OF THE HAUNTED (II,2538); TIC TAC DOUGH (II,2618); WAY OUT GAMES (II,2747); WINKY DINK AND YOU (I,4879); YOU'RE ON YOUR OWN (I,4968)

BARRY, PETER
JANE WYMAN PRESENTS THE FIRESIDE THEATER (I,2345)

BARRY, PHILIP
THE ALCOA HOUR (I,108); DOWN HOME (II,731); THE MANY LOVES OF ARTHUR (II,1625); MEET CORLISS ARCHER (I,2981); SANCTUARY OF FEAR

(II,2252); THE SILENT FORCE (I,4047)

BARRY, WESLEY
WILD BILL HICKOK (I,4861)

BARSOCCHINI, PETER
THE MERV GRIFFIN SHOW (II,1688)

BARTHOLOMEW, FRED
AS THE WORLD TURNS (II,110); SEARCH FOR TOMORROW (II,2284)

BARTLETT, JUANITA
THE GREATEST AMERICAN HERO (II,1060); NO MAN'S LAND (II,1855); THE QUEST (II,2098); THE ROCKFORD FILES (II,2197); SCARECROW AND MRS. KING (II,2268); STONE (II,2471); TENSPEED AND BROWN SHOE (II,2562)

BARTLETT, RICHARD H
CIMARRON CITY (I,953); RIVERBOAT (I,3797)

BARTLETT, WALTER E
COUNTRY GALAXY OF STARS (II,587)

BARTLEY, ANTHONY
ASSIGNMENT: FOREIGN LEGION (I,300)

BARTON, FRANKLIN
HAWAII FIVE-O (II,1110); OFF THE WALL (II,1879); RANSOM FOR ALICE (II,2113); SEVENTH AVENUE (II,2309)

BARTON, JOHN
PITFALL (II,2047)

BARZYK, FRED
JEAN SHEPHERD'S AMERICA (I,2354)

BASCH, CHARLES
VERSATILE VARIETIES (I,4713)

BASER, MICHAEL
9 TO 5 (II,1852)

BASKIN, JOHN
CRAZY LIKE A FOX (II,601); DEAR TEACHER (II,653); FISHERMAN'S WHARF (II,865); FOR MEMBERS ONLY (II,905); HIGH FIVE (II,1143)

BASS, JULES
'TWAS THE NIGHT BEFORE CHRISTMAS (II,2677); THE BEATLES (I,380); THE CONEHEADS (II,567); THE EASTER BUNNY IS COMIN' TO TOWN (II,753); FROSTY THE SNOWMAN (II,934); FROSTY'S WINTER WONDERLAND (II,935); HERE COMES PETER COTTONTAIL (II,1132); JACK FROST (II,1263); THE JACKSON FIVE (I,2326); KID POWER (I,2532); THE LEPRECHAUN'S CHRISTMAS GOLD (II,1461); THE LITTLE DRUMMER BOY (II,1485); LITTLE DRUMMER BOY, BOOK II (II,1486); THE NEW ADVENTURES OF PINOCCHIO (I,3253); THE OSMONDS (I,3410); PINOCCHIO'S CHRISTMAS (II,2045); THE RELUCTANT DRAGON AND MR. TOAD (I,3762); RUDOLPH THE RED-NOSED REINDEER (II,2227); RUDOLPH'S SHINY NEW YEAR (II,2228); SANTA CLAUS IS COMIN' TO TOWN (II,2259); SMOKEY THE BEAR SHOW (I,4094); TOMFOOLERY (I,4551); THE YEAR WITHOUT A SANTA CLAUS (II,2846)

BASS, STANLEY
JOHNNY BELINDA (I,2418)

BASSLER, ROBERT
BUCKSKIN (I,753); THE CALIFORNIANS (I,795)

BAST, WILLIAM
THE HAMPTONS (II,1076); TUCKER'S WITCH (II,2667)

BATCHELOR, JOY
DODO—THE KID FROM OUTER SPACE (I,1324); TOMFOOLERY (I,4551)

BATES, RICHARD
MARRIED ALIVE (I,2923); THE PRIME OF MISS JEAN BRODIE (II,2082)

BATTISTA, THOMAS M
BONKERS! (II,348)

BAUER, FRANK
ALCOA PREMIERE (I,109); ALCOA/GOODYEAR THEATER (I,107); WICHITA TOWN (I,4857)

BAUM, FRANK
M STATION: HAWAII (II,1568)

BAUM, FRED
HAWAII FIVE-O (II,1110)

BAUMES, WILFRED LLOYD
CALL HOLME (II,412); HONEYMOON SUITE (I,2108); WONDER WOMAN (SERIES 1) (II,2828); WONDER WOMAN (SERIES 2) (II,2829)

BAUMRUCKER, ALLAN
THE KEANE BROTHERS SHOW (II,1371); PUMPBOYS AND DINETTES ON TELEVISION (II,2090); VANITIES (II,2720)

BAUXABAUM, JAMES M
FLIPPER (I,1604)

BAXTER, BRUCE
DOCTOR SNUGGLES (II,687)

BAYER, WOLFGANG
WILD WILD WORLD OF ANIMALS (I,4864)

BEAN, JACK
MITZI (I,3071); MITZI AND A HUNDRED GUYS (II,1710); MITZI'S SECOND SPECIAL (I,3074); MITZI. . . ROARIN' IN THE 20S (II,1711); MITZI. . . WHAT'S HOT, WHAT'S NOT (II,1712); MITZI. . . ZINGS INTO SPRING (II,1713); MITZI. . . THE FIRST TIME (I,3073); MITZI: A TRIBUTE TO THE AMERICAN HOUSEWIFE (I,3072)

BEARDE, CHRIS
THE ANDY WILLIAMS SHOW (I,210); BOB HOPE SPECIAL: BOB HOPE'S ALL-STAR COMEDY SPECIAL FROM AUSTRALIA (II,300); BOB HOPE SPECIAL: BOB HOPE'S CHRISTMAS PARTY (II,307); THE BOBBY VINTON SHOW (II,345); THE CHEAP SHOW (II,491); COS (II,576); THE GONG SHOW (II,1026); THE GONG SHOW (II,1027); THE HALF-HOUR COMEDY HOUR (II,1074); THE HUDSON BROTHERS RAZZLE DAZZLE COMEDY SHOW (II,1201); THE HUDSON BROTHERS SHOW (II,1202); THE KEN BERRY WOW SHOW (I,2524); LI'L ABNER (I,2702); THE OSMONDS SPECIAL (II,1929); PUTTIN' ON THE HITS (II,2093); THE SMOTHERS BROTHERS COMEDY HOUR (I,4095); THE SONNY AND CHER COMEDY HOUR (II,2400); THE SONNY COMEDY REVUE (II,2403); THE STANLEY SIEGEL SHOW (II,2433); THAT'S MY MAMA (II,2580); TOP TEN (II,2648); WACKO (II,2733)

BEATON, ALEX
THE BLACK SHEEP SQUADRON (II,262); BOSTON AND KILBRIDE (II,356); DOCTOR SCORPION (II,685); DOCTOR STRANGE (II,688); THE DUKES (II,740); THE GREATEST AMERICAN HERO (II,1060); THE GYPSY WARRIORS (II,1070); HARRY O (II,1098); HARRY O (II,1099); KUNG FU (II,1416); LEGMEN (II,1458); MRS R.—DEATH AMONG FRIENDS (II,1759); THE NIGHT RIDER (II,1848); NIGHTSIDE (II,1851); RICHIE BROCKELMAN, PRIVATE EYE (II,2167); STONE (II,2470); STONE (II,2471); TENSPEED AND BROWN SHOE (II,2562)

BEATON, DOUGLAS
MOONLIGHT (II,1730)

BEATTS, ANNE
SQUARE PEGS (II,2431)

BEAUDINE JR, WILLIAM
LASSIE (I,2622); LASSIE: THE NEW BEGINNING (II,1433); TIMMY AND LASSIE (I,4520)

BEAUMONT, ALAN
THE ARLENE FRANCIS SHOW (I,258); A TIME TO LIVE (I,4510)

BEAVERS, JACKIE
TONY THE PONY (II,2641)

BEBAN, RICHARD
IT'S ONLY HUMAN (II,1257)

BECKER, FRED
PAROLE (I,3460)

BECKER, TERRY
BENDER (II,204); THE LAST HURRAH (I,2626); RIDING FOR THE PONY EXPRESS (II,2172); SAVAGE: IN THE ORIENT (II,2264)

BECKER, VERNON P
THE CHARLIE CHAPLIN COMEDY THEATRE (I,912); THE JERRY LESTER SHOW (I,2360); TOOTSIE HIPPODROME (I,4575)

BECKETT, KEITH
THE BENNY HILL SHOW (II,207)

BECKWITH, AARON
WHAT GAP? (I,4808)

BEDSON, SUSAN
AS THE WORLD TURNS (II,110)

BEETSON, FRANK
HIGHWAY HONEYS (II,1151)

BEGG, JIM
THE KID WITH THE BROKEN HALO (II,1387)

BEHAR, JOE
KOVACS ON THE CORNER (I,2582); LET'S MAKE A DEAL (II,1462)

BEHR, FELICIA
ALL MY CHILDREN (II,39); RYAN'S HOPE (II,2234)

BEICH, ALBERT
KENTUCKY JONES (I,2526)

BELANGER, PAUL
THROUGH THE CRYSTAL BALL (I,4494)

BELCHER, JO ANNE
A.K.A. PABLO (II,28)

BELKIN, GARY
SUGAR TIME (II,2489)

BELL, ALAN J W
THE HITCHHIKER'S GUIDE TO THE GALAXY (II,1158)

BELL, CHRISTOPHER
SHEENA EASTON—LIVE AT THE PALACE (II,2324)

BELL, DAVID
THE BENNY HILL SHOW (II,207); WHITNEY AND THE ROBOT (II,2789)

BELL, JANE UPTON
BOB HOPE SPECIAL: BOB HOPE'S STAND UP AND CHEER FOR THE NATIONAL FOOTBALL LEAGUE'S 60TH YEAR (II,316)

BELL, WILLIAM
THE YOUNG AND THE RESTLESS (II,2862)

BELLAMY, EARL
THE PIGEON (I,3596)

BELLAND, BRUCE
THE CROSS-WITS (II,605); KNOCKOUT (II,1403); THE NEW TRUTH OR CONSEQUENCES (II,1832)

BELLISARIO, DONALD P
AIRWOLF (II,27); BATTLESTAR GALACTICA (II,181); THE BLACK SHEEP SQUADRON (II,262); MAGNUM, P.I. (II,1601); QUINCY, M. E. (II,2102); THE RETURN OF LUTHER GILLIS (II,2137); STONE (II,2471); TALES OF THE GOLD MONKEY (II,2537)

BELOIN, EDMUND
FAMILY AFFAIR (I,1519); SO THIS IS HOLLYWOOD (I,4105); THE THIN MAN (I,4446)

BELOUS, PAUL M
THE OUTLAWS (II,1936); PEN 'N' INC. (II,1992)

BELSON, JERRY
EVIL ROY SLADE (I,1485); HEY, LANDLORD (I,2039); MIXED NUTS (II,1714); THE MURDOCKS AND THE MCCLAYS (I,3163); THE ODD COUPLE (II,1875); SHERIFF WHO? (I,4009); YOUNG GUY CHRISTIAN (II,2864)

BELTH, DON
MAKE A WISH (I,2838)

BENCIVENGO, RICHARD
GRAFFITI ROCK (II,1049)

BENEDETTO, JULIO D
THE ALAN KING SHOW (I,100)

BENEDICT, MARTHA
THE MAGIC OF DAVID COPPERFIELD (II,1593)

BENEDICT, RICHARD
MISSION: IMPOSSIBLE (I,3067)

BENEDICT, ROBERT
FEELING GOOD (I,1559); THE GREAT AMERICAN DREAM MACHINE (I,1871)

BENEDICT, TONY
PANDAMONIUM (II,1948)

BENJAMIN, BURTON
20TH CENTURY (I,4640); CRUNCH AND DES (I,1108); WORLD WAR I (I,4919); YOU ARE THERE (I,4930)

BENKOIL, MAURY
FOUL UPS, BLEEPS AND BLUNDERS (II,911)

BENNETT, DANNY
ON THE TOWN WITH TONY BENNETT (II,1895)

BENNETT, FRED W
THE ABC AFTERSCHOOL SPECIAL (II,1)

BENNETT, HARVE
ALEX AND THE DOBERMAN GANG (II,32); THE AMERICAN GIRLS (II,72); THE BIONIC WOMAN (II,255); FROM HERE TO ETERNITY (II,932); FROM HERE TO ETERNITY (II,933); THE GEMINI MAN (II,960); THE GEMINI MAN (II,961); GO WEST YOUNG GIRL (II,1013); THE INVISIBLE MAN (II,1242); THE INVISIBLE MAN (II,1243); THE LEGEND OF THE GOLDEN GUN (II,1455); THE MOD SQUAD (I,3078); NICK AND THE DOBERMANS (II,1843); THE POWERS OF MATTHEW STAR (II,2075); RICH MAN, POOR MAN—BOOK I (II,2161); SALVAGE (II,2244); SALVAGE 1 (II,2245); THE SIX-MILLION-DOLLAR MAN (II,2372); A WOMAN CALLED GOLDA (II,2818)

BENNETT, MEG
THE YOUNG AND THE RESTLESS (II,2862)

BENNETT, ROBERT
SHANNON (II,2314)

BENNETT, RONNI
THE BARBARA WALTERS SPECIAL (II,145)

BENNETT, STEPHANIE
THE EVERLY BROTHERS REUNION CONCERT (II,791)

BENNEWITZ, BETTINA
ROCK 'N' ROLL: THE FIRST 25 YEARS (II,2189)

BENNY, JACK
THE GISELE MACKENZIE SHOW (I,1812)

BENOFF, MAC
LIFE WITH LUIGI (I,2692)

BENSFIELD, DICK
ANOTHER MAN'S SHOES (II,96); HELLO, LARRY (II,1128); ONE DAY AT A TIME (II,1900); T.L.C. (II,2525)

BENSON, HUGH
THE BLUE AND THE GRAY (II,273); THE EYES OF CHARLES SAND (I,1497); FATHER KNOWS BEST: HOME FOR CHRISTMAS (II,839); FATHER KNOWS BEST: THE FATHER KNOWS BEST REUNION (II,840); GOLIATH AWAITS (II,1025); HANK (I,1931); HART TO HART (II,1102); THE OUTCASTS (I,3425); POLICE STORY (II,2062); THE SECOND HUNDRED YEARS (I,3959)

BENSON, JAY
AT EASE (II,116); LUCAS TANNER (II,1556); THE NIGHTENGALES (II,1850); THE STREETS (II,2479)

BENSON, LEON
FLIPPER (I,1604); HARBOURMASTER (I,1946); RIPCORD (I,3793); SEA HUNT (I,3948); THE WEST POINT STORY (I,4796)

BENSON, RAY
JOSEPH SCHILDKRAUT PRESENTS (I,2449)

BENTON, DOUGLAS
FEMALE INSTINCT (I,1564); THE GIRL FROM U.N.C.L.E. (I,1808); HEC RAMSEY (II,1121); MAGNUM, P.I. (II,1601); OWEN MARSHALL: COUNSELOR AT LAW (I,3436); POLICE WOMAN (II,2063); PUNCH AND JODY (II,2091); THE RETURN OF LUTHER GILLIS (II,2137)

BENZA, JACK
THE ADAMS CHRONICLES (II,8)

BERCOVICI, ERIC
ASSIGNMENT: MUNICH (I,302); ASSIGNMENT: VIENNA (I,304); CHICAGO STORY (II,506); THE CHICAGO STORY (II,507); JESSIE (II,1287); MCCLAIN'S LAW (II,1659); SHOGUN (II,2339); WASHINGTON: BEHIND CLOSED DOORS (II,2744)

BERCOVICI, JULIA
ROAD TO REALITY (I,3800)

BERG, CHERNEY
THE GOLDBERGS (I,1836)

BERG, DICK
ALCOA PREMIERE (I,109); THE BOB HOPE CHRYSLER THEATER (I,502); CHECKMATE (I,919); FIREHOUSE (I,1579); FIREHOUSE (II,859); THE FLIERS (I,1596); HEAT OF ANGER (I,1992); JOHNNY BELINDA (I,2418); THE MARTIAN CHRONICLES (II,1635); THE MUSIC MAKER (I,3173); THE WORD (II,2833)

BERGER, RICHARD
ANYTHING GOES (I,237); THE CHESTERFIELD SUPPER CLUB (I,923); MUSICAL COMEDY TIME (I,3183)

BERGER, ROBERT
THE FIRM (II,860); THE FOUR OF US (II,914); LAND OF HOPE (II,1424) RX FOR THE DEFENSE (I,3878); THE STRANGE REPORT (I,4248)

BERGMAN, ALAN
KID GLOVES (I,2531); THAT MCMAHON'S HERE AGAIN (I,4419)

BERGMAN, DAVID
LIVE FROM THE LONE STAR (II,1501); SOAP FACTORY DISCO (II,2393)

BERGMAN, DORIS
CINDERELLA (II,526)

BERGMAN, TED
D.H.O. (I,1249); GOOD OL' BOYS (II,1035); LOVE THY NEIGHBOR (I,2781); MISS WINSLOW AND SON (II,1705)

BERK, HOWARD
THE CONTENDER (II,571)

BERK, MICHAEL
MANIMAL (II,1622)

BERKE, LESTER
B.J. AND THE BEAR (II,125); QUINCY, M. E. (II,2102)

BERKE, WILLIAM
THE GOLDBERGS (I,1835); THE GOLDBERGS (I,1836); I SPY (I,2178)

BERKLEY, DEANNE
THE DAY THE WOMEN GOT EVEN (II,628) THE FAMILY TREE (II,820); THE SIX OF US (II,2369); VALENTINE MAGIC ON LOVE ISLAND (II,2716)

BERLE, MILTON
THE MILTON BERLE SHOW (I,3047); THE MILTON BERLE SHOW (I,3048); THE MILTON

BERLE SPECIAL (I,3051); TEXACO STAR THEATER (I,4411)

BERLINER, RALPH
BARBAPAPA (II,142)

BERMAN, HENRY
GETAWAY CAR (I,1783); NIGHT PROWL (I,3291); THE STAR MAKER (I,4193)

BERMAN, MARTIN
HOUR MAGAZINE (II,1193)

BERMAN, MARY
AMERICA ALIVE! (II,67)

BERMAN, MONTY
THE ADVENTURER (I,43); THE BARON (I,360); THE CHAMPIONS (I,896); DEPARTMENT S (I,1232); MY PARTNER THE GHOST (I,3199)

BERNARD, ALAN
THE ANDY WILLIAMS SHOW (I,210); THE CAPTAIN AND TENNILLE (II,429); THE JOHN DAVIDSON CHRISTMAS SHOW (II,1307)

BERNARD, BARRY
CHILDREN'S SKETCH BOOK (I,941)

BERNARD, GLEN
THE BIG TOP (I,435)

BERNARD, OLIVER
THE JOHN DAVIDSON SHOW (II,1309)

BERNDS, EDWARD
THE HANK MCCUNE SHOW (I,1932)

BERNER, IRENE
GETTING READY (II,984)

BERNHARD, HARVEY
FRANK SINATRA JR. WITH FAMILY AND FRIENDS (I,1663)

BERNS, LARRY
HOW TO (I,2145); OUR MISS BROOKS (I,3415); YES YES NANETTE (I,4927)

BERNS, SEYMOUR
THE IMPERIAL GRAND BAND (II,1224); THE LITTLEST HOBO (II,1500); NICK AND NORA (II,1842); THE QUIZ KIDS (II,2103); THE RED SKELTON SHOW (I,3755); A SALUTE TO STAN LAUREL (I,3892); SEARCH AND RESCUE: THE ALPHA TEAM (II,2283); THE TROUBLE WITH TRACY (I,4613)

BERNS, WILLIAM
LADIES BE SEATED (I,2598)

BERNSEN, HARRY
THE ABC AFTERSCHOOL SPECIAL (II,1)

BERNSTEIN, GARY
ANYTHING FOR MONEY (II,99)

BERNSTEIN, HARRY
THE AWAKENING LAND (II,122)

BERNSTEIN, JAY
BRING 'EM BACK ALIVE (II,377); MICKEY SPILLANE'S MARGIN FOR MURDER (II,1691); MICKEY SPILLANE'S MIKE HAMMER (II,1692); MICKEY SPILLANE'S MIKE HAMMER: MORE THAN MURDER (II,1693); MICKEY SPILLANE'S MIKE HAMMER: MURDER ME, MURDER YOU (II,1694); MORE WILD WILD WEST (II,1733); THE WILD WILD WEST REVISITED (II,2800)

BERNSTEIN, JERRY
THE UGLIEST GIRL IN TOWN (I,4663)

BERNSTEIN, RICK
THE 416TH (II,913); THE STEVE LANDESBERG TELEVISION SHOW (II,2458)

BERNSTEIN, STU
TELEVISION INSIDE AND OUT (II,2552)

BERNSTEIN, WALTER
SPARROW (II,2411)

BERRY, TIM
POOR RICHARD (II,2065)

BERTOLUCCI, GIOVANNI
MARCO POLO (II,1626)

BERZNER, JOHN
EARTH, WIND & FIRE IN CONCERT (II,750)

BESSADA, MILAND
SECOND CITY TELEVISION (II,2287)

BIBAS, FRANK P
THE VEIL (I,4709)

BICKLEY, WILLIAM S
HAPPY DAYS (II,1084); OUT OF THE BLUE (II,1934); THE PARTRIDGE FAMILY (II,1962); PLEASE STAND BY (II,2057); WHAT'S HAPPENING!! (II,2769)

BIDDELL, STEVE
BRONCO (I,745)

BIEN, WALTER
BARNEY AND ME (I,358); THAT'S MY MAMA (II,2580); TOPPER RETURNS (I,4583)

BIENER, TOM
THE ABC AFTERSCHOOL SPECIAL (II,1); THE DONNA FARGO SHOW (II,710); FLO (II,891)

BILL, RICHARD
WOMAN WITH A PAST (I,4889)

BILLETT, STU
GARROWAY (I,1736); IT'S ANYBODY'S GUESS (II,1255); ONE IN A MILLION (I,3379); THE PEOPLE'S COURT (II,2001); SO YOU THINK YOU GOT TROUBLES?! (II,2391); SPLIT SECOND (II,2428); THREE FOR THE MONEY (II,2606)

BILLINGSLEY, SHERMAN
THE STORK CLUB (I,4236)

BILLITERA, SALVATORE
JOHNNY SOKKO AND HIS FLYING ROBOT (I,2436)

BILSON, BRUCE
THE BAILEYS OF BALBOA (I,334)

BILSON, GEORGE
CAPTAIN MIDNIGHT (I,833)

BINDER, STEVE
AMERICA (I,164); THE BARRY MANILOW SPECIAL (II,155); BLONDES VS. BRUNETTES (II,271); DEBBY BOONE. . . ONE STEP CLOSER (II,655); DIANA (II,667); DOROTHY HAMILL'S CORNER OF THE SKY (II,721); ELVIS (I,1435); A LAST LAUGH AT THE 60'S (I,2627); THE LESLIE UGGAMS SHOW (I,2660); LUCY IN LONDON (I,2789); MAC DAVIS CHRISTMAS SPECIAL. . . WHEN I GROW UP (II,1574); THE MAC DAVIS SHOW (II,1576); OLIVIA (II,1883); PETULA (I,3572); SHIELDS AND YARNELL (II,2328); A SPECIAL EDDIE RABBITT (II,2415)

BING, MACK
HOLIDAY U.S.A. (I,2078); UNCLE CROC'S BLOCK (II,2702)

BINNS, BRONWYN
AGAINST THE WIND (II,24)

BIRCH, PETER
CAPTAIN KANGAROO (II,434); CHRISTMAS LEGEND OF NASHVILLE (II,522)

BIRCH, YVONNE KING
THE KING FAMILY CELEBRATE THANKSGIVING (I,2545); THE KING FAMILY CHRISTMAS SPECIAL (I,2546); THE KING FAMILY IN ALASKA (I,2547); THE KING FAMILY IN HAWAII (I,2548); THE KING FAMILY OCTOBER SPECIAL (I,2551); THE KING FAMILY SHOW (I,2552); THE KING FAMILY SPECIAL (I,2553); THE KING FAMILY VALENTINE'S DAY SPECIAL (I,2554)

BIRGLIA, RICHARD
CANDID CAMERA (II,420)

BIRNBAUM, BOB
EGAN (I,1419); THE PHOENIX (II,2036); SHIRLEY (II,2330)

BIRNBAUM, ROGER
RYAN'S FOUR (II,2233)

BIRTH, JURGEN
WILD WILD WORLD OF ANIMALS (I,4864)

BISCHOFF, SAMUEL
FOR THE DEFENSE (I,1624)

BISHOP, BENJAMIN
BLACK BEAUTY (II,261)

BISHOP, MARGE
ACTION AUTOGRAPHS (I,24)

BIXBY, BILL
GOODNIGHT, BEANTOWN (II,1041)

BLACH, ALBERT
THE ALAN DALE SHOW (I,96)

BLACK, JOHN D F
THE FUZZ BROTHERS (I,1722); A SHADOW IN THE STREETS (II,2312); STAR TREK (II,2440); WONDER WOMAN (PILOT 1) (II,2826)

BLACKBURN, NORMAN
CIRCUS BOY (I,959)

BLACTON, JENNIE
MOTHER AND ME, M.D. (II,1740)

BLAIR, GARY
SEYMOUR PRESENTS (I,3990)

BLAIR, JOCK
THE SULLIVANS (II,2490)

BLAIR, LEONARD
THE BRIGHTER DAY (I,728); VALIANT LADY (I,4695)

BLAIR, WENDY
THE ROPERS (II,2214); THREE'S A CROWD (II,2613)

BLAKE, ROBERT
JOE DANCER: THE BIG BLACK PILL (II,1300); JOE DANCER: THE MONKEY MISSION (II,1301); JOE DANCER: MURDER ONE, DANCER 0 (II,1302)

BLANCHARD, MITCHELL
RED SKELTON'S FUNNY FACES (II,2124)

BLATT, DANIEL H
BORN TO THE WIND (II,354); INSPECTOR PEREZ (II,1237); MICKEY SPILLANE'S MIKE HAMMER: MORE THAN MURDER (II,1693); THREE EYES (II,2605); V: THE FINAL BATTLE (II,2714); V: THE SERIES (II,2715)

BLAUBUT, DON
JOHNNY OLSEN'S RUMPUS ROOM (I,2433)

BLEES, ROBERT
BONANZA (II,347); BUS STOP (I,778); COMBAT (I,1011); COUNTY GENERAL (I,1076); KNIGHT'S GAMBIT (I,2576); OFF WE GO (I,3338); PROJECT UFO (II,2088)

BLEYER, BOB
SONGS FOR SALE (I,4124)

BLINN, WILLIAM
THE AMERICAN DREAM (II,71); THE LAZARUS SYNDROME (II,1451); THE MACKENZIES OF PARADISE COVE (II,1585); THE NEW LAND (II,1820); THE ROOKIES (II,2208); STICKIN' TOGETHER (II,2465)

BLOCK, PAUL
THE GEORGE JONES SPECIAL (II,975); ROCK-N-AMERICA (II,2195)

BLOODWORTH, LINDA
DRIBBLE (II,736); FILTHY RICH (II,856); LONDON AND DAVIS IN NEW YORK (II,1512)

BLOOM, GEORGE
STOCKARD CHANNING IN JUST FRIENDS (II,2467); WELCOME BACK, KOTTER (II,2761)

BLOOM, HAROLD JACK
THE D.A.: MURDER ONE (I,1122); HEC RAMSEY (II,1121)

BLOOMBERG, RON
9 TO 5 (II,1852); FIRST TIME, SECOND TIME (II,863)

BLUE, MARY JO
HOLLYWOOD STARS' SCREEN TESTS (II,1165); TV'S BLOOPERS AND PRACTICAL JOKES (II,2675)

BLUEL, RICHARD
CATCH 22 (I,871); THE GALLANT MEN (I,1728); GOLIATH AWAITS (II,1025); THE GREEN HORNET (I,1886); THE IMPOSTER (II,1225); THE MISADVENTURES OF SHERIFF LOBO (II,1704); TEMPLE HOUSTON (I,4392); THE WESTWIND (II,2766)

BLUM, DEBORAH
IN SEARCH OF. . . (II,1226); THAT'S INCREDIBLE! (II,2578)

BLYE, ALLAN
THE ANDY WILLIAMS SHOW (I,210); BIZARRE (II,258); BIZARRE (II,259); THE BOBBY VINTON SHOW (II,345); THE HUDSON BROTHERS RAZZLE DAZZLE COMEDY SHOW (II,1201); THE HUDSON BROTHERS SHOW (II,1202); JOEY AND DAD (II,1306); THE KEN BERRY WOW SHOW (I,2524); LI'L ABNER (I,2702); LOLA (II,1508); LOLA (II,1509); LOLA (II,1510); LOLA (II,1511); THE OSMONDS SPECIAL (II,1929); THE REDD FOXX COMEDY HOUR (II,2126); THE SMOTHERS BROTHERS COMEDY HOUR (I,4095); THE SONNY AND CHER COMEDY HOUR (II,2400); THE SONNY COMEDY REVUE (II,2403); THAT'S MY MAMA (II,2580); VAN DYKE AND COMPANY (II,2718); VAN DYKE AND COMPANY (II,2719)

BLYE, GARRY
DOLLY IN CONCERT (II,700)

BOBRICK, SAM
THE LATE FALL, EARLY SUMMER BERT CONVY SHOW (II,1441); THIS WEEK IN NEMTIN (I,4465); THE TIM CONWAY COMEDY HOUR (I,4501)

BOCHCO, STEVEN
BAY CITY BLUES (II,186); EVERY STRAY DOG AND KID (II,792); GRIFF (I,1888); HILL STREET BLUES (II,1154); THE INVISIBLE MAN (II,1242); THE INVISIBLE MAN (II,1243); PARIS (II,1957); RICHIE BROCKELMAN, PRIVATE EYE (II,2167); RICHIE BROCKELMAN: MISSING 24 HOURS (II,2168)

BOGART, PAUL
THE ADAMS CHRONICLES (II,8); HANSEL AND GRETEL (I,1934); HAWK (I,1974); MAMA MALONE (II,1609); SHIRLEY TEMPLE'S STORYBOOK: THE LEGEND OF SLEEPY HOLLOW (I,4018); YOU CAN'T TAKE IT WITH YOU (II,2857)

BOGIE, DUANE C
ALL CREATURES GREAT AND SMALL (I,129); THE BORROWERS (I,693); BRIEF ENCOUNTER (I,724); THE GATHERING STORM (I,1741); THE RIVALRY (I,3796); THE SMALL MIRACLE (I,4090)

BOHEM, ENDRE
RAWHIDE (I,3727)

BOLEN, LIN
FARRELL: FOR THE PEOPLE (II,834); GOLDEN GATE (II,1020); STUMPERS (II,2485); W.E.B. (II,2732)

BOLOGNA, JOSEPH
CALUCCI'S DEPARTMENT (I,805); GOOD PENNY (II,1036); LOVERS AND OTHER STRANGERS (II,1550)

BOMBECK, ERMA
MAGGIE (II,1589)

BONADUCE, JOSEPH
CALIFORNIA FEVER (II,411); MARIE (II,1628)

BOND, DENNIS M
THE PAUL WILLIAMS SHOW (II,1984)

BONI, JOHN
THE BAD NEWS BEARS (II,134); COMEDY OF HORRORS (II,557); SPACE FORCE (II,2407)

BONIS, HERB
COUNTY GENERAL (I,1076); THE JOHN GARY SHOW (I,2411); MUSICAL COMEDY TONIGHT (II,1777)

BONNER, MARY S
ANOTHER WORLD (II,97); TEXAS (II,2566)

BONOW, RAYSA
AMERICA ALIVE! (II,67)

BONSALL, SHULL
CRUSADER RABBIT (I,1109)

BOOKER, BOB
THE BEST LITTLE SPECIAL IN TEXAS (II,215); CHARO (II,488); COTTON CLUB '75 (II,580); THE DAVID STEINBERG SHOW (I,1179); FOUL UPS, BLEEPS AND BLUNDERS (II,911); THE PAUL LYNDE HALLOWEEN SPECIAL (II,1981); THE WAYNE NEWTON SPECIAL (II,2749); THE WORLD'S FUNNIEST COMMERCIAL GOOFS (II,2841)

BOORAREN, HENDRICK
JUST LIKE A WOMAN (I,2496)

BOOTH, PHILIP
BRIEF PAUSE FOR MURDER (I,725); THE MARSHAL OF GUNSIGHT PASS (I,2925); SHORTY (I,4025)

BORACK, CARL
KEY TORTUGA (II,1383)

BORDA, PAULO
FALCON'S GOLD (II,809)

BORGE, VICTOR
VICTOR BORGE'S COMEDY IN MUSIC III (I,4725)

BORMIS, GEORGE
THE ADAMS CHRONICLES (II,8)

BOROWITZ, ANDY
DREAMS (II,735)

BOSCO, WALLACE
BRAVE EAGLE (I,709)

BOSTON, JOE
B.J. AND THE BEAR (II,125); THE HARDY BOYS MYSTERIES (II,1090); THE MASTER (II,1648); THE MISADVENTURES OF SHERIFF LOBO (II,1704); THE NANCY DREW MYSTERIES (II,1789); SWORD OF JUSTICE (II,2521)

BOSUSTOW, STEPHEN
GERALD MCBOING-BOING (I,1778)

BOTNICK, BRUCE
KENNY LOGGINS IN CONCERT (II,1380)

BOWEN, JERRY
THE INA RAY HUTTON SHOW (I,2207)

BOWER, DALLAS
THE ADVENTURES OF SIR LANCELOT (I,76)

BOWERS, WILLIAM
MOBILE ONE (II,1717); MOBILE TWO (II,1718)

BOWMAN, CHUCK
240-ROBERT (II,2689); THE BLACK SHEEP SQUADRON (II,262); CASSIE AND COMPANY (II,455); HUNTER (II,1206); THE INCREDIBLE HULK (II,1232); THE ROUSTERS (II,2220); TENSPEED AND BROWN SHOE (II,2562); V (II,2713); YOUNG MAVERICK (II,2867)

BOWMAN, ROGER
KUDA BUX, HINDU MYSTIC (I,2593)

BOX, SYDNEY
WHITE HUNTER (I,4845)

BOYD, WILLIAM
HOPALONG CASSIDY (I,2118)

BOYER, CHARLES
FOUR STAR PLAYHOUSE (I,1652)

BOYETT, ROBERT L
ANGIE (II,80); FOUL PLAY (II,910); GOODTIME GIRLS (II,1042); JOANIE LOVES CHACHI (II,1295); OUT OF THE BLUE (II,1934)

BOYETT, WILLIAM
BOSOM BUDDIES (II,355); FOUL PLAY (II,910)

BOYLE, DONALD R
BIG FOOT AND WILD BOY (II,237); THE KROFFT SUPERSHOW (II,1412); THE KROFFT SUPERSHOW II (II,1413); MANIMAL (II,1622); THE SIX-MILLION-DOLLAR MAN (II,2372)

BOYRIVEN, PATRICK
V: THE FINAL BATTLE (II,2714)

BRADDOCK, STAN
PERRY COMO'S CHRISTMAS IN NEW YORK (II,2011)

BRADEMAN, BILL
QUICK AND QUIET (II,2099); TWO THE HARD WAY (II,2695)

BRADEN, EDDIE
THE MONTE CARLO SHOW (II,1726)

BRADFORD, HANK
DETECTIVE SCHOOL (II,661); SUGAR TIME (II,2489)

BRADFORD, JOHNNY
HERE'S EDIE (I,2030); THE JUDY GARLAND SHOW (I,2473); THE JUDY GARLAND SHOW (II,1349); A VERY SPECIAL OCCASION (I,4716)

BRADLEY, IAN
PRISONER: CELL BLOCK H (II,2085)

BRADY, BEN
THE KEN MURRAY SHOW (I,2525); OH, THOSE BELLS! (I,3345); THE OUTER LIMITS (I,3426); PERRY MASON (II,2025); THE RED BUTTONS SHOW (I,3750); THE RED SKELTON SHOW (I,3755)

BRAFF, DAVID
HOME ROOM (II,1172)

BRAND, JACK
ACTION AUTOGRAPHS (I,24)

BRAND, JOSHUA
ST. ELSEWHERE (II,2432)

BRAND, TONY
PEOPLE ARE FUNNY (II,1996)

BRANDERN, RUBEN
STRANGE PLACES (I,4247)

BRANDMAN, MICHAEL
BAREFOOT IN THE PARK (II,151); BLOCKHEADS (II,270); HERE IT IS, BURLESQUE! (II,1133); PAJAMA TOPS (II,1942)

BRANDON, JULIE
THE HOMEMADE COMEDY SPECIAL (II,1173)

BRANDON, PAUL TREVA
GOODNIGHT, BEANTOWN (II,1041)

BRANDSTEIN, EVE
E/R (II,748)

BRANIGAN, HUGH
DOUGH RE MI (I,1363)

BRANTON, MICHAEL
BACKSTAGE PASS (II,132)

BRANTON, RALPH
THE HORACE HEIDT SHOW (I,2120)

BRAO, LYNN FARR
WE GOT IT MADE (II,2752)

BRASSELLE, KEEFE
THE CARA WILLIAMS SHOW (I,843); THE KEEFE BRASSELLE SHOW (I,2514); THE REPORTER (I,3771)

BRAVERMAN, MICHAEL
QUINCY, M. E. (II,2102); THE RETURN OF MARCUS WELBY, M.D (II,2138)

BRAZIL, SCOTT
HILL STREET BLUES (II,1154)

BRECHER, IRVING
THE LIFE OF RILEY (I,2685); THE PEOPLE'S CHOICE (I,3521)

BRECKNER, BOB
THE PENDULUM (I,3511)

BREGMAN, BUDDY
AIN'T MISBEHAVIN' (II,25); THE GREAT AMERICAN MUSIC CELEBRATION (II,1052); OH MADELINE (II,1880)

BREGMAN, MARTIN
THE FOUR SEASONS (II,915); S*H*E (II,2235)

BRENNAN, ANDREW
THE OVER-THE-HILL GANG RIDES AGAIN (I,3432)

BRENNAN, BILL
EARN YOUR VACATION (I,1394); THE JOHNNY CARSON SHOW (I,2422)

BRENNER, NANCY
NOT NECESSARILY THE NEWS (II,1863)

BRENNER, NATHAN
MEN AT WORK IN CONCERT (II,1682)

BRENNER, ROBERT
QUIZZING THE NEWS (I,3713)

BRENT, JASON G
COME WITH ME—LAINIE KAZAN (I,1015)

BRESLER, JERRY
WHERE THE ACTION IS (I,4828); WHERE'S RAYMOND? (I,4837)

BRESLIN, HERBERT
PAVAROTTI AND FRIENDS (II,1985)

BRESLOW, LOU
DAMON RUNYON THEATER (I,1129)

BRESLOW, MARC
CARD SHARKS (II,438)

BREWER, JAMESON
BATTLE OF THE PLANETS (II,179)

BREZ, ETHEL
CASTLE ROCK (II,456)

BREZ, MEL
CASTLE ROCK (II,456)

BREZNER, LARRY
THE BILLY CRYSTAL COMEDY HOUR (II,248); BILLY CRYSTAL: A COMIC'S LINE (II,249); CHEERS (II,494); GOOD TIME HARRY (II,1037); STAR OF THE FAMILY (II,2436)

BRICKELL, BETH
A RAINY DAY (II,2110)

BRICKEN, JULES
THE 20TH CENTURY-FOX HOUR (I,4642); FORD THEATER HOUR (I,1635); JANE WYMAN PRESENTS THE FIRESIDE THEATER (I,2345); MIRACLE ON 34TH STREET (I,3056); RIVERBOAT (I,3797); SCHLITZ PLAYHOUSE OF STARS (I,3936)

BRIDGES, GINGER
HERE'S RICHARD (II,1137)

BRIERE, MARY S
TV'S BLOOPERS AND PRACTICAL JOKES (II,2675)

BRIGGLE, STOCKTON
CAPITOL (II,426)

BRIGGS, RICHARD
CUTTER TO HOUSTON (II,612); HAVING BABIES II (II,1108); UNCLE CROC'S BLOCK (II,2702)

BRIGHT, KEVIN
GEORGE BURNS' HOW TO LIVE TO BE 100 (II,972); THE MAGIC OF DAVID COPPERFIELD (II,1593)

BRILL, FRANK
THE JOHN DAVIDSON SHOW (II,1310); SIEGFRIED AND ROY (II,2350); THAT'S TV (II,2581)

BRILL, RICHARD
DATELINE: HOLLYWOOD (I,1167); STAND UP AND BE COUNTED (I,4187)

BRILLSTEIN, BERNIE
BUCKSHOT (II,385); BUFFALO BILL (II,387); THE BURNS AND SCHREIBER COMEDY HOUR (I,768); THE BURNS AND SCHREIBER COMEDY HOUR (I,769); JUMP (II,1356); OPEN ALL NIGHT (II,1914); SHOW BUSINESS (II,2343); SITCOM (II,2367)

BRINCKERHOFF, BURT
YOU ARE THE JURY (II,2854)

BRINGNOLO, GENEVA
CAR CARE CENTRAL (II,436)

BRINKLEY, DON
EXECUTIVE SUITE (II,796); MEDICAL CENTER (II,1675); TRAPPER JOHN, M.D. (II,2654)

BRISKIN, FRED
THE ADVENTURES OF RIN TIN TIN (I,73); TALES OF THE TEXAS RANGERS (I,4349); TOM, DICK, AND HARRY (I,4536)

BRISKIN, JERRY
MANHUNT (I,2890); SHANNON (I,3998); THE YOUNG LAWYERS (I,4941)

BRISKIN, MORT
OFFICIAL DETECTIVE (I,3339); SHERIFF OF COCHISE (I,4008); U.S. MARSHAL (I,4686)

BRITT, PONSONBY
THE BULLWINKLE SHOW (I,761)

BROD, SID
THE BING CROSBY SHOW (I,462)

BRODAX, AL
BLONDIE (I,487); COOL MCCOOL (I,1045); HELLO DERE (I,2010); OUTRAGEOUS OPINIONS (I,3429); POPEYE THE SAILOR (I,3649)

BRODAX, PAUL
KING FEATURES TRILOGY (I,2555)

BRODER, DICK
THE MARILYN MCCOO AND BILLY DAVIS JR. SHOW (II,1630)

BRODKIN, HERBERT
ABC ALBUM (I,4); THE ALCOA HOUR (I,108); BRENNER (I,719); CHARLIE WILD, PRIVATE DETECTIVE (I,914); COLONEL HUMPHREY J. FLACK (I,1004); CORONET BLUE (I,1055); THE DEFENDER (I,1219); THE DEFENDERS (I,1220); THE ELGIN HOUR (I,1428); ESPIONAGE (I,1464); THE FIRM (II,860); FOR THE PEOPLE (I,1626); THE FOUR OF US (II,914); THE HAPPENERS (I,1935); JAMIE (I,2333); JET FIGHTER (I,2374); JUSTICE (I,2498); LAND OF HOPE (II,1424); LIGHT'S OUT (I,2700); THE MOTOROLA TELEVISION HOUR (I,3112); MR. GLENCANNON TAKES ALL (I,3139); THE NURSES (I,3318); ONE-EYED JACKS ARE WILD (I,3374); THE PEOPLE NEXT DOOR (I,3519); RX FOR THE DEFENSE (I,3878); SHANE (I,3995); SKETCHBOOK (I,4072); STUDIO ONE (I,4268)

BRODSKY, SHELDON
CAN YOU TOP THIS? (I,817); THE NEW SOUPY SALES SHOW (II,1829)

BRODY, LARRY
AUTOMAN (II,120); THE FALL GUY (II,811); PARTNERS IN CRIME (II,1961); POLICE STORY (II,2062)

BROGGER, FREDERICK
THE MONROES (I,3089); TOM AND JOANN (II,2630); THE WORLD BEYOND (II,2835)

BROIDY, ELI
TOOTSIE HIPPODROME (I,4575)

BROIDY, WILLIAM F
WILD BILL HICKOK (I,4861)

BROOKS, JACK
ESTHER WILLIAMS AT CYPRESS GARDENS (I,1465)

BROOKS, JAMES L
THE ASSOCIATES (II,113); FRIENDS AND LOVERS (II,930); LOU GRANT (II,1526); THE MARY TYLER MOORE SHOW (II,1640); PAUL SAND IN FRIENDS AND LOVERS (II,1982); RHODA (II,2151); TAXI (II,2546)

BROOKS, MEL
WHEN THINGS WERE ROTTEN (II,2782)

BROTMAN, JOYCE
THE HARDY BOYS MYSTERIES (II,1090)

BROTT, LOU
TRAFFIC COURT (I,4590)

BROUGH, CHRISOPHER
GOING BANANAS (II,1015)

BROUGH, JOANNE
OUR FAMILY BUSINESS (II,1931); THIS IS KATE BENNETT (II,2597)

BROWAR, HARRY
THE ADDAMS FAMILY (II,11)

BROWAR, HERBERT W
MR. ED (I,3137); MY BOY GOOGIE (I,3185); PINE LAKE LODGE (I,3597)

BROWER, MITCHELL
NEWMAN'S DRUGSTORE (II,1837)

BROWLOW, KEVIN
HOLLYWOOD (II,1160)

BROWN JR, WILLIAM H
THE GOODYEAR REVUE (I,1856); SHIRLEY TEMPLE'S STORYBOOK (I,4017)

BROWN, ANDREW
JENNIE: LADY RANDOLPH CHURCHILL (II,1277); KENNEDY (II,1377); THE ROCK FOLLIES (II,2192)

BROWN, CHARLOTTE
BUMPERS (II,394); RHODA (II,2151)

BROWN, DAVE
DOLLAR A SECOND (I,1326); WINGO (I,4878)

BROWN, DICK
CAPTAIN FATHOM (I,830); CLUTCH CARGO (I,988); THE NEW THREE STOOGES (I,3274); SPACE ANGEL (I,4140)

BROWN, GREG
SQUARE PEGS (II,2431)

BROWN, HENRY
CAPTAIN Z-RO (I,839)

BROWN, HY
HIS HONOR, HOMER BELL (I,2063); TWELVE STAR SALUTE (I,4638)

BROWN, JAMES H
FOR LOVE AND HONOR (II,903); JOE FORRESTER (II,1303); THE QUEST (II,2097); SEVEN BRIDES FOR SEVEN BROTHERS (II,2307); THIS IS KATE BENNETT (II,2597); THE TURNING POINT OF JIM MALLOY (II,2670)

BROWN, JOE HARRY
CROSSROADS (I,1105)

BROWN, JOHN S
RADIO PICTURE SHOW (II,2104)

BROWN, MENDE
THE EVIL TOUCH (I,1486)

BROWN, PETER
YOU ASKED FOR IT (II,2855)

BROWN, RAY
1915 (II,1853)

BROWN, STEVE
CAGNEY AND LACEY (II,409)

BROWN, WILLIAM
STUDIO ONE (I,4268)

BROWNE, HOWARD
DESTRY (I,1241)

BROWNE, JERRY
FAMILY NIGHT WITH HORACE HEIDT (I,1523); LIVE LIKE A MILLIONAIRE (I,2728); SWIFT SHOW WAGON (I,4315)

BROWNING, KIRK
DAMN YANKEES (I,1128)

BROWNSTEIN, PAUL
BUDDY HACKETT—LIVE AND UNCENSORED (II,386)

BRUCE, DUSTY
JEFF'S COLLIE (I,2356)

BRUNNER, BOB
THE BAD NEWS BEARS (II,134); BROTHERS AND SISTERS (II,382); SISTER TERRI (II,2366); WORKING STIFFS (II,2834)

BRYANT, BARBARA
MANDY'S GRANDMOTHER (II,1618)

BRYCE, JOHN
60 YEARS OF SEDUCTION (II,2373); THE AVENGERS (II,121)

BRYNE, JOE
THE EVERLY BROTHERS SHOW (I,1479); THE YELLOW ROSE (II,2847)

BRYNNER, YUL
LIFE WITH SNARKY PARKER (I,2693)

BUCHANAN, JAMES DAVID
BEACH PATROL (II,192); CHARLIE'S ANGELS (II,486); CHOPPER ONE (II,515); FIREHOUSE (II,859); JIGSAW JOHN (II,1288); Q. E. D. (II,2094); SUPERCOPS (II,2502)

BUCHWALD, SAM
CASPER AND FRIENDS (I,867)

BUCK, JULES
O.S.S. (I,3411)

BUCKNER, BRAD
FIT FOR A KING (II,866); FOREVER FERNWOOD (II,909); HIGHCLIFFE MANOR (II,1150); MARY HARTMAN, MARY HARTMAN (II,1638); SCARECROW AND MRS. KING (II,2268); WHY US? (II,2796)

BUCKRIDGE, KENNETH
MISS SUSAN (I,3065)

BUFFUM, RAY
THE ART FORD SHOW (I,275); BROADWAY OPEN HOUSE (I,736); MYSTERIES OF CHINATOWN (I,3209); TAG THE GAG (I,4330)

BUGGENHEIM, CHARLES
FEARLESS FOSDICK (I,1554)

BULL, SHELDON
BEANE'S OF BOSTON (II,194); NEWHART (II,1835)

BULLOCK, HARVEY
ALICE (II,33); BIG JOHN, LITTLE JOHN (II,240); MAN IN THE MIDDLE (I,2870); MCDUFF, THE TALKING DOG (II,1662); THE MONSTER SQUAD (II,1725); THE RED HAND GANG (II,2122)

BUNETTA, FRANK
THE ADVENTURES OF OKY DOKY (I,70); DARK OF NIGHT (I,1156); JACKIE GLEASON AND HIS AMERICAN SCENE MAGAZINE (I,2321)

BUNIM, MARY-ELLIS
AS THE WORLD TURNS (II,110)

BUNIN, HOPE
FOODINI THE GREAT (I,1619); LUCKY PUP (I,2788)

BUNIN, MOREY
FOODINI THE GREAT (I,1619); LUCKY PUP (I,2788)

BURDITT, GEORGE
LOVE AND LEARN (II,1529); THREE'S COMPANY (II,2614)

BURKE, CAROLINE
THREE STEPS TO HEAVEN (I,4485)

BURKE, PAMELA
THE TOMORROW SHOW (II,2635)

BURMESTER, BRUCE

JACKPOT (II,1266); SHOOT FOR THE STARS (II,2340); WINNING STREAK (II,2809)

BURNETT, WALTER
AN EVENING WITH THE STATLER BROTHERS (II,790)

BURNS, ALLAN
THE DUCK FACTORY (II,738); FRIENDS AND LOVERS (II,930); LOU GRANT (II,1526); THE MARY TYLER MOORE SHOW (II,1640); PAUL SAND IN FRIENDS AND LOVERS (II,1982); RHODA (II,2151)

BURNS, BONNIE
THE JACKSONS (II,1267); SWEENEY TODD (II,2512); TWIGS (II,2684)

BURNS, FRANK
THE NEW ADVENTURES OF MARTIN KANE (I,3251)

BURNS, GEORGE
MCNAB'S LAB (I,2972); MONA MCCLUSKEY (I,3085); NO TIME FOR SERGEANTS (I,3300); WENDY AND ME (I,4793)

BURNS, JACK
BONKERS! (II,348); THE FLIP WILSON COMEDY SPECIAL (II,886); THE FLIP WILSON SPECIAL (I,1603); THE FLIP WILSON SPECIAL (II,887); THE FLIP WILSON SPECIAL (II,888); THE FLIP WILSON SPECIAL (II,889); HAPPY DAYS (I,1942); THE MUPPET SHOW (II,1766); THE PAUL LYNDE COMEDY HOUR (II,1977); WE'VE GOT EACH OTHER (II,2757)

BURNS, STAN
LANCELOT LINK, SECRET CHIMP (I,2609)

BURNS, VERNON
THE THIRD MAN (I,4450)

BURR, EUGENE
THE DUPONT SHOW OF THE WEEK (I,1388); FROM THESE ROOTS (I,1688)

BURRELL, GAIL
ALL MY CHILDREN (II,39); RITUALS (II,2182)

BURROWS, ABE
76 MEN AND PEGGY LEE (I,3989); ABE BURROWS ALMANAC (I,9); HOW TO SUCCEED IN BUSINESS WITHOUT REALLY TRYING (II,1197); THE REVLON REVUE (I,3782)

BURROWS, JAMES
CHEERS (II,494)

BURROWS, JOHN
TARGET: THE CORRUPTERS (I,4360)

BURRUD, BILL
THE AMERICAN WEST (I,174); ANIMALS ARE THE FUNNIEST PEOPLE (II,81)

BURT, CHRIS
REILLY, ACE OF SPIES (II,2129)

BURTON, AL
CHARLES IN CHARGE (II,482); MALIBU U. (I,2849); ROMP (I,3838); WHERE THE GIRLS ARE (I,4830)

BUSH, WARREN V
MAKE MINE RED, WHITE, AND BLUE (I,2841)

BUSHNELL, ANTHONY
THE ADVENTURES OF SIR FRANCIS DRAKE (I,75)

BUSS, FRANCES
HOLD IT PLEASE (I,2071); IT'S A GIFT (I,2255); KING'S PARTY LINE (I,2562); THE MARCH OF DIMES FASHION SHOW (I,2901); THE MARCH OF DIMES FASHION SHOW (I,2902); VANITY FAIR (I,4702); WHAT'S IT WORTH? (I,4818)

BUTLER, ROBERT
BLACK BART (II,260); N.Y.P.D. (I,3321); ONE NIGHT BAND (II,1909); REMINGTON STEELE (II,2130)

BUXTON, FRANK
HOT DOG (I,2122); SUNDAY FUNNIES (II,2493)

BUZZELL, EDWARD
THE FABULOUS SYCAMORES (I,1505); TEXACO STAR THEATER (I,4412)

BYRNE, JOE
THE BASTARD/KENT FAMILY CHRONICLES (II,159); THE HOYT AXTON SHOW (II,1200); INSPECTOR PEREZ (II,1237); THE JOHNNY CASH SHOW (I,2424); THE PAUL LYNDE HALLOWEEN SPECIAL (II,1981); STEAMBATH (II,2451)

BYRNES, JIM
THE BUFFALO SOLDIERS (II,388); THE BUSTERS (II,401); THE MACAHANS (II,1583); ROYCE (II,2224); WILD TIMES (II,2799)

BYRON, EDWARD
MR. DISTRICT ATTORNEY (I,3136); WONDERFUL JOHN ACTION (I,4892)

BYRON, PAUL
DIONE LUCAS' COOKING SCHOOL (I,1288)

BYRON, WARD
THE AT HOME SHOW (I,308); DAYDREAMING WITH LARAINE (I,1184); THE FRANCES LANGFORD-DON AMECHE SHOW (I,1655); NIGHT EDITOR (I,3286); PAUL WHITEMAN'S SATURDAY NIGHT REVUE (I,3490)

BYWATERS, TOM
MAKE A WISH (I,2838)

CACCITOTTI, TONY
FARRELL: FOR THE PEOPLE (II,834)

CADDIGAN, JAMES L
CAFE DE PARIS (I,791); CAPTAIN VIDEO AND HIS VIDEO RANGERS (I,838); FAMILY GENIUS (I,1521); FEDERAL AGENT (I,1557); HANDS OF MURDER (I,1929); LET'S TAKE A TRIP (I,2668); PROGRAM PLAYHOUSE (I,3679); THE TIMID SOUL (I,4519)

CAFFEY, MICHAEL
EMERGENCY PLUS FOUR (II,774)

CAFFEY, RICHARD
BUCK ROGERS IN THE 25TH CENTURY (II,383); CENTENNIAL (II,477); GARRISON'S GORILLAS (I,1735); HEAVEN ON EARTH (II,1120); PARIS 7000 (I,3459); THE SURVIVORS (I,4301)

CAHAN, GEORGE M
COWBOY IN AFRICA (I,1085); HOLLYWOOD THEATER TIME (I,2096); IT'S ABOUT TIME (I,2263); THE SILENT SERVICE (I,4049); TALES OF THE VIKINGS (I,4351); UNION PACIFIC (I,4676)

CAHN, SAMMY
THE BING CROSBY SPECIAL (I,472); THE FRANK SINATRA SHOW (I,1666); THE FRANK SINATRA TIMEX SHOW (I,1668); THE FRANK SINATRA TIMEX SHOW (I,1667)

CAIRNCROSS, WILLIAM
HAWAIIAN HEAT (II,1111); QUINCY, M. E. (II,2102)

CALANDRA, VINCE
THE JOHN DAVIDSON SHOW (II,1310); THICKE OF THE NIGHT (II,2587)

CALDWELL, STEPHEN
ARCHER—FUGITIVE FROM THE EMPIRE (II,103); GAVILAN (II,959)

CALEB, RUTH
OPPENHEIMER (II,1919)

CALHOUN, ROBERT
THE GUIDING LIGHT (II,1064); TEXAS (II,2566)

CALHOUN, RORY
THE TEXAN (I,4413)

CALIHAN JR, WILLIAM
THE ETHEL BARRYMORE THEATER (I,1467)

CALLENDER, COLIN
NICHOLAS NICKLEBY (II,1840)

CALLISTER, LIZ
BACKSTAGE PASS (II,132)

CALLNER, MARTY
DIANA ROSS IN CONCERT (II,668); FLEETWOOD MAC IN CONCERT (II,880); THE PEE WEE HERMAN SHOW (II,1988); THE RICH LITTLE SPECIAL (II,2155); STEVIE NICKS IN CONCERT (II,2463)

CALLO, JOSEPH F
MUGGSY (II,1761)

CALVELLI, JOSEPH
MR. NOVAK (I,3145)

CALVERT, FRED
EMERGENCY PLUS FOUR (II,774); I AM THE GREATEST: THE ADVENTURES OF MUHAMMED ALI (II,1211); NANNY AND THE PROFESSOR AND THE PHANTOM OF THE CIRCUS (I,3226); WINKY DINK AND YOU (I,4880)

CAMBOU, DON
THE LOVE REPORT (II,1543)

CAMBRIA, CATHY A
KATE AND ALLIE (II,1365)

CAMBRIDGE, BOB
ROCK-N-AMERICA (II,2195)

CAMP, CAROLYN
BENJI, ZAX AND THE ALIEN PRINCE (II,205)

CAMPBELL, BOB
MATINEE AT THE BIJOU (II,1651)

CAMPBELL, BRUCE
BILL COSBY DOES HIS OWN THING (I,441); A SPECIAL BILL COSBY SPECIAL (I,4147)

CAMPBELL, HOWARD
MR. ED (I,3137)

CAMPBELL, KEN

THE ED SULLIVAN SHOW (I,1401)

CAMPBELL, NORMAN
THE ANDY WILLIAMS CHRISTMAS SHOW (I,199); THE ANDY WILLIAMS CHRISTMAS SHOW (II,77); BING CROSBY'S WHITE CHRISTMAS (II,253); THE FURST FAMILY OF WASHINGTON (I,1718); THE GOLDDIGGERS IN LONDON (I,1837)

CANADY, RAY
OPRYLAND USA—1975 (II,1920)

CANFIELD, CURTIS
MASTERPIECE PLAYHOUSE (I,2942)

CANFIELD, HERMAN
DRAGNET (I,1369)

CANFIELD, HOMER
CRIME AND PUNISHMENT (I,1095); MAYOR OF HOLLYWOOD (I,2962)

CANNELL, STEPHEN J
THE A-TEAM (II,119); THE BLACK SHEEP SQUADRON (II,262); BOSTON AND KILBRIDE (II,356); DOCTOR SCORPION (II,685); THE DUKES (II,740); FOUR EYES (II,912); THE GREATEST AMERICAN HERO (II,1060); THE GYPSY WARRIORS (II,1070); HARDCASTLE AND MCCORMICK (II,1089); HUNTER (II,1206); THE NIGHT RIDER (II,1848); NIGHTSIDE (II,1851); THE QUEST (II,2098); RICHIE BROCKELMAN, PRIVATE EYE (II,2167); RICHIE BROCKELMAN: MISSING 24 HOURS (II,2168); RIPTIDE (II,2178); THE ROCKFORD FILES (II,2196); THE ROCKFORD FILES (II,2197); THE ROUSTERS (II,2220); STONE (II,2470); STONE (II,2471); TENSPEED AND BROWN SHOE (II,2562); TOMA (II,2634)

CANNON, ROBERT
GERALD MCBOING-BOING (I,1778)

CANTOR, EDDIE
THE EDDIE CANTOR COMEDY THEATER (I,1407)

CAPICE, PHILIP
THE BLUE KNIGHT (II,276); BRAVO TWO (II,368); BUNCO (II,395); DALLAS (II,614); EIGHT IS ENOUGH (II,762); ERIC (I,1451); FLATBUSH (II,877); HUNTER (II,1204); HUNTER (II,1205); A MAN CALLED INTREPID (II,1612); MARRIED: THE FIRST YEAR (II,1634); STUDS LONIGAN (II,2484); TWO GUYS FROM MUCK (II,2690); TWO MARRIAGES (II,2691); THE WILDS OF TEN THOUSAND ISLANDS (II,2803)

CAPLIN, CHARLES
FASHIONS ON PARADE (I,1538)

CARDEA, FRANK
BRING 'EM BACK ALIVE (II,377); CRAZY LIKE A FOX (II,601); HARD KNOCKS (II,1086); O'MALLEY (II,1872); SAWYER AND FINN (II,2265)

CARDONA, ROBERT
CRIMES OF PASSION (II,603)

CAREW, TOPPER
THE RIGHTEOUS APPLES (II,2174)

CARLBERG, LARS-OWEN
SCENES FROM A MARRIAGE (II,2272)

CARLIN, BRENDA
GEORGE CARLIN AT CARNEGIE HALL (II,974)

CARLIN, ROGER
THE MIGHTY HERCULES (I,3034)

CARLIN, STEVE
THE $100,000 BIG SURPRISE (I,3378); THE $128,000 QUESTION (II,1903); THE BIG SURPRISE (I,433); HOW DO YOU RATE? (I,2144); MADE IN AMERICA (I,2806); ROOTIE KAZOOTIE (I,3844); STOP THE MUSIC (I,4232); TOP DOLLAR (I,4577); YOU ASKED FOR IT (I,4931); YOU'RE IN THE PICTURE (I,4965)

CARLINER, MARK
THE AFRICAN QUEEN (II,18); FLYING HIGH (II,894); FLYING HIGH (II,895); LOCAL 306 (II,1505); MAUREEN (II,1656); MIXED NUTS (II,1714); NOWHERE TO HIDE (II,1868); OVER AND OUT (II,1938); THE PHOENIX (II,2035); THE PHOENIX (II,2036); RENDEZVOUS HOTEL (II,2131); WILDER AND WILDER (II,2802)

CARLTON, LEONARD
TOM CORBETT, SPACE CADET (I,4535)

CARMAN, BOB
JACK PAAR TONIGHT (I,2320)

CARMODY, DON
STRANGE TRUE STORIES (II,2476)

CARNEY, JACK
ARTHUR GODFREY AND FRIENDS (I,278); THE ARTHUR GODFREY SHOW (I,283); ARTHUR GODFREY'S TALENT SCOUTS (I,291); GIVE AND TAKE (I,1814)

CARNEY, ROBERT
JOHNNY GUITAR (I,2427)

CAROLEO, DINA
WONDER GIRL (I,4891)

CARON, GLEN GORDON
BREAKING AWAY (II,371); CONCRETE BEAT (II,562); REMINGTON STEELE (II,2130)

CARP, DAVID
ARCHER (II,102)

CARR, ALLAN
ANN-MARGRET OLSSON (II,84); ANN-MARGRET SMITH (II,85); ANN-MARGRET. . . RHINESTONE COWGIRL (II,87); ANN-MARGRET: FROM HOLLYWOOD WITH LOVE (I,217); ANN-MARGRET—WHEN YOU'RE SMILING (I,219); DON'T CALL ME MAMA ANYMORE (I,1350); FROM HOLLYWOOD WITH LOVE: THE ANN-MARGRET SPECIAL (I,1687)

CARRA, LAWRENCE
PULITZER PRIZE PLAYHOUSE (I,3692)

CARRINGTON, ROBERT
JOHNNY BLUE (II,1321)

CARROLL JR, BOB
ALICE (II,33); DOROTHY (II,717); MR. T. AND TINA (II,1758); PRIVATE BENJAMIN (II,2087)

CARROLL, MONROE
RYAN'S HOPE (II,2234); SALTY (II,2241)

CARROLL, RICH
HERNDON AND ME (II,1139)

CARROTHERS, RICHARD
CHARLEY'S AUNT (II,483)

CARRUTHERS, BILL
THE GIRL IN MY LIFE (II,996); GIVE-N-TAKE (II,1003); I LOVE LIBERTY (II,1215); JUST FRIENDS (I,2495); THE KING FAMILY IN WASHINGTON (I,2549); THE KING FAMILY JUNE SPECIAL (I,2550); THE NEIGHBORS (II,1801); OPERATION: ENTERTAINMENT (I,3401); PAT PAULSEN'S HALF A COMEDY HOUR (I,3473); PRESS YOUR LUCK (II,2078); SECOND CHANCE (II,2286); THIS IS YOUR LIFE (I,4461); A TRIBUTE TO "MR. TELEVISION" MILTON BERLE (II,2658); ULTRA QUIZ (II,2701); WHODUNIT? (II,2794); YOU DON'T SAY (II,2858)

CARSEY, MARCY
CALLAHAN (II,414); THE COSBY SHOW (II,578); I DO, I DON'T (II,1212); OH MADELINE (II,1880)

CARSON, HELEN
DON MCNEILL'S DINNER CLUB (I,1336)

CARSON, WHITEY
TOOTSIE HIPPODROME (I,4575)

CARTER, LYNDA
LYNDA CARTER: BODY AND SOUL (II,1565)

CARUSO, DEE
THE RED SKELTON SHOW (I,3756); WHAT'S HAPPENING!! (II,2769)

CARUSO, FRED
TODAY'S FBI (II,2628)

CARUSO, JAMES
THE VIDEO GAME (II,2727)

CASE, CARROLL
FRONTIER (I,1695)

CASE, WALT
ALL-STAR SECRETS (II,52); THE GUINNESS GAME (II,1066); THE NEW NEWLYWED GAME (II,1824); RHYME AND REASON (II,2152); STUMPERS (II,2485); THE TONI TENNILLE SHOW (II,2636); WE DARE YOU! (II,2751); YOU BET YOUR LIFE (II,2856); YOUR HIT PARADE (II,2872)

CASHMAN, ED
TEXACO STAR THEATER (I,4411)

CASSIDY, ALAN
THE ROUSTERS (II,2220)

CASSON, PHILIP
PICCADILLY PALACE (I,3589)

CASTLE, WILLIAM
CIRCLE OF FEAR (I,958); GHOST STORY (I,1788); MEN OF ANNAPOLIS (I,3008)

CASTRONOVA, T J
TALES FROM THE DARKSIDE (II,2534); TALES FROM THE DARKSIDE (II,2535)

CATES, GILBERT
CIRCUS LIONS, TIGERS AND MELISSAS TOO (II,528);

COUNTRY NIGHT OF STARS (II,592); COUNTRY NIGHT OF STARS II (II,593); COUNTRY STARS OF THE 70S (II,594); ELVIS REMEMBERED: NASHVILLE TO HOLLYWOOD (II,772); FAME (I,1517); FIFTY YEARS OF COUNTRY MUSIC (II,853); HAGGIS BAGGIS (I,1917); HAVE I GOT A CHRISTMAS FOR YOU (II,1105); I'VE GOT A SECRET (I,2283); OFF CAMPUS (II,1877); SKINFLINT (II,2377); STUBBY PRINGLE'S CHRISTMAS (II,2483)

CATES, JOSEPH
THE $64,000 QUESTION (I,4071); ACCENT ON LOVE (I,16); ALL COMMERCIALS—A STEVE MARTIN SPECIAL (II,37); ANN-MARGRET—WHEN YOU'RE SMILING (I,219); ANNIE, THE WOMAN IN THE LIFE OF A MAN (I,226); THE BACHELOR (I,325); BUCK ROGERS IN THE 25TH CENTURY (I,751); CAVALCADE OF STARS (I,877); CIRCUS LIONS, TIGERS AND MELISSAS TOO (II,528); COLISEUM (I,1000); COMEDY TONIGHT (I,1027); THE COMEDY ZONE (II,559); COS: THE BILL COSBY COMEDY SPECIAL (II,577); COUNTRY COMES HOME (II,585); COUNTRY COMES HOME (II,586); COUNTRY MUSIC HIT PARADE (I,1069); COUNTRY MUSIC HIT PARADE (II,588); COUNTRY MUSIC HIT PARADE (II,589); COUNTRY NIGHT OF STARS (II,592); COUNTRY NIGHT OF STARS II (II,593); COUNTRY STARS OF THE 70S (II,594); ELVIS REMEMBERED: NASHVILLE TO HOLLYWOOD (II,772); FAME (I,1517); FIFTY YEARS OF COUNTRY MUSIC (II,853); FIVE STARS IN SPRINGTIME (I,1590); THE GENE KELLY PONTIAC SPECIAL (I,1750); GEORGE M! (I,1772); GEORGE M! (II,976); HAVE I GOT A CHRISTMAS FOR YOU (II,1105); HIGH BUTTON SHOES (I,2050); HIPPODROME (I,2060); I BELIEVE IN MUSIC (I,2165); INTERNATIONAL SHOWTIME (I,2225); JACK LEMMON IN 'S WONDERFUL, 'S MARVELOUS, 'S GERSHWIN (I,2313); JACK LEMMON—GET HAPPY (I,2312); JOHNNY CASH AND FRIENDS (II,1322); JOHNNY CASH AND THE COUNTRY GIRLS (II,1323); JOHNNY CASH CHRISTMAS 1983 (II,1324); THE JOHNNY CASH CHRISTMAS SPECIAL (II,1325); THE JOHNNY CASH CHRISTMAS SPECIAL (II,1326); A JOHNNY CASH CHRISTMAS (II,1327); A JOHNNY CASH CHRISTMAS (II,1328); THE JOHNNY CASH SHOW (I,2424); THE JOHNNY CASH SPRING SPECIAL (II,1329); JOHNNY CASH'S AMERICA (II,1330); JOHNNY CASH: CHRISTMAS IN SCOTLAND (II,1332); JOHNNY CASH: CHRISTMAS ON THE ROAD (II,1333); JOHNNY CASH: COWBOY HEROES (II,1334); JOHNNY CASH: THE FIRST 25 YEARS (II,1336); JOHNNY CASH—A MERRY MEMPHIS CHRISTMAS (II,1331); THE MAGIC OF DAVID COPPERFIELD (II,1593); THE MAGIC OF DAVID COPPERFIELD (II,1594); MAGIC WITH THE STARS (II,1599); THE ROBERT KLEIN SHOW (II,2185); THE ROBERT KLEIN SHOW (II,2186); ROY ACUFF—50 YEARS THE KING OF COUNTRY MUSIC (II,2223); SKINFLINT (II,2377); STEVE MARTIN: COMEDY IS NOT PRETTY (II,2462); STOP THE MUSIC (I,4232); THE VICTOR BORGE SHOW (I,4719); THE VICTOR BORGE SHOW (I,4720)

CAULDWELL, DAVE
LOOK AT US (II,1515)

CAULEY, HARRY
ANGIE (II,80)

CAWLEY, BOB
THE ANITA BRYANT SPECTACULAR (II,82); PETER MARSHALL SALUTES THE BIG BANDS (II,2028)

CAZABON, JOHN
VANITY FAIR (I,4702)

CAZAU, CANDI
THE $128,000 QUESTION (II,1903); TALES OF THE HAUNTED (II,2538)

CELLCIO, JACK
EVENING AT THE MOULIN ROUGE (II,787)

CENTURY VIDEO CORPORATION
MR. MOON'S MAGIC CIRCUS (II,1753)

CERNY, MICHAEL
THE LOVE REPORT (II,1542)

CERRA, HAL
THE ROOTS OF ROCK 'N' ROLL (II,2212)

CESANA, RENZO
THE CONTINENTAL (I,1042)

CHAFFEY, DON
FANTASY ISLAND (II,829)

CHAIS, PAMELA
THE NATURAL LOOK (II,1799)

CHALOM, MARK
THE UNCLE FLOYD SHOW (II,2703)

CHALOPIN, JEAN
THE GET ALONG GANG (II,980); INSPECTOR GADGET (II,1236); KIDD VIDEO (II,1388); THE LITTLES (II,1499); POLE POSITION (II,2060); WOLF ROCK TV (II,2815)

CHAMBERLAIN, CRAN
YOU ASKED FOR IT (I,4931)

CHAMBERS, ANNE
WORLD WAR II: G.I. DIARY (II,2840)

CHAMBERS, ERNEST
20TH CENTURY FOLLIES (I,4641); BARBARA MANDRELL AND THE MANDRELL SISTERS (II,143); BARRY MANILOW—ONE VOICE (II,156); BATTLE OF THE BEAT (II,161); A BEDTIME STORY (I,384); THE BOBBY DARIN AMUSEMENT COMPANY (I,672); THE CAPTAIN AND TENNILLE SONGBOOK (II,432); CAROL CHANNING AND PEARL BAILEY ON BROADWAY (I,850); CAROL CHANNING PROUDLY PRESENTS THE SEVEN DEADLY SINS (I,851); CHARO (II,488); DICK VAN DYKE MEETS BILL COSBY (I,1274); THE DONNA SUMMER SPECIAL (II,711); THE DORIS MARY ANNE KAPPELHOFF SPECIAL (I,1356); FRANCIS ALBERT SINATRA DOES HIS THING (I,1658); FRIENDS AND NABORS (I,1684); HOW TO HANDLE A WOMAN (I,2146); THE JOHN GARY SHOW (I,2411); THE LESLIE UGGAMS SHOW (II,2660); THE LIVING END (I,2731); LOOK WHAT THEY'VE DONE TO MY SONG (II,1517); LOVE NEST (II,1541); LOVE, SIDNEY (II,1547); NEIL SEDAKA STEPPIN' OUT (II,1804); PRESENTING SUSAN ANTON (II,2077); PUMPBOYS AND DINETTES ON TELEVISION (II,2090); THE RAINBOW GIRL (II,2108); THE SECOND BARRY MANILOW SPECIAL (II,2285); THE SMOTHERS BROTHERS COMEDY HOUR (I,4095); THE THIRD BARRY MANILOW SPECIAL (II,2590); TONY ORLANDO AND DAWN (II,2639); VER-R-R-RY INTERESTING (I,4712); WHAT'S IT ALL ABOUT WORLD? (I,4816); ZERO INTELLIGENCE (II,2876)

CHAMBERS, EVERETT
B.A.D. CATS (II,123); COLUMBO (II,556); FUTURE COP (II,945); JOHNNY STACCATO (I,2437); THE LLOYD BRIDGES SHOW (I,2733); LUCAN (II,1554); MONTY NASH (I,3097); NERO WOLFE (II,1805); PARTNERS IN CRIME (II,1961); PEYTON PLACE (I,3574); THEY ONLY COME OUT AT NIGHT (II,2586); TURNOVER SMITH (II,2671); TWIN DETECTIVES (II,2687)

CHAMPION, GOWER
AN EVENING WITH JULIE ANDREWS AND HARRY BELAFONTE (I,1476); THE JULIE ANDREWS SPECIAL (I,2483); THE MARY MARTIN SHOW (I,2934)

CHAMPION, JOHN
ALCOA PREMIERE (I,109); LARAMIE (I,2614); SEVEN AGAINST THE SEA (I,3983)

CHAN, DANIEL
THE LEGEND OF THE GOLDEN GUN (II,1455)

CHAN, EDYTH
THE DAVID LETTERMAN SHOW (II,624)

CHAPIN, DOUG
MISSING PIECES (II,1706)

CHAPLIN, SAUL
THE GENE KELLY PONTIAC SPECIAL (I,1750)

CHAPMAN, BEN
SHOGUN (II,2339)

CHAPMAN, RICHARD
ALEX AND THE DOBERMAN GANG (II,32); LEGMEN (II,1458); NICK AND THE DOBERMANS (II,1843); THE SEAL (II,2282); SHADOW OF SAM PENNY (II,2313); SIMON AND SIMON (II,2357)

CHARLES, GEORGE
MUSIC ON ICE (I,3175); TONIGHT WITH BELAFONTE (I,4566)

CHARLES, GLEN
THE BOB NEWHART SHOW (II,342); CHEERS (II,494); TAXI (II,2546)

CHARLES, LES
THE BOB NEWHART SHOW (II,342); CHEERS (II,494); TAXI (II,2546)

CHARLES, RAY

THE BING CROSBY SHOW (I,468); THE BING CROSBY SHOW (I,469); THE KRAFT MUSIC HALL (I,2587)

CHARMOLI, TONY
FEATHERTOP (I,1556); THE SPECIAL GENTRY ONE (I,4148)

CHARNIN, MARTIN
ANNIE AND THE HOODS (II,91); ANNIE, THE WOMAN IN THE LIFE OF A MAN (I,226); GEORGE M! (I,1772); GEORGE M! (II,976); JACK LEMMON IN 'S WONDERFUL, 'S MARVELOUS, 'S GERSHWIN (I,2313); JACK LEMMON—GET HAPPY (I,2312)

CHASE, CARROLL
PUBLIC DEFENDER (I,3687); RACKET SQUAD (I,3714); SUGARFOOT (I,4277); THE TEXAN (I,4413)

CHASE, DAVID
MOONLIGHT (II,1730); PALMS PRECINCT (II,1946); THE ROCKFORD FILES (II,2197)

CHASTAIN, BILL
MUSICAL CHAIRS (II,1776)

CHAULY, JOHN C
THE FURTHER ADVENTURES OF WALLY BROWN (II,944)

CHENAULT, ROBERT
THE ABC AFTERSCHOOL SPECIAL (II,1); LITTLE LULU (II,1493); SHAZAM! (II,2319)

CHERCOVER, MURRAY
HALF THE GEORGE KIRBY COMEDY HOUR (I,1920); MIRROR, MIRROR, OFF THE WALL (I,3059)

CHERMAK, CY
AMY PRENTISS (II,74); THE BARBARY COAST (II,146); CHIPS (II,511); THE CIRCLE FAMILY (II,527); CONVOY (I,1043); THE DOCTORS (I,1323); FORCE SEVEN (II,908); IRONSIDE (II,1246); MITCHELL AND WOODS (II,1709); THE NIGHT STALKER (II,1849); RIDDLE AT 24000 (II,2171)

CHERONES, TOM
IN SECURITY (II,1227); OLD FRIENDS (II,1881); SUZANNE PLESHETTE IS MAGGIE BRIGGS (II,2509); THE TWO OF US (II,2693)

CHERRY, STANLEY Z
THE KIDS FROM C.A.P.E.R. (II,1391)

CHERTOK, JACK
CAVALCADE OF AMERICA (I,875); JOHNNY MIDNIGHT (I,2431); THE LAWLESS YEARS (I,2641); THE LONE RANGER (I,2740); THE MISS AND THE MISSILES (I,3062); MY FAVORITE MARTIAN (I,3189); MY LIVING DOLL (I,3195); PRIVATE SECRETARY (I,3672); SKY KING (I,4077); STEVE DONOVAN, WESTERN MARSHAL (I,4225)

CHESLER, LEWIS
CRYSTAL GAYLE IN CONCERT (II,608); THE HITCHHIKER (II,1157); RED SKELTON: A ROYAL PERFORMANCE (II,2125)

CHEVILLAT, DICK
IT'S A GREAT LIFE (I,2256); THE JIM BACKUS SHOW—HOT OFF THE WIRE (I,2379); THE MCGONIGLE (I,2968); WAR CORRESPONDENT (I,4767)

CHILDS, TED
SPECIAL BRANCH (II,2414); THE SWEENEY (II,2513)

CHOMSKY, MARVIN
RUN, BUDDY, RUN (I,3870)

CHOTZINOFF, SAMUEL
THE ALCOA HOUR (I,108); AMAHL AND THE NIGHT VISITORS (I,148); ROBERT MONTGOMERY PRESENTS YOUR LUCKY STRIKE THEATER (I,3809)

CHRISTENSEN, DON
THE BANG-SHANG LALAPALOOZA SHOW (II,140); BATMAN AND THE SUPER SEVEN (II,160); THE FABULOUS FUNNIES (II,802); FLASH GORDON (II,874); FLASH GORDON—THE GREATEST ADVENTURE OF ALL (II,875); THE NEW ADVENTURES OF BATMAN (II,1812); THE NEW ADVENTURES OF GILLIGAN (II,1813); THE NEW ADVENTURES OF MIGHTY MOUSE AND HECKLE AND JECKLE (II,1814); SPORT BILLY (II,2429); SUPER WITCH (II,2501); THE TARZAN/LONE RANGER/ZORRO ADVENTURE HOUR (II,2543); UNCLE CROC'S BLOCK (II,2702); THE YOUNG SENTINELS (II,2871)

CHRISTIAN, HARRY
THE WAYNE KING SHOW (I,4779)

CHRISTIANSEN, ROBERT
THE FABULOUS DR. FABLE (I,1500); THERE'S ALWAYS ROOM (II,2583); WISHMAN (II,2811)

CHRISTIANSON, CORNELL
THE PAPER CHASE: THE SECOND YEAR (II,1950)

CHRISTIE, HOWARD
THE ADAM MACKENZIE STORY (I,30); CHARLIE WOOSTER—OUTLAW (I,915); THE DOG TROOP (I,1325); THE ELISABETH MCQUEENEY STORY (I,1429); THE VIRGINIAN (I,4738); WAGON TRAIN (I,4747)

CICOTTI, BRUNO
THE CAPTAIN AND TENNILLE SONGBOOK (II,432)

CINADER, ROBERT A
CHASE (I,917); CONDOMINIUM (II,566); EMERGENCY! (II,775); EMERGENCY! (II,776); ESCAPE (I,1461); KNIGHT RIDER (II,1402); PINE CANYON IS BURNING (II,2040); QUINCY, M. E. (II,2102); THE REBELS (II,2121); THE SEEKERS (II,2298); SIERRA (II,2351)

CLAIR, DICK
MAMA'S FAMILY (II,1610); A NEW KIND OF FAMILY (II,1819); SOAP (II,2392)

CLAIR, JOE
THE GABBY HAYES SHOW (I,1723)

CLAMAN, JULIAN
ASSIGNMENT: MANHUNT (I,301); THE DUPONT SHOW OF THE WEEK (I,1388); HAVE GUN—WILL TRAVEL (I,1968); JAMIE (I,2334)

CLAMPETT, BOB
THE BEANY AND CECIL SHOW (I,374)

CLAPP, NICHOLAS
THE KIDS FROM FAME (II,1392)

CLARK, COLBERT
ANNIE OAKLEY (I,225); TALES OF THE TEXAS RANGERS (I,4349)

CLARK, DICK
ABC'S SILVER ANNIVERSARY CELEBRATION (II,2); AMERICAN BANDSTAND (II,70); ANIMALS ARE THE FUNNIEST PEOPLE (II,81); THE CAPTAIN AND TENNILLE (II,429); CELEBRITIES: WHERE ARE THEY NOW? (II,462); THE DAVID SOUL AND FRIENDS SPECIAL (II,626); DICK CLARK PRESENTS THE ROCK AND ROLL YEARS (I,1265); DICK CLARK PRESENTS THE ROCK AND ROLL YEARS (I,1266); DICK CLARK'S GOOD OL' DAYS: FROM BOBBY SOX TO BIKINIS (II,671); DICK CLARK'S LIVE WEDNESDAY (II,672); FRANKIE AND ANNETTE: THE SECOND TIME AROUND (II,917); THE FUNNIEST JOKE I EVER HEARD (II,939); THE FUNNIEST JOKE I EVER HEARD (II,940); GABRIEL KAPLAN PRESENTS THE FUTURE STARS (II,951); GET IT TOGETHER (I,1780); THE HALF-HOUR COMEDY HOUR (II,1074); HAPPENING '68 (I,1936); HOLLYWOOD STARS' SCREEN TESTS (II,1165); HOLLYWOOD'S PRIVATE HOME MOVIES (II,1167); HOLLYWOOD'S PRIVATE HOME MOVIES II (II,1168); I'VE HAD IT UP TO HERE (II,1221); THE IMPERIAL GRAND BAND (II,1224); INSIDE AMERICA (II,1234); IT'S HAPPENING (I,2269); THE JOHN DAVIDSON CHRISTMAS SHOW (II,1307); THE JOHN DAVIDSON SHOW (II,1309); THE LOU RAWLS SPECIAL (II,1527); MEN WHO RATE A "10" (II,1684); MR. & MS. AND THE BANDSTAND MURDERS (II,1746); MR. & MS. AND THE MAGIC STUDIO MYSTERY (II,1747); THE NATALIE COLE SPECIAL (II,1794); OPRYLAND: NIGHT OF STARS AND FUTURE STARS (II,1921); PUTTIN' ON THE HITS (II,2093); THE SENSATIONAL, SHOCKING, WONDERFUL, WACKY 70S (II,2302); SORORITY '62 (II,2405); THANKSGIVING REUNION WITH THE PARTRIDGE FAMILY AND MY THREE SONS (II,2569); TV'S BLOOPERS AND PRACTICAL JOKES (II,2675); ULTRA QUIZ (II,2701); VALENTINE MAGIC ON LOVE ISLAND (II,2716); WHATEVER BECAME OF. . . ? (II,2772); WHATEVER BECAME OF. . .? (II,2773); WOLF ROCK TV (II,2815); YOU ARE THE JURY (II,2854)

CLARK, JOHN L
THE PLAINCLOTHESMAN (I,3614)

CLARK, R A
PUTTIN' ON THE HITS (II,2093)

CLARK, ROBIN S
FLYING HIGH (II,895); HIGH SCHOOL, U.S.A. (II,1148)

CLARK, RON
ACE CRAWFORD, PRIVATE EYE (II,6); HOT L BALTIMORE (II,1187); THIS WEEK IN NEMTIN (I,4465); THE TIM CONWAY COMEDY HOUR (I,4501)

CLARK, VERNON E
HIGHWAY PATROL (I,2058); STARR, FIRST BASEMAN (I,4204)

CLARK, VIC
EASY DOES IT. . . STARRING FRANKIE AVALON (II,754)

CLARKE, CECIL
BRIEF ENCOUNTER (I,724); THE DOUBLE KILL (I,1360); AN ECHO OF THERESA (I,1399); THE EYES HAVE IT (I,1496); THE FEAR IS SPREADING (I,1553); A KILLER IN EVERY CORNER (I,2538); KILLER WITH TWO FACES (I,2539); KISS KISS, KILL KILL (I,2566); LADY KILLER (I,2603); LONG DAY'S JOURNEY INTO NIGHT (I,2746); MIRROR OF DECEPTION (I,3060); MURDER IS A ONE-ACT PLAY (I,3160); MURDER MOTEL (I,3161); MURDER ON THE MIDNIGHT EXPRESS (I,3162); ONE DEADLY OWNER (I,3373); ONLY A SCREAM AWAY (I,3393); A PLACE TO DIE (I,3612); POSSESSION (I,3655); SIGN IT DEATH (I,4046); SLEEPWALKER (I,4084); SOMEONE AT THE TOP OF THE STAIRS (I,4114); SPELL OF EVIL (I,4152); THE STRAUSS FAMILY (I,4253); TERROR FROM WITHIN (I,4404)

CLAROL, JOHN
THE PLAINCLOTHESMAN (I,3614)

CLASTER, JOHN
THE GREAT SPACE COASTER (II,1059)

CLAVELL, JAMES
SHOGUN (II,2339)

CLAVER, BOB
CAPTAIN KANGAROO (II,434); ENSIGN O'TOOLE (I,1448); GETTING TOGETHER (I,1784); GIDGET (I,1795); THE GIRL WITH SOMETHING EXTRA (II,999); HERE COME THE BRIDES (I,2024); THE INTERNS (I,2226); THE JIMMIE RODGERS SHOW (I,2382); MRS. G. GOES TO COLLEGE (I,3155); OCCASIONAL WIFE (I,3325); OVER AND OUT (II,1938); THE PARTRIDGE FAMILY (II,1962)

CLAVER, PHYLLIS
WOMAN'S PAGE (II,2822)

CLAXTON, WILLIAM F
THE HIGH CHAPARRAL (I,2051); LITTLE HOUSE ON THE PRAIRIE (II,1487)

CLAYMAN, ROBERT J
THE BAXTERS (II,183)

CLAYTON, BOB
PASS THE BUCK (II,1964)

CLEARY, DANNY
A SPECIAL OLIVIA NEWTON-JOHN (II,2418)

CLEARY, JACK
THE G.E. COLLEGE BOWL (I,1745)

CLEARY, JOHN P
ALUMNI FUN (I,145)

CLEMENS, BRIAN
THE AVENGERS (II,121); ESCAPADE (II,781); THE NEW AVENGERS (II,1816)

CLEMENTS JR, CALVIN
MATT HOUSTON (II,1654)

CLEMENTS, ALICE
THE CHILDREN'S HOUR (I,939)

CLEMENTS, CALVIN
BUCK ROGERS IN THE 25TH CENTURY (II,384); DOCTOR KILDARE (I,1315)

CLEMENTS, STEVE
HOUR MAGAZINE (II,1193)

CLEMMER, RICHARD
ONE MAN'S FAMILY (I,3383)

CLEMONS, WILLIAM TELL
HIDDEN TREASURE (I,2045)

CLEWS, COLIN
THE ENGELBERT HUMPERDINCK SHOW (I,1444); THE ENGELBERT HUMPERDINCK SHOW (I,1445); THE ENGELBERT HUMPERDINCK SPECIAL (I,1446); THE ENGELBERT HUMPERDINCK SPECIAL (I,1447); THE JOHN DAVIDSON SHOW (I,2409); THE LIBERACE SHOW (I,2677)

CLINE, EDWARD F
FIREBALL FUN FOR ALL (I,1577)

CLOKEY, ARTHUR
DAVY AND GOLIATH (I,1180)

CLOKEY, BOB
GUMBY (I,1905)

CLOSE, R DAVID
THE PALACE (II,1943)

CLOUD, HAMILTON
WELCOME TO THE FUN ZONE (II,2763)

CLUTTERCHALK, GRAHAM
PADDINGTON BEAR (II,1941)

COBLENZ, WALTER
APPLE'S WAY (II,101)

COCKRELL, PAUL
THE LOVE REPORT (II,1542); THE LOVE REPORT (II,1543)

COE, FRED
BONINO (I,685); CHEVROLET ON BROADWAY (I,925); THE CHEVROLET TELE-THEATER (I,926); DATELINE (I,1165); THE DOW HOUR OF GREAT MYSTERIES (I,1365); THE FARMER'S DAUGHTER (I,1534); FOR YOUR PLEASURE (I,1627); THE FOUR POSTER (I,1651); GIRL ABOUT TOWN (I,1806); THE GOLDEN AGE OF TELEVISION (II,1016); THE GREAT IMPERSONATION (I,1876); HEAVENS TO BETSY (I,1997); INTERMEZZO (I,2223); THE KING AND MRS. CANDLE (I,2544); THE LAST WAR (I,2630); LIGHT'S OUT (I,2699); MASTERPIECE PLAYHOUSE (I,2942); THE MICHELE LEE SHOW (II,1690); MR. MERGENTHWIRKER'S LOBBLIES (I,3144); MR. PEEPERS (I,3147); NOTORIOUS (I,3314); OF MEN OF WOMEN (I,3331); OF MEN OF WOMEN (I,3332); OUR TOWN (I,3419); THE PARADINE CASE (I,3453); PETER PAN (I,3566); THE PETRIFIED FOREST (I,3570); PHILCO TELEVISION PLAYHOUSE (I,3583); PLAYHOUSE 90 (I,3623); PLAYWRIGHTS '56 (I,3627); PRODUCERS SHOWCASE (I,3676); REBECCA (I,3745); REUNION IN VIENNA (I,3780); SKIN OF OUR TEETH (I,4073); SPELLBOUND (I,4153); THE SPIRAL STAIRCASE (I,4160); TONIGHT AT 8:30 (I,4556); THE WOMEN (I,4890)

COE, LIZ
FINDER OF LOST LOVES (II,857)

COFFEY, MARGARET
DANGER ZONE (I,1137)

COFOD, FRANKLIN
VOLTRON—DEFENDER OF THE UNIVERSE (II,2729)

COGAN, DAVID J
ANNIE AND THE HOODS (II,91)

COHAN, MARTIN
THE BOB CRANE SHOW (II,280); THE BOB NEWHART SHOW (II,342); DIFF'RENT STROKES (II,674); FLYING HIGH (II,894); MAUREEN (II,1656); SHEILA (II,2325); SUGAR TIME (II,2489); THE TED KNIGHT SHOW (II,2550); WHO'S THE BOSS? (II,2792)

COHAN, PHIL
THE GUY MITCHELL SHOW (I,1913)

COHEN, ALBERT J
THE ANN SOTHERN SHOW (I,220)

COHEN, ALEXANDER H
CBS: ON THE AIR (II,460); HELLZAPOPPIN (I,2011); I'M A FAN (I,2192); MARLENE DIETRICH: I WISH YOU LOVE (I,2918); NIGHT OF 100 STARS (II,1846); A WORLD OF LOVE (I,4914)

COHEN, ERIC
THE 416TH (II,913); AT EASE (II,116); GABRIEL KAPLAN PRESENTS THE SMALL EVENT (II,952); GUN SHY (II,1068); LAVERNE AND SHIRLEY (II,1446); UPTOWN SATURDAY NIGHT (II,2711); WELCOME BACK, KOTTER (II,2761); WHACKED OUT (II,2767); ZORRO AND SON (II,2878)

COHEN, HAROLD
THE EVERLY BROTHERS SHOW (I,1479); JIMMY DURANTE PRESENTS THE LENNON SISTERS HOUR (I,2387); THE JOHNNY CASH SHOW (I,2424); THE LENNON SISTERS SHOW (I,2656); SECOND CHANCE (I,3958)

COHEN, LAWRENCE J
AFTER GEORGE (II,19); EMPIRE (II,777); FOG (II,897); IS THERE A DOCTOR IN THE HOUSE (I,2241); MOMMA THE DETECTIVE (II,1721); THE SHAMEFUL SECRETS OF HASTINGS CORNERS (I,3994); STICK AROUND (II,2464)

COHEN, RALPH
KENNY & DOLLY: A CHRISTMAS TO REMEMBER (II,1378)

COHEN, RONALD M
CALL TO GLORY (II,413); FEEL THE HEAT (II,848)

COHEN, STUART
DUSTY (II,745); THE GANGSTER CHRONICLES (II,957); GAVILAN (II,959); KATE LOVES A MYSTERY (II,1367)

COHN, BRUCE
YESTERYEAR (II,2851)

COLBERT, STANLEY
THE GREATEST MAN ON EARTH (I,1882); THE GREATEST SHOW ON EARTH (I,1883); OUR MAN FLINT: DEAD ON TARGET (II,1932)

COLE, BUDDY
CLUB OASIS (I,984)

COLE, CLAY
PEOPLE (II,1995)

COLE, JOHN J
THE RETURN OF CHARLIE CHAN (II,2136)

COLE, RENATE
PADDINGTON BEAR (II,1941)

COLE, SIDNEY
THE ADVENTURES OF BLACK BEAUTY (I,51); THE ADVENTURES OF ROBIN HOOD (I,74); THE BUCCANEERS (I,749); DANGER MAN (I,1136); THE HIGHWAYMAN (I,2059); MAN IN A SUITCASE (I,2868); SECRET AGENT (I,3962); THE SWORD OF FREEDOM (I,4326)

COLE, TOM
CELEBRITY SWEEPSTAKES (II,476); LIE DETECTOR (II,1467)

COLEMAN, COLONEL W
PROJECT UFO (II,2088)

COLEMAN, HERBERT
ALFRED HITCHCOCK PRESENTS (I,115); WHISPERING SMITH (I,4842)

COLEN, NOREEN
STUMPERS (II,2485)

COLFAX, KEN
JUMP (II,1356)

COLLA, RICHARD A
SARGE: THE BADGE OR THE CROSS (I,3915)

COLLEARY, BOB
BENSON (II,208)

COLLERAN, BILL
THE BING CROSBY SPECIAL (I,471); THE BING CROSBY SPECIAL (I,472); A DATE WITH DEBBIE (I,1161); THE FRANK SINATRA SHOW (I,1666); THE FRANK SINATRA TIMEX SHOW (I,1668); HOLIDAY U.S.A. (I,2078); THE POLLY BERGEN SHOW (I,3643)

COLLIER, CHET
THE BAXTERS (II,184)

COLLINS, RICHARD
THE BOB HOPE CHRYSLER THEATER (I,502); BONANZA (II,347); THE CONTENDER (II,571); THE FAMILY HOLVAK (II,817); FIREHOUSE (II,859); LITTLE WOMEN (II,1498); THE OREGON TRAIL (II,1923); THE RHINEMANN EXCHANGE (II,2150); SARA (II,2260)

COLLINS, TED
THE KATE SMITH EVENING HOUR (I,2507); THE KATE SMITH HOUR (I,2508); THE KATE SMITH SHOW (I,2509); THE KATE SMITH SHOW (I,2510); MATINEE IN NEW YORK (I,2946)

COLLINSON, ROBERT
WORLD WAR I (I,4919)

COLLYER, CLAYTON 'BUD'
BEAT THE CLOCK (I,377)

COLMAN, HENRY
DINNER DATE (I,1286); THE LOVE BOAT (II,1535); THE LOVE BOAT II (II,1533); THE LOVE BOAT III (II,1534)

COLODNY, LES
TOM, DICK, AND MARY (I,4537)

COLOERAN, BILL
THE JUDY GARLAND SHOW (I,2473); THE JUDY GARLAND SHOW (II,1349)

COLOMBY, HARRY
COMEDY OF HORRORS (II,557); MCNAMARA'S BAND (II,1669); REPORT TO MURPHY (II,2134); SHEEHY AND THE SUPREME MACHINE (II,2322); WORKING STIFFS (II,2834)

COMFORT, BOB
THE BEST OF TIMES (II,220); FIRST TIME, SECOND TIME (II,863); JUST OUR LUCK (II,1360); NICHOLS AND DYMES (II,1841); ONE NIGHT BAND (II,1909)

COMINOS, N H
SURVIVAL (I,4300)

COMO, PERRY
THE PERRY COMO CHRISTMAS SHOW (II,2002); PERRY PRESENTS (I,3549)

CONBOY, JOHN
AFTER HOURS: FROM JANICE, JOHN, MARY AND MICHAEL, WITH LOVE (II,20); AFTER HOURS: SINGIN', SWINGIN' AND ALL THAT JAZZ (II,22); CAPITOL (II,426); THE YOUNG AND THE RESTLESS (II,2862)

CONNELL, DAVE
CAPTAIN KANGAROO (II,434); MR. MAYOR (I,3143); SIGN-ON (II,2353)

CONNELLY, JOE
BLONDIE (I,487); BRINGING UP BUDDY (I,731); CALVIN AND THE COLONEL (I,806); THE GENERAL ELECTRIC THEATER (I,1753); GOING MY WAY (I,1833); HARRIS AGAINST THE WORLD (I,1956); ICHABOD AND ME (I,2183); KAREN (I,2505); LEAVE IT TO BEAVER (I,2648); THE MARGE AND GOWER CHAMPION SHOW (I,2907); MEET MR. MCNUTLEY (I,2989); MUNSTER, GO HOME! (I,3157); THE MUNSTERS (I,3158); PISTOLS 'N' PETTICOATS (I,3608)

CONNOLLY, KATHY
THE CHEAP SHOW (II,491); WACKO (II,2733)

CONNORS, MICHELE
JABBERWOCKY (II,1262)

CONRAD, JOAN
HARD KNOX (II,1087)

CONRAD, WILLIAM
77 SUNSET STRIP (I,3988); KLONDIKE (I,2573); THIS MAN DAWSON (I,4463); TURNOVER SMITH (II,2671)

CONROY, DAVID
VANITY FAIR (I,4701); WAR AND PEACE (I,4765)

CONVERSE, TONY
BUTTERFLIES (II,405); THE DICK CAVETT SHOW (I,1264); GOSSIP (II,1044); THE MANIONS OF AMERICA (II,1623)

CONVY, BERT
THE CELEBRITY FOOTBALL CLASSIC (II,474)

CONWAY, JAMES L
GREATEST HEROES OF THE BIBLE (II,1062)

COOK, BRUCE W
ENTERTAINMENT TONIGHT (II,780)

COOK, DON
THE AL MORGAN SHOW (I,94)

COOK, FIELDER
BRIGADOON (I,726); KRAFT TELEVISION THEATER (I,2592); THE PHILADELPHIA STORY (I,3582); A STRING OF BEADS (I,4263)

COOK, PAUL
USA COUNTRY MUSIC (II,591)

COOLEY, LEE
THE BIG RECORD (I,430); THE GUY MITCHELL SHOW (I,1912); THE LANNY ROSS SHOW (I,2612); THE PERRY COMO SHOW (I,3531); THE PERRY COMO SHOW (I,3532); RED GOOSE KID'S SPECTACULAR (I,3751); THE SWIFT SHOW (I,4316); TEX AND JINX (I,4406); TV'S TOP TUNES (I,4634); TV'S TOP TUNES (I,4635)

COON, GENE L
IT TAKES A THIEF (I,2250); STAR TREK (II,2440); THE WILD WILD WEST (I,4863)

COONEY, JOAN GANZ
SESAME STREET (I,3982)

COOPER, EDWARD
THE YOUNG AND THE RESTLESS (II,2862)

COOPER, HAL
AND THEY ALL LIVED HAPPILY EVER AFTER (II,75); THE ASTRONAUTS (II,114); DID YOU HEAR ABOUT JOSH AND KELLY?! (II,673); FOR BETTER OR WORSE (I,1621); GIMME A BREAK (II,995); KING OF THE ROAD (II,1396); LOVE, SIDNEY (II,1547); MAGIC COTTAGE (I,2815); MR. & MRS. & MR. (II,1744); PHYL AND MIKHY (II,2037); POTTSVILLE (II,2072)

COOPER, JACKIE
CHARLIE ANGELO (I,906); DOCTOR DAN (II,684); HENNESSEY (I,2013); THE PEOPLE'S CHOICE (I,3521)

COOPER, JOHN
THE DEVIL'S WEB (I,1248); A KILLER IN EVERY CORNER (I,2538); KILLER WITH TWO FACES (I,2539); THE KILLING GAME (I,2540); MURDER ON THE MIDNIGHT EXPRESS (I,3162); SCREAMER (I,3945); TYCOON: THE STORY OF A WOMAN (II,2697)

COOPER, KAREN
WHACKED OUT (II,2767)

COOPER, LESTER
CALIFORNIA GIRL (I,794); MAKE A WISH (I,2838)

COOPER, SHELDON

AT THE MOVIES (II,117)

COOPER, WYLLIS
ESCAPE (I,1459); STAGE 13 (I,4183); VOLUME ONE (I,4742)

COOPERMAN, ALVIN
AIN'T MISBEHAVIN' (II,25); DAMN YANKEES (I,1128); THE SEEKERS (I,3974); SHIRLEY TEMPLE'S STORYBOOK (I,4017); THE UNTOUCHABLES (I,4682)

COOPERSMITH, JEROME
JOHNNY JUPITER (I,2429)

COOTE, BERNARD
IVANHOE (I,2282)

COPLEY, ANDI
THE JOE PISCOPO SPECIAL (II,1304)

COPPERFIELD, DAVID
THE MAGIC OF DAVID COPPERFIELD (II,1594); MAGIC WITH THE STARS (II,1599)

CORBETT, PATRICK
THE LOVE REPORT (II,1542)

CORCORAN, KEVIN
HERBIE, THE LOVE BUG (II,1131); ZORRO AND SON (II,2878)

CORDAY, BARBARA
THE BOUNDER (II,357); REGGIE (II,2127); USED CARS (II,2712)

CORDAY, BETTY
DAYS OF OUR LIVES (II,629)

CORDAY, KEN
DAYS OF OUR LIVES (II,629)

CORDAY, MRS TED
DAYS OF OUR LIVES (II,629)

CORDAY, TED
AS THE WORLD TURNS (II,110)

CORDERY, HOWARD
MR. MERGENTHWIRKER'S LOBBLIES (I,3144)

COREA, NICHOLAS
ARCHER—FUGITIVE FROM THE EMPIRE (II,103); GAVILAN (II,959); THE INCREDIBLE HULK (II,1232); THE RENEGADES (II,2133)

CORMAN, GENE
A WOMAN CALLED GOLDA (II,2818)

CORMAN, ROGER
THE GEORGIA PEACHES (II,979)

CORNELIUS, DON
SOUL TRAIN (I,4131)

CORNELL, JOHN
THE PAUL HOGAN SHOW (II,1975)

CORNELL, WENDY
A CHRISTMAS CAROL (II,517)

CORRELL, CHARLES
THE AMOS AND ANDY SHOW (I,179)

CORRELL, RICHARD
AT YOUR SERVICE (II,118); HAROLD LLOYD'S WORLD OF COMEDY (II,1093)

CORRIGAN, WILLIAM
HOLLYWOOD OPENING NIGHT (I,2087)

COSBY, BILL
THE BILL COSBY SHOW (I,442); THE BILL COSBY SPECIAL, OR? (I,443); FAT ALBERT AND THE COSBY KIDS (II,835)

COSSETTE, PIERRE
100 YEARS OF GOLDEN HITS (II,1905); THE ANDY WILLIAMS SHOW (II,78); BOWZER (II,358); THE DIONNE WARWICK SPECIAL (I,1289); THE GLEN CAMPBELL MUSIC SHOW (II,1006); THE GLEN CAMPBELL MUSIC SHOW (II,1007); JOHNNY MANN'S STAND UP AND CHEER (I,2430); THE KEANE BROTHERS SHOW (II,1371); MOVIN' (I,3117); SALUTE (II,2242); SAMMY AND COMPANY (II,2248); SHA NA NA (II,2311); STAND UP AND CHEER (I,4188)

COSTELLO, PAT
THE ABBOTT AND COSTELLO SHOW (I,2); I'M THE LAW (I,2195)

COSTELLO, ROBERT
THE ADAMS CHRONICLES (II,8); THE ARMSTRONG CIRCLE THEATER (I,260); DARK SHADOWS (I,1157); THE NURSES (I,3318); THE PATTY DUKE SHOW (I,3483); RYAN'S HOPE (II,2234); VANITY FAIR (I,4700)

COTLER, GORDON
LANIGAN'S RABBI (II,1428); SANCTUARY OF FEAR (II,2252); SING ALONG WITH MITCH (I,4057)

COTTLE, BOB
THE RUFF AND REDDY SHOW (I,3866)

COULTER, DOUG
BROADWAY OPEN HOUSE (I,736); SOUND OFF TIME (I,4135)

COURLEY, BUD
THE BULLWINKLE SHOW (I,761)

COURNEYA, JERRY
ADVENTURES OF NOAH BEERY JR. (I,69)

COURTLAND, JEROME
GIDGET GROWS UP (I,1797); MOVIN' ON (I,3118); NANCY (I,3221)

COURTNEY, ALAN D
THE GEORGE SEGAL SHOW (II,977)

COVAN, WILL
THE PAUL WINCHELL AND JERRY MAHONEY SHOW (I,3492)

COVINGTON, TREADWELL
THE BEAGLES (I,372); KING LEONARDO AND HIS SHORT SUBJECTS (I,2557); TENNESSEE TUXEDO AND HIS TALES (I,4403)

COWAN, GEOFFREY
THE QUIZ KIDS (II,2103)

COWAN, LOUIS
BALANCE YOUR BUDGET (I,336); THE BERT PARKS SHOW (I,400); THE BILL GOODWIN SHOW (I,447); THE COMEBACK STORY (I,1017); COSMOPOLITAN THEATER (I,1057); DOWN YOU GO (I,1367); THE QUIZ KIDS (I,3712); SING IT AGAIN (I,4059); STOP THE MUSIC (I,4232)

COX, DICK
HIGH HOPES (II,1144)

COX, GARRY
THE JOKER'S WILD (II,1341)

CRADDOCK, RON
PENMARRIC (II,1993)

CRAGG, STEPHEN
THE BOYS IN BLUE (II,359); SEVEN BRIDES FOR SEVEN BROTHERS (II,2307)

CRAIG, KEN
THE WAYNE KING SHOW (I,4779)

CRAIG, WALTER
THOSE TWO (I,4470)

CRAIS, ROBERT
THE MISSISSIPPI (II,1707)

CRAMER, DOUGLAS S
ALOHA PARADISE (II,58); AT EASE (II,115); B.A.D. CATS (II,123); BRIDGET LOVES BERNIE (I,722); CALL HER MOM (I,796); CALL HOLME (II,412); CASINO (II,452); FINDER OF LOST LOVES (II,857); THE FRENCH ATLANTIC AFFAIR (II,924); FRIENDS (II,927); GLITTER (II,1009); HONEYMOON SUITE (I,2108); HOTEL (II,1192); JOE AND SONS (II,1296); JOE AND SONS (II,1297); KATE BLISS AND THE TICKER TAPE KID (II,1366); KEEPING UP WITH THE JONESES (II,1374); LADIES IN BLUE (II,1422); THE LOVE BOAT (II,1535); THE LOVE BOAT I (II,1532); THE LOVE BOAT II (II,1533); THE LOVE BOAT III (II,1534); MAN ON A STRING (I,2877); MASSARATI AND THE BRAIN (II,1646); MATT HOUSTON (II,1654); MOVIN' ON (I,3118); MRS R.—DEATH AMONG FRIENDS (II,1759); MURDER CAN HURT YOU! (II,1768); THE POWER WITHIN (II,2073); THE SAN PEDRO BEACH BUMS (II,2250); THE SAN PEDRO BUMS (II,2251); SCARED SILLY (II,2269); SHOOTING STARS (II,2341); STRIKE FORCE (II,2480); VEGAS (II,2723); VEGAS (II,2724); VELVET (II,2725); VENICE MEDICAL (II,2726); WAIKIKI (II,2734); WEDNESDAY NIGHT OUT (II,2760); WHERE'S THE FIRE? (II,2785); WILD AND WOOLEY (II,2798); THE WILD WOMEN OF CHASTITY GULCH (II,2801); WONDER WOMAN (PILOT 2) (II,2827); WONDER WOMAN (SERIES 1) (II,2828); WONDER WOMAN (SERIES 2) (II,2829)

CRAMER, JOE L
KINGSTON: CONFIDENTIAL (II,1398); THE SIX-MILLION-DOLLAR MAN (II,2372); SWITCH (II,2519)

CRAMER, NED
ERNIE IN KOVACSLAND (I,1454); IT'S TIME FOR ERNIE (I,2277); KOVACS ON THE CORNER (I,2582)

CRAMOY, MICHAEL
THE RAY KNIGHT REVUE (I,3733)

CRANE, BARRY
THE MAGICIAN (I,2823); THE MAGICIAN (I,2824)

CRAVER, WILLIAM
THE DAIN CURSE (II,613); STUNT SEVEN (II,2486)

CRAWFORD, HECTOR
ALL THE RIVERS RUN (II,43)

CRAWFORD, HENRY
AGAINST THE WIND (II,24); HOMICIDE (I,2101); HUNTER (I,2161); THE SULLIVANS (II,2490); A TOWN LIKE ALICE

(II,2652)

CRAWFORD, IAN
ALL THE RIVERS RUN (II,43)

CRAWFORD, ROBERT L
WHAT REALLY HAPPENED TO THE CLASS OF '65? (II,2768)

CREDLE, GARY
THE MYSTERIOUS TWO (II,1783)

CRENESSE, PIERRE
INVITATION TO PARIS (I,2233)

CRENNA, RICHARD
MAKE ROOM FOR GRANDDADDY (I,2845)

CREWS, COLIN
PICCADILLY PALACE (I,3589)

CRICHTON, ROBIN
IN SEARCH OF. . . (II,1226)

CRIPPEN, FRED
SKYHAWKS (I,4079)

CROMMIE, KAREN
THE FANTASTIC FUNNIES (II,825)

CROMPTON, CYNTHIA
THE LOVE REPORT (II,1542)

CRONYN, HUME
ACTOR'S STUDIO (I,28); THE FOUR POSTER (I,1651) THE MARRIAGE (I,2922)

CROOLE, BARRY
TOP OF THE WORLD (II,2646)

CROSBY, GEORGE E
CENTENNIAL (II,477); THE JERK, TOO (II,1282)

CROSLAND JR, ALAN
O'HARA, UNITED STATES TREASURY (I,3346)

CROSS, JAMES
GAMES PEOPLE PLAY (II,956); THE GOLDEN AGE OF TELEVISION (II,1016); THE GOLDEN AGE OF TELEVISION (II,1017); THE GOLDEN AGE OF TELEVISION (II,1018); THE GOLDEN AGE OF TELEVISION (II,1019); LIFESTYLES OF THE RICH AND FAMOUS (II,1474)

CROSS, PERRY
ARCHIE (II,104); BE OUR GUEST (I,370); CAN YOU TOP THIS? (I,817); COUNTY FAIR (I,1075); THE ERNIE KOVACS SHOW (I,1455); HOLLYWOOD TALENT SCOUTS (I,2094); THE ICE PALACE (I,2182); THE JERRY LEWIS SHOW (I,2369); THE NEW SOUPY SALES SHOW (II,1829); THE NIGHT OF CHRISTMAS (I,3290); SONJA HENIE'S HOLIDAY ON ICE (I,4125); THE SOUPY SALES SHOW (I,4138); TOP OF THE MONTH (I,4578); THE WIZARD OF ODDS (I,4886)

CROTHERS, A J
GOODNIGHT, BEANTOWN (II,1041)

CROTTY, A BURKE
CELANESE THEATER (I,881); HOLLYWOOD STAR REVUE (I,2093)

CROWE, CHRISTOPHER
B.J. AND THE BEAR (II,125); B.J. AND THE BEAR (II,127); DARKROOM (II,619); THE HARDY BOYS MYSTERIES (II,1090)

CROWLEY, MART
HART TO HART (II,1102)

CRUM, JIM
WE'RE MOVIN' (II,2756)

CRYSTAL, BOB
SALE OF THE CENTURY (II,2239)

CULLEN, BILL
WHY? (I,4856)

CULVER, FELIX
JESSIE (II,1287)

CUMMINGS, IRVING
FURY (I,1721); THUNDER (II,2617)

CUNLIFFE, DAVID
FLAMBARDS (II,869)

CURRAN, MIMI
TOM COTTLE: UP CLOSE (II,2631)

CURTIS, DAN
THE BIG EASY (II,234); DARK SHADOWS (I,1157); FRANKENSTEIN (I,1671); IN THE DEAD OF NIGHT (I,2202); THE LONG DAYS OF SUMMER (II,1514); MELVIN PURVIS: G-MAN (II,1679); THE NIGHT STALKER (I,3292); THE NIGHT STRANGLER (I,3293); THE NORLISS TAPES (I,3305); THE PICTURE OF DORIAN GRAY (I,3592); THE STRANGE CASE OF DR. JEKYLL AND MR. HYDE (I,4244); SUPERTRAIN (II,2504); THE WINDS OF WAR (II,2807)

CURTIS, DAVID
RAQUEL (I,3725)

CURTIS, TOM R
SERGEANT PRESTON OF THE YUKON (I,3980)

CUSCUNA, SUSAN
THAT'S CAT (II,2576)

CUTLER, STAN
ME AND MAXX (II,1672); THE ROLLERGIRLS (II,2203); SHIPSHAPE (II,2329)

CUTTS, JOHN
ALLISON SIDNEY HARRISON (II,54); CHICAGO STORY (II,506); THE CHICAGO STORY (II,507); THE FITZPATRICKS (II,868); GAVILAN (II,959); THE GIRL, THE GOLD WATCH AND DYNAMITE (II,1001); TOP OF THE HILL (II,2645)

D'AMORE, BUCK
NOW YOU SEE IT (II,1867)

D'ANGELO, WILLIAM P
ALICE (II,33); BAREFOOT IN THE PARK (I,355); BIG JOHN, LITTLE JOHN (II,240); DOMINIC'S DREAM (II,704); THE KAREN VALENTINE SHOW (I,2506); MCDUFF, THE TALKING DOG (II,1662); THE MISADVENTURES OF SHERIFF LOBO (II,1704); THE MONSTER SQUAD (II,1725); THE RED HAND GANG (II,2122); ROOM 222 (I,3843); RUN RUN, JOE (II,2229); WEBSTER (II,2758); THE WESTWIND (II,2766); THE YOUNG LAWYERS (I,4940)

D'ANTONI, PHILIP
IN TANDEM (II,1228); MELINA MERCOURI'S GREECE (I,2997); MOVIN' ON (II,1743); MR. INSIDE/MR. OUTSIDE (II,1750); THE RUBBER GUN SQUAD (II,2225); SOPHIA LOREN IN ROME (I,4129)

DASILVA, DAVID
FAME (II,812)

DASILVA, HOWARD
WALTER FORTUNE (I,4758)

DACKOW, JOSEPH
THE OUTLAWS (I,3427); TEMPLE HOUSTON (I,4392)

DAHLMAN, LOU
THE ALAN DALE SHOW (I,96); DOORWAY TO FAME (I,1354)

DALRYMPLE, JEAN
REUNION IN VIENNA (I,3780)

DALY, JOHN
DATELINE (I,1165)

DAMES, BOB
CONDO (II,565)

DAMES, ROD
MR. & MRS. & MR. (II,1744)

DAMSKER, GARY
THE COPS AND ROBIN (II,573); FUTURE COP (II,945); FUTURE COP (II,946)

DANA, BILL
CAR WASH (II,437); DON KNOTTS NICE CLEAN, DECENT, WHOLESOME HOUR (I,1333); THE MILTON BERLE SHOW (I,3049)

DANDEE, GREG
MATINEE AT THE BIJOU (II,1651)

DANIEL, JAY
THE BILLION DOLLAR THREAT (II,246); BORDER PALS (II,352); CONCRETE BEAT (II,562); COREY: FOR THE PEOPLE (II,575); EISCHIED (II,763); FOR LOVERS ONLY (II,904); KISS ME, KILL ME (II,1400); ONCE UPON A SPY (II,1898)

DANIEL, ROD
THE DUCK FACTORY (II,738); HARPER VALLEY (II,1094); KUDZU (II,1415); WKRP IN CINCINNATI (II,2814)

DANIELS, ALYSON
LIFESTYLES OF THE RICH AND FAMOUS (II,1474)

DANIELS, MARC
FORD THEATER HOUR (I,1635); THE IMOGENE COCA SHOW (I,2198); JANE EYRE (I,2338); THE LIFE AND TIMES OF EDDIE ROBERTS (II,1468); THE NASH AIRFLYTE THEATER (I,3227); SAINTS AND SINNERS (I,3887); SKINFLINT (II,2377)

DANIELS, STAN
THE ASSOCIATES (II,113); THE BETTY WHITE SHOW (II,224); DOC (II,682); DOC (II,683); DON'T CALL US (II,709); THE MARY TYLER MOORE SHOW (II,1640); MR. SMITH (II,1755); PHYLLIS (II,2038); TAXI (II,2546)

DANOFF, BILL
ALL IN THE FAMILY (II,38)

DANSKA, DOLORES
PEOPLE (II,1995)

DANZIG, JERRY
CRIME SYNDICATED (I,1099); OPEN HOUSE (I,3394); SURE AS FATE (I,4297); WESTINGHOUSE SUMMER THEATER (I,4805)

DANZIGER, EDWARD J
ADVENTURE THEATER (I,42); THE MAN FROM INTERPOL (I,2866); MARK SABER (I,2914); RICHARD THE LION HEART (I,3786); SABER OF

LONDON (I,3879); THE VISE (I,4739)

DANZIGER, HARRY LEE
ADVENTURE THEATER (I,42); THE MAN FROM INTERPOL (I,2866); MARK SABER (I,2914); RICHARD THE LION HEART (I,3786); SABER OF LONDON (I,3879); THE VISE (I,4739)

DARE, DANNY
THE JACK CARTER SHOW (I,2306)

DARLEY, CHRIS
ULTRA QUIZ (II,2701)

DARLEY, DICK
CLUB OASIS (I,985); THE MICKEY MOUSE CLUB (I,3025); THE ROSEMARY CLOONEY SHOW (I,3848); SPACE PATROL (I,4144); THE SPIKE JONES SHOW (I,4156)

DAVENPORT, BILL
MAGGIE (II,1589)

DAVID, ALLAN
CELEBRITY BILLIARDS (I,882)

DAVID, JOHN
AIRWOLF (II,27); CASSIE AND COMPANY (II,455); PALMS PRECINCT (II,1946)

DAVID, MACK
COLT .45 (I,1010)

DAVIDSON, MARTIN
FLATBUSH/AVENUE J (II,878)

DAVIDSON, WILLIAM
ADVENTURES IN RAINBOW COUNTRY (I,47); THE STARLOST (I,4203)

DAVIES, JOHN
FAWLTY TOWERS (II,843); GOOD NEIGHBORS (II,1034); MONTY PYTHON'S FLYING CIRCUS (II,1729)

DAVIES, ROGER
OLIVIA NEWTON-JOHN IN CONCERT (II,1884)

DAVIS JR, OWEN
THE CHEVROLET TELE-THEATER (I,926)

DAVIS JR, SAMMY
THE SAMMY DAVIS JR. SPECIAL (I,3899)

DAVIS, ANN
THE ADVENTURES OF CYCLONE MALONE (I,57)

DAVIS, BILL
HEE HAW (II,1123); THE JACKSONS (II,1267)

DAVIS, DAVID
THE BOB NEWHART SHOW (II,342); BUMPERS (II,394); RHODA (II,2151); SHEPHERD'S FLOCK (I,4007); TAXI (II,2546)

DAVIS, DONALD
ACTOR'S STUDIO (I,28); ART AND MRS. BOTTLE (I,271); THE FRONT PAGE (I,1691); THE FUNNIEST JOKE I EVER HEARD (II,940); OUT OF OUR MINDS (II,1933); OUT THERE (I,3424); PRUDENTIAL FAMILY PLAYHOUSE (I,3683) THE SWAN (I,4312)

DAVIS, EDDIE
HARBOURMASTER (I,1946); THE MAN CALLED X (I,2860); TARGET (I,4359)

DAVIS, ELIAS
ALMOST HEAVEN (II,57); DELTA HOUSE (II,658); GOODNIGHT, BEANTOWN (II,1041); HOLLYWOOD HIGH (II,1163); THE SHOW MUST GO ON (II,2344); STEAMBATH (II,2451)

DAVIS, GARETH
BICENTENNIAL MINUTES (II,229); FLAMINGO ROAD (II,872); FREE COUNTRY (II,921); GOOD TIME HARRY (II,1037); THE OLDEST LIVING GRADUATE (II,1882); RAQUEL (II,2114); REMINGTON STEELE (II,2130)

DAVIS, HAL
BYLINE—BETTY FURNESS (I,785); TREASURE UNLIMITED (I,4603)

DAVIS, JERRY
THE BARBARA EDEN SHOW (I,346); BEWITCHED (I,418); THE COP AND THE KID (II,572); FUNNY FACE (I,1711); THE GOOD GUYS (I,1847); HARPER VALLEY (II,1094); HOUSE CALLS (II,1194); THE MURDOCKS AND THE MCCLAYS (I,3163); ONE MORE TRY (II,1908); ROSETTI AND RYAN (II,2216); ROSETTI AND RYAN: MEN WHO LOVE WOMEN (II,2217); SEMI-TOUGH (II,2300); SURFSIDE 6 (I,4299); THAT GIRL (II,2570)

DAVIS, JIM
THE ROARING TWENTIES (I,3805)

DAVIS, JOHN
BILLBOARD'S DISCO PARTY (II,245); THE GOODIES (II,1040); TAKE FIVE WITH STILLER AND MEARA (II,2530)

DAVIS, LEE
DIAGNOSIS: UNKNOWN (I,1251)

DAVIS, LUTHER
THE DOUBLE LIFE OF HENRY PHYFE (I,1361)

DAVIS, MADELYN
ALICE (II,33); DOROTHY (II,717); MR. T. AND TINA (II,1758); PRIVATE BENJAMIN (II,2087)

DAVIS, MICHAEL P
LET ME TELL YOU ABOUT A SONG (I,2661)

DAVIS, NELSON
THE NEW $100,000 NAME THAT TUNE (II,1810)

DAVIS, OWEN
BURLESQUE (I,764)

DAVIS, ROMMIE
VANITIES (II,2720)

DAVIS, SUNNI
DAYTIME (II,630)

DAWES, SKIP
PAUL WHITEMAN'S TV TEEN CLUB (I,3491)

DAWSON, CATHY
FAMILY FEUD (II,816)

DAWSON, GORDON
BRET MAVERICK (II,374)

DAY, DORIS
THE DORIS DAY SHOW (I,1355)

DE LOS SANTOS, NANCY
AT THE MOVIES (II,117); SNEAK PREVIEWS (II,2388)

DE'ANGELO, CARLO
CHARLIE WILD, PRIVATE DETECTIVE (I,914)

DEBELLIS, JOHN
THE JOE PISCOPO SPECIAL (II,1304)

DEBLASIO, EDWARD
PARIS (II,1957); POLICE WOMAN (II,2063)

DECORDOVA, FRED
DECEMBER BRIDE (I,1215); THE GEORGE BURNS AND GRACIE ALLEN SHOW (I,1763); THE GEORGE GOBEL SHOW (I,1768); THE JACK BENNY PROGRAM (I,2294); THE JACK BENNY SPECIAL (I,2296); JACK BENNY'S BIRTHDAY SPECIAL (I,2299); A LOVE LETTER TO JACK BENNY (II,1540); MR. ADAMS AND EVE (I,3121); THE SMOTHERS BROTHERS SHOW (I,4096); THE TONIGHT SHOW STARRING JOHNNY CARSON (II,2638)

DEFARIA, WALT
THE BEACH GIRLS (II,191) THE BORROWERS (I,693); TRAVELS WITH CHARLEY (I,4597); VALLEY FORGE (I,4698); THE WONDERFUL WORLD OF PIZZAZZ (I,4901)

DEGUERE, PHILIP
THE BLACK SHEEP SQUADRON (II,262); DOCTOR STRANGE (II,688); SHADOW OF SAM PENNY (II,2313); SIMON AND SIMON (II,2357); WHIZ KIDS (II,2790)

DEKAY, JIM
DAVID NIVEN'S WORLD (II,625)

DELEON, DICK
THE DIAHANN CARROLL SHOW (II,665)

DELEON, ROBERT
THE DIAHANN CARROLL SHOW (II,665)

DELUCA, RUDY
CALUCCI'S DEPARTMENT (I,805); THE INVESTIGATORS (II,1241) PEEPING TIMES (II,1990); STOPWATCH: THIRTY MINUTES OF INVESTIGATIVE TICKING (II,2474)

DEMET, PETER
THE DICK TRACY SHOW (I,1272)

DENOIA, NICK
UNICORN TALES (II,2706)

DEPATIE, DAVID
THE ABC AFTERSCHOOL SPECIAL (II,1); BAGGY PANTS AND THE NITWITS (II,135); BAILEY'S COMETS (I,333); THE BARKLEYS (I,356); DENNIS THE MENACE: MAYDAY FOR MOTHER (II,660); DOCTOR DOLITTLE (I,1309); DUNGEONS AND DRAGONS (II,744); THE FANTASTIC FOUR (II,824); G.I. JOE: A REAL AMERICAN HERO (II,948); HERE COMES THE GRUMP (I,2027); THE HOUNDCATS (I,2132); MEATBALLS AND SPAGHETTI (II,1674); THE ODDBALL COUPLE (II,1876); PANDAMONIUM (II,1948); THE PINK PANTHER (II,2042); PINK PANTHER AND SONS (II,2043); RETURN TO THE PLANET OF THE APES (II,2145); SPIDER-MAN AND HIS AMAZING FRIENDS (II,2425); SPIDER-WOMAN (II,2426); SUPER PRESIDENT (I,4291); THE SUPER SIX (I,4292); WHAT'S NEW MR. MAGOO? (II,2770)

DEPEW, JOSEPH
THE BEVERLY HILLBILLIES (I,417)

DEROY, RICHARD
PEYTON PLACE (I,3574)

DESOUZA, STEVEN E
V: THE SERIES (II,2715)

DEWITT, JOHN J
THAT TEEN SHOW (II,2573)

DEWOLFE, THOMAS
THE FIRM (II,860)

DEA, JOE
BACKSTAGE PASS (II,132)

DEAN, BARTON
CARLTON YOUR DOORMAN (II,440)

DEBIN, DAVID
STOCKARD CHANNING IN JUST FRIENDS (II,2467)

DEL GRANDE, LOUIS
KING OF KENSINGTON (II,1395); SEEING THINGS (II,2297)

DEL PRET, LOU
LORETTA LYNN IN THE BIG APPLE (II,1520)

DEL VALLE, JAIME
THE LINE-UP (I,2707)

DEL'ORO, ISOLA
THE ADVENTURES OF LONG JOHN SILVER (I,67)

DEL, JACK
THE KING FAMILY CELEBRATE THANKSGIVING (I,2545)

DELLA CIOPPA, GUY
HIPPODROME (I,2060); THE RED SKELTON SHOW (I,3756)

DELLINGER, ROBERT L
SOMETHING ELSE (I,4117)

DELMAINE, BARRY
SHIRLEY'S WORLD (I,4019)

DELMAR, MICHAEL
THE POPPY IS ALSO A FLOWER (I,3650)

DELSENER, RON
GOOD VIBRATIONS FROM CENTRAL PARK (I,1854); GOOD VIBRATIONS FROM CENTRAL PARK (I,1855)

DEMAS, CAROLE
A MAGIC GARDEN CHRISTMAS (II,1590)

DENNIS, HAL
LAUREL AND HARDY LAUGHTOONS (II,1445)

DENNIS, WILLIAM
26 MEN (I,4648)

DENOFF, SAM
BIG EDDIE (II,235); BIG EDDIE (II,236); THE BILL COSBY SPECIAL (I,444); THE BOYS (II,360); THE CONFESSIONS OF DICK VAN DYKE (II,568); DICK VAN DYKE AND THE OTHER WOMAN, MARY TYLER MOORE (I,1273); THE FIRST NINE MONTHS ARE THE HARDEST (I,1584); THE FUNNY SIDE (I,1714); THE FUNNY SIDE (I,1715); GOOD MORNING WORLD (I,1850); THE LATE FALL, EARLY SUMMER BERT CONVY SHOW (II,1441); THE MAN WHO CAME TO DINNER (I,2881); THE MONTEFUSCOS (II,1727); MY WIFE NEXT DOOR (II,1780); ON OUR OWN (II,1892); PURE GOLDIE (I,3696); THE SUNSHINE BOYS (II,2496); THAT GIRL (II,2570); TURNABOUT (II,2669)

DENOVE, JACK
CAVALCADE OF AMERICA (I,875); SUDDEN SILENCE (I,4275); THIS IS YOUR MUSIC (I,4462)

DEPPELER, RUSSELL
MEN AT WORK IN CONCERT (II,1682)

DEPPERSCHMIDT, EVE
BLONDES VS. BRUNETTES (II,271)

DERMAN, LOU
ALL IN THE FAMILY (II,38)

DERRIS, STEVE
THE POINTER SISTERS (II,2059)

DEUBEL, ROBERT
THOSE FABULOUS CLOWNS (II,2602)

DEVENNEY, SCOTT
MATINEE AT THE BIJOU (II,1651)

DEVLIN, DON
CONVOY (I,1043)

DEW, SIMON CHRISTOPHER
THE LITTLEST HOBO (II,1500)

DEWAR, STEVE
LORNE GREENE'S NEW WILDERNESS (II,1523)

DEXTER, BRAD
SKAG (II,2374)

DEZAKI, OSAMU
MIGHTY ORBOTS (II,1697)

DIBENEDETTO, JULIO
CANDID CAMERA (I,819); THE JIMMY DEAN SHOW (I,2385)

DIBONA, VIN
ENTERTAINMENT TONIGHT (II,780); ON STAGE AMERICA (II,1893)

DIMISSA JR, E V
FANTASY (II,828); THE LOVE REPORT (II,1542); THE MIKE DOUGLAS SHOW (II,1698); THE REGIS PHILBIN SHOW (II,2128); THICKE OF THE NIGHT (II,2587)

DIMISSA, ERNIE
THICKE OF THE NIGHT (II,2587)

DIAL, BILL
HARPER VALLEY (II,1094); LEGMEN (II,1458); LOBO (II,1504); WKRP IN CINCINNATI (II,2814)

DIAMOND, JANIS
I AM THE GREATEST: THE ADVENTURES OF MUHAMMED ALI (II,1211)

DICKER, DOROTHY
THE DANGEROUS CHRISTMAS OF RED RIDING HOOD (I,1139)

DIETRICK, GARTH
TIMEX ALL-STAR JAZZ SHOW III (I,4517)

DILLON, RITA
AND THEY ALL LIVED HAPPILY EVER AFTER (II,75); APPLE PIE (II,100); THE ASTRONAUTS (II,114); BOB & CAROL & TED & ALICE (I,495); HELLO, LARRY (II,1128); IN THE BEGINNING (II,1229); JO'S COUSINS (II,1293); LIVING IN PARADISE (II,1503); MR. & MRS. & MR. (II,1744); P.O.P. (II,1939); THE PARKERS (II,1959); PHYL AND MIKHY (II,2037); REX HARRISON PRESENTS SHORT STORIES OF LOVE (II,2148); TWO THE HARD WAY (II,2695); YOUNG LIVES (II,2866)

DILLON, SEAN
SPARRING PARTNERS (I,4145)

DIMITRI, RICHARD
ROOSEVELT AND TRUMAN (II,2209)

DINEHART, ALAN
ABE BURROWS ALMANAC (I,9); THE JEAN CARROLL SHOW (I,2353)

DINGWELL, ALLEN
THE ORCHID AWARD (I,3406)

DINSMORE, RICK
KID TALK (I,2533)

DIPEGO, GERALD
THE SHERIFF AND THE ASTRONAUT (II,2326)

DISMUKES, CHARLES
BARETTA (II,152); QUINCY, M. E. (II,2102)

DISNEY, WALT
ADVENTURES IN DAIRYLAND (I,44); ANNETTE (I,222); BORDER COLLIE (I,688); CORKY AND WHITE SHADOW (I,1049); DANIEL BOONE (I,1141); ELFEGO BACA (I,1427); THE FURTHER ADVENTURES OF SPIN AND MARTY (I,1720); THE HARDY BOYS AND THE MYSTERY OF GHOST FARM (I,1950); THE HARDY BOYS AND THE MYSTERY OF THE APPLEGATE TREASURE (I,1951); THE MICKEY MOUSE CLUB (I,3025); THE NEW ADVENTURES OF SPIN AND MARTY (I,3254); THE SAGA OF ANDY BURNETT (I,3883); THE SECRET OF MYSTERY LAKE (I,3967); SPIN AND MARTY (I,4158); SWAMP FOX (I,4311); TEXAS JOHN SLAUGHTER (I,4414); WHAT I WANT TO BE (I,4812); ZORRO (I,4982)

DOBBINS, SHELLEY
THE BIG SHOWDOWN (II,244)

DOBKIN, LAWRENCE
O'HARA, UNITED STATES TREASURY (I,3346); TEMPLE HOUSTON (I,4392)

DOBSON, BRIDGET
SANTA BARBARA (II,2258)

DOBSON, JEANNE
FOUL UPS, BLEEPS AND BLUNDERS (II,911); ROUGHCUTS (II,2218)

DOBSON, JEROME
SANTA BARBARA (II,2258)

DODGE, FRANK
SEARCH FOR TOMORROW (II,2284)

DOFF, RED
THE MANY SIDES OF MICKEY ROONEY (I,2899)

DOHANOS, PETER
STUBBY PRINGLE'S CHRISTMAS (II,2483)

DOHERTY, JAMES
SAM (II,2246); SAM (II,2247)

DONALD, JONATHAN
WILD WILD WORLD OF ANIMALS (I,4864)

DONALD, TOM
A WOMAN OF SUBSTANCE (II,2820)

DONALLY, ANTHONY
THE MARTIAN CHRONICLES (II,1635)

DONIGER, WALTER
THE SURVIVORS (I,4301)

DONNELLY, PAUL
SAM (II,2246); SAM (II,2247)

DONNER, JILL SHERMAN
CUTTER TO HOUSTON (II,612)

DONOHUE, JACK
THE DEAN MARTIN SHOW (I,1196); THE DEAN MARTIN SHOW (I,1197); THE DEAN MARTIN SHOW (I,1198); THE DEAN MARTIN SHOW (I,1199); THE DEAN MARTIN SHOW (I,1200); THE DICK VAN DYKE SPECIAL (I,1276); THE HOOFER (I,2113); THE MANY SIDES OF MICKEY ROONEY (I,2899); THE MICKEY ROONEY SHOW (I,3027); MY SON THE DOCTOR (I,3203); THE RED SKELTON REVUE (I,3754)

DONOVAN, HENRY
COWBOY G-MEN (I,1084)

DONOVAN, TOM
GENERAL HOSPITAL (II,964)

DORAN, PHIL
CARTER COUNTRY (II,449); JENNIFER SLEPT HERE (II,1278); TOO CLOSE FOR COMFORT (II,2642)

DORE, BONNY
THE HALF-HOUR COMEDY HOUR (II,1074); THE KROFFT KOMEDY HOUR (II,1411); MYSTERIES, MYTHS AND LEGENDS (II,1782)

DORFMAN, SID
GOOD TIMES (II,1038); ONE IN A MILLION (II,1906); WHERE THERE'S SMOKEY (I,4832)

DORSO, RICHARD
CROWN THEATER WITH GLORIA SWANSON (I,1107); THE DORIS DAY SHOW (I,1355); P.O.P. (II,1939)

DORTORT, DAVID
BONANZA (II,347); THE CHISHOLMS (II,512); THE COWBOYS (II,598); THE HIGH CHAPARRAL (I,2051); HUNTER'S MOON (II,1207); THE RESTLESS GUN (I,3773); SAM HILL (I,3895)

DOTY, DENIS E
THE RETURN OF MARCUS WELBY, M.D (II,2138)

DOUGHERTY, COLIN
MATINEE AT THE BIJOU (II,1651)

DOUGLAS, JACK
MAN TO WOMAN (I,2879)

DOUGLAS, MILTON
CAVALCADE OF BANDS (I,876); FRONT ROW CENTER (I,1693); SPRING HOLIDAY (I,4168)

DOUGLAS, PAULETTE
TELEVISION INSIDE AND OUT (II,2552)

DOUGLAS, ROBERT
ALFRED HITCHCOCK PRESENTS (I,115); COURT-MARTIAL (I,1080)

DOUMANIAN, JEAN
BOB & RAY & JANE, LARAINE & GILDA (II,279); SATURDAY NIGHT LIVE (II,2261)

DOWDEN, JANE
THE BOBBY GOLDSBORO SHOW (I,673)

DOWNING, STEPHEN
KNIGHT RIDER (II,1402)

DOZIER, ROBERT
THE CONTENDER (II,571); HARRY O (II,1099); INSPECTOR PEREZ (II,1237); SWEEPSTAKES (II,2514)

DOZIER, WILLIAM
BATMAN (I,366); BEN HECHT'S TALES OF THE CITY (I,395); DANGER (I,1134); THE GREEN HORNET (I,1886); THE LONER (I,2745); PENTAGON U.S.A. (I,3514); ROD BROWN OF THE ROCKET RANGERS (I,3825); THE TAMMY GRIMES SHOW (I,4358)

DRACKOW, JOSEPH
GUNSMOKE (II,1069)

DRAGON, DARYL
THE CAPTAIN AND TENNILLE SONGBOOK (II,432)

DRAKE, BRIGIT JENSEN
ALL IN THE FAMILY (II,38); IT'S NOT EASY (II,1256); ME AND MRS. C. (II,1673)

DRAMOND, VERN
YOU ARE THERE (I,4930)

DREBEN, STAN
FUNNY YOU SHOULD ASK (I,1717)

DREIFUSS, ARTHUR
SECRET FILE, U.S.A. (I,3963)

DRESNER, HAL
THE HARVEY KORMAN SHOW (II,1103); THE HARVEY KORMAN SHOW (II,1104); HUSBANDS AND WIVES (II,1208); HUSBANDS, WIVES AND LOVERS (II,1209); JOE AND VALERIE (II,1299); POOR RICHARD (II,2065)

DREYFUSS, LAUREN
SECOND EDITION (II,2288)

DRISCOLL, ROBERT M
HIGH HOPES (II,1144)

DRISKILL, BILL
THE FEATHER AND FATHER GANG (II,845); INSTITUTE FOR REVENGE (II,1239); KATE LOVES A MYSTERY (II,1367); PARTNERS IN CRIME (II,1961)

DRUCE, OLGA
CAPTAIN VIDEO AND HIS VIDEO RANGERS (I,838)

DRYER, SHERMAN H
THE GAYELORD HAUSER SHOW (I,1744); MASTERS OF MAGIC (I,2943)

DUMONT, DR ALLEN B
DENNIS JAMES SPORTS PARADE (I,1229); TELEVISION ROOF (I,4382)

DUBIN, CHARLES S
CINDERELLA (I,956); LUCAS TANNER (II,1556); THE NUT HOUSE (I,3320)

DUBIN, RONALD N
HOW DO YOU RATE? (I,2144)

DUBOV, PAUL
SHIRLEY (II,2330)

DUBROW, BURT
KIDS ARE PEOPLE TOO (II,1390)

DUCLON, DAVID
THE JEFFERSONS (II,1276); LAVERNE AND SHIRLEY (II,1446); MAKIN' IT (II,1603); PUNKY BREWSTER (II,2092); SILVER SPOONS (II,2355); THE TED KNIGHT SHOW (II,2550); WORKING STIFFS (II,2834)

DUCOVNY, ALLEN
CAPTAIN BILLY'S MISSISSIPPI MUSIC HALL (I,826); THE NEW ADVENTURES OF SUPERMAN (I,3255); THE PETER AND MARY SHOW (I,3561); TOM CORBETT, SPACE CADET (I,4535)

DUDLEY, DON
FOUL UPS, BLEEPS AND BLUNDERS (II,911)

DUDLEY, PAUL
THE FRANK SINATRA SHOW (I,1665)

DUDNEY, KEN
BARBARA MANDRELL—THE LADY IS A CHAMP (II,144)

DUERR, ED
THE ALDRICH FAMILY (I,111)

DUFAU, OSCAR
THE LITTLE RASCALS (II,1494); THE RICHIE RICH SHOW (II,2169); SPACE STARS (II,2408)

DUFF, GORDON
BUICK ELECTRA PLAYHOUSE (I,760); JUSTICE (I,2499); THE MILLION DOLLAR INCIDENT (I,3045); PHILCO TELEVISION PLAYHOUSE (I,3583); STUDIO ONE (I,4268); TWENTY-FOUR HOURS IN A WOMAN'S LIFE (I,4644)

DUFF, HOWARD
MR. ADAMS AND EVE (I,3121); THE TEENAGE IDOL (I,4375)

DUFF, JAMES
THE VIRGINIAN (I,4738)

DUFF, WARREN
MARKHAM (I,2917); THREE FOR DANGER (I,4475)

DUGAN, JOHN T
MR. NOVAK (I,3145)

DUGGAN, BARBARA
LOVING (II,1552)

DUKE, MAURICE
THE MUSIC SHOP (I,3177); SABU AND THE MAGIC RING (I,3881)

DUMM, J RICKLEY
THE DUKES (II,740); MAGNUM, P.I. (II,1601); PALMS PRECINCT (II,1946); QUINCY, M. E. (II,2102); RIPTIDE (II,2178); STONE (II,2471)

DUNAVAN, PAT
CHRIS AND THE MAGICAL DRIP (II,516)

DUNAWAY, DON CARLOS
THE DUKES (II,740); STONE (II,2470)

DUNGAN, FRANK
MR. MOM (II,1752)

DUNLAP, DICK
KRAFT TELEVISION THEATER (I,2592)

DUNLAP, REG
THE BOBBY GOLDSBORO SHOW (I,673); DOLLY (II,699);

LOUISE MANDRELL: DIAMONDS, GOLD AND PLATINUM (II,1528); THE MARTY ROBBINS SPOTLITE (II,1636); POP! GOES THE COUNTRY (II,2067)

DUNN, CAMILLA
ALMOST ANYTHING GOES (II,56)

DUNN, J RICKLEY
THE NIGHT RIDER (II,1848)

DUNN, RICHARD
BRIGHT PROMISE (I,727)

DUNNE, BARRY
SUGARFOOT (I,4277)

DUNNE, BRUCE
BRONCO (I,745)

DUNNE, DOMINICK
ADVENTURES IN PARADISE (I,46)

DUNNE, JAMES
JOANIE LOVES CHACHI (II,1295)

DUNNE, PETER
BULBA (II,392); KING'S CROSSING (II,1397); KNOTS LANDING (II,1404)

DUNPHY, JERRY
JERRY VISITS (I,2373)

DUPUY, JUDY
BILLY BOONE AND COUSIN KIBB (I,450)

DURHAM, EARL
ALL-STAR SECRETS (II,52); PEOPLE DO THE CRAZIEST THINGS (II,1997); PEOPLE DO THE CRAZIEST THINGS (II,1998); THE TONI TENNILLE SHOW (II,2636); YOU BET YOUR LIFE (II,2856)

DUTTON, MIKE
DOLLAR A SECOND (I,1326); THE STORK CLUB (I,4236)

DUVALL, SHELLEY
FAERIE TALE THEATER (II,807)

DYKERS, REAR ADMIRAL THOMAS
THE SILENT SERVICE (I,4049)

DYKSTRA, JOHN
BATTLESTAR GALACTICA (II,181)

DYLAN, ELLIE
THE LOVE REPORT (II,1542)

DYSON, FRANKLYN
CHESTER THE PUP (I,922)

EAGLE, DAVID J
THE GOLDEN AGE OF TELEVISION (II,1016); THE GOLDEN AGE OF TELEVISION (II,1017); THE GOLDEN AGE OF TELEVISION (II,1018); THE GOLDEN AGE OF TELEVISION (II,1019)

EARLL, ROBERT
THE POWERS OF MATTHEW STAR (II,2075); WIZARDS AND WARRIORS (II,2813)

EARNSHAW, FENTON
77 SUNSET STRIP (I,3988); THE RELUCTANT SPY (I,3763)

EATON, ANTHONY
DAVID BOWIE—SERIOUS MOONLIGHT (II,622); THE GEORGE JONES SPECIAL (II,975); GLEN CAMPBELL AND FRIENDS: THE SILVER ANNIVERSARY (II,1005); USA COUNTRY MUSIC (II,591)

EBB, FRED
GOLDIE AND LIZA TOGETHER (II,1022)

EBERSOL, DICK
THE MIDNIGHT SPECIAL (II,1696); SATURDAY NIGHT LIVE (II,2261)

ECCLESINE, STEVEN
CLOSE TIES (II,545)

ECKLER, RICHARD
THE GOODYEAR REVUE (I,1856); PAUL WHITEMAN'S SATURDAY NIGHT REVUE (I,3490)

ECKSTEIN, GEORGE
79 PARK AVENUE (II,2310); BANACEK (II,138); BANACEK: DETOUR TO NOWHERE (I,339); DUFFY (II,739); HEAVEN ON EARTH (II,1119); LOVE, SIDNEY (II,1547); MASADA (II,1642); THE MASK OF MARCELLA (I,2937); THE MISSISSIPPI (II,1707); THE NAME OF THE GAME (I,3217); THE RHINEMANN EXCHANGE (II,2150); SARA (II,2260); SIDNEY SHORR (II,2349); SUNSHINE (II,2495); TRAVIS MCGEE (II,2656); WILD ABOUT HARRY (II,2797)

ECONOMOU, MICHAEL
THE FOUR SEASONS (II,915)

EDELMAN, LOUIS F
THE ADVENTURES OF JIM BOWIE (I,63); ALONG THE BARBARY COAST (I,144); THE ANDY GRIFFITH SHOW (I,193); THE BARBARA STANWYCK THEATER (I,350); THE CALIFORNIANS (I,795); THE JOEY BISHOP SHOW (I,2404); JOSEPHINE LITTLE: ADVENTURES IN HAPPINESS (I,2450); JOSEPHINE LITTLE: DRAGON BY THE TAIL (I,2451); JOSEPHINE LITTLE: THE MIRACULOUS JOURNEY OFTADPOLE CHAN (I,2452); THE LIFE AND LEGEND OF WYATT EARP (I,2681); MAKE ROOM FOR DADDY (I,2843)

EDELSTEIN, RICK
CUTTER TO HOUSTON (II,612)

EDEN, ELIZABETH
TMT (II,2622)

EDENS, ROGER
MERMAN ON BROADWAY (I,3011)

EDGARTON, JUSTIN
THE JOKER'S WILD (II,1341)

EDWARDS, BLAKE
THE BOSTON TERRIER (I,696); THE BOSTON TERRIER (I,697); CITY DETECTIVE (I,966); JULIE AND DICK IN COVENT GARDEN (II,1352); JULIE ON SESAME STREET (I,2485); JULIE! (I,2477); PETER GUNN (I,3562)

EDWARDS, RALPH
THE $100,000 NAME THAT TUNE (II,1902); THE CROSS-WITS (II,605); DEALER'S CHOICE (II,635); END OF THE RAINBOW (I,1443); IT COULD BE YOU (I,2245); KNOCKOUT (II,1403); NAME THAT TUNE (II,1786); NAME THAT TUNE (II,1787); THE NEW TRUTH OR CONSEQUENCES (II,1832); THE PEOPLE'S COURT (II,2001); SO YOU THINK YOU GOT TROUBLES?! (II,2391); THIS IS YOUR LIFE (II,2598); TRUTH OR CONSEQUENCES (I,4619); TRUTH OR CONSEQUENCES (I,4620); WIDE COUNTRY (I,4858); YOU DON'T SAY (II,2859)

EGAN, JOHN
ROAD OF LIFE (I,3798)

EGAN, MARK
ALICE (II,33)

EGAN, SAM
AUTOMAN (II,120); QUINCY, M. E. (II,2102)

EGAN, TOM
LISA, BRIGHT AND DARK (I,2711); THICKER THAN WATER (I,4445); WARNING SHOT (I,4770)

EGAN, WILLIAM
TV'S FUNNIEST GAME SHOW MOMENTS (II,2676)

EHRLICH, KEN
CHRISTMAS SPECIAL. . . WITH LOVE, MAC DAVIS (II,523); FAME (II,812); MAC DAVIS 10TH ANNIVERSARY SPECIAL: I STILL BELIEVE IN MUSIC (II,1571); THE MAC DAVIS CHRISTMAS SPECIAL (II,1573); MAC DAVIS SPECIAL: THE MUSIC OF CHRISTMAS (II,1578); MAC DAVIS—I'LL BE HOME FOR CHRISTMAS (II,1580); RINGO (II,2176)

EINHORN, LAWRENCE
KIDS ARE PEOPLE TOO (II,1390); RONA LOOKS AT JAMES, MICHAEL, ELLIOTT, AND BURT (II,2205); RONA LOOKS AT RAQUEL, LIZA, CHER AND ANN-MARGRET (II,2207); TELEVISION INSIDE AND OUT (II,2552); THAT'S HOLLYWOOD (II,2577); THAT'S MY LINE (II,2579)

EINSTEIN, ALAN
ANGIE (II,80)

EINSTEIN, BOB
BIZARRE (II,258); BIZARRE (II,259); JOEY AND DAD (II,1306); LOLA (II,1508); LOLA (II,1509); LOLA (II,1510); LOLA (II,1511); THE REDD FOXX COMEDY HOUR (II,2126); VAN DYKE AND COMPANY (II,2718); VAN DYKE AND COMPANY (II,2719)

EISEN, GAIL
THE MIKE WALLACE PROFILES (II,1699)

EISENBERG, JERRY
FANG FACE (II,822); GOLDIE GOLD AND ACTION JACK (II,1023); THE HEATHCLIFF AND DINGBAT SHOW (II,1117); MEATBALLS AND SPAGHETTI (II,1674); PANDAMONIUM (II,1948); THE PLASTICMAN COMEDY/ADVENTURE SHOW (II,2051); THUNDARR THE BARBARIAN (II,2616)

EISENBERG, NAT
THE MAGIC CLOWN (I,2814)

EISENSTOCK, ALAN
BLUE JEANS (II,275); HIGH SCHOOL, U.S.A. (II,1148); HIGH SCHOOL, U.S.A. (II,1149)

EISNER, JANE
A NEW KIND OF FAMILY (II,1819)

ELBES, GEORGE
TWENTY QUESTIONS (I,4647)

ELDER, ANN
I'D RATHER BE CALM (II,1216); LITTLE LULU (II,1493); THE MARILYN MCCOO AND BILLY DAVIS JR. SHOW (II,1630); THAT'S MY LINE (II,2579)

ELIAS, MICHAEL

ALL'S FAIR (II,46); BLACK BART (II,260); SZYSZNYK (II,2523)

ELINSON, JACK
A.K.A. PABLO (II,28); THE ACADEMY (II,3); THE ACADEMY II (II,4); THE DORIS DAY SHOW (I,1355); THE FACTS OF LIFE (II,805); FACTS OF LIFE: THE FACTS OF LIFE GOES TO PARIS (II,806); JO'S COUSINS (II,1293); JOE'S WORLD (II,1305); MAKE ROOM FOR DADDY (I,2843); ONE DAY AT A TIME (II,1900); P.O.P. (II,1939); THE PARKERS (II,1959); RUN, BUDDY, RUN (I,3870)

ELISCU, COLONEL WILLIAM
O.S.S. (I,3411)

ELKINS, HILLARD
DEADLY GAME (II,632)

ELLENBOGEN, ERIC
THOSE WONDERFUL TV GAME SHOWS (II,2603)

ELLIOT, MICHAEL
V: THE SERIES (II,2715)

ELLIOTT, JACK
100 YEARS OF AMERICA'S POPULAR MUSIC (II,1904)

ELLIOTT, W C
THE B.B. BEEGLE SHOW (II,124)

ELLIS, ANTHONY
BLACK SADDLE (I,480)

ELLIS, BRIAN
WALTER CRONKITE'S UNIVERSE (II,2739)

ELLIS, MINERVA
THE SEEKING HEART (I,3975)

ELLIS, RALPH C
ADVENTURES IN RAINBOW COUNTRY (I,47)

ELLIS-BUNIN, MARY
SEARCH FOR TOMORROW (II,2284)

ELLISON, BOB
ALAN KING LOOKS BACK IN ANGER—A REVIEW OF 1972 (I,99); THE ALAN KING SPECIAL (I,103); ANGIE (II,80); THE BETTY WHITE SHOW (II,224); THE BURNS AND SCHREIBER COMEDY HOUR (I,768); THE BURNS AND SCHREIBER COMEDY HOUR (I,769); A COUPLE OF DONS (I,1077); COUSINS (II,595); IN SECURITY (II,1227); THE MAC DAVIS SHOW (II,1575); NUMBER 96 (II,1869); RHODA (II,2151); THE RICHARD PRYOR SPECIAL (II,2164); THE RICHARD PRYOR SPECIAL? (II,2165); THE SINGLE LIFE (II,2364); STRUCK BY LIGHTNING (II,2482); TOO GOOD TO BE TRUE (II,2643); THE WONDERFUL WORLD OF AGGRAVATION (I,4894); YOUR PLACE OR MINE? (II,2874)

ELLSWORTH, WHITNEY
THE ADVENTURES OF SUPERMAN (I,77)

ELMAN, IRVING
BEN CASEY (I,394); THE ELEVENTH HOUR (I,1425); MATT LINCOLN (I,2949)

ELSEY, IAN
LOVE IN A COLD CLIMATE (II,1539)

EMERY, BOB
THE SMALL FRY CLUB (I,4089)

EMERY, KAY
THE SMALL FRY CLUB (I,4089)

EMERY, RALPH
THE MARTY ROBBINS SPOTLITE (II,1636)

ENDERS, ROBERT J
BEST OF THE POST (I,408); THEY WENT THATAWAY (I,4443)

ENDLER, ESTELLE
RODNEY DANGERFIELD SHOW: I CAN'T TAKE IT NO MORE (II,2198); RODNEY DANGERFIELD SPECIAL: IT'S NOT EASY BEIN' ME (II,2199)

ENGEL, FRED L
THE HANDS OF CORMAC JOYCE (I,1927); KILLER BY NIGHT (I,2537)

ENGEL, PETER
THE 416TH (II,913); HOW TO SURVIVE A MARRIAGE (II,1198); THE ICE PALACE (I,2182); THE PAUL WILLIAMS SHOW (II,1984); SIROTA'S COURT (II,2365)

ENGLANDER, ROGER
LET'S TAKE A TRIP (I,2668)

ENGLE, HARRISON
LEGENDS OF THE WEST: TRUTH AND TALL TALES (II,1457)

ENGLE, LARRY
OMNI: THE NEW FRONTIER (II,1889)

ENGLISH, JOHN
SOLDIERS OF FORTUNE (I,4111)

ENGLUND, GEORGE
AN HOUR WITH ROBERT GOULET (I,2137)

ENRIGHT, DAN
ALL ABOUT FACES (I,126); BACK THE FACT (I,327); BERT CONVY SPECIAL—THERE'S A MEETING HERE TONIGHT (II,210); BREAK THE BANK (II,369); BULLSEYE (II,393); FAITH BALDWIN'S THEATER OF ROMANCE (I,1515); THE HOLLYWOOD CONNECTION (II,1161); HOT POTATO (II,1188); JOKER! JOKER!! JOKER!!! (II,1340); THE JOKER'S WILD (II,1341); JUVENILE JURY (I,2501); LIFE BEGINS AT EIGHTY (I,2682); OH, BABY! (I,3340); PLAY THE PERCENTAGES (II,2053); TALES OF THE HAUNTED (II,2538); TIC TAC DOUGH (II,2618); WAY OUT GAMES (II,2747); WINKY DINK AND YOU (I,4879); YOU'RE ON YOUR OWN (I,4968)

ENYEDA, DIANA
POWERHOUSE (II,2074)

EPAND, LEN
A HOT SUMMER NIGHT WITH DONNA (II,1190)

EPSTEIN, ALLEN
DOC ELLIOT (I,1305); DRIBBLE (II,736)

EPSTEIN, DONALD
BLANKETY BLANKS (II,266)

EPSTEIN, JON
ADVICE TO THE LOVELORN (II,16); THE CONTENDER (II,571); THE GOSSIP COLUMNIST (II,1045); HARBOURMASTER (I,1946); MCMILLAN (II,1666); MCMILLAN AND WIFE (II,1667); THE OUTCASTS (I,3425); OWEN MARSHALL: COUNSELOR AT LAW (I,3435); PARTNERS IN CRIME (I,3461); RICH MAN, POOR MAN—BOOK I (II,2161); RICH MAN, POOR MAN—BOOK II (II,2162); SCENE OF THE CRIME (II,2271); TARZAN (I,4361); TENAFLY (I,4395); WHAT REALLY HAPPENED TO THE CLASS OF '65? (II,2768); THE YOUNG REBELS (I,4944)

EPSTEIN, MEL
BROKEN ARROW (I,744); MAN WITHOUT A GUN (I,2884)

ERICKSON, ROD
THE BOB SMITH SHOW (I,668); THE FRED WARING SHOW (I,1679)

ERLICHMAN, MARTIN
BARBRA STREISAND AND OTHER MUSICAL INSTRUMENTS (I,353); BARBRA STREISAND: A HAPPENING IN CENTRAL PARK (I,352); THE BELLE OF 14TH STREET (I,390); THE CHEVY CHASE SHOW (II,505); MY NAME IS BARBRA (I,3198)

ERSKINE, CHESTER
TV READERS DIGEST (I,4630)

ERWIN JR, STU
THE ED SULLIVAN SHOW (I,1401)

ERWIN, STU
THE BUSTERS (II,401)

ESTIN, KEN
SHAPING UP (II,2316)

EUBANKS, BOB
ALL-STAR SECRETS (II,52); THE TONI TENNILLE SHOW (II,2636); YOU BET YOUR LIFE (II,2856)

EUSTIS, RICH
THE CARPENTERS (II,444); DEAN MARTIN PRESENTS MUSIC COUNTRY, U.S.A. (I,1192); DINAH IN SEARCH OF THE IDEAL MAN (I,1278); THE DOROTHY HAMILL WINTER CARNIVAL SPECIAL (II,720); FATHER O FATHER (II,842); THE GLEN CAMPBELL GOODTIME HOUR (I,1820); GOOBER AND THE TRUCKERS' PARADISE (II,1029); THE JERRY REED WHEN YOU'RE HOT, YOU'RE HOT HOUR (I,2372); THE JIM STAFFORD SHOW (II,1291); THE JOHN BYNER COMEDY HOUR (I,2407); JOHN DENVER ROCKY MOUNTAIN CHRISTMAS (II,1314); THE JOHN DENVER SPECIAL (II,1315); THE JOHN DENVER SPECIAL (II,1316); JOHN DENVER—THANK GOD I'M A COUNTRY BOY (II,1317); THE PAUL LYNDE COMEDY HOUR (II,1978); THE RICH LITTLE SHOW (II,2153); THE RICH LITTLE SHOW (II,2154); SZYSZNYK (II,2523)

EVANS, MARK
THE ADVENTURES OF LONG JOHN SILVER (I,67)

EVANS, MAURICE
ALICE IN WONDERLAND (I,120); THE CORN IS GREEN (I,1050); THE DEVIL'S DISCIPLE (I,1247); DREAM GIRL (I,1374); THE GOOD FAIRY (I,1846); TAMING OF THE SHREW (I,4356)

EVANS, TREVOR
CELEBRITY COOKS (II,472)

EVENS, SAMUEL H
BOBO HOBO AND HIS TRAVELING TROUPE (I,678)

FACTOR, ALAN JAY
THE NEXT STEP BEYOND (II,1839)

FAICHNEY, JAMES
AIR POWER (I,92)

FAIMAN, PETER
THE DON LANE SHOW (II,707)

FAIRBANKS, JERRY
CRUSADER RABBIT (I,1109); FRONT PAGE DETECTIVE (I,1689); PUBLIC PROSECUTOR (I,3689)

FAIRE, SANDRA
THE JOYCE DAVIDSON SHOW (I,2459)

FALSEY, JOHN
ST. ELSEWHERE (II,2432)

FARAGHAN, J E
CHICAGOLAND MYSTERY PLAYERS (I,936); THE MUSIC SHOW (I,3178)

FARBER, MEL
YOU'RE GONNA LOVE IT HERE (II,2860)

FARIA, WALT DE
THE RIVALRY (I,3796)

FARMER, GENE
THAT'S MY MAMA (II,2580)

FARR, GORDON
THE BOB NEWHART SHOW (II,342); HONEYMOON SUITE (I,2108); KEEPING UP WITH THE JONESES (II,1374); THE LOVE BOAT (II,1535); MAKE YOUR OWN KIND OF MUSIC (I,2847); WE GOT IT MADE (II,2752)

FARR, LYNNE
THE BOB NEWHART SHOW (II,342); THE LOVE BOAT (II,1535)

FARRELL, DICK
THE HANK MCCUNE SHOW (I,1932)

FARRELL, MARTY
THE MAC DAVIS CHRISTMAS SPECIAL (II,1572); THE MAC DAVIS SPECIAL (II,1577)

FARREN, JACK
WINKY DINK AND YOU (I,4879)

FASHKO, ROBERT
SUSPICION (I,4309)

FASS, GEORGE
THE THREE MUSKETEERS (I,4480)

FATES, GIL
THE EARL WRIGHTSON SHOW (I,1393); FAYE EMERSON'S WONDERFUL TOWN (I,1549); I'VE GOT A SECRET (II,1220); MAKE THE CONNECTION (I,2846); SCRAPBOOK JR. EDITION (I,3944); TO TELL THE TRUTH (II,2626); TV'S FUNNIEST GAME SHOW MOMENTS (II,2676); TWO FOR THE MONEY (I,4654); VANITY FAIR (I,4702); WHAT'S MY LINE? (I,4820); WHAT'S THE STORY? (I,4821); WINNER TAKE ALL (I,4881)

FAUSSETT, HUDSON
THE PAUL WINCHELL AND JERRY MAHONEY SHOW (I,3492); TIC TAC DOUGH (I,4498)

FAY JR, SHELDON
TRAVELS WITH CHARLEY (I,4597)

FAY, DEIDRE
ALL TOGETHER NOW (II,45)

FAY, PAT
GROWING PAYNES (I,1891)

FEDDERSON, DON
DATE WITH THE ANGELS (I,1164); FAMILY AFFAIR (I,1519); THE FESS PARKER SHOW (II,850); LIFE WITH ELIZABETH (I,2689); MEMORIES WITH LAWRENCE WELK (II,1681); THE MILLIONAIRE (I,3046); THE MILLIONAIRE (II,1700); MURDOCK'S GANG (I,3164); MY THREE SONS (I,3205); THE SMITH FAMILY (I,4092); TO ROME WITH LOVE (I,4526); WHO DO YOU TRUST? (I,4846)

FEEGAY, PAUL
THE MOREY AMSTERDAM SHOW (I,3100)

FEGAN, JOHN
NERO WOLFE (II,1806)

FEHRLE, PHIL
BEACH PATROL (II,192); EIGHT IS ENOUGH (II,762)

FEIN, DAVID
ALL ABOUT FACES (I,126); ANYTHING FOR MONEY (II,99); BULLSEYE (II,393); CELEBRITY CHARADES (II,470); FRED WARING: WAY BACK HOME (I,1680); THAT'S HOLLYWOOD (II,2577)

FEIN, IRVING
EVERYTHING YOU ALWAYS WANTED TO KNOW ABOUT JACK BENNY AND WERE AFRAID TO ASK (I,1482); GEORGE BURNS AND OTHER SEX SYMBOLS (II,966); GEORGE BURNS CELEBRATES 80 YEARS IN SHOW BUSINESS (II,967); GEORGE BURNS EARLY, EARLY, EARLY, CHRISTMAS SHOW (II,968); GEORGE BURNS IN NASHVILLE?? (II,969); THE GEORGE BURNS SPECIAL (II,970); GEORGE BURNS' 100TH BIRTHDAY PARTY (II,971); GEORGE BURNS' HOW TO LIVE TO BE 100 (II,972); I LOVE HER ANYWAY! (II,1214); THE JACK BENNY CHRISTMAS SPECIAL (I,2289); THE JACK BENNY PROGRAM (I,2294); THE JACK BENNY SPECIAL (I,2296); JACK BENNY'S 20TH ANNIVERSARY TV SPECIAL (I,2303); JACK BENNY'S BIRTHDAY SPECIAL (I,2299); JACK BENNY'S NEW LOOK (I,2301); JACK BENNY'S SECOND FAREWELL SHOW (I,2302); A LOVE LETTER TO JACK BENNY (II,1540)

FEINER, BEN
THE 20TH CENTURY-FOX HOUR (I,4642); LIFE WITH FATHER (I,2690); THE VALIANT YEARS (I,4696)

FEINER, BEN

FEINER, RICHARD
LAUREL AND HARDY LAUGHTOONS (II,1445)

FEIST, FELIX
THE CALIFORNIANS (I,795); HARBOURMASTER (I,1946); PEYTON PLACE (I,3574)

FEKE, STEVE
IT'S ANYBODY'S GUESS (II,1255)

FELD, IRVIN
SIEGFRIED AND ROY (II,2350)

FELD, KENNETH
SIEGFRIED AND ROY (II,2350)

FELDKAMP, FRED
MARCH OF TIME THROUGH THE YEARS (I,2903)

FELDMAN, CHESTER
CARD SHARKS (II,438); CHILD'S PLAY (II,510); I'VE GOT A SECRET (I,2283); I'VE GOT A SECRET (II,1220); JOHNNY CARSON DISCOVERS CYPRESS GARDENS (I,2419); TRIVIA TRAP (II,2659)

FELDMAN, DICK
THE WORLD'S FUNNIEST COMMERCIAL GOOFS (II,2841)

FELDMAN, EDWARD H
ALONE AT LAST (II,59); THE ANDROS TARGETS (II,76); ANGEL (I,216); THE BROTHERS (I,748); THE DORIS DAY SHOW (I,1355); ETHEL IS AN ELEPHANT (II,783); FAIR EXCHANGE (I,1512); FLAMINGO ROAD (II,871); FLAMINGO ROAD (II,872); GLYNIS (I,1829); GREAT DAY (II,1054); HOGAN'S HEROES (I,2069); JERRY (II,1283); JESSIE (II,1287); THE LUCILLE BALL COMEDY HOUR (I,2783); MURDOCK'S GANG (I,3164); MY WIVES JANE (I,3207); PAPA G.I. (I,3451); THE QUEEN AND I (I,3704); SPENCER'S PILOTS (II,2422)

FELDMAN, GENE
THE HORROR OF IT ALL (II,1181)

FELDMAN, PHIL
POWDERKEG (I,3657)

FELDMAN, SANFORD
BREAKAWAY (II,370)

FELL, DIANA
THE $1.98 BEAUTY SHOW (II,698); THE GONG SHOW (II,1026)

FELLOWS, ARTHUR
CANNON (I,820); THE GIRL, THE GOLD WATCH AND DYNAMITE (II,1001); THE GIRL, THE GOLD WATCH AND EVERYTHING (II,1002); KEY TORTUGA (II,1383); THE STREETS OF SAN FRANCISCO (I,4260); THE STREETS OF SAN FRANCISCO (II,2478); STRUCK BY LIGHTNING (II,2482); TOP OF THE HILL (II,2645); TRAVIS LOGAN, D.A. (I,4598)

FELLOWS, ROBERT
ALCOA/GOODYEAR THEATER (I,107)

FELSCHER, HOWARD
FAMILY FEUD (II,816); HE SAID, SHE SAID (I,1986); PASSWORD (II,1965); PASSWORD PLUS (II,1966); SHOWOFFS (II,2345); TIC TAC DOUGH (I,4498)

FELTHAM, KERRY
SHOGUN (II,2339)

FELTON, HAPPY
IT'S A HIT (I,2257)

FELTON, NORMAN
BAFFLED (I,332); CRISIS (I,1101); DOCTOR KILDARE (I,1315); THE ELEVENTH HOUR (I,1425); EXECUTIVE SUITE (II,796);

GHOSTBREAKER (I,1789); THE GIRL FROM U.N.C.L.E. (I,1808); THE GOOK FAMILY (I,1858); HAWKINS (I,1976); HAWKINS ON MURDER (I,1978); JERICHO (I,2358); JUDY SPLINTERS (I,2475); THE LADY DIED AT MIDNIGHT (I,2601); THE LIEUTENANT (I,2680); THE MAN FROM U.N.C.L.E. (I,2867); THE MOONGLOW AFFAIR (I,3099); PECK'S BAD GIRL (I,3503); THE PSYCHIATRIST (I,3686); THE PSYCHIATRIST: GOD BLESS THE CHILDREN (I,3685); PURSUIT (I,3700); THE STRANGE REPORT (I,4248); STUD'S PLACE (I,4271); STUDIO ONE (I,4268); THE UNITED STATES STEEL HOUR (I,4677)

FENADY, ANDREW J
BRANDED (I,707); HONDO (I,2103); THE REBEL (I,3746); THE STRANGER (I,4251)

FENNELL, ALBERT
THE AVENGERS (II,121); THE NEW AVENGERS (II,1816)

FENNELLY, VINCENT M
THE ALCOA HOUR (I,108); ALCOA/GOODYEAR THEATER (I,107); THE BOUNTY HUNTER (I,698); THE DAVID NIVEN THEATER (I,1178); GENTRY'S PEOPLE (I,1762); IF YOU KNEW TOMORROW (I,2187); JOHNNY RISK (I,2435); JUSTICE OF THE PEACE (I,2500); MAGGIE MALONE (I,2812); RAWHIDE (I,3727); STAGECOACH WEST (I,4185); TRACKDOWN (I,4589)

FENTON, MILDRED
FUN FOR THE MONEY (I,1708)

FERBER, MEL
A BELL FOR ADANO (I,388); CELEBRITY CHALLENGE OF THE SEXES (II,464); HERB SHRINER TIME (I,2020)

FERGUSON, ANDREW B
THE GREAT SPACE COASTER (II,1059)

FERGUSON, WILLIAM
GOOD OL' BOYS (II,1035)

FERRER, JOSE
CYRANO DE BERGERAC (I,1119)

FERRI, LUCY
GENERAL HOSPITAL (II,964)

FERRIN, FRANK
ANDY'S GANG (I,214)

FERRIS, BARBARA
THE CANDID CAMERA SPECIAL (II,422)

FERRO, BETH
THE PRICE IS RIGHT (I,3664)

FETTER, TED
ONE MAN SHOW (I,3381); YOUR HIT PARADE (I,4951)

FHERLE, PHIL
LADIES IN BLUE (II,1422); VEGAS (II,2724)

FICKS, BILL
THE AMES BROTHERS SHOW (I,178)

FIEDLER, PETER
THIS WAS AMERICA (II,2599)

FIELD, FERN
THE BAXTERS (II,183)

FIELD, TED
MARLO AND THE MAGIC MOVIE MACHINE (II,1631)

FIELDER, RICHARD
GEORGE WASHINGTON (II,978); SEVEN BRIDES FOR SEVEN BROTHERS (II,2307)

FIELDS, JOSEPH
WONDERFUL TOWN (I,4893)

FIELDS, RICHARD
PAT BENATAR IN CONCERT (II,1967)

FILERMAN, MICHAEL
BEHIND THE SCREEN (II,200); FALCON CREST (II,808); FLAMINGO ROAD (II,871); FLAMINGO ROAD (II,872); JOSHUA'S WORLD (II,1344); KING'S CROSSING (II,1397); KNOTS LANDING (II,1404); SECRETS OF MIDLAND HEIGHTS (II,2296); SPRAGGUE (II,2430); WILLOW B: WOMEN IN PRISON (II,2804)

FILLOUS, ROBERT
TURN OF FATE (I,4623)

FINCH, EDWARD
IT'S ROCK AND ROLL (II,1258)

FINCH, JAMES P
POLE POSITION (II,2060)

FINE, MORT
BEARCATS! (I,375); BERT D'ANGELO/SUPERSTAR (II,211); I SPY (I,2179); THE MOST DEADLY GAME (I,3106)

FINE, SYLVIA
AN HOUR WITH DANNY KAYE (I,2136)

FINK, MARK
ARCHIE BUNKER'S PLACE (II,105); GREAT DAY (II,1054)

FINKEL, BOB
THE ALL-STAR SALUTE TO MOTHER'S DAY (II,51); AND HERE'S THE SHOW (I,187); THE ANDY WILLIAMS SHOW (I,205); THE BEAUTIFUL PHYLLIS DILLER SHOW (I,381); BING CROSBY AND HIS FRIENDS (I,456); BING CROSBY AND HIS FRIENDS (II,251); BING CROSBY AND THE SOUNDS OF CHRISTMAS (I,457); BING CROSBY'S SUN VALLEY CHRISTMAS SHOW (I,475); BING CROSBY—COOLING IT (I,459); BING CROSBY—COOLING IT (I,460); THE BOB CUMMINGS SHOW (I,500); BRIEF ENCOUNTER (I,723); THE CHEVY CHASE SHOW (II,505); CHRISTMAS WITH THE BING CROSBYS (I,950); CHRISTMAS WITH THE BING CROSBYS (II,524); CIRCUS OF THE STARS (II,530); CIRCUS OF THE STARS (II,531); CIRCUS OF THE STARS (II,532); CIRCUS OF THE STARS (II,533); CIRCUS OF THE STARS (II,534); CIRCUS OF THE STARS (II,535); CIRCUS OF THE STARS (II,536); CIRCUS OF THE STARS (II,537); CIRCUS OF THE STARS (II,538); ELVIS (I,1435); THE JACKIE GLEASON SPECIAL (I,2325); THE JERRY LEWIS SHOW (I,2370); THE JIMMY MCNICHOL SPECIAL (II,1292); THE JOHN DAVIDSON CHRISTMAS SHOW (II,1308); JOHN DENVER AND THE MUPPETS: A CHRISTMAS TOGETHER (II,1313); THE MANY MOODS OF PERRY COMO (I,2897); MICKIE FINN'S (I,3029); THE MUHAMMED ALI VARIETY SPECIAL (II,1762); MYSTERIES OF CHINATOWN (I,3209); THE PEARL BAILEY SHOW (I,3501); THE PEOPLE'S CHOICE (I,3521); THE PERRY COMO WINTER SHOW (I,3543); THE PERRY COMO WINTER SHOW (I,3544); PERRY COMO'S SPRINGTIME SPECIAL (II,2023); PERRY COMO'S WINTER SHOW (I,3546); PERRY COMO'S WINTER SHOW (I,3547); PERRY COMO'S WINTER SHOW (I,3548); PRIDE OF THE FAMILY (I,3667); A SALUTE TO TELEVISION'S 25TH ANNIVERSARY (I,3893); TED KNIGHT MUSICAL COMEDY VARIETY SPECIAL SPECIAL (II,2549); THE WAYNE NEWTON SPECIAL (II,2750); THE WORLD OF ENTERTAINMENT (II,2837)

FINKLEHOFF, FRED
LOLLIPOP LOUIE (I,2738)

FINNEGAN, BILL
BIG HAWAII (II,238); CALLAHAN (II,414); THE CENTURY TURNS (I,890); D.H.O. (I,1249); DANGER IN PARADISE (II,617); DEAD MAN ON THE RUN (II,631); HAWAII FIVE-O (II,1110); SPENCER'S PILOTS (II,2421); STRANGER IN OUR HOUSE (II,2477); VALENTINE MAGIC ON LOVE ISLAND (II,2716)

FINNEGAN, PAT
DANGER IN PARADISE (II,617); STRANGER IN OUR HOUSE (II,2477); VALENTINE MAGIC ON LOVE ISLAND (II,2716)

FIRKIN, REX
UPSTAIRS, DOWNSTAIRS (II,2709)

FISCHER, BILLY
LACE (II,1418)

FISCHER, GEOFFREY
BRET MAVERICK (II,374)

FISCHER, PETER S
BLACK BEAUTY (II,261); CHARLIE COBB: NICE NIGHT FOR A HANGING (II,485); DARKROOM (II,619); THE EDDIE CAPRA MYSTERIES (II,758); ELLERY QUEEN (II,766); GRIFF (I,1888); MURDER, SHE WROTE (II,1770); ONCE AN EAGLE (II,1897); RICHIE BROCKELMAN, PRIVATE EYE (II,2167)

FISH, JOANNE
THE LOVE REPORT (II,1542)

FISHBURN, ALAN
BON VOYAGE (I,683)

FISHER, ART
33 1/3 REVOLUTIONS PER MONKEE (I,4451); CHER. . . SPECIAL (II,499); DONNY AND MARIE (II,712); MONSANTO PRESENTS MANCINI (I,3092); MONSANTO PRESENTS MANCINI (I,3093); MONSANTO PRESENTS MANCINI (I,3094); NEIL DIAMOND SPECIAL: I'M GLAD YOU'RE HERE WITH ME TONIGHT (II,1803); THE OSMOND BROTHERS SPECIAL (II,1925); SIEGFRIED AND ROY (II,2350); THE SUZANNE SOMERS SPECIAL (II,2510); THE SUZANNE SOMERS SPECIAL (II,2511); THAT'S TV (II,2581)

FISHER, BILL
THAT GOOD OLD NASHVILLE MUSIC (II,2571)

FISHER, BOB
HOORAY FOR LOVE (I,2116); MICKEY (I,3023)

FISHER, CHARLES
AS THE WORLD TURNS (II,110); THE EDGE OF NIGHT (II,760)

FISHER, HANNAH
THE FOUR JUST MEN (I,1650)

FISHER, MICHAEL
THE ASPHALT COWBOY (II,111); FANTASY ISLAND (II,829); MATT HOUSTON (II,1654); RETURN TO FANTASY ISLAND (II,2144); STRIKE FORCE (II,2480)

FISHER, STANFORD H
MARLO AND THE MAGIC MOVIE MACHINE (II,1631)

FISHER, STEVE
LUKE AND THE TENDERFOOT (I,2792)

FISHER, TERRY LOUISE
CAGNEY AND LACEY (II,409); CUTTER TO HOUSTON (II,612)

FISHMAN, DOUG
THE FUN FACTORY (II,938)

FISHMAN, ED
DEALER'S CHOICE (II,635); THE DIAMOND HEAD GAME (II,666); THE FUN FACTORY (II,938)

FITE, BEVERLY
ALKALI IKE (I,123)

FITZGERALD, JOHN
A FINE ROMANCE (II,858)

FITZGERALD, PRUDENCE
A FAMILY AFFAIR (II,814)

FITZPATRICK, CATHERINE
CAMELOT (II,416)

FITZSIMONS, CHARLES B
CASABLANCA (II,450); CONSPIRACY OF TERROR (II,569); COVER GIRLS (II,596); GOODNIGHT, BEANTOWN (II,1041); MASSARATI AND THE BRAIN (II,1646); MATT HELM (II,1653); NANNY AND THE PROFESSOR (I,3225); OH, NURSE! (I,3343); RIKER (II,2175); WHAT'S UP DOC? (II,2771); WONDER WOMAN (SERIES 2) (II,2829); THE WONDERFUL WORLD OF PHILIP MALLEY (II,2831)

FLACK, TIM
CALLAHAN (II,414)

FLANNERY, JUDY
COSMOS (II,579)

FLATT, ERNEST
THE JIMMIE RODGERS SHOW (I,2383)

FLATTERY, ED
YOU ASKED FOR IT (II,2855)

FLATTERY, PAUL
BRITT EKLAND'S JUKE BOX (II,378)

FLAUM, MARSHALL
20TH CENTURY (I,4640); BOB HOPE SPECIAL: BOB HOPE IN WHO MAKES THE WORLD LAUGH, PART 2 (II,286); BOB HOPE SPECIAL: BOB HOPE IN WHO MAKES THE WORLD LAUGH? (II,287); BOGART (I,679); HOLLYWOOD: THE SELZNICK YEARS (I,2095); LIFE'S MOST EMBARRASSING MOMENTS II (II,1471); PLAYBOY'S 25TH ANNIVERSARY CELEBRATION (II,2054); RIPLEY'S BELIEVE IT OR NOT (II,2177)

FLAUM, THEA
SNEAK PREVIEWS (II,2388)

FLEETWOOD, MICK
FLEETWOOD MAC IN CONCERT (II,880)

FLEISCHER, MAX
THE BETTY BOOP SHOW (I,411)

FLEISCHMAN, STEPHEN
ONE NIGHT STANDS (I,3387)

FLEISHER, CAROL L
THE SINGING COWBOYS RIDE AGAIN (II,2363); THOSE AMAZING ANIMALS (II,2601)

FLETCHER, ROBERT
LADIES AND GENTLEMAN. . . BOB NEWHART (II,1420)

FLICKER, THEODORE J
BARNEY MILLER (II,154)

FLINN, WILLIAM
FAME (II,812)

FLOREA, JOHN
NOT FOR HIRE (I,3308); SEA HUNT (I,3948)

FLORIMBI, DAVID
FINDER OF LOST LOVES (II,857)

FLORIN, FABRICE
BACKSTAGE PASS (II,132)

FLOTHOW, RUDOLPH
THE NEW ADVENTURES OF CHARLIE CHAN (I,3249); RAMAR OF THE JUNGLE (I,3718)

FOGARTY, JACK V
JESSIE (II,1287); T.J. HOOKER (II,2524)

FOLB, JAY
FOOT IN THE DOOR (II,902); FREEBIE AND THE BEAN (II,922); THE THREE WIVES OF DAVID WHEELER (II,2611)

FOLEY JR, GEORGE F
TALES OF TOMORROW (I,4352)

FOLEY, GEORGE
THE MARY MARGARET MCBRIDE SHOW (I,2933)

FONDA, JAMES
ALCOA/GOODYEAR THEATER (I,107); THE AMOS AND ANDY SHOW (I,179); BROCK CALLAHAN (I,741); DENNIS THE MENACE (I,1231); HAZEL (I,1982); STRATEGIC AIR COMMAND (I,4252); YOU ARE THERE (I,4929)

FONDA, JANE
9 TO 5 (II,1852)

FORBES, DOUGLAS C
DAVID BOWIE—SERIOUS MOONLIGHT (II,622)

FORCELLEDO, BETH
YOUR NEW DAY (II,2873)

FORD, DON
THE AMERICAN WEST OF JOHN FORD (I,173)

FORD, MONTGOMERY
THE EGG AND I (I,1420)

FORD, SENATOR EDWARD
CAN YOU TOP THIS? (I,816)

FORDYCE, IAN
APPOINTMENT WITH A KILLER (I,245); CRIMES OF PASSION (II,603); CRY TERROR! (I,1112); DEATH IN DEEP WATER (I,1206); DEATH IN SMALL DOSES (I,1207); DIAL A DEADLY NUMBER (I,1253); THE DOUBLE KILL (I,1360); FATHER BROWN (II,836); THE FEAR IS SPREADING (I,1553); IF IT'S A MAN, HANG UP (I,2186); IN THE STEPS OF A DEAD MAN (I,2205); LOOK BACK IN DARKNESS (I,2752); MELODY OF HATE (I,2999); MIRROR OF DECEPTION (I,3060); MURDER MOTEL (I,3161); THE NEXT VICTIM (I,3281); SLEEPWALKER (I,4084); TERROR FROM WITHIN (I,4404)

FORREST, ARTHUR
THE JERRY LEWIS SHOW (II,1284)

FORREST, ROBERT
CONTEST CARNIVAL (I,1040); COUNTY GENERAL (I,1076); IT'S WHAT'S HAPPENING, BABY! (I,2278)

FORRESTER, LARRY
LADIES IN BLUE (II,1422); VEGAS (II,2724)

FORSYTH, ED
STARSTRUCK (II,2446)

FORSYTHE, DON
MR. CHIPS (I,3132)

FORSYTHE, JOHN
THE FIRST HUNDRED YEARS (I,1582)

FORTIS, CHERIE
THE NEW SHOW (II,1828)

FORWARD, ROBERT H
CLINIC ON 18TH STREET (II,544); THE D.A. (I,1123); THE D.A.: CONSPIRACY TO KILL (I,1121); THE D.A.: MURDER ONE (I,1122)

FOSTER, BILL
FULL HOUSE (II,936); THE JERRY LEWIS SHOW (I,2370); PARADISE (II,1955); THE SINGERS (I,4060)

FOSTER, BOB
CHICAGO STORY (II,506); THE NEW MAVERICK (II,1822)

FOSTER, DICK
THE ARTHUR GODFREY SPECIAL (II,109); LINDSAY WAGNER—ANOTHER SIDE OF ME (II,1483); PEGGY FLEMING AT SUN VALLEY (I,3507); PEGGY FLEMING VISITS THE SOVIET UNION (I,3509); PERRY COMO'S HAWAIIAN HOLIDAY (II,2017); PERRY COMO'S LAKE TAHOE HOLIDAY (II,2018); TO EUROPE WITH LOVE (I,4525)

FOSTER, GEORGE
CHARO (II,488); COTTON CLUB '75 (II,580); THE DAVID STEINBERG SHOW (I,1179); MAKE ME LAUGH (I,2839); MAKE ME LAUGH (II,1602); THE PAUL LYNDE HALLOWEEN SPECIAL (II,1981); THE WAYNE NEWTON SPECIAL (II,2749)

FOSTER, HARRY
CHEYENNE (I,931)

FOSTER, HARVE

CROWN THEATER WITH GLORIA SWANSON (I,1107); PUBLIC DEFENDER (I,3687)

FOSTER, LEWIS
THE ADVENTURES OF JIM BOWIE (I,63)

FOSTER, ROBERT
ALL THAT GLITTERS (II,41); KATE MCSHANE (II,1368); KNIGHT RIDER (II,1402)

FOURNIER, RICK
NOWHERE TO HIDE (II,1868)

FOURNIER, RIFF
THE JOE NAMATH SHOW (I,2400)

FOWLER, BRUCE
LAND OF THE GIANTS (I,2611)

FOWLKES, MICHAEL
THE PARAGON OF COMEDY (II,1956)

FOX JR, FRED
HAPPY DAYS (II,1084); IT'S YOUR MOVE (II,1259); JOANIE LOVES CHACHI (II,1295)

FOX, BEN
FOREST RANGER (I,1641); WHIPLASH (I,4839)

FOX, CARL
JUBILEE U.S.A. (I,2461)

FOX, M BERNARD
WATERFRONT (I,4774)

FOX, SONNY
THE GOLDEN AGE OF TELEVISION (II,1016); THE GOLDEN AGE OF TELEVISION (II,1017); THE GOLDEN AGE OF TELEVISION (II,1018); THE GOLDEN AGE OF TELEVISION (II,1019); THE MYSTERIOUS TWO (II,1783)

FOXX, REDD
MY BUDDY (II,1778)

FRAMER, WALT
THE BIG PAYOFF (I,428); DOUBLE OR NOTHING (I,1362); FOR LOVE OR MONEY (I,1622); THE GREATEST MAN ON EARTH (I,1882); STRIKE IT RICH (I,4262)

FRANCHINI, BOB
SCROOGE'S ROCK 'N' ROLL CHRISTMAS (II,2279)

FRANCIS, CEDRIC
COLT .45 (I,1010)

FRANCK, DIETER
WILD WILD WORLD OF ANIMALS (I,4864)

FRAND, HARVEY
CALIFORNIA FEVER (II,411)

FRANK, BARRY
BATTLE OF THE NETWORK STARS (II,178)

FRANK, ETHEL
MATINEE THEATER (I,2947)

FRANK, GAIL
JABBERWOCKY (II,1262)

FRANK, JERRY
SORORITY '62 (II,2405); STAND UP AND CHEER (I,4188); A TRIBUTE TO "MR. TELEVISION" MILTON BERLE (II,2658)

FRANK, RICHARD H
ENTERTAINMENT TONIGHT (II,780); PLAZA SUITE (II,2056)

FRANK, SANDY
YOU ASKED FOR IT (II,2855)

FRANK, SARAH
TWIGS (II,2684)

FRANK, YASHA
PINOCCHIO (I,3602)

FRANKEL, ERNIE
CONCRETE COWBOYS (II,563); THE CONCRETE COWBOYS (II,564); THE COUNTRY MUSIC MURDERS (II,590); EBONY, IVORY AND JADE (II,755); MOVIN' ON (II,1743); NASHVILLE 99 (II,1791); THE NEW ADVENTURES OF PERRY MASON (I,3252); YOUNG DAN'L BOONE (II,2863)

FRANKENHEIMER, JOHN
THE TURN OF THE SCREW (I,4624)

FRANKLIN, JEFF
BOSOM BUDDIES (II,355); FM TV (II,896); HOT W.A.C.S. (II,1191); YOUNG HEARTS (II,2865)

FRANKOVICH, M J
BOB & CAROL & TED & ALICE (I,495); STATE FAIR (II,2448)

FRANKOVICH, PETER
STATE FAIR (II,2448)

FRASER, BOB
CONDO (II,565); MR. & MRS. & MR. (II,1744)

FRASER, FORREST
THE VIRGINIA GRAHAM SHOW (I,4736)

FRASER, NORA
HERE'S RICHARD (II,1136); HERE'S RICHARD (II,1137); THE RICHARD SIMMONS SHOW (II,2166)

FRASER, WENDY
PEOPLE DO THE CRAZIEST THINGS (II,1997)

FRASER, WOODY
AMERICA ALIVE! (II,67); HERE'S RICHARD (II,1136); HERE'S RICHARD (II,1137); IT ONLY HURTS WHEN YOU LAUGH (II,1251); THE KRYPTON FACTOR (II,1414); LEAVE IT TO THE WOMEN (II,1454); LIFE'S MOST EMBARRASSING MOMENTS (II,1470); LIFE'S MOST EMBARRASSING MOMENTS III (II,1472); LIFE'S MOST EMBARRASSING MOMENTS IV (II,1473); THE MIKE DOUGLAS SHOW (II,1698); PEOPLE DO THE CRAZIEST THINGS (II,1998); PEOPLE TO PEOPLE (II,2000); PERSONAL AND CONFIDENTIAL (II,2026); THE RICHARD SIMMONS SHOW (II,2166); SUCCESS: IT CAN BE YOURS (II,2488); THAT'S INCREDIBLE! (II,2578); THIS MORNING (I,4464); THOSE AMAZING ANIMALS (II,2601)

FRAZER-JONES, PETER
THE BENNY HILL SHOW (II,207); GEORGE AND MILDRED (II,965); MAN ABOUT THE HOUSE (II,1611); ROBIN'S NEST (II,2187)

FRAZIER, MICHAEL
LENA HORNE: THE LADY AND HER MUSIC (II,1460)

FRAZIER, RONALD
BEST OF THE WEST (II,218); DREAMS (II,735); JUST OUR LUCK (II,1360); SQUARE PEGS (II,2431)

FREBERG, STAN
TIME FOR BEANY (I,4505)

FREED, ALAN
THE ROCK N' ROLL SHOW (I,3820)

FREED, ARTHUR
HOLLYWOOD MELODY (I,2085)

FREED, FRED
THE LAST WORD (I,2631); THE RIGHT MAN (I,3790)

FREED, LORNE
ANYTHING YOU CAN DO (I,239)

FREED, RON
THE CHERYL LADD SPECIAL (II,501)

FREEDLEY, VINTON
SHOWTIME, U.S.A. (I,4038); TALENT JACKPOT (I,4339)

FREEDMAN, ALBERT Z
YOUR FUNNY FUNNY FILMS (I,4950)

FREEDMAN, GERALD
THE WAY THEY WERE (II,2748)

FREEDMAN, HY
YOUR FUNNY FUNNY FILMS (I,4950)

FREEDMAN, JERROLD
THE LAW ENFORCERS (I,2638); THE PSYCHIATRIST (I,3686)

FREEDMAN, LOUIS
BICENTENNIAL MINUTES (II,229); CAPTAIN SAFARI OF THE JUNGLE PATROL (I,836); THE DUPONT SHOW OF THE WEEK (I,1388)

FREEDMAN, STEVE
THE GIRL IN MY LIFE (II,996)

FREEMAN, ALBERT
TWENTY-ONE (I,4645)

FREEMAN, DEVERY
THE ANN SOTHERN SHOW (I,220); HARRIS AGAINST THE WORLD (I,1956); PETE AND GLADYS (I,3558); SIX GUNS FOR DONEGAN (I,4065)

FREEMAN, EVERETT
ALCOA PREMIERE (I,109); BACHELOR FATHER (I,323); THE FIRST HUNDRED YEARS (I,1582); I LOVE MY DOCTOR (I,2172); IVY LEAGUE (I,2286); THE PRUITTS OF SOUTHAMPTON (I,3684)

FREEMAN, FRED
AFTER GEORGE (II,19); EMPIRE (II,777); FOG (II,897); INSIDE O.U.T. (II,1235); THE SHAMEFUL SECRETS OF HASTINGS CORNERS (I,3994); STICK AROUND (II,2464)

FREEMAN, JOAN
WALTER CRONKITE'S UNIVERSE (II,2739)

FREEMAN, JOEL D
THE FUZZ BROTHERS (I,1722)

FREEMAN, LAWRENCE J
INSIDE O.U.T. (II,1235)

FREEMAN, LEONARD
FLOYD GIBBONS, REPORTER (I,1608); HAWAII FIVE-O (I,1972); HAWAII FIVE-O (II,1110); MEN AT LAW (I,3004); ROUTE 66 (I,3852); THE SEEKERS (I,3973); THE SEEKERS (I,3974); THE UNTOUCHABLES (I,4682)

FREEMAN, PAUL

THE CHISHOLMS (II,512); THE CHISHOLMS (II,513); THE OUTSIDER (I,3431); RUN FOR YOUR LIFE (I,3871); THE YELLOW ROSE (II,2847)

FREEMAN, SETH
LOU GRANT (II,1526); THINGS ARE LOOKING UP (II,2588)

FREER, RANDALL
DEALER'S CHOICE (II,635); THE DIAMOND HEAD GAME (II,666); THE FUN FACTORY (II,938)

FREIBERGER, FRED
BEYOND WESTWORLD (II,227); BIG SHAMUS, LITTLE SHAMUS (II,242); KORG: 70,000 B.C. (II,1408); A MAN CALLED SHENANDOAH (I,2859); THE SIX-MILLION-DOLLAR MAN (II,2372); SPACE: 1999 (II,2409); STAR TREK (II,2440); THE UNTOUCHABLES (I,4682); THE WILD WILD WEST (I,4863)

FREILICH, JEFF
BOONE (II,351); GALACTICA 1980 (II,953)

FRELENG, FRIZ
THE ABC AFTERSCHOOL SPECIAL (II,1); BAGGY PANTS AND THE NITWITS (II,135); BAILEY'S COMETS (I,333); THE BARKLEYS (I,356); THE BUGS BUNNY/ROAD RUNNER SHOW (II,391); DENNIS THE MENACE: MAYDAY FOR MOTHER (II,660); DOCTOR DOLITTLE (I,1309); THE FANTASTIC FOUR (II,824); HERE COMES THE GRUMP (I,2027); THE HOUNDCATS (I,2132); THE ODDBALL COUPLE (II,1876); THE PINK PANTHER (II,2042); RETURN TO THE PLANET OF THE APES (II,2145); SPIDER-WOMAN (II,2426); SUPER PRESIDENT (I,4291); THE SUPER SIX (I,4292); WHAT'S NEW MR. MAGOO? (II,2770)

FRIEBERG, HARDIE
15 WITH FAYE (I,1570)

FRIEDBERG, BILLY
CAMP RUNAMUCK (I,811); HARRY'S GIRLS (I,1959)

FRIEDBERG, WILLIAM
PETER LOVES MARY (I,3565)

FRIEDKIN, DAVID
BEARCATS! (I,375); CHASE (I,917); I SPY (I,2179); THE MOST DEADLY GAME (I,3106)

FRIEDLAND, CYNTHIA
FM TV (II,896)

FRIEDLANDER, LEONARD
THE GREAT SPACE COASTER (II,1059); HIZZONER (II,1159); THE LIFE AND TIMES OF EDDIE ROBERTS (II,1468); TV'S BLOOPERS AND PRACTICAL JOKES (II,2675)

FRIEDLANDER, SYLVIA
FORD STAR REVUE (I,1632)

FRIEDMAN, CHARLES
THE COLGATE COMEDY HOUR (I,997); FORD FESTIVAL (I,1629); RUGGLES OF RED GAP (I,3867)

FRIEDMAN, HAL
DAGMAR'S CANTEEN (I,1125)

FRIEDMAN, SEYMOUR
IVANHOE (I,2282)

FRIEDMAN, STEPHEN
CAMOUFLAGE (II,417); RHYME AND REASON (II,2152)

FRIEND, NOREEN
THE RICHARD SIMMONS SHOW (II,2166)

FRIENDLY, ANDY
ENTERTAINMENT TONIGHT (II,780); SPEAK UP AMERICA (II,2412); THIS IS YOUR LIFE (II,2598); TOM SNYDER'S CELEBRITY SPOTLIGHT (II,2632); THE WORLD OF ENTERTAINMENT (II,2837)

FRIENDLY, ED
BACKSTAIRS AT THE WHITE HOUSE (II,133); JUDGEMENT DAY (II,1348); LITTLE HOUSE ON THE PRAIRIE (II,1487); THE YOUNG PIONEERS (II,2868); THE YOUNG PIONEERS (II,2869); THE YOUNG PIONEERS' CHRISTMAS (II,2870)

FRIENDLY, FRED W
BACK THE FACT (I,327); LIFE BEGINS AT EIGHTY (I,2682); WHO SAID THAT? (I,4849)

FRIENDLY, SUZY
ALMOST ANYTHING GOES (II,56); ROCK PALACE (II,2193)

FRIES, CHARLES
THE ADDAMS FAMILY (II,13); THE AMAZING SPIDER-MAN (II,63); BABY, I'M BACK! (II,130); FULL HOUSE (II,936); HERE WE GO AGAIN (I,2028); THE HOME FRONT (II,1171); HUMAN FEELINGS (II,1203); THE MARTIAN CHRONICLES (II,1635); THE MICHELE LEE SHOW (II,1690); MURDER INK (II,1769); THE NORLISS TAPES (I,3305); PUNCH AND JODY (II,2091); THIS BETTER BE IT (II,2595); TWIN DETECTIVES (II,2687); THE WORD (II,2833)

FRIMBERG, HAL
YOUR SURPRISE STORE (I,4961)

FRITZ, KEN
THE MUSIC SCENE (I,3176); PAT PAULSEN FOR PRESIDENT (I,3472)

FRITZ, LESLIE
THE LOVEBIRDS (II,1548)

FROMKESS, LEON
THE COUNT OF MONTE CRISTO (I,1058); FURY (I,1721); GRUEN GUILD PLAYHOUSE (I,1892); THE NEW ADVENTURES OF CHARLIE CHAN (I,3249); NEW YORK CONFIDENTIAL (I,3276); RAMAR OF THE JUNGLE (I,3718); SAILOR OF FORTUNE (I,3885)

FROMKIN, PAUL
SUSAN'S SHOW (I,4303)

FROST, DAVID
THE DAVID FROST REVUE (I,1176); DYNASTY (II,747); HEADLINERS WITH DAVID FROST (II,1114); PEEPING TIMES (II,1990)

FROUG, WILLIAM
ADAM'S RIB (I,32); ALCOA/GOODYEAR THEATER (I,107); BEWITCHED (I,418); GILLIGAN'S ISLAND (II,990); MR. NOVAK (I,3145); SQUADRON (I,4170); THE TWILIGHT ZONE (I,4651)

FRYE, BEN
THE DUNNINGER SHOW (I,1385)

FRYE, WILLIAM
AMANDA FALLON (I,152); BACHELOR PARTY (I,324); THE BOB CUMMINGS SHOW (I,500); BRIDGET LOVES BERNIE (I,722); THE DEPUTY (I,1234); FOUR STAR PLAYHOUSE (I,1652); THE GENERAL ELECTRIC THEATER (I,1753); THE HALLS OF IVY (I,1923); JOHNNY STACCATO (I,2437); THE SLOWEST GUN IN THE WEST (I,4088); SUSPICION (I,4309); THRILLER (I,4492)

FRYER, ROBERT
WONDERFUL TOWN (I,4893)

FUCHS, MICHAEL
CAMELOT (II,416); DIANA ROSS IN CONCERT (II,668); JOHNNY CASH'S AMERICA (II,1330)

FUCHS, RICHARD
EVENING AT THE IMPROV (II,786)

FUJIOKA, YUTAKA
MIGHTY ORBOTS (II,1697)

FULLER, SAM
THE ALL-STAR REVUE (I,138); THE COLGATE COMEDY HOUR (I,997)

FULLERTON, SANDI
SOAP WORLD (II,2394)

FUNT, ALLEN
CANDID CAMERA (I,818); CANDID CAMERA (I,819); CANDID CAMERA (II,420); CANDID CAMERA LOOKS AT THE DIFFERENCE BETWEEN MEN AND WOMEN (II,421); THE CANDID CAMERA SPECIAL (II,422); IT'S ONLY HUMAN (II,1257); TELL IT TO THE CAMERA (I,4386)

FURIA JR, JOHN
APPLE'S WAY (II,101); THE BLUE KNIGHT (II,277); GIBBSVILLE (II,988); THE HEALERS (II,1115); THE HUSTLER OF MUSCLE BEACH (II,1210); KUNG FU (II,1416); RAGE OF ANGELS (II,2107); SWEEPSTAKES (II,2514)

GAFFNEY, RICKY
JUST MEN (II,1359)

GAGNON, DENNIS
CARD SHARKS (II,438)

GALAS, JUNE
FAMILY TIES (II,819); MAKING THE GRADE (II,1606)

GALBRAITH, JOHN
MATINEE AT THE BIJOU (II,1651)

GALE, BOB
USED CARS (II,2712)

GALE, GEORGE
DAKTARI (I,1127); JAMBO (I,2330)

GALEN, FRANK
MEET MILLIE (I,2988)

GALLACCIO, GEORGE
MACKENZIE (II,1584); THE OMEGA FACTOR (II,1888)

GALLAGHER, DAN
KOVACS UNLIMITED (I,2583)

GALLERY, MICHELE
THINGS ARE LOOKING UP (II,2588)

GALLIN, SANDY
CHER. . . SPECIAL (II,499); KENNY & DOLLY: A CHRISTMAS TO REMEMBER

(II,1378); MAC DAVIS 10TH ANNIVERSARY SPECIAL: I STILL BELIEVE IN MUSIC (II,1571); THE MAC DAVIS CHRISTMAS SPECIAL (II,1572); THE MAC DAVIS SHOW (II,1575); THE MAC DAVIS SPECIAL (II,1577); MAC DAVIS SPECIAL: THE MUSIC OF CHRISTMAS (II,1578); MAC DAVIS. . . I BELIEVE IN CHRISTMAS (II,1581); MAC DAVIS. . . SOUNDS LIKE HOME (II,1582); MAC DAVIS—I'LL BE HOME FOR CHRISTMAS (II,1580); PAUL LYNDE AT THE MOVIES (II,1976); THE PAUL LYNDE COMEDY HOUR (II,1978); THE PAUL LYNDE COMEDY HOUR (II,1979); PAUL LYNDE GOES M-A-A-A-AD (II,1980); THE PAUL LYNDE HALLOWEEN SPECIAL (II,1981); A SPECIAL OLIVIA NEWTON-JOHN (II,2418)

GALLO, LEW
ALOHA PARADISE (II,58); ANOTHER APRIL (II,93); HANDLE WITH CARE (II,1077); HAVING BABIES I (II,1107); HECK'S ANGELS (II,1122); LACY AND THE MISSISSIPPI QUEEN (II,1419); LUCAN (II,1554); MARY HARTMAN, MARY HARTMAN (II,1638); MICKEY SPILLANE'S MIKE HAMMER: MORE THAN MURDER (II,1693); MICKEY SPILLANE'S MIKE HAMMER: MURDER ME, MURDER YOU (II,1694); NOBODY'S PERFECT (II,1858)

GALLO, LILLIAN
THE NATURAL LOOK (II,1799)

GALLU, SAM
BEHIND CLOSED DOORS (I,385); THE BLUE ANGELS (I,490); NAVY LOG (I,3233); U.S. BORDER PATROL (I,4685)

GALVIN, BILL
AN EVENING WITH THE STATLER BROTHERS (II,790)

GAMMIE, BILL
MARINELAND CARNIVAL (I,2912)

GAMSON, MITCHELL
NUMBER 96 (II,1869)

GANAWAY, ALBERT
COUNTRY MUSIC CARAVAN (I,1066)

GANIS, SID
HEROES AND SIDEKICKS—INDIANA JONES AND THE TEMPLE OF DOOM (II,1140)

GANNON, JOE
ARCHIE BUNKER'S PLACE (II,105); GLORIA COMES HOME (II,1011)

GANTMAN, JOSEPH
BIG BOB JOHNSON AND HIS FANTASTIC SPEED CIRCUS (II,231); THE DUKES OF HAZZARD (II,742); MISSION: IMPOSSIBLE (I,3067); YOUNG DR. KILDARE (I,4937)

GANZ, JEFFREY
THE BAD NEWS BEARS (II,134); FOUL PLAY (II,910); MAKIN' IT (II,1603); SISTER TERRI (II,2366)

GANZ, LOWELL
BUSTING LOOSE (II,402); FLATFOOTS (II,879); FOUL PLAY (II,910); THE FURTHER ADVENTURES OF WALLY BROWN (II,944); HAPPY DAYS (II,1084); HERNDON AND ME (II,1139); JOANIE LOVES CHACHI (II,1295); LAVERNE AND SHIRLEY (II,1446); THE LOVEBIRDS (II,1548); MAKIN' IT (II,1603); THE RITA MORENO SHOW (II,2181); THE TED KNIGHT SHOW (II,2550)

GARA, MICHAEL
DANCE FEVER (II,615)

GARBAB, MARY ANN
BATTLE OF THE NETWORK STARS (II,163); BATTLE OF THE NETWORK STARS (II,164); BATTLE OF THE NETWORK STARS (II,165); BATTLE OF THE NETWORK STARS (II,166); BATTLE OF THE NETWORK STARS (II,167); BATTLE OF THE NETWORK STARS (II,168); BATTLE OF THE NETWORK STARS (II,169); BATTLE OF THE NETWORK STARS (II,170); BATTLE OF THE NETWORK STARS (II,171); BATTLE OF THE NETWORK STARS (II,172); BATTLE OF THE NETWORK STARS (II,173); BATTLE OF THE NETWORK STARS (II,174); BATTLE OF THE NETWORK STARS (II,175); BATTLE OF THE NETWORK STARS (II,176); BATTLE OF THE NETWORK STARS (II,177)

GARCIA, DAVID
THE TEXAS RANGERS (II,2567)

GARDEN, BILL
SHOW OF THE YEAR (I,4031)

GARDNER, ARTHUR
THE BIG VALLEY (I,437); THE DETECTIVES (I,1245); HONEY WEST (I,2105); THE INDIAN (I,2210); LAW OF THE PLAINSMAN (I,2639); THE RIFLEMAN (I,3789); WHICH WAY'D THEY GO? (I,4838)

GARDNER, DAVID
QUENTIN DERGENS, M.P. (I,3707)

GARDNER, GERALD
THE RED SKELTON SHOW (I,3756); THE RED SKELTON SHOW (I,3755)

GARDNER, JANE
THE LAST LEAF (II,1437)

GARDNER, MICHAEL
CHERYL LADD. . . LOOKING BACK—SOUVENIRS (II,502)

GARDNER, RICK
THE OTHER BROADWAY (II,1930)

GAREN, SCOTT
60 YEARS OF SEDUCTION (II,2373); TELEVISION'S GREATEST COMMERCIALS (II,2553); TELEVISION'S GREATEST COMMERCIALS II (II,2554); TELEVISION'S GREATEST COMMERCIALS III (II,2555); TELEVISION'S GREATEST COMMERCIALS IV (II,2556)

GARGIULO, MIKE
THE ALL-AMERICAN COLLEGE COMEDY SHOW (II,47); BOB HOPE SPECIAL: BOB HOPE'S ALL-STAR COMEDY BIRTHDAY PARTY AT WEST POINT (II,298); THE DAVID FROST REVUE (I,1176); LORETTA LYNN: THE LADY. . .THE LEGEND (II,1521); THE SOUND AND THE SCENE (I,4132); THAT'S MY LINE (II,2579)

GARNET, BILL
BATTLE OF THE NETWORK STARS (II,178)

GAROFALO, TONY
THE REGIS PHILBIN SHOW (II,2128)

GARRETT, LILA
BABY, I'M BACK! (II,130); GETTING THERE (II,985); INSTANT FAMILY (II,1238); THIS BETTER BE IT (II,2595)

GARRISON, GREG
THE CONNIE FRANCIS SHOW (I,1039); A COUNTRY HAPPENING (I,1065); DEAN MARTIN AT THE WILD ANIMAL PARK (II,636); THE DEAN MARTIN CELEBRITY ROAST (II,637); THE DEAN MARTIN CELEBRITY ROAST (II,638); THE DEAN MARTIN CELEBRITY ROAST (II,639); THE DEAN MARTIN CHRISTMAS SPECIAL (II,640); DEAN MARTIN IN LONDON (II,641); DEAN MARTIN PRESENTS MUSIC COUNTRY, U.S.A. (I,1192); DEAN MARTIN PRESENTS THE GOLDDIGGERS (I,1193); DEAN MARTIN PRESENTS THE GOLDDIGGERS (I,1194); THE DEAN MARTIN SHOW (I,1201); DEAN MARTIN'S CALIFORNIA CHRISTMAS (II,642); DEAN MARTIN'S CELEBRITY ROAST (II,643); DEAN MARTIN'S CHRISTMAS AT SEA WORLD (II,644); DEAN MARTIN'S CHRISTMAS IN CALIFORNIA (II,645); DEAN MARTIN'S COMEDY CLASSICS (II,646); DEAN MARTIN'S COMEDY WORLD (II,647); DEAN MARTIN'S RED HOT SCANDALS OF 1926 (II,648); DEAN MARTIN'S RED HOT SCANDALS PART 2 (II,649); DEAN'S PLACE (II,650); DOM DELUISE AND FRIENDS (II,701); DOM DELUISE AND FRIENDS, PART 2 (II,702); THE FIRST 50 YEARS (II,862); GENE KELLY'S WONDERFUL WORLD OF GIRLS (I,1752); THE GOLDDIGGERS (I,1838); THE JONATHAN WINTERS SHOW (I,2441); THE JONATHAN WINTERS SHOW (I,2442); THE KATE SMITH EVENING HOUR (I,2507); THE KEEFE BRASSELLE SHOW (I,2514); LADIES AND GENTLEMAN. . . BOB NEWHART (II,1420); LADIES AND GENTLEMAN. . . BOB NEWHART, PART II (II,1421); LADIES BE SEATED (I,2598); THE MARTY FELDMAN COMEDY MACHINE (I,2929); THE POWDER ROOM (I,3656); THE ROWAN AND MARTIN SHOW (I,3854); THE SID CAESAR SHOW (I,4043); STAND BY FOR CRIME (I,4186); STANDARD OIL ANNIVERSARY SHOW (I,4189); THE VIC DAMONE SHOW (I,4718); WHAT'S UP, AMERICA? (I,4824); THE WONDERFUL WORLD OF GIRLS (I,4898)

GARRISON, MICHAEL
THE INVESTIGATORS (I,2231); THE WILD WILD WEST (I,4863)

GARRY, AL
THE MAGIC CLOWN (I,2814)

GARSON, HENRY
FAMILY AFFAIR (I,1519)

GARVER, LLOYD
FAMILY TIES (II,819); MAKING THE GRADE (II,1606)

GARVES, JOHN

ASSIGNMENT: VIENNA (I,304)

GAST, HAROLD
CASABLANCA (II,450); JUDD, FOR THE DEFENSE (I,2463); LUCAN (II,1554); MCNAUGHTON'S DAUGHTER (II,1670); MEN AT LAW (I,3004); MOST WANTED (II,1739); THE STOREFRONT LAWYERS (I,4233)

GATWARD, JAMES
STAR MAIDENS (II,2435)

GATZIMOS, BILL
THE CRYSTAL GAYLE SPECIAL (II,609)

GAUNT, JOHN
THE BABY SITTER (I,321)

GAYNOR, JOHN
BUCK ROGERS IN THE 25TH CENTURY (II,383)

GEER, HAL
THE BUGS BUNNY/ROAD RUNNER SHOW (II,391); PORKY PIG AND FRIENDS (I,3651)

GEIGER, GEORGE
FLO (II,891)

GELBART, LARRY
IF I LOVE YOU, AM I TRAPPED FOREVER? (II,1222); KAREN (II,1363); M*A*S*H (II,1569); ROLL OUT! (I,3833); UNITED STATES (II,2708)

GELLER, AUDREY
DUPONT SHOW OF THE MONTH (I,1387)

GELLER, BRUCE
BRONK (II,379); THE JUDGE (I,2468); MISSION: IMPOSSIBLE (I,3067); RAWHIDE (I,3727); SUPERCOPS (II,2502)

GELLER, PHYLLIS
CAT ON A HOT TIN ROOF (II,457)

GERAGHTY, MAURICE
CAVALCADE OF AMERICA (I,875)

GERARD, ANDREW
LOW MAN ON THE TOTEM POLE (I,2782)

GERARD, ERICA
WE'RE MOVIN' (II,2756)

GERARD, MERWIN
KATE LOVES A MYSTERY (II,1367)

GERBER, CHARLES
THAT TEEN SHOW (II,2573)

GERBER, DAVID
BEULAH LAND (II,226); THE BILLION DOLLAR THREAT (II,246); BORN FREE (I,691); BORN FREE (II,353); THE BOYS IN BLUE (II,359); CADE'S COUNTY (I,787); COVER GIRLS (II,596); DAVID CASSIDY—MAN UNDERCOVER (II,623); DOCTORS' PRIVATE LIVES (II,692); DOCTORS' PRIVATE LIVES (II,693); EISCHIED (II,763); FOR LOVE AND HONOR (II,903); GEORGE WASHINGTON (II,978); THE GHOST AND MRS. MUIR (I,1786); GIBBSVILLE (II,988); JARRETT (I,2349); JESSIE (II,1287); JOE FORRESTER (II,1303); KANGAROOS IN THE KITCHEN (II,1362); THE LAST DAYS OF POMPEII (II,1436); MEDICAL STORY (II,1676); NAKIA (II,1784); NANNY AND THE PROFESSOR (I,3225); NEEDLES AND PINS (I,3246); NUTS AND BOLTS (II,1871); OH, NURSE! (I,3343); ONCE UPON A SPY (II,1898); THE POLICE STORY (I,3638); POLICE STORY (II,2062); POLICE WOMAN (II,2063); POLICE WOMAN: THE GAMBLE (II,2064); PRUDENCE AND THE CHIEF (I,3682); QUARK (II,2095); THE QUEST (II,2096); THE QUEST (II,2097); REVENGE OF THE GRAY GANG (II,2146); RIKER (II,2175); SEVEN BRIDES FOR SEVEN BROTHERS (II,2307); TO SIR, WITH LOVE (II,2624); TODAY'S FBI (II,2628); THE TURNING POINT OF JIM MALLOY (II,2670); WALKING TALL (II,2738); THE WONDERFUL WORLD OF PHILIP MALLEY (II,2831)

GERBER, FRED
HUNTER (II,1205)

GERNHARD, PHIL
THE JIM STAFFORD SHOW (II,1291)

GERRITY, WILLIAM C
THE DAIN CURSE (II,613); FOR THE PEOPLE (I,1626)

GERRY, ROGER
MELODY STREET (I,3000); SHADOW OF THE CLOAK (I,3992); STAGE A NUMBER (I,4176); THE STRAWHATTERS (I,4257)

GERSON, DON
THE DORIS DAY SHOW (I,1355); THE DORIS MARY ANNE KAPPELHOFF SPECIAL (I,1356)

GERSON, HANK
THE DEAN JONES SHOW (I,1191)

GERSTEIN, BERNARD
SHOWTIME, U.S.A. (I,4038)

GERTZ, FRANK
THE BOBBY VAN AND ELAINE JOYCE SHOW (I,677); SALT AND PEPE (II,2240)

GETCHELL, FRED
THE ABC AFTERSCHOOL SPECIAL (II,1)

GETHERS, STEVEN
THE BING CROSBY SHOW (I,467); THE FARMER'S DAUGHTER (I,1533)

GETZ, ROBERT
SEARCH FOR TOMORROW (II,2284)

GEWIRTZ, HOWARD
DOMESTIC LIFE (II,703); MR. SUCCESS (II,1756); TAXI (II,2546)

GEWIRTZ, IRA
SEMI-TOUGH (II,2300)

GIBBON, SAM
CAPTAIN KANGAROO (II,434); MR. MAYOR (I,3143)

GIBBS, JOHN
JOHNNY OLSEN'S RUMPUS ROOM (I,2433); TELLER OF TALES (I,4388); THE TROUBLESHOOTERS (I,4614)

GIBNEY, JACK
SUPER CIRCUS (I,4287); U.S. TREASURY SALUTES (I,4689)

GIBSON, CARL
RED SKELTON'S FUNNY FACES (II,2124); THE STEVE ALLEN COMEDY HOUR (II,2454)

GIBSON, KEN
MINSKY'S FOLLIES (II,1703); STAR CHART (II,2434)

GILBEL, ROGER
BUTTERFLIES (II,405)

GILBERT, ALAN
KEEP TALKING (I,2517); LET'S MAKE A DEAL (II,1464); MASQUERADE PARTY (II,1645); WINDOW SHOPPING (I,4875)

GILBERT, BRUCE
9 TO 5 (II,1852)

GILBERT, PAUL
DANCE FEVER (II,615); WHEEL OF FORTUNE (II,2775)

GILL, DAVID
HOLLYWOOD (II,1160)

GILL, MICHAEL
AMERICA (I,165)

GILLESPIE, HENRY
MUSIC HALL AMERICA (II,1773)

GILLESPIE, RITA
IT'S WHAT'S HAPPENING, BABY! (I,2278)

GILLETTE, BILL
THE FRED WARING SHOW (I,1679)

GILLETTE, ROLAND
THE FRED WARING SHOW (I,1679)

GILLIAT, LESLIE
COLONEL MARCH OF SCOTLAND YARD (I,1005)

GILLIGAN, JOHN
SURVIVAL (I,4300)

GILLIS, ANNE
WHO SAID THAT? (I,4849)

GILLOTT, NICK
ELLIS ISLAND (II,768); MASTER OF THE GAME (II,1647)

GILMER, ROBERT
ALL THAT GLITTERS (II,41); HAWAIIAN HEAT (II,1111)

GILROY, JOHN
HAPPY ENDINGS (II,1085); HEADLINERS WITH DAVID FROST (II,1114); THE ROBERT KLEIN SHOW (II,2186)

GIMBEL, ROGER
1968 HOLLYWOOD STARS OF TOMORROW (I,3297); GOSSIP (II,1043); GOSSIP (II,1044); THE MANIONS OF AMERICA (II,1623); MONTE CARLO, C'EST LA ROSE (I,3095); REPORT TO MURPHY (II,2134); WITH LOVE, SOPHIA (I,4884)

GINGOLD, DAN
CHRIS AND THE MAGICAL DRIP (II,516)

GINSBERG, MARLA
ANYTHING FOR MONEY (II,99)

GIOBUS, DOROTHY J
THE FOUR OF US (II,914)

GIRARD, BERNARD
COUNTERPOINT (I,1059); CROWN THEATER WITH GLORIA SWANSON (I,1107); MR. GARLUND (I,3138); WALTER FORTUNE (I,4758)

GITLIN, IRVING
PUREX SPECIAL FOR WOMEN (I,3698)

GITTERMAN, N GAYLE
THE WHIRLYBIRDS (I,4841)

GLAISTER, GERARD
SECRET ARMY (II,2290)

GLASS, KAREN
PAT BENATAR IN CONCERT (II,1967)

GLASSER, DICK
WHERE THE ACTION IS (I,4828)

GLATTERS, WOLFGANG
HIGH SCHOOL, U.S.A. (II,1148)

GLAZE, BILL
JULIE AND DICK IN COVENT GARDEN (II,1352)

GLAZER, BARRY
AMERICAN BANDSTAND (II,70)

GLEASON, MICHAEL
FOOLS, FEMALES AND FUN: I'VE GOTTA BE ME (II,899); FOOLS, FEMALES AND FUN: IS THERE A DOCTOR IN THE HOUSE? (II,900); FOOLS, FEMALES AND FUN: WHAT ABOUT THAT ONE? (II,901); FORCE FIVE (II,907); THE GOSSIP COLUMNIST (II,1045); MCCLOUD (II,1660); THE OREGON TRAIL (II,1922); THE OREGON TRAIL (II,1923); PARIS 7000 (I,3459); REMINGTON STEELE (II,2130); RICH MAN, POOR MAN—BOOK II (II,2162); THE SIX MILLION DOLLAR MAN (I,4067); THE SIX-MILLION-DOLLAR MAN (II,2372); SONS AND DAUGHTERS (II,2404)

GLENN, BILL
ROMANCE THEATER (II,2204)

GLICKSMAN, FRANK
THE BOBBY SHERMAN SPECIAL (I,676); THE CRIME CLUB (I,1096); FAMILY IN BLUE (II,818); HAGEN (II,1071); THE LEGEND OF CUSTER (I,2652); THE LONG HOT SUMMER (I,2747); MEDICAL CENTER (II,1675); TRAPPER JOHN, M.D. (II,2654); TWELVE O'CLOCK HIGH (I,4637); U.M.C. (I,4666); WOMAN ON THE RUN (II,2821)

GLOVER, BOB
THE NEW VOICE (II,1833)

GLUCKSMAN, ERNEST D
ALL THINGS BRIGHT AND BEAUTIFUL (I,141); THE ALL-STAR REVUE (I,138); THE COLGATE COMEDY HOUR (I,997); THE DONALD O'CONNOR SHOW (I,1344); THE FIFTH DIMENSION TRAVELING SHOW (I,1572); THE IMOGENE COCA SHOW (I,2198); JACK CARTER AND COMPANY (I,2305); THE JAZZ SINGER (I,2351); THE JERRY LEWIS SHOW (I,2362); THE JERRY LEWIS SHOW (I,2364); THE JERRY LEWIS SHOW (I,2367); THE JERRY LEWIS SHOW (I,2368); THE JERRY LEWIS SHOW (I,2369); THE SATURDAY NIGHT REVUE (I,3921); SHOW BIZ (I,4027)

GODFREY, ALAN
BARETTA (II,152); JOE DANCER: THE BIG BLACK PILL (II,1300); JOE DANCER: THE MONKEY MISSION (II,1301); JOE DANCER: MURDER ONE, DANCER 0 (II,1302); THE POWER WITHIN (II,2073); VEGAS (II,2724); WINTER KILL (II,2810)

GOFF, IVAN
CHARLIE'S ANGELS (II,487); THE KILLER WHO WOULDN'T DIE (II,1393); LOGAN'S RUN (II,1507); MANNIX (II,1624); NERO WOLFE (II,1806); TIME EXPRESS (II,2620)

GOFF-DAVIS, ANNABELLE
ADDIE AND THE KING OF HEARTS (I,35)

GOGGIN, RICHARD
STUMP THE AUTHORS (I,4272)

GOLAND, ALAN
COUNTERATTACK: CRIME IN AMERICA (II,581)

GOLD, BERNIE
THE JO STAFFORD SHOW (I,2392)

GOLD, BERT
THE FEATHER AND FATHER GANG (II,845); INSTITUTE FOR REVENGE (II,1239)

GOLD, DONALD
SUPERTRAIN (II,2505)

GOLD, MALCOLM
THE ROOTS OF ROCK 'N' ROLL (II,2212); TWIGGY'S JUKE BOX (II,2682)

GOLD, MARLENA
YOU ASKED FOR IT (II,2855)

GOLD, MARTIN J
PEOPLE DO THE CRAZIEST THINGS (II,1998)

GOLDBERG, GARY DAVID
FAMILY TIES (II,819); FAMOUS LIVES (II,821); THE LAST RESORT (II,1440); LOU GRANT (II,1526); MAKING THE GRADE (II,1606); THE TONY RANDALL SHOW (II,2640)

GOLDBERG, JACK
THE DUNNINGER SHOW (I,1385)

GOLDBERG, JOANN
THE BARBARA WALTERS SPECIAL (II,145); FAMOUS LIVES (II,821)

GOLDBERG, LEONARD
THE BAIT (I,335); BEACH PATROL (II,192); BLUE JEANS (II,275); CHARLIE'S ANGELS (II,486); CHARLIE'S ANGELS (II,487); CHOPPER ONE (II,515); THE DAUGHTERS OF JOSHUA CABE (I,1169); DELTA COUNTY, U.S.A. (II,657); FAMILY (II,813); FANTASY ISLAND (II,829); FANTASY ISLAND (II,830); FIREHOUSE (II,859); GAVILAN (II,959); HARD KNOCKS (II,1086); HART TO HART (II,1101); HART TO HART (II,1102); THE LETTERS (I,2670); LETTERS FROM THREE LOVERS (I,2671); THE NEW DAUGHTERS OF JOSHUA CABE (I,3261); THE OATH: 33 HOURS IN THE LIFE OF GOD (II,1873); THE OATH: THE SAD AND LONELY SUNDAYS (II,1874); PAPER DOLLS (II,1951); PAPER DOLLS (II,1952); RETURN TO FANTASY ISLAND (II,2144); THE ROOKIES (II,2208); S.W.A.T. (II,2236); STARSKY AND HUTCH (II,2444); STARSKY AND HUTCH (II,2445); T.J. HOOKER (II,2524); TONI'S BOYS (II,2637); WHEN THE WHISTLE BLOWS (II,2781)

GOLDBERG, LOU
TED MACK AND THE ORIGINAL AMATEUR HOUR (I,4370)

GOLDBERG, MADELYN
THE MORECAMBE AND WISE SHOW (II,1734)

GOLDBERG, RUTH
THE $1.98 BEAUTY SHOW (II,698)

GOLDEN, MYRON
FOR BETTER OR WORSE (I,1621); SEARCH FOR TOMORROW (II,2284)

GOLDEN, ROBERT
LASSIE (I,2622); TIMMY AND LASSIE (I,4520)

GOLDFARB, PETER
FELLINI: A DIRECTOR'S NOTEBOOK (I,1562)

GOLDHAMER, JOHN E
ENTERTAINMENT TONIGHT (II,780)

GOLDMAN, ROBERT
THE SEVEN LIVELY ARTS (I,3987)

GOLDSMITH, JERRY
THE LAS VEGAS SHOW (I,2619)

GOLDSTEIN, HARRIET
FOR RICHER, FOR POORER (II,906); LOVERS AND FRIENDS (II,1549)

GOLDSTEIN, JEFFREY
DREAM HOUSE (II,734)

GOLDSTEIN, JERRY
DOUG HENNING'S WORLD OF MAGIC I (II,725); DOUG HENNING'S WORLD OF MAGIC II (II,726); DOUG HENNING'S WORLD OF MAGIC III (II,727); DOUG HENNING: MAGIC ON BROADWAY (II,730); THE RICH LITTLE SHOW (II,2153); THE RICH LITTLE SHOW (II,2154); RICH LITTLE'S WASHINGTON FOLLIES (II,2159)

GOLDSTEIN, LEE
DREAM HOUSE (II,734)

GOLDSTEIN, MARTIN
CAPTAIN AMERICA (II,427); CAPTAIN AMERICA (II,428)

GOLDSTEIN, RICHARD
THE FACTS (II,804)

GOLDSTONE, DUKE
CAPTAIN DAVID GRIEF (I,829)

GOLDSTONE, JULES
THE COURT OF LAST RESORT (I,1082); THE SECRET LIFE OF JOHN MONROE (I,3966)

GOLDSTONE, RICHARD
ADVENTURES IN PARADISE (I,46); PEYTON PLACE (I,3574)

GOLDWYN JR, SAMUEL
THE UNEXPLAINED (I,4675)

GOLOD, JERRY
TALES FROM THE DARKSIDE (II,2534); TALES FROM THE DARKSIDE (II,2535)

GOLSMITH, KAREN MACK

TOO GOOD TO BE TRUE (II,2643)

GOMBERG, SY
BENDER (II,204); THE LAW AND MR. JONES (I,2637)

GOMEZ, DAN
WHITNEY AND THE ROBOT (II,2789)

GOOD, JACK
33 1/3 REVOLUTIONS PER MONKEE (I,4451); ANDY WILLIAMS MAGIC LANTERN SHOW COMPANY (I,202); ANDY'S LOVE CONCERT (I,215); CONSTANTINOPLE (II,570); MARY'S INCREDIBLE DREAM (II,1641)

GOODCHILD, PETER
OPPENHEIMER (II,1919)

GOODMAN, DANIEL R
THE AMAZING SPIDER-MAN (II,63)

GOODMAN, DONALD
TV'S BLOOPERS AND PRACTICAL JOKES (II,2675)

GOODMAN, GEORGE
MR. INSIDE/MR. OUTSIDE (II,1750)

GOODMAN, HAL
THE GOOD OLD DAYS (I,1851); MARGIE (I,2909)

GOODMAN, JASON
THE ROOTS OF ROCK 'N' ROLL (II,2212)

GOODMAN, JOSEPH
INSIDE O.U.T. (II,1235)

GOODMAN, MARTIN
THE JOAN EDWARDS SHOW (I,2396); MENASHA THE MAGNIFICENT (I,3010); THROUGH WENDY'S WINDOW (I,4495)

GOODSON, JONATHAN
CARD SHARKS (II,438); CHILD'S PLAY (II,510); THAT'S MY LINE (II,2579); TRIVIA TRAP (II,2659)

GOODSON, MARK
BEAT THE CLOCK (I,377); THE BETTER SEX (II,223); BY POPULAR DEMAND (I,784); CALL MY BLUFF (I,800); CHOOSE UP SIDES (I,945); FAMILY FEUD (II,816); GET THE MESSAGE (I,1782); I'VE GOT A SECRET (I,2284); I'VE GOT A SECRET (II,1220); IT'S NEWS TO ME (I,2272); JEFFERSON DRUM (I,2355); JUDGE FOR YOURSELF (I,2466); MADE IN AMERICA (I,2806); MAKE THE CONNECTION (I,2846); THE MATCH GAME (I,2944); THE MATCH GAME (II,1649); THE MATCH GAME—HOLLYWOOD SQUARES HOUR (II,1650); MINDREADERS (II,1702); THE NAME'S THE SAME (I,3220); NOTHING BUT THE TRUTH (I,3313); NOW YOU SEE IT (II,1867); ONE HAPPY FAMILY (I,3375); THE PRICE IS RIGHT (I,3665); THE RICHARD BOONE SHOW (I,3784); SHOWOFFS (II,2345); SPLIT PERSONALITY (I,4161); STOP THE MUSIC (I,4232); TATTLETALES (II,2545); THAT'S MY LINE (II,2579); TO TELL THE TRUTH (I,4527); TO TELL THE TRUTH (I,4528); TV'S FUNNIEST GAME SHOW MOMENTS (II,2676); TWO FOR THE MONEY (I,4654); THE WEB (I,4784); WINNER TAKE ALL (I,4881)

GOODWIN, BOB
BIG BOB JOHNSON AND HIS FANTASTIC SPEED CIRCUS (II,231)

GOODWIN, R W
JESSIE (II,1287)

GORDON, AL
SHIPSHAPE (II,2329)

GORDON, CHARLES
JUST OUR LUCK (II,1360); LONE STAR (II,1513); THE RENEGADES (II,2132); THE RENEGADES (II,2133); THE STREETS (II,2479)

GORDON, J ARNOLD
THE PALACE (II,1943)

GORDON, LAWRENCE
DOG AND CAT (II,695); DOG AND CAT (II,696); HOME COOKIN' (II,1170); LACY AND THE MISSISSIPPI QUEEN (II,1419); LONE STAR (II,1513); MATT HOUSTON (II,1654); THE NIGHTENGALES (II,1850); THE RENEGADES (II,2132); THE RENEGADES (II,2133); THE STREETS (II,2479); STUNTS UNLIMITED (II,2487)

GORDON, LES
JUST OUR LUCK (II,1360)

GORDON, MARGIE
A NEW KIND OF FAMILY (II,1819)

GORDON, MARK
OMNI: THE NEW FRONTIER (II,1889)

GORDON, MAX
THE FRANK SINATRA SHOW (I,1664); THE WOMEN (I,4890)

GORDON, MURRAY
YOU ASKED FOR IT (II,2855)

GORDON, RICHARD
A COUPLE OF JOES (I,1078)

GORDON, ROBERT
SCHOOL HOUSE (I,3937)

GORDON, SHEP
BLONDIE (II,272)

GORDON, STEVE
GOOD TIME HARRY (II,1037); THE PRACTICE (II,2076)

GORDY JR, BERRY
DIANA (I,1258)

GORFRESNSCO, GAIL
FIGHT BACK: WITH DAVID HOROWITZ (II,854)

GORING, MARIUS
THE ADVENTURES OF THE SCARLET PIMPERNEL (I,80)

GOSCH, MARTIN
THE COLLEGE BOWL (I,1001); OFF THE RECORD (I,3336); TONIGHT ON BROADWAY (I,4559)

GOSDEN, FREEMAN
THE AMOS AND ANDY SHOW (I,179)

GOSS, TRACY
ANYTHING FOR MONEY (II,99)

GOTTFRIED, HOWARD
CALUCCI'S DEPARTMENT (I,805)

GOTTLIEB, ALEX
THE ABBOTT AND COSTELLO SHOW (I,2); DEAR PHOEBE (I,1204); THE GALE STORM SHOW (I,1725); JOE AND MABEL (I,2398); LOVE THAT JILL (I,2780); THE PIGEON (I,3596); THE TAMMY GRIMES SHOW (I,4358)

GOTTLIEB, ANDREW
HAWAII FIVE-O (II,1110)

GOTTLIEB, CARL
THE NEW LORENZO MUSIC SHOW (II,1821)

GOTTLIEB, HERBERT
MAKE A FACE (I,2837)

GOTTLIEB, LESTER
THE BIG RECORD (I,430); FRANKIE LAINE TIME (I,1676); KEEP TALKING (I,2517); THE PAUL WINCHELL SHOW (I,3493); ROBERT Q'S MATINEE (I,3812); THE SHOW GOES ON (I,4030); SING IT AGAIN (I,4059)

GOTTLIEB, LINDA
THE ABC AFTERSCHOOL SPECIAL (II,1)

GOTTLIEB, RICHARD
THE $100,000 NAME THAT TUNE (II,1902); THE ABC AFTERSCHOOL SPECIAL (II,1); THIS IS YOUR LIFE (I,4461)

GOULD, BERNIE
I HEAR AMERICA SINGING (I,2168); THE JUDY GARLAND SHOW (I,2470); THE SKY'S THE LIMIT (I,4081)

GOULD, CLIFF
DEATH RAY 2000 (II,654); SCARECROW AND MRS. KING (II,2268); THE STREETS OF SAN FRANCISCO (II,2478); UNIT 4 (II,2707)

GOULD, DIANA
THE ROCK RAINBOW (II,2194)

GOULET, ROBERT
KISS ME, KATE (I,2569)

GOUVA, KEVIN
THAT TEEN SHOW (II,2573)

GOUVIEA, SHELLEY
ALIVE AND WELL (II,36)

GOVONS, MARCIA
BAREFOOT IN THE PARK (II,151); NIGHT COURT (II,1844); PLAZA SUITE (II,2056); THE RAINMAKER (II,2109)

GOWEN, BUD
THE EDGE OF NIGHT (II,760)

GOWER, SAMMY
CHRISTMAS SPECIAL. . . WITH LOVE, MAC DAVIS (II,523)

GOWERS, BRUCE
BRITT EKLAND'S JUKE BOX (II,378)

GRACE, MARTHA
WHAT A NIGHTMARE, CHARLIE BROWN (I,4806)

GRAD, SAM
INDEMNITY (I,2209)

GRAFE, ROBERT D
THE VALIANT YEARS (I,4696)

GRAHAM, BILL
THE BOBBY GOLDSBORO SHOW (I,673); DOLLY (II,699); THE MARTY ROBBINS SPOTLITE (II,1636); POP! GOES THE COUNTRY (II,2067)

GRAHAM, ED
LINUS THE LIONHEARTED (I,2709)

GRAHAM, JOE
INDEMNITY (I,2209)

GRAHAM, LOUIS
TED MACK'S FAMILY HOUR (I,4371); TED MACK'S MATINEE (I,4372)

GRAJEDA, WAYNE
PEOPLE DO THE CRAZIEST THINGS (II,1998)

GRANET, BERT
THE DESILU PLAYHOUSE (I,1237); THE LORETTA YOUNG THEATER (I,2756); THE LUCILLE BALL-DESI ARNAZ SHOW (I,2784); MEETING AT APALACHIN (I,2994); THE TIME ELEMENT (I,4504); THE UNTOUCHABLES (I,4681)

GRANOFF, BUDD
THE BOBBY VINTON SHOW (II,345); LEAVE IT TO THE WOMEN (II,1454); TREASURE HUNT (II,2657); YOUR HIT PARADE (II,2872)

GRANT, ARMAND
KANGAROOS IN THE KITCHEN (II,1362); LIKE MAGIC (II,1476); THE PAT BOONE SHOW (I,3471); THE VIN SCULLY SHOW (I,4734); WORDS AND MUSIC (I,4909)

GRANT, GIL
THE POWERS OF MATTHEW STAR (II,2075)

GRANT, LYLE
BULLSEYE (II,393)

GRANT, MERRILL
THE 36 MOST BEAUTIFUL GIRLS IN TEXAS (II,2594); THE CELEBRITY FOOTBALL CLASSIC (II,474); THE FUTURE: WHAT'S NEXT (II,947); KATE AND ALLIE (II,1365); THE KRYPTON FACTOR (II,1414); PEOPLE TO PEOPLE (II,2000); THAT'S INCREDIBLE! (II,2578); THOSE AMAZING ANIMALS (II,2601); THE WORLD'S FUNNIEST COMMERCIAL GOOFS (II,2841)

GRANT, MICKEY
GUILTY OR INNOCENT (II,1065)

GRANT, NORMAN
THE MAGIC SLATE (I,2820)

GRANT, PERRY
ANOTHER MAN'S SHOES (II,96); HELLO, LARRY (II,1128); ONE DAY AT A TIME (II,1900); T.L.C. (II,2525)

GRANVILLE, BONITA
TIMMY AND LASSIE (I,4520)

GRASS, CLANCY
THE PAUL ANKA SHOW (II,1972)

GRAUMAN, WALTER
ADAMS OF EAGLE LAKE (II,10); BARE ESSENCE (II,150); BLUE LIGHT (I,491); THE FELONY SQUAD (I,1563); MYSTERY AND MRS. (I,3210); THE NEW BREED (I,3258); THE SILENT FORCE (I,4047); THEY CALL IT MURDER (I,4441)

GRAVES, ELEANOR
HAROLD LLOYD'S WORLD OF COMEDY (II,1093)

GRAY, CAROL
BUFFALO BILL (II,387)

GRAY, IRVING
THE MILTON BERLE SHOW (I,3048); SHOW OF THE YEAR (I,4031)

GRAY, LOUIS
THE ADVENTURES OF CHAMPION (I,54); ANNIE OAKLEY (I,225); BUFFALO BILL JR. (I,755); THE GENE AUTRY SHOW (I,1748); THE RANGE RIDER (I,3720)

GRAY, MARV
THE JOE PYNE SHOW (I,2402)

GRAY, NORM
AT EASE (II,116)

GRAY, STEPHANIE B
WHITNEY AND THE ROBOT (II,2789)

GREEN, CARL
VALIANT LADY (I,4695)

GREEN, CAROL
LEGENDS OF THE SCREEN (II,1456)

GREEN, DOUGLAS
HAWAII FIVE-O (II,1110); HAWAIIAN HEAT (II,1111); MAGNUM, P.I. (II,1601)

GREEN, HAROLD
CASEY JONES (I,865)

GREEN, JAMES
UNCLE JOHNNY COONS (I,4668)

GREEN, JIM
DRIBBLE (II,736); TOURIST (II,2651)

GREEN, JOHN B
ONE HUNDRED GRAND (I,3377)

GREEN, JULES
THE STEVE ALLEN SHOW (I,4219)

GREEN, KATHERINE
ANOTHER MAN'S SHOES (II,96); IT'S YOUR MOVE (II,1259); ONE DAY AT A TIME (II,1900)

GREEN, KENNY
RADIO PICTURE SHOW (II,2104)

GREEN, MORT
MAKE ME LAUGH (I,2839); MITZI AND A HUNDRED GUYS (II,1710); MY HERO (I,3193); VAUDEVILLE (II,2722); YES, VIRGINIA, THERE IS A SANTA CLAUS (II,2849)

GREEN, WILLIAM
THE DOOLEY BROTHERS (II,715)

GREENBERG, AXEL
THIS IS YOUR LIFE (I,4461)

GREENBERG, BRENDA
AS THE WORLD TURNS (II,110)

GREENBERG, EARL
FANTASY (II,828); LIFE'S MOST EMBARRASSING MOMENTS (II,1470)

GREENBERG, HAROLD
A MAN CALLED INTREPID (II,1612)

GREENBERG, JUDY
PRINCESS (II,2084)

GREENBERG, RON
THE BIG SHOWDOWN (II,244); BULLSEYE (II,393); DREAM HOUSE (I,1376); JOKER! JOKER!! JOKER!!! (II,1340); THE JOKER'S WILD (II,1341); PLAY THE PERCENTAGES (II,2053); THE POP 'N' ROCKER GAME (II,2066); SALE OF THE CENTURY (I,3888); TIC TAC DOUGH (II,2618); THE WHO, WHAT OR WHERE GAME (I,4850)

GREENBERG, SANFORD
WAIT UNTIL DARK (II,2735)

GREENE, CHARLES
LORNE GREENE'S NEW WILDERNESS (II,1523)

GREENE, DAVID
LUCAN (II,1553); LUCAN (II,1554)

GREENE, F SHERWIN
SPACE: 1999 (II,2409)

GREENE, HAROLD
DOCTOR MIKE (I,1316); JOHNNY NIGHTHAWK (I,2432); JUNGLE JIM (I,2491)

GREENE, JOHN L
BLONDIE (I,486)

GREENE, LORNE
LORNE GREENE'S NEW WILDERNESS (II,1523)

GREENE, STAN
TIC TAC DOUGH (I,4498)

GREENE, TOM
ALL THAT GLITTERS (II,41); KNIGHT RIDER (II,1402)

GREENGRASS, KEN
THE ABC AFTERSCHOOL SPECIAL (II,1); DEAR ALEX AND ANNIE (II,651)

GREENWALD, ROBERT
DELTA COUNTY, U.S.A. (II,657); THE TEXAS RANGERS (II,2567)

GREENWOOD, JACK
FROM A BIRD'S EYE VIEW (I,1686)

GREER, BILL
GOODNIGHT, BEANTOWN (II,1041); HOUSE CALLS (II,1194)

GREER, KATHY
GOODNIGHT, BEANTOWN (II,1041); HOUSE CALLS (II,1194)

GREGORY, ARTHUR
HOUSE CALLS (II,1194)

GREGORY, DAN
CELEBRITY BOWLING (I,883); CAMELOT (II,416)

GREGORY, PAUL
CRESCENDO (I,1094); DUPONT SHOW OF THE MONTH (I,1387); THREE FOR TONIGHT (I,4478)

GRENGER, DEREK
BRIDESHEAD REVISITED (II,375)

GREY, VIRGINIA
EVERYTHING YOU EVER WANTED TO KNOW ABOUT MONSTERS . . . BUT WERE AFRAID! (II,794)

GREYHOSKY, BABS
FOUR EYES (II,912); RIPTIDE (II,2178); THE ROUSTERS (II,2220)

GRIEF, LESLIE
THE HALF-HOUR COMEDY HOUR (II,1074)

GRIES, TOM
HUNTER (II,1204); THE RAT PATROL (I,3726)

GRIFFIN, TOM
G.I. JOE: A REAL AMERICAN HERO (II,948); THE GREAT SPACE COASTER (II,1059); THE TRANSFORMERS (II,2653)

GRIFFITH, ANDY
MAYBERRY R.F.D. (I,2961)

GRIFFITHS, DAVID
THE COMEDY ZONE (II,559); SCENES FROM A MARRIAGE (II,2272)

GRILIKES, MICHAEL
CBS CARTOON THEATER (I,878)

GRILLO, BASIL
REBOUND (I,3747)

GRIMALDI, GIAN
ALL THAT GLITTERS (II,41); CONDOMINIUM (II,566); JOE DANCER: MURDER ONE, DANCER 0 (II,1302); PINE CANYON IS BURNING (II,2040); THE SEEKERS (II,2298)

GRIMALDI, ROBERT
KNIGHT RIDER (II,1402)

GRIMM, JULIET
BERGERAC (II,209)

GRINBERG, SHERMAN
BOGART (I,679); GREATEST HEADLINES OF THE CENTURY (I,1881); SURVIVAL (I,4300)

GRISIER, PHIL
GOTTA SING, GOTTA DANCE (II,1046)

GROBERMANN, JEFFREY
THE PAUL ANKA SHOW (II,1972)

GRODIN, CHARLES
PARADISE (II,1955)

GROSS, JACK
CURTAIN CALL THEATER (I,1115); THE LONE WOLF (I,2742)

GROSS, LLOYD
BIG TOWN (I,436); FREEDOM RINGS (I,1682); LUCKY PUP (I,2788); MASQUERADE PARTY (I,2941); WHAT'S MY LINE? (I,4820)

GROSSMAN, BUDD
ALMOST AMERICAN (II,55); DIFF'RENT STROKES (II,674); THREE'S COMPANY (II,2614)

GROSSMAN, HOWARD K
THE BEACH BOYS 20TH ANNIVERSARY (II,188); THE BEACH BOYS IN CONCERT (II,189)

GROSSMAN, STEPHEN
STEPHANIE (II,2453)

GROSSMAN, TERRY
CONDO (II,565)

GROSSO, SONNY
BAKER'S DOZEN (II,136); THE DAIN CURSE (II,613); JOHNNY GARAGE (II,1337); THE RUBBER GUN SQUAD (II,2225)

GROVES, HERMAN
SWORD OF JUSTICE (II,2521); VEGAS (II,2724)

GRUBER, FRANK
SHOTGUN SLADE (I,4026)

GUARNIERI, TERRI
NEVER AGAIN (II,1808)

GUBER, PETER
DREAMS (II,735)

GUCCIONE, BOB
OMNI: THE NEW FRONTIER (II,1889)

GUEDEL, JOHN
ANYBODY CAN PLAY (I,234); THE ART LINKLETTER SHOW (I,276); ART LINKLETTER'S HOUSE PARTY (I,277); FOR BETTER OR WORSE (I,1621); THE LID'S OFF (I,2678); LIFE WITH LINKLETTER (I,2691); PEOPLE ARE FUNNY (I,3518); YOU BET YOUR LIFE (I,4932)

GUENETTE, ROBERT
THE ABC AFTERSCHOOL SPECIAL (II,1); COUNTERATTACK: CRIME IN AMERICA (II,581); HEROES AND SIDEKICKS—INDIANA JONES AND THE TEMPLE OF DOOM (II,1140); LIE DETECTOR (II,1467); WOMEN WHO RATE A "10" (II,2825); YOU ASKED FOR IT (II,2855)

GUILD, LEO
NUMBER 13 DEMON STREET (I,3316)

GUNN, JOSEPH
AIRWOLF (II,27)

GUNTHER, LEE
DUNGEONS AND DRAGONS (II,744); JIM HENSON'S MUPPET BABIES (II,1289); SPIDER-MAN AND HIS AMAZING FRIENDS (II,2425); SPIDER-WOMAN (II,2426); THE TRANSFORMERS (II,2653)

GUNTZELMAN, DAN
CASS MALLOY (II,454); GLORIA (II,1010); OFF THE RACK (II,1878)

GURA, PHYLLIS
MAKE A WISH (I,2838)

GUSS, JACK
CHANNING (I,900); NURSE (II,1870)

GUSS, JOHN
SWITCH (II,2519)

GUTMAN, RUTH
BRAIN GAMES (II,366)

GUZMAN, CLAUDIO
THE GOOD LIFE (I,1848); THE GOOD LIFE (II,1033); I DREAM OF JEANNIE (I,2167)

GWENLAN, GARETH
BUTTERFLIES (II,404); THE FALL AND RISE OF REGINALD PERRIN (II,810); TO THE MANOR BORN (II,2627)

GYGAX, GARY
DUNGEONS AND DRAGONS (II,744)

HAAS, NANCY
THE STANLEY SIEGEL SHOW (II,2433)

HACK, MICHAEL
TURBO-TEEN (II,2668)

HACK, MONROE
SOUND OFF TIME (I,4135)

HACKETT, BUDDY
BUDDY HACKETT—LIVE AND UNCENSORED (II,386)

HAFFNER, CRAIG
EYE ON HOLLYWOOD (II,798); THE LOVE REPORT (II,1542); THE LOVE REPORT (II,1543)

HAFNER, LES
THE DENNIS O'KEEFE SHOW (I,1230)

HAFT, DAVID
STEVE CANYON (I,4224)

HAGAN, CHET
BARBI BENTON SPECIAL: A BARBI DOLL FOR CHRISTMAS (II,148); CHRISTMAS LEGEND OF NASHVILLE (II,522); COUNTRY COMES HOME (II,585); COUNTRY COMES HOME (II,586); COUNTRY MUSIC HIT PARADE (II,589); COUNTRY NIGHT OF STARS (II,592); COUNTRY NIGHT OF STARS II (II,593); COUNTRY STARS OF THE 70S (II,594); ELVIS REMEMBERED: NASHVILLE TO HOLLYWOOD (II,772); FIFTY YEARS OF COUNTRY MUSIC (II,853); THE FIRST 50 YEARS (II,862); JOHNNY CASH AND FRIENDS (II,1322); JOHNNY CASH AND THE COUNTRY GIRLS (II,1323); THE JOHNNY CASH CHRISTMAS SPECIAL (II,1326); A JOHNNY CASH CHRISTMAS (II,1328); THE JOHNNY CASH SHOW (I,2424); THE JOHNNY CASH SPRING SPECIAL (II,1329); JOHNNY CASH'S AMERICA (II,1330); JOHNNY CASH: SPRING FEVER (II,1335); JOHNNY CASH: THE FIRST 25 YEARS (II,1336); ROY ACUFF—50 YEARS THE KING OF COUNTRY MUSIC (II,2223)

HAGGOTT, JOHN
OUT THERE (I,3424); ROD BROWN OF THE ROCKET RANGERS (I,3825)

HAHN, PHIL
THE FUNNIEST JOKE I EVER HEARD (II,939); THE FUNNIEST JOKE I EVER HEARD (II,940); THE OSMOND FAMILY SHOW (II,1927)

HAIGHT, GEORGE
MEET MCGRAW (I,2984)

HAINES-STILES, GEOFFREY
COSMOS (II,579)

HAKIM, ANDRE
PARIS PRECINCT (I,3458)

HALAS, JOHN
DODO—THE KID FROM OUTER SPACE (I,1324); TOMFOOLERY (I,4551)

HALE, LEE
DEAN MARTIN AT THE WILD ANIMAL PARK (II,636); THE DEAN MARTIN CELEBRITY ROAST (II,637); THE DEAN MARTIN CELEBRITY ROAST (II,638); THE DEAN MARTIN CELEBRITY ROAST (II,639); THE DEAN MARTIN CHRISTMAS SPECIAL (II,640); DEAN MARTIN IN LONDON (II,641); DEAN MARTIN'S CELEBRITY ROAST (II,643); DEAN MARTIN'S CHRISTMAS AT SEA WORLD (II,644); DEAN MARTIN'S CHRISTMAS IN CALIFORNIA (II,645); DEAN MARTIN'S COMEDY CLASSICS (II,646); DEAN MARTIN'S RED HOT SCANDALS OF 1926 (II,648); DEAN MARTIN'S RED HOT SCANDALS PART 2 (II,649); DOM DELUISE AND FRIENDS (II,701); DOM DELUISE AND FRIENDS, PART 2 (II,702); THE FIRST 50 YEARS (II,862); LADIES AND GENTLEMAN. . . BOB NEWHART (II,1420); LADIES AND GENTLEMAN. . . BOB NEWHART, PART II (II,1421)

HALEE, ROY
SIMON AND GARFUNKEL IN CONCERT (II,2356)

HALEY JR, JACK
BEAT OF THE BRASS (I,376); BIOGRAPHY (I,476); BOB HOPE SPECIAL: BOB HOPE'S WORLD OF COMEDY (II,323); A FUNNY THING HAPPENED ON THE WAY TO HOLLYWOOD (I,1716); HOLLYWOOD AND THE STARS (I,2081); THE JACK PAAR SPECIAL (I,2319); THE MAC DAVIS CHRISTMAS SPECIAL (II,1572); THE MAC

DAVIS SPECIAL (II,1577); MILTON BERLE'S MAD MAD MAD WORLD OF COMEDY (II,1701); MONTE CARLO, C'EST LA ROSE (I,3095); MOVIN' WITH NANCY (I,3119); RIPLEY'S BELIEVE IT OR NOT (II,2177); THAT'S HOLLYWOOD (II,2577); WITH LOVE, SOPHIA (I,4884); THE WORLD OF JAMES BOND (I,4913)

HALEY, ALEX
PALMERSTOWN U.S.A. (II,1945)

HALL, DON
DUSTY'S TREEHOUSE (I,1391)

HALL, JOHN FLEMING
DAVID NIVEN'S WORLD (II,625)

HALL, JOSEPH L
THE MAGIC GARDEN (II,1591)

HALL, MONTY
IT'S ANYBODY'S GUESS (II,1255); LET'S MAKE A DEAL (II,1462); LET'S MAKE A DEAL (II,1463); MASQUERADE PARTY (II,1645); THE MCLEAN STEVENSON SHOW (II,1665); YOUR FIRST IMPRESSION (I,4949)

HALLECK, VANCE
TELE PUN (I,4376)

HALLENBECK, DARRELL
MCNAMARA'S BAND (II,1668); MCNAMARA'S BAND (II,1669)

HALLER, DANIEL
O'HARA, UNITED STATES TREASURY (I,3346)

HALLIDAY, RICHARD
ANNIE GET YOUR GUN (I,223); MAGIC WITH MARY MARTIN (I,2821); MUSIC WITH MARY MARTIN (I,3180); PETER PAN (I,3566)

HALLIN, RONNY
HAPPY DAYS (II,1084); HERNDON AND ME (II,1139); JOANIE LOVES CHACHI (II,1295)

HALMI, ROBERT
THE BEST OF FRIENDS (II,216); BUSH DOCTOR (II,400); NURSE (II,1870); PEKING ENCOUNTER (II,1991); WILSON'S REWARD (II,2805)

HALSEY, JIM
THE GEORGE JONES SPECIAL (II,975)

HALSEY, SHERMAN
USA COUNTRY MUSIC (II,591)

HAMEL, ALAN
MANTRAP (I,2893); THE SUZANNE SOMERS SPECIAL (II,2510); THE SUZANNE SOMERS SPECIAL (II,2511)

HAMILTON, BOB
MOBILE MEDICS (II,1716)

HAMILTON, JOE
6 RMS RIV VU (II,2371); CALAMITY JANE (I,793); CAROL AND COMPANY (I,847); CAROL BURNETT AND COMPANY (II,441); THE CAROL BURNETT SHOW (II,443); THE ENTERTAINERS (I,1449); EUNICE (II,784); THE GARRY MOORE SHOW (I,1739); THE GARRY MOORE SHOW (I,1740); THE GINGER ROGERS SHOW (I,1804); THE JIMMIE RODGERS SHOW (I,2383); JULIE AND CAROL AT CARNEGIE HALL (I,2478); JULIE AND CAROL AT LINCOLN CENTER (I,2479); MAMA'S FAMILY (II,1610); ONCE UPON A MATTRESS (I,3371); THE PRIMARY ENGLISH CLASS (II,2081); THE SAMMY DAVIS JR. SHOW (I,3898); SILLS AND BURNETT AT THE MET (II,2354); THE SMOTHERS BROTHERS SHOW (II,2384); THE TIM CONWAY SHOW (II,2619); THE TIM CONWAY SPECIAL (I,4503); TWIGS (II,2683); UNCLE TIM WANTS YOU! (II,2704)

HAMILTON, LANCE
BLITHE SPIRIT (I,484); TOGETHER WITH MUSIC (I,4532)

HAMILTON, NANCY
WE'RE MOVIN' (II,2756)

HAMILTON, WILLIAM
ETHEL AND ALBERT (I,1466)

HAMLIN, JOHN
THE 36 MOST BEAUTIFUL GIRLS IN TEXAS (II,2594); BOB HOPE SPECIAL: HAPPY BIRTHDAY, BOB! (II,326)

HAMMEL, THOMAS
THE GEORGIA PEACHES (II,979)

HAMMERSTEIN, TED
BROADWAY TO HOLLYWOOD (I,739)

HAMMOND, THOMAS
WE INTERRUPT THIS SEASON (I,4780)

HAMNER, EARL
APPLE'S WAY (II,101); BOONE (II,351); FALCON CREST (II,808); JOSHUA'S WORLD (II,1344); THE WALTONS (II,2740); THE WALTONS: A DAY FOR THANKS ON WALTON'S MOUNTAIN (EPISODE 3) (II,2741); THE WALTONS: MOTHER'S DAY ON WALTON'S MOUNTAIN (EPISODE 2) (II,2742); THE WALTONS: WEDDING ON WALTON'S MOUNTAIN (EPISODE 1) (II,2743); THE YOUNG PIONEERS (II,2869)

HAMNER, ROBERT
MALIBU (II,1607); MICKEY SPILLANE'S MARGIN FOR MURDER (II,1691); S.W.A.T. (II,2236)

HANDLEY, ALAN
THE ANDY GRIFFITH—DON KNOTTS—JIMNABORS SHOW (I,191); THE ANDY WILLIAMS CHRISTMAS SHOW (I,198); THE ANDY WILLIAMS SPECIAL (I,213); CAROL CHANNING AND 101 MEN (I,849); THE COMICS (I,1028); DANNY THOMAS GOES COUNTRY AND WESTERN (I,1146); THE DANNY THOMAS HOUR (I,1147); THE DANNY THOMAS SPECIAL (I,1150); THE DANNY THOMAS SPECIAL (I,1151); THE DANNY THOMAS SPECIAL (I,1152); DANNY THOMAS: AMERICA I LOVE YOU (I,1145); DANNY THOMAS: THE ROAD TO LEBANON (I,1153); DEBBIE REYNOLDS AND THE SOUND OF CHILDREN (I,1213); THE DINAH SHORE SHOW (I,1280); GUYS 'N' GEISHAS (I,1914); JIMMY DURANTE MEETS THE LIVELY ARTS (I,2386); THE JULIE ANDREWS SHOW (I,2481); THE JULIE ANDREWS SPECIAL (I,2482); THE MICKIE FINNS FINALLY PRESENT HOW THE WEST WAS LOST (II,1695); PORTRAIT OF PETULA (I,3654); THE ROYAL FOLLIES OF 1933 (I,3862); THE SID CAESAR SHOW (I,4042); SVENGALI AND THE BLONDE (I,4310); THE WALTER WINCHELL SHOW (I,4760); THE WONDERFUL WORLD OF BURLESQUE I (I,4895); THE WONDERFUL WORLD OF BURLESQUE II (I,4896); THE WONDERFUL WORLD OF BURLESQUE III (I,4897)

HANLEY, JIM
STRANGE TRUE STORIES (II,2476)

HANNA, WILLIAM
THE ABBOTT AND COSTELLO CARTOON SHOW (I,1); THE ABC AFTERSCHOOL SPECIAL (II,1); THE ADDAMS FAMILY (II,12); THE ADVENTURES OF GULLIVER (I,60); THE ADVENTURES OF JONNY QUEST (I,64); THE ALL-NEW POPEYE HOUR (II,49); THE AMAZING CHAN AND THE CHAN CLAN (I,157); THE ATOM ANT/SECRET SQUIRREL SHOW (I,311); THE BANANA SPLITS ADVENTURE HOUR (I,340); BIRDMAN (II,256); BIRDMAN AND THE GALAXY TRIO (I,478); THE BISKITTS (II,257); BUFORD AND THE GHOST (II,389); BUTCH CASSIDY AND THE SUNDANCE KIDS (I,781); THE C.B. BEARS (II,406); CAPTAIN CAVEMAN AND THE TEEN ANGELS (II,433); CASPER AND THE ANGELS (II,453); THE CATTANOOGA CATS (I,874); THE CLUE CLUB (II,546); DASTARDLY AND MUTTLEY IN THEIR FLYING MACHINES (I,1159); DEVLIN (II,663); THE DRAK PACK (II,733); THE DUKES (II,741); THE FANTASTIC FOUR (I,1529); THE FLINTSTONE COMEDY HOUR (II,881); FLINTSTONE FAMILY ADVENTURES (II,882); THE FLINTSTONE FUNNIES (II,883); THE FLINTSTONES (II,884); THE FLINTSTONES (II,885); FONZ AND THE HAPPY DAYS GANG (II,898); FRANKENSTEIN JR. AND THE IMPOSSIBLES (I,1672); FRED AND BARNEY MEET THE SHMOO (II,918); FRED AND BARNEY MEET THE THING (II,919); FRED FLINTSTONE AND FRIENDS (II,920); THE FUNKY PHANTOM (I,1709); THE GALAXY GOOFUPS (II,954); THE GARY COLEMAN SHOW (II,958); GIDGET MAKES THE WRONG CONNECTION (I,1798); GODZILLA (II,1014); GOOBER AND THE GHOST CHASERS (II,1028); THE GREAT GRAPE APE SHOW (II,1056); THE HARLEM GLOBETROTTERS (I,1953); HELP! IT'S THE HAIR BEAR BUNCH (I,2012); THE HERCULOIDS (I,2023); HERE COME THE STARS (I,2026); HONG KONG PHOOEY (II,1180); THE HUCKLEBERRY HOUND SHOW (I,2155); INCH HIGH, PRIVATE EYE (II,1231); JABBERJAW (II,1261); JACK AND THE BEANSTALK (I,2288); JEANNIE (II,1275); THE JETSONS (I,2376); JOKEBOOK (II,1339); JOSIE AND THE PUSSYCATS (I,2453); JOSIE AND THE PUSSYCATS IN OUTER SPACE (I,2454); KORG: 70,000 B.C. (II,1408); THE KWICKY KOALA SHOW (II,1417); LAVERNE AND SHIRLEY IN

THE ARMY (II,1448); LAVERNE AND SHIRLEY WITH THE FONZ (II,1449); LIPPY THE LION (I,2710); THE LITTLE RASCALS (II,1494); LOST IN SPACE (I,2759); THE MAGILLA GORILLA SHOW (I,2825); MOBY DICK AND THE MIGHTY MIGHTOR (I,3077); MONCHHICHIS (II,1722); MORK AND MINDY (II,1736); MOTOR MOUSE (I,3111); THE NEW ADVENTURES OF HUCKLEBERRY FINN (I,3250); THE NEW FRED AND BARNEY SHOW (II,1817); THE NEW SHMOO (II,1827); THE NEW SUPER FRIENDS HOUR (II,1830); PAC-MAN (II,1940); PARTRIDGE FAMILY: 2200 A.D. (II,1963); PEBBLES AND BAMM BAMM (II,1987); THE PERILS OF PENELOPE PITSTOP (I,3525); THE PETER POTAMUS SHOW (I,3568); PINK PANTHER AND SONS (II,2043); THE QUICK DRAW MCGRAW SHOW (I,3710); THE RICHIE RICH SHOW (II,2169); THE ROMAN HOLIDAYS (I,3835); THE RUFF AND REDDY SHOW (I,3866); SAMPSON AND GOLIATH (I,3903); SCOOBY'S ALL-STAR LAFF-A-LYMPICS (II,2273); SCOOBY-DOO AND SCRAPPY-DOO (II,2274); THE SCOOBY-DOO AND SCRAPPY-DOO SHOW (II,2275); SCOOBY-DOO, WHERE ARE YOU? (II,2276); THE SCOOBY-DOO/DYNOMUTT HOUR (II,2277); SEALAB 2020 (I,3949); SHAZZAN! (I,4002); THE SHIRT TALES (II,2337); THE SKATEBIRDS (II,2375); THE SMURFS (II,2385); SNORKS (II,2390); SPACE GHOST (I,4141); SPACE KIDDETTES (I,4143); SPACE STARS (II,2408); SPEED BUGGY (II,2419); SUPER FRIENDS (II,2497); THE SUPER GLOBETROTTERS (II,2498); THESE ARE THE DAYS (II,2585); THE THREE ROBONIC STOOGES (II,2609); THE TOM AND JERRY SHOW (I,4534); TOP CAT (I,4576); TOUCHE TURTLE (I,4586); TROLLKINS (II,2660); VALLEY OF THE DINOSAURS (II,2717); THE WACKY RACES (I,4745); WAIT TIL YOUR FATHER GETS HOME (I,4748); WALLY GATOR (I,4751); WHEELIE AND THE CHOPPER BUNCH (II,2776); WHERE'S HUDDLES (I,4836); THE WORLD'S GREATEST SUPER HEROES (II,2842); YOGI BEAR (I,4928); YOGI'S GANG (II,2852); YOGI'S SPACE RACE (II,2853)

HANSEN, PETER
THE MORECAMBE AND WISE SHOW (II,1734); WORLD WAR II: G.I. DIARY (II,2840)

HANSON, DAVID E
BATTLE OF THE PLANETS (II,179); OMNI: THE NEW FRONTIER (II,1889)

HANSON, J E
TALES OF THE RED CABOOSE (I,4347)

HARBACH, WALTER O
THE BING CROSBY SHOW (I,463)

HARBACH, WILLIAM
BING CROSBY AND CAROL BURNETT—TOGETHER AGAIN FOR THE FIRST TIME (I,454); THE BING CROSBY SHOW (I,464); BOB HOPE SPECIAL: BOB HOPE'S PINK PANTHER THANKSGIVING GALA (II,312); BOB HOPE SPECIAL: BOB HOPE PRESENTS A CELEBRATION WITH STARS OF COMEDY AND MUSIC (II,340); THE DON ADAMS SPECIAL: HOORAY FOR HOLLYWOOD (I,1329); THE DON KNOTTS SHOW (I,1334); GLENN MILLER TIME (I,1824); THE HOLLYWOOD PALACE (I,2088); HONEY WEST (I,2105); THE JULIE ANDREWS HOUR (I,2480); THE MILTON BERLE SHOW (I,3049); THE MILTON BERLE SPECIAL (I,3050); THE NEW STEVE ALLEN SHOW (I,3271); THE STEVE ALLEN COMEDY HOUR (II,2454); THE STEVE ALLEN SHOW (I,4219); THE STEVE ALLEN SHOW (I,4220); SWING OUT, SWEET LAND (II,2515)

HARBERT, JULIE
PEOPLE DO THE CRAZIEST THINGS (II,1998)

HARDIE, SEAN
NOT THE NINE O'CLOCK NEWS (II,1864)

HARDING, MALCOLM
CENTENNIAL (II,477); FALCON CREST (II,808); THE YELLOW ROSE (II,2847)

HARDY, ALAN
ALL THE RIVERS RUN (II,43)

HARDY, JOHN
PRISONERS OF THE LOST UNIVERSE (II,2086)

HARDY, JOSEPH
JAMES AT 15 (II,1270); JAMES AT 15 (II,1271); LOVE OF LIFE (I,2774); TAXI (I,4365)

HARGROVE, DEAN
COLUMBO (II,556); CUTTER (I,1117); DEAR DETECTIVE (II,652); THE FAMILY HOLVAK (II,817); THE GREATEST GIFT (II,1061); MCCLOUD (II,1660); THE NAME OF THE GAME (I,3217); RANSOM FOR A DEAD MAN (I,3722); RETURN OF THE WORLD'S GREATEST DETECTIVE (II,2142)

HARLAN, JOHN
YOU DON'T SAY (II,2858)

HARLIB, MATT
THE MARTHA WRIGHT SHOW (I,2927)

HARMON, BILL
MCKEEVER AND THE COLONEL (I,2970); THE TOM EWELL SHOW (I,4538)

HARMON, LARRY
BOZO THE CLOWN (I,702); POPEYE THE SAILOR (I,3649)

HARMON, SANDRA
REAL LIFE STORIES (II,2117)

HARMON, WILLIAM
BLONDIE (I,486); THE DUKE (I,1381); THE SOFT TOUCH (I,4107)

HARPER, GREG
THE $128,000 QUESTION (II,1903)

HARRIS, HARRY
THE TOM SWIFT AND LINDA CRAIG MYSTERY HOUR (II,2633); WANTED: DEAD OR ALIVE (I,4764)

HARRIS, JED
BILLY ROSE'S PLAYBILL (I,452)

HARRIS, JEFF
ALL-AMERICAN PIE (II,48); ALMOST ANYTHING GOES (II,56); COMEDY NEWS (I,1020); COMEDY NEWS II (I,1021); DETECTIVE SCHOOL (II,661); THE EVERLY BROTHERS SHOW (I,1479); FREEMAN (II,923); IN TROUBLE (II,1230); IT ONLY HURTS WHEN YOU LAUGH (II,1251); JIMMY DURANTE PRESENTS THE LENNON SISTERS HOUR (I,2387); JOE AND SONS (II,1296); JOE AND SONS (II,1297); THE LENNON SISTERS SHOW (I,2656); MA AND PA (II,1570); MCNAMARA'S BAND (II,1668); MCNAMARA'S BAND (II,1669); THE NEW OPERATION PETTICOAT (II,1826); PAT BOONE IN HOLLYWOOD (I,3469); SHEEHY AND THE SUPREME MACHINE (II,2322); SHOW BUSINESS SALUTE TO MILTON BERLE (I,4029); THE SHOW MUST GO ON (II,2344); THE STEVE ALLEN SHOW (I,4222); THAT SECOND THING ON ABC (II,2572); THAT THING ON ABC (II,2574)

HARRIS, KAREN
THE INCREDIBLE HULK (II,1232); SHANNON (II,2314)

HARRIS, KEN
THE JIM NABORS SHOW (II,1290); THE ROOTS OF ROCK 'N' ROLL (II,2212)

HARRIS, LESLIE
NAVY LOG (I,3233)

HARRIS, MARY
THE BERT PARKS SHOW (I,400); THE BRIGHTER DAY (I,728)

HARRIS, PAUL
IT'S ROCK AND ROLL (II,1258); THE JO STAFFORD SHOW (I,2392)

HARRIS, ROBERT
BARETTA (II,152)

HARRIS, SHERMAN
BOSS LADY (I,694); THE LONE RANGER (I,2740)

HARRIS, STAN
BLONDIE (II,272); DANCE FEVER (II,615); DOLLY IN CONCERT (II,700); GEORGE BURNS' 100TH BIRTHDAY PARTY (II,971); THE GISELE MACKENZIE SHOW (I,1813); HI, I'M GLEN CAMPBELL (II,1141); JACK BENNY'S 20TH ANNIVERSARY TV SPECIAL (I,2303); THE KENNY ROGERS SPECIAL (II,1382); LYNDA CARTER'S CELEBRATION (II,1563); LYNDA CARTER: ENCORE (II,1566); LYNDA CARTER: STREET LIGHTS (II,1567); THE MANCINI GENERATION (I,2885); THE MELBA MOORE-CLIFTON DAVIS SHOW (I,2996); THE MIDNIGHT SPECIAL (II,1696); THE MUSIC SCENE (I,3176); THE OSMOND BROTHERS SHOW (I,3409); A SPECIAL KENNY ROGERS (II,2417); THAT'S LIFE (I,4426)

HARRIS, SUSAN
BENSON (II,208); DAUGHTERS (II,620); LOVES ME, LOVES ME NOT (II,1551); SOAP (II,2392)

HARRISON, JERRY
BRENDA STARR, REPORTER (II,373); HOLLYWOOD TEEN (II,1166); ROCK 'N' ROLL: THE FIRST 25 YEARS (II,2189); THE ROOTS OF ROCK 'N' ROLL (II,2212); SINATRA: CONCERT FOR THE AMERICAS (II,2361); STEVE ALLEN'S LAUGH-

BACK (II,2455); THE WOLFMAN JACK RADIO SHOW (II,2816)

HARRISON, JOAN
ALFRED HITCHCOCK PRESENTS (I,115); JANET DEAN, REGISTERED NURSE (I,2348); JOURNEY TO THE UNKNOWN (I,2457); THE MOST DEADLY GAME (I,3106)

HARRISON, PAUL
I HEAR AMERICA SINGING (I,2168); THE MARGE AND GOWER CHAMPION SHOW (I,2907); OUR MAN HIGGINS (I,3414); THE SKY'S THE LIMIT (I,4081); WRANGLER (I,4921)

HARSBURGH, PATRICK
THE A-TEAM (II,119); HARDCASTLE AND MCCORMICK (II,1089)

HART, BOB
ROCK-N-AMERICA (II,2195); SHEENA EASTON—LIVE AT THE PALACE (II,2324)

HART, BRUCE
THE DICK CAVETT SHOW (II,669); HOT HERO SANDWICH (II,1186)

HART, CAROLE
THE DICK CAVETT SHOW (II,669); HOT HERO SANDWICH (II,1186); MARLO THOMAS AND FRIENDS IN FREE TO BE. . . YOU AND ME (II,1632)

HART, SAM
BONNIE AND THE FRANKLINS (II,349)

HART, TERRY
JENNIFER SLEPT HERE (II,1278); JOANIE LOVES CHACHI (II,1295)

HART, WALTER
ETHEL AND ALBERT (I,1466)

HARTLEY, BILL
CIRCUS OF THE 21ST CENTURY (II,529)

HARTMANN, EDMUND
THE EVE ARDEN SHOW (I,1470); FAMILY AFFAIR (I,1519); MY FAVORITE HUSBAND (I,3188); MY THREE SONS (I,3205); THE SMITH FAMILY (I,4092); TO ROME WITH LOVE (I,4526)

HARTUNG, ROBERT
EAGLE IN A CAGE (I,1392); A PUNT, A PASS, AND A PRAYER (I,3694)

HARWOOD, RICHARD S
GOOD PENNY (II,1036)

HATOS, STEFAN
CHAIN LETTER (I,892); IT COULD BE YOU (I,2245); IT PAYS TO BE MARRIED (I,2249); IT'S ANYBODY'S GUESS (II,1255); LET'S MAKE A DEAL (II,1462); LET'S MAKE A DEAL (II,1463); LET'S MAKE A DEAL (II,1464); MASQUERADE PARTY (II,1645); PANHANDLE PETE AND JENNIFER (I,3446); THERE'S ONE IN EVERY FAMILY (I,4439); THREE FOR THE MONEY (II,2606); UNCLE MISTLETOE AND HIS ADVENTURES (I,4669)

HAUCK, CHARLIE
APPLE PIE (II,100); BACK TOGETHER (II,131); BRANAGAN AND MAPES (II,367); A DOG'S LIFE (II,697); HARRY'S BATTLES (II,1100); HUSBANDS, WIVES AND LOVERS (II,1209); MR. DUGAN (II,1749); SUZANNE PLESHETTE IS MAGGIE BRIGGS (II,2509); THE TWO OF US (II,2693)

HAUSER, RICK
THE SCARLET LETTER (II,2270)

HAWKESWORTH, JOHN
Q. E. D. (II,2094); THE TALE OF BEATRIX POTTER (II,2533); UPSTAIRS, DOWNSTAIRS (II,2709)

HAWKINS, JOHN
THE COWBOYS (II,598); LITTLE HOUSE ON THE PRAIRIE (II,1487)

HAWKINS, RICK
PUNKY BREWSTER (II,2092)

HAWKSWORTH, JOHN
THE FLAME TREES OF THIKA (II,870)

HAWN, GOLDIE
GOLDIE AND KIDS: LISTEN TO ME (II,1021)

HAYDEN, JEFFREY
THE BILLY BEAN SHOW (I,449); HOW THE WEST WAS WON (II,1196); SANTA BARBARA (II,2258)

HAYDEN, RUSSELL
26 MEN (I,4648); JUDGE ROY BEAN (I,2467)

HAYES, DARLENE
THE PHIL DONAHUE SHOW (II,2032)

HAYES, JEFFREY
FINDER OF LOST LOVES (II,857); T.J. HOOKER (II,2524); VEGAS (II,2724)

HAYES, JOHN MICHAEL
NEVADA SMITH (II,1807)

HAYES, PETER LIND
WHEN TELEVISION WAS LIVE (II,2779)

HAYWARD, CHRIS
BARNEY MILLER (II,154); NOBODY'S PERFECT (II,1858); THE TEXAS WHEELERS (II,2568)

HAYWARD, LELAND
ANYTHING GOES (I,238); A BELL FOR ADANO (I,388); THE FABULOUS 50S (I,1501); THE FORD 50TH ANNIVERSARY SHOW (I,1630); THE GERSHWIN YEARS (I,1779); SATURDAY'S CHILDREN (I,3923); THAT WAS THE WEEK THAT WAS (I,4422); THAT WAS THE WEEK THAT WAS (I,4423)

HAYWOOD, BRUCE
THE MAGIC OF DAVID COPPERFIELD (II,1595)

HAZE, JONATHAN
CATALINA C-LAB (II,458)

HAZZARD, KAREN
HIGH HOPES (II,1144)

HEATH, BOB
YOU ASKED FOR IT (II,2855)

HEATH, LAURENCE
CALL TO DANGER (I,804); KHAN! (II,1384); THE MAGICIAN (I,2823); MISSION: IMPOSSIBLE (I,3067)

HEATTER, MERRILL
THE $64,000 QUESTION (I,4071); THE AMATEUR'S GUIDE TO LOVE (I,155); BATTLESTARS (II,182); BEDTIME STORIES (II,199); THE CELEBRITY GAME (I,884); FANTASY (II,828); GAMBIT (II,955); HIGH ROLLERS (II,1147); THE HOLLYWOOD SQUARES (II,1164); HOT SEAT (I,2125); LAS VEGAS GAMBIT (II,1430); THE MAGNIFICENT MARBLE MACHINE (II,1600); P.D.Q. (I,3497); PEOPLE WILL TALK (I,3520); SHOWDOWN (I,4034); THE STORYBOOK SQUARES (I,4242); TEMPTATION (I,4393); TO SAY THE LEAST (II,2623); VIDEO VILLAGE (I,4731)

HECHT, JOEL
HIT MAN (II,1156)

HECHT, KEN
MAKE YOUR OWN KIND OF MUSIC (I,2847)

HEDTON, JOHN
THE ALICE PEARCE SHOW (I,121)

HEERMANCE, RICHARD
THE WESTERNER (I,4801)

HEFNER, HUGH
PLAYBOY'S 25TH ANNIVERSARY CELEBRATION (II,2054); PLAYBOY'S PLAYMATE PARTY (II,2055); A SHADOW IN THE STREETS (II,2312)

HEIDER, FREDERICK
ARTHUR MURRAY'S DANCE PARTY (I,293); THE BELL TELEPHONE HOUR (I,389); THE BILLY DANIELS SHOW (I,451); MUSIC FOR A SPRING NIGHT (I,3169); MUSIC FOR A SUMMER NIGHT (I,3170); OPERA VS. JAZZ (I,3399); VOICE OF FIRESTONE (I,4741)

HEIDT, HORACE
FAMILY NIGHT WITH HORACE HEIDT (I,1523)

HEILWEIL, DAVID
MACREEDY'S WOMAN (I,2804)

HEIN, JACK
THE ERNIE KOVACS SHOW (I,1455)

HEINEMANN, GEORGE
CACTUS JIM (I,786); LITTLE WOMEN (I,2724); MUGGSY (II,1761)

HEINZ, JAMES
ENOS (II,779); HAVING BABIES III (II,1109); HAWAII FIVE-O (II,1110); JULIE FARR, M.D. (II,1354); KHAN! (II,1384); SIX PACK (II,2370)

HEISCH, GLEN
MR. MAGOO (I,3142)

HELBURN, THERESA
GREAT CATHERINE (I,1873); STAGE DOOR (I,4177)

HELFER, RALPH
MR. SMITH (II,1755)

HELFOTT, DANIEL
GOTTA SING, GOTTA DANCE (II,1046)

HELLER, FRANKLIN
THE AT HOME SHOW (I,308); THE AT LIBERTY CLUB (I,309); THE CLIFF EDWARDS SHOW (I,975); THE EDDIE ALBERT SHOW (I,1406); MASLAND AT HOME (I,2939); THE TRAP (I,4593); THE WEB (I,4784)

HELLER, GEORGANNE ALDRICH
FALCON'S GOLD (II,809)

HELLER, JOEL
EVERYTHING YOU EVER WANTED TO KNOW ABOUT MONSTERS . . . BUT WERE AFRAID! (II,794)

HELLER, PAUL
FALCON'S GOLD (II,809); WAIT UNTIL DARK (II,2735)

HELLER, ROSILYN
CELEBRITY (II,463)

HELPERN, STEVE
ALIAS SMITH AND JONES (I,118); THE LAWYERS (I,2646)

HEMANZA, GERALD
GEORGE CARLIN AT CARNEGIE HALL (II,974)

HEMION, DWIGHT
ANN-MARGRET OLSSON (II,84); ANN-MARGRET SMITH (II,85); ANN-MARGRET'S HOLLYWOOD MOVIE GIRLS (II,86); ANN-MARGRET. . . RHINESTONE COWGIRL (II,87); ANNE MURRAY'S CARIBBEAN CRUISE (II,88); ANNE MURRAY'S WINTER CARNIVAL. . . FROM QUEBEC (II,90); BARBRA STREISAND AND OTHER MUSICAL INSTRUMENTS (I,353); BARYSHNIKOV ON BROADWAY (II,158); BEN VEREEN—HIS ROOTS (II,202); BETTE MIDLER—OL' RED HAIR IS BACK (II,222); BING CROSBY'S MERRIE OLDE CHRISTMAS (II,252); BURT BACHARACH IN SHANGRI-LA (I,773); THE BURT BACHARACH SPECIAL (I,775); THE BURT BACHARACH SPECIAL (I,776); BURT BACHARACH! (I,771); BURT BACHARACH: CLOSE TO YOU (I,772); BURT BACHARACH—OPUS NO. 3 (I,774); THE CHERYL LADD SPECIAL (II,501); CHRISTMAS IN WASHINGTON (II,520); CHRISTMAS IN WASHINGTON (II,521); DICK CAVETT'S BACKLOT USA (II,670); DISNEYLAND'S 25TH ANNIVERSARY (II,681); DOROTHY HAMILL IN ROMEO & JULIET ON ICE (II,718); THE DOROTHY HAMILL SPECIAL (II,719); THE EDDIE RABBITT SPECIAL (II,759); EVENING AT THE MOULIN ROUGE (II,787); THE FRANK SINATRA SHOW (I,1667); FRANK SINATRA—A MAN AND HIS MUSIC (I,1661); GLEN CAMPBELL . . . DOWN HOME—DOWN UNDER (II,1008); THE GLEN CAMPBELL SPECIAL: THE MUSICAL WEST (I,1821); GOLDIE AND KIDS: LISTEN TO ME (II,1021); HERB ALBERT AND THE TIJUANA BRASS (I,2019); HERB ALPERT AND THE TIJUANA BRASS (II,1130); JAMES PAUL MCCARTNEY (I,2332); JULIE ON SESAME STREET (I,2485); THE KLOWNS (I,2574); THE KOPYCATS (I,2580); THE KRAFT 75TH ANNIVERSARY SPECIAL (II,1409); THE KRAFT MUSIC HALL (I,2588); KRAFT SALUTES WALT DISNEY WORLD'S 10TH ANNIVERSARY (II,1410); LARRY GATLIN AND THE GATLIN BROTHERS (II,1429); LAS VEGAS PALACE OF STARS (II,1431); LINDA IN WONDERLAND (II,1481); LUCY COMES TO NASHVILLE (II,1561); MAC DAVIS CHRISTMAS SPECIAL. . . WHEN I GROW UP (II,1574); THE MAC DAVIS SHOW (II,1576); MAC DAVIS'S CHRISTMAS ODYSSEY: TWO THOUSAND AND TEN (II,1579); MAC DAVIS. . . SOUNDS LIKE HOME (II,1582); THE MAGICAL MUSIC OF BURT BACHARACH (I,2822); MARLO THOMAS IN ACTS OF LOVE—AND OTHER COMEDIES (I,2919); MERRY CHRISTMAS FROM THE CROSBYS (II,1685); MERRY CHRISTMAS FROM THE GRAND OLE OPRY (II,1686); THE NEIL DIAMOND SPECIAL (II,1802); THE NEW CHRISTY MINSTRELS SHOW (I,3259); ON STAGE AMERICA (II,1893); PAVAROTTI AND FRIENDS (II,1985); PETER PAN (II,2030); THE ROGER MILLER SHOW (I,3830); ROYAL VARIETY PERFORMANCE (I,3865); SHEENA EASTON, ACT 1 (II,2323); SHIRLEY MACLAINE. . . EVERY LITTLE MOVEMENT (II,2334); A SPECIAL ANNE MURRAY CHRISTMAS (II,2413); STEVE AND EYDIE CELEBRATE IRVING BERLIN (II,2456); STEVE AND EYDIE: OUR LOVE IS HERE TO STAY (II,2457); STEVE LAWRENCE AND EYDIE GORME FROM THIS MOMENT ON. . . COLE PORTER (II,2459); THREE GIRLS THREE (II,2608); TIN PAN ALLEY TODAY (I,4521); THE TONY BENNETT SHOW (I,4570); UPTOWN (II,2710); THE VERY FIRST GLEN CAMPBELL SPECIAL (I,4714); ZERO HOUR (I,4980)

HENDRICKS, BRUCE
SQUARE PEGS (II,2431)

HENDRICKSON, FRED
SUSPENSE (I,4306)

HENERSON, JAMES S
BOB & CAROL & TED & ALICE (I,495); FOR LOVERS ONLY (II,904); IRENE (II,1245)

HENLEY, DON
DEBBY BOONE. . . THE SAME OLD BRAND NEW ME (II,656)

HENNESSEY, DENNIS
CHARLEY'S AUNT (II,483)

HENNING, PAUL
THE ALAN YOUNG SHOW (I,104); THE BEVERLY HILLBILLIES (I,417); CAROL (I,846); GREEN ACRES (I,1884); LOVE THAT BOB (I,2779); PETTICOAT JUNCTION (I,3571); THE R.C.A. VICTOR SHOW (I,3737); THE RETURN OF THE BEVERLY HILLBILLIES (II,2139); THE SECRET WORLD OF KIDS (I,3970); WHERE'S RAYMOND? (I,4837)

HENRY, BABETTE
BUCK ROGERS IN THE 25TH CENTURY (I,751)

HENRY, BOB
AFTER HOURS: GETTING TO KNOW US (II,21); ANDY WILLIAMS KALEIDOSCOPE COMPANY (I,201); THE ANDY WILLIAMS SHOW (I,205); THE CAPTAIN AND TENNILLE (II,429); THE CARPENTERS (II,445); THE CARPENTERS AT CHRISTMAS (II,446); THE CARPENTERS. . . SPACE ENCOUNTERS (II,447); THE DINAH SHORE CHEVY SHOW (I,1279); DO IT YOURSELF (I,1299); FELICIANO—VERY SPECIAL (I,1560); THE FLIP WILSON SHOW (I,1602); THE GIRL IN MY LIFE (II,996); GLADYS KNIGHT AND THE PIPS (II,1004); THE GLEN CAMPBELL MUSIC SHOW (II,1006); THE GLEN CAMPBELL MUSIC SHOW (II,1007); GOOD EVENING, CAPTAIN (II,1031); JONATHAN WINTERS PRESENTS 200 YEARS OF AMERICAN HUMOR (II,1342); KEEP U.S. BEAUTIFUL (I,2519); LAS VEGAS PALACE OF STARS (II,1431); MITZI (I,3071); MOVIN' (I,3117); THE NAT KING COLE SHOW (I,3230); THE PERRY COMO CHRISTMAS SHOW (I,3529); PERRY COMO'S MUSIC FROM HOLLYWOOD (II,2019); POETRY AND MUSIC (I,3632); THE ROY ROGERS AND DALE EVANS SHOW (I,3858); THE ROY ROGERS SHOW (I,3859); SALUTE (II,2242)

HENRY, BUCK
CAPTAIN NICE (I,835)

HENRY, FRED
DATE WITH THE ANGELS (I,1164); FAMILY AFFAIR (I,1519); THE MILLIONAIRE (I,3046); THE MILTON BERLE SHOW (I,3047); MY THREE SONS (I,3205)

HENRY, JOSEPH
KHAN! (II,1384)

HENRY, MIKE
YOU DON'T SAY (II,2858)

HENRY, NORMAN
O.K. CRACKERBY (I,3348); WEEKEND (I,4788)

HENSON, CAROL
SECOND CITY TELEVISION (II,2287)

HENSON, JIM
THE FANTASTIC MISS PIGGY SHOW (II,827); FRAGGLE ROCK (II,916); FROG PRINCE (I,1685); JIM HENSON'S MUPPET BABIES (II,1289); THE MUPPET SHOW (II,1766); THE MUPPETS GO TO THE MOVIES (II,1767); TALES FROM MUPPETLAND (I,4344)

HENSON, NANCY
CHER—A CELEBRATION AT CAESAR'S PALACE (II,498); THE MAC DAVIS SHOW (II,1576)

HERBERT, LUDWIG G
CRIME AND PUNISHMENT (I,1095)

HERLWELL, DAVID
SUSPENSE (I,4305)

HERMAN, MATTHEW N
ROCK 'N' ROLL: THE FIRST 25 YEARS (II,2189)

HERRIDGE, ROBERT
FOLK SOUND U.S.A. (I,1614); KRAFT TELEVISION THEATER (I,2592); THE ROBERT HERRIDGE THEATER (I,3808); THE SEVEN LIVELY ARTS (I,3987)

HERRMANN, HARRY
BELIEVE IT OR NOT (I,387); THE HARTMANS (I,1960); KRAFT TELEVISION THEATER (I,2592); STAR TONIGHT (I,4199)

HERZBERG, JACK
DAYS OF OUR LIVES (II,629); THE EVERGLADES (I,1478); HIGHWAY PATROL (I,2058)

HESSLER, GORDON
CONVOY (I,1043); DIAGNOSIS: DANGER (I,1250)

HEWITT, DON
SINATRA (I,4053)

HEYES, DOUGLAS
THE BARBARY COAST (II,146); THE BARBARY COAST (II,147); BEARCATS! (I,375); BRAVO DUKE (I,711); CIRCUS BOY (I,959); POWDERKEG (I,3657)

HEYNE, NORMAN
ACROBAT RANCH (I,20)

HEYWARD, ANDY
THE GET ALONG GANG (II,980); INSPECTOR GADGET (II,1236); KIDD VIDEO (II,1388); THE LITTLES (II,1499); POLE POSITION (II,2060); WOLF ROCK TV (II,2815)

HEYWARD, DEKE
THE DICK CLARK SATURDAY NIGHT BEECHNUT SHOW (I,1267)

HICKMAN, DARRYL
LOVE OF LIFE (I,2774); SIDE BY SIDE (II,2347); A YEAR AT THE TOP (II,2844); A YEAR AT THE TOP (II,2845)

HICKMAN, HUDSON
HOTEL (II,1192)

HICKOX, S BRYAN
BABY, I'M BACK! (II,130); JAKE'S WAY (II,1269); THOU SHALT NOT KILL (II,2604); WIZARDS AND WARRIORS (II,2813)

HICKS, TOM
BETTY CROCKER STAR MATINEE (I,412); WELCOME TRAVELERS (I,4791)

HIKEN, NAT
THE BALLAD OF LOUIE THE LOUSE (I,337); CAR 54, WHERE ARE YOU? (I,842); CAROL + 2 (I,853); THE NEW PHIL SILVERS SHOW (I,3269); PHIL SILVERS IN NEW YORK (I,3577); THE SLOWEST GUN IN THE WEST (I,4088); SUMMER IN NEW YORK (I,4279)

HILL, CHARLES
PERSON TO PERSON (I,3550)

HILL, JAMES B
MIDWESTERN HAYRIDE (I,3033)

HILL, JERRY
TREASURE UNLIMITED (I,4603)

HILL, KIMBERLY
ALL TOGETHER NOW (II,45)

HILL, LEONARD
HIGH SCHOOL, U.S.A. (II,1148); HIGH SCHOOL, U.S.A. (II,1149); TUCKER'S WITCH (II,2667)

HILL, LYLE B
SOMERSET (I,4115)

HILL, MICHAEL
ALL-STAR SECRETS (II,52); THE TONI TENNILLE SHOW (II,2636); WE DARE YOU! (II,2751); YOU BET YOUR LIFE (II,2856)

HILL, REG
CAPTAIN SCARLET AND THE MYSTERONS (I,837); THE PROTECTORS (I,3681); THUNDERBIRDS (I,4497); U.F.O. (I,4662)

HILL, SETH
IN SEARCH OF. . . (II,1226)

HILL, THOMAS A
THE NEW ZOO REVUE (I,3277); SIGMUND AND THE SEA MONSTERS (II,2352)

HILL, WILLIAM
COURT-MARTIAL (I,1080); THE LAST DAYS OF POMPEII (II,1436); LITTLE LORD FAUNTLEROY (II,1492)

HILLIER, BILL
SOAP WORLD (II,2394); THE WORLD OF PEOPLE (II,2839)

HILLIER, DAVID
SHEENA EASTON—LIVE AT THE PALACE (II,2324)

HILMER, DAVID
ANYTHING FOR MONEY (II,99)

HIMES, CAROL
FAMILY TIES (II,819)

HINCHCLIFFE, PHILIP
NANCY ASTOR (II,1788); TARGET (II,2541)

HINDERSTEIN, HOWARD
THE LATE FALL, EARLY SUMMER BERT CONVY SHOW (II,1441)

HINDS, ANTHONY
JOURNEY TO THE UNKNOWN (I,2457)

HINRIO, ADACHI
MARINE BOY (I,2911)

HIRSCH, JAMES G
FOR LOVERS ONLY (II,904); THE INCREDIBLE HULK (II,1232); IRENE (II,1245); KINGSTON: CONFIDENTIAL (II,1398)

HIRSCH, JANIS
DOUBLE TROUBLE (II,723); DOUBLE TROUBLE (II,724)

HIRSCH, MICHAEL
THE EDISON TWINS (II,761)

HIRSCHFIELD, JIM
CAPTAIN KANGAROO (II,434)

HIRSCHMAN, HERBERT
DOCTOR KILDARE (I,1315); THE DOCTORS (I,1323); THE ELEVENTH HOUR (I,1425); ERIC (I,1451); ESPIONAGE (I,1464); HONG KONG (I,2112); HUMAN FEELINGS (II,1203); THE MEN FROM SHILOH (I,3005); MISTRAL'S DAUGHTER (II,1708); PLANET OF THE APES (II,2049); REX HARRISON PRESENTS SHORT STORIES OF LOVE (II,2148); SCALPLOCK (I,3930); THE SCARLET LETTER (II,2270); THE TWILIGHT ZONE (I,4651); THE VIRGINIAN (I,4738); THE WACKIEST SHIP IN THE ARMY (I,4744); THE ZOO GANG (II,2877)

HITCHCOCK, ALFRED
ALFRED HITCHCOCK PRESENTS (I,115); SUSPICION (I,4309)

HITCHCOCK, BILL
THE VAL DOONICAN SHOW (I,4692)

HITTLEMAN, CARL
THE ADVENTURES OF CYCLONE MALONE (I,57)

HITZIG, RUPERT
ALAN KING'S FINAL WARNING (II,29); ALAN KING'S SECOND ANNUAL FINAL WARNING (II,30); ALAN KING'S THIRD ANNUAL FINAL WARNING!! (II,31); HAPPY ENDINGS (II,1085); IVAN THE TERRIBLE (II,1260); JOYCE AND BARBARA: FOR ADULTS ONLY (I,2458); LOVE, LIFE, LIBERTY & LUNCH (II,1544); PLAYBOY AFTER DARK (I,3620); SATURDAY NIGHT LIVE WITH HOWARD COSELL (II,2262); THE WONDERFUL WORLD OF JONATHAN WINTERS (I,4900)

HOAG, BOB
HAROLD LLOYD'S WORLD OF COMEDY (II,1093)

HOBIN, BILL
THE BIG TIME (I,434); THE BOBBY GOLDSBORO SHOW (I,673); THE CBS NEWCOMERS (I,879); THE CHOCOLATE SOLDIER (I,944); CLOWN ALLEY (I,979); A CONNECTICUT YANKEE (I,1038); DEAREST ENEMY (I,1205); THE DESERT SONG (I,1236); FANFARE (I,1526); FEELING GOOD (I,1559); THE FRED WARING SHOW (I,1679); GOOD TIMES (I,1853); THE GREAT WALTZ (I,1878); HEAVEN WILL PROTECT THE WORKING GIRL (I,1996); HEIDI (I,2004); HOLIDAY (I,2074); THE JUDY GARLAND SHOW (I,2473); THE JUDY GARLAND SHOW (II,1349); KALEIDOSCOPE (I,2504); MAKE MINE RED, WHITE, AND BLUE (I,2841); MARCO POLO (I,2904); THE MAURICE CHEVALIER SHOW (I,2954); MAX LIEBMAN PRESENTS (I,2958); MONTY HALL'S VARIETY HOUR (II,1728); THE MUSIC OF GERSHWIN (I,3174); NAUGHTY MARIETTA (I,3232); PANORAMA (I,3448); PARIS IN THE SPRINGTIME (I,3457); THE RED SKELTON SHOW (I,3755); SATINS AND SPURS (I,3917); SING ALONG WITH MITCH (I,4057); SPOTLIGHT (I,4162); STEP ON THE GAS (I,4215); SUNDAY IN TOWN (I,4285); THE TIM CONWAY COMEDY HOUR (I,4501); TIMEX ALL-STAR JAZZ SHOW II (I,4516); TIMEX ALL-STAR JAZZ SHOW IV (I,4518); VARIETY (I,4703); YOUR HIT PARADE (II,2872)

HOBLIT, GREGORY
BAY CITY BLUES (II,186); EVERY STRAY DOG AND KID (II,792); HILL STREET BLUES (II,1154); PARIS (II,1957)

HOBSON, JAMES
THE LAWRENCE WELK SHOW (I,2643); THE LAWRENCE WELK SHOW (I,2644); MEMORIES WITH LAWRENCE WELK (II,1681)

HODGE, AL
CAPTAIN VIDEO AND HIS VIDEO RANGERS (I,838)

HODGES, MARK
THE GIRL FROM U.N.C.L.E. (I,1808)

HOEY, MARK A
THE LAZARUS SYNDROME (II,1451)

HOEY, MICHAEL A
THE FITZPATRICKS (II,868)

HOFFE, ARTHUR
ALWAYS APRIL (I,147); THE ANN SOTHERN SHOW (I,220); PANDORA AND FRIEND (I,3445)

THE BOB HOPE SHOW (I,530); THE BOB HOPE SHOW (I,531); THE BOB HOPE SHOW (I,532); THE BOB HOPE SHOW (I,533); THE BOB HOPE SHOW (I,534); THE BOB HOPE SHOW (I,535); THE BOB HOPE SHOW (I,536); THE BOB HOPE SHOW (I,537); THE BOB HOPE SHOW (I,538); THE BOB HOPE SHOW (I,539); THE BOB HOPE SHOW (I,540); THE BOB HOPE SHOW (I,541); THE BOB HOPE SHOW (I,542); THE BOB HOPE SHOW (I,543); THE BOB HOPE SHOW (I,546); THE BOB HOPE SHOW (I,547); THE BOB HOPE SHOW (I,548); THE BOB HOPE SHOW (I,549); THE BOB HOPE SHOW (I,550); THE BOB HOPE SHOW (I,551); THE BOB HOPE SHOW (I,552); THE BOB HOPE SHOW (I,560); THE BOB HOPE SHOW (I,561); THE BOB HOPE SHOW (I,562); THE BOB HOPE SHOW (I,563); THE BOB HOPE SHOW (I,564); THE BOB HOPE SHOW (I,565); THE BOB HOPE SHOW (I,567); THE BOB HOPE SHOW (I,568); THE BOB HOPE SHOW (I,569); THE BOB HOPE SHOW (I,581); THE BOB HOPE SHOW (I,583); THE BOB HOPE SHOW (I,584); THE BOB HOPE SHOW (I,585); THE BOB HOPE SHOW (I,586); THE BOB HOPE SHOW (I,588); THE BOB HOPE SHOW (I,589); THE BOB HOPE SHOW (I,590); THE BOB HOPE SHOW (I,591); THE BOB HOPE SHOW (I,592); THE BOB HOPE SHOW (I,593); THE BOB HOPE SHOW (I,594); THE BOB HOPE SHOW (I,595); THE BOB HOPE SHOW (I,596); THE BOB HOPE SHOW (I,597); THE BOB HOPE SHOW (I,601); ROBERTA (I,3815); SOUND OFF TIME (I,4135)

HOPE, LINDA
BOB HOPE SPECIAL: BOB HOPE FOR PRESIDENT (II,282); BOB HOPE SPECIAL: BOB HOPE IN THE STAR-MAKERS (II,285); BOB HOPE SPECIAL: BOB HOPE ON CAMPUS (II,290); BOB HOPE SPECIAL: BOB HOPE'S 30TH ANNIVERSARY TV SPECIAL (II,293); BOB HOPE SPECIAL: BOB HOPE'S ALL-STAR BIRTHDAY PARTY (II,295); BOB HOPE SPECIAL: BOB HOPE'S ALL-STAR COMEDY BIRTHDAY PARTY (II,297); BOB HOPE SPECIAL: BOB HOPE'S ALL-STAR LOOK AT TV'S PRIME TIME WARS (II,303); BOB HOPE SPECIAL: BOB HOPE'S FUNNY VALENTINE (II,309); BOB HOPE SPECIAL: BOB HOPE'S OVERSEAS CHRISTMAS TOURS (II,311); BOB HOPE SPECIAL: BOB HOPE—HOPE, WOMEN AND SONG (II,324); BOB HOPE SPECIAL: THE BOB HOPE CHRISTMAS SPECIAL (II,330); BOB HOPE SPECIAL: THE BOB HOPE CHRISTMAS SPECIAL (II,331); JOE AND VALERIE (II,1298); JOE AND VALERIE (II,1299)

HOPLIN, MERT
THE $64,000 QUESTION (I,4071)

HOPPLE, HARRY
IT PAYS TO BE MARRIED (I,2249)

HOPPS, NORMAN
CARTER COUNTRY (II,449); FAMILY BUSINESS (II,815); MY BUDDY (II,1778); TOO CLOSE FOR COMFORT (II,2642)

HORGAN, SUSAN
AS THE WORLD TURNS (II,110)

HORL, ROY
THE CROSS-WITS (II,605); FACE THE MUSIC (II,803); NAME THAT TUNE (II,1786); NAME THAT TUNE (II,1787); THE NEW $100,000 NAME THAT TUNE (II,1810); YOU ASKED FOR IT (II,2855)

HORNER, CHUCK
THE ALL-AMERICAN COLLEGE COMEDY SHOW (II,47)

HOROWITZ, DAVID
FIGHT BACK: WITH DAVID HOROWITZ (II,854)

HOROWITZ, EDWARD
PEOPLE DO THE CRAZIEST THINGS (II,1998)

HORVATH, JOHN
WATCH ME (II,2745)

HORVITZ, LOUIS J
FERNWOOD 2-NIGHT (II,849)

HORWICH, FRANCES
DING DONG SCHOOL (I,1285)

HORWITZ, EDWARD R
PEOPLE TO PEOPLE (II,2000)

HORWITZ, HOWIE
77 SUNSET STRIP (I,3988); BANACEK (II,138); BARETTA (II,152); BATMAN (I,366); THE IMMORTAL (I,2197); MAVERICK (II,1657); REDIGO (I,3758); STRANDED (II,2475)

HOTCHNER, A E
BUICK ELECTRA PLAYHOUSE (I,760)

HOUGH, STAN
PLANET OF THE APES (II,2049)

HOUGHTON, A E
MAN WITH A CAMERA (I,2883); YANCY DERRINGER (I,4925)

HOUGHTON, BUCK
BLUE LIGHT (I,491); DYNASTY (II,747); EXECUTIVE SUITE (II,796); HARRY O (II,1099); THE HOME FRONT (II,1171); MR. BEVIS (I,3128); THE RICHARD BOONE SHOW (I,3784); THE TWILIGHT ZONE (I,4651)

HOUSE, JOHN B
AIR SUPPLY IN HAWAII (II,26)

HOUSEMAN, JOHN
GREAT ADVENTURE (I,1868); PLAYHOUSE 90 (I,3623)

HOVIS, LARRY
ANYTHING FOR MONEY (II,99); THE LIAR'S CLUB (II,1466)

HOWARD, AL
SALE OF THE CENTURY (I,3888); SALE OF THE CENTURY (II,2239)

HOWARD, BOB
GENERAL ELECTRIC'S ALL-STAR ANNIVERSARY (II,963)

HOWARD, CY
FAIR EXCHANGE (I,1512); HARRIGAN AND SON (I,1955); LIFE WITH LUIGI (I,2692); MICKEY AND THE CONTESSA (I,3024); MY FRIEND IRMA (I,3191); THAT'S MY BOY (I,4427)

HOWARD, DICK
USA COUNTRY MUSIC (II,591)

HOWARD, LINDA
THE $1.98 BEAUTY SHOW (II,698); THE GONG SHOW (II,1026); LEAVE IT TO THE WOMEN (II,1454)

HOWARD, RANCE
THROUGH THE MAGIC PYRAMID (II,2615)

HOWARD, RICK
THE GEORGE JONES SPECIAL (II,975)

HOWARD, RON
LITTLE SHOTS (II,1495); SKYWARD CHRISTMAS (II,2378); THROUGH THE MAGIC PYRAMID (II,2615)

HOWARD, SANDY
MACK AND MYER FOR HIRE (I,2802)

HOWARD, TERRY
MAKE A WISH (I,2838)

HOWARD, TOM
IT PAYS TO BE IGNORANT (I,2248)

HOWELL, WILLIAM
AS THE WORLD TURNS (II,110)

HOYEN, DONN
LOOK AT US (II,1515)

HUBER, HAROLD
I COVER TIMES SQUARE (I,2166)

HUBER, LARRY
SATURDAY SUPERCADE (II,2263); TURBO-TEEN (II,2668)

HUBERMAN, ARNOLD H
CANDIDA (II,423)

HUBLEY, JOHN
MR. MAGOO (I,3142)

HUCKER, WALTER
AROUND THE WORLD IN 80 DAYS (I,263)

HUDDLESTON, DAVID
HIZZONER (II,1159)

HUDSON, GARY
HIGHWAY HONEYS (II,1151)

HUDSON, HAL
ACTION (I,23); ALCOA PREMIERE (I,109); BLACK SADDLE (I,480); DOC HOLLIDAY (I,1307); THE LONER (I,2744); THE MAN FROM DENVER (I,2863); THE MAN FROM EVERYWHERE (I,2864); THE SHARPSHOOTER (I,4001); THE WESTERNER (I,4801); WINCHESTER (I,4872); ZANE GREY THEATER (I,4979)

HUDSON, WILLIAM
HARD CASE (I,1947)

HUFANE, MARK
ROCK PALACE (II,2193)

HUGGINS, ROY
77 SUNSET STRIP (I,3988); ALIAS SMITH AND JONES (I,118); ANYTHING FOR MONEY (I,236); BLUE THUNDER (II,278); CAPTAINS AND THE KINGS (II,435); CHEYENNE (I,931); CITY OF ANGELS (II,540); COLT .45 (I,1010); COOL MILLION (I,1046); THE GINGER ROGERS SHOW (I,1805); GIRL ON THE RUN (I,1809); THE GREEN FELT JUNGLE (I,1885); HAZARD'S PEOPLE (II,1112); THE JORDAN CHANCE (II,1343); KING'S ROW (I,2563); THE LAWYERS (I,2646); MAVERICK (II,1657); THE OUTSIDER (I,3430); THE OUTSIDER (I,3431); RAPTURE AT TWO-FORTY (I,3724); RUN FOR YOUR LIFE (I,3871); THE SOUND OF ANGER (I,4133);

TOMA (I,4547); TOMA (II,2634); THE VIRGINIAN (I,4738); WHEELS (II,2777); THE WHOLE WORLD IS WATCHING (I,4851)

HUGHES, GORDON B
THE CHARLIE FARRELL SHOW (I,913)

HUGHES, TERRY
THE CHEVY CHASE NATIONAL HUMOR TEST (II,504); EMPIRE (II,777); REPORT TO MURPHY (II,2134); THE TWO RONNIES (II,2694)

HULL, JOHN
RAMAR OF THE JUNGLE (I,3718)

HULL, SHELLY
FINDER OF LOST LOVES (II,857); THE MONK (I,3087); THE OVER-THE-HILL GANG RIDES AGAIN (I,3432); THE PIGEON (I,3596); STRIKE FORCE (II,2480); TATE (I,4364); THE WILD WOMEN OF CHASTITY GULCH (II,2801); WINDOWS (I,4876)

HUMPHREY, JACK
KING OF KENSINGTON (II,1395)

HUNCHER, GEORGE
SINATRA: CONCERT FOR THE AMERICAS (II,2361)

HUNT, GARY
TOP 40 VIDEOS (II,2644)

HUNT, PETER H
ADAM'S RIB (I,32); BUS STOP (II,399); SHERLOCK HOLMES (II,2327)

HUNTER, BARBARA
THE NEW PRICE IS RIGHT (I,3270); THE PRICE IS RIGHT (I,3665); THE PRICE IS RIGHT (II,2079)

HUNTER, BLAKE
DIFF'RENT STROKES (II,674); WHO'S THE BOSS? (II,2792); WKRP IN CINCINNATI (II,2814)

HUNTER, KEVIN
THE NATALIE COLE SPECIAL (II,1794)

HUNTER, LEW
THE YELLOW ROSE (II,2847)

HUNTER, ROSS
THE LIVES OF JENNY DOLAN (II,1502); THE MONEYCHANGERS (II,1724)

HUNTER, STAN
THE DANCE SHOW (II,616)

HURD, MARC
DUSTY'S TREEHOUSE (I,1391)

HURDLE, JACK
AMERICA'S GREATEST BANDS (I,177); CAVALCADE OF STARS (I,877); THE COLGATE COMEDY HOUR (I,997); THE HONEYMOONERS (I,2110); THE HONEYMOONERS (II,1175); THE JACKIE GLEASON SHOW (I,2322)

HUROK, SOL
FESTIVAL OF MUSIC (I,1567); ROMEO AND JULIET (I,3837); SLEEPING BEAUTY (I,4083)

HUSKY, RICK
CHARLIE'S ANGELS (II,486); MANDRAKE (II,1617); THE RENEGADES (II,2132); THE ROOKIES (II,2208); T.J. HOOKER (II,2524); WHAT REALLY HAPPENED TO THE CLASS OF '65? (II,2768); WHEN THE WHISTLE BLOWS (II,2781)

HUSON, PAUL
THE HAMPTONS (II,1076)

HUTSON, LEE
THE LONG DAYS OF SUMMER (II,1514)

HUTTON, DOUG
STAR CHART (II,2434)

HYDE, DONALD
THE FLYING DOCTOR (I,1610); GLENCANNON (I,1822); THE O. HENRY PLAYHOUSE (I,3322)

HYER, BILLY
HOLLYWOOD JUNIOR CIRCUS (I,2084)

HYMAN, STEPHANIE
GIGGLESNORT HOTEL (II,989)

HYSLOP, AL
CAPTAIN KANGAROO (II,434); MR. MAYOR (I,3143)

IACOFANO, TIM
TMT (II,2622)

IDELSON, BILL
ANNA AND THE KING (I,221); BEANE'S OF BOSTON (II,194); THE MONTEFUSCOS (II,1727)

IENO, MUKARTSUBO
WOLF ROCK TV (II,2815)

IKEUCHI, TAT
MIGHTY ORBOTS (II,1697)

ILLES, ROBERT
DOUBLE TROUBLE (II,723); FLO (II,891); SILVER SPOONS (II,2355)

ILOS, GEORGE
AMERICA'S TOP TEN (II,68)

ILSON, SAUL
20TH CENTURY FOLLIES (I,4641); THE BEATRICE ARTHUR SPECIAL (II,196); A BEDTIME STORY (I,384); THE BILLY CRYSTAL COMEDY HOUR (II,248); THE BOBBY DARIN AMUSEMENT COMPANY (I,672); CAROL CHANNING AND PEARL BAILEY ON BROADWAY (I,850); CAROL CHANNING PROUDLY PRESENTS THE SEVEN DEADLY SINS (I,851); CHARO (II,488); DICK VAN DYKE MEETS BILL COSBY (I,1274); THE DORIS MARY ANNE KAPPELHOFF SPECIAL (I,1356); FOR MEMBERS ONLY (II,905); FRANCIS ALBERT SINATRA DOES HIS THING (I,1658); FRIENDS AND NABORS (I,1684); HOW TO HANDLE A WOMAN (I,2146); THE JOHN GARY SHOW (I,2411); THE LESLIE UGGAMS SHOW (I,2660); THE LIVING END (I,2731); LOVE NEST (II,1541); LYNDA CARTER'S SPECIAL (II,1564); NEIL SEDAKA STEPPIN' OUT (II,1804); RICH LITTLE'S WASHINGTON FOLLIES (II,2159); THE SMOTHERS BROTHERS COMEDY HOUR (I,4095); THERE GOES THE NEIGHBORHOOD (II,2582); TONY ORLANDO AND DAWN (II,2639); TV FUNNIES (II,2672); VER-R-R-RY INTERESTING (I,4712); WHAT'S IT ALL ABOUT WORLD? (I,4816); ZERO INTELLIGENCE (II,2876)

IMPARTO, TOM
HOT (II,1184)

INADA, NOBUO
MIGHTY ORBOTS (II,1697)

INCH, KEVIN
REMINGTON STEELE (II,2130)

INGALLS, DON
FANTASY ISLAND (II,829); HARD KNOCKS (II,1086); HAVE GUN—WILL TRAVEL (I,1968); KINGSTON: CONFIDENTIAL (II,1398); SERPICO (II,2306); THE TRAVELS OF JAIMIE MCPHEETERS (I,4596)

INGSTER, BARRY
THE ALASKANS (I,106); THE ROARING TWENTIES (I,3805)

INGSTER, BORIS
CIMARRON CITY (I,953); THE MAN FROM U.N.C.L.E. (I,2867)

INTERNATIONAL TELEVISION PRODUCTIONS
POPI (II,2068)

IRVING, CHARLES
THE HAPPY TIME (I,1943); KITTY FOYLE (I,2571); SEARCH FOR TOMORROW (II,2284); THE SECRET JURY (I,3965); THAT WONDERFUL GUY (I,4425)

IRVING, JULES
THE ART OF CRIME (II,108); THE DETECTIVE (II,662); LOOSE CHANGE (II,1518); WHAT REALLY HAPPENED TO THE CLASS OF '65? (II,2768)

IRVING, RICHARD
THE ART OF CRIME (II,108); CONFIDENTIALLY YOURS (I,1035); CORONADO 9 (I,1053); COURT-MARTIAL (I,1080); CUTTER (I,1117); EXO-MAN (II,797); FRONTIER CIRCUS (I,1697); THE INVESTIGATORS (I,2231); JOHNNY BLUE (II,1321); THE LAST DAYS OF POMPEII (II,1436); MASADA (II,1642); MIKE HAMMER, DETECTIVE (I,3042); THE NAME OF THE GAME (I,3217); PRESCRIPTION: MURDER (I,3660); QUINCY, M. E. (II,2102); RANSOM FOR A DEAD MAN (I,3722); SAN FRANCISCO INTERNATIONAL AIRPORT (I,3905); SEVENTH AVENUE (II,2309); THE SIX-MILLION-DOLLAR MAN (II,2372); STATE TROOPER (I,4214); THE SWORD (I,4325); THE VIRGINIAN (I,4738); WHAT REALLY HAPPENED TO THE CLASS OF '65? (II,2768)

IRVINGS, RAY
FOUL PLAY (II,910)

IRWIN, CAROL
CLAUDIA: THE STORY OF A MARRIAGE (I,972); THE GIRLS (I,1811); I REMEMBER MAMA (I,2176); STAGE DOOR (I,4179); YOUNG DR. MALONE (I,4938)

ISAACS, CHARLES
AN EVENING WITH JIMMY DURANTE (I,1475); FULL SPEED ANYWHERE (I,1704); THE GISELE MACKENZIE SHOW (I,1812); GIVE MY REGARDS TO BROADWAY (I,1815); HEY, JEANNIE! (I,2038); THE REAL MCCOYS (I,3741); THE TYCOON (I,4660)

ISAACS, DAVID
AFTERMASH (II,23); CHARACTERS (II,481); CHEERS (II,494)

ISCOVE, ROB
DOROTHY HAMILL IN ROMEO & JULIET ON ICE (II,718)

ISENBERG, GERALD I
THE BUREAU (II,397); FAME (II,812); HAVING BABIES I (II,1107); HOLLYWOOD HIGH (II,1162); HOLLYWOOD HIGH (II,1163); JUDGE DEE IN THE MONASTERY MURDERS (I,2465); JULIE FARR, M.D. (II,1354); THE SUPER (I,4293)

ISHGURO, KOTCHI
POLE POSITION (II,2060)

ISRAEL, NEAL
MARIE (II,1629); TWILIGHT THEATER (II,2685); TWILIGHT THEATER II (II,2686)

JACK, DEL
THE KING FAMILY CHRISTMAS SPECIAL (I,2546); THE KING FAMILY IN ALASKA (I,2547); THE KING FAMILY IN HAWAII (I,2548); THE KING FAMILY OCTOBER SPECIAL (I,2551); THE KING FAMILY SHOW (I,2552); THE KING FAMILY SPECIAL (I,2553); THE KING FAMILY VALENTINE'S DAY SPECIAL (I,2554); MR. WIZARD (I,3154)

JACKS, ROBERT L
EIGHT IS ENOUGH (II,762); THE MAN WHO NEVER WAS (I,2882); MORE WILD WILD WEST (II,1733); STATE FAIR (II,2448); THREE COINS IN THE FOUNTAIN (I,4473); THE WALTONS (II,2740); THE WILD WILD WEST REVISITED (II,2800); THE YOUNG PIONEERS (II,2869)

JACKSON, BILL
GIGGLESNORT HOTEL (II,989)

JACKSON, CORNWELL
THE NEW ADVENTURES OF PERRY MASON (I,3252)

JACKSON, DAVID E
CAMP WILDERNESS (II,419)

JACKSON, FELIX
BROADWAY (I,733); CIMARRON CITY (I,953); DON'S MUSICAL PLAYHOUSE (I,1348); THE GUARDSMAN (I,1893); THE RESTLESS GUN (I,3774); SCHLITZ PLAYHOUSE OF STARS (I,3936); STAGE DOOR (I,4178); STUDIO ONE (I,4268)

JACKSON, GAIL PATRICK
PERRY MASON (II,2025)

JACKSON, JOE
THE JACKSONS (II,1267)

JACKSON, KATE
TOPPER (II,2649)

JACKSON, RILEY
FRONT PAGE DETECTIVE (I,1689)

JACOBS, ARTHUR
BARNEY AND ME (I,358); THE LONE RANGER (I,2741); TOPPER RETURNS (I,4583)

JACOBS, DAVID
BEHIND THE SCREEN (II,200); KNOTS LANDING (II,1404); LACE (II,1418); MARRIED: THE FIRST YEAR (II,1634); SECRETS OF MIDLAND HEIGHTS (II,2296)

JACOBS, JOHN
ORSON WELLES' GREAT MYSTERIES (I,3408)

JACOBS, MICHAEL
CHARLES IN CHARGE (II,482)

JACOBS, PAUL
IT'S ALWAYS JAN (I,2264)

JACOBS, RONALD
CITY VS. COUNTRY (I,968); THE DICK VAN DYKE SHOW (I,1275); FEATHERSTONE'S NEST (II,847); GOMER PYLE, U.S.M.C. (I,1843); MAKE ROOM FOR DADDY (I,2843); MAKE ROOM FOR GRANDDADDY (I,2844); SAMURAI (II,2249); STARTING FRESH (II,2447)

JACOBSON, JAKE
THE LOVE REPORT (II,1542)

JACOBSON, LARRY
BAKER'S DOZEN (II,136); JOHNNY GARAGE (II,1337)

JACOBSON, LAWRENCE
COMEBACK (I,1016)

JACOBSON, NORMAN
LLOYD BRIDGES WATER WORLD (I,2734)

JACOBSON, STAN
THE B.B. BEEGLE SHOW (II,124); THE HUDSON BROTHERS RAZZLE DAZZLE COMEDY SHOW (II,1201); THE HUDSON BROTHERS SHOW (II,1202); JOEY AND DAD (II,1306); THE JOHNNY CASH SHOW (I,2424); VIVA VALDEZ (II,2728)

JACOBY, COLEMAN
THE HALLOWEEN THAT ALMOST WASN'T (II,1075)

JACOBY, FRANK
A GUEST IN YOUR HOUSE (I,1898)

JAEGER, KOBI
SALTY (II,2241)

JAFFE, BARRY
TOP 40 VIDEOS (II,2644)

JAFFE, CHARLES
THE TED BESSELL SHOW (I,4369)

JAFFE, HENRY
THE BELL TELEPHONE HOUR (I,389); CHERYL LADD. . . LOOKING BACK—SOUVENIRS (II,502); CHERYL LADD: SCENES FROM A SPECIAL (II,503); COTTON CLUB '75 (II,580); DINAH AND FRIENDS (II,675); DINAH AND HER NEW BEST FRIENDS (II,676); DINAH IN SEARCH OF THE IDEAL MAN (I,1278); DINAH! (II,677); DINAH'S PLACE (II,678); EMILY, EMILY (I,1438); A SALUTE TO STAN LAUREL (I,3892); SHIRLEY TEMPLE'S STORYBOOK (I,4017); THE WAYNE NEWTON SPECIAL (II,2749)

JAFFE, MICHAEL
EMILY, EMILY (I,1438)

JAFFE, SAM
JUBILEE (II,1347)

JAFFE, SAUL
OUR TOWN (I,3421)

JAFFEE, STANLEY
MARINE BOY (I,2911)

JAMISON, MARSHALL
THE FABULOUS 50S (I,1501); THE GERSHWIN YEARS (I,1779); THE GOLDEN AGE OF TELEVISION (II,1019); THAT WAS THE WEEK THAT WAS (I,4423)

JAMPEL, CARL
MOTHERS DAY (I,3109); SEVEN KEYS (I,3985)

JANAVER, DIANE H
SUPER PASSWORD (II,2499)

JANES, ROBERT
THE AMAZING SPIDER-MAN (II,63); CHARLIE'S ANGELS (II,486); THE FALL GUY (II,811); LOTTERY (II,1525); TONI'S BOYS (II,2637); VOYAGERS (II,2730); WAIKIKI (II,2734)

JANIS, HAL
SID CAESAR INVITES YOU (I,4040)

JANIS, PAULA
A MAGIC GARDEN CHRISTMAS (II,1590)

JARVIS, BOB
WAYNE AND SHUSTER TAKE AN AFFECTIONATE LOOK AT. . . (I,4778)

JARVIS, IRV
THE MAGIC GARDEN (II,1591)

JARVIS, LUCY
THE BARBARA WALTERS SPECIAL (II,145)

JENKINS, GORDON
MANHATTAN TOWER (I,2889)

JENKS, GARY
THE PARENT GAME (I,3455)

JENNINGS, ROBERT
CHANCE OF A LIFETIME (I,898)

JENSEN-DRAKE, BRIGID
GIMME A BREAK (II,995)

JENSON, ERIC
THE ADVENTURES OF CHAMPION (I,54); BUFFALO BILL JR. (I,755) THE BUFFALO BILLY SHOW (I,756)

JESSELL, RAY
BERT CONVY SPECIAL—THERE'S A MEETING HERE TONIGHT (II,210); THE JACKSONS (II,1267)

JEWISON, NORMAN
THE BROADWAY OF LERNER AND LOEWE (I,735); THE JUDY GARLAND SHOW (I,2472); THE JUDY GARLAND SHOW (I,2473)

JOACHIMS, JOHN
THE B.B. BEEGLE SHOW (II,124)

JOELSON, BEN
THE COP AND THE KID (II,572); GLITTER (II,1009); IT'S YOUR MOVE (I,2281); THE LOVE BOAT (II,1535); MADHOUSE 90 (I,2807); PICTURE THIS (I,3593)

JOFFE, CHARLES H
CHEERS (II,494); GOOD TIME HARRY (II,1037); STAR OF THE FAMILY (II,2436); THE WOODY ALLEN SPECIAL (I,4904)

JOHN, BOBBIE
SEVEN KEYS (I,3985)

JOHNSON JR, DENIS
PRISONERS OF THE LOST UNIVERSE (II,2086)

JOHNSON, BIFF
GREATEST HEROES OF THE BIBLE (II,1062); MICKEY SPILLANE'S MARGIN FOR MURDER (II,1691)

JOHNSON, BOB A
THE ABC AFTERSCHOOL SPECIAL (II,1)

JOHNSON, BRUCE
ALICE (II,33); ALICE (II,34); ANGIE (II,80); ARNIE (I,261); BLANSKY'S BEAUTIES (II,267); GOMER PYLE, U.S.M.C. (I,1843); GOOD MORNING WORLD (I,1850); HOT W.A.C.S. (II,1191); THE JIM NABORS HOUR (I,2381); THE LITTLE PEOPLE (I,2716); LITTLE SHOTS (II,1495); MORK AND MINDY (II,1735); THE NEW ODD COUPLE (II,1825); THE NEW TEMPERATURES RISING SHOW (I,3273); QUARK (II,2095); SCALPELS (II,2266); SIERRA (II,2351); WEBSTER (II,2758)

JOHNSON, CHAS FLOYD
BRET MAVERICK (II,374); HELLINGER'S LAW (II,1127); MAGNUM, P.I. (II,1601); THE RETURN OF LUTHER GILLIS (II,2137); THE ROCKFORD FILES (II,2197); SIMON AND SIMON (II,2357)

JOHNSON, COSLOUGH
THE HUDSON BROTHERS RAZZLE DAZZLE COMEDY SHOW (II,1201); THE HUDSON BROTHERS SHOW (II,1202); JOEY AND DAD (II,1306); WACKO (II,2733)

JOHNSON, DENNIS
MARIE (II,1628)

JOHNSON, GARY
SCRABBLE (II,2278)

JOHNSON, JERRY
YOU ASKED FOR IT (II,2855)

JOHNSON, KENNETH
ALAN KING IN LAS VEGAS, PART I (I,97); ALAN KING IN LAS VEGAS, PART II (I,98); THE BIONIC WOMAN (II,255); THE CURSE OF DRACULA (II,611); HOLLYWOOD'S TALKING (I,2097); HOT PURSUIT (II,1189); THE INCREDIBLE HULK (II,1232); JUVENILE JURY (I,2501); THE REEL GAME (I,3759); THE SECRET EMPIRE (II,2291); THE SIX-MILLION-DOLLAR MAN (II,2372); STOP SUSAN WILLIAMS (II,2473); V (II,2713)

JOHNSON, LAURENCE
NOT FOR WOMEN ONLY (I,3310)

JOHNSON, MARK
DINER (II,679)

JOHNSON, MICKI
THE BEST LITTLE SPECIAL IN TEXAS (II,215)

JOHNSON, MONICA
LAVERNE AND SHIRLEY (II,1446); THE PLANT FAMILY (II,2050)

JOHNSTON, THOMAS
BEAUTY AND THE BEAST (I,382)

JOLLEY, NORMAN
CIMARRON CITY (I,953); FREEBIE AND THE BEAN (II,922); IRONSIDE (II,1246); RIVERBOAT (I,3797)

JONES, A ERIC
THE WORLD OF PEOPLE (II,2839)

JONES, CHARLOTTE SCHIFF
PEOPLE (II,1995)

JONES, CHUCK
CURIOSITY SHOP (I,1114); OFF TO SEE THE WIZARD (I,3337); THE TOM AND JERRY SHOW (I,4534)

JONES, CLARK
ANNIE GET YOUR GUN (I,224); THE PATRICE MUNSEL SHOW (I,3474); PEGGY FLEMING AT MADISON SQUARE GARDEN (I,3506); RAINBOW OF STARS (I,3717); TONY BENNETT IN WAIKIKI (I,4568)

JONES, CLAYLENE
BOONE (II,351); THE WALTONS: A DAY FOR THANKS ON WALTON'S MOUNTAIN (EPISODE 3) (II,2741); THE WALTONS: MOTHER'S DAY ON WALTON'S MOUNTAIN (EPISODE 2) (II,2742); THE WALTONS: WEDDING ON WALTON'S MOUNTAIN (EPISODE 1) (II,2743)

JONES, EUGENE S
THE WORLD OF MAURICE CHEVALIER (I,4916)

JONES, GARY
CELEBRITY CONCERTS (II,471); CELEBRITY REVUE (II,475)

JONES, GEOFFREY
FAITH BALDWIN'S THEATER OF ROMANCE (I,1515)

JONES, HILDY
PARADE OF STARS (II,1954)

JONES, IAN
AGAINST THE WIND (II,24)

JONES, MARK
RUBIK, THE AMAZING CUBE (II,2226)

JONES, MARTHA
PEOPLE TO PEOPLE (II,2000)

JONES, MARTIN
SHOW BUSINESS, INC. (I,4028)

JONES, NANCY
THE JOE PISCOPO SPECIAL (II,1304); WHEEL OF FORTUNE (II,2775)

JONES, PATRICIA
REPORT TO MURPHY (II,2134)

JONES, PHILIP
TOP OF THE WORLD (II,2646)

JONES, QUINCY
DUKE ELLINGTON. . . WE LOVE YOU MADLY (I,1380)

JORDAN JR, JIM
FRANKIE CARLE TIME (I,1674)

JORDAN, GLENN
FRIENDS (II,927)

JORDAN, TERRY
THE BEST OF SULLIVAN (II,217)

JORDAN, TONY
THE CRYSTAL GAYLE SPECIAL (II,609); THE LILY TOMLIN SHOW (I,2706)

JOSEFSBERG, MILT
ALL IN THE FAMILY (II,38); ARCHIE BUNKER'S PLACE (II,105); BUTTERFLIES (II,405); HOT W.A.C.S. (II,1191); THE JOEY BISHOP SHOW (I,2403); JOHNNY COME LATELY (I,2425); LAVERNE AND SHIRLEY (II,1446)

JOSLYN, JOHN
ROUGHCUTS (II,2218)

JOYCE, BERNADETTE
CONDOMINIUM (II,566)

JULIAN, ARTHUR
GIMME A BREAK (II,995); LOVE THY NEIGHBOR (I,2781); REAR GUARD (II,2120); THE SON-IN-LAW (II,2398); THE TWO OF US (I,4658)

JURIST, ED
DOTTO (I,1357); THE FLYING NUN (I,1611); HAWAIIAN EYE (I,1973); HELLO DERE (I,2010); ROOM FOR ONE MORE (I,3842); WHAT'S IT FOR? (I,4817); YOU'RE ONLY YOUNG TWICE (I,4970)

JURSIK, PETER
BACKSTAGE PASS (II,132)

JURWICH, DON
FONZ AND THE HAPPY DAYS GANG (II,898); G.I. JOE: A REAL AMERICAN HERO (II,948); THE RICHIE RICH SHOW (II,2169); SCOOBY'S ALL-STAR LAFF-A-LYMPICS (II,2273); SCOOBY-DOO AND SCRAPPY-DOO (II,2274); THE SCOOBY-DOO AND SCRAPPY-DOO SHOW (II,2275); SUPER FRIENDS (II,2497)

JUSTMAN, ROBERT
ASSIGNMENT: MUNICH (I,302); ASSIGNMENT: VIENNA (I,304); THE NEW PEOPLE (I,3268); THE MAN FROM ATLANTIS (II,1614); MCCLAIN'S LAW (II,1659); ON TRIAL (II,1896); PLANET EARTH (I,3615); SEARCH (I,3951); STAR TREK (II,2440); THEN CAME BRONSON (I,4435); THEN CAME BRONSON (I,4436)

KADISH, BEN
BATTLES: THE MURDER THAT WOULDN'T DIE (II,180); FITZ AND BONES (II,867); THE HARDY BOYS MYSTERIES (II,1090); SWEEPSTAKES (II,2514); TERROR AT ALCATRAZ (II,2563)

KADISON, ELLIS
CHALK ONE UP FOR JOHNNY (I,893)

KAGAN, MICHAEL
STEPHANIE (II,2453); THE STEVE LANDESBERG TELEVISION SHOW (II,2458)

KAHAN, EDWARD C
MARTIN KANE, PRIVATE EYE (I,2928)

KAHAN, JUDITH
LOVE, NATALIE (II,1545)

KAHN, BERNIE
JOE AND VALERIE (II,1298)

KAHN, JOAN
RIPLEY'S BELIEVE IT OR NOT (II,2177)

KALISH, AUSTIN
AMERICA 2100 (II,66); CARTER COUNTRY (II,449); GHOST OF A CHANCE (II,987); GOOD HEAVENS (II,1032); GOOD TIMES (II,1038); KANGAROOS IN THE KITCHEN (II,1362); OUT OF THE BLUE (II,1934); RENDEZVOUS HOTEL (II,2131); TOO CLOSE FOR COMFORT (II,2642); WILDER AND WILDER (II,2802)

KALISH, BRUCE
LANDON, LANDON & LANDON (II,1426)

KALISH, CHUCK
DEAR TEACHER (II,653); UNITED STATES (II,2708); THE WAYNE NEWTON SPECIAL (II,2750); THE WORLD OF ENTERTAINMENT (II,2837)

KALISH, IRMA
AMERICA 2100 (II,66); CARTER COUNTRY (II,449); FOOT IN THE DOOR (II,902); GHOST OF A CHANCE (II,987); GOOD HEAVENS (II,1032); GOOD TIMES (II,1038); KANGAROOS IN THE KITCHEN (II,1362); OH MADELINE (II,1880); OUT OF THE BLUE (II,1934); RENDEZVOUS HOTEL (II,2131); TOO CLOSE FOR COMFORT (II,2642); WILDER AND WILDER (II,2802)

KALLIS, STANLEY
THE ADVENTURES OF NICK CARTER (I,68); THE DANNY THOMAS HOUR (I,1147); FARADAY AND COMPANY (II,833); HAWAII FIVE-O (II,1110); JIGSAW (I,2377); KISS ME, KILL ME (II,1400); MAN ON THE MOVE (I,2878); THE MANIONS OF AMERICA (II,1623); MISSION: IMPOSSIBLE (I,3067); THE MISSISSIPPI (II,1707); POLICE STORY (II,2062); POLICE WOMAN: THE GAMBLE (II,2064); SHERLOCK HOLMES: THE HOUND OF THE BASKERVILLES (I,4011); TWO MARRIAGES (II,2691); WASHINGTON: BEHIND CLOSED DOORS (II,2744)

KALMER, KEITH
DICK TRACY (I,1271)

KAMINSKY, ANN
WHAT'S MY LINE? (I,4820)

KAMINSKY, BOB
BETTE MIDLER—ART OR BUST (II,221)

KAMMERMAN, ROY
THE BOOK OF LISTS (II,350); FISH (II,864); SO YOU THINK YOU GOT TROUBLES?! (II,2391); WEDDING PARTY (I,4786)

KAMPMANN, STEVEN
WKRP IN CINCINNATI (II,2814)

KANE, ARNOLD
HIS AND HERS (II,1155); HONEYMOON SUITE (I,2108); KEEPING UP WITH THE JONESES (II,1374); MAKE YOUR OWN KIND OF MUSIC (I,2847); ONE IN A MILLION (II,1906); PRIVATE BENJAMIN (II,2087)

KANE, BOB
COOL MCCOOL (I,1045)

KANE, BRUCE
TWO THE HARD WAY (II,2695); THE WAVERLY WONDERS (II,2746)

KANE, JOEL
DOBIE GILLIS (I,1302); THE JOHN FORSYTHE SHOW (I,2410)

KANEKO, MITSURU
POLE POSITION (II,2060)

KANTER, HAL
ALL IN THE FAMILY (II,38); BOB HOPE SPECIAL: BOB HOPE IN "JOYS" (II,284); CAP'N AHAB (I,824); CHICO AND THE MAN (II,508); DOWN HOME (I,1366); THE JIMMY STEWART SHOW (I,2391); JULIA (I,2476); THE MUSIC MART (II,1774); THE REASON NOBODY HARDLY EVER SEEN A FAT OUTLAW IN THE OLD WEST IS AS FOLLOWS: (I,3744); SALLY AND SAM (I,3891); THREE COINS IN THE FOUNTAIN (I,4473); THREE ON AN ISLAND (I,4483); VALENTINE'S DAY (I,4693); WHAT'S UP DOC? (II,2771)

KANTOR, RON
WE'RE DANCIN' (II,2755)

KAPLAN, BARRY
THE GEORGE JONES SPECIAL (II,975)

KAPLAN, BORIS
87TH PRECINCT (I,1422)

KAPLAN, CONSTANCE
SQUARE PEGS (II,2431)

KAPLAN, DICK
FLEETWOOD MAC IN CONCERT (II,880)

KAPLAN, E JACK
KUDZU (II,1415)

KAPLAN, GABRIEL
GABRIEL KAPLAN PRESENTS THE SMALL EVENT (II,952)

KAPLAN, VIC
BILLY CRYSTAL: A COMIC'S LINE (II,249); FRIDAYS (II,926); THE HELEN REDDY SPECIAL (II,1126); KENNY LOGGINS IN CONCERT (II,1380)

KAPPES, DAVID
THE ABC AFTERSCHOOL SPECIAL (II,1); TALES OF THE UNEXPECTED (II,2540)

KAPRALL, BO
THE BAY CITY AMUSEMENT COMPANY (II,185); FRIENDS (II,927)

KARLEN, BERNARD E
I'D LIKE TO SEE (I,2184)

KARLSON, PHIL
ALEXANDER THE GREAT (I,113)

KARMAN, JANICE
ALVIN AND THE CHIPMUNKS (II,61)

KARN, STELLA
THE MARY MARGARET MCBRIDE SHOW (I,2933)

KARP, DAVID
THE DANGEROUS DAYS OF KIOWA JONES (I,1140); HAWKINS (I,1976); HAWKINS ON MURDER (I,1978)

KARPF, MERRILL
THE BEST CHRISTMAS PAGEANT EVER (II,212)

KARRAS, ALEX
THE ABC AFTERSCHOOL SPECIAL (II,1)

KARWITZ, SHERMAN
BARYSHNIKOV IN HOLLYWOOD (II,157); BARYSHNIKOV ON BROADWAY (II,158)

KASEM, CASEY
AMERICA'S TOP TEN (II,68); PORTRAIT OF A LEGEND (II,2070)

KASHA, LAWRENCE
ANOTHER APRIL (II,93); APPLAUSE (I,244); BUSTING LOOSE (II,402); KOMEDY TONITE (II,1407); ROSENTHAL AND JONES (II,2215); TWO GUYS FROM MUCK (II,2690); WILLOW B: WOMEN IN PRISON (II,2804)

KASOFF, SY
PAT BOONE IN HOLLYWOOD (I,3469)

KASS, ELLEN
SWEENEY TODD (II,2512)

KASSEL, VIRGINIA
THE ADAMS CHRONICLES (II,8)

KATLEMAN, HARRIS
ALEX AND THE DOBERMAN GANG (II,32); THE AMERICAN GIRLS (II,72); THE DON RICKLES SHOW (I,1341); FROM HERE TO ETERNITY (II,932); FROM HERE TO ETERNITY (II,933); GO WEST YOUNG GIRL (II,1013); THE LEGEND OF THE GOLDEN GUN (II,1455); NICK AND THE DOBERMANS (II,1843); SALVAGE (II,2244); SALVAGE 1 (II,2245)

KATZ, ALLAN
ADAMS HOUSE (II,9); CHER (II,495); GOODBYE DOESN'T MEAN FOREVER (II,1039); M*A*S*H (II,1569); THE NATIONAL SNOOP (II,1797); RHODA (II,2151); WE'LL GET BY (II,2753)

KATZ, ARTHUR JOEL
FOR THE PEOPLE (I,1626); NORMAN CORWIN PRESENTS (I,3307); THE NURSES (I,3318)

KATZ, HOWARD
CELEBRITY CHALLENGE OF THE SEXES (II,464); CELEBRITY CHALLENGE OF THE SEXES 1 (II,465); CELEBRITY CHALLENGE OF THE SEXES 2 (II,466); CELEBRITY CHALLENGE OF THE SEXES 3 (II,467); CELEBRITY CHALLENGE OF THE SEXES 4 (II,468); CELEBRITY CHALLENGE OF THE SEXES 5 (II,469); GAMES PEOPLE PLAY (II,956); THE SUNDAY GAMES (II,2494)

KATZ, LENNIE
CHRISTMAS SPECIAL. . . WITH LOVE, MAC DAVIS (II,523)

KATZ, MARTY
CATALINA C-LAB (II,458)

KATZ, PETER
KAZ (II,1370); A MAN CALLED INTREPID (II,1612); NAKIA (II,1784); PEE WEE KING'S FLYING RANCH (I,3505); THE TURNING POINT OF JIM MALLOY (II,2670)

KATZ, RAYMOND
CHER. . . SPECIAL (II,499); DONNY AND MARIE (II,712); THE DONNY AND MARIE OSMOND SHOW (II,714); KOMEDY TONITE (II,1407); MAC DAVIS 10TH ANNIVERSARY SPECIAL: I STILL BELIEVE IN MUSIC (II,1571); THE MAC DAVIS CHRISTMAS SPECIAL (II,1573); MAC DAVIS. . . I BELIEVE IN CHRISTMAS (II,1581); MAC DAVIS. . . SOUNDS LIKE HOME (II,1582); MAC DAVIS—I'LL BE HOME FOR CHRISTMAS (II,1580); THE OSMONDS SPECIAL (II,1929); PAUL LYNDE AT THE MOVIES (II,1976); THE PAUL LYNDE COMEDY HOUR (II,1978); THE PAUL LYNDE COMEDY HOUR (II,1979); PAUL LYNDE GOES M-A-A-A-AD (II,1980); THE PAUL LYNDE HALLOWEEN

SPECIAL (II,1981)

KATZ, SANDY
A SPECIAL OLIVIA NEWTON-JOHN (II,2418)

KATZIN, LEE H
MISSION: IMPOSSIBLE (I,3067)

KATZKA, GABRIEL
ELLIS ISLAND (II,768)

KATZMAN, LEONARD
DALLAS (II,614); DIRTY SALLY (II,680); THE FANTASTIC JOURNEY (II,826); GUNSMOKE (II,1069); HAWAII FIVE-O (II,1110); LOGAN'S RUN (II,1507); PETROCELLI (II,2031); THE WILD WILD WEST (I,4863)

KAUFF, PETER
BETTE MIDLER—ART OR BUST (II,221); RINGO (II,2176)

KAUFMAN, DWANA
THE PEE WEE HERMAN SHOW (II,1988)

KAUFMAN, JOSEPH
THE ADVENTURES OF LONG JOHN SILVER (I,67)

KAUFMAN, KENNETH
THE BEST LEGS IN 8TH GRADE (II,214)

KAUFMAN, LEONARD B
THE AFRICAN QUEEN (II,18); ARCHER (II,102); BEYOND WESTWORLD (II,227); DAKTARI (I,1127); HAWAII FIVE-O (II,1110); KEEPER OF THE WILD (II,1373); MR. AND MRS. COP (II,1748); O'HARA, UNITED STATES TREASURY (I,3346); O'HARA, UNITED STATES TREASURY: OPERATION COBRA (I,3347); PRIVATE BENJAMIN (II,2087); SAM (II,2247); SCRUPLES (II,2280); SCRUPLES (II,2281); TIME EXPRESS (II,2620)

KAUFMAN, ROBERT
HERE WE GO AGAIN (I,2028); THE UGLIEST GIRL IN TOWN (I,4663)

KAY, JOE
SINATRA: THE MAN AND HIS MUSIC (II,2362)

KAY, MONTE
THE FLIP WILSON COMEDY SPECIAL (II,886); THE FLIP WILSON SPECIAL (I,1603); THE FLIP WILSON SPECIAL (II,887); THE FLIP WILSON SPECIAL (II,888); THE FLIP WILSON SPECIAL (II,889); FLIP WILSON. . . OF COURSE (II,890); THE HELEN REDDY SHOW (I,2008)

KAY, ROGER
THE ALDRICH FAMILY (I,111); BATTLE OF THE AGES (I,367)

KAYDEN, WILLIAM
BICENTENNIAL MINUTES (II,229); COMPUTERCIDE (II,561); THE COPS AND ROBIN (II,573); CRAZY TIMES (II,602); ON THE GO (I,3361); OPERATION ENTERTAINMENT (I,3400)

KAYE, SYLVIA FINE
MUSICAL COMEDY TONIGHT (II,1777)

KAYNE, ARNOLD
TURNABOUT (II,2669)

KEANE, EDWARD
THE BOB SMITH SHOW (I,668)

KEANE, REBECCA HUNT
THE DANCE SHOW (II,616)

KEARNEY, GENE
SWITCH (II,2519)

KEATS, ROBERT
SCALPELS (II,2266)

KEBBE, CHARLES
THE LILLI PALMER SHOW (I,2703); THE LILLI PALMER THEATER (I,2704)

KEEFE, DAN
TMT (II,2622)

KEEFE, PETER
VOLTRON—DEFENDER OF THE UNIVERSE (II,2729)

KEEGAN, TERRY
THE GIRL, THE GOLD WATCH AND DYNAMITE (II,1001); THE GIRL, THE GOLD WATCH AND EVERYTHING (II,1002); KEY TORTUGA (II,1383); STRUCK BY LIGHTNING (II,2482); TOP OF THE HILL (II,2645)

KEESHAN, BOB
CAPTAIN KANGAROO (II,434); GOOD EVENING, CAPTAIN (II,1031); WAKE UP (II,2736)

KEETON, KATHY
OMNI: THE NEW FRONTIER (II,1889)

KEIDEL, DALE
THE MADHOUSE BRIGADE (II,1588)

KEITH, HAL
MR. PEEPERS (I,3147); VILLAGE BARN (I,4733)

KELLARD, RICK
THE BEST OF TIMES (II,220); FIRST TIME, SECOND TIME (II,863); JUST OUR LUCK (II,1360); NICHOLS AND DYMES (II,1841); ONE NIGHT BAND (II,1909)

KELLER, DAVID
ON STAGE AMERICA (II,1893)

KELLER, EYTHAN
ANYTHING FOR MONEY (II,99); FOUL UPS, BLEEPS AND BLUNDERS (II,911); THAT'S HOLLYWOOD (II,2577)

KELLER, MAX
STRANGER IN OUR HOUSE (II,2477)

KELLER, MICHAEL
STRANGER IN OUR HOUSE (II,2477)

KELLER, SHELDON
BOB HOPE SPECIAL: BOB HOPE IN THE STAR-MAKERS (II,285); BOB HOPE SPECIAL: BOB HOPE'S ALL-STAR COMEDY SPECTACULAR FROM LAKE TAHOE (II,301); BOB HOPE SPECIAL: BOB HOPE'S ALL-STAR COMEDY TRIBUTE TO VAUDEVILLE (II,302); BOB HOPE SPECIAL: BOB HOPE'S ALL-STAR TRIBUTE TO THE PALACE THEATER (II,305); BOB HOPE SPECIAL: BOB HOPE'S CHRISTMAS SPECIAL (II,308); BOB HOPE SPECIAL: THE BOB HOPE COMEDY SPECIAL (II,332); CBS SALUTES LUCY—THE FIRST 25 YEARS (II,459); HIZZONER (II,1159); HOUSE CALLS (II,1194); THE JONATHAN WINTERS SHOW (I,2443); THE ODD COUPLE (II,1875)

KELLEY, DON
THE WOLFMAN JACK SHOW (II,2817)

KELLEY, WILLIAM
KEY TORTUGA (II,1383)

KELLIN, ROY
FOUR STAR PLAYHOUSE (I,1652)

KELLNER, JAMIE
20 MINUTE WORKOUT (II,2680); KICKS (II,1385)

KELLY, APRIL
9 TO 5 (II,1852); LOVE, SIDNEY (II,1547); TEACHERS ONLY (II,2548)

KELLY, GENE
AT YOUR SERVICE (I,310); JACK AND THE BEANSTALK (I,2288)

KEMP, BARRY
NEWHART (II,1835)

KEMP, HAL
THE ROWAN AND MARTIN SHOW (I,3854)

KEMPLEY, WALTER
HAPPY DAYS (II,1084); THE MERV GRIFFIN SHOW (I,3016); THE UGILY FAMILY (II,2699)

KENASTON, JACK
UNK AND ANDY (I,4679)

KENNARD, DAVID
COSMOS (II,579)

KENNEDY, BURT
THE ROUNDERS (I,3851); SIDEKICKS (II,2348)

KENNEDY, MICK
WE'RE DANCIN' (II,2755)

KENNEDY, RICHARD
THE JEAN ARTHUR SHOW (I,2352)

KENNEY, H WESLEY
DAYS OF OUR LIVES (II,629); THE YOUNG AND THE RESTLESS (II,2862)

KENNINGS, ROBERT
THINK FAST (I,4447)

KENNY, CHRIS
PHILIP MARLOWE, PRIVATE EYE (II,2033)

KENT, ALAN
A COUPLE OF JOES (I,1078)

KENT, STEVEN
CAPITOL (II,426); SANTA BARBARA (II,2258)

KENWITH, HERBERT
DIFF'RENT STROKES (II,674); JOE'S WORLD (II,1305)

KENWORTHY, DUNCAN
FRAGGLE ROCK (II,916)

KEREW, DIANA
THE ABC AFTERSCHOOL SPECIAL (II,1); TOM AND JOANN (II,2630); THE WORLD OF DARKNESS (II,2836)

KERN, JAMES V
THE ALL-STAR REVUE (I,138)

KERR, TREENA
THE GALLOPING GOURMET (I,1730)

KESSLER, HARRY
HARBOURMASTER (I,1946)

KESSLER, HENRY
LOCK UP (I,2737); THE WEST POINT STORY (I,4796)

KEYES, CHIP
BLISS (II,269)

KEYES, DOUG
BLISS (II,269)

KEYES, FREEMAN
THE RED SKELTON SHOW (I,3755)

KEYES, PAUL W
BOB HOPE SPECIAL: A QUARTER CENTURY OF BOB HOPE ON TELEVISION (II,281); BOB HOPE SPECIAL: BOB HOPE'S BICENTENNIAL STAR SPANGLED SPECTACULAR (II,306); THE DON RICKLES SHOW (II,708); GENERAL ELECTRIC'S ALL-STAR ANNIVERSARY (II,963); THE ROWAN AND MARTIN REPORT (II,2221); THE ROWAN AND MARTIN SPECIAL (I,3855); ROWAN AND MARTIN'S LAUGH-IN (I,3856); SINATRA AND FRIENDS (II,2358); SINATRA: THE MAN AND HIS MUSIC (II,2362); SINATRA—THE FIRST 40 YEARS (II,2359); SWING OUT, SWEET LAND (II,2515); TAKE ONE STARRING JONATHAN WINTERS (II,2532); THE VIN SCULLY SHOW (I,4734)

KIBBEE, DON
ALAN KING IN LAS VEGAS, PART I (I,97); ALAN KING IN LAS VEGAS, PART II (I,98); CIRCUS OF THE STARS (II,530); CIRCUS OF THE STARS (II,531); CIRCUS OF THE STARS (II,532); CIRCUS OF THE STARS (II,533); CIRCUS OF THE STARS (II,534); CIRCUS OF THE STARS (II,535); CIRCUS OF THE STARS (II,536); CIRCUS OF THE STARS (II,537); MEN AT LAW (I,3004)

KIBBEE, ROLAND
A.E.S. HUDSON STREET (II,17); THE BOB NEWHART SHOW (I,666); BROCK'S LAST CASE (I,742); COLUMBO (II,556); DEAR DETECTIVE (II,652); THE DEPUTY (I,1234); DIAGNOSIS: DANGER (I,1250); THE FAMILY HOLVAK (II,817); MADIGAN (I,2808); MCCOY (II,1661); RETURN OF THE WORLD'S GREATEST DETECTIVE (II,2142); THE STOREFRONT LAWYERS (I,4233); TENNESSEE ERNIE FORD MEETS KING ARTHUR (I,4397)

KIESER, ELLWOOD
IT CAN'T HAPPEN TO ME (II,1250); JOSIE (II,1345); THE JUGGLER OF NOTRE DAME (II,1350); LEADFOOT (II,1452); PRINCESS (II,2084); THE TROUBLE WITH GRANDPA (II,2662); WHEN, JENNY? WHEN (II,2783)

KIGER, BOB
STARSTRUCK (II,2446)

KILEY, TIM
THE KLOWNS (I,2574)

KILLAN, FRANK
TIN PAN ALLEY TV (I,4522)

KILLIAM, PAUL
CAPTAIN BILLY'S MISSISSIPPI MUSIC HALL (I,826); MOVIE MUSEUM (I,3115); SILENT'S, PLEASE (I,4050)

KIMBALL, WARD
THE MOUSE FACTORY (I,3113)

KIMMEL, JESS
DOLLAR A SECOND (I,1326); MIKE AND BUFF (I,3037)

KIMMEL, JOE
THE RED BUTTONS SHOW (I,3750)

KINBERG, JUD
QUINCY, M. E. (II,2102)

KINBERG, KIN
THE SEVEN LIVELY ARTS (I,3987)

KING, ALAN
ALAN KING LOOKS BACK IN ANGER—A REVIEW OF 1972 (I,99); THE ALAN KING SHOW (I,102); THE ALAN KING SPECIAL (I,103); ALAN KING'S FINAL WARNING (II,29); ALAN KING'S SECOND ANNUAL FINAL WARNING (II,30); ALAN KING'S THIRD ANNUAL FINAL WARNING!! (II,31); THE CORNER BAR (I,1051); THE CORNER BAR (I,1052); HAPPY ENDINGS (II,1085); IVAN THE TERRIBLE (II,1260); LOVE, LIFE, LIBERTY & LUNCH (II,1544); THE WONDERFUL WORLD OF AGGRAVATION (I,4894)

KING, ARCHER
SWEENEY TODD (II,2512)

KING, FRANK
MAYA (I,2960)

KING, LYNWOOD
ANOTHER LIFE (II,95)

KING, MAURICE
MAYA (I,2960)

KING, MURRAY
THE ADVENTURES OF BLINKEY (I,52)

KING, PAUL
CODE NAME: DIAMOND HEAD (II,549); MOST WANTED (II,1739)

KING, TIM
BIG SHAMUS, LITTLE SHAMUS (II,242); FORCE SEVEN (II,908)

KING, WALTER WOLFE
MYSTERY AND MRS. (I,3210)

KINGSLEY, DOROTHY
DEBBIE REYNOLDS AND THE SOUND OF CHILDREN (I,1213)

KINNEY, JACK
POPEYE THE SAILOR (I,3649)

KINOY, ERNEST
THE HAPPENERS (I,1935)

KIPNESS, JOSEPH
APPLAUSE (I,244)

KIRENTS, MILTON
DRAW ME A LAUGH (I,1372)

KIRGO, DIANA
REGGIE (II,2127); TEACHERS ONLY (II,2548)

KIRGO, GEORGE
ANOTHER DAY (II,94)

KIRGO, JULIE
REGGIE (II,2127)

KIRK, ROB
THE LOVE REPORT (II,1542); THE LOVE REPORT (II,1543)

KIRKLAND, DENNIS
THE BENNY HILL SHOW (II,207); THE UNEXPURGATED BENNY HILL SHOW (II,2705)

KIRNICK, WILLIAM
IN SEARCH OF. . . (II,1226)

KIRSCHNER, CAROLYN
FOR LOVERS ONLY (II,904)

KIRSHNER, DON
DON KIRSHNER'S ROCK CONCERT (I,1332); THE KIDS FROM C.A.P.E.R. (II,1391); ROXY PAGE (II,2222)

KISER, TONY
THE CONTENDER (II,571)

KISSINGER, DICK
THE FLORIAN ZABACH SHOW (I,1606)

KIVSCH, TERI
SUPER PASSWORD (II,2499)

KIYOSHI, TOSHITSUGA
WOLF ROCK TV (II,2815)

KLANE, ROBERT
CAMP GRIZZLY (II,418); MR. & MRS. DRACULA (II,1745); WHERE'S POPPA? (II,2784)

KLEIMAN, HARLAN P
CLOSE TIES (II,545)

KLEIN, DENNIS
BUFFALO BILL (II,387)

KLEIN, HAROLD J
CELEBRITY BILLIARDS (I,882)

KLEIN, HERBERT
THE LONE RANGER (I,2741)

KLEIN, LARRY
AMERICAN BANDSTAND (II,70); THE GOOD OLD DAYS (I,1851); INSIDE AMERICA (II,1234); MARGIE (I,2909); SHOWDOWN (I,4034)

KLEIN, MAL
MEET MARCEL MARCEAU (I,2983)

KLEIN, MARTY
JOHNNY CASH AND THE COUNTRY GIRLS (II,1323); JOHNNY CASH CHRISTMAS 1983 (II,1324); THE JOHNNY CASH CHRISTMAS SPECIAL (II,1326); A JOHNNY CASH CHRISTMAS (II,1328); THE JOHNNY CASH SPRING SPECIAL (II,1329); JOHNNY CASH'S AMERICA (II,1330); JOHNNY CASH: CHRISTMAS IN SCOTLAND (II,1332); JOHNNY CASH: CHRISTMAS ON THE ROAD (II,1333); JOHNNY CASH: COWBOY HEROES (II,1334); JOHNNY CASH: THE FIRST 25 YEARS (II,1336); JOHNNY CASH—A MERRY MEMPHIS CHRISTMAS (II,1331)

KLEIN, PAUL
THE DAY THE WOMEN GOT EVEN (II,628); VALENTINE MAGIC ON LOVE ISLAND (II,2716)

KLEINERMAN, ISAAC
20TH CENTURY (I,4640); WORLD WAR I (I,4919)

KLEINSCHMITT, CARL
FUNNY FACE (I,1711); KAREN (II,1363); PETE 'N' TILLIE (II,2027); PRYOR'S PLACE (II,2089)

KLINE, RICHARD
HOLLYWOOD'S TALKING (I,2097); SOAP WORLD (II,2394)

KLINE, ROBERT
OUTRAGEOUS OPINIONS (I,3429); THE SOUND AND THE SCENE (I,4132); THE STARLOST (I,4203)

KLINE, STEVE
TUCKER'S WITCH (II,2667)

KLINE, SUSAN COOKE

THE STARLOST (I,4203)

KLING, WOODY
ALL IN THE FAMILY (II,38); HELLO, LARRY (II,1128); HEREAFTER (II,1138); SANFORD ARMS (II,2257)

KLOSS, ALLEN
HOT POTATO (II,1188)

KLOSS, BUD
ALL MY CHILDREN (II,39); TEXAS (II,2566)

KLUGERMAN, IRA
POWERHOUSE (II,2074)

KLYNN, HERBERT
THE ALVIN SHOW (I,146); CURIOSITY SHOP (I,1114)

KNEELAND, TED
DAYDREAMING WITH LARAINE (I,1184); HOLD THAT CAMERA (I,2072)

KNEITEL, SEYMOUR
CARTOONSVILLE (I,858); CASPER AND FRIENDS (I,867)

KNELMAN, P K
CAGNEY AND LACEY (II,409); THIS GIRL FOR HIRE (II,2596)

KNIGHT, PAUL
THE ADVENTURES OF BLACK BEAUTY (I,51)

KNIGHT, VIVA
ALL THAT GLITTERS (II,42); EVERYDAY (II,793); GOOD TIMES (II,1038); MARY HARTMAN, MARY HARTMAN (II,1638)

KNOPF, CHRISTOPHER
CIMARRON STRIP (I,954)

KNOPF, EDWIN H
RENDEZVOUS (I,3766)

KNORP, ARTHUR
TEXACO STAR THEATER (I,4411)

KNOX, HAROLD
DANGEROUS ASSIGNMENT (I,1138)

KOBE, GAIL
THE EDGE OF NIGHT (II,760); THE GUIDING LIGHT (II,1064); TEXAS (II,2566)

KOCH, HOWARD W
MAGNAVOX PRESENTS FRANK SINATRA (I,2826); MIAMI UNDERCOVER (I,3020); TELLY. . . WHO LOVES YA, BABY? (II,2558)

KOCHMANN, WOLF
MINSKY'S FOLLIES (II,1703)

KOCHOFF, CHRIS
JOE AND SONS (II,1297)

KOENIG, DENNIS
AFTERMASH (II,23); M*A*S*H (II,1569)

KOGAN, ARNIE
DONNY AND MARIE (II,712)

KOHAN, BUZ
100 YEARS OF GOLDEN HITS (II,1905); THE ARTHUR GODFREY SPECIAL (I,288); ARTHUR GODFREY'S PORTABLE ELECTRIC MEDICINE SHOW (I,290); BING CROSBY AND THE SOUNDS OF CHRISTMAS (I,457); BING CROSBY'S CHRISTMAS SHOW (I,474); BING CROSBY'S SUN VALLEY CHRISTMAS SHOW (I,475); BING CROSBY—COOLING IT (I,459); BING CROSBY—COOLING IT (I,460); CHRISTMAS WITH THE BING CROSBYS (I,950); CHRISTMAS WITH THE BING CROSBYS (II,524); A COUNTRY CHRISTMAS (II,582); A COUNTRY CHRISTMAS (II,583); A COUNTRY CHRISTMAS (II,584); DISNEYLAND'S 25TH ANNIVERSARY (II,681); GENE KELLY. . . AN AMERICAN IN PASADENA (II,962); THE KEANE BROTHERS SHOW (II,1371); KRAFT SALUTES WALT DISNEY WORLD'S 10TH ANNIVERSARY (II,1410); LOOKING BACK (I,2754); OUR PLACE (I,3416); PERRY COMO'S WINTER SHOW (I,3546); SHIRLEY MACLAINE: ILLUSIONS (II,2336)

KOHRS, GEORGE
SHE'S IN THE ARMY NOW (II,2320)

KOLBE, WINRICH
BATTLESTAR GALACTICA (II,181)

KOMACK, JAMES
9 TO 5 (II,1852); ANOTHER DAY (II,94); ARCHIE (II,104); THE ARCHIE SITUATION COMEDY MUSICAL VARIETY SHOW (II,106); CHICO AND THE MAN (II,508); THE COURTSHIP OF EDDIE'S FATHER (I,1083); GET SMART (II,983); LADY LUCK (I,2604); ME AND MAXX (II,1672); MR. ROBERTS (I,3149); MR. T. AND TINA (II,1758); THE ROLLERGIRLS (II,2203); SHIPSHAPE (II,2329); SUGAR TIME (II,2489); WELCOME BACK, KOTTER (II,2761); WHATEVER HAPPENED TO DOBIE GILLIS? (II,2774)

KONDOLF, GEORGE
THE UNITED STATES STEEL HOUR (I,4677)

KONIGSBERG, FRANK
BIG CITY BOYS (II,232); BING CROSBY'S MERRIE OLDE CHRISTMAS (II,252); BING CROSBY'S WHITE CHRISTMAS (II,253); BING!. . . A 50TH ANNIVERSARY GALA (II,254); BREAKING AWAY (II,371); CALLING DR. STORM, M.D. (II,415); DOROTHY (II,717); ELLIS ISLAND (II,768); GENE KELLY. . . AN AMERICAN IN PASADENA (II,962); HARDCASE (II,1088); HIS AND HERS (II,1155); IT'S NOT EASY (II,1256); PEARL (II,1986); RITUALS (II,2182); YOU'RE GONNA LOVE IT HERE (II,2860)

KONNER, LARRY
I GAVE AT THE OFFICE (II,1213)

KOPELMAN, JEAN
THE MATCH GAME (I,2944)

KOPLAN, HARRY
THRILL SEEKERS (I,4491)

KOREY, STANLEY
MR. & MRS. DRACULA (II,1745)

KORTNER, PETER
THE FARMER'S DAUGHTER (I,1533); THE JOHN FORSYTHE SHOW (I,2410); PURSUIT (I,3700)

KOSLOW, IRA
LINDA RONSTADT IN CONCERT (II,1482)

KOSOFKY, JOEL
CAPTAIN KANGAROO (II,434); WAKE UP (II,2736)

KOSS, ALAN
TIC TAC DOUGH (II,2618)

KOVACS, BELA
SPACE PATROL (I,4144)

KOVACS, ERNIE
THE COMEDY OF ERNIE KOVACS (I,1022); THE ERNIE KOVACS SPECIAL (I,1456); THE NEW ERNIE KOVACS SHOW (I,3264)

KOWALSKI, BERNARD L
BARETTA (II,152); THE JUDGE (I,2468); RAWHIDE (I,3727)

KOZAK, ELLIOTT
BOB HOPE SPECIAL: BOB HOPE'S SUPER BIRTHDAY SPECIAL (II,319); BOB HOPE SPECIAL: BOB HOPE'S HILARIOUS UNREHEARSED ANTICS OF THE STARS (II,341); SHAUGHNESSEY (II,2318)

KOZOLL, ELLIOTT
BOB HOPE SPECIAL: HO HO HOPE'S JOLLY CHRISTMAS HOUR (II,327)

KOZOLL, MICHAEL
HILL STREET BLUES (II,1154)

KRAGEN, KEN
JUST FRIENDS (I,2495); KENNY & DOLLY: A CHRISTMAS TO REMEMBER (II,1378); KENNY ROGERS IN CONCERT (II,1381); THE KENNY ROGERS SPECIAL (II,1382); ON LOCATION WITH RICH LITTLE (II,1891); PAT PAULSEN FOR PRESIDENT (I,3472); THE RICH LITTLE SPECIAL (II,2156); RICH LITTLE'S ROBIN HOOD (II,2158); A SPECIAL KENNY ROGERS (II,2417); THREE FOR TAHITI (I,4476)

KRAIKE, MICHAEL
DAMON RUNYON THEATER (I,1129); THE DEPUTY (I,1234)

KRAMER, KENYON
FOUL UPS, BLEEPS AND BLUNDERS (II,911)

KRAMER, LEE
OLIVIA (II,1883); THE OLIVIA NEWTON-JOHN SHOW (II,1885); OLIVIA NEWTON-JOHN'S HOLLYWOOD NIGHTS (II,1886)

KRAMER, STANLEY
GUESS WHO'S COMING TO DINNER? (II,1063)

KRANTZ, STEVE
MISTRAL'S DAUGHTER (II,1708); ROCKET ROBIN HOOD (I,3821)

KRANZE, DON
EAST SIDE/WEST SIDE (I,1396)

KRASNE, PHILIP
CURTAIN CALL THEATER (I,1115); THE LONE WOLF (I,2742)

KRAUSS, ELLEN M
GOTTA SING, GOTTA DANCE (II,1046)

KRAUSS, PERRY
MARY HARTMAN, MARY HARTMAN (II,1638)

KRCITSEK, HOWARD B
SWINGIN' TOGETHER (I,4320)

KRIDOS, LISA

ALIVE AND WELL (II,36)

KRINSKI, SANDY
BABY, I'M BACK! (II,130)

KROFFT, MARTY
ANSON AND LORRIE (II,98); BARBARA MANDRELL AND THE MANDRELL SISTERS (II,143); THE BAY CITY ROLLERS SHOW (II,187); BIG FOOT AND WILD BOY (II,237); THE BRADY BUNCH HOUR (II,363); THE BUGALOOS (I,757); THE CRACKER BROTHERS (II,599); DOCTOR SHRINKER (II,686); DONNY AND MARIE (II,712); THE DONNY AND MARIE OSMOND SHOW (II,714); ELECTRA WOMAN AND DYNA GIRL (II,764); FAR OUT SPACE NUTS (II,831); H.R. PUFNSTUF (I,2154); THE KROFFT KOMEDY HOUR (II,1411); THE KROFFT SUPERSHOW (II,1412); THE KROFFT SUPERSHOW II (II,1413); LAND OF THE LOST (II,1425); LIDSVILLE (I,2679); THE LOST SAUCER (II,1524); MAGIC MONGO (II,1592); PINK LADY (II,2041); PRYOR'S PLACE (II,2089); SIGMUND AND THE SEA MONSTERS (II,2352); WONDERBUG (II,2830)

KROFFT, SID
ANSON AND LORRIE (II,98); BARBARA MANDRELL AND THE MANDRELL SISTERS (II,143); THE BAY CITY ROLLERS SHOW (II,187); BIG FOOT AND WILD BOY (II,237); THE BRADY BUNCH HOUR (II,363); THE BUGALOOS (I,757); THE CRACKER BROTHERS (II,599); DOCTOR SHRINKER (II,686); DONNY AND MARIE (II,712); THE DONNY AND MARIE OSMOND SHOW (II,714); ELECTRA WOMAN AND DYNA GIRL (II,764); FAR OUT SPACE NUTS (II,831); H.R. PUFNSTUF (I,2154); THE KROFFT KOMEDY HOUR (II,1411); THE KROFFT SUPERSHOW (II,1412); THE KROFFT SUPERSHOW II (II,1413); LAND OF THE LOST (II,1425); LIDSVILLE (I,2679); THE LOST SAUCER (II,1524); MAGIC MONGO (II,1592); PINK LADY (II,2041); PRYOR'S PLACE (II,2089); SIGMUND AND THE SEA MONSTERS (II,2352); WONDERBUG (II,2830)

KROKOW, HOWARD
THE FACTS (II,804)

KROLIK, DICK
MARCH OF TIME THROUGH THE YEARS (I,2903)

KRONICK, WILLIAM
PLIMPTON! DID YOU HEAR THE ONE ABOUT. . .? (I,3629); PLIMPTON! SHOWDOWN AT RIO LOBO (I,3630); PLIMPTON! THE MAN ON THE FLYING TRAPEZE (I,3631)

KRONMAN, HARRY
PROFESSIONAL FATHER (I,3677)

KROST, BARRY
MISSING PIECES (II,1706)

KRUEGER, ROB
THE HOMEMADE COMEDY SPECIAL (II,1173)

KRUMHOLTZ, CHESTER
DOCTOR SIMON LOCKE (I,1317); KOJAK (II,1406); POLICE SURGEON (I,3639)

KUBICHAN, JON
THE ABC AFTERSCHOOL SPECIAL (II,1); LAND OF THE LOST (II,1425); ROOM 222 (I,3843)

KUGEL, CARL
MODESTY BLAISE (II,1719)

KUHL, CALVIN
LIFE WITH LUIGI (I,2692); THE LUX VIDEO THEATER (I,2795)

KUHN, THOMAS
ALICE (II,33); THE AWAKENING LAND (II,122)

KUKOFF, BERNIE
ALL-AMERICAN PIE (II,48); ALMOST ANYTHING GOES (II,56); COMEDY NEWS (I,1020); COMEDY NEWS II (I,1021); DETECTIVE SCHOOL (II,661); THE EVERLY BROTHERS SHOW (I,1479); FREEMAN (II,923); IN TROUBLE (II,1230); JIMMY DURANTE PRESENTS THE LENNON SISTERS HOUR (I,2387); JOE AND SONS (II,1296); JOE AND SONS (II,1297); THE LENNON SISTERS SHOW (I,2656); MCNAMARA'S BAND (II,1668); MCNAMARA'S BAND (II,1669); THE NEW OPERATION PETTICOAT (II,1826); PAT BOONE IN HOLLYWOOD (I,3469); REGGIE (II,2127); SHEEHY AND THE SUPREME MACHINE (II,2322); SHOW BUSINESS SALUTE TO MILTON BERLE (I,4029); THE SHOW MUST GO ON (II,2344); THAT SECOND THING ON ABC (II,2572); THAT THING ON ABC (II,2574); TUCKER'S WITCH (II,2667)

KULIK, BUZZ
CAMPO 44 (I,813); COREY: FOR THE PEOPLE (II,575); THE FEATHER AND FATHER GANG (II,845); THE FEATHER AND FATHER GANG: NEVER CON A KILLER (II,846); FROM HERE TO ETERNITY (II,932); GEORGE WASHINGTON (II,978); KENTUCKY JONES (I,2526); MATT HELM (II,1652); THE SCOTT MUSIC HALL (I,3943); A STORM IN SUMMER (I,4237); WARNING SHOT (I,4770)

KULLER, SID
THE COLGATE COMEDY HOUR (I,997)

KUNEY, JACK
ARCHY AND MEHITABEL (I,254); ONE, TWO, THREE—GO! (I,3390)

KUPFER, MARVIN
THE SIX O'CLOCK FOLLIES (II,2368)

KUPPIN, LARRY
THE KID WITH THE BROKEN HALO (II,1387)

KURI, JOHN
SKYWARD CHRISTMAS (II,2378)

KURTZFIELD, STEPHEN
HIGH PERFORMANCE (II,1145)

KUSHNER, DONALD
AUTOMAN (II,120); FIRST & TEN (II,861)

KUSHNICK, JERROLD H
BEN VEREEN—HIS ROOTS (II,202)

KWARTIN, LESLIE
THE GUIDING LIGHT (II,1064)

KWESKIN, ROBERT
SAY WHEN (I,3929)

KYNE, TERRY
THE CHEAP SHOW (II,491)

LACOCK, SUZANNE
PEOPLE DO THE CRAZIEST THINGS (II,1998)

LAMONTE, LOU
CAT ON A HOT TIN ROOF (II,457)

LATOURETTE, FRANK
THE D.A.'S MAN (I,1158); MEDIC (I,2976); PEOPLE (I,3517)

LABELLA, VINCENZO
MARCO POLO (II,1626); MOSES THE LAWGIVER (II,1737)

LABINE, CLAIRE
RYAN'S HOPE (II,2234)

LACEY, JACK
BRAVE EAGLE (I,709); THE ROY ROGERS SHOW (I,3859)

LACHMAN, BOB
THE BOB HOPE SHOW (I,658)

LACHMAN, BRAD
THE DON HO SHOW (II,706); THE GIRL IN MY LIFE (II,996); MADAME'S PLACE (II,1587); SOLID GOLD (II,2395); SOLID GOLD '79 (II,2396); SOLID GOLD HITS (II,2397)

LACHMAN, DAVID
RUSSIAN ROULETTE (I,3875)

LACHMAN, MORT
ALL IN THE FAMILY (II,38); ARCHIE BUNKER'S PLACE (II,105); BABY MAKES FIVE (II,129); THE BOB HOPE SHOW (I,613); THE BOB HOPE SHOW (I,628); THE BOB HOPE SHOW (I,636); THE BOB HOPE SHOW (I,637); THE BOB HOPE SHOW (I,638); THE BOB HOPE SHOW (I,639); THE BOB HOPE SHOW (I,640); THE BOB HOPE SHOW (I,641); THE BOB HOPE SHOW (I,642); THE BOB HOPE SHOW (I,647); THE BOB HOPE SHOW (I,648); THE BOB HOPE SHOW (I,649); THE BOB HOPE SHOW (I,657); THE BOB HOPE SHOW (I,659); THE BOB HOPE SHOW (I,660); THE BOB HOPE SHOW (I,661); THE BOB HOPE SHOW (I,662); THE BOB HOPE SHOW (I,663); BOB HOPE SPECIAL: BOB HOPE ON CAMPUS (II,289); BOB HOPE SPECIAL: BOB HOPE PRESENTS THE STARS OF TOMORROW (II,292); BOB HOPE SPECIAL: THE BOB HOPE SPECIAL (II,333); BOB HOPE SPECIAL: THE BOB HOPE SPECIAL (II,334); THE BOOK OF LISTS (II,350); GIMME A BREAK (II,995); GOSSIP (II,1043); HAVE GIRLS—WILL TRAVEL (I,1967); HER SCHOOL FOR BACHELORS (I,2018); IN THE BEGINNING (II,1229); INSTANT FAMILY (II,1238); IT ONLY HURTS WHEN YOU LAUGH (II,1251); KATE AND ALLIE (II,1365); MAX (II,1658); NO SOAP, RADIO (II,1856); NOT IN FRONT OF THE KIDS (II,1861); ONE DAY AT A TIME (II,1900); SANFORD (II,2255); SHE'S WITH ME (II,2321); SPENCER (II,2420); SUTTERS BAY (II,2508)

LADD, CHERYL
CHERYL LADD. . . LOOKING BACK—SOUVENIRS (II,502)

LADD, HANK
OH, THOSE BELLS! (I,3345)

LADD, JOHNNY
THE PAUL HOGAN SHOW (II,1975)

LAEMMLE, NINA
SARA (II,2260)

LAFFERTY, PERRY
76 MEN AND PEGGY LEE (I,3989); ACRES AND PAINS (I,19); THE ANDY WILLIAMS SHOW (I,204); ARTHUR GODFREY AND THE SOUNDS OF NEW YORK (I,280); ARTHUR GODFREY IN HOLLYWOOD (I,281); ARTHUR GODFREY LOVES ANIMALS (I,282); BIG HAWAII (II,238); THE BIG PARTY FOR REVLON (I,427); CRUNCH AND DES (I,1108); THE DANNY KAYE SHOW (I,1143); EDDIE AND HERBERT (II,757); THE HAZEL BISHOP SHOW (I,1983); KAY KYSER'S KOLLEGE OF MUSICAL KNOWLEDGE (I,2512); LOOK OUT WORLD (II,1516); THE MARY TYLER MOORE COMEDY HOUR (II,1639); THE REVLON REVUE (I,3782); STAR OF THE FAMILY (I,4194); THE VICTOR BORGE SHOW (I,4721); YOUR HIT PARADE (I,4952)

LAFFERTY, PETER
DANGER IN PARADISE (II,617)

LAIBSON, MICHAEL
AS THE WORLD TURNS (II,110)

LAIRD, JACK
AMANDA FALLON (I,151); AMANDA FALLON (I,152); BEN CASEY (I,394); THE BOB HOPE CHRYSLER THEATER (I,502); BRILLIANT BENJAMIN BOGGS (I,730); CHANNING (I,900); CODE NAME: HERACLITUS (I,990); DOCTORS HOSPITAL (II,691); THE GANGSTER CHRONICLES (II,957); GUILTY OR NOT GUILTY (I,1903); HELLINGER'S LAW (II,1127); KOJAK (II,1406); THE LAW ENFORCERS (I,2638); NIGHT GALLERY (I,3287); ONE OF OUR OWN (II,1910); READY AND WILLING (II,2115); THE RETURN OF CHARLIE CHAN (II,2136); RUSSIAN ROULETTE (I,3875); SECOND LOOK (I,3960); SWITCH (II,2519); TESTIMONY OF TWO MEN (II,2564); WHAT REALLY HAPPENED TO THE CLASS OF '65? (II,2768)

LAKIN, RITA
EXECUTIVE SUITE (II,796); THE HOME FRONT (II,1171)

LAKSO, EDWARD J
CHARLIE'S ANGELS (II,486)

LAMAS, FERNANDO
SAMURAI (II,2249)

LAMB, DUNCAN
SEEING THINGS (II,2297)

LAMBERT, HUGH
MOVIN' WITH NANCY ON STAGE (I,3120)

LAMBERT, VERITY
THE FLAME TREES OF THIKA (II,870); REILLY, ACE OF SPIES (II,2129)

LAMMERS, PAUL
FROM THESE ROOTS (I,1688)

LAMMI, ED
ROMANCE THEATER (II,2204)

LAMOND, BILL
THE FALL GUY (II,811); HOTEL (II,1192)

LAMOND, JO
THE FALL GUY (II,811); HOTEL (II,1192)

LAMOUR, MARIANNE
DAVID NIVEN'S WORLD (II,625)

LANDAU, EDIE
DEADLY GAME (II,632); SEPARATE TABLES (II,2304)

LANDAU, ELY
DEADLY GAME (II,632); SEPARATE TABLES (II,2304)

LANDAU, RICHARD
THE SIX-MILLION-DOLLAR MAN (II,2372); THE WILD WILD WEST (I,4863)

LANDERS, PAUL
BOSTON BLACKIE (I,695); THE LONE RANGER (I,2740); O'HARA, UNITED STATES TREASURY (I,3346)

LANDIS, JAN
INCREDIBLE KIDS AND COMPANY (II,1233)

LANDIS, JOE
ABOUT FACES (I,12); THE LIBERACE SHOW (I,2674); PLACE THE FACE (I,3611); TAKE A GOOD LOOK (I,4332); TAKE A GOOD LOOK (II,2529); WHO DO YOU TRUST? (I,4846)

LANDON, JOE
WALTER FORTUNE (I,4758)

LANDON, MICHAEL
FATHER MURPHY (II,841); HIGHWAY TO HEAVEN (II,1152); LITTLE HOUSE ON THE PRAIRIE (II,1487); LITTLE HOUSE: A NEW BEGINNING (II,1488); LITTLE HOUSE: BLESS ALL THE DEAR CHILDREN (II,1489); LITTLE HOUSE: LOOK BACK TO YESTERDAY (II,1490); LITTLE HOUSE: THE LAST FAREWELL (II,1491)

LANDRY, RON
FLO (II,891)

LANDSBERG, CLEVE
THE PAPER CHASE: THE SECOND YEAR (II,1950)

LANDSBERG, DAVID
SECOND EDITION (II,2288)

LANDSBERG, KLAUS
SANDY DREAMS (I,3909)

LANDSBURG, ALAN
THE ABC AFTERSCHOOL SPECIAL (II,1); ALAN KING IN LAS VEGAS, PART I (I,97); ALAN KING IN LAS VEGAS, PART II (I,98); BIOGRAPHY (I,476); THE CHISHOLMS (II,512); THE CHISHOLMS (II,513); THE FUTURE: WHAT'S NEXT (II,947); IN SEARCH OF. . . (II,1226); IT ONLY HURTS WHEN YOU LAUGH (II,1251); IT WAS A VERY GOOD YEAR (I,2253); THE KIDS FROM C.A.P.E.R. (II,1391); THE KRYPTON FACTOR (II,1414); LIFE'S MOST EMBARRASSING MOMENTS (II,1470); LIFE'S MOST EMBARRASSING MOMENTS II (II,1471); LIFE'S MOST EMBARRASSING MOMENTS III (II,1472); LIFE'S MOST EMBARRASSING MOMENTS IV (II,1473); MEN IN CRISIS (I,3006); MIRROR, MIRROR, OFF THE WALL (I,3059); THE MYSTERIOUS TWO (II,1783); NO HOLDS BARRED (II,1854); PEOPLE DO THE CRAZIEST THINGS (II,1997); PEOPLE DO THE CRAZIEST THINGS (II,1998); PEOPLE TO PEOPLE (II,2000); PERSONAL AND CONFIDENTIAL (II,2026); SUCCESS: IT CAN BE YOURS (II,2488); THAT'S INCREDIBLE! (II,2578); THOSE AMAZING ANIMALS (II,2601); THE WORLD'S FUNNIEST COMMERCIAL GOOFS (II,2841)

LANE, BRIAN ALAN
COVER UP (II,597)

LANE, CARLA
BUTTERFLIES (II,405)

LANE, PETER
WORD FOR WORD (I,4907)

LANE, ROB
THE ALL-AMERICAN COLLEGE COMEDY SHOW (II,47)

LANFIELD, SIDNEY
MY DARLING JUDGE (I,3186)

LANG, OTTO
THE 20TH CENTURY-FOX HOUR (I,4642); LAURA (I,2635); WORLD OF GIANTS (I,4912)

LANG, RICHARD
HARRY O (II,1098); SHOOTING STARS (II,2341); VELVET (II,2725)

LANG, ROBERT
IN SEARCH OF. . . (II,1226)

LANGNER, LAWRENCE
GREAT CATHERINE (I,1873)

LANSBURY, BRUCE
BANJO HACKETT: ROAMIN' FREE (II,141); BELL, BOOK AND CANDLE (II,201); BUCK ROGERS IN THE 25TH CENTURY (II,383); ESCAPE (I,1460); THE FANTASTIC JOURNEY (II,826); THE LONER (I,2745); MISSION: IMPOSSIBLE (I,3067); MOBILE MEDICS (II,1716); THE POWERS OF MATTHEW STAR (II,2075); ROGER AND HARRY (II,2200); THE WILD WILD WEST (I,4863)

LANSBURY, EDGAR
CORONET BLUE (I,1055)

LANTZ, LOUIS
STORY THEATER (I,4240)

LANTZ, WALTER
THE WOODY WOODPECKER SHOW (I,4906)

LARKIN, JIM
THE MADHOUSE BRIGADE (II,1588)

LARKIN, JOHN
M SQUAD (I,2799)

LARKIN, RITA
FLAMINGO ROAD (II,872)

LARSEN, LARRY
KENNY LOGGINS IN CONCERT (II,1380)

LARSEN, WILLIAM
FRANCIS ALBERT SINATRA DOES HIS THING (I,1658)

LARSON, BOB
BRENDA STARR (II,372)

LARSON, CHARLES
CADE'S COUNTY (I,787); THE CRIME CLUB (I,1096); THE FBI (I,1551); THE INTERNS (I,2226); NAKIA (II,1785)

LARSON, GLEN A
ALIAS SMITH AND JONES (I,118); ALIAS SMITH AND JONES (I,119); ALL THAT GLITTERS (II,41); AUTOMAN (II,120); B.J. AND THE BEAR (II,125); B.J. AND THE BEAR (II,126); B.J. AND THE BEAR (II,127); BATTLES: THE MURDER THAT WOULDN'T DIE (II,180); BATTLESTAR GALACTICA (II,181); BENNY AND BARNEY: LAS VEGAS UNDERCOVER (II,206); BUCK ROGERS IN THE 25TH CENTURY (II,383); COVER UP (II,597); THE EYES OF TEXAS (II,799); THE EYES OF TEXAS II (II,800); THE FALL GUY (II,811); FOOLS, FEMALES AND FUN: I'VE GOTTA BE ME (II,899); FOOLS, FEMALES AND FUN: IS THERE A DOCTOR IN THE HOUSE? (II,900); FOOLS, FEMALES AND FUN: WHAT ABOUT THAT ONE? (II,901); GALACTICA 1980 (II,953); GET CHRISTIE LOVE! (II,981); THE HARDY BOYS MYSTERIES (II,1090); HOW DO I KILL A THIEF—LET ME COUNT THE WAYS (II,1195); THE ISLANDER (II,1249); KNIGHT RIDER (II,1402); LOBO (II,1504); MAGNUM, P.I. (II,1601); MANIMAL (II,1622); MASQUERADE (II,1644); MCCLOUD (II,1660); THE MISADVENTURES OF SHERIFF LOBO (II,1704); THE NANCY DREW MYSTERIES (II,1789); NIGHTSIDE (II,1851); QUINCY, M. E. (II,2102); ROOSTER (II,2210); THE SIX MILLION DOLLAR MAN (I,4067); THE SIX-MILLION-DOLLAR MAN (II,2372); SWITCH (II,2519); SWITCH (II,2520); SWORD OF JUSTICE (II,2521); TERROR AT ALCATRAZ (II,2563); TRAUMA CENTER (II,2655)

LATHAM, LARRY
SNORKS (II,2390)

LATHAM, MICHAEL
TEN WHO DARED (II,2559)

LAUBERT, PATRICK
INSPECTOR GADGET (II,1236)

LAURENCE, MICHAEL
RETURN TO EDEN (II,2143)

LAVEN, ARNOLD
THE BIG VALLEY (I,437); THE DETECTIVES (I,1245); HONEY WEST (I,2105); LAW OF THE PLAINSMAN (I,2639); THE RIFLEMAN (I,3789); WHICH WAY'D THEY GO? (I,4838)

LAVENHEIM, ROBERT
LOTTERY (II,1525)

LAVERY JR, EMMET G
NERO WOLFE (II,1805); SERPICO (II,2306)

LAW, GENE
HOW'S YOUR MOTHER-IN-LAW? (I,2148)

LAW, LINDSAY
YOU CAN'T TAKE IT WITH YOU (II,2857)

LAWRENCE, ANTHONY
THE PHOENIX (II,2035)

LAWRENCE, DAVID
THAT'S HOLLYWOOD (II,2577); TOURIST (II,2651)

LAWRENCE, NANCY
THE PHOENIX (II,2035)

LAWRENCE, PHIL
DEAR ALEX AND ANNIE (II,651)

LAWRENCE, ROBERT
THE MARVEL SUPER HEROES (I,2930); SPIDER-MAN (I,4154)

LAWRENCE, SHELLEY
THE WORLD OF PEOPLE (II,2839)

LAWRENCE, STEVE
STEVE AND EYDIE CELEBRATE IRVING BERLIN (II,2456); STEVE AND EYDIE. . . ON STAGE (I,4223)

LAYBOURNE, KIT
BRAIN GAMES (II,366)

LAYTON, JERRY
ADVENTURES OF COLONEL FLACK (I,56); BRIGHT PROMISE (I,727); CUT (I,1116); MODERN ROMANCES (I,3079); MR. ARSENIC (I,3126); ROCKY KING, INSIDE DETECTIVE (I,3824)

LAYTON, JOE
. . .AND DEBBIE MAKES SIX (I,185); BARBRA STREISAND AND OTHER MUSICAL INSTRUMENTS (I,353); THE BELLE OF 14TH STREET (I,390); AN EVENING WITH DIANA ROSS (II,788); THE HANNA-BARBERA HAPPINESS HOUR (II,1079); THE JACK JONES SPECIAL (I,2310); MAC DAVIS. . . I BELIEVE IN CHRISTMAS (II,1581); PAUL LYNDE AT THE MOVIES (II,1976); THE PAUL LYNDE COMEDY HOUR (II,1979); REALLY RAQUEL (II,2119); A SPECIAL OLIVIA NEWTON-JOHN (II,2418)

LAZAR, DAVID
THE MUPPET SHOW (II,1766); THE MUPPETS GO TO THE MOVIES (II,1767)

LEFRENAIS, IAN
MY WIFE NEXT DOOR (II,1781)

LEGRAND, HUDSON
FOUL UPS, BLEEPS AND BLUNDERS (II,911)

LEVIEN, JACK
THE GATHERING STORM (I,1741)

LEWELLEN, JOHN
THE QUIZ KIDS (I,3712)

LEACH, ROBIN
LIFESTYLES OF THE RICH AND FAMOUS (II,1474)

LEACOCK, PHILIP
BAFFLED (I,332); CIMARRON STRIP (I,954); GUNSMOKE (II,1069); HAWAII FIVE-O (II,1110); THE NEW LAND (II,1820); THE WILD WILD WEST (I,4863)

LEAF, PAUL
DISAPPEARANCE OF AIMEE (I,1290)

LEAR, GEORGE
THE GIRL FROM U.N.C.L.E. (I,1808)

LEAR, NORMAN
A.K.A. PABLO (II,28); ALL IN THE FAMILY (II,38); THE ANDY WILLIAMS SPECIAL (I,211); BAND OF GOLD (I,341); THE BAXTERS (II,183); THE DANNY KAYE SHOW (I,1144); FOREVER FERNWOOD (II,909); GOOD TIMES (II,1038); HENRY FONDA AND THE FAMILY (I,2015); HOT L BALTIMORE (II,1187); I LOVE LIBERTY (II,1215); KING OF THE ROAD (II,1396); THE MARTHA RAYE SHOW (I,2926); MARY HARTMAN, MARY HARTMAN (II,1638); MAUDE (II,1655); MR. DUGAN (II,1749); THE NANCY WALKER SHOW (II,1790); ONE DAY AT A TIME (II,1900); P.O.P. (II,1939); PALMERSTOWN U.S.A. (II,1945); SANFORD AND SON (II,2256); A YEAR AT THE TOP (II,2844); A YEAR AT THE TOP (II,2845)

LEAVITT, RON
THE BAD NEWS BEARS (II,134); IT'S YOUR MOVE (II,1259); THE JEFFERSONS (II,1276); SILVER SPOONS (II,2355)

LEDDING, ED
DYNASTY (II,746); SUMMER (II,2491)

LEDER, HERB
THE LAMBS GAMBOL (I,2607); TRY AND DO IT (I,4621); WHO SAID THAT? (I,4849)

LEDER, REUBEN
MAGNUM, P.I. (II,1601); THE RETURN OF LUTHER GILLIS (II,2137)

LEDUC, MARCEL
THE ERROL FLYNN THEATER (I,1458)

LEE, BILL
ABC'S SILVER ANNIVERSARY CELEBRATION (II,2); THE CAPTAIN AND TENNILLE IN HAWAII (II,430); THE CAPTAIN AND TENNILLE IN NEW ORLEANS (II,431); DICK CLARK PRESENTS THE ROCK AND ROLL YEARS (I,1265); DICK CLARK PRESENTS THE ROCK AND ROLL YEARS (I,1266); DICK CLARK'S GOOD OL' DAYS: FROM BOBBY SOX TO BIKINIS (II,671); DICK CLARK'S LIVE WEDNESDAY (II,672); EASY DOES IT. . . STARRING FRANKIE AVALON (II,754); FRIDAYS (II,926); GABRIEL KAPLAN PRESENTS THE FUTURE STARS (II,951); THE HELEN REDDY SPECIAL (II,1126); THE JOHN DAVIDSON CHRISTMAS SHOW (II,1307); THE JOHN DAVIDSON SHOW (II,1309); THE LOU RAWLS SPECIAL (II,1527); MANTRAP (I,2893); ROBERTA FLACK. . . THE FIRST TIME EVER (I,3817)

LEE, JAMES
RAFFERTY (II,2105)

LEE, JOANNA
MULLIGAN'S STEW (II,1763); MULLIGAN'S STEW (II,1764); SEARCH FOR TOMORROW (II,2284)

LEE, PAT
NOT NECESSARILY THE NEWS (II,1862); NOT NECESSARILY THE NEWS (II,1863)

LEEDS, HOWARD
ALMOST AMERICAN (II,55); THE BILL DANA SHOW (I,446); THE BRADY BUNCH (II,362); THE BUCCANEERS (I,749); DIFF'RENT STROKES (II,674); GEORGE GOBEL PRESENTS (I,1766); THE GHOST AND MRS. MUIR (I,1786); GRADY (II,1047); KELLY'S KIDS (I,2523); THE LAS VEGAS SHOW (I,2619); LI'L ABNER (I,2701); MY LIVING DOLL (I,3195)

LEEDS, ROBERT
PROJECT UFO (II,2088); SAM (II,2247)

LEESON, MICHAEL
THE ASSOCIATES (II,113); MIXED NUTS (II,1714); THERE'S ALWAYS ROOM (II,2583); YOUNG GUY CHRISTIAN (II,2864)

LEETCH, TOM
THE ADVENTURES OF POLLYANNA (II,14); GUN SHY (II,1068)

LEFCOURT, PETER
BULBA (II,392); THE DEVLIN CONNECTION (II,664); RIVKIN: BOUNTY HUNTER (II,2183); SCARECROW AND MRS. KING (II,2268)

LEFFERTS, GEORGE
BREAKING POINT (I,717); PUREX SPECIAL FOR WOMEN (I,3698); TEACHER, TEACHER (I,4366)

LEFFERTY, PERRY
GOOD OL' BOYS (II,1035)

LEFTWICH, ALEX
MAKE MINE MUSIC (I,2840)

LEHMAN, LEW
SEARCH AND RESCUE: THE ALPHA TEAM (II,2283)

LEHR, GEORGE
THE AMERICAN GIRLS (II,72); BRACKEN'S WORLD (I,703)

LEHR, MILTON
THE RAY ANTHONY SHOW (I,3729); THE SWINGING SCENE OF RAY ANTHONY (I,4323)

LEIGHTON, JIM
IT'S A MAN'S WORLD (I,2258)

LEMKIN, MARC
TOP 40 VIDEOS (II,2644)

LENNARD, LAURIE
BACKSTAGE PASS (II,132)

LENNON, JOHN
JOHN LENNON AND YOKO ONO PRESENT THE ONE-TO-ONE CONCERT (I,2414)

LENNY, RIPPS
13 THIRTEENTH AVENUE (II,2592)

LENOX, JOHN THOMAS
THE BRADY BRIDES (II,361); THE BRADY GIRLS GET MARRIED (II,364); BUS STOP (II,399); BUSTING LOOSE (II,402); STRUCK BY LIGHTNING (II,2482); TUCKER'S WITCH (II,2667)

LEO, MALCOLM
BOB HOPE SPECIAL: BOB HOPE'S OVERSEAS CHRISTMAS TOURS (II,311); BOB HOPE SPECIAL: BOB HOPE—HOPE, WOMEN AND SONG (II,324); E.T. & FRIENDS—MAGICAL MOVIE VISITORS (II,749); PRIME TIMES (II,2083)

LEOB, THOMAS
ETHEL AND ALBERT (I,1466)

LEONARD, BERT
NAKED CITY (I,3215)

LEONARD, ERNEST
MY FRIEND TONY (I,3192)

LEONARD, HERBERT B
THE ADVENTURES OF RIN TIN TIN (I,73); BREAKING AWAY (II,371); THE CATCHER (I,872); CIRCUS BOY (I,959); LADIES' MAN (II,1423); NAKED CITY (I,3216); NIGHTSIDE (I,3296); POPI (II,2068); POPI (II,2069); RESCUE 8 (I,3772); ROUTE 66 (I,3852); SPARROW (II,2410); SPARROW (II,2411); STARSTRUCK (II,2446); TALES OF THE 77TH BENGAL LANCERS (I,4348); TALLAHASSEE 7000 (I,4355)

LEONARD, SHELDON
ACCIDENTAL FAMILY (I,17); ACES UP (II,7); THE ANDY GRIFFITH SHOW (I,192); THE ANDY GRIFFITH SHOW (I,193); THE BILL DANA SHOW (I,446); THE DICK VAN DYKE SHOW (I,1275); THE DON RICKLES SHOW (I,1342); EVERYTHING HAPPENS TO ME (I,1481); GOMER PYLE, U.S.M.C. (I,1843); GOOD MORNING WORLD (I,1850); I SPY (I,2179); JEFF'S COLLIE (I,2356); MAKE ROOM FOR DADDY (I,2843); MY FRIEND TONY (I,3192); MY WORLD . . . AND WELCOME TO IT (I,3208); PATRICK STONE (I,3476); SHIRLEY'S WORLD (I,4019); SINGLES (I,4062); TOP SECRET (II,2647)

LEONE, ANTHONY
OUTLAW LADY (II,1935); WONDER GIRL (I,4891)

LEONETTI, CINDY
LITTLE LULU (II,1493)

LEOSCH, MARGARET
BENJI, ZAX AND THE ALIEN PRINCE (II,205)

LEPINER, MICHAEL
THE BEST LEGS IN 8TH GRADE (II,214)

LESCHIN, DEBORAH
MAKIN' IT (II,1603); SUZANNE PLESHETTE IS MAGGIE BRIGGS (II,2509)

LESLIE, ALEEN
A DATE WITH JUDY (I,1162)

LESSER, GENE
THE MCLEAN STEVENSON SHOW (II,1664)

LESSER, JULIAN
BOLD JOURNEY (I,680)

LESSER, SOL
TARZAN AND THE TRAPPERS (I,4362)

LESTER, LORRAINE
BOBO HOBO AND HIS TRAVELING TROUPE (I,678)

LETO, ROSEANNE
CHARLES IN CHARGE (II,482)

LETTS, BARRY
DOCTOR WHO (II,689); GREAT EXPECTATIONS (II,1055)

LEVANT, BRIAN
THE BAD NEWS BEARS (II,134); HAPPY DAYS (II,1084); HOT W.A.C.S. (II,1191); MORK AND MINDY (II,1735)

LEVEE, MICHAEL
THE SUNSHINE BOYS (II,2496)

LEVENBACK, PAULA
NEVER AGAIN (II,1808)

LEVEY, JAY
WELCOME TO THE FUN ZONE (II,2763)

LEVI, ALAN J
AIRWOLF (II,27); THE INVISIBLE WOMAN (II,1244); VOYAGERS (II,2730)

LEVIN, BIGGIE
STUD'S PLACE (I,4271)

LEVIN, CINDY
FRIENDS (II,927)

LEVIN, JACK
TELEVISION'S GREATEST COMMERCIALS II (II,2554)

LEVIN, LISSA
GLORIA (II,1010)

LEVIN, MIKE
HENRY MORGAN'S GREAT TALENT HUNT (I,2017)

LEVIN, PHIL
INTERNATIONAL SHOWTIME (I,2225)

LEVINE, DICK
FLOOR SHOW (I,1605)

LEVINE, JOSEPH E
HERCULES (I,2022)

LEVINE, KEN
AFTERMASH (II,23); CHARACTERS (II,481); CHEERS (II,494)

LEVINE, PHILIP
MISTER JERICO (I,3070)

LEVINSON, BARRY
DINER (II,679); THE INVESTIGATORS (II,1241); PEEPING TIMES (II,1990); STOPWATCH: THIRTY MINUTES OF INVESTIGATIVE TICKING (II,2474)

LEVINSON, DAVID
CHARLIE'S ANGELS (II,486); FORCE FIVE (II,907); GAVILAN (II,959); HART TO HART (II,1102); HERNANDEZ, HOUSTON P.D. (I,2034); MRS. COLUMBO (II,1760); SARGE (I,3914); SARGE: THE BADGE OR THE CROSS (I,3915); THE SENATOR (I,3976); SENIOR YEAR (II,2301); SONS AND DAUGHTERS (II,2404); TOP SECRET (II,2647)

LEVINSON, LEE
SKEEZER (II,2376)

LEVINSON, MARK
ROCK-N-AMERICA (II,2195)

LEVINSON, RICHARD
CHARLIE COBB: NICE NIGHT FOR A HANGING (II,485); COLUMBO (II,556); ELLERY QUEEN (II,766); ELLERY QUEEN: TOO MANY SUSPECTS (II,767); THE JUDGE AND JAKE WYLER (I,2464); MURDER, SHE WROTE (II,1770); THE NAME OF THE GAME (I,3217); PARTNERS IN CRIME (I,3461); SAVAGE (I,3924); TENAFLY (I,4395); TENAFLY (II,2560)

LEVITAS, WILLARD
MY NAME IS BARBRA (I,3198)

LEVITOW, ABE
CURIOSITY SHOP (I,1114); OFF TO SEE THE WIZARD (I,3337)

LEVITT, ALAN J
MISS WINSLOW AND SON (II,1705); YOU'RE JUST LIKE YOUR FATHER (II,2861)

LEVITT, GENE
BARNABY JONES (II,153); COMBAT (I,1011); HAWAII FIVE-O (II,1110); THE OUTSIDER (I,3431); PROJECT UFO (II,2088); S.W.A.T. (II,2236)

LEVITT, SAUL

I COVER TIMES SQUARE (I,2166)

LEVY, DAVID
THE ADDAMS FAMILY (II,11); THE ADDAMS FAMILY (II,13); FACE THE MUSIC (II,803); THE PRUITTS OF SOUTHAMPTON (I,3684); SARGE (I,3914); YOU ASKED FOR IT (II,2855)

LEVY, FRANKLIN
THE LAST HURRAH (I,2626); RETURN ENGAGEMENT (I,3775)

LEVY, JERRY
HAL LINDEN'S BIG APPLE (II,1072)

LEVY, JULES
THE BIG VALLEY (I,437); THE DETECTIVES (I,1245); HONEY WEST (I,2105); THE INDIAN (I,2210); LAW OF THE PLAINSMAN (I,2639); THE RIFLEMAN (I,3789); WHICH WAY'D THEY GO? (I,4838)

LEVY, PARKE
DECEMBER BRIDE (I,1215); MANY HAPPY RETURNS (I,2895); PETE AND GLADYS (I,3558)

LEVY, RALPH
THE ALAN YOUNG SHOW (I,104); A CHRISTMAS CAROL (I,947); DENNIS JAMES' CARNIVAL (I,1227); FROM A BIRD'S EYE VIEW (I,1686); THE GENERAL FOODS 25TH ANNIVERSARY SHOW (I,1756); THE GEORGE BURNS AND GRACIE ALLEN SHOW (I,1763); THE JACK BENNY HOUR (I,2291); THE JACK BENNY HOUR (I,2292); THE JACK BENNY HOUR (I,2293); THE JACK BENNY PROGRAM (I,2294); THE JACK BENNY SPECIAL (I,2295); JACK BENNY WITH GUEST STARS (I,2298); THE MISSUS GOES A SHOPPING (I,3069); QUADRANGLE (I,3703); TIME OUT FOR GINGER (I,4508)

LEWIN, CHARLES
THE AMAZING POLGAR (I,160)

LEWIN, ROBERT
BRACKEN'S WORLD (I,703); THE MAN FROM ATLANTIS (II,1615); THE PAPER CHASE (II,1949)

LEWINE, RICHARD
ALL IN FUN (I,130); THE BIL BAIRD SHOW (I,439); BLITHE SPIRIT (I,484); THE BLUE ANGEL (I,489); THE DANGEROUS CHRISTMAS OF RED RIDING HOOD (I,1139); DUPONT SHOW OF THE MONTH (I,1387); HOOTENANNY (I,2117); JUNIOR MISS (I,2493); LET'S CELEBRATE (I,2663); MUSIC '55 (I,3168); MY NAME IS BARBRA (I,3198); ON THE FLIP SIDE (I,3360); ONE MAN SHOW (I,3381); PINOCCHIO (I,3603); RODGERS AND HART TODAY (I,3829); SUMMERTIME U.S.A. (I,4283); THERE'S ONE IN EVERY FAMILY (I,4439); TOGETHER WITH MUSIC (I,4532)

LEWIS, AL
THE GEORGE GOBEL SHOW (I,1768); ONE HAPPY FAMILY (I,3375)

LEWIS, ARTHUR
THE ASPHALT JUNGLE (I,297); BRENNER (I,719); THE NURSES (I,3318)

LEWIS, BILL
THE ADVENTURES OF OZZIE AND HARRIET (I,71)

LEWIS, CHRISTOPHER
TALES OF THE UNEXPECTED (II,2540)

LEWIS, DAVID P
FARAWAY HILL (I,1532); STORIES IN ONE CAMERA (I,4234)

LEWIS, DICK
QUICK AS A FLASH (I,3709)

LEWIS, DRAPER
IT WAS A VERY GOOD YEAR (I,2253); THAT'S HOLLYWOOD (II,2577); THOSE AMAZING ANIMALS (II,2601)

LEWIS, EDWARD
SCHLITZ PLAYHOUSE OF STARS (I,3936); A TALE OF WELLS FARGO (I,4338); THE THORN BIRDS (II,2600)

LEWIS, ELLIOTT
THE COURT OF LAST RESORT (I,1082); THE LUCY SHOW (I,2791); MANHATTAN TOWER (I,2889); THE MOTHERS-IN-LAW (I,3110); THIS MAN DAWSON (I,4463); THE TWO OF US (I,4658)

LEWIS, JEFFREY
BAY CITY BLUES (II,186)

LEWIS, JERRY
THE JERRY LEWIS SHOW (I,2363)

LEWIS, JUDY
TEXAS (II,2566)

LEWIS, LEONARD
FLAMBARDS (II,869)

LEWIS, LESTER
ALL ABOARD (I,124); HOLLYWOOD OFF BEAT (I,2086); HOLLYWOOD SCREEN TEST (I,2089); MEET BETTY FURNESS (I,2978); MOVIELAND QUIZ (I,3116); PENTHOUSE PARTY (I,3515); WHO'S WHOSE? (I,4855)

LEWIS, MARCIA
THE BEST LEGS IN 8TH GRADE (II,214)

LEWIS, MARLO
THE BIL BAIRD SHOW (I,439); THE ED SULLIVAN SHOW (I,1401); FUN FOR 51 (I,1707)

LEWIS, R J
THE RENEGADES (II,2133)

LEWIS, RICHARD
ALCOA PREMIERE (I,109); BLIND DATE (I,482); CIMARRON CITY (I,953); THE CRUSADER (I,1110); IT'S A SMALL WORLD (I,2260); JAKE'S WAY (II,1269); LARAMIE (I,2614); M SQUAD (I,2799); MAMA MALONE (II,1609); MARKHAM (I,2916); NEW GIRL IN HIS LIFE (I,3265); RIVERBOAT (I,3797); SECRET AGENT (I,3961); SUSPICION (I,4309); TAKE A CHANCE (I,4331); THINK PRETTY (I,4448); TWO GIRLS NAMED SMITH (I,4656); WAGON TRAIN (I,4747); WHISPERING SMITH (I,4842); WHO'S THERE? (I,4854)

LEWIS, ROBERT
BARETTA (II,152)

LEWIS, THERESA
THE BRIGHTER DAY (I,728)

LEWIS, TOM
A LETTER TO LORETTA (I,2669); THE LORETTA YOUNG THEATER (I,2756)

LEWIS, WARREN
CHEVRON HALL OF STARS (I,927); CHICAGO 212 (I,935); HOORAY FOR HOLLYWOOD (I,2114); I AND CLAUDIE (I,2164); MAN WITH A CAMERA (I,2883); MEET MCGRAW (I,2985); STAGE 7 (I,4181); TERRY AND THE PIRATES (I,4405); WIRE SERVICE (I,4883); YANCY DERRINGER (I,4925)

LEWMAN AND REVUE PRODUCTIONS
THE DOCTOR WAS A LADY (I,1318)

LIBOW, MORT
STATE FAIR USA (II,2449); STATE FAIR USA (II,2450)

LICHTMAN, JIM
SO YOU WANT TO LEAD A BAND (I,4106)

LIEBER, ERIC
JOHN RITTER: BEING OF SOUND MIND AND BODY (II,1319); LEGENDS OF THE WEST: TRUTH AND TALL TALES (II,1457); THE LOVE CONNECTION (II,1536); THE MIKE DOUGLAS CHRISTMAS SPECIAL (I,3040); OMNIBUS (II,1890); SAMMY AND COMPANY (II,2248); SHOW BUSINESS (II,2343)

LIEBMAN, MAX
THE ADMIRAL BROADWAY REVUE (I,37); AMERICAN COWBOY (I,169); BABES IN TOYLAND (I,318); BEST FOOT FORWARD (I,401); THE BIG TIME (I,434); THE CHOCOLATE SOLDIER (I,944); A CONNECTICUT YANKEE (I,1038); DEAREST ENEMY (I,1205); THE DESERT SONG (I,1236); FANFARE (I,1526); GOOD TIMES (I,1853); THE GREAT WALTZ (I,1878); HEAVEN WILL PROTECT THE WORKING GIRL (I,1996); HEIDI (I,2004); HOLIDAY (I,2074); KALEIDOSCOPE (I,2504); LADY IN THE DARK (I,2602); MARCO POLO (I,2904); THE MAURICE CHEVALIER SHOW (I,2953); THE MAURICE CHEVALIER SHOW (I,2954); MAX LIEBMAN PRESENTS (I,2958); THE MERRY WIDOW (I,3012); THE MUSIC OF GERSHWIN (I,3174); NAUGHTY MARIETTA (I,3232); PANORAMA (I,3448); PARIS IN THE SPRINGTIME (I,3457); PRIVATE EYE, PRIVATE EYE (I,3671); SATINS AND SPURS (I,3917); SPOTLIGHT (I,4162); STANLEY (I,4190); STEP ON THE GAS (I,4215); SUNDAY IN TOWN (I,4285); VARIETY (I,4703); YOUR SHOW OF SHOWS (I,4957); YOUR SHOW OF SHOWS (II,2875)

LIESMORE, MARTIN
THE PALLISERS (II,1944)

LINDER, MICHAEL
THE LOVE REPORT (II,1542)

LINDHEIM, RICHARD
B.J. AND THE BEAR (II,125); B.J. AND THE BEAR (II,126)

LINDSAY, JEANNE
HOT WHEELS (I,2126); SKYHAWKS (I,4079)

LINDSAY-HOGG, MICHAEL
BRIDESHEAD REVISITED (II,375)

LINK, WILLIAM
CHARLIE COBB: NICE NIGHT FOR A HANGING (II,485); COLUMBO (II,556); ELLERY QUEEN (II,766); ELLERY QUEEN: TOO MANY SUSPECTS (II,767); THE JUDGE AND JAKE WYLER (I,2464); MURDER, SHE WROTE (II,1770); THE NAME OF THE GAME (I,3217); PARTNERS IN CRIME (I,3461); SAVAGE (I,3924); TENAFLY (I,4395); TENAFLY (II,2560)

LINKE, RICHARD O
ADAMS OF EAGLE LAKE (II,10); THE ANDY GRIFFITH SHOW (I,192); ANDY GRIFFITH'S UPTOWN-DOWNTOWN SHOW (I,194); THE ANDY GRIFFITH—DON KNOTTS—JIMNABORS SHOW (I,191); THE DEADLY GAME (II,633); FRIENDS AND NABORS (I,1684); GIRL FRIENDS AND NABORS (I,1807); THE GIRL IN THE EMPTY GRAVE (II,997); HEADMASTER (I,1990); LOOKING BACK (I,2754); MAYBERRY R.F.D. (I,2961); THE NEW ANDY GRIFFITH SHOW (I,3256); WINTER KILL (II,2810)

LINKROUM, RICHARD
THE ALAN YOUNG SHOW (I,104); HOME (I,2099); HOW TO (I,2145)

LIPMAN, ARLENE
TOP OF THE WORLD (II,2646)

LIPP, DON
THE BIG SHOWDOWN (II,244); THE MONEY MAZE (II,1723)

LIPSKY, MARK
BWANA MICHAEL OF AFRICA (I,783); THE DAYTON ALLEN SHOW (I,1186); ROD ROCKET (I,3827)

LIPSTONE, GREGORY
PEOPLE DO THE CRAZIEST THINGS (II,1998)

LIPTON, JAMES
BOB HOPE SPECIAL: BOB HOPE'S ALL-STAR BIRTHDAY AT ANNAPOLIS (II,294); BOB HOPE SPECIAL: BOB HOPE'S ALL-STAR COMEDY BIRTHDAY PARTY AT WEST POINT (II,298); BOB HOPE SPECIAL: BOB HOPE'S SUPER BIRTHDAY SPECIAL (II,319); BOB HOPE SPECIAL: HAPPY BIRTHDAY, BOB! (II,325); BOB HOPE SPECIAL: HAPPY BIRTHDAY, BOB! (II,326); LORETTA LYNN IN THE BIG APPLE (II,1520); LORETTA LYNN: THE LADY. . .THE LEGEND (II,1521)

LISANBY, CHARLES
THE ROCK RAINBOW (II,2194)

LISI, TOM
THE FIGHTING NIGHTINGALES (II,855)

LITTLE, RICH
ON LOCATION WITH RICH LITTLE (II,1891); THE RICH LITTLE SPECIAL (II,2156); RICH LITTLE'S ROBIN HOOD (II,2158)

LITTLEWOOD, YVONNE
PERRY COMO'S OLDE ENGLISH CHRISTMAS (II,2020); PETULA (I,3572)

LIVESAY, KENNETH
YOUNG LIVES (II,2866)

LIVINGSTON, NORMAN
BATTLE OF THE AGES (I,367)

LLEWELLYN, DOUG
ROUGHCUTS (II,2218)

LLOYD, DAVID
AT YOUR SERVICE (II,118); BEST OF THE WEST (II,218); NOT UNTIL TODAY (II,1865); YOUR PLACE OR MINE? (II,2874)

LLOYD, EDLYNE
SANTA BARBARA (II,2258)

LLOYD, EUAN
THE POPPY IS ALSO A FLOWER (I,3650)

LLOYD, JEREMY
BEANE'S OF BOSTON (II,194)

LLOYD, JOHN
THE HITCHHIKER'S GUIDE TO THE GALAXY (II,1158); NOT THE NINE O'CLOCK NEWS (II,1864)

LLOYD, MICHAEL
JOANIE LOVES CHACHI (II,1295)

LLOYD, NORMAN
ALFRED HITCHCOCK PRESENTS (I,115); TALES OF THE UNEXPECTED (II,2540)

LOBUE, J D
SOAP (II,2392)

LOCKE, ALBERT
THE LONDON PALLADIUM (I,2739)

LOCKE, PETER
AUTOMAN (II,120); FIRST & TEN (II,861); THE INVESTIGATORS (II,1241); LOVE AT FIRST SIGHT (II,1530); LOVE AT FIRST SIGHT (II,1531); STOCKARD CHANNING IN JUST FRIENDS (II,2467); STOPWATCH: THIRTY MINUTES OF INVESTIGATIVE TICKING (II,2474)

LOCKHART, WARREN L
THE BORROWERS (I,693); HAPPY ANNIVERSARY, CHARLIE BROWN (I,1939); YOU'RE A GOOD MAN, CHARLIE BROWN (I,4962)

LOCKWOOD, GREY
JOHNNY CARSON PRESENTS THE SUN CITY SCANDALS (I,2420)

LOESCH, MARGARET
GOING BANANAS (II,1015); JIM HENSON'S MUPPET BABIES (II,1289); THE TRANSFORMERS (II,2653)

LOEW, BOB
ONCE UPON A TUNE (I,3372)

LOMAN, MICHAEL
LADIES' MAN (II,1423)

LOMBARD, ROBERT
LINDA RONSTADT IN CONCERT (II,1482)

LONDON, JERRY
CHIEFS (II,509); THE DORIS DAY SHOW (I,1355); ELLIS ISLAND (II,768); HOTEL (II,1192)

LONDON, JOHN
THE LORETTA YOUNG SHOW (I,2755); THE LORETTA YOUNG THEATER (I,2756)

LONDON, MARC
SINATRA AND FRIENDS (II,2358)

LONDON, RICK
BILLY JOEL—A TV FIRST (II,250)

LONG, ROBERT
REAL PEOPLE (II,2118)

LONGSTREET, HARRY
HOT PURSUIT (II,1189); TRAUMA CENTER (II,2655); VOYAGERS (II,2730)

LONGSTREET, RENEE
HOT PURSUIT (II,1189); TRAUMA CENTER (II,2655); VOYAGERS (II,2730)

LOOS, MARY
YANCY DERRINGER (I,4925)

LOPEZ, ALBERT
CROWN THEATER WITH GLORIA SWANSON (I,1107)

LORD, JACK
M STATION: HAWAII (II,1568)

LORD, PHILLIPS H
THE BLACK ROBE (I,479); GANGBUSTERS (I,1733)

LORD, STEPHEN
JOHNNY RINGO (I,2434); ZANE GREY THEATER (I,4979)

LOREN, JAMES
MITZI'S SECOND SPECIAL (I,3074)

LORIE, EUGENE
STORY THEATER (I,4240)

LORIN, WILL
SEARCH AND RESCUE: THE ALPHA TEAM (II,2283)

LORING, LYNN
GLITTER (II,1009); THE RETURN OF THE MOD SQUAD (I,3776)

LORRIMER, VERE
BLAKE'S SEVEN (II,264)

LOSSO, ERNEST
JAMES AT 15 (II,1270); LOVES ME, LOVES ME NOT (II,1551); NAKIA (II,1785); THE PAPER CHASE: THE SECOND YEAR (II,1950)

LOTTERBY, SYDNEY
BUTTERFLIES (II,404); LAST OF THE SUMMER WINE (II,1439); OPEN ALL HOURS (II,1913); YES MINISTER (II,2848)

LOUBERT, PATRICK
THE EDISON TWINS (II,761)

LOUIS, R J
LONE STAR (II,1513)

LOUNSBERRY, DAN
THE BELL TELEPHONE HOUR (I,389); YOUR HIT PARADE (I,4951)

LOURIE, MILES
THE BARRY MANILOW SPECIAL (II,155); THE SECOND BARRY MANILOW SPECIAL (II,2285); THE THIRD BARRY MANILOW SPECIAL (II,2590)

LOUVELLO, SAM
THE HEE HAW HONEYS (II,1124); THE HEE HAW HONEYS (II,1125)

LOVE, HARRY
JOKEBOOK (II,1339)

LOVENHEIM, ROBERT
GABE AND WALKER (II,950); NEWMAN'S DRUGSTORE (II,1837); THE SHERIFF AND THE ASTRONAUT (II,2326)

LOVETON, JOHN W
MR. AND MRS. NORTH (I,3124); MR. AND MRS. NORTH (I,3125); TOPPER (I,4582)

LOVULLO, SAM
HEE HAW (II,1123); THE NASHVILLE PALACE (II,1792); THE NASHVILLE PALACE (II,1793)

LOVY, ALEX
THE ALL-NEW POPEYE HOUR (II,49); THE AMAZING CHAN AND THE CHAN CLAN (I,157); BUTCH CASSIDY AND THE SUNDANCE KIDS (I,781); CAPTAIN CAVEMAN AND THE TEEN ANGELS (II,433); CASPER AND THE ANGELS (II,453); THE FANTASTIC FOUR (I,1529); THE FLINTSTONE FUNNIES (II,883); THE FLINTSTONES (II,885); FRED AND BARNEY MEET THE THING (II,919); JOSIE AND THE PUSSYCATS (I,2453); MOTOR MOUSE (I,3111); THE NEW SHMOO (II,1827); SCOOBY'S ALL-STAR LAFF-A-LYMPICS (II,2273); SCOOBY-DOO, WHERE ARE YOU? (II,2276); THE SUPER GLOBETROTTERS (II,2498); WHERE'S HUDDLES (I,4836)

LOWDEN, FRASER
CASANOVA (II,451)

LOWE, DAVID
LOVE STORY (I,2776); WHAT'S THE STORY? (I,4821)

LOWE, JAMES
TONY THE PONY (II,2641)

LOWEI, BOB
THE S. S. HOLIDAY (I,4173)

LOWEN, BARRY
LUCAN (II,1553); LUCAN (II,1554)

LOWENSTEIN, ABRAHAM
ESCAPE (II,782)

LOWENSTEIN, AL
MR. SUCCESS (II,1756); REPORT TO MURPHY (II,2134)

LOWEY, LAWRENCE
WOMEN OF RUSSIA (II,2824)

LUBER, BERNARD
THE ADVENTURES OF SUPERMAN (I,77)

LUBIN, ARTHUR
MR. ED (I,3137); PINE LAKE LODGE (I,3597)

LUCAS, ALEX
MICKEY SPILLANE'S MARGIN FOR MURDER (II,1691)

LUCAS, JOHN MEREDYTH
BEYOND WESTWORLD (II,227); STAR TREK (II,2440)

LUCAS, JONATHAN
DEAN MARTIN'S CHRISTMAS AT SEA WORLD (II,644)

LUDLAM, STUART
PETER HUNTER, PRIVATE EYE (I,3563)

LUDWIG, JERRY
ASSIGNMENT: MUNICH (I,302); ASSIGNMENT: VIENNA (I,304); BUNCO (II,395); FOR LOVE AND HONOR (II,903); JESSICA NOVAK (II,1286)

LUDWIG, JULIAN
HOLLYWOOD AND THE STARS (I,2081); LORNE GREENE'S LAST OF THE WILD (II,1522)

LUEDECKE, WENZEL
FLASH GORDON (I,1593)

LUFT, SID
THE JUDY GARLAND SHOW (I,2470); THE JUDY GARLAND SHOW (I,2471); THE JUDY GARLAND SHOW (II,1349)

LUMET, SIDNEY
MR. BROADWAY (I,3130)

LUNDINGTON, ALLEN
MATTY'S FUNDAY FUNNIES (I,2951)

LUNDQUIST, PAMELA
THE HOMEMADE COMEDY SPECIAL (II,1173)

LUNNEY, DAVID
ONCE UPON A TIME. . . IS NOW THE STORY OF PRINCESS GRACE (II,1899)

LUPO, FRANK
THE A-TEAM (II,119); FOUR EYES (II,912); GALACTICA 1980 (II,953); THE GREATEST AMERICAN HERO (II,1060); HUNTER (II,1206); LOBO (II,1504); THE QUEST (II,2098); RIPTIDE (II,2178)

LUSITANA, DONNA E
THE LOVE REPORT (II,1542); THE LOVE REPORT (II,1543)

LUTZ, DOUGLAS
COLISEUM (I,1000)

LUTZ, SAM
THE LAWRENCE WELK SHOW (I,2643); THE LAWRENCE WELK SHOW (I,2644); LAWRENCE WELK'S TOP TUNES AND NEW TALENT (I,2645); MEMORIES WITH LAWRENCE WELK (II,1681)

LYDON, JAMES
ROLL OUT! (I,3833); TEMPLE HOUSTON (I,4392)

LYLES, A C
A CHRISTMAS FOR BOOMER (II,518); FLATFOOTS (II,879); HERE'S BOOMER (II,1134)

LYON, EARL
THE TALES OF WELLS FARGO (I,4353)

LYON, RONALD
RIPLEY'S BELIEVE IT OR NOT (II,2177)

LYONS, KENNY
JACK CARTER AND COMPANY (I,2305)

LYONS, RICHARD E
THE DAUGHTERS OF JOSHUA CABE (I,1169); THE DAUGHTERS OF JOSHUA CABE RETURN (I,1170); ROUGHNECKS (II,2219)

LYONS, ROBERT E
KATE BLISS AND THE TICKER TAPE KID (II,1366)

MACANDREW, JACK
EVENING AT THE IMPROV (II,786)

MACCALMAN, KRISTY
FOUL UPS, BLEEPS AND BLUNDERS (II,911)

MACDONALD, DAVID
THE FLYING DOCTOR (I,1610)

MACDONNELL, NORMAN
GUNSMOKE (II,1069)

MACDOUGALL, RONALD
FAME IS THE NAME OF THE GAME (I,1518)

MACDOWELL, NORMAN
THE VIRGINIAN (I,4738)

MACLENNAN, IAN
LET'S MAKE A DEAL (II,1463); PITFALL (II,2047)

MACNAUGHTON, IAN
MONTY PYTHON'S FLYING CIRCUS (II,1729)

MACK, DICK
I MARRIED JOAN (I,2174)

MACK, HELEN
MEET CORLISS ARCHER (I,2979)

MACK, KAREN
CASS MALLOY (II,454); THIS IS KATE BENNETT (II,2597)

MACK, TED
TED MACK AND THE ORIGINAL AMATEUR HOUR (I,4370)

MACNEE, ROBERT F
EVENING AT THE IMPROV (II,786)

MADDEN, JERRY
ALICE (II,33); DOROTHY (II,717); STEAMBATH (II,2451); YOU CAN'T TAKE IT WITH YOU (II,2857)

MAFFEO, GAYLE
HIS AND HERS (II,1155); OPEN ALL NIGHT (II,1914)

MAFFEO, NEIL T
APPLE'S WAY (II,101); THE HUSTLER OF MUSCLE BEACH (II,1210)

MAGISTRETTI, PAUL
THE OUTSIDE MAN (II,1937)

MAHIN, JOHN LEE
RIVAK, THE BARBARIAN (I,3795)

MAIBAUM, RICHARD
JARRETT (I,2349); MAISIE (I,2833)

MAJORS, CAROL
ELKE SOMMER'S WORLD OF SPEED AND BEAUTY (II,765)

MAJORS, LEE
THE FALL GUY (II,811); HOW DO I KILL A THIEF—LET ME COUNT THE WAYS (II,1195)

MAJORS, VERN
ELKE SOMMER'S WORLD OF SPEED AND BEAUTY (II,765)

MALLETT, DAVID
THE KENNY EVERETT VIDEO SHOW (II,1379)

MALLEY, HOWARD
HOT HERO SANDWICH (II,1186); SHOW BUSINESS (II,2343); YOUNG HEARTS (II,2865)

MALMUTH, DAN
FOUL UPS, BLEEPS AND BLUNDERS (II,911)

MALONE, ADRIAN
COSMOS (II,579)

MALONE, JOEL
SATAN'S WAITIN' (I,3916); THE WHISTLER (I,4843)

MALONE, NANCY
THE BIONIC WOMAN (II,255); HANDLE WITH CARE (II,1077)

MALONEY, DAVID
BLAKE'S SEVEN (II,264); DAY OF THE TRIFFIDS (II,627)

MALTZ, RENEE
KANGAROOS IN THE KITCHEN (II,1362)

MANBY, ROBERT

LENA HORNE: THE LADY AND HER MUSIC (II,1460)

MANDEL, ALAN
COPS (I,1047)

MANDELKER, PHILIP
BETWEEN THE LINES (II,225); THE FITZPATRICKS (II,868); HIGH SCHOOL, U.S.A. (II,1148); THE POSSESSED (II,2071); TUCKER'S WITCH (II,2667)

MANDRELL, BARBARA
BARBARA MANDRELL—THE LADY IS A CHAMP (II,144)

MANDUKE, JOE
FIREHOUSE (I,1579); FIREHOUSE (II,859)

MANETTA, JOE
MO AND JO (II,1715)

MANHEIM, MANNIE
GEORGE JESSEL'S SHOW BUSINESS (I,1770)

MANHEIM, MICHAEL
WHATEVER HAPPENED TO DOBIE GILLIS? (II,2774)

MANILOW, BARRY
BARRY MANILOW—ONE VOICE (II,156); THE SECOND BARRY MANILOW SPECIAL (II,2285); THE THIRD BARRY MANILOW SPECIAL (II,2590)

MANINGS, ALLAN
THE DAVID FROST REVUE (I,1176); GABE AND WALKER (II,950); GOOD TIMES (II,1038); NUMBER 96 (II,1869); ONE DAY AT A TIME (II,1900); ROXY PAGE (II,2222)

MANKIEWICZ, DON
LANIGAN'S RABBI (II,1428)

MANKIEWICZ, JOSEPH L
CAROL FOR ANOTHER CHRISTMAS (I,852)

MANN, ABBY
MEDICAL STORY (II,1676); SKAG (II,2374)

MANN, DANIEL
TELLER OF TALES (I,4388); THICKE OF THE NIGHT (II,2587)

MANN, MICHAEL
MIAMI VICE (II,1689)

MANN, ROBERT
BOSS LADY (I,694)

MANNERS, SAM
BREAKING AWAY (II,371); HARDCASE (II,1088); PEARL (II,1986); ROUTE 66 (I,3852); SPARROW (II,2410)

MANNES, ELENA
WALTER CRONKITE'S UNIVERSE (II,2739)

MANO, SUSAN
A HOT SUMMER NIGHT WITH DONNA (II,1190)

MANOFF, BILL
MAGGIE (I,2810)

MANSFIELD, IRVING
ARTHUR GODFREY'S TALENT SCOUTS (I,291); CELEBRITY TALENT SCOUTS (I,886); DICK CLARK'S WORLD OF TALENT (I,1268); FACE THE FACTS (I,1508); JACQUELINE SUSANN'S VALLEY OF THE DOLLS, 1981 (II,1268); THE MOREY AMSTERDAM SHOW (I,3101); NIGHT EDITOR (I,3286); ON BROADWAY TONIGHT (I,3355); THE SAM LEVENSON SHOW (I,3897); STOP ME IF YOU HEARD THIS ONE (I,4231); THE STORK CLUB (I,4236); TAKE A GOOD LOOK (I,4332); TAKE A GOOD LOOK (II,2529); TALENT SEARCH (I,4342); THIS IS SHOW BUSINESS (I,4459)

MANSFIELD, MIKE
ELTON JOHN IN CENTRAL PARK (II,769); TEDDY PENDERGRASS IN CONCERT (II,2551); TWIGGY'S JUKE BOX (II,2682)

MANSON, DAVID
THE WORD (II,2833)

MANTLEY, JOHN
BUCK ROGERS IN THE 25TH CENTURY (II,384); DIRTY SALLY (II,680); GUNSMOKE (II,1069); HOW THE WEST WAS WON (II,1196); THE MACAHANS (II,1583); THE WILD WILD WEST (I,4863)

MANULIS, JOHN BARD
THE COMEDY ZONE (II,559)

MANULIS, MARTIN
ARSENIC AND OLD LACE (I,268); THE BLUE MEN (I,492); CHIEFS (II,509); CLIMAX! (I,976); CRIME PHOTOGRAPHER (I,1097); DOBIE GILLIS (I,1302); FIVE FINGERS (I,1586); THE GOLDEN AGE OF TELEVISION (II,1018); THE GUARDSMAN (I,1893); JAMES AT 15 (II,1270); JAMES AT 15 (II,1271); THE MAN WHO CAME TO DINNER (I,2880); PANAMA HATTIE (I,3444); THE PHILADELPHIA STORY (I,3581); PLAYHOUSE 90 (I,3623); THE ROYAL FAMILY (I,3861); THE SHOWOFF (I,4035); SUSPENSE (I,4305)

MAPES, JACQUES
THE LIVES OF JENNY DOLAN (II,1502); THE MONEYCHANGERS (II,1724)

MARCH, ALEX
KRAFT TELEVISION THEATER (I,2592); PENTAGON U.S.A. (I,3514); SWISS FAMILY ROBINSON (I,4324)

MARCHETTA, CAMILLE
NURSE (II,1870); SCRUPLES (II,2281)

MARCIONE, GENE
BLUE JEANS (II,275); C.P.O. SHARKEY (II,407); CHARACTERS (II,481); GHOST OF A CHANCE (II,987); GOING BANANAS (II,1015); GREAT DAY (II,1053); HOT L BALTIMORE (II,1187); JACKIE AND DARLENE (II,1265); KANGAROOS IN THE KITCHEN (II,1362); MAKING IT (II,1605); NUTS AND BOLTS (II,1871); PIPER'S PETS (II,2046); POTTSVILLE (II,2072); THE RAG BUSINESS (II,2106); TEACHERS ONLY (II,2547); WENDY HOOPER—U. S. ARMY (II,2764); YOU'RE JUST LIKE YOUR FATHER (II,2861)

MARCUS, ANNE
FALCON CREST (II,808); THE LIFE AND TIMES OF EDDIE ROBERTS (II,1468)

MARCUS, DEVERE
COMPLETELY OFF THE WALL (II,560)

MARCUS, ELLIS
THE LIFE AND TIMES OF EDDIE ROBERTS (II,1468)

MARCUS, LARRY
ONE STEP BEYOND (I,3388)

MARGOLIN, ARNOLD
THE DOOLEY BROTHERS (II,715); GOING PLACES (I,1834); HARRY AND MAGGIE (II,1097); THE MCLEAN STEVENSON SHOW (II,1665); THE ORPHAN AND THE DUDE (II,1924); PRIVATE BENJAMIN (II,2087); WALKIN' WALTER (II,2737)

MARGOLIS, JEFF
PERRY COMO'S CHRISTMAS IN ENGLAND (II,2008); PERRY COMO'S CHRISTMAS IN NEW YORK (II,2011)

MARGULIES, STAN
MEN OF THE DRAGON (II,1683); ROOTS (II,2211); ROOTS: THE NEXT GENERATIONS (II,2213); THE THORN BIRDS (II,2600)

MARION, CHARLES
THUNDER (II,2617)

MARK, DAN
COCAINE AND BLUE EYES (II,548); HIGH FIVE (II,1143)

MARKELL, BOB
BICENTENNIAL MINUTES (II,229); THE DAIN CURSE (II,613); N.Y.P.D. (I,3321)

MARKES, LARRY
WE'LL TAKE MANHATTAN (I,4792)

MARKHAM, DICK
THAT'S MY LINE (II,2579)

MARKLE, FLETCHER
THE BARRETTS OF WIMPOLE STREET (I,361); FORD TELEVISION THEATER (I,1634); FRONT ROW CENTER (I,1694); JOHNNY BELINDA (I,2415); LIFE WITH FATHER (I,2690); STUDIO ONE (I,4268); THAT'S HOLLYWOOD (II,2577); THRILLER (I,4492); TONIGHT IN HAVANA (I,4557); YOUNG IN HEART (I,4939)

MARKOWITZ, GARY
UNITED STATES (II,2708)

MARKOWITZ, RIFF
FUNNY FACES (II,941); GEORGE BURNS: AN HOUR OF JOKES AND SONGS (II,973); THE HILARIOUS HOUSE OF FRIGHTENSTEIN (II,1153); THE HITCHHIKER (II,1157); RED SKELTON'S CHRISTMAS DINNER (II,2123); RED SKELTON: A ROYAL PERFORMANCE (II,2125); THE WOLFMAN JACK SHOW (II,2817)

MARKS, ARTHUR
PERRY MASON (II,2025)

MARKS, BRAD
WHEN THE WEST WAS FUN: A WESTERN REUNION (II,2780)

MARKS, DENNIS
SPIDER-MAN AND HIS AMAZING FRIENDS (II,2425)

MARKS, HILLIARD
THE JACK BENNY PROGRAM (I,2294)

MARKS, JOAN
CHRIS AND THE MAGICAL DRIP (II,516)

MARKS, LAWRENCE
DOC (II,683); MAKING IT (II,1605); YOU'RE JUST LIKE YOUR FATHER (II,2861)

MARKS, SHERMAN
THE BERT PARKS SHOW (I,400); THE BILL GOODWIN SHOW (I,447); COSMOPOLITAN THEATER (I,1057); THE PAUL WINCHELL AND JERRY MAHONEY SHOW (I,3492); TONI TWIN TIME (I,4554)

MARKS, SUSAN
FIGHT BACK: WITH DAVID HOROWITZ (II,854)

MARLOWE, ANNE
TELLER OF TALES (I,4388)

MARLOWE, HARVEY
ABC-POWERS CHARM SCHOOL (I,5); DETECT AND COLLECT (I,1242); HOTEL BROADWAY (I,2127); THE IREENE WICKER SHOW (I,2234); TELETIPS ON LOVELINESS (I,4380)

MARMELSTEIN, LINDA
THE ABC AFTERSCHOOL SPECIAL (II,1); LITTLE VIC (II,1496); THE SECRET OF CHARLES DICKENS (II,2293)

MARMER, MIKE
LANCELOT LINK, SECRET CHIMP (I,2609)

MARON, LINDA
ALIVE AND WELL (II,36)

MARRERO, FRANK
THE NEW VOICE (II,1833)

MARSH, LINDA
THE FACTS OF LIFE (II,805); FACTS OF LIFE: THE FACTS OF LIFE GOES TO PARIS (II,806)

MARSH, SY
THE NBC FOLLIES (I,3241)

MARSHALL, ALEXANDER
THE MADHOUSE BRIGADE (II,1588)

MARSHALL, FRANK
HEROES AND SIDEKICKS—INDIANA JONES AND THE TEMPLE OF DOOM (II,1140)

MARSHALL, GARRY K
ANGIE (II,80); BEANE'S OF BOSTON (II,194); BLANSKY'S BEAUTIES (II,267); DOMINIC'S DREAM (II,704); EVIL ROY SLADE (I,1485); HAPPY DAYS (II,1084); HERNDON AND ME (II,1139); HEY, LANDLORD (I,2039); JOANIE LOVES CHACHI (II,1295); LAVERNE AND SHIRLEY (II,1446); LEGS (II,1459); THE LITTLE PEOPLE (I,2716); ME AND THE CHIMP (I,2974); MORK AND MINDY (II,1735); THE MURDOCKS AND THE MCCLAYS (I,3163); THE NEW ODD COUPLE (II,1825); THE ODD COUPLE (II,1875); SHERIFF WHO? (I,4009); WALKIN' WALTER (II,2737); WHO'S WATCHING THE KIDS? (II,2793); WIVES (II,2812)

MARSHALL, NEIL
THE BLUE JEAN NETWORK (II,274); CELEBRITY SWEEPSTAKES (II,476); MUSIC CENTRAL (II,1771); NATIONAL LAMPOON'S HOT FLASHES (II,1795); THE PETER MARSHALL VARIETY SHOW (II,2029); THE WIZARD OF ODDS (I,4886)

MARSHALL, PETER
PETER MARSHALL SALUTES THE BIG BANDS (II,2028)

MARSHALL, SIDNEY
CITY BENEATH THE SEA (I,965); THE NEW ADVENTURES OF CHARLIE CHAN (I,3249)

MARSHALL, STEVE
CASS MALLOY (II,454); GLORIA (II,1010); OFF THE RACK (II,1878)

MARSHALL, TONY
BEANE'S OF BOSTON (II,194); BLANSKY'S BEAUTIES (II,267); DOMINIC'S DREAM (II,704); HAPPY DAYS (II,1084); LAVERNE AND SHIRLEY (II,1446); MORK AND MINDY (II,1735); THE NEW ODD COUPLE (II,1825); THE ODD COUPLE (II,1875); WHO'S WATCHING THE KIDS? (II,2793); WIVES (II,2812)

MARSON, BRUCE
GOOD DAY! (II,1030); SUMMER SOLSTICE (II,2492)

MARTI, JILL
CANDIDA (II,423)

MARTIN, ANTHONY S
KEY WEST (I,2528)

MARTIN, CHARLES
PHILIP MORRIS PLAYHOUSE (I,3585); THE TELLTALE CLUE (I,4389)

MARTIN, CRAIG
THE OSMOND FAMILY SHOW (II,1927)

MARTIN, DICK
ROWAN AND MARTIN BITE THE HAND THAT FEEDS THEM (I,3853)

MARTIN, PAUL
THE VIDEO GAME (II,2727)

MARTIN, QUINN
ALL ABOUT BARBARA (I,125); BANYON (I,344); BARNABY JONES (II,153); BERT D'ANGELO/SUPERSTAR (II,211); CANNON (I,820); CANNON (II,424); CARIBE (II,439); THE CITY (II,541); CODE NAME: DIAMOND HEAD (II,549); CRISIS (I,1102); CROSSFIRE (II,606); DAN AUGUST (I,1130); THE FBI (I,1551); THE FUGITIVE (I,1701); THE HOUSE ON GREENAPPLE ROAD (I,2140); INTERTECT (I,2228); THE INVADERS (I,2229); A MAN CALLED SLOANE (II,1613); MANHUNTER (II,1620); MANHUNTER (II,1621); MOST WANTED (II,1738); MOST WANTED (II,1739); THE NEW BREED (I,3258); THE RUNAWAYS (II,2231); THE STREETS OF SAN FRANCISCO (I,4260); THE STREETS OF SAN FRANCISCO (II,2478); TALES OF THE UNEXPECTED (II,2539); TRAVIS LOGAN, D.A. (I,4598); TWELVE O'CLOCK HIGH (I,4637); THE UNTOUCHABLES (I,4681); WINNER TAKE ALL (II,2808)

MARTIN, STEVE
DOMESTIC LIFE (II,703); THE JERK, TOO (II,1282); STEVE MARTIN'S THE WINDS OF WHOOPIE (II,2461); TWILIGHT THEATER (II,2685); TWILIGHT THEATER II (II,2686)

MARTIN, TOBY
BIG CITY COMEDY (II,233); LAUGH TRAX (II,1443); MEL AND SUSAN TOGETHER (II,1677); ROCK COMEDY (II,2191); ROCK PALACE (II,2193)

MARTIN, VIRGINIA
THE MAGIC GARDEN (II,1591)

MARTINELLI, JOHN A
MARCO POLO (II,1626)

MARTINSON, LESLIE H
THE ROY ROGERS SHOW (I,3859)

MARX, ARTHUR
MICKEY (I,3023)

MARX, MARVIN
THE BETTY HUTTON SHOW (I,413); THAT'S LIFE (I,4426)

MARX, SAMUEL
THE 20TH CENTURY-FOX HOUR (I,4642); DECEMBER BRIDE (I,1215); THE THIN MAN (I,4446); THOSE WHITING GIRLS (I,4471)

MASCHIO, MICHAEL
NO MAN'S LAND (II,1855); SCARECROW AND MRS. KING (II,2268)

MASIUS, JOHN
ST. ELSEWHERE (II,2432)

MASLANSKY, PAUL
THE GUN AND THE PULPIT (II,1067)

MASON, PAUL
ANOTHER DAY (II,94); BRENDA STARR (II,372); CHIPS (II,511); GET CHRISTIE LOVE! (II,981); IRONSIDE (II,1246); IT TAKES A THIEF (I,2250); MANIMAL (II,1622); MCMILLAN AND WIFE (II,1667); ONCE UPON A DEAD MAN (I,3369); SAVAGE (I,3924); WHATEVER HAPPENED TO DOBIE GILLIS? (II,2774)

MASSARI, MARK
RONA LOOKS AT KATE, PENNY, TONI, AND CINDY (II,2206); TAKE MY ADVICE (II,2531)

MASTERS, SUSAN
BLOCKHEADS (II,270)

MASTERSON, JOHN
BRIDE AND GROOM (I,721); LIVE LIKE A MILLIONAIRE (I,2728); SEARCH FOR BEAUTY (I,3953)

MASTERSON, PAUL
AUCTION AIRE (I,313)

MATHESON, MARGARET
KENNEDY (II,1377)

MAUZER, MERRILL
TOM COTTLE: UP CLOSE (II,2631)

MAXIM, ERNEST
THE MORECAMBE AND WISE SHOW (II,1734)

MAXWELL, ROBERT
THE ADVENTURES OF SUPERMAN (I,77); CANNONBALL (I,822); THE DEFENDERS (I,1220); FATHER OF THE BRIDE (I,1543); JEFF'S COLLIE (I,2356); NATIONAL VELVET (I,3231)

MAXWELL-SMITH, JOHN
HOT POTATO (II,1188)

MAY, PHIL
DISNEYLAND'S 25TH ANNIVERSARY (II,681); KRAFT SALUTES WALT DISNEY WORLD'S 10TH ANNIVERSARY (II,1410); THE MOUSEKETEERS REUNION (II,1742)

MAY, WAYLEEN
TWILIGHT THEATER II (II,2686)

MAYBERRY, ROBERT
FASHION MAGIC (I,1536)

MAYBERRY, RUSS
THE BLACK SHEEP SQUADRON (II,262)

MAYER, GERALD
O'HARA, UNITED STATES TREASURY (I,3346); POLICE SURGEON (I,3639); THE SWISS FAMILY ROBINSON (II,2518)

MAYER, JERRY
THE ACADEMY (II,3); THE ACADEMY II (II,4); BROTHERS AND SISTERS (II,382); THE FACTS OF LIFE (II,805); FACTS OF LIFE: THE FACTS OF LIFE GOES TO PARIS (II,806); FAY (II,844); JO'S COUSINS (II,1293); TABITHA (II,2527); TABITHA (II,2528)

MAYER, PAUL
RYAN'S HOPE (II,2234)

MAYER, PHIL
RIPLEY'S BELIEVE IT OR NOT (II,2177); THE VIRGINIA GRAHAM SHOW (I,4736)

MAYO, NICK
WITNESS (I,4885)

MAZZUCA, JOE
BLACKSTAR (II,263); THE TARZAN/LONE RANGER/ZORRO ADVENTURE HOUR (II,2543)

MCADAMS, JACK
THE GANGSTER CHRONICLES (II,957)

MCADAMS, JAMES
THE EDDIE CAPRA MYSTERIES (II,758); THE GANGSTER CHRONICLES (II,957); HELLINGER'S LAW (II,1127); KATE LOVES A MYSTERY (II,1367); KOJAK (II,1406); MRS. COLUMBO (II,1760); THE ROAD WEST (I,3801); SHANNON (II,2314); THE VIRGINIAN (I,4738)

MCAFFE, WES
THE JAMES BOYS (II,1272); LET'S CELEBRATE (I,2662); NANNY AND THE PROFESSOR (I,3225); THE PHIL SILVERS ARROW SHOW (I,3576); TELE-VARIETIES (I,4377)

MCAULIFFE, JODI
TOM COTTLE: UP CLOSE (II,2631)

MCAVITY, TOM
MEET CORLISS ARCHER (I,2979)

MCAVOY, JEAN
BEN VEREEN. . . COMIN' AT YA (II,203)

MCCALLUM, JOHN
SKIPPY, THE BUSH KANGAROO (I,4076)

MCCANN, ELIZABETH
PIAF (II,2039)

MCCARTHY, KEVIN
JEOPARDY (II,1280)

MCCLEAN, MICHAEL
THE MONSTER SQUAD (II,1725) SCARECROW AND MRS. KING (II,2268)

MCCLEERY, ALBERT
CAMEO THEATER (I,808); COSMOPOLITAN THEATER (I,1057); THE FURTHER ADVENTURES OF ELLERY QUEEN (I,1719); JANE WYMAN PRESENTS THE FIRESIDE THEATER (I,2345); HAMLET (I,1924); LITTLE WOMEN (I,2721); KING RICHARD II (I,2559); MACBETH (I,2800); MASTERPIECE PLAYHOUSE (I,2942); MATINEE THEATER (I,2947); MIDNIGHT MYSTERY (I,3030)

MCCONNOR, VINCENT
THE BARRETTS OF WIMPOLE STREET (I,361); LIFE WITH FATHER (I,2690); THE WEB (I,4785)

MCCORMACK, JAN
THE SQUARE WORLD OF ED BUTLER (I,4171)

MCCORMICK, PATRICK
THE ABC AFTERSCHOOL SPECIAL (II,1)

MCCOY, JACK
GLAMOUR GIRL (I,1817)

MCCRAY, KENT
FATHER MURPHY (II,841); HIGHWAY TO HEAVEN (II,1152); LITTLE HOUSE ON THE PRAIRIE (II,1487); LITTLE HOUSE: A NEW BEGINNING (II,1488); LITTLE HOUSE: BLESS ALL THE DEAR CHILDREN (II,1489); LITTLE HOUSE: LOOK BACK TO YESTERDAY (II,1490); LITTLE HOUSE: THE LAST FAREWELL (II,1491)

MCCULLOUGH, ROBERT L
FALCON CREST (II,808)

MCCUNE, HANK
THE HANK MCCUNE SHOW (I,1932)

MCCUTCHEN, BILL
IKE (II,1223); SCARECROW AND MRS. KING (II,2268); SKEEZER (II,2376)

MCDERMOTT, TOM
CONCERNING MISS MARLOWE (I,1032); THE DICK POWELL SHOW (I,1269); LAS VEGAS PALACE OF STARS (II,1431); LASSIE: THE NEW BEGINNING (II,1433); PURSUE AND DESTROY (I,3699)

MCDONALD, KATHY
BACKSTAGE PASS (II,132)

MCDOUGALL, IAN
20 MINUTE WORKOUT (II,2680); THE EDISON TWINS (II,761)

MCEDWARDS, JACK
MR. LUCKY (I,3141); PETER GUNN (I,3562)

MCELROY, HAL
RETURN TO EDEN (II,2143)

MCEUEN, JOHN
STEVE MARTIN: COMEDY IS NOT PRETTY (II,2462)

MCEVEETY, BERNARD
CIMARRON STRIP (I,954)

MCGEE, JAMES
TALES FROM THE DARKSIDE (II,2535)

MCGEEHAN, JACK
THE ED SULLIVAN SHOW (I,1401)

MCGHAN, JOHN
THE DANCE SHOW (II,616)

MCGINN, JIM
THE BEAN SHOW (I,373)

MCGOOHAN, PATRICK
THE PRISONER (I,3670)

MCGOWAN, DARRELL
DEATH VALLEY DAYS (I,1211); THE LITTLEST HOBO (I,2727); THE SILENT SERVICE (I,4049); SKY KING (I,4077)

MCGOWAN, STUART
THE LITTLEST HOBO (I,2727)

MCGRADY, PHYLLIS
SHOW BUSINESS (II,2343)

MCGUIRE, DON
DON'T CALL ME CHARLIE (I,1349); HENNESSEY (I,2013); MCGHEE (I,2967)

MCINTOSH, BOB
THE ASSASSINATION RUN (II,112); RUNNING BLIND (II,2232)

MCINTOSH, EZRA
FAIRMEADOWS, U.S.A. (I,1514)

MCKAY, BRUCE
THE TOMORROW SHOW (II,2635)

MCKAY, JIM
THE REAL MCKAY (I,3742)

MCKEAND, CAROL EVAN
CASSIE AND COMPANY (II,455); FAMILY (II,813); RYAN'S FOUR (II,2233)

MCKEAND, MARTIN
WHEN HAVOC STRUCK (II,2778)

MCKEAND, NIGEL
CASSIE AND COMPANY (II,455); FAMILY (II,813); THE FAMILY TREE (II,820); NORMA RAE (II,1860); RYAN'S FOUR (II,2233); THE SIX OF US (II,2369)

MCKENNA, WILLIAM
THE POINTER SISTERS (II,2059)

MCKEOWN, ALLAN
MY WIFE NEXT DOOR (II,1781)

MCKINLEY, BARRY
DOWN YOU GO (I,1367); SUSAN'S SHOW (I,4303)

MCKNIGHT, TOM
THE LIFE OF RILEY (I,2686); MAISIE (I,2833); MCKEEVER AND THE COLONEL (I,2970); OH, THOSE BELLS! (I,3345); PONY EXPRESS (I,3645); THE ROARING TWENTIES (I,3805); TOO MANY SERGEANTS (I,4573)

MCLAIRD, ARTHUR
BIG FOOT AND WILD BOY (II,237); RIVKIN: BOUNTY HUNTER (II,2183)

MCLAREN, IVOR
DON MCNEILL'S TV CLUB (I,1337)

MCLEOD, VIC
BELIEVE IT OR NOT (I,387); BROADWAY OPEN HOUSE (I,736); THE CHEVROLET TELE-THEATER (I,926); THE R.C.A. THANKSGIVING SHOW (I,3736); SO YOU WANT TO LEAD A BAND (I,4106); WELCOME ABOARD (I,4789)

MCMAHON, JENNA
MAMA'S FAMILY (II,1610); A NEW KIND OF FAMILY (II,1819); SOAP (II,2392)

MCMAHON, JOHN J

THE STOCKERS (II,2469); TELEVISION'S GREATEST COMMERCIALS (II,2553); TELEVISION'S GREATEST COMMERCIALS III (II,2555); TELEVISION'S GREATEST COMMERCIALS IV (II,2556); TV'S BLOOPERS AND PRACTICAL JOKES (II,2675); WHY US? (II,2796)

MCMILLEN, PATRICIA
THE PHIL DONAHUE SHOW (II,2032)

MCNALLY, TERRENCE
MAMA MALONE (II,1609)

MCNEELY, JERRY
BREAKING AWAY (II,371); LUCAS TANNER (II,1555); THREE FOR THE ROAD (II,2607)

MCPHEE, DANIEL
LEGMEN (II,1458); THE SEAL (II,2282)

MCPHIE, JERRY
DIANA (I,1258); GOSSIP (II,1044); I'VE HAD IT UP TO HERE (II,1221); THE JERRY REED WHEN YOU'RE HOT, YOU'RE HOT HOUR (I,2372); THE JOHN BYNER COMEDY HOUR (I,2407); LILY (I,2705); MARIE (II,1629); MEL AND SUSAN TOGETHER (II,1677); THE OSMOND FAMILY CHRISTMAS SPECIAL (II,1926); THE OSMOND FAMILY THANKSGIVING SPECIAL (II,1928); TOO CLOSE FOR COMFORT (II,2642)

MCRAE, JOHN
JANE EYRE (II,1273); LITTLE WOMEN (I,2723)

MCRAVEN, DALE
9 TO 5 (II,1852); ALMOST HEAVEN (II,57); ANGIE (II,80); THE BETTY WHITE SHOW (II,224); HOME ROOM (II,1172); MORK AND MINDY (II,1735); THE TEXAS WHEELERS (II,2568); YOUR PLACE OR MINE? (II,2874)

MEADOW, HERB
THE MAN FROM BLACKHAWK (I,2861)

MEAGHER, KEVIN
ANYTHING FOR MONEY (II,99)

MEDANN, ARLYNE
TONY ORLANDO AND DAWN (II,2639)

MEDLINSKY, HARVEY
PLAZA SUITE (II,2056)

MEEGAN, JACK
THE ED SULLIVAN SHOW (I,1401)

MEIER, DON
WELCOME TRAVELERS (I,4791)

MELAMED, LOU
THE SHOW GOES ON (I,4030)

MELCHER, TERRY
THE DORIS DAY SHOW (I,1355); THE DORIS MARY ANNE KAPPELHOFF SPECIAL (I,1356)

MELENDEZ, BILL
BE MY VALENTINE, CHARLIE BROWN (I,369); THE CHARLIE BROWN AND SNOOPY SHOW (II,484); A CHARLIE BROWN CHRISTMAS (I,908); A CHARLIE BROWN THANKSGIVING (I,909); CHARLIE BROWN'S ALL STARS (I,910); HE'S YOUR DOG, CHARLIE BROWN (I,2037); IS THIS GOODBYE, CHARLIE BROWN? (I,2242); IT WAS A SHORT SUMMER, CHARLIE BROWN (I,2252); IT'S AN ADVENTURE, CHARLIE BROWN (I,2266); IT'S ARBOR DAY, CHARLIE BROWN (I,2267); IT'S FLASHBEAGLE, CHARLIE BROWN (I,2268); IT'S MAGIC, CHARLIE BROWN (I,2271); IT'S THE EASTER BEAGLE, CHARLIE BROWN (I,2275); IT'S THE GREAT PUMPKIN, CHARLIE BROWN (I,2276); IT'S YOUR FIRST KISS, CHARLIE BROWN (I,2280); LIFE IS A CIRCUS, CHARLIE BROWN (I,2683); PLAY IT AGAIN, CHARLIE BROWN (I,3617); SHE'S A GOOD SKATE, CHARLIE BROWN (I,4012); SNOOPY'S GETTING MARRIED, CHARLIE BROWN (I,4102); SOMEDAY YOU'LL FIND HER, CHARLIE BROWN (I,4113); THERE'S NO TIME FOR LOVE, CHARLIE BROWN (I,4438); WHAT A NIGHTMARE, CHARLIE BROWN (I,4806); WHAT HAVE WE LEARNED, CHARLIE BROWN? (I,4810); YES, VIRGINIA, THERE IS A SANTA CLAUS (II,2849); YOU'RE A GOOD SPORT, CHARLIE BROWN (I,4963); YOU'RE IN LOVE, CHARLIE BROWN (I,4964); YOU'RE NOT ELECTED, CHARLIE BROWN (I,4967); YOU'RE THE GREATEST, CHARLIE BROWN (I,4973)

MELLING, RICK
THE BEACH BOYS IN CONCERT (II,189)

MELMAN, JEFF
LAVERNE AND SHIRLEY (II,1446); NIGHT COURT (II,1844)

MELMAN, JERRY
MAKING THE GRADE (II,1606)

MELNICK, DAN
THE DESPERATE HOURS (I,1239); THE GENERATION GAP (I,1759); HE AND SHE (I,1985); JERICHO (I,2358); A LAST LAUGH AT THE 60'S (I,2627); MR. BROADWAY (I,3131); N.Y.P.D. (I,3321); RUN, BUDDY, RUN (I,3870)

MEMISHIAR, STEPHEN E
GRAFFITI ROCK (II,1049)

MENDELSOHN, DEBORAH
IT'S NOT EASY (II,1256)

MENDELSOHN, JACK
SHIPSHAPE (II,2329)

MENDELSON, BILL
IT'S AN ADVENTURE, CHARLIE BROWN (I,2266)

MENDELSON, LEE
BE MY VALENTINE, CHARLIE BROWN (I,369); THE CHARLIE BROWN AND SNOOPY SHOW (II,484); A CHARLIE BROWN CHRISTMAS (I,908); A CHARLIE BROWN THANKSGIVING (I,909); CHARLIE BROWN'S ALL STARS (I,910); EIGHT IS ENOUGH (II,762); THE FABULOUS FUNNIES (II,801); THE FANTASTIC FUNNIES (II,825); HAPPY ANNIVERSARY, CHARLIE BROWN (I,1939); HE'S YOUR DOG, CHARLIE BROWN (I,2037); HOT DOG (I,2122); IS THIS GOODBYE, CHARLIE BROWN? (I,2242); IT WAS A SHORT SUMMER, CHARLIE BROWN (I,2252); IT'S A MYSTERY, CHARLIE BROWN (I,2259); IT'S ARBOR DAY, CHARLIE BROWN (I,2267); IT'S FLASHBEAGLE, CHARLIE BROWN (I,2268); IT'S MAGIC, CHARLIE BROWN (I,2271); IT'S THE EASTER BEAGLE, CHARLIE BROWN (I,2275); IT'S THE GREAT PUMPKIN, CHARLIE BROWN (I,2276); IT'S YOUR FIRST KISS, CHARLIE BROWN (I,2280); LIFE IS A CIRCUS, CHARLIE BROWN (I,2683); PLAY IT AGAIN, CHARLIE BROWN (I,3617); THE ROD MCKUEN SPECIAL (I,3826); SHE'S A GOOD SKATE, CHARLIE BROWN (I,4012); SNOOPY'S GETTING MARRIED, CHARLIE BROWN (I,4102); SOMEDAY YOU'LL FIND HER, CHARLIE BROWN (I,4113); SUNDAY FUNNIES (II,2493); THERE'S NO TIME FOR LOVE, CHARLIE BROWN (I,4438); TRAVELS WITH CHARLEY (I,4597); WHAT A NIGHTMARE, CHARLIE BROWN (I,4806); WHAT HAVE WE LEARNED, CHARLIE BROWN? (I,4810); THE WONDERFUL WORLD OF PIZZAZZ (I,4901); YOU ASKED FOR IT (II,2855); YOU'RE A GOOD MAN, CHARLIE BROWN (I,4962); YOU'RE A GOOD SPORT, CHARLIE BROWN (I,4963); YOU'RE IN LOVE, CHARLIE BROWN (I,4964); YOU'RE NOT ELECTED, CHARLIE BROWN (I,4967); YOU'RE THE GREATEST, CHARLIE BROWN (I,4973)

MENKIN, LAWRENCE
ONE MAN'S EXPERIENCE (I,3382); ONE WOMAN'S EXPERIENCE (I,3391); ROCKY KING, INSIDE DETECTIVE (I,3824); THE TIMID SOUL (I,4519)

MENTEER, GARY
AMERICA 2100 (II,66); HAPPY DAYS (II,1084); LAVERNE AND SHIRLEY (II,1446); PUNKY BREWSTER (II,2092); WHO'S WATCHING THE KIDS? (II,2793)

MERL, DONALD
INCREDIBLE KIDS AND COMPANY (II,1233)

MERLIS, GEORGE
ENTERTAINMENT TONIGHT (II,780)

MERLIS, IRIS
PAJAMA TOPS (II,1942)

MERMAN, DOC
THE BETTY HUTTON SHOW (I,413)

MERRICK, DAVID
SEMI-TOUGH (II,2299); SEMI-TOUGH (II,2300)

MERRICK, MIKE
CAMELOT (II,416)

MERSON, MARC
ANDROCLES AND THE LION (I,189); THE DAVID FROST REVUE (I,1176); DUSTY (II,745); FULL HOUSE (II,937); HICKEY VS. ANYBODY (II,1142); JESSICA NOVAK (II,1286); KAZ (II,1370); MARRIAGE IS ALIVE AND WELL (II,1633); ME AND DUCKY (II,1671); MO AND JO (II,1715); OFF THE RACK (II,1878); RIDING HIGH (II,2173); SIDE BY SIDE

(II,2346); SUSAN AND SAM (II,2506); THE WAVERLY WONDERS (II,2746); WE'LL GET BY (II,2753); WE'LL GET BY (II,2754)

MESHEKOFF, MICHAEL
DANTE (I,1155); DRAGNET (I,1369); GENERAL ELECTRIC TRUE (I,1754); THE MAN FROM GALVESTON (I,2865)

MESSINA, CEDRIC
MOLL FLANDERS (II,1720)

METCALFE, BURT
AFTERMASH (II,23); KAREN (II,1363); M*A*S*H (II,1569); MAKING IT (II,1605)

METRO MEDIA VIDEO PRODUCTIONS INC
I'D RATHER BE CALM (II,1216)

METTER, ALAN
STEVE MARTIN'S THE WINDS OF WHOOPIE (II,2461)

METZGER, MIKE
BLANK CHECK (II,265); CAMOUFLAGE (II,417); THE DATING GAME (I,1168); HOW'S YOUR MOTHER-IN-LAW? (I,2148); THE NEW TREASURE HUNT (I,3275); THE NEW TREASURE HUNT (II,1831); THE NEWLYWED GAME (II,1836); THREE'S A CROWD (II,2612); TREASURE HUNT (II,2657)

MEYER, STANLEY
DRAGNET (I,1369)

MEYERSON, PETER
HEAVEN ON EARTH (II,1120)

MEYROWITZ, BOB
BETTE MIDLER—ART OR BUST (II,221); RINGO (II,2176)

MICHAEL, TIM T
PRIMUS (I,3668)

MICHAELS, FAITH
THAT TEEN SHOW (II,2573)

MICHAELS, LORNE
THE BEACH BOYS SPECIAL (II,190); BOB & RAY & JANE, LARAINE & GILDA (II,279); THE CONEHEADS (II,567); FLIP WILSON. . . OF COURSE (II,890); LILY (II,1477); THE LILY TOMLIN SPECIAL (II,1479); THE NEW SHOW (II,1828); THE PAUL SIMON SPECIAL (II,1983); RANDY NEWMAN AT THE ODEON (II,2111); SATURDAY NIGHT LIVE (II,2261); SIMON AND GARFUNKEL IN CONCERT (II,2356); STEVE MARTIN'S BEST SHOW EVER (II,2460)

MICHAELSON, CHARLES
CAPSULE MYSTERIES (I,825)

MICHON, JACK
DREAMS (II,735)

MILKIS, EDWARD K
ANGIE (II,80); BLANSKY'S BEAUTIES (II,267); BOSOM BUDDIES (II,355); FEEL THE HEAT (II,848); FOUL PLAY (II,910); GOODTIME GIRLS (II,1042); HAPPY DAYS (II,1084); LAVERNE AND SHIRLEY (II,1446); MAKIN' IT (II,1603); NIGHT GAMES (II,1845); OUT OF THE BLUE (II,1934); PETROCELLI (II,2031); WALKIN' WALTER (II,2737)

MILLER, ALAN
BIFF BAKER, U.S.A. (I,420); SECRETS OF THE OLD BAILEY (I,3971); SUSPICION (I,4309)

MILLER, DIANE K
THE SCARLET LETTER (II,2270)

MILLER, E A
FLY AWAY HOME (II,893)

MILLER, FRANK R
THE JOHN DAVIDSON SHOW (II,1310)

MILLER, GLENN
THE HORACE HEIDT SHOW (I,2120)

MILLER, HARVEY
ANNA AND THE KING (I,221); THE ODD COUPLE (II,1875); SIROTA'S COURT (II,2365)

MILLER, HERMAN
KUNG FU (II,1416); THE MAN FROM ATLANTIS (II,1615)

MILLER, JACK
ARCHER (II,102); CAPTAIN KANGAROO (II,434); LASSIE: THE NEW BEGINNING (II,1433); MAGIC MIDWAY (I,2818); PIP THE PIPER (I,3606)

MILLER, JAMES M
WHAT REALLY HAPPENED TO THE CLASS OF '65? (II,2768)

MILLER, JENNIFER
PAPER DOLLS (II,1952)

MILLER, KARYL GELD
FIRST TIME, SECOND TIME (II,863)

MILLER, LEE
ANIMALS ARE THE FUNNIEST PEOPLE (II,81); THE BRADY BUNCH HOUR (II,363); CBS: ON THE AIR (II,460); CHER (II,495); CHER (II,496); CHER AND OTHER FANTASIES (II,497); DON'T CALL ME MAMA ANYMORE (I,1350); FLATBUSH/AVENUE J (II,878); HOLLYWOOD STARS' SCREEN TESTS (II,1165); INSIDE AMERICA (II,1234); LADIES' MAN (II,1423); MITZI...THE FIRST TIME (I,3073); MITZI: A TRIBUTE TO THE AMERICAN HOUSEWIFE (I,3072); MUSIC HALL AMERICA (II,1773); NEVER SAY NEVER (II,1809); TV'S BLOOPERS AND PRACTICAL JOKES (II,2675)

MILLER, MARVIN
THE BILL COSBY SHOW (I,442); THE BILL COSBY SPECIAL, OR? (I,443); STRUCK BY LIGHTNING (II,2482)

MILLER, MICHAEL
THE LOVE REPORT (II,1542); THE LOVE REPORT (II,1543); SUPER PASSWORD (II,2499)

MILLER, MORTON NEAL
SENSE OF HUMOR (II,2303)

MILLER, PETER
THE LITTLEST HOBO (II,1500)

MILLER, RON
DANIEL BOONE (I,1141); THE NEW MICKEY MOUSE CLUB (II,1823)

MILLER, SIDNEY
THE DONALD O'CONNOR SHOW (I,1345); WHAT'S HAPPENING!! (II,2769)

MILLER, STEPHEN
ARCHIE BUNKER'S PLACE (II,105); GREAT DAY (II,1054); THE LOVE REPORT (II,1542)

MILLER, THOMAS L
ANGIE (II,80); BLANSKY'S BEAUTIES (II,267); BOSOM BUDDIES (II,355); EGAN (I,1419); FOUL PLAY (II,910); GOODTIME GIRLS (II,1042); HAPPY DAYS (II,1084); JOANIE LOVES CHACHI (II,1295); LAVERNE AND SHIRLEY (II,1446); MAKIN' IT (II,1603); NIGHT GAMES (II,1845); OUT OF THE BLUE (II,1934); PETROCELLI (II,2031); WALKIN' WALTER (II,2737)

MILLER, WALTER C
100 YEARS OF GOLDEN HITS (II,1905); BOWZER (II,358); COUNTRY MUSIC HIT PARADE (I,1069); DOUG HENNING'S WORLD OF MAGIC I (II,725); DOUG HENNING'S WORLD OF MAGIC II (II,726); DOUG HENNING'S WORLD OF MAGIC III (II,727); DOUG HENNING'S WORLD OF MAGIC IV (II,728); DOUG HENNING'S WORLD OF MAGIC V (II,729); DOUG HENNING: MAGIC ON BROADWAY (II,730); ELECTRA WOMAN AND DYNA GIRL (II,764); GEORGE BURNS AND OTHER SEX SYMBOLS (II,966); GEORGE BURNS CELEBRATES 80 YEARS IN SHOW BUSINESS (II,967); GEORGE BURNS EARLY, EARLY, EARLY, CHRISTMAS SHOW (II,968); GEORGE BURNS IN NASHVILLE?? (II,969); GEORGE BURNS' HOW TO LIVE TO BE 100 (II,972); JOHNNY CASH AND FRIENDS (II,1322); THE JOHNNY CASH SHOW (I,2424); THE OSMOND FAMILY CHRISTMAS SPECIAL (II,1926); THE OSMOND FAMILY THANKSGIVING SPECIAL (II,1928); SHA NA NA (II,2311); TENNESSEE ERNIE FORD'S WHITE CHRISTMAS (I,4402)

MILLER, WINSTON
IT TAKES A THIEF (I,2250); LITTLE HOUSE ON THE PRAIRIE (II,1487); THE VIRGINIAN (I,4738)

MILLING, WILLIAM P
UNICORN TALES (II,2706)

MILLS, TED
GARROWAY AT LARGE (I,1737); MAURICE CHEVALIER'S PARIS (I,2955); THE PATTI PAGE SHOW (I,3482); THE SATURDAY NIGHT REVUE (I,3921); SHORT STORY THEATER (I,4024)

MILTON, RICHARD
THE CURSE OF DRACULA (II,611); THE SECRET EMPIRE (II,2291); STOP SUSAN WILLIAMS (II,2473)

MINER, LESLIE
JOAN BAEZ (II,1294)

MINER, PETER
DARK SHADOWS (I,1157)

MINER, WORTHINGTON
AUNT JENNY'S REAL LIFE STORIES (I,314); CURTAIN CALL THEATER (I,1115); FRONTIER (I,1695); THE GOLDBERGS (I,1835); LITTLE WOMEN (I,2720); MEDIC (I,2976); MR. I. MAGINATION (I,3140); STUDIO ONE (I,4268); WESLEY (I,4795)

MINOFF, MARVIN

THE GUINNESS GAME (II,1066); PEEPING TIMES (II,1990); WOMEN WHO RATE A "10" (II,2825)

MINOFF, PHYLLIS
THE SECRET OF CHARLES DICKENS (II,2293)

MINOR, MARCIA
THE NASHVILLE PALACE (II,1792)

MINTZ, LARRY
BLUE JEANS (II,275); HIGH SCHOOL, U.S.A. (II,1148); HIGH SCHOOL, U.S.A. (II,1149)

MINTZ, RICK
A.K.A. PABLO (II,28)

MINTZ, ROBERT
THE CABOT CONNECTION (II,408); THE FEATHER AND FATHER GANG (II,845); THE FRENCH ATLANTIC AFFAIR (II,924); GOOD HEAVENS (II,1032)

MIRAMS, ROGER
ADVENTURES OF THE SEASPRAY (I,82); WOOBINDA—ANIMAL DOCTOR (II,2832)

MIRELL, LEON
SHINDIG (I,4013)

MIRISCH, ANDREW
LEGMEN (II,1458)

MIRISCH, WALTER
WICHITA TOWN (I,4857)

MIRKIN, LAWRENCE
FRAGGLE ROCK (II,916)

MISCH, DAVID
CALLAHAN (II,414)

MISCHER, DON
ALAN KING'S THIRD ANNUAL FINAL WARNING!! (II,31); THE BARBARA WALTERS SPECIAL (II,145); BARYSHNIKOV IN HOLLYWOOD (II,157); THE BEST OF TIMES (II,219); BOB HOPE SPECIAL: HAPPY BIRTHDAY, BOB! (II,325); CHERYL LADD: SCENES FROM A SPECIAL (II,503); AN EVENING WITH ROBIN WILLIAMS (II,789); FAMOUS LIVES (II,821); GOLDIE AND LIZA TOGETHER (II,1022); JUMP (II,1356); LOVE, SEX. . . AND MARRIAGE (II,1546); LYNDA CARTER: BODY AND SOUL (II,1565); THE PHIL DONAHUE SHOW (II,2032); SHIRLEY MACLAINE: ILLUSIONS (II,2336)

MISHKIN, PHIL
FREE COUNTRY (II,921); THE ODD COUPLE (II,1875); SONNY BOY (II,2402); THE TV SHOW (II,2673)

MISROCK, HENRY
LADIES BEFORE GENTLEMEN (I,2599)

MITCHELL, BILL
PRESS YOUR LUCK (II,2078)

MITCHELL, COLEMAN
ANNIE FLYNN (II,92); GIMME A BREAK (II,995); MAX (II,1658); SPENCER (II,2420); WHO'S ON CALL? (II,2791)

MITCHELL, DAVE
LOOK AT US (II,1515)

MITCHELL, GORDON
GOOD TIMES (II,1038); HARPER VALLEY PTA (II,1095)

MITCHLLL, SCOEY
ME AND MRS. C. (II,1673)

MITTLEMAN, RICK
ARNIE (I,261); YOU ASKED FOR IT (I,4931)

MITZ, RICK
P.O.P. (II,1939)

MIZIKER, RON
THE MOUSEKETEERS REUNION (II,1742)

MOESSINGER, DAVID
BLUE THUNDER (II,278); QUINCY, M. E. (II,2102)

MOFFITT, JOHN
FRIDAYS (II,926); GOOD VIBRATIONS FROM CENTRAL PARK (I,1854); THE HELEN REDDY SPECIAL (II,1126); NOT NECESSARILY THE NEWS (II,1862); NOT NECESSARILY THE NEWS (II,1863)

MOGER, STANLEY H
YOUR NEW DAY (II,2873)

MOIR, JAMES
KELLY MONTEITH (II,1375)

MONASH, PAUL
BRADDOCK (I,704); CAIN'S HUNDRED (I,792); JUDD, FOR THE DEFENSE (I,2463); PEYTON PLACE (I,3574)

MONET, GARY
STRIPPERS (II,2481); THOSE FABULOUS CLOWNS (II,2602)

MONNICKENDAM, FREDDY
SNORKS (II,2390)

MONROE, ROBERT
THE DAUGHTERS OF JOSHUA CABE (I,1169); THEY ONLY COME OUT AT NIGHT (II,2586)

MONTAGNE, EDWARD J
BROADSIDE (I,732); DELTA HOUSE (II,658); ELLERY QUEEN: DON'T LOOK BEHIND YOU (I,1434); THE HUNTER (I,2162); THE MAN AGAINST CRIME (I,2853); MCHALE'S NAVY (I,2969); THE MISADVENTURES OF SHERIFF LOBO (II,1704); THE MUNSTERS' REVENGE (II,1765); NOBODY'S PERFECT (II,1858); THE PHIL SILVERS PONTIAC SPECIAL: KEEP IN STEP (I,3579); THE PHIL SILVERS SHOW (I,3580); QUINCY, M. E. (II,2102); SPIDER-MAN (II,2424); THE TALL MAN (I,4354); A VERY MISSING PERSON (I,4715)

MONTGOMERY, GARTH
ARSENIC AND OLD LACE (I,267); FORD TELEVISION THEATER (I,1634); LITTLE WOMEN (I,2719); MAGNAVOX THEATER (I,2827)

MONTGOMERY, ROBERT
EYE WITNESS (I,1494); ROBERT MONTGOMERY PRESENTS YOUR LUCKY STRIKE THEATER (I,3809)

MONTGOMERY, ROY
SHOW BIZ (I,4027)

MONTY, GLORIA
GENERAL HOSPITAL (II,964); THE HAMPTONS (II,1076)

MOONJEAN, HANK
BEAUTY AND THE BEAST (I,382)

MOORE, ARTHUR
COLONEL STOOPNAGLE'S STOOP (I,1006)

MOORE, BRIDGET
THE KENNY EVERETT VIDEO SHOW (II,1379)

MOORE, CARROLL
DOC (II,683)

MOORE, DICK
DICK TRACY (I,1271)

MOORE, JEAN ANNE
ROLL OF THUNDER, HEAR MY CRY (II,2202)

MOORE, KAREN
SPRAGGUE (II,2430)

MOORE, LOU
A CHRISTMAS CAROL (II,517)

MOORE, THOMAS W
DISAPPEARANCE OF AIMEE (I,1290); KUDZU (II,1415); ROLL OF THUNDER, HEAR MY CRY (II,2202)

MOORE, TOM
LADIES BE SEATED (I,2598)

MORDENTE, TONY
DON'T CALL ME MAMA ANYMORE (I,1350); LEGS (II,1459)

MORGAN, BREWSTER
THE ADVENTURES OF THE SEA HAWK (I,81); CURTAIN CALL THEATER (I,1115); DOCTOR HUDSON'S SECRET JOURNAL (I,1312)

MORGAN, CHRISTOPHER
BEULAH LAND (II,226); COLORADO C.I. (II,555); GLITTER (II,1009); HUNTER (II,1205); MEDICAL STORY (II,1676); THE MISSISSIPPI (II,1707); THE POLICE STORY (I,3638); POLICE STORY (II,2062); THE QUEST (II,2096); QUINCY, M. E. (II,2102); RIKER (II,2175); TODAY'S FBI (II,2628); W.E.B. (II,2732)

MORGAN, JIM
WHO DO YOU TRUST? (I,4846)

MORGAN, LEO
THE ALL-STAR REVUE (I,138); AS CAESAR SEES IT (I,294); CAESAR'S HOUR (I,790); E.S.P. (I,1462); E.S.P. (I,1463); MARRIAGE—HANDLE WITH CARE (I,2921); THE RED BUTTONS SHOW (I,3750); THE SID CAESAR SPECIAL (I,4044); TIPTOE THROUGH TV (I,4524); VARIETY: THE WORLD OF SHOW BIZ (I,4704)

MORGAN, MONTY
GIRL TALK (I,1810)

MORGAN, NORMAN
FULL CIRCLE (I,1702)

MORGAN, SHEP
TOP OF THE WORLD (II,2646)

MORHEIM, LOU
THE BIG VALLEY (I,437); THE IMMORTAL (I,2196); THE IMMORTAL (I,2197); MADAME SIN (I,2805)

MORIARITY, JAY
CHECKING IN (II,492); THE JEFFERSONS (II,1276)

MORRA, BUDDY
THE BILLY CRYSTAL COMEDY HOUR (II,248); BILLY CRYSTAL: A COMIC'S LINE (II,249); CHEERS (II,494); STAR OF THE FAMILY (II,2436)

MORRIS, HOWARD
THE CORNER BAR (I,1051); THE CORNER BAR (I,1052);

THE MANY FACES OF COMEDY (I,2894)

MORRIS, JOHN
MIDWESTERN HAYRIDE (I,3033)

MORRIS, LINDA
ALICE (II,33); JUST OUR LUCK (II,1360); PRIVATE BENJAMIN (II,2087)

MORRIS, MALCOLM
TOP OF THE WORLD (II,2646)

MORRIS, MICHAEL
CHICO AND THE MAN (II,508); WE'LL TAKE MANHATTAN (I,4792)

MORRIS, ROBERT
KENNEDY (II,1377)

MORRIS, STEWART
THE SHIRLEY BASSEY SPECIAL (II,2331); SOMETHING SPECIAL (I,4118)

MORRISON, AL
FIRST LOVE (I,1583)

MORROW, BILL
THE BING CROSBY SHOW (I,462); THE BING CROSBY SPECIAL (I,470); THE EDSEL SHOW (I,1417)

MORROW, TIM
TIN PAN ALLEY TV (I,4522)

MORSE JR, TERRY
SERGEANT T.K. YU (II,2305); THE SKATEBIRDS (II,2375); WISHMAN (II,2811)

MORSE, CARLETON E
MIXED DOUBLES (I,3075); ONE MAN'S FAMILY (I,3383)

MORSE, HERB
GO LUCKY (I,1832); SING IT AGAIN (I,4059); SONGS FOR SALE (I,4124)

MORTON, GARY
BUNGLE ABBEY (II,396); CBS SALUTES LUCY—THE FIRST 25 YEARS (II,459); HAPPY ANNIVERSARY AND GOODBYE (II,1082); HERE'S LUCY (II,1135); A LUCILLE BALL SPECIAL STARRING LUCILLE BALL AND DEAN MARTIN (II,1557); A LUCILLE BALL SPECIAL STARRING LUCILLE BALL AND JACKIE GLEASON (II,1558); THE LUCILLE BALL SPECIAL (II,1559); A LUCILLE BALL SPECIAL: WHAT NOW CATHERINE CURTIS? (II,1560); THE LUCY SHOW (I,2791)

MORTON, ROBERT
LATE NIGHT WITH DAVID LETTERMAN (II,1442)

MORWOOD, WILLIAM
THE GENERAL ELECTRIC THEATER (I,1753)

MOSER, JAMES E
BEN CASEY (I,394)

MOSES, HARRY
THE MIKE WALLACE PROFILES (II,1699)

MOSHER, BOB
BRINGING UP BUDDY (I,731); CALVIN AND THE COLONEL (I,806); THE GENERAL ELECTRIC THEATER (I,1753); GOING MY WAY (I,1833); ICHABOD AND ME (I,2183); LEAVE IT TO BEAVER (I,2648); THE MARGE AND GOWER CHAMPION SHOW (I,2907); MAYBERRY R.F.D. (I,2961); MEET MR. MCNUTLEY (I,2989); MUNSTER, GO HOME! (I,3157); THE MUNSTERS (I,3158)

MOSHER, JOE
KAREN (I,2505)

MOSMAN, JACK
HERB SHRINER TIME (I,2020)

MOSS, BOB
THE CHESTERFIELD SUPPER CLUB (I,923)

MOSS, CHARLES
QUICK AS A FLASH (I,3709)

MOSS, FRANK L
THE LIFE AND LEGEND OF WYATT EARP (I,2681)

MOSS, JEFFREY B
THE DRUNKARD (II,737)

MOSS, PAUL E
OLD NICKERBOCKER MUSIC HALL (I,3351); THE TIMES SQUARE STORY (I,4514)

MOSS, VIVIAN R
OMNI: THE NEW FRONTIER (II,1889)

MOUNT, MICHAEL
GIMME A BREAK (II,995); NOT IN FRONT OF THE KIDS (II,1861); SHE'S WITH ME (II,2321); SPENCER (II,2420); SUTTERS BAY (II,2508)

MOYE, MICHAEL
IT'S YOUR MOVE (II,1259); SILVER SPOONS (II,2355)

MOYNIHAN, GEORGE
MERV GRIFFIN'S SIDEWALKS OF NEW ENGLAND (I,3017); MERV GRIFFIN'S ST. PATRICK'S DAY SPECIAL (I,3018)

MUIR, E ROGER
THE AT LIBERTY CLUB (I,309); THE GABBY HAYES SHOW (I,1723); HOWDY DOODY (I,2151); JUDY SPLINTERS (I,2475); MATCHES 'N' MATES (I,2945); THE MOHAWK SHOWROOM (I,3080); THE NEW HOWDY DOODY SHOW (II,1818); PAY CARDS (I,3496); THE SHARI LEWIS SHOW (I,4000); SPIN-OFF (II,2427); SUPER PAY CARDS (II,2500)

MULCAHY, JACK
HOLLYWOOD AND THE STARS (I,2081)

MULLANEY, SCOTT
OLIVIA NEWTON-JOHN IN CONCERT (II,1884); THE OLIVIA NEWTON-JOHN SHOW (II,1885); OLIVIA NEWTON-JOHN: LET'S BE PHYSICAL (II,1887)

MULLER, ROMEO
DOROTHY IN THE LAND OF OZ (II,722)

MULLIGAN, BERT
THE BOB HOPE CHRYSLER THEATER (I,502)I; CHECKING IN (II,492)

MULLIGAN, GERALD
LATE NIGHT WITH DAVID LETTERMAN (II,1442)

MULLIGAN, JIM
AT EASE (II,115); FLO (II,891); IN THE BEGINNING (II,1229); M*A*S*H (II,1569); SCARED SILLY (II,2269); VENICE MEDICAL (II,2726)

MULLIGAN, MIKE
THE JEFFERSONS (II,1276)

MULLIGAN, ROBERT
MEMBER OF THE WEDDING (I,3002); THE MOON AND SIXPENCE (I,3098)

MULLIGAN, VERN
ELKE SOMMER'S WORLD OF SPEED AND BEAUTY (II,765)

MULLIN, JOHN
CAR CARE CENTRAL (II,436)

MUMFORD, THAD
THE DUCK FACTORY (II,738); M*A*S*H (II,1569)

MUNHALL, JACK
JACQUES FRAY'S MUSIC ROOM (I,2329)

MUNTNER, SIMON
THE AMERICAN GIRLS (II,72); THE SAN PEDRO BEACH BUMS (II,2250); T.J. HOOKER (II,2524)

MURPHY, BOB
THE MERV GRIFFIN SHOW (II,1688)

MURPHY, RICHARD
THE HARDY BOYS (I,1948)

MURRAY, ARTHUR
ARTHUR MURRAY'S DANCE PARTY (I,293)

MURRAY, KEN
EL COYOTE (I,1423); I HEAR AMERICA SINGING (I,2168); THE KEN MURRAY SHOW (I,2525); WHERE WERE YOU? (I,4834)

MURRAY, WARREN
THE BRADY BRIDES (II,361)

MURROW, EDWARD R
PERSON TO PERSON (I,3550)

MUSIC, LORENZO
THE BOB NEWHART SHOW (II,342); CARLTON YOUR DOORMAN (II,440); FRIENDS (II,928); THE LORENZO AND HENRIETTA MUSIC SHOW (II,1519); THE NEW LORENZO MUSIC SHOW (II,1821); RHODA (II,2151)

MUSIELLO, FRANK
THE PETER LIND HAYES SHOW (I,3564)

MYERSON, PETER
WELCOME BACK, KOTTER (II,2761)

MYHREIM, ROBERT
CAPTAIN KANGAROO (II,434)

MYMAN, ROBERT
JOHN RITTER: BEING OF SOUND MIND AND BODY (II,1319)

NBC RADIO CITY TELEVISION TEST STATION
THE ADVENTURE OF THE THREE GARRIDEBS (I,40)

NAAR, JOSEPH T
SCHLITZ PLAYHOUSE OF STARS (I,3936); STARSKY AND HUTCH (II,2444); STARSKY AND HUTCH (II,2445); STRIKE FORCE (II,2480)

NADEL, ARTHUR H
FAT ALBERT AND THE COSBY KIDS (II,835); HE-MAN AND THE MASTERS OF THE UNIVERSE (II,1113); ISIS (II,1248); JASON OF STAR COMMAND (II,1274); THE KID SUPER POWER HOUR WITH SHAZAM (II,1386); LAW OF THE PLAINSMAN (I,2639); SHAZAM! (II,2319); SPACE ACADEMY (II,2406)

NADLER, MARTY
LAVERNE AND SHIRLEY (II,1446); LEGS (II,1459); WHO'S WATCHING THE

KIDS? (II,2793)

NAFIE, CAROL LEE
THE REGIS PHILBIN SHOW (II,2128)

NARDINO, GARY
AT YOUR SERVICE (II,118); BROTHERS (II,380)

NARDO, PATRICIA
BETWEEN THE LINES (II,225); COUSINS (II,595); IT'S NOT EASY (II,1256)

NARZ, JACK
BEAT THE CLOCK (II,195)

NASH, DAVID
BARBARA MANDRELL—THE LADY IS A CHAMP (II,144)

NASHT, JOHN
CROWN THEATER WITH GLORIA SWANSON (I,1107)

NASSOUR, EDWARD
SHEENA, QUEEN OF THE JUNGLE (I,4003)

NASSOUR, WILLIAM
SHEENA, QUEEN OF THE JUNGLE (I,4003)

NAUD, TOM
THE GIRL IN MY LIFE (II,996)

NAYLOR, JACK
OPENING NIGHT: U.S.A. (I,3397)

NAZARRO, RAY
FURY (I,1721)

NEAME, CHRISTOPHER
THE FLAME TREES OF THIKA (II,870); Q. E. D. (II,2094)

NECESSARY, GARY
LOOK AT US (II,1515); REAL PEOPLE (II,2118); SALUTE TO LADY LIBERTY (II,2243); SPEAK UP AMERICA (II,2412)

NEDERLANDER, JAMES M
LENA HORNE: THE LADY AND HER MUSIC (II,1460)

NEECE, RANDY
HIT MAN (II,1156)

NEIGHER, GEOFFREY
ANNIE FLYNN (II,92); GIMME A BREAK (II,995); MAX (II,1658); SPENCER (II,2420); WHO'S ON CALL? (II,2791)

NEILSON, JAMES
O'HARA, UNITED STATES TREASURY (I,3346)

NELSON, A J
POPI (II,2069)

NELSON, CHRISTOPHER
THE GREATEST AMERICAN HERO (II,1060); NO MAN'S LAND (II,1855); SCARECROW AND MRS. KING (II,2268)

NELSON, DAVID
OZZIE'S GIRLS (I,3440); THE ROCK 'N ROLL SHOW (II,2188)

NELSON, DOLPH
ANYBODY CAN PLAY (I,234)

NELSON, DON
BRIDGET LOVES BERNIE (I,722); CRASH ISLAND (II,600); FRANKIE AND ANNETTE: THE SECOND TIME AROUND (II,917); THE MUNSTERS' REVENGE (II,1765)

NELSON, ERIK
BACKSTAGE PASS (II,132)

NELSON, GARY
THE BOYS IN BLUE (II,359); REVENGE OF THE GRAY GANG (II,2146)

NELSON, GEORGE R
THE MOHAWK SHOWROOM (I,3080)

NELSON, JOHN
BRIDE AND GROOM (I,721); LIVE LIKE A MILLIONAIRE (I,2728)

NELSON, OLIVER
THE PARAGON OF COMEDY (II,1956)

NELSON, OZZIE
THE ADVENTURES OF OZZIE AND HARRIET (I,71); OZZIE'S GIRLS (I,3439)

NELSON, PETER
THE BAIT (I,335); GET CHRISTIE LOVE! (II,981); GET CHRISTIE LOVE! (II,982)

NELSON, PHYLLIS
ME AND MAXX (II,1672); THE SHAPE OF THINGS (II,2315)

NELSON, PORTIA
THE SQUARE WORLD OF ED BUTLER (I,4171)

NELSON, RALPH
THE ARMSTRONG CIRCLE THEATER (I,260); AUCTION AIRE (I,313); THE DICK POWELL SHOW (I,1269); DUPONT SHOW OF THE MONTH (I,1387); THE DUPONT SHOW OF THE WEEK (I,1388); I REMEMBER MAMA (I,2176)

NELSON, RAYMOND E
ROAR OF THE RAILS (I,3803); TELEVISION FASHION FAIR (I,4381)

NEMEROVSLEI, JEFFREY
BACKSTAGE PASS (II,132)

NETTER, DOUGLAS
THE BUFFALO SOLDIERS (II,388); THE CHEROKEE TRAIL (II,500); ROUGHNECKS (II,2219); WILD TIMES (II,2799)

NEUFELD, MACE
THE AMERICAN DREAM (II,71); THE CAPTAIN AND TENNILLE (II,429); EAST OF EDEN (II,752); THE MAGIC PLANET (II,1598); QUARK (II,2095)

NEUFELD, SIGMUND
HAWKEYE AND THE LAST OF THE MOHICANS (I,1975)

NEUFIELD, STANLEY
THE CATCHER (I,872)

NEUMAN, E JACK
KATE MCSHANE (II,1368); KATE MCSHANE (II,1369); MR. NOVAK (I,3145); NIGHT GAMES (II,1845); SAM BENEDICT (I,3894); STAT! (I,4213)

NEUSTEIN, JOSEPH
SUPER PASSWORD (II,2499)

NEVENS, SUE
13 QUEENS BOULEVARD (II,2591); THE FACTS OF LIFE (II,805); JOE'S WORLD (II,1305); WHAT'S HAPPENING!! (II,2769)

NEVINS, ROY
THE ROOTS OF ROCK 'N' ROLL (II,2212)

NEWLAND, JOHN
THE MAN WHO NEVER WAS (I,2882)

NEWMAN, ALAN
LETTERS TO LAUGH-IN (I,2672)

NEWMAN, CARROLL
FAMILY (II,813); THE FAMILY TREE (II,820); TWO MARRIAGES (II,2691); WHEN THE WHISTLE BLOWS (II,2781)

NEWMAN, CY
THE JACQUELINE SUSANN SHOW (I,2327)

NEWMAN, RICK
PAT BENATAR IN CONCERT (II,1967)

NEWMAN, SUSAN KENDALL
CANDIDA (II,423)

NEWTON, RICHARD
CONCRETE COWBOYS (II,563); THE CONCRETE COWBOYS (II,564); THE FELONY SQUAD (I,1563); HAWAII FIVE-O (II,1110); HONEY WEST (I,2105)

NICHOLL, DON
ALL IN THE FAMILY (II,38); THE DUMPLINGS (II,743); GOOD TIMES (II,1038); THE JEFFERSONS (II,1276); KINGSTON: CONFIDENTIAL (II,1398); THE ROPERS (II,2214); THREE'S COMPANY (II,2614)

NICHOLL, PAUL
THREE'S A CROWD (II,2613)

NICHOLS, CHARLES
THE ADVENTURES OF GULLIVER (I,60); THE ADVENTURES OF JONNY QUEST (I,64); THE CATTANOOGA CATS (I,874); DASTARDLY AND MUTTLEY IN THEIR FLYING MACHINES (I,1159); THE FLINTSTONE COMEDY HOUR (II,881); THE FUNKY PHANTOM (I,1709); THE HARLEM GLOBETROTTERS (I,1953); HELP! IT'S THE HAIR BEAR BUNCH (I,2012); SPACE STARS (II,2408)

NICHOLS, J DANIEL
HAWAIIAN HEAT (II,1111)

NICHOLS, MIKE
FAMILY (II,813)

NICHOLS, WILLIAM
THE DUPONT SHOW OF THE WEEK (I,1388); THE PETRIFIED FOREST (I,3570); TALENT SEARCH (I,4342)

NICHOLSON, ERWIN
THE EDGE OF NIGHT (II,760)

NICHOLSON, NICK
MATCHES 'N' MATES (I,2945); THE NEW HOWDY DOODY SHOW (II,1818); PAY CARDS (I,3496); SPIN-OFF (II,2427); SUPER PAY CARDS (II,2500)

NICKELL, PAUL
GIANT IN A HURRY (I,1790); THE MAN AGAINST CRIME (I,2853); STUDIO ONE (I,4268)

NICKSAY, DAVID
CALL TO GLORY (II,413)

NICOLELLA, JOHN
MIAMI VICE (II,1689)

NIGRO-CHACON, GIOVANI
THAT'S CAT (II,2576)

NILES, FRED
UNCLE MISTLETOE AND HIS ADVENTURES (I,4669)

NILES, WENDELL
YOUR ALL AMERICAN COLLEGE SHOW (I,4946)

NISS, STANLEY
DANGER (I,1134); FBI CODE 98 (I,1550); HAWAIIAN EYE (I,1973); JERICHO (I,2358)

NIVEN, DAVID
FOUR STAR PLAYHOUSE (I,1652)

NOAH, LYLA OLIVER
THE PEOPLE'S COURT (II,2001)

NOAH, ROBERT
THE AMATEUR'S GUIDE TO LOVE (I,155); BATTLESTARS (II,182); BEDTIME STORIES (II,199); GAMBIT (II,955); HIGH ROLLERS (II,1147); LAS VEGAS GAMBIT (II,1430); THE MAGNIFICENT MARBLE MACHINE (II,1600); SCRABBLE (II,2278); TIC TAC DOUGH (I,4498); TO SAY THE LEAST (II,2623); TWENTY-ONE (I,4645)

NOBLE, PAUL
CINDERELLA (II,526)

NODELLA, BURT
GET SMART (II,983); THE TIM CONWAY SHOW (I,4502); WINTER KILL (II,2810)

NOONOO, BOB
FOUL UPS, BLEEPS AND BLUNDERS (II,911)

NORELL, MICHAEL
ALOHA PARADISE (II,58); FEATHERSTONE'S NEST (II,847); WHAT'S UP DOC? (II,2771)

NORMAN, JANE
PIXANNE (II,2048)

NORTH, MIKE
BRAVE EAGLE (I,709)

NORTH, STEVEN
COMEBACK (I,1016)

NORTHRUP, STUART
RICH LITTLE'S ROBIN HOOD (II,2158)

NOVELLO, DON
SECOND CITY TELEVISION (II,2287)

NOVIS, DOROTHY
THE ADVENTURES OF CYCLONE MALONE (I,57)

NOXON, NICHOLAS L
DEAR MR. GABLE (I,1203)

NUGENT, NELLE
PIAF (II,2039)

NUNN, TERRY
LOOK AT US (II,1515)

NYREN, DAVE
SOLDIER PARADE (I,4109)

O'BRIEN, ANN
LORNE GREENE'S NEW WILDERNESS (II,1523)

O'BRIEN, DEAN
V: THE FINAL BATTLE (II,2714); V: THE SERIES (II,2715)

O'BRIEN, JOEL
DRAW TO WIN (I,1373)

O'BRIEN, LIAM
THE MISSISSIPPI (II,1707); POLICE STORY (II,2062)

O'BRIEN, MIMI
BODY LANGUAGE (II,346); CHILD'S PLAY (II,510); MINDREADERS (II,1702); TO TELL THE TRUTH (II,2626)

O'BRIEN, WILLIAM
HAWAII FIVE-O (II,1110)

O'CONNELL, DAVID J
BATTLESTAR GALACTICA (II,181); BUCK ROGERS IN THE 25TH CENTURY (II,383); BUCK ROGERS IN THE 25TH CENTURY (II,384); THE CHADWICK FAMILY (II,478); COOL MILLION (I,1046); GALACTICA 1980 (II,953); KINGSTON: THE POWER PLAY (II,1399); LANIGAN'S RABBI (II,1428); MARCUS WELBY, M.D. (I,2905); MARCUS WELBY, M.D. (II,1627); THE MASK OF MARCELLA (I,2937); MR. TERRIFIC (I,3152); OPERATION PETTICOAT (II,1916); OPERATION PETTICOAT (II,1917)

O'CONNOR, CARROLL
THE BANANA COMPANY (II,139); BENDER (II,204); BRONK (II,379); GLORIA COMES HOME (II,1011)

O'CONNOR, DICK
RUN RUN, JOE (II,2229)

O'CONNOR, DONALD
THE DONALD O'CONNOR SHOW (I,1343)

O'CONNOR, FRANK
I AND CLAUDIE (I,2164); THE NATIVITY (II,1798); THE SNOW GOOSE (I,4103)

O'CONNOR, RICHARD L
CELEBRITY (II,463); KORG: 70,000 B.C. (II,1408)

O'DONOUGH, MICHAEL
SATURDAY NIGHT LIVE (II,2261)

O'DONOVAN, DANNY
AIR SUPPLY IN HAWAII (II,26); ELTON JOHN IN CENTRAL PARK (II,769)

O'FALLON, JAMES
CALL TO GLORY (II,413)

O'GORMAN, NICK
THE COSBY SHOW (II,578); THE NEW SHOW (II,1828)

O'HERLIHY, MICHAEL
BACKSTAIRS AT THE WHITE HOUSE (II,133)

O'KEEFE, JIM
THE NEW ODD COUPLE (II,1825)

O'KEEFE, WINSTON
ALCOA/GOODYEAR THEATER (I,107); ALL IN THE FAMILY (I,133); COOGAN'S REWARD (I,1044); DEAR MOM, DEAR DAD (I,1202); DENNIS THE MENACE (I,1231); DRAW TO WIN (I,1373); FORD TELEVISION THEATER (I,1634); GRINDL (I,1889); I REMEMBER CAVIAR (I,2175); MATINEE THEATER (I,2947); MR. TUTT (I,3153); SHOWTIME, U.S.A. (I,4038); THE VIRGINIAN (I,4737)

O'KELLY, JEFFREY
DOCTOR SNUGGLES (II,687)

O'KUN, LAN
THE JUGGLER OF NOTRE DAME (II,1350); THE LITTLEST ANGEL (I,2726)

O'NEILL, ROBERT F
B.J. AND THE BEAR (II,125); B.J. AND THE BEAR (II,126); DARKROOM (II,619); THE EYES OF TEXAS II (II,800); THE GEMINI MAN (II,961); GRIFF (I,1888); THE INVISIBLE MAN (II,1242); THE LAST CONVERTIBLE (II,1435); MISSION: IMPOSSIBLE (I,3067); MURDER, SHE WROTE (II,1770); QUINCY, M. E. (II,2102); WHEELS (II,2777); WOMEN IN WHITE (II,2823)

OCCHIOGROSSO, JIM
BACKSTAGE PASS (II,132)

OCHS, ACE
FACE THE MUSIC (I,1509)

OHLIGER III, JOSEPH
BACKSTAGE PASS (II,132)

OHLMEYER, DON
BATTLE OF THE NETWORK STARS (II,163); BATTLE OF THE NETWORK STARS (II,164); BATTLE OF THE NETWORK STARS (II,165); BATTLE OF THE NETWORK STARS (II,166); BATTLE OF THE NETWORK STARS (II,167); BATTLE OF THE NETWORK STARS (II,168); BATTLE OF THE NETWORK STARS (II,169); BATTLE OF THE NETWORK STARS (II,170); BATTLE OF THE NETWORK STARS (II,171); BATTLE OF THE NETWORK STARS (II,172); BATTLE OF THE NETWORK STARS (II,173); BATTLE OF THE NETWORK STARS (II,174); BATTLE OF THE NETWORK STARS (II,175); BATTLE OF THE NETWORK STARS (II,176); BATTLE OF THE NETWORK STARS (II,177); GAMES PEOPLE PLAY (II,956); PEOPLE ARE FUNNY (II,1996); THE SUNDAY GAMES (II,2494)

OLEK, HENRY
SCENE OF THE CRIME (II,2271)

OLIN, MARILYN
THE ABC AFTERSCHOOL SPECIAL (II,1); KIDS ARE PEOPLE TOO (II,1390)

OLIVER, GORDON
THE BOB HOPE CHRYSLER THEATER (I,502); CODE NAME: HERACLITUS (I,990); GUILTY OR NOT GUILTY (I,1903); IT TAKES A THIEF (I,2250); MR. LUCKY (I,3141); PETER GUNN (I,3562); PROFILES IN COURAGE (I,3678); THE REASON NOBODY HARDLY EVER SEEN A FAT OUTLAW IN THE OLD WEST IS AS FOLLOWS: (I,3744); THE SEVEN LITTLE FOYS (I,3986)

OLIVER, SHERLING
WREN'S NEST (I,4922)

OLSEN, FRED
TAKE TWO (I,4337)

OLSHEIN, MARILYN
IN TROUBLE (II,1230)

ONO, YOKO
JOHN LENNON AND YOKO ONO PRESENT THE ONE-TO-ONE CONCERT (I,2414)

OPIE, WIN
THE CLIFFWOOD AVENUE KIDS (II,543)

OPPENHEIMER, DAVID
HERE'S EDIE (I,2030)

OPPENHEIMER, DEBORAH
PAJAMA TOPS (II,1942)

OPPENHEIMER, JESS
ANGEL (I,216); THE BIG BRAIN (I,424); THE DEBBIE REYNOLDS SHOW (I,1214); FOR LOVE OR $$$ (I,1623); THE GENERAL MOTORS 50TH ANNIVERSARY SHOW

(I,1758); GET SMART (II,983); GLYNIS (I,1829); HERE'S HOLLYWOOD (I,2031); HIDE AND SEEK (I,2047); I LOVE LUCY (I,2171); THE LUCILLE BALL COMEDY HOUR (I,2783); THE ROSALIND RUSSELL SHOW (I,3845); THE THIRD COMMANDMENT (I,4449)

OPPENHEIMER, PIER
HERE'S HOLLYWOOD (I,2031)

ORENSTEIN, BERNIE
13 QUEENS BOULEVARD (II,2591); THE BEAUTIFUL PHYLLIS DILLER SHOW (I,381); CARTER COUNTRY (II,449); CONDO (II,565); DOUBLE TROUBLE (II,723); E/R (II,748); GO! (I,1830); GRADY (II,1047); THE NEW DICK VAN DYKE SHOW (I,3262); THE NEW DICK VAN DYKE SHOW (I,3263); ONE IN A MILLION (II,1906); ONE OF THE BOYS (II,1911); THE PERRY COMO WINTER SHOW (I,3543); SANFORD AND SON (II,2256); SANFORD ARMS (II,2257); SUPER COMEDY BOWL 2 (I,4289); THAT GIRL (II,2570); THIS WEEK IN NEMTIN (I,4465); A TOUCH OF GRACE (I,4585); WHAT'S HAPPENING!! (II,2769)

ORENSTEIN, BOB
PERRY COMO'S WINTER SHOW (I,3547)

ORIAND, JOHN
SAN FRANCISCO INTERNATIONAL AIRPORT (I,3906)

ORILIO, JOSEPH
CASPER AND FRIENDS (I,867); FELIX THE CAT (I,1561); THE MIGHTY HERCULES (I,3034)

ORINGER, BARRY
THE HUSTLER OF MUSCLE BEACH (II,1210); RAGE OF ANGELS (II,2107); SERPICO (II,2306)

ORMANDY, TRACEY
WHO'S THE BOSS? (II,2792)

ORR, PAUL
THE JACK PAAR SHOW (I,2317); THE JOEY BISHOP SHOW (I,2405)

ORR, WILLIAM T
77 SUNSET STRIP (I,3988); THE ALASKANS (I,106); BOURBON STREET BEAT (I,699); BRONCO (I,745); CHEYENNE (I,931); COLT .45 (I,1010); THE DAKOTAS (I,1126); F TROOP (I,1499); GIRL ON THE RUN (I,1809); HANK (I,1931); HAWAIIAN EYE (I,1973); LAWMAN (I,2642); MAVERICK (II,1657); NO TIME FOR SERGEANTS (I,3300); ROOM FOR ONE MORE (I,3842); SUGARFOOT (I,4277); SURFSIDE 6 (I,4299); THE TEXAN (I,4413)

ORSATTI, VIC
THE TEXAN (I,4413)

OSBORN, ANDREW
THE GATHERING STORM (I,1741)

OSGROVE, JOHN
REAL PEOPLE (II,2118)

OSHIMA, FAYE
BRANAGAN AND MAPES (II,367); OLD FRIENDS (II,1881); SCROOGE'S ROCK 'N' ROLL CHRISTMAS (II,2279)

OSMOND BROTHERS, THE
THE DONNA FARGO SHOW (II,710); DONNY AND MARIE (II,712); THE DONNY AND MARIE CHRISTMAS SPECIAL (II,713); MEL AND SUSAN TOGETHER (II,1677); THE OSMOND BROTHERS SPECIAL (II,1925); THE OSMOND FAMILY SHOW (II,1927)

OSMOND, ALAN
THE DONNY AND MARIE CHRISTMAS SPECIAL (II,713); MARIE (II,1629); THE OSMOND FAMILY SHOW (II,1927); THE OSMOND FAMILY THANKSGIVING SPECIAL (II,1928)

OSMOND, JAY
MARIE (II,1629); THE OSMOND FAMILY CHRISTMAS SPECIAL (II,1926)

OSMOND, MERRILL
THE OSMOND FAMILY THANKSGIVING SPECIAL (II,1928)

OSMOND, WAYNE
THE OSMOND FAMILY CHRISTMAS SPECIAL (II,1926)

OSTERMAN, LESTER
THE LITTLEST ANGEL (I,2726)

OSTROFF, MANNING
THE COLGATE COMEDY HOUR (I,997)

OWEN, TONY
THE BARBARA RUSH SHOW (I,349); THE DONNA REED SHOW (I,1347); HIS MODEL WIFE (I,2064)

OWENS, JIM
BARBI BENTON SPECIAL: A BARBI DOLL FOR CHRISTMAS (II,148); COUNTRY GALAXY OF STARS (II,587); AN EVENING WITH THE STATLER BROTHERS (II,790); JERRY REED AND SPECIAL FRIENDS (II,1285); LOUISE MANDRELL: DIAMONDS, GOLD AND PLATINUM (II,1528); MUSIC CITY NEWS TOP COUNTRY HITS OF THE YEAR (II,1772)

OXFORD, RONALD
WINDOW SHADE REVUE (I,4874)

PAAR, JACK
JACK PAAR AND A FUNNY THING HAPPENED EVERYWHERE (I,2314); JACK PAAR PRESENTS (I,2315); THE WONDERFUL WORLD OF JACK PAAR (I,4899)

PACKER, PETER
THE 20TH CENTURY-FOX HOUR (I,4642); LAW OF THE PLAINSMAN (I,2639); MAN WITHOUT A GUN (I,2884); THE MARRIAGE BROKER (I,2920); MY FRIEND FLICKA (I,3190)

PADILLA, MARIA
THE MISSISSIPPI (II,1707); SIDNEY SHORR (II,2349)

PAGET, DENNIS
ON THE TOWN WITH TONY BENNETT (II,1895)

PAIGE, GEORGE
MEN AT WORK IN CONCERT (II,1682)

PALADINO, PHIL
THE SQUARE WORLD OF ED BUTLER (I,4171)

PALASH, HARVEY
CAR CARE CENTRAL (II,436)

PALEY, IRVING
BLONDIE (I,487); ICHABOD AND ME (I,2183); PISTOLS 'N' PETTICOATS (I,3608)

PALLEY, NYLES
CHEZ PAREE REVUE (I,932)

PALMER, P K
DICK TRACY (I,1271)

PALMER, PATRICIA FASS
ALL TOGETHER NOW (II,45); ANOTHER MAN'S SHOES (II,96); CHECKING IN (II,492); HELLO, LARRY (II,1128); HIGHCLIFFE MANOR (II,1150); ONE DAY AT A TIME (II,1900); A YEAR AT THE TOP (II,2844)

PALMER, PAUL
OUT OF OUR MINDS (II,1933)

PALMER, PETER S
TV'S BLOOPERS AND PRACTICAL JOKES (II,2675); WHY US? (II,2796)

PALTROW, BRUCE
BIG CITY BOYS (II,232); ST. ELSEWHERE (II,2432); THE WHITE SHADOW (II,2788)

PAOLANTONIO, BILL
PEOPLE DO THE CRAZIEST THINGS (II,1998)

PAPAZIAN, ROBERT
TOPPER (II,2649); TROUBLE IN HIGH TIMBER COUNTRY (II,2661); THE YEAGERS (II,2843)

PAPP, FRANK
THE ALDRICH FAMILY (I,111)

PARENT, GAIL
I'D RATHER BE CALM (II,1216); SHEILA (II,2325); THE SMOTHERS BROTHERS SHOW (II,2384); THREE GIRLS THREE (II,2608)

PARETS, HAL
THE WACKY WORLD OF JONATHAN WINTERS (I,4746)

PARIS, GEORGE
THE GREAT AMERICAN MUSIC CELEBRATION (II,1052); RETURN TO PEYTON PLACE (I,3779); ULTRA QUIZ (II,2701)

PARIS, JERRY
HAPPY DAYS (II,1084)

PARK, BEN
HAWKINS FALLS, POPULATION 6200 (I,1977); STUD'S PLACE (I,4271); THOSE ENDEARING YOUNG CHARMS (I,4469); THE TRACER (I,4588)

PARKER, BENJAMIN R
THE ANSWER MAN (I,231)

PARKER, JAMES
HARRY AND MAGGIE (II,1097); LOVE, SIDNEY (II,1547); THE ORPHAN AND THE DUDE (II,1924); STAR OF THE FAMILY (II,2436)

PARKER, JIM
THE BOYS (I,701); GOING PLACES (I,1834)

PARKER, KIT
MATINEE AT THE BIJOU (II,1651)

PARKER, ROD
ALL'S FAIR (II,46); AND THEY ALL LIVED HAPPILY EVER AFTER (II,75); THE ASTRONAUTS (II,114); DID YOU HEAR ABOUT JOSH AND KELLY?! (II,673); THE DOM DELUISE SHOW (I,1328); GIMME A BREAK (II,995);

HOT L BALTIMORE (II,1187); KING OF THE ROAD (II,1396); LOVE, SIDNEY (II,1547); MAUDE (II,1655); MR. & MRS. & MR. (II,1744); THE NANCY WALKER SHOW (II,1790); PHYL AND MIKHY (II,2037); POTTSVILLE (II,2072); THE POWDER ROOM (I,3656); SNAVELY (II,2387)

PARKINSON, BOB
THIS IS YOUR LIFE (II,2598); THE WORLD OF ENTERTAINMENT (II,2837)

PARKS, HILDY
NIGHT OF 100 STARS (II,1846)

PARKS, WILLIAM
THOSE FABULOUS CLOWNS (II,2602)

PARLAN, STAN
TAXI (I,4365)

PARNELL, BARRY
FOUL UPS, BLEEPS AND BLUNDERS (II,911)

PARRIOTT, JAMES D
ALEX AND THE DOBERMAN GANG (II,32); THE BIONIC WOMAN (II,255); FITZ AND BONES (II,867); HAWAIIAN HEAT (II,1111); THE INCREDIBLE HULK (II,1232); THE LEGEND OF THE GOLDEN GUN (II,1455); NICK AND THE DOBERMANS (II,1843); THE SEAL (II,2282); VOYAGERS (II,2730)

PARSLOW, PHILIP
DYNASTY (II,746)

PARSON, CHARLES J
NOT FOR PUBLICATION (I,3309)

PARSON, FRANK
EMPIRE (I,1439)

PARSONNET, MARION
THE DOCTOR (I,1320); HOLLYWOOD OFF BEAT (I,2086)

PARSONS, LINDSLEY
THE FILES OF JEFFERY JONES (I,1576); THE GRAY GHOST (I,1867)

PASCHOLL, AL
THIS IS YOUR LIFE (I,4461)

PASETTA, MARTY
ARTHUR GODFREY'S PORTABLE ELECTRIC MEDICINE SHOW (I,290); BING!. . . A 50TH ANNIVERSARY GALA (II,254); BURNETT "DISCOVERS" DOMINGO (II,398); CELEBRATION: THE AMERICAN SPIRIT (II,461); CHRISTMAS IN DISNEYLAND (II,519); A COUNTRY CHRISTMAS (II,582); A COUNTRY CHRISTMAS (II,583); A COUNTRY CHRISTMAS (II,584); DEBBY BOONE. . . THE SAME OLD BRAND NEW ME (II,656); THE DON ADAMS SCREEN TEST (II,705); GENE KELLY. . . AN AMERICAN IN PASADENA (II,962); HAPPY BIRTHDAY, AMERICA (II,1083); JOHN SCHNEIDER'S CHRISTMAS HOLIDAY (II,1320); THE MONTE CARLO SHOW (II,1726); THE NOONDAY SHOW (II,1859); PAUL ANKA IN MONTE CARLO (II,1971); PAUL ANKA. . . MUSIC MY WAY (II,1973); A SALUTE TO TELEVISION'S 25TH ANNIVERSARY (I,3893); SANDY IN DISNEYLAND (II,2254); STEVE AND EYDIE. . . ON STAGE (I,4223); SUPER COMEDY BOWL 1 (I,4288); SUPER COMEDY BOWL 2 (I,4289); TEXACO STAR THEATER: OPENING NIGHT (II,2565)

PASQUALE, NICHOLAS
OUTLAW LADY (II,1935)

PATCHETT, TOM
THE BOB NEWHART SHOW (II,342); BUFFALO BILL (II,387); THE CHOPPED LIVER BROTHERS (II,514); MARY (II,1637); OPEN ALL NIGHT (II,1914); THE TONY RANDALL SHOW (II,2640); WE'VE GOT EACH OTHER (II,2757)

PATTERSON, RAY
SPIDER-MAN (I,4154); YOGI'S SPACE RACE (II,2853)

PATTON, PHIL
LADIES BE SEATED (I,2598); SUPER CIRCUS (I,4287)

PAUL, BYRON
THE CONFESSIONS OF DICK VAN DYKE (II,568); DICK VAN DYKE AND THE OTHER WOMAN, MARY TYLER MOORE (I,1273); DICK VAN DYKE MEETS BILL COSBY (I,1274); THE DICK VAN DYKE SPECIAL (I,1276); THE DICK VAN DYKE SPECIAL (I,1277); JANE FROMAN'S U.S.A. CANTEEN (I,2341); THE JULIUS LAROSA SHOW (I,2486); THE NANCY DUSSAULT SHOW (I,3222); THE NEW DICK VAN DYKE SHOW (I,3262); THE NEW DICK VAN DYKE SHOW (I,3263); OF ALL THINGS (I,3330); PRIZE PERFORMANCE (I,3673); SONG SNAPSHOTS ON A SUMMER HOLIDAY (I,4122); VAN DYKE AND COMPANY (II,2718); VAN DYKE AND COMPANY (II,2719)

PAUL, NORMAN
THE DORIS DAY SHOW (I,1355); GOOD TIMES (II,1038); MARIE (II,1628); ONE DAY AT A TIME (II,1900)

PAUL, ROD
LANIGAN'S RABBI (II,1427)

PAUL, SAM
FIFTH OF JULY (II,851)

PAYNE, JOHN
O'CONNOR'S OCEAN (I,3326); THE RESTLESS GUN (I,3773)

PEACH, PETER
TENNESSEE TUXEDO AND HIS TALES (I,4403)

PEARL, MICHAEL
BATTLE OF THE NETWORK STARS (II,163); BATTLE OF THE NETWORK STARS (II,164); BATTLE OF THE NETWORK STARS (II,165); BATTLE OF THE NETWORK STARS (II,166); BATTLE OF THE NETWORK STARS (II,167); BATTLE OF THE NETWORK STARS (II,168); BATTLE OF THE NETWORK STARS (II,169); BATTLE OF THE NETWORK STARS (II,170); BATTLE OF THE NETWORK STARS (II,171); BATTLE OF THE NETWORK STARS (II,172); BATTLE OF THE NETWORK STARS (II,173); BATTLE OF THE NETWORK STARS (II,174); BATTLE OF THE NETWORK STARS (II,175); BATTLE OF THE NETWORK STARS (II,176); BATTLE OF THE NETWORK STARS (II,177)

PECK, JAMES
CONFIDENTIAL FILE (I,1033)

PECK, STEPHEN
IN SEARCH OF. . . (II,1226)

PECKINPAH, DAVID
HIGH PERFORMANCE (II,1145)

PECKINPAH, SAM
THE WESTERNER (I,4801)

PEDERSON, ROD
ED MCMAHON AND HIS FRIENDS. . . DISCOVER WET AT CYPRESS GARDENS (I,1400); FIRST & TEN (II,861)

PEEPLES, SAMUEL A
OVERLAND TRAIL (I,3433); THE TALL MAN (I,4354)

PELANNE, CLAUDE
THIS WAS AMERICA (II,2599); TOM COTTLE: UP CLOSE (II,2631)

PENN, ARTHUR
FLESH AND BLOOD (I,1595); PLAYHOUSE 90 (I,3623)

PENN, LEO
THE ADVENTURES OF OZZIE AND HARRIET (I,71)

PEPPIATT, FRANK
BARBARA MANDRELL AND THE MANDRELL SISTERS (II,143); BOBBIE GENTRY'S HAPPINESS HOUR (I,671); BOBBIE GENTRY'S HAPPINESS HOUR (II,343); THE BRASS ARE COMING (I,708); DON KNOTTS NICE CLEAN, DECENT, WHOLESOME HOUR (I,1333); EVERYTHING GOES (I,1480); THE FUNNY WORLD OF FRED & BUNNI (II,943); THE HARLEM GLOBETROTTERS POPCORN MACHINE (II,1092); HECK'S ANGELS (II,1122); HEE HAW (II,1123); THE JONATHAN WINTERS SHOW (I,2443); THE JUD STRUNK SHOW (I,2462); KEEP ON TRUCKIN' (II,1372); THE OSMOND BROTHERS SHOW (I,3409); SHIELDS AND YARNELL (II,2328); THE STEVE ALLEN COMEDY HOUR (II,2454); WHO'S AFRAID OF MOTHER GOOSE? (I,4852)

PERIN, RICHARD
THE CORAL JUNGLE (II,574)

PERKINS, JODY
WALTER CRONKITE'S UNIVERSE (II,2739)

PERL, ARNOLD
DEADLINE (I,1188); EAST SIDE/WEST SIDE (I,1396)

PERLMUTTER, A H
BEDROOMS (II,198); THE GREAT AMERICAN DREAM MACHINE (I,1871)

PERLOWE, BOB
THE HOMEMADE COMEDY SPECIAL (II,1173)

PERRET, GENE
THAT'S TV (II,2581); THREE'S COMPANY (II,2614); THE TIM CONWAY SHOW (II,2619); WELCOME BACK, KOTTER (II,2761)

PERRIN, DOMINIQUE
CIRCUS OF THE STARS (II,530); CIRCUS OF THE STARS (II,531); CIRCUS OF THE STARS (II,532); CIRCUS OF THE STARS (II,533); CIRCUS OF THE STARS (II,534); CIRCUS OF THE STARS (II,535); CIRCUS OF THE STARS (II,536); CIRCUS OF THE STARS (II,537); MEN AT LAW (I,3004); THE

STOREFRONT LAWYERS (I,4233)

PERRIN, NAT
THE ADDAMS FAMILY (II,11); THE BEACHCOMBER (I,371); BURLESQUE (I,765); HOW TO MARRY A MILLIONAIRE (I,2147); THE JOHNNY CARSON SHOW (I,2422); THE RED SKELTON SHOW (I,3755)

PERRY, ALFRED
HONEY WEST (I,2105)

PERRY, DICK
THE PAUL DIXON SHOW (I,3487)

PERRY, FRANK
A CHRISTMAS MEMORY (I,948); THANKSGIVING VISITOR (I,4417)

PERSKY, BILL
BIG EDDIE (II,235); BIG EDDIE (II,236); THE BILL COSBY SPECIAL (I,444); BOBBY PARKER AND COMPANY (II,344); THE BOYS (II,360); THE CONFESSIONS OF DICK VAN DYKE (II,568); DICK VAN DYKE AND THE OTHER WOMAN, MARY TYLER MOORE (I,1273); THE FIRST NINE MONTHS ARE THE HARDEST (I,1584); THE FUNNY SIDE (I,1714); THE FUNNY SIDE (I,1715); GOOD MORNING WORLD (I,1850); HOW TO SURVIVE THE 70S AND MAYBE EVEN BUMP INTO HAPPINESS (II,1199); KATE AND ALLIE (II,1365); THE MAN WHO CAME TO DINNER (I,2881); THE MONTEFUSCOS (II,1727); MY WIFE NEXT DOOR (II,1780); PURE GOLDIE (I,3696); THAT GIRL (II,2570); THREE FOR TAHITI (I,4476)

PERSON, PATT
TV FUNNIES (II,2672)

PETERS, JON
DREAMS (II,735)

PETERS, MARGIE
THE FACTS OF LIFE (II,805); FACTS OF LIFE: THE FACTS OF LIFE GOES TO PARIS (II,806)

PETERSON, BONNIE
THAT'S HOLLYWOOD (II,2577)

PETERSON, CASANDRA
ELVIRA'S MOVIE MACABRE (II,771)

PETERSON, EDGAR
THE AMAZING MR. MALONE (I,159); PULITZER PRIZE PLAYHOUSE (I,3692); WORLD OF GIANTS (I,4912)

PETERSON, EDWARD
THE AMAZING MR. MALONE (I,159)

PETERSON, ROD
THE WALTONS (II,2740)

PETERSON, ROGER
I'VE GOT A SECRET (I,2283)

PETITCLERC, DENNE
SHANE (I,3995)

PETRANTO, RUSS
BETWEEN THE LINES (II,225)

PETRASHEVITCH, H M
SUSPENSE THEATER ON THE AIR (II,2507)

PETRASHEVITCH, VICTOR
SUSPENSE THEATER ON THE AIR (II,2507)

PETRIE, DANIEL
STUD'S PLACE (I,4271)

PETTER, WILL
COLISEUM (I,1000)

PETTIT, RICHARD
THE LID'S OFF (I,2678)

PETTUS, KEN
MATT HELM (II,1653)

PEVERALL, JOHN
THE FAR PAVILIONS (II,832)

PEWOLAR, JAMES
MR. WIZARD (I,3154)

PEYSER, JOHN
B.J. AND THE BEAR (II,125); B.J. AND THE BEAR (II,127); THE EYES OF TEXAS (II,799); GIANT IN A HURRY (I,1790); SWITCH (II,2519)

PHALEN, CHARLES
HOT WHEELS (I,2126); SKYHAWKS (I,4079)

PHARES, FRANK
N.O.P.D. (I,3303)

PHELPS, STUART W
WHAT'S THIS SONG? (I,4822)

PHILBIN, JACK
AMERICA'S GREATEST BANDS (I,177); THE BIG RECORD (I,430); THE BIG SELL (I,431); THE DUPONT SHOW OF THE WEEK (I,1388); THE HONEYMOONERS (I,2110); THE HONEYMOONERS (I,2111); THE HONEYMOONERS (II,1175); THE HONEYMOONERS CHRISTMAS SPECIAL (II,1176); THE HONEYMOONERS CHRISTMAS (II,1177); THE HONEYMOONERS SECOND HONEYMOON (II,1178); THE HONEYMOONERS VALENTINE SPECIAL (II,1179); THE JACKIE GLEASON SHOW (I,2323); THE JACKIE GLEASON SPECIAL (I,2324); THE JACKIE GLEASON SPECIAL (I,2325); THE KATE SMITH SHOW (I,2510); KEEFE BRASSELLE'S VARIETY GARDEN (I,2515); ONCE UPON A CHRISTMAS TIME (I,3368); YOU'RE IN THE PICTURE (I,4965)

PHILIPP, GODFREY
PRISONER: CELL BLOCK H (II,2085)

PHILLIPS, IRNA
THESE ARE MY CHILDREN (I,4440)

PHILLIPS, JACKIE
THE LOVE REPORT (II,1542)

PHILLIPS, KATHY
BULLSEYE (II,393)

PHILLIPS, MARK
BULLSEYE (II,393)

PHILLIPS, RANDY
ROCK PALACE (II,2193)

PHILLIPS, WILLIAM F
THE EARTHLINGS (II,751); EVERY STRAY DOG AND KID (II,792); HAWAII FIVE-O (II,1110); THE NIGHT RIDER (II,1848); THE OUTSIDE MAN (II,1937); RICHIE BROCKELMAN: MISSING 24 HOURS (II,2168); ROYCE (II,2224); THREE FOR THE ROAD (II,2607); TRAUMA CENTER (II,2655)

PHINNEY, DAVID
BATTLESTAR GALACTICA (II,181); COVER UP (II,597)

PICARD, PAUL R
CALIFORNIA FEVER (II,411); THE DUKES OF HAZZARD (II,742); ENOS (II,779); HIGH PERFORMANCE (II,1145); THE HOYT AXTON SHOW (II,1200); MELVIN PURVIS: G-MAN (II,1679); SCRUPLES (II,2281)

PIECH, PETER
THE BULLWINKLE SHOW (I,761); KING LEONARDO AND HIS SHORT SUBJECTS (I,2557)

PIERCE, ED
CHARGE ACCOUNT (I,903); NBC SATURDAY PROM (I,3245)

PIERCE, FRANK
MATT HELM (II,1653); MATT LINCOLN (I,2949)

PIERSON, ARTHUR
THE AFFAIRS OF CHINA SMITH (I,87); WATERFRONT (I,4774)

PIERSON, FRANK
HAVE GUN—WILL TRAVEL (I,1968); NICHOLS (I,3283)

PILL, JEFFREY
IN SEARCH OF. . . (II,1226)

PINA, BIANCA
THE $100,000 NAME THAT TUNE (II,1902)

PINCUS, IRVING
THE ADVENTURES OF ELLERY QUEEN (I,59); EDDIE (I,1405); MR. I. MAGINATION (I,3140); THE REAL MCCOYS (I,3741)

PINCUS, NORMAN
THE ADVENTURES OF ELLERY QUEEN (I,59); MR. I. MAGINATION (I,3140); THE REAL MCCOYS (I,3741)

PINGITORE, CARL
FROM HERE TO ETERNITY (II,933); POLICE STORY (II,2062); THE POWERS OF MATTHEW STAR (II,2075)

PINN, IRENE
LILY (I,2705); LILY (II,1477); THE LILY TOMLIN SHOW (I,2706); THE LILY TOMLIN SPECIAL (II,1479)

PINO, BIANCO
THICKE OF THE NIGHT (II,2587)

PINTOFF, ERNEST
THE KOWBOYS (I,2584); OCCASIONAL WIFE (I,3325)

PIOLI, JUDY
GOODTIME GIRLS (II,1042); I'M A BIG GIRL NOW (II,1218)

PIROSH, ROBERT
LARAMIE (I,2614)

PISCOPO, JOE
THE JOE PISCOPO SPECIAL (II,1304)

PITLIK, NOAM
9 TO 5 (II,1852); BARNEY MILLER (II,154); NOW WE'RE COOKIN' (II,1866)

PITTMAN, DEBBIE
INCREDIBLE KIDS AND COMPANY (II,1233)

PITTMAN, FRANK
BAT MASTERSON (I,363); THE GREAT GILDERSLEEVE (I,1875); TOMBSTONE TERRITORY (I,4550)

PITTS, CY
THE FRED WARING SHOW (I,1679)

PIVAL, STANLEY
REHEARSAL CALL (I,3761)

PIZZO, ANGELO
THE TOM SWIFT AND LINDA CRAIG MYSTERY HOUR (II,2633)

PLACE, LOU
DOC CORKLE (I,1304)

PLAYDON, PAUL
BATTLESTAR GALACTICA (II,181); THE NIGHT STALKER (II,1849); SWITCH (II,2519)

PLAYLOW, CLARK
SKY KING (I,4077)

PLEVEN, PAT
INTERNATIONAL SHOWTIME (I,2225)

PLOWDEN, JULIAN
HORNBLOWER (I,2121)

PODELL, RIC
BROTHERS (II,381)

PODMORE, BILL
CORONATION STREET (I,1054)

POINDEXTER, H R
ON THE ROCKS (II,1894)

POLACHEK, CHARLES
CLEAR HORIZON (I,973); THE EDGE OF NIGHT (II,760); VOICE OF FIRESTONE (I,4741)

POLINANI, KATINA
THE PAUL HOGAN SHOW (II,1975)

POLITO, ROSALIE
OUTLAW LADY (II,1935)

POLL, MARTIN
THE DAIN CURSE (II,613); HEAD OF THE FAMILY (I,1988); STUNT SEVEN (II,2486)

POLLOCK, DAVID
ALMOST HEAVEN (II,57); DELTA HOUSE (II,658); GOODNIGHT, BEANTOWN (II,1041); HOLLYWOOD HIGH (II,1163); THE SHOW MUST GO ON (II,2344); STEAMBATH (II,2451); THAT'S MY MAMA (II,2580)

POLSON, BETH
THE BARBARA WALTERS SPECIAL (II,145)

POMANSANOFF, ALEX
AN AMERICAN PORTRAIT (II,73); IN SEARCH OF. . . (II,1226); THAT'S INCREDIBLE! (II,2578)

POMERANTZ, EARL
BEST OF THE WEST (II,218); THE COSBY SHOW (II,578)

POMMER, JOHN E
BEN CASEY (I,394)

POMPIAN, PAUL
HEAR NO EVIL (II,1116)

POOLE, DUANE
GLITTER (II,1009); LAVERNE AND SHIRLEY IN THE ARMY (II,1448)

POOR, PETER
20TH CENTURY (I,4640)

POPE, JOE
SKYWARD CHRISTMAS (II,2378)

POPPE, HARRY H
THE LONE RANGER (I,2740); MY LIVING DOLL (I,3195)

PORTER, WILLIAM
THE LOVE BOAT (II,1535)

POSS, STANLEY
MEET YOUR COVER GIRL (I,2992)

POST, MIKE
MAC DAVIS CHRISTMAS SPECIAL. . . WHEN I GROW UP (II,1574); THE MAC DAVIS SHOW (II,1576)

POTTASCH, ALAN
PERSONALITY PUZZLE (I,3554)

POTTER, ALLEN
AS THE WORLD TURNS (II,110); THE BRIGHTER DAY (I,728); THE GUIDING LIGHT (II,1064); OUR PRIVATE WORLD (I,3417)

POTTER, ANDY
THE LIFE OF RILEY (I,2684)

POTTER, PETER
PETER POTTER'S JUKE BOX JURY (I,3569)

POTTER, STANLEY C
AS THE WORLD TURNS (II,110)

POULIOT, STEPHEN
HOT (II,1183); HOT (II,1184); JULIE ANDREWS: ONE STEP INTO SPRING (II,1353); LEAPIN' LIZARDS, IT'S LIBERACE (II,1453); LYNDA CARTER: ENCORE (II,1566); PERRY COMO IN LAS VEGAS (II,2003); PERRY COMO'S BAHAMA HOLIDAY (II,2006); PERRY COMO'S CHRISTMAS IN AUSTRIA (II,2007); PERRY COMO'S CHRISTMAS IN MEXICO (II,2009); PERRY COMO'S CHRISTMAS IN NEW MEXICO (II,2010); PERRY COMO'S CHRISTMAS IN PARIS (II,2012); PERRY COMO'S CHRISTMAS IN THE HOLY LAND (II,2013); PERRY COMO'S EASTER IN GUADALAJARA (II,2015); PERRY COMO'S EASTER BY THE SEA (II,2014); PERRY COMO'S FRENCH-CANADIAN CHRISTMAS (II,2016); PERRY COMO'S SPRING IN NEW ORLEANS (II,2021); PERRY COMO'S SPRING IN SAN FRANCISCO (II,2022)

POWELL, DICK
330 INDEPENDENCE S.W. (I,4486); THE BOSTON TERRIER (I,696); COLOSSUS (I,1009); THE DICK POWELL SHOW (I,1269); FOUR STAR PLAYHOUSE (I,1652); THE JUNE ALLYSON SHOW (I,2488); THE LOSERS (I,2757); LUXURY LINER (I,2796); SAFARI (I,3882); SAVAGE SUNDAY (I,3926)

POWELL, JONATHAN
PRIDE AND PREJUDICE (II,2080); SMILEY'S PEOPLE (II,2383); THERESE RAQUIN (II,2584); TINKER, TAILOR, SOLDIER, SPY (II,2621); THE WOMAN IN WHITE (II,2819)

POWELL, NORMAN S
THE BOB CRANE SHOW (II,280); FLATBUSH (II,877); RAFFERTY (II,2105); SALVAGE (II,2244); SALVAGE 1 (II,2245); THE THREE WIVES OF DAVID WHEELER (II,2611); WASHINGTON: BEHIND CLOSED DOORS (II,2744)

POWER, JULES
MR. DICKENS OF LONDON (I,3135)

POWERS, CHARLES
MY TRUE STORY (I,3206); WELCOME TRAVELERS (I,4791)

PRAGER, STANLEY
KIBBE HATES FINCH (I,2530)

PRAISER, IAN
DOMESTIC LIFE (II,703); MR. SUCCESS (II,1756); TAXI (II,2546)

PRATT, CHARLES
THE HAMPTONS (II,1076)

PRATT, JAMES
ELFEGO BACA (I,1427); THE SAGA OF ANDY BURNETT (I,3883); SWAMP FOX (I,4311); TEXAS JOHN SLAUGHTER (I,4414)

PRECHT, ROBERT H
THE BEST OF SULLIVAN (II,217); CALUCCI'S DEPARTMENT (I,805); CHANGE AT 125TH STREET (II,480); CLOWNAROUND (I,980); CRYSTAL (II,607); THE CRYSTAL GAYLE SPECIAL (II,609); THE ED SULLIVAN SHOW (I,1401); ED SULLIVAN'S BROADWAY (I,1402); THE FIFTH DIMENSION SPECIAL: AN ODYSSEY IN THE COSMIC UNIVERSE OF PETER MAX (I,1571); FUNNY PAPERS (I,1713); THE LILY TOMLIN SHOW (I,2706); MOVIN' WITH NANCY ON STAGE (I,3120); SEE AMERICA WITH ED SULLIVAN (I,3972); SONNY BOY (II,2402); THE SOUPY SALES SHOW (I,4138); THREE FOR THE GIRLS (I,4477)

PREMINGER, MICHAEL
BROTHERS (II,381)

PREMINGER, OTTO
TONIGHT AT 8:30 (I,4556)

PRESCOTT, NORM
THE ARCHIE COMEDY HOUR (I,252); ARCHIE'S TV FUNNIES (I,253); ARK II (II,107); THE BANG-SHANG LALAPALOOZA SHOW (II,140); BATMAN AND THE SUPER SEVEN (II,160); BLACKSTAR (II,263); THE BRADY KIDS (II,365); EVERYTHING'S ARCHIE (I,1483); THE FABULOUS FUNNIES (II,802); FANTASTIC VOYAGE (I,1530); FAT ALBERT AND THE COSBY KIDS (II,835); FLASH GORDON (II,874); FLASH GORDON—THE GREATEST ADVENTURE OF ALL (II,875); GILLIGAN'S PLANET (II,994); THE HARDY BOYS (I,1949); ISIS (II,1248); JASON OF STAR COMMAND (II,1274); JOURNEY TO THE CENTER OF THE EARTH (I,2456); THE KID SUPER POWER HOUR WITH SHAZAM (II,1386); LASSIE'S RESCUE RANGERS (II,1432); MISSION MAGIC (I,3068); MY FAVORITE MARTIANS (II,1779); THE NEW ADVENTURES OF BATMAN (II,1812); THE NEW ADVENTURES OF GILLIGAN (II,1813); THE NEW ADVENTURES OF MIGHTY MOUSE AND HECKLE AND JECKLE (II,1814); THE NEW ADVENTURES OF SUPERMAN (I,3255); RUN RUN, JOE (II,2229); SABRINA, THE TEENAGE WITCH (I,3880); THE SECRET LIVES OF WALDO KITTY (II,2292); SHAZAM! (II,2319); SPACE ACADEMY (II,2406); SPORT BILLY (II,2429); STAR TREK (II,2439); SUPER WITCH (II,2501); THE TARZAN/LONE RANGER/ZORRO ADVENTURE HOUR (II,2543);

TARZAN: LORD OF THE JUNGLE (II,2544); TEACHER'S PET (I,4367); THE TOM AND JERRY COMEDY SHOW (II,2629); THE U.S. OF ARCHIE (I,4687); U.S. OF ARCHIE (II,2698); UNCLE CROC'S BLOCK (II,2702); THE WESTWIND (II,2766); WILL THE REAL JERRY LEWIS PLEASE SIT DOWN? (I,4866); THE YOUNG SENTINELS (II,2871)

PRICE, FRANK
ALIAS SMITH AND JONES (I,119); THE CITY (I,969); CONVOY (I,1043); I LOVE A MYSTERY (I,2170); IRONSIDE (II,1246); IT TAKES A THIEF (I,2250); SAN FRANCISCO INTERNATIONAL AIRPORT (I,3906); THE TALL MAN (I,4354); THE VIRGINIAN (I,4738)

PRICE, JUDY
AMERICAN BANDSTAND (II,70)

PRICE, KENNY
AMERICA ALIVE! (II,67); ON STAGE AMERICA (II,1893)

PRICE, SHERWOOD
THE PROTECTORS (I,3681)

PRITZKER, STEVE
FISH (II,864); FRIENDS (II,928); PAUL SAND IN FRIENDS AND LOVERS (II,1982); THE SINGLE LIFE (II,2364); STRUCK BY LIGHTNING (II,2482)

PROCKTER, BERNARD
THE BIG STORY (I,432); MAN BEHIND THE BADGE (I,2857); SHORT, SHORT DRAMA (I,4023); TREASURY MEN IN ACTION (I,4604)

PROFT, PAT
BUCKSHOT (II,385); MARIE (II,1629)

PROSER, MONTE
HOLIDAY HOTEL (I,2075)

PUCK, LARRY
ARTHUR GODFREY TIME (I,289)

PURCELL, CLIE
THE DANCE SHOW (II,616)

PURCELL, SARAH
REAL KIDS (II,2116)

PURDY, HALL
ALL AROUND TOWN (I,128)

PURDY, JOHN
SINATRA: CONCERT FOR THE AMERICAS (II,2361)

PURDY, RAI
ROBERT Q'S MATINEE (I,3812)

PUTTERMAN, ZEV
THE INTERNATIONAL ANIMATION FESTIVAL (II,1240)

PYE, CHRIS
YOU ASKED FOR IT (II,2855)

QUIGLEY, BOB
THE AMATEUR'S GUIDE TO LOVE (I,155); BEDTIME STORIES (II,199); THE CELEBRITY GAME (I,884); GAMBIT (II,955); HIGH ROLLERS (II,1147); THE HOLLYWOOD SQUARES (II,1164); HOT SEAT (I,2125); LAS VEGAS GAMBIT (II,1430); THE MAGNIFICENT MARBLE MACHINE (II,1600); ON YOUR ACCOUNT (I,3364); P.D.Q. (I,3497); PEOPLE WILL TALK (I,3520); SHOWDOWN (I,4034); THE STORYBOOK SQUARES (I,4242); TEMPTATION (I,4393); TO SAY THE LEAST (II,2623); VIDEO VILLAGE (I,4731)

QUILL, JOHN E
CHIEFS (II,509); ME AND MAXX (II,1672); ONE MORE TRY (II,1908)

QUIMBY, FRED
THE TOM AND JERRY SHOW (I,4534)

QUIN, TONY
ROCK-N-AMERICA (II,2195)

QUINE, RICHARD
THE JEAN ARTHUR SHOW (I,2352)

QUINLAN, DORIS
ONE LIFE TO LIVE (II,1907); WHODUNIT? (II,2794); YOUNG DR. MALONE (I,4938)

QUINN, STANLEY
KRAFT TELEVISION THEATER (I,2592)

QUINT, GEORGE
BYLINE—BETTY FURNESS (I,785)

RKO VIDEO GROUP
LENA HORNE: THE LADY AND HER MUSIC (II,1460)

RABIN, AL
DAYS OF OUR LIVES (II,629)

RABWIN, PAUL
FORCE SEVEN (II,908)

RACHMIL, LEWIS
LIFE WITH VIRGINIA (I,2695); MEN INTO SPACE (I,3007)

RACKIN, MARTIN
NEVADA SMITH (II,1807); RIVAK, THE BARBARIAN (I,3795)

RACKMIL, GLADYS
THE COMEDY ZONE (II,559)

RADIN, PAUL
BORN FREE (I,691); BORN FREE (II,353)

RADY, SIMON
HOWDY DOODY (I,2151)

RAE, DAVID C
WHEN HAVOC STRUCK (II,2778)

RAFELSON, RALPH
THE MONKEES (I,3088)

RAFKIN, ALAN
ME AND THE CHIMP (I,2974); ONE DAY AT A TIME (II,1900); THE SUPER (I,4293); WE GOT IT MADE (II,2752)

RAISBECK, ROBERT
THE RUGGLES (I,3868)

RALLING, CHRISTOPHER
SEARCH FOR THE NILE (I,3954)

RALSTON, GILBERT A
CAPTAIN GALLANT OF THE FOREIGN LEGION (I,831); CAVALCADE OF AMERICA (I,875); THE GENERAL ELECTRIC THEATER (I,1753); HIGH ADVENTURE WITH LOWELL THOMAS (I,2048); SUSPENSE (I,4306); YOUR JEWELER'S SHOWCASE (I,4953)

RALSTON, RUDY
LAST STAGECOACH WEST (I,2629); STORIES OF THE CENTURY (I,4235)

RAMBEAU, LEONARD T
ANNE MURRAY'S CARIBBEAN CRUISE (II,88); ANNE MURRAY'S WINTER CARNIVAL. . . FROM QUEBEC (II,90)

RAMIS, HAROLD
RODNEY DANGERFIELD SPECIAL: IT'S NOT EASY BEIN' ME (II,2199)

RANDALL, BOB
KATE AND ALLIE (II,1365)

RANDALL, DICK
THE JACQUELINE SUSANN SHOW (I,2327)

RANDALL, RIC
CHIPS (II,511)

RANK, J ARTHUR
INTERPOL CALLING (I,2227)

RANKIN JR, ARTHUR
'TWAS THE NIGHT BEFORE CHRISTMAS (II,2677); THE BEATLES (I,380); THE CONEHEADS (II,567); THE EASTER BUNNY IS COMIN' TO TOWN (II,753); FROSTY THE SNOWMAN (II,934); FROSTY'S WINTER WONDERLAND (II,935); HERE COMES PETER COTTONTAIL (II,1132); JACK FROST (II,1263); THE JACKSON FIVE (I,2326); KID POWER (I,2532); THE KING KONG SHOW (I,2556); THE LEPRECHAUN'S CHRISTMAS GOLD (II,1461); THE LITTLE DRUMMER BOY (II,1485); LITTLE DRUMMER BOY, BOOK II (II,1486); THE NEW ADVENTURES OF PINOCCHIO (I,3253); THE OSMONDS (I,3410); PINOCCHIO'S CHRISTMAS (II,2045); THE RELUCTANT DRAGON AND MR. TOAD (I,3762); RUDOLPH THE RED-NOSED REINDEER (II,2227); RUDOLPH'S SHINY NEW YEAR (II,2228); SANTA CLAUS IS COMIN' TO TOWN (II,2259); SMOKEY THE BEAR SHOW (I,4094); TOMFOOLERY (I,4551); THE YEAR WITHOUT A SANTA CLAUS (II,2846)

RANSOHOFF, MARTIN
CO-ED FEVER (II,547)

RANTELS, DAVID
ALL THE WAY HOME (II,44)

RAPF, MATTHEW
BEN CASEY (I,394); DOCTORS HOSPITAL (II,691); DOCTORS' PRIVATE LIVES (II,692); EISCHIED (II,763); THE GANGSTER CHRONICLES (II,957); IRON HORSE (I,2238); JEFFERSON DRUM (I,2355); KOJAK (II,1406); A LETTER TO LORETTA (I,2669); ONE OF OUR OWN (II,1910); SLATTERY'S PEOPLE (I,4082); SWITCH (II,2519); TWO FACES WEST (I,4653); THE YOUNG LAWYERS (I,4941)

RAPOPORT, I C
BORN TO THE WIND (II,354)

RAPP, PHILIP
THE ADVENTURES OF HIRAM HOLIDAY (I,61); MIMI (I,3054)

RAPPAPORT, JOHN
M*A*S*H (II,1569); SECOND EDITION (II,2288); THREE TIMES DALEY (II,2610)

RAPPAPORT, MICHELLE
PAPER DOLLS (II,1951); PAPER DOLLS (II,1952)

RASKIN, CAROLYN
ARNOLD'S CLOSET REVIEW (I,262); BIG CITY COMEDY (II,233); BOB HOPE SPECIAL: BOB HOPE'S ROAD TO HOLLYWOOD (II,313); BOB HOPE SPECIAL: BOB HOPE'S STAR-STUDDED SPOOF OF THE NEW TV SEASON— G-RATED—WITH GLAMOUR, GLITTER & GAGS (II,317); BOB HOPE SPECIAL: BOB HOPE'S WICKI-WACKY SPECIAL FROM WAIKIKI (II,321); THE DIAHANN CARROLL SHOW (I,1252); DINAH AND HER NEW BEST FRIENDS (II,676); THE DINAH SHORE SPECIAL—LIKE HEP (I,1282); DINAH! (II,677); FRANK SINATRA—A MAN AND HIS MUSIC (I,1661); THE HELEN REDDY SHOW (I,2008); HELLZAPOPPIN (I,2011); HERB ALBERT AND THE TIJUANA BRASS (I,2019); THE JIM NABORS SHOW (II,1290); LAUGH TRAX (II,1443); ONE MORE TIME (I,3385); ONE MORE TIME (I,3386); ROCK COMEDY (II,2191); ROCK PALACE (II,2193); ROWAN AND MARTIN'S LAUGH-IN (I,3856); THE SHAPE OF THINGS (I,3999); SINATRA (I,4054); STATE FAIR USA (II,2449); STATE FAIR USA (II,2450); THE TENNESSEE ERNIE FORD SPECIAL (I,4401)

RASKY, HARRY
THE LEGEND OF SILENT NIGHT (I,2655); PERSPECTIVE ON GREATNESS (I,3555)

RAUCH, MICHAEL
BUSH DOCTOR (II,400); NURSE (II,1870)

RAUCH, PAUL
ANOTHER WORLD (II,97); FOR RICHER, FOR POORER (II,906); LOVERS AND FRIENDS (II,1549); TEXAS (II,2566)

RAUESO, VIC
ALICE (II,33); JUST OUR LUCK (II,1360); PRIVATE BENJAMIN (II,2087)

RAVIER, CLAUDE
BLONDES VS. BRUNETTES (II,271)

RAYEL, JACK
THE DEVIL'S DISCIPLE (I,1247); ENTERTAINMENT —1955 (I,1450); HOLIDAY IN LAS VEGAS (I,2076); ONE TOUCH OF VENUS (I,3389); TONIGHT AT 8:30 (I,4556)

RAYMOND, CHARLES
ALL THE WAY HOME (II,44); THE BETTY WHITE SHOW (II,224); BROTHERS (II,381); DOUBLE TROUBLE (II,724); FIRST TIME, SECOND TIME (II,863); GOODBYE DOESN'T MEAN FOREVER (II,1039); MOTHER AND ME, M.D. (II,1740); NOT UNTIL TODAY (II,1865); A ROCK AND A HARD PLACE (II,2190); WHITE AND RENO (II,2787)

READ, JOHN
CAPTAIN SCARLET AND THE MYSTERONS (I,837)

READICK, JERRY
BEDTIME STORIES (II,199)

RECHT, LORI
LEGENDS OF THE SCREEN (II,1456)

REDA, LOU
THE BLUE AND THE GRAY (II,273)

REDACK, JAY
BATTLESTARS (II,182); THE HOLLYWOOD SQUARES (II,1164); THE STORYBOOK SQUARES (I,4242)

REDDY, JOHN
BRIDE AND GROOM (I,721); LIVE LIKE A MILLIONAIRE (I,2728)

REDFORD, KEN
HOMEMAKER'S EXCHANGE (I,2100); MESSING PRIZE PARTY (I,3019); THE TED STEELE SHOW (I,4373); THE WARREN HULL SHOW (I,4771)

REED, DICK
HOT WHEELS (I,2126); SKYHAWKS (I,4079)

REED, LEWIS
DAN RAVEN (I,1131)

REED, ROLAND
BEULAH (I,416); ROCKY JONES, SPACE RANGER (I,3823); SABER OF LONDON (I,3879); TROUBLE WITH FATHER (I,4610)

REEDER, TOM
BENSON (II,208)

REEVE, GEOFFREY
THE FAR PAVILIONS (II,832)

REEVES, CHARLES
DICK CLARK'S WORLD OF TALENT (I,1268)

REEVES, GEORGE
BIG JOHN (II,239); CRAZY TIMES (II,602)

REEVES, JEROME R
ALLEN LUDDEN'S GALLERY (I,143)

REGAS, JACK
THE BAY CITY ROLLERS SHOW (II,187); DOCTOR SHRINKER (II,686); LOUISE MANDRELL: DIAMONDS, GOLD AND PLATINUM (II,1528); MAGIC MONGO (II,1592)

REID, DAVID
THE STRAUSS FAMILY (I,4253)

REID, DON
ANYTHING YOU CAN DO (I,239); DREAM HOUSE (I,1376); DREAM HOUSE (II,734)

REID, JOHN
ELTON JOHN IN CENTRAL PARK (II,769); ELTON JOHN: IN CONCERT (II,770)

REIG, JUNE
LITTLE WOMEN (I,2724)

REIKER, DONALD
REPORT TO MURPHY (II,2134)

REILLY, HOWARD
HOUR GLASS (I,2133); THE KEN MURRAY SHOW (I,2525)

REILLY, JACK
THE CORAL JUNGLE (II,574); ENTERTAINMENT TONIGHT (II,780); THE MIKE DOUGLAS SHOW (II,1698)

REINER, CARL
THE DICK VAN DYKE SHOW (I,1275); FLANNERY AND QUILT (II,873); GOOD HEAVENS (II,1032); GOOD MORNING WORLD (I,1850); THE NEW DICK VAN DYKE SHOW (I,3262)

REINER, ROB
FREE COUNTRY (II,921); SONNY BOY (II,2402); THE TV SHOW (II,2673)

REISMAN, DEL
BRACKEN'S WORLD (I,703); HOTEL DE PAREE (I,2129)

REISMAN, MARK
THICKE OF THE NIGHT (II,2587)

REISNER, ALLEN
O'HARA, UNITED STATES TREASURY (I,3346)

REITMAN, IVAN
DELTA HOUSE (II,658)

RELYEA, ROBERT E
THE AWAKENING LAND (II,122); EVEL KNIEVEL (II,785); HIGH RISK (II,1146)

RENAN, SHELDON
THE INTERNATIONAL ANIMATION FESTIVAL (II,1240)

REO, DON
CHER (II,495); LANDON, LANDON & LANDON (II,1426); M*A*S*H (II,1569); PRIVATE BENJAMIN (II,2087); RHODA (II,2151); A ROCK AND A HARD PLACE (II,2190); SUMMER (II,2491); WE'LL GET BY (II,2753); WHITE AND RENO (II,2787); WIZARDS AND WARRIORS (II,2813)

RESSEN, EDWARD
PLAYBOY'S 25TH ANNIVERSARY CELEBRATION (II,2054)

REUBENS, PAUL
THE PEE WEE HERMAN SHOW (II,1988)

REY, LOUISE KING
THE KING FAMILY CELEBRATE THANKSGIVING (I,2545); THE KING FAMILY CHRISTMAS SPECIAL (I,2546); THE KING FAMILY IN ALASKA (I,2547); THE KING FAMILY IN HAWAII (I,2548); THE KING FAMILY OCTOBER SPECIAL (I,2551); THE KING FAMILY SHOW (I,2552); THE KING FAMILY SPECIAL (I,2553); THE KING FAMILY VALENTINE'S DAY SPECIAL (I,2554)

REYNOLDS, BURT
COTTON CLUB '75 (II,580)

REYNOLDS, GENE
ALFRED OF THE AMAZON (I,116); ANNA AND THE KING (I,221); THE GHOST AND MRS. MUIR (I,1786); IF I LOVE YOU, AM I TRAPPED FOREVER? (II,1222); KAREN (II,1363); LOU GRANT (II,1526); M*A*S*H (II,1569); PEOPLE LIKE US (II,1999); ROLL OUT! (I,3833); ROOM 222 (I,3843); SOUTHERN FRIED (I,4139)

REYNOLDS, JIM
PAY CARDS (I,3496)

REYNOLDS, SHELDON
DICK AND THE DUCHESS (I,1263); FOREIGN INTRIGUE: CROSS CURRENT (I,1637); FOREIGN INTRIGUE: DATELINE EUROPE (I,1638); FOREIGN INTRIGUE: OVERSEAS ADVENTURES (I,1639); SHERLOCK HOLMES (I,4010)

REYNOLDS, STUART
SKIP TAYLOR (I,4075)

RHEINSTEIN, FRED
THE MELTING POT (II,1678)

RHINE, LARRY

JOE'S WORLD (II,1305); NUTS AND BOLTS (II,1871); SANFORD (II,2255)

RHINEHART, JOHN
THE PEOPLE'S COURT (II,2001)

RHODES, JACK
CELEBRITY REVUE (II,475); SECOND CITY TELEVISION (II,2287)

RHODES, MICHAEL
CANNON: THE RETURN OF FRANK CANNON (II,425); CUTTER TO HOUSTON (II,612); DELVECCHIO (II,659); ELLERY QUEEN (II,766); IT CAN'T HAPPEN TO ME (II,1250); JOSIE (II,1345); THE JUGGLER OF NOTRE DAME (II,1350); LEADFOOT (II,1452); LOOSE CHANGE (II,1518); THE NEW OPERATION PETTICOAT (II,1826); PRINCESS (II,2084); QUICK AND QUIET (II,2099); THE TROUBLE WITH GRANDPA (II,2662); TURNABOUT (II,2669); WHEN, JENNY? WHEN (II,2783)

RICCA, ERNEST
LOVE OF LIFE (I,2774)

RICH JR, JAMES
MINSKY'S FOLLIES (II,1703)

RICH, DAVID LOWELL
BRIDGER (II,376)

RICH, ELAINE
CHARLIE'S ANGELS (II,486); DYNASTY (II,746); FINDER OF LOST LOVES (II,857); HOTEL (II,1192); STRIKE FORCE (II,2480)

RICH, JOHN
ALL IN THE FAMILY (II,38); BENSON (II,208); BILLY (II,247); CHARO AND THE SERGEANT (II,489); CONDO (II,565); GRANDPA MAX (II,1051); I'LL NEVER FORGET WHAT'S HER NAME (II,1217); MOTHER, JUGGS AND SPEED (II,1741); ON THE ROCKS (II,1894); SLEZAK AND SON (I,4086)

RICH, LEE
APPLE'S WAY (II,101); BIG SHAMUS, LITTLE SHAMUS (II,242); THE BLUE KNIGHT (II,276); THE BLUE KNIGHT (II,277); BOONE (II,351); BRAVO TWO (II,368); BUNCO (II,395); CONSPIRACY OF TERROR (II,569); DALLAS (II,614); DOC ELLIOT (I,1305); DOC ELLIOT (I,1306); DUSTY (II,745); EIGHT IS ENOUGH (II,762); ERIC (I,1451); FLAMINGO ROAD (II,871); FLAMINGO ROAD (II,872); THE GOOD LIFE (I,1848); THE GOOD LIFE (II,1033); HEY, LANDLORD (I,2039); HUNTER (II,1204); HUNTER (II,1205); JOSHUA'S WORLD (II,1344); KAZ (II,1370); KING'S CROSSING (II,1397); KNOTS LANDING (II,1404); MAKING IT (II,1605); A MAN CALLED INTREPID (II,1612); MARRIAGE IS ALIVE AND WELL (II,1633); MARRIED: THE FIRST YEAR (II,1634); ME AND DUCKY (II,1671); OUR FAMILY BUSINESS (II,1931); PEOPLE LIKE US (II,1999); REWARD (II,2147); THE RUNAWAY BARGE (II,2230); SECRETS OF MIDLAND HEIGHTS (II,2296); SKAG (II,2374); STUDS LONIGAN (II,2484); THIS IS KATE BENNETT (II,2597); TWO GUYS FROM MUCK (II,2690); THE WALTONS (II,2740); THE WALTONS: A DAY FOR THANKS ON WALTON'S MOUNTAIN (EPISODE 3) (II,2741); THE WALTONS: MOTHER'S DAY ON WALTON'S MOUNTAIN (EPISODE 2) (II,2742); THE WALTONS: WEDDING ON WALTON'S MOUNTAIN (EPISODE 1) (II,2743); THE WAVERLY WONDERS (II,2746); THE WILDS OF TEN THOUSAND ISLANDS (II,2803); WILLOW B: WOMEN IN PRISON (II,2804); YOU'RE JUST LIKE YOUR FATHER (II,2861); THE YOUNG PIONEERS (II,2869)

RICH, TED
KOSTA AND HIS FAMILY (I,2581)

RICHARD, ROGERS
JESSIE (II,1287)

RICHARDS, ARTHUR
AS THE WORLD TURNS (II,110)

RICHARDS, LLOYD
THE UNTOUCHABLES (I,4682)

RICHARDSON, BOB
DUNGEONS AND DRAGONS (II,744); MEATBALLS AND SPAGHETTI (II,1674)

RICHARDSON, DONALD
I REMEMBER MAMA (I,2176)

RICHARDSON, ED
CIRCUS OF THE 21ST CENTURY (II,529); THE LITTLEST HOBO (II,1500)

RICHARDSON, MARGARET
ARTHUR GODFREY AND HIS UKULELE (I,279); THE ARTHUR GODFREY SHOW (I,283)

RICHARDSON, RON
JIM HENSON'S MUPPET BABIES (II,1289)

RICHE, WENDY
NEVER AGAIN (II,1808)

RICHETTA, DON
BENSON (II,208); I'M A BIG GIRL NOW (II,1218)

RICHMAN, STELLA
JENNIE: LADY RANDOLPH CHURCHILL (II,1277)

RICHMOND, BILL
DOUBLE TROUBLE (II,724); THE JERRY LEWIS SHOW (II,1284); NO SOAP, RADIO (II,1856); THAT'S TV (II,2581); THREE'S COMPANY (II,2614); THE TIM CONWAY SHOW (II,2619); WELCOME BACK, KOTTER (II,2761); WIZARDS AND WARRIORS (II,2813)

RICKABAUGH, KIMBERLY
TV'S BLOOPERS AND PRACTICAL JOKES (II,2675)

RICKEY, FRED
SORRY, WRONG NUMBER (I,4130)

RICKEY, PATRICIA
FAME (I,1517); THE FLIP WILSON SHOW (I,1602); HAVE I GOT A CHRISTMAS FOR YOU (II,1105); JENNIFER SLEPT HERE (II,1278); KEEP U.S. BEAUTIFUL (I,2519); LOVE, NATALIE (II,1545); MR. MOM (II,1752); WHERE'S POPPA? (II,2784)

RIDDINGTON, KEN
THE CITADEL (II,539); A HORSEMAN RIDING BY (II,1182)

RIDDLE, SAM
ALL-STAR ANYTHING GOES (II,50); ANDY WILLIAMS' EARLY NEW ENGLAND CHRISTMAS (II,79); BATTLE OF THE LAS VEGAS SHOW GIRLS (II,162); HOLLYWOOD A GO GO (I,2080); HOLLYWOOD TEEN (II,1166); LET'S GO GO (I,2665); STAR SEARCH (II,2437); STAR SEARCH (II,2438); TOUCH OF GOLD '75 (II,2650)

RILEY, THOMAS
GOLDEN WINDOWS (I,1841); WHEN THE NIGHTINGALE SANG IN BERKELEY SQUARE (I,4827)

RING, BILL
TALENT VARIETIES (I,4343)

RINGEL, JOHN E
YOUR NEW DAY (II,2873)

RINTELS, DAVID
THE MEMBER OF THE WEDDING (II,1680); MR. ROBERTS (II,1754); WASHINGTON: BEHIND CLOSED DOORS (II,2744)

RIPLEY, ARTHUR
CAVALCADE OF AMERICA (I,875); THE GENERAL ELECTRIC THEATER (I,1753)

RIPORT, LAWRENCE
THE RAY ANTHONY SHOW (I,3729); THE SWINGING SCENE OF RAY ANTHONY (I,4323)

RIPPER, JOHN
ON STAGE AMERICA (II,1893)

RIPPS, LEONARD
BOSOM BUDDIES (II,355); NOW WE'RE COOKIN' (II,1866)

RIPPS, MARTIN
THREE'S A CROWD (II,2613)

RIPPY, BILLY J
MUSIC WORLD (II,1775)

RISKIN, RALPH
BRIDGET LOVES BERNIE (I,722); THE COURTSHIP OF EDDIE'S FATHER (I,1083); THE DUKES OF HAZZARD (II,742); GIDGET'S SUMMER REUNION (I,1799)

RISS, SHELDON
BILLBOARD'S DISCO PARTY (II,245)

RISSIEN, EDWARD L
BIG BOB JOHNSON AND HIS FANTASTIC SPEED CIRCUS (II,231)

RITCHIE, MICHAEL
PROFILES IN COURAGE (I,3678)

RITT, MARTIN
DANGER (I,1134); TELLER OF TALES (I,4388)

RITTS, PAUL
THE BIG TOP (I,435)

RITZ, JAMES
LOOK WHAT THEY'VE DONE TO MY SONG (II,1517)

RIVERS, LUCILLE
THE LUCILLE RIVERS SHOW (I,2785)

RIVKIN, ALLEN
THE TROUBLESHOOTERS (I,4614)

RIVKIN, M J
MIRROR, MIRROR, OFF THE WALL (I,3059); FRANK SINATRA JR. WITH FAMILY

AND FRIENDS (I,1663)

RIZZO, TONY
MR. BLACK (I,3129)

ROACH JR, HAL
ALIAS MIKE HERCULES (I,117); BLONDIE (I,486); THE CHILDREN'S HOUR (I,940); CODE 3 (I,993); THE COUNT OF MONTE CRISTO (I,1058); DUFFY'S TAVERN (I,1379); FOREST RANGER (I,1641); THE GALE STORM SHOW (I,1725); LOVE THAT JILL (I,2780); MCGARRY AND ME (I,2966); MY LITTLE MARGIE (I,3194); PASSPORT TO DANGER (I,3467); PUBLIC DEFENDER (I,3687); RACKET SQUAD (I,3714); TELEPHONE TIME (I,4379); TROUBLE WITH FATHER (I,4610); THE VEIL (I,4709)

ROACH, HAL
COMEDY CAPERS (I,1018); ELFEGO BACA (I,1426); LAUREL AND HARDY LAUGHTOONS (II,1445); THE MISCHIEF MAKERS (I,3061)

ROBBINS, DICK
THE LID'S OFF (I,2678)

ROBERTS, BEN
CHARLIE'S ANGELS (II,487); THE KILLER WHO WOULDN'T DIE (II,1393); LOGAN'S RUN (II,1507); MANNIX (II,1624); NERO WOLFE (II,1806); TIME EXPRESS (II,2620)

ROBERTS, PAMELA
SECOND CITY TELEVISION (II,2287)

ROBERTS, RUTH
THE LORETTA YOUNG THEATER (I,2756)

ROBERTS, WILLIAM
THE DONNA REED SHOW (I,1347)

ROBERTSON, CAROL
THE TALE OF BEATRIX POTTER (II,2533)

ROBERTSON, HARRY
PRISONERS OF THE LOST UNIVERSE (II,2086)

ROBERTSON, MARC
HOT CITY DISCO (II,1185)

ROBERTSON, STANLEY G
HARRIS AND COMPANY (II,1096)

ROBERTSON, TOM
I'M SOOO UGLY (II,1219); THE TROUBLE WITH MOTHER (II,2663)

ROBINS, JOHN
THE BENNY HILL SHOW (II,207)

ROBINS, VALERIE
SHEENA EASTON—LIVE AT THE PALACE (II,2324)

ROBINSON, CHARLES
BIG TOWN (I,436)

ROBINSON, HUBBELL
87TH PRECINCT (I,1422); ACADEMY AWARD SONGS (I,13); ARSENIC AND OLD LACE (I,270); CHRISTMAS STARTIME WITH LEONARD BERNSTEIN (I,949); THE DEAN MARTIN SHOW (I,1198); THE DEAN MARTIN SHOW (I,1199); FUN FAIR (I,1706); GEORGE BURNS IN THE BIG TIME (I,1764); HAWK (I,1974); MEET CYD CHARISSE (I,2982); MY THREE ANGELS (I,3204); PLAYHOUSE 90 (I,3623); THE ROSALIND RUSSELL SHOW (I,3845); THE SECRET WORLD OF KIDS (I,3970); SING ALONG WITH MITCH (I,4056); SO HELP ME, APHRODITE (I,4104); SOLDIERS IN GREASEPAINT (I,4110); STARTIME (I,4212); THE SWINGIN' SINGIN' YEARS (I,4319); THE SWINGIN' YEARS (I,4321); THE TALENT SCOUTS PROGRAM (I,4341); TENNESSEE ERNIE FORD MEETS KING ARTHUR (I,4397); THRILLER (I,4492); THE TURN OF THE SCREW (I,4624); THE WONDERFUL WORLD OF JACK PAAR (I,4899)

ROBINSON, JEROME
MAN BEHIND THE BADGE (I,2857); POLICE STORY (I,3635)

ROBINSON, JOEL
CASABLANCA (I,860)

ROBINSON, JOHN
ACAPULCO (I,15); KING OF DIAMONDS (I,2558); WANTED: DEAD OR ALIVE (I,4764)

ROBINSON, LEE
SKIPPY, THE BUSH KANGAROO (I,4076)

ROBINSON, SALLY
KING'S CROSSING (II,1397); SECRETS OF MIDLAND HEIGHTS (II,2296)

RODDENBERRY, GENE
333 MONTGOMERY (I,4487); THE ALCOA HOUR (I,108); ASSIGNMENT: EARTH (I,299); GENESIS II (I,1760); THE LIEUTENANT (I,2680); PLANET EARTH (I,3615); POLICE STORY (I,3636); STAR TREK (II,2440)

RODGERS, MARK
240-ROBERT (II,2689); DAVID CASSIDY—MAN UNDERCOVER (II,623); THE GANGSTER CHRONICLES (II,957); JOE FORRESTER (II,1303); MASQUERADE (II,1644); MCCLAIN'S LAW (II,1659); OPERATION: RUNAWAY (II,1918); THE QUEST (II,2097); WONDER WOMAN (SERIES 2) (II,2829)

RODIN, GIL
THE FRED ASTAIRE SHOW (I,1677); THE SWINGIN' SINGIN' YEARS (I,4319); THE SWINGIN' YEARS (I,4321); WAYNE AND SHUSTER TAKE AN AFFECTIONATE LOOK AT. . . (I,4778)

RODNEY, EUGENE B
FATHER KNOWS BEST (II,838); WINDOW ON MAIN STREET (I,4873)

ROGERS, AL
BERT CONVY SPECIAL—THERE'S A MEETING HERE TONIGHT (II,210); THE CARPENTERS (II,444); DEAN MARTIN PRESENTS MUSIC COUNTRY, U.S.A. (I,1192); DINAH IN SEARCH OF THE IDEAL MAN (I,1278); THE DOROTHY HAMILL WINTER CARNIVAL SPECIAL (II,720); FATHER O FATHER (II,842); THE GLEN CAMPBELL GOODTIME HOUR (I,1820); THE JERRY REED WHEN YOU'RE HOT, YOU'RE HOT HOUR (I,2372); THE JIM STAFFORD SHOW (II,1291); THE JOHN BYNER COMEDY HOUR (I,2407); JOHN DENVER ROCKY MOUNTAIN CHRISTMAS (II,1314); THE JOHN DENVER SPECIAL (II,1315); THE JOHN DENVER SPECIAL (II,1316); JOHN DENVER—THANK GOD I'M A COUNTRY BOY (II,1317); THE PAUL LYNDE COMEDY HOUR (II,1978); THE RICH LITTLE SHOW (II,2153); THE RICH LITTLE SHOW (II,2154); THE STARLAND VOCAL BAND (II,2441); STOCKARD CHANNING IN JUST FRIENDS (II,2467)

ROGERS, FRED
MR. ROGERS' NEIGHBORHOOD (I,3150)

ROGERS, HERB
CHARLEY'S AUNT (II,483)

ROGERS, JAIME
BEN VEREEN. . . COMIN' AT YA (II,203)

ROGERS, JOHN W
MISSION: IMPOSSIBLE (I,3067)

ROGERS, KENNY
KENNY ROGERS IN CONCERT (II,1381)

ROGERS, PETER
CARRY ON LAUGHING (II,448)

ROGERS, ROY
THE ROY ROGERS SHOW (I,3859)

ROGOSIN, JOEL
77 SUNSET STRIP (I,3988); ALL THAT GLITTERS (II,41); BIG ROSE (II,241); THE BLUE KNIGHT (II,276); CALL TO GLORY (II,413); CIRCLE OF FEAR (I,958); THE DOCTORS (I,1323); GHOST STORY (I,1788); IRONSIDE (II,1246); JULIE FARR, M.D. (II,1354); KNIGHT RIDER (II,1402); LONGSTREET (I,2749); MAGNUM, P.I. (II,1601); MR. MERLIN (II,1751); SURFSIDE 6 (I,4299); WELCOME TO PARADISE (II,2762)

ROGOW, STAN
FAME (II,812)

ROLAND, ROY
THE LIFE AND LEGEND OF WYATT EARP (I,2681)

ROLAND, WILL
ARTHUR GODFREY TIME (I,289)

ROLEY, SUTTON
TARGET (I,4359)

ROLFE, SAM H
BIG SHAMUS, LITTLE SHAMUS (II,242); THE DELPHI BUREAU (I,1224); THE DELPHI BUREAU (I,1225); DUNDEE AND THE CULHANE (I,1383); THE ELEVENTH HOUR (I,1425); ENIGMA (II,778); KAZ (II,1370); THE MAN FROM U.N.C.L.E. (I,2867); MANHUNTER (II,1621)

ROLLINS, JACK
THE DAVID LETTERMAN SHOW (II,624); LATE NIGHT WITH DAVID LETTERMAN (II,1442); THE ROBERT KLEIN SHOW (II,2185); THE WOODY ALLEN SPECIAL (I,4904)

ROMAN, PHIL
IT'S A MYSTERY, CHARLIE BROWN (I,2259)

ROME, HARRY
COKE TIME WITH EDDIE FISHER (I,996)

ROMERO, GEORGE A

TALES FROM THE DARKSIDE (II,2534); TALES FROM THE DARKSIDE (II,2535)

ROMINI, CHARLES
ODYSSEY (I,3329)

ROMM, HAROLD
THE ORCHID AWARD (I,3406)

ROPOLO, ED
THE NEW MICKEY MOUSE CLUB (II,1823)

ROSE, DICK
THE GLORIA SWANSON HOUR (I,1827)

ROSE, JACK
THE GOOD GUYS (I,1847)

ROSE, SI
THE BUGALOOS (I,757); THE JEAN ARTHUR SHOW (I,2352); THE LOST SAUCER (II,1524); MCHALE'S NAVY (I,2969); OPERATION PETTICOAT (II,1917); SIGMUND AND THE SEA MONSTERS (II,2352)

ROSEMOND, PERRY
BIZARRE (II,259); THE COMEDY SHOP (II,558); DICK CLARK'S LIVE WEDNESDAY (II,672); KING OF KENSINGTON (II,1395); THE MAD MAD MAD MAD WORLD OF THE SUPER BOWL (II,1586); PEOPLE ARE FUNNY (II,1996)

ROSEMONT, NORMAN
BIG BEND COUNTRY (II,230); BRIGADOON (I,726); CAROUSEL (I,856); AN HOUR WITH ROBERT GOULET (I,2137); KISMET (I,2565); KISS ME, KATE (I,2569); LITTLE LORD FAUNTLEROY (II,1492); MASTER OF THE GAME (II,1647); MIRACLE ON 34TH STREET (I,3058); VARIETY (II,2721)

ROSEN, ARNE
ALFRED OF THE AMAZON (I,116); THE BOBBY VAN AND ELAINE JOYCE SHOW (I,677); THE CAROL BURNETT SHOW (II,443); DON RICKLES—ALIVE AND KICKING (I,1340); GET SMART (II,983); HE AND SHE (I,1985); HOTEL 90 (I,2130); THE MAC DAVIS SHOW (II,1575); OF THEE I SING (I,3334); SNAFU (II,2386)

ROSEN, BURT
ALL-STAR SWING FESTIVAL (II,53); THE ANN-MARGRET SHOW (I,218); ANN-MARGRET: FROM HOLLYWOOD WITH LOVE (I,217); THE BARBARA MCNAIR SHOW (I,348); THE BIG BAND AND ALL THAT JAZZ (I,422); THE BOBBY SHERMAN SPECIAL (I,676); DR JEKYLL AND MR. HYDE (I,1368); THE FIFTH DIMENSION TRAVELING SHOW (I,1572); FROM HOLLYWOOD WITH LOVE: THE ANN-MARGRET SPECIAL (I,1687); HALF THE GEORGE KIRBY COMEDY HOUR (I,1920); RAQUEL (I,3725); ROLLIN' ON THE RIVER (I,3834); SAGA OF SONORA (I,3884); SPECIAL LONDON BRIDGE SPECIAL (I,4150); STORY THEATER (I,4241); VAUDEVILLE (II,2722); WHERE THE GIRLS ARE (I,4830); YES, VIRGINIA, THERE IS A SANTA CLAUS (II,2849)

ROSEN, EDWARD
HERE COME THE STARS (I,2026)

ROSEN, FRED
CELEBRITY TIME (I,887); THE EYES HAVE IT (I,1495)

ROSEN, JERRY
THE BENNY RUBIN SHOW (I,398)

ROSEN, LARRY
ALONE AT LAST (II,59); ALONE AT LAST (II,60); THE ANDROS TARGETS (II,76); ETHEL IS AN ELEPHANT (II,783); JENNIFER SLEPT HERE (II,1278); MR. MERLIN (II,1751); PALS (II,1947); SPENCER'S PILOTS (II,2421); SPENCER'S PILOTS (II,2422); TEACHERS ONLY (II,2548); WEDNESDAY NIGHT OUT (II,2760)

ROSEN, MILT
THE DANNY THOMAS SPECIAL (I,1152); DANNY THOMAS: AMERICA I LOVE YOU (I,1145); ESTHER WILLIAMS AT CYPRESS GARDENS (I,1465)

ROSEN, NEIL
ME AND MAXX (II,1672); OFF THE WALL (II,1879); SHIPSHAPE (II,2329)

ROSEN, PAUL
IT'S A BUSINESS (I,2254)

ROSEN, ROBERT L
DOROTHY IN THE LAND OF OZ (II,722); GILLIGAN'S ISLAND (II,990); IT'S ABOUT TIME (I,2263)

ROSEN, SHELLEY L
FOUL UPS, BLEEPS AND BLUNDERS (II,911)

ROSEN, SY
BABY MAKES FIVE (II,129); THE BOOK OF LISTS (II,350); HANGING IN (II,1078); THE JEFFERSONS (II,1276); NOT IN FRONT OF THE KIDS (II,1861); SANFORD (II,2255); SHE'S WITH ME (II,2321); SPENCER (II,2420); SUTTERS BAY (II,2508)

ROSENBERG, AARON
DANIEL BOONE (I,1142)

ROSENBERG, ARNIE
IT'S ROCK AND ROLL (II,1258)

ROSENBERG, DICK
CAGNEY AND LACEY (II,409)

ROSENBERG, E J
330 INDEPENDENCE S.W. (I,4486); HAPPY (I,1938); THE PEOPLE'S CHOICE (I,3521)

ROSENBERG, EDGAR
CAROL FOR ANOTHER CHRISTMAS (I,852); HUSBANDS AND WIVES (II,1208); HUSBANDS, WIVES AND LOVERS (II,1209); THE POPPY IS ALSO A FLOWER (I,3650); THAT SHOW STARRING JOAN RIVERS (I,4421)

ROSENBERG, FRANK P
ARREST AND TRIAL (I,264); BELLE STARR (I,391); THE BOB HOPE CHRYSLER THEATER (I,502); THE LONELY WIZARD (I,2743); PAPA SAID NO (I,3452); SUSPICION (I,4309); THE TROUBLESHOOTERS (I,4614); WON'T IT EVER BE MORNING? (I,4902)

ROSENBERG, JOHN
TALES OF THE UNEXPECTED (II,2540)

ROSENBERG, LEONARD
AMANDA'S (II,62); NEVER SAY NEVER (II,1809)

ROSENBERG, MANNY
GUESS WHAT? (I,1895)

ROSENBERG, META
BRET MAVERICK (II,374); COPS (I,1047); THE NEW MAVERICK (II,1822); NICHOLS (I,3283); THE ROCKFORD FILES (II,2197)

ROSENBERG, RICK
THE FABULOUS DR. FABLE (I,1500); THERE'S ALWAYS ROOM (II,2583); WISHMAN (II,2811)

ROSENBERG, STUART
DECOY (I,1217); HEAD OF THE FAMILY (I,1988)

ROSENBLOOM, DICK

ROSENBLOOM, RICHARD M
240-ROBERT (II,2688); APPLAUSE (I,244); ARK II (II,107); CAGNEY AND LACEY (II,409); CAGNEY AND LACEY (II,410); EDDIE AND HERBERT (II,757); ISIS (II,1248); LOOK OUT WORLD (II,1516); MONTY NASH (I,3097); UNCLE CROC'S BLOCK (II,2702); THE WESTWIND (II,2766)

ROSENBLUM, ARTHUR
THE DUNNINGER SHOW (I,1385)

ROSENBROOK, JEB
STEAMBATH (II,2451); THE YELLOW ROSE (II,2847)

ROSENSTOCK, RICHARD
OH MADELINE (II,1880)

ROSENTHAL, EVERETT
ANYONE CAN WIN (I,235); THE BIG STORY (I,432); DECOY (I,1217); TREASURY MEN IN ACTION (I,4604)

ROSENZWEIG, BARNEY
THE AMERICAN DREAM (II,71); CAGNEY AND LACEY (II,409); CAGNEY AND LACEY (II,410); CHARLIE'S ANGELS (II,486); DANIEL BOONE (I,1142); EAST OF EDEN (II,752); MEN OF THE DRAGON (II,1683); MODESTY BLAISE (II,1719); THIS GIRL FOR HIRE (II,2596)

ROSETTI, RICHARD
GRANDPA GOES TO WASHINGTON (II,1050); KATE MCSHANE (II,1369)

ROSNER, RICK
240-ROBERT (II,2688); 240-ROBERT (II,2689); CHIPS (II,511); GARROWAY (I,1736); JUST MEN (II,1359); LOTTERY (II,1525)

ROSS, BOB
THE ANDY GRIFFITH SHOW (I,192); MAYBERRY R.F.D. (I,2961)

ROSS, CHARLES
CAVALCADE OF BANDS (I,876)

ROSS, DARRELL
MATINEE THEATER (I,2947)

ROSS, DEBORAH A
SHERLOCK HOLMES (II,2327)

ROSS, FRANK

SALLY (I,3889); SALLY (I,3890)

ROSS, GREGG
YOU ASKED FOR IT (II,2855)

ROSS, JERRY
GRADY (II,1047); HARPER VALLEY PTA (II,1095)

ROSS, JON
THE NEW TRUTH OR CONSEQUENCES (II,1832)

ROSS, MICHAEL
THE ALAN KING SHOW (I,101); ALL IN THE FAMILY (II,38); COMEDY IS KING (I,1019); THE DUMPLINGS (II,743); THE JEFFERSONS (II,1276); THE ROPERS (II,2214); SALVAGE (II,2244); SALVAGE 1 (II,2245) THREE'S A CROWD (II,2613); THREE'S COMPANY (II,2614)

ROSS, ROSALIND
SWINGING COUNTRY (I,4322)

ROSS, STANLEY RALPH
FLO'S PLACE (II,892); THE KALLIKAKS (II,1361); LOCAL 306 (II,1505)

ROSS-LEMING, EUGENIE
FIT FOR A KING (II,866); FOREVER FERNWOOD (II,909); HIGHCLIFFE MANOR (II,1150); MARY HARTMAN, MARY HARTMAN (II,1638); SCARECROW AND MRS. KING (II,2268); WHY US? (II,2796)

ROSTEN, IRWIN
DEAR MR. GABLE (I,1203); RIPLEY'S BELIEVE IT OR NOT (II,2177)

ROTH, LEON
FASHIONS ON PARADE (I,1538)

ROTH, LYNN
THE PAPER CHASE: THE SECOND YEAR (II,1950)

ROTH, PAUL
THE COMEDY SHOP (II,558)

ROTH, PAULA A
GOODTIME GIRLS (II,1042)

ROTH, RON
HEAT OF ANGER (I,1992)

ROTHCHILD, JANE
THE $20,000 PYRAMID (II,2681)

ROTHENBERGER, JOSEPH
ANOTHER WORLD (II,97)

ROTHMAN, BERNARD
ALL-STAR SWING FESTIVAL (II,53); AMERICA, YOU'RE ON (II,69); THE BIG BAND AND ALL THAT JAZZ (I,422); BUCKSHOT (II,385); BURT REYNOLDS' LATE SHOW (I,777); CELEBRITY DAREDEVILS (II,473); DIANA (I,1258); G.I.'S (II,949); THE GEORGE BURNS SPECIAL (II,970); HALF THE GEORGE KIRBY COMEDY HOUR (I,1920); THE LIBERACE SHOW (I,2677); THE MANHATTAN TRANSFER (II,1619); PAT BOONE AND FAMILY (II,1968); THE PAT BOONE AND FAMILY CHRISTMAS SPECIAL (II,1969); THE PAT BOONE AND FAMILY EASTER SPECIAL (II,1970); PINOCCHIO (II,2044); SANDY IN DISNEYLAND (II,2254); SHA NA NA (II,2311); THE SPECIAL GENTRY ONE (I,4148); VIVA VALDEZ (II,2728)

ROTHMAN, FRED
BURT AND THE GIRLS (I,770)

ROTHMAN, MARK
BUSTING LOOSE (II,402); THE FURTHER ADVENTURES OF WALLY BROWN (II,944); LAVERNE AND SHIRLEY (II,1446); THE LOVEBIRDS (II,1548); MAKIN' IT (II,1603); THE NEW ODD COUPLE (II,1825); THE RITA MORENO SHOW (II,2181); THE TED KNIGHT SHOW (II,2550)

ROTHSCHILD, HARRY S
RAMAR OF THE JUNGLE (I,3718)

ROTHSTEIN, FREDYA
EMERALD POINT, N.A.S. (II,773)

ROTHSTEIN, RICHARD
THE HITCHHIKER (II,1157)

ROTMAN, KEITH
FALCON'S GOLD (II,809)

ROUNDTREE, MARTHA
LEAVE IT TO THE GIRLS (I,2650)

ROUSSEAU, WILLIAM
THE JANE WYMAN THEATER (I,2346)

ROWAN, DAN
ROWAN AND MARTIN BITE THE HAND THAT FEEDS THEM (I,3853)

ROWE, ARTHUR
THE BIONIC WOMAN (II,255); FANTASY ISLAND (II,829); HARD KNOCKS (II,1086)

ROWE, ROBERT
PLAY YOUR HUNCH (I,3619); SPLIT PERSONALITY (I,4161)

ROWE, RUPERT S
SAY WHEN (I,3929)

ROY, FRAN
FOUL UPS, BLEEPS AND BLUNDERS (II,911)

ROYAL, JACK
ALICE IN WONDERLAND (I,120)

ROYCE, BILL
PEOPLE TO PEOPLE (II,2000); TELEVISION INSIDE AND OUT (II,2552)

RUBEN, AARON
THE ANDY GRIFFITH SHOW (I,192); ANDY GRIFFITH'S UPTOWN-DOWNTOWN SHOW (I,194); C.P.O. SHARKEY (II,407); CHARO AND THE SERGEANT (II,489); THE DON KNOTTS SPECIAL (I,1335); GOMER PYLE, U.S.M.C. (I,1843); GRANDPA MAX (II,1051); GREAT DAY (II,1053); JACKIE AND DARLENE (II,1265); THE MCLEAN STEVENSON SHOW (II,1664); THE NEW ANDY GRIFFITH SHOW (I,3256); THE PHIL SILVERS SHOW (I,3580); PIPER'S PETS (II,2046); THE RAG BUSINESS (II,2106); THE STOCKARD CHANNING SHOW (II,2468); TEACHERS ONLY (II,2547); TROUBLE WITH RICHARD (I,4612); WENDY HOOPER—U. S. ARMY (II,2764)

RUBEN, ALBERT
FOR THE PEOPLE (I,1626)

RUBEN, GLADYS
SANDY DREAMS (I,3909)

RUBICHAN, JOHN
CHECKMATE (I,919)

RUBIN, EDDIE
THE BOB CUMMINGS SHOW (I,500)

RUBIN, JACK
BREAK THE $250,000 BANK (I,713)

RUBIN, JOE
BREAK THE BANK (I,712)

RUBIN, ROBERT H
DREAM HOUSE (II,734); JEOPARDY (II,1279)

RUBIN, RONALD
THE BANANA COMPANY (II,139); JAMES AT 15 (II,1270); PALMERSTOWN U.S.A. (II,1945); ROOM 222 (I,3843); SHIRLEY'S WORLD (I,4019); TO SIR, WITH LOVE (II,2624)

RUBIN, SELMA
IRENE (II,1245)

RUBIN, STANLEY
BAND OF GOLD (I,341); BRACKEN'S WORLD (I,703); CHANNING (I,900); EXECUTIVE SUITE (II,796); THE FREEWHEELERS (I,1683); THE GENERAL ELECTRIC THEATER (I,1753); THE GHOST AND MRS. MUIR (I,1786); THE GRADUATION DRESS (I,1864); HOTEL DE PAREE (I,2129); THE MAN AND THE CITY (I,2856); MR. O'MALLEY (I,3146); PECK'S BAD GIRL (I,3503); THE SEARCH (I,3950); STORY THEATER (I,4240)

RUBIN, SYD
THE FLORIAN ZABACH SHOW (I,1606)

RUBINSTEIN, PAULETTE
SCENES FROM A MARRIAGE (II,2272)

RUBINSTEIN, RICHARD
TALES FROM THE DARKSIDE (II,2534); TALES FROM THE DARKSIDE (II,2535)

RUBINSTEIN, SHERMAN
GUILTY OR INNOCENT (II,1065)

RUBUCHUTAN, J REMI
THE PAPER CHASE: THE SECOND YEAR (II,1950)

RUBY, JOE
ALVIN AND THE CHIPMUNKS (II,61); BEAUTY AND THE BEAST (II,197); DRAGON'S LAIR (II,732); FANG FACE (II,822); THE HEATHCLIFF AND DINGBAT SHOW (II,1117); THE HEATHCLIFF AND MARMADUKE SHOW (II,1118); MORK AND MINDY (II,1736); MR. T (II,1757); THE PLASTICMAN COMEDY/ADVENTURE SHOW (II,2051); RUBIK, THE AMAZING CUBE (II,2226); SATURDAY SUPERCADE (II,2263); THUNDARR THE BARBARIAN (II,2616); TURBO-TEEN (II,2668)

RUBY, KEN
GOLDIE GOLD AND ACTION JACK (II,1023)

RUCKER, ELLEN
FROM CLEVELAND (II,931)

RUDD, LOUIS
CAESAR AND CLEOPATRA (I,789)

RUDDY, ALBERT S
THE STOCKERS (II,2469)

RUDOLPH, LOUIS
TRUE LIFE STORIES (II,2665)

RUSH, ARTHUR
BRAVE EAGLE (I,709)

RUSH, HERMAN
CELEBRATION: THE AMERICAN SPIRIT (II,461); D.H.O. (I,1249); REAR GUARD (II,2120); THAT WAS THE YEAR THAT WAS (II,2575)

RUSH, NORMAN
LOVE THY NEIGHBOR (I,2781)

RUSKIN, COBY
THE ALL-STAR COMEDY SHOW (I,137); THE ALL-STAR REVUE (I,138); SAMMY KAYE'S MUSIC FROM MANHATTAN (I,3902); SO YOU WANT TO LEAD A BAND (I,4106); STAR OF THE FAMILY (I,4194)

RUSSELL, BRIAN
CHERYL LADD: SCENES FROM A SPECIAL (II,503)

RUSSELL, CHARLES
ALFRED HITCHCOCK PRESENTS (I,115); BLITHE SPIRIT (I,484); CAIN'S HUNDRED (I,792); CRIME PHOTOGRAPHER (I,1097); DANGER (I,1134); JUDD, FOR THE DEFENSE (I,2463); NAKED CITY (I,3216); PURSUIT (I,3700); SPARROW (II,2410); TOGETHER WITH MUSIC (I,4532); THE UNTOUCHABLES (I,4682); YOU ARE THERE (I,4929)

RUSSO, AARON
BETTE MIDLER—OL' RED HAIR IS BACK (II,222); THE MANHATTAN TRANSFER (II,1619)

RUTT, STEVEN
MUSIC WORLD (II,1775)

RYAN, MARIANNE
WOMAN'S PAGE (II,2822)

RYAN, MARILYN
RIPLEY'S BELIEVE IT OR NOT (II,2177)

SABAN, HAIM
KIDD VIDEO (II,1388)

SABAROFF, ROBERT
THE NEW PEOPLE (I,3268); THEN CAME BRONSON (I,4435)

SABLE, SANDRA
SECOND EDITION (II,2288)

SACHNOFF, MARC
BLONDES VS. BRUNETTES (II,271)

SACK, TED
ODYSSEY (I,3329)

SACKHEIM, WILLIAM
ALCOA/GOODYEAR THEATER (I,107); ANOTHER DAY, ANOTHER DOLLAR (I,228); DEADLOCK (I,1190); DELVECCHIO (II,659); DIAL HOT LINE (I,1254); EMPIRE (I,1439); THE FLYING NUN (I,1611); GIDGET (I,1795); HOW'S BUSINESS? (I,2153); MALLORY: CIRCUMSTANTIAL EVIDENCE (II,1608); NIGHT GALLERY (I,3288); ONCE AN EAGLE (II,1897); REX HARRISON PRESENTS SHORT STORIES OF LOVE (II,2148); THE SENATOR (I,3976); THE SLIGHTLY FALLEN ANGEL (I,4087)

SACKS, ALAN
BEST FRIENDS (II,213); CHICO AND THE MAN (II,508); MURDER INK (II,1769); WELCOME BACK, KOTTER (II,2761)

SADLIER, MICHAEL
AGGIE (I,90)

SAGAL, BORIS
THE CLIFF DWELLERS (I,974); THE RUNAWAY BARGE (II,2230); T.H.E. CAT (I,4430)

SAGE, LIZ
PUNKY BREWSTER (II,2092)

SAHLINS, BERNARD
SECOND CITY TELEVISION (II,2287)

SAKAI, RICHARD
SHAPING UP (II,2316); TAXI (II,2546)

SAKIN, LEO
THE 2,000 YEAR OLD MAN (II,2696)

SAKS, SOL
AN APARTMENT IN ROME (I,242); OUT OF THE BLUE (I,3422)

SALAMON, OTTO
THE DAY THE WOMEN GOT EVEN (II,628); INSTITUTE FOR REVENGE (II,1239)

SALE, RICHARD
MR. BELVEDERE (I,3127); YANCY DERRINGER (I,4925)

SALEM, FRANK
THE KEN MURRAY SHOW (I,2525)

SALKIN, LEO
THE ALVIN SHOW (I,146)

SALKOW, SIDNEY
SIMON LASH (I,4052); THIS IS ALICE (I,4453)

SALKOWITZ, SY
TODAY'S FBI (II,2628)

SALLAN, BRUCE J
STEELTOWN (II,2452)

SALOMON, HENRY
VICTORY AT SEA (I,4730)

SALTER, HARRY
NAME THAT TUNE (I,3218); YOURS FOR A SONG (I,4975)

SALTER, KEN
TAKE MY ADVICE (II,2531)

SALTZMAN, BERT
TALES OF THE UNEXPECTED (II,2540)

SALTZMAN, PHILIP
THE ALIENS ARE COMING (II,35); BARE ESSENCE (II,149); BARNABY JONES (II,153); COLORADO C.I. (II,555); CROSSFIRE (II,606); DEATH RAY 2000 (II,654); ESCAPADE (II,781); THE FBI (I,1551); THE FELONY SQUAD (I,1563); FREEBIE AND THE BEAN (II,922); INTERTECT (I,2228); THE KILLIN' COUSIN (II,1394); A MAN CALLED SLOANE (II,1613); UNIT 4 (II,2707)

SALZER, ALBERT
HIGH PERFORMANCE (II,1145)

SALZMAN, DAVID
EVERYDAY (II,793); THE PETER MARSHALL VARIETY SHOW (II,2029)

SAMISH, ADRIAN
THE ATOM SQUAD (I,312); BARNABY JONES (II,153); CANNON (I,820); CRISIS (I,1102); DAN AUGUST (I,1130); THE HOUSE ON GREENAPPLE ROAD (I,2140); MANHUNTER (II,1620); NORTHWEST PASSAGE (I,3306); THE STREETS OF SAN FRANCISCO (I,4260); THE STREETS OF SAN FRANCISCO (II,2478); THREE STEPS TO HEAVEN (I,4485); A TIME TO LIVE (I,4510); TRAVIS LOGAN, D.A. (I,4598)

SAMMON, ROBERT
PERSON TO PERSON (I,3550)

SAMUELS, RON
LINDSAY WAGNER—ANOTHER SIDE OF ME (II,1483); LYNDA CARTER'S CELEBRATION (II,1563); LYNDA CARTER'S SPECIAL (II,1564); LYNDA CARTER: ENCORE (II,1566); LYNDA CARTER: STREET LIGHTS (II,1567); SCRUPLES (II,2280)

SAMUELSON, PAUL
THE CURSE OF DRACULA (II,611); THE SECRET EMPIRE (II,2291); STOP SUSAN WILLIAMS (II,2473)

SAND, BARRY
THE DAVID LETTERMAN SHOW (II,624); THE FIGHTING NIGHTINGALES (II,855); KLEIN TIME (II,1401); LATE NIGHT WITH DAVID LETTERMAN (II,1442)

SAND, BOB
100 CENTRE STREET (II,1901); FRIENDS (II,927); MADAME'S PLACE (II,1587)

SANDEFUR, B W
THE CURSE OF DRACULA (II,611); ENOS (II,779); THE HARDY BOYS MYSTERIES (II,1090); HAVING BABIES III (II,1109); HAWAII FIVE-O (II,1110); JULIE FARR, M.D. (II,1354); THE LEGEND OF THE GOLDEN GUN (II,1455); LITTLE HOUSE ON THE PRAIRIE (II,1487); THE NANCY DREW MYSTERIES (II,1789); THE SECRET EMPIRE (II,2291); STOP SUSAN WILLIAMS (II,2473)

SANDER, LOUIS D
THE LIBERACE SHOW (I,2675)

SANDERS, TERRY
THE LEGEND OF MARILYN MONROE (I,2654)

SANDLER, ALLAN
LANCELOT LINK, SECRET CHIMP (I,2609)

SANDLER, ROBERT
THE LIBERACE SHOW (I,2675)

SANDRICH, JAY
CAPTAIN NICE (I,835); GET SMART (II,983); THE HERO (I,2035)

SANFORD, GERALD
A MAN CALLED SLOANE (II,1613); WHEN THE WHISTLE BLOWS (II,2781)

SANFORD, HERB
THE GARRY MOORE SHOW (I,1738)

SANFORD, WENDY
CITY HOSPITAL (I,967)

SANGSTER, JIMMY
THE COUNTRY MUSIC MURDERS (II,590); EBONY, IVORY AND JADE (II,755); YOUNG DAN'L BOONE (II,2863)

SANIEE, DELIA GRAVEL
GRAFFITI ROCK (II,1049)

SANTA CROCE, ANTHONY
TALES FROM THE DARKSIDE (II,2535)

SANTLEY, JOSEPH
HEY MULLIGAN (I,2040)

SAPERSTEIN, HENRY
THE DICK TRACY SHOW (I,1272); THE FAMOUS ADVENTURES OF MR. MAGOO (I,1524); MR. MAGOO (I,3142)

SAPHIER, JAMES
FUN FOR THE MONEY (I,1708); IT PAYS TO BE MARRIED (I,2249)

SAPHIER, PETER
THE FOUR SEASONS (II,915)

SAPINSLEY, ALVIN
SHANNON (II,2314)

SARGENT, HERB
ALAN KING LOOKS BACK IN ANGER—A REVIEW OF 1972 (I,99); THE ALAN KING SPECIAL (I,103); THE GEORGE SEGAL SHOW (II,977); LILY (I,2705); LOVE, LIFE, LIBERTY & LUNCH (II,1544); THE NEWS IS THE NEWS (II,1838); THE WONDERFUL WORLD OF AGGRAVATION (I,4894)

SARGENT, JOSEPH
LONGSTREET (I,2750)

SARIEGO, RALPH
ALEX AND THE DOBERMAN GANG (II,32); THE BIONIC WOMAN (II,255); SALVAGE 1 (II,2245); THE SEAL (II,2282)

SARNOFF, THOMAS
HOLLYWOOD THEATER TIME (I,2096)

SAROYAN, HANK
ROBERTA FLACK. . . THE FIRST TIME EVER (I,3817)

SATENSTEIN, FRANK
BY POPULAR DEMAND (I,784)

SATLOF, RON
THE AMAZING SPIDER-MAN (II,63); BENNY AND BARNEY: LAS VEGAS UNDERCOVER (II,206); THE FALL GUY (II,811); GET CHRISTIE LOVE! (II,981); MCCLOUD (II,1660); WHAT REALLY HAPPENED TO THE CLASS OF '65? (II,2768)

SAUBER, HARRY
MEET THE GIRLS: THE SHAPE, THE FACE, AND THE BRAIN (I,2990)

SAUDEK, ROBERT
THE BAT (I,364); THE BURNING COURT (I,766); THE CAT AND THE CANARY (I,868); THE DATCHET DIAMONDS (I,1160); FOUR FOR TONIGHT (I,1648); INN OF THE FLYING DRAGON (I,2214); OMNIBUS (I,3353); PROFILES IN COURAGE (I,3678); THE WOMAN IN WHITE (I,4887)

SAUER, ERNEST
DAYTIME (II,630)

SAUNDERS, HERM
ADAM-12 (I,31); F TROOP (I,1499); WENDY AND ME (I,4793)

SAVADORE, LAURENCE D
THE ABC AFTERSCHOOL SPECIAL (II,1); ALAN KING IN LAS VEGAS, PART I (I,97); ALAN KING IN LAS VEGAS, PART II (I,98)

SAVAGE, JOHN
OMNI: THE NEW FRONTIER (II,1889)

SAVAGE, PAUL
THE NEW DAUGHTERS OF JOSHUA CABE (I,3261)

SAVENICK, PHILIP
THAT'S HOLLYWOOD (II,2577)

SAVITZ, CHARLOTTE
CAPITOL (II,426); ROMANCE THEATER (II,2204)

SAVORY, GERALD
LOVE IN A COLD CLIMATE (II,1539)

SAWYER, TOM
HIGH PERFORMANCE (II,1145)

SBARDELLATI, JAMES
THE GEORGIA PEACHES (II,979)

SCALEM, JIM
JOAN BAEZ (II,1294)

SCANDORE, JOSEPH
A COUPLE OF DONS (I,1077); THE DON RICKLES SHOW (I,1341); THE DON RICKLES SHOW (I,1342); THE DON RICKLES SHOW (II,708); DON RICKLES—ALIVE AND KICKING (I,1340); RICKLES (II,2170)

SCARPITTA, GUY
DOBIE GILLIS (I,1302)

SCHAEFER, ARMAND
ANNIE OAKLEY (I,225); BUFFALO BILL JR. (I,755); CAVALCADE OF AMERICA (I,875); THE GENE AUTRY SHOW (I,1748); THE RANGE RIDER (I,3720)

SCHAEFER, GEORGE
ABE LINCOLN IN ILLINOIS (I,11); THE ADMIRABLE CRICHTON (I,36); AH!, WILDERNESS (I,91); ALCOA PREMIERE (I,109); ANASTASIA (I,182); ARSENIC AND OLD LACE (I,269); BAREFOOT IN ATHENS (I,354); BERKELEY SQUARE (I,399); THE BEST CHRISTMAS PAGEANT EVER (II,212); BLITHE SPIRIT (I,485); BORN YESTERDAY (I,692); CAPTAIN BRASSBOUND'S CONVERSION (I,828); THE CRADLE SONG (I,1088); A CRY OF ANGELS (I,1111); CYRANO DE BERGERAC (I,1120); DIAL "M" FOR MURDER (I,1256); A DOLL'S HOUSE (I,1327); EAGLE IN A CAGE (I,1392); ELIZABETH THE QUEEN (I,1431); THE FANTASTICKS (I,1531); THE FILE ON DEVLIN (I,1575); GIDEON (I,1792); GIFT OF THE MAGI (I,1800); GIVE US BARABBAS! (I,1816); THE GOLDEN AGE OF TELEVISION (II,1017); GOLDEN CHILD (I,1839); THE GREEN PASTURES (I,1887); THE HALLMARK HALL OF FAME CHRISTMAS FESTIVAL (I,1921); HAMLET (I,1925); HANS BRINKER OR THE SILVER SKATES (I,1933); THE HOLY TERROR (I,2098); INHERIT THE WIND (I,2213); JANE EYRE (I,2339); JOHNNY BELINDA (I,2416); THE JOKE AND THE VALLEY (I,2438); KISS ME, KATE (I,2568); LAMP AT MIDNIGHT (I,2608); THE LARK (I,2616); THE LITTLE FOXES (I,2712); LITTLE MOON OF ALBAN (I,2714); LITTLE MOON OF ALBAN (I,2715); LOVE STORY (I,2778); MACBETH (I,2801); THE MAGNIFICENT YANKEE (I,2829); MAN AND SUPERMAN (I,2854); MY FAIR LADY (I,3187); ON BORROWED TIME (I,3354); OUR TOWN (I,3421); THE PATRIOTS (I,3477); SANDBURG'S LINCOLN (I,3907); SHANGRI-LA (I,3996); SOLDIER IN LOVE (I,4108); ST. JOAN (I,4175); THE TEAHOUSE OF THE AUGUST MOON (I,4368); THE TEMPEST (I,4391); THERE SHALL BE NO NIGHT (I,4437); TIME REMEMBERED (I,4509); TWELFTH NIGHT (I,4636); VICTORIA REGINA (I,4728); WINTERSET (I,4882); THE YEOMAN OF THE GUARD (I,4926)

SCHAFFEL, HAL
MAUREEN (II,1656); MOMMA THE DETECTIVE (II,1721)

SCHAFFNER, FRANKLIN
STUDIO ONE (I,4268); THE WIDE OPEN DOOR (I,4859)

SCHARLACH, ED
CHICO AND THE MAN (II,508); LEWIS AND CLARK (II,1465); THE MARILYN MCCOO AND BILLY DAVIS JR. SHOW (II,1630); MORK AND MINDY (II,1735)

SCHARY, DORY
SOUND OF THE 60S (I,4134)

SCHECK, GEORGE
THE ALAN DALE SHOW (I,96); DOORWAY TO FAME (I,1354); HAPPY BIRTHDAY (I,1940); THE JACQUELINE SUSANN SHOW (I,2327)

SCHEERER, ROBERT
THE ANDY WILLIAMS SHOW (II,78); BARBRA STREISAND: A HAPPENING IN CENTRAL PARK (I,352); CHANGING SCENE (I,899); A COUPLE OF DONS (I,1077); DISNEY WORLD —A GALA OPENING—DISNEYLAND EAST (I,1292); THE DOCTORS (I,1323); FRANK SINATRA: A MAN AND HIS MUSIC (I,1662); THE PEGGY FLEMING SHOW (I,3508); THE ROCK RAINBOW (II,2194); THE SHARI LEWIS SHOW (I,4000)

SCHEIMER, LOU
THE ARCHIE COMEDY HOUR (I,252); ARCHIE'S TV FUNNIES (I,253); ARK II (II,107); THE BANG-SHANG LALAPALOOZA SHOW (II,140); BATMAN AND THE SUPER SEVEN (II,160); BLACKSTAR (II,263); THE BRADY KIDS (II,365); EVERYTHING'S ARCHIE (I,1483); THE FABULOUS FUNNIES (II,802); FANTASTIC VOYAGE (I,1530); FAT ALBERT AND THE COSBY KIDS (II,835); FLASH GORDON (II,874); FLASH GORDON—THE GREATEST ADVENTURE OF ALL (II,875); GILLIGAN'S PLANET (II,994); THE HARDY BOYS (I,1949); HE-MAN AND THE MASTERS OF THE UNIVERSE (II,1113); ISIS (II,1248); JASON OF STAR COMMAND (II,1274); JOURNEY TO THE CENTER OF THE EARTH (I,2456); THE KID SUPER POWER HOUR

WITH SHAZAM (II,1386); LASSIE'S RESCUE RANGERS (II,1432); MISSION MAGIC (I,3068); MY FAVORITE MARTIANS (II,1779); THE NEW ADVENTURES OF BATMAN (II,1812); THE NEW ADVENTURES OF GILLIGAN (II,1813); THE NEW ADVENTURES OF MIGHTY MOUSE AND HECKLE AND JECKLE (II,1814); THE NEW ADVENTURES OF SUPERMAN (I,3255); ROD ROCKET (I,3827); RUN RUN, JOE (II,2229); SABRINA, THE TEENAGE WITCH (I,3880); THE SECRET LIVES OF WALDO KITTY (II,2292); SHAZAM! (II,2319); SPACE ACADEMY (II,2406); SPORT BILLY (II,2429); STAR TREK (II,2439); SUPER WITCH (II,2501); THE TARZAN/LONE RANGER/ZORRO ADVENTURE HOUR (II,2543); TARZAN: LORD OF THE JUNGLE (II,2544); TEACHER'S PET (I,4367); THE TOM AND JERRY COMEDY SHOW (II,2629); THE U.S. OF ARCHIE (I,4687); U.S. OF ARCHIE (II,2698); UNCLE CROC'S BLOCK (II,2702); THE WESTWIND (II,2766); WILL THE REAL JERRY LEWIS PLEASE SIT DOWN? (I,4866); THE YOUNG SENTINELS (II,2871)

SCHENCK, AUBREY
MIAMI UNDERCOVER (I,3020)

SCHENCK, GEORGE
BRING 'EM BACK ALIVE (II,377); CRAZY LIKE A FOX (II,601); O'MALLEY (II,1872); SAWYER AND FINN (II,2265)

SCHERICK, EDGAR J
JEREMIAH OF JACOB'S NECK (II,1281); THOU SHALT NOT KILL (II,2604)

SCHERMER, JEROME
THE DAKOTAS (I,1126); LAWMAN (I,2642); THE ROARING TWENTIES (I,3805)

SCHERMER, JULES
THE VIRGINIAN (I,4738)

SCHIEK, ELLIOTT
PRIVATE BENJAMIN (II,2087)

SCHIFF, SAM
THE WORLD OF MISTER SWEENY (I,4917)

SCHILLER, BOB
ALL'S FAIR (II,46); BUT MOTHER! (II,403); THE GOOD GUYS (I,1847); LIVING IN PARADISE (II,1503); MAUDE (II,1655); SIDE BY SIDE (II,2346); W*A*L*T*E*R (II,2731)

SCHILLER, CRAIG
THE BIONIC WOMAN (II,255)

SCHILLER, LAWRENCE
COME WITH ME—LAINIE KAZAN (I,1015)

SCHILLER, WILTON
BEN CASEY (I,394); DOCTOR SIMON LOCKE (I,1317); THE FUGITIVE (I,1701); POLICE SURGEON (I,3639)

SCHILZ, TED
DANIEL BOONE (I,1142)

SCHIRMER, GUS
THE SANDY DUNCAN SHOW (II,2253)

SCHLAMMUE, THOMAS
BETTE MIDLER—ART OR BUST (II,221)

SCHLATTER, GEORGE
ARNOLD'S CLOSET REVIEW (I,262); THE BEST OF TIMES (II,219); CHER (II,495); CHER (II,496); CHER AND OTHER FANTASIES (II,497); THE COLGATE COMEDY HOUR (I,998); THE DANNY THOMAS TV FAMILY REUNION (I,1154); DIANA ROSS AND THE SUPREMES AND THE TEMPTATIONS ON BROADWAY (I,1260); DORIS DAY TODAY (II,716); THE FABULOUS FUNNIES (II,801); GOLDIE AND LIZA TOGETHER (II,1022); THE GOLDIE HAWN SPECIAL (II,1024); JOHN DENVER AND FRIEND (II,1311); JOHN DENVER AND THE LADIES (II,1312); THE JUDY GARLAND SHOW (I,2473); THE JUDY GARLAND SHOW (II,1349); LAUGH-IN (II,1444); THE LISA HARTMAN SHOW (II,1484); LOOK AT US (II,1515); NBC FOLLIES OF 1965 (I,3240); THE NEW BILL COSBY SHOW (I,3257); ONE MORE TIME (I,3385); ONE MORE TIME (I,3386); RADIO CITY MUSIC HALL AT CHRISTMAS TIME (I,3715); REAL KIDS (II,2116); REAL PEOPLE (II,2118); ROWAN AND MARTIN'S LAUGH-IN (I,3856); ROWAN AND MARTIN'S LAUGH-IN (I,3857); SALUTE TO LADY LIBERTY (II,2243); THE SHAPE OF THINGS (I,3999); THE SHAPE OF THINGS (II,2315); THE SHIRLEY MACLAINE SPECIAL: WHERE DO WE GO FROM HERE? (II,2333); SPEAK UP AMERICA (II,2412); THE STEVE LAWRENCE SHOW (I,4227); THE TENNESSEE ERNIE FORD SPECIAL (I,4401); TEXACO STAR PARADE I (I,4409); TEXACO STAR PARADE II (I,4410); TURN ON (I,4625); VICTOR BORGE'S 20TH ANNIVERSARY SHOW (I,4727)

SCHLESINGER, LEON
THE BUGS BUNNY SHOW (II,390); THE ROAD RUNNER SHOW (I,3799)

SCHLITT, ROBERT
THE BLUE KNIGHT (II,276); THE LAZARUS SYNDROME (II,1451)

SCHLOES, BARRY
CARTOON TELETALES (I,857)

SCHLOM, HERMAN
MY FRIEND FLICKA (I,3190)

SCHLOW, STEPHEN
SUMMER SOLSTICE (II,2492)

SCHMERER, JAMES
THE DELPHI BUREAU (I,1225); THE HIGH CHAPARRAL (I,2051)

SCHMIDT, LARS
HEDDA GABLER (I,2001); THE HUMAN VOICE (I,2160); TWENTY-FOUR HOURS IN A WOMAN'S LIFE (I,4644)

SCHMITT, ANNE MARIE
THE $20,000 PYRAMID (II,2681); THE $25,000 PYRAMID (II,2679); THE LOVE EXPERTS (II,1537); THE NEW $25,000 PYRAMID (II,1811)

SCHNEIDER, ANDREW
THE INCREDIBLE HULK (II,1232); MAGNUM, P.I. (II,1601); MASQUERADE (II,1644)

SCHNEIDER, DICK
THE STARS AND STRIPES SHOW (I,4206); STARS AND STRIPES SHOW (II,2442); THE STARS AND STRIPES SHOW (II,2443)

SCHNEIDER, JOHN
JOHN SCHNEIDER'S CHRISTMAS HOLIDAY (II,1320)

SCHNEIDER, SASCHA
HILL STREET BLUES (II,1154)

SCHNER, CHARLES H
THE WEB (I,4785)

SCHNUR, JEROME
WHO PAYS? (I,4848)

SCHOCK, JEFF
BILLY JOEL—A TV FIRST (II,250)

SCHOENBAUM, DONALD
A CHRISTMAS CAROL (II,517)

SCHOENMAN, ELLIOT
AMANDA'S (II,62)

SCHONE, VIRGINIA
VANITY FAIR (I,4702)

SCHOTZ, ERIC
THE LOVE REPORT (II,1542)

SCHREDER, CAROL
CALL TO GLORY (II,413)

SCHREIBMAN, MYRIL A
THE GIRL, THE GOLD WATCH AND EVERYTHING (II,1002)

SCHROCK, LARRY
KIDS ARE PEOPLE TOO (II,1390)

SCHUBERT, BERNARD L
BLIND DATE (I,482); MR. AND MRS. NORTH (I,3125); MUSICAL COMEDY TIME (I,3183); TOPPER (I,4582); WHITE HUNTER (I,4845)

SCHUFFMAN, DAN
IT'S POLKA TIME (I,2273); OH, KAY! (I,3342)

SCHULER, ROBERT
THE PATRICE MUNSEL SHOW (I,3474)

SCHULKE, JAMES A
THE HAPPY TIME (I,1943)

SCHULMAN, ROGER
DEAR TEACHER (II,653); FISHERMAN'S WHARF (II,865); FOR MEMBERS ONLY (II,905); HIGH FIVE (II,1143)

SCHULTZ, CHARLES H
BELAFONTE, NEW YORK (I,386); STUDIO ONE (I,4268)

SCHULTZ, CHRIS
HARRY AND LENA (I,1957); THE JUDY GARLAND SHOW (I,2472); TALES OF THE UNEXPECTED (II,2540)

SCHULTZ, GLORIA
EARTH, WIND & FIRE IN CONCERT (II,750)

SCHULTZ, MICHAEL
EARTH, WIND & FIRE IN CONCERT (II,750)

SCHUMACHER, JOEL
NOW WE'RE COOKIN' (II,1866)

SCHUMMEL, CLARENCE

LUCKY PUP (I,2788)

SCHUTEK, DOUG
THE JOE NAMATH SHOW (I,2400)

SCHWAB, LAWRENCE
THE CLOCK (I,978); STOP ME IF YOU HEARD THIS ONE (I,4231); WHAT WILL THEY THINK OF NEXT? (I,4813)

SCHWARTZ, AL
CELEBRITIES: WHERE ARE THEY NOW? (II,462); CELEBRITY CHARADES (II,470); DICK CLARK'S GOOD OL' DAYS: FROM BOBBY SOX TO BIKINIS (II,671); FAR OUT SPACE NUTS (II,831); HOLLYWOOD'S PRIVATE HOME MOVIES (II,1167); HOLLYWOOD'S PRIVATE HOME MOVIES II (II,1168); I'VE HAD IT UP TO HERE (II,1221); INSIDE AMERICA (II,1234); MEN WHO RATE A "10" (II,1684); OPRYLAND: NIGHT OF STARS AND FUTURE STARS (II,1921); THE SENSATIONAL, SHOCKING, WONDERFUL, WACKY 70S (II,2302); THANKSGIVING REUNION WITH THE PARTRIDGE FAMILY AND MY THREE SONS (II,2569); TV'S BLOOPERS AND PRACTICAL JOKES (II,2675); WHATEVER BECAME OF. . .? (II,2772); WHATEVER BECAME OF. . .? (II,2773); WONDERBUG (II,2830); YOU ARE THE JURY (II,2854)

SCHWARTZ, ARTHUR
INSIDE U.S.A. WITH CHEVROLET (I,2220); TWENTIETH CENTURY (I,4639)

SCHWARTZ, BARBARA
YOU ARE THERE (I,4930)

SCHWARTZ, BILL
THE ABC AFTERSCHOOL SPECIAL (II,1)

SCHWARTZ, BOB
WATCH ME (II,2745)

SCHWARTZ, DOUGLAS
MANIMAL (II,1622)

SCHWARTZ, ELROY
DUSTY'S TRAIL (I,1390); GILLIGAN'S ISLAND: THE HARLEM GLOBETROTTERS ON GILLIGAN'S ISLAND (II,993)

SCHWARTZ, LLOYD J
BIG JOHN, LITTLE JOHN (II,240); THE BRADY BRIDES (II,361); THE BRADY BUNCH (II,362); THE BRADY GIRLS GET MARRIED (II,364); GILLIGAN'S ISLAND: RESCUE FROM GILLIGAN'S ISLAND (II,991); GILLIGAN'S ISLAND: THE CASTAWAYS ON GILLIGAN'S ISLAND (II,992); GILLIGAN'S ISLAND: THE HARLEM GLOBETROTTERS ON GILLIGAN'S ISLAND (II,993); THE INVISIBLE WOMAN (II,1244); KELLY'S KIDS (I,2523); SCAMPS (II,2267); WHAT'S HAPPENING!! (II,2769)

SCHWARTZ, MURRAY
THE MERV GRIFFIN SHOW (II,1688)

SCHWARTZ, SHERWOOD
BIG JOHN, LITTLE JOHN (II,240); THE BRADY BRIDES (II,361); THE BRADY BUNCH (II,362); THE BRADY GIRLS GET MARRIED (II,364); THE BRADY KIDS (II,365); DUSTY'S TRAIL (I,1390); GILLIGAN'S ISLAND (II,990); GILLIGAN'S ISLAND: RESCUE FROM GILLIGAN'S ISLAND (II,991); GILLIGAN'S ISLAND: THE CASTAWAYS ON GILLIGAN'S ISLAND (II,992); HARPER VALLEY PTA (II,1095); THE INVISIBLE WOMAN (II,1244); IT'S ABOUT TIME (I,2263); KELLY'S KIDS (I,2523); SCAMPS (II,2267); TEACHER'S PET (I,4367)

SCHWARTZBERG, LOUIS
PEN 'N' INC. (II,1992)

SCHWIMMER, STAN
GIDGET (I,1795); HERE COME THE BRIDES (I,2024); HERE WE GO AGAIN (I,2028); LOVE ON A ROOFTOP (I,2775)

SCHWIMMER, WALTER
THE CISCO KID (I,961)

SCIBETTA, JACK
YOUNG MR. BOBBIN (I,4943)

SCIBETTA, JOSEPH
THE ALDRICH FAMILY (I,111); HONESTLY, CELESTE! (I,2104)

SCIPIO, ALFRED D
BEAT OF THE BRASS (I,376)

SCOFFIELD, JOE
BOB HOPE SPECIAL: THE BOB HOPE SPECIAL (II,338)

SCOFFIELD, JON
SUNDAY NIGHT AT THE LONDON PALLADIUM (I,4286); THIS IS TOM JONES (I,4460); THE TOM JONES CHRISTMAS SPECIAL (I,4539); THE TOM JONES SPECIAL (I,4540); THE TOM JONES SPECIAL (I,4541); THE TOM JONES SPECIAL (I,4542); THE TOM JONES SPECIAL (I,4543); THE TOM JONES SPECIAL (I,4544); THE TOM JONES SPECIAL (I,4545)

SCOFIELD, IAN
SPOTLIGHT (I,4163)

SCOTT, ART
THE BISKITTS (II,257); BUFORD AND THE GHOST (II,389); CASPER AND THE ANGELS (II,453); THE DRAK PACK (II,733); FLINTSTONE FAMILY ADVENTURES (II,882); FONZ AND THE HAPPY DAYS GANG (II,898); FRED AND BARNEY MEET THE SHMOO (II,918); THE GALAXY GOOFUPS (II,954); THE GARY COLEMAN SHOW (II,958); THE KWICKY KOALA SHOW (II,1417); LAVERNE AND SHIRLEY IN THE ARMY (II,1448); LAVERNE AND SHIRLEY WITH THE FONZ (II,1449); LOST IN SPACE (I,2759); THE NEW FRED AND BARNEY SHOW (II,1817); THE NEW SHMOO (II,1827); THE RICHIE RICH SHOW (II,2169); THE SCOOBY-DOO AND SCRAPPY-DOO SHOW (II,2275); THE SUPER GLOBETROTTERS (II,2498); YOGI'S SPACE RACE (II,2853)

SCOTT, ASHMEAD
HERB SHRINER TIME (I,2020)

SCOTT, BILL
THE BULLWINKLE SHOW (I,761); THE DUDLEY DO-RIGHT SHOW (I,1378); FRACTURED FLICKERS (I,1653); GERALD MCBOING-BOING (I,1778); THE NUT HOUSE (I,3320); ROCKY AND HIS FRIENDS (I,3822)

SCOTT, EDWARD
THE YOUNG AND THE RESTLESS (II,2862)

SCOTT, FRANCES
VERSATILE VARIETIES (I,4713)

SCOTT, PETER GRAHAM
THE ONEDIN LINE (II,1912); QUILLER: NIGHT OF THE FATHER (II,2100); QUILLER: PRICE OF VIOLENCE (II,2101)

SCOTT, RITA
ANN-MARGRET'S HOLLYWOOD MOVIE GIRLS (II,86); BOB HOPE SPECIAL: BOB HOPE'S WOMEN I LOVE—BEAUTIFUL BUT FUNNY (II,322); DINAH AND HER NEW BEST FRIENDS (II,676); THE HARLEM GLOBETROTTERS POPCORN MACHINE (II,1092); THE KRAFT 75TH ANNIVERSARY SPECIAL (II,1409); LARRY GATLIN AND THE GATLIN BROTHERS (II,1429); LAS VEGAS PALACE OF STARS (II,1431); THE STEVE LANDESBERG TELEVISION SHOW (II,2458)

SCOTT, ROGER
HERE'S RICHARD (II,1137)

SCOTTI, TONY
AMERICA'S TOP TEN (II,68); THE EDDIE RABBITT SPECIAL (II,759); THE JIM STAFFORD SHOW (II,1291); PORTRAIT OF A LEGEND (II,2070); A SPECIAL EDDIE RABBITT (II,2415); WE'RE DANCIN' (II,2755)

SCULLY, JOE
ARTHUR GODFREY IN HOLLYWOOD (I,281)

SEARLES, BARBARA
THE LOST SAUCER (II,1524)

SEARLS, BOB
MEET MARCEL MARCEAU (I,2983)

SEARS, FRAN
THE ABC AFTERSCHOOL SPECIAL (II,1); HERE'S BOOMER (II,1134)

SEARS, FRANK
HONEST AL'S A-OK USED CAR AND TRAILER RENTAL TIGERS (II,1174)

SEAWELL, MIMI
MISS WINSLOW AND SON (II,1705)

SEDAWIE, NORMAN
RICH LITTLE'S A CHRISTMAS CAROL (II,2157); RICH LITTLE—COME LAUGH WITH ME (II,2160); THE SMOTHERS BROTHERS SHOW (I,4097); THE SMOTHERS ORGANIC PRIME TIME SPACE RIDE (I,4098)

SEDIER, CHRISTOPHER
THE BOYS IN BLUE (II,359)

SEEGER, HAL
BATFINK (I,365); THE MILTON THE MONSTER CARTOON SHOW (I,3053); OUT OF THE INKWELL (I,3423); PLANET PATROL (I,3616)

SEEGER, SUSAN
9 TO 5 (II,1852); A GIRL'S LIFE (II,1000); THE JAMES BOYS (II,1272)

SEGAL, ALEX
THE ALCOA HOUR (I,108); ALCOA PREMIERE (I,109); THE BORIS KARLOFF MYSTERY PLAYHOUSE (I,690); CELANESE THEATER (I,881)

SEGALL, STUART
HUNTER (II,1206)

SEGUIN, LISA
THE PRIME OF MISS JEAN BRODIE (II,2082)

SEID, ART
MY FRIEND TONY (I,3192); THE NEW ADVENTURES OF PERRY MASON (I,3252); PERRY MASON (II,2025); YOUNG DAN'L BOONE (II,2863)

SEIDEL, ARTHUR
HOT PURSUIT (II,1189)

SEIDELMAN, ARTHUR ALLAN
THE ABC AFTERSCHOOL SPECIAL (II,1)

SEIFERT, JACK
SUNDAY FUNNIES (II,2493)

SEINFIELD, JOHN
TALES OF THE TEXAS RANGERS (I,4349)

SEITNER, ROBERT M
NIGHT PARTNERS (II,1847)

SEITZ, CHRIS
WISHMAN (II,2811)

SELANDER, LESLEY
COWBOY G-MEN (I,1084)

SELDEN, ALBERT
GIFT OF THE MAGI (I,1800); LITTLE WOMEN (I,2722)

SELDEN, JACK
FEATHER YOUR NEST (I,1555)

SELDEN, WALTER
CITY HOSPITAL (I,967); THE DOOR WITH NO NAME (I,1352); DOORWAY TO DANGER (I,1353)

SELF, ED
CODE R (II,551); EMERGENCY! (II,775); QUICK AND QUIET (II,2099); THE RANGERS (II,2112); SIERRA (II,2351); TWO THE HARD WAY (II,2695)

SELF, WILLIAM
BUS STOP (I,778); COMBAT (I,1011); COMMAND PERFORMANCE (I,1029); HOTEL DE PAREE (I,2129); SCHLITZ PLAYHOUSE OF STARS (I,3936); STATE FAIR (II,2448)

SELIG, ANDREW
AIN'T MISBEHAVIN' (II,25); THE BOUNDER (II,357); FILTHY RICH (II,856); LEWIS AND CLARK (II,1465)

SELIGMAN, JAN
BABY MAKES FIVE (II,129)

SELIGMAN, MICHAEL
ANNE MURRAY'S WINTER CARNIVAL. . . FROM QUEBEC (II,90); EVENING AT THE MOULIN ROUGE (II,787); THE MONTE CARLO SHOW (II,1726)

SELIGMAN, SELIG J
ACCUSED (I,18); COMBAT (I,1011); DAY IN COURT (I,1182); GARRISON'S GORILLAS (I,1735); MICKEY (I,3023); SHINDIG (I,4013)

SELLIER JR, CHARLES E
GREATEST HEROES OF THE BIBLE (II,1062); THE LIFE AND TIMES OF GRIZZLY ADAMS (II,1469)

SELZNICK, DAVID O
LIGHT'S DIAMOND JUBILEE (I,2698)

SENNETT, MACK
COMEDY CAPERS (I,1018)

SERLING, ROD
THE TWILIGHT ZONE (I,4651)

SERTNER, ROBERT
WHACKED OUT (II,2767)

SEVANO, NICK
THE GLEN CAMPBELL GOODTIME HOUR (I,1820); HI, I'M GLEN CAMPBELL (II,1141); THE JOHN BYNER COMEDY HOUR (I,2407)

SHAFER, BETH HILL
BLISS (II,269)

SHAFFEL, JOSEF
THE APARTMENT HOUSE (I,241); STRAIGHTAWAY (I,4243)

SHALINS, BERNARD
THE YESTERDAY SHOW (II,2850)

SHALINS, JANE
THE YESTERDAY SHOW (II,2850)

SHALLECK, ALAN J
PIXANNE (II,2048)

SHAMBERG, MICHAEL
GOSSIP (II,1043); GOSSIP (II,1044); THE TV TV SHOW (II,2674)

SHANE, MAXWELL
CHECKMATE (I,919); THRILLER (I,4492)

SHANKMAN, NED
TED KNIGHT MUSICAL COMEDY VARIETY SPECIAL SPECIAL (II,2549); THE TED KNIGHT SHOW (II,2550)

SHANKS, ANN
AMERICAN LIFESTYLE (I,170)

SHANKS, BOB
THE MERV GRIFFIN SHOW (I,3014); THE MERV GRIFFIN SHOW (I,3016); MERV GRIFFIN'S SIDEWALKS OF NEW ENGLAND (I,3017); MERV GRIFFIN'S ST. PATRICK'S DAY SPECIAL (I,3018); SMALL WORLD (II,2381)

SHAPARIO, MARILYN
BONNIE AND THE FRANKLINS (II,349)

SHAPIRO, ARNOLD
THE SINGING COWBOYS RIDE AGAIN (II,2363)

SHAPIRO, ESTHER
DYNASTY (II,746); EMERALD POINT, N.A.S. (II,773)

SHAPIRO, GEORGE
BUCKSHOT (II,385); LEWIS AND CLARK (II,1465); YOUNG HEARTS (II,2865)

SHAPIRO, KEN
MONTY HALL'S VARIETY HOUR (II,1728)

SHAPIRO, MICKEY
FLEETWOOD MAC IN CONCERT (II,880)

SHAPIRO, RICHARD
DYNASTY (II,746); EMERALD POINT, N.A.S. (II,773)

SHAPIRO, STANLEY
MCGARRY AND ME (I,2966); WHERE'S RAYMOND? (I,4837)

SHAPIRO, STUART
FM TV (II,896)

SHARNIK, JOHN
WORLD WAR I (I,4919)

SHARP, PHIL
THE CARA WILLIAMS SHOW (I,843)

SHARPE, DON
CHICAGO 212 (I,935); FOUR STAR PLAYHOUSE (I,1652); MAN WITH A CAMERA (I,2883); MEET MCGRAW (I,2985); SHEENA, QUEEN OF THE JUNGLE (I,4003); WIRE SERVICE (I,4883); YANCY DERRINGER (I,4925)

SHAVELSON, MELVILLE
ELKE (I,1432); FATHER ON TRIAL (I,1544); IKE (II,1223)

SHAW, DAVID
SHANE (I,3995)

SHAW, FRANK
BLACK BART (II,260); CO-ED FEVER (II,547)

SHAW, GARY
100 CENTRE STREET (II,1901)

SHAW, JEROME
THE PAUL WINCHELL SHOW (I,3493); VICTOR BORGE'S COMEDY IN MUSIC I (I,4723); VICTOR BORGE'S COMEDY IN MUSIC II (I,4724)

SHAW, LOU
BEYOND WESTWORLD (II,227); THE FALL GUY (II,811); FITZ AND BONES (II,867); PLEASURE COVE (II,2058); QUINCY, M. E. (II,2102); TERROR AT ALCATRAZ (II,2563)

SHAYNE, ALAN
ADDIE AND THE KING OF HEARTS (I,35); THE EASTER PROMISE (I,1397); A HATFUL OF RAIN (I,1964); THE THANKSGIVING TREASURE (I,4416)

SHAYNE, BOB
AT EASE (II,116); COVER UP (II,597); FITZ AND BONES (II,867); WHIZ KIDS (II,2790)

SHAYON, ROBERT LEWIS
THE BIG STORY (I,432); HANDLE WITH CARE (I,1926)

SHEA, JACK
THE BOB HOPE SHOW (I,582); THE GLEN CAMPBELL GOODTIME HOUR (I,1820); THE JEFFERSONS (II,1276); THE JERRY LEWIS SHOW (I,2365); ROXY PAGE (II,2222); SLITHER (II,2379)

SHEAR, BARRY
THE ERNIE KOVACS SHOW (I,1455); THE GIRL FROM U.N.C.L.E. (I,1808); THE HAZEL SCOTT SHOW (I,1984); HERE'S EDIE (I,2030); HOORAY FOR HOLLYWOOD (I,2114); IT'S WHAT'S HAPPENING, BABY! (I,2278); JIMMY HUGHES, ROOKIE COP (I,2390); THE LIVELY ONES (I,2729); NOT FOR PUBLICATION (I,3309); RICKLES (II,2170); S.W.A.T. (II,2236); THE XAVIER CUGAT SHOW (I,4924)

SHEARER, HANNAH L
CONDOMINIUM (II,566); EMERGENCY! (II,776); KNIGHT RIDER (II,1402); PINE

CANYON IS BURNING (II,2040); QUINCY, M. E. (II,2102)

SHEARER, HARRY
THE TV SHOW (II,2673)

SHEEHAN, TONY
BARNEY MILLER (II,154)

SHEFFLER, MARC
LEWIS AND CLARK (II,1465)

SHELDON, JAMES
THE WEST POINT STORY (I,4796)

SHELDON, LES
THE BUFFALO SOLDIERS (II,388); CALIFORNIA FEVER (II,411); HARDCASTLE AND MCCORMICK (II,1089); HARPER VALLEY (II,1094); NICHOLS AND DYMES (II,1841); THE OUTLAWS (II,1936)

SHELDON, SIDNEY
I DREAM OF JEANNIE (I,2167); NANCY (I,3221); RAGE OF ANGELS (II,2107)

SHELLEY, KATHLEEN
FLAMINGO ROAD (II,872)

SHENSON, WALTER
THE EVE ARDEN SHOW (I,1471)

SHEPARD, GERALD S
BOBBY PARKER AND COMPANY (II,344)

SHERICK, EDGAR J
BORN TO THE WIND (II,354)

SHERMAN, ALLAN
PHIL SILVERS ON BROADWAY (I,3578); THE STEVE ALLEN SHOW (I,4221); YOUR SURPRISE PACKAGE (I,4960)

SHERMAN, GEORGE
DANIEL BOONE (I,1142); GENTLE BEN (I,1761)

SHERMAN, HARRY
SHE'S IN THE ARMY NOW (II,2320); STUDS LONIGAN (II,2484)

SHERMAN, JILL
THE INCREDIBLE HULK (II,1232); VOYAGERS (II,2730)

SHERMAN, ROBERT
BARNABY JONES (II,153); THE BETTER SEX (II,223); BODY LANGUAGE (II,346); FREEBIE AND THE BEAN (II,922); THE KILLIN' COUSIN (II,1394); THE MATCH GAME—HOLLYWOOD SQUARES HOUR (II,1650); PASSWORD PLUS (II,1966); SUPER PASSWORD (II,2499); TATTLETALES (II,2545)

SHERMAN, SI
JUBILEE U.S.A. (I,2461)

SHERRIN, NED
SONG BY SONG (II,2399)

SHERWIN, WALLY
MEET MARCEL MARCEAU (I,2983)

SHESLOW, STUART
THE BEST OF TIMES (II,220); THE STREETS (II,2479)

SHEVELOVE, BURT
ART CARNEY MEETS PETER AND THE WOLF (I,272); ART CARNEY MEETS THE SORCERER'S APPRENTICE (I,273); THE BIL BAIRD SHOW (I,439); THE BLUE ANGEL (I,489); THE CHEVROLET GOLDEN ANNIVERSARY SHOW (I,924); JUDY GARLAND AND HER GUESTS, PHIL SILVERS AND ROBERT GOULET (I,2469); OPENING NIGHT (I,3395); TEXACO COMMAND PERFORMANCE (I,4408); THAT WAS THE YEAR THAT WAS (II,2575)

SHIN, NELSON
THE TRANSFORMERS (II,2653)

SHIRLEY, LLOYD
THE SWEENEY (II,2513)

SHIRLEY, PAUL
SKYHAWKS (I,4079)

SHIVAS, MARK
CASANOVA (II,451); THE SIX WIVES OF HENRY VIII (I,4068)

SHIVELY, PAUL
HOT WHEELS (I,2126)

SHOENMAN, ELLIOT
CALUCCI'S DEPARTMENT (I,805); I LOVE HER ANYWAY! (II,1214); NEVER SAY NEVER (II,1809); OLD FRIENDS (II,1881)

SHOR, MITZI
BUCKSHOT (II,385)

SHORR, FREDERICK
LAREDO (I,2615)

SHOTEL, BARBARA
THAT'S INCREDIBLE! (II,2578)

SHPETNER, STAN
HEAVEN HELP US (I,1995); KODIAK (II,1405); THE SIXTH SENSE (I,4069); SWEET, SWEET RACHEL (I,4313); YOU'RE ONLY YOUNG TWICE (I,4971)

SHRIBMAN, JOSEPH S
THE ROSEMARY CLOONEY SHOW (I,3847); THE ROSEMARY CLOONEY SHOW (I,3848)

SHRIVER, MARIA
PORTRAIT OF A LEGEND (II,2070)

SHUBB, JASON
IT'S NOT EASY (II,1256)

SHULMAN, ROGER
CRAZY LIKE A FOX (II,601)

SHYER, CHARLES
COPS (I,1047)

SICHEL, JOHN
THE CARNATION KILLER (I,845); COLOR HIM DEAD (I,1007); COLOR HIM DEAD (II,554); COME OUT, COME OUT, WHEREVER YOU ARE (I,1014); AN ECHO OF THERESA (I,1399); THE EYES HAVE IT (I,1496); FILE IT UNDER FEAR (I,1574); I'M THE GIRL HE WANTS TO KILL (I,2194); KISS KISS, KILL KILL (I,2566); LADY KILLER (I,2603); MURDER IS A ONE-ACT PLAY (I,3160); NOT GUILTY! (I,3311); ONCE THE KILLING STARTS (I,3367); ONE DEADLY OWNER (I,3373); ONLY A SCREAM AWAY (I,3393); A PLACE TO DIE (I,3612); POSSESSION (I,3655); THE SAVAGE CURSE (I,3925); SIGN IT DEATH (I,4046); SOMEONE AT THE TOP OF THE STAIRS (I,4114); SPELL OF EVIL (I,4152)

SIDARIS, ARLENE
THE NANCY DREW MYSTERIES (II,1789)

SIDES, PATRICIA
MAKE A WISH (I,2838)

SIDNEY, GEORGE
WHO HAS SEEN THE WIND? (I,4847)

SIEGEL, DON
BALL FOUR (II,137); CELEBRITY CHARADES (II,470); CONVOY (I,1043); THE FOUR SEASONS (II,915); THE LEGEND OF JESSE JAMES (I,2653); TWO OF A KIND: GEORGE BURNS AND JOHN DENVER (II,2692)

SIEGEL, LIONEL E
THE AMAZING SPIDER-MAN (II,63); EXO-MAN (II,797); FROM HERE TO ETERNITY (II,933); SAVAGE: IN THE ORIENT (II,2264); THE SIX-MILLION-DOLLAR MAN (II,2372); STUNTS UNLIMITED (II,2487); THE ULTIMATE IMPOSTER (II,2700)

SIEGEL, NORMAN
SUPERTRAIN (II,2505)

SIEGLER, BILL
A CHRISTMAS CAROL (II,517); PIAF (II,2039)

SIEGLMAN, JAN
NOT IN FRONT OF THE KIDS (II,1861)

SIEGMAN, JOE
CELEBRITY BOWLING (I,883); THE COMEDY SHOP (II,558); THE WOLFMAN JACK RADIO SHOW (II,2816)

SIGNORELLI, JANIS
RANDY NEWMAN AT THE ODEON (II,2111)

SIGNORELLI, JAY
SIMON AND GARFUNKEL IN CONCERT (II,2356)

SILLIPHANT, STIRLING
CALLING DR. STORM, M.D. (II,415); FLY AWAY HOME (II,893); LONGSTREET (I,2749); LONGSTREET (I,2750); THE NEW HEALERS (I,3266); PEARL (II,1986); WELCOME TO PARADISE (II,2762)

SILLS, GREG
THE HOMEMADE COMEDY SPECIAL (II,1173); THE PEE WEE HERMAN SHOW (II,1988); STEVIE NICKS IN CONCERT (II,2463)

SILLS, STEPHANIE
ALL THAT GLITTERS (II,42); BIG CITY BOYS (II,232)

SILLS, THEODORE B
THE ROBBINS NEST (I,3806)

SILVER, ARTHUR
THE BAD NEWS BEARS (II,134); BRONCO (I,745); BROTHERS AND SISTERS (II,382); CHEYENNE (I,931); FLATFOOTS (II,879); SISTER TERRI (II,2366); WORKING STIFFS (II,2834)

SILVER, JOSEPH
DANIEL BOONE (I,1142)

SILVER, ROY
BILL COSBY DOES HIS OWN THING (I,441); A SPECIAL BILL COSBY SPECIAL (I,4147)

SILVER, STU
BROTHERS (II,380); STAR OF THE FAMILY (II,2436); WEBSTER (II,2758)

SILVERMAN, DON
THE JERRY LESTER SPECIAL (I,2361); THE UGILY FAMILY (II,2699); WHO'S WATCHING THE KIDS? (II,2793)

SILVERMAN, FRED

BIG JOHN (II,239); FARRELL: FOR THE PEOPLE (II,834); GREAT DAY (II,1054); THE LOVE REPORT (II,1542); THE LOVE REPORT (II,1543); MEATBALLS AND SPAGHETTI (II,1674); MIGHTY ORBOTS (II,1697); PANDAMONIUM (II,1948); THICKE OF THE NIGHT (II,2587); WE GOT IT MADE (II,2752)

SILVERMAN, SHANNON
SQUARE PEGS (II,2431)

SILVERS, HERB
MR. MOON'S MAGIC CIRCUS (II,1753)

SIMMONS, ED
AN AMATEUR'S GUIDE TO LOVE (I,154); THE CAROL BURNETT SHOW (II,443); HOTEL 90 (I,2130); THE KELLY MONTEITH SHOW (II,1376); MAMA'S FAMILY (II,1610); RODNEY DANGERFIELD SHOW: I CAN'T TAKE IT NO MORE (II,2198); ROWAN AND MARTIN BITE THE HAND THAT FEEDS THEM (I,3853); WELCOME BACK, KOTTER (II,2761); WHERE'S EVERETT? (I,4835)

SIMMONS, GARNER
V: THE SERIES (II,2715)

SIMMONS, JEFF
STATE FAIR USA (II,2449); STATE FAIR USA (II,2450)

SIMMONS, JIM
GREATEST HEROES OF THE BIBLE (II,1062)

SIMMONS, MATTY
DELTA HOUSE (II,658); NATIONAL LAMPOON'S HOT FLASHES (II,1795); NATIONAL LAMPOON'S TWO REELERS (II,1796)

SIMMONS, RICHARD ALAN
ALCOA/GOODYEAR THEATER (I,107); BANYON (I,344); BANYON (I,345); COLOSSUS (I,1009); COLUMBO (II,556); THE DICK POWELL SHOW (I,1269); LASSITER (I,2625); LASSITER (II,1434); LOCK, STOCK, AND BARREL (I,2736); MRS. COLUMBO (II,1760); THE TRIALS OF O'BRIEN (I,4605)

SIMON, AL
21 BEACON STREET (I,4646); THE BEVERLY HILLBILLIES (I,417); THE GEORGE BURNS AND GRACIE ALLEN SHOW (I,1763); HONEYMOON SUITE (I,2108); I MARRIED JOAN (I,2174); KEEPING UP WITH THE JONESES (II,1374); THE LORENZO AND HENRIETTA MUSIC SHOW (II,1519); LOVE THAT BOB (I,2779); MR. ED (I,3137); NO WARNING (I,3301); OZZIE'S GIRLS (I,3440); PANIC! (I,3447); PETER LOVES MARY (I,3565); PETTICOAT JUNCTION (I,3571); PINE LAKE LODGE (I,3597); THE RETURN OF THE BEVERLY HILLBILLIES (II,2139)

SIMON, ART
PURE GOLDIE (I,3696)

SIMON, DANNY
THE TROUBLE WITH PEOPLE (I,4611)

SIMON, JEFF
REAL PEOPLE (II,2118)

SIMON, JOHN
THE REPORTER (I,3771)

SIMON, JOSEPH
GILLIGAN'S PLANET (II,994); LOVE THAT BOB (I,2779)

SIMON, JUDY
THE PARAGON OF COMEDY (II,1956)

SIMON, LISA
SESAME STREET (I,3982)

SIMON, NEIL
THE TROUBLE WITH PEOPLE (I,4611)

SIMON, SAM
SHAPING UP (II,2316)

SIMONS, ED
THE MARTHA RAYE SHOW (I,2926)

SIMPSON, GARRY
THE ED WYNN SHOW (I,1404); IN TOWN TODAY (I,2206); PHILCO TELEVISION PLAYHOUSE (I,3583)

SIMPSON, GEORGE
THE ARMSTRONG CIRCLE THEATER (I,260)

SIMPSON, O J
COCAINE AND BLUE EYES (II,548); HIGH FIVE (II,1143)

SINATRA, FRANK
SINATRA (I,4054)

SINCLAIR, JEFF
YESTERYEAR (II,2851)

SINGER, ABBY
ST. ELSEWHERE (II,2432)

SINGER, AL
BE OUR GUEST (I,370); NAME THAT TUNE (I,3218); YOUR SURPRISE PACKAGE (I,4960)

SINGER, ARTHUR
DECOY (I,1217)

SINGER, DULCY
SESAME STREET (I,3982)

SINGER, GEORGE
MIGHTY ORBOTS (II,1697); MONCHHICHIS (II,1722)

SINGER, RAY
IT'S A GREAT LIFE (I,2256); THE JIM BACKUS SHOW—HOT OFF THE WIRE (I,2379); THE MCGONIGLE (I,2968); WAR CORRESPONDENT (I,4767)

SINGER, ROBERT
DOG AND CAT (II,695); DOG AND CAT (II,696); FREEBIE AND THE BEAN (II,922); LACY AND THE MISSISSIPPI QUEEN (II,1419); MICKEY SPILLANE'S MIKE HAMMER (II,1692); MICKEY SPILLANE'S MIKE HAMMER: MORE THAN MURDER (II,1693); THE NIGHT STRANGLER (I,3293); THREE EYES (II,2605); V: THE FINAL BATTLE (II,2714); V: THE SERIES (II,2715)

SINGER, SAM
THE ADVENTURES OF POW WOW (I,72); COURAGEOUS CAT (I,1079); PADDY THE PELICAN (I,3442)

SINGER, SHERI
THE PHIL DONAHUE SHOW (II,2032)

SINGLETON, RALPH S
STUNT SEVEN (II,2486)

SINSEL, DOUGLAS
BEDROOMS (II,198)

SIODMAK, CURT
NUMBER 13 DEMON STREET (I,3316)

SIRULNUCK, SID
SOMERSET (I,4115)

SISK, ROBERT
THE LIFE AND LEGEND OF WYATT EARP (I,2681)

SISTROM, JOSEPH
THE GENERAL ELECTRIC THEATER (I,1753); MARKHAM (I,2917)

SITOWITZ, HAL
FOUL PLAY (II,910); THE ROOKIES (II,2208)

SKELTON, RED
CLOWN ALLEY (I,979)

SKINNER, CHARLES E
SERGEANT PRESTON OF THE YUKON (I,3980)

SKOTCH, ED
MUSIC IN VELVET (I,3172); PENTHOUSE SONATA (I,3516)

SKUTCH, IRA
HE SAID, SHE SAID (I,1986); THE MATCH GAME (II,1649); PLAY YOUR HUNCH (I,3619); TATTLETALES (II,2545)

SLADKUS, PAUL
LOVE OF LIFE (I,2774)

SLATE, LEN
THE GIRL IN THE EMPTY GRAVE (II,997)

SLATER, BARNEY
MACKENZIE'S RAIDERS (I,2803)

SLOA, RICHARD
WAKE UP (II,2736)

SLOAN, ALAN P
MYSTERIES, MYTHS AND LEGENDS (II,1782); ON TRIAL (II,1896); ORSON WELLES' GREAT MYSTERIES (I,3408); SPECIAL EDITION (II,2416)

SLOAN, HARRY
THE KID WITH THE BROKEN HALO (II,1387)

SLOAN, MICHAEL
B.J. AND THE BEAR (II,125); B.J. AND THE BEAR (II,126); THE DEVLIN CONNECTION (II,664); THE EYES OF TEXAS (II,799); THE EYES OF TEXAS II (II,800); THE HARDY BOYS MYSTERIES (II,1090); THE MASTER (II,1648); RETURN OF THE MAN FROM U.N.C.L.E.: THE 15 YEARS LATER AFFAIR (II,2140); SWORD OF JUSTICE (II,2521)

SLOANE, ROBERT
TREASURY MEN IN ACTION (I,4604)

SMALL, EDGAR
THE PSYCHIATRIST: GOD BLESS THE CHILDREN (I,3685)

SMALL, FLORENCE
HOWDY DOODY AND FRIENDS (I,2152); THE SECRET WAR OF JACKIE'S GIRLS (II,2294); THE SHARI LEWIS SHOW (II,2317)

SMALL, JON
BILLY JOEL—A TV FIRST (II,250)

SMARDEN, ED
SKYHAWKS (I,4079)

SMART, BRIAN
RADIO PICTURE SHOW (II,2104)

SMART, RALPH
THE ADVENTURES OF WILLIAM TELL (I,85); THE

INVISIBLE MAN (I,2232); MAN OF THE WORLD (I,2875); SECRET AGENT (I,3962); WORLD OF GIANTS (I,4912)

SMIDT, BURR
THE LITTLEST ANGEL (I,2726)

SMIGHT, JACK
VICTOR BORGE'S COMEDY IN MUSIC IV (I,4726)

SMITH, APRIL
CAGNEY AND LACEY (II,409)

SMITH, BERNIE
TELL IT TO GROUCHO (I,4385)

SMITH, BEVERLY
PORTIA FACES LIFE (I,3653)

SMITH, BILLY RAY
THE MANIONS OF AMERICA (II,1623)

SMITH, CHRISTINE
A HOT SUMMER NIGHT WITH DONNA (II,1190)

SMITH, CHRISTOPHER
OLIVIA NEWTON-JOHN IN CONCERT (II,1884)

SMITH, CLEO
HERE'S LUCY (II,1135)

SMITH, CLIVE A
THE EDISON TWINS (II,761)

SMITH, DEREK
CELEBRITY COOKS (II,472)

SMITH, GARY
ANN-MARGRET OLSSON (II,84); ANN-MARGRET SMITH (II,85); ANN-MARGRET'S HOLLYWOOD MOVIE GIRLS (II,86); ANN-MARGRET. . . RHINESTONE COWGIRL (II,87); ANNE MURRAY'S CARIBBEAN CRUISE (II,88); ANNE MURRAY'S WINTER CARNIVAL. . . FROM QUEBEC (II,90); BARBRA STREISAND AND OTHER MUSICAL INSTRUMENTS (I,353); BARYSHNIKOV ON BROADWAY (II,158); BEN VEREEN—HIS ROOTS (II,202); BETTE MIDLER—OL' RED HAIR IS BACK (II,222); BING CROSBY'S MERRIE OLDE CHRISTMAS (II,252); BURT BACHARACH IN SHANGRI-LA (I,773); THE BURT BACHARACH SPECIAL (I,775); THE BURT BACHARACH SPECIAL (I,776); BURT BACHARACH! (I,771); BURT BACHARACH: CLOSE TO YOU (I,772); BURT BACHARACH—OPUS NO. 3 (I,774); THE CHERYL LADD SPECIAL (II,501); CHRISTMAS IN WASHINGTON (II,520); CHRISTMAS IN WASHINGTON (II,521); DICK CAVETT'S BACKLOT USA (II,670); DISNEYLAND'S 25TH ANNIVERSARY (II,681); DOROTHY HAMILL IN ROMEO & JULIET ON ICE (II,718); THE DOROTHY HAMILL SPECIAL (II,719); THE EDDIE RABBITT SPECIAL (II,759); EVENING AT THE MOULIN ROUGE (II,787); THE FRANK SINATRA SHOW (I,1667); FUNNY GIRL TO FUNNY LADY (II,942); GLEN CAMPBELL . . . DOWN HOME—DOWN UNDER (II,1008); THE GLEN CAMPBELL SPECIAL: THE MUSICAL WEST (I,1821); GOLDIE AND KIDS: LISTEN TO ME (II,1021); HERB ALBERT AND THE TIJUANA BRASS (I,2019); HERB ALPERT AND THE TIJUANA BRASS (II,1130); HULLABALOO (I,2156); JAMES PAUL MCCARTNEY (I,2332); JUBILEE (II,1347); THE JUDY GARLAND SHOW (I,2473); THE JUDY GARLAND SHOW (II,1349); JULIE ON SESAME STREET (I,2485); THE KLOWNS (I,2574); THE KOPYCATS (I,2580); THE KRAFT 75TH ANNIVERSARY SPECIAL (II,1409); THE KRAFT MUSIC HALL (I,2588); KRAFT SALUTES WALT DISNEY WORLD'S 10TH ANNIVERSARY (II,1410); LARRY GATLIN AND THE GATLIN BROTHERS (II,1429); LAS VEGAS PALACE OF STARS (II,1431); LINDA IN WONDERLAND (II,1481); LUCY COMES TO NASHVILLE (II,1561); MAC DAVIS CHRISTMAS SPECIAL. . . WHEN I GROW UP (II,1574); THE MAC DAVIS SHOW (II,1576); MAC DAVIS'S CHRISTMAS ODYSSEY: TWO THOUSAND AND TEN (II,1579); MAC DAVIS. . . SOUNDS LIKE HOME (II,1582); THE MAGICAL MUSIC OF BURT BACHARACH (I,2822); MARLO THOMAS IN ACTS OF LOVE—AND OTHER COMEDIES (I,2919); MERRY CHRISTMAS FROM THE CROSBYS (II,1685); MERRY CHRISTMAS FROM THE GRAND OLE OPRY (II,1686); THE NEIL DIAMOND SPECIAL (II,1802); THE NEW CHRISTY MINSTRELS SHOW (I,3259); ON STAGE AMERICA (II,1893); PAVAROTTI AND FRIENDS (II,1985); PETER PAN (II,2030); PETULA (I,3573); THE ROGER MILLER SHOW (I,3830); ROYAL VARIETY PERFORMANCE (I,3865); THE SANDY DUNCAN SHOW (II,2253); SHEENA EASTON, ACT 1 (II,2323); SHIRLEY MACLAINE. . . EVERY LITTLE MOVEMENT (II,2334); A SPECIAL ANNE MURRAY CHRISTMAS (II,2413); STEVE AND EYDIE CELEBRATE IRVING BERLIN (II,2456); STEVE AND EYDIE: OUR LOVE IS HERE TO STAY (II,2457); STEVE LAWRENCE AND EYDIE GORME FROM THIS MOMENT ON. . . COLE PORTER (II,2459); THREE GIRLS THREE (II,2608); TIN PAN ALLEY TODAY (I,4521); THE TONY BENNETT SHOW (I,4570); UPTOWN (II,2710); THE VERY FIRST GLEN CAMPBELL SPECIAL (I,4714); ZERO HOUR (I,4980)

SMITH, HERBERT
IVANHOE (I,2282)

SMITH, LEE ALLEN
STARS AND STRIPES SHOW (II,2442); THE STARS AND STRIPES SHOW (II,2443); THE STARS AND STRIPES SHOW (I,4206); THE STARS AND STRIPES SHOW (I,4207); THE STARS AND STRIPES SHOW (I,4208)

SMITH, LOU
THE WOLFMAN JACK RADIO SHOW (II,2816)

SMITH, PETE
PETE SMITH SPECIALTIES (I,3560)

SMITH, RICHARD
NO SOAP, RADIO (II,1856)

SMITH, ROBERT
THE FEMININE TOUCH (I,1565); JOHN CONTE'S LITTLE SHOW (I,2408)

SMITH, ROGER
ANN-MARGRET OLSSON (II,84); ANN-MARGRET SMITH (II,85); ANN-MARGRET. . . RHINESTONE COWGIRL (II,87); ANN-MARGRET: FROM HOLLYWOOD WITH LOVE (I,217); ANN-MARGRET—WHEN YOU'RE SMILING (I,219); FROM HOLLYWOOD WITH LOVE: THE ANN-MARGRET SPECIAL (I,1687)

SMITH, ROY A
THE BOOTS RANDOLPH SHOW (I,687); MUSIC HALL AMERICA (II,1773); THE NASHVILLE SOUND OF BOOTS RANDOLPH (I,3229); TENNESSEE ERNIE FORD'S WHITE CHRISTMAS (I,4402)

SMITH, SID
THE ALL-STAR REVUE (I,138); CIRCUS OF THE STARS (II,533); ELIZABETH TAYLOR IN LONDON (I,1430); THE MAGIC OF DAVID COPPERFIELD (II,1595); OPENING NIGHT: U.S.A. (I,3397)

SMITH, STEVE
OUT OF OUR MINDS (II,1933)

SMITH, TOM
YOU ARE THE JURY (II,2854)

SMOTHERS, TOM
PAT PAULSEN FOR PRESIDENT (I,3472); THE RETURN OF THE SMOTHERS BROTHERS (I,3777); THE SUMMER SMOTHERS BROTHERS SHOW (I,4281)

SNALL, MICHAEL
THE AMAZING YEARS OF CINEMA (II,64)

SNEED, SHERMAN
LENA HORNE: THE LADY AND HER MUSIC (II,1460)

SNEIDER, CARYN
ALL-AMERICAN PIE (II,48); THE COSBY SHOW (II,578); DETECTIVE SCHOOL (II,661); I DO, I DON'T (II,1212); IN TROUBLE (II,1230); THE NEW OPERATION PETTICOAT (II,1826); OH MADELINE (II,1880); THAT SECOND THING ON ABC (II,2572); THAT THING ON ABC (II,2574)

SNOWDEN, ALEX
SCOTLAND YARD (I,3941)

SNYDER, CARLYN
NATIONAL LAMPOON'S TWO REELERS (II,1796)

SNYDER, KEN
HOT WHEELS (I,2126); SKYHAWKS (I,4079)

SOBEL, JACK
JOHNNY MATHIS IN THE CANADIAN ROCKIES (II,1338); MONSANTO NIGHT PRESENTS BURL IVES (I,3090); MONSANTO NIGHT PRESENTS JOSE FELICIANO (I,3091); TONY BENNETT IN LONDON (I,4567); TONY BENNETT IN WAIKIKI (I,4568)

SOBOL, EDWARD
ABE LINCOLN IN ILLINOIS (I,10); THE ALFRED AND JOSEPHINE BUTLER SHOW (I,114); BLITHE SPIRIT (I,483); BROTHER RAT (I,746); COBINA (I,989); THE COLGATE COMEDY HOUR (I,997); THE ETHEL WATERS SHOW (I,1468); FAMILY HONOR (I,1522); THE

FORTUNE HUNTER (I,1646); THE GEORGE GOBEL SHOW (I,1767); THE GEORGE ROSS SHOW (I,1775); HAY FEVER (I,1979); IREENE WICKER SINGS (I,2236); THE JACK COLE DANCERS (I,2308); JANE EYRE (I,2337); THE LAWRENCE WELK SHOW (I,2643); LAWRENCE WELK'S TOP TUNES AND NEW TALENT (I,2645); THE LUCY MONROE SHOW (I,2790); MARIE AND GEORGE (I,2910); MEMORIES WITH LAWRENCE WELK (II,1681); THE MILKY WAY (I,3044); NIGHT CLUB (I,3284); PINKY AND COMPANY (I,3599); PLAY THE GAME (I,3618); THE RAY PERKINS REVUE (I,3735); SONG AND DANCE (I,4120); TEXACO STAR THEATER (I,4411); TOTAL ECLIPSE (I,4584); THE VALIANT (I,4697); WORLD'S FAIR BEAUTY SHOW (I,4920)

SOLLEY, RAY
SNEAK PREVIEWS (II,2388)

SOLMS, KENNY
BURNETT "DISCOVERS" DOMINGO (II,398); THE HOMEMADE COMEDY SPECIAL (II,1173); THE SMOTHERS BROTHERS SHOW (II,2384); TEXACO STAR THEATER: OPENING NIGHT (II,2565); THREE GIRLS THREE (II,2608)

SOLOMAN, AUBREY
THAT'S HOLLYWOOD (II,2577)

SOLOMAN, MARK
ALICE (II,33)

SOLOMON, LEO
LEAVE IT TO LARRY (I,2649)

SOLOMON, LINDA
THE FURTHER ADVENTURES OF WALLY BROWN (II,944)

SOLOMON, PAUL
THE MERV GRIFFIN SHOW (II,1688)

SOLOMON, SUE
PEOPLE (II,1995)

SOLOW, EUGENE
THE ADVENTURES OF THE SEA HAWK (I,81); CURTAIN CALL THEATER (I,1115); DOCTOR HUDSON'S SECRET JOURNAL (I,1312)

SOLOW, HERBERT F
THE DON RICKLES SHOW (II,708); THE MAN FROM ATLANTIS (II,1614); THE MAN FROM ATLANTIS (II,1615); MCLAREN'S RIDERS (II,1663); THEN CAME BRONSON (I,4435)

SOLT, ANDREW
BOB HOPE SPECIAL: BOB HOPE'S OVERSEAS CHRISTMAS TOURS (II,311); BOB HOPE SPECIAL: BOB HOPE—HOPE, WOMEN AND SONG (II,324); E.T. & FRIENDS—MAGICAL MOVIE VISITORS (II,749); PRIME TIMES (II,2083); THOSE WONDERFUL TV GAME SHOWS (II,2603)

SOMACH, DENNIS
THE ROCK 'N ROLL SHOW (II,2188)

SOMLYO, ROY
I'M A FAN (I,2192)

SOMMERS, JAY
CAROL (I,846); DADDY'S GIRL (I,1124); DOC (I,1303); GREEN ACRES (I,1884); HOWDY (I,2150); PIONEER SPIRIT (I,3604)

SONNENBERG, MARTIN
POWERHOUSE (II,2074)

SONNTAG, JACK
FAMILY IN BLUE (II,818)

SONTAG, DAVID
THE LAS VEGAS SHOW (I,2619)

SOTKIN, MARC
I'M A BIG GIRL NOW (II,1218); LAVERNE AND SHIRLEY (II,1446); WORKING STIFFS (II,2834)

SOWARDS, JACK
HAGEN (II,1071)

SPAFFORD, ROBERT
THE ORIENT EXPRESS (I,3407)

SPALLA, RICK
ENCHANTED LANDS (I,1440)

SPAN, AL
GUESS AGAIN (I,1894); THE JACK PAAR SHOW (I,2316); SONGS FOR SALE (I,4124); STAGE SHOW (I,4182)

SPANGLER, LARRY
THE JOE NAMATH SHOW (I,2400)

SPARBER, ISADORE
CASPER AND FRIENDS (I,867)

SPARGER, REX
SCROOGE'S ROCK 'N' ROLL CHRISTMAS (II,2279)

SPARKS, ROBERT
333 MONTGOMERY (I,4487); BACHELOR FATHER (I,323); THE EVE ARDEN SHOW (I,1470); HAVE GUN—WILL TRAVEL (I,1968); THE LINE-UP (I,2707); MUSIC U.S.A. (I,3179); THE TRAVELS OF JAIMIE MCPHEETERS (I,4596)

SPAULDING, KENNY
THE SCOEY MITCHLLL SHOW (I,3939)

SPEAKMAN, ROGER
HIT MAN (II,1156)

SPEAR, CHARLES
ROCKY KING, INSIDE DETECTIVE (I,3824)

SPEARS, JOE
GOLDIE GOLD AND ACTION JACK (II,1023)

SPEARS, KEN
ALVIN AND THE CHIPMUNKS (II,61); BEAUTY AND THE BEAST (II,197); DRAGON'S LAIR (II,732); FANG FACE (II,822); THE HEATHCLIFF AND DINGBAT SHOW (II,1117); THE HEATHCLIFF AND MARMADUKE SHOW (II,1118); MORK AND MINDY (II,1736); MR. T (II,1757); THE PLASTICMAN COMEDY/ADVENTURE SHOW (II,2051); RUBIK, THE AMAZING CUBE (II,2226); SATURDAY SUPERCADE (II,2263); THUNDARR THE BARBARIAN (II,2616); TURBO-TEEN (II,2668)

SPECTORSKY, BROOKE
THE DANCE SHOW (II,616)

SPEER, KATHY
CONDO (II,565)

SPELLING, AARON
ALOHA PARADISE (II,58); AMOS BURKE, SECRET AGENT (I,180); AMOS BURKE: WHO KILLED JULIE GREER? (I,181); AT EASE (II,115); B.A.D. CATS (II,123); THE BAIT (I,335); BEACH PATROL (II,192); BURKE'S LAW (I,762); CASINO (II,452); CHARLIE'S ANGELS (II,486); CHARLIE'S ANGELS (II,487); CHOPPER ONE (II,515); DANIEL BOONE (I,1142); THE DANNY THOMAS HOUR (I,1147); THE DAUGHTERS OF JOSHUA CABE (I,1169); THE DICK POWELL SHOW (I,1269); FAMILY (II,813); FANTASY ISLAND (II,829); FANTASY ISLAND (II,830); FINDER OF LOST LOVES (II,857); FIREHOUSE (II,859); THE FRENCH ATLANTIC AFFAIR (II,924); FRIENDS (II,927); GLITTER (II,1009); THE GUNS OF WILL SONNETT (I,1907); HARD KNOCKS (II,1086); HART TO HART (II,1101); HART TO HART (II,1102); HONEY WEST: WHO KILLED THE JACKPOT? (I,2106); HOTEL (II,1192); JOHNNY RINGO (I,2434); KATE BLISS AND THE TICKER TAPE KID (II,1366); LADIES IN BLUE (II,1422); LAST OF THE PRIVATE EYES (I,2628); THE LETTERS (I,2670); LETTERS FROM THREE LOVERS (I,2671); THE LLOYD BRIDGES SHOW (I,2733); THE LOSERS (I,2757); THE LOVE BOAT (II,1535); THE LOVE BOAT II (II,1533); THE LOVE BOAT III (II,1534); LUXURY LINER (I,2796); MAKE ROOM FOR GRANDDADDY (I,2844); MASSARATI AND THE BRAIN (II,1646); MATT HOUSTON (II,1654); THE MOD SQUAD (I,3078); THE MONK (I,3087); THE MOST DEADLY GAME (I,3106); MURDER CAN HURT YOU! (II,1768); THE NEW DAUGHTERS OF JOSHUA CABE (I,3261); THE NEW PEOPLE (I,3268); THE NIGHT STALKER (I,3292); THE NIGHT STRANGLER (I,3293); THE OATH: 33 HOURS IN THE LIFE OF GOD (II,1873); THE OATH: THE SAD AND LONELY SUNDAYS (II,1874); THE OVER-THE-HILL GANG RIDES AGAIN (I,3432); THE PIGEON (I,3596); THE POWER WITHIN (II,2073); RANGO (I,3721); THE RETURN OF THE MOD SQUAD (I,3776); RETURN TO FANTASY ISLAND (II,2144); THE ROOKIES (I,3840); THE ROOKIES (II,2208); S.W.A.T. (II,2236); THE SAN PEDRO BEACH BUMS (II,2250); THE SAN PEDRO BUMS (II,2251); SCARED SILLY (II,2269); SHOOTING STARS (II,2341); STARSKY AND HUTCH (II,2444); STARSKY AND HUTCH (II,2445); STRIKE FORCE (II,2480); T.J. HOOKER (II,2524); TONI'S BOYS (II,2637); TWO FOR THE MONEY (I,4655); VEGAS (II,2723); VEGAS (II,2724); VELVET (II,2725); VENICE MEDICAL (II,2726); WAIKIKI (II,2734); WILD AND WOOLEY (II,2798); THE WILD WOMEN OF CHASTITY GULCH (II,2801); THE YOUNG REBELS (I,4944); YUMA (I,4976); ZANE GREY THEATER (I,4979)

SPENCER, DENIS
THE BENNY HILL SHOW (II,207)

SPIDEL, SANDI
THE NEW $100,000 NAME THAT TUNE (II,1810)

SPIELVOGEL, DON
BATTLE OF THE LAS VEGAS SHOW GIRLS (II,162)

SPIER, WILLIAM
CHRYSLER MEDALLION THEATER (I,951); THE CLOCK (I,978); OMNIBUS (I,3353); WESTINGHOUSE SUMMER THEATER (I,4805); WILLY (I,4868); WILLY (I,4869)

SPINA, JACK
DEBBY BOONE. . . THE SAME OLD BRAND NEW ME (II,656)

SPINNER, ANTHONY
BARETTA (II,152); CANNON (II,424); CARIBE (II,439); THE DAKOTAS (I,1126); DAN AUGUST (I,1130); THE FBI (I,1551); THE LAST NINJA (II,1438); THE MAN FROM U.N.C.L.E. (I,2867); READY FOR THE PEOPLE (I,3739); RETURN OF THE SAINT (II,2141); ROGER AND HARRY (II,2200); SUPERTRAIN (II,2504)

SPIVA, TOM
GENTLE BEN (I,1761)

SPOSA, LOU
CHANCE OF A LIFETIME (I,898)

SPOTA, GEORGE
JONATHAN WINTERS PRESENTS 200 YEARS OF AMERICAN HUMOR (II,1342); WILL ROGERS' U.S.A. (I,4865); THE WONDERFUL WORLD OF JONATHAN WINTERS (I,4900)

ST JOHNS, RICHARD
JESSIE (II,1287); MCCLAIN'S LAW (II,1659)

STABLE, JOE
THE JERRY LEWIS SHOW (II,1284)

STAGER, MARTIN
HARRY'S BATTLES (II,1100)

STAGG, JERRY
BEN BLUE'S BROTHERS (I,393); HIGH ROAD (I,2054); TELEPHONE TIME (I,4379); THE TEXAN (I,4413)

STAMBLER, ROBERT
THE BEACHCOMBER (I,371); BRAVO TWO (II,368); CRISIS IN SUN VALLEY (II,604); THE DEADLY TRIANGLE (II,634); DOCTORS' PRIVATE LIVES (II,693); GRANDPA GOES TO WASHINGTON (II,1050); HAWAII FIVE-O (II,1110); THE IMPOSTER (II,1225); KATE MCSHANE (II,1368); PAPER MOON (II,1953); POOR DEVIL (I,3646); RISKO (II,2180); SUPERTRAIN (II,2504); SUPERTRAIN (II,2505); TABITHA (II,2527); TABITHA (II,2528)

STAMMER, NEWT
STAR OF THE FAMILY (I,4194)

STANDER, ARTHUR
IT'S ALWAYS JAN (I,2264)

STANFORD, TONY
PAUL WHITEMAN'S SATURDAY NIGHT REVUE (I,3490)

STANLEY, HAL
MR. SMITH GOES TO WASHINGTON (I,3151); THE PIED PIPER OF HAMELIN (I,3595)

STANLEY, JACKSON
THE BIG GAME (I,426)

STANLEY, JAMES
THE BELLE OF 14TH STREET (I,390); THE DANGEROUS CHRISTMAS OF RED RIDING HOOD (I,1139); FRED WARING: WAY BACK HOME (I,1680); MARLO THOMAS IN ACTS OF LOVE—AND OTHER COMEDIES (I,2919); ROLLIN' ON THE RIVER (I,3834)

STANLEY, JEROME
ESCAPE (I,1461); UNION PACIFIC (I,4676)

STANLEY, JOSEPH
THE ALL-STAR REVUE (I,138); THE ALL-STAR SUMMER REVUE (I,139); THE PATTI PAGE SHOW (I,3479); THE R.C.A. VICTOR SHOW (I,3737)

STANSON, MILTON
THE JERRY LESTER SHOW (I,2360); TOOTSIE HIPPODROME (I,4575)

STAPLETON, TERRY
ALL THE RIVERS RUN (II,43)

STAR, MICHAEL
QUINCY, M. E. (II,2102)

STARETSKI, JOSEPH
THREE'S A CROWD (II,2613); THREE'S COMPANY (II,2614)

STARGER, MARTIN
LEGENDS OF THE WEST: TRUTH AND TALL TALES (II,1457); MY WIFE NEXT DOOR (II,1781); OMNIBUS (II,1890); THE RETURN OF MARCUS WELBY, M.D (II,2138); THE ROCK RAINBOW (II,2194); SANCTUARY OF FEAR (II,2252); THE SON-IN-LAW (II,2398); THE TWO OF US (II,2693); WESTSIDE MEDICAL (II,2765); WHODUNIT? (II,2794)

STARK, ART
HOBBY LOBBY (I,2068); NAME THAT TUNE (I,3218); WEDDING PARTY (I,4786); WHO DO YOU TRUST? (I,4846)

STARK, RAY
FUNNY GIRL TO FUNNY LADY (II,942)

STARK, WILBUR
ADVENTURES OF COLONEL FLACK (I,56); THE BROTHERS BRANNAGAN (I,747); CRIME WITH FATHER (I,1100); CUT (I,1116); MODERN ROMANCES (I,3079); NEWSSTAND THEATER (I,3279); TRUE STORY (I,4617)

STARK, WILLIAM
THE 25TH MAN (MS) (II,2678); CHASE (I,917); THE LOG OF THE BLACK PEARL (II,1506)

STARKEY, DEWEY
IT'S ALWAYS JAN (I,2264)

STARR, BEN
DIFF'RENT STROKES (II,674); TOM, DICK, AND HARRY (I,4536)

STARR, IRVING
FORD TELEVISION THEATER (I,1634); THE MILTON BERLE SPECIAL (I,3052)

STARTZ, JANE
CHARLES IN CHARGE (II,482)

STEARNS, JOHNNY
FAYE AND SKITCH (I,1548); MAKE ME LAUGH (I,2839); MUSIC BINGO (I,3165)

STECK, JACK
PAUL WHITEMAN'S TV TEEN CLUB (I,3491)

STEELE, BARBARA
THE WINDS OF WAR (II,2807)

STEELE, BOB
A WOMAN TO REMEMBER (I,4888)

STEELE, TOM
ALL NIGHT RADIO (II,40)

STEENBERG, RICHARD
THE TROUBLESHOOTERS (I,4614)

STEFAN, BUD
SANDY DREAMS (I,3909)

STEFANO, JOSEPH
THE OUTER LIMITS (I,3426); THE UNKNOWN (I,4680)

STEFFNER, WAYNE
YOU ASKED FOR IT (I,4931)

STEIN, JACK
PRESENTING SUSAN ANTON (II,2077)

STEIN, JAMES
DOUBLE TROUBLE (II,723); FLO (II,891); SILVER SPOONS (II,2355)

STEIN, JEFF
MR. MOM (II,1752)

STEIN, JOEL
GIVE-N-TAKE (II,1003); THE NEIGHBORS (II,1801); SECOND CHANCE (II,2286); SOLID GOLD HITS (II,2397); WHODUNIT? (II,2794)

STEIN, PHIL
BELAFONTE, NEW YORK (I,386); TONIGHT WITH BELAFONTE (I,4566)

STEIN, WILLIE
THE $128,000 QUESTION (II,1903); SPIN-OFF (II,2427)

STEINBERG, ALAN
RINGO (II,2176)

STEINBERG, BARRY
BOONE (II,351); FALCON CREST (II,808); KING'S CROSSING (II,1397)

STEINBERG, NORMAN
100 CENTRE STREET (II,1901); ADVENTURING WITH THE CHOPPER (II,15); THE BAY CITY AMUSEMENT COMPANY (II,185); IN THE BEGINNING (II,1229); ROOSEVELT AND TRUMAN (II,2209); THE SIX O'CLOCK FOLLIES (II,2368); WHEN THINGS WERE ROTTEN (II,2782)

STEINBERG, SCOTT
THICKE OF THE NIGHT (II,2587); THE WOLFMAN JACK RADIO SHOW (II,2816)

STEINBERG, SHARI
PEOPLE DO THE CRAZIEST THINGS (II,1998)

STEINBERG, ZIGGY
THE JERK, TOO (II,1282)

STEINER, AARON
THE DENNIS JAMES SHOW (I,1228)

STEINHAUER, ROBERT
THE INCREDIBLE HULK (II,1232); LEGMEN (II,1458); VOYAGERS (II,2730)

STEINMAN, DANNY
SEPARATE TABLES (II,2304)

STELOFF, SKIP
LORNE GREENE'S LAST OF THE WILD (II,1522)

STENTZ, CHRISTIE
NOT NECESSARILY THE NEWS (II,1863)

STEPHENS, JOHN G
BUCK ROGERS IN THE 25TH CENTURY (II,384); THE GANGSTER CHRONICLES (II,957); HOW THE WEST WAS WON (II,1196); MAGNUM, P.I. (II,1601); SHADOW OF SAM PENNY (II,2313); SIMON AND SIMON (II,2357); TALES OF THE GOLD MONKEY (II,2537); THREE FOR THE ROAD (II,2607); WHIZ KIDS (II,2790); WONDER WOMAN (PILOT 1) (II,2826)

STERN, BERT
TWIGGY IN NEW YORK (I,4649)

STERN, JOSEPH
THE BIG EASY (II,234); CAGNEY AND LACEY (II,409); THE LONG DAYS OF SUMMER (II,1514); THE TOM SWIFT AND LINDA CRAIG MYSTERY HOUR (II,2633)

STERN, LEONARD B
BROCK'S LAST CASE (I,742); CAR WASH (II,437); COLUMBO (II,556); DIANA (I,1259); FARADAY AND COMPANY (II,833); FEMALE INSTINCT (I,1564); GET SMART (II,983); THE GOOD GUYS (I,1847); THE GOVERNOR AND J.J. (I,1863); HE AND SHE (I,1985); THE HERO (I,2035); HOLMES AND YOYO (II,1169); I'M DICKENS...HE'S FENSTER (I,2193); KOSTA AND HIS FAMILY (I,2581); LANIGAN'S RABBI (II,1427); LANIGAN'S RABBI (II,1428); MCMILLAN (II,1666); MCMILLAN AND WIFE (II,1667); ONCE UPON A DEAD MAN (I,3369); OPERATION PETTICOAT (II,1916); OPERATION PETTICOAT (II,1917); PHILLIP AND BARBARA (II,2034); ROSETTI AND RYAN (II,2216); ROSETTI AND RYAN: MEN WHO LOVE WOMEN (II,2217); RUN, BUDDY, RUN (I,3870); SNAFU (II,2386); THE SNOOP SISTERS (II,2389); THREE TIMES DALEY (II,2610); WILD ABOUT HARRY (II,2797); WINDOWS, DOORS AND KEYHOLES (II,2806)

STERN, ROBIN J
PEOPLE ARE FUNNY (II,1996)

STERN, SANDOR
DOC ELLIOT (I,1306) TRUE GRIT (II,2664);

STEVENS JR, GEORGE
CHRISTMAS IN WASHINGTON (II,520); CHRISTMAS IN WASHINGTON (II,521)

STEVENS, ANDREW
TOPPER (II,2649)

STEVENS, BOB
THE BOB HOWARD SHOW (I,665)

STEVENS, GARY
TWENTY QUESTIONS (I,4647)

STEVENS, JEREMY
THICKE OF THE NIGHT (II,2587)

STEVENS, LESLIE
BATTLESTAR GALACTICA (II,181); BORDER TOWN (I,689); BUCK ROGERS IN THE 25TH CENTURY (II,383); FRANK MERRIWELL (I,1659); THE GEMINI MAN (II,960); THE INVISIBLE MAN (II,1242); MCCLOUD (II,1660); MCCLOUD: WHO KILLED MISS U.S.A.? (I,2965); THE MEN FROM SHILOH (I,3005); THE NAME OF THE GAME (I,3217); THE OUTER LIMITS (I,3426); PROBE (I,3674); SEARCH (I,3951); STONESTREET: WHO KILLED THE CENTERFOLD MODEL? (II,2472); STONEY BURKE (I,4229); TACK REYNOLDS (I,4329); THE VIRGINIAN (I,4738); WEAPONS MAN (I,4783)

STEVENS, MARK
BIG TOWN (I,436); MICHAEL SHAYNE, DETECTIVE (I,3022)

STEVENS, MIKE
ADVICE TO THE LOVELORN (II,16); SCENE OF THE CRIME (II,2271)

STEVENS, ROBERT
THE MASK (I,2938); ROMANCE (I,3836); STARLIGHT THEATER (I,4201); SUSPENSE (I,4305)

STEWART, BOB
THE $20,000 PYRAMID (II,2681); THE $25,000 PYRAMID (II,2679); BLANKETY BLANKS (II,266); CHAIN REACTION (II,479); THE FACE IS FAMILIAR (I,1506); GO (II,1012); JACKPOT (II,1266); JOHNNY CARSON DISCOVERS CYPRESS GARDENS (I,2419); THE LOVE EXPERTS (II,1537); THE NEW $25,000 PYRAMID (II,1811); PASSWORD (I,3468); PERSONALITY (I,3553); THE PRICE IS RIGHT (I,3664); SHOOT FOR THE STARS (II,2340); WINNING STREAK (II,2809)

STEWART, CHARLES
ADAMS OF EAGLE LAKE (II,10); JOE'S WORLD (II,1305); THE JOEY BISHOP SHOW (I,2403); THE JOEY BISHOP SHOW (I,2404); MAKE ROOM FOR DADDY (I,2843); MAYBERRY R.F.D. (I,2961); PETTICOAT JUNCTION (I,3571)

STEWART, HERBERT F
CROSSROADS (I,1105)

STEWART, HILARY
ELTON JOHN IN CENTRAL PARK (II,769)

STEWART, LARRY
THRILL SEEKERS (I,4491)

STEWART, MALCOLM
JOHNNY BELINDA (I,2418)

STEWART, MEL
THE FUTURE: WHAT'S NEXT (II,947)

STEWART, PAUL
CONFLICT (I,1036)

STEWART, ROBERT BANKS
BERGERAC (II,209); SHOESTRING (II,2338)

STEWART, SANDE
CHAIN REACTION (II,479); GO (II,1012); PASS THE BUCK (II,1964)

STEWART, TOM
THE LOVE REPORT (II,1542)

STEWART, WILLIAM
BLESS THIS HOUSE (II,268); FATHER DEAR FATHER (II,837)

STIGWOOD, ROBERT
ALMOST ANYTHING GOES (II,56)

STILES, NORMAN
THE BAD NEWS BEARS (II,134); BROTHERS (II,381); SPACE FORCE (II,2407)

STILLMAN, ROBERT
FLIGHT (I,1597); PONY EXPRESS (I,3645); SKYFIGHTERS (I,4078)

STIVERS, BOB
THE ALL-STAR SALUTE TO MOTHER'S DAY (II,51); BID N' BUY (I,419); CELEBRITY DAREDEVILS (II,473); CIRCUS OF THE STARS (II,530); CIRCUS OF THE STARS (II,531); CIRCUS OF THE STARS (II,532); CIRCUS OF THE STARS (II,533); CIRCUS OF THE STARS (II,534); CIRCUS OF THE STARS (II,535); CIRCUS OF THE STARS (II,536); CIRCUS OF THE STARS (II,537); CIRCUS OF THE STARS (II,538); DINAH'S PLACE (II,678); MEN AT LAW (I,3004); THE STOREFRONT LAWYERS (I,4233)

STIZEL, MOLLY
THE RICH LITTLE SPECIAL (II,2155)

STODDARD, MICHAEL
NATIONAL LAMPOON'S TWO REELERS (II,1796)

STOKES, MICHAEL
THE CRACKER BROTHERS (II,599); WELCOME TO THE FUN ZONE (II,2763)

STOKEY, MIKE
ARMCHAIR DETECTIVE (I,259); PANTOMIME QUIZ (I,3449); STUMP THE STARS (I,4273); STUMP THE STARS (I,4274)

STOLFI, ROBERT
A FINE ROMANCE (II,858); FOOT IN THE DOOR (II,902)

STOLNITZ, ART
JEREMIAH OF JACOB'S NECK (II,1281); THE LIFE AND TIMES OF GRIZZLY ADAMS (II,1469)

STONE, CLIFF
THE TENNESSEE ERNIE FORD SHOW (I,4399)

STONE, EZRA
THE ALL-STAR REVUE (I,138); FIREBALL FUN FOR ALL (I,1577); THE HATHAWAYS (I,1965); TEXACO COMMAND PERFORMANCE (I,4407)

STONE, HAROLD
JULIA (I,2476)

STONE, JON
CAPTAIN KANGAROO (II,434); THE MUPPET SHOW (II,1766); SESAME STREET (I,3982)

STONE, KATHRYN
KIDS ARE PEOPLE TOO (II,1390)

STONE, MARSHALL
TIMEX ALL-STAR JAZZ SHOW III (I,4517)

STONE, MARTIN
THE GABBY HAYES SHOW (I,1723); HOWDY DOODY (I,2151); JOHNNY JUPITER (I,2429); SUPER CIRCUS (I,4287)

STONE, PAULA
PAULA STONE'S TOY SHOP (I,3495)

STONE, SID

RODNEY DANGERFIELD SHOW: I CAN'T TAKE IT NO MORE (II,2198)

STONEMAN, RUSSELL
BARNABY JONES (II,153); BRENNER (I,719); CANNON (II,424); PLAYHOUSE 90 (I,3623); THE STREETS OF SAN FRANCISCO (II,2478)

STONER, DOUGLAS
AMAZING BUT TRUE! (I,156); BELIEVE IT OR NOT (I,387)

STRAHORN, LEE
THE FRED WARING SHOW (I,1679)

STRANGIS, GREG
BULBA (II,392); EIGHT IS ENOUGH (II,762); HARPER VALLEY (II,1094); RIVKIN: BOUNTY HUNTER (II,2183); SHIRLEY (II,2330)

STRANGIS, SAM
HARPER VALLEY (II,1094); I GAVE AT THE OFFICE (II,1213); SEARCH AND RESCUE: THE ALPHA TEAM (II,2283); THE SIX-MILLION-DOLLAR MAN (II,2372)

STRATTON, CALEB
PERSPECTIVE ON GREATNESS (I,3555)

STRAUSS, HERB
WINGO (I,4878)

STRAUSS, TERRY
THE WORLD OF PEOPLE (II,2839)

STREAM, JAY
THE WAYNE NEWTON SPECIAL (II,2749)

STRICHMAN, LARRY
REX REED'S MOVIE GUIDE (II,2149)

STRICK, PHILIP
THE AMAZING YEARS OF CINEMA (II,64)

STRICKLAND, LORETTA
THE MIDNIGHT SPECIAL (II,1696)

STROCK, HERBERT L
THE CASES OF EDDIE DRAKE (I,863); HARBOR COMMAND (I,1945)

STRONG, JOHN
EVEL KNIEVEL (II,785)

STROOT, MIKE
YOU ASKED FOR IT (II,2855)

STRUD, GERALD
THE DOCTORS (II,694)

STUART, BILL
THE VAUGHN MONROE SHOW (I,4708)

STUART, JOSEPH
THE DOCTORS (II,694); LOVING (II,1552); ONE LIFE TO LIVE (II,1907)

STUART, LEN
SECOND CITY TELEVISION (II,2287)

STUART, MALCOLM
THE HOME FRONT (II,1171); HUMAN FEELINGS (II,1203)

STUART, MARK
THE BENNY HILL SHOW (II,207)

STUART, MEL
RIPLEY'S BELIEVE IT OR NOT (II,2177); SOPHIA! (I,4128); THE STORY OF— (I,4239); THOSE AMAZING ANIMALS (II,2601)

STUHL, JOSEPH
DOWN YOU GO (I,1367)

STUMP, KEN
ARCHIE BUNKER'S PLACE (II,105); A DOG'S LIFE (II,697); HANGING IN (II,1078); MR. DUGAN (II,1749)

STYNE, JULE
ANYTHING GOES (I,238); PANAMA HATTIE (I,3444)

SUBOTSKY, MILTON
THE MARTIAN CHRONICLES (II,1635)

SUDDLESTON, KEN
STEVE MARTIN'S THE WINDS OF WHOOPIE (II,2461); TWILIGHT THEATER (II,2685); TWILIGHT THEATER II (II,2686)

SUGARMAN, BURT
CELEBRITY SWEEPSTAKES (II,476); THE DIONNE WARWICK SPECIAL (I,1289); HI, I'M GLEN CAMPBELL (II,1141); JOHNNY MANN'S STAND UP AND CHEER (I,2430); THE JOKER'S WILD (II,1341); THE MANCINI GENERATION (I,2885); THE MIDNIGHT SPECIAL (II,1696); MOVIN' (I,3117); THE RICHARD PRYOR SHOW (II,2163); THE RICHARD PRYOR SPECIAL (II,2164); THE RICHARD PRYOR SPECIAL? (II,2165); STAND UP AND CHEER (I,4188); TAKE MY ADVICE (II,2531); WHEW! (II,2786); THE WIZARD OF ODDS (I,4886)

SUGERMAN, ANDREW
MANDY'S GRANDMOTHER (II,1618)

SULLIVAN, ED
THE ED SULLIVAN SHOW (I,1401)

SULTAN, ARNE
FAMILY BUSINESS (II,815); THE GOVERNOR AND J.J. (I,1863); HE AND SHE (I,1985); HOLMES AND YOYO (II,1169); THE JONES BOYS (I,2446); LAUGHS FOR SALE (I,2634); MY LUCKY PENNY (I,3196); NOBODY'S PERFECT (II,1858); THE PARTNERS (I,3462); POOR DEVIL (I,3646); POPI (II,2069); THE SANDY DUNCAN SHOW (I,3910); TOO CLOSE FOR COMFORT (II,2642); WINDOWS, DOORS AND KEYHOLES (II,2806)

SUNGA, GEORGE
THE DUMPLINGS (II,743); GOOD TIMES (II,1038); HOTEL 90 (I,2130); THE JEFFERSONS (II,1276); MAKE MINE RED, WHITE, AND BLUE (I,2841); NAKIA (II,1785); OF THEE I SING (I,3334); THE ROPERS (II,2214); ROWAN AND MARTIN BITE THE HAND THAT FEEDS THEM (I,3853); THE SUMMER SMOTHERS BROTHERS SHOW (I,4281); THREE'S A CROWD (II,2613); THREE'S COMPANY (II,2614)

SUNSHINE, MADELINE
WEBSTER (II,2758)

SUNSHINE, STEVE
WEBSTER (II,2758)

SURGALL, JON
THE SECRET WAR OF JACKIE'S GIRLS (II,2294)

SUSSAN, HERBERT
GUY LOMBARDO'S DIAMOND JUBILLE (I,1911)

SUSSKIND, DAVID
ALICE (II,33); ALICE (II,34); ALL CREATURES GREAT AND SMALL (I,129); ALL THE WAY HOME (I,140); APPOINTMENT WITH ADVENTURE (I,246); THE ART CARNEY SHOW (I,274); THE BELLS OF ST. MARY'S (I,392); THE BEST OF ANYTHING (I,404); CAESAR AND CLEOPATRA (I,789); CALL ME BACK (I,798); THE COUNTRY GIRL (I,1064); CROWN MATRIMONIAL (I,1106); DEATH OF A SALESMAN (I,1208); THE DEVIL AND DANIEL WEBSTER (I,1246); DIAL "M" FOR MURDER (I,1255); THE DIARY OF ANNE FRANK (I,1262); DUPONT SHOW OF THE MONTH (I,1387); THE DUPONT SHOW OF THE WEEK (I,1388); EAST SIDE/WEST SIDE (I,1396); FULL MOON OVER BROOKLYN (I,1703); THE GLASS MENAGERIE (I,1819); GOOD COMPANY (I,1844); HANSEL AND GRETEL (I,1934); HARVEY (I,1961); HARVEY (I,1962); A HATFUL OF RAIN (I,1964); HEDDA GABLER (I,2001); THE HEIRESS (I,2006); HOORAY FOR LOVE (I,2115); THE HUMAN VOICE (I,2160); JOHNNY BELINDA (I,2417); JOYCE AND BARBARA: FOR ADULTS ONLY (I,2458); JUSTICE (I,2499); KRAFT TELEVISION THEATER (I,2592); LAURA (I,2636); LITTLE WOMEN (I,2722); LOVERS AND OTHER STRANGERS (II,1550); THE MAN IN THE DOG SUIT (I,2869); MEET ME IN ST. LOUIS (I,2986); MEMBER OF THE WEDDING (I,3002); MIRACLE ON 34TH STREET (I,3057); MR. BROADWAY (I,3131); NINOTCHKA (I,3299); OF MICE AND MEN (I,3333); OLDSMOBILE MUSIC THEATER (I,3352); ON OUR OWN (II,1892); OUR TOWN (I,3420); PEOPLE (II,1995); THE PHILADELPHIA STORY (I,3582); THE PICTURE OF DORIAN GRAY (I,3591); THE POWER AND THE GLORY (I,3658); THE PRICE (I,3666); RUN, BUDDY, RUN (I,3870); THE STRAWBERRY BLONDE (I,4254); TEN LITTLE INDIANS (I,4394); THREE IN ONE (I,4479); THE THREE MUSKETEERS (I,4481); TOM AND JOANN (II,2630); TOO YOUNG TO GO STEADY (I,4574); TRUMAN AT POTSDAM (I,4618); VICTORY (I,4729); WAY OUT (I,4777); WITNESS (I,4885); THE WORLD BEYOND (II,2835); THE WORLD OF DARKNESS (II,2836); THE WORLD OF MAGIC (II,2838)

SUSSKIND, MURRAY
GIFT OF THE MAGI (I,1800); WITNESS (I,4885)

SUTHERLAND, EDDIE
INTERNATIONAL DETECTIVE (I,2224); MARTIN KANE, PRIVATE EYE (I,2928)

SUTHERLAND, HAL
WILL THE REAL JERRY LEWIS PLEASE SIT DOWN? (I,4866)

SWACKHAMER, E W
GIDGET GETS MARRIED (I,1796); IS THERE A DOCTOR IN THE HOUSE? (II,1247); LOVE ON A ROOFTOP

(I,2775); MR. DEEDS GOES TO TOWN (I,3134); THE OUTCASTS (I,3425); THREE FOR TAHITI (I,4476); UNDER THE YUM YUM TREE (I,4671)

SWALE, TOM
THE BRADY BUNCH HOUR (II,363); GLITTER (II,1009)

SWEENEY, BOB
ALONE AT LAST (II,59); THE ANDROS TARGETS (II,76); THE BAILEYS OF BALBOA (I,334); THE DORIS DAY SHOW (I,1355); ETHEL IS AN ELEPHANT (II,783); HAWAII FIVE-O (II,1110); HEY TEACHER (I,2041); SPENCER'S PILOTS (II,2421); SPENCER'S PILOTS (II,2422)

SWEENEY, TERENCE
THE JUGGLER OF NOTRE DAME (II,1350)

SWENSON, CHARLES
DOROTHY IN THE LAND OF OZ (II,722)

SWERDLOFF, HELAINE
PEOPLE TO PEOPLE (II,2000)

SWERLING JR, JO
THE A-TEAM (II,119); ALIAS SMITH AND JONES (I,118); BARETTA (II,152); THE BOB HOPE CHRYSLER THEATER (I,502); CAPTAINS AND THE KINGS (II,435); CITY OF ANGELS (II,540); COOL MILLION (I,1046); FOUR EYES (II,912); THE GREATEST AMERICAN HERO (II,1060); THE GREEN FELT JUNGLE (I,1885); HARDCASTLE AND MCCORMICK (II,1089); HAZARD'S PEOPLE (II,1112); HUNTER (II,1206); THE JORDAN CHANCE (II,1343); THE LAST CONVERTIBLE (II,1435); THE LAWYERS (I,2646); LOBO (II,1504); THE QUEST (II,2098); RAPTURE AT TWO-FORTY (I,3724); RIPTIDE (II,2178); THE ROCKFORD FILES (II,2196); THE ROUSTERS (II,2220); RUN FOR YOUR LIFE (I,3871); SAM HILL: WHO KILLED THE MYSTERIOUS MR. FOSTER? (I,3896); TOMA (I,4547); TOMA (II,2634); THE WHOLE WORLD IS WATCHING (I,4851)

SWIFT, DAVID
ARNIE (I,261); CAMP RUNAMUCK (I,811); NORBY (I,3304); POOR MR. CAMPBELL (I,3647)

SWIK, GEORGE
CODE RED (II,552)

SWOPE JR, HERBERT
THE CLOCK (I,978); FIVE FINGERS (I,1586); LIGHT'S OUT (I,2699); VOICE OF FIRESTONE (I,4741)

SWOPE, MEL
BANJO HACKETT: ROAMIN' FREE (II,141); CALIFORNIA FEVER (II,411); DAVID CASSIDY—MAN UNDERCOVER (II,623); FAME (II,812); THE GIRL WITH SOMETHING EXTRA (II,999); GOOD HEAVENS (II,1032); MIAMI VICE (II,1689); THE PARTRIDGE FAMILY (II,1962); PLEASURE COVE (II,2058); POLICE STORY (II,2062); WALKING TALL (II,2738)

SYATT, STEVE
THE JOYCE JILLSON SHOW (II,1346)

SYLVESTER, WARD
33 1/3 REVOLUTIONS PER MONKEE (I,4451); THE BOBBY SHERMAN SPECIAL (I,676); GETTING TOGETHER (I,1784); LEGENDS OF THE SCREEN (II,1456); THE MONKEES (I,3088)

SYMONDS, HUGH
CULTURE CLUB IN CONCERT (II,610)

SYNES, BOB
DREAM HOUSE (II,734); LET'S MAKE A DEAL (II,1464); THE MAGNIFICENT MARBLE MACHINE (II,1600)

SZUREK, SAM
LEGENDS OF THE SCREEN (II,1456)

SZWARC, JEANNOT
CODE NAME: HERACLITUS (I,990)

TAFFNER, DON
A FINE ROMANCE (II,858); FOOT IN THE DOOR (II,902); MISS WINSLOW AND SON (II,1705)

TAHSE, MARTIN
THE ABC AFTERSCHOOL SPECIAL (II,1)

TAKAMOTO, IWAO
THE ABC AFTERSCHOOL SPECIAL (II,1); THE ADDAMS FAMILY (II,12); THE ALL-NEW POPEYE HOUR (II,49); THE C.B. BEARS (II,406); THE CLUE CLUB (II,546); DEVLIN (II,663); FRED FLINTSTONE AND FRIENDS (II,920); GIDGET MAKES THE WRONG CONNECTION (I,1798); GOOBER AND THE GHOST CHASERS (II,1028); THE GREAT GRAPE APE SHOW (II,1056); THE HARLEM GLOBETROTTERS (I,1953); HONG KONG PHOOEY (II,1180); INCH HIGH, PRIVATE EYE (II,1231); JABBERJAW (II,1261); JEANNIE (II,1275); LOST IN SPACE (I,2759); THE NEW SUPER FRIENDS HOUR (II,1830); PARTRIDGE FAMILY: 2200 A.D. (II,1963); PEBBLES AND BAMM BAMM (II,1987); THE ROMAN HOLIDAYS (I,3835); SCOOBY'S ALL-STAR LAFF-A-LYMPICS (II,2273); SCOOBY-DOO, WHERE ARE YOU? (II,2276); THE SCOOBY-DOO/DYNOMUTT HOUR (II,2277); SEALAB 2020 (I,3949); SPEED BUGGY (II,2419); SUPER FRIENDS (II,2497); THESE ARE THE DAYS (II,2585); VALLEY OF THE DINOSAURS (II,2717); WHEELIE AND THE CHOPPER BUNCH (II,2776); THE WORLD'S GREATEST SUPER HEROES (II,2842); YOGI'S GANG (II,2852); YOGI'S SPACE RACE (II,2853)

TALSKY, RON
REALLY RAQUEL (II,2119)

TAMBER, SID
THE BIG PAYOFF (I,428)

TAMPLIN, ROBERT
THE ANDY WILLIAMS SHOW (I,204); BE OUR GUEST (I,370); THE BIG PARTY FOR REVLON (I,427); THE DANNY KAYE SHOW (I,1143); HOLLYWOOD TALENT SCOUTS (I,2094); THE KELLY MONTEITH SHOW (II,1376); THE LIBERACE SHOW (I,2677); YOUR HIT PARADE (I,4952)

TANG, CECILE
PEKING ENCOUNTER (II,1991)

TAPLIN, JONATHAN
FAERIE TALE THEATER (II,807)

TAPPER, OLIVIA
JEAN SHEPHERD'S AMERICA (I,2354)

TAPPLINHER, SYLVAN
ANYONE CAN WIN (I,235)

TARITERO, JOE
THE GREAT MYSTERIES OF HOLLYWOOD (II,1058)

TARSES, JAY
THE BOB NEWHART SHOW (II,342); BUFFALO BILL (II,387); THE CHOPPED LIVER BROTHERS (II,514); MARY (II,1637); OPEN ALL NIGHT (II,1914); SITCOM (II,2367); THE TONY RANDALL SHOW (II,2640); WE'VE GOT EACH OTHER (II,2757)

TATASHORE, FRED
BURT AND THE GIRLS (I,770); BURT REYNOLDS' LATE SHOW (I,777); DINAH AND FRIENDS (II,675); DINAH! (II,677); THE KRYPTON FACTOR (II,1414); THE REGIS PHILBIN SHOW (II,2128); WEDDING PARTY (I,4786); WOMAN'S PAGE (II,2822)

TATELMAN, HARRY
77 SUNSET STRIP (I,3988); THE ALASKANS (I,106); BOURBON STREET BEAT (I,699); THE CODE OF JONATHAN WEST (I,992); COLT .45 (I,1010); COME A RUNNING (I,1013); JIGSAW (I,2377); THE PSYCHIATRIST (I,3686); SUGARFOOT (I,4277)

TATOR, JOEL
THE TOMORROW SHOW (II,2635)

TAYLER, GEORGE
THE SWEENEY (II,2513)

TAYLOR, ART
CHER—A CELEBRATION AT CAESAR'S PALACE (II,498)

TAYLOR, BRUCE
T.L.C. (II,2525)

TAYLOR, DON
OCTAVIUS AND ME (I,3327); PURSUE AND DESTROY (I,3699)

TAYLOR, GAY
WINGO (I,4878)

TAYLOR, GEORGE
SPECIAL BRANCH (II,2414)

TAYLOR, JACK
THE PAUL DIXON SHOW (I,3487); THE PAUL DIXON SHOW (II,1974)

TAYLOR, JERI
BLUE THUNDER (II,278); QUINCY, M. E. (II,2102)

TAYLOR, PHILIP JOHN
LANDON, LANDON & LANDON (II,1426)

TAYLOR, RENEE
BEDROOMS (II,198); CALUCCI'S DEPARTMENT (I,805); LOVERS AND OTHER STRANGERS (II,1550)

TAYLOR, TALUS
BARBAPAPA (II,142)

TAYMOUR, LAWRENCE

TIBBLES, GEORGE
THE FESS PARKER SHOW (II,850); HELLO, LARRY (II,1128); MY THREE SONS (I,3205)

TIGHE, JANET
DEAN MARTIN IN LONDON (II,641)

TILLSTROM, BURR
KUKLA, FRAN, AND OLLIE (I,2594)

TILSON, ANNETTE
BARBAPAPA (II,142)

TIMMINS, REUBEN
COURAGEOUS CAT (I,1079)

TINKER, GRANT
CONSTANTINOPLE (II,570)

TINKER, MARK
OPERATING ROOM (II,1915); ST. ELSEWHERE (II,2432); THE WHITE SHADOW (II,2788)

TISCH, STEVE
CALL TO GLORY (II,413)

TISCHLER, BOB
SATURDAY NIGHT LIVE (II,2261)

TIVERS, CYNTHIA
PEOPLE TO PEOPLE (II,2000)

TOBYANSEN, JOHN
THE MEMORY GAME (I,3003)

TODDRE JR, RALPH
THE WORLD OF PEOPLE (II,2839)

TODMAN, BILL
BEAT THE CLOCK (I,377); THE BETTER SEX (II,223); BY POPULAR DEMAND (I,784); CALL MY BLUFF (I,800); CHOOSE UP SIDES (I,945); FAMILY FEUD (II,816); GET THE MESSAGE (I,1782); I'VE GOT A SECRET (I,2284); I'VE GOT A SECRET (II,1220); IT'S NEWS TO ME (I,2272); JEFFERSON DRUM (I,2355); JUDGE FOR YOURSELF (I,2466); MADE IN AMERICA (I,2806); MAKE THE CONNECTION (I,2846); THE MATCH GAME (I,2944); THE MATCH GAME (II,1649); MINDREADERS (II,1702); THE NAME'S THE SAME (I,3220); NOTHING BUT THE TRUTH (I,3313); NOW YOU SEE IT (II,1867); ONE HAPPY FAMILY (I,3375); THE PRICE IS RIGHT (I,3665); THE RICHARD BOONE SHOW (I,3784); SHOWOFFS (II,2345); SPLIT PERSONALITY (I,4161); TATTLETALES (II,2545); THAT'S MY LINE (II,2579); TO TELL THE TRUTH (I,4527); TO TELL THE TRUTH (I,4528); TWO FOR THE MONEY (I,4654); THE WEB (I,4784); WINNER TAKE ALL (I,4881)

TOKAR, NORMAN
ADVENTURES OF A MODEL (I,50); BIG DADDY (I,425); THE CLAUDETTE COLBERT SHOW (I,971); HIS MODEL WIFE (I,2064); MY FAVORITE HUSBAND (I,3188); MY SISTER HANK (I,3201); THE TAB HUNTER SHOW (I,4328)

TOKATYAN, DIAN
ALLISON SIDNEY HARRISON (II,54)

TOKATYAN, LEON
ALLISON SIDNEY HARRISON (II,54)

TOKOFSKY, JERRY
THE SECOND TIME AROUND (II,2289)

TOKOYAMA, MITSUTERU
THE EIGHTH MAN (I,1421); PRINCE PLANET (I,3669)

TOLKIN, MEL
JOE'S WORLD (II,1305); NUTS AND BOLTS (II,1871); SANFORD (II,2255)

TOMBLIN, DAVID
THE PRISONER (I,3670)

TOMLILSON, TOMMY
YOU ASKED FOR IT (I,4931)

TOMLIN, LILY
LILY—SOLD OUT (II,1480)

TOMPKINS, GEORGE
WE DARE YOU! (II,2751)

TORME, MEL
IT WAS A VERY GOOD YEAR (I,2253); THE SINGERS (I,4060)

TOROKVEI, PETER
WKRP IN CINCINNATI (II,2814)

TORS, IVAN
THE AQUANAUTS (I,250); DAKTARI (I,1127); DANNY AND THE MERMAID (II,618); FLIPPER (I,1604); GENTLE BEN (I,1761); JAMBO (I,2330); MALIBU RUN (I,2848); MAN AND THE CHALLENGE (I,2855); PRIMUS (I,3668); RIPCORD (I,3793); SCIENCE FICTION THEATER (I,3938); SEA HUNT (I,3948); TIGER! TIGER! (I,4499)

TOUB JR, WARNER
MR. ADAMS AND EVE (I,3121)

TOUCHSTONE, PEGGY
FACE THE MUSIC (II,803); THE NEW $100,000 NAME THAT TUNE (II,1810); YOU ASKED FOR IT (II,2855)

TOURTELLOT, ARTHUR
MARCH OF TIME THROUGH THE YEARS (I,2903)

TOWERS, HARRY ALAN
DIAL 999 (I,1257); THE NEW ADVENTURES OF MARTIN KANE (I,3251)

TRACH, EDWARD
SEARCH FOR TOMORROW (II,2284)

TRAPNELL, CHARLES
THE ALASKANS (I,106); LAWMAN (I,2642)

TREBEK, ALEX
JEOPARDY (II,1280)

TRENDLE, GEORGE W
THE LONE RANGER (I,2740)

TRENT, JOHN
QUENTIN DERGENS, M.P. (I,3707)

TRICKER, GEORGE
ME AND MAXX (II,1672); OFF THE WALL (II,1879); THE ROLLERGIRLS (II,2203); SHIPSHAPE (II,2329)

TRIKILIS, MICHAEL
PLAYBOY'S PLAYMATE PARTY (II,2055)

TRODD, KENITH
PENNIES FROM HEAVEN (II,1994)

TROWBRIDGE, WILLIAM
RONA LOOKS AT JAMES, MICHAEL, ELLIOTT, AND BURT (II,2205); RONA LOOKS AT KATE, PENNY, TONI, AND CINDY (II,2206); RONA LOOKS AT RAQUEL, LIZA, CHER AND ANN-MARGRET (II,2207)

TRUEBLOOD, GUERDON
BRAVO TWO (II,368)

TRUMBULL, DOUG
THE STARLOST (I,4203)

TSUNEDE, SACHIKO
MIGHTY ORBOTS (II,1697)

TUCHNER, MICHAEL
PEEK-A-BOO: THE ONE AND ONLY PHYLLIS DIXEY (II,1989)

TUCK, CECIL
THE GLEN CAMPBELL GOODTIME HOUR (I,1820)

TUCKER, LARRY
ALONE AT LAST (II,60); JENNIFER SLEPT HERE (II,1278); MR. MERLIN (II,1751); PALS (II,1947); TEACHERS ONLY (II,2548)

TUGEND, HARRY
DO NOT DISTURB (I,1300); THE GENERAL ELECTRIC THEATER (I,1753); I WAS A BLOODHOUND (I,2180); THE RAY MILLAND SHOW (I,3734); RUSSIAN ROULETTE (I,3875)

TULCHIN, HAL
JUST LIKE A WOMAN (I,2496)

TURBOW, SANDRA
MR. MOON'S MAGIC CIRCUS (II,1753)

TURMAN, LAWRENCE
GET CHRISTIE LOVE! (II,981); GET CHRISTIE LOVE! (II,982)

TURNER, ANN
AMERICA (I,165)

TURNER, ARNOLD
HARRIS AND COMPANY (II,1096)

TURNER, JOHN ANTHONY
DOCTOR WHO—THE FIVE DOCTORS (II,690)

TURNER, LLOYD
GOOD TIMES (II,1038)

TURNER, SKIP
WONDER GIRL (I,4891)

TURPIN, GEORGE
THE DORIS DAY SHOW (I,1355)

TURTELTAUB, SAUL
13 QUEENS BOULEVARD (II,2591); THE BEAUTIFUL PHYLLIS DILLER SHOW (I,381); CARTER COUNTRY (II,449); CONDO (II,565); DOUBLE TROUBLE (II,723); E/R (II,748); GO! (I,1830); GRADY (II,1047); THE NEW DICK VAN DYKE SHOW (I,3262); THE NEW DICK VAN DYKE SHOW (I,3263); ONE IN A MILLION (II,1906); ONE OF THE BOYS (II,1911); THE PERRY COMO WINTER SHOW (I,3543); PERRY COMO'S WINTER SHOW (I,3547); SANFORD AND SON (II,2256); SANFORD ARMS (II,2257); SUPER COMEDY BOWL 2 (I,4289); THAT GIRL (II,2570); THIS WEEK IN NEMTIN (I,4465); A TOUCH OF GRACE (I,4585); WHAT'S HAPPENING!! (II,2769)

TUSH, PAUL
HAL LINDEN'S BIG APPLE (II,1072)

TUTTLE, MARK
THE BEVERLY HILLBILLIES (I,417); BLACK BART (II,260)

TUZEE, JOHN
ED MCMAHON AND HIS FRIENDS. . . DISCOVER WET AT CYPRESS GARDENS (I,1400)

TWAIN, NORMAN
IT'S A BIRD, IT'S A PLANE, IT'S SUPERMAN (II,1253); VANITIES (II,2720)

UFFNER, BETH
THE PETER MARSHALL VARIETY SHOW (II,2029)

UGER, ALAN
THE FIGHTING NIGHTINGALES (II,855); KLEIN TIME (II,1401); STEPHANIE (II,2453); THE STEVE LANDESBERG TELEVISION SHOW (II,2458)

ULLMAN, FURTH
MANHATTAN TOWER (I,2889)

ULMAN, RICHARD H
COLONEL BLEEP (I,1002)

UNGER, BABE
LAWBREAKER (I,2640)

UNGER, GUSTAF
NUMBER 13 DEMON STREET (I,3316)

UNGER, MAURICE
DOCTOR CHRISTIAN (I,1308); RIPCORD (I,3793); TARZAN (I,4361); THE WEST POINT STORY (I,4796)

URBANO, CARL
THE FLINTSTONE FUNNIES (II,883)

URBISCI, ROCCO
ALAN KING'S SECOND ANNUAL FINAL WARNING (II,30); ALL NIGHT RADIO (II,40); BILLY CRYSTAL: A COMIC'S LINE (II,249); DELLA (I,1222); FROM CLEVELAND (II,931); THE JOE PISCOPO SPECIAL (II,1304); THE KENNY ROGERS SPECIAL (II,1382); LILY FOR PRESIDENT (II,1478); LILY—SOLD OUT (II,1480); THE PETER MARSHALL VARIETY SHOW (II,2029); THE RICHARD PRYOR SHOW (II,2163); A SPECIAL KENNY ROGERS (II,2417)

URBONT, JACQUES
JOHN DENVER: MUSIC AND THE MOUNTAINS (II,1318)

USDEN, PAMELA
TV'S FUNNIEST GAME SHOW MOMENTS (II,2676)

UTT, KENNETH
BAKER'S DOZEN (II,136); THE DEFENDERS (I,1220); LIZA WITH A Z (I,2732)

VAIL, LESTER
THE ALDRICH FAMILY (I,111)

VAJDA, ZELI
THE ROCK 'N ROLL SHOW (II,2188)

VALENTA, LEONARD
THE MASK (I,2938); WINDOWS (I,4876)

VALENTE, RENEE
FATHER KNOWS BEST: HOME FOR CHRISTMAS (II,839); JACQUELINE SUSANN'S VALLEY OF THE DOLLS, 1981 (II,1268); MASQUERADE (II,1644)

VAN AMBURG, TOM
EYE ON HOLLYWOOD (II,798)

VAN ATTA, DON
13 THIRTEENTH AVENUE (II,2592); BOSOM BUDDIES (II,355); BROTHERS (II,380); A COUNTRY HAPPENING (I,1065); HARPER VALLEY, U.S.A. (I,1954); THE HARVEY KORMAN SHOW (II,1103); THE HARVEY KORMAN SHOW (II,1104); HOME COOKIN' (II,1170); HUSBANDS AND WIVES (II,1208); HUSBANDS, WIVES AND LOVERS (II,1209); MADAME'S PLACE (II,1587); THE MCLEAN STEVENSON SHOW (II,1665); ME AND MAXX (II,1672); MISS WINSLOW AND SON (II,1705); THE MONTEFUSCOS (II,1727); MR. SUCCESS (II,1756); A NEW KIND OF FAMILY (II,1819); THE RETURN OF THE SMOTHERS BROTHERS (I,3777)

VAN HEUSEN, JIMMY
THE FRANK SINATRA SHOW (I,1666); THE FRANK SINATRA TIMEX SHOW (I,1668); THE FRANK SINATRA TIMEX SHOW (I,1667)

VAN KEUREN, SIDNEY
ALIAS MIKE HERCULES (I,117)

VAN NOYE, GEORGE
COMPLETELY OFF THE WALL (II,560); THINGS WE DID LAST SUMMER (II,2589)

VAN SCOYK, ROBERT
FLYING HIGH (II,894); RAFFERTY (II,2105); THREE EYES (II,2605); YOUNG MAVERICK (II,2867)

VANCE, DENNIS
JULIE AND DICK IN COVENT GARDEN (II,1352)

VANCE, LEIGH
BRONK (II,379); THE PHOENIX (II,2036); SWITCH (II,2519)

VANDA, CHARLES
ACTION IN THE AFTERNOON (I,25); THE BIG TOP (I,435); CONTEST CARNIVAL (I,1040); THE M AND M CANDY CARNIVAL (I,2797)

VANDEROOT, PHILIP
THE PEOPLE'S COURT (II,2001)

VANOFF, NICK
THE BIG SHOW (II,243); BING CROSBY AND CAROL BURNETT—TOGETHER AGAIN FOR THE FIRST TIME (I,454); THE BING CROSBY CHRISTMAS SHOW (I,458); THE BING CROSBY SHOW (I,465); THE BING CROSBY SHOW (I,466); CAMP GRIZZLY (II,418); THE DON ADAMS SPECIAL: HOORAY FOR HOLLYWOOD (I,1329); THE DON KNOTTS SHOW (I,1334); THE HOLLYWOOD PALACE (I,2088); THE JULIE ANDREWS HOUR (I,2480); JUST POLLY AND ME (I,2497); THE KING FAMILY SHOW (I,2552); THE KRAFT MUSIC HALL (I,2587); MERRY CHRISTMAS. . . WITH LOVE, JULIE (II,1687); THE MILTON BERLE SHOW (I,3049); ON STAGE AMERICA (II,1893); THE PERRY COMO CHRISTMAS SHOW (II,2002); THE PERRY COMO SUNSHINE SHOW (II,2005); PERRY COMO'S SUMMER OF '74 (II,2024); THE SONNY AND CHER SHOW (II,2401); SWING OUT, SWEET LAND (II,2515)

VASSER, DAVID
YOU ASKED FOR IT (II,2855)

VAUGHN, DONNA
ALL MY CHILDREN (II,39)

VEILLER, ANTHONY
TUGBOAT ANNIE (I,4622)

VEITH, SANDY
LOVE, SIDNEY (II,1547); NEVER AGAIN (II,1808)

VEJAR, MICHAEL
THE ASPHALT COWBOY (II,111)

VELIE, LESTER
TARGET: THE CORRUPTERS (I,4360)

VENZA, JAC
FIFTH OF JULY (II,851); MUSIC '55 (I,3168)

VERONA, STEPHEN
FLATBUSH/AVENUE J (II,878)

VERTUE, BERYL
ALL-STAR ANYTHING GOES (II,50); ALMOST ANYTHING GOES (II,56); BEACON HILL (II,193); JUNIOR ALMOST ANYTHING GOES (II,1357); PAROLE (II,1960); THE PRIME OF MISS JEAN BRODIE (II,2082)

VICKERS, DEBORAH
TOM SNYDER'S CELEBRITY SPOTLIGHT (II,2632)

VICTOR, DAVID
THE 13TH DAY: THE STORY OF ESTHER (II,2593); THE CHADWICK FAMILY (II,478); DOCTOR KILDARE (I,1315); DUNDEE AND THE CULHANE (I,1383); GRIFF (I,1888); JERICHO (I,2358); KINGSTON: CONFIDENTIAL (II,1398); KINGSTON: THE POWER PLAY (II,1399); LITTLE WOMEN (II,1497); LITTLE WOMEN (II,1498); LUCAS TANNER (II,1555); LUCAS TANNER (II,1556); THE MAN AND THE CITY (I,2856); MARCUS WELBY, M.D. (I,2905); MARCUS WELBY, M.D. (II,1627); MCNAUGHTON'S DAUGHTER (II,1670); OWEN MARSHALL: COUNSELOR AT LAW (I,3435); OWEN MARSHALL: COUNSELOR AT LAW (I,3436); RYAN'S FOUR (II,2233); STRANDED (II,2475); WOMEN IN WHITE (II,2823)

VINCENT, CHICK
FEDERAL AGENT (I,1557)

VINCENT, E DUKE
ALOHA PARADISE (II,58); ARNIE (I,261); AT EASE (II,115); B.A.D. CATS (II,123); CASINO (II,452); DYNASTY (II,746); EDDIE AND HERBERT (II,757); FINDER OF LOST LOVES (II,857); THE FRENCH ATLANTIC AFFAIR (II,924); FRIENDS (II,927); GLITTER (II,1009); GOMER PYLE, U.S.M.C. (I,1843); HOTEL (II,1192); THE JIM NABORS HOUR (I,2381); LADIES IN BLUE (II,1422); THE LITTLE PEOPLE (I,2716); MASSARATI AND THE BRAIN (II,1646); MATT HOUSTON (II,1654); MURDER CAN HURT YOU! (II,1768); THE NEW TEMPERATURES RISING SHOW (I,3273); THE POWER WITHIN (II,2073); THE RETURN OF THE MOD SQUAD (I,3776); SALT AND PEPE (II,2240); THE SAN PEDRO BEACH BUMS (II,2250); THE SAN PEDRO

BUMS (II,2251); SCARED SILLY (II,2269); SHOOTING STARS (II,2341); STRIKE FORCE (II,2480); VEGAS (II,2723); VEGAS (II,2724); VELVET (II,2725); VENICE MEDICAL (II,2726); WAIKIKI (II,2734); THE WILD WOMEN OF CHASTITY GULCH (II,2801)

VINNEDGE, SYD
AMERICA'S TOP TEN (II,68); BOB HOPE SPECIAL: THE BOB HOPE SPECIAL (II,338); THE EDDIE RABBITT SPECIAL (II,759); PORTRAIT OF A LEGEND (II,2070); A SPECIAL EDDIE RABBITT (II,2415); WE'RE DANCIN' (II,2755)

VINSON, GLORIA
SANFORD (II,2255)

VISCH, JOOP
DOCTOR SNUGGLES (II,687)

VITALIE, CARL
THE OREGON TRAIL (II,1923)

VIVINO, FLOYD
THE UNCLE FLOYD SHOW (II,2703)

VOIGHT, JURGEN
WILD WILD WORLD OF ANIMALS (I,4864)

VON SOOSTEN, JOHN
THE DANCE SHOW (II,616)

VON ZERNECK, FRANK
DELTA COUNTY, U.S.A. (II,657); FLATBED ANNIE AND SWEETIEPIE: LADY TRUCKERS (II,876); NIGHT PARTNERS (II,1847); THE TEXAS RANGERS (II,2567)

VOSBURGH, GEORGE
50 GRAND SLAM (II,852); SALE OF THE CENTURY (II,2239)

WADE, HARKER
AUTOMAN (II,120); COVER UP (II,597); THE FALL GUY (II,811); KNIGHT RIDER (II,1402); MANIMAL (II,1622); MASQUERADE (II,1644); ROOSTER (II,2210)

WADE, WARREN
THE R.C.A. THANKSGIVING SHOW (I,3736)

WAGLIN, ED
THE HONEYMOONERS CHRISTMAS SPECIAL (II,1176); THE HONEYMOONERS CHRISTMAS (II,1177); THE HONEYMOONERS SECOND HONEYMOON (II,1178); THE HONEYMOONERS VALENTINE SPECIAL (II,1179)

WAGNER, JANE
LILY (II,1477); LILY FOR PRESIDENT (II,1478); THE LILY TOMLIN SPECIAL (II,1479); LILY—SOLD OUT (II,1480)

WAGNER, RAYMOND
A WALK IN THE NIGHT (I,4750); WON'T IT EVER BE MORNING? (I,4902)

WAGNER, ROBERT
MADAME SIN (I,2805)

WAIGNER, PAUL
BICENTENNIAL MINUTES (II,229); FULL HOUSE (II,937); JESSICA NOVAK (II,1286); JESSIE (II,1287); MARRIAGE IS ALIVE AND WELL (II,1633); THE MEMBER OF THE WEDDING (II,1680); MR. ROBERTS (II,1754); SECRETS OF MIDLAND HEIGHTS (II,2296); STAR OF THE FAMILY (II,2436)

WAITE, RALPH
THE MISSISSIPPI (II,1707)

WALD, BOB
FOR LOVE OR MONEY (I,1622)

WALD, JEFF
THE HELEN REDDY SPECIAL (II,1126)

WALDMAN, TOM
THE BOSTON TERRIER (I,697); CLUB OASIS (I,985); THE SPIKE JONES SHOW (I,4156)

WALDRON, GY
THE DUKES OF HAZZARD (II,742); ENOS (II,779); SIX PACK (II,2370)

WALKER, BILL
THE LOVE REPORT (II,1542); THAT'S MY LINE (II,2579); THICKE OF THE NIGHT (II,2587)

WALKER, ELBERT
QUEEN FOR A DAY (I,3705)

WALKER, FRED
LENA HORNE: THE LADY AND HER MUSIC (II,1460)

WALKER, SUSAN
E.T. & FRIENDS—MAGICAL MOVIE VISITORS (II,749); LEAVE IT TO THE WOMEN (II,1454); THAT'S HOLLYWOOD (II,2577)

WALL, ANTHONY
THE EVERLY BROTHERS REUNION CONCERT (II,791)

WALLACE, ART
ADVENTURES IN PARADISE (I,46)

WALLACE, DON
THE EDGE OF NIGHT (II,760); RETURN TO PEYTON PLACE (I,3779)

WALLACE, EARL W
WILD AND WOOLEY (II,2798)

WALLACE, MARC
THE LOVE REPORT (II,1542)

WALLACE, RICK
BAY CITY BLUES (II,186)

WALLENSTEIN, JOSEPH B
HOTEL (II,1192); KNOTS LANDING (II,1404); SECRETS OF MIDLAND HEIGHTS (II,2296); SPRAGGUE (II,2430)

WALLER, ALBERT
THE ABC AFTERSCHOOL SPECIAL (II,1); FEELING GOOD (I,1559)

WALLERSTEIN, HERB
CALL HER MOM (I,796)

WALLING, ERNEST
THE CLOCK (I,978); LIGHT'S OUT (I,2699); MARGE AND JEFF (I,2908); MARY KAY AND JOHNNY (I,2932); THIS IS CHARLES LAUGHTON (I,4455)

WALLISER, BLAIR
THE SINGING LADY (I,4061)

WALSH, BILL
ADVENTURES IN DAIRYLAND (I,44); ANNETTE (I,222); BORDER COLLIE (I,688); CORKY AND WHITE SHADOW (I,1049); DAVY CROCKETT (I,1181); THE FURTHER ADVENTURES OF SPIN AND MARTY (I,1720); THE HARDY BOYS AND THE MYSTERY OF GHOST FARM (I,1950); THE HARDY BOYS AND THE MYSTERY OF THE APPLEGATE TREASURE (I,1951); THE MICKEY MOUSE CLUB (I,3025); THE NEW ADVENTURES OF SPIN AND MARTY (I,3254); ONE HOUR IN WONDERLAND (I,3376); THE SECRET OF MYSTERY LAKE (I,3967); SPIN AND MARTY (I,4158); THE WALT DISNEY CHRISTMAS SHOW (I,4755); WHAT I WANT TO BE (I,4812)

WALSH, TOM
WE DARE YOU! (II,2751)

WALTON, KIP
GET IT TOGETHER (I,1780); HOT CITY DISCO (II,1185); JUNIOR ALMOST ANYTHING GOES (II,1357); KICKS (II,1385); TOUCH OF GOLD '75 (II,2650)

WALWIN, KENT
RED SKELTON: A ROYAL PERFORMANCE (II,2125)

WALZ, GENE
RUTH LYONS 50 CLUB (I,3876)

WANAMAKER, SAM
LANCER (I,2610)

WANG, GENE
THE TRIALS OF O'BRIEN (I,4605)

WANG, GLEN
PHILIP MARLOWE (I,3584)

WARD, AL C
BETTY CROCKER STAR MATINEE (I,412); MEDICAL CENTER (II,1675); THE MONROES (I,3089)

WARD, BILL
THE LONDON PALLADIUM (I,2739); SUNDAY NIGHT AT THE LONDON PALLADIUM (I,4286)

WARD, BOB
THE NUT HOUSE (I,3320)

WARD, JAY
ADVENTURES OF HOPPITY HOPPER FROM FOGGY BOGG (I,62); THE BULLWINKLE SHOW (I,761); CRUSADER RABBIT (I,1109); THE DUDLEY DO-RIGHT SHOW (I,1378); FRACTURED FLICKERS (I,1653); GEORGE OF THE JUNGLE (I,1773); ROCKY AND HIS FRIENDS (I,3822)

WARD, JONATHAN
SHIRLEY MACLAINE: ILLUSIONS (II,2336); WALTER CRONKITE'S UNIVERSE (II,2739)

WARD, PHYLLIS
YOU ASKED FOR IT (II,2855)

WARE, CLYDE
AIRWOLF (II,27)

WARNER, MORT
HONEY WEST (I,2105)

WARREN, BOB
DANNY THOMAS GOES COUNTRY AND WESTERN (I,1146)

WARREN, CHARLES
GUNSLINGER (I,1908); GUNSMOKE (II,1069); OUR PRIVATE WORLD (I,3417); RAWHIDE (I,3727)

WARREN, ED
HOT CITY DISCO (II,1185)

WARREN, IAN
A WOMAN OF SUBSTANCE (II,2820)

WARREN, LEE
THE PINKY LEE SHOW (I,3601)

WARREN, LEWIS
DUPONT SHOW OF THE MONTH (I,1387)

WARREN, MICHAEL
OUT OF THE BLUE (II,1934); THE PARTRIDGE FAMILY (II,1962); PLEASE STAND BY (II,2057); WHAT'S HAPPENING!! (II,2769)

WARREN, RALPH
THE ALDRICH FAMILY (I,111); THE THORNTON SHOW (I,4468)

WARREN, ROD
ALAN KING'S THIRD ANNUAL FINAL WARNING!! (II,31); I'VE HAD IT UP TO HERE (II,1221); THE LISA HARTMAN SHOW (II,1484)

WARRENRATH, REINALD
DING DONG SCHOOL (I,1285)

WASHBURN, JIM
THIS IS YOUR LIFE (I,4461)

WASHINGTON, SCREECH
PEOPLE ARE FUNNY (II,1996)

WASSERMAN, AL
20TH CENTURY (I,4640); A LOOK AT THE LIGHT SIDE (I,2751)

WATERS, BILL
CIRCUS OF THE STARS (II,530); CIRCUS OF THE STARS (II,531); CIRCUS OF THE STARS (II,532); CIRCUS OF THE STARS (II,533); CIRCUS OF THE STARS (II,534); CIRCUS OF THE STARS (II,535); CIRCUS OF THE STARS (II,536); CIRCUS OF THE STARS (II,537); MEN AT LAW (I,3004); THE STOREFRONT LAWYERS (I,4233)

WATERS, ED
JESSIE (II,1287); THE MISSISSIPPI (II,1707)

WATERS, JOE
FANNY BY GASLIGHT (II,823)

WATERSON, HARRY
BIG CITY BOYS (II,232); CONDO (II,565); MITZI. . . ROARIN' IN THE 20S (II,1711)

WATERSON, HERB
THE MAGIC OF MARK WILSON (II,1596)

WATSON, JACK
BATTLE OF THE BEAT (II,161); THE DON KNOTTS SPECIAL (I,1335); GIRL FRIENDS AND NABORS (I,1807); LOOK WHAT THEY'VE DONE TO MY SONG (II,1517); MONTY HALL'S VARIETY HOUR (II,1728); THE RAINBOW GIRL (II,2108)

WATSON, REG
PRISONER: CELL BLOCK H (II,2085)

WATTS, BILLY
THE DICK CAVETT SHOW (II,669)

WATTS, NIGEL
THE MASTER (II,1648); RETURN OF THE MAN FROM U.N.C.L.E.: THE 15 YEARS LATER AFFAIR (II,2140)

WATTS, TESSA
CULTURE CLUB IN CONCERT (II,610)

WATTS, WILLIAM
TAKE FIVE WITH STILLER AND MEARA (II,2530)

WAUGH, IRVING
OPRYLAND USA—1975 (II,1920)

WAYNE, FRANK
BEAT THE CLOCK (I,378); BEAT THE CLOCK (II,195); THE DON RICKLES SHOW (I,1341); THE JACKIE GLEASON SHOW (I,2323); THE NEW PRICE IS RIGHT (I,3270); NOW YOU SEE IT (II,1867); PASSWORD (I,3468); PASSWORD (II,1965); THE PRICE IS RIGHT (I,3665); THE PRICE IS RIGHT (II,2079)

WAYNE, PAUL
DOC (II,682); LOVE AND LEARN (II,1529)

WAYNE, PHILIP
THE NEW PRICE IS RIGHT (I,3270); THE PRICE IS RIGHT (I,3665); THE PRICE IS RIGHT (II,2079)

WAYNE, RICHARD
WOMEN OF RUSSIA (II,2824)

WAYNE, RONALD
AWAY WE GO (I,317); THE BIG SELL (I,431); THE HONEYMOONERS (I,2111); THE JACKIE GLEASON SPECIAL (I,2324); THE KEEFE BRASSELLE SHOW (I,2514); THE NEW HOWDY DOODY SHOW (II,1818)

WEAVER, RICK
MAGNUM, P.I. (II,1601); TALES OF THE GOLD MONKEY (II,2537)

WEAVER, RON
RITUALS (II,2182)

WEAVER, SYLVESTER 'PAT'
COMEDY NEWS (I,1020); THE GARRY MOORE SHOW (I,1740); MAKE ME LAUGH (I,2839)

WEBB, GORDON A
THE DEADLY GAME (II,633); THE GIRL IN THE EMPTY GRAVE (II,997)

WEBB, JACK
THE 25TH MAN (MS) (II,2678); 77 SUNSET STRIP (I,3988); THE CENTURY TURNS (I,890); CHASE (I,917); CLINIC ON 18TH STREET (II,544); THE D.A. (I,1123); DRAGNET (I,1369); DRAGNET (I,1370); DRAGNET (I,1371); EMERGENCY! (II,775); ESCAPE (I,1461); GENERAL ELECTRIC TRUE (I,1754); THE LOG OF THE BLACK PEARL (II,1506); THE MAN FROM GALVESTON (I,2865); MOBILE ONE (II,1717); MOBILE TWO (II,1718); NOAH'S ARK (I,3302); O'HARA, UNITED STATES TREASURY (I,3346); O'HARA, UNITED STATES TREASURY: OPERATION COBRA (I,3347); PETE KELLY'S BLUES (I,3559); PROJECT UFO (II,2088); THE RANGERS (II,2112); SAM (II,2246); SAM (II,2247)

WEBB, WILLIAM
MR. ADAMS AND EVE (I,3121)

WEBER, FRANK
BILLY JOEL—A TV FIRST (II,250)

WEBER, JOHN
RED GOOSE KID'S SPECTACULAR (I,3751)

WEBER, JULIAN
NATIONAL LAMPOON'S HOT FLASHES (II,1795)

WEBSTER, NICHOLAS
ESCAPE (II,782)

WEBSTER, SKIP
FANTASY ISLAND (II,829); THE ROOKIES (II,2208)

WEEGE, REINHOLD
BARNEY MILLER (II,154); THE EARTHLINGS (II,751); NIGHT COURT (II,1844); PARK PLACE (II,1958); SAINT PETER (II,2238)

WEGHER, BARBARA
IN SEARCH OF. . . (II,1226)

WEINBERGER, ED
THE ASSOCIATES (II,113); BACHELOR AT LAW (I,322); THE BETTY WHITE SHOW (II,224); DOC (II,682); DOC (II,683); DON'T CALL US (II,709); THE MARY TYLER MOORE SHOW (II,1640); MR. SMITH (II,1755); PHYLLIS (II,2038); TAXI (II,2546); WHAT'S UP, AMERICA? (I,4824); WHAT'S UP? (I,4823)

WEINBERGER, MICHAEL
IT'S NOT EASY (II,1256)

WEINER, DON
HOT (II,1183)

WEINFELD, ANDRE
RAQUEL (II,2114)

WEINGART, MARK
THE RAT PATROL (I,3726)

WEINGARTEN, VICTOR
ANYWHERE, U.S.A. (I,240)

WEINROTT, LES
HOLD 'ER NEWT (I,2070)

WEINSTEIN, HANNAH
THE ADVENTURES OF ROBIN HOOD (I,74); THE ADVENTURES OF SIR LANCELOT (I,76); THE BUCCANEERS (I,749)

WEINSTOCK, KEN
BLONDIE (II,272); BUDDY HACKETT—LIVE AND UNCENSORED (II,386); CELEBRITY CHALLENGE OF THE SEXES 1 (II,465); CELEBRITY CHALLENGE OF THE SEXES 2 (II,466); CELEBRITY CHALLENGE OF THE SEXES 3 (II,467); CELEBRITY CHALLENGE OF THE SEXES 4 (II,468); CELEBRITY CHALLENGE OF THE SEXES 5 (II,469)

WEINTHEL, ARTHUR
THE B.B. BEEGLE SHOW (II,124); CIRCUS OF THE 21ST CENTURY (II,529)

WEINTHORN, MICHAEL
FAMILY TIES (II,819)

WEINTRAUB, CHARLES
THE SKY'S THE LIMIT (I,4081)

WEINTRAUB, JERRY
BLUE JEANS (II,275); THE CARPENTERS (II,444); THE CARPENTERS (II,445); THE CARPENTERS AT CHRISTMAS (II,446); THE CARPENTERS. . . SPACE ENCOUNTERS (II,447); DINER (II,679); THE DOROTHY HAMILL SPECIAL (II,719); THE DOROTHY HAMILL WINTER CARNIVAL SPECIAL (II,720); DOROTHY HAMILL'S

CORNER OF THE SKY (II,721); FATHER O FATHER (II,842); THE JIMMY MCNICHOL SPECIAL (II,1292); JOHN DENVER AND FRIEND (II,1311); JOHN DENVER AND THE LADIES (II,1312); JOHN DENVER ROCKY MOUNTAIN CHRISTMAS (II,1314); THE JOHN DENVER SPECIAL (II,1315); THE JOHN DENVER SPECIAL (II,1316); JOHN DENVER—THANK GOD I'M A COUNTRY BOY (II,1317); KING OF THE ROAD (II,1396); THE NEIL DIAMOND SPECIAL (II,1802); NEIL DIAMOND SPECIAL: I'M GLAD YOU'RE HERE WITH ME TONIGHT (II,1803); PAT BOONE AND FAMILY (II,1968); THE PAT BOONE AND FAMILY EASTER SPECIAL (II,1970); POOR RICHARD (II,2065); SINATRA—THE MAIN EVENT (II,2360); THE STARLAND VOCAL BAND (II,2441); SZYSZNYK (II,2523); TWO OF A KIND: GEORGE BURNS AND JOHN DENVER (II,2692); WHEN THE WHISTLE BLOWS (II,2781)

WEIS, DON
FANTASY ISLAND (II,829)

WEISBAR, DAVID
THE LEGEND OF CUSTER (I,2652)

WEISBATH, MICHAEL
PALMERSTOWN U.S.A. (II,1945)

WEISKOPF, BOB
ALL'S FAIR (II,46); BUT MOTHER! (II,403); THE GOOD GUYS (I,1847); LIVING IN PARADISE (II,1503); MAUDE (II,1655); SIDE BY SIDE (II,2346); W*A*L*T*E*R (II,2731)

WEISKOPF, KIM
9 TO 5 (II,1852)

WEISS, ADRIAN
THE CHUCKLE HEADS (I,952); CRAIG KENNEDY, CRIMINOLOGIST (I,1089)

WEISS, ARTHUR
SWISS FAMILY ROBINSON (II,2517)

WEISS, BILL
THE ASTRONAUT SHOW (I,307); DEPUTY DAWG (I,1233); THE HECKLE AND JECKLE SHOW (I,1999); THE HECTOR HEATHCOTE SHOW (I,2000); THE MIGHTY MOUSE PLAYHOUSE (I,3035); TOM TERRIFIC (I,4546)

WEISS, BOB
POLICE SQUAD! (II,2061)

WEISS, CHUCK
THE DOCTORS (II,694)

WEISS, DORI
HIGH SCHOOL, U.S.A. (II,1148); HIGH SCHOOL, U.S.A. (II,1149)

WEISS, PEPPER
PADDINGTON BEAR (II,1941)

WEISS, ROBERT K
WELCOME TO THE FUN ZONE (II,2763)

WEITZ, BARRY
THE CABOT CONNECTION (II,408); CRISIS IN SUN VALLEY (II,604); IN TANDEM (II,1228)

WEITZ, BRUCE
THE DEADLY TRIANGLE (II,634)

WELCH, KEN
HAL LINDEN'S BIG APPLE (II,1072); THE HANNA-BARBERA HAPPINESS HOUR (II,1079); KRAFT SUMMER MUSIC HALL (I,2590); LINDA IN WONDERLAND (II,1481); MAC DAVIS. . . I BELIEVE IN CHRISTMAS (II,1581); PAUL LYNDE AT THE MOVIES (II,1976); THE PAUL LYNDE COMEDY HOUR (II,1979); PAUL LYNDE GOES M-A-A-A-AD (II,1980)

WELCH, MITZI
BONNIE AND THE FRANKLINS (II,349); HAL LINDEN'S BIG APPLE (II,1072); THE HANNA-BARBERA HAPPINESS HOUR (II,1079); LINDA IN WONDERLAND (II,1481); MAC DAVIS. . . I BELIEVE IN CHRISTMAS (II,1581); PAUL LYNDE AT THE MOVIES (II,1976); THE PAUL LYNDE COMEDY HOUR (II,1979); PAUL LYNDE GOES M-A-A-A-AD (II,1980)

WELCH, ROBERT
OPERATION ENTERTAINMENT (I,3400); THREE WISHES (I,4488)

WELK, LARRY
THE LAWRENCE WELK CHRISTMAS SPECIAL (II,1450)

WELKER, MARY ANN
THE WORLD OF PEOPLE (II,2839); YOU ASKED FOR IT (II,2855)

WELLES, ORSON
FOUNTAIN OF YOUTH (I,1647)

WELLS, BOB
JULIE—MY FAVORITE THINGS (II,1355)

WELLS, DONNA COX
WOMEN OF RUSSIA (II,2824)

WELLS, HALLSTEAD
THE MASK (I,2938)

WELLS, HERBERT
THE VICTOR BORGE SPECIAL (I,4722)

WELLS, RICHARD J
COSMOS (II,579)

WELLS, ROBERT
THE GENE KELLY SHOW (I,1751); THE JANE POWELL SHOW (I,2344); THE MAN IN THE MOON (I,2871); SHIRLEY MACLAINE: IF THEY COULD SEE ME NOW (II,2335)

WELTMAN, PHILIP
ACE CRAWFORD, PRIVATE EYE (II,6)

WENDELL, JOHN
FOR RICHER, FOR POORER (II,906); LOVERS AND FRIENDS (II,1549)

WENDKOS, PAUL
79 PARK AVENUE (II,2310)

WENIG, PATRICIA
CAPITOL (II,426); DAYS OF OUR LIVES (II,629); THE YOUNG AND THE RESTLESS (II,2862)

WENLAND, BURT
STUMP THE STARS (I,4274)

WERNER, MARTIN
SEEING THINGS (II,2297)

WERNER, MORT
THE DOODLES WEAVER SHOW (I,1351)

WERNER, STEVE
RUBIK, THE AMAZING CUBE (II,2226)

WERNER, TOM
CALLAHAN (II,414); THE COSBY SHOW (II,578); I DO, I DON'T (II,1212); OH MADELINE (II,1880)

WERNICK, SANDY
JUMP (II,1356)

WESSON, DICK
THE JOHN FORSYTHE SHOW (I,2410); MY SISTER EILEEN (I,3200); PETTICOAT JUNCTION (I,3571); TAMMY (I,4357)

WEST, BERNIE
ALL IN THE FAMILY (II,38); THE DUMPLINGS (II,743); GOOD TIMES (II,1038); THE JEFFERSONS (II,1276); THE ROPERS (II,2214); THREE'S A CROWD (II,2613); THREE'S COMPANY (II,2614)

WEST, BROOKS
THE EVE ARDEN SHOW (I,1470)

WEST, HOWARD
BUCKSHOT (II,385); LEWIS AND CLARK (II,1465); YOUNG HEARTS (II,2865)

WEST, NORRIS
TELEVISION SYMPHONY (I,4384)

WEST, PAUL
PLEASE DON'T EAT THE DAISIES (I,3628)

WESTCOTT, TED
IT'S A HIT (I,2257)

WEYMAN, RONALD
QUENTIN DERGENS, M.P. (I,3707)

WHARMBY, TONY
THE SEVEN DIALS MYSTERY (II,2308); WHY DIDN'T THEY ASK EVANS? (II,2795)

WHEDON, TOM
BENSON (II,208); MAGGIE (II,1589)

WHITE, ANDY
BAT MASTERSON (I,363); COWBOY IN AFRICA (I,1085); GENTLE BEN (I,1761); IN SEARCH OF. . . (II,1226); THE LONER (I,2745); PRIMUS (I,3668); REDIGO (I,3758); THE RUNAWAYS (II,2231); TIGER! TIGER! (I,4499); TOMBSTONE TERRITORY (I,4550); THE WALTONS (II,2740); THE WILDS OF TEN THOUSAND ISLANDS (II,2803); YOUNG MAVERICK (II,2867)

WHITE, EDWARD J
FRONTIER DOCTOR (I,1698)

WHITE, LARRY
ACE (II,5); THE BLUE AND THE GRAY (II,273); THE FEATHER AND FATHER GANG (II,845); THE FEATHER AND FATHER GANG: NEVER CON A KILLER (II,846); FILTHY RICH (II,856); GOLIATH AWAITS (II,1025); GUESS WHAT? (I,1895); INTERNATIONAL SHOWTIME (I,2225); ON YOUR WAY (I,3366); THE PINKY LEE SHOW (I,3601); RISKO (II,2180); SWING INTO SPRING (I,4317); TIMEX ALL-STAR JAZZ SHOW I (I,4515); TIMEX ALL-STAR JAZZ SHOW II (I,4516); TIMEX ALL-STAR JAZZ SHOW III (I,4517); TIMEX ALL-STAR JAZZ SHOW IV (I,4518); THE VICTOR BORGE SHOW (I,4719); THE VICTOR BORGE

SHOW (I,4720)

WHITE, LEONARD
THE AVENGERS (II,121); CHARLIE WILD, PRIVATE DETECTIVE (I,914)

WHITE, MAURICE
EARTH, WIND & FIRE IN CONCERT (II,750)

WHITE, SAM
MY FRIEND FLICKA (I,3190); THE OUTER LIMITS (I,3426); PERRY MASON (II,2025)

WHITEMAN, PAUL
PAUL WHITEMAN'S TV TEEN CLUB (I,3491)

WHITFIELD, PAM
KIDS 2 KIDS (II,1389); THE LOVE REPORT (II,1542)

WHITLEY, PATRICK
CASTLE ROCK (II,456); SECOND CITY TELEVISION (II,2287)

WHITMAN, DAN
THE MAGIC LAND OF ALLAKAZAM (I,2817)

WHITMAN, JOHN
DOMESTIC LIFE (II,703); SEMI-TOUGH (II,2300)

WHITMORE, DEAN
BRIEF ENCOUNTER (I,723); JOHNNY MANN'S STAND UP AND CHEER (I,2430); MICKIE FINN'S (I,3029); SHINDIG (I,4013)

WHITMORE, PETER
DAVE ALLEN AT LARGE (II,621)

WHITNEY, DOUG
THE RED SKELTON REVUE (I,3754)

WHITTLE, JOHN
GETTING THERE (II,985)

WHITZ, BARRY
MOVIN' ON (II,1743)

WHORF, RICHARD
THE ANN SOTHERN SHOW (I,220); MY FRIEND IRMA (I,3191); PANDORA AND FRIEND (I,3445); THE TAMMY GRIMES SHOW (I,4358)

WICK, CHARLES
FRANCES LANGFORD PRESENTS (I,1656); THE FRANCES LANGFORD SHOW (I,1657)

WICKES, DAVID
PHILIP MARLOWE, PRIVATE EYE (II,2033)

WIDLEY, DOUG
GODZILLA (II,1014)

WILCOX, DAN
THE DUCK FACTORY (II,738); M*A*S*H (II,1569)

WILDER, JOHN
THE BASTARD/KENT FAMILY CHRONICLES (II,159); CENTENNIAL (II,477); THE CITY (II,541); THE DEVLIN CONNECTION (II,664); MOST WANTED (II,1738); MOST WANTED (II,1739); PARIS 7000 (I,3459); THE STREETS OF SAN FRANCISCO (II,2478); TALES OF THE UNEXPECTED (II,2539); WINNER TAKE ALL (II,2808); THE YELLOW ROSE (II,2847)

WILDER, MYLES
THE DUKES OF HAZZARD (II,742)

WILES, GORDON
THE NEW ZOO REVUE (I,3277)

WILK, RUTH
THAT'S OUR SHERMAN (I,4429)

WILKS, WENDELL
THE BAXTERS (II,184); CELEBRITY CONCERTS (II,471); THE PALACE (II,1943)

WILLIAMS, ANDY
THE ANDY WILLIAMS CHRISTMAS SHOW (I,198); THE ANDY WILLIAMS CHRISTMAS SHOW (I,199); THE ANDY WILLIAMS CHRISTMAS SHOW (II,77); ANDY WILLIAMS KALEIDOSCOPE COMPANY (I,201); ANDY WILLIAMS MAGIC LANTERN SHOW COMPANY (I,202); THE ANDY WILLIAMS SHOW (I,210); THE ANDY WILLIAMS SPECIAL (I,213); ANDY WILLIAMS' EARLY NEW ENGLAND CHRISTMAS (II,79); ANDY'S LOVE CONCERT (I,215)

WILLIAMS, ANSON
LITTLE SHOTS (II,1495); SKYWARD CHRISTMAS (II,2378)

WILLIAMS, BILL
THE BOOTS RANDOLPH SHOW (I,687); THE NASHVILLE SOUND OF BOOTS RANDOLPH (I,3229); TENNESSEE ERNIE FORD'S WHITE CHRISTMAS (I,4402)

WILLIAMS, DEBBIE
PORTRAIT OF A LEGEND (II,2070)

WILLIAMS, DON
LARAMIE (I,2614)

WILLIAMS, GRAHAM
DOCTOR WHO (II,689)

WILLIAMS, JACK
THE SEVEN DIALS MYSTERY (II,2308); WHY DIDN'T THEY ASK EVANS? (II,2795)

WILLIAMS, JAN
BEYOND WITCH MOUNTAIN (II,228); SMALL AND FRYE (II,2380)

WILLIAMS, LYNETTE
JEOPARDY (II,1279)

WILLIAMS, MIKE
LARAMIE (I,2614)

WILLIAMS, NORMAN
THE ERROL FLYNN THEATER (I,1458)

WILLIAMS, RENEE
SHORT STORY THEATER (I,4024)

WILLIAMS, ROBERT
BAREFOOT IN THE PARK (I,355)

WILLIAMS, TOM
ADAM-12 (I,31)

WILLIAMSON, WARREN
THE GIRL IN MY LIFE (II,996)

WILLIGER, DAVID
WHEEL OF FORTUNE (II,2775)

WILLINGHAM, WILLIAM
WHISPERING SMITH (I,4842)

WILLIS, JACK
THE GREAT AMERICAN DREAM MACHINE (I,1871)

WILLMORE, JOE
AS THE WORLD TURNS (II,110); THE GUIDING LIGHT (II,1064)

WILLOCK, DAVE
DAVE AND CHARLEY (I,1171)

WILSON, ANTHONY
COMPUTERCIDE (II,561); FOLLOW THE SUN (I,1617); FUTURE COP (II,945); FUTURE COP (II,946); THE IMMORTAL (I,2197); PAPER MOON (II,1953)

WILSON, CAREY
LIGHT'S DIAMOND JUBILEE (I,2698)

WILSON, DANIEL
THE ABC AFTERSCHOOL SPECIAL (II,1); THE BARBARA WALTERS SPECIAL (II,145); HENRY WINKLER MEETS WILLIAM SHAKESPEARE (II,1129); HERE'S BOOMER (II,1134); HONEST AL'S A-OK USED CAR AND TRAILER RENTAL TIGERS (II,1174); LITTLE VIC (II,1496); MR. DICKENS OF LONDON (I,3135); THE SECRET OF CHARLES DICKENS (II,2293)

WILSON, DENNIE MAIN
ROGER DOESN'T LIVE HERE ANYMORE (II,2201)

WILSON, DONALD
THE FORSYTE SAGA (I,1645)

WILSON, DOUG
BATTLE OF THE NETWORK STARS (II,163); BATTLE OF THE NETWORK STARS (II,164); BATTLE OF THE NETWORK STARS (II,165); BATTLE OF THE NETWORK STARS (II,166); BATTLE OF THE NETWORK STARS (II,167); BATTLE OF THE NETWORK STARS (II,168); BATTLE OF THE NETWORK STARS (II,169); BATTLE OF THE NETWORK STARS (II,170); BATTLE OF THE NETWORK STARS (II,171); BATTLE OF THE NETWORK STARS (II,172); BATTLE OF THE NETWORK STARS (II,173); BATTLE OF THE NETWORK STARS (II,174); BATTLE OF THE NETWORK STARS (II,175); BATTLE OF THE NETWORK STARS (II,176); BATTLE OF THE NETWORK STARS (II,177)

WILSON, FLIP
THE HELEN REDDY SHOW (I,2008)

WILSON, HUGH
THE TONY RANDALL SHOW (II,2640); WKRP IN CINCINNATI (II,2814)

WILSON, IRV
THAT WAS THE YEAR THAT WAS (II,2575)

WILSON, J DONALD
SABER OF LONDON (I,3879)

WILSON, JOHN C
BUICK CIRCUS HOUR (I,759)

WILSON, MARK
THE MAGIC LAND OF ALLAKAZAM (I,2817)

WILSON, MEREDITH
TEXACO STAR PARADE I (I,4409); TEXACO STAR PARADE II (I,4410)

WILSON, MICHAEL
I DO, I DON'T (II,1212)

WILSON, RONALD
THE PALLISERS (II,1944)

WILSON, TONY
THE COPS AND ROBIN (II,573)

WINCHELL, PAUL

JERRY MAHONEY'S CLUB HOUSE (I,2371); TOYLAND EXPRESS (I,4587)

WINDSOR, ROY
HOTEL COSMOPOLITAN (I,2128); LOVE OF LIFE (I,2774); THE PUBLIC LIFE OF CLIFF NORTON (I,3688)

WINDUST, BRETAIGNE
CLIMAX! (I,976); MY THREE ANGELS (I,3204)

WINER, ELIHU
STORY THEATER (I,4240)

WINKLER, HENRY
THE ABC AFTERSCHOOL SPECIAL (II,1); GABE AND WALKER (II,950); RYAN'S FOUR (II,2233)

WINSTON, SUSAN
AMERICA ALIVE! (II,67)

WINTER, GARY
GALACTICA 1980 (II,953); RIPTIDE (II,2178)

WINTERS, DAVID
ALL-STAR SWING FESTIVAL (II,53); THE ANN-MARGRET SHOW (I,218); ANN-MARGRET: FROM HOLLYWOOD WITH LOVE (I,217); THE BARBARA MCNAIR SHOW (I,348); THE BIG BAND AND ALL THAT JAZZ (I,422); FROM HOLLYWOOD WITH LOVE: THE ANN-MARGRET SPECIAL (I,1687); HALF THE GEORGE KIRBY COMEDY HOUR (I,1920); THE LISA HARTMAN SHOW (II,1484); RAQUEL (I,3725); ROLLIN' ON THE RIVER (I,3834); SAGA OF SONORA (I,3884); SPECIAL LONDON BRIDGE SPECIAL (I,4150); STORY THEATER (I,4241)

WINTERS, KAREN
THAT'S INCREDIBLE! (II,2578)

WINTERS, RENA
THE ANITA BRYANT SPECTACULAR (II,82)

WINTHER, JORN
ALL MY CHILDREN (II,39); THE BARBARA MCNAIR SHOW (I,348); BREAKAWAY (II,370); RITUALS (II,2182)

WINTLE, JULIAN
THE AVENGERS (II,121); MADAME SIN (I,2805); MISTER JERICO (I,3070)

WIPPER, JACK
THE MAGIC LAND OF ALLAKAZAM (I,2817)

WISBAR, FRANK
CURTAIN CALL THEATER (I,1115); FIRESIDE THEATER (I,1580); JANE WYMAN PRESENTS THE FIRESIDE THEATER (I,2345); MAN OF THE COMSTOCK (I,2873)

WISE, MIKE
THE LAST HURRAH (I,2626); RETURN ENGAGEMENT (I,3775)

WISEMAN, BUD
ONE DAY AT A TIME (II,1900)

WISENFELD, SUZANNE
MISTRAL'S DAUGHTER (II,1708)

WISER, BUD
ONE DAY AT A TIME (II,1900); SEMI-TOUGH (II,2299); WHO'S THE BOSS? (II,2792)

WITT, PAUL JUNGER
BENSON (II,208); BOBBY JO AND THE BIG APPLE GOODTIME BAND (I,674); CONDO (II,565); DAUGHTERS (II,620); FAY (II,844); GETTING TOGETHER (I,1784); THE GUN AND THE PULPIT (II,1067); HERE COME THE BRIDES (I,2024); HIGH RISK (II,1146); I'M A BIG GIRL NOW (II,1218); IT TAKES TWO (II,1252); IT'S A LIVING (II,1254); A KNIGHT IN SHINING ARMOUR (I,2575); THE LETTERS (I,2670); LOVES ME, LOVES ME NOT (II,1551); MAKING A LIVING (II,1604); THE PRACTICE (II,2076); THE ROOKIES (II,2208); SOAP (II,2392); TROUBLE IN HIGH TIMBER COUNTRY (II,2661); THE YEAGERS (II,2843)

WOHL, IRA
OMNI: THE NEW FRONTIER (II,1889)

WOHL, JACK
ALL-STAR SWING FESTIVAL (II,53); AMERICA, YOU'RE ON (II,69); THE BIG BAND AND ALL THAT JAZZ (I,422); BUCKSHOT (II,385); BURT AND THE GIRLS (I,770); BURT REYNOLDS' LATE SHOW (I,777); G.I.'S (II,949); THE GEORGE BURNS SPECIAL (II,970); HALF THE GEORGE KIRBY COMEDY HOUR (I,1920); THE MANHATTAN TRANSFER (II,1619); PAT BOONE AND FAMILY (II,1968); THE PAT BOONE AND FAMILY CHRISTMAS SPECIAL (II,1969); THE PAT BOONE AND FAMILY EASTER SPECIAL (II,1970); PINOCCHIO (II,2044); SANDY IN DISNEYLAND (II,2254); SHA NA NA (II,2311); VIVA VALDEZ (II,2728)

WOLAS, EVA
CLIMAX! (I,976); THE JANE WYMAN THEATER (I,2346); MUSIC U.S.A. (I,3179)

WOLF, ED
BREAK THE BANK (I,712); THE O'NEILLS (I,3392); PICK AND PAT (I,3590)

WOLF, FRED
DOROTHY IN THE LAND OF OZ (II,722)

WOLF, GEORGE
E.S.P. (I,1462); E.S.P. (I,1463)

WOLF, HERB
KEEP TALKING (I,2517); MASQUERADE PARTY (I,2941); PENNY TO A MILLION (I,3512)

WOLF, JOHN
ORSON WELLES' GREAT MYSTERIES (I,3408)

WOLFBERG, LEE
THE DON ADAMS SPECIAL: HOORAY FOR HOLLYWOOD (I,1329)

WOLFE, DIGBY
THE FABULOUS FORDIES (I,1502); FOL-DE-ROL (I,1613); THE PEAPICKER IN PICCADILLY (I,3500); TURN ON (I,4625)

WOLFE, KEN
BATTLE OF THE NETWORK STARS (II,163); BATTLE OF THE NETWORK STARS (II,164); BATTLE OF THE NETWORK STARS (II,165); BATTLE OF THE NETWORK STARS (II,166); BATTLE OF THE NETWORK STARS (II,167); BATTLE OF THE NETWORK STARS (II,168); BATTLE OF THE NETWORK STARS (II,169); BATTLE OF THE NETWORK STARS (II,170); BATTLE OF THE NETWORK STARS (II,171); BATTLE OF THE NETWORK STARS (II,172); BATTLE OF THE NETWORK STARS (II,173); BATTLE OF THE NETWORK STARS (II,174); BATTLE OF THE NETWORK STARS (II,175); BATTLE OF THE NETWORK STARS (II,176); BATTLE OF THE NETWORK STARS (II,177)

WOLFF, PERRY
AIR POWER (I,92)

WOLFSON, P J
I MARRIED JOAN (I,2174)

WOLLNER, TOM
CODE R (II,551)

WOLPER, DAVID L
BIOGRAPHY (I,476); CASABLANCA (II,450); GET CHRISTIE LOVE! (II,981); HOLLYWOOD AND THE STARS (I,2081); THE LEGEND OF MARILYN MONROE (I,2654); MAKE MINE RED, WHITE, AND BLUE (I,2841); ROOTS (II,2211); ROOTS: THE NEXT GENERATIONS (II,2213); SANDBURG'S LINCOLN (I,3907); SMALL WORLD (II,2381); THE THORN BIRDS (II,2600)

WOLPERT, JAY
HIT MAN (II,1156); THE PRICE IS RIGHT (I,3665); WHEW! (II,2786)

WOLPERT, JOHN
FIT FOR A KING (II,866)

WOLPERT, STUART
ALL TOGETHER NOW (II,45)

WOLTERSTORFF, ROBERT
THE OUTLAWS (II,1936); PEN 'N' INC. (II,1992)

WOMACK, STEVE
COUNTRY GALAXY OF STARS (II,587); MUSIC CITY NEWS TOP COUNTRY HITS OF THE YEAR (II,1772)

WONDERS, RALPH
THE ROY ROGERS AND DALE EVANS SHOW (I,3858)

WOOD, BARRY
54TH STREET REVUE (I,1573); AMERICA PAUSES FOR SPRINGTIME (I,166); AMERICA PAUSES FOR THE MERRY MONTH OF MAY (I,167); BACKSTAGE WITH BARRY WOOD (I,329); THE BELL TELEPHONE HOUR (I,389); THE CHEVROLET GOLDEN ANNIVERSARY SHOW (I,924); HOTPOINT HOLIDAY (I,2131); THE KATE SMITH EVENING HOUR (I,2507); THE KATE SMITH HOUR (I,2508); KOBB'S CORNER (I,2577); MANHATTAN SHOWCASE (I,2888); THE MARCH OF DIMES BENEFIT SHOW (I,2900); THE MOREY AMSTERDAM SHOW (I,3100); THE MOREY AMSTERDAM SHOW (I,3101); PLACES PLEASE (I,3613); THE ROBERT Q. LEWIS CHRISTMAS SHOW (I,3811); THE SONNY KENDIS SHOW (I,4127); STOP ME IF YOU HEARD THIS ONE (I,4231); STRICTLY FOR LAUGHS (I,4261); SUGAR HILL TIMES

(I,4276); THRU THE CRYSTAL BALL (I,4496); UKULELE IKE (I,4664)

WOOD, BOB
THE CHEAP SHOW (II,491); WACKO (II,2733)

WOOD, DEE DEE
THE LIGHT FANTASTIC, OR HOW TO TELL YOUR PAST, PRESENT AND MAYBE YOUR FUTURE THROUGH SOCIAL DANCING (I,2696)

WOOD, DOUGLAS
WORLD WAR I (I,4919)

WOOD, LANA
MICKEY SPILLANE'S MIKE HAMMER: MURDER ME, MURDER YOU (II,1694)

WOOD, PETER
COUNTERATTACK: CRIME IN AMERICA (II,581)

WOOD, ROBERT D
THE PLANT FAMILY (II,2050)

WOOD, WILLIAM
BIG HAWAII (II,238)

WOODFIELD, WILLIAM READ
SAN FRANCISCO INTERNATIONAL AIRPORT (I,3905); SAN FRANCISCO INTERNATIONAL AIRPORT (I,3906); SHAFT (I,3993)

WOODWARD, MARVIN
COURAGEOUS CAT (I,1079)

WOOLER, MIKE
HOLLYWOOD (II,1160)

WOOLERY, GERRY
THE DUCK FACTORY (II,738)

WOOLERY, TED
THE DUCK FACTORY (II,738)

WOOLF, LARRY
THE QUIZ KIDS (I,3712)

WOOLF, SIR JOHN
TALES OF THE UNEXPECTED (II,2540)

WORTH, MARVIN
THE JONES BOYS (I,2446); MY LUCKY PENNY (I,3196); WHERE'S POPPA? (II,2784)

WRATHER, BONITA GRANVILLE
LASSIE (I,2622); LASSIE (I,2623); LASSIE (I,2624)

WRATHER, JACK
BOSS LADY (I,694); THE FOUR JUST MEN (I,1650); INTERPOL CALLING (I,2227); JEFF'S COLLIE (I,2356); LASSIE (I,2622); LASSIE (I,2623); LASSIE (I,2624); TIMMY AND LASSIE (I,4520)

WRIGHT, HERBERT J
THROUGH THE MAGIC PYRAMID (II,2615)

WRIGHT, KAY
THE DUKES (II,741); MONCHHICHIS (II,1722); PAC-MAN (II,1940); PINK PANTHER AND SONS (II,2043); THE SHIRT TALES (II,2337); THE SMURFS (II,2385); TROLLKINS (II,2660)

WRIGHT, ROBERT
6 RMS RIV VU (II,2371); THE JIMMIE RODGERS SHOW (I,2383); THE KELLY MONTEITH SHOW (II,1376); ONCE UPON A MATTRESS (I,3371); STARTIME (I,4212); THE TIM CONWAY SHOW (II,2619); TWIGS (II,2683)

WRIGHT, WILLIAM H
THE BARBARA STANWYCK THEATER (I,350)

WRIGHT, WYNN
BARNEY BLAKE, POLICE REPORTER (I,359)

WUERGLER, MIKE
THE NEW MICKEY MOUSE CLUB (II,1823)

WYMAN, HAP
GILLIGAN'S ISLAND: THE HARLEM GLOBETROTTERS ON GILLIGAN'S ISLAND (II,993)

WYNN, BOB
THE 36 MOST BEAUTIFUL GIRLS IN TEXAS (II,2594); THE ANDY GRIFFITH—DON KNOTTS—JIMNABORS SHOW (I,191); BOB HOPE SPECIAL: BOB HOPE'S ALL-STAR BIRTHDAY PARTY (II,295); BOB HOPE SPECIAL: HAPPY BIRTHDAY, BOB! (II,326); CAROL CHANNING AND 101 MEN (I,849); DANNY THOMAS LOOKS AT YESTERDAY, TODAY AND TOMORROW (I,1148); THE DANNY THOMAS SPECIAL (I,1151); DANNY THOMAS: THE ROAD TO LEBANON (I,1153); DEBBIE REYNOLDS AND THE SOUND OF CHILDREN (I,1213); THE FABULOUS FORDIES (I,1502); GUYS 'N' GEISHAS (I,1914); JIMMY DURANTE MEETS THE LIVELY ARTS (I,2386); LIGHTS, CAMERA, MONTY! (II,1475); LOOK AT US (II,1515); THE NBC FOLLIES (I,3241); THE NBC FOLLIES (I,3242); OPRYLAND U.S.A. (I,3405); OPRYLAND USA—1975 (II,1920); THE PEAPICKER IN PICCADILLY (I,3500); PORTRAIT OF PETULA (I,3654); REAL KIDS (II,2116); REAL PEOPLE (II,2118); SALUTE TO LADY LIBERTY (II,2243); THE SHAPE OF THINGS (II,2315); SPEAK UP AMERICA (II,2412); TENNESSEE ERNIE'S NASHVILLE-MOSCOW EXPRESS (II,2561)

YAGEMAN, BILL
THE GIRL IN MY LIFE (II,996); IT TAKES TWO (I,2251); THE LIAR'S CLUB (II,1466); YOU DON'T SAY (I,4934); YOU DON'T SAY (II,2859)

YANOK, GEORGE
ARCHIE (II,104); THE ARCHIE SITUATION COMEDY MUSICAL VARIETY SHOW (II,106); FLATBUSH (II,877); GO WEST YOUNG GIRL (II,1013); HAPPY DAYS (I,1942); THE KALLIKAKS (II,1361); THE STOCKARD CHANNING SHOW (II,2468); TABITHA (II,2528); TEACHERS ONLY (II,2547); TOO CLOSE FOR COMFORT (II,2642); WELCOME BACK, KOTTER (II,2761)

YARBROUGH, JEAN
THE ABBOTT AND COSTELLO SHOW (I,2); I'M THE LAW (I,2195)

YARNELL, DAVID
THE CANDID CAMERA SPECIAL (II,422); DON KIRSHNER'S ROCK CONCERT (I,1332); LIFE'S MOST EMBARRASSING MOMENTS III (II,1472); LIFE'S MOST EMBARRASSING MOMENTS IV (II,1473); THE POP 'N' ROCKER GAME (II,2066); THAT'S INCREDIBLE! (II,2578); THOSE AMAZING ANIMALS (II,2601); WHAT GAP? (I,4808)

YATES, KEN
KENNY & DOLLY: A CHRISTMAS TO REMEMBER (II,1378); KENNY ROGERS IN CONCERT (II,1381); ON LOCATION WITH RICH LITTLE (II,1891); THE RICH LITTLE SPECIAL (II,2156)

YATES, PETER
BREAKING AWAY (II,371)

YATES, WILLIAM ROBERT
THE ADVENTURES OF POLLYANNA (II,14); BEYOND WITCH MOUNTAIN (II,228); GUN SHY (II,1068); HERBIE, THE LOVE BUG (II,1131); THE MOUSEKETEERS REUNION (II,1742); OPERATION: RUNAWAY (II,1918); THE RUNAWAYS (II,2231); SMALL AND FRYE (II,2380); THE STREETS OF SAN FRANCISCO (II,2478); TALES OF THE APPLE DUMPLING GANG (II,2536); TALES OF THE UNEXPECTED (II,2539); ZORRO AND SON (II,2878)

YENTOB, ALAN
THE EVERLY BROTHERS REUNION CONCERT (II,791)

YERKOVITCH, ANTHONY
HILL STREET BLUES (II,1154); MIAMI VICE (II,1689)

YORKIN, BUD
13 QUEENS BOULEVARD (II,2591); THE ANDY WILLIAMS CHRISTMAS SHOW (I,196); THE ANDY WILLIAMS SHOW (I,208); THE ANDY WILLIAMS SPECIAL (I,211); THE ANDY WILLIAMS SPECIAL (I,212); ANOTHER EVENING WITH FRED ASTAIRE (I,229); CARTER COUNTRY (II,449); THE COLGATE COMEDY HOUR (I,997); THE DANNY KAYE SHOW (I,1144); DUKE ELLINGTON. . . WE LOVE YOU MADLY (I,1380); AN EVENING WITH CAROL CHANNING (I,1473); AN EVENING WITH FRED ASTAIRE (I,1474); THE GEORGE GOBEL SHOW (I,1767); HENRY FONDA AND THE FAMILY (I,2015); I LOVE LIBERTY (II,1215); THE JACK BENNY HOUR (I,2290); THE MANY SIDES OF DON RICKLES (I,2898); OPERATION GREASEPAINT (I,3402); ROBERT YOUNG AND THE FAMILY (I,3813); SANFORD AND SON (II,2256); SANFORD ARMS (II,2257); THE SOLDIERS (I,4112); THE TENNESSEE ERNIE FORD SHOW (I,4398); THE TONY MARTIN SHOW (I,4572); WHAT'S HAPPENING!! (II,2769)

YOSHIDA, TATSUO
SPEED RACER (I,4151)

YOUNG, BILL
THE HOYT AXTON SHOW (II,1200)

YOUNG, COLLIER
HORNBLOWER (I,2121); IRONSIDE (I,2240); MR. ADAMS AND EVE (I,3121); THE NEXT STEP BEYOND (II,1839); ONE STEP BEYOND (I,3388); THE ROGUES (I,3832); THE WILD WILD WEST (I,4863)

YOUNG, JAMES
GENERAL HOSPITAL (II,964)

YOUNG, ROBERT
FATHER KNOWS BEST (II,838)

YOUNG, WILLIAM
THE HOYT AXTON SHOW (II,1200)

YOUNGSTEIN, MAX E
THE DANGEROUS DAYS OF KIOWA JONES (I,1140); THE DIAHANN CARROLL SHOW (II,665)

ZACHARIAS, STEVE
ME AND DUCKY (II,1671); SCALPELS (II,2266); THE WAVERLY WONDERS (II,2746)

ZACHARY, BEULAH
KUKLA, FRAN, AND OLLIE (I,2594); THE KUKLAPOLITAN EASTER SHOW (I,2595)

ZAGURY, BOB
BRIGITTE BARDOT (I,729)

ZAMORA, RUDY
KIDD VIDEO (II,1388)

ZANETOS, DEAN
HAWAIIAN HEAT (II,1111)

ZANETOS, FRANK
GALACTICA 1980 (II,953)

ZARCOFF, MORT
IT TAKES A THIEF (I,2250); THE MISADVENTURES OF SHERIFF LOBO (II,1704)

ZAVADA, ERVIN
THE ABC AFTERSCHOOL SPECIAL (II,1); RYAN'S FOUR (II,2233); STEELTOWN (II,2452)

ZAYLOR, JUDITH
ON STAGE AMERICA (II,1893)

ZEITMAN, JEROME
THE CHEERLEADERS (II,493); I LOVE HER ANYWAY! (II,1214); THE STARLOST (I,4203)

ZENO, NORMAN
THE ALL-STAR REVUE (I,138)

ZIARTO, BRUNO
FACE THE MUSIC (II,803)

ZIFFREN, JOHN
PAPER DOLLS (II,1952)

ZIMBALIST, AL
TARZAN THE APE MAN (I,4363)

ZIMBALIST, DONALD
TARZAN THE APE MAN (I,4363)

ZIMMER, JON
THE CARA WILLIAMS SHOW (I,843)

ZIMMERMAN, MORT
REX REED'S MOVIE GUIDE (II,2149)

ZINBERG, MICHAEL
THE BOB NEWHART SHOW (II,342); THE CHOPPED LIVER BROTHERS (II,514); A GIRL'S LIFE (II,1000); THE JAMES BOYS (II,1272); MOTHER AND ME, M.D. (II,1740); NOT UNTIL TODAY (II,1865); W*A*L*T*E*R (II,2731); THE YELLOW ROSE (II,2847)

ZIPPER, HARRY
LIKE MAGIC (II,1476); PINOCCHIO (II,2044)

ZIRATO, BRUNO
TO TELL THE TRUTH (I,4528); TO TELL THE TRUTH (II,2625)

ZIV, FREDERIC A
YOUR FAVORITE STORY (I,4948)

ZOUSMER, JESSE
THE DUPONT SHOW OF THE WEEK (I,1388); PERSON TO PERSON (I,3550)

ZUKER, DAVID
POLICE SQUAD! (II,2061)

ZUKER, JERRY
POLICE SQUAD! (II,2061)

ZURBURGY, SHAUNA
ON STAGE AMERICA (II,1893)

ZWICK, EDWARD
FAMILY (II,813)

ZWICK, JOEL
BROTHERS (II,380); HOT W.A.C.S. (II,1191); IT'S A LIVING (II,1254); THE NEW ODD COUPLE (II,1825)

Directors

ABBE, DERWIN
HIS HONOR, HOMER BELL (I,2063)

ABBOTT, GEORGE
U.S. ROYAL SHOWCASE (I,4688)

ABBOTT, NORMAN
ALICE (II,33); ANGIE (II,80); THE BAD NEWS BEARS (II,134); BLONDIE (I,487); BOB HOPE SPECIAL: BOB HOPE'S SPRING FLING OF COMEDY AND GLAMOUR (II,315); BOB HOPE SPECIAL: BOB HOPE'S STARS OVER TEXAS (II,318); CALIFORNIA FEVER (II,411); DANNY AND THE MERMAID (II,618); DANNY THOMAS LOOKS AT YESTERDAY, TODAY AND TOMORROW (I,1148); DENNIS THE MENACE (I,1231); THE DON KNOTTS SHOW (I,1334); EVERYTHING YOU ALWAYS WANTED TO KNOW ABOUT JACK BENNY AND WERE AFRAID TO ASK (I,1482); FATHER KNOWS BEST: THE FATHER KNOWS BEST REUNION (II,840); FISH (II,864); GET SMART (II,983); THE GHOST BUSTERS (II,986); ICHABOD AND ME (I,2183); THE JACK BENNY PROGRAM (I,2294); JACK BENNY'S FIRST FAREWELL SHOW (I,2300); JACK BENNY'S NEW LOOK (I,2301); JACK BENNY'S SECOND FAREWELL SHOW (I,2302); LEAVE IT TO BEAVER (I,2648); A LOVE LETTER TO JACK BENNY (II,1540); MCHALE'S NAVY (I,2969); THE MUNSTERS (I,3158); MUSIC U.S.A. (I,3179); NANNY AND THE PROFESSOR (I,3225); NOBODY'S PERFECT (II,1858); OH, THOSE BELLS! (I,3345); OPERATION PETTICOAT (II,1917); ROOM FOR ONE MORE (I,3842); SANFORD AND SON (II,2256); STAR SEARCH (II,2437); STAR SEARCH (II,2438); THAT'S MY MAMA (II,2580); WELCOME BACK, KOTTER (II,2761); WORKING STIFFS (II,2834)

ABBOTT, PHILIP
THE FBI (I,1551)

ABDO, NICK
BROTHERS AND SISTERS (II,382)

ABEL, RUDY
NATIONAL VELVET (I,3231)

ABRAHAM, MAURICE
BIZARRE (II,259)

ABRAHAMS, DERWIN
THE CISCO KID (I,961)

ABRAHAMS, JIM
POLICE SQUAD! (II,2061)

ABRAMS, ROBERT
DOCTORS HOSPITAL (II,691)

ABROMS, EDWARD
ALIAS SMITH AND JONES (I,118); APPLE'S WAY (II,101); CANNON (II,424); CHIPS (II,511); THE CHISHOLMS (II,513); COLUMBO (II,556); DAVID CASSIDY—MAN UNDERCOVER (II,623); DOC ELLIOT (I,1306); DOCTORS HOSPITAL (II,691); DOCTORS' PRIVATE LIVES (II,692); THE EDDIE CAPRA MYSTERIES (II,758); THE FEATHER AND FATHER GANG (II,845); GET CHRISTIE LOVE! (II,981); GRIFF (I,1888); THE HARDY BOYS MYSTERIES (II,1090); HAWAII FIVE-O (II,1110); THE IMPOSTER (II,1225); KOJAK (II,1406); THE MAN FROM ATLANTIS (II,1615); MCMILLAN AND WIFE (II,1667); MRS. COLUMBO (II,1760); NERO WOLFE (II,1806); NIGHT GALLERY (I,3287); POLICE STORY (II,2062); SALVAGE 1 (II,2245); THE SIX-MILLION-DOLLAR MAN (II,2372); SWITCH (II,2519)

ACOMBA, DAVID
THE MAGIC PLANET (II,1598); PERRY COMO'S EASTER BY THE SEA (II,2014)

ADAMS, DON
THE DON ADAMS SCREEN TEST (II,705); THE PARTNERS (I,3462)

ADAMS, LEE V
ROUGHCUTS (II,2218)

ADAMS, M CLAY
VICTORY AT SEA (I,4730)

ADAMS, ROBERT
THE CHAMBER MUSIC SOCIETY OF LOWER BASIN STREET (I,894)

ADAMSON, ED
WANTED: DEAD OR ALIVE (I,4764)

ADDISS, JUSTUS
CROSSROADS (I,1105); THE LORETTA YOUNG THEATER (I,2756); LOST IN SPACE (I,2758); RAWHIDE (I,3727); SAINTS AND SINNERS (I,3887); THE TWILIGHT ZONE (I,4651); VOYAGE TO THE BOTTOM OF THE SEA (I,4743); THE WILD WILD WEST (I,4863)

ADELSON, GARY
EIGHT IS ENOUGH (II,762)

ADREON, FRANCIS
CHEYENNE (I,931); COLT .45 (I,1010); SUGARFOOT (I,4277)

ADREON, FRANKLIN
THE ADVENTURES OF DR. FU MANCHU (I,58); COMMANDO CODY (I,1030)

AINSWORTH, JAMES
HAWAIIAN EYE (I,1973)

ALBERT, LOU
MR. CHIPS (I,3132); SUPER PAY CARDS (II,2500)

ALDA, ALAN
6 RMS RIV VU (II,2371); HICKEY VS. ANYBODY (II,1142); M*A*S*H (II,1569); MARLO THOMAS AND FRIENDS IN FREE TO BE. . . YOU AND ME (II,1632)

ALECK, JIMMY
ROCK-N-AMERICA (II,2195)

ALESIA, FRANK
LAVERNE AND SHIRLEY (II,1446)

ALEXANDER, A
Q.T. HUSH (I,3702)

ALEXANDER, DAVID
F TROOP (I,1499); GET SMART (II,983); THE GHOST AND MRS. MUIR (I,1786); MAGGIE BROWN (I,2811); THE MAN WHO CAME TO DINNER (I,2880); THE MUNSTERS (I,3158); MY FAVORITE MARTIAN (I,3189); NANNY AND THE PROFESSOR (I,3225); PANAMA HATTIE (I,3444); THE PHILADELPHIA STORY (I,3581); PLEASE DON'T EAT THE DAISIES (I,3628); THE REAL MCCOYS (I,3741); STAR TREK (II,2440); TELLER OF TALES (I,4388)

ALEXANDER, HAL
GENERAL HOSPITAL (II,964); MARY HARTMAN, MARY HARTMAN (II,1638); WHAT'S

HAPPENING!! (II,2769)

ALEXANDER, KIRK
MERV GRIFFIN'S SIDEWALKS OF NEW ENGLAND (I,3017); MERV GRIFFIN'S ST. PATRICK'S DAY SPECIAL (I,3018)

ALLEN, COREY
BRONK (II,379); CANNON: THE RETURN OF FRANK CANNON (II,425); CAPITOL (II,426); CHICAGO STORY (II,506); EXECUTIVE SUITE (II,796); THE FAMILY HOLVAK (II,817); HAWAII FIVE-O (II,1110); THE HIGH CHAPARRAL (I,2051); JESSIE (II,1287); KATE MCSHANE (II,1368); LEGMEN (II,1458); LOBO (II,1504); LUCAN (II,1554); MATT HOUSTON (II,1654); MCCLAIN'S LAW (II,1659); MOST WANTED (II,1739); MOVIN' ON (II,1743); MURDER, SHE WROTE (II,1770); THE PAPER CHASE: THE SECOND YEAR (II,1950); POLICE STORY (II,2062); POLICE WOMAN (II,2063); THE POWERS OF MATTHEW STAR (II,2075); THE QUEST (II,2097); QUINCY, M. E. (II,2102); THE ROCKFORD FILES (II,2197); SCARECROW AND MRS. KING (II,2268); SIMON AND SIMON (II,2357); STONE (II,2471); T.J. HOOKER (II,2524); TRAPPER JOHN, M.D. (II,2654); TUCKER'S WITCH (II,2667); WHIZ KIDS (II,2790)

ALLEN, CRAIG
DUNNINGER AND WINCHELL (I,1384); THE PAUL WINCHELL AND JERRY MAHONEY SHOW (I,3492); STARS IN KHAKI AND BLUE (I,4209); THE VAUGHN MONROE SHOW (I,4708)

ALLEN, GROVER J
DON MCNEILL'S TV CLUB (I,1337)

ALLEN, IRWIN
CITY BENEATH THE SEA (I,965); LAND OF THE GIANTS (I,2611); THE TIME TUNNEL (I,4511); VOYAGE TO THE BOTTOM OF THE SEA (I,4743)

ALLEN, LEWIS
THE 20TH CENTURY-FOX HOUR (I,4642); THE BIG VALLEY (I,437); BONANZA (II,347); BURKE'S LAW (I,762); DAN AUGUST (I,1130); THE DETECTIVES (I,1245); THE DICK POWELL SHOW (I,1269); THE FUGITIVE (I,1701); THE INVADERS (I,2229); LITTLE HOUSE ON THE PRAIRIE (II,1487); MAGGIE MALONE (I,2812); THE MARRIAGE BROKER (I,2920); MISSION: IMPOSSIBLE (I,3067); MY FRIEND TONY (I,3192); PERRY MASON (II,2025); THE RIFLEMAN (I,3789); THE ROGUES (I,3832); ROUTE 66 (I,3852); SCREEN DIRECTOR'S PLAYHOUSE (I,3946); SUDDEN SILENCE (I,4275); THE SURVIVORS (I,4301); TARGET: THE CORRUPTERS (I,4360)

ALLEN, WOODY
THE WOODY ALLEN SPECIAL (I,4904)

ALLMAN, JOHN
POWERHOUSE (II,2074)

ALTER, PAUL
BEAT THE CLOCK (I,378); BEAT THE CLOCK (II,195); THE BETTER SEX (II,223); BODY LANGUAGE (II,346); CARD SHARKS (II,438); FAMILY FEUD (II,816); FEATHER YOUR NEST (I,1555); HE SAID, SHE SAID (I,1986); THE PRICE IS RIGHT (I,3664); SHOWOFFS (II,2345); TATTLETALES (II,2545); TO TELL THE TRUTH (I,4528); TREASURE ISLE (I,4601); TV'S FUNNIEST GAME SHOW MOMENTS (II,2676)

ALTMAN, ROBERT
ALFRED HITCHCOCK PRESENTS (I,115); BRONCO (I,745); BUS STOP (I,778); COMBAT (I,1011); COUNTY GENERAL (I,1076); THE DESILU PLAYHOUSE (I,1237); THE GALLANT MEN (I,1728); LAWMAN (I,2642); M SQUAD (I,2799); MAVERICK (II,1657); THE MILLIONAIRE (I,3046); PETER GUNN (I,3562); THE ROARING TWENTIES (I,3805); ROUTE 66 (I,3852); SAM HILL (I,3895); SUGARFOOT (I,4277); SURFSIDE 6 (I,4299); A WALK IN THE NIGHT (I,4750)

ALTON, ROBERT
YOU'RE THE TOP (I,4974)

ALTRICE, WALT
ANOTHER LIFE (II,95)

ALZMANN, WALTER
THE WALTONS (II,2740)

AMATEAU, ROD
THE BOB HOPE CHRYSLER THEATER (I,502); THE CHARLIE FARRELL SHOW (I,913); DOBIE GILLIS (I,1302); THE DUKES OF HAZZARD (II,742); ENOS (II,779); THE GEORGE BURNS AND GRACIE ALLEN SHOW (I,1763); HIGH SCHOOL, U.S.A. (II,1148); HIGHWAY HONEYS (II,1151); MAGGIE (I,2810); MR. ED (I,3137); MY MOTHER THE CAR (I,3197); O.K. CRACKERBY (I,3348); THE PATTY DUKE SHOW (I,3483); PETER LOVES MARY (I,3565); PRIVATE SECRETARY (I,3672); SIX PACK (II,2370); SUPERTRAIN (II,2504); WEEKEND (I,4788); WHERE THERE'S SMOKEY (I,4832)

AMENTO, PINO
ALL THE RIVERS RUN (II,43)

AMOS, DICK
THE NEW MICKEY MOUSE CLUB (II,1823); SANTA BARBARA (II,2258)

AMYES, JULIAN
DANGER MAN (I,1136); GREAT EXPECTATIONS (II,1055); SECRET AGENT (I,3962)

ANDERSON, BRUCE
THE LILLI PALMER SHOW (I,2703); THE LILLI PALMER THEATER (I,2704)

ANDERSON, GERRY
U.F.O. (I,4662)

ANDERSON, LINDSAY
THE ADVENTURES OF ROBIN HOOD (I,74)

ANDERSON, MICHAEL
THE MARTIAN CHRONICLES (II,1635)

ANDRE, JACQUES
THE ED SULLIVAN SHOW (I,1401)

ANDREOLI, FRANK
CELEBRITY COOKS (II,472)

ANGUS, ALLAN
IT WAS A VERY GOOD YEAR (I,2253); THE JOHNNY CASH SHOW (I,2424)

ANNAKIN, KEN
HUNTER'S MOON (II,1207); INSTITUTE FOR REVENGE (II,1239)

ANSARA, MICHAEL
I DREAM OF JEANNIE (I,2167)

ANSPAUGH, DAVID
HILL STREET BLUES (II,1154); THE LAST LEAF (II,1437); ST. ELSEWHERE (II,2432)

ANTHONY, JOSEPH
BRENNER (I,719)

ANTONACCI, GREG
IT TAKES TWO (II,1252); MAKING A LIVING (II,1604); REGGIE (II,2127); WORKING STIFFS (II,2834)

ANTONIO, LOU
AMY PRENTISS (II,74); BANACEK (II,138); BOSTON AND KILBRIDE (II,356); THE CONTENDER (II,571); DELVECCHIO (II,659); FOOLS, FEMALES AND FUN: I'VE GOTTA BE ME (II,899); FOOLS, FEMALES AND FUN: IS THERE A DOCTOR IN THE HOUSE? (II,900); GABE AND WALKER (II,950); GENTLE BEN (I,1761); THE GIRL IN THE EMPTY GRAVE (II,997); GRIFF (I,1888); THE GYPSY WARRIORS (II,1070); HEAVEN ON EARTH (II,1120); LANIGAN'S RABBI (II,1427); LANIGAN'S RABBI (II,1428); MCCLOUD (II,1660); MCMILLAN (II,1666); MCMILLAN AND WIFE (II,1667); OWEN MARSHALL: COUNSELOR AT LAW (I,3435); RICH MAN, POOR MAN—BOOK II (II,2162); THE ROCKFORD FILES (II,2197); SONS AND DAUGHTERS (II,2404); THREE FOR THE ROAD (II,2607)

ANTOON, A J
HEREAFTER (II,1138)

APPEL, DON
THE BLUE ANGEL (I,489); PLEASE DON'T EAT THE DAISIES (I,3628); THE VAUGHN MONROE SHOW (I,4707); THE VIC DAMONE SHOW (I,4717)

APPEL, STANLEY
KELLY MONTEITH (II,1375)

ARCHINBAUD, GEORGE
THE ADVENTURES OF CHAMPION (I,54); ANNIE OAKLEY (I,225); BUFFALO BILL JR. (I,755); THE GENE AUTRY SHOW (I,1748); I'M THE LAW (I,2195); THE RANGE RIDER (I,3720)

ARDOLINO, EMIL
FAERIE TALE THEATER (II,807)

ARKIN, ALAN
FAY (II,844); TWIGS (II,2683)

ARKUS, ALLAN
SUMMER (II,2491)

ARKWRIGHT, BOB
FAT ALBERT AND THE COSBY KIDS (II,835)

ARLISS, LESLIE
THE NEW ADVENTURES OF CHARLIE CHAN (I,3249)

ARMSTRONG, JOHN
BLACKSTAR (II,263); FAT ALBERT AND THE COSBY KIDS (II,835); THE KID SUPER POWER HOUR WITH SHAZAM (II,1386); THE NEW ADVENTURES OF MIGHTY MOUSE AND HECKLE AND

JECKLE (II,1814); THE TARZAN/LONE RANGER/ZORRO ADVENTURE HOUR (II,2543)

ARMSTRONG, MORIA
THE ONEDIN LINE (II,1912)

ARNAZ, DESI
THE LUCILLE BALL-DESI ARNAZ SHOW (I,2784); THE MILTON BERLE SPECIAL (I,3052)

ARNER, GWEN
DYNASTY (II,746); FALCON CREST (II,808); FAME (II,812); FAMILY (II,813); KING'S CROSSING (II,1397); NURSE (II,1870); THE PAPER CHASE (II,1949); THE WALTONS (II,2740); THE WALTONS: MOTHER'S DAY ON WALTON'S MOUNTAIN (EPISODE 2) (II,2742)

ARNOLD, BOB
THAT GIRL (II,2570)

ARNOLD, DANNY
BARNEY MILLER (II,154); MY WORLD . . . AND WELCOME TO IT (I,3208); THE REAL MCCOYS (I,3741); SOMEWHERE IN ITALY. . . COMPANY B (I,4119)

ARNOLD, JACK
ALIAS SMITH AND JONES (I,118); THE BIONIC WOMAN (II,255); THE BRADY BUNCH (II,362); BUCK ROGERS IN THE 25TH CENTURY (II,384); THE ELEVENTH HOUR (I,1425); ELLERY QUEEN (II,766); GILLIGAN'S ISLAND (II,990); THE HOUSE NEXT DOOR (I,2139); IT TAKES A THIEF (I,2250); IT'S ABOUT TIME (I,2263); THE LOVE BOAT (II,1535); MCCLOUD (II,1660); MCNAUGHTON'S DAUGHTER (II,1670); THE MEN FROM SHILOH (I,3005); THE MISADVENTURES OF SHERIFF LOBO (II,1704); MR. LUCKY (I,3141); MR. TERRIFIC (I,3152); THE NANCY DREW MYSTERIES (II,1789); NANNY AND THE PROFESSOR (I,3225); OH, THOSE BELLS! (I,3345); PERRY MASON (II,2025); PETER GUNN (I,3562); RAWHIDE (I,3727); THE SAN PEDRO BEACH BUMS (II,2250); SCIENCE FICTION THEATER (I,3938); THINK PRETTY (I,4448); THE VIRGINIAN (I,4738); WONDER WOMAN (SERIES 2) (II,2829)

ARSENAULT, RAY
THE BAXTERS (II,184)

ARTHUR, GRAEME
PRISONER: CELL BLOCK H (II,2085)

ARTHUR, KAREN
BOONE (II,351); CAGNEY AND LACEY (II,409); EMERALD POINT, N.A.S. (II,773); HART TO HART (II,1102); REMINGTON STEELE (II,2130); RETURN TO EDEN (II,2143); RICH MAN, POOR MAN—BOOK II (II,2162); TWO MARRIAGES (II,2691)

ARTHUR, ROBERT
FUNNY PAPERS (I,1713)

ASHER, ROBERT
THE AVENGERS (II,121); THE BARON (I,360); THE PRISONER (I,3670); THE SAINT (II,2237)

ASHER, WILLIAM
ALICE (II,33); THE BAD NEWS BEARS (II,134); BEWITCHED (I,418); CHARLEY'S AUNT (II,483); A CHRISTMAS FOR BOOMER (II,518); CRAZY LIKE A FOX (II,601); THE DOCTOR WAS A LADY (I,1318); THE DUKES OF HAZZARD (II,742); FLATBUSH (II,877); FOUL PLAY (II,910); GIDGET (I,1795); HARPER VALLEY (II,1094); HARPER VALLEY PTA (II,1095); I LOVE LUCY (I,2171); LUCAS TANNER (II,1556); MICKEY AND THE CONTESSA (I,3024); MR. BEVIS (I,3128); OPERATION PETTICOAT (II,1917); THE PAUL LYNDE SHOW (I,3489); PRIVATE BENJAMIN (II,2087); SHORT STORY THEATER (I,4024); TABITHA (II,2526); TEMPERATURES RISING (I,4390); WILLY (I,4868)

ASHLEY, PAUL
THE ADVENTURES OF BLINKEY (I,52)

ASKEY, DAVID
DOCTOR IN THE HOUSE (I,1313); NO—HONESTLY (II,1857)

ASTIN, JOHN
CHIPS (II,511); ETHEL IS AN ELEPHANT (II,783); GETTING THERE (II,985); HOLMES AND YOYO (II,1169); JUST OUR LUCK (II,1360); MCMILLAN AND WIFE (II,1667); MR. MERLIN (II,1751); NIGHT GALLERY (I,3287); OPERATION PETTICOAT (II,1916); OPERATION PETTICOAT (II,1917); ROSETTI AND RYAN (II,2216); ROSETTI AND RYAN: MEN WHO LOVE WOMEN (II,2217); SCARED SILLY (II,2269); TWO GUYS FROM MUCK (II,2690)

ATKINSON, KEITH
B.J. AND THE BEAR (II,125); THE NANCY DREW MYSTERIES (II,1789)

ATTAWAY, PATTI
THE DANCE SHOW (II,616)

AUERBACH, LARRY
ALL MY CHILDREN (II,39); LOVE OF LIFE (I,2774); MR. WIZARD (I,3154)

AUSTIN, RAY
THE AVENGERS (II,121); B.J. AND THE BEAR (II,125); BARNABY JONES (II,153); THE BARON (I,360); DEPARTMENT S (I,1232); FROM HERE TO ETERNITY (II,933); HART TO HART (II,1102); HOUSE CALLS (II,1194); MAGNUM, P.I. (II,1601); A MAN CALLED SLOANE (II,1613); THE MASTER (II,1648); THE MISSISSIPPI (II,1707); THE NEW AVENGERS (II,1816); RETURN OF THE MAN FROM U.N.C.L.E.: THE 15 YEARS LATER AFFAIR (II,2140); RETURN OF THE SAINT (II,2141); THE SAINT (II,2237); SALVAGE 1 (II,2245); SHIRLEY'S WORLD (I,4019); SIMON AND SIMON (II,2357); SPACE: 1999 (II,2409); THE STRANGE REPORT (I,4248); SWORD OF JUSTICE (II,2521); TALES OF THE GOLD MONKEY (II,2537); THE UGLIEST GIRL IN TOWN (I,4663); VEGAS (II,2724); W.E.B. (II,2732)

AUSTIN, RONALD
CHARLIE'S ANGELS (II,486)

AVEDON, RICHARD
THE JUDY GARLAND SHOW (I,2471)

AVERBACK, HY
9 TO 5 (II,1852); ADVENTURING WITH THE CHOPPER (II,15); ALONE AT LAST (II,59); AT EASE (II,115); A BEDTIME STORY (I,384); BELL, BOOK AND CANDLE (II,201); BIG EDDIE (II,236); THE BROTHERS (I,748); THE BUREAU (II,397); BURKE'S LAW (I,762); BUS STOP (I,778); COLUMBO (II,556); THE DICK POWELL SHOW (I,1269); THE DON RICKLES SHOW (I,1342); DON RICKLES—ALIVE AND KICKING (I,1340); EDDIE (I,1405); ENSIGN O'TOOLE (I,1448); THE FLYING NUN (I,1611); THE FOUR SEASONS (II,915); FREEBIE AND THE BEAN (II,922); FRIENDS (II,928); THE GIRL, THE GOLD WATCH AND DYNAMITE (II,1001); THE GOOD LIFE (II,1033); HENNESSEY (I,2013); THE JIMMY DURANTE SHOW (I,2389); KAREN (II,1363); KEEPING AN EYE ON DENISE (I,2520); LOOK OUT WORLD (II,1516); THE LOVE BOAT II (II,1533); M*A*S*H (II,1569); THE MAN FROM U.N.C.L.E. (I,2867); MATT HOUSTON (II,1654); MCCLOUD (II,1660); MCMILLAN AND WIFE (II,1667); MO AND JO (II,1715); MRS. G. GOES TO COLLEGE (I,3155); NEEDLES AND PINS (I,3246); THE NEW MAVERICK (II,1822); NEWMAN'S DRUGSTORE (II,1837); THE NIGHT RIDER (II,1848); PEARL (II,1986); POPI (II,2069); QUARK (II,2095); THE REAL MCCOYS (I,3741); RICHIE BROCKELMAN, PRIVATE EYE (II,2167); RICHIE BROCKELMAN: MISSING 24 HOURS (II,2168); THE ROGUES (I,3832); THE RUBBER GUN SQUAD (II,2225); SHAUGHNESSEY (II,2318); SHE'S IN THE ARMY NOW (II,2320); SYBIL (I,4327); THREE FOR THE ROAD (II,2607); TOPPER RETURNS (I,4583); TWO'S COMPANY (I,4659); VENICE MEDICAL (II,2726); YOUNG MAVERICK (II,2867)

AVILDSEN, JOHN
MURDER INK (II,1769)

AWKIAN, ARAM
ONE NIGHT STANDS (I,3387)

AXNESS, JOHN
PEOPLE ARE FUNNY (II,1996)

BADHAM, JOHN
CANNON (II,424); COOL MILLION (I,1046); THE DOCTORS (I,1323); NIGHT GALLERY (I,3287); OWEN MARSHALL: COUNSELOR AT LAW (I,3435); POLICE STORY (II,2062); SARGE (I,3914); THE SENATOR (I,3976); THE SIXTH SENSE (I,4069); THE STREETS OF SAN FRANCISCO (II,2478); SUNSHINE (II,2495)

BADIYI, REZA S
240-ROBERT (II,2689); BANYON (I,344); BARETTA (II,152); BOONE (II,351); BRONK (II,379); CAGNEY AND LACEY (II,409); THE DORIS DAY SHOW (I,1355); EIGHT IS ENOUGH (II,762); THE EYES OF CHARLES SAND (I,1497); FARADAY AND COMPANY (II,833); GET SMART (II,983); THE GOOD GUYS (I,1847); THE GOVERNOR AND J.J. (I,1863); HART TO HART (II,1102); HAWAII FIVE-O

(II,1110); HE AND SHE (I,1985); HOLMES AND YOYO (II,1169); THE INCREDIBLE HULK (II,1232); INSIDE O.U.T. (II,1235); JIGSAW JOHN (II,1288); JOE DANCER: THE BIG BLACK PILL (II,1300); JOE DANCER: MURDER ONE, DANCER 0 (II,1302); JOE FORRESTER (II,1303); KATE LOVES A MYSTERY (II,1367); THE MAN FROM ATLANTIS (II,1615); MANNIX (II,1624); THE MARY TYLER MOORE SHOW (II,1640); MISSION: IMPOSSIBLE (I,3067); MODESTY BLAISE (II,1719); N.Y.P.D. (I,3321); THE PHOENIX (II,2036); POLICE SQUAD! (II,2061); THE ROCKFORD FILES (II,2197); S.W.A.T. (II,2236); SERPICO (II,2306); THE SIX-MILLION-DOLLAR MAN (II,2372); STOP SUSAN WILLIAMS (II,2473); SWITCH (II,2519); TENSPEED AND BROWN SHOE (II,2562); TRAUMA CENTER (II,2655)

BAER, DONALD
FITZ AND BONES (II,867)

BAIL, CHUCK
CHIPS (II,511); MANIMAL (II,1622)

BAIN, BILL
THE AVENGERS (II,121)

BAKER, JACK
THE AMES BROTHERS SHOW (I,178); HERE'S LUCY (II,1135)

BAKER, RAY
DEPARTMENT S (I,1232)

BAKER, ROY WARD
THE AVENGERS (II,121); THE BARON (I,360); THE CHAMPIONS (I,896); THE FLAME TREES OF THIKA (II,870); THE PERSUADERS! (I,3556); Q. E. D. (II,2094); THE SAINT (II,2237); THE STRANGE REPORT (I,4248)

BAKSHI, RALPH
CASPER AND FRIENDS (I,867)

BAKSHI, ROBERT
SPIDER-MAN (I,4154)

BALABAN, BOB
TALES FROM THE DARKSIDE (II,2534)

BALCH, JACK
THE HARTMANS (I,1960)

BALDWIN, GERALD
THE BULLWINKLE SHOW (I,761); THE FAMOUS ADVENTURES OF MR. MAGOO (I,1524)

BALDWIN, PETER
13 QUEENS BOULEVARD (II,2591); 9 TO 5 (II,1852); ALONE AT LAST (II,60); BENSON (II,208); THE BOB NEWHART SHOW (II,342); THE BRADY BRIDES (II,361); THE BRADY BUNCH (II,362); THE BRADY GIRLS GET MARRIED (II,364); C.P.O. SHARKEY (II,407); CARTER COUNTRY (II,449); CHICO AND THE MAN (II,508); THE DICK VAN DYKE SHOW (I,1275); DINAH AND HER NEW BEST FRIENDS (II,676); THE DORIS DAY SHOW (I,1355); THE DUCK FACTORY (II,738); GILLIGAN'S ISLAND: THE HARLEM GLOBETROTTERS ON GILLIGAN'S ISLAND (II,993); GOODNIGHT, BEANTOWN (II,1041); GOSSIP (II,1044); GREAT DAY (II,1053); GUN SHY (II,1068); HOLLYWOOD HIGH (II,1162); JOE AND SONS (II,1296); JOE AND SONS (II,1297); THE LIVING END (I,2731); THE LOVEBIRDS (II,1548); MAKING IT (II,1605); THE MICHELE LEE SHOW (II,1690); ONE IN A MILLION (II,1906); ONE OF THE BOYS (II,1911); OUT OF THE BLUE (II,1934); PLEASE DON'T EAT THE DAISIES (I,3628); SANFORD AND SON (II,2256); SPACE FORCE (II,2407); TEACHERS ONLY (II,2547); ZORRO AND SON (II,2878)

BALL, LUCILLE
BUNGLE ABBEY (II,396)

BALLARD, CHRISTINE
ALICE (II,33)

BALSER, ROBERT
THE JACKSON FIVE (I,2326)

BANNER, BOB
THE DINAH SHORE SHOW (I,1280); THE DINAH SHORE SHOW (I,1281); THE FRED WARING SHOW (I,1679); THE GINGER ROGERS SHOW (I,1804); ROSALINDA (I,3846)

BANNON, FRED C
COMMANDO CODY (I,1030)

BAR-DAVID, MAURICE
LAVERNE AND SHIRLEY (II,1446)

BARBERA, JOSEPH
THE ABBOTT AND COSTELLO CARTOON SHOW (I,1); THE ADVENTURES OF GULLIVER (I,60); THE ADVENTURES OF JONNY QUEST (I,64); THE AMAZING CHAN AND THE CHAN CLAN (I,157); THE ATOM ANT/SECRET SQUIRREL SHOW (I,311); BIRDMAN (II,256); BIRDMAN AND THE GALAXY TRIO (I,478); BUTCH CASSIDY AND THE SUNDANCE KIDS (I,781); THE CATTANOOGA CATS (I,874); DASTARDLY AND MUTTLEY IN THEIR FLYING MACHINES (I,1159); THE FANTASTIC FOUR (I,1529); THE FLINTSTONE COMEDY HOUR (II,881); THE FLINTSTONES (II,884); FRANKENSTEIN JR. AND THE IMPOSSIBLES (I,1672); THE FUNKY PHANTOM (I,1709); THE HARLEM GLOBETROTTERS (I,1953); HELP! IT'S THE HAIR BEAR BUNCH (I,2012); THE HERCULOIDS (I,2023); THE HUCKLEBERRY HOUND SHOW (I,2155); THE JETSONS (I,2376); JOKEBOOK (II,1339); JOSIE AND THE PUSSYCATS (I,2453); JOSIE AND THE PUSSYCATS IN OUTER SPACE (I,2454); LIPPY THE LION (I,2710); THE MAGILLA GORILLA SHOW (I,2825); MOBY DICK AND THE MIGHTY MIGHTOR (I,3077); MOTOR MOUSE (I,3111); THE NEW SUPER FRIENDS HOUR (II,1830); THE PERILS OF PENELOPE PITSTOP (I,3525); THE PETER POTAMUS SHOW (I,3568); SAMPSON AND GOLIATH (I,3903); SCOOBY-DOO, WHERE ARE YOU? (II,2276); THE SCOOBY-DOO/DYNOMUTT HOUR (II,2277); SHAZZAN! (I,4002); SPACE GHOST (I,4141); SPACE KIDDETTES (I,4143); THE TOM AND JERRY SHOW (I,4534); TOP CAT (I,4576); TOUCHE TURTLE (I,4586); THE WACKY RACES (I,4745); WAIT TIL YOUR FATHER GETS HOME (I,4748); WALLY GATOR (I,4751); WHERE'S HUDDLES (I,4836); THE WORLD'S GREATEST SUPER HEROES (II,2842); YOGI BEAR (I,4928)

BARE, RICHARD L
77 SUNSET STRIP (I,3988); ADVENTURES IN PARADISE (I,46); ALIAS SMITH AND JONES (I,118); BROKEN ARROW (I,744); BUS STOP (I,778); CASABLANCA (I,860); CHEYENNE (I,931); COLT .45 (I,1010); THE DAKOTAS (I,1126); DOC CORKLE (I,1304); THE DONNA REED SHOW (I,1347); FARADAY AND COMPANY (II,833); THE GALLANT MEN (I,1728); GIRL ON THE RUN (I,1809); GREEN ACRES (I,1884); LAWMAN (I,2642); MAN AGAINST CRIME (I,2852); MAVERICK (II,1657); THE RIFLEMAN (I,3789); ROUTE 66 (I,3852); RUN FOR YOUR LIFE (I,3871); SO THIS IS HOLLYWOOD (I,4105); SUGARFOOT (I,4277); TOPPER (I,4582); THE TWILIGHT ZONE (I,4651); THE WESTWIND (II,2766); YOU'RE ONLY YOUNG ONCE (I,4969)

BARKER, ALEX
CANDIDA (II,423)

BARLOW, ROGER
CONFIDENTIAL FILE (I,1033)

BARNES, BOB
THE ADVENTURES OF SUPERMAN (I,77)

BARNES, HOWARD G
ARTHUR MURRAY'S DANCE PARTY (I,293)

BARNHART, DON
BENSON (II,208); MADAME'S PLACE (II,1587)

BARNHIZER, DAVID
THE DICK CAVETT SHOW (I,1264); THE MIKE DOUGLAS CHRISTMAS SPECIAL (I,3040); THIS MORNING (I,4464)

BARNUM, PETE
THE ALL-STAR REVUE (I,138); THE ALL-STAR SUMMER REVUE (I,139); ANYTHING GOES (I,238)

BARON, ALLEN
AUTOMAN (II,120); BARNABY JONES (II,153); BRONK (II,379); CHARLIE'S ANGELS (II,486); THE DUKES OF HAZZARD (II,742); GRANDPA GOES TO WASHINGTON (II,1050); GRIFF (I,1888); HOUSE CALLS (II,1194); THE LOVE BOAT (II,1535); LUCAS TANNER (II,1556); MR. ROBERTS (I,3149); MY WORLD . . . AND WELCOME TO IT (I,3208); THE NIGHT STALKER (II,1849); ROOM 222 (I,3843); THE SAN PEDRO BEACH BUMS (II,2250); SCARECROW AND MRS. KING (II,2268); THE SIXTH SENSE (I,4069); SURFSIDE 6 (I,4299); WHEN THE WHISTLE BLOWS (II,2781)

BARON, MICKEY
NIGHT EDITOR (I,3286)

BARRET, EARL
TOO CLOSE FOR COMFORT (II,2642)

BARRON, ARTHUR
THE ABC AFTERSCHOOL SPECIAL (II,1)

BARRY, BRUCE
AS THE WORLD TURNS (II,110); THE GUIDING LIGHT (II,1064)

BARRY, CHRISTOPHER
DOCTOR WHO (II,689)

BARSCH, ARTHUR
DEPUTY DAWG (I,1233)

BARSHA, LEON
BOLD JOURNEY (I,680)

BARTH, CARL
THE FBI (I,1551)

BARTLETT, RICHARD H
HAWAIIAN EYE (I,1973); WAGON TRAIN (I,4747)

BARTON, CHARLES
THE AMOS AND ANDY SHOW (I,179); CAMP RUNAMUCK (I,811); DENNIS THE MENACE (I,1231); FAMILY AFFAIR (I,1519); HAZEL (I,1982); MEET MR. MCNUTLEY (I,2989); MEET THE GIRLS: THE SHAPE, THE FACE, AND THE BRAIN (I,2990); THE MUNSTERS (I,3158); OH, THOSE BELLS! (I,3345); PETTICOAT JUNCTION (I,3571); THE SMOTHERS BROTHERS SHOW (I,4096); SPIN AND MARTY (I,4158); TUGBOAT ANNIE (I,4622)

BARTSCH, ARTHUR
THE ASTRONAUT SHOW (I,307)

BARZYK, FRED
JEAN SHEPHERD'S AMERICA (I,2354)

BASINSKI, DAVE
WE'RE MOVIN' (II,2756)

BASS, JULES
'TWAS THE NIGHT BEFORE CHRISTMAS (II,2677); THE BEATLES (I,380); THE CONEHEADS (II,567); THE EASTER BUNNY IS COMIN' TO TOWN (II,753); FROSTY THE SNOWMAN (II,934); FROSTY'S WINTER WONDERLAND (II,935); HERE COMES PETER COTTONTAIL (II,1132); JACK FROST (II,1263); THE KING KONG SHOW (I,2556); THE LEPRECHAUN'S CHRISTMAS GOLD (II,1461); THE LITTLE DRUMMER BOY (II,1485); LITTLE DRUMMER BOY, BOOK II (II,1486); PINOCCHIO'S CHRISTMAS (II,2045); THE RELUCTANT DRAGON AND MR. TOAD (I,3762); RUDOLPH'S SHINY NEW YEAR (II,2228); SANTA CLAUS IS COMIN' TO TOWN (II,2259); THE YEAR WITHOUT A SANTA CLAUS (II,2846)

BASS, TONY
THE OSMONDS (I,3410)

BATCHELDER, WARREN
THE TRANSFORMERS (II,2653)

BATES, RICHARD
THE PRIME OF MISS JEAN BRODIE (II,2082)

BAUR, FRANK
BLACK SADDLE (I,480); JOHNNY RINGO (I,2434); THE RAT PATROL (I,3726)

BAUR, FRED
SURFSIDE 6 (I,4299)

BAXLEY, PAUL
THE DUKES OF HAZZARD (II,742)

BEATTY, ROGER
CAROL BURNETT AND COMPANY (II,441); MAMA'S FAMILY (II,1610); THE PRIMARY ENGLISH CLASS (II,2081); THE TIM CONWAY SHOW (II,2619)

BEAUDINE, WILLIAM
BROKEN ARROW (I,744); LASSIE (I,2622); THE MICKEY MOUSE CLUB (I,3025); NAKED CITY (I,3215); ROCKY JONES, SPACE RANGER (I,3823); THUNDER (II,2617); TIMMY AND LASSIE (I,4520); TV READERS DIGEST (I,4630)

BEAUMONT, ALAN
A TIME TO LIVE (I,4510)

BEAUMONT, CHARLES
PRIVATE BENJAMIN (II,2087)

BEAUMONT, GABRIELLE
DYNASTY (II,746); FLAMINGO ROAD (II,872); GLITTER (II,1009); THE GREATEST AMERICAN HERO (II,1060); HART TO HART (II,1102); M*A*S*H (II,1569); PRIVATE BENJAMIN (II,2087); SECRETS OF MIDLAND HEIGHTS (II,2296); TRAUMA CENTER (II,2655); VEGAS (II,2724); THE WALTONS (II,2740); ZORRO AND SON (II,2878)

BEAUMONT, HUGH
LEAVE IT TO BEAVER (I,2648)

BECKER, JOHN F
ANYWHERE, U.S.A. (I,240)

BECKER, ROBERT
EMERALD POINT, N.A.S. (II,773)

BECKER, TERRY
M*A*S*H (II,1569); MISSION: IMPOSSIBLE (I,3067); THE MOD SQUAD (I,3078)

BECKETT, KEITH
THE BENNY HILL SHOW (II,207)

BEEBE, FORD
THE ADVENTURES OF CHAMPION (I,54)

BEHAR, GEORGE
TAKE A GOOD LOOK (I,4332); TAKE A GOOD LOOK (II,2529)

BEHAR, HAL
DELLA (I,1222)

BEHAR, HANK
TAKE MY ADVICE (II,2531); THREE FOR THE MONEY (II,2606)

BEHAR, JOSEPH
CHAIN LETTER (I,892); CLEAR HORIZON (I,973); THE COMEDY OF ERNIE KOVACS (I,1022); DAYS OF OUR LIVES (II,629); FACE THE FACTS (I,1508); IT'S ANYBODY'S GUESS (II,1255); KOVACS ON THE CORNER (I,2582); LET'S MAKE A DEAL (II,1462); LET'S MAKE A DEAL (II,1464); PEOPLE WILL TALK (I,3520)

BEHAR, LEE
MASQUERADE PARTY (II,1645)

BEIER, CARL
MAGNAVOX THEATER (I,2827)

BELANGER, PAUL
THROUGH THE CRYSTAL BALL (I,4494)

BELL, ALAN J W
THE HITCHHIKER'S GUIDE TO THE GALAXY (II,1158)

BELL, DAVID
THE BENNY HILL SHOW (II,207); ELTON JOHN: IN CONCERT (II,770); U.F.O. (I,4662)

BELL, JOHN
OUT OF OUR MINDS (II,1933); SECOND CITY TELEVISION (II,2287)

BELL, MICHAEL
KIDD VIDEO (II,1388)

BELLAMY, BRUCE
THE HERO (I,2035)

BELLAMY, EARL
THE ADVENTURES OF RIN TIN TIN (I,73); ALFRED HITCHCOCK PRESENTS (I,115); THE ANDY GRIFFITH SHOW (I,192); BACHELOR FATHER (I,323); CHIPS (II,511); CODE RED (II,552); DANIEL BOONE (I,1142); THE DESILU PLAYHOUSE (I,1237); FANTASY ISLAND (II,829); THE FBI (I,1551); THE FIRST HUNDRED YEARS (I,1582); FUTURE COP (II,945); GET SMART (II,983); GETAWAY CAR (I,1783); GILLIGAN'S ISLAND: THE CASTAWAYS ON GILLIGAN'S ISLAND (II,992); HART TO HART (II,1102); I SPY (I,2179); ISIS (II,1248); A KNIGHT IN SHINING ARMOUR (I,2575); LARAMIE (I,2614); LAREDO (I,2615); LEAVE IT TO BEAVER (I,2648); THE LONE RANGER (I,2740); M SQUAD (I,2799); M*A*S*H (II,1569); MATT HELM (II,1653); MCHALE'S NAVY (I,2969); MEDICAL CENTER (II,1675); THE MOD SQUAD (I,3078); THE MONROES (I,3089); MUNSTER, GO HOME! (I,3157); THE MUNSTERS (I,3158); MY FRIEND TONY (I,3192); MY THREE SONS (I,3205); O'CONNOR'S OCEAN (I,3326); THE PARTNERS (I,3462); PERRY MASON (II,2025); THE PIGEON (I,3596); THE QUEST (II,2097); RAWHIDE (I,3727); S.W.A.T. (II,2236); THE SAN PEDRO BEACH BUMS (II,2250); SCIENCE FICTION THEATER (I,3938); SERGEANT PRESTON OF THE YUKON (I,3980); SHERIFF OF COCHISE (I,4008); THE SIXTH SENSE (I,4069); SOLDIERS OF FORTUNE (I,4111); STARSKY AND HUTCH (II,2444); STRANDED (II,2475); TARZAN (I,4361); TO ROME WITH LOVE (I,4526); TRAPPER JOHN, M.D. (II,2654); U.S. MARSHAL (I,4686); VALENTINE MAGIC ON LOVE ISLAND (II,2716); THE VIRGINIAN (I,4738); WAGON TRAIN (I,4747); YOUNG DAN'L BOONE (II,2863)

BELLISARIO, DONALD P
AIRWOLF (II,27); BATTLESTAR GALACTICA (II,181); MAGNUM, P.I. (II,1601)

BELSON, JERRY
COPS (I,1047); THE ODD COUPLE (II,1875); PETE 'N' TILLIE (II,2027); RHODA (II,2151)

BEMILLER, BOB
POPEYE THE SAILOR (I,3649)

BENDER, BOB
THE GREATEST AMERICAN HERO (II,1060); QUINCY, M. E. (II,2102)

BENDER, JACK
THE AMERICAN DREAM (II,71); BREAKING AWAY (II,371); EIGHT IS ENOUGH

(II,762); FALCON CREST (II,808); FAME (II,812); HIGH SCHOOL, U.S.A. (II,1149); KING'S CROSSING (II,1397); THE PAPER CHASE (II,1949); THE PAPER CHASE: THE SECOND YEAR (II,1950)

BENEDEK, LASLO
ALFRED HITCHCOCK PRESENTS (I,115); COMBAT (I,1011); THE FUGITIVE (I,1701); NAKED CITY (I,3216); THE OUTER LIMITS (I,3426); PERRY MASON (II,2025); RAWHIDE (I,3727); THRILLER (I,4492); THE UNTOUCHABLES (I,4682); VOYAGE TO THE BOTTOM OF THE SEA (I,4743)

BENEDICT, RICHARD
CHARLIE'S ANGELS (II,486); CODE R (II,551); COMBAT (I,1011); DAN AUGUST (I,1130); THE DOCTORS (I,1323); DOCTORS' PRIVATE LIVES (II,692); THE HARDY BOYS MYSTERIES (II,1090); HAWAII FIVE-O (II,1110); HAWAIIAN EYE (I,1973); HAWK (I,1974); THE INVADERS (I,2229); LAREDO (I,2615); LAWMAN (I,2642); THE LAWYERS (I,2646); THE MAN FROM ATLANTIS (II,1615); MATT HELM (II,1653); MAVERICK (II,1657); MEDICAL STORY (II,1676); THE MEN FROM SHILOH (I,3005); MISSION: IMPOSSIBLE (I,3067); THE NANCY DREW MYSTERIES (II,1789); NIGHT GALLERY (I,3287); THE PARTNERS (I,3462); POLICE STORY (II,2062); QUINCY, M. E. (II,2102); THE ROARING TWENTIES (I,3805); THE ROOKIES (II,2208); RUN FOR YOUR LIFE (I,3871); S.W.A.T. (II,2236); SAINTS AND SINNERS (I,3887); SAN FRANCISCO INTERNATIONAL AIRPORT (I,3905); SURFSIDE 6 (I,4299); THE VIRGINIAN (I,4738)

BENJAMIN, RICHARD
SEMI-TOUGH (II,2300); WHERE'S POPPA? (II,2784)

BENNETT, CHARLES
CAVALCADE OF AMERICA (I,875); THE NEW ADVENTURES OF CHARLIE CHAN (I,3249);

BENNETT, DEREK
UPSTAIRS, DOWNSTAIRS (II,2709)

BENNETT, MURRAY
I COVER TIMES SQUARE (I,2166)

BENNETT, RICHARD
THE ABC AFTERSCHOOL SPECIAL (II,1); APPLE'S WAY (II,101); CAPITOL (II,426); ENOS (II,779); HARPER VALLEY PTA (II,1095); LUCAS TANNER (II,1556); THE WALTONS (II,2740)

BENNETT, RODNEY
DOCTOR WHO (II,689)

BENNEWITZ, RICK
CAPITOL (II,426); THE DONNA FARGO SHOW (II,710); LAND OF THE LOST (II,1425); SANTA BARBARA (II,2258); THE STARLAND VOCAL BAND (II,2441); STOCKARD CHANNING IN JUST FRIENDS (II,2467)

BENNINGTON, WILLIAM
OPERATION ENTERTAINMENT (I,3400); STUMP THE STARS (I,4273)

BENOFF, MAC
LIFE WITH LUIGI (I,2692)

BENSON, JAMES
THE EDDIE CAPRA MYSTERIES (II,758); SALVAGE 1 (II,2245)

BENSON, JAY
LUCAS TANNER (II,1556)

BENSON, LEON
BEN CASEY (I,394); BONANZA (II,347); THE ELEVENTH HOUR (I,1425); FLIPPER (I,1604); THE HIGH CHAPARRAL (I,2051); THE OUTER LIMITS (I,3426); OWEN MARSHALL: COUNSELOR AT LAW (I,3435); THE RAT PATROL (I,3726); SCIENCE FICTION THEATER (I,3938); SEA HUNT (I,3948); SUNSHINE (II,2495); THE WILD WILD WEST (I,4863)

BENTON, DOUGLAS
HEC RAMSEY (II,1121); POLICE WOMAN (II,2063)

BERBER, OLLIE
15 WITH FAYE (I,1570)

BERGER, RICHARD
ANYTHING GOES (I,237); FORD STAR REVUE (I,1632); MUSICAL COMEDY TIME (I,3183);

BERGMAN, ALAN
KID GLOVES (I,2531); THE M AND M CANDY CARNIVAL (I,2797); THAT MCMAHON'S HERE AGAIN (I,4419)

BERGMAN, DAVID
IT'S ROCK AND ROLL (II,1258)

BERGMAN, INGMAR
SCENES FROM A MARRIAGE (II,2272)

BERGMANN, ALAN
BARNEY MILLER (II,154); FAMILY TIES (II,819); FLYING HIGH (II,894); GOODNIGHT, BEANTOWN (II,1041); HARPER VALLEY PTA (II,1095); HOUSE CALLS (II,1194); JUST OUR LUCK (II,1360); OPERATION PETTICOAT (II,1917); PRIVATE BENJAMIN (II,2087)

BERK, HOWARD
GET CHRISTIE LOVE! (II,981)

BERKE, LESTER
QUINCY, M. E. (II,2102)

BERKE, WILLIAM
ANNIE OAKLEY (I,225); THE GENE AUTRY SHOW (I,1748); THE GOLDBERGS (I,1835); THE GOLDBERGS (I,1836); THE HUNTER (I,2162); I SPY (I,2178)

BERLE, MILTON
THE MILTON BERLE SHOW (I,3047)

BERLIN, ABBY
THE ANN SOTHERN SHOW (I,220); BACHELOR FATHER (I,323); THE LIFE OF RILEY (I,2686); OH, THOSE BELLS! (I,3345)

BERNARD, BARRY
THE CHEVROLET TELE-THEATER (I,926); CHILDREN'S SKETCH BOOK (I,941)

BERNDS, EDWARD
COLT .45 (I,1010); THE HANK MCCUNE SHOW (I,1932); THE NEW THREE STOOGES (I,3274)

BERNDS, ERNIE
SUGARFOOT (I,4277)

BERNHARDI, LEE H
FISH (II,864); GABRIEL KAPLAN PRESENTS THE FUTURE STARS (II,951); JERRY REED AND SPECIAL FRIENDS (II,1285); THE JIM STAFFORD SHOW (II,1291); THE MONTE CARLO SHOW (II,1726); MUSIC CITY NEWS TOP COUNTRY HITS OF THE YEAR (II,1772); MUSIC HALL AMERICA (II,1773); ON OUR OWN (II,1892); ONE OF THE BOYS (II,1911); THE RICH LITTLE SHOW (II,2153); WE'VE GOT EACH OTHER (II,2757); WHAT'S HAPPENING!! (II,2769)

BERNHARDT, MELVIN
ANOTHER WORLD (II,97); MR. ROBERTS (II,1754)

BERNS, SEYMOUR
BILL COSBY DOES HIS OWN THING (I,441); BING CROSBY AND HIS FRIENDS (I,455); BURLESQUE (I,765); THE EDSEL SHOW (I,1417); FRANKIE LAINE TIME (I,1676); HOLIDAY LODGE (I,2077); HOWDY (I,2150); THE IMPERIAL GRAND BAND (II,1224); THE JACK BENNY PROGRAM (I,2294); THE JOHNNY CARSON SHOW (I,2422); THE MUNSTERS (I,3158); NICK AND NORA (II,1842); THE RED SKELTON CHEVY SPECIAL (I,3753); THE RED SKELTON TIMEX SPECIAL (I,3757); A SALUTE TO STAN LAUREL (I,3892); A SPECIAL BILL COSBY SPECIAL (I,4147); THE TROUBLE WITH TRACY (I,4613)

BERNSTEIN, JERROLD S
THE FARMER'S DAUGHTER (I,1533); THE FLYING NUN (I,1611); GIDGET (I,1795); HERE COME THE BRIDES (I,2024); NANNY AND THE PROFESSOR (I,3225); THE UGLIEST GIRL IN TOWN (I,4663)

BESSADA, MILAD
SECOND CITY TELEVISION (II,2287)

BESSELL, TED
THAT GIRL (II,2570)

BEST, JAMES
THE DUKES OF HAZZARD (II,742)

BETTMAN, GIL
B.J. AND THE BEAR (II,125); B.J. AND THE BEAR (II,126); THE FALL GUY (II,811); KNIGHT RIDER (II,1402)

BIBERMAN, ABNER
ALCOA/GOODYEAR THEATER (I,107); BEN CASEY (I,394); COLT .45 (I,1010); THE DOCTORS (I,1323); THE FUGITIVE (I,1701); HAWAII FIVE-O (II,1110); IRONSIDE (II,1246); MAVERICK (II,1657); THE MEN FROM SHILOH (I,3005); RUN FOR YOUR LIFE (I,3871); TEMPLE HOUSTON (I,4392); TIGHTROPE (I,4500); THE UNTOUCHABLES (I,4682); THE VIRGINIAN (I,4738)

BIELAK, ROBERT
MATT HOUSTON (II,1654)

BILL, TONY
FAERIE TALE THEATER (II,807); FULL HOUSE (II,937)

BILLETT, STU
GARROWAY (I,1736)

BILLINGTON, KEVIN
ECHOES OF THE SIXTIES (II,756); ONCE UPON A TIME. . . IS NOW THE STORY OF PRINCESS GRACE (II,1899)

BILSON, BRUCE
ALICE (II,33); ALOHA PARADISE (II,58); B.J. AND THE BEAR (II,125); B.J. AND THE BEAR (II,127); THE BAD NEWS BEARS (II,134); THE BANANA COMPANY (II,139); BAREFOOT IN THE PARK (I,355); BARNEY AND ME (I,358); BARNEY MILLER (II,154); THE BRADY BUNCH (II,362); BRING 'EM BACK ALIVE (II,377); CAMP RUNAMUCK (I,811); CAROL (I,846); DEAD MAN ON THE RUN (II,631); DUFFY (II,739); THE EYES OF TEXAS (II,799); THE EYES OF TEXAS II (II,800); THE FALL GUY (II,811); THE FEATHER AND FATHER GANG (II,845); THE FESS PARKER SHOW (II,850); FINDER OF LOST LOVES (II,857); GET SMART (II,983); THE GHOST AND MRS. MUIR (I,1786); GIDGET (I,1795); GIDGET'S SUMMER REUNION (I,1799); THE HALLOWEEN THAT ALMOST WASN'T (II,1075); HARPER VALLEY PTA (II,1095); HAWAII FIVE-O (II,1110); THE HERO (I,2035); HOTEL (II,1192); HOUSE CALLS (II,1194); HUNTER (II,1205); JUST OUR LUCK (II,1360); KNIGHT RIDER (II,1402); THE MISADVENTURES OF SHERIFF LOBO (II,1704); NANNY AND THE PROFESSOR (I,3225); THE NEW ADVENTURES OF HUCKLEBERRY FINN (I,3250); THE NEW DAUGHTERS OF JOSHUA CABE (I,3261); THE PATTY DUKE SHOW (I,3483); PLEASE DON'T EAT THE DAISIES (I,3628); PLEASURE COVE (II,2058); POPI (II,2069); THE POWERS OF MATTHEW STAR (II,2075); PRIVATE BENJAMIN (II,2087); QUARK (II,2095); THE RENEGADES (II,2133); THE ROOKIES (II,2208); S.W.A.T. (II,2236); SKAG (II,2374); SPENCER'S PILOTS (II,2422); TABITHA (II,2527); WONDER WOMAN (SERIES 1) (II,2828)

BINDER, STEVE
AMERICA (I,164); THE BARRY MANILOW SPECIAL (II,155); THE BIG SHOW (II,243); BLONDES VS. BRUNETTES (II,271); DEBBY BOONE. . . ONE STEP CLOSER (II,655); DIANA (II,667); DON KIRSHNER'S ROCK CONCERT (I,1332); DOROTHY HAMILL'S CORNER OF THE SKY (II,721); ELVIS (I,1435); IT'S ABOUT TIME (I,2263); A LAST LAUGH AT THE 60'S (I,2627); LEGENDS OF THE WEST: TRUTH AND TALL TALES (II,1457); THE LESLIE UGGAMS SHOW (I,2660); LOLA (II,1509); LUCY IN LONDON (I,2789); THE MAC DAVIS CHRISTMAS SPECIAL (II,1572); MAC DAVIS CHRISTMAS SPECIAL. . . WHEN I GROW UP (II,1574); THE MAC DAVIS SHOW (II,1576); THE MAC DAVIS SPECIAL (II,1577); MAC DAVIS. . . I BELIEVE IN CHRISTMAS (II,1581); OLIVIA (II,1883); PETULA (I,3572); SHIELDS AND YARNELL (II,2328); A SPECIAL EDDIE RABBITT (II,2415); THE STEVE ALLEN SHOW (I,4221)

BING, MACK
THE BEAN SHOW (I,373); THE GARRY MOORE SHOW (I,1740); MARY HARTMAN, MARY HARTMAN (II,1638); THE SMOTHERS BROTHERS SHOW (II,2384); UNCLE CROC'S BLOCK (II,2702)

BINKLEY, BAYRON
THE MARTY ROBBINS SPOTLITE (II,1636)

BINYON JR, CONRAD
HAWAIIAN EYE (I,1973)

BINYON, CLAUDE
SCREEN DIRECTOR'S PLAYHOUSE (I,3946)

BIRCH, PETER
CAPTAIN KANGAROO (II,434); THE JACK PAAR SHOW (I,2316); MR. MAYOR (I,3143); SONG SNAPSHOTS ON A SUMMER HOLIDAY (I,4122)

BIRNBAUM, BOB
THE ODD COUPLE (II,1875); SHIRLEY (II,2330)

BISHOP, TERRY
THE ADVENTURES OF ROBIN HOOD (I,74); DANGER MAN (I,1136); SECRET AGENT (I,3962)

BIXBY, BILL
THE BARBARY COAST (II,146); THE BARBARY COAST (II,147); BERT D'ANGELO/SUPERSTAR (II,211); THE BEST OF TIMES (II,220); CHARLIE'S ANGELS (II,486); THE COURTSHIP OF EDDIE'S FATHER (I,1083); DREAMS (II,735); GOODNIGHT, BEANTOWN (II,1041); HERBIE, THE LOVE BUG (II,1131); KATE MCSHANE (II,1368); MANNIX (II,1624); THE MANY LOVES OF ARTHUR (II,1625); MR. MERLIN (II,1751); THE OREGON TRAIL (II,1923); RICH MAN, POOR MAN—BOOK II (II,2162); SPENCER'S PILOTS (II,2422); W*A*L*T*E*R (II,2731); WIZARDS AND WARRIORS (II,2813)

BLACK, JOHN D F
CHARLIE'S ANGELS (II,486)

BLACK, NOEL
AMY PRENTISS (II,74); BIG HAWAII (II,238); LANIGAN'S RABBI (II,1428); MCCLOUD (II,1660); MULLIGAN'S STEW (II,1763); MULLIGAN'S STEW (II,1764); QUINCY, M. E. (II,2102); SWITCH (II,2519); THE WORLD BEYOND (II,2835)

BLAIR, GEORGE
SABU AND THE MAGIC RING (I,3881); WANTED: DEAD OR ALIVE (I,4764)

BLAKE, DARROL
THE ONEDIN LINE (II,1912)

BLAKE, GERALD
BLAKE'S SEVEN (II,264); DOCTOR WHO (II,689); THE OMEGA FACTOR (II,1888); THE ONEDIN LINE (II,1912)

BLAKE, ROBERT
BARETTA (II,152)

BLANCHARD, JOHN
RED SKELTON'S FUNNY FACES (II,2124); SECOND CITY TELEVISION (II,2287)

BLANK, TOM
THE ABC AFTERSCHOOL SPECIAL (II,1); THE AMAZING SPIDER-MAN (II,63); HARRIS AND COMPANY (II,1096)

BLECKNER, JEFF
THE ABC AFTERSCHOOL SPECIAL (II,1); BRET MAVERICK (II,374); DOC (II,682); THE FAMILY TREE (II,820); THE GUIDING LIGHT (II,1064); THE HARVEY KORMAN SHOW (II,1104); HENRY WINKLER MEETS WILLIAM SHAKESPEARE (II,1129); HILL STREET BLUES (II,1154); HONEST AL'S A-OK USED CAR AND TRAILER RENTAL TIGERS (II,1174); KING'S CROSSING (II,1397); KNOTS LANDING (II,1404); MR. DUGAN (II,1749); REMINGTON STEELE (II,2130); RYAN'S FOUR (II,2233); THE STOCKARD CHANNING SHOW (II,2468); THINGS ARE LOOKING UP (II,2588); TRAPPER JOHN, M.D. (II,2654); WILLOW B: WOMEN IN PRISON (II,2804)

BLEES, ROBERT
PROJECT UFO (II,2088)

BLEYER, BOB
SING IT AGAIN (I,4059); SONGS FOR SALE (I,4124)

BLOCH, HUNTER
ELKE SOMMER'S WORLD OF SPEED AND BEAUTY (II,765)

BLOCH, JOHN
TONIGHT AT 8:30 (I,4556)

BLOOM, ARTHUR
WALTER CRONKITE'S UNIVERSE (II,2739)

BLOOM, JEFFREY
DARKROOM (II,619)

BLOOMFIELD, GEORGE
FRAGGLE ROCK (II,916); SECOND CITY TELEVISION (II,2287)

BLUM, BOB
ARCHY AND MEHITABEL (I,254)

BLUNK, FIDELIS
SEARCH FOR BEAUTY (I,3953)

BLUNT, MICHAEL
DOCTOR WHO (II,689)

BOATMAN, BOB
HEE HAW (II,1123); THE HEE HAW HONEYS (II,1125)

BODE, LENORE
A MAGIC GARDEN CHRISTMAS (II,1590)

BODIE, WILLIAM
ACTION IN THE AFTERNOON (I,25)

BOETTICHER, BUDD
77 SUNSET STRIP (I,3988); ALIAS MIKE HERCULES (I,117); MAVERICK (II,1657); PUBLIC DEFENDER (I,3687); THE RIFLEMAN (I,3789); ZANE GREY THEATER (I,4979)

BOGART, PAUL
THE ADAMS CHRONICLES (II,8); ALICE (II,33); ALICE (II,34); ALL IN THE FAMILY (II,38); ARCHIE BUNKER'S PLACE (II,105); THE BOB HOPE CHRYSLER THEATER (I,502); CAROUSEL (I,856);

CORONET BLUE (I,1055); THE COUNTRY GIRL (I,1064); THE DEFENDERS (I,1220); THE DUMPLINGS (II,743); THE EASTER PROMISE (I,1397); GET SMART (II,983); HANSEL AND GRETEL (I,1934); HIGHER AND HIGHER, ATTORNEYS AT LAW (I,2057); THE HOUSE WITHOUT A CHRISTMAS TREE (I,2143); JOHNNY BELINDA (I,2417); KISS ME, KATE (I,2569); LOU GRANT (II,1526); MAMA MALONE (II,1609); THE PICTURE OF DORIAN GRAY (I,3591); SHIRLEY TEMPLE'S STORYBOOK: THE LEGEND OF SLEEPY HOLLOW (I,4018); SPELLBOUND (I,4153); TEN LITTLE INDIANS (I,4394); THE THANKSGIVING TREASURE (I,4416); YOU CAN'T TAKE IT WITH YOU (II,2857)

BOHL, DON
THE PRICE IS RIGHT (I,3664)

BOLE, CLIFF
THE AMAZING SPIDER-MAN (II,63); THE ASPHALT COWBOY (II,111); B.J. AND THE BEAR (II,125); CHARLIE'S ANGELS (II,486); EMERGENCY! (II,775); THE FALL GUY (II,811); FANTASY ISLAND (II,829); FINDER OF LOST LOVES (II,857); GAVILAN (II,959); MATT HOUSTON (II,1654); THE SIX-MILLION-DOLLAR MAN (II,2372); STRIKE FORCE (II,2480); T.J. HOOKER (II,2524); TIME EXPRESS (II,2620); VEGAS (II,2724)

BOLOGNA, JOSEPH
BEDROOMS (II,198); GOOD PENNY (II,1036)

BONDELLI, PHIL
240-ROBERT (II,2688); THE BIONIC WOMAN (II,255); CHARLIE'S ANGELS (II,486); CHIPS (II,511); CHOPPER ONE (II,515); CODE R (II,551); CODE RED (II,552); COVER UP (II,597); FANTASY ISLAND (II,829); GET CHRISTIE LOVE! (II,981); MASQUERADE (II,1644); THE ROOKIES (II,2208); S.W.A.T. (II,2236); THE SIX-MILLION-DOLLAR MAN (II,2372); T.J. HOOKER (II,2524); VEGAS (II,2724); WALKING TALL (II,2738)

BONERZ, PETER
9 TO 5 (II,1852); APPLE PIE (II,100); ARCHIE BUNKER'S PLACE (II,105); BACK TOGETHER (II,131); THE BOB NEWHART SHOW (II,342); A DOG'S LIFE (II,697); E/R (II,748); FATHER O FATHER (II,842); G.I.'S (II,949); GOOD HEAVENS (II,1032); HIGH FIVE (II,1143); IN SECURITY (II,1227); IT'S YOUR MOVE (II,1259); LOVE, NATALIE (II,1545); PARK PLACE (II,1958); PAUL SAND IN FRIENDS AND LOVERS (II,1982); SHEILA (II,2325); SUZANNE PLESHETTE IS MAGGIE BRIGGS (II,2509); SZYSZNYK (II,2523); THE TWO OF US (II,2693)

BONI, JOHN
ROCK-N-AMERICA (II,2195)

BONNER, FRANK
FAMILY TIES (II,819); WKRP IN CINCINNATI (II,2814)

BOOKE, SORRELL
THE DUKES OF HAZZARD (II,742)

BOONE, RICHARD
ARENA (I,256); HAVE GUN—WILL TRAVEL (I,1968)

BOOTH, PHILIP
BRIEF PAUSE FOR MURDER (I,725); THE MARSHAL OF GUNSIGHT PASS (I,2925); SHORTY (I,4025)

BORCHERT, RUDOLPH
THE GREATEST AMERICAN HERO (II,1060)

BORZAGE, FRANK
SCREEN DIRECTOR'S PLAYHOUSE (I,3946)

BOWAB, JOHN
13 THIRTEENTH AVENUE (II,2592); THE BAXTERS (II,183); BENSON (II,208); BOSOM BUDDIES (II,355); BROTHERS AND SISTERS (II,382); DOUBLE TROUBLE (II,723); THE FACTS OF LIFE (II,805); FOOT IN THE DOOR (II,902); I'M A BIG GIRL NOW (II,1218); IT TAKES TWO (II,1252); JENNIFER SLEPT HERE (II,1278); JOE'S WORLD (II,1305); MAKING A LIVING (II,1604); THE PARKERS (II,1959); REGGIE (II,2127); REPORT TO MURPHY (II,2134); ROMANCE THEATER (II,2204); SMALL AND FRYE (II,2380); TEACHERS ONLY (II,2548); WHY US? (II,2796)

BOWAN, ROSS
BIG JOHN, LITTLE JOHN (II,240)

BOWEN, GARY
THE DOCTORS (II,694)

BOWER, ROGER
TWENTY QUESTIONS (I,4647)

BOWKER, BOB
FROM CLEVELAND (II,931); THE STEVE ALLEN COMEDY HOUR (II,2454)

BOWMAN, CHUCK
THE A-TEAM (II,119); THE GREATEST AMERICAN HERO (II,1060); THE INCREDIBLE HULK (II,1232); TENSPEED AND BROWN SHOE (II,2562)

BOWMAN, ROSS
THE PARTRIDGE FAMILY (II,1962)

BOYLE, DONALD R
BIG FOOT AND WILD BOY (II,237)

BOYLE, JEFF
THE UNCLE FLOYD SHOW (II,2703)

BOYRIVAN, PATRICK
THE INCREDIBLE HULK (II,1232)

BOYT, JOHN
TIN PAN ALLEY TV (I,4522)

BRADFORD, JOHNNY
A VERY SPECIAL OCCASION (I,4716)

BRAHM, JOHN
THE 20TH CENTURY-FOX HOUR (I,4642); ALCOA PREMIERE (I,109); ALFRED HITCHCOCK PRESENTS (I,115); CAVALCADE OF AMERICA (I,875); THE DEFENDERS (I,1220); THE GIRL FROM U.N.C.L.E. (I,1808); LAURA (I,2635); M SQUAD (I,2799); THE MAN FROM U.N.C.L.E. (I,2867); MEDIC (I,2976); NAKED CITY (I,3215); NAKED CITY (I,3216); SCREEN DIRECTOR'S PLAYHOUSE (I,3946); SHANE (I,3995); THRILLER (I,4492); THE VIRGINIAN (I,4738)

BRAIN, DAVE
THE TRANSFORMERS (II,2653)

BRANCHI, BRUNO
INSPECTOR GADGET (II,1236)

BRAND, TONY
THE WALTONS (II,2740)

BRANDERN, RUBEN
STRANGE PLACES (I,4247)

BRASHER, GARY
THE WORLD OF PEOPLE (II,2839)

BRAVERMAN, CHUCK
DREAMS (II,735)

BRAYNE, WILLIAM
RUNNING BLIND (II,2232)

BREAUX, MARC
THE BING CROSBY SHOW (I,468); THE BING CROSBY SHOW (I,469); CAROL + 2 (I,853); THE CONFESSIONS OF DICK VAN DYKE (II,568); DEBBIE REYNOLDS AND THE SOUND OF CHILDREN (I,1213); THE DINAH SHORE SPECIAL—LIKE HEP (I,1282); THE HOLLYWOOD PALACE (I,2088); THE LIGHT FANTASTIC, OR HOW TO TELL YOUR PAST, PRESENT AND MAYBE YOUR FUTURE THROUGH SOCIAL DANCING (I,2696)

BREGMAN, BUDDY
THE WACKY WORLD OF JONATHAN WINTERS (I,4746)

BRESLOW, MARC
CARD SHARKS (II,438); THE MATCH GAME (II,1649); THE MATCH GAME—HOLLYWOOD SQUARES HOUR (II,1650); THE NEW PRICE IS RIGHT (I,3270); NOW YOU SEE IT (II,1867); THE PRICE IS RIGHT (I,3665); THE PRICE IS RIGHT (II,2079); THE SCOEY MITCHLLL SHOW (I,3939); TRIVIA TRAP (II,2659)

BRETHERTON, HOWARD
THE ADVENTURES OF SUPERMAN (I,77)

BRIAN, DAVE
JIM HENSON'S MUPPET BABIES (II,1289)

BRIANT, MICHAEL E
BLAKE'S SEVEN (II,264); DOCTOR WHO (II,689)

BRICKELL, BETH
A RAINY DAY (II,2110)

BRICKEN, JAMES
THE ANSWER MAN (I,231)

BRICKEN, JULES
THE 20TH CENTURY-FOX HOUR (I,4642); MARKHAM (I,2916); NAKED CITY (I,3216); THE RAY MILLAND SHOW (I,3734)

BRIDGES, ALAN
BRIEF ENCOUNTER (I,724); CROWN MATRIMONIAL (I,1106)

BRIGGLE, STOCKTON
ALICE (II,33)

BRINCKERHOFF, BURT
9 TO 5 (II,1852); ANOTHER DAY (II,94); BARETTA (II,152); THE BETTY WHITE SHOW (II,224); CAGNEY AND LACEY (II,409); CALUCCI'S

DEPARTMENT (I,805); THE DAY THE WOMEN GOT EVEN (II,628); DOC (II,683); DYNASTY (II,746); THE HAMPTONS (II,1076); HART TO HART (II,1102); HOLLYWOOD HIGH (II,1163); HOT L BALTIMORE (II,1187); HOW TO SUCCEED IN BUSINESS WITHOUT REALLY TRYING (II,1197); JOE DANCER: THE MONKEY MISSION (II,1301); LOU GRANT (II,1526); LOVERS AND OTHER STRANGERS (II,1550); THE MONTEFUSCOS (II,1727); OFF CAMPUS (II,1877); REMINGTON STEELE (II,2130); THE ROLLERGIRLS (II,2203); STEPHANIE (II,2453); STUBBY PRINGLE'S CHRISTMAS (II,2483); THAT'S MY MAMA (II,2580); THE THREE WIVES OF DAVID WHEELER (II,2611); TRAUMA CENTER (II,2655); TWO MARRIAGES (II,2691); YOU ARE THE JURY (II,2854)

BRINCKERHOFF, PETER
AS THE WORLD TURNS (II,110)

BRINKLEY, BYRON
THAT GOOD OLD NASHVILLE MUSIC (II,2571)

BRODERICK, JAMES
FAMILY (II,813)

BRODY, LARRY
BARNABY JONES (II,153)

BROOK, TRICIA
REAL LIFE STORIES (II,2117)

BROOKS, BOB
SPACE: 1999 (II,2409)

BROOKS, GEORGE E
HART TO HART (II,1102)

BROOKS, GEORGE W
CHARLIE'S ANGELS (II,486); FANTASY ISLAND (II,829)

BROWLOW, KEVIN
HOLLYWOOD (II,1160)

BROWN JR, WILLIAM H
THE GOODYEAR REVUE (I,1856)

BROWN, ARVIN
CLOSE TIES (II,545)

BROWN, CHARLOTTE
ARCHIE BUNKER'S PLACE (II,105); BETWEEN THE LINES (II,225); IT'S NOT EASY (II,1256); SECOND EDITION (II,2288)

BROWN, DAVID
AS CAESAR SEES IT (I,294); HAWKINS FALLS, POPULATION 6200 (I,1977); SHORT STORY THEATER (I,4024); THE WAYNE KING SHOW (I,4779)

BROWN, DICK
CLUTCH CARGO (I,988); SPACE ANGEL (I,4140)

BROWN, GARRY
BENSON (II,208)

BROWN, GEORG STANFORD
CAGNEY AND LACEY (II,409); CALL TO GLORY (II,413); CHARLIE'S ANGELS (II,486); DYNASTY (II,746); THE FALL GUY (II,811); FAME (II,812); FAMILY (II,813); FINDER OF LOST LOVES (II,857); THE FITZPATRICKS (II,868); THE GREATEST AMERICAN HERO (II,1060); HILL STREET BLUES (II,1154); THE LAZARUS SYNDROME (II,1451); LOU GRANT (II,1526); LUCAN (II,1554); MIAMI VICE (II,1689); PALMERSTOWN U.S.A. (II,1945); PARIS (II,1957); POLICE SQUAD! (II,2061); THE ROOKIES (II,2208); ROOTS: THE NEXT GENERATIONS (II,2213); STARSKY AND HUTCH (II,2444); TENSPEED AND BROWN SHOE (II,2562); TRAUMA CENTER (II,2655)

BROWN, MENDE
BATTLE OF THE AGES (I,367)

BROWN, PHILIP
COLONEL MARCH OF SCOTLAND YARD (I,1005)

BROWN, WILLIAM H
PAUL WHITEMAN'S SATURDAY NIGHT REVUE (I,3490)

BROWNING, KIRK
THE ALCOA HOUR (I,108); AMAHL AND THE NIGHT VISITORS (I,148); CYRANO DE BERGERAC (I,1119); DAMN YANKEES (I,1128); FIFTH OF JULY (II,851); HARRY AND LENA (I,1957); HOLIDAY IN LAS VEGAS (I,2076); ONCE UPON A CHRISTMAS TIME (I,3368); THE ROSALIND RUSSELL SHOW (I,3845)

BROWNING, RICOU
FLIPPER (I,1604); GENTLE BEN (I,1761); JAMBO (I,2330); PRIMUS (I,3668)

BRUCE, JOHN
TALES OF THE UNEXPECTED (II,2540); THE WOMAN IN WHITE (II,2819)

BRUCKNER, WILLIAM
THE LORETTA YOUNG THEATER (I,2756)

BRUNO, JOE
HOT WHEELS (I,2126); SKYHAWKS (I,4079)

BRYNNER, YUL
DANGER (I,1134); LIFE WITH SNARKY PARKER (I,2693); SURE AS FATE (I,4297)

BUCKHANTZ, ABNER
THE DAKOTAS (I,1126)

BUCKRIDGE, KENNETH
MISS SUSAN (I,3065)

BUNETTA, FRANK
THE BIG SELL (I,431); CAVALCADE OF BANDS (I,876); DARK OF NIGHT (I,1156); THE HONEYMOONERS (I,2111); IT'S A BUSINESS (I,2254); JACKIE GLEASON AND HIS AMERICAN SCENE MAGAZINE (I,2321); THE JACKIE GLEASON SHOW (I,2323); THE JACKIE GLEASON SPECIAL (I,2324); JOHNNY JUPITER (I,2429); SID CAESAR INVITES YOU (I,4040); THE SID CAESAR SPECIAL (I,4044); THE STRAWHATTERS (I,4257)

BUNIN, HOPE
FOODINI THE GREAT (I,1619)

BUNIN, MOREY
FOODINI THE GREAT (I,1619)

BURDEAU, GEORGE
PERSONAL AND CONFIDENTIAL (II,2026)

BURGER, RICHARD
MARRIED ALIVE (I,2923)

BURGESS, BRIAN
CAPTAIN SCARLET AND THE MYSTERONS (I,837)

BURKE, MELVILLE
THE MAN AGAINST CRIME (I,2853)

BURMESTER, BRUCE
CHAIN REACTION (II,479); THE DAVID LETTERMAN SHOW (II,624); GO (II,1012); THE LOVE EXPERTS (II,1537); THE NEW $25,000 PYRAMID (II,1811); SO YOU THINK YOU GOT TROUBLES?! (II,2391)

BURNESS, PETER
THE BULLWINKLE SHOW (I,761); MR. MAGOO (I,3142); ROCKY AND HIS FRIENDS (I,3822)

BURNLEY, FRED
SEARCH FOR THE NILE (I,3954)

BURNS, STAN
DEAN MARTIN'S RED HOT SCANDALS PART 2 (II,649)

BURR, WALTER
THE NEW ADVENTURES OF HUCKLEBERRY FINN (I,3250)

BURROWS, ABE
76 MEN AND PEGGY LEE (I,3989); O.K. CRACKERBY (I,3348)

BURROWS, BIL
ALL ABOUT FACES (I,126)

BURROWS, JAMES
THE ASSOCIATES (II,113); AT YOUR SERVICE (II,118); BEST OF THE WEST (II,218); THE BETTY WHITE SHOW (II,224); THE BOB NEWHART SHOW (II,342); BUMPERS (II,394); BUSTING LOOSE (II,402); BUTTERFLIES (II,405); CALLING DR. STORM, M.D. (II,415); CHEERS (II,494); EVERY STRAY DOG AND KID (II,792); FAY (II,844); FREE COUNTRY (II,921); GOOD TIME HARRY (II,1037); GOODBYE DOESN'T MEAN FOREVER (II,1039); HUSBANDS, WIVES AND LOVERS (II,1209); LAVERNE AND SHIRLEY (II,1446); LOU GRANT (II,1526); THE MARY TYLER MOORE SHOW (II,1640); NIGHT COURT (II,1844); ON OUR OWN (II,1892); P.O.P. (II,1939); PAUL SAND IN FRIENDS AND LOVERS (II,1982); PHYLLIS (II,2038); THE PLANT FAMILY (II,2050); RHODA (II,2151); ROOSEVELT AND TRUMAN (II,2209); THE STOCKARD CHANNING SHOW (II,2468); SZYSZNYK (II,2523); TAXI (II,2546); THE TED KNIGHT SHOW (II,2550); THE TONY RANDALL SHOW (II,2640); WE'VE GOT EACH OTHER (II,2757); YOUR PLACE OR MINE? (II,2874)

BURTON, AL
ROMP (I,3838)

BUSHELL, ANTHONY
DANGER MAN (I,1136); SECRET AGENT (I,3962)

BUSHNELL JR, WILLIAM
THE WALTONS (II,2740)

BUSS, FRANCES
HOLD IT PLEASE (I,2071); IT'S A GIFT (I,2255); KING'S PARTY LINE (I,2562); VANITY FAIR (I,4702); WHAT'S IT WORTH? (I,4818); THE WHISTLING WIZARD (I,4844);

WINNER TAKE ALL (I,4881)

BUSTANY, DON
THE BOB NEWHART SHOW (II,342)

BUTLER, DAVID
BELLE STARR (I,391); BRINGING UP BUDDY (I,731); THE DEPUTY (I,1234); I LOVE MY DOCTOR (I,2172); ICHABOD AND ME (I,2183); LEAVE IT TO BEAVER (I,2648); M SQUAD (I,2799); RUSSIAN ROULETTE (I,3875); SCREEN DIRECTOR'S PLAYHOUSE (I,3946); WAGON TRAIN (I,4747)

BUTLER, ROBERT
BATMAN (I,366); BEN CASEY (I,394); BLACK BART (II,260); CIMARRON STRIP (I,954); COLUMBO (II,556); CONCRETE BEAT (II,562); THE DEFENDERS (I,1220); THE DETECTIVES (I,1245); THE DICK POWELL SHOW (I,1269); THE DICK VAN DYKE SHOW (I,1275); DOC ELLIOT (I,1305); THE FUGITIVE (I,1701); GUNSMOKE (II,1069); HAVE GUN—WILL TRAVEL (I,1968); HAWAII FIVE-O (II,1110); HILL STREET BLUES (II,1154); HOGAN'S HEROES (I,2069); I SPY (I,2179); THE INVADERS (I,2229); KUNG FU (II,1416); LACY AND THE MISSISSIPPI QUEEN (II,1419); LANCER (I,2610); MISSION: IMPOSSIBLE (I,3067); MR. ROBERTS (I,3149); MRS. G. GOES TO COLLEGE (I,3155); N.Y.P.D. (I,3321); ONE NIGHT BAND (II,1909); REMINGTON STEELE (II,2130); RUN FOR YOUR LIFE (I,3871); THE SEEKERS (I,3974); SHANE (I,3995); THE UNTOUCHABLES (I,4682); THE WALTONS (II,2740)

BUTTERWORTH, KENT
FAT ALBERT AND THE COSBY KIDS (II,835); PANDAMONIUM (II,1948)

BUXTON, FRANK
HAPPY DAYS (II,1084); MORK AND MINDY (II,1735); THE ODD COUPLE (II,1875); SUNDAY FUNNIES (II,2493)

BUZZELL, EDWARD
THE FABULOUS SYCAMORES (I,1505); TEXACO STAR THEATER (I,4412)

BYRNE, GEORGE
HOLLYWOOD JUNIOR CIRCUS (I,2084)

BYRON, PAUL
DIONE LUCAS' COOKING SCHOOL (I,1288); THIS IS SHOW BUSINESS (I,4459)

CABRERA, PABLO
THE NEW VOICE (II,1833)

CADDIGAN, JAMES L
THE ALAN DALE SHOW (I,96); FAMILY GENIUS (I,1521)

CAFFEY, MICHAEL
THE AMAZING SPIDER-MAN (II,63); B.J. AND THE BEAR (II,125); BARNABY JONES (II,153); BERT D'ANGELO/SUPERSTAR (II,211); BUCK ROGERS IN THE 25TH CENTURY (II,383); CANNON (II,424); CHICAGO STORY (II,506); CHIPS (II,511); COMBAT (I,1011); DAN AUGUST (I,1130); THE DOCTORS (I,1323); EMERGENCY PLUS FOUR (II,774); ENOS (II,779); FREEBIE AND THE BEAN (II,922); THE GEMINI MAN (II,960); GRANDPA GOES TO WASHINGTON (II,1050); HAGEN (II,1071); HAWAII FIVE-O (II,1110); HERE'S BOOMER (II,1134); HIGH PERFORMANCE (II,1145); IT TAKES A THIEF (I,2250); KINGSTON: CONFIDENTIAL (II,1398); LANCER (I,2610); LEGMEN (II,1458); LOGAN'S RUN (II,1507); MANHUNTER (II,1621); MANNIX (II,1624); THE MASTER (II,1648); MEDICAL CENTER (II,1675); THE MOD SQUAD (I,3078); THE NANCY DREW MYSTERIES (II,1789); THE NIGHT STALKER (II,1849); PARIS 7000 (I,3459); SEARCH (I,3951); SERPICO (II,2306); THE SURVIVORS (I,4301); SWORD OF JUSTICE (II,2521); TIME EXPRESS (II,2620); TRAPPER JOHN, M.D. (II,2654); TWO MARRIAGES (II,2691); THE WILD WILD WEST (I,4863); WONDER WOMAN (SERIES 2) (II,2829)

CAHAN, GEORGE M
BOSTON BLACKIE (I,695); THE BRADY BUNCH (II,362); COWBOY G-MEN (I,1084); GILLIGAN'S ISLAND (II,990); IT'S ABOUT TIME (I,2263); MEDIC (I,2976); THE RUGGLES (I,3868); THE SILENT SERVICE (I,4049); THE UNEXPECTED (I,4674); WILLY (I,4869)

CAIRNCROSS, WILLIAM
QUINCY, M. E. (II,2102)

CALABRESE, PETER
BEN VEREEN. . . COMIN' AT YA (II,203); THE MIKE DOUGLAS SHOW (II,1698); TONY ORLANDO AND DAWN (II,2639)

CALDWELL, DAVE
THE SHAPE OF THINGS (II,2315); SPEAK UP AMERICA (II,2412)

CALDWELL, ROBERT
ST. ELSEWHERE (II,2432)

CALLNER, MARTY
CAMELOT (II,416); DIANA ROSS IN CONCERT (II,668); FLEETWOOD MAC IN CONCERT (II,880); THE GEORGE JONES SPECIAL (II,975); HERE IT IS, BURLESQUE! (II,1133); PAT BENATAR IN CONCERT (II,1967); THE PEE WEE HERMAN SHOW (II,1988); THE RICH LITTLE SPECIAL (II,2155); STEVIE NICKS IN CONCERT (II,2463); USA COUNTRY MUSIC (II,591)

CALPIN, ORESTES
PLANET PATROL (I,3616)

CALVERT, FRED
I AM THE GREATEST: THE ADVENTURES OF MUHAMMED ALI (II,1211)

CAMBERN, DONN
SOMETHING ELSE (I,4117)

CAMFIELD, DOUGLAS
BLAKE'S SEVEN (II,264); DOCTOR WHO (II,689)

CAMP, JOE
BENJI, ZAX AND THE ALIEN PRINCE (II,205)

CAMPANELLA JR, ROY
HAWAIIAN HEAT (II,1111); SIMON AND SIMON (II,2357)

CAMPBELL, MARTIN
BERGERAC (II,209)

CAMPBELL, NORMAN
THE ANDY WILLIAMS CHRISTMAS SHOW (I,199); THE ANDY WILLIAMS CHRISTMAS SHOW (II,77); THE ANDY WILLIAMS SHOW (I,210); BING CROSBY'S WHITE CHRISTMAS (II,253); AN EVENING WITH DIANA ROSS (II,788); FRAGGLE ROCK (II,916); THE FURST FAMILY OF WASHINGTON (I,1718); THE GERSHWIN YEARS (I,1779); THE GOLDDIGGERS IN LONDON (I,1837); THE JOHN DAVIDSON CHRISTMAS SHOW (II,1307); THE LIBERACE SHOW (I,2677); ONE DAY AT A TIME (II,1900); A SPECIAL OLIVIA NEWTON-JOHN (II,2418)

CAMPBELL, RON
THE SMURFS (II,2385)

CANNELL, STEPHEN J
ADAM-12 (I,31); CHASE (I,917); JIGSAW (I,2377); THE ROCKFORD FILES (II,2197); STONE (II,2471); TENSPEED AND BROWN SHOE (II,2562)

CANNON, ROBERT
MR. MAGOO (I,3142)

CARDONA, ROBERT D
CRIMES OF PASSION (II,603); DEATH IN SMALL DOSES (I,1207); MURDER IS A ONE-ACT PLAY (I,3160); NOT GUILTY! (I,3311)

CAREY, CHRISTOPHER
GIDGET (I,1795)

CARL, ANTHONY
STEVE ALLEN'S LAUGH-BACK (II,2455)

CARLSON, RICHARD
THE DETECTIVES (I,1245); THE LORETTA YOUNG THEATER (I,2756); THRILLER (I,4492)

CARLSON, ROBERT
ALCOA PREMIERE (I,109)

CARNEY, JACK
ARTHUR GODFREY'S TALENT SCOUTS (I,291)

CARNEY, JOE
BOZO THE CLOWN (I,702)

CAROLEO, DINA
WONDER GIRL (I,4891)

CARR, FRED
ACTOR'S STUDIO (I,28); BLIND DATE (I,482); PICK AND PAT (I,3590)

CARR, MICHAEL
WOMEN OF RUSSIA (II,2824)

CARR, THOMAS
THE ADVENTURES OF CHAMPION (I,54); THE ADVENTURES OF SUPERMAN (I,77); ANNIE OAKLEY (I,225); BUFFALO BILL JR. (I,755); CHEYENNE (I,931); THE DETECTIVES (I,1245); THE DICK POWELL SHOW (I,1269); THE GENE AUTRY SHOW (I,1748); GENTRY'S PEOPLE (I,1762); THE GUNS OF WILL SONNETT (I,1907); IF YOU KNEW TOMORROW (I,2187); LARAMIE (I,2614); THE RANGE RIDER (I,3720); RAWHIDE (I,3727); STAGECOACH WEST (I,4185); TURN OF FATE (I,4623); THE VIRGINIAN (I,4738); WANTED:

DEAD OR ALIVE (I,4764)

CARRADINE, DAVID
KUNG FU (II,1416)

CARRERE, LEON
CHARLIE'S ANGELS (II,486)

CARRUTHERS, BILL
ALL-STAR SECRETS (II,52); BIZARRE (II,258); I LOVE LIBERTY (II,1215); JUST FRIENDS (I,2495); THE KING FAMILY IN WASHINGTON (I,2549); THE KING FAMILY JUNE SPECIAL (I,2550); THE NEIGHBORS (II,1801); OPERATION: ENTERTAINMENT (I,3401); PRESS YOUR LUCK (II,2078); SINATRA—THE MAIN EVENT (II,2360); A TRIBUTE TO "MR. TELEVISION" MILTON BERLE (II,2658); ULTRA QUIZ (II,2701); WHEW! (II,2786); WHODUNIT? (II,2794)

CARSON, DICK
BUCKSHOT (II,385); THE DON RICKLES SHOW (I,1341); THE MERV GRIFFIN SHOW (I,3016); THE MERV GRIFFIN SHOW (II,1688); WHEEL OF FORTUNE (II,2775)

CARSON, HELEN
DON MCNEILL'S DINNER CLUB (I,1336); STUMP THE AUTHORS (I,4272)

CARSON, JEAN-PHILIPPE
HIGH ADVENTURE WITH LOWELL THOMAS (I,2048)

CARTER, JACK
HERE'S LUCY (II,1135)

CARTER, JOHN
BARNABY JONES (II,153)

CARTER, LES
CHARLIE'S ANGELS (II,486)

CARTER, PETER
A MAN CALLED INTREPID (II,1612); SEARCH AND RESCUE: THE ALPHA TEAM (II,2283); SHANNON (II,2314); THE SWISS FAMILY ROBINSON (II,2518)

CARTER, THOMAS
CALL TO GLORY (II,413); FAME (II,812); HILL STREET BLUES (II,1154); MIAMI VICE (II,1689); REMINGTON STEELE (II,2130); ST. ELSEWHERE (II,2432); TRAUMA CENTER (II,2655); THE WHITE SHADOW (II,2788)

CARUSO, JAMES
THE VIDEO GAME (II,2727)

CASE, BRAD
BAGGY PANTS AND THE NITWITS (II,135); BAILEY'S COMETS (I,333); THE BARKLEYS (I,356); THE FANTASTIC FOUR (II,824); THE HOUNDCATS (I,2132); JIM HENSON'S MUPPET BABIES (II,1289); PANDAMONIUM (II,1948)

CASE, DALE
THE 2,000 YEAR OLD MAN (II,2696)

CASSELLI, LAMAR
THE G.E. COLLEGE BOWL (I,1745); WHAT'S IN A WORD? (I,4815)

CASSON, PHILIP
THE MUPPET SHOW (II,1766); PICCADILLY PALACE (I,3589)

CASTLE, WILLIAM
MEN OF ANNAPOLIS (I,3008)

CATES, GILBERT
CIRCUS LIONS, TIGERS AND MELISSAS TOO (II,528); JOHNNY CASH CHRISTMAS 1983 (II,1324); JOHNNY CASH: CHRISTMAS ON THE ROAD (II,1333)

CATES, JOSEPH
ACCENT ON LOVE (I,16); THE BACHELOR (I,325); CIRCUS LIONS, TIGERS AND MELISSAS TOO (II,528); THE COMEDY ZONE (II,559); FIVE STARS IN SPRINGTIME (I,1590); THE GENE KELLY PONTIAC SPECIAL (I,1750); HIGH BUTTON SHOES (I,2050); JOHNNY CASH: CHRISTMAS ON THE ROAD (II,1333); STEVE MARTIN: COMEDY IS NOT PRETTY (II,2462); THE VICTOR BORGE SHOW (I,4719); THE VICTOR BORGE SHOW (I,4720)

CATES, RON
YOU ASKED FOR IT (II,2855)

CAULDWELL, DAVE
LOOK AT US (II,1515); REAL KIDS (II,2116); REAL PEOPLE (II,2118)

CAVALIER, JOE
BELIEVE IT OR NOT (I,387); BROADWAY OPEN HOUSE (I,736); MARY KAY AND JOHNNY (I,2932)

CAWLEY, BOB
PETER MARSHALL SALUTES THE BIG BANDS (II,2028)

CHAFFEY, DON
CASINO (II,452); CHARLIE'S ANGELS (II,486); FINDER OF LOST LOVES (II,857); GAVILAN (II,959); HOTEL (II,1192); IVANHOE (I,2282); LASSIE: THE NEW BEGINNING (II,1433); LOTTERY (II,1525); MATT HOUSTON (II,1654); THE NEW ADVENTURES OF CHARLIE CHAN (I,3249); THE PRISONER (I,3670); THE RENEGADES (II,2133); RIDING FOR THE PONY EXPRESS (II,2172); STRIKE FORCE (II,2480); T.J. HOOKER (II,2524); VEGAS (II,2724)

CHAIVAS, LOUIS
BLACKSTAR (II,263)

CHAMBERS, JEFF
ANGIE (II,80); BEST OF THE WEST (II,218); GOOD TIME HARRY (II,1037); MORK AND MINDY (II,1735); OUT OF THE BLUE (II,1934)

CHAMPION, GOWER
ACCENT ON LOVE (I,16); THE DATCHET DIAMONDS (I,1160); AN EVENING WITH JULIE ANDREWS AND HARRY BELAFONTE (I,1476); THE JULIE ANDREWS SPECIAL (I,2483); THE MARY MARTIN SHOW (I,2934); SCREEN DIRECTOR'S PLAYHOUSE (I,3946); THREE FOR TONIGHT (I,4478)

CHARMOLI, TONY
THE ANTHONY NEWLEY SHOW (I,233); THE BIG SHOW (II,243); BOB HOPE SPECIAL: BOB HOPE'S ALL-STAR SUPER BOWL PARTY (II,304); BOB HOPE SPECIAL: BOB HOPE'S PINK PANTHER THANKSGIVING GALA (II,312); BOB HOPE SPECIAL: BOB HOPE PRESENTS A CELEBRATION WITH STARS OF COMEDY AND MUSIC (II,340); BONNIE AND THE FRANKLINS (II,349); THE BUGALOOS (I,757); THE CAPTAIN AND TENNILLE (II,429); CENTER STAGE: CYD CHARISSE (I,889); CIRCUS OF THE STARS (II,530); CIRCUS OF THE STARS (II,531); CIRCUS OF THE STARS (II,533); CIRCUS OF THE STARS (II,534); CIRCUS OF THE STARS (II,535); CIRCUS OF THE STARS (II,536); CIRCUS OF THE STARS (II,537); DORIS DAY TODAY (II,716); THE EDDIE FISHER SPECIAL (I,1410); FOL-DE-ROL (I,1613); THE GIRL IN MY LIFE (II,996); GLADYS KNIGHT AND THE PIPS (II,1004); THE JANE MORGAN SHOW (I,2342); THE JIMMY MCNICHOL SPECIAL (II,1292); THE JOHN DAVIDSON CHRISTMAS SHOW (II,1308); JOHN DENVER AND THE MUPPETS: A CHRISTMAS TOGETHER (II,1313); THE KEANE BROTHERS SHOW (II,1371); THE KING FAMILY CELEBRATE THANKSGIVING (I,2545); THE KING FAMILY CHRISTMAS SPECIAL (I,2546); THE KING FAMILY IN ALASKA (I,2547); THE KING FAMILY IN HAWAII (I,2548); THE KING FAMILY OCTOBER SPECIAL (I,2551); THE KING FAMILY SHOW (I,2552); THE KING FAMILY SPECIAL (I,2553); THE KING FAMILY VALENTINE'S DAY SPECIAL (I,2554); LEAPIN' LIZARDS, IT'S LIBERACE (II,1453); LIDSVILLE (I,2679); MITZI AND A HUNDRED GUYS (II,1710); MITZI. . . ROARIN' IN THE 20S (II,1711); MITZI. . . WHAT'S HOT, WHAT'S NOT (II,1712); MITZI. . . ZINGS INTO SPRING (II,1713); MITZI: A TRIBUTE TO THE AMERICAN HOUSEWIFE (I,3072); MUSICAL COMEDY TONIGHT (II,1777); ON STAGE: PHIL SILVERS (I,3357); THE PAUL LYNDE COMEDY HOUR (II,1977); THE PERRY COMO WINTER SHOW (I,3544); PERRY COMO'S WINTER SHOW (I,3548); THE POLLY BERGEN SPECIAL (I,3644); RAQUEL (II,2114); THE SHIRLEY BASSEY SHOW (I,4014); THE SHIRLEY BASSEY SHOW (I,4015); THE SHIRLEY MACLAINE SPECIAL: WHERE DO WE GO FROM HERE? (II,2333); THE SPECIAL GENTRY ONE (I,4148); STAR SEARCH (II,2438); THE TRINI LOPEZ SHOW (I,4606); THE WAYNE NEWTON SPECIAL (II,2750)

CHARNIN, MARTIN
ANNIE AND THE HOODS (II,91); GEORGE M! (I,1772); JACK LEMMON IN 'S WONDERFUL, 'S MARVELOUS, 'S GERSHWIN (I,2313)

CHATTINGTON, BARRY
THE NATIVITY (II,1798)

CHEN, DAVID FU-YUNG
MR. ROGERS' NEIGHBORHOOD (I,3150)

CHENAULT, ROBERT
SHAZAM! (II,2319)

CHERRY, STANLEY Z
BACHELOR FATHER (I,323); THE BEACH GIRLS (II,191); THE DICK VAN DYKE SHOW (I,1275); DOBIE GILLIS (I,1302); GILLIGAN'S ISLAND

(II,990); THE KIDS FROM C.A.P.E.R. (II,1391); MCKEEVER AND THE COLONEL (I,2970); PLEASE DON'T EAT THE DAISIES (I,3628); TOO MANY SERGEANTS (I,4573)

CHESTNUT, BILL
TRUTH OR CONSEQUENCES (I,4620)

CHICKEY, ANTHONY
MAUDE (II,1655)

CHINIQUY, GERRY
DUNGEONS AND DRAGONS (II,744); G.I. JOE: A REAL AMERICAN HERO (II,948); JIM HENSON'S MUPPET BABIES (II,1289); MEATBALLS AND SPAGHETTI (II,1674); THE PINK PANTHER (II,2042); THE TRANSFORMERS (II,2653)

CHODERKER, GEORGE
THE $128,000 QUESTION (II,1903); PASSWORD PLUS (II,1966); SUPER PASSWORD (II,2499)

CHOMSKY, MARVIN
BIG HAWAII (II,238); DANGER IN PARADISE (II,617); THE DOCTORS (I,1323); HAWAII FIVE-O (II,1110); KATE MCSHANE (II,1368); KATE MCSHANE (II,1369); THE MAGICIAN (I,2823); THE MAGICIAN (I,2824); MISSION: IMPOSSIBLE (I,3067); THE NAME OF THE GAME (I,3217); ROOTS (II,2211); STAR TREK (II,2440); THEN CAME BRONSON (I,4435); THE WILD WILD WEST (I,4863)

CHOMYN, JOSEPH K
SOMERSET (I,4115)

CHRISTENSEN, DON
FAT ALBERT AND THE COSBY KIDS (II,835); MISSION MAGIC (I,3068); THE SECRET LIVES OF WALDO KITTY (II,2292); TARZAN: LORD OF THE JUNGLE (II,2544); THE TOM AND JERRY COMEDY SHOW (II,2629)

CHRISTMAN, BOB
COMEDY NEWS II (I,1021)

CHRISTOPHER, WILLIAM
AFTERMASH (II,23)

CHUDNOW, BYRON
ALEX AND THE DOBERMAN GANG (II,32)

CHULAY, JOHN
THE BOB NEWHART SHOW (II,342)

CIAPPESSONI, PAUL
A HORSEMAN RIDING BY (II,1182); THE ONEDIN LINE (II,1912)

CIRKER, IRA
ANOTHER WORLD (II,97)

CISNEY, MARCELLA
WOMAN WITH A PAST (I,4889)

CLAAR, JOHN
EARN YOUR VACATION (I,1394); LIFE WITH FATHER (I,2690); MEET CORLISS ARCHER (I,2979)

CLAIR, DICK
MAMA'S FAMILY (II,1610)

CLAMPETT, BOB
THE BEANY AND CECIL SHOW (I,374); TIME FOR BEANY (I,4505)

CLARK, BARRY
MYSTERIES, MYTHS AND LEGENDS (II,1782)

CLARK, GLORYETTE
RIPTIDE (II,2178)

CLARK, JAMES
BATMAN (I,366); BONANZA (II,347); BUICK ELECTRA PLAYHOUSE (I,760); LASSIE (I,2622); THE MONROES (I,3089); MY FRIEND FLICKA (I,3190); TIMMY AND LASSIE (I,4520); VOYAGE TO THE BOTTOM OF THE SEA (I,4743); THE WILD WILD WEST (I,4863)

CLARK, LEE
THE MICKEY MOUSE CLUB (I,3025)

CLARK, STEVE
THE FAMOUS ADVENTURES OF MR. MAGOO (I,1524); SPIDER-MAN AND HIS AMAZING FRIENDS (II,2425)

CLARKE, JOHN
HOUSE CALLS (II,1194)

CLARKE, LAWRENCE GORDON
FLAMBARDS (II,869)

CLAVELL, JAMES
THE DETECTIVES (I,1245); THE RIFLEMAN (I,3789)

CLAVER, BOB
ALL'S FAIR (II,46); AT EASE (II,116); AUTOMAN (II,120); CALIFORNIA FEVER (II,411); CAT BALLOU (I,870); DAUGHTERS (II,620); DOC (II,682); THE DUKES OF HAZZARD (II,742); ENSIGN O'TOOLE (I,1448); THE FACTS OF LIFE (II,805); THE FARMER'S DAUGHTER (I,1533); GIDGET (I,1795); GLORIA (II,1010); HERE COME THE BRIDES (I,2024); HOUSE CALLS (II,1194); JOE AND VALERIE (II,1298); JOE AND VALERIE (II,1299); THE LOVE BOAT (II,1535); MORK AND MINDY (II,1735); NO SOAP, RADIO (II,1856); OVER AND OUT (II,1938); PALS (II,1947); THE PARTRIDGE FAMILY (II,1962); PAUL SAND IN FRIENDS AND LOVERS (II,1982); THE POWERS OF MATTHEW STAR (II,2075); RHODA (II,2151); THE SHAMEFUL SECRETS OF HASTINGS CORNERS (I,3994); STARTING FRESH (II,2447); THE TEXAS WHEELERS (II,2568); WEBSTER (II,2758); WELCOME BACK, KOTTER (II,2761); YOUNG MAVERICK (II,2867)

CLAXTON, WILLIAM F
BLACK SADDLE (I,480); THE BLUE KNIGHT (II,276); BONANZA (II,347); BRET MAVERICK (II,374); DALLAS (II,614); THE DESILU PLAYHOUSE (I,1237); THE DETECTIVES (I,1245); FAME (II,812); FATHER MURPHY (II,841); HERE COME THE BRIDES (I,2024); THE HIGH CHAPARRAL (I,2051); LAW OF THE PLAINSMAN (I,2639); LITTLE HOUSE ON THE PRAIRIE (II,1487); MAVERICK (II,1657); THE ROOKIES (II,2208); ROUTE 66 (I,3852); SARA (II,2260); SHIRLEY (II,2330); THEN CAME BRONSON (I,4435); THRILLER (I,4492); THE TWILIGHT ZONE (I,4651)

CLEGG, TOM
RETURN OF THE SAINT (II,2141); SPACE: 1999 (II,2409)

CLEMENT, DICK
ON THE ROCKS (II,1894)

CLEMENTS, CALVIN
JIGSAW (I,2377)

CLEWS, COLIN
THE ENGELBERT HUMPERDINCK SPECIAL (I,1446)

CLIFFORD, GRAEME
BARNABY JONES (II,153); THE NEW AVENGERS (II,1816)

CLINE, EDWARD F
FIREBALL FUN FOR ALL (I,1577)

CLOUSE, ROBERT
THE MASTER (II,1648)

COBB, LEE J
TENNESSEE ERNIE FORD MEETS KING ARTHUR (I,4397)

COBURN, JAMES
THE ROCKFORD FILES (II,2197)

COE, BOB
LAVERNE AND SHIRLEY IN THE ARMY (II,1448)

COE, FRED
ALL THE WAY HOME (I,140); THE CLOCK (I,978); DATELINE (I,1165); FOR YOUR PLEASURE (I,1627); GREAT CATHERINE (I,1873); HEAVENS TO BETSY (I,1997); THE LAST WAR (I,2630); LIGHT'S OUT (I,2699); MASTERPIECE PLAYHOUSE (I,2942); MR. MERGENTHWIRKER'S LOBBLIES (I,3144); PHILCO TELEVISION PLAYHOUSE (I,3583)

COFFY, JACK
ALL MY CHILDREN (II,39); SOMERSET (I,4115)

COFOD, FRANKLIN
VOLTRON—DEFENDER OF THE UNIVERSE (II,2729)

COHAN, MARTY
DOC (II,682); RHODA (II,2151); THE TED KNIGHT SHOW (II,2550)

COHEN, LAWRENCE J
MOMMA THE DETECTIVE (II,1721)

COHN, BRUCE
YESTERYEAR (II,2851)

COKE, CYRIL
THE ONEDIN LINE (II,1912); PRIDE AND PREJUDICE (II,2080)

COLASANTO, NICHOLAS
COLUMBO (II,556); THE FELONY SQUAD (I,1563); HEC RAMSEY (II,1121); HERE COME THE BRIDES (I,2024); LOBO (II,1504); LOGAN'S RUN (II,1507); S.W.A.T. (II,2236); TODAY'S FBI (II,2628)

COLE, MARCUS
PRISONER: CELL BLOCK H (II,2085)

COLE, TOM
YOU DON'T SAY (II,2858); YOU DON'T SAY (II,2859)

COLEMAN, HENRY
DINNER DATE (I,1286)

COLEMAN, HERBERT
PLEASE DON'T EAT THE DAISIES (I,3628)

COLLA, RICHARD
BATTLESTAR GALACTICA (II,181); CHIPS (II,511); COVER UP (II,597); IRONSIDE (II,1246); JAKE'S WAY (II,1269); MCCLOUD (II,1660); MCCLOUD: WHO KILLED MISS U.S.A.? (I,2965); MIAMI VICE (II,1689); MURDER, SHE WROTE (II,1770); SARGE (I,3914); SARGE: THE BADGE OR THE CROSS (I,3915); SHANNON (II,2314); TENAFLY (II,2560); TENAFLY (I,4395); TRAPPER JOHN, M.D. (II,2654); THE WHOLE WORLD IS WATCHING (I,4851); WIZARDS AND WARRIORS (II,2813)

COLLERAN, BILL
THE BING CROSBY SPECIAL (I,471); THE BING CROSBY SPECIAL (I,472); CRESCENDO (I,1094); A DATE WITH DEBBIE (I,1161); THE FRANK SINATRA SHOW (I,1666); THE FRANK SINATRA TIMEX SHOW (I,1668); HOLIDAY U.S.A. (I,2078); THE POLLY BERGEN SHOW (I,3643)

COLLINS, GEOFF
LAVERNE AND SHIRLEY WITH THE FONZ (II,1449)

COLLINS, RICHARD
ROUTE 66 (I,3852)

COLLINS, ROBERT
DAN AUGUST (I,1130); THE DOCTORS (I,1323); MEDICAL STORY (II,1676); OUR FAMILY BUSINESS (II,1931); PAT PAULSEN FOR PRESIDENT (I,3472); SERPICO (II,2306); STEELTOWN (II,2452)

COMFORT, LANCE
IVANHOE (I,2282)

COMPTON, BENNETT
WHITE HUNTER (I,4845)

COMPTON, RICHARD
HARDCASTLE AND MCCORMICK (II,1089); WILD TIMES (II,2799)

CONNELL, TOM
A NEW KIND OF FAMILY (II,1819)

CONNOR, KEVIN
GOLIATH AWAITS (II,1025); HART TO HART (II,1102); MASTER OF THE GAME (II,1647); MISTRAL'S DAUGHTER (II,1708); REMINGTON STEELE (II,2130); SPACE: 1999 (II,2409); WIZARDS AND WARRIORS (II,2813)

CONNORS III, TOM
THE BIONIC WOMAN (II,255); THE FALL GUY (II,811); THE SIX-MILLION-DOLLAR MAN (II,2372)

CONRAD, ROBERT
THE BLACK SHEEP SQUADRON (II,262); THE DUKES (II,740); A MAN CALLED SLOANE (II,1613)

CONRAD, WILLIAM
77 SUNSET STRIP (I,3988); GENERAL ELECTRIC TRUE (I,1754); GUNSMOKE (II,1069); KLONDIKE (I,2573); THE MAN FROM GALVESTON (I,2865); NAKED CITY (I,3216); THE NAME OF THE GAME (I,3217); ROUTE 66 (I,3852); SAINTS AND SINNERS (I,3887); TARGET: THE CORRUPTERS (I,4360); TEMPLE HOUSTON (I,4392)

CONWAY, CURT
JAMIE (I,2333)

CONWAY, GARY
PRISONER: CELL BLOCK H (II,2085)

CONWAY, JAMES L
GREATEST HEROES OF THE BIBLE (II,1062); THE LIFE AND TIMES OF GRIZZLY ADAMS (II,1469); MATT HOUSTON (II,1654)

CONWAY, TIM
THE PRIMARY ENGLISH CLASS (II,2081)

COOK, ALAN
QUINCY, M. E. (II,2102)

COOK, DON
THE AL MORGAN SHOW (I,94); CHEZ PAREE REVUE (I,932)

COOK, FIELDER
BEACON HILL (II,193); BEAUTY AND THE BEAST (I,382); BEN CASEY (I,394); BRIGADOON (I,726); THE DEFENDERS (I,1220); DOWN HOME (II,731); THE ELEVENTH HOUR (I,1425); THE FARMER'S DAUGHTER (I,1534); THE HANDS OF CORMAC JOYCE (I,1927); HARVEY (I,1962); MIRACLE ON 34TH STREET (I,3058); MIRROR, MIRROR, OFF THE WALL (I,3059); MR. ROBERTS (I,3149); THE PHILADELPHIA STORY (I,3582); THE PRICE (I,3666); THE RIVALRY (I,3796); SAM HILL: WHO KILLED THE MYSTERIOUS MR. FOSTER? (I,3896); A STRING OF BEADS (I,4263); TEACHER, TEACHER (I,4366); VALLEY FORGE (I,4698); THE WALTONS (II,2740)

COOKE, ALAN
CAPITOL (II,426); DEAR DETECTIVE (II,652); FLAMINGO ROAD (II,872); GAVILAN (II,959); HARPER VALLEY (II,1094); HARPER VALLEY PTA (II,1095); HART TO HART (II,1102); HOUSE CALLS (II,1194); MURDER, SHE WROTE (II,1770); PRIVATE BENJAMIN (II,2087)

COOLEY, LEE
THE PERRY COMO SHOW (I,3531); THE PERRY COMO SHOW (I,3532); TV'S TOP TUNES (I,4635)

COON, GENE L
LAREDO (I,2615); THE MEN FROM SHILOH (I,3005); RAWHIDE (I,3727); THE VIRGINIAN (I,4738); WAGON TRAIN (I,4747)

COOPER, HAL
ALL'S FAIR (II,46); AND THEY ALL LIVED HAPPILY EVER AFTER (II,75); APPLE'S WAY (II,101); THE ASTRONAUTS (II,114); BOBBY JO AND THE BIG APPLE GOODTIME BAND (I,674); THE BRADY BUNCH (II,362); THE BRIGHTER DAY (I,728); THE COURTSHIP OF EDDIE'S FATHER (I,1083); THE DICK VAN DYKE SHOW (I,1275); DID YOU HEAR ABOUT JOSH AND KELLY?! (II,673); THE DUMPLINGS (II,743); A FINE ROMANCE (II,858); FOR BETTER OR WORSE (I,1621); FREE COUNTRY (II,921); FREEMAN (II,923); FUNNY FACE (I,1711); GIDGET (I,1795); GILLIGAN'S ISLAND (II,990); GIMME A BREAK (II,995); HAZEL (I,1982); HOT L BALTIMORE (II,1187); I DREAM OF JEANNIE (I,2167); JERRY (II,1283); KING OF THE ROAD (II,1396); LOVE THY NEIGHBOR (I,2781); LOVE, SIDNEY (II,1547); MAUDE (II,1655); MCNAMARA'S BAND (II,1668); MR. & MRS. & MR. (II,1744); MY WORLD . . . AND WELCOME TO IT (I,3208); N.Y.P.D. (I,3321); THE NANCY WALKER SHOW (II,1790); NEVER AGAIN (II,1808); THE ODD COUPLE (II,1875); ONE DAY AT A TIME (II,1900); PHYL AND MIKHY (II,2037); PORTIA FACES LIFE (I,3653); POTTSVILLE (II,2072); REAR GUARD (II,2120); ROOM 222 (I,3843); SNAVELY (II,2387); THAT GIRL (II,2570); TWO THE HARD WAY (II,2695); YOUR SURPRISE PACKAGE (I,4960)

COOPER, JACKIE
THE BLACK SHEEP SQUADRON (II,262); DOCTOR DAN (II,684); FAMILY IN BLUE (II,818); THE FEATHER AND FATHER GANG (II,845); GLITTER (II,1009); HAVING BABIES III (II,1109); HENNESSEY (I,2013); JESSIE (II,1287); KEEP THE FAITH (I,2518); LOU GRANT (II,1526); M*A*S*H (II,1569); MCMILLAN (II,1666); MOBILE ONE (II,1717); MOONLIGHT (II,1730); PARIS (II,1957); THE PEOPLE'S CHOICE (I,3521); QUINCY, M. E. (II,2102); SNAFU (II,2386); THE TEXAS WHEELERS (II,2568); TRAPPER JOHN, M.D. (II,2654); THE WHITE SHADOW (II,2788)

COOPER, JOHN
MURDER ON THE MIDNIGHT EXPRESS (I,3162); POSSESSION (I,3655); TYCOON: THE STORY OF A WOMAN (II,2697)

COOPER, LESTER
MAKE A WISH (I,2838)

COOPER, WYLLIS
ESCAPE (I,1459); STAGE 13 (I,4183); VOLUME ONE (I,4742)

COPELAN, JODIE
THE BETTY HUTTON SHOW (I,413); SKY KING (I,4077)

COPELAND, ALAN
THE WILD WILD WEST (I,4863)

COPPERFIELD, DAVID
THE MAGIC OF DAVID COPPERFIELD (II,1593); THE MAGIC OF DAVID COPPERFIELD (II,1595)

CORBETT, PATRICK
HIGH HOPES (II,1144)

CORDAY, TED
THE GUIDING LIGHT (II,1064)

CORDERY, HOWARD
DRAW ME A LAUGH (I,1372); THE RAY KNIGHT REVUE (I,3733); THAT'S OUR SHERMAN (I,4429)

COREA, NICHOLAS
ARCHER—FUGITIVE FROM THE EMPIRE (II,103); THE RENEGADES (II,2133)

COREY, ALLEN
STONE (II,2470)

COREY, JEFF
ALIAS SMITH AND JONES (I,118); THE BOB NEWHART SHOW (II,342); HAWKINS (I,1976); POLICE STORY (II,2062); THE PSYCHIATRIST

(I,3686); THE SIXTH SENSE (I,4069)

CORRIGAN, WILLIAM
HEDDA HOPPER'S HOLLYWOOD (I,2002); LIGHT'S OUT (I,2699); LITTLE WOMEN (I,2722); MASTERPIECE PLAYHOUSE (I,2942); MIRACLE ON 34TH STREET (I,3057); THE STRAWBERRY BLONDE (I,4254)

COSBY, BILL
THE BILL COSBY SHOW (I,442)

COSTALANO, NICHOLAS
FITZ AND BONES (II,867)

COSTALANO, RICHARD
THE CONTENDER (II,571)

COTTEN, JOSEPH
PETER HUNTER, PRIVATE EYE (I,3563)

COURTLAND, JEROME
DALLAS (II,614); DYNASTY (II,746); FANTASY ISLAND (II,829); HAWAIIAN EYE (I,1973); HOTEL (II,1192); THE LOVE BOAT (II,1535); MATT HOUSTON (II,1654)

COWARD, NOEL
BLITHE SPIRIT (I,484); TOGETHER WITH MUSIC (I,4532)

COX, JIM
THE MAC DAVIS SHOW (II,1576)

COX, NELL
9 TO 5 (II,1852); M*A*S*H (II,1569); THE WALTONS (II,2740)

COYLE, HARRY
GUESS WHAT? (I,1895); PANTOMIME QUIZ (I,3449)

CRABTREE, ARTHUR
THE ADVENTURES OF SIR LANCELOT (I,76); COLONEL MARCH OF SCOTLAND YARD (I,1005); IVANHOE (I,2282)

CRAFT, JOHN
JANE EYRE (II,1273)

CRAIN, WILLIAM
THE ROOKIES (II,2208)

CRAMER, NED
ERNIE IN KOVACSLAND (I,1454); IT'S TIME FOR ERNIE (I,2277); KOVACS ON THE CORNER (I,2582); KOVACS UNLIMITED (I,2583)

CRAMER, THOMAS
NOT NECESSARILY THE NEWS (II,1862)

CRANDELL, DAVID
COSMOPOLITAN THEATER (I,1057)

CRANE, BARRY
THE BIONIC WOMAN (II,255); BUCK ROGERS IN THE 25TH CENTURY (II,384); CARIBE (II,439); CHIPS (II,511); DALLAS (II,614); THE DEVLIN CONNECTION (II,664); THE FANTASTIC JOURNEY (II,826); FLAMINGO ROAD (II,872); GALACTICA 1980 (II,953); HAWAII FIVE-O (II,1110); HUNTER (II,1205); THE INCREDIBLE HULK (II,1232); KUNG FU (II,1416); THE LAZARUS SYNDROME (II,1451); LOTTERY (II,1525); THE MAGICIAN (I,2823); THE MAN FROM ATLANTIS (II,1615); MISSION: IMPOSSIBLE (I,3067); POLICE STORY (II,2062); POLICE WOMAN (II,2063); THE POWERS OF MATTHEW STAR (II,2075); RAFFERTY (II,2105); SEVEN BRIDES FOR SEVEN BROTHERS (II,2307); SHERLOCK HOLMES: THE HOUND OF THE BASKERVILLES (I,4011); THE SIX-MILLION-DOLLAR MAN (II,2372); THE STREETS OF SAN FRANCISCO (II,2478); SUPERTRAIN (II,2504); SUPERTRAIN (II,2505); THREE FOR THE ROAD (II,2607); TODAY'S FBI (II,2628); TRAPPER JOHN, M.D. (II,2654); TRAUMA CENTER (II,2655); WHIZ KIDS (II,2790); WONDER WOMAN (SERIES 1) (II,2828)

CRANE, PETER
B.J. AND THE BEAR (II,126); COVER UP (II,597); DARKROOM (II,619); THE FALL GUY (II,811); KNIGHT RIDER (II,1402); THE MASTER (II,1648)

CRANE, WILLIAM
S.W.A.T. (II,2236)

CRANNEY, JON
A CHRISTMAS CAROL (II,517)

CRANSTON, BARRY
HIGH HOPES (II,1144)

CRAVEN, WES
STRANGER IN OUR HOUSE (II,2477)

CRAWFORD, DENNY
THE OSMOND FAMILY SHOW (II,1927)

CRENNA, RICHARD
ALLISON SIDNEY HARRISON (II,54); THE ANDY GRIFFITH SHOW (I,192); CAP'N AHAB (I,824); THE CHEERLEADERS (II,493); GRANDPA GOES TO WASHINGTON (II,1050); THE HOYT AXTON SHOW (II,1200); LOU GRANT (II,1526); MARIE (II,1628); NO TIME FOR SERGEANTS (I,3300); THE REAL MCCOYS (I,3741); ROSETTI AND RYAN (II,2216); TURNABOUT (II,2669); WENDY AND ME (I,4793)

CREWS, COLIN
PICCADILLY PALACE (I,3589)

CRICHTON, CHARLES
THE ADVENTURES OF BLACK BEAUTY (I,51); THE AVENGERS (II,121); MAN IN A SUITCASE (I,2868); RETURN OF THE SAINT (II,2141); SHIRLEY'S WORLD (I,4019); SPACE: 1999 (II,2409); THE STRANGE REPORT (I,4248)

CRICHTON, DON
THE CHAMPIONS (I,896)

CRICHTON, ROBIN
IN SEARCH OF. . . (II,1226)

CRIPPEN, FRED
HOT WHEELS (I,2126); SKYHAWKS (I,4079)

CRISTOFER, MICHAEL
CANDIDA (II,423)

CROFT, PETER
HIPPODROME (I,2060)

CRONYN, HUME
ACTOR'S STUDIO (I,28)

CROPP, BEN
THE CORAL JUNGLE (II,574)

CROSLAND JR, ALAN
77 SUNSET STRIP (I,3988); ADAM-12 (I,31); ALCOA PREMIERE (I,109); ALFRED HITCHCOCK PRESENTS (I,115); BEN CASEY (I,394); THE BIONIC WOMAN (II,255); BONANZA (II,347); CHASE (I,917); CHEYENNE (I,931); COLT .45 (I,1010); THE D.A. (I,1123); EMERGENCY! (II,775); GENERAL ELECTRIC TRUE (I,1754); LAWMAN (I,2642); MAVERICK (II,1657); MR. LUCKY (I,3141); OF THIS TIME, OF THAT PLACE (I,3335); THE OUTER LIMITS (I,3426); PETER GUNN (I,3562); RAWHIDE (I,3727); THE SECRET EMPIRE (II,2291); SERGEANT PRESTON OF THE YUKON (I,3980); THE SIX-MILLION-DOLLAR MAN (II,2372); THE VIRGINIAN (I,4738); VOYAGE TO THE BOTTOM OF THE SEA (I,4743); WONDER WOMAN (SERIES 1) (II,2828); WONDER WOMAN (SERIES 2) (II,2829)

CROSS, JIM
GAMES PEOPLE PLAY (II,956); LIFESTYLES OF THE RICH AND FAMOUS (II,1474); THE SUNDAY GAMES (II,2494)

CROSS, PERRY
THE ART LINKLETTER SHOW (I,276); THE ERNIE KOVACS SHOW (I,1455)

CROTTY, BUCK
HOLLYWOOD STAR REVUE (I,2093)

CRUM, JIM
WE'RE MOVIN' (II,2756); YOUR NEW DAY (II,2873)

CSIKI, TONY
FERNWOOD 2-NIGHT (II,849); MAUDE (II,1655); SORORITY '62 (II,2405)

CUDDINGTON, CHRIS
THE DRAK PACK (II,733)

CULHANE, JAMES
CARTOONSVILLE (I,858)

CULHANE, SHAMUS
THE MILTON THE MONSTER CARTOON SHOW (I,3053); PLANET PATROL (I,3616)

CULLINGHAM, MARK
CASANOVA (II,451)

CULP, ROBERT
THE GREATEST AMERICAN HERO (II,1060); I SPY (I,2179)

CUMMING, FIONA
BLAKE'S SEVEN (II,264); THE OMEGA FACTOR (II,1888)

CUMMINGS, DREW
YOU ASKED FOR IT (II,2855)

CUNLIFFE, DAVID
THE ONEDIN LINE (II,1912)

CURTIS, DAN
DARK SHADOWS (I,1157); IN THE DEAD OF NIGHT (I,2202); THE LONG DAYS OF SUMMER (II,1514); MELVIN PURVIS: G-MAN (II,1679); THE NIGHT STRANGLER (I,3293); THE NORLISS TAPES (I,3305); SUPERTRAIN (II,2504); THE WINDS OF WAR (II,2807)

CURY, IVAN
BARBI BENTON SPECIAL: A BARBI DOLL FOR CHRISTMAS (II,148); COUNTRY NIGHT OF STARS (II,592); COUNTRY NIGHT OF STARS II (II,593); ELVIS REMEMBERED: NASHVILLE TO HOLLYWOOD (II,772); TAKE FIVE WITH STILLER AND MEARA (II,2530)

D'ANGELO, WILLIAM P
ALICE (II,33); MCDUFF, THE TALKING DOG (II,1662); THE MISADVENTURES OF SHERIFF LOBO (II,1704); PRIVATE BENJAMIN (II,2087); THE RED HAND GANG (II,2122); TURNABOUT (II,2669); WEBSTER (II,2758)

D'ARCY, WILLIAM
GILLIGAN'S ISLAND (II,990)

D'USSEAU, LORING
YOUR FUNNY FUNNY FILMS (I,4950)

DADARIO, JEAN
ALL MY CHILDREN (II,39)

DAGONE, ROBERT
IT'S ONLY HUMAN (II,1257)

DAHL, MARJA
BLACKSTAR (II,263); THE NEW ADVENTURES OF MIGHTY MOUSE AND HECKLE AND JECKLE (II,1814); THE TARZAN/LONE RANGER/ZORRO ADVENTURE HOUR (II,2543)

DAHL, MARSHALL
THE KID SUPER POWER HOUR WITH SHAZAM (II,1386)

DAILEY, BOB
PERSON TO PERSON (I,3550)

DAILEY, DAN
KOSTA AND HIS FAMILY (I,2581); THE ODD COUPLE (II,1875)

DALEN, ZALE
THE EDISON TWINS (II,761)

DAMES, ROD
BENSON (II,208)

DAMSKI, MEL
THE AMERICAN DREAM (II,71); THE ANDROS TARGETS (II,76); BARNABY JONES (II,153); THE BIONIC WOMAN (II,255); JULIE FARR, M.D. (II,1354); KAZ (II,1370); M*A*S*H (II,1569); MAKING THE GRADE (II,1606)

DANIEL, ROD
THE DUCK FACTORY (II,738); FILTHY RICH (II,856); FIRST & TEN (II,861); HARPER VALLEY (II,1094); KUDZU (II,1415); MAGNUM, P.I. (II,1601); NEWHART (II,1835); NICHOLS AND DYMES (II,1841); PARTNERS IN CRIME (II,1961); POOR RICHARD (II,2065); A ROCK AND A HARD PLACE (II,2190); TEACHERS ONLY (II,2548); TUCKER'S WITCH (II,2667); WKRP IN CINCINNATI (II,2814)

DANIELEWSKI, TAD
MUGGSY (II,1761)

DANIELS, DANNY
GENE KELLY'S WONDERFUL WORLD OF GIRLS (I,1752); THE WONDERFUL WORLD OF GIRLS (I,4898)

DANIELS, GLENN
MUSIC WORLD (II,1775)

DANIELS, MARC
ALICE (II,33); AMANDA'S (II,62); THE ANDROS TARGETS (II,76); APPLE'S WAY (II,101); ARSENIC AND OLD LACE (I,267); ASSIGNMENT: EARTH (I,299); BEN CASEY (I,394); BONANZA (II,347); BRANDED (I,707); BURKE'S LAW (I,762); CANNON (II,424); CO-ED FEVER (II,547); COLT .45 (I,1010); THE DICK POWELL SHOW (I,1269); DOCTORS' PRIVATE LIVES (II,692); THE DORIS DAY SHOW (I,1355); EIGHT IS ENOUGH (II,762); EMILY, EMILY (I,1438); FAME (I,1517); FAME (II,812); FATHER KNOWS BEST: HOME FOR CHRISTMAS (II,839); FLO (II,891); GIBBSVILLE (II,988); THE GOLDBERGS (I,1836); THE GOLDBERGS (I,1835); GUN SHY (II,1068); GUNSMOKE (II,1069); HAVE I GOT A CHRISTMAS FOR YOU (II,1105); THE HEIRESS (I,2006); HERE WE GO AGAIN (I,2028); HUSBANDS, WIVES AND LOVERS (II,1209); I LOVE LUCY (I,2171); I MARRIED JOAN (I,2174); THE IMOGENE COCA SHOW (I,2198); JAMES AT 15 (II,1270); JANE EYRE (I,2338); JIGSAW (I,2377); LAST OF THE PRIVATE EYES (I,2628); THE LIFE AND TIMES OF EDDIE ROBERTS (II,1468); LITTLE WOMEN (I,2719); THE LUCILLE BALL SPECIAL (II,1559); THE MAN FROM ATLANTIS (II,1615); THE MAN FROM U.N.C.L.E. (I,2867); MARCUS WELBY, M.D. (II,1627); THE MEN FROM SHILOH (I,3005); MISSION: IMPOSSIBLE (I,3067); MY HERO (I,3193); OWEN MARSHALL: COUNSELOR AT LAW (I,3435); PARIS 7000 (I,3459); PLANET EARTH (I,3615); THE POWER AND THE GLORY (I,3658); PRIVATE BENJAMIN (II,2087); PRUDENCE AND THE CHIEF (I,3682); SAINTS AND SINNERS (I,3887); SEARCH (I,3951); SHANE (I,3995); SKINFLINT (II,2377); SOLDIERS OF FORTUNE (I,4111); SPENCER'S PILOTS (II,2422); STAR TREK (II,2440); THE SURVIVORS (I,4301); TOMA (II,2634); VEGAS (II,2724); WHERE'S RAYMOND? (I,4837); WITNESS (I,4885)

DANIELS, NORMAN
THE DORIS DAY SHOW (I,1355)

DANIELS, STAN
MR. SMITH (II,1755)

DANTE, JOE
POLICE SQUAD! (II,2061)

DANTON, RAY
THE AMERICAN DREAM (II,71); BENDER (II,204); CAGNEY AND LACEY (II,409); FEEL THE HEAT (II,848); THE INCREDIBLE HULK (II,1232); QUINCY, M. E. (II,2102); SWITCH (II,2519); TALES OF THE UNEXPECTED (II,2540)

DARE, DANNY
HOW TO MARRY A MILLIONAIRE (I,2147)

DARLEY, CHRIS
ALL-STAR SECRETS (II,52); THE GIRL IN MY LIFE (II,996); THE LIAR'S CLUB (II,1466); SCRABBLE (II,2278); SECOND CHANCE (II,2286); YOU BET YOUR LIFE (II,2856)

DARLEY, DICK
CLUB OASIS (I,985); GEORGE BURNS IN THE BIG TIME (I,1764); THE ROSEMARY CLOONEY SHOW (I,3847); THE ROSEMARY CLOONEY SHOW (I,3848); SIGMUND AND THE SEA MONSTERS (II,2352); SPACE PATROL (I,4144); THE SPIKE JONES SHOW (I,4156); WHEW! (II,2786)

DARLING, JOAN
A.K.A. PABLO (II,28); DOC (II,682); DOC (II,683); THE FAMILY TREE (II,820); HIZZONER (II,1159); M*A*S*H (II,1569); MARY HARTMAN, MARY HARTMAN (II,1638); THE MARY TYLER MOORE SHOW (II,1640); PHYLLIS (II,2038); RHODA (II,2151); RICH MAN, POOR MAN—BOOK II (II,2162); STOCKARD CHANNING IN JUST FRIENDS (II,2467)

DATER, TOM
LEGENDS OF THE SCREEN (II,1456)

DAUGHERTY, HERSCHEL
ALCOA PREMIERE (I,109); ALFRED HITCHCOCK PRESENTS (I,115); APPLE'S WAY (II,101); BRACKEN'S WORLD (I,703); CANNON (II,424); CIMARRON STRIP (I,954); THE DEPUTY (I,1234); DO NOT DISTURB (I,1300); THE DOG TROOP (I,1325); EMERGENCY! (II,775); THE GIRL FROM U.N.C.L.E. (I,1808); THE GRADUATION DRESS (I,1864); GUNSMOKE (II,1069); HEC RAMSEY (II,1121); HERE COME THE BRIDES (I,2024); THE HIGH CHAPARRAL (I,2051); MARCUS WELBY, M.D. (II,1627); MY THREE SONS (I,3205); PETROCELLI (II,2031); POLICE WOMAN (II,2063); THE RAT PATROL (I,3726); RAWHIDE (I,3727); SHANE (I,3995); THE SLOWEST GUN IN THE WEST (I,4088); SOLDIERS OF FORTUNE (I,4111); STAR TREK (II,2440); THE TEXAS WHEELERS (II,2568); THREE FOR THE ROAD (II,2607); THRILLER (I,4492); THE TIME TUNNEL (I,4511); THE VIRGINIAN (I,4738); WAGON TRAIN (I,4747); YES YES NANETTE (I,4927)

DAVID, BILL
TV'S BLOOPERS AND PRACTICAL JOKES (II,2675)

DAVID, HUGH
THE PALLISERS (II,1944)

DAVID, JERRY
THE UGLIEST GIRL IN TOWN (I,4663)

DAVIDSON, ERIC
THE OMEGA FACTOR (II,1888)

DAVIDSON, MARTIN
CALL TO GLORY (II,413); FLATBUSH/AVENUE J (II,878)

DAVIDSON, WILLIAM
ADVENTURES IN RAINBOW COUNTRY (I,47)

DAVIES, GARETH
CRIMES OF PASSION (II,603)

DAVIES, HOWARD
PIAF (II,2039)

DAVIES, JOHN
GOOD NEIGHBORS (II,1034); MONTY PYTHON'S FLYING CIRCUS (II,1729); WAR AND PEACE (I,4765); WHY DIDN'T THEY ASK EVANS? (II,2795)

DAVIS, ART
THE PINK PANTHER (II,2042); THE SUPER SIX (I,4292)

DAVIS, BARRY
OPPENHEIMER (II,1919); WAIT UNTIL DARK (II,2735)

DAVIS, BILL

BOBBIE GENTRY'S HAPPINESS HOUR (I,671); BOBBIE GENTRY'S HAPPINESS HOUR (II,343); THE BOBBY VAN AND ELAINE JOYCE SHOW (I,677); THE BRASS ARE COMING (I,708); THE CARPENTERS (II,444); THE CELEBRITY FOOTBALL CLASSIC (II,474); CHER (II,495); THE DOROTHY HAMILL WINTER CARNIVAL SPECIAL (II,720); THE FLIP WILSON SPECIAL (I,1603); THE FLIP WILSON SPECIAL (II,888); THE FUNNY WORLD OF FRED & BUNNI (II,943); GABRIEL KAPLAN PRESENTS THE SMALL EVENT (II,952); GLEN CAMPBELL AND FRIENDS: THE SILVER ANNIVERSARY (II,1005); HEE HAW (II,1123); THE JACKSONS (II,1267); JIMMY DURANTE PRESENTS THE LENNON SISTERS HOUR (I,2387); JOHN DENVER AND FRIEND (II,1311); JOHN DENVER ROCKY MOUNTAIN CHRISTMAS (II,1314); THE JOHN DENVER SPECIAL (II,1315); THE JOHN DENVER SPECIAL (II,1316); JOHNNY CASH—A MERRY MEMPHIS CHRISTMAS (II,1331); THE JUD STRUNK SHOW (I,2462); THE JULIE ANDREWS HOUR (I,2480); THE KID SUPER POWER HOUR WITH SHAZAM (II,1386); LILY (I,2705); LILY FOR PRESIDENT (II,1478); LILY—SOLD OUT (II,1480); MARLO THOMAS AND FRIENDS IN FREE TO BE. . . YOU AND ME (II,1632); MERRY CHRISTMAS. . . WITH LOVE, JULIE (II,1687); THE NASHVILLE PALACE (II,1792); THE NATIONAL SNOOP (II,1797); OMNIBUS (II,1890); OPRYLAND: NIGHT OF STARS AND FUTURE STARS (II,1921); THE PAUL ANKA SHOW (II,1972); RODGERS AND HART TODAY (I,3829); SINATRA AND FRIENDS (II,2358); UPTOWN SATURDAY NIGHT (II,2711); VARIETY (II,2721); YOU CAN'T DO THAT ON TELEVISION (I,4933)

DAVIS, CHARLES R
GREATEST HEROES OF THE BIBLE (II,1062);

DAVIS, DAVID
FALCON'S GOLD (II,809)

DAVIS, DESMOND
LITTLE LORD FAUNTLEROY (II,1492); THE NEW AVENGERS (II,1816)

DAVIS, DONALD
ART AND MRS. BOTTLE (I,271); THE FUNNIEST JOKE I EVER HEARD (II,940); PRUDENTIAL FAMILY PLAYHOUSE (I,3683); THE REDD FOXX COMEDY HOUR (II,2126); THE WOLFMAN JACK SHOW (II,2817)

DAVIS, EDDIE
ADVENTURES OF THE SEASPRAY (I,82); BOSTON BLACKIE (I,695); THE CISCO KID (I,961); THE EDDIE CANTOR COMEDY THEATER (I,1407); MEET CORLISS ARCHER (I,2980); THE RAT PATROL (I,3726); TARGET (I,4359)

DAVIS, JERRY
BEWITCHED (I,418); THAT GIRL (II,2570)

DAVIS, MANNING
THE MIGHTY MOUSE PLAYHOUSE (I,3035)

DAVIS, MICHAEL P
LET ME TELL YOU ABOUT A SONG (I,2661)

DAVIS, ROBERT
DOG AND CAT (II,696)

DAVIS, WILLIAM
HULLABALOO (I,2156)

DAWSON, EARL
ALUMNI FUN (I,145)

DAY, ERNEST
THE NEW AVENGERS (II,1816)

DAY, LINDA
AFTER GEORGE (II,19); ARCHIE BUNKER'S PLACE (II,105); GIMME A BREAK (II,995); I'D RATHER BE CALM (II,1216); IT'S YOUR MOVE (II,1259); STAR OF THE FAMILY (II,2436); TEACHERS ONLY (II,2548); TOO CLOSE FOR COMFORT (II,2642); WKRP IN CINCINNATI (II,2814)

DAY, ROBERT
THE ADVENTURES OF POLLYANNA (II,14); THE AVENGERS (II,121); BANYON (I,344); BANYON (I,345); BEYOND WITCH MOUNTAIN (II,228); BRACKEN'S WORLD (I,703); CIRCLE OF FEAR (I,958); DALLAS (II,614); THE FBI (I,1551); GHOST STORY (I,1788); HAVING BABIES I (II,1107); THE HIGHWAYMAN (I,2059); THE HOUSE ON GREENAPPLE ROAD (I,2140); THE INVADERS (I,2229); KINGSTON: CONFIDENTIAL (II,1398); KINGSTON: THE POWER PLAY (II,1399); KODIAK (II,1405); LOGAN'S RUN (II,1507); LONDON AND DAVIS IN NEW YORK (II,1512); LUCAN (II,1554); MCCLOUD (II,1660); THE NAME OF THE GAME (I,3217); THE NEW ADVENTURES OF PERRY MASON (I,3252); OF MEN OF WOMEN (I,3332); SCRUPLES (II,2281); THE SENATOR (I,3976); THE SIXTH SENSE (I,4069); THE STREETS OF SAN FRANCISCO (II,2478); SUNSHINE (II,2495); SWITCH (II,2520); TENAFLY (II,2560); TWIN DETECTIVES (II,2687); WINNER TAKE ALL (II,2808)

DAYTON, DANNY
HERE'S LUCY (II,1135)

DEBOSIO, GIANFRANCO
MOSES THE LAWGIVER (II,1737)

DECAPRIO, AL
THE FRED WARING SHOW (I,1679); THE JOE NAMATH SHOW (I,2400); THE PHIL SILVERS PONTIAC SPECIAL: KEEP IN STEP (I,3579); THE PHIL SILVERS SHOW (I,3580); THAT SHOW STARRING JOAN RIVERS (I,4421); TROUBLE WITH RICHARD (I,4612)

DECORDOVA, FRED
THE BARBARA RUSH SHOW (I,349); THE BING CROSBY SPECIAL (I,470); BLITHE SPIRIT (I,484); THE DONNA REED SHOW (I,1347); THE DORIS DAY SHOW (I,1355); THE GEORGE BURNS AND GRACIE ALLEN SHOW (I,1763); THE GEORGE GOBEL SHOW (I,1768); HAVE GIRLS—WILL TRAVEL (I,1967); HER SCHOOL FOR BACHELORS (I,2018); THE JACK BENNY PROGRAM (I,2294); THE JACK BENNY SPECIAL (I,2296); JACK BENNY'S BIRTHDAY SPECIAL (I,2299); MR. ADAMS AND EVE (I,3121); MR. BELVEDERE (I,3127); MY THREE SONS (I,3205); THE SMOTHERS BROTHERS SHOW (I,4096); TO ROME WITH LOVE (I,4526)

DECOTIS, FRANCO
OUT OF OUR MINDS (II,1933)

DEFARIA, WALT
TRAVELS WITH CHARLEY (I,4597); THE WONDERFUL WORLD OF PIZZAZZ (I,4901)

DEGUERE, PHILIP
THE BLACK SHEEP SQUADRON (II,262); DOCTOR STRANGE (II,688)

DEHUFF, TOM
GAY NINETIES REVUE (I,1743)

DEKAY, JIM
DAVID NIVEN'S WORLD (II,625)

DELACEY, PHILLIPE
PANTOMIME QUIZ (I,3449)

DELUCA, RUDY
PEEPING TIMES (II,1990); STOPWATCH: THIRTY MINUTES OF INVESTIGATIVE TICKING (II,2474)

DEMORAES, RON
FANTASY (II,828); THE JOHN DAVIDSON SHOW (II,1310); THICKE OF THE NIGHT (II,2587)

DEPEW, JOSEPH
THE BEVERLY HILLBILLIES (I,417)

DESOUZA, STEVEN E
KNIGHT RIDER (II,1402)

DEVALLY JR, RAY
LAVERNE AND SHIRLEY (II,1446); WHO'S WATCHING THE KIDS? (II,2793)

DEVITO, DANNY
TAXI (II,2546)

DEARDEN, BASIL
THE PERSUADERS! (I,3556)

DEBHARDT, STEVE
JOHN LENNON AND YOKO ONO PRESENT THE ONE-TO-ONE CONCERT (I,2414)

DEIN, EDWARD
THE BLACK SHEEP SQUADRON (II,262); BRONCO (I,745); HAWAIIAN EYE (I,1973); THE ROARING TWENTIES (I,3805); THE WILD WILD WEST (I,4863)

DEITCH, GENE
TOM TERRIFIC (I,4546)

DEITCH, GEORGE
THE TOM AND JERRY SHOW (I,4534)

DESMOND, JOHN
THE BARBARA WALTERS SPECIAL (II,145); CASTLE ROCK (II,456); FEELING GOOD (I,1559); HOTEL COSMOPOLITAN (I,2128)

DETIEGE, DAVID
THE BUGS BUNNY/ROAD RUNNER SHOW (II,391); THE SUPER SIX (I,4292)

DETOTH, ANDRE
77 SUNSET STRIP (I,3988); BOURBON STREET BEAT (I,699); BRONCO (I,745)

HAWAIIAN EYE (I,1973); MAVERICK (II,1657); THE WESTERNER (I,4801)

DEUBEL, ROBERT
STRIPPERS (II,2481)

DEW, EDWARD
SERGEANT PRESTON OF THE YUKON (I,3980)

DEXTER, MAURY
FATHER MURPHY (II,841); LITTLE HOUSE ON THE PRAIRIE (II,1487); LITTLE HOUSE: A NEW BEGINNING (II,1488)

DEYRIES, BERNARD
INSPECTOR GADGET (II,1236); KIDD VIDEO (II,1388); THE LITTLES (II,1499); POLE POSITION (II,2060)

DIBENEDETTO, JULIO
THE ALAN KING SHOW (I,100); CANDID CAMERA (I,819)

DINOVI, GENE
CHANGING SCENE (I,899)

DIAMOND, VERN
EVERYTHING YOU EVER WANTED TO KNOW ABOUT MONSTERS . . . BUT WERE AFRAID! (II,794)

DICKSON, PAUL
THE CHAMPIONS (I,896); DEPARTMENT S (I,1232)

DIELHENN, ART
PUNKY BREWSTER (II,2092)

DIETRICK, GARTH
CANDID CAMERA (I,819); HOOTENANNY (I,2117); MATINEE IN NEW YORK (I,2946); ONE IN A MILLION (I,3379); TIMEX ALL-STAR JAZZ SHOW III (I,4517)

DILLON, SEAN
THE BILLY BEAN SHOW (I,449); JACK CARTER AND COMPANY (I,2305); SPARRING PARTNERS (I,4145)

DINEHART, ALAN
THE ALAN YOUNG SHOW (I,104); BATTLE OF THE PLANETS (II,179); CELEBRITY TIME (I,887); THE JEAN CARROLL SHOW (I,2353); MEET CORLISS ARCHER (I,2979); TURBO-TEEN (II,2668)

DISKIN, MARSHALL
BREAK THE BANK (I,712); THE COLLEGE BOWL (I,1001); FUN FOR THE MONEY (I,1708); THE STORK CLUB (I,4236)

DIXON, IVAN
THE A-TEAM (II,119); AIRWOLF (II,27); APPLE'S WAY (II,101); THE BILL COSBY SHOW (I,442); THE BILL COSBY SPECIAL, OR? (I,443); THE BLACK SHEEP SQUADRON (II,262); BLUE THUNDER (II,278); BRET MAVERICK (II,374); COUNTERATTACK: CRIME IN AMERICA (II,581); DELVECCHIO (II,659); GET CHRISTIE LOVE! (II,981); THE GREATEST AMERICAN HERO (II,1060); HARRIS AND COMPANY (II,1096); HOGAN'S HEROES (I,2069); KHAN! (II,1384); LEGMEN (II,1458); MAGNUM, P.I. (II,1601); THE NANCY DREW MYSTERIES (II,1789); NICHOLS (I,3283); PALMERSTOWN U.S.A. (II,1945); PALMS PRECINCT (II,1946); THE RIGHTEOUS APPLES (II,2174); THE ROCKFORD FILES (II,2197); SHAFT (I,3993); TALES OF THE GOLD MONKEY (II,2537); TENSPEED AND BROWN SHOE (II,2562); THE WALTONS (II,2740); WONDER WOMAN (SERIES 2) (II,2829)

DOBKIN, LAWRENCE
77 SUNSET STRIP (I,3988); ADAMS OF EAGLE LAKE (II,10); BARNABY JONES (II,153); THE BIG VALLEY (I,437); BRENDA STARR, REPORTER (II,373); BURKE'S LAW (I,762); CANNON (II,424); CARIBE (II,439); CHARLIE'S ANGELS (II,486); CONCRETE COWBOYS (II,563); THE DETECTIVES (I,1245); DOCTOR KILDARE (I,1315); THE DONNA REED SHOW (I,1347); DYNASTY (II,746); EMERGENCY! (II,775); ENSIGN O'TOOLE (I,1448); THE FALL GUY (II,811); FANTASY ISLAND (II,829); THE FITZPATRICKS (II,868); FREEBIE AND THE BEAN (II,922); THE FUGITIVE (I,1701); INTERTECT (I,2228); THE INVADERS (I,2229); MANHUNTER (II,1621); MAVERICK (II,1657); MOVIN' ON (II,1743); MR. ROBERTS (I,3149); THE MUNSTERS (I,3158); N.Y.P.D. (I,3321); NASHVILLE 99 (II,1791); THE RAT PATROL (I,3726); THE RIFLEMAN (I,3789); STAR TREK (II,2440); STRIKE FORCE (II,2480); VEGAS (II,2724); THE WALTONS (II,2740); WHAT REALLY HAPPENED TO THE CLASS OF '65? (II,2768); THE WILD WILD WEST (I,4863)

DOHENY, LAWRENCE
ADAM-12 (I,31); BIG HAWAII (II,238); THE BIG STORY (I,432); THE BLACK SHEEP SQUADRON (II,262); CHARLIE'S ANGELS (II,486); DOCTORS HOSPITAL (II,691); THE DUKES (II,740); THE GREATEST AMERICAN HERO (II,1060); MAGNUM, P.I. (II,1601); NAKED CITY (I,3216); QUINCY, M. E. (II,2102); THE ROCKFORD FILES (II,2197); THE SIX-MILLION-DOLLAR MAN (II,2372); SWITCH (II,2519); VEGAS (II,2724)

DOLLINGER, STEVE
AIRWOLF (II,27)

DONALD, JONATHAN
CALIFORNIA GIRL (I,794)

DONEHUE, VINCENT J
ANNIE GET YOUR GUN (I,223); THE GABBY HAYES SHOW (I,1723); MAGIC WITH MARY MARTIN (I,2821); MUSIC WITH MARY MARTIN (I,3180); PLAYWRIGHTS '56 (I,3627); REUNION IN VIENNA (I,3780); SKIN OF OUR TEETH (I,4073); THE WOMEN (I,4890)

DONER, RICHARD
THE SIX-MILLION-DOLLAR MAN (II,2372)

DONIGER, WALTER
BRADDOCK (I,704); CROWN THEATER WITH GLORIA SWANSON (I,1107); DELVECCHIO (II,659); THE DETECTIVES (I,1245); THE DICK POWELL SHOW (I,1269); GRIFF (I,1888); KUNG FU (II,1416); MARCUS WELBY, M.D. (II,1627); MCCLOUD (II,1660); OWEN MARSHALL: COUNSELOR AT LAW (I,3435); PEYTON PLACE (I,3574); SHANNON (II,2314); SQUADRON (I,4170); THE SURVIVORS (I,4301); SWITCH (II,2519); TARGET: THE CORRUPTERS (I,4360)

DONNELLY, DENNIS
THE A-TEAM (II,119); ADAM-12 (I,31); THE AMAZING SPIDER-MAN (II,63); BARNABY JONES (II,153); BLUE THUNDER (II,278); CALIFORNIA FEVER (II,411); CHARLIE'S ANGELS (II,486); THE D.A. (I,1123); DALLAS (II,614); EMERGENCY! (II,775); ENOS (II,779); FLYING HIGH (II,894); THE HARDY BOYS MYSTERIES (II,1090); HART TO HART (II,1102); HIGH PERFORMANCE (II,1145); THE MAN FROM ATLANTIS (II,1615); MARCUS WELBY, M.D. (II,1627); MOBILE ONE (II,1717); PROJECT UFO (II,2088); THE ROUSTERS (II,2220); SUPERTRAIN (II,2505); VEGAS (II,2724); WHIZ KIDS (II,2790)

DONNELLY, EDDIE
THE MIGHTY MOUSE PLAYHOUSE (I,3035)

DONNER, CLIVE
DANGER MAN (I,1136); SECRET AGENT (I,3962)

DONNER, RICHARD
THE BOB HOPE CHRYSLER THEATER (I,502); CIRCLE OF FEAR (I,958); THE DETECTIVES (I,1245); THE FBI (I,1551); THE FELONY SQUAD (I,1563); THE FUGITIVE (I,1701); GET SMART (II,983); GHOST STORY (I,1788); HERNANDEZ, HOUSTON P.D. (I,2034); KOJAK (II,1406); THE LORETTA YOUNG THEATER (I,2756); LUCAS TANNER (II,1555); LUCAS TANNER (II,1556); THE MAN FROM U.N.C.L.E. (I,2867); NIGHTSIDE (I,3296); PERRY MASON (II,2025); ROUTE 66 (I,3852); SENIOR YEAR (II,2301); A SHADOW IN THE STREETS (II,2312); SONS AND DAUGHTERS (II,2404); STAT! (I,4213); WANTED: DEAD OR ALIVE (I,4764); THE WILD WILD WEST (I,4863)

DONOHUE, JACK
ALL ABOUT BARBARA (I,125); ANDY GRIFFITH'S UPTOWN-DOWNTOWN SHOW (I,194); THE BRADY BUNCH (II,362); CHICO AND THE MAN (II,508); CITY VS. COUNTRY (I,968); THE DEAN MARTIN SHOW (I,1196); THE DEAN MARTIN SHOW (I,1197); THE DEAN MARTIN SHOW (I,1198); THE DEAN MARTIN SHOW (I,1199); THE DEAN MARTIN SHOW (I,1200); THE DICK VAN DYKE SPECIAL (I,1276); THE DON KNOTTS SPECIAL (I,1335); FAYE EMERSON'S WONDERFUL TOWN (I,1549); THE FRANK SINATRA SHOW (I,1664); GIRL FRIENDS AND NABORS (I,1807); HAPPY ANNIVERSARY AND GOODBYE (II,1082); HERE'S LUCY (II,1135); THE HOOFER (I,2113); THE JIM NABORS HOUR (I,2381); LOOKING BACK (I,2754); THE LUCILLE BALL COMEDY HOUR (I,2783); A LUCILLE BALL SPECIAL STARRING LUCILLE BALL AND DEAN MARTIN (II,1557); LUCY MOVES TO NBC (II,1562); THE MANY SIDES OF MICKEY ROONEY (I,2899);

THE MICKEY ROONEY SHOW (I,3027); MISS WINSLOW AND SON (II,1705); THE MUSIC MART (II,1774); THE RED SKELTON REVUE (I,3754)

DONOVAN, TOM
BUICK ELECTRA PLAYHOUSE (I,760); CALL ME BACK (I,798); THE DEVIL AND DANIEL WEBSTER (I,1246); HAWK (I,1974); NINOTCHKA (I,3299); OUR PRIVATE WORLD (I,3417); A PUNT, A PASS, AND A PRAYER (I,3694); SATURDAY'S CHILDREN (I,3923); THREE IN ONE (I,4479); THE THREE MUSKETEERS (I,4481); VANITY FAIR (I,4700)

DORFMAN, STANLEY
CELEBRITY CONCERTS (II,471); CELEBRITY REVUE (II,475); WACKO (II,2733)

DORSEY, JOHN
THE $1.98 BEAUTY SHOW (II,698); THE $100,000 NAME THAT TUNE (II,1902); CAMOUFLAGE (II,417); THE CHUCK BARRIS RAH-RAH SHOW (II,525); THE DATING GAME (I,1168); THE GAME GAME (I,1732); GIVE-N-TAKE (II,1003); THE GONG SHOW (II,1026); THE GONG SHOW (II,1027); HIT MAN (II,1156); THE JERRY LEWIS SHOW (I,2369); NAME THAT TUNE (II,1787); THE NEW NEWLYWED GAME (II,1824); THE NEW TREASURE HUNT (I,3275); THE NEW TREASURE HUNT (II,1831); THE NEWLYWED GAME (II,1836); THE PARENT GAME (I,3455); RHYME AND REASON (II,2152); THREE'S A CROWD (II,2612); TREASURE HUNT (II,2657)

DOUGLAS, GORDON
NEVADA SMITH (II,1807)

DOUGLAS, MILTON
FRONT ROW CENTER (I,1693)

DOUGLAS, PAULETTE
TELEVISION INSIDE AND OUT (II,2552)

DOUGLAS, ROBERT
77 SUNSET STRIP (I,3988); ALFRED HITCHCOCK PRESENTS (I,115); BARETTA (II,152); BARNABY JONES (II,153); BIG HAWAII (II,238); CANNON (II,424); CITY OF ANGELS (II,540); DAN AUGUST (I,1130); FAME (II,812); THE FBI (I,1551); THE FUGITIVE (I,1701); FUTURE COP (II,945); HOUSE CALLS (II,1194); HUNTER (II,1205); LOST IN SPACE (I,2758); THE MAN FROM ATLANTIS (II,1615); MAVERICK (II,1657); MEDICAL CENTER (II,1675); THE MONROES (I,3089); NOBODY'S PERFECT (II,1858); THE ROARING TWENTIES (I,3805); SHAZAM! (II,2319); SURFSIDE 6 (I,4299); TRAPPER JOHN, M.D. (II,2654)

DOYLE, BOB
A COUPLE OF JOES (I,1078); QUIZZING THE NEWS (I,3713)

DOYLE, DAVID
CHARLIE'S ANGELS (II,486)

DRAGOTI, STANLEY
MCCOY (II,1661)

DRAKE, JIM
ALICE (II,33); AMERICA 2-NIGHT (II,65); BABY MAKES FIVE (II,129); THE BAXTERS (II,183); BUFFALO BILL (II,387); COMPLETELY OFF THE WALL (II,560); DOMESTIC LIFE (II,703); DOUBLE TROUBLE (II,723); DOUBLE TROUBLE (II,724); FEATHERSTONE'S NEST (II,847); FERNWOOD 2-NIGHT (II,849); FOREVER FERNWOOD (II,909); GIMME A BREAK (II,995); IT'S YOUR MOVE (II,1259); JOE'S WORLD (II,1305); THE LIFE AND TIMES OF EDDIE ROBERTS (II,1468); MARY HARTMAN, MARY HARTMAN (II,1638); NEWHART (II,1835); PLEASE STAND BY (II,2057); SANFORD (II,2255); SECOND CITY TELEVISION (II,2287); WE GOT IT MADE (II,2752); WHO'S THE BOSS? (II,2792); A YEAR AT THE TOP (II,2845)

DRAKE, OLIVER
COLT .45 (I,1010); SKY KING (I,4077)

DRANKO, BOB
THE SMURFS (II,2385)

DREW, DI
1915 (II,1853)

DRIVAS, ROBERT
ANGIE (II,80); THE SECOND TIME AROUND (II,2289); STOCKARD CHANNING IN JUST FRIENDS (II,2467); THE UGILY FAMILY (II,2699)

DRYHURST, MICHAEL
BAFFLED (I,332)

DUFELL, PETER
THE AVENGERS (II,121)

DUBIN, CHARLES S
ACE CRAWFORD, PRIVATE EYE (II,6); AMANDA'S (II,62); BANYON (I,344); BARETTA (II,152); THE BIG VALLEY (I,437); THE BLUE KNIGHT (II,276); BORN TO THE WIND (II,354); BRACKEN'S WORLD (I,703); CHARLIE'S ANGELS (II,486); CINDERELLA (I,956); COOL MILLION (I,1046); CRIME WITH FATHER (I,1100); THE DEADLY TRIANGLE (II,634); THE DEFENDERS (I,1220); DRIBBLE (II,736); ELLERY QUEEN (II,766); EXECUTIVE SUITE (II,796); THE FOUR SEASONS (II,915); THE GENERAL MOTORS 50TH ANNIVERSARY SHOW (I,1758); HAWAII FIVE-O (II,1110); HERBIE, THE LOVE BUG (II,1131); HOLIDAY (I,2074); IRONSIDE (II,1246); JENNIFER SLEPT HERE (II,1278); JIGSAW JOHN (II,1288); JUDY GARLAND AND HER GUESTS, PHIL SILVERS AND ROBERT GOULET (I,2469); KOJAK (II,1406); KUNG FU (II,1416); LOU GRANT (II,1526); LUCAS TANNER (II,1556); M*A*S*H (II,1569); THE MAN AND THE CITY (I,2856); THE MAN FROM ATLANTIS (II,1615); THE MANIONS OF AMERICA (II,1623); MEDICAL CENTER (II,1675); MURDOCK'S GANG (I,3164); NEVER SAY NEVER (II,1809); THE NIGHTENGALES (II,1850); THE NUT HOUSE (I,3320); OF MEN OF WOMEN (I,3331); OPERA VS. JAZZ (I,3399); OWEN MARSHALL: COUNSELOR AT LAW (I,3435); PARTNERS IN CRIME (II,1961); PULITZER PRIZE PLAYHOUSE (I,3692); ROOTS: THE NEXT GENERATIONS (II,2213); SUPERTRAIN (II,2504); TABITHA (II,2528); TARZAN (I,4361); TEACHERS ONLY (II,2548); TEXACO COMMAND PERFORMANCE (I,4407); TOMA (II,2634); TOPPER (II,2649); TWO GIRLS NAMED SMITH (I,4656); THE VIRGINIAN (I,4738); THE WILDS OF TEN THOUSAND ISLANDS (II,2803)

DUBIN, JAY
BILLY JOEL—A TV FIRST (II,250); THE JOE PISCOPO SPECIAL (II,1304)

DUCHOWNY, ROGER
FRIENDS (II,927); THE KIDS FROM C.A.P.E.R. (II,1391); THE LOVE BOAT (II,1535); MCNAMARA'S BAND (II,1669); MURDER CAN HURT YOU! (II,1768); OF MEN OF WOMEN (I,3332); THE PARTRIDGE FAMILY (II,1962)

DUDLEY, PHILIP
A FAMILY AFFAIR (II,814); A HORSEMAN RIDING BY (II,1182)

DUFAU, CARL
THE FLINTSTONES (II,885)

DUFAU, OSCAR
THE DUKES (II,741); MONCHHICHIS (II,1722); PINK PANTHER AND SONS (II,2043); SCOOBY-DOO AND SCRAPPY-DOO (II,2274); SUPER FRIENDS (II,2497)

DUFAU, RAY
THE BISKITTS (II,257)

DUFF, GORDON
CHEVROLET ON BROADWAY (I,925); THE CHEVROLET TELE-THEATER (I,926)

DUFF, HOWARD
CAMP RUNAMUCK (I,811); THE FELONY SQUAD (I,1563)

DUFFELL, PETER
THE FAR PAVILIONS (II,832); FROM A BIRD'S EYE VIEW (I,1686); MAN IN A SUITCASE (I,2868)

DUFFELL, ROBERT
THE UGLIEST GIRL IN TOWN (I,4663)

DUFFY, PATRICK
DALLAS (II,614)

DUKE, BILL
EMERALD POINT, N.A.S. (II,773); FALCON CREST (II,808); FLAMINGO ROAD (II,872); HUNTER (II,1206); KNOTS LANDING (II,1404)

DUKE, DARYL
BANACEK (II,138); CIRCLE OF FEAR (I,958); COOL MILLION (I,1046); THE DOCTORS (I,1323); GHOST STORY (I,1788); THE LAW ENFORCERS (I,2638); NIGHT GALLERY (I,3287); THE PSYCHIATRIST (I,3686); THE PSYCHIATRIST: GOD BLESS THE CHILDREN (I,3685); THE RETURN OF CHARLIE CHAN (II,2136); THE SENATOR (I,3976); SLITHER (II,2379); THEY ONLY COME OUT AT NIGHT (II,2586); THE THORN BIRDS (II,2600)

DUMAS, JOHN
THE HARDY BOYS MYSTERIES (II,1090)

DUNAVAN, PAT
CHRIS AND THE MAGICAL DRIP (II,516)

DUNLAP, PAUL
POWERHOUSE (II,2074)

DUNLAP, REG
CHRISTMAS LEGEND OF NASHVILLE (II,522)

DUNLAP, RICHARD
AS THE WORLD TURNS (II,110); THE FRANK SINATRA TIMEX SHOW (I,1667); THE KATE SMITH SHOW (I,2511); MITZI'S SECOND SPECIAL (I,3074); THE YOUNG AND THE RESTLESS (II,2862)

DUNLOP, FRANK
CAMELOT (II,416)

DUNN, DAVID
STUMP THE STARS (I,4274)

DUNN, MARION
THE GALLOPING GOURMET (I,1730)

DWAN, ALLAN
IT'S ALWAYS SUNDAY (I,2265); SCREEN DIRECTOR'S PLAYHOUSE (I,3946)

DWAN, ROBERT
YOU BET YOUR LIFE (I,4932)

DYSON, FRANKLIN
KUDA BUX, HINDU MYSTIC (I,2593)

DYSON, FRANKLYN
CHESTER THE PUP (I,922)

EASON, MIKE
SERGEANT PRESTON OF THE YUKON (I,3980)

EAST, PHILIP
PRISONER: CELL BLOCK H (II,2085)

EASTMAN, ALLAN
THE LITTLEST HOBO (II,1500)

EASTMAN, CARL
FEDERAL AGENT (I,1557)

EATON, LEO
CAPTAIN SCARLET AND THE MYSTERONS (I,837)

EBERLE, BOB
THE JACQUELINE SUSANN SHOW (I,2327); MATINEE IN NEW YORK (I,2946)

EBI, EARL
KAY KYSER'S KOLLEGE OF MUSICAL KNOWLEDGE (I,2512)

EDELSTEIN, RICK
BOB & CAROL & TED & ALICE (I,495); THE BOB NEWHART SHOW (II,342); THE CORNER BAR (I,1051); MARCUS WELBY, M.D. (II,1627); MR. T. AND TINA (II,1758)

EDWARDS, BLAKE
THE DICK POWELL SHOW (I,1269); JULIE AND DICK IN COVENT GARDEN (II,1352); JULIE! (I,2477); JULIE—MY FAVORITE THINGS (II,1355); PETER GUNN (I,3562)

EDWARDS, JOHN
THE AMAZING YEARS OF CINEMA (II,64)

EDWARDS, VINCENT
B.J. AND THE BEAR (II,125); BATTLESTAR GALACTICA (II,181); BEN CASEY (I,394); DAVID CASSIDY—MAN UNDERCOVER (II,623); THE FALL GUY (II,811); FANTASY ISLAND (II,829); GALACTICA 1980 (II,953); THE HARDY BOYS MYSTERIES (II,1090); POLICE STORY (II,2062)

EGGART, HARRY
THE GUIDING LIGHT (II,1064)

EINHORN, LAWRENCE
DEAR ALEX AND ANNIE (II,651); KIDS ARE PEOPLE TOO (II,1390); RONA LOOKS AT JAMES, MICHAEL, ELLIOTT, AND BURT (II,2205); RONA LOOKS AT RAQUEL, LIZA, CHER AND ANN-MARGRET (II,2207); TELEVISION INSIDE AND OUT (II,2552); THAT'S MY LINE (II,2579)

EISENBERG, NAT
THE MAGIC CLOWN (I,2814)

EISENSTEIN, HAROLD
JERRY MAHONEY'S CLUB HOUSE (I,2371)

ELCAR, DANA
THE BLACK SHEEP SQUADRON (II,262); THE DUKES (II,740)

ELIKANN, LARRY
THE ABC AFTERSCHOOL SPECIAL (II,1); THE AMERICAN DREAM (II,71); BARNABY JONES (II,153); EISCHIED (II,763); EMERALD POINT, N.A.S. (II,773); FALCON CREST (II,808); FLAMINGO ROAD (II,872); GRANDPA GOES TO WASHINGTON (II,1050); HERE'S BOOMER (II,1134); KING'S CROSSING (II,1397); KNOTS LANDING (II,1404); MCCLAIN'S LAW (II,1659); MR. MERLIN (II,1751); PALMERSTOWN U.S.A. (II,1945); THE PAPER CHASE (II,1949); REMINGTON STEELE (II,2130); SPRAGGUE (II,2430); WESTSIDE MEDICAL (II,2765)

ELIZONDO, HECTOR
A.K.A. PABLO (II,28)

ELLIOT, DAPHNE
THE BIG STORY (I,432)

ELLIOTT, DAVID
THUNDERBIRDS (I,4497)

ELLIOTT, MICHAEL
THE GLASS MENAGERIE (I,1819)

ELLIOTT, WILLIAM
AIN'T MISBEHAVIN' (II,25)

ELTERMAN, JUDE
THE FACTS OF LIFE (II,805)

ENGLISH, JOHN
THE ADVENTURES OF CHAMPION (I,54); ANNIE OAKLEY (I,225); BLACK SADDLE (I,480); BUFFALO BILL JR. (I,755); JOHNNY RINGO (I,2434); THE MAN FROM DENVER (I,2863); MY FRIEND FLICKA (I,3190); SOLDIERS OF FORTUNE (I,4111); THRILLER (I,4492); WAGON TRAIN (I,4747); ZANE GREY THEATER (I,4979)

ENGLISH, RAY
LASSIE (I,2622)

ENRICO, ROBERT
THE TWILIGHT ZONE (I,4651)

ENRIGHT, DAN
ALL ABOUT FACES (I,126); THE JOE DIMAGGIO SHOW (I,2399); OH, BABY! (I,3340)

EPSTEIN, JON
MARCUS WELBY, M.D. (II,1627)

ERMAN, JOHN
BRACKEN'S WORLD (I,703); FAMILY (II,813); THE GHOST AND MRS. MUIR (I,1786); GOOD HEAVENS (II,1032); KAREN (II,1363); LETTERS FROM THREE LOVERS (I,2671); MY FAVORITE MARTIAN (I,3189); THE NEW LAND (II,1820); THE OUTER LIMITS (I,3426); PLEASE DON'T EAT THE DAISIES (I,3628); ROOTS (II,2211); ROOTS: THE NEXT GENERATIONS (II,2213); THAT GIRL (II,2570)

EUSTIS, RICH
GOOBER AND THE TRUCKERS' PARADISE (II,1029)

EVANS, BRUCE
SWITCH (II,2519)

EVANS, GRAHAM
TALES OF THE UNEXPECTED (II,2540)

EVANS, JERRY
LOVE OF LIFE (I,2774); RYAN'S HOPE (II,2234)

EVANS, OSMOND
THE ALVIN SHOW (I,146)

EVANS, TREVOR
ON LOCATION WITH RICH LITTLE (II,1891); THE RICH LITTLE SPECIAL (II,2156); RICH LITTLE'S A CHRISTMAS CAROL (II,2157); RICH LITTLE'S ROBIN HOOD (II,2158); RICH LITTLE—COME LAUGH WITH ME (II,2160)

EVERETT, CHAD
MEDICAL CENTER (II,1675)

FAILACE, SAL
THE BULLWINKLE SHOW (I,761)

FAIMAN, PETER
THE DON LANE SHOW (II,707); THE PAUL HOGAN SHOW (II,1975)

FALCON, ELLEN
BUFFALO BILL (II,387); REGGIE (II,2127); SANTA BARBARA (II,2258)

FALCON, ERROL
THE NEW HOWDY DOODY SHOW (II,1818)

FALK, HARRY
ADVICE TO THE LOVELORN (II,16); ALIAS SMITH AND JONES (I,118); THE ANDROS TARGETS (II,76); BERT D'ANGELO/SUPERSTAR (II,211); BEULAH LAND (II,226); BIG HAWAII (II,238); CANNON (II,424); CARIBE (II,439); CENTENNIAL (II,477); CHARLIE'S ANGELS (II,486); THE CONTENDER (II,571); THE COURTSHIP OF EDDIE'S FATHER (I,1083); THE DORIS DAY SHOW (I,1355); EMERALD POINT, N.A.S. (II,773); GOOD OL' BOYS (II,1035); HEAR NO EVIL (II,1116); HOW THE WEST WAS WON (II,1196); JIGSAW JOHN (II,1288); MAGNUM, P.I. (II,1601); MANDRAKE (II,1617); MCMILLAN AND WIFE (II,1667); MEN OF THE DRAGON (II,1683); OWEN MARSHALL: COUNSELOR AT LAW (I,3435); PARTNERS IN CRIME (II,1961); THE PARTRIDGE FAMILY (II,1962); RICH MAN, POOR MAN—BOOK I (II,2161); THE ROOKIES (II,2208); ROSETTI AND RYAN (II,2216); S.W.A.T. (II,2236); THE STREETS OF SAN FRANCISCO (II,2478); T.J. HOOKER (II,2524); TALES OF THE UNEXPECTED (II,2539); THAT GIRL (II,2570); THE TIM CONWAY SHOW (I,4502); VEGAS (II,2724); WHAT REALLY HAPPENED TO THE

CLASS OF '65? (II,2768); THE WONDERFUL WORLD OF PHILIP MALLEY (II,2831); THE YELLOW ROSE (II,2847)

FARALLA, WILLIAM DARIO
BLACK SADDLE (I,480); HARD CASE (I,1947); JOHNNY RINGO (I,2434); ROUTE 66 (I,3852)

FARGO, JAMES
TALES OF THE GOLD MONKEY (II,2537)

FARIS, JAY
LAWMAN (I,2642)

FARNHAM, BRIAN
THE ROCK FOLLIES (II,2192)

FARR, GORDON
THE LOVE BOAT (II,1535)

FARR, JAMIE
AFTERMASH (II,23); M*A*S*H (II,1569)

FARRAR, ANTHONY
TV'S TOP TUNES (I,4634)

FARRELL, MIKE
M*A*S*H (II,1569)

FAULL, BRIAN
PRISONER: CELL BLOCK H (II,2085)

FAY JR, SHELDON
TRAVELS WITH CHARLEY (I,4597)

FAY, PAT
CAPTAIN VIDEO AND HIS VIDEO RANGERS (I,838); GROWING PAYNES (I,1891); KIDS AND COMPANY (I,2534); THE S. S. HOLIDAY (I,4173)

FEIST, FELIX
ADVENTURES IN PARADISE (I,46); THE OUTER LIMITS (I,3426); SEA HUNT (I,3948); VOYAGE TO THE BOTTOM OF THE SEA (I,4743); ZANE GREY THEATER (I,4979)

FELDKAMP, FRED
MARCH OF TIME THROUGH THE YEARS (I,2903)

FELDMAN, DICK
ART CARNEY MEETS PETER AND THE WOLF (I,272); THE GREAT SPACE COASTER (II,1059); HAL LINDEN'S BIG APPLE (II,1072); THE WORLD'S FUNNIEST COMMERCIAL GOOFS (II,2841)

FELDMAN, EDWARD H
AFTERMASH (II,23); AT EASE (II,115); BAKER'S DOZEN (II,136); THE CHEAP DETECTIVE (II,490); FLAMINGO ROAD (II,872); MY WIVES JANE (I,3207); RAFFERTY (II,2105); SMALL AND FRYE (II,2380)

FELDMAN, GENE
THE HORROR OF IT ALL (II,1181)

FELDMAN, MARTY
WHEN THINGS WERE ROTTEN (II,2782)

FELLINI, FEDERICO
FELLINI: A DIRECTOR'S NOTEBOOK (I,1562)

FELTON, NORMAN
THE ALCOA HOUR (I,108); CRISIS (I,1101); THE GOOK FAMILY (I,1858); ROBERT MONTGOMERY PRESENTS YOUR LUCKY STRIKE THEATER (I,3809); STUD'S PLACE (I,4271); THESE ARE MY CHILDREN (I,4440)

FENADY, GEORG
B.J. AND THE BEAR (II,126); CHIPS (II,511); CODE R (II,551); CODE RED (II,552); COMBAT (I,1011); EMERGENCY! (II,776); KNIGHT RIDER (II,1402); MANIMAL (II,1622); MISSION: IMPOSSIBLE (I,3067); QUINCY, M. E. (II,2102); THE TEXAS RANGERS (II,2567)

FERBER, MEL
ALIAS SMITH AND JONES (I,118); ALICE (II,33); BEACON HILL (II,193); A BELL FOR ADANO (I,388); C.P.O. SHARKEY (II,407); CANDID CAMERA (I,819); FLATBUSH (II,877); FOLK SOUND U.S.A. (I,1614); FOOT IN THE DOOR (II,902); HAPPY DAYS (II,1084); HOUSE CALLS (II,1194); KAREN (II,1363); LOVE NEST (II,1541); LOVE, SIDNEY (II,1547); THE MARY TYLER MOORE SHOW (II,1640); MCMILLAN AND WIFE (II,1667); THE MISADVENTURES OF SHERIFF LOBO (II,1704); THE ODD COUPLE (II,1875); SHIRLEY (II,2330); SIROTA'S COURT (II,2365); SMALL AND FRYE (II,2380); WONDERFUL TOWN (I,4893)

FERGUSON, MICHAEL
FLAMBARDS (II,869)

FERRARO, DOLORES
DIFF'RENT STROKES (II,674); THE FACTS OF LIFE (II,805); WKRP IN CINCINNATI (II,2814)

FERRER, MEL
FALCON CREST (II,808); THE FARMER'S DAUGHTER (I,1533)

FERRIN, FRANK
ANDY'S GANG (I,214)

FIELD, JAMES
AMERICA 2-NIGHT (II,65); THE NEW MICKEY MOUSE CLUB (II,1823); TOM COTTLE: UP CLOSE (II,2631)

FIELD, RON
APPLAUSE (I,244); ED SULLIVAN'S BROADWAY (I,1402); THE MANHATTAN TRANSFER (II,1619); PINOCCHIO (II,2044)

FIELDS, PETER ALLAN
MCCLOUD (II,1660)

FINKEL, BOB
THE ALCOA HOUR (I,108); AND HERE'S THE SHOW (I,187); CIRCUS OF THE STARS (II,532); MCMILLAN (II,1666) MCMILLAN AND WIFE (II,1667) MYSTERY AND MRS. (I,3210); THE ORCHID AWARD (I,3406) THE PEOPLE'S CHOICE (I,3521)

FINNERMAN, GERALD PERRY
SALVAGE 1 (II,2245)

FISHER, ART
33 1/3 REVOLUTIONS PER MONKEE (I,4451); ANDY WILLIAMS MAGIC LANTERN SHOW COMPANY (I,202); ANDY'S LOVE CONCERT (I,215); ANN-MARGRET—WHEN YOU'RE SMILING (I,219); BING CROSBY'S SUN VALLEY CHRISTMAS SHOW (I,475); BING CROSBY—COOLING IT (I,459); BURT AND THE GIRLS (I,770); BURT REYNOLDS' LATE SHOW (I,777); CHER (II,495); CHER (II,496); CHER AND OTHER FANTASIES (II,497); CHER. . . SPECIAL (II,499); THE CHEVY CHASE SHOW (II,505); THE CRACKER BROTHERS (II,599); DONNY AND MARIE (II,712); THE DONNY AND MARIE OSMOND SHOW (II,714); THE FANTASTIC JOURNEY (II,826); HAPPY DAYS (II,1084); THE HUDSON BROTHERS SHOW (II,1202); THE KROFFT SUPERSHOW (II,1412); LINDSAY WAGNER—ANOTHER SIDE OF ME (II,1483); MONSANTO PRESENTS MANCINI (I,3092); MONSANTO PRESENTS MANCINI (I,3093); MONSANTO PRESENTS MANCINI (I,3094); NEIL DIAMOND SPECIAL: I'M GLAD YOU'RE HERE WITH ME TONIGHT (II,1803); NEIL SEDAKA STEPPIN' OUT (II,1804); THE OSMOND BROTHERS SPECIAL (II,1925); PINK LADY (II,2041); PORTRAIT OF A LEGEND (II,2070); SERPICO (II,2306); SIEGFRIED AND ROY (II,2350); THE SONNY AND CHER COMEDY HOUR (II,2400); THE SONNY COMEDY REVUE (II,2403); THE SUZANNE SOMERS SPECIAL (II,2510); THE SUZANNE SOMERS SPECIAL (II,2511); THAT'S TV (II,2581); VAN DYKE AND COMPANY (II,2718); WONDERBUG (II,2830)

FISHER, CHARLES
ON YOUR ACCOUNT (I,3364)

FISHER, TERENCE
COLONEL MARCH OF SCOTLAND YARD (I,1005)

FITZWATER, WILLIAM
PRISONER: CELL BLOCK H (II,2085)

FLAUM, MARSHALL
BOGART (I,679); HOLLYWOOD: THE SELZNICK YEARS (I,2095); PLAYBOY'S 25TH ANNIVERSARY CELEBRATION (II,2054)

FLEISCHER, DAVE
POPEYE THE SAILOR (I,3648)

FLEMYNG, GORDON
THE AVENGERS (II,121); THE SAINT (II,2237)

FLETCHER, JAMES
SPORT BILLY (II,2429)

FLICKER, THEODORE J
THE ANDY GRIFFITH SHOW (I,192); ANN IN BLUE (II,83); BANACEK (II,138); BANYON (I,344); BARNEY MILLER (II,154); THE DICK VAN DYKE SHOW (I,1275); HERE WE GO AGAIN (I,2028); NIGHT GALLERY (I,3287); THE ROGUES (I,3832); THE STREETS OF SAN FRANCISCO (II,2478)

FLOCKER, JAMES T
CAMP WILDERNESS (II,419)

FLOREA, JOHN
240-ROBERT (II,2688); THE BARBARY COAST (II,146); BONANZA (II,347); CHIPS (II,511); DANIEL BOONE (I,1142); THE DUKES OF HAZZARD (II,742); FOR LOVE AND HONOR (II,903); GENTLE BEN (I,1761); HONEY WEST (I,2105); IRONSIDE (II,1246); M SQUAD (I,2799); MISSION: IMPOSSIBLE (I,3067); OUTPOST (I,3428); PRIMUS (I,3668); SAM (II,2247); SEA HUNT (I,3948); SEVEN BRIDES

FOR SEVEN BROTHERS (II,2307); TEMPLE HOUSTON (I,4392); TODAY'S FBI (II,2628); THE VIRGINIAN (I,4738); WAGON TRAIN (I,4747); WALKING TALL (II,2738)

FLOREY, ROBERT
ADVENTURES IN PARADISE (I,46); ALCOA PREMIERE (I,109); THE BARBARA STANWYCK THEATER (I,350); CHECKMATE (I,919); THE DICK POWELL SHOW (I,1269); A LETTER TO LORETTA (I,2669); THE LORETTA YOUNG THEATER (I,2756); THE OUTER LIMITS (I,3426); THE RESTLESS GUN (I,3774); THRILLER (I,4492); TURN OF FATE (I,4623); THE UNTOUCHABLES (I,4682); THE WALT DISNEY CHRISTMAS SHOW (I,4755)

FLYNN, JOHN
LONE STAR (II,1513)

FOGARTY, KEN
PRISONER: CELL BLOCK H (II,2085)

FOGEL, JERRY
TALES OF THE BLACK CAT (I,4345)

FOLB, JAY
FREEBIE AND THE BEAN (II,922)

FORBES, BRYAN
PHILIP MARLOWE, PRIVATE EYE (II,2033)

FORBES, GEORGE
STARTIME (I,4212)

FORD, JOHN
ALCOA PREMIERE (I,109); SCREEN DIRECTOR'S PLAYHOUSE (I,3946); WAGON TRAIN (I,4747)

FORD, PHILIP
THE ADVENTURES OF SUPERMAN (I,77); JEFF'S COLLIE (I,2356)

FORDYCE, IAN
DIAL A DEADLY NUMBER (I,1253); THE DOUBLE KILL (I,1360); THE ENGELBERT HUMPERDINCK SHOW (I,1444); THE ENGELBERT HUMPERDINCK SHOW (I,1445); THE ENGELBERT HUMPERDINCK SPECIAL (I,1447)

FORDYCE, JOHN
ONE DEADLY OWNER (I,3373)

FORREST, ARTHUR
BATTLE OF THE LAS VEGAS SHOW GIRLS (II,162); FOUL UPS, BLEEPS AND BLUNDERS (II,911); HOLLYWOOD TEEN (II,1166); THE JERRY LEWIS SHOW (II,1284); JOYCE AND BARBARA: FOR ADULTS ONLY (I,2458); KNOCKOUT (II,1403); THE KRYPTON FACTOR (II,1414); LEAVE IT TO THE WOMEN (II,1454); THE MIKE DOUGLAS SHOW (II,1698); THE NEW TRUTH OR CONSEQUENCES (II,1832); NO HOLDS BARRED (II,1854); PEOPLE DO THE CRAZIEST THINGS (II,1997); PEOPLE DO THE CRAZIEST THINGS (II,1998); PEOPLE TO PEOPLE (II,2000); THE POP 'N' ROCKER GAME (II,2066); ROCK 'N' ROLL: THE FIRST 25 YEARS (II,2189); THE ROOTS OF ROCK 'N' ROLL (II,2212); SUCCESS: IT CAN BE YOURS (II,2488); THAT WAS THE YEAR THAT WAS (I,4424); THAT'S INCREDIBLE! (II,2578); THIS WILL BE THE YEAR THAT WILL BE (I,4466); THE WORLD'S FUNNIEST COMMERCIAL GOOFS (II,2841)

FOSSE, BOB
LIZA WITH A Z (I,2732)

FOSTER, BILL
APPLAUSE (I,244); BENSON (II,208); BERT CONVY SPECIAL—THERE'S A MEETING HERE TONIGHT (II,210); BUCKSHOT (II,385); DICK VAN DYKE MEETS BILL COSBY (I,1274); THE DON RICKLES SHOW (II,708); THE DORIS MARY ANNE KAPPELHOFF SPECIAL (I,1356); THE FLIP WILSON SPECIAL (II,887); FULL HOUSE (II,936); THE JACKIE GLEASON SPECIAL (I,2325); THE JERRY LEWIS SHOW (I,2370); JUST MEN (II,1359); THE LIAR'S CLUB (II,1466); LIKE MAGIC (II,1476); MAGIC MONGO (II,1592); PARADISE (II,1955); PLAYBOY AFTER DARK (I,3620); THE ROWAN AND MARTIN SPECIAL (I,3855); ROWAN AND MARTIN'S LAUGH-IN (I,3857); THE SINGERS (I,4060); THE SMOTHERS BROTHERS SHOW (II,2384); STAND UP AND CHEER (I,4188); SUGAR TIME (II,2489); THIS WEEK IN NEMTIN (I,4465); TONY ORLANDO AND DAWN (II,2639); VER-R-R-RY INTERESTING (I,4712); THE VIN SCULLY SHOW (I,4734)

FOSTER, DICK
PERRY COMO'S HAWAIIAN HOLIDAY (II,2017)

FOSTER, GEORGE SPENTON
DOCTOR WHO (II,689)

FOSTER, HARVE
CROWN THEATER WITH GLORIA SWANSON (I,1107); DUFFY'S TAVERN (I,1379); EXPERT WITNESS (I,1488); PUBLIC DEFENDER (I,3687)

FOSTER, LEWIS
CAVALCADE OF AMERICA (I,875); DANIEL BOONE (I,1141); THE SAGA OF ANDY BURNETT (I,3883); SWAMP FOX (I,4311)

FOSTER, NORMAN
BATMAN (I,366); CHICAGO 212 (I,935); DAVY CROCKETT (I,1181); ELFEGO BACA (I,1427); THE LORETTA YOUNG THEATER (I,2756); THE MONROES (I,3089); THE SMOTHERS BROTHERS SHOW (I,4096); ZORRO (I,4982)

FOSTER, ROBERT
KNIGHT RIDER (II,1402); PERSPECTIVE ON GREATNESS (I,3555)

FOURNIER, CLAUDE
THE NEW AVENGERS (II,1816)

FOWLER, NAT
TALES OF THE RED CABOOSE (I,4347)

FOX, BRUCE
THE GREAT MOVIE COWBOYS (II,1057)

FOX, VAN B
HERE'S HOLLYWOOD (I,2031)

FOXE, GEORGE
PEOPLE ARE FUNNY (I,3518)

FRANCHINI, BOB
SCROOGE'S ROCK 'N' ROLL CHRISTMAS (II,2279)

FRANCHOT, DICK
BRIGHT PROMISE (I,727)

FRANCIS, FREDDIE
MAN IN A SUITCASE (I,2868); STAR MAIDENS (II,2435)

FRANK, SELIG
THE FACTS OF LIFE (II,805)

FRANK, YASHA
PINOCCHIO (I,3602)

FRANKEL, CYRIL
THE AVENGERS (II,121); THE BARON (I,360); THE CHAMPIONS (I,896); DEPARTMENT S (I,1232); MY PARTNER THE GHOST (I,3199); RETURN OF THE SAINT (II,2141); U.F.O. (I,4662)

FRANKENHEIMER, JOHN
THE BLUE MEN (I,492); BUICK ELECTRA PLAYHOUSE (I,760); THE GOLDEN AGE OF TELEVISION (II,1018); THE RAINMAKER (II,2109); THE TURN OF THE SCREW (I,4624)

FRANKLIN, JIM
THE GOODIES (II,1040)

FRASER, WOODY
LIFE'S MOST EMBARRASSING MOMENTS (II,1470)

FRAWLEY, JAMES
ALL THAT GLITTERS (II,42); BULBA (II,392); COLUMBO (II,556); THE DEVLIN CONNECTION (II,664); THE EDDIE CAPRA MYSTERIES (II,758); FAERIE TALE THEATER (II,807); MAGNUM, P.I. (II,1601); MICKEY SPILLANE'S MIKE HAMMER (II,1692); THE MONKEES (I,3088); MR. MERLIN (II,1751); THE ORPHAN AND THE DUDE (II,1924); THE OUTLAWS (II,1936); SCARECROW AND MRS. KING (II,2268); TALES OF THE GOLD MONKEY (II,2537); THE TEXAS WHEELERS (II,2568); THAT GIRL (II,2570); THE UGLIEST GIRL IN TOWN (I,4663); WISHMAN (II,2811); WIZARDS AND WARRIORS (II,2813)

FRAZEN, STANLEY
DANGER ZONE (I,1137)

FRAZER-JONES, PETER
THE BENNY HILL SHOW (II,207); GEORGE AND MILDRED (II,965); MAN ABOUT THE HOUSE (II,1611); ROBIN'S NEST (II,2187)

FREA, PETER
WILD WILD WORLD OF ANIMALS (I,4864)

FREED, LORNE
THE JERRY LESTER SPECIAL (I,2361)

FREEDMAN, JERROLD
KOJAK (II,1406); NIGHT GALLERY (I,3287); THE PSYCHIATRIST (I,3686); THE SENATOR (I,3976)

FREEDMAN, LOUIS
CAPTAIN SAFARI OF THE JUNGLE PATROL (I,836)

FREEMAN, DEVERY
THE 20TH CENTURY-FOX HOUR (I,4642)

FREEMAN, PAUL
NOT FOR WOMEN ONLY (I,3310)

FREEMAN, SETH
LOU GRANT (II,1526)

FREGONESE, HUGO
THE THREE MUSKETEERS (I,4480)

FRELENG, FRIZ
THE BUGS BUNNY/ROAD RUNNER SHOW (II,391); PORKY PIG AND FRIENDS (I,3651); SYLVESTER AND TWEETY (II,2522)

FRELENG, I
THE ROAD RUNNER SHOW (I,3799)

FRENCH, VICTOR
BUCK ROGERS IN THE 25TH CENTURY (II,384); DALLAS (II,614); GUNSMOKE (II,1069); HIGHWAY TO HEAVEN (II,1152); LITTLE HOUSE ON THE PRAIRIE (II,1487); LITTLE HOUSE: A NEW BEGINNING (II,1488); LITTLE HOUSE: BLESS ALL THE DEAR CHILDREN (II,1489); LITTLE HOUSE: LOOK BACK TO YESTERDAY (II,1490)

FREND, CHARLES
DANGER MAN (I,1136); MAN IN A SUITCASE (I,2868); SECRET AGENT (I,3962)

FRIEBERG, RICHARD
THE LIFE AND TIMES OF GRIZZLY ADAMS (II,1469)

FRIEDKIN, DAVID
ALFRED HITCHCOCK PRESENTS (I,115); BERT D'ANGELO/SUPERSTAR (II,211); THE DICK POWELL SHOW (I,1269); DOCTORS HOSPITAL (II,691); GET CHRISTIE LOVE! (II,981); HAWAII FIVE-O (II,1110); I SPY (I,2179); IRONSIDE (II,1246); KATE MCSHANE (II,1368); KOJAK (II,1406); THE SNOOP SISTERS (II,2389); THE VIRGINIAN (I,4738)

FRIEDMAN, CHARLES
THE COLGATE COMEDY HOUR (I,997); FORD FESTIVAL (I,1629)

FRIEDMAN, ED
THE ANDROS TARGETS (II,76); BATMAN AND THE SUPER SEVEN (II,160); THE FABULOUS FUNNIES (II,802); FAT ALBERT AND THE COSBY KIDS (II,835); FLASH GORDON (II,874); THE KID SUPER POWER HOUR WITH SHAZAM (II,1386); SPORT BILLY (II,2429)

FRIEDMAN, JEFF
THE UNCLE FLOYD SHOW (II,2703)

FRIEDMAN, KIM
13 QUEENS BOULEVARD (II,2591); ALICE (II,33); FAMILY (II,813); FOR MEMBERS ONLY (II,905); GOODNIGHT, BEANTOWN (II,1041); KNOTS LANDING (II,1404); MARY HARTMAN, MARY HARTMAN (II,1638); NURSE (II,1870); SQUARE PEGS (II,2431)

FRIEND, MARTIN
BERGERAC (II,209); THE ONEDIN LINE (II,1912)

FRIEND, ROBERT L
THE MONROES (I,3089); THEY WENT THATAWAY (I,4443)

FRISTOE, ALLEN
AS THE WORLD TURNS (II,110); THE EDGE OF NIGHT (II,760); ONE LIFE TO LIVE (II,1907)

FRITSCH, GENE
BRONCO (I,745); LAWMAN (I,2642); SURFSIDE 6 (I,4299)

FUEST, ROBERT
THE ABC AFTERSCHOOL SPECIAL (II,1); THE AVENGERS (II,121)

FULLER, SAMUEL
330 INDEPENDENCE S.W. (I,4486); THE DICK POWELL SHOW (I,1269); THE VIRGINIAN (I,4738)

FULLILOVE, ERIC
SKIPPY, THE BUSH KANGAROO (I,4076)

FULMER, DAVE
CAPTAIN Z-RO (I,839)

FUNK, DAN
PEOPLE DO THE CRAZIEST THINGS (II,1997)

FUNT, ALLEN
CANDID CAMERA (I,818); CANDID CAMERA LOOKS AT THE DIFFERENCE BETWEEN MEN AND WOMEN (II,421); IT'S ONLY HUMAN (II,1257)

FUSARI, LOU
THE MAGNIFICENT MARBLE MACHINE (II,1600)

GABA, LESTER
THE MARCH OF DIMES FASHION SHOW (I,2901)

GAGE, JACK
THE EGG AND I (I,1420); THE NEW ADVENTURES OF CHARLIE CHAN (I,3249); SHERLOCK HOLMES (I,4010); YOU ARE THERE (I,4929)

GAIL, MAX
BARNEY MILLER (II,154); WHIZ KIDS (II,2790)

GALBRAITH, SCOTT
MATINEE AT THE BIJOU (II,1651)

GALLACCIO, GEORGE
THE OMEGA FACTOR (II,1888)

GALLAGHER, ELIZABETH
A MAN CALLED SLOANE (II,1613)

GANAWAY, ALBERT
COUNTRY MUSIC CARAVAN (I,1066)

GANGNEBIN, DEBRA
WHEN TELEVISION WAS LIVE (II,2779)

GANNON, EVERETT
THE FIRST HUNDRED YEARS (I,1581)

GANNON, JOE
ARCHIE BUNKER'S PLACE (II,105)

GANZ, JEFFREY
THE BAD NEWS BEARS (II,134)

GANZ, LOWELL
ANGIE (II,80); THE BAD NEWS BEARS (II,134); BROTHERS AND SISTERS (II,382); THE FURTHER ADVENTURES OF WALLY BROWN (II,944); JOANIE LOVES CHACHI (II,1295); MAKIN' IT (II,1603)

GANZER, ALVIN
ALCOA/GOODYEAR THEATER (I,107); THE AMERICAN GIRLS (II,72); BROKEN ARROW (I,744); CASABLANCA (I,860); CIMARRON STRIP (I,954); DAVID CASSIDY—MAN UNDERCOVER (II,623); THE DETECTIVES (I,1245); HAWAIIAN EYE (I,1973); JOE FORRESTER (II,1303); LARAMIE (I,2614); LOST IN SPACE (I,2758); THE MAN FROM U.N.C.L.E. (I,2867); THE NAME OF THE GAME (I,3217); PARIS (II,1957); PLEASE DON'T EAT THE DAISIES (I,3628); POLICE WOMAN (II,2063); THE ROOKIES (II,2208); ROUTE 66 (I,3852); TEMPLE HOUSTON (I,4392); TURN OF FATE (I,4623); THE WILD WILD WEST (I,4863)

GARDNER, HERB
CANDID CAMERA (II,420)

GARDNER, RICK
THE EVERLY BROTHERS REUNION CONCERT (II,791); THE OTHER BROADWAY (II,1930)

GAREN, SCOTT
TELEVISION'S GREATEST COMMERCIALS (II,2553); TELEVISION'S GREATEST COMMERCIALS II (II,2554); TELEVISION'S GREATEST COMMERCIALS III (II,2555); TELEVISION'S GREATEST COMMERCIALS IV (II,2556)

GARFEIN, JACK
THE MARRIAGE (I,2922); WINDOWS (I,4876)

GARGIULO, MIKE
THE $20,000 PYRAMID (II,2681); THE $25,000 PYRAMID (II,2679); THE ALL-AMERICAN COLLEGE COMEDY SHOW (II,47); BLANKETY BLANKS (II,266); THE DAVID FROST REVUE (I,1176); JACKPOT (II,1266); THE NEW $25,000 PYRAMID (II,1811); PASS THE BUCK (II,1964); PLAY YOUR HUNCH (I,3619); SALE OF THE CENTURY (I,3888); SHOOT FOR THE STARS (II,2340); THE SOUND AND THE SCENE (I,4132); THAT'S MY LINE (II,2579); VICTOR BORGE'S 20TH ANNIVERSARY SHOW (I,4727); WEDDING PARTY (I,4786); WINNING STREAK (II,2809)

GARLAND, PATRICK
THE SNOW GOOSE (I,4103)

GARNER, JAMES
MAVERICK (II,1657); THE ROCKFORD FILES (II,2197)

GARNETT, TAY
ALCOA/GOODYEAR THEATER (I,107); THE BEACHCOMBER (I,371); BONANZA (II,347); THE DEPUTY (I,1234); GUNSMOKE (II,1069); LARAMIE (I,2614); NAKED CITY (I,3216); PLEASE DON'T EAT THE DAISIES (I,3628); RAWHIDE (I,3727); SCREEN DIRECTOR'S PLAYHOUSE (I,3946); TURN OF FATE (I,4623); THE UNTOUCHABLES (I,4682); WAGON TRAIN (I,4747)

GARRETT, LILA
ARCHIE BUNKER'S PLACE (II,105); BABY MAKES FIVE (II,129); SPENCER (II,2420)

GARRISON, GREG
ASTAIRE TIME (I,305); BACHELOR FATHER (I,323); THE BALLAD OF LOUIE THE LOUSE (I,337); BON VOYAGE (I,683); THE CONNIE FRANCIS SHOW (I,1039); THE DANNY KAYE SHOW (I,1143); DEAN MARTIN AT THE WILD ANIMAL PARK (II,636); THE DEAN MARTIN CELEBRITY

ROAST (II,637); THE DEAN MARTIN CELEBRITY ROAST (II,638); THE DEAN MARTIN CELEBRITY ROAST (II,639); THE DEAN MARTIN CHRISTMAS SPECIAL (II,640); DEAN MARTIN IN LONDON (II,641); DEAN MARTIN PRESENTS THE GOLDDIGGERS (I,1193); DEAN MARTIN PRESENTS THE GOLDDIGGERS (I,1194); THE DEAN MARTIN SHOW (I,1201); DEAN MARTIN'S CALIFORNIA CHRISTMAS (II,642); DEAN MARTIN'S CELEBRITY ROAST (II,643); DEAN MARTIN'S CHRISTMAS IN CALIFORNIA (II,645); DEAN MARTIN'S COMEDY CLASSICS (II,646); DEAN MARTIN'S RED HOT SCANDALS OF 1926 (II,648); DEAN MARTIN'S RED HOT SCANDALS PART 2 (II,649); DEAN'S PLACE (II,650); DOM DELUISE AND FRIENDS (II,701); DOM DELUISE AND FRIENDS, PART 2 (II,702); THE DONALD O'CONNOR SHOW (I,1343); THE FIRST 50 YEARS (II,862); JACK PAAR PRESENTS (I,2315); THE JONATHAN WINTERS SHOW (I,2441); THE JONATHAN WINTERS SHOW (I,2442); THE KATE SMITH EVENING HOUR (I,2507); THE KATE SMITH HOUR (I,2508); THE KEEFE BRASSELLE SHOW (I,2514); LADIES AND GENTLEMAN... BOB NEWHART (II,1420); LADIES AND GENTLEMAN... BOB NEWHART, PART II (II,1421); LADIES BE SEATED (I,2598); MERMAN ON BROADWAY (I,3011); THE MILTON BERLE SHOW (I,3047); THE MILTON BERLE SPECIAL (I,3051); PHIL SILVERS IN NEW YORK (I,3577); THE ROWAN AND MARTIN SHOW (I,3854); THE SID CAESAR SHOW (I,4043); STANDARD OIL ANNIVERSARY SHOW (I,4189); SUMMER IN NEW YORK (I,4279); TEXACO STAR THEATER (I,4411); UNCLE MISTLETOE AND HIS ADVENTURES (I,4669); THE VIC DAMONE SHOW (I,4718); WASHINGTON SQUARE (I,4772); WHAT'S UP, AMERICA? (I,4824); THE WONDERFUL WORLD OF JACK PAAR (I,4899)

GARRISON, PAUL
THE CASES OF EDDIE DRAKE (I,863)

GARVECHA, BOB
AT THE MOVIES (II,117)

GARY, SAM
THREE'S COMPANY (II,2614)

GATES, JIM
THAT'S MY LINE (II,2579)

GATWARD, JAMES
STAR MAIDENS (II,2435)

GAUNT, JOHN
THE ADVENTURES OF CYCLONE MALONE (I,57); THE BABY SITTER (I,321)

GAYNES, GEORGE
WKRP IN CINCINNATI (II,2814)

GAZZARA, BEN
COLUMBO (II,556); THE NAME OF THE GAME (I,3217); RUN FOR YOUR LIFE (I,3871)

GEISEL, DAVID
THE ENTERTAINERS (I,1449); THE GARRY MOORE SHOW (I,1739); ON BROADWAY TONIGHT (I,3355); SWING INTO SPRING (I,4317)

GELBART, LARRY
AFTERMASH (II,23)

GENNARO, PETER
WHO'S AFRAID OF MOTHER GOOSE? (I,4852)

GENUS, KARL
PUREX SPECIAL FOR WOMEN (I,3698); THE ROBERT HERRIDGE THEATER (I,3808)

GERAGHTY, MAURY
THE VIRGINIAN (I,4738)

GERBER, DAVID
THE PRIME OF MISS JEAN BRODIE (II,2082)

GERONIMI, CLYDE
THE MICKEY MOUSE CLUB (I,3025); SPIDER-MAN (I,4154)

GERSON, HAL
THE BERT PARKS SHOW (I,400); THE FRANK SINATRA SHOW (I,1665); UKULELE IKE (I,4664)

GERSTAD, HARRY
THE ADVENTURES OF SUPERMAN (I,77)

GIBBS, JOHN
DUNGEONS AND DRAGONS (II,744); G.I. JOE: A REAL AMERICAN HERO (II,948); JIM HENSON'S MUPPET BABIES (II,1289); MEATBALLS AND SPAGHETTI (II,1674); SPIDER-MAN AND HIS AMAZING FRIENDS (II,2425); THE TRANSFORMERS (II,2653)

GIBSON, ALAN
SLEEPWALKER (I,4084); A WOMAN CALLED GOLDA (II,2818)

GIFFORD, GRAEME
FAERIE TALE THEATER (II,807)

GILBERT, KENNETH C
BARNABY JONES (II,153); THE BIONIC WOMAN (II,255); CANNON (II,424); THE INCREDIBLE HULK (II,1232); NATIONAL LAMPOON'S TWO REELERS (II,1796); PALMERSTOWN U.S.A. (II,1945); THE PAPER CHASE (II,1949); THE RUNAWAYS (II,2231); THE STREETS OF SAN FRANCISCO (II,2478)

GILBERT, RICHARD
DOCTOR SIMON LOCKE (I,1317); THE NEW AVENGERS (II,1816)

GILBERT, ROLAND "BILL"
CAPTAIN BILLY'S MISSISSIPPI MUSIC HALL (I,826)

GILES, DAVID
THE FORSYTE SAGA (I,1645); THE STRAUSS FAMILY (I,4253); VANITY FAIR (I,4701)

GILL, DAVID
HOLLYWOOD (II,1160)

GILLESPIE, RITA
CONSTANTINOPLE (II,570)

GILLETTE, ROLAND
THE FRED WARING SHOW (I,1679); WINNER TAKE ALL (I,4881)

GILLING, JOHN
THE CHAMPIONS (I,896); DEPARTMENT S (I,1232); THE SAINT (II,2237)

GILONA, MICHAEL
THE GUIDING LIGHT (II,1064)

GILROY, FRANK D
NERO WOLFE (II,1805); THE TURNING POINT OF JIM MALLOY (II,2670)

GIRALDI, BOB
KENNY & DOLLY: A CHRISTMAS TO REMEMBER (II,1378)

GIRARD, BERNARD
ADVENTURES IN PARADISE (I,46); ALCOA PREMIERE (I,109); ALFRED HITCHCOCK PRESENTS (I,115); COUNTERPOINT (I,1059); CROWN THEATER WITH GLORIA SWANSON (I,1107); THE LONE WOLF (I,2742); M SQUAD (I,2799); MEDIC (I,2976); RAWHIDE (I,3727); SEVEN AGAINST THE SEA (I,3983); THE SIXTH SENSE (I,4069); THE VIRGINIAN (I,4738); WAGON TRAIN (I,4747); WALTER FORTUNE (I,4758); YOU ARE THERE (I,4929)

GIST, ROBERT
BEN CASEY (I,394); THE DICK POWELL SHOW (I,1269); HAWAII FIVE-O (II,1110); MISSION: IMPOSSIBLE (I,3067); N.Y.P.D. (I,3321); NAKED CITY (I,3216); PETER GUNN (I,3562); ROUTE 66 (I,3852); STAR TREK (II,2440); THE UNTOUCHABLES (I,4682); THE VIRGINIAN (I,4738)

GLASER, PAUL MICHAEL
MIAMI VICE (II,1689); STARSKY AND HUTCH (II,2444)

GLAZER, BARRY
AMERICAN BANDSTAND (II,70); BUCKSHOT (II,385); DICK CLARK PRESENTS THE ROCK AND ROLL YEARS (I,1266); THE JIM NABORS SHOW (II,1290); THE JOHN DAVIDSON SHOW (II,1309); THE LOU RAWLS SPECIAL (II,1527); MEN WHO RATE A "10" (II,1684)

GLEASON, JACKIE
THE HONEYMOONERS CHRISTMAS SPECIAL (II,1176); THE HONEYMOONERS CHRISTMAS (II,1177); THE HONEYMOONERS SECOND HONEYMOON (II,1178); THE HONEYMOONERS VALENTINE SPECIAL (II,1179)

GLENISTER, JOHN
CASANOVA (II,451)

GLENN, WILLIAM E
AFTER HOURS: FROM JANICE, JOHN, MARY AND MICHAEL, WITH LOVE (II,20); AFTER HOURS: SINGIN', SWINGIN' AND ALL THAT JAZZ (II,22); BEHIND THE SCREEN (II,200); THE YOUNG AND THE RESTLESS (II,2862)

GLUCKSMAN, ERNEST D
THE ALL-STAR REVUE (I,138); THE COLGATE COMEDY HOUR (I,997); THE DONALD O'CONNOR SHOW (I,1344); THE IMOGENE COCA SHOW (I,2198)

GODDARD, JIM
KENNEDY (II,1377); NICHOLAS NICKLEBY (II,1840); REILLY, ACE OF SPIES (II,2129)

GODFREY, PETER
THE 20TH CENTURY-FOX HOUR (I,4642); CURTAIN CALL THEATER (I,1115); TURN OF FATE (I,4623)

GOE, BOB
LAVERNE AND SHIRLEY WITH THE FONZ (II,1449); THE LITTLE RASCALS (II,1494); PAC-MAN (II,1940); PINK PANTHER AND SONS (II,2043); THE RICHIE RICH SHOW (II,2169)

GOGGIN, RICHARD
MYSTERIES OF CHINATOWN (I,3209)

GOLD, JEFF
FITZ AND BONES (II,867)

GOLDBERG, BETTY
THE WHITE SHADOW (II,2788)

GOLDBERG, GARY DAVID
MAKING THE GRADE (II,1606)

GOLDEN, MURRAY
AMOS BURKE, SECRET AGENT (I,180); APPLE'S WAY (II,101); THE FLYING NUN (I,1611); GET SMART (II,983); MANNIX (II,1624); MEDICAL CENTER (II,1675); THE MEN FROM SHILOH (I,3005); MISSION: IMPOSSIBLE (I,3067); SIGMUND AND THE SEA MONSTERS (II,2352); THE SMOTHERS BROTHERS SHOW (I,4096); STAR TREK (II,2440); TABITHA (II,2528); THE TIME TUNNEL (I,4511); TRAPPER JOHN, M.D. (II,2654); WANTED: DEAD OR ALIVE (I,4764)

GOLDEN, TOM
PLANET PATROL (I,3616)

GOLDSTEIN, JEFFREY
ANYTHING FOR MONEY (II,99); DREAM HOUSE (II,734); JEOPARDY (II,1279); THE MEMORY GAME (I,3003); STUMPERS (II,2485)

GOLDSTEIN, JERRY
DOUG HENNING'S WORLD OF MAGIC III (II,727)

GOLDSTEIN, RICHARD
THE FACTS (II,804)

GOLDSTONE, DUKE
THE FLORIAN ZABACH SHOW (I,1607); THE FRANKIE LAINE SHOW (I,1675); THE LIBERACE SHOW (I,2675)

GOLDSTONE, JAMES
AMOS BURKE, SECRET AGENT (I,180); THE BOB HOPE CHRYSLER THEATER (I,502); CODE NAME: HERACLITUS (I,990); DENNIS THE MENACE (I,1231); DOCTOR KILDARE (I,1315); THE ELEVENTH HOUR (I,1425); ERIC (I,1451); THE FUGITIVE (I,1701); IRONSIDE (I,2240); THE OATH: THE SAD AND LONELY SUNDAYS (II,1874); THE OUTER LIMITS (I,3426); PERRY MASON (II,2025); ROUTE 66 (I,3852); SCALPLOCK (I,3930); SEA HUNT (I,3948); THE SENATOR (I,3976); STAR TREK (II,2440); STUDS LONIGAN (II,2484); VOYAGE TO THE BOTTOM OF THE SEA (I,4743)

GOMAVITZ, LEWIS
THE KUKLAPOLITAN EASTER SHOW (I,2595)

GOMEZ, DANNY
SALUTE TO LADY LIBERTY (II,2243)

GOODE, MARK
THE KING FAMILY IN WASHINGTON (I,2549); THE KING FAMILY JUNE SPECIAL (I,2550); PAT PAULSEN'S HALF A COMEDY HOUR (I,3473)

GOODE, RICHARD
THE BUFFALO BILLY SHOW (I,756); FOR YOUR PLEASURE (I,1627)

GOODHEART, JOANNE
THE EDGE OF NIGHT (II,760)

GOODMAN, MARSH
KIDD VIDEO (II,1388)

GOODMAN, ROGER
BATTLE OF THE NETWORK STARS (II,163); BATTLE OF THE NETWORK STARS (II,164); BATTLE OF THE NETWORK STARS (II,165); BATTLE OF THE NETWORK STARS (II,166); BATTLE OF THE NETWORK STARS (II,167); BATTLE OF THE NETWORK STARS (II,168); BATTLE OF THE NETWORK STARS (II,169); BATTLE OF THE NETWORK STARS (II,170); BATTLE OF THE NETWORK STARS (II,171); BATTLE OF THE NETWORK STARS (II,172); BATTLE OF THE NETWORK STARS (II,173); BATTLE OF THE NETWORK STARS (II,174); BATTLE OF THE NETWORK STARS (II,175); BATTLE OF THE NETWORK STARS (II,176); BATTLE OF THE NETWORK STARS (II,177)

GOODRUM, SKIP
INCREDIBLE KIDS AND COMPANY (II,1233)

GOODSON, JOSEPH
I DREAM OF JEANNIE (I,2167)

GOODWIN, AL
WHERE'S RAYMOND? (I,4837)

GOODWIN, DERRICK
DOCTOR WHO (II,689)

GOODWINS, LESLIE
77 SUNSET STRIP (I,3988); THE ALASKANS (I,106); THE BING CROSBY SHOW (I,462); BLONDIE (I,486); BRONCO (I,745); THE CISCO KID (I,961); F TROOP (I,1499); GILLIGAN'S ISLAND (II,990); HAWAIIAN EYE (I,1973); HEY, JEANNIE! (I,2038); IT'S ABOUT TIME (I,2263); THE LIFE OF RILEY (I,2685); MY FAVORITE MARTIAN (I,3189); THE NEW ADVENTURES OF CHARLIE CHAN (I,3249); SUGARFOOT (I,4277); SURFSIDE 6 (I,4299); TAMMY (I,4357); TOPPER (I,4582)

GOORIAN, LENNIE
THE PAUL DIXON SHOW (I,3487)

GORALL, LESLIE
ARTHUR MURRAY'S DANCE PARTY (I,293)

GORDON, DAN
CASPER AND FRIENDS (I,867)

GORDON, GEORGE
THE BISKITTS (II,257); CASPER AND THE ANGELS (II,453); FLINTSTONE FAMILY ADVENTURES (II,882); THE FLINTSTONE FUNNIES (II,883); FONZ AND THE HAPPY DAYS GANG (II,898); FRED AND BARNEY MEET THE SHMOO (II,918); FRED AND BARNEY MEET THE THING (II,919); THE GARY COLEMAN SHOW (II,958); THE KWICKY KOALA SHOW (II,1417); LAVERNE AND SHIRLEY IN THE ARMY (II,1448); LAVERNE AND SHIRLEY WITH THE FONZ (II,1449); THE LITTLE RASCALS (II,1494); MONCHHICHIS (II,1722); THE NEW SHMOO (II,1827); PAC-MAN (II,1940); THE RICHIE RICH SHOW (II,2169); SCOOBY-DOO AND SCRAPPY-DOO (II,2274); THE SHIRT TALES (II,2337); THE SMURFS (II,2385); SPACE STARS (II,2408); SUPER FRIENDS (II,2497); THE SUPER GLOBETROTTERS (II,2498); TROLLKINS (II,2660)

GORDON, JIM
THE DONALD O'CONNOR SHOW (I,1345)

GORDON, LEO
ADAM-12 (I,31)

GORDON, MARK
GOOD TIME HARRY (II,1037)

GORDON, RICHARD
THE ALASKANS (I,106)

GORDON, ROBERT
THE BIG STORY (I,432); LAW OF THE PLAINSMAN (I,2639); MY FRIEND FLICKA (I,3190); SCHOOL HOUSE (I,3937)

GORDON, STEVE
GOOD TIME HARRY (II,1037)

GOSCH, MARTIN
TONIGHT ON BROADWAY (I,4559)

GOTTLIEB, CARL
DELTA HOUSE (II,658); THE MUSIC SCENE (I,3176)

GOTTLIEB, RICHARD
THE $100,000 NAME THAT TUNE (II,1902); THE CROSS-WITS (II,605); NAME THAT TUNE (II,1786); THE NEW TRUTH OR CONSEQUENCES (II,1832); THE PEOPLE'S COURT (II,2001); THIS IS YOUR LIFE (I,4461); THIS IS YOUR LIFE (II,2598)

GOULD, CHARLES
THE ADVENTURES OF RIN TIN TIN (I,73)

GOULD, GEORGE
ROD BROWN OF THE ROCKET RANGERS (I,3825); TOM CORBETT, SPACE CADET (I,4535)

GOURNEE, HAL
THE DAVID LETTERMAN SHOW (II,624)

GOWERS, BRUCE
THE ALL-STAR SALUTE TO MOTHER'S DAY (II,51); BILLY CRYSTAL: A COMIC'S LINE (II,249); BRITT EKLAND'S JUKE BOX (II,378); GLEN CAMPBELL AND FRIENDS: THE SILVER ANNIVERSARY (II,1005); HEADLINERS WITH DAVID FROST (II,1114); MEN AT WORK IN CONCERT (II,1682); SHOW BUSINESS (II,2343); THIS IS YOUR LIFE (II,2598); THE TONI TENNILLE SHOW (II,2636); WE DARE YOU! (II,2751); THE WORLD OF ENTERTAINMENT (II,2837)

GRAHAM, WILLIAM A
ADAMSBURG, U.S.A. (I,33); BATMAN (I,366); THE BIG VALLEY (I,437); THE CAT

AND THE CANARY (I,868); CHECKMATE (I,919); THE DICK POWELL SHOW (I,1269); THE FBI (I,1551); FOUR FOR TONIGHT (I,1648); THE FUGITIVE (I,1701); GET CHRISTIE LOVE! (II,981); GET CHRISTIE LOVE! (II,982); IRONSIDE (II,1246); JIGSAW (I,2377); THE LAST NINJA (II,1438); MAN ON THE MOVE (I,2878); MEDICAL CENTER (II,1675); MR. INSIDE/MR. OUTSIDE (II,1750); NAKED CITY (I,3216); THE NAME OF THE GAME (I,3217); THE POLICE STORY (I,3638); RAPTURE AT TWO-FORTY (I,3724); THE ROGUES (I,3832); ROUTE 66 (I,3852); RUN FOR YOUR LIFE (I,3871); SAINTS AND SINNERS (I,3887); SOUND OF THE 60S (I,4134); SWISS FAMILY ROBINSON (I,4324); THEN CAME BRONSON (I,4435); THEN CAME BRONSON (I,4436); THREE FOR DANGER (I,4475); WITNESS (I,4885)

GRANITO, LIVIA
FULL CIRCLE (I,1702)

GRANT, BRIAN
A HOT SUMMER NIGHT WITH DONNA (II,1190); OLIVIA NEWTON-JOHN IN CONCERT (II,1884); OLIVIA NEWTON-JOHN: LET'S BE PHYSICAL (II,1887)

GRANT, LEE
THE SHAPE OF THINGS (I,3999)

GRASSHOFF, ALEX
THE ABC AFTERSCHOOL SPECIAL (II,1); THE BARBARY COAST (II,146); COUNTERATTACK: CRIME IN AMERICA (II,581); FRANK SINATRA JR. WITH FAMILY AND FRIENDS (I,1663); GET CHRISTIE LOVE! (II,981); MOVIN' ON (II,1743); THE NIGHT STALKER (II,1849); RAFFERTY (II,2105); THE ROCKFORD FILES (II,2197); TOMA (II,2634)

GRAUMAN, WALTER
ADAMS OF EAGLE LAKE (II,10); ALCOA/GOODYEAR THEATER (I,107); ANOTHER DAY, ANOTHER DOLLAR (I,228); BARE ESSENCE (II,149); BARE ESSENCE (II,150); BARNABY JONES (II,153); THE BOSTON TERRIER (I,697); COLT .45 (I,1010); DAN AUGUST (I,1130); THE ELEVENTH HOUR (I,1425); THE FELONY SQUAD (I,1563); FORCE FIVE (II,907); THE FUGITIVE (I,1701); HOW'S BUSINESS? (I,2153); JACQUELINE SUSANN'S VALLEY OF THE DOLLS, 1981 (II,1268); KNIGHT'S GAMBIT (I,2576); MANHUNTER (II,1620); MANHUNTER (II,1621); MOST WANTED (II,1738); MOST WANTED (II,1739); NAKED CITY (I,3216); OPERATION: RUNAWAY (II,1918); THE OUTER LIMITS (I,3426); PETER GUNN (I,3562); THE ROGUES (I,3832); ROUTE 66 (I,3852); SCENE OF THE CRIME (II,2271); THE STREETS OF SAN FRANCISCO (I,4260); THE STREETS OF SAN FRANCISCO (II,2478); TALES OF THE UNEXPECTED (II,2539); THEY CALL IT MURDER (I,4441); TOP OF THE HILL (II,2645); THE UNTOUCHABLES (I,4682); V: THE SERIES (II,2715)

GRAY, JOHN
POWERHOUSE (II,2074)

GRAY, MILT
DUNGEONS AND DRAGONS (II,744); PANDAMONIUM (II,1948)

GRAY, NORM
BUSTING LOOSE (II,402); THE ODD COUPLE (II,1875)

GREEDY, ALLAN
MISSION: IMPOSSIBLE (I,3067)

GREEN, BRUCE SETH
AUTOMAN (II,120); KNIGHT RIDER (II,1402); MAGNUM, P.I. (II,1601); V: THE SERIES (II,2715)

GREEN, RICHARD
PROJECT UFO (II,2088)

GREENBERG, ED
ARCHY AND MEHITABEL (I,254); ROBERTA (I,3815)

GREENE, DAVID
THE ALCOA HOUR (I,108); CORONET BLUE (I,1055); THE DEFENDERS (I,1220); ELLERY QUEEN (II,766); ELLERY QUEEN: TOO MANY SUSPECTS (II,767); THE HAPPENERS (I,1935); LUCAN (II,1553); LUCAN (II,1554); MADAME SIN (I,2805); THE PEOPLE NEXT DOOR (I,3519); THE PERSUADERS! (I,3556); RICH MAN, POOR MAN—BOOK I (II,2161); ROOTS (II,2211); THE STRANGE REPORT (I,4248)

GREENE, DOUGLAS
HAWAII FIVE-O (II,1110)

GREENE, MILTON
THE NEW ZOO REVUE (I,3277)

GREENWALD, ROBERT
FLATBED ANNIE AND SWEETIEPIE: LADY TRUCKERS (II,876)

GREGEEN, PETER
THE ONEDIN LINE (II,1912)

GRIES, TOM
ALCOA/GOODYEAR THEATER (I,107); BATMAN (I,366); THE BOB HOPE CHRYSLER THEATER (I,502); CALL TO DANGER (I,804); CHECKMATE (I,919); COMBAT (I,1011); THE DEFENDERS (I,1220); THE DETECTIVES (I,1245); THE HEALERS (II,1115); HUNTER (II,1204); THE MONROES (I,3089); THE RAT PATROL (I,3726); THE RIFLEMAN (I,3789); ROUTE 66 (I,3852); TURN OF FATE (I,4623); VOYAGE TO THE BOTTOM OF THE SEA (I,4743); THE WESTERNER (I,4801)

GRIEVE, KEN
CORONATION STREET (I,1054); THE OMEGA FACTOR (II,1888)

GRIFFIN, KEN
RED SKELTON: A ROYAL PERFORMANCE (II,2125)

GRILLO, GARY
SIMON AND SIMON (II,2357)

GRIMWADE, PETER
THE OMEGA FACTOR (II,1888)

GRINT, ALAN
A HORSEMAN RIDING BY (II,1182)

GRODIN, CHARLES
PARADISE (II,1955)

GROSS, LLOYD
BREAK THE $250,000 BANK (I,713); I'VE GOT A SECRET (II,1220); LUCKY PUP (I,2788); MAKE A FACE (I,2837); THE MEL TORME SHOW (I,2995); TO TELL THE TRUTH (I,4528); TO TELL THE TRUTH (II,2625); TO TELL THE TRUTH (II,2626)

GROSSAN, MARK
PEOPLE DO THE CRAZIEST THINGS (II,1998)

GROSSMAN, DAVID
PEOPLE ARE FUNNY (II,1996)

GROSSMAN, MARK
PEOPLE DO THE CRAZIEST THINGS (II,1997); PERSONAL AND CONFIDENTIAL (II,2026)

GRUENBERG, AXEL
THE UNEXPECTED (I,4674)

GUENETTE, ROBERT
HEROES AND SIDEKICKS—INDIANA JONES AND THE TEMPLE OF DOOM (II,1140)

GUEST, VAL
THE PERSUADERS! (I,3556); SPACE: 1999 (II,2409)

GUILFOYLE, PAUL
COLT .45 (I,1010); LAWMAN (I,2642); PUBLIC DEFENDER (I,3687); SUGARFOOT (I,4277)

GUILLERMIN, JOHN
AGGIE (I,90)

GUNTZELMAN, DAN
WKRP IN CINCINNATI (II,2814)

GURNEE, HAL
A FUNNY THING HAPPENED ON THE WAY TO HOLLYWOOD (I,1716); THE GARRY MOORE SHOW (I,1740); JACK PAAR AND A FUNNY THING HAPPENED EVERYWHERE (I,2314); THE JACK PAAR SPECIAL (I,2319); THE JOEY BISHOP SHOW (I,2405); LATE NIGHT WITH DAVID LETTERMAN (II,1442)

GURNEE, PAUL
JACK PAAR TONIGHT (I,2320)

GUTIERREZ, GERALD
THE COMEDY ZONE (II,559)

GUY, TONY
THE OSMONDS (I,3410)

GUZMAN, CLAUDIO
CALIFORNIA FEVER (II,411); THE DESILU PLAYHOUSE (I,1237); THE DESILU REVUE (I,1238); THE DICK VAN DYKE SHOW (I,1275); FAR OUT SPACE NUTS (II,831); FOR LOVERS ONLY (II,904); THE GOOD LIFE (I,1848); HARPER VALLEY PTA (II,1095); HERE'S BOOMER (II,1134); I DREAM OF JEANNIE (I,2167); THE PATTY DUKE SHOW (I,3483); THREE FOR THE ROAD (II,2607); THE TWO OF US (I,4658)

GWENLAN, GARETH
BUTTERFLIES (II,404); TO THE MANOR BORN (II,2627)

GYLLENHAAL, STEPHEN
THE ABC AFTERSCHOOL SPECIAL (II,1)

HAAS, CHARLES
77 SUNSET STRIP (I,3988); THE ALASKANS (I,106); ALFRED HITCHCOCK PRESENTS (I,115); BROKEN ARROW (I,744); BURKE'S

LAW (I,762); BUS STOP (I,778); THE DETECTIVES (I,1245); THE DICK POWELL SHOW (I,1269); THE HARDY BOYS AND THE MYSTERY OF THE APPLEGATETREASURE (I,1951); HAWAIIAN EYE (I,1973); LEAVE IT TO BEAVER (I,2648); MAVERICK (II,1657); THE MICKEY MOUSE CLUB (I,3025); THE NEW ADVENTURES OF CHARLIE CHAN (I,3249); THE OUTER LIMITS (I,3426); PERRY MASON (II,2025); THE ROARING TWENTIES (I,3805); ROUTE 66 (I,3852); SATAN'S WAITIN' (I,3916); STORY THEATER (I,4240); SURFSIDE 6 (I,4299)

HAAS, CLARK
CAPTAIN FATHOM (I,830)

HACKBORN, BOB
FRAGGLE ROCK (II,916); RICH LITTLE'S ROBIN HOOD (II,2158)

HAFFNER, CRAIG
EYE ON HOLLYWOOD (II,798)

HAGGARD, PIERS
PENNIES FROM HEAVEN (II,1994)

HAGMAN, LARRY
DALLAS (II,614); I DREAM OF JEANNIE (I,2167)

HAGMANN, STUART
BRONK (II,379); MANNIX (II,1624); MISSION: IMPOSSIBLE (I,3067); N.Y.P.D. (I,3321); SPARROW (II,2410)

HAHN, PHIL
THE FUNNIEST JOKE I EVER HEARD (II,939)

HAINES, RANDA
THE FAMILY TREE (II,820); FOR LOVE AND HONOR (II,903); HILL STREET BLUES (II,1154); KNOTS LANDING (II,1404); TUCKER'S WITCH (II,2667)

HALDANE, DON
THE SWISS FAMILY ROBINSON (II,2518)

HALE, BILLY
LACE (II,1418)

HALE, JEFF
JIM HENSON'S MUPPET BABIES (II,1289); THE TRANSFORMERS (II,2653)

HALE, LEE
DEAN MARTIN'S CHRISTMAS AT SEA WORLD (II,644);

HALE, WILLIAM
77 SUNSET STRIP (I,3988); BARNABY JONES (II,153); THE BOB HOPE CHRYSLER THEATER (I,502); BOURBON STREET BEAT (I,699); BRONCO (I,745); CARIBE (II,439); CHEYENNE (I,931); COLT .45 (I,1010); CROSSFIRE (II,606); THE FBI (I,1551); THE FELONY SQUAD (I,1563); HAWAII FIVE-O (II,1110); HAWAIIAN EYE (I,1973); THE INVADERS (I,2229); JUDD, FOR THE DEFENSE (I,2463); LANCER (I,2610); THE PAPER CHASE (II,1949); RUN FOR YOUR LIFE (I,3871); THE STREETS OF SAN FRANCISCO (II,2478); SUGARFOOT (I,4277)

HALEY JR, ARTHUR
RIPLEY'S BELIEVE IT OR NOT (II,2177)

HALEY JR, JACK
BEAT OF THE BRASS (I,376); BIOGRAPHY (I,476); BOB HOPE SPECIAL: BOB HOPE'S WORLD OF COMEDY (II,323); HOLLYWOOD AND THE STARS (I,2081); MOVIN' WITH NANCY (I,3119); THE WORLD OF JAMES BOND (I,4913)

HALL, JOSEPH L
THE MAGIC GARDEN (II,1591)

HALL, NORMAN
THE DOCTORS (II,694); ONE LIFE TO LIVE (II,1907)

HALLECK, VANCE
TELE PUN (I,4376); THIS IS CHARLES LAUGHTON (I,4455)

HALLENBACK, DANIEL
THE MAN FROM U.N.C.L.E. (I,2867)

HALLER, DANIEL
THE 25TH MAN (MS) (II,2678); B.J. AND THE BEAR (II,126); BANYON (I,344); BATTLESTAR GALACTICA (II,181); BLACK BEAUTY (II,261); THE BLUE KNIGHT (II,276); BUCK ROGERS IN THE 25TH CENTURY (II,383); CHARLIE'S ANGELS (II,486); THE FALL GUY (II,811); GALACTICA 1980 (II,953); THE GEORGIA PEACHES (II,979); THE HARDY BOYS MYSTERIES (II,1090); HIGH PERFORMANCE (II,1145); HOW DO I KILL A THIEF—LET ME COUNT THE WAYS (II,1195); IRONSIDE (II,1246); KHAN! (II,1384); KNIGHT RIDER (II,1402); KOJAK (II,1406); LOBO (II,1504); THE MAN AND THE CITY (I,2856); MANIMAL (II,1622); MCNAUGHTON'S DAUGHTER (II,1670); THE MISADVENTURES OF SHERIFF LOBO (II,1704); OWEN MARSHALL: COUNSELOR AT LAW (I,3435); QUINCY, M. E. (II,2102); ROSETTI AND RYAN (II,2216); SARA (II,2260); THE SIXTH SENSE (I,4069); SUNSHINE (II,2495); SWORD OF JUSTICE (II,2521); TOMA (II,2634); WALKING TALL (II,2738); WELCOME TO PARADISE (II,2762)

HALLER, DAVID
MICKEY SPILLANE'S MARGIN FOR MURDER (II,1691)

HALORAN, JACK
THE DEAN MARTIN SHOW (I,1201)

HALVORSON, GARY
PIAF (II,2039); SHERLOCK HOLMES (II,2327); VANITIES (II,2720)

HAMILTON, JOE
JULIE AND CAROL AT CARNEGIE HALL (I,2478)

HAMILTON, MICHAEL
WHIZ KIDS (II,2790)

HANDLEY, ALAN
THE ABC AFTERSCHOOL SPECIAL (II,1); THE ANDY GRIFFITH—DON KNOTTS—JIMNABORS SHOW (I,191); THE ANDY WILLIAMS CHRISTMAS SHOW (I,198); THE ANDY WILLIAMS SPECIAL (I,213); CAROL CHANNING AND 101 MEN (I,849); COKE TIME WITH EDDIE FISHER (I,996); THE COMICS (I,1028); DANNY THOMAS GOES COUNTRY AND WESTERN (I,1146); THE DANNY THOMAS SPECIAL (I,1150); THE DANNY THOMAS SPECIAL (I,1151); THE DANNY THOMAS SPECIAL (I,1152); THE DANNY THOMAS TV FAMILY REUNION (I,1154); DANNY THOMAS: AMERICA I LOVE YOU (I,1145); DANNY THOMAS: THE ROAD TO LEBANON (I,1153); THE DINAH SHORE SHOW (I,1280); ESTHER WILLIAMS AT CYPRESS GARDENS (I,1465); GUYS 'N' GEISHAS (I,1914); JIMMY DURANTE MEETS THE LIVELY ARTS (I,2386); THE JULIE ANDREWS SHOW (I,2481); THE JULIE ANDREWS SPECIAL (I,2482); THE MICKIE FINNS FINALLY PRESENT HOW THE WEST WAS LOST (II,1695); PORTRAIT OF PETULA (I,3654); THE ROYAL FOLLIES OF 1933 (I,3862); THE SID CAESAR SHOW (I,4042); SVENGALI AND THE BLONDE (I,4310); THE WALTER WINCHELL SHOW (I,4760); THE WONDERFUL WORLD OF BURLESQUE I (I,4895); THE WONDERFUL WORLD OF BURLESQUE II (I,4896); THE WONDERFUL WORLD OF BURLESQUE III (I,4897); THE WOODY ALLEN SPECIAL (I,4904)

HANDLEY, DREW
ME AND MRS. C. (II,1673)

HANLEY, JIM
STRANGE TRUE STORIES (II,2476)

HANNA, WILLIAM
THE ABBOTT AND COSTELLO CARTOON SHOW (I,1); THE ADVENTURES OF GULLIVER (I,60); THE ADVENTURES OF JONNY QUEST (I,64); THE AMAZING CHAN AND THE CHAN CLAN (I,157); THE ATOM ANT/SECRET SQUIRREL SHOW (I,311); BIRDMAN (II,256); BIRDMAN AND THE GALAXY TRIO (I,478); BUTCH CASSIDY AND THE SUNDANCE KIDS (I,781); THE CATTANOOGA CATS (I,874); DASTARDLY AND MUTTLEY IN THEIR FLYING MACHINES (I,1159); THE FANTASTIC FOUR (I,1529); THE FLINTSTONE COMEDY HOUR (II,881); THE FLINTSTONES (II,884); FRANKENSTEIN JR. AND THE IMPOSSIBLES (I,1672); THE FUNKY PHANTOM (I,1709); THE HARLEM GLOBETROTTERS (I,1953); HELP! IT'S THE HAIR BEAR BUNCH (I,2012); THE HERCULOIDS (I,2023); THE HUCKLEBERRY HOUND SHOW (I,2155); THE JETSONS (I,2376); JOKEBOOK (II,1339); JOSIE AND THE PUSSYCATS (I,2453); JOSIE AND THE PUSSYCATS IN OUTER SPACE (I,2454); LIPPY THE LION (I,2710); THE MAGILLA GORILLA SHOW (I,2825); MOBY DICK AND THE MIGHTY MIGHTOR (I,3077); MOTOR MOUSE (I,3111); THE NEW SUPER FRIENDS HOUR (II,1830); THE PERILS OF PENELOPE PITSTOP (I,3525); THE PETER POTAMUS SHOW (I,3568); SAMPSON AND GOLIATH (I,3903); SCOOBY-DOO, WHERE ARE YOU? (II,2276); THE SCOOBY-DOO/DYNOMUTT HOUR (II,2277); SHAZZAN! (I,4002); SPACE GHOST (I,4141); SPACE KIDDETTES (I,4143); THE TOM AND JERRY SHOW (I,4534); TOP CAT (I,4576); TOUCHE TURTLE (I,4586); THE WACKY RACES (I,4745);

WAIT TIL YOUR FATHER GETS HOME (I,4748); WALLY GATOR (I,4751); WHERE'S HUDDLES (I,4836); THE WORLD'S GREATEST SUPER HEROES (II,2842); YOGI BEAR (I,4928)

HANNAM, KEN
THE ASSASSINATION RUN (II,112); DAY OF THE TRIFFIDS (II,627); THE ONEDIN LINE (II,1912)

HANSON, DAVID E
BATTLE OF THE PLANETS (II,179)

HANSON, VALERIE
CRIMES OF PASSION (II,603)

HARBACH, WALTER O
THE BING CROSBY SHOW (I,463)

HARBACH, WILLIAM O
THE BING CROSBY SHOW (I,464)

HARDING, RICHMOND
THE AVENGERS (II,121); SAM (II,2247)

HARDWICKE, MARY
JERRY REED AND SPECIAL FRIENDS (II,1285); THAT'S INCREDIBLE! (II,2578); THOSE AMAZING ANIMALS (II,2601)

HARDWICKE, SIR CEDRIC
MR. GLENCANNON TAKES ALL (I,3139); SKETCHBOOK (I,4072)

HARDY, JOSEPH
ADDIE AND THE KING OF HEARTS (I,35); EXECUTIVE SUITE (II,796); JAMES AT 15 (II,1270); JAMES AT 15 (II,1271); THE LILY TOMLIN SHOW (I,2706); THE PAPER CHASE (II,1949); RETURN ENGAGEMENT (I,3775); TAXI (I,4365); TOMA (II,2634); TWO MARRIAGES (II,2691)

HARDY, ROD
PRISONER: CELL BLOCK H (II,2085)

HARGROVE, DEAN
DEAR DETECTIVE (II,652); RETURN OF THE WORLD'S GREATEST DETECTIVE (II,2142)

HARLIB, MATTHEW
RED GOOSE KID'S SPECTACULAR (I,3751); STRIKE IT RICH (I,4262)

HARMON, DAVID P
IT'S ABOUT TIME (I,2263)

HARMS, FRANK
THE REAL MCKAY (I,3742)

HARRELL, CHARLES
CHANCE OF A LIFETIME (I,898); THE PLAINCLOTHESMAN (I,3614); THINK FAST (I,4447)

HARRINGTON, CURTIS
BARETTA (II,152); CHARLIE'S ANGELS (II,486); DYNASTY (II,746); HOTEL (II,1192); LUCAN (II,1554); MULLIGAN'S STEW (II,1764); SWORD OF JUSTICE (II,2521); VEGAS (II,2724); WONDER WOMAN (SERIES 2) (II,2829)

HARRIS, HARRY
ADAM-12 (I,31); APPLE'S WAY (II,101); BOONE (II,351); BRANDED (I,707); CAGNEY AND LACEY (II,409); THE D.A. (I,1123); DALLAS (II,614); DOC ELLIOT (I,1306); EIGHT IS ENOUGH (II,762); FALCON CREST (II,808); FAME (II,812); GIBBSVILLE (II,988); GUNSMOKE (II,1069); THE HIGH CHAPARRAL (I,2051); THE HOME FRONT (II,1171); HOTEL (II,1192); HUNTER (II,1205); KUNG FU (II,1416); LAND OF THE GIANTS (I,2611); LOST IN SPACE (I,2758); THE MACKENZIES OF PARADISE COVE (II,1585); THE MEN FROM SHILOH (I,3005); THE MISSISSIPPI (II,1707); NAKED CITY (I,3216); NURSE (II,1870); RAWHIDE (I,3727); RIVKIN: BOUNTY HUNTER (II,2183); SCAMPS (II,2267); STAGECOACH WEST (I,4185); SWISS FAMILY ROBINSON (II,2516); SWISS FAMILY ROBINSON (II,2517); THE TIME TUNNEL (I,4511); THE TOM SWIFT AND LINDA CRAIG MYSTERY HOUR (II,2633); THE VIRGINIAN (I,4738); VOYAGE TO THE BOTTOM OF THE SEA (I,4743); THE WALTONS (II,2740); THE WALTONS: A DAY FOR THANKS ON WALTON'S MOUNTAIN (EPISODE 3) (II,2741); THE YOUNG PIONEERS (II,2869)

HARRIS, JEANNE
EASY ACES (I,1398)

HARRIS, JEFF
ALL-AMERICAN PIE (II,48); DETECTIVE SCHOOL (II,661); GOSSIP (II,1043); IT ONLY HURTS WHEN YOU LAUGH (II,1251); THE SHOW MUST GO ON (II,2344)

HARRIS, PAUL
LIVE FROM THE LONE STAR (II,1501)

HARRIS, PETER
BONKERS! (II,348); THE MUPPET SHOW (II,1766); THE MUPPETS GO TO THE MOVIES (II,1767)

HARRIS, RON
20 MINUTE WORKOUT (II,2680)

HARRIS, STAN
100 YEARS OF AMERICA'S POPULAR MUSIC (II,1904); 20TH CENTURY FOLLIES (I,4641); BLONDIE (II,272); BOB HOPE SPECIAL: BOB HOPE'S SALUTE TO NASA—25 YEARS OF REACHING FOR THE STARS (II,314); CHRISTMAS SPECIAL. . . WITH LOVE, MAC DAVIS (II,523); COUNTRY STARS OF THE 70S (II,594); CRYSTAL GAYLE IN CONCERT (II,608); DANCE FEVER (II,615); DOLLY IN CONCERT (II,700); DUKE ELLINGTON. . . WE LOVE YOU MADLY (I,1380); FANFARE (I,1528); FRIENDS AND NABORS (I,1684); THE GARRY MOORE SHOW (I,1739); GEORGE BURNS' 100TH BIRTHDAY PARTY (II,971); THE GISELE MACKENZIE SHOW (I,1813); HI, I'M GLEN CAMPBELL (II,1141); JACK BENNY'S 20TH ANNIVERSARY TV SPECIAL (I,2303); THE JOHN GARY SHOW (I,2411); JOHNNY CASH'S AMERICA (II,1330); JOHNNY CASH: CHRISTMAS IN SCOTLAND (II,1332); JOHNNY CASH: COWBOY HEROES (II,1334); LYNDA CARTER'S CELEBRATION (II,1563); LYNDA CARTER'S SPECIAL (II,1564); LYNDA CARTER: ENCORE (II,1566); LYNDA CARTER: STREET LIGHTS (II,1567); MAKE YOUR OWN KIND OF MUSIC (I,2847); THE MANCINI GENERATION (I,2885); THE MELBA MOORE-CLIFTON DAVIS SHOW (I,2996); THE MILTON BERLE SHOW (I,3049); THE MUSIC SCENE (I,3176); THE NASHVILLE PALACE (II,1793); THE OSMOND BROTHERS SHOW (I,3409); RICH LITTLE'S WASHINGTON FOLLIES (II,2159); THE ROBERT KLEIN SHOW (II,2185); THE SMOTHERS BROTHERS COMEDY HOUR (I,4095); A SPECIAL KENNY ROGERS (II,2417); THE STEVE LAWRENCE SHOW (I,4227); SWING OUT, SWEET LAND (II,2515); THAT'S LIFE (I,4426); TWO OF A KIND: GEORGE BURNS AND JOHN DENVER (II,2692)

HARRISON, CLAY
COUNTERATTACK: CRIME IN AMERICA (II,581); PEOPLE DO THE CRAZIEST THINGS (II,1998)

HARRISON, GREGORY
TRAPPER JOHN, M.D. (II,2654)

HARRISON, PAUL
THE ANN SOTHERN SHOW (I,220); I HEAR AMERICA SINGING (I,2168); THE JUDY GARLAND SHOW (I,2470); THE SKY'S THE LIMIT (I,4081); THE UNTOUCHABLES (I,4682)

HARRISON, SAM
MATINEE AT THE BIJOU (II,1651)

HARRISON, TERRY
LAVERNE AND SHIRLEY IN THE ARMY (II,1448); LAVERNE AND SHIRLEY WITH THE FONZ (II,1449); THE LITTLE RASCALS (II,1494); PAC-MAN (II,1940); THE RICHIE RICH SHOW (II,2169)

HART, HARVEY
ALFRED HITCHCOCK PRESENTS (I,115); THE ALIENS ARE COMING (II,35); BEN CASEY (I,394); THE BOB HOPE CHRYSLER THEATER (I,502); THE CITY (II,541); COLUMBO (II,556); DAN AUGUST (I,1130); EAST OF EDEN (II,752); HERE COME THE BRIDES (I,2024); LAREDO (I,2615); MANNIX (II,1624); MASSARATI AND THE BRAIN (II,1646); MASTER OF THE GAME (II,1647); MR. AND MRS. COP (II,1748); MY FRIEND TONY (I,3192); THE NAME OF THE GAME (I,3217); STAR TREK (II,2440); THE STARLOST (I,4203); THE SURVIVORS (I,4301); THIS IS KATE BENNETT (II,2597); W.E.B. (II,2732); THE WILD WILD WEST (I,4863); WOMAN ON THE RUN (II,2821); THE YELLOW ROSE (II,2847); THE YOUNG LAWYERS (I,4940); THE YOUNG LAWYERS (I,4941)

HART, TERRY
JOANIE LOVES CHACHI (II,1295)

HART, WALTER
ETHEL AND ALBERT (I,1466); THE GOLDBERGS (I,1835)

HARTUNG, ROBERT

GOLDEN CHILD (I,1839)

HARVEY JR, HARRY
MANNIX (II,1624)

HARVEY, ANTHONY
DISAPPEARANCE OF AIMEE (I,1290)

HARWOOD, RICHARD S
BABY, I'M BACK! (II,130); BOB HOPE SPECIAL: BOB HOPE'S STAND UP AND CHEER FOR THE NATIONAL FOOTBALL LEAGUE'S 60TH YEAR (II,316); GIMME A BREAK (II,995); GOOD PENNY (II,1036); HARPER VALLEY (II,1094); LOBO (II,1504); THE MISADVENTURES OF SHERIFF LOBO (II,1704); STOP SUSAN WILLIAMS (II,2473); SZYSZNYK (II,2523); WHAT'S HAPPENING!! (II,2769)

HASKIN, BYRON
ALCOA/GOODYEAR THEATER (I,107); THE NEW ADVENTURES OF HUCKLEBERRY FINN (I,3250); THE OUTER LIMITS (I,3426); SCREEN DIRECTOR'S PLAYHOUSE (I,3946); SECRET AGENT (I,3961)

HATHCOCK, BOB
THE GARY COLEMAN SHOW (II,958); LAVERNE AND SHIRLEY WITH THE FONZ (II,1449); THE LITTLE RASCALS (II,1494); PAC-MAN (II,1940); THE RICHIE RICH SHOW (II,2169); THE SHIRT TALES (II,2337); THE SMURFS (II,2385)

HATOS, STEFAN
IT PAYS TO BE MARRIED (I,2249); THERE'S ONE IN EVERY FAMILY (I,4439)

HAUSER, RICK
FROM HERE TO ETERNITY (II,933); THE SCARLET LETTER (II,2270)

HAVIGNA, NICK
AFTERMASH (II,23); ALICE (II,33); ANOTHER DAY (II,94); ARCHIE BUNKER'S PLACE (II,105); BABY, I'M BACK! (II,130); BALL FOUR (II,137); BEACON HILL (II,193); EMERALD POINT, N.A.S. (II,773); FLAMINGO ROAD (II,872); FLO (II,891); GHOST OF A CHANCE (II,987); HARPER VALLEY (II,1094); HIGHCLIFFE MANOR (II,1150); HOUSE CALLS (II,1194); KING'S CROSSING (II,1397); KNOTS LANDING (II,1404); LOVE AT FIRST SIGHT (II,1531); THE PAPER CHASE: THE SECOND YEAR (II,1950); REMINGTON STEELE (II,2130); UNITED STATES (II,2708)

HAWLEY, MARK
VERSATILE VARIETIES (I,4713); WHO SAID THAT? (I,4849)

HAYDEN, JEFFREY
77 SUNSET STRIP (I,3988); THE CHOCOLATE SOLDIER (I,944); THE CURSE OF DRACULA (II,611); THE DICK POWELL SHOW (I,1269); THE DONNA REED SHOW (I,1347); EMERALD POINT, N.A.S. (II,773); FROM HERE TO ETERNITY (II,933); HAWAIIAN HEAT (II,1111); THE INCREDIBLE HULK (II,1232); JESSICA NOVAK (II,1286); JULIE FARR, M.D. (II,1354); KNIGHT RIDER (II,1402); LADY IN THE DARK (I,2602); LEAVE IT TO BEAVER (I,2648); THE LORETTA YOUNG THEATER (I,2756); MAGNUM, P.I. (II,1601); THE MISSISSIPPI (II,1707); MR. MERLIN (II,1751); NO TIME FOR SERGEANTS (I,3300); PALMERSTOWN U.S.A. (II,1945); PLEASE DON'T EAT THE DAISIES (I,3628); THE POWERS OF MATTHEW STAR (II,2075); QUINCY, M. E. (II,2102); ROUTE 66 (I,3852); THE RUNAWAYS (II,2231); SHANE (I,3995); SPACE ACADEMY (II,2406); SURFSIDE 6 (I,4299); THAT GIRL (II,2570)

HAYDN, RICHARD
BACHELOR PARTY (I,324)

HAYERS, SIDNEY
THE AVENGERS (II,121); CONDOMINIUM (II,566); COVER UP (II,597); THE FALL GUY (II,811); THE FAMILY TREE (II,820); FITZ AND BONES (II,867); GALACTICA 1980 (II,953); THE GREATEST AMERICAN HERO (II,1060); THE HARDY BOYS MYSTERIES (II,1090); KNIGHT RIDER (II,1402); THE LAST CONVERTIBLE (II,1435); LOBO (II,1504); MAGNUM, P.I. (II,1601); MANIMAL (II,1622); MASQUERADE (II,1644); THE MASTER (II,1648); MISTER JERICO (I,3070); THE NEW AVENGERS (II,1816); THE PERSUADERS! (I,3556); PHILIP MARLOWE, PRIVATE EYE (II,2033); REMINGTON STEELE (II,2130); SAVAGE: IN THE ORIENT (II,2264); THE SEEKERS (II,2298); THE STRANGE REPORT (I,4248); TERROR AT ALCATRAZ (II,2563); THE ZOO GANG (II,2877)

HAYES, JEFFREY
LEGMEN (II,1458)

HAYS, BILL
FILE IT UNDER FEAR (I,1574); LADY KILLER (I,2603); THE TALE OF BEATRIX POTTER (II,2533)

HEATH, BOB
YOU ASKED FOR IT (II,2855)

HEATON, JEAN
THE EARL WRIGHTSON SHOW (I,1393)

HEDTON, JOHN
THE ALICE PEARCE SHOW (I,121)

HEFFRON, RICHARD T
BANACEK (II,138); THE LAWYERS (I,2646); THE ROCKFORD FILES (II,2196); THE ROCKFORD FILES (II,2197); TOMA (I,4547); TOMA (II,2634); TRUE GRIT (II,2664); V: THE FINAL BATTLE (II,2714)

HEGYES, ROBERT
WELCOME BACK, KOTTER (II,2761)

HEIN, JACK
THE ERNIE KOVACS SHOW (I,1455); TAG THE GAG (I,4330); TED MACK'S FAMILY HOUR (I,4371)

HEISLER, STUART
CHEYENNE (I,931); THE DAKOTAS (I,1126); MAVERICK (II,1657); THE ROARING TWENTIES (I,3805); SCREEN DIRECTOR'S PLAYHOUSE (I,3946); THE VIRGINIAN (I,4738)

HELFER, RALPH
MR. SMITH (II,1755)

HELLER, FRANKLIN
THE AT LIBERTY CLUB (I,309); THE CLIFF EDWARDS SHOW (I,975); THE FRONT PAGE (I,1691); I'VE GOT A SECRET (I,2283); MANHATTAN SHOWCASE (I,2888)

HELLER, JACK
MARY HARTMAN, MARY HARTMAN (II,1638)

HELLINGS, SIMON
PRISONER: CELL BLOCK H (II,2085)

HELLSTROM, GUNNAR
DALLAS (II,614); GUNSMOKE (II,1069); THE POWERS OF MATTHEW STAR (II,2075); THE WILD WILD WEST (I,4863)

HELM, JACK
BROADWAY OPEN HOUSE (I,736)

HEMION, DWIGHT
ANN-MARGRET OLSSON (II,84); ANN-MARGRET SMITH (II,85); ANN-MARGRET'S HOLLYWOOD MOVIE GIRLS (II,86); ANN-MARGRET. . . RHINESTONE COWGIRL (II,87); ANNE MURRAY'S CARIBBEAN CRUISE (II,88); ANNE MURRAY'S WINTER CARNIVAL. . . FROM QUEBEC (II,90); BARBRA STREISAND AND OTHER MUSICAL INSTRUMENTS (I,353); BARYSHNIKOV ON BROADWAY (II,158); BEN VEREEN—HIS ROOTS (II,202); BETTE MIDLER—OL' RED HAIR IS BACK (II,222); BING CROSBY'S MERRIE OLDE CHRISTMAS (II,252); BURT BACHARACH IN SHANGRI-LA (I,773); THE BURT BACHARACH SPECIAL (I,775); THE BURT BACHARACH SPECIAL (I,776); BURT BACHARACH! (I,771); BURT BACHARACH: CLOSE TO YOU (I,772); BURT BACHARACH—OPUS NO. 3 (I,774); THE CHERYL LADD SPECIAL (II,501); CHRISTMAS IN WASHINGTON (II,520); DICK CAVETT'S BACKLOT USA (II,670); DISNEYLAND'S 25TH ANNIVERSARY (II,681); THE DOROTHY HAMILL SPECIAL (II,719); THE EDDIE RABBITT SPECIAL (II,759); THE FRANK SINATRA SHOW (I,1667); FRANK SINATRA—A MAN AND HIS MUSIC (I,1661); FUNNY GIRL TO FUNNY LADY (II,942); GLEN CAMPBELL . . . DOWN HOME—DOWN UNDER (II,1008); THE GLEN CAMPBELL SPECIAL: THE MUSICAL WEST (I,1821); GOLDIE AND KIDS: LISTEN TO ME (II,1021); HERB ALBERT AND THE TIJUANA BRASS (I,2019); HERB ALPERT AND THE TIJUANA BRASS (II,1130); HONEY WEST (I,2105); JAMES PAUL MCCARTNEY (I,2332); JUBILEE (II,1347); JULIE ON SESAME STREET (I,2485); THE KOPYCATS (I,2580); THE KRAFT 75TH ANNIVERSARY SPECIAL (II,1409); THE KRAFT MUSIC HALL (I,2588); KRAFT SALUTES WALT DISNEY WORLD'S 10TH ANNIVERSARY (II,1410); LARRY GATLIN AND THE GATLIN BROTHERS (II,1429);

LINDA IN WONDERLAND (II,1481); LUCY COMES TO NASHVILLE (II,1561); MAC DAVIS'S CHRISTMAS ODYSSEY: TWO THOUSAND AND TEN (II,1579); MAC DAVIS. . . SOUNDS LIKE HOME (II,1582); THE MAGICAL MUSIC OF BURT BACHARACH (I,2822); MARLO THOMAS IN ACTS OF LOVE—AND OTHER COMEDIES (I,2919); MERRY CHRISTMAS FROM THE CROSBYS (II,1685); MERRY CHRISTMAS FROM THE GRAND OLE OPRY (II,1686); MY NAME IS BARBRA (I,3198); THE NEIL DIAMOND SPECIAL (II,1802); THE NEW CHRISTY MINSTRELS SHOW (I,3259); THE NEW STEVE ALLEN SHOW (I,3271); ON STAGE AMERICA (II,1893); PAVAROTTI AND FRIENDS (II,1985); PETER PAN (II,2030); PETULA (I,3573); THE ROGER MILLER SHOW (I,3830); ROOTIE KAZOOTIE (I,3844); ROYAL VARIETY PERFORMANCE (I,3865); THE SANDY DUNCAN SHOW (II,2253); SHEENA EASTON, ACT 1 (II,2323); SHIRLEY MACLAINE. . . EVERY LITTLE MOVEMENT (II,2334); A SPECIAL ANNE MURRAY CHRISTMAS (II,2413); THE STEVE ALLEN SHOW (I,4219); THE STEVE ALLEN SHOW (I,4220); STEVE AND EYDIE CELEBRATE IRVING BERLIN (II,2456); STEVE AND EYDIE: OUR LOVE IS HERE TO STAY (II,2457); STEVE LAWRENCE AND EYDIE GORME FROM THIS MOMENT ON. . . COLE PORTER (II,2459); TEXACO STAR PARADE II (I,4410); TIMEX ALL-STAR JAZZ SHOW I (I,4515); TIN PAN ALLEY TODAY (I,4521); THE TONY BENNETT SHOW (I,4570); UPTOWN (II,2710); THE VERY FIRST GLEN CAMPBELL SPECIAL (I,4714); ZERO HOUR (I,4980)

HEMION, MAC
ALMOST ANYTHING GOES (II,56)

HEMMINGS, DAVID
AIRWOLF (II,27); HAWAIIAN HEAT (II,1111)

HENNING, PAUL
PETTICOAT JUNCTION (I,3571)

HENRIED, PAUL
ALCOA/GOODYEAR THEATER (I,107); ALFRED HITCHCOCK PRESENTS (I,115); THE BIG VALLEY (I,437); BONANZA (II,347); BOURBON STREET BEAT (I,699); BRACKEN'S WORLD (I,703); HAWK (I,1974); JOHNNY RINGO (I,2434); THE MAN AND THE CITY (I,2856); MAVERICK (II,1657); SUGARFOOT (I,4277); THRILLER (I,4492); THE VIRGINIAN (I,4738)

HENRY, BABETTE
BUCK ROGERS IN THE 25TH CENTURY (I,751); CARTOON TELETALES (I,857); JACQUES FRAY'S MUSIC ROOM (I,2329); THE ROBBINS NEST (I,3806); THE SINGING LADY (I,4061); THAT WONDERFUL GUY (I,4425)

HENRY, BOB
AFTER HOURS: GETTING TO KNOW US (II,21); ANDY WILLIAMS KALEIDOSCOPE COMPANY (I,201); BARBARA MANDRELL AND THE MANDRELL SISTERS (II,143); THE CAPTAIN AND TENNILLE (II,429); THE CARPENTERS (II,445); THE CARPENTERS AT CHRISTMAS (II,446); THE CARPENTERS. . . SPACE ENCOUNTERS (II,447); THE DINAH SHORE CHEVY SHOW (I,1279); FELICIANO—VERY SPECIAL (I,1560); THE FLIP WILSON SHOW (I,1602); THE GLEN CAMPBELL MUSIC SHOW (II,1006); THE GLEN CAMPBELL MUSIC SHOW (II,1007); GO! (I,1830); GOOD EVENING, CAPTAIN (II,1031); THE JACK BENNY CHRISTMAS SPECIAL (I,2289); JONATHAN WINTERS PRESENTS 200 YEARS OF AMERICAN HUMOR (II,1342); KISMET (I,2565); LAS VEGAS PALACE OF STARS (II,1431); MITZI (I,3071); MOVIN' (I,3117); THE NAT KING COLE SHOW (I,3230); THE PERRY COMO CHRISTMAS SHOW (I,3529); PERRY COMO'S MUSIC FROM HOLLYWOOD (II,2019); POETRY AND MUSIC (I,3632); THE ROY ROGERS AND DALE EVANS SHOW (I,3858); SALUTE (II,2242)

HENSON, JIM
THE FANTASTIC MISS PIGGY SHOW (II,827); FRAGGLE ROCK (II,916); FROG PRINCE (I,1685)

HERBERT, HENRY
BERGERAC (II,209)

HERMAN, HARVEY
LITTLE VIC (II,1496)

HERMAN, KEN
THE NEW ERNIE KOVACS SHOW (I,3264)

HERRMANN, HARRY
BELIEVE IT OR NOT (I,387)

HERTS, KEN
NUMBER 13 DEMON STREET (I,3316)

HESSEMAN, HOWARD
WKRP IN CINCINNATI (II,2814)

HESSLER, GORDON
ALFRED HITCHCOCK PRESENTS (I,115); AMY PRENTISS (II,74); THE BLUE KNIGHT (II,276); THE BOB HOPE CHRYSLER THEATER (I,502); CHIPS (II,511); KUNG FU (II,1416); LITTLE WOMEN (II,1498); THE MASTER (II,1648); THE NIGHT STALKER (II,1849); RUN FOR YOUR LIFE (I,3871); SARA (II,2260); THE SECRET WAR OF JACKIE'S GIRLS (II,2294); SHANNON (II,2314); SPENCER'S PILOTS (II,2422); TALES OF THE HAUNTED (II,2538); WONDER WOMAN (SERIES 2) (II,2829)

HESSLER, STUART
LAWMAN (I,2642); RAWHIDE (I,3727)

HEYDRON, NANCY
TOO CLOSE FOR COMFORT (II,2642)

HEYES, DOUGLAS
77 SUNSET STRIP (I,3988); THE ADVENTURES OF RIN TIN TIN (I,73); ALFRED HITCHCOCK PRESENTS (I,115); ALIAS SMITH AND JONES (I,118); BARETTA (II,152); THE BOB HOPE CHRYSLER THEATER (I,502); BRAVO DUKE (I,711); CAPTAINS AND THE KINGS (II,435); CHECKMATE (I,919); CHEYENNE (I,931); CITY OF ANGELS (II,540); COLT .45 (I,1010); THE DESILU PLAYHOUSE (I,1237); THE FRENCH ATLANTIC AFFAIR (II,924); LARAMIE (I,2614); MAVERICK (II,1657); MCCLOUD (II,1660); NAKED CITY (I,3215); NIGHT GALLERY (I,3287); POWDERKEG (I,3657); SIX GUNS FOR DONEGAN (I,4065); SWITCH (II,2519); THRILLER (I,4492); THE TWILIGHT ZONE (I,4651)

HIATT, MICHAEL
HART TO HART (II,1102)

HIBBS, JESSE
THE ALASKANS (I,106); BRONCO (I,745); THE FBI (I,1551); THE FUGITIVE (I,1701); HAWAIIAN EYE (I,1973); THE INVADERS (I,2229); LARAMIE (I,2614); PERRY MASON (II,2025); THE RAT PATROL (I,3726); RAWHIDE (I,3727); WAGON TRAIN (I,4747); THE WILD WILD WEST (I,4863)

HICKOX, DOUGLAS
MISTRAL'S DAUGHTER (II,1708); THE PHOENIX (II,2035); THE PHOENIX (II,2036)

HIKEN, NAT
THE ALL-STAR REVUE (I,138); CAR 54, WHERE ARE YOU? (I,842)

HILBERMAN, DAVID
THE SMURFS (II,2385)

HILL, BEN
THE LID'S OFF (I,2678)

HILL, CHARLES N
ONE, TWO, THREE—GO! (I,3390); PERSON TO PERSON (I,3550)

HILL, DENNIS
SKIPPY, THE BUSH KANGAROO (I,4076)

HILL, HEATHER
LOVE OF LIFE (I,2774)

HILL, HERBERT
AS THE WORLD TURNS (II,110)

HILL, JACKSON
PAROLE (I,3460)

HILL, JAMES
THE AVENGERS (II,121); THE NEW AVENGERS (II,1816)

HILL, JEFF
JIM HENSON'S MUPPET BABIES (II,1289)

HILLER, ARTHUR
BEN CASEY (I,394); THE DESILU PLAYHOUSE (I,1237); THE DETECTIVES (I,1245); THE DICK POWELL SHOW (I,1269); NAKED CITY (I,3216); PERRY MASON (II,2025); THE RIFLEMAN (I,3789); ROUTE 66 (I,3852); STARR, FIRST BASEMAN (I,4204); TARGET: THE CORRUPTERS (I,4360); THRILLER (I,4492); WAGON TRAIN (I,4747)

HILLIE, ED
FRANCES LANGFORD PRESENTS (I,1656)

HILLIER, DAVID
SHEENA EASTON—LIVE AT THE PALACE (II,2324)

HILLMAN, DON
MATINEE IN NEW YORK (I,2946)

HILLYER, LAMBERT
THE CISCO KID (I,961)

HILMER, DAVE
CIRCUS OF THE STARS (II,530); CIRCUS OF THE STARS (II,531); CIRCUS OF THE STARS (II,533); CIRCUS OF THE STARS (II,534); CIRCUS OF THE STARS (II,535); CIRCUS OF THE STARS (II,536); CIRCUS OF THE STARS (II,537)

HILTON, ARTHUR DAVID
LAW OF THE PLAINSMAN (I,2639); WALDO (I,4749)

HILTON, LAWRENCE
WONDER GIRL (I,4891)

HIRSCHFELD, GERALD
MR. SMITH (II,1755)

HIRSCHFIELD, JIM
WAKE UP (II,2736)

HIRSCHMAN, HERBERT
THE ALCOA HOUR (I,108); THE FELONY SQUAD (I,1563); MEET CORLISS ARCHER (I,2981); THE MEN FROM SHILOH (I,3005); NERO WOLFE (II,1806); THE WALTONS (II,2740)

HIRSEN, STEVE
ENTERTAINMENT TONIGHT (II,780)

HITCHCOCK, ALFRED
ALFRED HITCHCOCK PRESENTS (I,115)

HIVELY JR, JACK B
THE LIFE AND TIMES OF GRIZZLY ADAMS (II,1469)

HIVELY, JACK B
EIGHT IS ENOUGH (II,762); GREATEST HEROES OF THE BIBLE (II,1062); LASSIE (I,2622); TIMMY AND LASSIE (I,4520)

HOBAN, CHARLES W
NO SOAP, RADIO (II,1856)

HOBBS, JOHN B
BUTTERFLIES (II,404); ROGER DOESN'T LIVE HERE ANYMORE (II,2201)

HOBIN, BILL
ALICE (II,33); AMERICA, YOU'RE ON (II,69); AMERICAN COWBOY (I,169); THE ARTHUR GODFREY SHOW (I,284); THE BILL COSBY SPECIAL (I,444); BOB HOPE SPECIAL: BOB HOPE ON CAMPUS (II,290); THE BOBBY GOLDSBORO SHOW (I,673); THE CBS NEWCOMERS (I,879); CLOWN ALLEY (I,979); A CONNECTICUT YANKEE (I,1038); THE DAVID STEINBERG SHOW (I,1179); DOCTOR SHRINKER (II,686); THE FLIP WILSON SPECIAL (II,889); THE FRED WARING SHOW (I,1679); GARROWAY AT LARGE (I,1737); THE GEORGE BURNS SPECIAL (II,970); THE JACK JONES SPECIAL (I,2310); THE LATE FALL, EARLY SUMMER BERT CONVY SHOW (II,1441); MAKE MINE RED, WHITE, AND BLUE (I,2841); THE MANY FACES OF COMEDY (I,2894); MAUDE (II,1655); THE MCLEAN STEVENSON SHOW (II,1665); MONTY HALL'S VARIETY HOUR (II,1728); NO SOAP, RADIO (II,1856); PHIL SILVERS ON BROADWAY (I,3578); THE SID CAESAR, IMOGENE COCA, CARL REINER, HOWARD MORRIS SPECIAL (I,4039); SING ALONG WITH MITCH (I,4057); STEP ON THE GAS (I,4215); STICK AROUND (II,2464); SUGAR TIME (II,2489); TEXACO STAR PARADE I (I,4409); THREE'S COMPANY (II,2614); THE TIM CONWAY COMEDY HOUR (I,4501); TIMEX ALL-STAR JAZZ SHOW II (I,4516); TIMEX ALL-STAR JAZZ SHOW IV (I,4518); YOUR HIT PARADE (II,2872); YOUR SHOW OF SHOWS (II,2875)

HOBLIT, GREGORY
BAY CITY BLUES (II,186); HILL STREET BLUES (II,1154)

HOBSON, JAMES
THE LAWRENCE WELK SHOW (I,2643); THE LAWRENCE WELK SHOW (I,2644); LAWRENCE WELK'S TOP TUNES AND NEW TALENT (I,2645); MEMORIES WITH LAWRENCE WELK (II,1681)

HODGES, MIKE
MISSING PIECES (II,1706)

HOEY, MICHAEL
FAME (II,812)

HOFFMAN, BERNIE
BATTLE OF THE NETWORK STARS (II,178); CELEBRITY CHALLENGE OF THE SEXES (II,464)

HOFFMAN, ELWOOD
TROUBLE INC. (I,4609)

HOFFMAN, HERMAN
ALCOA PREMIERE (I,109); BRENNER (I,719); COMBAT (I,1011); LUKE AND THE TENDERFOOT (I,2792); THE MONSTER SQUAD (II,1725); ROOM 222 (I,3843); RUN RUN, JOE (II,2229); SEA HUNT (I,3948); TEMPLE HOUSTON (I,4392)

HOFFMAN, SHERRELL
ALL MY CHILDREN (II,39)

HOFSISS, JACK
FOR RICHER, FOR POORER (II,906); LOVERS AND FRIENDS (II,1549); THE OLDEST LIVING GRADUATE (II,1882)

HOLCH, ARTHUR
CALIFORNIA GIRL (I,794)

HOLCOMB, ROD
THE AMERICAN GIRLS (II,72); B.J. AND THE BEAR (II,125); BATTLESTAR GALACTICA (II,181); BEYOND WESTWORLD (II,227); BIG SHAMUS, LITTLE SHAMUS (II,242); BRET MAVERICK (II,374); CAPTAIN AMERICA (II,427); CHICAGO STORY (II,506); THE DEVLIN CONNECTION (II,664); FANTASY ISLAND (II,829); THE GREATEST AMERICAN HERO (II,1060); HILL STREET BLUES (II,1154); MOONLIGHT (II,1730); NO MAN'S LAND (II,1855); THE QUEST (II,2098); QUINCY, M. E. (II,2102); SCARECROW AND MRS. KING (II,2268); SCENE OF THE CRIME (II,2271); THE SIX-MILLION-DOLLAR MAN (II,2372); TENSPEED AND BROWN SHOE (II,2562)

HOLLOWAY, JEAN
WHERE WERE YOU? (I,4834)

HOLT, SETH
DANGER MAN (I,1136); SECRET AGENT (I,3962)

HOOD, RANDAL
THE COURTSHIP OF EDDIE'S FATHER (I,1083)

HOOKS, KEVIN
CUTTER TO HOUSTON (II,612); ST. ELSEWHERE (II,2432)

HOPE, BOB
THE BOB HOPE SHOW (I,582)

HOPPER, JERRY
THE ADDAMS FAMILY (II,11); ALFRED HITCHCOCK PRESENTS (I,115); AMOS BURKE, SECRET AGENT (I,180); BURKE'S LAW (I,762); CHEYENNE (I,931); THE FUGITIVE (I,1701); GILLIGAN'S ISLAND (II,990); HAVE GUN—WILL TRAVEL (I,1968); HONEY WEST: WHO KILLED THE JACKPOT? (I,2106); HOORAY FOR LOVE (I,2116); IT'S A SMALL WORLD (I,2260); IT'S ABOUT TIME (I,2263); M SQUAD (I,2799); MR. LUCKY (I,3141); NAKED CITY (I,3215); NEW GIRL IN HIS LIFE (I,3265); PERRY MASON (II,2025); THE RIFLEMAN (I,3789); THE TIME TUNNEL (I,4511); THE UNTOUCHABLES (I,4682); VOYAGE TO THE BOTTOM OF THE SEA (I,4743); WAGON TRAIN (I,4747); YES YES NANETTE (I,4927)

HORAN, DON
REX REED'S MOVIE GUIDE (II,2149)

HORN, LEONARD J
ALFRED HITCHCOCK PRESENTS (I,115); THE BAIT (I,335); BRANDED (I,707); CHALK ONE UP FOR JOHNNY (I,893); THE DEFENDERS (I,1220); THE ELEVENTH HOUR (I,1425); FARADAY AND COMPANY (II,833); HAWK (I,1974); IRONSIDE (II,1246); KINCAID (I,2543); THE LAWYERS (I,2646); MCMILLAN AND WIFE (II,1667); MISSION: IMPOSSIBLE (I,3067); NAKIA (II,1784); THE OUTER LIMITS (I,3426); POLICE WOMAN (II,2063); THE ROOKIES (II,2208); ROUTE 66 (I,3852); THE SNOOP SISTERS (II,2389); THE UNTOUCHABLES (I,4682); VOYAGE TO THE BOTTOM OF THE SEA (I,4743); WONDER WOMAN (PILOT 2) (II,2827); WONDER WOMAN (SERIES 1) (II,2828)

HORNBACK, LEE
THE PAUL DIXON SHOW (I,3487); THE PAUL DIXON SHOW (II,1974)

HORNER, HARRY
CURTAIN CALL THEATER (I,1115); GUNSMOKE (II,1069); TV READERS DIGEST (I,4630)

HORTON, HERB
PAUL WHITEMAN'S TV TEEN CLUB (I,3491)

HORVATH, JOHN
WATCH ME (II,2745)

HORVITZ, LOUIS J
ALL-STAR ANYTHING GOES (II,50); THE CANDID CAMERA SPECIAL (II,422); EVERYDAY (II,793); FERNWOOD 2-NIGHT (II,849); SOLID GOLD (II,2395); SOLID GOLD '79 (II,2396); SOLID GOLD HITS (II,2397)

HOSTETTLER, JOSEPH

THE BOOTS RANDOLPH SHOW (I,687); TENNESSEE ERNIE FORD'S WHITE CHRISTMAS (I,4402)

HOUGH, JOHN
THE AVENGERS (II,121); THE NEW AVENGERS (II,1816); THE ZOO GANG (II,2877)

HOWARD, AL
JUDY SPLINTERS (I,2475)

HOWARD, KAREN
THOSE AMAZING ANIMALS (II,2601)

HOWARD, RON
LITTLE SHOTS (II,1495); THROUGH THE MAGIC PYRAMID (II,2615)

HOWE, JAMES
CHECKMATE (I,919)

HOWELL, WILLIAM
DREAM GIRL OF '67 (I,1375); FULL CIRCLE (I,1702)

HSU, VICTOR
RIPTIDE (II,2178); ST. ELSEWHERE (II,2432)

HUBLEY, JOHN
MR. MAGOO (I,3142)

HUGGINS, ROY
THE LAWYERS (I,2646); THE VIRGINIAN (I,4738)

HUGHES, BILL
PRISONER: CELL BLOCK H (II,2085)

HUGHES, DEL
ALL MY CHILDREN (II,39); THE BRIGHTER DAY (I,728)

HUGHES, TERRY
EMPIRE (II,777); MR. MOM (II,1752); REPORT TO MURPHY (II,2134); SQUARE PEGS (II,2431); SWEENEY TODD (II,2512); THE TWO RONNIES (II,2694)

HULL, SHELLY
THE STORK CLUB (I,4236)

HULTGREN, BOB
HOWDY DOODY (I,2151); THE SHARI LEWIS SHOW (I,4000)

HUMBERSTONE, BRUCE
THE SMOTHERS BROTHERS SHOW (I,4096)

HUNT, ED
GREATEST HEROES OF THE BIBLE (II,1062)

HUNT, PETER
THE LAST DAYS OF POMPEII (II,1436)

HUNT, PETER H
ADAM'S RIB (I,32); BRING 'EM BACK ALIVE (II,377); BUS STOP (II,399); ELLERY QUEEN (II,766); FLYING HIGH (II,894); FLYING HIGH (II,895); IVAN THE TERRIBLE (II,1260); THE KAREN VALENTINE SHOW (I,2506); LUCAN (II,1554); MASQUERADE (II,1644); MIXED NUTS (II,1714); NUTS AND BOLTS (II,1871); THE PERSUADERS! (I,3556); PHILIP MARLOWE, PRIVATE EYE (II,2033); QUARK (II,2095); RENDEZVOUS HOTEL (II,2131); SAWYER AND FINN (II,2265); SHERLOCK HOLMES (II,2327); SKEEZER (II,2376); TUCKER'S WITCH (II,2667); WHEN THINGS WERE ROTTEN (II,2782); WILDER AND WILDER (II,2802)

HURDLE, JACK
THE COLGATE COMEDY HOUR (I,997); THE TIMID SOUL (I,4519)

HURTZ, BILL
THE BULLWINKLE SHOW (I,761); MR. MAGOO (I,3142); ROCKY AND HIS FRIENDS (I,3822)

HUTCHINSON, TOM
TEX AND JINX (I,4406)

HUTTEN, BILL
LAVERNE AND SHIRLEY WITH THE FONZ (II,1449); THE LITTLE RASCALS (II,1494); PAC-MAN (II,1940); THE RICHIE RICH SHOW (II,2169); THE SHIRT TALES (II,2337); THE SMURFS (II,2385)

HYAMS, NESSA
MARY HARTMAN, MARY HARTMAN (II,1638)

IACOFANO, TIM
TMT (II,2622)

IDLE, ERIC
FAERIE TALE THEATER (II,807)

IMPASTATA, DAVID
SWITCH (II,2519)

IRVIN, JOHN
TINKER, TAILOR, SOLDIER, SPY (II,2621)

IRVING, CHARLES
SEARCH FOR TOMORROW (II,2284)

IRVING, JULES
THE DETECTIVE (II,662); THE JORDAN CHANCE (II,1343); LOOSE CHANGE (II,1518); WHAT REALLY HAPPENED TO THE CLASS OF '65? (II,2768)

IRVING, RICHARD
THE ART OF CRIME (II,108); CHECKMATE (I,919); CHIPS (II,511); CONFIDENTIALLY YOURS (I,1035); CUTTER (I,1117); EXO-MAN (II,797); GRUEN GUILD PLAYHOUSE (I,1892); HOLLYWOOD OPENING NIGHT (I,2087); JOHNNY BLUE (II,1321); LAREDO (I,2615); THE NAME OF THE GAME (I,3217); PRESCRIPTION: MURDER (I,3660); RANSOM FOR A DEAD MAN (I,3722); SEVENTH AVENUE (II,2309); SOLDIERS OF FORTUNE (I,4111); THE SWORD (I,4325); TERRY AND THE PIRATES (I,4405); WHAT REALLY HAPPENED TO THE CLASS OF '65? (II,2768)

IRWIN, CAROL
CLAUDIA: THE STORY OF A MARRIAGE (I,972)

ISCOVE, ROB
DOROTHY HAMILL IN ROMEO & JULIET ON ICE (II,718); MARY (II,1637); PAJAMA TOPS (II,1942)

ISENBERG, GERALD I
JULIE FARR, M.D. (II,1354)

ISHLI, CHRIS
MR. MAGOO (I,3142)

IZZARD, BRYAN
DOCTOR IN THE HOUSE (I,1313)

JACKMAN, FRED
THE ADVENTURES OF RIN TIN TIN (I,73); MEMORIES WITH LAWRENCE WELK (II,1681); ROUTE 66 (I,3852)

JACKMAN, MICHELE MATIS
THE LAWRENCE WELK CHRISTMAS SPECIAL (II,1450)

JACKSON, PAT
THE PRISONER (I,3670)

JACOBER, WARREN
THE DOODLES WEAVER SHOW (I,1351); SOUND OFF TIME (I,4135); THROUGH WENDY'S WINDOW (I,4495)

JACOBS, DAVID
KNOTS LANDING (II,1404)

JACOBSON, JAKE
THE LOVE REPORT (II,1543)

JACOBSON, STAN
THE B.B. BEEGLE SHOW (II,124); THE HUDSON BROTHERS RAZZLE DAZZLE COMEDY SHOW (II,1201); THE YESTERDAY SHOW (II,2850)

JACOBSON, TOM
THE NASHVILLE SOUND OF BOOTS RANDOLPH (I,3229)

JACOBY, COLEMAN
THOSE FABULOUS CLOWNS (II,2602)

JACOBY, FRANK
A GUEST IN YOUR HOUSE (I,1898)

JAIMES, SAM
THE CHARLIE BROWN AND SNOOPY SHOW (II,484); IT'S AN ADVENTURE, CHARLIE BROWN (I,2266)

JAMES, GABRIELLE ALICE
LAVERNE AND SHIRLEY (II,1446)

JAMES, LUTHER
THE BILL COSBY SHOW (I,442); THE COURTSHIP OF EDDIE'S FATHER (I,1083)

JAMES, TERRY
CAMELOT (II,416)

JAMESON, JERRY
HAWAII FIVE-O (II,1110); IRONSIDE (II,1246); THE LIVES OF JENNY DOLAN (II,1502); MCCLOUD (II,1660); THE MOD SQUAD (I,3078); THE ROOKIES (II,2208); SEARCH (I,3951); THE SIX-MILLION-DOLLAR MAN (II,2372); THIS GIRL FOR HIRE (II,2596)

JANES, IAN
THE SULLIVANS (II,2490)

JARRETT, CHARLES
THE STRANGE CASE OF DR. JEKYLL AND MR. HYDE (I,4244)

JARVIS, BOB
WAYNE AND SHUSTER TAKE AN AFFECTIONATE LOOK AT. . . (I,4778)

JARVIS, IRV
THE MAGIC GARDEN (II,1591)

JASON, GEORGE
MARLO AND THE MAGIC MOVIE MACHINE (II,1631)

JASON, LEIGH
SCIENCE FICTION THEATER (I,3938)

JASON, WILL
THE BIG STORY (I,432)

JEFFRIES, PETER
THE CITADEL (II,539); CRIMES OF PASSION (II,603); AN ECHO OF THERESA (I,1399); FANNY BY GASLIGHT (II,823); ONLY A SCREAM AWAY (I,3393); A PLACE TO DIE (I,3612)

JENKINS, ROGER
THE AVENGERS (II,121); THE ONEDIN LINE (II,1912)

JENNETTE, JIM
BATTLE OF THE NETWORK STARS (II,171); BATTLE OF THE NETWORK STARS (II,163); BATTLE OF THE NETWORK STARS (II,164); BATTLE OF THE NETWORK STARS (II,165); BATTLE OF THE NETWORK STARS (II,166); BATTLE OF THE NETWORK STARS (II,167); BATTLE OF THE NETWORK STARS (II,168); BATTLE OF THE NETWORK STARS (II,169); BATTLE OF THE NETWORK STARS (II,170); BATTLE OF THE NETWORK STARS (II,172); BATTLE OF THE NETWORK STARS (II,173); BATTLE OF THE NETWORK STARS (II,174); BATTLE OF THE NETWORK STARS (II,175); BATTLE OF THE NETWORK STARS (II,176); BATTLE OF THE NETWORK STARS (II,177)

JEWISON, NORMAN
BELAFONTE, NEW YORK (I,386); THE BIG PARTY FOR REVLON (I,427); THE BROADWAY OF LERNER AND LOEWE (I,735); THE FABULOUS 50S (I,1501); AN HOUR WITH DANNY KAYE (I,2136); THE JUDY GARLAND SHOW (I,2472); THE MILLION DOLLAR INCIDENT (I,3045); TONIGHT WITH BELAFONTE (I,4566); YOUR HIT PARADE (I,4952)

JOHNSON, JIM
DUSTY'S TREEHOUSE (I,1391)

JOHNSON, KEN
ADAM-12 (I,31); ALAN KING IN LAS VEGAS, PART I (I,97); ALAN KING IN LAS VEGAS, PART II (I,98); ALAN KING LOOKS BACK IN ANGER—A REVIEW OF 1972 (I,99); THE ALAN KING SPECIAL (I,103); THE BIONIC WOMAN (II,255); THE CURSE OF DRACULA (II,611); HOLLYWOOD'S TALKING (I,2097); HOT PURSUIT (II,1189); THE SECRET EMPIRE (II,2291); STOP SUSAN WILLIAMS (II,2473); V (II,2713);

JOHNSON, KEVIN
THE JIMMIE RODGERS SHOW (I,2382); THE KATE SMITH HOUR (I,2508); WINGO (I,4878)

JOHNSON, LAMONT
ANGEL (I,216); BUS STOP (I,778); CALL TO DANGER (I,803); CIMARRON STRIP (I,954); CORONET BLUE (I,1055); DEADLOCK (I,1190); THE DEFENDERS (I,1220); FAERIE TALE THEATER (II,807); HAVE GUN—WILL TRAVEL (I,1968); JOHNNY RINGO (I,2434); MIDNIGHT MYSTERY (I,3030); NAKED CITY (I,3216); PETER GUNN (I,3562); THE RIFLEMAN (I,3789); THE SEARCH (I,3950); THE TWILIGHT ZONE (I,4651); THE VIRGINIAN (I,4737)

JOHNSON, STERLING
ANDY WILLIAMS' EARLY NEW ENGLAND CHRISTMAS (II,79); THE ARTHUR GODFREY SPECIAL (II,109); THE BARBARA WALTERS SPECIAL (II,145); BLONDES VS. BRUNETTES (II,271); CRYSTAL (II,607); PEGGY FLEMING AT SUN VALLEY (I,3507); PEGGY FLEMING VISITS THE SOVIET UNION (I,3509); PERRY COMO'S BAHAMA HOLIDAY (II,2006); PERRY COMO'S CHRISTMAS IN MEXICO (II,2009); PERRY COMO'S CHRISTMAS IN NEW MEXICO (II,2010); PERRY COMO'S CHRISTMAS IN THE HOLY LAND (II,2013); PERRY COMO'S LAKE TAHOE HOLIDAY (II,2018); PERRY COMO'S SPRING IN SAN FRANCISCO (II,2022); TO EUROPE WITH LOVE (I,4525)

JOLLEY, STAN
TODAY'S FBI (II,2628)

JONES, CHUCK
THE BUGS BUNNY/ROAD RUNNER SHOW (II,391); PORKY PIG AND FRIENDS (I,3651); THE ROAD RUNNER SHOW (I,3799); SYLVESTER AND TWEETY (II,2522); THE TOM AND JERRY SHOW (I,4534)

JONES, CLARK
6 RMS RIV VU (II,2371); ANNIE GET YOUR GUN (I,224); CAROL AND COMPANY (I,847); CAROL CHANNING AND PEARL BAILEY ON BROADWAY (I,850); CAROL CHANNING PROUDLY PRESENTS THE SEVEN DEADLY SINS (I,851); CBS: ON THE AIR (II,460); THE DICK CAVETT SHOW (II,669); THE FIRST NINE MONTHS ARE THE HARDEST (I,1584); THE FORD 50TH ANNIVERSARY SHOW (I,1630); THE FOUR POSTER (I,1651); FRANCIS ALBERT SINATRA DOES HIS THING (I,1658); THE FUNNY SIDE (I,1714); THE FUNNY SIDE (I,1715); HELLZAPOPPIN (I,2011); AN HOUR WITH ROBERT GOULET (I,2137); I'M A FAN (I,2192); JOHNNY MATHIS IN THE CANADIAN ROCKIES (II,1338); THE KRAFT MUSIC HALL (I,2587); MARLENE DIETRICH: I WISH YOU LOVE (I,2918); THE MOHAWK SHOWROOM (I,3080); MONSANTO NIGHT PRESENTS BURL IVES (I,3090); MONSANTO NIGHT PRESENTS JOSE FELICIANO (I,3091); NIGHT OF 100 STARS (II,1846); PARADE OF STARS (II,1954); PEGGY FLEMING AT MADISON SQUARE GARDEN (I,3506); RAINBOW OF STARS (I,3717); RUGGLES OF RED GAP (I,3867); THE SAMMY DAVIS JR. SHOW (I,3898); SINATRA: THE MAN AND HIS MUSIC (II,2362); SINATRA—THE FIRST 40 YEARS (II,2359); SLEEPING BEAUTY (I,4083); TONY BENNETT IN LONDON (I,4567); TONY BENNETT IN WAIKIKI (I,4568); TWIGS (II,2683); THE WAY THEY WERE (II,2748); WILL ROGERS' U.S.A. (I,4865); A WORLD OF LOVE (I,4914); YOUR HIT PARADE (I,4951)

JONES, DAVID
TALES OF THE GOLD MONKEY (II,2537)

JONES, EUGENE S
THE WORLD OF MAURICE CHEVALIER (I,4916)

JONES, HARMON
BEN CASEY (I,394); THE MONROES (I,3089); RAWHIDE (I,3727); THE VIRGINIAN (I,4738); VOYAGE TO THE BOTTOM OF THE SEA (I,4743)

JONES, JAMES CELLAN
CAESAR AND CLEOPATRA (I,789); THE FOUR OF US (II,914); JENNIE: LADY RANDOLPH CHURCHILL (II,1277)

JONES, KIRK
QUENTIN DERGENS, M.P. (I,3707)

JONES, PENBERRY
DARK SHADOWS (I,1157)

JORDAN, GLENN
DELTA COUNTY, U.S.A. (II,657); FAMILY (II,813); FRANKENSTEIN (I,1671); FRIENDS (II,927); THE OATH: 33 HOURS IN THE LIFE OF GOD (II,1873); THE PICTURE OF DORIAN GRAY (I,3592); YOUNG DR. KILDARE (I,4937)

JORDAN, JIM
THE BOB HOPE SHOW (I,542); THE BOB HOPE SHOW (I,543); THE BOB HOPE SHOW (I,560); THE BOB HOPE SHOW (I,561); FRANKIE CARLE TIME (I,1674); THE GREAT GILDERSLEEVE (I,1875); WALTER FORTUNE (I,4758)

JORDAN, LARRY
THE MOONMAN CONNECTION (II,1731)

JORDAN, MAURY
YOU ASKED FOR IT (II,2855)

JOY, RON
SOMETHING ELSE (I,4117)

JOYCE, JIMMY
THE BING CROSBY SPECIAL (I,471)

JUHLAN, DALE
TAKE TWO (I,4337)

JULIANO, JOSEPH R
HOLLYWOOD HISTORAMA (I,2083)

JUMP, GORDON
WKRP IN CINCINNATI (II,2814)

JURAN, JERRY
LOST IN SPACE (I,2758)

JURAN, NATHAN
CROSSROADS (I,1105); DANIEL BOONE (I,1142); MY FRIEND FLICKA (I,3190); THE TIME TUNNEL (I,4511); VOYAGE TO THE BOTTOM OF THE SEA (I,4743)

JURGENSON, WILLIAM K
FLYING HIGH (II,894); M*A*S*H (II,1569)

JURKOSKI, TOM
JABBERWOCKY (II,1262)

JURWICH, DON
SPIDER-MAN AND HIS AMAZING FRIENDS (II,2425)

KACHIVAS, LOU
FAT ALBERT AND THE COSBY KIDS (II,835); FLASH GORDON (II,874); SPORT BILLY (II,2429)

KAGAN, JEREMY PAUL
COLUMBO (II,556); THE DOCTORS (I,1323); FAERIE TALE THEATER (II,807); JUDGE DEE IN THE MONASTERY MURDERS (I,2465); NICHOLS (I,3283)

KAHN, EDDIE
ONE MAN'S FAMILY (I,3383)

KAHN, RICHARD
SKY KING (I,4077)

KAHN, STEVE
SOAP FACTORY DISCO (II,2393)

KAMM, LARRY
BATTLE OF THE NETWORK STARS (II,163); BATTLE OF THE NETWORK STARS (II,164); BATTLE OF THE NETWORK STARS (II,165); BATTLE OF THE NETWORK STARS (II,166); BATTLE OF THE NETWORK STARS (II,167); BATTLE OF THE NETWORK STARS (II,168); BATTLE OF THE NETWORK STARS (II,169); BATTLE OF THE NETWORK STARS (II,170); BATTLE OF THE NETWORK STARS (II,171); BATTLE OF THE NETWORK STARS (II,172); BATTLE OF THE NETWORK STARS (II,173); BATTLE OF THE NETWORK STARS (II,174); BATTLE OF THE NETWORK STARS (II,175); BATTLE OF THE NETWORK STARS (II,176); BATTLE OF THE NETWORK STARS (II,177)

KANE, BRUCE
HAWAIIAN EYE (I,1973)

KANE, DENNIS
DARK SHADOWS (I,1157)

KANE, JOSEPH
BROKEN ARROW (I,744); LAST STAGECOACH WEST (I,2629); RAWHIDE (I,3727)

KANE, MICHAEL J
A BEDTIME STORY (I,384); CAN YOU TOP THIS? (I,817)

KANIN, GARSON
ALFRED HITCHCOCK PRESENTS (I,115); BORN YESTERDAY (I,692)

KANTER, HAL
DOWN HOME (I,1366); THE GEORGE GOBEL SHOW (I,1767); THE JIMMY STEWART SHOW (I,2391); JULIA (I,2476); THE REASON NOBODY HARDLY EVER SEEN A FAT OUTLAW IN THE OLD WEST IS AS FOLLOWS: (I,3744); THREE COINS IN THE FOUNTAIN (I,4473)

KANTOR, RON
CELEBRITY CHARADES (II,470); EVENING AT THE IMPROV (II,786); EVERYTHING GOES (I,1480); THE GUINNESS GAME (II,1066); THE HEE HAW HONEYS (II,1124); THE KALLIKAKS (II,1361); WE'RE DANCIN' (II,2755)

KAPLAN, HENRY
ALL MY CHILDREN (II,39); DARK SHADOWS (I,1157); THE MAN IN THE DOG SUIT (I,2869)

KAPLAN, JONATHAN
THE HUSTLER OF MUSCLE BEACH (II,1210)

KARLEN, BERNARD E
I'D LIKE TO SEE (I,2184)

KARLSON, PHIL
ALEXANDER THE GREAT (I,113); THE UNTOUCHABLES (I,4681)

KARN, BILL
DANGEROUS ASSIGNMENT (I,1138)

KARTUN, ALLAN
BREAKAWAY (II,370)

KASS, PETER
THE RED BUTTONS SHOW (I,3750)

KATTIN, STEVE
SPECIAL EDITION (II,2416)

KATZ, PETER
PEE WEE KING'S FLYING RANCH (I,3505)

KATZ, ROBERT
PEOPLE ARE FUNNY (II,1996)

KATZIN, LEE H
AUTOMAN (II,120); THE BASTARD/KENT FAMILY CHRONICLES (II,159); BONANZA (II,347); BRANDED (I,707); CHICAGO STORY (II,506); DEATH RAY 2000 (II,654); THE DEVLIN CONNECTION (II,664); THE FELONY SQUAD (I,1563); FORCE SEVEN (II,908); HARDCASE (II,1088); IT TAKES A THIEF (I,2250); THE MAN FROM ATLANTIS (II,1614); THE MAN FROM ATLANTIS (II,1615); MANNIX (II,1624); MCCLAIN'S LAW (II,1659); MCLAREN'S RIDERS (II,1663); MCMILLAN AND WIFE (II,1667); MIAMI VICE (II,1689); THE MISSISSIPPI (II,1707); THE MOD SQUAD (I,3078); PARTNERS IN CRIME (II,1961); POLICE STORY (II,2062); THE QUEST (II,2096); THE RAT PATROL (I,3726); RAWHIDE (I,3727); SAMURAI (II,2249); SPACE: 1999 (II,2409); THE STRANGER (I,4251); THE WILD WILD WEST (I,4863); THE YELLOW ROSE (II,2847)

KATZMAN, LEONARD
DALLAS (II,614); DIRTY SALLY (II,680)

KAUFMAN, LEONARD B
JAMBO (I,2330)

KAY, GILBERT L
PERRY MASON (II,2025)

KAY, ROGER
THE ALDRICH FAMILY (I,111); BATTLE OF THE AGES (I,367); NAKED CITY (I,3216); ROUTE 66 (I,3852); THE UNTOUCHABLES (I,4682)

KEAN, E ARTHUR
CHICAGO STORY (II,506); POLICE STORY (II,2062); POLICE WOMAN (II,2063); RIKER (II,2175)

KEARNEY, GENE
EISCHIED (II,763); NIGHT GALLERY (I,3287)

KEEFE
THE BEACH BOYS IN CONCERT (II,189); CULTURE CLUB IN CONCERT (II,610)

KEIDEL, DALE
THE MADHOUSE BRIGADE (II,1588)

KEITH, GERREN
ALMOST AMERICAN (II,55); DIFF'RENT STROKES (II,674); GOOD TIMES (II,1038); GRADY (II,1047); THE MARILYN MCCOO AND BILLY DAVIS JR. SHOW (II,1630); MY BUDDY (II,1778)

KEITH, HAL
MR. PEEPERS (I,3147); THE PHIL SILVERS ARROW SHOW (I,3576); VILLAGE BARN (I,4733); YOUR SHOW OF SHOWS (II,2875)

KELADA, ASAAD
THE ACADEMY (II,3); THE ACADEMY II (II,4); BABY, I'M BACK! (II,130); THE FACTS OF LIFE (II,805); FACTS OF LIFE: THE FACTS OF LIFE GOES TO PARIS (II,806); FAMILY TIES (II,819); FIRST TIME, SECOND TIME (II,863); JO'S COUSINS (II,1293); THE LAST RESORT (II,1440); NIGHT COURT (II,1844); PHYLLIS (II,2038); REPORT TO MURPHY (II,2134); RHODA (II,2151); SAINT PETER (II,2238); THE TONY RANDALL SHOW (II,2640); THE TWO OF US (II,2693); WKRP IN CINCINNATI (II,2814)

KELLER, EYTHAN
FOUL UPS, BLEEPS AND BLUNDERS (II,911)

KELLER, HARRY
COMMANDO CODY (I,1030); THE DICK POWELL SHOW (I,1269); THE LORETTA YOUNG THEATER (I,2756); TAMMY (I,4357); TEXAS JOHN SLAUGHTER (I,4414)

KELLER, LEW
THE BULLWINKLE SHOW (I,761)

KELLER, SHELDON
PAUL LYNDE AT THE MOVIES (II,1976)

KELLETT, BOB
SPACE: 1999 (II,2409)

KELLJAN, ROBERT
BEACH PATROL (II,192); BRING 'EM BACK ALIVE (II,377); CASSIE AND COMPANY (II,455); CHARLIE'S ANGELS (II,486); DOG AND CAT (II,695); THE DUKES OF HAZZARD (II,742); EISCHIED (II,763); FAME (II,812); HILL STREET BLUES (II,1154); JULIE FARR, M.D. (II,1354); KAZ (II,1370); MOVIN' ON (II,1743); NERO WOLFE (II,1806); RIKER (II,2175); STARSKY AND HUTCH (II,2444); VEGAS (II,2724); WONDER WOMAN (SERIES 2) (II,2829)

KELLMAN, BARNET
FOR RICHER, FOR POORER (II,906)

KELLY, CHRISTOPHER
BUDDY HACKETT—LIVE AND UNCENSORED (II,386)

KELLY, GENE
AT YOUR SERVICE (I,310); JACK AND THE BEANSTALK (I,2288)

KELLY, KEVIN
LOVERS AND FRIENDS (II,1549); TEXAS (II,2566)

KELLY, SHELBY
BLACKSTAR (II,263); THE KID SUPER POWER HOUR WITH SHAZAM (II,1386); THE TARZAN/LONE RANGER/ZORRO ADVENTURE HOUR (II,2543)

KELTON, RICHARD
ENSIGN O'TOOLE (I,1448)

KENASTON, JACK
UNK AND ANDY (I,4679)

KENIETEL, SEYMOUR
CARTOONSVILLE (I,858); POPEYE THE SAILOR (I,3648); POPEYE THE SAILOR (I,3649)

KENNARD, DAVID
COSMOS (II,579)

KENNEDY, BURT
COMBAT (I,1011); THE CONCRETE COWBOYS (II,564); HOW THE WEST WAS WON (II,1196); KATE BLISS AND THE TICKER TAPE KID (II,1366); LAWMAN (I,2642); MAGNUM, P.I. (II,1601); MORE WILD WILD WEST (II,1733);

THE RHINEMANN EXCHANGE (II,2150); SIDEKICKS (II,2348); SIMON AND SIMON (II,2357); THE VIRGINIAN (I,4738); THE WILD WILD WEST REVISITED (II,2800); THE YELLOW ROSE (II,2847)

KENNEDY, JOHN
ROBERTA (I,3816)

KENNEY, H WESLEY
THE ABC AFTERSCHOOL SPECIAL (II,1); ALL IN THE FAMILY (II,38); BIG JOHN, LITTLE JOHN (II,240); FAR OUT SPACE NUTS (II,831); FILTHY RICH (II,856); LADIES' MAN (II,1423); ROSENTHAL AND JONES (II,2215); SIDE BY SIDE (II,2347)

KENT, GORDON
THE HEATHCLIFF AND MARMADUKE SHOW (II,1118)

KENTON, ERLE C
CROSSROADS (I,1105); PUBLIC DEFENDER (I,3687)

KENWITH, HERBERT
ALL THAT GLITTERS (II,42); ALOHA PARADISE (II,58); BOSOM BUDDIES (II,355); DEAR TEACHER (II,653); DIFF'RENT STROKES (II,674); FISHERMAN'S WHARF (II,865); FLATFOOTS (II,879); GIMME A BREAK (II,995); GOOD TIMES (II,1038); GRANDPA GOES TO WASHINGTON (II,1050); HERE'S BOOMER (II,1134); HERE'S LUCY (II,1135); HOME COOKIN' (II,1170); JOE'S WORLD (II,1305); MAN IN THE MIDDLE (I,2870); ME AND MAXX (II,1672); MR. MERLIN (II,1751); A NEW KIND OF FAMILY (II,1819); ONE DAY AT A TIME (II,1900); THE PARTRIDGE FAMILY (II,1962); POPI (II,2069); PRIVATE BENJAMIN (II,2087); THE RAINBOW GIRL (II,2108); STAR TREK (II,2440); STRANGE PARADISE (I,4246); THAT'S MY MAMA (II,2580); TIGER! TIGER! (I,4499); VALIANT LADY (I,4695)

KENYON, SANDY
ONE DAY AT A TIME (II,1900)

KERN, JAMES V
THE 20TH CENTURY-FOX HOUR (I,4642); THE ALL-STAR REVUE (I,138); THE ANN SOTHERN SHOW (I,220); THE DONNA REED SHOW (I,1347); HEY, JEANNIE! (I,2038); JOHNNY COME LATELY (I,2425); THE MILLIONAIRE (I,3046); MY FAVORITE MARTIAN (I,3189); MY THREE SONS (I,3205); PETE AND GLADYS (I,3558); SUGARFOOT (I,4277); TOPPER (I,4582)

KERN, JEROME
BOURBON STREET BEAT (I,699); ROOM FOR ONE MORE (I,3842)

KERSHNER, IRVIN
BEN CASEY (I,394); CONFIDENTIAL FILE (I,1033); NAKED CITY (I,3216)

KERVEN, CLAUDE
THE ABC AFTERSCHOOL SPECIAL (II,1)

KESLER, HENRY
I LED THREE LIVES (I,2169)

KESSLER, BRUCE
240-ROBERT (II,2688); THE A-TEAM (II,119); ALIAS SMITH AND JONES (I,118); B.J. AND THE BEAR (II,125); BARETTA (II,152); BARNABY JONES (II,153); BORDER PALS (II,352); CALIFORNIA FEVER (II,411); CHICAGO STORY (II,506); CHIPS (II,511); CODE R (II,551); ENOS (II,779); THE FALL GUY (II,811); FOUR EYES (II,912); FREEBIE AND THE BEAN (II,922); GET CHRISTIE LOVE! (II,981); THE GREATEST AMERICAN HERO (II,1060); HARDCASTLE AND MCCORMICK (II,1089); HAWAIIAN HEAT (II,1111); HUNTER (II,1206); I DREAM OF JEANNIE (I,2167); IRONSIDE (II,1246); KNIGHT RIDER (II,1402); LOBO (II,1504); MARCUS WELBY, M.D. (II,1627); THE MASTER (II,1648); MCCLAIN'S LAW (II,1659); MCCLOUD (II,1660); MISSION: IMPOSSIBLE (I,3067); THE MONKEES (I,3088); THE QUEST (II,2098); RIPTIDE (II,2178); SWITCH (II,2519)

KETCHUM, DAVE
WHO'S WATCHING THE KIDS? (II,2793)

KIBBEE, DON
CIRCUS OF THE STARS (II,532)

KIBBEE, ROLAND
ALFRED HITCHCOCK PRESENTS (I,115); IT TAKES A THIEF (I,2250)

KIDD, MICHAEL
LAVERNE AND SHIRLEY (II,1446)

KILEY, TIM
THE BILLY CRYSTAL COMEDY HOUR (II,248); BOB HOPE SPECIAL: BOB HOPE IN WHO MAKES THE WORLD LAUGH, PART 2 (II,286); BOB HOPE SPECIAL: BOB HOPE'S ALL-STAR BIRTHDAY AT ANNAPOLIS (II,294); BOB HOPE SPECIAL: BOB HOPE'S ALL-STAR COMEDY BIRTHDAY PARTY AT WEST POINT (II,298); BOB HOPE SPECIAL: BOB HOPE'S FUNNY VALENTINE (II,309); BOB HOPE SPECIAL: BOB HOPE'S WICKI-WACKY SPECIAL FROM WAIKIKI (II,321); THE BOOK OF LISTS (II,350); THE ED SULLIVAN SHOW (I,1401); THE FLIP WILSON SHOW (I,1602); FLIP WILSON. . . OF COURSE (II,890); THE GOLDDIGGERS (I,1838); THE HELEN REDDY SHOW (I,2008); KEEP U.S. BEAUTIFUL (I,2519); THE KLOWNS (I,2574); THE LESLIE UGGAMS SHOW (I,2660); THE LISA HARTMAN SHOW (II,1484); LORETTA LYNN: THE LADY. . .THE LEGEND (II,1521); THE MAC DAVIS SHOW (II,1575); THE MAD MAD MAD MAD WORLD OF THE SUPER BOWL (II,1586); THE NATALIE COLE SPECIAL (II,1794); PAUL SAND IN FRIENDS AND LOVERS (II,1982); THE RETURN OF THE SMOTHERS BROTHERS (I,3777); SINATRA (I,4054); SOMETHING ELSE (I,4117); THE SONNY AND CHER SHOW (II,2401); STAR SEARCH (II,2438); THAT THING ON ABC (II,2574); THREE GIRLS THREE (II,2608); THE WAY THEY WERE (II,2748); WE'LL GET BY (II,2753)

KIMBALL, JOHN
ALVIN AND THE CHIPMUNKS (II,61); GOLDIE GOLD AND ACTION JACK (II,1023); THE HEATHCLIFF AND DINGBAT SHOW (II,1117); THE HEATHCLIFF AND MARMADUKE SHOW (II,1118); MORK AND MINDY (II,1736); THE PLASTICMAN COMEDY/ADVENTURE SHOW (II,2051); RUBIK, THE AMAZING CUBE (II,2226); SATURDAY SUPERCADE (II,2263); THUNDARR THE BARBARIAN (II,2616); TURBO-TEEN (II,2668)

KIMBALL, WARD
THE MOUSE FACTORY (I,3113)

KIMMEL, HENRY
ON STAGE AMERICA (II,1893)

KING, CHARLIE
CAR CARE CENTRAL (II,436)

KING, DONALD RAY
AMERICA ALIVE! (II,67); KIDS ARE PEOPLE TOO (II,1390); THE MIKE DOUGLAS SHOW (II,1698)

KING, DURNIE
SOMETHING ELSE (I,4117)

KING, LOUIS
TURN OF FATE (I,4623)

KING, LYNN
THE MAGIC SLATE (I,2820)

KING, LYNWOOD
MARLO AND THE MAGIC MOVIE MACHINE (II,1631); MUSICAL CHAIRS (II,1776); NBC SATURDAY PROM (I,3245); WE INTERRUPT THIS SEASON (I,4780)

KINNEY, JACK
POPEYE THE SAILOR (I,3649)

KINON, RICHARD
ALOHA PARADISE (II,58); AN APARTMENT IN ROME (I,242); THE BARBARA EDEN SHOW (I,346); BEWITCHED (I,418); THE BOB NEWHART SHOW (II,342); BURKE'S LAW (I,762); CAPTAIN NICE (I,835); CROSSROADS (I,1105); THE DEAN JONES SHOW (I,1191); THE DESILU PLAYHOUSE (I,1237); DYNASTY (II,746); FAMILY (II,813); THE FARMER'S DAUGHTER (I,1533); FAY (II,844); HECK'S ANGELS (II,1122); HERE COME THE BRIDES (I,2024); HOLMES AND YOYO (II,1169); HOTEL (II,1192); THE LOVE BOAT (II,1535); THE LOVE BOAT I (II,1532); THE LOVE BOAT III (II,1534); LOVES ME, LOVES ME NOT (II,1551); MR. ROBERTS (I,3149); MRS. G. GOES TO COLLEGE (I,3155); THE PARTRIDGE FAMILY (II,1962); PLEASE DON'T EAT THE DAISIES (I,3628); THE ROGUES (I,3832); ROLL OUT! (I,3833); THE SMOTHERS BROTHERS SHOW (I,4096); THIS BETTER BE IT (II,2595); WESTSIDE MEDICAL (II,2765); WONDER WOMAN (SERIES 1) (II,2828)

KIRKLAND, DENNIS
THE BENNY HILL SHOW (II,207); THE UNEXPURGATED BENNY HILL SHOW (II,2705)

KJELLIN, ALF
ALFRED HITCHCOCK PRESENTS (I,115); BARNABY JONES (II,153); CANNON (II,424); CASSIE AND COMPANY (II,455); CODE R (II,551); COLUMBO (II,556); DAVID CASSIDY—MAN

UNDERCOVER (II,623); DOCTOR KILDARE (I,1315); THE DOCTORS (I,1323); DYNASTY (II,746); FAMILY (II,813); THE FAMILY HOLVAK (II,817); FINDER OF LOST LOVES (II,857); GIBBSVILLE (II,988); HAWAII FIVE-O (II,1110); HOW THE WEST WAS WON (II,1196); IRONSIDE (II,1246); JOE FORRESTER (II,1303); LITTLE HOUSE ON THE PRAIRIE (II,1487); THE MAN FROM U.N.C.L.E. (I,2867); MANNIX (II,1624); MCMILLAN AND WIFE (II,1667); MISSION: IMPOSSIBLE (I,3067); PARIS (II,1957); PLANET OF THE APES (II,2049); POLICE WOMAN (II,2063); THE SIX-MILLION-DOLLAR MAN (II,2372); THE SIXTH SENSE (I,4069); SWITCH (II,2519); TRAPPER JOHN, M.D. (II,2654); VEGAS (II,2724); WALKING TALL (II,2738); THE WALTONS (II,2740); WHIZ KIDS (II,2790); THE YOUNG PIONEERS (II,2869)

KLAUSSEN, RAY
BING CROSBY AND THE SOUNDS OF CHRISTMAS (I,457);

KLEIN, DENNIS
BUFFALO BILL (II,387); FOREVER FERNWOOD (II,909); LAVERNE AND SHIRLEY (II,1446); MARY HARTMAN, MARY HARTMAN (II,1638); REPORT TO MURPHY (II,2134)

KLEIN, HERBERT
THE ABC AFTERSCHOOL SPECIAL (II,1)

KLEISER, RANDAL
FAMILY (II,813); LUCAS TANNER (II,1556); MARCUS WELBY, M.D. (II,1627); THE ROOKIES (II,2208); STARSKY AND HUTCH (II,2444)

KLINE, RICHARD S
BLANK CHECK (II,265); BREAK THE BANK (II,369); BULLSEYE (II,393); THE HOLLYWOOD CONNECTION (II,1161); HOT POTATO (II,1188); JOKER! JOKER!! JOKER!!! (II,1340); THE JOKER'S WILD (II,1341); PLAY THE PERCENTAGES (II,2053); SOAP WORLD (II,2394); TIC TAC DOUGH (II,2618); WAY OUT GAMES (II,2747)

KNORR, ARTHUR
SHOW OF THE YEAR (I,4031)

KNOWLES, BERNARD
THE ADVENTURES OF ROBIN HOOD (I,74); COLONEL MARCH OF SCOTLAND YARD (I,1005); IVANHOE (I,2282)

KOCH, HARRY
CHEYENNE (I,931)

KOCH, HOWARD W
JOHNNY RINGO (I,2434); MIAMI UNDERCOVER (I,3020); THE UNTOUCHABLES (I,4682)

KOLBE, WINRICH
ALL THAT GLITTERS (II,41); AUTOMAN (II,120); BARNABY JONES (II,153); CHIPS (II,511); THE FALL GUY (II,811); GAVILAN (II,959); KNIGHT RIDER (II,1402); MAGNUM, P.I. (II,1601); A MAN CALLED SLOANE (II,1613); SCARECROW AND MRS. KING (II,2268); STONE (II,2471); SWORD OF JUSTICE (II,2521); TALES OF THE GOLD MONKEY (II,2537); VOYAGERS (II,2730); THE YEAGERS (II,2843)

KOLSHURD, DON
JABBERWOCKY (II,1262)

KOMACK, JAMES
77 SUNSET STRIP (I,3988); ANOTHER DAY (II,94); CHICO AND THE MAN (II,508); THE COURTSHIP OF EDDIE'S FATHER (I,1083); THE DICK VAN DYKE SHOW (I,1275); ENSIGN O'TOOLE (I,1448); LADY LUCK (I,2604); ME AND MAXX (II,1672); MR. ROBERTS (I,3149); MR. T. AND TINA (II,1758); MY FAVORITE MARTIAN (I,3189); THE ROLLERGIRLS (II,2203); SHIPSHAPE (II,2329); STAR TREK (II,2440); WELCOME BACK, KOTTER (II,2761); WHATEVER HAPPENED TO DOBIE GILLIS? (II,2774)

KOPPERSTEIN, HARRY
THE STANLEY SIEGEL SHOW (II,2433)

KORDOS, RICHARD S
TWIGS (II,2684)

KORMAN, HARVEY
THE CRACKER BROTHERS (II,599); EUNICE (II,784); MAMA'S FAMILY (II,1610)

KORTNER, PETER
THE FARMER'S DAUGHTER (I,1533)

KOTCHEFF, TED
THE DESPERATE HOURS (I,1239); THE HUMAN VOICE (I,2160); OF MICE AND MEN (I,3333); RX FOR THE DEFENSE (I,3878)

KOVACS, ERNIE
THE COMEDY OF ERNIE KOVACS (I,1022); THE ERNIE KOVACS SPECIAL (I,1456); THE NEW ERNIE KOVACS SHOW (I,3264)

KOWALSKI, BERNARD L
B.A.D. CATS (II,123); BANACEK (II,138); BARETTA (II,152); BROKEN ARROW (I,744); CASSIE AND COMPANY (II,455); COLUMBO (II,556); THE DEVLIN CONNECTION (II,664); THE DICK POWELL SHOW (I,1269); HOT PURSUIT (II,1189); IN TANDEM (II,1228); JOHNNY BLUE (II,1321); THE JUDGE (I,2468); KNIGHT RIDER (II,1402); MAGNUM, P.I. (II,1601); MICKEY SPILLANE'S MIKE HAMMER (II,1692); MITCHELL AND WOODS (II,1709); THE MONROES (I,3089); MOVIN' ON (II,1743); THE NEW HEALERS (I,3266); NICK AND THE DOBERMANS (II,1843); NIGHTSIDE (II,1851); PERRY MASON (II,2025); RAWHIDE (I,3727); RISKO (II,2180); THE ROCKFORD FILES (II,2197); SEVEN BRIDES FOR SEVEN BROTHERS (II,2307); SIMON AND SIMON (II,2357); THE STREETS OF SAN FRANCISCO (II,2478); SUNSHINE (II,2495); SUPERCOPS (II,2502); TENAFLY (II,2560); TURNOVER SMITH (II,2671); TWO FOR THE MONEY (I,4655); THE WESTERNER (I,4801); WHIZ KIDS (II,2790); THE WILD WILD WEST (I,4863)

KRABER, TONY
OFF THE RECORD (I,3336)

KRAMER, STANLEY
GUESS WHO'S COMING TO DINNER? (II,1063)

KRAMER, THOMAS
NOT NECESSARILY THE NEWS (II,1863)

KRANZE, DONALD
GUY LOMBARDO'S DIAMOND JUBILLE (I,1911)

KRASNY, PAUL
THE ADVENTURES OF NICK CARTER (I,68); BIG ROSE (II,241); THE BLUE KNIGHT (II,276); BRING 'EM BACK ALIVE (II,377); BRONK (II,379); CATALINA C-LAB (II,458); CENTENNIAL (II,477); CHIPS (II,511); CLINIC ON 18TH STREET (II,544); CRAZY LIKE A FOX (II,601); THE D.A. (I,1123); THE D.A.: CONSPIRACY TO KILL (I,1121); FLY AWAY HOME (II,893); GAVILAN (II,959); HART TO HART (II,1102); THE ISLANDER (II,1249); JIGSAW JOHN (II,1288); THE LETTERS (I,2670); LOGAN'S RUN (II,1507); LUCAS TANNER (II,1556); THE MAN FROM ATLANTIS (II,1615); MANNIX (II,1624); MCCLAIN'S LAW (II,1659); MICKEY SPILLANE'S MIKE HAMMER (II,1692); MISSION: IMPOSSIBLE (I,3067); MOBILE MEDICS (II,1716); MOBILE ONE (II,1717); THE POWERS OF MATTHEW STAR (II,2075); QUINCY, M. E. (II,2102); SIMON AND SIMON (II,2357); SWITCH (II,2519); WIZARDS AND WARRIORS (II,2813)

KROEGER, BARRY
STAGE DOOR (I,4179)

KRONICK, WILLIAM
PLIMPTON! DID YOU HEAR THE ONE ABOUT. . .? (I,3629); PLIMPTON! SHOWDOWN AT RIO LOBO (I,3630); PLIMPTON! THE MAN ON THE FLYING TRAPEZE (I,3631)

KROWN, DICK
THE NEW MICKEY MOUSE CLUB (II,1823)

KUGEL, CARL
EIGHT IS ENOUGH (II,762); FANTASY ISLAND (II,829); THE PAPER CHASE (II,1949)

KUKOFF, BERNIE
ALL-AMERICAN PIE (II,48); DETECTIVE SCHOOL (II,661); GOSSIP (II,1043); IN TROUBLE (II,1230); MA AND PA (II,1570); THE NEW OPERATION PETTICOAT (II,1826); REGGIE (II,2127)

KULIK, BUZZ
CAMPO 44 (I,813); THE CASE AGAINST PAUL RYKER (I,861); COREY: FOR THE PEOPLE (II,575); THE DEFENDERS (I,1220); THE DESILU PLAYHOUSE (I,1237); THE DICK POWELL SHOW (I,1269); THE FEATHER AND FATHER GANG (II,845); THE FEATHER AND FATHER GANG: NEVER CON A KILLER (II,846); FORD STAR REVUE (I,1632); FROM HERE TO ETERNITY (II,932); GEORGE WASHINGTON (II,978); HAVE GUN—WILL TRAVEL (I,1968); KAY KYSER'S KOLLEGE OF MUSICAL KNOWLEDGE (I,2512); THE MAN WHO CAME TO DINNER (I,2881); MATT HELM (II,1652); NAKED CITY (I,3216); OWEN

MARSHALL: COUNSELOR AT LAW (I,3435); OWEN MARSHALL: COUNSELOR AT LAW (I,3436); RAGE OF ANGELS (II,2107); RAWHIDE (I,3727); READY FOR THE PEOPLE (I,3739); SAINTS AND SINNERS (I,3887); SAVAGE SUNDAY (I,3926); THE SCOTT MUSIC HALL (I,3943); A STORM IN SUMMER (I,4237); THE TWILIGHT ZONE (I,4651); WARNING SHOT (I,4770)

KULLER, SID
THE COLGATE COMEDY HOUR (I,997)

KUPCINET, JERRY
ENTERTAINMENT TONIGHT (II,780); HERE'S RICHARD (II,1136); HERE'S RICHARD (II,1137); LIFE'S MOST EMBARRASSING MOMENTS II (II,1471); LIFE'S MOST EMBARRASSING MOMENTS III (II,1472); LIFE'S MOST EMBARRASSING MOMENTS IV (II,1473); PERSONAL AND CONFIDENTIAL (II,2026); THE RICHARD SIMMONS SHOW (II,2166)

KUPFER, MARVIN
AMERICA 2-NIGHT (II,65); THE PAPER CHASE (II,1949)

KUWAHARA, ROBERT
THE ASTRONAUT SHOW (I,307)

KYNE, TERRY
THE CHEAP SHOW (II,491); THE DIAMOND HEAD GAME (II,666); THE GONG SHOW (II,1026); THE JOHN DAVIDSON SHOW (II,1310); NAME THAT TUNE (II,1786); NAME THAT TUNE (II,1787); THE RED SKELTON SHOW (I,3756); THICKE OF THE NIGHT (II,2587); THE WIZARD OF ODDS (I,4886)

LAHENDRO, BOB
ARCHIE BUNKER'S PLACE (II,105); CHANGE AT 125TH STREET (II,480); DETECTIVE SCHOOL (II,661); FISH (II,864); FLO (II,891); FRANKIE AND ANNETTE: THE SECOND TIME AROUND (II,917); GOING BANANAS (II,1015); HOT L BALTIMORE (II,1187); THE KALLIKAKS (II,1361); LEWIS AND CLARK (II,1465); OFF THE WALL (II,1879); THE RIGHTEOUS APPLES (II,2174); SONNY BOY (II,2402); STOCKARD CHANNING IN JUST FRIENDS (II,2467); THAT'S MY MAMA (II,2580); THREE FOR THE GIRLS (I,4477); WELCOME BACK, KOTTER (II,2761); WONDERBUG (II,2830)

LACHMAN, ED
TEXACO STAR THEATER (I,4411)

LACHMAN, MORT
THE BOB HOPE SHOW (I,562); THE BOB HOPE SHOW (I,613); THAT'S MY MAMA (II,2580)

LAFFERTY, PERRY
76 MEN AND PEGGY LEE (I,3989); ACRES AND PAINS (I,19); ARTHUR GODFREY AND THE SOUNDS OF NEW YORK (I,280); ARTHUR GODFREY IN HOLLYWOOD (I,281); ARTHUR GODFREY LOVES ANIMALS (I,282); HIDDEN TREASURE (I,2045); NAME THAT TUNE (I,3218); RAWHIDE (I,3727); ROBERT MONTGOMERY PRESENTS YOUR LUCKY STRIKE THEATER (I,3809); THE VICTOR BORGE SHOW (I,4721)

LAGOMARSINO, RON
ANOTHER WORLD (II,97)

LAIDMAN, HARVEY S
AIRWOLF (II,27); THE BLUE KNIGHT (II,276); CHICAGO STORY (II,506); THE DUKES OF HAZZARD (II,742); EIGHT IS ENOUGH (II,762); EISCHIED (II,763); FALCON CREST (II,808); FAMILY (II,813); THE FITZPATRICKS (II,868); FOR LOVE AND HONOR (II,903); THE INCREDIBLE HULK (II,1232); KINGSTON: CONFIDENTIAL (II,1398); KNIGHT RIDER (II,1402); KNOTS LANDING (II,1404); THE LAZARUS SYNDROME (II,1451); LOBO (II,1504); THE MAN FROM ATLANTIS (II,1615); THE PAPER CHASE (II,1949); SEVEN BRIDES FOR SEVEN BROTHERS (II,2307); TALES OF THE GOLD MONKEY (II,2537); TODAY'S FBI (II,2628); TUCKER'S WITCH (II,2667); WALKING TALL (II,2738); THE WALTONS (II,2740)

LAIRD, JACK
AMANDA FALLON (I,152); THE BOB HOPE CHRYSLER THEATER (I,502); BRONCO (I,745); HAVE GUN—WILL TRAVEL (I,1968); M SQUAD (I,2799); NIGHT GALLERY (I,3287); THE SEVEN LITTLE FOYS (I,3986)

LAIRD, MARLENA
DOUBLE TROUBLE (II,723); GENERAL HOSPITAL (II,964); LAVERNE AND SHIRLEY (II,1446); THE RIGHTEOUS APPLES (II,2174); A YEAR AT THE TOP (II,2844); YOUNG LIVES (II,2866)

LALLY, BOB
DOCTOR SHRINKER (II,686); IT'S YOUR MOVE (II,1259); THE JEFFERSONS (II,1276); LAND OF THE LOST (II,1425); THE LORENZO AND HENRIETTA MUSIC SHOW (II,1519); MARY HARTMAN, MARY HARTMAN (II,1638); SILVER SPOONS (II,2355)

LAMAS, FERNANDO
ALIAS SMITH AND JONES (I,118); THE AMAZING SPIDER-MAN (II,63); CODE RED (II,552); FALCON CREST (II,808); FLAMINGO ROAD (II,872); THE HARDY BOYS MYSTERIES (II,1090); HOUSE CALLS (II,1194); THE LAW ENFORCERS (I,2638); THE LAWYERS (I,2646); THE MAN AND THE CITY (I,2856); MANNIX (II,1624); THE ROOKIES (II,2208); RUN FOR YOUR LIFE (I,3871); S.W.A.T. (II,2236); SECRETS OF MIDLAND HEIGHTS (II,2296); THE SENATOR (I,3976); STARSKY AND HUTCH (II,2444); SWITCH (II,2519)

LAMBERT, ELEANOR
THE MARCH OF DIMES FASHION SHOW (I,2901); THE MARCH OF DIMES FASHION SHOW (I,2902)

LAMBERT, HUGH
DEAN MARTIN'S CHRISTMAS IN CALIFORNIA (II,645)

LAMBRECHT, IRV
ON THE GO (I,3361)

LAMKIN, KEN
KAZ (II,1370)

LAMMERS, PAUL
ANOTHER WORLD (II,97); AS THE WORLD TURNS (II,110); BEACON HILL (II,193)

LAMONT, CHARLES
ANNETTE (I,222)

LAMORE, MARSH
BATMAN AND THE SUPER SEVEN (II,160); BLACKSTAR (II,263); THE FABULOUS FUNNIES (II,802); FAT ALBERT AND THE COSBY KIDS (II,835); FLASH GORDON (II,874); FLASH GORDON—THE GREATEST ADVENTURE OF ALL (II,875); THE KID SUPER POWER HOUR WITH SHAZAM (II,1386); THE NEW ADVENTURES OF MIGHTY MOUSE AND HECKLE AND JECKLE (II,1814); SPORT BILLY (II,2429); THE TARZAN/LONE RANGER/ZORRO ADVENTURE HOUR (II,2543)

LAMPBERT, HUGH
DINAH AND HER NEW BEST FRIENDS (II,676)

LANCIT, LARRY
PLAY IT AGAIN, UNCLE SAM (II,2052)

LANDERS, HARRY
BEN CASEY (I,394)

LANDERS, LEW
THE ADVENTURES OF RIN TIN TIN (I,73); THE ADVENTURES OF SUPERMAN (I,77); BRONCO (I,745); CHEYENNE (I,931); COLT .45 (I,1010); THE FILES OF JEFFERY JONES (I,1576); TERRY AND THE PIRATES (I,4405); TOPPER (I,4582)

LANDERS, PAUL
77 SUNSET STRIP (I,3988); BLONDIE (I,486); BONANZA (II,347); BOSTON BLACKIE (I,695); BRONCO (I,745); CHEYENNE (I,931); THE CISCO KID (I,961); CROSSROADS (I,1105); THE DAKOTAS (I,1126); THE DETECTIVES (I,1245); FLIPPER (I,1604); FOREST RANGER (I,1641); HAWAIIAN EYE (I,1973); THE LONE RANGER (I,2740); MAVERICK (II,1657); THE RELUCTANT SPY (I,3763); THE RIFLEMAN (I,3789); SKY KING (I,4077); SUGARFOOT (I,4277)

LANDERS, ROBERT
SURFSIDE 6 (I,4299)

LANDIS, JOE
ABOUT FACES (I,12); THE LIBERACE SHOW (I,2674); THE TENNESSEE ERNIE FORD SHOW (I,4398)

LANDON, MICHAEL
BONANZA (II,347); FATHER MURPHY (II,841); HIGHWAY TO HEAVEN (II,1152); LITTLE HOUSE ON THE PRAIRIE (II,1487); LITTLE HOUSE: A NEW BEGINNING (II,1488); LITTLE HOUSE: THE LAST FAREWELL (II,1491)

LANDROT, PHILIPPE
INSPECTOR GADGET (II,1236)

LANDSBURG, ALAN
BIOGRAPHY (I,476); MEN IN CRISIS (I,3006); REAL LIFE STORIES (II,2117)

LANE, DAVID
CAPTAIN SCARLET AND THE MYSTERONS (I,837); THUNDERBIRDS (I,4497); U.F.O. (I,4662)

LANFIELD, SIDNEY
THE ADDAMS FAMILY (II,11); BRINGING UP BUDDY (I,731);

BURKE'S LAW (I,762); THE DEPUTY (I,1234); I WAS A BLOODHOUND (I,2180); ICHABOD AND ME (I,2183); M SQUAD (I,2799); MCHALE'S NAVY (I,2969); MY DARLING JUDGE (I,3186); NO TIME FOR SERGEANTS (I,3300); WAGON TRAIN (I,4747)

LANG, OTTO
THE DICK POWELL SHOW (I,1269); THE FELONY SQUAD (I,1563)

LANG, RICHARD
CHARLIE'S ANGELS (II,486); DOCTOR SCORPION (II,685); FANTASY ISLAND (II,830); HARRY O (II,1099); HAWAIIAN EYE (I,1973); KUNG FU (II,1416); MATT HOUSTON (II,1654); SHOOTING STARS (II,2341); THE STREETS OF SAN FRANCISCO (II,2478); STRIKE FORCE (II,2480); TALES OF THE UNEXPECTED (II,2539); VEGAS (II,2723); VELVET (II,2725); THE WORD (II,2833)

LANGE, HARRY
SHAZAM! (II,2319)

LANGE, TED
THE FALL GUY (II,811); THE LOVE BOAT (II,1535)

LANGTON, SIMON
SMILEY'S PEOPLE (II,2383); THERESE RAQUIN (II,2584)

LARGE, BRIAN
CINDERELLA (II,526)

LARRIVA, RUDY
ALVIN AND THE CHIPMUNKS (II,61); THE ALVIN SHOW (I,146); BEAUTY AND THE BEAST (II,197); FANG FACE (II,822); GOLDIE GOLD AND ACTION JACK (II,1023); THE HEATHCLIFF AND DINGBAT SHOW (II,1117); MISSION MAGIC (I,3068); MORK AND MINDY (II,1736); MR. MAGOO (I,3142); MR. T (II,1757); MY FAVORITE MARTIANS (II,1779); THE NEW ADVENTURES OF BATMAN (II,1812); THE NEW ADVENTURES OF GILLIGAN (II,1813); THE PLASTICMAN COMEDY/ADVENTURE SHOW (II,2051); RUBIK, THE AMAZING CUBE (II,2226); THE SECRET LIVES OF WALDO KITTY (II,2292); THUNDARR THE BARBARIAN (II,2616); THE U.S. OF ARCHIE (I,4687); U.S. OF ARCHIE (II,2698)

LARRY, SHELDON
EMERALD POINT, N.A.S. (II,773); KING OF KENSINGTON (II,1395); KNOTS LANDING (II,1404); REMINGTON STEELE (II,2130); THE SECRET OF CHARLES DICKENS (II,2293)

LARSON, CHARLES
THE FBI (I,1551)

LARSON, GLEN A
GET CHRISTIE LOVE! (II,981); MCCLOUD (II,1660); THE SIX-MILLION-DOLLAR MAN (II,2372)

LATEEF, AHMET
BATTLESTAR GALACTICA (II,181)

LATHAM, LARRY
THE SMURFS (II,2385)

LATHAN, STAN
THE ABC AFTERSCHOOL SPECIAL (II,1); BOONE (II,351); BREAKING AWAY (II,371); EIGHT IS ENOUGH (II,762); FALCON CREST (II,808); FAME (II,812); FLAMINGO ROAD (II,872); THE FLIP WILSON COMEDY SPECIAL (II,886); MIAMI VICE (II,1689); THE MUHAMMED ALI VARIETY SPECIAL (II,1762); REMINGTON STEELE (II,2130); THE RIGHTEOUS APPLES (II,2174); SHIRLEY (II,2330); THAT'S MY MAMA (II,2580); THE WALTONS (II,2740)

LAVEN, ARNOLD
THE A-TEAM (II,119); ALFRED HITCHCOCK PRESENTS (I,115); ALIAS SMITH AND JONES (I,118); THE BIG VALLEY (I,437); CASSIE AND COMPANY (II,455); CIRCLE OF FEAR (I,958); DAN AUGUST (I,1130); DELVECCHIO (II,659); THE DETECTIVES (I,1245); DOG AND CAT (II,696); EIGHT IS ENOUGH (II,762); FREEBIE AND THE BEAN (II,922); FRIENDS (II,927); GHOST STORY (I,1788); THE GREATEST AMERICAN HERO (II,1060); GRIFF (I,1888); GUNSMOKE (II,1069); HILL STREET BLUES (II,1154); HUNTER (II,1206); THE INDIAN (I,2210); ISIS (II,1248); MANNIX (II,1624); MARCUS WELBY, M.D. (II,1627); MICKEY SPILLANE'S MIKE HAMMER (II,1692); PLANET OF THE APES (II,2049); RAFFERTY (II,2105); RICHIE BROCKELMAN, PRIVATE EYE (II,2167); THE RIFLEMAN (I,3789); RIPTIDE (II,2178); THE ROUSTERS (II,2220); THE SHARPSHOOTER (I,4001); SHAZAM! (II,2319); THE SIX-MILLION-DOLLAR MAN (II,2372); TENSPEED AND BROWN SHOE (II,2562); TIME EXPRESS (II,2620); TURNABOUT (II,2669); WAGON TRAIN (I,4747); WHICH WAY'D THEY GO? (I,4838); THE YEAGERS (II,2843)

LAVIN, LINDA
ALICE (II,33)

LAW, GENE
HERE'S HOLLYWOOD (I,2031); ON THE GO (I,3361)

LAWRENCE, MARK
BRONCO (I,745); LAWMAN (I,2642); THE ROARING TWENTIES (I,3805)

LAWRENCE, QUENTIN
THE AVENGERS (II,121); THE INVISIBLE MAN (I,2232)

LAWRENCE, VERNON
SONG BY SONG (II,2399)

LAYTON, JOE
ANDROCLES AND THE LION (I,189); THE BELLE OF 14TH STREET (I,390); THE HANNA-BARBERA HAPPINESS HOUR (II,1079); THE LITTLEST ANGEL (I,2726); ON THE FLIP SIDE (I,3360); PAUL LYNDE AT THE MOVIES (II,1976); PAUL LYNDE GOES M-A-A-A-AD (II,1980)

LEBORG, REGINALD
TIGHTROPE (I,4500)

LEACOCK, PHILIP
ALFRED HITCHCOCK PRESENTS (I,115); APPLE'S WAY (II,101); BORN TO THE WIND (II,354); BUCK ROGERS IN THE 25TH CENTURY (II,383); THE DAUGHTERS OF JOSHUA CABE (I,1169); THE DEFENDERS (I,1220); DIRTY SALLY (II,680); DYNASTY (II,746); EIGHT IS ENOUGH (II,762); FAMILY (II,813); FANTASY ISLAND (II,829); THE FBI (I,1551); FINDER OF LOST LOVES (II,857); GUNSMOKE (II,1069); HAWAII FIVE-O (II,1110); HEAVEN ON EARTH (II,1119); HOTEL (II,1192); KATE LOVES A MYSTERY (II,1367); KEY WEST (I,2528); LITTLE WOMEN (II,1498); THE MEN FROM SHILOH (I,3005); THE MOD SQUAD (I,3078); THE NEW LAND (II,1820); NURSE (II,1870); THE PAPER CHASE (II,1949); ROUTE 66 (I,3852); SWEEPSTAKES (II,2514); THE VIRGINIAN (I,4738); THE WALTONS (II,2740); WILD AND WOOLEY (II,2798); THE WILD WOMEN OF CHASTITY GULCH (II,2801)

LEADER, ANTON
DANIEL BOONE (I,1142); GET SMART (II,983); GILLIGAN'S ISLAND (II,990); HAWAII FIVE-O (II,1110); I SPY (I,2179); IRONSIDE (II,1246); IT TAKES A THIEF (I,2250); LAWMAN (I,2642); LEAVE IT TO BEAVER (I,2648); MR. LUCKY (I,3141); NICHOLS (I,3283); THE O. HENRY PLAYHOUSE (I,3322); PERRY MASON (II,2025); RAWHIDE (I,3727); SOLDIERS OF FORTUNE (I,4111); STAR TREK (II,2440); SUGARFOOT (I,4277); TARZAN (I,4361); THE TWILIGHT ZONE (I,4651); THE VIRGINIAN (I,4738)

LEAF, PAUL
HERE'S BOOMER (II,1134); LOU GRANT (II,1526); TOP SECRET (II,2647)

LEAR, NORMAN
THE MARTHA RAYE SHOW (I,2926)

LEAVER, DON
APPOINTMENT WITH A KILLER (I,245); THE AVENGERS (II,121); BERGERAC (II,209); THE KILLING GAME (I,2540)

LEBORG, ROBERT
BOURBON STREET BEAT (I,699)

LEDERMAN, ROSS
ANNIE OAKLEY (I,225); THE GENE AUTRY SHOW (I,1748)

LEEDS, HERBERT
THE LIFE OF RILEY (I,2685)

LEEDS, ROBERT
77 SUNSET STRIP (I,3988); THE BEVERLY HILLBILLIES (I,417); GENERAL ELECTRIC TRUE (I,1754); JOHNNY GUITAR (I,2427); PROJECT UFO (II,2088); THE RETURN OF THE BEVERLY HILLBILLIES (II,2139); SAM (II,2247)

LEFTWICH, ALEX
FLOOR SHOW (I,1605); MAKE MINE MUSIC (I,2840); THE SHOW GOES ON (I,4030); THIS IS SHOW BUSINESS (I,4459); WHO'S WHOSE? (I,4855)

LEFTWICH, ED
THE CONTINENTAL (I,1042); LIVE LIKE A MILLIONAIRE (I,2728)

LEHMAN, BOB
MARINELAND CARNIVAL (I,2912); MARINELAND CARNIVAL (I,2913)

LEHR, GEORGE
POLICE WOMAN (II,2063)

LEHR, MILTON
THE RAY ANTHONY SHOW (I,3729); THE SWINGING SCENE OF RAY ANTHONY (I,4323)

LEIGHTON, RICK
THE FLORIAN ZABACH SHOW (I,1606)

LEISEN, MITCHELL
ADVENTURES IN PARADISE (I,46); THRILLER (I,4492); THE TWILIGHT ZONE (I,4651); WAGON TRAIN (I,4747)

LEITH, BARRY
PADDINGTON BEAR (II,1941)

LEMONT, JOHN
THE ERROL FLYNN THEATER (I,1458)

LENDERMAN, ROSS
CAPTAIN MIDNIGHT (I,833)

LENOX, JOHN THOMAS
BUSTING LOOSE (II,402); LAVERNE AND SHIRLEY (II,1446); WHO'S WATCHING THE KIDS? (II,2793)

LEO, MALCOLM
BOB HOPE SPECIAL: BOB HOPE'S OVERSEAS CHRISTMAS TOURS (II,311); E.T. & FRIENDS—MAGICAL MOVIE VISITORS (II,749)

LEONARD, HERBERT B
ROUTE 66 (I,3852)

LEONARD, SHELDON
ACES UP (II,7); THE ANDY GRIFFITH SHOW (I,193); THE DICK VAN DYKE SHOW (I,1275); EVERYTHING HAPPENS TO ME (I,1481); IT'S ALWAYS JAN (I,2264); JEFF'S COLLIE (I,2356); MAKE MORE ROOM FOR DADDY (I,2842); MAKE ROOM FOR DADDY (I,2843); MY FAVORITE MARTIAN (I,3189); MY WORLD . . . AND WELCOME TO IT (I,3208); PATRICK STONE (I,3476); THE REAL MCCOYS (I,3741); SINGLES (I,4062)

LEONARDI, ART
THE SUPER SIX (I,4292)

LEONARDI, BOB
DENNIS THE MENACE: MAYDAY FOR MOTHER (II,660)

LERSKOVITZ, MARSHALL
FAMILY (II,813)

LESHAY, JERRY
AMERICA 2-NIGHT (II,65)

LESIAN, MITCHELL
THE GIRL FROM U.N.C.L.E. (I,1808)

LESSAC, MICHAEL
9 TO 5 (II,1852); DOMESTIC LIFE (II,703); OLD FRIENDS (II,1881); SHAPING UP (II,2316); TAXI (II,2546)

LESSING, NORMAN
VEGAS (II,2724)

LETCH, CHRISTOPHER
THE HITCHHIKER (II,1157)

LETTS, BARRY
DOCTOR WHO (II,689)

LEVENS, PHIL
DOLLAR A SECOND (I,1326)

LEVI, ALAN J
AIRWOLF (II,27); ARNOLD'S CLOSET REVIEW (I,262); THE BIONIC WOMAN (II,255); THE GEMINI MAN (II,960); THE GEMINI MAN (II,961); GO WEST YOUNG GIRL (II,1013); THE INCREDIBLE HULK (II,1232); THE INVISIBLE MAN (II,1242); THE INVISIBLE WOMAN (II,1244); JUDGEMENT DAY (II,1348); THE LEGEND OF THE GOLDEN GUN (II,1455); LETTERS TO LAUGH-IN (I,2672); MAGNUM, P.I. (II,1601); A MAN CALLED SLOANE (II,1613); SCRUPLES (II,2280); SIMON AND SIMON (II,2357); TALES OF THE GOLD MONKEY (II,2537); TIME EXPRESS (II,2620); VOYAGERS (II,2730); WHAT REALLY HAPPENED TO THE CLASS OF '65? (II,2768)

LEVIN, DAN
VANITY FAIR (I,4702)

LEVIN, PETER
AFTERMASH (II,23); ANOTHER WORLD (II,97); BEACON HILL (II,193); BOONE (II,351); CAGNEY AND LACEY (II,409); CALL TO GLORY (II,413); EMERALD POINT, N.A.S. (II,773); FAME (II,812); JAMES AT 15 (II,1270); JOSHUA'S WORLD (II,1344); KAZ (II,1370); KNOTS LANDING (II,1404); LOU GRANT (II,1526); LOVERS AND FRIENDS (II,1549); PALMERSTOWN U.S.A. (II,1945); RYAN'S FOUR (II,2233); SEVEN BRIDES FOR SEVEN BROTHERS (II,2307); TRAUMA CENTER (II,2655); TWO MARRIAGES (II,2691)

LEVIN, ROBERT
MARRIED: THE FIRST YEAR (II,1634); THE PAPER CHASE (II,1949)

LEVINSON, BARRY
DINER (II,679); THE INVESTIGATORS (II,1241); PEEPING TIMES (II,1990)

LEVINSON, RICHARD
BURKE'S LAW (I,762); THE FUGITIVE (I,1701)

LEVITOW, ABE
THE FAMOUS ADVENTURES OF MR. MAGOO (I,1524); OFF TO SEE THE WIZARD (I,3337)

LEVITT, GENE
ALIAS SMITH AND JONES (I,118); ALIAS SMITH AND JONES (I,119); COOL MILLION (I,1046); GET CHRISTIE LOVE! (II,981); THE LAWYERS (I,2646); LOBO (II,1504); THE MASK OF MARCELLA (I,2937); THE NIGHT STALKER (II,1849); S.W.A.T. (II,2236); TENAFLY (II,2560)

LEVITT, GEORGE
GIBBSVILLE (II,988)

LEVY, LAWRENCE
LOTTERY (II,1525)

LEVY, RALPH
54TH STREET REVUE (I,1573); THE ALAN YOUNG SHOW (I,104); CELEBRITY TIME (I,887); A CHRISTMAS CAROL (I,947); DENNIS JAMES' CARNIVAL (I,1227); DETECTIVE SCHOOL (II,661); THE ED WYNN SHOW (I,1403); THE EYES HAVE IT (I,1495); FROM A BIRD'S EYE VIEW (I,1686); THE GENERAL FOODS 25TH ANNIVERSARY SHOW (I,1756); THE GEORGE BURNS AND GRACIE ALLEN SHOW (I,1763); GREEN ACRES (I,1884); HAWAII FIVE-O (II,1110); THE JACK BENNY HOUR (I,2291); THE JACK BENNY HOUR (I,2292); THE JACK BENNY HOUR (I,2293); THE JACK BENNY PROGRAM (I,2294); THE JACK BENNY SPECIAL (I,2295); JACK BENNY WITH GUEST STARS (I,2298); THE MISSUS GOES A SHOPPING (I,3069); MY BOY GOOGIE (I,3185); PETTICOAT JUNCTION (I,3571); PLACES PLEASE (I,3613); QUADRANGLE (I,3703); SHIRLEY'S WORLD (I,4019); TIME OUT FOR GINGER (I,4508); TRAPPER JOHN, M.D. (II,2654)

LEWIS, AL
MANY HAPPY RETURNS (I,2895); OUR MISS BROOKS (I,3415)

LEWIS, CHRISTOPHER
TALES OF THE UNEXPECTED (II,2540)

LEWIS, DAVID P
BARNEY BLAKE, POLICE REPORTER (I,359); FARAWAY HILL (I,1532); LINDA RONSTADT IN CONCERT (II,1482); STORIES IN ONE CAMERA (I,4234)

LEWIS, ELLIOT
THE BILL COSBY SHOW (I,442); PETTICOAT JUNCTION (I,3571)

LEWIS, JERRY
THE DOCTORS (I,1323); THE JERRY LEWIS SHOW (I,2363)

LEWIS, JOSEPH H
ALCOA/GOODYEAR THEATER (I,107); THE BIG VALLEY (I,437); BRANDED (I,707); THE DEFENDERS (I,1220); THE DETECTIVES (I,1245); THE DICK POWELL SHOW (I,1269); THE INVESTIGATORS (I,2231); THE RIFLEMAN (I,3789)

LEWIS, LEONARD
FLAMBARDS (II,869)

LEWIS, PETER
BALL FOUR (II,137)

LEWIS, ROBERT MICHAEL
CASABLANCA (II,450); COMPUTERCIDE (II,561); GRIFF (I,1888); HARRY O (II,1099); THE INVISIBLE MAN (II,1242); THE INVISIBLE MAN (II,1243); KUNG FU (II,1416); MARRIED: THE FIRST YEAR (II,1634); MCMILLAN AND WIFE (II,1667); THE MOD SQUAD (I,3078); S*H*E (II,2235); SECRETS OF MIDLAND HEIGHTS (II,2296)

LEWITT, GENE
THE YOUNG LAWYERS (I,4941)

LEYTES, JERRY
THE ALASKANS (I,106)

LEYTES, JOSEF
THE DICK POWELL SHOW (I,1269); VOYAGE TO THE BOTTOM OF THE SEA (I,4743)

LIBERMAN, ROBERT
THE ABC AFTERSCHOOL SPECIAL (II,1)

LIBERTI, JOHN
THE INCREDIBLE HULK (II,1232)

LIEBENHEINER, WOLFGANG
MARK TWAIN'S TOM SAWYER-HUCKLEBERRY

FINN (I,2915)

LIEBER, KEN
THE TONY RANDALL SHOW (II,2640)

LIEBMAN, MAX
THE ADMIRAL BROADWAY REVUE (I,37); BABES IN TOYLAND (I,318); BEST FOOT FORWARD (I,401); THE BIG TIME (I,434); A CONNECTICUT YANKEE (I,1038); DEAREST ENEMY (I,1205); THE DESERT SONG (I,1236); GOOD TIMES (I,1853); THE GREAT WALTZ (I,1878); HEAVEN WILL PROTECT THE WORKING GIRL (I,1996); HEIDI (I,2004); HOLIDAY (I,2074); KALEIDOSCOPE (I,2504); MARCO POLO (I,2904); THE MAURICE CHEVALIER SHOW (I,2953); THE MAURICE CHEVALIER SHOW (I,2954); THE MERRY WIDOW (I,3012); THE MUSIC OF GERSHWIN (I,3174); NAUGHTY MARIETTA (I,3232); PANORAMA (I,3448); PARIS IN THE SPRINGTIME (I,3457); STANLEY (I,4190); SUNDAY IN TOWN (I,4285); VARIETY (I,4703); YOUR SHOW OF SHOWS (I,4957)

LILLEY, JOE
THE BING CROSBY CHRISTMAS SHOW (I,458);

LINDSAY-HOGG, MICHAEL
RANDY NEWMAN AT THE ODEON (II,2111); SIMON AND GARFUNKEL IN CONCERT (II,2356)

LINK, WILLIAM
BURKE'S LAW (I,762); THE FUGITIVE (I,1701)

LINKROUM, RICHARD
THE ALAN YOUNG SHOW (I,104); FREEDOM RINGS (I,1682); HOW TO (I,2145); THE JACK BENNY PROGRAM (I,2294)

LIOTTA, CHARLES W
ON OUR OWN (II,1892); OPENING NIGHT: U.S.A. (I,3397); SANFORD AND SON (II,2256)

LISHAWA, CHET
THE DANCE SHOW (II,616)

LITTLEWOOD, YVONNE
PERRY COMO'S OLDE ENGLISH CHRISTMAS (II,2020)

LITVACK, JOHN
AS THE WORLD TURNS (II,110)

LIVINGSTON, CHARLES
THE LONE RANGER (I,2740); SERGEANT PRESTON OF THE YUKON (I,3980)

LIVINGSTON, ROBERT H
MAUDE (II,1655)

LLOYD, NORMAN
ALCOA PREMIERE (I,109); COLUMBO (II,556)

LO-RE, JOE
SOAP FACTORY DISCO (II,2393)

LOBIANCO, TONY
THE DUKES (II,740); KAZ (II,1370)

LOBL, VICTOR
CAGNEY AND LACEY (II,409); THE DUCK FACTORY (II,738); HERE'S BOOMER (II,1134); KING'S CROSSING (II,1397); PARIS (II,1957); ST. ELSEWHERE (II,2432); TALES OF THE GOLD MONKEY (II,2537); TRAUMA CENTER (II,2655); USED CARS (II,2712); V: THE SERIES (II,2715); THE WHITE SHADOW (II,2788)

LOBUE, J D
ALL'S FAIR (II,46); AMANDA'S (II,62); BLUE JEANS (II,275); CASS MALLOY (II,454); THE COMEDY ZONE (II,559); FOOT IN THE DOOR (II,902); OH MADELINE (II,1880); SOAP (II,2392); SOUL TRAIN (I,4131); STOCKARD CHANNING IN JUST FRIENDS (II,2467); TOO GOOD TO BE TRUE (II,2643)

LOCHHEAD, LEE
ALICE (II,33); BARNEY MILLER (II,154); CHARLES IN CHARGE (II,482); FLO (II,891); TOO CLOSE FOR COMFORT (II,2642)

LOCKE, ALBERT
THE LONDON PALLADIUM (I,2739)

LOCKE, DICK
STUD'S PLACE (I,4271)

LOCKE, RICK
MAGIC MONGO (II,1592)

LOCKWOOD, GREY
THE ALL-STAR REVUE (I,138); ALL-STAR SWING FESTIVAL (II,53); THE BIG BAND AND ALL THAT JAZZ (I,422); BING CROSBY AND CAROL BURNETT—TOGETHER AGAIN FOR THE FIRST TIME (I,454); THE BOBBY DARIN AMUSEMENT COMPANY (I,672); DIANA ROSS AND THE SUPREMES AND THE TEMPTATIONS ON BROADWAY (I,1260); THE DICK CLARK SATURDAY NIGHT BEECHNUT SHOW (I,1267); DICK CLARK'S WORLD OF TALENT (I,1268); THE DON ADAMS SPECIAL: HOORAY FOR HOLLYWOOD (I,1329); DON KNOTTS NICE CLEAN, DECENT, WHOLESOME HOUR (I,1333); FOUR FOR TONIGHT (I,1648); THE HOLLYWOOD PALACE (I,2088); JOHNNY CARSON PRESENTS THE SUN CITY SCANDALS (I,2420); THE LENNON SISTERS SHOW (I,2656); MUSIC ON ICE (I,3175); THE PERRY COMO SHOW (I,3532); SHOW BUSINESS SALUTE TO MILTON BERLE (I,4029); TAKE A CHANCE (I,4331); TALENT SEARCH (I,4342); WONDERFUL JOHN ACTION (I,4892)

LOEMAN, DICK
THE LONDON PALLADIUM (I,2739)

LOFGREN, DENNIS
COUNTERATTACK: CRIME IN AMERICA (II,581)

LOGGIA, ROBERT
MAGNUM, P.I. (II,1601)

LONDON, JERRY
THE BOB NEWHART SHOW (II,342); THE BRADY BUNCH (II,362); THE CHICAGO STORY (II,507); CHIEFS (II,509); COVER GIRLS (II,596); DELVECCHIO (II,659); ELLIS ISLAND (II,768); ESCAPADE (II,781); THE FEATHER AND FATHER GANG (II,845); THE FLYING NUN (I,1611); HAPPY DAYS (II,1084); HARRY O (II,1099); HERE WE GO AGAIN (I,2028); HOTEL (II,1192); JOE FORRESTER (II,1303); KOJAK (II,1406); LUCAS TANNER (II,1556); MA AND PA (II,1570); MARCUS WELBY, M.D. (II,1627); PETROCELLI (II,2031); THE ROCKFORD FILES (II,2197); SHOGUN (II,2339); WHEELS (II,2777); WOMEN IN WHITE (II,2823); THE WORLD OF DARKNESS (II,2836)

LONG, BOB
YOU ASKED FOR IT (II,2855)

LONG, RICHARD
THE BIG VALLEY (I,437)

LORD, JACK
HAWAII FIVE-O (II,1110); M STATION: HAWAII (II,1568)

LORD, STEPHEN
THE LORETTA YOUNG THEATER (I,2756)

LORRIMER, VERE
BLAKE'S SEVEN (II,264)

LOSSO, ERNEST
BEWITCHED (I,418); JAMES AT 15 (II,1270); THE PARTRIDGE FAMILY (II,1962)

LOTTERBY, SYDNEY
LAST OF THE SUMMER WINE (II,1439); OPEN ALL HOURS (II,1913)

LOUDIN, BOB
AMERICA ALIVE! (II,67); BEDTIME STORIES (II,199); THE REGIS PHILBIN SHOW (II,2128)

LOVE, HARRY
INSPECTOR GADGET (II,1236)

LOVE, JAMES A
LLOYD BRIDGES WATER WORLD (I,2734)

LOVELL, ANTHONY
THE ABC AFTERSCHOOL SPECIAL (II,1)

LOVEY, ALEX
THE FAMOUS ADVENTURES OF MR. MAGOO (I,1524); THE WOODY WOODPECKER SHOW (I,4906)

LOWE JR, DAVID
TV'S BLOOPERS AND PRACTICAL JOKES (II,2675)

LOWE, DAVID
LIFE BEGINS AT EIGHTY (I,2682)

LOWE, JAMES
TONY THE PONY (II,2641)

LOWRY, DICK
BARNABY JONES (II,153); BUCK ROGERS IN THE 25TH CENTURY (II,383)

LUBIN, ARTHUR
77 SUNSET STRIP (I,3988); BONANZA (II,347); BRONCO (I,745); CHEYENNE (I,931); THE DEPUTY (I,1234); THE DESILU PLAYHOUSE (I,1237); MR. ED (I,3137); LITTLE LULU (II,1493); MAVERICK (II,1657); YOU'RE ONLY YOUNG TWICE (I,4970)

LUBOFF, NORMAN
THE NIGHT OF CHRISTMAS (I,3290)

LUCAS, JOHN MEREDYTH
BEN CASEY (I,394); THE FUGITIVE (I,1701); THE INVADERS (I,2229); MEDIC (I,2976); NIGHT GALLERY (I,3287); PLANET OF THE APES (II,2049); THE SIX-

MILLION-DOLLAR MAN (II,2372); STAR TREK (II,2440)

LUCAS, JONATHAN
A COUNTRY HAPPENING (I,1065); DATELINE: HOLLYWOOD (I,1167); DEAN MARTIN'S COMEDY CLASSICS (II,646); THE GOLDDIGGERS (I,1838); HARPER VALLEY, U.S.A. (I,1954); THE MICKEY MOUSE CLUB (I,3025); THE POWDER ROOM (I,3656); WHAT'S UP? (I,4823)

LUDWIG, EDWARD
BRANDED (I,707); MAISIE (I,2833)

LUMET, SIDNEY
DANGER (I,1134); MR. BROADWAY (I,3130); THE SHOWOFF (I,4035); STAGE DOOR (I,4178); YOU ARE THERE (I,4929)

LUPINO, IDA
77 SUNSET STRIP (I,3988); ALFRED HITCHCOCK PRESENTS (I,115); THE BIG VALLEY (I,437); THE FUGITIVE (I,1701); THE GHOST AND MRS. MUIR (I,1786); GILLIGAN'S ISLAND (II,990); HAVE GUN—WILL TRAVEL (I,1968); HOLLOWAY'S DAUGHTERS (I,2079); MR. ADAMS AND EVE (I,3121); PLEASE DON'T EAT THE DAISIES (I,3628); THE RIFLEMAN (I,3789); THE ROGUES (I,3832); SCREEN DIRECTOR'S PLAYHOUSE (I,3946); THE TEENAGE IDOL (I,4375); THRILLER (I,4492); THE TWILIGHT ZONE (I,4651); THE UNTOUCHABLES (I,4682); THE VIRGINIAN (I,4738)

LUSK, DON
PINK PANTHER AND SONS (II,2043)

LYDON, JAMES
THE SIX-MILLION-DOLLAR MAN (II,2372)

LYMAN III, JOHN B
DAVE AND CHARLEY (I,1171); MAYOR OF HOLLYWOOD (I,2962)

LYMAN, JOHN
THE PINKY LEE SHOW (I,3601)

LYNCH, PAUL
DARKROOM (II,619); VOYAGERS (II,2730)

LYNN, ROBERT
CAPTAIN SCARLET AND THE MYSTERONS (I,837); SPACE: 1999 (II,2409)

LYON, FRANCIS D
THE LONER (I,2744); PERRY MASON (II,2025)

MACDEARMON, DAVID
PETER GUNN (I,3562); THAT GIRL (II,2570)

MACDONALD, DAVID
THE FLYING DOCTOR (I,1610); IVANHOE (I,2282); THE VEIL (I,4709)

MACDONALD, FRANK
THE WHISTLER (I,4843)

MACDOUGALL, DON
TARGET: THE CORRUPTERS (I,4360)

MACKENDRICK, ALEXANDER
THE DEFENDERS (I,1220)

MACKENZIE, WILL
13 QUEENS BOULEVARD (II,2591); AFTERMASH (II,23); ALL TOGETHER NOW (II,45); ARCHIE BUNKER'S PLACE (II,105); BEST OF THE WEST (II,218); BOSOM BUDDIES (II,355); BROTHERS (II,381); BROTHERS AND SISTERS (II,382); CHARACTERS (II,481); DOMESTIC LIFE (II,703); FAMILY TIES (II,819); GIMME A BREAK (II,995); HIZZONER (II,1159); I DO, I DON'T (II,1212); LEWIS AND CLARK (II,1465); A NEW KIND OF FAMILY (II,1819); NEWHART (II,1835); OPEN ALL NIGHT (II,1914); SHE'S WITH ME (II,2321); STOCKARD CHANNING IN JUST FRIENDS (II,2467); THE STOCKARD CHANNING SHOW (II,2468); T.L.C. (II,2525); TAXI (II,2546); TOO CLOSE FOR COMFORT (II,2642); UNITED STATES (II,2708); WE'VE GOT EACH OTHER (II,2757); WKRP IN CINCINNATI (II,2814)

MACNAUGHTON, IAN
MONTY PYTHON'S FLYING CIRCUS (II,1729)

MACDONALD, FRANK
THE RANGE RIDER (I,3720)

MACKAY, HARPER
THE WONDERFUL WORLD OF BURLESQUE II (I,4896)

MACKIE, BOB
THE HANNA-BARBERA HAPPINESS HOUR (II,1079)

MADDEN, LEE
SOMETHING ELSE (I,4117)

MAGAR, GUY
BLUE THUNDER (II,278); CAPITOL (II,426); COVER UP (II,597); HARDCASTLE AND MCCORMICK (II,1089); HUNTER (II,1206); PARTNERS IN CRIME (II,1961); THE POWERS OF MATTHEW STAR (II,2075); RIPTIDE (II,2178)

MAGWOOD, HOWARD
CBS CARTOON THEATER (I,878)

MAGYAR, DESZO
TALES OF THE UNEXPECTED (II,2540)

MAJORS, CAROL
ELKE SOMMER'S WORLD OF SPEED AND BEAUTY (II,765)

MAJORS, JASON
ELKE SOMMER'S WORLD OF SPEED AND BEAUTY (II,765)

MAJORS, LEE
THE SIX-MILLION-DOLLAR MAN (II,2372)

MALBY JR, RICHARD
AIN'T MISBEHAVIN' (II,25)

MALLET, DAVID
DAVID BOWIE—SERIOUS MOONLIGHT (II,622); THE KENNY EVERETT VIDEO SHOW (II,1379)

MALLORY, ED
RITUALS (II,2182)

MALMUTH, BRUCE
THE ABC AFTERSCHOOL SPECIAL (II,1)

MALONE, ADRIAN
COSMOS (II,579)

MALONEY, DAVID
BLAKE'S SEVEN (II,264); DOCTOR WHO (II,689)

MANDUKE, JOSEPH
THE AMAZING SPIDER-MAN (II,63); BARNABY JONES (II,153); DALLAS (II,614); FALCON CREST (II,808); HARDCASTLE AND MCCORMICK (II,1089); HARRY O (II,1099); HAWAII FIVE-O (II,1110); KNOTS LANDING (II,1404)

MANKIEWICZ, JOSEPH L
CAROL FOR ANOTHER CHRISTMAS (I,852)

MANKIEWICZ, TOM
HART TO HART (II,1101); HART TO HART (II,1102)

MANN, DANIEL
HOW THE WEST WAS WON (II,1196); THE LEGEND OF SILENT NIGHT (I,2655)

MANN, DELBERT
ALL THE WAY HOME (II,44); THE GOLDEN AGE OF TELEVISION (II,1016); JANE EYRE (I,2339); THE MEMBER OF THE WEDDING (II,1680); OUR TOWN (I,3419); THE PETRIFIED FOREST (I,3570); PHILCO TELEVISION PLAYHOUSE (I,3583); PLAYWRIGHTS '56 (I,3627); TOM AND JOANN (II,2630)

MANN, JOHNNY
THE NUT HOUSE (I,3320);

MANNERS, KIM
AUTOMAN (II,120); CHARLIE'S ANGELS (II,486); HAWAIIAN HEAT (II,1111); MATT HOUSTON (II,1654)

MANSFIELD, MIKE
AIR SUPPLY IN HAWAII (II,26); ELTON JOHN IN CENTRAL PARK (II,769); THE ROOTS OF ROCK 'N' ROLL (II,2212); TEDDY PENDERGRASS IN CONCERT (II,2551); TWIGGY'S JUKE BOX (II,2682)

MANTOOTH, RANDOLPH
EMERGENCY! (II,775)

MARCEL, TERRY
PRISONERS OF THE LOST UNIVERSE (II,2086)

MARCH, ALEX
BARETTA (II,152); BARNEY MILLER (II,154); BEN CASEY (I,394); BRONK (II,379); CASSIE AND COMPANY (II,455); THE COP AND THE KID (II,572); DALLAS (II,614); THE DANGEROUS DAYS OF KIOWA JONES (I,1140); FIREHOUSE (I,1579); HAGEN (II,1071); HART TO HART (II,1102); HEC RAMSEY (II,1121); HOUSE CALLS (II,1194); JOE FORRESTER (II,1303); JUDD, FOR THE DEFENSE (I,2463); MADIGAN (I,2808); MCCLOUD (II,1660); THE MISSISSIPPI (II,1707); N.Y.P.D. (I,3321); NAKED CITY (I,3216); NURSE (II,1870); THE PAPER CHASE (II,1949); THE PARADINE CASE (I,3453); PARIS (II,1957); POLICE STORY (II,2062); QUINCY, M. E. (II,2102); THE RETURN OF CAPTAIN NEMO (II,2135); ROSETTI AND RYAN (II,2216); SERPICO (II,2306); SHANE (I,3995); TRAPPER JOHN, M.D. (II,2654); TURNABOUT (II,2669); THE UNTOUCHABLES (I,4682); VOYAGE TO THE BOTTOM OF THE SEA (I,4743); W.E.B. (II,2732)

MARCH, HAL
CAMP RUNAMUCK (I,811); F TROOP (I,1499)

MARCLARE, WILLIAM
THE PLAINCLOTHESMAN (I,3614)

MARCUS, SID
BAGGY PANTS AND THE NITWITS (II,135); BAILEY'S COMETS (I,333); THE BARKLEYS (I,356); COURAGEOUS CAT (I,1079); G.I. JOE: A REAL AMERICAN HERO (II,948); THE HOUNDCATS (I,2132); SPIDER-MAN (I,4154); SPIDER-MAN AND HIS AMAZING FRIENDS (II,2425); WHAT'S NEW MR. MAGOO? (II,2770); THE WOODY WOODPECKER SHOW (I,4906)

MARDEN, JOHN
SCENES FROM A MARRIAGE (II,2272)

MARGOLIN, ARNOLD
THE BOYS (I,701); THAT'S MY MAMA (II,2580); WALKIN' WALTER (II,2737)

MARGOLIN, STUART
BRET MAVERICK (II,374); CRAZY LIKE A FOX (II,601); THE HARDY BOYS MYSTERIES (II,1090); THE LOVE BOAT (II,1535); PHYLLIS (II,2038); SARA (II,2260); THE TEXAS WHEELERS (II,2568); WONDER WOMAN (SERIES 1) (II,2828); YOUNG GUY CHRISTIAN (II,2864)

MARGOLIS, JEFF
60 YEARS OF SEDUCTION (II,2373); BATTLE OF THE BEAT (II,161); THE BEATRICE ARTHUR SPECIAL (II,196); THE CAPTAIN AND TENNILLE SONGBOOK (II,432); COS (II,576); DICK CLARK'S GOOD OL' DAYS: FROM BOBBY SOX TO BIKINIS (II,671); DINAH AND HER NEW BEST FRIENDS (II,676); THE DON HO SHOW (II,706); JULIE ANDREWS: ONE STEP INTO SPRING (II,1353); KENNY ROGERS IN CONCERT (II,1381); THE MAGIC OF DAVID COPPERFIELD (II,1593); MARIE (II,1629); THE OLIVIA NEWTON-JOHN SHOW (II,1885); OLIVIA NEWTON-JOHN'S HOLLYWOOD NIGHTS (II,1886); THE OSMONDS SPECIAL (II,1929); PERRY COMO'S CHRISTMAS IN ENGLAND (II,2008); PERRY COMO'S CHRISTMAS IN NEW YORK (II,2011); PERRY COMO'S EASTER IN GUADALAJARA (II,2015); PERRY COMO'S FRENCH-CANADIAN CHRISTMAS (II,2016); THE PETER MARSHALL VARIETY SHOW (II,2029); PRESENTING SUSAN ANTON (II,2077); PUMPBOYS AND DINETTES ON TELEVISION (II,2090); PUTTIN' ON THE HITS (II,2093); RINGO (II,2176); THE SENSATIONAL, SHOCKING, WONDERFUL, WACKY 70S (II,2302); TONY ORLANDO AND DAWN (II,2639); TV FUNNIES (II,2672); TWILIGHT THEATER II (II,2686); WHATEVER BECAME OF. . . ? (II,2772)

MARIE, ALBERT
BUICK ELECTRA PLAYHOUSE (I,760)

MARKLE, FLETCHER
FRONT ROW CENTER (I,1694); JOHNNY BELINDA (I,2415); TONIGHT IN HAVANA (I,4557); YOUNG IN HEART (I,4939)

MARKOWITZ, RIFF
THE HILARIOUS HOUSE OF FRIGHTENSTEIN (II,1153)

MARKOWITZ, ROBERT
SERPICO (II,2306)

MARKS, ARTHUR
THE DUKES OF HAZZARD (II,742); PERRY MASON (II,2025); STARSKY AND HUTCH (II,2444); YOUNG DAN'L BOONE (II,2863)

MARKS, SHERMAN
BALANCE YOUR BUDGET (I,336); THE BILL GOODWIN SHOW (I,447); COSMOPOLITAN THEATER (I,1057); THE FREEWHEELERS (I,1683); THE GHOST AND MRS. MUIR (I,1786); HIDE AND SEEK (I,2047); MANY HAPPY RETURNS (I,2895); MR. O'MALLEY (I,3146); OUT OF THE BLUE (I,3422); PAPA G.I. (I,3451); THE WILD WILD WEST (I,4863)

MARLOWE, HARVEY
ABC-POWERS CHARM SCHOOL (I,5); DETECT AND COLLECT (I,1242); HOTEL BROADWAY (I,2127); THE IREENE WICKER SHOW (I,2234); TELETIPS ON LOVELINESS (I,4380)

MARQUETTE, DESMOND
THE CHILDREN'S HOUR (I,939)

MARQUEZ, PAUL
GETTING READY (II,984)

MARRERO, FRANK
THE NEW VOICE (II,1833)

MARSHALL, GARRY K
BLANSKY'S BEAUTIES (II,267); DOMINIC'S DREAM (II,704); HERNDON AND ME (II,1139); MORK AND MINDY (II,1735)

MARSHALL, GEORGE
DANIEL BOONE (I,1142); HEC RAMSEY (II,1121); HERE'S LUCY (II,1135); SCREEN DIRECTOR'S PLAYHOUSE (I,3946)

MARSHALL, LEWIS
THE ODDBALL COUPLE (II,1876)

MARSHALL, PENNY
LAVERNE AND SHIRLEY (II,1446); WORKING STIFFS (II,2834)

MARTIN, ANDREW
FLIPPER (I,1604)

MARTIN, DICK
ARCHIE BUNKER'S PLACE (II,105); THE BOB NEWHART SHOW (II,342); BROTHERS AND SISTERS (II,382); FAMILY TIES (II,819); GOODNIGHT, BEANTOWN (II,1041); HOUSE CALLS (II,1194); NEWHART (II,1835); SEMI-TOUGH (II,2300); THERE GOES THE NEIGHBORHOOD (II,2582); THE WAVERLY WONDERS (II,2746); WHITE AND RENO (II,2787)

MARTIN, RICCI
ON STAGE AMERICA (II,1893)

MARTIN, ROSS
HERE'S LUCY (II,1135)

MARTIN, SOBEY
BOSTON BLACKIE (I,695); CAVALCADE OF AMERICA (I,875); THE CISCO KID (I,961); GUNSMOKE (II,1069); LANCER (I,2610); LAND OF THE GIANTS (I,2611); LOST IN SPACE (I,2758); THE MILLIONAIRE (I,3046); PUBLIC DEFENDER (I,3687); RAWHIDE (I,3727); STORY THEATER (I,4240); THE TIME TUNNEL (I,4511); THE UNEXPECTED (I,4674); VOYAGE TO THE BOTTOM OF THE SEA (I,4743)

MARTINSON, LESLIE H
77 SUNSET STRIP (I,3988); THE ALASKANS (I,106); ALIAS SMITH AND JONES (I,118); ALICE (II,33); BARNABY JONES (II,153); BATMAN (I,366); BIG FOOT AND WILD BOY (II,237); BIG SHAMUS, LITTLE SHAMUS (II,242); BOURBON STREET BEAT (I,699); THE BRADY BUNCH (II,362); BRONCO (I,745); BUCK ROGERS IN THE 25TH CENTURY (II,383); CANNON (II,424); CHEYENNE (I,931); CHIPS (II,511); CODE R (II,551); COLT .45 (I,1010); CROSSROADS (I,1105); DALLAS (II,614); EIGHT IS ENOUGH (II,762); FBI CODE 98 (I,1550); THE GALLANT MEN (I,1728); GHOST STORY (I,1788); GILLIGAN'S ISLAND: RESCUE FROM GILLIGAN'S ISLAND (II,991); HARPER VALLEY (II,1094); HAWAIIAN EYE (I,1973); THE KID WITH THE BROKEN HALO (II,1387); LAWMAN (I,2642); MANHUNTER (II,1621); MANIMAL (II,1622); MANNIX (II,1624); MAVERICK (II,1657); THE MISADVENTURES OF SHERIFF LOBO (II,1704); MOST WANTED (II,1739); MR. ROBERTS (I,3149); NO TIME FOR SERGEANTS (I,3300); THE POWERS OF MATTHEW STAR (II,2075); QUINCY, M. E. (II,2102); THE ROARING TWENTIES (I,3805); ROOM 222 (I,3843); ROOM FOR ONE MORE (I,3842); RUN FOR YOUR LIFE (I,3871); SMALL AND FRYE (II,2380); SUGARFOOT (I,4277); SURFSIDE 6 (I,4299); TEMPLE HOUSTON (I,4392); WONDER WOMAN (SERIES 2) (II,2829); YOUNG MAVERICK (II,2867)

MARTINUS, DEREK
BLAKE'S SEVEN (II,264); PENMARRIC (II,1993)

MARTON, ANDREW
MAURICE CHEVALIER'S PARIS (I,2955)

MASON, MARSHALL W
FIFTH OF JULY (II,851)

MASSARI, DICK
MADHOUSE 90 (I,2807)

MASSARI, MARK
RONA LOOKS AT KATE, PENNY, TONI, AND CINDY (II,2206)

MASSELL, ROBERT
THE FRANCES LANGFORD-DON AMECHE SHOW (I,1655); SOLDIER PARADE (I,4109)

MASTERSON, JOHN
BRIDE AND GROOM (I,721)

MASTROGEORGE, HARRY

FROM HERE TO ETERNITY (II,933); HART TO HART (II,1102); THE MARY TYLER MOORE SHOW (II,1640); PHYLLIS (II,2038)

MATE, RUDOLPH
THE LORETTA YOUNG THEATER (I,2756); RIVAK, THE BARBARIAN (I,3795)

MATHLON, EDDIE
BRIGITTE BARDOT (I,729)

MATT, DICK
ED MCMAHON AND HIS FRIENDS. . . DISCOVER WET AT CYPRESS GARDENS (I,1400)

MAUZER, MERRILL
REAL PEOPLE (II,2118)

MAXWELL, CHARLES
THE HANK MCCUNE SHOW (I,1932)

MAXWELL, PETER
THE INVISIBLE MAN (I,2232)

MAYBERRY, ROBERT
FASHION MAGIC (I,1536)

MAYBERRY, RUSS
ALIAS SMITH AND JONES (I,118); BARETTA (II,152); THE BLACK SHEEP SQUADRON (II,262); THE BRADY BUNCH (II,362); BRONK (II,379); THE CIRCLE FAMILY (II,527); THE FALL GUY (II,811); FOOLS, FEMALES AND FUN: WHAT ABOUT THAT ONE? (II,901); HARRY O (II,1099); HERE WE GO AGAIN (I,2028); IRONSIDE (II,1246); KAZ (II,1370); KOJAK (II,1406); MANIMAL (II,1622); MARRIAGE IS ALIVE AND WELL (II,1633); MCCLOUD (II,1660); THE MISSISSIPPI (II,1707); THE OUTSIDE MAN (II,1937); THE PARTRIDGE FAMILY (II,1962); PROBE (I,3674); THE REBELS (II,2121); THE ROCKFORD FILES (II,2197); ROOSTER (II,2210); SCARECROW AND MRS. KING (II,2268); SEARCH (I,3951); SEVENTH AVENUE (II,2309); SIDNEY SHORR (II,2349); THE SIX MILLION DOLLAR MAN (I,4067); STONESTREET: WHO KILLED THE CENTERFOLD MODEL? (II,2472); THAT GIRL (II,2570); A VERY MISSING PERSON (I,4715); THE VIRGINIAN (I,4738)

MAYER, GERALD
ADVENTURES IN PARADISE (I,46); BEN CASEY (I,394); BRACKEN'S WORLD (I,703); BRENNER (I,719); CIMARRON STRIP (I,954); DAN AUGUST (I,1130); THE DEFENDERS (I,1220); THE FUGITIVE (I,1701); HAVE GUN—WILL TRAVEL (I,1968); HUNTER (II,1205); THE INVADERS (I,2229); JUDD, FOR THE DEFENSE (I,2463); LOGAN'S RUN (II,1507); LOU GRANT (II,1526); MANNIX (II,1624); THE MILLIONAIRE (I,3046); MISSION: IMPOSSIBLE (I,3067); NERO WOLFE (II,1806); NICHOLS (I,3283); THE PERSUADERS! (I,3556); PRIVATE SECRETARY (I,3672); QUINCY, M. E. (II,2102); SHANE (I,3995); SHIRLEY (II,2330); THE SWISS FAMILY ROBINSON (II,2518); SWITCH (II,2519); THRILLER (I,4492); THE VIRGINIAN (I,4738); VOYAGE TO THE BOTTOM OF THE SEA (I,4743); WESTSIDE MEDICAL (II,2765)

MAYNE, LENNIE
DOCTOR WHO (II,689)

MAZUER, MERRILL
PEOPLE (II,1995)

MCBAIN, KENNY
MACKENZIE (II,1584); THE OMEGA FACTOR (II,1888)

MCCABE, NORM
JIM HENSON'S MUPPET BABIES (II,1289); PANDAMONIUM (II,1948); RUBIK, THE AMAZING CUBE (II,2226); THE TRANSFORMERS (II,2653)

MCCAIN, ROD
COSMOS (II,579)

MCCAMMON, ROBERT R
DARKROOM (II,619)

MCCAREY, LEO
MEET THE GOVERNOR (I,2991); SCREEN DIRECTOR'S PLAYHOUSE (I,3946); TOM AND JERRY (I,4533)

MCCARTHY, DESMOND
BLAKE'S SEVEN (II,264)

MCCARTHY, WILLIAM
THE ADVENTURES OF CHAMPION (I,54); BUFFALO BILL JR. (I,755)

MCCLEERY, ALBERT
COSMOPOLITAN THEATER (I,1057); JANE WYMAN PRESENTS THE FIRESIDE THEATER (I,2345); MASTERPIECE PLAYHOUSE (I,2942)

MCCONNELL, THOMAS
A.K.A. PABLO (II,28); OPEN ALL NIGHT (II,1914)

MCCOWAN, GEORGE
BANACEK (II,138); BARNABY JONES (II,153); BRING 'EM BACK ALIVE (II,377); CANNON (I,820); CANNON (II,424); CHARLIE'S ANGELS (II,486); DAN AUGUST (I,1130); FANTASY ISLAND (II,829); THE FELONY SQUAD (I,1563); THE INVADERS (I,2229); THE LOVE BOAT (II,1535); MANHUNTER (II,1621); THE MOD SQUAD (I,3078); THE MONK (I,3087); NERO WOLFE (II,1806); THE OVER-THE-HILL GANG RIDES AGAIN (I,3432); THE RETURN OF THE MOD SQUAD (I,3776); RETURN TO FANTASY ISLAND (II,2144); S.W.A.T. (II,2236); SEARCH (I,3951); SEEING THINGS (II,2297); THE STARLOST (I,4203); STARSKY AND HUTCH (II,2444); THE STREETS OF SAN FRANCISCO (II,2478); THE UGLIEST GIRL IN TOWN (I,4663); VEGAS (II,2724)

MCCOY, SID
SOUL TRAIN (I,4131)

MCCUE, RICHARD
HOW TO SURVIVE A MARRIAGE (II,1198)

MCCULLOUGH, ANDREW
77 SUNSET STRIP (I,3988); ALCOA/GOODYEAR THEATER (I,107); BRINGING UP BUDDY (I,731); THE DONNA REED SHOW (I,1347); HAWAIIAN EYE (I,1973); LEAVE IT TO BEAVER (I,2648); LIFE WITH VIRGINIA (I,2695); MR. I. MAGINATION (I,3140); OUT THERE (I,3424); ROOM FOR ONE MORE (I,3842); THE SMOTHERS BROTHERS SHOW (I,4096); THREE WISHES (I,4488); TURN OF FATE (I,4623); THE UNTOUCHABLES (I,4682)

MCDEARMON, DAVID
GILLIGAN'S ISLAND (II,990)

MCDERMOTT, TOM
THE LAMBS GAMBOL (I,2607)

MCDONALD, FRANK
THE ADVENTURES OF CHAMPION (I,54); ANNIE OAKLEY (I,225); BROKEN ARROW (I,744); BUFFALO BILL JR. (I,755); FLIPPER (I,1604); THE GENE AUTRY SHOW (I,1748); GET SMART (II,983); MEET MCGRAW (I,2984); PONY EXPRESS (I,3645)

MCDONALD, JOHN
COUNTERATTACK: CRIME IN AMERICA (II,581)

MCDONALD, LEROY
THE WHITE SHADOW (II,2788)

MCDONALD, TOM
MR. MAGOO (I,3142)

MCDONOUGH, DICK
50 GRAND SLAM (II,852); THE BOB HOPE SHOW (I,628); BOB HOPE SPECIAL: A QUARTER CENTURY OF BOB HOPE ON TELEVISION (II,281); BOB HOPE SPECIAL: BOB HOPE IN "JOYS" (II,284); BOB HOPE SPECIAL: BOB HOPE ON CAMPUS (II,289); BOB HOPE SPECIAL: BOB HOPE'S ALL-STAR COMEDY SPECIAL FROM AUSTRALIA (II,300); BOB HOPE SPECIAL: BOB HOPE'S ALL-STAR COMEDY SPECTACULAR FROM LAKE TAHOE (II,301); BOB HOPE SPECIAL: BOB HOPE'S ALL-STAR COMEDY TRIBUTE TO VAUDEVILLE (II,302); BOB HOPE SPECIAL: BOB HOPE'S ALL-STAR TRIBUTE TO THE PALACE THEATER (II,305); BOB HOPE SPECIAL: BOB HOPE'S BICENTENNIAL STAR SPANGLED SPECTACULAR (II,306); BOB HOPE SPECIAL: BOB HOPE'S CHRISTMAS PARTY (II,307); BOB HOPE SPECIAL: BOB HOPE'S CHRISTMAS SPECIAL (II,308); BOB HOPE SPECIAL: THE BOB HOPE COMEDY SPECIAL (II,332); BOB HOPE SPECIAL: THE BOB HOPE SPECIAL (II,333); BOB HOPE SPECIAL: THE BOB HOPE SPECIAL FROM PALM SPRINGS (II,339); CELEBRITY SWEEPSTAKES (II,476); DO IT YOURSELF (I,1299); DON RICKLES—ALIVE AND KICKING (I,1340); THE DUKE (I,1381); AN EVENING WITH JIMMY DURANTE (I,1475); FAMILY NIGHT WITH HORACE HEIDT (I,1523); GENERAL ELECTRIC'S ALL-STAR ANNIVERSARY (II,963); GIVE MY REGARDS TO BROADWAY (I,1815); IT TAKES TWO (I,2251); THE LIAR'S CLUB (II,1466); ROBERTA (I,3815); ROBERTA (I,3816); SHOW BIZ (I,4027); TAKE ONE STARRING JONATHAN WINTERS (II,2532)

MCDONOUGH, RICHARD
THE GEORGE GOBEL SHOW (I,1768)

MCDOUGALL, DON
THE ALCOA HOUR (I,108); ANNIE OAKLEY (I,225); THE BARBARY COAST (II,146); THE BIONIC WOMAN (II,255); THE BLUE KNIGHT (II,276); BONANZA (II,347); BUFFALO BILL JR. (I,755); CHIPS (II,511); DALLAS (II,614); THE DUKES OF HAZZARD (II,742); GHOST STORY (I,1788); THE HARDY BOYS MYSTERIES (II,1090); IRONSIDE (II,1246); JOHNNY RISK (I,2435); JUNGLE JIM (I,2491); JUSTICE OF THE PEACE (I,2500); KINGSTON: CONFIDENTIAL (II,1398); LANCER (I,2610); MANNIX (II,1624); THE MEN FROM SHILOH (I,3005); NASHVILLE 99 (II,1791); THE NIGHT STALKER (II,1849); PLANET OF THE APES (II,2049); THE RANGE RIDER (I,3720); RAWHIDE (I,3727); STAGECOACH WEST (I,4185); STAR TREK (II,2440); SWISS FAMILY ROBINSON (II,2517); T.H.E. CAT (I,4430); TURN OF FATE (I,4623); THE VIRGINIAN (I,4738); WONDER WOMAN (SERIES 2) (II,2829); YOUNG DAN'L BOONE (II,2863); YOUNG MAVERICK (II,2867)

MCDOUGALL, DOUGLAS
ALCOA/GOODYEAR THEATER (I,107)

MCDUFFIE, BRIAN
PRISONER: CELL BLOCK H (II,2085)

MCEVEETY, BERNARD
THE A-TEAM (II,119); B.J. AND THE BEAR (II,125); BANACEK (II,138); THE BIG VALLEY (I,437); BLUE THUNDER (II,278); BRANDED (I,707); BUCK ROGERS IN THE 25TH CENTURY (II,384); CENTENNIAL (II,477); CHARLIE'S ANGELS (II,486); CIMARRON STRIP (I,954); COMBAT (I,1011); DAVID CASSIDY—MAN UNDERCOVER (II,623); THE DUKES OF HAZZARD (II,742); EIGHT IS ENOUGH (II,762); ENOS (II,779); FOR LOVE AND HONOR (II,903); GUNSMOKE (II,1069); HAWAII FIVE-O (II,1110); HOT PURSUIT (II,1189); HOW THE WEST WAS WON (II,1196); THE INCREDIBLE HULK (II,1232); KAZ (II,1370); KILLER BY NIGHT (I,2537); KNIGHT RIDER (II,1402); LAREDO (I,2615); THE MACAHANS (II,1583); MANHUNTER (II,1621); MARCUS WELBY, M.D. (II,1627); PETROCELLI (II,2031); PLANET OF THE APES (II,2049); THE QUEST (II,2097); RAWHIDE (I,3727); ROUGHNECKS (II,2219); S.W.A.T. (II,2236); THREE FOR THE ROAD (II,2607); TRAPPER JOHN, M.D. (II,2654); VEGAS (II,2724); THE VIRGINIAN (I,4738); VOYAGERS (II,2730); THE WALTONS (II,2740); THE WILD WILD WEST (I,4863); YOUNG MAVERICK (II,2867)

MCEVEETY, VINCENT
BRANDED (I,707); BUCK ROGERS IN THE 25TH CENTURY (II,384); THE BUFFALO SOLDIERS (II,388); THE BUSTERS (II,401); CIMARRON STRIP (I,954); DALLAS (II,614); DIRTY SALLY (II,680); EIGHT IS ENOUGH (II,762); THE FANTASTIC JOURNEY (II,826); FUTURE COP (II,945); THE GANGSTER CHRONICLES (II,957); GUNSMOKE (II,1069); HERBIE, THE LOVE BUG (II,1131); HOTEL (II,1192); HOW THE WEST WAS WON (II,1196); LOTTERY (II,1525); MCCLAIN'S LAW (II,1659); THE NIGHT STALKER (II,1849); PERRY MASON (II,2025); PETROCELLI (II,2031); POLICE STORY (I,3636); THE ROCKFORD FILES (II,2197); SEVEN BRIDES FOR SEVEN BROTHERS (II,2307); SHADOW OF SAM PENNY (II,2313); SKYWARD CHRISTMAS (II,2378); STAR TREK (II,2440); THE UNTOUCHABLES (I,4682); WHIZ KIDS (II,2790); THE WILD WILD WEST (I,4863); WONDER WOMAN (PILOT 1) (II,2826)

MCGOOHAN, PATRICK
COLUMBO (II,556); THE PRISONER (I,3670); RAFFERTY (II,2105); SECRET AGENT (I,3962)

MCGOWAN, STUART E
THE SILENT SERVICE (I,4049); SKY KING (I,4077)

MCGUIRE, DON
CHARLIE ANGELO (I,906); MCGHEE (I,2967)

MCKAYLE, DONALD
GOOD TIMES (II,1038)

MCKEE, WILLIAM
MR. WIZARD (I,3154)

MCKENNEY, BYRAM
CAFE DE PARIS (I,791)

MCKIMSON, BOB
BAGGY PANTS AND THE NITWITS (II,135); BAILEY'S COMETS (I,333); THE BARKLEYS (I,356); THE BUGS BUNNY/ROAD RUNNER SHOW (II,391); THE HOUNDCATS (I,2132); PORKY PIG AND FRIENDS (I,3651); SYLVESTER AND TWEETY (II,2522); WHAT'S NEW MR. MAGOO? (II,2770)

MCKINNON, BOB
THE POINTER SISTERS (II,2059)

MCLAGLEN, ANDREW V
BANACEK (II,138); BANJO HACKETT: ROAMIN' FREE (II,141); THE BLUE AND THE GRAY (II,273); CODE R (II,551); THE FANTASTIC JOURNEY (II,826); GUNSMOKE (II,1069); HAVE GUN—WILL TRAVEL (I,1968); HEC RAMSEY (II,1121); THE LOG OF THE BLACK PEARL (II,1506); PERRY MASON (II,2025); RAWHIDE (I,3727); ROYCE (II,2224); TRAVIS MCGEE (II,2656); THE VIRGINIAN (I,4738)

MCLEAN, MICHAEL
VEGAS (II,2724)

MCLEOD, NELSON
THE GINGER ROGERS SHOW (I,1805)

MCLEOD, NORMAN Z
BEN BLUE'S BROTHERS (I,393); THE LIFE OF VERNON HATHAWAY (I,2687); SCREEN DIRECTOR'S PLAYHOUSE (I,3946)

MCLEOD, VIC
BURLESQUE (I,764); SO YOU WANT TO LEAD A BAND (I,4106); WELCOME ABOARD (I,4789)

MCMAHON, JENNA
MAMA'S FAMILY (II,1610)

MCMURRAY, LINDA
LAVERNE AND SHIRLEY (II,1446)

MCNEELY, JERRY
LUCAS TANNER (II,1556); OWEN MARSHALL: COUNSELOR AT LAW (I,3435); PARIS (II,1957)

MCPHERSON, CHUCK
THE INCREDIBLE HULK (II,1232)

MCPHERSON, ROBERT
DARKROOM (II,619)

MCPHILLIPS, HUGH
THE DOCTORS (II,694)

MCRAVEN, DALE
9 TO 5 (II,1852)

MCSWAIN, GINNY
KIDD VIDEO (II,1388); POLE POSITION (II,2060)

MCWHINNIE, DONALD
LOVE IN A COLD CLIMATE (II,1539); MOLL FLANDERS (II,1720)

MCKIMSON, BOB
THE FAMOUS ADVENTURES OF MR. MAGOO (I,1524)

MEDAK, PETER
FAERIE TALE THEATER (II,807); HART TO HART (II,1102); OF MEN OF WOMEN (I,3331); THE PERSUADERS! (I,3556); REMINGTON STEELE (II,2130); RETURN OF THE SAINT (II,2141); SPACE: 1999 (II,2409); THE STRANGE REPORT (I,4248)

MEDFORD, DON
ALFRED HITCHCOCK PRESENTS (I,115); BARETTA (II,152); CANNON (II,424); CIMARRON STRIP (I,954); CITY OF ANGELS (II,540); COLOSSUS (I,1009); DAVID CASSIDY—MAN UNDERCOVER (II,623); DECOY (I,1217); THE DETECTIVES (I,1245); THE DICK POWELL SHOW (I,1269); DOCTOR KILDARE (I,1315); DUSTY (II,745); DYNASTY (II,746); EMERALD POINT, N.A.S. (II,773); THE FALL GUY (II,811); THE FBI (I,1551); FOR LOVE AND HONOR (II,903); THE FUGITIVE (I,1701); THE FUZZ BROTHERS (I,1722); GHOSTBREAKER (I,1789); THE INVADERS (I,2229); JESSICA NOVAK (II,1286); JESSIE (II,1287); KATE LOVES A MYSTERY (II,1367); KAZ (II,1370); LANCER (I,2610); M SQUAD (I,2799); THE MAN FROM U.N.C.L.E. (I,2867); MOST WANTED (II,1739); MRS. COLUMBO (II,1760); THE RIFLEMAN (I,3789); THE STREETS OF SAN FRANCISCO (II,2478); TARGET: THE CORRUPTERS (I,4360); TRAUMA CENTER (II,2655); THE TWILIGHT ZONE (I,4651); TWO MARRIAGES (II,2691); THE UNTOUCHABLES (I,4682)

MEDLINSKY, HARVEY
ANGIE (II,80); BAREFOOT IN THE PARK (II,151); THE BETTY WHITE SHOW (II,224); MORK AND MINDY (II,1735); PLAZA SUITE (II,2056); THE TONY RANDALL SHOW

(II,2640); WE'VE GOT EACH OTHER (II,2757)

MEIER, DON
GARROWAY AT LARGE (I,1737)

MELENDEZ, BILL
THE CHARLIE BROWN AND SNOOPY SHOW (II,484); A CHARLIE BROWN CHRISTMAS (I,908); A CHARLIE BROWN THANKSGIVING (I,909); CHARLIE BROWN'S ALL STARS (I,910); HE'S YOUR DOG, CHARLIE BROWN (I,2037); IT WAS A SHORT SUMMER, CHARLIE BROWN (I,2252); IT'S AN ADVENTURE, CHARLIE BROWN (I,2266); IT'S FLASHBEAGLE, CHARLIE BROWN (I,2268); IT'S THE GREAT PUMPKIN, CHARLIE BROWN (I,2276); PLAY IT AGAIN, CHARLIE BROWN (I,3617); SNOOPY'S GETTING MARRIED, CHARLIE BROWN (I,4102); THERE'S NO TIME FOR LOVE, CHARLIE BROWN (I,4438); WHAT A NIGHTMARE, CHARLIE BROWN (I,4806); WHAT HAVE WE LEARNED, CHARLIE BROWN? (I,4810); YES, VIRGINIA, THERE IS A SANTA CLAUS (II,2849); YOU'RE IN LOVE, CHARLIE BROWN (I,4964); YOU'RE NOT ELECTED, CHARLIE BROWN (I,4967)

MELER, DON
THE QUIZ KIDS (I,3712)

MELLING, RICHARD
THE BEACH BOYS 20TH ANNIVERSARY (II,188)

MELMAN, JEFFREY
MAKING THE GRADE (II,1606); NIGHT COURT (II,1844); PARK PLACE (II,1958)

MENDELSON, LEE
THE FANTASTIC FUNNIES (II,825); HAPPY ANNIVERSARY, CHARLIE BROWN (I,1939)

MENKIN, LAWRENCE
HANDS OF MURDER (I,1929); ONE MAN'S EXPERIENCE (I,3382); ONE WOMAN'S EXPERIENCE (I,3391)

MERRILL, KEITH
THE CHEROKEE TRAIL (II,500)

METCALFE, BURT
AFTERMASH (II,23); M*A*S*H (II,1569)

METTER, ALAN
STEVE MARTIN'S THE WINDS OF WHOOPIE (II,2461)

MICHAELS, RICHARD
BEWITCHED (I,418); BIG HAWAII (II,238); THE BRADY BUNCH (II,362); CHARLIE COBB: NICE NIGHT FOR A HANGING (II,485); DELVECCHIO (II,659); THE FLYING NUN (I,1611); HAVING BABIES II (II,1108); JESSIE (II,1287); KELLY'S KIDS (I,2523); ONCE AN EAGLE (II,1897); ROOM 222 (I,3843)

MILESTONE, LEWIS
HAVE GUN—WILL TRAVEL (I,1968)

MILLAND, RAY
THE DICK POWELL SHOW (I,1269)

MILLAR, STUART
FAMILY (II,813); M*A*S*H (II,1569)

MILLER, ALLEN F
FRANK MERRIWELL (I,1659)

MILLER, BARBARA
MAKE A WISH (I,2838);

MILLER, GEORGE
AGAINST THE WIND (II,24); ALL THE RIVERS RUN (II,43)

MILLER, HARVEY
BUSTING LOOSE (II,402); THE ODD COUPLE (II,1875)

MILLER, J PHILIP
MUGGSY (II,1761)

MILLER, JAMES M
WHAT REALLY HAPPENED TO THE CLASS OF '65? (II,2768)

MILLER, JAY
NOT FOR WOMEN ONLY (I,3310)

MILLER, JOHN E
ALIVE AND WELL (II,36)

MILLER, JONATHAN
BLAKE'S SEVEN (II,264)

MILLER, LEE
ANIMALS ARE THE FUNNIEST PEOPLE (II,81); HOLLYWOOD STARS' SCREEN TESTS (II,1165); INSIDE AMERICA (II,1234)

MILLER, PAUL
THE BEST OF SULLIVAN (II,217); BUCKSHOT (II,385); A CHRISTMAS CAROL (II,517); THE LOVE CONNECTION (II,1536); MADAME'S PLACE (II,1587); THE WOLFMAN JACK RADIO SHOW (II,2816)

MILLER, PETER
BERGERAC (II,209)

MILLER, ROBERT ELLIS
ALCOA/GOODYEAR THEATER (I,107); AMOS BURKE: WHO KILLED JULIE GREER? (I,181); BEN CASEY (I,394); THE BOB HOPE CHRYSLER THEATER (I,502); BURKE'S LAW (I,762); THE DESILU PLAYHOUSE (I,1237); THE DICK POWELL SHOW (I,1269); THE DONNA REED SHOW (I,1347); THE HERO (I,2035); THE HUMAN COMEDY (I,2158); M SQUAD (I,2799); THE MUSIC MAKER (I,3173); THE ROGUES (I,3832); ROUTE 66 (I,3852); THE SLIGHTLY FALLEN ANGEL (I,4087)

MILLER, SHARRON
THE LIFE AND TIMES OF GRIZZLY ADAMS (II,1469)

MILLER, SIDNEY
THE ADDAMS FAMILY (II,11); THE ANN SOTHERN SHOW (I,220); THE APARTMENT HOUSE (I,241); THE GENE KELLY SHOW (I,1751); GET SMART (II,983); THE MICKEY MOUSE CLUB (I,3025); MY FAVORITE MARTIAN (I,3189); PLEASE DON'T EAT THE DAISIES (I,3628); THE REAL MCCOYS (I,3741); RODNEY DANGERFIELD SHOW: I CAN'T TAKE IT NO MORE (II,2198); THE SKATEBIRDS (II,2375); THE SMOTHERS BROTHERS SHOW (I,4096); TAMMY (I,4357); THAT GIRL (II,2570); TIGHTROPE (I,4500)

MILLER, STEPHEN
THE LOVE REPORT (II,1543)

MILLER, WALTER C
100 YEARS OF GOLDEN HITS (II,1905); THE ALAN KING SHOW (I,101); THE ALAN KING SHOW (I,102); ALL COMMERCIALS—A STEVE MARTIN SPECIAL (II,37); ANNIE, THE WOMAN IN THE LIFE OF A MAN (I,226); THE BELLE OF 14TH STREET (I,390); THE BIG SHOW (II,243); THE BORROWERS (I,693); BOWZER (II,358); CIRCUS OF THE STARS (II,538); COMEDY IS KING (I,1019); COMEDY TONIGHT (I,1027); COS: THE BILL COSBY COMEDY SPECIAL (II,577); COUNTRY COMES HOME (II,585); COUNTRY COMES HOME (II,586); COUNTRY MUSIC HIT PARADE (I,1069); COUNTRY MUSIC HIT PARADE (II,588); COUNTRY MUSIC HIT PARADE (II,589); DETECTIVE SCHOOL (II,661); THE DONNY AND MARIE CHRISTMAS SPECIAL (II,713); DOUG HENNING'S WORLD OF MAGIC I (II,725); DOUG HENNING'S WORLD OF MAGIC II (II,726); DOUG HENNING'S WORLD OF MAGIC IV (II,728); DOUG HENNING'S WORLD OF MAGIC V (II,729); DOUG HENNING: MAGIC ON BROADWAY (II,730); ELECTRA WOMAN AND DYNA GIRL (II,764); FAR OUT SPACE NUTS (II,831); FIFTY YEARS OF COUNTRY MUSIC (II,853); THE FUN FACTORY (II,938); GEORGE BURNS AND OTHER SEX SYMBOLS (II,966); GEORGE BURNS CELEBRATES 80 YEARS IN SHOW BUSINESS (II,967); GEORGE BURNS EARLY, EARLY, EARLY, CHRISTMAS SHOW (II,968); GEORGE BURNS IN NASHVILLE?? (II,969); GEORGE M! (I,1772); GEORGE M! (II,976); HANGING IN (II,1078); I BELIEVE IN MUSIC (I,2165); JACK LEMMON IN 'S WONDERFUL, 'S MARVELOUS, 'S GERSHWIN (I,2313); JOHN DENVER: MUSIC AND THE MOUNTAINS (II,1318); JOHN DENVER—THANK GOD I'M A COUNTRY BOY (II,1317); JOHNNY CASH AND FRIENDS (II,1322); JOHNNY CASH AND THE COUNTRY GIRLS (II,1323); THE JOHNNY CASH CHRISTMAS SPECIAL (II,1325); THE JOHNNY CASH CHRISTMAS SPECIAL (II,1326); A JOHNNY CASH CHRISTMAS (II,1327); A JOHNNY CASH CHRISTMAS (II,1328); THE JOHNNY CASH SHOW (I,2424); THE JOHNNY CASH SPRING SPECIAL (II,1329); JOHNNY CASH: SPRING FEVER (II,1335); JOHNNY CASH: THE FIRST 25 YEARS (II,1336); THE MAC DAVIS CHRISTMAS SPECIAL (II,1573); MAC DAVIS SPECIAL: THE MUSIC OF CHRISTMAS (II,1578); MAGIC WITH THE STARS (II,1599); THE OSMOND FAMILY CHRISTMAS SPECIAL (II,1926); THE OSMOND FAMILY SHOW (II,1927); THE OSMOND FAMILY THANKSGIVING SPECIAL (II,1928); THE RICH LITTLE SHOW (II,2154); RODNEY DANGERFIELD SPECIAL: IT'S NOT EASY BEIN' ME (II,2199); ROY ACUFF—50 YEARS THE KING OF COUNTRY MUSIC

(II,2223); SHA NA NA (II,2311); SINATRA: CONCERT FOR THE AMERICAS (II,2361); WHEN THE WEST WAS FUN: A WESTERN REUNION (II,2780); THE WORLD OF MAGIC (II,2838); YOU'RE A GOOD MAN, CHARLIE BROWN (I,4962)

MILLS, TED
THE SATURDAY NIGHT REVUE (I,3921)

MILTON, RICHARD
THE CURSE OF DRACULA (II,611); THE INCREDIBLE HULK (II,1232); MARCUS WELBY, M.D. (II,1627)

MINER, ALLEN H
THE CATCHER (I,872); THE DESILU PLAYHOUSE (I,1237); THE ELISABETH MCQUEENEY STORY (I,1429); M SQUAD (I,2799); MACREEDY'S WOMAN (I,2804); THE STAR MAKER (I,4193); TIGHTROPE (I,4500); THE UNTOUCHABLES (I,4682); WAGON TRAIN (I,4747)

MINER, LESLIE
JOAN BAEZ (II,1294)

MINER, TONY
FACE THE MUSIC (I,1509)

MINER, WORTHINGTON
AUNT JENNY'S REAL LIFE STORIES (I,314); FACE THE MUSIC (I,1509); MEDIC (I,2976)

MINNIX, BRUCE
HIGH HOPES (II,1144); SOMERSET (I,4115); TEXAS (II,2566)

MINOR, PETER
ONE LIFE TO LIVE (II,1907)

MINOR, PHILIP
HOUSE CALLS (II,1194)

MISCHER, BOB
OMNIBUS (II,1890)

MISCHER, DON
AIN'T MISBEHAVIN' (II,25); ALAN KING'S THIRD ANNUAL FINAL WARNING!! (II,31); THE BARBARA WALTERS SPECIAL (II,145); BARYSHNIKOV IN HOLLYWOOD (II,157); THE BEST OF TIMES (II,219); BOB HOPE SPECIAL: BOB HOPE GOES TO COLLEGE (II,283); BOB HOPE SPECIAL: HAPPY BIRTHDAY, BOB! (II,325); CHERYL LADD. . . LOOKING BACK—SOUVENIRS (II,502); CHERYL LADD: SCENES FROM A SPECIAL (II,503); THE DONNA SUMMER SPECIAL (II,711); AN EVENING WITH ROBIN WILLIAMS (II,789); FAMOUS LIVES (II,821); GOLDIE AND LIZA TOGETHER (II,1022); THE GOLDIE HAWN SPECIAL (II,1024); JOHN DENVER AND THE LADIES (II,1312); JUMP (II,1356); KLEIN TIME (II,1401); LAUGH-IN (II,1444); LYNDA CARTER: BODY AND SOUL (II,1565); SATURDAY NIGHT LIVE WITH HOWARD COSELL (II,2262); SHIRLEY MACLAINE: ILLUSIONS (II,2336); SMALL WORLD (II,2381); THAT WAS THE YEAR THAT WAS (II,2575); THE THIRD BARRY MANILOW SPECIAL (II,2590)

MISELOW, ALVIN
DREAM HOUSE (I,1376)

MITCHELL, JOHN W
FIGHT BACK: WITH DAVID HOROWITZ (II,854); WHITNEY AND THE ROBOT (II,2789)

MITCHELL, SUZANNE
THE 36 MOST BEAUTIFUL GIRLS IN TEXAS (II,2594)

MITCHLLL, SCOEY
ME AND MRS. C. (II,1673)

MOATES, BILL
MR. ROGERS' NEIGHBORHOOD (I,3150)

MODER, DICK
CHARLIE'S ANGELS (II,486); THE DUKES OF HAZZARD (II,742); THE GREATEST AMERICAN HERO (II,1060); KEEPER OF THE WILD (II,1373); KINGSTON: CONFIDENTIAL (II,1398); LASSIE (I,2622); LOST IN SPACE (I,2758); PROJECT UFO (II,2088); SAM (II,2247); THE SIX-MILLION-DOLLAR MAN (II,2372); STARSKY AND HUTCH (II,2444); SWITCH (II,2519); TIMMY AND LASSIE (I,4520); WONDER WOMAN (SERIES 2) (II,2829)

MODER, MIKE
THE WILD WILD WEST (I,4863)

MODER, RICHARD
JOHNNY RINGO (I,2434); ZANE GREY THEATER (I,4979)

MOESSINGER, DAVID
ALIAS SMITH AND JONES (I,118); APPLE'S WAY (II,101); BUCK ROGERS IN THE 25TH CENTURY (II,383); EIGHT IS ENOUGH (II,762); THE FANTASTIC JOURNEY (II,826); IT CAN'T HAPPEN TO ME (II,1250); KAZ (II,1370); THE LAWYERS (I,2646); THE MAN FROM ATLANTIS (II,1615); MARCUS WELBY, M.D. (II,1627); MOBILE ONE (II,1717); MOBILE TWO (II,1718); POLICE WOMAN (II,2063); QUINCY, M. E. (II,2102); SUPERTRAIN (II,2504)

MOFFATT, PETER
COLOR HIM DEAD (I,1007); COLOR HIM DEAD (II,554); CRIMES OF PASSION (II,603); DOCTOR WHO—THE FIVE DOCTORS (II,690)

MOFFITT, JOHN
THE BURNS AND SCHREIBER COMEDY HOUR (I,768); THE CAPTAIN AND TENNILLE IN HAWAII (II,430); THE CAPTAIN AND TENNILLE IN NEW ORLEANS (II,431); CLOWNAROUND (I,980); DEAN MARTIN'S COMEDY WORLD (II,647); DICK CLARK'S LIVE WEDNESDAY (II,672); DON'T CALL ME MAMA ANYMORE (I,1350); EASY DOES IT. . . STARRING FRANKIE AVALON (II,754); THE FIFTH DIMENSION SPECIAL: AN ODYSSEY IN THE COSMIC UNIVERSE OF PETER MAX (I,1571); FRIDAYS (II,926); FUNNY PAPERS (I,1713); THE GEORGE SEGAL SHOW (II,977); GOOD VIBRATIONS FROM CENTRAL PARK (I,1854); THE HELEN REDDY SPECIAL (II,1126); THE ICE PALACE (I,2182); LILY (II,1477); LOLA (II,1508); LOLA (II,1510); LOLA (II,1511); THE MCLEAN STEVENSON SHOW (II,1664); MITZI...THE FIRST TIME (I,3073); MOVIN' WITH NANCY ON STAGE (I,3120); NOT NECESSARILY THE NEWS (II,1862); NOT NECESSARILY THE NEWS (II,1863); THE PERRY COMO WINTER SHOW (I,3543); PERRY COMO'S WINTER SHOW (I,3547); THE RICHARD PRYOR SHOW (II,2163); THE RICHARD PRYOR SPECIAL (II,2164); THE RICHARD PRYOR SPECIAL? (II,2165); SAMMY AND COMPANY (II,2248); THE SOUPY SALES SHOW (I,4138); VAN DYKE AND COMPANY (II,2719)

MONTAGNE, EDWARD J
THE MAN AGAINST CRIME (I,2853); MCHALE'S NAVY (I,2969)

MONTAINO, GENNARO
BARNEY MILLER (II,154)

MONTALBAN, RICARDO
FANTASY ISLAND (II,829)

MONTALDO, GIULIANO
MARCO POLO (II,1626)

MONTGOMERY, LEONARD
MUSIC WORLD (II,1775)

MONTGOMERY, ROY
ESTHER WILLIAMS AT CYPRESS GARDENS (I,1465)

MONTGOMERY, THOMAS
DOBIE GILLIS (I,1302)

MONTY, GLORIA
BRIGHT PROMISE (I,727); GENERAL HOSPITAL (II,964); ONE DAY AT A TIME (II,1900)

MOODY, LAURENCE
BERGERAC (II,209)

MOORE, IRVING J
77 SUNSET STRIP (I,3988); THE ANDROS TARGETS (II,76); BONANZA (II,347); DALLAS (II,614); DIRTY SALLY (II,680); THE DORIS DAY SHOW (I,1355); DYNASTY (II,746); EIGHT IS ENOUGH (II,762); THE FANTASTIC JOURNEY (II,826); THE GREEN FELT JUNGLE (I,1885); GUNSMOKE (II,1069); HAWAIIAN EYE (I,1973); HERE COME THE BRIDES (I,2024); HOGAN'S HEROES (I,2069); KORG: 70,000 B.C. (II,1408); LAWMAN (I,2642); THE LIFE AND TIMES OF GRIZZLY ADAMS (II,1469); LOGAN'S RUN (II,1507); LOST IN SPACE (I,2758); MAVERICK (II,1657); PERRY MASON (II,2025); PETROCELLI (II,2031); THE QUEST (II,2097); THE ROARING TWENTIES (I,3805); SURFSIDE 6 (I,4299); SWEEPSTAKES (II,2514); TEMPLE HOUSTON (I,4392); TIGHTROPE (I,4500); THE WILD WILD WEST (I,4863); THE YOUNG PIONEERS (II,2869)

MOORE, KINGMAN T
KOBB'S CORNER (I,2577); LIGHT'S OUT (I,2699); MANHATTAN SHOWCASE (I,2888); THE MARCH OF DIMES BENEFIT SHOW (I,2900); THE MOREY AMSTERDAM SHOW (I,3100); THE MOREY AMSTERDAM SHOW (I,3101)

MOORE, MARY TYLER

THE MARY TYLER MOORE SHOW (II,1640)

MOORE, ROBERT
ANNIE FLYNN (II,92); DOC (II,683); DON'T CALL US (II,709); FIT FOR A KING (II,866); HAPPY ENDINGS (II,1085); IT'S NOT EASY (II,1256); THE NATURAL LOOK (II,1799); THE SUNSHINE BOYS (II,2496); THERE'S ALWAYS ROOM (II,2583); WILD ABOUT HARRY (II,2797)

MOORE, ROGER
THE PERSUADERS! (I,3556); THE SAINT (II,2237)

MOORE, TOM
FRIDAYS (II,926)

MORDENTE, TONY
ANGIE (II,80); THE ASSOCIATES (II,113); BAKER'S DOZEN (II,136); BENSON (II,208); THE BRADY BRIDES (II,361); BROTHERS AND SISTERS (II,382); BUSTING LOOSE (II,402); COUSINS (II,595); FAMILY TIES (II,819); FLATBUSH (II,877); FLO (II,891); GOODTIME GIRLS (II,1042); HARDCASTLE AND MCCORMICK (II,1089); THE HARLEM GLOBETROTTERS POPCORN MACHINE (II,1092); HOT L BALTIMORE (II,1187); I'M A BIG GIRL NOW (II,1218); KEEP ON TRUCKIN' (II,1372); LOVE, SIDNEY (II,1547); M*A*S*H (II,1569); THE NEW LORENZO MUSIC SHOW (II,1821); NOBODY'S PERFECT (II,1858); THE PRACTICE (II,2076); PRIVATE BENJAMIN (II,2087); RHODA (II,2151); THE RITA MORENO SHOW (II,2181); ROMANCE THEATER (II,2204); THE STEVE LANDESBERG TELEVISION SHOW (II,2458); THAT SECOND THING ON ABC (II,2572); THREE GIRLS THREE (II,2608); THE TONY RANDALL SHOW (II,2640); WE'VE GOT EACH OTHER (II,2757); WHO'S ON CALL? (II,2791); YOUNG HEARTS (II,2865)

MORGAN, ANDREW
BLAKE'S SEVEN (II,264)

MORGAN, EDWARD
UNCLE MISTLETOE AND HIS ADVENTURES (I,4669)

MORGAN, HARRY
ADAM-12 (I,31); THE D.A. (I,1123); HEC RAMSEY (II,1121); M*A*S*H (II,1569)

MORGAN, MONTY
GIRL TALK (I,1810)

MORRIS, HOWARD
ALOHA PARADISE (II,58); THE BEATRICE ARTHUR SPECIAL (II,196); THE DICK VAN DYKE SHOW (I,1275); THE GOOD OLD DAYS (I,1851); HOGAN'S HEROES (I,2069); LAVERNE AND SHIRLEY (II,1446); THE LOVE BOAT (II,1535); MIGHTY ORBOTS (II,1697); ONE DAY AT A TIME (II,1900); PLEASE DON'T EAT THE DAISIES (I,3628); PRIVATE BENJAMIN (II,2087); TURBO-TEEN (II,2668)

MORRIS, JEREMIAH
FISH (II,864); A NEW KIND OF FAMILY (II,1819); QUINCY, M. E. (II,2102); SZYSZNYK (II,2523); TOO CLOSE FOR COMFORT (II,2642)

MORRIS, JOHN
MIDWESTERN HAYRIDE (I,3033)

MORRIS, MALCOLM
TOP OF THE WORLD (II,2646)

MORRIS, MARTIN
ONE MAN SHOW (I,3381); THE ROWAN AND MARTIN REPORT (II,2221); THOSE AMAZING ANIMALS (II,2601)

MORRIS, MICHAEL
PLEASE DON'T EAT THE DAISIES (I,3628)

MORRIS, RICHARD
THE LORETTA YOUNG THEATER (I,2756)

MORRIS, STEWART
THE SHIRLEY BASSEY SPECIAL (II,2331)

MORRISON, ROBERT
HAWAII FIVE-O (II,1110)

MORROW, VIC
COMBAT (I,1011); QUINCY, M. E. (II,2102)

MORSE, BARRY
THE FUGITIVE (I,1701); MR. DICKENS OF LONDON (I,3135)

MORSE, CARLETON E
MIXED DOUBLES (I,3075)

MORSE, HOLLINGSWORTH
ADAM-12 (I,31); BROKEN ARROW (I,744); CRASH ISLAND (II,600); THE D.A. (I,1123); DELTA HOUSE (II,658); THE DUKES OF HAZZARD (II,742); ENOS (II,779); F TROOP (I,1499); FLIPPER (I,1604); THE GHOST AND MRS. MUIR (I,1786); H.R. PUFNSTUF (I,2154); HERE COME THE STARS (I,2026); ISIS (II,1248); LARAMIE (I,2614); LUCAN (II,1554); MARCUS WELBY, M.D. (II,1627); MCHALE'S NAVY (I,2969); THE MEN FROM SHILOH (I,3005); MULLIGAN'S STEW (II,1764); THE NEW ADVENTURES OF HUCKLEBERRY FINN (I,3250); THE NEW OPERATION PETTICOAT (II,1826); NO TIME FOR SERGEANTS (I,3300); OPERATION PETTICOAT (II,1917); THE OREGON TRAIL (II,1923); PLEASE DON'T EAT THE DAISIES (I,3628); THE RAT PATROL (I,3726); ROCKY JONES, SPACE RANGER (I,3823); SHAZAM! (II,2319); THE SKATEBIRDS (II,2375); THE VIRGINIAN (I,4738); WATERFRONT (I,4774); WHIZ KIDS (II,2790)

MORSE, JOHN H
THE LONE RANGER (I,2740); SKY KING (I,4077); SONGS FOR SALE (I,4124); ZORRO (I,4982)

MORSE, ROBERT
PAUL SAND IN FRIENDS AND LOVERS (II,1982)

MOSES, GILBERT
PALMERSTOWN U.S.A. (II,1945); ROOTS (II,2211)

MOSES, HARRY
THE MIKE WALLACE PROFILES (II,1699)

MOSS, JEFFREY B
THE DRUNKARD (II,737)

MOXEY, JOHN LLEWELLYN
THE AVENGERS (II,121); THE BARON (I,360); CHARLIE'S ANGELS (II,486); CHARLIE'S ANGELS (II,487); CIRCLE OF FEAR (I,958); CONSPIRACY OF TERROR (II,569); DIAL "M" FOR MURDER (I,1255); EBONY, IVORY AND JADE (II,755); ESCAPE (I,1460); GENESIS II (I,1760); GHOST STORY (I,1788); A HATFUL OF RAIN (I,1964); HAWAIIAN HEAT (II,1111); JUDD, FOR THE DEFENSE (I,2463); KUNG FU (II,1416); LAURA (I,2636); LEGMEN (II,1458); MANNIX (II,1624); MASQUERADE (II,1644); MIAMI VICE (II,1689); MISSION: IMPOSSIBLE (I,3067); MURDER, SHE WROTE (II,1770); THE NAME OF THE GAME (I,3217); THE NIGHT STALKER (I,3292); THE POWER WITHIN (II,2073); THE RETURN OF LUTHER GILLIS (II,2137); THE SAINT (II,2237); SAN FRANCISCO INTERNATIONAL AIRPORT (I,3905); SAN FRANCISCO INTERNATIONAL AIRPORT (I,3906); SANCTUARY OF FEAR (II,2252); SCARECROW AND MRS. KING (II,2268); SHAFT (I,3993)

MUIR, ROGER
THE AT LIBERTY CLUB (I,309); THE BOB SMITH SHOW (I,668)

MULLEN, MARJORIE
THE MARY TYLER MOORE SHOW (II,1640)

MULLIGAN, ROBERT
THE ALCOA HOUR (I,108); THE DEFENDER (I,1219); MEMBER OF THE WEDDING (I,3002); THE MOON AND SIXPENCE (I,3098); VICTOR BORGE'S COMEDY IN MUSIC I (I,4723); VICTOR BORGE'S COMEDY IN MUSIC II (I,4724)

MUNROE, PAUL
THE JACK CARTER SHOW (I,2306)

MUNTNER, SIMON
KAREN (II,1363)

MURPHY, JACK
GOOD COMPANY (I,1844)

MURPHY, RALPH FRANCIS
BROKEN ARROW (I,744); CROSSROADS (I,1105); DOBIE GILLIS (I,1302); THE MCGONIGLE (I,2968); MEDIC (I,2976); MR. AND MRS. NORTH (I,3124)

MURRAY, JOHN
THE DENNIS O'KEEFE SHOW (I,1230); GILLIGAN'S ISLAND (II,990)

MURRAY, KEN
EL COYOTE (I,1423); WHERE WERE YOU? (I,4834)

MUSCARELLA, STEVE
SANTA BARBARA (II,2258)

MYERSON, ALAN
THE BAD NEWS BEARS (II,134); THE BOB NEWHART SHOW (II,342); THE BRADY BRIDES (II,361); BUSTING LOOSE (II,402); CALIFORNIA FEVER (II,411); CAR WASH (II,437); CUTTER TO HOUSTON (II,612); DELTA HOUSE (II,658); THE DICK CAVETT SHOW (II,669); FLYING HIGH (II,894); FOUL PLAY (II,910); GUN SHY (II,1068); HARPER VALLEY (II,1094); HARPER VALLEY PTA (II,1095); THE HARVEY KORMAN SHOW (II,1103); THE

HARVEY KORMAN SHOW (II,1104); HUSBANDS, WIVES AND LOVERS (II,1209); KNIGHT RIDER (II,1402); THE KROFFT KOMEDY HOUR (II,1411); LAVERNE AND SHIRLEY (II,1446); LEWIS AND CLARK (II,1465); THE LOVE BOAT I (II,1532); THE MASTER (II,1648); THE MCLEAN STEVENSON SHOW (II,1665); MR. MERLIN (II,1751); A NEW KIND OF FAMILY (II,1819); PLEASE STAND BY (II,2057); PRIVATE BENJAMIN (II,2087); SHIRLEY (II,2330); SZYSZNYK (II,2523); THE TV TV SHOW (II,2674); VOYAGERS (II,2730); A YEAR AT THE TOP (II,2845); ZORRO AND SON (II,2878)

MYHRUM, ROBERT
AS THE WORLD TURNS (II,110); HOW TO SURVIVE A MARRIAGE (II,1198); RYAN'S HOPE (II,2234); SESAME STREET (I,3982)

NADEL, ARTHUR H
THE BIG VALLEY (I,437); ISIS (II,1248); JASON OF STAR COMMAND (II,1274); THE KID SUPER POWER HOUR WITH SHAZAM (II,1386); THE RIFLEMAN (I,3789); SHAZAM! (II,2319); SPACE ACADEMY (II,2406)

NAGY, IVAN
CAPTAIN AMERICA (II,428); DELVECCHIO (II,659); THE HITCHHIKER (II,1157); ONCE UPON A SPY (II,1898); THE POWERS OF MATTHEW STAR (II,2075); STARSKY AND HUTCH (II,2444)

NARIZZANO, SILVIO
TWENTY-FOUR HOURS IN A WOMAN'S LIFE (I,4644)

NASELLA, JIM
SQUARE PEGS (II,2431)

NATHAN, MARC
THE UNCLE FLOYD SHOW (II,2703)

NAZZARO, RAY
ANNIE OAKLEY (I,225); THE GENE AUTRY SHOW (I,1748)

NEEDHAM, HAL
THE STOCKERS (II,2469); STUNTS UNLIMITED (II,2487)

NEILSON, JAMES
ADAM-12 (I,31); ADVENTURES IN PARADISE (I,46); ALFRED HITCHCOCK PRESENTS (I,115); BRACKEN'S WORLD (I,703); CIRCLE OF FEAR (I,958); FATHER KNOWS BEST (II,838); HAVE GUN—WILL TRAVEL (I,1968); IRONSIDE (II,1246); LAW OF THE PLAINSMAN (I,2639); LEAVE IT TO BEAVER (I,2648); M SQUAD (I,2799); NIGHT PROWL (I,3291); PARIS 7000 (I,3459); THE VIRGINIAN (I,4738); WAGON TRAIN (I,4747); WE'LL TAKE MANHATTAN (I,4792)

NELSON, CRAIG RICHARD
SQUARE PEGS (II,2431)

NELSON, DAVID
GOODNIGHT, BEANTOWN (II,1041)

NELSON, DICK
THE FBI (I,1551); SLATTERY'S PEOPLE (I,4082)

NELSON, GARY
ACE (II,5); ADAM'S RIB (I,32); THE BOYS IN BLUE (II,359); BRACKEN'S WORLD (I,703); CALL HOLME (II,412); THE COP AND THE KID (II,572); CRAZY LIKE A FOX (II,601); FARADAY AND COMPANY (II,833); FOR LOVE AND HONOR (II,903); GET SMART (II,983); GOMER PYLE, U.S.M.C. (I,1843); THE GOOD GUYS (I,1847); HAVE GUN—WILL TRAVEL (I,1968); HAWAII FIVE-O (II,1110); THE HERO (I,2035); HUNTER (II,1205); MCCLAIN'S LAW (II,1659); MCMILLAN AND WIFE (II,1667); MEDICAL STORY (II,1676); MICKE Y SPILLANE'S MIKE HAMMER: MORE THAN MURDER (II,1693); MICKEY SPILLANE'S MIKE HAMMER: MURDER ME, MURDER YOU (II,1694); THE MOD SQUAD (I,3078); THE OUTER LIMITS (I,3426); THE PARTNERS (I,3462); THE PATTY DUKE SHOW (I,3483); PLEASE DON'T EAT THE DAISIES (I,3628); POLICE STORY (II,2062); REVENGE OF THE GRAY GANG (II,2146); SEVEN BRIDES FOR SEVEN BROTHERS (II,2307); SHANE (I,3995); TOMA (II,2634); WASHINGTON: BEHIND CLOSED DOORS (II,2744)

NELSON, GENE
THE BAD NEWS BEARS (II,134); BURKE'S LAW (I,762); THE CARA WILLIAMS SHOW (I,843); DAN AUGUST (I,1130); FANTASY ISLAND (II,829); THE FARMER'S DAUGHTER (I,1533); THE FBI (I,1551); THE FELONY SQUAD (I,1563); GET CHRISTIE LOVE! (II,981); I AND CLAUDIE (I,2164); I DREAM OF JEANNIE (I,2167); LANCER (I,2610); THE LETTERS (I,2670); MCNAUGHTON'S DAUGHTER (II,1670); THE MOD SQUAD (I,3078); THE NEW OPERATION PETTICOAT (II,1826); SALVAGE 1 (II,2245); THE SAN PEDRO BEACH BUMS (II,2250); SHIRLEY (II,2330); WHERE'S EVERETT? (I,4835)

NELSON, MERVYN
LEAVE IT TO LARRY (I,2649)

NELSON, OZZIE
ADAM-12 (I,31); THE ADVENTURES OF OZZIE AND HARRIET (I,71); THE D.A. (I,1123); OZZIE'S GIRLS (I,3439); OZZIE'S GIRLS (I,3440)

NELSON, RALPH
AUCTION AIRE (I,313); CELEBRITY TIME (I,887); CHRYSLER MEDALLION THEATER (I,951); THE DESILU PLAYHOUSE (I,1237); THE DICK POWELL SHOW (I,1269); THE FARMER'S DAUGHTER (I,1533); THE HAPPY TIME (I,1943); HOLD THAT CAMERA (I,2072); I REMEMBER MAMA (I,2176); THE JAZZ SINGER (I,2351); JUNIOR MISS (I,2493); JUSTICE (I,2498); SAY IT WITH ACTING (I,3928); SHOW BUSINESS, INC. (I,4028)

NELSON, RAYMOND E
TELEVISION FASHION FAIR (I,4381)

NEUFELD JR, SIGMUND
240-ROBERT (II,2688); BUCK ROGERS IN THE 25TH CENTURY (II,383); THE CHISHOLMS (II,513); CITY OF ANGELS (II,540); THE EDDIE CAPRA MYSTERIES (II,758); FINDER OF LOST LOVES (II,857); FLYING HIGH (II,894); GALACTICA 1980 (II,953); HARPER VALLEY PTA (II,1095); HERE'S BOOMER (II,1134); THE INCREDIBLE HULK (II,1232); THE INVISIBLE MAN (II,1242); KATE LOVES A MYSTERY (II,1367); KOJAK (II,1406); PARTNERS IN CRIME (II,1961); PROJECT UFO (II,2088); SCARECROW AND MRS. KING (II,2268); SERPICO (II,2306); SIMON AND SIMON (II,2357); SWITCH (II,2519); T.J. HOOKER (II,2524); THUNDER (II,2617); VOYAGERS (II,2730); WONDER WOMAN (SERIES 2) (II,2829)

NEUFELD, SAM
HAWKEYE AND THE LAST OF THE MOHICANS (I,1975)

NEUFELD, SIGMUND
HAWKEYE AND THE LAST OF THE MOHICANS (I,1975)

NEUMAN, ALAN
DAGMAR'S CANTEEN (I,1125); FAIRMEADOWS, U.S.A. (I,1514); MENASHA THE MAGNIFICENT (I,3010)

NEWLAND, JOHN
BACHELOR FATHER (I,323); DANIEL BOONE (I,1142); THE DEFENDERS (I,1220); EXECUTIVE SUITE (II,796); THE FAMILY HOLVAK (II,817); FANTASY ISLAND (II,829); FLAMINGO ROAD (II,872); HARRY O (II,1099); HORNBLOWER (I,2121); LITTLE WOMEN (II,1498); THE LORETTA YOUNG THEATER (I,2756); LOW MAN ON THE TOTEM POLE (I,2782); THE MAN FROM U.N.C.L.E. (I,2867); NAKED CITY (I,3216); THE NEXT STEP BEYOND (II,1839); NIGHT GALLERY (I,3287); ONE STEP BEYOND (I,3388); POLICE WOMAN (II,2063); THE ROGUES (I,3832); ROUTE 66 (I,3852); THE SIXTH SENSE (I,4069); STAR TREK (II,2440); THE SURVIVORS (I,4301); SWITCH (II,2519); TARGET: THE CORRUPTERS (I,4360); THRILLER (I,4492); WHIZ KIDS (II,2790); WONDER WOMAN (SERIES 2) (II,2829); THE YOUNG LAWYERS (I,4941)

NEWMAN, JOSEPH M
ALFRED HITCHCOCK PRESENTS (I,115); THE DESILU PLAYHOUSE (I,1237); MEETING AT APALACHIN (I,2994); TARZAN THE APE MAN (I,4363)

NEWTON, RICHARD
EMERGENCY! (II,775); THE MOD SQUAD (I,3078); THE SMOTHERS BROTHERS SHOW (I,4096)

NICHOL, PAUL
TIMMY AND LASSIE (I,4520)

NICHOLS, CHARLES A
THE ABC AFTERSCHOOL SPECIAL (II,1); THE ADDAMS FAMILY (II,12); ALVIN AND THE CHIPMUNKS (II,61); THE AMAZING CHAN AND THE CHAN CLAN (I,157); THE C.B. BEARS (II,406); CAPTAIN CAVEMAN AND THE TEEN ANGELS (II,433); THE CLUE CLUB (II,546); DEVLIN (II,663); DRAGON'S LAIR (II,732); FRED FLINTSTONE AND FRIENDS (II,920); GIDGET MAKES THE WRONG CONNECTION (I,1798);

GOOBER AND THE GHOST CHASERS (II,1028); THE GREAT GRAPE APE SHOW (II,1056); THE HEATHCLIFF AND DINGBAT SHOW (II,1117); THE HEATHCLIFF AND MARMADUKE SHOW (II,1118); HONG KONG PHOOEY (II,1180); INCH HIGH, PRIVATE EYE (II,1231); JABBERJAW (II,1261); JEANNIE (II,1275); JOSIE AND THE PUSSYCATS (I,2453); JOSIE AND THE PUSSYCATS IN OUTER SPACE (I,2454); LOST IN SPACE (I,2759); THE NEW ADVENTURES OF HUCKLEBERRY FINN (I,3250); THE NEW FRED AND BARNEY SHOW (II,1817); PARTRIDGE FAMILY: 2200 A.D. (II,1963); PEBBLES AND BAMM BAMM (II,1987); THE PLASTICMAN COMEDY/ADVENTURE SHOW (II,2051); THE ROMAN HOLIDAYS (I,3835); SATURDAY SUPERCADE (II,2263); SCOOBY'S ALL-STAR LAFF-A-LYMPICS (II,2273); SCOOBY-DOO AND SCRAPPY-DOO (II,2274); SCOOBY-DOO, WHERE ARE YOU? (II,2276); THE SCOOBY-DOO/DYNOMUTT HOUR (II,2277); SEALAB 2020 (I,3949); THE SKATEBIRDS (II,2375); SPEED BUGGY (II,2419); SUPER FRIENDS (II,2497); THESE ARE THE DAYS (II,2585); THE THREE ROBONIC STOOGES (II,2609); THUNDARR THE BARBARIAN (II,2616); VALLEY OF THE DINOSAURS (II,2717); WHEELIE AND THE CHOPPER BUNCH (II,2776); YOGI'S GANG (II,2852)

NICHOLSON, MEREDITH
THE HORACE HEIDT SHOW (I,2120)

NICKELL, PAUL
77 SUNSET STRIP (I,3988); THE ARTHUR GODFREY SHOW (I,283); THE BAT (I,364); BEN CASEY (I,394); CHARLIE WILD, PRIVATE DETECTIVE (I,914); THE FARMER'S DAUGHTER (I,1533); GIANT IN A HURRY (I,1790); THE GUARDSMAN (I,1893); THE LADY DIED AT MIDNIGHT (I,2601); LASSIE (I,2622); THE MAN AGAINST CRIME (I,2853); NAKED CITY (I,3216); ROAD TO REALITY (I,3800); THE ROYAL FAMILY (I,3861); THE WOMAN IN WHITE (I,4887)

NICOL, ALEX
DANIEL BOONE (I,1142); THE WILD WILD WEST (I,4863)

NIELSEN, JAMES
FOR THE DEFENSE (I,1624); THE FUGITIVE (I,1701); THE RAY MILLAND SHOW (I,3734); THE RIFLEMAN (I,3789)

NIGGEMEYER, AL
MAKE A WISH (I,2838)

NIGRO, BOB
THE LIFE AND TIMES OF EDDIE ROBERTS (II,1468)

NIGRO, GIOVANNI
MARY HARTMAN, MARY HARTMAN (II,1638)

NILES, DAVID
EVENING AT THE MOULIN ROUGE (II,787)

NIMOY, LEONARD
NIGHT GALLERY (I,3287); THE POWERS OF MATTHEW STAR (II,2075); T.J. HOOKER (II,2524)

NIVER, JAMES
THE DICK VAN DYKE SHOW (I,1275)

NOCKS, ARNEE
CAPTAIN VIDEO AND HIS VIDEO RANGERS (I,838)

NORMAN, LESLIE
THE AVENGERS (II,121); THE BARON (I,360); DEPARTMENT S (I,1232); THE PERSUADERS! (I,3556); RETURN OF THE SAINT (II,2141); THE SAINT (II,2237)

NORMAN, MARC
THE WHITE SHADOW (II,2788)

NORTON, CHARLES
GILLIGAN'S ISLAND (II,990)

NOSSECK, NOEL
NIGHT PARTNERS (II,1847)

NOYERS, ELI
BRAIN GAMES (II,366)

NUGENT, EDDIE
ARTHUR MURRAY'S DANCE PARTY (I,293); BLIND DATE (I,482); MR. ARSENIC (I,3126); PERSONALITY PUZZLE (I,3554)

NYBY II, CHRISTIAN I
240-ROBERT (II,2688); THE A-TEAM (II,119); ADAM-12 (I,31); B.J. AND THE BEAR (II,125); B.J. AND THE BEAR (II,126); BATTLESTAR GALACTICA (II,181); BONANZA (II,347); CASSIE AND COMPANY (II,455); CHASE (I,917); CHICAGO STORY (II,506); CHIPS (II,511); CODE RED (II,552); THE DEVLIN CONNECTION (II,664); EMERGENCY! (II,775); THE FALL GUY (II,811); THE FBI (I,1551); THE GREATEST AMERICAN HERO (II,1060); GUNSMOKE (II,1069); IRONSIDE (II,1246); KINGSTON: CONFIDENTIAL (II,1398); KNIGHT RIDER (II,1402); KOJAK (II,1406); KORG: 70,000 B.C. (II,1408); LASSIE (I,2622); LOBO (II,1504); MCCLAIN'S LAW (II,1659); MEDICAL CENTER (II,1675); THE MISADVENTURES OF SHERIFF LOBO (II,1704); THE NANCY DREW MYSTERIES (II,1789); THE PARTNERS (I,3462); PINE CANYON IS BURNING (II,2040); THE RANGERS (II,2112); RIPTIDE (II,2178); THE ROCKFORD FILES (II,2197); SCARECROW AND MRS. KING (II,2268); SIMON AND SIMON (II,2357); SWISS FAMILY ROBINSON (II,2517); SWORD OF JUSTICE (II,2521); TALES OF THE GOLD MONKEY (II,2537); TIMMY AND LASSIE (I,4520); THE WESTWIND (II,2766); WHEN THE WHISTLE BLOWS (II,2781)

NYBY, CHRISTIAN I
B.J. AND THE BEAR (II,126); CHASE (I,917); CROSSROADS (I,1105); EMERGENCY! (II,775); I SPY (I,2179); LANCER (I,2610); LIGHT'S DIAMOND JUBILEE (I,2698); THE LONE WOLF (I,2742); MY FRIEND TONY (I,3192); PERRY MASON (II,2025); PRIVATE SECRETARY (I,3672); RAWHIDE (I,3727); THE TWILIGHT ZONE (I,4651); WAGON TRAIN (I,4747); WAR CORRESPONDENT (I,4767); ZANE GREY THEATER (I,4979)

O'BRIEN, DAVE
THE JONES BOYS (I,2446)

O'BRIEN, JOEL
DRAW TO WIN (I,1373)

O'CONNELLY, JIM
THE SAINT (II,2237)

O'CONNOR, CARROLL
ARCHIE BUNKER'S PLACE (II,105); GLORIA COMES HOME (II,1011)

O'CURRAN, CHARLES
SATINS AND SPURS (I,3917)

O'HARA, GERRY
THE AVENGERS (II,121)

O'HERLIHY, MICHAEL
THE A-TEAM (II,119); BACKSTAIRS AT THE WHITE HOUSE (II,133); BLUE THUNDER (II,278); BRET MAVERICK (II,374); BRONCO (I,745); ENIGMA (II,778); EVEL KNIEVEL (II,785); THE FALL GUY (II,811); THE FBI (I,1551); THE GUNS OF WILL SONNETT (I,1907); GUNSMOKE (II,1069); HAWAII FIVE-O (II,1110); HAWAIIAN EYE (I,1973); KISS ME, KILL ME (II,1400); LOGAN'S RUN (II,1507); MAGNUM, P.I. (II,1601); THE MAN FROM ATLANTIS (II,1615); MANNIX (II,1624); MAVERICK (II,1657); MCCLAIN'S LAW (II,1659); MEDICAL CENTER (II,1675); MISSION: IMPOSSIBLE (I,3067); NERO WOLFE (II,1806); THE NEW ADVENTURES OF PERRY MASON (I,3252); O'MALLEY (II,1872); THE QUEST (II,2097); RAWHIDE (I,3727); SEVEN BRIDES FOR SEVEN BROTHERS (II,2307); SONS AND DAUGHTERS (II,2404); STAR TREK (II,2440); THE STREETS OF SAN FRANCISCO (II,2478); SURFSIDE 6 (I,4299); TODAY'S FBI (II,2628); THE YOUNG PIONEERS (II,2868); THE YOUNG PIONEERS' CHRISTMAS (II,2870)

O'LOUGHLIN, GERALD S
THE ROOKIES (II,2208)

O'NEAL, PATRICK
WILSON'S REWARD (II,2805)

O'RIORDAN, SHAUN
THE DEVIL'S WEB (I,1248); THE EYES HAVE IT (I,1496); I'M THE GIRL HE WANTS TO KILL (I,2194); IF IT'S A MAN, HANG UP (I,2186); IN THE STEPS OF A DEAD MAN (I,2205); SCREAMER (I,3945); SIGN IT DEATH (I,4046)

O'STEEN, SAM
HIGH RISK (II,1146)

OCHS, ACE
OPEN HOUSE (I,3394)

OFFNER, MORTIMER
A DATE WITH JUDY (I,1162)

OLDEN, GEORG
THE MOD SQUAD (I,3078)

OLINSKY, JOEL
EMERGENCY! (II,775)

OLIVER, SUSAN
M*A*S*H (II,1569)

OLSEN, STAN
THE LITTLEST HOBO (II,1500)

OPIE, WIN
THE CLIFFWOOD AVENUE KIDS (II,543); THE LAS VEGAS

SHOW (I,2619); PAT BOONE IN HOLLYWOOD (I,3469); THE STEVE ALLEN SHOW (I,4222)

OPPENHEIMER, JESS
FOR LOVE OR $$$ (I,1623)

ORILIO, JOSEPH
THE MIGHTY HERCULES (I,3034)

ORMEROD, JAMES
DEATH IN DEEP WATER (I,1206); THE NEXT VICTIM (I,3281); TERROR FROM WITHIN (I,4404)

ORNITZ, ARTHUR J
MAKE A WISH (I,2838)

ORR, MURRAY
THE NEW ERNIE KOVACS SHOW (I,3264)

ORR, WAYNE
THE PARAGON OF COMEDY (II,1956)

ORSATTI, FRANK
THE INCREDIBLE HULK (II,1232)

OSANI, HIRIO
MARINE BOY (I,2911)

OSBORNE, JOHN
TOO CLOSE FOR COMFORT (II,2642)

OSCARRUDOLPH
MCHALE'S NAVY (I,2969)

OSHINS, JACKIE
THE RED BUTTONS SHOW (I,3750)

OSWALD, GERD
THE 20TH CENTURY-FOX HOUR (I,4642); ADVENTURES IN PARADISE (I,46); BLACK SADDLE (I,480); BONANZA (II,347); DANIEL BOONE (I,1142); THE FELONY SQUAD (I,1563); THE FUGITIVE (I,1701); GENTLE BEN (I,1761); IT TAKES A THIEF (I,2250); NICHOLS (I,3283); THE OUTER LIMITS (I,3426); PERRY MASON (II,2025); RAWHIDE (I,3727); SHANE (I,3995); STAR TREK (II,2440); TEMPLE HOUSTON (I,4392); THE UNKNOWN (I,4680)

OWEN, CLIFF
THE AVENGERS (II,121)

OWENS, CLIFFORD
MELINA MERCOURI'S GREECE (I,2997)

OXFORD, RONALD
WINDOW SHADE REVUE (I,4874)

OYSTER, DAVID
COSMOS (II,579)

PACELLI, FRANK
DAYS OF OUR LIVES (II,629); RETURN TO PEYTON PLACE (I,3779); THE YOUNG AND THE RESTLESS (II,2862)

PAGE, ANTHONY
THE ADAMS CHRONICLES (II,8); JOHNNY BELINDA (I,2418)

PALMER, P K
DICK TRACY (I,1271)

PALTROW, BRUCE
OPERATING ROOM (II,1915); ST. ELSEWHERE (II,2432); THE WHITE SHADOW (II,2788); YOU'RE GONNA LOVE IT HERE (II,2860)

PAPP, FRANK
THE ALDRICH FAMILY (I,111)

PARIS, JERRY
BAREFOOT IN THE PARK (I,355); BEANE'S OF BOSTON (II,194); BEST FRIENDS (II,213); BLANSKY'S BEAUTIES (II,267); CALL HER MOM (I,796); CAT BALLOU (I,869); THE DICK VAN DYKE SHOW (I,1275); THE DICK VAN DYKE SPECIAL (I,1277); EVIL ROY SLADE (I,1485); THE FARMER'S DAUGHTER (I,1533); HAPPY DAYS (II,1084); HERE'S LUCY (II,1135); KAREN (II,1363); KEEPING UP WITH THE JONESES (II,1374); LAVERNE AND SHIRLEY (II,1446); MCCLOUD (II,1660); THE MUNSTERS (I,3158); THE ODD COUPLE (II,1875); THE PARTRIDGE FAMILY (II,1962); SHERIFF WHO? (I,4009); SISTER TERRI (II,2366); THE TED KNIGHT SHOW (II,2550); THICKER THAN WATER (I,4445); WEDNESDAY NIGHT OUT (II,2760); WHEN THINGS WERE ROTTEN (II,2782); WHERE'S THE FIRE? (II,2785)

PARK, BEN
THE PUBLIC LIFE OF CLIFF NORTON (I,3688); STUD'S PLACE (I,4271); THOSE ENDEARING YOUNG CHARMS (I,4469)

PARKER, FESS
DANIEL BOONE (I,1142)

PARKER, JOE
THE UNTOUCHABLES (I,4682)

PARMELEE, TED
THE BULLWINKLE SHOW (I,761); ROCKY AND HIS FRIENDS (I,3822)

PARONE, EDWARD
CASSIE AND COMPANY (II,455); FAMILY (II,813); THE FAMILY TREE (II,820); JULIE FARR, M.D. (II,1354); KNOTS LANDING (II,1404); NORMA RAE (II,1860); PAPER DOLLS (II,1952); RYAN'S FOUR (II,2233); THE SIX OF US (II,2369); SKAG (II,2374); WHAT REALLY HAPPENED TO THE CLASS OF '65? (II,2768); WHEN THE WHISTLE BLOWS (II,2781)

PARR, DAN
PADDINGTON BEAR (II,1941)

PARRIOTT, JAMES D
THE AMERICAN GIRLS (II,72); FROM HERE TO ETERNITY (II,933); VOYAGERS (II,2730)

PASETTA, MARTY
50 GRAND SLAM (II,852); THE ARTHUR GODFREY SPECIAL (I,288); ARTHUR GODFREY'S PORTABLE ELECTRIC MEDICINE SHOW (I,290); BING CROSBY AND HIS FRIENDS (I,456); BING CROSBY AND HIS FRIENDS (II,251); BING CROSBY'S CHRISTMAS SHOW (I,474); BING CROSBY—COOLING IT (I,460); BING!. . . A 50TH ANNIVERSARY GALA (II,254); BLANK CHECK (II,265); BURNETT "DISCOVERS" DOMINGO (II,398); CELEBRATION: THE AMERICAN SPIRIT (II,461); CHRISTMAS IN DISNEYLAND (II,519); CHRISTMAS WITH THE BING CROSBYS (I,950); CHRISTMAS WITH THE BING CROSBYS (II,524); A COUNTRY CHRISTMAS (II,582); A COUNTRY CHRISTMAS (II,583); A COUNTRY CHRISTMAS (II,584); DEBBY BOONE. . . THE SAME OLD BRAND NEW ME (II,656); THE DON ADAMS SCREEN TEST (II,705); THE EVERLY BROTHERS SHOW (I,1479); GENE KELLY. . . AN AMERICAN IN PASADENA (II,962); HAPPY BIRTHDAY, AMERICA (II,1083); THE HOMEMADE COMEDY SPECIAL (II,1173); HOW TO HANDLE A WOMAN (I,2146); JOHN SCHNEIDER'S CHRISTMAS HOLIDAY (II,1320); MAGNAVOX PRESENTS FRANK SINATRA (I,2826); THE MANY MOODS OF PERRY COMO (I,2897); THE NOONDAY SHOW (II,1859); ONE MORE TIME (I,3386); PAUL ANKA IN MONTE CARLO (II,1971); PAUL ANKA. . . MUSIC MY WAY (II,1973); PERRY COMO'S WINTER SHOW (I,3546); PURE GOLDIE (I,3696); THE REEL GAME (I,3759); THE ROD MCKUEN SPECIAL (I,3826); SAGA OF SONORA (I,3884); A SALUTE TO TELEVISION'S 25TH ANNIVERSARY (I,3893); SANDY IN DISNEYLAND (II,2254); STEVE AND EYDIE. . . ON STAGE (I,4223); STUMPERS (II,2485); THE SUMMER SMOTHERS BROTHERS SHOW (I,4281); SUPER COMEDY BOWL 1 (I,4288); SUPER COMEDY BOWL 2 (I,4289); TELLY. . . WHO LOVES YA, BABY? (II,2558); TEMPTATION (I,4393); TEXACO STAR THEATER: OPENING NIGHT (II,2565)

PASQUIN, JOHN
ALICE (II,33); FAMILY TIES (II,819); GIMME A BREAK (II,995); IT'S YOUR MOVE (II,1259); TEXAS (II,2566)

PASSER, IVAN
FAERIE TALE THEATER (II,807)

PATAKI, MICHAEL
THE NANCY DREW MYSTERIES (II,1789)

PATCHETT, TOM
THE BEST LEGS IN 8TH GRADE (II,214); BUFFALO BILL (II,387); I GAVE AT THE OFFICE (II,1213); OPEN ALL NIGHT (II,1914); SITCOM (II,2367)

PATILLO, ALAN
THUNDERBIRDS (I,4497)

PATRICK, DENNIS
FOR BETTER OR WORSE (I,1621)

PATTERSON, ARNY
GODZILLA (II,1014)

PATTERSON, DON
PINK PANTHER AND SONS (II,2043)

PATTERSON, JOHN
BRET MAVERICK (II,374); BUCK ROGERS IN THE 25TH CENTURY (II,384); CAGNEY AND LACEY (II,409); CUTTER TO HOUSTON (II,612); EMERALD POINT, N.A.S. (II,773); FOR LOVE AND HONOR (II,903); FOUL PLAY (II,910); HART TO HART (II,1102); MICKEY SPILLANE'S MIKE HAMMER (II,1692); THE MISSISSIPPI (II,1707); PROJECT UFO (II,2088); THE ROCKFORD FILES (II,2197); RYAN'S FOUR (II,2233); SEVEN BRIDES FOR SEVEN BROTHERS (II,2307); TENSPEED AND BROWN

SHOE (II,2562); TODAY'S FBI (II,2628)

PATTERSON, LAURA
PERSONAL AND CONFIDENTIAL (II,2026)

PATTERSON, RAY
THE ALL-NEW POPEYE HOUR (II,49); THE BISKITTS (II,257); BUFORD AND THE GHOST (II,389); CASPER AND THE ANGELS (II,453); FLINTSTONE FAMILY ADVENTURES (II,882); THE FLINTSTONE FUNNIES (II,883); THE FLINTSTONES (II,885); FONZ AND THE HAPPY DAYS GANG (II,898); FRED AND BARNEY MEET THE SHMOO (II,918); FRED AND BARNEY MEET THE THING (II,919); THE GALAXY GOOFUPS (II,954); THE GARY COLEMAN SHOW (II,958); LAVERNE AND SHIRLEY WITH THE FONZ (II,1449); THE LITTLE RASCALS (II,1494); MONCHHICHIS (II,1722); THE NEW FRED AND BARNEY SHOW (II,1817); PAC-MAN (II,1940); PINK PANTHER AND SONS (II,2043); THE RICHIE RICH SHOW (II,2169); SCOOBY-DOO AND SCRAPPY-DOO (II,2274); THE SCOOBY-DOO AND SCRAPPY-DOO SHOW (II,2275); THE SHIRT TALES (II,2337); THE SMURFS (II,2385); SNORKS (II,2390); SPIDER-MAN (I,4154); SUPER FRIENDS (II,2497); THE SUPER GLOBETROTTERS (II,2498)

PATTINSON, MICHAEL
PRISONER: CELL BLOCK H (II,2085)

PAUL, BYRON
BEN CASEY (I,394); COMBAT (I,1011); DANGER (I,1134); THE GUY MITCHELL SHOW (I,1912); THE JULIUS LAROSA SHOW (I,2486); MR. I. MAGINATION (I,3140); MY FAVORITE MARTIAN (I,3189); OUT THERE (I,3424); PRIZE PERFORMANCE (I,3673); THE SAM LEVENSON SHOW (I,3897); SEE AMERICA WITH ED SULLIVAN (I,3972); SONG SNAPSHOTS ON A SUMMER HOLIDAY (I,4122); SPRING HOLIDAY (I,4168); THE TRAP (I,4593); YOU ARE THERE (I,4929)

PAUL, GEORGE
TOM SNYDER'S CELEBRITY SPOTLIGHT (II,2632); THE TOMORROW SHOW (II,2635)

PAYNE, JERRY
THE CROSS-WITS (II,605)

PEACH, PETER
CHRISTMAS IN WASHINGTON (II,521)

PECHIN, CHRIS
PERSONAL AND CONFIDENTIAL (II,2026); THAT'S INCREDIBLE! (II,2578)

PECK, STEPHEN
IN SEARCH OF. . . (II,1226)

PECKINPAH, SAM
THE BOB HOPE CHRYSLER THEATER (I,502); BROKEN ARROW (I,744); THE DICK POWELL SHOW (I,1269); KLONDIKE (I,2573); THE RIFLEMAN (I,3789); ROUTE 66 (I,3852); THE WESTERNER (I,4801); WINCHESTER (I,4872)

PEEL, SPENCER
BAGGY PANTS AND THE NITWITS (II,135); BAILEY'S COMETS (I,333); THE BARKLEYS (I,356); THE HOUNDCATS (I,2132); WHAT'S NEW MR. MAGOO? (II,2770)

PENN, ARTHUR
FLESH AND BLOOD (I,1595); THE KING AND MRS. CANDLE (I,2544); PLAYWRIGHTS '56 (I,3627)

PENN, LEO
THE 13TH DAY: THE STORY OF ESTHER (II,2593); 77 SUNSET STRIP (I,3988); ALFRED HITCHCOCK PRESENTS (I,115); BARNABY JONES (II,153); BEN CASEY (I,394); THE BIONIC WOMAN (II,255); THE BLUE KNIGHT (II,276); BOONE (II,351); BRET MAVERICK (II,374); CAGNEY AND LACEY (II,409); CANNON (II,424); CIRCLE OF FEAR (I,958); CONCRETE COWBOYS (II,563); THE COUNTRY MUSIC MURDERS (II,590); DOCTORS HOSPITAL (II,691); EISCHIED (II,763); FATHER MURPHY (II,841); THE FUGITIVE (I,1701); THE GANGSTER CHRONICLES (II,957); THE GIRL FROM U.N.C.L.E. (I,1808); GUNSMOKE (II,1069); HART TO HART (II,1102); HAWAII FIVE-O (II,1110); HELLINGER'S LAW (II,1127); I SPY (I,2179); IRONSIDE (II,1246); JUDD, FOR THE DEFENSE (I,2463); KATE LOVES A MYSTERY (II,1367); KAZ (II,1370); LANCER (I,2610); LITTLE HOUSE ON THE PRAIRIE (II,1487); LITTLE WOMEN (II,1498); LUCAS TANNER (II,1556); MAGNUM, P.I. (II,1601); THE MAN FROM U.N.C.L.E. (I,2867); MANNIX (II,1624); MARCUS WELBY, M.D. (II,1627); MICKEY SPILLANE'S MIKE HAMMER (II,1692); THE MISSISSIPPI (II,1707); MOVIN' ON (II,1743); MR. MERLIN (II,1751); THE NEW ADVENTURES OF PERRY MASON (I,3252); OWEN MARSHALL: COUNSELOR AT LAW (I,3435); PAPER DOLLS (II,1952); REMINGTON STEELE (II,2130); RUN FOR YOUR LIFE (I,3871); SARA (II,2260); STAR TREK (II,2440); SWITCH (II,2519); TESTIMONY OF TWO MEN (II,2564); TRAPPER JOHN, M.D. (II,2654); THE VIRGINIAN (I,4738); WESTSIDE MEDICAL (II,2765); WHAT REALLY HAPPENED TO THE CLASS OF '65? (II,2768)

PEPPERMAN, RICHARD
THE EDGE OF NIGHT (II,760)

PEREZ, MANNY
THE PLASTICMAN COMEDY/ADVENTURE SHOW (II,2051)

PEREZ, PHIL
LAVERNE AND SHIRLEY (II,1446)

PERONE, HAL
THE AD-LIBBERS (I,29)

PERRIN, DOMINIQUE
CIRCUS OF THE STARS (II,532)

PERRIN, NAT
THE ADDAMS FAMILY (II,11)

PERRIS, ANTHONY
THE LITTLEST HOBO (II,1500)

PERRY, ALAN
CAPTAIN SCARLET AND THE MYSTERONS (I,837); U.F.O. (I,4662)

PERRY, FRANK
A CHRISTMAS MEMORY (I,948); SKAG (II,2374); THANKSGIVING VISITOR (I,4417)

PERSKY, BILL
ALAN KING'S FINAL WARNING (II,29); ALICE (II,33); ALMOST HEAVEN (II,57); BAKER'S DOZEN (II,136); THE BETTY WHITE SHOW (II,224); BIG CITY BOYS (II,232); BIG EDDIE (II,235); BOBBY PARKER AND COMPANY (II,344); THE BOYS (II,360); COMEDY OF HORRORS (II,557); FILTHY RICH (II,856); HERE WE GO AGAIN (I,2028); HOW TO SURVIVE THE 70S AND MAYBE EVEN BUMP INTO HAPPINESS (II,1199); HUSBANDS AND WIVES (II,1208); HUSBANDS, WIVES AND LOVERS (II,1209); JOE AND VALERIE (II,1298); JOHNNY GARAGE (II,1337); KATE AND ALLIE (II,1365); LOVE AT FIRST SIGHT (II,1530); ME AND DUCKY (II,1671); THE MONTEFUSCOS (II,1727); MY WIFE NEXT DOOR (II,1780); MY WIFE NEXT DOOR (II,1781); NOT IN FRONT OF THE KIDS (II,1861); THE PRACTICE (II,2076); RISE AND SHINE (II,2179); SEMI-TOUGH (II,2299); SEMI-TOUGH (II,2300); THE SINGLE LIFE (II,2364); SPENCER (II,2420); SUTTERS BAY (II,2508); THE TED BESSELL SHOW (I,4369); THAT GIRL (II,2570); THE WAVERLY WONDERS (II,2746); WELCOME BACK, KOTTER (II,2761); WHO'S THE BOSS? (II,2792)

PETCHFORD, PEITA
PRISONER: CELL BLOCK H (II,2085)

PETERS, BARBARA
BOONE (II,351); LOTTERY (II,1525); MATT HOUSTON (II,1654); THE POWERS OF MATTHEW STAR (II,2075); REMINGTON STEELE (II,2130); THE RENEGADES (II,2133)

PETERS, MICHAEL
FAME (II,812)

PETERSON, EDGAR
THE AMAZING MR. MALONE (I,159)

PETERSON, EDWARD
THE AMAZING MR. MALONE (I,159)

PETERSON, RAY
THE NEW SHMOO (II,1827)

PETRANTO, RUSS
THE BEST OF SULLIVAN (II,217); THE CRYSTAL GAYLE SPECIAL (II,609); INSTANT FAMILY (II,1238); JACKIE AND DARLENE (II,1265); MR. & MS. AND THE BANDSTAND MURDERS (II,1746); MR. & MS. AND THE MAGIC STUDIO MYSTERY (II,1747); THE PERRY COMO SPRINGTIME SPECIAL (II,2004); PIPER'S PETS (II,2046); THE RAG BUSINESS (II,2106); SANFORD AND SON (II,2256); THE STEVE ALLEN COMEDY HOUR (II,2454); TOO CLOSE FOR COMFORT

(II,2642)

PETRASHEVITCH, VICTOR
SUSPENSE THEATER ON THE AIR (II,2507)

PETRIE, DANIEL
THE ALCOA HOUR (I,108); ASSIGNMENT: MANHUNT (I,301); THE CENTURY TURNS (I,890); THE CITY (I,969); THE DEFENDERS (I,1220); THE GOLDEN AGE OF TELEVISION (II,1019); THE GUN AND THE PULPIT (II,1067); HEC RAMSEY (II,1121); IRONSIDE (II,1246); THE LAWYERS (I,2646); THE MAN AND THE CITY (I,2856); MARCUS WELBY, M.D. (II,1627); MCMILLAN AND WIFE (II,1667); MEDICAL CENTER (II,1675); N.Y.P.D. (I,3321); PANHANDLE PETE AND JENNIFER (I,3446); SAN FRANCISCO INTERNATIONAL AIRPORT (I,3905); SHORT, SHORT DRAMA (I,4023); VICTORY (I,4729)

PEVNEY, JOSEPH
ADAM-12 (I,31); ALCOA PREMIERE (I,109); ALFRED HITCHCOCK PRESENTS (I,115); BEN CASEY (I,394); EMERGENCY! (II,775); EXECUTIVE SUITE (II,796); FATHER MURPHY (II,841); GOING MY WAY (I,1833); GRANDPA GOES TO WASHINGTON (II,1050); HAGEN (II,1071); THE HARDY BOYS MYSTERIES (II,1090); THE HIGH CHAPARRAL (I,2051); THE INCREDIBLE HULK (II,1232); LANIGAN'S RABBI (II,1428); LUCAN (II,1554); MARCUS WELBY, M.D. (II,1627); MEDICAL CENTER (II,1675); MOBILE ONE (II,1717); THE MUNSTERS (I,3158); THE NAME OF THE GAME (I,3217); THE NANCY DREW MYSTERIES (II,1789); PALMERSTOWN U.S.A. (II,1945); THE PAPER CHASE: THE SECOND YEAR (II,1950); THE ROCKFORD FILES (II,2197); THE ROUSTERS (II,2220); SARA (II,2260); SEARCH (I,3951); THE SECRET EMPIRE (II,2291); STAR TREK (II,2440); SWORD OF JUSTICE (II,2521); TRAPPER JOHN, M.D. (II,2654); THE VIRGINIAN (I,4738); WAGON TRAIN (I,4747)

PEYSER, JOHN
240-ROBERT (II,2689); ALL IN FUN (I,130); THE AMERICAN GIRLS (II,72); AMOS BURKE, SECRET AGENT (I,180); THE BLACK SHEEP SQUADRON (II,262); BRONK (II,379); CANDID CAMERA (I,819); CASABLANCA (I,860); CHARLIE'S ANGELS (II,486); COMBAT (I,1011); CRIME SYNDICATED (I,1099); THE DICK POWELL SHOW (I,1269); GENERAL ELECTRIC TRUE (I,1754); GIANT IN A HURRY (I,1790); HANDLE WITH CARE (I,1926); HAWAII FIVE-O (II,1110); KATE MCSHANE (II,1368); KHAN! (II,1384); LAW OF THE PLAINSMAN (I,2639); M SQUAD (I,2799); MAN BEHIND THE BADGE (I,2857); THE MAN FROM U.N.C.L.E. (I,2867); MANNIX (II,1624); NAKED CITY (I,3216); PERRY MASON (II,2025); QUINCY, M. E. (II,2102); THE RAT PATROL (I,3726); ROUTE 66 (I,3852); SAINTS AND SINNERS (I,3887); SERGEANT PRESTON OF THE YUKON (I,3980); STARLIGHT THEATER (I,4201); STUNT SEVEN (II,2486); SWISS FAMILY ROBINSON (II,2517); SWITCH (II,2519); TALES OF THE UNEXPECTED (II,2540); TARGET: THE CORRUPTERS (I,4360); THE UNTOUCHABLES (I,4682); THE VIRGINIAN (I,4738); WESTINGHOUSE SUMMER THEATER (I,4805)

PFLEGHAR, MICHAEL
FRANK SINATRA: A MAN AND HIS MUSIC (I,1662); MONTE CARLO, C'EST LA ROSE (I,3095); WITH LOVE, SOPHIA (I,4884)

PHELPS, STUART W
LIE DETECTOR (II,1467); PASSWORD (II,1965); SHOWDOWN (I,4034); TRUTH OR CONSEQUENCES (I,4620); WHAT'S THIS SONG? (I,4822); YOU DON'T SAY (I,4934)

PHILIPP, GODFREY
PRISONER: CELL BLOCK H (II,2085)

PHILIPS, LEE
THE AMERICAN GIRLS (II,72); THE ANDY GRIFFITH SHOW (I,192); BRACKEN'S WORLD (I,703); CRAZY TIMES (II,602); THE DICK VAN DYKE SHOW (I,1275); DYNASTY (II,747); THE GHOST AND MRS. MUIR (I,1786); GIDGET (I,1795); GOING PLACES (I,1834); INSPECTOR PEREZ (II,1237); LOTTERY (II,1525); M*A*S*H (II,1569); THE MAN AND THE CITY (I,2856); MEDICAL CENTER (II,1675); MY WORLD . . . AND WELCOME TO IT (I,3208); OF MEN OF WOMEN (I,3332); THE PRACTICE (II,2076); RIDING HIGH (II,2173); ROOM 222 (I,3843); SALVAGE (II,2244); SALVAGE 1 (II,2245); THE SURVIVORS (I,4301); THE WALTONS (II,2740); THE WALTONS: WEDDING ON WALTON'S MOUNTAIN (EPISODE 1) (II,2743)

PHINNEY, DAVID
B.J. AND THE BEAR (II,126); BUCK ROGERS IN THE 25TH CENTURY (II,383); BUCK ROGERS IN THE 25TH CENTURY (II,384)

PICERNI, CHARLES
BLUE THUNDER (II,278); GAVILAN (II,959); MATT HOUSTON (II,1654); STRIKE FORCE (II,2480); T.J. HOOKER (II,2524); VEGAS (II,2724)

PICERNI, PAUL
GAVILAN (II,959)

PIERCE, LARRY
THE MONROES (I,3089)

PIERSON, ARTHUR
TERRY AND THE PIRATES (I,4405)

PIERSON, FRANK R
THE DOCTORS (I,1323); HAVE GUN—WILL TRAVEL (I,1968)

PINE, ROBERT
CHIPS (II,511)

PINSKER, SETH
EIGHT IS ENOUGH (II,762)

PINTOFF, ERNEST
BARNABY JONES (II,153); THE BIONIC WOMAN (II,255); BRAVO TWO (II,368); CALL TO GLORY (II,413); CODE RED (II,552); THE DUKES OF HAZZARD (II,742); EMERALD POINT, N.A.S. (II,773); THE FEATHER AND FATHER GANG (II,845); HAWAII FIVE-O (II,1110); HUMAN FEELINGS (II,1203); JAMES AT 15 (II,1270); KNOTS LANDING (II,1404); KOJAK (II,1406); THE KOWBOYS (I,2584); SPENCER'S PILOTS (II,2422); SWEEPSTAKES (II,2514); VOYAGERS (II,2730); THE WHITE SHADOW (II,2788); YOUNG DAN'L BOONE (II,2863)

PIPER, BILL
THE LOVE REPORT (II,1543)

PITLIK, NOAM
9 TO 5 (II,1852); A.E.S. HUDSON STREET (II,17); ANOTHER MAN'S SHOES (II,96); BARNEY MILLER (II,154); THE BETTY WHITE SHOW (II,224); BRANAGAN AND MAPES (II,367); FISH (II,864); LOVES ME, LOVES ME NOT (II,1551); NOW WE'RE COOKIN' (II,1866); ON OUR OWN (II,1892); ONE DAY AT A TIME (II,1900); ONE MORE TRY (II,1908); PHYLLIS (II,2038); THE PRACTICE (II,2076); TAXI (II,2546); THAT'S MY MAMA (II,2580); YOU'RE JUST LIKE YOUR FATHER (II,2861)

PITTMAN, MICHAEL
77 SUNSET STRIP (I,3988); CHEYENNE (I,931); COLT .45 (I,1010); LAWMAN (I,2642); SUGARFOOT (I,4277); SURFSIDE 6 (I,4299)

PITTMAN, MONTGOMERY
COME A RUNNING (I,1013); LUKE AND THE TENDERFOOT (I,2792); THE RIFLEMAN (I,3789); THE TWILIGHT ZONE (I,4651)

PLACE, LOU
DOC CORKLE (I,1304)

PLAXTON, GARRY
KING OF KENSINGTON (II,1395)

PLAYLOW, CLARK
SKY KING (I,4077)

PLESHETTE, JOHN
KNOTS LANDING (II,1404)

POGOSTIN, S LEE
THE BOB HOPE CHRYSLER THEATER (I,502)

POLACHECK, CHARLES
COLONEL STOOPNAGLE'S STOOP (I,1006)

POLK, JACKSON
THE MOONMAN CONNECTION (II,1731)

POLLACK, SYDNEY
ALFRED HITCHCOCK PRESENTS (I,115); BEN CASEY (I,394); THE BOB HOPE CHRYSLER THEATER (I,502); DIAGNOSIS: DANGER (I,1250); THE FLIERS (I,1596); SECOND LOOK (I,3960); TARGET: THE CORRUPTERS (I,4360)

POLLARD, DICK
THE JOE PYNE SHOW (I,2402)

POLONSKY, HENRY
BLOCKHEADS (II,270)

POMANSANOFF, ALEX
AN AMERICAN PORTRAIT (II,73)

PORTER, DON
GIDGET (I,1795)

POST, TED
THE 20TH CENTURY-FOX HOUR (I,4642); ALCOA PREMIERE (I,109); ARK II (II,107); B.A.D. CATS (II,123); BARETTA (II,152); BEYOND WESTWORLD (II,227); CAGNEY AND LACEY (II,410); CHECKMATE (I,919); COLUMBO (II,556); COMBAT (I,1011); THE DEFENDERS (I,1220); THE DESILU PLAYHOUSE (I,1237); THE DETECTIVES (I,1245); GUNSMOKE (II,1069); MEDIC (I,2976); PERRY MASON (II,2025); PEYTON PLACE (I,3574); RAWHIDE (I,3727); RICH MAN, POOR MAN—BOOK I (II,2161); ROUTE 66 (I,3852); SCREEN DIRECTOR'S PLAYHOUSE (I,3946); SECOND CHANCE (I,3957); THRILLER (I,4492); THE VIRGINIAN (I,4738); WAGON TRAIN (I,4747); WHEN, JENNY? WHEN (II,2783); YUMA (I,4976)

POTTER, H C
SCREEN DIRECTOR'S PLAYHOUSE (I,3946)

POTTER, PETER
THE STRAUSS FAMILY (I,4253)

POULIOT, STEPHEN
PERRY COMO'S CHRISTMAS IN AUSTRIA (II,2007); PERRY COMO'S SPRING IN NEW ORLEANS (II,2021)

POWELL, HOMER
BARNEY MILLER (II,154); THAT GIRL (II,2570)

POWELL, MICHAEL
THE DEFENDERS (I,1220)

POWELL, NORMAN S
THE BIG VALLEY (I,437); THE BOB CRANE SHOW (II,280)

POWERS, CHARLES
FAITH BALDWIN'S THEATER OF ROMANCE (I,1515); LIFE BEGINS AT EIGHTY (I,2682); MY TRUE STORY (I,3206); WREN'S NEST (I,4922)

POWERS, DAVE
THE CAROL BURNETT SHOW (II,443); THE JIMMIE RODGERS SHOW (I,2383); JOHN RITTER: BEING OF SOUND MIND AND BODY (II,1319); JULIE AND CAROL AT LINCOLN CENTER (I,2479); THE KELLY MONTEITH SHOW (II,1376); OF THEE I SING (I,3334); ONCE UPON A MATTRESS (I,3371); THE ROPERS (II,2214); SILLS AND BURNETT AT THE MET (II,2354); THREE'S A CROWD (II,2613); THREE'S COMPANY (II,2614); THE TIM CONWAY SPECIAL (I,4503); UNCLE TIM WANTS YOU! (II,2704)

PRAGER, STANLEY
CAR 54, WHERE ARE YOU? (I,842); KIBBE HATES FINCH (I,2530)

PRATT, HAWLEY
THE PINK PANTHER (II,2042); THE SUPER SIX (I,4292)

PREECE, LARRY
BATMAN (I,366); BRANDED (I,707); THE FELONY SQUAD (I,1563); THE HARDY BOYS (I,1948)

PREECE, MICHAEL
ACE CRAWFORD, PRIVATE EYE (II,6); B.J. AND THE BEAR (II,125); B.J. AND THE BEAR (II,126); BARNABY JONES (II,153); THE BIONIC WOMAN (II,255); BOONE (II,351); DALLAS (II,614); DOG AND CAT (II,696); FALCON CREST (II,808); FLAMINGO ROAD (II,872); FREEBIE AND THE BEAN (II,922); GREAT DAY (II,1054); HUNTER (II,1206); THE KILLIN' COUSIN (II,1394); LOGAN'S RUN (II,1507); A MAN CALLED SLOANE (II,1613); MICKEY SPILLANE'S MIKE HAMMER (II,1692); OPERATION: RUNAWAY (II,1918); THE RUNAWAYS (II,2231); SARA (II,2260); SHIRLEY (II,2330); THE STREETS OF SAN FRANCISCO (II,2478); T.J. HOOKER (II,2524); WHEN THE WHISTLE BLOWS (II,2781)

PREMINGER, OTTO
TONIGHT AT 8:30 (I,4556)

PRESSMAN, DAVID
CORONET BLUE (I,1055); N.Y.P.D. (I,3321); ONE LIFE TO LIVE (II,1907); THE SWAN (I,4312); TREASURY MEN IN ACTION (I,4604)

PREVIN, STEVE
SHERLOCK HOLMES (I,4010)

PRICE, ERIC
SKIPPY, THE BUSH KANGAROO (I,4076)

PRICE, NICK
DOCTOR SNUGGLES (II,687)

PRICE, PATON
THE SMOTHERS BROTHERS SHOW (I,4096)

PRICE, PAUL
77 SUNSET STRIP (I,3988)

PRINCE, HAROLD
SWEENEY TODD (II,2512)

PRINGLE, JULIAN
PRISONER: CELL BLOCK H (II,2085)

PROUDFOOT, DAVID SULLIVAN
THE ONEDIN LINE (II,1912)

PULTZ, ALAN
DAYS OF OUR LIVES (II,629); GENERAL HOSPITAL (II,964); RETURN TO PEYTON PLACE (I,3779)

PURDY, HALL
ALL AROUND TOWN (I,128)

PURDY, RAI
I'LL BUY THAT (I,2190); TAKE A GUESS (I,4333)

PYLE, DENVER
DIRTY SALLY (II,680); THE DORIS DAY SHOW (I,1355); THE DUKES OF HAZZARD (II,742)

QUINE, RICHARD
CATCH 22 (I,871); COLUMBO (II,556); HEC RAMSEY (II,1121); HEY MULLIGAN (I,2040); MCCOY (II,1661); PROJECT UFO (II,2088)

QUINLAN, BOB
BING CROSBY AND HIS FRIENDS (I,455)

QUINN, BOBBY
JOHNNY CARSON'S REPERTORY COMPANY IN AN EVENING OF COMEDY (I,2423); THE TONIGHT SHOW STARRING JOHNNY CARSON (II,2638)

QUINTERO, JOSE
OUR TOWN (I,3420); WINDOWS (I,4876)

RABB, ELLIS
MA AND PA (II,1570)

RABIN, AL
DAYS OF OUR LIVES (II,629)

RACHIBAS, LEW
THE KID SUPER POWER HOUR WITH SHAZAM (II,1386)

RACHINS, ALAN
PARIS (II,1957)

RADNITZ, BRAD
CALL TO GLORY (II,413)

RADY, SIMON
HOWDY DOODY (I,2151)

RAE, DAVID C
WHEN HAVOC STRUCK (II,2778)

RAFELSON, BOB
FAERIE TALE THEATER (II,807)

RAFELSON, RALPH
THE MONKEES (I,3088)

RAFKIN, ALAN
THE ANDY GRIFFITH SHOW (I,192); ANOTHER APRIL (II,93); BLANSKY'S BEAUTIES (II,267); THE BOB NEWHART SHOW (II,342); THE CARA WILLIAMS SHOW (I,843); CHARLES IN CHARGE (II,482); THE COURTSHIP OF EDDIE'S FATHER (I,1083); DADDY'S GIRL (I,1124); THE GOOD GUYS (I,1847); THE GOVERNOR AND J.J. (I,1863); HANDLE WITH CARE (II,1077); HANGING IN (II,1078); HARRY'S BATTLES (II,1100); HERE WE GO AGAIN (I,2028); I DREAM OF JEANNIE (I,2167); LAVERNE AND SHIRLEY (II,1446); LEGS (II,1459); LIVING IN PARADISE (II,1503); LOCAL 306 (II,1505); THE LOVE BOAT (II,1535); ME AND THE CHIMP (I,2974); MY FAVORITE MARTIAN (I,3189); MY WORLD . . . AND WELCOME TO IT (I,3208); THE NANCY WALKER SHOW (II,1790); ONE DAY AT A TIME (II,1900); PAUL SAND IN FRIENDS AND LOVERS (II,1982); RHODA (II,2151); SANFORD AND SON (II,2256); THE SUPER (I,4293); THAT'S MY MAMA (II,2580); THE TIM CONWAY SHOW (I,4502); VIVA VALDEZ (II,2728); WE GOT IT MADE (II,2752); WHAT'S HAPPENING!! (II,2769); A YEAR AT THE TOP (II,2844)

RAFKINKSON, JAN S
THAT'S INCREDIBLE! (II,2578)

RAINBOLT, BILL
THE LIAR'S CLUB (II,1466); THE SQUARE WORLD OF ED BUTLER (I,4171)

RAINBOLT, ROBERT
AMERICA'S TOP TEN (II,68)

RAKOFF, ALAN
THE NEW ADVENTURES OF CHARLIE CHAN (I,3249)

RALLING, CHRISTOPHER
SEARCH FOR THE NILE (I,3954)

RANKIN JR, ARTHUR
'TWAS THE NIGHT BEFORE CHRISTMAS (II,2677); THE BEATLES (I,380); THE CONEHEADS (II,567); THE EASTER BUNNY IS COMIN' TO TOWN (II,753); FROSTY

THE SNOWMAN (II,934); FROSTY'S WINTER WONDERLAND (II,935); HERE COMES PETER COTTONTAIL (II,1132); JACK FROST (II,1263); THE LEPRECHAUN'S CHRISTMAS GOLD (II,1461); THE LITTLE DRUMMER BOY (II,1485); LITTLE DRUMMER BOY, BOOK II (II,1486); THE OSMONDS (I,3410); PINOCCHIO'S CHRISTMAS (II,2045); THE RELUCTANT DRAGON AND MR. TOAD (I,3762); RUDOLPH'S SHINY NEW YEAR (II,2228); SANTA CLAUS IS COMIN' TO TOWN (II,2259); THE YEAR WITHOUT A SANTA CLAUS (II,2846)

RAPOPORT, I C
BORN TO THE WIND (II,354); THOU SHALT NOT KILL (II,2604)

RAPP, PHILIP
MIMI (I,3054); TOPPER (I,4582)

RASINSKI, CONNIE
THE ASTRONAUT SHOW (I,307); DEPUTY DAWG (I,1233); THE MIGHTY MOUSE PLAYHOUSE (I,3035)

RASKIN, CAROLYN
LAUGH TRAX (II,1443); ROCK COMEDY (II,2191); THE SHAPE OF THINGS (I,3999); STATE FAIR USA (II,2449); STATE FAIR USA (II,2450)

RASKY, HARRY
THE LEGEND OF SILENT NIGHT (I,2655); PERSPECTIVE ON GREATNESS (I,3555)

RAWLINS, PHIL
THE HIGH CHAPARRAL (I,2051)

RAY, TOM
DUNGEONS AND DRAGONS (II,744); G.I. JOE: A REAL AMERICAN HERO (II,948); JIM HENSON'S MUPPET BABIES (II,1289); MEATBALLS AND SPAGHETTI (II,1674); THE TRANSFORMERS (II,2653)

RAYEL, JACK
ENTERTAINMENT —1955 (I,1450)

REA, BEN
THE ONEDIN LINE (II,1912)

REARDON, JOHN
THE ADVENTURES OF BLACK BEAUTY (I,51)

REDFORD, KEN
HOMEMAKER'S EXCHANGE (I,2100); IT PAYS TO BE IGNORANT (I,2248); MESSING PRIZE PARTY (I,3019); THE TED STEELE SHOW (I,4373); THE WARREN HULL SHOW (I,4771)

REDMAN, SCOTT
GUILTY OR INNOCENT (II,1065)

REED, BILL
MISSION MAGIC (I,3068); MY FAVORITE MARTIANS (II,1779); THE NEW ADVENTURES OF GILLIGAN (II,1813); THE U.S. OF ARCHIE (I,4687); U.S. OF ARCHIE (II,2698)

REED, DEAN
THE PENDULUM (I,3511)

REED, ROBERT
THE BRADY BUNCH (II,362)

REEVES, GEORGE
THE ADVENTURES OF SUPERMAN (I,77)

REGAN, DARYL
LET'S MAKE A DEAL (II,1463)

REGAS, JACK
BARBARA MANDRELL AND THE MANDRELL SISTERS (II,143); BARBARA MANDRELL—THE LADY IS A CHAMP (II,144); THE BAY CITY ROLLERS SHOW (II,187); THE BEST LITTLE SPECIAL IN TEXAS (II,215); THE BRADY BUNCH HOUR (II,363); CHARO (II,488); DANCE FEVER (II,615); DINAH IN SEARCH OF THE IDEAL MAN (I,1278); DOCTOR SHRINKER (II,686); THE DON HO SHOW (II,706); FOUL UPS, BLEEPS AND BLUNDERS (II,911); HONEYMOON SUITE (I,2108); IT'S A BIRD, IT'S A PLANE, IT'S SUPERMAN (II,1253); THE JERRY REED WHEN YOU'RE HOT, YOU'RE HOT HOUR (I,2372); THE JOHN BYNER COMEDY HOUR (I,2407); THE KROFFT KOMEDY HOUR (II,1411); THE KROFFT SUPERSHOW II (II,1413); THE LOST SAUCER (II,1524); LOUISE MANDRELL: DIAMONDS, GOLD AND PLATINUM (II,1528); THE LOVE BOAT (II,1535); MAGIC MONGO (II,1592); MEL AND SUSAN TOGETHER (II,1677); THE PAT BOONE AND FAMILY CHRISTMAS SPECIAL (II,1969); THE PAT BOONE AND FAMILY EASTER SPECIAL (II,1970); PERRY COMO'S SPRINGTIME SPECIAL (II,2023); PLAYBOY'S PLAYMATE PARTY (II,2055); ROCK PALACE (II,2193)

REHBERG, EDDIE
THE NEW THREE STOOGES (I,3274)

REHR, DARYL
YOU ASKED FOR IT (II,2855)

REICHENBACH, FRANCOIS
BRIGITTE BARDOT (I,729)

REID, DAVID
THE STRAUSS FAMILY (I,4253)

REID, MAX
PERSONAL AND CONFIDENTIAL (II,2026)

REINER, CARL
FLANNERY AND QUILT (II,873); GOOD HEAVENS (II,1032); GOOD MORNING WORLD (I,1850); A TOUCH OF GRACE (I,4585)

REINER, ROB
SONNY BOY (II,2402)

REISNER, ALLEN
BARNABY JONES (II,153); BAY CITY BLUES (II,186); BEN CASEY (I,394); BRACKEN'S WORLD (I,703); BRANDED (I,707); CANNON (II,424); CAPTAINS AND THE KINGS (II,435); CITY HOSPITAL (I,967); THE COPS AND ROBIN (II,573); THE ELEVENTH HOUR (I,1425); FOR LOVE AND HONOR (II,903); GOING MY WAY (I,1833); HARDCASTLE AND MCCORMICK (II,1089); HAWAII FIVE-O (II,1110); THE HIGH CHAPARRAL (I,2051); I SPY (I,2179); IRONSIDE (II,1246); KOJAK (II,1406); LANCER (I,2610); LEAVE IT TO LARRY (I,2649); LEGMEN (II,1458); MANHUNTER (II,1621); THE MISSISSIPPI (II,1707); MURDER, SHE WROTE (II,1770); NIGHT GALLERY (I,3287); OWEN MARSHALL: COUNSELOR AT LAW (I,3435); PANTOMIME QUIZ (I,3449); RAWHIDE (I,3727); ROUTE 66 (I,3852); SAN FRANCISCO INTERNATIONAL AIRPORT (I,3905); SKAG (II,2374); TALES OF THE UNEXPECTED (II,2539); THE TIME ELEMENT (I,4504); THE TWILIGHT ZONE (I,4651); THE UNTOUCHABLES (I,4682)

REISTER, FRED
THE LOVE REPORT (II,1542); THE LOVE REPORT (II,1543)

REO, DON
LANDON, LANDON & LANDON (II,1426)

REY, ALEJANDRO
THE FACTS OF LIFE (II,805); FOREVER FERNWOOD (II,909)

REYNOLDS, BURT
HAWK (I,1974)

REYNOLDS, GENE
77 SUNSET STRIP (I,3988); ALFRED HITCHCOCK PRESENTS (I,115); BLISS (II,269); CAPTAIN NICE (I,835); THE CARA WILLIAMS SHOW (I,843); THE DUCK FACTORY (II,738); F TROOP (I,1499); THE FARMER'S DAUGHTER (I,1533); THE FITZPATRICKS (II,868); THE GHOST AND MRS. MUIR (I,1786); HOGAN'S HEROES (I,2069); IF I LOVE YOU, AM I TRAPPED FOREVER? (II,1222); KAREN (II,1363); LEAVE IT TO BEAVER (I,2648); LOU GRANT (II,1526); M*A*S*H (II,1569); MR. ROBERTS (I,3149); THE MUNSTERS (I,3158); MY THREE SONS (I,3205); PEOPLE LIKE US (II,1999); ROLL OUT! (I,3833); ROOM 222 (I,3843); ROOM FOR ONE MORE (I,3842); SOUTHERN FRIED (I,4139); SWINGIN' TOGETHER (I,4320); WANTED: DEAD OR ALIVE (I,4764); WENDY AND ME (I,4793)

REYNOLDS, HAL
OUTLAW LADY (II,1935)

REYNOLDS, JIM
PAY CARDS (I,3496)

REYNOLDS, SHELDON
DANGER (I,1134); THE DESILU PLAYHOUSE (I,1237); INN OF THE FLYING DRAGON (I,2214); SHERLOCK HOLMES (I,4010); SOPHIA LOREN IN ROME (I,4129)

RHEINSTEIN, FRED
THE MELTING POT (II,1678)

RHODER, MICHAEL
BAY CITY BLUES (II,186)

RHODES, MICHAEL
JOSIE (II,1345); THE JUGGLER OF NOTRE DAME (II,1350); LEADFOOT (II,1452); PRINCESS (II,2084); THE TROUBLE WITH GRANDPA (II,2662)

RHONE, ALLAN
FAIRMEADOWS, U.S.A. (I,1514)

RICH, DAVID LOWELL
77 SUNSET STRIP (I,3988); ALCOA PREMIERE (I,109); ALFRED HITCHCOCK PRESENTS (I,115); THE

AMAZING POLGAR (I,160); ARTHUR GODFREY'S TALENT SCOUTS (I,291); ASSIGNMENT: MUNICH (I,302); THE BARBARA STANWYCK THEATER (I,350); BEN CASEY (I,394); BIG TOWN (I,436); BLACK SADDLE (I,480); THE BOB HOPE CHRYSLER THEATER (I,502); BRIDGER (II,376); BROCK'S LAST CASE (I,742); THE CHADWICK FAMILY (II,478); CIRCLE OF FEAR (I,958); THE CRIME CLUB (I,1096); THE DAUGHTERS OF JOSHUA CABE RETURN (I,1170); GHOST STORY (I,1788); THE GIRLS (I,1811); IRONSIDE (II,1246); JOHNNY RINGO (I,2434); THE JUDGE AND JAKE WYLER (I,2464); LAW OF THE PLAINSMAN (I,2639); LITTLE WOMEN (II,1497); M SQUAD (I,2799); MARCUS WELBY, M.D. (I,2905); MARCUS WELBY, M.D. (II,1627); MISSION: IMPOSSIBLE (I,3067); NAKED CITY (I,3216); OWEN MARSHALL: COUNSELOR AT LAW (I,3435); PETER GUNN (I,3562); POLICE STORY (I,3635); RANSOM FOR ALICE (II,2113); ROUTE 66 (I,3852); STATE FAIR (II,2448); SWISS FAMILY ROBINSON (II,2517); THE VIRGINIAN (I,4737); WAGON TRAIN (I,4747); WON'T IT EVER BE MORNING? (I,4902); ZANE GREY THEATER (I,4979)

RICH, JOHN
ADVENTURES OF COLONEL FLACK (I,56); ALL IN THE FAMILY (II,38); AMANDA'S (II,62); BENSON (II,208); BILLY (II,247); THE BRADY BUNCH (II,362); CHARO AND THE SERGEANT (II,489); CONDO (II,565); THE DICK VAN DYKE SHOW (I,1275); DOROTHY (II,717); GILLIGAN'S ISLAND (II,990); GOMER PYLE, U.S.M.C. (I,1843); GRANDPA MAX (II,1051); I MARRIED JOAN (I,2174); I'LL NEVER FORGET WHAT'S HER NAME (II,1217); MCNAB'S LAB (I,2972); MOTHER, JUGGS AND SPEED (II,1741); MR. ED (I,3137); MY WORLD . . . AND WELCOME TO IT (I,3208); NEWHART (II,1835); ON THE ROCKS (II,1894); OUR MISS BROOKS (I,3415); PINE LAKE LODGE (I,3597); SCREEN DIRECTOR'S PLAYHOUSE (I,3946); SLEZAK AND SON (I,4086); THAT GIRL (II,2570); THE TWILIGHT ZONE (I,4651); WHERE'S RAYMOND? (I,4837)

RICHARDS, LLOYD
ROOTS: THE NEXT GENERATIONS (II,2213)

RICHARDS, PENNINGTON
DANGER MAN (I,1136); THE INVISIBLE MAN (I,2232); IVANHOE (I,2282); SECRET AGENT (I,3962)

RICHARDSON, BOB
DENNIS THE MENACE: MAYDAY FOR MOTHER (II,660); SPIDER-MAN AND HIS AMAZING FRIENDS (II,2425); SPIDER-WOMAN (II,2426)

RICHARDSON, DON
THE ADVENTURES OF ELLERY QUEEN (I,59); ARNIE (I,261); BILLY BOONE AND COUSIN KIBB (I,450); BONANZA (II,347); A DATE WITH JUDY (I,1162); THE DEFENDERS (I,1220); EMERGENCY! (II,775); THE HIGH CHAPARRAL (I,2051); I REMEMBER MAMA (I,2176); LANCER (I,2610); LOST IN SPACE (I,2758); MISSION: IMPOSSIBLE (I,3067); THE MONTEFUSCOS (II,1727); MR. I. MAGINATION (I,3140); MY LUCKY PENNY (I,3196); MY SON THE DOCTOR (I,3203); ONE DAY AT A TIME (II,1900); THE OREGON TRAIL (II,1923); THE VIRGINIAN (I,4738)

RICKEY, FRED
FLOOR SHOW (I,1605); MR. I. MAGINATION (I,3140); OPEN HOUSE (I,3394); RUTHIE ON THE TELEPHONE (I,3877); SORRY, WRONG NUMBER (I,4130); THE STORK CLUB (I,4236)

RIDGE, MARY
BLAKE'S SEVEN (II,264)

RIESNER, ALLEN
THE DESILU PLAYHOUSE (I,1237)

RIGSBY, GORDON
MY THREE ANGELS (I,3204); THE PAT BOONE SHOW (I,3471)

RILEY, THOMAS
WHEN THE NIGHTINGALE SANG IN BERKELEY SQUARE (I,4827)

RING, BILL
TALENT VARIETIES (I,4343)

RIPLEY, ALAN
COLT .45 (I,1010)

RIPLEY, ARTHUR
YOUR JEWELER'S SHOWCASE (I,4953)

RIPP, HEINO
THE NEW SHOW (II,1828)

RIPPEY, BOB
HOWDY DOODY (I,2151)

RISGBY, GORDON
YOU'RE GONNA LOVE IT HERE (II,2860)

RISKIN, RALPH
THE DUKES OF HAZZARD (II,742)

RITCHIE, MICHAEL
THE BIG VALLEY (I,437); THE BOB HOPE CHRYSLER THEATER (I,502); THE DETECTIVES (I,1245); DOCTOR KILDARE (I,1315); THE FELONY SQUAD (I,1563); THE OUTSIDER (I,3430); RUN FOR YOUR LIFE (I,3871); THE SOUND OF ANGER (I,4133)

RITELIS, VIKTORS
THE ONEDIN LINE (II,1912); QUILLER: NIGHT OF THE FATHER (II,2100)

RITT, MARTIN
THE ALCOA HOUR (I,108); TELLER OF TALES (I,4388)

RITTER, HOWARD L
SANTA BARBARA (II,2258)

RIZZO, ANTHONY
ACTION AUTOGRAPHS (I,24); MR. BLACK (I,3129); PADDY THE PELICAN (I,3442)

ROBBIE, SEYMOUR
THE ANDROS TARGETS (II,76); ART CARNEY MEETS THE SORCERER'S APPRENTICE (I,273); BARNABY JONES (II,153); BE OUR GUEST (I,370); BEWITCHED (I,418); BID N' BUY (I,419); BIG HAWAII (II,238); THE BILL COSBY SHOW (I,442); CANNON (II,424); DAN AUGUST (I,1130); THE ELEVENTH HOUR (I,1425); ELLERY QUEEN (II,766); F TROOP (I,1499); THE FBI (I,1551); THE FEATHER AND FATHER GANG (II,845); THE FELONY SQUAD (I,1563); HAGEN (II,1071); HART TO HART (II,1102); THE HIGH CHAPARRAL (I,2051); HONEY WEST (I,2105); HOW DO YOU RATE? (I,2144); JACKIE GLEASON AND HIS AMERICAN SCENE MAGAZINE (I,2321); KATE LOVES A MYSTERY (II,1367); LOST IN SPACE (I,2758); MANHUNTER (II,1621); MANNIX (II,1624); MISSION: IMPOSSIBLE (I,3067); THE MOD SQUAD (I,3078); MR. ROBERTS (I,3149); MURDER, SHE WROTE (II,1770); THE NAME OF THE GAME (I,3217); NEWSSTAND THEATER (I,3279); THE NIGHT STALKER (II,1849); NURSE (II,1870); THE PAPER CHASE (II,1949); POLICE STORY (II,2062); REMINGTON STEELE (II,2130); ROOM 222 (I,3843); SARGE (I,3914); THE STREETS OF SAN FRANCISCO (II,2478); SWITCH (II,2519); TRAPPER JOHN, M.D. (II,2654); THE VIRGINIAN (I,4738); WHAT REALLY HAPPENED TO THE CLASS OF '65? (II,2768); WONDER WOMAN (SERIES 2) (II,2829); YOU'RE IN THE PICTURE (I,4965)

ROBBINS, JEROME
PETER PAN (I,3566)

ROBERTS, PENNANT
BLAKE'S SEVEN (II,264); DOCTOR WHO (II,689)

ROBERTSON, CHRIS
BARNABY JONES (II,153)

ROBERTSON, TOM
I'M SOOO UGLY (II,1219); THE TROUBLE WITH MOTHER (II,2663)

ROBINS, JOHN
THE BENNY HILL SHOW (II,207); THE MARTY FELDMAN COMEDY MACHINE (I,2929); NO SOAP, RADIO (II,1856); ONE DAY AT A TIME (II,1900); THE SHANI WALLIS SHOW (I,3997); TOP TEN (II,2648); THE UGLIEST GIRL IN TOWN (I,4663)

ROBINSON, CHRIS
CANNON (II,424)

RODGERS, DOUGLAS
LIFE BEGINS AT EIGHTY (I,2682)

RODGERS, MARK
JIGSAW (I,2377)

ROEMER, LARRY
RUDOLPH THE RED-NOSED REINDEER (II,2227)

ROGELL, ALBERT S
THE 20TH CENTURY-FOX HOUR (I,4642); BROKEN ARROW (I,744); MY FRIEND FLICKA (I,3190)

ROGERS, ART
THE DANCE SHOW (II,616)

ROGERS, DOUG
THE ALL-STAR REVUE (I,138); BEST OF THE WEST (II,218); THE BETTY WHITE SHOW (II,224); DIFF'RENT STROKES (II,674); HELLO, LARRY (II,1128); I'M A BIG GIRL NOW (II,1218); IN THE BEGINNING (II,1229); LEWIS

AND CLARK (II,1465); THE MARY TYLER MOORE SHOW (II,1640); THE MOHAWK SHOWROOM (I,3080); MR. & MRS. DRACULA (II,1745); ON OUR OWN (II,1892); ONE OF THE BOYS (II,1911); PHYLLIS (II,2038); RHODA (II,2151)

ROGERS, HUGH
IT PAYS TO BE IGNORANT (I,2248); MASTERS OF MAGIC (I,2943)

ROGERS, JAIME
MARY'S INCREDIBLE DREAM (II,1641)

ROGERS, WAYNE
HOUSE CALLS (II,1194)

ROHRER, SUSAN
ANOTHER LIFE (II,95)

ROLAND, HERB
KING OF KENSINGTON (II,1395)

ROLEY, SUTTON
240-ROBERT (II,2688); 77 SUNSET STRIP (I,3988); B.A.D. CATS (II,123); BARETTA (II,152); THE BIG VALLEY (I,437); BRONK (II,379); CHARLIE'S ANGELS (II,486); COMBAT (I,1011); THE CONTENDER (II,571); THE DETECTIVES (I,1245); HAWAII FIVE-O (II,1110); THE INVADERS (I,2229); LOST IN SPACE (I,2758); LUCAN (II,1554); THE MAN FROM U.N.C.L.E. (I,2867); MANNIX (II,1624); MISSION: IMPOSSIBLE (I,3067); THE RAT PATROL (I,3726); RIKER (II,2175); S.W.A.T. (II,2236); SAINTS AND SINNERS (I,3887); STARSKY AND HUTCH (II,2444); SWEET, SWEET RACHEL (I,4313); VEGAS (II,2724); VOYAGE TO THE BOTTOM OF THE SEA (I,4743); WAGON TRAIN (I,4747)

ROMAN, PHIL
BE MY VALENTINE, CHARLIE BROWN (I,369); IS THIS GOODBYE, CHARLIE BROWN? (I,2242); IT'S A MYSTERY, CHARLIE BROWN (I,2259); IT'S AN ADVENTURE, CHARLIE BROWN (I,2266); IT'S ARBOR DAY, CHARLIE BROWN (I,2267); IT'S MAGIC, CHARLIE BROWN (I,2271); IT'S THE EASTER BEAGLE, CHARLIE BROWN (I,2275); IT'S YOUR FIRST KISS, CHARLIE BROWN (I,2280); LIFE IS A CIRCUS, CHARLIE BROWN (I,2683); SHE'S A GOOD SKATE, CHARLIE BROWN (I,4012); SOMEDAY YOU'LL FIND HER, CHARLIE BROWN (I,4113); WHAT A NIGHTMARE, CHARLIE BROWN (I,4806); YOU'RE A GOOD SPORT, CHARLIE BROWN (I,4963); YOU'RE THE GREATEST, CHARLIE BROWN (I,4973)

RONDEAU, CHARLES R
77 SUNSET STRIP (I,3988); ADVENTURES IN PARADISE (I,46); ALFRED OF THE AMAZON (I,116); B.J. AND THE BEAR (II,125); BAREFOOT IN THE PARK (I,355); BARETTA (II,152); BATMAN (I,366); BEN CASEY (I,394); BIG FOOT AND WILD BOY (II,237); BONANZA (II,347); BOURBON STREET BEAT (I,699); CALIFORNIA FEVER (II,411); CIMARRON STRIP (I,954); THE DAKOTAS (I,1126); F TROOP (I,1499); THE GALLANT MEN (I,1728); THE GEMINI MAN (II,960); HAWAIIAN EYE (I,1973); JIGSAW JOHN (II,1288); MAVERICK (II,1657); THE MISADVENTURES OF SHERIFF LOBO (II,1704); THE MUNSTERS (I,3158); THE MURDOCKS AND THE MCCLAYS (I,3163); NO TIME FOR SERGEANTS (I,3300); THE ODD COUPLE (II,1875); THE PARTRIDGE FAMILY (II,1962); PERRY MASON (II,2025); RAWHIDE (I,3727); THE RED HAND GANG (II,2122); THE ROARING TWENTIES (I,3805); ROOM 222 (I,3843); ROOM FOR ONE MORE (I,3842); RUN RUN, JOE (II,2229); SURFSIDE 6 (I,4299); TABITHA (II,2528); TEMPLE HOUSTON (I,4392); VOYAGE TO THE BOTTOM OF THE SEA (I,4743); THE WILD WILD WEST (I,4863); WONDER WOMAN (SERIES 1) (II,2828)

RONDELL, RIC
240-ROBERT (II,2688); BRONK (II,379)

ROSE, DICK
THE GLORIA SWANSON HOUR (I,1827)

ROSEMOND, PERRY
ABC'S SILVER ANNIVERSARY CELEBRATION (II,2); ANSON AND LORRIE (II,98); BIZARRE (II,259); THE COMEDY SHOP (II,558); THE DAVID SOUL AND FRIENDS SPECIAL (II,626); DEAN MARTIN PRESENTS MUSIC COUNTRY, U.S.A. (I,1192); DICK CLARK'S LIVE WEDNESDAY (II,672); DONNY AND MARIE (II,712); FRAGGLE ROCK (II,916); GOOD TIMES (II,1038); I'VE HAD IT UP TO HERE (II,1221); THE KEN BERRY WOW SHOW (I,2524); OUT OF OUR MINDS (II,1933); PAT BOONE AND FAMILY (II,1968); PEOPLE ARE FUNNY (II,1996); THANKSGIVING REUNION WITH THE PARTRIDGE FAMILY AND MY THREE SONS (II,2569); TWILIGHT THEATER (II,2685)

ROSENBAUM, ROBERT
BEWITCHED (I,418); CAMP RUNAMUCK (I,811); HERE COME THE BRIDES (I,2024)

ROSENBERG, META
THE ROCKFORD FILES (II,2197)

ROSENBERG, STUART
ALFRED HITCHCOCK PRESENTS (I,115); BEN CASEY (I,394); THE BIG STORY (I,432); THE BOB HOPE CHRYSLER THEATER (I,502); BUS STOP (I,778); DECOY (I,1217); THE DEFENDERS (I,1220); FAME IS THE NAME OF THE GAME (I,1518); NAKED CITY (I,3215); NAKED CITY (I,3216); THE NAME OF THE GAME (I,3217); RAWHIDE (I,3727); RUN FOR YOUR LIFE (I,3871); THE UNTOUCHABLES (I,4682)

ROSENBLUM, RALPH
BREAKING AWAY (II,371); SUMMER SOLSTICE (II,2492)

ROSENTHAL, RICK
DARKROOM (II,619); SECRETS OF MIDLAND HEIGHTS (II,2296)

ROSS, DICK
THE BARBARA MCNAIR AND DUKE ELLINGTON SPECIAL (I,347); THE CYD CHARISSE SPECIAL (I,1118); THE JULIE LONDON SPECIAL (I,2484); THE PEGGY LEE SPECIAL (I,3510); THE ROBERT GOULET SPECIAL (I,3807)

ROSS, HERB
FOUR FOR TONIGHT (I,1648)

ROSS, MICHAEL
SIMON AND SIMON (II,2357); THREE'S COMPANY (II,2614)

ROSS, NORMAN
CINDERELLA (II,526)

ROSTEN, IRWIN
RIPLEY'S BELIEVE IT OR NOT (II,2177)

ROYAL, THOMAS
GOTTA SING, GOTTA DANCE (II,1046)

RUBEN, AARON
HEADMASTER (I,1990); THE PHIL SILVERS PONTIAC SPECIAL: KEEP IN STEP (I,3579)

RUBEN, GLADYS
SANDY DREAMS (I,3909)

RUBEN, JOSEPH
BREAKING AWAY (II,371)

RUBIN, JACK
THE O'NEILLS (I,3392)

RUDD, AL
SUSAN'S SHOW (I,4303)

RUDOLPH, OSCAR
ALL IN THE FAMILY (I,133); THE ANN SOTHERN SHOW (I,220); BATMAN (I,366); THE BRADY BUNCH (II,362); DOCTOR MIKE (I,1316); THE DONNA REED SHOW (I,1347); THE GHOST AND MRS. MUIR (I,1786); GIDGET (I,1795); HERE WE GO AGAIN (I,2028); THE HUNTER (I,2162); JOHNNY NIGHTHAWK (I,2432); THE LONE RANGER (I,2740); MY FAVORITE MARTIAN (I,3189); MY SISTER EILEEN (I,3200); PRIVATE SECRETARY (I,3672); TIGHTROPE (I,4500); TOM, DICK, AND HARRY (I,4536)

RUMBAUGH, JOHN
STEVE ALLEN'S LAUGH-BACK (II,2455)

RUSKIN, COBY
THE ALL-STAR COMEDY SHOW (I,137); THE ALL-STAR REVUE (I,138); THE ANDY GRIFFITH SHOW (I,192); THE BILL COSBY SHOW (I,442); THE BOB NEWHART SHOW (I,666); THE DORIS DAY SHOW (I,1355); GEORGE GOBEL PRESENTS (I,1766); GOMER PYLE, U.S.M.C. (I,1843); HERE'S LUCY (II,1135); JULIA (I,2476); JUST POLLY AND ME (I,2497); LI'L ABNER (I,2701); SAMMY KAYE'S MUSIC FROM MANHATTAN (I,3902); SO YOU WANT TO LEAD A BAND (I,4106); WHEN THINGS WERE ROTTEN (II,2782)

RUSSELL, CHARLES
YOU ARE THERE (I,4929)

RUSSELL, PADDY
DOCTOR WHO (II,689); LITTLE WOMEN (I,2723); THE OMEGA FACTOR (II,1888)

RUSSELL, WILLIAM D
THE 20TH CENTURY-FOX HOUR (I,4642); BOSS LADY (I,694); DENNIS THE MENACE (I,1231); THE DONNA REED SHOW (I,1347); FAMILY

AFFAIR (I,1519); THE FARMER'S DAUGHTER (I,1533); FATHER KNOWS BEST (II,838); HAZEL (I,1982); YOU ARE THERE (I,4929)

RYDELL, MARK
BEN CASEY (I,394); FAMILY (II,813); GUNSMOKE (II,1069); I SPY (I,2179); THE WILD WILD WEST (I,4863)

RYDER, EDDIE
THE BOB NEWHART SHOW (II,342)

SAGAL, BORIS
77 SUNSET STRIP (I,3988); ADVENTURES IN PARADISE (I,46); THE ALCOA HOUR (I,108); ALFRED HITCHCOCK PRESENTS (I,115); AMY PRENTISS (II,74); THE AWAKENING LAND (II,122); CIMARRON STRIP (I,954); THE CLIFF DWELLERS (I,974); COLUMBO (II,556); COMBAT (I,1011); THE D.A.: MURDER ONE (I,1122); THE DEFENDERS (I,1220); THE GREATEST GIFT (II,1061); GRIFF (I,1888); HITCHED (I,2065); IKE (II,1223); IRONSIDE (II,1246); JUDD, FOR THE DEFENSE (I,2463); MADIGAN (I,2808); MALLORY: CIRCUMSTANTIAL EVIDENCE (II,1608); THE MAN FROM U.N.C.L.E. (I,2867); MANHATTAN TOWER (I,2889); MASADA (II,1642); MCCLOUD (II,1660); MEDICAL CENTER (II,1675); THE MONEYCHANGERS (II,1724); MR. LUCKY (I,3141); MRS. COLUMBO (II,1760); NAKED CITY (I,3216); NIGHT GALLERY (I,3288); NIGHT GALLERY (I,3287); THE OREGON TRAIL (II,1922); PETER GUNN (I,3562); REBECCA (I,3745); RICH MAN, POOR MAN—BOOK I (II,2161); THE RUNAWAY BARGE (II,2230); SAN FRANCISCO INTERNATIONAL AIRPORT (I,3905); THE SNOOP SISTERS (II,2389); THE SPIRAL STAIRCASE (I,4160); T.H.E. CAT (I,4430); THREE FOR THE ROAD (II,2607); THE TWILIGHT ZONE (I,4651); U.M.C. (I,4666)

SAKAI, RICHARD
TAXI (II,2546)

SAKIN, LEO
THE 2,000 YEAR OLD MAN (II,2696)

SAKS, GENE
LOVE, SEX. . . AND MARRIAGE (II,1546)

SALKOW, SIDNEY
SIMON LASH (I,4052); THE SOFT TOUCH (I,4107)

SALKOW, SY
77 SUNSET STRIP (I,3988); BRONCO (I,745); HAWAIIAN EYE (I,1973); THE ROARING TWENTIES (I,3805); SURFSIDE 6 (I,4299)

SALLIN, ROBERT
THE MISSISSIPPI (II,1707)

SALTZMAN, BERT
MUGGSY (II,1761); TALES OF THE UNEXPECTED (II,2540)

SAMETH, JACK
THE GREAT AMERICAN DREAM MACHINE (I,1871)

SAMMON, ROBERT
PERSON TO PERSON (I,3550)

SAMPSON, JACK
CIRCUS OF THE 21ST CENTURY (II,529)

SAMPSON, PADDY
LENA HORNE: THE LADY AND HER MUSIC (II,1460)

SAN FERNANDO, MANUEL
JOHNNY SOKKO AND HIS FLYING ROBOT (I,2436)

SANDERS, DENNIS
THE AMERICAN WEST OF JOHN FORD (I,173); THE DEFENDERS (I,1220); HAVE GUN—WILL TRAVEL (I,1968); NAKED CITY (I,3216); ROUTE 66 (I,3852)

SANDERS, TERRY
THE KIDS FROM FAME (II,1392); THE LEGEND OF MARILYN MONROE (I,2654)

SANDRICH, JAY
ADAMS HOUSE (II,9); BACHELOR AT LAW (I,322); BALL FOUR (II,137); BEACON HILL (II,193); BENSON (II,208); THE BILL COSBY SHOW (I,442); THE BOB NEWHART SHOW (II,342); CAPTAIN NICE (I,835); THE COSBY SHOW (II,578); THE EARTHLINGS (II,751); FOG (II,897); THE FOUR SEASONS (II,915); FRIENDS AND LOVERS (II,930); THE GHOST AND MRS. MUIR (I,1786); HARRY AND MAGGIE (II,1097); HE AND SHE (I,1985); HERE WE GO AGAIN (I,2028); THE HERO (I,2035); IT TAKES TWO (II,1252); IT'S A LIVING (II,1254); LAVERNE AND SHIRLEY (II,1446); THE LILY TOMLIN SPECIAL (II,1479); LOU GRANT (II,1526); LOVE, SIDNEY (II,1547); LOVES ME, LOVES ME NOT (II,1551); MAKING A LIVING (II,1604); THE MARY TYLER MOORE SHOW (II,1640); MAUREEN (II,1656); NIGHT COURT (II,1844); THE ODD COUPLE (II,1875); OFF THE RACK (II,1878); PAUL SAND IN FRIENDS AND LOVERS (II,1982); SIDE BY SIDE (II,2346); SOAP (II,2392); STOCKARD CHANNING IN JUST FRIENDS (II,2467); THE STOCKARD CHANNING SHOW (II,2468); SUSAN AND SAM (II,2506); THAT GIRL (II,2570); THREE TIMES DALEY (II,2610); TO SIR, WITH LOVE (II,2624); THE TONY RANDALL SHOW (II,2640); WE'LL GET BY (II,2753); WE'LL GET BY (II,2754); WIVES (II,2812); WKRP IN CINCINNATI (II,2814)

SANDRICH, MARK
77 SUNSET STRIP (I,3988)

SANDRICH, MICHAEL
HAWAIIAN EYE (I,1973); LAWMAN (I,2642)

SANDWICK, DICK
DAYS OF OUR LIVES (II,629); THE JOAN EDWARDS SHOW (I,2396); NIGHT EDITOR (I,3286); ROCKY KING, INSIDE DETECTIVE (I,3824); TWENTY QUESTIONS (I,4647)

SANFORD, ARLENE
CAPITOL (II,426)

SANFORD, GERALD
THE FBI (I,1551)

SANGSTER, JIMMY
BANACEK (II,138); CANNON (II,424); FARADAY AND COMPANY (II,833); GHOST STORY (I,1788); IRONSIDE (II,1246); A MAN CALLED SLOANE (II,1613); MCCLOUD (II,1660); RIPLEY'S BELIEVE IT OR NOT (II,2177)

SANIEE, CLARK
GRAFFITI ROCK (II,1049)

SANTOS, STEVEN J
GEORGE CARLIN AT CARNEGIE HALL (II,974)

SARAFIAN, RICHARD
THE AFRICAN QUEEN (II,18); BATMAN (I,366); BEN CASEY (I,394); THE BIG VALLEY (I,437); BRONCO (I,745); BUS STOP (I,778); CHEYENNE (I,931); CIMARRON STRIP (I,954); THE DAKOTAS (I,1126); DOCTOR KILDARE (I,1315); THE GALLANT MEN (I,1728) THE GANGSTER CHRONICLES (II,957); THE GIRL FROM U.N.C.L.E. (I,1808); GUNSMOKE (II,1069); HAWAIIAN EYE (I,1973); THE HIGH CHAPARRAL (I,2051); I SPY (I,2179); LAWMAN (I,2642); MAVERICK (II,1657); ONE OF OUR OWN (II,1910); THE ROARING TWENTIES (I,3805); SHANNON (II,2314); SURFSIDE 6 (I,4299); THE WILD WILD WEST (I,4863)

SARGENT, JOSEPH
GUNSMOKE (II,1069); THE IMMORTAL (I,2196); THE INVADERS (I,2229); LONGSTREET (I,2750); THE MAN FROM U.N.C.L.E. (I,2867); MAN ON A STRING (I,2877); THE MANIONS OF AMERICA (II,1623); THE MOONGLOW AFFAIR (I,3099); STAR TREK (II,2440)

SAROYAN, HANK
JIM HENSON'S MUPPET BABIES (II,1289)

SASDY, PETER
RETURN OF THE SAINT (II,2141)

SATENSTEIN, FRANK
AMERICA'S GREATEST BANDS (I,177); CAVALCADE OF STARS (I,877); THE HONEYMOONERS (I,2110); THE HONEYMOONERS (II,1175); THE JACKIE GLEASON SHOW (I,2322); THE KEN MURRAY SHOW (I,2525); YOUR SURPRISE STORE (I,4961)

SATLOF, RON
THE A-TEAM (II,119); THE AMAZING SPIDER-MAN (II,63); BARNABY JONES (II,153); BATTLES: THE MURDER THAT WOULDN'T DIE (II,180); BENNY AND BARNEY: LAS VEGAS UNDERCOVER (II,206); THE BEST OF FRIENDS (II,216); BUSH DOCTOR (II,400); THE EDDIE CAPRA MYSTERIES (II,758); THE FALL GUY (II,811); FLAMINGO ROAD (II,872); FROM HERE TO ETERNITY (II,933); GALACTICA 1980 (II,953); HARDCASTLE AND MCCORMICK (II,1089); THE HARDY BOYS MYSTERIES (II,1090); HUNTER (II,1206); NERO WOLFE (II,1806); THE POWERS OF MATTHEW STAR (II,2075); QUINCY, M. E. (II,2102); RIPTIDE (II,2178); THE ROUSTERS (II,2220); SALVAGE 1 (II,2245); TONI'S BOYS (II,2637); VOYAGERS (II,2730); WAIKIKI (II,2734); WHAT REALLY HAPPENED TO THE CLASS OF '65? (II,2768)

SAUER, ERNEST
DAYTIME (II,630)

SAUNDERS, DESMOND
CAPTAIN SCARLET AND THE MYSTERONS (I,837); THUNDERBIRDS (I,4497)

SAVALAS, TELLY
KOJAK (II,1406)

SCALEM, JAMES
THE INTERNATIONAL ANIMATION FESTIVAL (II,1240); JOAN BAEZ (II,1294)

SCANLAN, JOSEPH L
LAND OF THE LOST (II,1425); ON TRIAL (II,1896); OUR MAN FLINT: DEAD ON TARGET (II,1932); THE STARLOST (I,4203)

SCARIZA, VINCENT
BILLBOARD'S DISCO PARTY (II,245)

SCARPITTA, GUY
DOBIE GILLIS (I,1302); PETTICOAT JUNCTION (I,3571)

SCARZA, VINCENT
ON THE TOWN WITH TONY BENNETT (II,1895)

SCAVULLO, FRANCESCO
THE CRYSTAL GAYLE SPECIAL (II,609)

SCHAEFER, GEORGE
ABE LINCOLN IN ILLINOIS (I,11); THE ADMIRABLE CRICHTON (I,36); AH!, WILDERNESS (I,91); ALICE IN WONDERLAND (I,120); ANASTASIA (I,182); ARSENIC AND OLD LACE (I,269); BAREFOOT IN ATHENS (I,354); BARRY MANILOW—ONE VOICE (II,156); BERKELEY SQUARE (I,399); THE BEST CHRISTMAS PAGEANT EVER (II,212); BLITHE SPIRIT (I,485); CAPTAIN BRASSBOUND'S CONVERSION (I,828); THE CORN IS GREEN (I,1050); THE CRADLE SONG (I,1088); A CRY OF ANGELS (I,1111); CYRANO DE BERGERAC (I,1120); DEADLY GAME (II,632); THE DEVIL'S DISCIPLE (I,1247); DIAL "M" FOR MURDER (I,1256); A DOLL'S HOUSE (I,1327); DREAM GIRL (I,1374); EAGLE IN A CAGE (I,1392); ELIZABETH THE QUEEN (I,1431); THE FANTASTICKS (I,1531); THE FILE ON DEVLIN (I,1575); GIDEON (I,1792); GIFT OF THE MAGI (I,1800); GIVE US BARABBAS! (I,1816); THE GOLDEN AGE OF TELEVISION (II,1017); THE GOOD FAIRY (I,1846); THE GREEN PASTURES (I,1887); THE HALLMARK HALL OF FAME CHRISTMAS FESTIVAL (I,1921); HAMLET (I,1924); HANS BRINKER OR THE SILVER SKATES (I,1933); HARVEY (I,1961); THE HOLY TERROR (I,2098); INHERIT THE WIND (I,2213); JOHNNY BELINDA (I,2416); THE JOKE AND THE VALLEY (I,2438); KING RICHARD II (I,2559); KISS ME, KATE (I,2568); LAMP AT MIDNIGHT (I,2608); LAND OF HOPE (II,1424); THE LARK (I,2616); THE LITTLE FOXES (I,2712); LITTLE MOON OF ALBAN (I,2714); LITTLE MOON OF ALBAN (I,2715); MACBETH (I,2800); MACBETH (I,2801); THE MAGNIFICENT YANKEE (I,2829); MAN AND SUPERMAN (I,2854); MEET ME IN ST. LOUIS (I,2986); MY FAIR LADY (I,3187); ON BORROWED TIME (I,3354); ONE TOUCH OF VENUS (I,3389); OUR TOWN (I,3421); THE PATRIOTS (I,3477); SANDBURG'S LINCOLN (I,3907); THE SECOND BARRY MANILOW SPECIAL (II,2285); SHANGRI-LA (I,3996); SOLDIER IN LOVE (I,4108); ST. JOAN (I,4175); TAMING OF THE SHREW (I,4356); THE TEAHOUSE OF THE AUGUST MOON (I,4368); THE TEMPEST (I,4391); THERE SHALL BE NO NIGHT (I,4437); TIME REMEMBERED (I,4509); TRUMAN AT POTSDAM (I,4618); TWELFTH NIGHT (I,4636); VICTORIA REGINA (I,4728); WINTERSET (I,4882); THE YEOMAN OF THE GUARD (I,4926)

SCHAETZLE, BUD
KIDD VIDEO (II,1388)

SCHAFFNER, FRANKLIN
BROADWAY (I,733); THE DEFENDERS (I,1220); ONE-EYED JACKS ARE WILD (I,3374); PERSON TO PERSON (I,3550); WESLEY (I,4795); THE WIDE OPEN DOOR (I,4859)

SCHEERER, ROBERT
THE ANDY WILLIAMS SHOW (II,78); ARCHIE (II,104); ARSENIC AND OLD LACE (I,270); BARBRA STREISAND: A HAPPENING IN CENTRAL PARK (I,352); THE BLUE KNIGHT (II,276); CHANGING SCENE (I,899); A COUPLE OF DONS (I,1077); THE DANNY KAYE SHOW (I,1143); DISNEY WORLD —A GALA OPENING—DISNEYLAND EAST (I,1292); EDDIE AND HERBERT (II,757); FAME (II,812); THE FRED ASTAIRE SHOW (I,1677); THE FUTURE: WHAT'S NEXT (II,947); GLENN MILLER TIME (I,1824); GLITTER (II,1009); HAWAII FIVE-O (II,1110); HOTEL (II,1192); HOTEL 90 (I,2130); IRONSIDE (II,1246); KATE MCSHANE (II,1368); KHAN! (II,1384); THE LOVE BOAT (II,1535); THE MAN AND THE CITY (I,2856); THE MARY TYLER MOORE COMEDY HOUR (II,1639); THE NIGHT STALKER (II,1849); NUMBER 96 (II,1869); THE PEGGY FLEMING SHOW (I,3508); POOR DEVIL (I,3646); THE ROCK RAINBOW (II,2194); THE SHARI LEWIS SHOW (I,4000); SHIRLEY MACLAINE: IF THEY COULD SEE ME NOW (II,2335); YOU ASKED FOR IT (II,2855)

SCHELLERY, HENNING S
GREATEST HEROES OF THE BIBLE (II,1062)

SCHILLER, LAWRENCE
COME WITH ME—LAINIE KAZAN (I,1015)

SCHIMMEL, CLARENCE
THE GARRY MOORE SHOW (I,1738)

SCHLAMMUE, THOMAS
BETTE MIDLER—ART OR BUST (II,221)

SCHLATTER, GEORGE
DIANA ROSS AND THE SUPREMES AND THE TEMPTATIONS ON BROADWAY (I,1260); RADIO CITY MUSIC HALL AT CHRISTMAS TIME (I,3715)

SCHLESINGER, JOHN
SEPARATE TABLES (II,2304)

SCHMAH JR, LARRY
THE BENNY RUBIN SHOW (I,398)

SCHMIDT, ERNIE
BLACKSTAR (II,263); FAT ALBERT AND THE COSBY KIDS (II,835); THE KID SUPER POWER HOUR WITH SHAZAM (II,1386); THE NEW ADVENTURES OF MIGHTY MOUSE AND HECKLE AND JECKLE (II,1814); THE TARZAN/LONE RANGER/ZORRO ADVENTURE HOUR (II,2543)

SCHNEIDER, ALAN
SKIN OF OUR TEETH (I,4073)

SCHNEIDER, DICK
THE $128,000 QUESTION (II,1903); THE BIG SHOWDOWN (II,244); HENRY MORGAN'S GREAT TALENT HUNT (I,2017); JEOPARDY (II,1280); THE MOHAWK SHOWROOM (I,3080); THE QUIZ KIDS (II,2103); THE RUFF AND REDDY SHOW (I,3866); THE STARS AND STRIPES SHOW (I,4207); THE STARS AND STRIPES SHOW (I,4208); THE WHO, WHAT OR WHERE GAME (I,4850)

SCHNEIDER, MARCIA KUYPER
SING ALONG WITH MITCH (I,4057)

SCHNUR, JEROME
GO LUCKY (I,1832); GUESS AGAIN (I,1894); MAKE THE CONNECTION (I,2846); THE NAME'S THE SAME (I,3220); ROBERT Q'S MATINEE (I,3812); WHO PAYS? (I,4848)

SCHNURR, LARRY
THE PALACE (II,1943)

SCHOLEM, LEE
THE ADVENTURES OF SUPERMAN (I,77)

SCHOLZ-CONWAY, JOHN
KILLER WITH TWO FACES (I,2539); MELODY OF HATE (I,2999); MIRROR OF DECEPTION (I,3060); ONCE THE KILLING STARTS (I,3367)

SCHOTZ, ERIC
EYE ON HOLLYWOOD (II,798)

SCHUFFMAN, DAN
IT'S POLKA TIME (I,2273); OH, KAY! (I,3342)

SCHULTZ, MICHAEL
BARETTA (II,152); CHANGE AT 125TH STREET (II,480); EARTH, WIND & FIRE IN CONCERT (II,750); THE JERK, TOO (II,1282); MOVIN' ON (II,1743); THE ROCKFORD FILES (II,2197); TOMA (II,2634)

SCHULZ, BOB
FALCON'S GOLD (II,809)

SCHUTLE, TED
THE MAGIC RANCH (I,2819)

SCHWAB, LAWRENCE
JANE WYMAN PRESENTS THE FIRESIDE THEATER (I,2345); THE KATE SMITH HOUR (I,2508); WHAT WILL THEY THINK OF NEXT? (I,4813)

SCHWARTZ, AL
CELEBRITIES: WHERE ARE THEY NOW? (II,462); FAR OUT SPACE NUTS (II,831); HOLLYWOOD'S PRIVATE HOME MOVIES (II,1167); HOLLYWOOD'S PRIVATE HOME MOVIES II (II,1168); WELCOME BACK, KOTTER (II,2761); WHATEVER BECAME OF. . .? (II,2773)

SCHWARTZ, ARTHUR
TWENTIETH CENTURY (I,4639)

SCHWARTZ, BOB
CANDID CAMERA (II,420); SEARCH FOR TOMORROW (II,2284); SPIN-OFF (II,2427)

SCHWARTZ, JEAN-CLAUDE
INVITATION TO PARIS (I,2233)

SCHWARTZ, LEW
FRED WARING: WAY BACK HOME (I,1680)

SCHWARTZ, LLOYD J
THE BRADY BUNCH (II,362)

SCIBETTA, JOSEPH
DON'S MUSICAL PLAYHOUSE (I,1348); HONESTLY, CELESTE! (I,2104)

SCINTO, ROBERT
LOVING (II,1552)

SCITO, ROBERT
THE DOCTORS (II,694)

SCOFFIELD, JON
BOB HOPE SPECIAL: THE BOB HOPE SPECIAL (II,338); THE ROCK FOLLIES (II,2192); SUNDAY NIGHT AT THE LONDON PALLADIUM (I,4286); THIS IS TOM JONES (I,4460); THE TOM JONES CHRISTMAS SPECIAL (I,4539); THE TOM JONES SPECIAL (I,4540); THE TOM JONES SPECIAL (I,4541); THE TOM JONES SPECIAL (I,4542); THE TOM JONES SPECIAL (I,4543); THE TOM JONES SPECIAL (I,4544); THE TOM JONES SPECIAL (I,4545)

SCOFIELD, IAN
SPOTLIGHT (I,4163)

SCOTT, ASHMEAD
HERB SHRINER TIME (I,2020)

SCOTT, BILL
GEORGE OF THE JUNGLE (I,1773)

SCOTT, JACK
VAUDEVILLE (II,2722)

SCOTT, OZ
THE JEFFERSONS (II,1276); MAX (II,1658); THE MISSISSIPPI (II,1707)

SCOTT, PETER GRAHAM
THE AVENGERS (II,121); DANGER MAN (I,1136); THE ONEDIN LINE (II,1912); QUILLER: PRICE OF VIOLENCE (II,2101); SECRET AGENT (I,3962)

SCURTI, JIM
THAT TEEN SHOW (II,2573)

SEDAWIE, NORMAN
THE CATERINA VALENTE SHOW (I,873); ON STAGE WITH BARBARA MCNAIR (I,3358); THE SMOTHERS BROTHERS SHOW (I,4097); THE SMOTHERS ORGANIC PRIME TIME SPACE RIDE (I,4098)

SEDWICK, JOHN
AS THE WORLD TURNS (II,110); DARK SHADOWS (I,1157); THE EDGE OF NIGHT (II,760)

SEEGER, HAL
BATFINK (I,365)

SEGAL, ALEX
THE ALCOA HOUR (I,108); ALCOA PREMIERE (I,109); THE BILLY DANIELS SHOW (I,451); THE BOB HOPE CHRYSLER THEATER (I,502); THE BORIS KARLOFF MYSTERY PLAYHOUSE (I,690); CELANESE THEATER (I,881); DEATH OF A SALESMAN (I,1208); THE DIARY OF ANNE FRANK (I,1262); HEDDA GABLER (I,2001); PENTHOUSE PARTY (I,3515); PULITZER PRIZE PLAYHOUSE (I,3692); RICH MAN, POOR MAN—BOOK II (II,2162); VOLUME ONE (I,4742)

SEIDELMAN, ARTHUR ALAN
BAY CITY BLUES (II,186); CALL TO GLORY (II,413); COVER UP (II,597); PAPER DOLLS (II,1952); ROMANCE THEATER (II,2204)

SEIFERT, KIM
THE DANCE SHOW (II,616)

SEITER, WILLIAM
THE 20TH CENTURY-FOX HOUR (I,4642); THE ALASKANS (I,106); SCREEN DIRECTOR'S PLAYHOUSE (I,3946)

SEITZ JR, GEORGE B
THE LONE RANGER (I,2740)

SELANDER, LESLEY
COWBOY G-MEN (I,1084); JEFF'S COLLIE (I,2356); LARAMIE (I,2614)

SELIGMAN, SELIG J
DAY IN COURT (I,1182)

SENENSKY, RALPH
BANYON (I,344); BARNABY JONES (II,153); BIG BEND COUNTRY (II,230); THE BIG VALLEY (I,437); THE BLUE KNIGHT (II,276); CASABLANCA (II,450); CITY OF ANGELS (II,540); THE COURTSHIP OF EDDIE'S FATHER (I,1083); DAN AUGUST (I,1130); DOCTOR KILDARE (I,1315); DYNASTY (II,746); THE FAMILY HOLVAK (II,817); THE FBI (I,1551); THE FUGITIVE (I,1701); HART TO HART (II,1102); THE HIGH CHAPARRAL (I,2051); I SPY (I,2179); IRONSIDE (II,1246); JEREMIAH OF JACOB'S NECK (II,1281); LOU GRANT (II,1526); MANNIX (II,1624); MEDICAL CENTER (II,1675); MEDICAL STORY (II,1676); NAKED CITY (I,3216); PAPER DOLLS (II,1952); THE PARTRIDGE FAMILY (II,1962); PLANET OF THE APES (II,2049); POLICE STORY (II,2062); RIPLEY'S BELIEVE IT OR NOT (II,2177); THE ROOKIES (II,2208); ROUTE 66 (I,3852); SONS AND DAUGHTERS (II,2404); STAR TREK (II,2440); THREE FOR THE ROAD (II,2607); TRAPPER JOHN, M.D. (II,2654); THE WALTONS (II,2740); WESTSIDE MEDICAL (II,2765); THE WILD WILD WEST (I,4863); YOUNG MAVERICK (II,2867)

SENNA, LORRAINE
KNOTS LANDING (II,1404)

SERF, JOSEPH
THE PRISONER (I,3670)

SEYMOUR, ROBBIE
ROOM FOR ONE MORE (I,3842)

SGARRO, NICHOLAS
BRING 'EM BACK ALIVE (II,377); CAGNEY AND LACEY (II,409); CHARLIE'S ANGELS (II,486); CHIPS (II,511); DELTA HOUSE (II,658); THE EDDIE CAPRA MYSTERIES (II,758); EISCHIED (II,763); EMERALD POINT, N.A.S. (II,773); FAME (II,812); FLAMINGO ROAD (II,872); FLYING HIGH (II,894); THE GANGSTER CHRONICLES (II,957); KNOTS LANDING (II,1404); KOJAK (II,1406); THE MAN WITH THE POWER (II,1616); NURSE (II,1870); THE RAT PATROL (I,3726); SCARECROW AND MRS. KING (II,2268); TODAY'S FBI (II,2628)

SHAIN, CARL
THE GHOST AND MRS. MUIR (I,1786); OZMOE (I,3437)

SHALLAT, LEE
FAMILY TIES (II,819)

SHALLECK, ALAN J
PIXANNE (II,2048)

SHANE, MAXWELL
THE VIRGINIAN (I,4738)

SHAPIRO, KEN
TV'S BLOOPERS AND PRACTICAL JOKES (II,2675)

SHAPIRO, MEL
DOC (II,682); ON OUR OWN (II,1892); PHYLLIS (II,2038)

SHAPIRO, PAUL
THE EDISON TWINS (II,761)

SHAPIRO, STANLEY
THE R.C.A. VICTOR SHOW (I,3737)

SHAPIRO, STUART
FM TV (II,896)

SHAPLEN, ROBERT
THE HUNTER (I,2162)

SHARP, DON
THE AVENGERS (II,121); Q. E. D. (II,2094); A WOMAN OF SUBSTANCE (II,2820)

SHARP, JON
ARCHIE BUNKER'S PLACE (II,105); I'M A BIG GIRL NOW (II,1218); STAR OF THE FAMILY (II,2436); TOO CLOSE FOR COMFORT (II,2642)

SHATNER, WILLIAM
T.J. HOOKER (II,2524)

SHAVELSON, MELVILLE
ELKE (I,1432); FATHER ON TRIAL (I,1544); IKE (II,1223); TRUE LIFE STORIES (II,2665)

SHAW, DAVID
SHANE (I,3995)

SHAW, J EDWARD
ANNE MURRAY'S LADIES' NIGHT (II,89); FUNNY FACES (II,941)

SHAW, JEROME
BATTLESTARS (II,182); GAMBIT (II,955); GEORGE BURNS: AN HOUR OF JOKES AND SONGS (II,973); HIGH ROLLERS (II,1147); THE HOLLYWOOD SQUARES (II,1164); THE KATE SMITH SHOW (I,2510); KEEFE BRASSELLE'S VARIETY

GARDEN (I,2515); KEEP TALKING (I,2517); LAS VEGAS GAMBIT (II,1430); THE MAGNIFICENT MARBLE MACHINE (II,1600); MARRIAGE—HANDLE WITH CARE (I,2921); THE NAMEDROPPERS (I,3219); THE PAUL WINCHELL SHOW (I,3493); SALE OF THE CENTURY (II,2239); SUMMERTIME U.S.A. (I,4283); TIPTOE THROUGH TV (I,4524); TO SAY THE LEAST (II,2623); TOGETHER WITH MUSIC (I,4532); THE TOMMY HUNTER SHOW (I,4552); VARIETY: THE WORLD OF SHOW BIZ (I,4704); WINKY DINK AND YOU (I,4879)

SHAW, JOSEPH
THE AMATEUR'S GUIDE TO LOVE (I,155)

SHAW, LOU
THE SHIRT TALES (II,2337)

SHEA, JACK
APPLE'S WAY (II,101); THE BOB HOPE SHOW (I,549); THE BOB HOPE SHOW (I,552); THE BOB HOPE SHOW (I,553); THE BOB HOPE SHOW (I,554); THE BOB HOPE SHOW (I,555); THE BOB HOPE SHOW (I,556); THE BOB HOPE SHOW (I,558); THE BOB HOPE SHOW (I,559); THE BOB HOPE SHOW (I,581); THE BOB HOPE SHOW (I,583); THE BOB HOPE SHOW (I,584); THE BOB HOPE SHOW (I,585); THE BOB HOPE SHOW (I,586); THE BOB HOPE SHOW (I,588); THE BOB HOPE SHOW (I,589); THE BOB HOPE SHOW (I,590); THE BOB HOPE SHOW (I,591); THE BOB HOPE SHOW (I,592); THE BOB HOPE SHOW (I,603); THE BOB HOPE SHOW (I,604); THE BOB HOPE SHOW (I,606); THE BOB HOPE SHOW (I,607); BUT MOTHER! (II,403); CHECKING IN (II,492); THE GLEN CAMPBELL GOODTIME HOUR (I,1820); GOOD TIMES (II,1038); THE GRADY NUTT SHOW (II,1048); IN THE BEGINNING (II,1229); THE JEFFERSONS (II,1276); THE JERRY LEWIS SHOW (I,2363); THE JERRY LEWIS SHOW (I,2364); THE JERRY LEWIS SHOW (I,2365); THE JERRY LEWIS SHOW (I,2367); THE JERRY LEWIS SHOW (I,2368); KANGAROOS IN THE KITCHEN (II,1362); KATE MCSHANE (II,1368); LOVE AND LEARN (II,1529); MALIBU U. (I,2849); MASON (II,1643); MR. SUCCESS (II,1756); PEN 'N' INC. (II,1992); PUNKY BREWSTER (II,2092); THE ROPERS (II,2214); ROXY PAGE (II,2222); SALT AND PEPE (II,2240); SILVER SPOONS (II,2355); THREE'S COMPANY (II,2614); THE WALTONS (II,2740); WE'LL GET BY (II,2753); WENDY HOOPER—U. S. ARMY (II,2764); WHAT'S HAPPENING!! (II,2769); ZERO INTELLIGENCE (II,2876)

SHEAR, BARRY
ALIAS SMITH AND JONES (I,118); ALL THINGS BRIGHT AND BEAUTIFUL (I,141); THE BEAUTIFUL PHYLLIS DILLER SHOW (I,381); THE BILLION DOLLAR THREAT (II,246); THE BLACK SHEEP SQUADRON (II,262); CITY OF ANGELS (II,540); COOL MILLION (I,1046); DANIEL BOONE (I,1142); ELLERY QUEEN: DON'T LOOK BEHIND YOU (I,1434); THE ERNIE KOVACS SHOW (I,1455); THE FEATHER AND FATHER GANG (II,845); THE GIRL FROM U.N.C.L.E. (I,1808); HAWAII FIVE-O (II,1110); THE HAZEL SCOTT SHOW (I,1984); HERE'S EDIE (I,2030); HOORAY FOR HOLLYWOOD (I,2114); IRONSIDE (II,1246); IT TAKES A THIEF (I,2250); IT'S WHAT'S HAPPENING, BABY! (I,2278); THE JANE POWELL SHOW (I,2344); JARRETT (I,2349); JIGSAW (I,2377); JIMMY HUGHES, ROOKIE COP (I,2390); JOE FORRESTER (II,1303); JOSEPH SCHILDKRAUT PRESENTS (I,2449); JULIA (I,2476); THE LIVELY ONES (I,2729); LUCAN (II,1554); THE MAN FROM U.N.C.L.E. (I,2867); THE MAN IN THE MOON (I,2871); MCMILLAN AND WIFE (II,1667); MELODY STREET (I,3000); THE MILTON BERLE SPECIAL (I,3050); THE MOD SQUAD (I,3078); THE MUSIC SHOP (I,3177); THE NAME OF THE GAME (I,3217); NIGHT GALLERY (I,3288); NOT FOR PUBLICATION (I,3309); POLICE STORY (II,2062); POLICE WOMAN (II,2063); PRIVATE EYE, PRIVATE EYE (I,3671); PUNCH AND JODY (II,2091); RICKLES (II,2170); THE SAN PEDRO BEACH BUMS (II,2250); THE SAN PEDRO BUMS (II,2251); SEARCH (I,3951); SHINDIG (I,4013); THE SIXTH SENSE (I,4069); STARSKY AND HUTCH (II,2445); THE STREETS OF SAN FRANCISCO (II,2478); THE SWINGIN' SINGIN' YEARS (I,4319); THE SWINGIN' YEARS (I,4321); TAKE A GOOD LOOK (I,4332); TAKE A GOOD LOOK (II,2529); TARZAN (I,4361); THE VICTOR BORGE SPECIAL (I,4722); THE XAVIER CUGAT SHOW (I,4924)

SHEEHAN, TONY
BARNEY MILLER (II,154)

SHELDON, JAMES
240-ROBERT (II,2689); ALCOA PREMIERE (I,109); ALFRED HITCHCOCK PRESENTS (I,115); ALICE (II,33); APPLE'S WAY (II,101); THE BARRETTS OF WIMPOLE STREET (I,361); BATMAN (I,366); CAGNEY AND LACEY (II,409); CODE RED (II,552); THE CONTENDER (II,571); THE DEFENDERS (I,1220); THE DESILU PLAYHOUSE (I,1237); DOC ELLIOT (I,1306); DOC HOLLIDAY (I,1307); THE DUKES OF HAZZARD (II,742); THE EDDIE ALBERT SHOW (I,1406); ELLERY QUEEN (II,766); FAMILY (II,813); FAMILY AFFAIR (I,1519); THE FELONY SQUAD (I,1563); FLYING HIGH (II,894); THE FUGITIVE (I,1701); GIDGET GROWS UP (I,1797); GOOD HEAVENS (II,1032); THE GOSSIP COLUMNIST (II,1045); IRONSIDE (II,1246); JAMES AT 15 (II,1270); JESSIE (II,1287); KNOTS LANDING (II,1404); LOTTERY (II,1525); THE LOVE BOAT (II,1535); M*A*S*H (II,1569); THE MAN FROM U.N.C.L.E. (I,2867); MARCUS WELBY, M.D. (II,1627); THE MARGE AND GOWER CHAMPION SHOW (I,2907); MCCLAIN'S LAW (II,1659); MCDUFF, THE TALKING DOG (II,1662); MCMILLAN (II,1666); MCMILLAN AND WIFE (II,1667); THE MILLIONAIRE (I,3046); THE MISADVENTURES OF SHERIFF LOBO (II,1704); MODERN ROMANCES (I,3079); THE MONSTER SQUAD (II,1725); MR. T. AND TINA (II,1758); MY WORLD . . . AND WELCOME TO IT (I,3208); NAKED CITY (I,3216); NURSE (II,1870); OWEN MARSHALL: COUNSELOR AT LAW (I,3435); PARTNERS IN CRIME (II,1961); THE PATTY DUKE SHOW (I,3483); PERRY MASON (II,2025); PETROCELLI (II,2031); POPI (II,2069); ROBERT MONTGOMERY PRESENTS YOUR LUCKY STRIKE THEATER (I,3809); ROOM 222 (I,3843); ROUTE 66 (I,3852); THE SECRET LIFE OF JOHN MONROE (I,3966); SEVEN BRIDES FOR SEVEN BROTHERS (II,2307); SMALL AND FRYE (II,2380); SWEEPSTAKES (II,2514); THAT GIRL (II,2570); THERE'S ONE IN EVERY FAMILY (I,4439); TO ROME WITH LOVE (I,4526); TODAY'S FBI (II,2628); THE TWILIGHT ZONE (I,4651); THE VIRGINIAN (I,4738); WHAT REALLY HAPPENED TO THE CLASS OF '65? (II,2768); WHIZ KIDS (II,2790)

SHELLEY, JOSHUA
GET SMART (II,983); N.Y.P.D. (I,3321); POPI (II,2069); ROSETTI AND RYAN (II,2216); WHEN THINGS WERE ROTTEN (II,2782)

SHELTON, GILBERT
THE A-TEAM (II,119)

SHEPARD, GERALD
THE MONKEES (I,3088)

SHERMAN, GARY
THE MYSTERIOUS TWO (II,1783); THE STREETS (II,2479)

SHERMAN, GEORGE
GENTLE BEN (I,1761); MOBILE ONE (II,1717); NAKED CITY (I,3215); NAKED CITY (I,3216); NASHVILLE 99 (II,1791); SAM (II,2247); SCREEN DIRECTOR'S PLAYHOUSE (I,3946); STRATEGIC AIR COMMAND (I,4252)

SHERMAN, ROBERT
BARNABY JONES (II,153); THE WALTONS (II,2740)

SHERMAN, SAM
BICENTENNIAL MINUTES (II,229)

SHERMAN, VINCENT
77 SUNSET STRIP (I,3988); ALIAS SMITH AND JONES (I,118); BARETTA (II,152); DOCTORS HOSPITAL (II,691); EXECUTIVE SUITE (II,796); THE FAMILY HOLVAK (II,817); HAGEN (II,1071); THE LAST HURRAH (I,2626); THE LAWYERS (I,2646); MEDICAL CENTER (II,1675); SALLY AND SAM (I,3891); SIMON AND SIMON (II,2357); THREE ON AN ISLAND (I,4483); TROUBLE IN HIGH TIMBER COUNTRY (II,2661); WESTSIDE MEDICAL (II,2765)

SHERRY, ERNEST
COMEDY NEWS (I,1020); DINAH'S PLACE (II,678); MIKE AND PEARL (I,3038); THE MIKE DOUGLAS SHOW (II,1698)

SHERWIN, WALLY
MEET MARCEL MARCEAU (I,2983)

SHEVELOVE, BURT
THE ART CARNEY SHOW (I,274); THE BEST OF ANYTHING (I,404); THE BIL BAIRD SHOW (I,439); HOORAY FOR LOVE (I,2115); I'LL BUY THAT (I,2190); THE RIGHT MAN (I,3790); TEXACO COMMAND PERFORMANCE (I,4408)

SHILTON, GILBERT
BLUE THUNDER (II,278); V: THE SERIES (II,2715)

SHIMOKAWA, GARY
ALICE (II,33); ANOTHER DAY (II,94); ARCHIE BUNKER'S PLACE (II,105); THE BAY CITY AMUSEMENT COMPANY (II,185); FISH (II,864); GLORIA COMES HOME (II,1011); MR. T (II,1757); THE ROLLERGIRLS (II,2203); SHIPSHAPE (II,2329); THE SON-IN-LAW (II,2398); WELCOME BACK, KOTTER (II,2761); WHATEVER HAPPENED TO DOBIE GILLIS? (II,2774)

SHIN, NELSON
DUNGEONS AND DRAGONS (II,744); G.I. JOE: A REAL AMERICAN HERO (II,948); PANDAMONIUM (II,1948); SPIDER-MAN AND HIS AMAZING FRIENDS (II,2425)

SHOLEM, LEE
77 SUNSET STRIP (I,3988); BRONCO (I,745); CHEYENNE (I,931); COLT .45 (I,1010); LAWMAN (I,2642); MAVERICK (II,1657); SUGARFOOT (I,4277)

SICHEL, JOHN
COME OUT, COME OUT, WHEREVER YOU ARE (I,1014); KISS KISS, KILL KILL (I,2566); THE SAVAGE CURSE (I,3925); SOMEONE AT THE TOP OF THE STAIRS (I,4114); SPELL OF EVIL (I,4152)

SIDARIS, ANDREW
THE MAGIC LAND OF ALLAKAZAM (I,2817); THE NANCY DREW MYSTERIES (II,1789)

SIDNEY, GEORGE
WHO HAS SEEN THE WIND? (I,4847)

SIDNEY, ROBERT
THE BETTY HUTTON SHOW (I,413); DEAN MARTIN'S RED HOT SCANDALS OF 1926 (II,648)

SIEGEL, DON
BROCK CALLAHAN (I,741); BUS STOP (I,778)

SILLS, THEODORE
DECOY (I,1217); THE PETER AND MARY SHOW (I,3561)

SILVER, ARTHUR
WORKING STIFFS (II,2834)

SILVER, STUART
ONE LIFE TO LIVE (II,1907)

SILVERSTEIN, ELLIOT
ALCOA/GOODYEAR THEATER (I,107); THE DEFENDERS (I,1220); THE DICK POWELL SHOW (I,1269); THE FIRM (II,860); HAVE GUN—WILL TRAVEL (I,1968); NAKED CITY (I,3216); ROUTE 66 (I,3852); THE TWILIGHT ZONE (I,4651); THE WESTERNER (I,4801)

SIMMEL, EDWARD C
CROWN THEATER WITH GLORIA SWANSON (I,1107)

SIMMONS, GRANT
SPIDER-MAN (I,4154)

SIMON, DANNY
THE TROUBLE WITH PEOPLE (I,4611)

SIMON, JEFF
SALUTE TO LADY LIBERTY (II,2243); YOU ASKED FOR IT (II,2855)

SIMON, JIM
THE SHIRT TALES (II,2337)

SIMON, LISA
SESAME STREET (I,3982)

SIMON, ROBERT
OLD NICKERBOCKER MUSIC HALL (I,3351); THE TIMES SQUARE STORY (I,4514)

SIMON, SUSAN
ALL MY CHILDREN (II,39)

SIMPSON, GARRY
BARNEY BLAKE, POLICE REPORTER (I,359); THE CHEVROLET TELE-THEATER (I,926); MARY KAY AND JOHNNY (I,2932); THE MARY MARGARET MCBRIDE SHOW (I,2933); PHILCO TELEVISION PLAYHOUSE (I,3583); TONIGHT IN SAMARKAND (I,4558); WHO SAID THAT? (I,4849); THE WORLD OF MISTER SWEENY (I,4917)

SINCLAIR, RICHARD
77 SUNSET STRIP (I,3988); THE ALASKANS (I,106); BOURBON STREET BEAT (I,699); BRONCO (I,745); LAWMAN (I,2642); SURFSIDE 6 (I,4299)

SINCLAIR, ROBERT
ALCOA/GOODYEAR THEATER (I,107); THE DESILU PLAYHOUSE (I,1237)

SINGELTARY, TONY
GIMME A BREAK (II,995)

SINGER, ALEXANDER
ALIAS SMITH AND JONES (I,118); THE BOB HOPE CHRYSLER THEATER (I,502); BUNCO (II,395); CAGNEY AND LACEY (II,409); CASSIE AND COMPANY (II,455); CIRCLE OF FEAR (I,958); DALLAS (II,614); DAVID CASSIDY—MAN UNDERCOVER (II,623); EMERALD POINT, N.A.S. (II,773); THE FUGITIVE (I,1701); GHOST STORY (I,1788); GIBBSVILLE (II,988); HILL STREET BLUES (II,1154); JESSICA NOVAK (II,1286); JOE FORRESTER (II,1303); KNOTS LANDING (II,1404); LANCER (I,2610); LOGAN'S RUN (II,1507); LOST IN SPACE (I,2758); LOU GRANT (II,1526); LUCAS TANNER (II,1556); THE MAGICIAN (I,2823); MISSION: IMPOSSIBLE (I,3067); THE MONKEES (I,3088); THE NAME OF THE GAME (I,3217); PEARL (II,1986); POLICE WOMAN (II,2063); RAFFERTY (II,2105); REMINGTON STEELE (II,2130); THE RETURN OF MARCUS WELBY, M.D (II,2138); THE ROCKFORD FILES (II,2197); SECRETS OF MIDLAND HEIGHTS (II,2296); SHAFT (I,3993); WALKING TALL (II,2738)

SINGER, GEORGE
HOT WHEELS (I,2126); SKYHAWKS (I,4079); THE SMURFS (II,2385)

SINGLETARY, TONY
BENSON (II,208); CHARLES IN CHARGE (II,482); IT'S NOT EASY (II,1256); IT'S YOUR MOVE (II,1259); THE JEFFERSONS (II,1276); ONE DAY AT A TIME (II,1900); OPEN ALL NIGHT (II,1914); TEACHERS ONLY (II,2548)

SIODMAK, CURT
NUMBER 13 DEMON STREET (I,3316)

SKINNER, CHARLES E
THE BIG STORY (I,432); SERGEANT PRESTON OF THE YUKON (I,3980)

SKOTCH, ED
FUN FOR THE MONEY (I,1708); MUSIC IN VELVET (I,3172); PENTHOUSE SONATA (I,3516); SUPER CIRCUS (I,4287)

SKUTCH, IRA
CHILD'S PLAY (II,510); GIRL ABOUT TOWN (I,1806); MINDREADERS (II,1702)

SLAPEZYNSKI, RICHARD
AROUND THE WORLD IN 80 DAYS (I,263)

SLATE, LEN
THE DEADLY GAME (II,633); STRIKE FORCE (II,2480)

SLATE, WILLIAM
THOSE TWO (I,4470)

SLATER, WILLIAM
THE ONEDIN LINE (II,1912)

SLOANE, EVERETT
THE LORETTA YOUNG THEATER (I,2756)

SMART, BRIAN
RADIO PICTURE SHOW (II,2104)

SMART, RALPH
THE ADVENTURES OF ROBIN HOOD (I,74); THE ADVENTURES OF SIR LANCELOT (I,76); DANGER MAN (I,1136); THE INVISIBLE MAN (I,2232); SECRET AGENT (I,3962)

SMIGHT, JACK
ALFRED HITCHCOCK PRESENTS (I,115); BANACEK (II,138); BANACEK: DETOUR TO NOWHERE (I,339); COLUMBO (II,556); THE DEFENDERS (I,1220); FULL MOON OVER BROOKLYN (I,1703); MADIGAN (I,2808); MCCLOUD (II,1660); NAKED CITY (I,3216); NOTORIOUS (I,3314); OLDSMOBILE MUSIC THEATER (I,3352); PARTNERS IN CRIME (I,3461); ROLL OF THUNDER, HEAR MY CRY (II,2202); ROUTE 66 (I,3852); THE THIRD COMMANDMENT (I,4449); THE TWILIGHT ZONE (I,4651); VICTOR BORGE'S COMEDY IN MUSIC III (I,4725); VICTOR BORGE'S COMEDY IN MUSIC IV (I,4726)

SMIRNOFF, NICK
MR. MERLIN (II,1751)

SMITH, ARLANDO
IT'S YOUR MOVE (II,1259)

SMITH, BARNABY
THE S. S. HOLIDAY (I,4173)

SMITH, BERNIE
YOU BET YOUR LIFE (I,4932)

SMITH, DAN

BUCKSHOT (II,385); DEALER'S CHOICE (II,635); ED SULLIVAN'S BROADWAY (I,1402); THE LILY TOMLIN SHOW (I,2706); SIGN-ON (II,2353)

SMITH, JACK
MUSIC FOR A SPRING NIGHT (I,3169); MUSIC FOR A SUMMER NIGHT (I,3170);

SMITH, JAN DAINLEY
THE LITTLEST HOBO (II,1500)

SMITH, OZ
GIMME A BREAK (II,995)

SMITH, PAUL
THE DORIS DAY SHOW (I,1355); ONE HOUR IN WONDERLAND (I,3376); THE WOODY WOODPECKER SHOW (I,4906)

SMITH, SID
AMERICA PAUSES FOR SPRINGTIME (I,166); AMERICA PAUSES FOR THE MERRY MONTH OF MAY (I,167); BOB HOPE SPECIAL: BOB HOPE FOR PRESIDENT (II,282); BOB HOPE SPECIAL: BOB HOPE IN THE STAR-MAKERS (II,285); BOB HOPE SPECIAL: BOB HOPE IN WHO MAKES THE WORLD LAUGH? (II,287); BOB HOPE SPECIAL: BOB HOPE LAUGHS WITH THE MOVIE AWARDS (II,288); BOB HOPE SPECIAL: BOB HOPE PRESENTS THE STARS OF TOMORROW (II,292); BOB HOPE SPECIAL: BOB HOPE'S MERRY CHRISTMAS SHOW (II,310); BOB HOPE SPECIAL: BOB HOPE'S SUPER BIRTHDAY SPECIAL (II,319); BOB HOPE SPECIAL: BOB HOPE'S USO CHRISTMAS IN BEIRUT (II,320); BOB HOPE SPECIAL: HO HO HOPE'S JOLLY CHRISTMAS HOUR (II,327); BOB HOPE SPECIAL: THE BOB HOPE CHRISTMAS SPECIAL (II,328); BOB HOPE SPECIAL: THE BOB HOPE CHRISTMAS SPECIAL (II,330); BOB HOPE SPECIAL: THE BOB HOPE CHRISTMAS SPECIAL (II,331); BOB HOPE SPECIAL: THE BOB HOPE SPECIAL (II,334); BOB HOPE SPECIAL: BOB HOPE'S HILARIOUS UNREHEARSED ANTICS OF THE STARS (II,341); CBS: ON THE AIR (II,460); THE CHEVROLET GOLDEN ANNIVERSARY SHOW (I,924); CIRCUS OF THE STARS (II,530); CIRCUS OF THE STARS (II,531); CIRCUS OF THE STARS (II,534); CIRCUS OF THE STARS (II,535); CIRCUS OF THE STARS (II,536); CIRCUS OF THE STARS (II,537); THE DANGEROUS CHRISTMAS OF RED RIDING HOOD (I,1139); THE DUKE (I,1381); ELIZABETH TAYLOR IN LONDON (I,1430); LITTLE WOMEN (I,2724); LORETTA LYNN IN THE BIG APPLE (II,1520); THE MAGIC OF DAVID COPPERFIELD (II,1594); NBC FOLLIES OF 1965 (I,3240); OPENING NIGHT (I,3395); OPENING NIGHT: U.S.A. (I,3397); THE PAUL LYNDE COMEDY HOUR (II,1978); THE PAUL LYNDE HALLOWEEN SPECIAL (II,1981); PINOCCHIO (I,3603); PINOCCHIO (II,2044); TED KNIGHT MUSICAL COMEDY VARIETY SPECIAL SPECIAL (II,2549)

SMITH, SIDNEY
MUGGSY (II,1761)

SMITHEE, ALAN
OMNIBUS (II,1890); PAPER DOLLS (II,1952)

SOBEL, MARK
HERE'S BOOMER (II,1134)

SOBOL, EDWARD
ABE LINCOLN IN ILLINOIS (I,10); THE ALFRED AND JOSEPHINE BUTLER SHOW (I,114); BLITHE SPIRIT (I,483); BROTHER RAT (I,746); THE COLGATE COMEDY HOUR (I,997); FAMILY HONOR (I,1522); THE FORTUNE HUNTER (I,1646); THE GEORGE ROSS SHOW (I,1775); HAY FEVER (I,1979); HOUR GLASS (I,2133); IREENE WICKER SINGS (I,2236); JANE EYRE (I,2337); LET'S CELEBRATE (I,2662); THE LUCY MONROE SHOW (I,2790); THE MILKY WAY (I,3044); NIGHT CLUB (I,3284); PINKY AND COMPANY (I,3599); PLAY THE GAME (I,3618); THE RAY PERKINS REVUE (I,3735); SONG AND DANCE (I,4120); STAGE DOOR (I,4177); TELE-VARIETIES (I,4377); TEXACO STAR THEATER (I,4411); TOTAL ECLIPSE (I,4584); THE VALIANT (I,4697); WORLD'S FAIR BEAUTY SHOW (I,4920)

SOGARD, PHIL
GENERAL HOSPITAL (II,964)

SOLD, JACK
SPARROW (II,2411)

SOLKON, SIDNEY
THE ADDAMS FAMILY (II,11)

SOLOMON, ED
MY FAVORITE MARTIANS (II,1779)

SOLT, ANDREW
BOB HOPE SPECIAL: BOB HOPE'S OVERSEAS CHRISTMAS TOURS (II,311); E.T. & FRIENDS—MAGICAL MOVIE VISITORS (II,749); THOSE WONDERFUL TV GAME SHOWS (II,2603)

SOUL, DAVID
STARSKY AND HUTCH (II,2444)

SOVA, PETER
WOMEN OF RUSSIA (II,2824)

SPARBER, I
POPEYE THE SAILOR (I,3648)

SPARR, ROBERT
THE ALASKANS (I,106); BRONCO (I,745); THE GALLANT MEN (I,1728); HAWAIIAN EYE (I,1973); LASSIE (I,2622); LAWMAN (I,2642); THE RAT PATROL (I,3726); THE ROARING TWENTIES (I,3805); ROOM FOR ONE MORE (I,3842); STAR TREK (II,2440); SURFSIDE 6 (I,4299); TIMMY AND LASSIE (I,4520); THE TWILIGHT ZONE (I,4651); VOYAGE TO THE BOTTOM OF THE SEA (I,4743); THE WILD WILD WEST (I,4863)

SPECTOR, IRV
HOT WHEELS (I,2126); PORKY PIG AND FRIENDS (I,3651); SKYHAWKS (I,4079)

SPEIGEL, ED
RIPLEY'S BELIEVE IT OR NOT (II,2177)

SPELLING, AARON
THE SMOTHERS BROTHERS SHOW (I,4096)

SPENCE, LEIGH
PRISONER: CELL BLOCK H (II,2085)

SPENTON-FOSTER, GEORGE
BLAKE'S SEVEN (II,264)

SPIELBERG, STEVEN
COLUMBO (II,556); NIGHT GALLERY (I,3288); NIGHT GALLERY (I,3287); OWEN MARSHALL: COUNSELOR AT LAW (I,3435); THE PSYCHIATRIST (I,3686); SAVAGE (I,3924)

SPIER, WILLIAM
THE CLOCK (I,978); WILLY (I,4869)

SPIERS, BOB
FAWLTY TOWERS (II,843)

SPOSA, LOU
ARTHUR MURRAY'S DANCE PARTY (I,293); CHANCE OF A LIFETIME (I,898); THE DENNIS JAMES SHOW (I,1228); E.S.P. (I,1462); E.S.P. (I,1463)

SPOTS, GARY
MATINEE AT THE BIJOU (II,1651)

SPOTTISWOODE, ROGER
THE RENEGADES (II,2132)

SPRINGSTEEN, BUD
RAWHIDE (I,3727)

SPRINGSTEEN, R G
BONANZA (II,347); GENTLE BEN (I,1761); GUNSMOKE (II,1069); THE HARDY BOYS AND THE MYSTERY OF GHOST FARM (I,1950); THE HARDY BOYS AND THE MYSTERY OF THE APPLEGATETREASURE (I,1951); THE MONROES (I,3089)

SQUIRES, PAUL
THE BLUE JEAN NETWORK (II,274); MUSIC CENTRAL (II,1771)

STADER, PAUL
THE RETURN OF CAPTAIN NEMO (II,2135)

STAMOS, NICHOLAS
RITUALS (II,2182); WKRP IN CINCINNATI (II,2814)

STANFORD, DONALD
ALCOA PREMIERE (I,109)

STANLEY, JOSEPH
THE ALL-STAR REVUE (I,138); THE ALL-STAR SUMMER REVUE (I,139); THE PATTI PAGE SHOW (I,3479); THE R.C.A. VICTOR SHOW (I,3737)

STANLEY, PAUL
ADVENTURES IN PARADISE (I,46); THE ALCOA HOUR (I,108); BARETTA (II,152); BEYOND WESTWORLD (II,227); THE BIG BRAIN (I,424); THE BOB HOPE CHRYSLER THEATER (I,502); CANNON (II,424); CARIBE (II,439); CHARLIE'S ANGELS (II,486); COMBAT (I,1011); CRISIS IN SUN VALLEY (II,604); DEAR DETECTIVE (II,652); THE DEFENDERS (I,1220); THE DESILU PLAYHOUSE (I,1237); DOG AND CAT (II,696); THE FALL GUY (II,811); FITZ AND BONES (II,867); FLAMINGO ROAD (II,872); FOR LOVE AND HONOR (II,903); GHOST STORY (I,1788); GRANDPA GOES TO WASHINGTON (II,1050); GUNSMOKE (II,1069); HAWAII FIVE-O (II,1110); THE

HIGH CHAPARRAL (I,2051); KING'S CROSSING (II,1397); KNIGHT RIDER (II,1402); KOJAK (II,1406); LAREDO (I,2615); LOST IN SPACE (I,2758); MANHUNTER (II,1621); MEDICAL CENTER (II,1675); MISSION: IMPOSSIBLE (I,3067); NAKED CITY (I,3216); THE OUTER LIMITS (I,3426); PETER GUNN (I,3562); PETROCELLI (II,2031); THE RAT PATROL (I,3726); RICH MAN, POOR MAN—BOOK I (II,2161); RICH MAN, POOR MAN—BOOK II (II,2162); ROUTE 66 (I,3852); SEARCH (I,3951); SERGEANT T.K. YU (II,2305); SERPICO (II,2306); STONE (II,2471); THE STREETS OF SAN FRANCISCO (II,2478); TARZAN (I,4361); THREE EYES (II,2605); THE TIME TUNNEL (I,4511); THE ULTIMATE IMPOSTER (II,2700); THE UNTOUCHABLES (I,4682); VEGAS (II,2724); THE VIRGINIAN (I,4738); VOYAGERS (II,2730); WESTSIDE MEDICAL (II,2765); THE WILD WILD WEST (I,4863)

STARBUCK, JAMES
SING ALONG WITH MITCH (I,4057)

STARK, HERBERT
BONANZA (II,347); IN SEARCH OF. . . (II,1226)

STARK, NED
SEARCH FOR TOMORROW (II,2284)

STARK, WILBUR
CUT (I,1116)

STARRETT, JACK
BIG BOB JOHNSON AND HIS FANTASTIC SPEED CIRCUS (II,231); THE DUKES OF HAZZARD (II,742); EISCHIED (II,763); HILL STREET BLUES (II,1154); A MAN CALLED SLOANE (II,1613); NOWHERE TO HIDE (II,1868); PARIS (II,1957); PLANET OF THE APES (II,2049); ROGER AND HARRY (II,2200); STARSKY AND HUTCH (II,2444); WHAT REALLY HAPPENED TO THE CLASS OF '65? (II,2768)

STATES, BILL
THE SECRET JURY (I,3965)

STAUB, RALPH
WHERE WERE YOU? (I,4834)

STEARNS, JOHNNY
FAYE AND SKITCH (I,1548); MAKE ME LAUGH (I,2839)

STEELE, BOB
A WOMAN TO REMEMBER (I,4888)

STEELE, MICHAEL
THE BOBBY VINTON SHOW (II,345); HALF THE GEORGE KIRBY COMEDY HOUR (I,1920); ROLLIN' ON THE RIVER (I,3834)

STEEN, CORT
BYLINE—BETTY FURNESS (I,785); CITY HOSPITAL (I,967)

STEFAN, BUD
SANDY DREAMS (I,3909)

STEIN, JOEL
THE NEIGHBORS (II,1801)

STEINBERG, DENNIS
THE DUMPLINGS (II,743)

STEINBERG, NORMAN
THE BAY CITY AMUSEMENT COMPANY (II,185)

STEINBERG, SCOTT
TWIGS (II,2684)

STEINMETZ, DENNIS
THE ADDAMS FAMILY (II,13); ALLEN LUDDEN'S GALLERY (I,143); BARNEY MILLER (II,154); BROTHERS AND SISTERS (II,382); DOC (II,682); FISH (II,864); THE KALLIKAKS (II,1361); LAND OF THE LOST (II,1425); MR. T. AND TINA (II,1758); THE PAUL WILLIAMS SHOW (II,1984); WHAT'S HAPPENING!! (II,2769); THE YOUNG AND THE RESTLESS (II,2862)

STEPHANI, FREDERICK
MY FRIEND FLICKA (I,3190)

STERN, BERT
TWIGGY IN NEW YORK (I,4649)

STERN, DON
THE AMATEUR'S GUIDE TO LOVE (I,155)

STERN, LEONARD B
DIANA (I,1259); FEMALE INSTINCT (I,1564); THE GOVERNOR AND J.J. (I,1863); HE AND SHE (I,1985); HOLMES AND YOYO (II,1169); LANIGAN'S RABBI (II,1428); ONCE UPON A DEAD MAN (I,3369); PARTNERS IN CRIME (II,1961); PHILLIP AND BARBARA (II,2034); THE SNOOP SISTERS (II,2389); WINDOWS, DOORS AND KEYHOLES (II,2806)

STERN, SANDOR
CUTTER TO HOUSTON (II,612)

STERN, STEVEN H
CAMP GRIZZLY (II,418); DOCTORS' PRIVATE LIVES (II,693); DOG AND CAT (II,696); JESSICA NOVAK (II,1286); LOGAN'S RUN (II,1507); QUINCY, M. E. (II,2102); STILL THE BEAVER (II,2466)

STEVENS JR, GEORGE
PEOPLE (I,3517); PETER GUNN (I,3562)

STEVENS, BOB
THE BOB HOWARD SHOW (I,665); THE TIMES SQUARE STORY (I,4514)

STEVENS, DAVID
A TOWN LIKE ALICE (II,2652)

STEVENS, LESLIE
I LOVE A MYSTERY (I,2170); THE INVISIBLE MAN (II,1242); THE OUTER LIMITS (I,3426); TACK REYNOLDS (I,4329)

STEVENS, MARK
BIG TOWN (I,436); MICHAEL SHAYNE, DETECTIVE (I,3022)

STEVENS, ROBERT
ALFRED HITCHCOCK PRESENTS (I,115); APPOINTMENT WITH ADVENTURE (I,246); ARTHUR GODFREY'S TALENT SCOUTS (I,291); BEN HECHT'S TALES OF THE CITY (I,395); CORONET BLUE (I,1055); THE DEFENDERS (I,1220); ROMANCE (I,3836); SUSPENSE (I,4305); THE TWILIGHT ZONE (I,4651)

STEVENSON, ROBERT
THE 20TH CENTURY-FOX HOUR (I,4642); CAVALCADE OF AMERICA (I,875); MIRACLE ON 34TH STREET (I,3056); THIS IS YOUR MUSIC (I,4462)

STEWART, DOUGLAS
THE PSYCHIATRIST (I,3686)

STEWART, IRA
MR. ED (I,3137)

STEWART, LARRY
THE AMAZING SPIDER-MAN (II,63); THE BIONIC WOMAN (II,255); BUCK ROGERS IN THE 25TH CENTURY (II,383); CHARLIE'S ANGELS (II,486); HERE'S BOOMER (II,1134); HUNTER (II,1206); THE INCREDIBLE HULK (II,1232); JOHNNY RINGO (I,2434); SWORD OF JUSTICE (II,2521); THRILL SEEKERS (I,4491); THE WALTONS (II,2740)

STEWART, NORMAN
DOCTOR WHO (II,689); THE OMEGA FACTOR (II,1888)

STEWART, PAUL
THE DEFENDERS (I,1220); MAVERICK (II,1657)

STEWART, ROBERT
THE HUNTER (I,2162)

STEWART, WILLIAM
BLESS THIS HOUSE (II,268); FATHER DEAR FATHER (II,837)

STIERS, DAVID OGDEN
M*A*S*H (II,1569)

STIX, JOHN
WINDOWS (I,4876)

STONE, ANDREW
SCREEN DIRECTOR'S PLAYHOUSE (I,3946)

STONE, CORDELLA
THE GOLDEN AGE OF TELEVISION (II,1016); THE GOLDEN AGE OF TELEVISION (II,1017); THE GOLDEN AGE OF TELEVISION (II,1018); THE GOLDEN AGE OF TELEVISION (II,1019)

STONE, EZRA
THE ALL-STAR REVUE (I,138); ANGEL (I,216); FIREBALL FUN FOR ALL (I,1577); I MARRIED JOAN (I,2174); JOE AND MABEL (I,2398); LOST IN SPACE (I,2758); SOUND OFF TIME (I,4135); SPACE ACADEMY (II,2406); TAMMY (I,4357); TIME FOR ELIZABETH (I,4506)

STONE, PAULA
PAULA STONE'S TOY SHOP (I,3495)

STORCH, WOLFGANG
STAR MAIDENS (II,2435)

STORM, HOWARD
AMANDA'S (II,62); ANGIE (II,80); BEST OF THE WEST (II,218); BUSTING LOOSE (II,402); DOC (II,682); FAMILY BUSINESS (II,815); FERNWOOD 2-NIGHT (II,849); FISH (II,864); GIMME A BREAK (II,995); GOODTIME GIRLS (II,1042); HIZZONER (II,1159); JOANIE LOVES CHACHI (II,1295); THE KROFFT KOMEDY HOUR (II,1411); LAVERNE AND SHIRLEY (II,1446); MORK AND MINDY (II,1735); MR. MERLIN (II,1751); PLEASE STAND BY (II,2057); PRIME TIMES (II,2083); RHODA (II,2151); SHEEHY AND THE SUPREME MACHINE (II,2322); STAR OF THE FAMILY (II,2436); SUGAR TIME (II,2489); THE TED KNIGHT SHOW (II,2550); TOO

CLOSE FOR COMFORT (II,2642)

STRANGIS, SAM
BATMAN (I,366)

STRENGE, WALTER
MY LITTLE MARGIE (I,3194); WATERFRONT (I,4774)

STROCK, HERBERT L
77 SUNSET STRIP (I,3988); THE ALASKANS (I,106); BRONCO (I,745); CHEYENNE (I,931); COLT .45 (I,1010); DOCTOR CHRISTIAN (I,1308); HARBOR COMMAND (I,1945); I LED THREE LIVES (I,2169); SUGARFOOT (I,4277); THE VEIL (I,4709)

STRONG, JAY
VERSATILE VARIETIES (I,4713)

STROOT, MIKE
YOU ASKED FOR IT (II,2855)

STROUD, RICHARD
NANCY ASTOR (II,1788)

STUART, JOSEPH
THE DOCTORS (II,694)

STUART, MEL
BRENDA STARR (II,372); CASABLANCA (II,450); THE CHISHOLMS (II,512); THE CHISHOLMS (II,513); GET CHRISTIE LOVE! (II,981); HERE WE GO AGAIN (I,2028); RIPLEY'S BELIEVE IT OR NOT (II,2177); SOPHIA! (I,4128)

STUART, PETER
RIPLEY'S BELIEVE IT OR NOT (II,2177)

STURRIDGE, CHARLES
BRIDESHEAD REVISITED (II,375); CORONATION STREET (I,1054)

SUGERMAN, ANDREW
MANDY'S GRANDMOTHER (II,1618)

SULLIVAN, JACK
DARK SHADOWS (I,1157); ONE LIFE TO LIVE (II,1907)

SULLIVAN, SEAN DHU
DARK SHADOWS (I,1157)

SULTAN, ARNE
THE GOVERNOR AND J.J. (I,1863)

SUMMERS, JEREMY
INTERNATIONAL DETECTIVE (I,2224); MY PARTNER THE GHOST (I,3199); THE SAINT (II,2237); TOURIST (II,2651); U.F.O. (I,4662)

SUSSAN, HERBERT
BY POPULAR DEMAND (I,784); HOTPOINT HOLIDAY (I,2131); THE KEN MURRAY SHOW (I,2525); MEET YOUR COVER GIRL (I,2992)

SUTHERLAND, EDDIE
THE BLACK ROBE (I,479); INTERNATIONAL DETECTIVE (I,2224)

SUTHERLAND, HAL
THE ARCHIE COMEDY HOUR (I,252); ARCHIE'S TV FUNNIES (I,253); THE BRADY KIDS (II,365); EVERYTHING'S ARCHIE (I,1483); FANTASTIC VOYAGE (I,1530); FAT ALBERT AND THE COSBY KIDS (II,835); FLASH GORDON (II,874); GILLIGAN'S PLANET (II,994); HE-MAN AND THE MASTERS OF THE UNIVERSE (II,1113); JOURNEY TO THE CENTER OF THE EARTH (I,2456); LASSIE'S RESCUE RANGERS (II,1432); MISSION MAGIC (I,3068); MY FAVORITE MARTIANS (II,1779); THE NEW ADVENTURES OF BATMAN (II,1812); THE NEW ADVENTURES OF SUPERMAN (I,3255); SABRINA, THE TEENAGE WITCH (I,3880); STAR TREK (II,2439); TEACHER'S PET (I,4367); THE YOUNG SENTINELS (II,2871)

SWACKHAMER, E W
BEWITCHED (I,418); BRING 'EM BACK ALIVE (II,377); THE CABOT CONNECTION (II,408); CHOPPER ONE (II,515); COCAINE AND BLUE EYES (II,548); THE DAIN CURSE (II,613); EIGHT IS ENOUGH (II,762); FAMILY (II,813); THE FLYING NUN (I,1611); FOUL PLAY (II,910); GIDGET (I,1795); GIDGET GETS MARRIED (I,1796); HAZEL (I,1982); HERE COME THE BRIDES (I,2024); I DREAM OF JEANNIE (I,2167); IS THERE A DOCTOR IN THE HOUSE? (II,1247); M*A*S*H (II,1569); MALIBU (II,1607); MCCLOUD (II,1660); MCMILLAN AND WIFE (II,1667); MEDICAL CENTER (II,1675); MOBILE ONE (II,1717); MOVIN' ON (I,3118); THE NANCY DREW MYSTERIES (II,1789); ONCE AN EAGLE (II,1897); OWEN MARSHALL: COUNSELOR AT LAW (I,3435); THE PARTRIDGE FAMILY (II,1962); PEKING ENCOUNTER (II,1991); POLICE STORY (II,2062); POPI (II,2068); POPI (II,2069); QUINCY, M. E. (II,2102); REWARD (II,2147); THE ROOKIES (II,2208); THE ROUSTERS (II,2220); THE SHERIFF AND THE ASTRONAUT (II,2326); SPIDER-MAN (II,2424); SWEEPSTAKES (II,2514); TALES OF THE APPLE DUMPLING GANG (II,2536); TENSPEED AND BROWN SHOE (II,2562); THREE FOR TAHITI (I,4476); THE UGLIEST GIRL IN TOWN (I,4663); UNDER THE YUM YUM TREE (I,4671); WHAT'S UP DOC? (II,2771)

SWANSON, GLEN
DINAH AND FRIENDS (II,675); DINAH! (II,677); HOUR MAGAZINE (II,1193); MAKE ME LAUGH (II,1602)

SWANSON, JOAN
JIM HENSON'S MUPPET BABIES (II,1289)

SWEENEY, BOB
ALOHA PARADISE (II,58); THE ANDROS TARGETS (II,76); THE ANDY GRIFFITH SHOW (I,192); AT EASE (II,115); BRING 'EM BACK ALIVE (II,377); THE COURTSHIP OF EDDIE'S FATHER (I,1083); CRAZY LIKE A FOX (II,601); THE DORIS DAY SHOW (I,1355); THE DUKES OF HAZZARD (II,742); DYNASTY (II,746); FANTASY ISLAND (II,829); FINDER OF LOST LOVES (II,857); FLAMINGO ROAD (II,872); GOODNIGHT, BEANTOWN (II,1041); HEY TEACHER (I,2041); JUST OUR LUCK (II,1360); THE LOVE BOAT (II,1535); OH, NURSE! (I,3343); PRIVATE BENJAMIN (II,2087); THE SIX O'CLOCK FOLLIES (II,2368); SPENCER'S PILOTS (II,2421); STRIKE FORCE (II,2480); TRAPPER JOHN, M.D. (II,2654)

SWENSON, CHARLES
CARLTON YOUR DOORMAN (II,440); DOROTHY IN THE LAND OF OZ (II,722)

SWENSON, MURAKAMI WOLF
CARLTON YOUR DOORMAN (II,440)

SWERLING JR, JO
THE LAST CONVERTIBLE (II,1435)

SWIFT, DAVID
ARNIE (I,261); BARNEY MILLER (II,154); CAMP RUNAMUCK (I,811); POOR MR. CAMPBELL (I,3647); THE RIFLEMAN (I,3789); SWEEPSTAKES (II,2514)

SWIFT, LELA
DARK SHADOWS (I,1157); THE HOUSE ON HIGH STREET (I,2141); LITTLE WOMEN (I,2720); PUREX SPECIAL FOR WOMEN (I,3698); RYAN'S HOPE (II,2234); WESTINGHOUSE SUMMER THEATER (I,4805)

SWIT, LORETTA
M*A*S*H (II,1569)

SWITZER, MICHAEL
POWERHOUSE (II,2074)

SWOPE JR, HERBERT
ARSENIC AND OLD LACE (I,268); THE CLOCK (I,978); ROBERT MONTGOMERY PRESENTS YOUR LUCKY STRIKE THEATER (I,3809)

SWOPE, MEL
FAME (II,812); GOOD HEAVENS (II,1032); THE PARTRIDGE FAMILY (II,1962)

SYDOW, JACK
ANNIE GET YOUR GUN (I,224)

SYKES, PETER
THE AVENGERS (II,121)

SYNES, BOB
HOT SEAT (I,2125); SPLIT SECOND (II,2428)

SZWARC, JEANNOT
ALIAS SMITH AND JONES (I,118); BARETTA (II,152); THE BLACK SHEEP SQUADRON (II,262); CODE NAME: DIAMOND HEAD (II,549); HAZARD'S PEOPLE (II,1112); IRONSIDE (II,1246); LISA, BRIGHT AND DARK (I,2711); THE MAN AND THE CITY (I,2856); MARCUS WELBY, M.D. (II,1627); THE MEN FROM SHILOH (I,3005); THE NEW ADVENTURES OF PERRY MASON (I,3252); NIGHT GALLERY (I,3287); PARIS 7000 (I,3459); THE ROCKFORD FILES (II,2197); SARGE (I,3914); THE SIX-MILLION-DOLLAR MAN (II,2372); THE SMALL MIRACLE (I,4090); TOMA (II,2634); TRUE LIFE STORIES (II,2665); THE VIRGINIAN (I,4738)

TAKAMOTO, IWAO
THE HANNA-BARBERA HAPPINESS HOUR (II,1079)

TALMADGE, RICHARD
EL COYOTE (I,1423)

TAPPLINHER, SYLVAN
ANYONE CAN WIN (I,235)

TARRANT, ALAN

CARRY ON LAUGHING (II,448)

TASHLIN, FRANK
THE FRANCES LANGFORD SHOW (I,1657)

TATOR, JOEL
THE TOMORROW SHOW (II,2635)

TATOR, JOYCE
KIDS 2 KIDS (II,1389)

TAYBACK, VIC
ALICE (II,33)

TAYLOR JR, JACK GAMMON
SCARECROW AND MRS. KING (II,2268)

TAYLOR, ART
CHER—A CELEBRATION AT CAESAR'S PALACE (II,498)

TAYLOR, DON
ALCOA PREMIERE (I,109); ALCOA/GOODYEAR THEATER (I,107); ALFRED HITCHCOCK PRESENTS (I,115); AMANDA FALLON (I,151); AMOS BURKE, SECRET AGENT (I,180); THE BIG VALLEY (I,437); BURKE'S LAW (I,762); CHECKMATE (I,919); COOGAN'S REWARD (I,1044); DEAR MOM, DEAR DAD (I,1202); DENNIS THE MENACE (I,1231); THE DICK POWELL SHOW (I,1269); THE EVE ARDEN SHOW (I,1471); THE FARMER'S DAUGHTER (I,1533); FULL SPEED ANYWHERE (I,1704); HEAT OF ANGER (I,1992); I REMEMBER CAVIAR (I,2175); JOHNNY RINGO (I,2434); M SQUAD (I,2799); MANNIX (II,1624); MANY HAPPY RETURNS (I,2895); MOBILE ONE (II,1717); THE MOD SQUAD (I,3078); THE NAME OF THE GAME (I,3217); NIGHT GALLERY (I,3287); NIGHT GAMES (II,1845); OCTAVIUS AND ME (I,3327); PURSUE AND DESTROY (I,3699); THE RIFLEMAN (I,3789); THE WILD WILD WEST (I,4863); YOU'RE ONLY YOUNG TWICE (I,4971)

TAYLOR, GIL
DEPARTMENT S (I,1232)

TAYLOR, JUD
THE BIG EASY (II,234); CAPTAIN NICE (I,835); THE DOCTORS (I,1323); EGAN (I,1419); THE FELONY SQUAD (I,1563); FUTURE COP (II,946); THE GIRL FROM U.N.C.L.E. (I,1808); HAWKINS (I,1976); HAWKINS ON MURDER (I,1978); LOU GRANT (II,1526); MEDICAL CENTER (II,1675); THE ROOKIES (I,3840); SARA (II,2260); SHANE (I,3995); STAR TREK (II,2440); TENAFLY (II,2560); WINTER KILL (II,2810); THE YOUNG LAWYERS (I,4941)

TAYLOR, MALCOLM
A KILLER IN EVERY CORNER (I,2538); MURDER MOTEL (I,3161)

TAYLOR, RENEE
BEDROOMS (II,198)

TEAGUE, LEWIS
BARNABY JONES (II,153); LADIES IN BLUE (II,1422); A MAN CALLED SLOANE (II,1613); RIKER (II,2175); VEGAS (II,2724)

TEDESCO, LOU
THE FACE IS FAMILIAR (I,1506); FACE THE MUSIC (II,803); JOHNNY CARSON DISCOVERS CYPRESS GARDENS (I,2419); THE NEW SOUPY SALES SHOW (II,1829); SCROOGE'S ROCK 'N' ROLL CHRISTMAS (II,2279); TOP OF THE MONTH (I,4578); YOU ASKED FOR IT (II,2855)

TEICHMANN, HOWARD
SHOWTIME, U.S.A. (I,4038)

TELFORD, FRANK
THE DUNNINGER SHOW (I,1385); INDEMNITY (I,2209); PULITZER PRIZE PLAYHOUSE (I,3692)

TEMPLETON, GEORGE
CITIZEN SOLDIER (I,963)

TENDLAR, DAVE
THE ASTRONAUT SHOW (I,307); CARTOONSVILLE (I,858); CASPER AND FRIENDS (I,867); DEPUTY DAWG (I,1233); THE HECTOR HEATHCOTE SHOW (I,2000); THE MIGHTY MOUSE PLAYHOUSE (I,3035)

TETZLAFF, TED
SCREEN DIRECTOR'S PLAYHOUSE (I,3946)

TEWKSBURY, PETER
ALCOA PREMIERE (I,109); FATHER KNOWS BEST (II,838); THE FITZPATRICKS (II,868); THE MISS AND THE MISSILES (I,3062); MISS STEWART, SIR (I,3064); NANNY AND THE PROFESSOR (I,3225); THE PEOPLE'S CHOICE (I,3521); SECOND CHANCE (I,3958); SHEPHERD'S FLOCK (I,4007)

THAU, LEON
PRISONER: CELL BLOCK H (II,2085)

THEILL, WILLIAM
THE LONE RANGER (I,2740)

THEOBALD, GEOFF
CELEBRITY REVUE (II,475); LET'S MAKE A DEAL (II,1463); MANTRAP (I,2893); PITFALL (II,2047)

THOMAS, DANNY
MAKE ROOM FOR GRANDDADDY (I,2844)

THOMAS, DAVE
SANDY (I,3908)

THOMAS, LARRY
ELVIRA'S MOVIE MACABRE (II,771)

THOMAS, RICHARD
THE WALTONS (II,2740)

THOMPSON, BOB
CHICAGO STORY (II,506); GENERAL HOSPITAL (II,964)

THOMPSON, CHRIS
1915 (II,1853); BOSOM BUDDIES (II,355)

THOMPSON, DAN
G.I. JOE: A REAL AMERICAN HERO (II,948); THE TRANSFORMERS (II,2653)

THOMPSON, DON
THE NEW AVENGERS (II,1816)

THOMPSON, J LEE
THE BLUE KNIGHT (II,277); CODE RED (II,552); CODE RED (II,553)

THOMPSON, ROBERT
THE ABC AFTERSCHOOL SPECIAL (II,1); DYNASTY (II,746); THE GREATEST AMERICAN HERO (II,1060); HILL STREET BLUES (II,1154); MAGNUM, P.I. (II,1601); THE PAPER CHASE (II,1949); QUINCY, M. E. (II,2102); THE RENEGADES (II,2133); SEVEN BRIDES FOR SEVEN BROTHERS (II,2307)

THORNE, WORLEY
THE PAPER CHASE (II,1949)

THORP, MAUREEN
KIDS ARE PEOPLE TOO (II,1390)

THORPE, JERRY
THE 20TH CENTURY-FOX HOUR (I,4642); CHICAGO STORY (II,506); THE DESILU PLAYHOUSE (I,1237); DIAL HOT LINE (I,1254); HARRY O (II,1098); HARRY O (II,1099); THE LAZARUS SYNDROME (II,1451); LOCK, STOCK, AND BARREL (I,2736); THE MACKENZIES OF PARADISE COVE (II,1585); MANY HAPPY RETURNS (I,2895); MR. TUTT (I,3153); OF MEN OF WOMEN (I,3331); THE POSSESSED (II,2071); RAFFERTY (II,2105); SMILE JENNY, YOU'RE DEAD (II,2382); STICKIN' TOGETHER (II,2465)

THRASH, BILL
THE STARS AND STRIPES SHOW (I,4206); STARS AND STRIPES SHOW (II,2442); THE STARS AND STRIPES SHOW (II,2443)

THRONSON, ROBERT
THE BARON (I,360); TYCOON: THE STORY OF A WOMAN (II,2697)

THURMAN, GLYNN R
DYNASTY (II,746)

TIBBLES, DOUGLAS
THE DORIS DAY SHOW (I,1355)

TIGHE, KEVIN
EMERGENCY! (II,775)

TINGLING, JAMES
PUBLIC DEFENDER (I,3687)

TINKER, MARK
100 CENTRE STREET (II,1901); MAKING THE GRADE (II,1606); ST. ELSEWHERE (II,2432); THE WHITE SHADOW (II,2788)

TINNEY JR, JOSEPH
THE BIG TOP (I,435)

TOKAR, NORMAN
ADVENTURES OF A MODEL (I,50); BIG DADDY (I,425); THE CHICAGO TEDDY BEARS (I,934); THE CLAUDETTE COLBERT SHOW (I,971); THE DONNA REED SHOW (I,1347); THE DORIS DAY SHOW (I,1355); HIS MODEL WIFE (I,2064); IT'S ALWAYS JAN (I,2264); LEAVE IT TO BEAVER (I,2648); MCGARRY AND ME (I,2966); MY SISTER HANK (I,3201); NAKED CITY (I,3215); THE TAB HUNTER SHOW (I,4328); YOUNG MR. BOBBIN (I,4943)

TOMALIN, LES
MAGIC COTTAGE (I,2815)

TOMBLIN, DAVID
SPACE: 1999 (II,2409); U.F.O. (I,4662)

TORKLESON, PETER H
THE MONKEES (I,3088)

TOTTEN, ROBERT
THE DAKOTAS (I,1126); DOC ELLIOT (I,1306); ENOS (II,779); THE FITZPATRICKS (II,868); GUNSMOKE (II,1069); HAWAIIAN EYE (I,1973); KUNG FU (II,1416); MISSION: IMPOSSIBLE (I,3067); THE

MONROES (I,3089); TEMPLE HOUSTON (I,4392); THE TEXAS WHEELERS (II,2568); TODAY'S FBI (II,2628); THE VIRGINIAN (I,4738)

TOUBER, SELWIN
THE TENNESSEE ERNIE FORD SHOW (I,4399)

TOURNEUR, JACQUES
ADVENTURES IN PARADISE (I,46); THE ALASKANS (I,106); THE BARBARA STANWYCK THEATER (I,350); BONANZA (II,347); THE CODE OF JONATHAN WEST (I,992)

TOWNSEND, BUD
THE BEACHCOMBER (I,371)

TOWSLEY, DON
FAT ALBERT AND THE COSBY KIDS (II,835); FLASH GORDON (II,874); MISSION MAGIC (I,3068); MY FAVORITE MARTIANS (II,1779); THE NEW ADVENTURES OF BATMAN (II,1812); THE NEW ADVENTURES OF GILLIGAN (II,1813); THE U.S. OF ARCHIE (I,4687); U.S. OF ARCHIE (II,2698)

TOYNTON, IAN
BERGERAC (II,209)

TRACY, JOHN
ANGIE (II,80); BOSOM BUDDIES (II,355); I LOVE HER ANYWAY! (II,1214); IRENE (II,1245); IT'S A LIVING (II,1254); IT'S NOT EASY (II,1256); JOANIE LOVES CHACHI (II,1295); LADIES' MAN (II,1423); MAGGIE (II,1589); MAKIN' IT (II,1603); THE NEW MICKEY MOUSE CLUB (II,1823); THE NEW ODD COUPLE (II,1825); OUT OF THE BLUE (II,1934); SCALPELS (II,2266); SPENCER (II,2420); WHO'S THE BOSS? (II,2792)

TRBOVICH, TOM
THE ARCHIE SITUATION COMEDY MUSICAL VARIETY SHOW (II,106); BABY MAKES FIVE (II,129); BOSOM BUDDIES (II,355); THE CHEVY CHASE NATIONAL HUMOR TEST (II,504); CIRCUS OF THE STARS (II,530); CIRCUS OF THE STARS (II,531); CIRCUS OF THE STARS (II,533); CIRCUS OF THE STARS (II,534); CIRCUS OF THE STARS (II,535); CIRCUS OF THE STARS (II,536); CIRCUS OF THE STARS (II,537); DREAMS (II,735); THE FUN FACTORY (II,938); THE HALF-HOUR COMEDY HOUR (II,1074); HOT HERO SANDWICH (II,1186); JOANIE LOVES CHACHI (II,1295); LAVERNE AND SHIRLEY (II,1446); MAC DAVIS 10TH ANNIVERSARY SPECIAL: I STILL BELIEVE IN MUSIC (II,1571); MAC DAVIS—I'LL BE HOME FOR CHRISTMAS (II,1580); THE MANHATTAN TRANSFER (II,1619); THE MIDNIGHT SPECIAL (II,1696); THE MOUSEKETEERS REUNION (II,1742); NATIONAL LAMPOON'S HOT FLASHES (II,1795); THE ROBERT KLEIN SHOW (II,2186); SIROTA'S COURT (II,2365); SUGAR TIME (II,2489); THE TV SHOW (II,2673); WEDDING DAY (II,2759)

TRENT, JOHN
D.H.O. (I,1249); RED SKELTON'S CHRISTMAS DINNER (II,2123)

TRIKONIS, GUS
FLAMINGO ROAD (II,871)

TRONSON, ROBERT
THE CARNATION KILLER (I,845); CRY TERROR! (I,1112); FATHER BROWN (II,836); LOOK BACK IN DARKNESS (I,2752)

TRUMAN, MICHAEL
DANGER MAN (I,1136); SECRET AGENT (I,3962)

TUBBS, CARL
GIGGLESNORT HOTEL (II,989)

TUCHNER, MICHAEL
PAROLE (II,1960); PEEK-A-BOO: THE ONE AND ONLY PHYLLIS DIXEY (II,1989); TALES OF THE UNEXPECTED (II,2540)

TUCKER, ROGER
BERGERAC (II,209)

TULCHIN, HAL
JUST LIKE A WOMAN (I,2496); THE SAMMY DAVIS JR. SPECIAL (I,3899)

TULCHIN, HAROLD
THE WAYNE NEWTON SPECIAL (II,2749)

TURNER, ANN
AMERICA (I,165)

TURNER, BILL
DOLLY (II,699); POP! GOES THE COUNTRY (II,2067)

TURNER, GIL
MR. MAGOO (I,3142)

TURNER, KEN
CAPTAIN SCARLET AND THE MYSTERONS (I,837); U.F.O. (I,4662)

TURNER, SKIP
WONDER GIRL (I,4891)

TURNER, STEVE
THE DAVID SOUL AND FRIENDS SPECIAL (II,626)

TURRELL, SAUL J
SILENT'S, PLEASE (I,4050)

TURTELTAUB, SAUL
THAT GIRL (II,2570)

TUTTLE, FRANK
SCREEN DIRECTOR'S PLAYHOUSE (I,3946)

TWAIN, NORMAN
VANITIES (II,2720)

TYLER, JAMES
PLANET PATROL (I,3616)

TYNE, BUDDY
THE 416TH (II,913); THE BOB CRANE SHOW (II,280); BROTHERS AND SISTERS (II,382); FRIENDS (II,927); THE LOVE BOAT (II,1535)

TYNE, GEORGE
THE BOB NEWHART SHOW (II,342); THE BRADY BUNCH (II,362); THE FIGHTING NIGHTINGALES (II,855); GRANDPA GOES TO WASHINGTON (II,1050); JAMES AT 15 (II,1270); THE LOVE BOAT (II,1535); M*A*S*H (II,1569); MISS WINSLOW AND SON (II,1705); THE PRACTICE (II,2076); THE SAN PEDRO BEACH BUMS (II,2250); TABITHA (II,2528)

TYNE, JAMES
HAPPY DAYS (II,1084)

TYRRELL, LOUIS
MAGIC MIDWAY (I,2818)

TYTLA, BILL
CASPER AND FRIENDS (I,867); THE MIGHTY MOUSE PLAYHOUSE (I,3035)

ULLMAN, JOHN
ACTION IN THE AFTERNOON (I,25)

URBANO, CARL
THE ALL-NEW POPEYE HOUR (II,49); THE BISKITTS (II,257); CASPER AND THE ANGELS (II,453); FLINTSTONE FAMILY ADVENTURES (II,882); THE FLINTSTONES (II,885); FRED AND BARNEY MEET THE THING (II,919); THE GARY COLEMAN SHOW (II,958); THE KWICKY KOALA SHOW (II,1417); LAVERNE AND SHIRLEY IN THE ARMY (II,1448); LAVERNE AND SHIRLEY WITH THE FONZ (II,1449); THE LITTLE RASCALS (II,1494); MONCHHICHIS (II,1722); THE NEW FRED AND BARNEY SHOW (II,1817); PAC-MAN (II,1940); PINK PANTHER AND SONS (II,2043); THE RICHIE RICH SHOW (II,2169); SCOOBY-DOO AND SCRAPPY-DOO (II,2274); THE SHIRT TALES (II,2337); THE SMURFS (II,2385); SPACE STARS (II,2408); SUPER FRIENDS (II,2497); THE SUPER GLOBETROTTERS (II,2498); TROLLKINS (II,2660)

URBANO, TONY
BARBARA MANDRELL AND THE MANDRELL SISTERS (II,143)

URBISCI, ROCCO
ALAN KING'S SECOND ANNUAL FINAL WARNING (II,30); ALL NIGHT RADIO (II,40); THE KENNY ROGERS SPECIAL (II,1382)

USTINOV, PETER
LOVE, LIFE, LIBERTY & LUNCH (II,1544)

VAIL, LESTER
THE ALDRICH FAMILY (I,111)

VAJDA, ZELI
THE ROCK 'N ROLL SHOW (II,2188)

VALENTA, LEONARD
AS THE WORLD TURNS (II,110); BEAT THE CLOCK (I,377); CHARLIE WILD, PRIVATE DETECTIVE (I,914); THE GUIDING LIGHT (II,1064); MARGE AND JEFF (I,2908); THE SONNY KENDIS SHOW (I,4127); STRICTLY FOR LAUGHS (I,4261); WINDOWS (I,4876)

VAN ATTA, DON
BOSOM BUDDIES (II,355); ME AND MAXX (II,1672)

VAN DYKE, DICK
THE NANCY DUSSAULT SHOW (I,3222)

VAN HEUSEN, JIMMY
THE BING CROSBY SPECIAL (I,471)

VAN KUIKEN, RALPH
THE UNCLE FLOYD SHOW (II,2703)

VAN RYKER, PATRICIA
CALIFORNIA FEVER (II,411); THE POWERS OF MATTHEW STAR (II,2075); THE RETURN OF MARCUS WELBY, M.D (II,2138); V: THE SERIES (II,2715); WONDER WOMAN (SERIES 2) (II,2829)

VAN DEN ECKER, BEAU
GREEN ACRES (I,1884); HAWAII FIVE-O (II,1110)

VANCE, DENNIS
THE FEAR IS SPREADING (I,1553)

VANOFF, NICK
THE BING CROSBY SHOW (I,465); THE BING CROSBY SHOW (I,466); THE PERRY COMO CHRISTMAS SHOW (II,2002); THE PERRY COMO SUNSHINE SHOW (II,2005); PERRY COMO'S SUMMER OF '74 (II,2024)

VARDY, MIKE
THE CITADEL (II,539)

VARNEL, MAX
SKIPPY, THE BUSH KANGAROO (I,4076)

VASIR, DAVID
THOSE AMAZING ANIMALS (II,2601)

VASSAR, DAVID
YOU ASKED FOR IT (II,2855)

VAUGHN, ROBERT
POLICE WOMAN (II,2063)

VESOTO, BRUNO
CHICAGOLAND MYSTERY PLAYERS (I,936)

VEJAR, MICHAEL
THE A-TEAM (II,119); BLUE THUNDER (II,278); BRING 'EM BACK ALIVE (II,377); CUTTER TO HOUSTON (II,612); FANTASY ISLAND (II,829); HAWAIIAN HEAT (II,1111); HOT PURSUIT (II,1189); THE INCREDIBLE HULK (II,1232); LOTTERY (II,1525); MAGNUM, P.I. (II,1601); MATT HOUSTON (II,1654); QUINCY, M. E. (II,2102); SCARECROW AND MRS. KING (II,2268); SIMON AND SIMON (II,2357); TALES OF THE GOLD MONKEY (II,2537)

VEJAR, RUDY
ROMANCE THEATER (II,2204); THE YOUNG AND THE RESTLESS (II,2862)

VERDI, TONY
THE SHARI LEWIS SHOW (II,2317)

VERNA, TONY
CELEBRITY DAREDEVILS (II,473)

VIDOR, KING
LIGHT'S DIAMOND JUBILEE (I,2698)

VIOLA, AL
POPI (II,2069); STARSTRUCK (II,2446)

VITELLO, ART
G.I. JOE: A REAL AMERICAN HERO (II,948); PANDAMONIUM (II,1948); SPIDER-MAN AND HIS AMAZING FRIENDS (II,2425)

VOGEL, VIRGIL W
AIRWOLF (II,27); AMOS BURKE, SECRET AGENT (I,180); BARNABY JONES (II,153); BERT D'ANGELO/SUPERSTAR (II,211); BEULAH LAND (II,226); THE BIG VALLEY (I,437); BONANZA (II,347); CANNON (II,424); CARIBE (II,439); CENTENNIAL (II,477); COLORADO C.I. (II,555); DAN AUGUST (I,1130); DAVID CASSIDY—MAN UNDERCOVER (II,623); THE FANTASTIC JOURNEY (II,826); THE FBI (I,1551); HERE COME THE BRIDES (I,2024); THE HIGH CHAPARRAL (I,2051); KNIGHT RIDER (II,1402); LARAMIE (I,2614); LOTTERY (II,1525); M SQUAD (I,2799); MAGNUM, P.I. (II,1601); THE MAN FROM ATLANTIS (II,1615); MAVERICK (II,1657); MCCLAIN'S LAW (II,1659); MISSION: IMPOSSIBLE (I,3067); MOST WANTED (II,1739); NURSE (II,1870); THE OREGON TRAIL (II,1923); POLICE WOMAN (II,2063); SKAG (II,2374); STARSKY AND HUTCH (II,2444); THE STREETS OF SAN FRANCISCO (II,2478); TALES OF THE GOLD MONKEY (II,2537); TODAY'S FBI (II,2628); UNIT 4 (II,2707); VOYAGERS (II,2730)

VOIGHT, DON
SNEAK PREVIEWS (II,2388)

WAGGNER, GEORGE
77 SUNSET STRIP (I,3988); THE ALASKANS (I,106); ARROYO (I,266); BATMAN (I,366); BRONCO (I,745); CHEYENNE (I,931); COLT .45 (I,1010); HAWAIIAN EYE (I,1973); LAWMAN (I,2642); MAVERICK (II,1657); THE ROARING TWENTIES (I,3805); SCREEN DIRECTOR'S PLAYHOUSE (I,3946); SUGARFOOT (I,4277); SURFSIDE 6 (I,4299); WAGON TRAIN (I,4747)

WAGGONER, CHUCK
SNEAK PREVIEWS (II,2388)

WAGLIN, ED
THE DOM DELUISE SHOW (I,1328)

WAGNER, MARIA
AS THE WORLD TURNS (II,110)

WAITE, RALPH
THE WALTONS (II,2740)

WAKERELL, TINA
PENMARRIC (II,1993)

WALKER, JAMES
JIM HENSON'S MUPPET BABIES (II,1289); THE TRANSFORMERS (II,2653)

WALKER, JOHN
THE BISKITTS (II,257); MONCHHICHIS (II,1722); THE TRANSFORMERS (II,2653)

WALKER, NANCY
13 QUEENS BOULEVARD (II,2591); RHODA (II,2151)

WALKER, ROBERT G
THE ADVENTURES OF CHAMPION (I,54); THE ADVENTURES OF RIN TIN TIN (I,73); ANNIE OAKLEY (I,225); BUFFALO BILL JR. (I,755); THE RANGE RIDER (I,3720); RESCUE 8 (I,3772); THE WESTERNER (I,4800)

WALLACE, DON
WHODUNIT? (II,2794)

WALLACE, RICK
CALL TO GLORY (II,413)

WALLENSTEIN, JOSEPH B
KNOTS LANDING (II,1404)

WALLER, ALBERT
THE ABC AFTERSCHOOL SPECIAL (II,1)

WALLERSTEIN, HERB
THE BARBARY COAST (II,146); HAPPY DAYS (II,1084); HERE COME THE BRIDES (I,2024); MULLIGAN'S STEW (II,1764); THE NEW ADVENTURES OF PERRY MASON (I,3252); THE OREGON TRAIL (II,1923); THE PARTRIDGE FAMILY (II,1962); QUINCY, M. E. (II,2102); STAR TREK (II,2440); TABITHA (II,2528); THE WILD WILD WEST (I,4863); WONDER WOMAN (SERIES 1) (II,2828)

WALLERSTEIN, ROWE
EIGHT IS ENOUGH (II,762)

WALLIS, ALAN
DOCTOR IN THE HOUSE (I,1313)

WALRDON, GY
THE DUKES OF HAZZARD (II,742)

WALSH, BOB
MR. ROGERS' NEIGHBORHOOD (I,3150)

WALTERS, CHARLES
A LUCILLE BALL SPECIAL STARRING LUCILLE BALL AND JACKIE GLEASON (II,1558); A LUCILLE BALL SPECIAL: WHAT NOW CATHERINE CURTIS? (II,1560)

WALTON, KIP
ALMOST ANYTHING GOES (II,56); BOB HOPE SPECIAL: BOB HOPE'S 30TH ANNIVERSARY TV SPECIAL (II,293); BOB HOPE SPECIAL: BOB HOPE'S ALL-STAR COMEDY BIRTHDAY PARTY (II,297); BOB HOPE SPECIAL: BOB HOPE'S ALL-STAR COMEDY LOOK AT THE FALL SEASON: IT'S STILL FREE AND WORTH IT! (II,299); BOB HOPE SPECIAL: BOB HOPE'S ALL-STAR LOOK AT TV'S PRIME TIME WARS (II,303); BOB HOPE SPECIAL: BOB HOPE'S ROAD TO HOLLYWOOD (II,313); BOB HOPE SPECIAL: BOB HOPE'S STAR-STUDDED SPOOF OF THE NEW TV SEASON— G RATED—WITH GLAMOUR, GLITTER & GAGS (II,317); BOB HOPE SPECIAL: BOB HOPE'S WOMEN I LOVE—BEAUTIFUL BUT FUNNY (II,322); BOB HOPE SPECIAL: BOB HOPE—HOPE, WOMEN AND SONG (II,324); BOB HOPE SPECIAL: THE BOB HOPE CHRISTMAS SPECIAL (II,329); DIANA (I,1258); THE GEORGE KIRBY SPECIAL (I,1771); GET IT TOGETHER (I,1780); HOT CITY DISCO (II,1185); JUNIOR ALMOST ANYTHING GOES (II,1357); KICKS (II,1385); THE NEW $100,000 NAME THAT TUNE (II,1810); PERRY COMO IN LAS VEGAS (II,2003); PERRY COMO'S CHRISTMAS IN PARIS (II,2012); THE SMOKEY ROBINSON SHOW (I,4093); WOMAN'S PAGE (II,2822)

WANAMAKER, SAM
CIMARRON STRIP (I,954); COLUMBO (II,556); CORONET BLUE (I,1055); DAVID CASSIDY—MAN UNDERCOVER (II,623); THE DEFENDERS (I,1220); HART TO HART (II,1102); HAWK (I,1974); LANCER (I,2610); LASSITER (I,2625); LASSITER (II,1434); MRS. COLUMBO (II,1760); RETURN OF THE SAINT (II,2141)

WARD, JOHN
BARETTA (II,152)

WARD, PHYLLIS
YOU ASKED FOR IT (II,2855)

WARD, RICHARD
OZMOE (I,3437)

WARD, SCOTT
THE EARTHA KITT SHOW (I,1395)

WARE, CLYDE
ALFRED HITCHCOCK PRESENTS (I,115); GUNSMOKE (II,1069)

WARNICK, CLAY
YOUR HIT PARADE (I,4952); YOUR SHOW OF SHOWS (I,4957); YOUR SHOW OF SHOWS (II,2875)

WARREN, CHARLES MARQUIS
GUNSMOKE (II,1069); RAWHIDE (I,3727)

WARREN, FRANKLIN
ARTHUR MURRAY'S DANCE PARTY (I,293)

WARREN, MARK
BABY, I'M BACK! (II,130); BIG CITY COMEDY (II,233); COTTON CLUB '75 (II,580); THE DIAHANN CARROLL SHOW (I,1252); DICK CLARK PRESENTS THE ROCK AND ROLL YEARS (I,1265); DICK CLARK PRESENTS THE ROCK AND ROLL YEARS (I,1266); DINAH AND HER NEW BEST FRIENDS (II,676); GET CHRISTIE LOVE! (II,981); JOEY AND DAD (II,1306); KOMEDY TONITE (II,1407); THE NEW BILL COSBY SHOW (I,3257); ROWAN AND MARTIN'S LAUGH-IN (I,3856); TURN ON (I,4625); WHAT'S HAPPENING!! (II,2769); THE WOLFMAN JACK SHOW (II,2817); WOMEN WHO RATE A "10" (II,2825)

WARREN, MICHAEL
THE DIAHANN CARROLL SHOW (II,665); FISH (II,864); SANFORD AND SON (II,2256)

WARREN, RALPH
ACTOR'S STUDIO (I,28); THE FRED WARING SHOW (I,1679); MOVIELAND QUIZ (I,3116); ON THE CORNER (I,3359); STOP THE MUSIC (I,4232); THE THORNTON SHOW (I,4468)

WASHBURN, JIM
THE HANNA-BARBERA HAPPINESS HOUR (II,1079); MILTON BERLE'S MAD MAD MAD WORLD OF COMEDY (II,1701); THE PAUL LYNDE COMEDY HOUR (II,1979)

WASHMAN, TOM
THE TOM AND JERRY SHOW (I,4534)

WASSERMAN, AL
A LOOK AT THE LIGHT SIDE (I,2751)

WATERS, BILL
CIRCUS OF THE STARS (II,532)

WATERSON, HERB
THE MAGIC OF MARK WILSON (II,1596)

WATT, MICHAEL
MINSKY'S FOLLIES (II,1703); STAR CHART (II,2434)

WEAVER, GEOFFREY
DAVID NIVEN'S WORLD (II,625)

WEBB, GORDON
PERRY MASON (II,2025)

WEBB, JACK
ADAM-12 (I,31); CHASE (I,917); THE D.A. (I,1123); DRAGNET (I,1369); DRAGNET (I,1370); DRAGNET (I,1371); EMERGENCY! (II,775); GENERAL ELECTRIC TRUE (I,1754); NOAH'S ARK (I,3302); O'HARA, UNITED STATES TREASURY: OPERATION COBRA (I,3347); PETE KELLY'S BLUES (I,3559); SAM (II,2246)

WEBSTER, NICHOLAS
APPLE'S WAY (II,101); THE BIG VALLEY (I,437); THE CHISHOLMS (II,513); THE FBI (I,1551); RIPLEY'S BELIEVE IT OR NOT (II,2177)

WEDKOS, PAUL
BURKE'S LAW (I,762)

WEED, GENE
MR. MOON'S MAGIC CIRCUS (II,1753)

WEEGE, REINHOLD
NIGHT COURT (II,1844)

WEINBERG, DICK
AMERICA 2-NIGHT (II,65); THE JERRY LEWIS SHOW (I,2362)

WEINBERGER, ED
BEST OF THE WEST (II,218); MR. SMITH (II,1755); TAXI (II,2546)

WEINER, RON
HOT (II,1183); HOT (II,1184); THE PHIL DONAHUE SHOW (II,2032)

WEIS, DON
ALFRED HITCHCOCK PRESENTS (I,115); THE ANDROS TARGETS (II,76); THE BARBARY COAST (II,146); BARETTA (II,152); BATMAN (I,366); THE BOB HOPE CHRYSLER THEATER (I,502); BRING 'EM BACK ALIVE (II,377); BURKE'S LAW (I,762); CAGNEY AND LACEY (II,409); CASABLANCA (I,860); CHARLIE'S ANGELS (II,486); CHECKMATE (I,919); CHIPS (II,511); CODE RED (II,552); COMMAND PERFORMANCE (I,1029); DEAR PHOEBE (I,1204); DELTA HOUSE (II,658); THE DOOLEY BROTHERS (II,715); FANTASY ISLAND (II,829); FLAMINGO ROAD (II,872); FLO'S PLACE (II,892); HAPPY DAYS (II,1084); HARD KNOCKS (II,1086); HARRY O (II,1099); HAWAII FIVE-O (II,1110); HEAD OF THE FAMILY (I,1988); IRONSIDE (II,1246); IT TAKES A THIEF (I,2250); THE JACK BENNY PROGRAM (I,2294); KINGSTON: CONFIDENTIAL (II,1398); LOLLIPOP LOUIE (I,2738); LOTTERY (II,1525); THE LOVE BOAT (II,1535); M*A*S*H (II,1569); MANNIX (II,1624); MATT HELM (II,1653); MCKEEVER AND THE COLONEL (I,2970); THE MILLIONAIRE (II,1700); THE MUNSTERS' REVENGE (II,1765); THE NIGHT STALKER (II,1849); OFF WE GO (I,3338); PAPA SAID NO (I,3452); PARIS 7000 (I,3459); THE PATTY DUKE SHOW (I,3483); PLANET OF THE APES (II,2049); QUICK AND QUIET (II,2099); REMINGTON STEELE (II,2130); RIDDLE AT 24000 (II,2171); ROLL OUT! (I,3833); THE SAN PEDRO BEACH BUMS (II,2250); SECRETS OF THE OLD BAILEY (I,3971); THE SIX O'CLOCK FOLLIES (II,2368); SKIP TAYLOR (I,4075); SPENCER'S PILOTS (II,2422); STARSKY AND HUTCH (II,2444); THE SURVIVORS (I,4301); T.J. HOOKER (II,2524); THE VIRGINIAN (I,4738); WAGON TRAIN (I,4747)

WEIS, GARY
THE BEACH BOYS SPECIAL (II,190); THINGS WE DID LAST SUMMER (II,2589)

WEISMAN, BRUCE
BATTLE OF THE NETWORK STARS (II,163); BATTLE OF THE NETWORK STARS (II,164); BATTLE OF THE NETWORK STARS (II,165); BATTLE OF THE NETWORK STARS (II,166); BATTLE OF THE NETWORK STARS (II,167); BATTLE OF THE NETWORK STARS (II,168); BATTLE OF THE NETWORK STARS (II,169); BATTLE OF THE NETWORK STARS (II,170); BATTLE OF THE NETWORK STARS (II,171); BATTLE OF THE NETWORK STARS (II,172); BATTLE OF THE NETWORK STARS (II,173); BATTLE OF THE NETWORK STARS (II,174); BATTLE OF THE NETWORK STARS (II,175); BATTLE OF THE NETWORK STARS (II,176); BATTLE OF THE NETWORK STARS (II,177)

WEISMAN, SAM
THE BOUNDER (II,357); DOMESTIC LIFE (II,703); FAMILY TIES (II,819); HIS AND HERS (II,1155); WHO'S THE BOSS? (II,2792)

WEISS, SAM
G.I. JOE: A REAL AMERICAN HERO (II,948)

WEIST, GEORGE
LADIES BE SEATED (I,2598)

WELLES, ORSON
FOUNTAIN OF YOUTH (I,1647)

WELLS, GEORGE
THE FABULOUS DR. FABLE (I,1500)

WELLS, RICHARD
COSMOS (II,579); THE LOVE BOAT (II,1535)

WELLS, ROBERT
THE DIONNE WARWICK SPECIAL (I,1289)

WENDKOS, PAUL
333 MONTGOMERY (I,4487); 79 PARK AVENUE (II,2310); THE ALCOA HOUR (I,108); ALCOA/GOODYEAR THEATER (I,107); BEN CASEY (I,394); BIG JOHN (II,239); THE BIG VALLEY (I,437); BOONE (II,351); CELEBRITY (II,463); CRISIS (I,1102); THE DELPHI BUREAU (I,1224); THE DETECTIVES (I,1245); THE DICK POWELL SHOW (I,1269); FARRELL: FOR THE PEOPLE (II,834); GOLDEN GATE (II,1020); HAGEN (II,1071); HARRY O (II,1099); HAWAII FIVE-O (I,1972); HONEY WEST (I,2105); I SPY (I,2179); THE INVADERS (I,2229); LAW OF THE PLAINSMAN (I,2639); MEDICAL STORY (II,1676); MRS R.—DEATH AMONG FRIENDS (II,1759); NAKED CITY (I,3216); THE RIFLEMAN (I,3789); ROUTE 66 (I,3852); SAINTS AND SINNERS (I,3887); TRAVIS LOGAN, D.A. (I,4598); THE UNTOUCHABLES (I,4682);

THE WILD WILD WEST (I,4863)

WENMAN, DIANA
ALL MY CHILDREN (II,39)

WERNER, PETER
CALL TO GLORY (II,413); FAMILY (II,813); HARD KNOX (II,1087)

WESSON, DICK
READY AND WILLING (II,2115); THE SECRET WORLD OF KIDS (I,3970)

WESTCOTT, TED
IT'S A HIT (I,2257)

WETZLER, GWEN
BATMAN AND THE SUPER SEVEN (II,160); THE FABULOUS FUNNIES (II,802); FLASH GORDON (II,874); FLASH GORDON—THE GREATEST ADVENTURE OF ALL (II,875); THE NEW ADVENTURES OF BATMAN (II,1812)

WEYMAN, ANDREW
THE EDGE OF NIGHT (II,760); TEXAS (II,2566)

WEYMAN, ROBERT
LOVING (II,1552)

WHARF, RICHARD
ALWAYS APRIL (I,147)

WHARMBY, TONY
THE SEVEN DIALS MYSTERY (II,2308); WHY DIDN'T THEY ASK EVANS? (II,2795)

WHATHAM, CLAUDE
ALL CREATURES GREAT AND SMALL (I,129)

WHEELER, DAVID
THE WALTONS (II,2740)

WHEELER, DONNA
NEWHART (II,1835)

WHELAN, KEN
FOR LOVE OR MONEY (I,1622); OF ALL THINGS (I,3330)

WHELAN, TIM
CAVALCADE OF AMERICA (I,875); CROSSROADS (I,1105)

WHITE, JOSHUA
DELTA HOUSE (II,658); LOOK WHAT THEY'VE DONE TO MY SONG (II,1517); THAT'S CAT (II,2576)

WHITEHEAD, WILLIAM
SIMON AND SIMON (II,2357)

WHITESELL, JOHN
THE GUIDING LIGHT (II,1064); TEXAS (II,2566)

WHITING, JUD
MIKE AND BUFF (I,3037)

WHITMAN, JACK
THE DUKES OF HAZZARD (II,742); THE YELLOW ROSE (II,2847)

WHITMORE, DEAN
1968 HOLLYWOOD STARS OF TOMORROW (I,3297); AWAY WE GO (I,317); BRIEF ENCOUNTER (I,723); COLISEUM (I,1000); DICK VAN DYKE AND THE OTHER WOMAN, MARY TYLER MOORE (I,1273); THE DICK VAN DYKE SPECIAL (I,1277); FEATHERTOP (I,1556); JOHNNY MANN'S STAND UP AND CHEER (I,2430); THE JUDY GARLAND SHOW (I,2473); THE JUDY GARLAND SHOW (II,1349); MICKIE FINN'S (I,3029); THE NIGHT OF CHRISTMAS (I,3290); THE PEARL BAILEY SHOW (I,3501); THE TENNESSEE ERNIE FORD SPECIAL (I,4401); THE VIKKI CARR SHOW (I,4732)

WHITMORE, PETER
DAVE ALLEN AT LARGE (II,621)

WHITNEY, WILLIAM
LARAMIE (I,2614); THE WILD WILD WEST (I,4863); ZORRO (I,4982)

WHORF, RICHARD
THE ANN SOTHERN SHOW (I,220); THE BARBARA STANWYCK THEATER (I,350); CANNON (II,424); HEAVEN HELP US (I,1995); IVY LEAGUE (I,2286); LAW OF THE PLAINSMAN (I,2639); MY FRIEND IRMA (I,3191); PANDORA AND FRIEND (I,3445); THE UNTOUCHABLES (I,4682); THE WILD WILD WEST (I,4863)

WIARD, WILLIAM
BERT D'ANGELO/SUPERSTAR (II,211); THE BLACK SHEEP SQUADRON (II,262); BONANZA (II,347); BRET MAVERICK (II,374); CANNON (II,424); CARIBE (II,439); CHASE (I,917); DANIEL BOONE (I,1142); THE DORIS DAY SHOW (I,1355); THE EDDIE CAPRA MYSTERIES (II,758); THE GIRL, THE GOLD WATCH AND EVERYTHING (II,1002); THE HIGH CHAPARRAL (I,2051); JESSIE (II,1287); KEY TORTUGA (II,1383); M*A*S*H (II,1569); THE MISSISSIPPI (II,1707); THE MONROES (I,3089); MOST WANTED (II,1739); MR. ROBERTS (I,3149); OPERATION: RUNAWAY (II,1918); THE QUEST (II,2098); THE ROCKFORD FILES (II,2197); ROLL OUT! (I,3833); ROOM 222 (I,3843); SARA (II,2260); SCARECROW AND MRS. KING (II,2268); THE SEAL (II,2282); SEARCH (I,3951); THE STREETS OF SAN FRANCISCO (II,2478)

WICKES, DAVID
THE HITCHHIKER (II,1157); PHILIP MARLOWE, PRIVATE EYE (II,2033); TARGET (II,2541)

WIESEN, BERNARD
JULIA (I,2476)

WIESMAN, BERNARD
THE BRADY BUNCH (II,362)

WILCOX, LARRY
CHIPS (II,511)

WILDER, DOUG
RETURN TO THE PLANET OF THE APES (II,2145)

WILES, GEORGE
BIG FOOT AND WILD BOY (II,237)

WILES, GORDON
AN AMATEUR'S GUIDE TO LOVE (I,154); BIG JOHN, LITTLE JOHN (II,240); THE BOBBY DARIN AMUSEMENT COMPANY (I,672); THE COLGATE COMEDY HOUR (I,998); THE FABULOUS FUNNIES (II,801); LAND OF THE LOST (II,1425); LI'L ABNER (I,2702); MCDUFF, THE TALKING DOG (II,1662); ROWAN AND MARTIN BITE THE HAND THAT FEEDS THEM (I,3853); ROWAN AND MARTIN'S LAUGH-IN (I,3856); WHERE THE GIRLS ARE (I,4830)

WILK, ROBERT CLARK
OMNI: THE NEW FRONTIER (II,1889)

WILLIAMS, ALLEN
LOU GRANT (II,1526)

WILLIAMS, CINDY
LAVERNE AND SHIRLEY (II,1446)

WILLIAMS, DICK
THE BIG SHOW (II,243); FAVORITE SONGS (I,1545)

WILLIAMS, ROBIN
MORK AND MINDY (II,1735)

WILSON, ANDY
TOP 40 VIDEOS (II,2644)

WILSON, BILL
NOT THE NINE O'CLOCK NEWS (II,1864)

WILSON, DAVE
BOB & RAY & JANE, LARAINE & GILDA (II,279); THE CORNER BAR (I,1052); JACK LEMMON—GET HAPPY (I,2312); LET'S CELEBRATE (I,2663); THE NEWS IS THE NEWS (II,1838); THE PAUL SIMON SPECIAL (II,1983); REALLY RAQUEL (II,2119); SATURDAY NIGHT LIVE (II,2261); STEVE MARTIN'S BEST SHOW EVER (II,2460); THE TROUBLE WITH PEOPLE (I,4611); THE WONDERFUL WORLD OF AGGRAVATION (I,4894)

WILSON, HUGH
THE CHOPPED LIVER BROTHERS (II,514); THE TONY RANDALL SHOW (II,2640); WKRP IN CINCINNATI (II,2814)

WILSON, RICHARD
MEDIC (I,2976)

WILSON, RONALD
THE PALLISERS (II,1944)

WILSON, ROY
THE SHIRT TALES (II,2337)

WINBURN, RANDY
AMERICA 2-NIGHT (II,65); FOREVER FERNWOOD (II,909); IN THE BEGINNING (II,1229)

WINCER, SIMON
PRISONER: CELL BLOCK H (II,2085)

WINDUST, BRETAIGNE
LEAVE IT TO BEAVER (I,2648); MY THREE ANGELS (I,3204); THE PIED PIPER OF HAMELIN (I,3595)

WINER, HARRY
THE ABC AFTERSCHOOL SPECIAL (II,1); CALLAHAN (II,414); THE DUCK FACTORY (II,738); GOODNIGHT, BEANTOWN (II,1041); HART TO HART (II,1102); MR. MERLIN (II,1751); PAPER DOLLS (II,1952); STONE (II,2471); TENSPEED AND BROWN SHOE (II,2562); TUCKER'S WITCH (II,2667); TWO MARRIAGES (II,2691)

WINES, WILLIAM
THEY STAND ACCUSED (I,4442)

WINGER, SIMON
AGAINST THE WIND (II,24)

WINKLER, HENRY

JOANIE LOVES CHACHI (II,1295)

WINSTON, RONALD
INTERMEZZO (I,2223); THE TWILIGHT ZONE (I,4651)

WINTER, JACK
LAVERNE AND SHIRLEY (II,1446); THE ODD COUPLE (II,1875)

WINTERS, DAVID
THE ANN-MARGRET SHOW (I,218); ANN-MARGRET: FROM HOLLYWOOD WITH LOVE (I,217); DR JEKYLL AND MR. HYDE (I,1368); FROM HOLLYWOOD WITH LOVE: THE ANN-MARGRET SPECIAL (I,1687); THE MONKEES (I,3088); RAQUEL (I,3725); SPECIAL LONDON BRIDGE SPECIAL (I,4150); WHERE THE GIRLS ARE (I,4830)

WINTHER, JORN
THE BARBARA MCNAIR SHOW (I,348); THE BOBBY SHERMAN SPECIAL (I,676); THE FIFTH DIMENSION TRAVELING SHOW (I,1572); FUNNY YOU SHOULD ASK (I,1717); THE JONATHAN WINTERS SHOW (I,2443); KRAFT SUMMER MUSIC HALL (I,2590); RITUALS (II,2182); SHINDIG (I,4013); THE SONNY AND CHER COMEDY HOUR (II,2400); STORY THEATER (I,4241); SWINGING COUNTRY (I,4322)

WISBAR, FRANK
CURTAIN CALL THEATER (I,1115); FIRESIDE THEATER (I,1580); JANE WYMAN PRESENTS THE FIRESIDE THEATER (I,2345); MAN OF THE COMSTOCK (I,2873)

WISE, HERBERT
THE GATHERING STORM (I,1741); TALES OF THE UNEXPECTED (II,2540)

WITNEY, WILLIAM
THE ADVENTURES OF DR. FU MANCHU (I,58); ALFRED HITCHCOCK PRESENTS (I,115); BONANZA (II,347); BRANDED (I,707); DANIEL BOONE (I,1142); THE HIGH CHAPARRAL (I,2051); KODIAK (II,1405); LAREDO (I,2615); RESCUE 8 (I,3772); SKY KING (I,4077); TARZAN (I,4361); THE VIRGINIAN (I,4738); WAGON TRAIN (I,4747)

WITT, PAUL JUNGER
HERE COME THE BRIDES (I,2024); THE PARTRIDGE FAMILY (II,1962)

WOLF, FRED
CARLTON YOUR DOORMAN (II,440); MARLO THOMAS AND FRIENDS IN FREE TO BE. . . YOU AND ME (II,1632)

WOLFF, ART
ANOTHER WORLD (II,97); MARY HARTMAN, MARY HARTMAN (II,1638)

WOMACK, STEVE
COUNTRY GALAXY OF STARS (II,587); AN EVENING WITH THE STATLER BROTHERS (II,790)

WONG, ARNIE
G.I. JOE: A REAL AMERICAN HERO (II,948)

WOOD, BARRY
ALKALI IKE (I,123)

WOOD, IVOR
PADDINGTON BEAR (II,1941)

WOOD, PETER
HAMLET (I,1925); LONG DAY'S JOURNEY INTO NIGHT (I,2746)

WOOD, PRESTON
DUNNINGER AND WINCHELL (I,1384); G.E. GUEST HOUSE (I,1746); HOLIDAY HOTEL (I,2075)

WOODWARD, JOANNE
FAMILY (II,813); SENSE OF HUMOR (II,2303)

WOODWARD, MARVIN
COURAGEOUS CAT (I,1079)

WOPAT, TOM
THE DUKES OF HAZZARD (II,742)

WORSLEY JR, WALLACE
FLASH GORDON (I,1593)

WRANGLE, BASIL
THE HORACE HEIDT SHOW (I,2120)

WRAY, JOHN
THE ED SULLIVAN SHOW (I,1401); FUN FOR 51 (I,1707); STAR OF THE FAMILY (I,4194); SUGAR HILL TIMES (I,4276)

WRIGHT, KAY
BATMAN AND THE SUPER SEVEN (II,160); THE FABULOUS FUNNIES (II,802); FAT ALBERT AND THE COSBY KIDS (II,835); FLASH GORDON (II,874); FLASH GORDON—THE GREATEST ADVENTURE OF ALL (II,875); SPORT BILLY (II,2429)

WRIGHT, ROBERT
STARTIME (I,4212)

WULF, JOHN
FORMBY'S ANTIQUE WORKSHOP (I,1644)

WYMAN, STEVE
RYAN'S HOPE (II,2234)

WYNN, BOB
THE 36 MOST BEAUTIFUL GIRLS IN TEXAS (II,2594); BOB HOPE SPECIAL: BOB HOPE'S ALL-STAR BIRTHDAY PARTY (II,295); BOB HOPE SPECIAL: HAPPY BIRTHDAY, BOB! (II,326); THE FABULOUS FORDIES (I,1502); LIGHTS, CAMERA, MONTY! (II,1475); THE NBC FOLLIES (I,3241); THE NBC FOLLIES (I,3242); OPRYLAND U.S.A. (I,3405); OPRYLAND USA—1975 (II,1920); SALUTE TO LADY LIBERTY (II,2243); SAM (II,2247); TENNESSEE ERNIE'S NASHVILLE-MOSCOW EXPRESS (II,2561)

WYSE, BILL
GOOBER AND THE TRUCKERS' PARADISE (II,1029)

YAGEMANN, BILL
I'LL BET (I,2189)

YARBROUGH, JAMES P
THE BEACHCOMBER (I,371)

YARBROUGH, JEAN
THE ABBOTT AND COSTELLO SHOW (I,2); THE ADDAMS FAMILY (II,11); CHEYENNE (I,931); THE GUNS OF WILL SONNETT (I,1907); HELLO DERE (I,2010); I'M THE LAW (I,2195); THE LIFE OF RILEY (I,2686); MCHALE'S NAVY (I,2969); THE SILENT SERVICE (I,4049); SKYFIGHTERS (I,4078)

YATES, HAL
BLONDIE (I,486); MY LITTLE MARGIE (I,3194)

YORKIN, BUD
THE ANDY WILLIAMS CHRISTMAS SHOW (I,196); THE ANDY WILLIAMS SHOW (I,208); THE ANDY WILLIAMS SPECIAL (I,211); THE ANDY WILLIAMS SPECIAL (I,212); ANOTHER EVENING WITH FRED ASTAIRE (I,229); BAND OF GOLD (I,341); CARTER COUNTRY (II,449); THE COLGATE COMEDY HOUR (I,997); THE DANNY KAYE SHOW (I,1144); AN EVENING WITH CAROL CHANNING (I,1473); AN EVENING WITH FRED ASTAIRE (I,1474); HENRY FONDA AND THE FAMILY (I,2015); THE JACK BENNY HOUR (I,2290); THE MANY SIDES OF DON RICKLES (I,2898); MAUDE (II,1655); OPERATION GREASEPAINT (I,3402); ROBERT YOUNG AND THE FAMILY (I,3813); THE SOLDIERS (I,4112); THE TENNESSEE ERNIE FORD SHOW (I,4398); THE TONY MARTIN SHOW (I,4572); WHAT'S HAPPENING!! (II,2769)

YOSHIDA, TATSUO
SPEED RACER (I,4151)

YOUNG, EVAN
ANOTHER LIFE (II,95)

YOUNG, ROGER
HARDCASTLE AND MCCORMICK (II,1089); KNOTS LANDING (II,1404); LEGMEN (II,1458); LOU GRANT (II,1526); MAGNUM, P.I. (II,1601)

YOUNG, TERENCE
THE POPPY IS ALSO A FLOWER (I,3650)

YUKI, TOSHI
INSPECTOR GADGET (II,1236)

YUST, LARRY
TESTIMONY OF TWO MEN (II,2564)

ZAMORA, RUDY
THE BISKITTS (II,257); FLINTSTONE FAMILY ADVENTURES (II,882); THE FLINTSTONE FUNNIES (II,883); FONZ AND THE HAPPY DAYS GANG (II,898); THE GARY COLEMAN SHOW (II,958); THE KWICKY KOALA SHOW (II,1417); LAVERNE AND SHIRLEY IN THE ARMY (II,1448); LAVERNE AND SHIRLEY WITH THE FONZ (II,1449); THE LITTLE RASCALS (II,1494); MONCHHICHIS (II,1722); PAC-MAN (II,1940); THE RICHIE RICH SHOW (II,2169); THE SMURFS (II,2385); SPACE STARS (II,2408); TROLLKINS (II,2660)

ZASLOVE, ALAN
THE ALVIN SHOW (I,146)

ZAVADA, ERVIN
JERRY VISITS (I,2373)

ZEAMER, MICHAEL
THE ARTHUR GODFREY SHOW (I,285); THE ARTHUR GODFREY SHOW (I,286); BRIDE AND GROOM (I,721); HONEY WEST (I,2105)

ZIMMERMAN, GAIL
ONE LIFE TO LIVE (II,1907)

ZINBERG, MICHAEL
9 TO 5 (II,1852); THE BOB NEWHART SHOW (II,342); FAMILY TIES (II,819); A GIRL'S LIFE (II,1000); HOME ROOM (II,1172); THE JAMES BOYS (II,1272); LOU GRANT (II,1526); MOTHER AND ME, M.D. (II,1740); NEWHART (II,1835); NOT UNTIL TODAY (II,1865); TAXI (II,2546); THE TONY RANDALL SHOW (II,2640); WE'VE GOT EACH OTHER (II,2757); WHACKED OUT (II,2767); THE WHITE SHADOW (II,2788); WKRP IN CINCINNATI (II,2814)

ZUKER, DAVID
POLICE SQUAD! (II,2061)

ZUKER, JERRY
POLICE SQUAD! (II,2061)

ZUKOR, LOU
BATMAN AND THE SUPER SEVEN (II,160); BLACKSTAR (II,263); THE FABULOUS FUNNIES (II,802); FAT ALBERT AND THE COSBY KIDS (II,835); FLASH GORDON (II,874); HE-MAN AND THE MASTERS OF THE UNIVERSE (II,1113); THE KID SUPER POWER HOUR WITH SHAZAM (II,1386); MY FAVORITE MARTIANS (II,1779); THE NEW ADVENTURES OF BATMAN (II,1812); THE NEW ADVENTURES OF GILLIGAN (II,1813); THE NEW ADVENTURES OF MIGHTY MOUSE AND HECKLE AND JECKLE (II,1814); SPORT BILLY (II,2429); THE TARZAN/LONE RANGER/ZORRO ADVENTURE HOUR (II,2543); THE U.S. OF ARCHIE (I,4687); U.S. OF ARCHIE (II,2698)

ZWICK, EDWARD
PAPER DOLLS (II,1951)

ZWICK, JOEL
AMERICA 2100 (II,66); BOSOM BUDDIES (II,355); BROTHERS (II,380); BUSTING LOOSE (II,402); GOODTIME GIRLS (II,1042); HOT W.A.C.S. (II,1191); IT'S A LIVING (II,1254); JOANIE LOVES CHACHI (II,1295); LAVERNE AND SHIRLEY (II,1446); MAKIN' IT (II,1603); MORK AND MINDY (II,1735); THE NEW ODD COUPLE (II,1825); STAR OF THE FAMILY (II,2436); STRUCK BY LIGHTNING (II,2482); THE TED KNIGHT SHOW (II,2550); WEBSTER (II,2758)

Writers

ABATEMARCO, FRANK
CAGNEY AND LACEY (II,409); EISCHIED (II,763); MCCLAIN'S LAW (II,1659); MICKEY SPILLANE'S MIKE HAMMER (II,1692)

ABBOTT, CHRIS
FATHER MURPHY (II,841); LITTLE HOUSE ON THE PRAIRIE (II,1487); LITTLE HOUSE: A NEW BEGINNING (II,1488); LITTLE HOUSE: BLESS ALL THE DEAR CHILDREN (II,1489);; MAGNUM, P.I. (II,1601) NURSE (II,1870)

ABBOTT, NORMAN
THAT'S MY MAMA (II,2580)

ABEL, ROBERT
SOPHIA! (I,4128)

ABELL, JIM
DON'T CALL ME MAMA ANYMORE (I,1350); MADHOUSE 90 (I,2807); TOP OF THE MONTH (I,4578)

ABELSON, DANNY
NATIONAL LAMPOON'S HOT FLASHES (II,1795)

ABRAHAM, CYRIL
THE ONEDIN LINE (II,1912)

ABRAHAMS, JIM
POLICE SQUAD! (II,2061)

ABRAMS, LEON
COUNTERPOINT (I,1059); CROWN THEATER WITH GLORIA SWANSON (I,1107)

ABRAMS, MICHAEL
ANN-MARGRET SMITH (II,85)

ACE, GOODMAN
THE BIG PARTY FOR REVLON (I,427); EASY ACES (I,1398); THE GENERAL FOODS 25TH ANNIVERSARY SHOW (I,1756); THE KRAFT MUSIC HALL (I,2587); LADIES AND GENTLEMAN. . . BOB NEWHART (II,1420); THE MILTON BERLE SHOW (I,3047); THE PERRY COMO CHRISTMAS SHOW (I,3526); THE PERRY COMO CHRISTMAS SHOW (I,3528); THE PERRY COMO SHOW (I,3532); THE PERRY COMO THANKSGIVING SPECIAL (I,3540); RUTHIE ON THE TELEPHONE (I,3877); TEXACO STAR THEATER (I,4411); THE TROUBLE WITH TRACY (I,4613)

ADAIR, TOM
THE FARMER'S DAUGHTER (I,1533); HOLIDAY HOTEL (I,2075); THE MUNSTERS (I,3158); MY THREE SONS (I,3205); THE ROSEMARY CLOONEY SHOW (I,3847); THE TENNESSEE ERNIE FORD SHOW (I,4398)

ADAMS, CLAUDIA
CAGNEY AND LACEY (II,409); KNOTS LANDING (II,1404)

ADAMS, DON
THE DON ADAMS SPECIAL: HOORAY FOR HOLLYWOOD (I,1329)

ADAMS, DOUGLAS
THE HITCHHIKER'S GUIDE TO THE GALAXY (II,1158)

ADAMS, MICHAEL
THE FARMER'S DAUGHTER (I,1533); PLEASE DON'T EAT THE DAISIES (I,3628)

ADAMS, RICHARD
EVEL KNIEVEL (II,785); HAWAII FIVE-O (II,1110)

ADAMS, STANLEY
THE R.C.A. VICTOR SHOW (I,3737); STAR TREK (II,2440)

ADAMS, STEVE
DONNY AND MARIE (II,712)

ADAMSON, ED
BANYON (I,345); MANNIX (II,1624); PLEASE DON'T EAT THE DAISIES (I,3628); RAWHIDE (I,3727); THE RIFLEMAN (I,3789); THE UNTOUCHABLES (I,4682); THE WILD WILD WEST (I,4863)

ADDLESON, FLORENCE
FAMILY (II,813)

ADEKMAN, ALAN
WORKING STIFFS (II,2834)

ADELL, ILUNGA
SANFORD AND SON (II,2256); THAT'S MY MAMA (II,2580)

ADELMAN, BARRY
ALL NIGHT RADIO (II,40); THE FUNNY WORLD OF FRED & BUNNI (II,943); THE GRADY NUTT SHOW (II,1048); THE HEE HAW HONEYS (II,1124); THE JOHN DAVIDSON SHOW (II,1309); THE NASHVILLE PALACE (II,1792); PEOPLE ARE FUNNY (II,1996); SHIELDS AND YARNELL (II,2328); TV'S BLOOPERS AND PRACTICAL JOKES (II,2675)

ADELMAN, JERRY
DOCTOR CHRISTIAN (I,1308); FOREVER FERNWOOD (II,909); IT'S ABOUT TIME (I,2263); MEET CORLISS ARCHER (I,2979); PRIVATE SECRETARY (I,3672)

ADELMAN, SYBIL
CHARO (II,488); LILY (II,1477)

ADLER, DAVID
THE ANDY GRIFFITH SHOW (I,192); MY WORLD . . . AND WELCOME TO IT (I,3208)

ADLER, DICK
ENTERTAINMENT TONIGHT (II,780)

ADLER, EDWARD
N.Y.P.D. (I,3321)

ADLER, FELIX
THE ABBOTT AND COSTELLO SHOW (I,2)

ADLER, MARJORIE
MATINEE THEATER (I,2947)

AGNEW, JOHN
EIGHT IS ENOUGH (II,762)

AHERNS, LYN
DEAR ALEX AND ANNIE (II,651)

AHLERS, DAVID
TONIGHT IN HAVANA (I,4557)

AIDEKMAN, AL
IT'S YOUR MOVE (II,1259); THE JEFFERSONS (II,1276); LAVERNE AND SHIRLEY (II,1446); THE TED KNIGHT SHOW (II,2550)

AIDEM, MONTY
GENERAL ELECTRIC'S ALL-STAR ANNIVERSARY (II,963); THICKE OF THE NIGHT (II,2587); TOP TEN (II,2648)

AINOB, JACK
HOWDY DOODY AND FRIENDS (I,2152)

AKINS, ZOE
SCREEN DIRECTOR'S PLAYHOUSE (I,3946)

ALBRECHT, HOWARD
20TH CENTURY FOLLIES (I,4641); BOB HOPE SPECIAL: BOB HOPE'S ALL-STAR COMEDY SPECTACULAR FROM LAKE TAHOE (II,301); BOB HOPE SPECIAL: BOB HOPE'S ALL-STAR COMEDY TRIBUTE TO VAUDEVILLE (II,302); BOB HOPE SPECIAL: BOB HOPE'S ALL-STAR TRIBUTE TO THE PALACE THEATER (II,305); BOB HOPE SPECIAL: BOB HOPE'S CHRISTMAS SPECIAL (II,308); BOB HOPE SPECIAL: THE BOB HOPE SPECIAL FROM PALM SPRINGS (II,339); THE BOBBY DARIN AMUSEMENT COMPANY (I,672); THE DEAN MARTIN CELEBRITY ROAST (II,637); THE DEAN MARTIN CELEBRITY ROAST (II,638); THE DEAN MARTIN CELEBRITY ROAST (II,639); DEAN MARTIN'S CELEBRITY ROAST (II,643); DOM DELUISE AND FRIENDS, PART 2 (II,702); HELLZAPOPPIN (I,2011); HOW TO HANDLE A WOMAN (I,2146); THE JONATHAN WINTERS SHOW (I,2443); LADIES AND GENTLEMAN. . . BOB NEWHART (II,1420); LADIES AND GENTLEMAN. . . BOB NEWHART, PART II (II,1421); THE LESLIE UGGAMS SHOW (I,2660); THE LOVE BOAT (II,1535); THE MANY FACES OF COMEDY (I,2894); THE NBC FOLLIES (I,3241); THE NBC FOLLIES (I,3242); THE ODD COUPLE (II,1875); OUT OF THE BLUE (II,1934); THE PAUL LYNDE HALLOWEEN SPECIAL (II,1981); SZYSZNYK (II,2523); THREE'S COMPANY (II,2614); VOLTRON—DEFENDER OF THE UNIVERSE (II,2729); THE WAYNE NEWTON SPECIAL (II,2749)

ALBRECHT, JOIE
TELEVISION'S GREATEST COMMERCIALS III (II,2555); TELEVISION'S GREATEST COMMERCIALS IV (II,2556)

ALBRECHT, RICHARD
THE LOVE BOAT (II,1535); PAUL LYNDE AT THE MOVIES (II,1976)

ALCORN, R W
CITIZEN SOLDIER (I,963)

ALCROFT, JAMIE
ALL NIGHT RADIO (II,40)

ALDA, ALAN
THE FOUR SEASONS (II,915); HICKEY VS. ANYBODY (II,1142); M*A*S*H (II,1569); SUSAN AND SAM (II,2506); WE'LL GET BY (II,2753); WE'LL GET BY (II,2754)

ALDEN, JEROME
BARBAPAPA (II,142)

ALDEN, KAY
THE YOUNG AND THE RESTLESS (II,2862)

ALDER, EDWARD
THE ANDROS TARGETS (II,76)

ALDREDGE, DAWN
FLYING HIGH (II,895); HANDLE WITH CARE (II,1077); THE LOVE BOAT I (II,1532); THE LOVE BOAT II (II,1533); THE ROWAN AND MARTIN REPORT (II,2221); WHO'S THE BOSS? (II,2792)

ALDRICH, DAVID
STUDIO ONE (I,4268)

ALDRICH, HANK
THE HUNTER (I,2162)

ALDRICH, KEITH
THE HUNTER (I,2162)

ALDRIDGE, VIRGINIA
KNIGHT RIDER (II,1402); NURSE (II,1870)

ALEXANDER, ALTON
THE MAGIC GARDEN (II,1591)

ALEXANDER, BARRY
THE HARDY BOYS MYSTERIES (II,1090)

ALEXANDER, E NICK
240-ROBERT (II,2689); BARETTA (II,152); CASSIE AND COMPANY (II,455); THE FALL GUY (II,811); HUNTER (II,1206); VEGAS (II,2724)

ALEXANDER, LARRY
20TH CENTURY FOLLIES (I,4641); THE ANN-MARGRET SHOW (I,218); BARNABY JONES (II,153); BERT D'ANGELO/SUPERSTAR (II,211); BRONK (II,379); THE FLYING NUN (I,1611); GET CHRISTIE LOVE! (II,981); THE KING FAMILY IN WASHINGTON (I,2549); THE KING FAMILY JUNE SPECIAL (I,2550); MANHUNTER (II,1621); MATT HOUSTON (II,1654); PARIS (II,1957); RAQUEL (I,3725); THE RETURN OF CAPTAIN NEMO (II,2135); S.W.A.T. (II,2236); STAND UP AND CHEER (I,4188); WESTSIDE MEDICAL (II,2765)

ALEXANDER, LES
NO SOAP, RADIO (II,1856)

ALEXANDER, PATRICK
STUDIO ONE (I,4268)

ALEXANDER, ROD
BEN CASEY (I,394)

ALEXANDER, RONALD
LIFE WITH VIRGINIA (I,2695); THE MAN WHO CAME TO DINNER (I,2880); PANAMA HATTIE (I,3444); THE ROYAL FAMILY (I,3861); THE SHOWOFF (I,4035)

ALEY, ALBERT
ALCOA PREMIERE (I,109); THE FBI (I,1551); HAWAII FIVE-O (II,1110); RAWHIDE (I,3727); THE RIFLEMAN (I,3789); SEVEN AGAINST THE SEA (I,3983); YOUNG DAN'L BOONE (II,2863)

ALFIERI, RICHARD
I LOVE LIBERTY (II,1215)

ALLAN, BOB
BATTLESTARS (II,182)

ALLAN, JERRY
WOMEN OF RUSSIA (II,2824)

ALLARDICE, JAMES
THE FARMER'S DAUGHTER (I,1533); THE GEORGE GOBEL SHOW (I,1767); THE MUNSTERS (I,3158); MY THREE SONS (I,3205); THE TENNESSEE ERNIE FORD SHOW (I,4398)

ALLEN, AL
THE ABC AFTERSCHOOL SPECIAL (II,1)

ALLEN, CHRIS
THE BUFFALO BILLY SHOW (I,756); THE MARTY FELDMAN COMEDY MACHINE (I,2929)

ALLEN, DAVE
DAVE ALLEN AT LARGE (II,621)

ALLEN, FREDERICK LEWIS
THE FORD 50TH ANNIVERSARY SHOW (I,1630)

ALLEN, JAMES
BACHELOR FATHER (I,323); THE PINKY LEE SHOW (I,3601)

ALLEN, JAY PRESSON
THE BORROWERS (I,693)

ALLEN, R S
ALICE (II,33); FROM A BIRD'S EYE VIEW (I,1686); I SPY (I,2179); THE LOVE BOAT (II,1535); MR. TERRIFIC (I,3152); MY WORLD . . . AND WELCOME TO IT (I,3208); THE WONDERFUL WORLD OF BURLESQUE II (I,4896)

ALLEN, RAY
THE ANN SOTHERN SHOW (I,220); BACHELOR FATHER (I,323); CAVALCADE OF STARS (I,877); DOBIE GILLIS (I,1302); MAN IN THE MIDDLE (I,2870); MEDIC (I,2976); THE PINKY LEE SHOW (I,3601); THE WALTER WINCHELL SHOW (I,4760)

ALLEN, SIAN BARBARA
BARETTA (II,152)

ALLEN, STEVE
THE NEW STEVE ALLEN SHOW (I,3271); THE STEVE ALLEN COMEDY HOUR (II,2454); THE STEVE ALLEN SHOW (I,4219); THE STEVE ALLEN SHOW (I,4220); THE STEVE ALLEN SHOW (I,4221)

ALLEN, SYLVIA
REGGIE (II,2127)

ALLEN, VALERIE
VEGAS (II,2724)

ALLEN, WOODY
HOORAY FOR LOVE (I,2115); THE SID CAESAR SHOW (I,4042); THE WOODY ALLEN SPECIAL (I,4904); YOUR SHOW OF SHOWS (I,4957)

ALLENGREN, ERNIE
PALMERSTOWN U.S.A. (II,1945)

ALLIOTTE, JOHN
THE GREAT GILDERSLEEVE (I,1875)

ALLISON, JUDITH
PRIVATE BENJAMIN (II,2087); WIZARDS AND WARRIORS (II,2813)

ALLISON, LISA
WONDER GIRL (I,4891)

ALLYN, BARBARA
ALOHA PARADISE (II,58); TO SAY THE LEAST (II,2623)

ALPERT, ARTHUR M
WE INTERRUPT THIS SEASON (I,4780)

ALSBERG, ARTHUR
BACHELOR FATHER (I,323); CRASH ISLAND (II,600); THE DORIS DAY SHOW (I,1355); FRANKIE AND ANNETTE: THE SECOND TIME AROUND (II,917); THE GHOST AND MRS. MUIR (I,1786); HERBIE, THE LOVE BUG (II,1131); AN HOUR WITH ROBERT GOULET (I,2137); I DREAM OF JEANNIE (I,2167); THE MUNSTERS' REVENGE (II,1765); NANNY AND THE PROFESSOR AND THE PHANTOM OF THE CIRCUS

(I,3226); NEW GIRL IN HIS LIFE (I,3265)

ALTMAN, ROBERT
COMBAT (I,1011)

ALTMAN, WILIAM A
STUDIO ONE (I,4268)

AMATEAU, ROD
BORDER PALS (II,352); DOBIE GILLIS (I,1302); THE FALL GUY (II,811); HIGHWAY HONEYS (II,1151); WEEKEND (I,4788); WHERE THERE'S SMOKEY (I,4832)

AMBANDOS, JAMES
ONE, TWO, THREE—GO! (I,3390)

AMBLER, ERIC
ALFRED HITCHCOCK PRESENTS (I,115); SUSPICION (I,4309)

AMOS, JON
THE LESLIE UGGAMS SHOW (I,2660)

AMSTERDAM, MOREY
THE MOREY AMSTERDAM SHOW (I,3101)

ANASTI, TEDD
THE NEW MICKEY MOUSE CLUB (II,1823)

ANDERS, PATRICIA
OUTLAW LADY (II,1935)

ANDERSON, ALEXANDER
CRUSADER RABBIT (I,1109)

ANDERSON, BETH
WONDER GIRL (I,4891)

ANDERSON, CLIVE
NOT THE NINE O'CLOCK NEWS (II,1864)

ANDERSON, DAVID
WALTER CRONKITE'S UNIVERSE (II,2739)

ANDERSON, GEORGE
CHICAGOLAND MYSTERY PLAYERS (I,936); THE MAGIC RANCH (I,2819); PADDY THE PELICAN (I,3442); SANDY STRONG (I,3911)

ANDERSON, GERRY
CAPTAIN SCARLET AND THE MYSTERONS (I,837); THUNDERBIRDS (I,4497); U.F.O. (I,4662)

ANDERSON, HESPER
MARCUS WELBY, M.D. (II,1627); MEDICAL CENTER (II,1675)

ANDERSON, JOSEF
CALL TO GLORY (II,413); THE YELLOW ROSE (II,2847)

ANDERSON, MAXWELL
THE ALCOA HOUR (I,108); A CHRISTMAS CAROL (I,947)

ANDERSON, SYLVIA
CAPTAIN SCARLET AND THE MYSTERONS (I,837); THUNDERBIRDS (I,4497); U.F.O. (I,4662)

ANDRASKO, SCOTT
WONDER GIRL (I,4891)

ANDREWS, BETTY
NAKED CITY (I,3216); WAGON TRAIN (I,4747)

ANDREWS, CHARLES E
GARROWAY AT LARGE (I,1737); MARINELAND CARNIVAL (I,2912); MARINELAND CARNIVAL (I,2913); STUD'S PLACE (I,4271)

ANDREWS, CHRIS
THE WALTONS (II,2740)

ANDREWS, DANA RENO
THE FITZPATRICKS (II,868)

ANDREWS, ROBERT HARDY
GOING MY WAY (I,1833); THRILLER (I,4492)

ANFUSO, GUILLIO
ZANE GREY THEATER (I,4979)

ANGELL, BUCKLEY
RAWHIDE (I,3727)

ANGELL, DAVID
ARCHIE BUNKER'S PLACE (II,105); CHEERS (II,494)

ANGELOS, BILL
ANN-MARGRET: FROM HOLLYWOOD WITH LOVE (I,217); THE ARTHUR GODFREY SPECIAL (I,288); ARTHUR GODFREY'S PORTABLE ELECTRIC MEDICINE SHOW (I,290); BING CROSBY AND THE SOUNDS OF CHRISTMAS (I,457); BING CROSBY'S CHRISTMAS SHOW (I,474); BING CROSBY'S SUN VALLEY CHRISTMAS SHOW (I,475); BING CROSBY—COOLING IT (I,459); BING CROSBY—COOLING IT (I,460); CAROL AND COMPANY (I,847); CHRISTMAS WITH THE BING CROSBYS (I,950); CHRISTMAS WITH THE BING CROSBYS (II,524); FANFARE (I,1528); FROM HOLLYWOOD WITH LOVE: THE ANN-MARGRET SPECIAL (I,1687); THE KEEFE BRASSELLE SHOW (I,2514); THE KRAFT MUSIC HALL (I,2587); LOOKING BACK (I,2754); THE MANY MOODS OF PERRY COMO (I,2897); THE ODD COUPLE (II,1875); OUR PLACE (I,3416); THE PEARL BAILEY SHOW (I,3501); THE PERRY COMO CHRISTMAS SHOW (I,3529); PERRY COMO'S WINTER SHOW (I,3546); THE SAMMY DAVIS JR. SHOW (I,3898)

ANHALT, EDNA
BELLE STARR (I,391); PAPA SAID NO (I,3452); PLAYHOUSE 90 (I,3623)

ANHALT, EDWARD
NOWHERE TO HIDE (II,1868)

ANTONACCI, GREG
BUSTING LOOSE (II,402); IT TAKES TWO (II,1252); IT'S A LIVING (II,1254); MAKING A LIVING (II,1604); WEBSTER (II,2758)

ANTROBUS, JOHN
THE MARTY FELDMAN COMEDY MACHINE (I,2929)

APELIS, LIJANA
SUSPENSE THEATER ON THE AIR (II,2507)

APPEL, DON
APPLE'S WAY (II,101); MEDICAL CENTER (II,1675)

APPET, LEAH
FOR LOVE AND HONOR (II,903)

APPLEBAUM, COREY
BUCK ROGERS IN THE 25TH CENTURY (II,383); CHARLIE'S ANGELS (II,486)

APSTEIN, THEODORE
BAFFLED (I,332); BEN CASEY (I,394); BERKELEY SQUARE (I,399); CHANNING (I,900); JOHNNY BELINDA (I,2416); MATINEE THEATER (I,2947); STUDIO ONE (I,4268); SUSPICION (I,4309); TIME REMEMBERED (I,4509); THE TIME TUNNEL (I,4511); THE WALTONS (II,2740)

ARANGO, DOUGLAS
THE BOB NEWHART SHOW (II,342); CARTER COUNTRY (II,449); FAMILY BUSINESS (II,815); SANFORD (II,2255); THE TIM CONWAY COMEDY HOUR (I,4501); TONY ORLANDO AND DAWN (II,2639); TOO CLOSE FOR COMFORT (II,2642)

ARATOW, PAUL
SPORT BILLY (II,2429)

ARBOGAST, ROBERT L
THE NUT HOUSE (I,3320); PAT PAULSEN'S HALF A COMEDY HOUR (I,3473); TURN ON (I,4625)

ARCHIBALD, DOTTIE
THE SHAPE OF THINGS (II,2315)

ARENT, ARTHUR
THE CORN IS GREEN (I,1050)

ARKWRIGHT, BOB
GILLIGAN'S PLANET (II,994)

ARLISS, LESLIE
THE INVISIBLE MAN (I,2232)

ARMEN, MARGARET
BARNABY JONES (II,153); THE BIG VALLEY (I,437); EMERALD POINT, N.A.S. (II,773); FLAMINGO ROAD (II,872); MR. NOVAK (I,3145); THE RIFLEMAN (I,3789); STAR TREK (II,2439); STAR TREK (II,2440); WONDER WOMAN (SERIES 1) (II,2828); YOUNG DAN'L BOONE (II,2863)

ARMER, ALAN A
THE INVADERS (I,2229)

ARMOR, JOYCE
EMERALD POINT, N.A.S. (II,773); THE LOVE BOAT (II,1535); REMINGTON STEELE (II,2130); THE TONY RANDALL SHOW (II,2640); WKRP IN CINCINNATI (II,2814)

ARMSTRONG, BILL
PEOPLE DO THE CRAZIEST THINGS (II,1997); PEOPLE DO THE CRAZIEST THINGS (II,1998)

ARMUS, BURTON
AIRWOLF (II,27); BRONK (II,379); CASSIE AND COMPANY (II,455); THE FALL GUY (II,811); KOJAK (II,1406); PARIS (II,1957); VEGAS (II,2724)

ARNOLD, BUDDY
20TH CENTURY FOLLIES (I,4641); THE DOM DELUISE SHOW (I,1328); THE MILTON BERLE SHOW (I,3047); THE MILTON BERLE SHOW (I,3049); WASHINGTON SQUARE (I,4772)

ARNOLD, DANNY
ANN IN BLUE (II,83); BARNEY MILLER (II,154); MCHALE'S NAVY (I,2969); THE ROSEMARY CLOONEY SHOW (I,3848); SOMEWHERE IN ITALY. . . COMPANY B (I,4119); THE TENNESSEE ERNIE FORD SHOW (I,4398)

ARNOLD, ELLIOTT
BONANZA (II,347); RAWHIDE (I,3727)

ARNOLD, NICK
BAKER'S DOZEN (II,136); CHER (II,495); CHER (II,496); DEAN MARTIN'S CELEBRITY ROAST (II,643); JENNIFER SLEPT HERE (II,1278); LOVE AT FIRST SIGHT (II,1530); LOVE AT FIRST SIGHT (II,1531); A NEW KIND OF FAMILY (II,1819); PRIVATE BENJAMIN (II,2087); SMALL AND FRYE (II,2380); WHACKED OUT (II,2767); WHATEVER HAPPENED TO DOBIE GILLIS? (II,2774); ZORRO AND SON (II,2878)

ARNOTT, BOB
THE 36 MOST BEAUTIFUL GIRLS IN TEXAS (II,2594); AMERICA, YOU'RE ON (II,69); THE ANDY WILLIAMS SHOW (I,210); BARYSHNIKOV IN HOLLYWOOD (II,157); THE BEST OF TIMES (II,219); BOB HOPE SPECIAL: HAPPY BIRTHDAY, BOB! (II,326); DINAH AND HER NEW BEST FRIENDS (II,676); THE DONNY AND MARIE OSMOND SHOW (II,714); THE FUTURE: WHAT'S NEXT (II,947); HOLLYWOOD'S PRIVATE HOME MOVIES (II,1167); HOLLYWOOD'S PRIVATE HOME MOVIES II (II,1168); THE HUDSON BROTHERS RAZZLE DAZZLE COMEDY SHOW (II,1201); THE KEANE BROTHERS SHOW (II,1371); LIGHTS, CAMERA, MONTY! (II,1475); LYNDA CARTER: BODY AND SOUL (II,1565); MARIE (II,1629); OPRYLAND USA—1975 (II,1920); THE OSMOND FAMILY CHRISTMAS SPECIAL (II,1926); THE OSMOND FAMILY THANKSGIVING SPECIAL (II,1928); SALT AND PEPE (II,2240); THE SHAPE OF THINGS (II,2315); THE SMOTHERS BROTHERS SHOW (I,4097); THE SMOTHERS ORGANIC PRIME TIME SPACE RIDE (I,4098); THE SONNY AND CHER COMEDY HOUR (II,2400); THE SONNY AND CHER SHOW (II,2401); THE SONNY COMEDY REVUE (II,2403); THE WORLD'S FUNNIEST COMMERCIAL GOOFS (II,2841)

ARNSTEIN, LARRY
EIGHT IS ENOUGH (II,762); NOT NECESSARILY THE NEWS (II,1863)

AROESTS, JEAN LISETTE
STAR TREK (II,2440)

ARQUETTE, CLIFF
DAVE AND CHARLEY (I,1171); DO IT YOURSELF (I,1299)

ARQUETTE, LEWIS
THE LORENZO AND HENRIETTA MUSIC SHOW (II,1519); PRIME TIMES (II,2083)

ARRIGHI, MEL
MCCLOUD (II,1660)

ARRONS, RUTH
JACK CARTER AND COMPANY (I,2305)

ARTHUR, GEORGE LEE
REMINGTON STEELE (II,2130)

ARTHUR, ROBERT
ABC'S SILVER ANNIVERSARY CELEBRATION (II,2); ALFRED HITCHCOCK PRESENTS (I,115); THE BEST OF SULLIVAN (II,217); CELEBRITIES: WHERE ARE THEY NOW? (II,462); THE DAVID SOUL AND FRIENDS SPECIAL (II,626); INSIDE AMERICA (II,1234); MR. & MS. AND THE BANDSTAND MURDERS (II,1746); MR. & MS. AND THE MAGIC STUDIO MYSTERY (II,1747); THE NATALIE COLE SPECIAL (II,1794); OPRYLAND: NIGHT OF STARS AND FUTURE STARS (II,1921); THRILLER (I,4492); WHATEVER BECAME OF. . .? (II,2772); WHATEVER BECAME OF. . .? (II,2773)

ARTHUR, ROBERT ALAN
DUPONT SHOW OF THE MONTH (I,1387); MR. PEEPERS (I,3147); PLAYHOUSE 90 (I,3623); STUDIO ONE (I,4268)

ARTHUR, SID
HAPPY DAYS (II,1084)

ARTHUR, VICTOR
ANNIE OAKLEY (I,225)

ASAAL, DANIEL
ST. ELSEWHERE (II,2432)

ASH, ROD
FAERIE TALE THEATER (II,807); NOT NECESSARILY THE NEWS (II,1862); NOT NECESSARILY THE NEWS (II,1863)

ASHER, INEZ
THE CLAUDETTE COLBERT SHOW (I,971)

ASHER, TONY
SKYHAWKS (I,4079)

ASHTON, BRAD
THE UGLIEST GIRL IN TOWN (I,4663)

ASHTON, DAVID
FROM A BIRD'S EYE VIEW (I,1686)

ASINOF, ELIOT
CHANNING (I,900); DUPONT SHOW OF THE MONTH (I,1387)

ASSAEL, DAVID
CHICAGO STORY (II,506)

ATKINS, GEORGE
THE BULLWINKLE SHOW (I,761); THE DON HO SHOW (II,706); THE GET ALONG GANG (II,980); JOE AND SONS (II,1296); MADAME'S PLACE (II,1587); THE NUT HOUSE (I,3320); THREE'S COMPANY (II,2614)

ATKINSON, BUDDY
FAR OUT SPACE NUTS (II,831); THE JIM NABORS HOUR (I,2381); PARTRIDGE FAMILY: 2200 A.D. (II,1963)

ATKINSON, ROWAN
NOT THE NINE O'CLOCK NEWS (II,1864)

AUBRY, DANIEL
THE RAT PATROL (I,3726)

AUCHINCLOSS, GORDON
THE MATT DENNIS SHOW (I,2948)

AUERBACH, ARNOLD
TEXACO STAR THEATER (I,4411)

AUGUST, HELEN
BEWITCHED (I,418); THE BIONIC WOMAN (II,255); THE DONNA REED SHOW (I,1347); SAINTS AND SINNERS (I,3887)

AUGUST, KEN
THE IMPOSTER (II,1225)

AUGUST, TOM
ALCOA/GOODYEAR THEATER (I,107); BEWITCHED (I,418); THE BIONIC WOMAN (II,255); THE DONNA REED SHOW (I,1347); SAINTS AND SINNERS (I,3887)

AUSTIN, RONALD
BEACH PATROL (II,192); CANNON: THE RETURN OF FRANK CANNON (II,425); CHARLIE'S ANGELS (II,486); CHOPPER ONE (II,515); Q. E. D. (II,2094)

AVEDON, BARBARA
THE BARBARA RUSH SHOW (I,349); BEWITCHED (I,418); CAGNEY AND LACEY (II,409); CAGNEY AND LACEY (II,410); FISH (II,864); GIDGET (I,1795)

AXE, RONALD
THE HERO (I,2035)

AXELROD, DAVID
THE ALAN KING SHOW (I,102); THE BIG SHOW (II,243); DEAN MARTIN'S CELEBRITY ROAST (II,643); HOT HERO SANDWICH (II,1186); I'VE HAD IT UP TO HERE (II,1221); THE MIKE DOUGLAS CHRISTMAS SPECIAL (I,3040); THE ROBERT KLEIN SHOW (II,2186); THE SHOW MUST GO ON (II,2344); THAT WAS THE YEAR THAT WAS (I,4424); THIS MORNING (I,4464)

AXELROD, GEORGE
54TH STREET REVUE (I,1573); ALL IN FUN (I,130)

AXELROD, JOHN
RAQUEL (I,3725)

AYKROYD, DAN
THE BEACH BOYS SPECIAL (II,190); EVERYTHING GOES (I,1480)

AYLESWORTH, JOHN
THE BRASS ARE COMING (I,708); THE FUNNY WORLD OF FRED & BUNNI (II,943); THE GISELE MACKENZIE SHOW (I,1813); THE GRADY NUTT SHOW (II,1048); THE HARLEM GLOBETROTTERS POPCORN MACHINE (II,1092); HECK'S ANGELS (II,1122); THE HEE HAW HONEYS (II,1124); HERB ALBERT AND THE TIJUANA BRASS (I,2019); HULLABALOO (I,2156); THE JACKIE GLEASON SPECIAL (I,2325); THE JONATHAN WINTERS SHOW (I,2443); THE JUD STRUNK SHOW (I,2462); THE JULIE ANDREWS HOUR (I,2480); KEEP ON TRUCKIN' (II,1372); THE KOPYCATS (I,2580); THE KRAFT MUSIC HALL (I,2587); THE KRAFT MUSIC HALL (I,2588); MONTE CARLO, C'EST LA ROSE (I,3095); THE NASHVILLE PALACE (II,1792); THE PERRY COMO SUNSHINE SHOW (II,2005); SHIELDS AND YARNELL (II,2328); THE SONNY AND CHER SHOW (II,2401); TIN PAN ALLEY TODAY (I,4521); ZERO HOUR (I,4980)

AYRES, GENE
THE FLINTSTONE FUNNIES (II,883)

AYRES, GERALD
FAERIE TALE THEATER (II,807)

BABBIN, JACQUELINE
DUPONT SHOW OF THE MONTH (I,1387); HARVEY (I,1961); HARVEY (I,1962); THE HEIRESS (I,2006); MEMBER OF THE WEDDING (I,3002); OUR TOWN (I,3420); THE PHILADELPHIA STORY (I,3582); THE PICTURE OF DORIAN GRAY (I,3591)

BABCOCK, DWIGHT
THE ADVENTURES OF SUPERMAN (I,77); SKY KING (I,4077)

BACAL, JOHN
THE GREAT SPACE COASTER (II,1059)

BACHAR, KAREN
ALL'S FAIR (II,46)

BACKMAN, LEE
THE ADVENTURES OF SUPERMAN (I,77)

BACON, JAMES
THE GREAT MYSTERIES OF HOLLYWOOD (II,1058)

BACOS, CATHERINE
HART TO HART (II,1102); THE LOVE BOAT (II,1535)

BAEHR, NICHOLAS E
DAN AUGUST (I,1130); DELVECCHIO (II,659); MCCLOUD (II,1660)

BAER, ART
THE ANDY GRIFFITH SHOW (I,192); THE ANN SOTHERN SHOW (I,220); ARNIE (I,261); THE GARRY MOORE SHOW (I,1740); GLITTER (II,1009); GOOD TIMES (II,1038); THE JIM NABORS HOUR (I,2381); THE ODD COUPLE (II,1875); THE VICTOR BORGE SHOW (I,4719); THE VICTOR BORGE SHOW (I,4720)

BAER, JILL
FINDER OF LOST LOVES (II,857); THE LOVE BOAT (II,1535); TOO CLOSE FOR COMFORT (II,2642)

BAER, NORMAN
THE GREAT MYSTERIES OF HOLLYWOOD (II,1058)

BAER, RICHARD
ADAM'S RIB (I,32); BARNEY MILLER (II,154); THE FOUR SEASONS (II,915); KAREN (II,1363); LEAVE IT TO BEAVER (I,2648); THE MUNSTERS (I,3158); NUMBER 96 (II,1869)

BAGEN, TOM
STARSKY AND HUTCH (II,2444); SWITCH (II,2519)

BAGLEY, DESMOND
RUNNING BLIND (II,2232)

BAGNI, GWEN
BACKSTAIRS AT THE WHITE HOUSE (II,133); BURKE'S LAW (I,762); EIGHT IS ENOUGH (II,762); HONEY WEST: WHO KILLED THE JACKPOT? (I,2106); MEET MCGRAW (I,2984); TERRY AND THE PIRATES (I,4405); WONDER WOMAN (SERIES 1) (II,2828)

BAGNI, JOHN
MEET MCGRAW (I,2984); TERRY AND THE PIRATES (I,4405)

BAILEY, ANN HOWARD
APPOINTMENT WITH ADVENTURE (I,246); BEACON HILL (II,193)

BAILEY, JOSEPH
THE GREAT SPACE COASTER (II,1059); HOT HERO SANDWICH (II,1186); THE ROBERT KLEIN SHOW (II,2185)

BAILEY, PEARL
THE PEARL BAILEY SHOW (I,3501)

BAILEY, PHILIP
THE MUPPET SHOW (II,1766)

BAINE, SEAN
THE 25TH MAN (MS) (II,2678); EISCHIED (II,763); KOJAK (II,1406); THE REBELS (II,2121)

BAIRD, BIL
THE BIL BAIRD SHOW (I,439); LIFE WITH SNARKY PARKER (I,2693)

BAKER, ALAN
PERRY COMO'S SPRING IN NEW ORLEANS (II,2021)

BAKER, BOB
BERGERAC (II,209); DOCTOR WHO (II,689)

BAKER, DOROTHY
PLAYHOUSE 90 (I,3623)

BAKER, ELLIOT
LACE (II,1418); MALIBU (II,1607); ROBERT MONTGOMERY PRESENTS YOUR LUCKY STRIKE THEATER (I,3809)

BAKER, HARRIET
RICKLES (II,2170)

BAKER, HERBERT
100 YEARS OF AMERICA'S POPULAR MUSIC (II,1904); ANYTHING GOES (I,238); BING CROSBY'S WHITE CHRISTMAS (II,253); THE DANNY KAYE SHOW (I,1143); THE DEAN MARTIN SHOW (I,1196); THE DEAN MARTIN SHOW (I,1197); THE DEAN MARTIN SHOW (I,1198); THE DEAN MARTIN SHOW (I,1199); THE DEAN MARTIN SHOW (I,1200); THE FLIP WILSON SHOW (I,1602); GLADYS KNIGHT AND THE PIPS (II,1004); THE JOHN DAVIDSON CHRISTMAS SHOW (II,1308); KEEP U.S. BEAUTIFUL (I,2519); MOVIN' (I,3117); THE MUHAMMED ALI VARIETY SPECIAL (II,1762); THE PERRY COMO CHRISTMAS SHOW (II,2002); PERRY COMO'S SPRINGTIME SPECIAL (II,2023); PINOCCHIO (II,2044); THE SOUPY SALES SHOW (I,4138); TED KNIGHT MUSICAL COMEDY VARIETY SPECIAL SPECIAL (II,2549); THE WAYNE NEWTON SPECIAL (II,2750); THE WORLD OF ENTERTAINMENT (II,2837); YOU'RE THE TOP (I,4974)

BAKER, HOWARD
PLAYHOUSE 90 (I,3623)

BAKER, LYNN
WHIZ KIDS (II,2790)

BAKER, MELVILLE C
THE SWAN (I,4312)

BAKER, ROD
240-ROBERT (II,2688); HAWAII FIVE-O (II,1110); WONDER WOMAN (SERIES 2) (II,2829)

BAKER, RONNIE
THE TWO RONNIES (II,2694)

BALDWIN, EARL
PRIVATE SECRETARY (I,3672); RAWHIDE (I,3727)

BALKIN, DAVID
COVER UP (II,597); THE TEXAS RANGERS (II,2567)

BALLINGER, WILLIAM
BONANZA (II,347); MR. BLACK (I,3129); THE NIGHT STALKER (II,1849); THE OUTER LIMITS (I,3426); TIGHTROPE (I,4500)

BALLUCK, DON
CHOPPER ONE (II,515); HAWAII FIVE-O (II,1110); WILD TIMES (II,2799)

BALMAGIA, LARRY
9 TO 5 (II,1852); BARNEY MILLER (II,154); JENNIFER SLEPT HERE (II,1278); M*A*S*H (II,1569); MAKING A LIVING (II,1604); NIGHT COURT (II,1844); RHODA (II,2151); THREE'S COMPANY (II,2614)

BALTER, ALLAN
THE BARBARY COAST (II,146); THE MAN WITH THE POWER (II,1616); THE OUTER LIMITS (I,3426); THE POWERS OF MATTHEW STAR (II,2075); SAN FRANCISCO INTERNATIONAL AIRPORT (I,3906); THE TIME TUNNEL (I,4511); VOYAGE TO THE BOTTOM OF THE SEA (I,4743)

BALZER, GEORGE
THE DON KNOTTS SHOW (I,1334); THE GENERAL FOODS 25TH ANNIVERSARY SHOW (I,1756); THE JACK BENNY CHRISTMAS SPECIAL (I,2289); THE JACK BENNY HOUR (I,2290); THE JACK BENNY HOUR (I,2291); THE JACK BENNY HOUR (I,2292); THE JACK BENNY HOUR (I,2293); THE JACK BENNY PROGRAM (I,2294); THE JACK BENNY SPECIAL (I,2295); JACK BENNY WITH GUEST STARS (I,2298); JACK BENNY'S BIRTHDAY SPECIAL (I,2299)

BANKS, C J
SHE'S WITH ME (II,2321)

BANNICK, LISA
DOMESTIC LIFE (II,703); FAMILY TIES (II,819); OH MADELINE (II,1880)

BANTA, GLORIA
THE ABC AFTERSCHOOL SPECIAL (II,1); COUSINS (II,595); LILY (II,1477); THE MARY TYLER MOORE SHOW (II,1640); RHODA (II,2151)

BAR-DAVID, S
HEC RAMSEY (II,1121)

BARASCH, NORMAN
AMERICAN COWBOY (I,169); BENSON (II,208); BING CROSBY AND HIS FRIENDS (I,456); THE BURNS AND SCHREIBER COMEDY HOUR (I,768); THE BURNS AND SCHREIBER COMEDY HOUR (I,769); THE CONFESSIONS OF DICK VAN DYKE (II,568); DOC (II,682); FISH (II,864); THE FUNNY SIDE (I,1714); THE FUNNY SIDE (I,1715); THE GUY MITCHELL SHOW (I,1912); THE NEW ODD COUPLE (II,1825); ONE OF THE BOYS (II,1911); RHODA (II,2151); SEMI-TOUGH (II,2300); THE WONDERFUL WORLD OF AGGRAVATION (I,4894)

BARBASH, BOB
ALCOA PREMIERE (I,109); ALCOA/GOODYEAR THEATER (I,107); THE DESILU PLAYHOUSE (I,1237); THE DICK POWELL SHOW (I,1269); KINCAID (I,2543); MANHUNTER (II,1621); PLAYHOUSE 90 (I,3623); TURN OF FATE (I,4623); THE WILD WILD WEST (I,4863)

BARBERA, JOSEPH
THE ADVENTURES OF JONNY QUEST (I,64)

BARBOUR, JOHN
REAL PEOPLE (II,2118); SPEAK UP AMERICA (II,2412)

BARE, RICHARD
F TROOP (I,1499); POOR DEVIL (I,3646)

BARER, MARSHALL
ONCE UPON A MATTRESS (I,3371)

BARKIN, HASKELL
THE FLINTSTONE FUNNIES (II,883); THE LOVE BOAT (II,1535)

BARKLEY, DIANNE
THIS MORNING (I,4464)

BARLOW, DAVID
SEEING THINGS (II,2297)

BARLOW, MICHAEL
FAMILY (II,813)

BARMAK, IRA
DREAM GIRL OF '67 (I,1375)

BARNES, BILLY
...AND DEBBIE MAKES SIX (I,185); CHER (II,496); DIANA ROSS AND THE SUPREMES AND THE TEMPTATIONS ON BROADWAY (I,1260)

BARNES, CREIGHTON
THE SMURFS (II,2385)

BARNES, DALLAS
THE BLUE KNIGHT (II,276); EISCHIED (II,763); T.J. HOOKER (II,2524); VEGAS (II,2724); WONDER WOMAN (SERIES 2) (II,2829)

BARNES, JOANNA
T.J. HOOKER (II,2524)

BARNES, JUDY
STAR TREK (II,2440)

BARNES, KATHLEEN
BUCK ROGERS IN THE 25TH CENTURY (II,383); ISIS (II,1248); WONDER WOMAN (SERIES 2) (II,2829)

BARNETT, JACKIE
JIMMY DURANTE MEETS THE LIVELY ARTS (I,2386)

BARON, ALEXANDER
A HORSEMAN RIDING BY (II,1182)

BARON, PATRICIA
SIMON AND SIMON (II,2357)

BARR, BILL
BEAT THE CLOCK (I,378); SALE OF THE CENTURY (II,2239)

BARR, MARLENE
THE ODD COUPLE (II,1875)

BARRET, EARL
BATMAN (I,366); I SPY (I,2179); POOR DEVIL (I,3646); POPI (II,2069); THE SAN PEDRO BEACH BUMS (II,2250); TOO CLOSE FOR COMFORT (II,2642); VIVA VALDEZ (II,2728); THE WILD WILD WEST (I,4863); WINDOWS, DOORS AND KEYHOLES (II,2806)

BARRETT, BILL
THE ALCOA HOUR (I,108); FULL CIRCLE (I,1702); THOSE ENDEARING YOUNG CHARMS (I,4469)

BARRETT, JAMES LEE
THE AWAKENING LAND (II,122); BIG JOHN (II,239); KRAFT TELEVISION THEATER (I,2592); STUBBY PRINGLE'S CHRISTMAS (II,2483); YOU ARE THE JURY (II,2854)

BARRETT, JOHN
PAT PAULSEN FOR PRESIDENT (I,3472); RETURN TO THE PLANET OF THE APES (II,2145); THE SMOTHERS ORGANIC PRIME TIME SPACE RIDE (I,4098)

BARRETT, STEFFI
PETER GUNN (I,3562)

BARRETT, TONY
AMOS BURKE, SECRET AGENT (I,180); BURKE'S LAW (I,762); THE DICK POWELL SHOW (I,1269); THE MONK (I,3087); MR. LUCKY (I,3141); THE NEW ADVENTURES OF CHARLIE CHAN (I,3249); O'CONNOR'S OCEAN (I,3326); PETER GUNN (I,3562); THE ROGUES (I,3832); TIGHTROPE (I,4500)

BARRINGTON, DAVID
PRISONER: CELL BLOCK H (II,2085)

BARRINGTON, LOWELL
CLIMAX! (I,976); THE DESILU PLAYHOUSE (I,1237)

BARRIS, ALEX
THE BARBARA MCNAIR SHOW (I,348); THE DORIS MARY ANNE KAPPELHOFF SPECIAL (I,1356); THE PALACE (II,1943); ROLLIN' ON THE RIVER (I,3834)

BARRON, ARTHUR
THE ABC AFTERSCHOOL SPECIAL (II,1)

BARRON, BOB
THE WILD WILD WEST (I,4863)

BARRON, FRED
BETWEEN THE LINES (II,225); THE BOUNDER (II,357)

BARRON, JEFFREY
ALAN KING'S SECOND ANNUAL FINAL WARNING (II,30); AMERICA, YOU'RE ON (II,69); THE BEATRICE ARTHUR SPECIAL (II,196); BOB HOPE SPECIAL: BOB HOPE IN "JOYS" (II,284); BOB HOPE SPECIAL: BOB HOPE'S ALL-STAR COMEDY LOOK AT THE FALL SEASON: IT'S STILL FREE AND WORTH IT! (II,299); BOB HOPE SPECIAL: BOB HOPE'S ALL-STAR COMEDY SPECTACULAR FROM LAKE TAHOE (II,301); BOB HOPE SPECIAL: BOB HOPE'S ALL-STAR COMEDY TRIBUTE TO VAUDEVILLE (II,302); BOB HOPE SPECIAL: BOB HOPE'S BICENTENNIAL STAR SPANGLED SPECTACULAR (II,306); BOB HOPE SPECIAL: BOB HOPE'S CHRISTMAS PARTY (II,307); BOB HOPE SPECIAL: BOB HOPE'S WORLD OF COMEDY (II,323); BOB HOPE SPECIAL: THE BOB HOPE COMEDY SPECIAL (II,332); CHANGING SCENE (I,899); CHARO (II,488); DEAN MARTIN'S CELEBRITY ROAST (II,643); GENERAL ELECTRIC'S ALL-STAR ANNIVERSARY (II,963); LYNDA CARTER'S CELEBRATION (II,1563); LYNDA CARTER'S SPECIAL (II,1564); LYNDA CARTER: STREET LIGHTS (II,1567); LYNDA CARTER: ENCORE (II,1566); THE MANHATTAN TRANSFER (II,1619); THE RED SKELTON SHOW (I,3756); RICH LITTLE'S WASHINGTON FOLLIES (II,2159); SOLID GOLD (II,2395); THAT'S TV (II,2581); TWILIGHT THEATER (II,2685); TWILIGHT THEATER II (II,2686); THE WORLD'S FUNNIEST COMMERCIAL GOOFS (II,2841)

BARROWS, ROBERT
THE FUGITIVE (I,1701)

BARRY JR, PHILIP
THE PHILADELPHIA STORY (I,3581)

BARRY, MITCHELL
YESTERYEAR (II,2851)

BARRY, BRUCE
AS THE WORLD TURNS (II,110)

BARRY, J J
STOPWATCH: THIRTY MINUTES OF INVESTIGATIVE TICKING (II,2474)

BARRY, JOE
MATINEE THEATER (I,2947)

BARRY, MIKE
CHER (II,495)

BARRY, PAMELA
PADDINGTON BEAR (II,1941)

BARRY, PETER
BRANDED (I,707); THE MAN FROM U.N.C.L.E. (I,2867); THE NEW ADVENTURES OF CHARLIE CHAN (I,3249); STUDIO ONE (I,4268)

BARRY, PHILIP
THE ALCOA HOUR (I,108)

BARRY, R Z
HIGH PERFORMANCE (II,1145)

BARSOCCHINI, PETER
THE MERV GRIFFIN SHOW (II,1688)

BART, WILLIAM
SHANE (I,3995)

BARTLETT, JUANITA
THE GREATEST AMERICAN HERO (II,1060); LITTLE HOUSE ON THE PRAIRIE (II,1487); THE NEW MAVERICK (II,1822); NO MAN'S LAND (II,1855); PLANET EARTH (I,3615); THE QUEST (II,2098); THE ROCKFORD FILES (II,2197); SCARECROW AND MRS. KING (II,2268)

BARTLETT, SY
SUSPICION (I,4309)

BARTON, DAN
THE SQUARE WORLD OF ED BUTLER (I,4171)

BARTON, FRANKLIN
ROBERT MONTGOMERY PRESENTS YOUR LUCKY STRIKE THEATER (I,3809)

BARTON, JOHN
HAMLET (I,1925)

BARTON, KATHRYN

KUNG FU (II,1416)

BARWICK, TONY
CAPTAIN SCARLET AND THE MYSTERONS (I,837); THUNDERBIRDS (I,4497); U.F.O. (I,4662)

BASER, MICHAEL S
9 TO 5 (II,1852); AUTOMAN (II,120); CARTER COUNTRY (II,449); THREE'S COMPANY (II,2614)

BASKIN, JOHN
ALL IN THE FAMILY (II,38); CRAZY LIKE A FOX (II,601); DEAR TEACHER (II,653); FISHERMAN'S WHARF (II,865); FOR MEMBERS ONLY (II,905); GOOD TIMES (II,1038); HIGH FIVE (II,1143)

BASKIN, SUSAN
REMINGTON STEELE (II,2130)

BASS, ARTHUR
FANTASY ISLAND (II,829)

BASSETT, BRUCE
KRAFT TELEVISION THEATER (I,2592)

BAST, WILLIAM
APPLE'S WAY (II,101); BEN CASEY (I,394); COMBAT (I,1011); THE FUGITIVE (I,1701); THE HAMPTONS (II,1076); HAWAII FIVE-O (II,1110); HONEYMOON SUITE (I,2108); MANHUNTER (II,1621); THE NEW ADVENTURES OF PERRY MASON (I,3252); THE OUTER LIMITS (I,3426); PRIVATE SECRETARY (I,3672); THE ROGUES (I,3832); TUCKER'S WITCH (II,2667); THE WALTONS (II,2740)

BASTA, ANTHONY
ROUTE 66 (I,3852)

BASTONE, RON
THE NEW MICKEY MOUSE CLUB (II,1823)

BATCHELOR, RUTH
BOB HOPE SPECIAL: BOB HOPE IN "JOYS" (II,284)

BATES, RUSSELL
STAR TREK (II,2439)

BATTES, JOHN
THE SMURFS (II,2385)

BATTLE, JOHN TUCKER
BONANZA (II,347)

BAUBLITZ, BOB
HOUSE CALLS (II,1194); LEWIS AND CLARK (II,1465); THREE'S COMPANY (II,2614)

BAUM, THOMAS
THE ABC AFTERSCHOOL SPECIAL (II,1)

BAUMAN, PAUL
SPARROW (II,2410)

BAUMER, MARIE
THRILLER (I,4492)

BAXT, GEORGE
THE STRAWBERRY BLONDE (I,4254); THE THREE MUSKETEERS (I,4481)

BAXTER, CAROL
THE INCREDIBLE HULK (II,1232)

BAXTER, CLAYTON
RENDEZVOUS HOTEL (II,2131)

BAXTER, GEORGE
BEACON HILL (II,193)

BEARD, HENRY
THE FANTASTIC MISS PIGGY SHOW (II,827)

BEARDE, CHRIS
ANDY WILLIAMS MAGIC LANTERN SHOW COMPANY (I,202); THE ANDY WILLIAMS SHOW (I,210); ANDY'S LOVE CONCERT (I,215); ARNOLD'S CLOSET REVIEW (I,262); BARNEY AND ME (I,358); BOB HOPE SPECIAL: BOB HOPE'S ALL-STAR COMEDY SPECIAL FROM AUSTRALIA (II,300); THE DINAH SHORE SPECIAL—LIKE HEP (I,1282); ELVIS (I,1435); THE HALF-HOUR COMEDY HOUR (II,1074); THE HUDSON BROTHERS SHOW (II,1202); THE KEN BERRY WOW SHOW (I,2524); LI'L ABNER (I,2702); THE OSMONDS SPECIAL (II,1929); ROMP (I,3838); SANDY (I,3908); THE SMOTHERS BROTHERS COMEDY HOUR (I,4095); THE SONNY AND CHER COMEDY HOUR (II,2400); THE SONNY COMEDY REVUE (II,2403); WHERE THE GIRLS ARE (I,4830)

BEATTS, ANNE
SQUARE PEGS (II,2431)

BEATTY, ROGER
THE BOB NEWHART SHOW (II,342); THE CAROL BURNETT SHOW (II,443); THE PRIMARY ENGLISH CLASS (II,2081); UNCLE TIM WANTS YOU! (II,2704)

BEAUCHAMP, D D
CROWN THEATER WITH GLORIA SWANSON (I,1107)

BEAUMONT, CHARLES
ALCOA/GOODYEAR THEATER (I,107); ALFRED HITCHCOCK PRESENTS (I,115); NAKED CITY (I,3216); ONE STEP BEYOND (I,3388); THRILLER (I,4492); THE TWILIGHT ZONE (I,4651)

BEAUMONT, CHRISTOPHER
FAME (II,812)

BEAUMONT, DAN
BACHELOR FATHER (I,323); THE GHOST AND MRS. MUIR (I,1786); PLEASE DON'T EAT THE DAISIES (I,3628)

BEAUMONT, HUGH
LEAVE IT TO BEAVER (I,2648)

BEAVER, FLOYD
TURN OF FATE (I,4623)

BECK, GEORGE
DOBIE GILLIS (I,1302); MAKE ROOM FOR DADDY (I,2843)

BECKER, BOB
THE NBC FOLLIES (I,3241)

BECKERMAN, GEORGE
BLONDES VS. BRUNETTES (II,271)

BECKETT, ANNE
THE ADVENTURES OF POLLYANNA (II,14); SARA (II,2260)

BEDISTEIN, BOB
CHECKING IN (II,492)

BEECHER, ELIZABETH
ANNIE OAKLEY (I,225)

BEGEL, CINDY
JOANIE LOVES CHACHI (II,1295); LAVERNE AND SHIRLEY (II,1446); MORK AND MINDY (II,1735); OH MADELINE (II,1880)

BEGG, JIM
PARTRIDGE FAMILY: 2200 A.D. (II,1963)

BEGGS, RUSSEL
BEN BLUE'S BROTHERS (I,393)

BEHR, IRA
BRET MAVERICK (II,374); BRING 'EM BACK ALIVE (II,377); JESSICA NOVAK (II,1286); SEVEN BRIDES FOR SEVEN BROTHERS (II,2307)

BEHRSTOCK, ARTHUR
COLONEL MARCH OF SCOTLAND YARD (I,1005)

BEICH, ALBERT
AMOS BURKE, SECRET AGENT (I,180); BURKE'S LAW (I,762); THE DICK POWELL SHOW (I,1269); THE FLYING NUN (I,1611)

BEICH, ALFRED
MANNIX (II,1624)

BELGARD, ARNOLD
BONANZA (II,347); FIRESIDE THEATER (I,1580)

BELICHMAN, NORMAN
YOU ASKED FOR IT (II,2855)

BELKER, GARY
THE CHEVROLET GOLDEN ANNIVERSARY SHOW (I,924)

BELKIN, GARY
ANNIE AND THE HOODS (II,91); THE CAROL BURNETT SHOW (II,443); COMEDY TONIGHT (I,1027); THE DORIS MARY ANNE KAPPELHOFF SPECIAL (I,1356); FUNNY PAPERS (I,1713); THE GARRY MOORE SHOW (I,1740); SESAME STREET (I,3982); SHA NA NA (II,2311); SHIPSHAPE (II,2329); SINATRA: CONCERT FOR THE AMERICAS (II,2361); THE STEVE LAWRENCE SHOW (I,4227); SUGAR TIME (II,2489); THAT WAS THE YEAR THAT WAS (I,4424); THIS WILL BE THE YEAR THAT WILL BE (I,4466); THREE'S COMPANY (II,2614)

BELL, ALISTAIR
BERGERAC (II,209); LITTLE WOMEN (I,2723)

BELL, B GERALD
GET CHRISTIE LOVE! (II,981); MCCLOUD (II,1660)

BELL, BOB
SWINGING COUNTRY (I,4322); U.F.O. (I,4662)

BELL, WALTER
FALCON'S GOLD (II,809)

BELL, WILLIAM J
THE YOUNG AND THE RESTLESS (II,2862)

BELLAK, GEORGE
THE ANDROS TARGETS (II,76); EAST SIDE/WEST SIDE (I,1396); HIDDEN TREASURE (I,2045); N.Y.P.D. (I,3321); NURSE (II,1870); THE STREETS OF SAN FRANCISCO (II,2478); THRILLER (I,4492)

BELLAMY, EARL
HART TO HART (II,1102)

BELLAND, BRUCE
THE FITZPATRICKS (II,868); KNIGHT RIDER (II,1402)

BELLEM, ROBERT
THE ADVENTURES OF SUPERMAN (I,77)

BELLER, CARL
THE HALLMARK HALL OF FAME CHRISTMAS FESTIVAL (I,1921)

BELLISARIO, DONALD P
AIRWOLF (II,27); BATTLESTAR GALACTICA (II,181); KOJAK (II,1406); MAGNUM, P.I. (II,1601); TALES OF THE GOLD MONKEY (II,2537)

BELLWOOD, ROBERT
ANNIE, THE WOMAN IN THE LIFE OF A MAN (I,226)

BELOIN, EDMUND
THE DEAN JONES SHOW (I,1191); MY THREE SONS (I,3205)

BELOUS, PAUL M
THE OUTLAWS (II,1936); PEN 'N' INC. (II,1992)

BELSON, JERRY
BAREFOOT IN THE PARK (I,355); COPS (I,1047); THE DANNY THOMAS SPECIAL (I,1150); DANNY THOMAS: THE ROAD TO LEBANON (I,1153); THE DICK VAN DYKE SPECIAL (I,1277); EVIL ROY SLADE (I,1485); I SPY (I,2179); MIXED NUTS (II,1714); THE MURDOCKS AND THE MCCLAYS (I,3163); THE NEW ODD COUPLE (II,1825); SHERIFF WHO? (I,4009); THE TEXAS WHEELERS (II,2568); THINK PRETTY (I,4448); YOUNG GUY CHRISTIAN (II,2864)

BELUSHI, JOHN
THE BEACH BOYS SPECIAL (II,190)

BENCHLEY, ROBERT
JEREMIAH OF JACOB'S NECK (II,1281)

BENDER, KAY
TRAPPER JOHN, M.D. (II,2654)

BENDER, RUSS
THE HANK MCCUNE SHOW (I,1932)

BENDEROTH, MICHAEL
THE GEORGIA PEACHES (II,979)

BENDIX, MICHAEL
KEY TORTUGA (II,1383)

BENEDEK, BARBARA
CONDO (II,565); I'M A BIG GIRL NOW (II,1218); MAKING A LIVING (II,1604)

BENEDICT, TONY
THE JETSONS (I,2376)

BENEST, GLENN
STRANGER IN OUR HOUSE (II,2477)

BENJAMIN, BURTON
THE HUNTER (I,2162); KRAFT TELEVISION THEATER (I,2592); ROBERT MONTGOMERY PRESENTS YOUR LUCKY STRIKE THEATER (I,3809)

BENNETT, CHARLES
THE DICK POWELL SHOW (I,1269); LAND OF THE GIANTS (I,2611)

BENNETT, MEG
THE YOUNG AND THE RESTLESS (II,2862)

BENNETT, MURRAY
ROBERT MONTGOMERY PRESENTS YOUR LUCKY STRIKE THEATER (I,3809)

BENNETT, ROBERT RUSSELL
COMBAT (I,1011)

BENNETT, RUTH
FAMILY TIES (II,819); LAVERNE AND SHIRLEY (II,1446)

BENOFF, MAX
GEORGE BURNS IN THE BIG TIME (I,1764); LIFE WITH LUIGI (I,2692)

BENSFIELD, DICK
THE ADVENTURES OF OZZIE AND HARRIET (I,71); THE ANDY GRIFFITH SHOW (I,192); ANOTHER MAN'S SHOES (II,96); BIG DADDY (I,425); GOOD TIMES (II,1038); HAPPY DAYS (II,1084); HELLO, LARRY (II,1128); I DREAM OF JEANNIE (I,2167); MAUDE (II,1655); ONE DAY AT A TIME (II,1900); THE PARTRIDGE FAMILY (II,1962); POPI (II,2069); THREE FOR THE ROAD (II,2607)

BENSON, JESSICA
BRANDED (I,707)

BENSON, MARTIN
ONE STEP BEYOND (I,3388)

BENSON, SALLY
HANS BRINKER OR THE SILVER SKATES (I,1933); ROBERT MONTGOMERY PRESENTS YOUR LUCKY STRIKE THEATER (I,3809)

BENSON, SARI
GLITTER (II,1009)

BENTINE, MICHAEL
COMEDY IS KING (I,1019)

BENTON, DANIEL KING
DYNASTY (II,746)

BENTON, DOUGLAS
HEC RAMSEY (II,1121)

BERCOVICI, ERIC
ASSIGNMENT: MUNICH (I,302); THE CHICAGO STORY (II,507); I SPY (I,2179); JESSIE (II,1287); MCCLAIN'S LAW (II,1659); SHOGUN (II,2339); TOP OF THE HILL (II,2645); WASHINGTON: BEHIND CLOSED DOORS (II,2744)

BERDIS, BERT
THE TIM CONWAY SHOW (II,2619)

BERENBERG, BENEDICT
IVANHOE (I,2282)

BERG, CHERNEY
THE GOLDBERGS (I,1835); THE GOLDBERGS (I,1836); MRS. G. GOES TO COLLEGE (I,3155)

BERG, DAVID
TV FUNNIES (II,2672)

BERG, DICK
PLAYHOUSE 90 (I,3623); ROBERT MONTGOMERY PRESENTS YOUR LUCKY STRIKE THEATER (I,3809); STUDIO ONE (I,4268); THE WORD (II,2833)

BERG, GERTRUDE
THE GOLDBERGS (I,1835); THE GOLDBERGS (I,1836); MRS. G. GOES TO COLLEGE (I,3155)

BERG, HAMILTON
BRING 'EM BACK ALIVE (II,377)

BERG, JAMES
JUST OUR LUCK (II,1360)

BERG, KEN
FAME (II,812); HART TO HART (II,1102)

BERG, LEE
LAW OF THE PLAINSMAN (I,2639)

BERGAN, TED
THE SOUND AND THE SCENE (I,4132)

BERGER, HARVEY
BOB HOPE SPECIAL: BOB HOPE'S STAND UP AND CHEER FOR THE NATIONAL FOOTBALL LEAGUE'S 60TH YEAR (II,316); GEORGE BURNS' HOW TO LIVE TO BE 100 (II,972)

BERGMAN, ALAN
THAT MCMAHON'S HERE AGAIN (I,4419)

BERGMAN, PETER
THE STARLAND VOCAL BAND (II,2441)

BERGMAN, TED
GIMME A BREAK (II,995); GOOD OL' BOYS (II,1035); MADHOUSE 90 (I,2807); MAX (II,1658); THE MUNSTERS (I,3158); SANFORD (II,2255); SANFORD AND SON (II,2256); STAND UP AND CHEER (I,4188)

BERK, HOWARD
ADVICE TO THE LOVELORN (II,16); THE BARBARY COAST (II,146); COLUMBO (II,556); THE CONTENDER (II,571); THE DEVLIN CONNECTION (II,664); THE FALL GUY (II,811); MCMILLAN AND WIFE (II,1667); MRS. COLUMBO (II,1760); THE ROCKFORD FILES (II,2197)

BERK, MICHAEL
MANIMAL (II,1622)

BERKE, LESTER
QUINCY, M. E. (II,2102)

BERKELEY, MARTIN
WAGON TRAIN (I,4747)

BERLIN, PETER
THE GUINNESS GAME (II,1066)

BERMAN, HERB
JIGSAW (I,2377); S.W.A.T. (II,2236); WONDER WOMAN (SERIES 1) (II,2828)

BERNARD, IAN
LOOK WHAT THEY'VE DONE TO MY SONG (II,1517); SAM (II,2247)

BERNSTEIN, ARMYAN
FAMILY (II,813)

BERNSTEIN, ELLIOT
MAUDE (II,1655)

BERNSTEIN, WALTER
SPARROW (II,2410); SPARROW (II,2411)

BERTRAM, JOHN
BIZARRE (II,259)

BEST, HUGH
ACTION IN THE AFTERNOON (I,25)

BEZZERIDES, A I
SCREEN DIRECTOR'S PLAYHOUSE (I,3946)

BICKLEY, WILLIAM S
GOODTIME GIRLS (II,1042); HAPPY DAYS (II,1084); JOANIE LOVES CHACHI

(II,1295); THE KROFFT KOMEDY HOUR (II,1411); LAVERNE AND SHIRLEY (II,1446); OUT OF THE BLUE (II,1934)

BIELAK, ROBERT
KNOTS LANDING (II,1404); MATT HOUSTON (II,1654)

BIEN, WALTER N
THAT'S MY MAMA (II,2580)

BIENER, TOM
AT EASE (II,115); FLO (II,891)

BIERCE, AMBROSE
THE TWILIGHT ZONE (I,4651)

BIGELOW, JOE
THE HOLLYWOOD PALACE (I,2088); JOHNNY CARSON PRESENTS THE SUN CITY SCANDALS (I,2420); THE JOHNNY CARSON SHOW (I,2422); MERMAN ON BROADWAY (I,3011); MUSIC ON ICE (I,3175)

BIHELLER, ROBERT S
CHIPS (II,511)

BINDER, MIKE
HOT (II,1184)

BINDER, STEVE
THE BARRY MANILOW SPECIAL (II,155); DEBBY BOONE. . . ONE STEP CLOSER (II,655); DIANA (II,667); OLIVIA (II,1883); A SPECIAL EDDIE RABBITT (II,2415)

BINGHAM, CHARLOTTE
NO—HONESTLY (II,1857); ROBIN'S NEST (II,2187)

BINNS, BRONWYN
AGAINST THE WIND (II,24)

BIRCH, TONY
THE TWO RONNIES (II,2694)

BIRD, CORDWAINER
THE STARLOST (I,4203)

BIRD, MICHAEL J
THE ONEDIN LINE (II,1912); QUILLER: PRICE OF VIOLENCE (II,2101)

BIRKIN, ANDREW
PETER PAN (II,2030)

BIRNBAUM, STUART
THE CHEVY CHASE SHOW (II,505); THE JIM STAFFORD SHOW (II,1291); LILY (II,1477); THE REDD FOXX COMEDY HOUR (II,2126)

BIRNBEYER, EDMUND
HIGH ADVENTURE WITH LOWELL THOMAS (I,2048)

BISCHOF, LARRY
FATHER MURPHY (II,841); THE WALTONS (II,2740)

BISHOP, JIM
I'VE HAD IT UP TO HERE (II,1221)

BISHOP, JOHN
THE COMEDY ZONE (II,559)

BISHOP, MEL
RICH LITTLE'S ROBIN HOOD (II,2158)

BISHOP, RON
COMBAT (I,1011); FATHER MURPHY (II,841); HOW THE WEST WAS WON (II,1196); KUNG FU (II,1416); THE ROGUES (I,3832)

BIXBY, JEROME
STAR TREK (II,2440); THE TWILIGHT ZONE (I,4651)

BLACH, ALBERT
THE ALAN DALE SHOW (I,96)

BLACK, IAN STUART
THE CHAMPIONS (I,896); DANGER MAN (I,1136); THE INVISIBLE MAN (I,2232); SECRET AGENT (I,3962)

BLACK, JOHN D F
CHARLIE'S ANGELS (II,486); THE FBI (I,1551); THE FUGITIVE (I,1701); THE FUZZ BROTHERS (I,1722); HERE'S BOOMER (II,1134); A SHADOW IN THE STREETS (II,2312); STAR TREK (II,2440); THE UNTOUCHABLES (I,4682); WONDER WOMAN (PILOT 1) (II,2826)

BLACK, KAREN
TRAUMA CENTER (II,2655)

BLACK, STEPHEN
ALOHA PARADISE (II,58); ANOTHER DAY (II,94); EMERALD POINT, N.A.S. (II,773); FALCON CREST (II,808); FLAMINGO ROAD (II,872); HUSBANDS, WIVES AND LOVERS (II,1209); LOVE, SIDNEY (II,1547); PAPER DOLLS (II,1952)

BLACK, WALTER
THE FARMER'S DAUGHTER (I,1533); HAWAII FIVE-O (II,1110); RAWHIDE (I,3727); THE ROGUES (I,3832); S.W.A.T. (II,2236)

BLACKBURN, BRYAN
BARBARA MANDRELL AND THE MANDRELL SISTERS (II,143); BURT BACHARACH: CLOSE TO YOU (I,772); THE ENGELBERT HUMPERDINCK SHOW (I,1444); THE ENGELBERT HUMPERDINCK SHOW (I,1445); THE ENGELBERT HUMPERDINCK SPECIAL (I,1446); THE ENGELBERT HUMPERDINCK SPECIAL (I,1447); THE KOPYCATS (I,2580); THE MAGICAL MUSIC OF BURT BACHARACH (I,2822); MARIE (II,1629); PAUL LYNDE AT THE MOVIES (II,1976); THE VAL DOONICAN SHOW (I,4692)

BLACKBURN, CLARICE
ALL MY CHILDREN (II,39)

BLACKBURN, TOM
DAVY CROCKETT (I,1181); FOREST RANGER (I,1641)

BLACKTON, JENNIE
MOTHER AND ME, M.D. (II,1740); ON OUR OWN (II,1892)

BLAIL, KENDALL J
COCAINE AND BLUE EYES (II,548)

BLAINE, SEAN
BERT D'ANGELO/SUPERSTAR (II,211)

BLAKEMORE, MICHAEL
LONG DAY'S JOURNEY INTO NIGHT (I,2746)

BLAND, ROBIN
DOCTOR WHO (II,689)

BLANE, RALPH
TEXACO COMMAND PERFORMANCE (I,4408)

BLANKS, WILLIAM W
THE BLUE KNIGHT (II,276)

BLATT, JERRY
BETTE MIDLER—ART OR BUST (II,221); BETTE MIDLER—OL' RED HAIR IS BACK (II,222)

BLATTY, BILL
ON THE GO (I,3361)

BLAZER, GEORGE
THE BEAUTIFUL PHYLLIS DILLER SHOW (I,381)

BLECHMAN, COREY
THE CHISHOLMS (II,513)

BLEES, ROBERT
ALFRED HITCHCOCK PRESENTS (I,115); CLIMAX! (I,976); COLUMBO (II,556); THE DESILU PLAYHOUSE (I,1237); FLAMINGO ROAD (II,872); GIDGET'S SUMMER REUNION (I,1799); PETER GUNN (I,3562)

BLEICHMAN, NORMAN
GAMES PEOPLE PLAY (II,956); LIE DETECTOR (II,1467); PEOPLE ARE FUNNY (II,1996); THE SUNDAY GAMES (II,2494)

BLINN, WILLIAM
THE AMERICAN DREAM (II,71); EIGHT IS ENOUGH (II,762); FAME (II,812); HUNTER (II,1204); THE INVADERS (I,2229); A MAN CALLED INTREPID (II,1612); MY FAVORITE MARTIAN (I,3189); THE ROOKIES (I,3840); THE ROOKIES (II,2208); ROOTS (II,2211); SHANE (I,3995); STARSKY AND HUTCH (II,2445); STICKIN' TOGETHER (II,2465)

BLISH, JAMES
CAPTAIN VIDEO AND HIS VIDEO RANGERS (I,838)

BLITZER, BARRY
THE ANN SOTHERN SHOW (I,220); ARNIE (I,261); HOT L BALTIMORE (II,1187); THE JIMMIE RODGERS SHOW (I,2383); THE LOVE BOAT (II,1535); MCHALE'S NAVY (I,2969); MCKEEVER AND THE COLONEL (I,2970); PARTRIDGE FAMILY: 2200 A.D. (II,1963); THE PHIL SILVERS SHOW (I,3580); THE RICH LITTLE SHOW (II,2153); THE RICH LITTLE SHOW (II,2154); TOO MANY SERGEANTS (I,4573); TRAUMA CENTER (II,2655); WHEN THINGS WERE ROTTEN (II,2782)

BLOCH, JOHN
THE AMAZING SPIDER-MAN (II,63); THE INVADERS (I,2229); THE MAN FROM U.N.C.L.E. (I,2867)

BLOCH, ROBERT
ALFRED HITCHCOCK PRESENTS (I,115); DARKROOM (II,619); THE RETURN OF CAPTAIN NEMO (II,2135); STAR TREK (II,2440); THRILLER (I,4492)

BLOCK, HAL
THE KEN MURRAY SHOW (I,2525); THE VAUGHN MONROE SHOW (I,4708)

BLOODWORTH, JAMES
FOUR STAR PLAYHOUSE (I,1652)

BLOODWORTH, LINDA
DRIBBLE (II,736); FILTHY RICH (II,856); LONDON AND DAVIS IN NEW YORK (II,1512); M*A*S*H (II,1569); OVER AND OUT (II,1938); RHODA (II,2151)

BLOOM, CAROLE
SCENE OF THE CRIME (II,2271)

BLOOM, CHARLES
BURT AND THE GIRLS (I,770)

BLOOM, GEORGE
9 TO 5 (II,1852); ALAN KING'S FINAL WARNING (II,29); THE BIG SHOW (II,243); THE BILLY CRYSTAL COMEDY HOUR (II,248); BING CROSBY AND HIS FRIENDS (II,251); THE BOBBY VAN AND ELAINE JOYCE SHOW (I,677); CARTER COUNTRY (II,449); CHER (II,495); THE DEAN MARTIN SHOW (I,1201); THE INCREDIBLE HULK (II,1232); THE JULIE ANDREWS HOUR (I,2480); KEEP ON TRUCKIN' (II,1372); LOVE, SIDNEY (II,1547); MERRY CHRISTMAS. . . WITH LOVE, JULIE (II,1687); THE POWDER ROOM (I,3656); SANDY IN DISNEYLAND (II,2254); WELCOME TO THE FUN ZONE (II,2763); WHAT'S UP? (I,4823)

BLOOM, GEORGE ARTHUR
PHYLLIS (II,2038); THE TRANSFORMERS (II,2653)

BLOOM, HAROLD JACK
BONANZA (II,347); THE CENTURY TURNS (I,890); THE D.A.: MURDER ONE (I,1122); HEC RAMSEY (II,1121); THE LOG OF THE BLACK PEARL (II,1506); PLAYHOUSE 90 (I,3623); SHERLOCK HOLMES (I,4010); THE TIME TUNNEL (I,4511)

BLOOM, JEFFREY
DARKROOM (II,619); SCENE OF THE CRIME (II,2271); STARSKY AND HUTCH (II,2444); STRIKE FORCE (II,2480)

BLOOM, WILLIAM
BATTLE OF THE PLANETS (II,179)

BLOOMBERG, BEVERLY
ARCHIE (II,104); CHIPS (II,511); LADIES' MAN (II,1423); WELCOME BACK, KOTTER (II,2761)

BLOOMBERG, RON
HUSBANDS, WIVES AND LOVERS (II,1209); ONE DAY AT A TIME (II,1900)

BLOOMBERG, STUART
ABC'S SILVER ANNIVERSARY CELEBRATION (II,2)

BLOOMFIELD, ROBERT
PERRY MASON (II,2025); PURSUIT (I,3700); RAWHIDE (I,3727); TURN OF FATE (I,4623); VOYAGE TO THE BOTTOM OF THE SEA (I,4743)

BLOOMSTEIN, HENRY
HUMAN FEELINGS (II,1203)

BLOOMSTEIN, PAUL
RIPLEY'S BELIEVE IT OR NOT (II,2177)

BLUEL, RICHARD
BARETTA (II,152); GOLIATH AWAITS (II,1025); THE MISADVENTURES OF SHERIFF LOBO (II,1704); QUINCY, M. E. (II,2102); TODAY'S FBI (II,2628)

BLUESTEIN, STEVE
THE BRADY BUNCH HOUR (II,363)

BLUM, BOB
AMERICA ALIVE! (II,67)

BLUM, EDWIN
THE MAN FROM U.N.C.L.E. (I,2867)

BLYE, ALLAN
ANDY WILLIAMS KALEIDOSCOPE COMPANY (I,201); THE ANDY WILLIAMS SHOW (I,210); BARNEY AND ME (I,358); BIZARRE (II,258); ELVIS (I,1435); THE HUDSON BROTHERS SHOW (II,1202); JOEY AND DAD (II,1306); THE KEN BERRY WOW SHOW (I,2524); LI'L ABNER (I,2702); LOLA (II,1508); LOLA (II,1509); LOLA (II,1510); LOLA (II,1511); THE OSMONDS SPECIAL (II,1929); PAT PAULSEN FOR PRESIDENT (I,3472); PETULA (I,3572); THE REDD FOXX COMEDY HOUR (II,2126); THE SMOTHERS BROTHERS COMEDY HOUR (I,4095); THE SONNY AND CHER COMEDY HOUR (II,2400); THE SONNY COMEDY REVUE (II,2403); THE SUMMER SMOTHERS BROTHERS SHOW (I,4281); VAN DYKE AND COMPANY (II,2718); VAN DYKE AND COMPANY (II,2719)

BLYTHE, BRUCE W
PEGGY FLEMING AT MADISON SQUARE GARDEN (I,3506)

BOARDMAN, ERIC
THE TIM CONWAY SHOW (II,2619); THE YESTERDAY SHOW (II,2850)

BOARDMAN, TRUE
PERRY MASON (II,2025)

BOARDMAN, WILLIAM
THE NEWS IS THE NEWS (II,1838)

BOBRICK, SAM
EDDIE AND HERBERT (II,757); THE HERO (I,2035); HOW TO SURVIVE THE 70S AND MAYBE EVEN BUMP INTO HAPPINESS (II,1199); THE LATE FALL, EARLY SUMMER BERT CONVY SHOW (II,1441); QUICK AND QUIET (II,2099); SINGLES (I,4062); THIS WEEK IN NEMTIN (I,4465); THE TIM CONWAY COMEDY HOUR (I,4501)

BOCHCO, STEVEN
BAY CITY BLUES (II,186); COLUMBO (II,556); DELVECCHIO (II,659); GRIFF (I,1888); HILL STREET BLUES (II,1154); THE INVISIBLE MAN (II,1243); MCMILLAN (II,1666); MCMILLAN AND WIFE (II,1667); PARIS (II,1957); RICHIE BROCKELMAN: MISSING 24 HOURS (II,2168)

BOCK, JERRY
THE JOKE AND THE VALLEY (I,2438)

BOCKMAN, DAN
GABE AND WALKER (II,950)

BODE, WILLIAM T
STUDIO ONE (I,4268)

BOEHM, ANDRE
HAWKEYE AND THE LAST OF THE MOHICANS (I,1975)

BOEHM, SYDNEY
NAKED CITY (I,3216)

BOEHME, JAN
THIS IS YOUR LIFE (I,4461)

BOGARDUS, ROBERT
ROAR OF THE RAILS (I,3803)

BOGERT, VINCENT
THE GARRY MOORE SHOW (I,1739); THE GARRY MOORE SHOW (I,1740); LEAVE IT TO LARRY (I,2649)

BOHEM, ENDRE
THE CISCO KID (I,961); MEDIC (I,2976); RAWHIDE (I,3727)

BOLOGNA, JOSEPH
BEDROOMS (II,198); GOOD PENNY (II,1036); LOVERS AND OTHER STRANGERS (II,1550); A LUCILLE BALL SPECIAL STARRING LUCILLE BALL AND JACKIE GLEASON (II,1558); MARLO THOMAS IN ACTS OF LOVE—AND OTHER COMEDIES (I,2919); PARADISE (II,1955)

BOMBECK, ERMA
MAGGIE (II,1589)

BONADUCE, ANTHONY
THE FLINTSTONE FUNNIES (II,883)

BONADUCE, CELIA
THE FLINTSTONE FUNNIES (II,883)

BONADUCE, JOHN
FISH (II,864); MAUDE (II,1655)

BONADUCE, JOSEPH
THE ANDY GRIFFITH SHOW (I,192); APPLE'S WAY (II,101); CALIFORNIA FEVER (II,411); JOE'S WORLD (II,1305); LITTLE HOUSE ON THE PRAIRIE (II,1487); MARIE (II,1628); THE WALTONS (II,2740)

BOND, DENNIS M
THE PAUL WILLIAMS SHOW (II,1984)

BOND, JULIAN
THE FAR PAVILIONS (II,832); PENMARRIC (II,1993)

BOND, NELSON
MR. MERGENTHWIRKER'S LOBBLIES (I,3144)

BONDELLI, PHIL
GET CHRISTIE LOVE! (II,981)

BONI, JOHN
ALAN KING LOOKS BACK IN ANGER—A REVIEW OF 1972 (I,99); THE ALAN KING SPECIAL (I,103); THE CAPTAIN AND TENNILLE (II,429); CHER (II,495); CHER (II,496); COMEDY OF HORRORS (II,557); THE FLIP WILSON SPECIAL (II,887); FLIP WILSON. . . OF COURSE (II,890); THE MANY FACES OF COMEDY (I,2894); ROCK-N-AMERICA (II,2195); SPACE FORCE (II,2407); THE STOCKARD CHANNING SHOW (II,2468); THREE'S COMPANY (II,2614); WHEN THINGS WERE ROTTEN (II,2782); THE WONDERFUL WORLD OF AGGRAVATION (I,4894)

BONI, TOM
AMERICA 2-NIGHT (II,65)

BONICELLI, VITTORIO
MOSES THE LAWGIVER (II,1737)

BONNACCORSI, JOHN
THE SMURFS (II,2385)

BONNET, JAMES
KNOTS LANDING (II,1404)

BOOK, JEFF
KIDD VIDEO (II,1388)

BOOKER, BOB
THE BEST LITTLE SPECIAL IN TEXAS (II,215); CHARO (II,488); COTTON CLUB '75 (II,580); FOUL UPS, BLEEPS AND BLUNDERS (II,911); THE GARRY MOORE SHOW (I,1740); THE LOVE BOAT (II,1535); THE NBC FOLLIES (I,3242); OPRYLAND U.S.A. (I,3405); THE WAYNE NEWTON SPECIAL (II,2749); THE WORLD'S FUNNIEST COMMERCIAL GOOFS (II,2841)

BOOKER, CHRISTOPHER
THAT WAS THE WEEK THAT WAS (I,4423)

BOOTH, CONNIE
FAWLTY TOWERS (II,843)

BOOTH, DOUG
THE FLINTSTONE FUNNIES (II,883); THE SMURFS (II,2385)

BOOTHE, JON
GEORGE WASHINGTON (II,978)

BOOSLER, ELAYNE
PRIME TIMES (II,2083); RODNEY DANGERFIELD SPECIAL: IT'S NOT EASY BEIN' ME (II,2199); THE SHAPE OF THINGS (II,2315)

BORCHERT, RUDOLPH
THE BIONIC WOMAN (II,255); CHIPS (II,511); THE DEVLIN CONNECTION (II,664); GET CHRISTIE LOVE! (II,981); THE GREATEST AMERICAN HERO (II,1060); MOBILE ONE (II,1717); THE NIGHT STALKER (II,1849); THE ROCKFORD FILES (II,2197); SWITCH (II,2519)

BORETZ, ALVIN
THE ALCOA HOUR (I,108); ALCOA PREMIERE (I,109); THE BIG STORY (I,432); THE BLUE MEN (I,492); COREY: FOR THE PEOPLE (II,575); CRISIS IN SUN VALLEY (II,604); D.H.O. (I,1249); KOJAK (II,1406); KRAFT TELEVISION THEATER (I,2592); LIGHT'S OUT (I,2700); MASTER OF THE GAME (II,1647); MATINEE THEATER (I,2947); SPIDER-MAN (II,2424)

BORGE, VICTOR
VICTOR BORGE'S 20TH ANNIVERSARY SHOW (I,4727); VICTOR BORGE'S COMEDY IN MUSIC I (I,4723); VICTOR BORGE'S COMEDY IN MUSIC II (I,4724); VICTOR BORGE'S COMEDY IN MUSIC III (I,4725); VICTOR BORGE'S COMEDY IN MUSIC IV (I,4726)

BORISOFF, NORMAN
RITUALS (II,2182)

BORMET, GARRY
THE MAGIC OF DAVID COPPERFIELD (II,1593); MAGIC WITH THE STARS (II,1599)

BORNSTEIN, JUDY
POWERHOUSE (II,2074)

BOROWITZ, ANDY
ARCHIE BUNKER'S PLACE (II,105); THE BOYS IN BLUE (II,359); DREAMS (II,735); THE FACTS OF LIFE (II,805); SQUARE PEGS (II,2431)

BORSTEIN, PAUL
COUNTERATTACK: CRIME IN AMERICA (II,581); HART TO HART (II,1102)

BORSTEIN, SHARON
HART TO HART (II,1102)

BOSSON, BARBARA
FAMILY (II,813)

BOSWELL, BOBBY
HAPPY DAYS (II,1084)

BOTTOMLEY, ROY
FROM A BIRD'S EYE VIEW (I,1686)

BOUCHER, CHRIS
BLAKE'S SEVEN (II,264); DOCTOR WHO (II,689)

BOULES, PAUL
THE GLASS MENAGERIE (I,1819)

BOURJAILY, VANCE
KRAFT TELEVISION THEATER (I,2592)

BOURNE, LARZ
PARTRIDGE FAMILY: 2200 A.D. (II,1963); SPEED BUGGY (II,2419); WHEELIE AND THE CHOPPER BUNCH (II,2776)

BOURNE, PETER
THE MAN FROM U.N.C.L.E. (I,2867)

BOWEN, JOHN
BRIEF ENCOUNTER (I,724)

BOWERS, WILLIAM
THE GUN AND THE PULPIT (II,1067); KATE BLISS AND THE TICKER TAPE KID (II,1366); MORE WILD WILD WEST (II,1733); SIDEKICKS (II,2348); THE WILD WILD WEST REVISITED (II,2800)

BOWMAN, CHUCK
THE INCREDIBLE HULK (II,1232)

BOWMAN, JOHN
THE COMEDY ZONE (II,559)

BOX, BILL
DEAN MARTIN AT THE WILD ANIMAL PARK (II,636); THE DEAN MARTIN CELEBRITY ROAST (II,637); THE DEAN MARTIN CELEBRITY ROAST (II,638); THE DEAN MARTIN CELEBRITY ROAST (II,639); THE DEAN MARTIN CHRISTMAS SPECIAL (II,640); DEAN MARTIN'S CHRISTMAS AT SEA WORLD (II,644); DEAN MARTIN'S COMEDY CLASSICS (II,646); DOM DELUISE AND FRIENDS (II,701); DOM DELUISE AND FRIENDS, PART 2 (II,702); LADIES AND GENTLEMAN. . . BOB NEWHART (II,1420); LADIES AND GENTLEMAN. . . BOB NEWHART, PART II (II,1421); SANFORD (II,2255); SHOW BUSINESS SALUTE TO MILTON BERLE (I,4029); THE VICTOR BORGE SPECIAL (I,4722)

BOYES, MALCOLM
LIFESTYLES OF THE RICH AND FAMOUS (II,1474)

BOYLE, DONALD R
BIG FOOT AND WILD BOY (II,237); CODE RED (II,552); MANIMAL (II,1622); SIMON AND SIMON (II,2357); T.J. HOOKER (II,2524); WALKING TALL (II,2738)

BRACKETT, LEE
ALFRED HITCHCOCK PRESENTS (I,115); THE ROCKFORD FILES (II,2197)

BRADBURY, RAY
ALCOA PREMIERE (I,109); ALFRED HITCHCOCK PRESENTS (I,115); THE TWILIGHT ZONE (I,4651)

BRADDOCK, LORRAINE
ALL MY CHILDREN (II,39)

BRADFORD, HANK
IT ONLY HURTS WHEN YOU LAUGH (II,1251); JOE AND SONS (II,1296); LEWIS AND CLARK (II,1465); NO HOLDS BARRED (II,1854); PRIVATE BENJAMIN (II,2087); SUGAR TIME (II,2489)

BRADFORD, JOHN
BENJI, ZAX AND THE ALIEN PRINCE (II,205); THE DINAH SHORE SHOW (I,1280); THE DIONNE WARWICK SPECIAL (I,1289); THE DUKES (II,741); THE FRANK SINATRA SHOW (I,1666); THE FRANK SINATRA TIMEX SHOW (I,1668); THE FRANK SINATRA TIMEX SHOW (I,1667); THE GLEN CAMPBELL GOODTIME HOUR (I,1820); THE GREAT AMERICAN MUSIC CELEBRATION (II,1052); THE JANE POWELL SHOW (I,2344); THE LIVELY ONES (I,2729); MCKEEVER AND THE COLONEL (I,2970); THE MICKIE FINNS FINALLY PRESENT HOW THE WEST WAS LOST (II,1695); THE PERRY COMO WINTER SHOW (I,3544); PERRY COMO'S WINTER SHOW (I,3548); A SALUTE TO TELEVISION'S 25TH ANNIVERSARY (I,3893); SHIRLEY MACLAINE: IF THEY COULD SEE ME NOW (II,2335); TOO MANY SERGEANTS (I,4573)

BRADLEY, IAN
PRISONER: CELL BLOCK H (II,2085)

BRADSHAW, BOOKER
COLUMBO (II,556); GIMME A BREAK (II,995); MCMILLAN AND WIFE (II,1667); ROLL OUT! (I,3833)

BRADY, BEN
THE KEN MURRAY SHOW (I,2525)

BRADY, FREDERIC
ALCOA/GOODYEAR THEATER (I,107); GETAWAY CAR (I,1783); TIGHTROPE (I,4500); TURN OF FATE (I,4623)

BRADY, LEO
STUDIO ONE (I,4268)

BRADY, MURRAY
VERSATILE VARIETIES (I,4713)

BRADY, RAY
WALTER CRONKITE'S UNIVERSE (II,2739)

BRADY, TERENCE
NO—HONESTLY (II,1857); ROBIN'S NEST (II,2187)

BRAFF, DAVID
THE FALL GUY (II,811); HOME ROOM (II,1172); KNIGHT RIDER (II,1402); TOO CLOSE FOR COMFORT (II,2642); V: THE SERIES (II,2715)

BRAHMS, CARYL
THAT WAS THE WEEK THAT WAS (I,4423)

BRAINARD, ED

THE KATE SMITH EVENING HOUR (I,2507); THE KATE SMITH HOUR (I,2508)

BRALVER, CHARLENE
240-ROBERT (II,2688)

BRAND, JOSHUA
ST. ELSEWHERE (II,2432)

BRANDEL, MARC
ALFRED HITCHCOCK PRESENTS (I,115); AMOS BURKE, SECRET AGENT (I,180); DANGER MAN (I,1136); FOUR STAR PLAYHOUSE (I,1652); FRONT ROW CENTER (I,1694); PLAYHOUSE 90 (I,3623); RETURN TO FANTASY ISLAND (II,2144); STUDIO ONE (I,4268); SUSPICION (I,4309); TURN OF FATE (I,4623)

BRANDMAN, MICHAEL
THE NOONDAY SHOW (II,1859)

BRAO, LYNNE FARR
WE GOT IT MADE (II,2752)

BRASHAR, JAIE
YOU ASKED FOR IT (II,2855)

BRASON, JOHN
SECRET ARMY (II,2290)

BRAUNSTEIN, GENE
LAVERNE AND SHIRLEY (II,1446)

BRAVERMAN, DAVID
GILLIGAN'S ISLAND (II,990); GREEN ACRES (I,1884)

BRAVERMAN, HERB
THE DUKE (I,1381)

BRAVERMAN, MARTIN
THE FACTS OF LIFE (II,805)

BRAVERMAN, MICHAEL
QUINCY, M. E. (II,2102); THE RETURN OF MARCUS WELBY, M.D (II,2138)

BRAWN, PATRICK
INTERNATIONAL DETECTIVE (I,2224)

BRECHER, IRVING
THE LIFE OF RILEY (I,2685); MEET ME IN ST. LOUIS (I,2986)

BRECHER, JIM
THAT SECOND THING ON ABC (II,2572)

BRECKMAN, ANDY
HOT HERO SANDWICH (II,1186); THE JOE PISCOPO SPECIAL (II,1304); LATE NIGHT WITH DAVID LETTERMAN (II,1442)

BREECHER, ELIZABETH
THE ADVENTURES OF CHAMPION (I,54); BUFFALO BILL JR. (I,755); THE CISCO KID (I,961); THE GENE AUTRY SHOW (I,1748)

BREECHER, MICHAEL
FILTHY RICH (II,856)

BREEN, RICHARD L
DRAGNET (I,1370)

BRENNAN, ROBERT
MATT HOUSTON (II,1654)

BRENNAN, TOM
FROM A BIRD'S EYE VIEW (I,1686)

BRENNER, ALFRED
BEN CASEY (I,394); THE DICK POWELL SHOW (I,1269); KRAFT TELEVISION THEATER (I,2592); MCMILLAN AND WIFE (II,1667); ONE STEP BEYOND (I,3388)

BRENNER, RAY
CHARLIE'S ANGELS (II,486); CODE RED (II,552); ENOS (II,779); HERE WE GO AGAIN (I,2028); INSTANT FAMILY (II,1238); THE JIMMIE RODGERS SHOW (I,2383); KOJAK (II,1406); MCHALE'S NAVY (I,2969); NURSE (II,1870); TRAPPER JOHN, M.D. (II,2654)

BRENNERT, ALAN
BUCK ROGERS IN THE 25TH CENTURY (II,383); LEGMEN (II,1458); THE MISSISSIPPI (II,1707); SIMON AND SIMON (II,2357); WONDER WOMAN (SERIES 2) (II,2829)

BREWER, JAMESON
THE ADDAMS FAMILY (II,11); ALCOA PREMIERE (I,109); ALCOA/GOODYEAR THEATER (I,107); BATTLE OF THE PLANETS (II,179); BRANDED (I,707); BURKE'S LAW (I,762); THE GUNS OF WILL SONNETT (I,1907); MARKHAM (I,2916); MR. LUCKY (I,3141); PAPA G.I. (I,3451); VOLTRON—DEFENDER OF THE UNIVERSE (II,2729)

BREZ, ETHEL
CASTLE ROCK (II,456); ROXY PAGE (II,2222)

BREZ, MEL
CASTLE ROCK (II,456); ROXY PAGE (II,2222)

BRICE, MONTY
THE BOB HOPE SHOW (I,533); THE BOB HOPE SHOW (I,534)

BRICKELL, BETH
A RAINY DAY (II,2110)

BRICKMAN, MARSHALL
JOHNNY CARSON'S REPERTORY COMPANY IN AN EVENING OF COMEDY (I,2423); OFF CAMPUS (II,1877); THE WOODY ALLEN SPECIAL (I,4904)

BRIDGES, GINGER
PERSONAL AND CONFIDENTIAL (II,2026)

BRIDGES, JAMES
ALFRED HITCHCOCK PRESENTS (I,115); THE PAPER CHASE (II,1949)

BRIGGS, BILL
THE KING FAMILY SHOW (I,2552)

BRIGGS, BUNNY
MIKE AND PEARL (I,3038)

BRINDLEY, MICHAEL
PRISONER: CELL BLOCK H (II,2085)

BRINKLEY, DON
BEN CASEY (I,394); THE CISCO KID (I,961); THE DESILU PLAYHOUSE (I,1237); FAMILY IN BLUE (II,818); THE FBI (I,1551); THE FELONY SQUAD (I,1563); THE FUGITIVE (I,1701); THE INVADERS (I,2229); MEDICAL CENTER (II,1675); PERRY MASON (II,2025); THE RAT PATROL (I,3726); TRAPPER JOHN, M.D. (II,2654); VOYAGE TO THE BOTTOM OF THE SEA (I,4743)

BROADLEY, PHILIP
BERGERAC (II,209); THE CHAMPIONS (I,896); DEPARTMENT S (I,1232)

BROCK, E M
CAPTAIN VIDEO AND HIS VIDEO RANGERS (I,838)

BRODERICK, GEORGE
CHICAGOLAND MYSTERY PLAYERS (I,936)

BRODERICK, PATRICIA
INTERMEZZO (I,2223)

BRODNEY, OSCAR
DANGER MAN (I,1136)

BRODY, LARRY
AUTOMAN (II,120); BARNABY JONES (II,153); CANNON (II,424); THE FALL GUY (II,811); FARRELL: FOR THE PEOPLE (II,834); HAWAII FIVE-O (II,1110); PARTNERS IN CRIME (II,1961); STAR TREK (II,2439)

BRODY, SCOTT
FAME (II,812)

BROMFIELD, VALRI
THE DAVID LETTERMAN SHOW (II,624); THE NEW SHOW (II,1828); THAT SECOND THING ON ABC (II,2572); THAT THING ON ABC (II,2574)

BRONDFIELD, JEROME
THE GEORGE SANDERS MYSTERY THEATER (I,1776);

BROOKE-TAYLOR, TIM
CAMBRIDGE CIRCUS (I,807); THE GOODIES (II,1040)

BROOKER, JOAN
HARPER VALLEY PTA (II,1095); JOANIE LOVES CHACHI (II,1295)

BROOKS, ALBERT
TURN ON (I,4625)

BROOKS, C ROBERT
EIGHT IS ENOUGH (II,762)

BROOKS, HINDI
APPLE'S WAY (II,101); EIGHT IS ENOUGH (II,762); FAME (II,812); FAMILY (II,813); FRONT ROW CENTER (I,1694); JESSIE (II,1287); LITTLE HOUSE ON THE PRAIRIE (II,1487); LOTTERY (II,1525); NURSE (II,1870); THE WALTONS (II,2740)

BROOKS, JACK
ESTHER WILLIAMS AT CYPRESS GARDENS (I,1465)

BROOKS, JAMES L
FRIENDS AND LOVERS (II,930); GOING PLACES (I,1834); MY FRIEND TONY (I,3192); THE NEW LORENZO MUSIC SHOW (II,1821); RHODA (II,2151)

BROOKS, MEL
THE 2,000 YEAR OLD MAN (II,2696); ACCENT ON LOVE (I,16); THE JERRY LEWIS SHOW (I,2365); THE MAN IN THE MOON (I,2871); MARRIAGE—HANDLE WITH CARE (I,2921); THE SID CAESAR SPECIAL (I,4044); THE SID CAESAR, IMOGENE COCA, CARL REINER, HOWARD MORRIS SPECIAL (I,4039); TIPTOE THROUGH TV (I,4524); VARIETY: THE WORLD OF SHOW BIZ (I,4704); WHEN THINGS WERE ROTTEN (II,2782); YOUR SHOW OF SHOWS (I,4957); ZERO HOUR (I,4980)

BROOKS, T E
JOHNNY RINGO (I,2434)

BROPHY, EDNA
THE LORETTA YOUNG THEATER (I,2756)

BROWLOW, KEVIN
HOLLYWOOD (II,1160)

BROWN, BILL
THE KEEFE BRASSELLE SHOW (I,2514)

BROWN, CHARLOTTE
THE BOB NEWHART SHOW (II,342); BUMPERS (II,394); HERE WE GO AGAIN (I,2028); MITZI: A TRIBUTE TO THE AMERICAN HOUSEWIFE (I,3072); THE PARTRIDGE FAMILY (II,1962); REALLY RAQUEL (II,2119); RHODA (II,2151)

BROWN, DAVID
FLO (II,891); TALES OF THE GOLD MONKEY (II,2537)

BROWN, DOUGHERTY
CLAUDIA: THE STORY OF A MARRIAGE (I,972)

BROWN, EARL
CHER (II,496); DIANA ROSS AND THE SUPREMES AND THE TEMPTATIONS ON BROADWAY (I,1260); DONNY AND MARIE (II,712); MARIE (II,1629); THE SONNY AND CHER COMEDY HOUR (II,2400)

BROWN, FREDRIC
STAR TREK (II,2440)

BROWN, GEORGE CARLTON
MCHALE'S NAVY (I,2969); THE PEOPLE'S CHOICE (I,3521)

BROWN, HARRY
COMBAT (I,1011)

BROWN, MARY KAY
THE TV TV SHOW (II,2674)

BROWN, RITA MAE
I LOVE LIBERTY (II,1215)

BROWN, STANLEY
ROBERT MONTGOMERY PRESENTS YOUR LUCKY STRIKE THEATER (I,3809)

BROWN, STEVE
CAGNEY AND LACEY (II,409); THIS GIRL FOR HIRE (II,2596)

BROWN, WALTER C
FOUR STAR PLAYHOUSE (I,1652)

BROWNE JR, ARTHUR
THE BIG VALLEY (I,437); LAW OF THE PLAINSMAN (I,2639); PLANET OF THE APES (II,2049); THE RIFLEMAN (I,3789); VOYAGE TO THE BOTTOM OF THE SEA (I,4743); WAGON TRAIN (I,4747) WHICH WAY'D THEY GO? (I,4838)

BROWNE, HOWARD
BEN CASEY (I,394); THE FUGITIVE (I,1701); THE GREEN FELT JUNGLE (I,1885)

BROWNE, L VIRGINIA
RITUALS (II,2182)

BRUCE, GEORGE
CROSSROADS (I,1105); PLAYHOUSE 90 (I,3623); TIGHTROPE (I,4500)

BRUCKMAN, CLYDE
THE ABBOTT AND COSTELLO SHOW (I,2)

BRUCKNER, BRAD
THE ROGUES (I,3832)

BRUCKNER, WILLIAM
CROWN THEATER WITH GLORIA SWANSON (I,1107); THE LONE RANGER (I,2740); THE LORETTA YOUNG THEATER (I,2756); SKY KING (I,4077)

BRUNNER, BOB
LOVE, SIDNEY (II,1547); THE ODD COUPLE (II,1875); PRIVATE BENJAMIN (II,2087); SISTER TERRI (II,2366)

BRYANT, MICHAEL
BUCK ROGERS IN THE 25TH CENTURY (II,383)

BRYNE, JOE
THE YELLOW ROSE (II,2847)

BRYNES, STU
TOM CORBETT, SPACE CADET (I,4535)

BUCHANAN, JAMES DAVID
BEACH PATROL (II,192); CANNON: THE RETURN OF FRANK CANNON (II,425); CHARLIE'S ANGELS (II,486); CHOPPER ONE (II,515); Q. E. D. (II,2094)

BUCHWALD, ART
THAT WAS THE YEAR THAT WAS (II,2575)

BUCK, CRAIG
BLUE THUNDER (II,278); HOT PURSUIT (II,1189); THE INCREDIBLE HULK (II,1232); SKYWARD CHRISTMAS (II,2378); TRAUMA CENTER (II,2655)

BUCK, FAUSTAUS
V: THE FINAL BATTLE (II,2714)

BUCK, PEARL S
THE ALCOA HOUR (I,108)

BUCK, RON
STARSKY AND HUTCH (II,2444)

BUCKNER, BRAD
FIT FOR A KING (II,866); HIGHCLIFFE MANOR (II,1150); THE LOVE BOAT III (II,1534); SCARECROW AND MRS. KING (II,2268); WHY US? (II,2796)

BUCKNER, ROBERT
AMOS BURKE, SECRET AGENT (I,180); TWENTIETH CENTURY (I,4639)

BUHLER, KITTY
TIGHTROPE (I,4500)

BUIT, LIONEL
THE RED SKELTON SHOW (I,3756)

BULL, SHELDON
BEANE'S OF BOSTON (II,194); THE BETTY WHITE SHOW (II,224); CARTER COUNTRY (II,449); CONDO (II,565); IT'S A LIVING (II,1254); M*A*S*H (II,1569); MAKING A LIVING (II,1604); NEWHART (II,1835)

BULLOCK, DON
BRET MAVERICK (II,374); LITTLE HOUSE ON THE PRAIRIE (II,1487); LITTLE HOUSE: A NEW BEGINNING (II,1488); LOTTERY (II,1525); THE ROOKIES (II,2208)

BULLOCK, HARVEY
ALICE (II,33); THE ANDY GRIFFITH SHOW (I,192); BIG JOHN, LITTLE JOHN (II,240); FROM A BIRD'S EYE VIEW (I,1686); I SPY (I,2179); THE JETSONS (I,2376); MAN IN THE MIDDLE (I,2870); MR. TERRIFIC (I,3152); MY WORLD . . . AND WELCOME TO IT (I,3208); THE WALTER WINCHELL SHOW (I,4760); THE WONDERFUL WORLD OF BURLESQUE II (I,4896)

BUNCH, CHRIS
THE A-TEAM (II,119); BUCK ROGERS IN THE 25TH CENTURY (II,383); CODE RED (II,552); GAVILAN (II,959); HUNTER (II,1206); THE INCREDIBLE HULK (II,1232); JESSIE (II,1287)

BURBRIDGE, BETTY
THE CISCO KID (I,961)

BURDICK, HAL
NIGHT EDITOR (I,3286)

BURDITT, GEORGE
THE ANDY WILLIAMS SHOW (I,210); DOC (II,682); THE HUDSON BROTHERS SHOW (II,1202); JOEY AND DAD (II,1306); THE KEN BERRY WOW SHOW (I,2524); LOLA (II,1508); LOLA (II,1510); LOLA (II,1511); LOVE AND LEARN (II,1529); THE SMOTHERS BROTHERS SHOW (I,4097); THE SONNY AND CHER COMEDY HOUR (II,2400); THE SONNY COMEDY REVUE (II,2403); THREE'S A CROWD (II,2613); THREE'S COMPANY (II,2614); TURN ON (I,4625); VAN DYKE AND COMPANY (II,2718); VAN DYKE AND COMPANY (II,2719)

BURDITT, JOYCE
THREE'S COMPANY (II,2614)

BURDITT, PAUL
BENSON (II,208)

BURGESS, ANTHONY
MOSES THE LAWGIVER (II,1737)

BURGESS, GRANVILLE
CAPITOL (II,426)

BURNETT, ALAN
THE SMURFS (II,2385)

BURNETT, JACK
THE JACK CARTER SHOW (I,2306)

BURNETT, W R
NAKED CITY (I,3216); THE UNTOUCHABLES (I,4682)

BURNHAM, ED
THE JEFFERSONS (II,1276); RITUALS (II,2182)

BURNHAM, JEREMY
THE AVENGERS (II,121)

BURNIER, JEANNINE
THE SONNY AND CHER SHOW (II,2401)

BURNS, ALLAN
THE DUCK FACTORY (II,738); FRIENDS AND LOVERS (II,930); LOU GRANT (II,1526); MY MOTHER THE CAR (I,3197); THE NEW LORENZO MUSIC SHOW (II,1821); THE NUT HOUSE (I,3320); RHODA (II,2151); SHEPHERD'S FLOCK (I,4007)

BURNS, GARY
BOWZER (II,358)

BURNS, GEORGE
GEORGE BURNS: AN HOUR OF JOKES AND SONGS (II,973)

BURNS, JACK
THE BURNS AND SCHREIBER COMEDY HOUR (I,768); THE BURNS AND SCHREIBER COMEDY HOUR (I,769); THE FLIP WILSON COMEDY

SPECIAL (II,886); THE FLIP WILSON SHOW (I,1602); FLIP WILSON. . . OF COURSE (II,890); THE HARLEM GLOBETROTTERS POPCORN MACHINE (II,1092); JOHN RITTER: BEING OF SOUND MIND AND BODY (II,1319); THE JUD STRUNK SHOW (I,2462); THE KOPYCATS (I,2580); THE KRAFT MUSIC HALL (I,2588); THE MELBA MOORE-CLIFTON DAVIS SHOW (I,2996); MERRY CHRISTMAS. . . WITH LOVE, JULIE (II,1687); THE MUPPET SHOW (II,1766); PETER PAN (II,2030); THE SANDY DUNCAN SHOW (II,2253); TIN PAN ALLEY TODAY (I,4521); VARIETY (II,2721); WE'VE GOT EACH OTHER (II,2757); ZERO HOUR (I,4980)

BURNS, JUDY
THE FBI (I,1551); KNIGHT RIDER (II,1402); THE POWERS OF MATTHEW STAR (II,2075); THE SIX-MILLION-DOLLAR MAN (II,2372); VEGAS (II,2724); WONDER WOMAN (SERIES 2) (II,2829)

BURNS, LARRY
MAKE ROOM FOR DADDY (I,2843)

BURNS, STAN
DEAN MARTIN'S CELEBRITY ROAST (II,643); DEAN MARTIN'S RED HOT SCANDALS OF 1926 (II,648); DEAN MARTIN'S RED HOT SCANDALS PART 2 (II,649); DEAN'S PLACE (II,650); GILLIGAN'S ISLAND (II,990); HONEY WEST (I,2105); I'VE HAD IT UP TO HERE (II,1221); KEEP U.S. BEAUTIFUL (I,2519); THE MAC DAVIS SHOW (II,1575); THE MILTON BERLE SHOW (I,3049); THE NEW STEVE ALLEN SHOW (I,3271); PEOPLE ARE FUNNY (II,1996); THE SHOW MUST GO ON (II,2344); THE SMOTHERS BROTHERS COMEDY HOUR (I,4095); THE STEVE ALLEN SHOW (I,4219); THE STEVE ALLEN SHOW (I,4220); THE STEVE ALLEN SHOW (I,4221); TED KNIGHT MUSICAL COMEDY VARIETY SPECIAL SPECIAL (II,2549); THE TIM CONWAY SPECIAL (I,4503)

BURNS, TIMOTHY
SIMON AND SIMON (II,2357)

BURNS, WILLIAM
THE GEORGE BURNS AND GRACIE ALLEN SHOW (I,1763); MCNAB'S LAB (I,2972); MR. ED (I,3137); NO TIME FOR SERGEANTS (I,3300)

BURNSTEIN, P G
BOWZER (II,358)

BURROWS, ABE
HOW TO SUCCEED IN BUSINESS WITHOUT REALLY TRYING (II,1197); O.K. CRACKERBY (I,3348)

BURSTON, BUD
THE DINAH SHORE CHEVY SHOW (I,1279); JERRY MAHONEY'S CLUB HOUSE (I,2371)

BURT, FRANK
NOAH'S ARK (I,3302); TERRY AND THE PIRATES (I,4405)

BURTON, AL
GO! (I,1830)

BURTON, FRANK
BILL COSBY DOES HIS OWN THING (I,441)

BURTON, JAY
76 MEN AND PEGGY LEE (I,3989); AWAY WE GO (I,317); THE BIG PARTY FOR REVLON (I,427); THE DEAN MARTIN CELEBRITY ROAST (II,637); THE DEAN MARTIN CELEBRITY ROAST (II,638); THE DEAN MARTIN CELEBRITY ROAST (II,639); THE DEAN MARTIN SHOW (I,1201); DENNIS JAMES' CARNIVAL (I,1227); DOM DELUISE AND FRIENDS (II,701); DOM DELUISE AND FRIENDS, PART 2 (II,702); THE DON HO SHOW (II,706); HECK'S ANGELS (II,1122); THE HOLLYWOOD PALACE (I,2088); THE JULIE ANDREWS HOUR (I,2480); THE KOPYCATS (I,2580); THE KRAFT MUSIC HALL (I,2587); MERRY CHRISTMAS. . . WITH LOVE, JULIE (II,1687); THE MILTON BERLE SHOW (I,3047); ON STAGE AMERICA (II,1893); THE PERRY COMO SHOW (I,3532); THE PERRY COMO SUNSHINE SHOW (II,2005); PERRY COMO'S SUMMER OF '74 (II,2024); THE STEVE ALLEN COMEDY HOUR (II,2454); TEXACO STAR THEATER (I,4411); VICTOR BORGE'S 20TH ANNIVERSARY SHOW (I,4727)

BUSTANY, JUDITH
WHO'S THE BOSS? (II,2792)

BUTLER, DAVID
THE ADVENTURES OF BLACK BEAUTY (I,51); MARCO POLO (II,1626); THE STRAUSS FAMILY (I,4253)

BUTLER, JOHN
ANNIE OAKLEY (I,225); THE NEW ADVENTURES OF CHARLIE CHAN (I,3249)

BUTLER, MICHAEL
THE BARBARY COAST (II,146); BARETTA (II,152); JOE DANCER: THE BIG BLACK PILL (II,1300); KATE MCSHANE (II,1368); NAKIA (II,1784)

BUTLER, NATHAN
STAR TREK (II,2440)

BUTTRAM, PAT
DANNY THOMAS GOES COUNTRY AND WESTERN (I,1146); THE JERRY REED WHEN YOU'RE HOT, YOU'RE HOT HOUR (I,2372)

BUX, BILL
JIMMY DURANTE PRESENTS THE LENNON SISTERS HOUR (I,2387)

BUXTON, FRANK
THE ODD COUPLE (II,1875)

BUZBY, ZANE
ALL NIGHT RADIO (II,40)

BYNER, JOHN
THE JOHN BYNER COMEDY HOUR (I,2407)

BYRNE, ERICA
MATT HOUSTON (II,1654)

BYRNE, JOSEPH
ZANE GREY THEATER (I,4979)

BYRNES, JIM
THE BUFFALO SOLDIERS (II,388); THE BUSTERS (II,401); HOW THE WEST WAS WON (II,1196); THE MACAHANS (II,1583); RANSOM FOR ALICE (II,2113); ROYCE (II,2224); THE STREETS OF SAN FRANCISCO (II,2478); THE WALTONS (II,2740); ZANE GREY THEATER (I,4979)

CADDIGAN, JAMES L
THE TIMID SOUL (I,4519)

CAFFEY, RICHARD
HEAVEN ON EARTH (II,1120)

CAGNEY, TIM
TEXAS (II,2566)

CAHN, DANN
FRANCES LANGFORD PRESENTS (I,1656)

CAIDIN, MARTIN
EXO-MAN (II,797)

CAILLOU, ALAN
THE FUGITIVE (I,1701); THE MAN FROM U.N.C.L.E. (I,2867)

CAIRNCROSS, WILLIAM
QUINCY, M. E. (II,2102)

CALDWELL, BARRY
PANDAMONIUM (II,1948)

CALDWELL, JOE
DARK SHADOWS (I,1157)

CALDWELL, JOHN
BATMAN (I,366)

CALICK, LOUIS
ARCHIE BUNKER'S PLACE (II,105)

CALLAHAN, GEORGE
THE CISCO KID (I,961)

CALLAN, HAROLD
TARGET: THE CORRUPTERS (I,4360)

CALVELLI, JOSEPH
CLINIC ON 18TH STREET (II,544); THE MARRIAGE BROKER (I,2920); MR. NOVAK (I,3145)

CALVELLI, MICHAEL
HEC RAMSEY (II,1121)

CALVERT, FRED
I AM THE GREATEST: THE ADVENTURES OF MUHAMMED ALI (II,1211)

CAMERON, PATSY
THE SMURFS (II,2385)

CAMP, RICHARD
HOT HERO SANDWICH (II,1186)

CAMPANELLIS, JACOVOS
MELINA MERCOURI'S GREECE (I,2997)

CAMPBELL, ARCHIE
HEE HAW (II,1123)

CAMPBELL, BOB
MATINEE AT THE BIJOU (II,1651)

CAMPBELL, DICK
THE BOB SMITH SHOW (I,668)

CAMPBELL, R WRIGHT
MEDIC (I,2976)

CANDY, JOHN
BIG CITY COMEDY (II,233); SECOND CITY TELEVISION (II,2287)

CANNAN, TONY
THE STREETS OF SAN FRANCISCO (II,2478)

CANNELL, STEPHEN J
THE A-TEAM (II,119); BARETTA (II,152); BOSTON AND KILBRIDE (II,356); COLUMBO (II,556); DOCTOR SCORPION (II,685); THE GREATEST AMERICAN HERO (II,1060); THE GYPSY

WARRIORS (II,1070); HARDCASTLE AND MCCORMICK (II,1089); HUNTER (II,1206); THE JORDAN CHANCE (II,1343); THE NIGHT RIDER (II,1848); NIGHTSIDE (II,1851); THE QUEST (II,2098); RICHIE BROCKELMAN: MISSING 24 HOURS (II,2168); RIPTIDE (II,2178); THE ROCKFORD FILES (II,2196); THE ROCKFORD FILES (II,2197); THE ROUSTERS (II,2220); STONE (II,2470); TOMA (II,2634)

CANNING, BILL
JEANNIE (II,1275)

CANNON, DORAN
KNOTS LANDING (II,1404)

CANNON, WILLIAM
GOING BANANAS (II,1015)

CANTER, WALT
VICTOR BORGE'S 20TH ANNIVERSARY SHOW (I,4727)

CAPOTE, TRUMAN
LAURA (I,2636)

CAPP, AL
LI'L ABNER (I,2701)

CAPPS, DENNIS
CHICAGO STORY (II,506); TALES OF THE GOLD MONKEY (II,2537)

CARABATSOS, STEVEN W
BEN CASEY (I,394); THE BIG VALLEY (I,437); STAR TREK (II,2440)

CARDEA, FRANK
BRING 'EM BACK ALIVE (II,377); CRAZY LIKE A FOX (II,601); HARD KNOCKS (II,1086); O'MALLEY (II,1872)

CARLIN, GEORGE
AWAY WE GO (I,317); GEORGE CARLIN AT CARNEGIE HALL (II,974)

CARLIN, STEVE
ROOTIE KAZOOTIE (I,3844)

CARLINER, MARK
THE PHOENIX (II,2036)

CARLINO, LEWIS JOHN
DOC ELLIOT (I,1305)

CARLSON, CRAIG
CAPITOL (II,426)

CARLSON, JANET
THE FARMER'S DAUGHTER (I,1533)

CARLSON, JIM
THE BIONIC WOMAN (II,255); THE BOOTS RANDOLPH SHOW (I,687); KIDD VIDEO (II,1388); THE NASHVILLE SOUND OF BOOTS RANDOLPH (I,3229); ROWAN AND MARTIN'S LAUGH-IN (I,3856); TENNESSEE ERNIE FORD'S WHITE CHRISTMAS (I,4402)

CARLSON, MICHAEL
BATTLESTAR GALACTICA (II,181)

CARLTON, SAM
GEORGE JESSEL'S SHOW BUSINESS (I,1770)

CARMBOU, DON
EYE ON HOLLYWOOD (II,798)

CARMODY, DON
STRANGE TRUE STORIES (II,2476)

CARMOSANT, DONALD
THE RIVALRY (I,3796); VALLEY FORGE (I,4698)

CARNECE, JESS
COMBAT (I,1011)

CARNEY, OTIS
JOHNNY GUITAR (I,2427); WAR CORRESPONDENT (I,4767); ZANE GREY THEATER (I,4979)

CARON, GLENN GORDON
BREAKING AWAY (II,371); CONCRETE BEAT (II,562); GOOD TIME HARRY (II,1037); REMINGTON STEELE (II,2130)

CAROTHERS, A J
BARNEY AND ME (I,358); GOODNIGHT, BEANTOWN (II,1041); MISS STEWART, SIR (I,3064); TOPPER RETURNS (I,4583)

CARP, DAVID
NERO WOLFE (II,1806)

CARPENTER, JOEL
IVANHOE (I,2282); NAKED CITY (I,3216); ROUTE 66 (I,3852)

CARPENTER, RICHARD
THE ADVENTURES OF BLACK BEAUTY (I,51); DOCTOR SNUGGLES (II,687)

CARR, ALAN
THE WALTONS (II,2740)

CARR, DICK
BATMAN (I,366); MACREEDY'S WOMAN (I,2804)

CARR, JOHN DICKSON
CLIMAX! (I,976)

CARR, RICHARD
CASINO (II,452); DAN AUGUST (I,1130); THE DICK POWELL SHOW (I,1269); FOUR STAR PLAYHOUSE (I,1652); LAST OF THE PRIVATE EYES (I,2628); MAGGIE MALONE (I,2812); THE OVER-THE-HILL GANG RIDES AGAIN (I,3432); THE SIX-MILLION-DOLLAR MAN (II,2372); VEGAS (II,2724); THE WALTONS (II,2740); WONDER WOMAN (SERIES 2) (II,2829)

CARREN, DAVID
THE POWERS OF MATTHEW STAR (II,2075)

CARREROW, JACK
DOMESTIC LIFE (II,703)

CARRINGTON, JOHN WILLIAM
CAPITOL (II,426)

CARRINGTON, JOYCE
CAPITOL (II,426)

CARRINGTON, ROBERT
HIGH RISK (II,1146); JOHNNY BLUE (II,1321)

CARROLL JR, BOB
ALL ABOUT BARBARA (I,125); THE DESILU PLAYHOUSE (I,1237); DOROTHY (II,717); HERE'S LUCY (II,1135); I LOVE LUCY (I,2171); THE LUCILLE BALL SPECIAL (II,1559)

CARROLL, CARROLL
THE JUDY GARLAND SHOW (I,2470)

CARROLL, ELLISON
THE PAPER CHASE (II,1949)

CARROLL, SIDNEY
BEACON HILL (II,193); THE COUNTRY GIRL (I,1064); DYNASTY (II,747); PLAYWRIGHTS '56 (I,3627); TRUMAN AT POTSDAM (I,4618)

CARROLL, TODD
DELTA HOUSE (II,658)

CARRUTHERS, BILL
PRESS YOUR LUCK (II,2078)

CARSEY, JOHN JAY
AMERICA, YOU'RE ON (II,69)

CARSON, JOHNNY
THE JOHNNY CARSON SHOW (I,2422); JOHNNY CARSON'S REPERTORY COMPANY IN AN EVENING OF COMEDY (I,2423)

CARSON, RUBEN
CHARO (II,488)

CARTER, LES
BARETTA (II,152); CHARLIE'S ANGELS (II,486); STRIKE FORCE (II,2480)

CARTER, RAY
THE ADMIRAL BROADWAY REVUE (I,37)

CARUSO, DEE
...AND DEBBIE MAKES SIX (I,185); CALL HOLME (II,412); THE DON ADAMS SCREEN TEST (II,705); THE DON ADAMS SPECIAL: HOORAY FOR HOLLYWOOD (I,1329); JACK BENNY'S BIRTHDAY SPECIAL (I,2299); THE JERRY LEWIS SHOW (I,2370); THE NBC FOLLIES (I,3242); THE SMOTHERS BROTHERS SHOW (I,4096); A SPECIAL OLIVIA NEWTON-JOHN (II,2418); WE INTERRUPT THIS SEASON (I,4780); WHERE'S THE FIRE? (II,2785)

CARUTHERS, MARK
NOT NECESSARILY THE NEWS (II,1862); NOT NECESSARILY THE NEWS (II,1863)

CASADY, CORT
BARBARA MANDRELL AND THE MANDRELL SISTERS (II,143); THE EDDIE RABBITT SPECIAL (II,759); A SPECIAL KENNY ROGERS (II,2417); STAR SEARCH (II,2438)

CASEY, PAUL
BARETTA (II,152)

CASEY, PETER
THE JEFFERSONS (II,1276)

CASS, RONNIE
SPECIAL LONDON BRIDGE SPECIAL (I,4150)

CASSIDY, ALAN
THE INCREDIBLE HULK (II,1232)

CASSIDY, BRIAN
EIGHT IS ENOUGH (II,762); FAME (II,812); HUNTER (II,1205); THE ROOKIES (II,2208)

CASTLE JR, NICK
PERRY COMO'S CHRISTMAS IN MEXICO (II,2009); PERRY COMO'S HAWAIIAN HOLIDAY (II,2017)

CASTRO, DAVID
LAUGH TRAX (II,1443)

CATES, JOSEPH
COUNTRY MUSIC HIT PARADE (II,589); FIFTY YEARS OF COUNTRY MUSIC (II,853); JOHNNY CASH AND FRIENDS (II,1322)

CAULEY, HARRY
THE BAXTERS (II,183); HUSBANDS, WIVES AND LOVERS (II,1209)

CAUTHERN, KEN
FAME (II,812)

CAVANAUGH, JAMES P
PLAYHOUSE 90 (I,3623); STUDIO ONE (I,4268); THRILLER (I,4492)

CAVELLA, JOSEPH
SONS AND DAUGHTERS (II,2404)

CAVETT, DICK
THE DICK CAVETT SHOW (II,669); THE JERRY LEWIS SHOW (I,2369); THE MERV GRIFFIN SHOW (I,3014)

CECIL, VANDER
CHIPS (II,511)

CERVI, BRUCE
THE A-TEAM (II,119); GAVILAN (II,959); SIMON AND SIMON (II,2357)

CESANA, RENZA
THE CONTINENTAL (I,1042)

CHAD, SHELDON
SEEING THINGS (II,2297)

CHAIS, PAMELA HERBERT
HAVING BABIES III (II,1109); HERE WE GO AGAIN (I,2028); LOVE, SIDNEY (II,1547); MAUDE (II,1655); PHYLLIS (II,2038); RHODA (II,2151); SCARECROW AND MRS. KING (II,2268)

CHALOPIN, JEAN
INSPECTOR GADGET (II,1236)

CHAMBERLAIN, ANNE
ROBERT MONTGOMERY PRESENTS YOUR LUCKY STRIKE THEATER (I,3809)

CHAMBERS, DAVID
BOSOM BUDDIES (II,355); THE FALL GUY (II,811); WE GOT IT MADE (II,2752)

CHAMBERS, ERNEST
20TH CENTURY FOLLIES (I,4641); BARBARA MANDRELL AND THE MANDRELL SISTERS (II,143); BARRY MANILOW—ONE VOICE (II,156); A BEDTIME STORY (I,384); THE BOB NEWHART SHOW (I,666); THE BOBBY DARIN AMUSEMENT COMPANY (I,672); THE CAPTAIN AND TENNILLE SONGBOOK (II,432); CAROL CHANNING AND PEARL BAILEY ON BROADWAY (I,850); CAROL CHANNING PROUDLY PRESENTS THE SEVEN DEADLY SINS (I,851); THE DANNY KAYE SHOW (I,1143); THE DONNA SUMMER SPECIAL (II,711); THE DORIS MARY ANNE KAPPELHOFF SPECIAL (I,1356); FRANCIS ALBERT SINATRA DOES HIS THING (I,1658); FRIENDS AND NABORS (I,1684); HOW TO HANDLE A WOMAN (I,2146); THE JOHN GARY SHOW (I,2411); THE LESLIE UGGAMS SHOW (I,2660); THE LIVING END (I,2731); NEIL SEDAKA STEPPIN' OUT (II,1804); PUMPBOYS AND DINETTES ON TELEVISION (II,2090); RUN, BUDDY, RUN (I,3870); THE SECOND BARRY MANILOW SPECIAL (II,2285); THE SMOTHERS BROTHERS COMEDY HOUR (I,4095); THE THIRD BARRY MANILOW SPECIAL (II,2590); TONY ORLANDO AND DAWN (II,2639); TWO'S COMPANY (I,4659); VER-R-R-RY INTERESTING (I,4712)

CHAMPION, MADELEN
THE TWILIGHT ZONE (I,4651)

CHAN, RAY
PANHANDLE PETE AND JENNIFER (I,3446); UNCLE MISTLETOE AND HIS ADVENTURES (I,4669)

CHANCELLOR, MARY
PERSONAL AND CONFIDENTIAL (II,2026)

CHANDLER, NORM
RITUALS (II,2182)

CHANTLER, DAVID
THE ADVENTURES OF SUPERMAN (I,77); JOHNNY RINGO (I,2434); NAKED CITY (I,3216); THIS IS YOUR MUSIC (I,4462)

CHANTLER, PEGGY
THE ADVENTURES OF SUPERMAN (I,77); COOGAN'S REWARD (I,1044); HAZEL (I,1982); I DREAM OF JEANNIE (I,2167); MAKE ROOM FOR DADDY (I,2843); PRIVATE SECRETARY (I,3672); THE SLIGHTLY FALLEN ANGEL (I,4087)

CHAPIN, ANNE
CLIMAX! (I,976)

CHAPLIN, PRESCOTT
THE ANN SOTHERN SHOW (I,220)

CHAPMAN, GRAHAM
DOCTOR IN THE HOUSE (I,1313); MONTY PYTHON'S FLYING CIRCUS (II,1729)

CHAPMAN, LEIGH
BURKE'S LAW (I,762); WHERE THE GIRLS ARE (I,4830); THE WILD WILD WEST (I,4863)

CHAPMAN, RICHARD
ALEX AND THE DOBERMAN GANG (II,32); LEGMEN (II,1458); NICK AND THE DOBERMANS (II,1843); THE SEAL (II,2282); SIMON AND SIMON (II,2357)

CHAPMAN, ROBIN
JANE EYRE (II,1273)

CHAPMAN, TOM
THE JACKSONS (II,1267); JOHN DENVER ROCKY MOUNTAIN CHRISTMAS (II,1314)

CHARLES, GLEN
THE BETTY WHITE SHOW (II,224); THE BOB NEWHART SHOW (II,342); CHEERS (II,494); DOC (II,682); THE MARY TYLER MOORE SHOW (II,1640); PHYLLIS (II,2038); TAXI (II,2546)

CHARLES, LES
THE BETTY WHITE SHOW (II,224); THE BOB NEWHART SHOW (II,342); CHEERS (II,494); DOC (II,682); THE MARY TYLER MOORE SHOW (II,1640); PHYLLIS (II,2038); TAXI (II,2546)

CHARNIN, MARTIN
ANNIE AND THE HOODS (II,91); GEORGE M! (I,1772); JACK LEMMON IN 'S WONDERFUL, 'S MARVELOUS, 'S GERSHWIN (I,2313)

CHASE, BORDEN
BONANZA (II,347); BRANDED (I,707); THE ROY ROGERS AND DALE EVANS SHOW (I,3858)

CHASE, CHEVY
THE CHEVY CHASE SHOW (II,505); THE PAUL SIMON SPECIAL (II,1983); THE SMOTHERS BROTHERS SHOW (II,2384)

CHASE, DAVID
MOONLIGHT (II,1730); THE NIGHT STALKER (II,1849); PALMS PRECINCT (II,1946); THE ROCKFORD FILES (II,2197); SWITCH (II,2519)

CHASE, FRANK
BONANZA (II,347); BRANDED (I,707); ROUTE 66 (I,3852)

CHASE, MEL
DEAN MARTIN'S CELEBRITY ROAST (II,643)

CHAYEFSKY, PADDY
THE GOLDEN AGE OF TELEVISION (II,1016)

CHEEVER, JOHN
PLAYHOUSE 90 (I,3623)

CHEHAK, LARRY
JENNIFER SLEPT HERE (II,1278)

CHEHAK, TAD
MR. MERLIN (II,1751)

CHEHAK, TOM
ALOHA PARADISE (II,58); CRAZY LIKE A FOX (II,601); MR. MERLIN (II,1751); STAR OF THE FAMILY (II,2436); TEACHERS ONLY (II,2548); THE TONY RANDALL SHOW (II,2640); WKRP IN CINCINNATI (II,2814)

CHENEY, J BENTON
THE CISCO KID (I,961)

CHERBAK, CYNTHIA A
SWITCH (II,2519)

CHERMAK, CY
DECOY (I,1217); FRONT ROW CENTER (I,1694); KRAFT TELEVISION THEATER (I,2592)

CHESLER, LEWIS
CRYSTAL GAYLE IN CONCERT (II,608); THE HITCHHIKER (II,1157)

CHESNEY, RONALD
THE RAG BUSINESS (II,2106)

CHESTER, COLBY
THE WALTONS (II,2740)

CHEVALIER, MAURICE
MAURICE CHEVALIER'S PARIS (I,2955)

CHEVIGNY, HECTOR
MR. AND MRS. NORTH (I,3124)

CHEVILLAT, DICK
THE ANN SOTHERN SHOW (I,220); THE DANNY THOMAS TV FAMILY REUNION (I,1154); GREEN ACRES (I,1884); HOWDY (I,2150); IT'S A GREAT LIFE (I,2256); THE MCGONIGLE (I,2968)

CHICOS, CATHY
ALL MY CHILDREN (II,39)

CHIN, ANNIE
ENTERTAINMENT TONIGHT (II,780)

CHINIQUY, RON
LITTLE HOUSE ON THE PRAIRIE (II,1487)

CHISOLM, DAVID
CALL TO GLORY (II,413)

CHOLDORON, JEROME
WONDERFUL TOWN (I,4893)

CHOMSKY, DAVID
QUINCY, M. E. (II,2102)

CHORPENNY, CHARLOTTE B
THE MAGIC SLATE (I,2820)

CHRISTENSEN, DON
THE FLINTSTONE FUNNIES (II,883)

CHRISTOPHER, A L
TODAY'S FBI (II,2628)

CHRISTOPHER, C R
THE HARDY BOYS MYSTERIES (II,1090)

CHRISTOPHER, KATHRYN
BATTLE OF THE BEAT (II,161)

CHURCH, GENE
QUINCY, M. E. (II,2102)

CIBBER, COLLEY
THE MISSISSIPPI (II,1707); THE YELLOW ROSE (II,2847)

CINADER, ROBERT A
PINE CANYON IS BURNING (II,2040); QUINCY, M. E. (II,2102); THE RANGERS (II,2112); THE REBELS (II,2121)

CLAIR, DICK
THE CAROL BURNETT SHOW (II,443); EUNICE (II,784); FLO (II,891); HOW TO HANDLE A WOMAN (I,2146); A LAST LAUGH AT THE 60'S (I,2627); MAMA'S FAMILY (II,1610); MAUDE (II,1655); A NEW KIND OF FAMILY (II,1819); SOAP (II,2392)

CLAIRE, ALLEN
FAME (II,812)

CLAIRE, LUDI
THE DESILU PLAYHOUSE (I,1237); DUPONT SHOW OF THE MONTH (I,1387)

CLAMAN, JULIAN
APPOINTMENT WITH ADVENTURE (I,246)

CLAMPETT, BOB
THE BUFFALO BILLY SHOW (I,756); TIME FOR BEANY (I,4505)

CLARK, BARRY
MYSTERIES, MYTHS AND LEGENDS (II,1782)

CLARK, DON
THE BARRY MANILOW SPECIAL (II,155)

CLARK, ELIZABETH
FAMILY (II,813); HAVING BABIES II (II,1108); SECRETS OF MIDLAND HEIGHTS (II,2296); SEVEN BRIDES FOR SEVEN BROTHERS (II,2307)

CLARK, GLORYETTE
THE ROCKFORD FILES (II,2197)

CLARK, HARRY
FATHER KNOWS BEST (II,838)

CLARK, JOE
THE JIMMIE RODGERS SHOW (I,2383)

CLARK, NANCY
THE ANN SOTHERN SHOW (I,220)

CLARK, OLIVER
KNOTS LANDING (II,1404)

CLARK, RON
ACE CRAWFORD, PRIVATE EYE (II,6); THE DANNY THOMAS SPECIAL (I,1151); HOT L BALTIMORE (II,1187); JACKIE GLEASON AND HIS AMERICAN SCENE MAGAZINE (I,2321); THE RICH LITTLE SHOW (II,2153); THE RICH LITTLE SHOW (II,2154); SINGLES (I,4062); THIS WEEK IN NEMTIN (I,4465); THE TIM CONWAY COMEDY HOUR (I,4501); VARIETY (II,2721)

CLARK, SUSAN
THE BARRY MANILOW SPECIAL (II,155)

CLARKE, KATHARINE
CAPITOL (II,426)

CLARKE, ROY
LAST OF THE SUMMER WINE (II,1439); OPEN ALL HOURS (II,1913)

CLARKE, T E B
FROM A BIRD'S EYE VIEW (I,1686); THE WIDE OPEN DOOR (I,4859)

CLARKE, WILLIAM KENDALL
MISS SUSAN (I,3065); THE POWER WITHIN (II,2073); ROBERT MONTGOMERY PRESENTS YOUR LUCKY STRIKE THEATER (I,3809); TONIGHT IN SAMARKAND (I,4558)

CLAUSER, SUZANNE
LITTLE WOMEN (II,1497)

CLAVER, BOB
GIDGET (I,1795)

CLAY, ROGER
MEET THE GIRLS: THE SHAPE, THE FACE, AND THE BRAIN (I,2990)

CLAYTON, ROBERT
THE CISCO KID (I,961)

CLEESE, JOHN
DOCTOR IN THE HOUSE (I,1313); FAWLTY TOWERS (II,843); MONTY PYTHON'S FLYING CIRCUS (II,1729)

CLEMENS, BRIAN
APPOINTMENT WITH A KILLER (I,245); THE AVENGERS (II,121); THE CARNATION KILLER (I,845); THE CHAMPIONS (I,896); COLOR HIM DEAD (I,1007); COLOR HIM DEAD (II,554); COME OUT, COME OUT, WHEREVER YOU ARE (I,1014); CRY TERROR! (I,1112); DANGER MAN (I,1136); DARKROOM (II,619); DEATH IN DEEP WATER (I,1206); THE DEVIL'S WEB (I,1248); DIAL A DEADLY NUMBER (I,1253); THE DOUBLE KILL (I,1360); AN ECHO OF THERESA (I,1399); ESCAPADE (II,781); FILE IT UNDER FEAR (I,1574); I'M THE GIRL HE WANTS TO KILL (I,2194); IF IT'S A MAN, HANG UP (I,2186); IN THE STEPS OF A DEAD MAN (I,2205); A KILLER IN EVERY CORNER (I,2538); KILLER WITH TWO FACES (I,2539); THE KILLING GAME (I,2540); KISS KISS, KILL KILL (I,2566); LADY KILLER (I,2603); MELODY OF HATE (I,2999); MIRROR OF DECEPTION (I,3060); MURDER IS A ONE-ACT PLAY (I,3160); MURDER MOTEL (I,3161); MURDER ON THE MIDNIGHT EXPRESS (I,3162); THE NEW AVENGERS (II,1816); THE NEXT VICTIM (I,3281); NOT GUILTY! (I,3311); ONCE THE KILLING STARTS (I,3367); ONE DEADLY OWNER (I,3373); POSSESSION (I,3655); SCREAMER (I,3945); SECRET AGENT (I,3962); SLEEPWALKER (I,4084); SOMEONE AT THE TOP OF THE STAIRS (I,4114)

CLEMENT, DICK
I'LL NEVER FORGET WHAT'S HER NAME (II,1217); MY WIFE NEXT DOOR (II,1781); ON THE ROCKS (II,1894); THERE GOES THE NEIGHBORHOOD (II,2582)

CLEMENT, GAIL
MY THREE SONS (I,3205)

CLEMENTS JR, CALVIN
APPLE'S WAY (II,101); FAMILY (II,813); THE FEATHER AND FATHER GANG (II,845); MATT HOUSTON (II,1654); MICKEY SPILLANE'S MARGIN FOR MURDER (II,1691); MOBILE ONE (II,1717); THE WALTONS (II,2740); WONDER WOMAN (SERIES 2) (II,2829)

CLEMENTS, CALVIN
ALFRED HITCHCOCK PRESENTS (I,115); BARNABY JONES (II,153); THE DESILU PLAYHOUSE (I,1237); DIRTY SALLY (II,680); THE FBI (I,1551); HOW THE WEST WAS WON (II,1196); LAW OF THE PLAINSMAN (I,2639); STRIKE FORCE (II,2480); THE WILD WILD WEST (I,4863)

CLEMENTS, NANCY
THE FLINTSTONE FUNNIES (II,883)

CLEVER, FRANK
BONANZA (II,347)

CLOTH, SHERMAN
THE ALAN KING SHOW (I,102)

CLUESS, CHRIS
I'VE HAD IT UP TO HERE (II,1221); MARIE (II,1629); SCARECROW AND MRS. KING (II,2268)

CLYNE, KEVIN
KIDD VIDEO (II,1388)

COATES, KEVIN
BATTLE OF THE PLANETS (II,179)

COATES, PAUL
CONFIDENTIAL FILE (I,1033)

COATES, ROBERT
ROBERT MONTGOMERY PRESENTS YOUR LUCKY STRIKE THEATER (I,3809)

COBB, TY
THE ADVENTURES OF CHAMPION (I,54); ANNIE OAKLEY (I,225); BUFFALO BILL JR. (I,755); THE GENE AUTRY SHOW (I,1748)

COBEN, SHERRY
KATE AND ALLIE (II,1365)

COCHRAN, JOSEPH
IVANHOE (I,2282); KRAFT TELEVISION THEATER (I,2592)

COCKRELL, FRANCIS
ALFRED HITCHCOCK PRESENTS (I,115); BATMAN (I,366); ONE STEP BEYOND (I,3388); THE OUTER LIMITS (I,3426); PERRY MASON

(II,2025); SUSPICION (I,4309)

COCKRELL, MARIAN
ALFRED HITCHCOCK PRESENTS (I,115); BATMAN (I,366)

COE, LIZ
FINDER OF LOST LOVES (II,857)

COFFEY, MARGARET
DANGER ZONE (I,1137)

COHAN, BUZ
THE ODD COUPLE (II,1875)

COHAN, MARTIN
ALMOST AMERICAN (II,55); THE BOB NEWHART SHOW (II,342); FLYING HIGH (II,895); MAUREEN (II,1656); THE PARTRIDGE FAMILY (II,1962); SILVER SPOONS (II,2355); WHO'S THE BOSS? (II,2792)

COHAN, RONALD M
THE AMERICAN DREAM (II,71); CALL TO GLORY (II,413); FEEL THE HEAT (II,848)

COHEN, A M
SHANE (I,3995)

COHEN, CHARLES
CHANNING (I,900)

COHEN, ERIC
THE 416TH (II,913); ARCHIE (II,104); THE ARCHIE SITUATION COMEDY MUSICAL VARIETY SHOW (II,106); AT EASE (II,116); GABRIEL KAPLAN PRESENTS THE SMALL EVENT (II,952); GUN SHY (II,1068); LAVERNE AND SHIRLEY (II,1446); PRIVATE BENJAMIN (II,2087); THE TWO OF US (II,2693); UPTOWN SATURDAY NIGHT (II,2711); WELCOME BACK, KOTTER (II,2761); WHACKED OUT (II,2767); ZORRO AND SON (II,2878)

COHEN, LAWRENCE J
APPLE PIE (II,100); BRANDED (I,707); CALLING DR. STORM, M.D. (II,415); COLUMBO (II,556); EMPIRE (II,777); FOG (II,897); THE GOOD LIFE (II,1033); KRAFT TELEVISION THEATER (I,2592); THE MASK OF MARCELLA (I,2937); MOMMA THE DETECTIVE (II,1721); THE SHAMEFUL SECRETS OF HASTINGS CORNERS (I,3994); SPARROW (II,2410); STICK AROUND (II,2464); STRUCK BY LIGHTNING (II,2482)

COHEN, M CHARLES
ROOTS (II,2211); SENIOR YEAR (II,2301)

COHEN, RANDY
LATE NIGHT WITH DAVID LETTERMAN (II,1442)

COHEN, ROY
FOUR STAR PLAYHOUSE (I,1652)

COHN, JEFF
TOO CLOSE FOR COMFORT (II,2642)

COHON, BARRY
THE CISCO KID (I,961)

COLE, ALLAN
THE A-TEAM (II,119); BUCK ROGERS IN THE 25TH CENTURY (II,383); CODE RED (II,552); GAVILAN (II,959); HUNTER (II,1206); THE INCREDIBLE HULK (II,1232); JESSIE (II,1287)

COLE, DANNY LEE
THE GREATEST AMERICAN HERO (II,1060)

COLE, DAVID
THE MAGIC OF DAVID COPPERFIELD (II,1595)

COLE, DENNIS LEE
TALES OF THE GOLD MONKEY (II,2537)

COLE, ROYAL
THE ADVENTURES OF SUPERMAN (I,77); THE CISCO KID (I,961)

COLEMAN, CY
SHIRLEY MACLAINE: IF THEY COULD SEE ME NOW (II,2335)

COLEMAN, JANET
SZYSZNYK (II,2523)

COLES, STEDMAN
INDEMNITY (I,2209)

COLIN, SID
SUNDAY NIGHT AT THE LONDON PALLADIUM (I,4286)

COLLEARY, BOB
A.E.S. HUDSON STREET (II,17); BENSON (II,208); M*A*S*H (II,1569); MR. MAYOR (I,3143)

COLLEARY, R J
I'M A BIG GIRL NOW (II,1218); LOVE, SIDNEY (II,1547)

COLLIER, JOHN
THE TWILIGHT ZONE (I,4651)

COLLINS, ANNE
BUCK ROGERS IN THE 25TH CENTURY (II,383); THE DEVLIN CONNECTION (II,664); FANTASY ISLAND (II,829); HAWAII FIVE-O (II,1110); VEGAS (II,2724); WONDER WOMAN (SERIES 2) (II,2829)

COLLINS, HAL
HONESTLY, CELESTE! (I,2104); LETTERS TO LAUGH-IN (I,2672)

COLLINS, JOHN
HAPPY DAYS (II,1084); MORK AND MINDY (II,1735)

COLLINS, NANCY
LEGENDS OF THE SCREEN (II,1456)

COLLINS, RICHARD
THE CONTENDER (II,571); IN TANDEM (II,1228); PLANET OF THE APES (II,2049); THE RHINEMANN EXCHANGE (II,2150); THE UNTOUCHABLES (I,4682); WAGON TRAIN (I,4747)

COLLINS, ROBERT
REMINGTON STEELE (II,2130)

COLLYER, DEREK
THE VAL DOONICAN SHOW (I,4692)

COLMAN, HENRY
THE LOVE BOAT (II,1535)

COLOMBY, HARRY
COMEDY OF HORRORS (II,557); MCNAMARA'S BAND (II,1669); SOMETHING ELSE (I,4117)

COLVIN, TONY
JUST OUR LUCK (II,1360)

COMBEST, PHIL
SIMON AND SIMON (II,2357); WHIZ KIDS (II,2790)

COMDEN, BETTY
APPLAUSE (I,244); BUICK CIRCUS HOUR (I,759); LET'S CELEBRATE (I,2663)

COMFORT, BOB
ALAN KING'S SECOND ANNUAL FINAL WARNING (II,30); THE BEST OF TIMES (II,220); BIG BOB JOHNSON AND HIS FANTASTIC SPEED CIRCUS (II,231); DINAH IN SEARCH OF THE IDEAL MAN (I,1278); FIRST TIME, SECOND TIME (II,863); THE JOHN BYNER COMEDY HOUR (I,2407); JUST OUR LUCK (II,1360); NICHOLS AND DYMES (II,1841); ONE NIGHT BAND (II,1909); A SPECIAL KENNY ROGERS (II,2417); WACKO (II,2733); THE WONDERFUL WORLD OF PHILIP MALLEY (II,2831)

COMICI, E L
KING'S CROSSING (II,1397)

COMICI, LUCIANO
THE POWERS OF MATTHEW STAR (II,2075)

COMPTON, SARA
SESAME STREET (I,3982)

CONATO, ANN MARIE
OUTLAW LADY (II,1935)

CONDE, ESTELLE
ZANE GREY THEATER (I,4979)

CONE, BOB
HOWDY DOODY (I,2151)

CONNELL, DAVE
MR. MAYOR (I,3143); SIGN-ON (II,2353)

CONNELL, JIM
BATTLESTAR GALACTICA (II,181)

CONNELLY, JOE
THE AMOS AND ANDY SHOW (I,179); BRINGING UP BUDDY (I,731); GOING MY WAY (I,1833); ICHABOD AND ME (I,2183); IT'S A SMALL WORLD (I,2260); LEAVE IT TO BEAVER (I,2648); THE MARGE AND GOWER CHAMPION SHOW (I,2907); MEET MR. MCNUTLEY (I,2989); MUNSTER, GO HOME! (I,3157); THE MUNSTERS (I,3158)

CONNOR, MICHAEL
GILLIGAN'S PLANET (II,994)

CONRAD, BARNABY
PLAYHOUSE 90 (I,3623)

CONRAD, EVELYN
THE HUNTER (I,2162)

CONSTANDUROS, DENIS
LITTLE WOMEN (I,2723)

CONVY, ANNE
IT'S NOT EASY (II,1256); LADIES' MAN (II,1423); THE STOCKARD CHANNING SHOW (II,2468)

CONWAY, BILL
THE TIM CONWAY COMEDY HOUR (I,4501)

CONWAY, DICK
THE CISCO KID (I,961); FAR OUT SPACE NUTS (II,831); HOW TO MARRY A MILLIONAIRE (I,2147); LEAVE IT TO BEAVER (I,2648); MAKE ROOM FOR DADDY (I,2843); THE MUNSTERS (I,3158); PARTRIDGE FAMILY: 2200 A.D. (II,1963)

CONWAY, JAMES L
MATT HOUSTON (II,1654)

CONWAY, TIM
ACE CRAWFORD, PRIVATE EYE (II,6); THE CAROL BURNETT SHOW (II,443); THE TIM CONWAY SHOW (II,2619);

UNCLE TIM WANTS YOU! (II,2704)

COOK, FIELDER
BEACON HILL (II,193)

COOK, JOE
THE GREAT SPACE COASTER (II,1059)

COOK, T S
AIRWOLF (II,27); BARETTA (II,152); THE PAPER CHASE (II,1949)

COOK, WHITFIELD
THE CLAUDETTE COLBERT SHOW (I,971); CLIMAX! (I,976); THE GEORGE SANDERS MYSTERY THEATER (I,1776); LIFE WITH FATHER (I,2690); PLAYHOUSE 90 (I,3623); YOUNG IN HEART (I,4939)

COOKE, BRIAN
FATHER DEAR FATHER (II,837); FOOT IN THE DOOR (II,902); GEORGE AND MILDRED (II,965); MAN ABOUT THE HOUSE (II,1611); ROBIN'S NEST (II,2187)

COOLEY, LEE
RED GOOSE KID'S SPECTACULAR (I,3751)

COON, GENE L
BONANZA (II,347); COMBAT (I,1011); KUNG FU (II,1416); MCHALE'S NAVY (I,2969); MEDIC (I,2976); MR. LUCKY (I,3141); RAWHIDE (I,3727); RESCUE 8 (I,3772); STAR TREK (II,2440); WAGON TRAIN (I,4747); THE WILD WILD WEST (I,4863)

COONS, HANNIBAL
THE ADDAMS FAMILY (II,11); HEY TEACHER (I,2041)

COOPER, DENNIS
THE ADVENTURES OF SUPERMAN (I,77); ST. ELSEWHERE (II,2432)

COOPER, DOROTHY FOOTE
FATHER KNOWS BEST (II,838); THE FLYING NUN (I,1611)

COOPER, HAL
THE GOOD GUYS (I,1847)

COOPER, HELEN
SCREEN DIRECTOR'S PLAYHOUSE (I,3946)

COOPER, IRVING
SCREEN DIRECTOR'S PLAYHOUSE (I,3946)

COOPER, LESTER
MAKE A WISH (I,2838)

COOPER, PAUL W
THE ABC AFTERSCHOOL SPECIAL (II,1); BOONE (II,351); FATHER MURPHY (II,841); HIGHWAY TO HEAVEN (II,1152); LITTLE HOUSE ON THE PRAIRIE (II,1487); LITTLE HOUSE: A NEW BEGINNING (II,1488); THE WALTONS (II,2740)

COOPER, WYLLIS
VOLUME ONE (I,4742)

COOPERMAN, ALVIN
CLOWNAROUND (I,980)

COOPERSMITH, JEROME
'TWAS THE NIGHT BEFORE CHRISTMAS (II,2677); THE ANDROS TARGETS (II,76); DECOY (I,1217); THE GABBY HAYES SHOW (I,1723); HAVE I GOT A CHRISTMAS FOR YOU (II,1105); HAWAII FIVE-O (II,1110); JOHNNY JUPITER (I,2429); KRAFT TELEVISION THEATER (I,2592); MR. INSIDE/MR. OUTSIDE (II,1750)

COPP, DEWITT
KRAFT TELEVISION THEATER (I,2592); SKIP TAYLOR (I,4075)

COPPEL, ALEX
ALFRED HITCHCOCK PRESENTS (I,115); TURN OF FATE (I,4623)

CORBETT, RONNIE
THE TWO RONNIES (II,2694)

CORCORAN, BOB
JIMMY HUGHES, ROOKIE COP (I,2390)

CORDAY, BARBARA
THE AMERICAN DREAM (II,71); CAGNEY AND LACEY (II,409); FISH (II,864)

CORDERY, HOWARD
THE RAY KNIGHT REVUE (I,3733)

COREA, NICHOLAS
AIRWOLF (II,27); ARCHER—FUGITIVE FROM THE EMPIRE (II,103); GAVILAN (II,959); THE INCREDIBLE HULK (II,1232); THE RENEGADES (II,2133)

COREN, SHERRY
HOT HERO SANDWICH (II,1186)

CORNELL, JOHN
THE PAUL HOGAN SHOW (II,1975)

CORRELL, CHARLES
THE AMOS AND ANDY SHOW (I,179)

CORRELL, RICH
HAPPY DAYS (II,1084); HERE'S BOOMER (II,1134)

CORRINGTON, ELAINE
ROBERT MONTGOMERY PRESENTS YOUR LUCKY STRIKE THEATER (I,3809)

CORWIN, NORMAN
NORMAN CORWIN PRESENTS (I,3307)

CORWIN, RONALD
EMERALD POINT, N.A.S. (II,773)

COSBY, BILL
THE BILL COSBY SPECIAL (I,444); THE BILL COSBY SPECIAL, OR? (I,443)

COSGROVE, MICHAEL
FOR BETTER OR WORSE (I,1621); THE LORETTA YOUNG THEATER (I,2756)

COSTIGAN, JAMES
THE CRADLE SONG (I,1088); A DOLL'S HOUSE (I,1327); DUPONT SHOW OF THE MONTH (I,1387); THE GOLDEN AGE OF TELEVISION (II,1017); THE LARK (I,2616); LITTLE MOON OF ALBAN (I,2714); LITTLE MOON OF ALBAN (I,2715); ON BORROWED TIME (I,3354); THE TURN OF THE SCREW (I,4624)

COTLER, GORDON
LANIGAN'S RABBI (II,1427); MCMILLAN AND WIFE (II,1667); MURDER INK (II,1769); ROSETTI AND RYAN: MEN WHO LOVE WOMEN (II,2217); SANCTUARY OF FEAR (II,2252)

COTTON, HELEN
STUDIO ONE (I,4268)

COWARD, NOEL
BLITHE SPIRIT (I,484); THREE IN ONE (I,4479); TONIGHT AT 8:30 (I,4556)

COWEN, ROGER
FAMILY (II,813)

COWEN, RONALD
KNOTS LANDING (II,1404)

COWGILL, LINDA
QUINCY, M. E. (II,2102)

COWLEY, WILLIAM
THE FARMER'S DAUGHTER (I,1533); HAZEL (I,1982); MAKE ROOM FOR DADDY (I,2843); PLEASE DON'T EAT THE DAISIES (I,3628); THE SLIGHTLY FALLEN ANGEL (I,4087)

COX, DAVID
HERE'S EDIE (I,2030)

COX, WILLIAM R
BONANZA (II,347); FIRESIDE THEATER (I,1580); HEC RAMSEY (II,1121); THE MAN FROM DENVER (I,2863); ROUTE 66 (I,3852); SCREEN DIRECTOR'S PLAYHOUSE (I,3946); TURN OF FATE (I,4623); WAGON TRAIN (I,4747)

COXE, GEORGE HARMON
KRAFT TELEVISION THEATER (I,2592)

COYLE, PAUL
BARNABY JONES (II,153)

CRABTREE, PAUL
STUDIO ONE (I,4268)

CRAIS, ROBERT
BARETTA (II,152); CAGNEY AND LACEY (II,409); CASSIE AND COMPANY (II,455); HILL STREET BLUES (II,1154); JOE DANCER: THE MONKEY MISSION (II,1301); QUINCY, M. E. (II,2102); RIKER (II,2175)

CRAMOY, MICHAEL
THE BARON (I,360); THE INVISIBLE MAN (I,2232); THE PRISONER (I,3670); THE RAY KNIGHT REVUE (I,3733)

CRANE, HARRY
ALAN KING'S FINAL WARNING (II,29); ALAN KING'S SECOND ANNUAL FINAL WARNING (II,30); BING CROSBY AND HIS FRIENDS (I,456); BING CROSBY AND HIS FRIENDS (II,251); CAVALCADE OF STARS (I,877); THE DEAN MARTIN SHOW (I,1201); DEAN MARTIN'S CELEBRITY ROAST (II,643); THE FRANK SINATRA SHOW (I,1665); THE JERRY LEWIS SHOW (I,2362); THE JERRY LEWIS SHOW (I,2363); THE JERRY LEWIS SHOW (I,2365); LAS VEGAS PALACE OF STARS (II,1431); THE MILTON BERLE SPECIAL (I,3050); PERRY COMO'S SPRING IN SAN FRANCISCO (II,2022); STEVE AND EYDIE CELEBRATE IRVING BERLIN (II,2456); STEVE AND EYDIE: OUR LOVE IS HERE TO STAY (II,2457); STEVE LAWRENCE AND EYDIE GORME FROM THIS MOMENT ON. . . COLE PORTER (II,2459); UPTOWN (II,2710)

CRANSTON, TOLLER
THE MAGIC PLANET (II,1598)

CRAVEN, RICHARD
WE INTERRUPT THIS SEASON (I,4780)

CRAWFORD, OLIVER
BEN CASEY (I,394); THE BLUE KNIGHT (II,276); BRONK (II,379); COUNTERPOINT (I,1059); THE FUGITIVE (I,1701); LAND OF THE GIANTS (I,2611); THE OUTER LIMITS (I,3426); RAWHIDE (I,3727); THE RIFLEMAN (I,3789); STAR TREK (II,2440); TERRY AND THE PIRATES (I,4405); VOYAGE TO THE BOTTOM OF THE SEA (I,4743); THE WILD WILD WEST (I,4863)

CRAYS, DURRELL ROYCE
THE ABC AFTERSCHOOL SPECIAL (II,1)

CREAN, ROBERT J
CORONET BLUE (I,1055); KRAFT TELEVISION THEATER (I,2592); N.Y.P.D. (I,3321)

CRETO, ALFRED
THE ALCOA HOUR (I,108)

CREWS, ROGER
RYAN'S HOPE (II,2234)

CREWSON, BILL
MR. ED (I,3137)

CRISP, N J
A FAMILY AFFAIR (II,814)

CRISWELL, KIMBER RICKENBAUGH
HOT HERO SANDWICH (II,1186)

CROCKER, BEN
THE ADVENTURES OF SUPERMAN (I,77)

CROCKER, CARTER
MADAME'S PLACE (II,1587)

CROCKER, JAMES
THE ROCKFORD FILES (II,2197); SIMON AND SIMON (II,2357); WHIZ KIDS (II,2790)

CROFT, DAVID
BEANE'S OF BOSTON (II,194)

CRONE, WILLIAM C
JIMMY HUGHES, ROOKIE COP (I,2390)

CRONIN, LEE
STAR TREK (II,2440)

CRONKITE, WALTER
WALTER CRONKITE'S UNIVERSE (II,2739)

CROSLAND, HARRY
WAKE UP (II,2736)

CROUSE, RUSSELL
ARSENIC AND OLD LACE (I,268)

CROW, FRANK
HERE COME THE STARS (I,2026); MY THREE SONS (I,3205)

CROWE, CHRISTOPHER
B.J. AND THE BEAR (II,125); B.J. AND THE BEAR (II,127); DARKROOM (II,619); THE NANCY DREW MYSTERIES (II,1789)

CROWLEY, WILLIAM
COOGAN'S REWARD (I,1044)

CRUCIS, JUDD
STAR TREK (II,2440)

CRUMP, MARTIN
THUNDERBIRDS (I,4497)

CRUTCHER, ROBERT RILEY
BEWITCHED (I,418); THE FREEWHEELERS (I,1683); HAZEL (I,1982); THREE WISHES (I,4488)

CRUTCHFIELD, LES
RAWHIDE (I,3727)

CRYSTAL, BILLY
THE BILLY CRYSTAL COMEDY HOUR (II,248); BILLY CRYSTAL: A COMIC'S LINE (II,249)

CRYSTAL, RICHARD
THE BILLY CRYSTAL COMEDY HOUR (II,248); LIFE'S MOST EMBARRASSING MOMENTS (II,1470); LIFE'S MOST EMBARRASSING MOMENTS II (II,1471); LIFE'S MOST EMBARRASSING MOMENTS III (II,1472); LIFE'S MOST EMBARRASSING MOMENTS IV (II,1473)

CUCCI, FRANK
THE ANDROS TARGETS (II,76)

CULLEN, DON
EVERYTHING GOES (I,1480)

CULP, ROBERT
THE GREATEST AMERICAN HERO (II,1060); I SPY (I,2179)

CULVER, CARMEN
FAMILY (II,813); THE FITZPATRICKS (II,868); THE LAST DAYS OF POMPEII (II,1436); THE THORN BIRDS (II,2600)

CULVER, FELIX
JESSIE (II,1287)

CULVER, JOHN
STAR TREK (II,2439)

CUMMING, TANITH
BLAKE'S SEVEN (II,264)

CUMMINGS, BOB
MY HERO (I,3193)

CUMMINS, DWIGHT
ANNIE OAKLEY (I,225)

CURA, JOHN S
SHOW BUSINESS (II,2343)

CURRAN, LEIGH
ST. ELSEWHERE (II,2432)

CURRAN, PETER
CAPTAIN SCARLET AND THE MYSTERONS (I,837)

CURRAN, VINCE
OFF THE RECORD (I,3336)

CURTIN, VALERIE
THE MARY TYLER MOORE SHOW (II,1640); PHYLLIS (II,2038)

CURTIS, DAN
IN THE DEAD OF NIGHT (I,2202)

CURTIS, JACK
BEN CASEY (I,394); THE RIFLEMAN (I,3789); TARGET: THE CORRUPTERS (I,4360); WAGON TRAIN (I,4747); THE WESTERNER (I,4801)

CURTIS, MARK
FAERIE TALE THEATER (II,807)

CURTIS, NATHANIEL
BRILLIANT BENJAMIN BOGGS (I,730); CROWN THEATER WITH GLORIA SWANSON (I,1107)

CURTIS, RICHARD
NOT THE NINE O'CLOCK NEWS (II,1864)

CUTHBERT, NEIL
ST. ELSEWHERE (II,2432)

CUTLER, STAN
9 TO 5 (II,1852); THE FARMER'S DAUGHTER (I,1533); ME AND MAXX (II,1672); SHIPSHAPE (II,2329); SUGAR TIME (II,2489); THE UGLIEST GIRL IN TOWN (I,4663)

D'AMOUR, LOUIS
COUNTERPOINT (I,1059)

D'AVRAY, BILL
FLAMINGO ROAD (II,872)

DASILVA, HOWARD
WALTER FORTUNE (I,4758)

DAFFAN, HAL
LOVE, SIDNEY (II,1547)

DAGENAIS, TOM
THE ALL-NEW POPEYE HOUR (II,49); THE FLINTSTONE FUNNIES (II,883); THE MOUSE FACTORY (I,3113)

DALES, ARTHUR
BEN CASEY (I,394); DAN AUGUST (I,1130); THE FBI (I,1551); I SPY (I,2179); LAW OF THE PLAINSMAN (I,2639); REX HARRISON PRESENTS SHORT STORIES OF LOVE (II,2148); THE SMALL MIRACLE (I,4090)

DALEY, BILL
DEAN MARTIN'S CELEBRITY ROAST (II,643); HERE'S BOOMER (II,1134); LADIES AND GENTLEMAN. . . BOB NEWHART (II,1420); TOO CLOSE FOR COMFORT (II,2642)

DALEY, MADLYN
PHYL AND MIKHY (II,2037)

DALLENBACK, WALTER
THE BLUE KNIGHT (II,276); HART TO HART (II,1102); MCCLAIN'S LAW (II,1659); THE ROCKFORD FILES (II,2197)

DALTON, WALLY
BABY, I'M BACK! (II,130); BARNEY MILLER (II,154); IT'S A LIVING (II,1254); LAVERNE AND SHIRLEY (II,1446); MUSIC HALL AMERICA (II,1773); SEMI-TOUGH (II,2300)

DALY, JONATHAN
ADAMS OF EAGLE LAKE (II,10)

DALZELL, BILL
MELODY STREET (I,3000)

DAMES, BOB
BENSON (II,208); CONDO (II,565)

DAN, LIN
THE OUTER LIMITS (I,3426)

DANA, BILL
DON KNOTTS NICE CLEAN, DECENT, WHOLESOME HOUR (I,1333); DONNY AND MARIE (II,712); THE LAS VEGAS SHOW (I,2619); A LOOK AT THE LIGHT SIDE (I,2751); SPEAK UP AMERICA (II,2412); WINDOWS, DOORS AND KEYHOLES (II,2806)

DANCH, BILL
LASSIE'S RESCUE RANGERS (II,1432); MISSION MAGIC (I,3068); MY FAVORITE MARTIANS (II,1779); SHAZAM! (II,2319)

DANDRIDGE, FRANK
BARETTA (II,152); THE INCREDIBLE HULK (II,1232); PALMERSTOWN U.S.A. (II,1945)

DANE, JEFF
ANOTHER LIFE (II,95)

DANGERFIELD, RODNEY
RODNEY DANGERFIELD SHOW: I CAN'T TAKE IT NO MORE (II,2198); RODNEY DANGERFIELD SPECIAL: IT'S NOT EASY BEIN' ME (II,2199)

DANIELS, CHARLES
MARCUS WELBY, M.D. (II,1627)

DANIELS, DARI
MR. SMITH (II,1755)

DANIELS, MARC
STAR TREK (II,2439)

DANIELS, STAN
BROTHERS (II,380); THE DEAN MARTIN SHOW (I,1201); DOM DELUISE AND FRIENDS (II,701); GENE KELLY'S WONDERFUL WORLD OF GIRLS (I,1752); THE GOLDDIGGERS IN LONDON (I,1837); JACK BENNY'S FIRST FAREWELL SHOW (I,2300); THE MARY TYLER MOORE SHOW (II,1640); PHYLLIS (II,2038); WHAT'S UP, AMERICA? (I,4824)

DANN, SAM
KRAFT TELEVISION THEATER (I,2592)

DANTON, RAY
MCCLOUD (II,1660)

DANUS, CHRISTIAN
THE NIGHTENGALES (II,1850)

DANUS, RICHARD
MATT HOUSTON (II,1654); THE ROOKIES (II,2208)

DANZIGER, DENNIS
AFTER GEORGE (II,19); EMPIRE (II,777); KATE AND ALLIE (II,1365); OPEN ALL NIGHT (II,1914)

DARBON, LESLIE
DEPARTMENT S (I,1232)

DARLING, JEAN
DINAH IN SEARCH OF THE IDEAL MAN (I,1278)

DAVENPORT, M L
SWISS FAMILY ROBINSON (I,4324)

DAVENPORT, WILLIAM
ALL IN THE FAMILY (II,38); APPLE PIE (II,100); GOOD TIMES (II,1038); HANGING IN (II,1078); MAKE ROOM FOR DADDY (I,2843); MAUDE (II,1655); MICKEY AND THE CONTESSA (I,3024); MRS. G. GOES TO COLLEGE (I,3155); PINE LAKE LODGE (I,3597); THE SON-IN-LAW (II,2398); THE TWO OF US (II,2693)

DAVID, MADLYN
VALENTINE MAGIC ON LOVE ISLAND (II,2716)

DAVID, PAUL
MR. LUCKY (I,3141); ONE STEP BEYOND (I,3388); WAGON TRAIN (I,4747)

DAVID, RICHARD HARDING
FIRESIDE THEATER (I,1580)

DAVIDSON, BILL
THE ART OF CRIME (II,108)

DAVIDSON, DAVID
THE ALCOA HOUR (I,108); KRAFT TELEVISION THEATER (I,2592); N.Y.P.D. (I,3321); SAINTS AND SINNERS (I,3887); STUDIO ONE (I,4268)

DAVIDSON, LEE
SWITCH (II,2519)

DAVIDSON, LEW
SECRET AGENT (I,3962)

DAVIDSON, MARTIN
THE BOB NEWHART SHOW (II,342); FLATBUSH/AVENUE J (II,878)

DAVIS, DAVID
BUMPERS (II,394); RHODA (II,2151); WEEKEND (I,4788)

DAVIS, DEBORAH
FANTASY ISLAND (II,829); THE INCREDIBLE HULK (II,1232); KNIGHT RIDER (II,1402)

DAVIS, DONALD
CYRANO DE BERGERAC (I,1119); THE FUNNIEST JOKE I EVER HEARD (II,939)

DAVIS, ELIAS
CARTER COUNTRY (II,449); THE DON KNOTTS SHOW (I,1334); DON RICKLES—ALIVE AND KICKING (I,1340); THE EVERLY BROTHERS SHOW (I,1479); GOODNIGHT, BEANTOWN (II,1041); HOLLYWOOD HIGH (II,1163); HOT L BALTIMORE (II,1187); THE LENNON SISTERS SHOW (I,2656); M*A*S*H (II,1569); PAUL SAND IN FRIENDS AND LOVERS (II,1982); ROBERT YOUNG AND THE FAMILY (I,3813); THE SHOW MUST GO ON (II,2344); THE SINGERS (I,4060); SUPER COMEDY BOWL 2 (I,4289); THAT'S MY MAMA (II,2580)

DAVIS, GEENA
BUFFALO BILL (II,387)

DAVIS, GERRY
DOCTOR WHO (II,689); VEGAS (II,2724)

DAVIS, JAMES P
SUSPICION (I,4309)

DAVIS, JEFFREY
THE LOVE BOAT (II,1535)

DAVIS, JERRY
ACE (II,5); ALCOA/GOODYEAR THEATER (I,107); BACHELOR FATHER (I,323); THE BARBARA EDEN SHOW (I,346); BARNEY MILLER (II,154); BIG EDDIE (II,236); THE FARMER'S DAUGHTER (I,1533); MCHALE'S NAVY (I,2969); THE NEW LORENZO MUSIC SHOW (II,1821); SYBIL (I,4327)

DAVIS, LUTHER
ARSENIC AND OLD LACE (I,270)

DAVIS, MADELYN
ALL ABOUT BARBARA (I,125); DOROTHY (II,717); HERE'S LUCY (II,1135); I LOVE LUCY (I,2171); THE LUCILLE BALL SPECIAL (II,1559)

DAVIS, MICHAEL P
LET ME TELL YOU ABOUT A SONG (I,2661)

DAVIS, PHIL
THE ANN SOTHERN SHOW (I,220); THE BOB NEWHART SHOW (II,342); DOBIE GILLIS (I,1302); DOC (II,683); FATHER KNOWS BEST (II,838); MR. ED (I,3137); PRIVATE SECRETARY (I,3672); TRUTH OR CONSEQUENCES (I,4620)

DAVIS, STAN
MAYOR OF HOLLYWOOD (I,2962); THE MILTON BERLE SHOW (I,3049); MILTON BERLE'S MAD MAD MAD WORLD OF COMEDY (II,1701); MY THREE SONS (I,3205); THE ROY ROGERS AND DALE EVANS SHOW (I,3858); SHOW BUSINESS SALUTE TO MILTON BERLE (I,4029)

DAVIS, TOM
BOB & RAY & JANE, LARAINE & GILDA (II,279); THE CONEHEADS (II,567); THE NEW SHOW (II,1828); THE PAUL SIMON SPECIAL (II,1983); STEVE MARTIN'S BEST SHOW EVER (II,2460)

DAVIS, VALENTINE
THE GINGER ROGERS SHOW (I,1805)

DAVIS, WALTER
RAPTURE AT TWO-FORTY (I,3724)

DAWKINS, JOHNNY
LOU GRANT (II,1526)

DAWSON, DEBORAH ZOE
TRAPPER JOHN, M.D. (II,2654)

DAWSON, GORDON
BRET MAVERICK (II,374); CATALINA C-LAB (II,458); LOU GRANT (II,1526); THE ROCKFORD FILES (II,2197); SERGEANT T.K. YU (II,2305)

DAY, DOROTHY
THE KATE SMITH EVENING HOUR (I,2507); THE KATE SMITH HOUR (I,2508)

DAY, GERRY
BARNABY JONES (II,153); A CHRISTMAS FOR BOOMER (II,518); FALCON CREST (II,808); FAMILY (II,813); HAWAII FIVE-O (II,1110); LITTLE HOUSE ON THE PRAIRIE (II,1487); MATT HELM (II,1653); THE NEW ADVENTURES OF PERRY MASON (I,3252); THE ROOKIES (II,2208); WAGON TRAIN (I,4747); WILLOW B: WOMEN IN PRISON (II,2804)

DAY, LINDA
TAXI (II,2546)

DAY, MAE
THE FARMER'S DAUGHTER (I,1533)

DAY, RUDY
FROM HERE TO ETERNITY (II,933)

DAYLE, LOREN
RESCUE 8 (I,3772)

DEBELLIS, JOHN
THE BILLY CRYSTAL COMEDY HOUR (II,248); THE JOE PISCOPO SPECIAL (II,1304)

DEBLASIO, EDWARD
DYNASTY (II,746); PARIS (II,1957); THE STREETS OF SAN FRANCISCO (II,2478)

DEBOSIO, GIANFRANCO
MOSES THE LAWGIVER (II,1737)

DECOURCEY, KEVIN
THE WILD WILD WEST (I,4863)

DEGAW, BOYCE
TELE PUN (I,4376)

DEGENNARO, CARI ANNE
WONDER GIRL (I,4891)

DEGENNARO, TRICIA
OUTLAW LADY (II,1935)

DEGUERE, PHILIP
BARETTA (II,152); THE BIONIC WOMAN (II,255); DOCTOR STRANGE (II,688); THE GYPSY WARRIORS (II,1070); THE LAST CONVERTIBLE (II,1435); SIMON AND SIMON (II,2357); WHIZ KIDS (II,2790)

DEHARTOG, JAN
THE FOUR POSTER (I,1651)

DEKOKER, RICHARD
THE GANGSTER CHRONICLES (II,957)

DEKOVEN, DAVID
MEET YOUR COVER GIRL (I,2992)

DELAURENTIS, RAYMOND
ST. ELSEWHERE (II,2432)

DELAURENTIS, ROBERT
BIG CITY BOYS (II,232); ST. ELSEWHERE (II,2432)

DEPRIEST, MARGARET
BEHIND THE SCREEN (II,200)

DEROY, RICHARD
79 PARK AVENUE (II,2310); ALCOA PREMIERE (I,109); BELL, BOOK AND CANDLE (II,201); DUFFY (II,739); THE FLYING NUN (I,1611); HART TO HART (II,1102); MR. NOVAK (I,3145); THE PARTRIDGE FAMILY (II,1962); THE RAT PATROL (I,3726); REMINGTON STEELE (II,2130); SONS AND DAUGHTERS (II,2404)

DESOUSA, EDWARD
FOUL PLAY (II,910)

DESOUZA, STEVEN E
KNIGHT RIDER (II,1402); THE POWERS OF MATTHEW STAR (II,2075); THE RENEGADES (II,2132); THE SIX-MILLION-DOLLAR MAN (II,2372)

DEWITT, JACK
CLIMAX! (I,976); CROSSROADS (I,1105)

DEAN, BARTON
CARLTON YOUR DOORMAN (II,440); NEWHART (II,1835); OH MADELINE (II,1880); TAXI (II,2546)

DEAN, FREEMAN
STARR, FIRST BASEMAN (I,4204)

DEBONO, JERRY
MARCUS WELBY, M.D. (II,1627); RAFFERTY (II,2105)

DEL GRANDE, LOUIS
SEEING THINGS (II,2297)

DEL, LANE
COVER UP (II,597)

DELLIGAN, WILLIAM
ALL MY CHILDREN (II,39)

DELLINGER, ROBERT
THE BLUE KNIGHT (II,276); THE FEATHER AND FATHER GANG (II,845)

DELOUS, PAUL M
THE INCREDIBLE HULK (II,1232)

DELPH, DAGNY
ROBERT MONTGOMERY PRESENTS YOUR LUCKY STRIKE THEATER (I,3809)

DELUCA, RUDY
ACE CRAWFORD, PRIVATE EYE (II,6); THE CAROL BURNETT SHOW (II,443); COMEDY NEWS II (I,1021); HOT L BALTIMORE (II,1187); THE INVESTIGATORS (II,1241); IT ONLY HURTS WHEN YOU LAUGH (II,1251); THE JOHN BYNER COMEDY HOUR (I,2407); THE MARTY FELDMAN COMEDY MACHINE (I,2929); NUTS AND BOLTS (II,1871); PEEPING TIMES (II,1990); THE RICH LITTLE SHOW (II,2153); THE RICH LITTLE SHOW (II,2154); STOPWATCH: THIRTY MINUTES OF INVESTIGATIVE TICKING (II,2474); THE TIM CONWAY COMEDY HOUR (I,4501)

DEMAS, CAROLE
A MAGIC GARDEN CHRISTMAS (II,1590)

DENIM, SUE
KEEP ON TRUCKIN' (II,1372)

DENKER, HENRY
GIVE US BARABBAS! (I,1816); KRAFT TELEVISION THEATER (I,2592)

DENMARK, WILTON
THE BIONIC WOMAN (II,255); THE SIX-MILLION-DOLLAR MAN (II,2372); WONDER WOMAN (SERIES 2) (II,2829)

DENNIS, ROBERT C
ALFRED HITCHCOCK PRESENTS (I,115); BARNABY JONES (II,153); BATMAN (I,366); CANNON (II,424); CARIBE (II,439); CHARLIE'S ANGELS (II,486); DAN AUGUST (I,1130); EXPERT WITNESS (I,1488); THE FBI (I,1551); THE FUGITIVE (I,1701); THE HARDY BOYS MYSTERIES (II,1090); HARRY O (II,1099); I SPY (I,2179); MANHUNTER (II,1621); THE OUTER LIMITS (I,3426); PERRY MASON (II,2025); RAFFERTY (II,2105); THE RELUCTANT SPY (I,3763); THE RETURN OF CAPTAIN NEMO (II,2135); THE RIFLEMAN (I,3789); THE SIX-MILLION-DOLLAR MAN (II,2372); TURN OF FATE (I,4623); THE UNTOUCHABLES (I,4682); THE WILD WILD WEST (I,4863)

DENOFF, MAC
WASHINGTON SQUARE (I,4772)

DENOFF, SAM
BIG EDDIE (II,235); THE BILL COSBY SPECIAL (I,444); THE BOYS (II,360); THE CONFESSIONS OF DICK VAN DYKE (II,568); DICK VAN DYKE AND THE OTHER WOMAN, MARY TYLER MOORE (I,1273); THE FABULOUS FUNNIES (II,801); THE FIRST NINE MONTHS ARE THE HARDEST (I,1584); THE FUNNY SIDE (I,1714); THE FUNNY SIDE (I,1715); GOOD MORNING WORLD (I,1850); THE JULIE ANDREWS SHOW (I,2481); THE LATE FALL, EARLY SUMMER BERT CONVY SHOW (II,1441); THE MAN WHO CAME TO DINNER (I,2881); MY WIFE NEXT DOOR (II,1780); ON OUR OWN (II,1892); PURE GOLDIE (I,3696); THE SID CAESAR, IMOGENE COCA, CARL REINER, HOWARD MORRIS SPECIAL (I,4039)

DERMAN, BILL
THE PINKY LEE SHOW (I,3601)

DERMAN, LOU
ALL IN THE FAMILY (II,38); GOOD TIMES (II,1038); HERE'S LUCY (II,1135); LIFE WITH LUIGI (I,2692); THE MILTON BERLE SPECIAL (I,3052); MR. ED (I,3137); PINE LAKE LODGE (I,3597)

DESPRES, LORAINE
CHIPS (II,511); DALLAS (II,614); DYNASTY (II,746)

DETIEGE, DAVID
THE ODDBALL COUPLE (II,1876)

DEUTSCH, HELEN
THE GENERAL MOTORS 50TH ANNIVERSARY SHOW (I,1758); JACK AND THE BEANSTALK (I,2287)

DEVENNEY, SCOTT
MATINEE AT THE BIJOU (II,1651)

DEWITT, COPP
ONE STEP BEYOND (I,3388)

DEXTER, JOY
THE FUGITIVE (I,1701)

DICENZO, CHARLES
EMERALD POINT, N.A.S. (II,773)

DIMAGGIO-WAGNER, MADELINE
LACY AND THE MISSISSIPPI QUEEN (II,1419)

DIMARCO, TONY
THE BIONIC WOMAN (II,255); BLANSKY'S BEAUTIES (II,267); LOTTERY (II,1525); ON OUR OWN (II,1892); WONDER WOMAN (SERIES 1) (II,2828); WONDER WOMAN (SERIES 2) (II,2829)

DIMISSA JR, E
THE LOVE REPORT (II,1542); THICKE OF THE NIGHT (II,2587)

DISTEFANO, DAN
SPORT BILLY (II,2429)

DIZENO, CHUCK
AS THE WORLD TURNS (II,110)

DIZENO, PAT
AS THE WORLD TURNS (II,110)

DIAL, BILL
HARPER VALLEY (II,1094); LEGMEN (II,1458); LOBO (II,1504); SIMON AND SIMON (II,2357); WKRP IN CINCINNATI (II,2814)

DIAL, WALTER ALLEN
THE TONY RANDALL SHOW (II,2640)

DIAMOND, DEE
KEEFE BRASSELLE'S VARIETY GARDEN (I,2515)

DIAMOND, JANIS
I AM THE GREATEST: THE ADVENTURES OF MUHAMMED ALI (II,1211)

DIAMOND, MEL
THE ANN SOTHERN SHOW (I,220); BACHELOR FATHER (I,323); THE DONALD O'CONNOR SHOW (I,1343);

JACK CARTER AND COMPANY (I,2305); JACKIE GLEASON AND HIS AMERICAN SCENE MAGAZINE (I,2321); THE JIMMY DURANTE SHOW (I,2389); THE KATE SMITH EVENING HOUR (I,2507); THE KATE SMITH HOUR (I,2508); LEAVE IT TO BEAVER (I,2648); THE STEVE LAWRENCE SHOW (I,4227)

DIAMOND, SELMA
AQUA VARIETIES (I,248); THE BIG PARTY FOR REVLON (I,427); THE KRAFT MUSIC HALL (I,2587); THE MILTON BERLE SHOW (I,3047); TEXACO STAR THEATER (I,4411)

DICK, PEGGY CHANTLER
THE FARMER'S DAUGHTER (I,1533); THE PARTRIDGE FAMILY (II,1962)

DICKS, TERRENCE
DOCTOR WHO (II,689); DOCTOR WHO—THE FIVE DOCTORS (II,690)

DIGGS, ELIZABETH
CLOSE TIES (II,545); ST. ELSEWHERE (II,2432)

DILANNO, GREGORY
KNIGHT RIDER (II,1402)

DILBERT, BERNARD
MAMA MALONE (II,1609)

DILNIK, GEORGE S
HERE'S BOOMER (II,1134)

DILNIK, JOHN F
HERE'S BOOMER (II,1134)

DIMITRI, RICHARD
GOING BANANAS (II,1015); ROOSEVELT AND TRUMAN (II,2209)

DIMSDALE, HOWARD
THE HEALERS (II,1115); MEDICAL CENTER (II,1675); PLANET OF THE APES (II,2049); THE SIX-MILLION-DOLLAR MAN (II,2372)

DINALLO, GREGORY S
FAMILY (II,813); HAWAIIAN HEAT (II,1111); HUNTER (II,1205); MRS. COLUMBO (II,1760); THE POWERS OF MATTHEW STAR (II,2075); THE RENEGADES (II,2133); SCARECROW AND MRS. KING (II,2268); THE SIX-MILLION-DOLLAR MAN (II,2372); WALKING TALL (II,2738)

DINEHART, ALAN
GILLIGAN'S ISLAND (II,990); IT'S ABOUT TIME (I,2263)

DINELLI, MEL
LAURA (I,2635); THE UNEXPLAINED (I,4675)

DINI, PAUL
SPORT BILLY (II,2429)

DINSDALE, SHIRLEY
JUDY SPLINTERS (I,2475)

DIPEGO, GERALD
THE MANY LOVES OF ARTHUR (II,1625); THE SHERIFF AND THE ASTRONAUT (II,2326)

DISHELL, WALTER
M*A*S*H (II,1569)

DISNEY, DORIS MILES
KRAFT TELEVISION THEATER (I,2592)

DITTMAN, DEB
SCRABBLE (II,2278)

DIVINE, JERRY
THE SIX-MILLION-DOLLAR MAN (II,2372)

DIXON, BUZZ
GOLDIE GOLD AND ACTION JACK (II,1023); THUNDARR THE BARBARIAN (II,2616)

DIXON, PAUL
THE WALTONS (II,2740)

DIXON, PETER
THE ADVENTURES OF SUPERMAN (I,77); FATHER MURPHY (II,841); LITTLE HOUSE ON THE PRAIRIE (II,1487); SKYHAWKS (I,4079)

DIXON, SARAH
THE WALTONS (II,2740)

DIZON, JESSE
THE LOVE BOAT (II,1535)

DOBBS, CHARLOTTE
HAPPY DAYS (II,1084); LAVERNE AND SHIRLEY (II,1446)

DOBSON, BRIDGET
AS THE WORLD TURNS (II,110); SANTA BARBARA (II,2258)

DOBSON, JEROME
AS THE WORLD TURNS (II,110); SANTA BARBARA (II,2258)

DOCHTERMANN, RUDY
240-ROBERT (II,2688); CHIPS (II,511); LOTTERY (II,1525)

DOHERTY, JAMES
BLUE THUNDER (II,278); SAM (II,2247)

DOHERTY, JOHN
THE BARBARY COAST (II,146)

DOLAN, HARRY
THE PARTRIDGE FAMILY (II,1962)

DOLINSKY, MYER
CANNON (II,424); THE FARMER'S DAUGHTER (I,1533); HAWAII FIVE-O (II,1110); STAR TREK (II,2440)

DOLMAN, BOB
LITTLE SHOTS (II,1495)

DOLMAN, RICHARD
WKRP IN CINCINNATI (II,2814)

DONALDSON, MARTIN
BROCK'S LAST CASE (I,742)

DONATAN, DOROTHY
BEN CASEY (I,394)

DONIGER, WALTER
CROWN THEATER WITH GLORIA SWANSON (I,1107); THE DICK POWELL SHOW (I,1269); JIGSAW (I,2377); PERRY MASON (II,2025); SQUADRON (I,4170); STORY THEATER (I,4240)

DONLEY, JOHN
DIFF'RENT STROKES (II,674); FLIP WILSON. . . OF COURSE (II,890); GOOD TIMES (II,1038)

DONNELL, KATHY
LACY AND THE MISSISSIPPI QUEEN (II,1419); THREE'S COMPANY (II,2614); THE TONY RANDALL SHOW (II,2640); WE'VE GOT EACH OTHER (II,2757)

DONNER, JILL SHERMAN
CUTTER TO HOUSTON (II,612)

DONOVAN, MARTIN
THE FARMER'S DAUGHTER (I,1533); HUSBANDS, WIVES AND LOVERS (II,1209); THE MARY TYLER MOORE SHOW (II,1640); MR. MERLIN (II,1751); TABITHA (II,2528); THE UGLIEST GIRL IN TOWN (I,4663)

DONOVAN, MICHAEL
THE RANGERS (II,2112); SAM (II,2247)

DORAN, PHIL
THE BILLY CRYSTAL COMEDY HOUR (II,248); THE BOB NEWHART SHOW (II,342); CARTER COUNTRY (II,449); FAMILY BUSINESS (II,815); THE MERV GRIFFIN SHOW (I,3016); TONY ORLANDO AND DAWN (II,2639); TOO CLOSE FOR COMFORT (II,2642)

DORFMAN, PHIL
MAMA MALONE (II,1609)

DORFMAN, SID
BIG EDDIE (II,236); THE BING CROSBY SHOW (I,466); THE GEORGE BURNS AND GRACIE ALLEN SHOW (I,1763); GOOD TIMES (II,1038); THE LIFE OF RILEY (I,2685); M*A*S*H (II,1569); ROLL OUT! (I,3833); THE RUBBER GUN SQUAD (II,2225); THREE FOR THE ROAD (II,2607); WHERE THERE'S SMOKEY (I,4832)

DORNFELD, IRIS
ALCOA PREMIERE (I,109)

DORTORT, DAVID
BONANZA (II,347); HUNTER'S MOON (II,1207); SAM HILL (I,3895); SCREEN DIRECTOR'S PLAYHOUSE (I,3946)

DOUGHERTY, DON
THE FLINTSTONE FUNNIES (II,883)

DOUGLAS, JACK
THE GEORGE GOBEL SHOW (I,1767); THE JACK PAAR SHOW (I,2316); JACK PAAR TONIGHT (I,2320); THE SOLDIERS (I,4112); VICTOR BORGE'S COMEDY IN MUSIC III (I,4725); THE WONDERFUL WORLD OF JACK PAAR (I,4899)

DOUGLAS, WARREN
MCCLOUD (II,1660); THIS IS YOUR LIFE (I,4461)

DOUMANIAN, JEAN
BOB & RAY & JANE, LARAINE & GILDA (II,279)

DOWD, EARL
THE DONNY AND MARIE OSMOND SHOW (II,714); EIGHT IS ENOUGH (II,762); FAR OUT SPACE NUTS (II,831)

DOWLING, CHET
ALICE (II,33); AMERICA, YOU'RE ON (II,69); BABY, I'M BACK! (II,130); BUSTING LOOSE (II,402); DON'T CALL ME MAMA ANYMORE (I,1350); DONNY AND MARIE (II,712); THE DONNY AND MARIE OSMOND SHOW (II,714); GIMME A BREAK (II,995); MADHOUSE 90 (I,2807); THE PAUL LYNDE COMEDY HOUR (II,1978); ROWAN AND MARTIN'S LAUGH-IN (I,3856); SPENCER (II,2420); TOP OF THE MONTH (I,4578); WE GOT IT MADE (II,2752)

DOWN, RENA

KING'S CROSSING (II,1397); KNOTS LANDING (II,1404); NURSE (II,1870)

DOWNES, BARRY
STARS AND STRIPES SHOW (II,2442); THE STARS AND STRIPES SHOW (II,2443)

DOWNEY, JAMES
THE NEW SHOW (II,1828); STEVE MARTIN'S BEST SHOW EVER (II,2460)

DOWNING, STEPHEN
FOR LOVE AND HONOR (II,903); FORCE SEVEN (II,908); MCCLAIN'S LAW (II,1659); MICKEY SPILLANE'S MIKE HAMMER: MORE THAN MURDER (II,1693); NERO WOLFE (II,1806); T.J. HOOKER (II,2524); WALKING TALL (II,2738)

DOYLE, GORDON
THE PAUL LYNDE COMEDY HOUR (II,1977)

DOYLE-MURRAY, BRIAN
THE CHEVY CHASE SHOW (II,505)

DOZER, DAVID
SZYSZNYK (II,2523)

DOZIER, ROBERT
BATMAN (I,366); THE CONTENDER (II,571); DAN AUGUST (I,1130); THE DEVLIN CONNECTION (II,664); FRONT ROW CENTER (I,1694); INSPECTOR PEREZ (II,1237); THRILLER (I,4492)

DRAKE, ERVIN
SONG SNAPSHOTS ON A SUMMER HOLIDAY (I,4122)

DRAKE, OLIVER
THE ADVENTURES OF SUPERMAN (I,77)

DRAKE, THOMAS
JIGSAW (I,2377)

DRANDEL, MARC
SECRET AGENT (I,3962)

DRATLER, JAY
BURKE'S LAW (I,762); NAKED CITY (I,3216)

DREBEN, STAN
THE BEAUTIFUL PHYLLIS DILLER SHOW (I,381); F TROOP (I,1499); THE FACTS OF LIFE (II,805); FUNNY YOU SHOULD ASK (I,1717); JOHNNY CARSON PRESENTS THE SUN CITY SCANDALS (I,2420); LETTERS TO LAUGH-IN (I,2672); MCHALE'S NAVY (I,2969); THE MILTON BERLE SPECIAL (I,3050); PAT BOONE IN HOLLYWOOD (I,3469); THE PAUL WINCHELL AND JERRY MAHONEY SHOW (I,3492); TAKE MY ADVICE (II,2531); YOU'RE ONLY YOUNG TWICE (I,4971); YOUR FUNNY FUNNY FILMS (I,4950)

DRESNER, HAL
CATCH 22 (I,871); THE HARVEY KORMAN SHOW (II,1103); HUSBANDS AND WIVES (II,1208); HUSBANDS, WIVES AND LOVERS (II,1209); JERRY (II,1283); POOR RICHARD (II,2065)

DREW, JOHN
PRISONER: CELL BLOCK H (II,2085)

DREW, RICK
THE PAUL ANKA SHOW (II,1972)

DREXLER, ROSALYN
LILY (I,2705)

DREYFUSS, LORIN
REGGIE (II,2127)

DREYFUSS, MICHAEL
THE ALCOA HOUR (I,108)

DRISCOLL, DONALD
THE ALCOA HOUR (I,108)

DRISKILL, WILLIAM
COLUMBO (II,556); THE FEATHER AND FATHER GANG: NEVER CON A KILLER (II,846); GRIFF (I,1888); INSTITUTE FOR REVENGE (II,1239); MCMILLAN AND WIFE (II,1667); PARTNERS IN CRIME (II,1961); RISKO (II,2180); THE SIX-MILLION-DOLLAR MAN (II,2372); TALES OF THE GOLD MONKEY (II,2537)

DRUE, CYNDY
THE ROCK 'N ROLL SHOW (II,2188)

DRUGSTER, PETER
HAROLD LLOYD'S WORLD OF COMEDY (II,1093)

DRUYAN, ANN
COSMOS (II,579)

DRYDEN, MACK
ALL NIGHT RADIO (II,40)

DUANE, DIANE
THE FLINTSTONE FUNNIES (II,883)

DUBOIS, DOOLES
WELCOME TO THE FUN ZONE (II,2763)

DUBOV, PAUL
BACKSTAIRS AT THE WHITE HOUSE (II,133); EIGHT IS ENOUGH (II,762); HONEY WEST: WHO KILLED THE JACKPOT? (I,2106); WONDER WOMAN (SERIES 1) (II,2828)

DUCLON, DAVID W
BUSTING LOOSE (II,402); DOUBLE TROUBLE (II,724); THE JEFFERSONS (II,1276); LAVERNE AND SHIRLEY (II,1446); THE ODD COUPLE (II,1875); PUNKY BREWSTER (II,2092); SILVER SPOONS (II,2355); WORKING STIFFS (II,2834)

DUCOMMUN, RICK
ROCK-N-AMERICA (II,2195)

DUDDY, LYNN
THE VAUGHN MONROE SHOW (I,4708)

DUDLEY, PAUL
THE FRANK SINATRA SHOW (I,1665)

DUFF, WARREN
AMOS BURKE, SECRET AGENT (I,180); DAN AUGUST (I,1130); THE FBI (I,1551); THE INVADERS (I,2229); THE ROGUES (I,3832); THREE FOR DANGER (I,4475)

DUFFY, ALBERT
COUNTERPOINT (I,1059); CROWN THEATER WITH GLORIA SWANSON (I,1107); THE LONE RANGER (I,2740)

DUGAN, JOHN T
APPLE'S WAY (II,101); THE BLUE KNIGHT (II,276); BONANZA (II,347); COLUMBO (II,556); THE COPS AND ROBIN (II,573); FATHER MURPHY (II,841); THE FEATHER AND FATHER GANG (II,845); KUNG FU (II,1416); LITTLE HOUSE ON THE PRAIRIE (II,1487)

DUKE, BILL
GOOD TIMES (II,1038)

DUKE, ROBIN
SECOND CITY TELEVISION (II,2287)

DUNAWAY, DON CARLOS
BARETTA (II,152); THE ROCKFORD FILES (II,2197)

DUNCAN, BOB
ALCOA/GOODYEAR THEATER (I,107); LOST IN SPACE (I,2758); MATINEE THEATER (I,2947); ONE STEP BEYOND (I,3388); THE TIME TUNNEL (I,4511)

DUNCAN, WANDA
ALCOA/GOODYEAR THEATER (I,107); LOST IN SPACE (I,2758); MATINEE THEATER (I,2947); ONE STEP BEYOND (I,3388); THE TIME TUNNEL (I,4511)

DUNGAN, FRANK
MR. MOM (II,1752)

DUNKEL, JOHN
ONE STEP BEYOND (I,3388); RAWHIDE (I,3727); WAGON TRAIN (I,4747); THE WALTONS (II,2740); THE WESTERNER (I,4801)

DUNLOP, PAT
FROM A BIRD'S EYE VIEW (I,1686); A MAN CALLED SLOANE (II,1613); VEGAS (II,2724)

DUNN, JOHN W
THE ODDBALL COUPLE (II,1876)

DUNNE, JAMES P
HAPPY DAYS (II,1084); JOANIE LOVES CHACHI (II,1295)

DUNNING, PHILIP
BROADWAY (I,733)

DUNSMUIR, TOM
AMERICA 2-NIGHT (II,65); THE CAPTAIN AND TENNILLE (II,429); I'VE HAD IT UP TO HERE (II,1221); SESAME STREET (I,3982)

DURANG, CHRISTOPHER
THE COMEDY ZONE (II,559)

DURHAM, EARL
PEOPLE DO THE CRAZIEST THINGS (II,1997); PEOPLE DO THE CRAZIEST THINGS (II,1998); THE TONI TENNILLE SHOW (II,2636)

DURKEE, WILLIAM F
THE LORETTA YOUNG THEATER (I,2756); PLAYWRIGHTS '56 (I,3627)

DUXBURY, LESLIE
CORONATION STREET (I,1054)

DWAN, ALLAN
IT'S ALWAYS SUNDAY (I,2265)

DWAN, ROBERT
THE GENERAL FOODS 25TH ANNIVERSARY SHOW (I,1756)

DWIGHT, CURTIS
THE WALTONS (II,2740)

DWORSKI, DAVID
ISIS (II,1248)

DWORSKI, SUSAN
ISIS (II,1248)

DYER, BILL
AN EVENING WITH DIANA ROSS (II,788); MAC DAVIS. . . I BELIEVE IN CHRISTMAS (II,1581)

DYKERS, REAR ADMIRAL THOMAS
THE SILENT SERVICE (I,4049)

DYNE, MICHAEL
THE FILE ON DEVLIN (I,1575); JANE EYRE (I,2338); KRAFT TELEVISION THEATER (I,2592); STUDIO ONE (I,4268); VICTORY (I,4729)

DYNE, ROBERT
DUPONT SHOW OF THE MONTH (I,1387)

DYSLIN, GEORGE
KRAFT TELEVISION THEATER (I,2592)

EARLL, ROBERT
CHARLIE'S ANGELS (II,486); HILL STREET BLUES (II,1154); KOJAK (II,1406); THE NIGHT STALKER (II,1849); THE POWERS OF MATTHEW STAR (II,2075); THE RENEGADES (II,2133); STARSKY AND HUTCH (II,2444); SWITCH (II,2519); VEGAS (II,2724)

EARLY, CHARLES M
SHERLOCK HOLMES (I,4010)

EARNSHAW, FENTON
THE SWORD (I,4325)

EATON, DONALD
HART TO HART (II,1102)

EBB, FRED
BARYSHNIKOV ON BROADWAY (II,158); GOLDIE AND LIZA TOGETHER (II,1022); LIZA WITH A Z (I,2732); MAGNAVOX PRESENTS FRANK SINATRA (I,2826); THREE FOR THE GIRLS (I,4477)

ECKSTEIN, GEORGE
THE FUGITIVE (I,1701); HEAVEN ON EARTH (II,1119); THE HOUSE ON GREENAPPLE ROAD (I,2140); THE INVADERS (I,2229); THE UNTOUCHABLES (I,4682)

EDDO, NANCY
JOANIE LOVES CHACHI (II,1295)

EDELSTEIN, RICK
CHARLIE'S ANGELS (II,486); CODE RED (II,552); CUTTER TO HOUSTON (II,612); LUCAN (II,1554); MITCHELL AND WOODS (II,1709); STARSKY AND HUTCH (II,2444)

EDGAR, DAVID
NICHOLAS NICKLEBY (II,1840)

EDGE, FRED
KRAFT TELEVISION THEATER (I,2592)

EDGERTON, JUSTIN
THE BIONIC WOMAN (II,255); THE INCREDIBLE HULK (II,1232)

EDSON, ERIC
TRAPPER JOHN, M.D. (II,2654)

EDWARDS, BLAKE
HEY MULLIGAN (I,2040); JULIE—MY FAVORITE THINGS (II,1355); THE MONK (I,3087); PETER GUNN (I,3562)

EDWARDS, CHARLES
TEXAS (II,2566)

EDWARDS, JAMES
THE DESILU PLAYHOUSE (I,1237)

EDWARDS, JOHN
THE AMAZING YEARS OF CINEMA (II,64)

EDWARDS, MICHAEL
THE WILD WILD WEST (I,4863)

EDWARDS, PAUL
HIGH PERFORMANCE (II,1145); KNIGHT RIDER (II,1402); QUINCY, M. E. (II,2102); SHANNON (II,2314); V: THE SERIES (II,2715); THE YELLOW ROSE (II,2847)

EFRON, MARSHALL
THE DICK CAVETT SHOW (II,669)

EGAN, MARK
ALICE (II,33)

EGAN, SAM
AUTOMAN (II,120); THE INCREDIBLE HULK (II,1232); MANIMAL (II,1622); QUINCY, M. E. (II,2102)

EGAN, TOM
LINDSAY WAGNER—ANOTHER SIDE OF ME (II,1483); NURSE (II,1870)

EGGENWEILER, ROBERT
FOUR STAR PLAYHOUSE (I,1652); A WALK IN THE NIGHT (I,4750)

EHRENMAN, HOWARD N
NAKED CITY (I,3216)

EHRIN, KERRY
FAME (II,812)

EHRLICH, KEN
MAC DAVIS 10TH ANNIVERSARY SPECIAL: I STILL BELIEVE IN MUSIC (II,1571); THE MAC DAVIS CHRISTMAS SPECIAL (II,1573); MAC DAVIS SPECIAL: THE MUSIC OF CHRISTMAS (II,1578); MAC DAVIS—I'LL BE HOME FOR CHRISTMAS (II,1580)

EHRLICH, MAX
THE BIG STORY (I,432); THE DICK POWELL SHOW (I,1269); HANDLE WITH CARE (I,1926); STAR TREK (II,2440); TARGET: THE CORRUPTERS (I,4360); THE UNTOUCHABLES (I,4682); THE WILD WILD WEST (I,4863)

EHRMANN, PAUL
CAGNEY AND LACEY (II,409); LOU GRANT (II,1526); SEVEN BRIDES FOR SEVEN BROTHERS (II,2307)

EINSTEIN, BOB
ANDY WILLIAMS MAGIC LANTERN SHOW COMPANY (I,202); ANDY'S LOVE CONCERT (I,215); BIZARRE (II,258); THE HUDSON BROTHERS SHOW (II,1202); JOEY AND DAD (II,1306); THE KEN BERRY WOW SHOW (I,2524); LOLA (II,1508); LOLA (II,1509); LOLA (II,1511); THE REDD FOXX COMEDY HOUR (II,2126); THE SMOTHERS BROTHERS COMEDY HOUR (I,4095); THE SMOTHERS BROTHERS SHOW (I,4097); THE SONNY AND CHER COMEDY HOUR (II,2400); THE SONNY COMEDY REVUE (II,2403); THE SUMMER SMOTHERS BROTHERS SHOW (I,4281); VAN DYKE AND COMPANY (II,2718); VAN DYKE AND COMPANY (II,2719)

EINSTEIN, CHARLES
LOU GRANT (II,1526); THE WONDERFUL WORLD OF PIZZAZZ (I,4901)

EISENHORN, ART
EISCHIED (II,763); THE GANGSTER CHRONICLES (II,957)

EISENSTOCK, ALAN
BLUE JEANS (II,275); FOOT IN THE DOOR (II,902); HIGH SCHOOL, U.S.A. (II,1148); KOJAK (II,1406); MORK AND MINDY (II,1735); WHAT'S HAPPENING!! (II,2769)

EISINGER, JO
DANGER MAN (I,1136); PHILIP MARLOWE, PRIVATE EYE (II,2033); THE POPPY IS ALSO A FLOWER (I,3650); SECRET AGENT (I,3962)

EISMAN, MARK
THE GREAT SPACE COASTER (II,1059)

EISMANN, BERNARD
WILSON'S REWARD (II,2805)

ELBING, PETER
E.T. & FRIENDS—MAGICAL MOVIE VISITORS (II,749); PRIME TIMES (II,2083); THE TV TV SHOW (II,2674)

ELDER III, LONNE
THOU SHALT NOT KILL (II,2604)

ELDER, ANN
THE FLIP WILSON SPECIAL (I,1603); THE FLIP WILSON SPECIAL (II,887); THE FLIP WILSON SPECIAL (II,888); THE FLIP WILSON SPECIAL (II,889); I'D RATHER BE CALM (II,1216); THE KING FAMILY SHOW (I,2552); LILY FOR PRESIDENT (II,1478); THE LILY TOMLIN SHOW (I,2706); THE LILY TOMLIN SPECIAL (II,1479); LITTLE LULU (II,1493); THE MANY MOODS OF PERRY COMO (I,2897); THE MARILYN MCCOO AND BILLY DAVIS JR. SHOW (II,1630); MITZI (I,3071); MITZI'S SECOND SPECIAL (I,3074); THE PAUL LYNDE COMEDY HOUR (II,1977); TAKE MY ADVICE (II,2531); TEXACO STAR THEATER: OPENING NIGHT (II,2565)

ELDREDGE, DAWN
SUGAR TIME (II,2489)

ELEWIN, ALBERT
TEXAS JOHN SLAUGHTER (I,4414)

ELIAS, CAROLINE
BREAKING AWAY (II,371); SECRETS OF MIDLAND HEIGHTS (II,2296)

ELIAS, MICHAEL
ALL'S FAIR (II,46); THE BILL COSBY SPECIAL, OR? (I,443); BLACK BART (II,260); CO-ED FEVER (II,547); THE FUNNY SIDE (I,1714); THE FUNNY SIDE (I,1715); THE GLEN CAMPBELL GOODTIME HOUR (I,1820); THE LESLIE UGGAMS SHOW (I,2660); THE MARY TYLER MOORE SHOW (II,1640); PAT PAULSEN'S HALF A COMEDY HOUR (I,3473); ROBERT YOUNG AND THE FAMILY (I,3813); SCARED SILLY (II,2269)

ELINSON, IZZY
THE ANDY GRIFFITH SHOW (I,192); THE ANN SOTHERN SHOW (I,220); DOBIE GILLIS (I,1302); MY SISTER EILEEN (I,3200); RUN, BUDDY, RUN (I,3870)

ELINSON, JACK
A.K.A. PABLO (II,28); THE ALL-STAR REVUE (I,138); THE ANDY GRIFFITH SHOW (I,192); ARNIE (I,261); DANNY THOMAS LOOKS AT YESTERDAY, TODAY AND TOMORROW (I,1148); THE DANNY THOMAS TV FAMILY REUNION (I,1154); THE DORIS DAY SHOW (I,1355); THE DUKE (I,1381); THE FACTS OF LIFE (II,805); FACTS OF LIFE: THE FACTS OF LIFE GOES TO PARIS (II,806); FULL SPEED ANYWHERE (I,1704); GOOD TIMES (II,1038); HEY, JEANNIE! (I,2038); JOE'S WORLD (II,1305); MAKE MORE ROOM FOR DADDY (I,2842); MAKE ROOM FOR DADDY (I,2843); MAKE ROOM FOR GRANDDADDY (I,2844); MY HERO (I,3193); THE PARKERS (II,1959); RUN, BUDDY, RUN (I,3870); TEXACO STAR THEATER (I,4412)

ELINSON, NORMAN
MANY HAPPY RETURNS (I,2895)

ELISON, IRVING
GIVE MY REGARDS TO BROADWAY (I,1815); JOHNNY COME LATELY (I,2425); MAKE ROOM FOR DADDY (I,2843)

ELKINS, SAM
KRAFT TELEVISION THEATER (I,2592); PERRY MASON (II,2025)

ELLIOT, BRUCE
FLASH GORDON (I,1593)

ELLIOT, JOHN
WAR IN THE AIR (I,4768)

ELLIOT, SUSAN
OLIVIA (II,1883)

ELLIOTT, BOB
BOB & RAY & JANE, LARAINE & GILDA (II,279); FROM CLEVELAND (II,931)

ELLIOTT, CHRIS
LATE NIGHT WITH DAVID LETTERMAN (II,1442)

ELLIOTT, PAUL
DOLLY (II,699); THE MARTY ROBBINS SPOTLITE (II,1636); ROMANCE THEATER (II,2204)

ELLIOTT, PEGGY
COMEDY IS KING (I,1019); DEAN MARTIN PRESENTS THE GOLDDIGGERS (I,1193); THE GHOST AND MRS. MUIR (I,1786); HAVING BABIES I (II,1107)

ELLIOTT, PETER
TALES OF THE GOLD MONKEY (II,2537)

ELLIOTT, SUMNER LOCKE
THE ALCOA HOUR (I,108); DUPONT SHOW OF THE MONTH (I,1387); HEDDA HOPPER'S HOLLYWOOD (I,2002); THE KING AND MRS. CANDLE (I,2544); KRAFT TELEVISION THEATER (I,2592); LITTLE WOMEN (I,2720); NOTORIOUS (I,3314); PLAYWRIGHTS '56 (I,3627); SPELLBOUND (I,4153); STUDIO ONE (I,4268); THE WOMEN (I,4890)

ELLIOTTE, JOHN
THE DONNA REED SHOW (I,1347); FATHER KNOWS BEST (II,838); PLEASE DON'T EAT THE DAISIES (I,3628)

ELLIS, ANTHONY
BLACK SADDLE (I,480); ZANE GREY THEATER (I,4979)

ELLIS, ARNOLD
NAKED CITY (I,3216)

ELLIS, EDWARD J
LEAVE IT TO BEAVER (I,2648)

ELLIS, SIDNEY
THE A-TEAM (II,119); B.J. AND THE BEAR (II,125); BARETTA (II,152); HUNTER (II,1206); MCCLOUD (II,1660); TODAY'S FBI (II,2628)

ELLISON, BOB
ALAN KING LOOKS BACK IN ANGER—A REVIEW OF 1972 (I,99); THE ALAN KING SPECIAL (I,103); ANNIE AND THE HOODS (II,91); BING CROSBY AND HIS FRIENDS (I,456); THE BURNS AND SCHREIBER COMEDY HOUR (I,768); THE BURNS AND SCHREIBER COMEDY HOUR (I,769); THE BURT BACHARACH SPECIAL (I,775); BURT BACHARACH: CLOSE TO YOU (I,772); A COUPLE OF DONS (I,1077); DOM DELUISE AND FRIENDS (II,701); THE DOM DELUISE SHOW (I,1328); I'M A FAN (I,2192); IN SECURITY (II,1227); JULIE AND CAROL AT LINCOLN CENTER (I,2479); THE JULIE ANDREWS HOUR (I,2480); JULIE ON SESAME STREET (I,2485); THE KLOWNS (I,2574); THE MARY TYLER MOORE SHOW (II,1640); PERRY COMO'S SUMMER OF '74 (II,2024); PETULA (I,3573); PHYLLIS (II,2038); THE RICHARD PRYOR SHOW (II,2163); THE RICHARD PRYOR SPECIAL (II,2164); THE RICHARD PRYOR SPECIAL? (II,2165); SUZANNE PLESHETTE IS MAGGIE BRIGGS (II,2509); THAT'S LIFE (I,4426); THE WONDERFUL WORLD OF AGGRAVATION (I,4894); YOUR PLACE OR MINE? (II,2874)

ELLISON, HARLAN
ALFRED HITCHCOCK PRESENTS (I,115); BURKE'S LAW (I,762); LOGAN'S RUN (II,1507); THE MAN FROM U.N.C.L.E. (I,2867); THE OUTER LIMITS (I,3426)

ELLS, GEORGE
THIS IS YOUR LIFE (II,2598)

ELLSWORTH, WHITNEY
THE ADVENTURES OF SUPERMAN (I,77)

ELMORE, GLORIA
THE MAN FROM U.N.C.L.E. (I,2867)

ELSON, JAMES
THE BING CROSBY SHOW (I,463); THE BING CROSBY SHOW (I,464)

ELSTAD, LINDA
THE AMERICAN DREAM (II,71); CALL TO GLORY (II,413); DALLAS (II,614); THE FAMILY TREE (II,820); QUINCY, M. E. (II,2102); RYAN'S FOUR (II,2233); TWO MARRIAGES (II,2691)

ELWYN, CHARLES
VALIANT LADY (I,4695)

EMERICK, LUCILLE
THE ADVENTURES OF BLINKEY (I,52)

EMERON, ROY
THE KENNY EVERETT VIDEO SHOW (II,1379)

EMMETT, ROBERT
THE BELLE OF 14TH STREET (I,390); THE DANGEROUS CHRISTMAS OF RED RIDING HOOD (I,1139); AN EVENING WITH JULIE ANDREWS AND HARRY BELAFONTE (I,1476); THE JACK JONES SPECIAL (I,2310); THE MAN IN THE DOG SUIT (I,2869); ON THE FLIP SIDE (I,3360); SATURDAY'S CHILDREN (I,3923)

ENDRY, JACK
THE SHOW MUST GO ON (II,2344)

ENGELBACH, DAVID
LOTTERY (II,1525)

ENGELHARDT, JIM
MR. MOON'S MAGIC CIRCUS (II,1753)

ENGLEMAN, RON
TWILIGHT THEATER II (II,2686)

ENGLISH, DIANE
CALL TO GLORY (II,413)

ENGLISH, JOHN
THE ADVENTURES OF CHAMPION (I,54); ANNIE OAKLEY (I,225); BUFFALO BILL JR. (I,755); THE GENE AUTRY SHOW (I,1748); THE RANGE RIDER (I,3720)

ENGLUND, KEN
SHOW BIZ (I,4027)

ENOCHS, KENNETH
THE CISCO KID (I,961); LEAVE IT TO BEAVER (I,2648); THE NEW ADVENTURES OF CHARLIE CHAN (I,3249)

EPHRON, NORA
ADAM'S RIB (I,32)

EPSTEIN, DONALD
THE CRYSTAL GAYLE SPECIAL (II,609); DOUG HENNING: MAGIC ON BROADWAY (II,730)

EPSTEIN, HERMAN
FOUR STAR PLAYHOUSE (I,1652); PERRY MASON (II,2025)

EPSTEIN, RON
NBC SATURDAY PROM (I,3245)

ERVIN, JUDY
BLANSKY'S BEAUTIES (II,267); LAVERNE AND SHIRLEY (II,1446)

ERVING, JIM
MAMA'S FAMILY (II,1610)

ERWIN, BILL
THE FLYING NUN (I,1611)

ERWIN, EDDIE
SKY KING (I,4077)

ERWIN, LEE
ALFRED HITCHCOCK PRESENTS (I,115); THE BIG VALLEY (I,437); CROSSROADS (I,1105); THE NEW ADVENTURES OF CHARLIE CHAN (I,3249); PLEASE DON'T EAT THE DAISIES (I,3628); STAR TREK (II,2440)

ERWIN, ROY
THE DESILU PLAYHOUSE (I,1237)

ERWIN, STAN
THE FLYING NUN (I,1611)

ESMONDE, JOHN
GOOD NEIGHBORS (II,1034)

ESSEX, EDWARD
THE FILE ON DEVLIN (I,1575)

ESSEX, HARRY
THE ALCOA HOUR (I,108); TARGET: THE CORRUPTERS (I,4360); THE UNTOUCHABLES (I,4682)

ESSON, ROBERT
MATINEE THEATER (I,2947)

ESTIN, KEN
CHEERS (II,494); GOODTIME GIRLS (II,1042); SHAPING UP (II,2316); TAXI (II,2546)

ESTRIDGE, ROBIN
THE ROGUES (I,3832)

ETTLINGER, DON
DECOY (I,1217); KRAFT TELEVISION THEATER (I,2592)

EUNSON, DALE
BAND OF GOLD (I,341); CLIMAX! (I,976); DIRTY SALLY (II,680); FRONT ROW CENTER (I,1694); LEAVE IT TO BEAVER (I,2648); LITTLE HOUSE ON THE PRAIRIE (II,1487); THE WALTONS (II,2740)

EUNSON, KATHERINE
BAND OF GOLD (I,341); CLIMAX! (I,976); LEAVE IT TO BEAVER (I,2648)

EUSTIS, RALPH
DINAH IN SEARCH OF THE IDEAL MAN (I,1278)

EUSTIS, RICH
THE CARPENTERS (II,444); A COUNTRY HAPPENING (I,1065); DEAN MARTIN PRESENTS THE GOLDDIGGERS (I,1193); DOM DELUISE AND FRIENDS (II,701); THE DOROTHY HAMILL WINTER CARNIVAL SPECIAL (II,720); FATHER O FATHER (II,842); GOOBER AND THE TRUCKERS' PARADISE (II,1029); HARPER VALLEY, U.S.A. (I,1954); THE JERRY REED WHEN YOU'RE HOT, YOU'RE HOT HOUR (I,2372); THE JIM STAFFORD SHOW (II,1291); THE JOHN BYNER COMEDY HOUR (I,2407); JOHN DENVER ROCKY MOUNTAIN CHRISTMAS (II,1314); THE JOHN DENVER SPECIAL (II,1315); THE JOHN DENVER SPECIAL (II,1316); JOHN DENVER—THANK GOD I'M A COUNTRY BOY (II,1317); THE ROWAN AND MARTIN SHOW (I,3854); SCARED SILLY (II,2269)

EVANIER, MARK
ANSON AND LORRIE (II,98); GOLDIE GOLD AND ACTION JACK (II,1023); THE HALF-HOUR COMEDY HOUR (II,1074); THE LITTLE RASCALS (II,1494); PRYOR'S PLACE (II,2089); THE RICHIE RICH SHOW (II,2169); THUNDARR THE BARBARIAN (II,2616)

EVANS, BARBARA
THE LOVE BOAT (II,1535)

EVANS, EDITH
TEX AND JINX (I,4406)

EVANS, TIM
THE DON LANE SHOW (II,707)

EVERETT, KENNY
THE KENNY EVERETT VIDEO SHOW (II,1379)

EVERING, JIM
NUTS AND BOLTS (II,1871)

EWING, DICK
MAMA'S FAMILY (II,1610)

EXTON, CLIVE
THE DESPERATE HOURS (I,1239); THE HUMAN VOICE (I,2160)

FABER, CHARLES
THE GEORGE SANDERS MYSTERY THEATER (I,1776);

FALLBERG, CARL
THE FLINTSTONE FUNNIES (II,883)

FALLEN, LUCILLE
AMERICAN COWBOY (I,169)

FALSEY, JOHN
ST. ELSEWHERE (II,2432)

FALVO, JOHN
JOHNNY RINGO (I,2434)

FANAROW, BARRY
BENSON (II,208)

FARHI, MORRIS
MAN IN A SUITCASE (I,2868); THE ONEDIN LINE (II,1912)

FARMER, GENE
REAL PEOPLE (II,2118); ROWAN AND MARTIN'S LAUGH-IN (I,3856); THE SHAPE OF THINGS (II,2315); SPEAK UP AMERICA (II,2412); THAT'S MY MAMA (II,2580); WHAT'S HAPPENING!! (II,2769)

FARQUHAR, RALPH
HAPPY DAYS (II,1084); THE NEW ODD COUPLE (II,1825)

FARR, GORDON
THE BEAUTIFUL PHYLLIS DILLER SHOW (I,381); THE BOB NEWHART SHOW (II,342); BOBBIE GENTRY'S HAPPINESS HOUR (I,671); BOBBIE GENTRY'S HAPPINESS HOUR (II,343); THE FUNNY SIDE (I,1715); HERE WE GO AGAIN (I,2028); KEEPING UP WITH THE JONESES (II,1374); THE OSMOND BROTHERS SHOW (I,3409); PAUL SAND IN FRIENDS AND LOVERS (II,1982); PETULA (I,3572); PORTRAIT OF PETULA (I,3654); SUPER COMEDY BOWL 1 (I,4288); SUPER COMEDY BOWL 2 (I,4289); WE GOT IT MADE (II,2752)

FARR, LYNNE
THE BOB NEWHART SHOW (II,342); BOBBIE GENTRY'S HAPPINESS HOUR (II,343)

FARRELL, HENRY
THE EYES OF CHARLES SAND (I,1497)

FARRELL, JUDY
FAME (II,812)

FARRELL, MARTY
THE ALAN KING SHOW (I,101); THE ALAN KING SHOW (I,102); THE ANDY WILLIAMS CHRISTMAS SHOW (I,198); THE ANDY WILLIAMS SPECIAL (I,213); ANN-MARGRET OLSSON (II,84); ANN-MARGRET'S HOLLYWOOD MOVIE GIRLS (II,86); BING CROSBY AND HIS FRIENDS (I,456); BURT BACHARACH IN SHANGRI-LA (I,773); THE BURT BACHARACH SPECIAL (I,775); THE BURT BACHARACH SPECIAL (I,776); BURT BACHARACH! (I,771); BURT BACHARACH—OPUS NO. 3 (I,774); CELEBRATION: THE AMERICAN SPIRIT (II,461); COMEDY IS KING (I,1019); DICK CAVETT'S BACKLOT USA (II,670); THE JOHN DENVER SPECIAL (II,1315); JUBILEE (II,1347); JULIE AND CAROL AT LINCOLN CENTER (I,2479); JULIE AND DICK IN COVENT GARDEN (II,1352); JULIE ON SESAME STREET (I,2485); THE KLOWNS (I,2574); THE KRAFT 75TH ANNIVERSARY SPECIAL (II,1409); LAS VEGAS PALACE OF STARS (II,1431); THE MAC DAVIS CHRISTMAS SPECIAL (II,1572); THE MAC DAVIS SPECIAL (II,1577); MAC DAVIS—I'LL BE HOME FOR CHRISTMAS (II,1580); SALUTE (II,2242); THE SANDY DUNCAN SHOW (II,2253); SPECIAL LONDON BRIDGE SPECIAL (I,4150); STEVE AND EYDIE: OUR LOVE IS HERE TO STAY (II,2457); A TRIBUTE TO "MR. TELEVISION" MILTON BERLE (II,2658); UPTOWN (II,2710); THE WAY THEY WERE (II,2748); THE WORLD OF MAGIC (II,2838)

FARRELL, MIKE
M*A*S*H (II,1569)

FASS, GEORGE
SHERLOCK HOLMES (I,4010); SIMON LASH (I,4052); THE THREE MUSKETEERS (I,4480)

FASS, GERTRUDE
FOUR STAR PLAYHOUSE (I,1652); SHERLOCK HOLMES (I,4010); SIMON LASH (I,4052); THE THREE MUSKETEERS (I,4480)

FAULKNER, GEORGE
CAVALCADE OF AMERICA (I,875)

FAULKNER, MARJORIE
KRAFT TELEVISION THEATER (I,2592)

FAULKNER, NANCY
THE INCREDIBLE HULK (II,1232)

FAULKNER, WILLIAM
THE GRADUATION DRESS (I,1864)

FAUST, ERICH
THE MAN FROM U.N.C.L.E. (I,2867)

FAUST, GILBERT S
ROBERT MONTGOMERY PRESENTS YOUR LUCKY STRIKE THEATER (I,3809)

FAY, DEIDRE
ALL TOGETHER NOW (II,45); DOUBLE TROUBLE (II,723); THE FACTS OF LIFE (II,805); FACTS OF LIFE: THE FACTS OF LIFE GOES TO PARIS (II,806); ONE DAY AT A TIME (II,1900)

FAY, WILLIAM
ALCOA PREMIERE (I,109); ALFRED HITCHCOCK PRESENTS (I,115); BELLE STARR (I,391); COMBAT (I,1011); GOING MY WAY (I,1833); THE MAN FROM U.N.C.L.E. (I,2867); WAGON TRAIN (I,4747)

FEELY, TERENCE
BERGERAC (II,209); DEATH IN SMALL DOSES (I,1207); THE EYES HAVE IT (I,1496); LOOK BACK IN DARKNESS (I,2752); MISTRAL'S DAUGHTER (II,1708); ONLY A SCREAM AWAY (I,3393); A PLACE TO DIE (I,3612); THE PRISONER (I,3670); RETURN OF THE SAINT (II,2141); ROBIN'S NEST (II,2187); THE

SAINT (II,2237); THE SAVAGE CURSE (I,3925); SIGN IT DEATH (I,4046); SPELL OF EVIL (I,4152)

FEIBLEMAN, PETER
COLUMBO (II,556)

FEIFFER, JULES
THE COMEDY ZONE (II,559); HAPPY ENDINGS (II,1085); THAT WAS THE YEAR THAT WAS (II,2575)

FEIGENBAUM, JOEL
KNOTS LANDING (II,1404)

FEINBERG, ROBERT
THE MISADVENTURES OF SHERIFF LOBO (II,1704)

FEIRSTEIN, BRUCE
THE BEST LEGS IN 8TH GRADE (II,214)

FEIST, AUBREY
IVANHOE (I,2282)

FELDMAN, CHESTER
JOHNNY CARSON DISCOVERS CYPRESS GARDENS (I,2419)

FELDMAN, GENE
THE HORROR OF IT ALL (II,1181); MATINEE THEATER (I,2947)

FELDMAN, MARTY
DOM DELUISE AND FRIENDS (II,701); FLANNERY AND QUILT (II,873); THE MARTY FELDMAN COMEDY MACHINE (I,2929)

FELDON, BARBARA
DINAH IN SEARCH OF THE IDEAL MAN (I,1278)

FELLINI, FEDERICO
FELLINI: A DIRECTOR'S NOTEBOOK (I,1562)

FELTON, DAVID
SQUARE PEGS (II,2431)

FENADY, ANDREW J
BRANDED (I,707)

FENNELL, ALAN
THUNDERBIRDS (I,4497); U.F.O. (I,4662)

FENNERTON, WILLIAM
ALCOA PREMIERE (I,109)

FENTON, FRANK
THE DANGEROUS DAYS OF KIOWA JONES (I,1140)

FERBER, BRUCE
BOSOM BUDDIES (II,355); HOUSE CALLS (II,1194); JENNIFER SLEPT HERE (II,1278); OH MADELINE (II,1880); STAR OF THE FAMILY (II,2436); WEBSTER (II,2758)

FERGUSON, WILLIAM
GOOD OL' BOYS (II,1035)

FERMAN, BENNY
THE DUKES (II,741)

FERMEN, CLIVE
THE DUKES (II,741)

FERRIER, GARY
THE CRACKER BROTHERS (II,599); THE JOHNNY CASH SHOW (I,2424); THE KEANE BROTHERS SHOW (II,1371); PRIVATE BENJAMIN (II,2087); SANDY IN DISNEYLAND (II,2254); VAN DYKE AND COMPANY (II,2719)

FERRIS, WALTER
KRAFT TELEVISION THEATER (I,2592)

FERRO, JEFFREY
9 TO 5 (II,1852); ARCHIE BUNKER'S PLACE (II,105); TRAPPER JOHN, M.D. (II,2654)

FERRO, MATHILDE
LEAVE IT TO BEAVER (I,2648)

FERRO, THEODORE
KRAFT TELEVISION THEATER (I,2592); LEAVE IT TO BEAVER (I,2648)

FESSIER, MICHAEL
CAP'N AHAB (I,824); I AND CLAUDIE (I,2164); SECRET AGENT (I,3961)

FIELD, BARBARA
A CHRISTMAS CAROL (II,517)

FIELD, BILLY
FAME (II,812)

FIELD, GUSTAVE
THE SIX-MILLION-DOLLAR MAN (II,2372)

FIELDER, JOHN
BLACK SADDLE (I,480)

FIELDER, PAT
BARETTA (II,152); BEN CASEY (I,394); CHARLIE'S ANGELS (II,486); GOLIATH AWAITS (II,1025); LAW OF THE PLAINSMAN (I,2639); THE MISADVENTURES OF SHERIFF LOBO (II,1704); QUINCY, M. E. (II,2102); THE RIFLEMAN (I,3789); SWISS FAMILY ROBINSON (II,2517); TODAY'S FBI (II,2628)

FIELDER, RICHARD
ALCOA PREMIERE (I,109); APPLE'S WAY (II,101); CHANNING (I,900); GEORGE WASHINGTON (II,978); HEC RAMSEY (II,1121); IVANHOE (I,2282); JAKE'S WAY (II,1269); MARCUS WELBY, M.D. (II,1627); RAWHIDE (I,3727); STATE FAIR (II,2448); THE WALTONS (II,2740); THE WORD (II,2833); ZANE GREY THEATER (I,4979)

FIELDER, ROBERT
STUDIO ONE (I,4268)

FIELDS, GREG
MADAME'S PLACE (II,1587); SOLID GOLD (II,2395); SOLID GOLD HITS (II,2397)

FIELDS, JOSEPH
WONDERFUL TOWN (I,4893)

FIELDS, PETER ALLEN
CASSIE AND COMPANY (II,455); DARKROOM (II,619); THE FBI (I,1551); GET CHRISTIE LOVE! (II,981); MADIGAN (I,2808); A MAN CALLED SLOANE (II,1613); THE MAN FROM U.N.C.L.E. (I,2867); THE RAT PATROL (I,3726); SWITCH (II,2519)

FIELDS, SIDNEY
THE ABBOTT AND COSTELLO SHOW (I,2)

FILERMAN, MICHAEL
WKRP IN CINCINNATI (II,2814)

FIMBERG, HAL
THE BING CROSBY SHOW (I,468); THE BING CROSBY SHOW (I,469); THE ROBBINS NEST (I,3806)

FINBOW, COLIN
THE AVENGERS (II,121)

FINCH, SCOTT
FROM A BIRD'S EYE VIEW (I,1686)

FINE, BOB
54TH STREET REVUE (I,1573)

FINE, MORT
ALFRED HITCHCOCK PRESENTS (I,115); BANACEK (II,138); BERT D'ANGELO/SUPERSTAR (II,211); THE DICK POWELL SHOW (I,1269); I SPY (I,2179); KOJAK (II,1406); MCCLOUD (II,1660); THE STREETS OF SAN FRANCISCO (II,2478)

FINE, SYLVIA
THE DANNY KAYE SHOW (I,1143); AN HOUR WITH DANNY KAYE (I,2136)

FINESTRA, CARMEN
JOHNNY CASH AND FRIENDS (II,1322); JOHNNY CASH AND THE COUNTRY GIRLS (II,1323); MEL AND SUSAN TOGETHER (II,1677); PRIME TIMES (II,2083); PUNKY BREWSTER (II,2092); TWILIGHT THEATER II (II,2686)

FINFER, JUNE
SENSE OF HUMOR (II,2303)

FINK, HARRY JULIAN
ARENA (I,256); BEN CASEY (I,394); THE DICK POWELL SHOW (I,1269); ZANE GREY THEATER (I,4979)

FINK, MARK
ARCHIE BUNKER'S PLACE (II,105); GREAT DAY (II,1054); THE LOVE BOAT (II,1535); THE MISADVENTURES OF SHERIFF LOBO (II,1704); UNCLE CROC'S BLOCK (II,2702)

FINKEL, BOB
THE JIMMY MCNICHOL SPECIAL (II,1292); THE MUHAMMED ALI VARIETY SPECIAL (II,1762); THE WAYNE NEWTON SPECIAL (II,2750)

FINKEL, MORT
MAGIC COTTAGE (I,2815)

FINKELMAN, KEN
CALLAHAN (II,414); EVERYTHING GOES (I,1480); VAN DYKE AND COMPANY (II,2719)

FINKLE, DAVID
THE SINGERS (I,4060)

FINKLEHOFF, FRED
LOLLIPOP LOUIE (I,2738); MEET ME IN ST. LOUIS (I,2986)

FINN, HERBERT
GILLIGAN'S ISLAND (II,990); THE HONEYMOONERS (I,2110); THE HONEYMOONERS (II,1175); IT'S ABOUT TIME (I,2263)

FINNEY, JACK
ALCOA PREMIERE (I,109)

FINNEY, SARA V
THE JEFFERSONS (II,1276)

FINNIGAN, TOM
PEOPLE ARE FUNNY (II,1996)

FISCH, JOE
PUNKY BREWSTER (II,2092)

FISCHER, GEOFFREY
HOTEL (II,1192)

FISCHER, PETER S
BARETTA (II,152); BLACK BEAUTY (II,261); CHARLIE COBB: NICE NIGHT FOR A HANGING (II,485); COLUMBO (II,556); DARKROOM (II,619); MCMILLAN AND WIFE (II,1667); MURDER, SHE WROTE (II,1770); ONCE AN EAGLE (II,1897)

FISCHMAN, RUEL

FANTASY ISLAND (II,829)

FISHER, ART
33 1/3 REVOLUTIONS PER MONKEE (I,4451); MONSANTO PRESENTS MANCINI (I,3092); MONSANTO PRESENTS MANCINI (I,3093); MONSANTO PRESENTS MANCINI (I,3094)

FISHER, BOB
THE ANN SOTHERN SHOW (I,220); BACHELOR FATHER (I,323); DOC CORKLE (I,1304); HOORAY FOR LOVE (I,2116); I DREAM OF JEANNIE (I,2167); THE PEOPLE'S CHOICE (I,3521)

FISHER, JIM
BIG CITY COMEDY (II,233); THE BOOK OF LISTS (II,350); I'VE HAD IT UP TO HERE (II,1221); LAUGH TRAX (II,1443); ROCK COMEDY (II,2191); TWILIGHT THEATER (II,2685)

FISHER, MARVE
THE SOLDIERS (I,4112)

FISHER, MICHAEL
THE ASPHALT COWBOY (II,111); BARNABY JONES (II,153); FINDER OF LOST LOVES (II,857); KHAN! (II,1384); MATT HELM (II,1653); MATT HOUSTON (II,1654); SHOOTING STARS (II,2341); STARSKY AND HUTCH (II,2444); STRIKE FORCE (II,2480)

FISHER, ROBERT
ALICE (II,33)

FISHER, STEVE
BRINGING UP BUDDY (I,731); CANNON (II,424); THE DICK POWELL SHOW (I,1269); FANTASY ISLAND (II,829); THE GEORGE SANDERS MYSTERY THEATER (I,1776); LUKE AND THE TENDERFOOT (I,2792); MCMILLAN AND WIFE (II,1667); MICHAEL SHAYNE, DETECTIVE (I,3022); ON OUR OWN (II,1892); S.W.A.T. (II,2236); STARSKY AND HUTCH (II,2444); SWITCH (II,2519); THE WILD WILD WEST (I,4863)

FISHER, TERRY LOUISE
CAGNEY AND LACEY (II,409); THE MISSISSIPPI (II,1707); THIS GIRL FOR HIRE (II,2596)

FISHKIN, JEREMY
FAERIE TALE THEATER (II,807)

FLAHERTY, JOE
SECOND CITY TELEVISION (II,2287)

FLANAGAN, THOMAS
STUDIO ONE (I,4268)

FLAUM, MARSHALL
BOB HOPE SPECIAL: BOB HOPE IN WHO MAKES THE WORLD LAUGH, PART 2 (II,286); BOB HOPE SPECIAL: BOB HOPE IN WHO MAKES THE WORLD LAUGH? (II,287); BOGART (I,679); COUNTERATTACK: CRIME IN AMERICA (II,581); HOLLYWOOD: THE SELZNICK YEARS (I,2095); PLAYBOY'S 25TH ANNIVERSARY CELEBRATION (II,2054); RIPLEY'S BELIEVE IT OR NOT (II,2177); YOU ASKED FOR IT (II,2855)

FLEISCHMAN, STEPHEN
ONE NIGHT STANDS (I,3387)

FLEISHER, CAROL L
THE SINGING COWBOYS RIDE AGAIN (II,2363)

FLEMING, ERIC
RAWHIDE (I,3727)

FLEMING, THOMAS
CROSSROADS (I,1105)

FLETCHER, JULIE
THE LOVE BOAT (II,1535)

FLETCHER, LAWRENCE
DECOY (I,1217)

FLETCHER, LOUISE
THE TWILIGHT ZONE (I,4651)

FLICKER, THEODORE J
BANACEK (II,138); BARNEY MILLER (II,154)

FLIPPEN, RUTH BROOKS
BEWITCHED (I,418); THE GHOST AND MRS. MUIR (I,1786); GIDGET (I,1795); MY WORLD . . . AND WELCOME TO IT (I,3208)

FLOCKER, JAMES T
CAMP WILDERNESS (II,419)

FLOREY, ROBERT
WAGON TRAIN (I,4747)

FLOWERS, WAYLAND
MADAME'S PLACE (II,1587)

FLYNN, DON
ONE OF THE BOYS (II,1911)

FLYNN, JOHN
LONE STAR (II,1513)

FLYNN, MIRIAM
THE TIM CONWAY SHOW (II,2619)

FOGARTY, JACK V
BARNABY JONES (II,153); CHARLIE'S ANGELS (II,486); A MAN CALLED SLOANE (II,1613); S.W.A.T. (II,2236); T.J. HOOKER (II,2524); TRAPPER JOHN, M.D. (II,2654)

FOGEL, RICK
THE LITTLE RASCALS (II,1494); THE RICHIE RICH SHOW (II,2169)

FOLB, JAY
BIG EDDIE (II,236); FOOT IN THE DOOR (II,902); FREEBIE AND THE BEAN (II,922); GOOD HEAVENS (II,1032); MAUDE (II,1655); THE THREE WIVES OF DAVID WHEELER (II,2611)

FOLEY, JOHN
PUMPBOYS AND DINETTES ON TELEVISION (II,2090)

FOLLET, JAMES
BLAKE'S SEVEN (II,264)

FOLLOITT, DORIA
ROBERT MONTGOMERY PRESENTS YOUR LUCKY STRIKE THEATER (I,3809)

FOLSOM, ALLAN
HART TO HART (II,1102)

FONTANA, DOROTHY C
THE BIG VALLEY (I,437); THE FANTASTIC JOURNEY (II,826); KUNG FU (II,1416); LOGAN'S RUN (II,1507); STAR TREK (II,2439); STAR TREK (II,2440); THE STREETS OF SAN FRANCISCO (II,2478); THE WALTONS (II,2740)

FONTANA, TOM
ST. ELSEWHERE (II,2432)

FOOTE, DOROTHY COOPER
GIDGET (I,1795); MY THREE SONS (I,3205)

FOOTE, HORTON
DUPONT SHOW OF THE MONTH (I,1387); PLAYHOUSE 90 (I,3623); PLAYWRIGHTS '56 (I,3627)

FORBES, C SCOTT
THE SAINT (II,2237)

FORD, DEACON
THE SAINT (II,2237)

FORD, DEREK
THE SAINT (II,2237)

FORD, FRANK
THE WOMAN IN WHITE (I,4887)

FORD, OLGA
GET CHRISTIE LOVE! (II,981)

FORER, MORT
HIGH HOPES (II,1144)

FORMAN, EDDIE
THE ABBOTT AND COSTELLO SHOW (I,2)

FORRESTER, LARRY
BERT D'ANGELO/SUPERSTAR (II,211); CANNON (II,424); CARIBE (II,439); FANTASY ISLAND (II,829); THE FLYING DOCTOR (I,1610); IVANHOE (I,2282); LADIES IN BLUE (II,1422); MATT HOUSTON (II,1654); SWITCH (II,2519); VEGAS (II,2724)

FORSYTH, DAWNING
THE COPS AND ROBIN (II,573)

FORSYTH, HOLLY
PANDAMONIUM (II,1948)

FORTE, VINCENT
HARBOR COMMAND (I,1945)

FORTIN, MAJORIE
THE CISCO KID (I,961)

FORTUNE, JOHN
ROGER DOESN'T LIVE HERE ANYMORE (II,2201)

FOSTER, BOB
CASABLANCA (II,450); CHICAGO STORY (II,506); TALES OF THE GOLD MONKEY (II,2537)

FOSTER, GEORGE
76 MEN AND PEGGY LEE (I,3989); ARTHUR GODFREY AND THE SOUNDS OF NEW YORK (I,280); THE BIG PARTY FOR REVLON (I,427); CHARO (II,488); COTTON CLUB '75 (II,580); THE GARRY MOORE SHOW (I,1740); THE NBC FOLLIES (I,3241); THE NBC FOLLIES (I,3242); OPRYLAND U.S.A. (I,3405); THE PERRY COMO SHOW (I,3532); RAINBOW OF STARS (I,3717); THE WAYNE NEWTON SPECIAL (II,2749)

FOSTER, LEWIS
SWAMP FOX (I,4311)

FOSTER, NORMAN
DAVY CROCKETT (I,1181); ELFEGO BACA (I,1427)

FOSTER, ROBERT
ALL THAT GLITTERS (II,41); COMPUTERCIDE (II,561); CONCRETE COWBOYS (II,563); THE HOYT AXTON SHOW (II,1200); KATE

MCSHANE (II,1368); KNIGHT RIDER (II,1402); KOJAK (II,1406)

FOSTER, ROSEMARY
MATINEE THEATER (I,2947); MR. I. MAGINATION (I,3140)

FOSTER, ROYAL
THE KEN MURRAY SHOW (I,2525)

FOSTER, WARREN
THE JETSONS (I,2376)

FOURNIER, GENE
WKRP IN CINCINNATI (II,2814)

FOURNIER, RIFT
BARETTA (II,152); CODE RED (II,552); THE JOE NAMATH SHOW (I,2400); KOJAK (II,1406); YOU ASKED FOR IT (II,2855)

FOURZON, SAM
GET CHRISTIE LOVE! (II,981)

FOWLER, KEITH
THE ADDAMS FAMILY (II,11); THE BEAUTIFUL PHYLLIS DILLER SHOW (I,381); THE FARMER'S DAUGHTER (I,1533); THE GEORGE BURNS AND GRACIE ALLEN SHOW (I,1763); THE GEORGE GOBEL SHOW (I,1768); JACKIE GLEASON AND HIS AMERICAN SCENE MAGAZINE (I,2321); LEAVE IT TO BEAVER (I,2648); PAT BOONE IN HOLLYWOOD (I,3469); TEXACO STAR PARADE I (I,4409); TEXACO STAR PARADE II (I,4410)

FOWLER, MARJORIE
THE WALTONS: WEDDING ON WALTON'S MOUNTAIN (EPISODE 1) (II,2743)

FOX JR, FRED
HAPPY DAYS (II,1084); HERE'S BOOMER (II,1134); IT'S YOUR MOVE (II,1259); JOANIE LOVES CHACHI (II,1295); WEBSTER (II,2758)

FOX, BEN
FOREST RANGER (I,1641)

FOX, BRUCE
BOB HOPE SPECIAL: BOB HOPE'S SPRING FLING OF COMEDY AND GLAMOUR (II,315)

FOX, FRED S
3 (II,0); THE ANDY GRIFFITH SHOW (I,192); THE ANN SOTHERN SHOW (I,220); BOB HOPE SPECIAL: BOB HOPE FOR PRESIDENT (II,282); BOB HOPE SPECIAL: BOB HOPE GOES TO COLLEGE (II,283); BOB HOPE SPECIAL: BOB HOPE IN THE STAR-MAKERS (II,285); BOB HOPE SPECIAL: BOB HOPE IN WHO MAKES THE WORLD LAUGH, PART 2 (II,286); BOB HOPE SPECIAL: BOB HOPE IN WHO MAKES THE WORLD LAUGH? (II,287); BOB HOPE SPECIAL: BOB HOPE LAUGHS WITH THE MOVIE AWARDS (II,288); BOB HOPE SPECIAL: BOB HOPE'S 30TH ANNIVERSARY TV SPECIAL (II,293); BOB HOPE SPECIAL: BOB HOPE'S ALL-STAR BIRTHDAY AT ANNAPOLIS (II,294); BOB HOPE SPECIAL: BOB HOPE'S ALL-STAR COMEDY BIRTHDAY PARTY (II,297); BOB HOPE SPECIAL: BOB HOPE'S ALL-STAR COMEDY BIRTHDAY PARTY AT WEST POINT (II,298); BOB HOPE SPECIAL: BOB HOPE'S ALL-STAR COMEDY LOOK AT THE FALL SEASON: IT'S STILL FREE AND WORTH IT! (II,299); BOB HOPE SPECIAL: BOB HOPE'S ALL-STAR LOOK AT TV'S PRIME TIME WARS (II,303); BOB HOPE SPECIAL: BOB HOPE'S ALL-STAR SUPER BOWL PARTY (II,304); BOB HOPE SPECIAL: BOB HOPE'S FUNNY VALENTINE (II,309); BOB HOPE SPECIAL: BOB HOPE'S MERRY CHRISTMAS SHOW (II,310); BOB HOPE SPECIAL: BOB HOPE'S OVERSEAS CHRISTMAS TOURS (II,311); BOB HOPE SPECIAL: BOB HOPE'S PINK PANTHER THANKSGIVING GALA (II,312); BOB HOPE SPECIAL: BOB HOPE'S ROAD TO HOLLYWOOD (II,313); BOB HOPE SPECIAL: BOB HOPE'S SALUTE TO NASA—25 YEARS OF REACHING FOR THE STARS (II,314); BOB HOPE SPECIAL: BOB HOPE'S SPRING FLING OF COMEDY AND GLAMOUR (II,315); BOB HOPE SPECIAL: BOB HOPE'S STAR-STUDDED SPOOF OF THE NEW TV SEASON— G RATED—WITH GLAMOUR, GLITTER & GAGS (II,317); BOB HOPE SPECIAL: BOB HOPE'S STAND UP AND CHEER FOR THE NATIONAL FOOTBALL LEAGUE'S 60TH YEAR (II,316); BOB HOPE SPECIAL: BOB HOPE'S SUPER BIRTHDAY SPECIAL (II,319); BOB HOPE SPECIAL: BOB HOPE'S USO CHRISTMAS IN BEIRUT (II,320); BOB HOPE SPECIAL: BOB HOPE'S WICKI-WACKY SPECIAL FROM WAIKIKI (II,321); BOB HOPE SPECIAL: BOB HOPE'S WOMEN I LOVE—BEAUTIFUL BUT FUNNY (II,322); BOB HOPE SPECIAL: BOB HOPE—HOPE, WOMEN AND SONG (II,324); BOB HOPE SPECIAL: BOB HOPE PRESENTS A CELEBRATION WITH STARS OF COMEDY AND MUSIC (II,340); BOB HOPE SPECIAL: HAPPY BIRTHDAY, BOB! (II,325); BOB HOPE SPECIAL: HO HO HOPE'S JOLLY CHRISTMAS HOUR (II,327); BOB HOPE SPECIAL: THE BOB HOPE CHRISTMAS SPECIAL (II,328); BOB HOPE SPECIAL: THE BOB HOPE CHRISTMAS SPECIAL (II,329); BOB HOPE SPECIAL: THE BOB HOPE CHRISTMAS SPECIAL (II,331); BOB HOPE SPECIAL: BOB HOPE'S HILARIOUS UNREHEARSED ANTICS OF THE STARS (II,341); BUNGLE ABBEY (II,396); THE CBS NEWCOMERS (I,879); DOBIE GILLIS (I,1302); F TROOP (I,1499); GEORGE BURNS AND OTHER SEX SYMBOLS (II,966); GEORGE BURNS CELEBRATES 80 YEARS IN SHOW BUSINESS (II,967); GEORGE BURNS EARLY, EARLY, EARLY, CHRISTMAS SHOW (II,968); GEORGE BURNS IN NASHVILLE?? (II,969); THE GEORGE BURNS SPECIAL (II,970); GEORGE BURNS' 100TH BIRTHDAY PARTY (II,971); GEORGE BURNS' HOW TO LIVE TO BE 100 (II,972); GIVE MY REGARDS TO BROADWAY (I,1815); HERE'S LUCY (II,1135); HEY, JEANNIE! (I,2038); THE LOVE BOAT (II,1535); THE MAN FROM EVERYWHERE (I,2864); MONTY HALL'S VARIETY HOUR (II,1728); MY SISTER EILEEN (I,3200); SIGMUND AND THE SEA MONSTERS (II,2352); THE TIM CONWAY COMEDY HOUR (I,4501); WHERE'S RAYMOND? (I,4837)

FOX, FREDERIC
BRANDED (I,707); JOHNNY RINGO (I,2434)

FOX, NORM CHANDLER
RITUALS (II,2182)

FOX, PAULA
MATINEE THEATER (I,2947); MIDNIGHT MYSTERY (I,3030); NAKED CITY (I,3216)

FOX, WILLIAM P
OFF WE GO (I,3338); SOUTHERN FRIED (I,4139)

FRANCHINA, BASILO
SOPHIA LOREN IN ROME (I,4129)

FRANCIS, EUGENE
THE GEORGE SANDERS MYSTERY THEATER (I,1776);

FRANCIS, FREDDIE
THE SAINT (II,2237)

FRANCIS, RALPH
CROSSROADS (I,1105)

FRANK, COLIN
ALL THE RIVERS RUN (II,43)

FRANK, DEBRA
9 TO 5 (II,1852); DIFF'RENT STROKES (II,674); HOUSE CALLS (II,1194)

FRANK, FREDERIC
STRATEGIC AIR COMMAND (I,4252)

FRANK, JERRY
SORORITY '62 (II,2405)

FRANK, PAT
PLAYHOUSE 90 (I,3623); TURN OF FATE (I,4623)

FRANK, SANDY
LATE NIGHT WITH DAVID LETTERMAN (II,1442)

FRANK, SID
ONCE UPON A TUNE (I,3372)

FRANK, YASHA
PINOCCHIO (I,3602)

FRANKEL, ERNIE
THE COUNTRY MUSIC MURDERS (II,590); I SPY (I,2179); MY FRIEND TONY (I,3192); THE NEW ADVENTURES OF PERRY MASON (I,3252)

FRANKEN, AL
BOB & RAY & JANE, LARAINE & GILDA (II,279); THE CONEHEADS (II,567); THE NEW SHOW (II,1828); STEVE MARTIN'S BEST SHOW EVER (II,2460)

FRANKLIN, DANIEL J
KNOTS LANDING (II,1404)

FRANKLIN, JEFF
CONDO (II,565); HOT W.A.C.S. (II,1191); YOUNG HEARTS (II,2865)

FRANKLIN, JIM
LAVERNE AND SHIRLEY (II,1446)

FRANKLIN, NANCY
LOVING (II,1552)

FRANKLIN, PAUL
ANNIE OAKLEY (I,225); TURN OF FATE (I,4623)

FRANZ, CAROLINE
ALL MY CHILDREN (II,39); AS THE WORLD TURNS (II,110)

FRASER, BOB
BENSON (II,208); CONDO (II,565)

FRASER, PRUDENCE
ALICE (II,33); WHO'S THE BOSS? (II,2792)

FRAUDENBERGER, DANIEL
TRAUMA CENTER (II,2655)

FRAZEE, STEVE
ZANE GREY THEATER (I,4979)

FREDERICK, PAUL
COMBAT (I,1011)

FREED, FRED
THE RIGHT MAN (I,3790)

FREEDMAN, BILL
RUN, BUDDY, RUN (I,3870)

FREEDMAN, DAVID
COUNTERPOINT (I,1059)

FREEDMAN, HY
YOUR FUNNY FUNNY FILMS (I,4950)

FREEDMAN, JERROLD
JIGSAW (I,2377); KOJAK (II,1406); THE PSYCHIATRIST: GOD BLESS THE CHILDREN (I,3685)

FREEDMAN, PAUL
LETTERS TO LAUGH-IN (I,2672)

FREEDMAN, SEYMOUR
THE DONNA REED SHOW (I,1347)

FREELANDER, HOWARD
WAKE UP (II,2736)

FREEMAN, ALLYN
HART TO HART (II,1102); M*A*S*H (II,1569)

FREEMAN, BEN
THE ADVENTURES OF SUPERMAN (I,77)

FREEMAN, BENEDICT
THE ANDY GRIFFITH SHOW (I,192); GIVE MY REGARDS TO BROADWAY (I,1815); PANDORA AND FRIEND (I,3445); SABU AND THE MAGIC RING (I,3881)

FREEMAN, BILL
IT'S ABOUT TIME (I,2263)

FREEMAN, BUD
BERT D'ANGELO/SUPERSTAR (II,211); CAGNEY AND LACEY (II,409); HAWAII FIVE-O (II,1110); LOU GRANT (II,1526); PALMERSTOWN U.S.A. (II,1945)

FREEMAN, DAVE
THE AVENGERS (II,121); CARRY ON LAUGHING (II,448); THE CATCHER (I,872); MADIGAN (I,2808); ROBIN'S NEST (II,2187)

FREEMAN, DEVERY
DOC CORKLE (I,1304)

FREEMAN, EVERETT
BACHELOR FATHER (I,323); THE FIRST HUNDRED YEARS (I,1582); I LOVE MY DOCTOR (I,2172); IVY LEAGUE (I,2286); THE MISS AND THE MISSILES (I,3062)

FREEMAN, FRED
THE ANDY GRIFFITH SHOW (I,192); APPLE PIE (II,100); CALLING DR. STORM, M.D. (II,415); EMPIRE (II,777); FOG (II,897); THE GOOD LIFE (II,1033); INSIDE O.U.T. (II,1235); JACKIE GLEASON AND HIS AMERICAN SCENE MAGAZINE (I,2321); THE SHAMEFUL SECRETS OF HASTINGS CORNERS (I,3994); STICK AROUND (II,2464); STRUCK BY LIGHTNING (II,2482)

FREEMAN, LAWRENCE J
INSIDE O.U.T. (II,1235)

FREEMAN, LEONARD
ALCOA/GOODYEAR THEATER (I,107); HAWAII FIVE-O (I,1972); ROUTE 66 (I,3852); TURN OF FATE (I,4623)

FREEMAN, MARION
THE BEACH GIRLS (II,191); HANDLE WITH CARE (II,1077); THE LOVE BOAT (II,1535); THE LOVE BOAT I (II,1532); THE ROWAN AND MARTIN REPORT (II,2221)

FREEMAN, MICKEY
MCHALE'S NAVY (I,2969)

FREEMAN, SETH
DOC (II,682); HUNTER (II,1205); LOU GRANT (II,1526); PHYLLIS (II,2038); RHODA (II,2151); THE WALTONS (II,2740)

FREIBERGER, FRED
BONANZA (II,347); THE DUKES OF HAZZARD (II,742); JOHNNY RISK (I,2435); LOST IN SPACE (I,2759); NURSE (II,1870); RAWHIDE (I,3727); S.W.A.T. (II,2236); TURN OF FATE (I,4623); VEGAS (II,2724); THE WILD WILD WEST (I,4863)

FREILICH, JEFF
FLAMINGO ROAD (II,872); GALACTICA 1980 (II,953)

FREIWALD, ERIC
THE ADVENTURES OF CHAMPION (I,54); ANNIE OAKLEY (I,225); BUFFALO BILL JR. (I,755); THE GENE AUTRY SHOW (I,1748)

FREIWALD, ROBERT
THE RANGE RIDER (I,3720)

FRENCH, HELEN
THE STARLOST (I,4203)

FRESCO, ROBERT
BONANZA (II,347); WAGON TRAIN (I,4747)

FREUDBERG, JUDY
SESAME STREET (I,3982)

FREUDENBERGER, DAVID
BLUE THUNDER (II,278)

FREUND, PHILIP
MATINEE THEATER (I,2947)

FRIEBERG, BILLY
BEST FOOT FORWARD (I,401)

FRIEBERGER, FRED
ALCOA/GOODYEAR THEATER (I,107)

FRIEBUS, FLORIDA
ALICE IN WONDERLAND (I,120)

FRIEDBERG, WILLIAM
THE ALL-STAR REVUE (I,138); BABES IN TOYLAND (I,318); THE CHOCOLATE SOLDIER (I,944); A CONNECTICUT YANKEE (I,1038); DEAREST ENEMY (I,1205); THE DESERT SONG (I,1236); THE GREAT WALTZ (I,1878); THE JIMMY DURANTE SHOW (I,2389); LADY IN THE DARK (I,2602); LEAVE IT TO LARRY (I,2649); MARCO POLO (I,2904); THE MERRY WIDOW (I,3012); NAUGHTY MARIETTA (I,3232); PARIS IN THE SPRINGTIME (I,3457); PHIL SILVERS IN NEW YORK (I,3577); THE PHIL SILVERS PONTIAC SPECIAL: KEEP IN STEP (I,3579); SATINS AND SPURS (I,3917); SUMMER IN NEW YORK (I,4279)

FRIEDGREN, JULIE
MAGNUM, P.I. (II,1601); TOO CLOSE FOR COMFORT (II,2642)

FRIEDKIN, DAVID
ALFRED HITCHCOCK PRESENTS (I,115); BURLESQUE (I,765); THE DICK POWELL SHOW (I,1269); I SPY (I,2179); JIGSAW (I,2377); KATE MCSHANE (II,1368)

FRIEDMAN, AL
ALOHA PARADISE (II,58); COLUMBO (II,556); STARSKY AND HUTCH (II,2444)

FRIEDMAN, BRUCE J
THE TED BESSELL SHOW (I,4369)

FRIEDMAN, KEN
JOHNNY MATHIS IN THE CANADIAN ROCKIES (II,1338); LADIES AND GENTLEMAN. . . BOB NEWHART (II,1420); THE MELBA MOORE-CLIFTON DAVIS SHOW (I,2996)

FRIEDMAN, PAUL
BEWITCHED (I,418)

FRIEDMAN, RICHARD
THE PHIL SILVERS SHOW (I,3580)

FRIEDMAN, RON
ACE CRAWFORD, PRIVATE EYE (II,6); ALL IN THE FAMILY (II,38); ALL'S FAIR (II,46); BEWITCHED (I,418); CHANGING SCENE (I,899); CHARLIE'S ANGELS (II,486); DICK VAN DYKE MEETS BILL COSBY (I,1274); THE DICK VAN DYKE SPECIAL (I,1276); THE FALL GUY (II,811); FANTASY ISLAND (II,829); G.I. JOE: A REAL AMERICAN HERO (II,948); THE GHOST AND MRS. MUIR (I,1786); GILLIGAN'S ISLAND (II,990); HARPER VALLEY (II,1094); I DREAM OF JEANNIE (I,2167); THE JONATHAN WINTERS SHOW (I,2443); LUCY IN LONDON (I,2789); MURDER CAN HURT YOU! (II,1768); MY FAVORITE MARTIAN (I,3189); THE ODD COUPLE (II,1875); THE PARTRIDGE FAMILY (II,1962); THE SAN PEDRO BEACH BUMS (II,2250); STRIKE FORCE (II,2480); THAT'S MY MAMA (II,2580); VEGAS (II,2724); VER-R-R-RY INTERESTING (I,4712); WONDER WOMAN (SERIES 1) (II,2828)

FRIEDMAN, SARA ANN
KNOTS LANDING (II,1404)

FRIEMAN, RICHARD
HARPER VALLEY PTA (II,1095)

FRIEND, JOHN EDWARD
ANASTASIA (I,182); ELIZABETH THE QUEEN (I,1431); THE TEMPEST (I,4391)

FRIENDLY, ANDY

THIS IS YOUR LIFE (II,2598)

FRISCHMAN, RUEL
THE POWERS OF MATTHEW STAR (II,2075)

FRISTOE, ALLAN
AS THE WORLD TURNS (II,110)

FRITZ, PADDY
THE PRISONER (I,3670)

FRITZELL, JAMES
M*A*S*H (II,1569); MAKE ROOM FOR DADDY (I,2843); THE MCLEAN STEVENSON SHOW (II,1664); MR. PEEPERS (I,3147); MRS. G. GOES TO COLLEGE (I,3155); SYBIL (I,4327)

FRITZHAND, JAMES
FLAMINGO ROAD (II,872); HOTEL (II,1192); TRAPPER JOHN, M.D. (II,2654)

FROEHLICH, BILL
MICKEY SPILLANE'S MIKE HAMMER (II,1692)

FROHMAN, LORNE
THE CRACKER BROTHERS (II,599); PRYOR'S PLACE (II,2089)

FROLICK, SY
THE AT LIBERTY CLUB (I,309)

FROLOV, DIANE
HOT PURSUIT (II,1189); THE INCREDIBLE HULK (II,1232); MAGNUM, P.I. (II,1601); THE MISSISSIPPI (II,1707); V: THE FINAL BATTLE (II,2714)

FROMKIN, PAUL
SUSAN'S SHOW (I,4303)

FROST, DAVID
THAT WAS THE WEEK THAT WAS (I,4423)

FROST, MARK
GAVILAN (II,959); HILL STREET BLUES (II,1154)

FROUG, WILLIAM
CHARLIE'S ANGELS (II,486); THE DICK POWELL SHOW (I,1269); JUDGEMENT DAY (II,1348); Q. E. D. (II,2094)

FRUNKES, ROY
SUSPENSE THEATER ON THE AIR (II,2507)

FRYE, WILLIAM
BACHELOR PARTY (I,324)

FUCHS, THOMAS
ESCAPE (II,782); RIPLEY'S BELIEVE IT OR NOT (II,2177); YOU ASKED FOR IT (II,2855)

FULLER, DEAN
ONCE UPON A MATTRESS (I,3371)

FULLER, LESTER
INTERNATIONAL DETECTIVE (I,2224)

FUNT, JULIAN
CITY HOSPITAL (I,967)

FURIA JR, JOHN
APPLE'S WAY (II,101); BONANZA (II,347); THE HEALERS (II,1115); HOTEL (II,1192); THE WALTONS (II,2740)

FURINI, FRANK
THE HAMPTONS (II,1076)

FURINO, MICHAEL
FLAMINGO ROAD (II,872)

FURTH, GEORGE
MA AND PA (II,1570); TWIGS (II,2683); TWIGS (II,2684)

FYNE, MARTIN
BURLESQUE (I,765)

GABRIEL, JUDY
HARPER VALLEY (II,1094); HARPER VALLEY PTA (II,1095)

GABRIELSEN, FRANK
STAGE DOOR (I,4179)

GADNEY, REG
KENNEDY (II,1377)

GAFFNEY, RICKY
JUST MEN (II,1359)

GAINES, BARRY
SPORT BILLY (II,2429)

GALAY, PETER
KEEP U.S. BEAUTIFUL (I,2519); TONY ORLANDO AND DAWN (II,2639)

GALBRAITH, JOHN
MATINEE AT THE BIJOU (II,1651)

GALE, BOB
USED CARS (II,2712)

GALEN, FRANK
MY FRIEND IRMA (I,3191)

GALLAGHER, BARBARA
LILY (II,1477); MAUDE (II,1655)

GALLAGHER, FARNSWORTH
GUN SHY (II,1068)

GALLAY, PETER
ALL'S FAIR (II,46); ARCHIE (II,104); BARBARA MANDRELL AND THE MANDRELL SISTERS (II,143); DEAN MARTIN'S CELEBRITY ROAST (II,643); DINAH AND HER NEW BEST FRIENDS (II,676); GABRIEL KAPLAN PRESENTS THE SMALL EVENT (II,952); GLADYS KNIGHT AND THE PIPS (II,1004); THE JIMMY MCNICHOL SPECIAL (II,1292); NOBODY'S PERFECT (II,1858); SHA NA NA (II,2311)

GALLERY, MICHELE
LOU GRANT (II,1526); THINGS ARE LOOKING UP (II,2588)

GALLICO, PAUL
THE SNOW GOOSE (I,4103)

GALVIN, BARRY
COUNTRY GALAXY OF STARS (II,587)

GALVIN, BILL
AN EVENING WITH THE STATLER BROTHERS (II,790); LOUISE MANDRELL: DIAMONDS, GOLD AND PLATINUM (II,1528); MUSIC CITY NEWS TOP COUNTRY HITS OF THE YEAR (II,1772)

GALVIN, PAT
COUNTRY GALAXY OF STARS (II,587); AN EVENING WITH THE STATLER BROTHERS (II,790); LOUISE MANDRELL: DIAMONDS, GOLD AND PLATINUM (II,1528); MUSIC CITY NEWS TOP COUNTRY HITS OF THE YEAR (II,1772)

GAMMIE, BILL
MARINELAND CARNIVAL (I,2912); MARINELAND CARNIVAL (I,2913)

GAMMILL, TOM
LATE NIGHT WITH DAVID LETTERMAN (II,1442)

GANNON, JOE
WHIZ KIDS (II,2790)

GANZ, LOWELL
BUSTING LOOSE (II,402); FLATFOOTS (II,879); HERE'S BOOMER (II,1134); HERNDON AND ME (II,1139); JOANIE LOVES CHACHI (II,1295); LAVERNE AND SHIRLEY (II,1446); THE LOVEBIRDS (II,1548); THE NEW ODD COUPLE (II,1825); THE ODD COUPLE (II,1875); PAUL SAND IN FRIENDS AND LOVERS (II,1982); THE RITA MORENO SHOW (II,2181); THE TED KNIGHT SHOW (II,2550); WALKIN' WALTER (II,2737)

GANZER, ALVIN
ROUTE 66 (I,3852)

GARDEN, GRAEME
THE GOODIES (II,1040)

GARDNER JR, ED
THE CISCO KID (I,961)

GARDNER, CHARLES S
STUDIO ONE (I,4268)

GARDNER, ERLE STANLEY
PERRY MASON (II,2025)

GARDNER, GERALD
...AND DEBBIE MAKES SIX (I,185); CALL HOLME (II,412); THE DON ADAMS SCREEN TEST (II,705); THE DON ADAMS SPECIAL: HOORAY FOR HOLLYWOOD (I,1329); THE JERRY LEWIS SHOW (I,2370); THE NBC FOLLIES (I,3242); THE SMOTHERS BROTHERS SHOW (I,4096); A SPECIAL OLIVIA NEWTON-JOHN (II,2418); WHERE'S THE FIRE? (II,2785)

GARDNER, HERB
HAPPY ENDINGS (II,1085); LOVE, LIFE, LIBERTY & LUNCH (II,1544)

GARDNER, JOAN
THE BUFFALO BILLY SHOW (I,756)

GARDNER, NANCY
THE ABC AFTERSCHOOL SPECIAL (II,1)

GARDNER, STEVEN
DECOY (I,1217)

GARDNER, TED
RAWHIDE (I,3727)

GAREN, LEO
T.J. HOOKER (II,2524)

GAREN, SCOTT
TELEVISION'S GREATEST COMMERCIALS III (II,2555); TELEVISION'S GREATEST COMMERCIALS IV (II,2556)

GARGAN, LESLIE
RIPLEY'S BELIEVE IT OR NOT (II,2177)

GARLAND, BOB
THE BURNS AND SCHREIBER COMEDY HOUR (I,768); THE BURNS AND SCHREIBER COMEDY HOUR (I,769); COMEDY NEWS II (I,1021); THE FUNNY SIDE (I,1715)

GARMAN, STEPHANIE
AT EASE (II,115); MR. MERLIN (II,1751); ONE DAY AT A TIME (II,1900)

GARMET, CHARLES
STUDIO ONE (I,4268)

GARNDER, GRAEME
DOCTOR IN THE HOUSE (I,1313)

GARNER, GERALD
JACK BENNY'S BIRTHDAY SPECIAL (I,2299)

GARNETT, GRACE
KRAFT TELEVISION THEATER (I,2592)

GAROFALO, TONY
DANCE FEVER (II,615); MADAME'S PLACE (II,1587)

GARREN, DAVID
BUCK ROGERS IN THE 25TH CENTURY (II,383)

GARRETT, DAVID
CAPITOL (II,426)

GARRETT, KELLY
THE TIM CONWAY SHOW (II,2619)

GARRETT, LILA
ANOTHER APRIL (II,93); GETTING THERE (II,985); HONEYMOON SUITE (I,2108); THE JULIE ANDREWS HOUR (I,2480); MY WORLD . . . AND WELCOME TO IT (I,3208); NEWMAN'S DRUGSTORE (II,1837); THE SMOTHERS BROTHERS SHOW (I,4096); THIS BETTER BE IT (II,2595); THE UGLIEST GIRL IN TOWN (I,4663)

GARRETT, ROGER
BOSOM BUDDIES (II,355); LAVERNE AND SHIRLEY (II,1446); WEBSTER (II,2758)

GARRISON, GREG
THE DEAN MARTIN CELEBRITY ROAST (II,637); THE DEAN MARTIN CELEBRITY ROAST (II,638); THE DEAN MARTIN CELEBRITY ROAST (II,639); DOM DELUISE AND FRIENDS, PART 2 (II,702)

GARRY, AL
THE MAGIC CLOWN (I,2814)

GARSON, HENRY
MY THREE SONS (I,3205)

GARVER, LLOYD
ALICE (II,33); THE BOB NEWHART SHOW (II,342); A COUPLE OF DONS (I,1077); FAMILY TIES (II,819); GOOD TIMES (II,1038); HAPPY DAYS (II,1084); HOLLYWOOD HIGH (II,1163); THE LAST RESORT (II,1440); MAKING THE GRADE (II,1606); NUMBER 96 (II,1869); THAT'S MY MAMA (II,2580); WE'VE GOT EACH OTHER (II,2757)

GARY, CAROL
WHAT'S HAPPENING!! (II,2769)

GASS, MARC
GOING BANANAS (II,1015)

GAST, HAROLD
CASABLANCA (II,450); THE ELEVENTH HOUR (I,1425); THE FELONY SQUAD (I,1563); I SPY (I,2179); MATINEE THEATER (I,2947); N.Y.P.D. (I,3321); THE SEEKERS (I,3973); THE SEEKERS (I,3974); SUSPICION (I,4309); THE UNTOUCHABLES (I,4682); A WOMAN CALLED GOLDA (II,2818)

GAUNT, JOHN
THE BABY SITTER (I,321)

GAY, JOHN
PLAYHOUSE 90 (I,3623)

GAYNOR, JOHN
WONDER WOMAN (SERIES 2) (II,2829)

GAYNOR, LARRY
SEEING THINGS (II,2297)

GAZZO, MICHAEL
A HATFUL OF RAIN (I,1964)

GEIGER, GEORGE
BABY, I'M BACK! (II,130); THE CAPTAIN AND TENNILLE (II,429); THE CARPENTERS (II,444); CODE RED (II,552); FLO (II,891); HAWAIIAN HEAT (II,1111); LEGMEN (II,1458); LOVE, SIDNEY (II,1547); MAGNUM, P.I. (II,1601); MR. MERLIN (II,1751); THE STARLAND VOCAL BAND (II,2441); SZYSZNYK (II,2523); TALES OF THE GOLD MONKEY (II,2537)

GEIGER, HUBERT
GOOD TIMES (II,1038)

GEIGER, MILTON
FOUR STAR PLAYHOUSE (I,1652); PERRY MASON (II,2025)

GEIGER, TOM
BABY, I'M BACK! (II,130)

GEISS, TONY
COMEDY TONIGHT (I,1027); COS (II,576); COS: THE BILL COSBY COMEDY SPECIAL (II,577); THE DICK CAVETT SHOW (II,669); THE ROBERT KLEIN SHOW (II,2185); SESAME STREET (I,3982); THAT WAS THE YEAR THAT WAS (I,4424); THAT WAS THE YEAR THAT WAS (II,2575); THIS WILL BE THE YEAR THAT WILL BE (I,4466); WE INTERRUPT THIS SEASON (I,4780)

GELBART, LARRY
AFTERMASH (II,23); THE ALL-STAR REVUE (I,138); THE ART CARNEY SHOW (I,274); THE BEST OF ANYTHING (I,404); THE CHEVROLET GOLDEN ANNIVERSARY SHOW (I,924); THE DANNY KAYE SHOW (I,1143); EDDIE (I,1405); FOUR FOR TONIGHT (I,1648); HONESTLY, CELESTE! (I,2104); HOORAY FOR LOVE (I,2115); IF I LOVE YOU, AM I TRAPPED FOREVER? (II,1222); JUDY GARLAND AND HER GUESTS, PHIL SILVERS AND ROBERT GOULET (I,2469); KAREN (II,1363); M*A*S*H (II,1569); THE MARTY FELDMAN COMEDY MACHINE (I,2929); MY WIVES JANE (I,3207); OPENING NIGHT (I,3395); THE RED BUTTONS SHOW (I,3750); RIDING HIGH (II,2173); ROLL OUT! (I,3833); THE ROSALIND RUSSELL SHOW (I,3845); SID CAESAR INVITES YOU (I,4040); THE SID CAESAR SHOW (I,4042); UNITED STATES (II,2708); YOUR SHOW OF SHOWS (I,4957)

GELD, KARYL
KAREN (II,1363); THE LILY TOMLIN SHOW (I,2706); THE MARY TYLER MOORE SHOW (II,1640); THE SANDY DUNCAN SHOW (II,2253)

GELLEN, AUDREY
DUPONT SHOW OF THE MONTH (I,1387); HARVEY (I,1961); HARVEY (I,1962); THE HEIRESS (I,2006); MEMBER OF THE WEDDING (I,3002); OUR TOWN (I,3420); THE PHILADELPHIA STORY (I,3582); THE PICTURE OF DORIAN GRAY (I,3591)

GELLER, BRUCE
BRONK (II,379); THE DICK POWELL SHOW (I,1269); THE JUDGE (I,2468); THE WESTERNER (I,4801)

GELLER, MATT
ALOHA PARADISE (II,58)

GELMAN, LAURIE
OH MADELINE (II,1880); OUT OF THE BLUE (II,1934); TOO CLOSE FOR COMFORT (II,2642)

GELMAN, MILTON
COMMAND PERFORMANCE (I,1029); INDEMNITY (I,2209); THE MAN FROM U.N.C.L.E. (I,2867); ROBERT MONTGOMERY PRESENTS YOUR LUCKY STRIKE THEATER (I,3809)

GEMS, PAM
PIAF (II,2039)

GENDOT, ADRIAN
DANGEROUS ASSIGNMENT (I,1138); SKY KING (I,4077)

GEOGHAN, JIM
EMPIRE (II,777)

GEORGE, CAROL
MUSICAL CHAIRS (II,1776)

GEORGE, GEORGE W
ALCOA/GOODYEAR THEATER (I,107); DEAR MOM, DEAR DAD (I,1202); THE RIFLEMAN (I,3789)

GEORGE, JUDITH
ALCOA/GOODYEAR THEATER (I,107); COMBAT (I,1011); DEAR MOM, DEAR DAD (I,1202); THE RIFLEMAN (I,3789)

GEORGE, ROBERT
CHARLIE'S ANGELS (II,486)

GERALD, SANFORD
GET CHRISTIE LOVE! (II,981)

GERAMO, MURIEL
BATTLE OF THE PLANETS (II,179)

GERARD, JOHN
THE INVESTIGATORS (I,2231)

GERARD, MERWIN
BEN CASEY (I,394); BRIDGER (II,376); THE FANTASTIC JOURNEY (II,826); KATE LOVES A MYSTERY (II,1367); MANNIX (II,1624); NERO WOLFE (II,1806); THE NEXT STEP BEYOND (II,1839); ONE STEP BEYOND (I,3388); ROBERT MONTGOMERY PRESENTS YOUR LUCKY STRIKE THEATER (I,3809)

GERBER, CLAIRE
THE SMURFS (II,2385)

GERBER, DAVID
BABY, I'M BACK! (II,130); LOTTERY (II,1525)

GERBER, STEVE
BEAUTY AND THE BEAST (II,197)

GERMANO, PETER
THE FUGITIVE (I,1701); THE TIME TUNNEL (I,4511)

GERROLD, DAVID
LOGAN'S RUN (II,1507); STAR TREK (II,2439); STAR TREK (II,2440)

GERSHE, LEONARD
PRIVATE SECRETARY (I,3672)

GERSHMAN, BEN

THE ANDY GRIFFITH SHOW (I,192); LEAVE IT TO BEAVER (I,2648); PLEASE DON'T EAT THE DAISIES (I,3628); RUN, BUDDY, RUN (I,3870)

GERSON, JACK
THE OMEGA FACTOR (II,1888)

GERSON, NOEL B
ROBERT MONTGOMERY PRESENTS YOUR LUCKY STRIKE THEATER (I,3809)

GERSTAD, JOHN
ONE TOUCH OF VENUS (I,3389)

GERZGHTY, GERALD
THE CISCO KID (I,961)

GESNER, CLARK
THE DICK CAVETT SHOW (II,669); MR. MAYOR (I,3143); WE INTERRUPT THIS SEASON (I,4780)

GETCHELL, ROBERT
ALICE (II,34)

GETHERS, PETER
KATE AND ALLIE (II,1365)

GETHERS, STEVEN
THE FARMER'S DAUGHTER (I,1533); PLAYHOUSE 90 (I,3623); A STRING OF BEADS (I,4263); A WOMAN CALLED GOLDA (II,2818)

GEWIRTZ, HOWARD
BOSOM BUDDIES (II,355); BUSTING LOOSE (II,402); DOMESTIC LIFE (II,703); MR. SUCCESS (II,1756); ON OUR OWN (II,1892); TAXI (II,2546)

GIBBONS, JOHN
THE LORENZO AND HENRIETTA MUSIC SHOW (II,1519)

GIBBS, ANN
THE FACTS OF LIFE (II,805); THE LOVE BOAT (II,1535); TEACHERS ONLY (II,2548); A YEAR AT THE TOP (II,2844)

GIBNEY, SHERIDAN
THE SIX-MILLION-DOLLAR MAN (II,2372)

GIBSON, CHANNING
FAMILY (II,813)

GIEGER, GEORGE
THE PAUL LYNDE COMEDY HOUR (II,1978)

GIELGUD, GWEN
PURSUIT (I,3700)

GIELGUD, IRWIN
PURSUIT (I,3700)

GIL, JOANE A
HUSBANDS, WIVES AND LOVERS (II,1209)

GILBERT, ALAN
MASQUERADE PARTY (II,1645)

GILBERT, CATHY
UNITED STATES (II,2708)

GILBERT, DORIS
THE ADVENTURES OF SUPERMAN (I,77); THE UNEXPECTED (I,4674)

GILBERT, GARY
BAKER'S DOZEN (II,136); E/R (II,748); JOHNNY GARAGE (II,1337)

GILBERT, WILLIAM
DEVLIN (II,663); HOWDY DOODY (I,2151); THE NEW HOWDY DOODY SHOW (II,1818); ROD BROWN OF THE ROCKET RANGERS (I,3825)

GILDEN, JERRY
ANYTHING FOR MONEY (II,99)

GILER, BERNIE
THE MAN FROM U.N.C.L.E. (I,2867); PLAYHOUSE 90 (I,3623); TIGHTROPE (I,4500); ZANE GREY THEATER (I,4979)

GILER, DAVID
HOLLYWOOD (II,1160); THE MAN FROM U.N.C.L.E. (I,2867)

GILES, MARK
MURDER, SHE WROTE (II,1770)

GILL JR, FRANK
THE PEOPLE'S CHOICE (I,3521)

GILL, FRANK
MCHALE'S NAVY (I,2969)

GILLARD, STUART
QUARK (II,2095); THE SONNY AND CHER SHOW (II,2401)

GILLEN, SASHA
THE BIG VALLEY (I,437)

GILLES, D B
LOVE, SIDNEY (II,1547)

GILLETTE, WILLIAM
SHERLOCK HOLMES (II,2327)

GILLIAM, TERRY
MONTY PYTHON'S FLYING CIRCUS (II,1729)

GILLIAND, DEBORAH
TRAPPER JOHN, M.D. (II,2654)

GILLIGAN, JOHN
DEPARTMENT S (I,1232); RIPLEY'S BELIEVE IT OR NOT (II,2177)

GILLIS, JACKSON
THE ADVENTURES OF SUPERMAN (I,77); CANNON (II,424); CARIBE (II,439); CODE RED (II,552); COLUMBO (II,556); THE FBI (I,1551); THE HARDY BOYS AND THE MYSTERY OF GHOST FARM (I,1950); THE HARDY BOYS AND THE MYSTERY OF THE APPLEGATE TREASURE (I,1951); I'M THE LAW (I,2195); LOST IN SPACE (I,2758); THE MAN FROM U.N.C.L.E. (I,2867); MANNIX (II,1624); PARIS (II,1957); PERRY MASON (II,2025); THE WILD WILD WEST (I,4863); WONDER WOMAN (SERIES 2) (II,2829)

GILMER, RICHARD
KNOTS LANDING (II,1404)

GILMER, ROBERT
ALL THAT GLITTERS (II,41); CHICAGO STORY (II,506); THE DEVLIN CONNECTION (II,664); HAWAIIAN HEAT (II,1111); MAGNUM, P.I. (II,1601)

GILROY, FRANK D
AMOS BURKE: WHO KILLED JULIE GREER? (I,181); THE DICK POWELL SHOW (I,1269); KRAFT TELEVISION THEATER (I,2592); MATINEE THEATER (I,2947); NERO WOLFE (II,1805); PLAYHOUSE 90 (I,3623); THE RIFLEMAN (I,3789); STUDIO ONE (I,4268); TEXAS JOHN SLAUGHTER (I,4414); THE TURNING POINT OF JIM MALLOY (II,2670)

GIMBEL, ROGER
GOSSIP (II,1044)

GINNES, ABRAM S
CASSIE AND COMPANY (II,455); DECOY (I,1217); EGAN (I,1419); NAKED CITY (I,3216)

GIOFFRE, MARISA
THE ABC AFTERSCHOOL SPECIAL (II,1)

GIPE, GEORGE
THE INVESTIGATORS (II,1241); STOPWATCH: THIRTY MINUTES OF INVESTIGATIVE TICKING (II,2474)

GIPSON, FRED
SCREEN DIRECTOR'S PLAYHOUSE (I,3946)

GIRARD, BERNARD
COUNTERPOINT (I,1059); CROWN THEATER WITH GLORIA SWANSON (I,1107); FRONT ROW CENTER (I,1694); THE LONE WOLF (I,2742)

GITTLAN, JOYCE
THE JEFFERSONS (II,1276); MARIE (II,1629)

GIVENS, BETTIE JEAN
THE MARSHAL OF GUNSIGHT PASS (I,2925)

GLAISTER, GERARD
SECRET ARMY (II,2290)

GLASCO, GORDON
FAMILY (II,813)

GLASS, SANDY
PHYLLIS (II,2038)

GLASS, SIDNEY A
ROOTS: THE NEXT GENERATIONS (II,2213)

GLAVIN, GEORGE
BRET MAVERICK (II,374)

GLEASON, MICHAEL
THE BIG VALLEY (I,437); FOOLS, FEMALES AND FUN: I'VE GOTTA BE ME (II,899); FOOLS, FEMALES AND FUN: IS THERE A DOCTOR IN THE HOUSE? (II,900); FOOLS, FEMALES AND FUN: WHAT ABOUT THAT ONE? (II,901); FORCE FIVE (II,907); THE GOSSIP COLUMNIST (II,1045); MY FAVORITE MARTIAN (I,3189); THE OREGON TRAIL (II,1922); REMINGTON STEELE (II,2130); SARA (II,2260); SONS AND DAUGHTERS (II,2404)

GLICKMAN, WILL
1968 HOLLYWOOD STARS OF TOMORROW (I,3297); THE ALL-STAR COMEDY SHOW (I,137); THE ALL-STAR SUMMER REVUE (I,139); BEST FOOT FORWARD (I,401); THE CHOCOLATE SOLDIER (I,944); A CONNECTICUT YANKEE (I,1038); THE DESERT SONG (I,1236); JUNIOR MISS (I,2493); KEEFE BRASSELLE'S VARIETY GARDEN (I,2515); THE MERRY WIDOW (I,3012); NAUGHTY MARIETTA (I,3232); PHIL SILVERS ON BROADWAY (I,3578); PRIVATE EYE, PRIVATE EYE (I,3671); THE RED BUTTONS SHOW (I,3750); SCHOOL HOUSE (I,3937)

GLUCK, CAROL WARNER
MATINEE THEATER (I,2947)

GLUCKSMAN, ERNEST
THE PHIL SILVERS ARROW SHOW (I,3576)

GLUCKSMAN, FRANK D
THE BOBBY SHERMAN SPECIAL (I,676)

GODFREY, ALAN
240-ROBERT (II,2689); BARETTA (II,152); BRONK (II,379); THE HARDY BOYS MYSTERIES (II,1090)

GOFF, IVAN
CHARLIE'S ANGELS (II,487); MANNIX (II,1624); PURSUE AND DESTROY (I,3699); THE ROGUES (I,3832); THREE FOR DANGER (I,4475)

GOLATO, CHRISTOPHER
I DREAM OF JEANNIE (I,2167)

GOLBERG, SUSAN
CUTTER TO HOUSTON (II,612)

GOLD, BARRY
ALICE (II,33)

GOLD, CAREY
AS THE WORLD TURNS (II,110)

GOLD, KARYL
MAUDE (II,1655)

GOLDBERG, BETTY
TRAPPER JOHN, M.D. (II,2654)

GOLDBERG, GARY DAVID
ALICE (II,33); THE BOB NEWHART SHOW (II,342); FAMILY TIES (II,819); THE LAST RESORT (II,1440); LOU GRANT (II,1526); THE TONY RANDALL SHOW (II,2640)

GOLDBERG, HERMAN
MATINEE THEATER (I,2947); STUDIO ONE (I,4268)

GOLDBERG, MARSHALL
DIFF'RENT STROKES (II,674); THE FOUR SEASONS (II,915); SCARECROW AND MRS. KING (II,2268)

GOLDBERG, MEL
THE BIG VALLEY (I,437); DAN AUGUST (I,1130); DECOY (I,1217); EAST SIDE/WEST SIDE (I,1396); KRAFT TELEVISION THEATER (I,2592); NAKED CITY (I,3216); SAINTS AND SINNERS (I,3887); THE SIX-MILLION-DOLLAR MAN (II,2372); STUDIO ONE (I,4268); SWITCH (II,2519); THRILLER (I,4492)

GOLDBERG, ROSE LEIMAN
LAND OF HOPE (II,1424)

GOLDEN, HAL
THE MILTON BERLE SPECIAL (I,3051)

GOLDEN, IRIS
TONY ORLANDO AND DAWN (II,2639)

GOLDEN, MYRON
THE MAGIC RANCH (I,2819)

GOLDEN, RAY
TEXACO STAR THEATER (I,4411)

GOLDMAN, GINA
GIMME A BREAK (II,995); KING'S CROSSING (II,1397); LOU GRANT (II,1526); MARIE (II,1629); PARTNERS IN CRIME (II,1961)

GOLDMAN, HAL
THE BEATRICE ARTHUR SPECIAL (II,196); THE BILLY CRYSTAL COMEDY HOUR (II,248); CAROL CHANNING PROUDLY PRESENTS THE SEVEN DEADLY SINS (I,851); THE DEAN MARTIN CELEBRITY ROAST (II,637); EVERYTHING YOU ALWAYS WANTED TO KNOW ABOUT JACK BENNY AND WERE AFRAID TO ASK (I,1482); GEORGE BURNS AND OTHER SEX SYMBOLS (II,966); GEORGE BURNS CELEBRATES 80 YEARS IN SHOW BUSINESS (II,967); GEORGE BURNS EARLY, EARLY, EARLY, CHRISTMAS SHOW (II,968); GEORGE BURNS IN NASHVILLE?? (II,969); GEORGE BURNS' 100TH BIRTHDAY PARTY (II,971); GEORGE BURNS' HOW TO LIVE TO BE 100 (II,972); THE JACK BENNY CHRISTMAS SPECIAL (I,2289); THE JACK BENNY HOUR (I,2290); THE JACK BENNY HOUR (I,2291); THE JACK BENNY HOUR (I,2292); THE JACK BENNY HOUR (I,2293); THE JACK BENNY SPECIAL (I,2296); JACK BENNY WITH GUEST STARS (I,2298); JACK BENNY'S 20TH ANNIVERSARY TV SPECIAL (I,2303); JACK BENNY'S FIRST FAREWELL SHOW (I,2300); JACK BENNY'S NEW LOOK (I,2301); JACK BENNY'S SECOND FAREWELL SHOW (I,2302); THE JERRY LEWIS SHOW (I,2370); LOOK OUT WORLD (II,1516); A LOVE LETTER TO JACK BENNY (II,1540); THE MILTON BERLE SHOW (I,3049); THE MILTON BERLE SPECIAL (I,3051); MY SON THE DOCTOR (I,3203); THE NBC FOLLIES (I,3241); THE RED SKELTON REVUE (I,3754); THE RED SKELTON TIMEX SPECIAL (I,3757); RICH LITTLE'S WASHINGTON FOLLIES (II,2159); THE SMOTHERS BROTHERS COMEDY HOUR (I,4095); THAT'S MY MAMA (II,2580); TONY ORLANDO AND DAWN (II,2639); TV FUNNIES (II,2672)

GOLDMAN, JESSE
THE GEORGE BURNS AND GRACIE ALLEN SHOW (I,1763)

GOLDMAN, LAWRENCE
THE FUGITIVE (I,1701)

GOLDMAN, PEGGY
FOREVER FERNWOOD (II,909); HOT PURSUIT (II,1189); LOTTERY (II,1525); MADAME'S PLACE (II,1587); V: THE FINAL BATTLE (II,2714)

GOLDMAN, ROBERT
THE PARADINE CASE (I,3453); THE SPIRAL STAIRCASE (I,4160)

GOLDRUP, RAY
LITTLE HOUSE ON THE PRAIRIE (II,1487)

GOLDSMITH, ANTHONY
RETURN OF THE SAINT (II,2141)

GOLDSMITH, CLIFFORD
THE ALDRICH FAMILY (I,111); LEAVE IT TO BEAVER (I,2648)

GOLDSMITH, GLORIA
MCMILLAN AND WIFE (II,1667); MEDICAL CENTER (II,1675)

GOLDSMITH, MARTIN M
THE TWILIGHT ZONE (I,4651)

GOLDSTEIN, GARY
PANDAMONIUM (II,1948)

GOLDSTEIN, JEFF
DREAM HOUSE (II,734)

GOLDSTEIN, JESSE
AND HERE'S THE SHOW (I,187); I MARRIED JOAN (I,2174); THE RED SKELTON SHOW (I,3755)

GOLDSTEIN, SHELLEY
LAVERNE AND SHIRLEY (II,1446)

GOLDSTONE, JAMES
IRONSIDE (I,2240)

GOLITZEN, ALEXANDER
MUNSTER, GO HOME! (I,3157)

GOLLANCE, RICHARD
KNOTS LANDING (II,1404)

GOMBERG, SY
BENDER (II,204); THE EVE ARDEN SHOW (I,1471); GOOD HEAVENS (II,1032)

GONZALEZ, GLORIA
THE DAY THE WOMEN GOT EVEN (II,628)

GOOD, JACK
33 1/3 REVOLUTIONS PER MONKEE (I,4451); CONSTANTINOPLE (II,570)

GOODIS, DAVID
ALFRED HITCHCOCK PRESENTS (I,115)

GOODMAN, AL
THE JACK BENNY SPECIAL (I,2295); THE MELBA MOORE-CLIFTON DAVIS SHOW (I,2996)

GOODMAN, BOSCO
HAPPY DAYS (II,1084)

GOODMAN, DAVID Z
THE DESILU PLAYHOUSE (I,1237); THE UNTOUCHABLES (I,4682)

GOODMAN, HAL
DICK VAN DYKE MEETS BILL COSBY (I,1274); THE DONALD O'CONNOR SHOW (I,1343); THE FLIP WILSON SHOW (I,1602); THE GOOD OLD DAYS (I,1851); HOLIDAY LODGE (I,2077); THE JACK BENNY PROGRAM (I,2294); THE JIM NABORS HOUR (I,2381); THE JULIE ANDREWS HOUR (I,2480); THE MELBA MOORE-CLIFTON DAVIS SHOW (I,2996); RUN, BUDDY, RUN (I,3870)

GOODMAN, HANNAH
THE PAUL WINCHELL AND JERRY MAHONEY SHOW (I,3492)

GOODMAN, RALPH
THE FLYING NUN (I,1611); MCHALE'S NAVY (I,2969)

GOODSON, JOSEPH A
ON OUR OWN (II,1892)

GOODWIN, NANCY
THE MARSHAL OF GUNSIGHT PASS (I,2925); TIN PAN ALLEY TV (I,4522)

GOODWIN, ROBERT L
THE BIG VALLEY (I,437)

GOODWIN, ROBERT W
MANHUNTER (II,1621)

GORDEN, ROWBY
THREE'S COMPANY (II,2614)

GORDON, AL
BARBARA MANDRELL AND THE MANDRELL SISTERS (II,143); CAROL CHANNING PROUDLY PRESENTS THE SEVEN DEADLY SINS (I,851); CARTER COUNTRY (II,449); DICK VAN DYKE MEETS BILL COSBY (I,1274); EVERYTHING YOU ALWAYS WANTED TO KNOW ABOUT JACK BENNY AND WERE AFRAID TO ASK (I,1482); THE JACK BENNY CHRISTMAS SPECIAL (I,2289); THE JACK BENNY HOUR (I,2290); THE JACK BENNY HOUR (I,2291); THE JACK BENNY HOUR (I,2292); THE JACK BENNY HOUR (I,2293); THE JACK BENNY PROGRAM (I,2294); THE JACK BENNY SPECIAL (I,2296); JACK BENNY WITH GUEST STARS (I,2298); JACK BENNY'S 20TH ANNIVERSARY TV SPECIAL (I,2303); JACK BENNY'S FIRST FAREWELL SHOW (I,2300); JACK BENNY'S NEW LOOK (I,2301); JACK BENNY'S SECOND FAREWELL SHOW (I,2302); THE JIM NABORS HOUR (I,2381); LOOK OUT WORLD (II,1516); THE NBC FOLLIES (I,3241); THE RED SKELTON REVUE (I,3754); RUN, BUDDY, RUN (I,3870); THE SMOTHERS BROTHERS COMEDY HOUR (I,4095); THAT'S MY MAMA (II,2580); THREE'S COMPANY (II,2614); TONY ORLANDO AND DAWN (II,2639)

GORDON, ARTHUR
THE LORETTA YOUNG THEATER (I,2756)

GORDON, BEVERLY
DOC (II,682)

GORDON, DAN
HIGHWAY TO HEAVEN (II,1152)

GORDON, HAL
THE JACK BENNY SPECIAL (I,2295)

GORDON, JACK
THREE'S COMPANY (II,2614)

GORDON, JILL
DIFF'RENT STROKES (II,674); PAPER DOLLS (II,1952)

GORDON, JUNE
DOUBLE TROUBLE (II,723)

GORDON, KURTZ
MATINEE THEATER (I,2947)

GORDON, LARRY
BURKE'S LAW (I,762)

GORDON, LEO V
BRAVO TWO (II,368); ENOS (II,779)

GORDON, SONNY
MEL AND SUSAN TOGETHER (II,1677)

GORDON, STEVE
BARNEY MILLER (II,154); GOOD TIME HARRY (II,1037); PAUL SAND IN FRIENDS AND LOVERS (II,1982)

GORDON, WILLIAM D
ALFRED HITCHCOCK PRESENTS (I,115); THE BARBARY COAST (II,146); THE CASE AGAINST PAUL RYKER (I,861); THE FUGITIVE (I,1701); THRILLER (I,4492)

GORE, CHRISTOPHER
FAME (II,812)

GORE, STEPHEN
THE FACTS OF LIFE (II,805)

GOREN, RICH
ROMP (I,3838)

GOREN, ROWBY
HART TO HART (II,1102); POLE POSITION (II,2060)

GORES, JOE
MICKEY SPILLANE'S MIKE HAMMER (II,1692); STRIKE FORCE (II,2480)

GORODETSKY, EDDIE
LATE NIGHT WITH DAVID LETTERMAN (II,1442)

GOROG, LASZLO
FOUR STAR PLAYHOUSE (I,1652)

GORTER, FRED DE
BATMAN (I,366)

GOSDEN, FREEMAN
THE AMOS AND ANDY SHOW (I,179)

GOTTLIEB, ALEX
DEAR PHOEBE (I,1204); HAVE GIRLS—WILL TRAVEL (I,1967); HER SCHOOL FOR BACHELORS (I,2018); THE SMOTHERS BROTHERS SHOW (I,4096); TIME FOR ELIZABETH (I,4506)

GOTTLIEB, CARL
CRISIS IN SUN VALLEY (II,604); THE DEADLY TRIANGLE (II,634); THE FLIP WILSON COMEDY SPECIAL (II,886); THE FLIP WILSON SPECIAL (I,1603); THE FLIP WILSON SPECIAL (II,887); THE FLIP WILSON SPECIAL (II,888); THE FLIP WILSON SPECIAL (II,889); FLIP WILSON. . . OF COURSE (II,890); THE NEW LORENZO MUSIC SHOW (II,1821); THE SUMMER SMOTHERS BROTHERS SHOW (I,4281)

GOTTLIEB, RICHARD
NAME THAT TUNE (II,1786); NAME THAT TUNE (II,1787)

GOTTLIEB, THEODORE
THE BILLY CRYSTAL COMEDY HOUR (II,248)

GOUGH, BILL
SEEING THINGS (II,2297)

GOULD, ALAN
OUT OF OUR MINDS (II,1933)

GOULD, BERNIE
ABOUT FACES (I,12); DOBIE GILLIS (I,1302); LET'S MAKE A DEAL (II,1464); MASQUERADE PARTY (II,1645); THE SKY'S THE LIMIT (I,4081); TUGBOAT ANNIE (I,4622)

GOULD, CLIFF
DEATH RAY 2000 (II,654); THE DEVLIN CONNECTION (II,664); THE KILLER WHO WOULDN'T DIE (II,1393); MCLAREN'S RIDERS (II,1663); RAWHIDE (I,3727); SCARECROW AND MRS. KING (II,2268); UNIT 4 (II,2707); WINNER TAKE ALL (II,2808)

GOULD, DIANA
KNOTS LANDING (II,1404)

GOULD, GEORGE
ROD BROWN OF THE ROCKET RANGERS (I,3825)

GOULD, HEYWOOD
DOG AND CAT (II,695); HAZARD'S PEOPLE (II,1112)

GOULD, ROBERT
ALFRED HITCHCOCK PRESENTS (I,115)

GOULDING, RAY
BOB & RAY & JANE, LARAINE & GILDA (II,279); FROM CLEVELAND (II,931)

GRACE, MICHAEL
KNOTS LANDING (II,1404)

GRADY, O
THE DUKES (II,741)

GRAFTON, SAMUEL
STUDIO ONE (I,4268)

GRAFTON, SUE
RHODA (II,2151); SEVEN BRIDES FOR SEVEN BROTHERS (II,2307)

GRAHAM, BILL
DOLLY (II,699); THE MARTY ROBBINS SPOTLITE (II,1636); POP! GOES THE COUNTRY (II,2067)

GRAHAM, IRENE
THE ELEVENTH HOUR (I,1425)

GRAHAM, IRVING
SONG SNAPSHOTS ON A SUMMER HOLIDAY (I,4122)

GRAHAM, JOHN
THE DUKES (II,741)

GRAHAM, RON
A LAST LAUGH AT THE 60'S (I,2627)

GRAHAM, RONNY
THE BRADY BUNCH HOUR (II,363); THE CARPENTERS (II,444); THE HUDSON BROTHERS SHOW (II,1202); M*A*S*H (II,1569); THE NEW BILL COSBY SHOW (I,3257); THE OSMOND FAMILY CHRISTMAS SPECIAL (II,1926); THE OSMOND FAMILY THANKSGIVING SPECIAL (II,1928); THE PAUL LYNDE HALLOWEEN SPECIAL (II,1981); THE SONNY COMEDY REVUE (II,2403)

GRANDY, FRED
THE LOVE BOAT (II,1535)

GRANDY, JAN
THE LOVE BOAT (II,1535)

GRANGER, PERCY
LOVING (II,1552)

GRANT, ARMAND
LIKE MAGIC (II,1476)

GRANT, GIL
OPERATION PETTICOAT (II,1917)

GRANT, LEE H
BARNEY MILLER (II,154); FAME (II,812); M*A*S*H (II,1569); SILVER SPOONS (II,2355)

GRANT, PERRY
THE ADVENTURES OF OZZIE AND HARRIET (I,71); THE ANDY GRIFFITH SHOW (I,192); ANOTHER MAN'S SHOES (II,96); BIG DADDY (I,425); GOOD TIMES (II,1038); HAPPY DAYS (II,1084); HELLO, LARRY (II,1128); I DREAM OF JEANNIE (I,2167); MAUDE (II,1655); ONE DAY AT A TIME (II,1900); THE PARTRIDGE FAMILY (II,1962); POPI (II,2069); THREE FOR THE ROAD (II,2607)

GRANT, STEVE
JOANIE LOVES CHACHI (II,1295)

GRANVILLE, FRANK

SHAZAM! (II,2319)

GRAY, CAROL
BUFFALO BILL (II,387); IRENE (II,1245); OPEN ALL NIGHT (II,1914); THE TONY RANDALL SHOW (II,2640)

GRAY, PATRICIA
THE LITTLEST ANGEL (I,2726)

GREEN, ADOLPH
APPLAUSE (I,244); BUICK CIRCUS HOUR (I,759); LET'S CELEBRATE (I,2663)

GREEN, CATHERINE
THE TWO OF US (II,2693)

GREEN, ERIC
FROM A BIRD'S EYE VIEW (I,1686)

GREEN, HOWARD
THE ADVENTURES OF SUPERMAN (I,77)

GREEN, JAN
THE LITTLE RASCALS (II,1494); THE RICHIE RICH SHOW (II,2169)

GREEN, KATHERINE
THE DUCK FACTORY (II,738); NEWHART (II,1835); THAT'S TV (II,2581)

GREEN, MAURY
YOU ASKED FOR IT (II,2855)

GREEN, MORT
THE PERRY COMO SHOW (I,3532); VAUDEVILLE (II,2722); YES, VIRGINIA, THERE IS A SANTA CLAUS (II,2849); YOUR HIT PARADE (II,2872)

GREEN, PATRICIA
THE MISSISSIPPI (II,1707); SHIRLEY (II,2330)

GREEN, PETER
COLONEL MARCH OF SCOTLAND YARD (I,1005)

GREEN, SAL
STARSKY AND HUTCH (II,2444)

GREEN, SID
THE ANTHONY NEWLEY SHOW (I,233); THE BILL COSBY SPECIAL, OR? (I,443); THE DON KNOTTS SHOW (I,1334); THE GOLDDIGGERS (I,1838); KEEP U.S. BEAUTIFUL (I,2519); PICCADILLY PALACE (I,3589)

GREENBAUM, EVERETT
AFTERMASH (II,23); BACHELOR PARTY (I,324); THE GEORGE GOBEL SHOW (I,1768); LOU GRANT (II,1526); M*A*S*H (II,1569); THE MCLEAN STEVENSON SHOW (II,1664); MR. PEEPERS (I,3147); MRS. G. GOES TO COLLEGE (I,3155); SYBIL (I,4327); UNITED STATES (II,2708); W*A*L*T*E*R (II,2731)

GREENBAUM, FRITZELL
BACHELOR PARTY (I,324)

GREENBAUM, SAM
AMANDA'S (II,62); BARBARA MANDRELL AND THE MANDRELL SISTERS (II,143); TOO CLOSE FOR COMFORT (II,2642)

GREENBERG, HENRY F
ALCOA PREMIERE (I,109); CLIMAX! (I,976); PETER GUNN (I,3562); THE UNTOUCHABLES (I,4682)

GREENBERG, JOAN
ON OUR OWN (II,1892)

GREENBERG, RON
SEMI-TOUGH (II,2299); SEMI-TOUGH (II,2300)

GREENBERG, STANLEY R
CORONET BLUE (I,1055); MAN IN A SUITCASE (I,2868); ROUTE 66 (I,3852)

GREENBERG, STEVEN
BUCK ROGERS IN THE 25TH CENTURY (II,383); CAGNEY AND LACEY (II,409); QUINCY, M. E. (II,2102)

GREENBURG, DAN
ADAM'S RIB (I,32)

GREENE, ADAM
THE NEWS IS THE NEWS (II,1838)

GREENE, CATHERINE
CHEERS (II,494)

GREENE, GRAHAM
DUPONT SHOW OF THE MONTH (I,1387)

GREENE, JOHN L
BLONDIE (I,486); MY FAVORITE MARTIAN (I,3189); NO TIME FOR SERGEANTS (I,3300); SCREEN DIRECTOR'S PLAYHOUSE (I,3946)

GREENE, JOHNNY
MY FRIEND IRMA (I,3191)

GREENE, KATHERINE
BOB HOPE SPECIAL: BOB HOPE'S ALL-STAR COMEDY SPECTACULAR FROM LAKE TAHOE (II,301); BOB HOPE SPECIAL: BOB HOPE'S ALL-STAR COMEDY TRIBUTE TO VAUDEVILLE (II,302); BOB HOPE SPECIAL: BOB HOPE'S BICENTENNIAL STAR SPANGLED SPECTACULAR (II,306); BOB HOPE SPECIAL: BOB HOPE'S WORLD OF COMEDY (II,323)

GREENE, MARGE
MARGE AND JEFF (I,2908)

GREENE, MORT
CLOWN ALLEY (I,979); THE RED SKELTON SHOW (I,3756)

GREENE, SHEP
LOU GRANT (II,1526)

GREENE, TOM
ALOHA PARADISE (II,58); BEWITCHED (I,418); KNIGHT RIDER (II,1402); THE POWERS OF MATTHEW STAR (II,2075)

GREENFIELD, BARRY
SHAZAM! (II,2319)

GREENLAND, SETH
A.K.A. PABLO (II,28)

GREENWALD, NANCY
THE WALTONS (II,2740)

GREER, BILL
GOODNIGHT, BEANTOWN (II,1041); HOUSE CALLS (II,1194)

GREER, KATHY
GOODNIGHT, BEANTOWN (II,1041); HOUSE CALLS (II,1194)

GREER, LUANSHYA
IN THE STEPS OF A DEAD MAN (I,2205)

GREER, MIKE
ROMP (I,3838)

GREGORY, DAVID
FACE THE FACTS (I,1508); HOOTENANNY (I,2117); ON BROADWAY TONIGHT (I,3355)

GREGORY, MAURICE
THE ADVENTURES OF CHAMPION (I,54); ANNIE OAKLEY (I,225); BUFFALO BILL JR. (I,755); THE GENE AUTRY SHOW (I,1748); THE RANGE RIDER (I,3720)

GREGSON, JACK
THE ASSASSINATION RUN (II,112)

GREGSON, RICHARD
TYCOON: THE STORY OF A WOMAN (II,2697)

GREY, JOHN
SKY KING (I,4077)

GREY, RICHARD
THE NEW ADVENTURES OF CHARLIE CHAN (I,3249); PERRY MASON (II,2025)

GREYHOSKY, BABS
THE A-TEAM (II,119); FOUR EYES (II,912); THE GREATEST AMERICAN HERO (II,1060); HUNTER (II,1206); MAGNUM, P.I. (II,1601); RIPTIDE (II,2178); THE ROUSTERS (II,2220)

GRIES, TOM
THE RAT PATROL (I,3726); SKY KING (I,4077); THE WESTERNER (I,4801)

GRIFFIN, MERV
THE MERV GRIFFIN SHOW (II,1688)

GRIFFITH, ANDY
LOOKING BACK (I,2754)

GRIMALDI, GIAN
EMERGENCY! (II,776)

GRODIN, CHARLES
LOVE, SEX. . . AND MARRIAGE (II,1546); THE PAUL SIMON SPECIAL (II,1983)

GROSS JR, JACK
GILLIGAN'S ISLAND (II,990)

GROSS, LARRY
MICKEY SPILLANE'S MIKE HAMMER (II,1692)

GROSS, MARJORIE
THE HALF-HOUR COMEDY HOUR (II,1074); SQUARE PEGS (II,2431)

GROSSMAN, BUDD
DANNY AND THE MERMAID (II,618); THE DORIS DAY SHOW (I,1355); THE FARMER'S DAUGHTER (I,1533); FULL HOUSE (II,936); GILLIGAN'S ISLAND (II,990); IT'S ABOUT TIME (I,2263); MAUDE (II,1655); RUN, BUDDY, RUN (I,3870); THREE'S A CROWD (II,2613); THREE'S COMPANY (II,2614)

GROSSMAN, DIXIE BROWN
THREE'S COMPANY (II,2614)

GROSSMAN, JAY
ANYTHING FOR MONEY (II,99); THE LATE FALL, EARLY SUMMER BERT CONVY SHOW (II,1441); THE LOVE BOAT (II,1535)

GROSSMAN, LYNN
ONE MORE TRY (II,1908)

GROSSMAN, TERRY
BENSON (II,208); CONDO (II,565)

GROSSWIENER, HARRY
YOUR SURPRISE STORE (I,4961)

GROVES, HERMAN
240-ROBERT (II,2689); THE BIONIC WOMAN (II,255); THE BLUE KNIGHT (II,276); BONANZA (II,347); THE CONTENDER (II,571); FANTASY ISLAND (II,829); HARRY O (II,1099); HAWAII FIVE-O (II,1110); RAWHIDE (I,3727); THE RIFLEMAN (I,3789); THE SEEKERS (I,3973); THE SEEKERS (I,3974); THE UNTOUCHABLES (I,4682)

GRUBER, FRANK
CLIMAX! (I,976)

GRUSKIN, JEROME
CLIMAX! (I,976); NAKED CITY (I,3216); ONE STEP BEYOND (I,3388)

GUEDEL, JOHN
PEOPLE ARE FUNNY (I,3518)

GUENETTE, ROBERT
COUNTERATTACK: CRIME IN AMERICA (II,581); HEROES AND SIDEKICKS—INDIANA JONES AND THE TEMPLE OF DOOM (II,1140)

GUERDAT, ANDY
ARCHIE BUNKER'S PLACE (II,105); HUSBANDS, WIVES AND LOVERS (II,1209)

GUEST, CHRISTOPHER
THE LILY TOMLIN SPECIAL (II,1479); PEEPING TIMES (II,1990)

GUILD, LEE
NUMBER 13 DEMON STREET (I,3316)

GUIMAN, RICHARD
MORK AND MINDY (II,1735)

GUINES, A S
GROWING PAYNES (I,1891)

GULLIETH, RICHARD
TEXAS (II,2566)

GUMAR, RICHARD
LEWIS AND CLARK (II,1465)

GUNN, JAMES
FIRESIDE THEATER (I,1580); THE INVESTIGATORS (I,2231)

GUNN, JOSEPH
AIRWOLF (II,27); CHIPS (II,511); EVERY STRAY DOG AND KID (II,792); MANIMAL (II,1622); MICKEY SPILLANE'S MIKE HAMMER (II,1692)

GUNTZELMAN, DAN
CASS MALLOY (II,454); GLORIA (II,1010); OFF THE RACK (II,1878); WKRP IN CINCINNATI (II,2814)

GURMAN, RICHARD
THE BRADY BRIDES (II,361); HAPPY DAYS (II,1084)

GURMAN, STEPHANIE
HAPPY DAYS (II,1084)

GUSS, JACK
79 PARK AVENUE (II,2310); CHANNING (I,900); HERE WE GO AGAIN (I,2028); MEDICAL CENTER (II,1675); SWITCH (II,2519); TRAPPER JOHN, M.D. (II,2654)

GUTHRIE, DAVID
THE PHOENIX (II,2036); THE SECRET WAR OF JACKIE'S GIRLS (II,2294)

GUTIERREZ, VINCENT
FATHER MURPHY (II,841); LITTLE HOUSE ON THE PRAIRIE (II,1487); LITTLE HOUSE: LOOK BACK TO YESTERDAY (II,1490)

GUYLAS, ELLEN
THE FOUR SEASONS (II,915); THREE'S COMPANY (II,2614)

HAAS, CHARLES
STORY THEATER (I,4240)

HAAS, ED
THE CBS NEWCOMERS (I,879); THE JERRY LEWIS SHOW (I,2370); THE JOHN BYNER COMEDY HOUR (I,2407); MAKE MINE RED, WHITE, AND BLUE (I,2841); THE MUNSTERS (I,3158); ROWAN AND MARTIN BITE THE HAND THAT FEEDS THEM (I,3853)

HABER, LOU
SHOW BIZ (I,4027)

HABERMAN, DON
THE LOVE BOAT (II,1535)

HACKADAY, HAL
THE JACKSON FIVE (I,2326); NBC SATURDAY PROM (I,3245)

HACKEL, DAVE
9 TO 5 (II,1852); HARPER VALLEY (II,1094); SHIRLEY (II,2330)

HACKETT, BOB
KOMEDY TONITE (II,1407)

HACKETT, BUDDY
BUDDY HACKETT—LIVE AND UNCENSORED (II,386)

HAFFNER, CRAIG
THE LOVE REPORT (II,1542); THE LOVE REPORT (II,1543)

HAFFNER, MACK
EYE ON HOLLYWOOD (II,798)

HAGAN, CHET
BARBI BENTON SPECIAL: A BARBI DOLL FOR CHRISTMAS (II,148) CHRISTMAS LEGEND OF NASHVILLE (II,522); COUNTRY COMES HOME (II,585); COUNTRY COMES HOME (II,586); COUNTRY MUSIC HIT PARADE (I,1069); COUNTRY MUSIC HIT PARADE (II,588); COUNTRY MUSIC HIT PARADE (II,589); COUNTRY NIGHT OF STARS (II,592); COUNTRY NIGHT OF STARS II (II,593); COUNTRY STARS OF THE 70S (II,594); ELVIS REMEMBERED: NASHVILLE TO HOLLYWOOD (II,772); FIFTY YEARS OF COUNTRY MUSIC (II,853); I BELIEVE IN MUSIC (I,2165); JOHNNY CASH AND FRIENDS (II,1322); JOHNNY CASH AND THE COUNTRY GIRLS (II,1323); THE JOHNNY CASH CHRISTMAS SPECIAL (II,1325); THE JOHNNY CASH CHRISTMAS SPECIAL (II,1326); A JOHNNY CASH CHRISTMAS (II,1328); THE JOHNNY CASH SPRING SPECIAL (II,1329); JOHNNY CASH: CHRISTMAS IN SCOTLAND (II,1332); JOHNNY CASH: COWBOY HEROES (II,1334); JOHNNY CASH: SPRING FEVER (II,1335); JOHNNY CASH: THE FIRST 25 YEARS (II,1336); LARRY GATLIN AND THE GATLIN BROTHERS (II,1429); MERRY CHRISTMAS FROM THE GRAND OLE OPRY (II,1686); ROY ACUFF—50 YEARS THE KING OF COUNTRY MUSIC (II,2223)

HAGGART, JOHN
A WOMAN TO REMEMBER (I,4888)

HAGGIS, PAUL
MR. MERLIN (II,1751)

HAHN, PHIL
ABC'S SILVER ANNIVERSARY CELEBRATION (II,2); AMERICA (I,164); THE ANDY WILLIAMS SHOW (I,210); BARBARA MANDRELL AND THE MANDRELL SISTERS (II,143); THE DAVID SOUL AND FRIENDS SPECIAL (II,626); DONNY AND MARIE (II,712); THE FIFTH DIMENSION SPECIAL: AN ODYSSEY IN THE COSMIC UNIVERSE OF PETER MAX (I,1571); THE FUNNIEST JOKE I EVER HEARD (II,940); G.I.'S (II,949); THE HALF-HOUR COMEDY HOUR (II,1074); HOW TO SURVIVE THE 70S AND MAYBE EVEN BUMP INTO HAPPINESS (II,1199); THE JOHN DAVIDSON CHRISTMAS SHOW (II,1307); THE JOHN DAVIDSON SHOW (II,1309); THE KEN BERRY WOW SHOW (I,2524); THE NASHVILLE PALACE (II,1792); SHOW BUSINESS (II,2343); THE SONNY AND CHER COMEDY HOUR (II,2400); THE SOUPY SALES SHOW (I,4138); THREE'S COMPANY (II,2614); ULTRA QUIZ (II,2701)

HAIGHT, GEORGE
THE ADDAMS FAMILY (II,11)

HAILEY, ARTHUR
THE ALCOA HOUR (I,108); KRAFT TELEVISION THEATER (I,2592)

HAILEY, OLIVER
MCMILLAN AND WIFE (II,1667); SIDNEY SHORR (II,2349)

HALAS, PAUL
DOCTOR SNUGGLES (II,687)

HALE, WILLIAM
BARNABY JONES (II,153)

HALEY JR, JACK
A FUNNY THING HAPPENED ON THE WAY TO HOLLYWOOD (I,1716)

HALEY, ALEX
PALMERSTOWN U.S.A. (II,1945)

HALEY, ELIZABETH
FAMILY (II,813)

HALEY, OLIVER
FAMILY (II,813)

HALFF, ROBERT
THE LONE RANGER (I,2740)

HALL, ADAM
QUILLER: NIGHT OF THE FATHER (II,2100)

HALL, BARBARA
CONDO (II,565); DREAMS (II,735); THE DUCK FACTORY (II,738); FAMILY TIES (II,819); IT TAKES TWO (II,1252); NEWHART (II,1835)

HALL, DON
YOU ASKED FOR IT (II,2855)

HALL, JAMES ANDREW
GREAT EXPECTATIONS (II,1055); TELEVISION'S GREATEST COMMERCIALS (II,2553)

HALL, KAREN
HILL STREET BLUES (II,1154); M*A*S*H (II,1569)

HALL, KIMBERLY

THE FACTS OF LIFE (II,805)

HALL, MARILYN
LIGHTS, CAMERA, MONTY! (II,1475)

HALL, NORMAN
ANNIE OAKLEY (I,225); TERRY AND THE PIRATES (I,4405)

HALL, RICH
THE DAVID LETTERMAN SHOW (II,624); NO SOAP, RADIO (II,1856); NOT NECESSARILY THE NEWS (II,1862); NOT NECESSARILY THE NEWS (II,1863)

HALL, SAM
DARK SHADOWS (I,1157); FRANKENSTEIN (I,1671)

HALL, WILLIS
ROBIN'S NEST (II,2187)

HALLIDAY, FRED
THE WONDERFUL WORLD OF JONATHAN WINTERS (I,4900)

HAMILL, PETE
NIGHTSIDE (I,3296)

HAMILTON, ANDY
NOT THE NINE O'CLOCK NEWS (II,1864)

HAMILTON, DICK
THE ADVENTURES OF SUPERMAN (I,77)

HAMILTON, ROBERT
BARE ESSENCE (II,149); BARE ESSENCE (II,150); CODE R (II,551); THE CONTENDER (II,571); THE FANTASTIC JOURNEY (II,826); MAGNUM, P.I. (II,1601); MCCLOUD (II,1660); MOBILE MEDICS (II,1716); SKEEZER (II,2376); WHEELS (II,2777)

HAMILTON, ROY
THE ADVENTURES OF SUPERMAN (I,77); THE CISCO KID (I,961)

HAMMER, BARBARA
THE ANN SOTHERN SHOW (I,220); SCREEN DIRECTOR'S PLAYHOUSE (I,3946)

HAMMER, PAMELA
TEXAS (II,2566)

HAMMIL, JOE
STUDIO ONE (I,4268)

HAMMOND, BRAD
THE WOLFMAN JACK SHOW (II,2817)

HAMNER JR, EARL
THE TWILIGHT ZONE (I,4651); THE WALTONS (II,2740)

HAMNER, BARBARA
HIS MODEL WIFE (I,2064)

HAMNER, EARL
BOONE (II,351); FALCON CREST (II,808); JOSHUA'S WORLD (II,1344); TIGER! TIGER! (I,4499)

HAMNER, GARY
RITUALS (II,2182)

HAMNER, LINDA
RITUALS (II,2182)

HAMNER, ROBERT
CANNON (II,424); HAVE GIRLS—WILL TRAVEL (I,1967); HER SCHOOL FOR BACHELORS (I,2018); LOST IN SPACE (I,2758); MANNIX (II,1624); MCCLOUD (II,1660); MY FRIEND TONY (I,3192); PLANET OF THE APES (II,2049); THE ROCKFORD FILES (II,2197); THE ROOKIES (II,2208); S.W.A.T. (II,2236); STAR TREK (II,2440); THE TIME TUNNEL (I,4511); VOYAGE TO THE BOTTOM OF THE SEA (I,4743); THE WILD WILD WEST (I,4863)

HAMNER, SCOTT
BOONE (II,351); FALCON CREST (II,808); KNOTS LANDING (II,1404)

HAMPTON, GEORGE
THE GET ALONG GANG (II,980)

HAMPTON, JAMES
BURT AND THE GIRLS (I,770); BURT REYNOLDS' LATE SHOW (I,777)

HAMPTON, ORVILLE
THE DUKES (II,741); HAWAII FIVE-O (II,1110); THE NEW ADVENTURES OF PERRY MASON (I,3252); SKY KING (I,4077)

HANALIS, BLANCHE
BIG BEND COUNTRY (II,230); LITTLE LORD FAUNTLEROY (II,1492); MY FAVORITE MARTIAN (I,3189); THE YOUNG PIONEERS (II,2868); THE YOUNG PIONEERS' CHRISTMAS (II,2870)

HANARAHAN, JACK
THE ANDY WILLIAMS SHOW (I,210)

HANDEL, LEO
HARBOR COMMAND (I,1945)

HANDEY, JACK
PRIME TIMES (II,2083)

HANDLEMAN, STANLEY MYRON
RODNEY DANGERFIELD SPECIAL: IT'S NOT EASY BEIN' ME (II,2199)

HANDLER, JANIS
KNIGHT RIDER (II,1402)

HANDLEY, DAVID
KATE AND ALLIE (II,1365)

HANDY, JACK
THE NEW SHOW (II,1828)

HANLEY, JIM
STRANGE TRUE STORIES (II,2476)

HANLEY, WILLIAM
CELEBRITY (II,463); FLESH AND BLOOD (I,1595); TESTIMONY OF TWO MEN (II,2564)

HANNA, MARK
THRILLER (I,4492)

HANNA, WILLIAM
THE ADVENTURES OF JONNY QUEST (I,64)

HANRAHAN, JACK
THE ALL-NEW POPEYE HOUR (II,49); AMERICA (I,164); THE BOBBY DARIN AMUSEMENT COMPANY (I,672); THE FIFTH DIMENSION SPECIAL: AN ODYSSEY IN THE COSMIC UNIVERSE OF PETER MAX (I,1571); THE LAS VEGAS SHOW (I,2619); MARCUS WELBY, M.D. (II,1627); THE SOUPY SALES SHOW (I,4138); SPORT BILLY (II,2429)

HANRAHAN, JOHN
LITTLE HOUSE ON THE PRAIRIE (II,1487); THE MOUSE FACTORY (I,3113)

HANSER, RICHARD
VICTORY AT SEA (I,4730)

HARDIN, ROBERT
SHANE (I,3995)

HARDWICK, MARK
PUMPBOYS AND DINETTES ON TELEVISION (II,2090)

HARDY, JACK
MARIE (II,1629)

HARE, AMORY
FOUR STAR PLAYHOUSE (I,1652)

HARGROVE, DEAN
COLUMBO (II,556); CUTTER (I,1117); FOR LOVE OR $$$ (I,1623); LADY LUCK (I,2604); MADIGAN (I,2808); THE MAN FROM U.N.C.L.E. (I,2867); MCCLOUD (II,1660); THE MOONGLOW AFFAIR (I,3099); RANSOM FOR A DEAD MAN (I,3722); THE RAT PATROL (I,3726); RETURN OF THE WORLD'S GREATEST DETECTIVE (II,2142); SNAVELY (II,2387)

HARGROVE, MARION
BEN BLUE'S BROTHERS (I,393); BRET MAVERICK (II,374); GIRL ON THE RUN (I,1809); I SPY (I,2179); JEANNIE (II,1275); MY WORLD . . . AND WELCOME TO IT (I,3208); THE ROGUES (I,3832); THE WALTONS (II,2740)

HARGROVE, ROY
MATINEE THEATER (I,2947)

HARMAN, BARRY
THE CAROL BURNETT SHOW (II,443); THE GREAT SPACE COASTER (II,1059); ONE OF THE BOYS (II,1911)

HARMANN, RICK
THE GREAT SPACE COASTER (II,1059)

HARMON, DAVID P
ALL IN THE FAMILY (II,38); B.A.D. CATS (II,123); CALL TO DANGER (I,803); GILLIGAN'S ISLAND (II,990); GILLIGAN'S ISLAND: RESCUE FROM GILLIGAN'S ISLAND (II,991); GILLIGAN'S ISLAND: THE HARLEM GLOBETROTTERS ON GILLIGAN'S ISLAND (II,993); HARPER VALLEY PTA (II,1095); KILLER BY NIGHT (I,2537); MANNIX (II,1624); MCCLOUD (II,1660); MR. TERRIFIC (I,3152); THE NEW ADVENTURES OF PERRY MASON (I,3252); THE RIFLEMAN (I,3789); S.W.A.T. (II,2236); STAR TREK (II,2440); STARSKY AND HUTCH (II,2444); TODAY'S FBI (II,2628); VEGAS (II,2724); YOUNG DAN'L BOONE (II,2863)

HARMON, LARRY
BOZO THE CLOWN (I,702)

HARRINGTON, PAT
THE FUNNY WORLD OF FRED & BUNNI (II,943); ONE DAY AT A TIME (II,1900)

HARRIS, ALF
BARNABY JONES (II,153); THE WILD WILD WEST (I,4863); YOUNG DAN'L BOONE (II,2863)

HARRIS, DONALD
NOBODY'S PERFECT (II,1858)

HARRIS, FRAN
SKY KING (I,4077)

HARRIS, HARRY
JACKIE GLEASON AND HIS AMERICAN SCENE MAGAZINE (I,2321)

HARRIS, HOWARD
THE COLLEGE BOWL (I,1001)

HARRIS, JEFF
ALL-AMERICAN PIE (II,48); ALMOST ANYTHING GOES (II,56); COMEDY NEWS II (I,1021); DETECTIVE SCHOOL (II,661); THE EVERLY BROTHERS SHOW (I,1479); FREEMAN (II,923); GOSSIP (II,1043); IN TROUBLE (II,1230); IT ONLY HURTS WHEN YOU LAUGH (II,1251); JIMMY DURANTE PRESENTS THE LENNON SISTERS HOUR (I,2387); JOE AND SONS (II,1296); JOE AND SONS (II,1297); THE LENNON SISTERS SHOW (I,2656); MCNAMARA'S BAND (II,1668); OPERATION: ENTERTAINMENT (I,3401); PAT BOONE IN HOLLYWOOD (I,3469); THE ROGER MILLER SHOW (I,3830); SHEEHY AND THE SUPREME MACHINE (II,2322); THE SHOW MUST GO ON (II,2344); THAT THING ON ABC (II,2574); THE VAL DOONICAN SHOW (I,4692)

HARRIS, KAREN
THE INCREDIBLE HULK (II,1232)

HARRIS, LARRY
SO YOU THINK YOU GOT TROUBLES?! (II,2391)

HARRIS, LEONARD
OMNIBUS (II,1890)

HARRIS, RICHARD
THE AVENGERS (II,121)

HARRIS, STAN
THE GISELE MACKENZIE SHOW (I,1813)

HARRIS, STEPHEN
DOCTOR WHO (II,689)

HARRIS, SUSAN
BENSON (II,208); DAUGHTERS (II,620); I'M A BIG GIRL NOW (II,1218); IT TAKES TWO (II,1252); MAUDE (II,1655); THE PARTRIDGE FAMILY (II,1962); SOAP (II,2392)

HARRISON, JERRY
BRENDA STARR, REPORTER (II,373)

HARRISON, PAUL
THE SKY'S THE LIMIT (I,4081)

HARRISON, SAMUEL B
FIRESIDE THEATER (I,1580)

HARRON, DON
HEE HAW (II,1123)

HARROWER, ELIZABETH
THE YOUNG AND THE RESTLESS (II,2862)

HARSBURGH, PATRICIA
THE A-TEAM (II,119)

HARSBURGH, PATRICK
THE GREATEST AMERICAN HERO (II,1060); HARDCASTLE AND MCCORMICK (II,1089)

HART, BRUCE
THE DICK CAVETT SHOW (II,669); HOT HERO SANDWICH (II,1186); THE WONDERFUL WORLD OF JONATHAN WINTERS (I,4900)

HART, CAROLE
THE DICK CAVETT SHOW (II,669); THE WONDERFUL WORLD OF JONATHAN WINTERS (I,4900)

HART, CHRIS
MARIE (II,1629)

HART, DON
BENSON (II,208)

HART, SAM
BONNIE AND THE FRANKLINS (II,349)

HART, STAN
HAL LINDEN'S BIG APPLE (II,1072); OH, NURSE! (I,3343); THE PAUL LYNDE COMEDY HOUR (II,1979); YOUR SURPRISE STORE (I,4961)

HART, TERRY
BOSOM BUDDIES (II,355); DEAN MARTIN'S CELEBRITY ROAST (II,643); THE DON RICKLES SHOW (II,708); HERE'S BOOMER (II,1134); THE MAD MAD MAD MAD WORLD OF THE SUPER BOWL (II,1586); ME AND MAXX (II,1672); THE ROWAN AND MARTIN REPORT (II,2221); SHIRLEY (II,2330); SUGAR TIME (II,2489); THAT SECOND THING ON ABC (II,2572); A YEAR AT THE TOP (II,2844)

HARTE, BRET
FIRESIDE THEATER (I,1580)

HARTFORD, JOHN
THE SUMMER SMOTHERS BROTHERS SHOW (I,4281)

HARTIG, HERBERT
THAT WAS THE YEAR THAT WAS (I,4424); THIS WILL BE THE YEAR THAT WILL BE (I,4466)

HARTIGAN, KEVIN
BABY, I'M BACK! (II,130)

HARTLEY, BILL
THE TOMMY HUNTER SHOW (I,4552)

HARTMAN, JAN
THE ABC AFTERSCHOOL SPECIAL (II,1)

HARTMAN, PHIL
THE PEE WEE HERMAN SHOW (II,1988)

HARTMAN, SHIRLEY
CAPITOL (II,426)

HARTMAN, TED
PRIVATE SECRETARY (I,3672)

HARTUNG, ROBERT
ABE LINCOLN IN ILLINOIS (I,11); THE ADMIRABLE CRICHTON (I,36); AH!, WILDERNESS (I,91); ARSENIC AND OLD LACE (I,269); BAREFOOT IN ATHENS (I,354); CYRANO DE BERGERAC (I,1120); THE FANTASTICKS (I,1531); GIDEON (I,1792); GOLDEN CHILD (I,1839); THE LITTLE FOXES (I,2712); THE MAGNIFICENT YANKEE (I,2829); THE PATRIOTS (I,3477); ST. JOAN (I,4175); THE TEAHOUSE OF THE AUGUST MOON (I,4368); VICTORIA REGINA (I,4728); WINTERSET (I,4882)

HARVEY, JACK
FIRESIDE THEATER (I,1580)

HARVEY, MICHAEL
PRISONER: CELL BLOCK H (II,2085)

HASLETT, ANTHONY
THE HIGHWAYMAN (I,2059)

HASLEY, WILLIAM
THE SMURFS (II,2385)

HASS, ED
AN AMATEUR'S GUIDE TO LOVE (I,154); THE DIAHANN CARROLL SHOW (I,1252)

HASTINGS, MICHAEL
SEARCH FOR THE NILE (I,3954)

HATCH, DAVID
CAMBRIDGE CIRCUS (I,807)

HATFIELD, CAROL
THE BOOK OF LISTS (II,350); THE FACTS (II,804); I'VE HAD IT UP TO HERE (II,1221); WHATEVER BECAME OF. . . ? (II,2772)

HATTMAN, STEVE
HARPER VALLEY (II,1094); HARPER VALLEY PTA (II,1095)

HAUCK, CHARLIE
APPLE PIE (II,100); THE ASSOCIATES (II,113); BACK TOGETHER (II,131); BRANAGAN AND MAPES (II,367); A DOG'S LIFE (II,697); THE FLIP WILSON SPECIAL (II,887); HANGING IN (II,1078); HARRY'S BATTLES (II,1100); HOT L BALTIMORE (II,1187); MAUDE (II,1655); MR. DUGAN (II,1749); SUZANNE PLESHETTE IS MAGGIE BRIGGS (II,2509); THAT'S MY MAMA (II,2580); THE TWO OF US (II,2693)

HAVEN, SUE
THE WONDERFUL WORLD OF AGGRAVATION (I,4894)

HAWES, TONY
THE ENGELBERT HUMPERDINCK SHOW (I,1444); THE ENGELBERT HUMPERDINCK SHOW (I,1445); THE ENGELBERT HUMPERDINCK SPECIAL (I,1446); THE KOPYCATS (I,2580)

HAWKESWORTH, JOHN
THE FLAME TREES OF THIKA (II,870); THE TALE OF BEATRIX POTTER (II,2533)

HAWKINS, JOHN
ALCOA PREMIERE (I,109); BONANZA (II,347); LITTLE HOUSE ON THE PRAIRIE (II,1487); MR. LUCKY (I,3141); STUDIO ONE (I,4268); VOYAGE TO THE BOTTOM OF THE SEA (I,4743)

HAWKINS, ODIE
PALMERSTOWN U.S.A. (II,1945)

HAWKINS, RICK
THE CAROL BURNETT SHOW (II,443); DOROTHY (II,717); THE KELLY MONTEITH SHOW (II,1376); THE LOVE BOAT III (II,1534); MAMA'S FAMILY (II,1610); PUNKY BREWSTER (II,2092)

HAWKINS, WARD
ALCOA PREMIERE (I,109); BONANZA (II,347); LITTLE HOUSE ON THE PRAIRIE (II,1487); STUDIO ONE (I,4268); VOYAGE TO THE BOTTOM OF THE SEA (I,4743)

HAWLEY, LOWELL S
THE LORETTA YOUNG THEATER (I,2756)

HAWLEY, LUCAS
THE LORETTA YOUNG THEATER (I,2756)

HAYES, ALFRED
ALFRED HITCHCOCK PRESENTS (I,115); LOGAN'S RUN (II,1507); MANNIX (II,1624); NERO WOLFE (II,1806)

HAYES, JEFFREY
VEGAS (II,2724)

HAYES, JOHN MICHAEL
ADAMS OF EAGLE LAKE (II,10); NEVADA SMITH (II,1807); WINTER KILL (II,2810)

HAYES, PETER LIND
WHEN TELEVISION WAS LIVE (II,2779)

HAYES, RAPHAEL
BEN CASEY (I,394)

HAYES, STEVE
CAPITOL (II,426); CONDOMINIUM (II,566); THE SEEKERS (II,2298); SEVEN BRIDES FOR SEVEN BROTHERS (II,2307)

HAYMES, BOB
THE JULIUS LAROSA SHOW (I,2486)

HAYTER, JAMES
FATHER BROWN (II,836)

HAYWARD, CHRIS
BARNEY MILLER (II,154); THE BULLWINKLE SHOW (I,761); MY MOTHER THE CAR (I,3197); NOBODY'S PERFECT (II,1858); THE NUT HOUSE (I,3320); SHEPHERD'S FLOCK (I,4007)

HAYWARD, DEKE
THE ERNIE KOVACS SHOW (I,1455)

HAYWARD, LILLIE
ANNETTE (I,222)

HAZARD, LAWRENCE
IVANHOE (I,2282)

HEATH, LAURENCE
CALL TO DANGER (I,804); CODE RED (II,552); CODE RED (II,553); THE INVADERS (I,2229); JACQUELINE SUSANN'S VALLEY OF THE DOLLS, 1981 (II,1268); KHAN! (II,1384); THE MAGICIAN (I,2824); MOST WANTED (II,1738); SEVENTH AVENUE (II,2309); STEELTOWN (II,2452); STUNTS UNLIMITED (II,2487)

HECHT, BEN
BEN HECHT'S TALES OF THE CITY (I,395); THE THIRD COMMANDMENT (I,4449)

HECHT, JOEL
HIT MAN (II,1156)

HECHT, KEN
THE DON KNOTTS SHOW (I,1334); GOOD TIMES (II,1038); LOVE, SIDNEY (II,1547); NOBODY'S PERFECT (II,1858); REGGIE (II,2127); THREE'S COMPANY (II,2614)

HECHT, TED
THE AMERICAN DREAM (II,71)

HECKMAN, DON
BIG SHAMUS, LITTLE SHAMUS (II,242)

HEDLEY, BILL
CAPTAIN SCARLET AND THE MYSTERONS (I,837)

HEGERTY, TOM
AGAINST THE WIND (II,24)

HEIDEMAN, LEONARD
BONANZA (II,347); SKYFIGHTERS (I,4078)

HEIFNER, JACK
VANITIES (II,2720)

HEINEMANN, ARTHUR
THE ABC AFTERSCHOOL SPECIAL (II,1); THE FBI (I,1551); RAFFERTY (II,2105); ROLL OF THUNDER, HEAR MY CRY (II,2202); STAR TREK (II,2440); SWISS FAMILY ROBINSON (II,2517)

HELLBERG, SANDY
THE LORENZO AND HENRIETTA MUSIC SHOW (II,1519)

HELLER, JOEL
EVERYTHING YOU EVER WANTED TO KNOW ABOUT MONSTERS . . . BUT WERE AFRAID! (II,794); MCHALE'S NAVY (I,2969)

HELLER, LUKAS
ALFRED HITCHCOCK PRESENTS (I,115)

HELM, HARVEY
THE GEORGE BURNS AND GRACIE ALLEN SHOW (I,1763)

HELPERIN, MICHAEL
B.J. AND THE BEAR (II,126)

HEMMINGS, DAVID
AIRWOLF (II,27)

HENDERSON, JIM
FOR BETTER OR WORSE (I,1621)

HENDERSON, SYDNEY
COMBAT (I,1011)

HENDLER, JAMES
MATT HOUSTON (II,1654)

HENDLER, JANICE
FARRELL: FOR THE PEOPLE (II,834); PARTNERS IN CRIME (II,1961)

HENDRA, TONY
WE INTERRUPT THIS SEASON (I,4780)

HENDRYX, SHIRL
COMBAT (I,1011); THE FBI (I,1551); IVANHOE (I,2282); LAND OF THE GIANTS (I,2611); THE WILD WILD WEST (I,4863)

HENERSON, JAMES
BEWITCHED (I,418); COMBAT (I,1011); DOCTORS' PRIVATE LIVES (II,693); THE FLYING NUN (I,1611); I DREAM OF JEANNIE (I,2167); MANNIX (II,1624); THE NANCY DREW MYSTERIES (II,1789); THE PARTRIDGE FAMILY (II,1962)

HENIEMANN, ARTHUR
LITTLE HOUSE ON THE PRAIRIE (II,1487)

HENINGER, JAN
SCRABBLE (II,2278)

HENNING, DOUG
DOUG HENNING'S WORLD OF MAGIC III (II,727); DOUG HENNING'S WORLD OF MAGIC IV (II,728); DOUG HENNING'S WORLD OF MAGIC V (II,729); DOUG HENNING: MAGIC ON BROADWAY (II,730)

HENNING, PAUL
THE ALL-STAR REVUE (I,138); THE ANDY GRIFFITH SHOW (I,192); THE RETURN OF THE BEVERLY HILLBILLIES (II,2139); THE SECRET WORLD OF KIDS (I,3970); WHERE'S RAYMOND? (I,4837)

HENRY, BOB
THE ROY ROGERS AND DALE EVANS SHOW (I,3858)

HENRY, BUCK
THE BEAN SHOW (I,373); CAPTAIN NICE (I,835); THE NEW SHOW (II,1828); QUARK (II,2095); THAT WAS THE YEAR THAT WAS (II,2575)

HENRY, GIG
THE BOB HOPE SHOW (I,562); THE BOB HOPE SHOW (I,613); THE BOB HOPE SHOW (I,636); THE BOB HOPE SHOW (I,637); THE BOB HOPE SHOW (I,642); BOB HOPE SPECIAL: A QUARTER CENTURY OF BOB HOPE ON TELEVISION (II,281); BOB HOPE SPECIAL: BOB HOPE FOR PRESIDENT (II,282); BOB HOPE SPECIAL: BOB HOPE GOES TO COLLEGE (II,283); BOB HOPE SPECIAL: BOB HOPE IN "JOYS" (II,284); BOB HOPE SPECIAL: BOB HOPE IN THE STAR-MAKERS (II,285); BOB HOPE SPECIAL: BOB HOPE IN WHO MAKES THE WORLD LAUGH? (II,287); BOB HOPE SPECIAL: BOB HOPE LAUGHS WITH THE MOVIE AWARDS (II,288); BOB HOPE SPECIAL: BOB HOPE PRESENTS THE STARS OF TOMORROW (II,292); BOB HOPE SPECIAL: BOB HOPE'S 30TH ANNIVERSARY TV SPECIAL (II,293); BOB HOPE SPECIAL: BOB HOPE'S ALL-STAR BIRTHDAY AT ANNAPOLIS (II,294); BOB HOPE SPECIAL: BOB HOPE'S ALL-STAR COMEDY BIRTHDAY PARTY (II,297); BOB HOPE SPECIAL: BOB HOPE'S ALL-STAR COMEDY BIRTHDAY PARTY AT WEST POINT (II,298); BOB HOPE SPECIAL: BOB HOPE'S ALL-STAR COMEDY LOOK AT THE FALL SEASON: IT'S STILL FREE AND WORTH IT! (II,299); BOB HOPE SPECIAL: BOB HOPE'S ALL-STAR COMEDY SPECIAL FROM AUSTRALIA (II,300); BOB HOPE SPECIAL: BOB HOPE'S ALL-STAR COMEDY SPECTACULAR FROM LAKE TAHOE (II,301); BOB HOPE SPECIAL: BOB HOPE'S ALL-STAR COMEDY TRIBUTE TO VAUDEVILLE (II,302); BOB HOPE SPECIAL: BOB HOPE'S ALL-STAR LOOK AT TV'S PRIME TIME WARS (II,303); BOB HOPE SPECIAL: BOB HOPE'S ALL-STAR SUPER BOWL PARTY (II,304); BOB HOPE SPECIAL: BOB HOPE'S ALL-STAR TRIBUTE TO THE PALACE THEATER (II,305); BOB HOPE SPECIAL: BOB HOPE'S CHRISTMAS PARTY (II,307); BOB HOPE SPECIAL: BOB HOPE'S CHRISTMAS SPECIAL (II,308); BOB HOPE SPECIAL: BOB HOPE'S FUNNY VALENTINE (II,309); BOB HOPE SPECIAL: BOB HOPE'S MERRY CHRISTMAS SHOW (II,310); BOB HOPE SPECIAL: BOB HOPE'S OVERSEAS CHRISTMAS TOURS (II,311); BOB HOPE SPECIAL: BOB HOPE'S PINK PANTHER THANKSGIVING GALA (II,312); BOB HOPE SPECIAL: BOB HOPE'S ROAD TO HOLLYWOOD (II,313); BOB HOPE SPECIAL: BOB HOPE'S SALUTE TO NASA—25 YEARS OF REACHING FOR THE STARS (II,314); BOB HOPE SPECIAL: BOB HOPE'S SPRING FLING OF COMEDY AND GLAMOUR (II,315); BOB HOPE SPECIAL: BOB HOPE'S STAND UP AND CHEER FOR THE NATIONAL FOOTBALL LEAGUE'S 60TH YEAR (II,316); BOB HOPE SPECIAL:

BOB HOPE'S STARS OVER TEXAS (II,318); BOB HOPE SPECIAL: BOB HOPE'S STAR-STUDDED SPOOF OF THE NEW TV SEASON— G RATED—WITH GLAMOUR, GLITTER AND GAGS (II,317); BOB HOPE SPECIAL: BOB HOPE'S SUPER BIRTHDAY SPECIAL (II,319); BOB HOPE SPECIAL: BOB HOPE'S WOMEN I LOVE—BEAUTIFUL BUT FUNNY (II,322); BOB HOPE SPECIAL: BOB HOPE'S WORLD OF COMEDY (II,323); BOB HOPE SPECIAL: BOB HOPE—HOPE, WOMEN AND SONG (II,324); BOB HOPE SPECIAL: BOB HOPE PRESENTS A CELEBRATION WITH STARS OF COMEDY AND MUSIC (II,340); BOB HOPE SPECIAL: HAPPY BIRTHDAY, BOB! (II,325); BOB HOPE SPECIAL: THE BOB HOPE CHRISTMAS SPECIAL (II,328); BOB HOPE SPECIAL: THE BOB HOPE CHRISTMAS SPECIAL (II,329); BOB HOPE SPECIAL: THE BOB HOPE CHRISTMAS SPECIAL (II,331); BOB HOPE SPECIAL: THE BOB HOPE COMEDY SPECIAL (II,332); BOB HOPE SPECIAL: THE BOB HOPE SPECIAL (II,333); BOB HOPE SPECIAL: THE BOB HOPE SPECIAL (II,334); BOB HOPE SPECIAL: THE BOB HOPE SPECIAL FROM PALM SPRINGS (II,339); BOB HOPE SPECIAL: BOB HOPE'S HILARIOUS UNREHEARSED ANTICS OF THE STARS (II,341); ROBERTA (I,3816)

HENSON, JIM
FROG PRINCE (I,1685); THE MUPPET SHOW (II,1766)

HEPWORTH, JOHN
PRISONER: CELL BLOCK H (II,2085)

HERBERT, TIM
NO SOAP, RADIO (II,1856)

HERMAN, MAXENE
9 TO 5 (II,1852)

HERNDON, VENABLE
THE BEST OF FRIENDS (II,216)

HERRIDGE, ROBERT
FOLK SOUND U.S.A. (I,1614); THE ROBERT HERRIDGE THEATER (I,3808); STUDIO ONE (I,4268)

HERROLD, BRO
ANOTHER LIFE (II,95); THE DRUNKARD (II,737)

HERSKOVITZ, MARSHALL
SEVEN BRIDES FOR SEVEN BROTHERS (II,2307)

HERTS, KENNETH
NUMBER 13 DEMON STREET (I,3316)

HERTUNG, ROBERT
MY FAIR LADY (I,3187)

HERTZOG, LAWRENCE
HART TO HART (II,1102); KATE LOVES A MYSTERY (II,1367)

HERVERLY, ROBERT
BARNABY JONES (II,153)

HERVEY, WINIFRED
BENSON (II,208); MORK AND MINDY (II,1735)

HERZIG, SIEGFRIED
PRIVATE SECRETARY (I,3672)

HESS, JOHN D
KAREN (II,1363); MEDICAL CENTER (II,1675); ONE DAY AT A TIME (II,1900)

HESSEMAN, HOWARD
WKRP IN CINCINNATI (II,2814)

HEUBSCH, EDWARD
STORY THEATER (I,4240)

HEUER, KENNETH
STUDIO ONE (I,4268)

HEVERLY, ROBERT
THE FBI (I,1551); MOST WANTED (II,1739); SWITCH (II,2519); THE WESTERNER (I,4801)

HEWITT, ARNOLD
THE FARMER'S DAUGHTER (I,1533)

HEYES JR, DOUGLAS
AUTOMAN (II,120); CHIPS (II,511); COVER UP (II,597)

HEYES, DOUGLAS
THE BARBARY COAST (II,146); THE BARBARY COAST (II,147); BRAVO DUKE (I,711); CAPTAINS AND THE KINGS (II,435); THE DESILU PLAYHOUSE (I,1237); THE FRENCH ATLANTIC AFFAIR (II,924); POWDERKEG (I,3657); THRILLER (I,4492)

HEYWARD, ANDREW
INSPECTOR GADGET (II,1236)

HEYWARD, LOUIS
TALES OF THE HAUNTED (II,2538)

HICKEY, PAMELA
THE GET ALONG GANG (II,980)

HICKMAN, DARRYL
THE LORETTA YOUNG THEATER (I,2756)

HICKMAN, HUDSON
THE LOVE BOAT (II,1535)

HIDER, ED
100 YEARS OF GOLDEN HITS (II,1905); AFTER HOURS: GETTING TO KNOW US (II,21); THE CAPTAIN AND TENNILLE (II,429); DEAN MARTIN'S CELEBRITY ROAST (II,643); DONNY AND MARIE (II,712); THE EDDIE RABBITT SPECIAL (II,759); MOVIN' (I,3117); THE ROWAN AND MARTIN REPORT (II,2221); THE SOUPY SALES SHOW (I,4138); TURN ON (I,4625)

HIGGINS, PAUL
DIFF'RENT STROKES (II,674)

HIGLEY, PHILO
THE ALCOA HOUR (I,108)

HIGNETT, SEAN
THE OMEGA FACTOR (II,1888)

HIKEN, NAT
THE ALL-STAR REVUE (I,138); THE BALLAD OF LOUIE THE LOUSE (I,337); BUICK CIRCUS HOUR (I,759); CAROL + 2 (I,853); THE GENERAL FOODS 25TH ANNIVERSARY SHOW (I,1756); JUST POLLY AND ME (I,2497); THE MILTON BERLE SHOW (I,3047); PHIL SILVERS IN NEW YORK (I,3577); THE PHIL SILVERS SHOW (I,3580); THE SLOWEST GUN IN THE WEST (I,4088); SUMMER IN NEW YORK (I,4279)

HILL, BENNY
THE BENNY HILL SHOW (II,207); THE UNEXPURGATED BENNY HILL SHOW (II,2705)

HILL, DAVE
THE BILL COSBY SPECIAL, OR? (I,443)

HILL, GERALD
WONDER GIRL (I,4891)

HILL, HEATHER
AS THE WORLD TURNS (II,110)

HILL, KIMBERLY
E/R (II,748); FAMILY TIES (II,819); IT'S A LIVING (II,1254); ONE DAY AT A TIME (II,1900)

HILL, MARGARET BROOKMAN
LOST IN SPACE (I,2758)

HILL, MAURY
THE GEORGE SANDERS MYSTERY THEATER (I,1776);

HILL, MICHAEL
THE TONI TENNILLE SHOW (II,2636); WE DARE YOU! (II,2751)

HILL, ROBERT
THE MAN FROM U.N.C.L.E. (I,2867); TARZAN THE APE MAN (I,4363)

HILL, WALTER
DOG AND CAT (II,695)

HILLER, JANNE
RITUALS (II,2182)

HILLIARD, ROBERT
THE BOBBY VAN AND ELAINE JOYCE SHOW (I,677); A COUPLE OF DONS (I,1077); THE DOM DELUISE SHOW (I,1328); THE HONEYMOONERS CHRISTMAS SPECIAL (II,1176); THE HONEYMOONERS CHRISTMAS (II,1177); THE HONEYMOONERS SECOND HONEYMOON (II,1178); THE HONEYMOONERS VALENTINE SPECIAL (II,1179); MCHALE'S NAVY (I,2969); THE POWDER ROOM (I,3656); SHAUGHNESSEY (II,2318); THE VICTOR BORGE SHOW (I,4719); THE VICTOR BORGE SHOW (I,4720); WHAT'S UP? (I,4823)

HILLS, DICK
THE ANTHONY NEWLEY SHOW (I,233); JULIE AND DICK IN COVENT GARDEN (II,1352); KEEP U.S. BEAUTIFUL (I,2519); PICCADILLY PALACE (I,3589)

HILLYARD, BOB
VALENTINE MAGIC ON LOVE ISLAND (II,2716)

HILSMAN, HOYT
SANTA BARBARA (II,2258)

HINDER, ED
THE BOBBY GOLDSBORO SHOW (I,673); THE CBS NEWCOMERS (I,879)

HINKLEY, BOB
THE LAS VEGAS SHOW (I,2619)

HINKLEY, DON
ALL'S FAIR (II,46); THE BOB NEWHART SHOW (I,666); DEAN MARTIN'S CELEBRITY ROAST (II,643); DINAH AND HER NEW BEST FRIENDS (II,676); FLIP WILSON. . . OF COURSE (II,890); THE HERO (I,2035); KEEP U.S. BEAUTIFUL (I,2519); THE

MAC DAVIS SHOW (II,1575); THE MUPPET SHOW (II,1766); OF THEE I SING (I,3334); TONY ORLANDO AND DAWN (II,2639)

HIRE, LOIS
THE LORETTA YOUNG THEATER (I,2756); MY THREE SONS (I,3205)

HIRSCH, JAMES G
THE BLUE KNIGHT (II,276); FOR LOVERS ONLY (II,904); THE INCREDIBLE HULK (II,1232); THE LETTERS (I,2670)

HIRSCH, JANIS
AFTERMASH (II,23); DOUBLE TROUBLE (II,724); DREAMS (II,735)

HIRSON, JANICE
SQUARE PEGS (II,2431)

HIRSON, ROGER O
A BELL FOR ADANO (I,388); NINOTCHKA (I,3299); VANITY FAIR (I,4700)

HITE, KATHLEEN
APPLE'S WAY (II,101); CROWN THEATER WITH GLORIA SWANSON (I,1107); THE DAUGHTERS OF JOSHUA CABE RETURN (I,1170); FALCON CREST (II,808); THE GUNS OF WILL SONNETT (I,1907); SUSPICION (I,4309); WAGON TRAIN (I,4747); THE WALTONS (II,2740); THE WALTONS: A DAY FOR THANKS ON WALTON'S MOUNTAIN (EPISODE 3) (II,2741)

HOANE, KEN
ROBIN'S NEST (II,2187)

HOBAN, GORDON
THE PAPER CHASE (II,1949)

HOBART JR, FENTON
BARETTA (II,152)

HOBERG, RICK
SUPER FRIENDS (II,2497)

HOCK, ALLISON
BLUE THUNDER (II,278); KNOTS LANDING (II,1404); T.J. HOOKER (II,2524)

HOCKNER, STEVEN
PAC-MAN (II,1940)

HODGE, MAX
BATMAN (I,366); CANNON (II,424); ENOS (II,779); PALMERSTOWN U.S.A. (II,1945); THE WALTONS (II,2740); THE WILD WILD WEST (I,4863)

HODGES, GEORGE
CHARLIE'S ANGELS (II,486)

HODGES, MIKE
MISSING PIECES (II,1706)

HOEY, MICHAEL A
THE FITZPATRICKS (II,868); GET CHRISTIE LOVE! (II,981); MCCLOUD (II,1660); THE RAT PATROL (I,3726)

HOFFE, ARTHUR
THE ANN SOTHERN SHOW (I,220)

HOFFMAN, CHARLES
BATMAN (I,366); THE ROGUES (I,3832)

HOFFMAN, ELWOOD
TROUBLE INC. (I,4609)

HOFFMAN, FREDERICK
GLORIA (II,1010)

HOFFMAN, JOSEPH
LEAVE IT TO BEAVER (I,2648); MY THREE SONS (I,3205); THE SMOTHERS BROTHERS SHOW (I,4096); TO SAY THE LEAST (II,2623)

HOFFMAN, SANDRA
REAL LIFE STORIES (II,2117)

HOGAN, PAUL
THE PAUL HOGAN SHOW (II,1975)

HOINGSBERG, GAIL
WORKING STIFFS (II,2834)

HOLCH, ARTHUR
CALIFORNIA GIRL (I,794)

HOLE JR, WILLIAM
SKY KING (I,4077)

HOLLAND, RANDY
THE YOUNG AND THE RESTLESS (II,2862)

HOLLANDER, MELODIE
MAUDE (II,1655)

HOLLIS, DICK
THE DON KNOTTS SHOW (I,1334); THE GOLDDIGGERS (I,1838)

HOLLOWAY, JEAN
ANDERSON AND COMPANY (I,188); BRINGING UP BUDDY (I,731); CLIMAX! (I,976); EL COYOTE (I,1423); THE FLYING NUN (I,1611); THE GHOST AND MRS. MUIR (I,1786); I HEAR AMERICA SINGING (I,2168); THE LORETTA YOUNG THEATER (I,2756); MARCUS WELBY, M.D. (II,1627); SCREEN DIRECTOR'S PLAYHOUSE (I,3946); WAGON TRAIN (I,4747); WHERE WERE YOU? (I,4834)

HOLMES, ROBERT
BLAKE'S SEVEN (II,264); DOCTOR WHO (II,689); THE SAINT (II,2237)

HOLT, FRITZ
TEXACO STAR THEATER: OPENING NIGHT (II,2565)

HOLT, ROBERT I
BARETTA (II,152); BARNABY JONES (II,153); BRONK (II,379); CANNON (II,424); CARIBE (II,439); CHOPPER ONE (II,515); EIGHT IS ENOUGH (II,762); THE MAN FROM U.N.C.L.E. (I,2867); MANHUNTER (II,1621); THE ROOKIES (II,2208); S.W.A.T. (II,2236); SAM (II,2247); THE STREETS OF SAN FRANCISCO (II,2478); YOUNG DAN'L BOONE (II,2863)

HONIGBERG, GAIL
ALICE (II,33)

HOOD, THOMAS
THE WALTONS (II,2740)

HOPCRAFT, ARTHUR
TINKER, TAILOR, SOLDIER, SPY (II,2621)

HOPE, BOB
RUSSIAN ROULETTE (I,3875)

HOPKINS, JOHN
OF MICE AND MEN (I,3333); SMILEY'S PEOPLE (II,2383)

HOPKINS, WILLIAM
KNOTS LANDING (II,1404); LOU GRANT (II,1526); THE PAPER CHASE (II,1949); STRIKE FORCE (II,2480)

HOPPS, NORMAN
MY BUDDY (II,1778)

HOROVITZ, ISRAEL
FULL HOUSE (II,937)

HOROWITZ, ANDREW
HAPPY DAYS (II,1084)

HOROWITZ, ARNOLD
DOBIE GILLIS (I,1302); FIREBALL FUN FOR ALL (I,1577); THE FLYING NUN (I,1611); GREEN ACRES (I,1884); I DREAM OF JEANNIE (I,2167); MCHALE'S NAVY (I,2969); THE PATTY DUKE SHOW (I,3483)

HOTCHNER, A E
BUICK ELECTRA PLAYHOUSE (I,760); KRAFT TELEVISION THEATER (I,2592); OF MEN OF WOMEN (I,3331); PLAYHOUSE 90 (I,3623); PLAYWRIGHTS '56 (I,3627)

HOUGHTON, JAMES
KNOTS LANDING (II,1404)

HOUGHTON, MONA
KNOTS LANDING (II,1404)

HOVIS, LARRY
MITZI (I,3071); MITZI'S SECOND SPECIAL (I,3074); ROWAN AND MARTIN'S LAUGH-IN (I,3857)

HOWARD, BOB
THE BEAN SHOW (I,373); BEAT THE CLOCK (I,378); THE DON RICKLES SHOW (II,708); A FUNNY THING HAPPENED ON THE WAY TO HOLLYWOOD (I,1716); GENERAL ELECTRIC'S ALL-STAR ANNIVERSARY (II,963); HAPPY DAYS (II,1084); JACK PAAR PRESENTS (I,2315); THE JACK PAAR SHOW (I,2317); THE JACK PAAR SPECIAL (I,2319); JACK PAAR TONIGHT (I,2320); THE JERRY LEWIS SHOW (I,2369); LIKE MAGIC (II,1476); THE VIN SCULLY SHOW (I,4734); THE WONDERFUL WORLD OF JACK PAAR (I,4899)

HOWARD, BRUCE
THE DUKES OF HAZZARD (II,742); FAR OUT SPACE NUTS (II,831); THE FLYING NUN (I,1611); GILLIGAN'S ISLAND (II,990); IT'S ABOUT TIME (I,2263); TOO CLOSE FOR COMFORT (II,2642); WHAT'S HAPPENING!! (II,2769)

HOWARD, CAL
TRUTH OR CONSEQUENCES (I,4620)

HOWARD, CY
MICKEY AND THE CONTESSA (I,3024); MUSIC U.S.A. (I,3179); MY FRIEND IRMA (I,3191)

HOWARD, MATTHEW
ALIAS SMITH AND JONES (I,118); ALIAS SMITH AND JONES (I,119)

HOWARD, RANCE
THE ANDY GRIFFITH SHOW (I,192); PARTRIDGE FAMILY: 2200 A.D. (II,1963); THROUGH THE MAGIC PYRAMID (II,2615)

HOWARD, RUTH
IT PAYS TO BE IGNORANT (I,2248)

HOWELL, CHRIS
GILLIGAN'S PLANET (II,994)

HOWELLS, J HARVEY
ROBERT MONTGOMERY PRESENTS YOUR LUCKY STRIKE THEATER (I,3809)

HOYLE, TREVOR
BLAKE'S SEVEN (II,264)

HUDIS, NORMAN
BUCK ROGERS IN THE 25TH CENTURY (II,384); CHIPS (II,511); THE FBI (I,1551); THE MAN FROM U.N.C.L.E. (I,2867); MARCUS WELBY, M.D. (II,1627); TOURIST (II,2651); TURN ON (I,4625); THE WILD WILD WEST (I,4863)

HUDSON, GARY
THE FALL GUY (II,811); HIGHWAY HONEYS (II,1151)

HUFF, JOHN
240-ROBERT (II,2688); THE BARBARY COAST (II,146); THE NIGHT STALKER (II,1849); TALES OF THE GOLD MONKEY (II,2537)

HUFFAKER, CLAIRE
BONANZA (II,347)

HUGGINS, ROY
HAZARD'S PEOPLE (II,1112); THE OUTSIDER (I,3430)

HUGHES, GORDON
ABOUT FACES (I,12); BLONDIE (I,486)

HUGHES, JOHN
DELTA HOUSE (II,658)

HUGHES, RUSSELL S
PERRY MASON (II,2025)

HUGUELY, JAY
MAGNUM, P.I. (II,1601); TALES OF THE GOLD MONKEY (II,2537)

HUME, CYRIL
THE INDIAN (I,2210); LAW OF THE PLAINSMAN (I,2639); THE RIFLEMAN (I,3789)

HUME, EDWARD
CANNON (I,820); PAROLE (II,1960); THE STREETS OF SAN FRANCISCO (I,4260); TOMA (I,4547)

HUME, PETER
TOMA (II,2634)

HUMPHREY, STEVEN
SEVEN BRIDES FOR SEVEN BROTHERS (II,2307)

HUMPHREYS, JOEL DON
THE INCREDIBLE HULK (II,1232)

HUMPHREYS, MARTHA
SPORT BILLY (II,2429)

HUNTER, BLAKE
DIFF'RENT STROKES (II,674); THE TONY RANDALL SHOW (II,2640); WHO'S THE BOSS? (II,2792); WKRP IN CINCINNATI (II,2814)

HUNTER, EVAN
ALFRED HITCHCOCK PRESENTS (I,115); THE CHISHOLMS (II,512); GUILTY OR NOT GUILTY (I,1903); OF MEN OF WOMEN (I,3332)

HUNTER, IAN MCLELLAN
THE BLUE AND THE GRAY (II,273); N.Y.P.D. (I,3321); THE STRANGE CASE OF DR. JEKYLL AND MR. HYDE (I,4244); ZERO HOUR (I,4980)

HUNTER, LEW
240-ROBERT (II,2688); CODE RED (II,552)

HUNTER, PAUL
NEWHART (II,1835); PRIVATE BENJAMIN (II,2087); TOO CLOSE FOR COMFORT (II,2642); WKRP IN CINCINNATI (II,2814)

HUNTER, STAN
THE DANCE SHOW (II,616)

HURDIS, NORMAN
THE 13TH DAY: THE STORY OF ESTHER (II,2593); MCCLOUD (II,1660)

HURLBUT, GLADYS
PRIVATE SECRETARY (I,3672)

HURLEY, JOSEPH
MR. DICKENS OF LONDON (I,3135)

HURSLEY, DORIS
BRIGHT PROMISE (I,727)

HURSLEY, FRANK
BRIGHT PROMISE (I,727)

HURT, WILLIAM
KRAFT TELEVISION THEATER (I,2592)

HURWITZ, DAVID
EIGHT IS ENOUGH (II,762); NOT NECESSARILY THE NEWS (II,1863)

HUSKY, RICK
CHARLIE'S ANGELS (II,486); HART TO HART (II,1102); MANDRAKE (II,1617); THE RENEGADES (II,2132); THE ROOKIES (II,2208); S.W.A.T. (II,2236); THE STREETS OF SAN FRANCISCO (II,2478); T.J. HOOKER (II,2524)

HUSON, PAUL
FAMILY (II,813); THE HAMPTONS (II,1076); TUCKER'S WITCH (II,2667)

HUSTON, JIMMY
FOUL PLAY (II,910)

HUSTON, LOU
THE ADDAMS FAMILY (II,11)

HUTCHINSON, THOMAS
THE ADVENTURE OF THE THREE GARRIDEBS (I,40)

HUTSON, LEE
THE BIG EASY (II,234); THE LONG DAYS OF SUMMER (II,1514)

IDELSON, BILL
BAREFOOT IN THE PARK (I,355); BEANE'S OF BOSTON (II,194); THE BETTY WHITE SHOW (II,224); THE BOB NEWHART SHOW (II,342); FISH (II,864); THE GHOST AND MRS. MUIR (I,1786); GUESS WHO'S COMING TO DINNER? (II,1063); THE HERO (I,2035)

IDLE, ERIC
FAERIE TALE THEATER (II,807); MONTY PYTHON'S FLYING CIRCUS (II,1729)

ILLES, ROBERT
AMERICA 2-NIGHT (II,65); THE CAPTAIN AND TENNILLE (II,429); THE CAROL BURNETT SHOW (II,443); THE CRACKER BROTHERS (II,599); DICK CLARK'S GOOD OL' DAYS: FROM BOBBY SOX TO BIKINIS (II,671); DOC (II,682); DOUBLE TROUBLE (II,724); FLO (II,891); JOE AND SONS (II,1296); JOE AND SONS (II,1297); JOEY AND DAD (II,1306); LILY (I,2705); THE LOU RAWLS SPECIAL (II,1527); THE LOVE BOAT I (II,1532); A NEW KIND OF FAMILY (II,1819); NO SOAP, RADIO (II,1856); PEEPING TIMES (II,1990); PRIVATE BENJAMIN (II,2087); SILVER SPOONS (II,2355); THE SMOTHERS BROTHERS SHOW (II,2384); VAN DYKE AND COMPANY (II,2718)

ILSON, SAUL
20TH CENTURY FOLLIES (I,4641); ARTHUR GODFREY AND THE SOUNDS OF NEW YORK (I,280); THE BEATRICE ARTHUR SPECIAL (II,196); A BEDTIME STORY (I,384); THE BING CROSBY SHOW (I,463); THE BING CROSBY SHOW (I,464); THE BOBBY DARIN AMUSEMENT COMPANY (I,672); CAROL CHANNING AND PEARL BAILEY ON BROADWAY (I,850); CAROL CHANNING PROUDLY PRESENTS THE SEVEN DEADLY SINS (I,851); THE DANNY KAYE SHOW (I,1143); THE DINAH SHORE CHEVY SHOW (I,1279); THE DORIS MARY ANNE KAPPELHOFF SPECIAL (I,1356); FRANCIS ALBERT SINATRA DOES HIS THING (I,1658); FRIENDS AND NABORS (I,1684); HOW TO HANDLE A WOMAN (I,2146); THE JOHN GARY SHOW (I,2411); THE LESLIE UGGAMS SHOW (I,2660); THE LIVING END (I,2731); NEIL SEDAKA STEPPIN' OUT (II,1804); RAINBOW OF STARS (I,3717); RICH LITTLE'S WASHINGTON FOLLIES (II,2159); THE SMOTHERS BROTHERS COMEDY HOUR (I,4095); TEXACO STAR PARADE I (I,4409); TEXACO STAR PARADE II (I,4410); TONY ORLANDO AND DAWN (II,2639); TV FUNNIES (II,2672); TWO'S COMPANY (I,4659); VER-R-R-RY INTERESTING (I,4712)

INGALLS, DON
THE BIG VALLEY (I,437); BONANZA (II,347); CAPTAIN AMERICA (II,427); DANGER MAN (I,1136); FANTASY ISLAND (II,829); A MAN CALLED SLOANE (II,1613); SECRET AGENT (I,3962); STAR TREK (II,2440)

INGRAM, GAIL
THE BIG STORY (I,432); ROBERT MONTGOMERY PRESENTS YOUR LUCKY STRIKE THEATER (I,3809)

INMAN, JIM
THE ABC AFTERSCHOOL SPECIAL (II,1); THE YOUNG AND THE RESTLESS (II,2862)

INSANA, TINO
POLICE SQUAD! (II,2061); WELCOME TO THE FUN ZONE (II,2763)

IOVER, MORT
THE STARLOST (I,4203)

IRELAND JR, JOHN
THE NANCY DREW MYSTERIES (II,1789)

ISAACS, CHARLES
THE ALL-STAR REVUE (I,138); BOB HOPE SPECIAL: BOB HOPE'S ALL-STAR COMEDY BIRTHDAY PARTY AT WEST POINT (II,298); THE BOBBY DARIN AMUSEMENT COMPANY (I,672); THE CBS NEWCOMERS (I,879); THE DUKE (I,1381); ENTERTAINMENT —1955 (I,1450); FULL SPEED ANYWHERE (I,1704); GIVE MY REGARDS TO BROADWAY (I,1815); HARPER VALLEY PTA (II,1095); THE IMPERIAL GRAND BAND

(II,1224); MONTY HALL'S VARIETY HOUR (II,1728); A SALUTE TO STAN LAUREL (I,3892); VICTOR BORGE'S 20TH ANNIVERSARY SHOW (I,4727)

ISAACS, DAVID
AFTERMASH (II,23); THE BAY CITY AMUSEMENT COMPANY (II,185); CHARACTERS (II,481); CHEERS (II,494); JOE AND SONS (II,1296); M*A*S*H (II,1569); OPEN ALL NIGHT (II,1914); THE TONY RANDALL SHOW (II,2640)

ISENBERG, CHARLES
THE ANN SOTHERN SHOW (I,220)

ISRAEL, CHARLES
LOOSE CHANGE (II,1518); SONS AND DAUGHTERS (II,2404)

ISRAEL, NEIL
THE MAC DAVIS SHOW (II,1576); RINGO (II,2176); TWILIGHT THEATER (II,2685); TWILIGHT THEATER II (II,2686)

JACK, MAX
THE AMERICAN DREAM (II,71); CALL TO GLORY (II,413)

JACK, MENDELSHON
LOLA (II,1508)

JACKER, CORINNE
LOOSE CHANGE (II,1518)

JACKS, ROBERT L
THE DICK POWELL SHOW (I,1269); LAST OF THE PRIVATE EYES (I,2628)

JACKSON, BILL
GIGGLESNORT HOTEL (II,989)

JACKSON, REGINALD
THE GUARDSMAN (I,1893)

JACOB, NORMAN
BEN CASEY (I,394); MATINEE THEATER (I,2947)

JACOBS, BOB
THE DIONNE WARWICK SPECIAL (I,1289)

JACOBS, DAVID
BEHIND THE SCREEN (II,200); CHICAGO STORY (II,506); FAMILY (II,813); KNOTS LANDING (II,1404); SECRETS OF MIDLAND HEIGHTS (II,2296)

JACOBS, GARY
BABY, I'M BACK! (II,130); BOWZER (II,358); THE DAVID LETTERMAN SHOW (II,624); FOOT IN THE DOOR (II,902); MAMA'S FAMILY (II,1610); ON STAGE AMERICA (II,1893); ONE DAY AT A TIME (II,1900); SUNDAY FUNNIES (II,2493); THICKE OF THE NIGHT (II,2587)

JACOBS, HARVEY
OF MEN OF WOMEN (I,3331); THAT WAS THE YEAR THAT WAS (II,2575)

JACOBS, JACK
THE ALCOA HOUR (I,108); THE GEORGE SANDERS MYSTERY THEATER (I,1776); GREATEST HEROES OF THE BIBLE (II,1062); MEDIC (I,2976)

JACOBS, MICHAEL
CHARLES IN CHARGE (II,482); NO SOAP, RADIO (II,1856)

JACOBS, SEAMAN
3 (II,0); THE ADDAMS FAMILY (II,11); BACHELOR FATHER (I,323); BOB HOPE SPECIAL: BOB HOPE FOR PRESIDENT (II,282); BOB HOPE SPECIAL: BOB HOPE GOES TO COLLEGE (II,283); BOB HOPE SPECIAL: BOB HOPE IN THE STAR-MAKERS (II,285); BOB HOPE SPECIAL: BOB HOPE IN WHO MAKES THE WORLD LAUGH, PART 2 (II,286); BOB HOPE SPECIAL: BOB HOPE IN WHO MAKES THE WORLD LAUGH? (II,287); BOB HOPE SPECIAL: BOB HOPE LAUGHS WITH THE MOVIE AWARDS (II,288); BOB HOPE SPECIAL: BOB HOPE'S 30TH ANNIVERSARY TV SPECIAL (II,293); BOB HOPE SPECIAL: BOB HOPE'S ALL-STAR BIRTHDAY AT ANNAPOLIS (II,294); BOB HOPE SPECIAL: BOB HOPE'S ALL-STAR COMEDY BIRTHDAY PARTY (II,297); BOB HOPE SPECIAL: BOB HOPE'S ALL-STAR COMEDY BIRTHDAY PARTY AT WEST POINT (II,298); BOB HOPE SPECIAL: BOB HOPE'S ALL-STAR COMEDY LOOK AT THE FALL SEASON: IT'S STILL FREE AND WORTH IT! (II,299); BOB HOPE SPECIAL: BOB HOPE'S ALL-STAR LOOK AT TV'S PRIME TIME WARS (II,303); BOB HOPE SPECIAL: BOB HOPE'S ALL-STAR SUPER BOWL PARTY (II,304); BOB HOPE SPECIAL: BOB HOPE'S FUNNY VALENTINE (II,309); BOB HOPE SPECIAL: BOB HOPE'S MERRY CHRISTMAS SHOW (II,310); BOB HOPE SPECIAL: BOB HOPE'S OVERSEAS CHRISTMAS TOURS (II,311); BOB HOPE SPECIAL: BOB HOPE'S PINK PANTHER THANKSGIVING GALA (II,312); BOB HOPE SPECIAL: BOB HOPE'S ROAD TO HOLLYWOOD (II,313); BOB HOPE SPECIAL: BOB HOPE'S SALUTE TO NASA—25 YEARS OF REACHING FOR THE STARS (II,314); BOB HOPE SPECIAL: BOB HOPE'S SPRING FLING OF COMEDY AND GLAMOUR (II,315); BOB HOPE SPECIAL: BOB HOPE'S STAND UP AND CHEER FOR THE NATIONAL FOOTBALL LEAGUE'S 60TH YEAR (II,316); BOB HOPE SPECIAL: BOB HOPE'S STAR-STUDDED SPOOF OF THE NEW TV SEASON— G RATED—WITH GLAMOUR, GLITTER AND GAGS (II,317); BOB HOPE SPECIAL: BOB HOPE'S SUPER BIRTHDAY SPECIAL (II,319); BOB HOPE SPECIAL: BOB HOPE'S USO CHRISTMAS IN BEIRUT (II,320); BOB HOPE SPECIAL: BOB HOPE'S WICKI-WACKY SPECIAL FROM WAIKIKI (II,321); BOB HOPE SPECIAL: BOB HOPE'S WOMEN I LOVE—BEAUTIFUL BUT FUNNY (II,322); BOB HOPE SPECIAL: BOB HOPE—HOPE, WOMEN AND SONG (II,324); BOB HOPE SPECIAL: BOB HOPE PRESENTS A CELEBRATION WITH STARS OF COMEDY AND MUSIC (II,340); BOB HOPE SPECIAL: HAPPY BIRTHDAY, BOB! (II,325); BOB HOPE SPECIAL: HO HO HOPE'S JOLLY CHRISTMAS HOUR (II,327); BOB HOPE SPECIAL: THE BOB HOPE CHRISTMAS SPECIAL (II,328); BOB HOPE SPECIAL: THE BOB HOPE CHRISTMAS SPECIAL (II,329); BOB HOPE SPECIAL: THE BOB HOPE CHRISTMAS SPECIAL (II,331); BOB HOPE SPECIAL: BOB HOPE'S HILARIOUS UNREHEARSED ANTICS OF THE STARS (II,341); BRINGING UP BUDDY (I,731); BUNGLE ABBEY (II,396); THE CBS NEWCOMERS (I,879); F TROOP (I,1499); GEORGE BURNS AND OTHER SEX SYMBOLS (II,966); GEORGE BURNS CELEBRATES 80 YEARS IN SHOW BUSINESS (II,967); GEORGE BURNS EARLY, EARLY, EARLY, CHRISTMAS SHOW (II,968); GEORGE BURNS IN NASHVILLE?? (II,969); THE GEORGE BURNS SPECIAL (II,970); GEORGE BURNS' 100TH BIRTHDAY PARTY (II,971); GEORGE BURNS' HOW TO LIVE TO BE 100 (II,972); HERE'S LUCY (II,1135); HOW TO MARRY A MILLIONAIRE (I,2147); THE JOHNNY CARSON SHOW (I,2422); THE LOVE BOAT (II,1535); MAUDE (II,1655); MONTY HALL'S VARIETY HOUR (II,1728); MR. ROBERTS (I,3149); MY FAVORITE MARTIAN (I,3189); NO TIME FOR SERGEANTS (I,3300); RUN, BUDDY, RUN (I,3870); SIGMUND AND THE SEA MONSTERS (II,2352); THE TIM CONWAY COMEDY HOUR (I,4501)

JACOBSON, BILL
THE KATE SMITH EVENING HOUR (I,2507); THE KATE SMITH HOUR (I,2508)

JACOBSON, DANNY
I GAVE AT THE OFFICE (II,1213); THE RAINBOW GIRL (II,2108); SOAP (II,2392)

JACOBSON, DAVID
FAME (II,812)

JACOBSON, STAN
THE JOHNNY CASH SHOW (I,2424)

JACOBSON, WILLIAM
BEST FOOT FORWARD (I,401)

JACOBUS, JERRY
THE FACTS OF LIFE (II,805)

JACOBY, COLEMAN
THE ALAN KING SHOW (I,100); THE BACHELOR (I,325); THE GARRY MOORE SHOW (I,1739); NBC FOLLIES OF 1965 (I,3240); THE PHIL SILVERS PONTIAC SPECIAL: KEEP IN STEP (I,3579); THE PHIL SILVERS SHOW (I,3580); STRIPPERS (II,2481); THIS MORNING (I,4464)

JAFFE, ANITA
ALL MY CHILDREN (II,39)

JAFFE, JEWEL
THE BAXTERS (II,184)

JAKOBY, DON
BLUE THUNDER (II,278)

JAMES, ALEX
THE STARLOST (I,4203)

JAMES, DONALD
THE AVENGERS (II,121); THE CHAMPIONS (I,896); THE SAINT (II,2237); U.F.O. (I,4662)

JAMES, ED
THE ADDAMS FAMILY (II,11); DOBIE GILLIS (I,1302); LEAVE IT TO BEAVER (I,2648); MR. ROBERTS (I,3149); MY FAVORITE MARTIAN (I,3189); RUN, BUDDY, RUN (I,3870)

JAMES, JOHN THOMAS

BARETTA (II,152); TOMA (II,2634)

JAMPEL, CAROL
THE PAUL WINCHELL AND JERRY MAHONEY SHOW (I,3492)

JANAVER, RICHARD
SCRABBLE (II,2278)

JANES, ROBERT
THE AMAZING SPIDER-MAN (II,63); BARETTA (II,152); BUSH DOCTOR (II,400); THE FALL GUY (II,811); HAWAII FIVE-O (II,1110); M STATION: HAWAII (II,1568); VOYAGERS (II,2730); WAIKIKI (II,2734)

JANIS, PAULA
A MAGIC GARDEN CHRISTMAS (II,1590)

JANSON, LEN
BENJI, ZAX AND THE ALIEN PRINCE (II,205); GOING BANANAS (II,1015); SPEED BUGGY (II,2419); TARZAN: LORD OF THE JUNGLE (II,2544); UNCLE CROC'S BLOCK (II,2702); WHEELIE AND THE CHOPPER BUNCH (II,2776)

JAY, ANTHONY
LORNE GREENE'S LAST OF THE WILD (II,1522); YES MINISTER (II,2848)

JEAN, AL
CHARLES IN CHARGE (II,482)

JEAN, MONIQUE
MATINEE THEATER (I,2947)

JENKIN, LEN
FAMILY (II,813); THE INCREDIBLE HULK (II,1232)

JENKIN, RAY
THE WOMAN IN WHITE (II,2819)

JENKINS, DAL
CHIPS (II,511)

JENKINS, GORDON
MANHATTAN TOWER (I,2889)

JENKYNS, CHRIS
THE BULLWINKLE SHOW (I,761)

JENSON, LEN
STAR TREK (II,2439)

JEROME, STUART
AMOS BURKE, SECRET AGENT (I,180); THE FUGITIVE (I,1701); THE NEW ADVENTURES OF CHARLIE CHAN (I,3249); THE ROGUES (I,3832)

JESSEL, GEORGE
GEORGE JESSEL'S SHOW BUSINESS (I,1770)

JESSEL, RAY
BERT CONVY SPECIAL—THERE'S A MEETING HERE TONIGHT (II,210); THE BILL COSBY SPECIAL, OR? (I,443); THE BOB NEWHART SHOW (II,342); THE CAPTAIN AND TENNILLE (II,429); THE CARPENTERS (II,444); DINAH IN SEARCH OF THE IDEAL MAN (I,1278); FLIP WILSON. . . OF COURSE (II,890); GENE KELLY'S WONDERFUL WORLD OF GIRLS (I,1752); THE GLEN CAMPBELL GOODTIME HOUR (I,1820); THE JACKSONS (II,1267); THE JERRY REED WHEN YOU'RE HOT, YOU'RE HOT HOUR (I,2372); THE JOHN DENVER SPECIAL (II,1315); THE JOHN DENVER SPECIAL (II,1316); THE LOVE BOAT (II,1535); THE RICH LITTLE SHOW (II,2153); THE RICH LITTLE SHOW (II,2154); THE SMOTHERS BROTHERS SHOW (II,2384)

JESSUP, RICHARD
ROUTE 66 (I,3852); TURNOVER SMITH (II,2671)

JOELSON, BEN
THE ANDY GRIFFITH SHOW (I,192); THE ANN SOTHERN SHOW (I,220); ARNIE (I,261); THE GARRY MOORE SHOW (I,1740); GLITTER (II,1009); GOOD TIMES (II,1038); HAPPY DAYS (II,1084); THE JIM NABORS HOUR (I,2381); THE LOVE BOAT (II,1535); THE ODD COUPLE (II,1875); THE VICTOR BORGE SHOW (I,4719); THE VICTOR BORGE SHOW (I,4720)

JOHNS, JONNIE
ALOHA PARADISE (II,58); THE JOHN DAVIDSON CHRISTMAS SHOW (II,1307)

JOHNS, VICTORIA
TRAPPER JOHN, M.D. (II,2654)

JOHNSON, ANDREW
LAVERNE AND SHIRLEY (II,1446); PAUL SAND IN FRIENDS AND LOVERS (II,1982); THE REDD FOXX COMEDY HOUR (II,2126)

JOHNSON, ARTE
VER-R-R-RY INTERESTING (I,4712)

JOHNSON, BRUCE
MORK AND MINDY (II,1735)

JOHNSON, CATHERINE
THE SMURFS (II,2385)

JOHNSON, CHARLES
S.W.A.T. (II,2236)

JOHNSON, COSLOUGH
AMERICA, YOU'RE ON (II,69); ARNOLD'S CLOSET REVIEW (I,262); THE DINAH SHORE SPECIAL—LIKE HEP (I,1282); THE GLEN CAMPBELL GOODTIME HOUR (I,1820); THE HUDSON BROTHERS RAZZLE DAZZLE COMEDY SHOW (II,1201); THE JERRY REED WHEN YOU'RE HOT, YOU'RE HOT HOUR (I,2372); THE JOHN BYNER COMEDY HOUR (I,2407); LI'L ABNER (I,2702); LIGHTS, CAMERA, MONTY! (II,1475); LIKE MAGIC (II,1476); OPERATION PETTICOAT (II,1917); THE PARTRIDGE FAMILY (II,1962); ROWAN AND MARTIN'S LAUGH-IN (I,3856); SHA NA NA (II,2311); THE SONNY AND CHER COMEDY HOUR (II,2400); THE SONNY AND CHER SHOW (II,2401); THE SONNY COMEDY REVUE (II,2403); SPORT BILLY (II,2429); VER-R-R-RY INTERESTING (I,4712); VOLTRON—DEFENDER OF THE UNIVERSE (II,2729)

JOHNSON, DON
MAN IN A SUITCASE (I,2868); WHERE'S RAYMOND? (I,4837)

JOHNSON, DOUGLAS
HAWKINS FALLS, POPULATION 6200 (I,1977); SKY KING (I,4077)

JOHNSON, GARY
SALE OF THE CENTURY (II,2239)

JOHNSON, GEORGE CLAYTON
KUNG FU (II,1416); ROUTE 66 (I,3852); STAR TREK (II,2440); THE TWILIGHT ZONE (I,4651)

JOHNSON, JAMES BURR
THE FUNNY WORLD OF FRED & BUNNI (II,943); ONE DAY AT A TIME (II,1900)

JOHNSON, JAY
MATINEE IN NEW YORK (I,2946); WOLF ROCK TV (II,2815)

JOHNSON, KAY
THE ABC AFTERSCHOOL SPECIAL (II,1)

JOHNSON, KENNETH
THE BIONIC WOMAN (II,255); GRIFF (I,1888); HOT PURSUIT (II,1189); THE INCREDIBLE HULK (II,1232); THE SIX-MILLION-DOLLAR MAN (II,2372); V (II,2713)

JOHNSON, MONICA
THE CHEERLEADERS (II,493); LAVERNE AND SHIRLEY (II,1446); THE MARY TYLER MOORE SHOW (II,1640); PAUL SAND IN FRIENDS AND LOVERS (II,1982); THE PLANT FAMILY (II,2050)

JOHNSON, NUNNALLY
ROBERT MONTGOMERY PRESENTS YOUR LUCKY STRIKE THEATER (I,3809)

JOHNSON, ROBERT
DIFF'RENT STROKES (II,674)

JOHNSON, STERLING
PERRY COMO'S SPRING IN SAN FRANCISCO (II,2022)

JOHNSON, VICTOTRIA
TRAPPER JOHN, M.D. (II,2654)

JOLLEY, NORMAN
THE ADAM MACKENZIE STORY (I,30); BARNABY JONES (II,153); THE DOG TROOP (I,1325); THE FBI (I,1551); THE KILLIN' COUSIN (II,1394); WAGON TRAIN (I,4747)

JONAS, THEODORE
THE SECRET WAR OF JACKIE'S GIRLS (II,2294)

JONES, DAVID SCOTT
BLONDES VS. BRUNETTES (II,271)

JONES, HILDY
PARADE OF STARS (II,1954)

JONES, IAN
AGAINST THE WIND (II,24)

JONES, J FRANKLIN
JACKIE GLEASON AND HIS AMERICAN SCENE MAGAZINE (I,2321)

JONES, JULIAN
Q. E. D. (II,2094)

JONES, KAREN
FOREVER FERNWOOD (II,909)

JONES, MARK
THE FALL GUY (II,811); LOBO (II,1504); THE POWERS OF MATTHEW STAR (II,2075); RIPTIDE (II,2178); THE ROUSTERS (II,2220)

JONES, MICHAEL
MEATBALLS AND SPAGHETTI (II,1674)

JONES, MORT

THE MISADVENTURES OF SHERIFF LOBO (II,1704)

JONES, PATRICIA
THE BOB NEWHART SHOW (II,342); THE LAST RESORT (II,1440); REPORT TO MURPHY (II,2134); THE TONY RANDALL SHOW (II,2640)

JONES, PRESTON
THE OLDEST LIVING GRADUATE (II,1882)

JONES, TRENT
RYAN'S HOPE (II,2234)

JONES, TREVOR
THE DON LANE SHOW (II,707)

JOSEFSBERG, MILT
ALL IN THE FAMILY (II,38); ARCHIE BUNKER'S PLACE (II,105); DOWN HOME (I,1366); HERE'S LUCY (II,1135); THE JACK BENNY PROGRAM (I,2294); THE JACK BENNY SPECIAL (I,2296); THE MILTON BERLE SPECIAL (I,3050); THE MILTON BERLE SPECIAL (I,3051)

JOSEPH, BRYAN
THE ANDY WILLIAMS SHOW (I,210); THE BOBBY DARIN AMUSEMENT COMPANY (I,672); FOOT IN THE DOOR (II,902); THE NEIGHBORS (II,1801); OPERATION PETTICOAT (II,1917); PERRY COMO'S LAKE TAHOE HOLIDAY (II,2018); STRUCK BY LIGHTNING (II,2482); TOO CLOSE FOR COMFORT (II,2642); TURN ON (I,4625)

JOSEPH, JACKIE
BARNABY JONES (II,153); FAMILY (II,813)

JOSEPH, JOHN
THE WALTONS (II,2740)

JOSEPH, ROBERT L
ISIS (II,1248); RAGE OF ANGELS (II,2107); THE WORD (II,2833)

JOWITT, D A
FIRESIDE THEATER (I,1580)

JOYCE, WILLIAM
THE ADVENTURES OF SUPERMAN (I,77)

JUHL, JERRY
FRAGGLE ROCK (II,916); THE MUPPET SHOW (II,1766); THE MUPPETS GO TO THE MOVIES (II,1767)

JULIAN, ARTHUR
APPLE PIE (II,100); AT EASE (II,115); DON RICKLES—ALIVE AND KICKING (I,1340); THE DORIS DAY SHOW (I,1355); F TROOP (I,1499); FAIR EXCHANGE (I,1512); THE FLYING NUN (I,1611); GIMME A BREAK (II,995); HANGING IN (II,1078); HAPPY ANNIVERSARY AND GOODBYE (II,1082); LOVE THY NEIGHBOR (I,2781); THE MAC DAVIS SHOW (II,1575); MAUDE (II,1655); REAR GUARD (II,2120); THE SON-IN-LAW (II,2398); THE TWO OF US (I,4658); THE TWO OF US (II,2693)

JUNKIN, HARRY W
DEPARTMENT S (I,1232); FRONT ROW CENTER (I,1694); KRAFT TELEVISION THEATER (I,2592); THE SAINT (II,2237); WAGON TRAIN (I,4747)

JURIST, ED
THE ALDRICH FAMILY (I,111); BEWITCHED (I,418); THE CARA WILLIAMS SHOW (I,843); DIFF'RENT STROKES (II,674); GIMME A BREAK (II,995); THE HOOFER (I,2113); THE PATTY DUKE SHOW (I,3483); TABITHA (II,2526)

JUSTICE, PATRICIA
SHANNON (II,2314)

KADISON, ELLIS
CHALK ONE UP FOR JOHNNY (I,893); CLIMAX! (I,976); THE INVADERS (I,2229); NAKED CITY (I,3216); THE RAT PATROL (I,3726)

KADREY, RICHARD
THE SMURFS (II,2385)

KAGAN, MICHAEL
BEN VEREEN—HIS ROOTS (II,202); THE BRADY BUNCH HOUR (II,363); CONDO (II,565); THE KROFFT KOMEDY HOUR (II,1411); MEL AND SUSAN TOGETHER (II,1677); MUSIC HALL AMERICA (II,1773); THE OSMOND BROTHERS SPECIAL (II,1925); STEPHANIE (II,2453); THE STEVE LANDESBERG TELEVISION SHOW (II,2458)

KAHAN, JUDITH
E/R (II,748); LOVE, NATALIE (II,1545)

KAHANE, JACKIE
BIZARRE (II,259)

KAHN, BERNARD M
THE BANANA COMPANY (II,139); BOB & CAROL & TED & ALICE (I,495); HERE WE GO AGAIN (I,2028); MY WORLD . . . AND WELCOME TO IT (I,3208); THE SMOTHERS BROTHERS SHOW (I,4096); TABITHA (II,2528); THESE ARE THE DAYS (II,2585); THE UGLIEST GIRL IN TOWN (I,4663)

KALCHEIM, LEE
ALL IN THE FAMILY (II,38); HENRY WINKLER MEETS WILLIAM SHAKESPEARE (II,1129); MARRIAGE IS ALIVE AND WELL (II,1633); THE PAPER CHASE: THE SECOND YEAR (II,1950); THE SECRET OF CHARLES DICKENS (II,2293); SHANNON (II,2314); ZERO INTELLIGENCE (II,2876)

KALDOR, ERIC
BRONK (II,379); THE BUREAU (II,397); GET CHRISTIE LOVE! (II,981); THE ROCKFORD FILES (II,2197)

KALISH, ARTHUR
F TROOP (I,1499)

KALISH, AUSTIN
ALL IN THE FAMILY (II,38); AMERICA 2100 (II,66); APPLE'S WAY (II,101); CARTER COUNTRY (II,449); THE FLYING NUN (I,1611); GHOST OF A CHANCE (II,987); GIDGET (I,1795); GOOD HEAVENS (II,1032); GOOD TIMES (II,1038); HERE WE GO AGAIN (I,2028); JEANNIE (II,1275); KANGAROOS IN THE KITCHEN (II,1362); LOVE NEST (II,1541); MASON (II,1643); MY FAVORITE MARTIAN (I,3189); MY THREE SONS (I,3205); OH MADELINE (II,1880); PLEASE DON'T EAT THE DAISIES (I,3628); RENDEZVOUS HOTEL (II,2131); SUPERCOPS (II,2502); TOO CLOSE FOR COMFORT (II,2642); WILDER AND WILDER (II,2802)

KALISH, BRUCE
THE INCREDIBLE HULK (II,1232); LANDON, LANDON & LANDON (II,1426); LOTTERY (II,1525); QUICK AND QUIET (II,2099); TOO CLOSE FOR COMFORT (II,2642)

KALISH, IRMA
ALL IN THE FAMILY (II,38); AMERICA 2100 (II,66); APPLE'S WAY (II,101); CARTER COUNTRY (II,449); F TROOP (I,1499); FOOT IN THE DOOR (II,902); GHOST OF A CHANCE (II,987); GIDGET (I,1795); GOOD HEAVENS (II,1032); GOOD TIMES (II,1038); HERE WE GO AGAIN (I,2028); JEANNIE (II,1275); KANGAROOS IN THE KITCHEN (II,1362); LOVE NEST (II,1541); MASON (II,1643); MY THREE SONS (I,3205); OH MADELINE (II,1880); PLEASE DON'T EAT THE DAISIES (I,3628); RENDEZVOUS HOTEL (II,2131); SUPERCOPS (II,2502); TOO CLOSE FOR COMFORT (II,2642); WILDER AND WILDER (II,2802)

KALLEN, LUCILLE
THE ADMIRAL BROADWAY REVUE (I,37); HOLIDAY IN LAS VEGAS (I,2076); THE IMOGENE COCA SHOW (I,2198); PRIVATE EYE, PRIVATE EYE (I,3671); WE INTERRUPT THIS SEASON (I,4780)

KALLIS, DANNY
SILVER SPOONS (II,2355)

KAMMERMAN, ROY
BIG EDDIE (II,236); THE BOOK OF LISTS (II,350); DOC (II,682); FISH (II,864); GILLIGAN'S ISLAND (II,990); A NEW KIND OF FAMILY (II,1819); SO YOU THINK YOU GOT TROUBLES?! (II,2391)

KAMP, IRENE
SANDBURG'S LINCOLN (I,3907)

KAMP, LUIS
SANDBURG'S LINCOLN (I,3907)

KAMPMANN, STEVEN
RODNEY DANGERFIELD SPECIAL: IT'S NOT EASY BEIN' ME (II,2199); WKRP IN CINCINNATI (II,2814)

KAMUIMA, KAZUO
SPACE GIANTS (I,4142)

KANDEL, RICHARD
CANNON (II,424)

KANDEL, STEPHEN
BANACEK (II,138); BATMAN (I,366); BRONK (II,379); CALIFORNIA FEVER (II,411); CHARLIE'S ANGELS (II,486); FOR LOVE AND HONOR (II,903); GIDGET (I,1795); HARRY O (II,1099); HAWAII FIVE-O (II,1110); I SPY (I,2179); JESSICA NOVAK (II,1286); A MAN CALLED SLOANE (II,1613); MANNIX (II,1624); MARCUS WELBY, M.D. (II,1627); MEDICAL CENTER (II,1675); MOST WANTED (II,1739); NERO WOLFE (II,1806); THE PAPER CHASE (II,1949); THE ROGUES (I,3832); SCALPLOCK (I,3930); STAR TREK (II,2439); STAR TREK (II,2440); SWITCH (II,2519); THE WILD WILD WEST (I,4863); WONDER WOMAN (SERIES 1) (II,2828); WONDER WOMAN (SERIES 2) (II,2829)

KANDER, JOHN
LIZA WITH A Z (I,2732); THREE FOR THE GIRLS (I,4477)

KANE, ARNOLD
THE BOB NEWHART SHOW (II,342); BOBBIE GENTRY'S HAPPINESS HOUR (I,671); BOBBIE GENTRY'S HAPPINESS HOUR (II,343); DICK VAN DYKE AND THE OTHER WOMAN, MARY TYLER MOORE (I,1273); THE FUNNY SIDE (I,1714); THE FUNNY SIDE (I,1715); HERE WE GO AGAIN (I,2028); HIS AND HERS (II,1155); THE JIM NABORS HOUR (I,2381); KEEPING UP WITH THE JONESES (II,1374); ON OUR OWN (II,1892); ONE DAY AT A TIME (II,1900); THE OSMOND BROTHERS SHOW (I,3409); PAUL SAND IN FRIENDS AND LOVERS (II,1982); PRIVATE BENJAMIN (II,2087); SUPER COMEDY BOWL 1 (I,4288); SUPER COMEDY BOWL 2 (I,4289); WE GOT IT MADE (II,2752)

KANE, BRUCE
THE BOB NEWHART SHOW (II,342); DOC (II,683); ON OUR OWN (II,1892); QUARK (II,2095); RHODA (II,2151); SUGAR TIME (II,2489); TWO THE HARD WAY (II,2695)

KANE, DAVID
WHAT GAP? (I,4808)

KANE, DORIAN
THE ALL-STAR SALUTE TO MOTHER'S DAY (II,51)

KANE, HENRY
THE HUNTER (I,2162)

KANE, JOEL
DOBIE GILLIS (I,1302); THE GHOST AND MRS. MUIR (I,1786); THE INVADERS (I,2229); IT'S ABOUT TIME (I,2263); THE ODDBALL COUPLE (II,1876); PEBBLES AND BAMM BAMM (II,1987); PLEASE DON'T EAT THE DAISIES (I,3628); SPEED BUGGY (II,2419); THE WILD WILD WEST (I,4863)

KANE, JOSEPH NATHAN
BREAK THE BANK (I,712)

KANIN, FAY
HEAT OF ANGER (I,1992)

KANIN, GARSON
BORN YESTERDAY (I,692)

KANTER, HAL
ALL IN THE FAMILY (II,38); ARTHUR GODFREY IN HOLLYWOOD (I,281); BEYOND WITCH MOUNTAIN (II,228); CAP'N AHAB (I,824); THE DANNY KAYE SHOW (I,1144); THE GEORGE GOBEL SHOW (I,1767); HAPPY BIRTHDAY, AMERICA (II,1083); AN HOUR WITH DANNY KAYE (I,2136); THE JIMMY STEWART SHOW (I,2391); THE REASON NOBODY HARDLY EVER SEEN A FAT OUTLAW IN THE OLD WEST IS AS FOLLOWS: (I,3744); SALLY AND SAM (I,3891); THREE ON AN ISLAND (I,4483)

KANTER, JEFF
CODE R (II,551); MEDICAL CENTER (II,1675); STARSKY AND HUTCH (II,2444)

KANTOR, LEONARD
CANNON (II,424); THE FUGITIVE (I,1701); HART TO HART (II,1102); MANHUNTER (II,1621); PURSUIT (I,3700); THE UNTOUCHABLES (I,4682)

KAPLAN, E J
FILTHY RICH (II,856); HOT L BALTIMORE (II,1187); HOTEL (II,1192); KUDZU (II,1415); RETURN TO THE PLANET OF THE APES (II,2145); WHEN THINGS WERE ROTTEN (II,2782); WORKING STIFFS (II,2834)

KAPLAN, GABRIEL
GABRIEL KAPLAN PRESENTS THE SMALL EVENT (II,952); LEWIS AND CLARK (II,1465); WELCOME BACK, KOTTER (II,2761)

KAPLAN, JACK
LETTERS TO LAUGH-IN (I,2672); THE RICH LITTLE SHOW (II,2153); THE RICH LITTLE SHOW (II,2154); SPEED BUGGY (II,2419); TURN ON (I,4625)

KAPP, DAVID
SKY KING (I,4077)

KAPRALL, BO
BOB HOPE SPECIAL: BOB HOPE'S CHRISTMAS PARTY (II,307); CHER (II,495); CHER (II,496); THE JIM STAFFORD SHOW (II,1291); LAVERNE AND SHIRLEY (II,1446); SOLID GOLD '79 (II,2396); THANKSGIVING REUNION WITH THE PARTRIDGE FAMILY AND MY THREE SONS (II,2569); THAT SECOND THING ON ABC (II,2572); WACKO (II,2733); WELCOME BACK, KOTTER (II,2761); WHEN THINGS WERE ROTTEN (II,2782)

KARLAN, PATRICIA
SCREEN DIRECTOR'S PLAYHOUSE (I,3946)

KARLAN, RICHARD
SCREEN DIRECTOR'S PLAYHOUSE (I,3946)

KARLEN, BERNARD E
I'D LIKE TO SEE (I,2184)

KARP, DAVID
THE ALCOA HOUR (I,108); ALCOA PREMIERE (I,109); HAWKINS ON MURDER (I,1978); I SPY (I,2179); THE MISSISSIPPI (II,1707); PLAYHOUSE 90 (I,3623); QUINCY, M. E. (II,2102); SAINTS AND SINNERS (I,3887); TARGET: THE CORRUPTERS (I,4360); THE UNTOUCHABLES (I,4682)

KARPF, ELEANOR
CAPITOL (II,426); KUNG FU (II,1416)

KARPF, STEPHEN
CAPITOL (II,426); KUNG FU (II,1416)

KASCIA, ANN
KATE LOVES A MYSTERY (II,1367)

KASIAC, MARYANN
HART TO HART (II,1102); HOTEL (II,1192); TALES OF THE GOLD MONKEY (II,2537); TOPPER (II,2649); TUCKER'S WITCH (II,2667)

KASS, JEROME
LETTERS FROM THREE LOVERS (I,2671)

KATKOV, NORMAN
BEN CASEY (I,394); BIG SHAMUS, LITTLE SHAMUS (II,242); FRONT ROW CENTER (I,1694); NURSE (II,1870); THE RETURN OF CAPTAIN NEMO (II,2135); STUDIO ONE (I,4268); THE WILD WILD WEST (I,4863)

KATZ, ALLAN
ADAMS HOUSE (II,9); CHER (II,496); COMEDY NEWS II (I,1021); GOODBYE DOESN'T MEAN FOREVER (II,1039); THE JACKIE GLEASON SPECIAL (I,2325); THE NATIONAL SNOOP (II,1797); ONE MORE TIME (I,3385); ONE MORE TIME (I,3386); RHODA (II,2151); THE ROWAN AND MARTIN SPECIAL (I,3855); ROWAN AND MARTIN'S LAUGH-IN (I,3856); SHOW BUSINESS SALUTE TO MILTON BERLE (I,4029); WE'LL GET BY (II,2753)

KATZ, HOWARD
GAMES PEOPLE PLAY (II,956)

KATZ, STEPHEN
HARDCASTLE AND MCCORMICK (II,1089); KNIGHT RIDER (II,1402); TALES OF THE GOLD MONKEY (II,2537)

KATZMAN, LEONARD
DALLAS (II,614); DIRTY SALLY (II,680); THE FANTASTIC JOURNEY (II,826); LOGAN'S RUN (II,1507); THE WILD WILD WEST (I,4863)

KAUFFMANN, MARSHALL
STARSKY AND HUTCH (II,2444)

KAUFMAN, LEONARD B
THE DUKES OF HAZZARD (II,742); ENOS (II,779); HAWAII FIVE-O (II,1110); KEEPER OF THE WILD (II,1373); PRIVATE BENJAMIN (II,2087)

KAUFMAN, ROBERT
THE BOB NEWHART SHOW (I,666); COMBAT (I,1011); HERE WE GO AGAIN (I,2028); OFF WE GO (I,3338); THE UGLIEST GIRL IN TOWN (I,4663)

KAWTER, JONATHAN
QUARK (II,2095)

KAY, DUSTY
EIGHT IS ENOUGH (II,762)

KAYDEN, TONY
THE ABC AFTERSCHOOL SPECIAL (II,1); LUCAN (II,1554); MORE WILD WILD WEST (II,1733); THE WALTONS (II,2740)

KAYE, SYLVIA FINE
MUSICAL COMEDY TONIGHT (II,1777)

KAZARIAN, ANNE
COUNTERPOINT (I,1059)

KEAN, E ARTHUR
CHICAGO STORY (II,506); THE FBI (I,1551); THE FUGITIVE (I,1701); HAWAII FIVE-O (II,1110); KATE LOVES A MYSTERY (II,1367)

KEAN, EDWARD
THE BOB SMITH SHOW (I,668); HOWDY DOODY (I,2151)

KEANE, ART
THE NUT HOUSE (I,3320)

KEANE, REBECCA HUNT
THE DANCE SHOW (II,616)

KEANE, ROBERT
BOB HOPE SPECIAL: BOB HOPE'S STARS OVER TEXAS (II,318); BOB HOPE SPECIAL: BOB HOPE'S STAR-STUDDED SPOOF OF THE NEW TV

SEASON— G RATED—WITH GLAMOUR, GLITTER AND GAGS (II,317); THE ROWAN AND MARTIN REPORT (II,2221)

KEARNEY, GENE
EISCHIED (II,763); KOJAK (II,1406); LOU GRANT (II,1526); THE RETURN OF CHARLIE CHAN (II,2136)

KEARNEY, MARK
FIRESIDE THEATER (I,1580)

KEATON, DONALD
OMNI: THE NEW FRONTIER (II,1889)

KEATS, ROBERT
QUARK (II,2095); SCALPELS (II,2266)

KEEL, CARLENE
FANTASY ISLAND (II,829)

KEEN, MARTIN
OUT OF OUR MINDS (II,1933)

KEENAN, WILLIAM
THE JACKSON FIVE (I,2326); LAVERNE AND SHIRLEY (II,1446); THE RELUCTANT DRAGON AND MR. TOAD (I,3762); THE YEAR WITHOUT A SANTA CLAUS (II,2846)

KEENE, BOB
BOB HOPE SPECIAL: BOB HOPE'S SPRING FLING OF COMEDY AND GLAMOUR (II,315)

KELBAUGH, RICK
BARETTA (II,152); BLUE THUNDER (II,278); STRIKE FORCE (II,2480); TODAY'S FBI (II,2628)

KELLARD, PHIL
ANDY WILLIAMS' EARLY NEW ENGLAND CHRISTMAS (II,79); PERRY COMO'S CHRISTMAS IN PARIS (II,2012); PUTTIN' ON THE HITS (II,2093); THICKE OF THE NIGHT (II,2587); THE WOLFMAN JACK RADIO SHOW (II,2816)

KELLARD, RICK
ALAN KING'S SECOND ANNUAL FINAL WARNING (II,30); THE BEST OF TIMES (II,220); BIG BOB JOHNSON AND HIS FANTASTIC SPEED CIRCUS (II,231); FIRST TIME, SECOND TIME (II,863); JUST OUR LUCK (II,1360); LOLA (II,1508); LOLA (II,1510); LOLA (II,1511); NICHOLS AND DYMES (II,1841); ONE NIGHT BAND (II,1909); ROCK-N-AMERICA (II,2195); A SPECIAL KENNY ROGERS (II,2417); WACKO (II,2733); THE WONDERFUL WORLD OF PHILIP MALLEY (II,2831)

KELLBAUGH, RICHARD
B.J. AND THE BEAR (II,125)

KELLER, CASEY
THE LOVE BOAT (II,1535)

KELLER, MAX
STRANGER IN OUR HOUSE (II,2477)

KELLER, ROBERT CASEY
PAUL LYNDE AT THE MOVIES (II,1976)

KELLER, SHELDON
THE ART CARNEY SHOW (I,274); THE BEST OF ANYTHING (I,404); BING CROSBY AND CAROL BURNETT—TOGETHER AGAIN FOR THE FIRST TIME (I,454); THE BING CROSBY SHOW (I,463); THE BING CROSBY SHOW (I,464); BOB HOPE SPECIAL: BOB HOPE'S ALL-STAR COMEDY SPECTACULAR FROM LAKE TAHOE (II,301); BOB HOPE SPECIAL: BOB HOPE'S ALL-STAR COMEDY TRIBUTE TO VAUDEVILLE (II,302); BOB HOPE SPECIAL: BOB HOPE'S ALL-STAR TRIBUTE TO THE PALACE THEATER (II,305); BOB HOPE SPECIAL: BOB HOPE'S CHRISTMAS SPECIAL (II,308); CBS SALUTES LUCY—THE FIRST 25 YEARS (II,459); THE DANNY KAYE SHOW (I,1143); DIANA ROSS AND THE SUPREMES AND THE TEMPTATIONS ON BROADWAY (I,1260); THE DINAH SHORE CHEVY SHOW (I,1279); THE ENGELBERT HUMPERDINCK SHOW (I,1444); THE ENGELBERT HUMPERDINCK SHOW (I,1445); THE ENGELBERT HUMPERDINCK SPECIAL (I,1446); THE ENGELBERT HUMPERDINCK SPECIAL (I,1447); FRANK SINATRA: A MAN AND HIS MUSIC (I,1662); FRANK SINATRA—A MAN AND HIS MUSIC (I,1661); A FUNNY THING HAPPENED ON THE WAY TO HOLLYWOOD (I,1716); GUYS 'N' GEISHAS (I,1914); HELLZAPOPPIN (I,2011); HOORAY FOR HOLLYWOOD (I,2114); THE JACK PAAR SPECIAL (I,2319); THE JONATHAN WINTERS SHOW (I,2443); A LUCILLE BALL SPECIAL: WHAT NOW CATHERINE CURTIS? (II,1560); THE NASHVILLE PALACE (II,1792); THE NIGHT OF CHRISTMAS (I,3290); PAUL LYNDE AT THE MOVIES (II,1976); THE PEAPICKER IN PICCADILLY (I,3500); PEOPLE ARE FUNNY (II,1996); PORTRAIT OF PETULA (I,3654); ROLL OUT! (I,3833); SINATRA (I,4054); THE TENNESSEE ERNIE FORD SPECIAL (I,4401); WITH LOVE, SOPHIA (I,4884); WOMEN WHO RATE A "10" (II,2825); THE WONDERFUL WORLD OF BURLESQUE III (I,4897)

KELLER, SHERYL
THE FLYING NUN (I,1611)

KELLEY, JOHN T
BONANZA (II,347); THE MUSIC MAKER (I,3173); REX HARRISON PRESENTS SHORT STORIES OF LOVE (II,2148)

KELLEY, WELBOURN
THE CHAMBER MUSIC SOCIETY OF LOWER BASIN STREET (I,894)

KELLEY, WILLIAM
HOW THE WEST WAS WON (II,1196); KEY TORTUGA (II,1383); ROUTE 66 (I,3852); THREE FOR THE ROAD (II,2607)

KELLNER, BILL STEIN
THE PEE WEE HERMAN SHOW (II,1988)

KELLOGG, MARJORIE
MATINEE THEATER (I,2947); MIDNIGHT MYSTERY (I,3030)

KELLY, APRIL
BABY, I'M BACK! (II,130); THE CAPTAIN AND TENNILLE (II,429); THE CARPENTERS (II,444); GOOBER AND THE TRUCKERS' PARADISE (II,1029); HOW TO SURVIVE THE 70S AND MAYBE EVEN BUMP INTO HAPPINESS (II,1199); THE JACKSONS (II,1267); THE JIM STAFFORD SHOW (II,1291); JOHN DENVER ROCKY MOUNTAIN CHRISTMAS (II,1314); LOVE, SIDNEY (II,1547); MORK AND MINDY (II,1735); PARTNERS IN CRIME (II,1961); THE PAUL LYNDE COMEDY HOUR (II,1978); THE STARLAND VOCAL BAND (II,2441); SZYSZNYK (II,2523); TEACHERS ONLY (II,2548); WE GOT IT MADE (II,2752); YOU ASKED FOR IT (II,2855)

KELLY, SHAWN
THE FACTS (II,804); NATIONAL LAMPOON'S HOT FLASHES (II,1795)

KELLY, TIM
MATINEE THEATER (I,2947)

KELSEY, BILL
MY FAVORITE MARTIAN (I,3189)

KELSEY, JACK
THE ALCOA HOUR (I,108); FOR BETTER OR WORSE (I,1621)

KELSEY, LYNNE
PRIVATE BENJAMIN (II,2087)

KELSO, EDMOND
THE CISCO KID (I,961)

KELTON, KEVIN
NO SOAP, RADIO (II,1856); TWILIGHT THEATER II (II,2686)

KEMP, BARRY
IN SECURITY (II,1227); NEWHART (II,1835); OUT OF THE BLUE (II,1934); SZYSZNYK (II,2523); TAXI (II,2546)

KEMPLEY, WALTER
LEGS (II,1459); SATURDAY NIGHT LIVE WITH HOWARD COSELL (II,2262); THE UGILY FAMILY (II,2699)

KENASTON, JACK
UNK AND ANDY (I,4679)

KENDAL, STEPHEN
LOTTERY (II,1525)

KENDALL, JO
CAMBRIDGE CIRCUS (I,807)

KENKER, HENRY
THE DESILU PLAYHOUSE (I,1237)

KENNEDY, BURT
COMBAT (I,1011)

KENNEDY, KATHLEEN
THE YELLOW ROSE (II,2847)

KENNEDY, VERN
THE TOMMY HUNTER SHOW (I,4552)

KENNEDY-MARTIN, TROY
REILLY, ACE OF SPIES (II,2129)

KENT, DAVID
MR. MERGENTHWIRKER'S LOBBLIES (I,3144)

KENT, MONA
WOMAN WITH A PAST (I,4889)

KENT, ROBERT E
THE WILD WILD WEST (I,4863)

KENYON, CURTIS
THE LONE RANGER (I,2740); PRIVATE SECRETARY (I,3672); RAWHIDE (I,3727);

SKY KING (I,4077); THE UNTOUCHABLES (I,4682); WAIKIKI (II,2734)

KERR, JEAN
THE GOOD FAIRY (I,1846)

KERR, WALTER T
THE BAT (I,364); THE DATCHET DIAMONDS (I,1160)

KERSHAW, JOHN
BERGERAC (II,209)

KERWIN, MICHAEL
THE BAXTERS (II,184)

KESSLER, BRUCE
CALIFORNIA FEVER (II,411); HUNTER (II,1206); WONDER WOMAN (SERIES 1) (II,2828)

KETCHAM, HANK
DENNIS THE MENACE: MAYDAY FOR MOTHER (II,660)

KETCHUM, DAVID
THE BIONIC WOMAN (II,255); BLANSKY'S BEAUTIES (II,267); CAPTAIN NICE (I,835); JEANNIE (II,1275); LOTTERY (II,1525); M*A*S*H (II,1569); ON OUR OWN (II,1892); SWITCH (II,2519); WONDER WOMAN (SERIES 1) (II,2828); WONDER WOMAN (SERIES 2) (II,2829)

KEYES, CHIP
ALOHA PARADISE (II,58); BEST OF THE WEST (II,218); BLISS (II,269); GIMME A BREAK (II,995); LADIES' MAN (II,1423)

KEYES, DOUG
ALOHA PARADISE (II,58); BEST OF THE WEST (II,218); BLISS (II,269); GIMME A BREAK (II,995); LADIES' MAN (II,1423)

KEYES, PAUL W
BOB HOPE SPECIAL: A QUARTER CENTURY OF BOB HOPE ON TELEVISION (II,281); BOB HOPE SPECIAL: BOB HOPE'S BICENTENNIAL STAR SPANGLED SPECTACULAR (II,306); THE DON RICKLES SHOW (II,708); GENERAL ELECTRIC'S ALL-STAR ANNIVERSARY (II,963); JACK PAAR PRESENTS (I,2315); THE JACK PAAR SHOW (I,2317); THE ROWAN AND MARTIN SHOW (I,3854); THE ROWAN AND MARTIN SPECIAL (I,3855); ROWAN AND MARTIN'S LAUGH-IN (I,3856); ROWAN AND MARTIN'S LAUGH-IN (I,3857); SINATRA AND FRIENDS (II,2358); SINATRA: THE MAN AND HIS MUSIC (II,2362); SINATRA—THE FIRST 40 YEARS (II,2359); SWING OUT, SWEET LAND (II,2515); TAKE ONE STARRING JONATHAN WINTERS (II,2532); THE WONDERFUL WORLD OF JACK PAAR (I,4899)

KEYS, WILLIAM
BORN TO THE WIND (II,354); BUCK ROGERS IN THE 25TH CENTURY (II,384); THE RETURN OF CAPTAIN NEMO (II,2135)

KIBBE, DANIEL
THE BIONIC WOMAN (II,255)

KIBBEE, ROLAND
A.E.S. HUDSON STREET (II,17); ALFRED HITCHCOCK PRESENTS (I,115); THE BOB NEWHART SHOW (I,666); COLUMBO (II,556); DIAGNOSIS: DANGER (I,1250); GUILTY OR NOT GUILTY (I,1903); MADIGAN (I,2808); RETURN OF THE WORLD'S GREATEST DETECTIVE (II,2142); SNAVELY (II,2387); TENNESSEE ERNIE FORD MEETS KING ARTHUR (I,4397); THE TENNESSEE ERNIE FORD SHOW (I,4398)

KILLIAM, PAUL
SILENT'S, PLEASE (I,4050)

KIMMEL, HAROLD
ANYTHING FOR MONEY (II,99)

KIMMEL, HENRY
ON STAGE AMERICA (II,1893)

KIMMELL, JOEL
THE FACTS OF LIFE (II,805); A YEAR AT THE TOP (II,2844)

KINCAIDE, KRISTAN
THE DUKES OF HAZZARD (II,742)

KINDLEY, JEFFREY
THE ABC AFTERSCHOOL SPECIAL (II,1)

KING, ALAN
THE ALAN KING SHOW (I,101); THE ALAN KING SHOW (I,102); THE ALAN KING SPECIAL (I,103); ALAN KING'S FINAL WARNING (II,29); ALAN KING'S SECOND ANNUAL FINAL WARNING (II,30); THE MANY FACES OF COMEDY (I,2894); THE WONDERFUL WORLD OF AGGRAVATION (I,4894)

KING, DUSTY
SHIRLEY (II,2330)

KING, JONATHAN
AMERICA, YOU'RE ON (II,69)

KING, PAUL
BLACK SADDLE (I,480); BONANZA (II,347); CODE NAME: DIAMOND HEAD (II,549)

KING, TOM
AS THE WORLD TURNS (II,110); FOR RICHER, FOR POORER (II,906)

KING, TONY
LOVERS AND FRIENDS (II,1549)

KINGBORN, DAVID J
HARD KNOX (II,1087)

KINGSBRIDGE, JOHN
STAR TREK (II,2440)

KINGSLEY, DOROTHY
DEBBIE REYNOLDS AND THE SOUND OF CHILDREN (I,1213)

KINGSLEY, EMILY PERL
SESAME STREET (I,3982)

KINGSLEY, SIDNEY
DUPONT SHOW OF THE MONTH (I,1387)

KINON, RICHARD
THE FLYING NUN (I,1611)

KINOY, ERNEST
THE ALCOA HOUR (I,108); ALCOA/GOODYEAR THEATER (I,107); BRIGADOON (I,726); CHANGE AT 125TH STREET (II,480); THE DESILU PLAYHOUSE (I,1237); THE FIRM (II,860); FRONT ROW CENTER (I,1694); THE HAPPENERS (I,1935); THE IMOGENE COCA SHOW (I,2198); NAKED CITY (I,3216); PINOCCHIO (I,3603); THE RIVALRY (I,3796); ROOTS (II,2211); ROOTS: THE NEXT GENERATIONS (II,2213); ROUTE 66 (I,3852); RX FOR THE DEFENSE (I,3878); SAINTS AND SINNERS (I,3887); SHANE (I,3995); STUDIO ONE (I,4268); SUSPICION (I,4309); THE UNTOUCHABLES (I,4682); VALLEY FORGE (I,4698)

KIPERBERG, HOWARD
BLOCKHEADS (II,270)

KIRGO, DIANA
ANOTHER DAY (II,94); ONE OF THE BOYS (II,1911); REGGIE (II,2127)

KIRGO, DONNA
ONE DAY AT A TIME (II,1900)

KIRGO, GEORGE
ADAM'S RIB (I,32); APPLE'S WAY (II,101); BRENDA STARR (II,372); THE CABOT CONNECTION (II,408); GET CHRISTIE LOVE! (II,982); THE KID WITH THE BROKEN HALO (II,1387); MASSARATI AND THE BRAIN (II,1646); MR. SMITH (II,1755); TOPPER (II,2649)

KIRGO, JULIE
ANOTHER DAY (II,94); ONE DAY AT A TIME (II,1900); ONE OF THE BOYS (II,1911); REGGIE (II,2127)

KIRKPATRICK, DICK
LLOYD BRIDGES WATER WORLD (I,2734)

KIRSCHBAUM, BRUCE
DONNY AND MARIE (II,712); PEOPLE ARE FUNNY (II,1996)

KITE, LESA
JOANIE LOVES CHACHI (II,1295); LAVERNE AND SHIRLEY (II,1446); MORK AND MINDY (II,1735); OH MADELINE (II,1880); TOO CLOSE FOR COMFORT (II,2642)

KLANE, ROBERT
ACES UP (II,7); THE BANANA COMPANY (II,139); CAMP GRIZZLY (II,418); M*A*S*H (II,1569); THE MICHELE LEE SHOW (II,1690); MR. & MRS. DRACULA (II,1745); ROSENTHAL AND JONES (II,2215)

KLASSON, JOANNA
HARRY O (II,1099)

KLAUBER, MARCEL
STUDIO ONE (I,4268)

KLEIN, DENNIS
BUFFALO BILL (II,387); DOC (II,682); FOL-DE-ROL (I,1613); REPORT TO MURPHY (II,2134)

KLEIN, LARRY
THE DONALD O'CONNOR SHOW (I,1343); THE FLIP WILSON SHOW (I,1602); THE GOOD OLD DAYS (I,1851); HOLIDAY LODGE (I,2077); THE JACK CARTER SHOW (I,2306); THE JERRY LEWIS SHOW (I,2370); THE JULIE ANDREWS HOUR (I,2480); THE MILTON BERLE SHOW (I,3049); MY SON THE DOCTOR (I,3203); THE RED SKELTON TIMEX SPECIAL (I,3757)

KLEIN, MARTY
THE MILTON BERLE SPECIAL (I,3051)

KLEIN, ROBERT
KLEIN TIME (II,1401); THE ROBERT KLEIN SHOW (II,2185); THE ROBERT KLEIN SHOW (II,2186)

KLEINER, HARRY
TARGET: THE CORRUPTERS (I,4360)

KLEINSCHMITT, CARL
THE BRADY BUNCH HOUR (II,363); FIRST & TEN (II,861); KAREN (II,1363); THE LOVE BOAT (II,1535); THE LOVE BOAT I (II,1532); THE LOVE BOAT II (II,1533); MY WORLD . . . AND WELCOME TO IT (I,3208); PETE 'N' TILLIE (II,2027)

KLENMAN, NORMAN
OUR MAN FLINT: DEAD ON TARGET (II,1932)

KLINE, STEVE
THE DUCK FACTORY (II,738); FAME (II,812); GOODNIGHT, BEANTOWN (II,1041); LOU GRANT (II,1526); MAKING THE GRADE (II,1606); TUCKER'S WITCH (II,2667)

KLINE, WAYNE
THE JACKSONS (II,1267); RICH LITTLE'S WASHINGTON FOLLIES (II,2159); THICKE OF THE NIGHT (II,2587)

KLING, HEYWOOD
BATFINK (I,365)

KLING, WOODY
THE BOBBY VAN AND ELAINE JOYCE SHOW (I,677); A COUPLE OF DONS (I,1077); HANDLE WITH CARE (II,1077); HEREAFTER (II,1138); HOT L BALTIMORE (II,1187); MAUDE (II,1655); OF THEE I SING (I,3334); PEBBLES AND BAMM BAMM (II,1987); SPEED BUGGY (II,2419)

KLINKER, ZENA
FRANCES LANGFORD PRESENTS (I,1656)

KLOVE, JANE
HAZEL (I,1982)

KLUGMAN, DEBORAH
QUINCY, M. E. (II,2102)

KLUGMAN, JACK
KRAFT TELEVISION THEATER (I,2592); QUINCY, M. E. (II,2102)

KNEE, DENNIS
RODNEY DANGERFIELD SHOW: I CAN'T TAKE IT NO MORE (II,2198)

KNEELAND, TED
HOLD THAT CAMERA (I,2072)

KNEUBUHL, JOHN
ALCOA PREMIERE (I,109); CLIMAX! (I,976); THE DESILU PLAYHOUSE (I,1237); DOCTOR MIKE (I,1316); THE FUGITIVE (I,1701); MARKHAM (I,2916); MEDIC (I,2976); SECRETS OF THE OLD BAILEY (I,3971); SUSPICION (I,4309); THRILLER (I,4492); THE WILD WILD WEST (I,4863)

KNIGHT, DAMON
CAPTAIN VIDEO AND HIS VIDEO RANGERS (I,838); THE TWILIGHT ZONE (I,4651)

KNIGHT, VIC
PETER GUNN (I,3562)

KNOPF, CHRISTOPHER
THE DICK POWELL SHOW (I,1269); TARGET: THE CORRUPTERS (I,4360); TURN OF FATE (I,4623)

KNOTT, FREDERICK
DIAL "M" FOR MURDER (I,1255); DIAL "M" FOR MURDER (I,1256)

KNOTTS, DON
DON KNOTTS NICE CLEAN, DECENT, WHOLESOME HOUR (I,1333)

KNUTZ, JOHN
TEXAS (II,2566)

KOCH, TOM
THE GEORGE GOBEL SHOW (I,1768); THE JONATHAN WINTERS SHOW (I,2443); MY WORLD . . . AND WELCOME TO IT (I,3208); PAT PAULSEN'S HALF A COMEDY HOUR (I,3473)

KOCH, TONI
HENRY FONDA AND THE FAMILY (I,2015)

KOENIG, DENNIS
AFTERMASH (II,23); BARNEY MILLER (II,154); M*A*S*H (II,1569); RHODA (II,2151); THREE'S COMPANY (II,2614)

KOENIG, WALTER
THE POWERS OF MATTHEW STAR (II,2075); STAR TREK (II,2439)

KOGAN, ANNIE
ACE CRAWFORD, PRIVATE EYE (II,6)

KOGEN, ARNIE
THE CAROL BURNETT SHOW (II,443); THE FUNNY SIDE (I,1715); THE JACKSONS (II,1267); THE JIM NABORS HOUR (I,2381); THE PRIMARY ENGLISH CLASS (II,2081); WE DARE YOU! (II,2751)

KOHAN, BUZ
100 YEARS OF GOLDEN HITS (II,1905); 60 YEARS OF SEDUCTION (II,2373); ANN-MARGRET SMITH (II,85); ANN-MARGRET'S HOLLYWOOD MOVIE GIRLS (II,86); ANN-MARGRET. . . RHINESTONE COWGIRL (II,87); ANN-MARGRET: FROM HOLLYWOOD WITH LOVE (I,217); THE ARTHUR GODFREY SPECIAL (I,288); ARTHUR GODFREY'S PORTABLE ELECTRIC MEDICINE SHOW (I,290); BARYSHNIKOV IN HOLLYWOOD (II,157); BEN VEREEN. . . COMIN' AT YA (II,203); BETTE MIDLER—OL' RED HAIR IS BACK (II,222); BING CROSBY AND THE SOUNDS OF CHRISTMAS (I,457); BING CROSBY'S CHRISTMAS SHOW (I,474); BING CROSBY'S MERRIE OLDE CHRISTMAS (II,252); BING CROSBY'S SUN VALLEY CHRISTMAS SHOW (I,475); BING CROSBY—COOLING IT (I,459); BING CROSBY—COOLING IT (I,460); BING!. . . A 50TH ANNIVERSARY GALA (II,254); BLONDES VS. BRUNETTES (II,271); CAROL AND COMPANY (I,847); CHER. . . SPECIAL (II,499); CHRISTMAS IN DISNEYLAND (II,519); CHRISTMAS IN WASHINGTON (II,520); CHRISTMAS IN WASHINGTON (II,521); CHRISTMAS WITH THE BING CROSBYS (I,950); CHRISTMAS WITH THE BING CROSBYS (II,524); A COUNTRY CHRISTMAS (II,582); A COUNTRY CHRISTMAS (II,583); A COUNTRY CHRISTMAS (II,584); DEBBY BOONE. . . THE SAME OLD BRAND NEW ME (II,656); DIANA (II,667); DICK CAVETT'S BACKLOT USA (II,670); DISNEYLAND'S 25TH ANNIVERSARY (II,681); THE DOROTHY HAMILL SPECIAL (II,719); DOUG HENNING'S WORLD OF MAGIC I (II,725); DOUG HENNING'S WORLD OF MAGIC II (II,726); DOUG HENNING'S WORLD OF MAGIC III (II,727); DOUG HENNING'S WORLD OF MAGIC IV (II,728); DOUG HENNING'S WORLD OF MAGIC V (II,729); FANFARE (I,1528); THE FANTASTIC MISS PIGGY SHOW (II,827); FROM HOLLYWOOD WITH LOVE: THE ANN-MARGRET SPECIAL (I,1687); GENE KELLY. . . AN AMERICAN IN PASADENA (II,962); GLEN CAMPBELL . . . DOWN HOME—DOWN UNDER (II,1008); GOLDIE AND KIDS: LISTEN TO ME (II,1021); GOOD EVENING, CAPTAIN (II,1031); JOHN SCHNEIDER'S CHRISTMAS HOLIDAY (II,1320); JUBILEE (II,1347); THE KEEFE BRASSELLE SHOW (I,2514); KENNY & DOLLY: A CHRISTMAS TO REMEMBER (II,1378); THE KRAFT 75TH ANNIVERSARY SPECIAL (II,1409); KRAFT SALUTES WALT DISNEY WORLD'S 10TH ANNIVERSARY (II,1410); LAVERNE AND SHIRLEY (II,1446); LOOKING BACK (I,2754); MAC DAVIS CHRISTMAS SPECIAL. . . WHEN I GROW UP (II,1574); MAC DAVIS'S CHRISTMAS ODYSSEY: TWO THOUSAND AND TEN (II,1579); MAC DAVIS. . . SOUNDS LIKE HOME (II,1582); THE MANY MOODS OF PERRY COMO (I,2897); MERRY CHRISTMAS FROM THE CROSBYS (II,1685); MERRY CHRISTMAS FROM THE GRAND OLE OPRY (II,1686); THE NEIL DIAMOND SPECIAL (II,1802); ON STAGE AMERICA (II,1893); OUR PLACE (I,3416); PAUL ANKA IN MONTE CARLO (II,1971); PAUL ANKA. . . MUSIC MY WAY (II,1973); PAVAROTTI AND FRIENDS (II,1985); THE PEARL BAILEY SHOW (I,3501); THE PERRY COMO CHRISTMAS SHOW (I,3529); PERRY COMO'S WINTER SHOW (I,3546); THE SAMMY DAVIS JR. SHOW (I,3898); SHEENA EASTON, ACT 1 (II,2323); SHIRLEY MACLAINE. . . EVERY LITTLE MOVEMENT (II,2334); SHIRLEY MACLAINE: ILLUSIONS (II,2336); STEVE LAWRENCE AND EYDIE GORME FROM THIS MOMENT ON. . . COLE PORTER (II,2459); TELEVISION'S GREATEST COMMERCIALS III (II,2555); TELEVISION'S GREATEST COMMERCIALS IV (II,2556); TELLY. . . WHO LOVES YA, BABY? (II,2558); TWO OF A KIND: GEORGE BURNS AND JOHN DENVER (II,2692)

KOHN, JOHN
THE ANN SOTHERN SHOW (I,220); COUNTERPOINT (I,1059); CROWN THEATER WITH GLORIA SWANSON (I,1107)

KOLB, KEN
BEN CASEY (I,394); BLACK SADDLE (I,480); CHANNING (I,900); MEDIC (I,2976); THE RIFLEMAN (I,3789); SARA (II,2260); THE WILD WILD WEST (I,4863)

KOLDOR, ERIC
SWITCH (II,2519)

KOLLE, RAY
PRISONER: CELL BLOCK H (II,2085)

KOMACK, JAMES
ME AND MAXX (II,1672); MY FAVORITE MARTIAN (I,3189)

KONG, EMILIE
SUPER FRIENDS (II,2497)

KONNER, LAWRENCE
FAMILY (II,813); LITTLE HOUSE ON THE PRAIRIE (II,1487)

KONNER, RONNIE
HART TO HART (II,1102)

KOPIT, ARTHUR
STARSTRUCK (II,2446)

KOPLAN, HARRY
THRILL SEEKERS (I,4491)

KORMAN, JESS
THE ALAN KING SHOW (I,102)

KORN, DAVID MICHAEL
KUNG FU (II,1416)

KORNICK, WILLIAM
COUNTERATTACK: CRIME IN AMERICA (II,581)

KORR, DAVID
SESAME STREET (I,3982)

KOSKOS, ANDREW
ST. ELSEWHERE (II,2432)

KOSTMAYER, JOHN
THE FOUR SEASONS (II,915); SIMON AND SIMON (II,2357)

KOTT, GARY
IT'S NOT EASY (II,1256); PUNKY BREWSTER (II,2092)

KOUT, WENDY
9 TO 5 (II,1852)

KOVACS, ERNIE
THE COMEDY OF ERNIE KOVACS (I,1022); ERNIE IN KOVACSLAND (I,1454); THE ERNIE KOVACS SHOW (I,1455); THE ERNIE KOVACS SPECIAL (I,1456); IT'S TIME FOR ERNIE (I,2277); KOVACS ON THE CORNER (I,2582); KOVACS UNLIMITED (I,2583); THE NEW ERNIE KOVACS SHOW (I,3264)

KOZOLL, MICHAEL
DELVECCHIO (II,659); HILL STREET BLUES (II,1154); MCCLOUD (II,1660); THE NIGHT STALKER (II,1849); SWITCH (II,2519); THREE FOR THE ROAD (II,2607)

KRAFTZMER, HERBERT
THAT WAS THE WEEK THAT WAS (I,4423)

KRAGEN, JINX
THE SMOTHERS BROTHERS SHOW (I,4096)

KRAMER, RICHARD
FAMILY (II,813); THE PAPER CHASE (II,1949)

KRAMER, THOMAS
NOT NECESSARILY THE NEWS (II,1863)

KREINBERG, STEVE
ARCHIE BUNKER'S PLACE (II,105); HUSBANDS, WIVES AND LOVERS (II,1209)

KREISMAN, STU
I'VE HAD IT UP TO HERE (II,1221); SCARECROW AND MRS. KING (II,2268)

KRESS, EARL
THE ODDBALL COUPLE (II,1876)

KRIBACH, AUGIE
CAMP WILDERNESS (II,419)

KRIMS, MILTON
THE OUTER LIMITS (I,3426); PERRY MASON (II,2025); WAGON TRAIN (I,4747)

KRINSKI, SANDY
ALICE (II,33); BABY, I'M BACK! (II,130); BUSTING LOOSE (II,402); THE DIONNE WARWICK SPECIAL (I,1289); DONNY AND MARIE (II,712); THE DONNY AND MARIE OSMOND SHOW (II,714); GETTING THERE (II,985); GIMME A BREAK (II,995); THE GLEN CAMPBELL GOODTIME HOUR (I,1820); THE JIMMIE RODGERS SHOW (I,2383); NEWMAN'S DRUGSTORE (II,1837); THE PAUL LYNDE COMEDY HOUR (II,1978); SPENCER (II,2420); WE GOT IT MADE (II,2752)

KRISTOFFERSON, KRIS
JUST FRIENDS (I,2495)

KRONICK, WILLIAM
PLIMPTON! DID YOU HEAR THE ONE ABOUT. . .? (I,3629); PLIMPTON! SHOWDOWN AT RIO LOBO (I,3630); PLIMPTON! THE MAN ON THE FLYING TRAPEZE (I,3631)

KRONMAN, HARRY
BLONDIE (I,486); THE FUGITIVE (I,1701); THE RAT PATROL (I,3726); THE UNTOUCHABLES (I,4682)

KROPF, CHRIS
THE RIFLEMAN (I,3789)

KRUMHOLZ, BONNIE ANN
FANTASY ISLAND (II,829)

KRUMHOLZ, CHESTER
BEN CASEY (I,394); CASABLANCA (II,450); DEADLOCK (I,1190); DOCTOR SIMON LOCKE (I,1317); DYNASTY (II,746); KOJAK (II,1406); MANNIX (II,1624) ONCE UPON A DEAD MAN (I,3369)

KRUSE, JOHN
RETURN OF THE SAINT (II,2141); THE SAINT (II,2237)

KRUTCHER, JACK
DO IT YOURSELF (I,1299)

KUKOFF, BERNIE
ALL-AMERICAN PIE (II,48); ALMOST ANYTHING GOES (II,56); COMEDY NEWS II (I,1021); DETECTIVE SCHOOL (II,661); THE EVERLY BROTHERS SHOW (I,1479); FREEMAN (II,923); GOSSIP (II,1043); IN TROUBLE (II,1230); JIMMY DURANTE PRESENTS THE LENNON SISTERS HOUR (I,2387); JOE AND SONS (II,1296); JOE AND SONS (II,1297); THE LAS VEGAS SHOW (I,2619); THE LENNON SISTERS SHOW (I,2656); MCNAMARA'S BAND (II,1668); OPERATION: ENTERTAINMENT (I,3401); PAT BOONE IN HOLLYWOOD (I,3469); REGGIE (II,2127); THE ROGER MILLER SHOW (I,3830); SHEEHY AND THE SUPREME MACHINE (II,2322); THE SHOW MUST GO ON (II,2344); THAT THING ON ABC (II,2574); TUCKER'S WITCH (II,2667); THE VAL DOONICAN SHOW (I,4692)

KULLER, SID
JACKIE GLEASON AND HIS AMERICAN SCENE MAGAZINE (I,2321)

KUPFER, MARVIN
KOJAK (II,1406); THE PAPER CHASE (II,1949); THE SIX O'CLOCK FOLLIES (II,2368); THE STREETS OF SAN FRANCISCO (II,2478)

KUPINBERG, HOWARD
THE POP 'N' ROCKER GAME (II,2066)

KURLAND, JOHN
VALENTINE MAGIC ON LOVE ISLAND (II,2716)

KUTZ, MAX
PRIVATE BENJAMIN (II,2087)

LAFRENAIS, IAN
I'LL NEVER FORGET WHAT'S HER NAME (II,1217); ON THE ROCKS (II,1894)

LAMOND, BILL
FINDER OF LOST LOVES (II,857)

LAMOND, JO
FINDER OF LOST LOVES (II,857)

LATOURETTE, FRANK
THE LONELY WIZARD (I,2743)

LABELLA, VINCENZO
MARCO POLO (II,1626)

LABINE, CLAIRE
RYAN'S HOPE (II,2234)

LACHMAN, BRAD
SOLID GOLD (II,2395); THE VIN SCULLY SHOW (I,4734)

LACHMAN, MORT
BABY, I'M BACK! (II,130); THE BOB HOPE SHOW (I,533); THE BOB HOPE SHOW (I,534); THE BOB HOPE SHOW (I,541); THE BOB HOPE SHOW (I,543); THE BOB HOPE SHOW (I,552); THE BOB HOPE SHOW (I,555); THE BOB HOPE SHOW (I,560); THE BOB HOPE SHOW (I,561); THE BOB HOPE SHOW (I,562); THE BOB HOPE SHOW (I,603); THE BOB HOPE SHOW (I,604); THE BOB HOPE SHOW (I,606); THE BOB HOPE SHOW (I,607); THE BOB HOPE SHOW (I,613); THE BOB HOPE SHOW (I,636); THE BOB HOPE SHOW (I,637); THE BOB HOPE SHOW (I,642); BOB HOPE SPECIAL: BOB HOPE ON CAMPUS (II,289); BOB HOPE SPECIAL: THE BOB HOPE SPECIAL (II,333); BOB HOPE SPECIAL: THE BOB HOPE SPECIAL (II,334); THE BOOK OF LISTS (II,350); THE COLGATE COMEDY HOUR (I,998); GIMME A BREAK (II,995); THE HANK MCCUNE SHOW (I,1932); INSTANT FAMILY (II,1238); NOT IN FRONT OF THE KIDS (II,1861); ROBERTA (I,3815); ROBERTA (I,3816); SUTTERS BAY (II,2508)

LAEMMLE, NINA
THE MAN FROM DENVER (I,2863)

LAFFAN, KEVIN B
MAN IN A SUITCASE (I,2868)

LAGER, MARTIN
ADVENTURES IN RAINBOW COUNTRY (I,47)

LARVEY, EMMETT
GOING MY WAY (I,1833)

LASCELLER, KENDREW
THE SMOTHERS BROTHERS SHOW (I,4097)

LASER, MARTIN
THE STARLOST (I,4203)

LASKOS, ANDREW
HOTEL (II,1192); REMINGTON STEELE (II,2130)

LASKY JR, JESSE
NAKED CITY (I,3215); PHILIP MARLOWE, PRIVATE EYE (II,2033)

LATHAM, LARRY
SUPER FRIENDS (II,2497)

LATIMER, JONATHAN
COLUMBO (II,556); PERRY MASON (II,2025)

LAUHER, BOB
IT TAKES TWO (I,2251); MALIBU U. (I,2849)

LAUREATE, NOBLE
THE FORSYTE SAGA (I,1645)

LAURENCE, MICHAEL
RETURN TO EDEN (II,2143)

LAURENTS, ARTHUR
THE LIGHT FANTASTIC, OR HOW TO TELL YOUR PAST, PRESENT AND MAYBE YOUR FUTURE THROUGH SOCIAL DANCING (I,2696)

LAVEN, ARNOLD
EIGHT IS ENOUGH (II,762)

LAVERY, EMMETT
SANDBURG'S LINCOLN (I,3907)

LAW, WARNER
MATINEE THEATER (I,2947)

LAWLOR, KATHLEEN
LOVING (II,1552)

LAWRENCE, ANTHONY
BEN CASEY (I,394); THE BLUE KNIGHT (II,276); BONANZA (II,347); HAWAII FIVE-O (II,1110); NAKED CITY (I,3216); ON TRIAL (II,1896); THE PHOENIX (II,2035); PLANET OF THE APES (II,2049); THE RAT PATROL (I,3726); ROUTE 66 (I,3852); STRANDED (II,2475); SWEET, SWEET RACHEL (I,4313)

LAWRENCE, EDDIE
THE VICTOR BORGE SHOW (I,4719); THE VICTOR BORGE SHOW (I,4720)

LAWRENCE, FANYA
IVANHOE (I,2282)

LAWRENCE, JEROME
SANDBURG'S LINCOLN (I,3907); SHANGRI-LA (I,3996)

LAWRENCE, JERRY
THE UNEXPECTED (I,4674)

LAWRENCE, MARK
THE ALICE PEARCE SHOW (I,121); JACK CARTER AND COMPANY (I,2305)

LAWRENCE, NANCY
THE PHOENIX (II,2035)

LAWRENCE, TONY
RAFFERTY (II,2105); SWISS FAMILY ROBINSON (II,2517)

LAWSON, CHARLES
SKY KING (I,4077)

LAWSON, STEVE
ST. ELSEWHERE (II,2432)

LAYTON, GEORGE
DOCTOR IN THE HOUSE (I,1313)

LAYTON, JERRY
ADVENTURES OF COLONEL FLACK (I,56)

LAZARUS, TOM
COLUMBO (II,556); HEAR NO EVIL (II,1116)

LECARRE, JOHN
SMILEY'S PEOPLE (II,2383)

LEFRENAIS, IAN
MY WIFE NEXT DOOR (II,1781)

LEGALLIENNE, EVA
ALICE IN WONDERLAND (I,120)

LEROSE, NICK
LAVERNE AND SHIRLEY (II,1446)

LESEUER, JOE
FOR RICHER, FOR POORER (II,906)

LEACH, ROBIN
LIFESTYLES OF THE RICH AND FAMOUS (II,1474)

LEAR, NORMAN
A.K.A. PABLO (II,28); THE ANDY WILLIAMS SPECIAL (I,211); THE DANNY KAYE SHOW (I,1144); HENRY FONDA AND THE FAMILY (I,2015); I LOVE LIBERTY (II,1215); P.O.P. (II,1939); THE TENNESSEE ERNIE FORD SHOW (I,4398)

LEAVITT, RON
IT'S YOUR MOVE (II,1259); THE JEFFERSONS (II,1276); SILVER SPOONS (II,2355); THE TED KNIGHT SHOW (II,2550)

LEBOWITZ, NEIL
MORK AND MINDY (II,1735); SILVER SPOONS (II,2355)

LECHT, ALLEN
NURSE (II,1870)

LEDEKA, MARIA
THE BRADY BRIDES (II,361); FOREVER FERNWOOD (II,909); PRIVATE BENJAMIN (II,2087); TOP TEN (II,2648)

LEDER, REUBEN
THE INCREDIBLE HULK (II,1232); MAGNUM, P.I. (II,1601); THE RETURN OF LUTHER GILLIS (II,2137); TALES OF THE GOLD MONKEY (II,2537)

LEDON, JOHN W
ANYTHING GOES (I,237)

LEE, BILL
ABC'S SILVER ANNIVERSARY CELEBRATION (II,2); MR. & MS. AND THE BANDSTAND MURDERS (II,1746); MR. & MS. AND THE MAGIC STUDIO MYSTERY (II,1747)

LEE, CHARLES
THE BOB HOPE SHOW (I,541); THE BOB HOPE SHOW (I,543); THE BOB HOPE SHOW (I,552); THE BOB HOPE SHOW (I,560); THE BOB HOPE SHOW (I,562); THE BOB HOPE SHOW (I,613); THE BOB HOPE SHOW (I,636); BOB HOPE SPECIAL: A QUARTER CENTURY OF BOB HOPE ON TELEVISION (II,281); BOB HOPE SPECIAL: BOB HOPE IN "JOYS" (II,284); BOB HOPE SPECIAL: BOB HOPE ON CAMPUS (II,289); BOB HOPE SPECIAL: BOB HOPE PRESENTS THE STARS OF TOMORROW (II,292); BOB HOPE SPECIAL: BOB HOPE'S ALL-STAR COMEDY SPECIAL FROM AUSTRALIA (II,300); BOB HOPE SPECIAL: BOB HOPE'S ALL-STAR COMEDY SPECTACULAR FROM LAKE TAHOE (II,301); BOB HOPE SPECIAL: BOB HOPE'S ALL-STAR COMEDY TRIBUTE TO VAUDEVILLE (II,302); BOB HOPE SPECIAL: BOB HOPE'S ALL-STAR TRIBUTE TO THE PALACE THEATER (II,305); BOB HOPE SPECIAL: BOB HOPE'S BICENTENNIAL STAR SPANGLED SPECTACULAR (II,306); BOB HOPE SPECIAL: BOB HOPE'S CHRISTMAS PARTY (II,307); BOB HOPE SPECIAL: BOB HOPE'S CHRISTMAS SPECIAL (II,308); BOB HOPE SPECIAL: BOB HOPE'S WORLD OF COMEDY (II,323); BOB HOPE SPECIAL: THE BOB HOPE COMEDY SPECIAL (II,332); BOB HOPE SPECIAL: THE BOB HOPE SPECIAL (II,333); BOB HOPE SPECIAL: THE BOB HOPE SPECIAL (II,334); BOB HOPE SPECIAL: THE BOB HOPE SPECIAL FROM PALM SPRINGS (II,339); ROBERTA (I,3815); ROBERTA (I,3816)

LEE, DAVID
CAPTAIN SCARLET AND THE MYSTERONS (I,837); THE JEFFERSONS (II,1276)

LEE, GILBERT
NUTS AND BOLTS (II,1871)

LEE, JAMES
ALCOA PREMIERE (I,109); THE DIARY OF ANNE FRANK (I,1262); THE HOLY TERROR (I,2098); THE LIVES OF JENNY DOLAN (II,1502); RAFFERTY (II,2105); ROOTS (II,2211); SCRUPLES (II,2280)

LEE, JOANNA
GIDGET (I,1795); GILLIGAN'S ISLAND (II,990); MULLIGAN'S STEW (II,1763); THE WALTONS (II,2740)

LEE, ROBERT E
SANDBURG'S LINCOLN (I,3907); SHANGRI-LA (I,3996); THE UNEXPECTED (I,4674)

LEE, SUSAN H
WE GOT IT MADE (II,2752)

LEE, VIRGINIA
THE HAZEL SCOTT SHOW (I,1984)

LEEDS, HOWARD
ALMOST AMERICAN (II,55); BACHELOR FATHER (I,323); BEWITCHED (I,418); THE BING CROSBY SHOW (I,466); DIFF'RENT STROKES (II,674); THE FIRST HUNDRED YEARS (I,1582); GEORGE GOBEL PRESENTS (I,1766); THE GEORGE GOBEL SHOW (I,1768); THE GHOST AND MRS. MUIR (I,1786); THE LAS VEGAS SHOW (I,2619); MCHALE'S NAVY (I,2969); THE MILTON BERLE SHOW (I,3049); PLEASE DON'T EAT THE DAISIES (I,3628); THE ROSEMARY CLOONEY SHOW (I,3848); SILVER SPOONS (II,2355); SWINGIN' TOGETHER (I,4320); TENNESSEE ERNIE'S NASHVILLE-MOSCOW EXPRESS (II,2561); THE UGLIEST GIRL IN TOWN (I,4663)

LEESON, MICHAEL
THE ASSOCIATES (II,113); BEST OF THE WEST (II,218); THE COSBY SHOW (II,578); THE MARY TYLER MOORE SHOW (II,1640); MIXED NUTS

(II,1714); PHYLLIS (II,2038); TAXI (II,2546); THERE'S ALWAYS ROOM (II,2583); YOUNG GUY CHRISTIAN (II,2864)

LEFCOURT, PETER
BULBA (II,392); CAGNEY AND LACEY (II,409); THE DEVLIN CONNECTION (II,664); EIGHT IS ENOUGH (II,762); KATE MCSHANE (II,1368); REMINGTON STEELE (II,2130); RIVKIN: BOUNTY HUNTER (II,2183); SCARECROW AND MRS. KING (II,2268); TWO MARRIAGES (II,2691)

LEFFERTS, GEORGE
FAMILY (II,813); SUSPICION (I,4309)

LEHR, WILSON
GIFT OF THE MAGI (I,1800); LITTLE WOMEN (I,2722)

LEIBMAN, NORM
THE BOB NEWHART SHOW (I,666); DIFF'RENT STROKES (II,674); THE JERRY LEWIS SHOW (I,2370)

LEIBMAN, RON
SIDE BY SIDE (II,2346)

LEIGH, CAROLYN
I'M A FAN (I,2192)

LEIGHT, WARREN
NATIONAL LAMPOON'S HOT FLASHES (II,1795)

LEIGHTON, JAMES
ALCOA PREMIERE (I,109)

LEIGHTON, TED
CASSIE AND COMPANY (II,455); COLUMBO (II,556); ELLERY QUEEN: DON'T LOOK BEHIND YOU (I,1434)

LEIZMAN, JOAN
THE PEE WEE HERMAN SHOW (II,1988)

LEMAN, JAN
THE NEW ADVENTURES OF CHARLIE CHAN (I,3249)

LEMAY, HARDING
LOVERS AND FRIENDS (II,1549)

LEMONT, JOHN
IVANHOE (I,2282)

LENARD, KAY
COMBAT (I,1011)

LENDER, CARYL
THE LAST LEAF (II,1437); THE WALTONS (II,2740)

LENNOX, JOHN THOMAS
ALIAS SMITH AND JONES (I,118)

LENSKI, ROBERT W
THE ALIENS ARE COMING (II,35); BARNABY JONES (II,153); BRONK (II,379); CANNON (II,424); CHIEFS (II,509); COLORADO C.I. (II,555); THE DAIN CURSE (II,613); THE FBI (I,1551); FREEBIE AND THE BEAN (II,922); MANHUNTER (II,1621); PLANET OF THE APES (II,2049); SOMETHING ELSE (I,4117)

LENZ, RICK
ALOHA PARADISE (II,58)

LEO, MALCOLM
E.T. & FRIENDS—MAGICAL MOVIE VISITORS (II,749)

LEOB, LEE
THE FARMER'S DAUGHTER (I,1533)

LEONARD, HUGH
FATHER BROWN (II,836)

LEONARD, KAY
BACHELOR FATHER (I,323); THRILLER (I,4492)

LEONARD, SHELDON
PATRICK STONE (I,3476)

LEONE, ANTHONY
WONDER GIRL (I,4891)

LEONETTI, CINDY
LITTLE LULU (II,1493)

LEOPOLD, GLENN
THE ALL-NEW POPEYE HOUR (II,49)

LEOPOLD, THOMAS
THE CHEVY CHASE SHOW (II,505); THE MARILYN MCCOO AND BILLY DAVIS JR. SHOW (II,1630); THE STEVE ALLEN COMEDY HOUR (II,2454)

LERNER, DAVID
BOSOM BUDDIES (II,355); BUSTING LOOSE (II,402); HOUSE CALLS (II,1194); JENNIFER SLEPT HERE (II,1278); OH MADELINE (II,1880); ON OUR OWN (II,1892); STAR OF THE FAMILY (II,2436); WEBSTER (II,2758)

LESCHIN, DEBORAH
BUSTING LOOSE (II,402); I'M A BIG GIRL NOW (II,1218); LAVERNE AND SHIRLEY (II,1446); MAKING A LIVING (II,1604); RHODA (II,2151)

LESHNER, MARTY
STAND UP AND CHEER (I,4188)

LESLIE, ALEEN
A DATE WITH JUDY (I,1162)

LESLIE, BETHEL
BARNABY JONES (II,153); A CHRISTMAS FOR BOOMER (II,518); FALCON CREST (II,808); FAMILY (II,813); MATT HELM (II,1653); THE ROOKIES (II,2208)

LESLIE, PHIL
THE ADDAMS FAMILY (II,11); THE FARMER'S DAUGHTER (I,1533); LEAVE IT TO BEAVER (I,2648)

LESSER, BUDD
CROWN THEATER WITH GLORIA SWANSON (I,1107); SKY KING (I,4077)

LESSING, NORMAN
BONANZA (II,347); CANNON (II,424); CARIBE (II,439); THE FBI (I,1551); THE FUGITIVE (I,1701); GREATEST HEROES OF THE BIBLE (II,1062); HAWAII FIVE-O (II,1110); NAKED CITY (I,3216); SHIRLEY TEMPLE'S STORYBOOK: THE LEGEND OF SLEEPY HOLLOW (I,4018)

LESTER, SELEG
THE CASE AGAINST PAUL RYKER (I,861); CLIMAX! (I,976); HAWAII FIVE-O (II,1110); NERO WOLFE (II,1806); THE OUTER LIMITS (I,3426); PERRY MASON (II,2025)

LETNEY, DON
THE BLUE KNIGHT (II,276)

LETTERMAN, DAVID
BOB HOPE SPECIAL: BOB HOPE'S ALL-STAR COMEDY SPECIAL FROM AUSTRALIA (II,300); THE DAVID LETTERMAN SHOW (II,624); LATE NIGHT WITH DAVID LETTERMAN (II,1442); THE PAUL LYNDE COMEDY HOUR (II,1978); THE STARLAND VOCAL BAND (II,2441)

LEVANT, BRIAN
HOT W.A.C.S. (II,1191); MORK AND MINDY (II,1735); STILL THE BEAVER (II,2466)

LEVENE, PHILIP
THE AVENGERS (II,121); THE FLYING DOCTOR (I,1610); MISTER JERICO (I,3070)

LEVENSON, SAM
THE SAM LEVENSON SHOW (I,3897)

LEVEY, JAY
WELCOME TO THE FUN ZONE (II,2763)

LEVI, BRUCE
GAVILAN (II,959)

LEVIN, BERNARD
THAT WAS THE WEEK THAT WAS (I,4423)

LEVIN, BIGGIE
STUD'S PLACE (I,4271)

LEVIN, JEFFREY
CAGNEY AND LACEY (II,409)

LEVIN, LISSA
GLORIA (II,1010); THREE'S A CROWD (II,2613); WKRP IN CINCINNATI (II,2814)

LEVIN, PETER
EMERALD POINT, N.A.S. (II,773)

LEVINE, KEN
AFTERMASH (II,23); THE BAY CITY AMUSEMENT COMPANY (II,185); CHARACTERS (II,481); CHEERS (II,494); JOE AND SONS (II,1296); M*A*S*H (II,1569); OPEN ALL NIGHT (II,1914); THE TONY RANDALL SHOW (II,2640)

LEVINE, LAURA
9 TO 5 (II,1852); THE BOB NEWHART SHOW (II,342); THE FLIP WILSON COMEDY SPECIAL (II,886); LAVERNE AND SHIRLEY (II,1446); THE PAUL LYNDE COMEDY HOUR (II,1977); PRIVATE BENJAMIN (II,2087); THREE'S COMPANY (II,2614); WE GOT IT MADE (II,2752); WHO'S ON CALL? (II,2791)

LEVINSON, BARRY
THE CAROL BURNETT SHOW (II,443); COMEDY NEWS II (I,1021); DINER (II,679); HOT L BALTIMORE (II,1187); THE INVESTIGATORS (II,1241); THE JOHN BYNER COMEDY HOUR (I,2407); THE MARTY FELDMAN COMEDY MACHINE (I,2929); PEEPING TIMES (II,1990); THE RICH LITTLE SHOW (II,2153); STOPWATCH: THIRTY MINUTES OF INVESTIGATIVE TICKING (II,2474); THE TIM CONWAY COMEDY HOUR (I,4501)

LEVINSON, DAVID
CHARLIE'S ANGELS (II,486); FORCE FIVE (II,907); HART TO HART (II,1102); LEAVE IT TO BEAVER (I,2648); SONS AND DAUGHTERS (II,2404); TOP SECRET (II,2647)

LEVINSON, LARRY
JOANIE LOVES CHACHI (II,1295)

LEVINSON, RICHARD

ALFRED HITCHCOCK PRESENTS (I,115); BURKE'S LAW (I,762); COLUMBO (II,556); ELLERY QUEEN: TOO MANY SUSPECTS (II,767); THE FUGITIVE (I,1701); JOHNNY RINGO (I,2434); MCCLOUD: WHO KILLED MISS U.S.A.? (I,2965); MCMILLAN AND WIFE (II,1667); PRESCRIPTION: MURDER (I,3660); SAM HILL: WHO KILLED THE MYSTERIOUS MR. FOSTER? (I,3896); SAVAGE (I,3924); TENAFLY (I,4395); THE WHOLE WORLD IS WATCHING (I,4851)

LEVITT, ALAN J
BARETTA (II,152); MAUDE (II,1655); THREE'S COMPANY (II,2614); YOU'RE JUST LIKE YOUR FATHER (II,2861)

LEVITT, GENE
COMBAT (I,1011); FANTASY ISLAND (II,830); GET CHRISTIE LOVE! (II,981); THE LORETTA YOUNG THEATER (I,2756)

LEVITT, SAUL
CLIMAX! (I,976); IVANHOE (I,2282); THE UNTOUCHABLES (I,4682)

LEVNISON, RICHARD
THE JUDGE AND JAKE WYLER (I,2464)

LEVY, EUGENE
FROM CLEVELAND (II,931); SECOND CITY TELEVISION (II,2287)

LEVY, LAWRENCE
FANTASY ISLAND (II,829)

LEVY, MELVIN
CHARLIE'S ANGELS (II,486)

LEVY, PARKE
MANY HAPPY RETURNS (I,2895); THE R.C.A. VICTOR SHOW (I,3737)

LEWIN, ALBERT
DIFF'RENT STROKES (II,674); DOCTOR CHRISTIAN (I,1308); GET CHRISTIE LOVE! (II,981); THE GHOST AND MRS. MUIR (I,1786); THE HOUSE NEXT DOOR (I,2139); MY FAVORITE MARTIAN (I,3189); RUSSIAN ROULETTE (I,3875)

LEWIN, ROBERT
CANNON (II,424); THE FBI (I,1551); THE FUGITIVE (I,1701); I SPY (I,2179); KUNG FU (II,1416); MCMILLAN AND WIFE (II,1667); RAWHIDE (I,3727); THE RIFLEMAN (I,3789); TODAY'S FBI (II,2628)

LEWIS, AL
THE EVE ARDEN SHOW (I,1471); THE GEORGE GOBEL SHOW (I,1768); MANY HAPPY RETURNS (I,2895); OUR MISS BROOKS (I,3415)

LEWIS, ANDY
BIG ROSE (II,241); CORONET BLUE (I,1055); TRAVIS LOGAN, D.A. (I,4598)

LEWIS, ARTHUR BERNARD
DALLAS (II,614); HAWAII FIVE-O (II,1110); UNCLE MISTLETOE AND HIS ADVENTURES (I,4669)

LEWIS, BILL
THE JERRY LEWIS SHOW (I,2370)

LEWIS, DAVID P
COLUMBO (II,556); FARAWAY HILL (I,1532); RAFFERTY (II,2105); ROLL OUT! (I,3833)

LEWIS, DRAPER
THE KIDS FROM FAME (II,1392); MUSIC FOR A SPRING NIGHT (I,3169); MUSIC FOR A SUMMER NIGHT (I,3170); WE DARE YOU! (II,2751)

LEWIS, JACK
THE CISCO KID (I,961); MATINEE THEATER (I,2947)

LEWIS, JEFFREY
BAY CITY BLUES (II,186); HILL STREET BLUES (II,1154)

LEWIS, JERRY
THE JERRY LEWIS SHOW (I,2369); THE JERRY LEWIS SHOW (I,2370)

LEWIS, RICHARD
THE LORENZO AND HENRIETTA MUSIC SHOW (II,1519); THE STEVE LANDESBERG TELEVISION SHOW (II,2458)

LEWIS, ROGER
T.J. HOOKER (II,2524)

LEWIS, SHARI
STAR TREK (II,2440)

LIBERMAN, LEO
CLIMAX! (I,976)

LIBOTT, ROBERT YALE
ALCOA PREMIERE (I,109); WAGON TRAIN (I,4747)

LICHTMAN, PAUL
KATE MCSHANE (II,1368)

LIDEKS, MARK
ARCHIE BUNKER'S PLACE (II,105)

LIEBER, ERIC
JOHN RITTER: BEING OF SOUND MIND AND BODY (II,1319); LEGENDS OF THE WEST: TRUTH AND TALL TALES (II,1457); THE NOONDAY SHOW (II,1859)

LIEBER, LESLIE
THE EDDIE ALBERT SHOW (I,1406)

LIEBERMAN, IRWIN
THE CISCO KID (I,961)

LIEBLING, HOWARD
ALICE (II,33)

LIEBMAN, MAX
LADY IN THE DARK (I,2602); SATINS AND SPURS (I,3917)

LIEBMAN, NORM
BARETTA (II,152); THE DEAN MARTIN SHOW (I,1201); GENE KELLY'S WONDERFUL WORLD OF GIRLS (I,1752); THE GOLDDIGGERS (I,1838); GOOD TIMES (II,1038); LAS VEGAS PALACE OF STARS (II,1431); THE MUNSTERS (I,3158); THE NIGHT STALKER (II,1849); TONY ORLANDO AND DAWN (II,2639); YOUNG MAVERICK (II,2867)

LIEBMAN, RON
DUSTY (II,745)

LIGERMAN, NAT
LET'S MAKE A DEAL (II,1464); MASQUERADE PARTY (II,1645)

LILES, MARCIA DURANT
MIKE AND BUFF (I,3037)

LINDER, MICHAEL
THE LOVE REPORT (II,1542); THE LOVE REPORT (II,1543)

LINDER, SUSAN JANE
DOUBLE TROUBLE (II,723)

LINDSAY, CYNTHIA
AT YOUR SERVICE (I,310); MY THREE SONS (I,3205)

LINDSAY, HOWARD
ARSENIC AND OLD LACE (I,268)

LINDSAY, KATHLEEN
STUDIO ONE (I,4268)

LINDSAY, ROBERT HOWARD
STUDIO ONE (I,4268)

LINDSEY, GEORGE
GOOBER AND THE TRUCKERS' PARADISE (II,1029)

LINK, WILLIAM
ALFRED HITCHCOCK PRESENTS (I,115); BURKE'S LAW (I,762); COLUMBO (II,556); ELLERY QUEEN: TOO MANY SUSPECTS (II,767); THE FUGITIVE (I,1701); JOHNNY RINGO (I,2434); THE JUDGE AND JAKE WYLER (I,2464); MCCLOUD: WHO KILLED MISS U.S.A.? (I,2965); MCMILLAN AND WIFE (II,1667); PRESCRIPTION: MURDER (I,3660); SAM HILL: WHO KILLED THE MYSTERIOUS MR. FOSTER? (I,3896); SAVAGE (I,3924); TENAFLY (I,4395); THE WHOLE WORLD IS WATCHING (I,4851)

LINTZ, PAULA
MR. MERLIN (II,1751)

LINZE, DEWEY
TREASURE UNLIMITED (I,4603)

LIPMAN, DANIEL
EMERALD POINT, N.A.S. (II,773); FAMILY (II,813); KNOTS LANDING (II,1404)

LIPP, FREDERICK L
FOUR STAR PLAYHOUSE (I,1652)

LIPSCOTT, ALAN
THE ANN SOTHERN SHOW (I,220); BACHELOR FATHER (I,323); DOC CORKLE (I,1304); HOORAY FOR LOVE (I,2116); LEAVE IT TO BEAVER (I,2648); THE LIFE OF RILEY (I,2685); THE PEOPLE'S CHOICE (I,3521)

LIPTON, JAMES
BOB HOPE SPECIAL: BOB HOPE'S ALL-STAR BIRTHDAY AT ANNAPOLIS (II,294); BOB HOPE SPECIAL: BOB HOPE'S ALL-STAR COMEDY BIRTHDAY PARTY AT WEST POINT (II,298); BOB HOPE SPECIAL: BOB HOPE'S SUPER BIRTHDAY SPECIAL (II,319); BOB HOPE SPECIAL: HAPPY BIRTHDAY, BOB! (II,325); BOB HOPE SPECIAL: HAPPY BIRTHDAY, BOB! (II,326); LORETTA LYNN IN THE BIG APPLE (II,1520); LORETTA LYNN: THE LADY. . .THE LEGEND (II,1521)

LIPTON, LEE
FIREBALL FUN FOR ALL (I,1577)

LISS, JOSEPH
THE WORLD OF MAURICE CHEVALIER (I,4916); THE WORLD OF SOPHIA LOREN (I,4918)

LISSON, MARK

MICKEY SPILLANE'S MIKE HAMMER (II,1692)

LITTLE JR, HERBERT
RAWHIDE (I,3727)

LITTLE, RICH
ON LOCATION WITH RICH LITTLE (II,1891); THE RICH LITTLE SPECIAL (II,2155); THE RICH LITTLE SPECIAL (II,2156); RICH LITTLE'S A CHRISTMAS CAROL (II,2157); RICH LITTLE'S ROBIN HOOD (II,2158); RICH LITTLE—COME LAUGH WITH ME (II,2160)

LITVACK, JOHN
AS THE WORLD TURNS (II,110)

LIVESAY, KENNETH
YOUNG LIVES (II,2866)

LIVINGSTON, ALAN W
ONE IN A MILLION (II,1906)

LIVINGSTON, HAROLD
BANACEK (II,138); THE BARBARY COAST (II,146); THE FEATHER AND FATHER GANG (II,845); FUTURE COP (II,945); PARTNERS IN CRIME (II,1961)

LIVINGSTONE, DOUGLAS
DAY OF THE TRIFFIDS (II,627)

LLEWELLYN, RICHARD
CROWN THEATER WITH GLORIA SWANSON (I,1107)

LLOYD, DAVID
THE ASSOCIATES (II,113); AT YOUR SERVICE (II,118); THE BETTY WHITE SHOW (II,224); BOB & CAROL & TED & ALICE (I,495); BROTHERS (II,380); CHEERS (II,494); DOC (II,682); DOC (II,683); DON'T CALL US (II,709); A FUNNY THING HAPPENED ON THE WAY TO HOLLYWOOD (I,1716); THE JACK PAAR SPECIAL (I,2319); JACK PAAR TONIGHT (I,2320); JOHNNY CARSON'S REPERTORY COMPANY IN AN EVENING OF COMEDY (I,2423); LOU GRANT (II,1526); THE MARY TYLER MOORE SHOW (II,1640); THE MERV GRIFFIN SHOW (I,3014); MR. SMITH (II,1755); NOT UNTIL TODAY (II,1865); NUMBER 96 (II,1869); PHYLLIS (II,2038); RHODA (II,2151); TAXI (II,2546); THE TONY RANDALL SHOW (II,2640); YOUR PLACE OR MINE? (II,2874)

LLOYD, JACK
MAKE MINE RED, WHITE, AND BLUE (I,2841); TONY BENNETT IN LONDON (I,4567); TONY BENNETT IN WAIKIKI (I,4568); YOUR HIT PARADE (II,2872)

LLOYD, JEREMY
BEANE'S OF BOSTON (II,194)

LLOYD, MICHAEL
SIGMUND AND THE SEA MONSTERS (II,2352); THREE'S A CROWD (II,2613)

LLOYD, STEPHEN
MATT HOUSTON (II,1654)

LOCKE, SAM
DEVLIN (II,663); GILLIGAN'S ISLAND (II,990); IT'S ABOUT TIME (I,2263); MCHALE'S NAVY (I,2969); THE PATTY DUKE SHOW (I,3483); THE RED BUTTONS SHOW (I,3750); THESE ARE THE DAYS (II,2585)

LOCKE, SARA
MY DARLING JUDGE (I,3186)

LOCKWOOD, PERCY
TELEVISION'S GREATEST COMMERCIALS II (II,2554)

LOEB, JOSEPH
THE HITCHHIKER (II,1157)

LOEB, LEE
SCREEN DIRECTOR'S PLAYHOUSE (I,3946)

LOGAN, BOB
LAUGH TRAX (II,1443)

LOMAN, MALCOLM
AMANDA'S (II,62)

LOMAN, MICHAEL
FISH (II,864); LADIES' MAN (II,1423); ON THE ROCKS (II,1894)

LONDON, BARBARA
ROMANCE THEATER (II,2204)

LONDON, JOHN
TWILIGHT THEATER II (II,2686)

LONDON, MARC
BOBBIE GENTRY'S HAPPINESS HOUR (I,671); BOBBIE GENTRY'S HAPPINESS HOUR (II,343); CELEBRATION: THE AMERICAN SPIRIT (II,461); THE DON RICKLES SHOW (II,708); DON'T CALL ME MAMA ANYMORE (I,1350); FRANK SINATRA JR. WITH FAMILY AND FRIENDS (I,1663); HARPER VALLEY PTA (II,1095); THE MAD MAD MAD MAD WORLD OF THE SUPER BOWL (II,1586); THE MUPPET SHOW (II,1766); THE ROWAN AND MARTIN REPORT (II,2221); THE ROWAN AND MARTIN SPECIAL (I,3855); SINATRA AND FRIENDS (II,2358); SMALL WORLD (II,2381); TAKE ONE STARRING JONATHAN WINTERS (II,2532)

LONG, SUMNER
APPLE'S WAY (II,101); DOBIE GILLIS (I,1302); THE FABULOUS SYCAMORES (I,1505); FATHER KNOWS BEST (II,838); LOTTERY (II,1525); WALDO (I,4749)

LONGET, CLAUDINE
THE ANDY WILLIAMS CHRISTMAS SHOW (I,199); THE ANDY WILLIAMS CHRISTMAS SHOW (II,77)

LONGSTREET, HARRY
FATHER MURPHY (II,841); HOT PURSUIT (II,1189); QUINCY, M. E. (II,2102); TRAUMA CENTER (II,2655); VOYAGERS (II,2730)

LONGSTREET, RENEE
FATHER MURPHY (II,841); HOT PURSUIT (II,1189); QUINCY, M. E. (II,2102); TRAUMA CENTER (II,2655); VOYAGERS (II,2730)

LONGSTREET, STEPHEN
HIGH BUTTON SHOES (I,2050)

LONSDALE, FREDERICK
MATINEE THEATER (I,2947)

LONSDALE, KATE
THE PALACE (II,1943)

LONSDALE, TED
THE PALACE (II,1943)

LOOS, RON
HOT (II,1183); HOT (II,1184)

LORD, MINDRET
THE LONE WOLF (I,2742)

LORD, PHILLIPS H
THE BLACK ROBE (I,479); GANGBUSTERS (I,1733)

LORD, STEPHEN
BANACEK (II,138); THE BARBARY COAST (II,146); CHIPS (II,511); CRAZY LIKE A FOX (II,601); JOHNNY RINGO (I,2434); MCCLOUD (II,1660); MOST WANTED (II,1739); THE NIGHT STALKER (II,1849); THE OUTER LIMITS (I,3426); THE ROGUES (I,3832); ZANE GREY THEATER (I,4979)

LORENZO, EDWARD D
THE WILD WILD WEST (I,4863)

LORIN, WILL
CASSIE AND COMPANY (II,455); DALLAS (II,614); KRAFT TELEVISION THEATER (I,2592); RIKER (II,2175); ROUTE 66 (I,3852)

LORING, GEORGE
THE BILLY BEAN SHOW (I,449)

LORRE, CHUCK
POLE POSITION (II,2060); WOLF ROCK TV (II,2815)

LOTTERBY, SYDNEY
YES MINISTER (II,2848)

LOUIE, CHARLES
VERSATILE VARIETIES (I,4713)

LOUIS, C E
SEVEN BRIDES FOR SEVEN BROTHERS (II,2307)

LOVY, ALEX
MEATBALLS AND SPAGHETTI (II,1674); PANDAMONIUM (II,1948)

LOWE, SHERMAN
THE CISCO KID (I,961); SKY KING (I,4077)

LOWTHER, GEORGE
MATINEE THEATER (I,2947)

LUCAROTTI, JOHN
THE AVENGERS (II,121); THE ONEDIN LINE (II,1912)

LUCAS, ANN
PRISONER: CELL BLOCK H (II,2085)

LUCAS, JOHN MEREDYTH
BEN CASEY (I,394); BURKE'S LAW (I,762); CAVALCADE OF AMERICA (I,875); CITY BENEATH THE SEA (I,965); THE FUGITIVE (I,1701); HARRY O (II,1099); KOJAK (II,1406); LOGAN'S RUN (II,1507); THE LORETTA YOUNG THEATER (I,2756); MANNIX (II,1624); MEDIC (I,2976); NERO WOLFE (II,1806); RAFFERTY (II,2105); THE SIX-MILLION-DOLLAR MAN (II,2372); STAR TREK (II,2440)

LUCAS, JONATHAN
THE DEAN MARTIN SHOW (I,1201)

LUCCI, FRANK
FIREHOUSE (I,1579)

LUDIN, JOHN
THE GET ALONG GANG (II,980)

LUDLAM, GEORGE

PETER HUNTER, PRIVATE EYE (I,3563)

LUDWIG, JERRY
ASSIGNMENT: MUNICH (I,302); BUNCO (II,395); FOR LOVE AND HONOR (II,903); I SPY (I,2179); JESSICA NOVAK (II,1286); RIKER (II,2175); SAMURAI (II,2249); TODAY'S FBI (II,2628)

LUGER, LOIS
THE GEORGIA PEACHES (II,979)

LUKAS, C W
ON THE TOWN WITH TONY BENNETT (II,1895)

LUKES, JACK
LAVERNE AND SHIRLEY (II,1446)

LUND, TRIG
KOVACS UNLIMITED (I,2583)

LUPINO, IDA
THRILLER (I,4492)

LUPO, FRANK
THE A-TEAM (II,119); B.J. AND THE BEAR (II,125); GALACTICA 1980 (II,953); THE GREATEST AMERICAN HERO (II,1060); HUNTER (II,1206); LOBO (II,1504); MAGNUM, P.I. (II,1601); THE QUEST (II,2098); RIPTIDE (II,2178)

LUSITANA, DONNA E
THE LOVE REPORT (II,1542); THE LOVE REPORT (II,1543)

LUSSIER, DANE
SCREEN DIRECTOR'S PLAYHOUSE (I,3946)

LUSSOR, DAVE
FRONT ROW CENTER (I,1694)

LUTZ, BILL
THE ADDAMS FAMILY (II,11); QUEEN FOR A DAY (I,3705)

LUTZ, TOM
THE JUD STRUNK SHOW (I,2462)

LUXLEY, JAY
THE PHIL SILVERS SHOW (I,3580)

LYLE, FRED
REMINGTON STEELE (II,2130)

LYNCH, PEG
ETHEL AND ALBERT (I,1466)

LYNDE, JACQUELINE
HAWAII FIVE-O (II,1110)

LYNDE, PAUL
THE PAUL LYNDE COMEDY HOUR (II,1977)

LYNDON, BARRE
THRILLER (I,4492)

LYNN, BILL
THE GOLDDIGGERS IN LONDON (I,1837)

LYNN, JONATHAN
CAMBRIDGE CIRCUS (I,807); DOCTOR IN THE HOUSE (I,1313); YES MINISTER (II,2848)

MABLEY, EDWARD
KRAFT TELEVISION THEATER (I,2592)

MACANDREW, JACK
EVENING AT THE IMPROV (II,786)

MACDONALD, PHILIP
THRILLER (I,4492)

MACDONNELL, BRUCE
COLISEUM (I,1000)

MACDOUGALL, RONALD
FAME IS THE NAME OF THE GAME (I,1518)

MACFARLAND, LOUELLA
THE ANN SOTHERN SHOW (I,220); HAZEL (I,1982); PRIVATE SECRETARY (I,3672); THE TEENAGE IDOL (I,4375)

MACKENZIE, JACK
NAKED CITY (I,3215)

MACKILLOP, ED
WESTSIDE MEDICAL (II,2765)

MACLANE, ROLAND
HOW TO MARRY A MILLIONAIRE (I,2147); IT'S ABOUT TIME (I,2263); LEAVE IT TO BEAVER (I,2648)

MACMAHON, JENNA
THE CAROL BURNETT SHOW (II,443)

MACAULAY, RICHARD
PERRY MASON (II,2025)

MACK, BURRELL
THE FACTS OF LIFE (II,805)

MACK, DICK
BLONDIE (I,486)

MACKIE, PHILIP
THERESE RAQUIN (II,2584)

MADDEN, PETER
MAN ON A STRING (I,2877)

MADDOX, BEN
BRADDOCK (I,704); NAKED CITY (I,3216); THE UNTOUCHABLES (I,4682)

MADDOX, LEE
ALL NIGHT RADIO (II,40); I'VE HAD IT UP TO HERE (II,1221)

MADRID III, LANCE
MAGNUM, P.I. (II,1601)

MADY, A J
GIDGET (I,1795)

MAGAR, GUY
BUCK ROGERS IN THE 25TH CENTURY (II,383)

MAGEE, JIM
THE JERRY LESTER SPECIAL (I,2361)

MAGISTRETTI, PAUL
BARETTA (II,152); FOR LOVE AND HONOR (II,903); SIMON AND SIMON (II,2357); TODAY'S FBI (II,2628); WHIZ KIDS (II,2790)

MAGISTRETTI, WILLIAM F
THE OUTSIDE MAN (II,1937)

MAHER, JAMES T
100 YEARS OF AMERICA'S POPULAR MUSIC (II,1904)

MAHIN, RICHARD
RIVAK, THE BARBARIAN (I,3795)

MAIBAUM, RICHARD
COMBAT (I,1011); JARRETT (I,2349); S*H*E (II,2235); WAGON TRAIN (I,4747)

MAIKIN, DAVID
ALL NIGHT RADIO (II,40)

MAINWARING, DANIEL
TARGET: THE CORRUPTERS (I,4360); THE WILD WILD WEST (I,4863)

MAITLAND, JULES
THE DICK POWELL SHOW (I,1269)

MAKOUL, RUDY
EL COYOTE (I,1423)

MALBY JR, RICHARD
AIN'T MISBEHAVIN' (II,25)

MALEX, BRYCE
THE LITTLE RASCALS (II,1494); THE RICHIE RICH SHOW (II,2169)

MALIANI, FELICIA
THE GET ALONG GANG (II,980)

MALIANI, MICHAEL
THE GET ALONG GANG (II,980)

MALKO, GEORGE
THE ABC AFTERSCHOOL SPECIAL (II,1)

MALLORS, PARKER
YOU ASKED FOR IT (II,2855)

MALMBERG, DAN
THIS IS YOUR LIFE (I,4461)

MALONE, HALSEY
SUSPICION (I,4309)

MALONE, JOEL
SATAN'S WAITIN' (I,3916)

MALTESE, MIKE
THE JETSONS (I,2376)

MALVIN, ARTIE
THE STEVE LAWRENCE SHOW (I,4227)

MANDEL, ALEX
LADY LUCK (I,2604)

MANDEL, BABALOO
BUSTING LOOSE (II,402); HERNDON AND ME (II,1139); LAVERNE AND SHIRLEY (II,1446)

MANDEL, BARBARA
THE FURTHER ADVENTURES OF WALLY BROWN (II,944)

MANDEL, HOWIE
WELCOME TO THE FUN ZONE (II,2763)

MANDEL, LORING
SANDBURG'S LINCOLN (I,3907); STUDIO ONE (I,4268); TOM AND JOANN (II,2630)

MANDEL, MARC
M*A*S*H (II,1569)

MANDEL, MEL
SKINFLINT (II,2377)

MANDELL, SID
CAMP RUNAMUCK (I,811)

MANGO, CARLO
THE NEW VOICE (II,1833)

MANHEIM, CHRIS
THE MISSISSIPPI (II,1707); SHIRLEY (II,2330)

MANHEIM, MANNIE
THE ALL-STAR REVUE (I,138); DO NOT DISTURB (I,1300); FRANCES LANGFORD PRESENTS (I,1656)

MANHOFF, BILL
THE ANN SOTHERN SHOW (I,220); THE DICK POWELL SHOW (I,1269); THE DOCTOR WAS A LADY (I,1318); LEAVE IT TO BEAVER (I,2648); LOW MAN ON THE TOTEM POLE (I,2782); MAGGIE BROWN (I,2811); MAKE ROOM FOR DADDY (I,2843); THE MILTON BERLE SHOW (I,3047); MY BOY GOOGIE (I,3185); MY WORLD . . . AND WELCOME TO IT (I,3208); THE PARTRIDGE FAMILY (II,1962); THE TENNESSEE ERNIE FORD SHOW (I,4398)

MANILOW, BARRY
THE BARRY MANILOW SPECIAL (II,155); BARRY MANILOW—ONE VOICE (II,156); THE SECOND BARRY MANILOW SPECIAL (II,2285); THE THIRD BARRY MANILOW SPECIAL (II,2590)

MANINGS, ALLAN
THE COMICS (I,1028); GABE AND WALKER (II,950); GOOD TIMES (II,1038); ICHABOD AND ME (I,2183); LEAVE IT TO BEAVER (I,2648); THE LILY TOMLIN SHOW (I,2706); PLEASE DON'T EAT THE DAISIES (I,3628); ROXY PAGE (II,2222); A SALUTE TO STAN LAUREL (I,3892); THE WONDERFUL WORLD OF BURLESQUE I (I,4895)

MANKIEWICZ, DON M
THE BAIT (I,335); IRONSIDE (I,2240); KRAFT TELEVISION THEATER (I,2592); LANIGAN'S RABBI (II,1427); MARCUS WELBY, M.D. (I,2905); MCMILLAN AND WIFE (II,1667); MURDER INK (II,1769); ON TRIAL (I,3362); ONE STEP BEYOND (I,3388); PLAYHOUSE 90 (I,3623); ROSETTI AND RYAN: MEN WHO LOVE WOMEN (II,2217); SANCTUARY OF FEAR (II,2252); SARGE: THE BADGE OR THE CROSS (I,3915); WHO HAS SEEN THE WIND? (I,4847)

MANKIEWICZ, TOM
BEAT OF THE BRASS (I,376); HART TO HART (II,1101); MOTHER, JUGGS AND SPEED (II,1741); MOVIN' WITH NANCY (I,3119)

MANN, ABBY
THE FIRST 50 YEARS (II,862); THE GREATEST GIFT (II,1061); MATINEE THEATER (I,2947); PLAYHOUSE 90 (I,3623); ROBERT MONTGOMERY PRESENTS YOUR LUCKY STRIKE THEATER (I,3809)

MANN, DANNY
EVENING AT THE IMPROV (II,786); THICKE OF THE NIGHT (II,2587)

MANN, MICHAEL
BRONK (II,379); STARSKY AND HUTCH (II,2444); VEGAS (II,2723); VEGAS (II,2724)

MANN, STANLEY
THE INVISIBLE MAN (I,2232)

MANNING, MONROE
THE ADVENTURES OF SUPERMAN (I,77); COUNTERPOINT (I,1059); CROWN THEATER WITH GLORIA SWANSON (I,1107)

MANOFF, WILLIAM
HEY, JEANNIE! (I,2038); MAGGIE (I,2810)

MANSFIELD, IRVING
THE MOREY AMSTERDAM SHOW (I,3101)

MANTLEY, JOHN
THE DESILU PLAYHOUSE (I,1237); DIRTY SALLY (II,680); HOW THE WEST WAS WON (II,1196); THE OUTER LIMITS (I,3426); RAWHIDE (I,3727); SIX GUNS FOR DONEGAN (I,4065); THE UNTOUCHABLES (I,4682)

MANZANO, SONIA
SESAME STREET (I,3982)

MAPES, JOANNA
GILLIGAN'S ISLAND (II,990)

MAPES, TERENCE
GILLIGAN'S ISLAND (II,990); THE LONE RANGER (I,2740); THE NEW ADVENTURES OF CHARLIE CHAN (I,3249)

MARCEL, TERRY
PRISONERS OF THE LOST UNIVERSE (II,2086)

MARCH, JACK
WESTSIDE MEDICAL (II,2765)

MARCHETTA, CAMILLE
DALLAS (II,614); LUCAN (II,1554); SCRUPLES (II,2281)

MARCUS, ANN
CAMP RUNAMUCK (I,811); FLAMINGO ROAD (II,872); KNOTS LANDING (II,1404); LETTERS FROM THREE LOVERS (I,2671)

MARCUS, BOB
GREEN ACRES (I,1884)

MARCUS, DIANA KOPALD
EMERALD POINT, N.A.S. (II,773)

MARCUS, ELLIS
BEN CASEY (I,394); FLAMINGO ROAD (II,872); KNOTS LANDING (II,1404); THE LETTERS (I,2670); LITTLE WOMEN (I,2719); THE MAN FROM U.N.C.L.E. (I,2867); MR. TUTT (I,3153); TARGET: THE CORRUPTERS (I,4360)

MARCUS, LARRY
ALCOA PREMIERE (I,109); CAVALCADE OF AMERICA (I,875); FOUR STAR PLAYHOUSE (I,1652); OF THIS TIME, OF THAT PLACE (I,3335); ON TRIAL (I,3362); ONE STEP BEYOND (I,3388); ROUTE 66 (I,3852); SUDDEN SILENCE (I,4275); WAGON TRAIN (I,4747); WON'T IT EVER BE MORNING? (I,4902)

MARCUS, LEAH
DYNASTY (II,746); FAME (II,812); PAPER DOLLS (II,1951); PAPER DOLLS (II,1952)

MARGOLIN, ARNOLD
THE DOOLEY BROTHERS (II,715); HARRY AND MAGGIE (II,1097); MR. TERRIFIC (I,3152); THE ORPHAN AND THE DUDE (II,1924); PRIVATE BENJAMIN (II,2087); THE SMOTHERS BROTHERS SHOW (I,4096)

MARGOLIS, HERBERT
ALIAS MIKE HERCULES (I,117); GILLIGAN'S ISLAND (II,990)

MARGOLIS, JACK S
DICK CLARK PRESENTS THE ROCK AND ROLL YEARS (I,1265); THE NUT HOUSE (I,3320); ROWAN AND MARTIN'S LAUGH-IN (I,3856)

MARGOLIS, PAUL
T.J. HOOKER (II,2524)

MARGULIES, HARRIET
KOJAK (II,1406)

MARION, CHARLES
DOBIE GILLIS (I,1302); THE UNTOUCHABLES (I,4682)

MARK, DAVID
A PUNT, A PASS, AND A PRAYER (I,3694)

MARKES, LARRY
THE ALAN KING SHOW (I,101); THE CARA WILLIAMS SHOW (I,843); COMEDY IS KING (I,1019); COS (II,576); COS: THE BILL COSBY COMEDY SPECIAL (II,577); JACK AND THE BEANSTALK (I,2288); THE JACK PAAR SHOW (I,2316); THE JETSONS (I,2376); THE JOHNNY CASH CHRISTMAS SPECIAL (II,1325); THE LESLIE UGGAMS SHOW (I,2660); THE MANHATTAN TRANSFER (II,1619); MCHALE'S NAVY (I,2969); THE NIGHT STALKER (II,1849); THE VICTOR BORGE SHOW (I,4719); THE VICTOR BORGE SHOW (I,4720); WE'LL TAKE MANHATTAN (I,4792); THE WONDERFUL WORLD OF PIZZAZZ (I,4901)

MARKHAM, EARL
FLASH GORDON (I,1593)

MARKHAM, PACY
TELEVISION'S GREATEST COMMERCIALS (II,2553)

MARKIN, CAROLE
KIDD VIDEO (II,1388)

MARKO, ZEKIAL
THE NIGHT STALKER (II,1849)

MARKOWITZ, BRAD
THE LOVE REPORT (II,1542)

MARKOWITZ, GARY
M*A*S*H (II,1569)

MARKOWITZ, MITCH
BEST OF THE WEST (II,218); BUFFALO BILL (II,387); THE FACTS OF LIFE (II,805); FOREVER FERNWOOD (II,909); M*A*S*H (II,1569); REPORT TO MURPHY (II,2134); TOO CLOSE FOR COMFORT (II,2642); VAN DYKE AND COMPANY (II,2719)

MARKOWITZ, RIFF
THE HILARIOUS HOUSE OF FRIGHTENSTEIN (II,1153)

MARKS, BILL
BRING 'EM BACK ALIVE (II,377); THE WILD WILD WEST (I,4863)

MARKS, DANIEL
THE STARS AND STRIPES SHOW (I,4206)

MARKS, DARA
ROUGHCUTS (II,2218)

MARKS, DENNIS
BATFINK (I,365); THE STARS AND STRIPES SHOW (I,4208)

MARKS, HERBERT
THE JACK BENNY SPECIAL (I,2296)

MARKS, HILLIARD
EVERYTHING YOU ALWAYS WANTED TO KNOW ABOUT JACK BENNY AND WERE AFRAID TO ASK (I,1482); THE HOLLYWOOD PALACE (I,2088); JACK BENNY'S 20TH ANNIVERSARY TV SPECIAL (I,2303); JACK BENNY'S BIRTHDAY SPECIAL (I,2299); JACK BENNY'S FIRST FAREWELL SHOW (I,2300); JACK BENNY'S NEW LOOK (I,2301); JACK BENNY'S SECOND FAREWELL SHOW (I,2302)

MARKS, JOAN
CHRIS AND THE MAGICAL DRIP (II,516)

MARKS, LARRY

DEAN MARTIN'S CELEBRITY ROAST (II,643); DICK CLARK'S WORLD OF TALENT (I,1268); JOHNNY CASH AND FRIENDS (II,1322); PAT PAULSEN'S HALF A COMEDY HOUR (I,3473); THE PHIL SILVERS ARROW SHOW (I,3576); TONY ORLANDO AND DAWN (II,2639)

MARKS, LAWRENCE
THE DORIS DAY SHOW (I,1355); I WAS A BLOODHOUND (I,2180); M*A*S*H (II,1569); MY WORLD . . . AND WELCOME TO IT (I,3208); PERRY MASON (II,2025); PHYLLIS (II,2038); READY AND WILLING (II,2115)

MARKS, LOUIS
DOCTOR WHO (II,689)

MARKS, SHERMAN
THE FREEWHEELERS (I,1683); TONI TWIN TIME (I,4554)

MARKSTEIN, GEORGE
THE PRISONER (I,3670)

MARKUS, JOHN
AMANDA'S (II,62); THE COSBY SHOW (II,578); GIMME A BREAK (II,995)

MARLAND, DOUGLAS
LOVING (II,1552)

MARLENS, NEAL
AMANDA'S (II,62); OH MADELINE (II,1880)

MARLOWE, DEREK
NANCY ASTOR (II,1788)

MARLOWE, JACK
THE WILD WILD WEST (I,4863)

MARMER, MIKE
DEAN MARTIN'S CELEBRITY ROAST (II,643); DEAN MARTIN'S RED HOT SCANDALS OF 1926 (II,648); DEAN MARTIN'S RED HOT SCANDALS PART 2 (II,649); DEAN'S PLACE (II,650); F TROOP (I,1499); THE FACTS OF LIFE (II,805); THE FIRST 50 YEARS (II,862); GILLIGAN'S ISLAND (II,990); KEEP U.S. BEAUTIFUL (I,2519); LADIES AND GENTLEMAN. . . BOB NEWHART (II,1420); LINDSAY WAGNER—ANOTHER SIDE OF ME (II,1483); THE LOVE BOAT (II,1535); THE MAC DAVIS SHOW (II,1575); ME AND MAXX (II,1672); THE MILTON BERLE SHOW (I,3049); NOBODY'S PERFECT (II,1858); THE SMOTHERS BROTHERS COMEDY HOUR (I,4095); THE STEVE ALLEN SHOW (I,4221); SUCCESS: IT CAN BE YOURS (II,2488); TED KNIGHT MUSICAL COMEDY VARIETY SPECIAL SPECIAL (II,2549); THE TIM CONWAY SPECIAL (I,4503); VICTOR BORGE'S 20TH ANNIVERSARY SHOW (I,4727)

MARMORSTEIN, MALCOLM
DARK SHADOWS (I,1157)

MAROKE, MERRILL
THE DAVID LETTERMAN SHOW (II,624); LATE NIGHT WITH DAVID LETTERMAN (II,1442); LOVE, NATALIE (II,1545); MAKING THE GRADE (II,1606); OPEN ALL NIGHT (II,1914); SHAPING UP (II,2316)

MARRIOTT, ANTHONY
FROM A BIRD'S EYE VIEW (I,1686)

MARRIS, WEBB
THE ROOKIES (II,2208)

MARSH, GWENDA
ALL THE RIVERS RUN (II,43)

MARSH, LINDA
13 QUEENS BOULEVARD (II,2591); THE FACTS OF LIFE (II,805); FACTS OF LIFE: THE FACTS OF LIFE GOES TO PARIS (II,806); JO'S COUSINS (II,1293)

MARSH, SID
JUST MEN (II,1359)

MARSHALL, EMILY
THE BOB NEWHART SHOW (II,342); NEWHART (II,1835)

MARSHALL, GARRY K
BAREFOOT IN THE PARK (I,355); THE DANNY THOMAS SPECIAL (I,1150); DANNY THOMAS: THE ROAD TO LEBANON (I,1153); THE DICK VAN DYKE SPECIAL (I,1277); DOMINIC'S DREAM (II,704); EVIL ROY SLADE (I,1485); I SPY (I,2179); THE MURDOCKS AND THE MCCLAYS (I,3163); THE ODD COUPLE (II,1875); SHERIFF WHO? (I,4009); THINK PRETTY (I,4448); WEDNESDAY NIGHT OUT (II,2760); WIVES (II,2812)

MARSHALL, GEORGE LEE
CRAZY LIKE A FOX (II,601); MICKEY SPILLANE'S MIKE HAMMER (II,1692)

MARSHALL, LEW
MEATBALLS AND SPAGHETTI (II,1674)

MARSHALL, NEAL
THE BOBBY DARIN AMUSEMENT COMPANY (I,672)

MARSHALL, ROGER
THE AVENGERS (II,121)

MARSHALL, SIDNEY
LAND OF THE GIANTS (I,2611); MEDIC (I,2976); THE NEW ADVENTURES OF CHARLIE CHAN (I,3249); VOYAGE TO THE BOTTOM OF THE SEA (I,4743)

MARSHALL, STEVE
CASS MALLOY (II,454); GLORIA (II,1010); OFF THE RACK (II,1878); WKRP IN CINCINNATI (II,2814)

MARTIN, AL
MATINEE THEATER (I,2947); MY FAVORITE MARTIAN (I,3189)

MARTIN, ANDREA
FROM CLEVELAND (II,931); SECOND CITY TELEVISION (II,2287)

MARTIN, ANN
CRYSTAL (II,607); TEACHERS ONLY (II,2548)

MARTIN, ANTHONY S
KEY WEST (I,2528)

MARTIN, DAVE
DOCTOR WHO (II,689)

MARTIN, DON
COUNTERPOINT (I,1059)

MARTIN, EDITH
FIRESIDE THEATER (I,1580)

MARTIN, EDWARD
STORY THEATER (I,4240)

MARTIN, IAN
APPOINTMENT WITH ADVENTURE (I,246); THE ONEDIN LINE (II,1912); STRANGE PARADISE (I,4246)

MARTIN, JEFF
LATE NIGHT WITH DAVID LETTERMAN (II,1442)

MARTIN, MADELYN
THE DESILU PLAYHOUSE (I,1237)

MARTIN, NORMAN
DEBBY BOONE. . . ONE STEP CLOSER (II,655); DINAH AND FRIENDS (II,675); DINAH! (II,677)

MARTIN, RICCI
ON STAGE AMERICA (II,1893)

MARTIN, STEVE
JOHN DENVER ROCKY MOUNTAIN CHRISTMAS (II,1314); THE KEN BERRY WOW SHOW (I,2524); PAT PAULSEN'S HALF A COMEDY HOUR (I,3473); THE SUMMER SMOTHERS BROTHERS SHOW (I,4281); VAN DYKE AND COMPANY (II,2718)

MARTIN, VIRGINIA
THE MAGIC GARDEN (II,1591)

MARTINEZ, AL
B.A.D. CATS (II,123); THEY ONLY COME OUT AT NIGHT (II,2586)

MARTINO, EDWARD
HART TO HART (II,1102)

MARVIN, MITZIE
KING'S CROSSING (II,1397); SEVEN BRIDES FOR SEVEN BROTHERS (II,2307)

MARX, ARTHUR
ALICE (II,33); DO NOT DISTURB (I,1300)

MARX, MARVIN
AS CAESAR SEES IT (I,294); CAVALCADE OF STARS (I,877); THE HONEYMOONERS (II,1175); THE JACK CARTER SHOW (I,2306); JACKIE GLEASON AND HIS AMERICAN SCENE MAGAZINE (I,2321); THE JACKIE GLEASON SHOW (I,2322); THE JACKIE GLEASON SPECIAL (I,2324); THAT'S LIFE (I,4426)

MARX, SAM
NAKED CITY (I,3216)

MASCHELLA, TOM
STARSKY AND HUTCH (II,2444)

MASCHLER, TIM
BRING 'EM BACK ALIVE (II,377); CRAZY LIKE A FOX (II,601); HAWAII FIVE-O (II,1110); KHAN! (II,1384); MAGNUM, P.I. (II,1601); MR. MERLIN (II,1751); THE NIGHT STALKER (II,1849)

MASCOTT, HOLLY
BLANSKY'S BEAUTIES (II,267); LAVERNE AND SHIRLEY (II,1446)

MASCOTT, LAWRENCE
CROWN THEATER WITH GLORIA SWANSON (I,1107)

MASIUS, JOHN
ST. ELSEWHERE (II,2432)

MASON, EDWARD J
INTERNATIONAL DETECTIVE (I,2224)

MASON, JUDI ANN
GOOD TIMES (II,1038)

MASON, PAUL
CHIPS (II,511); GET CHRISTIE LOVE! (II,981); MANIMAL

(II,1622)

MASOUKA, MARK
E/R (II,748)

MASS, AUDREY
CAESAR AND CLEOPATRA (I,789)

MASSELINK, BEN
CODE RED (II,552); THE FBI (I,1551); THE INCREDIBLE HULK (II,1232); S.W.A.T. (II,2236); STARSKY AND HUTCH (II,2444)

MASSELNIK, MICHAEL
BARNABY JONES (II,153)

MASSUOLKA, MARK
TOO CLOSE FOR COMFORT (II,2642)

MASTERS, SUSAN
THE NOONDAY SHOW (II,1859)

MASTROGEORGE, PAUL
THE NIGHT STALKER (II,1849)

MATHESON, RICHARD
THE A-TEAM (II,119); ENOS (II,779); THE INCREDIBLE HULK (II,1232); KNIGHT RIDER (II,1402); THE MARTIAN CHRONICLES (II,1635); THE NIGHT STALKER (I,3292); THE NIGHT STRANGLER (I,3293); THE POWERS OF MATTHEW STAR (II,2075); STAR TREK (II,2440); THE TWILIGHT ZONE (I,4651)

MATHEWS, PATRICK
240-ROBERT (II,2688); A MAN CALLED SLOANE (II,1613); QUINCY, M. E. (II,2102); WONDER WOMAN (SERIES 2) (II,2829)

MATHEWS, WALTER
THE JIMMIE RODGERS SHOW (I,2383)

MATHIAS, SONNY
KNIGHT RIDER (II,1402)

MATTHEWS, DOROTHY
CYRANO DE BERGERAC (I,1119)

MATTHEWS, MARK
FOUL UPS, BLEEPS AND BLUNDERS (II,911)

MATTHEWS, PATRICK
SWITCH (II,2519)

MAUER, MICHAEL
MR. MERLIN (II,1751)

MAULDIN, NAT
IT TAKES TWO (II,1252)

MAURER, NORMAN
SPEED BUGGY (II,2419)

MAURY, DICK
JACK CARTER AND COMPANY (I,2305); THE WONDERFUL WORLD OF JACK PAAR (I,4899)

MAUS, WILLIAM
BRANDED (I,707)

MAXFIELD, B R
TWO MARRIAGES (II,2691)

MAXWELL, ROBERT
THE ADVENTURES OF SUPERMAN (I,77)

MAXWELL-SMITH, MARK
THE CROSS-WITS (II,605)

MAY, BOB
JIGSAW (I,2377)

MAY, PHIL
KRAFT SALUTES WALT DISNEY WORLD'S 10TH ANNIVERSARY (II,1410)

MAYBERRY, ROBERT
WINDOW SHADE REVUE (I,4874)

MAYER, DAVID
THE GOLDDIGGERS IN LONDON (I,1837)

MAYER, JEFF
THE MANY MOODS OF PERRY COMO (I,2897); PERRY COMO'S WINTER SHOW (I,3546)

MAYER, JERRY
THE ACADEMY (II,3); THE ACADEMY II (II,4); BEWITCHED (I,418); THE BOB NEWHART SHOW (II,342); DON KNOTTS NICE CLEAN, DECENT, WHOLESOME HOUR (I,1333); THE FACTS OF LIFE (II,805); FACTS OF LIFE: THE FACTS OF LIFE GOES TO PARIS (II,806); FOL-DE-ROL (I,1613); HERE WE GO AGAIN (I,2028); THE MARY TYLER MOORE SHOW (II,1640); MITZI AND A HUNDRED GUYS (II,1710); MITZI. . . ROARIN' IN THE 20S (II,1711); MITZI. . . WHAT'S HOT, WHAT'S NOT (II,1712); MITZI. . . ZINGS INTO SPRING (II,1713); MITZI...THE FIRST TIME (I,3073); MITZI: A TRIBUTE TO THE AMERICAN HOUSEWIFE (I,3072); THE NEW BILL COSBY SHOW (I,3257); PHILLIP AND BARBARA (II,2034); TABITHA (II,2527); TABITHA (II,2528)

MAYES, WENDELL
KRAFT TELEVISION THEATER (I,2592); SAVAGE: IN THE ORIENT (II,2264)

MAYNARD, BIFF
THE JACKSONS (II,1267)

MAZARSKY, PHIL
THE CARA WILLIAMS SHOW (I,843)

MAZER, VIRGINIA
SUSPICION (I,4309)

MAZURSKY, PAUL
THE DANNY KAYE SHOW (I,1143); THE RIFLEMAN (I,3789)

MAZZA, PIETRA
SARA (II,2260)

MCADAMS, JAMES
KOJAK (II,1406)

MCCALL, MARY
GILLIGAN'S ISLAND (II,990); MAISIE (I,2833)

MCCALL, MITZI
IT'S NOT EASY (II,1256); LADIES' MAN (II,1423); THE STOCKARD CHANNING SHOW (II,2468)

MCCANN, CHUCK
THE DONNY AND MARIE OSMOND SHOW (II,714); FAR OUT SPACE NUTS (II,831)

MCCAREY, LEO
MEET THE GOVERNOR (I,2991); TOM AND JERRY (I,4533)

MCCAREY, MARY
SCREEN DIRECTOR'S PLAYHOUSE (I,3946)

MCCARTHY, IRVING J
THE DICK POWELL SHOW (I,1269)

MCCARTHY, NICK
THE OMEGA FACTOR (II,1888)

MCCAULEY, PAUL
KEEP U.S. BEAUTIFUL (I,2519)

MCCLAIN, JACK
BONANZA (II,347)

MCCLANE, ROLAND
GILLIGAN'S ISLAND (II,990)

MCCLEERY, WILLIAM
THE ALCOA HOUR (I,108); MATINEE THEATER (I,2947); MEET CORLISS ARCHER (I,2981)

MCCLELLAN, MAX
NURSE (II,1870)

MCCLOSKY, MARGARET
PRISONER: CELL BLOCK H (II,2085)

MCCLURG, EDIE
THE DAVID LETTERMAN SHOW (II,624); THE PARAGON OF COMEDY (II,1956); THE PEE WEE HERMAN SHOW (II,1988)

MCCONNACHIE, BRIAN
THE TV TV SHOW (II,2674)

MCCONNOR, VINCENT
THE BARRETTS OF WIMPOLE STREET (I,361)

MCCORMICK, BRIAN
RODNEY DANGERFIELD SPECIAL: IT'S NOT EASY BEIN' ME (II,2199)

MCCORMICK, JAN
THE NEIGHBORS (II,1801); SECOND CHANCE (II,2286)

MCCORMICK, NANCE
GOOD TIMES (II,1038)

MCCORMICK, PAT
BETTE MIDLER—OL' RED HAIR IS BACK (II,222); COMEDY NEWS II (I,1021); A COUPLE OF DONS (I,1077); THE CRACKER BROTHERS (II,599); THE DANNY THOMAS SPECIAL (I,1151); THE DICK VAN DYKE SPECIAL (I,1276); THE DON RICKLES SHOW (I,1341); THE FUNNY SIDE (I,1714); THE FUNNY SIDE (I,1715); JOHNNY CARSON DISCOVERS CYPRESS GARDENS (I,2419); THE JONATHAN WINTERS SHOW (I,2442); THE KRAFT MUSIC HALL (I,2588); LUCY IN LONDON (I,2789); THE MANY SIDES OF DON RICKLES (I,2898); THE MARTY FELDMAN COMEDY MACHINE (I,2929); THE MERV GRIFFIN SHOW (I,3014); THE NEW BILL COSBY SHOW (I,3257); THE RED SKELTON SHOW (I,3756); SHAUGHNESSEY (II,2318); TIN PAN ALLEY TODAY (I,4521); ZERO HOUR (I,4980)

MCCOY, DENNIS
THE GET ALONG GANG (II,980)

MCCOY, ELIZABETH
CLIMAX! (I,976)

MCCOY, ESTHER
THE LORETTA YOUNG THEATER (I,2756)

MCCOY, HOMER
CROWN THEATER WITH GLORIA SWANSON (I,1107)

MCCRACKEN, ELLEN
MATINEE THEATER (I,2947)

MCCULLEN, CARSON
THE MEMBER OF THE WEDDING (II,1680)

MCCULLOUGH, ROBERT L
B.J. AND THE BEAR (II,125); B.J. AND THE BEAR (II,126); THE BIONIC WOMAN (II,255); EIGHT IS ENOUGH (II,762); FALCON CREST (II,808); GALACTICA 1980 (II,953)

MCDONAGH, RICHARD
SHANE (I,3995)

MCDONALD, ERIC
THE OMEGA FACTOR (II,1888)

MCDONNELL, TERRENCE
BATTLESTAR GALACTICA (II,181); THE BIONIC WOMAN (II,255); KIDD VIDEO (II,1388)

MCFADDEN, CYRA
13 QUEENS BOULEVARD (II,2591)

MCGEEHAN, BERNICE
THE LAWRENCE WELK CHRISTMAS SPECIAL (II,1450)

MCGIBBON, DUNCAN
ARCHIE BUNKER'S PLACE (II,105)

MCGIBBON, SCOTT
DONNY AND MARIE (II,712); THE STEVE LANDESBERG TELEVISION SHOW (II,2458); THAT SECOND THING ON ABC (II,2572)

MCGINN, JIM
APPLE'S WAY (II,101); LEADFOOT (II,1452); PRINCESS (II,2084)

MCGLYNN, DON
KIDD VIDEO (II,1388)

MCGOOHAN, PATRICK
THE PRISONER (I,3670)

MCGOVERN, WILLIAM
ALCOA PREMIERE (I,109); BEN CASEY (I,394); MADIGAN (I,2808)

MCGREEVEY, JOHN
THE ANDY WILLIAMS CHRISTMAS SHOW (I,199); THE ANDY WILLIAMS CHRISTMAS SHOW (II,77); APPLE'S WAY (II,101); BLACK SADDLE (I,480); CLIMAX! (I,976); THE DESILU PLAYHOUSE (I,1237); DISAPPEARANCE OF AIMEE (I,1290); THE FARMER'S DAUGHTER (I,1533); THE FESS PARKER SHOW (II,850); THE FLYING NUN (I,1611); GIDGET (I,1795); GIDGET GETS MARRIED (I,1796); GIDGET GROWS UP (I,1797); HAZEL (I,1982); THE MILLIONAIRE (II,1700); MY THREE SONS (I,3205); PUNCH AND JODY (II,2091); THE RETURN OF MARCUS WELBY, M.D (II,2138); ROOTS: THE NEXT GENERATIONS (II,2213); ROUTE 66 (I,3852); SCREEN DIRECTOR'S PLAYHOUSE (I,3946); VOYAGE TO THE BOTTOM OF THE SEA (I,4743); WAGON TRAIN (I,4747); THE WALTONS (II,2740)

MCGREEVEY, MICHAEL
BOONE (II,351); FATHER MURPHY (II,841); NURSE (II,1870); PALMERSTOWN U.S.A. (II,1945); QUINCY, M. E. (II,2102); WONDER WOMAN (SERIES 2) (II,2829)

MCGUIRE, DON
CHARLIE ANGELO (I,906); FROM HERE TO ETERNITY (II,932); MCGHEE (I,2967)

MCILDOWNIE, GEORGE
THE POWERS OF MATTHEW STAR (II,2075)

MCKANE, JACKIE
HARPER VALLEY PTA (II,1095); HOUSE CALLS (II,1194)

MCKAY, ANDY
KOVACS UNLIMITED (I,2583)

MCKAY, BRIAN
CAGNEY AND LACEY (II,409); SONS AND DAUGHTERS (II,2404); VEGAS (II,2724); A WALK IN THE NIGHT (I,4750); WONDER WOMAN (SERIES 2) (II,2829)

MCKAY, KEVIN
ROBERT MONTGOMERY PRESENTS YOUR LUCKY STRIKE THEATER (I,3809)

MCKENNA, BERNARD
DOCTOR IN THE HOUSE (I,1313); ROBIN'S NEST (II,2187)

MCKNIGHT, FRED
QUINCY, M. E. (II,2102)

MCLANE, RONALD
MAKE ROOM FOR DADDY (I,2843)

MCLEOD, VIC
WELCOME ABOARD (I,4789)

MCLEVELY, JACK
CHARLIE'S ANGELS (II,486)

MCMAHON, ED
THAT MCMAHON'S HERE AGAIN (I,4419)

MCMAHON, JENNA
EUNICE (II,784); FLO (II,891); HOW TO HANDLE A WOMAN (I,2146); A LAST LAUGH AT THE 60'S (I,2627); MAMA'S FAMILY (II,1610); MAUDE (II,1655); A NEW KIND OF FAMILY (II,1819); SOAP (II,2392)

MCMAHON, PATRICK
HOT HERO SANDWICH (II,1186)

MCMANUS, KAY
CORONATION STREET (I,1054)

MCMANUS, MARSHALL
TWILIGHT THEATER II (II,2686)

MCMANUS, MIKE
MARIE (II,1629)

MCNALLY, TERRENCE
MAMA MALONE (II,1609)

MCNEELY, JERRY
LUCAS TANNER (II,1555); THE MAN FROM U.N.C.L.E. (I,2867); MARCUS WELBY, M.D. (II,1627); MCMILLAN (II,1666); OWEN MARSHALL: COUNSELOR AT LAW (I,3436); SANDBURG'S LINCOLN (I,3907); THREE FOR THE ROAD (II,2607); TRAUMA CENTER (II,2655); THE TWILIGHT ZONE (I,4651)

MCNEIL, STEVE
BONANZA (II,347); MR. LUCKY (I,3141)

MCPHERSON, STEPHEN
BUCK ROGERS IN THE 25TH CENTURY (II,384); THE GANGSTER CHRONICLES (II,957); THE LAST CONVERTIBLE (II,1435); THE RENEGADES (II,2133)

MCRAVEN, DALE
ALMOST HEAVEN (II,57); THE BETTY WHITE SHOW (II,224); HOME ROOM (II,1172); LAVERNE AND SHIRLEY (II,1446); MORK AND MINDY (II,1735); THE PARTRIDGE FAMILY (II,1962); THE TEXAS WHEELERS (II,2568)

MCKEAN, MICHAEL
LAVERNE AND SHIRLEY (II,1446)

MCKEAND, CAROL EVAN
APPLE'S WAY (II,101); CASSIE AND COMPANY (II,455); ERIC (I,1451); FAMILY (II,813); THE FAMILY TREE (II,820); NORMA RAE (II,1860); RYAN'S FOUR (II,2233); THE SIX OF US (II,2369); THE WALTONS (II,2740)

MCKEAND, NIGEL
ERIC (I,1451); FAMILY (II,813); THE WALTONS (II,2740)

MCKEE, ROBERT
KATE LOVES A MYSTERY (II,1367)

MCLEAN, RON
PRISONER: CELL BLOCK H (II,2085)

MEADOW, BARRY
ONE IN A MILLION (II,1906)

MEADOW, HERB
IN TANDEM (II,1228); MAN AGAINST CRIME (I,2852)

MEADOW, HERMAN
ROUTE 66 (I,3852)

MECCHI, IRENE
LILY—SOLD OUT (II,1480); SOLID GOLD (II,2395)

MEDFORD, HAROLD
MANNIX (II,1624)

MEDWAY, ELIZABETH
SOLID GOLD HITS (II,2397)

MEDWAY, LISA
BARBARA MANDRELL AND THE MANDRELL SISTERS (II,143); BLONDES VS. BRUNETTES (II,271)

MEEHAN, THOMAS
ANNIE AND THE HOODS (II,91); ANNIE, THE WOMAN IN THE LIFE OF A MAN (I,226); COMEDY TONIGHT (I,1027); COS (II,576); THE DICK CAVETT SHOW (II,669); THAT WAS THE YEAR THAT WAS (I,4424); THAT WAS THE YEAR THAT WAS (II,2575); THIS WILL BE THE YEAR THAT WILL BE (I,4466)

MEGDALL, TOM
FOR RICHER, FOR POORER (II,906)

MELCHOIR, IB
THE OUTER LIMITS (I,3426)

MELENDEZ, BILL
HAPPY ANNIVERSARY, CHARLIE BROWN (I,1939)

MELFI, LEONARD
MAMA MALONE (II,1609)

MELLIN, SHERMAN

GHOSTBREAKER (I,1789)

MELMAN, JEFF
NIGHT COURT (II,1844)

MELTZER, LOU
DAGMAR'S CANTEEN (I,1125); THE MOREY AMSTERDAM SHOW (I,3101)

MELTZER, NEWTON
APPOINTMENT WITH ADVENTURE (I,246)

MELVOIN, JEFF
REMINGTON STEELE (II,2130)

MELVOIN, MICHAEL
REMINGTON STEELE (II,2130)

MENDELSOHN, JACK
CARTER COUNTRY (II,449); DOC (II,682); THE JIM NABORS HOUR (I,2381); JOE AND SONS (II,1296); THE LITTLE RASCALS (II,1494); MEATBALLS AND SPAGHETTI (II,1674); PARTRIDGE FAMILY: 2200 A.D. (II,1963); THE RICHIE RICH SHOW (II,2169); SPEED BUGGY (II,2419); THREE'S COMPANY (II,2614)

MENDELSON, LEE
THE FABULOUS FUNNIES (II,801); THE FANTASTIC FUNNIES (II,825); SUNDAY FUNNIES (II,2493); THE WONDERFUL WORLD OF PIZZAZZ (I,4901)

MENKEN, JOHN
KUNG FU (II,1416)

MENOTTI, GIAN-CARLO
AMAHL AND THE NIGHT VISITORS (I,148); AMAHL AND THE NIGHT VISITORS (I,150)

MENTEER, GARY
JOANIE LOVES CHACHI (II,1295)

MENVILLE, CHUCK
BENJI, ZAX AND THE ALIEN PRINCE (II,205); GOING BANANAS (II,1015); SPEED BUGGY (II,2419); STAR TREK (II,2439); TARZAN: LORD OF THE JUNGLE (II,2544); UNCLE CROC'S BLOCK (II,2702); WHEELIE AND THE CHOPPER BUNCH (II,2776)

MENZIES, JAMES
THE FUGITIVE (I,1701); MY FAVORITE MARTIAN (I,3189); THE PAPER CHASE (II,1949); RIDING FOR THE PONY EXPRESS (II,2172); THE WALTONS (II,2740)

MENZIES, WILLIAM
COMBAT (I,1011)

MERCER, JACK
THE MILTON THE MONSTER CARTOON SHOW (I,3053)

MEREDITH, SEAN
THE YELLOW ROSE (II,2847)

MERL, JUDY
FALCON CREST (II,808); FLAMINGO ROAD (II,872); NIGHT PARTNERS (II,1847)

MERLIN, BARBARA
THE LIFE OF VERNON HATHAWAY (I,2687); THE LORETTA YOUNG THEATER (I,2756); SCREEN DIRECTOR'S PLAYHOUSE (I,3946)

MERLIN, MILTON
THE FUGITIVE (I,1701)

MERRICK, LEONARD
CROWN THEATER WITH GLORIA SWANSON (I,1107)

MERRILL, HOWARD
F TROOP (I,1499); THE LOVE BOAT (II,1535); MRS. G. GOES TO COLLEGE (I,3155)

MERRILL, KEITH
THE CHEROKEE TRAIL (II,500)

MERSON, MARC
DUSTY (II,745); RISE AND SHINE (II,2179)

MESSINA, DIANE
THE FACTS OF LIFE (II,805); HARPER VALLEY PTA (II,1095); WEBSTER (II,2758)

MESSINA, LOU
THE FACTS OF LIFE (II,805); HARPER VALLEY PTA (II,1095); WEBSTER (II,2758)

MESTON, JOHN
LITTLE HOUSE ON THE PRAIRIE (II,1487)

METCALFE, BURT
M*A*S*H (II,1569)

METCALFE, RORY
RYAN'S HOPE (II,2234)

METCALFE, STEVE
THE COMEDY ZONE (II,559)

METTER, ALAN
MONTY HALL'S VARIETY HOUR (II,1728)

METZGER, MIKE
CAMOUFLAGE (II,417); HOW'S YOUR MOTHER-IN-LAW? (I,2148); THE NEW TREASURE HUNT (II,1831); TREASURE HUNT (II,2657); WEDDING DAY (II,2759)

MEUGNIOT, WILL
SUPER FRIENDS (II,2497)

MEUNIER, JACQUES
BEULAH LAND (II,226)

MEYER, BOB
T.L.C. (II,2525)

MEYER, NICHOLAS
JUDGE DEE IN THE MONASTERY MURDERS (I,2465)

MEYER, RICH
A MAN CALLED SLOANE (II,1613)

MEYERS, HOWARD
DIFF'RENT STROKES (II,674)

MEYERS, JANET
CALIFORNIA FEVER (II,411)

MEYERSON, PETER
BEST FRIENDS (II,213); THE BOB NEWHART SHOW (II,342); CAPTAIN NICE (I,835); WE'LL GET BY (II,2753); WHATEVER HAPPENED TO DOBIE GILLIS? (II,2774)

MICHAEL, SANDRA
ROBERT MONTGOMERY PRESENTS YOUR LUCKY STRIKE THEATER (I,3809)

MICHAELIAN, KATHARYN
KHAN! (II,1384); KUNG FU (II,1416); LOGAN'S RUN (II,1507); SARA (II,2260)

MICHAELIAN, MICHAEL
BARNABY JONES (II,153); CHARLIE'S ANGELS (II,486); THE DUKES OF HAZZARD (II,742); THE FANTASTIC JOURNEY (II,826); KHAN! (II,1384); KUNG FU (II,1416); ROUGHNECKS (II,2219); SARA (II,2260); THE TOM SWIFT AND LINDA CRAIG MYSTERY HOUR (II,2633)

MICHAELS, LORNE
THE BEACH BOYS SPECIAL (II,190); THE BEAUTIFUL PHYLLIS DILLER SHOW (I,381); FLIP WILSON. . . OF COURSE (II,890); LILY (I,2705); LILY (II,1477); THE LILY TOMLIN SPECIAL (II,1479); THE NEW SHOW (II,1828); THE PAUL SIMON SPECIAL (II,1983); THE PERRY COMO WINTER SHOW (I,3544); PERRY COMO'S WINTER SHOW (I,3548)

MICHAELS, SIDNEY
CAROUSEL (I,856)

MICHELLE, STEVEN
THE BOOK OF LISTS (II,350); SUNDAY FUNNIES (II,2493); TV FUNNIES (II,2672)

MICKLIN, JOAN
FAERIE TALE THEATER (II,807)

MIDLER, BETTE
BETTE MIDLER—ART OR BUST (II,221); BETTE MIDLER—OL' RED HAIR IS BACK (II,222)

MILBURN, SUE
THE BIONIC WOMAN (II,255); BORN TO THE WIND (II,354); CHARLIE'S ANGELS (II,486); THE JACKSON FIVE (I,2326); JOHNNY BELINDA (I,2418); RAFFERTY (II,2105); THIS IS KATE BENNETT (II,2597)

MILCH, DAVID
BAY CITY BLUES (II,186); HILL STREET BLUES (II,1154)

MILES, HANK
OF ALL THINGS (I,3330)

MILES, LYNDA
AS THE WORLD TURNS (II,110)

MILIO, JIM
PRIME TIMES (II,2083)

MILITO, SEBASTIAN
QUINCY, M. E. (II,2102)

MILIUS, JOHN
MELVIN PURVIS: G-MAN (II,1679)

MILLANDER, CAREY
THE ROGUES (I,3832)

MILLARD, OSCAR
ALCOA PREMIERE (I,109); SECRETS OF THE OLD BAILEY (I,3971); THRILLER (I,4492)

MILLER, ARTHUR
FAME (I,1517); THE PRICE (I,3666)

MILLER, BARRY
ANIMALS ARE THE FUNNIEST PEOPLE (II,81)

MILLER, CAROLYN
THE ABC AFTERSCHOOL SPECIAL (II,1); POWERHOUSE (II,2074)

MILLER, CHRIS
RAWHIDE (I,3727); SQUARE PEGS (II,2431)

MILLER, ELIZABETH
THE BAXTERS (II,184)

MILLER, GARY
9 TO 5 (II,1852); BOSOM BUDDIES (II,355); DOMESTIC LIFE (II,703); GIMME A BREAK (II,995)

MILLER, HARVEY
BAREFOOT IN THE PARK (I,355); THE GHOST AND MRS. MUIR (I,1786)

MILLER, HERMAN
KUNG FU (II,1416)

MILLER, IRA
THE LORENZO AND HENRIETTA MUSIC SHOW (II,1519); WELCOME BACK, KOTTER (II,2761)

MILLER, JP
THE PEOPLE NEXT DOOR (I,3519); PLAYHOUSE 90 (I,3623)

MILLER, JACK
HOW THE WEST WAS WON (II,1196); LASSIE: THE NEW BEGINNING (II,1433); THREE FOR THE ROAD (II,2607)

MILLER, JAMES M
CASABLANCA (II,450); KOJAK (II,1406); THE MISSISSIPPI (II,1707); MOBILE TWO (II,1718); THE POWERS OF MATTHEW STAR (II,2075); TESTIMONY OF TWO MEN (II,2564)

MILLER, JASON
REWARD (II,2147)

MILLER, JENNIFER
LOOSE CHANGE (II,1518); PAPER DOLLS (II,1952); TESTIMONY OF TWO MEN (II,2564)

MILLER, JIM
KOJAK (II,1406)

MILLER, LARRY
SWINGING COUNTRY (I,4322)

MILLER, MARILYN
BARNEY MILLER (II,154); LILY (II,1477); THE MARY TYLER MOORE SHOW (II,1640); RHODA (II,2151)

MILLER, MARK
THE NEW ODD COUPLE (II,1825)

MILLER, MERLE
PLAYHOUSE 90 (I,3623)

MILLER, MORTON NEAL
SENSE OF HUMOR (II,2303)

MILLER, NANCY ANN
THE RENEGADES (II,2133)

MILLER, PAUL
DIFF'RENT STROKES (II,674)

MILLER, SIDNEY
THE BOBBY DARIN AMUSEMENT COMPANY (I,672); THE DONALD O'CONNOR SHOW (I,1343); THE GENE KELLY SHOW (I,1751)

MILLER, STEPHEN
ARCHIE BUNKER'S PLACE (II,105); FLO (II,891); GREAT DAY (II,1054); THE MISADVENTURES OF SHERIFF LOBO (II,1704)

MILLER, SUSAN
FAMILY (II,813)

MILLER, WALTER C
BOWZER (II,358); THE OSMOND FAMILY CHRISTMAS SPECIAL (II,1926); THE OSMOND FAMILY THANKSGIVING SPECIAL (II,1928)

MILLER, WINSTON
RAWHIDE (I,3727); SCREEN DIRECTOR'S PLAYHOUSE (I,3946)

MILLERICK, KERRY
ELKE SOMMER'S WORLD OF SPEED AND BEAUTY (II,765)

MILLIGAN, SPIKE
THE MARTY FELDMAN COMEDY MACHINE (I,2929)

MILLINAIRE, NICOLE
SHERLOCK HOLMES (I,4010)

MILLS, ED
ON THE GO (I,3361)

MILLS, ROBERT L
BOB HOPE SPECIAL: BOB HOPE FOR PRESIDENT (II,282); BOB HOPE SPECIAL: BOB HOPE GOES TO COLLEGE (II,283); BOB HOPE SPECIAL: BOB HOPE IN THE STAR-MAKERS (II,285); BOB HOPE SPECIAL: BOB HOPE IN WHO MAKES THE WORLD LAUGH, PART 2 (II,286); BOB HOPE SPECIAL: BOB HOPE IN WHO MAKES THE WORLD LAUGH? (II,287); BOB HOPE SPECIAL: BOB HOPE LAUGHS WITH THE MOVIE AWARDS (II,288); BOB HOPE SPECIAL: BOB HOPE'S 30TH ANNIVERSARY TV SPECIAL (II,293); BOB HOPE SPECIAL: BOB HOPE'S ALL-STAR BIRTHDAY AT ANNAPOLIS (II,294); BOB HOPE SPECIAL: BOB HOPE'S ALL-STAR COMEDY BIRTHDAY PARTY (II,297); BOB HOPE SPECIAL: BOB HOPE'S ALL-STAR COMEDY BIRTHDAY PARTY AT WEST POINT (II,298); BOB HOPE SPECIAL: BOB HOPE'S ALL-STAR COMEDY LOOK AT THE FALL SEASON: IT'S STILL FREE AND WORTH IT! (II,299); BOB HOPE SPECIAL: BOB HOPE'S ALL-STAR COMEDY SPECIAL FROM AUSTRALIA (II,300); BOB HOPE SPECIAL: BOB HOPE'S ALL-STAR LOOK AT TV'S PRIME TIME WARS (II,303); BOB HOPE SPECIAL: BOB HOPE'S ALL-STAR SUPER BOWL PARTY (II,304); BOB HOPE SPECIAL: BOB HOPE'S ALL-STAR TRIBUTE TO THE PALACE THEATER (II,305); BOB HOPE SPECIAL: BOB HOPE'S CHRISTMAS SPECIAL (II,308); BOB HOPE SPECIAL: BOB HOPE'S FUNNY VALENTINE (II,309); BOB HOPE SPECIAL: BOB HOPE'S MERRY CHRISTMAS SHOW (II,310); BOB HOPE SPECIAL: BOB HOPE'S OVERSEAS CHRISTMAS TOURS (II,311); BOB HOPE SPECIAL: BOB HOPE'S PINK PANTHER THANKSGIVING GALA (II,312); BOB HOPE SPECIAL: BOB HOPE'S ROAD TO HOLLYWOOD (II,313); BOB HOPE SPECIAL: BOB HOPE'S SALUTE TO NASA—25 YEARS OF REACHING FOR THE STARS (II,314); BOB HOPE SPECIAL: BOB HOPE'S SPRING FLING OF COMEDY AND GLAMOUR (II,315); BOB HOPE SPECIAL: BOB HOPE'S STAND UP AND CHEER FOR THE NATIONAL FOOTBALL LEAGUE'S 60TH YEAR (II,316); BOB HOPE SPECIAL: BOB HOPE'S STAR-STUDDED SPOOF OF THE NEW TV SEASON— G RATED—WITH GLAMOUR, GLITTER AND GAGS (II,317); BOB HOPE SPECIAL: BOB HOPE'S STARS OVER TEXAS (II,318); BOB HOPE SPECIAL: BOB HOPE'S SUPER BIRTHDAY SPECIAL (II,319); BOB HOPE SPECIAL: BOB HOPE'S USO CHRISTMAS IN BEIRUT (II,320); BOB HOPE SPECIAL: BOB HOPE'S WICKI-WACKY SPECIAL FROM WAIKIKI (II,321); BOB HOPE SPECIAL: BOB HOPE'S WOMEN I LOVE—BEAUTIFUL BUT FUNNY (II,322); BOB HOPE SPECIAL: BOB HOPE—HOPE, WOMEN AND SONG (II,324); BOB HOPE SPECIAL: BOB HOPE PRESENTS A CELEBRATION WITH STARS OF COMEDY AND MUSIC (II,340); BOB HOPE SPECIAL: HAPPY BIRTHDAY, BOB! (II,325); BOB HOPE SPECIAL: HO HO HOPE'S JOLLY CHRISTMAS HOUR (II,327); BOB HOPE SPECIAL: THE BOB HOPE CHRISTMAS SPECIAL (II,328); BOB HOPE SPECIAL: THE BOB HOPE CHRISTMAS SPECIAL (II,329); BOB HOPE SPECIAL: THE BOB HOPE CHRISTMAS SPECIAL (II,331); BOB HOPE SPECIAL: THE BOB HOPE SPECIAL FROM PALM SPRINGS (II,339); BOB HOPE SPECIAL: BOB HOPE'S HILARIOUS UNREHEARSED ANTICS OF THE STARS (II,341); DEAN MARTIN'S CELEBRITY ROAST (II,643)

MILLS, STUART
I'VE HAD IT UP TO HERE (II,1221)

MILLS, TED
MAURICE CHEVALIER'S PARIS (I,2955)

MILNER, JACK
THE INVADERS (I,2229)

MILTON, RICHARD
MARCUS WELBY, M.D. (II,1627)

MINER, ALLEN H
THE DESILU PLAYHOUSE (I,1237); THE DICK POWELL SHOW (I,1269); THE ELISABETH MCQUEENEY STORY (I,1429); WAGON TRAIN (I,4747)

MINGO, JACK
WELCOME TO THE FUN ZONE (II,2763)

MINOFF, PHIL
THE POLLY BERGEN SHOW (I,3643)

MINTNER, GORDON
BRIEF PAUSE FOR MURDER (I,725)

MINTZ, LARRY
BLUE JEANS (II,275); FOOT IN THE DOOR (II,902); HIGH SCHOOL, U.S.A. (II,1148); MORK AND MINDY (II,1735); WHAT'S HAPPENING!! (II,2769)

MINTZ, RICK
A.K.A. PABLO (II,28)

MINTZ, ROBERT
THE OUTER LIMITS (I,3426)

MIRKIN, DAVID
THREE'S A CROWD (II,2613); THREE'S COMPANY (II,2614)

MISCH, DAVID
CALLAHAN (II,414); MORK AND MINDY (II,1735)

MISCHEL, JOSEF
FIRESIDE THEATER (I,1580)

MISHKIN, PHIL
FREE COUNTRY (II,921); HAPPY DAYS (II,1084); HARPER VALLEY (II,1094); HARPER VALLEY PTA (II,1095); PAUL SAND IN FRIENDS AND LOVERS (II,1982); ROBERT YOUNG AND THE FAMILY (I,3813); SONNY BOY (II,2402); THE SUPER (I,4293); THREE FOR THE GIRLS (I,4477)

MISROCK, HENRY
MATINEE THEATER (I,2947)

MITCHELL, BERNARD
S.W.A.T. (II,2236)

MITCHELL, BILL
GIVE-N-TAKE (II,1003); THE NEIGHBORS (II,1801); SECOND CHANCE (II,2286)

MITCHELL, BOB
LAND OF THE GIANTS (I,2611); PERRY MASON (II,2025)

MITCHELL, CASEY T
PAPER DOLLS (II,1951); PAPER DOLLS (II,1952)

MITCHELL, CHICK
WE GOT IT MADE (II,2752)

MITCHELL, COLEMAN
ANNIE FLYNN (II,92); THE BOB NEWHART SHOW (II,342); GIMME A BREAK (II,995); PAUL SAND IN FRIENDS AND LOVERS (II,1982); RHODA (II,2151); SPENCER (II,2420)

MITCHELL, ESTHER
BUCK ROGERS IN THE 25TH CENTURY (II,384); CHIPS (II,511); COMBAT (I,1011); LAND OF THE GIANTS (I,2611); PERRY MASON (II,2025); S.W.A.T. (II,2236); SWISS FAMILY ROBINSON (II,2517)

MITCHELL, GORDON
AFTERMASH (II,23); GILLIGAN'S ISLAND: THE HARLEM GLOBETROTTERS ON GILLIGAN'S ISLAND (II,993); HARPER VALLEY PTA (II,1095); HONEYMOON SUITE (I,2108); JOE'S WORLD (II,1305); MORK AND MINDY (II,1735); THE PARTRIDGE FAMILY (II,1962)

MITCHELL, JULIAN
JENNIE: LADY RANDOLPH CHURCHILL (II,1277)

MITCHELL, ROBERT
BUCK ROGERS IN THE 25TH CENTURY (II,384); CHIPS (II,511); COMBAT (I,1011); SWISS FAMILY ROBINSON (II,2517)

MITCHLLL, SCOEY
ME AND MRS. C. (II,1673)

MITTLEMAN, RICK
BOB & CAROL & TED & ALICE (I,495); CHIPS (II,511); THE DORIS DAY SHOW (I,1355); ENOS (II,779); FREEBIE AND THE BEAN (II,922); GLITTER (II,1009); GOOD MORNING WORLD (I,1850); JENNIFER SLEPT HERE (II,1278); JUST FRIENDS (I,2495); THE LESLIE UGGAMS SHOW (I,2660); MY WORLD ... AND WELCOME TO IT (I,3208); THE ODD COUPLE (II,1875); ONE DAY AT A TIME (II,1900); THAT'S MY MAMA (II,2580); VAN DYKE AND COMPANY (II,2718); WHAT'S HAPPENING!! (II,2769)

MITZ, RICK
I LOVE LIBERTY (II,1215)

MOESSINGER, DAVID
APPLE'S WAY (II,101); BANACEK (II,138); BLUE THUNDER (II,278); MARCUS WELBY, M.D. (II,1627); MOBILE TWO (II,1718); QUINCY, M. E. (II,2102); THE WILD WILD WEST (I,4863)

MOFFETT, JORDAN
THE DUCK FACTORY (II,738); HARPER VALLEY PTA (II,1095); IT TAKES TWO (II,1252)

MOFFLY, JOE REB
STARSKY AND HUTCH (II,2444)

MOLL, ELICK
PLAYHOUSE 90 (I,3623)

MOLLIN, LARRY
BRET MAVERICK (II,374); CHIPS (II,511); THE CIRCLE FAMILY (II,527); CUTTER TO HOUSTON (II,612)

MONAGHAN, ED
COMEDY NEWS II (I,1021); THE ROWAN AND MARTIN REPORT (II,2221)

MONASH, PAUL
COLONEL MARCH OF SCOTLAND YARD (I,1005); THE DESILU PLAYHOUSE (I,1237); HANDLE WITH CARE (I,1926); KRAFT TELEVISION THEATER (I,2592); THE LONELY WIZARD (I,2743); PLAYHOUSE 90 (I,3623); STUDIO ONE (I,4268); THE UNTOUCHABLES (I,4681)

MONASTER, NATE
THE ALAN YOUNG SHOW (I,104); ALL IN THE FAMILY (I,133); BACHELOR FATHER (I,323); THE GHOST AND MRS. MUIR (I,1786); HEY, JEANNIE! (I,2038); I REMEMBER CAVIAR (I,2175); MEDICAL CENTER (II,1675); THE MILTON BERLE SHOW (I,3047); THE PHIL SILVERS SHOW (I,3580)

MONET, GABY
THE HALLOWEEN THAT ALMOST WASN'T (II,1075)

MONK, DEBRA
PUMPBOYS AND DINETTES ON TELEVISION (II,2090)

MONKS JR, JOHN
S.W.A.T. (II,2236)

MONTALDO, GIULIANO
MARCO POLO (II,1626)

MONTE, ERIC
GOOD TIMES (II,1038)

MONTEITH, KELLY
KELLY MONTEITH (II,1375); NO HOLDS BARRED (II,1854)

MONTGOMERY, DOREEN
DANGER MAN (I,1136); THE INVISIBLE MAN (I,2232); THE NEW ADVENTURES OF CHARLIE CHAN (I,3249); SECRET AGENT (I,3962)

MOODY, JOHN
THE PEE WEE HERMAN SHOW (II,1988)

MOONEY, PAUL
FREEMAN (II,923); THE RICHARD PRYOR SHOW (II,2163); THE RICHARD PRYOR SPECIAL (II,2164); THE RICHARD PRYOR SPECIAL? (II,2165)

MOORE JR, TOM
CHER (II,495); SUNDAY FUNNIES (II,2493)

MOORE, ARTHUR
COLONEL STOOPNAGLE'S STOOP (I,1006)

MOORE, CARROLL
AMERICAN COWBOY (I,169); BING CROSBY AND HIS FRIENDS (I,456); THE BURNS AND SCHREIBER COMEDY HOUR (I,768); THE BURNS AND SCHREIBER COMEDY HOUR (I,769); THE CONFESSIONS OF DICK VAN DYKE (II,568); DOC (II,682); THE FUNNY SIDE (I,1715); THE GUY MITCHELL SHOW (I,1912); RHODA (II,2151); THE WONDERFUL WORLD OF AGGRAVATION (I,4894)

MOORE, DON
ROD BROWN OF THE ROCKET RANGERS (I,3825)

MOORE, MIKE
THE GET ALONG GANG (II,980)

MOORE, TOM
ALAN KING'S FINAL WARNING (II,29); AMERICA 2-NIGHT (II,65); THE ANDY WILLIAMS SHOW (II,78); THE DIAHANN CARROLL SHOW (II,665); FERNWOOD 2-NIGHT (II,849); JOE'S WORLD (II,1305); THE STEVE ALLEN COMEDY HOUR (II,2454); THICKE OF THE NIGHT (II,2587); WHAT'S HAPPENING!! (II,2769)

MORAN, DEBORAH
SIMON AND SIMON (II,2357)

MORAN, PHIL
SANFORD (II,2255)

MORAN, VINCE
ALL THE RIVERS RUN (II,43)

MORANIS, RICK
SECOND CITY TELEVISION (II,2287)

MORELL, DENNIS M
WESTSIDE MEDICAL (II,2765)

MORGAN, BREWSTER
FIRESIDE THEATER (I,1580)

MORGAN, CASS
PUMPBOYS AND DINETTES ON TELEVISION (II,2090)

MORGAN, DENISE
PRISONER: CELL BLOCK H (II,2085)

MORGAN, DICK
TRAPPER JOHN, M.D. (II,2654)

MORGAN, ELAINE
THE ONEDIN LINE (II,1912)

MORGAN, GUY
INTERNATIONAL DETECTIVE (I,2224)

MORGAN, HENRY
VICTOR BORGE'S COMEDY IN MUSIC III (I,4725)

MORGAN, OWEN
DOG AND CAT (II,695)

MORGAN, RICHARD
BONANZA (II,347)

MORGAN, TONY
BARETTA (II,152)

MORGANSTERN, EDWARD
LETTERS TO LAUGH-IN (I,2672)

MORGANSTERN, HOWARD
PEBBLES AND BAMM BAMM (II,1987)

MORHAIM, JOSEPH
THE SAINT (II,2237)

MORHEIM, LOU
BRONK (II,379); GENERAL ELECTRIC TRUE (I,1754); THE OUTER LIMITS (I,3426); SHERLOCK HOLMES (I,4010)

MORIARTY, JAY
THE JEFFERSONS (II,1276); JOE AND SONS (II,1296); THAT'S MY MAMA (II,2580)

MORRIS, ALAN
WELCOME TO THE FUN ZONE (II,2763)

MORRIS, COLIN
THE GATHERING STORM (I,1741)

MORRIS, EDMUND
CHANNING (I,900); THE WILD WILD WEST (I,4863)

MORRIS, LINDA
ALICE (II,33); JUST OUR LUCK (II,1360)

MORRIS, MICHAEL
ALL IN THE FAMILY (II,38); BEWITCHED (I,418); THE CARA WILLIAMS SHOW (I,843); THE FLYING NUN (I,1611); THE GOLDBERGS (I,1835); THE GOLDBERGS (I,1836); JACK AND THE BEANSTALK (I,2288); MCHALE'S NAVY (I,2969); PERRY MASON (II,2025); PLEASE DON'T EAT THE DAISIES (I,3628); SECOND CHANCE (I,3958); WE'LL TAKE MANHATTAN (I,4792)

MORRIS, RICHARD
LATE NIGHT WITH DAVID LETTERMAN (II,1442); THE LORETTA YOUNG THEATER (I,2756); MOBILE ONE (II,1717); PRIVATE SECRETARY (I,3672)

MORRISON, GERRY
THE WAYNE KING SHOW (I,4779)

MORROW, BILL
THE BING CROSBY CHRISTMAS SHOW (I,458); THE BING CROSBY SHOW (I,462); THE BING CROSBY SHOW (I,465); THE BING CROSBY SHOW (I,466); THE BING CROSBY SHOW (I,468); THE BING CROSBY SHOW (I,469); THE BING CROSBY SPECIAL (I,470); THE BING CROSBY SPECIAL (I,471); THE BING CROSBY SPECIAL (I,472); THE EDSEL SHOW (I,1417)

MORSE, CARLETON E
MIXED DOUBLES (I,3075); ONE MAN'S FAMILY (I,3383)

MORSE, RAY
TOM CORBETT, SPACE CADET (I,4535)

MORSE, SID
THE ANDY GRIFFITH SHOW (I,192); BATTLE OF THE PLANETS (II,179); ISIS (II,1248); JEANNIE (II,1275); THE LOVE BOAT (II,1535)

MORTIMER, JOHN
BRIDESHEAD REVISITED (II,375); FATHER DEAR FATHER (II,837); GEORGE AND MILDRED (II,965); MAN ABOUT THE HOUSE (II,1611); MARRIED ALIVE (I,2923); ROBIN'S NEST (II,2187); TWENTY-FOUR HOURS IN A WOMAN'S LIFE (I,4644)

MORTON, HARRY
LETTERS TO LAUGH-IN (I,2672)

MORTON, JACK
FAMILY (II,813)

MOSEL, TAD
ALL THE WAY HOME (I,140); ALL THE WAY HOME (II,44); PLAYHOUSE 90 (I,3623); PLAYWRIGHTS '56 (I,3627); STUDIO ONE (I,4268)

MOSER, JAMES E
BEN CASEY (I,394); JOSIE (II,1345); MEDIC (I,2976); THE NATIVITY (II,1798); O'HARA, UNITED STATES TREASURY: OPERATION COBRA (I,3347)

MOSES, HARRY
THE MIKE WALLACE PROFILES (II,1699)

MOSHER, BOB
THE AMOS AND ANDY SHOW (I,179); BRINGING UP BUDDY (I,731); GOING MY WAY (I,1833); ICHABOD AND ME (I,2183); IT'S A SMALL WORLD (I,2260); LEAVE IT TO BEAVER (I,2648); THE MARGE AND GOWER CHAMPION SHOW (I,2907); MEET MR. MCNUTLEY (I,2989); MUNSTER, GO HOME! (I,3157); THE MUNSTERS (I,3158)

MOSS, FRANCIS
BUCK ROGERS IN THE 25TH CENTURY (II,384); WAGON TRAIN (I,4747)

MOSS, FRANK L
CROSSROADS (I,1105); FOUR STAR PLAYHOUSE (I,1652); ROUTE 66 (I,3852); THE WILD WILD WEST (I,4863)

MOSS, GENE
CHANGING SCENE (I,899); CURIOSITY SHOP (I,1114)

MOSS, JEFFREY
SESAME STREET (I,3982)

MOSS, VIVIAN R
OMNI: THE NEW FRONTIER (II,1889)

MOSS, WINSTON
THE FLIP WILSON SHOW (I,1602); THE JACKSONS (II,1267)

MOURNE, WILLIAM
KRAFT TELEVISION THEATER (I,2592); MATINEE THEATER (I,2947)

MOYE, MICHAEL
CHECKING IN (II,492); GOOD TIMES (II,1038); IT'S YOUR MOVE (II,1259); SILVER SPOONS (II,2355)

MUIR, JOHN
FROM A BIRD'S EYE VIEW (I,1686)

MULA, FRANK
LADIES AND GENTLEMAN. . . BOB NEWHART, PART II (II,1421); RICH LITTLE—COME LAUGH WITH ME (II,2160)

MULA, FRED
MADAME'S PLACE (II,1587)

MULCAHEY, PATRICK
LOVING (II,1552)

MULHEIM, HARRY
MIRACLE ON 34TH STREET (I,3057)

MULHOLLAND, JIM
CHER (II,495); JACK PAAR TONIGHT (I,2320)

MULLALLY, DONN
CROSSROADS (I,1105); FOR THE DEFENSE (I,1624); MANNIX (II,1624); THE WILD WILD WEST (I,4863)

MULLER, MARK
THE JEFFERSONS (II,1276)

MULLER, ROMEO
DOROTHY IN THE LAND OF OZ (II,722); THE EASTER BUNNY IS COMIN' TO TOWN (II,753); FROSTY THE SNOWMAN (II,934); FROSTY'S WINTER WONDERLAND (II,935); HERE COMES PETER COTTONTAIL (II,1132); IT'S A BIRD, IT'S A PLANE, IT'S SUPERMAN (II,1253); JACK FROST (II,1263); THE JACKSON FIVE (I,2326); THE KIDS FROM C.A.P.E.R. (II,1391); THE LEPRECHAUN'S CHRISTMAS GOLD (II,1461); THE LITTLE DRUMMER BOY (II,1485); LITTLE DRUMMER BOY, BOOK II (II,1486); PINOCCHIO'S CHRISTMAS (II,2045); THE RELUCTANT DRAGON AND MR. TOAD (I,3762); RUDOLPH THE RED-NOSED REINDEER (II,2227); RUDOLPH'S SHINY NEW YEAR (II,2228); SANTA CLAUS IS COMIN' TO TOWN (II,2259)

MULLIGAN, GERALD
LATE NIGHT WITH DAVID LETTERMAN (II,1442)

MULLIGAN, JIM
THE CARPENTERS (II,444); THE JACKSONS (II,1267); JOHN DENVER ROCKY MOUNTAIN CHRISTMAS (II,1314); THE KEN BERRY WOW SHOW (I,2524); THE LATE FALL, EARLY SUMMER BERT CONVY SHOW (II,1441); M*A*S*H (II,1569); THE PAUL LYNDE COMEDY HOUR (II,1978); PERRY COMO IN LAS VEGAS (II,2003); PERRY COMO'S EASTER BY THE SEA (II,2014); THE RICH LITTLE SHOW (II,2154); ROWAN AND MARTIN'S LAUGH-IN (I,3856); ROWAN AND MARTIN'S LAUGH-IN (I,3857); THE SMOTHERS BROTHERS SHOW (II,2384); SZYSZNYK (II,2523); VENICE MEDICAL (II,2726); WHEN THINGS WERE ROTTEN (II,2782)

MULLIGAN, MIKE
THE JEFFERSONS (II,1276); JOE AND SONS (II,1296); THAT'S MY MAMA (II,2580)

MUMFORD, THAD
BAY CITY BLUES (II,186); THE CAPTAIN AND TENNILLE (II,429); THE FLIP WILSON SPECIAL (I,1603); THE FLIP WILSON SPECIAL (II,887); THE FLIP WILSON SPECIAL (II,888); THE FLIP WILSON SPECIAL (II,889); FLIP WILSON. . . OF COURSE (II,890); KEEP ON TRUCKIN' (II,1372); M*A*S*H (II,1569); THE MAC DAVIS SHOW (II,1576); ROOTS: THE NEXT GENERATIONS (II,2213); WHAT'S HAPPENING!! (II,2769)

MUNISTERI, MARY RYAN
MANDY'S GRANDMOTHER (II,1618); RYAN'S HOPE (II,2234)

MUNTNER, SIMON
240-ROBERT (II,2688); ALICE (II,33); THE AMERICAN GIRLS (II,72); BIG EDDIE (II,236); THE DUKES OF HAZZARD (II,742); THE FEATHER AND FATHER GANG (II,845); KAREN (II,1363); KATE LOVES A MYSTERY (II,1367); KUNG FU (II,1416); PARTNERS IN CRIME (II,1961); PRIVATE BENJAMIN (II,2087); THE RUBBER GUN SQUAD (II,2225); THE SAN PEDRO BEACH BUMS (II,2250); T.J. HOOKER (II,2524); THAT'S MY MAMA (II,2580)

MURCOTT, JOEL

ALFRED HITCHCOCK PRESENTS (I,115); BARNABY JONES (II,153); CANNON (II,424); LITTLE HOUSE ON THE PRAIRIE (II,1487); NIGHT PROWL (I,3291); PERRY MASON (II,2025)

MURPHY, BOB
THE MERV GRIFFIN SHOW (II,1688)

MURPHY, GARY
THICKE OF THE NIGHT (II,2587)

MURPHY, MARILYN
SKY KING (I,4077)

MURPHY, MAUREEN
THE SHAPE OF THINGS (II,2315)

MURPHY, RALPH FRANCIS
CROSSROADS (I,1105)

MURPHY, RICHARD
THE HARDY BOYS (I,1948)

MURRAIS, WARREN
DIFF'RENT STROKES (II,674)

MURRAY, ARTHUR L
BEN CASEY (I,394)

MURRAY, BILL
THE TV TV SHOW (II,2674)

MURRAY, BRIAN DOYLE
RODNEY DANGERFIELD SPECIAL: IT'S NOT EASY BEIN' ME (II,2199)

MURRAY, DON
THE GHOST AND MRS. MUIR (I,1786)

MURRAY, GERALD
MATINEE THEATER (I,2947)

MURRAY, JOHN FENTON
THE ANDY GRIFFITH SHOW (I,192); GOOD TIMES (II,1038); MCHALE'S NAVY (I,2969); THE NEW SHOW (II,1828); OPERATION PETTICOAT (II,1917); PANDORA AND FRIEND (I,3445); SABU AND THE MAGIC RING (I,3881); SIGMUND AND THE SEA MONSTERS (II,2352)

MURRAY, LARRY
ANDY WILLIAMS MAGIC LANTERN SHOW COMPANY (I,202); ANDY'S LOVE CONCERT (I,215); JOHNNY CASH CHRISTMAS 1983 (II,1324); JOHNNY CASH: CHRISTMAS ON THE ROAD (II,1333)

MURRAY, TOM
BATTLE OF THE AGES (I,367); LIFE WITH SNARKY PARKER (I,2693); THE POLLY BERGEN SHOW (I,3643); SKY KING (I,4077)

MURRAY, WARREN
AN AMATEUR'S GUIDE TO LOVE (I,154); BLANSKY'S BEAUTIES (II,267); THE FACTS OF LIFE (II,805); HARPER VALLEY PTA (II,1095)

MUSANTE, TONY
TOMA (II,2634)

MUSHEIM, HARRY
OF MEN OF WOMEN (I,3332)

MUSIC, JERRY
THE LESLIE UGGAMS SHOW (I,2660)

MUSIC, LORENZO
CARLTON YOUR DOORMAN (II,440); FRIENDS (II,928); THE LORENZO AND HENRIETTA MUSIC SHOW (II,1519); THE NEW LORENZO MUSIC SHOW (II,1821); RHODA (II,2151)

MUSKAT, JOYCE
STAR TREK (II,2440)

MYER, BOB
THE FACTS OF LIFE (II,805)

MYER, GEORGE
THE NEW SHOW (II,1828)

MYER, MARIANNE
HOT HERO SANDWICH (II,1186)

MYERS, PAUL ERIC
FALCON CREST (II,808); FLAMINGO ROAD (II,872); NIGHT PARTNERS (II,1847)

MYERSON, ALAN
THE KROFFT KOMEDY HOUR (II,1411)

MYERSON, PETER
WELCOME BACK, KOTTER (II,2761)

MYHRAM, ROBERT
AS THE WORLD TURNS (II,110)

MYROW, DONNA
HART TO HART (II,1102)

MYROW, JEFF
CHARLIE'S ANGELS (II,486); HART TO HART (II,1102); VEGAS (II,2724)

NABLO, JAN
THE PAUL ANKA SHOW (II,1972)

NADEL, ARTHUR H
SHAZAM! (II,2319)

NADLER, MARTY
THE LAST RESORT (II,1440); LEGS (II,1459); LOVE, SIDNEY (II,1547); THE STEVE LANDESBERG TELEVISION SHOW (II,2458); SUNDAY FUNNIES (II,2493)

NALEVANSKY, STEVEN
STARSKY AND HUTCH (II,2444)

NAPOLEON, ART
THE DICK POWELL SHOW (I,1269); MEDIC (I,2976)

NAPOLEON, JO
THE DICK POWELL SHOW (I,1269); MEDIC (I,2976)

NARDO, PATRICIA
THE ABC AFTERSCHOOL SPECIAL (II,1); COUSINS (II,595); IT'S NOT EASY (II,1256); LILY (II,1477); THE MARY TYLER MOORE SHOW (II,1640); RHODA (II,2151)

NASCO, PETER
NERO WOLFE (II,1806)

NASCOTT, LAURENCE
BONANZA (II,347)

NASH, N RICHARD
CAVALCADE OF AMERICA (I,875); MATINEE THEATER (I,2947); THE RAINMAKER (II,2109)

NATHAN, MORT
ARCHIE BUNKER'S PLACE (II,105)

NATHAN, STEVEN
REGGIE (II,2127)

NATION, JOHN
MASTER OF THE GAME (II,1647)

NATION, TERRY
THE AVENGERS (II,121); THE BARON (I,360); BLAKE'S SEVEN (II,264); THE CHAMPIONS (I,896); COLOR HIM DEAD (I,1007); COLOR HIM DEAD (II,554); DEPARTMENT S (I,1232); DOCTOR WHO (II,689); THE SAINT (II,2237)

NEALE, L FORD
240-ROBERT (II,2688); THE BARBARY COAST (II,146); THE NIGHT STALKER (II,1849); TALES OF THE GOLD MONKEY (II,2537)

NEER, JUDI
EMERALD POINT, N.A.S. (II,773); THE LOVE BOAT (II,1535); REMINGTON STEELE (II,2130); THE TONY RANDALL SHOW (II,2640); WKRP IN CINCINNATI (II,2814)

NEIGHER, GEOFFREY
ANNIE FLYNN (II,92); THE BOB NEWHART SHOW (II,342); GIMME A BREAK (II,995); PAUL SAND IN FRIENDS AND LOVERS (II,1982); RHODA (II,2151); SPENCER (II,2420); WE GOT IT MADE (II,2752)

NEIGHER, STEPHEN
IT TAKES TWO (II,1252)

NEIMAN, IRVING GAYNOR
THE AFRICAN QUEEN (II,18); BEACON HILL (II,193); THE BELLS OF ST. MARY'S (I,392); FULL MOON OVER BROOKLYN (I,1703); HIGHER AND HIGHER, ATTORNEYS AT LAW (I,2057); MATINEE THEATER (I,2947); PLAYHOUSE 90 (I,3623)

NELBERT, SHELLEY
THE MARY TYLER MOORE SHOW (II,1640)

NELSON, CRAIG
THE JOHN BYNER COMEDY HOUR (I,2407); THE TIM CONWAY COMEDY HOUR (I,4501)

NELSON, DICK
ALIAS SMITH AND JONES (I,118); BARNABY JONES (II,153); CHIPS (II,511); DYNASTY (II,746); THE ELEVENTH HOUR (I,1425); FALCON CREST (II,808); THE FBI (I,1551); FREEBIE AND THE BEAN (II,922); KINGSTON: THE POWER PLAY (II,1399); A MAN CALLED SLOANE (II,1613); THE MAN FROM U.N.C.L.E. (I,2867); NERO WOLFE (II,1806); S.W.A.T. (II,2236); THE SOUND OF ANGER (I,4133); TARGET: THE CORRUPTERS (I,4360); THREE FOR THE ROAD (II,2607); WONDER WOMAN (SERIES 2) (II,2829)

NELSON, DON
THE ADVENTURES OF OZZIE AND HARRIET (I,71); BACHELOR FATHER (I,323); CRASH ISLAND (II,600); THE DORIS DAY SHOW (I,1355); FRANKIE AND ANNETTE: THE SECOND TIME AROUND (II,917); THE GHOST AND MRS. MUIR (I,1786); HERBIE, THE LOVE BUG (II,1131); THE MUNSTERS' REVENGE (II,1765); NANNY AND THE PROFESSOR AND THE PHANTOM OF THE CIRCUS (I,3226); THE SOFT TOUCH (I,4107)

NELSON, OZZIE
THE ADVENTURES OF OZZIE AND HARRIET (I,71); OZZIE'S GIRLS (I,3439); OZZIE'S GIRLS (I,3440)

NELSON, PORTIA
DEBBIE REYNOLDS AND THE SOUND OF CHILDREN (I,1213)

NELSON, RALPH
THE DESILU PLAYHOUSE (I,1237)

NELSON, RICHARD
KOJAK (II,1406)

NEMO, PHILIP
THE RAT PATROL (I,3726)

NEPUS, RIA
HAPPY DAYS (II,1084); JUST OUR LUCK (II,1360); LAVERNE AND SHIRLEY (II,1446)

NESLER, BOB
THE SMURFS (II,2385)

NESTEM, JOSEPH
HIT MAN (II,1156)

NEUMAN, E JACK
BONANZA (II,347); KATE MCSHANE (II,1369); MR. NOVAK (I,3145); NIGHT GAMES (II,1845); THE POLICE STORY (I,3638); STAT! (I,4213); THE TWILIGHT ZONE (I,4651); THE UNTOUCHABLES (I,4682); WAGON TRAIN (I,4747)

NEUMAN, MATT
THE JIM STAFFORD SHOW (II,1291); NOT NECESSARILY THE NEWS (II,1862); NOT NECESSARILY THE NEWS (II,1863); THE REDD FOXX COMEDY HOUR (II,2126)

NEUMAN, SAM
HAWAII FIVE-O (II,1110)

NEUMANN, SAM
THE OUTER LIMITS (I,3426)

NEUSTEIN, JOSEPH
THE SMURFS (II,2385); WHAT'S HAPPENING!! (II,2769)

NEWBOUND, LAURIE
ONE OF THE BOYS (II,1911)

NEWBURY, MICKEY
JUST FRIENDS (I,2495)

NEWHAFER, RICHARD
CANNON (II,424)

NEWHART, BOB
THE BOB NEWHART SHOW (I,666); LADIES AND GENTLEMAN. . . BOB NEWHART, PART II (II,1421)

NEWMAN, ANDREA
MACKENZIE (II,1584)

NEWMAN, CHRISTOPHER
ELLIS ISLAND (II,768)

NEWMAN, ELAINE
THE JEFFERSONS (II,1276); RITUALS (II,2182)

NEWMAN, MATT
LILY (II,1477)

NEWMAN, RICHARD
BONANZA (II,347); JOHNNY RINGO (I,2434)

NEWMAN, ROBERT
CITY HOSPITAL (I,967)

NEWMAN, SAMUEL
PERRY MASON (II,2025); THE WILD WILD WEST (I,4863)

NEWMAN, WALTER
THE DESILU PLAYHOUSE (I,1237)

NEWTON, DWIGHT
WAGON TRAIN (I,4747)

NIBLEY, SLOAN
NAKED CITY (I,3216); THE WESTERNER (I,4800)

NIBLEY, SLOEN
THE AMES BROTHERS SHOW (I,178)

NICHOL, B P
FRAGGLE ROCK (II,916)

NICHOLAS, WILLIAM
TWELFTH NIGHT (I,4636)

NICHOLL, DON
ALL IN THE FAMILY (II,38); THE DUMPLINGS (II,743); THREE'S A CROWD (II,2613); THREE'S COMPANY (II,2614)

NICHOLS, MIKE
JULIE AND CAROL AT CARNEGIE HALL (I,2478)

NICHOLS, WILLIAM
TAMING OF THE SHREW (I,4356); THREE FOR TONIGHT (I,4478); THE YEOMAN OF THE GUARD (I,4926)

NICHOLSON, NICK
THE NEW HOWDY DOODY SHOW (II,1818)

NICKERSON, IRA
BOB HOPE SPECIAL: BOB HOPE'S CHRISTMAS PARTY (II,307)

NICKOLSON, CHALESA
CAGNEY AND LACEY (II,409)

NIELSEN, HELEN
ALCOA PREMIERE (I,109); THE DICK POWELL SHOW (I,1269); PERRY MASON (II,2025)

NIEMACK, ROBERT
THE SINGING COWBOYS RIDE AGAIN (II,2363)

NIEMAN, IRVING GAYNOR
THE ANDROS TARGETS (II,76)

NIGRO-CHACON, GIOVANNI
ROMANCE THEATER (II,2204)

NISS, STANLEY
THE DESILU PLAYHOUSE (I,1237); FBI CODE 98 (I,1550); FOUR STAR PLAYHOUSE (I,1652); PERRY MASON (II,2025)

NIXON, AGNES
ALL MY CHILDREN (II,39)

NOAH, LYLA OLIVER
THE ALL-STAR SALUTE TO MOTHER'S DAY (II,51)

NOAH, PETER
BATTLESTARS (II,182); DREAM HOUSE (II,734); THE FACTS OF LIFE (II,805); ONE DAY AT A TIME (II,1900)

NOBBS, DAVID
THE TWO RONNIES (II,2694)

NOBLE, DON
SAM (II,2246)

NOLAN, WILLIAM
240-ROBERT (II,2689); THE NORLISS TAPES (I,3305)

NOONAN, JOHN FORD
THE COMEDY ZONE (II,559); ST. ELSEWHERE (II,2432)

NOONAN, MICHAEL
THE FLYING DOCTOR (I,1610)

NORELL, MICHAEL
ALOHA PARADISE (II,58); FEATHERSTONE'S NEST (II,847); THE LOVE BOAT III (II,1534); REVENGE OF THE GRAY GANG (II,2146); WHAT'S UP DOC? (II,2771)

NORRIS, PAMELA
IT'S YOUR MOVE (II,1259); THE NEWS IS THE NEWS (II,1838)

NORRIS, R HAMER
BONANZA (II,347)

NORTH, EDMUND
MURDOCK'S GANG (I,3164)

NORTON, ROGER
ROBIN'S NEST (II,2187)

NOVELLO, DON
BLONDES VS. BRUNETTES (II,271); THINGS WE DID LAST SUMMER (II,2589); VAN DYKE AND COMPANY (II,2719)

NOVIER, FRANCIS
THE SMURFS (II,2385)

NOWINSON, DAVID
THE CISCO KID (I,961)

NOXON, NICHOLAS L
DEAR MR. GABLE (I,1203)

NUGENT, FRANK
SCREEN DIRECTOR'S PLAYHOUSE (I,3946)

NUNIS, RICHARD
THE HUNTER (I,2162)

NYE, BUD
DOBIE GILLIS (I,1302); MY FAVORITE MARTIAN (I,3189)

O'BANNON, DAN
BLUE THUNDER (II,278)

O'BRIEN, BARRY
HAPPY DAYS (II,1084)

O'BRIEN, BOB
BURKE'S LAW (I,762); THE DICK POWELL SHOW (I,1269); THE DON RICKLES SHOW (II,708); HERE'S LUCY (II,1135); A LUCILLE BALL SPECIAL STARRING LUCILLE BALL AND DEAN MARTIN (II,1557); MAKE ROOM FOR DADDY (I,2843); MR. ED (I,3137); THE PAUL LYNDE COMEDY HOUR (II,1977)

O'BRIEN, CAROL
MR. NOVAK (I,3145)

O'BRIEN, DAVE
CLOWN ALLEY (I,979); THE RED SKELTON CHEVY SPECIAL (I,3753); THE RED SKELTON SHOW (I,3755)

O'BRIEN, LIAM
REX HARRISON PRESENTS SHORT STORIES OF LOVE (II,2148); TODAY'S FBI (II,2628)

O'BRIEN, PADDY
THE SAINT (II,2237)

O'BRIEN, ROBERT
THE DEAN MARTIN SHOW (I,1198); THE DEAN MARTIN SHOW (I,1199); JOHNNY COME LATELY (I,2425)

O'CASEY, SEAN
THREE IN ONE (I,4479)

O'CHRISTOPHER, C R
AIRWOLF (II,27)

O'CONNOR, CARROLL
THE LAST HURRAH (I,2626); THREE FOR THE GIRLS (I,4477)

O'DEA, JOHN
THE BIG VALLEY (I,437)

O'DONNELL, MARK
THE COMEDY ZONE (II,559)

O'DONNELL, STEVE
LATE NIGHT WITH DAVID LETTERMAN (II,1442)

O'DONNELL, TIM
GLORIA (II,1010)

O'FARRELL, WILLIAM
PERRY MASON (II,2025)

O'GRADY, TONY
THE BARON (I,360); THE INVISIBLE MAN (I,2232)

O'HALLORAN, GEORGE
THE ANN SOTHERN SHOW (I,220); CHESTER THE PUP (I,922); GILLIGAN'S ISLAND (II,990)

O'HARA, CATHERINE
FROM CLEVELAND (II,931); SECOND CITY TELEVISION (II,2287)

O'HARA, GERRY
BERGERAC (II,209)

O'HARA, MICHAEL
CHICAGO STORY (II,506)

O'KUN, LAN
HOWDY DOODY AND FRIENDS (I,2152); IT CAN'T HAPPEN TO ME (II,1250); THE JUGGLER OF NOTRE DAME (II,1350); THE LITTLEST ANGEL (I,2726); THE LOVE BOAT (II,1535); PHYL AND MIKHY (II,2037); THE SHARI LEWIS SHOW (I,4000); THE TROUBLE WITH GRANDPA (II,2662); WHEN, JENNY? WHEN (II,2783)

O'MAHONEY, MIKE
WOLF ROCK TV (II,2815)

O'MALLEY, DAVID
JOHN DENVER ROCKY MOUNTAIN CHRISTMAS (II,1314); MORK AND MINDY (II,1735)

O'NEAL, CHARLES
THE UNTOUCHABLES (I,4682)

O'NEILL, EUGENE
THREE IN ONE (I,4479)

O'SHEA, PEGGY
CAPITOL (II,426)

OAKES, GARY
TV FUNNIES (II,2672)

OBELL, DAVID
THE MUPPET SHOW (II,1766)

OBERMAN, MARGARET
SQUARE PEGS (II,2431)

OBOLER, RUTH
TELEVISION FASHION FAIR (I,4381)

OCHS, ACE
OPEN HOUSE (I,3394)

ODDIE, BILL
CAMBRIDGE CIRCUS (I,807); DOCTOR IN THE HOUSE (I,1313); THE GOODIES (II,1040)

OGLE, ROBERT
BAGGY PANTS AND THE NITWITS (II,135); THE ODDBALL COUPLE (II,1876); WHEELIE AND THE CHOPPER BUNCH (II,2776)

OHLMEYER, DON
GAMES PEOPLE PLAY (II,956)

OLEK, HENRY
SCENE OF THE CRIME (II,2271); SPRAGGUE (II,2430)

OLESKER, HARRY
KRAFT TELEVISION THEATER (I,2592)

OLESKER, JACK
THE GET ALONG GANG (II,980)

OLIANSKY, JOEL
KOJAK (II,1406); MALLORY: CIRCUMSTANTIAL EVIDENCE (II,1608); MASADA (II,1642)

OLIVER, RICHARD
STAR SEARCH (II,2437); STAR SEARCH (II,2438)

OLIVER, STAN
VOLTRON—DEFENDER OF THE UNIVERSE (II,2729)

OLSEN, BETTY
THE DICK CAVETT SHOW (II,669)

OLSON, DANA
LAVERNE AND SHIRLEY (II,1446)

OLSON, GLEN
240-ROBERT (II,2688); HAWAII FIVE-O (II,1110); WONDER WOMAN (SERIES 2) (II,2829)

OMACH, DANNY
THE ROCK 'N ROLL SHOW (II,2188)

OPIE, WIN
THE CLIFFWOOD AVENUE KIDS (II,543)

OPPENHEIMER, JESS
THE BIG BRAIN (I,424); THE FIRST 50 YEARS (II,862); HIDE AND SEEK (I,2047); I LOVE LUCY (I,2171)

ORDLEY, NANCY
LILY—SOLD OUT (II,1480)

ORENSTEIN, BERNIE
13 QUEENS BOULEVARD (II,2591); AWAY WE GO (I,317); THE BEAUTIFUL PHYLLIS DILLER SHOW (I,381); DOUBLE TROUBLE (II,723); E/R (II,748); GO! (I,1830); THE HOLLYWOOD PALACE (I,2088); ONE IN A MILLION (II,1906); ONE OF THE BOYS (II,1911); THE PERRY COMO WINTER SHOW (I,3543); PERRY COMO'S WINTER SHOW (I,3547); SANFORD AND SON (II,2256); SUPER COMEDY BOWL 1 (I,4288); SUPER COMEDY BOWL 2 (I,4289); THIS WEEK IN NEMTIN (I,4465); WHERE THE GIRLS ARE (I,4830)

ORGEL, LEE
BATMAN (I,366)

ORINGER, BARRY
BARNABY JONES (II,153); BEN CASEY (I,394); THE FELONY SQUAD (I,1563); THE FUGITIVE (I,1701); GENERAL ELECTRIC TRUE (I,1754); HOTEL (II,1192); I SPY (I,2179); THE INVADERS (I,2229); MADAME SIN (I,2805); MANNIX (II,1624); MEDICAL CENTER (II,1675); PLANET OF THE APES (II,2049)

ORKIN, DICK
THE TIM CONWAY SHOW (II,2619)

ORKIN, HARRY
ACRES AND PAINS (I,19); THE PHIL SILVERS SHOW (I,3580)

ORLOFF, ARTHUR F
CROWN THEATER WITH GLORIA SWANSON (I,1107)

ORLOFF, RICHARD
FLO (II,891); MAKING THE GRADE (II,1606); MAMA'S FAMILY (II,1610); REPORT TO MURPHY (II,2134); THREE'S COMPANY (II,2614)

ORME, GEOFFREY
IVANHOE (I,2282)

ORSINI, PAOLO
SHAZAM! (II,2319)

OSBORN JR, JOHN JAY
THE PAPER CHASE (II,1949); THE PAPER CHASE: THE SECOND YEAR (II,1950); SILVER SPOONS (II,2355)

OSBORN, PAUL
SCREEN DIRECTOR'S PLAYHOUSE (I,3946)

OSBORN, RON
GOODNIGHT, BEANTOWN (II,1041); TOO CLOSE FOR COMFORT (II,2642)

OSGOOD, CHARLES
EVERYTHING YOU EVER WANTED TO KNOW ABOUT MONSTERS . . . BUT WERE AFRAID! (II,794); WALTER CRONKITE'S UNIVERSE (II,2739)

OSMOND, CLIFF
THE HARDY BOYS MYSTERIES (II,1090)

OSTROFF, HOWARD
FISH (II,864); HOUSE CALLS (II,1194); HUSBANDS, WIVES AND LOVERS (II,1209); NUMBER 96 (II,1869)

OWEN, JOHN STEVEN
THE ASSOCIATES (II,113); THE BAXTERS (II,183); THE DUCK FACTORY (II,738); LADIES' MAN (II,1423); NEWHART (II,1835); SEMI-TOUGH (II,2300); THE SIX O'CLOCK FOLLIES (II,2368)

PAAR, JACK
A FUNNY THING HAPPENED ON THE WAY TO HOLLYWOOD (I,1716); JACK PAAR AND A FUNNY THING HAPPENED EVERYWHERE (I,2314); THE JACK PAAR SHOW (I,2316); THE JACK PAAR SHOW (I,2317); THE JACK PAAR SPECIAL (I,2319)

PACKARD, ED
NO TIME FOR SERGEANTS (I,3300); WENDY AND ME (I,4793)

PACKARD, ELON
BRANDED (I,707); THE GEORGE BURNS SPECIAL (II,970); GEORGE GOBEL PRESENTS (I,1766); GOOD TIMES (II,1038); THE MANY SIDES OF MICKEY ROONEY (I,2899); MAYOR OF HOLLYWOOD (I,2962); MCNAB'S LAB (I,2972); THE MICKEY ROONEY SHOW (I,3027); THE ROY ROGERS AND DALE EVANS SHOW (I,3858); RUN, BUDDY, RUN (I,3870); THE WILD WILD WEST (I,4863)

PACKER, JULIET
BRING 'EM BACK ALIVE (II,377); PALMERSTOWN U.S.A. (II,1945); THE WALTONS (II,2740); THE WALTONS: MOTHER'S DAY ON WALTON'S MOUNTAIN (EPISODE 2) (II,2742)

PACKER, PETER

LAND OF THE GIANTS (I,2611); LOST IN SPACE (I,2758)

PADDOCK, JIM
WKRP IN CINCINNATI (II,2814)

PADNICK, GLENN
DIFF'RENT STROKES (II,674); SILVER SPOONS (II,2355)

PAGANO, JO
NAKED CITY (I,3216); ROUTE 66 (I,3852); SAINTS AND SINNERS (I,3887)

PAGLIANO, JOANNE
FAMILY TIES (II,819); LOU GRANT (II,1526); REPORT TO MURPHY (II,2134)

PAINE, SIDNEY
MATINEE THEATER (I,2947)

PALEY, SARAH
THE NEW SHOW (II,1828)

PALILARO, JOANNE
BOONE (II,351)

PALIN, MICHAEL
MONTY PYTHON'S FLYING CIRCUS (II,1729)

PALLEY, NYLES
CHEZ PAREE REVUE (I,932)

PALMER, GENE
REAL KIDS (II,2116)

PALMER, P K
DICK TRACY (I,1271); JOHNNY RINGO (I,2434); PETER GUNN (I,3562)

PALMER, STUART
PERRY MASON (II,2025)

PALMER, TONY
FROM HERE TO ETERNITY (II,933)

PALTROW, BRUCE
OPERATING ROOM (II,1915); YOU'RE GONNA LOVE IT HERE (II,2860)

PALUMBO, GENE
RITUALS (II,2182)

PANAMA, NORMAN
GET CHRISTIE LOVE! (II,981)

PANICH, DAVID
CHER (II,495); CHER (II,496); DOM DELUISE AND FRIENDS (II,701); DON'T CALL ME MAMA ANYMORE (I,1350); FRANK SINATRA JR. WITH FAMILY AND FRIENDS (I,1663); GABRIEL KAPLAN PRESENTS THE SMALL EVENT (II,952); THE HUDSON BROTHERS SHOW (II,1202); REAL PEOPLE (II,2118); THE ROWAN AND MARTIN SPECIAL (I,3855); ROWAN AND MARTIN'S LAUGH-IN (I,3857); ROWAN AND MARTIN'S LAUGH-IN (I,3856); THE SONNY COMEDY REVUE (II,2403); SPEAK UP AMERICA (II,2412)

PANTER, NICOLE
THE PEE WEE HERMAN SHOW (II,1988)

PAOLANTONIO, BILL
PEOPLE DO THE CRAZIEST THINGS (II,1997); PEOPLE DO THE CRAZIEST THINGS (II,1998)

PARAGON, JOHN
THE PARAGON OF COMEDY (II,1956); THE PEE WEE HERMAN SHOW (II,1988)

PARENT, GAIL
ANN-MARGRET: FROM HOLLYWOOD WITH LOVE (I,217); ANNIE AND THE HOODS (II,91); BING CROSBY AND CAROL BURNETT—TOGETHER AGAIN FOR THE FIRST TIME (I,454); CALL HER MOM (I,796); FROM HOLLYWOOD WITH LOVE: THE ANN-MARGRET SPECIAL (I,1687); HELLZAPOPPIN (I,2011); I'D RATHER BE CALM (II,1216); THE MANY SIDES OF DON RICKLES (I,2898); RHODA (II,2151); SHEILA (II,2325); SILLS AND BURNETT AT THE MET (II,2354); THE SMOTHERS BROTHERS SHOW (II,2384); SONS AND DAUGHTERS (II,2404); THREE GIRLS THREE (II,2608)

PARIS, FRANK
BRANDED (I,707)

PARKER, G ROSS
THE GREAT SPACE COASTER (II,1059)

PARKER, JAMES
GIMME A BREAK (II,995); HANDLE WITH CARE (II,1077); HARRY AND MAGGIE (II,1097); THE INCREDIBLE HULK (II,1232); KOMEDY TONITE (II,1407); THE KROFFT SUPERSHOW (II,1412); THE KROFFT SUPERSHOW II (II,1413); THE ORPHAN AND THE DUDE (II,1924); THE OSMOND BROTHERS SPECIAL (II,1925); THE TEXAS WHEELERS (II,2568)

PARKER, JIM
MR. TERRIFIC (I,3152); THE SMOTHERS BROTHERS SHOW (I,4096)

PARKER, MONICA
EVENING AT THE IMPROV (II,786); WHO'S THE BOSS? (II,2792)

PARKER, RAY
CALIFORNIA FEVER (II,411); FAR OUT SPACE NUTS (II,831); PARTRIDGE FAMILY: 2200 A.D. (II,1963)

PARKER, ROD
ALL'S FAIR (II,46); AND THEY ALL LIVED HAPPILY EVER AFTER (II,75); THE ASTRONAUTS (II,114); THE DEAN MARTIN SHOW (I,1201); DID YOU HEAR ABOUT JOSH AND KELLY?! (II,673); THE DOM DELUISE SHOW (I,1328); HOT L BALTIMORE (II,1187); THE JACKIE GLEASON SPECIAL (I,2324); KING OF THE ROAD (II,1396); LOVE, SIDNEY (II,1547); MAUDE (II,1655); MR. & MRS. & MR. (II,1744); MR. DUGAN (II,1749); PHYL AND MIKHY (II,2037); POTTSVILLE (II,2072); THE POWDER ROOM (I,3656); THAT'S LIFE (I,4426); WHAT'S UP, AMERICA? (I,4824); WHAT'S UP? (I,4823)

PARKER, SKIP
OUTLAW LADY (II,1935)

PARKER, WILLIAM
NURSE (II,1870); THE WALTONS (II,2740)

PARKES, ROGER
BLAKE'S SEVEN (II,264); MAN IN A SUITCASE (I,2868); THE PRISONER (I,3670)

PARKS, HILDY
CBS: ON THE AIR (II,460); NIGHT OF 100 STARS (II,1846); A WORLD OF LOVE (I,4914)

PARR, LARRY
THE GET ALONG GANG (II,980)

PARRIOTT, JAMES D
ALEX AND THE DOBERMAN GANG (II,32); THE AMERICAN GIRLS (II,72); THE BIONIC WOMAN (II,255); FITZ AND BONES (II,867); FROM HERE TO ETERNITY (II,933); HAWAIIAN HEAT (II,1111); THE INCREDIBLE HULK (II,1232); THE LEGEND OF THE GOLDEN GUN (II,1455); NICK AND THE DOBERMANS (II,1843); THE SEAL (II,2282); THE SIX-MILLION-DOLLAR MAN (II,2372); VOYAGERS (II,2730)

PARRIOTT, SARA
VOYAGERS (II,2730)

PARSONNETT, MARION
BONANZA (II,347)

PARSONS, E M
BONANZA (II,347); READY FOR THE PEOPLE (I,3739)

PARTIZ, JACK
BATMAN (I,366); VOLTRON—DEFENDER OF THE UNIVERSE (II,2729)

PARTIZA, JACK
KRAFT TELEVISION THEATER (I,2592)

PASCAL, ERNEST
CONFIDENTIALLY YOURS (I,1035)

PASCAL, MILTON
I WAS A BLOODHOUND (I,2180); MAUDE (II,1655); PLEASE DON'T EAT THE DAISIES (I,3628); THESE ARE THE DAYS (II,2585); TUGBOAT ANNIE (I,4622)

PASHDAG, JOHN
TALES OF THE GOLD MONKEY (II,2537)

PASKO, MARTIN
BEAUTY AND THE BEAST (II,197); BUCK ROGERS IN THE 25TH CENTURY (II,383); GOLDIE GOLD AND ACTION JACK (II,1023); THUNDARR THE BARBARIAN (II,2616)

PASKY, STEVE
ENTERTAINMENT TONIGHT (II,780)

PATCHETT, TOM
THE ARTHUR GODFREY SPECIAL (I,288); ARTHUR GODFREY'S PORTABLE ELECTRIC MEDICINE SHOW (I,290); BING CROSBY'S SUN VALLEY CHRISTMAS SHOW (I,475); THE BOB NEWHART SHOW (II,342); BUFFALO BILL (II,387); THE CHOPPED LIVER BROTHERS (II,514); DIANA (I,1258); OPEN ALL NIGHT (II,1914); SITCOM (II,2367); THE TONY RANDALL SHOW (II,2640)

PATILLO, ALAN
CAPTAIN SCARLET AND THE MYSTERONS (I,837); THUNDERBIRDS (I,4497); U.F.O. (I,4662)

PATRICK, JACK
COUNTERPOINT (I,1059)

PATRICK, JOHN
THE SMALL MIRACLE (I,4090)

PATTERSON, DON
KOJAK (II,1406); STARSKY AND HUTCH (II,2444)

PATTERSON, HARVEY
THE TOMMY HUNTER SHOW (I,4552)

PATTERSON, ROBERT
COUNTERPOINT (I,1059); CROWN THEATER WITH GLORIA SWANSON (I,1107)

PAUL, NORMAN
ARNIE (I,261); DANNY THOMAS LOOKS AT YESTERDAY, TODAY AND TOMORROW (I,1148); THE DORIS DAY SHOW (I,1355); THE GEORGE BURNS AND GRACIE ALLEN SHOW (I,1763); GOOD TIMES (II,1038); LEAVE IT TO BEAVER (I,2648); MAKE MORE ROOM FOR DADDY (I,2842); MAKE ROOM FOR GRANDDADDY (I,2844); MANY HAPPY RETURNS (I,2895); MCNAB'S LAB (I,2972); MR. ED (I,3137); MY HERO (I,3193); NO TIME FOR SERGEANTS (I,3300); RUN, BUDDY, RUN (I,3870); WENDY AND ME (I,4793)

PAULSEN, DAVID
CHICAGO STORY (II,506); DALLAS (II,614)

PAXTON, JOHN
THE CODE OF JONATHAN WEST (I,992)

PAYNE, DANIEL
MATT HOUSTON (II,1654)

PAYNE, JERRY
THE CROSS-WITS (II,605)

PAYNE, JULIE
PRIME TIMES (II,2083)

PAYNE, PATRICIA
CAPTAIN AMERICA (II,428)

PEACOCK, WILBUR S
THE CISCO KID (I,961); SCREEN DIRECTOR'S PLAYHOUSE (I,3946)

PEARLMAN, B K
RYAN'S HOPE (II,2234)

PEARLMAN, RON
THE BARRY MANILOW SPECIAL (II,155); CHER (II,496); THE PAUL LYNDE HALLOWEEN SPECIAL (II,1981)

PEATTIE, NANCY
SEARCH FOR BEAUTY (I,3953)

PECK JR, CHARLES K
BEN CASEY (I,394)

PECK, JAMES
CONFIDENTIAL FILE (I,1033)

PECKINPAH, DAVID
GABE AND WALKER (II,950); HIGH PERFORMANCE (II,1145); YOUNG MAVERICK (II,2867)

PECKINPAH, SAM
THE DICK POWELL SHOW (I,1269); THE RIFLEMAN (I,3789); THE SHARPSHOOTER (I,4001); THE WESTERNER (I,4801); WINCHESTER (I,4872)

PEDDERSON, RON
ED MCMAHON AND HIS FRIENDS. . . DISCOVER WET AT CYPRESS GARDENS (I,1400)

PEDERSON, TED
GOLDIE GOLD AND ACTION JACK (II,1023); THE SMURFS (II,2385); SPORT BILLY (II,2429); THUNDARR THE BARBARIAN (II,2616)

PEEPLES, SAMUEL A
FLASH GORDON—THE GREATEST ADVENTURE OF ALL (II,875); RAWHIDE (I,3727); THE RIFLEMAN (I,3789); STAR TREK (II,2439); STAR TREK (II,2440)

PEETE, BOB
THE FACTS OF LIFE (II,805)

PEKKONEN, DONNA
HART TO HART (II,1102)

PELLETIER, LOUIS
KRAFT TELEVISION THEATER (I,2592); MR. O'MALLEY (I,3146); THE UNTOUCHABLES (I,4682); WILLY (I,4868); WILLY (I,4869)

PELUTSKY, BERT
THE ROCKFORD FILES (II,2197)

PENN, RICHARD
OH MADELINE (II,1880)

PEPPIATT, FRANK
BARBARA MANDRELL AND THE MANDRELL SISTERS (II,143); THE BRASS ARE COMING (I,708); THE FUNNY WORLD OF FRED & BUNNI (II,943); THE GISELE MACKENZIE SHOW (I,1813); THE HARLEM GLOBETROTTERS POPCORN MACHINE (II,1092); HECK'S ANGELS (II,1122); THE HEE HAW HONEYS (II,1124); HERB ALBERT AND THE TIJUANA BRASS (I,2019); HULLABALOO (I,2156); THE JACKIE GLEASON SPECIAL (I,2325); THE JONATHAN WINTERS SHOW (I,2443); THE JUD STRUNK SHOW (I,2462); KEEP ON TRUCKIN' (II,1372); THE KOPYCATS (I,2580); THE KRAFT MUSIC HALL (I,2587); THE KRAFT MUSIC HALL (I,2588); MONTE CARLO, C'EST LA ROSE (I,3095); THE PERRY COMO SUNSHINE SHOW (II,2005); SHIELDS AND YARNELL (II,2328); THE SONNY AND CHER SHOW (II,2401); TIN PAN ALLEY TODAY (I,4521); WHO'S AFRAID OF MOTHER GOOSE? (I,4852)

PERELMAN, S J
ELIZABETH TAYLOR IN LONDON (I,1430)

PERINE, PARKE
AUTOMAN (II,120); EIGHT IS ENOUGH (II,762); FAME (II,812); HUNTER (II,1205); MR. MERLIN (II,1751); STARSKY AND HUTCH (II,2444)

PERKINS, JODY
WALTER CRONKITE'S UNIVERSE (II,2739)

PERKUNY, ROBERT
HAPPY DAYS (II,1084)

PERL, ARNOLD
EAST SIDE/WEST SIDE (I,1396); NAKED CITY (I,3216)

PERLBERG, IRVING
ALIAS SMITH AND JONES (I,118); APPLE'S WAY (II,101); CANNON (II,424); COLUMBO (II,556); EISCHIED (II,763); THE FBI (I,1551); FOUL PLAY (II,910); HAWAII FIVE-O (II,1110); THE MISSISSIPPI (II,1707); PARIS (II,1957); THE ROOKIES (II,2208)

PERLMAN, HEIDI
CHEERS (II,494)

PERLMAN, RON
CHER (II,495)

PERLOVE, PAUL
M*A*S*H (II,1569); ONE DAY AT A TIME (II,1900)

PERLOWE, ROBERT
LAVERNE AND SHIRLEY (II,1446); NEWHART (II,1835)

PEROW, TOM
ON STAGE AMERICA (II,1893)

PERRET, GENE
BOB HOPE SPECIAL: BOB HOPE IN WHO MAKES THE WORLD LAUGH, PART 2 (II,286); BOB HOPE SPECIAL: BOB HOPE'S ALL-STAR BIRTHDAY AT ANNAPOLIS (II,294); BOB HOPE SPECIAL: BOB HOPE'S STAND UP AND CHEER FOR THE NATIONAL FOOTBALL LEAGUE'S 60TH YEAR (II,316); BOB HOPE SPECIAL: BOB HOPE'S SUPER BIRTHDAY SPECIAL (II,319); BOB HOPE SPECIAL: BOB HOPE'S USO CHRISTMAS IN BEIRUT (II,320); BOB HOPE SPECIAL: BOB HOPE'S WICKI-WACKY SPECIAL FROM WAIKIKI (II,321); BOB HOPE SPECIAL: HO HO HOPE'S JOLLY CHRISTMAS HOUR (II,327); BOB HOPE SPECIAL: THE BOB HOPE CHRISTMAS SPECIAL (II,329); THE CAROL BURNETT SHOW (II,443); COMEDY NEWS II (I,1021); GIMME A BREAK (II,995); HELLZAPOPPIN (I,2011); JOE AND SONS (II,1296); JOHN RITTER: BEING OF SOUND MIND AND BODY (II,1319); THE KELLY MONTEITH SHOW (II,1376); MAMA'S FAMILY (II,1610); THE NEW BILL COSBY SHOW (I,3257); PEEPING TIMES (II,1990); ROWAN AND MARTIN'S LAUGH-IN (I,3856); THE SHAPE OF THINGS (II,2315); THAT'S TV (II,2581); THE TIM CONWAY SHOW (II,2619); UNCLE TIM WANTS YOU! (II,2704)

PERRIN, SAM
THE DON KNOTTS SHOW (I,1334); THE GENERAL FOODS 25TH ANNIVERSARY SHOW (I,1756); THE JACK BENNY HOUR (I,2290); THE JACK BENNY HOUR (I,2291); THE JACK BENNY HOUR (I,2292); THE JACK BENNY HOUR (I,2293); THE JACK BENNY PROGRAM (I,2294); THE JACK BENNY SPECIAL (I,2295); JACK BENNY WITH GUEST STARS (I,2298); JACK BENNY'S BIRTHDAY SPECIAL (I,2299); JACK BENNY'S NEW LOOK (I,2301)

PERRY, DICK
THE PAUL DIXON SHOW (I,3487)

PERRY, ELEANOR
A CHRISTMAS MEMORY (I,948); THE HOUSE WITHOUT A CHRISTMAS TREE (I,2143); THE THANKSGIVING TREASURE (I,4416); THANKSGIVING VISITOR (I,4417)

PERRY, JOHN FENTON
PARTRIDGE FAMILY: 2200 A.D. (II,1963)

PERRY, JOYCE
EIGHT IS ENOUGH (II,762); FLAMINGO ROAD (II,872); NURSE (II,1870); STAR TREK (II,2439); THE WALTONS (II,2740)

PERRY, MARLENE
THE LOVE BOAT (II,1535)

PERSKY, BILL
ALAN KING'S FINAL WARNING (II,29); BIG EDDIE (II,235); THE BILL COSBY SPECIAL (I,444); BOBBY PARKER AND COMPANY

(II,344); THE BOYS (II,360); THE CONFESSIONS OF DICK VAN DYKE (II,568); DICK VAN DYKE AND THE OTHER WOMAN, MARY TYLER MOORE (I,1273); THE FABULOUS FUNNIES (II,801); THE FIRST NINE MONTHS ARE THE HARDEST (I,1584); THE FUNNY SIDE (I,1714); GOOD MORNING WORLD (I,1850); HOW TO SURVIVE THE 70S AND MAYBE EVEN BUMP INTO HAPPINESS (II,1199); THE JULIE ANDREWS SHOW (I,2481); THE MAN WHO CAME TO DINNER (I,2881); MY WIFE NEXT DOOR (II,1780); PURE GOLDIE (I,3696); THE SID CAESAR, IMOGENE COCA, CARL REINER, HOWARD MORRIS SPECIAL (I,4039)

PERTWEE, MICHAEL
RETURN OF THE SAINT (II,2141); THE SAINT (II,2237); SECRET AGENT (I,3962)

PERTWEE, WILLIAM
SUSPICION (I,4309)

PERZIGIAN, JERRY
THE JEFFERSONS (II,1276); THE KRAFT 75TH ANNIVERSARY SPECIAL (II,1409)

PETAL, MARION
THE TWILIGHT ZONE (I,4651)

PETERS, ERIC
BEN CASEY (I,394); THRILLER (I,4492)

PETERS, MARGIE
13 QUEENS BOULEVARD (II,2591); THE FACTS OF LIFE (II,805); FACTS OF LIFE: THE FACTS OF LIFE GOES TO PARIS (II,806); JO'S COUSINS (II,1293)

PETERSON, DON
OF MEN OF WOMEN (I,3331)

PETERSON, PAUL
LIFE WITH SNARKY PARKER (I,2693)

PETERSON, ROD
THE FITZPATRICKS (II,868); THE WALTONS (II,2740)

PETERSON, SHIRLEY
STUDIO ONE (I,4268)

PETERSON, TED
POLE POSITION (II,2060)

PETICLERE, DENNE BART
MEN OF THE DRAGON (II,1683); SHANE (I,3995); THEN CAME BRONSON (I,4436); THE WILD WILD WEST (I,4863)

PETITO, DAVID
THE NEW NEWLYWED GAME (II,1824)

PETRACCA, JOSEPH
ONE STEP BEYOND (I,3388); RAWHIDE (I,3727); THE UNTOUCHABLES (I,4682)

PETRAGNA, MICHAEL
SEVEN BRIDES FOR SEVEN BROTHERS (II,2307)

PETRAKIS, HARRY MARK
THE JUDGE (I,2468)

PETTE, BOB
DIFF'RENT STROKES (II,674)

PETTLER, PAMELA
CHARLES IN CHARGE (II,482)

PETTUS, KEN
THE ADVENTURES OF NICK CARTER (I,68); B.J. AND THE BEAR (II,126); THE BIG VALLEY (I,437); BRANDED (I,707); DEAD MAN ON THE RUN (II,631); HAWAII FIVE-O (II,1110); JIGSAW (I,2377); SHANNON (II,2314); VEGAS (II,2724); THE WILD WILD WEST (I,4863)

PEYSER JR, TONY
KOMEDY TONITE (II,1407)

PEYSER, ARNOLD
GILLIGAN'S ISLAND (II,990); MY FAVORITE MARTIAN (I,3189); THREE FOR THE ROAD (II,2607)

PEYSER, LARS GREEN
MIKE AND BUFF (I,3037)

PEYSER, LOIS
GILLIGAN'S ISLAND (II,990); MY FAVORITE MARTIAN (I,3189); OF ALL THINGS (I,3330); THREE FOR THE ROAD (II,2607)

PFANNER, JIM
ROUGHCUTS (II,2218)

PHILLIPS, ARNOLD
FIRESIDE THEATER (I,1580)

PHILLIPS, ARTHUR
THE ANN SOTHERN SHOW (I,220); CAVALCADE OF STARS (I,877); THE CBS NEWCOMERS (I,879); THE JERRY LEWIS SHOW (I,2362); THE JERRY LEWIS SHOW (I,2363); MAKE ROOM FOR DADDY (I,2843); MONTY HALL'S VARIETY HOUR (II,1728); THE RED SKELTON SHOW (I,3755)

PHILLIPS, BILL
SUMMER SOLSTICE (II,2492)

PHILLIPS, CAL
MYSTERY AND MRS. (I,3210)

PHILLIPS, LUCY
ANOTHER DAY (II,94)

PHILLIPS, MARGARET
MATINEE THEATER (I,2947)

PHILLIPS, ROGER
BENSON (II,208)

PHIPPS, THOMAS W
THE FARMER'S DAUGHTER (I,1534); LAURA (I,2636)

PICCRANO, GREG
THE DANCE SHOW (II,616)

PICKARD, JOHN
ADAMSBURG, U.S.A. (I,33); THE DICK POWELL SHOW (I,1269)

PIEFE, ALAN
MAGIC MIDWAY (I,2818)

PIERCE, ARTHUR
FANTASY ISLAND (II,829)

PIERCE, PAUL
ACTION IN THE AFTERNOON (I,25)

PIERSON, ARTHUR
TERRY AND THE PIRATES (I,4405)

PIERSON, FRANK
AMANDA FALLON (I,152); NAKED CITY (I,3216)

PILLER, GENE
HEAVENS TO BETSY (I,1997)

PILLER, MICHAEL
CAGNEY AND LACEY (II,409); LEGMEN (II,1458); SHADOW OF SAM PENNY (II,2313); SIMON AND SIMON (II,2357)

PINACH, DAVID
REAL KIDS (II,2116)

PINE, LESTER
BEN CASEY (I,394); THE BOSTON TERRIER (I,696); THE DICK POWELL SHOW (I,1269); DOBIE GILLIS (I,1302); FOL-DE-ROL (I,1613); MR. LUCKY (I,3141); NAKED CITY (I,3216); PETER GUNN (I,3562); POPI (II,2068); POPI (II,2069); ROUTE 66 (I,3852); TARGET: THE CORRUPTERS (I,4360); TURN ON (I,4625)

PINE, TINA
POPI (II,2068); POPI (II,2069)

PINSKER, JUDITH
RYAN'S HOPE (II,2234)

PIOLI, JUDY
CHARLES IN CHARGE (II,482); GOODTIME GIRLS (II,1042); I'M A BIG GIRL NOW (II,1218); LAVERNE AND SHIRLEY (II,1446); WEBSTER (II,2758)

PIROSH, ROBERT
ALEXANDER THE GREAT (I,113); COMBAT (I,1011); THE FUGITIVE (I,1701); THE GUNS OF WILL SONNETT (I,1907); THE HARDY BOYS MYSTERIES (II,1090); HAWAII FIVE-O (II,1110); SARA (II,2260); THE WALTONS (II,2740); THE YOUNG PIONEERS (II,2869)

PISCOPO, JOE
THE JOE PISCOPO SPECIAL (II,1304)

PISTOLE, GREGORY
THE UGILY FAMILY (II,2699)

PITTMAN, MONTGOMERY
THE RIFLEMAN (I,3789); THE TWILIGHT ZONE (I,4651)

PIZER, ELIZABETH
KNOTS LANDING (II,1404)

PLACE, MARY KAY
M*A*S*H (II,1569); THE MARY TYLER MOORE SHOW (II,1640); PHYLLIS (II,2038)

PLANT, MICHAEL
ONE STEP BEYOND (I,3388); THE VEIL (I,4709)

PLATER, ALAN
FLAMBARDS (II,869)

PLATT, KIN
THE MILTON THE MONSTER CARTOON SHOW (I,3053)

PLAYDON, PAUL
BANACEK (II,138); BATTLESTAR GALACTICA (II,181); COMBAT (I,1011); ESCAPE (I,1460); THE NIGHT STALKER (II,1849); SWITCH (II,2519); THE WILD WILD WEST (I,4863)

PLESHETTE, JOHN
KNOTS LANDING (II,1404); RYAN'S FOUR (II,2233)

PLIMPTON, GEORGE
PLIMPTON! DID YOU HEAR THE ONE ABOUT. . .? (I,3629); PLIMPTON! SHOWDOWN AT RIO LOBO (I,3630); PLIMPTON! THE MAN ON THE FLYING TRAPEZE (I,3631)

POCKRESS, LEE
THE MILTON BERLE SHOW (I,3047)

PODELL, RIC
BROTHERS (II,381)

POE, JAMES
THE DICK POWELL SHOW (I,1269)

POE, JEROME

COME A RUNNING (I,1013)

POGOSTIN, S LEE
THE DICK POWELL SHOW (I,1269); THE MOON AND SIXPENCE (I,3098)

POLIZZI, JOSEPH
GET CHRISTIE LOVE! (II,981); KOJAK (II,1406); MALLORY: CIRCUMSTANTIAL EVIDENCE (II,1608)

POLLOCK, DAVID
CARTER COUNTRY (II,449); THE DON KNOTTS SHOW (I,1334); DON RICKLES—ALIVE AND KICKING (I,1340); THE EVERLY BROTHERS SHOW (I,1479); GOODNIGHT, BEANTOWN (II,1041); HOLLYWOOD HIGH (II,1163); HOT L BALTIMORE (II,1187); THE LENNON SISTERS SHOW (I,2656); M*A*S*H (II,1569); PAUL SAND IN FRIENDS AND LOVERS (II,1982); ROBERT YOUNG AND THE FAMILY (I,3813); THE SHOW MUST GO ON (II,2344); THE SINGERS (I,4060); SUPER COMEDY BOWL 2 (I,4289); THAT'S MY MAMA (II,2580); WHAT'S HAPPENING!! (II,2769)

POLLOCK, EILEEN
APPOINTMENT WITH ADVENTURE (I,246); THE BRIGHTER DAY (I,728); CLIMAX! (I,976)

POLLOCK, MARION
ON THE GO (I,3361)

POLLOCK, ROBERT MASON
APPOINTMENT WITH ADVENTURE (I,246); THE BRIGHTER DAY (I,728); CLIMAX! (I,976); DYNASTY (II,746)

POLLOCK, RUTH
POWERHOUSE (II,2074)

POLLON, DAPHNE
LAUGH TRAX (II,1443)

POLONSKY, SUSAN
THE NOONDAY SHOW (II,1859)

POMERANTZ, EARL
BEST OF THE WEST (II,218); THE BETTY WHITE SHOW (II,224); CHEERS (II,494); THE COSBY SHOW (II,578); EVERYTHING GOES (I,1480); FLIP WILSON. . . OF COURSE (II,890); FREE COUNTRY (II,921); KEEP ON TRUCKIN' (II,1372); LILY (II,1477); THE LILY TOMLIN SPECIAL (II,1479); THE MARY TYLER MOORE SHOW (II,1640); NEWHART (II,1835); PHYLLIS (II,2038); RHODA (II,2151); TAXI (II,2546); THE TONY RANDALL SHOW (II,2640)

POMERANTZ, HART
THE BEAUTIFUL PHYLLIS DILLER SHOW (I,381)

POOLE, DUANE
THE B.B. BEEGLE SHOW (II,124); THE FALL GUY (II,811); HART TO HART (II,1102); THE KROFFT KOMEDY HOUR (II,1411); LOTTERY (II,1525)

POOLEY, OLAF
FALCON'S GOLD (II,809)

POPE, ELAINE
LILY FOR PRESIDENT (II,1478); LILY—SOLD OUT (II,1480); NOT NECESSARILY THE NEWS (II,1862); NOT NECESSARILY THE NEWS (II,1863)

POPP, M ANDRE
VICTOR BORGE'S COMEDY IN MUSIC IV (I,4726)

POPPE JR, HARRY
THE LONE RANGER (I,2740); SKY KING (I,4077)

PORTER, WILL
PEOPLE ARE FUNNY (II,1996)

PORTERFIELD, CHRISTOPHER
THE DICK CAVETT SHOW (II,669)

POST, HOWARD
BATTLE OF THE PLANETS (II,179)

POTTER, DENNIS
CASANOVA (II,451)

POTTER, ELLEN
DOUBLE TROUBLE (II,723)

POTTET, MICHELE
TEXAS (II,2566)

POULIOT, STEPHEN
PERRY COMO'S BAHAMA HOLIDAY (II,2006); PERRY COMO'S CHRISTMAS IN AUSTRIA (II,2007); PERRY COMO'S CHRISTMAS IN PARIS (II,2012); PERRY COMO'S CHRISTMAS IN THE HOLY LAND (II,2013); PERRY COMO'S EASTER IN GUADALAJARA (II,2015); PERRY COMO'S FRENCH-CANADIAN CHRISTMAS (II,2016); PERRY COMO'S SPRING IN SAN FRANCISCO (II,2022)

POWELL, LESTER
IVANHOE (I,2282)

POWELL, RICHARD
APPLE PIE (II,100); BABY, I'M BACK! (II,130); BIG EDDIE (II,236); THE CHEAP DETECTIVE (II,490); THE FARMER'S DAUGHTER (I,1533); THE LUCILLE BALL COMEDY HOUR (I,2783); THE MARY TYLER MOORE SHOW (II,1640)

POWELL, RURIC
U.F.O. (I,4662)

POWELL, WILLIAM D
BONANZA (II,347)

POWERS, KATHARYN MICHAELIAN
CHARLIE'S ANGELS (II,486); FALCON CREST (II,808); THE FANTASTIC JOURNEY (II,826); FANTASY ISLAND (II,829); THE FITZPATRICKS (II,868); HOW THE WEST WAS WON (II,1196); THE TOM SWIFT AND LINDA CRAIG MYSTERY HOUR (II,2633); TONI'S BOYS (II,2637); WONDER WOMAN (SERIES 2) (II,2829); THE YOUNG PIONEERS (II,2869)

PRAISER, IAN
BOSOM BUDDIES (II,355); DOMESTIC LIFE (II,703); MR. SUCCESS (II,1756); TAXI (II,2546); THREE'S COMPANY (II,2614)

PRASKINS, LEONARD
NAKED CITY (I,3216); WAGON TRAIN (I,4747)

PRAY, POOT
THE WHISTLING WIZARD (I,4844)

PREBBLE, JOHN
PENMARRIC (II,1993)

PRELUSKY, BURT
MCMILLAN AND WIFE (II,1667)

PREMINGER, MICHAEL
BROTHERS (II,381); STAR OF THE FAMILY (II,2436); THE STEVE LANDESBERG TELEVISION SHOW (II,2458); WEBSTER (II,2758)

PRENTICE, PATRICK
POWERHOUSE (II,2074)

PRESNELL JR, ROBERT
THE ELEVENTH HOUR (I,1425); THE GEORGE SANDERS MYSTERY THEATER (I,1776); MCCLOUD (II,1660); STUDIO ONE (I,4268); THE TWILIGHT ZONE (I,4651)

PRESNELL JR, RUSSELL
BANACEK (II,138)

PRESSBURGER, FRED
THE ABC AFTERSCHOOL SPECIAL (II,1)

PRICE, EUGENE
EMERALD POINT, N.A.S. (II,773); KOJAK (II,1406); MARCUS WELBY, M.D. (II,1627); RYAN'S HOPE (II,2234)

PRICE, PAUL
REGGIE (II,2127)

PRICE, ROBERT
PALMERSTOWN U.S.A. (II,1945)

PRICE, ROGER
THE DON KNOTTS SHOW (I,1334); KOSTA AND HIS FAMILY (I,2581)

PRICE, STANLEY
THE UGLIEST GIRL IN TOWN (I,4663)

PRIDEAUX, JAMES
RETURN ENGAGEMENT (I,3775)

PRINCE, EUGENE
CONCRETE COWBOYS (II,563)

PRINCE, PETER
OPPENHEIMER (II,1919)

PRINGLE, DON
ACTION IN THE AFTERNOON (I,25)

PRIOR, ALLAN
BLAKE'S SEVEN (II,264); THE ONEDIN LINE (II,1912)

PRITZKER, STEVE
FRIENDS (II,928); HERE WE GO AGAIN (I,2028); THE LOVE BOAT II (II,1533); THE PARTRIDGE FAMILY (II,1962); PAUL SAND IN FRIENDS AND LOVERS (II,1982); SILVER SPOONS (II,2355); TURN ON (I,4625)

PROCTOR, PHIL
THE STARLAND VOCAL BAND (II,2441)

PROCTOR, RICHARD
PEOPLE ARE FUNNY (II,1996)

PROFT, PAT
BOB HOPE SPECIAL: BOB HOPE'S CHRISTMAS PARTY (II,307); CHER (II,495); CHER (II,496); DETECTIVE SCHOOL (II,661); HIGH SCHOOL, U.S.A. (II,1149); THE JIM STAFFORD SHOW (II,1291); POLICE SQUAD! (II,2061); THE REDD FOXX COMEDY HOUR (II,2126); RINGO (II,2176); THE SMOTHERS BROTHERS SHOW (II,2384); TWILIGHT

THEATER II (II,2686); ULTRA QUIZ (II,2701); VAN DYKE AND COMPANY (II,2719); WELCOME BACK, KOTTER (II,2761); WHEN THINGS WERE ROTTEN (II,2782)

PROSS, MAX
THE NEW SHOW (II,1828)

PROVO, FRANK
THE DICK POWELL SHOW (I,1269)

PRUSS, NANCY
LET'S MAKE A DEAL (II,1464)

PRYOR, RICHARD
LILY (I,2705); THE LILY TOMLIN SHOW (I,2706); THE RICHARD PRYOR SHOW (II,2163); THE RICHARD PRYOR SPECIAL (II,2164); THE RICHARD PRYOR SPECIAL? (II,2165)

PUGH, MADELYN
I LOVE LUCY (I,2171)

PULLBROOK, VIOLET
DARKROOM (II,619)

PULMAN, JACK
JANE EYRE (I,2339); WAR AND PEACE (I,4765)

PUMPIAN, PAUL
BOB HOPE SPECIAL: BOB HOPE IN "JOYS" (II,284); DONNY AND MARIE (II,712); I'VE HAD IT UP TO HERE (II,1221); THE MAD MAD MAD MAD WORLD OF THE SUPER BOWL (II,1586); TV FUNNIES (II,2672)

PUMPIAN, PHIL
100 YEARS OF GOLDEN HITS (II,1905)

PURDHAM, EMILY
LAVERNE AND SHIRLEY (II,1446)

PURDUM, HERBERT R
CROSSROADS (I,1105); RAWHIDE (I,3727)

PURSER, DOROTHY
FOR RICHER, FOR POORER (II,906); TEXAS (II,2566)

PURSER, PHILIP
PEEK-A-BOO: THE ONE AND ONLY PHYLLIS DIXEY (II,1989)

PUTNAM, WILLIAM
LITTLE HOUSE ON THE PRAIRIE (II,1487); THE MISSISSIPPI (II,1707)

PYLE, DENVER
DIRTY SALLY (II,680)

PYNE, DANIEL
MIAMI VICE (II,1689)

QUICKSILVER, ELIZABETH
SECRETS OF MIDLAND HEIGHTS (II,2296)

QUILLAN, JOE
BING CROSBY AND HIS FRIENDS (I,455); OUR MISS BROOKS (I,3415)

QUINE, RICHARD
HEY MULLIGAN (I,2040)

QUINN, JERRY
ONE STEP BEYOND (I,3388)

QUINN, LOU
CELEBRITY TIME (I,887)

QUINN, SPENCER
COMEDY NEWS II (I,1021)

RABIN, ARTHUR
PHYLLIS (II,2038)

RABINOWITZ, BARRY
LAVERNE AND SHIRLEY (II,1446)

RACKIN, MARTIN
NEVADA SMITH (II,1807); RIVAK, THE BARBARIAN (I,3795)

RADANO, GENE
THE BLUE KNIGHT (II,276)

RADAR, WILLIAM C
ALL IN THE FAMILY (II,38)

RADCLIFF, SAMUEL
TEXAS (II,2566)

RADDICK, JOHN
DANGER MAN (I,1136)

RADER, PAUL
TEXAS (II,2566)

RADNITZ, BRAD
CALL TO GLORY (II,413); CANNON (II,424); THE COPS AND ROBIN (II,573); GET CHRISTIE LOVE! (II,981); GILLIGAN'S ISLAND (II,990); HARPER VALLEY PTA (II,1095); HEC RAMSEY (II,1121); IT'S ABOUT TIME (I,2263); MCMILLAN AND WIFE (II,1667); NURSE (II,1870)

RAGAWAY, MARTIN A
THE BILLY CRYSTAL COMEDY HOUR (II,248); THE CRYSTAL GAYLE SPECIAL (II,609); DEAN MARTIN'S CELEBRITY ROAST (II,643); DIANA (I,1258); F TROOP (I,1499); GIRL FRIENDS AND NABORS (I,1807); HERE'S LUCY (II,1135); THE JERRY LEWIS SHOW (I,2370); THE NIGHT OF CHRISTMAS (I,3290); THE PARTRIDGE FAMILY (II,1962); PRIVATE SECRETARY (I,3672); THE RED SKELTON REVUE (I,3754); THE RED SKELTON SHOW (I,3755); THE UGLIEST GIRL IN TOWN (I,4663)

RAINER, IRIS
CHER (II,495); CHER (II,496); THE JOHN DAVIDSON SHOW (II,1309); THE SONNY AND CHER SHOW (II,2401); SUGAR TIME (II,2489)

RAKER, FRED
THE CRACKER BROTHERS (II,599); NO SOAP, RADIO (II,1856)

RALEY, RON
SPECIAL EDITION (II,2416)

RALLY, PAUL
BENSON (II,208); THE DAVID LETTERMAN SHOW (II,624)

RALSTON, GILBERT
AMOS BURKE, SECRET AGENT (I,180); BEN CASEY (I,394); COMBAT (I,1011); I SPY (I,2179); NAKED CITY (I,3216); ROUTE 66 (I,3852); STAR TREK (II,2440); THE WILD WILD WEST (I,4863)

RAMBEAU, LOU
ALFRED HITCHCOCK PRESENTS (I,115)

RAMIS, HAROLD
RODNEY DANGERFIELD SPECIAL: IT'S NOT EASY BEIN' ME (II,2199)

RAMUS, AL
THE MAN FROM U.N.C.L.E. (I,2867)

RANDALL, BOB
6 RMS RIV VU (II,2371); ANNIE AND THE HOODS (II,91); KATE AND ALLIE (II,1365); MO AND JO (II,1715); ON OUR OWN (II,1892)

RANDOLPH, JOHN
SAM (II,2246)

RANKIN, ARTHUR
FISH (II,864)

RANNOW, JERRY
THE BAXTERS (II,184); HARPER VALLEY (II,1094); THE KAREN VALENTINE SHOW (I,2506); WELCOME BACK, KOTTER (II,2761)

RANNOW, JEWEL JAFFE
WELCOME BACK, KOTTER (II,2761)

RANSONE, ROBERT
SEVEN BRIDES FOR SEVEN BROTHERS (II,2307)

RAPF, MATTHEW
KOJAK (II,1406)

RAPHAELSON, SAMPSON
THE JAZZ SINGER (I,2351)

RAPOPORT, I C
BORN TO THE WIND (II,354); THOU SHALT NOT KILL (II,2604)

RAPP, JOEL
GILLIGAN'S ISLAND (II,990); IT'S ABOUT TIME (I,2263); MCHALE'S NAVY (I,2969); THE PATTY DUKE SHOW (I,3483)

RAPP, JOHN
THE BOB HOPE SHOW (I,533); THE BOB HOPE SHOW (I,534); THE BOB HOPE SHOW (I,541); THE BOB HOPE SHOW (I,543); THE BOB HOPE SHOW (I,552); THE BOB HOPE SHOW (I,555); THE BOB HOPE SHOW (I,560); THE BOB HOPE SHOW (I,561); THE BOB HOPE SHOW (I,562); THE BOB HOPE SHOW (I,604); THE BOB HOPE SHOW (I,606); THE BOB HOPE SHOW (I,607); THE BOB HOPE SHOW (I,613); THE EDDIE CANTOR COMEDY THEATER (I,1407); ROBERTA (I,3815)

RAPP, PHILIP
MIMI (I,3054)

RAPPAPORT, JOHN
ALL IN THE FAMILY (II,38); HELLZAPOPPIN (I,2011); THE LILY TOMLIN SHOW (I,2706); M*A*S*H (II,1569); THE ODD COUPLE (II,1875); ROWAN AND MARTIN'S LAUGH-IN (I,3856); SECOND EDITION (II,2288); THREE TIMES DALEY (II,2610)

RASCHELLA, CAROLE
CODE RED (II,552); LITTLE HOUSE ON THE PRAIRIE (II,1487)

RASCHELLA, MICHAEL
CODE RED (II,552); LITTLE HOUSE ON THE PRAIRIE (II,1487)

RASKIN, RICHARD
THE FALL GUY (II,811); FINDER OF LOST LOVES (II,857); HART TO HART (II,1102)

RASKY, HARRY
THE LEGEND OF SILENT NIGHT (I,2655)

RATCLIFFE, VIRGINIA
KRAFT TELEVISION THEATER (I,2592)

RATTIGAN, TERENCE

OF MEN OF WOMEN (I,3331); SEPARATE TABLES (II,2304)

RAUCHER, HERMAN
THE ALCOA HOUR (I,108)

RAUESO, VIC
ALICE (II,33); JUST OUR LUCK (II,1360)

RAVEN, SIMON
LOVE IN A COLD CLIMATE (II,1539); THE PALLISERS (II,1944)

RAY, FRANK
THE DON RICKLES SHOW (I,1341)

RAY, JEFF
THE A-TEAM (II,119); THE GREATEST AMERICAN HERO (II,1060); TALES OF THE GOLD MONKEY (II,2537)

RAY, MARC
20TH CENTURY FOLLIES (I,4641); THE ANN-MARGRET SHOW (I,218); BURT BACHARACH IN SHANGRI-LA (I,773); BURT BACHARACH—OPUS NO. 3 (I,774); THE KING FAMILY IN WASHINGTON (I,2549); THE KING FAMILY JUNE SPECIAL (I,2550); THE NEW MICKEY MOUSE CLUB (II,1823); SPECIAL LONDON BRIDGE SPECIAL (I,4150); STAND UP AND CHEER (I,4188)

RAYFIELD, DAVID
CHANNING (I,900); COLUMBO (II,556); THE FLIERS (I,1596); SECOND LOOK (I,3960)

RAYFIELD, HOWARD
THE ABC AFTERSCHOOL SPECIAL (II,1)

RAYMOND, CHARLES
MOTHER AND ME, M.D. (II,1740)

RAYMOND, JACK
THE FARMER'S DAUGHTER (I,1533); GILLIGAN'S ISLAND (II,990); THE NBC FOLLIES (I,3241); PLEASE DON'T EAT THE DAISIES (I,3628); SIGMUND AND THE SEA MONSTERS (II,2352); THE SMOTHERS BROTHERS SHOW (I,4096)

RAYNOR, WILLIAM
BACHELOR FATHER (I,323); THE DUKES OF HAZZARD (II,742); HELLO DERE (I,2010); MCHALE'S NAVY (I,2969); PARTRIDGE FAMILY: 2200 A.D. (II,1963); THE SAN PEDRO BEACH BUMS (II,2250); THE TIM CONWAY COMEDY HOUR (I,4501)

RAZNICK, DEBORAH
MORK AND MINDY (II,1735)

REACH, JAMES
OLD NICKERBOCKER MUSIC HALL (I,3351); THE TIMES SQUARE STORY (I,4514)

READ, ANTHONY
THE OMEGA FACTOR (II,1888)

REAGAN, DENNIS
THE LORENZO AND HENRIETTA MUSIC SHOW (II,1519)

REAVES, MICHAEL
GOING BANANAS (II,1015); ISIS (II,1248); POLE POSITION (II,2060); THE SMURFS (II,2385)

REBACK, KATHERINE
HUSBANDS, WIVES AND LOVERS (II,1209)

REDACK, JAY
BATTLESTARS (II,182)

REDMOND, MELBA
MATINEE THEATER (I,2947)

REED, BILL
GILLIGAN'S PLANET (II,994)

REED, GREG
BIZARRE (II,259)

REED, LEWIS
BONANZA (II,347); BURKE'S LAW (I,762); MR. LUCKY (I,3141); PETER GUNN (I,3562)

REED, LUTHER
MATINEE THEATER (I,2947)

REED, PHILLIP
MIAMI VICE (II,1689)

REED, REX
REX REED'S MOVIE GUIDE (II,2149)

REEDER, TOM
A.E.S. HUDSON STREET (II,17); BARNEY MILLER (II,154); BENSON (II,208); CHEERS (II,494); FISH (II,864); M*A*S*H (II,1569); NIGHT COURT (II,1844); PARK PLACE (II,1958); PHYL AND MIKHY (II,2037)

REENAN, WILLIAM
MR. MERLIN (II,1751)

REES, GARY
THE FOUR SEASONS (II,915); TOO CLOSE FOR COMFORT (II,2642)

REESE, RAY
GIVE-N-TAKE (II,1003); THE NEIGHBORS (II,1801); SECOND CHANCE (II,2286)

REEVES, GEORGE
CRAZY TIMES (II,602)

REHAK, BRIAN
THE INCREDIBLE HULK (II,1232); MOBILE ONE (II,1717)

REHAK, KANDY
MOBILE ONE (II,1717)

REID, DON
AN EVENING WITH THE STATLER BROTHERS (II,790)

REID, ELLIOTT
AFTERMASH (II,23)

REID, HOWARD
AN EVENING WITH THE STATLER BROTHERS (II,790)

REID, TIM
WKRP IN CINCINNATI (II,2814)

REIG, JUNE
LITTLE WOMEN (I,2724)

REIGER, JOHN
MAKING IT (II,1605)

REIKER, DONALD
THE BOB NEWHART SHOW (II,342); THE LAST RESORT (II,1440); REPORT TO MURPHY (II,2134); THE TONY RANDALL SHOW (II,2640)

REINER, CARL
THE 2,000 YEAR OLD MAN (II,2696); A DATE WITH DEBBIE (I,1161); FLANNERY AND QUILT (II,873); HEAD OF THE FAMILY (I,1988); THE NANCY DUSSAULT SHOW (I,3222); THE SID CAESAR, IMOGENE COCA, CARL REINER, HOWARD MORRIS SPECIAL (I,4039)

REINER, ROB
FREE COUNTRY (II,921); ROBERT YOUNG AND THE FAMILY (I,3813); ROMP (I,3838); SONNY BOY (II,2402); THE SUMMER SMOTHERS BROTHERS SHOW (I,4281); THE SUPER (I,4293); THREE FOR THE GIRLS (I,4477)

REINHART, RICH
BENSON (II,208); THE LAST RESORT (II,1440); SEMI-TOUGH (II,2300); TOO CLOSE FOR COMFORT (II,2642)

REINHART, RUTH
FAMILY TIES (II,819)

REISMAN JR, PHIL
THE ANDROS TARGETS (II,76); THE DEVIL AND DANIEL WEBSTER (I,1246); DUPONT SHOW OF THE MONTH (I,1387); HEDDA GABLER (I,2001); THE HUNTER (I,2162); KRAFT TELEVISION THEATER (I,2592); SANDBURG'S LINCOLN (I,3907); SHANE (I,3995); TEN LITTLE INDIANS (I,4394); A VERY MISSING PERSON (I,4715)

REISMAN, DEL
BANACEK (II,138); BORN TO THE WIND (II,354); BRONK (II,379); CAGNEY AND LACEY (II,409); CANNON (II,424); CHARLIE'S ANGELS (II,486); GENERAL ELECTRIC TRUE (I,1754); LOU GRANT (II,1526); PARIS (II,1957); THE STREETS OF SAN FRANCISCO (II,2478); THREE FOR THE ROAD (II,2607)

REISMAN, MARK
THICKE OF THE NIGHT (II,2587)

REISNER, DEAN
DOBIE GILLIS (I,1302); THE OUTER LIMITS (I,3426); RAWHIDE (I,3727); RICH MAN, POOR MAN—BOOK I (II,2161)

REISNER, DEL
THE MONEYCHANGERS (II,1724)

REISNER, JUNE
COMEDY IS KING (I,1019); COS (II,576); COS: THE BILL COSBY COMEDY SPECIAL (II,577)

REISNER, SHERRY
THIS WAS AMERICA (II,2599)

REISS, MICHAEL
CHARLES IN CHARGE (II,482)

REITZ, MATHIAS
THE ROOKIES (II,2208)

REMBACH, FRANK
THE SMOTHERS BROTHERS SHOW (I,4097)

REMPLETON, WILLIAM
FAIR EXCHANGE (I,1512)

RENNE, GEORGE
LITTLE HOUSE: BLESS ALL THE DEAR CHILDREN (II,1489)

RENO, JEFF
GOODNIGHT, BEANTOWN (II,1041); SILVER SPOONS (II,2355); TOO CLOSE FOR COMFORT (II,2642)

RENWICK, DAVID
NOT THE NINE O'CLOCK NEWS (II,1864)

REO, DON
CHER (II,496); CHER AND OTHER FANTASIES (II,497); COMEDY NEWS II (I,1021); DOUBLE TROUBLE (II,724); THE JACKIE GLEASON SPECIAL (I,2325); JIMMY

DURANTE PRESENTS THE LENNON SISTERS HOUR (I,2387); ONE MORE TIME (I,3385); ONE MORE TIME (I,3386); PRIVATE BENJAMIN (II,2087); RHODA (II,2151); A ROCK AND A HARD PLACE (II,2190); THE ROWAN AND MARTIN SPECIAL (I,3855); ROWAN AND MARTIN'S LAUGH-IN (I,3856); SHOW BUSINESS SALUTE TO MILTON BERLE (I,4029); SUMMER (II,2491); WHITE AND RENO (II,2787); WIZARDS AND WARRIORS (II,2813)

RESNICK, PATRICIA
CHER. . . SPECIAL (II,499); FAERIE TALE THEATER (II,807)

REUBENS, PAUL
THE PARAGON OF COMEDY (II,1956); THE PEE WEE HERMAN SHOW (II,1988)

REYNOLDS, AL
MRS. COLUMBO (II,1760)

REYNOLDS, GENE
LOU GRANT (II,1526)

REYNOLDS, JONATHAN
THAT WAS THE YEAR THAT WAS (II,2575)

REYNOLDS, REBECCA
THE FALL GUY (II,811)

REYNOLDS, SHELDON
INN OF THE FLYING DRAGON (I,2214); SHERLOCK HOLMES (I,4010); SOPHIA LOREN IN ROME (I,4129)

REZNICK, SIDNEY
THE ALL-STAR COMEDY SHOW (I,137); THE MANHATTAN TRANSFER (II,1619); THE ODD COUPLE (II,1875); THE VIN SCULLY SHOW (I,4734)

RHEA, FREDDY
MY THREE SONS (I,3205)

RHINE, LARRY
ALL IN THE FAMILY (II,38); GIMME A BREAK (II,995); HERE'S LUCY (II,1135); JOE'S WORLD (II,1305); MR. ED (I,3137); THE ODD COUPLE (II,1875); PRIVATE SECRETARY (I,3672); THE RED SKELTON SHOW (I,3755); SPEED BUGGY (II,2419)

RHINEHART, RICH
GLORIA (II,1010)

RHODES, MICHAEL
DELVECCHIO (II,659)

RHYMER, PAUL
THE GOOK FAMILY (I,1858)

RICE, CY
ABOUT FACES (I,12)

RICE, ED
THROUGH THE CRYSTAL BALL (I,4494)

RICE, SAMUEL
THE LONE RANGER (I,2740)

RICH, KENNY
PRIVATE BENJAMIN (II,2087)

RICH, STEVEN
ALCOA/GOODYEAR THEATER (I,107)

RICHARDS, ALUN
THE ONEDIN LINE (II,1912)

RICHARDS, CHET
STAR TREK (II,2440)

RICHARDS, CLAYTON
BUCK ROGERS IN THE 25TH CENTURY (II,383)

RICHARDS, JULES
STORIES IN ONE CAMERA (I,4234)

RICHARDS, LAWRENCE
HART TO HART (II,1102)

RICHARDS, LLOYD
KUNG FU (II,1416)

RICHARDS, MARC
THE GHOST BUSTERS (II,986); HELLZAPOPPIN (I,2011); THE KLOWNS (I,2574); MISSION MAGIC (I,3068); MY FAVORITE MARTIANS (II,1779); TEACHER'S PET (I,4367)

RICHARDS, MICHAEL
BUCK ROGERS IN THE 25TH CENTURY (II,383); LOGAN'S RUN (II,1507); STAR TREK (II,2440)

RICHARDS, RON
THE CRACKER BROTHERS (II,599); THE DAVID LETTERMAN SHOW (II,624); IT ONLY HURTS WHEN YOU LAUGH (II,1251); NO SOAP, RADIO (II,1856); NOT NECESSARILY THE NEWS (II,1862); NOT NECESSARILY THE NEWS (II,1863)

RICHARDSON, DON
THE RAT PATROL (I,3726)

RICHARDSON, JOE
ANNIE OAKLEY (I,225); THE GENE AUTRY SHOW (I,1748); THE LONE RANGER (I,2740); SKY KING (I,4077)

RICHLIN, MAURICE
MCGARRY AND ME (I,2966); THE TAB HUNTER SHOW (I,4328); WHERE'S RAYMOND? (I,4837)

RICHMAN, DON
THE MAN FROM U.N.C.L.E. (I,2867)

RICHMAN, JEFFREY
THE JEFFERSONS (II,1276); MARIE (II,1629)

RICHMOND, BILL
THE CAROL BURNETT SHOW (II,443); THE DIAHANN CARROLL SHOW (I,1252); HOTEL 90 (I,2130); THE JERRY LEWIS SHOW (I,2369); THE JERRY LEWIS SHOW (II,1284); JOE AND SONS (II,1296); JOHN RITTER: BEING OF SOUND MIND AND BODY (II,1319); THE KELLY MONTEITH SHOW (II,1376); NO SOAP, RADIO (II,1856); PEEPING TIMES (II,1990); ROWAN AND MARTIN BITE THE HAND THAT FEEDS THEM (I,3853); ROWAN AND MARTIN'S LAUGH-IN (I,3856); THE SINGERS (I,4060); THAT'S TV (II,2581); THE TIM CONWAY SHOW (II,2619); UNCLE TIM WANTS YOU! (II,2704); WIZARDS AND WARRIORS (II,2813)

RICHMOND, JANE
ALOHA PARADISE (II,58); ON OUR OWN (II,1892)

RICHMOND, STEPHEN
HART TO HART (II,1102)

RICHTER, W D
SLITHER (II,2379)

RICKEY, FRED
OPEN HOUSE (I,3394)

RICKEY, PATRICIA
WHERE'S POPPA? (II,2784)

RICKMAN, THOMAS
DELTA COUNTY, U.S.A. (II,657); HOME COOKIN' (II,1170)

RIDDLE, SAM
STAR SEARCH (II,2438)

RIDGEWAY, AGNES
FAIRMEADOWS, U.S.A. (I,1514)

RIED, DAVID
THE STRAUSS FAMILY (I,4253)

RIESNER, DEAN
BEN CASEY (I,394); THE MAN FROM GALVESTON (I,2865)

RIFKIN, LEO
HERE COME THE STARS (I,2026); MY FAVORITE MARTIAN (I,3189); THESE ARE THE DAYS (II,2585)

RIKER, DONALD
THE BOB NEWHART SHOW (II,342)

RILEY, JACK
THE DON RICKLES SHOW (I,1341); THE MANY SIDES OF DON RICKLES (I,2898); THE TIM CONWAY SHOW (II,2619)

RILEY, JAMES C
CAPITOL (II,426)

RILEY, LEN
GOOD TIMES (II,1038)

RILEY, NORD
YOU'RE ONLY YOUNG ONCE (I,4969)

RILEY, THOMAS
WHEN THE NIGHTINGALE SANG IN BERKELEY SQUARE (I,4827)

RIMMER, SHANE
CAPTAIN SCARLET AND THE MYSTERONS (I,837)

RING, PAUL
BONANZA (II,347); TARGET: THE CORRUPTERS (I,4360)

RINSLER, DENNIS
THE HALF-HOUR COMEDY HOUR (II,1074); MADAME'S PLACE (II,1587); SUNDAY FUNNIES (II,2493)

RINTELS, DAVID
THE CLIFF DWELLERS (I,974); THE INVADERS (I,2229); WASHINGTON: BEHIND CLOSED DOORS (II,2744)

RIPLEY, ARTHUR
CAVALCADE OF AMERICA (I,875)

RIPPS, LENNIE
13 THIRTEENTH AVENUE (II,2592); BOSOM BUDDIES (II,355); THE CAPTAIN AND TENNILLE (II,429); THE MAC DAVIS SHOW (II,1576); THE REDD FOXX COMEDY HOUR (II,2126); RODNEY DANGERFIELD SPECIAL: IT'S NOT EASY BEIN' ME (II,2199); SMALL AND FRYE (II,2380); VAN DYKE AND COMPANY (II,2719)

RIPS, MARTIN
BABY, I'M BACK! (II,130); THREE'S A CROWD (II,2613); THREE'S COMPANY (II,2614)

RISSIEN, EDWARD L
A SHADOW IN THE STREETS (II,2312)

RITCH, STEVEN
COMBAT (I,1011); TIGHTROPE (I,4500)

RITCHIE, BLAKE
MANNIX (II,1624)

RITTEL, HELEN
THE LORETTA YOUNG THEATER (I,2756)

RITZ, JAMES
GOOD TIMES (II,1038); LOOK WHAT THEY'VE DONE TO MY SONG (II,1517); THE MARILYN MCCOO AND BILLY DAVIS JR. SHOW (II,1630); NEIL SEDAKA STEPPIN' OUT (II,1804); THREE'S COMPANY (II,2614)

RIVERA, JOSE
A.K.A. PABLO (II,28)

RIVERS, JOAN
HUSBANDS AND WIVES (II,1208)

RIVKIN, ALLEN
PAPA SAID NO (I,3452); SAINTS AND SINNERS (I,3887)

ROACH, JULIAN
CORONATION STREET (I,1054)

ROBBINS, DICK
THE B.B. BEEGLE SHOW (II,124); BIG SHAMUS, LITTLE SHAMUS (II,242); THE KROFFT KOMEDY HOUR (II,1411); THE LID'S OFF (I,2678); THE LITTLE RASCALS (II,1494); THE RICHIE RICH SHOW (II,2169)

ROBERTS, A A
THE DESILU PLAYHOUSE (I,1237)

ROBERTS, BEN
CHARLIE'S ANGELS (II,487); MANNIX (II,1624); PURSUE AND DESTROY (I,3699); THE ROGUES (I,3832); THREE FOR DANGER (I,4475)

ROBERTS, ERIC
THE YOUNG AND THE RESTLESS (II,2862)

ROBERTS, JUAREZ
SAFARI (I,3882)

ROBERTS, MARK
EXPOSE (I,1491)

ROBERTS, MEADE
PLAYHOUSE 90 (I,3623)

ROBERTS, PAUL
DEVLIN (II,663)

ROBERTS, RUTH
HEAVENS TO BETSY (I,1997)

ROBERTS, SONYA
THE DESILU PLAYHOUSE (I,1237)

ROBERTS, STANLEY
BANACEK (II,138); KEEPING AN EYE ON DENISE (I,2520); PLAYHOUSE 90 (I,3623)

ROBERTS, WILLIAM
LIFE WITH FATHER (I,2690)

ROBERTSON, DONALD
THUNDERBIRDS (I,4497)

ROBERTSON, HARRY
PRISONERS OF THE LOST UNIVERSE (II,2086)

ROBERTSON, TOM
I'M SOOO UGLY (II,1219); THE TROUBLE WITH MOTHER (II,2663)

ROBIN, ALLEN
THE ERNIE KOVACS SHOW (I,1455)

ROBINS, JOHN
TOP TEN (II,2648)

ROBINSON, BARBARA
THE BEST CHRISTMAS PAGEANT EVER (II,212)

ROBINSON, DAVID
BEWITCHED (I,418); FIRESIDE THEATER (I,1580); FOL-DE-ROL (I,1613)

ROBINSON, JOHN
ALCOA/GOODYEAR THEATER (I,107); GENTRY'S PEOPLE (I,1762); JUSTICE OF THE PEACE (I,2500); MAGGIE MALONE (I,2812); TURN OF FATE (I,4623)

ROBINSON, MATT
THE COSBY SHOW (II,578); KOMEDY TONITE (II,1407); THE WALTONS (II,2740)

ROBINSON, PETER
THE WILD WILD WEST (I,4863)

ROBINSON, PHIL
TRAPPER JOHN, M.D. (II,2654)

ROBINSON, ROBIE
FLATBED ANNIE AND SWEETIEPIE: LADY TRUCKERS (II,876)

ROBINSON, SALLY
THE FAMILY TREE (II,820); KING'S CROSSING (II,1397); LOU GRANT (II,1526)

ROBS, INEZ
THE MARCH OF DIMES FASHION SHOW (I,2901)

ROBSON, MICHAEL
PENMARRIC (II,1993)

ROCK, GAIL
ADDIE AND THE KING OF HEARTS (I,35); THE EASTER PROMISE (I,1397)

RODDENBERRY, GENE
333 MONTGOMERY (I,4487); THE ALCOA HOUR (I,108); ASSIGNMENT: EARTH (I,299); GENERAL ELECTRIC TRUE (I,1754); GENESIS II (I,1760); NAKED CITY (I,3216); PLANET EARTH (I,3615); POLICE STORY (I,3636); STAR TREK (II,2440)

RODDICK, JOHN
SECRET AGENT (I,3962)

RODGERS, BOB
. . .AND DEBBIE MAKES SIX (I,185); THE ABC AFTERSCHOOL SPECIAL (II,1); BATMAN (I,366); THE BING CROSBY SHOW (I,466); CAMP RUNAMUCK (I,811); GILLIGAN'S ISLAND (II,990); THE PARTRIDGE FAMILY (II,1962)

RODGERS, MARK
COVER GIRLS (II,596); EISCHIED (II,763); THE FBI (I,1551); THE GANGSTER CHRONICLES (II,957); JIGSAW (I,2377); MASQUERADE (II,1644); MCCLAIN'S LAW (II,1659); SAVAGE (I,3924); T.J. HOOKER (II,2524); WONDER WOMAN (SERIES 2) (II,2829)

RODGERS, MARY
MUSIC WITH MARY MARTIN (I,3180)

RODMAN, HOWARD
ALCOA PREMIERE (I,109); ALCOA/GOODYEAR THEATER (I,107); ANOTHER DAY, ANOTHER DOLLAR (I,228); THE CITY (I,969); CONSPIRACY OF TERROR (II,569); THE DEVLIN CONNECTION (II,664); EXO-MAN (II,797); HARRY O (II,1098); HARRY O (II,1099); HOW'S BUSINESS? (I,2153); THE MAN FROM U.N.C.L.E. (I,2867); MATINEE THEATER (I,2947); MR. AND MRS. COP (II,1748); NAKED CITY (I,3216); ROUTE 66 (I,3852); RYAN'S FOUR (II,2233); SMILE JENNY, YOU'RE DEAD (II,2382); STUDIO ONE (I,4268); THRILLER (I,4492); TWO FOR THE MONEY (I,4655)

ROECA, SAM
ANNIE OAKLEY (I,225); HAWAII FIVE-O (II,1110); THE OUTER LIMITS (I,3426); RAWHIDE (I,3727); SABU AND THE MAGIC RING (I,3881); VOLTRON—DEFENDER OF THE UNIVERSE (II,2729)

ROGER, AGNES
THE FORD 50TH ANNIVERSARY SHOW (I,1630)

ROGERS, AL
BERT CONVY SPECIAL—THERE'S A MEETING HERE TONIGHT (II,210); THE CARPENTERS (II,444); A COUNTRY HAPPENING (I,1065); DEAN MARTIN PRESENTS THE GOLDDIGGERS (I,1193); DINAH IN SEARCH OF THE IDEAL MAN (I,1278); DOM DELUISE AND FRIENDS (II,701); THE DOROTHY HAMILL WINTER CARNIVAL SPECIAL (II,720); FATHER O FATHER (II,842); HARPER VALLEY, U.S.A. (I,1954); THE JERRY REED WHEN YOU'RE HOT, YOU'RE HOT HOUR (I,2372); THE JIM STAFFORD SHOW (II,1291); THE JOHN BYNER COMEDY HOUR (I,2407); JOHN DENVER ROCKY MOUNTAIN CHRISTMAS (II,1314); THE JOHN DENVER SPECIAL (II,1315); THE JOHN DENVER SPECIAL (II,1316); JOHN DENVER—THANK GOD I'M A COUNTRY BOY (II,1317); THE ROWAN AND MARTIN SHOW (I,3854)

ROGERS, JIM
THE DUKES OF HAZZARD (II,742); ENOS (II,779); GOOD TIMES (II,1038); THE LOVE BOAT (II,1535); ON OUR OWN (II,1892); THREE'S COMPANY (II,2614); TRAPPER JOHN, M.D. (II,2654)

ROGERS, ROSWELL
FATHER KNOWS BEST (II,838); THE HERO (I,2035)

ROGERS, SHORTY
B.A.D. CATS (II,123)

ROGERS, WAYNE
HOUSE CALLS (II,1194)

ROGOSIN, PAUL
KNIGHT RIDER (II,1402)

ROLEY, SUTTON
PARTNERS IN CRIME (II,1961)

ROLFE, SAM H
THE DELPHI BUREAU (I,1224); ENIGMA (II,778); JOHNNY NIGHTHAWK (I,2432); THE MAN FROM U.N.C.L.E. (I,2867); MANHUNTER (II,1620); MANHUNTER (II,1621); MATT HELM (II,1652); ROSETTI AND RYAN: MEN WHO LOVE WOMEN (II,2217); THEY CALL IT MURDER (I,4441); THE TWILIGHT ZONE (I,4651)

ROMAN, LARRY
TERRY AND THE PIRATES (I,4405)

ROMAN, MAURY
THE SUMMER SMOTHERS BROTHERS SHOW (I,4281)

ROMERO, GEORGE A
TALES FROM THE DARKSIDE (II,2534)

ROMINS, CHARLES
THE EDDIE ALBERT SHOW (I,1406)

ROONEY, ANDY
SINATRA (I,4053)

ROOS, AUDREY
THE CAT AND THE CANARY (I,868)

ROOS, DONALD PAUL
PAPER DOLLS (II,1952)

ROOS, KELLY
THE BURNING COURT (I,766)

ROOS, WILLIAM
THE CAT AND THE CANARY (I,868)

ROOT, DAVID BOEHMWELLS
FIRESIDE THEATER (I,1580)

ROOT, WELLS
COMBAT (I,1011); THE ROGUES (I,3832)

ROPER, CAROL
KNOTS LANDING (II,1404)

ROPES, BRADFORD
THE HUNTER (I,2162)

ROSATO, TONY
SECOND CITY TELEVISION (II,2287)

ROSE, DONALD
DIFF'RENT STROKES (II,674)

ROSE, JERRY
ARCHIE BUNKER'S PLACE (II,105); THE GOLDDIGGERS IN LONDON (I,1837)

ROSE, LARRY
JENNIFER SLEPT HERE (II,1278)

ROSE, MICKEY
ACE CRAWFORD, PRIVATE EYE (II,6); ARCHIE (II,104); CHANGING SCENE (I,899); CHARLIE'S ANGELS (II,486); COMEDY NEWS II (I,1021); THE FUNNY SIDE (I,1715); GOOD TIME HARRY (II,1037); IT ONLY HURTS WHEN YOU LAUGH (II,1251); THE LOVE BOAT (II,1535); THE SMOTHERS BROTHERS SHOW (II,2384); THE STOCKARD CHANNING SHOW (II,2468); VAN DYKE AND COMPANY (II,2719); THE WOODY ALLEN SPECIAL (I,4904)

ROSE, REGINALD
THE ALCOA HOUR (I,108); THE DEFENDER (I,1219); THE FOUR OF US (II,914); PLAYHOUSE 90 (I,3623); STUDIO ONE (I,4268); STUDS LONIGAN (II,2484); THE TWILIGHT ZONE (I,4651)

ROSE, SI
BACHELOR FATHER (I,323); BRINGING UP BUDDY (I,731); THE DUKES OF HAZZARD (II,742); FRANCES LANGFORD PRESENTS (I,1656); THE HANK MCCUNE SHOW (I,1932); HOW TO MARRY A MILLIONAIRE (I,2147); OPERATION PETTICOAT (II,1917); SIGMUND AND THE SEA MONSTERS (II,2352)

ROSEBROOK, JEB
THE YELLOW ROSE (II,2847)

ROSEN, ALAN
ARCHIE BUNKER'S PLACE (II,105); DIFF'RENT STROKES (II,674); SZYSZNYK (II,2523)

ROSEN, ARNE
THE ALAN KING SHOW (I,100); ALFRED OF THE AMAZON (I,116); THE BACHELOR (I,325); DON RICKLES—ALIVE AND KICKING (I,1340); THE GARRY MOORE SHOW (I,1739); HAPPY ANNIVERSARY AND GOODBYE (II,1082); HOTEL 90 (I,2130); THE JEAN CARROLL SHOW (I,2353); NBC FOLLIES OF 1965 (I,3240); OF THEE I SING (I,3334); THE PHIL SILVERS PONTIAC SPECIAL: KEEP IN STEP (I,3579); SNAFU (II,2386); THE TIM CONWAY SPECIAL (I,4503)

ROSEN, BURT
THE BOBBY SHERMAN SPECIAL (I,676)

ROSEN, DICK
LOOK WHAT THEY'VE DONE TO MY SONG (II,1517)

ROSEN, DONALD
THE BIG BAND AND ALL THAT JAZZ (I,422)

ROSEN, KENNETH
NAKED CITY (I,3216)

ROSEN, LARRY
ALONE AT LAST (II,60); MR. MERLIN (II,1751); PALS (II,1947); TEACHERS ONLY (II,2548)

ROSEN, MILT
...AND DEBBIE MAKES SIX (I,185); THE ANDY WILLIAMS CHRISTMAS SHOW (I,198); THE ANDY WILLIAMS SPECIAL (I,213); CAROL CHANNING AND 101 MEN (I,849); CHIPS (II,511); THE COLGATE COMEDY HOUR (I,998); DANNY THOMAS GOES COUNTRY AND WESTERN (I,1146); DANNY THOMAS: AMERICA I LOVE YOU (I,1145); DEAN MARTIN'S CELEBRITY ROAST (II,643); E/R (II,748); ENOS (II,779); ESTHER WILLIAMS AT CYPRESS GARDENS (I,1465); THE FLYING NUN (I,1611); HERE WE GO AGAIN (I,2028); JIMMY DURANTE MEETS THE LIVELY ARTS (I,2386); KATE MCSHANE (II,1368); LEWIS AND CLARK (II,1465); THE NBC FOLLIES (I,3241); PLEASE DON'T EAT THE DAISIES (I,3628); THE RED BUTTONS SHOW (I,3750); TALES OF THE GOLD MONKEY (II,2537); TOO CLOSE FOR COMFORT (II,2642); THE WALTER WINCHELL SHOW (I,4760)

ROSEN, NEIL
ARCHIE (II,104); DINAH AND HER NEW BEST FRIENDS (II,676); JOANIE LOVES CHACHI (II,1295); THE MAC DAVIS SHOW (II,1576); ME AND MAXX (II,1672); OFF THE WALL (II,1879); PUMPBOYS AND DINETTES ON TELEVISION (II,2090); THE ROLLERGIRLS (II,2203); SHIPSHAPE (II,2329); SUGAR TIME (II,2489); TONY ORLANDO AND DAWN (II,2639); WELCOME BACK, KOTTER (II,2761)

ROSEN, STU
DUSTY'S TREEHOUSE (I,1391)

ROSEN, SY
BABY MAKES FIVE (II,129); BEST OF THE WEST (II,218); THE BOB NEWHART SHOW (II,342); THE BOOK OF LISTS (II,350); FREE COUNTRY (II,921); GIMME A BREAK (II,995); HANGING IN (II,1078); MAUDE (II,1655); NOT IN FRONT OF THE KIDS (II,1861); RHODA (II,2151); SHE'S WITH ME (II,2321); SPENCER (II,2420); SUTTERS BAY (II,2508); THE TONY RANDALL SHOW (II,2640); WE'VE GOT EACH OTHER (II,2757)

ROSENBAUM, HENRY
DOG AND CAT (II,695); LONE STAR (II,1513)

ROSENBERG, FRED
LAUGH TRAX (II,1443)

ROSENBERG, PHILIP
BAKER'S DOZEN (II,136); NURSE (II,1870)

ROSENBERG, STUART
BAKER'S DOZEN (II,136)

ROSENBROOK, JEB
MIRACLE ON 34TH STREET (I,3058); TROUBLE IN HIGH TIMBER COUNTRY (II,2661); THE WALTONS (II,2740); THE YEAGERS (II,2843)

ROSENFARB, BOB
CUTTER TO HOUSTON (II,612); THE GET ALONG GANG (II,980)

ROSENFELD, MAX
KRAFT TELEVISION THEATER (I,2592)

ROSENSTOCK, RICHARD
LAVERNE AND SHIRLEY (II,1446); MORK AND MINDY (II,1735); OH MADELINE (II,1880); THE TED KNIGHT SHOW (II,2550)

ROSENTHAL, MARK
CASSIE AND COMPANY (II,455)

ROSENZWEIG, BARNEY
CAGNEY AND LACEY (II,409)

ROSIN, CHARLES
BREAKING AWAY (II,371); ST. ELSEWHERE (II,2432)

ROSIN, JAMES
QUINCY, M. E. (II,2102)

ROSNER, RICK
JUST MEN (II,1359)

ROSOFF, LOU
TERRY AND THE PIRATES (I,4405)

ROSS, ARTHUR
ALFRED HITCHCOCK PRESENTS (I,115); CROWN THEATER WITH GLORIA SWANSON (I,1107)

ROSS, BOB
THE AMOS AND ANDY SHOW (I,179); THE ANDY GRIFFITH SHOW (I,192); ICHABOD AND ME (I,2183); LEAVE IT TO BEAVER (I,2648)

ROSS, DAVID
SPECIAL LONDON BRIDGE SPECIAL (I,4150)

ROSS, DIANA
DIANA (II,667)

ROSS, DONALD
ALL-STAR SWING FESTIVAL (II,53); DINAH AND FRIENDS (II,675); DINAH! (II,677); GAVILAN (II,959); HARPER VALLEY (II,1094); HART TO HART (II,1102); HOUSE CALLS (II,1194); THE LOVE BOAT (II,1535)

ROSS, JEROME
THE ALCOA HOUR (I,108); APPOINTMENT WITH ADVENTURE (I,246); MATINEE THEATER (I,2947); NAKED CITY (I,3216); SOLDIER IN LOVE (I,4108); THE UNTOUCHABLES (I,4682)

ROSS, JERRY
BARNEY MILLER (II,154); THE COSBY SHOW (II,578); GIMME A BREAK (II,995); HARPER VALLEY PTA (II,1095); LEWIS AND CLARK (II,1465); MARCUS WELBY, M.D. (II,1627); THE NEW ODD COUPLE (II,1825); THE OUTER LIMITS (I,3426); THE PAPER CHASE (II,1949); SANFORD AND SON (II,2256); THAT'S MY MAMA (II,2580); WELCOME BACK, KOTTER (II,2761); WHAT'S HAPPENING!! (II,2769); YOUNG MAVERICK (II,2867)

ROSS, JIM
MAMA'S FAMILY (II,1610)

ROSS, JOEL
MATINEE THEATER (I,2947)

ROSS, MARC
TRAUMA CENTER (II,2655)

ROSS, MARTIN
THE BAXTERS (II,184)

ROSS, MAX
STEVE MARTIN'S BEST SHOW EVER (II,2460)

ROSS, MICHAEL
THE DUMPLINGS (II,743); THREE'S A CROWD (II,2613); THREE'S COMPANY (II,2614)

ROSS, MIKE LLOYD
THE AMERICAN GIRLS (II,72); SALVAGE (II,2244)

ROSS, SAM
MR. LUCKY (I,3141); NAKED CITY (I,3215); THE SIX-MILLION-DOLLAR MAN (II,2372)

ROSS, SHELLEY
PERSONAL AND CONFIDENTIAL (II,2026); SHOW BUSINESS (II,2343)

ROSS, SHIBOHN
THE BILLY CRYSTAL COMEDY HOUR (II,248)

ROSS, STANLEY RALPH
BANACEK (II,138); BATMAN (I,366); BEST FRIENDS (II,213); COLUMBO (II,556); FLO'S PLACE (II,892); THE FURST FAMILY OF WASHINGTON (I,1718); LOCAL 306 (II,1505); THE MAN FROM U.N.C.L.E. (I,2867); MITZI...THE FIRST TIME (I,3073); MRS R.—DEATH AMONG FRIENDS (II,1759); THE NEW BILL COSBY SHOW (I,3257); ON STAGE AMERICA (II,1893); THE TEXAS RANGERS (II,2567); WONDER WOMAN (PILOT 2) (II,2827)

ROSS-LEMING, EUGENIE
FIT FOR A KING (II,866); HIGHCLIFFE MANOR (II,1150); SCARECROW AND MRS. KING (II,2268); SECOND CITY TELEVISION (II,2287); WHY US? (II,2796)

ROSTEN, IRWIN
RIPLEY'S BELIEVE IT OR NOT (II,2177)

ROSWELL, MAGGIE
THE TIM CONWAY SHOW (II,2619)

ROTH, LYNN
HONEYMOON SUITE (I,2108); JOE AND SONS (II,1296); A LUCILLE BALL SPECIAL: WHAT NOW CATHERINE CURTIS? (II,1560); THAT WAS THE YEAR THAT WAS (II,2575); THIS BETTER BE IT (II,2595)

ROTH, MARTIN
BEYOND WESTWORLD (II,227); THE DUKES OF HAZZARD (II,742); HAWAII FIVE-O (II,1110); MATT HELM (II,1653); MEDICAL CENTER (II,1675); THREE'S COMPANY (II,2614)

ROTH, MARTY
AS CAESAR SEES IT (I,294); GLENN MILLER TIME (I,1824); MCHALE'S NAVY (I,2969); MY FAVORITE MARTIAN (I,3189)

ROTH, MONTY
I DREAM OF JEANNIE (I,2167)

ROTH, PAULA A
JOANIE LOVES CHACHI (II,1295); LAVERNE AND SHIRLEY (II,1446)

ROTH, SIMON
THE DUKES OF HAZZARD (II,742)

ROTHMAN, BERNARD
20TH CENTURY FOLLIES (I,4641); AMERICA, YOU'RE ON (II,69); BURT AND THE GIRLS (I,770); BURT REYNOLDS' LATE SHOW (I,777); A COUNTRY HAPPENING (I,1065); THE DEAN MARTIN SHOW (I,1201); DIANA (I,1258); FROM A BIRD'S EYE VIEW (I,1686); G.I.'S (II,949); GIRL FRIENDS AND NABORS (I,1807); HALF THE GEORGE KIRBY COMEDY HOUR (I,1920); HARPER VALLEY, U.S.A. (I,1954); THE MANHATTAN TRANSFER (II,1619); PAT BOONE AND FAMILY (II,1968); SAGA OF SONORA (I,3884); SANDY IN DISNEYLAND (II,2254); THE SPECIAL GENTRY ONE (I,4148)

ROTHMAN, MARK
BUSTING LOOSE (II,402); HAPPY DAYS (II,1084); IT'S A LIVING (II,1254); LAVERNE AND SHIRLEY (II,1446); THE LOVEBIRDS (II,1548); THE NEW ODD COUPLE (II,1825); THE ODD COUPLE (II,1875); PAUL SAND IN FRIENDS AND LOVERS (II,1982); THE RITA MORENO SHOW (II,2181); THE TED KNIGHT SHOW (II,2550); WALKIN' WALTER (II,2737)

ROUSSEAU, WILLIAM
CHICAGO 212 (I,935)

ROWE, ARTHUR
THE BIG VALLEY (I,437); THE BIONIC WOMAN (II,255); FANTASY ISLAND (II,829); THE NIGHT STALKER (II,1849); SWITCH (II,2519)

ROWE, NICHOLAS
BRANDED (I,707)

ROWLAND, MELODY
GLORIA (II,1010)

ROWLEY, LAURIE
NOT THE NINE O'CLOCK NEWS (II,1864)

ROYCE, BILL
PEOPLE TO PEOPLE (II,2000)

RUBEN, AARON
THE ANDY GRIFFITH SHOW (I,192); ANDY GRIFFITH'S UPTOWN-DOWNTOWN SHOW (I,194); THE ANDY GRIFFITH—DON KNOTTS—JIMNABORS SHOW (I,191); CHARO AND THE SERGEANT (II,489); THE DON KNOTTS SPECIAL (I,1335); GRANDPA MAX (II,1051); GREAT DAY (II,1053); JACKIE AND DARLENE (II,1265); THE MILTON BERLE SHOW (I,3047); PIPER'S PETS (II,2046); SANFORD AND SON (II,2256); THE STOCKARD CHANNING SHOW (II,2468); TEACHERS ONLY (II,2547); TEXACO STAR THEATER (I,4411); TROUBLE WITH RICHARD (I,4612); WENDY HOOPER—U. S. ARMY (II,2764)

RUBEN, ALBERT
THE BLUE KNIGHT (II,277); CORONET BLUE (I,1055); THE DESILU PLAYHOUSE (I,1237); KOJAK (II,1406); N.Y.P.D. (I,3321); ONE-EYED JACKS ARE WILD (I,3374); SHANNON (II,2314); THE STREETS OF SAN FRANCISCO (II,2478)

RUBEN, GLADYS
SANDY DREAMS (I,3909)

RUBIN, FRED
ARCHIE BUNKER'S PLACE (II,105); MAMA'S FAMILY (II,1610); SZYSZNYK (II,2523)

RUBIN, MANN
ALFRED HITCHCOCK PRESENTS (I,115); APPOINTMENT WITH ADVENTURE (I,246); BARNABY JONES (II,153); CLIMAX! (I,976); DYNASTY (II,746); FUTURE COP (II,945); LAND OF THE GIANTS (I,2611); LUCAN (II,1554); MANNIX (II,1624); THE RETURN OF CAPTAIN NEMO (II,2135); THE ROCKFORD FILES (II,2197); THE ROOKIES (II,2208); TODAY'S FBI (II,2628); WARNING SHOT (I,4770)

RUBIN, MARC
FAME (II,812); TUCKER'S WITCH (II,2667)

RUBIN, MICHAEL
TO SAY THE LEAST (II,2623)

RUBIN, RONALD
THE BANANA COMPANY (II,139); PALMERSTOWN U.S.A. (II,1945); TO SIR, WITH LOVE (II,2624)

RUBIN, STANLEY
STORY THEATER (I,4240)

RUBINOWITZ, BARRY
BUSTING LOOSE (II,402); THE TED KNIGHT SHOW (II,2550)

RUBINSTEIN, SCOTT
9 TO 5 (II,1852); HOUSE CALLS (II,1194)

RUBINTON, ERIC
TEXAS (II,2566)

RUBY, CLIFF
CODE RED (II,552); LOBO (II,1504)

RUBY, JOE
PEBBLES AND BAMM BAMM (II,1987)

RUDDY, JOHN
SEEING THINGS (II,2297)

RUDLEY, SARETT
THE DETECTIVE (II,662)

RUEGGER, TOM
THE DUKES (II,741); SPORT BILLY (II,2429)

RUGGIERO JR, ALFONSE
MIAMI VICE (II,1689)

RUSK, JIM
LILY (I,2705); THE LILY TOMLIN SHOW (I,2706); THE LILY TOMLIN SPECIAL (II,1479)

RUSO, R A
FROM HERE TO ETERNITY (II,933)

RUSSELL, A J
ART CARNEY MEETS PETER AND THE WOLF (I,272); ART CARNEY MEETS THE SORCERER'S APPRENTICE (I,273); THE BIG SELL (I,431); THE FABULOUS 50S (I,1501); THE GERSHWIN YEARS (I,1779); HOLIDAY U.S.A. (I,2078); THE HONEYMOONERS (I,2110); THE HONEYMOONERS (II,1175); LIFE WITH FATHER (I,2690); MERMAN ON BROADWAY (I,3011); THE MILLION DOLLAR INCIDENT (I,3045); ONCE UPON A CHRISTMAS TIME (I,3368)

RUSSELL, ANDY
THE JACKIE GLEASON SHOW (I,2322)

RUSSELL, GORDON
DARK SHADOWS (I,1157)

RUSSELL, JERRY
WOMEN WHO RATE A "10" (II,2825)

RUSSELL, PAMELA
THE MARY TYLER MOORE SHOW (II,1640)

RUSSELL, ROY
THE ONEDIN LINE (II,1912); THE SAINT (II,2237)

RUSSELL, W J
PLAYHOUSE 90 (I,3623)

RUSSNOW, MICHAEL
BARNEY MILLER (II,154); EMERALD POINT, N.A.S. (II,773); FAMILY TIES (II,819); THE HAMPTONS (II,1076); JUST OUR LUCK (II,1360); STRUCK BY LIGHTNING (II,2482); THE WALTONS (II,2740)

RYAN, ELAINE
LITTLE WOMEN (I,2721)

RYAN, JIM
LASSIE'S RESCUE RANGERS (II,1432); MISSION MAGIC (I,3068); MY FAVORITE MARTIANS (II,1779); SHAZAM! (II,2319)

RYAN, JOHN
MR. NOVAK (I,3145)

RYAN, PAMELA
CHIPS (II,511)

RYAN, TERRY
THE ANN SOTHERN SHOW (I,220); DOBIE GILLIS (I,1302); THE PHIL SILVERS PONTIAC SPECIAL: KEEP IN STEP (I,3579); THE PHIL SILVERS SHOW (I,3580); ROWAN AND MARTIN BITE THE HAND THAT FEEDS THEM (I,3853)

RYDER, EDDIE
THE DON RICKLES SHOW (I,1341)

RYERSON, ANN
THE YESTERDAY SHOW (II,2850)

RYF, ROBERT
DANGEROUS ASSIGNMENT (I,1138)

RYTON, ROYCE
CROWN MATRIMONIAL (I,1106)

SABATINO, TONY
CIRCUS OF THE STARS (II,538)

SACKHEIM, JERRY
THE NEW ADVENTURES OF CHARLIE CHAN (I,3249)

SACKHEIM, WILLIAM
DELVECCHIO (II,659); POOR MR. CAMPBELL (I,3647)

SACKS, ALAN
THE ROCK RAINBOW (II,2194)

SACKS, DOROTHY
SHOW BIZ (I,4027)

SAFFRON, JOHN
ALL MY CHILDREN (II,39); AS THE WORLD TURNS (II,110)

SAGAN, CARL
COSMOS (II,579)

SAGE, LIZ
THE CAROL BURNETT SHOW (II,443); DOROTHY (II,717); THE KELLY MONTEITH SHOW (II,1376); THE LOVE BOAT III (II,1534); MAMA'S FAMILY (II,1610); PUNKY BREWSTER (II,2092); RODNEY DANGERFIELD SHOW: I CAN'T TAKE IT NO MORE (II,2198)

SAIDY, FRED
BEST FOOT FORWARD (I,401)

SAILOR, CHARLES
BRONK (II,379); THE BUREAU (II,397); CHARLIE'S ANGELS (II,486); GET CHRISTIE LOVE! (II,981); THE ROCKFORD FILES (II,2197); SWITCH (II,2519)

SAINO, ART
MARCUS WELBY, M.D. (II,1627)

SAKETT, NANCY
GLITTER (II,1009)

SAKS, SOL
AN APARTMENT IN ROME (I,242); HEAVEN HELP US (I,1995); OUT OF THE BLUE (I,3422); THE SLIGHTLY FALLEN ANGEL (I,4087)

SALAMAN, LOUIS
NAKED CITY (I,3215)

SALAMON, OTTO
INSTITUTE FOR REVENGE (II,1239)

SALE, RICHARD
MR. BELVEDERE (I,3127)

SALE, VIRGINIA
WREN'S NEST (I,4922)

SALKOWITZ, SY
ALIAS SMITH AND JONES (I,118); JESSIE (II,1287); MCCLOUD (II,1660); NAKED CITY (I,3216); NAKIA (II,1784); PERRY MASON (II,2025); RAWHIDE (I,3727); READY FOR THE PEOPLE (I,3739); TODAY'S FBI (II,2628); THE UNTOUCHABLES (I,4682)

SALOMON, HENRY
VICTORY AT SEA (I,4730)

SALTZMAN, PHILIP
ALCOA/GOODYEAR THEATER (I,107); CROSSFIRE (II,606); THE FUGITIVE (I,1701); INTERTECT (I,2228); TURN OF FATE (I,4623); THE WILD WILD WEST (I,4863)

SAMUELS, JEFF
LIFESTYLES OF THE RICH AND FAMOUS (II,1474)

SAND, BARRY
COMEDY TONIGHT (I,1027); THE FIGHTING NIGHTINGALES (II,855); KLEIN TIME (II,1401)

SAND, BOB
100 CENTRE STREET (II,1901); THE CAPTAIN AND TENNILLE (II,429); LAVERNE AND SHIRLEY (II,1446); MADAME'S PLACE (II,1587); MCNAMARA'S BAND (II,1669); THE NEW TREASURE HUNT (II,1831); SUNDAY FUNNIES (II,2493); THANKSGIVING REUNION WITH THE PARTRIDGE FAMILY AND MY THREE SONS (II,2569); THAT SECOND THING ON ABC (II,2572)

SANDBURG, DON
SHIELDS AND YARNELL (II,2328)

SANDEFER, DALTON
THE ALL-NEW POPEYE HOUR (II,49); WHEELIE AND THE CHOPPER BUNCH (II,2776)

SANDEFUR, B W
THE AMAZING SPIDER-MAN (II,63); BEYOND WITCH MOUNTAIN (II,228); BRING 'EM BACK ALIVE (II,377); CHARLIE'S ANGELS (II,486); CODE RED (II,552); ENOS (II,779); LITTLE HOUSE ON THE PRAIRIE (II,1487); LITTLE HOUSE: A NEW BEGINNING (II,1488); A MAN CALLED SLOANE (II,1613); MICKEY SPILLANE'S MIKE HAMMER (II,1692); VOYAGERS (II,2730)

SANDERS, RICHARD
WKRP IN CINCINNATI (II,2814)

SANDERS, TERRY
THE LEGEND OF MARILYN MONROE (I,2654)

SANDLER, ELLEN
AFTER GEORGE (II,19); EMPIRE (II,777); KATE AND ALLIE (II,1365); OPEN ALL NIGHT (II,1914)

SANDLER, JESSE
ROUTE 66 (I,3852)

SANDOR, ANNA
SEEING THINGS (II,2297)

SANDOZ, HENRY
SHERLOCK HOLMES (I,4010)

SANDRICH, JAY
THE GOOD GUYS (I,1847)

SANDS, LANA
HART TO HART (II,1102)

SANDY, JOE
THE MAN FROM U.N.C.L.E. (I,2867)

SANFORD, ARLENE
ON OUR OWN (II,1892)

SANFORD, CHARLES
FEDERAL AGENT (I,1557)

SANFORD, DONALD S
BONANZA (II,347); THE FBI (I,1551); THE OUTER LIMITS (I,3426); PERRY MASON (II,2025); WAGON TRAIN (I,4747)

SANFORD, DOUGLAS
THRILLER (I,4492)

SANFORD, GERALD
BARNABY JONES (II,153); CANNON (II,424); CHIPS (II,511); ENOS (II,779); THE FBI (I,1551); A MAN CALLED

SLOANE (II,1613); NERO WOLFE (II,1806); THE STRANGER (I,4251); T.J. HOOKER (II,2524)

SANGSTER, JIMMY
BANACEK (II,138); THE BILLION DOLLAR THREAT (II,246); CONCRETE COWBOYS (II,563); THE CONCRETE COWBOYS (II,564); THE COUNTRY MUSIC MURDERS (II,590); EBONY, IVORY AND JADE (II,755); A MAN CALLED SLOANE (II,1613); MCCLOUD (II,1660); THE NIGHT STALKER (II,1849); ONCE UPON A SPY (II,1898); RIPLEY'S BELIEVE IT OR NOT (II,2177); WONDER WOMAN (SERIES 1) (II,2828)

SANGSTER, MARGARET
MY TRUE STORY (I,3206)

SANS, NANCY
SESAME STREET (I,3982)

SANTICERO, LUIS
SESAME STREET (I,3982)

SANTILLO, CINDY
THE TED KNIGHT SHOW (II,2550)

SAPINSLEY, ALVIN
THE ALCOA HOUR (I,108); CANNON (II,424); CODE NAME: HERACLITUS (I,990); THE FRONT PAGE (I,1691); HAWAII FIVE-O (II,1110); THE HITCHHIKER (II,1157); THE MAN FROM U.N.C.L.E. (I,2867); NICK AND NORA (II,1842); ROBERT MONTGOMERY PRESENTS YOUR LUCKY STRIKE THEATER (I,3809); ROGER AND HARRY (II,2200); SPENCER'S PILOTS (II,2421); THE UNTOUCHABLES (I,4682)

SARACENO, CAROL
HART TO HART (II,1102)

SARASOHN, CAROL
THE BEST OF TIMES (II,219)

SARASOHN, LANE
THE BEST OF TIMES (II,219); THE BOOK OF LISTS (II,350); THE FACTS (II,804); I'VE HAD IT UP TO HERE (II,1221)

SARDY, FRED
BABES IN TOYLAND (I,318)

SARECKY, BARNEY
TERRY AND THE PIRATES (I,4405)

SARGENT, ALVIN
ALFRED HITCHCOCK PRESENTS (I,115); BEN CASEY (I,394); CORONET BLUE (I,1055); NAKED CITY (I,3216); ROUTE 66 (I,3852); THE SEARCH (I,3950)

SARGENT, HERB
ALAN KING LOOKS BACK IN ANGER—A REVIEW OF 1972 (I,99); THE ALAN KING SPECIAL (I,103); ANNIE, THE WOMAN IN THE LIFE OF A MAN (I,226); THE BING CROSBY SHOW (I,463); THE BING CROSBY SHOW (I,464); BURT BACHARACH! (I,771); BURT BACHARACH: CLOSE TO YOU (I,772); FUNNY GIRL TO FUNNY LADY (II,942); THE GEORGE SEGAL SHOW (II,977); HONEY WEST (I,2105); LILY (I,2705); THE MAGICAL MUSIC OF BURT BACHARACH (I,2822); THE MANY FACES OF COMEDY (I,2894); THE MILTON BERLE SPECIAL (I,3050); THE NEW STEVE ALLEN SHOW (I,3271); THE NEWS IS THE NEWS (II,1838); PETULA (I,3573); THE STEVE ALLEN SHOW (I,4219); THE STEVE ALLEN SHOW (I,4220); THE WONDERFUL WORLD OF AGGRAVATION (I,4894)

SAROYAN, WILLIAM
THE HUMAN COMEDY (I,2158); SCREEN DIRECTOR'S PLAYHOUSE (I,3946)

SAUDER, PETER
THE EDISON TWINS (II,761)

SAUNDERS, GLORIA
MACREEDY'S WOMAN (I,2804)

SAUNDERS, TOM
NO SOAP, RADIO (II,1856)

SAVAGE, PAUL
THE CHISHOLMS (II,513); THE DESILU PLAYHOUSE (I,1237); JESSIE (II,1287); THE NEW DAUGHTERS OF JOSHUA CABE (I,3261); THE STREETS OF SAN FRANCISCO (II,2478); T.J. HOOKER (II,2524); TODAY'S FBI (II,2628); WAGON TRAIN (I,4747); WALKING TALL (II,2738); THE WALTONS (II,2740); THE YELLOW ROSE (II,2847)

SAVAGE, PHILIP
THE DAUGHTERS OF JOSHUA CABE (I,1169)

SAVETH, PHIL
THE PAUL ANKA SHOW (II,1972)

SAWYER, TOM
B.J. AND THE BEAR (II,126); BRING 'EM BACK ALIVE (II,377); CALIFORNIA FEVER (II,411); THE CARPENTERS. . . SPACE ENCOUNTERS (II,447); GAVILAN (II,959); HOW TO SURVIVE THE 70S AND MAYBE EVEN BUMP INTO HAPPINESS (II,1199); THE MASTER (II,1648); PAT BOONE AND FAMILY (II,1968); SCARECROW AND MRS. KING (II,2268); WHIZ KIDS (II,2790); WONDER WOMAN (SERIES 2) (II,2829)

SCHAEFER, GEORGE
ALCOA PREMIERE (I,109)

SCHAEFER, JACK
THEY WENT THATAWAY (I,4443)

SCHAEFER, ROBERT
THE ADVENTURES OF CHAMPION (I,54); ANNIE OAKLEY (I,225); BUFFALO BILL JR. (I,755); THE GENE AUTRY SHOW (I,1748); THE RANGE RIDER (I,3720)

SCHAETZLE, BUD
KIDD VIDEO (II,1388)

SCHARFMAN, MORT
CARTER COUNTRY (II,449); CHER (II,495); DINAH AND HER NEW BEST FRIENDS (II,676); DONNY AND MARIE (II,712); THE MANY FACES OF COMEDY (I,2894); PERRY COMO'S SUMMER OF '74 (II,2024); THE WOLFMAN JACK SHOW (II,2817)

SCHARLACH, ED
DEAN MARTIN PRESENTS THE GOLDDIGGERS (I,1193); THE DEAN MARTIN SHOW (I,1201); THE GHOST AND MRS. MUIR (I,1786); LEWIS AND CLARK (II,1465); THE MARILYN MCCOO AND BILLY DAVIS JR. SHOW (II,1630); MORK AND MINDY (II,1735); NEIL SEDAKA STEPPIN' OUT (II,1804); THE PETER MARSHALL VARIETY SHOW (II,2029); THAT'S MY MAMA (II,2580)

SCHARY, DORY
SOUND OF THE 60S (I,4134)

SCHAUER, MITCHELL
PAC-MAN (II,1940)

SCHEFF, MICHAEL
DARKROOM (II,619); HART TO HART (II,1102)

SCHENCK, GEORGE
BRANDED (I,707); BRING 'EM BACK ALIVE (II,377); CRAZY LIKE A FOX (II,601); HARD KNOCKS (II,1086); SAWYER AND FINN (II,2265); THE WILD WILD WEST (I,4863)

SCHIBI, MARGARET
BEHIND THE SCREEN (II,200); FLAMINGO ROAD (II,872)

SCHICKEL, RICHARD
THE CORAL JUNGLE (II,574)

SCHIFF, MICHAEL
TOPPER (II,2649)

SCHILITT, ROBERT
LOU GRANT (II,1526)

SCHILLER, BOB
ALL IN THE FAMILY (II,38); THE ANN SOTHERN SHOW (I,220); ARCHIE BUNKER'S PLACE (II,105); THE BEAUTIFUL PHYLLIS DILLER SHOW (I,381); BUT MOTHER! (II,403); CHECKING IN (II,492); THE DESILU PLAYHOUSE (I,1237); THE DESILU REVUE (I,1238); THE FLIP WILSON SHOW (I,1602); LIVING IN PARADISE (II,1503); THE LUCILLE BALL-DESI ARNAZ SHOW (I,2784); MAUDE (II,1655); SANFORD (II,2255); SIDE BY SIDE (II,2346); W*A*L*T*E*R (II,2731); YOU'RE ONLY YOUNG TWICE (I,4970)

SCHILLER, FRED
THE VEIL (I,4709)

SCHILLER, LYDIA
THE LORETTA YOUNG THEATER (I,2756)

SCHILLER, WILTON
THE ADVENTURES OF SUPERMAN (I,77); BEN CASEY (I,394); CAPTAIN AMERICA (II,428); DOCTOR SIMON LOCKE (I,1317); THE FUGITIVE (I,1701); JOHNNY RINGO (I,2434)

SCHIMMEL, JOHN
PUMPBOYS AND DINETTES ON TELEVISION (II,2090)

SCHISGAL, MURRAY
LOVE, LIFE, LIBERTY & LUNCH (II,1544)

SCHLATTER, GEORGE
CHER (II,495); DORIS DAY TODAY (II,716); THE FABULOUS FUNNIES (II,801); JOHN DENVER AND FRIEND (II,1311); JOHN DENVER AND THE LADIES (II,1312); REAL KIDS (II,2116); REAL PEOPLE (II,2118); ROWAN AND MARTIN'S LAUGH-IN (I,3857); SALUTE TO LADY LIBERTY (II,2243); THE SHAPE OF THINGS (II,2315); SPEAK UP AMERICA (II,2412)

SCHLITT, ROBERT
CHICAGO STORY (II,506); EMERALD POINT, N.A.S. (II,773); KUNG FU (II,1416); Q. E. D. (II,2094)

SCHMERER, JAMES

240-ROBERT (II,2688); CODE R (II,551); ISIS (II,1248); LOGAN'S RUN (II,1507); STAR TREK (II,2439)

SCHMIDT, WILLIAM
KNIGHT RIDER (II,1402)

SCHNECK, GEORGE
O'MALLEY (II,1872)

SCHNEIDER, ANDREW
THE INCREDIBLE HULK (II,1232); MAGNUM, P.I. (II,1601); MASQUERADE (II,1644); TALES OF THE GOLD MONKEY (II,2537)

SCHNEIDER, MARGARET
KING'S CROSSING (II,1397); MARCUS WELBY, M.D. (II,1627); THE SIX-MILLION-DOLLAR MAN (II,2372)

SCHNEIDER, PAUL
BUCK ROGERS IN THE 25TH CENTURY (II,384); THE FBI (I,1551); KING'S CROSSING (II,1397); MARCUS WELBY, M.D. (II,1627); THE SIX-MILLION-DOLLAR MAN (II,2372); STAR TREK (II,2440)

SCHOENFELD, BERNARD C
CAVALCADE OF AMERICA (I,875); MADIGAN (I,2808); MANNIX (II,1624); PETER GUNN (I,3562); THE TWILIGHT ZONE (I,4651)

SCHOENMAN, ELLIOT
AMANDA'S (II,62)

SCHONE, VIRGINIA
OF ALL THINGS (I,3330)

SCHRANK, JOSEPH
CINDERELLA (I,956)

SCHREDER, CAROL
CALL TO GLORY (II,413)

SCHREIBER, AVERY
THE BURNS AND SCHREIBER COMEDY HOUR (I,768); THE BURNS AND SCHREIBER COMEDY HOUR (I,769); ZERO HOUR (I,4980)

SCHROCK, RAYMOND L
THE CISCO KID (I,961)

SCHULBERG, BUDD
FIRESIDE THEATER (I,1580)

SCHULER, ANNE
TRAPPER JOHN, M.D. (II,2654)

SCHULMAN, ARNOLD
THE GOLDEN AGE OF TELEVISION (II,1019)

SCHULMAN, ROGER
DEAR TEACHER (II,653); FISHERMAN'S WHARF (II,865); FOR MEMBERS ONLY (II,905); HIGH FIVE (II,1143); THREE'S COMPANY (II,2614)

SCHULZ, CHARLES M
BE MY VALENTINE, CHARLIE BROWN (I,369); THE CHARLIE BROWN AND SNOOPY SHOW (II,484); A CHARLIE BROWN CHRISTMAS (I,908); A CHARLIE BROWN THANKSGIVING (I,909); CHARLIE BROWN'S ALL STARS (I,910); HE'S YOUR DOG, CHARLIE BROWN (I,2037); IT WAS A SHORT SUMMER, CHARLIE BROWN (I,2252); IT'S A MYSTERY, CHARLIE BROWN (I,2259); IT'S AN ADVENTURE, CHARLIE BROWN (I,2266); IT'S ARBOR DAY, CHARLIE BROWN (I,2267); IT'S FLASHBEAGLE, CHARLIE BROWN (I,2268); IT'S MAGIC, CHARLIE BROWN (I,2271); IT'S THE EASTER BEAGLE, CHARLIE BROWN (I,2275); IT'S THE GREAT PUMPKIN, CHARLIE BROWN (I,2276); IT'S YOUR FIRST KISS, CHARLIE BROWN (I,2280); LIFE IS A CIRCUS, CHARLIE BROWN (I,2683); PLAY IT AGAIN, CHARLIE BROWN (I,3617); SHE'S A GOOD SKATE, CHARLIE BROWN (I,4012); SNOOPY'S GETTING MARRIED, CHARLIE BROWN (I,4102); SOMEDAY YOU'LL FIND HER, CHARLIE BROWN (I,4113); THERE'S NO TIME FOR LOVE, CHARLIE BROWN (I,4438); WHAT A NIGHTMARE, CHARLIE BROWN (I,4806); WHAT HAVE WE LEARNED, CHARLIE BROWN? (I,4810); YOU'RE A GOOD SPORT, CHARLIE BROWN (I,4963); YOU'RE IN LOVE, CHARLIE BROWN (I,4964); YOU'RE NOT ELECTED, CHARLIE BROWN (I,4967); YOU'RE THE GREATEST, CHARLIE BROWN (I,4973)

SCHUMACHER, JOEL
NOW WE'RE COOKIN' (II,1866)

SCHUMAN, HOWARD
THE ROCK FOLLIES (II,2192)

SCHUMATE, HAROLD
CAVALCADE OF AMERICA (I,875)

SCHUSTER, ROSE
SQUARE PEGS (II,2431)

SCHWARTZ, AL
A CONNECTICUT YANKEE (I,1038); GILLIGAN'S ISLAND (II,990); GILLIGAN'S ISLAND: RESCUE FROM GILLIGAN'S ISLAND (II,991); GILLIGAN'S ISLAND: THE CASTAWAYS ON GILLIGAN'S ISLAND (II,992); GILLIGAN'S ISLAND: THE HARLEM GLOBETROTTERS ON GILLIGAN'S ISLAND (II,993); HERE'S LUCY (II,1135); JOE'S WORLD (II,1305); THE MILTON BERLE SHOW (I,3047); THE RED SKELTON CHEVY SPECIAL (I,3753); WASHINGTON SQUARE (I,4772)

SCHWARTZ, BILL
KOJAK (II,1406)

SCHWARTZ, DAVID R
ACRES AND PAINS (I,19); THE ALAN YOUNG SHOW (I,104)

SCHWARTZ, DOUGLAS
MANIMAL (II,1622)

SCHWARTZ, ELROY
ENOS (II,779); GILLIGAN'S ISLAND (II,990); GILLIGAN'S ISLAND: RESCUE FROM GILLIGAN'S ISLAND (II,991); GILLIGAN'S ISLAND: THE CASTAWAYS ON GILLIGAN'S ISLAND (II,992); HARRY O (II,1099); IT'S ABOUT TIME (I,2263); MCHALE'S NAVY (I,2969); MOST WANTED (II,1739); MY FAVORITE MARTIAN (I,3189); MY THREE SONS (I,3205); THE SIX-MILLION-DOLLAR MAN (II,2372); WONDER WOMAN (SERIES 1) (II,2828); YOUR SURPRISE STORE (I,4961)

SCHWARTZ, JAN
KNIGHT RIDER (II,1402)

SCHWARTZ, LEW
CARRY ON LAUGHING (II,448); FROM A BIRD'S EYE VIEW (I,1686)

SCHWARTZ, LLOYD J
ALICE (II,33); BIG JOHN, LITTLE JOHN (II,240); THE BRADY BRIDES (II,361); THE BRADY GIRLS GET MARRIED (II,364); HARPER VALLEY PTA (II,1095); THE INVISIBLE WOMAN (II,1244); SCAMPS (II,2267)

SCHWARTZ, NANCY LYNN
WHEELS (II,2777)

SCHWARTZ, SHERWOOD
BIG JOHN, LITTLE JOHN (II,240); THE BRADY BRIDES (II,361); THE BRADY GIRLS GET MARRIED (II,364); GILLIGAN'S ISLAND (II,990); GILLIGAN'S ISLAND: RESCUE FROM GILLIGAN'S ISLAND (II,991); GILLIGAN'S ISLAND: THE CASTAWAYS ON GILLIGAN'S ISLAND (II,992); GILLIGAN'S ISLAND: THE HARLEM GLOBETROTTERS ON GILLIGAN'S ISLAND (II,993); HARPER VALLEY PTA (II,1095); I MARRIED JOAN (I,2174); THE INVISIBLE WOMAN (II,1244); IT'S ABOUT TIME (I,2263); KELLY'S KIDS (I,2523); THE RED SKELTON CHEVY SPECIAL (I,3753); THE RED SKELTON SHOW (I,3755); SCAMPS (II,2267)

SCHWARTZ, WILLIAM
THE INCREDIBLE HULK (II,1232)

SCHWEITZER, S S
BARETTA (II,152); BOONE (II,351); CANNON (II,424); CARIBE (II,439); THE FBI (I,1551); THE HANDS OF CORMAC JOYCE (I,1927); MANHUNTER (II,1621); MATINEE THEATER (I,2947); SARA (II,2260); STRIKE FORCE (II,2480); WONDER WOMAN (SERIES 2) (II,2829); THE WORD (II,2833)

SCOTT, ALLAN
BEN CASEY (I,394)

SCOTT, ASHMEAD
THE LIFE OF RILEY (I,2685)

SCOTT, DEBORAH K
DANCE FEVER (II,615)

SCOTT, ED
THE STEVE LAWRENCE SHOW (I,4227)

SCOTT, ERIC
ROUTE 66 (I,3852)

SCOTT, GARY
FAME (II,812)

SCOTT, HARRY LEE
DINAH IN SEARCH OF THE IDEAL MAN (I,1278); THE JOHN BYNER COMEDY HOUR (I,2407); THE JOHN DENVER SPECIAL (II,1316); REGGIE (II,2127)

SCOTT, JEFFREY
GOLDIE GOLD AND ACTION JACK (II,1023); LOBO (II,1504); MR. MERLIN (II,1751); THE POWERS OF MATTHEW STAR (II,2075); SUPER FRIENDS (II,2497); THUNDARR THE BARBARIAN (II,2616)

SCOTT, JOAN
THE WALTONS (II,2740)

SCOTT, JUD

THE DAY THE WOMEN GOT EVEN (II,628); MATT HOUSTON (II,1654); SEVEN BRIDES FOR SEVEN BROTHERS (II,2307)

SCOTT, MARC
THE GET ALONG GANG (II,980); POLE POSITION (II,2060)

SCOTT, PETER GRAHAM
THE ONEDIN LINE (II,1912)

SEARLS, HANK
THE FUGITIVE (I,1701); WHEELS (II,2777)

SEARS, FRANK
HONEST AL'S A-OK USED CAR AND TRAILER RENTAL TIGERS (II,1174)

SEE, CAROLYN
BARNABY JONES (II,153); FAMILY (II,813)

SEEGER, SUSAN
9 TO 5 (II,1852); A GIRL'S LIFE (II,1000); I'M A BIG GIRL NOW (II,1218); THE JAMES BOYS (II,1272)

SEELEN, JERRY
THE MILTON BERLE SHOW (I,3047); MY FAVORITE MARTIAN (I,3189)

SEFF, BRIAN
THE PEE WEE HERMAN SHOW (II,1988)

SEGAL, FRED
THE BLUE KNIGHT (II,276)

SEGALL, DON
FISH (II,864); THE FOUR SEASONS (II,915); ONE IN A MILLION (I,3379)

SEIDELMAN, ARTHUR ALLAN
I LOVE LIBERTY (II,1215)

SEIGEL, DON
THE KRAFT 75TH ANNIVERSARY SPECIAL (II,1409)

SEITER, JOE
GIVE-N-TAKE (II,1003)

SEITZ JR, GEORGE B
THE LONE RANGER (I,2740)

SELBY, J E
BRANDED (I,707); RAWHIDE (I,3727)

SELDON, GEORGE
STUDIO ONE (I,4268)

SELF, ED
CODE R (II,551)

SELLARS, ARTHUR
MEL AND SUSAN TOGETHER (II,1677)

SELLER, THOMAS
RAWHIDE (I,3727); PRIVATE SECRETARY (I,3672); SKY KING (I,4077)

SEMPLE JR, LORENZO
THE ALCOA HOUR (I,108); BATMAN (I,366); BURKE'S LAW (I,762); KNIGHT'S GAMBIT (I,2576); THE RAT PATROL (I,3726); THE ROGUES (I,3832); YOU'RE ONLY YOUNG TWICE (I,4971)

SENF, CHARLES R
THE ANSWER MAN (I,231)

SENTNER, JOYCE
GIRL ABOUT TOWN (I,1806)

SERLING, ROD
APPOINTMENT WITH ADVENTURE (I,246); CAROL FOR ANOTHER CHRISTMAS (I,852); THE DESILU PLAYHOUSE (I,1237); THE GOLDEN AGE OF TELEVISION (II,1018); KRAFT TELEVISION THEATER (I,2592); MATINEE THEATER (I,2947); MR. BEVIS (I,3128); NIGHT GALLERY (I,3288); THE OATH: THE SAD AND LONELY SUNDAYS (II,1874); PLAYHOUSE 90 (I,3623); PURSUIT (I,3700); A STORM IN SUMMER (I,4237); STUDIO ONE (I,4268); THE TIME ELEMENT (I,4504); THE TWILIGHT ZONE (I,4651)

SETTLE, MIKE
THE EVERLY BROTHERS SHOW (I,1479)

SETTZER, DAVID
SOPHIA! (I,4128)

SEVAREID, MICHAEL
THE DUKES OF HAZZARD (II,742)

SEVINE, SAURA
THE SINGLE LIFE (II,2364)

SEVORG, JON
THE FITZPATRICKS (II,868); THE IMPOSTER (II,1225)

SEWALL, PHYLLIS
THE BAXTERS (II,184)

SEYMOUR, EDITH
CAVALCADE OF AMERICA (I,875)

SHALINS, BERNARD
THE YESTERDAY SHOW (II,2850)

SHAMBERG, MICHAEL
THE TV TV SHOW (II,2674)

SHAND, NEIL
KELLY MONTEITH (II,1375)

SHANE, MAXWELL
THRILLER (I,4492)

SHANER, JOHN
THE MAN FROM U.N.C.L.E. (I,2867)

SHANKS, BOB
MERV GRIFFIN'S SIDEWALKS OF NEW ENGLAND (I,3017); MERV GRIFFIN'S ST. PATRICK'S DAY SPECIAL (I,3018)

SHAPIRO, ARNOLD
THE SINGING COWBOYS RIDE AGAIN (II,2363)

SHAPIRO, DAN
HELLZAPOPPIN (I,2011); THE MILTON BERLE SPECIAL (I,3050); THE MILTON BERLE SPECIAL (I,3051)

SHAPIRO, ESTHER
DYNASTY (II,746); EMERALD POINT, N.A.S. (II,773)

SHAPIRO, KEN
MONTY HALL'S VARIETY HOUR (II,1728)

SHAPIRO, RICHARD
DYNASTY (II,746); EAST OF EDEN (II,752); EMERALD POINT, N.A.S. (II,773); LAND OF THE GIANTS (I,2611)

SHAPIRO, STANLEY
THE ALAN YOUNG SHOW (I,104); THE ALL-STAR REVUE (I,138); HEY, JEANNIE! (I,2038); MCGARRY AND ME (I,2966); THE TAB HUNTER SHOW (I,4328); WHERE'S RAYMOND? (I,4837)

SHAPLEN, ROBERT
THE HUNTER (I,2162)

SHARKEY, MIKE
THE NATIVITY (II,1798)

SHARNIK, JOHN
WORLD WAR I (I,4919)

SHARP, HENRY
THE ANN SOTHERN SHOW (I,220); THE MAN FROM U.N.C.L.E. (I,2867); THE WILD WILD WEST (I,4863)

SHARP, PHIL
ALL IN THE FAMILY (II,38); THE DONNA REED SHOW (I,1347); I MARRIED JOAN (I,2174); JOE'S WORLD (II,1305); MAUDE (II,1655); MY WORLD . . . AND WELCOME TO IT (I,3208)

SHATZ, TOM
HOT (II,1183)

SHAUGHNESSEY, ALFRED
THE SAINT (II,2237)

SHAVELSON, MELVILLE
FATHER ON TRIAL (I,1544); IKE (II,1223); MY WORLD . . . AND WELCOME TO IT (I,3208); THE SECRET LIFE OF JOHN MONROE (I,3966); THREE COINS IN THE FOUNTAIN (I,4473); TRUE LIFE STORIES (II,2665)

SHAW, DAVID
COUNTY GENERAL (I,1076); THE JUDGE AND JAKE WYLER (I,2464); MIRROR, MIRROR, OFF THE WALL (I,3059); THE MISSISSIPPI (II,1707); ONE-EYED JACKS ARE WILD (I,3374); OUR TOWN (I,3419); PARTNERS IN CRIME (I,3461); PLAYHOUSE 90 (I,3623); REUNION IN VIENNA (I,3780); RUGGLES OF RED GAP (I,3867); SHANE (I,3995); STUNT SEVEN (II,2486)

SHAW, DON
THE CITADEL (II,539)

SHAW, FRANK
THE BILL COSBY SPECIAL, OR? (I,443); BLACK BART (II,260); CO-ED FEVER (II,547); DINAH IN SEARCH OF THE IDEAL MAN (I,1278); THE GLEN CAMPBELL GOODTIME HOUR (I,1820); THE JERRY REED WHEN YOU'RE HOT, YOU'RE HOT HOUR (I,2372); THE LESLIE UGGAMS SHOW (I,2660); THE NEW BILL COSBY SHOW (I,3257); PAT PAULSEN'S HALF A COMEDY HOUR (I,3473); ROBERT YOUNG AND THE FAMILY (I,3813)

SHAW, GEORGE BERNARD
CANDIDA (II,423); GREAT CATHERINE (I,1873)

SHAW, LOU
ALCOA PREMIERE (I,109); BARNABY JONES (II,153); BEYOND WESTWORLD (II,227); BRANDED (I,707); COLUMBO (II,556); THE FALL GUY (II,811); FITZ AND BONES (II,867); GET CHRISTIE LOVE! (II,981); LOBO (II,1504); MCCLOUD (II,1660); NAKED CITY (I,3216); PEOPLE ARE FUNNY (I,3518); PLEASURE COVE (II,2058); RAWHIDE (I,3727); THE SIX-MILLION-DOLLAR MAN (II,2372); THE STAR MAKER (I,4193); SWITCH (II,2519); TERROR AT ALCATRAZ (II,2563); TURN OF FATE (I,4623); WAGON TRAIN (I,4747)

SHAW, PEGGY
ALCOA PREMIERE (I,109); NAKED CITY (I,3216); RAWHIDE (I,3727); THE STAR MAKER (I,4193); TURN OF FATE (I,4623); WAGON TRAIN (I,4747)

SHAW, RICK
BATTLE OF THE PLANETS (II,179); BATTLESTARS (II,182); THE LOVE BOAT (II,1535)

SHAW, ROBERT J
MATINEE THEATER (I,2947); OUR PRIVATE WORLD (I,3417); PERRY MASON (II,2025); ROBERT MONTGOMERY PRESENTS YOUR LUCKY STRIKE THEATER (I,3809)

SHAYNE, BOB
AT EASE (II,116); COVER UP (II,597); DINAH AND FRIENDS (II,675); DINAH! (II,677); FOUL PLAY (II,910); HART TO HART (II,1102); KNIGHT RIDER (II,1402); MAGNUM, P.I. (II,1601); SIMON AND SIMON (II,2357); THAT'S MY MAMA (II,2580); TONY ORLANDO AND DAWN (II,2639); WHIZ KIDS (II,2790)

SHEA, JACK
DEAN MARTIN'S CELEBRITY ROAST (II,643); SILVER SPOONS (II,2355)

SHEA, PAT
ARCHIE BUNKER'S PLACE (II,105); GLORIA COMES HOME (II,1011); LOU GRANT (II,1526)

SHEAR, BARRY
ALL THINGS BRIGHT AND BEAUTIFUL (I,141)

SHEARLES, BUCKY
EVERYTHING YOU ALWAYS WANTED TO KNOW ABOUT JACK BENNY AND WERE AFRAID TO ASK (I,1482)

SHEEHAN, TONY
A.E.S. HUDSON STREET (II,17); BARNEY MILLER (II,154)

SHEFFLER, MARC
LEWIS AND CLARK (II,1465)

SHEINER, MARY-DAVID
THE TONY RANDALL SHOW (II,2640); WE'VE GOT EACH OTHER (II,2757)

SHEKTER, MARK
THE ANDY WILLIAMS SHOW (I,210); DIANA (I,1258); HOW TO HANDLE A WOMAN (I,2146); THE JERRY REED WHEN YOU'RE HOT, YOU'RE HOT HOUR (I,2372)

SHELDON, LEE
CALIFORNIA FEVER (II,411); CHARLIE'S ANGELS (II,486); THE GREATEST AMERICAN HERO (II,1060); THE HARDY BOYS MYSTERIES (II,1090); THE NANCY DREW MYSTERIES (II,1789); NERO WOLFE (II,1806); TODAY'S FBI (II,2628); TUCKER'S WITCH (II,2667); WALKING TALL (II,2738)

SHELDON, LES
FATHER MURPHY (II,841)

SHELDON, SANFORD
THE FUNNY SIDE (I,1715); THE PAT BOONE SHOW (I,3471)

SHELDON, SIDNEY
ADVENTURES OF A MODEL (I,50); HART TO HART (II,1101); I DREAM OF JEANNIE (I,2167); THE PATTY DUKE SHOW (I,3483)

SHELLEY, KATHLEEN
EMERALD POINT, N.A.S. (II,773); FLAMINGO ROAD (II,872)

SHELLY, BRUCE
CAPTAIN NICE (I,835); EIGHT IS ENOUGH (II,762); FAME (II,812); FINDER OF LOST LOVES (II,857); JEANNIE (II,1275); LEAVE IT TO BEAVER (I,2648); THE LOVE BOAT (II,1535); M*A*S*H (II,1569); THE POWERS OF MATTHEW STAR (II,2075); RETURN TO THE PLANET OF THE APES (II,2145); SWITCH (II,2519); TWO MARRIAGES (II,2691); VOYAGERS (II,2730); WONDER WOMAN (SERIES 2) (II,2829)

SHELTON, JAMES
THE DOM DELUISE SHOW (I,1328)

SHENK, JERRY
THIS IS YOUR LIFE (I,4461)

SHEPERD, SCOTT
MATT HOUSTON (II,1654)

SHEPHERD, JEAN
JEAN SHEPHERD'S AMERICA (I,2354)

SHEPHERD, SANDRA
THE BAXTERS (II,183)

SHEPPARD, DAVID
THE LONE RANGER (I,2740); SKY KING (I,4077)

SHEPPARD, DON
PANDAMONIUM (II,1948); SUPER FRIENDS (II,2497)

SHERDEMAN, TED
HAZEL (I,1982); WAGON TRAIN (I,4747)

SHERER, MEL
LAVERNE AND SHIRLEY (II,1446)

SHERIDAN, ANN
THE NEW VOICE (II,1833)

SHERLOCK, JOHN
LOGAN'S RUN (II,1507)

SHERMAN, ALLAN
54TH STREET REVUE (I,1573); PHIL SILVERS ON BROADWAY (I,3578)

SHERMAN, CHARLES
THE BOB NEWHART SHOW (I,666); PHIL SILVERS IN NEW YORK (I,3577); SUMMER IN NEW YORK (I,4279)

SHERMAN, DON
THE BOBBY DARIN AMUSEMENT COMPANY (I,672); SWINGING COUNTRY (I,4322)

SHERMAN, GARY
THE MYSTERIOUS TWO (II,1783); THE STREETS (II,2479)

SHERMAN, GEORGE
THE SMOTHERS BROTHERS SHOW (I,4097)

SHERMAN, JILL
THE INCREDIBLE HULK (II,1232); VOYAGERS (II,2730)

SHERMAN, MARTIN
DON'T CALL ME MAMA ANYMORE (I,1350)

SHERMAN, ROBERT
BARNABY JONES (II,153); CANNON (II,424); THE INVADERS (I,2229); JOHNNY RINGO (I,2434); THE RAT PATROL (I,3726); RAWHIDE (I,3727); T.J. HOOKER (II,2524)

SHERMAN, STANFORD
THE MAN FROM U.N.C.L.E. (I,2867)

SHERMAN, TEDDI
BEN CASEY (I,394); JOHNNY RINGO (I,2434); LAW OF THE PLAINSMAN (I,2639)

SHERWIN, WALLY
MEET MARCEL MARCEAU (I,2983)

SHERWOOD, CHRISTOPHER
THE LEGEND OF SILENT NIGHT (I,2655)

SHERWOOD, HOPE
THE BRADY BRIDES (II,361)

SHERWOOD, ROBERT E
BACKBONE OF AMERICA (I,328)

SHEVELSON, MELVILLE
ELKE (I,1432)

SHEVIN, FRED
LEAVE IT TO BEAVER (I,2648)

SHIELDS, MEL
THE LOVE BOAT (II,1535)

SHIELDS, ROBERT
MAC DAVIS. . . I BELIEVE IN CHRISTMAS (II,1581)

SHIFF, ROBERT
ALOHA PARADISE (II,58)

SHINKAI, BILL
DIFF'RENT STROKES (II,674); THE FACTS OF LIFE (II,805)

SHIPMAN, BARRY
LAST STAGECOACH WEST (I,2629)

SHIPP, REUBEN
CAVALCADE OF STARS (I,877); THE LIFE OF RILEY (I,2685)

SHIRL, JIMMY
SONG SNAPSHOTS ON A SUMMER HOLIDAY (I,4122)

SHKENAZY, IRWIN
TERRY AND THE PIRATES (I,4405)

SHOEMAKER, EMILY
FAMILY (II,813); THE FAMILY TREE (II,820)

SHOENFELD, BERNARD
ALFRED HITCHCOCK PRESENTS (I,115); COMBAT (I,1011)

SHOENMAN, ELLIOT
I LOVE HER ANYWAY! (II,1214); NEVER SAY NEVER (II,1809); OLD FRIENDS (II,1881)

SHORE, WILMA
COUNTERPOINT (I,1059)

SHORR, BOB
THE STEVE ALLEN COMEDY HOUR (II,2454)

SHORT, LUKE
HARD CASE (I,1947)

SHORT, MARTIN
SECOND CITY TELEVISION (II,2287)

SHORT, MICHAEL
BIG CITY COMEDY (II,233); OUT OF OUR MINDS (II,1933)

SHRANK, JOSEPH
BRIEF ENCOUNTER (I,723)

SHUKEN, PHIL
CALAMITY JANE (I,793); SCREEN DIRECTOR'S PLAYHOUSE (I,3946); SHINDIG (I,4013)

SHULKIN, JOE
THE REDD FOXX COMEDY HOUR (II,2126)

SHULMAN, MAX
DOBIE GILLIS (I,1302)

SHULMAN, ROGER
ALL IN THE FAMILY (II,38); CRAZY LIKE A FOX (II,601); GOOD TIMES (II,1038)

SHUMATE, HAROLD
BONANZA (II,347); ZANE GREY THEATER (I,4979)

SHUSTER, FRANK
OPERATION GREASEPAINT (I,3402); WAYNE AND SHUSTER TAKE AN AFFECTIONATE LOOK AT. . . (I,4778)

SHYER, CHARLES
LADY LUCK (I,2604); THE PARTRIDGE FAMILY (II,1962)

SIAMIS, KORBY
BENSON (II,208); LOVE, SIDNEY (II,1547); MAGGIE (II,1589)

SIBER, BARRY
THE KENNY EVERETT VIDEO SHOW (II,1379)

SIBLEY, SHEILA
PRISONER: CELL BLOCK H (II,2085)

SIDNEY, ROBERT
DOM DELUISE AND FRIENDS (II,701)

SIDOMAK, CURT
FIRESIDE THEATER (I,1580)

SIEGEL, DONALD L
THE JEFFERSONS (II,1276); YOU ASKED FOR IT (II,2855)

SIEGEL, GEORGE S
DARKROOM (II,619)

SIEGEL, GERALD K
MURDER, SHE WROTE (II,1770)

SIEGEL, JEFF
THE SMURFS (II,2385)

SIEGEL, LARRY
THE CAROL BURNETT SHOW (II,443); OH, NURSE! (I,3343); ROWAN AND MARTIN'S LAUGH-IN (I,3856); THAT'S MY MAMA (II,2580); WHEN THINGS WERE ROTTEN (II,2782)

SIEGEL, LIONEL E
79 PARK AVENUE (II,2310); THE AMAZING SPIDER-MAN (II,63); BEN CASEY (I,394); THE BIONIC WOMAN (II,255); LISA, BRIGHT AND DARK (I,2711); MR. NOVAK (I,3145); THE SIX-MILLION-DOLLAR MAN (II,2372); THE ULTIMATE IMPOSTER (II,2700); THE WALTONS (II,2740)

SIEGEL, SANDRA
FALCON CREST (II,808); FINDER OF LOST LOVES (II,857)

SILBERG, BOB
LOOK WHAT THEY'VE DONE TO MY SONG (II,1517)

SILBERMAN, HARVEY
AMANDA'S (II,62)

SILLIPHANT, STIRLING
ALCOA/GOODYEAR THEATER (I,107); ALFRED HITCHCOCK PRESENTS (I,115); BROCK CALLAHAN (I,741); FLY AWAY HOME (II,893); GOLDEN GATE (II,1020); HARDCASE (II,1088); LONGSTREET (I,2750); MOVIN' ON (I,3118); NAKED CITY (I,3215); NAKED CITY (I,3216); THE NEW HEALERS (I,3266); PEARL (II,1986); PERRY MASON (II,2025); ROUTE 66 (I,3852); TIGHTROPE (I,4500); TRAVIS MCGEE (II,2656); WELCOME TO PARADISE (II,2762)

SILLS, PAUL
STORY THEATER (I,4241)

SILVER, ARTHUR
FLATFOOTS (II,879); HAPPY DAYS (II,1084); HERE'S BOOMER (II,1134); LAVERNE AND SHIRLEY (II,1446); SISTER TERRI (II,2366)

SILVER, BARRY
THE FUNNY WORLD OF FRED & BUNNI (II,943); THE GRADY NUTT SHOW (II,1048); THE HEE HAW HONEYS (II,1124); THE JOHN DAVIDSON SHOW (II,1309); SHIELDS AND YARNELL (II,2328)

SILVER, FRANELLE
THE CAROL BURNETT SHOW (II,443); DONNY AND MARIE (II,712); THE STEVE LANDESBERG TELEVISION SHOW (II,2458)

SILVER, PAT
PHILIP MARLOWE, PRIVATE EYE (II,2033)

SILVER, STU
IT'S A LIVING (II,1254); THE NEW ODD COUPLE (II,1825); SOAP (II,2392); STAR OF THE FAMILY (II,2436); WEBSTER (II,2758)

SILVER, SUSAN
THE BOB NEWHART SHOW (II,342); SQUARE PEGS (II,2431); WE'LL GET BY (II,2753)

SILVERMAN, DAVID
ALICE (II,33); ONE DAY AT A TIME (II,1900)

SILVERMAN, RON
THE WILD WILD WEST (I,4863)

SILVERMAN, STANLEY
GIANT IN A HURRY (I,1790); SKY KING (I,4077)

SILVERMAN, STUART
THE DUCK FACTORY (II,738); NEWHART (II,1835)

SILVERMAN, TREVA
CAPTAIN NICE (I,835); THE MARY TYLER MOORE SHOW (II,1640); OH, NURSE! (I,3343); WE INTERRUPT THIS SEASON (I,4780)

SILVERS, SID
THE MANY SIDES OF MICKEY ROONEY (I,2899); THE MICKEY ROONEY SHOW (I,3027)

SILVERSTONE, LOU
THE JACKSON FIVE (I,2326)

SILVERTON, DORIS
THE HAMPTONS (II,1076)

SIMMONS, ALBERT
ALICE THROUGH THE LOOKING GLASS (I,122); THE GEORGE GOBEL SHOW (I,1768)

SIMMONS, ED
AN AMATEUR'S GUIDE TO LOVE (I,154); THE CAROL BURNETT SHOW (II,443); HOLIDAY LODGE (I,2077); HOTEL 90 (I,2130); THE JERRY LEWIS SHOW (I,2370); KEEP THE FAITH (I,2518); THE KELLY MONTEITH SHOW (II,1376); THE MARTHA RAYE SHOW (I,2926); RODNEY DANGERFIELD SHOW: I CAN'T TAKE IT NO MORE (II,2198); ROWAN AND MARTIN BITE THE HAND THAT FEEDS THEM (I,3853); WHERE'S EVERETT? (I,4835)

SIMMONS, GARNER
V: THE SERIES (II,2715)

SIMMONS, KIRBY
THE WALTONS (II,2740)

SIMMONS, MATTY
DELTA HOUSE (II,658); NATIONAL LAMPOON'S TWO REELERS (II,1796)

SIMMONS, RICHARD ALAN
COLOSSUS (I,1009); THE DICK POWELL SHOW (I,1269); HITCHED (I,2065); LASSITER (I,2625); LASSITER (II,1434); THE LIVES OF JENNY DOLAN (II,1502); LOCK, STOCK, AND BARREL (I,2736)

SIMMS, JAY
THE BIG VALLEY (I,437); DIRTY SALLY (II,680)

SIMMS, MARLEY
ARCHIE BUNKER'S PLACE (II,105)

SIMMS, OLGA
SHAZAM! (II,2319)

SIMON, AL
PETER LOVES MARY (I,3565)

SIMON, DANNY
THE ANN SOTHERN SHOW (I,220); GIVE MY REGARDS TO BROADWAY (I,1815); THE JERRY LEWIS SHOW (I,2365); MCHALE'S NAVY (I,2969); MY THREE SONS (I,3205); THE PHIL SILVERS ARROW SHOW (I,3576); THE RED BUTTONS SHOW (I,3750); SID CAESAR INVITES YOU (I,4040)

SIMON, DOC
THE GARRY MOORE SHOW (I,1739); THE PHIL SILVERS ARROW SHOW (I,3576)

SIMON, GEORGE T
AMERICA'S GREATEST BANDS (I,177); TIMEX ALL-STAR JAZZ SHOW I (I,4515); TIMEX ALL-STAR JAZZ SHOW II (I,4516); TIMEX ALL-STAR JAZZ SHOW III (I,4517); TIMEX ALL-STAR JAZZ SHOW IV (I,4518)

SIMON, JIM
THE LITTLE RASCALS (II,1494); THE RICHIE RICH SHOW (II,2169)

SIMON, MAYO
THE MAN FROM ATLANTIS (II,1614); PLAYHOUSE 90 (I,3623); STUDIO ONE (I,4268)

SIMON, NEIL
BABES IN TOYLAND (I,318); BAREFOOT IN THE PARK (II,151); BEST FOOT FORWARD (I,401); THE CHOCOLATE SOLDIER (I,944); A CONNECTICUT YANKEE (I,1038); DEAREST ENEMY

(I,1205); THE DESERT SONG (I,1236); THE GREAT WALTZ (I,1878); HAPPY ENDINGS (II,1085); HOLIDAY IN LAS VEGAS (I,2076); THE JERRY LEWIS SHOW (I,2364); KIBBE HATES FINCH (I,2530); LOVE, LIFE, LIBERTY & LUNCH (II,1544); MARCO POLO (I,2904); THE MERRY WIDOW (I,3012); NAUGHTY MARIETTA (I,3232); PARIS IN THE SPRINGTIME (I,3457); THE PHIL SILVERS SHOW (I,3580); PLAZA SUITE (II,2056); SID CAESAR INVITES YOU (I,4040); THE SUNSHINE BOYS (II,2496); THE TROUBLE WITH PEOPLE (I,4611); YOUR SHOW OF SHOWS (I,4957)

SIMON, PAUL
THE PAUL SIMON SPECIAL (II,1983)

SIMON, SAM
CHEERS (II,494); SHAPING UP (II,2316); TAXI (II,2546)

SIMON, SUZY
ONE OF THE BOYS (II,1911)

SIMOUN, HENRI
THE DEVLIN CONNECTION (II,664)

SIMS, BURT
SKY KING (I,4077)

SINDLE, MARTIN
THIS WAS AMERICA (II,2599)

SINGER, AL
THE ALL-STAR REVUE (I,138)

SINGER, RAY
THE ANN SOTHERN SHOW (I,220); THE DANNY THOMAS SPECIAL (I,1151); THE DANNY THOMAS TV FAMILY REUNION (I,1154); GILLIGAN'S ISLAND (II,990); HERE'S LUCY (II,1135); IT'S A GREAT LIFE (I,2256); JOE'S WORLD (II,1305); THE MCGONIGLE (I,2968); RUN, BUDDY, RUN (I,3870)

SIODMAK, CURT
NUMBER 13 DEMON STREET (I,3316)

SISKO, SUSAN
THE TED KNIGHT SHOW (II,2550)

SISSON, ROSEMARY
THE ADVENTURES OF BLACK BEAUTY (I,51); THE MANIONS OF AMERICA (II,1623); MISTRAL'S DAUGHTER (II,1708); A TOWN LIKE ALICE (II,2652)

SITOWITZ, HAL
FOUL PLAY (II,910); THE LETTERS (I,2670); THE OATH: 33 HOURS IN THE LIFE OF GOD (II,1873); THE ROOKIES (II,2208)

SKELTON, RED
CLOWN ALLEY (I,979); THE RED SKELTON CHEVY SPECIAL (I,3753); THE RED SKELTON SHOW (I,3755); THE RED SKELTON SHOW (I,3756); THE RED SKELTON TIMEX SPECIAL (I,3757); RED SKELTON'S CHRISTMAS DINNER (II,2123); RED SKELTON'S FUNNY FACES (II,2124); RED SKELTON: A ROYAL PERFORMANCE (II,2125)

SKENE, ANTHONY
THE PRISONER (I,3670); THE STRAUSS FAMILY (I,4253)

SLADE, BERNARD
BOBBY JO AND THE BIG APPLE GOODTIME BAND (I,674); THE FLYING NUN (I,1611); GOOD HEAVENS (II,1032); IS THERE A DOCTOR IN THE HOUSE (I,2241); IS THERE A DOCTOR IN THE HOUSE? (II,1247); A KNIGHT IN SHINING ARMOUR (I,2575); THE PARTRIDGE FAMILY (II,1962); UNDER THE YUM YUM TREE (I,4671)

SLADE, MARK
THE ROOKIES (II,2208)

SLATE, LANE
THE AMERICAN GIRLS (II,72); THE DEADLY GAME (II,633); THE GIRL IN THE EMPTY GRAVE (II,997); STRIKE FORCE (II,2480); OUR FAMILY BUSINESS (II,1931)

SLATE, LARRY
RIDDLE AT 24000 (II,2171)

SLATER, BARNEY
COLUMBO (II,556); THE DESILU PLAYHOUSE (I,1237); JOHNNY NIGHTHAWK (I,2432); JOHNNY RINGO (I,2434); LOST IN SPACE (I,2758); THE RIFLEMAN (I,3789); THE ROGUES (I,3832); THE WILD WILD WEST (I,4863)

SLAVIN, GEORGE
BARNABY JONES (II,153); THE FLYING NUN (I,1611); HAWAII FIVE-O (II,1110); S.W.A.T. (II,2236); TURN OF FATE (I,4623); THE UNTOUCHABLES (I,4682)

SLESAR, HENRY
ALFRED HITCHCOCK PRESENTS (I,115); THE MAN FROM U.N.C.L.E. (I,2867); THE TWILIGHT ZONE (I,4651)

SLOAN, MICHAEL
B.J. AND THE BEAR (II,125); B.J. AND THE BEAR (II,126); BATTLES: THE MURDER THAT WOULDN'T DIE (II,180); BATTLESTAR GALACTICA (II,181); THE DEVLIN CONNECTION (II,664); THE HARDY BOYS MYSTERIES (II,1090); HARRY O (II,1099); THE MASTER (II,1648); MCCLOUD (II,1660); THE NANCY DREW MYSTERIES (II,1789); RETURN OF THE MAN FROM U.N.C.L.E.: THE 15 YEARS LATER AFFAIR (II,2140); SWITCH (II,2519)

SLOANE, ALLEN
330 INDEPENDENCE S.W. (I,4486); THE BIG STORY (I,432); CROSSROADS (I,1105); THE DICK POWELL SHOW (I,1269); EMILY, EMILY (I,1438); JOHNNY BELINDA (I,2417); TEACHER, TEACHER (I,4366); THEY WENT THATAWAY (I,4443)

SLOCUM, CHARLES
OF ALL THINGS (I,3330)

SLOCUM, FRANK
BARBI BENTON SPECIAL: A BARBI DOLL FOR CHRISTMAS (II,148); CIRCUS LIONS, TIGERS AND MELISSAS TOO (II,528); COUNTRY COMES HOME (II,585); COUNTRY COMES HOME (II,586); COUNTRY MUSIC HIT PARADE (II,589); COUNTRY NIGHT OF STARS (II,592); COUNTRY NIGHT OF STARS II (II,593); COUNTRY STARS OF THE 70S (II,594); FIFTY YEARS OF COUNTRY MUSIC (II,853); JOHNNY CASH AND FRIENDS (II,1322); JOHNNY CASH AND THE COUNTRY GIRLS (II,1323); THE JOHNNY CASH CHRISTMAS SPECIAL (II,1325); THE JOHNNY CASH CHRISTMAS SPECIAL (II,1326); A JOHNNY CASH CHRISTMAS (II,1327); A JOHNNY CASH CHRISTMAS (II,1328); THE JOHNNY CASH SPRING SPECIAL (II,1329); JOHNNY CASH'S AMERICA (II,1330); JOHNNY CASH: CHRISTMAS IN SCOTLAND (II,1332); JOHNNY CASH: COWBOY HEROES (II,1334); JOHNNY CASH: SPRING FEVER (II,1335); JOHNNY CASH: THE FIRST 25 YEARS (II,1336); JOHNNY CASH—A MERRY MEMPHIS CHRISTMAS (II,1331); THE MAGIC OF DAVID COPPERFIELD (II,1593); MAGIC WITH THE STARS (II,1599); ROY ACUFF—50 YEARS THE KING OF COUNTRY MUSIC (II,2223)

SLON, SIDNEY
INVITATION TO PARIS (I,2233)

SLOTE, LESLIE
COLONEL MARCH OF SCOTLAND YARD (I,1005); DUPONT SHOW OF THE MONTH (I,1387)

SMALL, ALEXANDER
THE QUIZ KIDS (II,2103)

SMALL, EMILE
TRAPPER JOHN, M.D. (II,2654)

SMALLWOOD, LOUIS
DIFF'RENT STROKES (II,674)

SMART, RALPH
THE CHAMPIONS (I,896); DANGER MAN (I,1136); THE INVISIBLE MAN (I,2232); SECRET AGENT (I,3962)

SMIGHT, JACK
MCCLOUD (II,1660)

SMILOW, DAVID
THE HUSTLER OF MUSCLE BEACH (II,1210); THE JACKSONS (II,1267)

SMITH, ANDREW
THE BOB NEWHART SHOW (II,342); THE MERV GRIFFIN SHOW (I,3016); THE NEWS IS THE NEWS (II,1838); NOT THE NINE O'CLOCK NEWS (II,1864); ON OUR OWN (II,1892)

SMITH, APRIL
CAGNEY AND LACEY (II,409); LOU GRANT (II,1526)

SMITH, BARBARA ELAINE
EIGHT IS ENOUGH (II,762); FAMILY (II,813)

SMITH, BERNIE
THE GENERAL FOODS 25TH ANNIVERSARY SHOW (I,1756)

SMITH, BOB
HOWDY DOODY (I,2151)

SMITH, CECIL
THE DICK POWELL SHOW (I,1269)

SMITH, CHARLES
RAWHIDE (I,3727); TURN OF FATE (I,4623)

SMITH, DUNCAN
REMINGTON STEELE (II,2130)

SMITH, F D
ALL ABOUT BARBARA (I,125)

SMITH, HANNAH

STUDIO ONE (I,4268)

SMITH, JOHN F
THE YOUNG AND THE RESTLESS (II,2862)

SMITH, MARTIN
THE ART OF CRIME (II,108)

SMITH, PAUL GERALD
GAY NINETIES REVUE (I,1743)

SMITH, PENROD
THE FBI (I,1551)

SMITH, ROBERT KIMMEL
SIDE BY SIDE (II,2347)

SMITH, S MARK
THE DEVIL'S DISCIPLE (I,1247); DREAM GIRL (I,1374)

SMITH, STEVE
OUT OF OUR MINDS (II,1933)

SMOOT, FRED
A SPECIAL OLIVIA NEWTON-JOHN (II,2418)

SMOTHERS, TOM
THE RETURN OF THE SMOTHERS BROTHERS (I,3777)

SMYRL, DAVID
BENSON (II,208)

SNEE, DENNIS
GAMES PEOPLE PLAY (II,956)

SNEIDER, VERN
THE ALCOA HOUR (I,108); ROBERT MONTGOMERY PRESENTS YOUR LUCKY STRIKE THEATER (I,3809)

SNELL, DAVID
RYAN'S HOPE (II,2234)

SNYDER, HOWARD
THE FRANK SINATRA SHOW (I,1665); I MARRIED JOAN (I,2174); TIME OUT FOR GINGER (I,4508)

SNYDER, HUGH
NATIONAL LAMPOON'S TWO REELERS (II,1796)

SOBELMAN, BORIS
THE MAN FROM U.N.C.L.E. (I,2867); STAR TREK (II,2440); THRILLER (I,4492)

SOBIESKI, CAROL
DIAL HOT LINE (I,1254); TWO MARRIAGES (II,2691)

SODERBERG, ROBERT
THE ALDRICH FAMILY (I,111); BATTLE OF THE AGES (I,367); CAVALCADE OF AMERICA (I,875); SANTA BARBARA (II,2258); SUSPICION (I,4309)

SOHL, JERRY
THE OUTER LIMITS (I,3426); STAR TREK (II,2440); TARGET: THE CORRUPTERS (I,4360)

SOLMS, KENNY
ANN-MARGRET: FROM HOLLYWOOD WITH LOVE (I,217); ANNIE AND THE HOODS (II,91); BING CROSBY AND CAROL BURNETT—TOGETHER AGAIN FOR THE FIRST TIME (I,454); BURNETT "DISCOVERS" DOMINGO (II,398); CALL HER MOM (I,796); FROM HOLLYWOOD WITH LOVE: THE ANN-MARGRET SPECIAL (I,1687); HELLZAPOPPIN (I,2011); THE HOMEMADE COMEDY SPECIAL (II,1173); JULIE ANDREWS: ONE STEP INTO SPRING (II,1353); THE MANY SIDES OF DON RICKLES (I,2898); SHEILA (II,2325); SILLS AND BURNETT AT THE MET (II,2354); THE SMOTHERS BROTHERS SHOW (II,2384); TEXACO STAR THEATER: OPENING NIGHT (II,2565); THREE GIRLS THREE (II,2608)

SOLOMAN, MARK
ALICE (II,33)

SOLOMON, AUBREY
CAGNEY AND LACEY (II,409); QUINCY, M. E. (II,2102)

SOLOMON, DAVID
HART TO HART (II,1102); PARIS (II,1957)

SOLOMON, LEO
THE ALAN YOUNG SHOW (I,104); THE ANDY GRIFFITH SHOW (I,192); LEAVE IT TO LARRY (I,2649)

SOLOMON, LOUIS
THE DINAH SHORE SPECIAL (I,1284); ELIZABETH TAYLOR IN LONDON (I,1430); OPENING NIGHT: U.S.A. (I,3397)

SOLOW, EUGENE
THE ADVENTURES OF SUPERMAN (I,77)

SOLT, ANDREW
E.T. & FRIENDS—MAGICAL MOVIE VISITORS (II,749); THOSE WONDERFUL TV GAME SHOWS (II,2603)

SOMMER, EDITH
THE ALDRICH FAMILY (I,111); BATTLE OF THE AGES (I,367); BURKE'S LAW (I,762)

SOMMERBY, JON
PENMARRIC (II,1993)

SOMMERS, JAY
THE ADVENTURES OF OZZIE AND HARRIET (I,71); BACHELOR FATHER (I,323); CAROL (I,846); DADDY'S GIRL (I,1124); GOOD TIMES (II,1038); GREEN ACRES (I,1884); HOT L BALTIMORE (II,1187); HOWDY (I,2150); MY FRIEND IRMA (I,3191) THE SOFT TOUCH (I,4107)

SONDHEIM, STEPHEN
THE FABULOUS 50S (I,1501)

SONNTAG, ROBERT
TRAPPER JOHN, M.D. (II,2654)

SONTAG, DAVID
THE PAPER CHASE: THE SECOND YEAR (II,1950)

SORECKY, BARNEY A
THE CISCO KID (I,961)

SOSIN, ALLEN
BATTLE OF THE PLANETS (II,179)

SOSSI, RONALD
THE RAT PATROL (I,3726)

SOTER, STEVE
COSMOS (II,579)

SOTKIN, MARC
BLANSKY'S BEAUTIES (II,267); CHARLES IN CHARGE (II,482); I'M A BIG GIRL NOW (II,1218); WORKING STIFFS (II,2834)

SOWARDS, JACK
BARNABY JONES (II,153); HAGEN (II,1071); THE STREETS OF SAN FRANCISCO (II,2478)

SPARGER, REX
SCROOGE'S ROCK 'N' ROLL CHRISTMAS (II,2279)

SPARKES, JOHN
TOMA (II,2634)

SPARR, QUENTIN
THE RAT PATROL (I,3726)

SPEARS, KEN
PEBBLES AND BAMM BAMM (II,1987)

SPEARS, ROBERT
FANTASY ISLAND (II,829)

SPEARS, STEPHEN
AFTER HOURS: GETTING TO KNOW US (II,21); THE CAPTAIN AND TENNILLE (II,429); THE CARPENTERS AT CHRISTMAS (II,446); THE CARPENTERS. . . SPACE ENCOUNTERS (II,447); THE FACTS OF LIFE (II,805); JONATHAN WINTERS PRESENTS 200 YEARS OF AMERICAN HUMOR (II,1342); ROWAN AND MARTIN'S LAUGH-IN (I,3856)

SPECHT, GEORGE
KUNG FU (II,1416)

SPECHT, ROBERT
240-ROBERT (II,2688); THE IMMORTAL (I,2196); THE OUTER LIMITS (I,3426); TWIN DETECTIVES (II,2687)

SPECTOR, WARREN
BLONDIE (I,486)

SPEER, CHARLES
KOBB'S CORNER (I,2577)

SPEER, KATHY
BENSON (II,208); CONDO (II,565)

SPELLING, AARON
THE DESILU PLAYHOUSE (I,1237); THE DICK POWELL SHOW (I,1269); DOC HOLLIDAY (I,1307); JOHNNY RINGO (I,2434); THE LONER (I,2744); PLAYHOUSE 90 (I,3623); THE ROOKIES (II,2208); WAGON TRAIN (I,4747); ZANE GREY THEATER (I,4979)

SPENCE, PETER
NOT THE NINE O'CLOCK NEWS (II,1864); TO THE MANOR BORN (II,2627)

SPENCER, FRANZ
THE HUNTER (I,2162)

SPENCER, LARRY
THE $1.98 BEAUTY SHOW (II,698); JENNIFER SLEPT HERE (II,1278)

SPEWACK, BELLA
KISS ME, KATE (I,2568); MR. BROADWAY (I,3130); MY THREE ANGELS (I,3204)

SPEWACK, SAM
KISS ME, KATE (I,2568); MR. BROADWAY (I,3130); MY THREE ANGELS (I,3204)

SPIEGEL, ED
THE DICK POWELL SHOW (I,1269)

SPIEGEL, LARRY
THE ABC AFTERSCHOOL SPECIAL (II,1); RETURN TO THE PLANET OF THE APES (II,2145)

SPIELMAN, ED
THE LAST NINJA (II,1438)

SPIER, WILLIAM
THE UNTOUCHABLES (I,4682); WILLY (I,4868); WILLY (I,4869)

SPIES, ADRIAN
THE ALCOA HOUR (I,108); ALCOA/GOODYEAR THEATER (I,107); CLIMAX!

(I,976); THE DESILU PLAYHOUSE (I,1237); THE DICK POWELL SHOW (I,1269); FRONT ROW CENTER (I,1694); MEETING AT APALACHIN (I,2994); ON YOUR ACCOUNT (I,3364); PURSUIT (I,3700); SAINTS AND SINNERS (I,3887); SAVAGE SUNDAY (I,3926); STAR TREK (II,2440); SUSPICION (I,4309); WAGON TRAIN (I,4747)

SPIES, VIRGINIA
FRONT ROW CENTER (I,1694)

SPIGELGASS, LEONARD
PLAYHOUSE 90 (I,3623)

SPINNER, ANTHONY
CANNON (II,424); CARIBE (II,439); CRISIS (I,1102); THE FBI (I,1551); KOJAK (II,1406); THE MAN FROM U.N.C.L.E. (I,2867); MATINEE THEATER (I,2947); NAKED CITY (I,3216)

SPINRAD, NORMAN
STAR TREK (II,2440)

SPIRO, HERBERT ABBOTT
THE UNTOUCHABLES (I,4682)

SPOONER, DENNIS
THE AVENGERS (II,121); THE BARON (I,360); BERGERAC (II,209); THE FEAR IS SPREADING (I,1553); THE NEW AVENGERS (II,1816); TERROR FROM WITHIN (I,4404); THUNDERBIRDS (I,4497)

SPROAT, RON
DARK SHADOWS (I,1157)

SPRUNG, SANDY
IT'S YOUR MOVE (II,1259); THE JEFFERSONS (II,1276)

ST CLAIR, ARTHUR
FIRESIDE THEATER (I,1580)

ST CLAIR, LEONARD
COUNTERPOINT (I,1059)

ST ELLIS, JOSEPH
CLIMAX! (I,976); LAND OF THE GIANTS (I,2611); THE ROGUES (I,3832); THE TIME TUNNEL (I,4511)

STAAHL, JIM
BIG CITY COMEDY (II,233); THE BOOK OF LISTS (II,350); I'VE HAD IT UP TO HERE (II,1221); LAUGH TRAX (II,1443); MARIE (II,1629); ROCK COMEDY (II,2191); TWILIGHT THEATER (II,2685); THE YESTERDAY SHOW (II,2850)

STADD, ARLENE
HAWAII FIVE-O (II,1110); JEANNIE (II,1275)

STADD, LEONARD
EIGHT IS ENOUGH (II,762); HAWAII FIVE-O (II,1110); JEANNIE (II,1275) THE TIME TUNNEL (I,4511);

STAFFORD, JIM
THE JIM STAFFORD SHOW (II,1291)

STAHL, C RAY
SKY KING (I,4077)

STALL, MARSHA
THE QUIZ KIDS (II,2103)

STAMBLER, MARTIN
POWERHOUSE (II,2074)

STAMBLER, ROBERT
PLEASE DON'T EAT THE DAISIES (I,3628)

STANDER, ARTHUR
THE ANDY GRIFFITH SHOW (I,192); THE ANDY GRIFFITH SHOW (I,193); EVERYTHING HAPPENS TO ME (I,1481); I MARRIED JOAN (I,2174); MAKE ROOM FOR DADDY (I,2843); MRS. G. GOES TO COLLEGE (I,3155)

STANDER, JERRY
TEXACO STAR THEATER (I,4412)

STANLEY, JACKSON
COUNTERPOINT (I,1059); CROWN THEATER WITH GLORIA SWANSON (I,1107); PEOPLE ARE FUNNY (I,3518)

STANLEY, LEE
BANACEK (II,138)

STANLEY, RALPH
AUNT JENNY'S REAL LIFE STORIES (I,314)

STANTON, JOHN
THE SAINT (II,2237)

STARETSKI, JOSEPH
BABY, I'M BACK! (II,130); THREE'S A CROWD (II,2613); THREE'S COMPANY (II,2614)

STARK, JAMES
PARTNERS IN CRIME (II,1961)

STARK, SHELDON
BATMAN (I,366); THE BIG STORY (I,432); CLIMAX! (I,976); THE FUGITIVE (I,1701); IVANHOE (I,2282); LAND OF THE GIANTS (I,2611); THE MAN FROM U.N.C.L.E. (I,2867); MATINEE THEATER (I,2947); QUINCY, M. E. (II,2102); THE RAT PATROL (I,3726); VOYAGE TO THE BOTTOM OF THE SEA (I,4743); THE WALTONS (II,2740)

STARKE, ROBERT
WONDER WOMAN (SERIES 2) (II,2829)

STARR, BEN
ALL IN THE FAMILY (II,38); ALMOST AMERICAN (II,55); BACHELOR FATHER (I,323); BOB HOPE SPECIAL: BOB HOPE IN "JOYS" (II,284); THE DICK POWELL SHOW (I,1269); DIFF'RENT STROKES (II,674); I MARRIED JOAN (I,2174); MISSION MAGIC (I,3068); MORK AND MINDY (II,1735); MR. ED (I,3137); MY FAVORITE MARTIAN (I,3189); MY FAVORITE MARTIANS (II,1779); MY SISTER HANK (I,3201); SILVER SPOONS (II,2355); TOM, DICK, AND HARRY (I,4536)

STEARNS, JOHNNY
MARY KAY AND JOHNNY (I,2932)

STEED, BEN
BLAKE'S SEVEN (II,264)

STEELE, AUSTIN
DAVE ALLEN AT LARGE (II,621)

STEEN, NANCY
DREAMS (II,735); MARIE (II,1629); POLICE SQUAD! (II,2061)

STEFAN, BUD
SANDY DREAMS (I,3909)

STEFANO, JOSEPH
THE OUTER LIMITS (I,3426); THE UNKNOWN (I,4680)

STEFFINS, ROY
CAPTAIN Z-RO (I,839)

STEIGER, JOEL
THE FAMILY TREE (II,820); FLAMINGO ROAD (II,872); REMINGTON STEELE (II,2130); RYAN'S FOUR (II,2233); SCARECROW AND MRS. KING (II,2268); TWO MARRIAGES (II,2691); THE YEAGERS (II,2843)

STEIN, HENRY
FLAMINGO ROAD (II,872)

STEIN, JAMES
AMERICA 2-NIGHT (II,65); THE CAPTAIN AND TENNILLE (II,429); THE CAROL BURNETT SHOW (II,443); THE CRACKER BROTHERS (II,599); DICK CLARK'S GOOD OL' DAYS: FROM BOBBY SOX TO BIKINIS (II,671); DOC (II,682); DOUBLE TROUBLE (II,724); JOE AND SONS (II,1296); JOE AND SONS (II,1297); JOEY AND DAD (II,1306); LILY (I,2705); THE LOU RAWLS SPECIAL (II,1527); THE LOVE BOAT I (II,1532); A NEW KIND OF FAMILY (II,1819); NO SOAP, RADIO (II,1856); PEEPING TIMES (II,1990); PRIVATE BENJAMIN (II,2087); SILVER SPOONS (II,2355); THE SMOTHERS BROTHERS SHOW (II,2384); VAN DYKE AND COMPANY (II,2718)

STEIN, JEFF
MR. MOM (II,1752)

STEIN, JOE
THE ALL-STAR SUMMER REVUE (I,139); AMERICAN COWBOY (I,169); A DATE WITH DEBBIE (I,1161); THE GINGER ROGERS SHOW (I,1804); PHIL SILVERS ON BROADWAY (I,3578); THE RED BUTTONS SHOW (I,3750); SCHOOL HOUSE (I,3937); SOLID GOLD HITS (II,2397)

STEIN, SOL
PERRY MASON (II,2025)

STEINBERG, DAVID
THE NOONDAY SHOW (II,1859); THE RETURN OF THE SMOTHERS BROTHERS (I,3777)

STEINBERG, NORMAN
ALAN KING LOOKS BACK IN ANGER—A REVIEW OF 1972 (I,99); COMEDY TONIGHT (I,1027); THE MANY FACES OF COMEDY (I,2894); ROOSEVELT AND TRUMAN (II,2209); THE SIX O'CLOCK FOLLIES (II,2368); THE WONDERFUL WORLD OF AGGRAVATION (I,4894)

STEINBERG, SCOTT
THE WOLFMAN JACK RADIO SHOW (II,2816)

STEINBERG, ZIGGY
THE BOB NEWHART SHOW (II,342); THE JERK, TOO (II,1282); LILY FOR PRESIDENT (II,1478); LILY—SOLD OUT (II,1480); THE MARY TYLER MOORE SHOW (II,1640); THE NOONDAY SHOW (II,1859); THREE'S COMPANY (II,2614)

STEINEM, GLORIA
THAT WAS THE YEAR THAT WAS (II,2575)

STEINER, FRED
HOTEL (II,1192)

STENGER, DICK
CLIMAX! (I,976); PERRY MASON (II,2025)

STEPHENS, GARN
ST. ELSEWHERE (II,2432); TRAPPER JOHN, M.D.

(II,2654)

STEPHENSON, EDWARD
MAUDE (II,1655)

STERN, ALLAN
THE WHISTLING WIZARD (I,4844)

STERN, HENRY
ALOHA PARADISE (II,58); ANOTHER DAY (II,94); EMERALD POINT, N.A.S. (II,773); FALCON CREST (II,808); HUSBANDS, WIVES AND LOVERS (II,1209); LOVE, SIDNEY (II,1547); PAPER DOLLS (II,1952)

STERN, JOHN
WISHMAN (II,2811)

STERN, JOSEPH
JUNIOR MISS (I,2493)

STERN, LEONARD B
FEMALE INSTINCT (I,1564); THE HERO (I,2035); THE HONEYMOONERS (II,1175); THE HONEYMOONERS (I,2110); THE JACKIE GLEASON SHOW (I,2322); MCMILLAN AND WIFE (II,1667); THE NEW STEVE ALLEN SHOW (I,3271); ONCE UPON A DEAD MAN (I,3369); OPERATION PETTICOAT (II,1916); OPERATION PETTICOAT (II,1917); PARTNERS IN CRIME (II,1961); THE PHIL SILVERS SHOW (I,3580); SNAFU (II,2386); THE STEVE ALLEN SHOW (I,4220); WINDOWS, DOORS AND KEYHOLES (II,2806)

STERN, MARTIN
AGGIE (I,90)

STERN, ROBERT
OH MADELINE (II,1880)

STERN, SANDOR
CUTTER TO HOUSTON (II,612); THE ROOKIES (II,2208); TRUE GRIT (II,2664)

STERN, STEVEN H
PETULA (I,3572)

STERN, STEWART
PLAYHOUSE 90 (I,3623)

STERNIN, ROBERT
ALICE (II,33); IT'S NOT EASY (II,1256); WHO'S THE BOSS? (II,2792)

STETTIN, MONTE
THE GEORGIA PEACHES (II,979)

STEVEN, ANTHONY
FANNY BY GASLIGHT (II,823)

STEVEN, JEREMY
PLAYBOY AFTER DARK (I,3620)

STEVENS JR, GEORGE
CHRISTMAS IN WASHINGTON (II,520)

STEVENS, BOB
GILLIGAN'S ISLAND (II,990); NIGHT COURT (II,1844)

STEVENS, JEREMY
ALAN KING'S FINAL WARNING (II,29); AMERICA 2-NIGHT (II,65); THE ANDY WILLIAMS SHOW (II,78); BARBARA MANDRELL AND THE MANDRELL SISTERS (II,143); THE DIAHANN CARROLL SHOW (II,665); JOE'S WORLD (II,1305); THE MAC DAVIS CHRISTMAS SPECIAL (II,1572); THE MAC DAVIS SPECIAL (II,1577); THE STEVE ALLEN COMEDY HOUR (II,2454); THICKE OF THE NIGHT (II,2587); WHAT'S HAPPENING!! (II,2769); THE WONDERFUL WORLD OF JONATHAN WINTERS (I,4900)

STEVENS, JIMMY
FERNWOOD 2-NIGHT (II,849)

STEVENS, LESLIE
BUCK ROGERS IN THE 25TH CENTURY (II,383); CRESCENDO (I,1094); FOUR STAR PLAYHOUSE (I,1652); FRANK MERRIWELL (I,1659); THE GEMINI MAN (II,961); I LOVE A MYSTERY (I,2170); MCCLOUD (II,1660); THE OUTER LIMITS (I,3426); PROBE (I,3674); STONESTREET: WHO KILLED THE CENTERFOLD MODEL? (II,2472); TACK REYNOLDS (I,4329); THE VIRGINIAN (I,4737); WEAPONS MAN (I,4783)

STEVENSON, GEORGE
CAVALCADE OF AMERICA (I,875)

STEVENSON, JOCELYN
FRAGGLE ROCK (II,916)

STEVENSON, JOHN
CORONATION STREET (I,1054)

STEVENSON, MCLEAN
M*A*S*H (II,1569); THE MCLEAN STEVENSON SHOW (II,1664); THE SUMMER SMOTHERS BROTHERS SHOW (I,4281)

STEVENSON, ROBERT
CAVALCADE OF AMERICA (I,875)

STEVIN, GEORGE
COMBAT (I,1011)

STEWART, BRUCE
THE ONEDIN LINE (II,1912)

STEWART, CHARLES
ADAMS OF EAGLE LAKE (II,10); THE ANDY GRIFFITH SHOW (I,192); HEY, JEANNIE! (I,2038); JOE'S WORLD (II,1305); MAKE ROOM FOR DADDY (I,2843)

STEWART, CHUCK
ALL IN THE FAMILY (II,38); THE DON KNOTTS SHOW (I,1334); THREE'S COMPANY (II,2614)

STEWART, DOROTHY
MATINEE THEATER (I,2947)

STEWART, FRED MUSTARD
ELLIS ISLAND (II,768)

STEWART, IAN SCOTT
U.F.O. (I,4662)

STEWART, LARRY
THRILL SEEKERS (I,4491)

STEWART, LYNNE
THE PEE WEE HERMAN SHOW (II,1988)

STEWART, MIKE
FIREBALL FUN FOR ALL (I,1577); THE POLLY BERGEN SHOW (I,3643); SID CAESAR INVITES YOU (I,4040)

STEWART, R J
REMINGTON STEELE (II,2130)

STEWART, ROBERT BANKS
THE AVENGERS (II,121); BERGERAC (II,209); DANGER MAN (I,1136); DOCTOR WHO (II,689); THE HUNTER (I,2162); SECRET AGENT (I,3962)

STEWART, SUSAN
HARPER VALLEY (II,1094); KNOTS LANDING (II,1404); ROMANCE THEATER (II,2204)

STEWART, WILLIAM L
LAND OF THE GIANTS (I,2611)

STILES, NORMAN
AMERICA 2-NIGHT (II,65); THE CAPTAIN AND TENNILLE (II,429); SESAME STREET (I,3982); SPACE FORCE (II,2407); WHEN THINGS WERE ROTTEN (II,2782)

STILLMAN, DEANN
SQUARE PEGS (II,2431)

STIRDIVANT, MARC
BARETTA (II,152)

STIVERS, BUNNY
CELEBRITY DAREDEVILS (II,473); CIRCUS OF THE STARS (II,538)

STIVERS, ROBERT
CELEBRITY DAREDEVILS (II,473)

STONE JR, N B
BONANZA (II,347); THE RIFLEMAN (I,3789)

STONE, ANDREW
SCREEN DIRECTOR'S PLAYHOUSE (I,3946)

STONE, ERIC
THE ELEVENTH HOUR (I,1425)

STONE, JON
JULIE ON SESAME STREET (I,2485); MR. MAYOR (I,3143); TALES FROM MUPPETLAND (I,4344)

STONE, JOSEPH
BLACK SADDLE (I,480); BONANZA (II,347); WAGON TRAIN (I,4747)

STONE, MARY
ADAM'S RIB (I,32)

STONE, N B
THE RESTLESS GUN (I,3774)

STONE, NOREEN
PALMERSTOWN U.S.A. (II,1945)

STONE, PAULA
PAULA STONE'S TOY SHOP (I,3495)

STONE, PAULINE
FOR BETTER OR WORSE (I,1621); THE LORETTA YOUNG THEATER (I,2756)

STONE, PETER
ADAM'S RIB (I,32); ANDROCLES AND THE LION (I,189); GHOSTBREAKER (I,1789); HAPPY ENDINGS (II,1085)

STONE, SID
JACKIE GLEASON AND HIS AMERICAN SCENE MAGAZINE (I,2321)

STONE, WALTER
THE BIG SELL (I,431); CAVALCADE OF STARS (I,877); THE HONEYMOONERS (I,2110); THE HONEYMOONERS (II,1175); THE HONEYMOONERS CHRISTMAS SPECIAL (II,1176); THE HONEYMOONERS CHRISTMAS (II,1177); THE HONEYMOONERS SECOND HONEYMOON (II,1178); THE

HONEYMOONERS VALENTINE SPECIAL (II,1179); THE JACKIE GLEASON SHOW (I,2322); THE JACKIE GLEASON SPECIAL (I,2324); THE JACKIE GLEASON SPECIAL (I,2325); THE MILLION DOLLAR INCIDENT (I,3045); THAT'S LIFE (I,4426)

STORM, HOWARD
THE PARTRIDGE FAMILY (II,1962); PEOPLE LIKE US (II,1999)

STRANGIS, GREG
EIGHT IS ENOUGH (II,762); FALCON CREST (II,808); HARPER VALLEY (II,1094); THE KAREN VALENTINE SHOW (I,2506); LAVERNE AND SHIRLEY (II,1446); SHIRLEY (II,2330); TWO MARRIAGES (II,2691)

STRASSFIELD, ADLE
THE TWILIGHT ZONE (I,4651)

STRATTON, BILL
THE BLUE KNIGHT (II,276); HAWAII FIVE-O (II,1110); MICKEY SPILLANE'S MIKE HAMMER: MORE THAN MURDER (II,1693); MICKEY SPILLANE'S MIKE HAMMER: MURDER ME, MURDER YOU (II,1694); THE ROCKFORD FILES (II,2197); STRIKE FORCE (II,2480); SWITCH (II,2519); VEGAS (II,2724)

STRAUSS, OSKAR
THE CHOCOLATE SOLDIER (I,944)

STRAUSS, RICHARD
TELE-VARIETIES (I,4377)

STRONG, JOHN
RETURN TO THE PLANET OF THE APES (II,2145)

STRUTTON, BILL
IVANHOE (I,2282)

STRYER, ADELE
TOO CLOSE FOR COMFORT (II,2642)

STRYER, BART
TOO CLOSE FOR COMFORT (II,2642)

STUART, JEFF
PAPER DOLLS (II,1952); TRAPPER JOHN, M.D. (II,2654)

STUART, LAURA
WONDER GIRL (I,4891)

STUART, MEL
RIPLEY'S BELIEVE IT OR NOT (II,2177); SOPHIA! (I,4128)

STURGEON, TED
ROD BROWN OF THE ROCKET RANGERS (I,3825); STAR TREK (II,2440)

STYLER, ADELE
DOC (II,683); JOE AND SONS (II,1296); JOE'S WORLD (II,1305); PAT BOONE AND FAMILY (II,1968); POPI (II,2069)

STYLER, BURT
DOC (II,683); DOCTOR CHRISTIAN (I,1308); THE FLYING NUN (I,1611); THE HOUSE NEXT DOOR (I,2139); JOE AND SONS (II,1296); JOE'S WORLD (II,1305); M*A*S*H (II,1569); MY FAVORITE MARTIAN (I,3189); PAT BOONE AND FAMILY (II,1968); POPI (II,2069); RUSSIAN ROULETTE (I,3875); TEXAS JOHN SLAUGHTER (I,4414)

STYLES, STUART
IT'S ABOUT TIME (I,2263)

SUBOTSKY, MILT
QUIZZING THE NEWS (I,3713)

SUDINOW, JACK
BAY CITY BLUES (II,186)

SUHOSKY, BOB
CODE R (II,551)

SULLIVAN, NORMAN
THE BOB HOPE SHOW (I,541); THE BOB HOPE SHOW (I,543); THE BOB HOPE SHOW (I,560); THE BOB HOPE SHOW (I,561); ROBERTA (I,3815)

SULLY, PAUL
THE GEORGE GOBEL SHOW (I,1768)

SULTAN, ARNE
BARNEY MILLER (II,154); CAPTAIN NICE (I,835); THE HERO (I,2035); THE JONES BOYS (I,2446); MY LUCKY PENNY (I,3196); NOBODY'S PERFECT (II,1858); POOR DEVIL (I,3646); POPI (II,2069); TOO CLOSE FOR COMFORT (II,2642); WINDOWS, DOORS AND KEYHOLES (II,2806)

SULTAN, RICK
THE FOUR SEASONS (II,915); TOO CLOSE FOR COMFORT (II,2642)

SUMMERS, JARON
THE INCREDIBLE HULK (II,1232)

SUMMERS, MARK
HIT MAN (II,1156)

SUNDELL, FRANK
THE NEW $100,000 NAME THAT TUNE (II,1810)

SUNGA, GEORGE
ROWAN AND MARTIN BITE THE HAND THAT FEEDS THEM (I,3853)

SUNSHINE, MADELYN
THE NEW ODD COUPLE (II,1825); STAR OF THE FAMILY (II,2436); WEBSTER (II,2758)

SUNSHINE, STEVE
THE NEW ODD COUPLE (II,1825); STAR OF THE FAMILY (II,2436); WEBSTER (II,2758)

SURGAL, JON
THE BAXTERS (II,183); HOWDY DOODY AND FRIENDS (I,2152)

SURNOW, JOEL
LEGMEN (II,1458); MIAMI VICE (II,1689); ST. ELSEWHERE (II,2432)

SURTEES, JAMES T
MANNIX (II,1624)

SUSSMAN, BRUCE
MUSICAL CHAIRS (II,1776)

SUSSMAN, SALLY
THE YOUNG AND THE RESTLESS (II,2862)

SUSTARSIC, STEPHEN
ONE DAY AT A TIME (II,1900)

SUTHERLAND, DEBORAH
TOP OF THE WORLD (II,2646)

SUTTON, MICHAEL
SQUARE PEGS (II,2431)

SVANOE, BILL
GRIFF (I,1888)

SWALE, TOM
THE FALL GUY (II,811); HART TO HART (II,1102); LOTTERY (II,1525)

SWANSON, HAROLD
ALCOA PREMIERE (I,109)

SWANSON, LOGAN
COMBAT (I,1011)

SWANSON, ROBERT E
HART TO HART (II,1102); HOTEL (II,1192); KOJAK (II,1406); MCMILLAN (II,1666); MURDER, SHE WROTE (II,1770); STARSKY AND HUTCH (II,2444); WALKING TALL (II,2738)

SWANTON, HAROLD
THE ALCOA HOUR (I,108); ALFRED HITCHCOCK PRESENTS (I,115); THE CHISHOLMS (II,513); HAWAII FIVE-O (II,1110); HEC RAMSEY (II,1121); LITTLE HOUSE ON THE PRAIRIE (II,1487); MR. TUTT (I,3153); PERRY MASON (II,2025); TURN OF FATE (I,4623); WAGON TRAIN (I,4747)

SWANTON, SCOTT
LITTLE HOUSE ON THE PRAIRIE (II,1487)

SWAYBILL, ROGER
ONE IN A MILLION (I,3379)

SWEENEY, RENEE
GETTING READY (II,984)

SWEENEY, TERRENCE A
BARNABY JONES (II,153)

SWERDLOFF, HELAINE
PEOPLE TO PEOPLE (II,2000)

SWERLING JR, JO
THE A-TEAM (II,119); HAZARD'S PEOPLE (II,1112)

SWIFT, DAVID
CAMP RUNAMUCK (I,811); COOGAN'S REWARD (I,1044); HOT L BALTIMORE (II,1187); JAMIE (I,2333); MR. PEEPERS (I,3147); PLAYHOUSE 90 (I,3623); RAWHIDE (I,3727); WAGON TRAIN (I,4747)

SWIMER, LARRY
THE FACTS OF LIFE (II,805)

SYLVESTER, ROBERT
NAKED CITY (I,3215)

SYLVESTER, STEVE
ANOTHER LIFE (II,95)

SYMINGTON, DONALD
MATINEE THEATER (I,2947)

SYNES, BOB
DREAM HOUSE (II,734)

SZOLLASI, THOMAS
THE INCREDIBLE HULK (II,1232); THE POWERS OF MATTHEW STAR (II,2075); REMINGTON STEELE (II,2130)

SZUREK, SAM
LEGENDS OF THE SCREEN (II,1456)

TABAT, JORDAN
THE PLANT FAMILY (II,2050)

TABMAN, AUDREY
DONNY AND MARIE (II,712)

TABOR, PAUL
COLONEL MARCH OF SCOTLAND YARD (I,1005)

TACKABERRY, JOHN
THE JUDY GARLAND SHOW (I,2470); TIME OUT FOR GINGER (I,4508)

TADMAN, AUBREY

9 TO 5 (II,1852); THE CRACKER BROTHERS (II,599); THE JOHNNY CASH SHOW (I,2424); THE KEANE BROTHERS SHOW (II,1371); PRIVATE BENJAMIN (II,2087); SANDY IN DISNEYLAND (II,2254); THAT'S TV (II,2581); VAN DYKE AND COMPANY (II,2719)

TAGGART, MILLEE
AS THE WORLD TURNS (II,110)

TAGGERT, BRIAN
V: THE FINAL BATTLE (II,2714)

TAIT, DON
COMBAT (I,1011); HERBIE, THE LOVE BUG (II,1131); MR. MERLIN (II,1751); SKY KING (I,4077)

TALLISMAN, DAVID
THE NEW MICKEY MOUSE CLUB (II,1823)

TAMS, VIC
THE DICK POWELL SHOW (I,1269)

TANACHUCK, NATHANIEL
BONANZA (II,347); BRANDED (I,707); WAGON TRAIN (I,4747)

TANG, CECILE
PEKING ENCOUNTER (II,1991)

TANNER, CHARLES
GILLIGAN'S ISLAND (II,990)

TANNER, JOHN
SHANE (I,3995)

TAPP, GORDIE
HEE HAW (II,1123)

TAPPIS, JOEL
HARPER VALLEY PTA (II,1095); HOUSE CALLS (II,1194)

TARAS, MARTIN
BATFINK (I,365)

TARCHER, JEREMY
STAR TREK (II,2440)

TARLOFF, ERIK
ALL IN THE FAMILY (II,38); HOUSE CALLS (II,1194); M*A*S*H (II,1569); MY WORLD . . . AND WELCOME TO IT (I,3208); THAT'S MY MAMA (II,2580); TOO CLOSE FOR COMFORT (II,2642); TRAPPER JOHN, M.D. (II,2654)

TARLOFF, FRANK
I MARRIED JOAN (I,2174)

TARSES, JAY
THE ARTHUR GODFREY SPECIAL (I,288); ARTHUR GODFREY'S PORTABLE ELECTRIC MEDICINE SHOW (I,290); BING CROSBY'S SUN VALLEY CHRISTMAS SHOW (I,475); THE BOB NEWHART SHOW (II,342); BUFFALO BILL (II,387); THE CHOPPED LIVER BROTHERS (II,514); DIANA (I,1258); OPEN ALL NIGHT (II,1914); SITCOM (II,2367); THE TONY RANDALL SHOW (II,2640)

TASH, MAX
FAME (II,812); WKRP IN CINCINNATI (II,2814)

TASHLIN, FRANK
THE FRANCES LANGFORD SHOW (I,1657)

TATARANOWICZ, THOMAS
PAC-MAN (II,1940)

TATASHORE, FRED
BURT AND THE GIRLS (I,770); BURT REYNOLDS' LATE SHOW (I,777)

TAUB, BILL
POPI (II,2069); THE POWERS OF MATTHEW STAR (II,2075)

TAWELL, GEORGE
STAR SEARCH (II,2438)

TAYLOR, BILL
BUCK ROGERS IN THE 25TH CENTURY (II,383); WONDER WOMAN (SERIES 2) (II,2829)

TAYLOR, BRUCE
DIFF'RENT STROKES (II,674); THE ROWAN AND MARTIN REPORT (II,2221); T.L.C. (II,2525); THE VIN SCULLY SHOW (I,4734)

TAYLOR, DAVID
KOJAK (II,1406); THE ROCKFORD FILES (II,2197); SWITCH (II,2519)

TAYLOR, DON
BURKE'S LAW (I,762); ROBIN'S NEST (II,2187)

TAYLOR, DOUGLAS
APPOINTMENT WITH ADVENTURE (I,246)

TAYLOR, GIL
DEPARTMENT S (I,1232)

TAYLOR, IRVING
AND HERE'S THE SHOW (I,187)

TAYLOR, JERI
BLUE THUNDER (II,278); THE INCREDIBLE HULK (II,1232); QUINCY, M. E. (II,2102)

TAYLOR, KURT
FLIP WILSON. . . OF COURSE (II,890); GOOD TIMES (II,1038); THE JEFFERSONS (II,1276); THE NEW ODD COUPLE (II,1825)

TAYLOR, LEIGH
THE REDD FOXX COMEDY HOUR (II,2126); THAT'S TV (II,2581)

TAYLOR, MARK
FOUR STAR PLAYHOUSE (I,1652)

TAYLOR, PAUL
YOU ASKED FOR IT (II,2855)

TAYLOR, PHILIP
THE BRADY BRIDES (II,361); THE INCREDIBLE HULK (II,1232); LANDON, LANDON & LANDON (II,1426); QUICK AND QUIET (II,2099)

TAYLOR, RAY
CHER (II,495); CHER (II,496); COMEDY NEWS II (I,1021); GABRIEL KAPLAN PRESENTS THE SMALL EVENT (II,952); GLEN CAMPBELL . . . DOWN HOME—DOWN UNDER (II,1008); THAT THING ON ABC (II,2574)

TAYLOR, RENEE
BEDROOMS (II,198); GOOD PENNY (II,1036); LOVERS AND OTHER STRANGERS (II,1550); A LUCILLE BALL SPECIAL STARRING LUCILLE BALL AND JACKIE GLEASON (II,1558); MARLO THOMAS IN ACTS OF LOVE—AND OTHER COMEDIES (I,2919); PARADISE (II,1955)

TAYLOR, SAMUEL W
THE BIG BRAIN (I,424); THE HAPPY TIME (I,1943); WESLEY (I,4795)

TAYLOR, WARREN
WOLF ROCK TV (II,2815)

TEED, TOM
CROSSROADS (I,1105)

TEEL, ROSS
HART TO HART (II,1102); HOTEL (II,1192); KOJAK (II,1406)

TEFFEL, BESS
STORY THEATER (I,4240)

TEGLAND, ARCHIE
THE MAN FROM U.N.C.L.E. (I,2867)

TEICHER, ROY
GOODTIME GIRLS (II,1042)

TEICHMANN, HOWARD
THE FORD 50TH ANNIVERSARY SHOW (I,1630); SHOWTIME, U.S.A. (I,4038); SLEZAK AND SON (I,4086)

TELFORD, FRANK
CHIPS (II,511); FATHER MURPHY (II,841); KNIGHT RIDER (II,1402); MANNIX (II,1624); MCCLAIN'S LAW (II,1659); QUINCY, M. E. (II,2102); THE ROOKIES (II,2208)

TEMPLETON, WILLIAM
THE DESILU PLAYHOUSE (I,1237)

TENNANT, DON
HOLD 'ER NEWT (I,2070)

TENNEY, MARTY
BIZARRE (II,259); LIFE'S MOST EMBARRASSING MOMENTS (II,1470); LIFE'S MOST EMBARRASSING MOMENTS II (II,1471); LIFE'S MOST EMBARRASSING MOMENTS III (II,1472); LIFE'S MOST EMBARRASSING MOMENTS IV (II,1473)

TENNILLE, TONI
THE CAPTAIN AND TENNILLE SONGBOOK (II,432)

TENNIS, CRAIG
ROCK-N-AMERICA (II,2195); TV FUNNIES (II,2672)

TENOWICH, TOM
THE DEAN MARTIN SHOW (I,1201); THE DOM DELUISE SHOW (I,1328); THE GOLDDIGGERS (I,1838); THE GOLDDIGGERS IN LONDON (I,1837); JACK BENNY'S FIRST FAREWELL SHOW (I,2300); LEWIS AND CLARK (II,1465); MORK AND MINDY (II,1735); PHYLLIS (II,2038); THAT'S MY MAMA (II,2580)

TERPILOFF, ANTHONY
RETURN OF THE SAINT (II,2141)

TERRANCE, MICHAEL
THE CHEROKEE TRAIL (II,500)

TESCHER, ROY
OH MADELINE (II,1880)

TESICH, STEVE
BREAKING AWAY (II,371)

TESSIER, MICHAEL
LOST IN SPACE (I,2758)

TEWKSBURY, PETER
ALCOA PREMIERE (I,109)

TEZUKA, OSAMU
SPACE GIANTS (I,4142)

THACKABERRY, JOHN
THE JACK BENNY PROGRAM (I,2294)

THAW, MORT
ROUTE 66 (I,3852); THE WALTONS (II,2740)

THICKE, ALAN
AMERICA 2-NIGHT (II,65); ANNE MURRAY'S CARIBBEAN CRUISE (II,88); ANNE MURRAY'S LADIES' NIGHT (II,89); ANNE MURRAY'S WINTER CARNIVAL. . . FROM QUEBEC (II,90); THE BARRY MANILOW SPECIAL (II,155); THE BOBBY DARIN AMUSEMENT COMPANY (I,672); THE BOBBY VINTON SHOW (II,345); FERNWOOD 2-NIGHT (II,849); THE FLIP WILSON SPECIAL (I,1603); THE FLIP WILSON SPECIAL (II,887); THE FLIP WILSON SPECIAL (II,888); THE FLIP WILSON SPECIAL (II,889); LOLA (II,1508); LOLA (II,1509); LOLA (II,1510); LOLA (II,1511); MAC DAVIS CHRISTMAS SPECIAL. . . WHEN I GROW UP (II,1574); OLIVIA (II,1883); THE OLIVIA NEWTON-JOHN SHOW (II,1885); OLIVIA NEWTON-JOHN'S HOLLYWOOD NIGHTS (II,1886); THE PAUL LYNDE COMEDY HOUR (II,1977); PLAY IT AGAIN, UNCLE SAM (II,2052); THE RICHARD PRYOR SHOW (II,2163); THE RICHARD PRYOR SPECIAL (II,2164); THE RICHARD PRYOR SPECIAL? (II,2165); THE SANDY DUNCAN SHOW (II,2253); A SPECIAL ANNE MURRAY CHRISTMAS (II,2413); THICKE OF THE NIGHT (II,2587)

THIEL, NICK
THE FALL GUY (II,811); HAWAIIAN HEAT (II,1111); HOME ROOM (II,1172); MAGNUM, P.I. (II,1601); VOYAGERS (II,2730)

THOMAS JR, LOWELL
HIGH ADVENTURE WITH LOWELL THOMAS (I,2048)

THOMAS, BARRY
THE ONEDIN LINE (II,1912)

THOMAS, BRANDON
PLAYHOUSE 90 (I,3623)

THOMAS, CHARLES
JIGSAW (I,2377)

THOMAS, DAVE
FROM CLEVELAND (II,931); THE NEW SHOW (II,1828); SECOND CITY TELEVISION (II,2287)

THOMAS, FRANKIE
TOM CORBETT, SPACE CADET (I,4535)

THOMAS, JERRY
MANNIX (II,1624); NAKED CITY (I,3216); THE WILD WILD WEST (I,4863)

THOMAS, LARRY
ELVIRA'S MOVIE MACABRE (II,771)

THOMAS, LESLIE
AS THE WORLD TURNS (II,110)

THOMAS, ROY
GOLDIE GOLD AND ACTION JACK (II,1023); THUNDARR THE BARBARIAN (II,2616)

THOMAS, TED
THE NEW ADVENTURES OF CHARLIE CHAN (I,3249)

THOMPSON, BERNIE
ONE DAY AT A TIME (II,1900)

THOMPSON, CHRIS
BLANSKY'S BEAUTIES (II,267); BOSOM BUDDIES (II,355); I DO, I DON'T (II,1212)

THOMPSON, CYNTHIA
THE LOVE BOAT (II,1535)

THOMPSON, GENE
APPLE'S WAY (II,101); CANNON (II,424); COLUMBO (II,556); THE FLYING NUN (I,1611); HERE WE GO AGAIN (I,2028); MARCUS WELBY, M.D. (II,1627); MY FAVORITE MARTIAN (I,3189); MY THREE SONS (I,3205); THE SMOTHERS BROTHERS SHOW (I,4096); SWITCH (II,2519); THESE ARE THE DAYS (II,2585)

THOMPSON, JACK
THE BEST LITTLE SPECIAL IN TEXAS (II,215)

THOMPSON, JAY
ONCE UPON A MATTRESS (I,3371)

THOMPSON, MARION
THE LORETTA YOUNG THEATER (I,2756)

THOMPSON, NEIL
DREAMS (II,735); POLICE SQUAD! (II,2061)

THOMPSON, PALMER
AMOS BURKE, SECRET AGENT (I,180); IF YOU KNEW TOMORROW (I,2187); LAW OF THE PLAINSMAN (I,2639); THE RIFLEMAN (I,3789); TARGET: THE CORRUPTERS (I,4360); TURN OF FATE (I,4623); THE UNTOUCHABLES (I,4682)

THOMPSON, ROBERT E
DEADLOCK (I,1190); KISS ME, KILL ME (II,1400); THE MAN FROM U.N.C.L.E. (I,2867); MAN ON THE MOVE (I,2878); MATINEE THEATER (I,2947); SHERLOCK HOLMES: THE HOUND OF THE BASKERVILLES (I,4011); WAGON TRAIN (I,4747)

THOMPSON, RON ALLEN
GOOD TIMES (II,1038)

THOMPSON, THOMAS
BONANZA (II,347); RAWHIDE (I,3727); WAGON TRAIN (I,4747)

THOMPSON, WALLACE
MIKE AND BUFF (I,3037)

THOR, LARRY
SKYHAWKS (I,4079)

THORNE, ROBERT
BARNABY JONES (II,153)

THORNE, WORLEY
CHARLIE'S ANGELS (II,486); WESTSIDE MEDICAL (II,2765)

THORNLEY, STEVEN
BRING 'EM BACK ALIVE (II,377)

THUNA, LEONORA
FAMILY (II,813); LOU GRANT (II,1526); THE NATURAL LOOK (II,1799)

THURMAN, JAMES F
CHANGING SCENE (I,899); CURIOSITY SHOP (I,1114); SIGN-ON (II,2353)

THWAYTES, JOY
THE ADVENTURES OF BLACK BEAUTY (I,51)

TIBBLES, DOUGLAS
THE DORIS DAY SHOW (I,1355); THE MUNSTERS (I,3158); MY THREE SONS (I,3205); RAQUEL (I,3725)

TIBBLES, GEORGE
THE ADDAMS FAMILY (II,13); THE APARTMENT HOUSE (I,241); BRINGING UP BUDDY (I,731); HELLO, LARRY (II,1128); LEAVE IT TO BEAVER (I,2648); MAUDE (II,1655); MUNSTER, GO HOME! (I,3157); THE MUNSTERS (I,3158); MY THREE SONS (I,3205); OCTAVIUS AND ME (I,3327)

TIEFER, GREGORY
BARETTA (II,152)

TILLER, TED
FREEDOM RINGS (I,1682); MR. I. MAGINATION (I,3140)

TILLSTROM, BURR
KUKLA, FRAN, AND OLLIE (I,2594)

TILSLEY, VINCENT
MAN IN A SUITCASE (I,2868); THE PRISONER (I,3670)

TINKER, JOHN
ST. ELSEWHERE (II,2432)

TISDALE, JAMES
THE INCREDIBLE HULK (II,1232); THE JACKSONS (II,1267)

TIVERS, CYNTHIA
PEOPLE TO PEOPLE (II,2000)

TOBIAS, JOHN
OF MEN OF WOMEN (I,3332)

TODD, CLIFF
THE GUNS OF WILL SONNETT (I,1907)

TODMAN, DAVID
THE DUKES (II,741)

TOKAR, NORMAN
THE ALDRICH FAMILY (I,111); ALL IN THE FAMILY (I,133); I REMEMBER CAVIAR (I,2175); YOU'RE ONLY YOUNG TWICE (I,4970)

TOKATYAN, DIANA
FOR LOVE AND HONOR (II,903)

TOKATYAN, LEON
ALLISON SIDNEY HARRISON (II,54); THE BIG STORY (I,432); DECOY (I,1217); FOR LOVE AND HONOR (II,903); KATE MCSHANE (II,1368); LOU GRANT (II,1526)

TOKOFSKY, JERRY
THE SECOND TIME AROUND (II,2289)

TOLKIN, MEL
ACCENT ON LOVE (I,16); THE ADMIRAL BROADWAY REVUE (I,37); ALL IN THE FAMILY (II,38); BACHELOR FATHER (I,323); THE BOB HOPE SHOW (I,636); BOB HOPE SPECIAL: BOB HOPE ON CAMPUS (II,289); BOB HOPE SPECIAL: BOB HOPE PRESENTS THE STARS OF TOMORROW (II,292); BOB HOPE SPECIAL: THE BOB HOPE SPECIAL (II,333); BOB HOPE SPECIAL: THE BOB HOPE SPECIAL (II,334); THE DANNY KAYE SHOW (I,1143); HOLIDAY IN LAS VEGAS (I,2076); THE JERRY LEWIS

SHOW (I,2364); THE JIMMY DURANTE SHOW (I,2389); JOE'S WORLD (II,1305); LOVE, SIDNEY (II,1547); MARRIAGE—HANDLE WITH CARE (I,2921); ROBERTA (I,3816); RUN, BUDDY, RUN (I,3870); THE SID CAESAR SPECIAL (I,4044); THE SID CAESAR, IMOGENE COCA, CARL REINER, HOWARD MORRIS SPECIAL (I,4039); TIPTOE THROUGH TV (I,4524); VARIETY: THE WORLD OF SHOW BIZ (I,4704); YOUR SHOW OF SHOWS (I,4957)

TOMARKEN, PETER
THOSE WONDERFUL TV GAME SHOWS (II,2603)

TOMBLIN, DAVID
THE PRISONER (I,3670); U.F.O. (I,4662)

TOMBRAGEL, MAURICE
DEVLIN (II,663)

TOMERLIN, JOHN
THE PICTURE OF DORIAN GRAY (I,3592); S.W.A.T. (II,2236); THRILLER (I,4492)

TOMLIN, GARY
TEXAS (II,2566)

TOMLIN, LILY
LILY (I,2705); LILY (II,1477); THE LILY TOMLIN SHOW (I,2706); THE LILY TOMLIN SPECIAL (II,1479); THE PAUL SIMON SPECIAL (II,1983)

TOMLINSON, TOMMY
SATAN'S WAITIN' (I,3916)

TOPLYN, JOE
LATE NIGHT WITH DAVID LETTERMAN (II,1442)

TORGAN, BARBARA
SHANE (I,3995)

TORME, MEL
MANNIX (II,1624); THE SINGERS (I,4060)

TOROKVEI, PETER
WKRP IN CINCINNATI (II,2814)

TORS, IVAN
LORNE GREENE'S LAST OF THE WILD (II,1522)

TOUCHSTONE, BOB
MY THREE SONS (I,3205)

TOWBIN, FRED
HERE'S BOOMER (II,1134); LOVE, SIDNEY (II,1547)

TOWNE, ROBERT
THE OUTER LIMITS (I,3426)

TOWNLEY, JACK
THE ABBOTT AND COSTELLO SHOW (I,2); THE ADVENTURES OF CHAMPION (I,54); ANNIE OAKLEY (I,225); BUFFALO BILL JR. (I,755); THE RANGE RIDER (I,3720)

TOWNSEND, LEO
PERRY MASON (II,2025); WAGON TRAIN (I,4747)

TRACY, EMILY
TOO GOOD TO BE TRUE (II,2643)

TRAVIS, NEIL
PLEASE DON'T EAT THE DAISIES (I,3628)

TRELZGER, LEO
FOUR STAR PLAYHOUSE (I,1652)

TREVEY, KEN
BANJO HACKETT: ROAMIN' FREE (II,141); BRANDED (I,707); CODE R (II,551); LOU GRANT (II,1526); MATT HOUSTON (II,1654); NAKED CITY (I,3216); SWISS FAMILY ROBINSON (II,2516); WAGON TRAIN (I,4747)

TRICKER, GEORGE
ARCHIE (II,104); BARBARA MANDRELL AND THE MANDRELL SISTERS (II,143); DINAH AND HER NEW BEST FRIENDS (II,676); THE JIM STAFFORD SHOW (II,1291); JOANIE LOVES CHACHI (II,1295); THE MAC DAVIS SHOW (II,1576); ME AND MAXX (II,1672); OFF THE WALL (II,1879); THE PETER MARSHALL VARIETY SHOW (II,2029); PUMPBOYS AND DINETTES ON TELEVISION (II,2090); THE ROLLERGIRLS (II,2203); SHIPSHAPE (II,2329); SUGAR TIME (II,2489); TONY ORLANDO AND DAWN (II,2639); WELCOME BACK, KOTTER (II,2761)

TRICKER, NEIL
CARTER COUNTRY (II,449)

TRIPP, PAUL
FREEDOM RINGS (I,1682); MR. I. MAGINATION (I,3140)

TRIVERS, BARRY
MANNIX (II,1624); NAKED CITY (I,3216); ROUTE 66 (I,3852); STAR TREK (II,2440)

TRIVERS, PAUL
STORY THEATER (I,4240)

TROGDON, MIRIAM
NEWHART (II,1835)

TRUEBLOOD, GUERDON
THE BASTARD/KENT FAMILY CHRONICLES (II,159); BRAVO TWO (II,368); THE DEVLIN CONNECTION (II,664); THE STREETS OF SAN FRANCISCO (II,2478)

TRUMBO, CHRISTOPHER
NAKIA (II,1784)

TUCK, CECIL
PAT PAULSEN FOR PRESIDENT (I,3472); ROLLIN' ON THE RIVER (I,3834); THE SMOTHERS BROTHERS SHOW (I,4097); THE SUMMER SMOTHERS BROTHERS SHOW (I,4281)

TUCKAHOE, PAUL
COLUMBO (II,556)

TUCKER, LARRY
ALONE AT LAST (II,59); ALONE AT LAST (II,60); THE CARA WILLIAMS SHOW (I,843); THE DANNY KAYE SHOW (I,1143); ETHEL IS AN ELEPHANT (II,783); JENNIFER SLEPT HERE (II,1278); MR. MERLIN (II,1751); PALS (II,1947); TEACHERS ONLY (II,2548)

TUCKER, REX
VANITY FAIR (I,4701)

TUNBERG, KARL
CANNON (II,424)

TUNBERG, WILLIAM
SCREEN DIRECTOR'S PLAYHOUSE (I,3946); THE WILD WILD WEST (I,4863)

TUNICK, IRV
COMBAT (I,1011)

TURKUS, BURTON
MR. ARSENIC (I,3126)

TURLEY, JACK
BANACEK (II,138); DAN AUGUST (I,1130); THE FBI (I,1551); THE FUGITIVE (I,1701); I SPY (I,2179); LAND OF THE GIANTS (I,2611); THE LOVE BOAT (II,1535); THE MAN FROM U.N.C.L.E. (I,2867); RAWHIDE (I,3727); ROUTE 66 (I,3852); THREE FOR THE ROAD (II,2607); VEGAS (II,2724)

TURNER, CONNIE
SZYSZNYK (II,2523)

TURNER, DAVID
THE ROADS TO FREEDOM (I,3802)

TURNER, DENNIS
DYNASTY (II,746)

TURNER, LLOYD
ALICE (II,33); ALL IN THE FAMILY (II,38); THE BULLWINKLE SHOW (I,761); HONEYMOON SUITE (I,2108); THE MISADVENTURES OF SHERIFF LOBO (II,1704); MORK AND MINDY (II,1735); THE NUT HOUSE (I,3320); THE PARTRIDGE FAMILY (II,1962)

TURNER, RICHARD
PLAYWRIGHTS '56 (I,3627)

TURNER, SKIP
WONDER GIRL (I,4891)

TURNEY, CATHERINE
ONE STEP BEYOND (I,3388)

TURRELL, SAUL J
SILENT'S, PLEASE (I,4050)

TURRENTINE, ROGERS
BRET MAVERICK (II,374); CAGNEY AND LACEY (II,409); THE MISSISSIPPI (II,1707); RIKER (II,2175); THE ROCKFORD FILES (II,2197); TODAY'S FBI (II,2628)

TURTELTAUB, SAUL
13 QUEENS BOULEVARD (II,2591); THE BEAUTIFUL PHYLLIS DILLER SHOW (I,381); DOUBLE TROUBLE (II,723); E/R (II,748); GO! (I,1830); ON BROADWAY TONIGHT (I,3355); ONE IN A MILLION (II,1906); ONE OF THE BOYS (II,1911); THE PAT BOONE SHOW (I,3471); THE PERRY COMO WINTER SHOW (I,3543); PERRY COMO'S WINTER SHOW (I,3547); SANFORD AND SON (II,2256); THE SHARI LEWIS SHOW (I,4000); THE STEVE LAWRENCE SHOW (I,4227); SUPER COMEDY BOWL 1 (I,4288); SUPER COMEDY BOWL 2 (I,4289); THIS WEEK IN NEMTIN (I,4465); WHERE THE GIRLS ARE (I,4830)

TUTTLE, MARK
PRIVATE BENJAMIN (II,2087); THREE'S COMPANY (II,2614)

TUZEE, JOAN
ED MCMAHON AND HIS FRIENDS. . . DISCOVER WET AT CYPRESS GARDENS (I,1400)

TWIST, JOHN
BONANZA (II,347)

TYLER, EDWARD
MEDIC (I,2976); SHOW BIZ (I,4027)

UDELL, ANN
ISIS (II,1248)

UDES, LEN

WOLF ROCK TV (II,2815)

UGER, ALAN
ALAN KING LOOKS BACK IN ANGER—A REVIEW OF 1972 (I,99); CONDO (II,565); FAMILY TIES (II,819); THE FIGHTING NIGHTINGALES (II,855); KLEIN TIME (II,1401); THE MANY FACES OF COMEDY (I,2894); STEPHANIE (II,2453); THE STEVE LANDESBERG TELEVISION SHOW (II,2458); THE WONDERFUL WORLD OF AGGRAVATION (I,4894)

ULLETT, NICK
WE INTERRUPT THIS SEASON (I,4780)

ULLMAN, DAN
THE FUGITIVE (I,1701); THE INCREDIBLE HULK (II,1232); THE INVADERS (I,2229); LAND OF THE GIANTS (I,2611); MANNIX (II,1624); MOBILE ONE (II,1717); THE OUTER LIMITS (I,3426); THE RAT PATROL (I,3726); THE WILD WILD WEST (I,4863) WONDER WOMAN (SERIES 2) (II,2829)

ULLMAN, ELWOOD
THE LONE RANGER (I,2740)

UNGER, FRANK
BONANZA (II,347)

UNSWORTH, S E
PRISONER: CELL BLOCK H (II,2085)

UPTON, GABRIELLE
BEN CASEY (I,394); ONE STEP BEYOND (I,3388)

UPTON, JOHN
PRISONER: CELL BLOCK H (II,2085)

URBACK, LESLIE
FIRESIDE THEATER (I,1580)

URBISCI, ROCCO
ALAN KING'S SECOND ANNUAL FINAL WARNING (II,30); ALL NIGHT RADIO (II,40); BILLY CRYSTAL: A COMIC'S LINE (II,249); FROM CLEVELAND (II,931); THE JERK, TOO (II,1282); THE JOE PISCOPO SPECIAL (II,1304); LILY FOR PRESIDENT (II,1478); LILY—SOLD OUT (II,1480); THE RICHARD PRYOR SHOW (II,2163); THE RICHARD PRYOR SPECIAL (II,2164); THE RICHARD PRYOR SPECIAL? (II,2165); A SPECIAL KENNY ROGERS (II,2417)

USTINOV, PETER
CRESCENDO (I,1094); LOVE, LIFE, LIBERTY & LUNCH (II,1544)

VACHY, RICHARD
BENSON (II,208)

VAIL, LAWRENCE
HELLINGER'S LAW (II,1127)

VALANCE, BRUCE
THE SUZANNE SOMERS SPECIAL (II,2511)

VALENTA, LEONARD
AS THE WORLD TURNS (II,110)

VAN DRUTEN, JOHN
SLEEPING BEAUTY (I,4083)

VAN DYKE, DICK
THE DICK VAN DYKE SPECIAL (I,1276); VAN DYKE AND COMPANY (II,2718); VAN DYKE AND COMPANY (II,2719)

VAN FLEET, DAVID
20 MINUTE WORKOUT (II,2680)

VAN HARTESVELDT, FRAN
LEAVE IT TO BEAVER (I,2648)

VAN HORNE, TONI
EIGHT IS ENOUGH (II,762)

VAN HORTESVELDT, FRAN
SKY KING (I,4077)

VAN LIEU, FELIX
IVANHOE (I,2282)

VAN MARTER, GEORGE
THE LONE RANGER (I,2740)

VAN PEEBLES, MELVIN
DOWN HOME (II,731)

VAN RONKEL, RIP
FRANCES LANGFORD PRESENTS (I,1656)

VAN SCOG, ROBERT
TALES OF THE APPLE DUMPLING GANG (II,2536)

VAN SCOYCK, JULIE
MAGNUM, P.I. (II,1601)

VAN SCOYK, ROBERT
ALL'S FAIR (II,46); ALWAYS APRIL (I,147); THE ANN SOTHERN SHOW (I,220); BABY MAKES FIVE (II,129); BANACEK (II,138); CORONET BLUE (I,1055); DELVECCHIO (II,659); DOBIE GILLIS (I,1302); EAST SIDE/WEST SIDE (I,1396); GAVILAN (II,959); HERNANDEZ, HOUSTON P.D. (I,2034); KRAFT TELEVISION THEATER (I,2592); LOVE, SIDNEY (II,1547); MARCUS WELBY, M.D. (II,1627); MURDER, SHE WROTE (II,1770); N.Y.P.D. (I,3321); RAFFERTY (II,2105); THREE EYES (II,2605); TRAUMA CENTER (II,2655); YOUNG MAVERICK (II,2867)

VAN WAGONER, JAMES
BONANZA (II,347)

VAN, DEBORAH
MAMA'S FAMILY (II,1610)

VANCE, BRIAN
THE SUZANNE SOMERS SPECIAL (II,2510)

VANCE, LEIGH
THE AVENGERS (II,121); FANTASY ISLAND (II,829); JIGSAW (I,2377); THE PHOENIX (II,2036); THE SAINT (II,2237); SWITCH (II,2519)

VANE, CHRISTOPHER
FINDER OF LOST LOVES (II,857); GOLDIE GOLD AND ACTION JACK (II,1023); THE LOVE BOAT (II,1535); THUNDARR THE BARBARIAN (II,2616); TOO CLOSE FOR COMFORT (II,2642)

VARELA, MIGDIA
THE FACTS OF LIFE (II,805); THE INCREDIBLE HULK (II,1232)

VAUGHN, JUANITA
GOING MY WAY (I,1833)

VEITH, SANDY
DIFF'RENT STROKES (II,674); HERE'S BOOMER (II,1134); LOVE, SIDNEY (II,1547); NEVER AGAIN (II,1808); A YEAR AT THE TOP (II,2844)

VENABLE, LYNN
THE TWILIGHT ZONE (I,4651)

VENDIG, IRVING
THREE STEPS TO HEAVEN (I,4485)

VERNEY, ANTONY
IVANHOE (I,2282)

VICTOR, DAVID
THE CHADWICK FAMILY (II,478); I'M THE LAW (I,2195); THE MAN FROM U.N.C.L.E. (I,2867); RAWHIDE (I,3727)

VIDAL, GORE
PLAYWRIGHTS '56 (I,3627); STAGE DOOR (I,4178)

VIGON, BARRY
I GAVE AT THE OFFICE (II,1213); JUST OUR LUCK (II,1360); THE RAINBOW GIRL (II,2108); SOAP (II,2392)

VILANCH, BILL
JOHN RITTER: BEING OF SOUND MIND AND BODY (II,1319)

VILANCH, BRUCE
THE BRADY BUNCH HOUR (II,363); DONNY AND MARIE (II,712)

VINCENT, E DUKE
GOOD MORNING WORLD (I,1850); LOOK OUT WORLD (II,1516); SALT AND PEPE (II,2240); THE SAN PEDRO BUMS (II,2251)

VINCENT, ED
OPEN ALL NIGHT (II,1914)

VINCENT, JOHN
KIDS 2 KIDS (II,1389)

VINCENT, PETER
DAVE ALLEN AT LARGE (II,621)

VINNERA, CHICK
VEGAS (II,2724)

VIOLA, AL
POPI (II,2069)

VIOLA, JAMES
THE BIONIC WOMAN (II,255)

VIOLA, JOE
MICKEY SPILLANE'S MIKE HAMMER (II,1692); T.J. HOOKER (II,2524)

VIOLETT, ELLEN M
REBECCA (I,3745); SHANE (I,3995); SKIN OF OUR TEETH (I,4073)

VIORST, JUDITH
ANNIE, THE WOMAN IN THE LIFE OF A MAN (I,226)

VIPPERMAN, LOUIS
MAGNUM, P.I. (II,1601)

VITTES, LOUIS
HAWKEYE AND THE LAST OF THE MOHICANS (I,1975); THE INVADERS (I,2229) MEDIC (I,2976); RAWHIDE (I,3727); THE WILD WILD WEST (I,4863)

VITTES, MICHAEL
LOU GRANT (II,1526)

VIVINO, FLOYD
THE UNCLE FLOYD SHOW (II,2703)

VLAHOS, JOHN
ROUTE 66 (I,3852)

VOLLAERTS, RIK
ADVENTURES OF COLONEL FLACK (I,56); BATMAN (I,366); THE BRIGHTER DAY (I,728) COMBAT (I,1011); DEVLIN (II,663); THE FABULOUS SYCAMORES (I,1505); LEAVE IT TO BEAVER (I,2648); MYSTERIES OF CHINATOWN

(I,3209); THE NEW ADVENTURES OF CHARLIE CHAN (I,3249); RAWHIDE (I,3727); STAR TREK (II,2440); VOYAGE TO THE BOTTOM OF THE SEA (I,4743); WAGON TRAIN (I,4747)

VON ZELL, HARRY
WAGON TRAIN (I,4747)

VOSBURGH, DICK
SUNDAY NIGHT AT THE LONDON PALLADIUM (I,4286)

VOSBURGH, MARCY
IT'S YOUR MOVE (II,1259); THE JEFFERSONS (II,1276)

VOWELL, DAVID
RIPLEY'S BELIEVE IT OR NOT (II,2177)

WAALSTEN, ROBERT
THE ALCOA HOUR (I,108)

WADE, SALLY
WHAT'S HAPPENING!! (II,2769)

WAGGNER, GEORGE
ARROYO (I,266); SCREEN DIRECTOR'S PLAYHOUSE (I,3946)

WAGNER, JANE
LILY (I,2705); LILY (II,1477); LILY FOR PRESIDENT (II,1478); THE LILY TOMLIN SHOW (I,2706); THE LILY TOMLIN SPECIAL (II,1479); LILY—SOLD OUT (II,1480)

WAGNER, MADELINE
FANTASY ISLAND (II,829); THE TONY RANDALL SHOW (II,2640); WE'VE GOT EACH OTHER (II,2757)

WAGNER, MICHAEL
THE BLUE KNIGHT (II,276); HILL STREET BLUES (II,1154); KOJAK (II,1406); THE ROCKFORD FILES (II,2197)

WAKEFIELD, DAN
JAMES AT 15 (II,1271)

WALD, MALVIN
THE ALCOA HOUR (I,108); CLIMAX! (I,976); COMBAT (I,1011); DOBIE GILLIS (I,1302); THE GEORGE SANDERS MYSTERY THEATER (I,1776); GREATEST HEROES OF THE BIBLE (II,1062); PERRY MASON (II,2025); PETER GUNN (I,3562); YOU ASKED FOR IT (II,2855)

WALDBOTT, HELEN
FIRESIDE THEATER (I,1580)

WALDMAN, FRANK
THE BOSTON TERRIER (I,697); I DREAM OF JEANNIE (I,2167); JEANNIE (II,1275); JULIE AND DICK IN COVENT GARDEN (II,1352); JULIE—MY FAVORITE THINGS (II,1355); MCHALE'S NAVY (I,2969); THE SWINGIN' SINGIN' YEARS (I,4319) THE SWINGIN' YEARS (I,4321); THE VAL DOONICAN SHOW (I,4692)

WALDMAN, FRED
THE ROGUES (I,3832)

WALDMAN, TOM
THE BOSTON TERRIER (I,697); I DREAM OF JEANNIE (I,2167); MCHALE'S NAVY (I,2969); PETER GUNN (I,3562); THE ROGUES (I,3832); THE ROSEMARY CLOONEY SHOW (I,3847); THE ROSEMARY CLOONEY SHOW (I,3848); THE SPIKE JONES SHOW (I,4156); THE SWINGIN' SINGIN' YEARS (I,4319); THE SWINGIN' YEARS (I,4321); THE VAL DOONICAN SHOW (I,4692)

WALDRON, GY
ENOS (II,779); SIX PACK (II,2370); A YEAR AT THE TOP (II,2845)

WALDRON, TOM
THE DEAN MARTIN CELEBRITY ROAST (II,637); THE DEAN MARTIN CELEBRITY ROAST (II,638); THE DEAN MARTIN CELEBRITY ROAST (II,639); DOM DELUISE AND FRIENDS, PART 2 (II,702)

WALKER, ALLAN
DAGMAR'S CANTEEN (I,1125)

WALKER, BILL
DINAH AND FRIENDS (II,675); DINAH! (II,677); THE JOHNNY CASH SHOW (I,2424); THE LID'S OFF (I,2678); THE LOVE REPORT (II,1542); ON THE GO (I,3361); THICKE OF THE NIGHT (II,2587)

WALKER, DAVID E
THE ALCOA HOUR (I,108)

WALKER, GERTRUDE
FRONT ROW CENTER (I,1694); THE NEW ADVENTURES OF CHARLIE CHAN (I,3249); SCREEN DIRECTOR'S PLAYHOUSE (I,3946)

WALKER, KEITH
THE NANCY DREW MYSTERIES (II,1789)

WALKER, MICHAEL
MAUDE (II,1655); VOLTRON—DEFENDER OF THE UNIVERSE (II,2729)

WALKER, SUSAN
PRIME TIMES (II,2083)

WALKER, TURNLEY
THE DICK POWELL SHOW (I,1269)

WALLACE, ART
ASSIGNMENT: EARTH (I,299); COMBAT (I,1011); CORONET BLUE (I,1055); THE FELONY SQUAD (I,1563); THE INVADERS (I,2229); KRAFT TELEVISION THEATER (I,2592); LITTLE VIC (II,1496); PLANET OF THE APES (II,2049); STAR TREK (II,2440); STUDIO ONE (I,4268); THE WORLD BEYOND (II,2835); THE WORLD OF DARKNESS (II,2836)

WALLACE, CHARLES
HARD CASE (I,1947); YUMA (I,4976)

WALLACE, DAVID
THE SMURFS (II,2385)

WALLACE, EARL W
THE BLUE KNIGHT (II,276); BRONK (II,379); HOW THE WEST WAS WON (II,1196); SHE'S IN THE ARMY NOW (II,2320); WILD AND WOOLEY (II,2798); THE WILD WOMEN OF CHASTITY GULCH (II,2801)

WALLACE, ELIZABETH
ALL MY CHILDREN (II,39)

WALLACE, HENRY
THE REDD FOXX COMEDY HOUR (II,2126)

WALLACE, IRVING
ALCOA/GOODYEAR THEATER (I,107); CLIMAX! (I,976)

WALLACE, MIKE
THE MIKE WALLACE PROFILES (II,1699)

WALLACE, ROBERT
ROBERT MONTGOMERY PRESENTS YOUR LUCKY STRIKE THEATER (I,3809)

WALLENGREN, E F
FALCON CREST (II,808); LITTLE HOUSE: A NEW BEGINNING (II,1488); THE WALTONS (II,2740)

WALLER, ALBERT
THE ABC AFTERSCHOOL SPECIAL (II,1)

WALMAN, MARION
THE STARLOST (I,4203)

WALSH, BILL
ONE HOUR IN WONDERLAND (I,3376); THE WALT DISNEY CHRISTMAS SHOW (I,4755)

WANG, GENE
THE GEORGE SANDERS MYSTERY THEATER (I,1776); THE NEW ADVENTURES OF CHARLIE CHAN (I,3249); PERRY MASON (II,2025)

WANN, JIM
PUMPBOYS AND DINETTES ON TELEVISION (II,2090)

WARD, AL C
BEN CASEY (I,394); BONANZA (II,347); CLIMAX! (I,976); THE FUGITIVE (I,1701); MEDICAL CENTER (II,1675); PERRY MASON (II,2025); PETER GUNN (I,3562); PLAYHOUSE 90 (I,3623); RAWHIDE (I,3727); TIGHTROPE (I,4500); U.M.C. (I,4666)

WARD, EDMUND
MAN IN A SUITCASE (I,2868)

WARD, JAY
THE BULLWINKLE SHOW (I,761); CRUSADER RABBIT (I,1109)

WARD, JONATHAN
WALTER CRONKITE'S UNIVERSE (II,2739)

WARDE, HARLAN
THE LORETTA YOUNG THEATER (I,2756)

WARE, CLYDE
AIRWOLF (II,27); ALFRED HITCHCOCK PRESENTS (I,115); THE LAST CONVERTIBLE (II,1435); RAWHIDE (I,3727)

WARE, LEO
ALCOA/GOODYEAR THEATER (I,107); TURN OF FATE (I,4623)

WARE, WALLACE
HUNTER (II,1205); NERO WOLFE (II,1806); Q. E. D. (II,2094)

WARGA, WAYNE
ENTERTAINMENT TONIGHT (II,780)

WARING, RICHARD
WILD ABOUT HARRY (II,2797)

WARNER, ALLYN
THE CAPTAIN AND TENNILLE SONGBOOK (II,432); STAR SEARCH (II,2437); STAR SEARCH (II,2438); THAT THING ON ABC (II,2574)

WARNER, CHUCK
OF ALL THINGS (I,3330)

WARNER, DARYL
THE ABC AFTERSCHOOL SPECIAL (II,1)

WARREN, MARC
THE HALF-HOUR COMEDY HOUR (II,1074); MADAME'S PLACE (II,1587); SUNDAY FUNNIES (II,2493)

WARREN, MICHAEL
GOODTIME GIRLS (II,1042); HAPPY DAYS (II,1084); JOANIE LOVES CHACHI (II,1295); LAVERNE AND SHIRLEY (II,1446); OUT OF THE BLUE (II,1934)

WARREN, ROD
100 YEARS OF GOLDEN HITS (II,1905); BETTE MIDLER—OL' RED HAIR IS BACK (II,222); BING CROSBY'S CHRISTMAS SHOW (I,474); BLONDES VS. BRUNETTES (II,271); THE CARPENTERS (II,445); CHER. . . SPECIAL (II,499); CHERYL LADD. . . LOOKING BACK—SOUVENIRS (II,502); DONNY AND MARIE (II,712); THE EDDIE RABBITT SPECIAL (II,759); FROM CLEVELAND (II,931); THE GLEN CAMPBELL MUSIC SHOW (II,1006); THE GLEN CAMPBELL MUSIC SHOW (II,1007); GOOD EVENING, CAPTAIN (II,1031); HOLLYWOOD STARS' SCREEN TESTS (II,1165); THE JIM STAFFORD SHOW (II,1291); LEGENDS OF THE WEST: TRUTH AND TALL TALES (II,1457); LILY FOR PRESIDENT (II,1478); THE LILY TOMLIN SHOW (I,2706); THE LILY TOMLIN SPECIAL (II,1479); LILY—SOLD OUT (II,1480); THE LISA HARTMAN SHOW (II,1484); LORETTA LYNN: THE LADY. . .THE LEGEND (II,1521); MAC DAVIS. . . SOUNDS LIKE HOME (II,1582); MEN WHO RATE A "10" (II,1684); NEIL DIAMOND SPECIAL: I'M GLAD YOU'RE HERE WITH ME TONIGHT (II,1803); PERRY COMO'S CHRISTMAS IN ENGLAND (II,2008); PERRY COMO'S CHRISTMAS IN NEW YORK (II,2011); PUMPBOYS AND DINETTES ON TELEVISION (II,2090); ROCK COMEDY (II,2191); SALUTE TO LADY LIBERTY (II,2243); THE SMOTHERS BROTHERS SHOW (II,2384); A SPECIAL EDDIE RABBITT (II,2415); TELEVISION INSIDE AND OUT (II,2552)

WASHAM, WISNER
ALL MY CHILDREN (II,39)

WASHER, WENDELL
PAC-MAN (II,1940)

WASSERMAN, AL
A LOOK AT THE LIGHT SIDE (I,2751)

WASSERMAN, DALE
THE ALCOA HOUR (I,108); CLIMAX! (I,976); DUPONT SHOW OF THE MONTH (I,1387); KRAFT TELEVISION THEATER (I,2592); THE POWER AND THE GLORY (I,3658); STUDIO ONE (I,4268)

WASSERSTEIN, WENDY
THE COMEDY ZONE (II,559)

WATERS, ED
BARETTA (II,152); CARIBE (II,439); THE FBI (I,1551); JESSIE (II,1287); JOE DANCER: MURDER ONE, DANCER 0 (II,1302); KUNG FU (II,1416); THE MISSISSIPPI (II,1707); TODAY'S FBI (II,2628)

WATKINS, JON
ROBIN'S NEST (II,2187)

WATSON, MICHAEL
THE ADVENTURES OF BLACK BEAUTY (I,51)

WAXMAN, MARK
THE YOUNG AND THE RESTLESS (II,2862)

WAYNE, FRANK
BEAT THE CLOCK (I,378)

WAYNE, JOHNNY
OPERATION GREASEPAINT (I,3402); WAYNE AND SHUSTER TAKE AN AFFECTIONATE LOOK AT. . . (I,4778)

WAYNE, PAUL
BENSON (II,208); BEWITCHED (I,418); THE FLYING NUN (I,1611); THE GHOST AND MRS. MUIR (I,1786); THE KEN BERRY WOW SHOW (I,2524); THE LESLIE UGGAMS SHOW (I,2660); LOVE AND LEARN (II,1529); MY WORLD . . . AND WELCOME TO IT (I,3208); PAT PAULSEN'S HALF A COMEDY HOUR (I,3473); THE SMOTHERS ORGANIC PRIME TIME SPACE RIDE (I,4098); THE SONNY AND CHER COMEDY HOUR (II,2400); THREE'S COMPANY (II,2614)

WAYNE, PHILIP
BEAT THE CLOCK (I,378)

WEBB, AMY
HARPER VALLEY (II,1094)

WEBB, JACK
NOAH'S ARK (I,3302)

WEBB, MALCOLM
HARPER VALLEY (II,1094)

WEBSTER, DIANA
RIPLEY'S BELIEVE IT OR NOT (II,2177)

WEBSTER, SKIP
CHARLIE'S ANGELS (II,486); FANTASY ISLAND (II,829); THE LOVE BOAT (II,1535); MATT HOUSTON (II,1654); THE PARTRIDGE FAMILY (II,1962); THE ROOKIES (II,2208); WONDER WOMAN (SERIES 1) (II,2828)

WEBSTER, TOM
MADHOUSE 90 (I,2807)

WEBSTER, TONY
BING CROSBY—COOLING IT (I,460); CALL ME BACK (I,798); DOC (II,682); THE JONATHAN WINTERS SHOW (I,2443); THE LOVE BOAT (II,1535); PHYLLIS (II,2038); THE STEVE LAWRENCE SHOW (I,4227)

WEDLOCK, HUGH
20TH CENTURY FOLLIES (I,4641); AS CAESAR SEES IT (I,294); THE COMICS (I,1028); EVERYTHING YOU ALWAYS WANTED TO KNOW ABOUT JACK BENNY AND WERE AFRAID TO ASK (I,1482); THE FRANK SINATRA SHOW (I,1665); I MARRIED JOAN (I,2174); JACK BENNY'S 20TH ANNIVERSARY TV SPECIAL (I,2303); JACK BENNY'S FIRST FAREWELL SHOW (I,2300); JACK BENNY'S NEW LOOK (I,2301); JACK BENNY'S SECOND FAREWELL SHOW (I,2302); JIMMY DURANTE PRESENTS THE LENNON SISTERS HOUR (I,2387); A LOVE LETTER TO JACK BENNY (II,1540); A SALUTE TO STAN LAUREL (I,3892); TIME OUT FOR GINGER (I,4508); THE WONDERFUL WORLD OF BURLESQUE I (I,4895)

WEEDEN, BILL
THE SINGERS (I,4060)

WEEDON, TOM
SUZANNE PLESHETTE IS MAGGIE BRIGGS (II,2509)

WEEGE, REINHOLD
BARNEY MILLER (II,154); THE EARTHLINGS (II,751); FISH (II,864); NIGHT COURT (II,1844); PARK PLACE (II,1958); SAINT PETER (II,2238)

WEIDMAN, JOHN
NATIONAL LAMPOON'S HOT FLASHES (II,1795); THE NEWS IS THE NEWS (II,1838)

WEINBERGER, ALAN
THE PAUL LYNDE COMEDY HOUR (II,1977)

WEINBERGER, ED
BACHELOR AT LAW (I,322); BILL COSBY DOES HIS OWN THING (I,441); BROTHERS (II,380); THE COSBY SHOW (II,578); DOCTOR DAN (II,684); JOHNNY CARSON PRESENTS THE SUN CITY SCANDALS (I,2420); THE MARY TYLER MOORE SHOW (II,1640); PHYLLIS (II,2038); WHAT'S UP, AMERICA? (I,4824)

WEINBERGER, MICHAEL
EIGHT IS ENOUGH (II,762); THE FLIP WILSON COMEDY SPECIAL (II,886); HARPER VALLEY (II,1094); HOLLYWOOD HIGH (II,1162); LADIES' MAN (II,1423); PRIVATE BENJAMIN (II,2087); TEACHERS ONLY (II,2547); THREE'S COMPANY (II,2614); WELCOME BACK, KOTTER (II,2761)

WEINER, ELLIS
NATIONAL LAMPOON'S HOT FLASHES (II,1795)

WEINER, REX
MIAMI VICE (II,1689)

WEINER, WILLARD
FOUR STAR PLAYHOUSE (I,1652)

WEINER-KONNER, RONNIE
BEHIND THE SCREEN (II,200)

WEINFELD, ANDRE
RAQUEL (II,2114)

WEINGART, MARK
THE FBI (I,1551); GOING MY WAY (I,1833); THE MAN FROM U.N.C.L.E. (I,2867); THE RAT PATROL (I,3726); THE STREETS OF SAN FRANCISCO (II,2478)

WEINGARTEN, ARTHUR
BRING 'EM BACK ALIVE (II,377); DAN AUGUST (I,1130); IT'S ABOUT TIME (I,2263); THE SIX-MILLION-DOLLAR MAN (II,2372); THE SMOTHERS BROTHERS SHOW (I,4096); T.J. HOOKER (II,2524); WONDER WOMAN (SERIES 2) (II,2829)

WEINK, CHRIS
DEAN MARTIN'S CELEBRITY ROAST (II,643)

WEINRIB, LENNIE
A SALUTE TO TELEVISION'S 25TH ANNIVERSARY (I,3893)

WEINSTAD, MICHAEL
BABY MAKES FIVE (II,129)

WEINSTEIN, HOWARD
STAR TREK (II,2439)

WEINSTEIN, JACK
TELEVISION'S GREATEST COMMERCIALS (II,2553)

WEINSTEIN, SOL
20TH CENTURY FOLLIES (I,4641); BOB HOPE SPECIAL: BOB HOPE'S ALL-STAR COMEDY SPECTACULAR FROM LAKE TAHOE (II,301); BOB HOPE SPECIAL: BOB HOPE'S ALL-STAR COMEDY TRIBUTE TO VAUDEVILLE (II,302); THE BOBBY DARIN AMUSEMENT COMPANY (I,672); THE DEAN MARTIN CELEBRITY ROAST (II,637); THE DEAN MARTIN CELEBRITY ROAST (II,638); THE DEAN MARTIN CELEBRITY ROAST (II,639); DEAN MARTIN'S CELEBRITY ROAST (II,643); HELLZAPOPPIN (I,2011); HOW TO HANDLE A WOMAN (I,2146); LADIES AND GENTLEMAN. . . BOB NEWHART (II,1420); LADIES AND GENTLEMAN. . . BOB NEWHART, PART II (II,1421); THE LOVE BOAT (II,1535); THE MANY FACES OF COMEDY (I,2894); THE NBC FOLLIES (I,3241); THE NBC FOLLIES (I,3242); THE ODD COUPLE (II,1875); THE PAUL LYNDE HALLOWEEN SPECIAL (II,1981); SZYSZNYK (II,2523); VOLTRON—DEFENDER OF THE UNIVERSE (II,2729); THE WAYNE NEWTON SPECIAL (II,2749)

WEINSTOCK, JACK
HOWDY DOODY (I,2151); ROD BROWN OF THE ROCKET RANGERS (I,3825)

WEINTHORN, MICHAEL
FAMILY TIES (II,819)

WEIR, DAVID
THE ONEDIN LINE (II,1912)

WEISBERG, BRENDA
FIRESIDE THEATER (I,1580)

WEISBERG, SHEILA JUDIS
THE TONY RANDALL SHOW (II,2640); WE'VE GOT EACH OTHER (II,2757)

WEISBURD, DAN E
LAVERNE AND SHIRLEY (II,1446); MEDICAL CENTER (II,1675); WESTSIDE MEDICAL (II,2765)

WEISKOPF, BOB
ALL IN THE FAMILY (II,38); THE ANN SOTHERN SHOW (I,220); ARCHIE BUNKER'S PLACE (II,105); THE BEAUTIFUL PHYLLIS DILLER SHOW (I,381); BUT MOTHER! (II,403); CHECKING IN (II,492); THE DESILU PLAYHOUSE (I,1237); THE DESILU REVUE (I,1238); THE FLIP WILSON SHOW (I,1602); IT'S A BUSINESS (I,2254); LIVING IN PARADISE (II,1503); THE LUCILLE BALL-DESI ARNAZ SHOW (I,2784); MAUDE (II,1655); SANFORD (II,2255); SIDE BY SIDE (II,2346); W*A*L*T*E*R (II,2731); YOU'RE ONLY YOUNG TWICE (I,4970)

WEISKOPF, KIM
9 TO 5 (II,1852); AUTOMAN (II,120); CARTER COUNTRY (II,449); THREE'S COMPANY (II,2614)

WEISMAN, MATTHEW
THE HITCHHIKER (II,1157)

WEISS, ARTHUR
THE FUGITIVE (I,1701); LAND OF THE GIANTS (I,2611); THE RIFLEMAN (I,3789); VOYAGE TO THE BOTTOM OF THE SEA (I,4743)

WEISS, FREDRIC
9 TO 5 (II,1852); ARCHIE BUNKER'S PLACE (II,105)

WEISS, HARRIET
ARCHIE BUNKER'S PLACE (II,105); BABY MAKES FIVE (II,129); GLORIA COMES HOME (II,1011); IT TAKES TWO (II,1252) LOU GRANT (II,1526)

WEISS, ROBERT K
WELCOME TO THE FUN ZONE (II,2763)

WEISSMAN, NORMAN
LLOYD BRIDGES WATER WORLD (I,2734)

WEITHORN, MICHAEL J
MAKING THE GRADE (II,1606)

WEITZMAN, HARVEY
BOB HOPE SPECIAL: BOB HOPE IN "JOYS" (II,284); CARTER COUNTRY (II,449); DONNY AND MARIE (II,712); JOEY AND DAD (II,1306); THE MAD MAD MAD MAD WORLD OF THE SUPER BOWL (II,1586)

WELCH, KEN
BING CROSBY AND HIS FRIENDS (II,251); JULIE AND CAROL AT CARNEGIE HALL (I,2478); LINDA IN WONDERLAND (II,1481)

WELCH, MITZI
BING CROSBY AND HIS FRIENDS (II,251); BONNIE AND THE FRANKLINS (II,349); BURNETT "DISCOVERS" DOMINGO (II,398); HAL LINDEN'S BIG APPLE (II,1072); LINDA IN WONDERLAND (II,1481)

WELCH, RAQUEL
RAQUEL (II,2114)

WELCH, WILLIAM
LAND OF THE GIANTS (I,2611); LOST IN SPACE (I,2758); SKY KING (I,4077); SWISS FAMILY ROBINSON (II,2517); THE TIME TUNNEL (I,4511); VOYAGE TO THE BOTTOM OF THE SEA (I,4743); THE WALTONS (II,2740)

WELDON, FAY
PRIDE AND PREJUDICE (II,2080)

WELDON, LOIS
DEBBY BOONE. . . ONE STEP CLOSER (II,655)

WELLES, HALSTED
ALCOA/GOODYEAR THEATER (I,107); ALFRED HITCHCOCK PRESENTS (I,115); THE GEORGE SANDERS MYSTERY THEATER (I,1776); JUSTICE (I,2498); KNIGHT'S GAMBIT (I,2576); KOJAK (II,1406); PLAYHOUSE 90 (I,3623); SUSPICION (I,4309); WAGON TRAIN (I,4747)

WELLES, ORSON
THE FIRST 50 YEARS (II,862); FOUNTAIN OF YOUTH (I,1647)

WELLESLEY, GORDON
INTERNATIONAL DETECTIVE (I,2224)

WELLMAN, MANLEY WADE
THE TWILIGHT ZONE (I,4651)

WELLS, GEORGE
THE FABULOUS DR. FABLE (I,1500)

WELLS, HALSTEAD
MANNIX (II,1624)

WELLS, MARY K
ALL MY CHILDREN (II,39)

WELLS, ROBERT
THE ANN-MARGRET SHOW (I,218); THE DINAH SHORE SHOW (I,1280); THE DIONNE WARWICK SPECIAL (I,1289); THE GENE KELLY SHOW (I,1751); THE JANE POWELL SHOW (I,2344); THE JIMMY MCNICHOL SPECIAL (II,1292); JULIE—MY FAVORITE THINGS (II,1355); THE PEGGY FLEMING SHOW (I,3508); THE PERRY COMO WINTER SHOW (I,3544); PERRY COMO'S WINTER SHOW (I,3548); THE MAN IN THE MOON (I,2871); A SALUTE TO TELEVISION'S 25TH ANNIVERSARY (I,3893); SHIRLEY MACLAINE: IF THEY COULD SEE ME NOW (II,2335)

WENDELL, SOL
DOC (II,682)

WENKLER-KONNER, RONNIE
YOUNG LIVES (II,2866)

WERRIS, SNAG
THE JACK CARTER SHOW (I,2306); SHOW BIZ (I,4027)

WESHNER, SKIP
OZMOE (I,3437)

WESSON, DICK
THESE ARE THE DAYS (II,2585)

WEST, BERNIE
ALL IN THE FAMILY (II,38); THE DUMPLINGS (II,743); THREE'S A CROWD (II,2613); THREE'S COMPANY (II,2614)

WEST, ELIOT
FOUR STAR PLAYHOUSE (I,1652); LOU GRANT (II,1526)

WEST, PAUL
ADVENTURES OF COLONEL FLACK (I,56); FATHER KNOWS BEST (II,838); FATHER KNOWS BEST: HOME FOR CHRISTMAS (II,839); FATHER KNOWS BEST: THE FATHER KNOWS BEST REUNION (II,840); THE GREAT GILDERSLEEVE (I,1875); MY THREE SONS (I,3205); PLEASE DON'T EAT THE DAISIES (I,3628); THE WALTONS (II,2740); THE WILDS OF TEN THOUSAND ISLANDS (II,2803)

WESTERBY, ROBERT
THE INVISIBLE MAN (I,2232)

WESTERSCHULTE, DICK
THE DOM DELUISE SHOW (I,1328); STAND UP AND CHEER (I,4188)

WESTHEIMER, DAVID
CAMPO 44 (I,813)

WHEATON, GLENN
FRANCES LANGFORD PRESENTS (I,1656); FRANK SINATRA—A MAN AND HIS MUSIC (I,1661); OPERATION ENTERTAINMENT (I,3400)

WHEDON, JOHN
54TH STREET REVUE (I,1573); THE ALCOA HOUR (I,108); THE ANDY GRIFFITH SHOW (I,192); THE DONNA REED SHOW (I,1347); KILROY (I,2541); KRAFT TELEVISION THEATER (I,2592); LEAVE IT TO BEAVER (I,2648)

WHEDON, TOM
THE ALAN KING SHOW (I,102); ALL'S FAIR (II,46); THE MIKE DOUGLAS CHRISTMAS SPECIAL (I,3040); MR. MAYOR (I,3143); TALES FROM MUPPETLAND (I,4344); THIS MORNING (I,4464); THE TWO OF US (II,2693)

WHEELER, HUGH
FEMALE INSTINCT (I,1564)

WHELPLEY, JOHN
TRAPPER JOHN, M.D. (II,2654)

WHITCOMB, DENNIS
THE MUNSTERS (I,3158)

WHITE, ANDY
THE BOB HOPE SHOW (I,637); THE BOB HOPE SHOW (I,642); FATHER KNOWS BEST (II,838); THE GREAT GILDERSLEEVE (I,1875); THE WALTONS (II,2740); THE WILDS OF TEN THOUSAND ISLANDS (II,2803)

WHITE, HOLLACE
AT EASE (II,115); HAPPY DAYS (II,1084); MR. MERLIN (II,1751); ONE DAY AT A TIME (II,1900)

WHITE, LESTER
THE BOB HOPE SHOW (I,533); THE BOB HOPE SHOW (I,534); THE BOB HOPE SHOW (I,541); THE BOB HOPE SHOW (I,543); THE BOB HOPE SHOW (I,552); THE BOB HOPE SHOW (I,555); THE BOB HOPE SHOW (I,560); THE BOB HOPE SHOW (I,561); THE BOB HOPE SHOW (I,562); THE BOB HOPE SHOW (I,603); THE BOB HOPE SHOW (I,604); THE BOB HOPE SHOW (I,606); THE BOB HOPE SHOW (I,607); THE BOB HOPE SHOW (I,613); THE BOB HOPE SHOW (I,636); BOB HOPE SPECIAL: BOB HOPE ON CAMPUS (II,289); BOB HOPE SPECIAL: THE BOB HOPE SPECIAL (II,333); THE EDDIE CANTOR COMEDY THEATER (I,1407); ROBERTA (I,3815); ROBERTA (I,3816)

WHITE, PHYLLIS
CANNON (II,424); JEANNIE (II,1275); MEDICAL CENTER (II,1675)

WHITE, ROBERT A
CANNON (II,424); THE CISCO KID (I,961); JEANNIE (II,1275); MEDICAL CENTER (II,1675); PERRY MASON (II,2025)

WHITE, STEVE
BOB HOPE SPECIAL: BOB HOPE ON CAMPUS (II,289); BOB HOPE SPECIAL: BOB HOPE PRESENTS THE STARS OF TOMORROW (II,292); BOB HOPE SPECIAL: THE BOB HOPE SPECIAL (II,333); BOB HOPE SPECIAL: THE BOB HOPE SPECIAL (II,334); MAURICE CHEVALIER'S PARIS (I,2955)

WHITEHEAD, WILLIAM
THE INCREDIBLE HULK (II,1232)

WHITEMORE, HUGH
ALL CREATURES GREAT AND SMALL (I,129); MOLL FLANDERS (II,1720)

WHITMORE, STANFORD
THE D.A.: CONSPIRACY TO KILL (I,1121); THE EYES OF CHARLES SAND (I,1497); THE FUGITIVE (I,1701); MADIGAN (I,2808); MCCLOUD: WHO KILLED MISS U.S.A.? (I,2965); THE MONEYCHANGERS (II,1724); THE RUNAWAY BARGE (II,2230); THE WILD WILD WEST (I,4863)

WHITTAKER, CLAIRE
THE FITZPATRICKS (II,868); THE WALTONS (II,2740)

WHITTERMORE, L H
BARETTA (II,152)

WHITTINGHAM, JACK
SECRET AGENT (I,3962)

WHUL, ROBERT
POLICE SQUAD! (II,2061)

WIBBERLY, LEONARD
THE HANDS OF CORMAC JOYCE (I,1927)

WICKER, IREENE
THE IREENE WICKER SHOW (I,2234); THE SINGING LADY (I,4061)

WICKES, DAVID
PHILIP MARLOWE, PRIVATE EYE (II,2033)

WICKLINE, MATT
LATE NIGHT WITH DAVID LETTERMAN (II,1442)

WIDLEY, DOUGLAS
THE ADVENTURES OF JONNY QUEST (I,64)

WIENER, WILLARD
SCREEN DIRECTOR'S PLAYHOUSE (I,3946)

WIENGARTEN, ARTHUR
THE WILD WILD WEST (I,4863)

WILBER, CAREY
BARNABY JONES (II,153); BONANZA (II,347); BRONK (II,379); CANNON (II,424); CAPTAIN VIDEO AND HIS VIDEO RANGERS (I,838); CARIBE (II,439); CODE R (II,551); DEVLIN (II,663); HAWAII FIVE-O (II,1110); LOST IN SPACE (I,2758); MANHUNTER (II,1621); MATT HELM (II,1653); RAWHIDE (I,3727); STAR TREK (II,2440); STUDIO ONE (I,4268); SWITCH (II,2519); TARGET: THE CORRUPTERS (I,4360); THE TIME TUNNEL (I,4511); THE UNTOUCHABLES (I,4682)

WILCOX, DAN
AMERICA 2-NIGHT (II,65); BAY CITY BLUES (II,186); M*A*S*H (II,1569); RISE AND SHINE (II,2179); ROOTS: THE NEXT GENERATIONS (II,2213)

WILDE, HAGAR
CLIMAX! (I,976); FRONT ROW CENTER (I,1694); PLAYHOUSE 90 (I,3623)

WILDER, FRANK
BRANDED (I,707)

WILDER, JOHN
CENTENNIAL (II,477); THE CITY (II,541); THE DEVLIN CONNECTION (II,664); THE YELLOW ROSE (II,2847)

WILDER, MARGARET BUELL
THE LORETTA YOUNG THEATER (I,2756)

WILDER, MYLES
BACHELOR FATHER (I,323); THE DUKES OF HAZZARD (II,742); HELLO DERE (I,2010); MCHALE'S NAVY (I,2969); RUN, BUDDY, RUN (I,3870); THE SAN PEDRO BEACH BUMS (II,2250); THE TIM CONWAY COMEDY HOUR (I,4501); WELCOME BACK, KOTTER (II,2761)

WILDER, REX
CAMP WILDERNESS (II,419)

WILDERBLOOD, PETER
FATHER BROWN (II,836)

WILE, SHELDON
HAWAII FIVE-O (II,1110)

WILES, JOHN
A HORSEMAN RIDING BY (II,1182)

WILEY, CARSON
STUDIO ONE (I,4268)

WILEY, MARTY
KIDD VIDEO (II,1388)

WILHELM, JEFF
THE BOB NEWHART SHOW (II,342); LOBO (II,1504); MAGNUM, P.I. (II,1601)

WILK, DIANE
AMANDA'S (II,62); IT TAKES TWO (II,1252)

WILK, MAX
54TH STREET REVUE (I,1573); THE BING CROSBY CHRISTMAS SHOW (I,458); THE BING CROSBY SHOW (I,465); THE IMOGENE COCA SHOW (I,2198); JONATHAN WINTERS PRESENTS 200 YEARS OF AMERICAN HUMOR (II,1342); MATINEE THEATER (I,2947); MELINA MERCOURI'S GREECE (I,2997); THAT'S OUR SHERMAN (I,4429); THE WONDERFUL WORLD OF JONATHAN WINTERS (I,4900)

WILKERSON, MARSHA
ROBERT MONTGOMERY PRESENTS YOUR LUCKY STRIKE THEATER (I,3809)

WILLENS, MICHELE
9 TO 5 (II,1852)

WILLENS, SHEL
BRET MAVERICK (II,374); HARDCASTLE AND MCCORMICK (II,1089); JESSIE (II,1287); THE MISSISSIPPI (II,1707)

WILLIAMS, ALLEN
FROM HERE TO ETERNITY (II,933)

WILLIAMS, DAVID
CAPTAIN SCARLET AND THE MYSTERONS (I,837)

WILLIAMS, DICK
THE STEVE LAWRENCE SHOW (I,4227)

WILLIAMS, JASTON
THE COMEDY ZONE (II,559)

WILLIAMS, LAWRENCE
DOBIE GILLIS (I,1302)

WILLIAMS, MAGGIE
DOBIE GILLIS (I,1302)

WILLIAMS, MASON
ANDY WILLIAMS KALEIDOSCOPE COMPANY (I,201); PAT PAULSEN FOR PRESIDENT (I,3472); PETULA

(I,3572); THE ROGER MILLER SHOW (I,3830); THE SMOTHERS BROTHERS COMEDY HOUR (I,4095); THE SMOTHERS BROTHERS SHOW (II,2384); THE SUMMER SMOTHERS BROTHERS SHOW (I,4281)

WILLIAMS, PAUL
THE PAUL WILLIAMS SHOW (II,1984); ROOSTER (II,2210)

WILLIAMS, ROBIN
AN EVENING WITH ROBIN WILLIAMS (II,789)

WILLIAMS, TENNESSEE
THE GLASS MENAGERIE (I,1819); KRAFT TELEVISION THEATER (I,2592)

WILLIAMSON, MARTHA
THE HOMEMADE COMEDY SPECIAL (II,1173)

WILLIAMSON, TONY
THE AVENGERS (II,121); DEPARTMENT S (I,1232)

WILLIS, TED
THE ADVENTURES OF BLACK BEAUTY (I,51)

WILLISON, BOB
THE MAGICAL MUSIC OF BURT BACHARACH (I,2822)

WILLOCK, DAVE
DAVE AND CHARLEY (I,1171); DO IT YOURSELF (I,1299)

WILSON, ANTHONY
BANACEK: DETOUR TO NOWHERE (I,339); COMBAT (I,1011); COMPUTERCIDE (II,561); THE FUGITIVE (I,1701); FUTURE COP (II,945); FUTURE COP (II,946); GIDGET (I,1795); THE INVADERS (I,2229); LAND OF THE GIANTS (I,2611); THE TWILIGHT ZONE (I,4651)

WILSON, DICK
THE LOVE REPORT (II,1543)

WILSON, DONALD
HORNBLOWER (I,2121)

WILSON, ELIZABETH
THE DICK POWELL SHOW (I,1269); DYNASTY (II,746); SARA (II,2260)

WILSON, FLIP
THE FLIP WILSON COMEDY SPECIAL (II,886); THE FLIP WILSON SHOW (I,1602); THE FLIP WILSON SPECIAL (I,1603); THE FLIP WILSON SPECIAL (II,887); THE FLIP WILSON SPECIAL (II,888); THE FLIP WILSON SPECIAL (II,889); FLIP WILSON. . . OF COURSE (II,890)

WILSON, FRANK
THE KATE SMITH EVENING HOUR (I,2507)

WILSON, HUGH
THE BOB NEWHART SHOW (II,342); THE CHOPPED LIVER BROTHERS (II,514); THE TONY RANDALL SHOW (II,2640); WKRP IN CINCINNATI (II,2814)

WILSON, LANFORD
FIFTH OF JULY (II,851); TAXI (I,4365)

WILSON, RALPH
THE BAXTERS (II,184)

WILSON, RAY
THE GABBY HAYES SHOW (I,1723)

WILSON, RICHARD
DYNASTY (II,746)

WILSON, ROY
PAC-MAN (II,1940)

WILSON, WARREN
ANNIE OAKLEY (I,225); THE CISCO KID (I,961); THE RANGE RIDER (I,3720); WAGON TRAIN (I,4747)

WILSON, WENDY
POWERHOUSE (II,2074)

WILTSE, DAVID
BEACON HILL (II,193); LADIES' MAN (II,1423)

WINCH, ARDEN
A HORSEMAN RIDING BY (II,1182)

WINCHELBERG, SHIMON
BRONK (II,379); DALLAS (II,614); DEVLIN (II,663); HAGEN (II,1071); HEC RAMSEY (II,1121); LOGAN'S RUN (II,1507); MANNIX (II,1624); NAKED CITY (I,3216); THE NEW ADVENTURES OF PERRY MASON (I,3252); THE PAPER CHASE (II,1949); ROUTE 66 (I,3852); STAR TREK (II,2440); TRAPPER JOHN, M.D. (II,2654); THE WILD WILD WEST (I,4863)

WINCHELL, PAUL
JERRY MAHONEY'S CLUB HOUSE (I,2371); THE PAUL WINCHELL AND JERRY MAHONEY SHOW (I,3492)

WINCHELL, WALTER
THE WALTER WINCHELL SHOW (I,4760)

WINDER, MICHAEL
THE AVENGERS (II,121); THE FEATHER AND FATHER GANG (II,845); HARRY O (II,1099); THE SAINT (II,2237)

WINDSOR, GABRIELLE
FIRESIDE THEATER (I,1580)

WINDSOR, ROY
ANOTHER LIFE (II,95); THE PUBLIC LIFE OF CLIFF NORTON (I,3688)

WINER, ELIHU
STORY THEATER (I,4240)

WINER, STEVE
LATE NIGHT WITH DAVID LETTERMAN (II,1442)

WINES, WILLIAM
THEY STAND ACCUSED (I,4442)

WINGARD, BILL
THE BOOTS RANDOLPH SHOW (I,687); HEE HAW (II,1123); THE NASHVILLE SOUND OF BOOTS RANDOLPH (I,3229); TENNESSEE ERNIE FORD'S WHITE CHRISTMAS (I,4402)

WINGREEN, JASON
THE WILD WILD WEST (I,4863)

WINKLER, HARRY
THE ADDAMS FAMILY (II,11); BATTLE OF THE PLANETS (II,179); THE GEORGE GOBEL SHOW (I,1767); THE GEORGE GOBEL SHOW (I,1768); HEY TEACHER (I,2041); THE MANY SIDES OF MICKEY ROONEY (I,2899); THE MICKEY ROONEY SHOW (I,3027); THE PARTRIDGE FAMILY (II,1962)

WINNICK, JERRY
THE FLIP WILSON SPECIAL (I,1603); THE FLIP WILSON SPECIAL (II,887); THE FLIP WILSON SPECIAL (II,888); THE FLIP WILSON SPECIAL (II,889); THE LOVE BOAT (II,1535); THE MANY FACES OF COMEDY (I,2894); ONE IN A MILLION (II,1906); TOO CLOSE FOR COMFORT (II,2642)

WINTER, JACK
HAPPY DAYS (II,1084); LAVERNE AND SHIRLEY (II,1446); THE MARY TYLER MOORE SHOW (II,1640); THE ODD COUPLE (II,1875); RHODA (II,2151)

WINTER, SUZETTE
THE HORROR OF IT ALL (II,1181)

WINTERS, JAN
RAWHIDE (I,3727)

WINTERS, JONATHAN
JONATHAN WINTERS PRESENTS 200 YEARS OF AMERICAN HUMOR (II,1342); UNCLE TIM WANTS YOU! (II,2704); THE WONDERFUL WORLD OF JONATHAN WINTERS (I,4900)

WINTERS, ROLAND
FOUR STAR PLAYHOUSE (I,1652)

WINTHROP, HARVEY
SHERLOCK HOLMES (I,4010)

WISBAR, FRANK
FIRESIDE THEATER (I,1580)

WISE, DAVID
BUCK ROGERS IN THE 25TH CENTURY (II,383); ISIS (II,1248); STAR TREK (II,2439); WONDER WOMAN (SERIES 2) (II,2829)

WISE, WILLIAM
SUSPICION (I,4309)

WISER, BUD
ALL'S FAIR (II,46); PAUL SAND IN FRIENDS AND LOVERS (II,1982); SEMI-TOUGH (II,2299); SEMI-TOUGH (II,2300); WHO'S THE BOSS? (II,2792)

WISHENGRAD, MORTON
SUSPICION (I,4309); THERE SHALL BE NO NIGHT (I,4437)

WITTINGHAM, JACK
DANGER MAN (I,1136)

WOHL, JACK
AMERICA, YOU'RE ON (II,69); BURT AND THE GIRLS (I,770); BURT REYNOLDS' LATE SHOW (I,777); THE DEAN MARTIN SHOW (I,1201); DIANA (I,1258); G.I.'S (II,949); HALF THE GEORGE KIRBY COMEDY HOUR (I,1920); THE MANHATTAN TRANSFER (II,1619); PAT BOONE AND FAMILY (II,1968); ROWAN AND MARTIN'S LAUGH-IN (I,3856); SAGA OF SONORA (I,3884); SANDY IN DISNEYLAND (II,2254)

WOLFE, DIGBY
CHER (II,495); CHER (II,496); CHER AND OTHER FANTASIES (II,497); THE COLGATE COMEDY HOUR (I,998); DORIS DAY TODAY (II,716); THE FABULOUS FORDIES (I,1502); FOL-DE-ROL (I,1613); THE GOLDIE HAWN SPECIAL (II,1024); JOHN DENVER AND FRIEND (II,1311); JOHN DENVER AND THE LADIES (II,1312); THE JONATHAN WINTERS SHOW (I,2443); LAUGH-IN (II,1444);

THE LESLIE UGGAMS SHOW (I,2660); THE PEAPICKER IN PICCADILLY (I,3500); ROWAN AND MARTIN'S LAUGH-IN (I,3857); THE SHIRLEY MACLAINE SPECIAL: WHERE DO WE GO FROM HERE? (II,2333); THE TENNESSEE ERNIE FORD SPECIAL (I,4401); THE WILD WILD WEST (I,4863)

WOLFE, GLEN
PERRY MASON (II,2025)

WOLFE, IAN
MAKING A LIVING (II,1604)

WOLFE, LAWRENCE
STAR TREK (II,2440)

WOLFE, RONALD
THE RAG BUSINESS (II,2106)

WOLFE, WINIFRED
HIGH HOPES (II,1144)

WOLFF, JURGEON
GLORIA (II,1010) THE INVESTIGATORS (II,1241); JENNIFER SLEPT HERE (II,1278); TOO CLOSE FOR COMFORT (II,2642)

WOLFF, PAUL
LITTLE HOUSE ON THE PRAIRIE (II,1487)

WOLFT, WILLIAM
TERRY AND THE PIRATES (I,4405)

WOLK, MAX
THE FABULOUS 50S (I,1501)

WOLLA, TOM
GIMME A BREAK (II,995)

WOLPERT, DAVE
GIMME A BREAK (II,995)

WOLPERT, JAY
HIT MAN (II,1156)

WOLPERT, ROLAND
NAKED CITY (I,3216); TABITHA (II,2528)

WOLPERT, STUART
ALL TOGETHER NOW (II,45); THE FACTS OF LIFE (II,805); FACTS OF LIFE: THE FACTS OF LIFE GOES TO PARIS (II,806); ONE DAY AT A TIME (II,1900)

WOLTERSTOFF, ROBERT
THE INCREDIBLE HULK (II,1232); THE MISADVENTURES OF SHERIFF LOBO (II,1704); THE OUTLAWS (II,1936); PEN 'N' INC. (II,1992)

WOOD, BOB
THE NUT HOUSE (I,3320)

WOOD, GENE
THE JERRY LESTER SPECIAL (I,2361)

WOOD, JACK
ALL MY CHILDREN (II,39)

WOOD, JOHN
PRISONER: CELL BLOCK H (II,2085)

WOOD, PETER
LONG DAY'S JOURNEY INTO NIGHT (I,2746)

WOOD, PRESTON
JESSICA NOVAK (II,1286); QUINCY, M. E. (II,2102); THE RANGERS (II,2112); THE RETURN OF CAPTAIN NEMO (II,2135); THE WILD WILD WEST (I,4863)

WOOD, WILLIAM
DANGER IN PARADISE (II,617); THE MARY TYLER MOORE SHOW (II,1640)

WOODALD, ANN
MR. MERLIN (II,1751)

WOODFIELD, WILLIAM REED
THE BARBARY COAST (II,146); MASQUERADE (II,1644); SAN FRANCISCO INTERNATIONAL AIRPORT (I,3906); THE TIME TUNNEL (I,4511); VOYAGE TO THE BOTTOM OF THE SEA (I,4743)

WOODHOUSE, MARTIN
THE AVENGERS (II,121)

WOODS, JACK
THE EDDIE ALBERT SHOW (I,1406)

WOODS, RAYMOND
WOMEN OF RUSSIA (II,2824)

WOODWARD, JOANNE
SENSE OF HUMOR (II,2303)

WOODY, RUSS
BENSON (II,208); WEBSTER (II,2758)

WOOLEY, SHEB
RAWHIDE (I,3727)

WOOLF, CHARLES
THE FARMER'S DAUGHTER (I,1533)

WOOLLEN, SUSAN
THE INCREDIBLE HULK (II,1232); THE MASTER (II,1648)

WOPERT, STUART
DOUBLE TROUBLE (II,723)

WORTH, MARTIN
THE ONEDIN LINE (II,1912)

WORTH, MARVIN
THE MILTON BERLE SHOW (I,3049); MY LUCKY PENNY (I,3196)

WORTHEN, WARREN J
BARETTA (II,152)

WORTHINGTON, DAVE
PRISONER: CELL BLOCK H (II,2085)

WOUK, HERMAN
THE WINDS OF WAR (II,2807)

WRAY, JOHN
TARGET: THE CORRUPTERS (I,4360)

WRIGHT, BOB
THE WILD WILD WEST (I,4863)

WRIGHT, HERBERT J
THROUGH THE MAGIC PYRAMID (II,2615)

WRIGHT, ROBERT VINCENT
BONANZA (II,347); LITTLE HOUSE ON THE PRAIRIE (II,1487); VOYAGE TO THE BOTTOM OF THE SEA (I,4743)

WRIGHT, ROGER
MASQUERADE PARTY (II,1645)

WRIGHT, TOM
THE OMEGA FACTOR (II,1888)

WRIGHT, WILLIAM H
AMOS BURKE, SECRET AGENT (I,180); THE DICK POWELL SHOW (I,1269)

WUHL, JACK
CHERYL LADD: SCENES FROM A SPECIAL (II,503)

WUHL, ROBERT
LEGENDS OF THE WEST: TRUTH AND TALL TALES (II,1457)

WYANT, SCOTT
HOT POTATO (II,1188)

WYCKOFF, GENE
BUCK ROGERS IN THE 25TH CENTURY (I,751)

WYLES, J DAVID
FAERIE TALE THEATER (II,807)

WYLIE, MAX
SKIP TAYLOR (I,4075)

WYMAN, DOUGLAS
FAMILY TIES (II,819); MR. SMITH (II,1755)

WYNN, BOB
MATT HOUSTON (II,1654); THE PEAPICKER IN PICCADILLY (I,3500)

WYNN, NED
VELVET (II,2725)

WYNN, TRACY KEENAN
THE QUEST (II,2096)

YAGERMAN, BILL
THE TENNESSEE ERNIE FORD SHOW (I,4398)

YAKUTIS, TOM
THE LITTLE RASCALS (II,1494); THE RICHIE RICH SHOW (II,2169)

YANOK, GEORGE
ARCHIE (II,104); THE ARCHIE SITUATION COMEDY MUSICAL VARIETY SHOW (II,106); THE BRASS ARE COMING (I,708); THE BURNS AND SCHREIBER COMEDY HOUR (I,768); THE BURNS AND SCHREIBER COMEDY HOUR (I,769); COMEDY NEWS II (I,1021); THE FLIP WILSON COMEDY SPECIAL (II,886); THE FLIP WILSON SPECIAL (I,1603); THE FLIP WILSON SPECIAL (II,888); THE FLIP WILSON SPECIAL (II,889); GLADYS KNIGHT AND THE PIPS (II,1004); GO WEST YOUNG GIRL (II,1013); HOW TO HANDLE A WOMAN (I,2146); THE JIMMIE RODGERS SHOW (I,2383); THE LILY TOMLIN SHOW (I,2706); THE LILY TOMLIN SPECIAL (II,1479); THE MANY FACES OF COMEDY (I,2894); THE MELBA MOORE-CLIFTON DAVIS SHOW (I,2996); THE PAUL LYNDE COMEDY HOUR (II,1977); PERRY COMO'S MUSIC FROM HOLLYWOOD (II,2019); THE STOCKARD CHANNING SHOW (II,2468); TABITHA (II,2528); TEACHERS ONLY (II,2547); THE WAYNE NEWTON SPECIAL (II,2749); WELCOME BACK, KOTTER (II,2761)

YARNELL, BRUCE
WHAT GAP? (I,4808)

YARO, BORIS
CODE R (II,551)

YASHA, FRANK
HANSEL AND GRETEL (I,1934)

YATES, BROCK
THE STOCKERS (II,2469)

YATES, ROBERT MALCOLM
WOMEN IN WHITE (II,2823)

YATES, WILLIAM ROBERT

ALEXANDER THE GREAT (I,113); COMBAT (I,1011)

YELDMAN, PETER
1915 (II,1853) ALL THE RIVERS RUN (II,43)

YELLEN, SHELDON
A CRY OF ANGELS (I,1111)

YELLEN, SHERMAN
BEAUTY AND THE BEAST (I,382); DR JEKYLL AND MR. HYDE (I,1368)

YERKOVICH, ANTHONY
240-ROBERT (II,2688); HART TO HART (II,1102); HILL STREET BLUES (II,1154); MIAMI VICE (II,1689); STARSKY AND HUTCH (II,2444)

YORK, TIFFANY
HAPPY DAYS (II,1084)

YORKIN, BUD
THE ANDY WILLIAMS SPECIAL (I,211); THE TONY MARTIN SHOW (I,4572)

YOUNG, ALAN
THE ALAN YOUNG SHOW (I,104)

YOUNG, BOB
THE FACTS OF LIFE (II,805); T.L.C. (II,2525)

YOUNG, COLLIER
THE DICK POWELL SHOW (I,1269); ONE STEP BEYOND (I,3388)

YOUNG, JEFF
DALLAS (II,614)

YOUNG, JOHN SCARET
THE FITZPATRICKS (II,868); THE POSSESSED (II,2071)

YOUNG, ROBERT MALCOLM
AMANDA FALLON (I,151); BEYOND WITCH MOUNTAIN (II,228); THE FBI (I,1551); HARRY O (II,1099); MARCUS WELBY, M.D. (II,1627); THE STREETS OF SAN FRANCISCO (II,2478); THE WALTONS (II,2740)

YURICK, PAUL
MASTER OF THE GAME (II,1647)

ZACHARIAS, STEVE
ME AND DUCKY (II,1671); QUARK (II,2095); SCALPELS (II,2266); TWO GUYS FROM MUCK (II,2690); VIVA VALDEZ (II,2728)

ZAFRAN, JACK
POWERHOUSE (II,2074)

ZAGOR, MICHAEL
BEN CASEY (I,394); THE FUGITIVE (I,1701); GENERAL ELECTRIC TRUE (I,1754); GOING PLACES (I,1834); THE MAN FROM GALVESTON (I,2865); TO SIR, WITH LOVE (II,2624); THE YOUNG LAWYERS (I,4940)

ZALMAN, SHELLEY
SEMI-TOUGH (II,2300)

ZAPPONI, BERNARDINO
MOSES THE LAWGIVER (II,1737)

ZARCOFF, MORT
SWITCH (II,2519)

ZATESLO, GEORGE
EMPIRE (II,777); GIDGET'S SUMMER REUNION (I,1799); THE GIRL, THE GOLD WATCH AND DYNAMITE (II,1001); THE GIRL, THE GOLD WATCH AND EVERYTHING (II,1002)

ZAVATINI, BOBBY
THE RENEGADES (II,2133)

ZEHREN, LEROY
THE ADVENTURES OF SUPERMAN (I,77)

ZEIGLER, TED
LI'L ABNER (I,2702); SHIELDS AND YARNELL (II,2328)

ZELINKA, SYDNEY
THE ALL-STAR COMEDY SHOW (I,137); THE BIG SELL (I,431); THE COLLEGE BOWL (I,1001); THE HERO (I,2035); THE HONEYMOONERS (I,2110); THE HONEYMOONERS (II,1175); THE JACKIE GLEASON SHOW (I,2322); THE JERRY LEWIS SHOW (I,2370); THE JIMMIE RODGERS SHOW (I,2383); KEEFE BRASSELLE'S VARIETY GARDEN (I,2515); MARRIAGE—HANDLE WITH CARE (I,2921); THE MILLION DOLLAR INCIDENT (I,3045); PRIVATE EYE, PRIVATE EYE (I,3671); THE SID CAESAR SPECIAL (I,4044); THE STEVE LAWRENCE SHOW (I,4227); TIPTOE THROUGH TV (I,4524); VARIETY: THE WORLD OF SHOW BIZ (I,4704)

ZELLMAN, SHELLEY
BABY, I'M BACK! (II,130); BARNEY MILLER (II,154); CHARLES IN CHARGE (II,482); DONNY AND MARIE (II,712); IT'S A LIVING (II,1254); NEWHART (II,1835); SPENCER (II,2420); THREE'S COMPANY (II,2614)

ZICREE, MARK SCOTT
THE SMURFS (II,2385)

ZIEGMAN, JERRY
BRANDED (I,707); CENTENNIAL (II,477); THE MISSISSIPPI (II,1707); THE YELLOW ROSE (II,2847)

ZIMM, MAURICE
PERRY MASON (II,2025)

ZIMMER, JON
LEAVE IT TO BEAVER (I,2648)

ZIMMERMAN, STAN
JUST OUR LUCK (II,1360)

ZINBERG, MICHAEL
THE BOB NEWHART SHOW (II,342); THE JAMES BOYS (II,1272)

ZITO, STEPHEN
MODESTY BLAISE (II,1719)

ZLOTOFF, DAVID
BRET MAVERICK (II,374)

ZLOTOFF, LEE
HILL STREET BLUES (II,1154); REMINGTON STEELE (II,2130)

ZODORON, JOHN
KATE BLISS AND THE TICKER TAPE KID (II,1366)

ZOLOTOW, MAURICE
SWING INTO SPRING (I,4317)

ZUKER, DAVID
POLICE SQUAD! (II,2061)

ZUKER, JERRY
POLICE SQUAD! (II,2061)

ZUMAN, BOB
RODNEY DANGERFIELD SPECIAL: IT'S NOT EASY BEIN' ME (II,2199)

ZWEIBEL, ALAN
THE BEACH BOYS SPECIAL (II,190); THE NEW SHOW (II,1828)

ZWERBECK, A MARTIN
KUNG FU (II,1416); THE RIFLEMAN (I,3789)

ZWICK, EDWARD
FAMILY (II,813)